T0310114

Myeloid Cells in Health and Disease

A SYNTHESIS

Myeloid Cells in Health and Disease

A SYNTHESIS

Edited by

Siamon Gordon

Emeritus Professor of Cellular Pathology
Sir William Dunn School of Pathology
Oxford, United Kingdom

Washington, DC

Library of Congress Cataloging-in-Publication Data

Names: Gordon, Siamon, editor.
Title: Myeloid cells in health and disease : a synthesis / edited by Siamon Gordon, Emeritus Professor of Cellular Pathology, Sir William Dunn School of Pathology, Oxford, United Kingdom.
Description: Washington, DC : ASM Press, [2017] | Includes index.
Identifiers: LCCN 2016057179 (print) | LCCN 2016059319 (ebook) | ISBN 9781555819187 (hardcover) | ISBN 9781555819194 (ebook)
Subjects: LCSH: Bone marrow cells. | Leucocytes. | Granulocytes. | Monocytes. | Phagocytes. | Hematopoietic stem cells. | Cellular immunity.
Classification: LCC QP92 .M945 2017 (print) | LCC QP92 (ebook) | DDC 616.07/98--dc23
LC record available at https://lccn.loc.gov/2016057179

Printed in the United States of America

10 9 8 7 6 5 4 3 2 1

Address editorial correspondence to
ASM Press, 1752 N St., N.W.,
Washington, DC 20036-2904, USA

Send orders to ASM Press, P.O. Box 605, Herndon, VA 20172, USA
Phone: 800-546-2416; 703-661-1593
Fax: 703-661-1501
E-mail: books@asmusa.org
Online: http://www.asmscience.org

Cover: Image shows primary human monocyte-derived macrophages ingesting IgG-opsonized red cells that were fixed and stained for F-actin with phalloidin. Pseudocoloring was used to grade the intensity of the F-actin staining (blue, lowest; yellow, highest). Courtesy of Daniel Schlam and Sergio Grinstein.

For Zan, Jim, and Ralph

Contents

Contributors

Shizuo Akira
Department of Host Defense
Research Institute for Microbial Diseases
Laboratory of Host Defense
WPI Immunology Frontier Research Center
Osaka University
Osaka 565-0871, Japan

Rob J. W. Arts
Department of Internal Medicine
Radboud University Medical Center
Nijmegen, The Netherlands

Jonathan M. Austyn
Nuffield Department of Surgical Sciences
University of Oxford
John Radcliffe Hospital
Oxford OX3 9DU, United Kingdom

Daniel R. Barreda
Department of Biological Sciences
Department of Agricultural, Food and Nutritional Science
University of Alberta, Edmonton
Alberta T6G 2P5, Canada

Concetta Beninati
Metchnikoff Laboratory
Department of Pediatric, Gynecological, Microbiological and Biomedical Sciences
University of Messina
Messina, Italy

MELANIE BENNETT
Roche Products Limited
Shire Park
Welwyn Garden City AL7 1TW, United Kingdom

VENETIA BIGLEY
Institute of Cellular Medicine
Newcastle University
Newcastle upon Tyne NE2 4HH, United Kingdom

CARMELO BIONDO
Metchnikoff Laboratory
Department of Pediatric, Gynecological, Microbiological and Biomedical Sciences
University of Messina
Messina, Italy

BARRY R. BLOOM
Harvard School of Public Health
Boston, MA 02115

EDUARDO BONAVITA
Humanitas Clinical Research Center
20089, Rozzano (Milano), Italy
Cancer Research UK Manchester Institute
The University of Manchester
Manchester M20 4QL, United Kingdom

STYLIANOS BOURNAZOS
Laboratory of Molecular Genetics and Immunology
The Rockefeller University
New York, NY 10065

CLINTON J. BRADFIELD
Howard Hughes Medical Institute
Chevy Chase, MD 20815
Yale Systems Biology Institute
Departments of Microbial Pathogenesis and Immunobiology
Yale University School of Medicine
New Haven, CT 06520

DONNA L. BRATTON
Departments of Pediatrics and Medicine
National Jewish Health, Denver
Departments of Immunology, Medicine, and Pediatrics
University of Colorado, Denver
Denver, CO 80231

W. JUNE BRICKEY
Lineberger Comprehensive Cancer Center
The University of North Carolina
Chapel Hill, NC 27599

VINCENZO BRONTE
Immunology Section
Department of Pathology and Diagnostics
University of Verona
37135, Verona, Italy

GORDON D. BROWN
MRC Centre for Medical Mycology
University of Aberdeen
Foresterhill, Aberdeen AB25 2ZD, United Kingdom

JENNA L. CASH
MRC Centre for Inflammation Research
The University of Edinburgh
The Queen's Medical Research Institute
Edinburgh EH16 4TJ, United Kingdom

MATILDA F. CHAN
Department of Ophthalmology
University of California, San Francisco
San Francisco, CA 94143

MARK P. CHAO
Institute for Stem Cell Biology and Regenerative Medicine
Ludwig Center for Cancer Stem Cell Research and Medicine
Stanford Cancer Institute
Stanford University School of Medicine
Stanford, CA 94305

CHEN CHEN
Department of Molecular and Cell Biology
University of California, Berkeley
Berkeley, CA 94720

AKANKSHA CHHABRA
Blood and Marrow Transplantation
Stanford University School of Medicine
Stanford, CA 94305

JONATHAN CHOU
Department of Anatomy
Department of Medicine
University of California, San Francisco
San Francisco, CA 94143

STEPHEN COBBOLD
Sir William Dunn School of Pathology
Oxford OX1 3RE, United Kingdom

MATTHEW COLLIN
Institute of Cellular Medicine
Newcastle University
Newcastle upon Tyne NE2 4HH, United Kingdom

PETER CRESSWELL
Howard Hughes Medical Institute
Department of Immunobiology
Yale University School of Medicine
New Haven, CT 06520

PAUL R. CROCKER
Division of Cell Signalling and Immunology
School of Life Sciences
University of Dundee
Dundee DD1 5EH, United Kingdom

JESMOND DALLI
Center for Experimental Therapeutics and Reperfusion Injury
Harvard Institutes of Medicine
Brigham and Women's Hospital and Harvard Medical School
Boston, MA 02115

JUSTIN F. DENISET
Department of Physiology and Pharmacology
Calvin, Phoebe, and Joan Snyder Institute for Chronic Diseases
University of Calgary
Calgary, AB T2N 4N1, Canada

FRANCESCO DE SANCTIS
Immunology Section
Department of Pathology and Diagnostics
University of Verona
37135, Verona, Italy

MICHAEL L. DUSTIN
Kennedy Institute of Rheumatology
Nuffield Department of Orthopedics, Rheumatology and Musculoskeletal Sciences
The University of Oxford
Headington, OX3 7FY, United Kingdom

LAURE EL CHAMY
Laboratoire de Génétique de la Drosophile et Virulence Microbienne, UR. EGFEM
Faculté des Sciences
Université Saint-Joseph de Beyrouth,
B.P. 17-5208 Mar Mikhaël
Beyrouth 1104 2020, Liban

MARC FELDMANN
Kennedy Institute of Rheumatology, NDORMS
University of Oxford
Botnar Research Centre
Headington, Oxford OX3 7LD, United Kingdom

MINGYE FENG
Institute for Stem Cell Biology and Regenerative Medicine
Ludwig Center for Cancer Stem Cell Research and Medicine
Stanford Cancer Institute
Stanford University School of Medicine
Stanford, CA 94305

NOAH FINE
Department of Dentistry
Matrix Dynamics Group
University of Toronto
Toronto, ON M5S 3E2, Canada

EDWARD A. FISHER
Departments of Medicine (Cardiology) and Cell Biology
New York University School of Medicine
New York, NY 10016

MARTIN F. FLAJNIK
Department of Microbiology and Immunology
University of Maryland Baltimore
Baltimore, MD 21201

GREGORY J. FONSECA
Department of Cellular and Molecular Medicine
University of California, San Diego
La Jolla, CA 92093

TOMAS GANZ
Departments of Medicine and Pathology
David Geffen School of Medicine
University of California, Los Angeles
Los Angeles, CA 90095

CECILIA GARLANDA
Humanitas Clinical Research Center
Humanitas University
20089, Rozzano (Milano), Italy

RYAN G. GAUDET
Howard Hughes Medical Institute
Chevy Chase, MD 20815
Yale Systems Biology Institute
Departments of Microbial Pathogenesis and Immunobiology
Yale University School of Medicine
New Haven, CT 06520

DEREK W. GILROY
Centre for Clinical Pharmacology and Therapeutics
Division of Medicine
University College London
London WC1 E6JJ, United Kingdom

CHRISTOPHER K. GLASS
Department of Cellular and Molecular Medicine
Department of Medicine
University of California, San Diego
La Jolla, CA 92093

MICHAEL GLOGAUER
Department of Dentistry
Matrix Dynamics Group
University of Toronto
Toronto, ON M5S 3E2, Canada

DAVID R. GREAVES
Sir William Dunn School of Pathology
University of Oxford
Oxford OX1 3RE, United Kingdom

GILLIAN M. GRIFFITHS
Cambridge Institute for Medical Research
Cambridge Biomedical Campus
Cambridge CB2 0XY, United Kingdom

SERGIO GRINSTEIN
Cell Biology Program
The Hospital for Sick Children
Toronto, Ontario M5G 1X8, Canada

HAMIDA HAMMAD
VIB Center for Inflammation Research
Ghent University
Department of Pulmonary Medicine
Ghent University Hospital
9000 Gent, Belgium

SIAVASH HASSANPOUR
Department of Dentistry
Matrix Dynamics Group
University of Toronto
Toronto, ON M5S 3E2, Canada

PETER M. HENSON
Departments of Pediatrics and Medicine
National Jewish Health, Denver
Departments of Immunology, Medicine, and Pediatrics
University of Colorado, Denver
Denver, CO 80231

JAMES P. HEWITSON
Department of Biology
University of York
York YO10 5DD, United Kingdom

DUNCAN HOWIE
Sir William Dunn School of Pathology
Oxford OX1 3RE, United Kingdom

MARK M. HUGHES
School of Biochemistry and Immunology
Trinity Biomedical Science Institute
Trinity College Dublin
Dublin 2, Ireland

DAVID A. HUME
University of Edinburgh
The Roslin Institute and Royal (Dick) School of Veterinary Studies
Midlothian EH25 9RG, United Kingdom

Asif J. Iqbal
Sir William Dunn School of Pathology
University of Oxford
Oxford OX1 3RE, United Kingdom

Lionel B. Ivashkiv
Arthritis and Tissue Degeneration Program and David Z. Rosensweig Genomics Center
Hospital for Special Surgery
Department of Medicine and Immunology and Microbial Pathogenesis Program
Weill Cornell Medical College
New York, NY 10021

Sébastien Jaillon
Humanitas Clinical Research Center
Humanitas University
20089, Rozzano (Milano), Italy

Claudia V. Jakubzick
Departments of Pediatrics and Medicine
National Jewish Health, Denver
Departments of Immunology, Medicine, and Pediatrics
University of Colorado, Denver
Denver, CO 80231

William J. Janssen
Departments of Pediatrics and Medicine
National Jewish Health, Denver
Departments of Immunology, Medicine, and Pediatrics
University of Colorado, Denver
Denver, CO 80231

Valentin Jaumouillé
Cell Biology Program
The Hospital for Sick Children
Toronto, Ontario M5G 1X8, Canada

Sushmita Jha
Department of Bio-Sciences and Bio-Engineering
Indian Institute of Technology Jodhpur
Rajasthan, 342011, India

Steffen Jung
Department of Immunology
The Weizmann Institute of Science
Rehovot 76100, Israel

A. Barry Kay
Inflammation, Repair, and Development
National Heart & Lung Institute
Imperial College
London SW3 2AZ, United Kingdom

Samira Khaliq
Department of Dentistry
Matrix Dynamics Group
University of Toronto
Toronto, ON M5S 3E2, Canada

Paul Kubes
Department of Physiology and Pharmacology
Department of Microbiology and Infectious Diseases
Calvin, Phoebe, and Joan Snyder Institute for Chronic Diseases
University of Calgary
Calgary, AB T2N 4N1, Canada

Thomas A. Kufer
Institute of Nutritional Medicine
Department of Immunology
University of Hohenheim
70593 Stuttgart, Germany

Bart N. Lambrecht
VIB Center for Inflammation Research
Ghent University
Department of Pulmonary Medicine
Ghent University Hospital
9000 Gent, Belgium

Toby Lawrence
Centre d'Immunologie de Marseille-Luminy
Aix Marseille Université
Inserm, CNRS
Marseille, France

Adam P. Levine
Division of Medicine
University College London
London WC1E 6JF, United Kingdom

Peter Libby
Division of Cardiovascular Medicine
Department of Medicine
Brigham and Women's Hospital
Harvard Medical School
Boston, MA 02115

Hsi-Hsien Lin
Department of Microbiology and Immunology
College of Medicine
Chang Gung University
Chang Gung Immunology Consortium and Department of Anatomic Pathology
Chang Gung Memorial Hospital-Linkou
Tao-Yuan, Taiwan

John D. MacMicking
Howard Hughes Medical Institute
Chevy Chase, MD 20815
Yale Systems Biology Institute
Departments of Microbial Pathogenesis and Immunobiology
Yale University School of Medicine
New Haven, CT 06520

Rick M. Maizels
Wellcome Trust Centre for Molecular Parasitology
Institute of Infection, Immunity and Inflammation
University of Glasgow
Glasgow G12 8TA, United Kingdom

Kevin J. Maloy
Sir William Dunn School of Pathology
Oxford, OX1 3RE, United Kingdom

Alberto Mantovani
Humanitas Clinical Research Center
Humanitas University
20089, Rozzano (Milano), Italy

Goran Marinkovic
Department of Immunology
The Weizmann Institute of Science
Rehovot 76100, Israel

Paul Martin
School of Biochemistry, Medical Sciences, University Walk
Bristol University
Bristol BS8 1TD, United Kingdom

Nicolas Matt
Université de Strasbourg
UPR 9022 du CNRS
Institut de Biologie Moléculaire et Cellulaire
Strasbourg Cedex 67084, France

Gabriel Mbalaviele
Musculoskeletal Research Center
Division of Bone and Mineral Diseases
Department of Medicine
Washington University School of Medicine
St. Louis, MO 63110

Anne F. McGettrick
School of Biochemistry and Immunology
Trinity Biomedical Science Institute
Trinity College Dublin
Dublin 2, Ireland

DONALD METCALF†
Cancer and Haematology Division
The Walter and Eliza Hall Institute of Medical Research
Parkville, Victoria 3052, Australia

ALEXANDER MILDNER
Department of Immunology
The Weizmann Institute of Science
Rehovot 76100, Israel

GABRIEL MITCHELL
Department of Molecular and Cell Biology
University of California, Berkeley
Berkeley, CA 94720

ROBERT L. MODLIN
David Geffen School of Medicine
University of California at Los Angeles
Los Angeles, CA 90095

CLAUDIA MONACO
Kennedy Institute of Rheumatology, NDORMS
University of Oxford
Botnar Research Centre
Headington, Oxford OX3 7LD, United Kingdom

MATTHIAS NAHRENDORF
Center for Systems Biology
Massachusetts General Hospital
Boston, MA 02114

JAGDEEP NANCHAHAL
Kennedy Institute of Rheumatology, NDORMS
University of Oxford
Botnar Research Centre
Headington, Oxford OX3 7LD, United Kingdom

HAROLD R. NEELY
Department of Microbiology and Immunology
University of Maryland Baltimore
Baltimore, MD 21201

MIHAI G. NETEA
Department of Internal Medicine
Radboud University Medical Center
Nijmegen, The Netherlands

NICOS A. NICOLA
The Walter and Eliza Hall Institute of Medical Research
Parkville, Victoria 3052, Australia

†Deceased.

GIULIA NIGRO
Institut Pasteur
Unité de Pathogénie Microbienne Moléculaire
75724 Paris, France
INSERM, U1202
75015 Paris, France

DEBORAH VEIS NOVACK
Musculoskeletal Research Center
Division of Bone and Mineral Diseases
Department of Medicine
Department of Pathology and Immunology
Washington University School of Medicine
St. Louis, MO 63110

LUKE A.J. O'NEILL
School of Biochemistry and Immunology
Trinity Biomedical Science Institute
Trinity College Dublin
Dublin 2, Ireland

WENDY W. PANG
Institute for Stem Cell Biology and Regenerative Medicine
Ludwig Center for Cancer Stem Cell Research and Medicine
Stanford Cancer Institute
Blood and Marrow Transplantation
Stanford University School of Medicine
Stanford, CA 94305

SUNG HO PARK
Arthritis and Tissue Degeneration Program and David Z. Rosensweig Genomics Center
Hospital for Special Surgery
New York, NY 10021

V. HUGH PERRY
Centre for Biological Sciences
University of Southampton
Southampton SO16 6YD, United Kingdom

EMMA K. PERSSON
VIB Center for Inflammation Research
Ghent University
9000 Gent, Belgium

DANIEL A. PORTNOY
Department of Molecular and Cell Biology
School of Public Health
University of California, Berkeley
Berkeley, CA 94720

JEFFREY V. RAVETCH
Laboratory of Molecular Genetics and Immunology
The Rockefeller University
New York, NY 10065

MICHAEL REHLI
University Hospital Regensburg
Department of Internal Medicine III
D-93047 Regensburg, Germany

JEAN-MARC REICHHART
Université de Strasbourg
UPR 9022 du CNRS
Institut de Biologie Moléculaire et Cellulaire
Strasbourg Cedex 67084, France

PAUL A. ROCHE
Experimental Immunology Branch
Center for Cancer Research
National Cancer Institute, NIH
Bethesda, MD 20892

MARC E. ROTHENBERG
Division of Allergy and Immunology
Cincinnati Children's Hospital Medical Center
University of Cincinnati College of Medicine
Cincinnati, OH 45229

THEODORE J. SANDERS
Sir William Dunn School of Pathology
Oxford, OX1 3RE, United Kingdom

PHILIPPE J. SANSONETTI
Institut Pasteur
Unité de Pathogénie Microbienne Moléculaire
75724 Paris, France
INSERM, U1202
75015 Paris, France
Collège de France
75005 Paris, France

TAKASHI SATOH
Department of Host Defense
Research Institute for Microbial Diseases
Laboratory of Host Defense
WPI Immunology Frontier Research Center
Osaka University
Osaka 565-0871, Japan

PETER J. SCHNORR
Institute for Stem Cell Biology and Regenerative Medicine
Ludwig Center for Cancer Stem Cell Research and Medicine
Stanford Cancer Institute
Stanford University School of Medicine
Stanford, CA 94305

ANTHONY W. SEGAL
Division of Medicine
University College London
London WC1E 6JF, United Kingdom

JASON S. SEIDMAN
Department of Cellular and Molecular Medicine
Biomedical Sciences Graduate Program
University of California, San Diego
La Jolla, CA 92093

JUN SEITA
Institute for Stem Cell Biology and Regenerative Medicine
Ludwig Center for Cancer Stem Cell Research and Medicine
Stanford Cancer Institute
Stanford University School of Medicine
Stanford, CA 94305

CHARLES SERHAN
Center for Experimental Therapeutics and Reperfusion Injury
Harvard Institutes of Medicine
Brigham and Women's Hospital and Harvard Medical School
Boston, MA 02115

MARTIN STACEY
School of Molecular and Cellular Biology
University of Leeds
Leeds, LS2 9JT, United Kingdom

KIM M. SUMMERS
University of Edinburgh
The Roslin Institute and Royal (Dick) School of Veterinary Studies
Midlothian EH25 9RG, United Kingdom

FILIP K. SWIRSKI
Center for Systems Biology
Massachusetts General Hospital
Boston, MA 02114

GIUSEPPE TETI
Metchnikoff Laboratory
Department of Pediatric, Gynecological, Microbiological and Biomedical Sciences
University of Messina
Messina, Italy

JENNY PAN-YUN TING
Lineberger Comprehensive Cancer Center
Department of Genetics
University of North Carolina
Chapel Hill, NC 27599

IRINA UDALOVA
Kennedy Institute of Rheumatology, NDORMS
University of Oxford
Botnar Research Centre
Headington, Oxford OX3 7LD, United Kingdom

STEFANO UGEL
Immunology Section
Department of Pathology and Diagnostics
University of Verona
37135, Verona, Italy

EMIL R. UNANUE
Department of Pathology and Immunology
Washington University School of Medicine
St. Louis, MO 63103

HERMAN WALDMANN
Sir William Dunn School of Pathology
Oxford OX1 3RE, United Kingdom

TAIA T. WANG
Laboratory of Molecular Genetics and Immunology
The Rockefeller University
New York, NY 10065

KIPP WEISKOPF
Department of Medicine
Brigham and Women's Hospital
Boston, MA 02115
Institute for Stem Cell Biology and Regenerative Medicine
Ludwig Center for Cancer Stem Cell Research and Medicine
Stanford Cancer Institute
Stanford University School of Medicine
Stanford, CA 94305

IRVING L. WEISSMAN
Institute for Stem Cell Biology and Regenerative Medicine
Ludwig Center for Cancer Stem Cell Research and Medicine
Stanford Cancer Institute
Stanford University School of Medicine
Stanford, CA 94305

TING WEN
Division of Allergy and Immunology
Cincinnati Children's Hospital Medical Center
University of Cincinnati College of Medicine
Cincinnati, OH 45229

ZENA WERB
Department of Anatomy
University of California, San Francisco
San Francisco, CA 94143

ULF YRLID
Department of Microbiology and Immunology
Institute of Biomedicine
University of Gothenburg
S-405 30 Gothenburg, Sweden

Preface

Two of the fathers of myeloid cell immunobiology, Elie Metchnikoff and Paul Ehrlich, joint Nobel laureates in 1908 for their contributions to cellular and humoral immunity, died in 1916 and 1915, respectively. The 2011 award to Bruce Beutler, Jules Hoffman, and the recently deceased Ralph Steinman served to demarcate a century of discovery in myeloid cells, no longer overshadowed by research into lymphoid cell function and adaptive immunity. It has therefore been an opportune time to take stock of a century of progress in myeloid cell biology, to gain a perspective on present knowledge, and to identify future needs. This is timely since increased specialization in molecular analysis has fragmented information gained for individual myeloid cell types. A comparative approach yields insights into cellular differentiation and activation, illuminates understanding of cell structure and function, and sharpens selectivity of therapeutic targeting. This volume compares polymorphonuclear and mononuclear leukocytes, granulocytic and secretory cells, and the effects of different tissue microenvironments on tissue specialization. Authors were invited to be analytic and integrative, to provide a perspective on their subject rather than excessive detail associated with acceleration of research.

The overall presentation is rooted in history. Gio Teti and colleagues deal with the life and work of Elie Metchnikoff; Barry Kay describes early studies by Paul Ehrlich of blood cell cytochemistry following developments in the dye industry. Donald Metcalf provided an account of the identification of colony-stimulating factors, using assays that he had practiced throughout his career. In a poignant covering letter, he stated that he was terminally ill and could not see to proofs of his final submission. This was overseen by his colleagues and I asked Nick Nicola to write an appreciation of his mentor and friend.

Topics and authors were selected to provide an overview of the remarkably broad range of disciplines and approaches used to investigate myeloid cell biology.

From the mid-19th century, these cells have contributed to the growth of hematology, immunology, microbiology, pathology, and pharmacology. Their origins and functions attracted developmental biologists and zoologists, using invertebrate and genetically tractable model organisms, including mice and humans. Enhanced imaging methods made it possible to dissect their ontogeny and role in physiology and disease, in addition to phagocytic clearance and host defense. Because of ready availability, myeloid cells have provided models to study general cellular structure and function, benefitting from growth in molecular cloning and gene expression analysis. Recent developments in stem cell biology grew from early advances in hematopoietic and cell culture methods. Monoclonal antibody technology provided antigen markers to define lineage relationships and phenotypic heterogeneity, within tissues as well as blood.

The table of contents is organized by grouping of topics, with some overlap of subjects in different chapters. The goal was to be illustrative, rather than encyclopaedic and individual chapters provide references for further details. Specialists from different disciplines, backgrounds, and countries have been brought together in a single volume, an additional form of synthesis. To appreciate where the field has arrived, I believe it is necessary to appreciate, very briefly, a few of the milestones along the way.

SETTING THE STAGE

Myeloid cells and their role in homeostasis, inflammation, and infection were discovered mainly by early investigators in Europe during the second half of the 19th century. As American bioscience research developed, by the mid-20th century, the laboratory of James Hirsch and Zanvil Cohn at the Rockefeller Institute, later University, became the Mecca of myeloid cell biologists. They were influenced by developments in cell biology of George Palade and Christian de Duve, applied to the study of granulocytes (polymorphonuclear leukocytes and eosinophils, especially), and of mononuclear monocytes, macrophages and, later, dendritic cells (DC), when Ralph Steinman joined the laboratory. The discovery in the mid-1970s of the potent antigen presentation activity of DC broadened the group's research interest to include antigen-specific lymphoid cells and their interactions with DC and macrophages in cell-mediated immunity. The discovery by Tim Mossman and Bob Coffman of Th1 and Th2 subdivisions of cellular immunity gave rise to corresponding distinct functions of macrophages activated by interferon gamma and interleukin 4/13.

During this period, research at many centres throughout the world, defined the recognition, killing, uptake, and digestion of microbes and altered host components by phagocytic leukocytes. Opsonic (FcR and complement receptors) and non-opsonic (scavenger and lectin-like) phagocytic recognition molecules were defined, as well as the toll-like sensors of innate cell activation. Phosphatidylserine is one of several recognition receptors for uptake of apoptotic cells. Autophagy has been recognized as an important counterpart to heterophagy, ending in a common lysosomal degradative pathway following cellular injury and infection. Cytosolic sensing of microbial constituents and of nucleic acids, and inflammasome activation of cytokine release became a topic of major interest. The induction and secretion of neutral proteinases, bioactive cytokines and chemokines, notably TNF alpha, arachidonate, oxygen, and nitrogen metabolites, established that myeloid leukocytes were secretory cells as well as phagocytes. Individual chapters build on many of these topics.

The development of monoclonal antibodies, such as F4/80 to a seven transmembrane adhesion GPCR, provided markers to detect tissue macrophages and, along with other antigen markers such as CD68, CD169, CD206, and CD163, defined macrophage heterogeneity in different tissues during development and throughout adult life. Resident and newly recruited cells were shown to differ in their fetal and adult origin and in function. Apart from inflammation and infection, myeloid cells play an important role in metabolic diseases such as atherosclerosis, as well as in malignancy. Several of the articles in this volume explore the regulation of gene expression by myeloid cells, in relation to chromatin structure.

CONCLUSION

The current explosion of genome-wide analysis of gene expression and the resultant growth of systems biology, have generated new insights into myeloid cell differentiation and activation, leading to the discovery of novel functions. However, a formidable challenge remains, for example, to investigate tissue macrophage functions *in situ*, and to reconstruct their heterogeneous phenotype *in vitro*. iPSC and CRISPR/Cas9 technology is proceeding apace, and has been increasingly applied to myeloid cells. The widely held notion of plasticity of differentiated myeloid cells is unsubstantiated at present, incorporating as it does, changes in population dynamics as well as modulation of the activity of individual cells. This has to be integrated with the study of metabolic regulation *in vivo*. The study of myeloid cell biology will no doubt create further opportunities for selective therapeutic interventions in the future.

Siamon Gordon

Acknowledgments

I am grateful to Amelia Molloy-Bland and Annette Pluddemann for their unstinting support throughout this project. At ASM Press I acknowledge encouragement by Greg Payne and Christine Charlip and the highly professional editorial assistance of Cathy Balogh, Lauren Luethy, and Megan Angelini at all stages of preparation of the manuscript.

About the Editor

Siamon Gordon was born in Cape Town, South Africa in 1938 and graduated in medicine at the University of Cape Town in 1961. After a year's introduction to immunology in the department of Rodney Porter at the Wright Fleming Institute in London, he moved to the Rockefeller University in New York in 1965, first as a postdoc with Alick Bearn in human genetics and then as a doctoral student with Zanvil Cohn. His thesis project used macrophage-melanoma cell heterokaryons to study cellular differentiation, leading to graduation in 1971. He continued researching macrophage immunobiology for the remainder of his academic career, first at Rockefeller, then at the University of Oxford, retiring as Glaxo Wellcome Professor of Cellular Pathology at the Sir William Dunn School of Pathology in 2008. Important mentors included Golda Selzer in South Africa, Zanvil Cohn and James Hirsch in New York, and Henry Harris in Oxford. Research interests in New York focused on phagocytosis and secretion of lysozyme and plasminogen activator. At Oxford his group produced monoclonal antibodies to characterize the function of myeloid differentiation antigens such as F/80, dectin-1, complement, scavenger, and lectin-like receptors. Modulation of macrophages by cytokines such as interleukin-4 led to the concept of an alternative pathway of macrophage activation. Returning to cell fusion, later studies demonstrated that macrophage giant cells are more than the sum of their parts, but represent a further stage of differentiation with specialized functions such as enhanced clearance. He was elected to fellowship of the Royal Society in 2007. After retiring from his academic post, he spent a year with Alan Sher and Giorgio Trinchieri at the NIH in Bethesda. He has continued to be active in promoting research in South Africa, as well as initiating an AIDS education project there. His hobbies include the history of macrophage research, with special interest in the life and work of Elie Metchnikoff.

Introduction and History

I

Myeloid Cells in Health and Disease: A Synthesis
Edited by Siamon Gordon

doi:10.1128/microbiolspec.MCHD-0032-2016

A. Barry Kay[1]

1 Paul Ehrlich and the Early History of Granulocytes

INTRODUCTION

Between 1878 and 1880, Paul Ehrlich (1854-1915), a medical student and then assistant physician at the Charité Hospital, Berlin, demonstrated, using acid and basic aniline coal tar dyes, that the different types of blood leukocytes could be distinguished on the basis of the staining properties of their granules (Fig. 1). Ehrlich's technique for staining blood films and his method of differential blood cell counting ended years of speculation regarding the classification of white cells. His discoveries were among the greatest advances in modern hematology, and the principles surrounding his methodology are applied to this day. Nevertheless, prior to Ehrlich there were several notable landmarks that led to a fuller understanding of white cells, their origin, and possible function. This chapter highlights many of the important achievements of Paul Ehrlich and others. It is not intended to be comprehensive. The priority of many of the observations described, particularly in pre-Ehrlich time, remains controversial to this day.

EARLY DESCRIPTIONS OF WHITE CELLS

Man has been fascinated by blood since prehistoric times. Blood as a life-giving substance, holding the body's vital forces, was believed by the ancients. The Romans drank the blood of their enemies thinking it would confer the courage of their vanquished foes. It was not until the advent of the compound microscope, around 1590, by Hans and Zacharias Janssen that the content of blood could actually be examined. In 1658, the Dutch naturalist Jan Swammerdam (1637-1680) was the first to observe red blood cells under the microscope. A few years later, in 1695, another Dutch microscopist, Antonie van Leeuwenhoek (1632-1723), described the size and shape of "red corpuscles" (1). White cells were paid little attention to, presumably because of their infrequency and transparency. In 1749, the French physician Joseph Lieutaud (1703-1780) observed white cells ("*globuli albicantes*") in postmortem material (2), and in the same year another French doctor, Jean-Baptiste de Senac (1693-1770), described pus cells ("*globules blanc du pus*") (3). However, it was the Englishman William Hewson (1739-1774) who, in 1773, made the first detailed study on blood and lymph (Fig. 2). He found that "colourless cells" or "central particles" (terms he used to describe white cells) were rare in comparison with red corpuscles (4). He surmised that these cells made their way into the lymphatic system via the thoracic duct. Hewson is credited with the discovery of the lymphocyte but was unable to

[1]Inflammation, Repair, and Development, National Heart & Lung Institute, Imperial College, London, SW3 2AZ, United Kingdom.

Über die specifischen Granulationen des Blutes*

P. Ehrlich

Die bei Weitem wichtigste dieser Körnungen ist die eosinophile oder α-Granulation, über welche ich schon am 17. Januar d. J. vor der Gesellschaft berichten konnte.* Die α-Granulation ist durch ihre Verwandtschaft zu der grossen Reihe der *sauren* Theerfarbstoffe charakterisirt, d. h. solchen, in denen wie im picrinsauren Ammon das färbende Princip eine Säure darstellt. Die Farbstoffe zerfallen, entsprechend

Figure 1 Paul Ehrlich in 1878 (from Hirsch and Hirsch [23]). The first description of the eosinophil appeared in the *Archive für Anatomie und Physiologie* in 1879. Reprinted from (26), with permission.

differentiate it from other white cells. His best-known contribution to medicine was the demonstration that fibrinogen leads to coagulation. He is often referred to as the Father of Hematology.

THE MID-19TH CENTURY

Much of the knowledge about white cells in the pre-Ehrlich era came from three distinct but overlapping areas, namely, the first descriptions of leukemia, theories on the nature of pus, and theories on the nature of inflammation. The introduction of the microscope into clinical diagnosis is largely attributed to Alfred Donné (1801-1878) (Fig. 2) of Paris, as is the description of the first case of leukemia (Fig. 3) (5, 6). In 1839, when assisting a colleague with another early recorded case of leukemia, Donné wrote:

> The blood you sent me shows a remarkable and most conspicuous change. More than half of the cells were mucous globules: you know that normal blood contains three types of cells 1) red cells, the essential globules of the blood 2) white cells or mucous cells: and 3) the small globules. It is the second variety which dominates so much, that, one wonders, knowing nothing about the clinical course, whether this blood does not contain pus. As you know the pus cells cannot yet be differentiated with accuracy from mucous cells.

Then in a later case he remarked:

> The blood of this patient showed such a number of white cells that I thought his blood was mixed with pus, but in the end was able to observe a clear-cut difference between these cells and white cells. I believe that the excess of white cells is due to an arrest of maturation of blood. From my theory on the origin of red blood cells, the overabundance of white blood cells should be the result of an arrest of development of intermediate cells.

As Xavier Thomas remarked in his excellent history of the disease (6), these were seminal observations since they linked leukemia with abnormal pathology for the first time.

By 1843, William Addison (1802-1881) had accurately observed that pus cells were derived from blood leukocytes that had passed through the capillary wall (7). John Hughes Bennett (1812-1875) (Fig. 4), an

Figure 2 (Left) William Hewson, the Father of Hematology. (Right) Alfred Donné, who introduced the microscope into clinical diagnosis and recorded one of the first descriptions of white cells in leukemia. Reprinted from (26), with permission.

Edinburgh physician and former pupil of Donné, is often credited with the discovery of leukemia. He published a detailed scientific description of the disease as a clinical entity in 1845 but believed that the blood vessels contained "thick, creamy pus." It was the famous German pathologist Rudolf Virchow (1821-1902) (Fig. 5) who recognized, shortly afterwards in 1847, with his own case, that the cells in leukemia are not pus but originate from blood (8). This established that the alteration in the blood was independent of inflammation.

For centuries, inflammation was considered as a separate disease. It was John Hunter (1728-1793) who suggested that inflammation was a response to injury (9). Early microscopists believed that red cells and "colourless cells" seeped from the small vessels into the tissues, where they provided nutrition by transforming into various tissue cells (the "intercalatory" or "corpuscular" theory). The concept of capillaries having walls with the movement of white cells from the vessels across the walls to the surrounding tissues was not established until the second half of the century. In 1863, Friedrich von Recklinghausen (Fig. 6), working as an assistant to Virchow at the Charité, showed that the amoeboid properties of white cells (first observed by Thomas Wharton Jones [see below]) were the result of both locomotion and contractility (10). When von Recklinghausen moved to the chair of pathology in Konigsberg, Virchow appointed Julius Cohnheim in his place (Fig. 6). It was Cohnheim who showed that leukocytes actually passed through the apparently intact walls of the capillaries by amoeboid movement (11).

The concept of different types of white cells based on their size, shape, and granularity was slow to evolve. Several investigators observed granulated cells that, from their diagrams, were almost certainly what Ehrlich was to name eosinophils based on his precise staining techniques. One of the earliest descriptions, if not the first, of "granule cells" in inflammatory exudates was made by the Belgium clinician and scientist Gottlieb (Théophile) Gluge (1812-1898) (Fig. 7) (12). Gluge was a pioneer medical researcher and personal physician to the king of Belgium. He was one of the first clinicians to examine diseased tissue microscopically. His major work was the *Atlas der Pathologischen Anatomie* (1843), in which he described "inflammatory globules," a term he preferred to "granular cells." These cells were observed not only in inflammatory exudates but also in colostrum and the ovary and resemble eosinophils (Fig. 7). Gluge believed that his inflammation corpuscles were derived from precursor material by a "second mode of cell formation" in which "small bodies appear" that

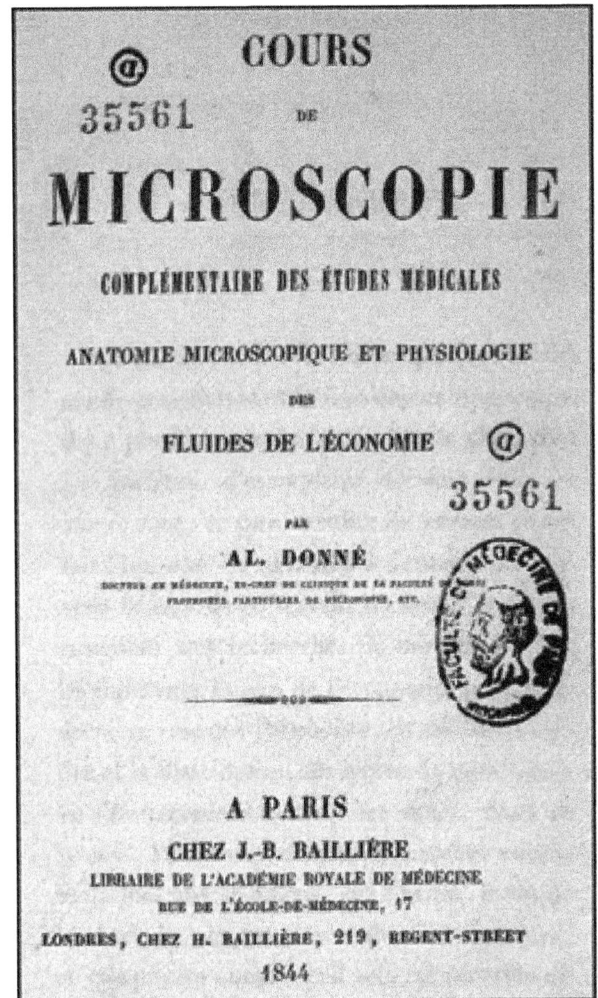

COURS
DE
MICROSCOPIE
COMPLÉMENTAIRE DES ÉTUDES MÉDICALES
ANATOMIE MICROSCOPIQUE ET PHYSIOLOGIE
DES
FLUIDES DE L'ÉCONOMIE

PAR
AL. DONNÉ

A PARIS
CHEZ J.-B. BAILLIÈRE
LIBRAIRE DE L'ACADÉMIE ROYALE DE MÉDECINE
RUE DE L'ÉCOLE-DE-MÉDECINE, 17
LONDRES, CHEZ H. BAILLIÈRE, 219, REGENT-STREET
1844

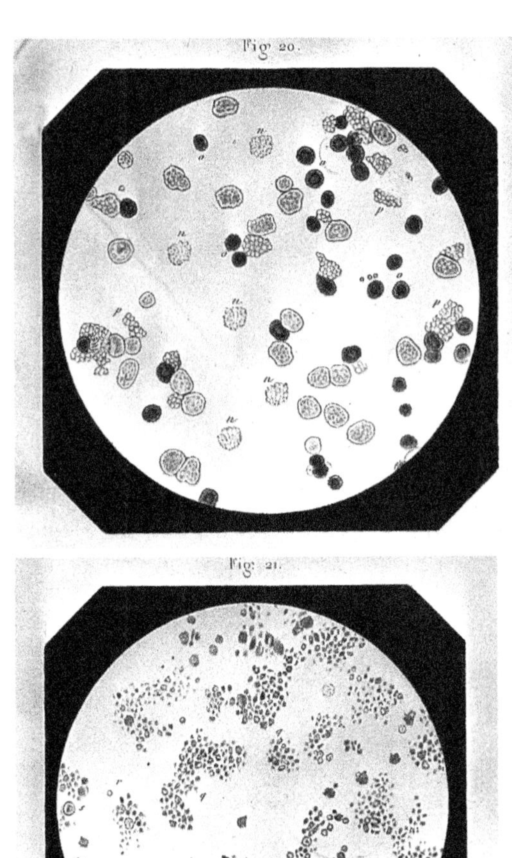

Figure 3 Alfred Donné's microscopic images of blood from a patient with leukemia. (Top right) Red and white blood cells from a leukemic patient. (Lower right) "Mucous globules," or white blood cells, from the same patient. The images were published in the supplement to *Cours de Microscopie* in 1845 (left). Reprinted from (26), with permission.

> resemble fat granules ... but also frequently may consist of protein or sometimes of pigment. These (granules) average in size from 1/500 to 1/400 of a millimetre in diameter, and become associated, in groups of from 10-40 or more ... the groups of granules are mulberry-like globules, measuring on average 1/30 of a millimetre, and in this condition are frequently observed.

In one illustration Gluge describes "Stasis and metamorphosis of blood corpuscles into inflammation-corpuscles." The drawing shows eosinophil-like cells (his "inflammation globules"), probable red cells, and empty cells (neutrophils?) together with venous stasis. Taken together, his various drawings of inflammation globules, which he believed essentially were large pus cells containing fat droplets, have the appearance of intact and partially degranulated eosinophils. This appears to be the first description of granular, eosinophil-like cells.

Gluge's work was recognized and extended by the eminent pathologist Julius Vogel (1814-1880) (Fig. 8). Vogel was a skillful microscopist and an expert in basic sciences. He was the professor of pathology at Göttingen, Germany, and published his pioneering book (also in 1843), *The Pathological Anatomy of the Human Body* (13). It was the first comprehensive general pathology book and contained numerous macroscopic and microscopic illustrations. The book was remarkable in that the bulk of the figures showed, for the first time, microscopic features of specific pathologic lesions (Fig. 8). Among the illustrations are cytological preparations of sputum and pleural and pericardial

John Hughes Bennett
(1812-1875)

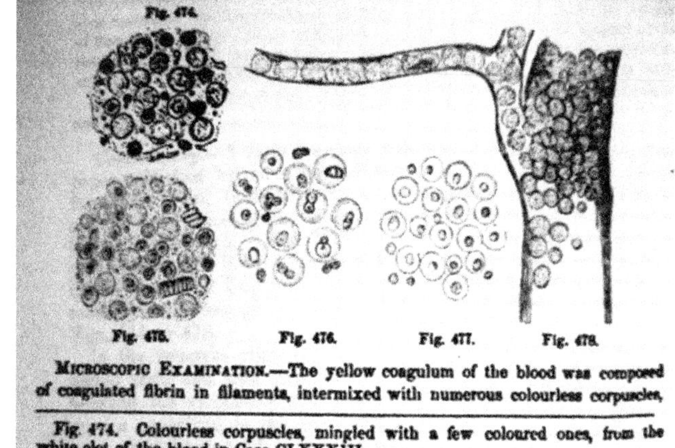

Fig. 474. Fig. 475. Fig. 476. Fig. 477. Fig. 478.

MICROSCOPIC EXAMINATION.—The yellow coagulum of the blood was composed of coagulated fibrin in filaments, intermixed with numerous colourless corpuscles,

Fig. 474. Colourless corpuscles, mingled with a few coloured ones, from the white clot of the blood in Case CLXXXIII.

Fig. 475. The same bodies, mingled with a larger number of yellow blood corpuscles in the red clot.

Fig. 476. Change produced on the colourless corpuscles on the addition of acetic acid, the yellow corpuscles being dissolved.

Fig. 477. Cells in the fluid squeezed from the lymphatic glands, after the addition of acetic acid.

Fig. 478. Blood-vessels giving off a capillary from the pia mater; the latter is seen filled with colourless corpuscles; the former partly with colourless, mingled with coloured corpuscles. 250 diam.

Figure 4 John Bennett published the first detailed description of leukemia. His illustrations show "colourless corpuscles" in the circulation. Reprinted from (26), with permission.

effusions of patients with bronchitis, tracheitis, and pneumonia. In these, Vogel described numerous granular cells that "are for the most part round, but sometimes elongated or even angular ... it is difficult to distinguish whether they (the granules) are on the surface or in the interior of the cell." Vogel believed that the granules consisted of fat (olein and margarine) since they dissolved in organic solvents. He also wrote: "when the granular cells have attained their full development, the cell-wall disappears, and the granules contained in the interior are liberated, and form larger and smaller heaps." The illustrations show cells with the clear morphological characteristics of eosinophils, both intact and in various stages of degranulation.

Another convincing and comprehensive morphological pre-Ehrlich study of granular blood cells was by the British physiologist and ophthalmologist Thomas Wharton Jones (1808-1891) (Fig. 9). Wharton Jones studied medicine in Edinburgh and for a few years was the assistant to the anatomist Robert Knox (of Burke and Hare fame). He left Edinburgh for Glasgow, where he studied ophthalmology and eventually settled in London, first as a lecturer at the Charing Cross Hospital and then as the professor of ophthalmic medicine at University College. He had a special interest in inflammation and the vasculature and in 1846 published "The Blood Corpuscle Considered in Its Different Phases of Development in the Animal Series" (14). In this paper he announced his discovery of "the spontaneous change in shape and other movements—so called amoeboid movements—of the colourless corpuscles in the blood of the skate, frog, and other animals including man." In the same article he described "finely granular and coarse granular blood-cells" in a large number of species including the lamprey, frog, fowl, horse, elephant, and man (Fig. 9). He speculated (incorrectly) that the coarse granular cells were a stage in development of the finely granular cells. The blood specimens were treated with water or dilute acetic acid in an attempt to distend the cell. This naturally dissolved the intracellular granules, but nevertheless his color drawings show clearly, particularly in higher mammals, that the coarsely granular cells had the typical appearance of eosinophils. Even with primitive optics (a Ross compound microscope with a one-eighth-inch objective), he was able to estimate the diameter of the coarse granule to be 1/25,000 of an inch.

FUNCTIONAL STUDIES

Several years after the description of granulated cells by Wharton Jones, Vogel, and Gluge, but almost 20 years

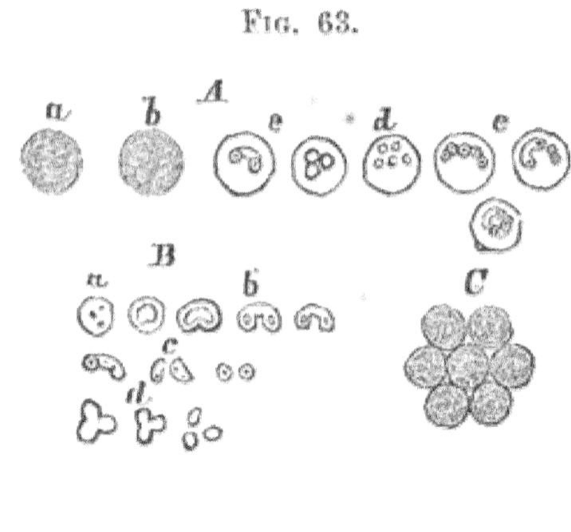

Rudolph Virchow (1821-1902)

Figure 5 Rudolf Virchow and an illustration taken from his volume *Cellular Pathology as Based upon Physiological and Pathological Histology* (7th American edition, Robert M. de Witt, New York, NY, 1856):

> Fig. 63. A. Pus-corpuscles, a fresh, b after the addition of a little water, c–e after treatment with acetic acid, the contents cleared up, the nuclei which were in process of division, or already divided, visible, at e with a slight depression on their surface. B. Nuclei of pus corpuscles in gonorrhoea; a simple nucleus with nucleoli, b incipient division, with depressions (by many held to be nuclei) on the surface of the nuclei, c progressive bi-partition, d tri-partition. C. Pus-corpuscles in their natural position with regard to one another. 500 diameters.

Reprinted from (26), with permission.

before Ehrlich developed his staining methods, Max Johann Sigismund Schultze (1825-1874) (Fig. 10) performed functional experiments on coarse granular cells (which were almost certainly eosinophils). Schultze was a distinguished German microscopic anatomist who was noted for his work on cell theory. He had studied medicine at Greifswald and Berlin and was appointed extraordinary professor at Halle in 1854 and later as professor of anatomy and histology and director of the Anatomical Institute at Bonn. In his study, published in 1865 ("The Heated Stage and Its Use in the Investigation of the Blood"), he described for the first time four different types of leukocytes, which we now recognize as the monocyte, lymphocyte, neutrophil, and eosinophil (15). With his warm stage microscope, he established that finely granular and coarsely granular human white blood cells moved and phagocytosed small particles (Fig. 10). His illustrations and descriptions of movement leave little doubt that these finely and coarsely granulated cells were neutrophils and eosinophils, respectively. He wrote:

> If one increases the temperature to that of the body a much quicker movement (of coarse granular white cells) develops with striking changes in form and a more rapid amoeba-like movement develops with movement from place to place as we have just described in the finely granular white cell.

He then went on to add particulate matter such as dyes and milk to the cell preparation and observed that

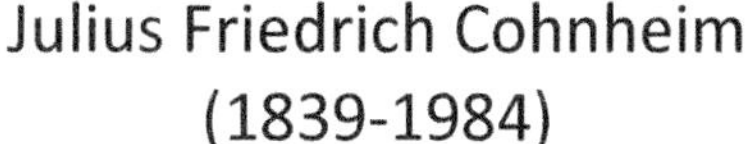

Friedrich Daniel von Recklinghausen
(1833-1910)

Figure 6 (Left) Julius Cohnheim demonstrated that leukocytes passed through the apparently intact walls of the capillaries by amoeboid movement. (Right) Friedrich von Recklinghausen showed that the amoeboid properties of white cells were the result of both locomotion and contractility. Reprinted from (26), with permission.

> The coarsely granular white cells do take up granules of dyes as I have observed with aniline blue, amongst others. It was striking to me that the liveliness of the movement (of the coarse granular cells) was obviously much reduced following the uptake of dye particles, an appearance that I often met with the finely granular cells.

According to Douglas Brewer (16), "the work of Schultze was more important than that of Ehrlich: not only did he distinguish four different types of white cells, but he demonstrated important differences in their movement and phagocytic abilities."

Max Schultze is also credited with the discovery of platelets, which he observed as part of his studies on white cells. Previously, in 1842, Alfred Donné had described a third cellular element in blood (the platelet) but mistook them for fat globules of chyle. However, it was Giulio Bizzozero (1846-1901) who showed that platelets adhered to fibrin and were an essential part of the clotting mechanism (17).

PAUL EHRLICH AND THE EARLY DISCOVERY OF CELLS OF THE MYELOID LINEAGE

In 1879, Paul Ehrlich published his technique for staining blood films and his method for differential blood cell counting using coal tar dyes. His use of stains was a landmark contribution and heralded modern studies on blood leukocytes. The work originated from his days as a medical student in Leipzig, where, in 1878, he submitted his dissertation on the use of aniline dyes for staining microscopic specimens (18). The earliest experiments were with tissue sections using the dye dahlia (monophenyl rosaniline). At first the results were unsatisfactory, and he only managed to achieve diffuse, patchy staining of the cells. However, when he reduced the staining intensity with dilute acetic acid ("decolorization"), this produced an intense blue-violet nuclear staining. He also observed that, with certain cell types such as plasma cells (discovered by his mentor Heinrich von Waldeyer-Hartz [1836-1921]), there was staining of the cytoplasmic granules. Ehrlich became fascinated with the concept of cytoplasmic "granulation" but realized it was important to distinguish true cytoplasmic granules from artifactual aggregates of cytoplasmic material. Using basic aniline dyes he was able to demonstrate that various connective tissue cells appeared to be packed with acidic granules, and so he proposed the name "mast cell" (from the German *mästen*, meaning "to fatten or force-feed") (18). He then turned his attention to the staining of blood cells with dyes. This was initially difficult since the chemical fixatives used

Gottlieb Gluge
(1812-1898)

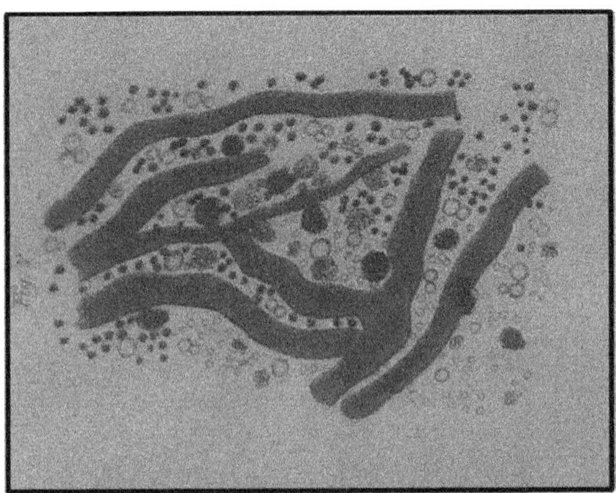

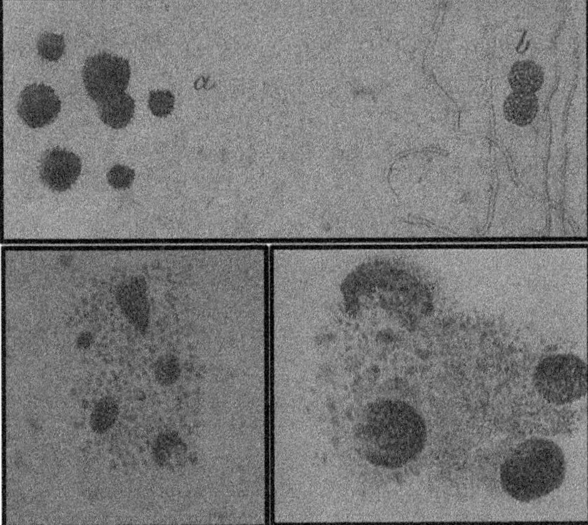

Figure 7 Gottlieb Gluge was possibly the first to describe granular cells, which he referred to as "inflammatory globules." He proposed that blood corpuscles "metamorphosed" into inflammation globules. The cells shown in the top right panel were believed to represent stages in this process. However, they have the appearance of intact or degranulated eosinophils. Further examples of Gluge's "inflammation corpuscles" are shown in the middle right and bottom panels. Reprinted from (26), with permission.

at the time destroyed intracellular granules. When he used a simple air-drying method on blood that had been spread on the slide as thinly as possible, he was able to observe "fine structural elements" in white blood cells. Further exposure of the air-dried preparations to heat gave even better staining clarity. Ehrlich was able to obtain a large number of acid and basic aniline coal tar dyes (courtesy of his pharmacist friend Herr Frank) and embarked on experiments which were to lead to the discovery of the eosinophil, neutrophil, basophil, and lymphocyte.

Initially his attention was taken largely with the eosinophil (originally called the "acidophil" since the granules stained with >30 of the acid dyes tried out). Surprisingly, the first published work on the eosinophil ("Contribution to Knowledge of Granulated Connective Tissue Cells and of Eosinophil Leukocytes," presented to the Physiological Society of Berlin on January 17, 1879) did not mention the cell at all (19), and so it was assumed that the section on the eosinophil had been deleted. It was in his next paper ("About the Specific Granulations of the Blood"), also based on a presentation to the Physiological Society of Berlin in 1879, that the eosinophil was mentioned for the first time (20). When discussing granule types in the white cells of vertebrates he wrote: "The most important of these granulations by far is the eosinophil or alpha granulation, about which I have already spoken before this society on the 17th January of this year. These alpha granules are characterised by their affinity for a wide

Julius Vogel
(1814-1880)

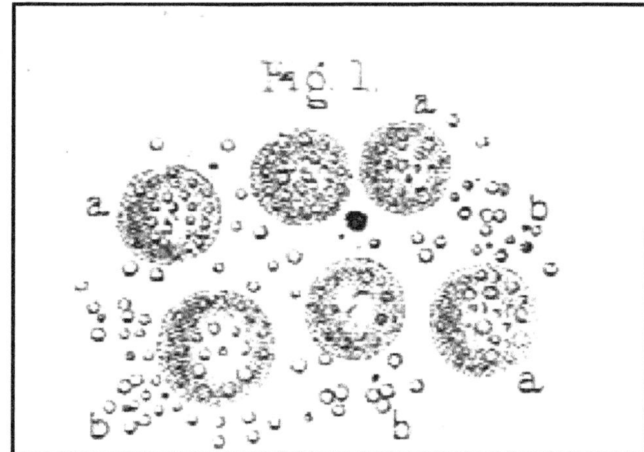

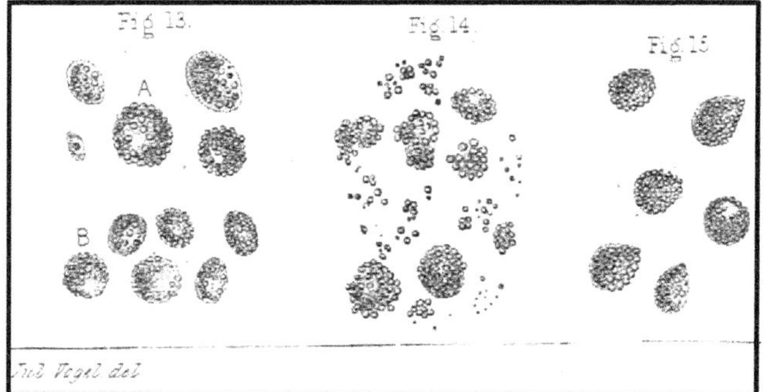

Figure 8 Julius Vogel made extensive drawings of granular cells in inflammatory exudates. The right panels are examples of Vogel's "pus corpuscles," which he also refers to as "granular cells" or "Gluge's compound inflammatory globules." They bear a striking resemblance to eosinophils. Reprinted from (13).

range of acid coal tar dyes." Ehrlich not only identified the eosinophil but described many of its features in detail and speculated, to the most part correctly, on its formation and function. For example, he noted that the alpha granules (more commonly referred to now as specific or crystalloid granules) were either round or had the shape of short rods with rounded ends, that there was variation in the number of nuclear lobes, and that the number of granules fluctuates from cell to cell. Ehrlich observed eosinophils in all species of animals he studied and showed that there were large numbers in the bone marrow and suggested that this was the site of their formation. Moreover, he noticed a second type of granule in these bone marrow-derived eosinophils that stained black with an eosin-indulin-glycerin stain. He called these "beta granules" and thought they probably represented a developmental stage in the formation of alpha granules. It would be reasonable to speculate that the beta granules were immature crystalloid granules (sometimes called primary granules), which are well documented in eosinophil myelocytes and appear blue or purple with differential stains such as Giemsa.

At a lecture he gave in 1900 at an International Congress of Medicine in Paris, Ehrlich speculated that leukocyte granules were secretory products and "extruded by the protoplasm which contained them" (21). He also suggested that "the phenomenon of eosinophilia is dependent on the circulation of a substance which has a chemotactic action on eosinophils, and which serves to release preformed eosinophils from the bone marrow into the blood." Thus Ehrlich described the cell and the staining properties of its granules, studied its distribution in various species and tissues as well as its morphology and formation in the bone marrow, and speculated correctly that the granule contents were secretory products. He studied the distribution of the cell in various species and in tissues and commented on the physiological aspects of the composition of the granule. He also described various causes of eosinophilia (i.e., asthma, various

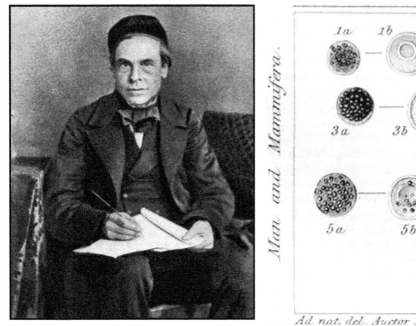

Figure 9 Thomas Wharton Jones made an extensive study of granular cells in a wide variety of species. His drawings show coarsely granular cells in the blood of human, horse, and elephant. Reprinted from (26), with permission.

skin diseases, helminths, postfebrile states, malignant tumors, as well as reactions to medications). Ehrlich also considered that myeloid and lymphocyte development was separate.

Having described the eosinophil in some detail, Ehrlich then turned his attention to the neutrophil. He demonstrated that the coal tar dyes could be divided into two general classes: basic and acidic for identifying granules in mast cells and eosinophils, respectively. However, he noticed that "most of the leukocytes display an affinity for neither acid or basic dyes in their granules." He then attempted to make neutral dyes, initially unsuccessfully because of solubility problems. Ehrlich was able to overcome the problem of precipitation by using a slight excess of acid and initially used a combination of methylene blue and acid fuchsin. He eventually developed triacid (orange G, acid fuchsin, and methyl green) to simultaneously identify acid, basic, and neutral granules (Fig. 11) (22). Ehrlich used the term "epsilon" for the neutrophil granule and established that they were not protein or fat deposits.

Using these fixation and staining methods Ehrlich was able to give a remarkably accurate picture of neutrophil development ("polynuclear cells develop by the progressive metamorphosis of the mononuclear elements").

The basophil leukocytes were also first observed by Ehrlich in 1891 in the blood of a patient with myeloid leukemia (23), and because of their resemblance to tissue mast cells, which he had previously described and differentiated from the plasma cells of Waldeyer, he called these "blood mast cells." Considerable controversy arose over the relationship between these two types of cells. Ehrlich's conviction that the blood mast cell and the tissue mast cell were distinct has now been amply substantiated since it is now recognized that, while both are of myeloid origin, they have different hematopoietic lineages.

Paul Ehrlich was a monumental figure in the fields of immunology, hematology, oncology, and chemotherapy (Fig. 1 and 11). His childhood was not especially remarkable. He was the son of a distiller, but his mother's cousin was the famous pathologist Carl Weigert, who inspired Ehrlich in the staining of microscopic tissue sections (24). As was the custom, Ehrlich attended many universities and after graduation became assistant physician to Professor Frerichs at the

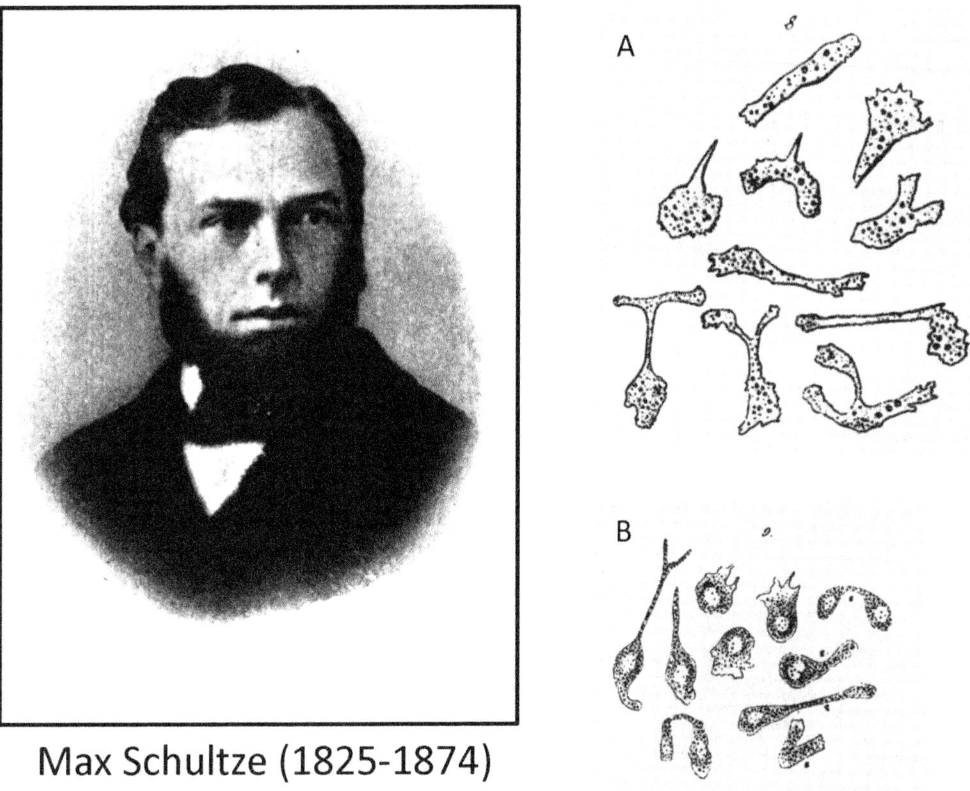

Figure 10 Max Schultze performed functional studies on finely (A) and coarsely (B) granular cells. The cells were observed on a warm stage at 38°C. Reprinted from (16), with permission.

Charité Hospital in Berlin. From the outset his heart was in research, although he managed a heavy clinical load. Frerichs was very supportive of Ehrlich and recognized his exceptional gifts. Tragically, Frerichs committed suicide in 1885 and was succeeded by Professor Gerhardt, who, in contrast, was unsympathetic to Ehrlich's research. Added to this, Ehrlich developed a persistent cough and found tubercle bacilli in his own sputum. He and his wife went to recuperate in Italy and Egypt, but on his return to Berlin in 1889 he was essentially out of a job. With help from his wife's parents, he set up his own laboratory in an apartment until Robert Koch secured a post for him at Moabit Hospital in Berlin. Later Koch invited Ehrlich to join the newly founded Institute for Infectious Diseases (Institut für Infektionskrankheiten) but with no official appointment or salary. In 1899, Ehrlich moved to Frankfurt am Main to the Institute of Experimental Therapy (Institut für experimentelle Therapie), and in 1906 he became the director of the Georg Speyer House in Frankfurt, a private research foundation affiliated with his institute.

In 1908, Ehrlich shared the Nobel Prize in Physiology and Medicine with Ilya Metchnikov (1845-1916) for their contributions to immunology. This wise decision by the Nobel committee in recognizing the contribution of both humoral and cellular immunity in the fight against infection had been preceded by years of rancor. Despite Ehrlich's work on the identification of white blood cells, he was, at the time, the leader of the humoral theory. The work of Robert Koch, Hans Buchner (1850-1902), and Emil von Behring (1854-1917) on antitoxins and the experiments of Richard Pfeiffer on the extracellular killing of *Vibrio cholerae* provided strong evidence in favor of antibody-mediated immunity without the requirement for cells. Metchnikov, on the other hand, working at the Institut Pasteur, had brilliantly defined the role of monocytes, macrophages, and neutrophils in inflammation and innate immunity and passionately argued his "theory of phagocytes." Reconciliation, at least in part, was eventually provided by Edward Almroth Wright (1861-1947), who discovered the phenomenon of opsonization, which he aptly defined as: "The body fluids modify

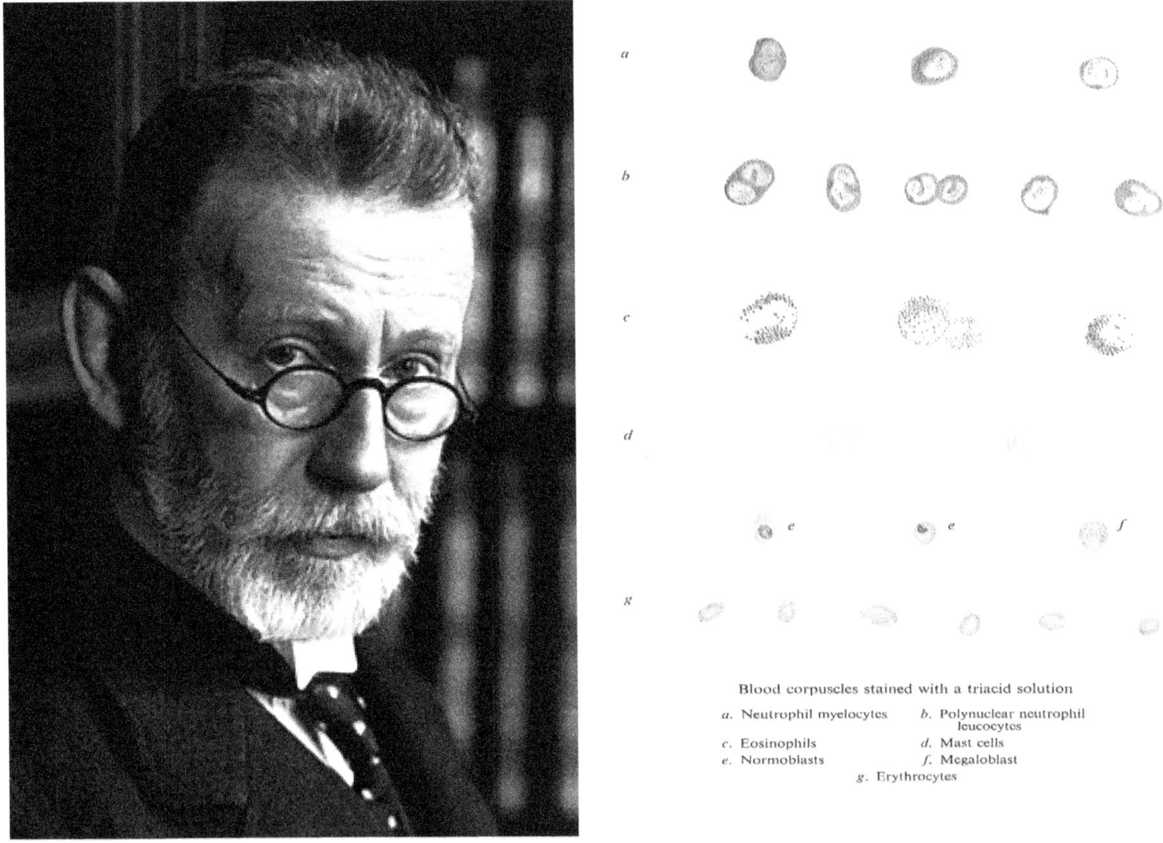

Figure 11 Paul Ehrlich in 1908. Color plate from *Histology of the Blood, Normal and Pathological* (22). Triacid (orange G, acid fuchsin, and methyl green) was used as a differential leukocyte stain. Reprinted from (26), with permission.

bacteria in a manner which renders them a ready prey to phagocytes" (25).

Citation. Kay AB. 2016. Paul Ehrlich and the early history of granulocytes. Microbiol Spectrum 4(4):MCHD-0032-2016.

References

1. **Van Leeuwenhoek A.** 1674. Microscopical observations from Mr. Leeuwenhoeck, about blood, milk, bones, the brain, spitle, cuticula, sweat, fatt, teares; communicated in two letters to the publisher. *Phil Trans* **9:**121–128.
2. **Lieutaud J.** 1749. *Elementa Physiologiae*, p 82–84. Amsterdam, The Netherlands. (Later translated and quoted in Dreyfus C, 1957, *Some Milestones in the History of Hematology*, p 11–12, Grune and Stratton, New York, NY.)
3. **de Senac JB.** 1749. *Traite de la Structure du Coeur, de son Action, et de ses Maladies.* Jacque Vincent, Paris, France.
4. **Hewson W.** 1774. *Experimental Inquiries, Part I. A Description of the Lymphatic System in the Human Subject and Other Animals*, p 30. J. Johnson, London, United Kingdom.
5. **Diamantis A, Magiorkinis E, Androutsos G.** 2009. Alfred Francois Donné (1801-78): a pioneer of microscopy, microbiology and haematology. *J Med Biogr* **17:**81–87.
6. **Thomas X.** 2013. First contributors in the history of leukemia. *World J Haematol* **2:**62–70.
7. **Addison W.** 1840-1841. Colourless globules in the buffy coat of the blood. *Lond Med Gaz* **27:**524–527.
8. **Virchow RL.** 1846. Weisses Blut und Milztumoren. *Med Zeitung* **14:**157–163.
9. **Rather LJ.** 1972. *Addison and the White Corpuscles: an Aspect of Nineteenth-Century Biology.* Wellcome Institute for the History of Medicine, London, United Kingdom.
10. **Von Recklinghausen FD.** 1863. Über Eiter- und Bindegewebskörperchen. *Virchows Arch Pathol Anat* **28:**157–197.
11. **Wohlgemuth B, Borte G.** 1989. The 150th birthday of Julius Cohnheim. *Z Arztl Fortbild (Jena)* **83:**743–745. (In German.)
12. **Gluge G.** 1843. *Atlas der Pathologischen Anatomie*, Mauke, Jena, Germany. (Translated from French by Blanchard and Lea, Philadelphia, PA, 1853.)

13. **Vogel J.** 1847. *The Pathological Anatomy of the Human Body* (Day GE, transl). Blanchard and Lea, Philadelphia, PA.
14. **Wharton Jones T.** 1846. The blood-corpuscle considered in its different phases of development in the animal series. Memoir I. Vertebrata. *Philos Trans R Soc Lond* **136:**63–87.
15. **Schultze MJ.** 1865. Ein heizbarer Objecttisch und seine Verwendung bei Untersuchungen des Blutes. *Arch Mikrosk Anat* **1:**1–42.
16. **Brewer DB.** 1994. Max Schultze and the living, moving, phagocytosing leucocytes: 1865. *Med Hist* **38:**91–101.
17. **Bizzozero G.** 1881. Su di un nuovo elemento morfologico del sangue dei mammiferi e sulla sua importanza nella trombosi e nella coagulazione. *Osserv Gazz Clin* **17:**785–787.
18. **Ehrlich P.** 1878. *Beiträge zyr Theorie und Praxis der histologischen Färbung. Doctoral thesis dissertation.* University of Leipzig, Leipzig, Germany.
19. **Ehrlich P.** 1879. Beiträge zur Kenntnis der granulirten Bindegewbszellen und der ecosinophilen Leukocythen. *Arch Anat Physiol (Leipzig)* **3:**166–169.
20. **Ehrlich P.** 1879. Über die specifischen Granulationen des Blutes. *Arch Anat Physiol (Leipzig)*571–579.
21. **Ehrlich P.** 1900. *La Leukocytose.* X11le Congres Internationale de Medicine, Paris 1900.
22. **Ehrlich P, Lazarus A.** 1900. *Histology of the Blood, Normal and Pathological* (Myers W, transl). Cambridge University Press, Cambridge, United Kingdom.
23. **Hirsch JG, Hirsch BI.** 1980. Paul Ehrlich and the discovery of the eosinophil, p 3–23. *In* Mahmoud AA, Austen KF (ed), *The Eosinophil in Health and Disease.* Grune and Stratton, New York, NY.
24. **Hüntelmann AC.** 2010. Legend of science. External constructions by the extended "family"—the biography of Paul Ehrlich. *InterDisciplines* **2:**13–36.
25. **Wright AE, Douglas SR, Sanderson JB.** 1903. An experimental investigation on the role of the blood fluids in connection with phagocytosis. *Proc R Soc Lond* **72:** 357–370.
26. **Kay AB.** 2015. The early history of the eosinophil. *Clin Exp Allergy* **45:**575–582.

Myeloid Cells in Health and Disease: A Synthesis
Edited by Siamon Gordon

doi:10.1128/microbiolspec.MCHD-0009-2015

Giuseppe Teti[1]
Carmelo Biondo[1]
Concetta Beninati[1]

2 The Phagocyte, Metchnikoff, and the Foundation of Immunology

INTRODUCTION

The life and work of Elie Metchnikoff are a rich source of inspiration to anybody interested in the biology and pathophysiology of myeloid cells. He made the fundamental discoveries that subsequently shaped the development of the field and that represent, still today, the basis of our knowledge. First and foremost, he defined these cells by their function (i.e., "phagocytosis"), a definition that suits better than any other designation the nature of these cells, including perhaps the term "myeloid" itself. Metchnikoff described for the first time a number of crucial features of phagocytic cells, including (i) phagocyte-mediated host protection; (ii) active internalization of live, in addition to dead, organisms; (iii) uptake of senescent or damaged host cells; (iv) destruction of internalized particles; (v) bacterial killing by virtue of enzymes ("cytases"); (vi) vacuolar acidification; (vii) distinction between microphages (polymorphonuclear leukocytes) and macrophages; (viii) inflammatory recruitment of phagocytes; (ix) chemotaxis; and (x) diapedesis. For these reasons, Metchnikoff is unanimously considered the founding father of the field of phagocyte biology.

However, since the ability of some cell types to actively engulf particulate material was observed in both invertebrates and vertebrates before Metchnikoff, he did not "discover" phagocytosis, as is sometimes mentioned in textbooks. His contribution to biology is far greater and extends beyond the field of phagocyte biology. By assigning to particle internalization the function of defending the host against noxious stimuli (this represented a new function relative to the previously recognized task of intracellular digestion), Metchnikoff envisioned for the first time the presence of an active body defense system and created the theoretical framework that led to the birth of immunology, an entirely new science. In this sense, Metchnikoff can be rightly viewed as the father of all immunological sciences and not only of innate immunity, cellular immunology, or phagocyte biology. Indeed, before Metchnikoff, immune phenomena were explained by "nonimmunological" mechanisms (1). For example, Louis Pasteur

[1]Metchnikoff Laboratory, Department of Pediatric, Gynecological, Microbiological and Biomedical Sciences, University of Messina, Messina, Italy.

believed that the resistance to infection he had observed in animals vaccinated with attenuated microbes was linked to the consumption by the latter of specific growth factors required for the reproduction of bacteria inside the body. Pasteur himself soon realized the fallacy of his interpretation when he observed growth of pathogens in the blood of immune animals (2). Metchnikoff's concept of immunity as an active body function derived directly from his embryological studies. Indeed, he brought a fresh biological perspective to the field of medical pathology, which made it possible to assign to inflammatory phenomena a new functional significance. In addition, the recognition properties of Metchnikoff's phagocyte fit surprisingly well with recent discoveries and modern models of the "immune self" (3–5). For example, rather than assigning to the immune system exclusively the function of eliminating nonself components (as others did for many years after him), Metchnikoff also endowed his phagocyte with the ability to check for the presence of unwanted or damaged endogenous components, i.e., to detect the "altered" self. We will review here the fascinating story of how myeloid cells and their evolutionary ancestors inspired Metchnikoff's theoretical achievements. The historical and philosophical aspects of his discoveries have been the subject of extensive research by Alfred Tauber and coworkers (6, 7), to whom we owe some of the concepts presented here. In addition, Metchnikoff's life and discoveries have been the subject of several excellent accounts (8–16).

FATE OF THE PHAGOCYTE THEORY

Metchnikoff became a famous scientist in the 1890s after his theory had struck popular imagination by depicting armies of phagocytes moving against infectious agents to destroy them and save the body from deadly diseases. His phagocyte theory provided a vivid representation of immunity at work and a simple explanation for the resistance of some individuals to contagious infections despite exposure during epidemics. Moreover, the powerful and eccentric personality of Metchnikoff, who was heavily influenced by Mitteleuropean 19th-century romanticism, has lent itself to a number of picturesque portraits. He was depicted sometimes as a "mad scientist" battling relentlessly to defend his theory from the attacks of his detractors (17). Indeed, his phagocyte or "cellular" theory (presented in 1883) came immediately under ferocious criticism even before alternative explanations were offered. This occurred only in the late 1880s with the formulation of the humoral theory of immunity, according to which soluble factors present in serum and secretions (later identified as antibodies)—and not cells—were exclusively responsible for immunity. Great German scientists, such as Emil von Behring, Richard Pfeiffer, and Paul Ehrlich, championed the humoral theory, while French immunologists took sides with Metchnikoff, who was working at the Institut Pasteur at the time. The debate took belligerent tones and was influenced by the heated atmosphere of nationalism that followed the Franco-Prussian War (1870–71). Finally, after Almroth Wright showed that humoral factors (i.e., opsonins) could increase the susceptibility of bacteria to phagocytosis (18), the 1908 Nobel Committee declared a sort of cease-fire by awarding the Nobel Prize jointly to Metchnikoff and Ehrlich. Clearly, cellular and humoral theories were not mutually exclusive. Important observations were made in the 1920s and '30s showing that blockade of the phagocyte system (which had been renamed the "reticuloendothelial system") by India ink or quartz particles reduced antibody formation (19–24). Despite this, the popularity of the cellular (or phagocyte) theory steadily declined during the 20th century as a result of difficulties in demonstrating immune specificity in the action of phagocytes. At the same time, biochemistry was taking the center stage in biology, and immunochemistry was increasingly successful in demonstrating the structural basis of antigen-antibody interactions. The success of "cellular immunology" in the '70s was mostly associated with excitement about the role of lymphocytes, and Metchnikoff's contributions continued to be relegated to history libraries. Only quite recently, after the rediscovery of the centrality of innate immunity in human health and disease and throughout evolution, was Metchnikoff's phagocyte vindicated as the initiator and orchestrator of immune responses. Yet Metchnikoff's image still suffers today from the same stereotype that originally made him famous as the vociferous protagonist of the humoral/cellular controversy. Not many realize that he was the first to bring into the context of medical sciences an evolutionary perspective that has had a major impact lasting until today. As a comparative embryologist, Metchnikoff had a different scientific background from that of Ehrlich and von Behring, who were physicians, or from that of other medically oriented, contemporary microbiologists and pathologists. Both biology and immunology owe a lot to Metchnikoff's original reinterpretation of Darwinian evolutionist principles: in recognizing the limits of a purely morphological approach, he focused on how a given activity (specifically, particle internalization) acquires new functional meanings through phylogenesis.

He concluded that, in higher animals, particle internalization had lost its evolutionarily ancient function (nutrition) to fulfill the new tasks of tissue remodeling and elimination of potentially noxious agents. Studying a conserved activity or structure (such as a gene) from the perspective of its functional adaptations during evolution would seem today an obvious strategy. However, few realize that Metchnikoff was the first to fully demonstrate the power of this approach. In addition, as an embryologist, he was quick to extrapolate his conclusions from phylogeny to ontogeny. For example, he noted that the same activity (phagocytosis) was used for different purposes (respectively, resorption of the tadpole tail and antimicrobial defenses) during metamorphosis and adult life (25). Metchnikoff would have been delighted, but by no means surprised, to learn about the different functions of the *Drosophila* Toll pathway in the embryo (establishment of dorsoventral patterning [26]) and in the adult fly (host defense against infection [27]). To him, Toll and Toll-like receptor activities would have represented the perfect articulation of the basic function of his phagocyte, namely, to shape the identity of life during individual development and to preserve it later.

METCHNIKOFF'S PERSONALITY

In the case of Metchnikoff, it is difficult to understand the nature of his discoveries without referring to his life and personality. For example, pessimistic or optimistic feelings about his personal life heavily influenced his scientific ideas. During the first half of his existence, he was often affected by misanthropy and depression and saw in living creatures—including himself—contrasting and disharmonious features that worried him. This was a different attitude from that of many naturalists of his time, who were inclined to admire the functional perfection of the organisms they were studying. The awareness of disharmony in nature generated a strong need in Metchnikoff to find some counterbalancing force that could be leveraged by science to ultimately solve the problems affecting mankind. Once he was convinced, after 1881, that he was on the right track toward defining such a harmonizing principle (which he later identified in the phagocyte), he became more optimistic and found new motivations for existence. Science was a religion to him and totally shaped his life. In this attitude he was not alone, and a similar romantic orientation can also be found in many of his contemporaries and in the scientists who preceded him, including Pasteur and his coworkers. Metchnikoff showed, from adolescence, a compulsory need to learn all he could about biology and nature using direct observation, books, visits to scientists, and attendance at scientific meetings throughout Europe. He continuously attempted to integrate the enormous body of knowledge he was acquiring into a unifying theory, and failure to find a satisfactory explanation for scientific evidence invariably threw him into a state of anxiety and despair. The need to resolve this state of unhappiness, which originated from frustration with his work, was the driving force behind the development of the phagocytosis theory, in which he finally found intellectual satisfaction. Since this achievement was his raison d'être, he always vigorously defended the theory against any criticism. It is not surprising that he wrote in 1913: "The controversy over phagocytosis could have killed me, or permanently weakened me sooner. Sometimes, (I remember such attacks of Lubarsch in 1889, and those of Pfeiffer in 1894) I was ready to get rid of life" (15).

Yet he managed to maintain a friendly and respectful attitude toward his critics, as shown by his letters to Ehrlich, who visited him at Institut Pasteur in 1903 (15). Satisfaction with his scientific work was a necessary, but insufficient, condition for Metchnikoff's happiness in life, and a second "requirement" had to be met, namely the love and dedication of a feminine figure, which he found first in his mother and then in his wives. His two suicide attempts were driven by the loss of his first wife and by a serious disease of his second. His second wife, Olga, gave him happiness and all the support he needed to carry on his work until the end of his life. Her biography of Metchnikoff is a pleasant read that vividly and faithfully pictures his complex personality and the mental processes behind his discoveries (28).

Elie Metchnikoff (also spelled Ilya Mechnikov) was born in 1845 in his family estate near the village of Ivanovka in the Governorate of Kharkov (Kharkiv), Little Russia (Ukraine), then a province of the Russian Empire. He was the last of five children. His father (Fig. 1) was a middle-class aristocrat of Moldavian descent and a retired Imperial Guard officer, who apparently had only a minor role in Elie's education. His mother, Emilia Lvovna née Nevakhovich (Fig. 1), the beautiful and intelligent daughter of a converted Jewish entrepreneur, had instead a great influence on Metchnikoff, who constantly referred to her even when an adult. Elie spent his childhood in the family estate. As a child, he was very active, sensitive, demanding, and manipulative (his mother defined his temperament as "neurotic"). As we will see, this personality persisted throughout his life: he often showed frustration with the smallest complication and had

Le père
Ilya Ivanovitch Metchnikoff
La mère
Emilia Lvovna Panasovka

Figure 1 Metchnikoff's parents. Reproduced from reference 10, with permission.

difficulties in coping with problems at work, such as academic restrictions and less-than-ideal research facilities, which led to resignation from his position several times during his life.

THE EMBRYOLOGIST

Metchnikoff was tutored at home under the attentive supervision of his mother, who loved him dearly and chose for him the best local teachers. As a child, he showed great interest in the animals and plants he observed in his native land and in illustrations. Elie entered the Kharkov Lycée (a school with progressively oriented teaching) when he was 11 and soon concentrated on natural history, botany, and geology, brilliantly completing his studies at 17. He was extremely intelligent, active, and talented, with a prodigious memory and imagination. In order to read the original works of German philosophers and scientists, he learned German when he was 14 and at 18 became acquainted with Darwin's *Origin of Species* and Rudolf Virchow's contributions to the cell theory. At 16, he published in the *Journal de Moscou* a critical review of a geology book written by a professor of Kharkov. After exiting the Lycée with a gold medal, Metchnikoff hastened to complete his studies at Kharkov University (which he disliked) and published, at 18, his first research article on *Vorticella*. In this paper, he compared the pseudopod, which functions as a stalk in this protozoan, with the vertebrate skeletal muscle (he concluded that there was no analogy, provoking a ferocious reaction from Professor Kuehne, a celebrated physiologist). At age 19, soon after completing his university studies, the young scientist felt a strong need to visit research laboratories throughout Europe. He visited the Universities of Giessen, Gottingen, and Munich and the marine stations of Heligoland Island and Naples, where he could find collections and fresh samples of different kinds of invertebrates. He met in Naples another young Russian, Alexander Kowalevsky, who was 5 years older and had already started studies in comparative embryology. His friendship with Kowalevsky was an important factor in Metchnikoff's decision to concentrate on invertebrate embryology, which he viewed as an ideal tool to identify similarities between different species. In 1865, while in Giessen, Metchnikoff observed particle internalization in protozoa (called "infusoria" at the time) and in the primitive gut of the flatworm *Geodesmus bilineatus*, confirming previous observations conducted by Lieberkuehn in sponges 10 years earlier. After these observations in Giessen, Metchnikoff continued his studies on the development of a wide range of invertebrate species, focusing on primary embryonic layers, and—notably—did not show any further interest in particle internalization. Only in the late 1870s, for reasons that will be apparent below, did he resume his interest in particle internalization, which ultimately inspired his great discovery.

In 1867, Metchnikoff returned to Russia and soon moved to St. Petersburg to defend his doctoral thesis on primary embryonic layers in invertebrates. He was also awarded there the Baer Prize, which he shared with Kowalevsky (Carl von Baer was an authoritative Estonian embryologist who taught from 1834 to 1862 in St. Petersburg). At age 22, Metchnikoff was appointed as docent at the University of Odessa, but he soon entered into conflict with academic authorities and moved to St. Petersburg University. In 1868, he again visited Italy (where he found Kowalevsky), touching Naples, Reggio Calabria, Messina, and Trieste. Back in St. Petersburg, he became fond of the young daughters of the professor of botany, Beketov, and started to conceive the rather peculiar idea of training one of them to conform to his feminine ideals, in order to subsequently marry her. Having failed, he married a friend of the Beketovs, Ludmila Vassilievna Fedorovitch, who was roughly as old as him and was already seriously ill with tuberculosis. Indeed, when they married in 1869, the bride could not walk or stand because of breathlessness and was taken to church on a chair. Metchnikoff's marriage was very unhappy because of the illness of his wife and serious financial difficulties. She died in Madeira in 1873, which left Metchnikoff in a state of deep depression and despair, resulting in a suicide attempt with opium. After recovering, Metchnikoff lived and worked in Odessa, where he continued to suffer from depression, pessimism, and misanthropy. He had joined the faculty of the University of Odessa in 1870 and continued to teach there until 1882. During his early life and until 1881, Metchnikoff was pessimistic not only about his life but also about the nature of humans in general. Moreover, as mentioned above, he saw a number of biological incongruities in many creatures. Luckily, in Odessa, he soon fell in love with one of the daughters of his neighbors, the beautiful 15-year-old Olga Belokopytova (Fig. 2 and 3), to whom he offered to give private lessons in zoology. He married Olga in 1875 and lived with her happily for the rest of his life. Olga immediately showed a strong devotion to her husband and helped him with his work by preparing illustrations and translating articles.

After marrying Olga, particularly from the late 1870s to the early 1880s, he started to take a novel evolutionary approach toward the solution of the problem at which he had worked for the first half of his scientific life, namely, the development of the primary embryonic layers in invertebrates. He focused on the function and fate of the mesoderm, which he believed to have a crucial role in gastrulation in invertebrates. In those years, a controversy had been going on between him and Ernst Haeckel concerning the hypothetical progenitor of multicellular organisms. Metchnikoff hypothesized that the first metazoan (which he named "*parenchymella*") was similar to the larvae of the most primitive invertebrates, the sponges, which are made up of a solid internal mass (parenchyma) of larger cells surrounded by smaller, externally flagellated cellular elements. Incidentally, the term "*parenchymella*" is still

Olga Nikolaevna Belokopitova
lycéenne jeune mariée (1874)

Figure 2 Olga Metchnikoff. Reproduced from reference 10, with permission.

Figure 3 Elie and Olga Metchnikoff. Reproduced from reference 15, with permission.

used today to designate the larvae of *Demospongiae*, the largest and most ancient class of sponges, or *Porifera*. Haeckel had proposed instead that the first multicellular organism was similar to the invaginated gastrulas that Kowalevsky had observed in primitive chordates. He called this hypothetical metazoan progenitor "gastrea." The controversy with Haeckel had a role in the genesis of the phagocyte theory, because it helped Metchnikoff reach the following conclusions: (i) the inner *parenchymella* mass contained mobile, or "wandering ameboid," cells originating from the mesoderm and capable of taking up particulate material; (ii) the wandering cells served as nutritive cells in animals without a gut, such as sponges; and (iii) in higher invertebrates the mesoderm gave rise to the wandering ameboid cells, the locomotion apparatus, and the circulatory system. The entoderm gave origin to a well-developed intestine capable of extracellular digestion and absorption.

In 1880, Olga had typhoid fever and Metchnikoff went through a stressful situation, as described in her book: "Though worn out with devoted nursing, he tried to make up the time lost to research and overworked himself, with the result that cardiac trouble was followed by fits of giddiness and unconquerable insomnia. He fell into such a state of neuroasthenia that, in 1881, he resolved in a moment of depression to do away with his life" (28).

Curiously, having decided to die, he thought that he could use that circumstance to solve a scientific problem, namely, to determine whether relapsing fever was transmissible with blood. According to Olga, he did this also to "spare his family from an obvious suicide." At any rate, he injected himself with the blood from a patient with relapsing fever, contracted the disease, and almost died. Strangely, after recovering, he underwent a sort of physical and psychological resurrection: the eye problems that had tortured him for most of his life suddenly disappeared and he found new vitality. This resulted in a period of intense work that led to his most important discovery, which occurred in Messina, and to the formulation of the phagocytosis theory.

THE MESSINA DISCOVERY

After Olga's parents died in 1881 and 1882, Metchnikoff took care of his wife's family and properties. He managed to sell Olga's share of her father's land and to leave in the hands of her older brother the care of the remaining part of the property. At the same time, he also got rid of his share of the land property near Ivanovka. In 1882, having resigned from the University of Odessa because of the political turmoil that followed the assassination of Czar Alexander II, he was finally free to realize his dream to reach the Sicilian town of Messina. He wrote about this town in 1908, after a terrible earthquake had destroyed it: "Thus it was in Messina that the great event of my scientific life took place. A zoologist until then, I suddenly became a pathologist. I entered a new road in which my later activity was to be exerted. It is with warm feeling that I evoke that distant past and with tenderness that I think of Messina, of which the terrible fate has deeply moved my heart" (28).

In her biography of Metchnikoff, Olga vividly describes their life in Messina (28):

> At Messina we settled in a suburb, the Ringo, on the quay of the Straits, in a small flat with a garden and a

splendid view over the sea. We did not have much room, and the laboratory had to be installed in the drawing-room, but, on the other hand, Elie only had to cross the quay in order to find the fisherman that provided him with the material needed for his researches and with whom he frequently went out sailing. Metchnikoff loved Messina, with its rich fauna and beautiful scenery. The splendid view of the sea and the calm outline of the Calabrian coast over the Straits delighted him.

The conceptual path that led Metchnikoff to the Messina discovery is deeply rooted in his embryological studies. As outlined above, Metchnikoff believed that nutrition in primitive invertebrates, such as sponges, occurred by intracellular digestion and was performed by mesodermic ameboid cells capable of ingesting food particles, moving around the body, and feeding other cells. In higher invertebrates, starting with the echinoderms (e.g., starfish and sea urchins), a well-developed intestine took up nutritional functions. What could then be the function of mobile "devouring" cells in the higher animals? These cells were no longer performing their original nutritive function, yet they retained their ability to ingest particulate material. While in Messina, Metchnikoff took advantage of the abundant and varied local marine fauna to undertake a systematic research program. First, he set out to confirm that the mesodermic ameboid cells could actively internalize particulate material and digest it: "I found it an easy matter to demonstrate that these elements seized foreign bodies of very varied nature by means of their living processes, and certain of these bodies underwent a true digestion within the amoeboid cells" (29).

Second, he sought to investigate the role of these cells during ontogenesis. After observing, in the larvae of *Synaptae* (a family of echinoderms), that the ameboid cells "accumulate and unite into masses" in the numerous organs that undergo atrophy during metamorphosis, he concluded that these cells had a causal role in tissue resorption during development (29). Notably, while in Messina, Metchnikoff began to study with great interest Ernst Ziegler's treatise on pathological anatomy (29), in order to gather more information on a process whose details were apparently unknown to him until quite recently: "some time before my departure from Messina, I listened to the reading of Cohnheim's treatise on General Pathology and I was struck by his description of the facts and of his theory on inflammation. The former, especially his description of the diapedesis of white corpuscles across the vessel wall seemed to me of momentous interest. His theory, on the other hand, appeared to be extremely vague and nebulous" (29).

The accumulation of white corpuscles in the extravascular space described by Ziegler and Cohnheim must have been reminiscent to Metchnikoff of the "masses" of ameboid cells that he was observing at the time in the atrophied organs of echinoderm larvae. These elements (i.e., his novel concern with inflammation and the congregation of ameboid cells in atrophied organs) led Metchnikoff to conceive a crucial experiment, which was performed in December 1882 and is vividly described in a famous account (28):

I remained alone with my microscope, observing the life in the mobile cells of a transparent star-fish larva, when a new thought suddenly flashed across my brain. It struck me that similar cells might serve in the defense of the organism against intruders. Feeling that there was in this something of surpassing interest, I felt so excited that I began striding up and down the room and even went to the seashore in order to collect my thoughts. I said to myself that, if my supposition was true, a splinter introduced in the body of a star-fish larva, devoid of blood vessels or of a central nervous system, should soon be surrounded by mobile cells as is to be observed in a man who runs a splinter into his finger. This was no sooner said than done.

There was a small garden in our dwelling, in which we had a few days previously organized a "Christmas tree" for the children on a little tangerine tree; I fetched from it a few rose thorns and introduced them at once under the skin of some beautiful star fish larvae as transparent as water.

I was too excited to sleep that night in the expectation of the results of my experiment, and very early the next morning I ascertained that it had fully succeeded.

That experiment formed the basis of the phagocyte theory, in the development of which I devoted the next twenty-five years of my life.

Was the phagocytosis theory the result of a spark of intuition or the logical development of a conceptual trajectory lasting several years? No doubt, the theory was based on a continuous intellectual effort going back to at least the mid-1870s. Yet the Messina studies reveal a conceptual shift that marked the beginning of his transition from zoology to pathology. Indeed, two novel elements are obvious in the December 1882 experiment: (i) the use of a *noxious*, as opposed to a *harmless*, stimulus (i.e., a rose thorn as opposed to colored particles) to provoke a response in the mobile cells; and (ii) the choice of cell *congregation* around the stimulus rather than particle *internalization* or *digestion* as the readout of such a response. Clearly, Metchnikoff applied to the research model he used at the time (direct observation of live invertebrate larvae) an experimental design used in vertebrates by Cohnheim, Ziegler, and other medical pathologists (induction of inflammation by splinters or croton oil). Metchnikoff and other naturalists before him had been using grains of carmine, indigo, or red blood cells for many years to observe particle internalization in invertebrates (Fig. 4). It never occurred to anybody that this process might represent a defense reaction against the particulate agents under study, and the phenomenon was always interpreted as serving a nutritional purpose or not explained at all. Now Metchnikoff addressed a different subject, inflammation, which was outside the realm of his own specialty. And he entered this new arena à la Metchnikoff. He immediately confuted the prevailing theory, championed by Cohnheim, that inflammation originated from a pathological process in blood vessels or, as others asserted, in nerve terminations. He showed (this is the third key element of the December 1882 experiment) that inflammation could be induced in animals, such as starfish larvae, which are devoid of vascular and central nervous systems. This paved the way for the new concept that cell accumulation during inflammation was an active leukocyte (or "ameboid cell") function, rather than the passive result of circulatory dynamics, as Cohnheim believed. Moreover, Metchnikoff's idea that inflammation was beneficial to the host was not prevalent at the time and was violently criticized. Pathologists were convinced that bacteria hijacked white blood cells to find further nourishment and to disseminate in the body. Virchow, who was visiting Messina at the time, viewed Metchnikoff's ideas favorably but warned him that his theory was in contrast with the prevailing opinion on the effects of inflammation.

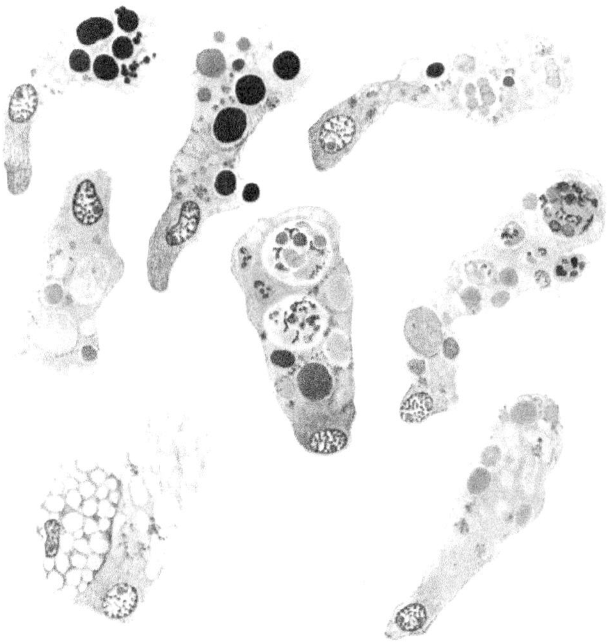

Figure 4 Columnar cells from a flatworm showing intracellular digestion in planariae. Reproduced from reference 44.

In the summer of 1883, Metchnikoff moved to Riva del Garda in northern Italy, where he wrote an article on his new ideas. Returning to Russia, Metchnikoff stopped in Vienna on the way and presented his theory to Claus, the local professor of zoology, who suggested the term "phagocyte" for the "devouring" cells, from the Greek *phagein* ("eat") and *kutos* ("hollow vessel or cell"). Ultimately, in 1883, Metchnikoff presented at a naturalists' meeting in Odessa his first paper on phagocytosis. When Metchnikoff's theory of inflammation came to the attention of professional pathologists, it provoked a violent and persistent reaction. The objections varied widely in nature, but the more difficult to answer came from reductionists (e.g., Baumgarten) claiming that Metchnikoff's theory lacked mechanistic physicochemical evidence and was based on vitalistic and teleological notions. In other words, to his critics, Metchnikoff had arbitrarily endowed the phagocyte with unexplained vitality and purpose to defend the body against infections. He managed, however, to successfully defend his theory by continuously providing new biological, if not physicochemical, evidence in favor of it.

THE PATHOLOGIST

The Messina discovery had a marked influence on Metchnikoff himself, who became more optimistic and set out to find further proof of his hypothesis. In his first efforts as a pathologist, Metchnikoff worked with the small freshwater crustacean *Daphnia*, which was the victim of a fungal infection under natural conditions. He observed first that spindle-shaped fungal spores penetrated the intestinal wall and reproduced in the body. He also noticed that the phagocytes of the crustacean attacked the fungal cells. He turned to anthrax bacilli and found that the phagocytes could not attack the more virulent strains (or the spore form of the bacterium), while they could destroy the less virulent ones.

After a difficult period in Odessa, where he was appointed director of an institute established in 1886 to carry out Pasteur's vaccine treatment of rabies and other diseases, in 1888 he resigned and traveled through Europe. In Paris he met Pasteur, who showed appreciation for his ideas. Pasteur invited Metchnikoff to join the Institut Pasteur, where he remained to work for the rest of his life (Fig. 5 to 8).

At Pasteur, Metchnikoff was engaged, during the 1890s, in work aimed at disproving the arguments of his critics and at confirming his theory of phagocyte-mediated immunity. He published two important books summarizing the work developed in this period: *Lectures on the Comparative Pathology of Inflammation* (1892) (30) and *Immunity in Infective Diseases* (1901) (29). In 1903, in collaboration with Emile Roux, he showed that syphilis could be transmitted to anthropoid monkeys (reproducible animal models of the disease were not available at the time) and could be prevented by the topical application of mercurials (e.g., calomel) after inoculation with infectious material.

Figure 6 Elie Metchnikoff and Alexandre Besredka, Institut Pasteur, 1914. Besredka was a medical doctor from Odessa who collaborated with Metchnikoff at the Institut Pasteur from 1897. Reproduced from reference 10, with permission.

These experiments were confirmed in 1906 on a volunteer, the medical student Paul Maisonneuve. This represented the first successful attempt to prevent syphilis after exposure and the beginning of chemotherapy for the disease, although the discovery was soon overshadowed by the introduction of Ehrlich's arsphenamine (Salvarsan) in 1910.

In those years, Metchnikoff was also attempting to extend the boundaries of his phagocyte theory in an effort to ameliorate the consequences of senescence. He proposed that toxins produced by the intestinal bacteria responsible for the putrefaction of food residues were absorbed into the body and damaged host cells. Phagocytes, in turn, in an effort to limit the consequence of chronic cell damage, were ultimately responsible for the body changes (including graying of hair) associated with senescence. He proposed that the process of senescence could be slowed down by a diet that could replace, at least in part, the endogenous gut flora, leading to a healthier state that he called "orthobiosis." He successfully promoted, to this end, the use of fermented milk products, such as yogurt, because of their high content in lactobacilli. Yogurt consumption thereafter became widely popular, giving rise to a new industry. From 1913, Metchnikoff suffered from several bouts of heart failure, which remained compensated until 1916. He died on July 16, 1916, in the apartment originally occupied by Pasteur at the Institut Pasteur. Milestones in the life of Metchnikoff are reported in Fig. 9.

Figure 5 Elie Metchnikoff at 46 years of age. Reproduced from reference 10, with permission.

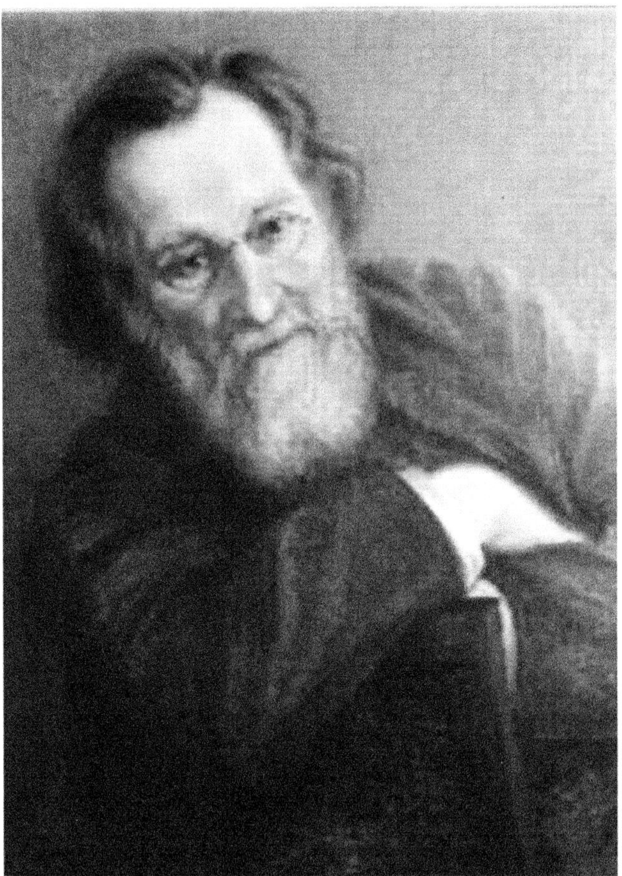

Figure 7 Elie Metchnikoff portrait painted by Olga. Reproduced from reference 15.

THE IMPORTANCE OF SELECTIVE EATING

In conclusion, Metchnikoff pioneered research on immunity, evolutionary developmental biology, aging, intestinal microbiota, and probiotics, to cite only a few research areas. However, his major scientific contribution lay in the novel approach he successfully used to answer basic biological questions. As an embryologist, he went to the heart of the matter and set out to identify the mechanisms that determine the shape and identity of living creatures. This was a formidable undertaking in an era in which the principles of modern genetics were still very far from being defined. Focusing on the role of primary embryonic layers in the development of invertebrate animals, he looked at how these structures "reinterpreted" their basic functions in different settings (e.g., in higher versus lower invertebrate physiology). In this approach, a crucial choice was to focus on nutrition, which Metchnikoff considered the most ancient and fundamental biological activity of all. It was at this point that the ameboid myeloid cell ancestors bewitched him. To him, they became not only an essential marker of the mesoderm but, more importantly, a tool to discern how an old function (nutrition by intracellular digestion) could be adapted to new needs during evolution. This was an extraordinary intuition and a new way to approach a biological problem. By observing them in a wide variety of higher animals, Metchnikoff realized that ameboid cells had the ability to move freely around the body and interact with other cell types, unbound by any obligations to perform a specific function. He believed that these "communication skills," which derived from their primitive activity of feeding other cells after intracellular digestion, enabled phagocytes to modulate the otherwise conflicting activity of other cell types, thereby integrating or harmonizing the function of different body components into a coherent "plan." The ameboid cells retained their primitive "eating" functions, but now with the new task of eliminating unwanted, damaged, or senescent cells of the body or tissue debris (in a function that Metchnikoff called "physiological inflammation"). This conclusion anticipated by more than a

Figure 8 Robert Koch visiting the Institut Pasteur, accompanied by Elie Metchnikoff (1904). Reproduced from reference 15, with permission.

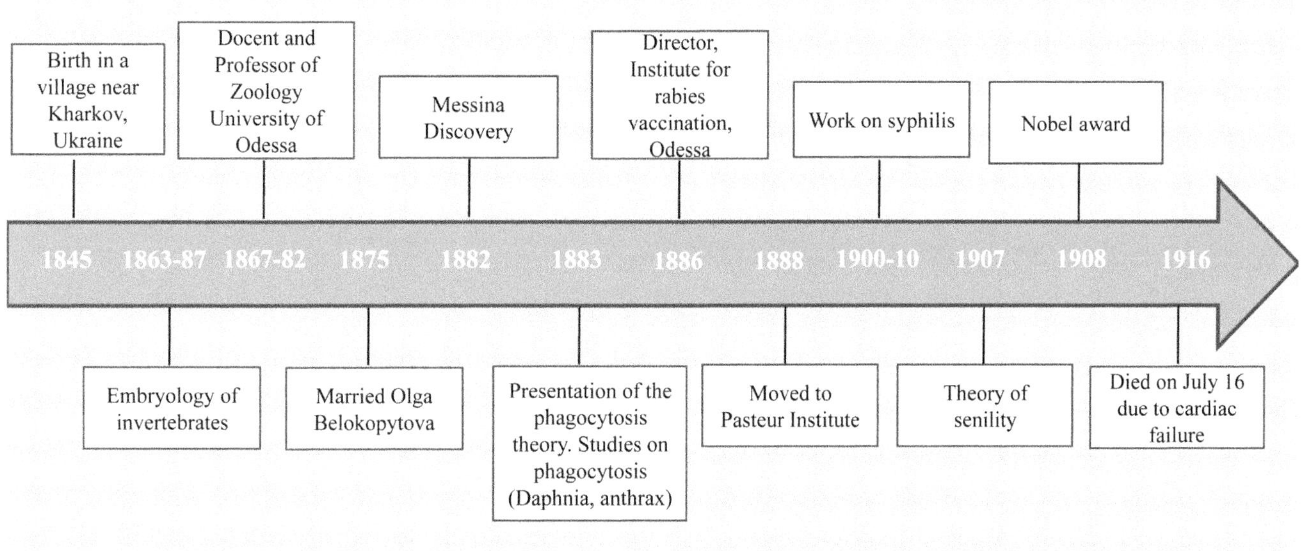

Figure 9 Milestones in the life of Elie Metchnikoff.

century our current knowledge of the ability of phagocytes to recognize and ingest apoptotic cells and, in general, of the ability of the innate immune system to detect damaged endogenous elements (31). Moreover, his proposal that phagocytes promote tissue trophism and growth has been confirmed by the ability of macrophages to stimulate angiogenesis, neuronal patterning, bone morphogenesis, metabolism, and wound healing (32). The role in the preservation of organism integrity assigned by Metchnikoff to the phagocyte is equally evident in its ability to reach infection sites and destroy pathogens (in a function he called "pathological inflammation"). This activity represented, to Metchnikoff, the adaptation of a "selective attack apparatus," to use Alfred Tauber's words (33), used by primitive animals to detect the presence of food, or, in other words, a transition from an "eat-to-feed" to an "eat-to-defend" function. Intriguingly, recent studies point to similarities and interconnections in the signal transduction pathways involved in sensing of nutrients and pathogens (34, 35), with potentially profound implications for the pathogenesis of chronic metabolic diseases (35). For example, classic sensors of microbial molecules, such as Toll-like receptors 2 and 4, can respond, in both macrophages and adipocytes, to the presence of nutritional lipids (36–39). It is also intriguing that macrophages and adipocytes show similar transcriptional profiles and ability to generate proinflammatory responses under certain conditions (41, 42). Moreover, autophagy can be triggered in response to nutrient starvation or invasion of the cytosol by intracellular pathogens (43). In summary, Metchnikoff's vision of the immune system and of myeloid cell biology is surprisingly modern, as indicated by recent discoveries. No doubt, his total commitment to science and his creativity in tackling fundamental biological questions will continue to provide a powerful source of inspiration.

Acknowledgments. The comments and suggestions of Siamon Gordon and Alfred Tauber are gratefully acknowledged.

Citation. Teti G, Biondo C, Beninati C. 2016. The phagocyte, Metchnikoff, and the foundation of immunology. Microbiol Spectrum 4(2):MCHD-0009-2015.

References

1. **Silverstein AM.** 1989. *History of Immunology*. Academic Press, San Diego, CA.
2. **Mechnikov I.** 1908. On the present state of the question of immunity in infectious diseases. *Nobel lecture*, December 11, 1908.
3. **Tauber AI.** 1994. *The Immune Self: Theory or Metaphor?* Cambridge University Press, Cambridge, United Kingdom.
4. **Tauber AI.** 1999. The elusive immune self: a case of category errors. *Perspect Biol Med* **42:**459–474.
5. **Tauber AI.** 1991. The immunological self: a centenary perspective. *Perspect Biol Med* **35:**74–86.
6. **Tauber AI, Chernyak L.** 1991. *Metchnikoff and the Origins of Immunology: from Metaphor to Theory.* Oxford University Press, Oxford, United Kingdom.
7. **Gourko H, Williamson DI, Tauber AI.** 2000. Introduction, p 3–32. *In* Gourko H, Williamson DI, Tauber AI (ed), *The Evolutionary Biology Papers of Elie Metchnikoff.* Kluwer Academic Publishers, Dordrecht, The Netherlands.
8. **Hirsch JG.** 1959. Immunity to infectious diseases: review of some concepts of Metchnikoff. *Bacteriol Rev* **23:**48–60.
9. **Vaughan RB.** 1965. The romantic rationalist: a study of Elie Metchnikoff. *Med Hist* **9:**201–215.
10. **Lépine P.** 1966. *Elie Metchnikoff et l'immunologie.* Pierre Seghers, Paris, France.
11. **Metchnikoff E.** 1959. *Souvenirs. Recueil d'articles autobiographiques,* pages, traduits du russe par L. Piatigorski et commentes par A. Gaissinovitch, Moscou.
12. **Besredka A.** 1921. *Histoire d'un idee.* Paris, France.
13. **Levaditi C.** 1945. Centenaire d'Elie Metchnikoff. *Presse Medicale* **30:**363–364.
14. **Ramon G.** 1945. Eloge d'Elie Metchnikoff à l'occasion du centième anniversaire de sa naissance. *Bull Acad Med* **129:**294–301.
15. **Cavaillon JM.** 2011. The historical milestones in the understanding of leukocyte biology initiated by Elie Metchnikoff. *J Leukoc Biol* **90:**413–424.
16. **Gordon S.** 2008. Elie Metchnikoff: father of natural immunity. *Eur J Immunol* **38:**3257–3264.
17. **De Kruif P.** 1954. *The Microbe Hunters*, 2nd ed. Harcourt, Brace and Jovanovich, San Diego, CA.
18. **Wright AE, Douglas SR, Sanderson JB.** 1903. An experimental investigation of the rôle of the blood fluids in connection with phagocytosis. *Proc R Soc Lond* **72:** 357–370.
19. **Jungeblut CW, Berlot JA.** 1926. The role of the reticulo–endothelial system in immunity. I. The role of the reticulo–endothelial system in the production of diphtheria antitoxin. *J Exp Med* **43:**613–622.
20. **Roberts EF.** 1929. The reticulo-endothelial system and antibody production. I. The appearance of antibody in the circulation. *J Immunol* **16:**137–149.
21. **Cannon P, Baer R, Sullivan TL, Webster JR.** 1929. The influence of blockade of the reticulo-endothelial system in the formation of antibodies. *J Immunol* **17:**441–463.
22. **Tuft L.** 1934. The effect of the reticulo-endothelial cell blockade upon antibody formation in rabbits. *J Immunol* **27:**63–80.
23. **Siegmund H.** 1922. Speicherung Dutch reticuloendothelien, cellulite Reaktion und Immunitat. *Klin Wehnschr* **1:**2566–2567.
24. **Elvidge AR.** 1933. The reticulo-endothelial system and the source of opsonin. *J Immunol* **24:**31–64.
25. **Metchnikoff E.** 2000. The struggle for existence between parts of the animal organism, p 207–216. *In* Gourko H, Williamson DI, Tauber AI (ed), *The Evolutionary Biology Papers of Elie Metchnikoff.* Kluwer Academic Publishers, Dordrecht, The Netherlands.
26. **Anderson KV, Jürgens G, Nüsslein-Volhard C.** 1985. Establishment of dorsal-ventral polarity in the Drosophila embryo: genetic studies on the role of the *Toll* gene product. *Cell* **42:**779–789.
27. **Lemaitre B, Nicolas E, Michaut L, Reichhart JM, Hoffmann JA.** 1996. The dorsoventral regulatory gene cassette *spätzle/Toll/cactus* controls the potent antifungal response in Drosophila adults. *Cell* **86:**973–983.
28. **Metchnikoff O.** 1924. *The Life of Elie Metchnikoff 1845–1916* (Lankester ER, transl). Constable, London, United Kingdom.
29. **Metchnikoff E.** 1905. *Immunity in Infective Diseases* (Binnie FG, transl). Cambridge University Press, Cambridge, United Kingdom.
30. **Metchnikoff E.** 1968. *Lectures on the Comparative Pathology of Inflammation* (Starling FA, Starling EH, transl). Dover, New York, NY.
31. **Seong SY, Matzinger P.** 2004. Hydrophobicity: an ancient damage-associated molecular pattern that initiates innate immune responses. *Nat Rev Immunol* **4:**469–478.
32. **Pollard JW.** 2009. Trophic macrophages in development and disease. *Nat Rev Immunol* **9:**259–270.
33. **Tauber AI.** 2003. Metchnikoff and the phagocytosis theory. *Nat Rev Mol Cell Biol* **4:**898–901.
34. **Hotamisligil GS, Erbay E.** 2008. Nutrient sensing and inflammation in metabolic diseases. *Nat Rev Immunol* **8:** 923–934.
35. **Efeyan A, Comb WC, Sabatini DM.** 2015. Nutrient-sensing mechanisms and pathways. *Nature* **517:**302–310.
36. **Shi H, Kokoeva MV, Inouye K, Tzameli I, Yin H, Flier JS.** 2006. TLR4 links innate immunity and fatty acid-induced insulin resistance. *J Clin Invest* **116:** 3015–3025.
37. **Tsukumo DM, Carvalho-Filho MA, Carvalheira JB, Prada PO, Hirabara SM, Schenka AA, Araújo EP, Vassallo J,**

Curi R, Velloso LA, Saad MJ. 2007. Loss-of-function mutation in Toll-like receptor 4 prevents diet-induced obesity and insulin resistance. *Diabetes* **56:**1986–1998.

38. **Davis JE, Gabler NK, Walker-Daniels J, Spurlock ME.** 2008. Tlr-4 deficiency selectively protects against obesity induced by diets high in saturated fat. *Obesity (Silver Spring)* **16:**1248–1255.

39. **Ajuwon KM, Spurlock ME.** 2005. Palmitate activates the NF-κB transcription factor and induces IL-6 and TNFα expression in 3T3-L1 adipocytes. *J Nutr* **135:**1841–1846.

40. **Akira S, Uematsu S, Takeuchi O.** 2006. Pathogen recognition and innate immunity. *Cell* **124:**783–801.

41. **Khazen W, M'bika JP, Tomkiewicz C, Benelli C, Chany C, Achour A, Forest C.** 2005. Expression of macrophage-selective markers in human and rodent adipocytes. *FEBS Lett* **579:**5631–5634.

42. **Chung S, Lapoint K, Martinez K, Kennedy A, Boysen Sandberg M, McIntosh MK.** 2006. Preadipocytes mediate lipopolysaccharide-induced inflammation and insulin resistance in primary cultures of newly differentiated human adipocytes. *Endocrinology* **147:**5340–5351.

43. **Levine B, Mizushima N, Virgin HW.** 2011. Autophagy in immunity and inflammation. *Nature* **469:**323–335.

44. **Arnold G.** 1909. Intra-cellular and general digestive processes in Planariae. *Quart J Micro Sci* **54:**207.

Myeloid Cells in Health and Disease: A Synthesis
Edited by Siamon Gordon
© 2017 American Society for Microbiology, Washington, DC
doi:10.1128/microbiolspec.MCHD-0004-2015

Donald Metcalf[1,†]

3 Growth and Differentiation Factors

THE CELLULAR ORIGIN OF MYELOID CELLS

Most hematologists accept that hematopoiesis is initiated by multipotential hematopoietic stem cells. These cells are regarded as being few in number but diverse in properties and as being usually noncycling in normal health. It is well documented that, when myeloid populations are damaged by irradiation or chemotherapy, hematopoietic repopulation is initiated by such stem cells.

Much less well documented is the assumption that hematopoiesis under basal conditions in healthy adults is also initiated and sustained by similar stem cells. On the contrary, recent evidence has become persuasive that, under basal conditions, hematopoiesis is initiated and sustained by a heterogeneous population of more-mature cells that are 100-fold more numerous than stem cells (1, 2). Importantly, these cells are able to be cultured and analyzed clonally *in vitro*. These cells are defined by their capacity to generate colonies composed of blast cells in semisolid cultures (3, 4). These colonies themselves contain blast colony-forming cells (BL-CFCs), indicating an ability for self-generation, and also committed progenitor cells in all lineages in numbers that greatly exceed those required to maintain normal hematopoiesis (Fig. 1).

The importance of recognizing that BL-CFCs are likely to be "stem cells" in normal hematopoiesis is that the regulation of BL-CFCs can readily be analyzed in clonal cultures. In contrast, little is known regarding what factors might influence the proliferative activity and differentiation commitment of the traditional stem cells, due to the necessity to monitor them indirectly by their *in vivo* repopulating capacity in severely myeloid-depleted recipients.

This discussion will commence with the discovery and properties of the factors controlling myeloid lineage-committed progenitor cells, some of the progeny of BL-CFCs.

THE DISCOVERY AND PROPERTIES OF THE CSFs

The discovery of the colony-stimulating factors (CSFs) arose from the accidental development of semisolid culture systems in which colonies of granulocytes and/or macrophages developed in cultures of murine bone marrow or spleen cells (5, 6). Such colonies were shown to be clones derived from single precursor cells, and these precursors are now known as lineage-committed progenitor cells.

In these cultures, cell division to form colonies was accompanied by maturation of the colony cells to recognizably mature granulocytic (neutrophil) or macro-

[1]Cancer and Haematology Division, The Walter and Eliza Hall Institute of Medical Research, Parkville, Victoria 3052, Australia. [†]Deceased.

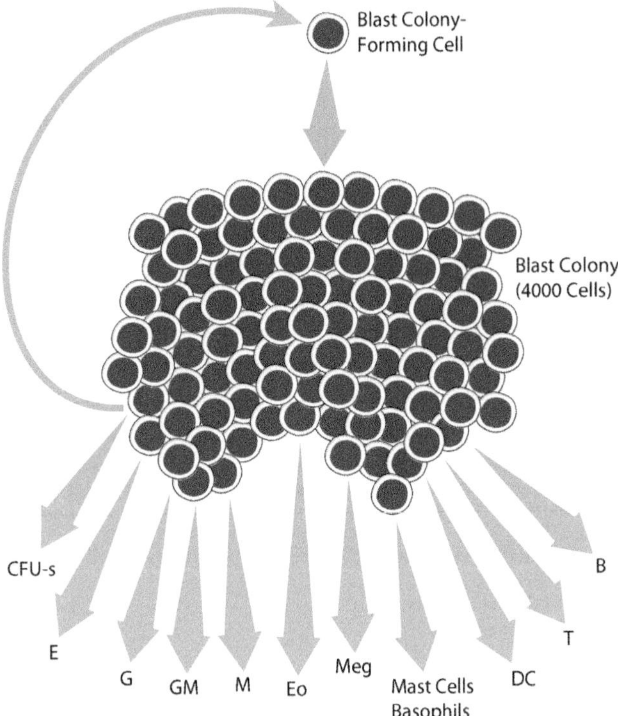

Figure 1 When stimulated in culture, BL-CFCs can each produce large colonies containing a wide variety of committed progenitor cells in different lineages. Present in many colonies are also BL-CFCs, indicating a capacity of the initiating cells to self-generate and to sustain the continuous production of maturing progeny cells. CFU-s, colony-forming unit-spleen; E, erythroid progenitors; G, granulocyte progenitors; GM, granulocyte-macrophage progenitors; M, macrophage progenitors; Eo, eosinophil progenitors; Meg, megakaryocyte progenitors; DC, dendritic cell progenitors; T, T lymphocyte progenitors; B, B lymphocyte progenitors.

phage cells. Cell division during colony formation was absolutely dependent on the addition to the cultures of certain cells, cell extracts, or medium conditioned by various cells or tissues. The operational term "colony-stimulating factor" was chosen for the active agent from these sources that was required for the proliferation of granulocytes or macrophages *in vitro*.

CSF was shown to be present in human serum or urine and to be produced *in vitro* by most mouse organs. Furthermore, CSF levels *in vivo* were elevated by infections or the injection of endotoxin (7). With this evidence suggesting that CSF might be a bona fide regulator controlling the proliferation of granulocytes and macrophages, efforts were commenced to purify CSF and to characterize its action. There followed a 15-year period during which CSFs were slowly purified from the richest available murine or human sources. Complicating this program was the eventual recognition that four distinct glycoprotein CSFs existed, some requiring extreme levels of purification of up to a millionfold.

The CSF glycoproteins range in molecular weight from 18 to 70 kDa; they were too large to synthesize accurately, and it was impractical to extract satisfactory amounts from even the richest tissue sources. This impasse was overcome by cloning cDNAs for each of the four murine and corresponding human CSFs. This was accomplished in a productive period between 1984 and 1986 (7).

As shown by the typical data on cultured murine bone marrow cells in Table 1, each CSF stimulates colony formation by characteristically different subsets of committed progenitor cells, no CSF having an action restricted to a single lineage. The range of action of each CSF is dictated by the distribution of specific membrane receptors on responding cells (7). Despite the extremely high specific activity of the CSFs, which typically stimulate maximal numbers of colonies to develop at concentrations of 10 ng/ml, CSF receptors number only ~300 to 500 per cell (8). The CSFs therefore have extremely high specific activity but are required constantly for continuing cell division. As shown in the example in Fig. 2, the CSF concentration used determines the number of colonies developing. This is due to the heterogeneity of committed progenitor cells, because some cells and their progeny require higher CSF concentrations than others to proliferate. The CSF concentration also determines the number of maturing progeny produced by individual progenitor cells during a defined period of incubation. These relationships result in the sigmoid dose-response curve seen in Fig. 2 between CSF concentrations and colony numbers—a relationship that was used as a monitoring system during the purification and cloning of the CSFs (7).

Multiple cell types and organs can produce granulocyte-macrophage CSF (GM-CSF) and G-CSF in low concentrations, but these can be profoundly and rapidly increased following stimulation by agents such as endotoxin (7). M-CSF is produced in higher concentrations, and levels tend to be less variable. Interleukin-3 (IL-3) is not detectable in the normal mouse or in murine organ-conditioned medium (9), although murine lymphoid cells readily produce IL-3 *in vitro* when stimulated by IL-2, foreign antigens, or mitogens (7).

The CSFs have some properties resembling hormones because they are secreted molecules acting at a distance, but differ radically in that multiple cell types are able to produce CSF. In addition, M-CSF can also be membrane displayed in a biologically active form, permitting local stimulation (10).

Table 1 Colony-stimulating activity for mouse bone marrow cells of CSFs and other cytokines[a]

Stimulus	Final concentration	No. of colonies[b]					
		Blast	G	GM	M	Eo	Meg
GM-CSF	10 ng/ml		19 ± 4	8 ± 3	22 ± 7	4 ± 1	
G-CSF	10 ng/ml		15 ± 5	0 ± 0	0 ± 0		
M-CSF	10 ng/ml		1 ± 1	2 ± 1	30 ± 5		
IL-3 (multi-CSF)	10 ng/ml	4 ± 2	16 ± 6	10 ± 5	10 ± 3	2 ± 1	6 ± 3
SCF	100 ng/ml	9 ± 2	20 ± 6	4 ± 3	1 ± 1		
IL-6 + SCF	100 + 100 ng/ml	13 ± 6	20 ± 6	10 ± 1	12 ± 5		6 ± 4
SCF + IL-3 + EPO	100 + 10 + 1 IU	8 ± 3	23 ± 7	18 ± 7	20 ± 5	3 ± 2	22 ± 6
IL-6	100 ng/ml		10 ± 2	0 ± 0	0 ± 0		
Saline	–		0 ± 0	0 ± 0	0 ± 0		

[a]Cultures contained 25,000 C57BL bone marrow cells, and colonies were scored from stained cultures after 7 days of incubation. Data are mean ± standard deviation from 3 mice.
[b]Blast, blast colonies; G, granulocyte colonies; GM, granulocyte-macrophage colonies; M, macrophage colonies; Eo, eosinophil colonies; Meg, megakaryocyte colonies.

Subsequent to the discovery of the CSFs and the parallel characterization and cloning of the erythroid-specific regulator erythropoietin (11), a number of factors were discovered that also had proliferative actions on granulocyte, macrophage, or eosinophil populations. These include IL-5, which is a selective proliferative stimulus for eosinophil colony formation (12), and IL-6, which has pleiotropic actions but also an ability to stimulate granulocyte colony formation *in vitro* that closely resembles that stimulated by G-CSF (13). There then followed the discovery of stem cell factor (SCF) (14), an agent with significant stimulating actions on the ancestral BL-CFCs as well as weaker direct actions on granulocyte colony formation (7). Finally, thrombopoietin (TPO) was discovered (15) as the dominant stimulator of megakaryocyte formation but was then shown to be an agent with important actions on stem and early progenitor cells of multiple lineages (16). In each case, the actions of these regulators are made possible by the membrane display of the appropriate specific receptors.

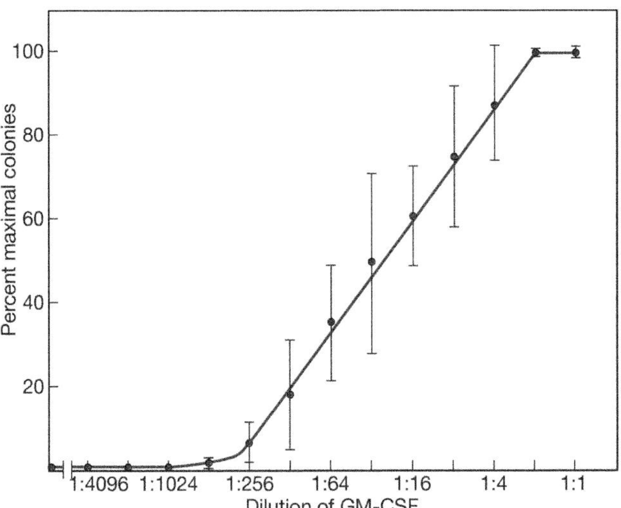

Figure 2 Colony formation in 7-day cultures of C57BL mouse bone marrow cells stimulated by GM-CSF. Final concentration of GM-CSF in the 1:1 cultures was 10 ng/ml. The colony numbers ± standard deviations were derived from cultures of five different bone marrow cell populations.

The multiplicity of agents able to stimulate the proliferation of cells in a particular lineage raised the criticism of possible redundancy in the regulatory system. This required analysis of the consequences of deletion of the gene encoding each regulator. No example of genuine redundancy has so far been discovered, but certain regulators clearly have dominant actions. From the gene deletion studies, G-CSF appears to be the major regulator of granulocyte formation (17), M-CSF for macrophage formation (18), and IL-5 for eosinophil production (19), while SCF and TPO have important actions on stem cell and early progenitor cell proliferation (16, 20).

In contrast to possible redundancy of hematopoietic growth factors, experiments have in fact shown that combinations of certain growth factors have powerful synergistic effects. Examples of such synergy include the combination of GM-CSF with M-CSF, which enhances granulocyte-macrophage colony formation (21), and particularly the combination of early-acting SCF with IL-6 or G-CSF, which strikingly increases blast colony and granulocyte colony formation (3, 4). Similarly, to stimulate maximum numbers of megakaryocyte colonies to develop, a combination of SCF + IL-3 + erythropoietin (EPO) is required (22). This design of the regulatory system also involves the local, simultaneous production by cells of more than one growth factor; e.g., bone shaft cells or spleen cells can produce multiple regulatory factors (7).

There is a quantitative discrepancy between responses elicited by the CSFs *in vivo* compared with their effects *in vitro*. G-CSF has the strongest effects *in vivo* in elevating neutrophil granulocyte levels but is the weakest CSF *in vitro*, stimulating relatively few colonies to develop, which are of very small size. Conversely, IL-3 is a strong CSF *in vitro*, but when injected *in vivo* the most evident effect is a mild stimulation of mast cell numbers (7).

The design for proliferative hematopoietic stimulation appears to involve cooperation between regulatory factors in order to achieve the desired cellular responses with the minimum effort in growth factor production. A striking example is the marked neutrophil response to the injection of G-CSF. This is entirely dependent on a synergistic action with SCF. In Steel mice, lacking SCF, or W^v mice, lacking SCF receptors, G-CSF elicits only very weak responses *in vivo* (23).

It is a misconception to label the CSFs or comparable growth factors as "lymphokines," or worse, to regard them as T-lymphocyte products. Certainly, T lymphocytes can produce many of these factors, but so can other cell types (9). IL-5 is commonly misrepresented in this context as a T-lymphocyte cytokine, but it is produced in detectable amounts by tissues not normally containing lymphocytes and even by organs from mice lacking T lymphocytes (24).

MULTIPLE ACTIONS OF THE CSFs

It was surprising, and initially disputed, that the CSFs were found to have multiple actions on responding populations. The CSFs certainly are mandatory proliferative stimuli but they also have additional actions on responding cells. The first such additional action noted was on the survival of progenitor cells and their maturing progeny. In the absence of CSFs, progenitor cells die by apoptosis (25). Some concluded from this that the real action of CSFs was merely to ensure cell survival, and then, given good health, progenitor cells would be able to proliferate without stimulation. With the elucidation of the signaling pathways activated by CSF stimulation, this interpretation was made most unlikely. The strongest single experiment disproving this hypothesis was the overexpression of Bcl-2 in a CSF-dependent cell line. These cells survived in the absence of CSF but did not undergo cell division (26). The cell survival action applies throughout the maturing cells of relevant lineages, and, for example, GM-CSF promotes increased cell survival of mature neutrophils, with lower CSF concentrations being more effective than those required to stimulate cell proliferation.

A further controversial aspect of CSF action was whether CSFs could dictate commitment of cells to particular lineages. Several arguments had previously supported the view that lineage commitment is a random event. The progenitor cell progeny of individual BL-CFCs certainly are extremely heterogeneous: no two blast colonies generate similar numbers or types of lineage-committed progeny (3, 4). Furthermore, many mutant mice have been developed with quantitative defects in blast cell and progenitor cell numbers, but no examples have been encountered in which commitment has been skewed to one lineage.

On the other hand, transfection of certain transcription factors can result in skewing of commitment in progenitor cells. Thus, overexpression of GATA-1 enhances erythroid commitment and PU.1 enhances commitment to form granulocyte and macrophage cells (27). Similarly, complete reversion of commitment is possible by transfection of multiple transcription factors (28), but it was unclear from these studies whether or not extrinsic regulatory factors can induce any of these changes.

However, using continuous bioimaging it has been shown that CSFs can instruct hematopoietic lineage choice (29). This has been confirmed and extended in studies showing that M-CSF can actively commit cells to enter the macrophage lineage (30). These studies provide proof in principle that CSFs can influence commitment. It remains likely, however, that multiple factors can act to determine lineage commitment, with growth factors merely being one such agent. This makes it difficult to document clear examples of such actions.

Another controversial action of the CSFs appeared to be that they could promote or accelerate maturation. At first sight, if CSFs have a concentration-dependent action in increasing progeny cell numbers, this must involve a delay in maturation to the postmitotic state; otherwise continued cell proliferation would not be possible. Indeed, it was reported that a CSF-dependent cell line with *bcl*-2-enhanced survival could produce maturing progeny in the absence of CSF (31). Despite these observations, specific regions have been identified in CSF receptors that are necessary to permit CSF action that results in the production of maturing progeny (32).

The least controversial actions of the CSFs are their capacity to stimulate the functional activity of mature neutrophils and macrophages. This has been demonstrated *in vitro* using assays ranging from phagocytosis and killing of microorganisms to production of molecules of biological importance. Again, this stimulation of functional activity is not an exclusive action of the CSFs; other agents can achieve the same end results, but CSFs remain prominent molecules whose actions are at times unique. Thus, deletion of GM-CSF results in hypofunctioning of alveolar macrophages that cannot be corrected by other agents. GM-CSF-deprived lung macrophages cannot process surfactant, and the consequent accumulation of surfactant results in the disease alveolar proteinosis (7).

When the actions of injected CSFs were first investigated in mice and humans, a surprising consequence of the actions of G-CSF and GM-CSF was the release of stem cells and progenitor cells from the bone marrow to the peripheral blood (33, 34). This has resulted in revolutionary changes in bone marrow transplantation because the CSF-elicited stem and progenitor cells are more numerous than can be collected by bone marrow aspiration. As a consequence, peripheral blood stem and progenitor cells have made transplantation easier, cheaper, safer, and more effective. The mechanisms of cell release remain only partially characterized. Release from the marrow can be achieved of cells not bearing G-CSF receptors. This reinforces evidence that release involves changes of cellular attachments with marrow stromal cells (35), with the consequent appreciation that CSF actions are not restricted to hematopoietic cells and that CSF-initiated changes can directly or indirectly affect marrow stromal cell function.

Model builders have always held that to achieve the sustained and stable production of cells it is insufficient to use stimulating factors alone. To avoid cycling of cell numbers, models indicate that stimulation needs to be balanced by inhibitory factors, possibly of equivalent specificity. Despite 40 years of exploration, no satisfactory extracellular inhibitors of granulocyte-macrophage populations have been uncovered.

What has been discovered is a family of intracellular factors whose transcription is activated by signaling from cytokine-activated membrane receptors (36). This family of proteins, known as the suppressors of cytokine signaling (SOCS) molecules, serve to modulate or terminate cytokine signaling by preventing phosphorylation of cytokine receptor cytoplasmic regions, inhibiting JAK tyrosine kinase activity and the consequent phosphorylation of STATs. This prevents dimerization of these molecules and their subsequent movement to the nucleus to activate transcription genes involved in proliferative responses. The SOCS system represents a finely tuned inhibitory system for restricting cytokine signaling.

The prototype SOCS1 molecule regulates signaling by gamma interferon, SOCS2 modulates signaling by growth hormone, and SOCS3 restricts proliferative responses to G-CSF and IL-6 family cytokines (37). In the absence of SOCS3, injection of normal doses of G-CSF into mice produces hind-limb paralysis within 3 to 4 days due to vast accumulations of neutrophils in the spinal canal, accompanied by cannonball aggregates of neutrophils in the liver, lung, and marrow (37).

This raises the subject of the possible effects of sustained excess CSF levels. Excessive stimulation by G-CSF eventually leads to massive accumulations of mature neutrophils in the marrow, lung, and liver. Excess levels of GM-CSF cause a variety of inflammatory lesions due to excessive functional activity of macrophages; excess levels of multi-CSF (IL-3) lead to accumulation of mast cells in a variety of tissues (7). Some of these effects have restricted clinical use of the CSFs to G-CSF and GM-CSF. Clinical trials of M-CSF were terminated because of thrombocytopenia, possibly due to macrophage activation, and clinical trials of IL-3 were terminated because of allergic-type reactions, possibly the consequence of mast cell accumulation and activation (7).

RELATED HEMATOPOIETIC LINEAGES

EPO has similar proliferative actions to those of the CSFs. Deletion of the EPO gene from conception leads to embryonic death due to failure of definitive erythropoiesis to develop in the fetal liver. Conversely, overexpression of EPO leads to polycythemia. Unlike the CSFs, EPO is inactive *in vivo* unless the carbohydrate moiety of the molecule is present. Glycosylated EPO has had wide clinical use in patients with anemia secondary to renal disease (38) and apparently in cyclists and other athletes wishing to enhance their physical performance by increasing hemoglobin levels.

The master regulator of megakaryocyte and platelet formation was revealed by the identification and cloning of TPO. TPO stimulates megakaryocyte colony formation *in vitro*, but its action is strongly enhanced by combination with SCF + IL-3. Deletion of the gene for TPO, or its receptor (Mpl), leads to profound thrombocytopenia (16, 39). TPO has proliferative actions on stem and progenitor cells in multiple lineages (16). However, its subsequent actions are restricted to the megakaryocytic lineage. Curiously, TPO is not necessary for the final stages of megakaryocyte or platelet formation (40).

CSF RECEPTORS

Specific membrane receptors exist for each CSF, and individual hematopoietic cells can coexpress all four types of CSF receptors, often together with receptors for SCF, IL-5, and IL-6.

Only a limited number of a few hundred receptors of each type exist on any one immature or mature cell in the granulocyte-macrophage lineage. There is little in common in the structure of these receptors apart from the exquisite specificity with which they bind their ligands and initiate intracytoplasmic signaling. Receptors for G-CSF and M-CSF are homodimers, but only those for M-CSF possess intrinsic kinase activity. Receptors for GM-CSF and IL-3 are heterodimers with unique α chains but a shared common signaling β chain. Additionally, in the mouse, IL-6 also shares receptor chains with several other ligand receptors (41).

Binding of the CSF to its receptor induces dimer formation or reorientation of the cytoplasmic regions of the receptors. This positions the receptor-bound JAK kinases so that they can phosphorylate each other and activate their tyrosine kinase activity for cytoplasmic regions of the receptor and subsequently STAT molecules. Dimeric STAT molecules can then transit to the nucleus to initiate events leading to activation of genes necessary for various responses.

The use of mutated receptors has allowed functionally distinct regions of receptor chains to be identified. The ability of CSFs to have multiple functional actions on responding cells is made possible by specific regions of the receptor that selectively activate different signaling molecules to result in such diverse responses as cell division, maturation, or functional activation (32, 42).

Following binding to a ligand such as CSF, internalization of the ligand-receptor complex rapidly follows, with degradation and destruction of the complex. Continued stimulation of a cell depends on continuous display of new receptors at the cell surface. These are newly synthesized receptors rather than receptors recycled from previous ligand-receptor complexes (41). This process is presumably under genetic control and is part of the gene complex involved in differentiation commitment of precursor cells to become lineage-specific progenitor cells.

PLEIOTROPIC ACTIONS OF THE HEMATOPOIETIC REGULATORS

Initially, regulators such as the CSFs were envisaged as exclusive regulators for specific subsets of hematopoietic populations. The notion of private regulators and their responding cells did have to accept that individual CSFs could influence more than one cell type, e.g., that GM-CSF was a proliferative factor for granulocytes, granulocyte-macrophages, macrophages, and eosinophil progenitor cells and their progeny. The two CSFs with relatively selective action also had some actions on other lineages. Thus, G-CSF can stimulate occasional macrophage colonies to develop, while M-CSF regularly stimulates the formation of some granulocytic and granulocyte-macrophage colonies (Table 1).

When GM-CSF was shown to be necessary for the *in vitro* development of dendritic cells (43), such cells could be accepted as closely related to the granulocyte-macrophage lineage. Further analysis of dendritic cells has shown that they can be the progeny of the same BL-CFCs that generate granulocytes, macrophages, and eosinophil progeny.

However, M-CSF was shown to also be a proliferative factor for placental trophoblast cells (44). Also disturbing for the notion of lineage-specific hematopoietic regulators was the demonstration that the CSF receptors were present on pain neurons in bone (45)—a puzzling combination of apparently unrelated biology that may explain why CSF injections frequently lead to bone pain.

The cytokine world was radically overturned by the discovery and characterization of leukemia inhibitory factor (LIF) (46) and IL-6, with their astounding range of actions on unrelated tissues. For example, LIF clearly had actions on some myeloid leukemic cells but was also shown to be necessary for the maintenance of totipotential stem cells and the implantation of blastocysts. LIF also had powerful actions on the sympathetic nervous system, lipid metabolism, pituitary function, and osteoblast biology, to mention only some of its actions. IL-6 has a similarly bizarre range of biological actions but was otherwise an unremarkable cytokine active on lymphoid and granulocytic populations (13) (Table 1).

It remains obscure what biological advantages are achieved by having regulator molecules with such a bizarre range of actions, but clearly there is a logic that refutes our present notions of organ biology. While other hematopoietic cytokines like the CSFs and erythropoietin may yet prove to have a similar disturbing range of additional actions, the practical fact remains that they function, and can be used clinically, as hematopoietic regulators with insignificant side effects that might result from such possible actions on other organ systems.

THE CSFs AND MYELOID LEUKEMIAS

When agar cultures were adapted to support the growth of human granulocyte-macrophage colonies, it was surprising to find that the proliferation *in vitro* of clonogenic chronic myeloid and acute myeloid leukemic cells was dependent on the same concentrations of CSF as were required for normal cells (Fig. 3) (47). If *in vitro* behavior is at all an indication of behavior *in vivo*, then, at the least, CSFs would be mandatory cofactors in the emergence and progressive growth of myeloid leukemic populations.

It was subsequently shown that transfection of GM-CSF or IL-3 to nonleukemic continuous cell lines resulted in immediate leukemic transformation (48). This was extended by studies in which cotransfection of a homeobox gene plus IL-3 also resulted in immediate leukemic transformation of normal bone marrow cells (49). Indeed, in a number of transformation studies it was observed that the final step in leukemic transformation commonly involved the acquired capacity by gene rearrangement of the cells to exhibit autocrine GM-CSF or IL-3 production (50).

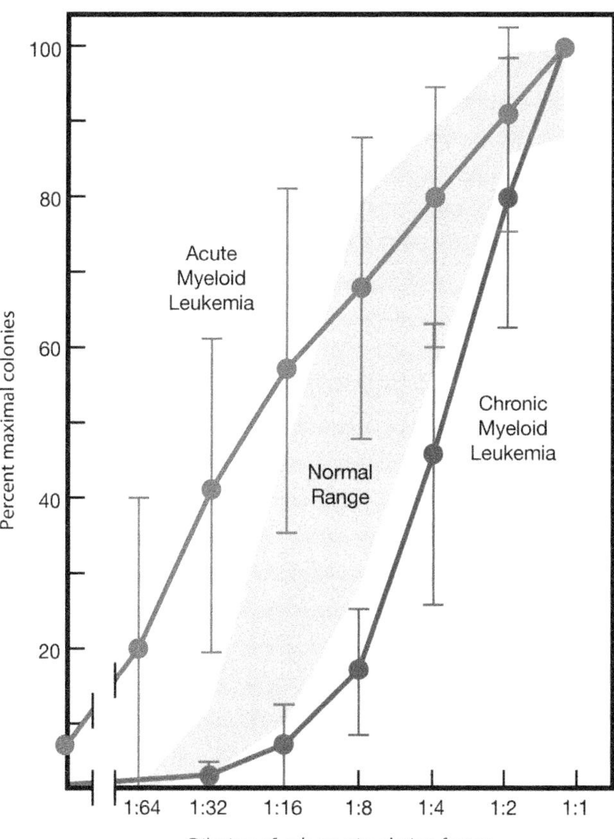

Figure 3 Colony formation by marrow cells from patients with acute or chronic myeloid leukemia using varying concentrations of material with colony-stimulating activity. The responsiveness of the leukemic cells to generate colonies (verified by karyotypic colony analysis) is broadly similar to that of normal human marrow cells, although some acute myeloid leukemia populations contained some cells capable of limited unstimulated proliferation.

The relevance of these findings for human myeloid leukemia remains unclear because autocrine CSF production based on rearrangement of the CSF genes has not been noted, although activating mutations of CSF receptors have occasionally been observed (51).

On balance, the studies suggest strongly that autocrine production of CSF and possibly other comparable growth factors can be one of the steps necessary for myeloid leukemia development.

UNRESOLVED PROBLEMS

Remarkable as the progress has been in understanding the biology of granulocyte-macrophage populations, many unresolved questions remain, and a few of these can be listed.

Are multipotential stem cells, as detected in myeloid-depressed recipients, responsible for the initiation and maintenance of hematopoietic populations in normal health or is this the role of BL-CFCs?

A recurrent and unresolved question in hematopoiesis is whether the generation of lineage-committed precursor cells is due to random chance or is significantly modulated by extrinsic regulatory factors. This is relevant for all cells maintaining bi- and multipotentiality and is most acute for stem cells and BL-CFCs. Is the generation of lineage-committed cells due to random chance or directed commitment? Do the cytokines induce commitment of precursor cells or merely facilitate cell division by already committed precursors? The problem seems impossible to resolve for stem cells, relying as they do on indirect *in vivo* readouts, but answers should be able to be obtained with BL-CFCs and their immediate progeny.

Other striking phenomena in need of molecular explanation are the superadditive synergistic responses induced by some growth factor combinations, for example, SCF + IL-6. Are there separate signaling pathways able to control cell division or is a common final pathway involved, and, if the latter, why are the combined responses superadditive?

The exact role of CSFs in the development of myeloid leukemia needs to be clarified. Why are myeloid leukemic cells so CSF dependent for cell proliferation *in vitro*? Does this indicate an oncogenic role for the CSFs, and why are current sequencing data failing to detect aberrations in the structure or regulation of the CSFs or their receptors?

A dominant unresolved feature of all hematopoietic subpopulations is their heterogeneity. What is the basis for this?

As judged *in vitro*, precursor cells have a greatly excessive capacity to generate lineage-committed progenitor cells. Does this indicate extensive local cell death in normal health, as is true for marrow neutrophils (52), or are the concentrations of stimuli being used *in vitro* far higher than are actually available *in vivo*?

Do all the cells in every lineage originate from BL-CFCs or do alternate pathways exist? Is the inability of lineage-committed progenitor cells to self-generate real or an *in vitro* artifact? Is lineage commitment irreversible or can reprogramming be induced by external regulatory factors as distinct from genetic manipulation?

Can segregation of subsets of pure stem cells, BL-CFCs, or lineage-committed progenitor cells be achieved by flow cytometry using more-selective monoclonal antibodies or will intrinsic heterogeneity always defeat efforts to produce pure, homogeneous populations?

Acknowledgments. *This work was supported by the Carden Fellowship Fund of the Cancer Council, Victoria; the National Health and the Medical Research Council (NHMRC), Canberra (Programs 461219 and 1016647); the National Institutes of Health (NIH), Bethesda (Grant No. CA22556); the NHMRC IRIISS (Independent Research Institutes Infrastructure Support Scheme) Grant No. 361646; and a Victorian State Government OIS (Operational Infrastructure Support) grant.*

DONALD METCALF (1929–2014)–A PERSONAL PERSPECTIVE ON HIS CONTRIBUTIONS TO MYELOID HEMATOPOIESIS.

Nicos A. Nicola, The Walter and Eliza Hall Institute of Medical Research, Parkville, Victoria 3052, Australia

Don Metcalf's work brought our understanding of myeloid hematopoiesis from a phenomenological description of the process to a hierarchical description of the cellular developmental tree and a molecular understanding of the cytokines/growth factors that regulate all aspects of myeloid cell production and function. In the process, his work laid the foundations for the current clinical uses of the CSFs in treating patients with cancer and neutropenic diseases as well as the potential uses of antibodies to CSFs or their receptors to treat autoimmune and inflammatory diseases.

Don was a driven individual. He never tired of emphasizing to his students and collaborators that the job of medical research was not to create elaborate and elegant scientific theories but to alleviate human suffering. His Scottish Presbyterian upbringing and schoolteacher parents taught him that hard work and discipline were the keys to success in life, and there was little that upset him more than laziness on the part of others. Perhaps the only thing that upset him more was overspeculation based on limited data. Don's papers are characterized by a high ratio of data and facts to speculation! Indeed, in later years he confessed to me that when he trawled back through his early papers to find the first mention of the potential clinical uses of CSFs (that were obvious to all of us) he was dismayed to find that there wasn't any.

There were several key events in Don's life. The first was his decision in 1950, during his medical degree, to enter a new research training course (B.Sc. in medical science) with Patrick de Burgh in the department of bacteriology at Sydney University. This experience taught him that his career should be in research, not medical practice, and he became interested in the leukemia-causing viruses. The second was the offer in 1953 from the Anti-Cancer Council of Victoria and the Walter and Eliza Hall Institute in Melbourne of the inaugural Carden Fellowship in cancer research. Don was to hold this fellowship for the next 60 years and to remain at the Institute throughout that time. The third was a sabbatical Don had with Jacob Furth at the Harvard Medical School in 1956, where his interest in hormonal imbalance as a cause of leukemia was initiated. The fourth, and most important, was the day in 1965 when his collaborator Ray Bradley from the University of Melbourne physiology department brought along agar petri dishes containing colonies growing from the plated bone marrow cells. In fact, Ray had been trying to grow AKR thymic lymphoma cells in agar and had used a bone marrow underlay in an attempt to stimulate their growth but had instead grown bone marrow colonies in the underlay. Don's instant recognition of the potential of this simple assay to elucidate the hormonal control of hematopoiesis was to set his course for the next several decades.

Because the agar assay for hematopoietic cell colony formation (independently codiscovered by Leo Sachs's group in Israel) required an overlay or underlay of feeder cells or medium conditioned by other cells, it was obvious that it could be used to identify the regulatory growth factors (CSFs). However, Don was concerned from the start that the feeder cells might be producing transforming viruses or merely providing essential nutrients or minerals lacking in the culture media being used to grow the colonies, and this fear stayed with him until the CSFs were finally purified and cloned. To allay these fears, he early on sought to demonstrate that CSF activity was present in mouse and human sera and urine and that it might be regulated by infection or in patients with leukemia. The early positive results gave him some comfort, but screens of various tissues, organs, and cell types showed that CSF was ubiquitously expressed, and this did not fit the then paradigm that growth factors and hormones would be produced only in specialized organs like the pancreas or liver. Today, of course, this makes perfect sense since an efficient response to infection requires all tissues to be able to respond rapidly with CSF production, but at the time the cynicism that met Don's fearless reporting of the facts was encapsulated by a question I heard put to him at a conference: "Have you purified toenail CSF yet?"

Don also recognized the power of the agar assay to define the developmental pathways and cell lineages from stem cells to individual blood cell types. Over the next decade, he and Greg Johnson developed clonal assays for progenitors of granulocytes, macrophages, eosinophils, erythroid cells, megakaryocytes, mast cells, T- and B-lymphoid cells, and multipotential cells. He also pioneered assays to detect self-renewal capacity of colony cells and to determine differentiation potentials by the use of paired daughter cells.

Don recognized that he did not have the skills himself to biochemically purify and clone the CSFs, so he went about recruiting protein chemists (Richard Stanley, Tony Burgess, myself, and Doug Hilton) and molecular biologists (Ashley Dunn, Nick Gough, and David Gearing), among others, to help him achieve his goals. It is a testament to his singular vision and drive that so many talented individuals were wholeheartedly converted to the cause.The purification of the CSFs was also to prove a decade-long endeavor. The difficulty lay first in the unprecedented high specific activity of the CSFs, meaning that even the richest sources contained minuscule amounts of protein. Second, the then available protein purification schemes were incapable of achieving the required millionfold purifications, and this had to await the development of high-performance liquid chromatography. Third, the submicrogram amounts of CSF being purified meant that as soon as near purity was achieved all the available protein would adsorb irreversibly to the glass or plastic test tubes. As each of these obstacles was overcome, M-CSF (now known as CSF-1), GM-CSF, G-CSF, and multi-CSF (now known as IL-3) were each purified and ultimately molecularly cloned.

As mentioned above, it was only then that sufficient quantities of the CSFs were available to test in mice and then humans. These were anxious times for Don because at the back of his mind he was always concerned that colony growth in agar might be artifactual. He was relieved and excited when G-CSF and GM-CSF were injected into mice and humans and elevated circulating granulocyte levels. Since neutropenia (low granulocyte numbers) was a dose-limiting toxicity of chemotherapy, the early clinical trials of G-CSF and GM-CSF were in cancer patients receiving various chemotherapeutic regimens and the results were encouraging in reducing the period of neutropenia. During these early trials in Melbourne (led by George Morstyn and Richard Fox), Don and Uli Duhrsen noticed a dramatic mobilization of hematopoietic colony-forming cells out of the bone marrow into the blood. This observation has led to the use of G-CSF-mobilized peripheral blood stem cell transplants replacing bone marrow transplants in modern cancer treatments. G-CSF has also found use in treating patients with rare chronic or cyclic neutropenias.

From his time with Jacob Furth, Don was interested in the concept that altered levels of CSFs might help to bring leukemic cell growth back into control. In the early 1980s, he used the mouse myeloid leukemic cell line WEHI-3B to search for factors that induced terminal differentiation and suppressed its growth (differentiation factor). Ultimately, differentiation factor was shown to be identical to G-CSF, but a second myeloid leukemic cell line used by Leo Sachs (M1) was unresponsive to G-CSF. Don and Doug Hilton used this cell line to purify and clone LIF, a factor that proved to be extraordinarily pleiotropic (including having the exact opposite action on embryonic stem cells—preventing differentiation and maintaining pluripotency). On the other side of the coin, Don wondered whether inappropriate expression of CSFs (especially autocrine production) could cause leukemic transformation. With Richard Lang and Tom Gonda, he showed that indeed the nonleukemic GM-CSF-dependent myeloid cell line FDCP-1 could be transformed to leukemic simply by infecting with a retrovirus that expressed GM-CSF.

Don, Tony Burgess, and I began studies to identify the cellular receptors for CSFs in the mid-1980s. Though it proved difficult to radiolabel CSFs while retaining biological activity, this was eventually achieved for all the CSFs and revealed that CSF receptors had a high affinity for their ligands but were usually present at low numbers (a few hundred per cell). Cell microscopic autoradiographs showed that CSF receptors were expressed selectively on cell types known to be responsive to each CSF, and this exquisite specificity suggested a way of screening for cDNA clones expressing each receptor. This was to use a human placental cDNA expression library, transfect pools into COS cells, incubate them with radioactive GM-CSF, and use microscopic autoradiography to detect positive cells. This proved successful, and its cloning ultimately allowed the generation of neutralizing monoclonal antibodies to the GM-CSF receptor. Since Don's earlier work with Mathew Vadas and Angel Lopez had shown that GM-CSF and G-CSF were also potent activators of the inflammatory actions of mature neutrophils and macrophages, the development of neutralizing antibodies to CSFs and their receptors is now being explored as a potential treatment for inflammatory and autoimmune diseases like rheumatoid arthritis.

Don was extremely gratified that his research had a direct effect on the health of millions of patients worldwide. For him this was the only reason for doing medical research. The lingering doubts about the CSFs, the inevitable frustrations at almost every turn, the (literally) backbreaking work schedule of 8- to 10-hour stints at his microscope were all made worthwhile by the eventual outcomes. In the process, his work fundamentally changed the field of molecular hematology forever.

Citation. Metcalf D. 2016. Growth and differentiation factors. Microbiol Spectrum 4(4):MCHD-0004-2015.

References

1. **Sun J, Ramos A, Chapman B, Johnnidis JB, Le L, Ho YJ, Klein A, Hofmann O, Camargo FD.** 2014. Clonal dynamics of native haematopoiesis. *Nature* **514:**322–327.
2. **Becher B, Schlitzer A, Chen J, Mair F, Sumatoh HR, Teng KW, Low D, Ruedl C, Riccardi-Castagnoli P, Poidinger M, Greter M, Ginhoux F, Newell EW.** 2014. High-dimensional analysis of the murine myeloid cell system. *Nat Immunol* **15:**1181–1189.
3. **Metcalf D, Greig KT, de Graaf CA, Loughran SJ, Alexander WS, Kauppi M, Hyland CD, Di Rago L, Mifsud S.** 2008. Two distinct types of murine blast colony-forming cells are multipotential hematopoietic precursors. *Proc Natl Acad Sci U S A* **105:**18501–18506.
4. **Metcalf D, Ng A, Mifsud S, Di Rago L.** 2010. Multipotential hematopoietic blast colony-forming cells exhibit delays in self-generation and lineage commitment. *Proc Natl Acad Sci U S A* **107:**16257–16261.
5. **Bradley TR, Metcalf D.** 1966. The growth of mouse bone marrow cells *in vitro*. *Aust J Exp Biol Med Sci* **44:**287–299.
6. **Ichikawa Y, Pluznik DH, Sachs L.** 1966. *In vitro* control of the development of macrophage and granulocyte colonies. *Proc Natl Acad Sci U S A* **56:**488–495.
7. **Metcalf D, Nicola NA.** 1995. *The Hemopoietic Colony-Stimulating Factors: From Biology to Clinical Applications.* Cambridge University Press, Cambridge, United Kingdom.
8. **Nicola NA, Peterson L, Hilton DJ, Metcalf D.** 1988. Cellular processing of murine colony-stimulating factor (Multi-CSF, GM-CSF, G-CSF) receptors by normal hemopoietic cells and cell lines. *Growth Factors* **1:**41–49.

9. **Metcalf D, Willson TA, Hilton DJ, Di Rago L, Mifsud S.** 1995. Production of hematopoietic regulatory factors in cultures of adult and fetal mouse organs: measurement by specific bioassays. *Leukemia* **9:**1556–1564.
10. **Stein J, Borzillo GV, Rettenmier CW.** 1990. Direct stimulation of cells expressing receptors for macrophage colony-stimulating factor (CSF-1) by a plasma membrane-bound precursor of human CSF-1. *Blood* **76:**1308–1314.
11. **Miyake T, Kung CK, Goldwasser E.** 1977. Purification of human erythropoietin. *J Biol Chem* **252:**5558–5564.
12. **Sanderson CJ.** 1992. Interleukin-5, eosinophils, and disease. *Blood* **79:**3101–3109.
13. **Tanaka T, Kishimoto T.** 2014. The biology and medical implications of interleukin-6. *Cancer Immunol Res* **2:** 288–294.
14. **Zsebo KM, Wypych J, McNiece IK, Lu HS, Smith KA, Karkare SB, Sachdev RK, Yuschenkoff VN, Birkett NC, Williams LR, Satyagal VN, Tung W, Bosselman RA, Mendiaz EA, Langley KE.** 1990. Identification, purification, and biological characterization of hematopoietic stem cell factor from buffalo rat liver-conditioned medium. *Cell* **63:**195–201.
15. **de Sauvage FJ, Hass PE, Spencer SD, Malloy BE, Gurney AL, Spencer SA, Darbonne WC, Henzel WJ, Wong SC, Kuang WJ, Oles KJ, Hultgren B, Solberg LA, Goeddel DV, Eaton DL.** 1994. Stimulation of megakaryocytopoiesis and thrombopoiesis by the c-Mpl ligand. *Nature* **369:**533–538.
16. **Alexander WS, Roberts AW, Nicola NA, Li R, Metcalf D.** 1996. Deficiencies in progenitor cells of multiple hematopoietic lineages and defective megakaryocytopoiesis in mice lacking the thrombopoietic receptor c-Mpl. *Blood* **87:**2162–2170.
17. **Lieschke GJ, Grail D, Hodgson G, Metcalf D, Stanley E, Cheers C, Fowler KJ, Basu S, Zhan YF, Dunn AR.** 1994. Mice lacking granulocyte colony-stimulating factor have chronic neutropenia, granulocyte and macrophage progenitor cell deficiency, and impaired neutrophil mobilization. *Blood* **84:**1737–1746.
18. **Wiktor-Jedrzejczak W, Ratajczak MZ, Ptasznik A, Sell KW, Ahmed-Ansari A, Ostertag W.** 1992. CSF-1 deficiency in the *op/op* mouse has differential effects on macrophage populations and differentiation stages. *Exp Hematol* **20:**1004–1010.
19. **Kopf M, Brombacher F, Hodgkin PD, Ramsay AJ, Milbourne EA, Dai WJ, Ovington KS, Behm CA, Köhler G, Young IG, Matthaei KI.** 1996. IL-5-deficient mice have a developmental defect in CD5$^+$ B-1 cells and lack eosinophilia but have normal antibody and cytotoxic T cell responses. *Immunity* **4:**15–24.
20. **Qian H, Buza-Vidas N, Hyland CD, Jensen CT, Antonchuk J, Månsson R, Thoren LA, Ekblom M, Alexander WS, Jacobsen SE.** 2007. Critical role of thrombopoietin in maintaining adult quiescent hematopoietic stem cells. *Cell Stem Cell* **1:**671–684.
21. **Metcalf D, Nicola NA.** 1992. The clonal proliferation of normal mouse hematopoietic cells: enhancement and suppression by colony-stimulating factor combinations. *Blood* **79:**2861–2866.
22. **Metcalf D, Di Rago L, Mifsud S.** 2002. Synergistic and inhibitory interactions in the *in vitro* control of murine megakaryocyte colony formation. *Stem Cells* **20:**552–560.
23. **Cynshi O, Satoh K, Shimonaka Y, Hattori K, Nomura H, Imai N, Hirashima K.** 1991. Reduced response to granulocyte colony-stimulating factor in *W/W^v* and *Sl/Sl^d* mice. *Leukemia* **5:**75–77.
24. **Ryan PJ, Willson T, Alexander WS, Di Rago L, Mifsud S, Metcalf D.** 2001. The multi-organ origin of interleukin-5 in the mouse. *Leukemia* **15:**1248–1255.
25. **Williams GT, Smith CA, Spooncer E, Dexter TM, Taylor DR.** 1990. Haemopoietic colony stimulating factors promote cell survival by suppressing apoptosis. *Nature* **343:** 76–79.
26. **Vaux DL, Cory S, Adams JM.** 1988. *Bcl-2* gene promotes haemopoietic cell survival and cooperates with *c-myc* to immortalize pre-B cells. *Nature* **335:**440–442.
27. **Graf T, Enver T.** 2009. Forcing cells to change lineages. *Nature* **462:**587–594.
28. **Riddell J, Gazit R, Garrison BS, Guo G, Saadatpour A, Mandal PK, Ebina W, Volchkov P, Yuan GC, Orkin SH, Rossi DJ.** 2014. Reprogramming committed murine blood cells to induced hematopoietic stem cells with defined factors. *Cell* **157:**549–564.
29. **Rieger MA, Hoppe PS, Smejkal BM, Eitelhuber AC, Schroeder T.** 2009. Hematopoietic cytokines can instruct lineage choice. *Science* **325:**217–218.
30. **Mossadegh-Keller N, Sarrazin S, Kandalla PK, Espinosa L, Stanley ER, Nutt SL, Moore J, Sieweke MH.** 2013. M-CSF instructs myeloid lineage fate in single haematopoietic stem cells. *Nature* **497:**239–243.
31. **Fairbairn LJ, Cowling GJ, Reipert BM, Dexter TM.** 1993. Suppression of apoptosis allows differentiation and development of a multipotent hemopoietic cell line in the absence of added growth factors. *Cell* **74:**823–832.
32. **Dong F, van Buitenen C, Pouwels K, Hoefsloot LH, Löwenberg B, Touw IP.** 1993. Distinct cytoplasmic regions of the human granulocyte colony-stimulating factor receptor involved in induction of proliferation and maturation. *Mol Cell Biol* **13:**7774–7781.
33. **Dührsen U, Villeval JL, Boyd J, Kannourakis G, Morstyn G, Metcalf D.** 1988. Effects of recombinant human granulocyte colony-stimulating factor on hematopoietic progenitor cells in cancer patients. *Blood* **72:**2074–2081.
34. **Socinski MA, Cannistra SA, Elias A, Antman KH, Schnipper L, Griffin JD.** 1988. Granulocyte-macrophage colony stimulating factor expands the circulating haemopoietic progenitor cell compartment in man. *Lancet* **1:** 1194–1198.
35. **Papayannopoulou T, Priestley GV, Bonig H, Nakamoto B.** 2003. The role of G-protein signaling in hematopoietic stem/progenitor cell mobilization. *Blood* **101:** 4739–4747.
36. **Krebs DL, Hilton DJ.** 2001. SOCS proteins: negative regulators of cytokine signaling. *Stem Cells* **19:**378–387.
37. **Croker BA, Metcalf D, Robb L, Wei W, Mifsud S, DiRago L, Cluse LA, Sutherland KD, Hartley L, Williams E, Zhang JG, Hilton DJ, Nicola NA, Alexander WS,**

Roberts AW. 2004. SOCS3 is a critical physiological negative regulator of G-CSF signaling and emergency granulopoiesis. *Immunity* **20:**153–165.

38. **Phrommintikul A, Haas SJ, Elsik M, Krum H.** 2007. Mortality and target haemoglobin concentrations in anaemic patients with chronic kidney disease treated with erythropoietin: a meta-analysis. *Lancet* **369:**381–388.

39. **de Sauvage FJ, Carver-Moore K, Luoh SM, Ryan A, Dowd M, Eaton DL, Moore MW.** 1996. Physiological regulation of early and late stages of megakaryocytopoiesis by thrombopoietin. *J Exp Med* **183:**651–656.

40. **Ng AP, Kauppi M, Metcalf D, Hyland CD, Josefsson EC, Lebois M, Zhang JG, Baldwin TM, Di Rago L, Hilton DJ, Alexander WS.** 2014. Mpl expression on megakaryocytes and platelets is dispensable for thrombopoiesis but essential to prevent myeloproliferation. *Proc Natl Acad Sci U S A* **111:**5884–5889.

41. **Nicola NA.** 1991. Structural and functional characteristics of receptors for colony-stimulating factors (CSFs), p 101–120. *In* Quesenberry PJ, Asano S, Saito K (ed), *Hemopoietic Growth Factors.* Excerpta Medica, Amsterdam, The Netherlands.

42. **Nicholson SE, Novak U, Ziegler SF, Layton JE.** 1995. Distinct regions of the granulocyte colony-stimulating factor receptor are required for tyrosine phosphorylation of the signaling molecules JAK2, Stat3, and p42, $p44^{MAPK}$. *Blood* **86:**3698–3704.

43. **Inaba K, Inaba M, Romani N, Aya H, Deguchi M, Ikehara S, Muramatsu S, Steinman RM.** 1992. Generation of large numbers of dendritic cells from mouse bone marrow cultures supplemented with granulocyte/macrophage colony-stimulating factor. *J Exp Med* **176:**1693–1702.

44. **Athanassakis I, Bleackley RC, Paetkau V, Guilbert L, Barr PJ, Wegmann TG.** 1987. The immunostimulatory effect of T cells and T cell lymphokines on murine fetally derived placental cells. *J Immunol* **138:**37–44.

45. **Schweizerhof M, Stösser S, Kurejova M, Njoo C, Gangadharan V, Agarwal N, Schmelz M, Bali KK, Michalski CW, Brugger S, Dickenson A, Simone DA, Kuner R.** 2009. Hematopoietic colony-stimulating factors mediate tumor-nerve interactions and bone cancer pain. *Nat Med* **15:**802–807.

46. **Hilton DJ, Gough NM.** 1991. Leukemia inhibitory factor: a biological perspective. *J Cell Biochem* **46:**21–26.

47. **Moore MA, Williams N, Metcalf D.** 1973. *In vitro* colony formation by normal and leukemic human hematopoietic cells: interaction between colony-forming and colony-stimulating cells. *J Natl Cancer Inst* **50:**591–602.

48. **Lang RA, Metcalf D, Gough NM, Dunn AR, Gonda TJ.** 1985. Expression of a hemopoietic growth factor cDNA in a factor-dependent cell line results in autonomous growth and tumorigenicity. *Cell* **43:**531–542.

49. **Perkins A, Kongsuwan K, Visvader J, Adams JM, Cory S.** 1990. Homeobox gene expression plus autocrine growth factor production elicits myeloid leukemia. *Proc Natl Acad Sci U S A* **87:**8398–8402.

50. **Metcalf D.** 2010. The colony-stimulating factors and cancer. *Nat Rev Cancer* **10:**425–434.

51. **Gonda TJ, D'Andrea RJ.** 1997. Activating mutations in cytokine receptors: implications for receptor function and role in disease. *Blood* **89:**355–369.

52. **Metcalf D, Roberts AW, Willson TA.** 1996. The regulation of hematopoiesis in *max 41* transgenic mice with sustained excess granulopoiesis. *Leukemia* **10:**311–320.

General Aspects

II

Myeloid Cells in Health and Disease: A Synthesis
Edited by Siamon Gordon

doi:10.1128/microbiolspec.MCHD-0007-2015

Daniel R. Barreda[1,2]
Harold R. Neely[3]
Martin F. Flajnik[3]

Evolution of Myeloid Cells

4

INTRODUCTION

The drive to clear dying cells is evident even in the deepest branches of the *Archaea* and *Bacteria* phylogeny, where colonial biofilms appeared as a mode to promote survival in diverse environments (1). These ancient colonies (earliest fossil record ~3.25 billion years ago) already displayed attributes of multicellular organism specialization, removing nonfunctional spent cells to recycle nutrients and maintain the integrity of the colony (1, 2). Multicellularity followed shortly after and introduced a requirement for removal of nonself (3, 4). Phagocytosis provided an elegant answer to both challenges, and has since served as a primary tool for cell turnover and removal of foreign invaders across all animal groups. It therefore provides a good stage to examine the evolution of myeloid cells through their contributions to homeostasis and host defenses (Fig. 1). This chapter first focuses on ancestral phagocytes and examines their progression from primarily homeostatic cells to multifaceted effectors and regulators of immunity. The literature provides some insight into macrophage and lower metazoan hemocyte function as far back as echinoderms and urochordates. Further examination of gene marker conservation (e.g., apoptotic genes) in sponges and other colonial organisms allows us to dig deeper to examine the factors that led to the phylogenetic origins of cell clearance mechanisms and their continued evolution across newly developing animal branches. Subsequently, we focus on key challenges encountered by higher vertebrate myeloid cells as they manage increasingly complex mechanisms of immunity while maintaining a strict balance between proinflammatory and homeostatic cellular responses. In one example, we examine the impact of specialization through the diverging contributions of macrophages and neutrophils. We then consider the continued specialization of the myeloid lineage through the eyes of the dendritic cell (DC), which, through antigen presentation, effectively integrated new adaptive features into well-established and robust innate mechanisms of immunity. Indeed, documenting the multiple facets that comprise the life history of myeloid cells across evolution would not be possible in a single chapter. However, by focusing on their origins as phagocytes, we can appreciate the continued struggle of a host to develop novel and effective strategies to combat invading pathogens while ensuring the continued maintenance of tissue integrity and homeostasis.

[1]Department of Biological Sciences, University of Alberta, Edmonton, Alberta T6G 2P5, Canada; [2]Department of Agricultural, Food and Nutritional Science, University of Alberta, Edmonton, Alberta T6G 2P5, Canada; [3]Department of Microbiology and Immunology, University of Maryland Baltimore, Baltimore, MD 21201.

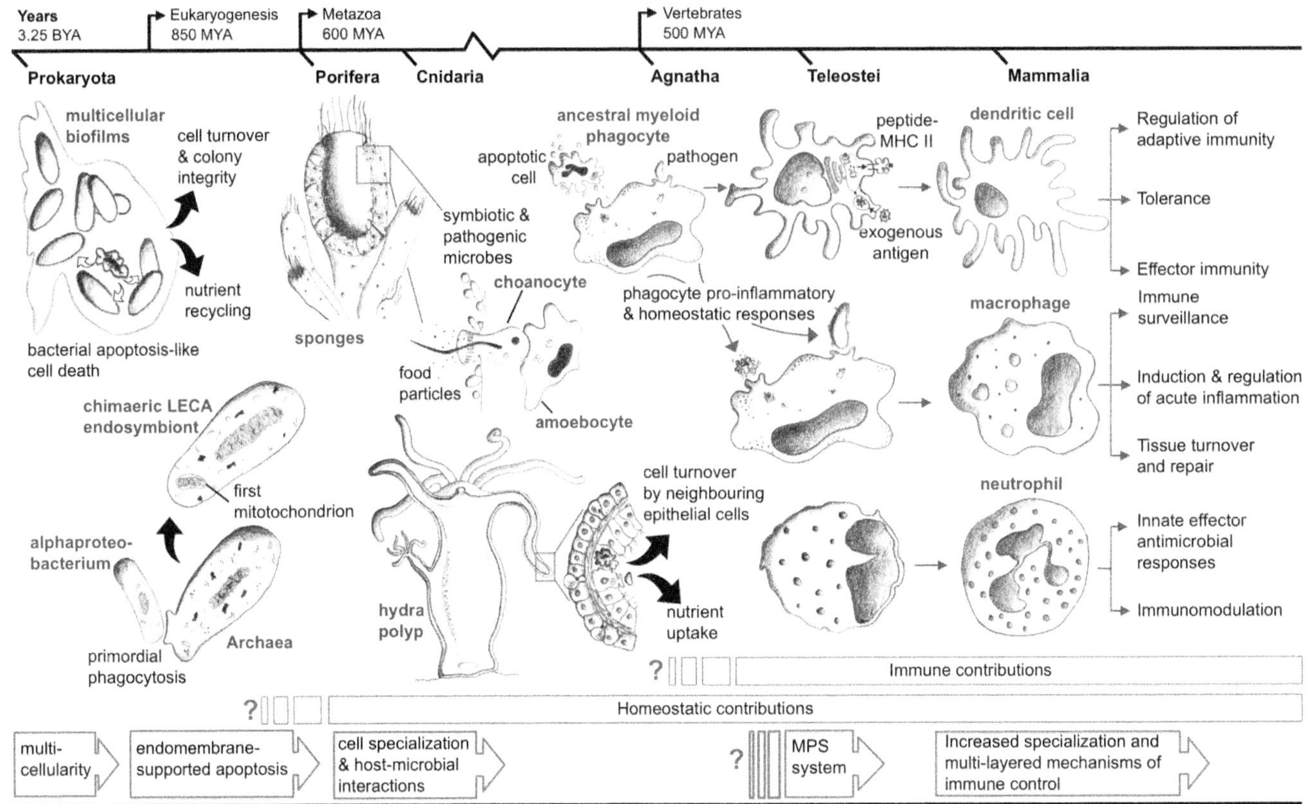

Figure 1 Contributions of phagocytosis to the evolution of myeloid cell function. Major events and key features identified in comparative models are highlighted. Metazoa refers to multicellular animals; invertebrates of the protostome lineage arose 600 MYA and deuterostomes ~500 MYA. Agnathans are jawless fish, and all other vertebrates have jaws (gnathostomes). Adaptive immunity is found only in the vertebrates, as well as the division of labor among myeloid cells that is well known in mammals. Refer to the text for details of each of the particular features described in the figure. MPS, mononuclear phagocyte system.

ANCESTRAL FOUNDATIONS FOR PROGRAMMED CELL DEATH FROM PROKARYOTES

Biofilms represent large, complex bacterial communities that have long shared some of the challenges exhibited as eukaryotes manage multicellular structures. Ecological opportunities within these colonies promote phenotypic diversity that enhances long-term population survival. As in eukaryotes, cell death allows for targeted turnover during development and removal of damaged cells following periods of stress (Fig. 1) (2, 5). The bacterial programmed cell death process is coordinated through the action of holin-like proteins derived from *cid* and *lrg* operons, which were originally identified in *Staphylococcus aureus* but are widely conserved across Gram-positive and Gram-negative bacteria as well as *Archaea* (2, 6). Functions for these proteins are analogous to the proapoptotic effector and antiapoptotic regulators of the B-cell lymphoma 2 (BCL-2) protein family (2). Their capacity to oligomerize in the bacterial membrane is also reminiscent of the oligomerization of effector BCL-2 proteins in the mitochondrial outer membrane, leading to membrane permeabilization and the release of cytochrome *c* during apoptosis (7). Additionally, similar to the manner by which eukaryotic proapoptotic BCL-2 effectors drive caspase activation and initiate the controlled demolition of cellular constituents, bacterial holin-like molecules activate peptidoglycan hydrolases that promote prokaryotic cell disassembly (6).

Other bacterial effectors including RecA, ClpXP, and BapE; the increased production of reactive oxygen species (ROS); and the SOS response further contribute to this process (8–10). These molecular events lead to downstream phenotypic features that are also consistent with a bacterial apoptosis-like cell death process based on observations of cell shrinkage, DNA fragmentation, chromosome condensation, extracellular exposure of phosphatidylserine, and membrane depolarization (8, 9, 11, 12). Indeed, analyses of the evolution of apoptosis regulatory networks suggest that these already displayed a significant level of complexity prior to the development of the metazoan line (13). As such, the seemingly altruistic behaviors of individual members of multicellular bacterial biofilms are simply a result of a cell turnover program that shares several features of modern eukaryotic apoptotic mechanisms, ultimately contributing to the maintenance of colony integrity through management of cellular constituents after death.

Despite the parallels between eukaryotic apoptosis and apoptotic-like mechanisms in bacteria, however, there is little evidence that the latter is the preferred mode of death if biofilm homeostasis is to be maintained. Further, the lytic nature of this mechanism among bacteria is central to the release of genomic DNA that incorporates into the biofilm matrix to enhance biofilm integrity (14–16). The result is a colony with remarkable structural, mechanical, and chemical properties that, among others, offers significant resistance to antimicrobial agents derived from other microorganisms in their environment or human intervention (17). This is markedly different from the classic inflammation paradigm of eukaryotes in which cell lysis events contribute to proinflammatory rather than homeostatic tissue repair outcomes, based on the release of intracellular constituents (Fig. 1). Thus, the mechanisms by which bacterial programmed cell death contributes to the integrity of the colony are quite distinct from those apoptotic-driven events that promote tissue repair and a return to homeostasis in multicellular eukaryotes.

THE BIRTH OF THE *EUKARYA* AND ENDOMEMBRANE CONTAINMENT FOR APOPTOTIC EVENTS

Phylogenetic analyses looking for conserved homologs coding for phagocytosis-relevant proteins identified the early presence of these among a subset of *Archaea*. Among others, these include actin nucleators that are monophyletic with eukaryotic players and further share unique structural features with the modern actin-related protein (ARP)-2/3 (18). The ARP2/3 complex is well established as a central regulator for the polymerization, organization, and recycling of actin filament networks (19). In phagocytosis, ARP2/3 is known to promote the formation of y-branched actin filament networks that provide the structural integrity and mechanical force for lamellipodial protrusions and extension of the plasma membrane around the target particle. Results from phylogenomic analyses also point to the near universality of the actin-centered functional core across the *Eukarya*, based on the conservation of nucleation-promoting factors WASP (Wiskott–Aldrich syndrome protein)/N-WASP (neural Wiskott–Aldrich syndrome protein) or WAVE (WASP-family verprolin-homologous protein)/SCAR (suppressor of cyclic AMP repressor) as well as the actin-binding proteins gelsolin, profilin, cofilin, formin, and coronin, which are involved in the remodeling of actin filaments (18). Their availability to the last eukaryotic common ancestor (LECA) would have offered the capacity to form branched-filament structures and networks during actin polymerization, an important feature for structural remodeling of the cell membrane that would then have facilitated bacterial uptake. This capacity for primordial phagocytosis appears to have been central to the establishment of endosymbionts, whereby eubacterial and archaebacterial prokaryotic ancestors merged to establish a hybrid parent line for today's eukaryotes (Fig. 1). The eubacterial target, presumably an alphaproteobacterium, subsequently became the first functional mitochondrion (20).

Despite continued debate over the morphological features of this archaeal ancestor and the relative efficiency of its phagocytic machinery, support for this dominant model outlining the origins of *Eukarya* can be found in cell biology, paleontology, and biochemical data sets (18, 20, 21). Additional features of contemporary eukaryotes that are proposed to have already been present in the LECA are well reviewed elsewhere (20, 22). Overall, the chimeric nature of this ancestral eukaryote offered new opportunities for further sophistication of the phagocytosis process beyond those which led to uptake of the bacterial endosymbiont. For example, phylogenomic analysis of the Ras superfamily of small GTPases indicates that this group of important actin polymerization regulators was of bacterial origin—both eukaryotic and archaeal members are embedded within bacterial branches, and the cluster that includes Rab-Ran-Ras-Rho GTPases was found to be overwhelmingly bacterial (18, 23). Thus, based on the symbiogenic model for the origin of the *Eukarya* described above, horizontal gene transfer would have allowed for acquisition of novel genes from this family of GTPases from the phagocytosed eubacterial endosymbiont, further enriching the archaeal parent genetic repertoire. As such, basal eukaryotes appear to have already possessed the capacity for phagocytosis and primordial mechanisms of programmed cell death that could help address early challenges associated with multicellularity (Fig. 1).

Additional features acquired during eukaryogenesis further strengthened the platform for the long-standing evolutionary role that apoptosis and phagocytes play in homeostatic control. Among others, ancestral eukaryotes also gained containment of programmed cell death molecular events within membrane envelopes, which prevented the release of intracellular constituents (24). The emergence of this intricate endomembrane system is unique to eukaryotes (25). Based on this model, the eubacterial endosymbiont that evolved into the modern mitochondrion appears to have retained the capacity for autolysis as a mechanism to trigger downstream programmed cell death pathways. However, in light of the development of an endomembrane system, autolytic events now took place without the release of intracellular components that could be construed as danger signals to surrounding cells. Since acquisition of mitochondria appears to predate the origin of all known extant eukaryotes, current support points to a broad application for this novel mode of controlled cell death from the earliest days of the *Eukarya*.

HOST-MICROBE INTERACTIONS AND CELLULAR SPECIALIZATION IN BASAL MULTICELLULAR METAZOANS

The sponges (phylum *Porifera*) are sessile, benthic aquatic organisms that represent the most basal clade of the extant multicellular animals (26). They lack specialized organs and a nervous system and thus must rely on the individual behavior of a limited number of cell types to manage their physiological requirements (27). For ancestral phagocytes, these contributions were based on meeting nutritional needs and managing emerging challenges with symbiotic and pathogenic microorganisms (Fig. 1). Consistent with this idea, carnivorous sponges such as *Asbestopluma hypogea* show the presence of specialized migrating cells that contribute to the digestion of disrupted macroprey (28). In other sponges, nutrients are derived from phagocytosis of free-living bacteria and other particles in the water column filtered through sponge choanocyte chambers (29, 30). As such, examples from the *Porifera* further support the early presence of the specialized cellular machinery necessary for chemotaxis, phagocytosis, and intracellular degradation as well as their application in basal metazoans. Still, further research would help determine whether these archaeocytes (also called amebocytes) and other sponge cells described to internalize food particles represent bona fide ancestral phagocytes that predate immune hemocytes and phagocytic myeloid cells of higher metazoans or an independent wave of cell specialization that permitted a shift in energy acquisition in the absence of a digestive cavity in deep-sea oligotrophic environments.

Notably, *Porifera* microbes comprise as much as 40% of the sponge volume, are well known to establish intimate relationships that range from mutualism to host-pathogen associations, and can be located in the host extracellular and intracellular spaces (31). This provides an opportunity to examine how early host-microbe associations may have led to the development of functional adaptations in basal metazoans and their interacting microorganisms. Not surprisingly, the most abundant bacterial morphotypes among microbial communities colonizing sponges display thickened cell walls, multiple membranes, and slime capsules, which appear to serve to counteract the phagocytic activity by sponge archaeocytes (32). Indeed, early analysis of discrimination among food bacteria (those utilized for nutrient acquisition) and bacterial symbionts showed selectivity in the recognition and uptake by sponge phagocytes (33). Whereas a large proportion of bacterial symbionts were seen to pass through the sponge and were expelled in the exhalant current, food bacteria were retained within the sponge via phagocytic uptake. This is an extraordinary feat given the immense filtering capacity of sponges, estimated at 24,000 liters of water per day for a 1-kg sponge—such volume would translate to 2.4×10^{13} bacteria based on an estimated 10^6 bacteria per ml of seawater (34).

The selectivity of *Porifera* hosts for microbial targets described above is further complemented by well-developed bacterial strategies to survive intracellular killing mechanisms. Most recently, for example, symbionts from the sponge *Cymbastela concentrica* were found to contain a genomic fragment that encoded four proteins whose closest known relatives are part of the sponge genome (35). Two of these eukaryotic-like ankyrin repeat proteins (ARPs) were found to interfere with phagosome development and acidification by blocking phagosome-lysosome fusion. Interestingly, ARPs are well-established tools for molecular mimicry of eukaryotic domains by bacterial pathogens like *Legionella pneumophila* and *Coxiella burnetii*, where they provide an effective mechanism to facilitate host infection and bacterial intracellular survival and proliferation (36–39). Thus, based on the characterization of modern sponges, both host and microbial contributions may have shaped the complexity of early phagocyte-microbe interactions in ancestral metazoans to determine if facultative or obligate symbiotic associations would promote colonization or whether effective rejection would result upon encounter of a pathogen.

DEVELOPMENT OF HOMEOSTATIC PATHWAYS BASED ON PHAGOCYTIC UPTAKE OF APOPTOTIC CELLS

Early work by Ellis and Horvitz in *Caenorhabditis elegans* identified genetic contributors with core regulatory roles in apoptotic events (40). *C. elegans ced-4* was found to encode a molecular chaperone necessary for the activation of *ced-3* and promotion of apoptosis (41). In contrast, *ced-9* antagonized the actions of CED-3 and CED-4 in cells that should survive (42). We now know that these genes represent homologs of the mammalian apoptotic protease-activating factor 1 (*Apaf-1*), interleukin-1β-converting enzyme (*ICE*), and proto-oncogene *bcl-2*, respectively.

More recent studies have extended the origin of this programmed cell death process into basal metazoans. Some of this work was described above. In addition, cnidarians like *Hydra* have helped significantly to define the molecular events associated with apoptosis and the role of phagocytes in its homeostatic contributions in pre-Bilaterians (Fig. 1). This interplay has been observed during its regulation of cell numbers when feeding, tissue remodeling, gametogenesis (oogenesis and spermatogenesis), regeneration, and allorecognition (24, 43, 44). In one example, apoptotic interstitial stem cells were rapidly internalized during normal growth and budding of *Hydra* polyps by neighboring epidermal and gastrodermal epithelial cells. In another, *Hydra* small nurse cells were internalized by surrounding oocytes during oogenesis. This process allowed for the efficient removal of spent cells and further supplied nourishment to developing polyps. The morphology of apoptotic *Hydra* cells is almost indistinguishable from that of higher metazoans (45). Further, the biochemical mechanisms mediating these events, the complexity of gene families contributing to them, as well as the intracellular localization and function of key players such as the caspase and BCL-2 family proteins show a significant degree of conservation with those of higher metazoans (24, 43, 46). Thus, the contribution of phagocytes to the internalization and removal of apoptotic cells—defined as efferocytosis—is well established in cnidarians and appears to represent a platform mechanism for maintenance of homeostasis that arose in basal metazoans. Notably, the apoptosis cell death machinery of *Hydra* displays a significantly higher level of complexity than that of *Caenorhabditis* and *Drosophila*. This suggests that the primordial establishment of this novel homeostatic strategy was followed by selective gene loss after the divergence of the last common eumetazoan ancestor toward the protostome clade rather than the previously suggested more dramatic evolution of the apoptotic machinery from worms to man based on the traditional Coelomata hypothesis that was in place at the time of early characterization of *C. elegans* apoptosis (47–50). This is consistent with examination of the apoptotic machinery in other "intermediate" metazoan clades such as the *Mollusca*, which also show a higher level of complexity than *Caenorhabditis* and *Drosophila* (51).

An additional point to highlight relates to the nature of phagocytes mediating the internalization of apoptotic cells in these basal metazoans. Despite the perceived dominance of mononuclear phagocytes for this uptake in mammals (tissue-resident macrophages and immature DCs) (52, 53), it appears that it was epithelial cells rather than bona fide professional phagocytes that were originally responsible for carrying out this process (Fig. 1) (45). This may point to a segregation of labor toward the mononuclear phagocyte system that later evolved as these cells acquired a central position as inducers and regulators of inflammatory processes in higher metazoans. Alternatively, given that most studies on efferocytosis have focused largely on immune cell fractions and that uptake and removal of apoptotic cells is highly efficient (1 to 2 h), we may find that adherent epithelial populations actually play a dominant (rather than supportive) role in the execution of homeostatic cell clearance programs, and that this remains at least partially true in higher vertebrates. Of course, these studies will require further improvements in our current capacity to detect and quantitate the uptake of apoptotic cells *in vivo* in a range of cells and comparative animal models.

Apoptotic cell recognition by myeloid phagocytes initiates discrete molecular programs aimed at the promotion of homeostasis. In higher vertebrates, we find both the active induction of anti-inflammatory, wound repair, and tolerogenic mechanisms as well as the active repression of proinflammatory programs. Contributing factors include interleukin-10 (IL-10), transforming growth factor β (TGF-β), platelet-activating factor, prostaglandin E_2, lipoxin A_4, lactoferrin, vascular endothelial growth factor, and others (54–56). For cytokines like IL-10 and TGF-β, and lipoxins like lipoxin A_4, this is mediated through a p38 mitogen-activated protein kinase-dependent mechanism and, at least in some cases, requires recognition of apoptotic cell phosphatidylserine (57–59). Increased production of the factors above is accompanied by active translational control of proinflammatory mediators like IL-1β, IL-6, IL-8, and tumor necrosis factor α (TNF-α) (56). This is at least partially linked to repression of Toll-like receptor (TLR)-dependent and -independent proinflammatory signaling cascades (56, 60). The production of ROS through the action of NADPH oxidase provides an additional regulatory node for the transition from a proinflammatory to an anti-inflammatory homeostatic one (Fig. 1). On the one hand, mammalian neutrophil ROS activate RhoA GTPase, a negative regulator of efferocytosis, in surrounding mature macrophages, maintaining a proinflammatory phenotype and causing a reduction in apoptotic cell uptake (61). Conversely, inhibition of ROS production and RhoA activity increases efferocytosis efficiency, reduces TNF-α production, inhibits recruitment of inflammatory leukocytes, and promotes TGF-β release (62). This is consistent with experiments in which the addition of exogenous apoptotic cells reduced neutrophil recruitment, promoted a shift in macrophages toward an anti-inflammatory phenotype, reduced ROS and proinflammatory cytokine production, and increased TGF-β expression (52, 63, 64). As such, phagocyte recognition of apoptotic cells offers key contributions for the modulation of inflammation from an innate immune mechanism that clearly discriminates between live and effete cells (65). Despite the complexity of these homeostatic programs, many of their components appear to have been already well established in pre-Bilaterian metazoans. One good example is the TGF-β superfamily signaling pathway, for which several components have already been identified, among others, in the ctenophore *Mnemiopsis leidyi*, the sponge *Amphimedon*, and the cnidarian *Trichoplax* (66). However, we must be careful to not attribute inflammation regulatory roles to these pathways too early since, as with other components of the immune system, members of this network saw their origins not in immunity but rather in the establishment of tolerance and tissue regeneration. Additional developmental functions described in both protostome and deuterostome models include tissue morphogenesis and dorsal-ventral patterning (67–69). Thus, it remains to be determined when exactly efferocytosis offered its first contributions to the control of proinflammatory immunity programs.

CORE CONTRIBUTIONS OF MYELOID CELLS TO THE BALANCE OF PROINFLAMMATORY AND HOMEOSTATIC RESPONSES IN HIGHER METAZOANS

Elie Metchnikoff's seminal identification of phagocytosis in starfish larvae and an appreciation of its broader significance across various animal groups (Daphnia, snails, rhinoceros beetles, osseous fishes, frogs, turtles, lizards, pigeons, guinea pigs, rabbits, and others) provided strong early support for the prominent role of phagocytes in host defense and homeostasis across evolution. Unlike Virchow and Cohnheim, whose concurrent theories focused on parenchymal and vascular responses, respectively, Metchnikoff saw leukocytes as the most active participants in an inflammatory response that was not a deleterious pathological process but one that reflected a host's continued drive to maintain its integrity (70, 71): "It is possible to state as a general principle that the mesodermic phagocytes, which originally (as in the sponges of our days) acted as digestive cells, retained their role to absorb the dead or weakened parts of the organism as much as different foreign intruders" (71). More recent work has expanded Metchnikoff's early work on basal deuterostomes. It is now known, for example, that at least three distinct phagocyte morphotypes contribute to surveillance and immune defenses in larval and adult stages of echinoderms like the purple sea urchin (discoidal cells, polygonal cells, and small phagocytes) (72). These, along with other echinoderm coelomocytes, further display nearly complete sets of homologs of vertebrate transcriptional regulators of myelopoiesis (e.g., PU.1/SpiB/SpiC/Ets) as well as key molecular players in pathogen recognition and uptake (e.g., greatly expanded gene families for TLRs [>210 genes], NOD-like receptors [NLRs; >200 genes], and scavenger receptors [>1,000 genes]). Importantly, although "the attacked organism defended itself against these little aggressors by all the means at their disposal," Metchnikoff also saw the internalization of neutrophils (also known as microphages) by phagocytic macrophages as a key step in the subsequent resolution of inflammation (71). Indeed, it is now well established that phagocytes serve as ground zero for vertebrate induction and control of divergent proinflammatory and homeostatic responses following internalization of exogenous particles. At the site of infection, pathogen engagement leads to rapid production of proinflammatory mediators including reactive oxygen and nitrogen species (73, 74); antimicrobial peptides (75); and cytokines like TNF-α, IL-1β, and gamma interferon (76). The goal is to elicit potent responses that will deal with the incoming threat rapidly to prevent systemic dispersal of pathogens and additional commitment of energy resources. In contrast, internalization of apoptotic cells initiates a shift toward resolution mechanisms that promote tissue repair and a return to homeostasis once the pathogen has been effectively cleared. This is marked by increases in the production of TGF-β1, IL-10, prostaglandin E_2, and platelet-activating factor (55, 77), combined with decreases in proinflammatory mediators, including

TNF-α, leukotriene C_4, thromboxane B_2, and the interleukins IL-6, IL-8, IL-12, IL-17, and IL-23 (52, 78). Consistent with Metchnikoff's early ideas, this reflects a host's continued drive to maintain its integrity and places myeloid phagocytes as central effectors and regulators of the inflammatory response. However, additional work is still required to determine whether this dichotomy is a feature unique to vertebrates or is also shared by invertebrate phagocytes.

Comparative examination of chordate myeloid populations shows the long-standing role that phagocytes have played in the induction and control of divergent proinflammatory and homeostatic responses. As in mammals, professional phagocytes (macrophages, monocytes, and neutrophils) are the primary contributors to the internalization of proinflammatory particles among teleost fish and agnathan hematopoietic leukocytes (79). Further, individual phagocytes of the jawless vertebrate *Petromyzon marinus* (sea lamprey), like those of teleost fish and mice, display the capacity for divergent proinflammatory and homeostatic responses following internalization of pathogens and apoptotic cells, respectively (63, 79). Despite these similarities, there are also preliminary indications of an evolutionary shift between the contributions of phagocytes to the induction versus the control of inflammation. When compared to their counterparts from jawless fish, teleost phagocytes showed a greater capacity for internalization of pathogen-derived particles and for effective induction of proinflammatory antimicrobial programs. At the same time, higher vertebrate phagocytes displayed a lower sensitivity to homeostatic signals derived from apoptotic cell internalization (79). Thus, an early dominance of phagocyte homeostatic programs in evolution appears to have been followed by a progressive shift toward increased robustness among proinflammatory antimicrobial responses (Fig. 1) (79). Of course, given the critical importance of inflammation control to the maintenance of tissue integrity and the management of precious metabolic resources, it would seem ill-serving to favor the induction of proinflammatory host defenses at the expense of inflammation control. Indeed, it appears that higher vertebrates have answered this dilemma through the establishment of new mechanisms of inflammation control among first-line phagocytes, including the novel capacity of mammalian neutrophils to internalize apoptotic cells, a feature that is not available to their teleost fish counterparts (63, 80). Thus, an expanding body of evidence shows that the contributions of mammalian neutrophils to inflammation control go well beyond their timely induction of apoptosis and subsequent removal by macrophages (81–83). An added division of labor now also sees important contributions to the control of inflammation beyond the myeloid professional phagocyte group (54, 82, 84, 85). However, as is often the case, increased complexity is not always without problems; a quick scan of the literature shows extensive documentation of inflammation-driven disease among biomedical journals (86). An interesting possibility is that this may partially stem from a shorter evolutionary "break-in" period for novel mechanisms of immune control that have only recently developed among the higher vertebrates.

LINKING INNATE AND ADAPTIVE IMMUNITY THROUGH DC SPECIALIZATION

With the appearance of adaptive immunity, the need for novel mechanisms of immune control in jawed vertebrates is obvious, not only with respect to the induction, regulation, and eventual resolution of inflammatory responses but also with respect to the specific (and context-dependent) activation of antigen-restricted lymphocytes. The necessarily low precursor frequency of antigen-restricted lymphocytes generated by germ line DNA rearrangement also necessitated specific anatomical locations (secondary lymphoid organs [SLOs]) in which both lymphocytes and antigen could be concentrated to facilitate the interaction of lymphocytes with cognate antigen. Additionally, a need arose for cells within the SLOs capable of acquiring and presenting antigen to lymphocytes, as well as cells outside the SLOs to acquire antigen and traffic it to the SLOs for presentation.

The term "dendritic cell" was coined by Steinman and Cohn (87) to describe and define a morphologically distinct subset of cells in the SLOs of mice, just over a century after the description of the archetypal DC, the Langerhans cell, in human skin (Fig. 1) (88, 89). In the years since their discovery, mammalian DCs have been extensively characterized, and multiple subtypes with unique functions, tissue distributions, and surface phenotypes have been described (90). Macrophages and DCs share multiple functional and phenotypic similarities (e.g., phagocytic capacity and major histocompatibility complex [MHC] class II expression) (Fig. 1), as well as (in the case of conventional DCs) a common progenitor/lineage (91). Despite the wealth of new and increasingly detailed information on DCs, or perhaps because of it, the line between DCs and macrophages has recently blurred (92). However, conventional DCs can be functionally defined as cells with both a canonical dendritic morphology and, more importantly, the ability to capture and present antigen to antigen-restricted lymphocytes for the purpose of eliciting a specific adaptive immune response. DCs with these properties have existed (at least) since the *Teleostei*, and likely since the appearance of Ig and TCR/peptide-MHC (pMHC)-based adaptive immunity in the cartilaginous fish. Additionally, DCs have been identified and studied with increasing detail in reptiles, amphibians, and birds. Herein, we hope to illustrate a progressive specialization in DC phenotype and function, alongside a retention of their ancestral and defining characteristics—MHC-II expression and the ability to elicit specific adaptive immune responses—since their first appearance in early jawed vertebrates.

Adaptive immunity based on germ line rearrangement of leucine-rich repeat gene segments, along with a T/B dichotomy, has recently been demonstrated in agnathans (93). However, as "there has been no sign of MHC I or II genes in animals older than the cartilaginous fish" (94), it seems unlikely that a canonical DC population (at least one partially defined by MHC-II expression) exists in the jawless vertebrates. Adaptive immunity based on rearranging Ig and TCR/pMHC arose in the jawed vertebrates, of which the *Chondrichthyes* is the oldest living group (cartilaginous fishes, i.e., chimeras, sharks, skates, and rays; common ancestor with humans ~500 million years ago [MYA]), as did the primordial SLO the spleen. Along with lymphocytes in the white pulp (WP) of the quiescent nurse shark (*Ginglymostoma cirratum*) spleen, we have described large cells expressing high levels of MHC-II and displaying dendritic processes (95). Immunization of adult nurse sharks with biotinylated bovine serum albumin resulted in an accumulation of immunogen in the splenic red pulp (RP) within 1 week; 4 weeks after immunization, the immunogen had localized to the splenic WP and was predominantly observed at the plasma membrane of large cells with dendritic processes (our unpublished observations). The only other putative DC population described in cartilaginous fish was identified in dogfish; Torroba and colleagues reported "irregular macrophages with long cell processes" in the hypothalamic ventricle (96), though their functional role in adaptive immunity was not addressed. Despite a lack of functional data for these putative DCs in the nurse shark, their predicted role in the induction of adaptive immunity is supported by the appearance of not only MHC-II but also CD83 (an Ig superfamily transmembrane protein associated with activated/differentiated DCs) (97) in the nurse shark genome.

Some of the most definitive recent work on DC evolution has been performed in teleost fish (common ancestor with humans ~400 MYA), particularly in the zebrafish. Employing a cell-sorting method based on a combination of a light-scatter profile characteristic of myelomonocytes and binding of peanut agglutinin, a lectin that binds Gal-β-1-3-GalNAc (a carbohydrate moiety present on, among other leukocytes, DCs), Lugo-Villarino and colleagues (98) isolated cells from zebrafish whole-kidney marrow and demonstrated that these cells are bona fide DCs by cellular morphology, expression of DC-associated transcripts (e.g., IL-12, MHC-II invariant chain), and induction of antigen-restricted T-cell proliferation. Further, they showed that these cells are present in peritoneal cavity, spleen, gut, thymus, and skin, though not in brain or liver (99), demonstrating that the zebrafish possesses not only cells with the phenotypic and functional characteristics of conventional DCs but also a tissue distribution pattern of these cells to some extent resembling that found in mammals.

In the Atlantic salmon (*Salmo salar* L.), a non-lymphocyte population of MHC-II$^+$ CD83$^+$ (demonstrated by reverse transcriptase PCR) cells with the ability to differentiate into cells with dendritic morphology has been identified (100). In the rainbow trout (*Oncorhynchus mykiss*), cells with dendritic morphology have been identified as well, and were also demonstrated to be MHC-II positive, able to stimulate T-cell proliferation in a mixed-lymphocyte reaction, migratory, and phagocytic and to have a gene expression profile of conventional DCs (e.g., TLR, B7 family, CD83, CD209/DC-SIGN). In addition, like the cells identified in the salmon, the trout DCs mature upon TLR simulation, increasing their expression of CD83 and MHC-II and elongating their dendritic processes (101). While these data do not formally identify a DC lineage distinct from that of macrophages, they do demonstrate a phenotypic and functional specialization strongly suggestive of canonical DCs in teleost fish.

In the spleen of the amphibian *Xenopus* (common ancestor with humans ~350 MYA), "large, mitotically active cells with abundant electron lucent cytoplasm, large hyperlobated nuclei and prominent nucleoli are found in the periphery of the splenic white pulp" (102). These cells, termed XL cells, possess long cytoplasmic processes and, upon immunization, position themselves in a discrete ring at the internal perimeter of the WP, just inside the Grenzschichtenmembran of Sterba (a double layer of cells surrounding the WP and forming a boundary between it and the RP). XL cells have been implicated in the trafficking of antigen from the RP into the WP and have been suggested to be capable of retaining native (i.e., unprocessed) antigen at their plasma membrane. Also upon immunization, native antigen is detectable at the internal perimeter of the WP, colocalizing with the immigrating/repositioning XL cells (103). Further, this antigen import into the WP was shown to be thymus dependent—immunogen was undetectable in the WP of thymectomized animals. With these data in mind, XL cells were proposed to be primitive follicular DCs.

Follicular dendritic cells (FDCs) were "originally identified by their striking morphology and their ability to trap immune complexes (ICs) of antigen and antibodies in B cell follicles" (104), and are instrumental in the initiation of humoral immune responses by virtue of their ability to retain and present native antigen, at their plasma membrane, to B cells. The precursor cells to FDCs (pre-FDCs) were recently identified as ubiquitous, perivascular mural cells expressing the platelet-derived growth factor receptor β (105), and as such these cells are of a nonhematopoietic lineage fundamentally distinct from both macrophages and conventional DCs. Nonetheless, we feel that their inclusion in this review is necessary; FDCs, or cells functionally comparable to FDCs in their ability to trap antigen/ICs at their plasma membrane for presentation to B lymphocytes, are as central to the initiation of adaptive, humoral immune responses as conventional DCs are to adaptive, cell-mediated immune responses. Additionally, the microarchitecture of the mammalian WP (characterized by a central arteriole surrounded by a periarteriolar sheath of T cells, with one or more B-cell follicles adjacent, and all bounded by a marginal zone comprising a unique subset of B cells and two distinct subpopulations of macrophages) is dependent on the lymphotoxin (LT)-$\alpha_1\beta_2$-dependent maturation of pre-FDCs into mature FDCs. Likewise, the establishment of germinal centers (GCs) in mammals is both LT dependent and FDC dependent.

We have shown that the XL cells are indeed capable of acquiring and retaining native antigen at their plasma membrane after thymus-dependent immunization with phycoerythrin in incomplete Freund adjuvant, suggesting the proposed FDC-like function. In addition, we have shown, by both flow cytometry and reverse transcriptase PCR, that the XL cells express high levels of MHC-II, which, along with the thymus dependence of their antigen transport into the WP, suggests classical DC function as well (our unpublished observations). We have therefore proposed that the XL cells, the DCs of the *Xenopus* spleen, perform "double duty," presenting both native, surface-bound antigen to B cells as well as pMHC antigen to T cells. However, our data do not provide evidence for or against a distinct lineage of DCs (separate from macrophages) in *Xenopus*. Nonetheless, the T-cell-dependent migration of the *Xenopus* XL cells into the WP, as well as their positioning within the B-cell follicle, demonstrate a functional maturation dependent on (and a dedication to) interaction with lymphocytes, suggestive of a specific role in the induction of adaptive immunity, distinct from their phagocytic capacity. Outside the spleen (the only SLO in *Xenopus*), DCs (as defined by morphology and MHC-II expression) have been identified in both the thymus and skin (106, 107). Furthermore, cells in the skin with classical dendritic morphology and expressing vimentin have been identified, suggesting that the specialization of cells analogous to Langerhans cells occurred at the latest in the amphibians (108).

A population of cells similar to the *Xenopus* XL cells has been described in another amphibian, the natterjack toad (*Bufo calamita*) (109). In this study, colloidal carbon particles injected into the dorsal lymph sac were first captured by macrophages and then transported into the RP, where cells with a similar morphology to *Xenopus* XL cells were observed. In a subsequent study (110), immunization with sheep red blood cells resulted in a similar appearance of antigen in the RP, followed by a progression of antigen into the WP, and also suggested that immunization resulted in the maturation/differentiation of monocytic cells into "giant, dendritic-like" cells (residing predominantly in the RP). As with the *Xenopus* XL cells, it is difficult to declare these cells distinct from macrophages; we hope that our current studies will help define the multiplicity versus differentiation state of APC lineages in amphibians and rekindle a general interest in the evolution of antigen presentation.

While reptiles (common ancestor with humans ~300 MYA) have been studied less extensively than amphibians, at least two examples of putative DCs have been described. In the WP of the Asiatic reticulated python (*Python reticulatus*) is a population of cells that, like the XL cells in *Xenopus*, localize to the periphery of the WP upon immunization and are also able to trap and retain surface immunogen (111). In the Caspian turtle (*Mauremys caspica*), Zapata and colleagues originally described "dendritic macrophages" present within the reticular network of the WP (112), and later identified DCs, characterized by dendritic morphology (with ultrastructural similarity to mammalian FDCs) and weak phagocytic capacity, in the inner region of the periellipsoidal lymphoid sheath of the WP (113). Notably, the splenic WP of *M. caspica* comprises a central arteriole surrounded by a periarteriolar sheath of Ig-negative lymphocytes (presumably T cells), which is in turn surrounded by a periellipsoidal sheath, the inner layer of which contains Ig-positive B cells and the described DCs. This is the earliest instance of splenic WP microarchitecture with remarkable similarity to the mammalian WP (though GCs have not been identified in any reptile).

Outside of mammals, DC subpopulations and lineages have been investigated most extensively in birds (common ancestor with humans ~200 MYA), particularly in the chicken *Gallus gallus*. Indeed, multiple subtypes of DCs (in multiple anatomical locations) have been identified and characterized in the chicken (114). Langerhans cells in the esophagus and skin, identified by their cellular morphology, positioning within the epidermis, and expression of vimentin and MHC-II, have been shown to be migratory from the epidermis to the dermal lymphoid nodules in response to hapten exposure. In the spleen, both ellipsoid-associated cells (similar to the cells described in the ellipsoid of the reptilian spleen) and their progeny, CD83-expressing interdigitating DCs, are found in intimate contact with WP lymphocytes. In the bursa of Fabricius, bursal secretory DCs are found. Each of these populations is $CD45^+$, derived from hematopoietic precursors.

In addition to the DC subpopulations described above, Wu and colleagues have established a method of generating chicken bone marrow-derived DCs *in vitro* after culture in a combination of recombinant chicken granulocyte-macrophage colony-stimulating factor and IL-4 (115). These cells display the canonical dendritic morphology, are moderately phagocytic, are capable of eliciting T-cell proliferation in a mixed-lymphocyte reaction, and express high levels of surface MHC-II and CD11c, along with moderate levels of CD40. Upon stimulation with either lipopolysaccharide or CD40L, their phagocytic capacity decreases, surface expression of CD83 is acquired, and transcription of genes associated with induction of a canonical Th1 response is induced. Additionally, expression of the chemokine receptor CCR6 decreases, concurrent with an increase in the expression of CCR7, indicating not only a functional maturation but also a propensity toward migration/repositioning after activation (116).

Within the WP of the chicken spleen, cells with the canonical morphology, surface phenotype, and function of mammalian FDCs have been identified. These chicken FDCs are stellate, express vascular cellular adhesion molecule-1 and intercellular adhesion molecule-1 (among other cellular adhesion molecules), stain as positive for surface immunoglobulin (IgM and IgG/Y), and have a demonstrated capacity to stimulate both B-cell proliferation and class switch recombination to IgG/Y without the ability to stimulate T-cell proliferation (117). The lineage of these cells is controversial; conflicting reports regarding their expression of both CD45 and MHC-II (positivity of each indicating a hematopoietic lineage) have been published. Regardless of the lineage of the chicken FDCs, the microarchitecture of the chicken WP, like that of the reptile WP, is remarkably similar to that of mammals. In addition, GCs (which in mammals are FDC dependent) form in the chicken spleen. The microarchitectural organization of the chicken WP, along with the presence of GCs, are surprising, given that the chicken genome has lost TNF-α, LT-α, and LT-β (118, 119), all of which are necessary for the establishment and maintenance of mammalian WP and GCs.

The earlier years of the study of macrophages and DCs naturally yielded differences between the two cell types; a relative paucity of reagents for distinguishing the two made a distinction between them a natural starting point for experimentation. With the technological advances in multiparameter flow cytometry (and the dramatic increase in monoclonal antibodies to leukocyte markers) and transcriptomic analysis, it was almost inevitable that more and more similarities between cells with a common progenitor would be found. Nonetheless, we feel that DCs are functionally distinct from macrophages in their propensity/dedication to the initiation of adaptive immune responses. With this almost purely functional distinction in mind, it is difficult to unequivocally identify the initial appearance of the DC—as a cell separate from the macrophage—in the evolution of adaptive immunity.

Citation. Barreda DR, Neely HR, Flajnik MF. 2016. Evolution of myeloid cells. Microbiol Spectrum 4(3):MCHD-0007-2015.

References

1. **Hall-Stoodley L, Costerton JW, Stoodley P.** 2004. Bacterial biofilms: from the natural environment to infectious diseases. *Nat Rev Microbiol* **2:**95–108.
2. **Bayles KW.** 2007. The biological role of death and lysis in biofilm development. *Nat Rev Microbiol* **5:** 721–726.
3. **Ereskovsky AV, Renard E, Borchiellini C.** 2013. Cellular and molecular processes leading to embryo formation in sponges: evidences for high conservation of processes throughout animal evolution. *Dev Genes Evol* **223:** 5–22.
4. **Nedelcu AM.** 2012. The evolution of self during the transition to multicellularity. *Adv Exp Med Biol* **738:** 14–30.
5. **Lewis K.** 2000. Programmed death in bacteria. *Microbiol Mol Biol Rev* **64:**503–514.

6. **Bayles KW.** 2014. Bacterial programmed cell death: making sense of a paradox. *Nat Rev Microbiol* **12:** 63–69.
7. **Ranjit DK, Endres JL, Bayles KW.** 2011. *Staphylococcus aureus* CidA and LrgA proteins exhibit holin-like properties. *J Bacteriol* **193:**2468–2476.
8. **Bos J, Yakhnina AA, Gitai Z.** 2012. BapE DNA endonuclease induces an apoptotic-like response to DNA damage in *Caulobacter*. *Proc Natl Acad Sci U S A* **109:** 18096–18101.
9. **Dwyer DJ, Camacho DM, Kohanski MA, Callura JM, Collins JJ.** 2012. Antibiotic-induced bacterial cell death exhibits physiological and biochemical hallmarks of apoptosis. *Mol Cell* **46:**561–572.
10. **Wadhawan S, Gautam S, Sharma A.** 2010. Metabolic stress-induced programmed cell death in *Xanthomonas*. *FEMS Microbiol Lett* **312:**176–183.
11. **Bidle KD, Falkowski PG.** 2004. Cell death in planktonic, photosynthetic microorganisms. *Nat Rev Microbiol* **2:**643–655.
12. **Hakansson AP, Roche-Hakansson H, Mossberg AK, Svanborg C.** 2011. Apoptosis-like death in bacteria induced by HAMLET, a human milk lipid-protein complex. *PLoS One* **6:**e17717. doi:10.1371/journal.pone.0017717.
13. **Zmasek CM, Godzik A.** 2013. Evolution of the animal apoptosis network. *Cold Spring Harb Perspect Biol* **5:** a008649. doi:10.1101/cshperspect.a008649.
14. **Conover MS, Mishra M, Deora R.** 2011. Extracellular DNA is essential for maintaining *Bordetella* biofilm integrity on abiotic surfaces and in the upper respiratory tract of mice. *PLoS One* **6:**e16861. doi:10.1371/journal.pone.0016861.
15. **Rice KC, Mann EE, Endres JL, Weiss EC, Cassat JE, Smeltzer MS, Bayles KW.** 2007. The *cidA* murein hydrolase regulator contributes to DNA release and biofilm development in *Staphylococcus aureus*. *Proc Natl Acad Sci U S A* **104:**8113–8118.
16. **Whitchurch CB, Tolker-Nielsen T, Ragas PC, Mattick JS.** 2002. Extracellular DNA required for bacterial biofilm formation. *Science* **295:**1487. doi:10.1126/science.295.5559.1487.
17. **Schultz D, Onuchic JN, Ben-Jacob E.** 2012. Turning death into creative force during biofilm engineering. *Proc Natl Acad Sci U S A* **109:**18633–18634.
18. **Yutin N, Wolf MY, Wolf YI, Koonin EV.** 2009. The origins of phagocytosis and eukaryogenesis. *Biol Direct* **4:**9. doi:10.1186/1745-6150-4-9.
19. **Goley ED, Welch MD.** 2006. The ARP2/3 complex: an actin nucleator comes of age. *Nat Rev Mol Cell Biol* **7:** 713–726.
20. **McInerney JO, O'Connell MJ, Pisani D.** 2014. The hybrid nature of the Eukaryota and a consilient view of life on Earth. *Nat Rev Microbiol* **12:**449–455.
21. **Martijn J, Ettema TJ.** 2013. From archaeon to eukaryote: the evolutionary dark ages of the eukaryotic cell. *Biochem Soc Trans* **41:**451–457.
22. **Poole AM, Neumann N.** 2011. Reconciling an archaeal origin of eukaryotes with engulfment: a biologically plausible update of the Eocyte hypothesis. *Res Microbiol* **162:**71–76.
23. **Eliáš M, Klimeš V.** 2012. Rho GTPases: deciphering the evolutionary history of a complex protein family. *Methods Mol Biol* **827:**13–34.
24. **David CN, Schmidt N, Schade M, Pauly B, Alexandrova O, Böttger A.** 2005. *Hydra* and the evolution of apoptosis. *Integr Comp Biol* **45:**631–638.
25. **Jékely G.** 2003. Small GTPases and the evolution of the eukaryotic cell. *BioEssays* **25:**1129–1138.
26. **Thacker RW, Díaz MC, Kerner A, Vignes-Lebbe R, Segerdell E, Haendel MA, Mungall CJ.** 2014. The Porifera Ontology (PORO): enhancing sponge systematics with an anatomy ontology. *J Biomed Semantics* **5:** 39. doi:10.1186/2041-1480-5-39.
27. **Bergquist PR.** 1978. *Sponges*. University of California Press, Berkeley, CA.
28. **Vacelet J, Duport E.** 2004. Prey capture and digestion in the carnivorous sponge *Asbestopluma hypogea* (Porifera: Demospongiae). *Zoomorphology* **123:**179–190.
29. **Wehrl M, Steinert M, Hentschel U.** 2007. Bacterial uptake by the marine sponge *Aplysina aerophoba*. *Microb Ecol* **53:**355–365.
30. **Hadas E, Shpigel M, Ilan M.** 2009. Particulate organic matter as a food source for a coral reef sponge. *J Exp Biol* **212:**3643–3650.
31. **Taylor MW, Radax R, Steger D, Wagner M.** 2007. Sponge-associated microorganisms: evolution, ecology, and biotechnological potential. *Microbiol Mol Biol Rev* **71:**295–347.
32. **Hentschel U, Hopke J, Horn M, Friedrich AB, Wagner M, Hacker J, Moore BS.** 2002. Molecular evidence for a uniform microbial community in sponges from different oceans. *Appl Environ Microbiol* **68:**4431–4440.
33. **Wilkinson CR, Garrone R, Vacelet J.** 1984. Marine sponges discriminate between food bacteria and bacterial symbionts: electron microscope autoradiography and *in situ* evidence. *Proc R Soc Lond B Biol Sci* **220:** 519–528.
34. **Vogel S.** 1977. Current-induced flow through living sponges in nature. *Proc Natl Acad Sci U S A* **74:**2069–2071.
35. **Nguyen MT, Liu M, Thomas T.** 2014. Ankyrin-repeat proteins from sponge symbionts modulate amoebal phagocytosis. *Mol Ecol* **23:**1635–1645.
36. **Habyarimana F, Al-Khodor S, Kalia A, Graham JE, Price CT, Garcia MT, Kwaik YA.** 2008. Role for the Ankyrin eukaryotic-like genes of *Legionella pneumophila* in parasitism of protozoan hosts and human macrophages. *Environ Microbiol* **10:**1460–1474.
37. **Pan X, Lührmann A, Satoh A, Laskowski-Arce MA, Roy CR.** 2008. Ankyrin repeat proteins comprise a diverse family of bacterial type IV effectors. *Science* **320:** 1651–1654.
38. **Price CT, Al-Khodor S, Al-Quadan T, Abu Kwaik Y.** 2010. Indispensable role for the eukaryotic-like ankyrin domains of the ankyrin B effector of *Legionella pneumophila* within macrophages and amoebae. *Infect Immun* **78:**2079–2088.

39. **van Schaik EJ, Chen C, Mertens K, Weber MM, Samuel JE.** 2013. Molecular pathogenesis of the obligate intracellular bacterium *Coxiella burnetii*. *Nat Rev Microbiol* **11:**561–573.
40. **Ellis HM, Horvitz HR.** 1986. Genetic control of programmed cell death in the nematode *C. elegans*. *Cell* **44:**817–829.
41. **Yuan J, Horvitz HR.** 1992. The *Caenorhabditis elegans* cell death gene *ced-4* encodes a novel protein and is expressed during the period of extensive programmed cell death. *Development* **116:**309–320.
42. **Hengartner MO, Ellis RE, Horvitz HR.** 1992. *Caenorhabditis elegans* gene *ced-9* protects cells from programmed cell death. *Nature* **356:**494–499.
43. **Reiter S, Crescenzi M, Galliot B, Buzgariu W.** 2012. *Hydra*, a versatile model to study the homeostatic and developmental functions of cell death. *Int J Dev Biol* **56:**593–604.
44. **Burnet FM.** 1971. "Self-recognition" in colonial marine forms and flowering plants in relation to the evolution of immunity. *Nature* **232:**230–235.
45. **Böttger A, Alexandrova O.** 2007. Programmed cell death in *Hydra*. *Semin Cancer Biol* **17:**134–146.
46. **Lasi M, David CN, Böttger A.** 2010. Apoptosis in pre-Bilaterians: *Hydra* as a model. *Apoptosis* **15:**269–278.
47. **Edgecombe GD, Giribet G, Dunn CW, Hejnol A, Kristensen RM, Neves RC, Rouse GW, Worsaae K, Sorensen MV.** 2011. Higher-level metazoan relationships: recent progress and remaining questions. *Org Divers Evol* **11:**151–172.
48. **Holland PW.** 1999. The future of evolutionary developmental biology. *Nature* **402**(Suppl)**:**C41–C44.
49. **Aravind L, Dixit VM, Koonin EV.** 2001. Apoptotic molecular machinery: vastly increased complexity in vertebrates revealed by genome comparisons. *Science* **291:**1279–1284.
50. **Vaux DL, Strasser A.** 1996. The molecular biology of apoptosis. *Proc Natl Acad Sci U S A* **93:**2239–2244.
51. **Sokolova IM.** 2009. Apoptosis in molluscan immune defense. *Invertebrate Surviv J* **6:**49–58.
52. **Poon IK, Lucas CD, Rossi AG, Ravichandran KS.** 2014. Apoptotic cell clearance: basic biology and therapeutic potential. *Nat Rev Immunol* **14:**166–180.
53. **Henson PM, Hume DA.** 2006. Apoptotic cell removal in development and tissue homeostasis. *Trends Immunol* **27:**244–250.
54. **Ortega-Gómez A, Perretti M, Soehnlein O.** 2013. Resolution of inflammation: an integrated view. *EMBO Mol Med* **5:**661–674.
55. **Fadok VA, Chimini G.** 2001. The phagocytosis of apoptotic cells. *Semin Immunol* **13:**365–372.
56. **Elliott MR, Ravichandran KS.** 2010. Clearance of apoptotic cells: implications in health and disease. *J Cell Biol* **189:**1059–1070.
57. **Chung EY, Liu J, Homma Y, Zhang Y, Brendolan A, Saggese M, Han J, Silverstein R, Selleri L, Ma X.** 2007. Interleukin-10 expression in macrophages during phagocytosis of apoptotic cells is mediated by homeodomain proteins Pbx1 and Prep-1. *Immunity* **27:**952–964.
58. **Hottz ED, Medeiros-de-Moraes IM, Vieira-de-Abreu A, de Assis EF, Vals-de-Souza R, Castro-Faria-Neto HC, Weyrich AS, Zimmerman GA, Bozza FA, Bozza PT.** 2014. Platelet activation and apoptosis modulate monocyte inflammatory responses in dengue. *J Immunol* **193:**1864–1872.
59. **Krönke G, Katzenbeisser J, Uderhardt S, Zaiss MM, Scholtysek C, Schabbauer G, Zarbock A, Koenders MI, Axmann R, Zwerina J, Baenckler HW, van den Berg W, Voll RE, Kühn H, Joosten LA, Schett G.** 2009. 12/15-Lipoxygenase counteracts inflammation and tissue damage in arthritis. *J Immunol* **183:**3383–3389.
60. **Pattabiraman G, Lidstone EA, Palasiewicz K, Cunningham BT, Ucker DS.** 2014. Recognition of apoptotic cells by viable cells is specific, ubiquitous, and species independent: analysis using photonic crystal biosensors. *Mol Biol Cell* **25:**1704–1714.
61. **McPhillips K, Janssen WJ, Ghosh M, Byrne A, Gardai S, Remigio L, Bratton DL, Kang JL, Henson P.** 2007. TNF-α inhibits macrophage clearance of apoptotic cells via cytosolic phospholipase A_2 and oxidant-dependent mechanisms. *J Immunol* **178:**8117–8126.
62. **Moon C, Lee YJ, Park HJ, Chong YH, Kang JL.** 2010. N-Acetylcysteine inhibits RhoA and promotes apoptotic cell clearance during intense lung inflammation. *Am J Respir Crit Care Med* **181:**374–387.
63. **Rieger AM, Konowalchuk JD, Grayfer L, Katzenback BA, Havixbeck JJ, Kiemele MD, Belosevic M, Barreda DR.** 2012. Fish and mammalian phagocytes differentially regulate pro-inflammatory and homeostatic responses in vivo. *PLoS One* **7:**e47070. doi:10.1371/journal.pone.0047070.
64. **Rieger AM, Havixbeck JJ, Belosevic M, Barreda DR.** 2015. Teleost soluble CSF-1R modulates cytokine profiles at an inflammatory site, and inhibits neutrophil chemotaxis, phagocytosis, and bacterial killing. *Dev Comp Immunol* **49:**259–266.
65. **Birge RB, Ucker DS.** 2008. Innate apoptotic immunity: the calming touch of death. *Cell Death Differ* **15:**1096–1102.
66. **Pang K, Ryan JF, Baxevanis AD, Martindale MQ.** 2011. Evolution of the TGF-β signaling pathway and its potential role in the ctenophore, *Mnemiopsis leidyi*. *PLoS One* **6:**e24152. doi:10.1371/journal.pone.0024152.
67. **De Robertis EM, Sasai Y.** 1996. A common plan for dorsoventral patterning in Bilateria. *Nature* **380:**37–40.
68. **Herpin A, Lelong C, Favrel P.** 2004. Transforming growth factor-β-related proteins: an ancestral and widespread superfamily of cytokines in metazoans. *Dev Comp Immunol* **28:**461–485.
69. **Kingsley DM.** 1994. The TGF-β superfamily: new members, new receptors, and new genetic tests of function in different organisms. *Genes Dev* **8:**133–146.
70. **Cavaillon JM.** 2011. The historical milestones in the understanding of leukocyte biology initiated by Elie Metchnikoff. *J Leukoc Biol* **90:**413–424.

71. **Metchnikoff E.** 1968; orig. 1905. *Immunity in infective diseases*. Johnson Reprint Corporation, New York and London.

72. **Smith LC, Ghosh J, Buckley KM, Clow LA, Dheilly NM, Haug T, Henson JH, Li C, Lun CM, Majeske AJ, Matranga V, Nair SV, Rast JP, Raftos DA, Roth M, Sacchi S, Schrankel CS, Stensvag K.** 2010. Echinoderm immunity, p 260–301. *In* Söderhäll K (ed), *Invertebrate Immunity*. Springer, New York, NY.

73. **West AP, Brodsky IE, Rahner C, Woo DK, Erdjument-Bromage H, Tempst P, Walsh MC, Choi Y, Shadel GS, Ghosh S.** 2011. TLR signalling augments macrophage bactericidal activity through mitochondrial ROS. *Nature* **472:**476–480.

74. **Underhill DM, Ozinsky A.** 2002. Phagocytosis of microbes: complexity in action. *Annu Rev Immunol* **20:** 825–852.

75. **Danilova N.** 2006. The evolution of immune mechanisms. *J Exp Zoolog B Mol Dev Evol* **306:**496–520.

76. **Jung HC, Eckmann L, Yang SK, Panja A, Fierer J, Morzycka-Wroblewska E, Kagnoff MF.** 1995. A distinct array of proinflammatory cytokines is expressed in human colon epithelial cells in response to bacterial invasion. *J Clin Invest* **95:**55–65.

77. **Maderna P, Godson C.** 2003. Phagocytosis of apoptotic cells and the resolution of inflammation. *Biochim Biophys Acta* **1639:**141–151.

78. **Henson PM, Bratton DL, Fadok VA.** 2001. Apoptotic cell removal. *Curr Biol* **11:**R795–R805.

79. **Havixbeck JJ, Rieger AM, Wong ME, Wilkie MP, Barreda DR.** 2014. Evolutionary conservation of divergent pro-inflammatory and homeostatic responses in lamprey phagocytes. *PLoS One* **9:**e86255. doi:10.1371/journal.pone.0086255.

80. **Esmann L, Idel C, Sarkar A, Hellberg L, Behnen M, Möller S, van Zandbergen G, Klinger M, Köhl J, Bussmeyer U, Solbach W, Laskay T.** 2010. Phagocytosis of apoptotic cells by neutrophil granulocytes: diminished proinflammatory neutrophil functions in the presence of apoptotic cells. *J Immunol* **184:**391–400.

81. **Devitt A, Marshall LJ.** 2011. The innate immune system and the clearance of apoptotic cells. *J Leukoc Biol* **90:** 447–457.

82. **Silva MT.** 2011. Macrophage phagocytosis of neutrophils at inflammatory/infectious foci: a cooperative mechanism in the control of infection and infectious inflammation. *J Leukoc Biol* **89:**675–683.

83. **Soehnlein O, Lindbom L.** 2010. Phagocyte partnership during the onset and resolution of inflammation. *Nat Rev Immunol* **10:**427–439.

84. **Mulero V, Sepulcre MP, Rainger GE, Buckley CD.** 2011. Editorial: neutrophils live on a two-way street. *J Leukoc Biol* **89:**645–647.

85. **Scapini P, Cassatella MA.** 2014. Social networking of human neutrophils within the immune system. *Blood* **124:**710–719.

86. **Weissmann G.** 2010. It's complicated: inflammation from Metchnikoff to Meryl Streep. *FASEB J* **24:**4129–4132.

87. **Steinman RM, Cohn ZA.** 1973. Identification of a novel cell type in peripheral lymphoid organs of mice. I. Morphology, quantitation, tissue distribution. *J Exp Med* **137:**1142–1162.

88. **Jolles S.** 2002. Paul Langerhans. *J Clin Pathol* **55:**243.

89. **Langerhans P.** 1868. Ueber die Nerven der menschlichen Haut. *Arch Pathol Anat Physiol Klin Med* **44:**325–337.

90. **Karmaus PW, Chi H.** 2014. Genetic dissection of dendritic cell homeostasis and function: lessons from cell type-specific gene ablation. *Cell Mol Life Sci* **71:**1893–1906.

91. **Satpathy AT, Wu X, Albring JC, Murphy KM.** 2012. Re(de)fining the dendritic cell lineage. *Nat Immunol* **13:** 1145–1154.

92. **Hume DA.** 2008. Macrophages as APC and the dendritic cell myth. *J Immunol* **181:**5829–5835.

93. **Flajnik MF.** 2014. Re-evaluation of the immunological Big Bang. *Curr Biol* **24:**R1060–R1065.

94. **Flajnik MF, Du Pasquier L.** 2004. Evolution of innate and adaptive immunity: can we draw a line? *Trends Immunol* **25:**640–644.

95. **Rumfelt LL, McKinney EC, Taylor E, Flajnik MF.** 2002. The development of primary and secondary lymphoid tissues in the nurse shark *Ginglymostoma cirratum*: B-cell zones precede dendritic cell immigration and T-cell zone formation during ontogeny of the spleen. *Scand J Immunol* **56:**130–148.

96. **Torroba M, Chiba A, Vicente A, Varas A, Sacedón R, Jimenez E, Honma Y, Zapata AG.** 1995. Macrophage-lymphocyte cell clusters in the hypothalamic ventricle of some elasmobranch fish: ultrastructural analysis and possible functional significance. *Anat Rec* **242:**400–410.

97. **Ohta Y, Landis E, Boulay T, Phillips RB, Collet B, Secombes CJ, Flajnik MF, Hansen JD.** 2004. Homologs of CD83 from elasmobranch and teleost fish. *J Immunol* **173:**4553–4560.

98. **Lugo-Villarino G, Balla KM, Stachura DL, Bañuelos K, Werneck MB, Traver D.** 2010. Identification of dendritic antigen-presenting cells in the zebrafish. *Proc Natl Acad Sci U S A* **107:**15850–15855.

99. **Wittamer V, Bertrand JY, Gutschow PW, Traver D.** 2011. Characterization of the mononuclear phagocyte system in zebrafish. *Blood* **117:**7126–7135.

100. **Haugland GT, Jordal AE, Wergeland HI.** 2012. Characterization of small, mononuclear blood cells from salmon having high phagocytic capacity and ability to differentiate into dendritic like cells. *PLoS One* **7:** e49260. doi:10.1371/journal.pone.0049260.

101. **Bassity E, Clark TG.** 2012. Functional identification of dendritic cells in the teleost model, rainbow trout (*Oncorhynchus mykiss*). *PLoS One* **7:**e33196. doi: 10.1371/journal.pone.0033196.

102. **Baldwin WM III, Cohen N.** 1981. A giant cell with dendritic cell properties in spleens of the anuran amphibian *Xenopus laevis*. *Dev Comp Immunol* **5:**461–473.

103. **Horton JD, Manning MJ.** 1974. Effect of early thymectomy on the cellular changes occuring in the spleen of the clawed toad following administration of soluble antigen. *Immunology* **26:**797–807.

104. **Aguzzi A, Kranich J, Krautler NJ.** 2014. Follicular dendritic cells: origin, phenotype, and function in health and disease. *Trends Immunol* **35:**105–113.

105. **Krautler NJ, Kana V, Kranich J, Tian Y, Perera D, Lemm D, Schwarz P, Armulik A, Browning JL, Tallquist M, Buch T, Oliveira-Martins JB, Zhu C, Hermann M, Wagner U, Brink R, Heikenwalder M, Aguzzi A.** 2012. Follicular dendritic cells emerge from ubiquitous perivascular precursors. *Cell* **150:**194–206.

106. **Du Pasquier L, Flajnik MF.** 1990. Expression of MHC class II antigens during *Xenopus* development. *Dev Immunol* **1:**85–95.

107. **Turpen JB, Smith PB.** 1986. Analysis of hemopoietic lineage of accessory cells in the developing thymus of *Xenopus laevis*. *J Immunol* **136:**412–421.

108. **Mescher AL, Wolf WL, Moseman EA, Hartman B, Harrison C, Nguyen E, Neff AW.** 2007. Cells of cutaneous immunity in *Xenopus*: studies during larval development and limb regeneration. *Dev Comp Immunol* **31:** 383–393.

109. **García Barrutia MS, Leceta J, Fonfría J, Garrido E, Zapata A.** 1983. Non-lymphoid cells of the anuran spleen: an ultrastructural study in the natterjack, *Bufo calamita*. *Am J Anat* **167:**83–94.

110. **García Barrutia MS, Villena A, Gomariz RP, Razquin B, Zapata A.** 1985. Ultrastructural changes in the spleen of the natterjack, *Bufo calamita*, after antigenic stimulation. *Cell Tissue Res* **239:**435–441.

111. **Kroese FG, Leceta J, Döpp EA, Herraez MP, Nieuwenhuis P, Zapata A.** 1985. Dendritic immune complex trapping cells in the spleen of the snake, *Python reticulatus*. *Dev Comp Immunol* **9:**641–652.

112. **Zapata A, Leceta J, Barrutia MG.** 1981. Ultrastructure of splenic white pulp of the turtle, *Mauremys caspica*. *Cell Tissue Res* **220:**845–855.

113. **Leceta J, Zapata AG.** 1991. White pulp compartments in the spleen of the turtle *Mauremys caspica*: a light-microscopic, electron-microscopic, and immunehistochemical study. *Cell Tissue Res* **266:**605–613.

114. **Oláh I, Nagy N.** 2013. Retrospection to discovery of bursal function and recognition of avian dendritic cells; past and present. *Dev Comp Immunol* **41:**310–315.

115. **Wu Z, Rothwell L, Young JR, Kaufman J, Butter C, Kaiser P.** 2010. Generation and characterization of chicken bone marrow-derived dendritic cells. *Immunology* **129:**133–145.

116. **Wu Z, Hu T, Kaiser P.** 2011. Chicken CCR6 and CCR7 are markers for immature and mature dendritic cells respectively. *Dev Comp Immunol* **35:**563–567.

117. **Del Cacho E, Gallego M, Lillehoj HS, López-Bernard F, Sánchez-Acedo C.** 2009. Avian follicular and interdigitating dendritic cells: isolation and morphologic, phenotypic, and functional analyses. *Vet Immunol Immunopathol* **129:**66–75.

118. **Kaiser P.** 2012. The long view: a bright past, a brighter future? Forty years of chicken immunology pre- and post-genome. *Avian Pathol* **41:**511–518.

119. **Magor KE, Miranzo Navarro D, Barber MR, Petkau K, Fleming-Canepa X, Blyth GA, Blaine AH.** 2013. Defense genes missing from the flight division. *Dev Comp Immunol* **41:**377–388.

Myeloid Cells in Health and Disease: A Synthesis
Edited by Siamon Gordon

doi:10.1128/microbiolspec.MCHD-0038-2016

Laure El Chamy[1]
Nicolas Matt[2]
Jean-Marc Reichhart[2]

5 Advances in Myeloid-Like Cell Origins and Functions in the Model Organism *Drosophila melanogaster*

INTRODUCTION

Innate immunity shields all metazoans against infections. Its main features, including sensing, signaling, and effector mechanisms, are conserved from invertebrates to vertebrates. The hallmark of innate immunity is its reliance on a limited set of non-clonally-distributed receptors, which detect signature molecules of microbial origin and activate subsequent effector mechanisms. This concept, coined by Charles Janeway in 1989 as the self-versus-microbial-nonself discrimination system, has opened a large field of research for the so-called pattern recognition receptors (PRRs) and their cognate microbial elicitors, the pathogen-associated molecular patterns (1). *Drosophila* has rapidly emerged as a particularly suitable model organism for this research. Indeed, like all invertebrates, *Drosophila* exclusively relies on an innate immune system, which fends off infections in highly contaminated environments. Most importantly, *Drosophila* has benefited from more than a century of laboratory-use experience, yielding a wide array of molecular and genetic tools. Investigations on the defense reactions in flies rapidly provided valuable insights into the evolutionary conservation between insects and mammals, including humans, of the signal transduction pathways that control the innate immune system (2). Most prominent is the seminal finding in 1996 of the chief role of the Toll signaling pathway in the control of fungal infections in *Drosophila* (3). This study paved the way for the identification of the first mammalian PRR, Toll-like receptor 4 (the launching member of the TLR family), and the understanding of the innate immune system's molecular mechanisms for sensing, signaling, and activation of adaptive immunity (4–6). Following more than 2 decades of in-depth analysis exploiting several infection models combined with genetic and genomic approaches, research on the

[1]Laboratoire de Génétique de la drosophile et virulence microbienne, UR. EGFEM, Faculté des Sciences, Université Saint-Joseph de Beyrouth, B.P. 17-5208 Mar Mikhaël Beyrouth 1104 2020, Liban; [2]Université de Strasbourg, UPR 9022 du CNRS, Institut de Biologie Moléculaire et Cellulaire, Strasbourg Cedex 67084, France.

Drosophila immune system revealed complex interconnected humoral and cellular processes, both of which show striking similarities with those of mammals. In this review, we provide a global view of the *Drosophila* host defense while drawing particular attention to the role of its monocyte-macrophage-like cells, the plasmatocytes. We provide general insights on the recent advances in *Drosophila* hematopoiesis and give a comprehensive summary on the so-far identified receptors involved in microbial detection, binding, and the ensuing internalization processes.

DROSOPHILA DEFENSE REACTIONS

As in all metazoans, the *Drosophila* innate immune system includes physical barriers together with local and systemic responses (Fig. 1). Physical barriers include the body cuticle, the peritrophic chitinous membrane lining the gut lumen and the underlying epithelia, which also harbor a potent local response involving the production of antimicrobial peptides (AMPs) and reactive oxygen species (ROS) (7, 8). Microorganisms getting access to the insect body cavity rapidly trigger a systemic immune response characterized mostly by the secretion of AMPs by the fat body cells (an organ equivalent to the mammalian liver) into the hemolymph (9). Two signaling pathways, the so-called Toll and IMD (immune deficiency) pathways, control the expression of AMPs and many other genes through the activation of members of the NF-κB family of transcription factors. Extensive research has now generated a fairly clear image of the Toll and IMD pathways, both of which share striking similarities with mammalian NF-κB signaling cascades. Namely, the Toll pathway is reminiscent of the myeloid differentiation primary-response protein 88 (MyD88)-dependent TLR/interleukin-1 receptor pathway, whereas the IMD signaling cascade is analogous to the tumor necrosis factor receptor (TNFR) and TIR-domain-containing adaptor protein inducing beta interferon (TRIF)-dependent TLR pathways (2, 10, 11). However, unlike the situation in mammals, microbial sensing upstream of the *Drosophila* NF-κB pathways is not mediated by the Toll receptor nor by its related proteins (the *Drosophila* genome encodes nine Toll proteins). This function is performed by canonical PRR members the peptidoglycan recognition proteins (PGRPs) and the glucan-binding proteins (GNBPs), which allow a discriminative activation of the two path-

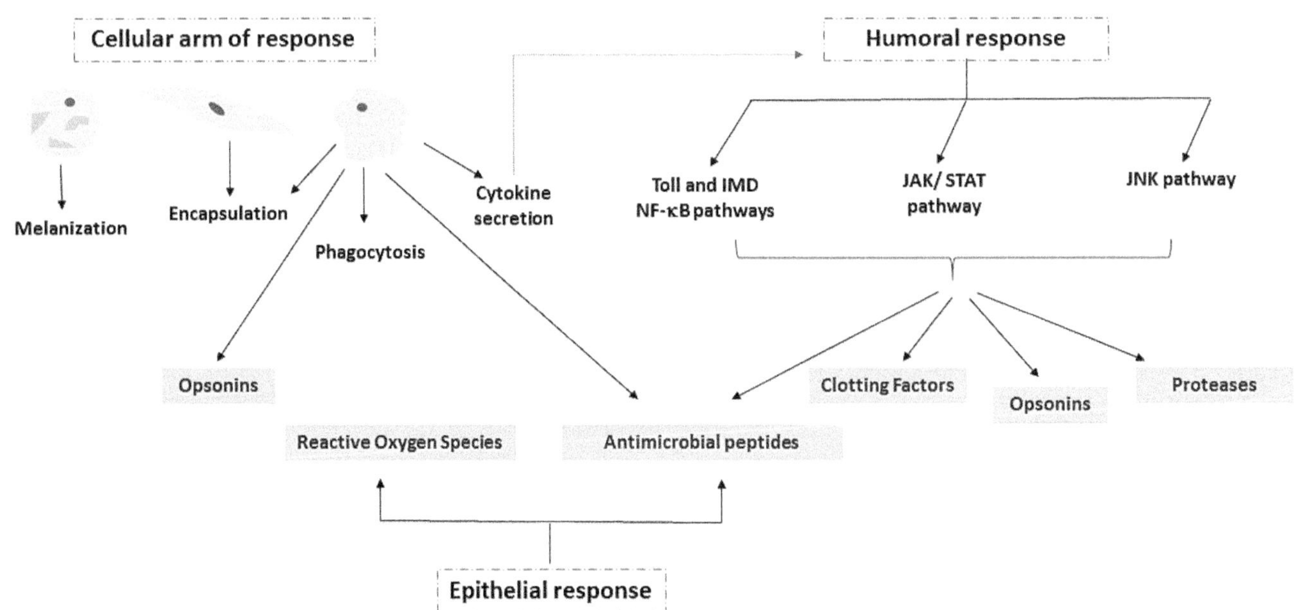

Figure 1 *Drosophila* immune reactions. *Drosophila* immune response comprises a local barrier response based on the secretion of AMPs and a fine-tuned oxidative response. Breaching of this barrier triggers a systemic humoral antimicrobial response as well as a cellular response (refer to text for a detailed description).

ways (12, 13). Precisely, the transmembrane PGRP-LC and the secreted/cytosolic PGRP-LE act upstream of the IMD pathway as receptors for the diaminopimelic acid-type peptidoglycan (PGN) that is commonly found in the membrane of Gram-negative bacteria and bacilli; whereas lysine-type PGN, which is a signature of most Gram-positive bacteria (except for bacilli), and β-(1,3)-glucans from fungal cell walls, trigger the Toll pathway following binding of a PRR complex including PGRP-SA, GNBP1, and PGRP-SD and the GNBP3 receptor, respectively. These soluble PRRs activate a cascade of serine proteases, analogous to the complement and coagulation cascades in mammals, which culminate in the maturation of the Toll endogenous ligand, Spätzle, by the Spätzle-processing enzyme (SPE) (14). The Toll pathway can also be activated independently of PRRs through the sensing of microbial proteolytic activities by the circulating host protease Persephone, also leading to activation of SPE (15, 16).

Microbial detection also triggers a (different) cascade of serine proteases, which culminates in the activation of prophenoloxidase (PPO), a key enzyme for converting phenols to quinones that polymerize to form melanin deposits around the microorganism and on the site of the injury. This melanization reaction is important for microbial killing during the cellular encapsulation response; it has a role in blood clotting and has been documented as conveying a signal to the fat body for the expression of an AMP-coding gene, *drosomycin*, upon a local reaction at the tracheal level (17–19).

The *Drosophila* immune response has been thoroughly investigated by transcriptomic and proteomic analyses. These analyses showed that, in addition to AMPs, many potential effectors and immune-related molecules are induced upon infection. Most of the transcriptomic profiles following bacterial and fungal infections were shown to be dependent on the Toll and IMD pathways. Nevertheless, two other conserved pathways, the c-Jun N-terminal protein kinase (JNK) and the JAK/STAT pathways, have been linked to the regulation of the septic injury response (20). Further inquiries have shown that the JNK pathway is associated with the tissue-repair and wound-healing response (21–24), whereas the JAK/STAT pathway is more likely involved in the stress-induced response (25–29). The *Drosophila* model has also been used to study reactions against viral infections (30, 31). The characterization of the transcriptional program triggered upon different viral infections suggested the contribution of the NF-κB and JAK/STAT pathways to the overall *Drosophila* viral immune reactions (32–40). In agreement with these data, flies mutant for genes involved in these signaling pathways exhibited a differential sensitivity to distinct viral infections (32, 37, 41–46). However, the sensing mechanisms driving these responses and the identity of the antiviral effectors are still largely unknown. Apoptosis and autophagy have also been described as part of the antiviral response in *Drosophila* (47–49). However, their contribution is limited to the restriction of a limited set of viral infections.

In addition to these virus-specific inducible responses, RNA interference (RNAi) provides the most potent and efficient response against infections by RNA as well as DNA viruses in *Drosophila* (37, 50–54). The RNase III enzyme Dicer-2 senses viral double-stranded RNAs and produces small interfering RNAs of which a guide strand is incorporated in the RNA-induced silencing complex to mediate sequence-specific degradation of the target viral RNAs (55–57). Interestingly, Dicer-2 is also involved in the induction of a part of the transcriptomic program activated upon *Drosophila* C virus infection. This function is dependent on its amino-terminal DExD/H box helicase domain, phylogenetically related to the helicase domain of the mammalian retinoic acid-inducible gene I (RIG-I)-like receptor family (58). These PRRs control interferon-induced gene expression, central to mammalian antiviral immunity (59). Thus, studies of the *Drosophila* antiviral response again reveal evolutionarily conserved aspects of sensing and signaling mechanisms from flies to mammals.

Drosophila immunity also relies on professional hemocytes that participate in the elimination of the intruders through phagocytosis or encapsulation but also by contributing to the humoral reactions. These functions will be discussed in the following sections.

DROSOPHILA HEMATOPOIESIS

The hematopoietic cell lineage of *Drosophila* includes three cell types referred to as hemocytes. These include the plasmatocytes, which are professional phagocytes and thus constitute the functional equivalent of mammalian monocyte-macrophages; the crystal cells (named after their crystalline inclusions), which are involved in microbial killing through the secretion of the proteolytic enzymes of the melanization cascade but also in blood coagulation as counterparts of mammalian platelets; and the lamellocytes, which only develop in parasitized larvae and are largely involved in the encapsulation of large targets (60–62).

During development from the embryonic to the larval stages, *Drosophila* hematopoiesis occurs in three waves, which give rise to a capital of hemocytes derived from two distinct lineages, embryonic or tissue hemocytes and lymph gland hemocytes. This situation is reminiscent of the dual origins of myeloid blood cells in vertebrates, which arise either from the self-renewal of tissue macrophages or from the sustainable proliferation and differentiation of hematopoietic progenitors (63). Both hemocyte lineages can be found persevering as a mixed population in the adult fly (64).

The first hematopoietic wave takes place during early embryogenesis when blood cell progenitors emanate from the procephalic mesoderm (64, 65) (Fig. 2). These progenitors give rise to a limited number of embryonic hemocytes, which differentiate mainly into plasmatocytes and into a few crystal cells. The embryonic plasmatocytes are required for embryonic development, notably for the engulfment of apoptotic cells and the morphogenesis of the central nervous system (65–71). The molecular cascades that control embryonic hematopoiesis are relatively well documented, and they pinpoint the involvement of transactivators such as GATA, Friend of GATA (FOG), and RUNX factors, which are also involved in mammalian hematopoiesis (61, 63, 71–78). After their differentiation, crystal cells remain around their points of origins and populate the proven-

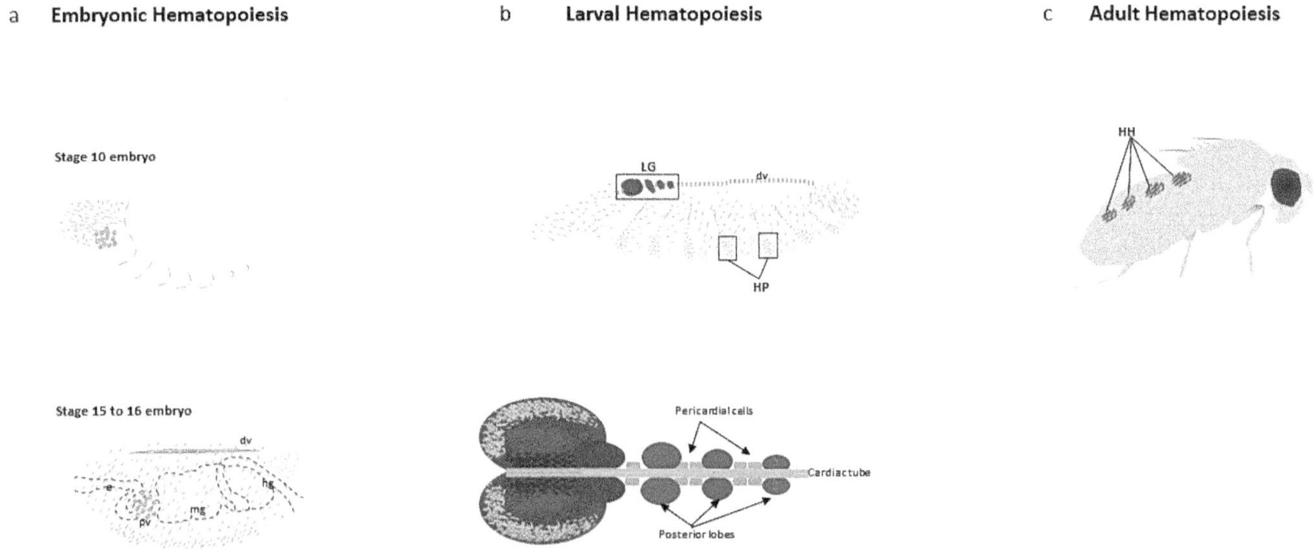

Figure 2 *Drosophila* hematopoiesis. Hematopoiesis in *Drosophila* starts during embryogenesis and continues till the adult stage. (a) Embryonic hematopoiesis. Schematic presentations of stage 10 and 16 *Drosophila* embryos with the head on the left and the ventral axis facing down. The gut is shown in dashed lines (dv, dorsal vessel; e, esophagus; pv, proventriculus; mg, midgut; hg, hindgut). *Drosophila* embryonic hemocyte progenitors, the prohemocytes, emanate from the procephalic mesoderm. After their differentiation, plasmatocytes (shown in blue) migrate to populate the whole embryo, whereas crystal cells (shown in yellow) remain around their points of origin and populate the proventriculus. (b) Larval hematopoiesis. Schematic presentation of a *Drosophila* third instar larva with the head at left and the ventral axis facing down (top). In the larvae, embryonic hemocytes proliferate within the hematopoietic pockets (HPs), giving rise to sessile hemocytes, which could differentiate into plasmatocytes but also crystal cells and lamellocytes in the case of parasitization. In the larvae, another center of hematopoiesis, the lymph gland (LG), originates from an anlage of the thoracic mesoderm and differentiates into four (to six) bilaterally paired lobes along the anterior part of the dorsal vessel. The cellular organization of the LG is shown (bottom). The primary lobes are indicated in red, the posterior signaling center (PSC) in purple, and the posterior lobes in blue. Within the primary lobes, core progenitor hemocytes are shown in purple, progenitors in dark blue, and plasmatocytes and crystal cells in light blue and yellow, respectively. LG-derived hemocytes are normally released at the beginning of pupariation only under immune challenge. A detailed description of signaling pathways controlling *Drosophila* hematopoiesis is reviewed in reference 62. (c) Adult hematopoiesis. Four hematopoietic hubs (HH) have been identified in the dorsal part of adult fly abdomen. These hubs enclose hematopoietic progenitors, derived from the third and fourth lobes of the LG, together with differentiated hemocytes of embryonic and larval origins.

triculus (a valve-like structure that allows the circulation of the food bowl from the esophagus to the midgut), whereas the embryonic phagocytic cells migrate to spread along the embryonic tissues (65, 73). In the larval stages, these cells are found along the dorsal vessel (a circulatory organ with heart-like function) and in the proventriculus but also in specialized microenvironments of the subepidermal layers of the body cavity, the hematopoietic pockets (HPs) (64, 79–81) (Fig. 2). These cells are commonly referred to as the sessile hemocytes, and they are the equivalent of mammalian tissue macrophages, which self-renew in their differentiated state (63, 82). The expansion of sessile hemocytes occurs during the second or larval hematopoietic wave within the HPs in response to trophic signals delivered by the peripheral nervous system, with which they colocalize (83, 84). Because of these findings, the transparent and genetically powerful *Drosophila* larva has been proposed as a model to study the role of the nervous system in the regulation of hematopoiesis (84). The sessile hemocyte clusters have first been defined as a bona fide hematopoietic tissue based on the observation that these cells could differentiate into plasmatocytes, but also, upon parasitization of *Drosophila* larvae with a wasp egg, into lamellocytes (79, 85). More recently, by employing cell-lineage tracing and live imaging, Leitão and Sucena have elegantly shown that the self-renewing plasmatocytes in the HPs could also give rise to crystal cells through a Notch signaling-dependent transdifferentiation process (86). Tissue macrophages mostly form resident clusters in HPs of first instar larvae, whereas an increasing number of these hemocytes circulate in the hemolymph of second and third instar larvae, a trend that culminates at the onset of metamorphosis in the pupal stage (87). These cells ensure all phagocytic and immune surveillance functions in the larvae and are complemented by lymph gland-derived hemocytes, which are normally released at the beginning of pupariation only under extreme immune challenge (63, 88–91). The latter hemocytes are the product of the third hematopoietic wave, which occurs in a specialized hematopoietic organ, the lymph gland.

The lymph gland progenitors originate from an anlage of the thoracic mesoderm and proliferate until the second larval instar to form four to six bilaterally paired lobes along the anterior part of the dorsal vessel (64, 78, 92) (Fig. 2). The lymph gland is also developmentally related to the dorsal vessel, as it arises from a common hemangioblast progenitor similarly to the situation in mammals, where endothelial and hematopoietic cells differentiate from a common precursor (84, 93). Therefore, as it also shares an origin with the excretory cell lineages, the ontogeny of the lymph gland in the *Drosophila* embryo has been compared to that of the aorta-gonad-mesonephros region of vertebrates (78, 84, 94). The differentiation of lymph gland hemocytes is detected starting from the third larval stage; it develops in a spatiotemporally organized manner (91, 95). The medullar zone of the most anterior primary lobes as well as the secondary lobes contain undifferentiated prohemocytes. A group of cells located at the posterior end of the primary lobes constitute a signaling center, referred to as the posterior signaling center (PSC), which plays a key role in the control of hematopoiesis (75, 96, 97). Differentiated cells, mostly plasmatocytes and a small proportion of crystal cells, are located in the cortical zone of the primary lobes (75). In the case of parasitization by wasp eggs, the differentiation of prohemocytes is shifted to the production of lamellocytes (85, 87, 95, 96).

The signaling pathways involved in the specification of lymph gland cells, the maintenance of pluripotent progenitor cells, and the differentiation of blood cells in the *Drosophila* lymph gland have been thoroughly investigated (62, 97–100). Findings in these fields indicate a global conservation of the molecular mechanisms that control the progenitor-based hematopoiesis from flies to vertebrates (63). The early specification of the lymph gland is under the control of the Notch signaling pathway, which is also involved in the differentiation of crystal cells (75, 93). The core population of the hematopoietic progenitors is defined by low expression of *collier*, the *Drosophila* ortholog of the vertebrate early B-cell factor-encoding gene (101). Collier, together with the Hox factor Antennapedia (Antp), is also required for the differentiation of the PSC (96, 97). The cardiac tube controls proliferation and clustering of the PSC cells through Slit/Robo signaling. Interestingly, several studies have also underlined the implication of Slit ligands and Robo receptors in mouse hematopoiesis. Thus, altogether these data emphasize a conserved role of the vascular system in regulating hematopoiesis from insects to vertebrates (102). Wingless (Wnt) signaling, which regulates the proliferation and the maintenance of the PSC cells, is also required for the autonomous maintenance in an undifferentiated state of progenitor cells of the medullary zone upon delivery of nutritional signals (103, 104). As for the differentiation of lymph gland progenitors, it is controlled by developmentally regulated ROS (105).

Intriguingly, it has recently been shown that olfactory signals are also involved in the fate determination of progenitor cells. Indeed, neurosecretory cells of the brain produce γ-aminobutyric acid (GABA), which binds $GABA_B$ receptors on progenitor cells, resulting in high cytosolic calcium that is necessary and sufficient for progenitor maintenance (106). The PSC also plays a crucial role in hemocyte differentiation, most probably by regulating the maturation of intermediate progenitors (107). In agreement with this hypothesis, it was shown that the inhibition of a Hedgehog signal in the PSC cells promotes the differentiation of hemocytes (96, 97, 100). Many signaling pathways, including the Toll, JAK/STAT, and Decapentaplegic (Dpp, the homolog of mammalian bone morphogenetic protein) pathways; the polycomb group gene multiple sex combs (*mxc*); and the transcription factor Zfrp8, have been implicated in controlling hemocyte proliferation and density as well as their differentiation (108–113). The lymph gland dissociates shortly after pupariation, thus releasing a mixture of fully or partially differentiated hemocytes of its primary and secondary lobes, respectively (91). Lymph gland dissociation and differentiation of hemocytes is accelerated in the case of an infection with the remarkable generation of lamellocytes in the case of wasp parasitization (87, 96, 114–116). Several signaling pathways, including the JAK/STAT and JNK pathways, are involved in the differentiation of plasmatocytes (89), whereas Collier is critical for the differentiation of lamellocytes (114).

Contrasting with the long-held belief that there is no hematopoietic activity in adult flies, a recent study by Ghosh et al. has uncovered hematopoietic niches or "hubs" at the dorsal side of the adult fly abdomen, where progenitor cells deriving from the third and fourth lobes of the lymph gland reside. Through a differential activation of Notch signaling, these progenitors give rise to crystal cells or plasmatocytes. The organization of the hematopoietic hubs, which comprise both hematopoietic progenitors and differentiated hemocytes of embryonic and larval origins, in addition to the differentiation process of their hematopoietic progenitors, draws additional similarities to vertebrate hematopoiesis in the bone marrow niches (117).

OVERVIEW OF HEMOCYTE IMMUNE FUNCTIONS

Drosophila hemocytes play a crucial role during embryonic development (70). They engulf apoptotic bodies and secrete proteins of the extracellular matrix required for the development of many tissues, including the central nervous system and the renal tubule (69, 71, 118–122). Despite previous assumptions that hemocytes are required for tissue remodeling in the pupae, studies based on the ablation of *Drosophila* blood cells clearly indicate that they are not implicated in metamorphosis. However, hemocytes are required to prevent bacterial-induced lethality at the pupal stage, which is rescued by raising the hemocyte-depleted flies under axenic conditions. These studies have also revealed the crucial role of plasmatocytes in controlling several bacterial infections, notably through their phagocytic activity (13, 70). Phagocytosis has also been associated with the antiviral response in *Drosophila.* Notably, phagocytes are required to eliminate apoptotic infected cells (49, 123). Advances in the field of phagocytosis in *Drosophila* will be detailed in a separate section.

Besides phagocytosis, encapsulation represents a potent *Drosophila* cellular response that is activated against large extracellular parasites. Encapsulation of wasp eggs involves the three types of *Drosophila* hemocytes, starting with the spread of plasmatocytes around the egg, hiding it from the circulation; followed by the recruitment of lamellocytes that form large adhesive masses; and finally the degranulation of crystal cells, activating the melanization reaction, thought to be toxic for the parasite. (For a detailed review of encapsulation, see references 61 and 124.) Other aspects of hemocytes' immune functions are linked to their cooperation for the onset of the humoral response and their coordinated migration during the immune response. Advances in these fields of research are summarized in the two following sections.

Cytokine Expression and Cooperation with the Systemic Humoral Response

The susceptibility of phagocyte-depleted flies to bacterial infections was aggravated in a loss-of-function mutant background for the Toll or IMD pathway. These observations are indicative of the cooperative roles of the cellular and humoral reactions in the *Drosophila* immune response and are in agreement with other studies that suppressed phagocytosis by saturating plasmatocytes with latex beads (13, 125, 126). Hemocytes have also been proposed to mediate the activation of the humoral systemic response through the secretion of cytokine-like molecules or other signaling molecules (such as nitric oxide) (28, 29, 127–132). The first evidence for such an interplay between hemocytes and the humoral response came from investigations on mutant larvae carrying *domeless* or *lethal 3 hematopoietic organ missing* [*l(3)hem*] mutations, which greatly reduce hemocyte counts (127, 133–135). Indeed, it was shown that the systemic expression of AMP-encoding genes in the fat body of larvae orally infected by the phytopathogenic bacterium *Erwinia carotovora* is abolished in *Dom* and *l(3)hem* mutants. However, it is important to mention that both mutations also affect cell proliferation and thus are pleiotropic. We therefore cannot exclude the possibility that this phenomenon could be due to other physiological processes. However, similar results were obtained in hemocyte-ablated larvae by targeted activation of apoptotic signals. In the same study, silencing of *spätzle* transcripts in hemocytes inhibited the expression of *drosomycin* in the fat body of infected larvae, indicating that hemocytes are required for the production of the cytokine-like Spätzle, ligand of the Toll receptor (128). Another study, investigating the mechanisms triggering JAK/STAT signaling in response to stress, has shown that hemocytes represent the source of Unpaired 3 (UPD3), the cytokine binding the JAK/STAT pathway receptor Domeless (26). Finally, it has been shown that a defect in the processing of phagocytosed bacteria in the *psidin* mutant (phacytocyte signaling impaired), which lacks the lysosomal protein Psidin, results in a reduced production of the AMP defensin by the fat body cells. This result led to the hypothesis that phagosome maturation might be involved in the initiation of the systemic immune response, although this hypothesis is challenged by the fact that the expression of all other AMP-encoding genes is unaffected (136).

Hemocytes are also involved in the melanization reaction, as they represent the main source of PPO in the larvae. Indeed, the *Drosophila* genome contains three PPO-encoding genes, of which PPO1 and -2 are primarily expressed in crystal cells and PPO3 in lamellocytes (137–140). The first two enzymes are activated by proteolytic cascades, including two so far identified phenoloxidase-activating enzymes (PAEs) or melanization proteases (MPs), namely MP1 and MP2 (also known as PAE1), whereas PPO3 is active in its zymogen form (138, 141, 142). The importance of the melanization reaction in the immune response was first proposed based on the phenotype of *Black cell* (*Bc*) mutant flies, which contain atypical crystal cells and show compromised melanization and reduced survival in infection (21, 133, 141–146). The *Bc* mutation was mapped to the *ppo1* gene (146). Mutation and knockdown experiments affecting the expression of the MP1, MP2, and PPO enzymes were shown to cause enhanced sensitivity of mutant flies to several bacterial and fungal infections, providing additional support for the role of the melanization reaction in the overall host defense reaction (140, 142, 147). The emergence of microbial resistance mechanisms targeting the melanization reaction provides further support for its importance in host defense in insects (17). Experimental evidence indicates that local melanization in *Drosophila* trachea relays a signal for the systemic activation of the Toll pathway in the fat body (19). The melanization reaction is also involved in the systemic wound response, through the production of ROS as by-products of a cascade activated by the serine protease Hayan in the hemolymph of wounded flies. This response is mediated by the activation of JNK signaling in neurons (148). Finally, some proteins secreted by hemocytes, such as hemolectin, were shown to be required for blood clotting (149–151).

Hemocyte Mobility and Inflammation

Drosophila hemocytes are highly motile; they move through directed migration during embryogenesis, in order to populate the embryo, but also to heal a wound upon tissue damage (152). *Drosophila* has thus been adopted as a model organism to study blood cell migration during the inflammatory process. Indeed, beyond the advantages of its sophisticated genetic manipulations, the transparent *Drosophila* embryo is particularly suited for live imaging as an optimized model to decipher the dynamics of the inflammatory response *in vivo* (153). Advances in this field have revealed a significant conservation of the molecular mechanisms driving cellular mobility in *Drosophila* and mammals (152). Notably, two distinct pathways control hemocytes' developmental and damage-induced migrations (154, 155). Hemocyte chemotaxis to wounds is driven by H_2O_2 gradients generated by the damaged cells via a calcium flash triggered at the wound site (155, 156). This chemotaxis is controlled by the conserved Src family kinase intracellular immunoreceptor tyrosine-based activation motifs (ITAM)-Syk signaling pathway, which is also implicated in the vertebrate adaptive immune response. It has thus been proposed that the mechanism of self-versus-nonself distinction by the mammalian adaptive immune receptors has evolved from an ancient mechanism of damage sensing (157). Moreover, phosphoinositide 3-kinase signaling is also required for hemocyte recruitment to damaged tissue, this situation being reminiscent of the same requirement for neutrophil chemotactic movements in mammals (124).

PHAGOCYTOSIS: DETECTION AND ENGULFMENT OF CARGO

Phagocytosis is an intricate cellular reaction that begins with receptor-mediated recognition of microbial agents or altered, mainly apoptotic, self-cells. This recognition event triggers the reorganization of the cellular cytoskeleton, leading to the formation of a highly complex organelle, the phagosome. Beyond its central role as a "phagocytic organ" where the enzymatic digestion of the internalized particles occurs, the phagosome is a crossroad for a multitude of molecules serving for the initiation of crucial signaling events for both innate and adaptive immunity (158).

Drosophila plasmatocytes have long served as a model for studying phagocytosis. This field of research has been especially facilitated by the use of the embryonic-derived Schneider 2 (S2) cell culture. These cells are highly phagocytic and readily amenable to high-throughput RNAi screens (159–161) that have therefore been adapted for the identification of host genes involved in the cellular response against *Escherichia coli* (162), *Staphylococcus aureus* (163, 164) *Mycobacterium fortuitum* (165), *Listeria monocytogenes* (166, 167), and *Candida albicans* (168). Furthermore, the genetic tractability of the *Drosophila* model has allowed for the evaluation of the role of several phagocytic machineries and genes involved in the immune response either *in vivo* or *ex vivo* (66, 169).

PHAGOCYTIC RECEPTORS

Drosophila membrane-bound phagocytic receptors belong to two major families (Table 1). The first is the family of scavenger receptors, whose members are structurally diverse but share the capacity of binding polyanionic ligands (158, 170). Croquemort was the first described receptor, mediating phagocytosis and clearance of apoptotic cells in *Drosophila* embryos (66, 171). This receptor was further shown to be involved in the uptake of *S. aureus*. These findings in *Drosophila* have led to the identification of CD36, the mammalian ortholog of Croquemort, which mediates binding of lipoteichoic acid and diacylated bacterial lipopeptides (163, 172). Peste, another member of the family of class B scavenger receptors, also exemplifies the high conservation of phagocytic receptors through evolution. Peste was identified as a receptor for *M. fortuitum* and *L. monocytogenes* in *Drosophila* S2 cells, and overexpression of its mammalian ortholog, SR-BI, conferred uptake of *M. fortuitum* to nonphagocytic cells (165, 166). Remarkably, besides phagocytosis, several members of the class B scavenger receptor family have been associated with physiological roles pertaining to their capability of lipoprotein binding. These functions include pheromone sensing, steroidogenesis, and carotenoid uptake (173, 174). Indeed, the *Drosophila* scavenger receptors were initially identified through their role in lipoprotein uptake, as is the case for their mammalian counterparts (175). Overall, these data underscore evolutionarily conserved functions of scavenger receptors for immune sensing but also for the regulation of lipoprotein-related physiological processes. The *Drosophila* genome also contains a class C scavenger receptor gene family, with no identified mammalian homologs, of which SR-CI was shown to partially contribute to the phagocytosis of *E. coli* and *S. aureus* in S2 cells (176).

The second family of phagocytic receptors described in *Drosophila* contains multiple epidermal growth factor (EGF)-like repeats (177–179). Three members of this family, namely, Eater (the founding member), Nimrod, and Draper, are major actors in the phagocytosis of Gram-positive bacteria (126, 177, 180–183). *Eater* was identified as a gene regulated by the *Drosophila* GATA factor Serpent, which is exclusively expressed in

Table 1 *Drosophila* phagocytic receptors and opsonins and their corresponding ligands

Receptors and opsonins	*D. melanogaster* receptors	Ligand(s)	Reference(s)
Scavenger receptors	Croquemort	Apoptotic cells, *S. aureus*	66, 163, 171
	Peste	*M. fortuitum*, *L. monocytogenes*	165, 166
	SR-CI	*S. aureus*, *E. coli*	176
EGF-like repeat-containing Nimrod receptors	Eater	*S. aureus*, *Enterococcus faecalis*, *Serratia marcescens*, *E. coli*	126, 177
	Nimrod C1	*S. aureus*, *E. coli*	178
	Draper	Apoptotic cells, *S. aureus*	182, 191
	SIMU (NimC4)	*S. aureus*	192
Opsonins	TEP2	*E. coli*	168
	TEP3	*S. aureus*	168
	TEP6	*C. albicans*	168
	Dscam		209
Others	αPS3/βν integrin heterodimer	Apoptotic cells, *S. aureus*	183, 198, 199
	PGRP-LC	*E. coli*	162

macrophages. Eater mutants are highly susceptible to infections by Gram-negative and Gram-positive bacteria (126, 177). Further characterization of Eater's binding properties indicated that this receptor readily binds Gram-positive but not Gram-negative bacteria, unless their membrane is previously disrupted by AMPs (180, 181). Interestingly, Eater is also required for enabling attachment of hemocytes in the sessile compartments (184). Nimrod, another member of the EGF-like repeat family receptors, has been proposed to function as an adhesion molecule (178). Like Eater, this receptor is exclusively expressed in plasmatocytes and is involved in the phagocytosis of the Gram-positive bacterium *S. aureus* (178). The third member of the family, Draper, is a homolog of the *Caenorhabditis elegans* apoptotic cell receptor CED-1 (185, 186). The Draper gene was identified as a target of the transcription factor Glial cells missing (Gcm), which controls the development of glial cells, the primary immune cells of the nervous system (187). In agreement with its expression on the surface of glial cells and macrophages, Draper was shown to be required for apoptotic neuron engulfment, axon pruning, and the clearance of severed axons upon injury or during developmental degeneration, as well as the elimination of embryonic apoptotic cells (188–191). Apoptotic corpse elimination involves an additional receptor, harboring four EGF-like repeats in its extracellular domain, Six-microns-under (SIMU), which acts upstream of Draper (192). Interestingly, two mammalian receptors homologous to Draper, namely multiple-EGF-like domain 10 (MEGF10) and MEGF12, also referred to as Jedi, have been shown to mediate the elimination of apoptotic corpses in developing mouse dorsal root ganglia (193–195). Like Draper, these mammalian receptors engage their intracellular ITAM domains, which become phosphorylated and interact with the tyrosine kinase Syk to promote phagocytosis (194, 196, 197). More recently, Draper was shown to be required for the uptake of *S. aureus* by binding to bacterial membrane lipoteichoic acid. This function of Draper is required for the survival of adult flies of septic infection by *S. aureus* (182). Whether the mammalian homologs of Draper are also involved in the clearance of bacteria during infections remains to be clarified. Draper has also been shown to be required for plasmatocyte chemotaxis to a wounded site through signaling by its intracellular ITAM motif (157). It is thus tempting to speculate that this double function of the phagocytic receptor Draper would allow an adapted and efficient cellular response to limit pathogenic invasions. Through its ability to sense altered damaged self, Draper drives hemocyte recruitment to the potential portal for pathogen entry, thus establishing a predisposed cellular arm to counter infectious microorganisms by phagocytosis.

Several studies have extended the list of *Drosophila* phagocytic receptors. Among these is the integrin αPS3/βν heterodimer, which mediates internalization of apoptotic cells and bacteria, namely, *S. aureus* (183, 198, 199). This receptor complex acts independently of Draper, thus defining a second path for internalization of cell corpses and bacteria in *Drosophila* (183, 198). This situation is reminiscent of the involvement of two pathways for the elimination of apoptotic cells in *C. elegans* and mammals (195, 200–202). Overall, these data highlight an evolutionarily conserved role of integrins in the phagocytic uptake of apoptotic cells. In addition to its critical role upstream of the IMD pathway, the PGRP-LC receptor has also been proposed to participate, although moderately, in the cellular response (162, 203, 204). Indeed, the silencing of the *pgrp-lc* gene led to a slight reduction in the phagocytic index of S2 cells upon their infection by *E. coli* (162).

In addition to cell-bound receptors, secreted proteins have been proposed to act like opsonins, facilitating the phagocytic uptake of microorganisms by *Drosophila* hemocytes. These include the secreted thioester-containing proteins (TEPs), which are related to mammalian α_2-macroglobulin and C3/C4/C5 complement factors. The *Drosophila* TEP family comprises six genes (*Tep1* to -6), which—apart from *Tep5*, which does not seem to be expressed—exhibit a specific expression profile in immunocompetent tissues such as hemocytes, fat body, and barrier epithelia. As with mammalian acute-phase proteins, the expression and secretion of these proteins is enhanced upon infection (137, 205, 206). The TEP function has been studied mostly in the model organism *Anopheles gambiae*, where TEP1 was shown to act as an opsonin, binding the surface of bacteria and promoting their phagocytosis (207). TEP1 also binds the surface of *Plasmodium berghei* ookinetes and promotes their killing, thus acting as a key molecule for the determination of the vectorial capacity in this mosquito species (208). In *Drosophila*, an RNAi screen performed in S2 cells identified TEP6 as an opsonin required for the phagocytosis of *C. albicans*. The same study showed that TEP2 and TEP3 are required for efficient phagocytosis of *E. coli* and *S. aureus*, respectively (168).

Finally, in addition to TEPs, secreted proteins encoded by the immunoglobulin superfamily gene member *Dscam* (Down syndrome cell adhesion molecule) have been proposed to mediate phagocytosis of microbes in *Drosophila* (209). These proteins have gained particular attention especially because they could develop receptor diversity through a mechanism of alternative splicing of the *Dscam* transcript. The *Dscam* gene was first described for its role in axon guidance in the development of the nervous system (210, 211). Interestingly, this gene was further shown to be expressed by hemocytes and fat body cells, and secreted Dscam protein isoforms were detected in the hemolymph. Dscam was shown to bind *E. coli*, and its specific knockdown in hemocytes impairs the phagocytic uptake of this bacterium (209). Moreover, Dscam isoforms were identified in a study exploring the phagosome proteome (212). Based on these data, and the estimation that ~18,000 different Dscam isoforms could be expressed by immunocompetent cells, it is tempting to speculate that the Dscam family could provide *Drosophila* with a large network of phagocytic receptors evocative of the function of vertebrate immunoglobulins. However, this hypothesis is challenged by the lack of any evidence for increased secretion of pathogen-specific Dscam isoforms after infection (213).

In addition to the identification of phagocytic receptors, RNAi screens exploring the cellular response in *Drosophila* S2 cells have identified several intracellular proteins essential for the engulfment of microorganisms and the ensuing phagocytic process (162, 165–168, 214–217). In this context, the *Drosophila* homolog of vacuolar protein sorting 35 (Vps35), a main actor in the retromer complex that controls receptor trafficking from the endosome to the Golgi, was shown to be required for scavenger receptor-mediated endocytosis (217). *Vps35* mutants exhibit a defective scavenger receptor-mediated ligand uptake in hemocytes. However, the localization of scavenger receptors at the plasma membrane is slightly increased in these mutants, thus indicating that the endocytic defect is due not to the scarcity of receptors but rather to the inhibition of receptor-mediated uptake. Hemocytes of *Vps35* mutants showed a considerable increase in actin filaments that were proven to be Rac1 dependent. Based on these results, it is postulated that Vps35 is required for scavenger receptor-mediated endocytosis through the regulation of Rac1-dependent actin polymerization (217). Phagosome constituents have been thoroughly explored through combined proteomic and systems biology approaches (212, 218). The accumulated data give further evidence for the high conservation of the phagocytic process during evolution (158, 161). The global functions of the identified genes are related to cytoskeleton remodeling, vesicle trafficking, endocytic machinery, as well as signaling. Interestingly, in some cases, the identification of *Drosophila* phagosomal proteins gave insights into the function of their mammalian orthologs. Most prominent among these cases is the involvement of the exocyst, the secretory vesicle complex, in microbial phagocytosis in *Drosophila* and mouse macrophages (212). Several screens in *Drosophila* have isolated components of the coatomer complex, coat protein complex I and II (COPI and COPII), as part of the phagosome complex (162, 164, 167). These results suggest a role for the coatomer in the maintenance of the membrane curvature for the establishment of the phagocytic cup. However, the hypothesis remains to be confirmed, as the COP complex could be associated with the membranes of endosomal or lysosomal vesicles that fuse with the phagosome during its maturation (158). Finally, a genetic analysis has shown that phagosomal maturation and the subsequent digestion of internalized bacteria in *Drosophila* require members of the evolutionarily conserved HOPS (homotypic fusion and vacuole protein sorting) complex. More precisely, a null mutation of the *Vps16* gene or a knockdown of *Vps33B* leads to a phenotype of high susceptibility to nonpathogenic bacteria in flies. Hemocytes of the *fob* mutant were shown to be unable to digest the phagocytosed bacteria due to a defect in the fusion of lysosomes with phagosomes (219).

CONCLUDING REMARKS

More than 2 decades of research on *Drosophila* hemocytes has revealed significant similarities between *Drosophila* and mammalian hematopoiesis and blood cell immune functions. These results emphasize the advantages of using this genetically powerful and yet simple model organism for elucidating the molecular and cellular processes of vertebrate blood cell development and macrophage functions. Notably, *Drosophila* is presently used as a model for the study of human leukemia. Indeed, as previously described, hemocyte development and differentiation involve highly conserved signaling pathways. Interestingly, mutants of some genes involved in these pathways have shown exacerbated hematopoietic cell proliferation that is sometimes associated with the development of melanotic tumors in the larval lymph gland (220). Moreover, several pathways, such as the JAK/STAT and Wnt pathways, have been linked to diverse cases of human leukemia or lymphoma (reviewed in reference 221). Thus, due to its powerful genetics and the relative simplicity of its genome compared to the highly redundant mammalian genomes, *Drosophila* is currently used as a screening model for genes involved in the development of blood cell-derived cancers.

As presented in this review, *Drosophila* hemocytes are multitasking cells that are required for proper embryogenesis and morphogenesis but also for the coordination of the immune response through the secretion of cytokines, as well as the elimination of deadly microorganisms by the means of phagocytosis. High-throughput genetic screens have allowed the accumulation of considerable data related to hemocyte immune functions. Prominent among these is the involvement of several phagocytic receptors for the clearance of apoptotic bodies and/or microorganisms. It has been proposed that the expansion of phagocytic receptors in *Drosophila* would cope with the high needs of morphogenesis in holometabolous insects but would also compensate for the absence of an adaptive immune response (158). Genetic screens and massive proteomics have allowed for the identification of several components of the phagosomal compartments in addition to phagocytic receptors. It remains to confirm these data by asserting the relevance of each of the elements in controlling the phagocytic process through genetic analysis *in vivo* and to translate them into a comprehensive view of the orchestrated events controlling the phagocytic uptake and clearance of microorganisms through live imaging techniques.

Finally, beyond their role in containing microbial infections, macrophages have long been suspected to be involved in the development of several human diseases associated with an inflammatory response, such as metabolic disorders as well as tumor development and progression. The analysis of the specific contribution of macrophages to such pathologies in mammalian systems is hampered by several factors, including the complexity of blood cell lineages and the overlapping effects of innate and adaptive immune responses, but also by limited technical facilities and time-consuming genetic manipulations. Therefore, these fields of research are currently expanding in the *Drosophila* model, which provides all required advantages for their development. In this context, Woodcock and colleagues have recently defined macrophages as the causative agent of impaired glucose homeostasis and a reduced life span following a chronic lipid-rich diet. This study, mimicking the effect of "obesity" in flies, has clearly shown that lipid excess is scavenged by the CD36 receptor homolog, Croquemort, which activates JNK signaling in macrophages, further driving their production of UPD3. Upon its secretion, this cytokine drives a systemic JAK/STAT response responsible for decreased insulin sensitivity and a compromised life span in flies (28). Other studies in *Drosophila* have put the macrophages at the front line of damage sensing and the consequent responses of tissue repair and cellular homeostasis (88). Indeed, *Drosophila* models featuring the development of solid tumors have highlighted the recruitment of hemocytes to the aberrant tissues marked by the disruption of the basement membrane (88, 222, 223). The same kind of observation was noted at wounded sites, drawing similarities between the immune responses to tumors and wounds in flies (29, 88, 154).

As in metabolic stress, tissue damage is translated in systemic JAK/STAT signaling through JNK-dependent secretion of the UPD cytokine. Interestingly, it was shown that hemocytes participate in the amplification of UPD expression and JAK/STAT signaling and are essential for the restriction of tumor growth (88). More recently, another study by Parisi and colleagues also evidenced the contribution of macrophages to tumor restriction through the secretion of other cytokines, notably Spätzle, which drives systemic Toll signaling, and Eiger, the *Drosophila* homolog of mammalian TNF-α, which triggers apoptotic death of tumor cells (132). Altogether, these findings provide evidence for the role of macrophages in cancer immunosurveillance. However, an inflammatory response driven by macrophages is also suspected to enhance tumor development. This was clearly shown in a *Drosophila* model of apoptosis-induced proliferation, in which macrophages were shown to be the sensors of ROS danger signals produced by undead cells. Activated phagocytes produce Eiger, which provides a proliferative signal to the aberrant cells through the JNK signaling cascade (224). These recent data highlight the central role of macrophages as sensors of danger signals, orchestrating cellular reactions and shaping the tissue damage-induced responses through the production of diverse cytokines. Another interesting finding in this perspective is the decisive role of macrophages in the control of intestinal stem cell proliferation and gut homeostasis, both in the case of a local oxidative stress or infection but also following a septic injury (29, 225). Macrophages were also shown to be involved in the establishment of gut dysplasia in aging flies (225). In all these cases, the hemocytes are recruited to the gut and were shown to deliver signaling molecules to activate stem cell division. These observations in *Drosophila* provide insights into the role of immune cells in the development and progression of gut-associated inflammatory diseases in humans. Given the high conservation of core signaling pathways between *Drosophila* and human, further investigations in this model organism will likely uncover essential mediators of inflammatory diseases and potential targets for therapeutic interventions.

Acknowledgments. *This work has been published under the framework of the LABEX: ANR-10-LABX-0036_NETRNA and benefits from a funding from the state managed by the French National Research Agency as part of the Investments for the Future program to J.-M.R. and N.M.; the Centre National de la Recherche Scientifique; and a European Research Council Advanced Grant (AdG_20090506 "Immudroso," to J.-M.R.). L.E.C's research is funded with support from the National Council for Scientific Research in Lebanon, the Research Council of the Saint-Joseph University of Beirut, the UNESCO-L'Oréal for women in Sciences Program and the Cooperation for the Evaluation and Development of Research program (CEDRE- 32942VD).*

Citation. El Chamy L, Matt N, Reichhart J-M. 2017. Advances in myeloid-like cell origins and functions in the model organism *Drosophila melanogaster.* Microbiol Spectrum 5(1):MCHD-0038-2016.

References

1. **Janeway CA Jr.** 1989. Approaching the asymptote? Evolution and revolution in immunology. *Cold Spring Harb Symp Quant Biol* **54**(Pt 1)**:**1–13.
2. **Ferrandon D, Imler JL, Hetru C, Hoffmann JA.** 2007. The *Drosophila* systemic immune response: sensing and signalling during bacterial and fungal infections. *Nat Rev Immunol* **7:**862–874.
3. **Lemaitre B, Nicolas E, Michaut L, Reichhart JM, Hoffmann JA.** 1996. The dorsoventral regulatory gene cassette *spätzle/Toll/cactus* controls the potent antifungal response in Drosophila adults. *Cell* **86:**973–983.
4. **Medzhitov R, Preston-Hurlburt P, Janeway CA Jr.** 1997. A human homologue of the *Drosophila* Toll protein signals activation of adaptive immunity. *Nature* **388:**394–397.
5. **Poltorak A, He X, Smirnova I, Liu MY, Van Huffel C, Du X, Birdwell D, Alejos E, Silva M, Galanos C, Freudenberg M, Ricciardi-Castagnoli P, Layton B, Beutler B.** 1998. Defective LPS signaling in C3H/HeJ and C57BL/10ScCr mice: mutations in *Tlr4* gene. *Science* **282:**2085–2088.
6. **Kawai T, Akira S.** 2011. Toll-like receptors and their crosstalk with other innate receptors in infection and immunity. *Immunity* **34:**637–650.
7. **Ferrandon D.** 2013. The complementary facets of epithelial host defenses in the genetic model organism *Drosophila melanogaster*: from resistance to resilience. *Curr Opin Immunol* **25:**59–70.
8. **El Chamy L, Matt N, Ntwasa M, Reichhart JM.** 2015. The multilayered innate immune defense of the gut. *Biomed J* **38:**276–284.
9. **Lemaitre B, Hoffmann J.** 2007. The host defense of *Drosophila melanogaster. Annu Rev Immunol* **25:**697–743.
10. **Valanne S, Wang JH, Rämet M.** 2011. The *Drosophila* Toll signaling pathway. *J Immunol* **186:**649–656.
11. **Myllymäki H, Valanne S, Rämet M.** 2014. The *Drosophila* Imd signaling pathway. *J Immunol* **192:**3455–3462.
12. **Royet J, Reichhart JM, Hoffmann JA.** 2005. Sensing and signaling during infection in *Drosophila. Curr Opin Immunol* **17:**11–17.
13. **Charroux B, Rival T, Narbonne-Reveau K, Royet J.** 2009. Bacterial detection by *Drosophila* peptidoglycan recognition proteins. *Microbes Infect* **11:**631–636.
14. **Jang IH, Chosa N, Kim SH, Nam HJ, Lemaitre B, Ochiai M, Kambris Z, Brun S, Hashimoto C, Ashida M, Brey PT, Lee WJ.** 2006. A Spätzle-processing enzyme required for Toll signaling activation in *Drosophila* innate immunity. *Dev Cell* **10:**45–55.
15. **El Chamy L, Leclerc V, Caldelari I, Reichhart JM.** 2008. Sensing of 'danger signals' and pathogen-associated molecular patterns defines binary signaling pathways 'upstream' of Toll. *Nat Immunol* **9:**1165–1170.
16. **Gottar M, Gobert V, Matskevich AA, Reichhart JM, Wang C, Butt TM, Belvin M, Hoffmann JA, Ferrandon D.** 2006. Dual detection of fungal infections in *Drosophila* via recognition of glucans and sensing of virulence factors. *Cell* **127:**1425–1437.
17. **Cerenius L, Kawabata S, Lee BL, Nonaka M, Soderhall K.** 2010. Proteolytic cascades and their involvement in invertebrate immunity. *Trends Biochem Sci* **35:**575–583.
18. **Tang H.** 2009. Regulation and function of the melanization reaction in *Drosophila. Fly (Austin)* **3:**105–111.
19. **Tang H, Kambris Z, Lemaitre B, Hashimoto C.** 2008. A serpin that regulates immune melanization in the respiratory system of *Drosophila. Dev Cell* **15:**617–626.
20. **Boutros M, Agaisse H, Perrimon N.** 2002. Sequential activation of signaling pathways during innate immune responses in *Drosophila. Dev Cell* **3:**711–722.
21. **Rämet M, Lanot R, Zachary D, Manfruelli P.** 2002. JNK signaling pathway is required for efficient wound healing in *Drosophila. Dev Biol* **241:**145–156.
22. **Galko MJ, Krasnow MA.** 2004. Cellular and genetic analysis of wound healing in *Drosophila* larvae. *PLoS Biol* **2:**E239. doi:10.1371/journal.pbio.0020239.
23. **Lesch C, Jo J, Wu Y, Fish GS, Galko MJ.** 2010. A targeted *UAS-RNAi* screen in *Drosophila* larvae identifies wound closure genes regulating distinct cellular processes. *Genetics* **186:**943–957.
24. **Kwon YC, Baek SH, Lee H, Choe KM.** 2010. Nonmuscle myosin II localization is regulated by JNK during *Drosophila* larval wound healing. *Biochem Biophys Res Commun* **393:**656–661.
25. **Ekengren S, Hultmark D.** 2001. A family of *Turandot*-related genes in the humoral stress response of *Drosophila. Biochem Biophys Res Commun* **284:**998–1003.
26. **Agaisse H, Petersen UM, Boutros M, Mathey-Prevot B, Perrimon N.** 2003. Signaling role of hemocytes in *Drosophila* JAK/STAT-dependent response to septic injury. *Dev Cell* **5:**441–450.
27. **Brun S, Vidal S, Spellman P, Takahashi K, Tricoire H, Lemaitre B.** 2006. The MAPKKK Mekk1 regulates the expression of *Turandot* stress genes in response to septic injury in *Drosophila. Genes Cells* **11:**397–407.
28. **Woodcock KJ, Kierdorf K, Pouchelon CA, Vivancos V, Dionne MS, Geissmann F.** 2015. Macrophage-derived *upd3* cytokine causes impaired glucose homeostasis and reduced lifespan in *Drosophila* fed a lipid-rich diet. *Immunity* **42:**133–144.

29. **Chakrabarti S, Dudzic JP, Li X, Collas EJ, Boquete JP, Lemaitre B.** 2016. Remote control of intestinal stem cell activity by haemocytes in *Drosophila*. *PLoS Genet* **12:** e1006089. doi:10.1371/journal.pgen.1006089.
30. **Martins N, Imler JL, Meignin C.** 2016. Discovery of novel targets for antivirals: learning from flies. *Curr Opin Virol* **20:**64–70.
31. **Xu J, Cherry S.** 2014. Viruses and antiviral immunity in *Drosophila*. *Dev Comp Immunol* **42:**67–84.
32. **Dostert C, Jouanguy E, Irving P, Troxler L, Galiana-Arnoux D, Hetru C, Hoffmann JA, Imler JL.** 2005. The Jak-STAT signaling pathway is required but not sufficient for the antiviral response of drosophila. *Nat Immunol* **6:**946–953.
33. **Carpenter J, Hutter S, Baines JF, Roller J, Saminadin-Peter SS, Parsch J, Jiggins FM.** 2009. The transcriptional response of *Drosophila melanogaster* to infection with the sigma virus (Rhabdoviridae). *PLoS One* **4:** e6838. doi:10.1371/journal.pone.0006838.
34. **Castorena KM, Stapleford KA, Miller DJ.** 2010. Complementary transcriptomic, lipidomic, and targeted functional genetic analyses in cultured *Drosophila* cells highlight the role of glycerophospholipid metabolism in Flock House virus RNA replication. *BMC Genomics* **11:**183. doi:10.1186/1471-2164-11-183.
35. **Mudiganti U, Hernandez R, Brown DT.** 2010. Insect response to alphavirus infection—establishment of alphavirus persistence in insect cells involves inhibition of viral polyprotein cleavage. *Virus Res* **150:**73–84.
36. **Xu J, Grant G, Sabin LR, Gordesky-Gold B, Yasunaga A, Tudor M, Cherry S.** 2012. Transcriptional pausing controls a rapid antiviral innate immune response in *Drosophila*. *Cell Host Microbe* **12:**531–543.
37. **Kemp C, Mueller S, Goto A, Barbier V, Paro S, Bonnay F, Dostert C, Troxler L, Hetru C, Meignin C, Pfeffer S, Hoffmann JA, Imler JL.** 2013. Broad RNA interference-mediated antiviral immunity and virus-specific inducible responses in *Drosophila*. *J Immunol* **190:**650–658.
38. **Cordes EJ, Licking-Murray KD, Carlson KA.** 2013. Differential gene expression related to Nora virus infection of *Drosophila melanogaster*. *Virus Res* **175:**95–100.
39. **Huang Z, Kingsolver MB, Avadhanula V, Hardy RW.** 2013. An antiviral role for antimicrobial peptides during the arthropod response to alphavirus replication. *J Virol* **87:**4272–4280.
40. **Lamiable O, Imler JL.** 2014. Induced antiviral innate immunity in *Drosophila*. *Curr Opin Microbiol* **20:** 62–68.
41. **Zambon RA, Nandakumar M, Vakharia VN, Wu LP.** 2005. The Toll pathway is important for an antiviral response in *Drosophila*. *Proc Natl Acad Sci U S A* **102:** 7257–7262.
42. **Avadhanula V, Weasner BP, Hardy GG, Kumar JP, Hardy RW.** 2009. A novel system for the launch of alphavirus RNA synthesis reveals a role for the Imd pathway in arthropod antiviral response. *PLoS Pathog* **5:**e1000582. doi:10.1371/journal.ppat.1000582.
43. **Costa A, Jan E, Sarnow P, Schneider D.** 2009. The Imd pathway is involved in antiviral immune responses in *Drosophila*. *PLoS One* **4:**e7436. doi:10.1371/journal.pone.0007436.
44. **Rancès E, Johnson TK, Popovici J, Iturbe-Ormaetxe I, Zakir T, Warr CG, O'Neill SL.** 2013. The Toll and Imd pathways are not required for *Wolbachia*-mediated dengue virus interference. *J Virol* **87:**11945–11949.
45. **Ferreira AG, Naylor H, Esteves SS, Pais IS, Martins NE, Teixeira L.** 2014. The Toll-dorsal pathway is required for resistance to viral oral infection in *Drosophila*. *PLoS Pathog* **10:**e1004507. doi:10.1371/journal.ppat.1004507.
46. **Merkling SH, Bronkhorst AW, Kramer JM, Overheul GJ, Schenck A, Van Rij RP.** 2015. The epigenetic regulator G9a mediates tolerance to RNA virus infection in *Drosophila*. *PLoS Pathog* **11:**e1004692. doi:10.1371/journal.ppat.1004692.
47. **Shelly S, Lukinova N, Bambina S, Berman A, Cherry S.** 2009. Autophagy is an essential component of *Drosophila* immunity against vesicular stomatitis virus. *Immunity* **30:**588–598.
48. **Liu B, Behura SK, Clem RJ, Schneemann A, Becnel J, Severson DW, Zhou L.** 2013. P53-mediated rapid induction of apoptosis conveys resistance to viral infection in *Drosophila melanogaster*. *PLoS Pathog* **9:** e1003137. doi:10.1371/journal.ppat.1003137.
49. **Nainu F, Tanaka Y, Shiratsuchi A, Nakanishi Y.** 2015. Protection of insects against viral infection by apoptosis-dependent phagocytosis. *J Immunol* **195:**5696–5706.
50. **Galiana-Arnoux D, Dostert C, Schneemann A, Hoffmann JA, Imler JL.** 2006. Essential function *in vivo* for Dicer-2 in host defense against RNA viruses in drosophila. *Nat Immunol* **7:**590–597.
51. **van Rij RP, Saleh MC, Berry B, Foo C, Houk A, Antoniewski C, Andino R.** 2006. The RNA silencing endonuclease Argonaute 2 mediates specific antiviral immunity in *Drosophila melanogaster*. *Genes Dev* **20:** 2985–2995.
52. **Wang XH, Aliyari R, Li WX, Li HW, Kim K, Carthew R, Atkinson P, Ding SW.** 2006. RNA interference directs innate immunity against viruses in adult *Drosophila*. *Science* **312:**452–454.
53. **Mueller S, Gausson V, Vodovar N, Deddouche S, Troxler L, Perot J, Pfeffer S, Hoffmann JA, Saleh MC, Imler JL.** 2010. RNAi-mediated immunity provides strong protection against the negative-strand RNA vesicular stomatitis virus in *Drosophila*. *Proc Natl Acad Sci U S A* **107:**19390–19395.
54. **Bronkhorst AW, van Cleef KW, Vodovar N, Ince IA, Blanc H, Vlak JM, Saleh MC, van Rij RP.** 2012. The DNA virus Invertebrate iridescent virus 6 is a target of the *Drosophila* RNAi machinery. *Proc Natl Acad Sci U S A* **109:**E3604–E3613. doi:10.1073/pnas.1207213109.
55. **Bernstein E, Caudy AA, Hammond SM, Hannon GJ.** 2001. Role for a bidentate ribonuclease in the initiation step of RNA interference. *Nature* **409:**363–366.
56. **Rand TA, Ginalski K, Grishin NV, Wang X.** 2004. Biochemical identification of Argonaute 2 as the sole protein required for RNA-induced silencing complex activity. *Proc Natl Acad Sci U S A* **101:**14385–14389.

57. **Okamura K, Ishizuka A, Siomi H, Siomi MC.** 2004. Distinct roles for Argonaute proteins in small RNA-directed RNA cleavage pathways. *Genes Dev* **18:**1655–1666.

58. **Deddouche S, Matt N, Budd A, Mueller S, Kemp C, Galiana-Arnoux D, Dostert C, Antoniewski C, Hoffmann JA, Imler JL.** 2008. The DExD/H-box helicase Dicer-2 mediates the induction of antiviral activity in drosophila. *Nat Immunol* **9:**1425–1432.

59. **Desmet CJ, Ishii KJ.** 2012. Nucleic acid sensing at the interface between innate and adaptive immunity in vaccination. *Nat Rev Immunol* **12:**479–491.

60. **Meister M.** 2004. Blood cells of *Drosophila*: cell lineages and role in host defence. *Curr Opin Immunol* **16:** 10–15.

61. **Crozatier M, Meister M.** 2007. *Drosophila* haematopoiesis. *Cell Microbiol* **9:**1117–1126.

62. **Letourneau M, Lapraz F, Sharma A, Vanzo N, Waltzer L, Crozatier M.** 2016. *Drosophila* hematopoiesis under normal conditions and in response to immune stress. *FEBS Lett* **590:**4034–4051.

63. **Gold KS, Bruckner K.** 2014. *Drosophila* as a model for the two myeloid blood cell systems in vertebrates. *Exp Hematol* **42:**717–727.

64. **Holz A, Bossinger B, Strasser T, Janning W, Klapper R.** 2003. The two origins of hemocytes in *Drosophila*. *Development* **130:**4955–4962.

65. **Tepass U, Fessler LI, Aziz A, Hartenstein V.** 1994. Embryonic origin of hemocytes and their relationship to cell death in *Drosophila*. *Development* **120:**1829–1837.

66. **Franc NC, Heitzler P, Ezekowitz RA, White K.** 1999. Requirement for Croquemort in phagocytosis of apoptotic cells in *Drosophila*. *Science* **284:**1991–1994.

67. **Franc NC.** 2002. Phagocytosis of apoptotic cells in mammals, *Caenorhabditis elegans* and *Drosophila melanogaster*: molecular mechanisms and physiological consequences. *Front Biosci* **7:**d1298–d1313.

68. **Sears HC, Kennedy CJ, Garrity PA.** 2003. Macrophage-mediated corpse engulfment is required for normal *Drosophila* CNS morphogenesis. *Development* **130:**3557–3565.

69. **Olofsson B, Page DT.** 2005. Condensation of the central nervous system in embryonic *Drosophila* is inhibited by blocking hemocyte migration or neural activity. *Dev Biol* **279:**233–243.

70. **Defaye A, Evans I, Crozatier M, Wood W, Lemaitre B, Leulier F.** 2009. Genetic ablation of *Drosophila* phagocytes reveals their contribution to both development and resistance to bacterial infection. *J Innate Immun* **1:**322–334.

71. **Wood W, Jacinto A.** 2007. *Drosophila melanogaster* embryonic haemocytes: masters of multitasking. *Nat Rev Mol Cell Biol* **8:**542–551.

72. **Terriente-Felix A, Li J, Collins S, Mulligan A, Reekie I, Bernard F, Krejci A, Bray S.** 2013. Notch cooperates with Lozenge/Runx to lock haemocytes into a differentiation programme. *Development* **140:**926–937.

73. **Lebestky T, Chang T, Hartenstein V, Banerjee U.** 2000. Specification of *Drosophila* hematopoietic lineage by conserved transcription factors. *Science* **288:**146–149.

74. **Bernardoni R, Vivancos V, Giangrande A.** 1997. *glide/gcm* is expressed and required in the scavenger cell lineage. *Dev Biol* **191:**118–130.

75. **Lebestky T, Jung SH, Banerjee U.** 2003. A Serrate-expressing signaling center controls *Drosophila* hematopoiesis. *Genes Dev* **17:**348–353.

76. **Muratoglu S, Garratt B, Hyman K, Gajewski K, Schulz RA, Fossett N.** 2006. Regulation of *Drosophila* Friend of GATA gene, *u-shaped*, during hematopoiesis: a direct role for Serpent and Lozenge. *Dev Biol* **296:**561–579.

77. **Bataille L, Auge B, Ferjoux G, Haenlin M, Waltzer L.** 2005. Resolving embryonic blood cell fate choice in *Drosophila*: interplay of GCM and RUNX factors. *Development* **132:**4635–4644.

78. **Evans CJ, Hartenstein V, Banerjee U.** 2003. Thicker than blood: conserved mechanisms in *Drosophila* and vertebrate hematopoiesis. *Dev Cell* **5:**673–690.

79. **Markus R, Laurinyecz B, Kurucz E, Honti V, Bajusz I, Sipos B, Somogyi K, Kronhamn J, Hultmark D, Ando I.** 2009. Sessile hemocytes as a hematopoietic compartment in *Drosophila melanogaster*. *Proc Natl Acad Sci U S A* **106:**4805–4809.

80. **Zaidman-Remy A, Regan JC, Brandao AS, Jacinto A.** 2012. The *Drosophila* larva as a tool to study gut-associated macrophages: PI3K regulates a discrete hemocyte population at the proventriculus. *Dev Comp Immunol* **36:**638–647.

81. **Stofanko M, Kwon SY, Badenhorst P.** 2008. A misexpression screen to identify regulators of *Drosophila* larval hemocyte development. *Genetics* **180:**253–267.

82. **Hashimoto D, Chow A, Noizat C, Teo P, Beasley MB, Leboeuf M, Becker CD, See P, Price J, Lucas D, Greter M, Mortha A, Boyer SW, Forsberg EC, Tanaka M, van Rooijen N, Garcia-Sastre A, Stanley ER, Ginhoux F, Frenette PS, Merad M.** 2013. Tissue-resident macrophages self-maintain locally throughout adult life with minimal contribution from circulating monocytes. *Immunity* **38:**792–804.

83. **Makhijani K, Alexander B, Tanaka T, Rulifson E, Bruckner K.** 2011. The peripheral nervous system supports blood cell homing and survival in the *Drosophila* larva. *Development* **138:**5379–5391.

84. **Makhijani K, Bruckner K.** 2012. Of blood cells and the nervous system: hematopoiesis in the *Drosophila* larva. *Fly (Austin)* **6:**254–260.

85. **Honti V, Csordas G, Markus R, Kurucz E, Jankovics F, Ando I.** 2010. Cell lineage tracing reveals the plasticity of the hemocyte lineages and of the hematopoietic compartments in *Drosophila melanogaster*. *Mol Immunol* **47:**1997–2004.

86. **Leitão AB, Sucena E.** 2015. *Drosophila* sessile hemocyte clusters are true hematopoietic tissues that regulate larval blood cell differentiation. *eLife* **4:**e06166. doi: 10.7554/eLife.06166.

87. **Lanot R, Zachary D, Holder F, Meister M.** 2001. Postembryonic hematopoiesis in *Drosophila*. *Dev Biol* **230:** 243–257.

88. **Pastor-Pareja JC, Wu M, Xu T.** 2008. An innate immune response of blood cells to tumors and tissue

damage in *Drosophila*. *Dis Model Mech* **1**:144–154; discussion 153.

89. **Zettervall CJ, Anderl I, Williams MJ, Palmer R, Kurucz E, Ando I, Hultmark D.** 2004. A directed screen for genes involved in *Drosophila* blood cell activation. *Proc Natl Acad Sci U S A* **101**:14192–14197.

90. **Babcock DT, Brock AR, Fish GS, Wang Y, Perrin L, Krasnow MA, Galko MJ.** 2008. Circulating blood cells function as a surveillance system for damaged tissue in *Drosophila* larvae. *Proc Natl Acad Sci U S A* **105**: 10017–10022.

91. **Grigorian M, Mandal L, Hartenstein V.** 2011. Hematopoiesis at the onset of metamorphosis: terminal differentiation and dissociation of the *Drosophila* lymph gland. *Dev Genes Evol* **221**:121–131.

92. **Rugendorff AE, Younossi-Hartenstein A, Hartenstein V.** 1994. Embryonic origin and differentiation of the *Drosophila* heart. *Roux's Arch Dev Bio* **203**:266–280.

93. **Mandal L, Banerjee U, Hartenstein V.** 2004. Evidence for a fruit fly hemangioblast and similarities between lymph-gland hematopoiesis in fruit fly and mammal aorta-gonadal-mesonephros mesoderm. *Nat Genet* **36**: 1019–1023.

94. **Hartenstein V.** 2006. Blood cells and blood cell development in the animal kingdom. *Annu Rev Cell Dev Biol* **22**:677–712.

95. **Jung SH, Evans CJ, Uemura C, Banerjee U.** 2005. The *Drosophila* lymph gland as a developmental model of hematopoiesis. *Development* **132**:2521–2533.

96. **Krzemień J, Dubois L, Makki R, Meister M, Vincent A, Crozatier M.** 2007. Control of blood cell homeostasis in *Drosophila* larvae by the posterior signalling centre. *Nature* **446**:325–328.

97. **Mandal L, Martinez-Agosto JA, Evans CJ, Hartenstein V, Banerjee U.** 2007. A Hedgehog- and Antennapedia-dependent niche maintains *Drosophila* haematopoietic precursors. *Nature* **446**:320–324.

98. **Crozatier M, Krzemień J, Vincent A.** 2007. The hematopoietic niche: a *Drosophila* model, at last. *Cell Cycle* **6**:1443–1444.

99. **Minakhina S, Steward R.** 2010. Hematopoietic stem cells in *Drosophila*. *Development* **137**:27–31.

100. **Mondal BC, Mukherjee T, Mandal L, Evans CJ, Sinenko SA, Martinez-Agosto JA, Banerjee U.** 2011. Interaction between differentiating cell- and niche-derived signals in hematopoietic progenitor maintenance. *Cell* **147**:1589–1600.

101. **Benmimoun B, Polesello C, Haenlin M, Waltzer L.** 2015. The EBF transcription factor Collier directly promotes *Drosophila* blood cell progenitor maintenance independently of the niche. *Proc Natl Acad Sci U S A* **112**:9052–9057.

102. **Morin-Poulard I, Sharma A, Louradour I, Vanzo N, Vincent A, Crozatier M.** 2016. Vascular control of the *Drosophila* haematopoietic microenvironment by Slit/Robo signalling. *Nat Commun* **7**:11634. doi:10.1038/ncomms11634.

103. **Sinenko SA, Mandal L, Martinez-Agosto JA, Banerjee U.** 2009. Dual role of Wingless signaling in stem-like hematopoietic precursor maintenance in *Drosophila*. *Dev Cell* **16**:756–763.

104. **Shim J, Mukherjee T, Banerjee U.** 2012. Direct sensing of systemic and nutritional signals by haematopoietic progenitors in *Drosophila*. *Nat Cell Biol* **14**:394–400.

105. **Owusu-Ansah E, Banerjee U.** 2009. Reactive oxygen species prime *Drosophila* haematopoietic progenitors for differentiation. *Nature* **461**:537–541.

106. **Shim J, Mukherjee T, Mondal BC, Liu T, Young GC, Wijewarnasuriya DP, Banerjee U.** 2013. Olfactory control of blood progenitor maintenance. *Cell* **155**: 1141–1153.

107. **Oyallon J, Vanzo N, Krzemień J, Morin-Poulard I, Vincent A, Crozatier M.** 2016. Two independent functions of Collier/Early B Cell Factor in the control of *Drosophila* blood cell homeostasis. *PLoS One* **11**: e0148978. doi:10.1371/journal.pone.0148978.

108. **Qiu P, Pan PC, Govind S.** 1998. A role for the *Drosophila* Toll/Cactus pathway in larval hematopoiesis. *Development* **125**:1909–1920.

109. **Myrick KV, Dearolf CR.** 2000. Hyperactivation of the *Drosophila* Hop Jak kinase causes the preferential overexpression of eIF1A transcripts in larval blood cells. *Gene* **244**:119–125.

110. **Minakhina S, Tan W, Steward R.** 2011. JAK/STAT and the GATA factor Pannier control hemocyte maturation and differentiation in Drosophila. *Dev Biol* **352**: 308–316.

111. **Pennetier D, Oyallon J, Morin-Poulard I, Dejean S, Vincent A, Crozatier M.** 2012. Size control of the *Drosophila* hematopoietic niche by bone morphogenetic protein signaling reveals parallels with mammals. *Proc Natl Acad Sci U S A* **109**:3389–3394.

112. **Remillieux-Leschelle N, Santamaria P, Randsholt NB.** 2002. Regulation of larval hematopoiesis in *Drosophila melanogaster*: a role for the *multi sex combs* gene. *Genetics* **162**:1259–1274.

113. **Minakhina S, Druzhinina M, Steward R.** 2007. *Zfrp8*, the *Drosophila* ortholog of *PDCD2*, functions in lymph gland development and controls cell proliferation. *Development* **134**:2387–2396.

114. **Crozatier M, Ubeda JM, Vincent A, Meister M.** 2004. Cellular immune response to parasitization in *Drosophila* requires the EBF orthologue Collier. *PLoS Biol* **2**: E196. doi:10.1371/journal.pbio.0020196.

115. **Rizki TM, Rizki RM.** 1992. Lamellocyte differentiation in *Drosophila* larvae parasitized by *Leptopilina*. *Dev Comp Immunol* **16**:103–110.

116. **Sinenko SA, Shim J, Banerjee U.** 2012. Oxidative stress in the haematopoietic niche regulates the cellular immune response in *Drosophila*. *EMBO Rep* **13**:83–89.

117. **Ghosh S, Singh A, Mandal S, Mandal L.** 2015. Active hematopoietic hubs in *Drosophila* adults generate hemocytes and contribute to immune response. *Dev Cell* **33**:478–488.

118. **Martinek N, Shahab J, Saathoff M, Ringuette M.** 2008. Haemocyte-derived SPARC is required for collagen-IV-dependent stability of basal laminae in *Drosophila* embryos. *J Cell Sci* **121**:1671–1680.

119. **Bunt S, Hooley C, Hu N, Scahill C, Weavers H, Skaer H.** 2010. Hemocyte-secreted type IV collagen enhances BMP signaling to guide renal tubule morphogenesis in *Drosophila*. *Dev Cell* **19:**296–306.

120. **Fessler JH, Fessler LI.** 1989. *Drosophila* extracellular matrix. *Annu Rev Cell Biol* **5:**309–339.

121. **Gullberg D, Fessler LI, Fessler JH.** 1994. Differentiation, extracellular matrix synthesis, and integrin assembly by *Drosophila* embryo cells cultured on vitronectin and laminin substrates. *Dev Dyn* **199:**116–128.

122. **Hortsch M, Olson A, Fishman S, Soneral SN, Marikar Y, Dong R, Jacobs JR.** 1998. The expression of MDP-1, a component of *Drosophila* embryonic basement membranes, is modulated by apoptotic cell death. *Int J Dev Biol* **42:**33–42.

123. **Lamiable O, Arnold J, de Faria IJ, Olmo RP, Bergami F, Meignin C, Hoffmann JA, Marques JT, Imler JL.** 2016. Analysis of the contribution of hemocytes and autophagy to *Drosophila* antiviral immunity. *J Virol* **90:** 5415–5426.

124. **Vlisidou I, Wood W.** 2015. *Drosophila* blood cells and their role in immune responses. *FEBS J* **282:**1368–1382.

125. **Elrod-Erickson M, Mishra S, Schneider D.** 2000. Interactions between the cellular and humoral immune responses in *Drosophila*. *Curr Biol* **10:**781–784.

126. **Nehme NT, Quintin J, Cho JH, Lee J, Lafarge MC, Kocks C, Ferrandon D.** 2011. Relative roles of the cellular and humoral responses in the *Drosophila* host defense against three Gram-positive bacterial infections. *PLoS One* **6:**e14743. doi:10.1371/journal.pone.0014743.

127. **Basset A, Khush RS, Braun A, Gardan L, Boccard F, Hoffmann JA, Lemaitre B.** 2000. The phytopathogenic bacteria *Erwinia carotovora* infects *Drosophila* and activates an immune response. *Proc Natl Acad Sci U S A* **97:**3376–3381.

128. **Shia AK, Glittenberg M, Thompson G, Weber AN, Reichhart JM, Ligoxygakis P.** 2009. Toll-dependent antimicrobial responses in *Drosophila* larval fat body require Spätzle secreted by haemocytes. *J Cell Sci* **122:** 4505–4515.

129. **Foley E, O'Farrell PH.** 2003. Nitric oxide contributes to induction of innate immune responses to gram-negative bacteria in *Drosophila*. *Genes Dev* **17:**115–125.

130. **Wu SC, Liao CW, Pan RL, Juang JL.** 2012. Infection-induced intestinal oxidative stress triggers organ-to-organ immunological communication in *Drosophila*. *Cell Host Microbe* **11:**410–417.

131. **Glittenberg MT, Kounatidis I, Christensen D, Kostov M, Kimber S, Roberts I, Ligoxygakis P.** 2011. Pathogen and host factors are needed to provoke a systemic host response to gastrointestinal infection of *Drosophila* larvae by *Candida albicans*. *Dis Model Mech* **4:**515–525.

132. **Parisi F, Stefanatos RK, Strathdee K, Yu Y, Vidal M.** 2014. Transformed epithelia trigger non-tissue-autonomous tumor suppressor response by adipocytes via activation of Toll and Eiger/TNF signaling. *Cell Rep* **6:**855–867.

133. **Braun A, Hoffmann JA, Meister M.** 1998. Analysis of the *Drosophila* host defense in *domino* mutant larvae, which are devoid of hemocytes. *Proc Natl Acad Sci U S A* **95:**14337–14342.

134. **Ruhf ML, Braun A, Papoulas O, Tamkun JW, Randsholt N, Meister M.** 2001. The *domino* gene of *Drosophila* encodes novel members of the SWI2/SNF2 family of DNA-dependent ATPases, which contribute to the silencing of homeotic genes. *Development* **128:**1429–1441.

135. **Gateff E.** 1994. Tumor suppressor and overgrowth suppressor genes of *Drosophila melanogaster*: developmental aspects. *Int J Dev Biol* **38:**565–590.

136. **Brennan CA, Delaney JR, Schneider DS, Anderson KV.** 2007. Psidin is required in *Drosophila* blood cells for both phagocytic degradation and immune activation of the fat body. *Curr Biol* **17:**67–72.

137. **Irving P, Ubeda JM, Doucet D, Troxler L, Lagueux M, Zachary D, Hoffmann JA, Hetru C, Meister M.** 2005. New insights into *Drosophila* larval haemocyte functions through genome-wide analysis. *Cell Microbiol* **7:** 335–350.

138. **Nam HJ, Jang IH, Asano T, Lee WJ.** 2008. Involvement of pro-phenoloxidase 3 in lamellocyte-mediated spontaneous melanization in *Drosophila*. *Mol Cells* **26:**606–610.

139. **Waltzer L, Ferjoux G, Bataille L, Haenlin M.** 2003. Cooperation between the GATA and RUNX factors Serpent and Lozenge during *Drosophila* hematopoiesis. *EMBO J* **22:**6516–6525.

140. **Dudzic JP, Kondo S, Ueda R, Bergman CM, Lemaitre B.** 2015. *Drosophila* innate immunity: regional and functional specialization of prophenoloxidases. *BMC Biol* **13:**81. doi:10.1186/s12915-015-0193-6.

141. **Leclerc V, Pelte N, El Chamy L, Martinelli C, Ligoxygakis P, Hoffmann JA, Reichhart JM.** 2006. Prophenoloxidase activation is not required for survival to microbial infections in *Drosophila*. *EMBO Rep* **7:**231–235.

142. **Tang H, Kambris Z, Lemaitre B, Hashimoto C.** 2006. Two proteases defining a melanization cascade in the immune system of *Drosophila*. *J Biol Chem* **281:**28097–28104.

143. **Rizki TM, Rizki RM, Bellotti RA.** 1985. Genetics of a *Drosophila* phenoloxidase. *Mol Gen Genet* **201:**7–13.

144. **Ayres JS, Schneider DS.** 2008. A signaling protease required for melanization in *Drosophila* affects resistance and tolerance of infections. *PLoS Biol* **6:**2764–2773.

145. **Lemaitre B, Kromer-Metzger E, Michaut L, Nicolas E, Meister M, Georgel P, Reichhart JM, Hoffmann JA.** 1995. A recessive mutation, immune deficiency (*imd*), defines two distinct control pathways in the *Drosophila* host defense. *Proc Natl Acad Sci U S A* **92:**9465–9469.

146. **Neyen C, Binggeli O, Roversi P, Bertin L, Sleiman MB, Lemaitre B.** 2015. The *Black cells* phenotype is caused by a point mutation in the *Drosophila pro-phenoloxidase 1* gene that triggers melanization and hematopoietic defects. *Dev Comp Immunol* **50:**166–174.

147. **Binggeli O, Neyen C, Poidevin M, Lemaitre B.** 2014. Prophenoloxidase activation is required for survival to microbial infections in *Drosophila*. *PLoS Pathog* **10:** e1004067. doi:10.1371/journal.ppat.1004067.

148. **Nam HJ, Jang IH, You H, Lee KA, Lee WJ.** 2012. Genetic evidence of a redox-dependent systemic wound response via Hayan protease-phenoloxidase system in *Drosophila. EMBO J* **31:**1253–1265.

149. **Scherfer C, Karlsson C, Loseva O, Bidla G, Goto A, Havemann J, Dushay MS, Theopold U.** 2004. Isolation and characterization of hemolymph clotting factors in *Drosophila melanogaster* by a pullout method. *Curr Biol* **14:**625–629.

150. **Goto A, Kadowaki T, Kitagawa Y.** 2003. *Drosophila* hemolectin gene is expressed in embryonic and larval hemocytes and its knock down causes bleeding defects. *Dev Biol* **264:**582–591.

151. **Goto A, Kumagai T, Kumagai C, Hirose J, Narita H, Mori H, Kadowaki T, Beck K, Kitagawa Y.** 2001. A *Drosophila* haemocyte-specific protein, hemolectin, similar to human von Willebrand factor. *Biochem J* **359:**99–108.

152. **Evans IR, Wood W.** 2014. *Drosophila* blood cell chemotaxis. *Curr Opin Cell Biol* **30:**1–8.

153. **Stramer B, Wood W, Galko MJ, Redd MJ, Jacinto A, Parkhurst SM, Martin P.** 2005. Live imaging of wound inflammation in *Drosophila* embryos reveals key roles for small GTPases during in vivo cell migration. *J Cell Biol* **168:**567–573.

154. **Wood W, Faria C, Jacinto A.** 2006. Distinct mechanisms regulate hemocyte chemotaxis during development and wound healing in *Drosophila melanogaster. J Cell Biol* **173:**405–416.

155. **Moreira S, Stramer B, Evans I, Wood W, Martin P.** 2010. Prioritization of competing damage and developmental signals by migrating macrophages in the *Drosophila* embryo. *Curr Biol* **20:**464–470.

156. **Razzell W, Evans IR, Martin P, Wood W.** 2013. Calcium flashes orchestrate the wound inflammatory response through DUOX activation and hydrogen peroxide release. *Curr Biol* **23:**424–429.

157. **Evans IR, Rodrigues FS, Armitage EL, Wood W.** 2015. Draper/CED-1 mediates an ancient damage response to control inflammatory blood cell migration in vivo. *Curr Biol* **25:**1606–1612.

158. **Stuart LM, Ezekowitz RA.** 2008. Phagocytosis and comparative innate immunity: learning on the fly. *Nat Rev Immunol* **8:**131–141.

159. **Schneider I.** 1972. Cell lines derived from late embryonic stages of *Drosophila melanogaster. J Embryol Exp Morphol* **27:**353–365.

160. **Cherry S.** 2008. Genomic RNAi screening in *Drosophila* S2 cells: what have we learned about host-pathogen interactions? *Curr Opin Microbiol* **11:**262–270.

161. **Ulvila J, Vanha-Aho LM, Ramet M.** 2011. *Drosophila* phagocytosis—still many unknowns under the surface. *APMIS* **119:**651–662.

162. **Ramet M, Manfruelli P, Pearson A, Mathey-Prevot B, Ezekowitz RA.** 2002. Functional genomic analysis of phagocytosis and identification of a *Drosophila* receptor for *E. coli. Nature* **416:**644–648.

163. **Stuart LM, Deng J, Silver JM, Takahashi K, Tseng AA, Hennessy EJ, Ezekowitz RA, Moore KJ.** 2005. Response to *Staphylococcus aureus* requires CD36-mediated phagocytosis triggered by the COOH-terminal cytoplasmic domain. *J Cell Biol* **170:**477–485.

164. **Stuart LM, Bell SA, Stewart CR, Silver JM, Richard J, Goss JL, Tseng AA, Zhang A, El Khoury JB, Moore KJ.** 2007. CD36 signals to the actin cytoskeleton and regulates microglial migration via a p130Cas complex. *J Biol Chem* **282:**27392–27401.

165. **Philips JA, Rubin EJ, Perrimon N.** 2005. *Drosophila* RNAi screen reveals CD36 family member required for mycobacterial infection. *Science* **309:**1251–1253.

166. **Agaisse H, Burrack LS, Philips JA, Rubin EJ, Perrimon N, Higgins DE.** 2005. Genome-wide RNAi screen for host factors required for intracellular bacterial infection. *Science* **309:**1248–1251.

167. **Cheng LW, Viala JP, Stuurman N, Wiedemann U, Vale RD, Portnoy DA.** 2005. Use of RNA interference in *Drosophila* S2 cells to identify host pathways controlling compartmentalization of an intracellular pathogen. *Proc Natl Acad Sci U S A* **102:**13646–13651.

168. **Stroschein-Stevenson SL, Foley E, O'Farrell PH, Johnson AD.** 2006. Identification of *Drosophila* gene products required for phagocytosis of *Candida albicans. PLoS Biol* **4:**e4. doi:10.1371/journal.pbio.0040004.

169. **Pearson AM, Baksa K, Ramet M, Protas M, McKee M, Brown D, Ezekowitz RA.** 2003. Identification of cytoskeletal regulatory proteins required for efficient phagocytosis in *Drosophila. Microbes Infect* **5:**815–824.

170. **Nichols Z, Vogt RG.** 2008. The SNMP/CD36 gene family in Diptera, Hymenoptera and Coleoptera: *Drosophila melanogaster*, *D. pseudoobscura*, *Anopheles gambiae*, *Aedes aegypti*, *Apis mellifera*, and *Tribolium castaneum. Insect Biochem Mol Biol* **38:**398–415.

171. **Franc NC, Dimarcq JL, Lagueux M, Hoffmann J, Ezekowitz RA.** 1996. Croquemort, a novel Drosophila hemocyte/macrophage receptor that recognizes apoptotic cells. *Immunity* **4:**431–443.

172. **Hoebe K, Georgel P, Rutschmann S, Du X, Mudd S, Crozat K, Sovath S, Shamel L, Hartung T, Zahringer U, Beutler B.** 2005. CD36 is a sensor of diacylglycerides. *Nature* **433:**523–527.

173. **Talamillo A, Herboso L, Pirone L, Pérez C, González M, Sánchez J, Mayor U, Lopitz-Otsoa F, Rodriguez MS, Sutherland JD, Barrio R.** 2013. Scavenger receptors mediate the role of SUMO and Ftz-f1 in *Drosophila* steroidogenesis. *PLoS Genet* **9:**e1003473. doi:10.1371/journal.pgen.1003473.

174. **Benton R, Vannice KS, Vosshall LB.** 2007. An essential role for a CD36-related receptor in pheromone detection in *Drosophila. Nature* **450:**289–293.

175. **Abrams JM, Lux A, Steller H, Krieger M.** 1992. Macrophages in *Drosophila* embryos and L2 cells exhibit scavenger receptor-mediated endocytosis. *Proc Natl Acad Sci U S A* **89:**10375–10379.

176. **Ramet M, Pearson A, Manfruelli P, Li X, Koziel H, Gobel V, Chung E, Krieger M, Ezekowitz RA.** 2001. *Drosophila* scavenger receptor CI is a pattern recognition receptor for bacteria. *Immunity* **15:**1027–1038.

177. Kocks C, Cho JH, Nehme N, Ulvila J, Pearson AM, Meister M, Strom C, Conto SL, Hetru C, Stuart LM, Stehle T, Hoffmann JA, Reichhart JM, Ferrandon D, Rämet M, Ezekowitz RA. 2005. Eater, a transmembrane protein mediating phagocytosis of bacterial pathogens in *Drosophila*. *Cell* **123**:335–346.

178. Kurucz E, Márkus R, Zsámboki J, Folkl-Medzihradszky K, Darula Z, Vilmos P, Udvardy A, Krausz I, Lukacsovich T, Gateff E, Zettervall CJ, Hultmark D, Andó I. 2007. Nimrod, a putative phagocytosis receptor with EGF repeats in *Drosophila* plasmatocytes. *Curr Biol* **17**:649–654.

179. Somogyi K, Sipos B, Penzes Z, Ando I. 2010. A conserved gene cluster as a putative functional unit in insect innate immunity. *FEBS Lett* **584**:4375–4378.

180. Chung YS, Kocks C. 2011. Recognition of pathogenic microbes by the *Drosophila* phagocytic pattern recognition receptor Eater. *J Biol Chem* **286**:26524–26532.

181. Chung YS, Kocks C. 2012. Phagocytosis of bacterial pathogens. *Fly (Austin)* **6**:21–25.

182. Hashimoto Y, Tabuchi Y, Sakurai K, Kutsuna M, Kurokawa K, Awasaki T, Sekimizu K, Nakanishi Y, Shiratsuchi A. 2009. Identification of lipoteichoic acid as a ligand for Draper in the phagocytosis of *Staphylococcus aureus* by *Drosophila* hemocytes. *J Immunol* **183**:7451–7460.

183. Shiratsuchi A, Mori T, Sakurai K, Nagaosa K, Sekimizu K, Lee BL, Nakanishi Y. 2012. Independent recognition of *Staphylococcus aureus* by two receptors for phagocytosis in *Drosophila*. *J Biol Chem* **287**:21663–21672.

184. Bretscher AJ, Honti V, Binggeli O, Burri O, Poidevin M, Kurucz E, Zsamboki J, Ando I, Lemaitre B. 2015. The Nimrod transmembrane receptor Eater is required for hemocyte attachment to the sessile compartment in *Drosophila melanogaster*. *Biol Open* **4**:355–363.

185. Gumienny TL, Brugnera E, Tosello-Trampont AC, Kinchen JM, Haney LB, Nishiwaki K, Walk SF, Nemergut ME, Macara IG, Francis R, Schedl T, Qin Y, Van Aelst L, Hengartner MO, Ravichandran KS. 2001. CED-12/ELMO, a novel member of the CrkII/Dock180/Rac pathway, is required for phagocytosis and cell migration. *Cell* **107**:27–41.

186. Gumienny TL, Hengartner MO. 2001. How the worm removes corpses: the nematode *C. elegans* as a model system to study engulfment. *Cell Death Differ* **8**:564–568.

187. Freeman MR, Delrow J, Kim J, Johnson E, Doe CQ. 2003. Unwrapping glial biology: Gcm target genes regulating glial development, diversification, and function. *Neuron* **38**:567–580.

188. Awasaki T, Tatsumi R, Takahashi K, Arai K, Nakanishi Y, Ueda R, Ito K. 2006. Essential role of the apoptotic cell engulfment genes *draper* and *ced-6* in programmed axon pruning during *Drosophila* metamorphosis. *Neuron* **50**:855–867.

189. MacDonald JM, Beach MG, Porpiglia E, Sheehan AE, Watts RJ, Freeman MR. 2006. The *Drosophila* cell corpse engulfment receptor Draper mediates glial clearance of severed axons. *Neuron* **50**:869–881.

190. Hoopfer ED, McLaughlin T, Watts RJ, Schuldiner O, O'Leary DD, Luo L. 2006. Wld^s protection distinguishes axon degeneration following injury from naturally occurring developmental pruning. *Neuron* **50**: 883–895.

191. Manaka J, Kuraishi T, Shiratsuchi A, Nakai Y, Higashida H, Henson P, Nakanishi Y. 2004. Draper-mediated and phosphatidylserine-independent phagocytosis of apoptotic cells by *Drosophila* hemocytes/macrophages. *J Biol Chem* **279**:48466–48476.

192. Kurant E, Axelrod S, Leaman D, Gaul U. 2008. Six-microns-under acts upstream of Draper in the glial phagocytosis of apoptotic neurons. *Cell* **133**:498–509.

193. Krivtsov AV, Rozov FN, Zinovyeva MV, Hendrikx PJ, Jiang Y, Visser JW, Belyavsky AV. 2007. Jedi—a novel transmembrane protein expressed in early hematopoietic cells. *J Cell Biochem* **101**:767–784.

194. Hamon Y, Trompier D, Ma Z, Venegas V, Pophillat M, Mignotte V, Zhou Z, Chimini G. 2006. Cooperation between engulfment receptors: the case of ABCA1 and MEGF10. *PLoS One* **1**:e120. doi:10.1371/journal.pone.0000120.

195. Wu HH, Bellmunt E, Scheib JL, Venegas V, Burkert C, Reichardt LF, Zhou Z, Farinas I, Carter BD. 2009. Glial precursors clear sensory neuron corpses during development via Jedi-1, an engulfment receptor. *Nat Neurosci* **12**:1534–1541.

196. Scheib JL, Sullivan CS, Carter BD. 2012. Jedi-1 and MEGF10 signal engulfment of apoptotic neurons through the tyrosine kinase Syk. *J Neurosci* **32**:13022–13031.

197. Ziegenfuss JS, Biswas R, Avery MA, Hong K, Sheehan AE, Yeung YG, Stanley ER, Freeman MR. 2008. Draper-dependent glial phagocytic activity is mediated by Src and Syk family kinase signalling. *Nature* **453**: 935–939.

198. Nagaosa K, Okada R, Nonaka S, Takeuchi K, Fujita Y, Miyasaka T, Manaka J, Ando I, Nakanishi Y. 2011. Integrin βν-mediated phagocytosis of apoptotic cells in *Drosophila* embryos. *J Biol Chem* **286**:25770–25777.

199. Nonaka S, Nagaosa K, Mori T, Shiratsuchi A, Nakanishi Y. 2013. Integrin αPS3/βν-mediated phagocytosis of apoptotic cells and bacteria in *Drosophila*. *J Biol Chem* **288**:10374–10380.

200. Lettre G, Hengartner MO. 2006. Developmental apoptosis in *C. elegans*: a complex CEDnario. *Nat Rev Mol Cell Biol* **7**:97–108.

201. Kinchen JM, Ravichandran KS. 2007. Journey to the grave: signaling events regulating removal of apoptotic cells. *J Cell Sci* **120**:2143–2149.

202. Hsu TY, Wu YC. 2010. Engulfment of apoptotic cells in *C. elegans* is mediated by integrin α/SRC signaling. *Curr Biol* **20**:477–486.

203. Gottar M, Gobert V, Michel T, Belvin M, Duyk G, Hoffmann JA, Ferrandon D, Royet J. 2002. The *Drosophila* immune response against Gram-negative bacteria is mediated by a peptidoglycan recognition protein. *Nature* **416**:640–644.

204. Choe KM, Werner T, Stöven S, Hultmark D, Anderson KV. 2002. Requirement for a peptidoglycan recognition

protein (PGRP) in Relish activation and antibacterial immune responses in *Drosophila. Science* **296:**359–362.

205. **Lagueux M, Perrodou E, Levashina EA, Capovilla M, Hoffmann JA.** 2000. Constitutive expression of a complement-like protein in Toll and JAK gain-of-function mutants of *Drosophila. Proc Natl Acad Sci U S A* **97:**11427–11432.

206. **Bou Aoun R, Hetru C, Troxler L, Doucet D, Ferrandon D, Matt N.** 2011. Analysis of thioester-containing proteins during the innate immune response of *Drosophila melanogaster. J Innate Immun* **3:**52–64.

207. **Levashina EA, Moita LF, Blandin S, Vriend G, Lagueux M, Kafatos FC.** 2001. Conserved role of a complement-like protein in phagocytosis revealed by dsRNA knockout in cultured cells of the mosquito, *Anopheles gambiae. Cell* **104:**709–718.

208. **Blandin S, Shiao SH, Moita LF, Janse CJ, Waters AP, Kafatos FC, Levashina EA.** 2004. Complement-like protein TEP1 is a determinant of vectorial capacity in the malaria vector *Anopheles gambiae. Cell* **116:**661–670.

209. **Watson FL, Puttmann-Holgado R, Thomas F, Lamar DL, Hughes M, Kondo M, Rebel VI, Schmucker D.** 2005. Extensive diversity of Ig-superfamily proteins in the immune system of insects. *Science* **309:**1874–1878.

210. **Schmucker D, Clemens JC, Shu H, Worby CA, Xiao J, Muda M, Dixon JE, Zipursky SL.** 2000. *Drosophila* Dscam is an axon guidance receptor exhibiting extraordinary molecular diversity. *Cell* **101:**671–684.

211. **Wojtowicz WM, Flanagan JJ, Millard SS, Zipursky SL, Clemens JC.** 2004. Alternative splicing of *Drosophila* Dscam generates axon guidance receptors that exhibit isoform-specific homophilic binding. *Cell* **118:** 619–633.

212. **Stuart LM, Boulais J, Charriere GM, Hennessy EJ, Brunet S, Jutras I, Goyette G, Rondeau C, Letarte S, Huang H, Ye P, Morales F, Kocks C, Bader JS, Desjardins M, Ezekowitz RA.** 2007. A systems biology analysis of the *Drosophila* phagosome. *Nature* **445:**95–101.

213. **Armitage SA, Sun W, You X, Kurtz J, Schmucker D, Chen W.** 2014. Quantitative profiling of *Drosophila melanogaster* Dscam1 isoforms reveals no changes in splicing after bacterial exposure. *PLoS One* **9:**e108660. doi:10.1371/journal.pone.0108660.

214. **Derre I, Pypaert M, Dautry-Varsat A, Agaisse H.** 2007. RNAi screen in *Drosophila* cells reveals the involvement of the Tom complex in *Chlamydia* infection. *PLoS Pathog* **3:**1446–1458.

215. **Koo IC, Ohol YM, Wu P, Morisaki JH, Cox JS, Brown EJ.** 2008. Role for lysosomal enzyme β-hexosaminidase in the control of mycobacteria infection. *Proc Natl Acad Sci U S A* **105:**710–715.

216. **Ulvila J, Vanha-aho LM, Kleino A, Vähä-Mäkilä M, Vuoksio M, Eskelinen S, Hultmark D, Kocks C, Hallman M, Parikka M, Rämet M.** 2011. Cofilin regulator 14-3-3ζ is an evolutionarily conserved protein required for phagocytosis and microbial resistance. *J Leukoc Biol* **89:**649–659.

217. **Korolchuk VI, Schütz MM, Gómez-Llorente C, Rocha J, Lansu NR, Collins SM, Wairkar YP, Robinson IM, O'Kane CJ.** 2007. *Drosophila* Vps35 function is necessary for normal endocytic trafficking and actin cytoskeleton organisation. *J Cell Sci* **120:**4367–4376.

218. **Charrière GM, Ip WE, Dejardin S, Boyer L, Sokolovska A, Cappillino MP, Cherayil BJ, Podolsky DK, Kobayashi KS, Silverman N, Lacy-Hulbert A, Stuart LM.** 2010. Identification of *Drosophila* Yin and PEPT2 as evolutionarily conserved phagosome-associated muramyl dipeptide transporters. *J Biol Chem* **285:**20147–20154.

219. **Akbar MA, Tracy C, Kahr WH, Krämer H.** 2011. The *full-of-bacteria* gene is required for phagosome maturation during immune defense in *Drosophila. J Cell Biol* **192:**383–390.

220. **Harrison DA, Binari R, Nahreini TS, Gilman M, Perrimon N.** 1995. Activation of a *Drosophila* Janus kinase (JAK) causes hematopoietic neoplasia and developmental defects. *EMBO J* **14:**2857–2865.

221. **Wang L, Kounatidis I, Ligoxygakis P.** 2014. *Drosophila* as a model to study the role of blood cells in inflammation, innate immunity and cancer. *Front Cell Infect Microbiol* **3:**113. doi:10.3389/fcimb.2013.00113.

222. **Srivastava A, Pastor-Pareja JC, Igaki T, Pagliarini R, Xu T.** 2007. Basement membrane remodeling is essential for *Drosophila* disc eversion and tumor invasion. *Proc Natl Acad Sci U S A* **104:**2721–2726.

223. **Hauling T, Krautz R, Markus R, Volkenhoff A, Kucerova L, Theopold U.** 2014. A *Drosophila* immune response against Ras-induced overgrowth. *Biol Open* **3:**250–260.

224. **Fogarty CE, Diwanji N, Lindblad JL, Tare M, Amcheslavsky A, Makhijani K, Bruckner K, Fan Y, Bergmann A.** 2016. Extracellular reactive oxygen species drive apoptosis-induced proliferation via *Drosophila* macrophages. *Curr Biol* **26:**575–584.

225. **Ayyaz A, Li H, Jasper H.** 2015. Haemocytes control stem cell activity in the *Drosophila* intestine. *Nat Cell Biol* **17:**736–748.

Myeloid Cells in Health and Disease: A Synthesis
Edited by Siamon Gordon

doi:10.1128/microbiolspec.MCHD-0015-2015

Matthew Collin[1]
Venetia Bigley[1]

6 Monocyte, Macrophage, and Dendritic Cell Development: the Human Perspective

INTRODUCTION

Monocytes, macrophages, and dendritic cells (DCs) are populations of myeloid mononuclear cells (MMCs) that provide critical sensing functions in innate immunity and a bridge to adaptive immunity through antigen presentation. They also perform important effector functions and contribute to chronic inflammation and healing. Collectively they have been described as the "mononuclear phagocyte system" (MPS) (1). As originally conceived, the MPS had a single blood-borne precursor, the monocyte. It is now appreciated that the development and homeostasis of MMCs is considerably more complex. In this chapter, we discuss the ontogeny of these diverse cells and reflect on ways in which ontogeny is linked to functional specialization, plasticity, and immune regulation.

MYTHS, CONTROVERSIES, AND OTHER DIFFICULTIES

Resolving the developmental relationships between monocytes, macrophages, and DCs and their respective roles in immunity are challenges that continue to attract strong opinion. For many years, the MPS model was widely accepted and both resident histiocytes and myeloid inflammatory infiltrates were presumed to be populations derived from monocytes (Fig. 1A). The ability of human monocytes to take on a wide range of phenotypes *in vitro* appeared to support this notion (2, 3).

DCs: a Hematopoietic Lineage

The discovery of DCs posed a problem for the classic MPS model (4): if they were included as members of the MPS, this implied monocyte origin, promulgating the view that DCs were just a type of specialized macrophage (5). If one accepted their uniqueness, then this suggested the existence of a distinct DC lineage that was not consistent with a unified MPS. The heritage of DCs took some years to emerge following their discovery, but it is now accepted by most that DCs emerge from the bone marrow as a defined hematopoietic lineage that is distinct from monocytes (6) (Fig. 1B).

While monocyte function overlaps the remit of DCs in inflammation, it is clear that in the steady state easily distinguishable populations of DCs and monocytes, or

[1]Institute of Cellular Medicine, Newcastle University, Newcastle upon Tyne NE2 4HH, United Kingdom.

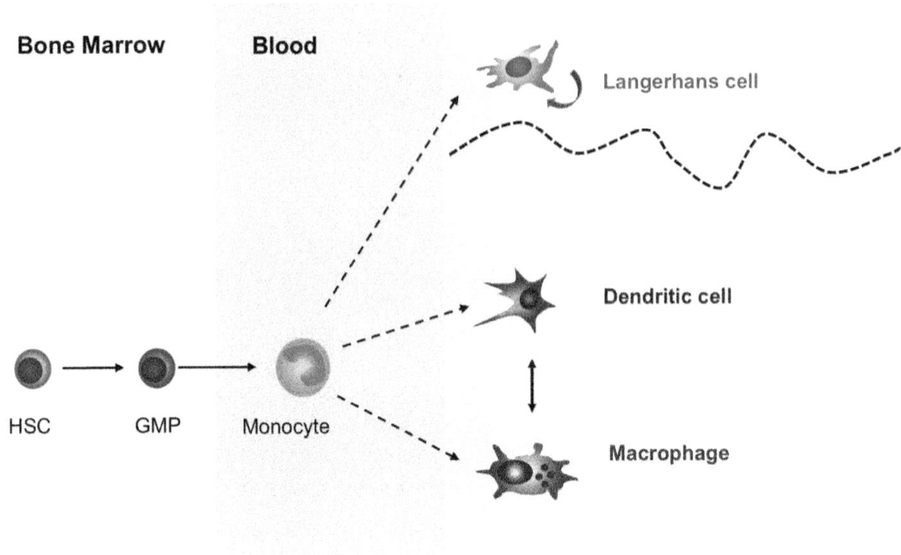

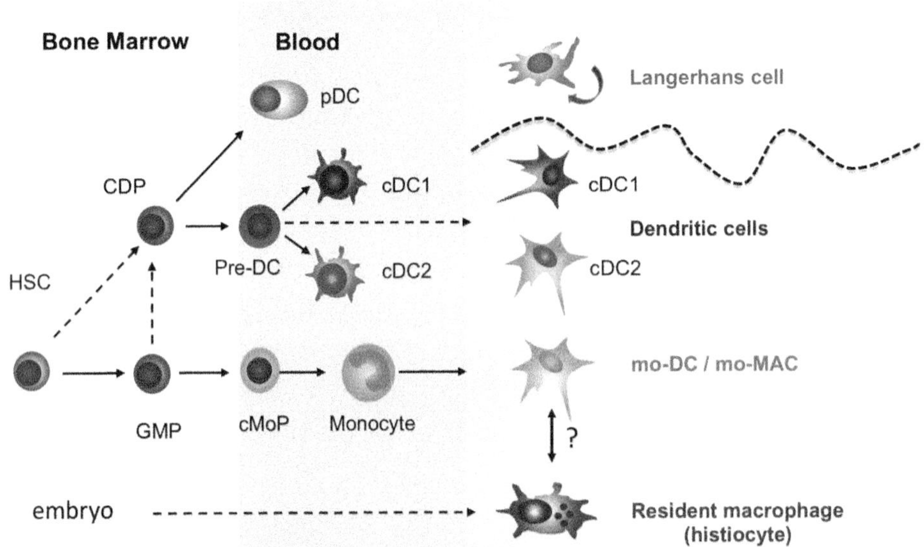

Figure 1 Steady-state MMC homeostasis. (A) Classical mononuclear phagocyte model showing the development of LCs, DCs, and macrophages from a common monocyte precursor. (B) Current model of DC, monocyte, and macrophage development in which DCs develop through a discrete hematopoietic lineage arising from a bone marrow-resident restricted DC precursor, the CDP. CDPs give rise to pDCs and a myeloid pre-DC population that differentiates into two myeloid DCs (cDC1 and cDC2). It is not known if the pre-DC or cDC1 and cDC2 circulating in the blood give rise to tissue populations of cDC1 and cDC2. Monocytes also contribute to steady-state populations of monocyte-derived macrophages and DCs especially in the skin, gut, and lung. These can be clearly distinguished from the CDP-derived DCs and resident macrophages. Resident macrophages are originally derived from prenatal hematopoiesis. It is unknown whether monocyte-derived cells can contribute to long-term resident macrophages. cMoP, common monocyte progenitor; MAC, macrophage; mo, monocyte.

their derivatives, exist in the blood and tissues of humans and other mammals (7–11). The significance of a distinct DC lineage in the induction of immunity is still incompletely understood, as although DCs are potent at stimulating naive T cells, they are relatively rare, especially during inflammation, when monocyte-derived, highly proficient antigen-presenting cells are rapidly formed (12). While monocyte function may be sufficient to resist infection at some level, DCs presumably improve the fitness of the immune system to respond rapidly, to generate anamnestic responses, or to combat a small number of highly virulent pathogens exerting strong selective pressure. The nonredundant role of DCs in infection and evidence of their tolerogenic functions are established in a number of mouse models (13, 14).

Primitive Macrophages

A second, major challenge to the MPS model was the more recent discovery that the majority of histiocytes, or tissue-resident macrophages, are extremely stable populations that do not require replenishment by monocytes or any other precursor (15, 16). In the absence of inflammatory challenge, microglia, Langerhans cells (LCs), Kupffer cells, and alveolar, peritoneal, and splenic macrophages persist from fetal or even embryonic hematopoiesis (12, 17, 18). Notable exceptions are the contributions of monocyte-derived cells to skin, gut, and interstitial lung macrophages (10, 19, 20). Significant import has been attached to the primitive origin of tissue macrophages, and recent studies in the heart suggest that they have a much more prominent role in repair than recently recruited monocyte-derived cells (21). It is not clear whether monocyte-independent routes of macrophage differentiation can also persist in some form after birth. Given that hematopoietic stem cells (HSCs) are detectable in the peripheral tissues of mice (22), it is hard to exclude the occurrence of a direct route of macrophage differentiation from an early hematopoietic precursor, residing in adult tissues.

PITFALLS IN THE STUDY OF MMCs

This narrative illustrates a number of generic challenges in the study of monocytes, macrophages, and DCs. First, it is important to distinguish between the steady state and inflammation. There is no *a priori* reason to suppose that inflammatory precursors and their progeny are related to steady-state populations of MMCs. Indeed, temporal and special separation of quiescent and inflammatory pathways of differentiation may provide critical aspects of immune regulation. Major questions remain unsolved, particularly in humans, concerning the functional differences between resident and recruited cells and their respective fates after inflammation has resolved. Second, overreliance on a small number of surface antigens to identify a particular cell type can be misleading. The use of CD11c and major histocompatibility complex class II (MHC-II) to define murine DCs has been criticized (5). In humans, it was already well known that CD11c and MHC-II are highly expressed by both monocytes and DCs and cannot be used to separate them in the blood, although CD11c defines different populations of dermal DCs and macrophages quite clearly in tissues (8). This issue has been partly circumvented by the discovery of more-restricted antigens such as the suite of "blood DC antigens," or BDCA-1 to -4, in humans (23, 24) and, more recently, new DC and macrophage signposts such as CLEC9A (C-type lectin domain family 9 member A), XCR1 (chemokine XC receptor 1), SIRPα (signal regulatory protein α), and MerTK (MER tyrosine kinase) identified by unbiased computational approaches to classification (25, 26). Third, the plasticity of MMCs to respond to environmental cues means that different anatomical sites are likely to induce specific phenotypes (27). Furthermore, this plasticity inflates the problems already alluded to that inflammation and the steady state are not equivalent and that individual surface antigens may be fickle. A cogent argument has been made to mitigate these problems by using ontogeny as the basis of MMC classification (28). This approach has many attractions, and although definitive ontological experiments are difficult to perform in humans, the concepts developed in mice can often translate to humans. As recent data show, the prenatal development of MMCs is an important dimension to consider, in addition to the lineages that arise later from definitive HSCs (12, 17).

THE REPERTOIRE OF HUMAN MONOCYTES, MACROPHAGES, AND DCs

Monocytes

Monocytes are the archetypal MMCs, constituting ~10% of human peripheral blood mononuclear cells. They are blood borne by convention, although recent studies in humans and mice describe steady-state populations of tissue monocytes or monocyte-derived cells (9, 19, 20). Monocytes have a plethora of functions in inflammation and rapidly differentiate into cells with DC-like and macrophage phenotypes that can be difficult to separate from resident populations. Recent transcriptomic profiling experiments highlight the complexity of monocyte differentiation potential (29).

Human monocytes are heterogeneous in their expression of CD14 and CD16 (Fc receptor III), and this has been used to divide them into subpopulations (30). CD14 and CD16 are both continuously distributed (Fig. 2), but the majority of monocytes are CD14$^+$ and are defined as classical or inflammatory monocytes. Most of the precursor functions defined *in vitro*, including the formation of monocyte-derived DCs and macrophages, arise from classical monocytes (29). Some studies ascribe distinct functions to "intermediate" CD14$^+$CD16$^+$ monocytes at the vertex of the two-dimensional plot. These have the highest MHC-II expression and cytokine production and may be derived by activation of classical monocytes.

Nonclassical CD16$^+$ monocytes also express high MHC-II and costimulatory antigens, leading some authors to regard them as DCs (24). However, transcriptional profiling and unsupervised hierarchical clustering is unequivocal that all monocytes cluster independently of DCs (9, 31). Within the CD16$^+$ subset a population expressing the antigen 6-sulfo LacNAc (SLAN-DCs) is reported to secrete large amounts of tumor necrosis factor α, interleukin-1β (IL-1β), and IL-12 and to respond rapidly to inflammatory stimuli (32).

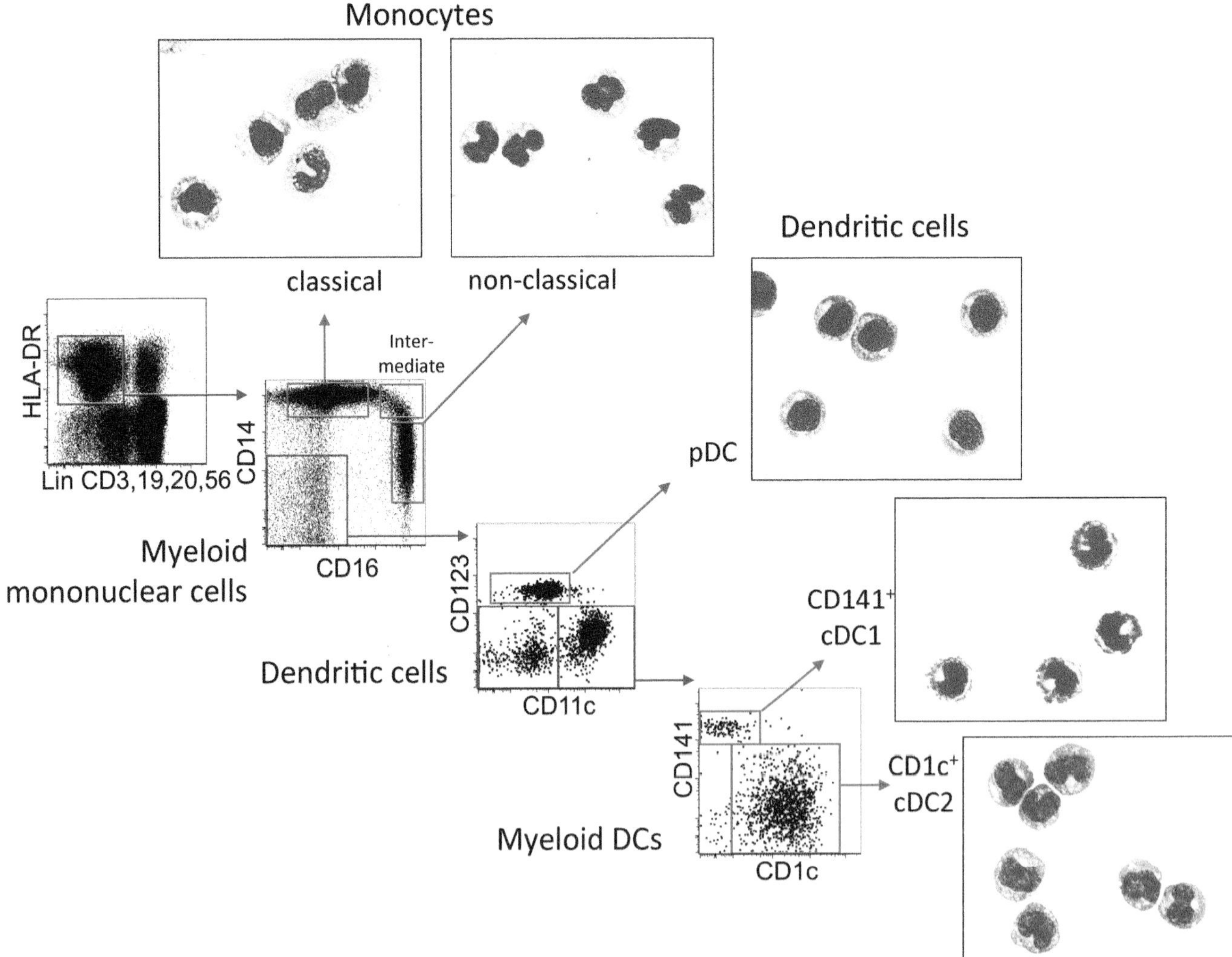

Figure 2 Human blood monocytes and DCs. Gating strategy to identify human monocytes and DCs in peripheral blood. Monocytes and DCs are all found in the HLA-DR$^+$, lineage-negative compartment. CD14 versus CD16 displays monocyte subsets and double-negative DC populations. These may be separated in a variety of ways. Here the markers CD123 and CD11c are used to define pDCs and myeloid DCs. The latter can be separated into cDC1 and cDC2 using CD141 and CD1c, respectively.

Other studies suggest that CD16$^+$ monocytes, including SLAN$^+$ cells, have low inflammatory activity and are homologous to murine Gr-1/Ly6C^{lo} "patrolling" monocytes (33). Inconsistencies may have arisen through different gating of the CD14 versus CD16 plot and inclusion or exclusion of the CD14$^+$CD16$^+$ intermediate cells, which have a highly activated phenotype. Anatomically, it appears that nonclassical monocytes are well positioned to infiltrate tissues or to control diapedesis across the endothelium; the importance of both of these potential functions remains to be tested in humans (34).

The origin of human nonclassical monocytes remains controversial. Kinetic studies in the mouse suggest that they are derived from classical monocytes (35), although this is recently disputed (36). Human CD16$^+$ monocytes also have higher expression of the nuclear factor Nur77, proposed to control murine nonclassical monocyte differentiation (37).

There is a prominent literature in humans describing abnormalities of monocyte development in disease, notably the appearance of "sick" monocytes with lower MHC-II expression and impaired antimicrobial functions (38). Variation in the CD14/CD16 ratio due to expansion or contraction of either subset is also associated with many different conditions. Of note, CD16$^+$ monocytes are highly sensitive to depletion following treatment with corticosteroids (39). The appearance of monocytic "myeloid-derived suppressor cells" that can inhibit T-cell responses is associated with cancer and chronic infection (40). These demographic changes may simply represent a "left shift" in myelopoiesis, but nonetheless this appears to provide immune regulation under conditions of stress or chronic inflammation.

Finally, heterogeneous expression of CD68, CD163, and CD202b (Tie2; angiopoietin receptor) has been noted among human monocytes. Tie2$^+$ monocytes are found throughout classical, intermediate, and nonclassical populations and are implicated in tumor angiogenesis (41).

Macrophages: Histiocytes Old and New

Macrophages are classically tissue-resident cells, or "histiocytes." Many different specialized populations of macrophages exist in different sites. Observations from human HSC transplantation and monocytopenic states suggest that as in mice, these are likely to be developmentally old cells that can persist for long periods without replenishment from the bone marrow (7, 42–44). The label "histiocyte" is perhaps still useful to denote these developmentally older populations from more recently recruited monocyte-derived cells. Location matters, and the tissue environment induces specific gene regulation in resident histiocytes, regardless of their origin (27).

Recent advances have led to much clearer definitions between long-term resident histiocytes, DCs, and more transient populations of monocyte-derived cells (Fig. 3). Macrophages are operationally distinguished from DCs by their much lower ability to simulate naive T cells (4). There are also many differences in gene expression (9, 25, 26). Useful markers in human skin such as CD163, factor XIIIA, and LYVE-1 (lymphatic vessel endothelial hyaluronan receptor 1) identify macrophage populations that are completely nonoverlapping with DCs, identified by CD11c or CD1c expression (8, 45). In addition to histiocytes and DCs, populations of monocyte-derived cells such as CD14$^+$ or "interstitial-type" DCs are also found in human skin. These were originally classified as DCs owing to their ability to migrate from explanted skin (46), but transcriptomic profiling shows close relationships with the human dermal macrophage and mouse monocyte-derived macrophages (9, 10). As predicted, these monocyte-derived macrophages are indeed poor allostimulators, produce IL-10, and can provide regulatory function (47). Confusingly, CD141 may also be induced on monocyte-derived cells, but they remain distinct from CD141$^+$ DCs particularly through the expression of CD14, SIRPα, and other markers (11, 47).

In vivo human data show that monocyte-derived macrophages are rapidly repopulated after HSC transplantation, in contrast to resident histiocytes (7, 9). Monocyte-derived cells are also rapidly turned over in murine tissues and may be found in afferent lymph (10, 19, 20). Together, these observations support a model that recently derived monocyte-macrophages are mobile, in contrast to the long-term histiocyte population. Immobility and the progressive accumulation of lipofuscin and pigment that gives macrophages autofluorescence in the flow cytometer are physical signs of longevity that are experimentally tractable to identify histiocytes and to separate them from monocyte-derived cells (7, 9).

One difference between mouse and human is that it has been difficult to demonstrate homeostatic proliferation in human macrophage populations. Observations in quiescent skin suggest that if proliferation does occur, it is less than one-tenth of the level observed in LCs (~2 to 3%) (7, 43). These observations need to be extended to inflammatory situations such as Th2 environments that have been observed to stimulate murine macrophage proliferation (48).

The relatedness of monocytes, monocyte-derived macrophages, and tissue histiocytes by unsupervised cluster-

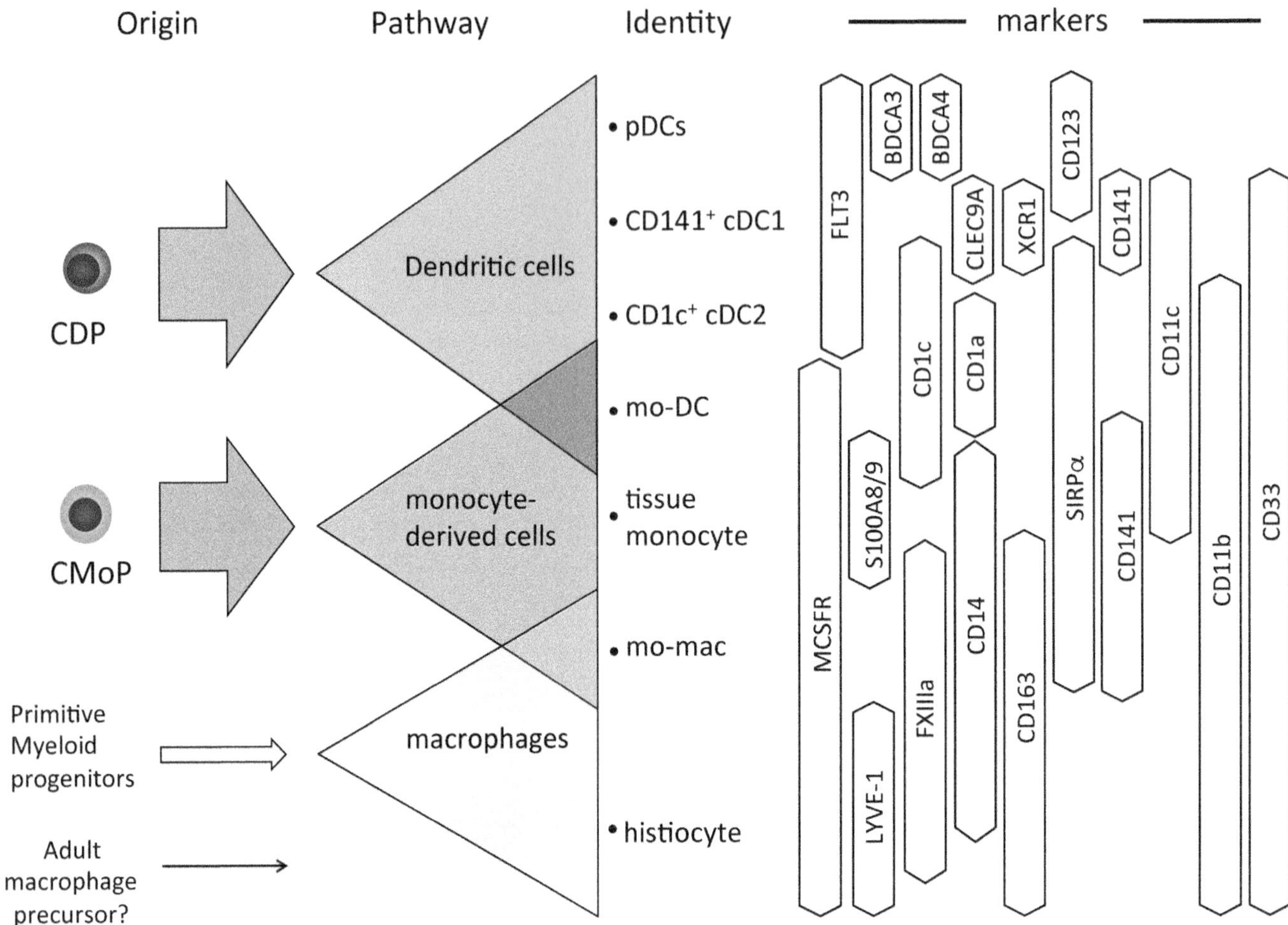

Figure 3 Summary of human DC, monocyte, and macrophage origins in relation to steady-state surface markers. The most distinctive DC phenotypes are displayed by pDCs and CD141$^+$ cDC1 cells. There is more overlap between cDC2 and monocyte-derived cells that may coexist in tissues. Monocyte-derived macrophages also share many markers with resident histiocytes. The relative size of arrows from the precursor populations indicates the estimated steady-state flux. CMoP, common monocyte progenitor; mac, macrophage; mo, monocyte.

ing of their gene expression signatures is reminiscent of the classic picture of the MPS as a monocyte-based system but also strong evidence of the separateness of the DC lineage, members of which cluster very distinctly (9, 31). However, care must be made in extrapolating these static similarities to population dynamics. The plasticity of monocytes allows them take on genuine DC properties during inflammation (49–51). We concur with others that in experimental work the term "monocyte-derived DC" should be used to describe such populations (12, 28, 52).

Dendritic Cells

DCs were first isolated from the mouse spleen as nonadherent branching cells that were 100-fold more potent at activating naive T cells compared with adherent macrophages (4). "Resident" populations of DCs that are derived from blood-borne precursors are found in lymphoid tissues, while "migratory" DCs found in peripheral tissues take up exogenous antigens and migrate to draining lymph nodes, performing a surveillance role (6).

For obvious logistical reasons the analysis of human lymphoid and tissue DCs has proceeded more slowly than the study of human blood. In human blood, three populations of DCs are readily described: plasmacytoid DCs (pDCs) and two populations of "myeloid" DCs (Fig. 3). The term "myeloid" is equivalent to "classical" or "conventional" in mice. Its usage in humans is quite specific and based on the expression of antigens typically seen on granulocytes or monocytes, including CD13, CD33, CD11b, and CD11c. These markers are lower or absent on pDCs, which are also morphologically and functionally distinct, providing a major source of type I interferon in response to viral infection (reviewed in reference 53). All DCs carry antigens typically associated with lymphoid differentiation, such as CD4 (54). MHC-II and CD11c are not useful for separating DCs from monocytes in humans, but the series of BDCA antigens is extremely powerful (23, 24). Myeloid DCs separate into a major CD1c$^+$ (BDCA-1) population and a minor CD141$^+$ (BDCA-3) population. The minor population is homologous with the mouse CD8$^+$/CD103$^+$ DC lineage, has high expression of XCR1 and CLEC9A, and has natural cross-presenting capacity (55, 56). In a recent nomenclature system this "conventional" DC lineage has been called cDC1, while human CD1c$^+$ DCs together with mouse conventional CD4$^+$ or CD11b$^+$ DCs are cDC2 (28). An alternative system is simply to name these cells CD141$^+$ myeloid DCs and CD1c$^+$ myeloid DCs, although, as previously noted, CD141$^+$ is quite a promiscuous antigen (30). It has been suggested that more-universal markers are XCR1 and CLEC9A for cDC1 and SIRPα for cDC2 (57). The conservation of pDCs is apparent in many different mammalian species, and in humans they express CD123, CD303 (BDCA-2; CLEC4C), and CD304 (BDCA-4; neuropilin).

In tissues it is possible to detect equivalent populations of migratory myeloid DCs. Lymphoid organs also contain CD1c$^+$ and CD141$^+$ DCs in addition to pDCs and LCs in skin-draining locations (58–61).

Langerin-Positive DCs

The expression of Langerin on non-LCs was first highlighted in the mouse by several groups working with models of inducible LC depletion, in which Langerin$^+$CD103$^+$EpCAM$^-$ DCs were observed in the complete absence of LCs (62). Langerin expression had already been observed in CD8$^+$ lymph node-resident DCs (63), and it became clear that Langerin is expressed by murine CD8$^+$/CD103$^+$ cDC1s. In contrast, a small component of CD1c$^+$ cDC2s express Langerin in human tissues *in vivo* (64). This population is distinct from LCs, and the human CD141$^+$ cDC1 population in the steady state, but may be a precursor of LCs after inflammation (65).

Langerhans Cells

LCs are an unusual population of migratory myeloid DCs situated in the epidermis and mucosae (66). They express Langerin and CD1a at high levels and contain Birbeck granules by electron microscopy (63). LCs can be stimulated to leave the epidermis, forming migratory DCs detectable in skin-draining lymph nodes. LCs appear to play multiple roles in immunity, including the presentation of lipid antigens (67), inactivation of pathogens (68), and induction of Th17 responses (69) and graft-versus-host disease (70), in addition to the maintenance of tolerance in the steady state (14).

LCs are derived from myeloid precursors during fetal life (71, 72) that differentiate under the influence of the colony-stimulating factor-1 receptor (CSF-1R; CD115), cytokine IL-34 (73, 74), and autocrine production of transforming growth factor β (75). Proliferation of epidermal LCs was first reported in humans (76), but elegant experiments in mice subsequently proved that self-renewal was sufficient to provide local homeostasis (77). LCs were the first example of an MMC that was not continuously replenished from the bone marrow. Limb transplantation and bone marrow failure syndromes confirm their independence of bone marrow-derived precursors in humans (43, 44, 78). Interestingly, it has been difficult to reproduce mouse data showing that LCs survive HSC transplantation, and most are replaced by donor cells within a few months, even with nonmyeloablative conditioning and in the absence of graft-versus-host disease (79, 80).

EXPERIMENTAL APPROACHES TO ANALYZING MYELOID DIFFERENTIATION

Hematopoiesis has provided one of the seminal model systems of developmental biology. The key concepts are that stem cells are self-renewing and pluripotent and that differentiation proceeds by a progressive restriction of cell potential, through multiple rounds of proliferation. The focus of almost all research has been to define the nodes in this process, the points at which cell potential bifurcates, through the isolation of clonogenic progenitors that give rise to daughter cells each with nonoverlapping developmental potential. Combined with flow cytometry, the major fruit of this labor is the purification of cells with progressively restricted potential, manifest by a number of *in vitro* or *in vivo* assays. There are a number of limitations of this approach that are pertinent to unraveling the development and differentiation of MMCs.

Not All Swans Are White

Experiments subjecting purified progenitors to any interrogation of their potential can only *rule in* a particular cell fate potential; they cannot exclude it. Even when a comparator population is used to verify the performance of a readout, a cell that fails in one assay may have the same potential under more sophisticated conditions, for example, the use of stromal cells or new growth factors. This problem has led to the appearance of a number of black swans over the years as new potentials are revealed through improvements in the sensitivity of readouts. Xenografting in immunodeficient mice is no exception, as the expression of cell potential is still limited by an arbitrary range of cross-reactive hematopoietins (81).

Uncertainty Principles *In Vitro*

A further problem in analyzing single cells with dual potential is that it may be difficult or impossible to find a set of conditions that allow simultaneous differentiation along more than one lineage. A progenitor may exist in one of several states of potential, but often this can only be interrogated by collapsing the potential to a singular outcome through defined culture conditions. Critically, attempts to define a "universal" readout are likely to increase the chance of missing a potential outcome and creating an unseen black swan. An interesting meta-problem of the same nature is that antibodies to growth factor receptors are frequently used to sort populations and can potentially interfere with growth factor signaling during the analysis of progeny. Thus, it has been suggested that high expression of CD115 used to isolate restricted DC progenitors may blunt the ability of these cells to develop into monocytes (5). Recent efforts to define DC progenitors in humans have all capitalized on growth factor expression to isolate prospective cell populations (82, 83).

Monocyte and DC Potential

The black swan and simultaneous differentiation problems both impede the study of monocyte and DC development in humans. First, conditions permissive for the development of human DCs *in vitro* and in animal models have only recently been defined (56, 82, 84); and second, the phenotypic readouts are subtle, and robust definitions of what constitutes a DC or monocyte-derived cell have been hard to formulate, particularly due to the extreme plasticity of monocytes in adopting DC-like phenotypes (51).

In terms of defining the "true" DC potential of an assay, the appearance of pDCs and XCR1$^+$CLEC9A$^+$ myeloid DCs (cDC1) is instrumental. This is because CD1c$^+$ myeloid DCs (cDC2) are close in phenotype to monocyte-derived DCs and it may be difficult to discern if cDC2 phenotype output is simply a reflection of monocyte potential; this is especially true when granulocyte-macrophage CSF (GM-CSF) and IL-4 are present, conditions that promote the formation of monocyte-derived DCs (85). It is necessary to perform quite detailed transcriptional profiling in order to ascertain the true lineage of culture outputs (82). The ability of CD1c$^+$ DCs to form LC-like cells in response to transforming growth factor β may be useful in distinguishing them from monocyte-derived DCs (65, 86).

THE MYELOID-BASED MODEL OF MOUSE MMC DIFFERENTIATION

Many critical details of murine MMC development have recently been elaborated. As alluded to previously, it is now appreciated that primitive myelopoiesis contributes almost exclusively to microglia, which arise directly from yolk sac macrophages, and that LCs and most tissue macrophages also have a primitive origin via fetal liver monocytes (12, 17, 18).

Definitive hematopoiesis continues to supply monocytes and DCs throughout the life of an animal, and unraveling a discrete pathway of DC development in the mouse has been a significant advance in the field of hematopoiesis (6). The model that has emerged is firmly myeloid based, in which granulocyte-macrophage progenitors (GMPs) give rise to macrophage and DC progenitors (MDPs), which have lost granulocyte potential (87, 88). MDPs in turn give rise to a monocyte precursor (35) and common DC precursor (CDP) (89, 90). The latter has restricted DC potential and is the key staging post of the DC lineage where monocyte potential should disappear. pDCs are said to arise directly from CDPs, but classical or conventional DCs descend from an intermediate pre-cDC population that can be found in the blood (91, 92) (Fig. 1B).

This model appeals to the reductionist by proposing sequential bifurcations of cell potential consistent with textbook models of hematopoiesis. However, a number of inconsistencies are emerging. The MDP has recently come under fresh scrutiny with evidence that its ability to generate pDCs and XCR1$^+$ cDC1s (markers of "true" DC potential) is much less than its monocyte potential and that it may still retain some ability to form granulocytes (93). The identity of the CDP as a clonogenic progenitor has also been questioned, with reports that populations originally classified as CDPs are primed to produce either classical DCs (cDC1 and cDC2) or pDCs according to their level of CLEC9A and CD115 expression (94, 95). Pre-DC function has also been observed in a pDC-like cell that is distinct from the pre-DC (96). Furthermore, the pre-DC itself shows evidence of precommitment to either cDC1 or cDC2 progeny (97). The concept of precommitment or lineage priming is not new to stem cell biology but is at variance with models based on discrete cell fate "decisions" at the nodes of clonogenic progenitors (98). Lentiviral barcoding reveals that lineage priming may bias the entire progeny of early progenitors, with a significant tranche apparently devoted to the generation of DCs (99). A final aspect of the myeloid-based model that has been difficult to assimilate with other mouse and human data is the lack of consideration for lymphoid pathways of DC development (100, 101). The original concept that pDCs might have a lymphoid heritage stems not only from their morphology and surface markers but also from the fact that rearranged IgH and T-cell receptor loci are found at high frequency (102, 103). Rather than "convergent development" from two different branches of hematopoiesis, it appears more likely that lymphoid and myeloid potential are conjoined until late stages of differentiation.

MYELOID AND LYMPHOID OR LYMPHOMYELOID?

The terms "myeloid" and "lymphoid" dominate hematopoiesis. "Myeloid" is used here in the broad sense of bone marrow derived, encompassing erythropoiesis, megakaryopoiesis, granulopoiesis, and the production of monocytes and macrophages; indeed anything that does not include small lymphoid cells. In the classical model of hematopoiesis, the earliest dichotomy occurs between myeloid and lymphoid, based on the description of common myeloid (erythro-mega-granulocyte-macrophage) and common lymphoid (B and T) progenitors (104, 105).

However, in assays that supported both myeloid and lymphoid differentiation, lymphomyeloid progenitors lacking erythro-megakaryocyte potential were observed, suggesting that the specification of lymphoid potential is a more continuous process (81, 106). Many groups have since described "lymphoid-primed multipotent progenitors" (LMPPs) in mice and humans, in support of this concept (85, 107) (Fig. 4). From an epigenetic standpoint, the gradual, rather than abrupt, loss of lymphoid potential concords with progressive methylation of myeloid regulatory loci (108). From the point of trying to decipher monocyte and DC development, these findings are absolutely key as it is no longer necessary to invoke an unorthodox blending of myeloid and lymphoid potential if instead they coexist. The DC embodies innate myeloid phenotype with an ability to interact with adaptive immune system lymphocytes, and difficulties in mapping its precise origin may be intimately linked to problems with the myeloid-lymphoid dichotomy of classical hematopoiesis. A graphical illustration of this is the structure of hematopoiesis derived from an unbiased analysis of transcription factor expression by the "known" hematopoietic lineages and their intermediates: this places the DC lineage as neither myeloid nor lymphoid but arising directly from HSCs (109).

HUMAN MONOCYTE AND DC DIFFERENTIATION

Serious analysis of human hematopoiesis began with *in vitro* colony-forming assays defining erythroid-, megakaryocytic-, granulocyte-, and "macrophage"-forming units, thus providing the first example of readouts that could interrogate either granulocyte or monocyte potential (110). CD34 is a key marker of the human progenitor/stem cell compartment. Most multilineage potential is contained within the CD38^{lo} fraction of CD34^{+} cells, while CD38^{+} cells contain committed cells. The CD38^{+}CD10^{+} fraction contains lymphoid potential, while GMPs fall in the CD38^{+} CD45RA^{+} fraction. The importance of CSF-1R in identifying progenitors with monocyte but not granulocyte potential was recognized early (111). This population potentially represents MDP-like cells within the GMP, but to our knowledge a human equivalent of the common monocyte progenitor has not been identified. In addition, within the GMP, a discrete population of CD123^{hi} (IL-3 receptor) cells was noted to give rise to pDCs (112). CD123^{hi} cells form a distinct peninsula of the GMP cloud that has been observed in a number of studies (113).

Upon *in vitro* exposure to Flt3 ligand, stem cell factor, GM-CSF, and stromal cells, the CD123^{hi} population of GMPs contains pDC and myeloid cDC1 and cDC2 potential but lacks monocyte potential (82). In this study, molecular profiling was required to differentiate between monocytes and cDC2 because CD1c^{+} cDC2s can acquire CD14 expression *in vitro* and CD14 monocytes can differentiate into CD1c^{+} cDC2-like cells. These results show that good DC potential resides in the CD123^{hi} population, but the statement that these cells are "DC restricted" should be qualified to the specified conditions. Under these conditions, GMPs generated both monocytes and DCs. However, strong monocytic differentiation with M-CSF was not tested, and it might therefore be argued that monocyte potential was not completely excluded from the CD123^{hi} fraction.

It was also previously shown that CD38^{+}CD10^{+} lymphoid progenitors have the ability to develop into

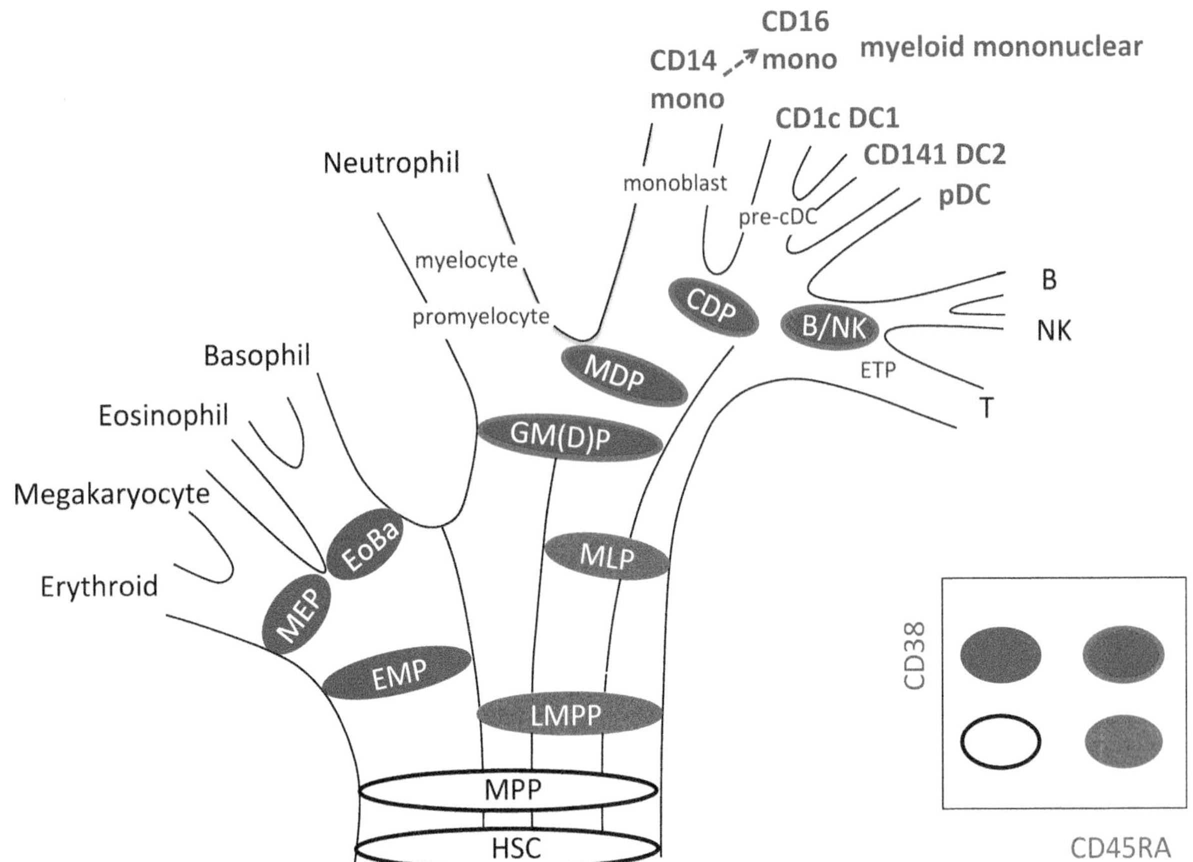

Figure 4 Revised structure of human hematopoiesis showing common lymphomyeloid origin of MMCs including DCs and monocytes. Color coding of progenitor populations is according to their expression of CD38 and CD45RA (inset). B/NK, B- and NK-cell progenitor; EMP, erythromyeloid progenitor; EoBa, eosinophil-basophil progenitor; ETP, early thymic progenitor; GM(D)P, granulocyte-monocyte (DC) progenitor; MEP, mega-erythroid progenitor; MLP, multi-lymphoid progenitor; MDP, monocyte-DC progenitor.

HLA-DR$^+$CD1a$^+$ DC-like cells and pDCs (100–102). More-recent studies suggest that the CD38$^-$CD45RA$^+$ LMPPs (or multi-lymphoid progenitors), which can give rise to monocytes, will also contain genuine DC potential (85). Whether these lymphoid pathways funnel into the CD123^{hi} population is unknown. Also, it is possible that accessory pathways of DC differentiation exist, as the authors who first described pDC potential in the CD123^{hi} population also noted LC-like potential in the CD123^{lo} GMP fraction (112); this may indicate the presence of CD1c$^+$ cDC2s (65, 86).

The same authors who defined a CDP also reported pre-cDCs in human blood by scrutinizing small populations of CD34$^-$, lineage-negative cells expressing Flt3, GM-CSF, and M-CSF receptors (83). The pre-cDCs that were defined did not express high CD123 and had no pDC potential. The transition between CD123^{hi} CDP and CD123^{lo} pre-cDC is not immediately obvious, and other transitional CD34$^-$, lineage-negative intermediates may exist.

Bone marrow transplantation and human DC deficiency states indicate that continual replenishment from blood-borne precursors is required for tissue populations of CD141$^+$ and CD1c$^+$ DCs (7, 43, 44). Comparison with the mouse suggests that these cells are derived from rare pre-DC populations although the potential precursor role of blood DCs has not been excluded (114).

MONOCYTE-DERIVED CELLS IN INFLAMMATION

The derivation of macrophages from monocytes *in vitro* has been studied for many years and was one of the founding observations leading to the MPS model. A "classical" highly immune-activated macrophage phenotype is seen after exposure to gamma interferon, while "alternative" activation with IL-4 was developed as a model of resident macrophage differentiation (3). These two phenotypes, known as M1 and M2, respectively, have been regarded as two poles of macrophage differentiation with contrasting functions. More-recent studies have highlighted the complexity of monocyte activation using a greater range of stimuli and employing transcriptomic analysis to derive modules of coregulated genes and their regulators (29). The isolation of primary human macrophages from different sites will now facilitate a more systematic comparison of *ex vivo* populations with *in vitro*-derived cells (7, 29, 115, 116).

The observation that monocytes could differentiate into myeloid DCs with the ability to stimulate naive T cells was a significant experimental breakthrough facilitating a great number of studies in humans (2). However, recognition of naturally occurring DCs *in vivo* has led to a reappraisal of the monocyte-derived DC as probably more closely linked with inflammatory monocyte function than to steady-state DC biology. Latterly, the importance of monocyte-derived DCs in resistance to infection has been fully recognized in the mouse, thus validating the human *in vitro* data (49, 50), and inflammatory DCs have been described in humans in several pathological settings (115–118).

In addition to monocyte-derived DCs, the direct recruitment of DC-lineage cells to inflammatory infiltrates remains to be investigated. The problem of separating monocyte-origin and DC-derived inflammatory cells is highlighted by *in vitro* studies showing that inflammation drives convergent transcriptional programs probably leading to very similar phenotypes (51). Indicators of recent monocyte origin include S100A8/9, recognized by antibody MAC387, a marker long used to describe "monocyte-macrophages" by pathologists (119).

Further information is also required to understand the repopulation of LCs following an inflammatory insult. In humans, LCs are replaced by bone marrow-derived cells after HSC transplantation and graft-versus-host disease (79, 80). A two-phase kinetic of LC recovery was observed many years ago in humans in serial skin biopsies of delayed-type hypersensitivity reactions (120). In mice, more detailed analysis shows an initial infiltration by classical monocytes that express low levels of Langerin (121) followed by a second wave of more long-lived LC precursors (122, 123). Langerin$^+$ cells may be derived from monocytes and CD34$^+$ progenitors *in vitro* (124–126). Recent experiments indicate that monocytes express only a low level of Langerin in comparison with CD1c$^+$ DCs, which rapidly form LC-like cells with Birbeck granules (65, 86). Together, these results suggest that the DC differentiation pathway may contribute to long-term LC repopulation following inflammation.

GENETIC CONTROL OF THE HUMAN MONOCYTE AND DC LINEAGES

The genetic control of human monocyte and DC development has been illuminated in a number of ways, using defined growth factors, knockdown of specific genes, and xenograft models (56, 102). Humans with genetic defects of monocyte and DC development have also been described (reviewed in reference 127). DC production from human progenitors is promoted by Flt3 ligand, GM-CSF, IL-4, and the presence of stromal cells (82, 84). Administration of Flt3 ligand to humans expands DCs *in vivo* (83). GM-CSF is implicated in DC development in mice. This function has not been thoroughly explored in humans, but blockade of GM-CSF signaling by spontaneously occurring autoantibodies (or rare receptor mutations) dramatically inhibits alveolar macrophage development, causing pulmonary alveolar proteinosis (128).

Human monocyte and DC development becomes severely impaired in humans with heterozygous *GATA2* mutation (129). GATA2 is required for stem cell self-renewal, and the hemizygous state preferentially affects mononuclear cell development, resulting in a combined DC, monocyte, B, and NK lymphoid (DCML) deficiency (130). The clinical picture of immunodeficiency is variable, but susceptibility to viral infections and mycobacteria is common (131). Flt3 ligand is hugely elevated, probably in response to a fundamental disturbance of stem cell equilibrium rather than as a result of DC deficiency. *Ikaros* mutation also causes a severe pancytopenia that includes monocytes and probably DCs (132). Neutropenia may involve genes implicated in MMC development in mice, such as *GFI1*, but monocytes and DCs have not been examined systematically in these disorders (133). Finally, WHIM (warts, hypogammaglobulinemia, infections, and myelokathexis) syndrome, due to gain-of-function mutation of *CXCR4*, causes neutropenia and global mononuclear deficiency by preventing leukocytes from leaving the marrow. Clinical manifestations of WHIM may overlap with the features of *GATA2* mutation, including warts and susceptibility to mycobacteria (134).

Biallelic *IRF8* mutation in humans was reported to cause a severe defect of monocytes, myeloid DCs, and pDCs, while a heterozygous state induced only changes in CD1c$^+$ DCs (44). A gene dosage effect may be at work because in the mouse there are substantial differences between the hypomorphic BXH2 mouse, which has a specific defect of cDC1, and the *IRF8* gene-deleted mouse, which shows multiple lineage deficiencies and expanded granulopoiesis. Monocytopenia appears to be a direct consequence of the IRF8 dependency of Kruppel-like factor 4 (KLF4) (135), while granulopoiesis is promoted by the failure of IRF8 to suppress CCAAT/enhancer-binding protein α (C/EBPα) activity (136). IRF8 also associates with IRF4, a transcription factor implicated in the development of CD1c DCs (cDC2) (137). The variable effect of heterozygous *IRF8* mutation on CD1c$^+$ DCs may be mediated through an IRF8/IRF4 interaction, but further clarification of this is required.

In cooperation with IRF8, the activator protein 1 (AP1) factor BATF3 plays a key role in DC development in mice and in human *in vitro* culture (56, 138). Within the DC lineage, the balance of basic helix-loop-helix protein E2-2 and its inhibitor ID2 is critical for the specification of pDCs or classical DCs from DC precursors (53). Mutation of E2-2 is the cause of Pitt-Hopkins syndrome, in which abnormal pDC development has been noted (139). The recently described transcription factor ZBTB46 is specifically expressed in myeloid CD1c$^+$ DCs and their murine counterparts (31). Although it is not required for DC development, ZBTB46 is very useful in lineage marking and lineage-specific DC depletion to demonstrate that cDC2s have specific roles in immunity (97, 140). Monocytes do not express ZBTB46 at rest, but it is induced when they differentiate into DCs (97).

CONCLUSIONS AND FUTURE PERSPECTIVES

The MPS provided a conceptual framework for understanding the development of MMCs for nearly half a century. It is becoming clear that monocytes are not the only axis of homeostasis and that long-lived macrophages and a specialized DC lineage contribute to tissue leukocyte populations. The mouse has been an indispensable model for the analysis of hematopoiesis and continues to provide many insights into the pathways of human monocyte and DC development. While the study of monocyte-derived populations *in vitro* has furnished many exciting experimental results, the true physiological context of this work remains incompletely understood but is likely to provide significant insights into inflammatory monocyte differentiation. A distinct DC lineage is well described in humans, but more studies are required to understand the nonredundant role of these cells in immunity and how they can be manipulated for therapeutic purposes.

In the near future, single-cell genomics will provide a new level of detail in the analysis of the progenitor cells and of the genomic and transcriptional architecture controlling cell fate (reviewed in reference 141). This is likely to revise classical models of hematopoiesis; rather than successive bifurcations in cell potential, the picture that is emerging is that differentiation is a graded process punctuated by many phenotypic states. Continuous phenotypic variables are likely to reflect gradients in cell potential, which may be in equilibrium and appear stochastic at given points. A further addition to the modern armamentarium is the use of deep phenotyping and unbiased hierarchical clustering to assist the identification of related populations (142). Intermediate populations may indeed harbor cells with intermediate potentials. The diversity of MMCs provides an important regulatory dimension in immunity and the study of their ontogeny endures as a priority area of research. Although many difficulties relating to their plasticity arise, the ability of progenitors to remain flexible in their potential is likely to be critical in allowing hematopoiesis and immunity to adapt to environmental challenges.

Acknowledgments. *The authors are grateful to M. A. Haniffa for critical review of the manuscript. V.B. is funded by Wellcome Intermediate Clinical Fellowship WT088555MA.*

Citation. Collin M, Bigley V. 2016. Monocyte, macrophage, and dendritic cell development: the human perspective. Microbiol Spectrum 4(5):MCHD-0015-2015.

References

1. **van Furth R, Cohn ZA, Hirsch JG, Humphrey JH, Spector WG, Langevoort HL.** 1972. The mononuclear phagocyte system: a new classification of macrophages, monocytes, and their precursor cells. *Bull World Health Organ* **46:**845–852.
2. **Sallusto F, Lanzavecchia A.** 1994. Efficient presentation of soluble antigen by cultured human dendritic cells is maintained by granulocyte/macrophage colony-stimulating factor plus interleukin 4 and downregulated by tumor necrosis factor α. *J Exp Med* **179:** 1109–1118.
3. **Murray PJ, Allen JE, Biswas SK, Fisher EA, Gilroy DW, Goerdt S, Gordon S, Hamilton JA, Ivashkiv LB, Lawrence T, Locati M, Mantovani A, Martinez FO, Mege JL, Mosser DM, Natoli G, Saeij JP, Schultze JL, Shirey KA, Sica A, Suttles J, Udalova I, van Ginderachter JA, Vogel SN, Wynn TA.** 2014. Macrophage activation and polarization: nomenclature and experimental guidelines. *Immunity* **41:**14–20.
4. **Steinman RM, Lustig DS, Cohn ZA.** 1974. Identification of a novel cell type in peripheral lymphoid organs

of mice. 3. Functional properties in vivo. *J Exp Med* **139:**1431–1445.

5. **Hume DA.** 2008. Macrophages as APC and the dendritic cell myth. *J Immunol* **181:**5829–5835.
6. **Merad M, Sathe P, Helft J, Miller J, Mortha A.** 2013. The dendritic cell lineage: ontogeny and function of dendritic cells and their subsets in the steady state and the inflamed setting. *Annu Rev Immunol* **31:**563–604.
7. **Haniffa M, Ginhoux F, Wang XN, Bigley V, Abel M, Dimmick I, Bullock S, Grisotto M, Booth T, Taub P, Hilkens C, Merad M, Collin M.** 2009. Differential rates of replacement of human dermal dendritic cells and macrophages during hematopoietic stem cell transplantation. *J Exp Med* **206:**371–385.
8. **Wang XN, McGovern N, Gunawan M, Richardson C, Windebank M, Siah TW, Lim HY, Fink K, Li JL, Ng LG, Ginhoux F, Angeli V, Collin M, Haniffa M.** 2014. A three-dimensional atlas of human dermal leukocytes, lymphatics, and blood vessels. *J Invest Dermatol* **134:** 965–974.
9. **McGovern N, Schlitzer A, Gunawan M, Jardine L, Shin A, Poyner E, Green K, Dickinson R, Wang XN, Low D, Best K, Covins S, Milne P, Pagan S, Aljefri K, Windebank M, Miranda-Saavedra D, Larbi A, Wasan PS, Duan K, Poidinger M, Bigley V, Ginhoux F, Collin M, Haniffa M.** 2014. Human dermal CD14$^+$ cells are a transient population of monocyte-derived macrophages. *Immunity* **41:**465–477.
10. **Tamoutounour S, Guilliams M, Montanana Sanchis F, Liu H, Terhorst D, Malosse C, Pollet E, Ardouin L, Luche H, Sanchez C, Dalod M, Malissen B, Henri S.** 2013. Origins and functional specialization of macrophages and of conventional and monocyte-derived dendritic cells in mouse skin. *Immunity* **39:**925–938.
11. **Watchmaker PB, Lahl K, Lee M, Baumjohann D, Morton J, Kim SJ, Zeng R, Dent A, Ansel KM, Diamond B, Hadeiba H, Butcher EC.** 2014. Comparative transcriptional and functional profiling defines conserved programs of intestinal DC differentiation in humans and mice. *Nat Immunol* **15:**98–108.
12. **Ginhoux F, Jung S.** 2014. Monocytes and macrophages: developmental pathways and tissue homeostasis. *Nat Rev Immunol* **14:**392–404.
13. **Bar-On L, Jung S.** 2010. Defining dendritic cells by conditional and constitutive cell ablation. *Immunol Rev* **234:**76–89.
14. **Seneschal J, Clark RA, Gehad A, Baecher-Allan CM, Kupper TS.** 2012. Human epidermal Langerhans cells maintain immune homeostasis in skin by activating skin resident regulatory T cells. *Immunity* **36:**873–884.
15. **Yona S, Kim KW, Wolf Y, Mildner A, Varol D, Breker M, Strauss-Ayali D, Viukov S, Guilliams M, Misharin A, Hume DA, Perlman H, Malissen B, Zelzer E, Jung S.** 2013. Fate mapping reveals origins and dynamics of monocytes and tissue macrophages under homeostasis. *Immunity* **38:**79–91.
16. **Hashimoto D, Chow A, Noizat C, Teo P, Beasley MB, Leboeuf M, Becker CD, See P, Price J, Lucas D, Greter M, Mortha A, Boyer SW, Forsberg EC, Tanaka M, van Rooijen N, García-Sastre A, Stanley ER, Ginhoux F, Frenette PS, Merad M.** 2013. Tissue-resident macrophages self-maintain locally throughout adult life with minimal contribution from circulating monocytes. *Immunity* **38:**792–804.
17. **Hoeffel G, Chen J, Lavin Y, Low D, Almeida FF, See P, Beaudin AE, Lum J, Low I, Forsberg EC, Poidinger M, Zolezzi F, Larbi A, Ng LG, Chan JK, Greter M, Becher B, Samokhvalov IM, Merad M, Ginhoux F.** 2015. C-Myb$^+$ erythro-myeloid progenitor-derived fetal monocytes give rise to adult tissue-resident macrophages. *Immunity* **42:**665–678.
18. **Gomez Perdiguero E, Klapproth K, Schulz C, Busch K, Azzoni E, Crozet L, Garner H, Trouillet C, de Bruijn MF, Geissmann F, Rodewald HR.** 2015. Tissue-resident macrophages originate from yolk-sac-derived erythro-myeloid progenitors. *Nature* **518:**547–551.
19. **Jakubzick C, Gautier EL, Gibbings SL, Sojka DK, Schlitzer A, Johnson TE, Ivanov S, Duan Q, Bala S, Condon T, van Rooijen N, Grainger JR, Belkaid Y, Ma'ayan A, Riches DW, Yokoyama WM, Ginhoux F, Henson PM, Randolph GJ.** 2013. Minimal differentiation of classical monocytes as they survey steady-state tissues and transport antigen to lymph nodes. *Immunity* **39:**599–610.
20. **Bain CC, Mowat AM.** 2014. Macrophages in intestinal homeostasis and inflammation. *Immunol Rev* **260:** 102–117.
21. **Lavine KJ, Epelman S, Uchida K, Weber KJ, Nichols CG, Schilling JD, Ornitz DM, Randolph GJ, Mann DL.** 2014. Distinct macrophage lineages contribute to disparate patterns of cardiac recovery and remodeling in the neonatal and adult heart. *Proc Natl Acad Sci U S A* **111:**16029–16034.
22. **Massberg S, Schaerli P, Knezevic-Maramica I, Köllnberger M, Tubo N, Moseman EA, Huff IV, Junt T, Wagers AJ, Mazo IB, von Andrian UH.** 2007. Immunosurveillance by hematopoietic progenitor cells trafficking through blood, lymph, and peripheral tissues. *Cell* **131:**994–1008.
23. **Dzionek A, Fuchs A, Schmidt P, Cremer S, Zysk M, Miltenyi S, Buck DW, Schmitz J.** 2000. BDCA-2, BDCA-3, and BDCA-4: three markers for distinct subsets of dendritic cells in human peripheral blood. *J Immunol* **165:**6037–6046.
24. **MacDonald KP, Munster DJ, Clark GJ, Dzionek A, Schmitz J, Hart DN.** 2002. Characterization of human blood dendritic cell subsets. *Blood* **100:**4512–4520.
25. **Miller JC, Brown BD, Shay T, Gautier EL, Jojic V, Cohain A, Pandey G, Leboeuf M, Elpek KG, Helft J, Hashimoto D, Chow A, Price J, Greter M, Bogunovic M, Bellemare-Pelletier A, Frenette PS, Randolph GJ, Turley SJ, Merad M, Immunological Genome Consortium.** 2012. Deciphering the transcriptional network of the dendritic cell lineage. *Nat Immunol* **13:**888–899.
26. **Gautier EL, Shay T, Miller J, Greter M, Jakubzick C, Ivanov S, Helft J, Chow A, Elpek KG, Gordonov S, Mazloom AR, Ma'ayan A, Chua WJ, Hansen TH, Turley SJ, Merad M, Randolph GJ; Immunological Genome Consortium.** 2012. Gene-expression profiles

and transcriptional regulatory pathways that underlie the identity and diversity of mouse tissue macrophages. *Nat Immunol* **13:**1118–1128.

27. **Lavin Y, Winter D, Blecher-Gonen R, David E, Keren-Shaul H, Merad M, Jung S, Amit I.** 2014. Tissue-resident macrophage enhancer landscapes are shaped by the local microenvironment. *Cell* **159:**1312–1326.
28. **Guilliams M, Ginhoux F, Jakubzick C, Naik SH, Onai N, Schraml BU, Segura E, Tussiwand R, Yona S.** 2014. Dendritic cells, monocytes and macrophages: a unified nomenclature based on ontogeny. *Nat Rev Immunol* **14:** 571–578.
29. **Xue J, Schmidt SV, Sander J, Draffehn A, Krebs W, Quester I, De Nardo D, Gohel TD, Emde M, Schmidleithner L, Ganesan H, Nino-Castro A, Mallmann MR, Labzin L, Theis H, Kraut M, Beyer M, Latz E, Freeman TC, Ulas T, Schultze JL.** 2014. Transcriptome-based network analysis reveals a spectrum model of human macrophage activation. *Immunity* **40:**274–288.
30. **Ziegler-Heitbrock L, Ancuta P, Crowe S, Dalod M, Grau V, Hart DN, Leenen PJ, Liu YJ, MacPherson G, Randolph GJ, Scherberich J, Schmitz J, Shortman K, Sozzani S, Strobl H, Zembala M, Austyn JM, Lutz MB.** 2010. Nomenclature of monocytes and dendritic cells in blood. *Blood* **116:**e74–e80. doi:10.1182/blood-2010-02-258558.
31. **Robbins SH, Walzer T, Dembélé D, Thibault C, Defays A, Bessou G, Xu H, Vivier E, Sellars M, Pierre P, Sharp FR, Chan S, Kastner P, Dalod M.** 2008. Novel insights into the relationships between dendritic cell subsets in human and mouse revealed by genome-wide expression profiling. *Genome Biol* **9:**R17. doi:10.1186/gb-2008-9-1-r17.
32. **Hänsel A, Günther C, Ingwersen J, Starke J, Schmitz M, Bachmann M, Meurer M, Rieber EP, Schäkel K.** 2011. Human slan (6-sulfo LacNAc) dendritic cells are inflammatory dermal dendritic cells in psoriasis and drive strong T_H17/T_H1 T-cell responses. *J Allergy Clin Immunol* **127:**787–794.e9. doi:10.1016/j.jaci.2010.12.009.
33. **Cros J, Cagnard N, Woollard K, Patey N, Zhang SY, Senechal B, Puel A, Biswas SK, Moshous D, Picard C, Jais JP, D'Cruz D, Casanova JL, Trouillet C, Geissmann F.** 2010. Human $CD14^{dim}$ monocytes patrol and sense nucleic acids and viruses via TLR7 and TLR8 receptors. *Immunity* **33:**375–386.
34. **Carlin LM, Stamatiades EG, Auffray C, Hanna RN, Glover L, Vizcay-Barrena G, Hedrick CC, Cook HT, Diebold S, Geissmann F.** 2013. *Nr4a1*-dependent $Ly6C^{low}$ monocytes monitor endothelial cells and orchestrate their disposal. *Cell* **153:**362–375.
35. **Hettinger J, Richards DM, Hansson J, Barra MM, Joschko AC, Krijgsveld J, Feuerer M.** 2013. Origin of monocytes and macrophages in a committed progenitor. *Nat Immunol* **14:**821–830.
36. **Hanna RN, Carlin LM, Hubbeling HG, Nackiewicz D, Green AM, Punt JA, Geissmann F, Hedrick CC.** 2011. The transcription factor NR4A1 (Nur77) controls bone marrow differentiation and the survival of $Ly6C^-$ monocytes. *Nat Immunol* **12:**778–785.
37. **Schmidl C, Renner K, Peter K, Eder R, Lassmann T, Balwierz PJ, Itoh M, Nagao-Sato S, Kawaji H, Carninci P, Suzuki H, Hayashizaki Y, Andreesen R, Hume DA, Hoffmann P, Forrest AR, Kreutz MP, Edinger M, Rehli M, FANTOM consortium.** 2014. Transcription and enhancer profiling in human monocyte subsets. *Blood* **123:**e90–e99. doi:10.1182/blood-2013-02-484188.
38. **Abeles RD, McPhail MJ, Sowter D, Antoniades CG, Vergis N, Vijay GK, Xystrakis E, Khamri W, Shawcross DL, Ma Y, Wendon JA, Vergani D.** 2012. CD14, CD16 and HLA-DR reliably identifies human monocytes and their subsets in the context of pathologically reduced HLA-DR expression by $CD14^{hi}/CD16^{neg}$ monocytes: expansion of $CD14^{hi}/CD16^{pos}$ and contraction of $CD14^{lo}/CD16^{pos}$ monocytes in acute liver failure. *Cytometry A* **81:**823–834.
39. **Fingerle-Rowson G, Angstwurm M, Andreesen R, Ziegler-Heitbrock HW.** 1998. Selective depletion of $CD14^+$ $CD16^+$ monocytes by glucocorticoid therapy. *Clin Exp Immunol* **112:**501–506.
40. **Damuzzo V, Pinton L, Desantis G, Solito S, Marigo I, Bronte V, Mandruzzato S.** 2015. Complexity and challenges in defining myeloid-derived suppressor cells. *Cytometry B Clin Cytom* **88:**77–91.
41. **Wong KL, Yeap WH, Tai JJ, Ong SM, Dang TM, Wong SC.** 2012. The three human monocyte subsets: implications for health and disease. *Immunol Res* **53:**41–57.
42. **Cogle CR, Yachnis AT, Laywell ED, Zander DS, Wingard JR, Steindler DA, Scott EW.** 2004. Bone marrow transdifferentiation in brain after transplantation: a retrospective study. *Lancet* **363:**1432–1437.
43. **Bigley V, Haniffa M, Doulatov S, Wang XN, Dickinson R, McGovern N, Jardine L, Pagan S, Dimmick I, Chua I, Wallis J, Lordan J, Morgan C, Kumararatne DS, Doffinger R, van der Burg M, van Dongen J, Cant A, Dick JE, Hambleton S, Collin M.** 2011. The human syndrome of dendritic cell, monocyte, B and NK lymphoid deficiency. *J Exp Med* **208:**227–234.
44. **Hambleton S, Salem S, Bustamante J, Bigley V, Boisson-Dupuis S, Azevedo J, Fortin A, Haniffa M, Ceron-Gutierrez L, Bacon CM, Menon G, Trouillet C, McDonald D, Carey P, Ginhoux F, Alsina L, Zumwalt TJ, Kong XF, Kumararatne D, Butler K, Hubeau M, Feinberg J, Al-Muhsen S, Cant A, Abel L, Chaussabel D, Doffinger R, Talesnik E, Grumach A, Duarte A, Abarca K, Moraes-Vasconcelos D, Burk D, Berghuis A, Geissmann F, Collin M, Casanova JL, Gros P.** 2011. *IRF8* mutations and human dendritic-cell immunodeficiency. *N Engl J Med* **365:**127–138.
45. **Zaba LC, Fuentes-Duculan J, Steinman RM, Krueger JG, Lowes MA.** 2007. Normal human dermis contains distinct populations of $CD11c^+BDCA-1^+$ dendritic cells and $CD163^+FXIIIA^+$ macrophages. *J Clin Invest* **117:** 2517–2525.
46. **Lenz A, Heine M, Schuler G, Romani N.** 1993. Human and murine dermis contain dendritic cells. Isolation by means of a novel method and phenotypical and functional characterization. *J Clin Invest* **92:**2587–2596.
47. **Chu CC, Ali N, Karagiannis P, Di Meglio P, Skowera A, Napolitano L, Barinaga G, Grys K, Sharif-Paghaleh**

E, Karagiannis SN, Peakman M, Lombardi G, Nestle FO. 2012. Resident CD141 (BDCA3)+ dendritic cells in human skin produce IL-10 and induce regulatory T cells that suppress skin inflammation. *J Exp Med* **209**:935–945.

48. **Jenkins SJ, Ruckerl D, Cook PC, Jones LH, Finkelman FD, van Rooijen N, MacDonald AS, Allen JE.** 2011. Local macrophage proliferation, rather than recruitment from the blood, is a signature of T_H2 inflammation. *Science* **332**:1284–1288.

49. **Cheong C, Matos I, Choi JH, Dandamudi DB, Shrestha E, Longhi MP, Jeffrey KL, Anthony RM, Kluger C, Nchinda G, Koh H, Rodriguez A, Idoyaga J, Pack M, Velinzon K, Park CG, Steinman RM.** 2010. Microbial stimulation fully differentiates monocytes to DC-SIGN/CD209+ dendritic cells for immune T cell areas. *Cell* **143**:416–429.

50. **León B, López-Bravo M, Ardavín C.** 2007. Monocyte-derived dendritic cells formed at the infection site control the induction of protective T helper 1 responses against *Leishmania*. *Immunity* **26**:519–531.

51. **Manh TP, Alexandre Y, Baranek T, Crozat K, Dalod M.** 2013. Plasmacytoid, conventional, and monocyte-derived dendritic cells undergo a profound and convergent genetic reprogramming during their maturation. *Eur J Immunol* **43**:1706–1715.

52. **Scott CL, Henri S, Guilliams M.** 2014. Mononuclear phagocytes of the intestine, the skin, and the lung. *Immunol Rev* **262**:9–24.

53. **Reizis B, Bunin A, Ghosh HS, Lewis KL, Sisirak V.** 2011. Plasmacytoid dendritic cells: recent progress and open questions. *Annu Rev Immunol* **29**:163–183.

54. **Jardine L, Barge D, Ames-Draycott A, Pagan S, Cookson S, Spickett G, Haniffa M, Collin M, Bigley V.** 2013. Rapid detection of dendritic cell and monocyte disorders using CD4 as a lineage marker of the human peripheral blood antigen-presenting cell compartment. *Front Immunol* **4**:495. doi:10.3389/fimmu.2013.00495.

55. **Crozat K, Guiton R, Contreras V, Feuillet V, Dutertre CA, Ventre E, Vu Manh TP, Baranek T, Storset AK, Marvel J, Boudinot P, Hosmalin A, Schwartz-Cornil I, Dalod M.** 2010. The XC chemokine receptor 1 is a conserved selective marker of mammalian cells homologous to mouse CD8α+ dendritic cells. *J Exp Med* **207**:1283–1292.

56. **Poulin LF, Reyal Y, Uronen-Hansson H, Schraml BU, Sancho D, Murphy KM, Håkansson UK, Moita LF, Agace WW, Bonnet D, Reis e Sousa C.** 2012. DNGR-1 is a specific and universal marker of mouse and human Batf3-dependent dendritic cells in lymphoid and nonlymphoid tissues. *Blood* **119**:6052–6062.

57. **Gurka S, Hartung E, Becker M, Kroczek RA.** 2015. Mouse conventional dendritic cells can be universally classified based on the mutually exclusive expression of XCR1 and SIRPα. *Front Immunol* **6**:35. doi:10.3389/fimmu.2015.00035.

58. **van de Ven R, van den Hout MF, Lindenberg JJ, Sluijter BJ, van Leeuwen PA, Lougheed SM, Meijer S, van den Tol MP, Scheper RJ, de Gruijl TD.** 2011. Characterization of four conventional dendritic cell subsets in human skin-draining lymph nodes in relation to T-cell activation. *Blood* **118**:2502–2510.

59. **Mittag D, Proietto AI, Loudovaris T, Mannering SI, Vremec D, Shortman K, Wu L, Harrison LC.** 2011. Human dendritic cell subsets from spleen and blood are similar in phenotype and function but modified by donor health status. *J Immunol* **186**:6207–6217.

60. **Haniffa M, Shin A, Bigley V, McGovern N, Teo P, See P, Wasan PS, Wang XN, Malinarich F, Malleret B, Larbi A, Tan P, Zhao H, Poidinger M, Pagan S, Cookson S, Dickinson R, Dimmick I, Jarrett RF, Renia L, Tam J, Song C, Connolly J, Chan JK, Gehring A, Bertoletti A, Collin M, Ginhoux F.** 2012. Human tissues contain $CD141^{hi}$ cross-presenting dendritic cells with functional homology to mouse CD103+ nonlymphoid dendritic cells. *Immunity* **37**:60–73.

61. **Segura E, Durand M, Amigorena S.** 2013. Similar antigen cross-presentation capacity and phagocytic functions in all freshly isolated human lymphoid organ-resident dendritic cells. *J Exp Med* **210**:1035–1047.

62. **Merad M, Ginhoux F, Collin M.** 2008. Origin, homeostasis and function of Langerhans cells and other langerin-expressing dendritic cells. *Nat Rev Immunol* **8**:935–947.

63. **Valladeau J, Ravel O, Dezutter-Dambuyant C, Moore K, Kleijmeer M, Liu Y, Duvert-Frances V, Vincent C, Schmitt D, Davoust J, Caux C, Lebecque S, Saeland S.** 2000. Langerin, a novel C-type lectin specific to Langerhans cells, is an endocytic receptor that induces the formation of Birbeck granules. *Immunity* **12**:71–81.

64. **Bigley V, McGovern N, Milne P, Dickinson R, Pagan S, Cookson S, Haniffa M, Collin M.** 2015. Langerin-expressing dendritic cells in human tissues are related to CD1c+ dendritic cells and distinct from Langerhans cells and $CD141^{high}$ XCR1+ dendritic cells. *J Leukoc Biol* **97**:627–634.

65. **Milne P, Bigley V, Gunawan M, Haniffa M, Collin M.** 2015. CD1c+ blood dendritic cells have Langerhans cell potential. *Blood* **125**:470–473.

66. **Romani N, Clausen BE, Stoitzner P.** 2010. Langerhans cells and more: langerin-expressing dendritic cell subsets in the skin. *Immunol Rev* **234**:120–141.

67. **Hunger RE, Sieling PA, Ochoa MT, Sugaya M, Burdick AE, Rea TH, Brennan PJ, Belisle JT, Blauvelt A, Porcelli SA, Modlin RL.** 2004. Langerhans cells utilize CD1a and langerin to efficiently present nonpeptide antigens to T cells. *J Clin Invest* **113**:701–708.

68. **de Witte L, Nabatov A, Pion M, Fluitsma D, de Jong MA, de Gruijl T, Piguet V, van Kooyk Y, Geijtenbeek TB.** 2007. Langerin is a natural barrier to HIV-1 transmission by Langerhans cells. *Nat Med* **13**:367–371.

69. **Igyártó BZ, Haley K, Ortner D, Bobr A, Gerami-Nejad M, Edelson BT, Zurawski SM, Malissen B, Zurawski G, Berman J, Kaplan DH.** 2011. Skin-resident murine dendritic cell subsets promote distinct and opposing antigen-specific T helper cell responses. *Immunity* **35**:260–272.

70. **Bennett CL, Fallah-Arani F, Conlan T, Trouillet C, Goold H, Chorro L, Flutter B, Means TK, Geissmann**

F, Chakraverty R. 2011. Langerhans cells regulate cutaneous injury by licensing CD8 effector cells recruited to the skin. *Blood* **117**:7063–7069.

71. Schuster C, Vaculik C, Fiala C, Meindl S, Brandt O, Imhof M, Stingl G, Eppel W, Elbe-Bürger A. 2009. HLA-DR$^+$ leukocytes acquire CD1 antigens in embryonic and fetal human skin and contain functional antigen-presenting cells. *J Exp Med* **206**:169–181.
72. Hoeffel G, Wang Y, Greter M, See P, Teo P, Malleret B, Leboeuf M, Low D, Oller G, Almeida F, Choy SH, Grisotto M, Renia L, Conway SJ, Stanley ER, Chan JK, Ng LG, Samokhvalov IM, Merad M, Ginhoux F. 2012. Adult Langerhans cells derive predominantly from embryonic fetal liver monocytes with a minor contribution of yolk sac-derived macrophages. *J Exp Med* **209**: 1167–1181.
73. Wang Y, Szretter KJ, Vermi W, Gilfillan S, Rossini C, Cella M, Barrow AD, Diamond MS, Colonna M. 2012. IL-34 is a tissue-restricted ligand of CSF1R required for the development of Langerhans cells and microglia. *Nat Immunol* **13**:753–760.
74. Greter M, Lelios I, Pelczar P, Hoeffel G, Price J, Leboeuf M, Kündig TM, Frei K, Ginhoux F, Merad M, Becher B. 2012. Stroma-derived interleukin-34 controls the development and maintenance of Langerhans cells and the maintenance of microglia. *Immunity* **37**: 1050–1060.
75. Borkowski TA, Letterio JJ, Farr AG, Udey MC. 1996. A role for endogenous transforming growth factor β1 in Langerhans cell biology: the skin of transforming growth factor β1 null mice is devoid of epidermal Langerhans cells. *J Exp Med* **184**:2417–2422.
76. Czernielewski JM, Demarchez M. 1987. Further evidence for the self-reproducing capacity of Langerhans cells in human skin. *J Invest Dermatol* **88**:17–20.
77. Merad M, Manz MG, Karsunky H, Wagers A, Peters W, Charo I, Weissman IL, Cyster JG, Engleman EG. 2002. Langerhans cells renew in the skin throughout life under steady-state conditions. *Nat Immunol* **3**:1135–1141.
78. Kanitakis J, Morelon E, Petruzzo P, Badet L, Dubernard JM. 2011. Self-renewal capacity of human epidermal Langerhans cells: observations made on a composite tissue allograft. *Exp Dermatol* **20**:145–146.
79. Collin MP, Hart DN, Jackson GH, Cook G, Cavet J, Mackinnon S, Middleton PG, Dickinson AM. 2006. The fate of human Langerhans cells in hematopoietic stem cell transplantation. *J Exp Med* **203**:27–33.
80. Mielcarek M, Kirkorian AY, Hackman RC, Price J, Storer BE, Wood BL, Leboeuf M, Bogunovic M, Storb R, Inamoto Y, Flowers ME, Martin PJ, Collin M, Merad M. 2014. Langerhans cell homeostasis and turnover after nonmyeloablative and myeloablative allogeneic hematopoietic cell transplantation. *Transplantation* **98**:563–568.
81. Doulatov S, Notta F, Laurenti E, Dick JE. 2012. Hematopoiesis: a human perspective. *Cell Stem Cell* **10**: 120–136.
82. Lee J, Breton G, Oliveira TY, Zhou YJ, Aljoufi A, Puhr S, Cameron MJ, Sékaly RP, Nussenzweig MC, Liu K. 2015. Restricted dendritic cell and monocyte progenitors in human cord blood and bone marrow. *J Exp Med* **212**:385–399.
83. Breton G, Lee J, Zhou YJ, Schreiber JJ, Keler T, Puhr S, Anandasabapathy N, Schlesinger S, Caskey M, Liu K, Nussenzweig MC. 2015. Circulating precursors of human CD1c$^+$ and CD141$^+$ dendritic cells. *J Exp Med* **212**:401–413.
84. Balan S, Ollion V, Colletti N, Chelbi R, Montanana-Sanchis F, Liu H, Vu Manh TP, Sanchez C, Savoret J, Perrot I, Doffin AC, Fossum E, Bechlian D, Chabannon C, Bogen B, Asselin-Paturel C, Shaw M, Soos T, Caux C, Valladeau-Guilemond J, Dalod M. 2014. Human XCR1$^+$ dendritic cells derived in vitro from CD34$^+$ progenitors closely resemble blood dendritic cells, including their adjuvant responsiveness, contrary to monocyte-derived dendritic cells. *J Immunol* **193**:1622–1635.
85. Doulatov S, Notta F, Eppert K, Nguyen LT, Ohashi PS, Dick JE. 2010. Revised map of the human progenitor hierarchy shows the origin of macrophages and dendritic cells in early lymphoid development. *Nat Immunol* **11**: 585–593.
86. Martínez-Cingolani C, Grandclaudon M, Jeanmougin M, Jouve M, Zollinger R, Soumelis V. 2014. Human blood BDCA-1 dendritic cells differentiate into Langerhans-like cells with thymic stromal lymphopoietin and TGF-β. *Blood* **124**:2411–2420.
87. Fogg DK, Sibon C, Miled C, Jung S, Aucouturier P, Littman DR, Cumano A, Geissmann F. 2006. A clonogenic bone marrow progenitor specific for macrophages and dendritic cells. *Science* **311**:83–87.
88. Waskow C, Liu K, Darrasse-Jèze G, Guermonprez P, Ginhoux F, Merad M, Shengelia T, Yao K, Nussenzweig M. 2008. The receptor tyrosine kinase Flt3 is required for dendritic cell development in peripheral lymphoid tissues. *Nat Immunol* **9**:676–683.
89. Naik SH, Sathe P, Park HY, Metcalf D, Proietto AI, Dakic A, Carotta S, O'Keeffe M, Bahlo M, Papenfuss A, Kwak JY, Wu L, Shortman K. 2007. Development of plasmacytoid and conventional dendritic cell subtypes from single precursor cells derived *in vitro* and *in vivo*. *Nat Immunol* **8**:1217–1226.
90. Onai N, Obata-Onai A, Schmid MA, Ohteki T, Jarrossay D, Manz MG. 2007. Identification of clonogenic common Flt3$^+$M-CSFR$^+$ plasmacytoid and conventional dendritic cell progenitors in mouse bone marrow. *Nat Immunol* **8**:1207–1216.
91. Naik SH, Metcalf D, van Nieuwenhuijze A, Wicks I, Wu L, O'Keeffe M, Shortman K. 2006. Intrasplenic steady-state dendritic cell precursors that are distinct from monocytes. *Nat Immunol* **7**:663–671.
92. Liu K, Victora GD, Schwickert TA, Guermonprez P, Meredith MM, Yao K, Chu FF, Randolph GJ, Rudensky AY, Nussenzweig M. 2009. In vivo analysis of dendritic cell development and homeostasis. *Science* **324**:392–397.
93. Sathe P, Metcalf D, Vremec D, Naik SH, Langdon WY, Huntington ND, Wu L, Shortman K. 2014. Lymphoid tissue and plasmacytoid dendritic cells and macrophages

do not share a common macrophage-dendritic cell-restricted progenitor. *Immunity* **41**:104–115.

94. Schraml BU, van Blijswijk J, Zelenay S, Whitney PG, Filby A, Acton SE, Rogers NC, Moncaut N, Carvajal JJ, Reis e Sousa C. 2013. Genetic tracing via DNGR-1 expression history defines dendritic cells as a hematopoietic lineage. *Cell* **154**:843–858.
95. Onai N, Kurabayashi K, Hosoi-Amaike M, Toyama-Sorimachi N, Matsushima K, Inaba K, Ohteki T. 2013. A clonogenic progenitor with prominent plasmacytoid dendritic cell developmental potential. *Immunity* **38**: 943–957.
96. Schlitzer A, Heiseke AF, Einwächter H, Reindl W, Schiemann M, Manta CP, See P, Niess JH, Suter T, Ginhoux F, Krug AB. 2012. Tissue-specific differentiation of a circulating CCR9$^-$ pDC-like common dendritic cell precursor. *Blood* **119**:6063–6071.
97. Satpathy AT, Kc W, Albring JC, Edelson BT, Kretzer NM, Bhattacharya D, Murphy TL, Murphy KM. 2012. *Zbtb46* expression distinguishes classical dendritic cells and their committed progenitors from other immune lineages. *J Exp Med* **209**:1135–1152.
98. Sanjuan-Pla A, Macaulay IC, Jensen CT, Woll PS, Luis TC, Mead A, Moore S, Carella C, Matsuoka S, Bouriez Jones T, Chowdhury O, Stenson L, Lutteropp M, Green JC, Facchini R, Boukarabila H, Grover A, Gambardella A, Thongjuea S, Carrelha J, Tarrant P, Atkinson D, Clark SA, Nerlov C, Jacobsen SE. 2013. Platelet-biased stem cells reside at the apex of the haematopoietic stem-cell hierarchy. *Nature* **502**:232–236.
99. Naik SH, Perié L, Swart E, Gerlach C, van Rooij N, de Boer RJ, Schumacher TN. 2013. Diverse and heritable lineage imprinting of early haematopoietic progenitors. *Nature* **496**:229–232.
100. Galy A, Travis M, Cen D, Chen B, Human T. 1995. Human T, B, natural killer, and dendritic cells arise from a common bone marrow progenitor cell subset. *Immunity* **3**:459–473.
101. Chicha L, Jarrossay D, Manz MG. 2004. Clonal type I interferon-producing and dendritic cell precursors are contained in both human lymphoid and myeloid progenitor populations. *J Exp Med* **200**:1519–1524.
102. Spits H, Couwenberg F, Bakker AQ, Weijer K, Uittenbogaart CH. 2000. Id2 and Id3 inhibit development of CD34$^+$ stem cells into predendritic cell (pre-DC)2 but not into pre-DC1: evidence for a lymphoid origin of pre-DC2. *J Exp Med* **192**:1775–1784.
103. Sathe P, Vremec D, Wu L, Corcoran L, Shortman K. 2013. Convergent differentiation: myeloid and lymphoid pathways to murine plasmacytoid dendritic cells. *Blood* **121**:11–19.
104. Kondo M, Weissman IL, Akashi K. 1997. Identification of clonogenic common lymphoid progenitors in mouse bone marrow. *Cell* **91**:661–672.
105. Akashi K, Traver D, Miyamoto T, Weissman IL. 2000. A clonogenic common myeloid progenitor that gives rise to all myeloid lineages. *Nature* **404**:193–197.
106. Kawamoto H, Ikawa T, Masuda K, Wada H, Katsura Y. 2010. A map for lineage restriction of progenitors during hematopoiesis: the essence of the myeloid-based model. *Immunol Rev* **238**:23–36.
107. Goardon N, Marchi E, Atzberger A, Quek L, Schuh A, Soneji S, Woll P, Mead A, Alford KA, Rout R, Chaudhury S, Gilkes A, Knapper S, Beldjord K, Begum S, Rose S, Geddes N, Griffiths M, Standen G, Sternberg A, Cavenagh J, Hunter H, Bowen D, Killick S, Robinson L, Price A, Macintyre E, Virgo P, Burnett A, Craddock C, Enver T, Jacobsen SE, Porcher C, Vyas P. 2011. Coexistence of LMPP-like and GMP-like leukemia stem cells in acute myeloid leukemia. *Cancer Cell* **19**:138–152.
108. Ji H, Ehrlich LI, Seita J, Murakami P, Doi A, Lindau P, Lee H, Aryee MJ, Irizarry RA, Kim K, Rossi DJ, Inlay MA, Serwold T, Karsunky H, Ho L, Daley GQ, Weissman IL, Feinberg AP. 2010. Comprehensive methylome map of lineage commitment from haematopoietic progenitors. *Nature* **467**:338–342.
109. Novershtern N, Subramanian A, Lawton LN, Mak RH, Haining WN, McConkey ME, Habib N, Yosef N, Chang CY, Shay T, Frampton GM, Drake AC, Leskov I, Nilsson B, Preffer F, Dombkowski D, Evans JW, Liefeld T, Smutko JS, Chen J, Friedman N, Young RA, Golub TR, Regev A, Ebert BL. 2011. Densely interconnected transcriptional circuits control cell states in human hematopoiesis. *Cell* **144**:296–309.
110. Metcalf D. 1997. The molecular control of granulocytes and macrophages. *Ciba Found Symp* **204**:40–50, discussion 50–56.
111. Olweus J, Thompson PA, Lund-Johansen F. 1996. Granulocytic and monocytic differentiation of CD34^{hi} cells is associated with distinct changes in the expression of the PU.1-regulated molecules, CD64 and macrophage colony-stimulating factor receptor. *Blood* **88**: 3741–3754.
112. Olweus J, BitMansour A, Warnke R, Thompson PA, Carballido J, Picker LJ, Lund-Johansen F. 1997. Dendritic cell ontogeny: a human dendritic cell lineage of myeloid origin. *Proc Natl Acad Sci U S A* **94**:12551–12556.
113. Manz MG, Miyamoto T, Akashi K, Weissman IL. 2002. Prospective isolation of human clonogenic common myeloid progenitors. *Proc Natl Acad Sci U S A* **99**: 11872–11877.
114. Collin M, McGovern N, Haniffa M. 2013. Human dendritic cell subsets. *Immunology* **140**:22–30.
115. Beitnes AC, Ráki M, Brottveit M, Lundin KE, Jahnsen FL, Sollid LM. 2012. Rapid accumulation of CD14$^+$ CD11c$^+$ dendritic cells in gut mucosa of celiac disease after *in vivo* gluten challenge. *PLoS One* **7**:e33556. doi: 10.1371/journal.pone.0033556.
116. Segura E, Touzot M, Bohineust A, Cappuccio A, Chiocchia G, Hosmalin A, Dalod M, Soumelis V, Amigorena S. 2013. Human inflammatory dendritic cells induce Th17 cell differentiation. *Immunity* **38**:336–348.
117. Wollenberg A, Mommaas M, Oppel T, Schottdorf EM, Günther S, Moderer M. 2002. Expression and function of the mannose receptor CD206 on epidermal dendritic cells in inflammatory skin diseases. *J Invest Dermatol* **118**:327–334.

118. Lowes MA, Chamian F, Abello MV, Fuentes-Duculan J, Lin SL, Nussbaum R, Novitskaya I, Carbonaro H, Cardinale I, Kikuchi T, Gilleaudeau P, Sullivan-Whalen M, Wittkowski KM, Papp K, Garovoy M, Dummer W, Steinman RM, Krueger JG. 2005. Increase in TNF-α and inducible nitric oxide synthase-expressing dendritic cells in psoriasis and reduction with efalizumab (anti-CD11a). *Proc Natl Acad Sci U S A* **102**:19057–19062.

119. Soulas C, Conerly C, Kim WK, Burdo TH, Alvarez X, Lackner AA, Williams KC. 2011. Recently infiltrating MAC387+ monocytes/macrophages a third macrophage population involved in SIV and HIV encephalitic lesion formation. *Am J Pathol* **178**:2121–2135.

120. Kaplan G, Nusrat A, Witmer MD, Nath I, Cohn ZA. 1987. Distribution and turnover of Langerhans cells during delayed immune responses in human skin. *J Exp Med* **165**:763–776.

121. Ginhoux F, Tacke F, Angeli V, Bogunovic M, Loubeau M, Dai XM, Stanley ER, Randolph GJ, Merad M. 2006. Langerhans cells arise from monocytes *in vivo*. *Nat Immunol* **7**:265–273.

122. Seré K, Baek JH, Ober-Blöbaum J, Müller-Newen G, Tacke F, Yokota Y, Zenke M, Hieronymus T. 2012. Two distinct types of Langerhans cells populate the skin during steady state and inflammation. *Immunity* **37**: 905–916.

123. Nagao K, Kobayashi T, Moro K, Ohyama M, Adachi T, Kitashima DY, Ueha S, Horiuchi K, Tanizaki H, Kabashima K, Kubo A, Cho YH, Clausen BE, Matsushima K, Suematsu M, Furtado GC, Lira SA, Farber JM, Udey MC, Amagai M. 2012. Stress-induced production of chemokines by hair follicles regulates the trafficking of dendritic cells in skin. *Nat Immunol* **13**:744–752.

124. Geissmann F, Prost C, Monnet JP, Dy M, Brousse N, Hermine O. 1998. Transforming growth factor β1, in the presence of granulocyte/macrophage colony-stimulating factor and interleukin 4, induces differentiation of human peripheral blood monocytes into dendritic Langerhans cells. *J Exp Med* **187**:961–966.

125. Caux C, Vanbervliet B, Massacrier C, Dezutter-Dambuyant C, de Saint-Vis B, Jacquet C, Yoneda K, Imamura S, Schmitt D, Banchereau J. 1996. CD34+ hematopoietic progenitors from human cord blood differentiate along two independent dendritic cell pathways in response to GM-CSF+TNFα. *J Exp Med* **184**:695–706.

126. Strunk D, Rappersberger K, Egger C, Strobl H, Krömer E, Elbe A, Maurer D, Stingl G. 1996. Generation of human dendritic cells/Langerhans cells from circulating CD34+ hematopoietic progenitor cells. *Blood* **87**:1292–1302.

127. Collin M, Bigley V, Haniffa M, Hambleton S. 2011. Human dendritic cell deficiency: the missing ID? *Nat Rev Immunol* **11**:575–583.

128. Sakagami T, Uchida K, Suzuki T, Carey BC, Wood RE, Wert SE, Whitsett JA, Trapnell BC, Luisetti M. 2009. Human GM-CSF autoantibodies and reproduction of pulmonary alveolar proteinosis. *N Engl J Med* **361**: 2679–2681.

129. Collin M, Dickinson R, Bigley V. 2015. Haematopoietic and immune defects associated with *GATA2* mutation. *Br J Haematol* **169**:173–187.

130. Dickinson RE, Milne P, Jardine L, Zandi S, Swierczek SI, McGovern N, Cookson S, Ferozepurwalla Z, Langridge A, Pagan S, Gennery A, Heiskanen-Kosma T, Hämäläinen S, Seppänen M, Helbert M, Tholouli E, Gambineri E, Reykdal S, Gottfreðsson M, Thaventhiran JE, Morris E, Hirschfield G, Richter AG, Jolles S, Bacon CM, Hambleton S, Haniffa M, Bryceson Y, Allen C, Prchal JT, Dick JE, Bigley V, Collin M. 2014. The evolution of cellular deficiency in *GATA2* mutation. *Blood* **123**:863–874.

131. Spinner MA, Sanchez LA, Hsu AP, Shaw PA, Zerbe CS, Calvo KR, Arthur DC, Gu W, Gould CM, Brewer CC, Cowen EW, Freeman AF, Olivier KN, Uzel G, Zelazny AM, Daub JR, Spalding CD, Claypool RJ, Giri NK, Alter BP, Mace EM, Orange JS, Cuellar-Rodriguez J, Hickstein DD, Holland SM. 2014. GATA2 deficiency: a protean disorder of hematopoiesis, lymphatics, and immunity. *Blood* **123**:809–821.

132. Goldman FD, Gurel Z, Al-Zubeidi D, Fried AJ, Icardi M, Song C, Dovat S. 2012. Congenital pancytopenia and absence of B lymphocytes in a neonate with a mutation in the Ikaros gene. *Pediatr Blood Cancer* **58**:591–597.

133. Boztug K, Klein C. 2011. Genetic etiologies of severe congenital neutropenia. *Curr Opin Pediatr* **23**:21–26.

134. Hernandez PA, Gorlin RJ, Lukens JN, Taniuchi S, Bohinjec J, Francois F, Klotman ME, Diaz GA. 2003. Mutations in the chemokine receptor gene *CXCR4* are associated with WHIM syndrome, a combined immunodeficiency disease. *Nat Genet* **34**:70–74.

135. Kurotaki D, Osato N, Nishiyama A, Yamamoto M, Ban T, Sato H, Nakabayashi J, Umehara M, Miyake N, Matsumoto N, Nakazawa M, Ozato K, Tamura T. 2013. Essential role of the IRF8-KLF4 transcription factor cascade in murine monocyte differentiation. *Blood* **121**:1839–1849.

136. Kurotaki D, Yamamoto M, Nishiyama A, Uno K, Ban T, Ichino M, Sasaki H, Matsunaga S, Yoshinari M, Ryo A, Nakazawa M, Ozato K, Tamura T. 2014. IRF8 inhibits C/EBPα activity to restrain mononuclear phagocyte progenitors from differentiating into neutrophils. *Nat Commun* **5**:4978. doi:10.1038/ncomms5978.

137. Schlitzer A, McGovern N, Teo P, Zelante T, Atarashi K, Low D, Ho AW, See P, Shin A, Wasan PS, Hoeffel G, Malleret B, Heiseke A, Chew S, Jardine L, Purvis HA, Hilkens CM, Tam J, Poidinger M, Stanley ER, Krug AB, Renia L, Sivasankar B, Ng LG, Collin M, Ricciardi-Castagnoli P, Honda K, Haniffa M, Ginhoux F. 2013. IRF4 transcription factor-dependent CD11b+ dendritic cells in human and mouse control mucosal IL-17 cytokine responses. *Immunity* **38**:970–983.

138. Tussiwand R, Lee WL, Murphy TL, Mashayekhi M, Kc W, Albring JC, Satpathy AT, Rotondo JA, Edelson BT, Kretzer NM, Wu X, Weiss LA, Glasmacher E, Li P, Liao W, Behnke M, Lam SS, Aurthur CT, Leonard WJ, Singh H, Stallings CL, Sibley LD, Schreiber RD, Murphy KM. 2012. Compensatory dendritic cell development mediated by BATF-IRF interactions. *Nature* **490**:502–507.

139. **Cisse B, Caton ML, Lehner M, Maeda T, Scheu S, Locksley R, Holmberg D, Zweier C, den Hollander NS, Kant SG, Holter W, Rauch A, Zhuang Y, Reizis B.** 2008. Transcription factor E2-2 is an essential and specific regulator of plasmacytoid dendritic cell development. *Cell* **135:**37–48.

140. **Meredith MM, Liu K, Darrasse-Jeze G, Kamphorst AO, Schreiber HA, Guermonprez P, Idoyaga J, Cheong C, Yao KH, Niec RE, Nussenzweig MC.** 2012. Expression of the zinc finger transcription factor zDC (Zbtb46, Btbd4) defines the classical dendritic cell lineage. *J Exp Med* **209:**1153–1165.

141. **Jaitin DA, Keren-Shaul H, Elefant N, Amit I.** 2015. Each cell counts: hematopoiesis and immunity research in the era of single cell genomics. *Semin Immunol* **27:**67–71.

142. **Becher B, Schlitzer A, Chen J, Mair F, Sumatoh HR, Teng KW, Low D, Ruedl C, Riccardi-Castagnoli P, Poidinger M, Greter M, Ginhoux F, Newell EW.** 2014. High-dimensional analysis of the murine myeloid cell system. *Nat Immunol* **15:**1181–1189.

Myeloid Cells in Health and Disease: A Synthesis
Edited by Siamon Gordon

doi:10.1128/microbiolspec.MCHD-0005-2015

William J. Janssen,[1] Donna L. Bratton,[1] Claudia V. Jakubzick,[1] and Peter M. Henson[1]

7 Myeloid Cell Turnover and Clearance

INTRODUCTION

Few, if any, individual cells survive throughout the life of the animal, an observation that sets up the critical concepts of cell life span, turnover, and removal and maintenance of homeostatic cell numbers. These issues are of special interest for understanding the properties of the myeloid cell lineage, which includes cells such as neutrophils, which may exhibit in the normal naive adult mammal the shortest life span of all but yet are maintained in relatively constant numbers within the circulation. However, our understanding of the underlying mechanisms for myeloid cell maintenance and removal is still substantially limited and also requires reexamination in light of new ideas about the ontogeny, characterization, and distribution of the myeloid cells in general. Accordingly, this essay will focus on the concepts and questions that, we argue, are in need of exploration, rather than providing a detailed review of what is a huge literature. By focusing on four of the myeloid-lineage cell types (neutrophils, monocytes, macrophages, and dendritic cells [DCs]), we will also be able to bring to the fore many of the key issues that characterize this set of questions.

Removal of cells implies cell death and destruction, with uptake into phagocytes and subsequent digestion within the endosomes. An exception would be loss at extracorporeal sites such as lung, gut, skin, etc., where the cells may be removed physically. Various forms of programmed cell death (PCD), often, but erroneously, subsumed under the term "apoptosis," lead to uptake. Clearly, unprogrammed cell death (often called necrosis) can generate dead cells and cell debris that are also generally removed by being engulfed by phagocytic cell engulfment. Key to these processes is the necessary recognition of the dying cell or its constituents by the phagocyte—unique forms of "self-recognition" that seem at first hand to defy the concepts of self/nonself that underpin how we usually think of "immunity." In addition, and possibly of significant importance, stimulated cells that are still "living" may also exhibit such recognition signals that lead to their removal while still active (see the section on neutrophils), thus serving a potential regulatory role at the level of whole, living cells.

This removal by endocytic uptake inevitably puts the emphasis on myeloid cells themselves, especially macrophages, as key instruments of the cell and debris clearance (a legacy of Metchnikov's phagocyte theories). However, it is increasingly clear that many nonmyeloid tissue cell types can, either endogenously or after appropriate stimulation, exhibit these endocytic

[1]Departments of Pediatrics and Medicine, National Jewish Health, Denver; Departments of Immunology, Medicine, and Pediatrics, University of Colorado, Denver, Denver, CO 80231.

functions, including the engulfment of whole cells up to 15 µm in diameter, i.e., clear-cut phagocytosis. This point is also exemplified by the extensive literature on intact apoptotic cell clearance in *Caenorhabditis elegans* carried out by near-neighbor tissue cells in the absence of macrophages. A primitive clearance function would be an obvious requirement for tissue development in multicellular organisms, especially evident in substantial metamorphic alterations at different organizational phases seen in numerous animal groups. Implications from some of the observations noted below emphasize the possible unique elements of these endocytic clearance functions for tissue or inflammatory cells that would be in keeping with their early metazoan evolutionary development. In the context of understanding the life history and functions of mammalian myeloid cells, especially in the normal resolution of inflammatory processes to restore tissue homeostasis, the mechanisms underlying their recognition and removal become critical, and even therapeutically targetable.

Accordingly, in this discussion we will first briefly address general issues of removal of cells and cell debris, which, as noted, also significantly involve a key function of phagocytic myeloid cells. Subsequent sections will focus on the four individual cell types, in each case with an emphasis on general points and concepts that, we suggest, are understudied and ripe for new experimental analysis.

RECOGNITION AND REMOVAL OF CELLS OR OF CELL DEBRIS

Before discussing turnover and removal of the different myeloid cell types, it is relevant to first consider the broader question of how dying, effete, or fragmented/disrupted cells are recognized and then ingested. This is a form of self-recognition that, for the most part, does not in itself initiate inflammatory, immunological, or defensive responses in the tissues and thus must be distinguished from more classical innate or adaptive immune system recognition processes. Strikingly, this normal cell removal (including of myeloid cells) is occurring all the time in very large numbers ($>10^{10}$ per day) in fashions that are essentially immunologically invisible.

Recognition

Extensive studies over the last 20 years have revealed numerous surface alterations that occur during PCD that contribute to recognition and removal of the dying cells. Many of these surface alterations are likely also to extend to cell debris if the cell undergoes disruption, either due to induction of externally induced "necrosis" or postapoptotic cytolysis (often termed secondary necrosis). A general concept here is that as a consequence of such changes on apoptotic/PCD cells, their removal is usually so efficient that (i) they are generally cleared before undergoing secondary necrosis and (ii) it is very hard to detect apoptotic cells in tissues unless the clearance mechanisms are disrupted or overloaded.

A key early change during PCD is the cell's inability to maintain the normal phospholipid asymmetry of the plasma membrane, resulting in the outward exposure of phosphatidylserines (PSs) from their normal location on the inner leaflet (1). Detection of this surface PS is an often-used assay for apoptosis and other forms of PCD and is also one of the most important recognition signals for normal removal of the cells. The PS species themselves may be in the form normally present on the inner leaflet or may become oxidized, leading to various sets of recognition structures. These may be direct receptors on the engulfing phagocyte (for example, brain-specific angiogenesis inhibitor-1 [BAI1], CD36, and some scavenger receptors) or various sets of "bridge" molecules (opsonins) that bind the PS on the dying cell or cell debris (e.g., protein S and milk fat globule-epidermal growth factor-factor 8 [MFGE8]) but also link to the surface of the phagocyte to initiate ingestion via different sets of receptors (Mer or αv integrins as examples). Notably, such "bridge" molecules may be produced by the phagocytes themselves (e.g., MFGE8 and thrombospondin-1) or may be present in the local environment (e.g., protein S produced by endothelial cells).

It is noteworthy that phospholipid asymmetry of the plasma membrane, and thus the normal inner leaflet distribution of PS, is energy dependent (see, for example, references 2 and 3). PS may be exposed on the outer leaflet as a potential clearance recognition ligand under many different circumstances, including inevitably when the plasma membrane is disrupted, i.e., during primary or secondary necrosis as well as in extrusion of numerous forms of exosomes, microparticles, or even enveloped viruses. A number of specific molecules involved in altering the distribution of phospholipids in/on the cell membrane have recently been characterized (4), as well as mechanisms for maintaining the asymmetry under normal circumstances. In addition to PS exposure, many other molecular and structural changes on the surface of cells undergoing PCD have been described, each potentially also leading to interaction with the phagocyte via either direct receptor ligation or relevant bridge molecules. Examples include calreticulin from the dying cell endoplasmic reticulum, externally exposed chromatin and DNA, alterations in surface charge, and very likely (but not studied) changes in glycosyl groups (glycocalyx). More complete consideration of these recognition processes is given elsewhere (5–7), but the overall picture indicates high levels of redundancy, further supported by the difficulty of experimentally blocking these systems *in vivo*.

Also implicit in these comments is the expectation that different cell types and different circumstances that render cells effete and ready for removal and/or induce frank PCD or necrosis are likely to generate different combinations and permutations of these recognition structures and removal processes. Unfortunately, this point has not always been considered in the past, leading to studies of cell removal using standardized "apoptotic" or "necrotic" cells that may have only minimal bearing on their removal *in vivo* as well as ignoring the contribution to clearance of various forms of cell "debris." These concepts apply also to normal myeloid cells that end their effective life spans in the blood or tissues, and is especially true for cells exposed to inflammatory or toxic environmental processes.

Removal

As noted, the most effective route of cell removal in vertebrates appears to be phagocytosis by various categories of tissue macrophages. Importantly, however, the cells that remove effete cells in the bone marrow are poorly categorized or understood, though also important in clearing excess leukocytes that have developed at this site but which are not called into the circulation. This substantial clearance mechanism in the bone marrow extends beyond removal of myeloid cells to, for example, clearance of extruded nuclear material from developing erythrocytes (8).

Ingestion of intact apoptotic cells (the normal process in homeostatic conditions) into macrophages appears in general to involve a somewhat unique phagocyte process that can be distinguished in a number of ways from "classical" phagocytosis mediated by normal innate or adaptive immunological recognition. The involved signal pathways for the apoptotic cell uptake are highly conserved between *C. elegans*, *Drosophila*, and mammals (9–11) and show a significant requirement for the GTPase Rac with, in fact, a balancing suppressive action from the GTPase Rho. It has been suggested that the process represents a version of stimulated macropinocytosis (similarly Rac requiring and similarly responsive to macropinocytosis inhibitors and generally resulting in early spacious phagosomes and concurrent uptake of surrounding fluid and its contents). In this case, the usual 200-nm size consideration for macropinocytosis is clearly overcome and there is also a requirement for initial tethering of the target cell to the phagocyte. To emphasize this distinction, we have termed the process "efferocytosis" (12, 13; Fig. 1). Not surprisingly, given the variability of the surface ligands and potential signaling responses, this concept of a primary uptake mechanism is likely oversimplified, and we have argued that the specific uptake mechanisms, as well as the disposition of the phagosome and its contents, are urgently in need of detailed investigation. Similarly, the increasing recognition that uptake of necrotic and disrupted cells (cell debris) is critical to understanding resolution of inflammation and normal tissue homeostasis requires investigation of the underlying mechanisms—to this point substantially understudied.

Neutral, Inflammosuppressive, and Immunosuppressive Responses to Cell Removal

The massive ongoing uptake and removal of circulating cells (including myeloid cells) is normally immunologically and inflammationally silent (as is the similar daily removal of a massive load of retinal outer segments in the eye [14]). Simplistically, this may be because the triggers for uptake (recognition signals) just do not stimulate proinflammatory responses in the phagocytosing cells. On the other hand, in many experimental systems, uptake of intact apoptotic cells has been shown to be actively anti-inflammatory and anti-immunogenic. This effect may in turn be mediated by intrinsic blockade of potential inflammatory responses (e.g., inhibition of proinflammatory transcription factors such as NF-κB) or by initiation of anti-inflammatory mediators such as transforming growth factor β or interleukin-10 (IL-10) that can act in autocrine or paracrine fashion to suppress responses of cells in the local environment. Examples of both possibilities abound (9, 11, 15, 16). Notably, most of these studies have focused on the effects of apoptotic cells in suppressing *stimulated* inflammatory or immunogenic responses in the macrophages or DCs rather than addressing the importance of mere lack of initiation of inflammatory responses when the phagocytes interact and ingest apoptotic cells. The potential difference here between efferocytosis and the phagocyte uptake of true foreign or opsonin-coated particles is more evidence of the specialized receptors and signaling involved in efferocytosis as well as in the impact of the actual uptake process on the general response of the ingesting cell. These issues are of substantial importance to host response processes in general and the questions underlying the initiation of autoimmunity.

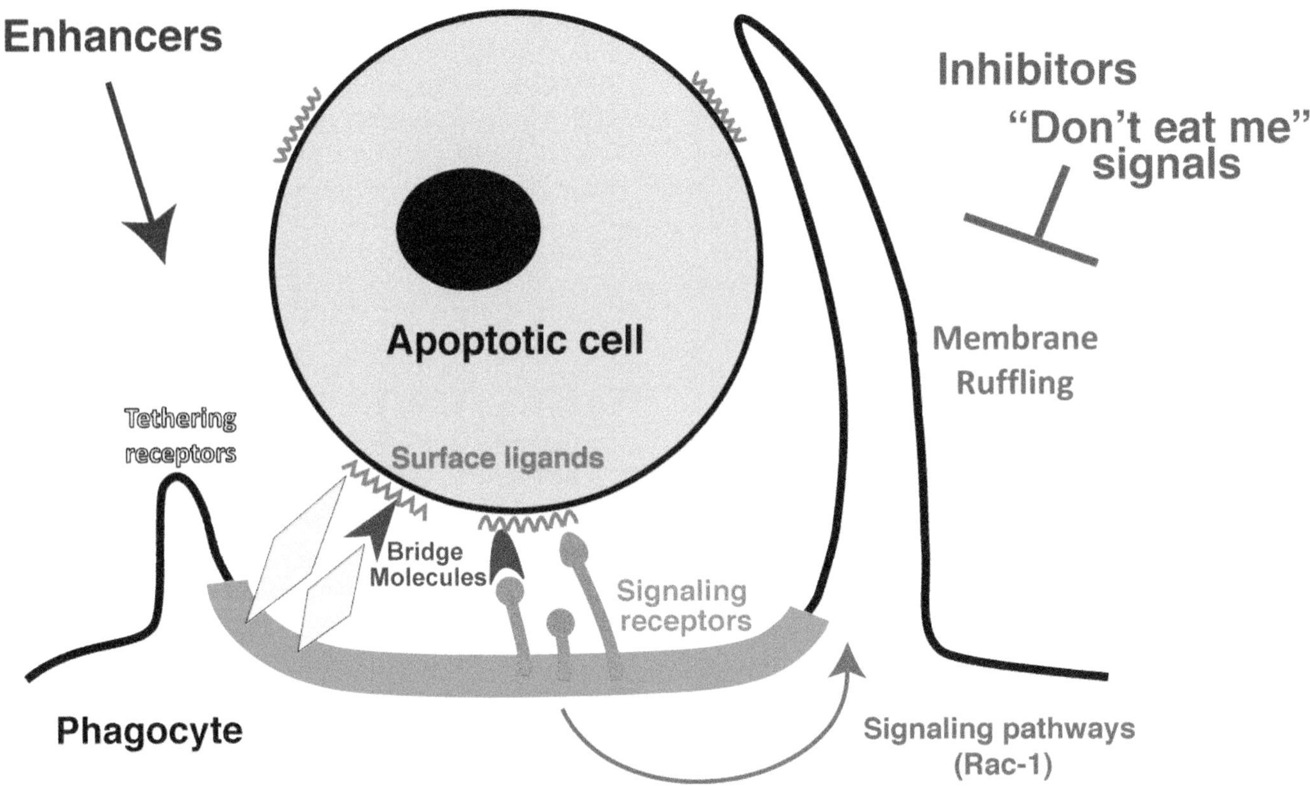

Figure 1 Mechanisms for recognition and uptake of apoptotic cells. Changes on the apoptotic cell surface, including exposure of PS and other normally internally located molecules, are recognized by surface receptors on the phagocyte, leading to tethering of the apoptotic cell and transduction of uptake signals. Bridge molecules (opsonins) in the environment or produced by the phagocyte may also recognize the apoptotic surface changes and also a different set of receptors on the phagocyte to initiate tethering and/or signaling. These processes are significantly redundant and also highly regulated by enhancing or inhibitory stimuli. Viable cells may also avoid removal by expressing "don't eat me" stimuli that block the recognition and/or uptake processes.

For example, there are numerous reports of associations between ineffective removal of apoptotic cells and induction of autoimmunity in both animals and man (see below). The implications are that if dying or effete cells are not removed (especially, for example, circulating granulocytes) they can undergo secondary necrosis, thereby releasing internal constituents such as chromatins or other nuclear contents that can initiate inflammation and immune responses via standard recognition and uptake as foreign stimuli, for example, involving Toll-like receptors (TLRs). On the other hand, there are increasing reasons to believe that at least some forms of non- or postapoptotic cell dissolution may in fact generate particles that can also be cleared efficiently without inflammatory or immunological sequelae.

These concepts led to ideas of a contrasting dualism between reactions to apoptotic versus necrotic cells and even the teleology of "good" versus "bad" cell death. This is unfortunate, since the situation is clearly much more complex and the responses of phagocytes to cells and cell debris are much more nuanced and balanced. More realistically, the complexity and redundancy of ligands on dying cells and receptors on the phagocytes are likely to generate a wide spectrum of effects that vary from nothing more than uptake itself, to induction of anti-inflammatory processes, to frank initiation of inflammation and, in the absence of appropriate immunological tolerance, to autoimmunity. As an example, macrophage responses to ligation of receptors associated with PS recognition versus those associated with calreticulin recognition were shown to induce anti-inflammatory versus proinflammatory responses, respectively (5). In real life, however, both PS and calreticulin are likely to be present on a given apoptotic cell (as are many other potential ligands) and a given macrophage likely expresses various receptors either for these or their bridge molecules and the capacity for variable inflammatory or anti-inflammatory signaling. The net effects are going to be in balance, and certainly make it extremely difficult to mimic *in vitro*. Additionally, as mentioned earlier, it has to be noted that all apoptotic/PCD cells are not the same. They differ substantially by cell type, mode of PCD initiation, stage of cell death, and the nature of the local environment. This is likely even more true of "necrotic" cells, even when this arises from apoptosis when the cells are not cleared (secondary necrosis) and even more in the case of various forms of cell debris in general. In passing, it should be noted that most studies of necrotic cells have involved *ex vivo* dissolution of cells by sonication, heat, or freeze-thaw cycles—processes not commonly seen *in vivo*.

Intracellular Fate of Ingested Cells

Basically, ingested PCD cells and cell debris are effectively and rapidly digested within the endosome and, as a consequence, are not commonly visible in tissue sections—i.e., the inability to detect apoptotic or dying cells does not disprove some ongoing level of cell removal (note reference 17 as an example). Macrophages are suggested to mediate rapid phagosomal maturation and content digestion compared to DCs, which may reflect the increased ability of the latter to present antigens to the adaptive immune system. Since DCs are also believed to ingest significantly through macropinocytic mechanisms and because this may itself result in slower endosomal maturation and digestive capability, in addition to immunological implications, a careful analysis of not only the uptake processes for dying and degraded cells but also their intracellular digestion would seem to be needed.

CIRCULATING NEUTROPHILS, MONOCYTES, AND DCs

Different cell types within the blood are normally maintained at relatively constant numbers. Since they all have finite life spans in the circulation, this implies control of source input (the bone marrow in mature mammals) and removal as well as some mechanism for sensing the circulating concentrations. For erythrocytes, the sensor appears to be local tissue oxygenation, with erythropoietin serving as a rheostat for bone marrow production and release. In the case of myeloid cells (for the purpose of this essay we will focus on neutrophils, monocytes, and DCs), the details are much less clear. The best-studied here are neutrophils, with less known about monocytes and even less about circulating human DCs.

Neutrophil Life Span and Cell Turnover in the Bloodstream

Studies in the 1960s using tagged cells suggested that mature neutrophils in the bloodstream have a life span measured in hours. While this has recently been challenged and might be extended to days (18), the cells are undoubtedly short-lived. A key question for all the circulating myeloid cells is whether their time in the circulation in normal circumstances, i.e., loss from the bloodstream, is driven by their emigration into the tissues or clearance in the three main removal organs: liver, spleen, and bone marrow. This latter fate also raises a significant question. Do the cells in the circulation undergo apoptosis and are then recognized and removed, or are they recognized in some way as old or "effete" and then induced to undergo PCD by interaction with cells such as macrophages in the clearance sites, or even endothelial cells in the vasculature?

Equally challenging is the question of how the circulating numbers of neutrophils are regulated. While certainly fluctuating, including in a circadian fashion, the numbers in normal, noninflammatory conditions remain within certain limits, and estimates of 10^9 neutrophils/kg are released and removed daily (19). Given the life span, this inevitably means rates of input and output are relatively constant under homeostatic conditions, and implies a sensor of some sort to maintain the consistency. Input into the circulation involves release of mature neutrophils from stores in the bone marrow and, upstream of this, production of the cells from hematopoietic precursors. Various candidates have been proposed for the mechanisms of sensing neutrophil numbers, though hard and fast conclusions are lacking.

One significant site of neutrophil removal from the blood is in the bone marrow itself, thereby setting up this as the possible site of sensing and control of the numbers in the blood in a so-called neutrophil "turnstile" driven by the IL-23/IL-17/granulocyte colony-stimulating factor (G-CSF) signaling axis (20). Investigation has shown that rhythmic release of neutrophils into the circulation (triggered by cycled exposure to light in mice) is followed by their return to the bone marrow and rhythmic reductions in the size and function of the hematopoietic niche (21). This suggests a rheostat for production that is turned down based on the numbers of returning neutrophils, and as return slows, production responds by ramping up again. The mechanism appears to depend on signals from engulfing bone marrow-resident macrophages. Alternatively, other investigators describe the active production of G-CSF by bone marrow macrophages in response to engulfment of returning neutrophils, thereby spurring neutrophil production and release to replace the cells that have just been removed (22). These data may certainly be reconciled, though as yet they have not been, as representing different phases and signals of the same finely tuned rhythmic process underlying neutrophil homeostasis.

While there is some increasing evidence for constituent migration of at least a few neutrophils into tissues (as well as possibly most eosinophils), nonetheless, in the absence of an inflammatory response (considered below), most of the constant daily removal of neutrophils occurs within the spleen (red pulp), liver (Kupffer cells in the sinusoids), or bone marrow—notably all sites with relatively low blood flow, perhaps a requirement for the attendant phagocytes to have enough time to interact with and ingest the target neutrophil. In keeping with this hypothesis, a role for functional β_2 integrins on the neutrophils themselves has been implicated as a requirement for their proper clearance (20).

As far as the role of macrophages in carrying out the actual neutrophil removal, their programming state is critical in determining their capacity for cell clearance. For example, activation of peroxisome proliferator-activated receptors (PPARs) and liver X receptors (LXRs) in macrophages has been shown in response to the engulfment of apoptotic cells, and in turn, their activation leads to expression of downstream targets promoting the uptake and disposal of apoptotic cells (23–25). Recent data show a special role for LXRα/LXRβ in neutrophil clearance and, in turn, in the maintenance of neutrophil homeostasis. The LXRs are expressed in bone marrow macrophages in an oscillatory fashion, and their expression and activation was shown to follow the wave of returning neutrophils. Genetic mutations of these receptors resulted in loss of the rhythmic modulation of the hematopoietic niche, suggesting a role within the neutrophil turnstile (21). Further, genetic loss of the LXRs was also shown to be key to neutrophil removal in the periphery: LXR-deficient mice showed increases in circulating neutrophil numbers, slower turnover, and accumulation in the spleen and liver under homeostatic conditions (26). Equally, perturbation of macrophage populations, either depleting their numbers from bone marrow and spleen (27) or genetically deleting macrophage receptors for the recognition of effete cells (26), similarly leads to neutrophilia, though inflammation inevitably results from these manipulations, making assessment of their role in homeostasis difficult.

As noted above, the signals on the neutrophil for removal are still not precisely defined, but are hypothesized to depend on the fate of the neutrophil, as part of its aging, death, or activation, rather than stochastic processes (15). Aging murine neutrophils are characterized as having relatively low CD62L and high CXCR4 and appear to interact with bone marrow macrophages, leading to their preferential engulfment in comparison to younger neutrophils (21). Whether the aged neutrophils are alive or undergoing death at the time of engulfment or whether death comes after ingestion has not been fully resolved, but experimental data exist to support both possibilities. As with most cell types, there are numerous possible pathways by which neutrophils undergo PCD (particularly relevant to inflammatory settings, discussed below), including standard, caspase-induced apoptosis. *Ex vivo*, without additional stimuli, neutrophils undergo apoptosis relatively rapidly and, as a consequence, expose surface markers, e.g., PSs, that are recognized by phagocytes and initiate their engulfment into phagosomes and eventual digestion. Finally, the unique role of oxidants from the neutrophil NADPH oxidase acting either directly (e.g., generation of oxidant-modified phospholipids) or indirectly via influencing the form of PCD (see below) deserves consideration in the generation of signals for engulfment.

Monocytes in the Circulation

The current concept is that there are two major monocyte types within the circulation. In the mouse these can be distinguished by expression of Ly6C, and in humans most readily by CD14 and CD16. The Ly6C^{lo} monocytes are considered to be largely restricted to the intravascular location and have been suggested to play roles of "patrolling" the vasculature (28). The luminal crawling of Ly6C^{lo} monocytes on the endothelium requires firm adhesion with lymphocyte function-associated antigen-1 (CD11ca/CD18; $\alpha_L\beta_2$) integrin and CX_3CR1 to intercellular adhesion molecule-1 and -2 and CX_3CL1 on endothelial cells (28, 29). On the other hand, circulating Ly6C^{hi} monocytes have the ability to extravasate into the tissues, both constitutively and in response to inflammatory conditions, with subsequent ability to either maintain their monocyte characteristics or mature in a number of different ways, including development into monocyte-derived macrophages or monocyte-derived DCs (30), even without inflammation, as has been noted in the gut, skin, and lung (31). Tracer studies using bromodeoxyuridine have shown that monocytes in the circulation have a life span of ~4 days for Ly6C^{hi} monocytes and 11 days for Ly6C^{lo} monocytes (31–33). Once again, the triggers for eventual removal of monocytes from the circulation are not clear, whether frank apoptosis or surface changes leading to recognition and/or stimulation of apoptosis or uptake.

Major sites for removal are likely macrophages in liver and spleen as well as phagocytic cells in the bone marrow, but this too requires detailed investigation. The proportion of the migratory Ly6C^{hi} monocytes that are lost from the circulation as a consequence of their extravasation versus clearance from the bloodstream directly is also not known. Notably, the Ly6C^{hi} monocytes that traffic into noninflamed tissues and bypass macrophage differentiation can subsequently migrate, as monocytes, down lymphatics to lymph nodes, with presumed properties that include local antigen presentation (33). Once in the lymph nodes, the final disposition of the monocytes, as for other myeloid cells gaining access to lymph nodes via the afferent lymphatics, may be of some importance and will be discussed in more detail below.

Dendritic Cells in the Circulation

Unlike mice, humans have fully differentiated DCs circulating in the blood, including those of the myeloid lineage. Although many transcriptional and functional parallels between murine tissue and human intravascular DCs have been assessed, the clearance and turnover rate of human DCs in the blood is currently unknown.

DENDRITIC CELLS

DC life span in lymphoid and nonlymphoid organs varies depending on the DC subtype and location examined. However, most emphasis to date has been on lymphoid DCs in the lymph nodes (34, 35). Nonetheless, bromodeoxyuridine studies have demonstrated that the life span of myeloid DCs in the skin is ~10 to 14 days, but how generally this applies to other tissues needs to be explored. Tissue-resident monocyte-derived DCs were previously hypothesized to require granulocyte-macrophage CSF (GM-CSF), as bone marrow-derived DCs *in vitro* require GM-CSF (36). However, GM-CSF-deficient mice demonstrated that monocyte-derived tumor necrosis factor-α- and inducible nitric oxide synthase-producing DCs did not require GM-CSF for development *in vivo* (37). Therefore, what mediators drive tissue-trafficking monocytes to either remain as monocytes or differentiate into macrophage-like or DC-like cells is currently unknown. The exact counterpart of this dichotomy is likely common to the human situation, though it may not be as clearly defined as in mice.

When activated to migrate from their tissue locations into the lymph node, it is generally recognized that the DCs are short-lived and undergo apoptosis and removal within 2 to 3 days. The stimuli/conditions driving this effect again need further investigation. Furthermore, inflammation in the tissue of origin did not substantially prolong the life span of migratory DCs within the draining lymph node (34). On the other hand, some studies have suggested that upon TLR stimulation Bcl-2/Bcl-xL is upregulated and DC survival is prolonged. In addition, in the absence of Bcl-2, DC turnover was increased (38). It has been hypothesized that the fast turnover rate of DCs functions to regulate antigen presentation to cognate T cells, ultimately limiting immune activation. Since DCs do not exit lymph nodes via efferent lymphatics, removal occurs within the T-cell zone of the lymph node, but which cell type performs this clearance function is unclear—possibly local DCs themselves but perhaps, more likely, medullary macrophages or stromal cells.

The recent emphasis on studies of parabiotic mice to address issues of myeloid cell origin and turnover suggests that this approach could help answer many of these questions. For example, longer-lived cells (B cells, T cells, and Ly6C^{lo} monocytes) reach equilibrium between both parabiotic partners. However, due to their shorter life spans, neutrophils and Ly6C^{hi} monocytes do not attain equilibrium in the blood even after 120 days of parabiosis (35). It was also shown that blood-derived DC precursors (pre-DCs) are rapidly cleared from circulation (residence time, <2 h) and do not fully equilibrate between the parabionts, i.e., behave similarly to circulating Ly6C^{hi} monocytes.

EFFECTS OF NEUTROPHIL AND MONOCYTE CELL EMIGRATION INTO TISSUES AND CLEARANCE FROM INFLAMMATORY LESIONS

Neutrophils in Inflamed Tissues

It has generally been assumed that neutrophils remain in the vasculature until called into inflammatory sites and that they are subsequently removed locally in the inflamed tissues, or if not so removed, remain as a major constituent of pus. Notably, relevant *in vivo* studies of their accumulation and fate are less clear, in part because of the difficulties in distinguishing neutrophil persistence from newly arriving cells in the face of normal efficient clearance of the dying cells. Further, there are increasing suggestions that neutrophils may under some inflammatory circumstances migrate back into the bloodstream (39) as well as down afferent lymphatics to gain access to, and accumulate in, lymph nodes (40). We will consider the issue of clearance of myeloid cells within the lymph nodes below in a separate section. Additional routes of clearance apply after migration onto epithelial surfaces. For example, within the inflamed alveolus, much of the postinflammatory neutrophil clearance likely occurs through phagocytic uptake and digestion by alveolar macrophages just as in tissues (41) (Fig. 2).

In other instances, neutrophils exit the body and escape phagocytic recycling. An interesting special case is the constant emigration of neutrophils to the mouth through the gingival crevice, with removal via the saliva (42). Similarly, emigration into the bronchi results in removal by the mucociliary escalator, and emigration through the colonic epithelium results in elimination via the gastrointestinal tract. While critical for host defense at these sites of the body that interface with the external environment, the constant loss of such neutrophils, arguably, in response to inflammatory signals of relatively low intensity, has yet to be quantified or accounted for in overall recycling models.

Importantly, whatever the actual life span of neutrophils in the circulation may turn out to be, various forms of inflammatory stimulation of neutrophils appear to delay the "spontaneous" development of their apoptotic pathways. This has mostly been studied *ex vivo*, where onset of "spontaneous" apoptosis without specific added stimuli usually occurs over 18 h. Stimulation of the cells with promigratory or proinflammatory stimuli such as chemokines or TLR agonists extends the time before apoptosis sets in (43). This has been suggested to have functional consequences in optimizing neutrophil-protective activities in the inflammatory process but with potentially negative consequences

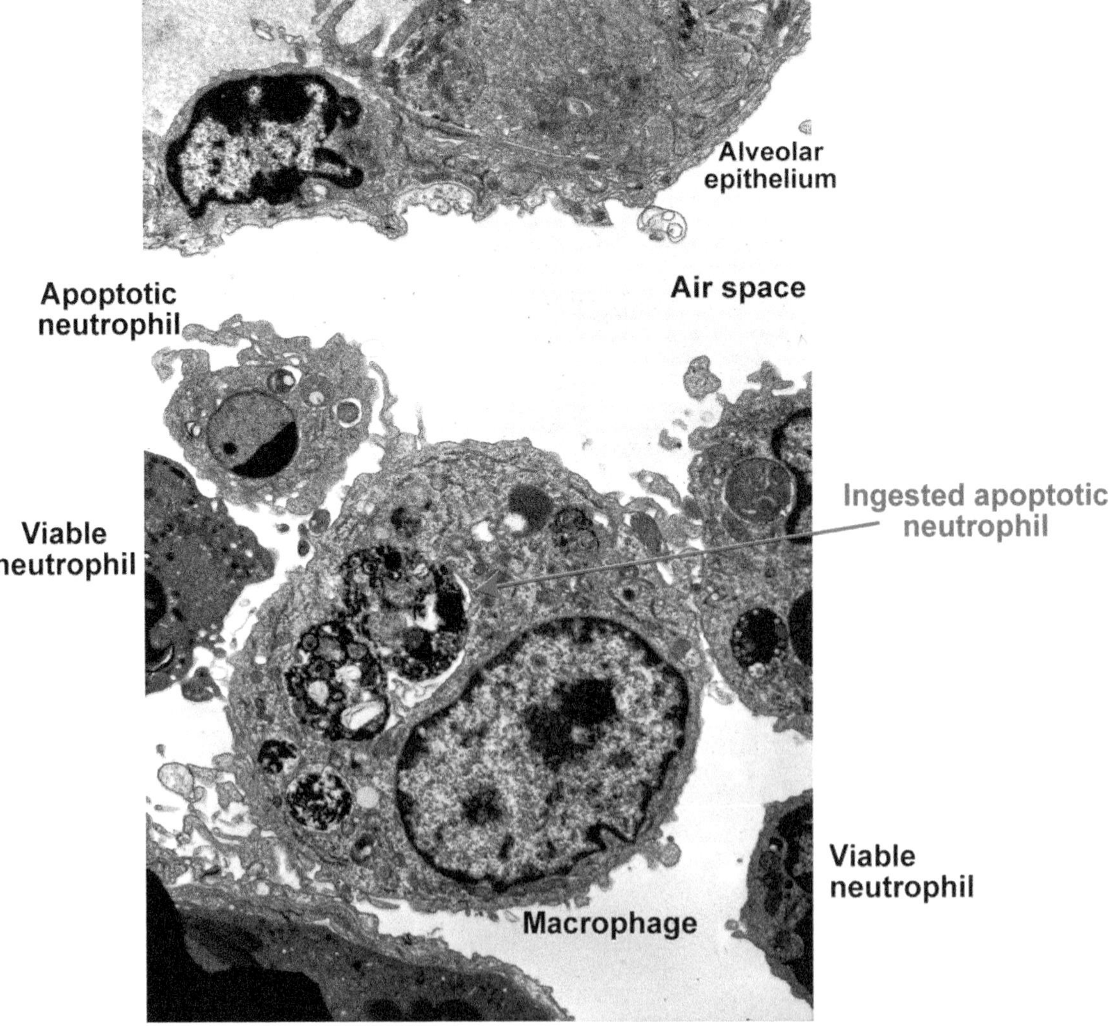

Figure 2 Uptake of apoptotic neutrophils by a macrophage in the alveolar air space during resolution of inflammation in the lung.

in disease states in which neutrophil life span is prolonged and inflammation persists.

Neutrophil apoptosis has been well studied *in vitro*, including that caused by both intrinsic as well as extrinsic signals, and in the important context of phagocytizing microbes (43). Standard caspase-mediated apoptotic signaling pathways, nuclear condensation, exposure of PS and other recognition ligands, etc., are all evident. Additionally, other signals, some unique to neutrophils and subverted from host defense activities, have also been described (15). Of some interest, neutrophils, though they do release microparticles, do not appear to generate the apoptotic blebs and larger apoptotic bodies that are seen in many parenchymal cell apoptotic processes. Equally, there is some indication that the cells remain intact for longer in the apoptotic state. Whether these are general features of myeloid cell PCD is not clear, as detailed study in monocytes and macrophages is less extensive. Teleological suggestions have been raised about the need to maintain cell membrane integrity throughout the clearance process in the face of the potentially injurious contents of these cells and their potential to cause tissue injury. Clearly, in circumstances in which removal of apoptotic neutrophils by macrophages is delayed, including a number of chronic disease states (see below), the cells undergo so-called postapoptotic or secondary necrosis, in which the integrity of the plasma membrane is lost and intracellular contents including a variety of danger-associated molecular patterns and proteases are released.

Of note, other forms of neutrophil PCD are increasingly recognized, each likely resulting in differing signals to the inflammatory milieu and consequences. Neutrophils harboring large vacuoles, likely representing autophagosomes, evident during infections and autoimmunity, are thought to undergo autophagy-associated death (44). Characteristics include mitochondrial swelling, nuclear condensation, and intact plasma membranes. Though precise mechanisms are yet to be elucidated, this form of death may represent a branch point from apoptosis in which caspase activation has been inhibited by survival signals, e.g., GM-CSF.

In addition, neutrophils undergo a unique form of "death," termed "NETosis," in which the nuclear membrane is disrupted, with the extrusion of chromatin both into the cytoplasm and eventually into the extracellular environment (45). Of note, the original descriptions of this process relied on stimulation of the neutrophils with the phorbol diester, phorbol myristate acetate, and criticisms of these observations have been levied accordingly. Recent studies using more-physiologic stimuli have shown that neutrophils can extrude nuclear DNA (mitochondrial DNA extrusion has also been described) while still clearly carrying on the business of being alive; they can still chemotax and actively phagocytose microbes (46). This cellular disruption is hypothesized to play an important role in host defense demonstrated both *in vitro* and *in vivo* (46). In the context of this essay, we clearly need also to think about mechanisms by which these disrupted cells and their contents are ultimately removed, and the inflammatory consequences of signals from their intracellular contents.

Reactive oxygen species almost inevitably play important roles in cell death pathways, and special mention should be made of the role of the NADPH oxidase in determining the fate of neutrophils. For instance, activation of the oxidase by tumor necrosis factor-α and integrin engagement appears to enhance apoptosis by enhancing mitochondrial membrane permeability, whereas high levels of oxidase activation are implicated in autophagy-associated death as well as NETosis (15, 43) Notably, mutations of the NADPH oxidase that render it nonfunctional are associated with prolonged survival of neutrophils (see below). Aside from playing a role in determining the PCD pathway, activation of the NADPH oxidase leads to the production of a modified PS, lysophosphatidylserine. This lipid accumulates both in neutrophils that are activated and fully functional as well as those aged in culture and apoptotic (47). It signals via a G protein-coupled receptor, G2A, on macrophages for engulfment of neutrophils, alive or dead (48, 49).

MONOCYTES AND MACROPHAGES

Origins

As discussed elsewhere, the current paradigm indicates two independent sources for macrophages: embryonic progenitors in the fetal liver and yolk sac and bone marrow-derived monocytes. Most so-called "tissue-resident" macrophages derive exclusively from the former, and colonize tissues during embryogenesis (or immediately after birth). Exceptions occur in barrier tissues, namely the lungs, skin, and gut, where mixed ontogenies may exist. For example, in the lungs, alveolar macrophages arise exclusively from embryonic precursors, whereas extraluminal macrophages (often referred to as "interstitial" macrophages) may arise from circulating monocytes (50, 51) or embryonic precursors (33). In comparison, macrophages in the dermis and gut appear to derive primarily from circulating monocytes (52, 53), Whether commensal microbiota play a role in monocyte recruitment and subsequent development at these barrier sites during homeostasis remains unclear. Certainly, during inflammatory processes the recruitment, maturation, programming, and fate of monocytes and macrophages change considerably. Accordingly, in the following paragraphs homeostasis and inflammation will be discussed separately.

Tissue-Resident Macrophages

Classic examples of embryonic-derived tissue macrophages include brain microglia, liver Kupffer cells, and alveolar macrophages. These macrophages replicate locally throughout postnatal life, and unless they are depleted or destroyed (by gamma radiation, clodronate liposomes, etc.), they are not replaced by monocytes from the circulation. While the mechanisms that implement and regulate either the removal or local self-renewal of these macrophages are not known, it is noteworthy that their normal life spans (i.e., in the absence of overt inflammation) are markedly different. For example, alveolar macrophages in the mouse have been shown to persist for at least 60 days and probably longer (52, 54), whereas resident peritoneal macrophages have considerably shorter life spans (~15 days) (our unpublished observations). The programming states (sometimes termed "phenotype," though we would argue that this term is misleading) of macrophages in these sites also vary considerably. Critically, both cell turnover rates and programming states are highly influenced by the local environment. To illustrate, the alveolar macrophage exists in an extraluminal compartment where it is bathed in a lipidic film at an air-liquid interface with ambient oxygen tension. In stark contrast, peritoneal macrophages exist in a near anoxic environment that is bounded by the mesothelium. Intriguingly, peritoneal macrophages transferred into the airspaces display cell surface molecules and functional properties characteristic of alveolar macrophages within a week (55).

Thus, the general concept of the macrophage as a cell that is highly adaptable to its environment (i.e., exhibiting enormous plasticity) is critical. Whatever the precise signals and processes involved, the environment is clearly paramount in determining macrophage life history. For the moment we must assume that the eventual death and removal of these resident cells likely will involve some form of PCD, but the triggers here for normal turnover are completely unknown, as to a significant degree are the removal processes. A little more is known for monocyte-derived macrophages that accumulate in tissues during the process of inflammation, and thus removal of these cells is discussed in more detail below and contrasted where possible with the resident cell type.

Inflammatory Macrophages

During inflammation, mononuclear phagocyte numbers increase dramatically. Two sources account for this—local replication of tissue-resident macrophages and recruitment of monocytes from the circulation. As noted above, some of the monocytes that emigrate into an inflammatory site undergo maturational changes that result in their legitimate consideration as macrophages. This includes a huge increase in cell constituent synthesis, resulting a much larger cell; higher cytoplasm-to-nuclear ratio; and increased numbers of mitochondria, lysosomes, and membranes, as well as a number of surface markers. As a consequence, in any inflammatory lesion there is a spectrum of cells in various stages of this maturation that are often lumped together under the general term "inflammatory monocytes." Importantly, the relative contribution provided to the total mononuclear phagocyte pool by tissue-resident macrophages versus recruited mononuclear phagocytes varies by anatomic site and is greatly influenced by the nature and magnitude of the stimulus. For example, administration of lipopolysaccharide into the lungs leads to massive recruitment of circulating monocytes yet induces only modest proliferation of resident alveolar macrophages (52). Conversely, pleural infection with helminthic organisms drives marked proliferation of resident pleural macrophages but only minimal recruitment of monocytes from the circulation (56). Additionally, as discussed at length elsewhere, the macrophages that exist in inflammatory conditions exhibit a variety of environment-induced programming states that change considerably over time and that are often associated in a contributory way with the progression of inflammation at its early stages and later with its resolution. Critically, as noted in Fig. 3, these monocyte-derived macrophages, as well as the increased numbers of monocytes accumulated in inflamed tissues, are eventually cleared as the inflammation resolves.

Resolution of Inflammation

As inflammation resolves, tissue macrophage and monocyte numbers return to homeostatic levels. The question of where the "inflammatory macrophages" go is long-standing and only partially answered. In basic terms, three potential fates exist: migration from the site; *in situ* death; and, in the case of tissues that contact the external environment (such as the respiratory tract, gut lumen, and skin), extracorporeal elimination (e.g., expectoration and sloughing). As described above, monocytes clearly have the ability to migrate from tissues using the lymphatics as a conduit (33). However, whether monocytes can also exit tissues by

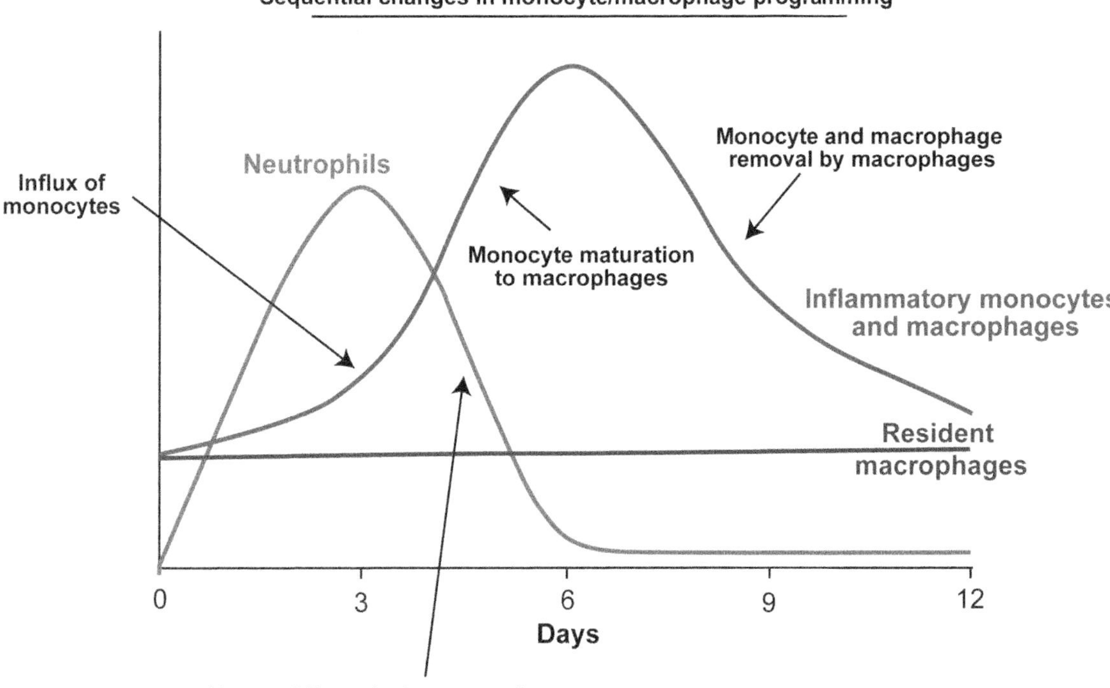

Figure 3 Time course of a standardized acute inflammatory response in the lung, showing accumulation and removal of neutrophils and monocyte/macrophages. The resident alveolar macrophages persist throughout the inflammation and do not substantially change in numbers. Some of the recruited Ly6C^{hi} monocytes mature into macrophages (and DCs), and some remain as monocytes. The macrophages undergo a variety of programming changes during the course of the inflammation, participating in first its initiation and then its resolution. The monocytes and macrophages are cleared as the inflammation wanes, mostly by undergoing PCD and engulfment, though some of the monocytes migrate to the local lymph nodes and, at this site, some of the cells may also be cleared physically up the airways by the mucociliary escalator.

direct egress back into the bloodstream remains unclear. That macrophages emigrate from the peritoneum via the lymphatics has also been reported (57, 58). As mentioned above, monocyte maturation occurs across a continuum, and it is likely that at least some of these emigrating macrophages represent inflammatory monocytes that have gained cell surface molecules typically associated with macrophages. *In situ* death of macrophages and subsequent efferocytic removal represents a clear mode of exit, and in some instances provides a critical means of host defense by eliminating ingested pathogens. Notably, in the peritoneum and lungs, macrophage death appears to be the dominant mechanism for macrophage elimination during resolving inflammation, with lymphatic egress playing only a minor role (52, 59). As mentioned above, numerous pathways for death exist, including apoptosis, autophagy, necroptosis, and frank necrosis—all of which may be operative in driving elimination of macrophages from inflammatory sites. Critically, both the mode of death and the frequency with which cells die are likely to be influenced by the local environment tissue-specific factors, and perhaps even the origin of the cell itself. In this regard, studies of resolving inflammation in the peritoneum (59) and lungs (52) demonstrate that recruited monocyte-derived macrophages undergo PCD, while at the same time resident macrophages (i.e., embryonic-derived macrophages) persist—even though both populations of cells exist in the same environment.

The developing theme of fundamental (transcription factor-dependent) differences between embryonic-derived self-renewing "resident" macrophages and the monocyte-derived inflammatory macrophages under consideration here may in fact be somewhat blurred and in need of more investigation. Thus, although study of resident alveolar macrophages shows their considerable ability to survive throughout the course of an acute inflammatory reaction, there is nonetheless some loss of these cells, which are eventually replaced, in part, apparently, from a proportion of the monocyte-derived cells (52). Intriguingly, gene expression comparison of these versus the persistent original "resident" macrophages shows that over time the monocyte-derived cells became almost identical to the resident cells, differing in only a few genes (60). This observation, which is likely mirrored at other sites, such as the peritoneum, makes a strong case for the influence of the environment in molding the character of the local macrophage, whatever its proximal origin.

The frequency of apoptotic macrophages that can be detected in resolving inflammatory lesions is generally low, suggesting their rapid removal and digestion. However, the question of what clears the dying macrophages remains largely unanswered. Both the remaining macrophages and nonprofessional phagocytes, such as epithelial cells, have the capacity to engulf dying cells, though quantitative studies describing their relative contributions are lacking. The answer will almost certainly be organ/site specific and may also vary with the inflammatory stimulus. In any case, as described above for neutrophils, the process appears to involve exposure of "eat me" signals on the dying cell's surface. Again, PS most likely plays a major role, although additional apoptotic cell-associated molecular patterns, including changes in surface charge and glycosylation and exposure of calreticulin on the cell surface, may also be operative. Notably, activated macrophages can express PS on their surfaces, even when they are not apoptotic (61), thereby suggesting that PS exposure alone is insufficient to drive engulfment and that loss of "don't eat me" signals (e.g., CD47 and CD31) on the surface of the dying cell may also be required. The many molecules used by macrophages to recognize dying cells are described in detail elsewhere (5–7). However, it must be noted that most of our knowledge derives from studies performed either in the steady state (as described above) or during the acute neutrophilic phase of inflammation. The local environment changes significantly as inflammation resolves, and it is therefore likely that phagocytic receptor expression and uptake mechanisms are also likely to change.

DISPOSITION OF MYELOID CELLS MIGRATING TO THE LYMPH NODES

One central theme in the removal of a number of myeloid cell types from tissues is migration/clearance down the lymphatics. Thus, as noted, this appears to be one mode of removal of macrophages from the inflamed peritoneum and is clearly relevant for disposition of tissue migratory DCs or constitutively migrating monocytes. However, this process only kicks the can down the street. Are the cells then removed in the local lymph node? Do some even pass on down the chain to further lymph nodes or even into the bloodstream via the lymphatic ducts? We suggest that this last pathway for removal seems unlikely since mature macrophages are of substantial size and will not pass readily through capillary beds. Basically, PCD and local uptake by phagocytes in the lymph nodes seems most likely for this secondary process of clearance from the tissues. As noted above, the life span of migratory myeloid DCs that have reached the lymph node is no more than a few days. However, reflecting a constant theme in this essay, the precise stimuli that initiate this, and the cells (macrophages, stromal cells, etc.) and recognition processes involved in their subsequent phagocytic removal, have not been clearly defined. Nonetheless, we must suppose this to be a quantitatively substantial clearance process with significant relevance for maintenance of tissue homeostasis.

DEFECTS AND DISEASE

To no great surprise, systems as essential as these to normal development and homeostasis are not immune from inherent or induced defects, and these can include loss-of-function as well as gain-of-function effects. In the case of myeloid cells, their additional role in the removal of damaged cells and debris leads to potential defects in both their own removal as well as their ability to mediate the removal process. A few examples will suffice to illustrate these points, but one can expect many others and a significant contribution to a wide variety of disease processes.

Defects in Removal of Inflammatory Myeloid Cells: COPD

In any inflammatory condition, inability to mediate the rapid and highly efficient removal of the accumulated neutrophils and monocytes would be expected to, and does, prolong the inflammation. Such effects can be caused by the effects of stimuli to the inflammatory cells that block their normal induction of PCD as well as by defects of the removal process itself. As noted above, neutrophil apoptosis can indeed be delayed by a host of potential inflammatory mediators as well as by stimulation of pattern recognition molecules from external or host-derived sources. We bring up chronic obstructive pulmonary disease (COPD) in this context because the long-term destruction of the lung tissue underlying this extremely common and important disease has been suggested to result from persistent low-grade inflammatory, especially neutrophil-driven, injury. Cigarette smoke, either directly or indirectly in a feed-forward inflammatory mediator-induced fashion, can prolong the life of neutrophils, leading to accumulation in the lung tissue. However, it has additionally become apparent that COPD is associated with defects in the ability of macrophages in this tissue to recognize and remove apoptotic cells (i.e., the neutrophils), thereby adding an additional component to persistence of inflammation and its contribution to tissue destruction (62–64). Fundamentally, this example leads one to raise a challenge to the general possibility of similar feedback loops contributing to many forms of chronic inflammation, whether the targets are neutrophils selectively or inflammatory myeloid (or lymphoid) cells more generally. In each case, the mechanisms underlying the defects may be different, and may be either external or internal (or both), but, we would argue, are worth further exploration in the hope of ameliorating the ongoing process.

Chronic Granulomatous Disease

Another example, in this case with a genetic underpinning, is chronic granulomatous disease. Various defects in subunits of the NADPH oxidase lead to loss of the oxidative burst (particularly relevant in neutrophils), with concomitant defects in bacterial and fungal host defense. However, in addition to problems with constant infections, the patients also exhibit persistent inflammatory and granulomatous problems (hence the name). Relevant to the current discussions, the defective neutrophil NADPH oxidase results in lack of normal changes on the cells' surface during inflammation, such as generation of lysophosphatidylserine, that usually render the cells more palatable for uptake and removal by macrophages (48, 65). In addition, the genetic defects are also associated with altered programing of the macrophages toward decreased ability to ingest apoptotic cells. Consequently, these two effects act together to result in defective neutrophil clearance and, thus, persistent inflammation. Notably, the effect is seen most clearly in the context of inflammation, i.e., when the NADPH oxidase is normally activated, and to a much lower extent at the level of normal neutrophil turnover in the blood.

Autoimmune Processes Including Systemic Lupus Erythematosus

An intriguing observation that pervades the studies of mechanisms and extent of apoptotic cell clearance is that defects in these processes, for example, loss of key receptors for apoptotic cell recognition and uptake, are frequently associated with generalized autoimmune disease. Examples include mice lacking the scavenger receptor CD36, bridge molecules MFGE8 and C1q, Mer tyrosine kinase, and G protein-coupled receptor kinase 6 (66), which normally enhances apoptotic cell engulfment through Rac1 activation. The development of autoimmunity in these mouse models coupled with observations of defective apoptotic cell clearance in human systemic lupus erythematosus, lead to the hypothesis that inefficient clearance of apoptotic cells (e.g., dying neutrophils) enables the release of intracellular constituents, which can then initiate autoimmune responses (67). Note also that chronic granulomatous disease patients and mice, again showing inefficient clearance of neutrophils, are also prone over time to develop autoimmunity (68, 69). The extent to which defects in normal cell removal are generally associated with autoimmune responses and the additional controls that may serve to limit this extent are of considerable interest and potential importance.

Hemophagocytic Lymphohistiocytosis

In the other direction, a group of syndromes often termed hemophagocytic lymphohistiocytosis is associated with *increased* phagocytic uptake and removal of blood cells, including myeloid cells, throughout the body (70, 71). These can be either developmental or associated with viral infections and are accompanied and presumably induced by very high levels of systemic cytokines, particularly gamma interferon. Specific mechanisms by which the leukocytes (and erythrocytes) are altered or become recognizable by the phagocytes are not clear. The macrophages involved in the ingestion are likely substantially overstimulated by the cytokine storm, but whether the loss of normal discrimination against uptake of blood leukocytes is solely at the level of the targets or also in the functions of the phagocytes is also unclear. While the clinical conditions themselves are rare, and animal models are limited (70), full exploration of the underlying mechanisms would seem to have potential for general understanding of the controls normally involved in leukocyte maintenance and removal.

IMPLICATIONS AND CONCLUSIONS—MAINTENANCE OF HOMEOSTASIS

We are consistently impressed by the extraordinary variety of forms and mechanisms by which cells can be induced to undergo programmed death, variety that is certainly manifest in the myeloid lineage as well as within other cell categories. This extensive repertoire extends also to the changes on the cell surface, or on cell debris that results, leading to recognition by phagocytes and mediating their removal. Increasingly it is also becoming apparent that these different ligands likely stimulate many different uptake mechanisms in the phagocytes. A sense of enormous redundancy results, suggesting the substantial importance to a multicellular organism of being efficient in its ability to remove cells no longer needed or undergoing aging, damage, abnormal function, or death, i.e., to maintain homeostasis. On the other hand, it also emphasizes the need to consider each cell type, situation, and environment separately, and explicitly, where possible, to study the removal processes *in situ* in the relevant environment *in vivo*.

Not surprisingly, increasing understanding of the processes of myeloid cell removal has led to consideration of therapeutic approaches to its enhancement, for example, in inflammation. Given the complexities noted above, these approaches are in their infancy and fraught with not only the redundancy problem but also contrasting concerns relating to the balance between protective versus damaging effects of inflammatory leukocytes. We note only two potential examples here (of many) to illustrate the possibilities as well as issues involved. In our own studies, stimulation of PPAR-γ with pioglitazone is showing intriguing promise in overcoming some of the defects in neutrophil removal seen in chronic granulomatous disease by altering both the palatability of the neutrophils for uptake as well as the effectiveness of their uptake by the macrophages (72). In another example, a flavone (wogonin) inducer of eosinophil apoptosis was recently reported to have attenuating effects on airway inflammation and allergic responses (73, 74). Other areas of investigation that would seem appropriate to this discussion and that have also perhaps not received the attention that would now be due are the effects of aging of the animal and human on the clearance processes and, at a cellular level, the effects of senescence of the cells (or precursors) themselves. In the case of the myeloid cells, this could apply both to the cells destined for removal as well as those doing the removal itself.

Acknowledgments. *This work was supported by NIH grants HL114381, HL109517, AI110408, HL34303, and HL115334.*

Citation. Janssen WJ, Bratton DL, Jakubzick CV, Henson PM. 2016. Myeloid cell turnover and clearance. Microbiol Spectrum 4(6):MCHD-0005-2015.

References

1. **Fadok VA, Voelker DR, Campbell PA, Cohen JJ, Bratton DL, Henson PM.** 1992. Exposure of phosphatidylserine on the surface of apoptotic lymphocytes triggers specific recognition and removal by macrophages. *J Immunol* **148**:2207–2216.
2. **Takatsu H, Tanaka G, Segawa K, Suzuki J, Nagata S, Nakayama K, Shin HW.** 2014. Phospholipid flippase activities and substrate specificities of human type IV P-type ATPases localized to the plasma membrane. *J Biol Chem* **289**:33543–33556.
3. **Suzuki J, Nagata S.** 2014. Phospholipid scrambling on the plasma membrane. *Methods Enzymol* **544**:381–393.
4. **Segawa K, Nagata S.** 2015. An apoptotic 'eat me' signal: phosphatidylserine exposure. *Trends Cell Biol* **25**:639–650.
5. **Gardai SJ, Bratton DL, Ogden CA, Henson PM.** 2006. Recognition ligands on apoptotic cells: a perspective. *J Leukoc Biol* **79**:896–903.
6. **Poon IK, Lucas CD, Rossi AG, Ravichandran KS.** 2014. Apoptotic cell clearance: basic biology and therapeutic potential. *Nat Rev Immunol* **14**:166–180.
7. **Ravichandran KS.** 2011. Beginnings of a good apoptotic meal: the find-me and eat-me signaling pathways. *Immunity* **35**:445–455.
8. **Yoshida H, Kawane K, Koike M, Mori Y, Uchiyama Y, Nagata S.** 2005. Phosphatidylserine-dependent engulfment by macrophages of nuclei from erythroid precursor cells. *Nature* **437**:754–758.
9. **Birge RB, Ucker DS.** 2008. Innate apoptotic immunity: the calming touch of death. *Cell Death Differ* **15**:1096–1102.
10. **D'mello V, Birge RB.** 2010. Apoptosis: conserved roles for integrins in clearance. *Curr Biol* **20**:R324–R327.
11. **Elliott MR, Ravichandran KS.** 2010. Clearance of apoptotic cells: implications in health and disease. *J Cell Biol* **189**:1059–1070.
12. **deCathelineau AM, Henson PM.** 2003. The final step in programmed cell death: phagocytes carry apoptotic cells to the grave. *Essays Biochem* **39**:105–117.
13. **Hoffmann PR, deCathelineau AM, Ogden CA, Leverrier Y, Bratton DL, Daleke DL, Ridley AJ, Fadok VA, Henson PM.** 2001. Phosphatidylserine (PS) induces PS receptor-mediated macropinocytosis and promotes clearance of apoptotic cells. *J Cell Biol* **155**:649–659.
14. **Mazzoni F, Safa H, Finnemann SC.** 2014. Understanding photoreceptor outer segment phagocytosis: use and utility of RPE cells in culture. *Exp Eye Res* **126**:51–60.
15. **Bratton DL, Henson PM.** 2011. Neutrophil clearance: when the party is over, clean-up begins. *Trends Immunol* **32**:350–357.
16. **Erwig LP, Henson PM.** 2007. Immunological consequences of apoptotic cell phagocytosis. *Am J Pathol* **171**:2–8.
17. **Scott RS, McMahon EJ, Pop SM, Reap EA, Caricchio R, Cohen PL, Earp HS, Matsushima GK.** 2001. Phagocytosis and clearance of apoptotic cells is mediated by MER. *Nature* **411**:207–211.
18. **Pillay J, den Braber I, Vrisekoop N, Kwast LM, de Boer RJ, Borghans JA, Tesselaar K, Koenderman L.** 2010. In vivo labeling with $^{2}H_2O$ reveals a human neutrophil lifespan of 5.4 days. *Blood* **116**:625–627.
19. **Summers C, Rankin SM, Condliffe AM, Singh N, Peters AM, Chilvers ER.** 2010. Neutrophil kinetics in health and disease. *Trends Immunol* **31**:318–324.

20. Stark MA, Huo Y, Burcin TL, Morris MA, Olson TS, Ley K. 2005. Phagocytosis of apoptotic neutrophils regulates granulopoiesis via IL-23 and IL-17. *Immunity* **22**:285–294.

21. Casanova-Acebes M, Pitaval C, Weiss LA, Nombela-Arrieta C, Chèvre R, A-González N, Kunisaki Y, Zhang D, van Rooijen N, Silberstein LE, Weber C, Nagasawa T, Frenette PS, Castrillo A, Hidalgo A. 2013. Rhythmic modulation of the hematopoietic niche through neutrophil clearance. *Cell* **153**:1025–1035.

22. Furze RC, Rankin SM. 2008. The role of the bone marrow in neutrophil clearance under homeostatic conditions in the mouse. *FASEB J* **22**:3111–3119.

23. A-Gonzalez N, Bensinger SJ, Hong C, Beceiro S, Bradley MN, Zelcer N, Deniz J, Ramirez C, Díaz M, Gallardo G, de Galarreta CR, Salazar J, Lopez F, Edwards P, Parks J, Andujar M, Tontonoz P, Castrillo A. 2009. Apoptotic cells promote their own clearance and immune tolerance through activation of the nuclear receptor LXR. *Immunity* **31**:245–258.

24. Majai G, Sarang Z, Csomós K, Zahuczky G, Fésüs L. 2007. PPARγ-dependent regulation of human macrophages in phagocytosis of apoptotic cells. *Eur J Immunol* **37**:1343–1354.

25. Mukundan L, Odegaard JI, Morel CR, Heredia JE, Mwangi JW, Ricardo-Gonzalez RR, Goh YP, Eagle AR, Dunn SE, Awakuni JU, Nguyen KD, Steinman L, Michie SA, Chawla A. 2009. PPAR-δ senses and orchestrates clearance of apoptotic cells to promote tolerance. *Nat Med* **15**:1266–1272.

26. Hong C, Kidani Y, A-Gonzalez N, Phung T, Ito A, Rong X, Ericson K, Mikkola H, Beaven SW, Miller LS, Shao WH, Cohen PL, Castrillo A, Tontonoz P, Bensinger SJ. 2012. Coordinate regulation of neutrophil homeostasis by liver X receptors in mice. *J Clin Invest* **122**:337–347.

27. Gordy C, Pua H, Sempowski GD, He YW. 2011. Regulation of steady-state neutrophil homeostasis by macrophages. *Blood* **117**:618–629.

28. Auffray C, Fogg D, Garfa M, Elain G, Join-Lambert O, Kayal S, Sarnacki S, Cumano A, Lauvau G, Geissmann F. 2007. Monitoring of blood vessels and tissues by a population of monocytes with patrolling behavior. *Science* **317**:666–670.

29. Carlin LM, Stamatiades EG, Auffray C, Hanna RN, Glover L, Vizcay-Barrena G, Hedrick CC, Cook HT, Diebold S, Geissmann F. 2013. *Nr4a1*-dependent $Ly6C^{low}$ monocytes monitor endothelial cells and orchestrate their disposal. *Cell* **153**:362–375.

30. Guilliams M, Ginhoux F, Jakubzick C, Naik SH, Onai N, Schraml BU, Segura E, Tussiwand R, Yona S. 2014. Dendritic cells, monocytes and macrophages: a unified nomenclature based on ontogeny. *Nat Rev Immunol* **14**:571–578.

31. Yona S, Kim KW, Wolf Y, Mildner A, Varol D, Breker M, Strauss-Ayali D, Viukov S, Guilliams M, Misharin A, Hume DA, Perlman H, Malissen B, Zelzer E, Jung S. 2013. Fate mapping reveals origins and dynamics of monocytes and tissue macrophages under homeostasis. *Immunity* **38**:79–91.

32. Hettinger J, Richards DM, Hansson J, Barra MM, Joschko AC, Krijgsveld J, Feuerer M. 2013. Origin of monocytes and macrophages in a committed progenitor. *Nat Immunol* **14**:821–830.

33. Jakubzick C, Gautier EL, Gibbings SL, Sojka DK, Schlitzer A, Johnson TE, Ivanov S, Duan Q, Bala S, Condon T, van Rooijen N, Grainger JR, Belkaid Y, Ma'ayan A, Riches DW, Yokoyama WM, Ginhoux F, Henson PM, Randolph GJ. 2013. Minimal differentiation of classical monocytes as they survey steady-state tissues and transport antigen to lymph nodes. *Immunity* **39**:599–610.

34. Kamath AT, Henri S, Battye F, Tough DF, Shortman K. 2002. Developmental kinetics and lifespan of dendritic cells in mouse lymphoid organs. *Blood* **100**:1734–1741.

35. Liu K, Waskow C, Liu X, Yao K, Hoh J, Nussenzweig M. 2007. Origin of dendritic cells in peripheral lymphoid organs of mice. *Nat Immunol* **8**:578–583.

36. Serbina NV, Salazar-Mather TP, Biron CA, Kuziel WA, Pamer EG. 2003. TNF/iNOS-producing dendritic cells mediate innate immune defense against bacterial infection. *Immunity* **19**:59–70.

37. Greter M, Helft J, Chow A, Hashimoto D, Mortha A, Agudo-Cantero J, Bogunovic M, Gautier EL, Miller J, Leboeuf M, Lu G, Aloman C, Brown BD, Pollard JW, Xiong H, Randolph GJ, Chipuk JE, Frenette PS, Merad M. 2012. GM-CSF controls nonlymphoid tissue dendritic cell homeostasis but is dispensable for the differentiation of inflammatory dendritic cells. *Immunity* **36**:1031–1046.

38. Chen M, Huang L, Shabier Z, Wang J. 2007. Regulation of the lifespan in dendritic cell subsets. *Mol Immunol* **44**:2558–2565.

39. Woodfin A, Voisin MB, Beyrau M, Colom B, Caille D, Diapouli FM, Nash GB, Chavakis T, Albelda SM, Rainger GE, Meda P, Imhof BA, Nourshargh S. 2011. The junctional adhesion molecule JAM-C regulates polarized transendothelial migration of neutrophils *in vivo*. *Nat Immunol* **12**:761–769.

40. Chtanova T, Schaeffer M, Han SJ, van Dooren GG, Nollmann M, Herzmark P, Chan SW, Satija H, Camfield K, Aaron H, Striepen B, Robey EA. 2008. Dynamics of neutrophil migration in lymph nodes during infection. *Immunity* **29**:487–496.

41. McCubbrey AL, Curtis JL. 2013. Efferocytosis and lung disease. *Chest* **143**:1750–1757.

42. Scott DA, Krauss J. 2012. Neutrophils in periodontal inflammation. *Front Oral Biol* **15**:56–83.

43. Geering B, Stoeckle C, Conus S, Simon HU. 2013. Living and dying for inflammation: neutrophils, eosinophils, basophils. *Trends Immunol* **34**:398–409.

44. Mihalache CC, Simon HU. 2012. Autophagy regulation in macrophages and neutrophils. *Exp Cell Res* **318**:1187–1192.

45. Brinkmann V, Reichard U, Goosmann C, Fauler B, Uhlemann Y, Weiss DS, Weinrauch Y, Zychlinsky A. 2004. Neutrophil extracellular traps kill bacteria. *Science* **303**:1532–1535.

46. Yipp BG, Petri B, Salina D, Jenne CN, Scott BN, Zbytnuik LD, Pittman K, Asaduzzaman M, Wu K,

Meijndert HC, Malawista SE, de Boisfleury Chevance A, Zhang K, Conly J, Kubes P. 2012. Infection-induced NETosis is a dynamic process involving neutrophil multitasking *in vivo*. *Nat Med* **18:**1386–1393.

47. Frasch SC, Berry KZ, Fernandez-Boyanapalli R, Jin HS, Leslie C, Henson PM, Murphy RC, Bratton DL. 2008. NADPH oxidase-dependent generation of lysophosphatidylserine enhances clearance of activated and dying neutrophils via G2A. *J Biol Chem* **283:**33736–33749.
48. Frasch SC, Fernandez-Boyanapalli RF, Berry KA, Murphy RC, Leslie CC, Nick JA, Henson PM, Bratton DL. 2013. Neutrophils regulate tissue neutrophilia in inflammation via the oxidant-modified lipid lysophosphatidylserine. *J Biol Chem* **288:**4583–4593.
49. Frasch SC, Fernandez-Boyanapalli RF, Berry KZ, Leslie CC, Bonventre JV, Murphy RC, Henson PM, Bratton DL. 2011. Signaling via macrophage G2A enhances efferocytosis of dying neutrophils by augmentation of Rac activity. *J Biol Chem* **286:**12108–12122.
50. Schulz C, Gomez Perdiguero E, Chorro L, Szabo-Rogers H, Cagnard N, Kierdorf K, Prinz M, Wu B, Jacobsen SE, Pollard JW, Frampton J, Liu KJ, Geissmann F. 2012. A lineage of myeloid cells independent of Myb and hematopoietic stem cells. *Science* **336:**86–90.
51. Scott CL, Henri S, Guilliams M. 2014. Mononuclear phagocytes of the intestine, the skin, and the lung. *Immunol Rev* **262:**9–24.
52. Janssen WJ, Barthel L, Muldrow A, Oberley-Deegan RE, Kearns MT, Jakubzick C, Henson PM. 2011. Fas determines differential fates of resident and recruited macrophages during resolution of acute lung injury. *Am J Respir Crit Care Med* **184:**547–560.
53. Bain CC, Bravo-Blas A, Scott CL, Gomez Perdiguero E, Geissmann F, Henri S, Malissen B, Osborne LC, Artis D, Mowat AM. 2014. Constant replenishment from circulating monocytes maintains the macrophage pool in the intestine of adult mice. *Nat Immunol* **15:**929–937.
54. Murphy J, Summer R, Wilson AA, Kotton DN, Fine A. 2008. The prolonged life-span of alveolar macrophages. *Am J Respir Cell Mol Biol* **38:**380–385.
55. Guth AM, Janssen WJ, Bosio CM, Crouch EC, Henson PM, Dow SW. 2009. Lung environment determines unique phenotype of alveolar macrophages. *Am J Physiol Lung Cell Mol Physiol* **296:**L936–L946.
56. Jenkins SJ, Ruckerl D, Cook PC, Jones LH, Finkelman FD, van Rooijen N, MacDonald AS, Allen JE. 2011. Local macrophage proliferation, rather than recruitment from the blood, is a signature of T_H2 inflammation. *Science* **332:**1284–1288.
57. Bellingan GJ, Caldwell H, Howie SE, Dransfield I, Haslett C. 1996. In vivo fate of the inflammatory macrophage during the resolution of inflammation: inflammatory macrophages do not die locally, but emigrate to the draining lymph nodes. *J Immunol* **157:**2577–2585.
58. Cao C, Lawrence DA, Strickland DK, Zhang L. 2005. A specific role of integrin Mac-1 in accelerated macrophage efflux to the lymphatics. *Blood* **106:**3234–3241.
59. Gautier EL, Ivanov S, Lesnik P, Randolph GJ. 2013. Local apoptosis mediates clearance of macrophages from resolving inflammation in mice. *Blood* **122:**2714–2722.
60. Gibbings SL, Goyal R, Desch AN, Leach SM, Prabagar M, Atif SM, Bratton DL, Janssen W, Jakubzick CV. 2015. Transcriptome analysis highlights the conserved difference between embryonic and postnatal-derived alveolar macrophages. *Blood* **126:**1357–1366.
61. Marguet D, Luciani MF, Moynault A, Williamson P, Chimini G. 1999. Engulfment of apoptotic cells involves the redistribution of membrane phosphatidylserine on phagocyte and prey. *Nat Cell Biol* **1:**454–456.
62. Petrusca DN, Gu Y, Adamowicz JJ, Rush NI, Hubbard WC, Smith PA, Berdyshev EV, Birukov KG, Lee CH, Tuder RM, Twigg HL III, Vandivier RW, Petrache I. 2010. Sphingolipid-mediated inhibition of apoptotic cell clearance by alveolar macrophages. *J Biol Chem* **285:**40322–40332.
63. Henson PM, Vandivier RW, Douglas IS. 2006. Cell death, remodeling, and repair in chronic obstructive pulmonary disease? *Proc Am Thorac Soc* **3:**713–717.
64. Hamon R, Homan CC, Tran HB, Mukaro VR, Lester SE, Roscioli E, Bosco MD, Murgia CM, Ackland ML, Jersmann HP, Lang C, Zalewski PD, Hodge SJ. 2014. Zinc and zinc transporters in macrophages and their roles in efferocytosis in COPD. *PLoS One* **9:**e110056. doi:10.1371/journal.pone.0110056.
65. Fernandez-Boyanapalli R, Frasch SC, Riches DW, Vandivier RW, Henson PM, Bratton DL. 2010. PPARγ activation normalizes resolution of acute sterile inflammation in murine chronic granulomatous disease. *Blood* **116:**4512–4522.
66. Nakaya M, Tajima M, Kosako H, Nakaya T, Hashimoto A, Watari K, Nishihara H, Ohba M, Komiya S, Tani N, Nishida M, Taniguchi H, Sato Y, Matsumoto M, Tsuda M, Kuroda M, Inoue K, Kurose H. 2013. GRK6 deficiency in mice causes autoimmune disease due to impaired apoptotic cell clearance. *Nat Commun* **4:**1532. doi:10.1038/ncomms2540.
67. Muñoz LE, Janko C, Schulze C, Schorn C, Sarter K, Schett G, Herrmann M. 2010. Autoimmunity and chronic inflammation—two clearance-related steps in the etiopathogenesis of SLE. *Autoimmun Rev* **10:**38–42.
68. Rupec RA, Petropoulou T, Belohradsky BH, Walchner M, Liese JG, Plewing G, Messer G. 2000. Lupus erythematosus tumidus and chronic discoid lupus erythematosus in carriers of X-linked chronic granulomatous disease. *Eur J Dermatol* **10:**184–189.
69. Sanford AN, Suriano AR, Herche D, Dietzmann K, Sullivan KE. 2006. Abnormal apoptosis in chronic granulomatous disease and autoantibody production characteristic of lupus. *Rheumatology (Oxford)* **45:**178–181.
70. Brisse E, Wouters CH, Matthys P. 2015. Hemophagocytic lymphohistiocytosis (HLH): a heterogeneous spectrum of cytokine-driven immune disorders. *Cytokine Growth Factor Rev* **26:**263–280.
71. Usmani GN, Woda BA, Newburger PE. 2013. Advances in understanding the pathogenesis of HLH. *Br J Haematol* **161:**609–622.
72. Fernandez-Boyanapalli RF, Falcone EL, Zerbe CS, Marciano BE, Frasch SC, Henson PM, Holland SM, Bratton DL. 2015. Impaired efferocytosis in human

chronic granulomatous disease is reversed by pioglitazone treatment. *J Allergy Clin Immunol* **136:**1399–1401.

73. **Lucas CD, Dorward DA, Sharma S, Rennie J, Felton JM, Alessandri AL, Duffin R, Schwarze J, Haslett C, Rossi AG.** 2015. Wogonin induces eosinophil apoptosis and attenuates allergic airway inflammation. *Am J Respir Crit Care Med* **191:**626–636.

74. **Persson C.** 2015. Drug-induced death of eosinophils. Promises and pitfalls. *Am J Respir Crit Care Med* **191:** 605–606.

Specialization

III

Myeloid Cells in Health and Disease: A Synthesis
Edited by Siamon Gordon

doi:10.1128/microbiolspec.MCHD-0024-2015

David A. Hume[1]
Kim M. Summers[1]
Michael Rehli[2]

8 Transcriptional Regulation and Macrophage Differentiation

THE CELLS OF THE MONONUCLEAR PHAGOCYTE SYSTEM

The mononuclear phagocyte system (MPS) was originally defined by van Furth and Cohn (1) as a family of cells of the innate immune system derived from hematopoietic progenitor cells under the influence of specific growth factors (2, 3). Differentiated cells of the MPS, monocytes and macrophages, are effectors of innate immunity, engulfing and killing pathogens. They are also needed for tissue repair and resolution of inflammation and for the generation of an appropriate acquired immune response. Their biology and differentiation have been reviewed by a number of authors (2, 4–8). The original definition of the MPS considered an essentially linear sequence from pluripotent progenitors, through committed myeloid progenitors shared with granulocytes, to promonocytes and blood monocytes, and thence to tissue macrophages (2, 4–8). Resident macrophages differ in function between tissues, and within tissues they occupy a specific niche (9). In some locations, for example, associated with epithelia, they clearly have individual identifiable territories that form a regular pattern (2, 3).

THE BIOLOGY OF BLOOD MONOCYTE SUBSETS

Monocytes in peripheral blood have been subdivided into subsets based on certain surface markers (4, 9–13). The seminal study in the area of monocyte subset function (14) segregated mouse peripheral blood monocytes based on their expression of chemokine receptors and behavior on adoptive transfer; those expressing CCR2 (and the marker Ly6C) were recruited to inflammatory sites, whereas those expressing CX3CR1 were selectively recruited to noninflammatory sites. Subsequent mouse studies have indicated that Ly6C^{hi} monocytes replenish the large resident macrophage population of the gastrointestinal tract (15–18) and patrol the extravascular space in many other organs (19). The more mature Ly6C^{lo} populations, which have a much longer half-life in the circulation, may perform a patrolling function in the circulation and sense nucleic acids and viruses (16, 20, 21). The subsets have also been referred to as "classical" (Ly6C^{hi} in mice or CD14^{hi} in humans) and "nonclassical" (Ly6C^{lo} in mice or CD16^{hi} in humans) under a proposed unifying nomenclature (11, 13). Given the lack of correlation between markers,

[1]University of Edinburgh, The Roslin Institute and Royal (Dick) School of Veterinary Studies, Midlothian EH25 9RG, United Kingdom;
[2]University Hospital Regensburg, Department of Internal Medicine III, D-93047 Regensburg, Germany.

and species differences, it is likely that these populations could be further subdivided into different and/or smaller subsets using additional markers. Rather than give the subsets names, it may be better to refer specifically to the markers used. Expression array profiling of human and mouse monocyte subsets defined by the markers described above (Ly6C and CD14/16) supported the idea that these subsets were functionally equivalent across species, although there were many species-specific expression differences (22). One major difference between mice and humans is the relative abundance of the subsets. In mice, the major populations are equally abundant, whereas in humans, the CD16^{hi} subset is a minor subpopulation that also varies widely with disease states. CD16 as a marker is also difficult to interpret since there is extensive copy number variation in the gene encoding this marker, *FCG3RB*, underlying inflammatory disease susceptibility in humans (23). In humans, an "intermediate" population (CD14^{int}, CD16^{int}) has been identified and attributed specialized functions (10–13). As discussed further below, the definition of the intermediate population depends somewhat on the position of gates on the fluorescence-activated cell sorter, and the use of other markers such as SLAN and TIE2 may produce distinct subsets (10, 12). Bovine monocytes have also been subdivided based on CD14 and CD16 as markers, with CD172a (SIRPA) and CD163 providing additional markers that varied between subpopulations (24). Pig peripheral blood monocytes, which were more human-like than were mouse macrophages in their overall gene expression profiles, could also be subdivided into roughly equal subsets, depending on reciprocal expression of CD14 and CD163, but appeared to lack a genuine "nonclassical" population (25). In rats, a separate marker, CD43, separates monocytes into two populations (26). As in the mouse, these populations differ in their expression of chemokine receptors CCR2 and CX3CR1 (26). CD43 expression also differs between the mouse monocyte subpopulations (27). Adoptive transfer and lineage trace studies in the mouse and rat confirmed that the mature (Ly6^{lo}CD43^{hi}) cells derive from the Ly6C^{hi}CD43^{lo} cells (26, 28, 29). The monocyte subpopulations are therefore likely to be the extremes of a differentiation series, in which the time spent in transition is relatively short.

THE DEVELOPMENT OF THE MPS

During embryonic development in the mouse, macrophages are first identifiable as a distinct cell type in the yolk sac, and from that location they migrate and infiltrate all of the tissues of the body. In mice, the first appearance occurs in Reichert's membrane (30). The yolk sac-derived macrophages are apparently generated as mature macrophages, without an obvious monocyte-like intermediate, and are highly motile, actively phagocytic, and proliferative (31–33). In outbred mice, these yolk sac-derived macrophages lacked detectable expression of the macrophage-specific transcription factor PU.1, and the null mutation of the *Spi1* gene encoding PU.1 did not compromise their development (32, 33). Subsequent studies demonstrated that the impact of the *Spi1* knockout mutation depends strongly on genetic background; in some strains the knockout is midgestation lethal, whereas in others it is myelodeficient at birth (34). The initial view, based on studies in the chick, was that yolk sac-derived macrophages are replaced later in development by the products of definitive hematopoiesis (35). This view was strongly supported by recent cellular transplantation studies in the chicken, where yolk sac cells injected into early embryos produced macrophages that were lost by the time of hatch, but bone marrow-derived cells produced long-term chimerism (36). Unexpectedly, this chimerism was restricted to the macrophage lineage, suggesting the existence of a macrophage-restricted progenitor cell in the bone marrow that has self-renewal capacity. There is also some evidence for such a progenitor in the mouse (37–39).

In the mouse, an emerging consensus is that populations of tissue macrophages, notably the microglia of the brain (40), the epidermal macrophages (Langerhans cells) of the skin (41, 42), and alveolar macrophages of the lung (43), are seeded from the yolk sac or fetal liver during development, and thereafter are maintained by self-renewal (reviewed in reference 38). However, this consensus depends on assumptions about the validity of the models used. Almost all of these studies use a single inbred mouse strain, C57BL/6, because most knockouts and conditional reporters have been made on this background. By contrast to the outbred mice, in the C57BL/6 line yolk sac macrophages were apparently dependent on PU.1, which was expressed in the yolk sac, and were independent of the expression of c-MYB, which is expressed in definitive hematopoietic stem cells (HSCs) (44). The view that the yolk sac is a major source of tissue macrophages in the adult has been extended to the point where blood monocytes are no longer believed to make a significant contribution to any tissue macrophage population (29, 44, 45) other than that of the gastrointestinal tract (46). The vast majority of adult tissue-resident macrophages in adult mouse liver, brain, skin, and lung were proposed to originate from yolk sac erythromyeloid progenitors, distinct from HSCs that colonize the fetal liver, with minimal replacement from HSC progeny (via blood monocytes) even by 1 year of age (47, 48).

Other authors, using different experimental models, have reached a different conclusion about the role of the yolk sac and embryonic macrophages in adult macrophage populations. Epelman et al. (49) reported that ~50% of macrophages in most organs, other than brain, were labeled with a conditional reporter driven by the *Flt3* promoter, which is expressed in HSCs. Neither *Flt3* mRNA nor the conditional reporter was detected in the yolk sac, and they were relatively low even in fetal liver. Sheng et al. (50), using a conditional reporter gene based on the *Kit* locus, which is expressed only in HSCs, also concluded that the large majority of tissue macrophages derive from definitive progenitors. On the basis of these data, one would conclude that definitive marrow-derived macrophages gradually replace those of embryonic origin. The most recent contribution to this emerging field made the interesting observation that the yolk sac-derived macrophages could be ablated with injection of antibody to macrophage colony-stimulating factor receptor (CSF-1R) into the embryo, and they were apparently replaced by distinct c-MYB-dependent fetal liver-derived monocytes by the end of gestation (51). These authors used inducible lineage trace markers to support the view that the large majority of tissue macrophages are derived from fetal liver-derived monocytes that seed tissue in the embryo and then self-renew. They reiterate the view that microglia are derived exclusively from the yolk sac macrophage population. However, in mice treated with anti-CSF-1R the microglia were repopulated, presumably from the monocytic source, by the time of birth. This finding may relate to much earlier observations, using outbred mice, that monocytes entered the brain, and especially the retina, and clearly transdifferentiated into microglia, around the time of birth (52, 53).

HOMEOSTASIS AND THE MACROPHAGE NICHE

As noted above, the distribution of macrophages in tissues is very regular; individual cells occupy a specific niche and seldom overlap processes with each other. Each of the models used to infer the origins and turnover of tissue macrophages in mice makes the assumption that the methods used do not disturb the steady state, and this assumption may also not be valid. The proliferation and differentiation of macrophages is controlled by CSF-1R, which is activated by two ligands, macrophage colony-stimulating factor (CSF-1) and interleukin-34 (IL-34) (7, 54). Administration of CSF-1 to mice promoted expansion of the monocyte pool and increased monocyte infiltration into tissues and tissue-resident macrophage proliferation (55–57). CSF-1 is cleared from the circulation through receptor-mediated endocytosis by the CSF-1 receptor, involving both the macrophages of the liver and spleen (58) and blood monocytes (29). Accordingly, blockade of the CSF-1R with a monoclonal antibody causes a massive increase in circulating CSF-1 (59). The central role of CSF-1 in macrophage homeostasis is evident from the CSF-1-deficient osteopetrotic ($Csf1^{op/op}$) mouse and toothless rat ($Csf1^{tl/tl}$), which have gross deficiencies of tissue macrophages and many pleiotropic consequences of that deficiency (60). Treatment with anti-CSF-1R antibody also depletes most tissue macrophage populations in adult mice, indicating that CSF-1 dependence is maintained throughout life. Anti-CSF-1R does not deplete monocytes (59), nor does it impact on monocytopoiesis (61), but it does prevent differentiation to form the $Ly6C^{lo}$ population (29, 59).

Based on the available data, there is an intrinsic CSF-1/CSF-1R feedback loop that controls monocyte numbers and differentiation, monocyte recruitment, and tissue macrophage numbers through the local and systemic availability of CSF-1. When an individual macrophage niche becomes vacant, CSF-1 is no longer consumed and the local CSF-1 concentration rises to promote either local proliferation or recruitment of blood monocytes. Studies of development and turnover of mononuclear phagocytes need to take account of this homeostatic loop. Inducible Cre reporters, bone marrow transplantation, parabiosis, and monocyte depletion/mutations are all very likely to alter the homeostatic balance by altering the availability of CSF-1 (17). For example, the estrogen analog tamoxifen is commonly used in lineage trace experiments. There is evidence that CSF-1 expression is regulated by estrogen and is elevated in ovariectomized mice and in pregnancy (62, 63). Furthermore, hematopoietic progenitor cells respond directly to estrogen, which modulates their differentiation in the presence of myeloid growth factors (64). When embryos are pulsed with tamoxifen to induce recombinase activity, the same treatment may induce or repress CSF-1 in the embryo or the mother and alter the relative contribution of yolk sac phagocytes. Once a macrophage niche has been occupied by a yolk sac-derived progenitor, it may be less available to cells of definitive origin. Similarly, the lack of any impact on tissue macrophage numbers when monocytes are depleted (for example, in MYB or CCR2 knockout mice) could be explained by the increased availability of circulating CSF-1 (29), leading to the compensatory proliferation of the tissue-resident macrophages.

TRANSCRIPTOMIC ANALYSIS OF THE RELATIONSHIP BETWEEN MACROPHAGES AND DENDRITIC CELLS

The functional definition of a macrophage is in the name, "big eater," coined by Elie Metchnikoff (65). One might expect that the set of genes required to make up a professional phagocyte would be coexpressed in macrophages, and this might also provide clues to the function of genes for which there is poor annotation. The principle of guilt by association has been confirmed many times in analysis of large data sets. Genes that are coexpressed at the mRNA level commonly encode protein products that participate in the same pathway or process (66–68). Since the pioneering efforts of Su et al. (69) to generate the SymAtlas (now BioGPS; http://biogps.org) from sets of microarray data from mouse and human tissues, there have been numerous gene expression "atlases" across multiple tissues and within tissues across cell types and developmental time. The network tool BioLayout *Express*^{3D} was developed to allow the visualization of coexpression relationships in large data sets (70, 71) and was used to dissect the mouse BioGPS data set (67) and subsequently in a meta-analysis of publicly available mouse (72) and human data (73) relating to hematopoietic differentiation and macrophage biology and in a preliminary pig expression atlas (66). BioLayout analysis was used to identify a large set of coexpressed genes that were most highly expressed in phagocytes and that encoded proteins associated with lysosomes, including all of the components of the vacuolar ATPase proton pump and lysosomal hydrolases. This lysosomal/endocytosis cluster of coexpressed genes included those encoding transcription factors such as PU.1 and C/EBP (CCAAT/enhancer-binding protein) that likely contribute to transcriptional control in macrophage differentiation, and indeed can drive differentiation to a phagocyte phenotype when expressed in fibroblasts (74). The promoters of these phagocyte-enriched genes, in common with those of known human lysosomal proteins (75), contained purine-rich motifs, binding sites for the macrophage transcription factor PU.1, and the recognition motifs for basic helix-loop-helix transcription factors of the microphthalmia transcription factor family: MITF, TFEB, TFEC, and TFE3. All four MITF family members are expressed in macrophages. TFEC is a macrophage-specific transcription factor and itself a PU.1 target gene (76). MITF interacts both physically and genetically with PU.1 (34) and can transactivate the promoter of the *ACP5* lysosomal enzyme gene (34).

By contrast to the clear coexpression of phagocyte-specific genes, the genes encoding surface antigen markers, recognized by monoclonal antibodies, that are commonly used to divide monocytes and macrophages into subpopulations, were not stringently coexpressed with either phagocytic function or with each other. The gene encoding the well-studied macrophage marker EMR1 (F4/80), for example, was in a very small coexpression cluster with the gene for a transcription factor, MAFB, that was subsequently shown to regulate its expression (77, 78). High expression of F4/80 was proposed as a marker for tissue macrophages that derive from yolk sac progenitors (44). The clear lack of correlation between surface markers is important because it means that we cannot predict the gene expression or surface antigen profile of individual cells based on the presence of any marker. Each of the proteins we regard as surface markers has a function and has its own intrinsic regulation.

One of the most prevalent uses of surface markers is the separation of macrophages from dendritic cells (DCs) and the segregation of both populations into subpopulations. The activation of T lymphocytes requires the presentation of the antigen on major histocompatibility complex (MHC) molecules on the surface of an antigen-presenting cell (APC). For some groups, the term "dendritic cell" has been merged with "antigen-presenting cell." The problem with this view is that antigen presentation is a regulated pathway, and the genes involved are not correlated with any other cellular function. For some significant time, the integrin CD11c (ITGAX) was considered a DC marker in mice, even though it was clearly expressed by many tissue macrophage populations (6, 79). In the cluster analysis of mouse BioGPS data, CD11c was not coexpressed with any other marker (67).

The original description of the classical DC by Steinman and Cohn made the clear distinction that "unlike macrophages, they do not appear to engage in active endocytosis" (80). Analysis of large transcriptomic data sets reveals that the *a priori* definition of cells as DCs currently groups together cells with very distinct transcriptional profiles (6, 17, 79, 81, 82). The Immunological Genome Consortium (ImmGen) produced transcriptomic profiles from a large number of different myeloid cell populations isolated from multiple mouse tissues based on cell surface marker expression, and proposed distinct expression signatures unique to DCs and macrophages (83, 84). Their analysis concluded that MHC class II (MHC-II) and the transcriptional regulator *CIITA* were part of a DC signature, which was predictable given the selection of MHC-II-negative macrophages for comparison. A separate analysis of the entire data set failed to identify coexpressed genes that correlated with the *a priori* definition of cells as macrophages or DCs (85). Even within this large myeloid-specific data set, there was no correlation between any of the cell surface receptors, such as CD11b, CD11c, F4/80, MHC-II, CD64, CSF-1R, and FLT3, that are commonly used to classify subpopulations or cell types of macrophages and DCs. As discussed in detail previously (6), this lack of correlation between markers is supported by both fluorescence-activated cell sorter profiling and immunohistochemistry, where the number of subpopulations that can be defined is a function of the number of markers examined.

With respect to the macrophage-DC distinction, meta-analysis of large microarray data sets from mouse and human (72, 73) produced the same conclusion as analysis of the ImmGen data; cells annotated as DCs clearly segregated into two groups. As one might have predicted based on the original definition of the DC, the expression cluster that divided all macrophage/DC-related networks into two clear classes was the cluster containing lysosomal/endocytosis genes described above. The classical DCs isolated as defined by Steinman had very low expression of all of the genes in this cluster, and lower expression of the putative regulators (PU.1, C/EBP, and MITF family), whereas there was a separate class of cells that included monocyte- and bone marrow-derived cells described as "DCs" and many MHC-II-positive myeloid cells from tissues, which were clearly capable of being active phagocytes based on their expression of the large endocytic cluster (67, 72, 81).

The only transcriptional signature associated with APC activity in both macrophages and "DCs" is a very small one, including MHC-II, CD74, and the regulator CIITA. Several recent reviews (82, 86, 87) have considered the transcriptional control in the development of the classical DC, and the BioGPS and ImmGen data confirm the separation of these cells from macrophages. The data also confirm some of the lineage-restricted transcription factors in mice that distinguish the classical, FLT3-dependent DC from the macrophage (86). The transcription factors that most clearly define classical DCs, such as BATF3 and ZBTB46, are transcriptional repressors. One of the functions of these transcription factors could be to block macrophage differentiation and expression of endocytic function (88).

The ImmGen data were also used to define a number of surface markers that positively identify "macrophages" as distinct from "DCs," including the Fc receptor CD64 and the signaling molecule MERTK (83). These two markers did not correlate with each other in our reanalysis of the ImmGen data. CD64 is a direct target of CSF-1 signaling in mice and has been considered a marker, alongside CD163, of M2 polarization (89). CD64 is absent from elicited peritoneal macrophages, whereas MERTK is very highly expressed and further induced by lipopolysaccharide (LPS) (http://biogps.org). Therefore, like CD11c discussed above, their expression is state specific, and they cannot be considered as markers for a definitive "macrophage" or "DC." The shortcomings of CD64 as a marker are illustrated by a study of dermal myeloid cells. CD64 was used to distinguish CD11b$^+$ macrophages from CD11b$^+$ DCs in the dermis, leading to the conclusion that only ~10% of the CD11b$^+$ myeloid cells in the dermis are macrophages (90, 91). The CD11b$^+$ "DCs" were monocyte derived and expressed most other macrophage "markers" (e.g., F4/80, CD68, and lysozyme) (90, 91). Alongside Langerhans cells of the epidermis, the vast majority of dermal myeloid cells expressed a *Csf1r*-EGFP (enhanced green fluorescent protein) transgene and were completely ablated by treatment of mice with anti-CSF-1R antibody (59). A recent proposal for a unified nomenclature that separates macrophages and DCs based on ontogeny and growth factor dependence would clearly separate the CD11b$^+$ DCs in skin, which are CSF-1 dependent, from classical DCs, which depend on FLT3 ligand (92). This distinction would also separate the putative classical DC subclasses in mouse, cDC1 and cDC2, which differ in expression of the chemokine receptor XCR1 (GPR5) and CD8 (86). Within the BioGPS data, the CD8$^+$ DCs express *Gpr5* at high levels and lack *Csf1r*. Plasmacytoid DCs also lack *Csf1r* mRNA. Conversely, the CD8$^-$ DCs lack GPR5 and expressed both *Csf1r* and *Flt3*, and detectable EMR1 (F4/80). It is not clear that the markers are expressed on the same cells within this population. What is clear is that the definition of cDC2s as distinct from macrophages is rather more contentious (86).

The separation of cell populations based on cell surface markers is clearly a mainstream technology in immunology. Objections to the use of markers are not solely semantic. The core problem lies with the concept of a marker. A marker is only useful if it actually predicts function. Markers expressed by T lymphocytes such as CD3, CD4, and CD8 are informative, and predictive, because they are required for the function of the cells in which they are expressed, and they are, in fact, very well correlated with expression of genes required for lineage-specific function. The endosome/lysosome gene cluster discussed above is also clearly linked to function. Any one of the genes present within that cluster has a predictive value as a marker; a cell that expresses it is likely to be phagocytic. On the other hand, CD11b and CD11c encode complement receptors that have no function in antigen presentation. They are clearly likely to be required for complement receptor function, but there is no reason *a priori* to suspect that they will be highly expressed in APCs. The obvious counterargument is that markers are useful; they can be used to purify populations of cells that are enriched for a particular function of interest. However, the expression data tell us that such populations are intrinsically heterogeneous.

GENETIC VARIATION AND MACROPHAGE BIOLOGY

Like the studies of macrophage ontogeny above, the ImmGen and BioGPS data sets and much of the literature on antigen presentation and DC/macrophage divergence derive from the C57BL/6 mouse. In mice, many tissue macrophages, and cells grown in CSF-1, lack MHC-II, whereas in most other species, including humans and pigs, monocytes and monocyte-derived macrophages (MDMs) are strongly MHC-II positive. Furthermore, the C57BL/6 mouse has only one MHC-II locus since H2-Ea is absent. This locus is inducible by LPS in macrophages from BALB/c mice (93). C57BL/6 macrophages do not express cathepsin E, which is needed for antigen processing (93). A gene promoter polymorphism in the arginine transporter, *Slc7a2*, means that C57BL/6 mice have very different rates of arginine metabolism than other strains (94). At least 65 genes distinguish the transcriptome of C57BL/6 macrophages absolutely from BALB/c macrophages (93), including the C1q components proposed as unique markers for microglia (95), which were much more highly expressed and inducible in C57BL/6 than in BALB/c mouse macrophages (93). Strain-specific variation in gene expression is also reflected in allele-specific methylation patterns in macrophages from an F1 cross between these strains (96), and there is similar diversity between other mouse strains (97, 98). This means that a view of the MPS or the macrophage transcriptome based on the biology of the C57BL/6 mouse may not be generalizable to all mice, let alone to other species.

TRANSCRIPTIONAL REGULATION AND MACROPHAGE DIFFERENTIATION

Whether or not tissue macrophages derive from definitive progenitors in the marrow, and must transit through a blood monocyte precursor, it is possible to isolate committed progenitor cells from bone marrow that express receptors for hematopoietic growth factors, and to use those growth factors to generate relatively pure populations of macrophages (in CSF-1), classical DCs (in FLT3L), or macrophages with APC activity (in granulocyte-macrophage CSF [GM-CSF]; otherwise called bone marrow-derived DCs). Some of the underlying transcriptional regulation of lineage commitment in mice has been inferred from large-scale transcriptomic profiling of purified progenitor cells (99–102). Similarly, there have been detailed transcriptomic studies of the differentiation of human blood monocytes to mature macrophages, or DCs, in the presence of CSF-1 or GM-CSF (103, 104). The process of macrophage differentiation can also be modeled to some extent using myeloid leukemia cell lines. The FANTOM4 (Functional Annotation of the Mammalian Genome) consortium used the THP-1 human monocytic leukemia line to study the cascade of transcriptional events associated with macrophage differentiation (105). Genome-scale 5′ rapid amplification of cDNA ends (5′ RACE; also known as cap analysis of gene expression [CAGE]) was used in parallel with microarrays to identify changes in promoter activity with time and to infer the transcription factors involved. One of the most interesting findings was the rapid downregulation of a large cohort of transcriptional regulators, exemplified by c-MYB, and the finding that small interfering RNA-mediated downmodulation of those factors was sufficient to drive the differentiation process (105).

The largest data sets available for human macrophage differentiation have come from the FANTOM5 consortium, which also utilized CAGE to identify the sets of promoters and enhancers utilized by hundreds of different cells and tissues (106, 107). These data sets include multiple samples of different human myeloid lineages, including multipotent progenitors ($CD34^+$ stem cells), common myeloid progenitors (CMPs), and granulocyte-macrophage progenitors (GMPs), three monocyte subsets defined by CD14 and CD16 expression, MDMs (grown in CSF-1), monocyte-derived “DCs” (MDCs; grown in GM-CSF plus IL-4), Langerhans cells (LC) and migratory DCs from skin lymphatics. Although there are subtle differences in experimental systems and purification protocols, the data complement and parallel two large data sets generated by the Blueprint consortium, one of which profiled transcriptional regulation in isolated human progenitor cells (99), and another which detailed the human monocyte response to CSF-1 and tolerization with LPS (108).

The expression profiles of myeloid lineage cells have been analyzed separately from the complete FANTOM5 data set (109) and data access provided on a separate portal at http://www.myeloidome.roslin.ed.ac.uk. This portal enables visualization of the transcription start sites and expression profiles in this set of myeloid cell types of any gene of interest, using the genome browser. It also summarizes expression profiles of long noncoding RNAs and microRNAs and cell type enhancers that correlate in their expression with promoters in the same genomic region.

Figure 1 shows the expression levels of key growth factor receptors in a macrophage-DC differentiation series extracted from the human FANTOM5 gene expression data. The combined total expression of these five growth factors was remarkably similar across all cell types, but the proportion of total expression contributed by each gene varied considerably. Consistent with the identity of the samples, *KIT* (receptor for stem cell factor), *FLT3*, *CSF2RB* (GM-CSFR), and *CSF3R* (G-CSFR) are expressed in progenitors, the CD34$^+$ multipotent stem cells (MSCs), the CMPs, and the GMPs. *Csf1r* is first expressed in the GMPs. Outside of the progenitor population, *Csf1r* and *FLT3* show opposite expression patterns; the classical DCs express *FLT3* and all other populations, including monocyte-derived "DCs" and Langerhans cells of the skin (sometimes regarded as DCs), express *Csf1r*. As expected, given the clinical use of G-CSF for stem cell mobilization, *CSF3R* is expressed in the CD34$^+$ progenitors, but interestingly, its expression is retained in CD14$^+$ monocytes and downregulated with maturation in response to CSF-1, varying inversely with *Csf1r* mRNA. The expression of *CSF3R* in monocytes and their progenitors may be related to the finding that G-CSF treatment of humans mobilizes a population of monocytes that retain expression of CD34 (110).

As we concluded from a meta-analysis of published human microarray data (73), the transcriptional

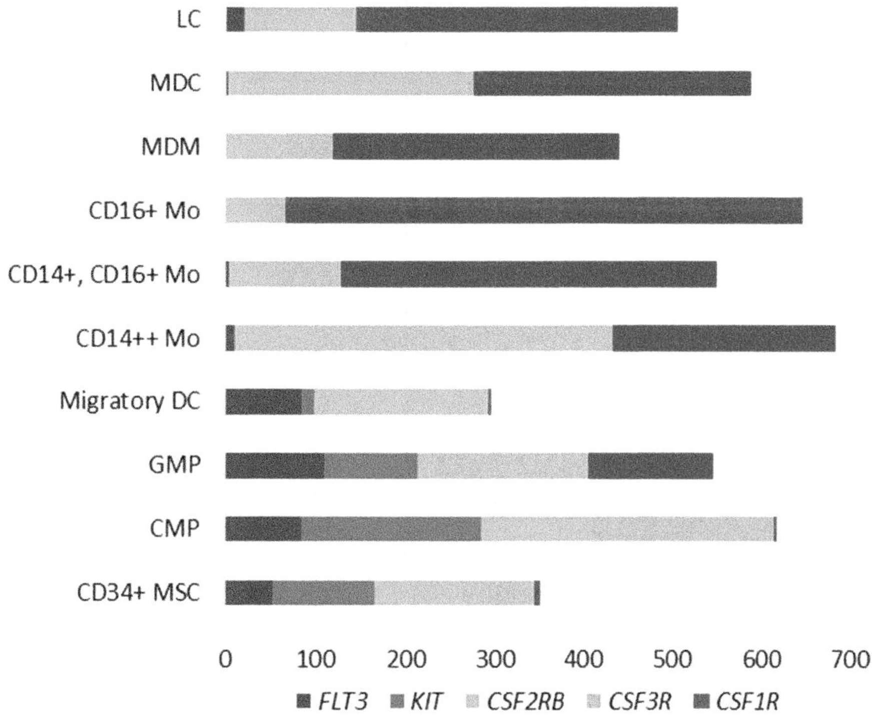

Figure 1 Expression of selected genes encoding myeloid-restricted growth factor receptors. Stacked bars show expression of each gene in the cell type (normalized tags per million) derived from FANTOM5 CAGE data for human cells (107). Cell types are presented in the order of maturation: CD34$^+$ MSCs; CMPs; GMPs; migratory DCs; CD14^{++} monocytes (CD14^{++} Mo); CD14$^+$CD16$^+$ monocytes (CD14$^+$, CD16$^+$ Mo); CD16$^+$ monocytes (CD16$^+$ Mo); MDMs (cultured in CSF-1); monocytes cultured in GM-CSF (MDCs); and migratory DCs from skin lymphatics (LC).

profiles of monocyte-derived populations grown in different growth factors clustered together, first and foremost because of their shared expression of phagocytic genes and some endocytic receptors (109). Many transcripts were more highly expressed in monocyte-derived macrophages grown in CSF-1 (*ACP5*, *CD163*, *C1Q*, tetraspanins, *IGF1*, *VEGFB*, scavenger receptors, and *CHI3L1*) or the monocytes grown in GM-CSF (*CD1A*, *-B*, *-C*, and *-E*; *DCSTAMP*; *CLEC4A*; *CLEC10A*; *CLEC4G*; and *TNFRSF11A*, also known as *RANK*). As in other data sets discussed above, the MHC-II genes were separately clustered because they are also expressed by monocytes and their committed progenitors. As expected, the monocyte-derived "DC" cells, grown in GM-CSF, maintained higher levels of the MHC-II cluster, whereas these genes were somewhat repressed in the monocytes grown in CSF-1.

The three human monocyte populations in the FANTOM5 data set were also subjected to detailed chromatin analysis, providing global maps of H3K4me1 (promoter) and H3K27ac (enhancer) locations that largely confirmed the CAGE data (111). The CAGE-based methodology also permitted identification of regulated enhancer activity, enabling, for example, the identification of active enhancers responsible for the differential expression of *CD14* in the monocyte subpopulations. Over all, the FANTOM5 data supported the view that the monocytes are a CSF-1-dependent differentiation series, with the intermediate monocytes showing intermediate expression of the vast majority of the hundreds of genes that showed differential expression between the two extremes of monocyte phenotype. This pattern includes *Csf1r* itself, which is 3-fold elevated in $CD16^{++}$ monocytes. The one exception, consistent with findings of others (reviewed in references 10–12), is ~2-fold elevation of a subset of MHC-II (HLA-DP, HLA-DQ, and HLA-DR) transcripts in the intermediate cells relative to $CD14^{++}$ and $CD16^{++}$ extremes, despite a progressive decline in expression of the regulator CIITA. If monocytes are a differentiation series, the selective increase in the intermediate population reported in many clinical settings probably represents a change in transit time between the states.

Aside from CD14 and CD16, two other markers have been proposed to provide functional delineation of subsets of human monocytes: TIE2 and SLAN (10, 12). The former encodes a tyrosine kinase receptor for angiopoietin-2, but it was not detected in any of the monocyte preparations in FANTOM5, where it was highly expressed as expected in endothelial cells. SLAN is an antigen formed by a novel O-linked 6-sulfo-LacNac modification of P-selectin glycoprotein ligand (PSGL1). PSGL1 is itself around 2 times more highly expressed in $CD16^{++}$ monocytes than $CD14^{++}$ monocytes. However, the enzyme most likely to be required for the modification, carbohydrate sulfotransferase 2, was undetectable in $CD14^{++}$ monocytes, induced in intermediate cells, and highly expressed in $CD16^{++}$ cells. One other marker, CD43, has been used in other species as a subset marker (10, 12), and the FANTOM5 data demonstrate that it is likely to provide a similar discrimination of the $CD16^{++}$ subset in humans.

The analysis of monocyte subsets by FANTOM5 identified the coregulation of genes required for glycolytic metabolism and the metabolic burst in the classical monocytes (111). Conversely, the $CD16^{++}$ monocytes were apparently more committed to oxidative metabolism and mitochondrial energy generation. It may be that the more mature, long-lived monocytes adapt with time to the aerobic environment of the bloodstream, where the "inflammatory" monocytes are adapted to enter the relatively low-oxygen environments of tissues and inflammatory sites. Among the genes strongly downregulated in the $CD16^{++}$ monocytes is the hypoxia-inducible transcription factor *HIF1A* and the glucose transporter *GLUT3* (*SLC2A3*).

About two-thirds of the annotated transcription factors in the genome can be detected in some state of differentiation or activation of the macrophage lineage. Most do not vary greatly in their expression across myeloid lineages. About 100 transcriptional regulators showed significant variation with differentiation in the populations shown in Fig. 1. An analysis of the monocyte cell types based on expression of 108 transcription factor and growth factor receptor genes was performed using BioLayout *Express*^{3D} (70). Eleven coexpression clusters containing three or more genes were identified. The patterns of these 11 clusters are shown in Fig. 2, and their profiles are summarized in Table 1. The largest cluster consisted of the set of transcription factor genes that was expressed in progenitors and downregulated with differentiation. All of these genes were highly expressed in C57BL/6 mouse progenitor cells and downregulated with differentiation (http://biogps.org). Many of them were also expressed constitutively by THP1 cells and downregulated in response to phorbol myristate acetate-induced differentiation (105).

Among these coregulated transcription factors, clusters 3 and 6 were expressed most highly in the classical monocytes and declined with maturation to the $CD16^{+}$ subset, and still further in response to differentiation *in vitro*. Conversely, cluster 4 included transcription factors such as *CEBPB*, *MAFB*, and *TFEB* that increased with monocyte differentiation and still

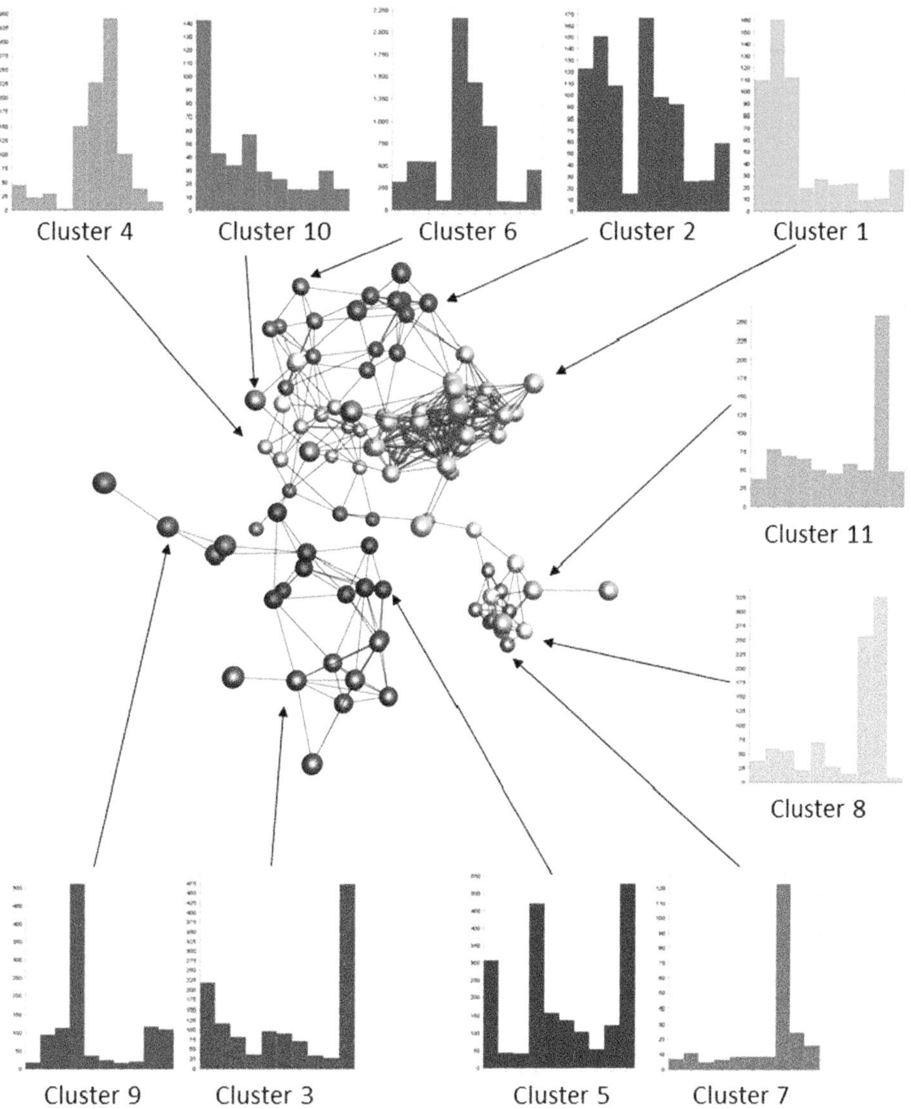

Figure 2 Network layout of 108 growth factor receptor and transcription factor genes in myeloid lineages. Nodes represent genes and edges correlation between expression patterns of genes at a Pearson correlation coefficient of 0.74 or greater. Nodes of the same color form a cluster. Histograms show the average expression pattern of genes within the cluster. *x* axis, cell type. Each column represents one cell type, presented in the order of maturation: CD34$^+$ mesenchymal stem cells; CMPs; GMPs; migratory DCs; CD14^{++} monocytes; CD14$^+$CD16$^+$ monocytes; CD16$^+$ monocytes; MDMs (cultured in CSF-1); monocytes cultured in GM-CSF; and migratory DCs from skin lymphatics. Column colors are the same as the nodes in the cluster. *y* axis, average expression of genes in the cluster (normalized tags per million) derived from FANTOM5 CAGE data for human cells (107).

further in response to CSF-1 *in vitro*. The results are broadly consistent with a recent review of monocyte differentiation by Huber et al. (112) but greatly extend the transcriptional network. In every case, the CD14$^+$CD16$^+$ cells have intermediate expression, again reinforcing the view that they are a differentiation intermediate between the CD14^{++}CD16$^-$ and CD14$^-$CD16^{++} subsets. A number of genes of interest have idiosyncratic expression patterns and are not included in clusters (not shown). For example, *BATF3*, the mouse DC regulator mentioned earlier, was expressed in human monocytes and increased markedly with matura-

Table 1 Coexpression of transcription factor genes in cells of myeloid lineages[a]

Cluster	Description	Transcription factor genes
1	High in precursors (FLT3⁺, KIT⁺)	*BACH2, DACH1, E2F1, E2F3, EGR1, ERG, ELF2, FOXP1, GFI1, GFIB, HOXA9, JUND, LMO2, MYB, MYC, MYCN, NFE2, PRDM8, RFX8, RUNX1, SOX12, TAL, TCF3, TCF4*
2	High in precursors and monocytes; down in MDMs, MDCs, and G-CSFR⁺	*ELF1(A), ETS2, FLI1, FRA2, KLF7, PER1, PER2, SMAD3*
3	Declines with maturation from MSCs through monocyte subsets to MDMs/MDCs; retained in LCs	*ATF3, FOSB, FRA1, IRF6, JUN, MAFF, NFATC1, NR4A1, NR4A2*
4	Highest in monocytes and MDMs, increased in CD16⁺ monocytes	*CEBPB, IRF7, KLF2, LMO2 (C), MAFB, NR1D1, POU2F2, TCF7L2, TFEB*
5	Highest in SCs, migratory DCs, and LCs; declines with monocyte maturation	*ARNT, BHLHE40, ETV3, ID2, NR4A3, PRDM1, REL, RELB, ZBTB46*
6	Highest in CD14⁺ monocytes, declines in CD16⁺, and suppressed in MDMs/MDCs	*CEBPD, FOS, HIF1A, JDP2, KLF4, MEF2C*
7	Induced in MDMs, low in all others	*BHLHE41, EGR2, ETV5, MAF, MITF, NR1H3 (LXRA), PPARD*
8	Induced in both MDMs and MDCs	*CREG1, PPARG, SNAI3, TFEC*
9	High in MDCs, also increased in MDCs and LCs	*ARNTL2, CIITA, IRF4, SPIB*
10	High in SCs, down in CMPs	*ETS1, SMAD7, SOX4*
11	High in MDCs (GM-CSFR⁺⁺)	*CEBPA, FOXQ1*

[a]Cluster numbers are derived from the BioLayout *Express*^{3D} analysis shown in Fig. 2. Transcription factors are those with high expression in the cluster. MDM, monocyte-derived macrophage (cultured in CSF-1); MDC, monocyte cultured in GM-CSF plus IL-4; LC, Langerhans cell. Migratory DCs are cells isolated from skin lymphatics.

tion to the CD16^{++} subset. It was maintained in monocytes grown in GM-CSF but ablated in those grown in CSF-1. In mice, the generation of the Ly6C^{-}, or nonclassical, monocyte subset depends on the CSF-1-responsive early response gene *Nr4a1* (*Nur77*) (113). The *Nr4a1* gene was a member of cluster 3, and it is likely that NR4A1 protein also performs some function in maturation in human monocytes.

CSf1R, THE ARCHETYPAL MACROPHAGE-SPECIFIC GENE

The CSF-1 receptor (*Csf1r*) locus has been studied in great detail to provide an understanding of the processes of macrophage differentiation. Expression of *Csf1r* mRNA is one of the earliest markers of macrophage lineage commitment from multipotent progenitors in mouse bone marrow, as well as the appearance of macrophages in the yolk sac (32, 114). In the mouse, CSF-1 was shown to direct myeloid lineage fate in single multipotent HSCs by activating the expression of transcription factor PU.1 (115). This finding appears incompatible with human and mouse transcriptomic data (both in FANTOM5 and in BioGPS), which indicate that *Spi1* (encoding PU.1) is already highly expressed in CD34^{+} MSCs, whereas *Csf1r* is first detectable in GMPs (see Fig. 1). Further, the TATA-less, macrophage-specific promoter of the *Csf1r* locus in both species contains multiple PU.1/ETS binding sites that are bound by PU.1 and required for maximal activity (116, 117), and PU.1 is required for *Csf1r* expression in progenitor cells (116). Hence, a model that places PU.1 downstream of CSF-1R signaling does not fit the available expression data. Tagoh et al. (114) purified mouse c-KIT-positive progenitors and confirmed that (i) they express c-MYB and PU.1 and (ii) they lack surface CSF-1R as well as *Csf1r* mRNA. These cells require a cocktail of IL-1, IL-3, and CSF-1 to drive macrophage differentiation. The critical event that permits the expression of *Csf1r* and lineage commitment is more likely loss of c-MYB rather than induction of PU.1 (118). Detailed studies of the chromatin structure of the mouse *Csf1r* locus during differentiation of progenitors (114) revealed the occupation of both purine-rich motifs and C/EBP binding sites in the *Csf1r* promoter in the immature precursors, consistent with expression of both PU.1 and CEBP family members in these cells. These findings indicated that regulatory elements elsewhere in the locus drive transcriptional activation of *Csf1r* transcription during differentiation. Multiple analyses of transcription factor binding site occupancy and epigenetic profiles in mouse and human provide a consistent view of potential regulatory elements in the *Csf1r* locus (104, 111, 119, 120), summarized in Fig. 3.

The purine-rich promoter architecture of *Csf1r* is shared by other myeloid-specific promoters, and basal activity requires cooperation between PU.1 and other ETS family transcription factors (121) as well as binding of Ewing sarcoma protein or the related FUS/TLS to the transcription start site (TSS) (122). Multiple ETS family members expressed in precursors and monocytes (ERG, ELF2, ELF1, ETS2, FLI1, ETV3, and ETV5; Table 1) are likely to fulfil the role of cooperating with PU.1 to activate purine-rich promoters. Conserved candidate enhancer sites upstream of the major macrophage *Csf1r* TSS, and within several introns, are also evident in chromatin analysis (104, 111, 119, 120), and the enhancer activity was confirmed based on bidirectional promoter activity in data from both human and mouse macrophages from the FANTOM5 consortium (106, 107). All mammalian *Csf1r* genes contain a conserved intronic enhancer referred as the Fms intronic regulatory element (FIRE) (123, 124). Tagoh et al. (114) demonstrated by *in vivo* footprinting that activation of mouse *Csf1r* transcription in c-KIT-positive precursors involves sequential occupation of multiple sites in the FIRE region. FIRE provides a remarkable cluster of highly conserved binding sites for many of the regu-

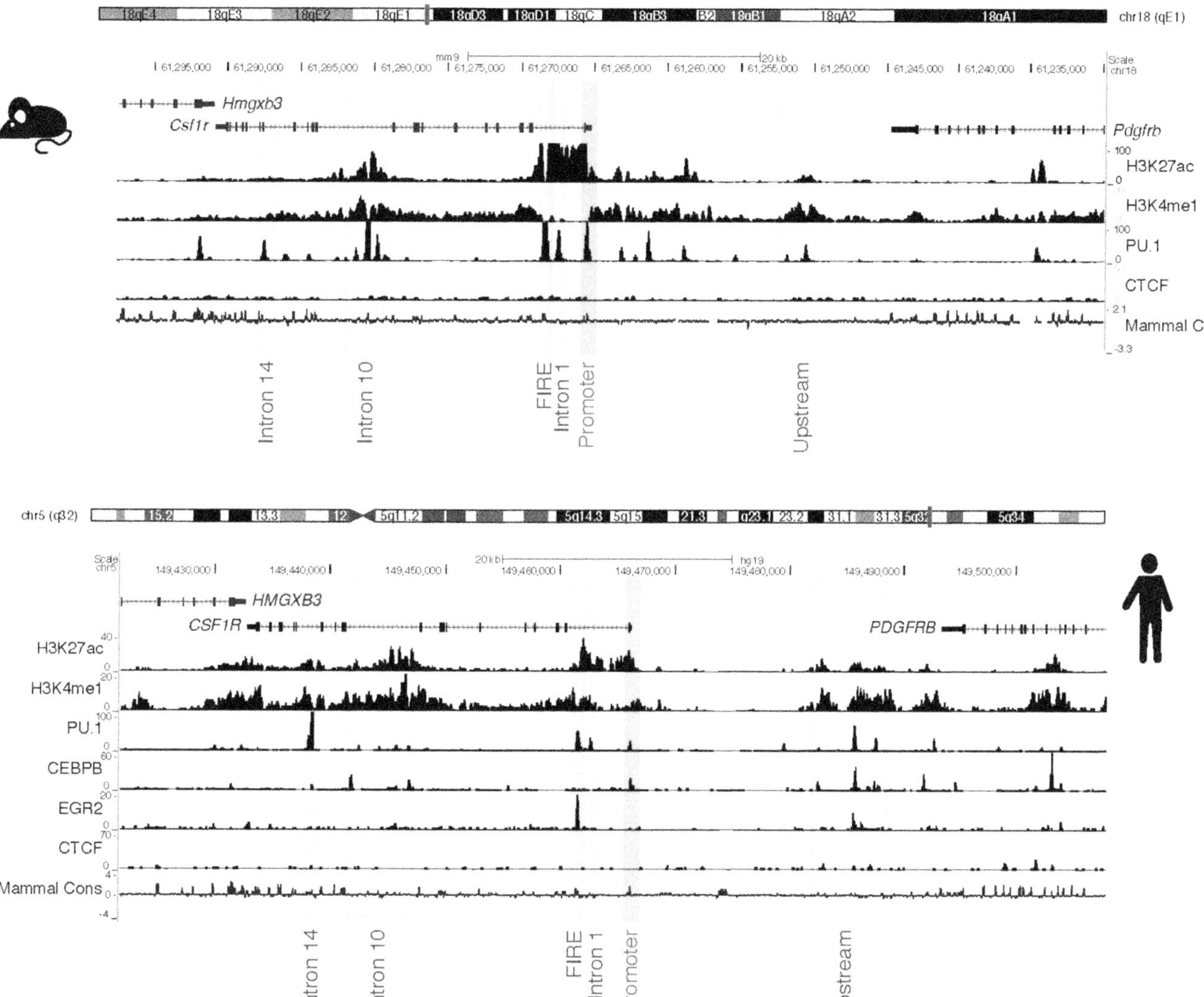

Figure 3 Chromatin architecture of the mouse and human *Csf1r* loci. Genome browser tracks of indicated histone modifications and transcription factors associated with enhancer elements are shown. The filled green box indicates the macrophage promoter. Boxes in blue identify intergenic and intragenic enhancer candidates. FIRE is represented by a filled blue box. Chip-Seq data sets that formed the basis of this figure for human macrophages are from derived from references 104 and 159. The mouse PU.1 track is derived from reference 136, and other mouse tracks from ENCODE.

lated transcription factors shown in Table 1, notably GC-rich elements likely bound by EGR1/EGR2 and KLF family members (104, 116, 125).

Csf1r reporter mice have been especially useful for studies of both CSF-1 and anti-CSF-1R treatments (55, 59, 126). A *Csf1r*-EGFP reporter containing the *Csf1r* promoter plus FIRE was detectable in all myeloid cells throughout the body, generating the MacGreen mouse (123) as well as constitutive and tamoxifen-inducible *Csf1r*-Cre transgenes used in lineage trace studies (44, 127). A 150-bp distal promoter element, containing a conserved AP1 site, is required for expression in mature macrophages and osteoclasts (128). The MacBlue transgenic line (*Csf1r*-ECFP [enhanced cyan fluorescent protein]) was generated using a promoter with this element deleted (129, 130). The reporter protein was expressed on macrophages as they appeared in the yolk sac and provided a striking picture of their abundance during embryonic development (30). Unexpectedly, the expression disappeared from the majority of tissue macrophages postnatally, whereas expression was retained in blood monocytes and their progenitors, and in the IL-34-dependent populations, the microglia and Langerhans cells in the skin (30). Even in the gut, where the macrophage population clearly derives from continuous renewal from monocytes (46), the *Csf1r*-ECFP transgene was extinguished in the majority of resident macrophages. The transgene therefore permits live imaging of monocyte trafficking (131).

Expression of the reporter genes based on the *Csf1r* locus requires the activity of FIRE (123), and the activity of FIRE is functionally conserved across mammalian species (132). FIRE is conserved also in the *Csf1r* locus of birds, and *Csf1r* reporter transgenic chickens have been produced in which all of the macrophages are labeled from the earliest yolk sac-derived cells (133). The chick provides an alternative model to study the origin of macrophages, and is the model in which the yolk sac-derived macrophages were first described (35, 134). Using transgenic reporters, we found that macrophages derived from the yolk sac are capable of extensive proliferation in response to CSF-1 *in vitro*, and in response to CSF-1 administration into the embryo. When transplanted to a nontransgenic recipient, they give rise to macrophages all over the body. However, these cells are not retained anywhere after hatching. By contrast, transplantation of cells from hatchling bone marrow into embryos prior to the onset of definitive hematopoiesis produced extensive macrophage chimerism in the embryo that was retained in mature adult birds, and which included regeneration of macrophage progenitors of donor origin in the marrow (36). We suggest that the trophic environment of the early embryo can only support the committed macrophage progenitors, perhaps similar to those of mouse (37). In keeping with the earliest findings using chick-quail yolk sac chimeras (134), these findings suggest that the contribution of the yolk sac to tissue macrophage populations is transient, and yolk sac-derived cells are replaced entirely by the progeny of definitive hematopoiesis.

TISSUE-SPECIFIC ADAPTATION AND ACTIVATION OF MACROPHAGES

Regardless of the embryonic origin, each population of tissue-resident macrophages adapts specifically to the environment in which it finds itself. Sample-to-sample comparison of the extensive ImmGen data set for the mouse demonstrated that each tissue macrophage population formed a distinct cluster (81). One recent study focused on the differentiation of microglia, the macrophages of the brain, and identified a marker set that distinguished these from blood monocytes and from tissue macrophages isolated from other locations (95). The BioGPS data, the FANTOM5 data, and the ImmGen data all include strongly validated microglial data sets. The former two confirm known microglial-enriched genes including *Aif1* (*Iba1*) and the scavenger receptors *Fcrls* and *Marco* but do not support others such as *C1q* and *Csf1r*, which are also highly expressed in other macrophage populations. A more extensive comparison combined expression profiling of isolated mouse tissue macrophage populations with extensive analysis of chromatin architecture (135, 136). Several of the genes that marked particular macrophage populations were transcription factors, notably *Sall1* and *Mef2c* in microglia, *Gata6* in peritoneal macrophages, *Lxra* in Kupffer cells, *Spic* in spleen, and *Runx3* in the gut (135). The expression of *Gata6* in mouse peritoneal macrophages appears to be required for their self-renewal and homeostasis; two groups separately reported on the select impact of myeloid-specific deletions of the gene in the resident peritoneal population (137, 138). The tissue-specific macrophage phenotypes correlated with the presence of H3K4Me1/2 enhancer profiles in their genomic vicinity. The role of environment in the acquisition of tissue-specific phenotypes was examined in bone marrow chimeras. The outcomes indicated not only that the tissue determines the phenotype but, *inter alia*, that monocytes derived from bone marrow progenitors can, and do, give rise to tissue-resident macrophages if the niche is available.

There is considerably less data available on tissue-specific macrophage development from other species.

However, the number of macrophages in most organs is sufficiently high that we can extract a macrophage signature without isolating the cells. Within the pig gene expression atlas, we profiled lung alveolar macrophages, blood monocytes, and bone marrow-derived macrophages alongside many different tissues (66). Comparative analysis revealed that alveolar macrophages express exceptionally high levels of a wide diversity of C-type lectin receptors (SIGLEC and CLEC family members) and Toll-like receptors (TLRs), presumably to deal with inhaled particles. This pattern is also evident in the many profiles of human alveolar macrophages that can be accessed through NCBI GEO. By contrast, in the wall of the gut we could detect a clear signature of the abundant lamina propria macrophage population, but the C-type lectins and TLRs were absent.

The apparent plasticity of tissue macrophages reflects, in part, their responsiveness to numerous different stimuli, some of which are tissue specific and others shared by locations or induced in response to environmental challenges. The generic term "activation" was originally applied to the ability of recruited macrophages to acquire microbicidal and tumoricidal activity in response to products of activated T cells. A number of groups have advocated subclassification of the activation states seen in recruited macrophages, broadly into M1 and M2 or classically activated and alternatively activated (5, 139–141). The M1 and M2 nomenclature links the state of activation of the macrophages to the activation of Th1 and Th2 lymphocytes, which in turn links them to the actions of gamma interferon (IFN-γ; originally known as macrophage-activating factor) (142) and IL-4. Interestingly, although it is not commonly regarded as a macrophage growth factor, IL-4 can promote macrophage proliferation in mice *in vivo* (57, 143).

The proposed M1/M2 dichotomy is not supported by genome-scale data (73). If there were coordinated M1 or M2 "regulons," we would expect to see correlated expression of the classical marker genes for the putative M1 and M2 phenotypes across different cellular states, and this was not observed. The utility of the M1/M2 concept across species is also not clear. The gene expression profiles of "classically activated" mouse, pig, and human macrophages are very different (144), with the pig being rather more human-like (145). Similarly, the IL-4-inducible alternatively activated profiles, and putative M2 markers, are poorly conserved between mouse and human. Transglutaminase 2 was proposed as the only well-conserved marker (146). Mosser and Edwards (147) suggested that macrophage activation may better be described as a spectrum, analogous to a color wheel. Xue et al. (148) confirmed the spectrum model with an extensive comparison of human monocyte-derived macrophages grown in CSF-1 or GM-CSF and exposed to numerous distinct stimuli including IFN-γ, IFN-β, IL-4/IL-13, IL-10, glucocorticoids, TLR agonists, tumor necrosis factor, and prostaglandin. While there was a broad dichotomy between IFN-γ (M1) and IL-4 (M2) directed states, addition of further stimuli segregated the transcriptional response into many separate modules, with 49 distinct coexpression clusters containing 27 to 884 genes per module (148). The simplistic view of macrophage activation that pervades the literature is based in part on simplified models in which individual agonists are studied individually, often at a single dose, a single time point, and in a single mouse strain and/or small subset of individuals. The problem with this approach is illustrated by several findings. IFN-γ is normally made by activated T cells alongside GM-CSF. As the sole stimulus, IFN-γ causes growth arrest, whereas when GM-CSF is present, it is a mitogen (149). On the matter of dose and mouse strain, the dose response to LPS differs at the single-gene level (some genes are induced only by higher doses), and profoundly between mouse strains (93, 98). And finally, in humans it has been shown that some 80% of genes in monocytes exhibit heritable differences in gene expression and there is a major divergence in the secondary induction of IFN-β target genes in response to LPS (150).

A secondary issue is the massive heterogeneity of gene expression in macrophages at the single-cell level. Single-cell analysis of LPS-inducible genes in macrophages demonstrated essentially bimodal variation between individual cells; genes are either induced or they are not (151). The authors in this case sought causal explanations for this variation based on covariance of transcription factor expression at the single-cell level. However, an alternative view is that transcriptional activation at the individual gene level is intrinsically probabilistic (152). Stochastic variation in gene expression may occur even at the single-allele level. Indeed, the LPS receptor, TLR4, is expressed from only one allele in individual cells, with an allele-counting mechanism similar to that of the X chromosome (153). This finding explains the semidominance of the *Tlr4* mutation in C3H/HeJ mice, since in heterozygotes 50% of cells express the nonfunctional protein. Conversely, when one allele is deleted, all cells express *Tlr4* (from the other allele). Furthermore, because of the complex feedback loops in stimulated cells, in which LPS rapidly induces inhibitors that block signaling and degrade induced mRNAs and proteins (154, 155), individual cells show an oscillating response over relatively short time frames, eventually reaching a new steady state (156).

Against this background of complexity and heterogeneity, Murray et al. (157) proposed a set of guidelines to be used when describing macrophage activation states. They propose that descriptions of macrophage activation in disease states accurately describe the system, the way populations were isolated, and then utilize markers to enhance the description. However, even this rigorous approach is confounded if each macrophage is unique, deploying the potential arsenal of host defense weaponry in distinct combinations that change with time. The plasticity and diversity of individual macrophages may be a necessary part of their innate immune function, enabling the generation of a combinatorial diversity that cannot readily be overcome by a single pathogenicity determinant. An alternative to cell-based models is to describe diseases in terms of interacting genes. For example, if one examines very large data sets of cancer expression arrays, it is possible to extract a set of coexpressed genes that includes many known phagocyte markers, and *inter alia* indicates that there is a common tumor-associated macrophage signature, regardless of tumor type, that fits neither M1 nor M2 profiles (85).

DATABASES, WEBSITES, AND THE FUTURE

The escalating amount of data on mononuclear phagocyte biology coming from genome-scale technologies now taxes the capacity of any individual to access all of the useful information about the regulation of their favorite gene. BioGPS is an example of a new era of more user-friendly portals, as is the Immunological Genome portal (http://www.immgen.org). We established the website http://www.macrophages.com as a community website for sharing access to macrophage-related genomic and other information (158), including the massive promoter-related data sets arising from the FANTOM projects. Macrophages.com also provides macrophage-related pathway annotation data and links to the growing InnateDB (http://www.innateDB.org), which curates molecular interactions among macrophage-expressed proteins. The website also contains a curated compendium of major reviews on macrophage biology and transcriptional regulation, many of which provide much more comprehensive coverage of subtopics than this brief overview. With completed genomes, we are seeing comparable data sets available for other species, including domestic pigs, chickens, sheep, and cattle, that will underpin more-rigorous studies of the evolution of innate immunity, and also the recognition that there is very substantial genetic variation within species that underlies disease susceptibility loci.

In summary, macrophages are a very diverse cell type, expressing a subset of surface markers and derived from both yolk sac and circulating monocytes. While useful information has been gained by studies of mice, primarily the C57BL/6 strain, these are not always generalizable to humans and other animals, and more-extensive studies of larger mammals such as pigs and sheep, as well as studies utilizing the flexibility and unprecedented developmental analysis of the chicken, will be needed to clarify the role, ontogeny, and differential function of the range of myeloid cells.

Acknowledgments. *The Roslin Institute is supported by Institute Strategic Programme Grants from the Biotechnology and Biological Sciences Research Council.*

Citation. Hume DA, Summers KM, Rehli M. 2016. Transcriptional regulation and macrophage differentiation. Microbiol Spectrum 4(3):MCHD-0024-2015.

References

1. **van Furth R, Cohn ZA.** 1968. The origin and kinetics of mononuclear phagocytes. *J Exp Med* **128:**415–435.
2. **Hume DA.** 2006. The mononuclear phagocyte system. *Curr Opin Immunol* **18:**49–53.
3. **Hume DA, Ross IL, Himes SR, Sasmono RT, Wells CA, Ravasi T.** 2002. The mononuclear phagocyte system revisited. *J Leukoc Biol* **72:**621–627.
4. **Geissmann F, Manz MG, Jung S, Sieweke MH, Merad M, Ley K.** 2010. Development of monocytes, macrophages, and dendritic cells. *Science* **327:**656–661.
5. **Gordon S, Taylor PR.** 2005. Monocyte and macrophage heterogeneity. *Nat Rev Immunol* **5:**953–964.
6. **Hume DA.** 2008. Differentiation and heterogeneity in the mononuclear phagocyte system. *Mucosal Immunol* **1:**432–441.
7. **Hume DA, MacDonald KP.** 2012. Therapeutic applications of macrophage colony-stimulating factor-1 (CSF-1) and antagonists of CSF-1 receptor (CSF-1R) signaling. *Blood* **119:**1810–1820.
8. **Wynn TA, Chawla A, Pollard JW.** 2013. Macrophage biology in development, homeostasis and disease. *Nature* **496:**445–455.
9. **Gordon S, Taylor PR.** 2005. Monocyte and macrophage heterogeneity. *Nat Rev Immunol* **5:**953–964.
10. **Wong KL, Yeap WH, Tai JJ, Ong SM, Dang TM, Wong SC.** 2012. The three human monocyte subsets: implications for health and disease. *Immunol Res* **53:**41–57.
11. **Ziegler-Heitbrock L.** 2014. Monocyte subsets in man and other species. *Cell Immunol* **289:**135–139.
12. **Ziegler-Heitbrock L.** 2015. Blood monocytes and their subsets: established features and open questions. *Front Immunol* **6:**423. doi:10.3389/fimmu.2015.00423.
13. **Ziegler-Heitbrock L, Ancuta P, Crowe S, Dalod M, Grau V, Hart DN, Leenen PJ, Liu YJ, MacPherson G,**

Randolph GJ, Scherberich J, Schmitz J, Shortman K, Sozzani S, Strobl H, Zembala M, Austyn JM, Lutz MB. 2010. Nomenclature of monocytes and dendritic cells in blood. *Blood* **116:**e74–e80. doi:10.1182/blood-2010-02-258558.

14. **Geissmann F, Jung S, Littman DR.** 2003. Blood monocytes consist of two principal subsets with distinct migratory properties. *Immunity* **19:**71–82.
15. **Bain CC, Mowat AM.** 2014. The monocyte-macrophage axis in the intestine. *Cell Immunol* **291:**41–48.
16. **Epelman S, Lavine KJ, Randolph GJ.** 2014. Origin and functions of tissue macrophages. *Immunity* **41:**21–35.
17. **Jenkins SJ, Hume DA.** 2014. Homeostasis in the mononuclear phagocyte system. *Trends Immunol* **35:**358–367.
18. **Zigmond E, Jung S.** 2013. Intestinal macrophages: well educated exceptions from the rule. *Trends Immunol* **34:** 162–168.
19. **Jakubzick C, Gautier EL, Gibbings SL, Sojka DK, Schlitzer A, Johnson TE, Ivanov S, Duan Q, Bala S, Condon T, van Rooijen N, Grainger JR, Belkaid Y, Ma'ayan A, Riches DW, Yokoyama WM, Ginhoux F, Henson PM, Randolph GJ.** 2013. Minimal differentiation of classical monocytes as they survey steady-state tissues and transport antigen to lymph nodes. *Immunity* **39:**599–610.
20. **Cros J, Cagnard N, Woollard K, Patey N, Zhang SY, Senechal B, Puel A, Biswas SK, Moshous D, Picard C, Jais JP, D'Cruz D, Casanova JL, Trouillet C, Geissmann F.** 2010. Human $CD14^{dim}$ monocytes patrol and sense nucleic acids and viruses via TLR7 and TLR8 receptors. *Immunity* **33:**375–386.
21. **Thomas G, Tacke R, Hedrick CC, Hanna RN.** 2015. Nonclassical patrolling monocyte function in the vasculature. *Arterioscler Thromb Vasc Biol* **35:**1306–1316.
22. **Ingersoll MA, Spanbroek R, Lottaz C, Gautier EL, Frankenberger M, Hoffmann R, Lang R, Haniffa M, Collin M, Tacke F, Habenicht AJ, Ziegler-Heitbrock L, Randolph GJ.** 2010. Comparison of gene expression profiles between human and mouse monocyte subsets. *Blood* **115:**e10–e19. doi:10.1182/blood-2009-07-235028.
23. **Hollox EJ, Hoh BP.** 2014. Human gene copy number variation and infectious disease. *Hum Genet* **133:** 1217–1233.
24. **Hussen J, Düvel A, Sandra O, Smith D, Sheldon IM, Zieger P, Schuberth HJ.** 2013. Phenotypic and functional heterogeneity of bovine blood monocytes. *PLoS One* **8:**e71502. doi:10.1371/journal.pone.0071502.
25. **Fairbairn L, Kapetanovic R, Beraldi D, Sester DP, Tuggle CK, Archibald AL, Hume DA.** 2013. Comparative analysis of monocyte subsets in the pig. *J Immunol* **190:**6389–6396.
26. **Yrlid U, Jenkins CD, MacPherson GG.** 2006. Relationships between distinct blood monocyte subsets and migrating intestinal lymph dendritic cells in vivo under steady-state conditions. *J Immunol* **176:**4155–4162.
27. **Burke B, Ahmad R, Staples KJ, Snowden R, Kadioglu A, Frankenberger M, Hume DA, Ziegler-Heitbrock L.** 2008. Increased TNF expression in $CD43^{++}$ murine blood monocytes. *Immunol Lett* **118:**142–147.
28. **Sunderkötter C, Nikolic T, Dillon MJ, Van Rooijen N, Stehling M, Drevets DA, Leenen PJ.** 2004. Subpopulations of mouse blood monocytes differ in maturation stage and inflammatory response. *J Immunol* **172:**4410–4417.
29. **Yona S, Kim KW, Wolf Y, Mildner A, Varol D, Breker M, Strauss-Ayali D, Viukov S, Guilliams M, Misharin A, Hume DA, Perlman H, Malissen B, Zelzer E, Jung S.** 2013. Fate mapping reveals origins and dynamics of monocytes and tissue macrophages under homeostasis. *Immunity* **38:**79–91.
30. **Sauter KA, Pridans C, Sehgal A, Bain CC, Scott C, Moffat L, Rojo R, Stutchfield BM, Davies CL, Donaldson DS, Renault K, McColl BW, Mowat AM, Serrels A, Frame MC, Mabbott NA, Hume DA.** 2014. The MacBlue binary transgene (csf1r-gal4VP16/UAS-ECFP) provides a novel marker for visualisation of subsets of monocytes, macrophages and dendritic cells and responsiveness to CSF1 administration. *PLoS One* **9:**e105429. doi:10.1371/journal.pone.0105429.
31. **Naito M, Yamamura F, Nishikawa S, Takahashi K.** 1989. Development, differentiation, and maturation of fetal mouse yolk sac macrophages in cultures. *J Leukoc Biol* **46:**1–10.
32. **Lichanska AM, Browne CM, Henkel GW, Murphy KM, Ostrowski MC, McKercher SR, Maki RA, Hume DA.** 1999. Differentiation of the mononuclear phagocyte system during mouse embryogenesis: the role of transcription factor PU.1. *Blood* **94:**127–138.
33. **Lichanska AM, Hume DA.** 2000. Origins and functions of phagocytes in the embryo. *Exp Hematol* **28:** 601–611.
34. **Luchin A, Suchting S, Merson T, Rosol TJ, Hume DA, Cassady AI, Ostrowski MC.** 2001. Genetic and physical interactions between microphthalmia transcription factor and PU.1 are necessary for osteoclast gene expression and differentiation. *J Biol Chem* **276:**36703–36710.
35. **Cuadros MA, Coltey P, Carmen Nieto M, Martin C.** 1992. Demonstration of a phagocytic cell system belonging to the hemopoietic lineage and originating from the yolk sac in the early avian embryo. *Development* **115:**157–168.
36. **Garceau V, Balic A, Garcia-Morales C, Sauter KA, McGrew MJ, Smith J, Vervelde L, Sherman A, Fuller TE, Oliphant T, Shelley JA, Tiwari R, Wilson TL, Chintoan-Uta C, Burt DW, Stevens MP, Sang HM, Hume DA.** 2015. The development and maintenance of the mononuclear phagocyte system of the chick is controlled by signals from the macrophage colony-stimulating factor receptor. *BMC Biol* **13:**12. doi:10.1186/s12915-015-0121-9.
37. **Hettinger J, Richards DM, Hansson J, Barra MM, Joschko AC, Krijgsveld J, Feuerer M.** 2013. Origin of monocytes and macrophages in a committed progenitor. *Nat Immunol* **14:**821–830.
38. **Swirski FK, Hilgendorf I, Robbins CS.** 2014. From proliferation to proliferation: monocyte lineage comes full circle. *Semin Immunopathol* **36:**137–148.

39. Yamamoto R, Morita Y, Ooehara J, Hamanaka S, Onodera M, Rudolph KL, Ema H, Nakauchi H. 2013. Clonal analysis unveils self-renewing lineage-restricted progenitors generated directly from hematopoietic stem cells. *Cell* **154:**1112–1126.

40. Ginhoux F, Greter M, Leboeuf M, Nandi S, See P, Gokhan S, Mehler MF, Conway SJ, Ng LG, Stanley ER, Samokhvalov IM, Merad M. 2010. Fate mapping analysis reveals that adult microglia derive from primitive macrophages. *Science* **330:**841–845.

41. Chorro L, Sarde A, Li M, Woollard KJ, Chambon P, Malissen B, Kissenpfennig A, Barbaroux JB, Groves R, Geissmann F. 2009. Langerhans cell (LC) proliferation mediates neonatal development, homeostasis, and inflammation-associated expansion of the epidermal LC network. *J Exp Med* **206:**3089–3100.

42. Hoeffel G, Wang Y, Greter M, See P, Teo P, Malleret B, Leboeuf M, Low D, Oller G, Almeida F, Choy SH, Grisotto M, Renia L, Conway SJ, Stanley ER, Chan JK, Ng LG, Samokhvalov IM, Merad M, Ginhoux F. 2012. Adult Langerhans cells derive predominantly from embryonic fetal liver monocytes with a minor contribution of yolk sac-derived macrophages. *J Exp Med* **209:** 1167–1181.

43. Guilliams M, De Kleer I, Henri S, Post S, Vanhoutte L, De Prijck S, Deswarte K, Malissen B, Hammad H, Lambrecht BN. 2013. Alveolar macrophages develop from fetal monocytes that differentiate into long-lived cells in the first week of life via GM-CSF. *J Exp Med* **210:**1977–1992.

44. Schulz C, Gomez Perdiguero E, Chorro L, Szabo-Rogers H, Cagnard N, Kierdorf K, Prinz M, Wu B, Jacobsen SE, Pollard JW, Frampton J, Liu KJ, Geissmann F. 2012. A lineage of myeloid cells independent of Myb and hematopoietic stem cells. *Science* **336:**86–90.

45. Hashimoto D, Chow A, Noizat C, Teo P, Beasley MB, Leboeuf M, Becker CD, See P, Price J, Lucas D, Greter M, Mortha A, Boyer SW, Forsberg EC, Tanaka M, van Rooijen N, García-Sastre A, Stanley ER, Ginhoux F, Frenette PS, Merad M. 2013. Tissue-resident macrophages self-maintain locally throughout adult life with minimal contribution from circulating monocytes. *Immunity* **38:**792–804.

46. Bain CC, Bravo-Blas A, Scott CL, Gomez Perdiguero E, Geissmann F, Henri S, Malissen B, Osborne LC, Artis D, Mowat AM. 2014. Constant replenishment from circulating monocytes maintains the macrophage pool in the intestine of adult mice. *Nat Immunol* **15:** 929–937.

47. Kierdorf K, Erny D, Goldmann T, Sander V, Schulz C, Perdiguero EG, Wieghofer P, Heinrich A, Riemke P, Hölscher C, Müller DN, Luckow B, Brocker T, Debowski K, Fritz G, Opdenakker G, Diefenbach A, Biber K, Heikenwalder M, Geissmann F, Rosenbauer F, Prinz M. 2013. Microglia emerge from erythromyeloid precursors via Pu.1- and Irf8-dependent pathways. *Nat Neurosci* **16:**273–280.

48. Gomez Perdiguero E, Klapproth K, Schulz C, Busch K, Azzoni E, Crozet L, Garner H, Trouillet C, de Bruijn MF, Geissmann F, Rodewald HR. 2015. Tissue-resident macrophages originate from yolk-sac-derived erythro-myeloid progenitors. *Nature* **518:**547–551.

49. Epelman S, Lavine KJ, Beaudin AE, Sojka DK, Carrero JA, Calderon B, Brija T, Gautier EL, Ivanov S, Satpathy AT, Schilling JD, Schwendener R, Sergin I, Razani B, Forsberg EC, Yokoyama WM, Unanue ER, Colonna M, Randolph GJ, Mann DL. 2014. Embryonic and adult-derived resident cardiac macrophages are maintained through distinct mechanisms at steady state and during inflammation. *Immunity* **40:**91–104.

50. Sheng J, Ruedl C, Karjalainen K. 2015. Most tissue-resident macrophages except microglia are derived from fetal hematopoietic stem cells. *Immunity* **43:**382–393.

51. Hoeffel G, Chen J, Lavin Y, Low D, Almeida FF, See P, Beaudin AE, Lum J, Low I, Forsberg EC, Poidinger M, Zolezzi F, Larbi A, Ng LG, Chan JK, Greter M, Becher B, Samokhvalov IM, Merad M, Ginhoux F. 2015. C-Myb$^+$ erythro-myeloid progenitor-derived fetal monocytes give rise to adult tissue-resident macrophages. *Immunity* **42:**665–678.

52. Hume DA, Perry VH, Gordon S. 1983. Immunohistochemical localization of a macrophage-specific antigen in developing mouse retina: phagocytosis of dying neurons and differentiation of microglial cells to form a regular array in the plexiform layers. *J Cell Biol* **97:** 253–257.

53. Perry VH, Hume DA, Gordon S. 1985. Immunohistochemical localization of macrophages and microglia in the adult and developing mouse brain. *Neuroscience* **15:** 313–326.

54. Nakamichi Y, Udagawa N, Takahashi N. 2013. IL-34 and CSF-1: similarities and differences. *J Bone Miner Metab* **31:**486–495.

55. Gow DJ, Sauter KA, Pridans C, Moffat L, Sehgal A, Stutchfield BM, Raza S, Beard PM, Tsai YT, Bainbridge G, Boner PL, Fici G, Garcia-Tapia D, Martin RA, Oliphant T, Shelly JA, Tiwari R, Wilson TL, Smith LB, Mabbott NA, Hume DA. 2014. Characterisation of a novel Fc conjugate of macrophage colony-stimulating factor. *Mol Ther* **22:**1580–1592.

56. Hume DA, Pavli P, Donahue RE, Fidler IJ. 1988. The effect of human recombinant macrophage colony-stimulating factor (CSF-1) on the murine mononuclear phagocyte system in vivo. *J Immunol* **141:**3405–3409.

57. Jenkins SJ, Ruckerl D, Thomas GD, Hewitson JP, Duncan S, Brombacher F, Maizels RM, Hume DA, Allen JE. 2013. IL-4 directly signals tissue-resident macrophages to proliferate beyond homeostatic levels controlled by CSF-1. *J Exp Med* **210:**2477–2491.

58. Bartocci A, Mastrogiannis DS, Migliorati G, Stockert RJ, Wolkoff AW, Stanley ER. 1987. Macrophages specifically regulate the concentration of their own growth factor in the circulation. *Proc Natl Acad Sci U S A* **84:** 6179–6183.

59. MacDonald KP, Palmer JS, Cronau S, Seppanen E, Olver S, Raffelt NC, Kuns R, Pettit AR, Clouston A, Wainwright B, Branstetter D, Smith J, Paxton RJ, Cerretti DP, Bonham L, Hill GR, Hume DA. 2010. An antibody against the colony-stimulating factor 1 recep-

tor depletes the resident subset of monocytes and tissue- and tumor-associated macrophages but does not inhibit inflammation. *Blood* **116:**3955–3963.

60. **Pollard JW.** 2009. Trophic macrophages in development and disease. *Nat Rev Immunol* **9:**259–270.
61. **Sudo T, Nishikawa S, Ogawa M, Kataoka H, Ohno N, Izawa A, Hayashi S, Nishikawa S.** 1995. Functional hierarchy of *c-kit* and *c-fms* in intramarrow production of CFU-M. *Oncogene* **11:**2469–2476.
62. **Yao GQ, Wu JJ, Troiano N, Zhu ML, Xiao XY, Insogna K.** 2012. Selective deletion of the membrane-bound colony stimulating factor 1 isoform leads to high bone mass but does not protect against estrogen-deficiency bone loss. *J Bone Miner Metab* **30:**408–418.
63. **De M, Sanford T, Wood GW.** 1993. Relationship between macrophage colony-stimulating factor production by uterine epithelial cells and accumulation and distribution of macrophages in the uterus of pregnant mice. *J Leukoc Biol* **53:**240–248.
64. **Carreras E, Turner S, Paharkova-Vatchkova V, Mao A, Dascher C, Kovats S.** 2008. Estradiol acts directly on bone marrow myeloid progenitors to differentially regulate GM-CSF or Flt3 ligand-mediated dendritic cell differentiation. *J Immunol* **180:**727–738.
65. **Gordon S.** 2008. Elie Metchnikoff: father of natural immunity. *Eur J Immunol* **38:**3257–3264.
66. **Freeman TC, Ivens A, Baillie JK, Beraldi D, Barnett MW, Dorward D, Downing A, Fairbairn L, Kapetanovic R, Raza S, Tomoiu A, Alberio R, Wu C, Su AI, Summers KM, Tuggle CK, Archibald AL, Hume DA.** 2012. A gene expression atlas of the domestic pig. *BMC Biol* **10:** 90. doi:10.1186/1741-7007-10-90.
67. **Hume DA, Summers KM, Raza S, Baillie JK, Freeman TC.** 2010. Functional clustering and lineage markers: insights into cellular differentiation and gene function from large-scale microarray studies of purified primary cell populations. *Genomics* **95:**328–338.
68. **van Dam S, Craig T, de Magalhães JP.** 2015. GeneFriends: a human RNA-seq-based gene and transcript co-expression database. *Nucleic Acids Res* **43**(Database issue):D1124–D1132.
69. **Su AI, Wiltshire T, Batalov S, Lapp H, Ching KA, Block D, Zhang J, Soden R, Hayakawa M, Kreiman G, Cooke MP, Walker JR, Hogenesch JB.** 2004. A gene atlas of the mouse and human protein-encoding transcriptomes. *Proc Natl Acad Sci U S A* **101:**6062–6067.
70. **Theocharidis A, van Dongen S, Enright AJ, Freeman TC.** 2009. Network visualization and analysis of gene expression data using BioLayout *Express*^{3D}. *Nat Protoc* **4:**1535–1550.
71. **Freeman TC, Goldovsky L, Brosch M, van Dongen S, Mazière P, Grocock RJ, Freilich S, Thornton J, Enright AJ.** 2007. Construction, visualisation, and clustering of transcription networks from microarray expression data. *PLoS Comput Biol* **3:**2032–2042.
72. **Mabbott NA, Kenneth Baillie J, Hume DA, Freeman TC.** 2010. Meta-analysis of lineage-specific gene expression signatures in mouse leukocyte populations. *Immunobiology* **215:**724–736.
73. **Mabbott NA, Baillie JK, Brown H, Freeman TC, Hume DA.** 2013. An expression atlas of human primary cells: inference of gene function from coexpression networks. *BMC Genomics* **14:**632. doi:10.1186/1471-2164-14-632.
74. **Feng R, Desbordes SC, Xie H, Tillo ES, Pixley F, Stanley ER, Graf T.** 2008. PU.1 and C/EBPα/β convert fibroblasts into macrophage-like cells. *Proc Natl Acad Sci U S A* **105:**6057–6062.
75. **Settembre C, Di Malta C, Polito VA, Garcia Arencibia M, Vetrini F, Erdin S, Erdin SU, Huynh T, Medina D, Colella P, Sardiello M, Rubinsztein DC, Ballabio A.** 2011. TFEB links autophagy to lysosomal biogenesis. *Science* **332:**1429–1433.
76. **Rehli M, Lichanska A, Cassady AI, Ostrowski MC, Hume DA.** 1999. TFEC is a macrophage-restricted member of the microphthalmia-TFE subfamily of basic helix-loop-helix leucine zipper transcription factors. *J Immunol* **162:**1559–1565.
77. **Moriguchi T, Hamada M, Morito N, Terunuma T, Hasegawa K, Zhang C, Yokomizo T, Esaki R, Kuroda E, Yoh K, Kudo T, Nagata M, Greaves DR, Engel JD, Yamamoto M, Takahashi S.** 2006. MafB is essential for renal development and F4/80 expression in macrophages. *Mol Cell Biol* **26:**5715–5727.
78. **Sarrazin S, Mossadegh-Keller N, Fukao T, Aziz A, Mourcin F, Vanhille L, Kelly Modis L, Kastner P, Chan S, Duprez E, Otto C, Sieweke MH.** 2009. MafB restricts M-CSF-dependent myeloid commitment divisions of hematopoietic stem cells. *Cell* **138:**300–313.
79. **Hume DA.** 2008. Macrophages as APC and the dendritic cell myth. *J Immunol* **181:**5829–5835.
80. **Geissmann F, Gordon S, Hume DA, Mowat AM, Randolph GJ.** 2010. Unravelling mononuclear phagocyte heterogeneity. *Nat Rev Immunol* **10:**453–460.
81. **Hume DA, Mabbott N, Raza S, Freeman TC.** 2013. Can DCs be distinguished from macrophages by molecular signatures? *Nat Immunol* **14:**187–189.
82. **Mildner A, Jung S.** 2014. Development and function of dendritic cell subsets. *Immunity* **40:**642–656.
83. **Gautier EL, Shay T, Miller J, Greter M, Jakubzick C, Ivanov S, Helft J, Chow A, Elpek KG, Gordonov S, Mazloom AR, Ma'ayan A, Chua WJ, Hansen TH, Turley SJ, Merad M, Randolph GJ, Immunological Genome Consortium.** 2012. Gene-expression profiles and transcriptional regulatory pathways that underlie the identity and diversity of mouse tissue macrophages. *Nat Immunol* **13:**1118–1128.
84. **Miller JC, Brown BD, Shay T, Gautier EL, Jojic V, Cohain A, Pandey G, Leboeuf M, Elpek KG, Helft J, Hashimoto D, Chow A, Price J, Greter M, Bogunovic M, Bellemare-Pelletier A, Frenette PS, Randolph GJ, Turley SJ, Merad M, Immunological Genome Consortium.** 2012. Deciphering the transcriptional network of the dendritic cell lineage. *Nat Immunol* **13:**888–899.
85. **Doig TN, Hume DA, Theocharidis T, Goodlad JR, Gregory CD, Freeman TC.** 2013. Coexpression analysis of large cancer datasets provides insight into the cellular phenotypes of the tumour microenvironment. *BMC Genomics* **14:**469. doi:10.1186/1471-2164-14-469.

86. Vu Manh TP, Bertho N, Hosmalin A, Schwartz-Cornil I, Dalod M. 2015. Investigating evolutionary conservation of dendritic cell subset identity and functions. *Front Immunol* 6:260. doi:10.3389/fimmu.2015.00260.

87. Merad M, Sathe P, Helft J, Miller J, Mortha A. 2013. The dendritic cell lineage: ontogeny and function of dendritic cells and their subsets in the steady state and the inflamed setting. *Annu Rev Immunol* 31:563–604.

88. Everitt AR, Clare S, Pertel T, John SP, Wash RS, Smith SE, Chin CR, Feeley EM, Sims JS, Adams DJ, Wise HM, Kane L, Goulding D, Digard P, Anttila V, Baillie JK, Walsh TS, Hume DA, Palotie A, Xue Y, Colonna V, Tyler-Smith C, Dunning J, Gordon SB, GenISIS Investigators, MOSAIC Investigators, Smyth RL, Openshaw PJ, Dougan G, Brass AL, Kellam P. 2012. IFITM3 restricts the morbidity and mortality associated with influenza. *Nature* 484:519–523.

89. Haegel H, Thioudellet C, Hallet R, Geist M, Menguy T, Le Pogam F, Marchand JB, Toh ML, Duong V, Calcei A, Settelen N, Preville X, Hennequi M, Grellier B, Ancian P, Rissanen J, Clayette P, Guillen C, Rooke R, Bonnefoy JY. 2013. A unique anti-CD115 monoclonal antibody which inhibits osteolysis and skews human monocyte differentiation from M2-polarized macrophages toward dendritic cells. *MAbs* 5:736–747.

90. Malissen B, Tamoutounour S, Henri S. 2014. The origins and functions of dendritic cells and macrophages in the skin. *Nat Rev Immunol* 14:417–428.

91. Tamoutounour S, Guilliams M, Montanana Sanchis F, Liu H, Terhorst D, Malosse C, Pollet E, Ardouin L, Luche H, Sanchez C, Dalod M, Malissen B, Henri S. 2013. Origins and functional specialization of macrophages and of conventional and monocyte-derived dendritic cells in mouse skin. *Immunity* 39:925–938.

92. Guilliams M, Ginhoux F, Jakubzick C, Naik SH, Onai N, Schraml BU, Segura E, Tussiwand R, Yona S. 2014. Dendritic cells, monocytes and macrophages: a unified nomenclature based on ontogeny. *Nat Rev Immunol* 14: 571–578.

93. Raza S, Barnett MW, Barnett-Itzhaki Z, Amit I, Hume DA, Freeman TC. 2014. Analysis of the transcriptional networks underpinning the activation of murine macrophages by inflammatory mediators. *J Leukoc Biol* 96: 167–183.

94. Sans-Fons MG, Yeramian A, Pereira-Lopes S, Santamaría-Babi LF, Modolell M, Lloberas J, Celada A. 2013. Arginine transport is impaired in C57Bl/6 mouse macrophages as a result of a deletion in the promoter of *Slc7a2* (CAT2), and susceptibility to *Leishmania* infection is reduced. *J Infect Dis* 207:1684–1693.

95. Butovsky O, Jedrychowski MP, Moore CS, Cialic R, Lanser AJ, Gabriely G, Koeglsperger T, Dake B, Wu PM, Doykan CE, Fanek Z, Liu L, Chen Z, Rothstein JD, Ransohoff RM, Gygi SP, Antel JP, Weiner HL. 2014. Identification of a unique TGF-β-dependent molecular and functional signature in microglia. *Nat Neurosci* 17:131–143.

96. Schilling E, El Chartouni C, Rehli M. 2009. Allele-specific DNA methylation in mouse strains is mainly determined by *cis*-acting sequences. *Genome Res* 19: 2028–2035.

97. Heinz S, Romanoski CE, Benner C, Allison KA, Kaikkonen MU, Orozco LD, Glass CK. 2013. Effect of natural genetic variation on enhancer selection and function. *Nature* 503:487–492.

98. Wells CA, Ravasi T, Faulkner GJ, Carninci P, Okazaki Y, Hayashizaki Y, Sweet M, Wainwright BJ, Hume DA. 2003. Genetic control of the innate immune response. *BMC Immunol* 4:5. doi:10.1186/1471-2172-4-5.

99. Chen L, Kostadima M, Martens JH, Canu G, Garcia SP, Turro E, Downes K, Macaulay IC, Bielczyk-Maczynska E, Coe S, Farrow S, Poudel P, Burden F, Jansen SB, Astle WJ, Attwood A, Bariana T, de Bono B, Breschi A, Chambers JC, BRIDGE Consortium, Choudry FA, Clarke L, Coupland P, van der Ent M, Erber WN, Jansen JH, Favier R, Fenech ME, Foad N, Freson K, van Geet C, Gomez K, Guigo R, Hampshire D, Kelly AM, Kerstens HH, Kooner JS, Laffan M, Lentaigne C, Labalette C, Martin T, Meacham S, Mumford A, Nürnberg S, Palumbo E, van der Reijden BA, Richardson D, Sammut SJ, Slodkowicz G, Tamuri AU, Vasquez L, Voss K, Watt S, Westbury S, Flicek P, Loos R, Goldman N, Bertone P, Read RJ, Richardson S, Cvejic A, Soranzo N, Ouwehand WH, Stunnenberg HG, Frontini M, Rendon A. 2014. Transcriptional diversity during lineage commitment of human blood progenitors. *Science* 345:1251033. doi: 10.1126/science.1251033.

100. Laurenti E, Doulatov S, Zandi S, Plumb I, Chen J, April C, Fan JB, Dick JE. 2013. The transcriptional architecture of early human hematopoiesis identifies multilevel control of lymphoid commitment. *Nat Immunol* 14: 756–763.

101. Novershtern N, Subramanian A, Lawton LN, Mak RH, Haining WN, McConkey ME, Habib N, Yosef N, Chang CY, Shay T, Frampton GM, Drake AC, Leskov I, Nilsson B, Preffer F, Dombkowski D, Evans JW, Liefeld T, Smutko JS, Chen J, Friedman N, Young RA, Golub TR, Regev A, Ebert BL. 2011. Densely interconnected transcriptional circuits control cell states in human hematopoiesis. *Cell* 144:296–309.

102. Qiao W, Wang W, Laurenti E, Turinsky AL, Wodak SJ, Bader GD, Dick JE, Zandstra PW. 2014. Intercellular network structure and regulatory motifs in the human hematopoietic system. *Mol Syst Biol* 10:741. doi: 10.15252/msb.20145141.

103. Martinez FO, Gordon S, Locati M, Mantovani A. 2006. Transcriptional profiling of the human monocyte-to-macrophage differentiation and polarization: new molecules and patterns of gene expression. *J Immunol* 177: 7303–7311.

104. Pham TH, Benner C, Lichtinger M, Schwarzfischer L, Hu Y, Andreesen R, Chen W, Rehli M. 2012. Dynamic epigenetic enhancer signatures reveal key transcription factors associated with monocytic differentiation states. *Blood* 119:e161–e171. doi:10.1182/blood-2012-01-402453.

105. FANTOM Consortium, Suzuki H, Forrest AR, van Nimwegen E, Daub CO, Balwierz PJ, Irvine KM, Lassmann T, Ravasi T, Hasegawa Y, de Hoon MJ, Katayama S, Schroder K, Carninci P, Tomaru Y, Kanamori-Katayama M, Kubosaki A, Akalin A, Ando

Y, Arner E, Asada M, Asahara H, Bailey T, Bajic VB, Bauer D, Beckhouse AG, Bertin N, Björkegren J, Brombacher F, Bulger E, Chalk AM, Chiba J, Cloonan N, Dawe A, Dostie J, Engström PG, Essack M, Faulkner GJ, Fink JL, Fredman D, Fujimori K, Furuno M, Gojobori T, Gough J, Grimmond SM, Gustafsson M, Hashimoto M, Hashimoto T, Hatakeyama M, Heinzel S, et al. 2009. The transcriptional network that controls growth arrest and differentiation in a human myeloid leukemia cell line. *Nat Genet* **41**:553–562.

106. Andersson R, Gebhard C, Miguel-Escalada I, Hoof I, Bornholdt J, Boyd M, Chen Y, Zhao X, Schmidl C, Suzuki T, Ntini E, Arner E, Valen E, Li K, Schwarzfischer L, Glatz D, Raithel J, Lilje B, Rapin N, Bagger FO, Jørgensen M, Andersen PR, Bertin N, Rackham O, Burroughs AM, Baillie JK, Ishizu Y, Shimizu Y, Furuhata E, Maeda S, Negishi Y, Mungall CJ, Meehan TF, Lassmann T, Itoh M, Kawaji H, Kondo N, Kawai J, Lennartsson A, Daub CO, Heutink P, Hume DA, Jensen TH, Suzuki H, Hayashizaki Y, Müller F, FANTOM Consortium, Forrest AR, Carninci P, Rehli M, Sandelin A. 2014. An atlas of active enhancers across human cell types and tissues. *Nature* **507**:455–461.

107. FANTOM Consortium and the RIKEN PMI and CLST (DGT), Forrest AR, Kawaji H, Rehli M, Baillie JK, de Hoon MJ, Haberle V, Lassmann T, Kulakovskiy IV, Lizio M, Itoh M, Andersson R, Mungall CJ, Meehan TF, Schmeier S, Bertin N, Jørgensen M, Dimont E, Arner E, Schmidl C, Schaefer U, Medvedeva YA, Plessy C, Vitezic M, Severin J, Semple C, Ishizu Y, Young RS, Francescatto M, Alam I, Albanese D, Altschuler GM, Arakawa T, Archer JA, Arner P, Babina M, Rennie S, Balwierz PJ, Beckhouse AG, Pradhan-Bhatt S, Blake JA, Blumenthal A, Bodega B, Bonetti A, Briggs J, Brombacher F, Burroughs AM, Califano A, Cannistraci CV, Carbajo D, et al. 2014. A promoter-level mammalian expression atlas. *Nature* **507**:462–470.

108. Saeed S, Quintin J, Kerstens HH, Rao NA, Aghajanirefah A, Matarese F, Cheng SC, Ratter J, Berentsen K, van der Ent MA, Sharifi N, Janssen-Megens EM, Ter Huurne M, Mandoli A, van Schaik T, Ng A, Burden F, Downes K, Frontini M, Kumar V, Giamarellos-Bourboulis EJ, Ouwehand WH, van der Meer JW, Joosten LA, Wijmenga C, Martens JH, Xavier RJ, Logie C, Netea MG, Stunnenberg HG. 2014. Epigenetic programming of monocyte-to-macrophage differentiation and trained innate immunity. *Science* **345**:1251086. doi:10.1126/science.1251086.

109. Joshi A, Pooley C, Freeman TC, Lennartsson A, Babina M, Schmidl C, Geijtenbeek T, the FANTOM Consortium, Michoel T, Severin J, Itoh M, Lassmann T, Kawaji H, Hayashizaki Y, Carninci P, Forrest AR, Rehli M, Hume DA. 2015. Transcription factor, promoter, and enhancer utilization in human myeloid cells. *J Leukoc Biol* doi:10.1189/jlb.6TA1014-477RR.

110. D'Aveni M, Rossignol J, Coman T, Sivakumaran S, Henderson S, Manzo T, Santos e Sousa P, Bruneau J, Fouquet G, Zavala F, Alegria-Prévot O, Garfa-Traoré M, Suarez F, Trebeden-Nègre H, Mohty M, Bennett CL, Chakraverty R, Hermine O, Rubio MT. 2015. G-CSF mobilizes CD34^{+} regulatory monocytes that inhibit graft-versus-host disease. *Sci Transl Med* **7**: 281ra42. doi:10.1126/scitranslmed.3010435.

111. Schmidl C, Renner K, Peter K, Eder R, Lassmann T, Balwierz PJ, Itoh M, Nagao-Sato S, Kawaji H, Carninci P, Suzuki H, Hayashizaki Y, Andreesen R, Hume DA, Hoffmann P, Forrest AR, Kreutz MP, Edinger M, Rehli M, FANTOM consortium. 2014. Transcription and enhancer profiling in human monocyte subsets. *Blood* **123**:e90–e99. doi:10.1182/blood-2013-02-484188.

112. Huber R, Pietsch D, Günther J, Welz B, Vogt N, Brand K. 2014. Regulation of monocyte differentiation by specific signaling modules and associated transcription factor networks. *Cell Mol Life Sci* **71**:63–92.

113. Carlin LM, Stamatiades EG, Auffray C, Hanna RN, Glover L, Vizcay-Barrena G, Hedrick CC, Cook HT, Diebold S, Geissmann F. 2013. *Nr4a1*-dependent Ly6C^{low} monocytes monitor endothelial cells and orchestrate their disposal. *Cell* **153**:362–375.

114. Tagoh H, Himes R, Clarke D, Leenen PJ, Riggs AD, Hume D, Bonifer C. 2002. Transcription factor complex formation and chromatin fine structure alterations at the murine c-fms (CSF-1 receptor) locus during maturation of myeloid precursor cells. *Genes Dev* **16**: 1721–1737.

115. Mossadegh-Keller N, Sarrazin S, Kandalla PK, Espinosa L, Stanley ER, Nutt SL, Moore J, Sieweke MH. 2013. M-CSF instructs myeloid lineage fate in single haematopoietic stem cells. *Nature* **497**:239–243.

116. Krysinska H, Hoogenkamp M, Ingram R, Wilson N, Tagoh H, Laslo P, Singh H, Bonifer C. 2007. A two-step, PU.1-dependent mechanism for developmentally regulated chromatin remodeling and transcription of the c-*fms* gene. *Mol Cell Biol* **27**:878–887.

117. Ross IL, Dunn TL, Yue X, Roy S, Barnett CJ, Hume DA. 1994. Comparison of the expression and function of the transcription factor PU.1 (Spi-1 proto-oncogene) between murine macrophages and B lymphocytes. *Oncogene* **9**:121–132.

118. Reddy MA, Yang BS, Yue X, Barnett CJ, Ross IL, Sweet MJ, Hume DA, Ostrowski MC. 1994. Opposing actions of c-*ets*/PU.1 and c-*myb* protooncogene products in regulating the macrophage-specific promoters of the human and mouse colony-stimulating factor-1 receptor (c-*fms*) genes. *J Exp Med* **180**:2309–2319.

119. Bonifer C, Hume DA. 2008. The transcriptional regulation of the colony-stimulating factor 1 receptor (*csf1r*) gene during hematopoiesis. *Front Biosci* **13**:549–560.

120. Heinz S, Benner C, Spann N, Bertolino E, Lin YC, Laslo P, Cheng JX, Murre C, Singh H, Glass CK. 2010. Simple combinations of lineage-determining transcription factors prime cis-regulatory elements required for macrophage and B cell identities. *Mol Cell* **38**:576–589.

121. Ross IL, Yue X, Ostrowski MC, Hume DA. 1998. Interaction between PU.1 and another Ets family transcription factor promotes macrophage-specific basal transcription initiation. *J Biol Chem* **273**:6662–6669.

122. Hume DA, Sasmono T, Himes SR, Sharma SM, Bronisz A, Constantin M, Ostrowski MC, Ross IL. 2008. The

Ewing sarcoma protein (EWS) binds directly to the proximal elements of the macrophage-specific promoter of the CSF-1 receptor (*csf1r*) gene. *J Immunol* **180**:6733–6742.

123. **Sasmono RT, Oceandy D, Pollard JW, Tong W, Pavli P, Wainwright BJ, Ostrowski MC, Himes SR, Hume DA.** 2003. A macrophage colony-stimulating factor receptor-green fluorescent protein transgene is expressed throughout the mononuclear phagocyte system of the mouse. *Blood* **101**:1155–1163.

124. **Sauter KA, Bouhlel MA, O'Neal J, Sester DP, Tagoh H, Ingram RM, Pridans C, Bonifer C, Hume DA.** 2013. The function of the conserved regulatory element within the second intron of the mammalian *Csf1r* locus. *PLoS One* **8**:e54935. doi:10.1371/journal.pone.0054935.

125. **Alder JK, Georgantas RW III, Hildreth RL, Kaplan IM, Morisot S, Yu X, McDevitt M, Civin CI.** 2008. Kruppel-like factor 4 is essential for inflammatory monocyte differentiation in vivo. *J Immunol* **180**:5645–5652.

126. **Sauter KA, Pridans C, Sehgal A, Tsai YT, Bradford BM, Raza S, Moffat L, Gow DJ, Beard PM, Mabbott NA, Smith LB, Hume DA.** 2014. Pleiotropic effects of extended blockade of CSF1R signaling in adult mice. *J Leukoc Biol* **96**:265–274.

127. **Deng L, Zhou JF, Sellers RS, Li JF, Nguyen AV, Wang Y, Orlofsky A, Liu Q, Hume DA, Pollard JW, Augenlicht L, Lin EY.** 2010. A novel mouse model of inflammatory bowel disease links mammalian target of rapamycin-dependent hyperproliferation of colonic epithelium to inflammation-associated tumorigenesis. *Am J Pathol* **176**:952–967.

128. **Ovchinnikov DA, DeBats CE, Sester DP, Sweet MJ, Hume DA.** 2010. A conserved distal segment of the mouse CSF-1 receptor promoter is required for maximal expression of a reporter gene in macrophages and osteoclasts of transgenic mice. *J Leukoc Biol* **87**:815–822.

129. **Ovchinnikov DA, van Zuylen WJ, DeBats CE, Alexander KA, Kellie S, Hume DA.** 2008. Expression of Gal4-dependent transgenes in cells of the mononuclear phagocyte system labeled with enhanced cyan fluorescent protein using *Csf1r*-Gal4VP16/UAS-ECFP double-transgenic mice. *J Leukoc Biol* **83**:430–433.

130. **van Zuylen WJ, Garceau V, Idris A, Schroder K, Irvine KM, Lattin JE, Ovchinnikov DA, Perkins AC, Cook AD, Hamilton JA, Hertzog PJ, Stacey KJ, Kellie S, Hume DA, Sweet MJ.** 2011. Macrophage activation and differentiation signals regulate Schlafen-4 gene expression: evidence for Schlafen-4 as a modulator of myelopoiesis. *PLoS One* **6**:e15723. doi:10.1371/journal.pone.0015723.

131. **Jacquelin S, Licata F, Dorgham K, Hermand P, Poupel L, Guyon E, Deterre P, Hume DA, Combadière C, Boissonnas A.** 2013. CX3CR1 reduces Ly6C^{high}-monocyte motility within and release from the bone marrow after chemotherapy in mice. *Blood* **122**:674–683.

132. **Pridans CE, Lillico S, Whitelaw CBA, Hume DA.** 2014. Lentiviral vectors containing mouse Csf1r control elements direct macrophage-restricted expression in multiple species of birds and mammals. *Mol Ther Methods Clin Dev* **1**:14010. doi:10.1038/mtm.2014.10.

133. **Balic A, Garcia-Morales C, Vervelde L, Gilhooley H, Sherman A, Garceau V, Gutowska MW, Burt DW, Kaiser P, Hume DA, Sang HM.** 2014. Visualisation of chicken macrophages using transgenic reporter genes: insights into the development of the avian macrophage lineage. *Development* **141**:3255–3265.

134. **Cuadros MA, Martin C, Coltey P, Almendros A, Navascués J.** 1993. First appearance, distribution, and origin of macrophages in the early development of the avian central nervous system. *J Comp Neurol* **330**:113–129.

135. **Lavin Y, Winter D, Blecher-Gonen R, David E, Keren-Shaul H, Merad M, Jung S, Amit I.** 2014. Tissue-resident macrophage enhancer landscapes are shaped by the local microenvironment. *Cell* **159**:1312–1326.

136. **Gosselin D, Link VM, Romanoski CE, Fonseca GJ, Eichenfield DZ, Spann NJ, Stender JD, Chun HB, Garner H, Geissmann F, Glass CK.** 2014. Environment drives selection and function of enhancers controlling tissue-specific macrophage identities. *Cell* **159**:1327–1340.

137. **Gautier EL, Ivanov S, Williams JW, Huang SC, Marcelin G, Fairfax K, Wang PL, Francis JS, Leone P, Wilson DB, Artyomov MN, Pearce EJ, Randolph GJ.** 2014. Gata6 regulates aspartoacylase expression in resident peritoneal macrophages and controls their survival. *J Exp Med* **211**:1525–1531.

138. **Rosas M, Davies LC, Giles PJ, Liao CT, Kharfan B, Stone TC, O'Donnell VB, Fraser DJ, Jones SA, Taylor PR.** 2014. The transcription factor Gata6 links tissue macrophage phenotype and proliferative renewal. *Science* **344**:645–648.

139. **Gordon S.** 2007. The macrophage: past, present and future. *Eur J Immunol* **37**(Suppl 1):S9–S17.

140. **Taylor PR, Gordon S.** 2003. Monocyte heterogeneity and innate immunity. *Immunity* **19**:2–4.

141. **Sica A, Mantovani A.** 2012. Macrophage plasticity and polarization: in vivo veritas. *J Clin Invest* **122**:787–795.

142. **Schroder K, Hertzog PJ, Ravasi T, Hume DA.** 2004. Interferon-γ: an overview of signals, mechanisms and functions. *J Leukoc Biol* **75**:163–189.

143. **Jenkins SJ, Ruckerl D, Cook PC, Jones LH, Finkelman FD, van Rooijen N, MacDonald AS, Allen JE.** 2011. Local macrophage proliferation, rather than recruitment from the blood, is a signature of T_H2 inflammation. *Science* **332**:1284–1288.

144. **Schroder K, Irvine KM, Taylor MS, Bokil NJ, Le Cao KA, Masterman K-A, Labzin LI, Semple CA, Kapetanovic R, Fairbairn L, Akalin A, Faulkner GJ, Baillie JK, Gongora M, Daub CO, Kawaji H, McLachlan GJ, Goldman N, Grimmond SM, Carninci P, Suzuki H, Hayashizaki Y, Lenhard B, Hume DA, Sweet MJ.** 2012. Conservation and divergence in Toll-like receptor 4-regulated gene expression in primary human versus mouse macrophages. *Proc Natl Acad Sci U S A* **109**:E944–E953. doi:10.1073/pnas.1110156109.

145. **Kapetanovic R, Fairbairn L, Beraldi D, Sester DP, Archibald AL, Tuggle CK, Hume DA.** 2012. Pig bone

marrow-derived macrophages resemble human macrophages in their response to bacterial lipopolysaccharide. *J Immunol* 188:3382–3394.

146. **Martinez FO, Helming L, Milde R, Varin A, Melgert BN, Draijer C, Thomas B, Fabbri M, Crawshaw A, Ho LP, Ten Hacken NH, Cobos Jiménez V, Kootstra NA, Hamann J, Greaves DR, Locati M, Mantovani A, Gordon S.** 2013. Genetic programs expressed in resting and IL-4 alternatively activated mouse and human macrophages: similarities and differences. *Blood* **121:**e57–e69. doi:10.1182/blood-2012-06-436212.

147. **Mosser DM, Edwards JP.** 2008. Exploring the full spectrum of macrophage activation. *Nat Rev Immunol* **8:** 958–969.

148. **Xue J, Schmidt SV, Sander J, Draffehn A, Krebs W, Quester I, De Nardo D, Gohel TD, Emde M, Schmidleithner L, Ganesan H, Nino-Castro A, Mallmann MR, Labzin L, Theis H, Kraut M, Beyer M, Latz E, Freeman TC, Ulas T, Schultze JL.** 2014. Transcriptome-based network analysis reveals a spectrum model of human macrophage activation. *Immunity* **40:**274–288.

149. **Breen FN, Hume DA, Weidemann MJ.** 1991. Interactions among granulocyte-macrophage colony-stimulating factor, macrophage colony-stimulating factor, and IFN-gamma lead to enhanced proliferation of murine macrophage progenitor cells. *J Immunol* **147:**1542–1547.

150. **Fairfax BP, Humburg P, Makino S, Naranbhai V, Wong D, Lau E, Jostins L, Plant K, Andrews R, McGee C, Knight JC.** 2014. Innate immune activity conditions the effect of regulatory variants upon monocyte gene expression. *Science* **343:**1246949. doi:10.1126/science.1246949.

151. **Shalek AK, Satija R, Adiconis X, Gertner RS, Gaublomme JT, Raychowdhury R, Schwartz S, Yosef N, Malboeuf C, Lu D, Trombetta JJ, Gennert D, Gnirke A, Goren A, Hacohen N, Levin JZ, Park H, Regev A.** 2013. Single-cell transcriptomics reveals bimodality in expression and splicing in immune cells. *Nature* **498:** 236–240.

152. **Ravasi T, Wells C, Forest A, Underhill DM, Wainwright BJ, Aderem A, Grimmond S, Hume DA.** 2002. Generation of diversity in the innate immune system: macrophage heterogeneity arises from gene-autonomous transcriptional probability of individual inducible genes. *J Immunol* **168:**44–50.

153. **Pereira JP, Girard R, Chaby R, Cumano A, Vieira P.** 2003. Monoallelic expression of the murine gene encoding Toll-like receptor 4. *Nat Immunol* **4:**464–470.

154. **Kondo T, Kawai T, Akira S.** 2012. Dissecting negative regulation of Toll-like receptor signaling. *Trends Immunol* **33:**449–458.

155. **Wells CA, Ravasi T, Hume DA.** 2005. Inflammation suppressor genes: please switch out all the lights. *J Leukoc Biol* **78:**9–13.

156. **Lee TK, Covert MW.** 2010. High-throughput, single-cell NF-κB dynamics. *Curr Opin Genet Dev* **20:**677–683.

157. **Murray PJ, Allen JE, Biswas SK, Fisher EA, Gilroy DW, Goerdt S, Gordon S, Hamilton JA, Ivashkiv LB, Lawrence T, Locati M, Mantovani A, Martinez FO, Mege JL, Mosser DM, Natoli G, Saeij JP, Schultze JL, Shirey KA, Sica A, Suttles J, Udalova I, van Ginderachter JA, Vogel SN, Wynn TA.** 2014. Macrophage activation and polarization: nomenclature and experimental guidelines. *Immunity* **41:**14–20.

158. **Robert C, Lu X, Law A, Freeman TC, Hume DA.** 2011. Macrophages.com: an on-line community resource for innate immunity research. *Immunobiology* **216:**1203–1211.

159. **Pham TH, Minderjahn J, Schmidl C, Hoffmeister H, Schmidhofer S, Chen W, Längst G, Benner C, Rehli M.** 2013. Mechanisms of *in vivo* binding site selection of the hematopoietic master transcription factor PU.1. *Nucleic Acids Res* **41:**6391–6402.

Myeloid Cells in Health and Disease: A Synthesis
Edited by Siamon Gordon

doi:10.1128/microbiolspec.MCHD-0033-2016

Alexander Mildner[1]
Goran Marinkovic[1]
Steffen Jung[1]

9 Murine Monocytes: Origins, Subsets, Fates, and Functions

Monocytes are a conserved population of leukocytes that are present in all vertebrates, with some evidence of a parallel cell population in fly hemolymph (1). Monocytes are defined by their location in the bloodstream, their phenotype and nuclear morphology, as well as by their characteristic gene and microRNA expression signatures (2–5). In mice, monocytes represent 4% of the nucleated cells in the blood, with considerable marginal pools in the spleen and lungs that can be mobilized on demand (6, 7). Within the blood, monocytes, and in particular the classical Ly6C$^+$ mouse subset, exhibit a characteristically short half-life of 20 h (8), akin to that of similar ephemer neutrophils (9).

MONOCYTE ORIGINS

Monocyte development in adult mice depends on colony-stimulating factor-1 (CSF-1; also known as macrophage CSF [M-CSF]), and mice deficient for this growth factor or its receptor, CSF-1R (also known as CD115 and M-CSFR), exhibit severe monocytopenia (9). Monocytes arise from myeloid precursor cells in primary lymphoid organs, including the fetal liver (10) and bone marrow (BM), during both embryonic and adult hematopoiesis. As opposed to adult monocytes, fetal liver monocytes were shown to proliferate and to be independent of CSF-1 (10). However, available information on these cells is scarce, and we will focus hence in the remainder of this review on monocytes in the adult.

A major breakthrough in our understanding of monocyte biology, and the organization of the mononuclear phagocyte system as a whole, was the identification of dedicated clonotypic precursor cells, termed monocyte/macrophage dendritic cell (DC) progenitors (MDPs), in mouse BM (11) (Fig. 1). MDPs are defined as lineage-negative (Lin$^-$) cells that express CD117 (also known as KIT), CD135 (also known as FLT3), and CSF-1R (Fig. 2A). Adoptive transfer of MDPs into BM cavities has established that these cells are bona fide monocyte precursors (12). As opposed to granulocyte/macrophage progenitors (GMPs), MDPs have been reported to lack the potential to generate neutrophils but are still multipotent, as they can give rise to plasmacytoid DCs (pDCs) and classical DCs (cDCs) (11–13). MDP differentiation toward monocytes includes a defined transient intermediate, the common monocyte progenitors (cMoPs) (14). cMoPs differ from MDPs by their lack

[1]Department of Immunology, The Weizmann Institute of Science, Rehovot 76100, Israel.

of CD135 expression, as well as by their loss of cDC or pDC potential (14). Adoptively transferred cMoPs have been shown to give rise to both Ly6C$^+$ and Ly6C$^-$ monocyte subsets, albeit with distinct kinetics (14). Of note, single-cell transcriptome profiling (15) und advanced multiparameter flow analysis (16) are likely to reveal heterogeneity within myeloid precursor populations, which might explain current controversies (17)

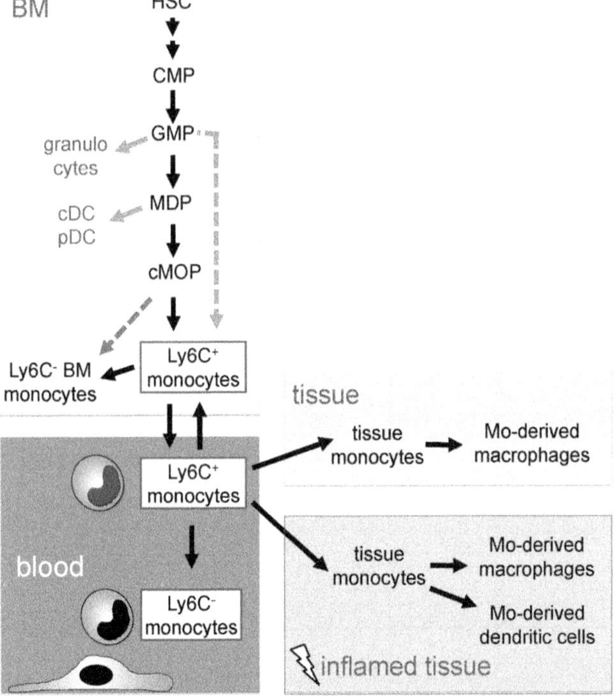

Figure 1 Schematic of murine monocyte development and monocyte fates. Ly6C$^+$ monocytes are continuously generated in the adult BM in steady state from hematopoietic stem cells (HSCs) via a sequence comprising common myeloid precursors (CMPs), GMPs, MDPs (11), and cMoPs (14). The BM also harbors a rare population of Ly6C$^-$ monocytes that could derive from Ly6C$^+$ monocytes or cMoPs, but the function of this subset is unknown. Ly6C$^+$ monocytes egress to the circulation, a process that requires CCR2 (18). Ly6C$^+$ blood monocytes have a half-life of 20 h and several potential fates: (i) differentiation into Ly6C$^-$ monocytes that patrol the endothelial surface of the vasculature; (ii) extravasation to selected tissues whose steady-state macrophage compartment requires maintenance by monocyte recruitment, such as the intestine (6, 7); and (iii) recruitment to sites of tissue injury, infection, wounds, tumors, and inflammation, where the cells differentiate depending on local cues and give rise to cells with macrophage or DC features. Ly6C$^+$ monocytes can also return from the circulation to the BM cavities (12), where their fate remains unclear. Of note, the scheme refers to monocyte development in steady state, while under challenge alternative developmental routes could be activated that might, for instance, bypass the MDP stage.

and could result in revisions of the current hematopoietic trees.

Steady-state adult monopoiesis is restricted to the BM, and egress of monocytes from BM has emerged as a critical checkpoint controlling the abundance of blood monocytes. Specifically, the Pamer group has shown that monocyte emigration from the BM requires the chemokine receptor CCR2, whose expression is restricted to Ly6C$^+$ cells (18). Toll-like receptor (TLR) ligands or peripheral bacterial infections rapidly induce expression of the CCR2 ligand CCL2 (monocyte chemoattractant protein-1) on BM mesenchymal cells and thereby boost CCR2-dependent monocyte egress (19, 20). Interestingly, steady-state abundance of Ly6C$^+$ but not Ly6C$^-$ monocytes in the circulation was reported to undergo diurnal oscillations (21), although the exact mechanism underlying this phenomenon remains to be elucidated.

Monocyte development has to date been studied mainly in steady state. Given the ample communication circuits between the periphery and the BM, monopoiesis is, however, likely to be affected by challenges. Peripheral inflammation and stress are documented to lead to increased monocyte numbers in the circulation presumably by impacting on hematopoiesis (22, 23). Acute elevation of "classical" monocyte numbers in the circulation, for instance, following exercise, is probably linked to the release of these cells from marginal pools (24), i.e., areas of reduced blood velocity from where the cells can be mobilized in a catecholamine-dependent fashion (25). Interestingly, though, emerging evidence suggests that the monocyte compartment is not only affected quantitatively, but can also be qualitatively changed. Thus, monocytes have been proposed to be preemptively educated in the BM to promote their tissue-specific function at sites of persistent challenge (26). Specifically, an intestinal *Toxoplasma* parasite infection was shown to lead to systemic interleukin-12 that triggered gamma interferon production by BM-resident natural killer (NK) cells, which in turn profoundly affected the gene expression profile of cMoPs and BM monocytes (26). Similar scenarios, combined with longer-lasting epigenetic alterations in monocyte precursors, may account for the reported intriguing phenomenon of "trained immunity" (27, 28), although the underlying mechanisms remain to be defined.

Given the central role of monocytes in inflammation, tissue restoration, and immunopathologies, the impact of peripheral challenges and chronic inflammation on the monocyte compartment clearly deserves further study and should profit from the recent advance in profiling techniques, including single-cell analysis

(29). Mechanisms that alter monocyte signatures could include mere acceleration of monopoiesis or the alteration of developmental routes (Fig. 1), but also the activation of alternative extramedullary sites for monocyte generation, as shown for an atherosclerosis model (30).

MONOCYTE SUBSETS

Heterogeneity within monocytes was first reported in 1989 by Ziegler-Heitbrock and colleagues for human blood (31), and the two main human monocyte subpopulations are currently defined as $CD14^{+}CD16^{-}$ and $CD14^{dim}CD16^{+}$ cells (32). In addition, human blood harbors an intriguing rare "intermediate" $CD14^{+}CD16^{+}$ monocyte population, whose differential abundance in individuals could have diagnostic value (32). Monocyte heterogeneity in mice was only defined more than a decade after its discovery in man. Specifically, flow cytometric analysis of a transgenic mouse strain that carried an insertion of a green fluorescent protein (GFP) reporter gene in the locus of the CX3CR1 chemokine receptor ($CX3CR1^{gfp}$ animals) (9) had revealed discrete monocyte subpopulations that could be discriminated according to GFP reporter intensity, as well as surface markers such as the chemokine receptor CCR2 and CD62L (L-selectin) (33, 34). The two main murine monocyte subsets, in unchallenged animals collectively defined as blood cells that express high levels of the CSF-1 receptor CD115, are currently phenotypically characterized as $CX3CR1^{med}CCR2^{+}CD62L^{+}Ly6C^{hi}$ ($Ly6C^{+}$) and $CX3CR1^{hi}CCR2^{-}CD62L^{-}Ly6C^{lo}$ ($Ly6C^{-}$) cells (33–35) (Fig. 2B). Gene expression profiling and functional studies established that $Ly6C^{+}$ and $Ly6C^{-}$ monocytes are equivalents of the human $CD14^{+}$ and $CD14^{dim}CD16^{+}$ monocyte subsets, respectively, although interesting distinctions remain (2, 4). Of note, as in the human, mouse monocytes also comprise an additional distinct transition state (36), which remains, however, less well characterized in terms of phenotype and function.

Animals transiently depleted of monocytes using clodronate liposomes show distinct reappearance kinetics of $Ly6C^{+}$ and $Ly6C^{-}$ monocytes. This suggests that $Ly6C^{+}$ monocytes might be precursors of $Ly6C^{-}$ cells (37). The identification of a putative intermediate $Ly6C^{mid}CCR7^{+}CCR8^{+}$ monocyte subset supported this notion (36), which was subsequently proven directly by adoptive transfer experiment. Thus, $Ly6C^{+}$ monocytes isolated from BM or spleen of $CX3CR1^{gfp}$ donor mice were shown to give upon intravenous transfer into wild-type recipients efficient rise to circulating $Ly6C^{-}$ cells (8, 12). Moreover, following adoptive transfer of MDPs or cMoPs, graft-derived $Ly6C^{-}$ cells arose, as compared to $Ly6C^{+}$ monocytes, with a considerable delay (14). Steady-state blood monocyte conversion was further supported by results of less invasive bromodeoxyuridine incorporation assays showing (i) that $Ly6C^{+}$ and $Ly6C^{-}$ monocytes display sequential label acquisition and (ii) that labeling of $Ly6C^{-}$ monocytes requires the presence of $Ly6C^{+}$ blood monocytes (8). Interestingly, the bromodeoxyuridine assay showed similar results for the $CD14^{+}$ and $CD14^{dim}CD16^{+}$ monocyte subsets of macaques (38), indicating that monocyte conversion is a conserved feature among rodents and primates. Of note, certain gene mutations, including the one affecting expression of CCR2, and the transcription factors (TFs) interferon regulatory factor 8 (IRF8) and Krüppel-like factor 4 (KLF4) were shown to selectively impair the generation of $Ly6C^{+}$ monocytes (18, 39–41) but seemingly did not affect the $Ly6C^{-}$ cells. This could be interpreted as evidence for the existence of a developmental pathway of $Ly6C^{-}$ monocytes that is independent of $Ly6C^{+}$ monocytes. However, fate mapping experiments have shown that absence of $Ly6C^{+}$ monocytes in CCR2-deficient animals results in a compensatory extension of the half-life of $Ly6C^{-}$ cells that could be linked to CSF-1 availability (8). Impairment of this cellular compartment could hence also be masked in other mutant mice. Conversely, the selective absence of $Ly6C^{-}$ monocytes in NR4A1-deficient animals has been interpreted as evidence for a dedicated developmental pathway of these cells (42). However, the NR4A1 deficiency could affect the generation of these cells from $Ly6C^{-}$ monocytes rather than a putative direct derivation from cMoPs. Collectively, the above studies and that of others (7) provide compelling evidence that the two main circulating murine monocyte subsets form in steady state an obligatory developmental sequence. Notably, the existence of an $Ly6C^{+}$-independent route from cMoPs or other precursors to $Ly6C^{-}$ monocytes, particularly in pathological settings, can currently not be excluded but awaits experimental evidence.

MOLECULAR DEFINITIONS OF MONOCYTES

Gene expression profiling allowed the alignment of murine and primate monocyte subsets and established mouse $Ly6C^{+}$ monocytes as equivalent of classical $CD14^{+}$ human monocytes (2, 4). Comparative transcriptome analysis of monocytes with other leukocytes, as well as tissue-resident mononuclear phagocytes, has furthermore highlighted the uniqueness of these cells (43, 44). The distinct monocyte expression profiles also

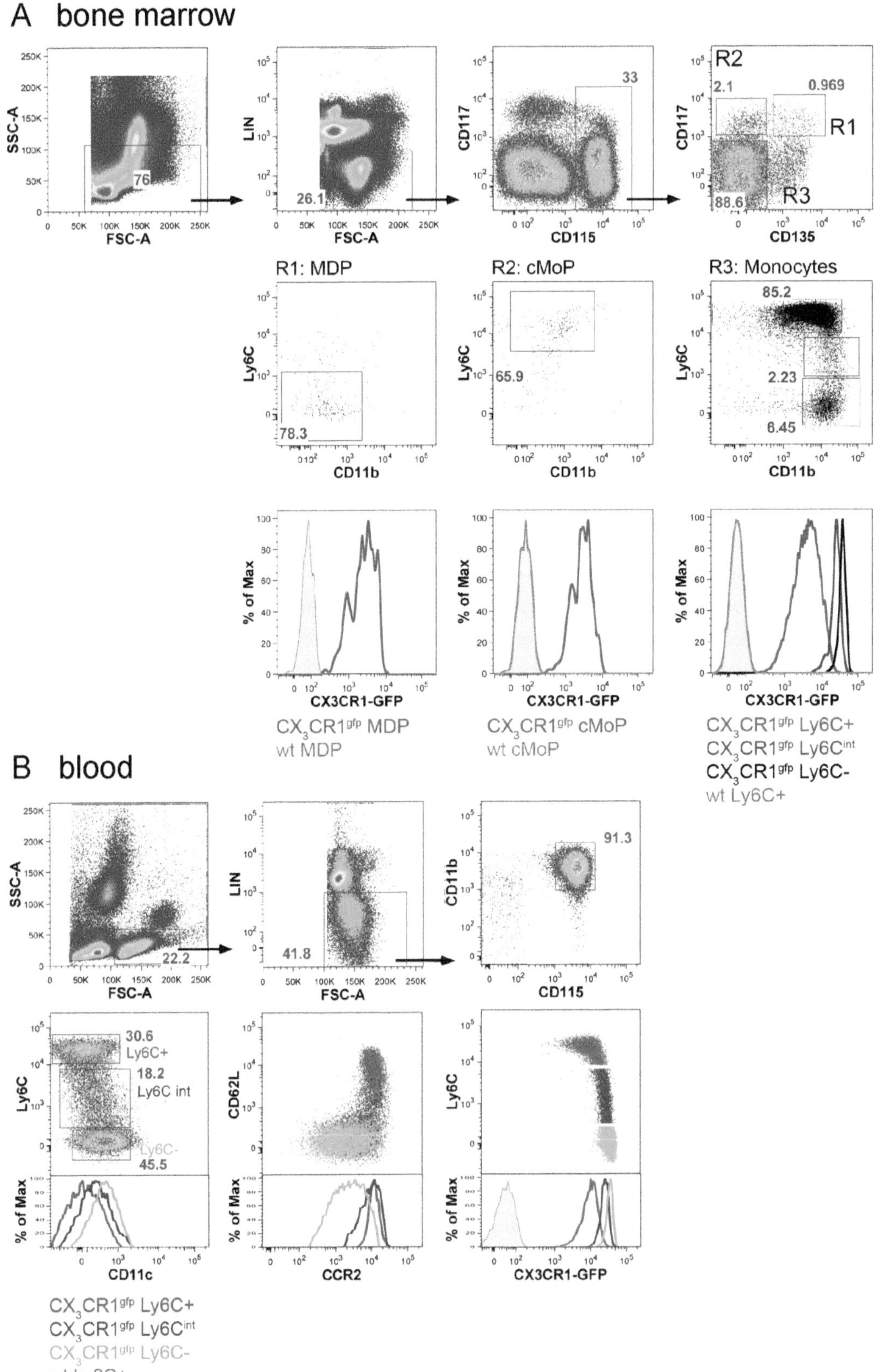
A bone marrow
SSC-A
FSC-A
76
LIN
26.1
CD117
CD115
33
R2
2.1
0.969
R1
R3
88.6
CD135
R1: MDP
R2: cMoP
R3: Monocytes
Ly6C
CD11b
78.3
65.9
85.2
2.23
6.45
% of Max
CX3CR1-GFP
CX3CR1gfp MDP
wt MDP
CX3CR1gfp cMoP
wt cMoP
CX3CR1gfp Ly6C+
CX3CR1gfp Ly6Cint
CX3CR1gfp Ly6C-
wt Ly6C+
B blood
22.2
41.8
CD11b
91.3
30.6
Ly6C+
18.2
Ly6C int
Ly6C-
45.5
CD62L
CCR2
CD11c
CX3CR1gfp Ly6C+
CX3CR1gfp Ly6Cint
CX3CR1gfp Ly6C-
wt Ly6C+

include small noncoding RNAs, or microRNAs (3, 5), that critically shape cell identities as posttranscriptional filters (45). Moreover, in line with the emerging notion that epigenetic signatures are imprinted by environment (46), Ly6C$^+$ monocytes also display distinct enhancer usage (44). Specifically, this study revealed a considerable overlap with respect to promoter activities between monocytes, neutrophils, and the seven tissue macrophages tested (82% of the total 10,806 promoters). On the other hand, monocytes shared only 27% of the total 30,976 putative revealed myeloid enhancers with neutrophils and macrophages. Consistent with the prominent expression of KLF4 by Ly6C$^+$ monocytes and earlier reports of its role as a monocyte regulator (40, 41), enhancer regions found unique to monocytes were enriched for DNA sequence motif bound by KLF4 (44). Future detailed analysis of monocytes, including Ly6C$^+$ cells, intermediates, and Ly6C$^-$ cells, using advanced genomic approaches such as indexing-first chromatin immunoprecipitation (47) and single-cell profiling (48) are likely to aid our appreciation of monocyte heterogeneity. Furthermore, analysis of alterations of the monocyte compartment following defined challenges should allow definition of signatures associated with pathology that can be translated into the human setting.

The identity and response repertoire of monocytes are defined by hierarchical and combinatorial action of a selected set of TFs (49). For monocytes this includes "pioneer" or "lineage-determining" TFs, such as PU1.1; cell-specific TFs, such as KLF4; and stimulus-responsive TFs, such as NF-κB (50). TFs or TF combinations that specifically regulate monocyte development remain incompletely understood, however, and most of the candidates also show effects on other hematopoietic lineages (49). Moreover, TFs such as IRF8, PU.1 (SFPI1), and NR4A1 (Nur77), as well as members of the NF-κB family, might also promote expansion and/or survival rather than providing only instructive cues (51).

The establishment of the requirement of a specific TF for monocyte development will arguably require its mutagenesis, preferentially restricted to monocytes. Notably, though, monocytes, and in particular the Ly6C$^+$ subset, are due to their extreme short half-life refractory to Cre/loxP-based mutagenesis (8, 52). In absence of suitable *in vitro* models that recapitulate monopoiesis, the elucidation of TF hierarchies governing monocyte development will require the establishment of novel experimental approaches. Once the regulatory elements that govern monocyte-specific gene expression are mapped in detail, this could include interference with TF binding to cell-specific enhancers. Below, we discuss selected TF deficiencies affecting monocyte development and maintenance that have emerged from the analysis of respective mouse mutants and can provide a framework for future studies. More comprehensive coverage of this topic and additional primary references can be found in recent reviews (49, 51, 53).

As master regulator of the myeloid lineage, the ETS family TF PU.1 (SFPI1) drives expression of CSF-1R receptor, which is essential for monocyte generation. Expression of low levels of PU.1 in GMPs induces granulopoiesis, whereas high PU.1 levels induce monopoiesis and expression of monocyte-associated TFs, including IRF8. IRF8-deficient mice lack Ly6C$^+$ monocytes and have reduced numbers of Ly6C$^-$ monocytes (39). Interestingly, IRF8 was shown to drive expression of KLF4,

Figure 2 Example flow cytometric analysis of BM-resident monocyte precursors and blood monocytes. (A) Analysis of myeloid BM compartment of C57BL/6 and CX3CR1^{gfp} C57BL/6 mouse (9). Blood cells were lysed with BD lysis buffer (#349202). Cells are gated for scatter, singlets, and Lin (Ter-119, B220, Ly6G, NK1.1, TCRγδ, CD4, CD8)-negative cells expressing CD115. This gate comprises CD135$^+$CD117$^+$ MDPs and CD135$^-$CD117$^+$ cMoPs and CD135$^-$CD117$^-$ monocytes. MDPs are CD11b$^-$Ly6C$^-$, while cMoPs are CD11b^{int}Ly6C$^+$. BM monocytes are CD11b$^+$ and can be subdivided according to Ly6C expression into major Ly6C$^+$ and minor Ly6C^{lo-neg} populations. Histogram shows CX3CR1-GFP reporter gene expression of MDPs, cMoPs, and monocytes, as compared to wild-type (WT) BM (gray). Note upregulation of reporter in Ly6C$^{int-neg}$ BM monocytes. (B) Analysis of blood monocyte compartment of C57BL/6 and CX3CR1^{gfp} C57BL/6 mouse (9). Blood cells were lysed with BD lysis buffer (#349202). Cells are gated for scatter (excluding sideward scatted [SSC] high neutrophils), singlets, and lineage negative according to Lin marker expression (B220, CD19, CD3, CD4, CD8, Ly6G, NK1.1, TCRγδ). Lin$^-$ cells comprise largely CD115$^+$CD11b$^+$ monocytes. "Ly6C$^+$ monocytes" (red) can be defined as Ly6C$^+$CD11c$^-$CCR2$^+$CD62L$^+$-CX3CR1^{int} and "Ly6C$^-$ monocytes" (green) can be defined as Ly6C$^-$CD11c^{int}CCR2$^-$CD62L$^-$ CX3CR1^{hi} cells. Note that "Ly6C^{int} monocytes" (blue) also show intermediate expression of the other markers, as seen in the respective histograms. This supports the notion that these cells are a transitional population. FSC, forward scatter.

previously shown to be critical for the generation of Ly6C$^+$ monocytes (40, 41). The only TF so far reported to be selectively required for the generation/survival of Ly6C$^-$ monocytes is the orphan nuclear receptor NR4A1 (Nur77) (42). Interestingly, NR4A1 is induced during the conversion of Ly6C$^+$ monocytes to Ly6C$^-$ cells (A. Mildner et al., in preparation). Moreover, NR4A1 is also expressed by tissue macrophages and was reported to be required for their polarization toward proinflammatory phenotypes (54) but also activities that curb inflammation (55). The above TFs are critical for monocyte homeostasis; of equal interest, but even less well understood, are factors governing discrete *in vivo* responses of monocyte subsets. The microRNA miR-146a, which is highly expressed in both mouse Ly6C$^+$ and human CD14$^+$ monocytes but not the Ly6C$^-$ and CD14^{dim}CD16$^+$ subset, was for instance reported to specifically control the response of Ly6C$^+$ monocytes during inflammation by directly targeting the noncanonical NF-κB/Rel family member Relb (3).

MONOCYTE FATES AND FUNCTIONS

Monocytes were long believed to constitute merely a transient precursor reservoir for tissue-resident mononuclear phagocytes. Classical labeling experiments with [^{3}H]thymidine by van Furth and Cohn that aimed at defining the kinetics of monocyte transit in the bloodstream revealed that these cells spend at most a few days in the circulation and are then mobilized to various tissues, including the inflamed peritoneum (56). Likewise, monocytes give efficient rise to macrophages and DC-like cells under conditions of inflammation, tissue injury, and infection (57–60). The notion that monocytes act mainly as precursors was further supported by the relative ease with which these cells can be differentiated into cells with macrophage or DC features in cultures driven by CSF-1 (M-CSF) and CSF-2 (granulocyte-macrophage CSF), respectively (61, 62); the validity of such culture-derived cells to serve as proxies for bona fide cells isolated from tissue has, however, recently been challenged and the cultures might be more complex than originally assumed (63–65). Rather, it is now firmly established that classical FLT3 ligand-dependent migratory DCs (cDCs) that harbor a characteristic and unique potential to stimulate naive T cells derive from dedicated BM precursors originating from MDPs but are independent of cMoPs or monocytes (66, 67) (Fig. 1). Conversely, many early studies had suggested that tissue macrophages, as diverse as liver Kupffer cells, microglia, as well as splenic and lung macrophages, are largely maintained through local proliferation (for primary references, please see reference 68). More-recent fate mapping experiments involving Cre transgenic mice and inducible reporter alleles have corroborated this notion. Furthermore, they revealed that most tissue-resident macrophage compartments are established prenatally and develop locally, alongside their respective host tissue and independent from each other (8, 69–72). Maintenance of these macrophages relies on their longevity and self-renewal and is independent of ongoing hematopoiesis (73).

Interestingly, though, specific tissues, including the gut, the skin, and the heart, show a replacement of embryonic macrophages by adult monocyte-derived cells that can either be quantitative close to birth (74, 75) or progress with time (74–77). The reason why monocyte-derived cells in these selected tissues can seed "tissue macrophage niches" that are otherwise restricted and how monocyte-derived cells can efficiently compete with the original embryonic populations remains unknown (78). Given the current examples, it could be linked to unique homeostatic challenges of these organs, such as microbiota exposure in gut and skin or continuous microtrauma induced by contraction in the heart. In fact, monocyte-derived cells among tissue macrophages can also be remnants of acute monocyte infiltrates associated with past challenges, as shown, for instance, following sterile peritonitis (8). In particular, the peritoneum seems to accumulate monocyte-derived cells alongside the embryonic-derived cells with time, although the extent of this phenomenon can vary among mouse strains (8, 79) and with gender (80). Together with the observation that monocyte-derived cells can permanently replace embryonic macrophage populations in lung and liver when the latter are compromised by a CSF-1R deficiency or experimentally ablated (81, 82), this establishes that monocytes can give rise to persisting tissue macrophages. Of note, gene expression profiling suggests so far that with time the monocyte-derived cells become virtually indistinguishable from the original populations (82, 83). However, the cells could differ on the epigenome level, including the abundance of "poised" enhancer loci (46), which could encode distinct response patterns to stimulation. It thus remains to be shown whether the cells are functionally identical.

Inflammation is a crucial component of the host defense against injury and infection. Prolonged and chronic inflammatory responses, however, are detrimental for the host and can cause considerable collateral damage (84). Accordingly, mechanisms have evolved that actively promote the resolution of inflammatory reaction and thereby efficiently prevent immunopathology (85). Tissue infiltration of Ly6C$^+$ monocytes is central to this regulatory circuit and a hallmark of sterile and infectious inflammation. With their striking plasticity, monocytes can contribute pro- or anti-inflammatory activities, depending on the context and time of their arrival at the site of injury (60). However, unless the original macrophage population is severely compromised (86), monocyte-derived cells disappear once the steady state is restored. This scenario has probably been best demonstrated in the murine model for multiple sclerosis, experimental autoimmune encephalomyelitis (EAE). In this paradigm, monocytes breach the compromised blood-brain barrier of the central nervous system and actively promote pathology (87–89). However, once acute inflammation is resolved, monocyte-derived cells disappear and the central nervous system macrophage compartment is restored to exclusively comprise yolk sac-derived microglia (89).

Given space limitations, we refer the interested reader for a comprehensive overview of the monocyte contributions to various pathologies, such as rheumatoid arthritis, colitis, atherosclerosis, and bacterial infections, to recent excellent reviews on the topic (20, 59). Another particularly intensely studied field is tumor settings where Ly6C$^+$ monocyte-derived tumor-associated macrophages (TAMs) can been discriminated from tissue-resident macrophages. In a mammary tumor model, for instance, it was shown that TAM development from Ly6C$^+$ monocytes required Notch signaling and ablation of monocyte-derived TAMs suppressed tumor growth (90). Monocyte-derived cells have been linked to myeloid suppressor activity; however, differential contributions of the different macrophages to cancer development clearly remain to be further defined (91). Highlighting the central role of Ly6C$^+$ monocytes in diseases, monocyte-targeted strategies have been shown to bear considerable therapeutic potential. Thus, acute ablation of Ly6C$^+$ monocytes from the circulation using an anti-CCR2 antibody regimen was shown to ameliorate experimentally induced colitis, arthritis, and neuroinflammation (88, 92, 93). Moreover, therapeutic CCR2 silencing in monocytes using short interfering RNA attenuated multiple pathologies, including infarct size after coronary artery occlusion and tumor progression (94). Finally, Ly6C$^+$ monocytes have recently been successfully redirected from the circulation to the spleen for clearance using negatively charged microparticles, and this monocyte neutralization reduced disease symptoms and promoted tissue repair in a number of tested disease models (95).

Of note, a recent study also reported on extravasated monocytes in noninflamed tissues that retain much of their monocytic character rather than differentiating to macrophages or DCs (35). However, these cells are short-lived and could be intermediates; hence their specific contribution remains to be further elucidated.

Sites of injury often harbor monocyte-derived cells that can be divided according to Ly6C expression. This was originally interpreted as evidence that both Ly6C$^+$ and Ly6C$^-$ monocytes extravasate and might even contribute distinct pro- and anti-inflammatory activities (96). Alternatively, Ly6C$^+$ monocytes could have lost Ly6C expression along with their differentiation into tissue cells (97), akin to their conversion into Ly6C$^-$ monocytes. Indeed, it is now generally accepted that monocytic tissue infiltrates comprise predominantly, if not exclusively, Ly6C$^+$ monocytes, or in humans the "classical" CD14$^+$ monocyte subset (98). These infiltrating cells are prominently equipped with chemokine receptors allowing recruitment, including CCR2 (20). As opposed to "classical" Ly6C$^+$ monocytes, Ly6C$^-$ monocytes lack potential to give rise to tissue-resident cells, such as intestinal macrophages (6), but their activities seem restricted to the vascular lumen.

Seminal work by the Geissmann group has shown that a fraction of the Ly6C$^-$ monocytes continuously patrol the vasculature of healthy tissues through lymphocyte function-associated antigen/intercellular adhesion molecule-dependent long-range crawling on the resting endothelium (99, 100). Likewise, human CD14^{dim}CD16$^+$ monocytes also showed patrolling activity when adoptively transferred into immunodeficient mice (2). Ly6C$^-$ monocytes display, with 2 days, a longer half-life than Ly6C$^+$ monocytes (8). Depending on the availability of CSF-1, which has, together with the alternative CSF-1R ligand interleukin-34 (101, 102), emerged as a critical factor controlling monocyte/macrophage maintenance (103, 104), the half-life of Ly6C$^-$ monocytes can moreover even be extended to almost 2 weeks (8). In further support of their macrophage identity, Ly6C$^-$ monocytes efficiently scavenge luminal microparticles in steady state (100) and were also reported to be critical for the removal of luminal vascular β-amyloid in a murine Alzheimer's disease model (105). Another key role of Ly6C$^-$ monocytes might be their surveillance of endothelial integrity. Following exposure to TLR7-mediated "danger" signals, endothelium was thus shown to induce the retention of Ly6C$^-$ monocytes. Simultaneously, the TLR7 ligand triggered monocytes to locally recruit neutrophils, which mediated focal endothelial necrosis, and the

Ly6C⁻ monocytes removed the cellular debris (100). Animals that display reduced numbers of patrolling Ly6C⁻ monocytes due to a missing CX3CR1 survival signal (106) or lack the cells altogether, such as NR4A1$^{-/-}$ mice (42), have not been reported to display spontaneous vascular abnormalities. However, they likely will be instrumental in deciphering additional physiological roles of these cells. Indeed, NR4A1$^{-/-}$ mice were shown to display increased lung metastasis in a tumor model, and transfer of wild-type Ly6C⁻ monocytes prevented tumor invasion of the lung (107). In this model, patrolling monocytes interacted with metastasizing tumor cells, scavenged tumor material from the lung vasculature, and promoted NK-cell recruitment and activation (107).

In-depth understanding of Ly6C⁻ monocytes clearly requires further investigation with intravital approaches. Their activities will likely be related to their prevalence in capillaries (100) and strategic positioning on the blood vessel wall. Ly6C⁻ monocytes thus highlight monocyte functions related to their specific location within the blood circulation. Intravascular functions of Ly6C⁺ monocytes, on the other hand, and interactions of the two monocyte subsets with other lymphoid and myeloid leukocytes remain poorly understood.

CONCLUDING REMARKS

Monocytes are defined mainly by their location in the bloodstream, and full appreciation of these fascinating cells will require understanding of their activities in this special compartment, including communication with other leukocytes and the vessel wall. In addition, Ly6C⁺ monocytes and their descendants have emerged as a third, highly plastic and dynamic cellular system that complements the two classical, tissue-resident mononuclear phagocyte compartments, i.e., macrophages and DCs, on demand. Given the central role of monocytes in homeostasis and pathology, multiparameter flow analysis and genomic profiling of these cells and their discrete subsets in human blood samples should be highly informative on the health state of individuals. Moreover, promising results from preclinical disease models suggest that monocytes represent attractive targets for therapeutic intervention for pathologies associated with acute and chronic inflammation.

Acknowledgments. *S.J. is supported by the European Research Council (340345).*

Citation. Mildner A, Marinkovic G, Jung S. 2016. Murine monocytes: origins, subsets, fates, and functions. Microbiol Spectrum 4(5): MCHD-0033-2016.

References

1. **Williams MJ.** 2007. *Drosophila* hemopoiesis and cellular immunity. *J Immunol* **178:**4711–4716.
2. **Cros J, Cagnard N, Woollard K, Patey N, Zhang SY, Senechal B, Puel A, Biswas SK, Moshous D, Picard C, Jais JP, D'Cruz D, Casanova JL, Trouillet C, Geissmann F.** 2010. Human CD14^{dim} monocytes patrol and sense nucleic acids and viruses via TLR7 and TLR8 receptors. *Immunity* **33:**375–386.
3. **Etzrodt M, Cortez-Retamozo V, Newton A, Zhao J, Ng A, Wildgruber M, Romero P, Wurdinger T, Xavier R, Geissmann F, Meylan E, Nahrendorf M, Swirski FK, Baltimore D, Weissleder R, Pittet MJ.** 2012. Regulation of monocyte functional heterogeneity by *miR-146a* and Relb. *Cell Rep* **1:**317–324.
4. **Ingersoll MA, Spanbroek R, Lottaz C, Gautier EL, Frankenberger M, Hoffmann R, Lang R, Haniffa M, Collin M, Tacke F, Habenicht AJ, Ziegler-Heitbrock L, Randolph GJ.** 2010. Comparison of gene expression profiles between human and mouse monocyte subsets. *Blood* **115:**e10–e19. doi:10.1182/blood-2009-07-235028.
5. **Mildner A, Chapnik E, Manor O, Yona S, Kim KW, Aychek T, Varol D, Beck G, Itzhaki ZB, Feldmesser E, Amit I, Hornstein E, Jung S.** 2013. Mononuclear phagocyte miRNome analysis identifies miR-142 as critical regulator of murine dendritic cell homeostasis. *Blood* **121:**1016–1027.
6. **Varol C, Vallon-Eberhard A, Elinav E, Aychek T, Shapira Y, Luche H, Fehling HJ, Hardt WD, Shakhar G, Jung S.** 2009. Intestinal lamina propria dendritic cell subsets have different origin and functions. *Immunity* **31:**502–512.
7. **MacDonald KP, Palmer JS, Cronau S, Seppanen E, Olver S, Raffelt NC, Kuns R, Pettit AR, Clouston A, Wainwright B, Branstetter D, Smith J, Paxton RJ, Cerretti DP, Bonham L, Hill GR, Hume DA.** 2010. An antibody against the colony-stimulating factor 1 receptor depletes the resident subset of monocytes and tissue- and tumor-associated macrophages but does not inhibit inflammation. *Blood* **116:**3955–3963.
8. **Yona S, Kim KW, Wolf Y, Mildner A, Varol D, Breker M, Strauss-Ayali D, Viukov S, Guilliams M, Misharin A, Hume DA, Perlman H, Malissen B, Zelzer E, Jung S.** 2013. Fate mapping reveals origins and dynamics of monocytes and tissue macrophages under homeostasis. *Immunity* **38:**79–91.
9. **Jung S, Aliberti J, Graemmel P, Sunshine MJ, Kreutzberg GW, Sher A, Littman DR.** 2000. Analysis of fractalkine receptor CX_3CR1 function by targeted deletion and green fluorescent protein reporter gene insertion. *Mol Cell Biol* **20:**4106–4114.
10. **Hoeffel G, Chen J, Lavin Y, Low D, Almeida FF, See P, Beaudin AE, Lum J, Low I, Forsberg EC, Poidinger M, Zolezzi F, Larbi A, Ng LG, Chan JK, Greter M, Becher B, Samokhvalov IM, Merad M, Ginhoux F.** 2015. C-Myb⁺ erythro-myeloid progenitor-derived fetal monocytes give rise to adult tissue-resident macrophages. *Immunity* **42:**665–678.

11. **Fogg DK, Sibon C, Miled C, Jung S, Aucouturier P, Littman DR, Cumano A, Geissmann F.** 2006. A clonogenic bone marrow progenitor specific for macrophages and dendritic cells. *Science* **311:**83–87.

12. **Varol C, Landsman L, Fogg DK, Greenshtein L, Gildor B, Margalit R, Kalchenko V, Geissmann F, Jung S.** 2007. Monocytes give rise to mucosal, but not splenic, conventional dendritic cells. *J Exp Med* **204:**171–180.

13. **Liu K, Victora GD, Schwickert TA, Guermonprez P, Meredith MM, Yao K, Chu FF, Randolph GJ, Rudensky AY, Nussenzweig M.** 2009. In vivo analysis of dendritic cell development and homeostasis. *Science* **324:**392–397.

14. **Hettinger J, Richards DM, Hansson J, Barra MM, Joschko AC, Krijgsveld J, Feuerer M.** 2013. Origin of monocytes and macrophages in a committed progenitor. *Nat Immunol* **14:**821–830.

15. **Paul F, Arkin Y, Giladi A, Jaitin DA, Kenigsberg E, Keren-Shaul H, Winter D, Lara-Astiaso D, Gury M, Weiner A, David E, Cohen N, Lauridsen FK, Haas S, Schlitzer A, Mildner A, Ginhoux F, Jung S, Trumpp A, Porse BT, Tanay A, Amit I.** 2015. Transcriptional heterogeneity and lineage commitment in myeloid progenitors. *Cell* **163:**1663–1677.

16. **Spitzer MH, Gherardini PF, Fragiadakis GK, Bhattacharya N, Yuan RT, Hotson AN, Finck R, Carmi Y, Zunder ER, Fantl WJ, Bendall SC, Engleman EG, Nolan GP.** 2015. IMMUNOLOGY. An interactive reference framework for modeling a dynamic immune system. *Science* **349:**1259425. doi:10.1126/science.1259425.

17. **Sathe P, Metcalf D, Vremec D, Naik SH, Langdon WY, Huntington ND, Wu L, Shortman K.** 2014. Lymphoid tissue and plasmacytoid dendritic cells and macrophages do not share a common macrophage-dendritic cell-restricted progenitor. *Immunity* **41:**104–115.

18. **Serbina NV, Pamer EG.** 2006. Monocyte emigration from bone marrow during bacterial infection requires signals mediated by chemokine receptor CCR2. *Nat Immunol* **7:**311–317.

19. **Shi C, Jia T, Mendez-Ferrer S, Hohl TM, Serbina NV, Lipuma L, Leiner I, Li MO, Frenette PS, Pamer EG.** 2011. Bone marrow mesenchymal stem and progenitor cells induce monocyte emigration in response to circulating Toll-like receptor ligands. *Immunity* **34:**590–601.

20. **Shi C, Pamer EG.** 2011. Monocyte recruitment during infection and inflammation. *Nat Rev Immunol* **11:** 762–774.

21. **Nguyen KD, Fentress SJ, Qiu Y, Yun K, Cox JS, Chawla A.** 2013. Circadian gene *Bmal1* regulates diurnal oscillations of Ly6C^{hi} inflammatory monocytes. *Science* **341:** 1483–1488.

22. **Griseri T, McKenzie BS, Schiering C, Powrie F.** 2012. Dysregulated hematopoietic stem and progenitor cell activity promotes interleukin-23-driven chronic intestinal inflammation. *Immunity* **37:**1116–1129.

23. **Heidt T, Sager HB, Courties G, Dutta P, Iwamoto Y, Zaltsman A, von Zur Muhlen C, Bode C, Fricchione GL, Denninger J, Lin CP, Vinegoni C, Libby P, Swirski FK, Weissleder R, Nahrendorf M.** 2014. Chronic variable stress activates hematopoietic stem cells. *Nat Med* **20:**754–758.

24. **Klonz A, Wonigeit K, Pabst R, Westermann J.** 1996. The marginal blood pool of the rat contains not only granulocytes, but also lymphocytes, NK-cells and monocytes: a second intravascular compartment, its cellular composition, adhesion molecule expression and interaction with the peripheral blood pool. *Scand J Immunol* **44:**461–469.

25. **Steppich B, Dayyani F, Gruber R, Lorenz R, Mack M, Ziegler-Heitbrock HW.** 2000. Selective mobilization of CD14^{+}CD16^{+} monocytes by exercise. *Am J Physiol Cell Physiol* **279:**C578–C586.

26. **Askenase MH, Han SJ, Byrd AL, Morais da Fonseca D, Bouladoux N, Wilhelm C, Konkel JE, Hand TW, Lacerda-Queiroz N, Su XZ, Trinchieri G, Grainger JR, Belkaid Y.** 2015. Bone-marrow-resident NK cells prime monocytes for regulatory function during infection. *Immunity* **42:**1130–1142.

27. **Netea MG, Quintin J, van der Meer JW.** 2011. Trained immunity: a memory for innate host defense. *Cell Host Microbe.* **9:**355–361.

28. **Saeed S, Quintin J, Kerstens HH, Rao NA, Aghajanirefah A, Matarese F, Cheng SC, Ratter J, Berentsen K, van der Ent MA, Sharifi N, Janssen-Megens EM, Ter Huurne M, Mandoli A, van Schaik T, Ng A, Burden F, Downes K, Frontini M, Kumar V, Giamarellos-Bourboulis EJ, Ouwehand WH, van der Meer JW, Joosten LA, Wijmenga C, Martens JH, Xavier RJ, Logie C, Netea MG, Stunnenberg HG.** 2014. Epigenetic programming of monocyte-to-macrophage differentiation and trained innate immunity. *Science* **345:**1251086. doi:10.1126/science.1251086.

29. **Jaitin DA, Kenigsberg E, Keren-Shaul H, Elefant N, Paul F, Zaretsky I, Mildner A, Cohen N, Jung S, Tanay A, Amit I.** 2014. Massively parallel single-cell RNA-seq for marker-free decomposition of tissues into cell types. *Science* **343:**776–779.

30. **Robbins CS, Chudnovskiy A, Rauch PJ, Figueiredo JL, Iwamoto Y, Gorbatov R, Etzrodt M, Weber GF, Ueno T, van Rooijen N, Mulligan-Kehoe MJ, Libby P, Nahrendorf M, Pittet MJ, Weissleder R, Swirski FK.** 2012. Extramedullary hematopoiesis generates Ly-6C^{high} monocytes that infiltrate atherosclerotic lesions. *Circulation* **125:** 364–374.

31. **Passlick B, Flieger D, Ziegler-Heitbrock HW.** 1989. Identification and characterization of a novel monocyte subpopulation in human peripheral blood. *Blood* **74:** 2527–2534.

32. **Ziegler-Heitbrock L.** 2015. Blood monocytes and their subsets: established features and open questions. *Front Immunol* **6:**423. doi:10.3389/fimmu.2015.00423.

33. **Palframan RT, Jung S, Cheng G, Weninger W, Luo Y, Dorf M, Littman DR, Rollins BJ, Zweerink H, Rot A, von Andrian UH.** 2001. Inflammatory chemokine transport and presentation in HEV: a remote control mechanism for monocyte recruitment to lymph nodes in inflamed tissues. *J Exp Med* **194:**1361–1373.

34. **Geissmann F, Jung S, Littman DR.** 2003. Blood monocytes consist of two principal subsets with distinct migratory properties. *Immunity* **19:**71–82.

35. **Jakubzick C, Gautier EL, Gibbings SL, Sojka DK, Schlitzer A, Johnson TE, Ivanov S, Duan Q, Bala S, Condon T, van Rooijen N, Grainger JR, Belkaid Y, Ma'ayan A, Riches DW, Yokoyama WM, Ginhoux F, Henson PM, Randolph GJ.** 2013. Minimal differentiation of classical monocytes as they survey steady-state tissues and transport antigen to lymph nodes. *Immunity* **39:**599–610.

36. **Qu C, Edwards EW, Tacke F, Angeli V, Llodrá J, Sanchez-Schmitz G, Garin A, Haque NS, Peters W, van Rooijen N, Sanchez-Torres C, Bromberg J, Charo IF, Jung S, Lira SA, Randolph GJ.** 2004. Role of CCR8 and other chemokine pathways in the migration of monocyte-derived dendritic cells to lymph nodes. *J Exp Med* **200:**1231–1241.

37. **Sunderkötter C, Nikolic T, Dillon MJ, Van Rooijen N, Stehling M, Drevets DA, Leenen PJ.** 2004. Subpopulations of mouse blood monocytes differ in maturation stage and inflammatory response. *J Immunol* **172:**4410–4417.

38. **Sugimoto C, Hasegawa A, Saito Y, Fukuyo Y, Chiu KB, Cai Y, Breed MW, Mori K, Roy CJ, Lackner AA, Kim WK, Didier ES, Kuroda MJ.** 2015. Differentiation kinetics of blood monocytes and dendritic cells in macaques: insights to understanding human myeloid cell development. *J Immunol* **195:**1774–1781.

39. **Kurotaki D, Osato N, Nishiyama A, Yamamoto M, Ban T, Sato H, Nakabayashi J, Umehara M, Miyake N, Matsumoto N, Nakazawa M, Ozato K, Tamura T.** 2013. Essential role of the IRF8-KLF4 transcription factor cascade in murine monocyte differentiation. *Blood* **121:**1839–1849.

40. **Feinberg MW, Wara AK, Cao Z, Lebedeva MA, Rosenbauer F, Iwasaki H, Hirai H, Katz JP, Haspel RL, Gray S, Akashi K, Segre J, Kaestner KH, Tenen DG, Jain MK.** 2007. The Kruppel-like factor KLF4 is a critical regulator of monocyte differentiation. *EMBO J* **26:** 4138–4148.

41. **Alder JK, Georgantas RW III, Hildreth RL, Kaplan IM, Morisot S, Yu X, McDevitt M, Civin CI.** 2008. Kruppel-like factor 4 is essential for inflammatory monocyte differentiation in vivo. *J Immunol* **180:**5645–5652.

42. **Hanna RN, Carlin LM, Hubbeling HG, Nackiewicz D, Green AM, Punt JA, Geissmann F, Hedrick CC.** 2011. The transcription factor NR4A1 (Nur77) controls bone marrow differentiation and the survival of Ly6C$^-$ monocytes. *Nat Immunol* **12:**778–785.

43. **Gautier EL, Shay T, Miller J, Greter M, Jakubzick C, Ivanov S, Helft J, Chow A, Elpek KG, Gordonov S, Mazloom AR, Ma'ayan A, Chua WJ, Hansen TH, Turley SJ, Merad M, Randolph GJ; Immunological Genome Consortium.** 2012. Gene-expression profiles and transcriptional regulatory pathways that underlie the identity and diversity of mouse tissue macrophages. *Nat Immunol* **13:**1118–1128.

44. **Lavin Y, Winter D, Blecher-Gonen R, David E, Keren-Shaul H, Merad M, Jung S, Amit I.** 2014. Tissue-resident macrophage enhancer landscapes are shaped by the local microenvironment. *Cell* **159:**1312–1326.

45. **Mehta A, Baltimore D.** 2016. MicroRNAs as regulatory elements in immune system logic. *Nat Rev Immunol* **16:** 279–294.

46. **Amit I, Winter DR, Jung S.** 2016. The role of the local environment and epigenetics in shaping macrophage identity and their effect on tissue homeostasis. *Nat Immunol* **17:**18–25.

47. **Lara-Astiaso D, Weiner A, Lorenzo-Vivas E, Zaretsky I, Jaitin DA, David E, Keren-Shaul H, Mildner A, Winter D, Jung S, Friedman N, Amit I.** 2014. Immunogenetics. Chromatin state dynamics during blood formation. *Science* **345:**943–949.

48. **Jaitin DA, Keren-Shaul H, Elefant N, Amit I.** 2015. Each cell counts: hematopoiesis and immunity research in the era of single cell genomics. *Semin Immunol* **27:** 67–71.

49. **Tussiwand R, Gautier EL.** 2015. Transcriptional regulation of mononuclear phagocyte development. *Front Immunol* **6:**533. doi:10.3389/fimmu.2015.00533.

50. **Glass CK, Natoli G.** 2016. Molecular control of activation and priming in macrophages. *Nat Immunol* **17:** 26–33.

51. **Friedman AD.** 2007. Transcriptional control of granulocyte and monocyte development. *Oncogene* **26:**6816–6828.

52. **Goldmann T, Wieghofer P, Müller PF, Wolf Y, Varol D, Yona S, Brendecke SM, Kierdorf K, Staszewski O, Datta M, Luedde T, Heikenwalder M, Jung S, Prinz M.** 2013. A new type of microglia gene targeting shows TAK1 to be pivotal in CNS autoimmune inflammation. *Nat Neurosci* **16:**1618–1626.

53. **Terry RL, Miller SD.** 2014. Molecular control of monocyte development. *Cell Immunol* **291:**16–21.

54. **Hanna RN, Shaked I, Hubbeling HG, Punt JA, Wu R, Herrley E, Zaugg C, Pei H, Geissmann F, Ley K, Hedrick CC.** 2012. NR4A1 (Nur77) deletion polarizes macrophages toward an inflammatory phenotype and increases atherosclerosis. *Circ Res* **110:**416–427.

55. **Shaked I, Hanna RN, Shaked H, Chodaczek G, Nowyhed HN, Tweet G, Tacke R, Basat AB, Mikulski Z, Togher S, Miller J, Blatchley A, Salek-Ardakani S, Darvas M, Kaikkonen MU, Thomas GD, Lai-Wing-Sun S, Rezk A, Bar-Or A, Glass CK, Bandukwala H, Hedrick CC.** 2015. Transcription factor Nr4a1 couples sympathetic and inflammatory cues in CNS-recruited macrophages to limit neuroinflammation. *Nat Immunol* **16:**1228–1234.

56. **van Furth R, Cohn ZA.** 1968. The origin and kinetics of mononuclear phagocytes. *J Exp Med* **128:**415–435.

57. **Serbina NV, Salazar-Mather TP, Biron CA, Kuziel WA, Pamer EG.** 2003. TNF/iNOS-producing dendritic cells mediate innate immune defense against bacterial infection. *Immunity* **19:**59–70.

58. **Segura E, Amigorena S.** 2013. Inflammatory dendritic cells in mice and humans. *Trends Immunol* **34:** 440–445.

59. **Wynn TA, Chawla A, Pollard JW.** 2013. Macrophage biology in development, homeostasis and disease. *Nature* **496:**445–455.

60. **Mildner A, Yona S, Jung S.** 2013. A close encounter of the third kind: monocyte-derived cells. *Adv Immunol* **120:**69–103.

61. **Murray PJ, Allen JE, Biswas SK, Fisher EA, Gilroy DW, Goerdt S, Gordon S, Hamilton JA, Ivashkiv LB, Lawrence T, Locati M, Mantovani A, Martinez FO, Mege JL, Mosser DM, Natoli G, Saeij JP, Schultze JL, Shirey KA, Sica A, Suttles J, Udalova I, van Ginderachter JA, Vogel SN, Wynn TA.** 2014. Macrophage activation and polarization: nomenclature and experimental guidelines. *Immunity* **41:**14–20.

62. **Sallusto F, Lanzavecchia A.** 1994. Efficient presentation of soluble antigen by cultured human dendritic cells is maintained by granulocyte/macrophage colony-stimulating factor plus interleukin 4 and downregulated by tumor necrosis factor α. *J Exp Med* **179:**1109–1118.

63. **Gosselin D, Link VM, Romanoski CE, Fonseca GJ, Eichenfield DZ, Spann NJ, Stender JD, Chun HB, Garner H, Geissmann F, Glass CK.** 2014. Environment drives selection and function of enhancers controlling tissue-specific macrophage identities. *Cell* **159:**1327–1340.

64. **Helft J, Böttcher JP, Chakravarty P, Zelenay S, Huotari J, Schraml BU, Goubau D, Reis E Sousa C.** 2016. Alive but confused: heterogeneity of CD11c[+] MHC class II[+] cells in GM-CSF mouse bone marrow cultures. *Immunity* **44:**3–4.

65. **Lacey DC, Achuthan A, Fleetwood AJ, Dinh H, Roiniotis J, Scholz GM, Chang MW, Beckman SK, Cook AD, Hamilton JA.** 2012. Defining GM-CSF- and macrophage-CSF-dependent macrophage responses by in vitro models. *J Immunol* **188:**5752–5765.

66. **Merad M, Sathe P, Helft J, Miller J, Mortha A.** 2013. The dendritic cell lineage: ontogeny and function of dendritic cells and their subsets in the steady state and the inflamed setting. *Annu Rev Immunol* **31:**563–604.

67. **Mildner A, Jung S.** 2014. Development and function of dendritic cell subsets. *Immunity* **40:**642–656.

68. **Yona S, Jung S.** 2010. Monocytes: subsets, origins, fates and functions. *Curr Opin Hematol* **17:**53–59.

69. **Ginhoux F, Greter M, Leboeuf M, Nandi S, See P, Gokhan S, Mehler MF, Conway SJ, Ng LG, Stanley ER, Samokhvalov IM, Merad M.** 2010. Supporting fate mapping analysis reveals that adult microglia derive from primitive macrophages. *Science* **330:**841–845.

70. **Schulz C, Gomez Perdiguero E, Chorro L, Szabo-Rogers H, Cagnard N, Kierdorf K, Prinz M, Wu B, Jacobsen SE, Pollard JW, Frampton J, Liu KJ, Geissmann F.** 2012. A lineage of myeloid cells independent of Myb and hematopoietic stem cells. *Science* **336:**86–90.

71. **Hashimoto D, Chow A, Noizat C, Teo P, Beasley MB, Leboeuf M, Becker CD, See P, Price J, Lucas D, Greter M, Mortha A, Boyer SW, Forsberg EC, Tanaka M, van Rooijen N, García-Sastre A, Stanley ER, Ginhoux F, Frenette PS, Merad M.** 2013. Tissue-resident macrophages self-maintain locally throughout adult life with minimal contribution from circulating monocytes. *Immunity* **38:**792–804.

72. **Gomez Perdiguero E, Klapproth K, Schulz C, Busch K, Azzoni E, Crozet L, Garner H, Trouillet C, de Bruijn MF, Geissmann F, Rodewald HR.** 2015. Tissue-resident macrophages originate from yolk-sac-derived erythro-myeloid progenitors. *Nature* **518:**547–551.

73. **Ginhoux F, Jung S.** 2014. Monocytes and macrophages: developmental pathways and tissue homeostasis. *Nat Rev Immunol* **14:**392–404.

74. **Bain CC, Bravo-Blas A, Scott CL, Gomez Perdiguero E, Geissmann F, Henri S, Malissen B, Osborne LC, Artis D, Mowat AM.** 2014. Constant replenishment from circulating monocytes maintains the macrophage pool in the intestine of adult mice. *Nat Immunol* **15:**929–937.

75. **Zigmond E, Jung S.** 2013. Intestinal macrophages: well educated exceptions from the rule. *Trends Immunol* **34:**162–168.

76. **Molawi K, Wolf Y, Kandalla PK, Favret J, Hagemeyer N, Frenzel K, Pinto AR, Klapproth K, Henri S, Malissen B, Rodewald HR, Rosenthal NA, Bajenoff M, Prinz M, Jung S, Sieweke MH.** 2014. Progressive replacement of embryo-derived cardiac macrophages with age. *J Exp Med* **11:**2151–2158.

77. **Malissen B, Tamoutounour S, Henri S.** 2014. The origins and functions of dendritic cells and macrophages in the skin. *Nat Rev Immunol* **14:**417–428.

78. **Ginhoux F, Guilliams M.** 2016. Tissue-resident macrophage ontogeny and homeostasis. *Immunity* **44:**439–449.

79. **Sheng J, Ruedl C, Karjalainen K.** 2015. Most tissue-resident macrophages except microglia are derived from fetal hematopoietic stem cells. *Immunity* **43:**382–393.

80. **Bain CC, Hawley CA, Garner H, Scott CL, Schridde A, Steers NJ, Mack M, Joshi A, Guilliams M, Mowat AM, Geissmann F, Jenkins SJ.** 2016. Long-lived self-renewing bone marrow-derived macrophages displace embryo-derived cells to inhabit adult serous cavities. *Nat Commun* **7:**s11852.

81. **Scott CL, Zheng F, De Baetselier P, Martens L, Saeys Y, De Prijck S, Lippens S, Abels C, Schoonooghe S, Raes G, Devoogdt N, Lambrecht BN, Beschin A, Guilliams M.** 2016. Bone marrow-derived monocytes give rise to self-renewing and fully differentiated Kupffer cells. *Nat Commun* **7:**10321. doi:10.1038/ncomms10321.

82. **van de Laar L, Saelens W, De Prijck S, Martens L, Scott CL, Van Isterdael G, Hoffmann E, Beyaert R, Saeys Y, Lambrecht BN, Guilliams M.** 2016. Yolk sac macrophages, fetal liver, and adult monocytes can colonize an empty niche and develop into functional tissue-resident macrophages. *Immunity* **44:**755–768.

83. **Bruttger J, Karram K, Wörtge S, Regen T, Marini F, Hoppmann N, Klein M, Blank T, Yona S, Wolf Y, Mack M, Pinteaux E, Müller W, Zipp F, Binder H, Bopp T, Prinz M, Jung S, Waisman A.** 2015. Genetic cell ablation reveals clusters of local self-renewing microglia in the mammalian central nervous system. *Immunity* **43:**92–106.

84. Nathan C, Ding A. 2010. Nonresolving inflammation. *Cell* 140:871–882.

85. Serhan CN, Brain SD, Buckley CD, Gilroy DW, Haslett C, O'Neill LA, Perretti M, Rossi AG, Wallace JL. 2007. Resolution of inflammation: state of the art, definitions and terms. *FASEB J* 21:325–332.

86. Blériot C, Dupuis T, Jouvion G, Eberl G, Disson O, Lecuit M. 2015. Liver-resident macrophage necroptosis orchestrates type 1 microbicidal inflammation and type-2-mediated tissue repair during bacterial infection. *Immunity* 42:145–158.

87. Yamasaki R, Lu H, Butovsky O, Ohno N, Rietsch AM, Cialic R, Wu PM, Doykan CE, Lin J, Cotleur AC, Kidd G, Zorlu MM, Sun N, Hu W, Liu L, Lee JC, Taylor SE, Uehlein L, Dixon D, Gu J, Floruta CM, Zhu M, Charo IF, Weiner HL, Ransohoff RM. 2014. Differential roles of microglia and monocytes in the inflamed central nervous system. *J Exp Med* 211:1533–1549.

88. Mildner A, Mack M, Schmidt H, Brück W, Djukic M, Zabel MD, Hille A, Priller J, Prinz M. 2009. CCR2 $^{+}$Ly-6C^{hi} monocytes are crucial for the effector phase of autoimmunity in the central nervous system. *Brain* 132: 2487–2500.

89. Ajami B, Bennett JL, Krieger C, McNagny KM, Rossi FMV. 2011. Infiltrating monocytes trigger EAE progression, but do not contribute to the resident microglia pool. *Nat Neurosci* 14:1142–1149.

90. Franklin RA, Liao W, Sarkar A, Kim MV, Bivona MR, Liu K, Pamer EG, Li MO. 2014. The cellular and molecular origin of tumor-associated macrophages. *Science* 344:921–925.

91. Lahmar Q, Keirsse J, Laoui D, Movahedi K, Van Overmeire E, Van Ginderachter JA. 2016. Tissue-resident versus monocyte-derived macrophages in the tumor microenvironment. *Biochim Biophys Acta* 1865: 23–34.

92. Brühl H, Cihak J, Plachý J, Kunz-Schughart L, Niedermeier M, Denzel A, Rodriguez Gomez M, Talke Y, Luckow B, Stangassinger M, Mack M. 2007. Targeting of Gr-1+,CCR2+ monocytes in collagen-induced arthritis. *Arthritis Rheum* 56:2975–2985.

93. Zigmond E, Varol C, Farache J, Elmaliah E, Satpathy AT, Friedlander G, Mack M, Shpigel N, Boneca IG, Murphy KM, Shakhar G, Halpern Z, Jung S. 2012. Ly6C^{hi} monocytes in the inflamed colon give rise to proinflammatory effector cells and migratory antigen presenting cells. *Immunity* 37:1076–1090.

94. Leuschner F, Dutta P, Gorbatov R, Novobrantseva TI, Donahoe JS, Courties G, Lee KM, Kim JI, Markmann JF, Marinelli B, Panizzi P, Lee WW, Iwamoto Y, Milstein S, Epstein-Barash H, Cantley W, Wong J, Cortez-Retamozo V, Newton A, Love K, Libby P, Pittet MJ, Swirski FK, Koteliansky V, Langer R, Weissleder R, Anderson DG, Nahrendorf M. 2011. Therapeutic siRNA silencing in inflammatory monocytes in mice. *Nat Biotechnol* 29:1005–1010.

95. Terry RL, Getts MT, Deffrasnes C, Müller M, van Vreden C, Ashhurst TM, Chami B, McCarthy D, Wu H, Ma J, Martin A, Shae LD, Witting P, Kansas GS, Kühn J, Hafezi W, Campbell IL, Reilly D, Say J, Brown L, White MY, Cordwell SJ, Chadban SJ, Thorp EB, Bao S, Miller SD, King NJ. 2014. Therapeutic inflammatory monocyte modulation using immune-modifying microparticles. *Science* 6:219ra7. doi:10.1126/scitranslmed.3007563.

96. Nahrendorf M, Swirski FK, Aikawa E, Stangenberg L, Wurdinger T, Figueiredo JL, Libby P, Weissleder R, Pittet MJ. 2007. The healing myocardium sequentially mobilizes two monocyte subsets with divergent and complementary functions. *J Exp Med* 204:3037–3047.

97. Arnold L, Henry A, Poron F, Baba-Amer Y, van Rooijen N, Plonquet A, Gherardi RK, Chazaud B. 2007. Inflammatory monocytes recruited after skeletal muscle injury switch into antiinflammatory macrophages to support myogenesis. *J Exp Med* 204:1057–1069.

98. McGovern N, Schlitzer A, Gunawan M, Jardine L, Shin A, Poyner E, Green K, Dickinson R, Wang XN, Low D, Best K, Covins S, Milne P, Pagan S, Aljefri K, Windebank M, Miranda-Saavedra D, Larbi A, Wasan PS, Duan K, Poidinger M, Bigley V, Ginhoux F, Collin M, Haniffa M. 2014. Human dermal CD14^{+} cells are a transient population of monocyte-derived macrophages. *Immunity* 41: 465–477.

99. Auffray C, Fogg D, Garfa M, Elain G, Join-Lambert O, Kayal S, Sarnacki S, Cumano A, Lauvau G, Geissmann F. 2007. Monitoring of blood vessels and tissues by a population of monocytes with patrolling behavior. *Science* 317:666–670.

100. Carlin LM, Stamatiades EG, Auffray C, Hanna RN, Glover L, Vizcay-Barrena G, Hedrick CC, Cook HT, Diebold S, Geissmann F. 2013. *Nr4a1*-dependent Ly6C^{low} monocytes monitor endothelial cells and orchestrate their disposal. *Cell* 153:362–375.

101. Wang Y, Szretter KJ, Vermi W, Gilfillan S, Rossini C, Cella M, Barrow AD, Diamond MS, Colonna M. 2012. IL-34 is a tissue-restricted ligand of CSF1R required for the development of Langerhans cells and microglia. *Nat Immunol* 13:753–760.

102. Greter M, Lelios I, Pelczar P, Hoeffel G, Price J, Leboeuf M, Kündig TM, Frei K, Ginhoux F, Merad M, Becher B. 2012. Stroma-derived interleukin-34 controls the development and maintenance of Langerhans cells and the maintenance of microglia. *Immunity* 37: 1050–1060.

103. Bartocci A, Mastrogiannis DS, Migliorati G, Stockert RJ, Wolkoff AW, Stanley ER. 1987. Macrophages specifically regulate the concentration of their own growth factor in the circulation. *Proc Natl Acad Sci U S A* 84: 6179–6183.

104. Elmore MR, Najafi AR, Koike MA, Dagher NN, Spangenberg EE, Rice RA, Kitazawa M, Matusow B, Nguyen H, West BL, Green KN. 2014. Colony-stimulating factor 1 receptor signaling is necessary for microglia viability, unmasking a microglia progenitor cell in the adult brain. *Neuron* 82:380–397.

105. Michaud JP, Bellavance MA, Préfontaine P, Rivest S. 2013. Real-time in vivo imaging reveals the ability of monocytes to clear vascular amyloid beta. *Cell Rep* 5: 646–653.

106. **Landsman L, Bar-On L, Zernecke A, Kim KW, Krauthgamer R, Shagdarsuren E, Lira SA, Weissman IL, Weber C, Jung S.** 2009. CX3CR1 is required for monocyte homeostasis and atherogenesis by promoting cell survival. *Blood* **113:**963–972.

107. **Hanna RN, Cekic C, Sag D, Tacke R, Thomas GD, Nowyhed H, Herrley E, Rasquinha N, McArdle S, Wu R, Peluso E, Metzger D, Ichinose H, Shaked I, Chodaczek G, Biswas SK, Hedrick CC.** 2015. Patrolling monocytes control tumor metastasis to the lung. *Science* **350:**985–990.

Myeloid Cells in Health and Disease: A Synthesis
Edited by Siamon Gordon

doi:10.1128/microbiolspec.MCHD-0046-2016

Jonathan M. Austyn[1]

10 Dendritic Cells in the Immune System—History, Lineages, Tissues, Tolerance, and Immunity

Ralph Steinman was posthumously awarded a Nobel Prize in 2011 "for his discovery of the dendritic cell and its role in adaptive immunity." He first coined the term "dendritic cell" in 1973 while working with Zanvil Cohn. His findings laid the foundations for a new field of immunology. This has enormous therapeutic potential for new vaccination strategies against cancer and infectious diseases. It also holds promise for new therapies for autoimmune disease, allergy, and transplantation reactions.

This review seeks to provide a conceptual framework for understanding dendritic cells (DCs). It focuses mainly on DCs of mice, and to a lesser extent humans, although a menagerie of other creatures will be mentioned. It deals mainly with DCs and "conventional" T cells in the adaptive immune system, but other, specialized "innate-like" T cells will be noted. A number of recent reviews (cited later) contain much additional information, including many molecular details not covered here.

The introduction to this review provides essential background information on the immune system and mechanisms of tolerance and immunity, particularly for nonspecialist readers. Further general information can be found in standard texts (e.g., references 1–4). With this background, the remaining sections may be read selectively or consecutively. To aid cross-referencing, the five central sections are cited in brief as "History," "Lineages," "Tissues," "Tolerance," and "Immunity." The review ends with a discussion and conclusion.

INTRODUCTION

A crucial function of the immune system is to provide defense against infection, or immunity. It is conventionally divided into a phylogenetically ancient innate immune system, which exists in all animals, and an adaptive immune system, which evolved later in jawed vertebrates (gnathostomes), including mice and humans. The immune system has been described as "the most efficient killing machine ever created." It is therefore essential that it does not attack its host and maintains a state of unresponsiveness, or "tolerance," to its own tissues. In addition, it must establish a state of mutualism with

[1]Nuffield Department of Surgical Sciences, University of Oxford, John Radcliffe Hospital, Oxford OX3 9DU, United Kingdom.

the beneficial populations of commensal organisms that populate the skin and tracts of mucosal tissues, and which have helped shape the evolution of the immunity system. DCs appear to have evolved around the time that adaptive immunity "suddenly" appeared. They play a fundamental role in linking the innate and adaptive immune systems together so they can work together synergistically. DCs are crucially important in tolerance and immunity.

Infection and Immunity

All living things, both plants and animals, must defend themselves against infection. The types of infectious agent that would infect and inhabit larger eukaryotic animals are the viruses; the prokaryotic bacteria; and the eukaryotic groups of fungi, single-celled protozoa, and multicellular helminths. To infect, they must cross the defensive barriers provided by host epithelia, including the external surfaces of the skin and the internal (but topologically external) mucosal surfaces. These preexisting barriers to infection have their own forms of constitutive defense, provided, for example, by antiviral and antimicrobial proteins such as defensins and cathelicidins. Having gained access to the internal milieu of the host, in order to survive and replicate, some infectious agents invade intracellular niches (typically viruses, intracellular bacteria, and protozoa), while others adopt an extracellular habit (commonly pyogenic bacteria and other protozoa). Other organisms survive and replicate either within or outside of the epithelial barriers, as exemplified by fungi colonizing the skin or helminths cohabiting within the mucosal tracts of the intestine. Hence, to survive, any multicellular animal must be able to eliminate or at least control pathogens from any class of infectious agent. In the case of humans, these range in size over 8 or more orders of magnitude from the smallest viruses to longest tapeworms, can be located both within and outside of host cells, and may have enormously complex and transformative life cycles (the malarial parasite, for example).

Innate Immunity

All plants can defend themselves against infection, although they do not possess an immune system as such. The innate immune system emerged with the dawning of the animal kingdom and particularly with the evolution of the first multicellular organisms. Innate immunity provides the earliest molecular and cellular, and often system-wide, mechanisms of defense that are rapidly triggered by infection. For example, phagocytes such as macrophages can capture and internalize many bacteria, which are then degraded. The soluble complement system can also rapidly be activated to coat microbes and help macrophages do this. Fundamentally, innate responses have several crucial functions. First, they may rapidly and efficiently eliminate infectious agents. In general, immunological mechanisms that effect, or bring about, the elimination or control of infection are termed effector mechanisms. Second, early innate responses typically cause local inflammation. This enables yet more molecular and cellular effectors to be recruited from the blood into local sites of infection, further amplifying their efficacy. Third, innate responses can lead to later system-wide modulations of organ function. These result, for example, in fever and "sickness behavior" through actions on the brain, and the so-called acute-phase response produced largely by the liver. In the latter case, extraordinarily high quantities of new molecules are pumped into the bloodstream. Many of these seem to be opsonins, which coat infectious agents, bind to complementary receptors on cells such as macrophages, and promote uptake and clearance. Finally, innate responses are absolutely essential for the subsequent initiation and regulation of adaptive immune responses (see below).

So-called pattern recognition receptors (PRRs) are essential for innate immunity. These cellular receptors enable different types of infectious agents to be sensed with a high degree of accuracy. This is because PRRs can recognize highly conserved molecular motifs on which infectious agents depend for their integrity, replication, and survival. These components, which cannot be synthesized by the host, are generally known as pathogen-associated molecular patterns (PAMPs). The PRRs also facilitate precise discrimination between classes of infectious agents, for example, by recognition of lipopolysaccharide as a component of Gram-negative bacterial cell walls, absent in Gram-positive species. Similar molecules also exist in soluble form and are sometimes termed pattern recognition molecules. They include molecules such as C1q and mannose-binding lectin of the complement system. These can bind directly to bacterial surfaces or to opsonic molecules attached to them, and trigger the so-called classical and lectin pathways, which both result in complement activation. Arguably, PRRs enable the innate immune system to discriminate perfectly between infectious nonself and healthy self.

Some PRRs, such as scavenger receptors and members of the C-type lectin receptor family, can promote the uptake and internalization of infectious agents through receptor-mediated endocytosis or phagocytosis, by macrophages, for example. Other PRRs, however, trigger rapid secretion of proinflammatory cytokines and/or type I interferons (IFN-I), which contribute to rapid initiation of inflammatory and antiviral responses, respectively. These PRRs include the Toll-like receptors (TLRs), associated with the plasma membrane and endocytic compartments of cells such as macrophages, and the RIG-I (retinoic acid-inducible gene I)-like and NOD (nucleotide oligomerization domain)-like receptors in the cytosol. PRRs are also known to sense cell or tissue damage and death, whether or not it results from infection. This is through their capacity to recognize so-called damage-associated molecular patterns (DAMPs). One example is extracellular ATP, which in fact acts as a DAMP in animals as well as plants. Hence, through such mechanisms, innate immunity can rapidly sense and discriminate between different types of "danger" that may be presented to the host and trigger the responses needed for defense and restoration of homeostasis.

The rapid initiation of innate responses critically depends on the different types of cells that are present at sites of infection, all of which express PRRs. These include epithelial cells of tissues such as skin and the mucosae (the gut, lung and airways, and urogenital tract). Under physiological conditions, most tissues also contain resident populations of macrophages and mast cells and, in many, recently described innate lymphoid cells (ILCs). Collectively, it may be convenient to think of these as "alarm cells" that sense danger and rapidly sound "alarm signals" that are provided by molecular alarmins. The latter include the proinflammatory cytokines and lipid mediators that induce local inflammation. Subsequently, large numbers of reinforcements are recruited from the blood to help fight the danger and repair the damage. These include the granulocytes—neutrophils, or basophils and eosinophils—and later monocytes, as well as complement components and "natural" antibodies.

The abundant cellular expression of multiple types of different PRRs that are "hard-wired" to induce rapid cellular responses, as well as soluble forms that can be quickly activated, ensures that innate responses to infection are extremely fast, generally large, and often (though not always) highly effective. DCs are also present in tissues. They too can detect infection, tissue damage, and cell death through their own expression of PRRs and by sensing of alarmins, and also contribute to innate responses. However, they also have pivotal functions in the adaptive immune system, in the relatively few species that have one.

Adaptive Immunity

The vast majority of animal species on the planet survive and continue to evolve with the innate immune system as their only means of defense against infection. And yet, around the time of the Cambrian explosion some 540 million years ago, the adaptive immune system evolved within a remarkably short period of geological time (5). This provides immunity that is based on two main types of lymphocytes, T cells and antibody-producing B cells. Each has its own specialized type of receptor that enables it to recognize components of infectious agents or antigens: T-cell receptors (TCRs), and B-cell receptors (BCRs) which can be secreted as antibodies. This new type of immunity, as recognized in humans, probably first evolved in jawed fish. Through the millennia, with the emergence of all other vertebrates, adaptive immunity has coopted the mechanisms of innate immunity, with which it coexists, for its own purposes. Recent evidence indicates that an alternative form of adaptive immunity originated even earlier in jawless fish (agnathans), such as lampreys and hagfish, but is apparently an evolutionary dead end. In this review, the term "adaptive immunity" will be used only for the type that exists in jawed vertebrates (gnathostomes).

Precisely why the adaptive immune system "suddenly" appeared remains an almost complete mystery. However, according to the 2R hypothesis, it may have been facilitated by two entire genome replications before or around the time of the Cambrian explosion; even a third round may have occurred in some bony fish. If so, it would have enabled the different classes of antigen receptors (BCRs and TCRs) to evolve on paralogous chromosomes from an ancestral gene locus. This form of immunity probably also depended on the incorporation into the genome of a transposon, perhaps from a virus, that encodes the recombinase-activating gene (*RAG*). (There are in fact two *RAG* genes in gnathostomes, but only one in agnathans.) *RAG* enables a remarkable genetic mechanism known as somatic recombination to operate in developing T cells and B cells (but in no other cell type). This generates the extremely diverse sets of TCRs and BCRs. These lymphocyte antigen receptors are clonally distributed such that, at any one time, all those expressed by any T cell or B cell are essentially the same. They are in fact encoded by germ line gene segments that can be joined by somatic recombination to form functional genes. Together with additional mechanisms of diversification, a quite astonishing total number of receptors could be generated from extremely small portions of the genome that would otherwise require billions or trillions more DNA.

T cells and B cells, and their respective receptors, have very different but complementary functions in adaptive immunity. BCRs recognize small portions (epitopes) of extracellular molecules that typically retain their native, three-dimensional conformations. Hence, antibodies can bind directly to antigens such as viruses and bacteria before they infect cells. In contrast, so-called conventional T cells express αβ TCRs. In general, αβ TCRs can recognize intracellular antigens such as viruses and bacteria that have infected or been internalized by host cells. In actual fact, they recognize peptides derived from antigens inside cells that have been bound to classical major histocompatibility complex (MHC) molecules, which enable their transport from intracellular locations up to the cell surface. Thus, αβ TCRs recognize peptide-MHC complexes but not free peptides alone, and are often said to be MHC restricted. Classical MHC molecules are expressed by host cells so that they can be continuously surveyed by T cells and recognized if they have become infected by infectious agents or otherwise contain them. Conventionally, host cells are said to present their antigens to T cells and are hence termed antigen-presenting cells (APCs); the process as a whole is known as antigen presentation. In this review, the terms "conventional" and "classical" (see above) will normally be omitted, but they are used here since, for example, some "unconventional" T cells express γδ TCRs instead and some of these can recognize other, "nonclassical" MHC molecules.

Overall, antigen recognition by lymphocytes is anticipatory in nature, since vast repertoires of antigen receptors are generated in any individual even before any infection has occurred. It is also highly discriminatory, since even the slightest molecular change in an epitope may abolish recognition. It is, however, highly promiscuous, in that any given receptor has the potential to cross-react with perhaps millions of different epitopes from unrelated antigens that just happen to have the same shape (6). (The same is actually also true of MHC molecules binding peptides.) Functionally, what this means is that there is an extremely high chance that the adaptive immune system will be able to recognize an antigen from any given infectious agent it may encounter in the future, but the frequency of lymphocytes that can recognize any given epitope is extremely low. Hence, in any adaptive immune response, only these extremely rare clones of lymphocytes that recognize antigen can participate. Originating from a hypothesis put forward decades ago, this is known as clonal selection (by antigen). These few clones first need to be activated, since naive (antigen-inexperienced) lymphocytes are resting and impotent. They also need to be greatly expanded in numbers through multiple rounds of proliferation (or clonal expansion) before they can reach a critical mass to function effectively. In addition, lymphocytes are in fact highly plastic cells that can generate different types of effector responses (see below) but need to be "instructed" as to which type they should produce, as well as where they should go. All this takes time. For these reasons, the first, primary adaptive response produced against any infectious agent appears to be extremely slow. As measured, for example, by the frequency of antigen-specific T cells and antibodies, it may take days to reach a peak, in contrast to innate responses measured in minutes or hours. These responses are, however, much faster if the same infection occurs again, because the adaptive immune system can generate an immunological memory of what it has previously encountered (see below). DCs have essential roles in all stages of adaptive immune responses, and we now know they also enable these responses to be initiated much more quickly than it might seem.

Evolution of lymphoid tissues

The evolution of adaptive immunity necessitated the coevolution of specialized lymphoid organs and tissues for generating lymphocytes and regulating their responses (5). These are conventionally divided into primary lymphoid organs, where lymphocytes develop, and secondary lymphoid organs (SLOs; this abbreviation also encompasses secondary lymphoid tissues), where the fully developed lymphocytes can be activated and regulated during the adaptive immune response. The former include the thymus and sometimes the bone marrow. The latter include the spleen and lymph nodes, together with specialized mucosa-associated lymphoid tissue (MALT) such as Peyer's patches in the small intestine. All jawed vertebrates have a thymus, and a related structure known as the thymoid appears to exist even in earlier agnathans. In some species, there are additional specialized primary organs where B cells are generated. In humans and mice, B cells undergo most of their development in bone marrow, but in chickens they develop in the bursa of Fabricius, which is where the term "B" cell originated. All gnathostomes also have a spleen, which has been retained throughout evolution from jawed fish to humans. Lymph nodes evolved much later and perhaps first originated in some (but not all) species of birds, though related aggregates of lymphoid tissue are present in reptiles and amphibians. However, Peyer's patches are absent in birds and appear to have evolved even more recently. Hence, while adaptive immunity arose within a very short period of geological time, the different lymphoid tissues have continued to evolve in both structure and most probably function. Notably, with time, lymphoid tissues also became increasingly organized. Typically, T cells became segregated in distinct T-cell zones separate from the B cells in follicles (and, in birds and mammals, within germinal centers, which can develop from them). DCs have been identified in primary lymphoid organs and SLOs of all species of jawed vertebrates that have been closely studied. This suggests that they have played important roles in adaptive immunity throughout its evolution.

Immunological tolerance

In innate immunity, the expression of PRRs that can recognize PAMPs and DAMPs enables precise discrimination between infectious nonself and normal self components. This situation is, however, completely different in adaptive immunity. Because the somatic recombination of antigen receptor gene segments during BCR and TCR generation is an inherently random process, any newly formed receptor might potentially be able to recognize a self-antigen. Therefore, during lymphocyte development, cells with newly created autoreactive antigen receptors that recognize self-antigens with a high affinity are typically deleted from the repertoire. Because this process occurs within the primary, or central, lymphoid organs (i.e., the thymus and bone marrow in mice and humans), it contributes to what is known as central tolerance. It is, however, not perfect. Hence, there are additional mechanisms that can suppress the function of lymphocytes with autoreactive receptors if they escape from the thymus and enter peripheral SLOs and nonlymphoid tissues (NLTs). This can involve deletion of the autoreactive cells or the imposition of a state of unresponsiveness, or anergy, prior to their removal. Importantly, however, specialized populations of T cells, known as regulatory T cells (Tregs), can be generated that actively suppress potentially damaging autoreactive responses in the periphery. Some are generated within the thymus itself and are known as natural or thymically induced Tregs (tTregs). Others are generated within peripheral tissues and may also be produced during adaptive responses. These are collectively known as induced or peripherally induced Tregs (pTregs). DCs may play crucial roles in both central tolerance and the induction and maintenance of peripheral tolerance.

T-cell responses

Innate immune responses are absolutely essential for the subsequent induction of adaptive immune responses that are mediated by populations of αβ T cells and conventional B cells. This is because antigen-specific lymphocytes that recognize their cognate epitopes need to be instructed as to its "context," in other words, whether the source of that antigen is "dangerous" or not. If it is, they can be stimulated to mount a response, while if not, they may develop into Tregs, become unresponsive, or be killed. In the case of T cells, DCs are essential for helping to place the antigen in its context. Importantly, this is because if DCs sense danger themselves, they induce the expression of specialized costimulatory molecules that help to trigger T-cell activation; if not, they don't. Hence, generally speaking, two different sets of signals need to be delivered to a T cell before it can respond: the first is delivered through the TCR, which recognizes its cognate peptide-MHC complex (derived from the antigen the DC has acquired); the second is delivered through costimulation (provided by the DC if it has sensed danger). A similar situation actually also applies to many B-cell responses. These can be greatly increased if the antigen they recognize is bound to the specialized C3d component, which can be generated by the complement system after it is activated during innate responses. The requirement for two signals for lymphocyte activation ensures that provided there is no danger, and hence no innate response, recognition of any self component in normal tissues helps to reinforce peripheral tolerance (see above).

While all conventional T cells express αβ TCRs, they can contribute to fundamentally different types of adaptive responses. There are two types of T cells, defined by expression of CD8 or CD4; two types of classical MHC molecules, class I and class II; and two main sources of peptides that can be loaded onto the latter, from intracellular and extracellular antigens, respectively. In general, $CD8^+$ T cells recognize peptide–MHC-I complexes after peptides have been generated in the cytosol from "endogenous" antigens, such as viruses that have infected cells. These cells can develop into cytotoxic T lymphocytes (CTLs), which can kill the cells they recognize. Because essentially all cells can express MHC-I molecules, $CD8^+$ T cells can therefore help to eliminate any host cell that has been infected. In contrast, $CD4^+$ T cells recognize peptide–MHC-II complexes after the peptides have been generated within endosomal compartments from "exogenous" antigens, such as bacteria that were phagocytosed. These cells can develop into effector cells and acquire the capacity to secrete hormone-like cytokines and deliver other molecular signals that stimulate the effector functions of the cell that they recognize (in other cases they function as Tregs, as noted above). The constitutive expression of MHC-II is restricted to a very few specialized cell types, such as DCs and B cells, although expression can be induced in others, such as epithelial cells, during adaptive responses. In general, therefore, these two types of T-cell responses are quite segregated. An exception to this rule is cross-presentation, whereby peptides from exogenous antigens can be rerouted for loading onto MHC-I molecules. Roughly speaking, this enables specialized populations of DCs, for example, to initiate $CD8^+$ T-cell responses even if they have not been infected. Conversely, peptides from endogenous antigens may be rerouted onto MHC-II molecules during the cellular process of autophagy, when a cell digests its own cytoplasmic organelles.

In general, so-called helper $CD4^+$ T-cell responses are absolutely essential for subsequent induction of other types of adaptive responses. For example, they are generally required to activate conventional B cells, which may then develop into antibody-secreting plasma cells. In this way $CD4^+$ T cells can help to stimulate robust antibody responses of different types, typically against protein antigens from infectious agents. Generally speaking, in mice and humans, these antibodies are usually IgG, IgA, or IgE for systemic, mucosal, or barrier defenses, respectively. (IgM is the first type of antibody that is secreted in all T-dependent [TD] responses, but IgD usually remains membrane bound, though it can also be secreted for defense.) In many cases, helper T cells are also essential for activation of $CD8^+$ T cells and their subsequent development into CTLs that can kill infected cells. These types of response are therefore termed helper-dependent or TD responses. (This is to distinguish them from T-independent responses that can also be produced by other specialized populations of innate-like B cells, not considered here.)

A central paradigm in current immunology is that $CD4^+$ T cells can adopt a relatively small number of characteristic polarized types of response in different infectious settings. The best known of these are mediated by so-called Th17, Th1, and Th2 cells, often referred to as $CD4^+$ T-cell subsets. For example, many infections caused by extracellular bacteria and fungi can induce Th17 responses, which recruit neutrophils to the site of infection and may stimulate the production of IgA by plasma cells. Alternatively, many viruses, intracellular bacteria, and protozoa more typically induce Th1 responses, which can activate potent antimicrobial mechanisms of macrophages and induce specific types of IgG response. In contrast, helminths (and perhaps hematogenous insects and arachnids biting the skin) usually elicit Th2 responses, which can recruit basophils and eosinophils to the site of infection, elicit IgE responses, and help to maintain or restore barrier defenses. In each setting, the particular combinations of effector mechanisms that are induced can then synergize to bring about a coordinated and integrated response that is best suited to eliminate or control the infectious agent in question.

Immunological memory

During primary T-cell responses, effector cells are generated and many of these migrate from SLOs to sites of infection within NLTs. Here the effector $CD4^+$ T cells and $CD8^+$ CTLs can contribute to the clearance or control of infection in different ways, noted above. However, during primary responses, memory T cells can also be generated, which remain long after the antigen has been eliminated and the bulk of effector cells that were generated have been cleared. The same is true for B-cell responses, which can also generate memory B cells. These antigen-specific clones of memory cells persist in much higher numbers than were present in the naive cell repertoires, and are more easily and rapidly reactivated. Hence, they enable the adaptive immune system to respond very rapidly to subsequent (secondary, tertiary, etc.) encounters with the same antigen, so much so that an infectious agent can be rapidly eliminated without causing any disease. Memory cells can provide protection against any pathogen to which they were initially induced and are long-lived cells. Hence, they provide immunological memory that can often last for a lifetime. (Long-lived plasma cells can also secrete protective antibodies for considerable periods of time.)

Immunological memory is a unique and specialized feature of the adaptive immune system and does not apply to innate immunity. However, other types of immune memory exist in even the most ancient animals, such as sponges, which lack an adaptive immune system. More recently the term "innate memory" has also been introduced to describe the phenomenon in which innate cells, such as macrophages, for example, may respond faster or differently against the same class of PAMPs to which they were initially exposed. However, this form of memory is not clonal and arguably might be better termed "innate imprinting." In this review, the term "immunological memory" will be used exclusively within the context of adaptive immune responses.

DCs may have the specialized capacity to trigger T-cell activation, over and above any other cell type, and hence are essential for initiating T-cell and TD responses. They also contribute directly to the effector functions of T cells by helping to polarize their responses. In addition, they may also play central roles in memory responses, not least because they may reactivate memory T cells if the antigen is encountered again. At a species level, natural selection has driven the evolution and diversification of the adaptive immune system. In turn, DCs may drive the evolution and diversification of adaptive responses required for tolerance and immunity.

DCs IN HISTORY

This section provides a brief historical introduction to the evolution of early ideas about DCs. It also highlights some sources of confusion that persist even today. To put this into some sort of context, Ralph Steinman's first paper on DCs was published in 1973 (7). Earlier studies had indicated that an "accessory cell" was required to induce T-cell and B-cell responses, and that APCs were somehow required to "process and present" antigens in a form that could be recognized by T cells. Then, in 1974, Zinkernagel and Doherty (8) first described the phenomenon of MHC restriction, for which they later received a Nobel Prize. This phenomenon is now understood as being due to the requirement for αβ TCRs to recognize peptide-MHC complexes. However, it was not until a decade or more later that the structure of the TCR and the gene segments that encoded it became known, and precisely how peptides bound to MHC molecules was understood. All of this was completely unknown in 1973. Moreover, at that time, it seemed most likely that macrophages were essential accessory cells and APCs for T-cell responses.

Lymphoid DCs and Interdigitating Cells

The original description (7) of a Steinman-Cohn DC was of a morphologically distinct, trace cell type that was noted among adherent cells cultured from mouse SLOs, such as spleen, lymph nodes, and Peyer's patches of the intestine. These cells, as originally defined, continually extended and contracted spiny cell extensions in culture that were termed "dendrites." They did not phagocytose particulate tracers, in contrast to the majority population of relatively sessile macrophages, which were avid phagocytes. Subsequently, DCs were discovered to be the most potent stimulators of T cell and T-cell-mediated immune responses, compared to other cell types tested (9–13). They seemed to have a unique capacity to initiate T-cell activation, thus acting as accessory cells. Moreover, through elaboration of cytokines such as interleukin-12 (IL-12), it seemed likely that they could also regulate the subsequent type of immunity that was induced (13). Hence, the original definition of a DC was a distinct cell type that could be isolated from SLOs and which was nonphagocytic but highly immunostimulatory.

Soon after their description, it was suggested that isolated DCs might be counterparts of the interdigitating cells. These had previously been identified within sections of lymphoid tissues, and particularly in the T-cell-rich zones of spleen and lymph nodes (14–16). It was only later realized that these cells were mostly liberated from lymphoid tissues after enzymatic digestion, for example, with collagenase. Hence, interdigitating cells appeared to be relatively fixed within the T-cell areas, in contrast to the apparently more free DCs first studied. These likely represented the nests of DCs that were later noted in more peripheral regions of tissues such as the marginal zones of spleen (17).

The development of monoclonal antibodies (MAbs) was published just 2 years after the initial description of DCs, and also rewarded with a Nobel Prize (18). This was crucial for further exploration of the phenotype and function of DCs in culture, and of their possible relationship to other cell types. It was soon appreciated that both isolated DCs and interdigitating cells expressed MHC-II molecules, and they were later found to express costimulatory molecules, thus helping to explain their capacity to initiate T-cell responses. Newly developed MAbs also showed that both cell types also expressed CD11c (part of the complement receptor type 4). However, they could be distinguished phenotypically since the majority of DCs that were readily liberated expressed 33D1 (the first MAb generated against DCs, later shown to recognize DCIR2 [DC inhibitory receptor 2]), whereas interdigitating cells expressed CD205 (a C-type lectin receptor) (17, 19).

The more general term "lymphoid DC" was often adopted for presumptive DCs isolated from or localized within SLOs. These also included cells within the medulla of the thymus, which, after isolation, were also found to be nonphagocytic but to have immunostimulatory function in culture (20, 21). However, early evidence suggested that these thymic DCs might be lymphoid in origin, deriving from the common lymphoid progenitor (CLP) rather than common myeloid progenitor (CMP) downstream of the hematopoietic stem cell. Matters were further complicated by the later finding that some DCs expressed the lymphoid markers CD4 (a subset of DCIR2$^+$ cells) or CD8 (generally interdigitating cells). Hence, the term "lymphoid" was used rather uneasily to reflect either or both a lymphoid tissue and/or a hematopoietic lymphoid origin.

Langerhans Cells and Interstitial DCs

Attention then turned to presumptive DCs in NLTs (this abbreviation also including "nonlymphoid organs"). The first detailed studies were of Langerhans cells (LCs) isolated from the epidermis of mouse skin (22, 23). These cells have a dendritic morphology within the tissue and form a dense network of cells in the suprabasal region of the epidermis. They also contain a characteristic organelle known as the Birbeck granule. After isolation from the skin, LCs were found to be weak or inactive as stimulators of T-cell responses *in vitro*. They were also shown to be phagocytic (24). During culture, however, the cells acquired potent immunostimulatory activity accompanied by profound phenotypic and functional changes, including increased expression of MHC-II molecules but loss of Birbeck granules and a markedly reduced capacity for phagocytosis. They thus came to resemble Steinman-Cohn DCs.

Working around the same time, other investigators had been studying a trace cell type that expressed MHC-II molecules in NLTs such as heart and kidney (25, 26). These cells, which became known as interstitial DCs, were identified in the connective tissue of all NLTs studied, except for the central nervous system. The impetus for such studies was the suggestion, made as early as 1957, that transplant rejection was initiated by a small number of "passenger leukocytes" that were carried over in the graft, while the bulk of the transplanted material was in fact nonimmunogenic. Consistent with this idea, when interstitial cells were isolated from vascularized organs such as heart and kidney, it was found that their phenotypes and functions changed during culture similarly to what had been observed for LCs (27).

The term "myeloid DC" was introduced by some workers to discriminate between the presumptive DCs that were isolated from or identified within NLTs and the lymphoid DCs and interdigitating cells noted above (28). The use of this term probably also originated from observations that DCs in NLTs, as well as myeloid cells such as monocytes and macrophages, expressed Fc receptors and CD11b. These cells did not, however, express CD4 or CD8, and later evidence indicated that they probably originated from myeloid progenitors downstream of the CMP. Unfortunately, the focus on shared myeloid markers to characterize DC populations has bedeviled the field and caused further confusion that, to some extent, has still not been fully resolved.

Afferent Lymph Veiled Cells and Blood DCs

Because of the developments above, the concept of a DC needed to encompass both a phagocytic, nonimmunostimulatory cell in NLTs as well as the Steinman-Cohn DC isolated from lymphoid tissues. Moreover, it appeared likely that one could develop into the other. This transition was termed "maturation," since cells in NLTs were considered to be relatively immature compared to mature cells in lymphoid tissues that had immunostimulatory activity. The concept of a DC also needed to be revised to encompass a migratory stage, as the cells apparently moved from peripheral tissues into SLOs, for example, via afferent lymph. In larger animals, it is possible to cannulate afferent lymphatics and to study migrating cells directly, but this is extremely difficult in smaller animals such as rodents. However, by surgically removing regional lymph nodes and allowing time for the lymphatics to reanastomose, one can obtain cells that would otherwise travel in afferent lymph after cannulation of the thoracic duct; this source is termed "pseudoafferent lymph" (29).

In culture, cells obtained from lymph or pseudoafferent lymph were found to have characteristic lamellipodiae, or "veils," and became known as veiled cells (16). These cells were found to be traveling in lymph derived from tissues such as gut, liver, and skin in a continuous manner (30, 31). In the latter case they resembled LCs since they contained Birbeck granules, and similar cells were also detected in regional lymph nodes (32, 33). (An early pioneer, Brigitte Balfour, who was related to a former British prime minister, cannulated the lymphatics of her own leg in order to obtain veiled cells, and made vast numbers of cinematographic recordings of their movements in culture that now fill a back room somewhere in London [S. C. Knight, personal communication]). Veiled cells were also found to stimulate T-cell responses in culture (30, 34). The concept that veiled cells represent an intermediate stage of DC trafficking from NLTs into SLOs was reinforced by other studies. For example, studies of skin explants and transplants enabled the direct visualization of presumptive LCs leaving the epidermis and migrating through dermal lymphatics out of the skin (35, 36). Moreover, while the numbers of such cells in lymph were generally quite low, the flux of these cells increased massively after infection or inflammation in the site from which they originated. This suggested that they may have important roles in generating immune responses within SLOs.

DCs had also been identified in human blood, although it was not clear if these might be en route to or migrating from NLTs (37). In early studies, the capacity of DCs to migrate via the blood into the spleen, where they homed to areas rich in T cells, was demonstrated after adoptive transfer of labeled DCs in mice (38, 39). Moreover, DCs were observed to disappear from cardiac tissue after mouse heart transplantation, and donor cells were then detected in recipient spleen, where they were visualized often in close association with recipient T cells (40). Stimuli promoting DC maturation and migration also started to be identified. These included proinflammatory cytokines such as IL-1 and tumor necrosis factor α (TNF-α) (41, 42) and microbial products such as bacterial lipopolysaccharide, which also induced potent immunostimulatory activity *in vitro* (43). Hence, the general concept evolved that, *in vivo*, infection and inflammation in NLTs stimulated the maturation and migration of DCs into SLOs to stimulate TD immunity.

Ex Vivo-Generated Bone Marrow-Derived and Monocyte-Derived DCs

DCs were clearly shown to have a hematopoietic origin and to originate from bone marrow of adult animals. It was discovered that cells resembling DCs could also be generated in culture from mouse and rat bone marrow progenitors, particularly after supplementation of cultures with granulocyte-macrophage colony-stimulating factor (GM-CSF; CSF-2) (44, 45). There were perhaps two main drivers for these studies. First, as a trace cell type, the isolation of "primary" DCs from tissues was laborious, time-consuming, and often costly (not just expensive) in terms of the large number of animals that were needed for even a single experiment. Second, the appreciation that DCs might be central to the initiation and regulation of T-cell-mediated immunity almost immediately suggested that they may have enormous therapeutic potential. It was subsequently shown that DCs could be generated from macaque and human bone marrow $CD34^+$ progenitors (46–48). At around the same time it was also discovered that cells resembling DCs could also be generated from human monocytes in culture (49, 50), although their differentiation into veiled cells in culture had in fact been reported some years earlier (51).

The discovery that it was possible to generate monocyte-derived DCs (moDCs) suggested that, *in vivo*, monocytes are bipotential cells that can differentiate into either macrophages or DCs with immunostimulatory activity. It was therefore proposed that monocytes might be recruited to peripheral sites of inflammation and infection, where they differentiated into moDCs before migrating to SLOs to stimulate T-cell-mediated immunity. Later work further reinforced this view, particularly after injection of particulates into the skin or the blood (see "Lineages"). But, from a therapeutic standpoint, these findings were critical. At last it became possible to initiate human clinical trials in which cancer patients received infusions of their own "*ex vivo*-generated" DCs that had been pulsed with tumor antigens and matured in one form or another. Thousands of patients have now been treated in a large number of clinical trials, which continue with varying degrees of success (52–54). (In some centers, perhaps a quarter of patients with end-stage [IV] melanoma who came off trial but continued to be treated with their DCs on an *ad hoc* basis have lived for many years afterwards, compared sadly to survival rates more in terms of months [C. Figdor and G. Schuler, personal communication].)

Plasmacytoid DCs

As early as the 1950s, pathologists had noted an unusual cell type within SLOs, particularly in disease settings, which was originally known as a plasmacytoid monocyte or plasmacytoid T cell. In 1997, it was discovered that if these cells were isolated from inflamed human tonsils and cultured with IL-3 and CD40 ligand (normally expressed by activated T cells *in vivo*), they developed immunostimulatory activity (55). They also appeared to be nonphagocytic and thus, functionally, came to resemble Steinman-Cohn DCs. In many other respects, however, they were very different: they appeared to circulate primarily in the blood (but not in lymph), had not been identified in many other tissues under physiological conditions, had a very different phenotype, were later shown to have the specialized capacity to secrete extremely high levels of IFN-I during viral infections, and additionally showed evidence of a possible lymphoid origin. Whether or not these cells should be considered as DCs is still debated by some investigators.

Immunostimulatory and Tolerogenic DCs

Many early studies focused particularly on the immunostimulatory properties of DCs. Parallel developments in the field, however, clearly demonstrated that at least some, if not all, of these cells might also contribute to tolerance. For example, increasing evidence showed that interdigitating DCs of the thymus were probably involved in the induction of central tolerance during T-cell development (despite their immunostimulatory function *in vitro*). Moreover, much later studies showed that veiled cells migrated constitutively from the gut in a relatively immature state. Some of them contained apoptotic cell contents, presumably derived from the tissue (56). It was therefore proposed that these cells might contribute to peripheral tolerance in regional lymphoid tissues. This paradigm was subsequently extended to DCs that migrated constitutively from all other NLTs in which they had been identified. Hence, DCs in different types of tissue appeared to have different cellular functions (e.g., phagocytosis versus immunostimulation); could contribute to tolerance and/or immunity depending on circumstances; and while some were clearly related (e.g., those migrating from one site to another), others (e.g., plasmacytoid DCs [pDCs]) seemed very different.

Follicular DCs Are Not "Dendritic Cells"

Another cell with dendritic morphology had been described within SLOs in 1965 (57, 58). Unlike interdigitating cells, however, these cells were identified within the germinal centers that develop from B-cell follicles during the course of adaptive immune responses. They were originally known as "antigen-retaining reticular cells." However, they were termed "follicular dendritic cells" by Steinman in 1978 (59), since at that time he believed they might be related to the Steinman-Cohn DCs he first described. Despite this early confusion, it is now absolutely clear that the follicular DC is a completely distinct cell type, in terms of both its stromal cell origin and functions. These cells retain native antigens on their cell surface and are involved in selection of B cells during the so-called germinal center reaction (60). Hence, in this review, the term "dendritic cell" specifically excludes follicular DCs.

Multiple DC Subsets

By the turn of the century, at least eight types of DCs or dendritic-like cells had been identified: LCs and interstitial DCs within peripheral NLTs (commonly termed myeloid DCs); afferent lymph veiled cells and blood DCs in the circulation; Steinman-Cohn DCs and interdigitating cells within the lymphoid tissues (sometimes termed lymphoid DCs); moDCs in peripheral and possibly lymphoid tissues; and pDCs. Nevertheless, there appeared to be a commonality between them in that each cell type was increasingly shown to contribute to both T-cell tolerance and T-cell-mediated immune responses. Even so, into the fifth decade after Steinman's first publication, the field of DC immunobiology appears to have become very complicated and often confused. It seems that an ever increasing number of "new DCs" are being described, particularly from phenotypic studies of cells from different tissues, some of which appear to have varying immunological functions depending on the experimental system that is tested. There is therefore understandable debate about if and how it might ever be possible to define a DC, and continuing controversy over the relationship between these and other cell types, and particularly macrophages, that seems well beyond semantics. This review focuses specifically on just a few types of DCs that seem quite distinctive, though others will be mentioned. This is to try to distinguish clearly between "the wood and the trees." Whether or not it leads only to a clearing in a forest remains to be seen; it may be the edge of the wood.

DCs IN LINEAGES

This review is based on the premise that five distinct lineages of DCs or DC-like cells have to date been identified with some clarity. These populations comprise two subsets of "classical" DCs, pDCs, moDCs, and LCs. Evidence that these are different lineages include the findings that (i) they can develop from distinct hematopoietic or myelopoietic progenitors, (ii) they require different sets of transcription factors and cytokines for development, (iii) the fully differentiated cells adopt distinct transcriptional programs that regulate their functions, and (iv) interconversion from one type to another has not been seen (61–68). In addition, (v) cells that seem to resemble these subsets can be generated from progenitors in culture (68–71) (see "History"), though much care is needed in extrapolating findings from these systems to the cells in tissues (72). In this section, some of the molecular and cellular specializations of each will be highlighted. Their distributions and behavior within tissues and their functions in tolerance and immunity are discussed in the following sections.

Fate-mapping studies have been particularly helpful in elucidating the developmental origins of different populations of DCs from progenitors. The techniques used have included adoptive cell transfer, cellular barcoding, expression of reporter genes, and genetic lineage tracing in the Cre-Lox system (73). However, any such technique is potentially associated with off-target effects. For example, the use of irradiated recipients in cell transfer studies may disturb normal developmental signals within a leukopenic environment, and virus-mediated transduction during barcoding may skew cell fates. Moreover, the promoters used may not be strictly cell specific. This includes CD11c, which is often considered to be a specific marker for all five populations of DCs, but is and can be expressed on other cell types. Genetic ablation studies, using gene-targeted knockout mice, have also been very informative (74) but are likewise potentially subject to off-target effects (75). These considerations become particularly important when discoveries of "new types" of DCs are reported. Very rare individuals with DC deficiency have also recently been identified (76). However, the underlying genetic defects also affect development or homeostasis of other cell lineages, so they are rather uninformative when trying to elucidate the precise origins and functions of DCs.

It is generally accepted that the pluripotent hematopoietic stem cell generates both a CMP and a CLP (61–63). Adoptive cell transfer studies have shown that both the CMP and CLP and other progenitors can at least give rise to classical DCs and pDCs in tissues under experimental conditions, and most likely during times of physiological stress. However, current evidence also indicates that, under normal steady-state and homeostatic conditions, the two subsets of classical DCs, pDCs, and moDCs derive from the CMP. In contrast, LCs are initially produced from primitive myelopoietic progenitors during embryonic development that also generate resident macrophage populations of some tissues (see below). According to one simplified, linear model, downstream progenitors of the CMP become progressively committed toward the generation of monocytes (which can subsequently develop into moDCs), pDCs, and, lastly, the two populations of classical DCs (77). In this type of model, the CMP generates a bipotential cell termed the macrophage-DC progenitor (MDP). In turn, this produces monocytes, perhaps via a common monocyte progenitor, with another bipotential cell called the common DC progenitor (CDP). The latter then generates pDCs, which complete their development in the bone marrow, together with a final bipotential cell termed the pre-DC. The pre-DC then migrates into the tissues, where it generates the two subsets of classical DCs. However, such schemes have been contested (78, 79). Apart from possible species differences that may exist, such differentiation pathways may well not conform to strictly linear models (80).

Classical DCs

The two types of classical DCs are crucially important for initiation of primary T-cell responses and make a substantial contribution to the regulation of adaptive immunity. These two subsets are as distinct from each other as are T cells and B cells (81). Classical DCs have been identified in most tissues studied closely, and all contain both types. However, these cells are absent from the parenchyma of the brain and the eye, though they do populate sites such as meninges and outer cornea. They are also absent from stratified squamous epithelia such as the epidermis of skin, which contains LCs instead. Both types of classical DCs have been identified in all warm-blooded species studied, including the mouse, rat, sheep, cow, pig, and human; the subsets in mouse and human seem extremely similar to each other at a transcriptional level (82, 83). Cells resembling classical DCs have also been identified in fish and amphibians (84–86), though further study is required to determine whether or not these are classical DCs and, if so, whether there are two subsets. Nevertheless, the evolutionary conservation of two distinct subsets of classical DCs in warm-blooded species at least suggests that they play fundamentally different roles in the immune system.

In mice, pre-DCs migrate from the bone marrow via the blood into the tissues, where, locally, they differentiate into mature cells. Both the pre-DC and its presumptive upstream CDP and MDP progenitors express the CX3CR1 chemokine (fractalkine) receptor, but expression is extinguished as they differentiate into the mature cells (87). Classical DCs express the transcription factor Zbtb46 (zDC; Btbd4), which helps to discriminate between these cells and moDCs, pDCs, and LCs (64). However, this is not a stable or definitive marker, as it is downregulated on DC maturation, can be induced in monocytes, and is or can be expressed in other cell types. The homeostatic proliferation of pre-DCs and/or the development of classical DCs also depend on the Flt3L growth factor and its receptor, CD135. In general, classical DCs have a short half-life of just a few days in tissues, and this is tightly regulated: experimentally decreasing their life span has detrimental effects on the induction of immunity, while increasing it can result in autoimmune manifestations (88).

In this review, for simplicity, the two types of classical DCs will be designated as DCI and DCII subsets, and these are intended as neutral terms. The pre-DC generates two resident populations in all SLOs, and probably the thymus. These will be termed cDCI and cDCII cells, an abbreviation (cDC) that is otherwise used for classical or conventional DCs. The pre-DC also generates two populations in NLTs, which, however, are only transiently resident because they can subsequently migrate to SLOs. To distinguish these cells from the resident populations in SLOs, they will be termed migratory mDCI and mDCII cells, an abbreviation (mDC) that is otherwise used for myeloid DCs. Although detailed studies have not been performed in all cases, the respective populations of cDCI and mDCI, and cDCII and mDCII, appear to be extremely similar in many respects. However, their phenotypes do vary depending on their tissue of localization (see below), presumably reflecting tissue-specific specializations (89).

In general, classical DCs in NLTs express relatively low levels of MHC-II molecules compared to some of their counterparts in SLOs. A common view is that DCs in NLTs can endocytose or phagocytose antigens and generate intracellular peptide–MHC-II complexes. (They may also retain intracellular antigenic epitopes for considerable periods of time, even after antigen has been eliminated [90].) In response to stimuli such as TLR agonists and proinflammatory mediators, they decrease further uptake, begin to express high levels of peptide–MHC-II molecules at the cell surface, and increase their expression of costimulatory molecules such as CD86. They also express CD40, which enables cross talk between DCs and activated T cells via CD40L. Together, these regulate complex, bidirectional cascades of costimulatory (and conversely, coinhibitory) molecular interactions. The related phenotypic and functional changes are commonly referred to as DC maturation.

The cells then migrate from the tissue into SLOs (91). Here, due to their new and increased expression of CCR7, they home to T-cell areas where the corresponding chemokines (CCL19 and CCL21) are produced (92, 93). Here these migratory DCs can activate antigen-specific T cells and trigger subsequent adaptive responses. An extreme view might hold that just one population of classical DCs could do the whole job itself. And yet there are two distinct subsets of migratory DCs in NLTs, and it is unlikely that the two resident subsets in SLOs are mere spectators. This strongly implies a marked division of labor between the two main types of classical DCs (DCI and DCII), and very close cooperation between all four in adaptive immune responses as a whole.

The DCI population

The DCI subset may have specialized capacities to present cell-associated antigens to CD8$^+$ T cells, and perhaps differentially to sense PAMPs and DAMPs from these cellular sources. Differentiation of DCI cells from the pre-DCs requires transcription factors such as IFN regulatory factor 8 (IRF8), Batf3 (which, however, is also expressed by DCII cells), and Id2. Mice with targeted deletions in the corresponding genes have profound deficiencies in DCI cells. As a general rule, DCI cells in mice express molecules such as CD8 in SLOs but CD103 and/or CD207 (Langerin) in NLTs, which facilitates phenotypic discrimination from DCII cells. The function of CD8, which is expressed as an $\alpha\alpha$ homodimer, in contrast to the $\alpha\beta$ heterodimer of conventional CD8$^+$ T cells, is unknown. (Perhaps future insights might come from consideration of other systems [94].) CD103 is the α_E integrin, which can combine with β_7 to form a receptor for E-cadherin, discussed later. The corresponding lineage in humans expresses CD141 (BDCA-3 [blood DC antigen 3]) or thrombomodulin (95, 96). This plays regulatory roles in both the complement and coagulation pathways, but its relevance to DC function is not clear. This aspect may be worthy of further study, especially since the complement and coagulation systems are so tightly linked.

Several examples of the specialized functions of the DCI population can be highlighted that seem to set them apart from DCII cells. First, DCI cells in mouse and human differentially express TLR3. This may enable them to sense double-stranded RNA produced during viral replication in infected cells, for example, after uptake of apoptotic or necrotic bodies. They also selectively express TLR11 and TLR5 in mouse and TLR10 in human (97, 98). Second, DCI cells differentially express receptors that can recognize DAMPs and/or capture and internalize dead and dying cells (99). For example, both mouse and human DCI cells express CD36, which is a receptor for apoptotic cells (and plasma thrombospondin) (100, 101). They also express CLEC9A (DC NK lectin group receptor-1 [DNGR-1]), which is a receptor for necrotic cells and recognizes F-actin as a DAMP (102). In mice, DCI cells also express CD205 (DEC-205), which is a receptor for uptake of both necrotic and apoptotic cells (103), as well as CD24, which can recognize high-mobility group box-1 (HMGB1) as a DAMP (104). Hence, these cells may be specialized for uptake of cell-associated antigens, and perhaps sensing of cell-associated DAMPs. Third, DCI cells express exceptionally high levels of components required for the peptide–MHC-I pathway and may constitutively cross-present exogenous peptides into the same route (105). Consequently, they seem particularly well specialized for antigen presentation to CD8$^+$ T cells. Fourth, DCI cells can secrete high levels of cytokines such as IL-12, which is associated with polarization of CD4$^+$ Th1 cells (13). Finally, in both mice and humans, DCI cells differentially express the chemokine receptor XCR1, which has been associated with homeostasis and tolerance (106, 107). The potential relevance of some of these specializations is discussed later (see "Tolerance" and, especially, "Immunity").

The DCII population

In contrast to the above, the DCII subset may be specialized to present extracellular antigens to CD4$^+$ T cells and perhaps differentially to sense extracellular PAMPs and soluble DAMPs. Differentiation of DCII cells requires a distinct set of transcription factors, which include Irf4 and RelB. Mice with targeted deletion of the corresponding genes have profound deficiencies in DCII cells. In mice and humans, these cells in probably all tissues express CD11b, which can associate with CD18 to form the complement receptor type 3 (CR3). However, expression of CD11b has proven a confounding factor in studies that have attempted to characterize DC populations in tissues since it is also expressed by monocytes, monocyte-derived cells (such as moDCs), and macrophages. Presumably the expression of CR3 (CD11b/CD18) facilitates uptake of complement-opsonized antigens by DCs and enhances antigen presentation. It seems generally presumed that this is also the case for CR4 (CD11d/CD18), which is expressed by all DC subsets. It remains possible, however, that these receptors may play other regulatory roles, and it is notable that, unusually, Fc receptors on DCs tend to be inhibitory in function (108). Mouse, rat, and human DCII cells, but not the DCI subset, also express CD172 (signal regulatory protein α [SIRPα]; see below), and this has been very helpful in distinguishing between these cell types, but again, this is not a definitive marker, as it can also be expressed by monocytes.

Several examples of the specialized functions of the DCII population can be highlighted that distinguish them from the DCI subset. First, mouse DCII cells selectively express TLR7 and TLR5 (109). These respectively may enable them to sense genomic RNA of viruses after uptake into endosomes, and perhaps the abnormal endosomal localization of host nucleic acids or chromatin as DAMPs (110), as well as bacterial flagellin at the cell surface. In contrast, DCII cells in human selectively express TLR2 and TLR4 (97), which, in addition to PAMPs, can recognize HMGB1 as a soluble DAMP (110). Second, mouse and human DCII cells selectively express CD172 (SIRPα), which is an inhibitory receptor that recognizes CD47 on host cells and prevents phagocytosis ("phagoptosis"). Interestingly, CD47 is also a receptor for thrombospondin. In addition, mouse DCII cells in SLOs express Clec4a4 (DCIR2; 33D1), which is an inhibitory receptor that has been implicated in maintenance of homeostatic tolerance (111). Third, the DCII population expresses high levels of all components required for the peptide–MHC-II pathway (105, 112), though cross-presentation may be inducible. Consequently, they may be specialized for antigen presentation to $CD4^+$ T cells. Finally, DCII cells have been more generally associated with the selective induction of Th17 and/or Th2 responses (113). Some of these specializations are discussed later (see "Immunity").

Plasmacytoid DCs

pDCs seem specialized to induce antiviral resistance and may regulate $CD4^+$ T-cell responses and perhaps contribute to immunological memory of $CD8^+$ T cells. Despite the original demonstration that pDCs can develop into immunostimulatory cells (55), subsequent studies have found that this capacity may be relatively limited compared to that of classical DCs. There seems to be little or no persuasive evidence that pDCs initiate T-cell responses *in vivo*. Unlike differentiation of classical DCs, the development of pDCs is dependent on the transcription factor E2-2 although it also requires Flt3L (65, 114, 115). While pDCs may normally be generated from the CMP under homeostatic conditions, they can also be generated from the CLP. Interestingly, and irrespective of myeloid or lymphoid origin, human pDCs express pre-Tα, mouse pDCs have rearranged IgH DJ segments, and both express RAG (116). These respective molecules are normally essential for formation of the pre-TCR and expression of the pre-BCR during intermediate stages of T- and B-cell differentiation, and for somatic recombination of both TCRs and BCRs. It has been proposed that this reflects the evolutionary origins of pDCs (116), a point that is further considered below.

In general, pDCs circulate in the blood and populate SLOs, which they enter via high endothelial venules (HEVs), and may also home to the thymus (see "Tolerance"). Small numbers of pDCs also populate mucosal tissues, within the lamina propria of the gut and lung and airways, for example, but appear to be excluded from other peripheral tissues under homeostatic conditions. Hence, migration of pDCs to the mucosal tissues could represent homing to lymphoid tissue within the gut wall, for example, or more generally migration in response to homeostatic inflammation (see "Tissues"). Such homing is dependent on expression of CCR9, although entry to SLOs is not. However, in pathological settings, pDCs can accumulate in large numbers in sites such as inflamed skin and within melanomas and ovarian and mammary carcinomas. Their roles in these sites are unclear.

Perhaps the best-known function of pDCs is the expression of exceptionally high levels of IFN-I in response to viral infections, and particularly of IFN-α. Both human and mouse pDCs selectively express TLR7 and TLR9, which enable them to sense genomic single-stranded RNA and DNA in endosomes. In fact, they may also sense host nucleic acids liberated during tissue damage and, in turn, facilitate wound healing through ancient (innate) mechanisms (117, 118). Nevertheless, IFN-I secretion is likely to be tightly regulated through a number of inhibitory receptors, such as Siglec-H (sialic acid-binding immunoglobulin-type lectin H) in mouse as well as CD303 (Clec4c; BDCA-2) and immunoglobulin-like transcript 7 (ILT7) in human, all of which regulate TLR responses (95, 119). Mouse pDCs also express tetherin (BST-2 [bone marrow stromal antigen 2]), which can physically bind live viruses to infected cells. In response to viral infection, the cells undergo autophagy, which may route endogenous viral antigens into the peptide–MHC-II pathway.

Through their capacity to secrete high levels of IFN-I, pDCs may play central roles in promoting antiviral resistance as well as regulating innate and adaptive responses. In the latter case, for example, at a molecular level, IFN-I can modulate functions of classical DCs by promoting extended antigen sampling of endosomal compartments for peptide–MHC-II loading (120); enhance cross-presentation and peptide–MHC-I loading; and regulate IL-12 expression, either to promote Th1 polarization or conversely to facilitate activation of CTLs (and innate-like NK cells). At the cellular level, IFN-I stimulates the differentiation of classical DCs from progenitors (121) and increases the maturation and migration of classical DCs from NLTs into SLOs. Furthermore, IFN-I decreases the generation of Tregs; increases the expression by $CD8^+$ T cells of perforin and granzymes, which are both important for CTL killing; and promotes the survival of memory $CD8^+$ T cells. Nevertheless, some of these proposed functions are inferences from the described roles of IFN-I in other settings, and, particularly since IFN-I is a large family of different classes, further study is required to confirm or refute a direct role of pDCs in some of them. Overall, however, pDCs may play regulatory roles in adaptive immunity either directly or by modulating the functions of other DC populations (122). This apparent synergy has, for example, been shown to be crucial for clearance of viral infections (123).

Monocyte-Derived DCs

moDCs may play particularly important roles in the effector phase of T-cell-mediated responses, and perhaps particularly those of $CD4^+$ T cells. It now seems clear that monocytes are bipotential cells that can home to the tissues and differentiate into moDCs or macrophages, though monocytes, as such, have also been identified within some tissues (124). Other populations of probable monocyte-derived cells, described as "Tip-DCs" and "LysoDCs," have also been reported (below, and see "Tissues"). Discrimination between moDCs, classical DCs (and particularly DCII cells), as well as monocyte-derived and other macrophages has proven highly problematic. In large part, this is because of the shared expression of molecules that can be further modulated during differentiation. For example, most or all express CD11c, and moDCs and mDCII cells both express MHC-II and CD11b. Nevertheless, a particularly useful marker of monocyte-derived macrophages in tissues is CX3CR1, since its expression increases to high levels during the differentiation of these cells from monocytes; it is also expressed by resident macrophage progenitors (125). In contrast, expression is lost during differentiation of pre-DCs into classical DCs. However, CX3CR1 is probably expressed at intermediate levels during development of moDCs in tissues, so this is not definitive. This molecule is the fractalkine receptor, which is presumed to facilitate cellular interactions with epithelial cells in tissues, though its precise functions remain unclear (126). In the mouse, many macrophage populations, including monocyte-derived populations as well as LCs, but not classical DCs, express the F4/80 molecule (127, 128). Importantly, CD64 (FcγR) also seems to be selectively expressed by moDCs but not classical DCs, thus helping to distinguish between them (108, 129).

Different monocyte subsets have been identified in the blood. In humans, those that express high levels of CD14, with or without CD16, are the likely equivalents of those that express high levels of Ly6C (Gr-1) in the mouse. Most evidence indicates that these cells can be recruited to inflammatory sites, where they have the potential to develop into moDCs or macrophages. Hence, this population has been termed inflammatory monocytes. In contrast, in physiological conditions, another human subset with high-level expression of CD16 but low CD14, which is the likely equivalent of those expressing low levels of Ly6C in mouse, are associated with the luminal surfaces of endothelial cells. Hence, they have been termed patrolling monocytes. It seems probable that both types of monocytes can be recruited to tissues under different circumstances and, depending on which and the particular microenvironment in which they find themselves, may "choose" alternative fates. Hence, they may develop into either moDCs or macrophages (130, 131). However, these tissue populations probably represent a continuum of different phenotypes and functions ranging from moDCs, LysoDCs, and Tip-DCs to macrophages (which are further subdivided by some into M1 or M2 types), and quite a lot more in between (132). Nevertheless, moDCs and macrophages, as commonly described, are probably very different in function.

There is much evidence that monocytes and/or monocyte-derived cells such as moDCs play important roles in the resolution of infections by bacteria and fungi; examples include infection with *Citrobacter* in the intestine and *Aspergillus* in the lungs (133–135). They have also been associated with the polarization of Th1 responses due at least in part to their secretion of IL-12, but also with Th2 responses in other settings such as allergic asthma (136). In addition, moDCs seem able to elicit "recall" responses by activating memory T cells, though their relative importance in this respect, compared to classical DCs, is still not clear (137). In fact, both moDCs and classical DCs most likely play complementary and synergistic roles during adaptive immune responses in general (132) (see "Immunity"). In addition, populations of Tip-DCs have been identified within the spleen during infections with intracellular pathogens such as *Listeria monocytogenes* or *Trypanosoma cruzi*. These cells can secrete TNF-α and express inducible nitric oxide synthase, which are required for synthesis of reactive nitrogen intermediates, but also stimulate primary T-cell responses in culture. They are most likely monocyte-derived cells but functionally appear to represent a transitional stage between moDCs and macrophages. This fate choice appears to depend in part on which TLR agonists might be sensed (138).

A particularly potent stimulus for the differentiation of monocytes into moDCs may be phagocytosis of particles at endothelial cell surfaces. For example, in a transmigration system *in vitro*, CD14$^+$ and/or CD16$^+$ monocytes were able to traverse endothelial cell monolayers into an artificial matrix where they internalized particles, after which they crossed back and developed into moDCs (139, 140). Furthermore, injection of particulates into mouse skin was observed to recruit monocytes into the site, which subsequently migrated into the regional lymph nodes and developed into particle-laden moDCs in the T-cell-rich paracortical regions (141). Earlier work had also demonstrated that injection of particulates into the bloodstream of rats led to their uptake by cells within the marginating pool in the sinusoids of the liver, where Kupffer cells reside (142–144). Subsequently, particle-laden cells were detected within the hepatic lymph, presumably en route to the regional lymph nodes. These findings raise the possibility that Kupffer cells initially phagocytosed the particles and perhaps then elaborated chemokines to recruit monocytes from the bloodstream, which, after phagocytosis, began to develop into moDCs. The cells that were subsequently detected in lymph nodes indeed had little phagocytic activity and could stimulate primary T-cell responses in culture, but additionally expressed CD103. Whether or not such expression was due to their microenvironment is not clear. However, taken together with the fact that the two main subsets of monocytes can migrate into tissues (above), it is possible that each could separately generate a distinct population of moDCs. While there seems to be no evidence to support this as yet, potential parallels with the classical DCI and DCII subsets should be obvious and worthy of closer scrutiny.

At least three different roles for moDCs in immunity may be envisaged. First, at some stages of differentiation, they may acquire effector functions, which contribute to the elimination of infectious agents in peripheral tissues, as seems likely for cells such as Tip-DCs. Second, moDCs may acquire peripheral antigens and migrate into SLOs to initiate primary T-cell responses. Evidence has been presented that, while the entry of monocytes into the tissues is CCR2 dependent, subsequent migration of moDCs into SLOs depends on expression of CCR7 and CCR8 (145). In this respect they could be viewed as a rapidly recruited source of immunostimulatory cells that assist the induction of adaptive immune responses, perhaps after classical DCs have migrated from the tissue and before repopulation from the pre-DC. Nevertheless, at present there seems to be little available evidence to support a major role of moDCs in the activation of primary T-cell responses. Finally, there is good evidence that moDCs may instead play essential roles in the effector phases of T-cell-mediated immune responses in NLTs (see "Immunity").

Langerhans Cells

Epidermal LCs were first identified in 1868, but even today their roles in immunological responses have not been clearly elucidated. The dermis of skin, like most other tissues, contains the two subsets of classical DCs. In contrast, the epidermis contains only LCs. Related cells have been identified in all other stratified squamous epithelia, such as those of the oropharynx, vagina, ectocervix, and anus. LCs are situated in a suprabasal position deep within the epidermis, where they express MHC-II and CD11c and, in the mouse, F4/80. Epidermal LCs and dermal DCI cells also express CD207 (Langerin). This is involved in the development of Birbeck granules, which are specialized endosomal compartments of unclear function that have, however, only been identified in LCs (see "Tissues"). Unlike the development of other DC subsets, that of mouse LCs requires transcription factors such as ID2 and RUNX3 and is also dependent on transforming growth factor β (TGF-β) (which can be used to generate LC-like cells from progenitors in culture [70]). In addition, LC development is dependent on the CSF-1 receptor (M-CSFR; CD115). However, LCs persist in the skin of CD115-deficient mice, most probably because they require IL-34, which is an alternative ligand for this receptor to which it binds with a higher affinity than CSF-1.

LCs may have important functions related to monitoring the integrity of the epidermal barrier and contributing to repair. For example, in mice, skin oils apparently diffusing through the epidermis after damage to the outermost stratum corneum may be presented by LCs to a subset of specialized, innate-like NKT cells (146). These in turn may elaborate cytokines such as IL-22 to induce keratinocyte proliferation, and hence effect repair. Though their cell bodies are situated deep within the epidermis, LCs can extend cytoplasmic processes upwards between the keratinocytes and appear able to sample antigens beneath the stratum corneum (147). There seems to be little available evidence supporting the idea that LCs play any major role in initiating T-cell responses. In contrast, they may induce or maintain tolerance within the skin, even in the presence of TLR agonists (148). In part this seems due to their capacity to maintain populations of neonatal Tregs that home to the skin after birth (149) (see "Tolerance"). LCs also appear to contribute to polarized $CD4^+$ T-cell responses of different types (Th17, Th1, and/or Th2) against antigens that may breach the uppermost layers of the skin or penetrate the epidermis. Other evidence tends to suggest that their contributions to $CD8^+$ T-cell responses may be indirect (150).

It is, however, important to note that there are very distinct differences between the skin of different species of jawed vertebrates, which all possess epidermal LCs. For example, obviously the skin of mammals has an outer keratinized layer that is coated with skin oils and generally surrounded by air (except, of course, for aquatic mammals such as whales). In fish and amphibian tadpoles, however, the skin is a mucosal tissue surrounded by an aqueous environment, similar to that of the mouse or human embryo developing in the amniotic sac. There are even major differences between mouse and human skin. Both contain T cells in the epidermis, but those in mouse are mostly unconventional γδ T cells (dendritic epidermal T cells), whereas most in human are conventional αβ memory T cells. Possibly, therefore, the epidermal LCs of different species may have evolved some different functions, and there may be immunological differences between LCs of mouse and human.

Unlike all other DC lineages discussed here, the epidermal LCs of skin first arise very early during embryonic development. In the mouse, primitive myelopoiesis begins in the yolk sac at around embryonic day 7 (E7) and primitive myelopoietic (nonmonocytic) progenitors seed the epidermis at around E8.5. Here they generate LCs, which form a self-renewing population (151, 152); they also generate the resident populations of macrophages in the brain and liver, the microglial and Kupffer cells (153). A second wave of LCs then populates the epidermis, which develop from fetal liver-derived monocytes at E16.5 after the commencement of definitive hematopoiesis. (The same progenitors may also generate a second population of Kupffer cells in the liver, populate the lamina propria of the lung with cells that develop into alveolar macrophages, and probably give rise to resident macrophages of many other tissues [154].) The fetal-derived epidermal LCs then undergo rapid proliferation after birth and come to represent the predominant, self-renewing population in adult skin under homeostatic conditions (151). Under inflammatory conditions, at least, epidermal LCs can be further supplemented by bone marrow-derived monocytes that home to the skin epidermis and develop into apparently indistinguishable cells. By contrast, corresponding populations in the oral mucosa appear to originate from pre-DCs, with a contribution from monocyte-derived cells, though they too express a very similar transcriptomic signature to those in skin (155).

While the LCs of adult mouse and human skin have been associated with tolerogenic and immune functions (see above), the functions of the yolk sac-derived LCs within the embryo are unknown. These cells populate the epidermis even before the endothelium develops at around E9 and later differentiates into lymphatic and venous components at E11.5; hence, these LCs are present in the skin before the blood and lymph circulatory systems have developed. Potentially they may have specialized homeostatic functions within this specialized "mucosal-like" environment (cf. fish and amphibian tadpoles above). Curiously, however, the establishment of this population coincides with development of the umbilical link between fetus and mother. The only immunological cells of relevance at this stage would seem to be maternally transferred lymphocytes. In principle at least, maternal T cells could conceivably gain access to fetal skin perhaps in response to the endogenous chemokine CXCL12 (stromal cell-derived factor 1 [SDF-1]). This plays important roles in the movement of embryonic cells during morphogenesis and is a ligand for the CXCR4 receptor on mature T cells. Fetal engraftment by maternal T cells has in fact been demonstrated in some cases of SCID, for example (156, 157). Such maternally transferred T cells could be detrimental to the fetus (through recognition of paternal antigens) or beneficial if they were Tregs. Hence, it is an open question whether embryonic epidermal LCs might be involved in the induction or maintenance of tolerance, through local interactions or even after migration from fetus to mother (158, 159).

Other "Subsets" of DCs

Numerous publications have postulated the existence of other presumptive DC subsets in addition to the five types discussed here. Perhaps the most persuasive (or at least suggestive) evidence is for a subdivision of mouse DCII cells in SLOs into two further subsets. This is based on their apparent dependence on the transcription factors Notch2 or klf4 for development (79). These putative subsets have also been associated with Th17 and Th2 responses, respectively. Moreover, there is undoubted phenotypic heterogeneity of DCII cells within mouse SLOs based on differential expression of molecules such as Clec4a4 (CDIR2; 33D1; see above),

CD4, and ESAM. As another example, subsets of "double-negative" DCs that do not express "selected" DCI or DCII markers have been reported in various tissues. These include the dermis of skin ($XCR1^-$ $CD11b^-$) (160), spleen ($CD4^-CD8^-$) (161), Peyer's patches ($CD8^-CD11b^-$) (162, 163), and mesenteric lymph nodes (MLNs) ($CD103^-CD11b^-$) (164). These double-negative cells have variably been reported to be IRF4 dependent, to secrete IL-12, and/or to express CD207 (Langerin) or CX3CR1. In addition, a double-positive population has been reported in gut ($CD103^+$ $CX3CR1^+$) (164), and there are undoubtedly other examples. However, the expression of different transcription factors can be induced or extinguished because of intrinsic developmental programs or extrinsic microenvironmental influences, and can also be modulated during cellular responses. It is therefore possible that at least some of the additional DC subsets that have been proposed, such as those noted above, might represent early stages during the development of progenitors, such as pre-DCs, or of monocytes adopting different cell fates within tissues (compare Tip-DCs and LysoDCs above). Given the general preoccupation with three main types of polarized $CD4^+$ T-cell responses (Th17, Th1, and Th2), it is perhaps not surprising that some might expect the existence of three dedicated subsets of classical DCs (see "Immunity"). There are almost certainly two subsets of classical DCs; there may be three; many more seem like fashion accessories.

DCs IN TISSUES

This section provides an overview of the dynamic anatomy of DCs within tissues. It focuses on their distributions in anatomical compartments, including their migration from NLTs to SLOs, and their behavior within microenvironments under physiological conditions and after inflammation or infection. Other local or migrant cell populations will be noted, with an emphasis on macrophage-like cells. Unless otherwise stated, two subsets of classical DCs are present in all, and moDCs may also develop under inflammatory conditions. DCs within the thymus are discussed in "Tolerance."

Steady State versus Homeostatic Inflammation

The term "steady state" is often used to describe conditions in tissues under physiological conditions in the absence of overt infection. However, the presence of commensal organisms on external stratified epithelia such as the epidermis of skin, and at luminal surfaces within mucosal tracts such as that of the gut, is continuously sensed by the immune system. Indeed, the microbiota also shape adaptive immunity, and a state of mutualism is established and maintained between them and their hosts (165–169). The microbiota induce a state of "homeostatic inflammation," also termed "tonic stimulation," which is physiological and tightly regulated and controlled (170). However, depending on the relative loads of commensals, the levels of homeostatic inflammation may vary, perhaps being highest in gut and lowest in organs and tissues such as heart and skeletal muscle, for example. This should be borne in mind when considering the relative complexity of DCs and other populations in different tissues. It is also a general, though not entirely accurate, rule that naive T cells are excluded from NLTs. However, populations of effector and/or memory T cells may or may not constitutively traffic into different tissues, which adds a further layer of complexity to the physiological conditions within each. Hence, in this review the term "homeostatic conditions" is often used for those in normal tissues in general, while the term "inflammation" relates to conditions that prevail after infection or tissue damage.

The Skin and Peripheral Lymph Nodes

Skin

The skin has two distinct anatomical compartments: the epidermis and the dermis, separated by a basement membrane. The outer epidermis is a stratified squamous epithelium that is avascular and lacks its own blood supply. The connective tissue of the dermis is penetrated by a rich network of blood vessels and capillaries as well as blind-ended, fenestrated capillaries; nerves also traverse the dermis, as in all tissues, and some extend into the epidermis. The extravascular fluid of the dermis drains into blind-ended lymphatics and flows as afferent lymph into regional lymph nodes, which are often connected to each other in chains. The skin contains one population of LCs within the epidermis and the two subsets of classical DCs, mDCI and mDCII, in the dermis; a third, double-negative population has also been reported (see "Lineages"). The functions of these DC populations are tightly knitted with those of other cell types that exist within the skin. Collectively, they play crucial roles during immune surveillance for pathogens (160, 171, 172) and continuously maintain states of mutualism with commensals (173) and tolerance to self-antigens of the tissue (174).

Epidermal LCs are localized with their cell bodies mostly positioned within the spinous layer (stratum spinosum) just above the basement membrane (stratum basale). They are intimately associated with keratinocytes, with which they closely interact (see "Lineages"). For example, they can deliver topically applied carcinogens to neighboring keratinocytes, which subsequently undergo malignant transformation, but this does not occur in LC-depleted skin (175). It has been claimed that each LC is separate and does not communicate directly with any other, but instead extends its dendritic processes around its "own group" of keratinocytes. These express PRRs, and their inflammatory responses may stimulate LC migration from the tissue. More-recent findings show that in fact the dendritic processes of LCs can extend upwards through the tight junctions of the keratinocytes in the uppermost layers, and just below the outermost stratum corneum, which comprises dead keratinocytes (147). Here they can sample antigens and contribute to the induction of immune responses against them. LCs were, for example, found to be essential for the induction of IgG against the exfoliative toxin of *Staphylococcus aureus* and could prophylactically protect mice from subsequent disease (176). In this respect, LCs may resemble the "antigen-sampling" cells within simple epithelia of other tissues such as the gut and airways (see below). However, LCs may have very limited ability to phagocytose bacteria (177). Whether LCs may also extend dendritic processes from the epidermis through the basement membrane below and into the dermis, perhaps to communicate directly or indirectly with dermal DCs, deserves further study.

Within the dermis, the mDCI and mDCII subsets reciprocally express CD103 and CD11b, respectively. Both dermal classical DC subsets are believed to undergo "homeostatic maturation" and hence migrate constitutively from the dermis into SLOs, where they may contribute to induction or maintenance of peripheral tolerance. The dermis also contains other populations such as mast cells close to the blood vessels and innate lymphoid cells. During infection and inflammation, there is a complex interplay between these and other cells in skin that ultimately leads to maturation of the classical DCs and migration to the regional peripheral lymph nodes (pLNs). During inflammation, monocytes may enter and develop into moDCs, and pDCs may also be present in pathological settings. It is, however, quite clear that dermal mDCI cells also express CD207 (178), although they appear not to possess Birbeck granules. This is an important point to note in studies that have endeavored to trace the migration of LCs into pLNs, for example (179). It also raises the issue, which does not yet seem to have been addressed, as to whether or not CD207 might also induce development of Birbeck granules in these cells within pLNs. This seems to be especially pertinent since, in contrast, earlier studies clearly documented the loss of Birbeck granules by LCs during maturation in culture (see "History"). In fact, other studies have also shown that after migration into the dermis, LCs can undergo apoptosis (150). The apoptotic contents were then acquired by DCs, particularly the mDCI subset, which subsequently migrated to the pLNs, where they (rather than LCs) stimulated antigen-specific responses. Hence, some immunological responses that have been associated with LCs may in fact not be induced directly by these cells. Perhaps it is time for a much closer examination of these issues.

Peripheral lymph nodes

The pLNs are specialized SLOs that monitor peripheral tissues and integrate the different types of information derived from them under homeostatic and inflammatory conditions. Subsequently, tolerogenic or immunogenic adaptive responses can be initiated within them. Lymph nodes are surrounded by a dense fibrous capsule that is penetrated by multiple afferent lymphatics leading from the tissues, such as dermis of skin, which during infection and inflammation contains antigens and proinflammatory mediators and chemokines. The afferent lymphatics also provide the route by which migrating DCs that have acquired antigens, as well as necrotic or apoptotic cells and debris, can enter pLNs and subsequently interact with T cells. Hence, three populations of peripheral DCs are likely to enter pLNs from skin: the epidermal LCs and both subsets of dermal DCs. Presumably, moDCs that develop in response to inflammation may also migrate to nodes via afferent lymphatics, although inflammatory monocytes may also enter directly from blood (see below).

After entering a pLN, the afferent lymph drains into a narrow subcapsular space immediately beneath the capsule and surrounding much of the underlying cortex and medulla. The "floor" of this space is populated by macrophages above the tissue below. The cortex is penetrated by trabeculae through which the lymph enters and is distributed through the lymphatic sinuses or medullary cords. These then drain into usually a single efferent lymphatic, and the lymph leaves the lymph node at the hilum. Ultimately, the efferent lymph is collected by the thoracic duct and drains back to the blood. Blood vessels enter the node at the hilum and run upwards mostly into the cortex (which may not be directly accessible to lymph). Here there are multiple arteriovenous communications surrounded by specialized endothelial cells that comprise the HEVs.

Lymphocytes enter the lymph node from the blood via the HEVs and generally segregate within the cortex. The B cells become localized to follicles, with T cells in the interfollicular areas between them. (In fact, entry of T cells into the lymph nodes via HEVs is regulated by DCs [180].) This latter region is distinct from the more superficial parafollicular areas adjacent to the trabeculae. Pre-DCs also enter the nodes from the blood via HEVs, as do monocytes and pDCs during inflammation (181).

Lymph nodes also contain several networks of reticular cells with specialized functions in different anatomical compartments (182). These include a fibroblastic reticular cell (FRC) network that extends from the subcapsular space, ramifies through the cortex, and connects to the HEV. The fibers of the FRC network contain a central core of collagen bundles that is surrounded by extracellular matrix components that are synthesized by the FRCs that closely surround them (183, 184). These structures are otherwise known as conduits, and have also been identified in the spleen and more recently the thymus (see "Tolerance"), although there may be molecular differences in their composition. At the distal ends of these pLN conduits, a specialized protein, PLVAP (plasmalemma vesicle-associated protein), of lymphatic endothelial cells forms diaphragms that act as molecular sieves (185). These permit small (<70-kDa) molecules to enter the conduits, from where they can be delivered deep into the cortex to the HEVs, but excludes all those that are larger. It is now clear that low-molecular-weight antigens and chemokines, and potentially cytokines, can pass directly through the central core of the conduits from peripheral regions such as the subcapsular sinus (186, 187). DCs remodel the FRC network over which they migrate, and both may promote the survival of T cells within the tissue (187–189). A population of CD11b$^+$ presumptive DCs is closely associated with the FRC network. Intermittent gaps, or "windows," between the FRCs enable these cells to insert dendritic processes into the conduits and directly to sample antigens (190). Whether these cells are in fact classical DCs, possibly resident cDCII cells, or moDCs requires further investigation. In addition, chemokines can also be delivered via conduits directly to HEVs, where, under inflammatory conditions, they may recruit monocytes from the blood into the node (181).

Under homeostatic conditions, it is likely that the majority of DCs in pLNs are resident populations, with relatively few immigrants present (191). More recently, detailed information on the localization of DCs in pLNs has been gleaned from studies using multiphoton microscopy and a more recent analytical technique termed histo-cytometry (192). These have visualized distinct subsets of DCs within different microenvironments of the pLNs. The resident cDCI subset is mostly localized centrally within the T-cell zones. In contrast, the cDCII subset resides within the medullary lymphatic zone close to the surrounding subcapsular sinus. Here, these cells have been observed to capture particles from the lymph (193). The mDCI subset also localizes deep within the paracortex, whereas the mDCII subset tends to be positioned in the outer paracortical regions. Interestingly, under homeostatic conditions, the presumptive DCII subsets were found to express retinoic acid and be capable of inducing Tregs (features actually associated with the converse subsets in gut MLNs) (194). However, in models of skin inflammation, it was shown that the migratory DCs subsequently homed into the outer cortex, in the T-cell zone but adjacent to B-cell follicles. In contrast, presumptive LCs were observed to migrate from the skin much more slowly into lymph nodes, and were later detected within the inner paracortex, separated from the migratory DCs (179).

It is clear, even from the few examples above, that lymph nodes are highly dynamic structures and that DCs within them perform an extremely complex choreography. Nevertheless, some functional consequences for the initiation of CD4$^+$ T-cell responses are becoming apparent (195). For example, it has been shown that antigens injected into the skin can be very rapidly acquired by resident lymph node DCs, which then induce the early stages of CD4$^+$ T-cell activation (196). However, robust T-cell responses were only induced after the entry of antigen-bearing DCs that migrated from the site of antigen injection much later. It may be that conduit-associated DCs that have sampled small antigens also mature in response to the chemokines (some of which have proinflammatory activities) these structures may deliver, though whether this also includes proinflammatory cytokines requires further study. What these and other studies highlight is the fact that T-cell activation occurs much faster after antigen delivery than was previously appreciated. Moreover, many CD4$^+$ T-cell responses may require serial engagement with different DC subsets to become fully effective (see below). Presumably, the cortex of the lymph nodes is relatively isolated and "sterile," much like within the thymus (see earlier). Otherwise it would seem that uncontrolled access of immunomodulatory agents to DCs would presumably further modulate their functions and lead to dysregulated T-cell responses; for example, functional cell-cell communication may need cytokine gradients to be established between them (197). This is an interesting aspect of SLO structure and function that has been relatively little investigated.

Typically, activation of many CD8$^+$ T-cell responses requires "help" delivered by CD4$^+$ T cells. Other studies have visualized the sequential interactions of T cells with DCs during helper-dependent CD8$^+$ T-cell activation (198, 199). These have revealed that a series of spatiotemporally distinct cellular interactions occur within different microenvironments of a lymph node, and that these are required to generate a robust CD8$^+$ T-cell response. They also highlight serial engagements between different subsets of DCs and CD4$^+$ and CD8$^+$ T cells. Such findings are reminiscent of an earlier model proposed for helper-dependent CD8$^+$ T-cell responses (200). It seems possible that antigen-specific CD4$^+$ T cells are first activated by migratory DCs. These activated T cells then home toward cDCI cells and stimulate ("license") them to increase their expression of costimulatory molecules sufficiently for CD8$^+$ T-cell activation (roughly speaking, this is a "higher" level than is required by CD4$^+$ T cells). These cDCI cells may then activate CD8$^+$ T cells that subsequently engage with them. However, what does not seem to have been well defined is the precise source of antigen for the cDCI cells, which seem relatively sequestered deep within the T zone. Interestingly, they may communicate with each other over large distances through the formation of "tunneling nanotubules" between them, particularly after CD40 ligation, which could be provided by activated T cells (201).

Studies of other cell populations within "steady-state" pLNs have demonstrated clear differences in cellular compartmentalization (183). For example, NK cells, specialized invariant NKT cells, and unconventional γδ T cells are present in the interfollicular regions and medullary lymphatic zones, to which mDCII and cDCII cells, respectively, home. However, elucidation of potential roles of these cells in modulating DC responses, and possibly subsequent αβ T-cell responses, awaits further investigation. Furthermore, different types of macrophages populate the subcapsular space and medullary sinuses. Two of these, for example, express CD169 (sialoadhesin) and F4/80, respectively, in mice. The CD169 macrophages appear to comprise part of the "firewall" preventing access of lymph-derived pathogens into downstream tissues since, if they are depleted, infectious agents can spread to other connecting lymph nodes (202). They have also been implicated in the generation of CTL responses (203). Potentially, too, small-molecular mediators produced by these macrophages could also be delivered into the tissue via conduits.

The GI Tract, MLNs, and GALTs

Lymph from large regions of the gastrointestinal (GI) tract probably drains to the MLNs, though some areas drain into other regional lymph nodes (see below). Throughout the small and large intestine there are also distinct types of gut-associated lymphoid tissues (GALTs) that play specialized roles in tolerance and immunity (204). Descending from the duodenum, the jejunum and ileum contain variable numbers of Peyer's patches, which, in humans, reach a peak in the third decade of life but then decline in number (205). These are embedded within the mucosal propria and submucosal tissue. They have specialized, sieve-like M cells within a follicle-associated epithelium, juxtaposed between the intestinal epithelial cells. Descending further into the cecum is the appendix, which may have specialized functions yet to be discovered. There are also increasing numbers of "solitary isolated lymphoid tissues," such as isolated lymphoid follicles, descending from the ileum to the colon. These most probably represent tertiary lymphoid structures. They have much less well-defined structures than SLOs, lacking defined follicles and T-cell zones, for example, and are not developmentally controlled but are locally induced by a subset of ILCs (206). DCs populate the lamina propria and submucosal tissue of the gut wall and GALTs.

GI tract

The relative distributions of the two subsets of classical DCs within the lamina propria and submucosa of the GI tract may differ (29, 207). Their relative proportions also differ between the small intestine and colon, with a predominance of cDCII cells in the former and of cDCI cells in the latter. These cells are generated from pre-DCs that home to the gut through expression of the $\alpha_4\beta_7$ integrin. Precisely how and where the homing of these progenitors is determined is not clear, and presumably different homing molecules may be required for migration to other NLTs. The two subsets of classical DCs in the gut have characteristic expression of CD11b (which is expressed by mDCII but not by mDCI cells), but both express CD103. The reason for this is not clear, though in other settings expression of CD103 can be induced by GM-CSF. Similar subsets have been identified in gut-derived pseudoafferent lymph in rats, indicating that they constitutively migrate into the MLNs (208). Under homeostatic conditions, the cDCI subset in afferent lymph was found to contain apoptotic cell inclusions, and it seems likely that these cells may contribute to tolerance induction against cell-derived components (see "Tolerance").

However, the flux of both subsets is dramatically increased in inflammatory settings.

In mice, the macrophage populations within the lamina propria of the gut most likely originate from fetal liver-derived monocytes, but are supplemented by bone marrow-derived monocytes after birth when the gut microbiota first becomes established (68). There is a general consensus that gut macrophages express high levels of CX3CR1, as well as CD11b. Hence, high-level expression of CX3CR1 or CD103 has been used to distinguish phenotypically between gut macrophage and DC populations, respectively, in a number of studies (87, 204, 209–212). However, variant populations, including a double-positive subset of cells that expresses both CX3CR1 and CD103, have also been described (106, 164, 211), and the situation is further complicated by the fact that moDCs can express intermediate levels of CX3CR1 (see "Lineages"). Overall, therefore, multiple phenotypic subsets of presumptive DCs (or macrophages) have been identified, particularly in the gut, it seems. Possible explanations for this have been noted (see "Lineages"). Nevertheless, there appear to be clear differences in the subanatomical localization of the major populations of presumptive CX3CR1^{+} macrophages and CD103^{+} classical DCs in the small intestine, and most likely their functions.

Under homeostatic conditions, CX3CR1^{+} cells have been identified within the gut epithelium. It is now generally accepted that these cells express tight junction proteins that enable them to penetrate the epithelial cell layer, through which they extend dendritic processes to sample luminal contents (212, 213). However, this appears to be an inducible behavior since it is enhanced by TLR-dependent responses of the epithelial cells (213) and does not occur in germ-free gnotobiotic mice, though whether it occurs at all has been questioned (214). Nevertheless, under these circumstances, the CX3CR1^{+} cells may contribute to a state of tolerance or unresponsiveness to innocuous soluble food antigens. It has been shown they can acquire such antigens from the lumen and transfer them to lamina propria CD103^{+} cells, with subsequent induction of Tregs and tolerance, presumably after migration of the latter to MLNs (210, 215). The transfer of antigens is dependent on the formation of gap junctions between the CX3CR1^{+} and CD103^{+} cells, since genetic ablation of connexin-43 prevented the generation of Tregs and induction of oral tolerance. It has also been shown that soluble antigens can be transferred directly through mucus-secreting goblet cells to CD103^{+} cells in the lamina propria (214).

The capacity of CX3CR1^{+} cells to promote tolerance induction under homeostatic conditions may be due to their apparently constitutive, anti-inflammatory properties. These cells seem relatively refractile to stimulation by TLR agonists, can secrete IL-10, and also inhibit T-cell proliferation in a contact-dependent manner (210). They also appear to be specialized for the uptake of soluble antigens, and may have little or no capacity to phagocytose particulates. It is generally accepted that these cells are sessile and do not migrate from the tissue to the MLNs, though contrary evidence has been put forward (216). However, during experimental *Salmonella* infection of the gut, the CX3CR1^{+} cells may secrete chemokines to recruit CD103^{+} cells from the lamina propria into the epithelium above the basement membrane (217). Here, the CD103^{+} cells were also observed to extend processes into the lumen, capture bacteria, and rapidly retract their processes. This behavior was associated with the subsequent induction of CD8^{+} antigen-specific T-cell proliferation within the MLNs. Under homeostatic conditions, rare CD103^{+} cells were also observed "patrolling" the epithelium (217). It is possible that, under these circumstance, these cells might also sample innocuous luminal contents, migrate to the MLNs, and facilitate tolerance induction, though this requires further investigation.

In the gut, the specialized anti-inflammatory subset of CX3CR1^{+} macrophages may be required to dampen DC responses and help maintain tolerance to innocuous food antigen; otherwise homeostatic inflammation might promote sufficient DC maturation to enable them to induce immunity, although this aspect deserves further investigation. It is quite evident, however, that tolerance needs to be maintained even during times of infection, when DCs become able to stimulate TD primary responses. It is known that CD103^{+} DCs can produce the vitamin A derivative retinoic acid and anti-inflammatory TGF-β, and can promote the generation of FoxP3^{+} Tregs within the MLNs (217–220). Locally, gut epithelial cells also produce thymic stromal lymphopoietin, which drives DCs toward a "regulatory" phenotype, and the latter can also produce indoleamine 2,3-dioxygenase, which depletes tryptophan and inhibits T-cell proliferation. In addition, the mDCI subset, through secretion of IL-12 and IL-15, can induce secretion of anti-inflammatory IFN-γ by T cells in the colon (221). Precisely how this balancing act is achieved remains a major puzzle in mucosal immunology and more generally in adaptive immunity as a whole.

Mesenteric lymph nodes

The MLNs that receive efferent lymph from the GI tract are developmentally distinct from the majority of "peripheral" lymph nodes that drain other tissues. They are described as mucosal lymph nodes, and also include the cervical lymph nodes that drain the nasopharynx, for example, and the sacral lymph nodes associated with the rectum. Development of the MLNs as well as Peyer's patches requires key members of the lymphotoxin and lymphotoxin receptor family, which are related to the TNF family and its receptors (222). Nevertheless, at a gross level, the overall structure of the MLNs is similar to that of pLNs. However, there are likely to be functional differences due to the additional influence of gut-derived immunomodulatory molecules draining from afferent lymph (above) and perhaps produced locally. These may also contribute to tolerance under homeostatic conditions, for example, by inducing pTregs (223) (see "Tolerance").

The MLNs play a central role in the induction of tolerance to food antigens, particularly in soluble form (224, 225). Oral tolerance cannot be established in mice that developmentally lack MLNs (as well as pLNs). The MLNs have been described as a firewall that helps to prevent commensals that may have invaded the gut lamina propria and traveled in lymph from entering the blood. Because of this, it is believed that the systemic immune system normally remains "ignorant" of the gut microbiota. Mice lacking MLNs develop fatal splenomegaly and lymphadenopathy. Likewise, the liver, which receives the entire output of intestinal venous blood through the portal vein, acts as an additional firewall to capture and clear gut commensals that may enter the blood in more-pathological settings (226). Damage to the liver impairs its function and can also lead to fatal infection with commensal organisms.

The MLNs contain the two resident subsets of classical DCs as well as corresponding subsets that migrate from the gut wall. As noted above, the migratory DCs may play important roles in the induction or maintenance of tolerance within the gut, although it is not clear if this is primarily against normal gut tissue antigens or against antigens associated with commensals or food. Certainly it seems likely that the induction of oral tolerance against food antigens is intimately associated with the migration of DCs to the gut since this is induced in a CCR7-dependent manner. During immunity to gut pathogens, the MLNs have been particularly associated with Th17 responses, which play particularly important protective roles in gut immunity, as well as Th1 responses. The MLNs also contain pDCs, though these appear to arrive from the blood rather than after traveling in lymph (227). Their precise contributions to tolerance and immunity are not precisely defined, though they are presumably important in responses against enteric viruses (228). It also seems likely that moDCs, developing during gut infections, migrate to MLNs and also contribute to T-cell responses.

Peyer's patches

Peyer's patches are specialized SLOs that monitor the intestinal lumen. They possess sieve-like M cells that are in direct contact with the lumen, and through which contents may be sampled (205). Below the follicle-associated epithelium, with its juxtaposed M cells, is the subepithelial dome (SED) region. Below this, the B cells and T cells are segregated into follicles (or, more usually, in this tissue, germinal centers [229]) and the interfollicular regions, respectively. Lymphatics do not appear to have been well described in Peyer's patches, but if they exist, they would presumably represent efferents that drain to the MLNs. Different subsets of DCs have been characterized within the Peyer's patches in generally discrete subanatomical compartments (230). The DCI cells occupy the interfollicular regions among the T cells, whereas DCII cells are located in the SED region. It seems likely that most or all are resident DC subsets (i.e., cDCI and cDCII) derived from the pre-DCs. Some contribution from migratory DCs derived from the gut lamina propria and/or submucosa is possible, but alternatively they may be excluded from this tissue. Within the interfollicular and SED regions, both pDCs and a subset of double-negative cells has also been identified that lacks expression of CX3CR1 and CD8 (see "Lineages"). The latter have not yet been identified within MLNs, suggesting that they do not migrate from the tissue, though relevant molecular expression could conceivably be induced if they do. Two populations of monocyte-derived cells have also been well characterized within the Peyer's patches (230, 231). One has been termed a lysozyme-containing DC (LysoDC) and is likely to represent a form of moDC (cf. Tip-DC earlier), while the other has been termed a lysozyme-containing macrophage.

It seems generally assumed that soluble molecules may traverse the M cells. If so, they do not seem to be involved in the induction of oral tolerance, since this is not impaired in mice with developmental deficits of Peyer's patches. What seems clear is that pathogenic bacteria and inert particulates can be captured by the LysoDCs that extend dendrites through the M-cell pores into the gut lumen (232). While these $CX3CR1^{+}$ cells somewhat resemble those described within the gut epithelium (above), they appear to be distinct in that they do not express IL-10 and can secrete proinflammatory cytokines in responses to TLR7 agonists. They have been described as short-lived cells and associated with induction of Th17 responses. In contrast, the cDCI subset can secrete IL-12 in response to pathogenic bacteria and may be involved in induction of Th1 responses. It is possible that the cDCII subset may contribute to the initiation of TD B-cell responses and production of protective IgA in response to gut pathogens. This may be in contrast to the IgA that is secreted in a T-independent manner and that plays essential roles in confining commensals to the gut lumen (233). Nevertheless, in other settings, the Peyer's patches may represent specialized inductive sites for Th2 responses, for example, in response to helminth infections. Collectively, therefore, DCs within the lamina propria and submucosa of the gut may facilitate the concomitant induction of tolerance to food antigens and polarized immune responses against pathogens. At the same time, those within Peyer's patches may additionally maintain responses that help to contain commensal organisms to the lumen, but induce differently polarized responses against other pathogens.

Other Mucosal Tissues and MALTs

Distinct subsets of DCs also populate the lamina propria of all other mucosal tissues and their respective MALTs. Here, the focus will be on the lung and airways and their associated MALTs, though the urogenital tract will be mentioned briefly.

Lung and airways

Within the lung and airways, the mDCI subset is localized close to or within the pulmonary epithelium, while the mDCII subset tends to be confined within the lamina propria; different populations of moDCs or macrophages have also been identified (234, 235). There is evidence that a subset of presumptive DCs within alveoli can extend dendrites into the airspace and sample antigens, presumably for subsequent induction of tolerance or immunity (236, 237). In contrast, the alveolar macrophages remain relatively sessile and apparently focus on noninflammatory clearance of foreign particulates. The precise subset concerned with airway sampling awaits further characterization. However, evidence has also been presented for a two-step mechanism by which presumptive DCs may be recruited to the epithelium from the lamina propria in a chemokine-dependent manner (237). This is reminiscent of that described within the intestine whereby $CX3CR1^+$ cells recruit $CD103^+$ DCs to sample particulates (see above). It may also help explain the apparent contribution of CX3CR1 cells to antiviral responses in the lung, for example (238). The mDCI subset expresses tight junction proteins (236), which may enable it to penetrate the epithelium for direct sampling of the airspace, though this has not been directly proven. This subset selectively transports apoptotic cell contents to the regional pLNs under homeostatic conditions (239). However, during enteric viral infections, the same subset has also been associated with cross-presentation of antigens and the induction of $CD8^+$ T-cell responses. Antigen-sampling cells have also been identified within human nasal mucosa (240). In addition, pDCs can be detected within the lamina propria of the airways, where they presumably contribute to antiviral responses through production of IFN-I, which has been shown also to protect classical DCs from viral infection (241). Finally, DCs have also been identified throughout the urogenital tract (242). It would perhaps not be surprising if some of these cells, too, had the capacity to sample luminal antigens similarly to those in other epithelial sites.

Waldeyer's ring, BALTs, and NALTs

The DCs of the lung and airways and of the urogenital tract can migrate into regional pLNs, which are unlike the MLNs associated with the gut. In addition, specialized lymphoid tissues are, or can be, associated with the nasopharynx and the airways. The tonsils are situated within Waldeyer's ring of the pharynx in humans; some animals also have additional tonsils. Mice, but not humans, have nasal-associated lymphoid tissue (NALT), which is developmentally controlled, like other SLOs. Furthermore, bronchial-associated lymphoid tissues (BALTs) may be present, particularly in pathological settings, though this might represent tertiary lymphoid tissue. Classical DCs have been identified in all these tissues, and in addition, pDCs may be present in large numbers in pathological circumstances. It is not yet clear whether or not classical DCs can contribute jointly to tolerance against innocuous antigens and the containment of commensals (particularly in the upper regions of the airways and nasopharynx), as well as immunity against pathogens, though this seems likely. Specialized mechanisms by which they may do so at these sites, however, awaits further study.

Fully Vascularized Tissues and the Spleen

Early studies demonstrated that DCs could be isolated from the blood, where they appeared to circulate in both relatively immature and mature forms (37). It was believed that the former represented DCs en route to the tissues, while the latter represented those migrating from these sites. However, this interpretation needs to be revisited in the light of our current understanding that classical DCs are generated locally in tissues from the pre-DCs.

The heart and nonmucosal tissues

Two subsets of classical DCs have been identified in all vascularized tissues that have been closely studied. Such tissues include the heart (243); cardiac valves and aorta (244, 245); skeletal muscle (178); kidney (246, 247); and tissues such as pancreas (248). Interestingly, antigen-sampling CD11c$^+$ cells have been described within the subintimal space of the aorta, where they appeared to probe the vascular lumen (245). The macrophage populations in at least some of these tissues, such as heart and arterial walls, originate from both yolk sac-derived progenitors and fetal liver monocytes, but may be supplemented from bone marrow-derived monocytes later in life (125, 249). It is generally presumed that DCs from the above sites can all traffic into regional pLNs. What is not at all clear is whether many can also travel in blood.

As discussed earlier (see "History"), there seems to be good evidence that DCs can migrate from the heart via the blood into the spleen and colocalize with CD4$^+$ T cells (38, 39, 250). Here, it was proposed, they initiated transplant rejection of cardiac allografts. And yet the heart also possesses abundant pericardial lymphatics (251), and it seems generally assumed that DCs migrate from the heart into the regional pLNs. It seems completely unknown why there should be two DC "outputs" from the heart, one via the blood and one via the lymph. However, an alternative (and speculative) explanation might be that DCs within different anatomical regions of the heart differentially migrate to the pLNs or spleen (for example, from the outer epicardium in the former case, but the myocardium and/or endocardium in the latter). It might also be the case that DCs from epithelial sites such as skin and mucosal tissues migrate preferentially via the afferent lymph to regional lymph nodes, while those from nonmucosal or fully vascularized sites migrate to the spleen. This seems to be an absolutely fundamental issue that does not appear to have been clearly raised, and almost certainly not addressed.

The migration of DCs from peripheral tissues via the blood is a really important aspect that deserves much further investigation. The reasons for this include the following. First, rare DCs can be detected in efferent and thoracic duct lymph, from which they might be transported, via the thoracic duct, into the blood; this would provide an alternative explanation for DCs migrating to spleen and other tissues, rather than direct entry into the blood (91, 252). Second, there is evidence that DCs from the oral submucosa may migrate to distant lymph nodes associated with the genital mucosa (253). Third, both DCI and DCII cells can migrate via the blood to the bone marrow, where they can elicit recall responses of memory T cells that appear to home to this tissue (252). Finally, cells that closely resemble the DCII subset can migrate from peripheral tissues into the thymus (see "Tolerance"). Hence, these issues are extremely important in terms of subsequent induction of both immunity and tolerance. There seems little doubt that further insights into the immunobiology of DCs may come from much more detailed examination of their migration patterns, and particularly via blood.

The spleen (and liver)

The spleen is an encapsulated, blood-filtering organ that lacks an afferent lymphatic supply and does not possess HEVs. It is anatomically divided into white pulp and red pulp. These are separated by the marginal zone, and in some species by a marginal sinus as well. Arterial blood enters the spleen via trabecular arteries, which then become "central arteries" surrounded by the so-called periarteriolar lymphoid sheaths (PALS) of the white pulp. Within the PALS, the B cells are segregated within follicles, surrounded by the T-cell zone. Small capillaries arise from the central vein and traverse the PALS to enter the marginal zone, which is generally considered the splenic equivalent of the subcapsular sinus of lymph nodes, though containing blood rather than afferent lymph. The lymphocytes enter the white pulp from the marginal zones, rather than via HEVs as in lymph nodes, before becoming segregated in their respective areas. The blood from the central veins and marginal zones is then collected into penicillar arterioles, which become capillaries entering the red pulp. The blood in some capillaries is transported directly through these vessels out of the tissue. However, others are open-ended and "dump" the blood into the red pulp, from where it travels through splenic cords before leaving the tissue. (As the blood enters the splenic cords, aging and effete erythrocytes that lack the elasticity to pass through are retained in the red pulp and eliminated by splenic macrophages.) There are, however, structural differences in the spleen between different species, and the functional implications of these are not well understood (254).

It is presumed that pre-DCs enter the tissue of the spleen via the marginal zone and generate the two subsets of classical DCs locally. In some studies in mice, these have been characterized according to their expression of CD8 or CD4, by DCI or DCII cells, respectively;

a double-negative population in other respects resembling the latter has also been described (see "Lineages"). It is generally believed that the cDCI subset is localized within the T-cell zones of the PALS. The spleen also has so-called bridging channels, which traverse the marginal sinus and connect the outer region of the white pulp to that of the red pulp. Though this has yet to be clearly demonstrated, these may represent counterparts to the medulla of the lymph nodes. If so, by analogy, the DCs within the bridging channels might be predicted to contain resident cDCII cells, which, in lymph nodes, are located within lymphatic sinuses from where they may sample the lymph. A subset of cells resembling the DCII subset is indeed present within the bridging channels, from where they may sample particulates from the blood (255). The localization of this subset has been shown to be dependent on a specific chemotactic receptor (EBI2), and these cells seem essential for antigen-specific $CD4^{+}$ T-cell activation and antibody responses (256). In addition, retinoic acid may be essential for development for some of these (Notch-dependent) DC cells (161). Much remains to be learned about the regulation of DC functions within the spleen in relation to their potential roles in tolerance and immunity, though some interesting insights have been obtained (257).

Under homeostatic conditions, "nests" of DCs were originally described penetrating the marginal zone (see "History"). These cells can phagocytose apoptotic cells and subsequently home into the T-cell zones, presumably for induction of tolerance (258). These findings are reminiscent of those showing similar nests of cells penetrating the subcapsular sinus of lymph nodes. It is tempting to speculate that they may represent migratory DCs that may transiently reside in these sites before migrating constitutively in small numbers into the T zones for induction of tolerance. Conceivably, however, after infection, their homing to these sites could be accelerated and they may acquire the capacity to activate T cells rapidly against antigens they captured from lymph. Within the marginal zone there are also specialized macrophage populations that include the same $CD169^{+}$ subset present in the subcapsular sinus of lymph nodes (above); if these are depleted, pathogens can spread systemically (202). In addition, fibroblastic reticular networks have been identified within the spleen, similar to those in nodes, including one ramifying through the white pulp from the marginal zone.

It might be expected that that the spleen is important for the induction of robust immunity against blood-borne antigens. However, genetically asplenic and splenectomized individuals have increased susceptibility only to encapsulated bacterial infections. This is most likely because of the absence or loss of a specialized subset of B-1 cells, and perhaps accessory marginal-zone macrophages, that can produce T-independent antibody responses against such pathogens. Defects of T-cell-dependent immunity have not generally been reported, presumably because such responses can be induced at other sites. Importantly, the venous blood from the spleen is transported via the hepatic arteries into the liver, where the two subsets of classical DCs also reside. Relatively little is known about the immunological functions of the liver, though presumably these DCs may migrate to regional lymph nodes by the same route that appears to be taken by presumptive moDCs that have captured particulates from the blood (143); it has even been suggested that the liver acts as a "biological concentrator" for blood DCs, to direct them to hepatic nodes (142). Nevertheless, in transplant settings the liver appears to be a relatively nonimmunogenic organ, so much so that liver allografts can be transplanted with little or no immunosuppression of the recipients. It has been suggested that this is due to its large bulk of poorly immunogenic hepatocytes and a relative paucity of DCs, though other factors contributing to this may yet be discovered.

Common Themes in Tissues

Some common themes emerge. The increasing number of reports of antigen-sampling cells in NLTs is provocative (see above). Epidermal LCs in skin sample antigens from under the stratum corneum impregnated with skin oils and surrounded by air (or water). Within mucosal tissues, $CX3CR1^{+}$ "macrophages" of the gut sample the lumen under the mucus, while perhaps related cells of the lung sample the airspaces under mucus or airway lubricants. Perhaps they exist in the urogenital tract. Within nonmucosal and fully vascularized tissues, similar cells in contact with blood appear to do likewise. Perhaps these are also present in other tissues, such as kidney, where they may sample filtrates. It might be speculated that the function of these macrophage-like populations may be to monitor "danger." If none is detected, they may regulate the populations of classical DCs that populate the dermis of skin, lamina propria of the mucosal tissues, and connective tissue of other organs and tissues. Hence, antigen-sampling LCs and macrophages may control the homeostatic maturation of classical DCs and/or induce regulatory functions that promote tolerance when the latter migrate into SLOs.

If danger is sensed by the antigen-sampling macrophage-like cells, they may recruit classical DCs, such as DCI cells, to sample its source for themselves. These then transport the antigen into SLOs to initiate protective immunity. Conceivably, the CDII cells are coopted if the danger spreads into the tissues. Hence, innocuous antigens such as from food or aerosols may generally result in tolerance, or perhaps "ignorance." Commensals may be contained by barrier defenses and homeostatic "background immunity," which includes T-independent production of IgA (these responses generally lack memory, to enable the microbiota to evolve). However, pathogens lead to the initiation of protective immunity, which may include TD IgA responses against commensals that invade the lamina propria. At the tissue and organ level, firewalls also exist. The MLNs and the liver, respectively, provide firewall protection for the lymph and blood derived from the gut. Likewise, the pLNs and the spleen may play comparable roles for the lymph and blood derived from all other peripheral tissues and organs. Presumably, related mechanisms may also be involved in protection of the brain, though much of this tissue seems very different (see below).

The Central Nervous System, Brain, and "Glymphatics"

It is generally accepted that DCs are absent from the parenchyma of the brain. Some CD11c$^+$ populations have been identified in isolated areas such as the pituitary (259), although these could well represent monocyte-derived cells rather than classical DCs. Moreover, there is no evidence that the abundant microglial cells that are present throughout are any more than relatively sessile resident macrophages that remain within the tissues. However, cells that closely resemble classical DCs have been identified in sites such as the meninges and the choroid plexus in contact with the cerebrospinal fluid. Early studies demonstrated that small tracers could be transported from the brain into the deep cervical lymph nodes, and were believed to do so via drainage through the cribriform plate. However, the more recent finding of the existence of a specialized dural lymphatic system (sometimes termed "glymphatics") that drains to these lymph nodes may provide an alternative explanation for these observations (260, 261). It may also provide a potential route by which DCs may migrate to SLOs, potentially to induce immunity against pathogens in the central nervous system, though this has yet to be studied in depth.

There are very close links between the central nervous system and the immune system. For example, it is quite clear that inflammation in peripheral tissues can be regulated via the hypothalamic-pituitary-adrenal axis (262). Furthermore, gut immunity can be regulated by the enteric nervous system and is controlled in part by the vagus nerve (263). Even the development of SLOs may be initiated by nerves that infiltrate tissues (222). Hence, the nervous system may regulate DC responses at least indirectly. However, both nociceptors (264) and neurotransmitters can directly influence DCs (265). Perhaps more remarkable, however, is the recent report that DCs (and macrophages) may be directly innervated in lymph nodes "by a mesh of filamentous neurofilament positive structures originating from single nerve fibers and covering each single APC similar to a glass fishing float" (266). There are also other highly provocative findings suggesting, for example, that nerve endings in spleen can form synapses with a subset of memory T cells that produce acetylcholine (262). Hence, investigation of the bipotential links between the central nervous system and the immune system, and of potential roles of DCs in them, may be a most fruitful field of study for the future.

DCs IN TOLERANCE

At a fundamental level, the immune system must discriminate with a high degree of precision between what is self and what is danger. In all settings, it is critical for survival of the host to maintain, and if necessary restore, the normal state of immunological unresponsiveness to self, or tolerance. (The converse state, immunological responsiveness to self, is, of course, termed autoreactivity.) Innate immunity is inherently self-tolerant. PRRs have evolved to recognize PAMPs and DAMPs but not normal components of self with exquisite sensitivity. (A pertinent example is the gene for TLR11, which is expressed in mice and can recognize protozoal profilin, but is a pseudogene in humans and not expressed because profilin-like molecules are synthesized [267, 268].) In contrast, a substantial proportion of the enormous repertoires of lymphocyte antigen receptors that are generated for adaptive immunity is inherently autoreactive. Moreover, nonfunctional receptors can be generated that are unable to recognize peptides in the context of MHC. This necessitates the existence of specialized mechanisms to promote and ensure the operational fitness of the total repertoire of TCRs and of the T-cell clones that express them (and similarly for BCRs and B cells).

T cells developing within the thymus are known as thymocytes. During their maturation, the developing thymocytes are subjected to the two critical selection processes of positive and negative selection, after which vast numbers will have been killed. Positive selection facilitates the further development of thymocytes that express TCRs that recognize peptide-MHC complexes, rather than either alone (though if the affinity is too high, they too are eliminated). It also controls whether the developing cells will eventually become mature $CD4^+$ or $CD8^+$ T cells. Negative selection involves the deletion through apoptosis of thymocytes that express TCRs that recognize self peptide-MHC complexes with a high affinity, and therefore contributes to tolerance. Within the thymus, there is also a third process whereby the fate of certain thymocytes, probably those of intermediate affinity, is redirected to generate populations of "thymus-derived" regulatory T cells (tTregs), which have counterregulatory (suppressive) functions (269, 270). These tTregs subsequently migrate from the thymus and seed the extrathymic tissues, where they contribute to the maintenance of peripheral tolerance. They are reinforced by additional populations of peripherally induced Tregs (pTregs) that are generated extrathymically and which most probably have high-affinity TCRs (271, 272). Hence, the relatively small proportion of remaining T cells that eventually emerge from the thymus have TCRs that may be able to recognize an epitope from an infectious agent that might be encountered in the future, purely by chance. (If any engineer were to design such a dodgy-looking system, they would be sacked on the spot.)

Central Tolerance

The thymus is an encapsulated and lobulated organ. Within the lobules, the tissue is broadly divided into an outer cortex and inner medulla by the corticomedullary junction, which is rich in blood vessels and afferent lymphatics. A conduit system has also been identified, akin to that described in SLOs (273). The conduits appear to originate at blood vessels, traverse the medulla, and may terminate at Hassall's corpuscles, which develop from medullary thymic epithelial cells (mTECs) after they lose Aire expression (see below); alternatively, they may originate at the latter and extend to the blood vessels. The early thymic precursors migrate to the thymus from the blood via the corticomedullary junction and home to the outer cortex. As these progenitors develop into mature T cells, they are believed to travel through the cortex into the medulla. Finally, mature T cells leave the thymus, enter the blood, and begin their lymphocyte recirculation between different SLOs.

The thymic cortex and medulla contain dense networks of highly specialized epithelial cells that nurture thymocyte development and play largely distinct roles in thymocyte selection. The cortical thymic epithelial cells (cTECs) contain a distinct form of the proteasome termed the thymoproteasome, which generates peptides for MHC-I loading. They also express specialized lysosomal proteases, cathepsin L and thymus-specific serine protease (TSSP), which generate peptides for MHC-II loading (274). Hence, cTECs express a unique spectrum of self peptide–MHC-I and peptide–MHC-II complexes that are not likely to be generated in other tissues. These positively select $CD8^+$ and $CD4^+$ thymocytes, respectively. Mice lacking thymoproteasomes or the specialized lysosomal proteases have deficiencies of $CD8^+$ and $CD4^+$ T cells, respectively.

In contrast, mTECs have the remarkable capacity to express an enormously diverse spectrum of otherwise tissue-specific antigens (TSAs) that are normally expressed only in peripheral (extrathymic) tissues. Such ectopic expression within the thymus is regulated, in whole or in part, by a transcriptional regulator termed the autoimmune regulator (Aire) and the transcription factor Fezf2, which induce the expression of distinct subsets of TSAs (275–282). Consequently, mTECs can generate tissue-specific peptide–MHC-I complexes that contribute to negative selection of $CD8^+$ thymocytes. In addition, mTECs may undergo macroautophagy and generate tissue-specific peptide–MHC-II complexes that may contribute to negative selection of $CD4^+$ thymocytes. Humans with mutations in Aire suffer from the devastating autoimmune disease autoimmune polyendocrinopathy-candidiasis ectodermal dystrophy (APECED), and different hypomorphic mutations may underlie many other autoimmune manifestations. Likewise, mice deficient in Aire manifest with autoimmune disease, although the condition is generally less severe than that in humans. In contrast, mice deficient in FezF2 also present with autoimmune disease, but typically affecting a different spectrum of tissues. Hence, Aire and Fezf2 control the expression of distinct subsets of TSAs.

DCs in central tolerance

The thymus contains distinct populations of DCs that are predominantly localized within the medulla, though some are also present in the cortex. Most attention has been paid to a resident cDCI population, which is most likely generated from the pre-DC progenitor, and mDCII and pDC populations, which appear to originate from extrathymic tissues and migrate to the thymus via the blood (283). The expression of Aire by mTECs also regulates the expression of chemokines that may recruit the three DC populations toward mTECs, namely XCR1 for DCI, CCL2 for DCII, and CCL25 for pDC (106, 279). While thymic DCs do not seem to express Aire, any or all of these populations may acquire ectopic TSAs, which are promiscuously expressed by the mTECs in their vicinity (284). In addition, B cells are present in the thymus, where they can be induced to express Aire and a different spectrum of TSAs in a cell context-dependent manner (285). Thymic macrophages appear primarily to be responsible for clearing the large number of thymocytes that undergo apoptosis, having failed positive or negative selection. Inflammatory cells are normally excluded from the thymic microenvironment. What seems clear is that thymic DCs of one type or another play essential roles in central tolerance (286), and that both they and the corresponding extrathymic populations have complementary roles in peripheral tolerance (see below) (287).

In mice, the expression of Aire in mTECs has been shown to be crucial for the development of a perinatal population of tTregs that persists into adult life (288). The TSAs that are generated presumably represent those that are expressed in subsets of normal tissues, or at least subsets of self components within them. It is also clear that Aire-dependent TSAs can be acquired from mTECs and presented by bone marrow-derived APCs (280, 284, 289). However, these respective cell types play distinct roles in shaping of the adult T-cell repertoire through both deletion of autoreactive T cells and the generation of distinct populations of tTregs. It has been estimated that approximately half of both the Aire-dependent deletion of autoreactive thymocytes and selection of tTregs may be controlled by bone marrow-derived APCs that acquire TSAs from mTECs (280). The APCs responsible for generation of tTregs are most likely cDCI cells (290), although they are present at much lower frequencies in the thymus of perinatal mice than adults. It has also been suggested that Aire induces apoptosis of mTECs, potentially providing an abundant source of antigens for the cross-presenting cDCI population. Direct presentation of TSAs by mTECs, primarily or exclusively, induces the deletion of autoreactive $CD8^+$ thymocytes. In contrast, through their additional costimulatory activities, which may be required for robust generation of tTregs, the cDCI cells may be particularly adept at controlling the generation of tTregs. In the periphery, these may regulate the functions of the mDCI population within normal tissues. What does not seem to have been explained is why the intrathymic cDCI subset, which would otherwise preferentially present antigens to $CD8^+$ T cells in the periphery, seems so important in selecting $CD4^+$ tTregs.

It has been suggested that $CD11c^+MHC\text{-}II^+$ cells, and most likely a population of mDCII cells as defined by expression of CD11b and/or CD172 (SIRPα), can traffic from peripheral tissues via the blood into the thymus. This comes from studies in mice that have used, for example, adoptive cell transfers (291, 292), bone marrow chimeras and culture systems (290), and parabiosis models (291, 293, 294). In general, however, it is not always possible to exclude the trafficking of progenitors such as pre-DCs rather than fully differentiated cells, particularly after transfer of cells subjected to *in vitro* manipulations such as expansion with Flt3L. What is clear is that, after intravenous injection of labeled soluble tracers and antigens, labeled cells resembling mDCII can subsequently be detected within the thymus (290, 292, 295). Here they appear to induce both deletion of autoreactive $CD4^+$ T cells and generation of $CD4^+$ tTregs (292, 295). Importantly, however, there may also be a large population of resident cDCII cells in the thymus, derived from the pre-DCs that also generate the cDCI population. This possibility is generally overlooked but was in fact acknowledged in a well-cited study that nevertheless classified these cells as being of extrathymic origin "for convenience" (293), and is consistent with findings of others (295). It has also been shown that cells resembling DCs are closely associated with recently described medullary conduits, though their phenotype was not fully explored (273). These findings are reminiscent of those described for SLOs, in which presumptive cDCII cells are closely juxtaposed to conduits from where they may sample small soluble molecules. The possibility that such molecules may also gain access to a resident cDCII population via thymic medullary conduits thus deserves further study.

Other studies have clearly shown that if particulate tracers too large to enter the thymus are injected intravenously, they can subsequently be detected in cells resembling the mDCII subset within the thymus (293, 296). However, this is also reminiscent of other studies that have documented the capture of such tracers by monocytes in peripheral tissues, and their subsequent differentiation into monocyte-derived DCs that traffic to the lymph nodes (141, 143). Interestingly, one study has documented the perivascular capture of a soluble tracer by cells that subsequently migrated in a CCR2-dependent manner into the thymic cortex (rather than medulla), where they remained in close proximity to blood vessels (292). However, CCR2-dependent migration is typically associated with monocyte migration and does not seem to have been described as important for migration of classical DCs. The thymus, similar to any other tissue, presumably requires defense against infection. This could therefore reflect a mechanism that might be involved in induction of protective (thymic) immunity, rather than tolerance, for example, after further trafficking through lymphatics into the regional lymphatics. This too deserves further investigation.

Additional evidence for trafficking of a tolerogenic presumptive DC population from peripheral tissues into the thymus has come from other studies. For example, after skin painting with a fluorescently labeled contact-sensitizing agent, labeled CD11c$^+$ cells were detected in the thymus, but their accumulation was inhibited by blockade of the α_4 integrin of very late antigen-4 (VLA-4), which therefore seems to plays a central role in their trafficking to the thymus (291). Furthermore, transgenic expression of a membrane-bound antigen exclusively in cardiomyocytes resulted in thymic deletion of antigen-specific CD4$^+$ T cells, and this too was prevented by similar blockade. The former could represent a peripheral DC population that was induced to migrate in response to "sterile" inflammation, perhaps for induction of intrathymic tolerance against "damaged" tissue antigens (possibly via induction of tTregs). In contrast, the latter may represent migration from a "steady-state" tissue for induction of deletional tolerance against normal tissue antigens. Further studies are required, however, to identify the precise cells involved and determine whether or not such differences exist.

It has also been found, using techniques noted above, that pDCs can migrate from the blood into the thymus. These cells may endocytose soluble tracers and antigens after intravenous or subcutaneous injection (291, 293, 295, 296) and migrate to the thymus, where they appear to delete antigen-specific CD4$^+$ T cells and induce CD4$^+$ tTregs (295, 296). The apparent capture by phagocytosis of intravenously or subcutaneously injected particulates by pDCs has also been demonstrated (296), with subsequent migration of particle-laden cells to the thymus in an α_4 integrin-dependent manner (296). The migration of pDCs into the thymus was shown to be dependent on CCR9. Interestingly, pDCs that were stimulated by TLR9 agonists appear to be excluded from the thymic microenvironment (296). If so, this might be a mechanism to prevent the transport of infectious viruses or microbes into the thymus. It has also been shown that traffic of adoptively transferred mDCII cells is much decreased after maturation in response to a TLR4 agonist (291). Further investigation is needed to establish whether this is a general mechanism to ensure that DCs can only traffic to the thymus under homeostatic conditions, but are prevented from doing so from infected tissues.

In summary, Aire and Fezf2 control the expression of TSAs by mTECs that subsequently induce deletion of autoreactive CD8$^+$ thymocytes. These TSAs can be acquired by thymic cDCI cells, which subsequently induce tTregs that may perhaps control the activity of the mDCI populations in the periphery. In addition, the cDCII and/or extrathymically derived mDCII populations, together with pDCs, may induce the deletion of CD4$^+$ T cells and the induction of tTregs against additional peripheral tissue antigens. Collectively, these mechanisms can lead to the deletion of newly generated autoreactive CD4$^+$ and CD8$^+$ thymocytes and generate a diverse spectrum of tTregs specific for peripheral tissue antigens. Some of these may possibly have roles in regulating the homeostatic maturation of DCs (see below) or in dampening their maturation at sites of homeostatic inflammation (see "Tissues").

Under physiological conditions, any DC expresses its own self peptide-MHC complexes, which represent the normal epitopes that can be generated from its own cellular and molecular components (i.e., those that make a DC a DC rather than any other cell type). It would therefore seem essential to ensure that autoreactive thymocytes that might be specific for such components, which might be termed DC-specific antigens (DCAs), are also deleted or regulated. For example, if the concept that classical DCs are essential for initiating primary TD responses is correct, as generally seems to be the case, then DCs would be able to readily activate any autoreactive T cells that were specific for their own DCAs. Potentially, this would be drastic, as it could ultimately result in elimination of all DCs from SLOs and NLTs. Hence, one could argue that, in addition to inducing tolerance to a diverse spectrum of TSAs, a crucial role of the thymus may be to induce tolerance to the specialized cells that can initiate primary T-cell-dependent responses, and particularly the thymic cDCI and cDCII subsets. In this respect, therefore, the landscape of immunostimulatory cells within the thymus may mirror that which exists in the periphery. A corollary of this hypothesis is that pDCs and/or B cells can also activate primary T-cell responses, which under certain circumstances is possible, or that they are present within the thymus for different functions (e.g., for presentation of peripheral or Aire-dependent antigens, respectively).

Peripheral Tolerance

Peripheral tissues contain different populations of Tregs (269, 297). The centrally generated tTregs home to these sites from the thymus and are supplemented by pTregs that are generated within the periphery; conversely, it seems likely that pTregs, as well as small numbers of naive T cells (298), may home back to the thymus, for reasons largely unknown. While tTregs develop from thymocytes during their differentiation, pTregs can develop from mature, naive T cells that have these regulatory functions "imposed" upon them. In both settings, the master transcription factor FoxP3 regulates their development and functions (299). Mice that lack expression of FoxP3 develop lymphoproliferative autoimmune disease but can be rescued if normal expression is restored. Rare humans lacking FoxP3 expression develop IPEX (immune dysregulation, polyendocrinopathy, enteropathy, X-linked). Typically, they suffer from severe gut and skin inflammation, as well as autoimmune-mediated damage to other tissues and organs such as the thyroid. (It is interesting that many autoimmune conditions resulting from defects in central tolerance involve the endocrine tissues; see below.)

The thymus is critically important particularly for generating tTregs around the perinatal period, which migrate to peripheral tissues. Those that were induced by thymic DCs may, in turn, be selected, expanded, and maintained by peripheral DCs that express TSAs (300). Neonatally thymectomized but not adult mice, as well as rare humans with thymic hypoplasia (DiGeorge syndrome), suffer from severe autoimmune disease caused by the early, or aberrant, populations of T cells they respectively develop. A major stimulus for recruitment of tTregs into some tissues is likely the early colonization with commensals. For example, a wave of Tregs is recruited into neonatal skin of mice in response to establishment of its microbiota (301). Peripheral tolerance in the intestine and lung is associated with an influx of tTregs, perhaps also in response to their distinct and diverse populations of commensals (302). Hence, it appears that tTregs may be particularly important in maintaining tolerance to the microbiota at such sites. However, other evidence indicates that the majority of pTregs in the intestine are generated against food antigens (303). Nevertheless, from TCR repertoire analysis it is quite clear that the populations of Tregs vary enormously between different tissues (thousands of different specificities have been detected in peripheral sites, compared to just hundreds in SLOs) (297). The largely Aire-dependent, and hence TSA-specific, population of tTregs might control the induction of autoimmune T-cell responses against TSAs that might be liberated, for example, during tissue damage caused by invading commensals. But if, as seems likely, classical DCs are crucial for T-cell activation, then this would need to be regulated at the level of DCs. Alternatively, or in addition, a hypothetical population of DCA-specific tTregs (above) might also regulate the responses of DCs that captured and expressed commensal or other antigens. (This is not, however, intended to imply that Tregs only act on DCs and may not also regulate the functions of other cell types.)

In contrast to skin and gut (see above), tTregs may not be recruited to tissues lacking commensals, such as the exocrine pancreas, to which conventional T cells may normally be "ignorant" (302). If so, this may pertain to other nonmucosal or fully vascularized organs and tissues, such as heart and skeletal muscle. Perhaps this necessitates additional mechanisms for the maintenance of tolerance under other circumstances, such as might be provided by pTregs in such sites. It is currently believed that the continuous, homeostatic maturation and migration of DCs from peripheral tissues into SLOs plays a central role in generating pTregs (see below). In addition, there is accumulating evidence that expression of FoxP3 and regulatory function can be induced or imposed on populations of effector and memory T cells (304, 305).

Aire can also be expressed in the periphery and may control expression of TSAs in a cell context-dependent manner, different from those generated by mTECs. So-called extrathymic Aire-expressing cells have been identified in mouse SLOs (277, 306). These are bone marrow-derived cells but seem to be distinct from DCs, although Aire expression has also been reported in splenic marginal-zone DCs (probably the cDCII subset) (307), as well as other cells such as CD14$^+$ monocytes (308). There is evidence that extrathymic Aire-expressing cells can directly inactivate, or perhaps anergize, CD4$^+$ T cells even under inflammatory conditions in the lymph node (rather than inducing deletional tolerance or Tregs) (277). Another transcription factor, deformed epidermal autoregulatory factor 1 (Deaf-1), also controls further ectopic TSA expression in the periphery (309). Deaf-1 is expressed by lymph node stromal cells (LNSCs) and/or FRCs and can also be expressed in the pancreas (310). Hence, it appears that Aire and Deaf-1 may regulate ectopic expression of distinct subsets of TSAs in the periphery (i.e., within both SLOs and at least some NLTs). Moreover, FRCs in lymph nodes may themselves delete autoreactive CD8$^+$ T cells that recognize the TSAs they express (309). Whether or not such antigens can also be acquired by peripheral DCs, perhaps for generation of pTregs, awaits further investigation. However, a converse pathway has also been demonstrated, in which LNSCs acquire peptide-MHC complexes from DCs and subsequently delete CD4$^+$ T cells (311). Whether they also strip other membrane molecules from DCs (or "cross-dress" in them [312]) remains to be seen, but could explain observations that LNSCs may stimulate CD4$^+$ T-cell responses (313).

DCs in peripheral tolerance

Within peripheral tissues, DCs can control the homeostatic expansion of Tregs. These may include the tTregs recruited to these sites that were originally generated by

DCs within the thymus (300). In turn, Tregs may control the maturation of DCs, which express tissue antigens and which might otherwise be sufficient for T-cell activation (314). Some of these interactions have been directly visualized in mouse lymph nodes, where, for example, clusters of Tregs and effector cells were found to be tightly associated with migratory DCs in paracortical regions (315). In general, persistent interactions between $CD4^+$ Tregs and DCs are essential for maintaining peripheral tolerance. If these are experimentally disrupted, for example, by preventing the expression of MHC-II on DCs, severe disease is induced (316). This can involve severe gut inflammation that is driven by commensals and can be prevented by their removal. It also leads to a failure to control $CD8^+$ T-cell tolerance and results in severe autoimmune tissue damage, not only in the intestine but also in sites such as the pancreas noted earlier (317). The normal control of $CD8^+$ T-cell responses seems to be regulated by Tregs, which most likely act at the level of helper $CD4^+$ T cells in a cytotoxic-T-lymphocyte-associated antigen 4 (CTLA-4)-dependent and/or programmed death 1 (PD-1)-dependent manner (318, 319). These and probably other coinhibitory interactions may also contribute to the capacity of Tregs to dampen DC responses sufficiently to maintain normal homeostasis, perhaps especially at sites of homeostatic inflammation driven by commensals. In addition, other mechanisms of regulation by Tregs may be through production of IL-10 and/or TGF-β, as well as perforin-mediated cytolysis of DCs. Nevertheless, there is likely to be a delicate balance between controlling the activation of DCs locally to prevent the induction of immunity and allowing their migration into SLOs to induce pTregs perhaps against the commensals and other antigens they acquired. What is not clear, however, is the relative importance of tTregs and pTregs in controlling these states (320). It would also seem critical to ensure that such regulatory interactions are normally maintained while immunological responses are being induced by other DCs at peripheral sites of infection.

Numerous studies have documented the constitutive migration of DCs from the tissues under normal, homeostatic conditions (29, 321). Both types of classical DCs migrate constitutively from tissues such as gut, skin, and liver in a relatively immature state. Most probably the DCI subset can transport apoptotic cell contents from the gut and the lung via the afferent lymph (56, 239). These cells express relatively low levels of costimulatory molecules that may be inadequate to initiate T-cell activation against the soluble or cell-derived antigens they present, but sufficient to drive the production of Tregs and/or anergy or deletion (88, 299, 322). The NF-κB transcription factor has been shown to be essential for both constitutive migration and the subsequent induction of tolerance against normal tissue antigens (321). There is also evidence from studies in mice lacking IRF4 that this transcription factor regulates the migration of cDCII cells from the tissues under both homeostatic and inflammatory conditions, as well as contributing to their development from progenitors and enhancing their MHC-II antigen presentation capacity (112). Such information is currently lacking for the mDCI population, which, in mice, expresses CD103. However, a natural ligand for CD103 is E-selectin, which plays an essential role in the turnover of apoptotic cells (323). In addition, disruption of E-cadherin-mediated intracellular interactions between mouse DCs in culture induces a transcriptionally distinct maturation process. During this, the cells increase expression of costimulatory molecules but fail to mature fully or to secrete proinflammatory cytokines (324). At least in part, this process is regulated by β-catenin, which can form complexes with E-cadherin and has also been shown to control the homeostatic migration of DCs from the intestine and the subsequent induction of pTregs. Hence, while constitutive migration of mDCII cells may be part of a normal, IRF4-dependent differentiation program, that of cDCI cells may be regulated by apoptotic turnover in tissues and/or through a CD103-dependent mechanism controlled by a different program.

If DCs are crucial for the induction and maintenance of tolerance, then at first sight one might predict that deletion of DCs, or selective populations of DCs, would lead to autoimmune disease. This has been demonstrated in one setting in which all $CD11c^+$ cells were deleted, with resultant induction of devastating autoimmune disease (325). In a related model, which did not, however, delete pDCs and LCs, myeloproliferative disease resulted instead (326). Other evidence is indirect; for example, intrathymic expression of XCL1 is controlled by Aire apparently to recruit DCI cells toward mTECs, and deficiency leads to autoimmune disease (107). Nevertheless, some have suggested that DCs play only minimal roles, if any, in tolerance. In part, this is because selective depletion of the DCI or DCII populations in IRF8- or IRF4-deficient mice, for example, does not result in autoimmune manifestations. Crucially, however, these manipulations may deplete not only the DC populations that are required for induction and maintenance of tolerance but also the very same populations that are required for induction of immunity. One possibility is that there is redundancy in immunological functions, such that depletion of one DC population spares the capacity to induce tolerance to overlapping antigens by the other. Alternatively, it is conceivable that each DC population induces tolerance to a distinct and selective repertoire of self-derived antigens, which is perfectly mirrored by those it naturally presents in the periphery.

DCs IN IMMUNITY

This section focuses, mainly at a cellular level, on the role of DCs and conventional αβ T cells in defense against infection. In particular, it deals with the functions of DCs in driving the different stages of adaptive immune responses. These stages are broadly divided into (i) sensing of infection and tissue damage, (ii) the priming phase, (iii) the effector phase, and (iv) the memory phase. All are, however, closely interlinked.

Sensing of Infection and Tissue Damage

Any tissue monitors itself, or is monitored, to sense infection and damage. The PRRs of epithelial cells in mucosal tissues are generally localized on the tissue-facing, abluminal surfaces. Hence, they preferentially recognize invaders of the lamina propria rather than commensals contained within the lumen. In addition to specialized populations of macrophages, which may play homeostatic rather than primary defense roles, most NLTs contain resident cells such as mast cells within the connective tissue or mucosae. Tissues such as the dermis of skin, the lamina propria of the gut and lung, and the liver additionally harbor different populations of ILCs (327–329). They are also present in SLOs such as the spleen, MLNs, and tonsils (330). ILCs are lymphoid cells in origin, deriving from the CLP, and they closely resemble T cells but produce innate-like responses and lack expression of TCRs. They include three different types that are classified as ILC1, ILC2, and ILC3 cells. In response to infection or tissue damage, these secrete polarized patterns of cytokines very similar to those produced by Th1, Th2, and Th17 cells, respectively. They also express the corresponding master transcription factors T-bet, GATA-3, and ROR-γ, respectively. Innate-like NK cells have also been assigned to the general class of ILCs and closely resemble CTLs in their cellular cytotoxicity and cytokine secretion profiles. ILCs may represent very ancient lineages that evolved before the emergence of conventional T cells and presumably classical MHC molecules. Their existence reinforces the common view that the immune system focuses on sensing general classes of infectious agents and generating characteristic immunological outputs to deal with each.

Within peripheral tissues, DCs coexist with diverse types of tissue-specific and resident cells and may be bathed in complement components (etc.) produced by resident macrophages; the latter indeed may also regulate DC functions (see "Tissues"). Collectively, however, these may constitute local immunological networks that monitor their environment for infection and damage, immediately mount and amplify innate effector responses, and rapidly trigger inflammation to recruit new effectors. The composition of the DC populations itself also varies between tissues: classical DCs, LCs, and/or pDCs under homeostatic conditions in some, with pDCs and/or moDCs during and after inflammation in these or others (the same will be true for regulatory, effector, and/or memory T cells). Any or all DCs may contribute to innate responses, but it may be the classical DCs that are also essential for initiating subsequent adaptive responses. First, each classical DC may sense alarmins, PAMPs, and/or DAMPs within its own tissue microenvironment. This type of information from many varied inputs is then likely to be integrated, and in turn, each cell may generate a finely tuned molecular and cellular response as an output. Second, any DC may internalize antigens (through pinocytosis, receptor-mediated endocytosis, phagocytosis, and/or macropinocytosis) and generate representative peptide-MHC complexes from them. Third, DCs then migrate from NLTs into SLOs, where they can select antigen-specific T cells, deliver costimulatory signals to activate them, and modulate the functions of activated T cells at least in part according to the information they received in their peripheral microenvironment.

Heterogeneity of DC responses

While the above may seem obvious to most in the field, it raises two very important and closely related issues that have not yet been adequately addressed. The first is the extent of heterogeneity of responses that can be generated by peripheral DCs, and the second is the resultant diversity of T-cell responses that are subsequently induced. In the first case, it seems likely that, at any point in time, DCs in peripheral tissues are at different stages of homeostatic maturation and therefore heterogeneous. Such heterogeneity may be reinforced by the particular microenvironment that any such DC senses during inflammatory responses. An important issue therefore is whether, in the latter case, all DCs are driven to homogeneity during maturation before the critical stage is reached when they egress from the tissue, or whether they migrate as a highly heterogeneous population (331). At a population level, core transcriptomic signatures of the two subsets of classical DCs and pDCs converge during maturation in response to viral infection, although other cell-specific characteristics are retained (331), but clearly this says nothing about diversity at a single-cell level. The issue of DC heterogeneity has only recently begun to be studied (in contrived experimental settings) but indicates that DCs can be diverse at the single-cell level (332, 333). It also seems likely that, during the priming phase, DCs may select T cells with TCRs of different affinity for the peptide-MHC complexes they present, and TCR affinity is directly linked to the subsequent (effector) responses of these T-cell clones (334). Moreover, because of the plasticity of T cells, they respond differentially depending on the extent of costimulatory and other signals they receive. Hence, it seems likely (though yet to be proven, or otherwise) that highly heterogeneous populations of migratory DCs can subsequently generate highly diverse populations of activated and, subsequently, effector T cells.

The Priming Phase

Primary T-cell responses are initiated within SLOs. There is abundant evidence that each subset of classical DCs has specialized capacities to activate naive, antigen-specific T cells (and hence TD responses), and this is essential for generating protective responses (see "Lineages" and "Tissues"). Other DC subsets can express MHC-II and costimulatory molecules, as may other lineages, such as B cells. It is, however, important to stress that expression of these molecules is also essential for the induction of Tregs (see "Tolerance") and may regulate the differentiation of effector and/or memory T cells. Moreover, the exuberant expression of MHC or costimulatory molecules that can be driven by various stimuli *in vitro* does not necessarily translate to the *in vivo* setting, where it may be more tightly regulated and controlled. The evidence that DCs can initiate primary T-cell responses *in vivo* seems strongest for classical DCs, perhaps less so for moDCs, and least convincing for LCs and pDCs.

Segregation of DC responses

Conventional T-cell responses are highly segregated. $CD8^+$ and $CD4^+$ T cells are restricted to recognizing peptide–MHC-I or peptide–MHC-II complexes, respectively. Consequently, recognition is focused toward the respective sources of antigen from which peptides can be generated and loaded onto each type of MHC molecule. Hence, $CD8^+$ T cells are focused toward recognition of intracellularly derived "endogenous" antigens (from the cytosol), while $CD4^+$ T cells recognize extracellularly derived "exogenous" antigens (which are endosomal). However, mere recognition is not enough. To respond, each subset of T cells requires recognition of peptide-MHC complexes together with costimulation that is provided by DCs particularly after they recognize PAMPs and/or DAMPs. Therefore, an extreme view might hold that responses of the two subsets of classical DCs may likewise be segregated. They may have evolved to acquire antigens and to sense DAMPs and PAMPs, preferentially from the corresponding sources that T cells recognize. For DCI cells, these include dead, dying, or damaged cells; apoptotic and necrotic bodies; and infectious agents contained within them (e.g., cytosolic viruses). In contrast, DCII cells may selectively acquire extracellular antigens (e.g., phagocytosed bacteria) and sense the PAMPs associated with them, as well as DAMPs that are released or produced in soluble form. Hence, each may present antigen to T cells and express essential costimulatory molecules to activate them only if they have concomitantly sensed the PAMPs and DAMPs within each respective context.

The relative segregation of discrete antigen presentation and costimulation pathways between the two subsets of classical DCs could promote and ensure the differential activation, and hence segregation, of $CD8^+$ versus $CD4^+$ T-cell responses. This segregation may be further reinforced by the rapid shutdown of further antigen sampling once DC maturation has commenced. It is, however, clear that DCI cells can capture exogenous antigens and quite possible that DCII cells acquire cell-derived components. Potentially this may facilitate subsequent interactions with activated $CD4^+$ and $CD8^+$ T cells, respectively (since activated T cells are no longer dependent on primary costimulation). In this way, DCII cells may selectively activate $CD4^+$ T cells and help to regulate the responses of activated $CD8^+$ T cells, and vice versa for DCI cells (see below). It is also possible that these (converse) interactions may induce early T-cell activation, but that robust T-cell responses require subsequent interactions with the other "default" subset (see "Tissues"). It should be noted that human DCs may also selectively express CD1 molecules, which bind lipids much as classical MHC molecules bind peptides. DCs may therefore differentially present lipid-CD1 complexes to populations of specialized innate-like T cells that recognize them (335), and may regulate or be regulated by them.

Integration of DC responses

The two subsets of classical DCs may function synergistically to enable the adaptive immune system to sense peripheral infection and tissue damage qualitatively, quantitatively, and temporarily, through a continuous stream of short-lived migratory DCs arriving in the SLOs (each of which may have a snapshot of prevailing conditions in the periphery). The mechanism for this would involve multiple sequential interactions between any given T cell and different DCs (see below). In turn, the adaptive immune system may be able to continually fine-tune its responses through many different immunological outputs. For example, the differential expression of TLR3 and CLEC9A (DNGR-1) by mDCI cells, coupled with that of TLR5 and TLR7 by mDCII cells, could ultimately permit discrimination between peripheral infection by an RNA virus or a flagellated bacterium; between the different stages of an RNA viral infection, through sensing of double-stranded RNA replicative intermediates or of genomic RNA before infection; and between cytopathic or noncytopathic infections, through sensing of apoptotic or necrotic bodies. Diverse PRRs for PAMPs and DAMPs are selectively, differentially, or commonly expressed by classical DCs (336). Hence, it is possible that the adaptive immune system may ultimately discriminate between even closely related infectious agents with exquisite selectivity and sensitivity. Taking this one step further, it seems possible that it may also clearly differentiate not just between apoptosis and necrosis but also between different forms of regulated necroptosis (337–340), and hence monitor different stages of peripheral tissue damage and healing.

Activation of T-cell responses

Migratory classical DCs have a relatively short life span within the SLOs, on the order of 1 to 3 days. The same is thought to be true for resident classical DCs, which are constantly regenerated from pre-DCs within these tissues. Migratory DCs die within SLOs, but presumably their contents could be internalized and recycled by other cells, including resident DCs, thus further modulating responses of the latter and perpetuating the T-cell response until the peripheral source of antigen is eliminated. It is estimated that naive lymphocytes transiently reside within SLOs for between 6 and 24 h before leaving to continue their recirculation between these tissues if not activated (183); $CD4^+$ T cells may dwell for the shortest times, $CD8^+$ T cells for longer, and B cells for the longest. The intricate architecture within the SLOs seems to have evolved to greatly increase the probability, and effectively ensure, that any antigen-specific T cell can be activated by a DC that expresses its cognate antigen as a peptide-MHC complex. In large part, this may result from the T cells and migratory DCs crawling (or gliding) along the fibrils of the FRC network rather than attempting to traverse the spaces between them (182, 341–345). In this way it is estimated that a DC can sample ~5,000 T cells every hour (183). When a T cell meets its cognate DC, they remain tightly associated for up to 16 to 20 h, though this can be as short as 3 to 4 h, during which time the T cell is activated. Subsequently, the activated T cell detaches from its priming DC but remains within the SLO for some time (see below).

The Effector Phase

Some of the complex choreographies that are executed by T cells and DCs within different microenvironments of SLOs during priming have been noted (see "Tissues"). It is quite clear that both activated $CD4^+$ and $CD8^+$ T cells can make further sequential interactions with other cognate DCs in the tissue. During these, they presumably start to express their effector programs, which may be serially modulated by the different sets of information they receive from successive DCs (for example, depending on the quantity and quality of costimulatory and/or coinhibitory signals each can deliver). The frequency of these interactions is further enhanced by the increased dwell times of activated T cells within the SLOs compared to their naive counterparts, which are controlled in part by IFN-I (such as might be secreted by pDCs) (344). T cells proliferate at an exceptionally fast rate compared to other eukaryotic cells, each cell cycle being completed within just 6 to 8 h. This leads to a rapid expansion of the antigen-specific clones, each of which may interact with migratory and/or resident DCs. In turn, the effector programs of each T cell may be further modulated depending on the particular microenvironments that migratory DCs experienced in the periphery and/or that resident DCs experience within the SLOs. Potentially, the responses of any DC itself may be modulated by small molecules delivered via conduits or mediators produced by monocytes (181), moDCs, or "swarms" of neutrophils (345) that enter SLOs (see "Tissues"). This may drive further diversification of T-cell responses, producing even more heterogeneity in early effector-T-cell populations.

Some T cells execute their effector functions within the SLOs. These include $CD4^+$ T cells that activate $CD8^+$ T cells and potentially drive the generation of CTLs locally or later within NLTs. They also include $CD4^+$ T cells that migrate to the borders of T-cell zones and the follicles where they activate antigen-specific B cells. Some $CD4^+$ T cells may alternatively differentiate into follicular helper T cells, which are involved in the germinal center reaction; precisely where this latter fate is decided is not entirely clear, though it may be controlled at least in part by moDCs (346). Other $CD4^+$ effector T cells may instead develop into memory T cells (see below). However, some $CD4^+$ T cells migrate from the SLOs and execute their effector functions instead within NLTs (see below). There is good evidence that homing of effector T cells to different NLTs can be dependent on where their antigen originated, and this is controlled at least in part by DCs; hence, effector T cells may selectively migrate to distinct sites such as the skin, gut, or lung (347–349).

Polarization of T-cell responses

For decades, many in the field have focused on polarized types of T-cell effector functions, such as Th17, Th1, and Th2 responses, which have been well studied at the cell population level (348). There has also been a tendency by some (though not all) to correlate these responses with different DC subsets that might induce them. This type of view originated from the initial description of the Th1-Th2 paradigm followed by many early studies (noted elsewhere [61]) showing that adoptive cell transfer of different DC populations, or selective targeting of antigens to one or the other, resulted in polarized Th1 or Th2 responses (105). This general concept was extended following the discovery of Th17 cells. It now seems to be believed by many that Th1 responses are typically induced by DCI cells, while Th17 and Th2 responses are more generally induced by DCII cells (whether or not of different subsets). Such notions indeed seem consistent with known functional specializations of these subsets (see "Lineages"). And yet there have been many contradictory findings, depending on the precise experimental system that is studied. Moreover, the description of apparently different polarized subsets, such as Th9 and Th22 cells that may play specialized roles in defense of skin or gut (349), is hard to reconcile with this general line of thinking unless further DC subsets are invoked. Furthermore, during natural infections of mostly outbred species such as human, such highly polarized patterns of T-cell behavior have been hard to find. It is, however, now absolutely clear that T cells are highly plastic and that, in any given response, $CD4^+$ T cells coexpressing different combinations of master transcription factors can be readily detected (350); responses of $CD8^+$ T cells are likewise highly diverse (351).

Rather than invoking a "one DC subset, one T-cell subset" view, an alternative is that, at the single-cell level, the effector function of any given T cell represents a summation of all the influences to which it has been subject during its short-term (but adequately long-lived) residence in an SLO and/or later in NLTs. In the simplest case, the initial activation of a $CD4^+$ T cell by a DCI cell followed by serial interactions with DCs of the same subset may bias responses toward Th1 polarization. A similar case might also be made for DCII cells and Th2 responses, while reciprocal interactions between the two DC subsets might result in others, such as Th17 responses. Nevertheless, the serial interactions of any given activated T cell with DCs are likely to be considerably more heterogeneous in nature. It therefore seems possible that, during the effector phase, many different permutations and combinations of effector functions are generated at a single-cell level (though at a population level they may seem biased). Hence, from this large effector cell pool, only some T-cell clones may have the precise sets of effector functions that can promote the elimination or control of any given infectious agent. This hypothesis predicts the existence of heterogeneity in antigen-specific effector T cells at a single-cell level.

Heterogeneity of T-cell effector responses

Recent studies have revealed a remarkable diversity of effector-T-cell functions at the single-cell level, even during apparently polarized responses to a defined infectious agent. Earlier work that studied individual $CD8^+$ T cells responding to a given antigen by multiplex analysis demonstrated considerable heterogeneity within the population (352, 353), though 14 different types of overall response were initially proposed. (Good luck to those who would seek 14 new DC subsets to drive these responses.) Recent high-dimensional, multiparametric techniques have reinforced these findings, but also tend to suggest that T-cell effector responses are even more heterogeneous and perhaps more closely resemble a continuum (354). If the above hypothesis is correct, then presumably an adequate diversity of effector functions can be generated for those clones that remain in the SLOs to execute their responses. Conceivably, however, the effector functions of those clones that migrate to NLTs, and their clonal progeny, could be further diversified in the periphery. If so, an essential cell type in this respect may be the moDC (136, 145, 153), though any APC might of course also contribute. Through their capacity to capture antigens within the periphery, moDCs are likely to express peptide-MHC complexes that can drive further proliferation of effector T cells for as long as any infection persists. They certainly appear to be particularly important in the effector phases of $CD4^+$ T-cell responses. Potentially, too, moDCs may be modulated by their microenvironments such that the population as a whole is highly diverse at a single-cell level (compare classical DCs above). While activated T cells are not dependent on primary costimulation, the expression of qualitatively and perhaps quantitatively different costimulatory (and/or coinhibitory) molecules by moDCs may further differentially modulate the effector functions of each T-cell clone with which they interact. If this is the case, then conceivably LCs may function similarly during T-cell effector phases within the skin and structurally similar sites. Possibly pDCs could also contribute to diversification within the mucosal tissues (and perhaps even at inflammatory sites and in cancers).

The Memory Phase

Memory T cells can persist for the lifetime of any mouse or human, though they decline in numbers with time. While much still remains to be discovered, it seems clear that $CD4^+$ memory T cells can persist in the apparent absence of the antigen to which they were induced, and that $CD4^+$ T cells are also essential for the generation and maintenance of $CD8^+$ memory T cells. However, the latter are better understood primarily because of the relative ease of studying their function in *in vitro* assays. Nevertheless, based on their phenotype and function, $CD4^+$ memory T cells have been divided into central memory (Tcm) and effector memory (Tem) subsets, while others that appear to reside in peripheral tissues are termed resident memory (Trm) cells (355). The former, Tcm cells, recirculate between SLOs and can rapidly reexpress molecules such as CD40L on restimulation; within SLOs they may be positioned most closely to potential sites of pathogen entry (356). In contrast, Tem cells tend to populate peripheral tissues and can rapidly reacquire effector cell functions such as polarized patterns of cytokine secretion. It has been suggested that Tcm cells can develop into Tem cells after restimulation.

There is still much to be learned about the factors that regulate the generation of $CD4^+$ and $CD8^+$ memory T cells. However, pDCs have been implicated particularly in this process. As noted earlier, these cells may play specialized roles in viral infections and perhaps particularly in the regulation of $CD8^+$ T-cell and CTL responses (see "Lineages"). As was also noted, pDCs can secrete high levels of IFN-I. Importantly, at the tissue level, it is known that IFN-I can prevent the egress of lymphocytes from SLOs (344). This latter example may be most relevant to the phenomenon of lymph node shutdown, which is observed soon after infection and can be induced in regional lymph nodes after injection of IFN-I into the skin (G. G. Macpherson, personal communication). The markedly increased cellularity of SLOs during this period may increase the frequency and duration of interaction between T cells and their cognate DCs, and possibly enhance the diversification of effector cells. Because $CD8^+$ T cells generally have a higher activation threshold than $CD4^+$ T cells (200), it is possible that this could also apply to their development into memory T cells. Hence, in principle, the capacity of pDCs to contribute to lymph node shutdown and extend intercellular communications could directly or indirectly contribute to robust CD8 memory-T-cell generation. Whether or not there might be segregation between moDCs and pDCs for generation of $CD4^+$ and $CD8^+$ memory T cells remains to be seen. The relative contributions of the different subsets to reactivation of memory cells during secondary responses is unclear. However, there is good evidence at least that classical DCs can elicit such recall responses, presumably in SLOs and also in bone marrow and NLTs (104, 357, 358).

DISCUSSION AND CONCLUSION

Collectively, the different DC subsets that are the focus of this review may constitute an integrated cellular network that is essential for both tolerance and immunity in the adaptive immune system. In this final section, two main areas will briefly be discussed. The first provides some speculations on the evolution of DCs. The second revisits concepts of the generation of diversity and clonal selection, but within the context of DCs driving tolerance and immunity.

DCs in Evolution

From an evolutionary standpoint, and if ontogeny indeed recapitulates phylogeny, it is tempting to speculate about the origins of DCs. The embryonic, yolk sac-derived population of LCs may be the most ancient of all. All animals have hematopoietic tissues that can generate cells, such as the wandering amoebocytes observed by Metchnikoff in starfish larvae or, later in evolution, blood-circulating cells such as monocytes (5). It is possible to envisage that cells such as amoebocytes might first have evolved into resident macrophages within the tissues, including LCs in the epidermis of skin. Potentially the earliest LCs, generated by a form of primitive myelopoiesis well before lymphoid cells evolved, may have played homeostatic roles in the skin of ancient fish, but might have used phagocytosis in defense. From extant species we know that LCs are present in all jawed fish that have been studied, but this does not seem known for agnathans. During the earliest stages of adaptive evolution in jawed fish, LCs may then have evolved into essential immunostimulatory cells for immune responses within skin-associated lymphoid tissues (359) (unless they could migrate to the spleen). The later LC populations successively derived from fetal liver- and bone marrow-derived monocytes, as observed in mice, may conceivably represent examples of evolutionary convergence. Certainly, as a population, all LCs seem homogeneous in skin at a transcriptional level (including the oral mucosal LCs that are generated from pre-DCs; see "Tissues").

Blood-circulating monocytes may have evolved before adaptive immunity. Monocytic cells can even be found in agnathans (360). They were presumably early inflammatory cells since their capacity to circulate meant that they could be rapidly recruited to sites of infection and damage. Here, their greater numbers might have helped to reinforce the defense and repair functions of the more ancient and sessile (amoebocyte-derived) macrophages and LCs. With the evolution of adaptive immunity, their functions may have also diverged to generate populations of moDCs as immunostimulatory cells. Again, their circulatory capacity may have enabled them now to migrate from the tissues into the newly evolved spleen. Hence, these may have represented early migratory DCs. In the natterjack toad (*Bufo calamita*), for example, migratory monocytes in the red pulp of spleen appear to differentiate into giant, dendritic-like cells after immunization (86).

Potentially, the phagocytic activity of early monocyte-derived cells may have enabled them to contribute particularly in early immunity against bacteria, which perhaps have the greatest diversity of PAMPs for any eukaryotic host. However, most viruses are too small to be phagocytosed and may have fewer PAMPs, in part because they can envelop themselves within host cell-derived membranes. This may have provided the evolutionary pressure for the emergence of pDCs. These cells can be produced from both myeloid and lymphoid lineages and have certain features of both. In particular, they contain RAG and partial TCR or BCR transcripts (116). This could suggest that pDCs evolved during the diversification of the myeloid and lymphoid lineages. It would therefore be interesting to investigate whether pDCs are also present in fish and amphibians, which this hypothesis would predict, in addition to warm-blooded animals such as rodents, pigs, and humans. The earliest functions of pDCs may have been for inducing antiviral resistance. However, these cells also had the capacity to circulate within the blood and therefore could travel to the spleen. Perhaps therefore they acquired specialized functions to facilitate $CD8^+$ T-cell responses against viruses. In contrast, the moDCs were presumably more specialized for $CD4^+$ T-cell responses against bacteria, and most likely protozoa and fungal spores since these can be readily phagocytosed. Potentially, therefore, moDCs and pDCs could have represented the earliest DC-like cells with diversified but synergistic functions that drove early $CD4^+$ and $CD8^+$ T-cell responses.

Classical DCs can apparently also be generated from both myeloid and lymphoid lineages *in vivo*. Those in lymphoid tissues can also express either CD4 or CD8, which are otherwise generally associated with conventional or specialized $\alpha\beta$ T-cell populations. However, they do not contain RAG or the abortive transcripts of pDCs noted above. It is therefore possible that they evolved even later than pDCs, as is also suggested by the apparently late appearance of the pre-DCs in simple linear models of differentiation (see "Lineages"). These two subsets of CDI and CDII cells might therefore be viewed as being "superimposed" upon the pDC and moDC populations that already existed, with LCs perhaps confined to local defense of the skin. Instead of focusing on particular types of infectious agents, the classical DC may have evolved primarily to sense PAMPs and DAMPs derived from intracellular and extracellular sources in general. This may have enabled a fine-tuning of protective T-cell responses against the respective infectious agents noted above. These DCs may also have facilitated more-robust responses to be generated against infectious agents such as helminths, which are generally difficult to eliminate and often cohabit with humans. Helminths are eukaryotes, and some can mask themselves in host proteins. Hence, they may express many fewer PAMPs than other types of infectious agents but can cause barrier damage that generates DAMPs. Sensing of these by classical DCs may have enhanced the sensing of "danger" and the capacity to induce T-cell responses in general. Interdigitating cells resembling classical DCs have also been identified in the splenic white pulp of different fish, including sharks (85), trout (84), and zebrafish (361). It is not, however, clear if these are classical DCs or monocyte-derived cells. Transcriptional comparisons with those of more recently evolved species may therefore provide insights into whether one subset of classical pre-DCs evolved before another, or whether they arose even later in the evolution of jawed vertebrates.

There are many ways to build an immune system (362–364). All jawed animals have skin and gut (and fish have gills, too), so it is important to endow them with local forms of defense at the very least (359, 365). The essential specialized tissues for adaptive immunity seem to be a thymus and spleen, but thereafter additional primary lymphoid tissues and SLOs can evolve in different species. The essential cellular building blocks are T cells, B cells to produce antibodies, and almost certainly DCs. The critical molecular building blocks include lymphocytes with highly diversified antigen receptor repertoires and MHC molecules. However, the genetic mechanisms for diversification can differ between species, including, for example, somatic recombination in humans and mice, gene conversion in chickens, and somatic hypermutation in sharks. And at the cellular and molecular level, some startling differences are also apparent. The cod, for example, lacks $CD4^{+}$ T cells and all components associated with the peptide–MHC-II pathway (366). This was presumably not an act of wanton vandalism, but due to evolutionary pressures that are still not understood (though a possible third round of genome duplication in some bony fish may be relevant [367]). Studies of DCs in this species may prove particularly interesting.

DCs in Generation of Diversity and Clonal Selection

Two cornerstones in our understanding of adaptive immunity were laid decades ago. The first was the clonal selection theory proposed separately by Burnet and Talmage. This envisaged the existence of a vast number of clonally distributed lymphocyte receptors that could potentially recognize almost any antigen in the universe. The second, originating from the later work of Tonegawa, was the mechanistic insight into precisely how this remarkable generation of diversity could be accomplished, through rearrangement of germ line gene segments accompanied by additional mechanisms of diversification. Since then, our thinking about both processes has been dominated by the role of antigen, in terms of both the diversification of antigen receptors and the subsequent selection of lymphocyte clones by antigen in immunological responses.

There are, in fact, two well-recognized systems for receptor diversification and clonal selection in adaptive immunity that have also been studied for decades. The first was as originally conceived, and occurs within primary lymphoid tissues that have evolved in all jawed vertebrates. Through whichever genetic mechanisms are used in the species, enormous repertoires of BCRs and TCRs can be created. Functional clones of naive B cells and T cells are selected in the thymus through their capacities to recognize self-antigen, in its native form for B cells and as self peptide-MHC complexes for T cells. This generates the primary repertoires of mature T cells and B cells that can subsequently be selected by foreign antigen for protective immunity. During T-cell development in the thymus, DCs may facilitate deletion of autoreactive thymocytes. Crucially, however, they may also select and shape the precise repertoires of tTregs that are subsequently required to regulate their functions in peripheral tissues. During this process it seems likely that they select tTregs with diverse, though moderate, affinities. In turn, the affinity of TCRs may regulate the precise effector functions of these cells (272). Hence, DCs may clonally select and diversify these tTreg populations. Moreover, because it is not possible to ensure that all representations of self are available within the thymus, DCs in normal peripheral tissues generate further repertoires of pTregs that regulate their own functions (and perhaps of course those of other APCs) to ensure the maintenance of tolerance. Again, the same considerations apply. Therefore, at these levels, DCs may be responsible for both clonal selection and generation of diversity of Tregs in tolerance.

The second well-recognized system for clonal diversification and receptor selection occurs within germinal centers of SLOs and acts exclusively on B cells (368). In this case, the generation of diversity occurs through somatic hypermutation, a mechanism that exists in all jawed vertebrates, although germinal centers only evolved later in birds and mammals. During this process, random point mutations are introduced into the variable regions of BCRs in clones of B cells that were selected by foreign antigen and activated in primary responses. This process massively increases diversification of their BCRs, and these diversified clones then undergo repeated rounds of iterative selection against foreign antigen. This is retained for considerable periods of time and displayed on the follicular DC network within the germinal center. This results in the selection of B-cell clones with high-affinity BCRs (different from those they initially expressed) and from which memory B cells can be subsequently generated; all other clones are eliminated. It has been estimated that in some cases the germinal center reaction may select for survival as few as one B-cell clone out of the vast numbers that were originally generated. For B cells, antigen recognition is inseparable from their primary effector functions. These are mediated by soluble forms of their BCRs, antibodies, that are secreted after differentiation into plasma cells. Hence, the germinal center reaction effectively both diversifies and clonally selects for enhanced B-cell effector functions.

For T cells, the situation is entirely different. There is no known equivalent of the germinal center reaction for T cells and no mechanism for clonal diversification of TCRs (since T cells are not subject to somatic hypermutation in birds and mammals). Moreover, the effector functions of T cells are completely unrelated to their antigen specificities, even though TCRs directionally target these toward other cell types (for example, during formation of immunological synapses between helper T cells and B cells, or between CTLs and infected cells). Instead, the effector functions of T cells are mediated by the specialized membrane molecules and soluble mediators they can express or secrete, and which modulate the functions of other cell types, to help, suppress, or otherwise regulate them or to induce apoptosis to kill. Hence, the only way in which it possible to generate diversity of T-cell responses is to select antigen-specific clones and differentially modify their effector responses (or to recruit others into the response for the same purposes). This diversification may originate during priming itself within the T-cell zones of SLOs. Even at this stage, DCs are likely to select clones of T cells with TCRs of different affinities that directly regulate their effector functions (334). Subsequently, the activated T cells execute their complex choreographies as they serially engage with DCs in the T-cell zone, just as B cells do with follicular DCs within their separate follicular compartment (though they may dance together at the borders).

Distinct populations of DCs may be centrally involved in clonal selection and generation of diversity during the different stages of T-cell or TD responses. The classical DCs (which may themselves be highly diversified) may select heterogeneous populations of antigen-specific T cells to be activated within SLOs. Consequently, they generate diverse populations of effector T cells within them. Within NLTs, moDCs (and perhaps LCs) may select from these clones and further diversify their functions during clonal expansion. Meanwhile, pDCs may modulate the functions of classical DCs in SLOs, and possibly moDCs in NLTs where they are present, in order to help drive further diversification of effector T cells (and perhaps recruit additional clones within SLOs). These processes may continue until protective T-cell clones are generated, at which point antigen can be eliminated. Selection and diversification of effector-T-cell clones by all DCs then inevitably ceases (although potentially DCs might subsequently drive other responses that help dampen the response). However, many or all of the early effector-T-cell clones that are generated in response to infection have the potential to develop into memory-T-cell clones. Some of these become long-lived cells that can provide immunological memory for a lifetime. Hence, if the same antigen is encountered again, these memory T cells can be rapidly reactivated by classical (and perhaps other) DCs to provide almost immediate protection. Hence, DCs may drive the evolution of adaptive responses that are central to immunity and tolerance (see above).

Conclusion

There is continuing debate about whether it is possible to define "dendritic cells," whether these are in fact "just another type of macrophage," or whether they should be included in the mononuclear phagocyte system. From the viewpoints expressed in this review, the five main lineages of DCs that are discussed appear to comprise, at least in part, an integrated DC network. If more are identified, perhaps these too may be accommodated within the general framework of ideas that has been presented. The individual functions of each DC subset seem diverse and specialized, as are those of macrophage populations within different tissues. Collectively they seem to be synergistic; any DC therefore may be essential for adaptive immunity or tolerance, but none by itself is sufficient. Whether or not DCs should be included in the mononuclear phagocyte system seems mainly a semantic argument, for example, depending on whether a cutoff point is taken above or below monocytes during development or if all are simply grouped together. It matters not. All such categorizations are anthropomorphic constructs designed to help our understanding of systems whose complexity may be beyond mere human comprehension. It is with this in mind that this review ends.

Acknowledgments. *Sincere thanks are due to the numerous academic colleagues, guests, students, and friends whose intellectual input over the years has helped shape the ideas presented in this review. It was great fortune and a privilege to work with Siamon Gordon as his first graduate student, and subsequently with Ralph Steinman as his first postdoctoral fellow within Zanvil Cohn's department. This review is offered in gratitude to Siamon and to the memory of Ralph and Zan.*

Citation. Austyn JM. 2016. Dendritic cells in the immune system—history, lineages, tissues, tolerance, and immunity. Microbiol Spectrum 4(6):MCHD-0046-2016.

References

1. **Sompayrak L.** 2016. *How the Immune System Works*, 5th ed. John Wiley & Sons, Chichester, United Kingdom.
2. **MacPherson GG, Austyn JM.** 2012. *Exploring Immunology: Concepts and Evidence*, 1st ed. Wiley-VCH Verlag, Weinheim, Germany.
3. **Murphy KM, Weaver CT.** 2016. *Janeway's Immunobiology*, 9th ed. Garland Science, Taylor & Francis Group, New York, NY.
4. **Paul WE (ed).** 2013. *Fundamental Immunology*, 7th ed. Lippincott Williams & Wilkins, Philadelphia, PA.

5. **Flajnik MF, Du Pasquier L.** 2013. Evolution of the immune system, p 67–128. *In* Paul WE (ed), *Fundamental Immunology*, 7th ed. Lippincott Williams & Wilkins, Philadelphia, PA.
6. **Sewell AK.** 2012. Why must T cells be cross-reactive? *Nat Rev Immunol* **12:**669–677.
7. **Steinman RM, Cohn ZA.** 1973. Identification of a novel cell type in peripheral lymphoid organs of mice. I. Morphology, quantitation, tissue distribution. *J Exp Med* **137:**1142–1162.
8. **Zinkernagel RM, Doherty PC.** 1974. Restriction of in vitro T cell-mediated cytotoxicity in lymphocytic choriomeningitis within a syngeneic or semiallogeneic system. *Nature* **248:**701–702.
9. **Steinman RM, Witmer MD.** 1978. Lymphoid dendritic cells are potent stimulators of the primary mixed leukocyte reaction in mice. *Proc Natl Acad Sci U S A* **75:** 5132–5136.
10. **Farrant J, Clark JC, Lee H, Knight SC, O'Brien J.** 1980. Conditions for measuring DNA synthesis in PHA stimulated human lymphocytes in 20 microliters hanging drops with various cell concentrations and periods of culture. *J Immunol Methods* **33:**301–312.
11. **Inaba K, Steinman RM, Van Voorhis WC, Muramatsu S.** 1983. Dendritic cells are critical accessory cells for thymus-dependent antibody responses in mouse and in man. *Proc Natl Acad Sci U S A* **80:**6041–6045.
12. **Austyn JM, Steinman RM, Weinstein DE, Granelli-Piperno A, Palladino MA.** 1983. Dendritic cells initiate a two-stage mechanism for T lymphocyte proliferation. *J Exp Med* **157:**1101–1115.
13. **Macatonia SE, Hosken NA, Litton M, Vieira P, Hsieh CS, Culpepper JA, Wysocka M, Trinchieri G, Murphy KM, O'Garra A.** 1995. Dendritic cells produce IL-12 and direct the development of Th1 cells from naive CD4+ T cells. *J Immunol* **154:**5071–5079.
14. **Veldman JE, Molenaar I, Keuning FJ.** 1978. Electron microscopy of cellular immunity reactions in B-cell deprived rabbits. Thymus derived antigen reactive cells, their micro-environment and progeny in the lymph node. *Virchows Arch B Cell Pathol* **28:**217–228.
15. **Veerman AJ.** 1974. On the interdigitating cells in the thymus-dependent area of the rat spleen: a relation between the mononuclear phagocyte system and T-lymphocytes. *Cell Tissue Res* **148:**247–257.
16. **Balfour BM, Drexhage HA, Kamperdijk EW, Hoefsmit EC.** 1981. Antigen-presenting cells, including Langerhans cells, veiled cells and interdigitating cells. *Ciba Found Symp* **84:**281–301.
17. **Metlay JP, Witmer-Pack MD, Agger R, Crowley MT, Lawless D, Steinman RM.** 1990. The distinct leukocyte integrins of mouse spleen dendritic cells as identified with new hamster monoclonal antibodies. *J Exp Med* **171:**1753–1771.
18. **Köhler G, Milstein C.** 1975. Continuous cultures of fused cells secreting antibody of predefined specificity. *Nature* **256:**495–497.
19. **Crowley M, Inaba K, Witmer-Pack M, Steinman RM.** 1989. The cell surface of mouse dendritic cells: FACS analyses of dendritic cells from different tissues including thymus. *Cell Immunol* **118:**108–125.
20. **Landry D, Lafontaine M, Cossette M, Barthélémy H, Chartrand C, Montplaisir S, Pelletier M.** 1988. Human thymic dendritic cells. Characterization, isolation and functional assays. *Immunology* **65:**135–142.
21. **Inaba K, Hosono M, Inaba M.** 1990. Thymic dendritic cells and B cells: isolation and function. *Int Rev Immunol* **6:**117–126.
22. **Schuler G, Romani N, Steinman RM.** 1985. A comparison of murine epidermal Langerhans cells with spleen dendritic cells. *J Invest Dermatol* **85**(Suppl)**:**99s–106s.
23. **Schuler G, Steinman RM.** 1985. Murine epidermal Langerhans cells mature into potent immunostimulatory dendritic cells in vitro. *J Exp Med* **161:**526–546.
24. **Reis e Sousa C, Stahl PD, Austyn JM.** 1993. Phagocytosis of antigens by Langerhans cells in vitro. *J Exp Med* **178:**509–519.
25. **Hart DN, Fabre JW.** 1981. Demonstration and characterization of Ia-positive dendritic cells in the interstitial connective tissues of rat heart and other tissues, but not brain. *J Exp Med* **154:**347–361.
26. **Hart DN, McKenzie JL.** 1990. Interstitial dendritic cells. *Int Rev Immunol* **6:**127–138.
27. **Austyn JM, Hankins DF, Larsen CP, Morris PJ, Rao AS, Roake JA.** 1994. Isolation and characterization of dendritic cells from mouse heart and kidney. *J Immunol* **152:**2401–2410.
28. **Steinman RM, Pack M, Inaba K.** 1997. Dendritic cells in the T-cell areas of lymphoid organs. *Immunol Rev* **156:**25–37.
29. **Milling S, Yrlid U, Cerovic V, MacPherson G.** 2010. Subsets of migrating intestinal dendritic cells. *Immunol Rev* **234:**259–267.
30. **Mason DW, Pugh CW, Webb M.** 1981. The rat mixed lymphocyte reaction: roles of a dendritic cell in intestinal lymph and T-cell subsets defined by monoclonal antibodies. *Immunology* **44:**75–87.
31. **Spry CJ, Pflug AJ, Janossy G, Humphrey JH.** 1980. Large mononuclear (veiled) cells like 'Ia-like' membrane antigens in human afferent lymph. *Clin Exp Immunol* **39:**750–755.
32. **Drexhage HA, Mullink H, de Groot J, Clarke J, Balfour BM.** 1979. A study of cells present in peripheral lymph of pigs with special reference to a type of cell resembling the Langerhans cell. *Cell Tissue Res* **202:** 407–430.
33. **Thorbecke GJ, Silberberg-Sinakin I, Flotte TJ.** 1980. Langerhans cells as macrophages in skin and lymphoid organs. *J Invest Dermatol* **75:**32–43.
34. **Knight SC, Balfour BM, O'Brien J, Buttifant L, Sumerska T, Clarke J.** 1982. Role of veiled cells in lymphocyte activation. *Eur J Immunol* **12:**1057–1060.
35. **Larsen CP, Steinman RM, Witmer-Pack M, Hankins DF, Morris PJ, Austyn JM.** 1990. Migration and maturation of Langerhans cells in skin transplants and explants. *J Exp Med* **172:**1483–1493.

36. **Stoitzner P, Pfaller K, Stössel H, Romani N.** 2002. A close-up view of migrating Langerhans cells in the skin. *J Invest Dermatol* **118:**117–125.

37. **O'Doherty U, Peng M, Gezelter S, Swiggard WJ, Betjes M, Bhardwaj N, Steinman RM.** 1994. Human blood contains two subsets of dendritic cells, one immunologically mature and the other immature. *Immunology* **82:** 487–493.

38. **Kupiec-Weglinski JW, Austyn JM, Morris PJ.** 1988. Migration patterns of dendritic cells in the mouse. Traffic from the blood, and T cell-dependent and -independent entry to lymphoid tissues. *J Exp Med* **167:**632–645.

39. **Austyn JM, Kupiec-Weglinski JW, Hankins DF, Morris PJ.** 1988. Migration patterns of dendritic cells in the mouse. Homing to T cell-dependent areas of spleen, and binding within marginal zone. *J Exp Med* **167:** 646–651.

40. **Larsen CP, Morris PJ, Austyn JM.** 1990. Migration of dendritic leukocytes from cardiac allografts into host spleens. A novel pathway for initiation of rejection. *J Exp Med* **171:**307–314.

41. **Cumberbatch M, Dearman RJ, Kimber I.** 1997. Interleukin 1 beta and the stimulation of Langerhans cell migration: comparisons with tumour necrosis factor alpha. *Arch Dermatol Res* **289:**277–284.

42. **Kimber I, Cumberbatch M.** 1992. Stimulation of Langerhans cell migration by tumor necrosis factor alpha (TNF-alpha). *J Invest Dermatol* **99:**48S–50S.

43. **Roake JA, Rao AS, Morris PJ, Larsen CP, Hankins DF, Austyn JM.** 1995. Dendritic cell loss from nonlymphoid tissues after systemic administration of lipopolysaccharide, tumor necrosis factor, and interleukin 1. *J Exp Med* **181:**2237–2247.

44. **Bowers WE, Berkowitz MR.** 1986. Differentiation of dendritic cells in cultures of rat bone marrow cells. *J Exp Med* **163:**872–883.

45. **Inaba K, Inaba M, Romani N, Aya H, Deguchi M, Ikehara S, Muramatsu S, Steinman RM.** 1992. Generation of large numbers of dendritic cells from mouse bone marrow cultures supplemented with granulocyte/macrophage colony-stimulating factor. *J Exp Med* **176:**1693–1702.

46. **Caux C, Dezutter-Dambuyant C, Schmitt D, Banchereau J.** 1992. GM-CSF and TNF-α cooperate in the generation of dendritic Langerhans cells. *Nature* **360:**258–261.

47. **Siena S, Di Nicola M, Bregni M, Mortarini R, Anichini A, Lombardi L, Ravagnani F, Parmiani G, Gianni AM.** 1995. Massive ex vivo generation of functional dendritic cells from mobilized $CD34^+$ blood progenitors for anticancer therapy. *Exp Hematol* **23:**1463–1471.

48. **O'Doherty U, Ignatius R, Bhardwaj N, Pope M.** 1997. Generation of monocyte-derived dendritic cells from precursors in rhesus macaque blood. *J Immunol Methods* **207:**185–194.

49. **Sallusto F, Lanzavecchia A.** 1994. Efficient presentation of soluble antigen by cultured human dendritic cells is maintained by granulocyte/macrophage colony-stimulating factor plus interleukin 4 and downregulated by tumor necrosis factor α. *J Exp Med* **179:** 1109–1118.

50. **Romani N, Reider D, Heuer M, Ebner S, Kämpgen E, Eibl B, Niederwieser D, Schuler G.** 1996. Generation of mature dendritic cells from human blood. An improved method with special regard to clinical applicability. *J Immunol Methods* **196:**137–151.

51. **Peters JH, Ruhl S, Friedrichs D.** 1987. Veiled accessory cells deduced from monocytes. *Immunobiology* **176:** 154–166.

52. **Palucka K, Banchereau J.** 2012. Cancer immunotherapy via dendritic cells. *Nat Rev Cancer* **12:**265–277.

53. **Elster JD, Krishnadas DK, Lucas KG.** 2016. Dendritic cell vaccines: a review of recent developments and their potential pediatric application. *Hum Vaccin Immunother* **12:**2232–2239.

54. **Kantoff PW, Higano CS, Shore ND, Berger ER, Small EJ, Penson DF, Redfern CH, Ferrari AC, Dreicer R, Sims RB, Xu Y, Frohlich MW, Schellhammer PF, IMPACT Study Investigators.** 2010. Sipuleucel-T immunotherapy for castration-resistant prostate cancer. *N Engl J Med* **363:**411–422.

55. **Grouard G, Rissoan MC, Filgueira L, Durand I, Banchereau J, Liu YJ.** 1997. The enigmatic plasmacytoid T cells develop into dendritic cells with interleukin (IL)-3 and CD40-ligand. *J Exp Med* **185:** 1101–1111.

56. **Huang FP, Platt N, Wykes M, Major JR, Powell TJ, Jenkins CD, MacPherson GG.** 2000. A discrete subpopulation of dendritic cells transports apoptotic intestinal epithelial cells to T cell areas of mesenteric lymph nodes. *J Exp Med* **191:**435–444.

57. **Tew JG, Phipps RP, Mandel TE.** 1980. The maintenance and regulation of the humoral immune response: persisting antigen and the role of follicular antigen-binding dendritic cells as accessory cells. *Immunol Rev* **53:**175–201.

58. **Aguzzi A, Kranich J, Krautler NJ.** 2014. Follicular dendritic cells: origin, phenotype, and function in health and disease. *Trends Immunol* **35:**105–113.

59. **Chen LL, Adams JC, Steinman RM.** 1978. Anatomy of germinal centers in mouse spleen, with special reference to "follicular dendritic cells." *J Cell Biol* **77:**148–164.

60. **Heesters BA, Myers RC, Carroll MC.** 2014. Follicular dendritic cells: dynamic antigen libraries. *Nat Rev Immunol* **14:**495–504.

61. **Pulendran B.** 2015. The varieties of immunological experience: of pathogens, stress, and dendritic cells. *Annu Rev Immunol* **33:**563–606.

62. **Merad M, Sathe P, Helft J, Miller J, Mortha A.** 2013. The dendritic cell lineage: ontogeny and function of dendritic cells and their subsets in the steady state and the inflamed setting. *Annu Rev Immunol* **31:**563–604.

63. **Miller JC, Brown BD, Shay T, Gautier EL, Jojic V, Cohain A, Pandey G, Leboeuf M, Elpek KG, Helft J, Hashimoto D, Chow A, Price J, Greter M, Bogunovic M, Bellemare-Pelletier A, Frenette PS, Randolph GJ, Turley SJ, Merad M, Immunological Genome Consortium.** 2012. Deciphering the transcriptional network of the dendritic cell lineage. *Nat Immunol* **13:** 888–899.

64. **Schraml BU, Reis e Sousa C.** 2015. Defining dendritic cells. *Curr Opin Immunol* **32:**13–20.

65. **Reizis B, Bunin A, Ghosh HS, Lewis KL, Sisirak V.** 2011. Plasmacytoid dendritic cells: recent progress and open questions. *Annu Rev Immunol* **29:**163–183.

66. **Ginhoux F, Merad M.** 2010. Ontogeny and homeostasis of Langerhans cells. *Immunol Cell Biol* **88:**387–392.

67. **Guilliams M, Ginhoux F, Jakubzick C, Naik SH, Onai N, Schraml BU, Segura E, Tussiwand R, Yona S.** 2014. Dendritic cells, monocytes and macrophages: a unified nomenclature based on ontogeny. *Nat Rev Immunol* **14:** 571–578.

68. **De Kleer I, Willems F, Lambrecht B, Goriely S.** 2014. Ontogeny of myeloid cells. *Front Immunol* **5:**423. doi: 10.3389/fimmu.2014.00423.

69. **Lee J, Breton G, Oliveira TYK, Zhou YJ, Aljoufi A, Puhr S, Cameron MJ, Sékaly RP, Nussenzweig MC, Liu K.** 2015. Restricted dendritic cell and monocyte progenitors in human cord blood and bone marrow. *J Exp Med* **212:**385–399.

70. **Strobl H, Bello-Fernandez C, Riedl E, Pickl WF, Majdic O, Lyman SD, Knapp W.** 1997. flt3 ligand in cooperation with transforming growth factor-β1 potentiates in vitro development of Langerhans-type dendritic cells and allows single-cell dendritic cell cluster formation under serum-free conditions. *Blood* **90:**1425–1434.

71. **Guo X, Zhou Y, Wu T, Zhu X, Lai W, Wu L.** 2016. Generation of mouse and human dendritic cells *in vitro*. *J Immunol Methods* **432:**24–29.

72. **Helft J, Böttcher J, Chakravarty P, Zelenay S, Huotari J, Schraml BU, Goubau D, Reis e Sousa C.** 2015. GM-CSF mouse bone marrow cultures comprise a heterogeneous population of $CD11c^+MHCII^+$ macrophages and dendritic cells. *Immunity* **42:**1197–1211.

73. **Poltorak MP, Schraml BU.** 2015. Fate mapping of dendritic cells. *Front Immunol* **6:**199. doi:10.3389/fimmu.2015.00199.

74. **Bar-On L, Jung S.** 2010. Defining dendritic cells by conditional and constitutive cell ablation. *Immunol Rev* **234:**76–89.

75. **Bennett CL, Clausen BE.** 2007. DC ablation in mice: promises, pitfalls, and challenges. *Trends Immunol* **28:** 525–531.

76. **Collin M, Bigley V, Haniffa M, Hambleton S.** 2011. Human dendritic cell deficiency: the missing ID? *Nat Rev Immunol* **11:**575–583.

77. **Liu K, Victora GD, Schwickert TA, Guermonprez P, Meredith MM, Yao K, Chu F-F, Randolph GJ, Rudensky AY, Nussenzweig M.** 2009. In vivo analysis of dendritic cell development and homeostasis. *Science* **324:**392–397.

78. **Sathe P, Metcalf D, Vremec D, Naik SH, Langdon WY, Huntington ND, Wu L, Shortman K.** 2014. Lymphoid tissue and plasmacytoid dendritic cells and macrophages do not share a common macrophage-dendritic cell-restricted progenitor. *Immunity* **41:**104–115.

79. **Murphy TL, Grajales-Reyes GE, Wu X, Tussiwand R, Briseño CG, Iwata A, Kretzer NM, Durai V, Murphy KM.** 2016. Transcriptional control of dendritic cell development. *Annu Rev Immunol* **34:**93–119.

80. **Naik SH, Perié L, Swart E, Gerlach C, van Rooij N, de Boer RJ, Schumacher TN.** 2013. Diverse and heritable lineage imprinting of early haematopoietic progenitors. *Nature* **496:**229–232.

81. **Liu K, Nussenzweig MC.** 2013. Dendritic cells, p 381–384. *In* Paul WE (ed), *Fundamental Immunology*, 7th ed. Lippincott Williams & Wilkins, Philadelphia, PA.

82. **Robbins SH, Walzer T, Dembélé D, Thibault C, Defays A, Bessou G, Xu H, Vivier E, Sellars M, Pierre P, Sharp FR, Chan S, Kastner P, Dalod M.** 2008. Novel insights into the relationships between dendritic cell subsets in human and mouse revealed by genome-wide expression profiling. *Genome Biol* **9:**R17. doi:10.1186/gb-2008-9-1-r17.

83. **Crozat K, Guiton R, Guilliams M, Henri S, Baranek T, Schwartz-Cornil I, Malissen B, Dalod M.** 2010. Comparative genomics as a tool to reveal functional equivalences between human and mouse dendritic cell subsets. *Immunol Rev* **234:**177–198.

84. **Granja AG, Leal E, Pignatelli J, Castro R, Abós B, Kato G, Fischer U, Tafalla C.** 2015. Identification of teleost skin $CD8\alpha^+$ dendritic-like cells, representing a potential common ancestor for mammalian cross-presenting dendritic cells. *J Immunol* **195:**1825–1837.

85. **Rumfelt LL, McKinney EC, Taylor E, Flajnik MF.** 2002. The development of primary and secondary lymphoid tissues in the nurse shark *Ginglymostoma cirratum*: B-cell zones precede dendritic cell immigration and T-cell zone formation during ontogeny of the spleen. *Scand J Immunol* **56:**130–148.

86. **García Barrutia MS, Villena A, Gomariz RP, Razquin B, Zapata A.** 1985. Ultrastructural changes in the spleen of the natterjack, *Bufo calamita*, after antigenic stimulation. *Cell Tissue Res* **239:**435–441.

87. **Bogunovic M, Ginhoux F, Helft J, Shang L, Hashimoto D, Greter M, Liu K, Jakubzick C, Ingersoll MA, Leboeuf M, Stanley ER, Nussenzweig M, Lira SA, Randolph GJ, Merad M.** 2009. Origin of the lamina propria dendritic cell network. *Immunity* **31:**513–525.

88. **Hammer GE, Ma A.** 2013. Molecular control of steady-state dendritic cell maturation and immune homeostasis. *Annu Rev Immunol* **31:**743–791.

89. **Jiao Z, Bedoui S, Brady JL, Walter A, Chopin M, Carrington EM, Sutherland RM, Nutt SL, Zhang Y, Ko H-J, Wu L, Lew AM, Zhan Y.** 2014. The closely related $CD103^+$ dendritic cells (DCs) and lymphoid-resident $CD8^+$ DCs differ in their inflammatory functions. *PLoS One* **9:**e91126. doi:10.1371/journal.pone.0091126.

90. **Li C, Buckwalter MR, Basu S, Garg M, Chang J, Srivastava PK.** 2012. Dendritic cells sequester antigenic epitopes for prolonged periods in the absence of antigen-encoding genetic information. *Proc Natl Acad Sci U S A* **109:**17543–17548.

91. **Randolph GJ, Ochando J, Partida-Sánchez S.** 2008. Migration of dendritic cell subsets and their precursors. *Annu Rev Immunol* **26:**293–316.

92. **Sánchez-Sánchez N, Riol-Blanco L, Rodríguez-Fernández JL.** 2006. The multiple personalities of the chemokine receptor CCR7 in dendritic cells. *J Immunol* **176:**5153–5159.

93. **Ohl L, Mohaupt M, Czeloth N, Hintzen G, Kiafard Z, Zwirner J, Blankenstein T, Henning G, Förster R.** 2004. CCR7 governs skin dendritic cell migration under inflammatory and steady-state conditions. *Immunity* **21:** 279–288.

94. **Smith TRF, Kumar V.** 2008. Revival of CD8$^+$ Treg-mediated suppression. *Trends Immunol* **29:**337–342.

95. **Dzionek A, Fuchs A, Schmidt P, Cremer S, Zysk M, Miltenyi S, Buck DW, Schmitz J.** 2000. BDCA-2, BDCA-3, and BDCA-4: three markers for distinct subsets of dendritic cells in human peripheral blood. *J Immunol* **165:**6037–6046.

96. **Adams TE, Huntington JA.** 2006. Thrombin-cofactor interactions: structural insights into regulatory mechanisms. *Arterioscler Thromb Vasc Biol* **26:**1738–1745.

97. **Chistiakov DA, Sobenin IA, Orekhov AN, Bobryshev YV.** 2015. Myeloid dendritic cells: development, functions, and role in atherosclerotic inflammation. *Immunobiology* **220:**833–844.

98. **Uematsu S, Fujimoto K, Jang MH, Yang B-G, Jung Y-J, Nishiyama M, Sato S, Tsujimura T, Yamamoto M, Yokota Y, Kiyono H, Miyasaka M, Ishii KJ, Akira S.** 2008. Regulation of humoral and cellular gut immunity by lamina propria dendritic cells expressing Toll-like receptor 5. *Nat Immunol* **9:**769–776.

99. **Iyoda T, Shimoyama S, Liu K, Omatsu Y, Akiyama Y, Maeda Y, Takahara K, Steinman RM, Inaba K.** 2002. The CD8$^+$ dendritic cell subset selectively endocytoses dying cells in culture and in vivo. *J Exp Med* **195:** 1289–1302.

100. **Savill J, Hogg N, Ren Y, Haslett C.** 1992. Thrombospondin cooperates with CD36 and the vitronectin receptor in macrophage recognition of neutrophils undergoing apoptosis. *J Clin Invest* **90:**1513–1522.

101. **Urban BC, Willcox N, Roberts DJ.** 2001. A role for CD36 in the regulation of dendritic cell function. *Proc Natl Acad Sci U S A* **98:**8750–8755.

102. **Hanč P, Fujii T, Iborra S, Yamada Y, Huotari J, Schulz O, Ahrens S, Kjær S, Way M, Sancho D, Namba K, Reis e Sousa C.** 2015. Structure of the complex of F-actin and DNGR-1, a C-type lectin receptor involved in dendritic cell cross-presentation of dead cell-associated antigens. *Immunity* **42:**839–849.

103. **Cao L, Shi X, Chang H, Zhang Q, He Y.** 2015. pH-dependent recognition of apoptotic and necrotic cells by the human dendritic cell receptor DEC205. *Proc Natl Acad Sci U S A* **112:**7237–7242.

104. **Kim TS, Gorski SA, Hahn S, Murphy KM, Braciale TJ.** 2014. Distinct dendritic cell subsets dictate the fate decision between effector and memory CD8$^+$ T cell differentiation by a CD24-dependent mechanism. *Immunity* **40:**400–413.

105. **Dudziak D, Kamphorst AO, Heidkamp GF, Buchholz VR, Trumpfheller C, Yamazaki S, Cheong C, Liu K, Lee H-W, Park CG, Steinman RM, Nussenzweig MC.** 2007. Differential antigen processing by dendritic cell subsets in vivo. *Science* **315:**107–111.

106. **Ohta T, Sugiyama M, Hemmi H, Yamazaki C, Okura S, Sasaki I, Fukuda Y, Orimo T, Ishii KJ, Hoshino K, Ginhoux F, Kaisho T.** 2016. Crucial roles of XCR1-expressing dendritic cells and the XCR1-XCL1 chemokine axis in intestinal immune homeostasis. *Sci Rep* **6:** 23505. doi:10.1038/srep23505.

107. **Lei Y, Ripen AM, Ishimaru N, Ohigashi I, Nagasawa T, Jeker LT, Bösl MR, Holländer GA, Hayashi Y, Malefyt RW, Nitta T, Takahama Y.** 2011. Aire-dependent production of XCL1 mediates medullary accumulation of thymic dendritic cells and contributes to regulatory T cell development. *J Exp Med* **208:**383–394.

108. **Guilliams M, Bruhns P, Saeys Y, Hammad H, Lambrecht BN.** 2014. The function of Fcγ receptors in dendritic cells and macrophages. *Nat Rev Immunol* **14:** 94–108.

109. **Atif SM, Uematsu S, Akira S, McSorley SJ.** 2014. CD103$^-$CD11b$^+$ dendritic cells regulate the sensitivity of CD4 T-cell responses to bacterial flagellin. *Mucosal Immunol* **7:**68–77.

110. **Iwasaki A, Medzhitov R.** 2015. Control of adaptive immunity by the innate immune system. *Nat Immunol* **16:** 343–353.

111. **Uto T, Fukaya T, Takagi H, Arimura K, Nakamura T, Kojima N, Malissen B, Sato K.** 2016. Clec4A4 is a regulatory receptor for dendritic cells that impairs inflammation and T-cell immunity. *Nat Commun* **7:**11273. doi:10.1038/ncomms11273.

112. **Vander Lugt B, Khan AA, Hackney JA, Agrawal S, Lesch J, Zhou M, Lee WP, Park S, Xu M, DeVoss J, Spooner CJ, Chalouni C, Delamarre L, Mellman I, Singh H.** 2014. Transcriptional programming of dendritic cells for enhanced MHC class II antigen presentation. *Nat Immunol* **15:**161–167.

113. **Persson EK, Uronen-Hansson H, Semmrich M, Rivollier A, Hägerbrand K, Marsal J, Gudjonsson S, Håkansson U, Reizis B, Kotarsky K, Agace WW.** 2013. IRF4 transcription-factor-dependent CD103$^+$CD11b$^+$ dendritic cells drive mucosal T helper 17 cell differentiation. *Immunity* **38:**958–969.

114. **Swiecki M, Colonna M.** 2015. The multifaceted biology of plasmacytoid dendritic cells. *Nat Rev Immunol* **15:** 471–485.

115. **Reizis B, Colonna M, Trinchieri G, Barrat F, Gilliet M.** 2011. Plasmacytoid dendritic cells: one-trick ponies or workhorses of the immune system? *Nat Rev Immunol* **11:**558–565.

116. **Shigematsu H, Reizis B, Iwasaki H, Mizuno S, Hu D, Traver D, Leder P, Sakaguchi N, Akashi K.** 2004. Plasmacytoid dendritic cells activate lymphoid-specific genetic programs irrespective of their cellular origin. *Immunity* **21:**43–53.

117. **Tian J, Avalos AM, Mao SY, Chen B, Senthil K, Wu H, Parroche P, Drabic S, Golenbock D, Sirois C, Hua J, An LL, Audoly L, La Rosa G, Bierhaus A, Naworth P, Marshak-Rothstein A, Crow MK, Fitzgerald KA, Latz

E, Kiener PA, Coyle AJ. 2007. Toll-like receptor 9-dependent activation by DNA-containing immune complexes is mediated by HMGB1 and RAGE. *Nat Immunol* **8**:487–496.

118. **Gregorio J, Meller S, Conrad C, Di Nardo A, Homey B, Lauerma A, Arai N, Gallo RL, Digiovanni J, Gilliet M.** 2010. Plasmacytoid dendritic cells sense skin injury and promote wound healing through type I interferons. *J Exp Med* **207**:2921–2930.

119. **Cao W, Rosen DB, Ito T, Bover L, Bao M, Watanabe G, Yao Z, Zhang L, Lanier LL, Liu YJ.** 2006. Plasmacytoid dendritic cell-specific receptor ILT7-FcεRIγ inhibits Toll-like receptor-induced interferon production. *J Exp Med* **203**:1399–1405.

120. **Simmons DP, Wearsch PA, Canaday DH, Meyerson HJ, Liu YC, Wang Y, Boom WH, Harding CV.** 2012. Type I IFN drives a distinctive dendritic cell maturation phenotype that allows continued class II MHC synthesis and antigen processing. *J Immunol* **188**:3116–3126.

121. **Watowich SS, Liu YJ.** 2010. Mechanisms regulating dendritic cell specification and development. *Immunol Rev* **238**:76–92.

122. **Guéry L, Hugues S.** 2013. Tolerogenic and activatory plasmacytoid dendritic cells in autoimmunity. *Front Immunol* **4**:59. doi:10.3389/fimmu.2013.00059.

123. **Cervantes-Barragan L, Lewis KL, Firner S, Thiel V, Hugues S, Reith W, Ludewig B, Reizis B.** 2012. Plasmacytoid dendritic cells control T-cell response to chronic viral infection. *Proc Natl Acad Sci U S A* **109**:3012–3017.

124. **Jakubzick C, Gautier EL, Gibbings SL, Sojka DK, Schlitzer A, Johnson TE, Ivanov S, Duan Q, Bala S, Condon T, van Rooijen N, Grainger JR, Belkaid Y, Ma'ayan A, Riches DWH, Yokoyama WM, Ginhoux F, Henson PM, Randolph GJ.** 2013. Minimal differentiation of classical monocytes as they survey steady-state tissues and transport antigen to lymph nodes. *Immunity* **39**:599–610.

125. **Ensan S, Li A, Besla R, Degousee N, Cosme J, Roufaiel M, Shikatani EA, El-Maklizi M, Williams JW, Robins L, Li C, Lewis B, Yun TJ, Lee JS, Wieghofer P, Khattar R, Farrokhi K, Byrne J, Ouzounian M, Zavitz CCJ, Levy GA, Bauer CMT, Libby P, Husain M, Swirski FK, Cheong C, Prinz M, Hilgendorf I, Randolph GJ, Epelman S, Gramolini AO, Cybulsky MI, Rubin BB, Robbins CS.** 2016. Self-renewing resident arterial macrophages arise from embryonic CX3CR1$^+$ precursors and circulating monocytes immediately after birth. *Nat Immunol* **17**:159–168.

126. **Jung S, Aliberti J, Graemmel P, Sunshine MJ, Kreutzberg GW, Sher A, Littman DR.** 2000. Analysis of fractalkine receptor CX_3CR1 function by targeted deletion and green fluorescent protein reporter gene insertion. *Mol Cell Biol* **20**:4106–4114.

127. **Austyn JM, Gordon S.** 1981. F4/80, a monoclonal antibody directed specifically against the mouse macrophage. *Eur J Immunol* **11**:805–815.

128. **Lin HH, Stacey M, Stein-Streilein J, Gordon S.** 2010. F4/80: the macrophage-specific adhesion-GPCR and its role in immunoregulation. *Adv Exp Med Biol* **706**:149–156.

129. **Tamoutounour S, Guilliams M, Montanana Sanchis F, Liu H, Terhorst D, Malosse C, Pollet E, Ardouin L, Luche H, Sanchez C, Dalod M, Malissen B, Henri S.** 2013. Origins and functional specialization of macrophages and of conventional and monocyte-derived dendritic cells in mouse skin. *Immunity* **39**:925–938.

130. **Geissmann F, Jung S, Littman DR.** 2003. Blood monocytes consist of two principal subsets with distinct migratory properties. *Immunity* **19**:71–82.

131. **Geissmann F, Manz MG, Jung S, Sieweke MH, Merad M, Ley K.** 2010. Development of monocytes, macrophages, and dendritic cells. *Science* **327**:656–661.

132. **Schlitzer A, McGovern N, Ginhoux F.** 2015. Dendritic cells and monocyte-derived cells: two complementary and integrated functional systems. *Semin Cell Dev Biol* **41**:9–22.

133. **Espinosa V, Jhingran A, Dutta O, Kasahara S, Donnelly R, Du P, Rosenfeld J, Leiner I, Chen C-C, Ron Y, Hohl TM, Rivera A.** 2014. Inflammatory monocytes orchestrate innate antifungal immunity in the lung. *PLoS Pathog* **10**:e1003940. doi:10.1371/journal.ppat.1003940.

134. **Lauvau G, Loke P, Hohl TM.** 2015. Monocyte-mediated defense against bacteria, fungi, and parasites. *Semin Immunol* **27**:397–409.

135. **Hohl TM, Rivera A, Lipuma L, Gallegos A, Shi C, Mack M, Pamer EG.** 2009. Inflammatory monocytes facilitate adaptive CD4 T cell responses during respiratory fungal infection. *Cell Host Microbe* **6**:470–481.

136. **Plantinga M, Guilliams M, Vanheerswynghels M, Deswarte K, Branco-Madeira F, Toussaint W, Vanhoutte L, Neyt K, Killeen N, Malissen B, Hammad H, Lambrecht BN.** 2013. Conventional and monocyte-derived CD11b$^+$ dendritic cells initiate and maintain T helper 2 cell-mediated immunity to house dust mite allergen. *Immunity* **38**:322–335.

137. **Soudja SM, Ruiz AL, Marie JC, Lauvau G.** 2012. Inflammatory monocytes activate memory CD8$^+$ T and innate NK lymphocytes independent of cognate antigen during microbial pathogen invasion. *Immunity* **37**:549–562.

138. **Rotta G, Edwards EW, Sangaletti S, Bennett C, Ronzoni S, Colombo MP, Steinman RM, Randolph GJ, Rescigno M.** 2003. Lipopolysaccharide or whole bacteria block the conversion of inflammatory monocytes into dendritic cells in vivo. *J Exp Med* **198**:1253–1263.

139. **Randolph GJ, Beaulieu S, Lebecque S, Steinman RM, Muller WA.** 1998. Differentiation of monocytes into dendritic cells in a model of transendothelial trafficking. *Science* **282**:480–483.

140. **Randolph GJ, Sanchez-Schmitz G, Liebman RM, Schäkel K.** 2002. The CD16$^+$ (FcγRIII$^+$) subset of human monocytes preferentially becomes migratory dendritic cells in a model tissue setting. *J Exp Med* **196**:517–527.

141. **Randolph GJ, Inaba K, Robbiani DF, Steinman RM, Muller WA.** 1999. Differentiation of phagocytic

monocytes into lymph node dendritic cells in vivo. *Immunity* **11**:753–761.

142. **Kudo S, Matsuno K, Ezaki T, Ogawa M.** 1997. A novel migration pathway for rat dendritic cells from the blood: hepatic sinusoids-lymph translocation. *J Exp Med* **185**:777–784.

143. **Matsuno K, Ezaki T, Kudo S, Uehara Y.** 1996. A life stage of particle-laden rat dendritic cells in vivo: their terminal division, active phagocytosis, and translocation from the liver to the draining lymph. *J Exp Med* **183**: 1865–1878.

144. **Yu B, Ueta H, Kitazawa Y, Tanaka T, Adachi K, Kimura H, Morita M, Sawanobori Y, Qian HX, Kodama T, Matsuno K.** 2012. Two immunogenic passenger dendritic cell subsets in the rat liver have distinct trafficking patterns and radiosensitivities. *Hepatology* **56**:1532–1545.

145. **Qu C, Brinck-Jensen N-S, Zang M, Chen K.** 2014. Monocyte-derived dendritic cells: targets as potent antigen-presenting cells for the design of vaccines against infectious diseases. *Int J Infect Dis* **19**:1–5.

146. **de Jong A, Cheng TY, Huang S, Gras S, Birkinshaw RW, Kasmar AG, Van Rhijn I, Peña-Cruz V, Ruan DT, Altman JD, Rossjohn J, Moody DB.** 2014. CD1a-autoreactive T cells recognize natural skin oils that function as headless antigens. *Nat Immunol* **15**: 177–185.

147. **Kubo A, Nagao K, Yokouchi M, Sasaki H, Amagai M.** 2009. External antigen uptake by Langerhans cells with reorganization of epidermal tight junction barriers. *J Exp Med* **206**:2937–2946.

148. **Flacher V, Tripp CH, Mairhofer DG, Steinman RM, Stoitzner P, Idoyaga J, Romani N.** 2014. Murine Langerin$^+$ dermal dendritic cells prime CD8$^+$ T cells while Langerhans cells induce cross-tolerance. *EMBO Mol Med* **6**:1191–1204.

149. **Seneschal J, Clark RA, Gehad A, Baecher-Allan CM, Kupper TS.** 2012. Human epidermal Langerhans cells maintain immune homeostasis in skin by activating skin resident regulatory T cells. *Immunity* **36**:873–884.

150. **Kim M, Truong NR, James V, Bosnjak L, Sandgren KJ, Harman AN, Nasr N, Bertram KM, Olbourne N, Sawleshwarkar S, McKinnon K, Cohen RC, Cunningham AL.** 2015. Relay of herpes simplex virus between Langerhans cells and dermal dendritic cells in human skin. *PLoS Pathog* **11**:e1004812. doi:10.1371/journal.ppat.1004812.

151. **Hoeffel G, Wang Y, Greter M, See P, Teo P, Malleret B, Leboeuf M, Low D, Oller G, Almeida F, Choy SHY, Grisotto M, Renia L, Conway SJ, Stanley ER, Chan JKY, Ng LG, Samokhvalov IM, Merad M, Ginhoux F.** 2012. Adult Langerhans cells derive predominantly from embryonic fetal liver monocytes with a minor contribution of yolk sac-derived macrophages. *J Exp Med* **209**:1167–1181.

152. **Schuster C, Mildner M, Mairhofer M, Bauer W, Fiala C, Prior M, Eppel W, Kolbus A, Tschachler E, Stingl G, Elbe-Bürger A.** 2014. Human embryonic epidermis contains a diverse Langerhans cell precursor pool. *Development* **141**:807–815.

153. **Ginhoux F, Jung S.** 2014. Monocytes and macrophages: developmental pathways and tissue homeostasis. *Nat Rev Immunol* **14**:392–404.

154. **Sheng J, Ruedl C, Karjalainen K.** 2015. Most tissue-resident macrophages except microglia are derived from fetal hematopoietic stem cells. *Immunity* **43**:382–393.

155. **Capucha T, Mizraji G, Segev H, Blecher-Gonen R, Winter D, Khalaileh A, Tabib Y, Attal T, Nassar M, Zelentsova K, Kisos H, Zenke M, Seré K, Hieronymus T, Burstyn-Cohen T, Amit I, Wilensky A, Hovav AH.** 2015. Distinct murine mucosal Langerhans cell subsets develop from pre-dendritic cells and monocytes. *Immunity* **43**:369–381.

156. **Liu C, Duffy B, Bednarski JJ, Calhoun C, Lay L, Rundblad B, Payton JE, Mohanakumar T.** 2016. Maternal T-cell engraftment interferes with human leukocyte antigen typing in severe combined immunodeficiency. *Am J Clin Pathol* **145**:251–257.

157. **Buckley RH, Schiff RI, Schiff SE, Markert ML, Williams LW, Harville TO, Roberts JL, Puck JM.** 1997. Human severe combined immunodeficiency: genetic, phenotypic, and functional diversity in one hundred eight infants. *J Pediatr* **130**:378–387.

158. **Dawe GS, Tan XW, Xiao ZC.** 2007. Cell migration from baby to mother. *Cell Adhes Migr* **1**:19–27.

159. **Nijagal A, Wegorzewska M, Jarvis E, Le T, Tang Q, MacKenzie TC.** 2011. Maternal T cells limit engraftment after in utero hematopoietic cell transplantation in mice. *J Clin Invest* **121**:582–592.

160. **Malissen B, Tamoutounour S, Henri S.** 2014. The origins and functions of dendritic cells and macrophages in the skin. *Nat Rev Immunol* **14**:417–428.

161. **Beijer MR, Molenaar R, Goverse G, Mebius RE, Kraal G, den Haan JM.** 2013. A crucial role for retinoic acid in the development of Notch-dependent murine splenic CD8$^-$CD4$^-$ and CD4$^+$ dendritic cells. *Eur J Immunol* **43**:1608–1616.

162. **Iwasaki A, Kelsall BL.** 2001. Unique functions of CD11b$^+$, CD8α^+, and double-negative Peyer's patch dendritic cells. *J Immunol* **166**:4884–4890.

163. **De Jesus M, Ostroff GR, Levitz SM, Bartling TR, Mantis NJ.** 2014. A population of Langerin-positive dendritic cells in murine Peyer's patches involved in sampling β-glucan microparticles. *PLoS One* **9**:e91002. doi:10.1371/journal.pone.0091002.

164. **Cerovic V, Bain CC, Mowat AM, Milling SW.** 2014. Intestinal macrophages and dendritic cells: what's the difference? *Trends Immunol* **35**:270–277.

165. **Belkaid Y, Segre JA.** 2014. Dialogue between skin microbiota and immunity. *Science* **346**:954–959.

166. **Belkaid Y, Naik S.** 2013. Compartmentalized and systemic control of tissue immunity by commensals. *Nat Immunol* **14**:646–653.

167. **Belkaid Y, Bouladoux N, Hand TW.** 2013. Effector and memory T cell responses to commensal bacteria. *Trends Immunol* **34**:299–306.

168. **Bessman NJ, Sonnenberg GF.** 2016. Emerging roles for antigen presentation in establishing host-microbiome symbiosis. *Immunol Rev* **272**:139–150.

169. **Cullender TC, Chassaing B, Janzon A, Kumar K, Muller CE, Werner JJ, Angenent LT, Bell ME, Hay AG, Peterson DA, Walter J, Vijay-Kumar M, Gewirtz AT, Ley RE.** 2013. Innate and adaptive immunity interact to quench microbiome flagellar motility in the gut. *Cell Host Microbe* **14**:571–581.

170. **Miyake K, Kaisho T.** 2014. Homeostatic inflammation in innate immunity. *Curr Opin Immunol* **30**:85–90.

171. **Heath WR, Carbone FR.** 2013. The skin-resident and migratory immune system in steady state and memory: innate lymphocytes, dendritic cells and T cells. *Nat Immunol* **14**:978–985.

172. **Nestle FO, Di Meglio P, Qin JZ, Nickoloff BJ.** 2009. Skin immune sentinels in health and disease. *Nat Rev Immunol* **9**:679–691.

173. **Naik S, Bouladoux N, Linehan JL, Han SJ, Harrison OJ, Wilhelm C, Conlan S, Himmelfarb S, Byrd AL, Deming C, Quinones M, Brenchley JM, Kong HH, Tussiwand R, Murphy KM, Merad M, Segre JA, Belkaid Y.** 2015. Commensal-dendritic-cell interaction specifies a unique protective skin immune signature. *Nature* **520**:104–108.

174. **Shklovskaya E, O'Sullivan BJ, Ng LG, Roediger B, Thomas R, Weninger W, Fazekas de St Groth B.** 2011. Langerhans cells are precommitted to immune tolerance induction. *Proc Natl Acad Sci U S A* **108**:18049–18054.

175. **Modi BG, Neustadter J, Binda E, Lewis J, Filler RB, Roberts SJ, Kwong BY, Reddy S, Overton JD, Galan A, Tigelaar R, Cai L, Fu P, Shlomchik M, Kaplan DH, Hayday A, Girardi M.** 2012. Langerhans cells facilitate epithelial DNA damage and squamous cell carcinoma. *Science* **335**:104–108.

176. **Ouchi T, Kubo A, Yokouchi M, Adachi T, Kobayashi T, Kitashima DY, Fujii H, Clausen BE, Koyasu S, Amagai M, Nagao K.** 2011. Langerhans cell antigen capture through tight junctions confers preemptive immunity in experimental staphylococcal scalded skin syndrome. *J Exp Med* **208**:2607–2613.

177. **van der Aar AM, Picavet DI, Muller FJ, de Boer L, van Capel TM, Zaat SA, Bos JD, Janssen H, George TC, Kapsenberg ML, van Ham SM, Teunissen MB, de Jong EC.** 2013. Langerhans cells favor skin flora tolerance through limited presentation of bacterial antigens and induction of regulatory T cells. *J Invest Dermatol* **133**:1240–1249.

178. **Langlet C, Tamoutounour S, Henri S, Luche H, Ardouin L, Grégoire C, Malissen B, Guilliams M.** 2012. CD64 expression distinguishes monocyte-derived and conventional dendritic cells and reveals their distinct role during intramuscular immunization. *J Immunol* **188**:1751–1760.

179. **Kissenpfennig A, Henri S, Dubois B, Laplace-Builhé C, Perrin P, Romani N, Tripp CH, Douillard P, Leserman L, Kaiserlian D, Saeland S, Davoust J, Malissen B.** 2005. Dynamics and function of Langerhans cells in vivo: dermal dendritic cells colonize lymph node areas distinct from slower migrating Langerhans cells. *Immunity* **22**:643–654.

180. **Moussion C, Girard JP.** 2011. Dendritic cells control lymphocyte entry to lymph nodes through high endothelial venules. *Nature* **479**:542–546.

181. **Palframan RT, Jung S, Cheng G, Weninger W, Luo Y, Dorf M, Littman DR, Rollins BJ, Zweerink H, Rot A, von Andrian UH.** 2001. Inflammatory chemokine transport and presentation in HEV: a remote control mechanism for monocyte recruitment to lymph nodes in inflamed tissues. *J Exp Med* **194**:1361–1373.

182. **Mueller SN, Germain RN.** 2009. Stromal cell contributions to the homeostasis and functionality of the immune system. *Nat Rev Immunol* **9**:618–629.

183. **Qi H, Kastenmüller W, Germain RN.** 2014. Spatiotemporal basis of innate and adaptive immunity in secondary lymphoid tissue. *Annu Rev Cell Dev Biol* **30**:141–167.

184. **Malhotra D, Fletcher AL, Turley SJ.** 2013. Stromal and hematopoietic cells in secondary lymphoid organs: partners in immunity. *Immunol Rev* **251**:160–176.

185. **Rantakari P, Auvinen K, Jäppinen N, Kapraali M, Valtonen J, Karikoski M, Gerke H, Iftakhar-E-Khuda I, Keuschnigg J, Umemoto E, Tohya K, Miyasaka M, Elima K, Jalkanen S, Salmi M.** 2015. The endothelial protein PLVAP in lymphatics controls the entry of lymphocytes and antigens into lymph nodes. *Nat Immunol* **16**:386–396.

186. **Gretz JE, Norbury CC, Anderson AO, Proudfoot AE, Shaw S.** 2000. Lymph-borne chemokines and other low molecular weight molecules reach high endothelial venules via specialized conduits while a functional barrier limits access to the lymphocyte microenvironments in lymph node cortex. *J Exp Med* **192**:1425–1440.

187. **Roozendaal R, Mempel TR, Pitcher LA, Gonzalez SF, Verschoor A, Mebius RE, von Andrian UH, Carroll MC.** 2009. Conduits mediate transport of low-molecular-weight antigen to lymph node follicles. *Immunity* **30**:264–276.

188. **Acton SE, Reis e Sousa C.** 2016. Dendritic cells in remodeling of lymph nodes during immune responses. *Immunol Rev* **271**:221–229.

189. **Acton SE, Farrugia AJ, Astarita JL, Mourão-Sá D, Jenkins RP, Nye E, Hooper S, van Blijswijk J, Rogers NC, Snelgrove KJ, Rosewell I, Moita LF, Stamp G, Turley SJ, Sahai E, Reis e Sousa C.** 2014. Dendritic cells control fibroblastic reticular network tension and lymph node expansion. *Nature* **514**:498–502.

190. **Sixt M, Kanazawa N, Selg M, Samson T, Roos G, Reinhardt DP, Pabst R, Lutz MB, Sorokin L.** 2005. The conduit system transports soluble antigens from the afferent lymph to resident dendritic cells in the T cell area of the lymph node. *Immunity* **22**:19–29.

191. **Jakubzick C, Bogunovic M, Bonito AJ, Kuan EL, Merad M, Randolph GJ.** 2008. Lymph-migrating, tissue-derived dendritic cells are minor constituents within steady-state lymph nodes. *J Exp Med* **205**:2839–2850.

192. **Gerner MY, Kastenmuller W, Ifrim I, Kabat J, Germain RN.** 2012. Histo-cytometry: a method for highly multiplex quantitative tissue imaging analysis applied to

dendritic cell subset microanatomy in lymph nodes. *Immunity* 37:364–376.

193. **Gerner MY, Torabi-Parizi P, Germain RN.** 2015. Strategically localized dendritic cells promote rapid T cell responses to lymph-borne particulate antigens. *Immunity* **42:**172–185.

194. **Guilliams M, Crozat K, Henri S, Tamoutounour S, Grenot P, Devilard E, de Bovis B, Alexopoulou L, Dalod M, Malissen B.** 2010. Skin-draining lymph nodes contain dermis-derived $CD103^-$ dendritic cells that constitutively produce retinoic acid and induce $Foxp3^+$ regulatory T cells. *Blood* **115:**1958–1968.

195. **Catron DM, Itano AA, Pape KA, Mueller DL, Jenkins MK.** 2004. Visualizing the first 50 hr of the primary immune response to a soluble antigen. *Immunity* **21:**341–347.

196. **Itano AA, McSorley SJ, Reinhardt RL, Ehst BD, Ingulli E, Rudensky AY, Jenkins MK.** 2003. Distinct dendritic cell populations sequentially present antigen to CD4 T cells and stimulate different aspects of cell-mediated immunity. *Immunity* **19:**47–57.

197. **Thurley K, Gerecht D, Friedmann E, Höfer T.** 2015. Three-dimensional gradients of cytokine signaling between T cells. *PLoS Comput Biol* **11:**e1004206. doi: 10.1371/journal.pcbi.1004206.

198. **Hor JL, Whitney PG, Zaid A, Brooks AG, Heath WR, Mueller SN.** 2015. Spatiotemporally distinct interactions with dendritic cell subsets facilitates $CD4^+$ and $CD8^+$ T cell activation to localized viral infection. *Immunity* **43:**554–565.

199. **Eickhoff S, Brewitz A, Gerner MY, Klauschen F, Komander K, Hemmi H, Garbi N, Kaisho T, Germain RN, Kastenmüller W.** 2015. Robust anti-viral immunity requires multiple distinct T cell-dendritic cell interactions. *Cell* **162:**1322–1337.

200. **Smith CM, Wilson NS, Waithman J, Villadangos JA, Carbone FR, Heath WR, Belz GT.** 2004. Cognate $CD4^+$ T cell licensing of dendritic cells in $CD8^+$ T cell immunity. *Nat Immunol* **5:**1143–1148.

201. **Zaccard CR, Watkins SC, Kalinski P, Fecek RJ, Yates AL, Salter RD, Ayyavoo V, Rinaldo CR, Mailliard RB.** 2015. CD40L induces functional tunneling nanotube networks exclusively in dendritic cells programmed by mediators of type 1 immunity. *J Immunol* **194:**1047–1056.

202. **Junt T, Scandella E, Ludewig B.** 2008. Form follows function: lymphoid tissue microarchitecture in antimicrobial immune defence. *Nat Rev Immunol* **8:**764–775.

203. **Bernhard CA, Ried C, Kochanek S, Brocker T.** 2015. $CD169^+$ macrophages are sufficient for priming of CTLs with specificities left out by cross-priming dendritic cells. *Proc Natl Acad Sci U S A* **112:**5461–5466.

204. **Bain CC, Mowat AM.** 2014. Macrophages in intestinal homeostasis and inflammation. *Immunol Rev* **260:** 102–117.

205. **Jung C, Hugot J-P, Barreau F.** 2010. Peyer's patches: the immune sensors of the intestine. *Int J Inflamm* **2010:**823710.

206. **Kruglov AA, Grivennikov SI, Kuprash DV, Winsauer C, Prepens S, Seleznik GM, Eberl G, Littman DR, Heikenwalder M, Tumanov AV, Nedospasov SA.** 2013. Nonredundant function of soluble LTα3 produced by innate lymphoid cells in intestinal homeostasis. *Science* **342:**1243–1246.

207. **Mowat AM, Agace WW.** 2014. Regional specialization within the intestinal immune system. *Nat Rev Immunol* **14:**667–685.

208. **Pugh CW, MacPherson GG, Steer HW.** 1983. Characterization of nonlymphoid cells derived from rat peripheral lymph. *J Exp Med* **157:**1758–1779.

209. **Bain CC, Scott CL, Uronen-Hansson H, Gudjonsson S, Jansson O, Grip O, Guilliams M, Malissen B, Agace WW, Mowat AM.** 2013. Resident and pro-inflammatory macrophages in the colon represent alternative context-dependent fates of the same $Ly6C^{hi}$ monocyte precursors. *Mucosal Immunol* **6:**498–510.

210. **Mazzini E, Massimiliano L, Penna G, Rescigno M.** 2014. Oral tolerance can be established via gap junction transfer of fed antigens from $CX3CR1^+$ macrophages to $CD103^+$ dendritic cells. *Immunity* **40:**248–261.

211. **Scott CL, Bain CC, Wright PB, Sichien D, Kotarsky K, Persson EK, Luda K, Guilliams M, Lambrecht BN, Agace WW, Milling SW, Mowat AM.** 2015. $CCR2^+$ $CD103^-$ intestinal dendritic cells develop from DC-committed precursors and induce interleukin-17 production by T cells. *Mucosal Immunol* **8:**327–339.

212. **Rescigno M, Urbano M, Valzasina B, Francolini M, Rotta G, Bonasio R, Granucci F, Kraehenbuhl JP, Ricciardi-Castagnoli P.** 2001. Dendritic cells express tight junction proteins and penetrate gut epithelial monolayers to sample bacteria. *Nat Immunol* **2:**361–367.

213. **Chieppa M, Rescigno M, Huang AYC, Germain RN.** 2006. Dynamic imaging of dendritic cell extension into the small bowel lumen in response to epithelial cell TLR engagement. *J Exp Med* **203:**2841–2852.

214. **McDole JR, Wheeler LW, McDonald KG, Wang B, Konjufca V, Knoop KA, Newberry RD, Miller MJ.** 2012. Goblet cells deliver luminal antigen to $CD103^+$ dendritic cells in the small intestine. *Nature* **483:**345–349.

215. **Shakhar G, Kolesnikov M.** 2014. Intestinal macrophages and DCs close the gap on tolerance. *Immunity* **40:**171–173.

216. **Diehl GE, Longman RS, Zhang J-X, Breart B, Galan C, Cuesta A, Schwab SR, Littman DR.** 2013. Microbiota restricts trafficking of bacteria to mesenteric lymph nodes by CX_3CR1^{hi} cells. *Nature* **494:**116–120.

217. **Farache J, Koren I, Milo I, Gurevich I, Kim K-W, Zigmond E, Furtado GC, Lira SA, Shakhar G.** 2013. Luminal bacteria recruit $CD103^+$ dendritic cells into the intestinal epithelium to sample bacterial antigens for presentation. *Immunity* **38:**581–595.

218. **Coombes JL, Siddiqui KRR, Arancibia-Cárcamo CV, Hall J, Sun CM, Belkaid Y, Powrie F.** 2007. A functionally specialized population of mucosal $CD103^+$ DCs induces $Foxp3^+$ regulatory T cells via a TGF-β and

retinoic acid-dependent mechanism. *J Exp Med* **204:** 1757–1764.

219. **Coombes JL, Powrie F.** 2008. Dendritic cells in intestinal immune regulation. *Nat Rev Immunol* **8:**435–446.

220. **Travis MA, Reizis B, Melton AC, Masteller E, Tang Q, Proctor JM, Wang Y, Bernstein X, Huang X, Reichardt LF, Bluestone JA, Sheppard D.** 2007. Loss of integrin $\alpha_v\beta_8$ on dendritic cells causes autoimmunity and colitis in mice. *Nature* **449:**361–365.

221. **Muzaki ARBM, Tetlak P, Sheng J, Loh SC, Setiagani YA, Poidinger M, Zolezzi F, Karjalainen K, Ruedl C.** 2016. Intestinal $CD103^+CD11b^-$ dendritic cells restrain colitis via IFN-γ-induced anti-inflammatory response in epithelial cells. *Mucosal Immunol* **9:**336–351.

222. **van de Pavert SA, Mebius RE.** 2010. New insights into the development of lymphoid tissues. *Nat Rev Immunol* **10:**664–674.

223. **Tanoue T, Atarashi K, Honda K.** 2016. Development and maintenance of intestinal regulatory T cells. *Nat Rev Immunol* **16:**295–309.

224. **Worbs T, Bode U, Yan S, Hoffmann MW, Hintzen G, Bernhardt G, Förster R, Pabst O.** 2006. Oral tolerance originates in the intestinal immune system and relies on antigen carriage by dendritic cells. *J Exp Med* **203:**519–527.

225. **Macpherson AJ, Smith K.** 2006. Mesenteric lymph nodes at the center of immune anatomy. *J Exp Med* **203:**497–500.

226. **Balmer ML, Slack E, de Gottardi A, Lawson MAE, Hapfelmeier S, Miele L, Grieco A, Van Vlierberghe H, Fahrner R, Patuto N, Bernsmeier C, Ronchi F, Wyss M, Stroka D, Dickgreber N, Heim MH, McCoy KD, Macpherson AJ.** 2014. The liver may act as a firewall mediating mutualism between the host and its gut commensal microbiota. *Sci Transl Med* **6:**237ra66. doi:10.1126/scitranslmed.3008618.

227. **Yrlid U, Cerovic V, Milling S, Jenkins CD, Zhang J, Crocker PR, Klavinskis LS, MacPherson GG.** 2006. Plasmacytoid dendritic cells do not migrate in intestinal or hepatic lymph. *J Immunol* **177:**6115–6121.

228. **Yang JY, Kim MS, Kim E, Cheon JH, Lee YS, Kim Y, Lee SH, Seo SU, Shin SH, Choi SS, Kim B, Chang SY, Ko HJ, Bae JW, Kweon MN.** 2016. Enteric viruses ameliorate gut inflammation via Toll-like receptor 3 and Toll-like receptor 7-mediated interferon-β production. *Immunity* **44:**889–900.

229. **Reboldi A, Cyster JG.** 2016. Peyer's patches: organizing B-cell responses at the intestinal frontier. *Immunol Rev* **271:**230–245.

230. **Bonnardel J, Da Silva C, Henri S, Tamoutounour S, Chasson L, Montañana-Sanchis F, Gorvel J-P, Lelouard H.** 2015. Innate and adaptive immune functions of Peyer's patch monocyte-derived cells. *Cell Rep* **11:**770–784.

231. **Bonnardel J, Da Silva C, Masse M, Montañana-Sanchis F, Gorvel JP, Lelouard H.** 2015. Gene expression profiling of the Peyer's patch mononuclear phagocyte system. *Genom Data* **5:**21–24.

232. **Lelouard H, Fallet M, de Bovis B, Méresse S, Gorvel J-P.** 2012. Peyer's patch dendritic cells sample antigens by extending dendrites through M cell-specific transcellular pores. *Gastroenterology* **142:**592–601.e3. doi:10.1053/j.gastro.2011.11.039.

233. **Macpherson AJ, McCoy KD.** 2015. Independence Day for IgA. *Immunity* **43:**416–418.

234. **Cook PC, MacDonald AS.** 2016. Dendritic cells in lung immunopathology. *Semin Immunopathol* **38:**449–460.

235. **Kopf M, Schneider C, Nobs SP.** 2015. The development and function of lung-resident macrophages and dendritic cells. *Nat Immunol* **16:**36–44.

236. **Sung S-SJ, Fu SM, Rose CE Jr, Gaskin F, Ju S-T, Beaty SR.** 2006. A major lung CD103 (α_E)-β_7 integrin-positive epithelial dendritic cell population expressing Langerin and tight junction proteins. *J Immunol* **176:**2161–2172.

237. **Thornton EE, Looney MR, Bose O, Sen D, Sheppard D, Locksley R, Huang X, Krummel MF.** 2012. Spatiotemporally separated antigen uptake by alveolar dendritic cells and airway presentation to T cells in the lung. *J Exp Med* **209:**1183–1199.

238. **Bonduelle O, Duffy D, Verrier B, Combadière C, Combadière B.** 2012. Cutting edge: protective effect of $CX3CR1^+$ dendritic cells in a vaccinia virus pulmonary infection model. *J Immunol* **188:**952–956.

239. **Desch AN, Randolph GJ, Murphy K, Gautier EL, Kedl RM, Lahoud MH, Caminschi I, Shortman K, Henson PM, Jakubzick CV.** 2011. $CD103^+$ pulmonary dendritic cells preferentially acquire and present apoptotic cell-associated antigen. *J Exp Med* **208:**1789–1797.

240. **Takano K, Kojima T, Go M, Murata M, Ichimiya S, Himi T, Sawada N.** 2005. HLA-DR- and CD11c-positive dendritic cells penetrate beyond well-developed epithelial tight junctions in human nasal mucosa of allergic rhinitis. *J Histochem Cytochem* **53:**611–619.

241. **Helft J, Manicassamy B, Guermonprez P, Hashimoto D, Silvin A, Agudo J, Brown BD, Schmolke M, Miller JC, Leboeuf M, Murphy KM, García-Sastre A, Merad M.** 2012. Cross-presenting $CD103^+$ dendritic cells are protected from influenza virus infection. *J Clin Invest* **122:**4037–4047.

242. **Abraham SN, Miao Y.** 2015. The nature of immune responses to urinary tract infections. *Nat Rev Immunol* **15:**655–663.

243. **Dieterlen MT, John K, Reichenspurner H, Mohr FW, Barten MJ.** 2016. Dendritic cells and their role in cardiovascular diseases: a view on human studies. *J Immunol Res* **2016:**5946807. doi:10.1155/2016/5946807.

244. **Busch M, Westhofen TC, Koch M, Lutz MB, Zernecke A.** 2014. Dendritic cell subset distributions in the aorta in healthy and atherosclerotic mice. *PLoS One* **9:**e88452. doi:10.1371/journal.pone.0088452.

245. **Choi JH, Do Y, Cheong C, Koh H, Boscardin SB, Oh YS, Bozzacco L, Trumpfheller C, Park CG, Steinman RM.** 2009. Identification of antigen-presenting dendritic cells in mouse aorta and cardiac valves. *J Exp Med* **206:**497–505.

246. **Rogers NM, Ferenbach DA, Isenberg JS, Thomson AW, Hughes J.** 2014. Dendritic cells and macrophages in the kidney: a spectrum of good and evil. *Nat Rev Nephrol* **10:**625–643.

247. **Gottschalk C, Kurts C.** 2015. The debate about dendritic cells and macrophages in the kidney. *Front Immunol* **6:**435. doi:10.3389/fimmu.2015.00435.

248. **Ferris ST, Carrero JA, Mohan JF, Calderon B, Murphy KM, Unanue ER.** 2014. A minor subset of Batf3-dependent antigen-presenting cells in islets of Langerhans is essential for the development of autoimmune diabetes. *Immunity* **41:**657–669.

249. **Epelman S, Lavine KJ, Beaudin AE, Sojka DK, Carrero JA, Calderon B, Brija T, Gautier EL, Ivanov S, Satpathy AT, Schilling JD, Schwendener R, Sergin I, Razani B, Forsberg EC, Yokoyama WM, Unanue ER, Colonna M, Randolph GJ, Mann DL.** 2014. Embryonic and adult-derived resident cardiac macrophages are maintained through distinct mechanisms at steady state and during inflammation. *Immunity* **40:**91–104.

250. **Larsen CP, Morris PJ, Austyn JM.** 1990. Donor dendritic leukocytes migrate from cardiac allografts into recipients' spleens. *Transplant Proc* **22:**1943–1944.

251. **Aspelund A, Robciuc MR, Karaman S, Makinen T, Alitalo K.** 2016. Lymphatic system in cardiovascular medicine. *Circ Res* **118:**515–530.

252. **Cavanagh LL, Bonasio R, Mazo IB, Halin C, Cheng G, van der Velden AW, Cariappa A, Chase C, Russell P, Starnbach MN, Koni PA, Pillai S, Weninger W, von Andrian UH.** 2005. Activation of bone marrow-resident memory T cells by circulating, antigen-bearing dendritic cells. *Nat Immunol* **6:**1029–1037.

253. **Hervouet C, Luci C, Bekri S, Juhel T, Bihl F, Braud VM, Czerkinsky C, Anjuère F.** 2014. Antigen-bearing dendritic cells from the sublingual mucosa recirculate to distant systemic lymphoid organs to prime mucosal CD8 T cells. *Mucosal Immunol* 7:280–291.

254. **Steiniger BS.** 2015. Human spleen microanatomy: why mice do not suffice. *Immunology* **145:**334–346.

255. **Yi T, Cyster JG.** 2013. EBI2-mediated bridging channel positioning supports splenic dendritic cell homeostasis and particulate antigen capture. *eLife* **2:**e00757. doi:10.7554/eLife.00757.

256. **Gatto D, Wood K, Caminschi I, Murphy-Durland D, Schofield P, Christ D, Karupiah G, Brink R.** 2013. The chemotactic receptor EBI2 regulates the homeostasis, localization and immunological function of splenic dendritic cells. *Nat Immunol* **14:**446–453.

257. **Yi T, Li J, Chen H, Wu J, An J, Xu Y, Hu Y, Lowell CA, Cyster JG.** 2015. Splenic dendritic cells survey red blood cells for missing self-CD47 to trigger adaptive immune responses. *Immunity* **43:**764–775.

258. **Qiu CH, Miyake Y, Kaise H, Kitamura H, Ohara O, Tanaka M.** 2009. Novel subset of CD8α^+ dendritic cells localized in the marginal zone is responsible for tolerance to cell-associated antigens. *J Immunol* **182:**4127–4136.

259. **Glennon E, Kaunzner UW, Gagnidze K, McEwen BS, Bulloch K.** 2015. Pituitary dendritic cells communicate immune pathogenic signals. *Brain Behav Immun* **50:**232–240.

260. **Louveau A, Smirnov I, Keyes TJ, Eccles JD, Rouhani SJ, Peske JD, Derecki NC, Castle D, Mandell JW, Lee KS, Harris TH, Kipnis J.** 2015. Structural and functional features of central nervous system lymphatic vessels. *Nature* **523:**337–341.

261. **Aspelund A, Antila S, Proulx ST, Karlsen TV, Karaman S, Detmar M, Wiig H, Alitalo K.** 2015. A dural lymphatic vascular system that drains brain interstitial fluid and macromolecules. *J Exp Med* **212:**991–999.

262. **Chavan SS, Tracey KJ.** 2013. Neurophysiologic reflex mechanisms in immunology, p 850–862. *In* Paul WE (ed), *Fundamental Immunology*, 7th ed. Lippincott Williams & Wilkins, Philadelphia, PA.

263. **Rhee SH, Pothoulakis C, Mayer EA.** 2009. Principles and clinical implications of the brain-gut-enteric microbiota axis. *Nat Rev Gastroenterol Hepatol* **6:**306–314.

264. **McMahon SB, La Russa F, Bennett DLH.** 2015. Crosstalk between the nociceptive and immune systems in host defence and disease. *Nat Rev Neurosci* **16:**389–402.

265. **Prado C, Contreras F, González H, Díaz P, Elgueta D, Barrientos M, Herrada AA, Lladser Á, Bernales S, Pacheco R.** 2012. Stimulation of dopamine receptor D5 expressed on dendritic cells potentiates Th17-mediated immunity. *J Immunol* **188:**3062–3070.

266. **Wülfing C, Günther HS.** 2015. Dendritic cells and macrophages neurally hard-wired in the lymph node. *Sci Rep* **5:**16866. doi:10.1038/srep16866.

267. **Yarovinsky F.** 2014. Innate immunity to *Toxoplasma gondii* infection. *Nat Rev Immunol* **14:**109–121.

268. **Balenga NA, Balenga NA.** 2007. Human TLR11 gene is repressed due to its probable interaction with profilin expressed in human. *Med Hypotheses* **68:**456.

269. **Abbas AK, Benoist C, Bluestone JA, Campbell DJ, Ghosh S, Hori S, Jiang S, Kuchroo VK, Mathis D, Roncarolo MG, Rudensky A, Sakaguchi S, Shevach EM, Vignali DAA, Ziegler SF.** 2013. Regulatory T cells: recommendations to simplify the nomenclature. *Nat Immunol* **14:**307–308.

270. **Caramalho Í, Nunes-Cabaço H, Foxall RB, Sousa AE.** 2015. Regulatory T-cell development in the human thymus. *Front Immunol* **6:**395. doi:10.3389/fimmu.2015.00395.

271. **Panduro M, Benoist C, Mathis D.** 2016. Tissue Tregs. *Annu Rev Immunol* **34:**609–633.

272. **Li MO, Rudensky AY.** 2016. T cell receptor signalling in the control of regulatory T cell differentiation and function. *Nat Rev Immunol* **16:**220–233.

273. **Drumea-Mirancea M, Wessels JT, Müller CA, Essl M, Eble JA, Tolosa E, Koch M, Reinhardt DP, Sixt M, Sorokin L, Stierhof YD, Schwarz H, Klein G.** 2006. Characterization of a conduit system containing laminin-5 in the human thymus: a potential transport system for small molecules. *J Cell Sci* **119:**1396–1405.

274. **Klein L, Kyewski B, Allen PM, Hogquist KA.** 2014. Positive and negative selection of the T cell repertoire:

what thymocytes see (and don't see). *Nat Rev Immunol* **14:**377–391.

275. **Kurd N, Robey EA.** 2016. T-cell selection in the thymus: a spatial and temporal perspective. *Immunol Rev* **271:**114–126.
276. **Takaba H, Morishita Y, Tomofuji Y, Danks L, Nitta T, Komatsu N, Kodama T, Takayanagi H.** 2015. Fezf2 orchestrates a thymic program of self-antigen expression for immune tolerance. *Cell* **163:**975–987.
277. **Gardner JM, Metzger TC, McMahon EJ, Au-Yeung BB, Krawisz AK, Lu W, Price JD, Johannes KP, Satpathy AT, Murphy KM, Tarbell KV, Weiss A, Anderson MS.** 2013. Extrathymic Aire-expressing cells are a distinct bone marrow-derived population that induce functional inactivation of $CD4^+$ T cells. *Immunity* **39:**560–572.
278. **Lucas B, McCarthy NI, Baik S, Cosway E, James KD, Parnell SM, White AJ, Jenkinson WE, Anderson G.** 2016. Control of the thymic medulla and its influence on αβT-cell development. *Immunol Rev* **271:**23–37.
279. **Perry JS, Hsieh CS.** 2016. Development of T-cell tolerance utilizes both cell-autonomous and cooperative presentation of self-antigen. *Immunol Rev* **271:**141–155.
280. **Perry JS, Lio CW, Kau AL, Nutsch K, Yang Z, Gordon JI, Murphy KM, Hsieh CS.** 2014. Distinct contributions of Aire and antigen-presenting-cell subsets to the generation of self-tolerance in the thymus. *Immunity* **41:** 414–426.
281. **Abramson J, Husebye ES.** 2016. Autoimmune regulator and self-tolerance—molecular and clinical aspects. *Immunol Rev* **271:**127–140.
282. **St-Pierre C, Brochu S, Vanegas JR, Dumont-Lagacé M, Lemieux S, Perreault C.** 2013. Transcriptome sequencing of neonatal thymic epithelial cells. *Sci Rep* **3:**1860. doi:10.1038/srep01860.
283. **Proietto AI, van Dommelen S, Wu L.** 2009. The impact of circulating dendritic cells on the development and differentiation of thymocytes. *Immunol Cell Biol* **87:** 39–45.
284. **Taniguchi RT, DeVoss JJ, Moon JJ, Sidney J, Sette A, Jenkins MK, Anderson MS.** 2012. Detection of an autoreactive T-cell population within the polyclonal repertoire that undergoes distinct autoimmune regulator (Aire)-mediated selection. *Proc Natl Acad Sci U S A* **109:**7847–7852.
285. **Yamano T, Nedjic J, Hinterberger M, Steinert M, Koser S, Pinto S, Gerdes N, Lutgens E, Ishimaru N, Busslinger M, Brors B, Kyewski B, Klein L.** 2015. Thymic B cells are licensed to present self antigens for central T cell tolerance induction. *Immunity* **42:**1048–1061.
286. **Oh J, Shin JS.** 2015. The role of dendritic cells in central tolerance. *Immune Netw* **15:**111–120.
287. **Osorio F, Fuentes C, López MN, Salazar-Onfray F, González FE.** 2015. Role of dendritic cells in the induction of lymphocyte tolerance. *Front Immunol* **6:**535. doi:10.3389/fimmu.2015.00535.
288. **Yang S, Fujikado N, Kolodin D, Benoist C, Mathis D.** 2015. Immune tolerance. Regulatory T cells generated early in life play a distinct role in maintaining self-tolerance. *Science* **348:**589–594.
289. **Hubert FX, Kinkel SA, Davey GM, Phipson B, Mueller SN, Liston A, Proietto AI, Cannon PZ, Forehan S, Smyth GK, Wu L, Goodnow CC, Carbone FR, Scott HS, Heath WR.** 2011. Aire regulates the transfer of antigen from mTECs to dendritic cells for induction of thymic tolerance. *Blood* **118:**2462–2472.
290. **Proietto AI, van Dommelen S, Zhou P, Rizzitelli A, D'Amico A, Steptoe RJ, Naik SH, Lahoud MH, Liu Y, Zheng P, Shortman K, Wu L.** 2008. Dendritic cells in the thymus contribute to T-regulatory cell induction. *Proc Natl Acad Sci U S A* **105:**19869–19874.
291. **Bonasio R, Scimone ML, Schaerli P, Grabie N, Lichtman AH, von Andrian UH.** 2006. Clonal deletion of thymocytes by circulating dendritic cells homing to the thymus. *Nat Immunol* **7:**1092–1100.
292. **Baba T, Nakamoto Y, Mukaida N.** 2009. Crucial contribution of thymic $Sirp\alpha^+$ conventional dendritic cells to central tolerance against blood-borne antigens in a CCR2-dependent manner. *J Immunol* **183:**3053–3063.
293. **Li J, Park J, Foss D, Goldschneider I.** 2009. Thymus-homing peripheral dendritic cells constitute two of the three major subsets of dendritic cells in the steady-state thymus. *J Exp Med* **206:**607–622.
294. **Donskoy E, Goldschneider I.** 2003. Two developmentally distinct populations of dendritic cells inhabit the adult mouse thymus: demonstration by differential importation of hematogenous precursors under steady state conditions. *J Immunol* **170:**3514–3521.
295. **Atibalentja DF, Murphy KM, Unanue ER.** 2011. Functional redundancy between thymic $CD8\alpha^+$ and $Sirp\alpha^+$ conventional dendritic cells in presentation of blood-derived lysozyme by MHC class II proteins. *J Immunol* **186:**1421–1431.
296. **Hadeiba H, Lahl K, Edalati A, Oderup C, Habtezion A, Pachynski R, Nguyen L, Ghodsi A, Adler S, Butcher EC.** 2012. Plasmacytoid dendritic cells transport peripheral antigens to the thymus to promote central tolerance. *Immunity* **36:**438–450.
297. **Yadav M, Stephan S, Bluestone JA.** 2013. Peripherally induced Tregs—role in immune homeostasis and autoimmunity. *Front Immunol* **4:**232. doi:10.3389/fimmu.2013.00232.
298. **Bosco N, Kirberg J, Ceredig R, Agenès F.** 2009. Peripheral T cells in the thymus: have they just lost their way or do they do something? *Immunol Cell Biol* **87:**50–57.
299. **Josefowicz SZ, Lu LF, Rudensky AY.** 2012. Regulatory T cells: mechanisms of differentiation and function. *Annu Rev Immunol* **30:**531–564.
300. **Leventhal DS, Gilmore DC, Berger JM, Nishi S, Lee V, Malchow S, Kline DE, Kline J, Vander Griend DJ, Huang H, Socci ND, Savage PA.** 2016. Dendritic cells coordinate the development and homeostasis of organ-specific regulatory T cells. *Immunity* **44:**847–859.
301. **Scharschmidt TC, Vasquez KS, Truong HA, Gearty SV, Pauli ML, Nosbaum A, Gratz IK, Otto M, Moon JJ, Liese J, Abbas AK, Fischbach MA, Rosenblum MD.**

2015. A wave of regulatory T cells into neonatal skin mediates tolerance to commensal microbes. *Immunity* **43**:1011–1021.

302. **Legoux FP, Lim J-B, Cauley AW, Dikiy S, Ertelt J, Mariani TJ, Sparwasser T, Way SS, Moon JJ.** 2015. CD4+ T cell tolerance to tissue-restricted self antigens is mediated by antigen-specific regulatory T cells rather than deletion. *Immunity* **43**:896–908.

303. **Kim KS, Hong SW, Han D, Yi J, Jung J, Yang BG, Lee JY, Lee M, Surh CD.** 2016. Dietary antigens limit mucosal immunity by inducing regulatory T cells in the small intestine. *Science* **351**:858–863.

304. **Rosenblum MD, Way SS, Abbas AK.** 2016. Regulatory T cell memory. *Nat Rev Immunol* **16**:90–101.

305. **Sanchez Rodriguez R, Pauli ML, Neuhaus IM, Yu SS, Arron ST, Harris HW, Yang SH, Anthony BA, Sverdrup FM, Krow-Lucal E, MacKenzie TC, Johnson DS, Meyer EH, Löhr A, Hsu A, Koo J, Liao W, Gupta R, Debbaneh MG, Butler D, Huynh M, Levin EC, Leon A, Hoffman WY, McGrath MH, Alvarado MD, Ludwig CH, Truong HA, Maurano MM, Gratz IK, Abbas AK, Rosenblum MD.** 2014. Memory regulatory T cells reside in human skin. *J Clin Invest* **124**:1027–1036.

306. **Eldershaw SA, Sansom DM, Narendran P.** 2011. Expression and function of the autoimmune regulator (Aire) gene in non-thymic tissue. *Clin Exp Immunol* **163**:296–308.

307. **Lindmark E, Chen Y, Georgoudaki AM, Dudziak D, Lindh E, Adams WC, Loré K, Winqvist O, Chambers BJ, Karlsson MC.** 2013. AIRE expressing marginal zone dendritic cells balances adaptive immunity and T-follicular helper cell recruitment. *J Autoimmun* **42**: 62–70.

308. **Suzuki E, Kobayashi Y, Kawano O, Endo K, Haneda H, Yukiue H, Sasaki H, Yano M, Maeda M, Fujii Y.** 2008. Expression of AIRE in thymocytes and peripheral lymphocytes. *Autoimmunity* **41**:133–139.

309. **Fletcher AL, Malhotra D, Turley SJ.** 2011. Lymph node stroma broaden the peripheral tolerance paradigm. *Trends Immunol* **32**:12–18.

310. **Yip L, Su L, Sheng D, Chang P, Atkinson M, Czesak M, Albert PR, Collier AR, Turley SJ, Fathman CG, Creusot RJ.** 2009. *Deaf1* isoforms control the expression of genes encoding peripheral tissue antigens in the pancreatic lymph nodes during type 1 diabetes. *Nat Immunol* **10**:1026–1033.

311. **Dubrot J, Duraes FV, Potin L, Capotosti F, Brighouse D, Suter T, LeibundGut-Landmann S, Garbi N, Reith W, Swartz MA, Hugues S.** 2014. Lymph node stromal cells acquire peptide-MHCII complexes from dendritic cells and induce antigen-specific CD4+ T cell tolerance. *J Exp Med* **211**:1153–1166.

312. **Campana S, De Pasquale C, Carrega P, Ferlazzo G, Bonaccorsi I.** 2015. Cross-dressing: an alternative mechanism for antigen presentation. *Immunol Lett* **168**: 349–354.

313. **Fletcher AL, Lukacs-Kornek V, Reynoso ED, Pinner SE, Bellemare-Pelletier A, Curry MS, Collier AR, Boyd RL, Turley SJ.** 2010. Lymph node fibroblastic reticular cells directly present peripheral tissue antigen under steady-state and inflammatory conditions. *J Exp Med* **207**:689–697.

314. **Scheinecker C, McHugh R, Shevach EM, Germain RN.** 2002. Constitutive presentation of a natural tissue autoantigen exclusively by dendritic cells in the draining lymph node. *J Exp Med* **196**:1079–1090.

315. **Liu Z, Gerner MY, Van Panhuys N, Levine AG, Rudensky AY, Germain RN.** 2015. Immune homeostasis enforced by co-localized effector and regulatory T cells. *Nature* **528**:225–230.

316. **Loschko J, Schreiber HA, Rieke GJ, Esterházy D, Meredith MM, Pedicord VA, Yao KH, Caballero S, Pamer EG, Mucida D, Nussenzweig MC.** 2016. Absence of MHC class II on cDCs results in microbial-dependent intestinal inflammation. *J Exp Med* **213**:517–534.

317. **Muth S, Schütze K, Schild H, Probst HC.** 2012. Release of dendritic cells from cognate CD4+ T-cell recognition results in impaired peripheral tolerance and fatal cytotoxic T-cell mediated autoimmunity. *Proc Natl Acad Sci U S A* **109**:9059–9064.

318. **Schildknecht A, Brauer S, Brenner C, Lahl K, Schild H, Sparwasser T, Probst HC, van den Broek M.** 2010. FoxP3+ regulatory T cells essentially contribute to peripheral CD8+ T-cell tolerance induced by steady-state dendritic cells. *Proc Natl Acad Sci U S A* **107**:199–203.

319. **Probst HC, McCoy K, Okazaki T, Honjo T, van den Broek M.** 2005. Resting dendritic cells induce peripheral CD8+ T cell tolerance through PD-1 and CTLA-4. *Nat Immunol* **6**:280–286.

320. **Shimoda M, Mmanywa F, Joshi SK, Li T, Miyake K, Pihkala J, Abbas JA, Koni PA.** 2006. Conditional ablation of MHC-II suggests an indirect role for MHC-II in regulatory CD4 T cell maintenance. *J Immunol* **176**: 6503–6511.

321. **Baratin M, Foray C, Demaria O, Habbeddine M, Pollet E, Maurizio J, Verthuy C, Davanture S, Azukizawa H, Flores-Langarica A, Dalod M, Lawrence T.** 2015. Homeostatic NF-κB signaling in steady-state migratory dendritic cells regulates immune homeostasis and tolerance. *Immunity* **42**:627–639.

322. **Suffner J, Hochweller K, Kühnle MC, Li X, Kroczek RA, Garbi N, Hämmerling GJ.** 2010. Dendritic cells support homeostatic expansion of Foxp3+ regulatory T cells in Foxp3. LuciDTR mice. *J Immunol* **184**:1810–1820.

323. **Lubkov V, Bar-Sagi D.** 2014. E-cadherin-mediated cell coupling is required for apoptotic cell extrusion. *Curr Biol* **24**:868–874.

324. **Jiang A, Bloom O, Ono S, Cui W, Unternaehrer J, Jiang S, Whitney JA, Connolly J, Banchereau J, Mellman I.** 2007. Disruption of E-cadherin-mediated adhesion induces a functionally distinct pathway of dendritic cell maturation. *Immunity* **27**:610–624.

325. **Ohnmacht C, Pullner A, King SBS, Drexler I, Meier S, Brocker T, Voehringer D.** 2009. Constitutive ablation of dendritic cells breaks self-tolerance of CD4 T cells

and results in spontaneous fatal autoimmunity. *J Exp Med* **206:**549–559.

326. **Birnberg T, Bar-On L, Sapoznikov A, Caton ML, Cervantes-Barragán L, Makia D, Krauthgamer R, Brenner O, Ludewig B, Brockschnieder D, Riethmacher D, Reizis B, Jung S.** 2008. Lack of conventional dendritic cells is compatible with normal development and T cell homeostasis, but causes myeloid proliferative syndrome. *Immunity* **29:**986–997.
327. **Eberl G, Colonna M, Di Santo JP, McKenzie ANJ.** 2015. Innate lymphoid cells: a new paradigm in immunology. *Science* **348:**aaa6566. doi:10.1126/science.aaa6566.
328. **McKenzie AN, Spits H, Eberl G.** 2014. Innate lymphoid cells in inflammation and immunity. *Immunity* **41:**366–374.
329. **Hazenberg MD, Spits H.** 2014. Human innate lymphoid cells. *Blood* **124:**700–709.
330. **Bar-Ephraïm YE, Mebius RE.** 2016. Innate lymphoid cells in secondary lymphoid organs. *Immunol Rev* **271:** 185–199.
331. **Manh TP, Alexandre Y, Baranek T, Crozat K, Dalod M.** 2013. Plasmacytoid, conventional, and monocyte-derived dendritic cells undergo a profound and convergent genetic reprogramming during their maturation. *Eur J Immunol* **43:**1706–1715.
332. **Shalek AK, Satija R, Adiconis X, Gertner RS, Gaublomme JT, Raychowdhury R, Schwartz S, Yosef N, Malboeuf C, Lu D, Trombetta JJ, Gennert D, Gnirke A, Goren A, Hacohen N, Levin JZ, Park H, Regev A.** 2013. Single-cell transcriptomics reveals bimodality in expression and splicing in immune cells. *Nature* **498:** 236–240.
333. **Shalek AK, Satija R, Shuga J, Trombetta JJ, Gennert D, Lu D, Chen P, Gertner RS, Gaublomme JT, Yosef N, Schwartz S, Fowler B, Weaver S, Wang J, Wang X, Ding R, Raychowdhury R, Friedman N, Hacohen N, Park H, May AP, Regev A.** 2014. Single-cell RNA-seq reveals dynamic paracrine control of cellular variation. *Nature* **510:**363–369.
334. **DuPage M, Bluestone JA.** 2016. Harnessing the plasticity of $CD4^+$ T cells to treat immune-mediated disease. *Nat Rev Immunol* **16:**149–163.
335. **Mori L, Lepore M, De Libero G.** 2016. The immunology of CD1- and MR1-restricted T cells. *Annu Rev Immunol* **34:**479–510.
336. **Wang D, Sun B, Feng M, Feng H, Gong W, Liu Q, Ge S.** 2015. Role of scavenger receptors in dendritic cell function. *Hum Immunol* **76:**442–446.
337. **Vanden Berghe T, Linkermann A, Jouan-Lanhouet S, Walczak H, Vandenabeele P.** 2014. Regulated necrosis: the expanding network of non-apoptotic cell death pathways. *Nat Rev Mol Cell Biol* **15:**135–147.
338. **Pasparakis M, Vandenabeele P.** 2015. Necroptosis and its role in inflammation. *Nature* **517:**311–320.
339. **Yang WS, Stockwell BR.** 2016. Ferroptosis: death by lipid peroxidation. *Trends Cell Biol* **26:**165–176.
340. **Bergsbaken T, Fink SL, Cookson BT.** 2009. Pyroptosis: host cell death and inflammation. *Nat Rev Microbiol* **7:** 99–109.
341. **Kumar V, Dasoveanu DC, Chyou S, Tzeng TC, Rozo C, Liang Y, Stohl W, Fu YX, Ruddle NH, Lu TT.** 2015. A dendritic-cell-stromal axis maintains immune responses in lymph nodes. *Immunity* **42:**719–730.
342. **Roozendaal R, Mebius RE.** 2011. Stromal cell-immune cell interactions. *Annu Rev Immunol* **29:**23–43.
343. **Turley SJ, Fletcher AL, Elpek KG.** 2010. The stromal and haematopoietic antigen-presenting cells that reside in secondary lymphoid organs. *Nat Rev Immunol* **10:** 813–825.
344. **Shiow LR, Rosen DB, Brdicková N, Xu Y, An J, Lanier LL, Cyster JG, Matloubian M.** 2006. CD69 acts downstream of interferon-α/β to inhibit $S1P_1$ and lymphocyte egress from lymphoid organs. *Nature* **440:**540–544.
345. **Hampton HR, Chtanova T.** 2016. The lymph node neutrophil. *Semin Immunol* **28:**129–136.
346. **Chakarov S, Fazilleau N.** 2014. Monocyte-derived dendritic cells promote T follicular helper cell differentiation. *EMBO Mol Med* **6:**590–603.
347. **Johansson-Lindbom B, Svensson M, Pabst O, Palmqvist C, Marquez G, Förster R, Agace WW.** 2005. Functional specialization of gut $CD103^+$ dendritic cells in the regulation of tissue-selective T cell homing. *J Exp Med* **202:** 1063–1073.
348. **Annunziato F, Romagnani C, Romagnani S.** 2015. The 3 major types of innate and adaptive cell-mediated effector immunity. *J Allergy Clin Immunol* **135:**626–635.
349. **Raphael I, Nalawade S, Eagar TN, Forsthuber TG.** 2015. T cell subsets and their signature cytokines in autoimmune and inflammatory diseases. *Cytokine* **74:** 5–17.
350. **Weinmann AS.** 2014. Roles for helper T cell lineage-specifying transcription factors in cellular specialization. *Adv Immunol* **124:**171–206.
351. **Arens R, Schoenberger SP.** 2010. Plasticity in programming of effector and memory CD8 T-cell formation. *Immunol Rev* **235:**190–205.
352. **Peixoto A, Evaristo C, Munitic I, Monteiro M, Charbit A, Rocha B, Veiga-Fernandes H.** 2007. CD8 single-cell gene coexpression reveals three different effector types present at distinct phases of the immune response. *J Exp Med* **204:**1193–1205.
353. **Monteiro M, Evaristo C, Legrand A, Nicoletti A, Rocha B.** 2007. Cartography of gene expression in CD8 single cells: novel $CCR7^-$ subsets suggest differentiation independent of CD45RA expression. *Blood* **109:**2863–2870.
354. **Buchholz VR, Schumacher TN, Busch DH.** 2016. T cell fate at the single-cell level. *Annu Rev Immunol* **34:** 65–92.
355. **Shin H, Iwasaki A.** 2013. Tissue-resident memory T cells. *Immunol Rev* **255:**165–181.
356. **Kastenmüller W, Brandes M, Wang Z, Herz J, Egen JG, Germain RN.** 2013. Peripheral prepositioning and local CXCL9 chemokine-mediated guidance orchestrate rapid memory $CD8^+$ T cell responses in the lymph node. *Immunity* **38:**502–513.

357. **Alexandre YO, Ghilas S, Sanchez C, Le Bon A, Crozat K, Dalod M.** 2016. XCR1[+] dendritic cells promote memory CD8[+] T cell recall upon secondary infections with *Listeria monocytogenes* or certain viruses. *J Exp Med* **213:**75–92.

358. **Wakim LM, Waithman J, van Rooijen N, Heath WR, Carbone FR.** 2008. Dendritic cell-induced memory T cell activation in nonlymphoid tissues. *Science* **319:** 198–202.

359. **Ángeles Esteban M.** 2012. An overview of the immunological defenses in fish skin. *ISRN Immunol* **2012:** 1–29.

360. **Han Q, Das S, Hirano M, Holland SJ, McCurley N, Guo P, Rosenberg CS, Boehm T, Cooper MD.** 2015. Characterization of lamprey IL-17 family members and their receptors. *J Immunol* **195:**5440–5451.

361. **Lugo-Villarino G, Balla KM, Stachura DL, Bañuelos K, Werneck MBF, Traver D.** 2010. Identification of dendritic antigen-presenting cells in the zebrafish. *Proc Natl Acad Sci U S A* **107:**15850–15855.

362. **Flajnik MF, Kasahara M.** 2010. Origin and evolution of the adaptive immune system: genetic events and selective pressures. *Nat Rev Genet* **11:**47–59.

363. **Hirano M.** 2015. Evolution of vertebrate adaptive immunity: immune cells and tissues, and AID/APOBEC cytidine deaminases. *BioEssays* **37:**877–887.

364. **Sun JC, Ugolini S, Vivier E.** 2014. Immunological memory within the innate immune system. *EMBO J* **33:** 1295–1303.

365. **Gross M, Salame TM, Jung S.** 2015. Guardians of the gut—murine intestinal macrophages and dendritic cells. *Front Immunol* **6:**254. doi:10.3389/fimmu.2015.00254.

366. **Star B, Nederbragt AJ, Jentoft S, Grimholt U, Malmstrøm M, Gregers TF, Rounge TB, Paulsen J, Solbakken MH, Sharma A, Wetten OF, Lanzén A, Winer R, Knight J, Vogel JH, Aken B, Andersen O, Lagesen K, Tooming-Klunderud A, Edvardsen RB, Tina KG, Espelund M, Nepal C, Previti C, Karlsen BO, Moum T, Skage M, Berg PR, Gjøen T, Kuhl H, Thorsen J, Malde K, Reinhardt R, Du L, Johansen SD, Searle S, Lien S, Nilsen F, Jonassen I, Omholt SW, Stenseth NC, Jakobsen KS.** 2011. The genome sequence of Atlantic cod reveals a unique immune system. *Nature* **477:**207–210.

367. **Kasahara M.** 2007. The 2R hypothesis: an update. *Curr Opin Immunol* **19:**547–552.

368. **De Silva NS, Klein U.** 2015. Dynamics of B cells in germinal centres. *Nat Rev Immunol* **15:**137–148.

369. **Dudda JC, Simon JC, Martin S.** 2004. Dendritic cell immunization route determines CD8[+] T cell trafficking to inflamed skin: role for tissue microenvironment and dendritic cells in establishment of T cell-homing subsets. *J Immunol* **172:**857–863.

Myeloid Cells in Health and Disease: A Synthesis
Edited by Siamon Gordon

doi:10.1128/microbiolspec.MCHD-0008-2015

Paul A. Roche[1]
Peter Cresswell[2]

11 Antigen Processing and Presentation Mechanisms in Myeloid Cells

INTRODUCTION TO ANTIGEN PROCESSING

Major histocompatibility complex class I molecules (MHC-I) and class II molecules (MHC-II) are transmembrane glycoproteins that share the property of binding short peptides that are produced by the cells that express them. The generation of peptides and their subsequent association with MHC molecules is referred to as antigen processing. Antigen processing by myeloid cells, particularly dendritic cells (DCs), and the presentation of antigen-derived peptides to CD4$^+$ and CD8$^+$ T cells by MHC-I and MHC-II expressed on these cells are critical steps for effective adaptive immune responses. However, the mechanisms involved in antigen processing for MHC-I and MHC-II are different (Fig. 1). For recognition by mature effector CD4$^+$ T cells MHC-II-associated peptides are generated and bind within the endolysosomal system, while for recognition by mature CD8$^+$ T cells MHC-I-associated peptides are generated in the cytosol from newly synthesized proteins and bind to MHC-I molecules in the endoplasmic reticulum (ER). For priming naive CD4$^+$ T cells, the MHC-II processing pathway used by DCs also relies on peptide generation and binding in the endolysosomal system. However, priming CD8$^+$ T cells requires endocytosis of antigens by the DCs followed by their transfer into the cytosol for proteolysis into peptides that ultimately bind to MHC-I molecules, a process known as cross-presentation or cross-priming. In this chapter we will discuss both general and myeloid-specific mechanisms of both MHC-I- and MHC-II-restricted antigen processing and presentation, phenomena that are intimately involved with the biosynthesis of the MHC glycoproteins.

OVERVIEW OF MHC-II-RESTRICTED ANTIGEN PROCESSING

MHC-II is constitutively expressed on a subset of cells termed professional antigen-presenting cells (APCs), which include most classes of DCs, B cells, and thymic epithelial cells. MHC-II expression is inducible, however, on most cell types, including monocytes and macrophages, most notably by gamma interferon (IFN-γ)-mediated activation.

[1]Experimental Immunology Branch, Center for Cancer Research, National Cancer Institute, NIH, Bethesda, MD 20892; [2]Howard Hughes Medical Institute, Department of Immunobiology, Yale University School of Medicine, New Haven, CT 06520.

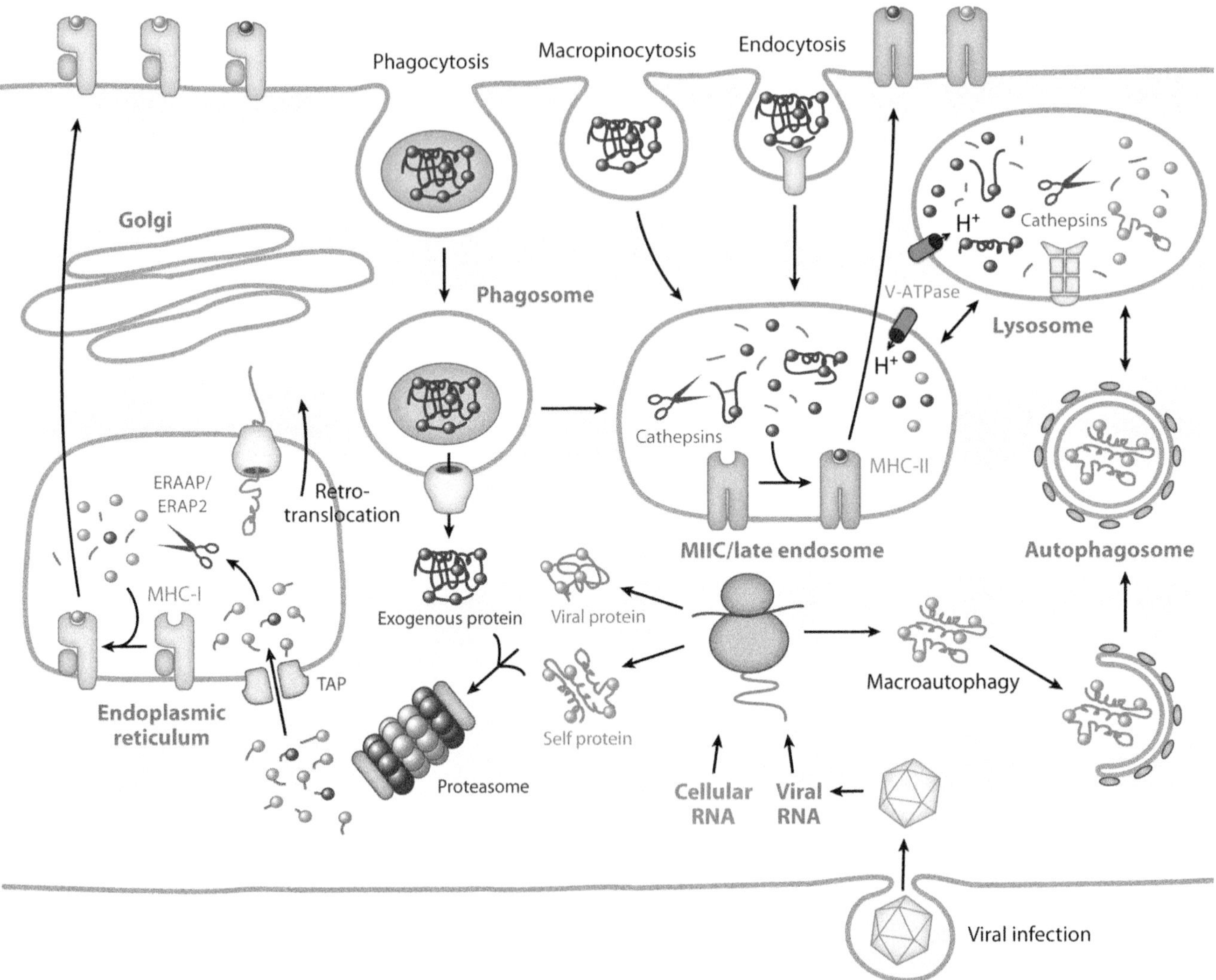

Figure 1 Overview of MHC-peptide complex biogenesis. Cytosolic proteins are degraded by the proteasome into small peptides that are imported into the lumen of the ER by TAP, where they bind to nascent MHC-I molecules. ER peptides can be trimmed to 8 to 10 residues by the action of ERAAP/ERAP1 and ERAP2. Fully assembled MHC-I–peptide complexes leave the ER and are delivered to the plasma membrane by recognition by $CD8^+$ T cells. Proteins internalized into endosomes by a variety of mechanisms are degraded into peptides in late endosomes rich in proteinases, classically called cathepsins, active at acidic pH. MHC-II molecules are transported to these compartments from the ER by virtue of its association with a chaperone termed the invariant chain (not shown). The MHC-II-positive compartment is indicated as MIIC/late endosome in the figure. Invariant chain is also proteolytically degraded in late endosomes, thereby making the MHC-II molecules available for peptide binding. Following a series of peptide-editing processes, immunodominant MHC-II–peptide complexes move to the plasma membrane for recognition by $CD4^+$ T cells. In specialized APCs, particularly DCs, proteins that enter the cell by endocytosis/phagocytosis are retrotranslocated into the cytosol for subsequent proteasomal degradation and binding to MHC-I in a process termed cross-presentation. The retrotranslocation mechanism is currently undefined, but here it is depicted as a channel responsible for ERAD that may be recruited to the phagosome from the ER. This hypothesis remains unproven. Reprinted from reference 32, with permission.

As discussed below, enhanced MHC-II biosynthesis or regulated degradation of MHC-II is an important way for APCs to focus their attention on pathogens that "alert" the immune system to an infection.

MHC-II binds peptides generated by proteolysis of antigens in endosomal/lysosomal "antigen-processing compartments." Antigens gain access to these compartments by various mechanisms, including receptor-mediated endocytosis, macropinocytosis, phagocytosis, and autophagy. MHC-II molecules, which consist of a heterodimer of transmembrane α and β subunits, gain access to these same compartments by association with an accessory protein termed the invariant chain (I_i) shortly after biosynthesis in the ER (Fig. 2). I_i provides three distinct functions for MHC-II: (i) it acts as a molecular chaperone and promotes proper folding and movement of the MHC-II–I_i complex from the ER through the Golgi apparatus (1, 2); (ii) it prevents peptides and unfolded proteins present in the ER from binding to the peptide-binding site on the nascent MHC-II molecule (3, 4); and (iii) it contains targeting signals in its cytoplasmic domain that direct the MHC-II–I_i complex to antigen-processing compartments (5, 6). The precise pathway taken by MHC-II–I_i complexes to access these compartments (i.e., whether the complexes are delivered directly into the endosomal pathway from the *trans*-Golgi network or whether they traffic to the plasma membrane and are then internalized) has been a matter of considerable debate (reviewed in 7). However, regardless of the pathway used, efficient movement of MHC-II into the late endocytic pathway depends on I_i association. The targeting signal in I_i consists of two dileucine-based internalization motifs (5, 6, 8). These motifs interact with clathrin-associated adaptor proteins to drive MHC-II–I_i complexes into the endocytic pathway (9, 10).

In principle, any endo/lysosomal compartment that generates antigenic peptides capable of binding to MHC-II can be considered an antigen-processing compartment, and MHC-II–peptide complexes can indeed be generated throughout the endocytic pathway (11). The findings that the MHC-II–I_i complex can enter the earliest of endosomes by endocytosis from the cell surface (12) and that all endosomes contain at least some proteinase activity (13) are consistent with the idea that MHC-II is available throughout the endocytic pathway for peptide loading.

MHC-II is not able to bind antigenic peptides until I_i is proteolytically degraded and dissociates from the MHC-II–I_i complex (3). The degradation of MHC-II-associated I_i occurs in a series of discrete steps catalyzed by different proteinases (14–16), leaving an I_i-derived polypeptide, termed CLIP (class II-associated invariant chain peptides) (17), associated with the MHC-II peptide-binding groove. CLIP is catalytically removed to make room for lysosomally generated peptides, including those derived from internalized antigens, by a homolog of MHC-II, termed HLA-DM in humans and H2-M or DM in mice (18). Newly synthesized DM traffics to antigen-processing compartments by clathrin-mediated endocytosis after arrival at the plasma membrane. Unlike I_i, however, the internalization motif on DM is tyrosine based and preferentially sorts DM to mature endosomal antigen-processing compartments (19). DM not only catalyzes CLIP release but also promotes the dissociation of MHC-II-bound peptides that possess an intrinsically fast off rate (20), thereby serving as a "peptide editor" for MHC-II to foster the generation of high-affinity immunodominant epitopes (21). Recent data have revealed that DM interacts with the MHC-II–CLIP complex near the P1 peptide-binding pocket on MHC-II and stabilizes an intermediate conformation of MHC-II that permits dissociation of weakly bound peptides (22, 23).

A second MHC-II homolog, called HLA-DO in humans and H2-O in mice (referred to here as DO), regulates the peptide-editing function of DM. DM binds tightly to DO in the ER and serves to escort DO to lysosome-like antigen-processing compartments (24). DO is expressed in both human and mouse B cells, thymic epithelial cells, and Langerhans cells and is present in all $CD11c^+$ spleen DC subsets in the mouse (25, 26). DO expression is suppressed during DC maturation (25, 27), while DM expression changes are modest (25, 26). Most published studies show that DO association suppresses DM activity (28). *In vitro* peptide-binding assays have demonstrated that DO inhibition of DM activity is pH dependent: at pH of >5.5, DO completely abrogates DM activity, but at the pH of most antigen-processing compartments (4.5 to 5.0), DO does not inhibit DM function (29). Whether this is due to pH-dependent dissociation of DO from DM or conformational alterations in the DO/DM complex remains undetermined.

OVERVIEW OF MHC-I RESTRICTED ANTIGEN PROCESSING

The pathways of MHC-I-restricted antigen processing are indicated in detail in Fig. 3. MHC-I presentation to effector $CD8^+$ T cells, or cytotoxic T lymphocytes, involves the generation of peptides from newly synthesized cytosolic proteins, including, for example, viral proteins produced during infection of a cell. These

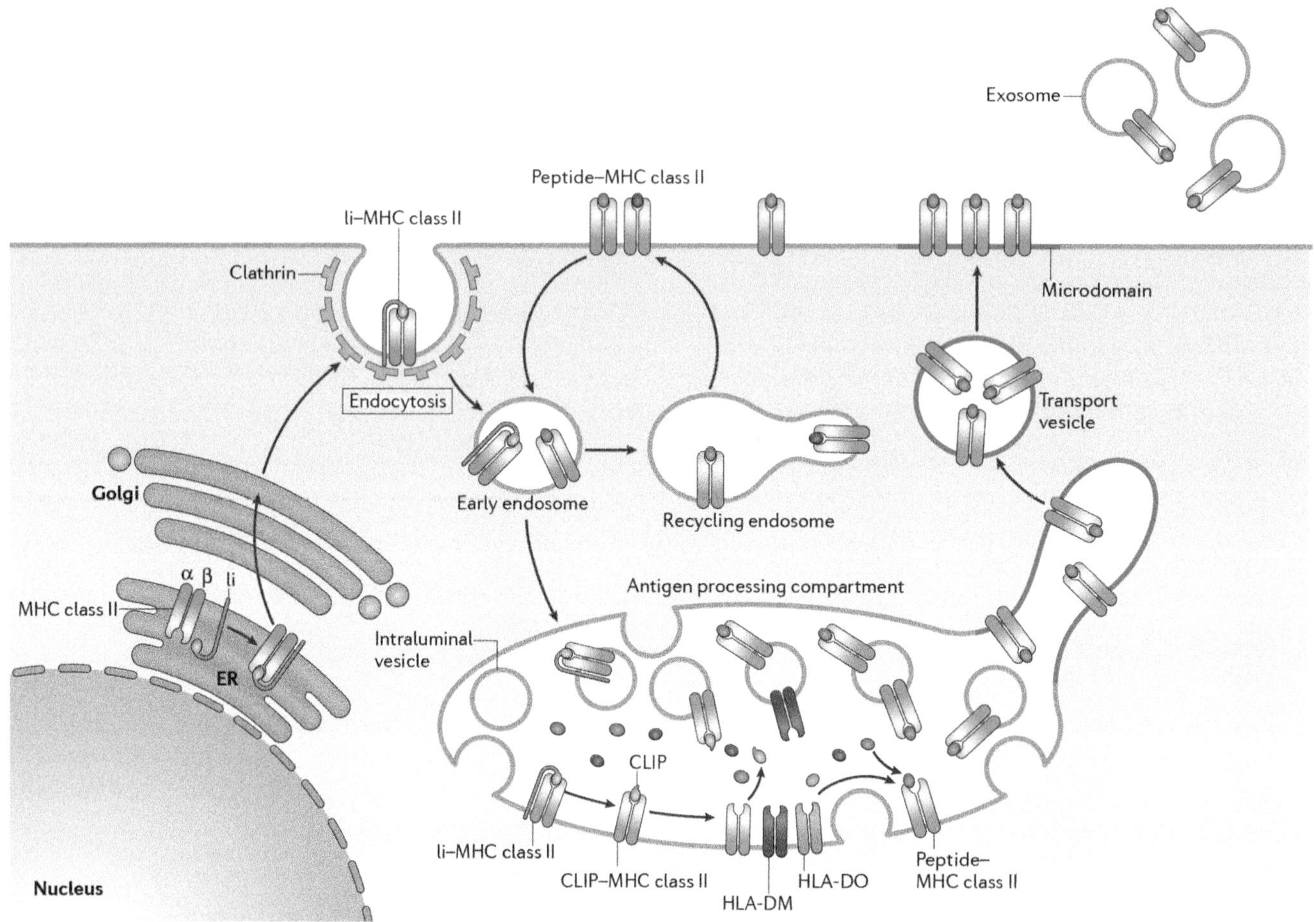

Figure 2 Biosynthesis of MHC-II–peptide complexes. MHC-II αβ dimers associate with I_i in the ER, and the assembled MHC-II–I_i complexes traffic through the Golgi apparatus and are delivered to the plasma membrane. The complexes are internalized by clathrin-mediated endocytosis and are transported to late endosomal multivesicular antigen-processing compartments. Some of these complexes sort onto the ILVs of these compartments, where sequential I_i proteolysis leads to persistence of a derived fragment (termed CLIP) in the MHC-II peptide-binding groove. CLIP is removed from CLIP–MHC-II complexes by DM molecules that are present on the ILV and limiting membrane of antigen-processing compartments, thereby allowing peptide binding onto nascent MHC-II. The activity of DM is regulated by DO; however, the mechanism of regulation remains unknown. It is likely that ILV-associated MHC-II is transferred to the limiting membrane and endo/lysosomal tubules that either directly fuse, or give rise to transport vesicles that fuse, with the plasma membrane. MHC-II–peptide association with lipid microdomains first occurs in antigen-processing compartments and allows clustering of MHC-II–peptide complexes on the cell surface. If an entire antigen-processing compartment fuses with the plasma membrane, the ILV can be released from the cell in the form of exosomes. Surface-expressed MHC-II–peptide complexes can internalize using a clathrin-independent endocytosis pathway and are targeted for lysosomal degradation or may be recycled back to the plasma membrane. Reprinted from reference 103, with permission.

proteins are degraded by the proteasome into peptides that, potentially after further processing by cytosolic aminopeptidases, are translocated into the ER by a dedicated ATP-dependent transporter, the transporter associated with antigen processing (TAP). TAP is composed of two MHC-encoded subunits, TAP1 and TAP2, and is a member of the ATP-binding cassette family of transporters (30). Once in the ER, the peptides can be further trimmed by ER-resident aminopeptidases, called ERAP1 (ERAAP in the mouse) and ERAP2 (absent from the mouse), to a length of 8 to 10 amino acids suitable for binding to newly synthesized MHC-I molecules (31).

MHC-I molecules are heterodimers consisting of a glycosylated transmembrane heavy chain of ~45 kDa, which is the polymorphic MHC-I gene product, and a small subunit of ~12 kDa called β_2-microglobulin (β_2m). The heavy chain-β_2m dimers fold and assemble in the ER with the assistance of a number of chaperones, but peptide binding occurs after incorporation of the assembled dimers into the peptide loading complex (PLC). The PLC consists of TAP, tapasin (a transmembrane glycoprotein also encoded in the MHC), a protein disulfide isomerase homolog called ERp57, and the soluble chaperone calreticulin (CRT). Stoichiometric analysis indicates that there are two tapasin molecules per PLC, each of which is permanently disulfide linked to an ERp57 molecule. MHC-I molecules interact directly with tapasin and also, via their N-linked glycans, with CRT (reviewed in 32).

CRT is a lectin with specificity for a single terminal glucose residue transiently present on the glycans of newly synthesized glycoproteins. Such glycoproteins are subjected to a folding cycle in which CRT (or the related chaperone calnexin) also cooperates with ERp57 via a glycan-independent, noncovalent interaction to facilitate their correct folding and disulfide bond formation (33). After dissociation of glycoproteins from CRT, the glycan is enzymatically deglucosylated. However, if the glycoprotein remains improperly folded, it can be reglucosylated by the enzyme UDP-glucose glycoprotein transferase-1 (UGT-1), allowing reentry into the folding cycle (34). The covalent association of ERp57 with tapasin in the PLC provides a secondary anchor via CRT to cooperatively maintain the association of newly synthesized MHC-I with the PLC in an adaptation of the normal glycoprotein folding cycle. UGT-1 is used to maintain monoglucosylation of MHC-I molecules that lack associated high-affinity peptides. CRT, ERp57, and UGT-1 are all required for optimal MHC-I peptide loading (35–37). This is even more dependent on tapasin, which has a similar peptide-editing role for MHC-I that DM has for MHC-II, promoting the association of high-affinity peptides at the expense of low-affinity ones (38–40). Our molecular understanding of how tapasin does this is less advanced, but when peptides of sufficiently high affinity are bound, the completed MHC-I–peptide complexes permanently dissociate from the PLC and are transported to the cell surface.

Cross-presentation, or cross-priming, involves the binding of peptides derived from extracellular antigens with MHC-I and the recognition of these complexes by naive $CD8^+$ T cells. Most data are consistent with a role for components of the conventional MHC-I processing pathway in cross-presentation (reviewed in 41); however, the precise cell biological mechanisms regulating this process are still not well understood. The most favored mechanism involves antigen internalization into endosomes, translocation of the antigens (or large fragments of them) from the endocytic pathway into the cytosol by an undetermined mechanism, and finally antigen proteolysis by proteasomal degradation. Cytosolically generated peptides are then translocated into either the ER, where they bind MHC-I molecules in a PLC-mediated fashion as in conventional MHC-I processing, or back into an endocytic or phagocytic compartment. Here they bind either to MHC-I molecules recycling between the plasma membrane and this compartment or to MHC-I molecules recruited to that compartment from the ER, along with PLC components. Some data in the literature argue that cross-presented peptides are generated by lysosomal proteolysis, much as they are for MHC-II. Data showing that DCs lacking the lysosomal enzyme cathepsin S are deficient in cross-presentation of certain antigens support this model (42). However, the mechanism underlying this observation is undefined, and the principle that cross-presented antigens undergo proteasomal processing in the cytosol prior to transport into MHC-I-containing compartments is generally accepted.

Although a number of cell types have been shown to be capable of cross-presentation *in vitro*, DCs are the major cell type that primes $CD8^+$ T-cell responses *in vivo* (41). Considerable evidence indicates that in mice a particular subset of DCs, characterized by expression of the surface molecule CD8α, is the dominant cross-priming cell (43). Curiously, surface expression of CD8α is not believed to have any functional significance in this process. Whether a dominant cross-priming DC subset exists in humans is less clear. Human DCs expressing the marker CD141, or BDCA3, have been suggested to be the homolog of $CD8^+$ mouse DCs (44, 45), but a recent study of human tonsillar DCs found that all subsets, identifiable by expression of a variety of surface markers, were competent to cross-present exogenous antigens via MHC-I (46).

DELIVERY OF ANTIGENS INTO ANTIGEN-PROCESSING COMPARTMENTS

Cross-presentation by MHC-I and successful antigen presentation by MHC-II share the requirement that

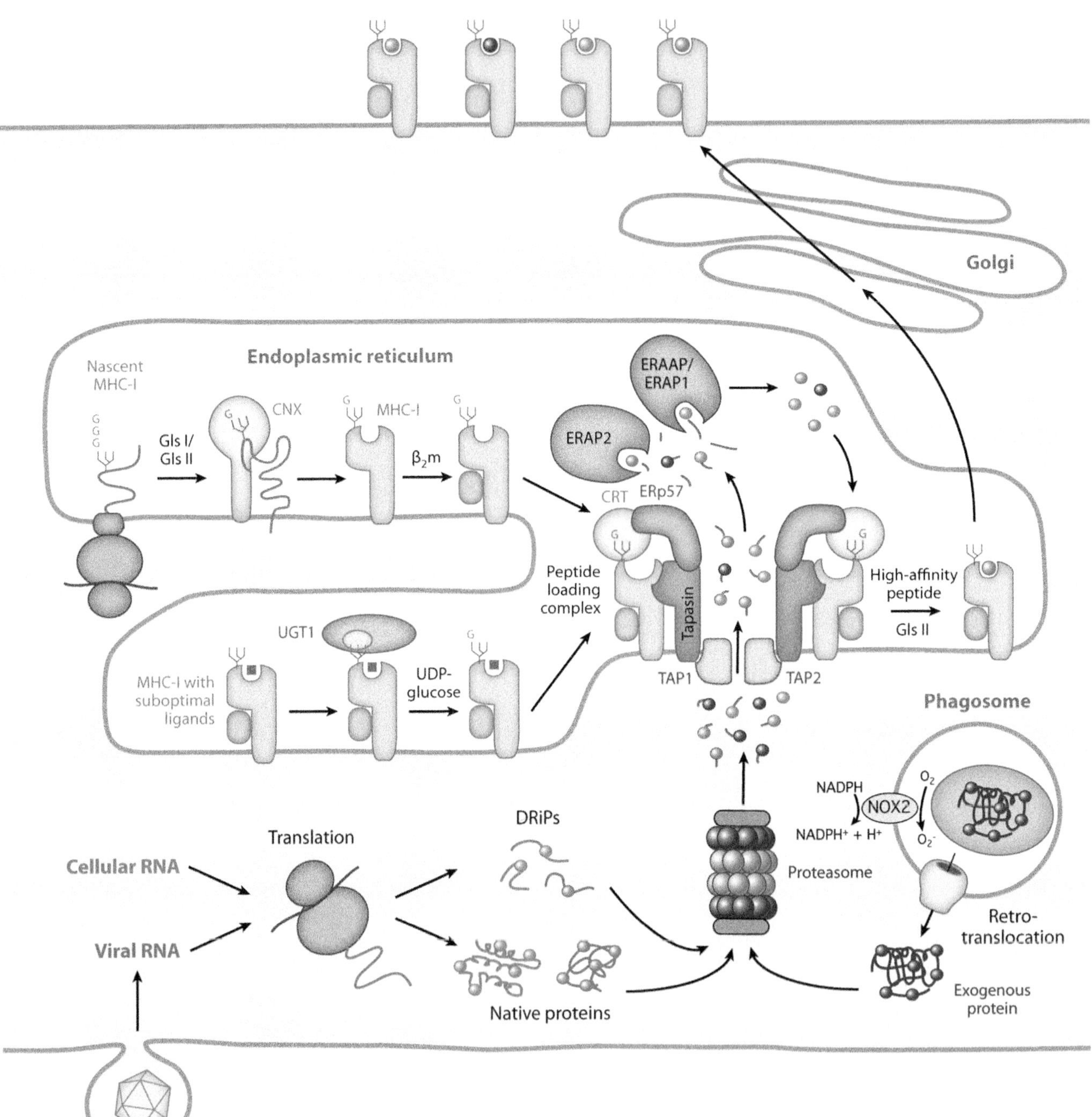

Golgi
Endoplasmic reticulum
Nascent MHC-I
Gls I/ Gls II
CNX
MHC-I
β2m
ERAAP/ ERAP1
ERAP2
CRT
ERp57
Peptide loading complex
Tapasin
TAP1
TAP2
High-affinity peptide
Gls II
UGT1
MHC-I with suboptimal ligands
UDP-glucose
Phagosome
NADPH
NOX2
NADPH+ + H+
O2
O2-
DRiPs
Translation
Cellular RNA
Viral RNA
Proteasome
Retro-translocation
Exogenous protein
Native proteins

protein antigens must gain access to the endocytic pathway. Here we discuss the various mechanisms used by DCs to mediate this process, illustrated in Fig. 4.

Macropinocytosis

One process used to capture extracellular material is macropinocytosis (47), an endocytic process that is responsible for the nonspecific uptake of extracellular material that can vary in size from small molecules to intact bacteria and protozoa. Macropinosomes are generated from plasma membrane ruffles that extend from the cell, fold back onto themselves, and then fuse with the plasma membrane. Macropinosomes ultimately fuse with early endosomes, delivering extracellular material to the endolysosomal pathway for antigen processing. Resting DCs are capable of internalizing large amounts of fluid by constitutive macropinocytosis (up to 2 fl/cell/min) (48), and this pathway is thought to represent a major mechanism of antigen acquisition by DCs.

Macropinocytosis is controlled by the Rho GTPase Cdc42 and Rac-mediated reorganization of the cortical actin cytoskeleton (49, 50). Activation of DCs *in vitro*, for example, by lipopolysaccharide, reduces active Cdc42 levels and profoundly suppresses macropinocytosis (49); however, some studies have shown that unlike their *in vitro*-activated counterparts, DCs activated *in vivo* retain the ability to internalize, process, and present soluble exogenous antigens to CD4 T cells (51–53). *In vitro* activation of macrophages does not alter their capacity for macropinocytosis (54); however, activation does reprogram the endocytic machinery from receptor-mediated phagocytosis to macropinocytosis (55), thereby increasing their ability to internalize and destroy infectious agents in an inflammatory environment.

Receptor-Mediated Endocytosis

APCs possess a variety of different surface receptors that mediate antigen internalization. Fcγ receptors on macrophages and DCs bind immune complexes and efficiently deliver them to antigen-processing compartments (52). DCs also possess lectin receptors, such as the mannose receptor and DEC-205, that recognize carbohydrate residues on self-proteins and some pathogens and target them for internalization via receptor-mediated phagocytosis. Conjugation of antigens to ligands for specific APC surface receptors can dramatically enhance the efficiency of processing and presentation to antigen-specific T cells (56). By following different endocytic routes, different receptors deliver their cargo to distinct classes of endosomes in DCs (57). For example, targeting antigens to the mannose receptor leads to their delivery to early endosomes (which can be useful for MHC-I cross-presentation), whereas targeting antigens to Fcγ receptors or DEC-205 leads to their delivery to late endosomes/prelysosomes for efficient antigen processing and presentation by MHC-II (52).

Phagocytosis

Perhaps the most important mechanism of antigen uptake in macrophages and DCs is phagocytosis. This process allows these cells to internalize a wide variety of insoluble particulate antigens including necrotic/apoptotic cells, bacteria, and viruses (58). Unlike nonspecific macropinocytosis, phagocytosis generally involves recognition of particles by specific phagocytic receptors on APCs. There are a wide variety of such receptors on DCs and macrophages, including diverse Fc receptors, complement receptors, and C-type lectin receptors. Although *in vitro* activation suppresses phagocytosis in DCs, *in vivo* activation does not significantly

Figure 3 MHC-I biosynthesis and peptide binding. The proteasome generates short antigenic peptides capable of binding to MHC-I molecules. These peptides are derived from native cytosolic proteins, defective ribosomal products (DRiPs), or, in the case of cross-presentation, exogenous proteins that enter the cell by phagocytosis and are translocated into the cytosol, either intact or as large proteolytic fragments. In cross-presenting mouse $CD8^+$ DCs, the presence of NOX2 on the phagosomal membrane neutralizes acidification and reduces proteolytic activity, preserving protein integrity. Nascent MHC-I heavy chains initially interact with the molecular chaperone calnexin (CNX) and, after binding β_2m, are recruited to the PLC by simultaneous noncovalent CRT interactions with a monoglucosylated N-linked glycan on the heavy chain and ERp57 disulfide linked to tapasin in the PLC. Peptide-free MHC-I molecules and those possessing suboptimal ligands are subject to a series of "editing" steps mediated by interaction with tapasin within the PLC as well as maintenance of the monoglucosylated N-linked glycan by the opposing actions of the enzymes glucosidase 2 (GlsII), which removes the terminal glucose residue, and UGT1, which adds back glucose to preserve the CRT interaction. MHC-I molecules containing high-affinity peptides ultimately leave the ER and are transported to the plasma membrane. Reprinted from reference 32, with permission.

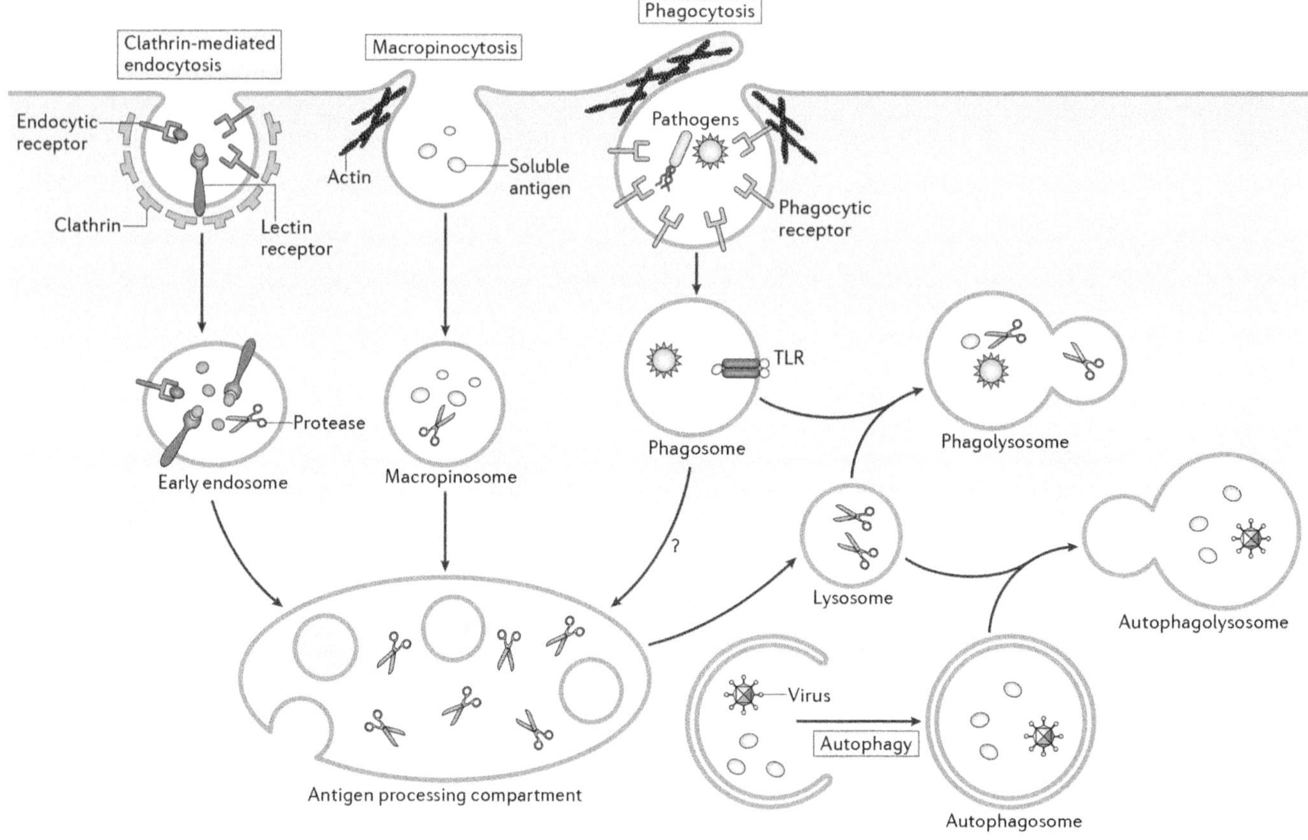

Figure 4 Pathways of antigen entry into the processing compartments of myeloid cells. Pathogens as well as soluble and particulate antigens access the endolysosomal pathway of antigen-processing cells by a variety of mechanisms. Clathrin-mediated endocytosis generally involves the binding of ligands to one of a variety of endocytic receptors that deliver endocytosed cargo to early endosomes. Macropinocytosis is a nonspecific form of endocytosis that involves actin-dependent membrane ruffling that leads to solute encapsulation in structures that give rise to macropinosomes. Like early endosomes, macropinosomes are not highly proteolytic and antigen degradation only occurs following their fusion with acidic late endosomal/lysosomal compartments containing lysosomal proteinases. Pathogens and large particles that possess specific binding sites for surface receptors are internalized by phagocytosis, an endocytic process that combines the features of macropinocytosis and receptor-mediated endocytosis. Phagosomes are not acidic nor proteinase rich; however, maturation of phagosomes by fusion with late endosomes or lysosomes gives rise to proteolytic phagolysosomes that degrade phagocytosed material. Autophagy also provides material for endolysosomal degradation by sequestering cytosol into a double-membrane encapsulated autophagosome that, like a conventional phagosome, undergoes maturation upon fusion with lysosomes to generate proteolytic autophagolysosomes. Reprinted from reference 103, with permission.

alter the ability of DCs to capture antigens by phagocytosis and stimulate antigen-specific CD4 T cells (52). Sustained phagocytosis after maturation could be important to generate MHC-II complexes with pathogen-derived peptides and perhaps for prolonging cross-presentation by MHC-I.

Phagocytosis requires large amounts of membrane to generate a developing phagosome. Proteomic analysis of phagosomes has revealed the presence of ER proteins on phagosomes (59). This initially led to the suggestion that the ER is a major source of membrane during phagocytosis, although more-recent data suggest that the amount of ER recruitment to the phagosome, while significant, is actually quite small. This observation also led to considerable speculation that the mechanisms responsible for ER-associated degradation (ERAD), the process by which misfolded proteins in the ER are translocated into the cytosol, are adapted for transfer from the phagosome to mediate cross-presentation. Although some components, such as the AAA-ATPase p97, do appear to be involved in both, evidence that the ERAD retrotranslocation apparatus is involved has been difficult to come by. Curiously, components of the ER-associated MHC-I PLC (including TAP and tapasin) are present on phagosome membranes, allowing the phagosome to function as a "surrogate ER" for peptide loading onto phagosome-associated MHC-I during cross-presentation (60–62). Although recent studies have shown that recycling surface MHC-I enters phagosomes (63), it remains to be determined how significantly these MHC-I molecules contribute to phagosome-dependent cross-presentation.

Initially phagosomes are minimally proteolytic and therefore do not generate large amounts of antigenic peptides. Internalized cargo is only degraded during the process of phagosome maturation, in which phagosomes fuse with late endosomes/lysosomes to generate phagolysosomes (58). The comparative lack of proteolysis within early phagosomes makes them the organelle of choice for mediating cross-presentation because premature degradation of internalized proteins can actually destroy potential MHC-I epitopes prior to antigen entry into the cytosol. More extensive proteolysis is critical for MHC-II function, however, and the fusion of a phagosome with a late endosomal MHC-II-positive compartment leads to formation of a hybrid organelle that possesses all the components necessary to generate MHC-II complexes with peptides derived from phagocytosed cargo. Phagosome maturation is stimulated by Toll-like receptor (TLR) signaling in macrophages and DCs (64, 65), providing these cells with a mechanism to increase MHC-II-restricted antigen processing during phagocytosis of pathogens bearing TLR ligands. Given the importance of antigen integrity for translocation from phagosomes, cross-presentation by MHC-I is actually reduced during phagosome acidification. For this reason, DCs that are specialized in cross-presentation have adopted mechanisms to control the proteolytic activity of phagosomes. Cross-presenting $CD8^+$ DCs recruit the NADPH oxidase NOX2 to the phagosomal membrane, a process that results in the alkalinization of the lumen of the phagosome by the reactive oxygen species generated by NOX2 (66, 67). This reduces the activity of cathepsins, which have acidic pH optima, thereby suppressing antigen proteolysis in DC phagosomes.

Autophagy

Autophagy is a process in which cytosol is encapsulated in a double-membrane structure termed an autophagosome (68). Like conventional phagosomes, autophagosomes are not highly proteolytic; however, fusion with a lysosome-like, MHC-II-positive antigen-processing compartment forms a hybrid autophagolysosome that contains all of the machinery required to degrade antigens and generate MHC-II–peptide complexes (69). Since the protein precursors of MHC-I-associated peptides are already cytosolic, autophagy may not be important for conventional MHC-I-restricted antigen processing. However, ~25% of MHC-II-associated peptides in DCs are derived from cytosolic and/or nuclear proteins, highlighting the importance of this pathway for MHC-II function (70). Genetic disruption of the process of autophagy severely compromises positive and negative selection of CD4 T cells by thymic epithelial cells (71, 72), pointing to a prominent role for autophagy in the function of APCs in the thymus.

APCs also possess an alternative autophagy pathway termed chaperone-mediated autophagy (CMA) (73). CMA is distinct from macroautophagy in a number of ways. Whereas macroautophagy is induced rapidly upon cell stress (such as nutrient deprivation) and wanes within 24 h, CMA increases as macroautophagy decreases. Unlike macroautophagy, CMA does not generate double-membrane autophagosomes, but instead results in the formation of a macromolecular complex containing the late endosome/lysosome-associated membrane protein LAMP2A and the heat shock proteins Hsc70 and Hsp90. This molecular translocation complex results in the delivery of cytosolic proteins into the endosome/lysosome lumen for degradation.

REGULATION OF PROTEOLYSIS IN ANTIGEN-PROCESSING COMPARTMENTS

Optimal MHC-II function requires proteolytic digestion of antigens in late endosomal/lysosomal antigen-processing compartments. However, a delicate balance must be maintained in APCs that allows the generation of immunodominant, antigenic peptides but does not result in their complete destruction (74). Lysosomal enzyme activity in DCs is ~50 times lower than it is in macrophages (75), and this leads (in part) to prolonged antigen retention and MHC-II stability in DCs compared to macrophages. Lysosomes are less acidic in DCs than in macrophages, in part because of reduced accumulation of the vacuolar ATPase (V-ATPase) that pumps protons into these compartments (76), thereby reducing their proteolytic activity. Similarly, phagosomes of DCs are less acidic (and less proteolytic) than phagosomes in macrophages. Recent work has shown that a major difference in the lysosomal and phagosomal properties of DCs and macrophages results from their differential expression and activation of transcription factor EB (TFEB), which is a master regulator of lysosomal function (77, 78). Transcription of a number of cathepsin genes, as well as genes encoding the subunits of V-ATPase, is regulated by TFEB, and DCs express significantly less TFEB than macrophages (104). Notably, $CD8^+$ DCs in the spleen, which are the primary mediators of cross-presentation, express significantly less TFEB than other DC subsets in the mouse. It is therefore not surprising that overexpressing TFEB in DCs results in a reduction of cross-presentation, while suppressing TFEB expression with a short hairpin RNA in macrophages allows them to effectively mediate cross-presentation. Reciprocal effects were observed on MHC-II function: MHC-II-restricted antigen processing was increased in DCs overexpressing TFEB while it was decreased in macrophages with reduced TFEB. As noted above, the selective association of the ROS-generating enzyme NOX2 with phagosomal membranes in $CD8^+$ DCs also increases the pH of developing phagosomes, thereby limiting antigen degradation and prolonging cross-presentation by MHC-I (66).

Immature DCs can retain intracellular antigens for extended periods of time, and acute stimulation of antigen-loaded DCs leads to rapid antigen degradation, the formation of MHC-II–peptide complexes, and their accumulation on the surface of the now activated DCs (79, 80). The DC activation process leads to increased association of the ATP-dependent vacuolar proton pump with antigen-processing compartments (76), increasing their acidification, and also induces the redistribution of cathepsins from conventional lysosomes into antigen-processing compartments (81). Taken together, these activation-induced changes promote the generation of MHC-II–peptide complexes in activating DCs that are required for effective antigen processing and presentation to CD4 T cells.

MOVEMENT OF MHC MOLECULES TO THE PLASMA MEMBRANE

Like most cargo internalized from the plasma membrane, internalized MHC-II–I_i complexes enter early endosomes and eventually sort into late endosomal antigen-processing compartments that have the properties of multivesicular bodies (MVBs). MHC-II–I_i complexes reside primarily on the intraluminal vesicles (ILVs) of MVBs in DCs (82); however, the signals present on the MHC-II–I_i complex that are required for sorting into these vesicles remains to be determined. It is likely that peptide loading onto MHC-II occurs when MHC-II is present on these ILVs, since MHC-II bound to the I_i degradation product CLIP (83) as well as other MHC-II–peptide complexes are readily observed on these internal membranes by immunoelectron microscopy (84). MHC-II–I_i, MHC-II–CLIP, and peptide-loaded MHC-II molecules are also found on the peripheral, limiting, membrane of MVBs, but it remains to be determined whether or not MHC-II–CLIP or peptide-loaded MHC-II are actually generated on these membranes. To be competent for insertion into the PM, the MHC-II must leave the ILV and be deposited into the limiting membrane of the MVB in a process that has been termed "back-fusion" (84). Whether back-fusion actually occurs remains unknown, and it is also unknown how MHC-II moves to the limiting membrane of the MVB for eventual transport to the plasma membrane.

When an intact MVB directly fuses with the plasma membrane, the MHC-II-bearing ILVs are released from the cell and these cell-free vesicles are termed exosomes (85). Exosomes are secreted from most cell types in the body, and DC-derived exosomes contain antigenic MHC-II–peptide complexes, as well as costimulatory and adhesion molecules that allow exosomes to function as "mini-APCs" that are capable of directly activating T cells or indirectly activating T cells (after acquisition by other APCs) (86, 87). While the physiological role of DC-derived exosomes remains unknown, data showing that engagement of DCs with $CD4^+$ T cells promotes exosome release (88) has led to the speculation that exosomes are able to help propagate T-cell activation.

The membrane transport pathways and molecular mechanisms that allow newly generated MHC-II–peptide complexes to move from intracellular antigen-processing compartments to the APC surface are poorly understood. Activation of DCs with TLR ligands (89) or interaction of antigen-loaded DCs with antigen-specific $CD4^+$ T cells (90) results in the formation of elongated tubules that emanate from antigen-processing compartments toward the DC plasma membrane (90). Whether tubules or vesicles derived from tubules are responsible for the direct delivery of MHC-II to the cell surface remains to be conclusively demonstrated. MHC-II-containing vesicles have been observed to fuse with the surface of MHC-II-expressing melanoma cells (91), and even in professional APCs, these vesicles travel in a stop-and-go pattern along microtubule tracks in an actin-dependent manner from antigen-processing compartments to the plasma membrane (92). More-recent studies have identified a variety of actin-based molecular motors and GTPases that regulate MHC-II transport to the plasma membrane in

DCs; however, the mechanisms used by these proteins to regulate vesicle movement are unknown. Once on the plasma membrane, MHC-II–peptide complexes are present in small microclusters (93). This has been attributed to the association of MHC-II–peptide complexes with lipid raft membrane microdomains (94), thereby locally concentrating small numbers of specific MHC-II–peptide complexes for efficient activation of $CD4^+$ T cells.

ROLE OF MHC-II BIOSYNTHESIS/ TURNOVER FOR APC FUNCTION

While all APCs can ultimately stimulate antigen-specific $CD4^+$ T cells, expression of MHC-II in resting and activated states differs among different APC subtypes. For example, MHC-II mRNA is expressed in resting B cells, thymic epithelial cells, and DCs, whereas, particularly in the mouse, monocytes and macrophages do not constitutively express MHC-II. However, treatment with IFN-γ promotes the expression of the class II transactivator (CIITA) that induces MHC-II transcription and protein expression in monocytes, macrophages, and other IFN-γ-responsive cells (95). Activation of DCs leads to a burst in MHC-II transcription and protein synthesis; however, this increase is short-lived and DC activation eventually leads to a profound reduction in MHC-II biosynthesis that has been observed both *in vitro* and *in vivo* (96, 97). Activation of either DCs or IFN-γ-treated macrophages with TLR ligands (such as lipopolysaccharide or CpG DNA) ultimately terminates CIITA expression and MHC-II synthesis (98, 99). Indeed, injection of the TLR ligand CpG into mice results in a near complete cessation of MHC-II biosynthesis within 16 h (97). This increase in MHC-II synthesis followed by a rapid decline serves to enhance the surface expression of MHC-II complexes with pathogen-derived peptides.

Under steady-state conditions, the continuous input of newly generated MHC-II–peptide complexes on the surface of DCs is accompanied by their rapid turnover. Without a mechanism to protect MHC-II from degradation, termination of MHC-II synthesis upon DC activation would be accompanied by a reduction in total MHC-II on the cell surface. The rapid turnover of MHC-II in immature DCs is mediated by ubiquitination of MHC-II by the E3 ubiquitin ligase March-I (100, 101). March-I is only expressed in immature DCs, and termination of March-I expression upon DC activation results in long-lived MHC-II–peptide complexes on the surface of activated DCs (100, 101). Taken together with the data showing regulated synthesis of MHC-II upon DC activation (96, 102), these findings have led to a widely accepted model in which DCs respond to activating pathogens by transiently increasing MHC-II synthesis and generating pathogen-derived MHC-II–peptide complexes that have enhanced stability on the surface of the pathogen-activated DC.

CONCLUDING REMARKS

As is clear from the foregoing narrative, antigen processing requires dedicated accessory components, such as tapasin and TAP for MHC-I and I_i and DM for MHC-II, that interact in sophisticated ways with evolutionarily ancient "housekeeping" functions that exist in all eukaryotic cells. These include proteasomal proteolysis in the cytosol (for MHC-I) and cathepsin-mediated proteolysis and pH control in the endocytic pathway (for MHC-II), as well as chaperone-mediated glycoprotein folding and assembly processes that are required to produce functional MHC molecules. These general housekeeping functions are adopted and modified in myeloid cells to work in concert with specific modulators of antigen processing to produce an optimal outcome for the immune system, namely, the efficient and appropriate generation of MHC-peptide complexes that result in effective T-cell immunity.

Citation. Roche PA, Cresswell P. 2016. Antigen processing and presentation mechanisms in myeloid cells. Microbiol Spectrum 4(3):MCHD-0008-2015.

References

1. **Anderson MS, Miller J.** 1992. Invariant chain can function as a chaperone protein for class II major histocompatibility complex molecules. *Proc Natl Acad Sci U S A* **89:**2282–2286.
2. **Elliott EA, Drake JR, Amigorena S, Elsemore J, Webster P, Mellman I, Flavell RA.** 1994. The invariant chain is required for intracellular transport and function of major histocompatibility complex class II molecules. *J Exp Med* **179:**681–694.
3. **Roche PA, Cresswell P.** 1990. Invariant chain association with HLA-DR molecules inhibits immunogenic peptide binding. *Nature* **345:**615–618.
4. **Teyton L, O'Sullivan D, Dickson PW, Lotteau V, Sette A, Fink P, Peterson PA.** 1990. Invariant chain distinguishes between the exogenous and endogenous antigen presentation pathways. *Nature* **348:**39–44.
5. **Bakke O, Dobberstein B.** 1990. MHC class II-associated invariant chain contains a sorting signal for endosomal compartments. *Cell* **63:**707–716.
6. **Lotteau V, Teyton L, Peleraux A, Nilsson T, Karlsson L, Schmid SL, Quaranta V, Peterson PA.** 1990. Intracellular transport of class II MHC molecules directed by invariant chain. *Nature* **348:**600–605.

7. **Hiltbold EM, Roche PA.** 2002. Trafficking of MHC class II molecules in the late secretory pathway. *Curr Opin Immunol* **14:**30–35.

8. **Pieters J, Bakke O, Dobberstein B.** 1993. The MHC class II-associated invariant chain contains two endosomal targeting signals within its cytoplasmic tail. *J Cell Sci* **106:**831–846.

9. **Dugast M, Toussaint H, Dousset C, Benaroch P.** 2005. AP2 clathrin adaptor complex, but not AP1, controls the access of the major histocompatibility complex (MHC) class II to endosomes. *J Biol Chem* **280:**19656–19664.

10. **McCormick PJ, Martina JA, Bonifacino JS.** 2005. Involvement of clathrin and AP-2 in the trafficking of MHC class II molecules to antigen-processing compartments. *Proc Natl Acad Sci U S A* **102:**7910–7915.

11. **Castellino F, Germain RN.** 1995. Extensive trafficking of MHC class II-invariant chain complexes in the endocytic pathway and appearance of peptide-loaded class II in multiple compartments. *Immunity* **2:**73–88.

12. **Roche PA, Teletski CL, Stang E, Bakke O, Long EO.** 1993. Cell surface HLA-DR-invariant chain complexes are targeted to endosomes by rapid internalization. *Proc Natl Acad Sci U S A* **90:**8581–8585.

13. **Tjelle TE, Brech A, Juvet LK, Griffiths G, Berg T.** 1996. Isolation and characterization of early endosomes, late endosomes and terminal lysosomes: their role in protein degradation. *J Cell Sci* **109**(Pt 12):2905–2914.

14. **Shi GP, Villadangos JA, Dranoff G, Small C, Gu L, Haley KJ, Riese R, Ploegh HL, Chapman HA.** 1999. Cathepsin S required for normal MHC class II peptide loading and germinal center development. *Immunity* **10:** 197–206.

15. **Nakagawa TY, Brissette WH, Lira PD, Griffiths RJ, Petrushova N, Stock J, McNeish JD, Eastman SE, Howard ED, Clarke SR, Rosloniec EF, Elliott EA, Rudensky AY.** 1999. Impaired invariant chain degradation and antigen presentation and diminished collagen-induced arthritis in cathepsin S null mice. *Immunity* **10:** 207–217.

16. **Manoury B, Mazzeo D, Li DN, Billson J, Loak K, Benaroch P, Watts C.** 2003. Asparagine endopeptidase can initiate the removal of the MHC class II invariant chain chaperone. *Immunity* **18:**489–498.

17. **Riberdy JM, Newcomb JR, Surman MJ, Barbosa JA, Cresswell P.** 1992. HLA-DR molecules from an antigen-processing mutant cell line are associated with invariant chain peptides. *Nature* **360:**474–477.

18. **Denzin LK, Cresswell P.** 1995. HLA-DM induces CLIP dissociation from MHC class II αβ dimers and facilitates peptide loading. *Cell* **82:**155–165.

19. **Marks MS, Roche PA, van Donselaar E, Woodruff L, Peters PJ, Bonifacino JS.** 1995. A lysosomal targeting signal in the cytoplasmic tail of the β chain directs HLA-DM to MHC class II compartments. *J Cell Biol* **131:**351–369.

20. **Kropshofer H, Vogt AB, Moldenhauer G, Hammer J, Blum JS, Hammerling GJ.** 1996. Editing of the HLA-DR-peptide repertoire by HLA-DM. *EMBO J* **15:**6144–6154.

21. **Sant AJ, Chaves FA, Jenks SA, Richards KA, Menges P, Weaver JM, Lazarski CA.** 2005. The relationship between immunodominance, DM editing, and the kinetic stability of MHC class II:peptide complexes. *Immunol Rev* **207:**261–278.

22. **Yin L, Stern LJ.** 2013. HLA-DM focuses on conformational flexibility around P1 pocket to catalyze peptide exchange. *Front Immunol* **4:**336. doi:10.3389/fimmu.2013.00336.

23. **Pos W, Sethi DK, Call MJ, Schulze MS, Anders AK, Pyrdol J, Wucherpfennig KW.** 2012. Crystal structure of the HLA-DM-HLA-DR1 complex defines mechanisms for rapid peptide selection. *Cell* **151:**1557–1568.

24. **Liljedahl M, Kuwana T, Fung-Leung WP, Jackson MR, Peterson PA, Karlsson L.** 1996. HLA-DO is a lysosomal resident which requires association with HLA-DM for efficient intracellular transport. *EMBO J* **15:**4817–4824.

25. **Chen X, Reed-Loisel LM, Karlsson L, Jensen PE.** 2006. H2-O expression in primary dendritic cells. *J Immunol* **176:**3548–3556.

26. **Fallas JL, Yi W, Draghi NA, O'Rourke HM, Denzin LK.** 2007. Expression patterns of H2-O in mouse B cells and dendritic cells correlate with cell function. *J Immunol* **178:**1488–1497.

27. **Hornell TM, Burster T, Jahnsen FL, Pashine A, Ochoa MT, Harding JJ, Macaubas C, Lee AW, Modlin RL, Mellins ED.** 2006. Human dendritic cell expression of HLA-DO is subset specific and regulated by maturation. *J Immunol* **176:**3536–3547.

28. **Denzin LK, Fallas JL, Prendes M, Yi W.** 2005. Right place, right time, right peptide: DO keeps DM focused. *Immunol Rev* **207:**279–292.

29. **Liljedahl M, Winqvist O, Surh CD, Wong P, Ngo K, Teyton L, Peterson PA, Brunmark A, Rudensky AY, Fung-Leung WP, Karlsson L.** 1998. Altered antigen presentation in mice lacking H2-O. *Immunity* **8:**233–243.

30. **Hinz A, Tampe R.** 2012. ABC transporters and immunity: mechanism of self-defense. *Biochemistry* **51:**4981–4989.

31. **Saveanu L, Carroll O, Lindo V, Del Val M, Lopez D, Lepelletier Y, Greer F, Schomburg L, Fruci D, Niedermann G, van Endert PM.** 2005. Concerted peptide trimming by human ERAP1 and ERAP2 aminopeptidase complexes in the endoplasmic reticulum. *Nat Immunol* **6:** 689–697.

32. **Blum JS, Wearsch PA, Cresswell P.** 2013. Pathways of antigen processing. *Annu Rev Immunol* **31:**443–473.

33. **Hebert DN, Garman SC, Molinari M.** 2005. The glycan code of the endoplasmic reticulum: asparagine-linked carbohydrates as protein maturation and quality-control tags. *Trends Cell Biol* **15:**364–370.

34. **D'Alessio C, Caramelo JJ, Parodi AJ.** 2010. UDP-GlC: glycoprotein glucosyltransferase-glucosidase II, the ying-yang of the ER quality control. *Seminars Cell Dev Biol* **21:**491–499.

35. **Gao B, Adhikari R, Howarth M, Nakamura K, Gold MC, Hill AB, Knee R, Michalak M, Elliott T.** 2002. Assembly and antigen-presenting function of MHC class I molecules in cells lacking the ER chaperone calreticulin. *Immunity* **16:**99–109.
36. **Garbi N, Tanaka S, Momburg F, Hammerling GJ.** 2006. Impaired assembly of the major histocompatibility complex class I peptide-loading complex in mice deficient in the oxidoreductase ERp57. *Nat Immunol* **7:** 93–102.
37. **Zhang W, Wearsch PA, Zhu Y, Leonhardt RM, Cresswell P.** 2011. A role for UDP-glucose glycoprotein glucosyltransferase in expression and quality control of MHC class I molecules. *Proc Natl Acad Sci U S A* **108:** 4956–4961.
38. **Williams AP, Peh CA, Purcell AW, McCluskey J, Elliott T.** 2002. Optimization of the MHC class I peptide cargo is dependent on tapasin. *Immunity* **16:**509–520.
39. **Howarth M, Williams A, Tolstrup AB, Elliott T.** 2004. Tapasin enhances MHC class I peptide presentation according to peptide half-life. *Proc Natl Acad Sci U S A* **101:**11737–11742.
40. **Wearsch PA, Cresswell P.** 2007. Selective loading of high-affinity peptides onto major histocompatibility complex class I molecules by the tapasin-ERp57 heterodimer. *Nat Immunol* **8:**873–881.
41. **Joffre OP, Segura E, Savina A, Amigorena S.** 2012. Cross-presentation by dendritic cells. *Nat Rev Immunol* **12:**557–569.
42. **Shen L, Sigal LJ, Boes M, Rock KL.** 2004. Important role of cathepsin S in generating peptides for TAP-independent MHC class I crosspresentation in vivo. *Immunity* **21:**155–165.
43. **den Haan JM, Lehar SM, Bevan MJ.** 2000. $CD8^+$ but not $CD8^-$ dendritic cells cross-prime cytotoxic T cells in vivo. *J Exp Med* **192:**1685–1696.
44. **Jongbloed SL, Kassianos AJ, McDonald KJ, Clark GJ, Ju X, Angel CE, Chen CJ, Dunbar PR, Wadley RB, Jeet V, Vulink AJ, Hart DN, Radford KJ.** 2010. Human $CD141^+$ $(BDCA\text{-}3)^+$ dendritic cells (DCs) represent a unique myeloid DC subset that cross-presents necrotic cell antigens. *J Exp Med* **207:**1247–1260.
45. **Poulin LF, Salio M, Griessinger E, Anjos-Afonso F, Craciun L, Chen JL, Keller AM, Joffre O, Zelenay S, Nye E, Le Moine A, Faure F, Donckier V, Sancho D, Cerundolo V, Bonnet D, Reis e Sousa C.** 2010. Characterization of human $DNGR\text{-}1^+$ $BDCA3^+$ leukocytes as putative equivalents of mouse $CD8\alpha^+$ dendritic cells. *J Exp Med* **207:**1261–1271.
46. **Segura E, Durand M, Amigorena S.** 2013. Similar antigen cross-presentation capacity and phagocytic functions in all freshly isolated human lymphoid organ-resident dendritic cells. *J Exp Med* **210:**1035–1047.
47. **Lim JP, Gleeson PA.** 2011. Macropinocytosis: an endocytic pathway for internalising large gulps. *Immunol Cell Biol* **89:**836–843.
48. **Norbury CC, Chambers BJ, Prescott AR, Ljunggren HG, Watts C.** 1997. Constitutive macropinocytosis allows TAP-dependent major histocompatibility complex class I presentation of exogenous soluble antigen by bone marrow-derived dendritic cells. *Eur J Immunol* **27:**280–288.
49. **Garrett WS, Chen LM, Kroschewski R, Ebersold M, Turley S, Trombetta S, Galan JE, Mellman I.** 2000. Developmental control of endocytosis in dendritic cells by Cdc42. *Cell* **102:**325–334.
50. **West MA, Prescott AR, Eskelinen EL, Ridley AJ, Watts C.** 2000. Rac is required for constitutive macropinocytosis by dendritic cells but does not control its downregulation. *Curr Biol* **10:**839–848.
51. **Ruedl C, Koebel P, Karjalainen K.** 2001. In vivo-matured Langerhans cells continue to take up and process native proteins unlike in vitro-matured counterparts. *J Immunol* **166:**7178–7182.
52. **Platt CD, Ma JK, Chalouni C, Ebersold M, Bou-Reslan H, Carano RA, Mellman I, Delamarre L.** 2010. Mature dendritic cells use endocytic receptors to capture and present antigens. *Proc Natl Acad Sci U S A* **107:** 4287–4292.
53. **Drutman SB, Trombetta ES.** 2010. Dendritic cells continue to capture and present antigens after maturation in vivo. *J Immunol* **185:**2140–2146.
54. **Jayachandran R, Sundaramurthy V, Combaluzier B, Mueller P, Korf H, Huygen K, Miyazaki T, Albrecht I, Massner J, Pieters J.** 2007. Survival of mycobacteria in macrophages is mediated by coronin 1-dependent activation of calcineurin. *Cell* **130:**37–50.
55. **Bosedasgupta S, Pieters J.** 2014. Inflammatory stimuli reprogram macrophage phagocytosis to macropinocytosis for the rapid elimination of pathogens. *PLoS Pathog* **10:** e1003879. doi:10.1371/journal.ppat.1003879.
56. **Bonifaz LC, Bonnyay DP, Charalambous A, Darguste DI, Fujii S, Soares H, Brimnes MK, Moltedo B, Moran TM, Steinman RM.** 2004. In vivo targeting of antigens to maturing dendritic cells via the DEC-205 receptor improves T cell vaccination. *J Exp Med* **199:**815–824.
57. **Chatterjee B, Smed-Sörensen A, Cohn L, Chalouni C, Vandlen R, Lee BC, Widger J, Keler T, Delamarre L, Mellman I.** 2012. Internalization and endosomal degradation of receptor-bound antigens regulate the efficiency of cross presentation by human dendritic cells. *Blood* **120:**2011–2020.
58. **Stuart LM, Ezekowitz RA.** 2005. Phagocytosis: elegant complexity. *Immunity* **22:**539–550.
59. **Gagnon E, Duclos S, Rondeau C, Chevet E, Cameron PH, Steele-Mortimer O, Paiement J, Bergeron JJ, Desjardins M.** 2002. Endoplasmic reticulum-mediated phagocytosis is a mechanism of entry into macrophages. *Cell* **110:**119–131.
60. **Ackerman AL, Kyritsis C, Tampe R, Cresswell P.** 2003. Early phagosomes in dendritic cells form a cellular compartment sufficient for cross presentation of exogenous antigens. *Proc Natl Acad Sci U S A* **100:** 12889–12894.
61. **Guermonprez P, Saveanu L, Kleijmeer M, Davoust J, Van Endert P, Amigorena S.** 2003. ER-phagosome fusion defines an MHC class I cross-presentation compartment in dendritic cells. *Nature* **425:**397–402.

62. **Houde M, Bertholet S, Gagnon E, Brunet S, Goyette G, Laplante A, Princiotta MF, Thibault P, Sacks D, Desjardins M.** 2003. Phagosomes are competent organelles for antigen cross-presentation. *Nature* **425**:402–406.

63. **Nair-Gupta P, Baccarini A, Tung N, Seyffer F, Florey O, Huang Y, Banerjee M, Overholtzer M, Roche PA, Tampe R, Brown BD, Amsen D, Whiteheart SW, Blander JM.** 2014. TLR signals induce phagosomal MHC-I delivery from the endosomal recycling compartment to allow cross-presentation. *Cell* **158**:506–521.

64. **Blander JM, Medzhitov R.** 2004. Regulation of phagosome maturation by signals from Toll-like receptors. *Science* **304**:1014–1018.

65. **Blander JM, Medzhitov R.** 2006. Toll-dependent selection of microbial antigens for presentation by dendritic cells. *Nature* **440**:808–812.

66. **Savina A, Jancic C, Hugues S, Guermonprez P, Vargas P, Moura IC, Lennon-Duménil AM, Seabra MC, Raposo G, Amigorena S.** 2006. NOX2 controls phagosomal pH to regulate antigen processing during cross-presentation by dendritic cells. *Cell* **126**:205–218.

67. **Savina A, Peres A, Cebrian I, Carmo N, Moita C, Hacohen N, Moita LF, Amigorena S.** 2009. The small GTPase Rac2 controls phagosomal alkalinization and antigen crosspresentation selectively in $CD8^+$ dendritic cells. *Immunity* **30**:544–555.

68. **Crotzer VL, Blum JS.** 2010. Autophagy and adaptive immunity. *Immunology* **131**:9–17.

69. **Schmid D, Pypaert M, Munz C.** 2007. Antigen-loading compartments for major histocompatibility complex class II molecules continuously receive input from autophagosomes. *Immunity* **26**:79–92.

70. **Adamopoulou E, Tenzer S, Hillen N, Klug P, Rota IA, Tietz S, Gebhardt M, Stevanovic S, Schild H, Tolosa E, Melms A, Stoeckle C.** 2013. Exploring the MHC-peptide matrix of central tolerance in the human thymus. *Nat Commun* **4**:2039. doi:10.1038/ncomms3039.

71. **Nedjic J, Aichinger M, Emmerich J, Mizushima N, Klein L.** 2008. Autophagy in thymic epithelium shapes the T-cell repertoire and is essential for tolerance. *Nature* **455**:396–400.

72. **Aichinger M, Wu C, Nedjic J, Klein L.** 2013. Macroautophagy substrates are loaded onto MHC class II of medullary thymic epithelial cells for central tolerance. *J Exp Med* **210**:287–300.

73. **Kaushik S, Cuervo AM.** 2012. Chaperone-mediated autophagy: a unique way to enter the lysosome world. *Trends Cell Biol* **22**:407–417.

74. **Manoury B, Mazzeo D, Fugger L, Viner N, Ponsford M, Streeter H, Mazza G, Wraith DC, Watts C.** 2002. Destructive processing by asparagine endopeptidase limits presentation of a dominant T cell epitope in MBP. *Nat Immunol* **3**:169–174.

75. **Delamarre L, Pack M, Chang H, Mellman I, Trombetta ES.** 2005. Differential lysosomal proteolysis in antigen-presenting cells determines antigen fate. *Science* **307**: 1630–1634.

76. **Trombetta ES, Ebersold M, Garrett W, Pypaert M, Mellman I.** 2003. Activation of lysosomal function during dendritic cell maturation. *Science* **299**:1400–1403.

77. **Sardiello M, Palmieri M, di Ronza A, Medina DL, Valenza M, Gennarino VA, Di Malta C, Donaudy F, Embrione V, Polishchuk RS, Banfi S, Parenti G, Cattaneo E, Ballabio A.** 2009. A gene network regulating lysosomal biogenesis and function. *Science* **325**: 473–477.

78. **Settembre C, Fraldi A, Medina DL, Ballabio A.** 2013. Signals from the lysosome: a control centre for cellular clearance and energy metabolism. *Nat Rev Mol Cell Biol* **14**:283–296.

79. **Inaba K, Turley S, Iyoda T, Yamaide F, Shimoyama S, Reis e Sousa C, Germain RN, Mellman I, Steinman RM.** 2000. The formation of immunogenic major histocompatibility complex class II-peptide ligands in lysosomal compartments of dendritic cells is regulated by inflammatory stimuli. *J Exp Med* **191**:927–936.

80. **Turley SJ, Inaba K, Garrett WS, Ebersold M, Unternaehrer J, Steinman RM, Mellman I.** 2000. Transport of peptide-MHC class II complexes in developing dendritic cells. *Science* **288**:522–527.

81. **Lautwein A, Burster T, Lennon-Duménil AM, Overkleeft HS, Weber E, Kalbacher H, Driessen C.** 2002. Inflammatory stimuli recruit cathepsin activity to late endosomal compartments in human dendritic cells. *Eur J Immunol* **32**:3348–3357.

82. **Kleijmeer MJ, Oorschot VM, Geuze HJ.** 1994. Human resident Langerhans cells display a lysosomal compartment enriched in MHC class II. *J Invest Dermatol* **103**: 516–523.

83. **Stang E, Guerra CB, Amaya M, Paterson Y, Bakke O, Mellins ED.** 1998. DR/CLIP (class II-associated invariant chain peptides) and DR/peptide complexes colocalize in prelysosomes in human B lymphoblastoid cells. *J Immunol* **160**:4696–4707.

84. **Kleijmeer M, Ramm G, Schuurhuis D, Griffith J, Rescigno M, Ricciardi-Castagnoli P, Rudensky AY, Ossendorp F, Melief CJ, Stoorvogel W, Geuze HJ.** 2001. Reorganization of multivesicular bodies regulates MHC class II antigen presentation by dendritic cells. *J Cell Biol* **155**:53–63.

85. **Denzer K, Kleijmeer MJ, Heijnen HF, Stoorvogel W, Geuze HJ.** 2000. Exosome: from internal vesicle of the multivesicular body to intercellular signaling device. *J Cell Sci* **113**(Pt 19):3365–3374.

86. **Zitvogel L, Regnault A, Lozier A, Wolfers J, Flament C, Tenza D, Ricciardi-Castagnoli P, Raposo G, Amigorena S.** 1998. Eradication of established murine tumors using a novel cell-free vaccine: dendritic cell-derived exosomes. *Nat Med* **4**:594–600.

87. **Thery C, Duban L, Segura E, Veron P, Lantz O, Amigorena S.** 2002. Indirect activation of naive $CD4^+$ T cells by dendritic cell-derived exosomes. *Nat Immunol* **3**:1156–1162.

88. **Buschow SI, Nolte-'t Hoen EN, van Niel G, Pols MS, ten Broeke T, Lauwen M, Ossendorp F, Melief CJ, Raposo G, Wubbolts R, Wauben MH, Stoorvogel W.** 2009. MHC II in dendritic cells is targeted to lysosomes

or T cell-induced exosomes via distinct multivesicular body pathways. *Traffic* **10:**1528–1542.

89. **Chow A, Toomre D, Garrett W, Mellman I.** 2002. Dendritic cell maturation triggers retrograde MHC class II transport from lysosomes to the plasma membrane. *Nature* **418:**988–994.

90. **Boes M, Cerny J, Massol R, Op den Brouw M, Kirchhausen T, Chen J, Ploegh HL.** 2002. T-cell engagement of dendritic cells rapidly rearranges MHC class II transport. *Nature* **418:**983–988.

91. **Wubbolts R, Fernandez-Borja M, Oomen L, Verwoerd D, Janssen H, Calafat J, Tulp A, Dusseljee S, Neefjes J.** 1996. Direct vesicular transport of MHC class II molecules from lysosomal structures to the cell surface. *J Cell Biol* **135:**611–622.

92. **Rocha N, Neefjes J.** 2008. MHC class II molecules on the move for successful antigen presentation. *EMBO J* **27:**1–5.

93. **Bosch B, Heipertz EL, Drake JR, Roche PA.** 2013. Major histocompatibility complex (MHC) class II-peptide complexes arrive at the plasma membrane in cholesterol-rich microclusters. *J Biol Chem* **288:**13236–13242.

94. **Anderson HA, Roche PA.** 2015. MHC class II association with lipid rafts on the antigen presenting cell surface. *Biochim Biophys Acta* **1853:**775–780.

95. **Reith W, LeibundGut-Landmann S, Waldburger JM.** 2005. Regulation of MHC class II gene expression by the class II transactivator. *Nat Rev Immunol* **5:**793–806.

96. **Cella M, Engering A, Pinet V, Pieters J, Lanzavecchia A.** 1997. Inflammatory stimuli induce accumulation of MHC class II complexes on dendritic cells. *Nature* **388:** 782–787.

97. **Young LJ, Wilson NS, Schnorrer P, Mount A, Lundie RJ, La Gruta NL, Crabb BS, Belz GT, Heath WR, Villadangos JA.** 2007. Dendritic cell preactivation impairs MHC class II presentation of vaccines and endogenous viral antigens. *Proc Natl Acad Sci U S A* **104:** 17753–17758.

98. **Landmann S, Muhlethaler-Mottet A, Bernasconi L, Suter T, Waldburger JM, Masternak K, Arrighi JF, Hauser C, Fontana A, Reith W.** 2001. Maturation of dendritic cells is accompanied by rapid transcriptional silencing of class II transactivator (CIITA) expression. *J Exp Med* **194:**379–391.

99. **Yao Y, Xu Q, Kwon MJ, Matta R, Liu Y, Hong SC, Chang CH.** 2006. ERK and p38 MAPK signaling pathways negatively regulate CIITA gene expression in dendritic cells and macrophages. *J Immunol* **177:**70–76.

100. **De Gassart A, Camosseto V, Thibodeau J, Ceppi M, Catalan N, Pierre P, Gatti E.** 2008. MHC class II stabilization at the surface of human dendritic cells is the result of maturation-dependent MARCH I down-regulation. *Proc Natl Acad Sci U S A* **105:**3491–3496.

101. **Walseng E, Furuta K, Bosch B, Weih KA, Matsuki Y, Bakke O, Ishido S, Roche PA.** 2010. Ubiquitination regulates MHC class II-peptide complex retention and degradation in dendritic cells. *Proc Natl Acad Sci U S A* **107:**20465–20470.

102. **Pierre P, Turley SJ, Gatti E, Hull M, Meltzer J, Mirza A, Inaba K, Steinman RM, Mellman I.** 1997. Developmental regulation of MHC class II transport in mouse dendritic cells. *Nature* **388:**787–792.

103. **Roche PA, Furuta K.** 2015. The ins and outs of MHC class II-mediated antigen processing and presentation. *Nat Rev Immunol* **15:**203–216.

104. **Samie M, Cresswell P.** 2015. The transcription factor TFEB acts as a molecular switch that regulates exogenous antigen-presentation pathways. *Nat Immunol* **16:** 729–736.

Myeloid Cells in Health and Disease: A Synthesis
Edited by Siamon Gordon

doi:10.1128/microbiolspec.MCHD-0003-2015

V. Hugh Perry[1]

Microglia

12

INTRODUCTION

Microglia are the resident macrophages of the brain parenchyma (1). Although it has long been known that microglia are of myeloid lineage, based on immunocytochemical detection of macrophage-restricted antigens (2), it has only relatively recently been shown, by fate mapping studies, that these cells are of yolk sac origin and enter the developing neuroepithelium of the central nervous system (CNS) in the embryo (3). They are present throughout the length of the neuraxis, characterized by their fine processes emanating from a small cell body, and each cell appears to occupy its own territory. The morphology of microglia and their territorial behavior is well illustrated in retina whole mounts (Fig. 1). The density and morphology of the microglia vary between distinct functional divisions of the CNS, with the lowest density found in the cerebellum and perhaps the highest density in the substantia nigra (4). These regional differences have been well studied in rodents, the most common experimental animal models, and although similar regional differences are seen in the human brain, there are some notable differences. In the rodent brain, the microglia are denser in gray matter than in white matter, while in the human brain, the microglia are denser in the large-fiber tracts that dominate the larger brain (5).

Microglia in the adult brain have been imaged in real time using two-photon confocal microscopy to view cells genetically engineered to express green fluorescent protein (GFP): the cell processes possess a remarkable degree of motility. The cell soma appears to remain in a relatively fixed position, while the processes palpate the surface of the neurons and glia in the surrounding microenvironment (6). It is estimated that a microglia cell can survey its entire territory in about 1 hour but a laser-induced microlesion can induce a rapid response from the processes of microglia as they orient toward and surround the lesion without necessitating movement of the cell soma (6). In the aging brain, the density of microglia is modestly increased; the regular lattice is still maintained, although less precisely than in young animals; and the rate of surveillance of the surrounding tissue is slower (7).

A striking feature of the microglia is the lack or low level of expression of many molecules typically expressed by other tissue macrophages: components of the CNS microenvironment keep these cells under tight phenotypic control, as is discussed below. While it is widely appreciated that all tissue macrophages differ one from another to some degree, studies of the transcriptome of the microglia reveal that they express species of mRNA distinct from other tissue macrophages. A growing body of evidence indicates that in the normal-health CNS, in homeostatic conditions, the microglia express not only fewer transcripts than other tissue macrophages but also molecules not expressed on other macrophages (8, 9). A transcriptome signature distinct for murine microglia in homeostasis is

[1]Centre for Biological Sciences, University of Southampton, Southampton SO16 6YD, United Kingdom.

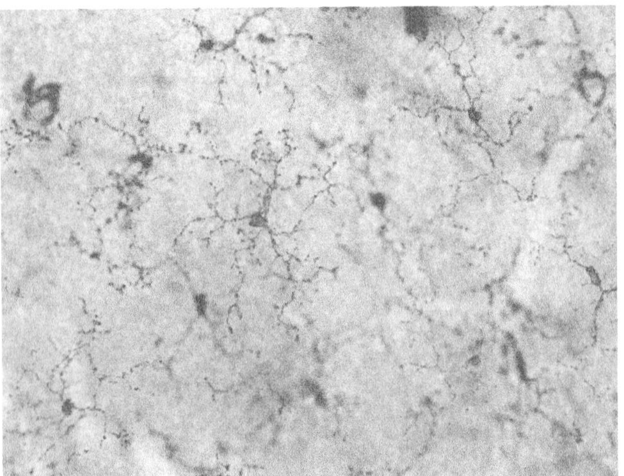

Figure 1 Microglia in the outer plexiform layer of the retina of the mouse, illustrating the delicate branching of the processes and the territory occupied by each cell. GFP-labeled cell from a MacGreen mouse with immunocytochemistry. Courtesy of Sallome Murinello.

beginning to be defined, with the unique expression of molecules such as the purinergic receptor P2Y12, Hexb, Gpr34, and others. Many of these molecules are also expressed in human fetal and adult microglia. Of particular note, given the use of cell lines and microglia isolated from the newborn murine brain, for studies of microglia function *in vitro*, the transcriptome of homeostatic microglia is not expressed in these conditions and the specialized phenotype of microglia is absent or lost.

The molecular characterization of the microglia phenotype reveals that there are numerous molecules that are involved in interactions with endogenous ligands in the local CNS microenvironment, and it is these ligand-receptor interactions that contribute to their downregulated phenotype. The microglia express on their cell surface a spectrum of receptors that have in common the property of inhibiting myeloid cell activation. For example, CD200R, C3XCR1, and CD33 all bind ligands expressed by neurons—CD200, C3XCL1, and sialic acid, respectively—that activate inhibitory signaling in the cytoplasm of the microglia (1 and references therein). These, and other molecules expressed in the CNS, regulate the microglia phenotype (Fig. 2). The microglia express receptors for several neurotransmitters, which may also contribute to the regulation of the microglia phenotype (10). A simple corollary of these observations is that if the microglia are tightly regulated by endogenous ligands in their local molecular environment, then degeneration or loss of neurons will lead to the activation of the microglia.

The microglia population is maintained by local proliferation, and although an early study suggested that their proliferation was lower than that of other tissue macrophage populations (11), recent studies suggest that ~0.5% or more of microglia are undergoing division at any moment (D. Gomez-Nicola, personal communication). A reevaluation of the data from bone marrow chimeras has shown that evidence demonstrating that microglia were replaced by bone marrow-derived monocytes was likely an artifact of the irradiation protocol (12). In the normal healthy CNS, few, if any, bone marrow-derived monocytes enter the CNS parenchyma, although the perivascular macrophages located in the perivascular space are replaced by monocytes in CCR2-dependent fashion. If the microglia proliferate throughout life but the density remains relatively constant, the suggestion is that there must be significant death among the adult microglia. This, however, has yet to be investigated in detail.

MICROGLIA HOMEOSTATIC FUNCTIONS IN THE ADULT CNS

The ubiquitous distribution and surveillance of the CNS parenchyma by microglia is consistent with the role of tissue macrophages being critical as the first line of defense against infection or injury. However, the spectrum of possible homeostatic functions of microglia in the normal healthy CNS remains ill defined and subject to much speculation and conjecture. Electron microscopy studies show that microglia processes appear to make contact with synapses within their territory and that the frequency of contacts may be modified by activity (13). It has been suggested that these transient contacts reflect the monitoring by microglia of the "health of the synapse." There is, however, little direct evidence for this, and other studies have shown that only a small percentage (3.5%) of synapses are contacted by microglia at any one time (14). Although microglia have the capacity to secrete molecules, such as glutamate, that influence neuronal activity, the physiological significance of these transient contacts remains to be properly evaluated, particularly with regard to the human CNS, where the microglia density in the synapse-rich neuropil is even less than that of the rodent.

As was noted above, the microglia density in the white matter of the human brain is greater than that seen in the rodent CNS, and this raises the question as to whether there is a homeostatic role for microglia in white matter tracts. In individuals with mutations in receptors expressed on microglia, and important in their

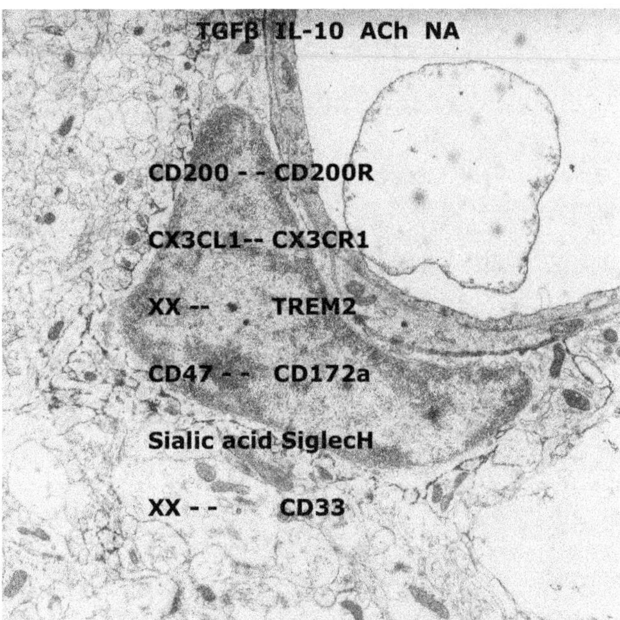

Figure 2 Electron micrograph of a microglia in the adult mouse cortex immunolabeled with F4/80. The thin rim of cytoplasm and the sparse rough endoplasmic reticulum point to the downregulated phenotype of these cells. Ligands expressed in the CNS (left-hand column) engage receptors expressed on the microglia (right-hand column) that inhibit their activation. Other soluble mediators, including the neurotransmitters acetylcholine (ACh) and noradrenaline (NA), are shown at the top of the figure.

phagocytic function (triggering receptor expressed on myeloid cells 2 [Trem2]) (15) or in their maintenance (colony-stimulating factor-1 receptor [CSF-1R]) (16), it is the white matter in which the most severe pathology is apparent. The secretion of exosomes by oligodendrocytes (17) and their removal by microglia may be an important function. A role for microglia in providing support for neurons has also been proposed (18), and it has been further suggested that microglia senescence and decline in this trophic function is a significant component of a number of chronic neurodegenerative diseases associated with the accumulation of misfolded proteins, such as amyloid-β (Aβ) in Alzheimer's disease (AD). As noted above, it is clear that, in aged rodents, microglia still survey their microenvironment and respond to local microlesions; hence, evidence that microglia senescence is critical to any disease state is not yet on a firm footing.

A relatively neglected component of microglia homeostatic function is the role that they play in communication between systemic inflammation and the CNS. Systemic inflammation, arising from infection or tissue damage, generates cytokines that communicate with the CNS by at least three routes (19). A neural route involves activation of receptors on sensory afferents of the vagus nerve, which innervates organs in the thoracic-abdominal cavity (20). The humoral route involves circulating cytokines in the blood that also communicate with the CNS. Circulating cytokines signal to macrophages in the circumventricular organs, such as the area postrema and subfornical organ, that line the ventricles, and these regions lack a patent blood-brain barrier. The macrophages in the circumventricular organs have a different morphology and phenotype compared to microglia and signals pass from the macrophages in the circumventricular organ into the brain parenchyma (21). The other major route of communication by circulating cytokines or other mediators involves the activation of the cerebral endothelial cells, which express Toll-like receptors to bind products of microorganisms and receptors for circulating cytokines. The activated endothelium, which expresses cyclooxygenase-1 (COX-1) and COX-2, generates prostaglandins, for example, PGE_2, which in turn signals to the perivascular macrophages and the microglia in the parenchyma (22). Following a peripheral challenge with lipopolysaccharide, there is rapid and widespread activation of the microglia, as can be shown *in vivo* using positron emission tomography imaging and ligand PK11195, which binds relatively selectively the 18-kDa translocator protein (TSPO) expressed by microglia (23). Microglia synthesize cytokines and other mediators, which in turn communicate with neurons in regions of the CNS important in the generation of fever and the behavioral changes associated with systemic inflammation: so-called sickness behaviors (19, 22). These signaling pathways between the periphery and the brain are not only an important part of how we adapt our behavior to deal with systemic infections and inflammation but also highlight the fact that there is intimate and ongoing communication between the immune system and the brain, which involves microglia: as discussed below, this may be of particular importance when there is ongoing disease in the CNS and accompanying activation of the microglia.

The communication between the immune system and the brain is not only one-way traffic from the periphery to the CNS. Systemic inflammation that activates pathways to the brain along the vagus, for example, leads to activation of the descending fibers of the vagus nerve, which in turn downregulate macrophage activation by the secretion of acetylcholine acting on the α7 acetylcholine receptor. The activation of circuits to the pituitary leads to the release of adrenocorticotropic hormone and release of glucocorticoids from the adrenals, which also modifies the systemic inflammatory response (20).

ROLE OF MICROGLIA DEVELOPMENT

The presence of macrophages in the developing CNS has long been recognized, although their yolk sac origin has only been recently established (3). There is a growing body of evidence to show that, during the late embryonic period and early postnatal life, microglia play an important role in the modeling of the CNS circuitry. Developmental or programmed cell death of neurons is a major part of CNS development, with as many as 50% of the cells generated in the embryo undergoing apoptosis before maturity. These apoptotic cells are phagocytosed by the immature microglia and their corpses rapidly removed (2). Recent evidence shows that the microglia may also be involved in more-subtle modifications of neuronal circuitry. Studies on the visual system of rodents have shown that selective pruning of axon terminals is an activity-dependent process, but it has now been shown that microglia play a role in the removal of these synapses by recognizing components of complement (C3) that have tagged selective synapses for removal (24). In the absence of C3, in C3 knockout mice, excess innervation of some target neurons of the incoming retinal ganglion cell axons persists. In the hippocampus, microglia have also been implicated in the removal of excess synapses. In mice lacking the receptor CX3CR1, there is a transient excess of synapses in the postnatal period, although the density returns to normal wild-type numbers by the end of postnatal development (25). There are a number of unanswered questions as to how the different components of activity—marking of the terminal by complement or other components and recognition by the microglia—are orchestrated. Despite these uncertainties, it has been suggested that deficits or increases in the removal of synapses by microglia may contribute to diverse neurodevelopmental disorders that lead to conditions such as autism spectrum disorder and schizophrenia, which is perhaps a step beyond the limits of what we know.

MICROGLIA IN PATHOLOGY

It has long been known that microglia are exquisitely sensitive to injury and disease of the CNS and can serve as a sensor of almost any form of disruption of homeostasis (26). We restrict our discussion here to role of microglia in multiple sclerosis, the archetypal inflammatory disease of the CNS, and chronic neurodegenerative diseases, in which the role of microglia has become a topic of growing interest.

Multiple Sclerosis

Multiple sclerosis is an inflammatory demyelinating disease of the CNS, characterized by local lesions or plaques, predominantly in white matter tracts, although cortical lesions are also present. In active lesions, there is a prominent T-cell infiltrate and large numbers of macrophages containing myelin debris; as the lesions evolve, so the infiltrate is confined to form a rim around a demyelinated central region. The inactive or chronic lesions, common in advanced disease, are areas of demyelination with few activated microglia or macrophages. The most common form of multiple sclerosis begins as a relapsing-remitting disease, which evolves into a secondary-progressive phase with little evidence of relapses. The pathology of MS is that of a highly complex, T-cell-driven disease with different T-cell subsets involved, and genome-wide association studies show that the immune system is critical to the etiology of the disease (27). To dissect the cellular and molecular events that underlie disease, investigators have used the animal model of multiple sclerosis experimental allergic encephalomyelitis (EAE). In this model and its variants, susceptible strains of rodents are immunized with components of myelin in adjuvant, leading to the development of clinical symptoms over the next 1 to 2 weeks. Depending on the particular model, this may be a transient limb paralysis with recovery or a more prolonged and chronic disease with varying degrees of relapse and remission. The initiation of the disease involves the activation of autoreactive T cells by perivascular macrophages as they cross the cerebral endothelium (28), which in turn leads to the activation of microglia and the recruitment of monocytes from the blood. To elucidate the different roles of microglia and monocyte-derived macrophages is difficult in postmortem human tissue owing to the lack of markers that definitively discriminate between the two subsets of myeloid cells. Evidence that blood-borne monocytes were critical in disease initiation was shown by the depletion of monocytes using peripherally delivered liposomes (29). The activated microglia alone appear to be insufficient for development of disease. More-recent studies have confirmed that it is the recruited monocytes that are critical in disease severity and progression. Using a parabiosis and irradiation combination, with the circulating cell population marked with GFP, Ajami and colleagues (30) showed that microglia proliferation occurred prior to monocyte entry but that it was the density of monocyte-derived macrophages that correlated with disease severity. As disease resolved, the monocyte-derived cells were reduced in number and did not appear to have the capacity to become resident microglia. Further evidence that the recruited monocyte-derived macrophages play a key role in the CNS pathology comes from studies using genetically engineered mice with GFP to label the resident microglia population, CX3CR1-GFP, and red fluorescent protein (RFP) to label the monocyte-derived macrophages, CCR2-RFP (31). In these mice, the induction of EAE enabled the distinction between the two myeloid populations, and with a combination of confocal microscopy and serial block-face scanning electron microscopy it was shown that the monocyte-derived macrophages play an active role in the stripping of myelin from the axon. The fine processes of these cells invade the nodes of Ranvier and appear to lift the myelin away from the axon plasma membrane before phagocytosis. The different color marking of the two populations also enabled gene expression analysis of the two types of myeloid cells. The monocyte-derived macrophages showed upregulation of genes associated with phagocytosis, calcium signaling, and autophagy, and although the microglia also upregulated genes associated with chemoattraction, cell movement, and motility, they also showed downregulation of genes of the naive homeostatic microglia signature and downregulation of an activated phenotype. The microglia and monocyte-derived macrophages have distinct expression profiles throughout the disease evolution and play distinct roles in disease pathogenesis.

Microglia in Chronic Neurodegeneration

In chronic neurodegenerative diseases of the CNS such as AD, Parkinson's disease, and prion diseases, the microglia take on an activated morphology (Fig. 3) and express diverse molecules not typically expressed by microglia in the healthy brain. The presence of these activated microglia is now referred to as neuroinflammation and is common to diverse degenerative diseases of the CNS (32). In contrast to the pathology of multiple sclerosis, cells of the myeloid lineage wholly dominate the neuroinflammation in chronic neurodegeneration. The possibility that these activated innate immune cells might play a role in AD progression arose from not only studies of the pathology of the AD brain but also epidemiological evidence that individuals taking anti-inflammatory medicine may be protected in some way from the onset or progression of the disease. Similar evidence has been obtained for neuroinflammation in Parkinson's disease. The strongest evidence that neuroinflammation is not simply a consequence of the pathology, the accumulation of amyloid, and the degeneration of neurons, but is a causal component of the disease in AD, comes from genome-wide association studies. There are a number of genes associated with innate immune cells, more specifically cells of the macrophage lineage, that confer risk of AD (33).

Investigation into the potential role of microglia in the brains of individuals with AD has generated an enormous amount of data but little clarity. The amyloid hypothesis, which dominates much thinking about the pathogenesis of AD, proposes that it is the

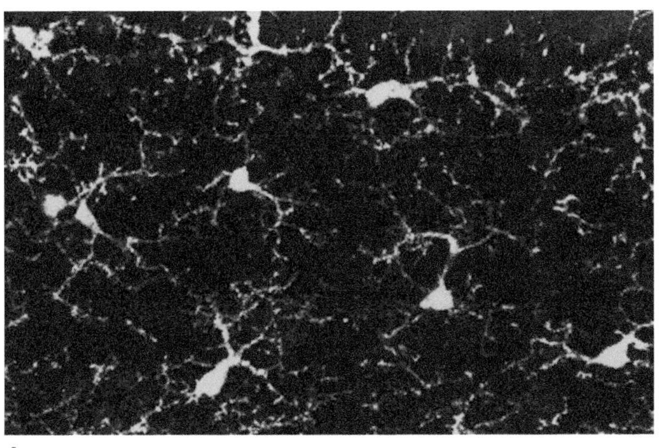
A

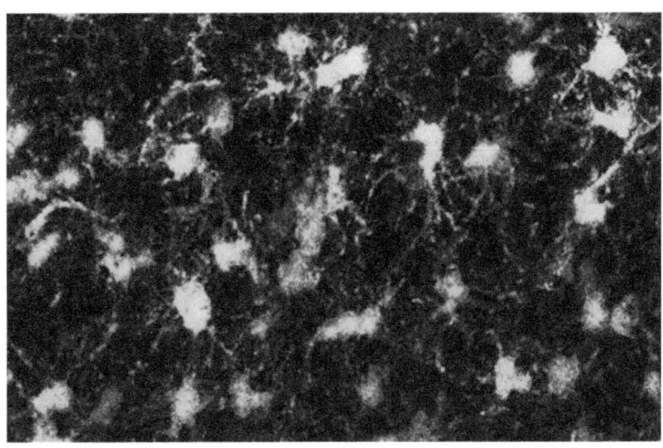
B

Figure 3 GFP-labeled microglia in the hippocampus of the normal adult mouse (**A**), and morphologically activated microglia in the hippocampus of a mouse with prion disease (**B**). Note the greater density of cells, larger cell bodies, and multiple processes of those in panel B compared to panel A. Courtesy of Diego Gomez-Nicola.

generation of neurotoxic amyloid species that drives the disease. Aβ fibrils are generated from the amyloid precursor protein (APP) expressed on neurons and cleaved to $A\beta_{1\text{-}40}$ and $A\beta_{1\text{-}42}$ peptides, which aggregate, particularly the latter, to form extracellular oligomers and the plaques characteristic of the disease. On the one hand, microglia may play a critical role in the removal of the misfolded Aβ fibrils and thus protect the neuron, or they may become activated by the presence of the amyloid fibrils and secrete mediators that contribute to neurodegeneration. Receptors that have been implicated in the phagocytosis of Aβ fibrils include, among others, CD14, CD33, CD47, CD204/SCARA1, CR3, Trem2, and the Toll-like receptors (32 and references therein). Engaging these receptors on microglia may trigger phagocytosis, or activation, or both. A significant problem for the interpretation of the details is that many studies on interactions between microglia and Aβ peptides are carried out *in vitro* when the cells no longer retain their microglia phenotype, as noted above. Often the concentrations of Aβ used do not mimic those *in vivo* in the human CNS, and furthermore, the Aβ is rarely if ever added to the microglia in a slowly rising concentration, such as would occur in a chronic neurodegenerative disease. It is hardly surprising that the rapid addition of Aβ peptides to cultured macrophages/microglia, a step function change in the concentration, leads to activation of the cells.

In vivo animal models generated to mimic aspects of AD pathology have relied on the generation of transgenic mice that significantly overexpress mutated forms of the APP found in humans with early-onset familial AD (APP transgenics) or mutated forms of enzymes, such as β-site amyloid precursor-cleaving enzyme (BACE) and PS1, involved in the cleavage of the Aβ peptides from APP (34). In these mice, the Aβ accumulates in plaques appearing at different ages depending on the particular transgenic construct. Associated with the raised Aβ and formation of plaques, the animals develop cognitive deficits, abnormal synaptic signaling, and, in the longer term, synaptic loss. In these transgenic mice, the microglia have an activated morphology, particularly in the region of the plaques, and express many of the receptors that would be expected to phagocytose the Aβ. Despite the activated morphology of these microglia, they do not have an overtly proinflammatory or anti-inflammatory phenotype but rather exist as cells expressing a broad spectrum of both pro- and anti-inflammatory molecules.

A direct route to study the role of microglia in the phagocytosis of Aβ plaques would be to deplete the microglia and follow the changes in plaques. Injection of ganciclovir into the cortex of APP transgenic mice crossed to CD11b-HSVTK mice led to the nearly complete ablation of the microglia in a focal region of the cortex; despite the absence of the microglia, the plaque size or number did not appear to have increased at a greater rate, suggesting that the microglia are not critical in plaque dynamics (35). Two-photon imaging experiments in APP transgenics crossed to mice with fluorescently labeled microglia have not resolved the role of microglia in Aβ clearance.

It appears that the microglia are indeed relatively reluctant to remove these amyloid plaques despite the expression of relevant phagocytic receptors. A number of approaches have been used to enhance the phagocytosis of amyloid by microglia. A simple but apparently effective approach is to inject endotoxin directly into the brain parenchyma, which will induce a proinflammatory and phagocytic phenotype in the microglia. The Aβ was removed, although the core of the plaques stained with Congo red persisted. Other approaches to enhance amyloid clearance include promoting the invasion of the CNS by monocyte-derived macrophages. In APP transgenics, the increase in the number of microglia or plaque-associated macrophages depends not on the invasion of the brain parenchyma by monocytes but rather on proliferation of microglia (12). When bone marrow chimeras are generated with whole-body irradiation and no shielding of the head from irradiation, monocyte-derived macrophages invade the brain and have been reported to aid in the clearance of plaques (36). If, as has been argued, the limited clearance of Aβ is because the microglia are not sufficiently active, then ameliorating endogenous inhibition of the microglia by deletion of interleukin-10 (IL-10) would be expected to reduce amyloid deposition and in turn mitigate cognitive dysfunction, as has been reported (37).

An important and interesting experiment, with regard to the role of the immune system in the regulation of amyloid deposition in the brain, was the demonstration that immunization of mice against Aβ leads to clearance of the amyloid plaques and an improvement in cognitive function (38). The systemic immunization generates antibodies to Aβ, and small amounts of the antibody gain access to the brain, where they decorate the amyloid and enhance its removal by the microglia. The rapid translation of this approach to the clinical setting and the demonstration that active and passive immunization can lead to amyloid clearance have generated considerable interest in this approach as a potential lead therapy in AD. The clinical trials to date have not yet yielded the improvements in cognitive function seen in animal models, but this has been attributed to the fact that the majority of trials to date have been initiated in patients with relatively advanced disease (39). The removal of the amyloid is not going to reverse the damage to the neuronal circuits. Other issues also remain to be resolved, such as how the antibody's access across the blood-brain barrier to the CNS can be improved, how the antibody interacts with the microglia without leading to microglia activation through Fc receptors, and whether careful design of antibody targets can circumvent the development of the unwanted imaging abnormalities that result with higher doses of antibody.

A major feature of all the neurodegenerative diseases of the CNS is the slow progressive degeneration and loss of neurons, and yet this is not recapitulated in most if not all of the APP transgenics. To investigate the microglial reaction to neurodegeneration, we have studied a mouse model of prion disease—a fatal and progressive degenerative disease that mimics the human disease, which is associated with the accumulation of a misfolded disease-related protein, PrP^{Sc}, derived from the normal cellular PrP^{C} expressed on neurons and other cells. In murine prion disease, there are numerous activated microglia, generated by the proliferation of the local resident cells, driven by CSF-1 and IL-34 (40). As the disease progresses, the activated microglia become more widespread, associated with the spread of the neurodegeneration. These activated microglia are associated with the expression of both pro- and anti-inflammatory mediators, of which some of the most abundant are transforming growth factor β, PGE_2, CCL2, and CSF-1. Although it has been unclear how this phenotype might contribute to or protect from disease progression, recent studies show that inhibition of signaling through the CSF-1R kinase pathway leads to a reduction in neuronal degeneration, delay in the onset of behavioral deficits, and extension of life (40). It appears that microglia are actively contributing to disease progression but by pathways that remain to be elucidated.

A significant difference between all of the animal models of chronic neurodegenerative disease and human disease is that the animals are kept in a tightly controlled, pathogen-free environment and are not exposed to the many factors that lead to systemic comorbidities typically found in aged humans. Old age is the single largest risk factor for the common neurodegenerative conditions such as AD and Parkinson's disease. Many of the risk factors now associated with AD, for example, smoking, obesity, and type 2 diabetes, are also associated with systemic inflammation (41). We thus explored how a systemic inflammatory challenge might impact on the microglia in the brains of mice with prion disease. A systemic challenge with lipopolysaccharide or poly(I:C), to mimic aspects of a systemic infection, switched the microglia to a more aggressive proinflammatory phenotype with the generation of potentially tissue-damaging cytokines. The systemic inflammatory challenges induced significantly exacerbated symptoms of sickness when compared to naive animals and accelerated the onset of behavioral deficits associated with disease progression (42). It appears that the microglia in the diseased brain are both more numerous and more sensitive to a secondary stimulus arising from the systemic challenge, and we have proposed that the microglia are "primed" by the ongoing neurodegeneration in a manner akin to that described for gamma interferon priming of macrophages *in vitro* (43). In the brains of animals with prion disease, although there are small numbers of recruited T cells, there is no detectable gamma interferon. The most likely candidate for priming the microglia is the significantly raised level of CSF-1 (44). The phenomenon of enhanced microglia responsiveness to systemic inflammation, or priming, is not confined to prion disease but has now been reported in numerous other animal models of both acute and chronic neurodegeneration and in the brains of aged animals (43 and references therein).

It is well known in the clinic that an acute infection may elicit the symptoms of delirium, acute mental deterioration, impaired cognition, and attention deficits in elderly individuals or those with underlying CNS disease (45). The experiments described above provide an explanation as to how primed microglia and the cytokines they secrete contribute to this condition. However, the results also suggest that systemic inflammation may contribute to chronic disease progression. We have studied a cohort of patients with AD and followed these patients for 6 months with a careful record of systemic events that may trigger a systemic inflammatory response and assessed blood markers for underlying chronic systemic inflammation. The patients with raised levels of tumor necrosis factor α (TNF-α) and one or more systemic infections in the 6-month period had a more rapid cognitive decline that those with neither (46). The patients with raised levels of TNF-α also experienced more depression, apathy, and anxiety—all symptoms of sickness behavior—than those with low levels of TNF-α. The data support the idea that systemic inflammation may drive disease progression via signaling of primed microglia and offer a route to modifying disease progression using reagents that target systemic inflammation.

SUMMARY

Since the first recognition in the 1980s (2) that microglia express myeloid-restricted molecules and are members of the myeloid lineage, the field of microglia biology has advanced rapidly. The microglia are involved in immune system-to-brain communication, illustrating the key role that these cells play in homeostasis following systemic disease in those with a healthy brain; however, this signaling by microglia in response to systemic disease may be maladaptive in individuals with a diseased brain and primed microglia. The active role of microglia in sculpting neuronal connectivity in the developing brain opens new avenues for investigation into the role of microglia in diverse neurodevelopmental disorders. The role of the microglia and other macrophages in the brain in chronic inflammatory and degenerative disease of the brain has come to the fore and again offers new opportunities for the development of interventions that may influence the onset and progression of these devastating diseases.

Acknowledgments. *Experimental work in the author's laboratory is supported by the Medical Research Council (UK), the Wellcome Trust, and Alzheimer's Research UK (ARUK). Thanks to Sallome Murinello for the image in Fig. 1 and Diego Gomez-Nicola for the images in Fig. 3.*

Citation. Perry VH. 2016. Microglia. Microbiol Spectrum 4(3):MCHD-0003-2015.

References

1. **Ransohoff RM, Perry VH.** 2009. Microglial physiology: unique stimuli, specialized responses. *Annu Rev Immunol* **27:**119–145.
2. **Perry VH, Hume DA, Gordon S.** 1985. Immunohistochemical localization of macrophages and microglia in the adult and developing mouse brain. *Neuroscience* **15:** 313–326.
3. **Ginhoux F, Greter M, Leboeuf M, Nandi S, See P, Gokhan S, Mehler MF, Conway SJ, Ng LG, Stanley ER, Samokhvalov IM, Merad M.** 2010. Fate mapping analysis reveals that adult microglia derive from primitive macrophages. *Science* **330:**841–845.
4. **Lawson LJ, Perry VH, Gordon S.** 1992. Turnover of resident microglia in the normal adult mouse brain. *Neuroscience* **48:**405–415.
5. **Mittelbronn M, Dietz K, Schluesener HJ, Meyermann R.** 2001. Local distribution of microglia in the normal adult human central nervous system differs by up to one order of magnitude. *Acta Neuropathol* **101:**249–255.
6. **Nimmerjahn A, Kirchhoff F, Helmchen F.** 2005. Resting microglial cells are highly dynamic surveillants of brain parenchyma in vivo. *Science* **308:**1314–1318.
7. **Hefendehl JK, Neher JJ, Sühs RB, Kohsaka S, Skodras A, Jucker M.** 2014. Homeostatic and injury-induced microglia behavior in the aging brain. *Aging Cell* **13:**60–69.
8. **Hickman SE, Kingery ND, Ohsumi TK, Borowsky ML, Wang LC, Means TK, El Khoury J.** 2013. The microglial sensome revealed by direct RNA sequencing. *Nat Neurosci* **16:**1896–1905.
9. **Butovsky O, Jedrychowski MP, Moore CS, Cialic R, Lanser AJ, Gabriely G, Koeglsperger T, Dake B, Wu PM, Doykan CE, Fanek Z, Liu L, Chen Z, Rothstein JD, Ransohoff RM, Gygi SP, Antel JP, Weiner HL.** 2014. Identification of a unique TGF-β-dependent molecular and functional signature in microglia. *Nat Neurosci* **17:** 131–143.
10. **Kettenmann H, Hanisch UK, Noda M, Verkhratsky A.** 2011. Physiology of microglia. *Physiol Rev* **91:**461–553.
11. **Lawson LJ, Perry VH, Gordon S.** 1992. Turnover of resident microglia in the normal adult mouse brain. *Neuroscience* **48:**405–415.
12. **Mildner A, Schlevogt B, Kierdorf K, Böttcher C, Erny D, Kummer MP, Quinn M, Brück W, Bechmann I, Heneka MT, Priller J, Prinz M.** 2011. Distinct and non-redundant roles of microglia and myeloid subsets in mouse models of Alzheimer's disease. *J Neurosci* **31:**11159–11171.
13. **Tremblay MÈ, Lowery RL, Majewska AK.** 2010. Microglial interactions with synapses are modulated by visual experience. *PLoS Biol* **8:**e1000527. doi:10.1371/journal.pbio.1000527.
14. **Sogn CJ, Puchades M, Gundersen V.** 2013. Rare contacts between synapses and microglial processes containing high levels of Iba1 and actin—a postembedding immunogold study in the healthy rat brain. *Eur J Neurosci* **38:** 2030–2040.
15. **Satoh J, Tabunoki H, Ishida T, Yagishita S, Jinnai K, Futamura N, Kobayashi M, Toyoshima I, Yoshioka T, Enomoto K, Arai N, Arima K.** 2011. Immunohistochemical characterization of microglia in Nasu-Hakola disease brains. *Neuropathology* **31:**363–375.
16. **Nicholson AM, Baker MC, Finch NA, Rutherford NJ, Wider C, Graff-Radford NR, Nelson PT, Clark HB, Wszolek ZK, Dickson DW, Knopman DS, Rademakers R.** 2013. *CSF1R* mutations link POLD and HDLS as a single disease entity. *Neurology* **80:**1033–1040.
17. **Krämer-Albers EM, Bretz N, Tenzer S, Winterstein C, Möbius W, Berger H, Nave KA, Schild H, Trotter J.** 2007. Oligodendrocytes secrete exosomes containing major myelin and stress-protective proteins: trophic support for axons? *Proteomics Clin Appl* **1:**1446–1461.
18. **Streit WJ, Xue QS.** 2014. Human CNS immune senescence and neurodegeneration. *Curr Opin Immunol* **29:** 93–96.
19. **Dantzer R, O'Connor JC, Freund GG, Johnson RW, Kelley KW.** 2008. From inflammation to sickness and depression: when the immune system subjugates the brain. *Nat Rev Neurosci* **9:**46–56.
20. **Tracey KJ.** 2007. Physiology and immunology of the cholinergic antiinflammatory pathway. *J Clin Invest* **117:** 289–296.
21. **Lacroix S, Feinstein D, Rivest S.** 1998. The bacterial endotoxin lipopolysaccharide has the ability to target the brain in upregulating its membrane CD14 receptor within specific cellular populations. *Brain Pathol* **8:**625–640.

22. **Teeling JL, Perry VH.** 2009. Systemic infection and inflammation in acute CNS injury and chronic neurodegeneration: underlying mechanisms. *Neuroscience* **158:** 1062–1073.

23. **Hannestad J, Gallezot JD, Schafbauer T, Lim K, Kloczynski T, Morris ED, Carson RE, Ding YS, Cosgrove KP.** 2012. Endotoxin-induced systemic inflammation activates microglia: [^{11}C]PBR28 positron emission tomography in nonhuman primates. *Neuroimage* **63:**232–239.

24. **Stevens B, Allen NJ, Vazquez LE, Howell GR, Christopherson KS, Nouri N, Micheva KD, Mehalow AK, Huberman AD, Stafford B, Sher A, Litke AM, Lambris JD, Smith SJ, John SW, Barres BA.** 2007. The classical complement cascade mediates CNS synapse elimination. *Cell* **131:**1164–1178.

25. **Paolicelli RC, Bolasco G, Pagani F, Maggi L, Scianni M, Panzanelli P, Giustetto M, Ferreira TA, Guiducci E, Dumas L, Ragozzino D, Gross CT.** 2011. Synaptic pruning by microglia is necessary for normal brain development. *Science* **333:**1456–1458.

26. **Kreutzberg GW.** 1996. Microglia: a sensor for pathological events in the CNS. *Trends Neurosci* **19:**312–318.

27. **Sawcer S, Franklin RJ, Ban M.** 2014. Multiple sclerosis genetics. *Lancet Neurol* **13:**700–709.

28. **Hickey WF, Kimura H.** 1988. Perivascular microglial cells of the CNS are bone marrow-derived and present antigen in vivo. *Science* **239:**290–292.

29. **Huitinga I, van Rooijen N, de Groot CJ, Uitdehaag BM, Dijkstra CD.** 1990. Suppression of experimental allergic encephalomyelitis in Lewis rats after elimination of macrophages. *J Exp Med* **172:**1025–1033.

30. **Ajami B, Bennett JL, Krieger C, McNagny KM, Rossi FM.** 2011. Infiltrating monocytes trigger EAE progression, but do not contribute to the resident microglia pool. *Nat Neurosci* **14:**1142–1149.

31. **Yamasaki R, Lu H, Butovsky O, Ohno N, Rietsch AM, Cialic R, Wu PM, Doykan CE, Lin J, Cotleur AC, Kidd G, Zorlu MM, Sun N, Hu W, Liu L, Lee JC, Taylor SE, Uehlein L, Dixon D, Gu J, Floruta CM, Zhu M, Charo IF, Weiner HL, Ransohoff RM.** 2014. Differential roles of microglia and monocytes in the inflamed central nervous system. *J Exp Med* **211:**1533–1549.

32. **Heneka MT, Carson MJ, El Khoury J, Landreth GE, Brosseron F, Feinstein DL, Jacobs AH, Wyss-Coray T, Vitorica J, Ransohoff RM, Herrup K, Frautschy SA, Finsen B, Brown GC, Verkhratsky A, Yamanaka K, Koistinaho J, Latz E, Halle A, Petzold GC, Town T, Morgan D, Shinohara ML, Perry VH, Holmes C, Bazan NG, Brooks DJ, Hunot S, Joseph B, Deigendesch N, Garaschuk O, Boddeke E, Dinarello CA, Breitner JC, Cole GM, Golenbock DT, Kummer MP.** 2015. Neuroinflammation in Alzheimer's disease. *Lancet Neurol* **14:** 388–405.

33. **Karch CM, Goate AM.** 2015. Alzheimer's disease risk genes and mechanisms of disease pathogenesis. *Biol Psychiatry* **77:**43–51.

34. **Webster SJ, Bachstetter AD, Nelson PT, Schmitt FA, Van Eldik LJ.** 2014. Using mice to model Alzheimer's dementia: an overview of the clinical disease and the preclinical behavioral changes in 10 mouse models. *Front Genet* **5:** 88. doi:10.3389/fgene.2014.00088.

35. **Grathwohl SA, Kälin RE, Bolmont T, Prokop S, Winkelmann G, Kaeser SA, Odenthal J, Radde R, Eldh T, Gandy S, Aguzzi A, Staufenbiel M, Mathews PM, Wolburg H, Heppner FL, Jucker M.** 2009. Formation and maintenance of Alzheimer's disease β-amyloid plaques in the absence of microglia. *Nat Neurosci* **12:**1361–1363.

36. **Simard AR, Soulet D, Gowing G, Julien JP, Rivest S.** 2006. Bone marrow-derived microglia play a critical role in restricting senile plaque formation in Alzheimer's disease. *Neuron* **49:**489–502.

37. **Guillot-Sestier MV, Doty KR, Gate D, Rodriguez J Jr, Leung BP, Rezai-Zadeh K, Town T.** 2015. *Il10* deficiency rebalances innate immunity to mitigate Alzheimer-like pathology. *Neuron* **85:**534–548.

38. **Schenk D, Barbour R, Dunn W, Gordon G, Grajeda H, Guido T, Hu K, Huang J, Johnson-Wood K, Khan K, Kholodenko D, Lee M, Liao Z, Lieberburg I, Motter R, Mutter L, Soriano F, Shopp G, Vasquez N, Vandevert C, Walker S, Wogulis M, Yednock T, Games D, Seubert P.** 1999. Immunization with amyloid-β attenuates Alzheimer-disease-like pathology in the PDAPP mouse. *Nature* **400:** 173–177.

39. **Karran E, Hardy J.** 2014. A critique of the drug discovery and phase 3 clinical programs targeting the amyloid hypothesis for Alzheimer disease. *Ann Neurol* **76:** 185–205.

40. **Gómez-Nicola D, Fransen NL, Suzzi S, Perry VH.** 2013. Regulation of microglial proliferation during chronic neurodegeneration. *J Neurosci* **33:**2481–2493.

41. **Deckers K, van Boxtel MP, Schiepers OJ, de Vugt M, Muñoz Sánchez JL, Anstey KJ, Brayne C, Dartigues JF, Engedal K, Kivipelto M, Ritchie K, Starr JM, Yaffe K, Irving K, Verhey FR, Köhler S.** 2015. Target risk factors for dementia prevention: a systematic review and Delphi consensus study on the evidence from observational studies. *Int J Geriatr Psychiatry* **30:**234–246.

42. **Cunningham C, Campion S, Lunnon K, Murray CL, Woods JF, Deacon RM, Rawlins JN, Perry VH.** 2009. Systemic inflammation induces acute behavioral and cognitive changes and accelerates neurodegenerative disease. *Biol Psychiatry* **65:**304–312.

43. **Perry VH, Holmes C.** 2014. Microglial priming in neurodegenerative disease. *Nat Rev Neurol* **10:**217–224.

44. **Paniagua RT, Chang A, Mariano MM, Stein EA, Wang Q, Lindstrom TM, Sharpe O, Roscow C, Ho PP, Lee DM, Robinson WH.** 2010. c-Fms-mediated differentiation and priming of monocyte lineage cells play a central role in autoimmune arthritis. *Arthritis Res Ther* **12:**R32. doi:10.1186/ar2940.

45. **Maclullich AM, Anand A, Davis DH, Jackson T, Barugh AJ, Hall RJ, Ferguson KJ, Meagher DJ, Cunningham C.** 2013. New horizons in the pathogenesis, assessment and management of delirium. *Age Ageing* **42:**667–674.

46. **Holmes C, Cunningham C, Zotova E, Woolford J, Dean C, Kerr S, Culliford D, Perry VH.** 2009. Systemic inflammation and disease progression in Alzheimer disease. *Neurology* **73:**768–774.

Myeloid Cells in Health and Disease: A Synthesis
Edited by Siamon Gordon

doi:10.1128/microbiolspec.MCHD-0011-2015

Deborah Veis Novack[1,2]
Gabriel Mbalaviele[1]

13 Osteoclasts—Key Players in Skeletal Health and Disease

INTRODUCTION

Although bone is one of the hardest tissues in the body, necessary for its structural and protective roles, this organ is not static. Bone matrix must be renewed over time in order to maintain its mechanical properties, and myeloid lineage cells called osteoclasts (OCs) are the specialized cells that perform this critical function. Since bone is the major storage site for calcium, OCs play an important role in the regulation of this signaling ion by releasing it from bone. In this process, OCs respond indirectly to calcium-regulating hormones such as parathyroid hormone and $1,25(OH)_2$ vitamin D_3. Growth factors such as insulin-like growth factor-1 (IGF-1) and transforming growth factor β (TGF-β) are also incorporated into bone matrix and released by OCs, affecting the coupling of bone formation to bone resorption and potentially targeting other cells in the microenvironment, such as metastatic tumors. Lastly, OCs retain features of other myeloid cells, such as antigen presentation and cytokine production, which afford them the potential to affect immune responses. Thus, the OC plays many roles in health and disease.

OCs are multinucleated cells formed by fusion of myeloid precursors. In normal circumstances, OCs are only found on bone surfaces, although cells with similar features can be found in association with some tumors, even outside of the bone. In contrast with the multinucleated giant cells found in granulomas such as in sarcoidosis, tuberculosis, or foreign body responses, OCs express tartrate-resistant acid phosphatase, the vitronectin receptor, $\alpha_v\beta_3$ integrin, and the calcitonin receptor. These polykaryons have a unique cytoskeletal organization and membrane polarization that allows them to isolate the bone-apposed extracellular space where bone resorption occurs. Not surprisingly, OC differentiation and function are tightly regulated and rely on many signaling pathways important to other immune cells, both innate and adaptive, providing both challenges and opportunities for their therapeutic targeting in disease states.

In this chapter, we will discuss the differentiation of OCs, a process known as osteoclastogenesis, and the relationship of OC precursors to other myeloid lineages. We will also describe their unique functional features, providing a glimpse into their unusual cell biology and interactions with other cells. Finally, we will address the pathophysiology of several diseases associated with bone loss, focusing on several key extracellular mediators dysregulated in osteolytic conditions.

[1]Musculoskeletal Research Center, Division of Bone and Mineral Diseases, Department of Medicine; [2]Department of Pathology and Immunology, Washington University School of Medicine, St. Louis, MO 63110.

OC PRECURSORS

Hematopoietic stem cells (HSCs) differentiate into all blood cell lineages (1), including the mononuclear phagocyte system from which OCs arise. Consistent with the conventional view that bone marrow niches are the primary sites of postnatal hematopoiesis (1), medullary fractions from humans and animal models are excellent sources of OC precursors. Indeed, unfractionated marrow leukocytes (2, 3, 4) or macrophages expanded *in vitro* in the presence of macrophage colony-stimulating factor (M-CSF) (5, 6) are commonly used to induce *in vitro* osteoclastogenesis under the influence of appropriate cues, especially M-CSF and receptor activator of NF-κB ligand (RANKL), as described below. OCs can also be differentiated from peripheral blood (7–9), spleen, and fetal liver (8, 10, 11). While early committed OC precursors express tartrate-resistant acid phosphatase (12) and $\alpha_v\beta_5$ integrin (13), late precursors and fully differentiated OCs upregulate cathepsin K (14, 15), calcitonin receptor (16), OC-associated receptor (OSCAR) (17), and $\alpha_v\beta_3$ integrin (18, 19).

Bone marrow macrophages expanded in M-CSF for several days are technically easy to obtain, the yields are quite high (up to 50 million from one adult mouse), and the extent of OC formation reflects the intrinsic differentiation potential of a relatively uniform population of precursors. However, in some disease states or abnormal/mutant genotypes, there are changes in the frequency of precursors rather than in the response of individual cells to differentiation cues (20). Use of unfractionated marrow cells rather than expanded macrophages can reveal alterations in early precursors and also prevents loss of effects due to the microenvironment *in vivo*, such as those stemming from hormonal effects or stromal cell support.

Much effort has gone into identifying specific precursor populations, but like many myeloid cell types, the OC precursor has defied a single definition. Hematopoietic cells from various developmental stages, including $CD34^+$ HSCs (21, 22), CD265 $(RANK)^+$ monocytes (23), and $CD14^+$ monocytes (9, 24, 25), possess the ability to undergo osteoclastogenesis. M-CSF-expanded marrow macrophages express CD11b (α_M integrin), CD265, and CD115 (M-CSF receptor/c-Fms) (8). The critical role of these last two receptors for OC differentiation will be discussed in detail below. The utility of CD11b as a marker of OC precursors and its role in osteoclastogenesis are more complex. Some groups have found that precursors expressing CD11b have high osteoclastogenic potential (8, 26), while others claim that the $CD11b^{lo/neg}$ population is more efficient (27, 28). Jacquin et al. found that $CD11b^{hi}$ bone marrow fractions generate OCs more slowly than a $CD3^-B220^-$ $CD11b^-$ fraction but with a similar efficiency (29). Conflicting results may come from differences in experimental techniques such as the time allowed for differentiation, the plasticity of the precursor populations, and the dynamic regulation of CD11b expression. Indeed, CD11b is upregulated by M-CSF and subsequently downregulated by RANKL (23, 29, 30). A recent study of CD11b-deficient mice showed that this molecule acts early as a negative regulator of OC differentiation and its engagement favors an inflammatory macrophage phenotype (31).

Several groups have tried to further refine the identity of OC precursors, with the expectation of improving cellular homogeneity and differentiation efficiency. CD117 (c-Kit) is expressed on all early hematopoietic progenitors, and cells expressing both CD115 and CD117 form OCs readily (29). This population can also form phagocytic macrophages and dendritic cells (28). Ly6C, a component of the Gr1 epitope found on monocytic myeloid-derived suppressor cells (MDSCs), was found on an OC precursor pool present in normal bone marrow but amplified in inflammatory arthritis (27). However, the population identified in this study remained heterogeneous for other markers such as CD11b and CD117. Others have shown that OCs can differentiate from $Gr1^+CD11b^+$ MDSCs in the context of cancer (32–34). Besides CD markers, OC precursors also display several additional surface molecules at some point during differentiation, including dendritic cell-specific transmembrane protein (DC-STAMP) (35), OC-STAMP (36), calcium-dependent cell-cell adhesion molecules, E-cadherin (37, 38) and protocadherin-7 (39), and a disintegrin and metalloproteinase (ADAM) family members, mainly ADAM8 (40). None of these is completely restricted to the OC lineage.

OC DIFFERENTIATION

The presence in the bloodstream of cells capable of OC differentiation is well documented (9, 21, 22, 25). However, it is still unclear whether bone marrow-derived myeloid cells in the adult bone marrow directly differentiate into OCs or enter the bloodstream before reentering the bone microenvironment to form OCs. It is possible that, rather than providing a source of homeostatic OCs for bone, circulating OC precursors represent a pool that can be recruited in pathological states such as inflammatory arthritis or heterotopic ossification. Irrespective of the potential systemic circulation of OC precursors, the conceptual framework is that within the bone microenvironment, OC precursors are generated in or near perivascular HSC niches and migrate toward the bone surface, where they terminally mature into OCs (Fig. 1). According to this concept, OC precursors may be lured to bone surfaces by gradients generated by bone matrix proteins and degradation products, calcium ions, lipid mediators such as sphingosine-1-phosphate (41–43), or chemokines such as CXCL12, which are produced by the osteoblast lineage cells (44), the main regulators of OC differentiation in homeostatic states.

Even before the identification of specific factors governing osteoclastogenesis, it was clear that stromal, osteoblast lineage cells could provide support for this process (45, 46). The essential role of M-CSF in the proliferation and survival of the OC lineage was uncovered in the early 1980s following an observation that hematopoietic progenitors from *op/op* mice formed normal monocyte/macrophage colonies only when exposed to colony-stimulating activity derived from wild-type stromal cells (47). This activity was assigned to M-CSF a decade later when it was discovered that *op/op* mice failed to produce this growth factor due to a mutation in the *M-csf* gene (48). Osteopetrosis in *op/op* mice stems from inadequate OC lineage commitment of macrophages, which are also diminished

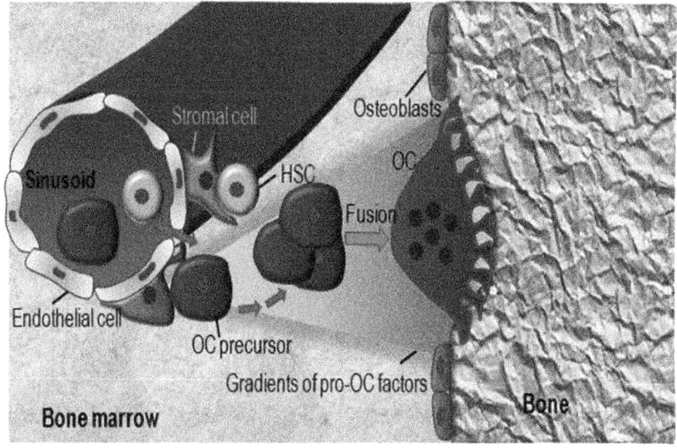

Figure 1 A model of OC differentiation. OCs differentiate from HSCs. The hematopoietic niche comprises endothelial cells and perivascular stromal cells, which exhibit mesenchymal stem cell (MSC) features. It is still unclear whether OC precursors directly differentiate into OCs or enter the bloodstream before reentering the bone microenvironment to form OCs. In any scenario, higher levels of chemoattractants toward bone surfaces, including bone ECM proteins, lipid mediators (e.g., sphingosine-1-phosphate), and ECM degradation products, create gradients that attract OC precursors to the hard tissue, where they fuse and complete the differentiation process. Conversely, higher levels of perivascular chemorepellents (not drawn for simplicity) may also contribute to the migration of OC precursors toward the endosteum.

in numbers (49). M-CSF is involved in the growth and survival of the OC lineage through regulation of numerous pathways such as the mitogen-activated protein kinases (MAPK), extracellular signal-regulated kinase (ERK), and Jun N-terminal kinase (JNK) (50), glycogen synthase kinase-3β/β-catenin (51), and mammalian target of rapamycin (mTOR), whose signaling intermediates include phosphatidylinositol 3-kinase (PI3K) and Akt (52). Moreover, M-CSF was more recently found to increase diacylglycerol levels through stimulation of phospholipase C-γ (PLC-γ), an event that causes c-Fos activation (53), thus suggesting that this growth factor can also induce OC differentiation signals. The M-CSF receptor c-Fms (CD115) can also be activated by newly identified interleukin-34 (IL-34) (54, 55), and mice lacking *c-Fms* show a more severe osteopetrotic phenotype compared to those with impaired M-CSF production (56). M-CSF, as well as IL-34, promotes the expression of RANK (CD265), whose downstream signaling pathways also drive differentiation (see below). Other factors such as granulocyte-CSF (57), vascular endothelial growth factor (58, 59), and hepatocyte growth factor (60) can also provide mitogenic and survival signals to the OC lineage, though with less efficiency compared to M-CSF.

RANKL is another essential factor required for the differentiation, survival, and function of OCs. Its discovery 2 decades ago was a breakthrough that reinvigorated the field of OC biology (61, 62). The findings that transgenic mice expressing high levels of RANKL develop osteoporosis whereas mice lacking RANKL or the decoy receptor osteoprotegerin (OPG) are osteopetrotic (63) and osteoporotic (64), respectively, underscore the central role of this pathway in OC development. Furthermore, rare cases of osteopetrosis have been ascribed to inactivation or deficiency of RANKL or RANK, its receptor (65). Deactivating mutations in OPG cause the high-turnover bone disorder known as juvenile Paget's disease, and activating mutations of RANK lead to a range of osteolytic conditions, including familial expansile osteolysis (66).

Monocytes expressing RANK, a tumor necrosis factor (TNF) receptor family member, exposed to RANKL exit the cell cycle and fuse to form multinucleated OCs. This process involves activation of various pathways including MAPKs (p38, JNK, ERK), mTOR, PI3K, NF-κB (canonical and alternative), and microphthalmia-associated transcription factor (MITF) (52, 67–69). Many of these pathways contribute to expression of nuclear factor of activated T cells, cytoplasmic 1 (NFATc1), the "master" osteoclastogenic transcription factor (70), whose activation also depends on a PLC-

$\gamma 2/Ca^{2+}$ pathway (71). As it is emerging for other signaling pathways, RANK signaling induces protein posttranslational modifications such as phosphorylation (71), ubiquitination (72), SUMOylation (73), and poly-ADP-ribosylation (our unpublished data). Developing evidence also indicates a crucial role of epigenetic mechanisms in RANKL control of OC differentiation (74, 75). Indeed, RANKL regulates the expression of several species of microRNA (76). DNA methyltransferase 3a is a DNA-modifying enzyme with profound effects on bone homeostasis, as mice lacking this protein develop a high bone mass phenotype due to defective osteoclastogenesis (77). RANKL also induces histone 3 (H3) lysine 4 trimethylation (H3K4me3, a mark of transcriptionally active chromatin) while attenuating H3K27me3 (a mark of silent chromatin) near the transcription start site of several genes encoding OC transcription factors, such as NFATc1 and NF-κB (78). Pharmacological interference with the recruitment of CBP to acetylated histones at the NFATc1 promoter impairs RANKL's osteoclastogenic effects (79). Accordingly, the deacetylase SIRT1, which removes acetyl groups from proteins, including histones, also modulates RANKL-induced OC formation (80, 81). Thus, RANKL regulation of OC development involves a plethora of signaling networks, including epigenetic mechanisms.

A third signal reported to control osteoclastogenesis emanates from immunoreceptor tyrosine-based activation motif (ITAM)-harboring adaptors such as DNAX activation protein of 12 kDa (DAP12) (82) and Fc receptor γ (FcRγ). These cell surface molecules have very small extracellular domains and interact with coreceptors TREM2 and Sirpβ1 (DAP12) or OSCAR and PIR-A (FcRγ) in order to signal to Ca^{2+}/NFATc1 through PLC-γ2 (83, 84). In mice, deletion of either ITAM protein has no effect on bone mass, while deletion of both causes severe osteopetrosis. While Mócsai et al. (83) showed normal expression of a number of markers of mature OCs (β_3 integrin, calcitonin receptor, cathepsin K), suggesting intact differentiation, Koga et al. (84) found a failure to induce Ca^{2+} oscillations and NFATc1 induction, and thus poor differentiation, in DAP12/FcRγ double-deficient cells. Complicating interpretation of the phenotype, DAP12/FcRγ-deficient mice lose bone very efficiently in response to ovariectomy in the long bones but not vertebrae (85). RBP-J, a transcription factor usually associated with Notch signaling, was shown to impose a requirement for ITAM-mediated costimulation on RANKL-induced osteoclastogenesis; in the absence of RBP-J, ITAM signaling was completely dispensable (86). Thus, a role for ITAM in OC differentiation may be site and/or context dependent. Other studies have indicated that the primary role of the ITAM-mediated signaling appears to be in the regulation of the OC cytoskeleton (discussed below), with little effect on differentiation (82, 87). Furthermore, loss of TREM2, the coreceptor for DAP12, increases osteoclastogenesis, via β-catenin, suggesting a Ca^{2+}/NFATc1-independent effect (51). Thus, the role of ITAM signaling in OC differentiation remains incompletely resolved.

Beyond RANKL, M-CSF, and the ligands for ITAM-associated receptors, the normal bone marrow microenvironment contains a wide variety of OC-regulating molecules, including cytokines, growth factors, hormones, and the recently reported danger signals, which function via the inflammasomes. These molecules act through or interact with M-CSF and RANKL, the most proximal signals in OC differentiation, to fine-tune pro- and anti-osteoclastogenic inputs, thereby ensuring bone homeostasis. For example, although TNF-α is an important inflammatory cytokine (discussed in detail below), mice lacking TNF-α have higher bone mass at baseline, indicating a modulatory role in normal conditions as well (88). Thus, it is the imbalance in the levels or actions of osteoclastogenic factors that causes bone pathology, as discussed below.

RECIPROCAL REGULATION OF OCs AND IMMUNE CELLS

Although osteoblast lineage cells have the strongest influence on OC development in homeostatic conditions, several immune cell types can influence this process too. Indeed, B cells (89) and activated T cells express RANKL (90–92), but ablation of RANKL expression from these cells does not affect bone mass at baseline through 7 months of age (89). Thus, lymphocyte-produced RANKL, as well as other cytokines, is likely important only in pathological bone loss. Nevertheless, in basal conditions, mice that lack T cells or B cells are osteopenic, at least in part due to reduced OPG production by B cells (93–96), indicating that lymphocytes play a role in bone homeostasis, acting to protect bone. There are many different subsets of T cells, and those that appear to have the dominant role in protecting bone are regulatory T cells. These T cells, which act to reduce effector T-cell activity, differentiate from naive $CD4^+$ T cells (Tregs) or $CD8^+$ T cells (Tcregs), and express the transcription factor Forkhead box protein 3 (FoxP3) and CD25. The $CD4^+$ Tregs have been more extensively studied, and in many contexts they suppress OC formation and inflammatory osteolysis (97–100). The $CD8^+$ Tcregs also inhibit OC differentiation and bone resorption *in vitro* and *in vivo* in murine models (101–103). Moreover, forced expression of FoxP3 hinders OC differentiation (99), whereas loss of function of this protein causes severe osteoporosis, owing to the hypersensitivity of OC precursor pools to M-CSF (20). Thus, regulatory T cells may limit bone resorption by affecting OCs at multiple stages.

Interestingly, the relationship between OCs and T cells appears to be bidirectional. Indeed, OCs express major histocompatibility complex classes I and II, costimulatory molecules CD80 and CD86, and are capable of presenting antigens. Cross-presentation of antigens by OCs leads naive $CD8^+$ T cells to become Tcregs, which in turn inhibit bone resorption (102, 103). OCs can also activate alloreactive $CD4^+$ T cells, and $CD8^+$ T cells to some extent, and suppress the ability of $CD4^+$ T cells and $CD8^+$ T cells to produce TNF-α and gamma interferon (IFN-γ), respectively (104, 105). A recent study shows that the suppressive function of OCs is inducible as it is triggered by activated T cell-derived IFN-γ and CD40 ligand (106). Whether the immune-modulating function of OCs is important at baseline, or only in disease states, remains an open question that is difficult to answer since there does not appear to be a unique molecule responsible and OC-specific conditional knockouts have not yet been studied.

MECHANISMS OF BONE RESORPTION

As noted earlier, terminal OC differentiation occurs on the bone surface. OC adhesion to bone is mediated mainly by the $\alpha_v\beta_3$ integrin, which recognizes the amino acid motif Arg-Gly-Asp contained in certain bone extracellular matrix (ECM) proteins, including vitronectin, osteopontin, bone sialoprotein, and denatured collagen I. Although the α_v subunit is expressed throughout differentiation, the β_3 subunit is potently upregulated downstream of RANKL and serves as a marker of OC differentiation. Accordingly, mice lacking the β_3 subunit develop high bone mass late in life due to defective OC adhesion to bone matrix, diminishing their resorptive activity (107). The bone resorption phase starts with dramatic morphological changes in OCs, a process referred to as OC polarization whereby a segment of the bone-facing plasma membrane expands and becomes highly convoluted (and known as ruffled border) whereas the portion toward the marrow space, the basolateral membrane, is enriched in ion transporters (Fig. 2). The ruffled border is formed by fusion of secretory lysosomes with the plasma membrane in a process controlled by the autophagy machinery (108). This specialized membrane domain is surrounded by actin microfilaments and actin-binding proteins such as talin (109), vinculin (110), kindlin 3 (111), myosin IIA (112, 113), and paxillin (113) organized as podosomes, which cluster to form the actin ring, a dense cytoskeletal structure that excludes all organelles, also called the sealing zone (67). Loss of any of these actin-associated proteins in the podosome diminishes bone resorption, although they have distinct roles in organizing the cytoskeleton.

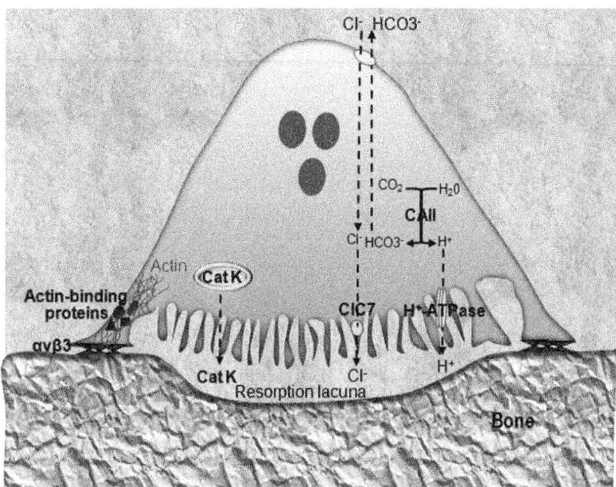

Figure 2 Key molecules involved in OC function. Loss of function of any of the depicted molecules causes osteopetrosis due to defective OC activity. OCs adhere to bone matrix proteins via integrin $\alpha_v\beta_3$ and are polarized such that the plasma membrane-facing bone is convoluted (ruffled) and contains the proton pump (v-ATPase) and Cl^- channel 7 (ClC7), whereas the basolateral membrane bears the HCO_3^-/Cl^- antiporter. Cytoplasmic carbonic anhydrase type II (CAII) generates the protons to be secreted into the resorption lacuna beneath the cell. This lacuna becomes isolated from the rest of the extracellular space by the tight adhesion of $\alpha_v\beta_3$ to the bone surface at the sealing zone. The cytoplasmic domain of β_3 recruits signaling proteins, which induce the association of actin with interacting partners (including talin, vinculin, kindlin, myosin IIA, and paxillin) and formation of an actin ring that defines the periphery of the ruffled membrane. Concerted action of ClC7 and v-ATPase produces a high concentration of HCl that acidifies the resorption lacuna, leading to the dissolution of the inorganic components of the bone matrix. Acidified cytoplasmic vesicles containing lysosomal enzymes such as cathepsin K (Cat K) are also transported toward the bone-apposed plasma membrane and, ultimately, the sealed resorption lacuna, where they digest the exposed matrix proteins.

During the process of cytoskeletal reorganization, activated β_3 recruits to its cytoplasmic domain a variety of signaling proteins, including tyrosine kinases c-Src and Syk, and guanine nucleotide exchange factors, primarily Vav3 (114), leading to activation of Rac and formation of the actin ring (115). Mice lacking c-Src, Syk, or Vav3 (114, 116, 117) exhibit higher bone mass due to defective resorption, which requires adhesion to bone, cell spreading, and actin ring formation (Fig. 2). ITAM signaling is required for these signaling events downstream of the integrin (118). Interestingly, however, the two ITAMs, DAP12 and FcRγ, do not work identically. FcRγ requires the presence of β_3 integrin to compensate for the absence of DAP12, while DAP12 can function with β_1 integrin to mediate cytoskeletal organization (118). OC cytoskeletal reorganization is also indirectly regulated by M-CSF, which through c-Fms activates β_3 integrin intracellularly, causing conformational changes to its extracellular ligand-binding region (inside-out activation). M-CSF stimulates phosphorylation of DAP12 by c-Src and its recruitment to the cytoplasmic tail of c-Fms, leading to activation of Syk and downstream signals to the cytoskeleton. This interaction with c-Fms is another example of differences between DAP12 and FcRγ, as the latter is not able to form this complex (82). In sum, $\alpha_v\beta_3$ integrin, c-Fms, and ITAM proteins work together to control the OC cytoskeleton, which mediates attachment to bone, migration along the surface, and formation of the actin ring to allow bone resorption.

OC polarization enables directed transport of acidified cytoplasmic vesicles toward the bone-apposed plasma membrane, i.e., at the ruffled border (119). The sealing zone creates an isolated extracellular microenvironment, enclosed by the cell and the bone surface, where bone resorption takes place in two steps. First, the acidification of the resorption lacuna leads to dissolution of the inorganic components of the bone matrix, including hydroxyapatite (119, 120). During this process, carbonic anhydrase type II generates protons and HCO_3^- by hydrating CO_2 (121). The protons are released into the resorptive compartment via an electrogenic proton pump mediated by vacuolar H^+-ATPase (v-H^+-ATPase), contained in vesicles that fuse with and expand the ruffled border (120, 122). The combined actions of the Cl^-/HCO_3^- anion exchanger and Cl^- channel (ClC-7) located on the basolateral membrane and ruffled membrane, respectively, ensure OC cytoplasmic pH neutrality as the Cl^- ions are released in the resorption lacuna. The most common forms of osteopetrosis in humans are caused by mutations in genes controlling acidification, including *TCIRG1*, *CLCN7*, and *OSTM1* (123), emphasizing the importance of this step. Other rare forms of osteopetrosis include mutations in PLEKHM1, which is involved is lysosomal trafficking (124, 125). HCl-mediated demineralization exposes the organic phase of the bone matrix, which is made up of ~95% type I collagen. The degradation of bone matrix proteins is carried out by secreted lysosomal enzymes (122), mainly the cysteine protease cathepsin K. Indeed, deficiency of cathepsin K in humans causes pycnodysostosis, a high-bone-mass disease that can be mimicked in mouse models (14, 15, 126). Additionally, there is evidence that neutral matrix metalloproteinases (MMPs), including MMP-2, -9, and -13, may also participate in bone resorption (127, 128).

Both the organic and inorganic degradation products from bone ECM are endocytosed by the ruffled membrane (129, 130), a trafficking process involving small GTPases and microtubules, and released from the basolateral membrane to the bloodstream. This process enables OCs to excrete degraded matrix components while digging deep into bone and maintaining an enclosed resorption site. Some of the collagen I degradation products, such as C-telopeptide of type I collagen (CTX-1), can be used as markers of bone resorption in patients (131) and animal models (4). The OC eventually detaches through yet unknown mechanisms, migrates, and reinitiates another bone matrix degradation cycle. One potential mechanism for sustained bone resorption may be provided by degradation products of the mineral phase functioning as crystalline danger-associated molecular patterns that activate pathways such as the inflammasomes, which are expressed by the OC lineage (4, 132–135), and may function as a positive feedback mechanism that amplifies bone resorption.

COUPLING OF BONE FORMATION TO BONE RESORPTION

In normal bone homeostasis, OCs remove old or damaged bone matrix and then osteoblasts replace the matrix to restore the bone to its original state in a process known as the bone remodeling cycle. Thus, it makes sense that OCs might send signals to osteoblasts to induce bone formation at sites of resorption. In fact, multiple molecules made or activated by OCs have the potential to enhance bone formation, acting on various stages of osteoblast differentiation. TGF-β and IGF-1 are both stored in bone matrix and released by its resorption by OCs, with the former acting to promote migration of mesenchymal stem cells to newly resorbed sites (136, 137). TGF-β also induces OCs to express CXCL16 and Wnt10b, which induce recruitment and mineralization by osteoblasts, respectively (138, 139). Sphingosine-1-phosphate (140) and collagen triple helix repeat containing 1 (CTHRC1) (141) have been proposed as possible secreted coupling factors that enhance bone formation. In contrast, semaphorin 4D produced by OCs appears to inhibit bone formation (142), and perhaps its role is analogous to that of OPG, which limits the OC-promoting actions of RANKL. An additional proposed pathway for OC-osteoblast coupling is via direct cell-cell interactions between ephrin B2 (on OCs) and EphB4 (on osteoblasts). These transmembrane proteins can signal bidirectionally and *in vitro* experiments show enhanced osteoblast activity (143), but the *in vivo* evidence for coupling is not strong (144). While more work is required to resolve the specific mechanisms involved, it is clear that OCs influence the differentiation and activity of osteoblasts *in vivo*.

PATHOLOGICAL BONE LOSS

Although other cells are able to secrete matrix-degrading enzymes, such as MMPs, it is generally accepted that OCs are the primary effectors of bone loss in osteolytic disease states. Furthermore, upregulation of the osteoclastogenic cytokine RANKL is a common feature in states of abnormal bone loss. Nevertheless, in most cases other cytokines and growth factors that interact directly with OCs or their precursors have powerful effects on OC differentiation and function in these conditions, with each disease featuring a different "cocktail" of factors. In fact, both pro- and anti-OC signals may be increased, such that the net effects on bone resorption represent the relative amounts of the various factors. Similarly, bone formation may be increased through the normal coupling mechanisms or decreased due to opposing effects of cytokines on osteoblasts versus OCs. Thus, although the mechanisms for bone loss are complex in most diseases, our discussion here will focus specifically on bone resorption.

By far the most common form of bone loss, osteoporosis, affects a large portion of the elderly population. Although the timing and progression of this disease differ in men and women, age-related changes occur in both sexes and can be largely attributed to effects of estrogen loss, as estrogen acts to maintain bone formation and limit resorption (145). None of the effects of androgen on bone mass are through direct effects on OCs (146). In osteoporosis, bone resorption outpaces formation, and drugs that inhibit OCs (such as bisphosphonates and denosumab, an anti-RANKL antibody) significantly reduce the risk of fracture in patients (147). Another feature of aging probably linked to low estrogen is an increase in chronic low-grade inflammation, so-called "inflamm-aging" (148, 149). This is characterized by a generalized increase in the number of myeloid cells and cytokine levels, but without overt tissue-specific inflammatory infiltrates. T cells may be an important source of osteoclastogenic cytokines, as T-cell production of TNF-α and RANKL is elevated in estrogen-deficient humans and mice (150, 151). Other age-related changes in the immune system, such as a failure of OCs to induce Tcregs in the context of estrogen loss, may also contribute to bone loss (102). As will be discussed below, inflammatory cytokines have potent effects on OCs, and although the rise in cytokines is likely mechanistically linked to estrogen loss, their effects on OCs occur via separate molecular pathways. Much of our current understanding of cellular and molecular mechanisms is based on models of ovariectomy in rodents, which causes acute loss of estrogen and rapid, sustained loss of bone mass, particularly at trabecular sites.

Autoimmune inflammatory diseases, most frequently rheumatoid arthritis, are associated with focal bone loss, and increased numbers of OCs are generally found at inflamed bone surfaces. Because inflammatory cytokines (especially TNF-α and IL-1) are often coregulated or induced in a cascade, activating both positive and negative feedback loops, there are many potential osteoclastogenic mediators present at the same time. Besides the cytokines with direct influence on OCs, others including IL-17, generated by Th17 cells, and IL-6, produced by many cell types, act mainly by increasing RANKL produced by local osteoblasts and other stromal cells such as synovial fibroblasts (152, 153). Many cytokine-blocking agents (biologics) are now in clinical use to treat inflammatory arthritic diseases and have shown efficacy in blocking the associated bone loss (152). Patients with rheumatoid arthritis also have a higher incidence of osteoporosis (154), and although anti-TNF-α therapies reduce generalized bone loss in these patients, recent studies suggest that the autoantibodies themselves may initiate bone loss prior to the onset of inflammation (155).

Several murine models of inflammatory arthritis have been useful in advancing our understanding of inflammatory bone loss, as they implicate the same cytokines (primarily RANKL, TNF-α, and IL-1) and cells (lymphocytes and myeloid cells). These include collagen-induced arthritis (156) and antigen-induced arthritis (157), both of which involve T-cell-mediated immune responses, recruitment of innate immune cells, and subsequent osteoclastic bone resorption localized to the joint (typically one knee) in which the antigen is injected. The serum transfer arthritis model (158) is a model of systemic antibody transfer and complement fixation that leads to inflammation and bone erosion in many peripheral joints, in a lymphocyte-independent manner. Antibodies to collagen II can also generate a similar form of arthritis (159). Another model that has been used extensively to study arthritis is the human TNF-α transgenic (hTNF-Tg) mouse (160, 161). In this purely cytokine-driven model, arthritis develops spontaneously and remains chronic for the life of the animal, which is a unique feature of this model. However, since cytokines often affect both the inflammatory component and the OCs, many experiments need to be interpreted with caution regarding the direct effects of particular pathways on the osteolytic component of disease. The clearest results regarding effects on OCs occur when bone loss can be evaluated in the presence of an intact inflammatory response (162–164). For example, NF-κB-inducing kinase (NIK)-deficient mice have reduced bone erosion in inflammatory arthritis models. Since they develop inflammation in the serum transfer model but not in the antigen-induced model, we can conclude that osteolysis is specifically impacted by the loss of NIK only from the serum transfer model. In the end, however, none of the animal models precisely replicates the human disease state, nor captures the heterogeneity of real patient populations.

Joint manifestations also occur in sterile autoinflammatory disorders (165), including neonatal-onset multisystem inflammatory disease (NOMID), the most severe form of cryopyrin-associated periodic syndromes. Cryopyrinopathy disorders are caused by activating mutations in *NLRP3*, resulting in systemic inflammation driven by IL-1β and IL-18 oversecretion (165). NOMID patients exhibit prominent skeletal malformations, including short stature and low bone mass (166). Abnormal IL-1 biosynthesis or signaling also causes joint complications in a variety of other autoinflammatory conditions, including TNF receptor-associated periodic syndrome, due to mutations in the *TNFRSF1A* gene, encoding the type 1 TNF receptor (167); familial Mediterranean fever, induced by mutated pyrin (168); systemic-onset juvenile idiopathic arthritis (169); and Blau syndrome, caused by *NOD2* mutations (170).

Another very common form of inflammatory bone loss is that associated with periodontitis, a bacteria-driven process. Despite differences in the immune effector cells between periodontitis and autoimmune diseases, the direct modulators of OCs are the same (RANKL, TNF-α, and IL-1) (171). Although periodontal bacteria may produce forms of lipopolysaccharide and other Toll-like receptor-activating ligands that could directly stimulate osteoclastogenesis (172), the strong induction of inflammatory cytokines is likely more important for osteolysis as cytokine blockade is effective, at least in animal models (173, 174).

Most types of solid tumors that readily metastasize to bone cause OC-mediated bone loss, also known as tumor-mediated osteolysis (175). Although there are complex interactions among tumor cells and the many cell types in the bone microenvironment that affect tumor growth, interactions with OCs have perhaps received the most attention because the clinical consequences of their activity, known as skeletal-related events (e.g., fracture and hypercalcemia), are very significant. Abundant data from animal models of osteolytic metastasis indicate that increased OC activity correlates with higher tumor burden (176, 177) while defects therein reduce it (178, 179). OC-inhibitory therapy with bisphosphonates or denosumab is now standard of care for many patients either at high risk or with clinically evident bone metastases, and these reduce skeletal-related events (180). Through a variety of mediators and interactions with osteoblast lineage cells, tumors increase local expression of RANKL and may decrease OPG as well (175). Other factors may include TGF-β, which is thought to be another mediator of osteolysis in the tumor microenvironment (discussed below). Additionally, tumors increase the number of MDSCs in the bone marrow, and in addition to their role in suppressing antitumor immune responses, these immature myeloid cells can serve as OC precursors (32–34).

Multiple myeloma, a highly osteolytic tumor of plasma B cells that originates in the bone marrow, secretes several factors that promote osteoclastogenesis, including Dickkopf-1 (DKK-1), macrophage inflammatory protein-1α (MIP-1α), and IL-6. DKK-1, which has inhibitory effects on osteoblasts, may also promote osteoclastogenesis through inhibition of β-catenin (181). MIP-1α can directly act on OC precursors to promote differentiation (182, 183). In contrast, the effects of IL-6 appear to be indirect, but its blockade may be useful in targeting tumor growth as well as osteolysis (184). Another interesting feature of multiple myeloma that has therapeutic implications is a commonality of signaling pathways in tumor cells and OCs. A current first-line therapy is proteasome inhibition such as with bortezomib, whose mechanisms of action include inhibition of NF-κB, p38, NFATc1, and TNF receptor-associated factor 6 (TRAF6), all pathways that impact osteoclastogenesis. In fact, studies have shown that these drugs have dual antitumor and antiresorptive activity (185–187). Recent studies have shown that activating mutations in the alternative NF-κB pathway are common in multiple myeloma and play a role in tumor survival (188, 189). Since this pathway is also important for OC formation and function (190–192), its inhibition could also represent a powerful therapeutic strategy (193).

MODULATORS OF OSTEOCLAST DIFFERENTIATION AND FUNCTION

In this section, we discuss several of the most important modulators of OCs involved in pathological bone loss. We have chosen only those that bind to receptors on OCs or their progenitors, and will not discuss those that are produced by other cells in the microenvironment, such as IL-17, IL-6, and oncostatin M, but act indirectly and induce expression of RANKL and the inflammatory mediators discussed below.

Estrogen

In vitro studies demonstrated that mature OCs express receptors for estrogen, especially ERα, and respond to this hormone by undergoing apoptosis (194, 195). One study using ERα conditional knockout mice showed that this is due to upregulation of FasL signaling (196). Further upstream, estrogen blocks differentiation of OC precursors in response to RANKL (197). It also affects the frequency of OC precursors, as more OCs can be generated from the bones of ovariectomized mice (198) and from peripheral blood of postmenopausal women (199) than from healthy controls. The relative importance of the estrogen effects on different stages of OC survival and differentiation is not clear. A recent three-dimensional analysis of *in vivo* resorption in ovariectomized rats showed an increase in the number of resorption sites (reflecting increased differentiation) with little change in the size of the lacunae (presumably a measure of survival of individual OCs), suggesting that perhaps early effects on differentiation that determine initiation are more important than those on OC life span (200).

Studies using a variety of mice with tissue-specific deletions of ERα demonstrate an interesting and perhaps unexpected difference in the mechanism by which estrogen controls resorption in the cortical and trabecular compartments in females. Deletion of ERα from OCs, either in progenitors using lysozyme M-Cre or more mature cells using cathepsin K-Cre, reveals bone loss only at trabecular sites but not in the cortex (196, 201). In contrast, deletion of ERα from osteoblast progenitors, with either Prx1-Cre or Osterix-Cre, leads to loss only of cortical bone (202) due to defects in bone formation and a secondary increase in endocortical resorption. The early and rapid bone loss seen at the onset of menopause is largely trabecular. Thus, despite the fact that ERα is deleted since conception in mouse models as opposed to menopausal estrogen decline in humans, these findings suggest that the direct inhibitory effects of estrogen on OCs are physiologically important (203). In addition to the direct effects of estrogen on bone cells, this hormone also acts as an immunosuppressive agent, reducing levels of TNF-α and IL-1β, whose osteoclastogenic actions are discussed below. Supporting these cytokines as significant mediators of bone loss *in vivo*, their removal or inhibition not only reduces osteoporosis in mouse models (204), but reduces bone turnover markers in women following acute estrogen withdrawal (205).

TNF-α

Even before the identification of RANKL as the "osteoclast differentiation factor," TNF-α was recognized for its ability to stimulate bone resorption *in vivo* and *in vitro* (206, 207). We now know that TNF-α promotes RANKL and M-CSF expression by osteoblast lineage cells (208, 209) and also acts directly on OC precursors (210–213). Since TNF-α and RANKL, and their respective receptors, are highly related, many of the signaling pathways activated by these two cytokines are very similar. Thus, it is not surprising that TNF-α can act with RANKL to promote OC differentiation (26, 214). However, the ability of TNF-α alone to induce osteoclastogenesis is quite low in most cases, due to subtle differences in signaling downstream of RANK and TNFR1. Indeed, RANK recruits TRAF3, allowing NIK-mediated processing of p100 to p52, thereby removing significant brakes on the alternative NF-κB signaling pathway (191). TNFR1 signaling increases expression of p100, but not its processing to p52, in wild-type cells. However, in the absence of TRAF3, TNF-α is able to induce OC formation (215). Another molecule that limits the ability of TNF-α to promote osteoclastogenesis is RBP-J. This transcription factor strongly suppresses differentiation downstream of TNF-α but not RANKL, and its removal allows robust osteolysis in response to TNF-α (216). In normal cells, small amounts of RANKL appear to counteract the negative regulation of TNF-α such that the action of the two cytokines is synergistic. However, in conditions in which RANKL is stringently excluded from the system, exposure of precursors to TNF-α prior to RANKL inhibits osteoclastogenesis (26).

IL-1

IL-1, present as α and β isoforms (from distinct gene products), is another cytokine abundantly produced in a number of inflammatory conditions. Although not structurally related to RANKL and TNF-α, IL-1, when bound to its receptor, IL-1R1, activates the same signaling pathways as the other two cytokines, including TRAF6/Src, NF-κB, ERK, and PI3K (217). IL-1 does not affect early stages of OC differentiation, likely due to low levels of expression of IL-1R1, which is upregulated by RANKL, but promotes multinucleation and terminal differentiation of RANKL-treated precursors, even without further RANKL exposure (208, 218, 219). RANKL seems to sensitize NFATc1 to induction by IL-1 (219). Intriguingly, one study showed that ectopic expression of IL-1R1 allows IL-1 to drive osteoclastogenesis, dependent on MITF, but not c-Fos or NFATc1 (220). Since RANKL is likely present in the inflammatory environments in which IL-1 acts *in vivo*, a significant portion of its osteolytic activity may be due to promotion of bone resorption by mature OCs rather than effects on early stages of differentiation. IL-1 appears to support actin ring formation and cell survival (221–223). The cytokine also promotes osteolysis indirectly via osteoblast lineage cells (208), in which IL-1 is an important mediator of TNF-α-induced RANKL expression.

IL-1β is initially produced in a pro-form that must be cleaved by caspase-1 to become mature and active, and this step is tightly controlled by the inflammasomes, including the NLR family, pyrin domain-containing 3 (NLRP3) inflammasome. As described above, activating mutations of NLRP3 cause NOMID as a consequence of systemic elevation of IL-1β (166). Although high inflammatory cytokine levels certainly contribute to bone loss (4), a mouse model expressing NOMID-mutated NLRP3 in committed OC progenitors lacks inflammation but still demonstrates osteopenia (132). Interestingly, hydroxyapatite particles released from bone activate the NLRP3 inflammasome in OC lineage cells (our unpublished data), and mature OCs show evidence of active inflammasomes in the absence of inflammatory stimuli (132). Thus, the NLRP3 inflammasome may amplify bone loss through IL-1β-dependent and -independent mechanisms.

TGF-β

TGF-β interacts directly with its receptor on OC progenitors and has significant osteoclastogenic effects. On early progenitors, TGF-β enhances lineage commitment through upregulation of SOCS (suppressors of cytokine signaling) (224) and RANK expression (225). These early effects may serve to limit differentiation down other lineages such as inflammatory macrophages (226). It also promotes signaling downstream of RANK via interactions between TRAF6, an adaptor for RANK, and Smad3, a transcription factor downstream of TGF-β receptor (227). A more recent study by the Tanaka group using genome-wide analysis of chromatin highlighted cooperativity of Smad2/3 with c-Fos in the promotion of NFATc1 expression (228). MITF, another transcription factor important for osteoclastogenesis, is also enhanced by TGF-β treatment (229), likely due to activation of p38 (230), providing further mechanisms for promotion of differentiation. However, TGF-β does not appear to promote resorption by mature OCs (231). In fact, we have found that stimulation of bone marrow macrophages with TGF-β for as little as 4 h is sufficient for the full pro-osteoclastogenic effects of this factor (our unpublished data).

Some early studies suggested inhibitory actions of TGF-β on OC differentiation, but these were all in complex coculture or organ culture models in which effects on OC differentiation were likely indirect (232, 233). TGF-β is synthesized by osteoblasts, stored as a latent protein in the bone matrix, and mobilized by OCs in an active form. Thus, TGF-β is thought to be an important player in the vicious cycle of osteolytic bone metastasis in which OC action enhances TGF-β release, feeding growth of tumor cells as well as differentiation of more OCs. In multiple mouse models, blockade of TGF-β signaling increases bone mass and reduces bone metastasis (234–236). Despite the promise shown in animal models for therapy of bone metastasis, the role of TGF-β in the context of tumors as well as inflammatory diseases is very complex (237). For example, TGF-β may be an important mediator of disseminated tumor cell dormancy such that its inhibition could enhance bone metastasis in some cases (238). Furthermore, the effects of TGF-β or its inhibition in disease settings is complicated by the universality of TGF-β receptor expression. For bone, osteoblasts also mount physiologically relevant responses to TGF-β, and these may be dominant to the effects on OCs, for example, in renal osteodystrophy (239). In this study, TGF-β neutralization limited downstream signaling in osteoblasts and osteocytes but not in OCs. In diseases such as Marfan syndrome, mutations in matrix proteins reduce the sequestration of TGF-β and increase free TGF-β levels, causing osteoblast lineage cells to secrete more RANKL, providing another mechanism for osteolytic effects of this factor (240–242). Aberrant activation of TGF-β may also play a role in the bone abnormalities in neurofibromatosis (243).

IFN-γ

OC precursors bear receptors for IFN-γ, and the direct effects of this cytokine produced by T cells and innate immune cells are strongly inhibitory for osteoclastogenesis in mice (244) and humans (245). IFN-γ acts on early OC precursors, but prior exposure to RANKL seems to abrogate the inhibitory effects (246). In models of inflammatory bone loss or tumor-mediated osteolysis, IFN-γ has an inhibitory role, and global deletion of its receptor exacerbates disease (244, 247, 248). However, in some contexts this cytokine may support OC activity. IFN-γ has been used with some success to improve bone resorption in patients with osteopetrosis, possibly through induction of superoxide production as a direct effect on OC precursors (249, 250). In conditions of estrogen deficiency, IFN-γ acts indirectly, causing T-cell proliferation and increased TNF-α and RANKL expression, which increase bone loss (251, 252). Thus, like TGF-β, the effects of IFN-γ on OCs may be very context dependent.

CONCLUSION

While members of the myeloid lineage, to which OCs belong, are specialized in the clearance of pathogens and cell debris, the function of OCs is to remove damaged or aged components of the bone ECM, which are then replaced by osteoblasts. These tightly controlled sequential responses ensure that appropriate bone mass and quality is maintained throughout life. Perturbation of this equilibrium in metabolic, inflammatory, and neoplastic diseases negatively impacts the skeleton. Despite the elusive identity of the most proximal OC precursors, considerable advances in the field have been achieved, including methods for efficient generation of OCs *in vitro* and elucidation of the molecular mechanisms that govern OC differentiation, survival, and function. OC biology is driven mainly by RANKL and M-CSF, but is modulated by a variety of autocrine, paracrine, and hormonal factors in health and disease conditions. Additionally, the active fields of costimulatory signaling and epigenetic effects have intersections with OC biology. OCs develop and act in a rich microenvironment, and they have important reciprocal interactions with other immune cells and mesenchymal cells that impact homeostasis and disease states. Despite these breakthroughs, we still do not understand how all of the various modulators of the osteoclastogenic process interact at baseline and in disease states. In particular, we lack insight into differences between responses of OCs at different sites (e.g., vertebrae versus long bones) or compartments (cancellous versus cortical) that are observed with systemic stimuli such as estrogen deprivation. Further work into the dynamics of the various myeloid populations that can function as OC progenitors may provide additional insight into pathological bone loss. Surprisingly, given that osteoporosis is an age-related disease, there is still much to be learned about changes in OCs with aging as well. Considering all the newly emerging tools for generating highly manipulable animal models, as well as our ability to use human samples, the coming years should provide answers to many of these questions.

Acknowledgments. *We wish to thank Roberta Faccio for critical reading of the manuscript. We have tried to cite primary research in most cases, but due to the large amount of research in this area, we are sure to have missed some important papers. We apologize in advance to authors who have been omitted. We are grateful for funding support from the National Institutes of Health: NIH/NIAMS RO1-AR052705 (to D.V.N.) and NIH/NIAMS RO1-AR064755 (to G.M.).*

Citation. Novack DV, Mbalaviele G. 2016. Osteoclasts—key players in skeletal health and disease. Microbiol Spectrum 4(3):MCHD-0011-2015.

References

1. **Mosaad YM.** 2014. Hematopoietic stem cells: an overview. *Transfus Apheresis Sci* **51:**68–82.
2. **Demulder A, Takahashi S, Singer FR, Hosking DJ, Roodman GD.** 1993. Abnormalities in osteoclast precursors and marrow accessory cells in Paget's disease. *Endocrinology* **133:**1978–1982.
3. **Demulder A, Suggs SV, Zsebo KM, Scarcez T, Roodman GD.** 1992. Effects of stem cell factor on osteoclast-like cell formation in long-term human marrow cultures. *J Bone Miner Res* **7:**1337–1344.
4. **Bonar SL, Brydges SD, Mueller JL, McGeough MD, Pena C, Chen D, Grimston SK, Hickman-Brecks CL, Ravindran S, McAlinden A, Novack DV, Kastner DL, Civitelli R, Hoffman HM, Mbalaviele G.** 2012. Constitutively activated NLRP3 inflammasome causes inflammation and abnormal skeletal development in mice. *PLoS One* **7:**e35979. doi:10.1371/journal.pone.0035979.
5. **Mediero A, Perez-Aso M, Cronstein BN.** 2014. Activation of EPAC1/2 is essential for osteoclast formation by modulating NFκB nuclear translocation and actin cytoskeleton rearrangements. *FASEB J* **28:**4901–4913.
6. **Xing L, Boyce B.** 2014. RANKL-based osteoclastogenic assays from murine bone marrow Cells, p 307–313. *In* Hilton MJ (ed), *Skeletal Development and Repair*, **vol 1130.** Humana Press, Totowa, NJ.
7. **Mabilleau G, Pascaretti-Grizon F, Baslé MF, Chappard D.** 2012. Depth and volume of resorption induced by osteoclasts generated in the presence of RANKL, TNF-alpha/IL-1 or LIGHT. *Cytokine* **57:**294–299.
8. **Li P, Schwarz EM, O'Keefe RJ, Ma L, Looney RJ, Ritchlin CT, Boyce BF, Xing L.** 2004. Systemic tumor necrosis factor α mediates an increase in peripheral $CD11b^{high}$ osteoclast precursors in tumor necrosis factor α-transgenic mice. *Arthritis Rheum* **50:**265–276.
9. **Henriksen K, Karsdal M, Taylor A, Tosh D, Coxon F.** 2012. Generation of human osteoclasts from peripheral blood, p 159–175. *In* Helfrich MH, Ralston SH (ed),

Bone Research Protocols, vol **816**. Humana Press, Totowa, NJ.

10. **Bradley E, Oursler M.** 2008. Osteoclast culture and resorption assays, p 19–35. *In* Westendorf J (ed), *Osteoporosis*, vol **455**. Humana Press, Totowa, NJ.
11. **Wang Y, Menendez A, Fong C, ElAlieh HZ, Chang W, Bikle DD.** 2014. Ephrin B2/EphB4 mediates the actions of IGF-I signaling in regulating endochondral bone formation. *J Bone Miner Res* **29**:1900–1913.
12. **Hayman AR, Jones SJ, Boyde A, Foster D, Colledge WH, Carlton MB, Evans MJ, Cox TM.** 1996. Mice lacking tartrate-resistant acid phosphatase (Acp 5) have disrupted endochondral ossification and mild osteopetrosis. *Development* **122**:3151–3162.
13. **Sago K, Teitelbaum SL, Venstrom K, Reichardt LF, Ross FP.** 1999. The integrin $\alpha_v\beta_5$ is expressed on avian osteoclast precursors and regulated by retinoic acid. *J Bone Miner Res* **14**:32–38.
14. **Saftig P, Hunziker E, Wehmeyer O, Jones S, Boyde A, Rommerskirch W, Moritz JD, Schu P, von Figura K.** 1998. Impaired osteoclastic bone resorption leads to osteopetrosis in cathepsin-K-deficient mice. *Proc Natl Acad Sci USA* **95**:13453–13458.
15. **Gowen M, Lazner F, Dodds R, Kapadia R, Feild J, Tavaria M, Bertoncello I, Drake F, Zavarselk S, Tellis I, Hertzog P, Debouck C, Kola I.** 1999. Cathepsin K knockout mice develop osteopetrosis due to a deficit in matrix degradation but not demineralization. *J Bone Miner Res* **14**:1654–1663.
16. **Hoff AO, Catala-Lehnen P, Thomas PM, Priemel M, Rueger JM, Nasonkin I, Bradley A, Hughes MR, Ordonez N, Cote GJ, Amling M, Gagel RF.** 2002. Increased bone mass is an unexpected phenotype associated with deletion of the calcitonin gene. *J Clin Invest* **110**:1849–1857.
17. **Kim N, Takami M, Rho J, Josien R, Choi Y.** 2002. A novel member of the leukocyte receptor complex regulates osteoclast differentiation. *J Exp Med* **195**:201–209.
18. **Sørensen MG, Henriksen K, Schaller S, Henriksen DB, Nielsen FC, Dziegiel MH, Karsdal MA.** 2007. Characterization of osteoclasts derived from CD14$^+$ monocytes isolated from peripheral blood. *J Bone Miner Metab* **25**: 36–45.
19. **McHugh KP, Hodivala-Dilke K, Zheng MH, Namba N, Lam J, Novack D, Feng X, Ross FP, Hynes RO, Teitelbaum SL.** 2000. Mice lacking β3 integrins are osteosclerotic because of dysfunctional osteoclasts. *J Clin Invest* **105**:433–440.
20. **Chen TH, Swarnkar G, Mbalaviele G, Abu-Amer Y.** 2015. Myeloid lineage skewing due to exacerbated NF-κB signaling facilitates osteopenia in Scurfy mice. *Cell Death Dis* **6**:e1723. doi:10.1038/cddis.2015.87.
21. **Mbalaviele G, Jaiswal N, Meng A, Cheng L, Bos CV, Thiede M.** 1999. Human mesenchymal stem cells promote human osteoclast differentiation from CD34$^+$ bone marrow hematopoietic progenitors. *Endocrinology* **140**: 3736–3743.
22. **Matayoshi A, Brown C, DiPersio JF, Haug J, Abu-Amer Y, Liapis H, Kuestner R, Pacifici R.** 1996. Human blood-mobilized hematopoietic precursors differentiate into osteoclasts in the absence of stromal cells. *Proc Natl Acad Sci USA* **93**:10785–10790.
23. **Muto A, Mizoguchi T, Udagawa N, Ito S, Kawahara I, Abiko Y, Arai A, Harada S, Kobayashi Y, Nakamichi Y, Penninger JM, Noguchi T, Takahashi N.** 2011. Lineage-committed osteoclast precursors circulate in blood and settle down into bone. *J Bone Miner Res* **26**: 2978–2990.
24. **Durand M, Komarova SV, Bhargava A, Trebec-Reynolds DP, Li K, Fiorino C, Maria O, Nabavi N, Manolson MF, Harrison RE, Dixon SJ, Sims SM, Mizianty MJ, Kurgan L, Haroun S, Boire G, de Fatima Lucena-Fernandes M, de Brum-Fernandes AJ.** 2013. Monocytes from patients with osteoarthritis display increased osteoclastogenesis and bone resorption: the In Vitro Osteoclast Differentiation in Arthritis study. *Arthritis Rheum* **65**:148–158.
25. **Hemingway F, Cheng X, Knowles HJ, Estrada FM, Gordon S, Athanasou NA.** 2011. In vitro generation of mature human osteoclasts. *Calcif Tissue Int* **89**:389–395.
26. **Lam J, Takeshita S, Barker JE, Kanagawa O, Ross FP, Teitelbaum SL.** 2000. TNF-α induces osteoclastogenesis by direct stimulation of macrophages exposed to permissive levels of RANK ligand. *J Clin Invest* **106**: 1481–1488.
27. **Charles JF, Hsu LY, Niemi EC, Weiss A, Aliprantis AO, Nakamura MC.** 2012. Inflammatory arthritis increases mouse osteoclast precursors with myeloid suppressor function. *J Clin Invest* **122**:4592–4605.
28. **Jacome-Galarza CE, Lee S-K, Lorenzo JA, Aguila HL.** 2013. Identification, characterization, and isolation of a common progenitor for osteoclasts, macrophages, and dendritic cells from murine bone marrow and periphery. *J Bone Miner Res* **28**:1203–1213.
29. **Jacquin C, Gran DE, Lee SK, Lorenzo JA, Aguila HL.** 2006. Identification of multiple osteoclast precursor populations in murine bone marrow. *J Bone Miner Res* **21**:67–77.
30. **Takahashi N, Udagawa N, Tanaka S, Murakami H, Owan I, Tamura T, Suda T.** 1994. Postmitotic osteoclast precursors are mononuclear cells which express macrophage-associated phenotypes. *Dev Biol* **163**:212–221.
31. **Park-Min KH, Lee EY, Moskowitz NK, Lim E, Lee SK, Lorenzo JA, Huang C, Melnick AM, Purdue PE, Goldring SR, Ivashkiv LB.** 2013. Negative regulation of osteoclast precursor differentiation by CD11b and β2 integrin-B-cell lymphoma 6 signaling. *J Bone Miner Res* **28**:135–149.
32. **Zhuang J, Zhang J, Lwin ST, Edwards JR, Edwards CM, Mundy GR, Yang X.** 2012. Osteoclasts in multiple myeloma are derived from Gr-1+CD11b+myeloid-derived suppressor cells. *PLoS One* **7**:e48871. doi: 10.1371/journal.pone.0048871.
33. **Sawant A, Deshane J, Jules J, Lee CM, Harris BA, Feng X, Ponnazhagan S.** 2013. Myeloid-derived suppressor cells function as novel osteoclast progenitors enhancing bone loss in breast cancer. *Cancer Res* **73**:672–682.

34. **Danilin S, Merkel AR, Johnson JR, Johnson RW, Edwards JR, Sterling JA.** 2012. Myeloid-derived suppressor cells expand during breast cancer progression and promote tumor-induced bone destruction. *OncoImmunology* **1:**1484–1494.

35. **Yagi M, Miyamoto T, Sawatani Y, Iwamoto K, Hosogane N, Fujita N, Morita K, Ninomiya K, Suzuki T, Miyamoto K, Oike Y, Takeya M, Toyama Y, Suda T.** 2005. DC-STAMP is essential for cell-cell fusion in osteoclasts and foreign body giant cells. *J Exp Med* **202:** 345–351.

36. **Miyamoto H, Suzuki T, Miyauchi Y, Iwasaki R, Kobayashi T, Sato Y, Miyamoto K, Hoshi H, Hashimoto K, Yoshida S, Hao W, Mori T, Kanagawa H, Katsuyama E, Fujie A, Morioka H, Matsumoto M, Chiba K, Takeya M, Toyama Y, Miyamoto T.** 2012. Osteoclast stimulatory transmembrane protein and dendritic cell-specific transmembrane protein cooperatively modulate cell-cell fusion to form osteoclasts and foreign body giant cells. *J Bone Miner Res* **27:**1289–1297.

37. **Mbalaviele G, Chen H, Boyce BF, Mundy GR, Yoneda T.** 1995. The role of cadherin in the generation of multinucleated osteoclasts from mononuclear precursors in murine marrow. *J Clin Invest* **95:**2757–2765.

38. **Van den Bossche J, Malissen B, Mantovani A, De Baetselier P, Van Ginderachter JA.** 2012. Regulation and function of the E-cadherin/catenin complex in cells of the monocyte-macrophage lineage and DCs. *Blood* **119:**1623–1633.

39. **Nakamura H, Nakashima T, Hayashi M, Izawa N, Yasui T, Aburatani H, Tanaka S, Takayanagi H.** 2014. Global epigenomic analysis indicates protocadherin-7 activates osteoclastogenesis by promoting cell-cell fusion. *Biochem Biophys Res Commun* **455:**305–311.

40. **Ishizuka H, García-Palacios V, Lu G, Subler MA, Zhang H, Boykin CS, Choi SJ, Zhao L, Patrene K, Galson DL, Blair HC, Hadi TM, Windle JJ, Kurihara N, Roodman GD.** 2011. ADAM8 enhances osteoclast precursor fusion and osteoclast formation in vitro and in vivo. *J Bone Miner Res* **26:**169–181.

41. **Ishii M, Egen JG, Klauschen F, Meier-Schellersheim M, Saeki Y, Vacher J, Proia RL, Germain RN.** 2009. Sphingosine-1-phosphate mobilizes osteoclast precursors and regulates bone homeostasis. *Nature* **458:**524–528.

42. **Ishii M, Kikuta J, Shimazu Y, Meier-Schellersheim M, Germain RN.** 2010. Chemorepulsion by blood S1P regulates osteoclast precursor mobilization and bone remodeling in vivo. *J Exp Med* **207:**2793–2798.

43. **Ishii M, Kikuta J.** 2013. Sphingosine-1-phosphate signaling controlling osteoclasts and bone homeostasis. *Biochim Biophys Acta* **1831:**223–227.

44. **Shahnazari M, Chu V, Wronski TJ, Nissenson RA, Halloran BP.** 2013. CXCL12/CXCR4 signaling in the osteoblast regulates the mesenchymal stem cell and osteoclast lineage populations. *FASEB J* **27:**3505–3513.

45. **Takahashi N, Akatsu T, Udagawa N, Sasaki T, Yamaguchi A, Moseley JM, Martin TJ, Suda T.** 1988. Osteoblastic cells are involved in osteoclast formation. *Endocrinology* **123:**2600–2602.

46. **Udagawa N, Takahashi N, Akatsu T, Tanaka H, Sasaki T, Nishihara T, Koga T, Martin TJ, Suda T.** 1990. Origin of osteoclasts: mature monocytes and macrophages are capable of differentiating into osteoclasts under a suitable microenvironment prepared by bone marrow-derived stromal cells. *Proc Natl Acad Sci USA* **87:** 7260–7264.

47. **Wiktor-Jedrzejczak WW, Ahmed A, Szczylik C, Skelly RR.** 1982. Hematological characterization of congenital osteopetrosis in *op/op* mouse. Possible mechanism for abnormal macrophage differentiation. *J Exp Med* **156:** 1516–1527.

48. **Yoshida H, Hayashi SI, Kunisada T, Ogawa M, Nishikawa S, Okamura H, Sudo T, Shultz LD, Nishikawa SI.** 1990. The murine mutation osteopetrosis is in the coding region of the macrophage colony stimulating factor gene. *Nature* **345:**442–444.

49. **Felix R, Cecchini MG, Hofstetter W, Elford PR, Stutzer A, Fleisch H.** 1990. Impairment of macrophage colony-stimulating factor production and lack of resident bone marrow macrophages in the osteopetrotic *op/op* mouse. *J Bone Miner Res* **5:**781–789.

50. **Stanley ER, Chitu V.** 2014. CSF-1 receptor signaling in myeloid cells. *Cold Spring Harb Perspect Biol* **6:** a021857. doi:10.1101/cshperspect.a021857.

51. **Otero K, Turnbull IR, Poliani PL, Vermi W, Cerutti E, Aoshi T, Tassi I, Takai T, Stanley SL, Miller M, Shaw AS, Colonna M.** 2009. Macrophage colony-stimulating factor induces the proliferation and survival of macrophages via a pathway involving DAP12 and β-catenin. *Nat Immunol* **10:**734–743.

52. **Glantschnig H, Fisher JE, Wesolowski G, Rodan GA, Reszka AA.** 2003. M-CSF, TNFα and RANK ligand promote osteoclast survival by signaling through mTOR/S6 kinase. *Cell Death Differ* **10:**1165–1177.

53. **Zamani A, Decker C, Cremasco V, Hughes L, Novack DV, Faccio R.** 2015. Diacylglycerol kinase ζ (DGKζ) is a critical regulator of bone homeostasis via modulation of c-Fos levels in osteoclasts. *J Bone Miner Res* **30:** 1852–1863.

54. **Baud'Huin M, Renault R, Charrier C, Riet A, Moreau A, Brion R, Gouin F, Duplomb L, Heymann D.** 2010. Interleukin-34 is expressed by giant cell tumours of bone and plays a key role in RANKL-induced osteoclastogenesis. *J Pathol* **221:**77–86.

55. **Chen Z, Buki K, Vääräniemi J, Gu G, Väänänen HK.** 2011. The critical role of IL-34 in osteoclastogenesis. *PLoS One* **6:**e18689. doi:10.1371/journal.pone.0018689.

56. **Li J, Chen K, Zhu L, Pollard JW.** 2006. Conditional deletion of the colony stimulating factor-1 receptor (*c-fms* proto-oncogene) in mice. *Genesis* **44:**328–335.

57. **Lee MS, Kim HS, Yeon JT, Choi SW, Chun CH, Kwak HB, Oh J.** 2009. GM-CSF regulates fusion of mononuclear osteoclasts into bone-resorbing osteoclasts by activating the Ras/ERK pathway. *J Immunol* **183:** 3390–3399.

58. **Niida S, Kaku M, Amano H, Yoshida H, Kataoka H, Nishikawa S, Tanne K, Maeda N, Nishikawa SI, Kodama H.** 1999. Vascular endothelial growth factor can substitute for macrophage colony-stimulating factor

in the support of osteoclastic bone resorption. *J Exp Med* **190:**293–298.

59. **Nakagawa M, Kaneda T, Arakawa T, Morita S, Sato T, Yomada T, Hanada K, Kumegawa M, Hakeda Y.** 2000. Vascular endothelial growth factor (VEGF) directly enhances osteoclastic bone resorption and survival of mature osteoclasts. *FEBS Lett* **473:**161–164.
60. **Adamopoulos IE, Xia Z, Lau YS, Athanasou NA.** 2006. Hepatocyte growth factor can substitute for M-CSF to support osteoclastogenesis. *Biochem Biophys Res Commun* **350:**478–483.
61. **Lacey DL, Timms E, Tan HL, Kelley MJ, Dunstan CR, Burgess T, Elliott R, Colombero A, Elliott G, Scully S, Hsu H, Sullivan J, Hawkins N, Davy E, Capparelli C, Eli A, Qian YX, Kaufman S, Sarosi I, Shalhoub V, Senaldi G, Guo J, Delaney J, Boyle WJ.** 1998. Osteoprotegerin ligand is a cytokine that regulates osteoclast differentiation and activation. *Cell* **93:**165–176.
62. **Yasuda H, Shima N, Nakagawa N, Yamaguchi K, Kinosaki M, Mochizuki S, Tomoyasu A, Yano K, Goto M, Murakami A, Tsuda E, Morinaga T, Higashio K, Udagawa N, Takahashi N, Suda T.** 1998. Osteoclast differentiation factor is a ligand for osteoprotegerin/osteoclastogenesis-inhibitory factor and is identical to TRANCE/RANKL. *Proc Natl Acad Sci USA* **95:**3597–3602.
63. **Kong YY, Yoshida H, Sarosi I, Tan HL, Timms E, Capparelli C, Morony S, Oliveira-dos-Santos AJ, Van G, Itie A, Khoo W, Wakeham A, Dunstan CR, Lacey DL, Mak TW, Boyle WJ, Penninger JM.** 1999. OPGL is a key regulator of osteoclastogenesis, lymphocyte development and lymph-node organogenesis. *Nature* **397:**315–323.
64. **Bucay N, Sarosi I, Dunstan CR, Morony S, Tarpley J, Capparelli C, Scully S, Tan HL, Xu W, Lacey DL, Boyle WJ, Simonet WS.** 1998. *osteoprotegerin*-deficient mice develop early onset osteoporosis and arterial calcification. *Genes Dev* **12:**1260–1268.
65. **Whyte MP, Tau C, McAlister WH, Zhang X, Novack DV, Preliasco V, Santini-Araujo E, Mumm S.** 2014. Juvenile Paget's disease with heterozygous duplication within *TNFRSF11A* encoding RANK. *Bone* **68:**153–161.
66. **Hughes AE, Ralston SH, Marken J, Bell C, MacPherson H, Wallace RG, van Hul W, Whyte MP, Nakatsuka K, Hovy L, Anderson DM.** 2000. Mutations in *TNFRSF11A*, affecting the signal peptide of RANK, cause familial expansile osteolysis. *Nat Genet* **24:**45–48.
67. **Novack DV, Teitelbaum SL.** 2008. The osteoclast: friend or foe? *Annu Rev Pathol* **3:**457–484.
68. **Smink JJ, Bégay V, Schoenmaker T, Sterneck E, de Vries TJ, Leutz A.** 2009. Transcription factor C/EBPβ isoform ratio regulates osteoclastogenesis through MafB. *EMBO J* **28:**1769–1781.
69. **Smink J, Tunn PU, Leutz A.** 2012. Rapamycin inhibits osteoclast formation in giant cell tumor of bone through the C/EBPβ-MafB axis. *J Mol Med Berl* **90:**25–30.
70. **Takayanagi H, Kim S, Koga T, Nishina H, Isshiki M, Yoshida H, Saiura A, Isobe M, Yokochi T, Inoue J, Wagner EF, Mak TW, Kodama T, Taniguchi T.** 2002. Induction and activation of the transcription factor NFATc1 (NFAT2) integrate RANKL signaling in terminal differentiation of osteoclasts. *Dev Cell* **3:**889–901.
71. **Mao D, Epple H, Uthgenannt B, Novack DV, Faccio R.** 2006. PLCγ2 regulates osteoclastogenesis via its interaction with ITAM proteins and GAB2. *J Clin Invest* **116:**2869–2879.
72. **Alhawagri M, Yamanaka Y, Ballard D, Oltz E, Abu-Amer Y.** 2012. Lysine392, a K63-linked ubiquitination site in NEMO, mediates inflammatory osteoclastogenesis and osteolysis. *J Orthop Res* **30:**554–560.
73. **Bronisz A, Carey HA, Godlewski J, Sif S, Ostrowski MC, Sharma SM.** 2014. The multifunctional protein fused in sarcoma (FUS) is a coactivator of microphthalmia-associated transcription factor (MITF). *J Biol Chem* **289:**326–334.
74. **Yasui T, Hirose J, Aburatani H, Tanaka S.** 2011. Epigenetic regulation of osteoclast differentiation. *Ann N Y Acad Sci* **1240:**7–13.
75. **Kim JH, Kim N.** 2014. Regulation of NFATc1 in osteoclast differentiation. *J Bone Metab* **21:**233–241.
76. **Mizoguchi F, Izu Y, Hayata T, Hemmi H, Nakashima K, Nakamura T, Kato S, Miyasaka N, Ezura Y, Noda M.** 2010. Osteoclast-specific Dicer gene deficiency suppresses osteoclastic bone resorption. *J Cell Biochem* **109:**866–875.
77. **Nishikawa K, Iwamoto Y, Kobayashi Y, Katsuoka F, Kawaguchi S, Tsujita T, Nakamura T, Kato S, Yamamoto M, Takayanagi H, Ishii M.** 2015. DNA methyltransferase 3a regulates osteoclast differentiation by coupling to an S-adenosylmethionine-producing metabolic pathway. *Nat Med* **21:**281–287.
78. **Yasui T, Hirose J, Tsutsumi S, Nakamura K, Aburatani H, Tanaka S.** 2011. Epigenetic regulation of osteoclast differentiation: possible involvement of Jmjd3 in the histone demethylation of *Nfatc1*. *J Bone Miner Res* **26:**2665–2671.
79. **Park-Min KH, Lim E, Lee MJ, Park SH, Giannopoulou E, Yarilina A, van der Meulen M, Zhao B, Smithers N, Witherington J, Lee K, Tak PP, Prinjha RK, Ivashkiv LB.** 2014. Inhibition of osteoclastogenesis and inflammatory bone resorption by targeting BET proteins and epigenetic regulation. *Nat Commun* **5:**5418. doi:10.1038/ncomms6418.
80. **Shakibaei M, Buhrmann C, Mobasheri A.** 2011. Resveratrol-mediated SIRT-1 interactions with p300 modulate receptor activator of NF-κB ligand (RANKL) activation of NF-κB signaling and inhibit osteoclastogenesis in bone-derived cells. *J Biol Chem* **286:**11492–11505.
81. **Hah YS, Cheon YH, Lim HS, Cho HY, Park BH, Ka SO, Lee YR, Jeong DW, Kim HO, Han MK, Lee SI.** 2014. Myeloid deletion of SIRT1 aggravates serum transfer arthritis in mice via nuclear factor-κB activation. *PLoS One* **9:**e87733. doi:10.1371/journal.pone.0087733.
82. **Zou W, Reeve JL, Liu Y, Teitelbaum SL, Ross FP.** 2008. DAP12 couples c-Fms activation to the osteoclast cytoskeleton by recruitment of Syk. *Mol Cell* **31:**422–431.

83. **Mócsai A, Humphrey MB, Van Ziffle JAG, Hu Y, Burghardt A, Spusta SC, Majumdar S, Lanier LL, Lowell CA, Nakamura MC.** 2004. The immunomodulatory adapter proteins DAP12 and Fc receptor γ-chain (FcRγ) regulate development of functional osteoclasts through the Syk tyrosine kinase. *Proc Natl Acad Sci USA* **101:**6158–6163.

84. **Koga T, Inui M, Inoue K, Kim S, Suematsu A, Kobayashi E, Iwata T, Ohnishi H, Matozaki T, Kodama T, Taniguchi T, Takayanagi H, Takai T.** 2004. Costimulatory signals mediated by the ITAM motif cooperate with RANKL for bone homeostasis. *Nature* **428:**758–763.

85. **Wu Y, Torchia J, Yao W, Lane NE, Lanier LL, Nakamura MC, Humphrey MB.** 2007. Bone microenvironment specific roles of ITAM adapter signaling during bone remodeling induced by acute estrogen-deficiency. *PLoS One* **2:**e586. doi:10.1371/journal.pone.0000586.

86. **Li S, Miller CH, Giannopoulou E, Hu X, Ivashkiv LB, Zhao B.** 2014. RBP-J imposes a requirement for ITAM-mediated costimulation of osteoclastogenesis. *J Clin Invest* **124:**5057–5073.

87. **Zou W, Teitelbaum SL.** 2015. Absence of Dap12 and the αvβ3 integrin causes severe osteopetrosis. *J Cell Biol* **208:**125–136.

88. **Li Y, Li A, Strait K, Zhang H, Nanes MS, Weitzmann MN.** 2007. Endogenous TNFα lowers maximum peak bone mass and inhibits osteoblastic Smad activation through NF-κB. *J Bone Miner Res* **22:**646–655.

89. **Onal M, Xiong J, Chen X, Thostenson JD, Almeida M, Manolagas SC, O'Brien CA.** 2012. Receptor activator of nuclear factor κB ligand (RANKL) protein expression by B lymphocytes contributes to ovariectomy-induced bone loss. *J Biol Chem* **287:**29851–29860.

90. **Kong YY, Feige U, Sarosi I, Bolon B, Tafuri A, Morony S, Capparelli C, Li J, Elliott R, McCabe S, Wong T, Campagnuolo G, Moran E, Bogoch ER, Van G, Nguyen LT, Ohashi PS, Lacey DL, Fish E, Boyle WJ, Penninger JM.** 1999. Activated T cells regulate bone loss and joint destruction in adjuvant arthritis through osteoprotegerin ligand. *Nature* **402:**304–309.

91. **Weitzmann MN, Cenci S, Rifas L, Haug J, Dipersio J, Pacifici R.** 2001. T cell activation induces human osteoclast formation via receptor activator of nuclear factor κB ligand-dependent and -independent mechanisms. *J Bone Miner Res* **16:**328–337.

92. **Horwood NJ, Kartsogiannis V, Quinn JM, Romas E, Martin TJ, Gillespie MT.** 1999. Activated T lymphocytes support osteoclast formation *in vitro*. *Biochem Biophys Res Commun* **265:**144–150.

93. **Lee SK, Kadono Y, Okada F, Jacquin C, Koczon-Jaremko B, Gronowicz G, Adams DJ, Aguila HL, Choi Y, Lorenzo JA.** 2006. T lymphocyte-deficient mice lose trabecular bone mass with ovariectomy. *J Bone Miner Res* **21:**1704–1712.

94. **Toraldo G, Roggia C, Qian WP, Pacifici R, Weitzmann MN.** 2003. IL-7 induces bone loss *in vivo* by induction of receptor activator of nuclear factor κB ligand and tumor necrosis factor α from T cells. *Proc Natl Acad Sci USA* **100:**125–130.

95. **Li Y, Li A, Yang X, Weitzmann MN.** 2007. Ovariectomy-induced bone loss occurs independently of B cells. *J Cell Biochem* **100:**1370–1375.

96. **Li Y, Toraldo G, Li A, Yang X, Zhang H, Qian W-P, Weitzmann MN.** 2007. B cells and T cells are critical for the preservation of bone homeostasis and attainment of peak bone mass in vivo. *Blood* **109:**3839–3848.

97. **Zaiss MM, Axmann R, Zwerina J, Polzer K, Gückel E, Skapenko A, Schulze-Koops H, Horwood N, Cope A, Schett G.** 2007. Treg cells suppress osteoclast formation: a new link between the immune system and bone. *Arthritis Rheum* **56:**4104–4112.

98. **Kelchtermans H, Geboes L, Mitera T, Huskens D, Leclercq G, Matthys P.** 2009. Activated $CD4^+CD25^+$ regulatory T cells inhibit osteoclastogenesis and collagen-induced arthritis. *Ann Rheum Dis* **68:**744–750.

99. **Zaiss MM, Sarter K, Hess A, Engelke K, Böhm C, Nimmerjahn F, Voll R, Schett G, David JP.** 2010. Increased bone density and resistance to ovariectomy-induced bone loss in FoxP3-transgenic mice based on impaired osteoclast differentiation. *Arthritis Rheum* **62:**2328–2338.

100. **Luo CY, Wang L, Sun C, Li DJ.** 2011. Estrogen enhances the functions of $CD4^+CD25^+Foxp3^+$ regulatory T cells that suppress osteoclast differentiation and bone resorption in vitro. *Cell Mol Immunol* **8:**50–58.

101. **Kiesel JR, Buchwald ZS, Aurora R.** 2009. Cross-presentation by osteoclasts induces FoxP3 in $CD8^+$ T cells. *J Immunol* **182:**5477–5487.

102. **Buchwald ZS, Kiesel JR, Yang C, DiPaolo R, Novack DV, Aurora R.** 2013. Osteoclast-induced $Foxp3^+$ CD8 T-cells limit bone loss in mice. *Bone* **56:**163–173.

103. **Buchwald ZS, Yang C, Nellore S, Shashkova EV, Davis JL, Cline A, Ko J, Novack DV, DiPaolo R, Aurora R.** 2015. A bone anabolic effect of RANKL in a murine model of osteoporosis mediated through $FoxP3^+$ CD8 T cells. *J Bone Miner Res* **30:**1508–1522.

104. **Grassi F, Manferdini C, Cattini L, Piacentini A, Gabusi E, Facchini A, Lisignoli G.** 2011. T cell suppression by osteoclasts in vitro. *J Cell Physiol* **226:**982–990.

105. **Li H, Hong S, Qian J, Zheng Y, Yang J, Yi Q.** 2010. Cross talk between the bone and immune systems: osteoclasts function as antigen-presenting cells and activate $CD4^+$ and $CD8^+$ T cells. *Blood* **116:**210–217.

106. **Li H, Lu Y, Qian J, Zheng Y, Zhang M, Bi E, He J, Liu Z, Xu J, Gao JY, Yi Q.** 2014. Human osteoclasts are inducible immunosuppressive cells in response to T cell-derived IFN-γ and CD40 ligand in vitro. *J Bone Miner Res* **29:**2666–2675.

107. **McHugh KP, Hodivala-Dilke K, Zheng MH, Namba N, Lam J, Novack D, Feng X, Ross FP, Hynes RO, Teitelbaum SL.** 2000. Mice lacking β3 integrins are osteosclerotic because of dysfunctional osteoclasts. *J Clin Invest* **105:**433–440.

108. **DeSelm CJ, Miller BC, Zou W, Beatty WL, van Meel E, Takahata Y, Klumperman J, Tooze SA, Teitelbaum SL, Virgin HW.** 2011. Autophagy proteins regulate the secretory component of osteoclastic bone resorption. *Dev Cell* **21:**966–974.

109. Zou W, Izawa T, Zhu T, Chappel J, Otero K, Monkley SJ, Critchley DR, Petrich BG, Morozov A, Ginsberg MH, Teitelbaum SL. 2013. Talin1 and Rap1 are critical for osteoclast function. *Mol Cell Biol* **33**:830–844.

110. Fukunaga T, Zou W, Warren JT, Teitelbaum SL. 2014. Vinculin regulates osteoclast function. *J Biol Chem* **289**: 13554–13564.

111. Schmidt S, Nakchbandi I, Ruppert R, Kawelke N, Hess MW, Pfaller K, Jurdic P, Fässler R, Moser M. 2011. Kindlin-3-mediated signaling from multiple integrin classes is required for osteoclast-mediated bone resorption. *J Cell Biol* **192**:883–897.

112. Krits I, Wysolmerski RB, Holliday LS, Lee BS. 2002. Differential localization of myosin II isoforms in resting and activated osteoclasts. *Calcif Tissue Int* **71**:530–538.

113. Zou W, DeSelm CJ, Broekelmann TJ, Mecham RP, Vande Pol S, Choi K, Teitelbaum SL. 2012. Paxillin contracts the osteoclast cytoskeleton. *J Bone Miner Res* **27**:2490–2500.

114. Faccio R, Teitelbaum SL, Fujikawa K, Chappel J, Zallone A, Tybulewicz VL, Ross FP, Swat W. 2005. Vav3 regulates osteoclast function and bone mass. *Nat Med* **11**:284–290.

115. Croke M, Ross FP, Korhonen M, Williams DA, Zou W, Teitelbaum SL. 2011. Rac deletion in osteoclasts causes severe osteopetrosis. *J Cell Sci* **124**:3811–3821.

116. Zou W, Croke M, Fukunaga T, Broekelmann TJ, Mecham RP, Teitelbaum SL. 2013. Zap70 inhibits Syk-mediated osteoclast function. *J Cell Biochem* **114**: 1871–1878.

117. Soriano P, Montgomery C, Geske R, Bradley A. 1991. Targeted disruption of the c-*src* proto-oncogene leads to osteopetrosis in mice. *Cell* **64**:693–702.

118. Zou W, Kitaura H, Reeve J, Long F, Tybulewicz VLJ, Shattil SJ, Ginsberg MH, Ross FP, Teitelbaum SL. 2007. Syk, c-Src, the αvβ3 integrin, and ITAM immunoreceptors, in concert, regulate osteoclastic bone resorption. *J Cell Biol* **176**:877–888.

119. Baron R, Neff L, Louvard D, Courtoy PJ. 1985. Cell-mediated extracellular acidification and bone resorption: evidence for a low pH in resorbing lacunae and localization of a 100-kD lysosomal membrane protein at the osteoclast ruffled border. *J Cell Biol* **101**: 2210–2222.

120. Vaes G. 1968. On the mechanisms of bone resorption: the action of parathyroid hormone on the excretion and synthesis of lysosomal enzymes and on the extracellular release of acid by bone cells. *J Cell Biol* **39**:676–697.

121. Gay CV, Schraer H, Anderson RE, Cao H. 1984. Current studies on the location and function of carbonic anhydrase in osteoclasts. *Ann N Y Acad Sci* **429**:473–478.

122. Baron R, Neff L, Brown W, Courtoy PJ, Louvard D, Farquhar MG. 1988. Polarized secretion of lysosomal enzymes: co-distribution of cation-independent mannose-6-phosphate receptors and lysosomal enzymes along the osteoclast exocytic pathway. *J Cell Biol* **106**:1863–1872.

123. Sobacchi C, Schulz A, Coxon FP, Villa A, Helfrich MH. 2013. Osteopetrosis: genetics, treatment and new insights into osteoclast function. *Nat Rev Endocrinol* **9**:522–536.

124. Van Wesenbeeck L, Odgren PR, Coxon FP, Frattini A, Moens P, Perdu B, MacKay CA, Van Hul E, Timmermans JP, Vanhoenacker F, Jacobs R, Peruzzi B, Teti A, Helfrich MH, Rogers MJ, Villa A, Van Hul W. 2007. Involvement of *PLEKHM1* in osteoclastic vesicular transport and osteopetrosis in *incisors absent* rats and humans. *J Clin Invest* **117**:919–930.

125. Ye S, Fowler TW, Pavlos NJ, Ng PY, Liang K, Feng Y, Zheng M, Kurten R, Manolagas SC, Zhao H. 2011. LIS1 regulates osteoclast formation and function through its interactions with dynein/dynactin and Plekhm1. *PLoS One* **6**:e27285. doi:10.1371/journal.pone.0027285.

126. Fujita Y, Nakata K, Yasui N, Matsui Y, Kataoka E, Hiroshima K, Shiba RI, Ochi T. 2000. Novel mutations of the cathepsin K gene in patients with pycnodysostosis and their characterization. *J Clin Endocrinol Metab* **85**: 425–431.

127. Andersen TL, del Carmen Ovejero M, Kirkegaard T, Lenhard T, Foged NT, Delaissé JM. 2004. A scrutiny of matrix metalloproteinases in osteoclasts: evidence for heterogeneity and for the presence of MMPs synthesized by other cells. *Bone* **35**:1107–1119.

128. Mosig RA, Dowling O, DiFeo A, Ramirez MC, Parker IC, Abe E, Diouri J, Aqeel AA, Wylie JD, Oblander SA, Madri J, Bianco P, Apte SS, Zaidi M, Doty SB, Majeska RJ, Schaffler MB, Martignetti JA. 2007. Loss of MMP-2 disrupts skeletal and craniofacial development and results in decreased bone mineralization, joint erosion and defects in osteoblast and osteoclast growth. *Hum Mol Genet* **16**:1113–1123.

129. Nesbitt SA, Horton MA. 1997. Trafficking of matrix collagens through bone-resorbing osteoclasts. *Science* **276**:266–269.

130. Salo J, Lehenkari P, Mulari M, Metsikkö K, Väänänen HK. 1997. Removal of osteoclast bone resorption products by transcytosis. *Science* **276**:270–273.

131. Kawana K, Takahashi M, Hoshino H, Kushida K. 2002. Comparison of serum and urinary C-terminal telopeptide of type I collagen in aging, menopause and osteoporosis. *Clin Chim Acta* **316**:109–115.

132. Qu C, Bonar SL, Hickman-Brecks CL, Abu-Amer S, McGeough MD, Peña CA, Broderick L, Yang C, Grimston SK, Kading J, Abu-Amer Y, Novack DV, Hoffman HM, Civitelli R, Mbalaviele G. 2015. NLRP3 mediates osteolysis through inflammation-dependent and -independent mechanisms. *FASEB J* **29**:1269–1279.

133. Burton L, Paget D, Binder NB, Bohnert K, Nestor BJ, Sculco TP, Santambrogio L, Ross FP, Goldring SR, Purdue PE. 2013. Orthopedic wear debris mediated inflammatory osteolysis is mediated in part by NALP3 inflammasome activation. *J Orthop Res* **31**:73–80.

134. Youm YH, Grant RW, McCabe LR, Albarado DC, Nguyen KY, Ravussin A, Pistell P, Newman S, Carter R, Laque A, Münzberg H, Rosen CJ, Ingram DK, Salbaum JM, Dixit VD. 2013. Canonical Nlrp3 inflammasome links systemic low-grade inflammation to functional decline in aging. *Cell Metab* **18**:519–532.

135. Scianaro R, Insalaco A, Bracci Laudiero L, De Vito R, Pezzullo M, Teti A, De Benedetti F, Prencipe G. 2014.

Deregulation of the IL-1β axis in chronic recurrent multifocal osteomyelitis. *Pediatr Rheumatol Online J* **12**:30–30.

136. **Tang Y, Wu X, Lei W, Pang L, Wan C, Shi Z, Zhao L, Nagy TR, Peng X, Hu J, Feng X, Van Hul W, Wan M, Cao X.** 2009. TGF-β1-induced migration of bone mesenchymal stem cells couples bone resorption with formation. *Nat Med* **15**:757–765.
137. **Xian L, Wu X, Pang L, Lou M, Rosen CJ, Qiu T, Crane J, Frassica F, Zhang L, Rodriguez JP, Jia X, Yakar S, Xuan S, Efstratiadis A, Wan M, Cao X.** 2012. Matrix IGF-1 maintains bone mass by activation of mTOR in mesenchymal stem cells. *Nat Med* **18**:1095–1101.
138. **Ota K, Quint P, Ruan M, Pederson L, Westendorf JJ, Khosla S, Oursler MJ.** 2013. TGF-β induces Wnt10b in osteoclasts from female mice to enhance coupling to osteoblasts. *Endocrinology* **154**:3745–3752.
139. **Ota K, Quint P, Weivoda MM, Ruan M, Pederson L, Westendorf JJ, Khosla S, Oursler MJ.** 2013. Transforming growth factor beta 1 induces CXCL16 and leukemia inhibitory factor expression in osteoclasts to modulate migration of osteoblast progenitors. *Bone* **57**:68–75.
140. **Lotinun S, Kiviranta R, Matsubara T, Alzate JA, Neff L, Lüth A, Koskivirta I, Kleuser B, Vacher J, Vuorio E, Horne WC, Baron R.** 2013. Osteoclast-specific cathepsin K deletion stimulates S1P-dependent bone formation. *J Clin Invest* **123**:666–681.
141. **Takeshita S, Fumoto T, Matsuoka K, Park KA, Aburatani H, Kato S, Ito M, Ikeda K.** 2013. Osteoclast-secreted CTHRC1 in the coupling of bone resorption to formation. *J Clin Invest* **123**:3914–3924.
142. **Negishi-Koga T, Shinohara M, Komatsu N, Bito H, Kodama T, Friedel RH, Takayanagi H.** 2011. Suppression of bone formation by osteoclastic expression of semaphorin 4D. *Nat Med* **17**:1473–1480.
143. **Irie N, Takada Y, Watanabe Y, Matsuzaki Y, Naruse C, Asano M, Iwakura Y, Suda T, Matsuo K.** 2009. Bidirectional signaling through ephrinA2-EphA2 enhances osteoclastogenesis and suppresses osteoblastogenesis. *J Biol Chem* **284**:14637–14644.
144. **Zhao C, Irie N, Takada Y, Shimoda K, Miyamoto T, Nishiwaki T, Suda T, Matsuo K.** 2006. Bidirectional ephrinB2-EphB4 signaling controls bone homeostasis. *Cell Metab* **4**:111–121.
145. **Cauley JA.** 2015. Estrogen and bone health in men and women. *Steroids* **99**(Pt A):11–15.
146. **Manolagas SC, O'Brien CA, Almeida M.** 2013. The role of estrogen and androgen receptors in bone health and disease. *Nat Rev Endocrinol* **9**:699–712.
147. **Andreopoulou P, Bockman RS.** 2015. Management of postmenopausal osteoporosis. *Annu Rev Med* **66**:329–342.
148. **Franceschi C, Campisi J.** 2014. Chronic inflammation (inflammaging) and its potential contribution to age-associated diseases. *J Gerontol A Biol Sci Med Sci* **69**: S4–S9.
149. **Sanguineti R, Puddu A, Mach F, Montecucco F, Viviani GL.** 2014. Advanced glycation end products play adverse proinflammatory activities in osteoporosis. *Mediators Inflamm* **975872**: doi:10.1155/2014/975872.
150. **D'Amelio P, Grimaldi A, Di Bella S, Brianza SZ, Cristofaro MA, Tamone C, Giribaldi G, Ulliers D, Pescarmona GP, Isaia G.** 2008. Estrogen deficiency increases osteoclastogenesis up-regulating T cells activity: a key mechanism in osteoporosis. *Bone* **43**:92–100.
151. **Cenci S, Weitzmann MN, Roggia C, Namba N, Novack D, Woodring J, Pacifici R.** 2000. Estrogen deficiency induces bone loss by enhancing T-cell production of TNF-α. *J Clin Invest* **106**:1229–1237.
152. **Koenders MI, van den Berg WB.** 2015. Novel therapeutic targets in rheumatoid arthritis. *Trends Pharmacol Sci* **36**:189–195.
153. **Tanaka T, Narazaki M, Kishimoto T.** 2014. IL-6 in inflammation, immunity, and disease. *Cold Spring Harb Perspect Biol* **6**:a016295. doi:10.1101/cshperspect.a016295.
154. **van Staa TP, Geusens P, Bijlsma JW, Leufkens HG, Cooper C.** 2006. Clinical assessment of the long-term risk of fracture in patients with rheumatoid arthritis. *Arthritis Rheum* **54**:3104–3112.
155. **Harre U, Georgess D, Bang H, Bozec A, Axmann R, Ossipova E, Jakobsson P-J, Baum W, Nimmerjahn F, Szarka E, Sarmay G, Krumbholz G, Neumann E, Toes R, Scherer HU, Catrina AI, Klareskog L, Jurdic P, Schett G.** 2012. Induction of osteoclastogenesis and bone loss by human autoantibodies against citrullinated vimentin. *J Clin Invest* **122**:1791–1802.
156. **Seki N, Sudo Y, Yoshioka T, Sugihara S, Fujitsu T, Sakuma S, Ogawa T, Hamaoka T, Senoh H, Fujiwara H.** 1988. Type II collagen-induced murine arthritis. I. Induction and perpetuation of arthritis require synergy between humoral and cell-mediated immunity. *J Immunol* **140**:1477–1484.
157. **Brackertz D, Mitchell GF, Mackay IR.** 1977. Antigen-induced arthritis in mice. I. Induction of arthritis in various strains of mice. *Arthritis Rheum* **20**:841–850.
158. **Korganow AS, Ji H, Mangialaio S, Duchatelle V, Pelanda R, Martin T, Degott C, Kikutani H, Rajewsky K, Pasquali JL, Benoist C, Mathis D.** 1999. From systemic T cell self-reactivity to organ-specific autoimmune disease via immunoglobulins. *Immunity* **10**:451–461.
159. **Khachigian LM.** 2006. Collagen antibody-induced arthritis. *Nat Protoc* **1**:2512–2516.
160. **Keffer J, Probert L, Cazlaris H, Georgopoulos S, Kaslaris E, Kioussis D, Kollias G.** 1991. Transgenic mice expressing human tumour necrosis factor: a predictive genetic model of arthritis. *EMBO J* **10**:4025–4031.
161. **Li P, Schwarz E.** 2003. The TNF-α transgenic mouse model of inflammatory arthritis. *Springer Semin Immunopathol* **25**:19–33.
162. **Mukai T, Gallant R, Ishida S, Kittaka M, Yoshitaka T, Fox DA, Morita Y, Nishida K, Rottapel R, Ueki Y.** 2015. Loss of SH3 domain-binding protein 2 function suppresses bone destruction in tumor necrosis factor-driven and collagen-induced arthritis in mice. *Arthritis Rheumatol* **67**:656–667.
163. **Aya K, Alhawagri M, Hagen-Stapleton A, Kitaura H, Kanagawa O, Novack DV.** 2005. NF-κB-inducing

kinase controls lymphocyte and osteoclast activities in inflammatory arthritis. *J Clin Invest* **115:**1848–1854.

164. **Cremasco V, Benasciutti E, Cella M, Kisseleva M, Croke M, Faccio R.** 2010. Phospholipase C gamma 2 is critical for development of a murine model of inflammatory arthritis by affecting actin dynamics in dendritic cells. *PLoS One* **5:**e8909. doi:10.1371/journal.pone.0008909.

165. **Masters SL, Simon A, Aksentijevich I, Kastner DL.** 2009. *Horror autoinflammaticus*: the molecular pathophysiology of autoinflammatory disease. *Annu Rev Immunol* **27:**621–668.

166. **Hill S, Namde M, Dwyer A, Poznanski A, Canna S, Goldbach-Mansky R.** 2007. Arthropathy of neonatal onset multisystem inflammatory disease (NOMID/CINCA). *Pediatr Radiol* **37:**145–152.

167. **Schoindre Y, Feydy A, Giraudet-Lequintrec JS, Kahan A, Allanore Y.** 2009. TNF receptor-associated periodic syndrome (TRAPS): a new cause of joint destruction? *Joint Bone Spine* **76:**567–569.

168. **Koca SS, Etem EO, Isik B, Yuce H, Ozgen M, Dag MS, Isik A.** 2010. Prevalence and significance of *MEFV* gene mutations in a cohort of patients with rheumatoid arthritis. *Joint Bone Spine* **77:**32–35.

169. **Lang BA, Schneider R, Reilly BJ, Silverman ED, Laxer RM.** 1995. Radiologic features of systemic onset juvenile rheumatoid arthritis. *J Rheumatol* **22:**168–173.

170. **Rosé CD, Pans S, Casteels I, Anton J, Bader-Meunier B, Brissaud P, Cimaz R, Espada G, Fernandez-Martin J, Hachulla E, Harjacek M, Khubchandani R, Mackensen F, Merino R, Naranjo A, Oliveira-Knupp S, Pajot C, Russo R, Thomée C, Vastert S, Wulffraat N, Arostegui JI, Foley KP, Bertin J, Wouters CH.** 2015. Blau syndrome: cross-sectional data from a multicentre study of clinical, radiological and functional outcomes. *Rheumatology (Oxford)* **54:**1008–1016.

171. **Chen B, Wu W, Sun W, Zhang Q, Yan F, Xiao Y.** 2014. RANKL expression in periodontal disease: where does RANKL come from? *BioMed Res Int* **731039:** doi: 10.1155/2014/731039.

172. **Zhang P, Liu J, Xu Q, Harber G, Feng X, Michalek SM, Katz J.** 2011. TLR2-dependent modulation of osteoclastogenesis by *Porphyromonas* gingivalis through differential induction of NFATc1 and NF-κB. *J Biol Chem* **286:**24159–24169.

173. **Assuma R, Oates T, Cochran D, Amar S, Graves DT.** 1998. IL-1 and TNF antagonists inhibit the inflammatory response and bone loss in experimental periodontitis. *J Immunol* **160:**403–409.

174. **Graves DT, Oskoui M, Voleinikova S, Naguib G, Cai S, Desta T, Kakouras A, Jiang Y.** 2001. Tumor necrosis factor modulates fibroblast apoptosis, PMN recruitment, and osteoclast formation in response to *P. gingivalis* infection. *J Dent Res* **80:**1875–1879.

175. **Weilbaecher KN, Guise TA, McCauley LK.** 2011. Cancer to bone: a fatal attraction. *Nat Rev Cancer* **11:** 411–425.

176. **Hirbe AC, Uluçkan O, Morgan EA, Eagleton MC, Prior JL, Piwnica-Worms D, Trinkaus K, Apicelli A, Weilbaecher K.** 2007. Granulocyte colony-stimulating factor enhances bone tumor growth in mice in an osteoclast-dependent manner. *Blood* **109:**3424–3431.

177. **Yang C, Davis JL, Zeng R, Vora P, Su X, Collins LI, Vangveravong S, Mach RH, Piwnica-Worms D, Weilbaecher KN, Faccio R, Novack DV.** 2013. Antagonism of inhibitor of apoptosis proteins increases bone metastasis via unexpected osteoclast activation. *Cancer Discov* **3:**212–223.

178. **Morony S, Capparelli C, Sarosi I, Lacey DL, Dunstan CR, Kostenuik PJ.** 2001. Osteoprotegerin inhibits osteolysis and decreases skeletal tumor burden in syngeneic and nude mouse models of experimental bone metastasis. *Cancer Res* **61:**4432–4436.

179. **Canon JR, Roudier M, Bryant R, Morony S, Stolina M, Kostenuik PJ, Dougall WC.** 2008. Inhibition of RANKL blocks skeletal tumor progression and improves survival in a mouse model of breast cancer bone metastasis. *Clin Exp Metastasis* **25:**119–129.

180. **Gampenrieder SP, Rinnerthaler G, Greil R.** 2014. Bone-targeted therapy in metastatic breast cancer—all well-established knowledge? *Breast Care (Basel)* **9:** 323–330.

181. **Otero K, Shinohara M, Zhao H, Cella M, Gilfillan S, Colucci A, Faccio R, Ross FP, Teitelbaum SL, Takayanagi H, Colonna M.** 2012. TREM2 and β-catenin regulate bone homeostasis by controlling the rate of osteoclastogenesis. *J Immunol* **188:**2612–2621.

182. **Oba Y, Lee JW, Ehrlich LA, Chung HY, Jelinek DF, Callander NS, Horuk R, Choi SJ, Roodman GD.** 2005. MIP-1α utilizes both CCR1 and CCR5 to induce osteoclast formation and increase adhesion of myeloma cells to marrow stromal cells. *Exp Hematol* **33:**272–278.

183. **Han JH, Choi SJ, Kurihara N, Koide M, Oba Y, Roodman GD.** 2001. Macrophage inflammatory protein-1α is an osteoclastogenic factor in myeloma that is independent of receptor activator of nuclear factor κB ligand. *Blood* **97:**3349–3353.

184. **Tawara K, Oxford JT, Jorcyk CL.** 2011. Clinical significance of interleukin (IL)-6 in cancer metastasis to bone: potential of anti-IL-6 therapies. *Cancer Manag Res* **3:** 177–189.

185. **Hurchla MA, Garcia-Gomez A, Hornick MC, Ocio EM, Li A, Blanco JF, Collins L, Kirk CJ, Piwnica-Worms D, Vij R, Tomasson MH, Pandiella A, San Miguel JF, Garayoa M, Weilbaecher KN.** 2013. The epoxyketone-based proteasome inhibitors carfilzomib and orally bioavailable oprozomib have anti-resorptive and bone-anabolic activity in addition to anti-myeloma effects. *Leukemia* **27:**430–440.

186. **von Metzler I, Krebbel H, Hecht M, Manz RA, Fleissner C, Mieth M, Kaiser M, Jakob C, Sterz J, Kleeberg L, Heider U, Sezer O.** 2007. Bortezomib inhibits human osteoclastogenesis. *Leukemia* **21:**2025–2034.

187. **Boissy P, Andersen TL, Lund T, Kupisiewicz K, Plesner T, Delaissé JM.** 2008. Pulse treatment with the proteasome inhibitor bortezomib inhibits osteoclast resorptive activity in clinically relevant conditions. *Leuk Res* **32:**1661–1668.

188. **Keats JJ, Fonseca R, Chesi M, Schop R, Baker A, Chng WJ, Van Wier S, Tiedemann R, Shi CX, Sebag M, Braggio E, Henry T, Zhu YX, Fogle H, Price-Troska T, Ahmann G, Mancini C, Brents LA, Kumar S, Greipp P, Dispenzieri A, Bryant B, Mulligan G, Bruhn L, Barrett M, Valdez R, Trent J, Stewart AK, Carpten J, Bergsagel PL.** 2007. Promiscuous mutations activate the non-canonical NF-κB pathway in multiple myeloma. *Cancer Cell* **12:**131–144.

189. **Annunziata CM, Davis RE, Demchenko Y, Bellamy W, Gabrea A, Zhan F, Lenz G, Hanamura I, Wright G, Xiao W, Dave S, Hurt EM, Tan B, Zhao H, Stephens O, Santra M, Williams DR, Dang L, Barlogie B, Shaughnessy JD Jr, Kuehl WM, Staudt LM.** 2007. Frequent engagement of the classical and alternative NF-κB pathways by diverse genetic abnormalities in multiple myeloma. *Cancer Cell* **12:**115–130.

190. **Vaira S, Johnson T, Hirbe AC, Alhawagri M, Anwisye I, Sammut B, O'Neal J, Zou W, Weilbaecher KN, Faccio R, Novack DV.** 2008. RelB is the NF-κB subunit downstream of NIK responsible for osteoclast differentiation. *Proc Natl Acad Sci USA* **105:**3897–3902.

191. **Novack DV, Yin L, Hagen-Stapleton A, Schreiber RD, Goeddel DV, Ross FP, Teitelbaum SL.** 2003. The IκB function of NF-κB2 p100 controls stimulated osteoclastogenesis. *J Exp Med* **198:**771–781.

192. **Yang C, McCoy K, Davis JL, Schmidt-Supprian M, Sasaki Y, Faccio R, Novack DV.** 2010. NIK stabilization in osteoclasts results in osteoporosis and enhanced inflammatory osteolysis. *PLoS One* 5:e15383. doi: 10.1371/journal.pone.0015383.

193. **Demchenko YN, Brents LA, Li Z, Bergsagel LP, McGee LR, Kuehl MW.** 2014. Novel inhibitors are cytotoxic for myeloma cells with NFkB inducing kinase-dependent activation of NFkB. *Oncotarget* **5:**4554–4566.

194. **Kameda T, Mano H, Yuasa T, Mori Y, Miyazawa K, Shiokawa M, Nakamaru Y, Hiroi E, Hiura K, Kameda A, Yang NN, Hakeda Y, Kumegawa M.** 1997. Estrogen inhibits bone resorption by directly inducing apoptosis of the bone-resorbing osteoclasts. *J Exp Med* **186:** 489–495.

195. **Hughes DE, Dai A, Tiffee JC, Li HH, Mundy GR, Boyce BF.** 1996. Estrogen promotes apoptosis of murine osteoclasts mediated by TGF-beta. *Nat Med* **2:** 1132–1136.

196. **Nakamura T, Imai Y, Matsumoto T, Sato S, Takeuchi K, Igarashi K, Harada Y, Azuma Y, Krust A, Yamamoto Y, Nishina H, Takeda S, Takayanagi H, Metzger D, Kanno J, Takaoka K, Martin TJ, Chambon P, Kato S.** 2007. Estrogen prevents bone loss via estrogen receptor α and induction of Fas ligand in osteoclasts. *Cell* **130:**811–823.

197. **Shevde NK, Bendixen AC, Dienger KM, Pike JW.** 2000. Estrogens suppress RANK ligand-induced osteoclast differentiation via a stromal cell independent mechanism involving c-Jun repression. *Proc Natl Acad Sci USA* **97:**7829–7834.

198. **Shevde N, Pike J.** 1996. Estrogen modulates the recruitment of myelopoietic cell progenitors in rat through a stromal cell-independent mechanism involving apoptosis. *Blood* 87:2683–2692.

199. **D'Amelio P, Grimaldi A, Pescarmona GP, Tamone C, Roato I, Isaia G.** 2004. Spontaneous osteoclast formation from peripheral blood mononuclear cells in postmenopausal osteoporosis. *FASEB J* **19:**410–412.

200. **Slyfield CR, Tkachenko EV, Wilson DL, Hernandez CJ.** 2012. Three-dimensional dynamic bone histomorphometry. *J Bone Miner Res* **27:**486–495.

201. **Martin-Millan M, Almeida M, Ambrogini E, Han L, Zhao H, Weinstein RS, Jilka RL, O'Brien CA, Manolagas SC.** 2010. The estrogen receptor-α in osteoclasts mediates the protective effects of estrogens on cancellous but not cortical bone. *Mol Endocrinol* **24:** 323–334.

202. **Almeida M, Iyer S, Martin-Millan M, Bartell SM, Han L, Ambrogini E, Onal M, Xiong J, Weinstein RS, Jilka RL, O'Brien CA, Manolagas SC.** 2013. Estrogen receptor-α signaling in osteoblast progenitors stimulates cortical bone accrual. *J Clin Invest* **123:**394–404.

203. **Khosla S.** 2010. Pathogenesis of osteoporosis. *Transl Endocrinol Metab* **1:**55–86.

204. **Roggia C, Gao Y, Cenci S, Weitzmann MN, Toraldo G, Isaia G, Pacifici R.** 2001. Up-regulation of TNF-producing T cells in the bone marrow: a key mechanism by which estrogen deficiency induces bone loss *in vivo*. *Proc Natl Acad Sci USA* **98:**13960–13965.

205. **Charatcharoenwitthaya N, Khosla S, Atkinson EJ, McCready LK, Riggs BL.** 2007. Effect of blockade of TNF-α and interleukin-1 action on bone resorption in early postmenopausal women. *J Bone Miner Res* **22:** 724–729.

206. **Johnson RA, Boyce BF, Mundy GR, Roodman GD.** 1989. Tumors producing human tumor necrosis factor induce hypercalcemia and osteoclastic bone resorption in nude mice. *Endocrinology* **124:**1424–1427.

207. **Pfeilschifter J, Chenu C, Bird A, Mundy GR, Roodman DG.** 1989. Interleukin-1 and tumor necrosis factor stimulate the formation of human osteoclastlike cells in vitro. *J Bone Miner Res* **4:**113–118.

208. **Wei S, Kitaura H, Zhou P, Ross FP, Teitelbaum SL.** 2005. IL-1 mediates TNF-induced osteoclastogenesis. *J Clin Invest* **115:**282–290.

209. **Kaplan DL, Eielson CM, Horowitz MC, Insogna KL, Weir EC.** 1996. Tumor necrosis factor-α induces transcription of the colony-stimulating factor-1 gene in murine osteoblasts. *J Cell Physiol* **168:**199–208.

210. **Azuma Y, Kaji K, Katogi R, Takeshita S, Kudo A.** 2000. Tumor necrosis factor-α induces differentiation of and bone resorption by osteoclasts. *J Biol Chem* **275:** 4858–4864.

211. **Kobayashi K, Takahashi N, Jimi E, Udagawa N, Takami M, Kotake S, Nakagawa N, Kinosaki M, Yamaguchi K, Shima N, Yasuda H, Morinaga T, Higashio K, Martin TJ, Suda T.** 2000. Tumor necrosis factor α stimulates osteoclast differentiation by a mechanism independent of the ODF/RANKL-RANK interaction. *J Exp Med* **191:**275–286.

212. **Kanazawa K, Kudo A.** 2005. TRAF2 is essential for TNF-α-induced osteoclastogenesis. *J Bone Miner Res* **20:**840–847.

213. **Kudo O, Fujikawa Y, Itonaga I, Sabokbar A, Torisu T, Athanasou NA.** 2002. Proinflammatory cytokine (TNFα/IL-1α) induction of human osteoclast formation. *J Pathol* **198:**220–227.

214. **O'Gradaigh D, Ireland D, Bord S, Compston JE.** 2004. Joint erosion in rheumatoid arthritis: interactions between tumour necrosis factor α, interleukin 1, and receptor activator of nuclear factor κB ligand (RANKL) regulate osteoclasts. *Ann Rheum Dis* **63:**354–359.

215. **Yao Z, Xing L, Boyce BF.** 2009. NF-κB p100 limits TNF-induced bone resorption in mice by a TRAF3-dependent mechanism. *J Clin Invest* **119:**3024–3034.

216. **Zhao B, Grimes SN, Li S, Hu X, Ivashkiv LB.** 2012. TNF-induced osteoclastogenesis and inflammatory bone resorption are inhibited by transcription factor RBP-J. *J Exp Med* **209:**319–334.

217. **Nakamura I, Jimi E.** 2006. Regulation of osteoclast differentiation and function by interleukin-1. *Vitam Horm* **74:**357–370.

218. **Ma T, Miyanishi K, Suen A, Epstein NJ, Tomita T, Smith RL, Goodman SB.** 2004. Human interleukin-1-induced murine osteoclastogenesis is dependent on RANKL, but independent of TNF-α. *Cytokine* **26:** 138–144.

219. **Jules J, Zhang P, Ashley JW, Wei S, Shi Z, Liu J, Michalek SM, Feng X.** 2012. Molecular basis of requirement of receptor activator of nuclear factor κB signaling for interleukin 1-mediated osteoclastogenesis. *J Biol Chem* **287:**15728–15738.

220. **Kim JH, Jin HM, Kim K, Song I, Youn BU, Matsuo K, Kim N.** 2009. The mechanism of osteoclast differentiation induced by IL-1. *J Immunol* **183:**1862–1870.

221. **Nakamura I, Kadono Y, Takayanagi H, Jimi E, Miyazaki T, Oda H, Nakamura K, Tanaka S, Rodan GA, Duong LT.** 2002. IL-1 regulates cytoskeletal organization in osteoclasts via TNF receptor-associated factor 6/c-Src complex. *J Immunol* **168:**5103–5109.

222. **Jimi E, Nakamura I, Ikebe T, Akiyama S, Takahashi N, Suda T.** 1998. Activation of NF-κB is involved in the survival of osteoclasts promoted by interleukin-1. *J Biol Chem* **273:**8799–8805.

223. **Jimi E, Nakamura I, Duong LT, Ikebe T, Takahashi N, Rodan GA, Suda T.** 1999. Interleukin 1 induces multinucleation and bone-resorbing activity of osteoclasts in the absence of osteoblasts/stromal cells. *Exp Cell Res* **247:**84–93.

224. **Fox SW, Haque SJ, Lovibond AC, Chambers TJ.** 2003. The possible role of TGF-β-induced suppressors of cytokine signaling expression in osteoclast/macrophage lineage commitment in vitro. *J Immunol* **170:**3679–3687.

225. **Yan T, Riggs BL, Boyle WJ, Khosla S.** 2001. Regulation of osteoclastogenesis and RANK expression by TGF-β1. *J Cell Biochem* **83:**320–325.

226. **Fox SW, Lovibond AC.** 2005. Current insights into the role of transforming growth factor-β in bone resorption. *Mol Cell Endocrinol* **243:**19–26.

227. **Yasui T, Kadono Y, Nakamura M, Oshima Y, Matsumoto T, Masuda H, Hirose J, Omata Y, Yasuda H, Imamura T, Nakamura K, Tanaka S.** 2011. Regulation of RANKL-induced osteoclastogenesis by TGF-β through molecular interaction between Smad3 and Traf6. *J Bone Miner Res* **26:**1447–1456.

228. **Omata Y, Yasui T, Hirose J, Izawa N, Imai Y, Matsumoto T, Masuda H, Tokuyama N, Nakamura S, Tsutsumi S, Yasuda H, Okamoto K, Takayanagi H, Hikita A, Imamura T, Matsuo K, Saito T, Kadono Y, Aburatani H, Tanaka S.** 2014. Genome-wide comprehensive analysis reveals critical cooperation between Smad and c-Fos in RANKL-induced osteoclastogenesis. *J Bone Miner Res* **30:**869–877.

229. **Asai K, Funaba M, Murakami M.** 2014. Enhancement of RANKL-induced MITF-E expression and osteoclastogenesis by TGF-β. *Cell Biochem Funct* **32:**401–409.

230. **Mansky KC, Sankar U, Han J, Ostrowski MC.** 2002. Microphthalmia transcription factor is a target of the p38 MAPK pathway in response to receptor activator of NF-κB ligand signaling. *J Biol Chem* **277:**11077–11083.

231. **Fuller K, Kirstein B, Chambers TJ.** 2006. Murine osteoclast formation and function: differential regulation by humoral agents. *Endocrinology* **147:**1979–1985.

232. **Mundy GR.** 1991. The effects of TGF-beta on bone. *Ciba Found Symp* **157:**137–143, discussion 143–151.

233. **Dieudonne SC, Foo P, van Zoelen EJ, Burger EH.** 1991. Inhibiting and stimulating effects of TGF-β_1 on osteoclastic bone resorption in fetal mouse bone organ cultures. *J Bone Miner Res* **6:**479–487.

234. **Edwards JR, Nyman JS, Lwin ST, Moore MM, Esparza J, O'Quinn EC, Hart AJ, Biswas S, Patil CA, Lonning S, Mahadevan-Jansen A, Mundy GR.** 2010. Inhibition of TGF-β signaling by 1D11 antibody treatment increases bone mass and quality in vivo. *J Bone Miner Res* **25:** 2419–2426.

235. **Mohammad KS, Chen CG, Balooch G, Stebbins E, McKenna CR, Davis H, Niewolna M, Peng XH, Nguyen DHN, Ionova-Martin SS, Bracey JW, Hogue WR, Wong DH, Ritchie RO, Suva LJ, Derynck R, Guise TA, Alliston T.** 2009. Pharmacologic inhibition of the TGF-β type I receptor kinase has anabolic and anti-catabolic effects on bone. *PLoS One* **4:**e5275. doi: 10.1371/journal.pone.0005275.

236. **Yin JJ, Selander K, Chirgwin JM, Dallas M, Grubbs BG, Wieser R, Massagué J, Mundy GR, Guise TA.** 1999. TGF-β signaling blockade inhibits PTHrP secretion by breast cancer cells and bone metastases development. *J Clin Invest* **103:**197–206.

237. **Massagué J.** 2012. TGFβ signalling in context. *Nat Rev Mol Cell Biol* **13:**616–630.

238. **Bragado P, Estrada Y, Parikh F, Krause S, Capobianco C, Farina HG, Schewe DM, Aguirre-Ghiso JA.** 2013. TGFβ2 dictates disseminated tumour cell fate in target organs through TGFβ-RIII and p38α/β signalling. *Nat Cell Biol* **15:**1351–1361.

239. **Liu S, Song W, Boulanger JH, Tang W, Sabbagh Y, Kelley B, Gotschall R, Ryan S, Phillips L, Malley K, Cao X, Xia TH, Zhen G, Cao X, Ling H, Dechow PC,**

Bellido TM, Ledbetter SR, Schiavi SC. 2014. Role of TGF-β in a mouse model of high turnover renal osteodystrophy. *J Bone Miner Res* **29:**1141–1157.

240. **Nistala H, Lee-Arteaga S, Smaldone S, Siciliano G, Ramirez F.** 2010. Extracellular microfibrils control osteoblast-supported osteoclastogenesis by restricting TGFβ stimulation of RANKL production. *J Biol Chem* **285:**34126–34133.

241. **Hayata T, Ezura Y, Asashima M, Nishinakamura R, Noda M.** 2015. *Dullard/Ctdnep1* regulates endochondral ossification via suppression of TGF-β signaling. *J Bone Miner Res* **30:**318–329.

242. **Craft CS, Broekelmann TJ, Zou W, Chappel JC, Teitelbaum SL, Mecham RP.** 2012. Oophorectomy-induced bone loss is attenuated in MAGP1-deficient mice. *J Cell Biochem* **113:**93–99.

243. **Rhodes SD, Wu X, He Y, Chen S, Yang H, Staser KW, Wang J, Zhang P, Jiang C, Yokota H, Dong R, Peng X, Yang X, Murthy S, Azhar M, Mohammad KS, Xu M, Guise TA, Yang FC.** 2013. Hyperactive transforming growth factor-β1 signaling potentiates skeletal defects in a neurofibromatosis type 1 mouse model. *J Bone Miner Res* **28:**2476–2489.

244. **Takayanagi H, Ogasawara K, Hida S, Chiba T, Murata S, Sato K, Takaoka A, Yokochi T, Oda H, Tanaka K, Nakamura K, Taniguchi T.** 2000. T-cell-mediated regulation of osteoclastogenesis by signalling cross-talk between RANKL and IFN-γ. *Nature* **408:**600–605.

245. **Ji JD, Park-Min KH, Shen Z, Fajardo RJ, Goldring SR, McHugh KP, Ivashkiv LB.** 2009. Inhibition of RANK expression and osteoclastogenesis by TLRs and IFN-γ in human osteoclast precursors. *J Immunol* **183:**7223–7233.

246. **Huang W, O'Keefe R, Schwarz E.** 2003. Exposure to receptor-activator of NFκB ligand renders pre-osteoclasts resistant to IFN-γ by inducing terminal differentiation. *Arthritis Res Ther* **5:**R49–R59.

247. **Vermeire K, Heremans H, Vandeputte M, Huang S, Billiau A, Matthys P.** 1997. Accelerated collagen-induced arthritis in IFN-γ receptor-deficient mice. *J Immunol* **158:**5507–5513.

248. **Xu Z, Hurchla MA, Deng H, Uluçkan O, Bu F, Berdy A, Eagleton MC, Heller EA, Floyd DH, Dirksen WP, Shu S, Tanaka Y, Fernandez SA, Rosol TJ, Weilbaecher KN.** 2009. Interferon-γ targets cancer cells and osteoclasts to prevent tumor-associated bone loss and bone metastases. *J Biol Chem* **284:**4658–4666.

249. **Key LL Jr, Rodriguiz RM, Willi SM, Wright NM, Hatcher HC, Eyre DR, Cure JK, Griffin PP, Ries WL.** 1995. Long-term treatment of osteopetrosis with recombinant human interferon gamma. *N Engl J Med* **332:**1594–1599.

250. **Madyastha PR, Yang S, Ries WL, Key LL Jr.** 2000. IFN-γ enhances osteoclast generation in cultures of peripheral blood from osteopetrotic patients and normalizes superoxide production. *J Interferon Cytokine Res* **20:**645–652.

251. **Cenci S, Toraldo G, Weitzmann MN, Roggia C, Gao Y, Qian WP, Sierra O, Pacifici R.** 2003. Estrogen deficiency induces bone loss by increasing T cell proliferation and lifespan through IFN-γ-induced class II transactivator. *Proc Natl Acad Sci USA* **100:**10405–10410.

252. **Gao Y, Grassi F, Ryan MR, Terauchi M, Page K, Yang X, Weitzmann MN, Pacifici R.** 2007. IFN-γ stimulates osteoclast formation and bone loss in vivo via antigen-driven T cell activation. *J Clin Invest* **117:**122–132.

Myeloid Cells in Health and Disease: A Synthesis
Edited by Siamon Gordon

doi:10.1128/microbiolspec.MCHD-0020-2015

Ting Wen[1]
Marc E. Rothenberg[1]

14 The Regulatory Function of Eosinophils

EOSINOPHILS ARE KEY INNATE REGULATOR/EFFECTOR CELLS

Eosinophils represent a minor component of circulating leukocytes and are generally considered to be terminally differentiated as postmitotic cells, yet it is now appreciated that they can be long-lived, multifaceted granulocytes involved in a variety of regulatory functions. Like other granulocytes, eosinophils develop and differentiate in the bone marrow. Under homeostasis, eosinophils are distributed in the blood, lung, thymus, uterus, adipose tissues, mammary gland, spleen and the lamina propria of the gastrointestinal (GI) tract (1), indicating a physiological function in each organ. Although eosinophils outside of the bone marrow are deemed as mature, recent evidence suggests the existence of multiple tissue-specific subtypes on the basis of distinct cell surface marker expression and functions (2–4). Driven by eosinophil-specific chemokines (primarily eotaxins) produced at baseline and markedly upregulated after a variety of stimuli (5), mature eosinophils are recruited from the circulation into their physiological locations and inflammatory sites, respectively. The cytokine interleukin-5 (IL-5), produced primarily by type 2 T helper (Th2) cells (6) and type 2 innate helper lymphoid cells (ILC2) (7), is crucial for eosinophil differentiation, priming, and survival (8). Conversely, eosinophils also serve as a source of a variety of cytokines and growth factors closely associated with multiple immunomodulatory functions to be discussed later. Through their vast cytokine arsenal and engagement of cell contact, eosinophils modulate immune responses through an array of interactive and orchestrated mechanisms, in *trans* and *cis* fashions, by cellular and humoral mediators, in both innate and adaptive immune responses. Recently, a burgeoning body of evidence has uncovered several underappreciated roles for eosinophils that could modulate both the adaptive and innate arms of immunity. An essential goal of this chapter is to summarize the role of eosinophils in physiological and inflammatory processes in human and small mammal models in order to identify novel pharmacological targets for specific disease management.

EOSINOPHILS INTERACTIVELY REGULATE MULTIPLE COMPONENTS OF ADAPTIVE IMMUNITY

Eosinophils Modulate Lymphocyte Recruitment and Homeostasis

The canonical theory for eosinophil recruitment into the GI and the lung tissue is highlighted by the "T helper-

[1]Division of Allergy and Immunology, Cincinnati Children's Hospital Medical Center, University of Cincinnati College of Medicine, Cincinnati, OH 45229.

Th2 cytokine-epithelium-eosinophil chemokine" axis (9), emphasizing the influence of lymphocytes on eosinophil recruitment. Specifically, antigen-experienced local Th2 cells produce the cytokine IL-5, which promotes eosinophil production, priming, and survival; and IL-13, which induces local cells to produce eosinophil-specific chemokines—the eotaxins—which attract circulating eosinophils into their niche. There are three eotaxins identified, namely, eotaxin 1, 2, and 3, all of which were shown to induce eosinophilia in asthma models and some human diseases. Meanwhile, the inflammatory environment (e.g., tumor necrosis factor α) prepares endothelial cells for adhesion. However, the ability of eosinophils to influence lymphocyte recruitment and activity was not appreciated until several recent studies provided alternative evidence showing a pronounced reduction of Th2 cytokines and effector T-cell recruitment in eosinophil-deficient mice, a deficit fully rescued by eosinophil reintroduction (10), suggesting that eosinophils are critical for T-cell homing in the lung. Indeed, in an *Il5*$^{-/-}$ background, the airway hyperreactivity and mucus production associated with experimental asthma were shown to be functions of both T-cell and eosinophil factors, as each alone was insufficient for induction of allergic airway inflammation (11). Therefore, at least in a rodent pulmonary system, eosinophils may interact with lymphocytes in a bidirectional manner rather than passively responding to chemotactic and priming signals. Consistent with this theory, in eosinophil-deficient ΔdblGATA-1 mice, reduced Peyer's patch development and T helper cell cytokine production was observed, highlighting the promoting role of eosinophils on lymphocyte homeostasis (10, 12, 13). Extending this paradigm to B cells, it has recently become appreciated that bone marrow eosinophils colocalize with plasma cells during their maturation, secrete the cytokines APRIL (a proliferation-inducing ligand) and IL-6, and contribute to the survival of bone marrow plasma cells, whose death is augmented in eosinophil-deficient ΔdblGATA-1 mice (14). In addition, the B-cell process of IgA class switching was recently found to be positively regulated by GI eosinophils in the intestinal tissue (12, 13). Finally, recent studies have uncovered a novel role for eosinophils in promoting B-cell proliferation upon eosinophil activation in mice and a positive correlation between blood eosinophil and B-cell counts in humans (15). In light of these findings, a systemic scanning of lymphocyte phenotypes and functions in eosinophil-deficient mice (not restricted to the pulmonary system) should be prioritized.

Eosinophils Behave as Antigen-Presenting Cells

Other than regulating lymphocyte recruitment and function, it is now appreciated that eosinophils have the capacity to present antigen to T cells. This topic stems from the original findings that granulocyte-macrophage colony-stimulating factor (GM-CSF)-treated eosinophils have a "nonprofessional" antigen presentation function *in vitro*, as shown by their capacity to induce antigen-specific T-cell clone proliferation (16). Eosinophils, after allergen exposure, express the machinery for antigen presentation and a full set of costimulation molecules, including major histocompatibility complex class II (MHC-II), CD80, CD86, CD9, CD28, and CD40, at the protein level (17, 18). Eosinophils labeled in the airway lumen migrate to the draining lymph node, reaching the T-cell proliferation zone, in a process that is independent of the eotaxin receptor CCR3. Moreover, these antigen-experienced eosinophils promote antigen-specific T-cell proliferation *ex vivo*, suggesting an antigen-presenting cell (APC) behavior. Antigen (ovalbumin [OVA])-loaded and GM-CSF-treated eosinophils, when instilled intratracheally, promote the proliferation of adoptively transferred OVA-specific T-cell clones, which is accompanied by T-cell CD69 upregulation and IL-4 production and T-cell/eosinophil colocalization in the draining lymph node (19). In a murine allergic asthma model, peripheral eosinophil recruitment into the lymph node is required for antigen-specific T-cell proliferation *in situ* (20). In addition to the pulmonary system, murine studies have also demonstrated the antigen-presenting capacity of eosinophils in threadworm (21) and fungus infections (22). Moreover, considering that eosinophils are potent cytokine producers and regulators of humoral immunity and are usually isolated from a population rich in professional APCs, the direct evidence for the physical T-cell/eosinophil interaction (e.g., by confocal or intravital microscopy) is needed. Conceivably, eosinophil-deficient or eosinophil-specific (23) MHC-II-deficient mice would serve as the best tools to substantiate this interaction.

Eosinophils, Alone or via Dendritic Cells, Drive Th2 Polarization

Eosinophils are capable of driving a Th2 response in multiple ways. A crucial Th2 characteristic of eosinophils is their capacity to produce canonical Th2 cytokines (IL-4, IL-5, and IL-13) upon stimulation (1, 24). In addition, eosinophils isolated from patients with asthma may sustain Th2 polarization by maintaining a high intracellular indoleamine 2,3-dioxygenase (IDO) level, with IDO being a Th2 differentiation regulator (25). In mice, eosinophils are required for dendritic cells (DCs) and T cells to initiate Th2 inflammation in the lung (20). Emerging evidence also suggests that eosinophils suppress Th17 and Th1 responses via DC regulation, which will be further discussed below.

As a messenger between innate and adaptive immunity, eosinophils direct cross talk with DCs. Although eosinophils themselves are able to present antigen to T cells *de novo*, the well-recognized positive regulation by eosinophils on professional APCs should be emphasized as well. Conventional myeloid DCs (mDCs, immunotyped as IL-12, Toll-like receptor 2 [TLR2], and TLR4 positive) can ingest the eosinophil granule protein major basic protein (MBP) *in vivo*, and the physical interaction of human blood eosinophils and DCs, revealed by confocal imaging, results in DC maturation in the presence of the bacterial pathogen-associated molecular pattern (PAMP) CpG-C (26). In general, the presence of either eosinophils or DCs in tumor tissue is an indicator of positive prognosis or negative prognosis, respectively, for allograft, further suggesting the role of eosinophils in antigen presentation, either alone or in combination with the DCs (27, 28). Furthermore, the eosinophil granule protein eosinophil-derived neurotoxin (EDN), which is a member of the RNase family, has been shown to be a chemotactic factor for mDCs (29), triggering mDC cytokine production (30). Moreover, EDN specifically binds to TLR2 on DCs as an endogenous ligand, triggering the MyD88-dependent pathway in TLR-transfected HEK293 cells (31). Following OVA immunization *in vivo*, coimmunization with EDN potentiates OVA-specific IgG1 (a Th2 Ig) production, but not IgG2 and IgG3 production. Splenocytes from $TLR2^{+/+}$ mice immunized with OVA in the presence of EDN primarily produce IL-5, IL-6, IL-10, and IL-13 (Th2 cytokine signature), whereas the $TLR2^{-/-}$ splenocytes (lacking the signal through EDN) primarily produce gamma interferon (31). Collectively, these findings indicate that specific eosinophil products can serve as Th2 adjuvants via DC regulation, at least partially contributing to the Th2-promoting role of eosinophils, and facilitate maturation of other immunocytes including the lineage commitment of eosinophils themselves (32), although more studies are needed to substantiate these novel observations.

In a murine OVA-induced asthma model, cellular trafficking of eosinophils into the lung lymphatic compartment is a prerequisite for mDC accumulation in draining lymph nodes (20) and, consequently, for allergen-specific effector T-cell proliferation. Importantly, this pulmonary Th2 enhancement through DC regulation is not MHC-II (classical antigen presentation complex) or CCR7 (classical lymph node homing signal) dependent (20). Whether the DC regulation is through eosinophil cytokine secretion or cellular interaction remains to be resolved. It should also be noted that most of the studies of DC regulation are confined to the lung and use allergic asthma models; the generalization of these processes in other tissues is ripe for investigation.

Regulation of T-Cell Development in the Thymus

The thymus is one of several organs where eosinophils are readily found under homeostatic conditions. Eosinophils migrate into the thymus during the neonatal period and wane after adolescence (33). Eosinophils are present in the thymus under homeostasis and have a unique $CD11c^{+}CD11b^{+}CD44^{hi}MHC\text{-}II^{lo}SSC^{hi}$ phenotype, distinct from circulating eosinophils. Thymus-bound eosinophils exhibit an activated phenotype, as shown by expression of several surface activation markers, including CD25, CD69, and mRNA for Th2 cytokines IL-4, IL-5, IL-13, and GM-CSF (34). Initial data suggest that eosinophils have a role in MHC-I-dependent negative selection and may induce T-cell apoptosis by free radicals to facilitate negative selection. Eosinophils in the thymus are IDO positive (25) and correlate with local Th2 cytokine levels (33), suggesting a Th1/Th2-regulating role within the thymus, possibly via DC regulation.

Taken together, the results show that eosinophils actively orchestrate a chain response in adaptive immunity at earlier points than previously appreciated, as evidenced by the lymphocyte deficiency and impaired local Th2 cytokine profile seen in eosinophil-deficient mice. Notably, eosinophils also possess a robust capacity to regulate a variety of cell types in the non-adaptive immune system, which will be discussed in the following sections.

EOSINOPHILS DIRECTLY PARTICIPATE IN INNATE RECOGNITION AND ACTIVELY REGULATE OTHER ASPECTS OF NON-ADAPTIVE IMMUNITY

Unique PAMP Receptor Repertoire for Specific Pathogen Recognition and Clearance

Despite early studies showing that eosinophils are able to engulf foreign pathogens, it is now generally accepted that they are not directly involved in cellular phagocytosis. Rather, eosinophils rapidly release (catapult) their mitochondrial DNA to confine bacterial infection in the GI mucosa (35). Furthermore, it was recently found that the eosinophil extracellular DNA trap formation after cytolysis is also accompanied by release of cell-free secretion-competent granules in an NADPH oxidase-dependent fashion (36), further strengthening the evidence of the pathogen-combating function of human eosinophils. In addition, evidence exists that eosinophils have antiparasitic activity, mediated by release of their cytotoxic granule proteins (37). The capacity of eosinophils to release antiparasitic mediators has been shown in multiple studies demonstrating the deposition of eosinophil granule proteins around parasites (38, 39). Gene deletion studies in mice have also indicated that disruption of eosinophil peroxidase (EPO) or MBP results in significantly higher worm burdens compared to wild-type mice (40). The key activating signal for promoting these responses is not yet agreed upon, but multiple lines of evidence suggest that functional Fc receptors on eosinophils (allowing mediation of antibody-dependent cellular cytotoxicity), complement receptors, and several receptors for PAMPs/damage-associated molecular patterns (DAMPs) may contribute (41, 42).

As to PAMP recognition, eosinophils express several functional TLRs, including TLR1, -2, -3, -5, -6, -7, and -9 (43, 44), and activation of these receptors by microbes and pathogens leads to intracellular signal activation and cytokine production by eosinophils (45). These findings not only imply that eosinophils may take part in PAMP recognition and subsequent defensive processes but also suggest a potential mechanism explaining the exacerbation of allergic inflammation by bacterial/viral infection (46). Notably, human data regarding eosinophil TLR2 and TLR4 expression exhibit a certain degree of heterogeneity associated with atopic and eosinophilia status, suggesting that the TLR pathway in eosinophils is functionally linked to the Th2 phenotype in humans (47).

Specifically regarding viral PAMP recognition, TLR7, which recognizes viral single-stranded RNA, is expressed in human eosinophils with intracellular signaling capacity (43). In addition to their proinflammatory roles in asthma, eosinophils are tightly associated with antiviral activity in a variety of systems. EDN has been shown to possess an inhibitory effect on HIV (48), and eosinophil-tropic IL-5 transgenic mice have a more rapid, MyD88-dependent clearance of respiratory syncytial virus compared to wild-type mice (49). Likewise, in pulmonary viral infection, eosinophils cooperate with macrophages to prevent the infection from spreading to uninfected epithelial cells (50), suggesting a positive role of eosinophils in combatting viral infections through TLRs. However, in the context of human rhinovirus, eosinophils enhanced the viral load by inhibiting epithelial interferon production (51), indicating the complexity of eosinophil functions in viral infections.

Currently, a large portion of our knowledge regarding eosinophil-viral interactions comes from respiratory syncytial virus. At a mechanistic level, it is not clear how the virus initiates eosinophilia and how the eosinophil-viral interactions are regulated. Further insights are expected by broadening the scope to other organs/systems (such as the GI), and investigating other types of viruses may provide further information, such as rotavirus infection in mouse and human GI tract (52).

Eosinophil: a Key Orchestrator of Asthma and Other Allergic Diseases

In addition to the Th2-tilted cytokine production and DC induction discussed earlier in the adaptive immunity section, eosinophils also contribute to asthma pathogenesis by serving as a key orchestrator. Indeed, genetic variants affecting eosinophilia, such as variants in *WDR36*, *ST2*, *IL33*, and *MYB*, have been associated with human asthma (53). Importantly, major eosinophil granule products are composed of four bioactive cytotoxic proteins, namely MBP, EPO, eosinophil cationic protein (ECP), and EDN. In human lung, eosinophil products can be detected in macrophage intracellular compartments, contributing to macrophage activation. EPO has been shown to positively regulate macrophage phagocytosis (54). MBP has been shown to disrupt barrier function, induce airway smooth muscle contraction, and elicit mast cell/basophil degranulation (55). As for the interaction with lung epithelium, emerging evidence suggests that eosinophils regulate the airway epithelial cytokine profile and permeability (51). Multiple lines of evidence also suggest that eosinophils take an active role in epithelial damage and basal membrane hyperproliferation (56, 57). Additionally, the observation that human eosinophils induce mucus production via epidermal growth factor (EGF) activation (58) underlines the interaction between eosinophils and asthmatic lung epithelium. Notably, with the vital role of mast cells in asthma increasingly revealed (59), the close interaction between eosinophils and mast cells is highly likely to be important as well, which will be discussed in the next subsection.

In human and murine asthmatic lung, eosinophils are generally recognized as a contributor to airway hyperreactivity. Eosinophils are sources of IL-5 and GM-CSF, thereby promoting their own survival in an autocrine fashion (60), but the major source of IL-5 is believed to be the Th2 cells or ILC2 cells. The causal relationship between asthma and eosinophils has been best elucidated by a series of anti-IL-5 studies in human asthma. Although early anti-IL-5 studies yielded controversial results, more recent studies focusing on severe eosinophilic asthmatic phenotypes, as well as sputum eosinophil count, clearly demonstrated the efficacy of humanized anti-IL-5 therapy (mepolizumab), representing an important treatment avenue in an area of unmet clinical need (61–63). From an independent development perspective, another anti-IL-5 humanized antibody, reslizumab, also convincingly demonstrated efficacy in severe asthma management, including improvements in lung function, significantly reducing the asthmatic exacerbation frequency in two independent phase 3 trials (64). Interestingly, the optimal effect was achieved in inadequately controlled asthma with elevated blood eosinophil counts. Of note, with the demonstrated efficacy during the late-phase development, both anti-IL-5 humanized antibodies have now been FDA approved for clinical management of asthma (65, 66). Human eosinophils have been shown to directly activate neutrophils by releasing ENA-78/CXCL5 (67). Additionally, MBP stimulates IL-8 secretion by neutrophils, with regulation occurring at both transcriptional and posttranscriptional levels (68). These can lead to neutrophilia and neutrophil activation, which is also a critical component of asthma (69). In the presence of GM-CSF, eosinophils express basal levels of Notch ligand (60, 70), capable of regulating both innate and adaptive immunity, including macrophage polarization (71) and cytokine-independent T-cell lineage commitment (72), respectively, suggesting another regulatory function of lung eosinophils in asthma.

Eosinophils were identified as the key factor for human asthma exacerbation (73) and lung connective tissue remodeling, which has been attributed to granule products such as MBP and cytokines such as transforming growth factor β (TGF-β). Recent research in patients with asthma indicates that eosinophils actively participate in the lung tissue fibrosis and remodeling, linking eosinophils to the potential etiology that commonly leads to progressive worsening of quality of life (74, 75). In addition to the well-documented fibrosis-promoting cytokine TGF-β produced by eosinophils, *in vitro* evidence indicates that the eosinophil product ECP induces migration of lung fibroblasts (76) and stimulates them to produce TGF-β *in vitro* (77). Interestingly, the tissue-remodeling capacity of GI eosinophils seems to be suppressed by their surface inhibitory receptors, such as CD172a (3), implying a promising potential to suppress this adverse effect.

Eosinophil-Mast Cell Cross Talk

As pivotal components of allergic hypersensitivity, eosinophils and mast cells mutually potentiate each other in several allergic disorders, such as asthma and eosinophilic esophagitis (EoE) (78). While mast cells support the survival and activation of eosinophils by secreting IL-5, eosinophil MBP directly activates mast cells and basophils, triggering the release of an arsenal of allergic mediators and cytokines, including histamine and tumor necrosis factor α. In the murine system, ECP and EPO also activate allergic mediator release from mast cells (1). Conversely, mast cells are the major sources of prostaglandin D_2 (PGD_2) (79, 80), a key inflammation mediator whose receptor, CRTH2/CD294, is robustly expressed on rodent and human eosinophils (81, 82). This transmembrane G protein-coupled receptor has the dual functions of eosinophil-activating receptor and chemotaxis receptor. Exposure of eosinophils to PGD_2 induces rapid morphological changes, intracellular calcium flux, chemotaxis, and cellular degranulation of human eosinophils. CRTH2 is also known to be expressed on Th2 cells, whose Th2 cytokines will promote eosinophilia as well (83). The eosinophil-mast cell interaction was highlighted by a recent clinical study showing that in human patients with the allergic GI disorder EoE, eosinophils are physically coupled with mast cells as assessed by immunohistochemistry (84) and are a major source of the mast cell-supporting cytokine IL-9 (85). *In vitro* ultrastructural evidence also indicated that eosinophils and mast cells form cell-cell contact and exchange their products for reciprocal activation (84, 86). Importantly, anti-IL-5 therapy concomitantly reduced both eosinophils and mast cell numbers (84), indicating that eosinophils promote mastocytosis in human allergic disease. Of note, in eosinophilic gastrointestinal disorders (EGIDs; e.g., EoE), despite the fact that eosinophil migration is relatively better understood, it remains a mystery how circulating mast cell precursors migrate into tissue and become mature/activated. The reciprocal eosinophil-mast cell potentiation may be critical to understanding the pathogenesis of allergic disease and developing new intervention platforms for human allergic disorders.

In summary, eosinophils serve as recognition cells of certain unique PAMPs, playing a vital role in innate defense against viral, parasitic, and bacterial infection.

Aside from the Th2 induction and T-cell/DC interactions, the regulating role of the eosinophil in asthma can be reflected by its multifaceted interaction with a myriad of cell types in the lung and by its robust tissue-remodeling capacity.

SOME NEW AND INTRIGUING AREAS OF RESEARCH ON EOSINOPHILS

Paradoxical Roles in Tissue Destruction and Repair

Eosinophils are equipped with a tissue damage-sensing system, including several histamine receptors (HR1, HR2, and HR4) (87, 88), which enables them to release multiple tissue-repairing molecules. Damaged epithelial cells from different tissue origins directly stimulate eosinophil secretion of TGF-β and fibroblast growth factor (FGF) (89). Indeed, the eosinophil is capable of producing TGF-β, TGF-α, EGF, (90), FGF (89), platelet-derived growth factor (91), and vascular endothelial growth factor (92), all of which have well-recognized beneficial roles in tissue repair (8). The tissue repair function of eosinophils was substantiated by a recent study showing that eosinophil IL-4 production is necessary for hepatocyte regeneration after hepatectomy or toxin injury, as this effect is abolished in eosinophil-deficient mice (93). Although it was proposed that eosinophils promote wound healing due to growth factor production (90, 94), *Il5*-overexpressing mice have a delayed wound healing due to augmented inflammatory responses and delayed matrix synthesis (95, 96), suggesting that the reparative function of eosinophils is tonically modulated.

Adipose Tissue-Residential Eosinophils Regulate Local Cytokine Milieu and Glucose Homeostasis

Recent studies indicated that eosinophils are tightly associated with alternatively activated macrophages (M2) (97, 98), whereas obesity is interpreted as an uncontrolled chronic inflammatory response associated with dysregulated macrophage populations in the adipose tissue (99). Interestingly, with eosinophils found as a residential cell type in the visceral adipose tissue under homeostasis, a recent study has uncovered nonredundant roles of eosinophils in adipose tissue macrophage polarization and glucose metabolism (100). In murine models, eosinophils have been proven to be critical for maintaining adipose tissue M2 macrophages as a major contributor of IL-4, a key factor for M2 polarization. In the absence of eosinophils, mice are prone to develop obesity, glucose intolerance, and insulin resistance (100). Notably, the eosinophilia induced by parasitic infection has been shown to enhance glucose tolerance, reinforcing a regulatory effect of eosinophils on metabolism. The field is awaiting studies performed in human adipose tissue demonstrating similar regulatory roles of eosinophils on macrophage subtypes and body metabolism.

Eosinophils May Contribute to Neurologic Symptoms by Regulating Neural Activity

Notably, several studies suggest a link between eosinophils and neuronal homeostasis. First, peripheral dorsal root ganglia and airway neurons produce eotaxins to chemoattract eosinophils into their niche (101, 102). Furthermore, eosinophils are capable of producing nerve growth factor and neurotrophins (e.g., NT-3), at the mRNA and protein level, to directly regulate neuronal activity. This constitutive activity can be further boosted by Fc receptor-mediated eosinophil activation (103). *In vitro*, both murine and human eosinophils promoted dorsal root ganglia neuron branching, an effect independent of physical contact (101) and a plausible explanation for the cutaneous nerve outgrowth and associated neurologic symptoms in atopic dermatitis. EGIDs have pronounced tissue-specific eosinophilia, with expression of several neurofilament elements being robustly upregulated in EoE (104). Notably, a vast majority of patients with EoE experience a certain degree of neuropathy and somatosensory alterations (105). Additionally, the neurotrophins present in the asthmatic lung promote the survival of airway eosinophils (106), forming a potential vicious cycle. It is therefore conceivable that eosinophils may directly contribute to these neuronal dysregulations. These observations may at least partially explain the neurologic hypersensitivity in a myriad of allergic diseases, such as asthma, atopic dermatitis, and allergic rhinitis (107, 108), and in some eosinophilic disorders, such as EGIDs (109, 110). Although the neural hyperplasia in atopic dermatitis has been well documented, the anomalous neurologic changes in EGIDs have yet to be characterized. In a guinea pig model of asthma, it is notable that eosinophils migrate into the nerves in a CCR3-dependent fashion (102) (especially the vagal nerve) and release MBP (111), which serves as a muscarinic receptor 2 antagonist (112) that enhances the acetylcholine release and thereby exacerbates bronchoconstriction. In human asthma and EoE, eosinophils localize near nerve endings with extracellular MBP adhered to the nerve endings, suggesting a neuronal regulatory role of eosinophils (113).

The Enigmatic Functions of GI Eosinophils

Rodent and human GI tissue harbors the largest reservoir of eosinophils compared to other anatomical compartments including blood and bone marrow. Although intestinal eosinophils are extensively present in the lamina propria of the full length of the GI tract, their function is little understood compared to eosinophils in other organs. From a limited number of studies, eosinophils in the GI compartment seem to adopt a unique surface expression profile to accommodate their GI-specific functions (2). For reasons that are not fully comprehended, GI eosinophil turnover rate is much slower than that of the lung and blood eosinophils, as assessed by bromodeoxyuridine incorporation (114). It was also found that signaling through the common γ chain receptor is necessary for the longer survival of GI eosinophils (114), but the functions of these cells still remain largely unknown. Recently, utilizing the eosinophil-deficient ΔdblGATA-1 and PHIL mice, multiple research groups showed that GI eosinophils promote the generation and production of IgA-producing plasma cells in the GI tract (12, 13). These studies also determined that eosinophils have novel and profound functions in the GI tract, such as promoting IgA class

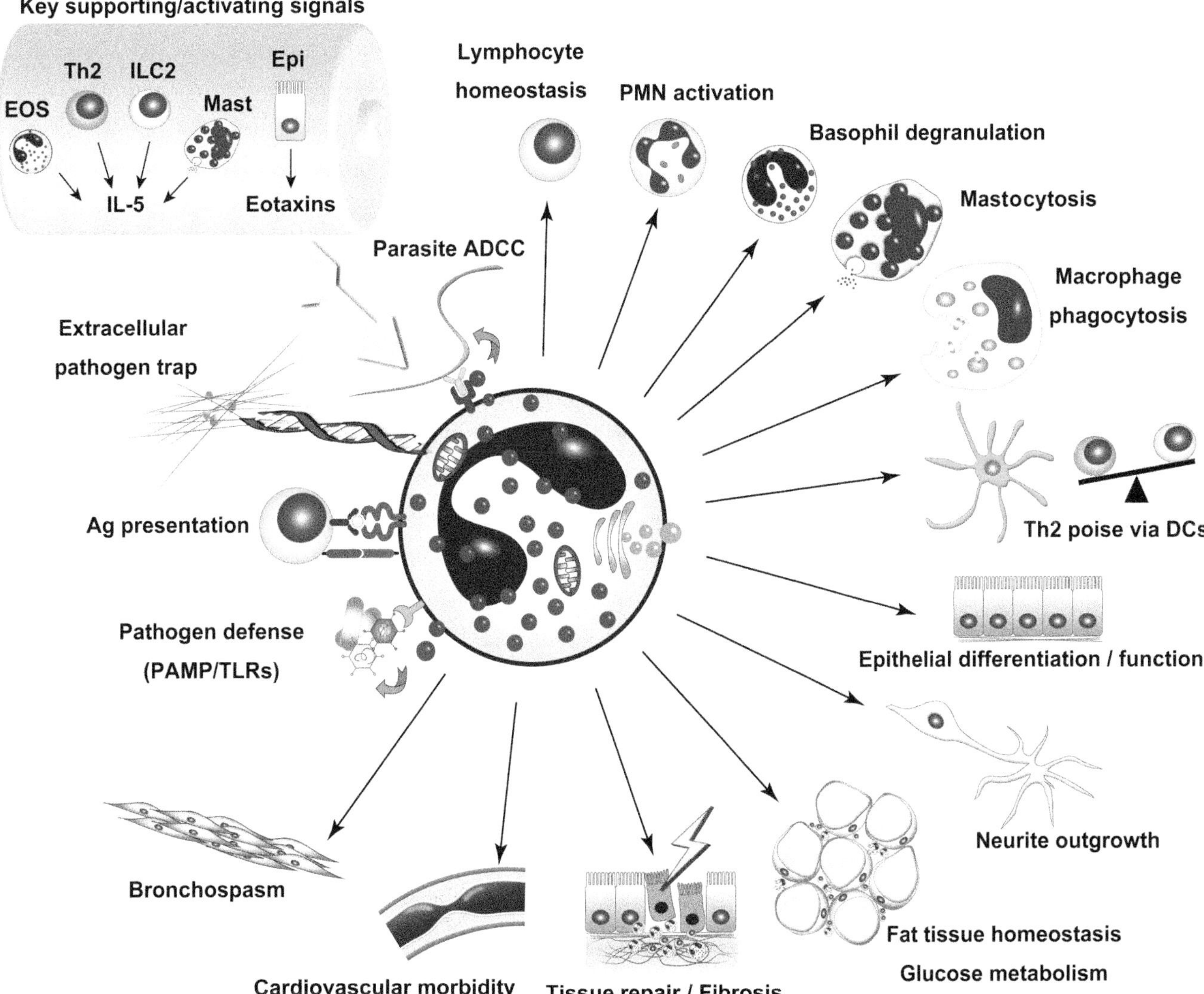

Figure 1 Schematic summary of eosinophil-tropic signaling and eosinophil cellular and humoral regulatory functions. EOS, eosinophils; Mast, mast cells; Epi, epithelium; ADCC, antibody-dependent cell-mediated cytotoxicity; Ag, antigen; PMN, polymorphonuclear leukocyte (neutrophil).

switching, enhancing intestinal mucus secretions, determining intestinal microbiota, and inducing the development of Peyer's patches (12, 13). The functional exploration of GI eosinophils has just started. With conditional, eosinophil lineage-deficient mice and genome-wide screening methods becoming increasingly available, key functions of GI eosinophils in homeostasis and disease contexts, such as EGIDs, are soon to be discovered.

PERSPECTIVES

With advancing technology and accumulating knowledge, the understanding of the eosinophil has changed from that of a simple postmitotic cell with limited functional capacity in parasite infection and allergy to a cell type that is actively involved in orchestrating a variety of mucosal and nonmucosal immune responses at baseline and during a variety of disease responses (see Fig. 1

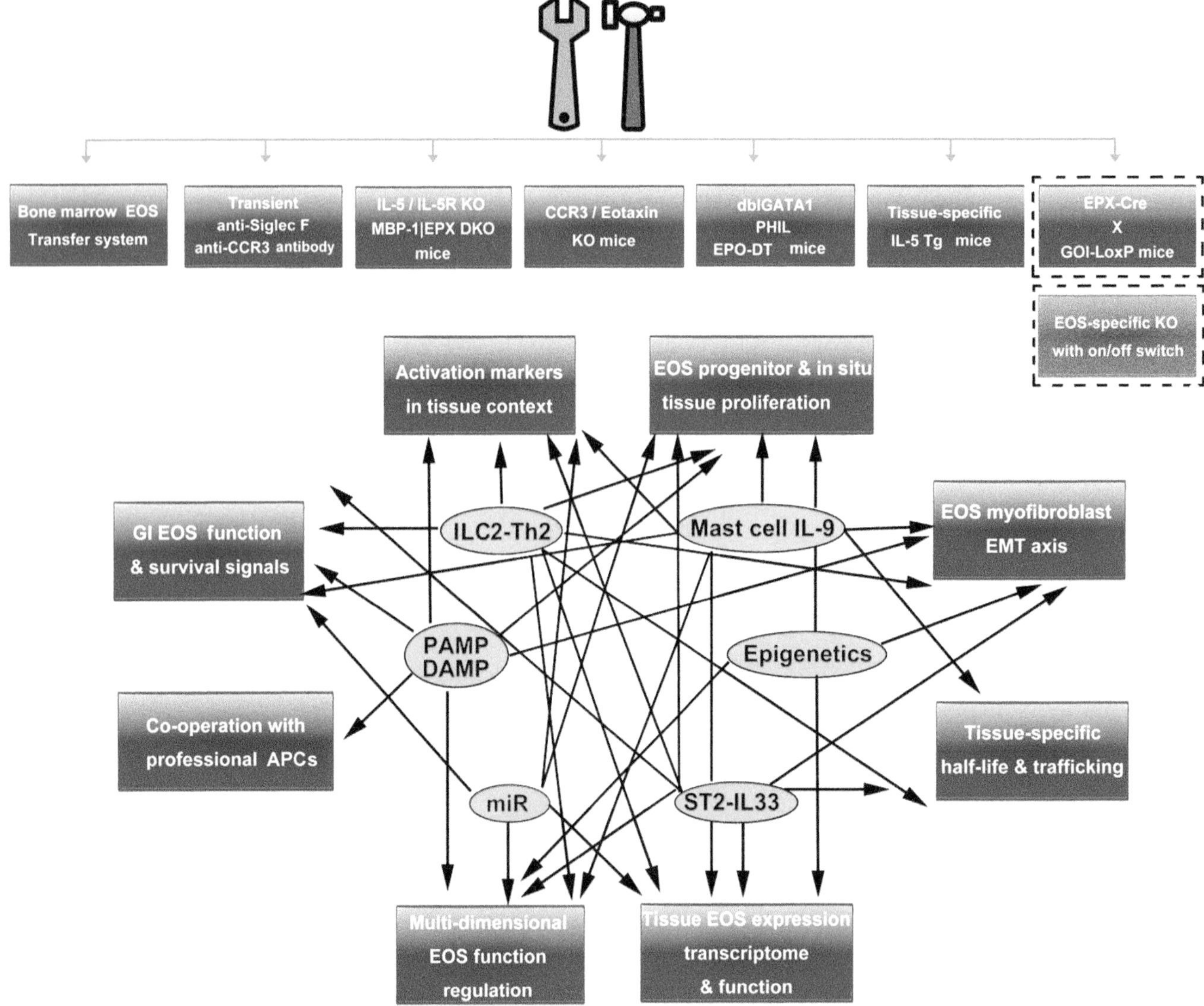

Figure 2 Research tool summary and questions to be answered. Although recent advances provide tremendous insight into the regulatory functions of eosinophils, important questions remain. With the increasing number of tools available (upper panel), progress in the listed areas (lower panel, blue boxes), in the context of key eosinophil regulation elements (ovals), will be interesting and crucial to understanding the still enigmatic function of eosinophils. EOS, eosinophils; EPO-DT, EPO-driven diphtheria toxin expression mice; EMT, epithelial-mesenchymal transition; miR, microRNA; GOI, gene of interest; Tg, transgenic; (D) KO, (double) knockout; APCs, antigen-presenting cells. Gene EPX encodes the protein product of eosinophil peroxidase (EPO).

for schematic summary illustration). With more novel functions of eosinophils being actively investigated and an increasing number of murine models available, rapid advances in further understanding this cell type are likely to occur in the near future (summarized in Fig. 2). The recent generation of EPO-diphtheria toxin eosinophil-depleted mice (115) and MBP-1 EPX-disrupted eosinophil progenitor-deficient mice (23) is expected to provide a complementary and independent means to assess the participation of eosinophils in a variety of responses. The interaction network between eosinophils and the recently identified IL-33–ST2 axis provides an opportunity to further understand the activation of this cell type in innate allergic responses. The interplay between eosinophils and newly identified ILC2s (116), which produce abundant amounts of IL-5 and primarily reside in nonlymphoid tissues (7), is a particularly promising new finding that may explain the key source of eosinophilopoietins and early involvement of eosinophils in acute injury responses. The recent generation of an eosinophil-specific gene disruption model, the EPO-driven Cre expression mouse (117), is likely to facilitate the investigation of underappreciated functions of eosinophils. As to advancing asthma management strategy, as illustrated by the success of the anti-IL-5 studies, eosinophil-suppressive therapy represents an effective and well-tolerated treatment that reduces the possibility of asthma exacerbation and will likely receive further approval in the market and more clinical and research attention. Collectively, these studies and the implications of their findings support the likelihood of an emerging and expanding body of data concerning the function and role of eosinophils in immunity.

Acknowledgments. *This work is supported by the NIH grants R37 A1045898, R01 AI083450, and P30 DK078392 (the Digestive Diseases Research Core Center in Cincinnati) and the CURED (Campaign Urging Research for Eosinophilic Disease) Foundation, the Food Allergy Research & Education (FARE), the Buckeye Foundation, and the American Partnership for Eosinophilic Disorders (APFED). We thank Shawna Hottinger for editorial assistance.*

Citation. Wen T, Rothenberg ME. 2016. The regulatory function of eosinophils. Microbiol Spectrum 4(5):MCHD-0020-2015.

References

1. **Rothenberg ME, Hogan SP.** 2006. The eosinophil. *Annu Rev Immunol* **24:**147–174.
2. **Wen T, Mingler MK, Blanchard C, Wahl B, Pabst O, Rothenberg ME.** 2012. The pan-B cell marker CD22 is expressed on gastrointestinal eosinophils and negatively regulates tissue eosinophilia. *J Immunol* **188:**1075–1082.
3. **Verjan Garcia N, Umemoto E, Saito Y, Yamasaki M, Hata E, Matozaki T, Murakami M, Jung YJ, Woo SY, Seoh JY, Jang MH, Aozasa K, Miyasaka M.** 2011. SIRPα/CD172a regulates eosinophil homeostasis. *J Immunol* **187:**2268–2277.
4. **Rose CE Jr, Lannigan JA, Kim P, Lee JJ, Fu SM, Sung SS.** 2010. Murine lung eosinophil activation and chemokine production in allergic airway inflammation. *Cell Mol Immunol* **7:**361–374.
5. **Conroy DM, Williams TJ.** 2001. Eotaxin and the attraction of eosinophils to the asthmatic lung. *Respir Res* **2:** 150–156.
6. **Upadhyaya B, Yin Y, Hill BJ, Douek DC, Prussin C.** 2011. Hierarchical IL-5 expression defines a subpopulation of highly differentiated human Th2 cells. *J Immunol* **187:**3111–3120.
7. **Nussbaum JC, Van Dyken SJ, von Moltke J, Cheng LE, Mohapatra A, Molofsky AB, Thornton EE, Krummel MF, Chawla A, Liang HE, Locksley RM.** 2013. Type 2 innate lymphoid cells control eosinophil homeostasis. *Nature* **502:**245–248.
8. **Hogan SP.** 2009. Functional role of eosinophils in gastrointestinal inflammation. *Immunol Allergy Clin North Am* **29:**129–140, xi.
9. **Munitz A, Brandt EB, Mingler M, Finkelman FD, Rothenberg ME.** 2008. Distinct roles for IL-13 and IL-4 via IL-13 receptor α1 and the type II IL-4 receptor in asthma pathogenesis. *Proc Natl Acad Sci U S A* **105:** 7240–7245.
10. **Jacobsen EA, Ochkur SI, Pero RS, Taranova AG, Protheroe CA, Colbert DC, Lee NA, Lee JJ.** 2008. Allergic pulmonary inflammation in mice is dependent on eosinophil-induced recruitment of effector T cells. *J Exp Med* **205:**699–710.
11. **Shen HH, Ochkur SI, McGarry MP, Crosby JR, Hines EM, Borchers MT, Wang H, Biechelle TL, O'Neill KR, Ansay TL, Colbert DC, Cormier SA, Justice JP, Lee NA, Lee JJ.** 2003. A causative relationship exists between eosinophils and the development of allergic pulmonary pathologies in the mouse. *J Immunol* **170:**3296–3305.
12. **Chu VT, Beller A, Rausch S, Strandmark J, Zanker M, Arbach O, Kruglov A, Berek C.** 2014. Eosinophils promote generation and maintenance of immunoglobulin-A-expressing plasma cells and contribute to gut immune homeostasis. *Immunity* **40:**582–593.
13. **Jung Y, Wen T, Mingler MK, Caldwell JM, Wang YH, Chaplin DD, Lee EH, Jang MH, Woo SY, Seoh JY, Miyasaka M, Rothenberg ME.** 2015. IL-1β in eosinophil-mediated small intestinal homeostasis and IgA production. *Mucosal Immunol* **8:**930–942.
14. **Chu VT, Fröhlich A, Steinhauser G, Scheel T, Roch T, Fillatreau S, Lee JJ, Löhning M, Berek C.** 2011. Eosinophils are required for the maintenance of plasma cells in the bone marrow. *Nat Immunol* **12:**151–159.
15. **Wong TW, Doyle AD, Lee JJ, Jelinek DF.** 2014. Eosinophils regulate peripheral B cell numbers in both mice and humans. *J Immunol* **192:**3548–3558.
16. **Del Pozo V, De Andrés B, Martín E, Cárdaba B, Fernández JC, Gallardo S, Tramón P, Leyva-Cobian F,**

Palomino P, Lahoz C. 1992. Eosinophil as antigen-presenting cell: activation of T cell clones and T cell hybridoma by eosinophils after antigen processing. *Eur J Immunol* **22:**1919–1925.

17. **Shi HZ, Humbles A, Gerard C, Jin Z, Weller PF.** 2000. Lymph node trafficking and antigen presentation by endobronchial eosinophils. *J Clin Invest* **105:**945–953.
18. **Akuthota P, Melo RC, Spencer LA, Weller PF.** 2012. MHC Class II and CD9 in human eosinophils localize to detergent-resistant membrane microdomains. *Am J Respir Cell Mol Biol* **46:**188–195.
19. **Wang HB, Ghiran I, Matthaei K, Weller PF.** 2007. Airway eosinophils: allergic inflammation recruited professional antigen-presenting cells. *J Immunol* **179:**7585–7592.
20. **Jacobsen EA, Zellner KR, Colbert D, Lee NA, Lee JJ.** 2011. Eosinophils regulate dendritic cells and Th2 pulmonary immune responses following allergen provocation. *J Immunol* **187:**6059–6068.
21. **Padigel UM, Hess JA, Lee JJ, Lok JB, Nolan TJ, Schad GA, Abraham D.** 2007. Eosinophils act as antigen-presenting cells to induce immunity to *Strongyloides stercoralis* in mice. *J Infect Dis* **196:**1844–1851.
22. **Garro AP, Chiapello LS, Baronetti JL, Masih DT.** 2011. Eosinophils elicit proliferation of naive and fungal-specific cells *in vivo* so enhancing a T helper type 1 cytokine profile in favour of a protective immune response against *Cryptococcus neoformans* infection. *Immunology* **134:**198–213.
23. **Doyle AD, Jacobsen EA, Ochkur SI, McGarry MP, Shim KG, Nguyen DT, Protheroe C, Colbert D, Kloeber J, Neely J, Shim KP, Dyer KD, Rosenberg HF, Lee JJ, Lee NA.** 2013. Expression of the secondary granule proteins major basic protein 1 (MBP-1) and eosinophil peroxidase (EPX) is required for eosinophilopoiesis in mice. *Blood* **122:**781–790.
24. **Spencer LA, Szela CT, Perez SA, Kirchhoffer CL, Neves JS, Radke AL, Weller PF.** 2009. Human eosinophils constitutively express multiple Th1, Th2, and immunoregulatory cytokines that are secreted rapidly and differentially. *J Leukoc Biol* **85:**117–123.
25. **Odemuyiwa SO, Ghahary A, Li Y, Puttagunta L, Lee JE, Musat-Marcu S, Ghahary A, Moqbel R.** 2004. Cutting edge: human eosinophils regulate T cell subset selection through indoleamine 2,3-dioxygenase. *J Immunol* **173:**5909–5913.
26. **Lotfi R, Lotze MT.** 2008. Eosinophils induce DC maturation, regulating immunity. *J Leukoc Biol* **83:** 456–460.
27. **Weir MR, Hall-Craggs M, Shen SY, Posner JN, Alongi SV, Dagher FJ, Sadler JH.** 1986. The prognostic value of the eosinophil in acute renal allograft rejection. *Transplantation* **41:**709–712.
28. **Lotfi R, Lee JJ, Lotze MT.** 2007. Eosinophilic granulocytes and damage-associated molecular pattern molecules (DAMPs): role in the inflammatory response within tumors. *J Immunother* **30:**16–28.
29. **Yang D, Rosenberg HF, Chen Q, Dyer KD, Kurosaka K, Oppenheim JJ.** 2003. Eosinophil-derived neurotoxin (EDN), an antimicrobial protein with chemotactic activities for dendritic cells. *Blood* **102:**3396–3403.
30. **Yang D, Chen Q, Rosenberg HF, Rybak SM, Newton DL, Wang ZY, Fu Q, Tchernev VT, Wang M, Schweitzer B, Kingsmore SF, Patel DD, Oppenheim JJ, Howard OM.** 2004. Human ribonuclease A superfamily members, eosinophil-derived neurotoxin and pancreatic ribonuclease, induce dendritic cell maturation and activation. *J Immunol* **173:**6134–6142.
31. **Yang D, Chen Q, Su SB, Zhang P, Kurosaka K, Caspi RR, Michalek SM, Rosenberg HF, Zhang N, Oppenheim JJ.** 2008. Eosinophil-derived neurotoxin acts as an alarmin to activate the TLR2-MyD88 signal pathway in dendritic cells and enhances Th2 immune responses. *J Exp Med* **205:**79–90.
32. **O'Connell AE, Hess JA, Santiago GA, Nolan TJ, Lok JB, Lee JJ, Abraham D.** 2011. Major basic protein from eosinophils and myeloperoxidase from neutrophils are required for protective immunity to *Strongyloides stercoralis* in mice. *Infect Immun* **79:**2770–2778.
33. **Tulic MK, Sly PD, Andrews D, Crook M, Davoine F, Odemuyiwa SO, Charles A, Hodder ML, Prescott SL, Holt PG, Moqbel R.** 2009. Thymic indoleamine 2,3-dioxygenase-positive eosinophils in young children: potential role in maturation of the naive immune system. *Am J Pathol* **175:**2043–2052.
34. **Throsby M, Herbelin A, Pléau JM, Dardenne M.** 2000. $CD11c^+$ eosinophils in the murine thymus: developmental regulation and recruitment upon MHC class I-restricted thymocyte deletion. *J Immunol* **165:**1965–1975.
35. **Yousefi S, Gold JA, Andina N, Lee JJ, Kelly AM, Kozlowski E, Schmid I, Straumann A, Reichenbach J, Gleich GJ, Simon HU.** 2008. Catapult-like release of mitochondrial DNA by eosinophils contributes to antibacterial defense. *Nat Med* **14:**949–953.
36. **Ueki S, Melo RC, Ghiran I, Spencer LA, Dvorak AM, Weller PF.** 2013. Eosinophil extracellular DNA trap cell death mediates lytic release of free secretion-competent eosinophil granules in humans. *Blood* **121:**2074–2083.
37. **Klion AD, Nutman TB.** 2004. The role of eosinophils in host defense against helminth parasites. *J Allergy Clin Immunol* **113:**30–37.
38. **Kephart GM, Gleich GJ, Connor DH, Gibson DW, Ackerman SJ.** 1984. Deposition of eosinophil granule major basic protein onto microfilariae of *Onchocerca volvulus* in the skin of patients treated with diethylcarbamazine. *Lab Invest* **50:**51–61.
39. **Mehlotra RK, Hall LR, Higgins AW, Dreshaj IA, Haxhiu MA, Kazura JW, Pearlman E.** 1998. Interleukin-12 suppresses filaria-induced pulmonary eosinophilia, deposition of major basic protein and airway hyperresponsiveness. *Parasite Immunol* **20:**455–462.
40. **Specht S, Saeftel M, Arndt M, Endl E, Dubben B, Lee NA, Lee JJ, Hoerauf A.** 2006. Lack of eosinophil peroxidase or major basic protein impairs defense against murine filarial infection. *Infect Immun* **74:**5236–5243.
41. **Bandeira-Melo C, Weller PF.** 2005. Mechanisms of eosinophil cytokine release. *Mem Inst Oswaldo Cruz* **100** (Suppl 1):73–81.

42. **Elsner J, Oppermann M, Kapp A.** 1996. Detection of C5a receptors on human eosinophils and inhibition of eosinophil effector functions by anti-C5a receptor (CD88) antibodies. *Eur J Immunol* **26:**1560–1564.

43. **Nagase H, Okugawa S, Ota Y, Yamaguchi M, Tomizawa H, Matsushima K, Ohta K, Yamamoto K, Hirai K.** 2003. Expression and function of Toll-like receptors in eosinophils: activation by Toll-like receptor 7 ligand. *J Immunol* **171:**3977–3982.

44. **Mansson A, Cardell LO.** 2009. Role of atopic status in Toll-like receptor (TLR)7- and TLR9-mediated activation of human eosinophils. *J Leukoc Biol* **85:**719–727.

45. **Wong CK, Cheung PF, Ip WK, Lam CW.** 2007. Intracellular signaling mechanisms regulating Toll-like receptor-mediated activation of eosinophils. *Am J Respir Cell Mol Biol* **37:**85–96.

46. **Guilbert TW, Denlinger LC.** 2010. Role of infection in the development and exacerbation of asthma. *Expert Rev Respir Med* **4:**71–83.

47. **Driss V, Legrand F, Hermann E, Loiseau S, Guerardel Y, Kremer L, Adam E, Woerly G, Dombrowicz D, Capron M.** 2009. TLR2-dependent eosinophil interactions with mycobacteria: role of α-defensins. *Blood* **113:** 3235–3244.

48. **Bedoya VI, Boasso A, Hardy AW, Rybak S, Shearer GM, Rugeles MT.** 2006. Ribonucleases in HIV type 1 inhibition: effect of recombinant RNases on infection of primary T cells and immune activation-induced RNase gene and protein expression. *AIDS Res Hum Retroviruses* **22:**897–907.

49. **Phipps S, Lam CE, Mahalingam S, Newhouse M, Ramirez R, Rosenberg HF, Foster PS, Matthaei KI.** 2007. Eosinophils contribute to innate antiviral immunity and promote clearance of respiratory syncytial virus. *Blood* **110:**1578–1586.

50. **Soukup JM, Becker S.** 2003. Role of monocytes and eosinophils in human respiratory syncytial virus infection in vitro. *Clin Immunol* **107:**178–185.

51. **Mathur SK, Fichtinger PS, Kelly JT, Lee WM, Gern JE, Jarjour NN.** 2013. Interaction between allergy and innate immunity: model for eosinophil regulation of epithelial cell interferon expression. *Ann Allergy Asthma Immunol* **111:**25–31.

52. **Knipping K, McNeal MM, Crienen A, van Amerongen G, Garssen J, Van't Land B.** 2011. A gastrointestinal rotavirus infection mouse model for immune modulation studies. *Virol J* **8:**109. doi:10.1186/1743-422X-8-109.

53. **Gudbjartsson DF, Bjornsdottir US, Halapi E, Helgadottir A, Sulem P, Jonsdottir GM, Thorleifsson G, Helgadottir H, Steinthorsdottir V, Stefansson H, Williams C, Hui J, Beilby J, Warrington NM, James A, Palmer LJ, Koppelman GH, Heinzmann A, Krueger M, Boezen HM, Wheatley A, Altmuller J, Shin HD, Uh ST, Cheong HS, Jonsdottir B, Gislason D, Park CS, Rasmussen LM, Porsbjerg C, Hansen JW, Backer V, Werge T, Janson C, Jönsson UB, Ng MC, Chan J, So WY, Ma R, Shah SH, Granger CB, Quyyumi AA, Levey AI, Vaccarino V, Reilly MP, Rader DJ, Williams MJ, van Rij AM, Jones GT, Trabetti E, Malerba G, Pignatti PF, Boner A, Pescollderungg L, Girelli D, Olivieri O, Martinelli N, Ludviksson BR, Ludviksdottir D, Eyjolfsson GI, Arnar D, Thorgeirsson G, Deichmann K, Thompson PJ, Wjst M, Hall IP, Postma DS, Gislason T, Gulcher J, Kong A, Jonsdottir I, Thorsteinsdottir U, Stefansson K.** 2009. Sequence variants affecting eosinophil numbers associate with asthma and myocardial infarction. *Nat Genet* **41:**342–347.

54. **Lefkowitz DL, Lincoln JA, Howard KR, Stuart R, Lefkowitz SS, Allen RC.** 1997. Macrophage-mediated candidacidal activity is augmented by exposure to eosinophil peroxidase: a paradigm for eosinophil-macrophage interaction. *Inflammation* **21:**159–172.

55. **Furuta GT, Nieuwenhuis EE, Karhausen J, Gleich G, Blumberg RS, Lee JJ, Ackerman SJ.** 2005. Eosinophils alter colonic epithelial barrier function: role for major basic protein. *Am J Physiol Gastrointest Liver Physiol* **289:**G890–G897.

56. **Hirata A, Motojima S, Fukuda T, Makino S.** 1996. Damage to respiratory epithelium by guinea-pig eosinophils stimulated with IgG-coated Sepharose beads. *Clin Exp Allergy* **26:**848–858.

57. **Ricciardolo FL, Di Stefano A, van Krieken JH, Sont JK, van Schadewijk A, Rabe KF, Donner CF, Hiemstra PS, Sterk PJ, Mauad T.** 2003. Proliferation and inflammation in bronchial epithelium after allergen in atopic asthmatics. *Clin Exp Allergy* **33:**905–911.

58. **Burgel PR, Lazarus SC, Tam DC, Ueki IF, Atabai K, Birch M, Nadel JA.** 2001. Human eosinophils induce mucin production in airway epithelial cells via epidermal growth factor receptor activation. *J Immunol* **167:** 5948–5954.

59. **Bradding P.** 2008. Asthma: eosinophil disease, mast cell disease, or both? *Allergy Asthma Clin Immunol* **4:** 84–90.

60. **Radke AL, Reynolds LE, Melo RC, Dvorak AM, Weller PF, Spencer LA.** 2009. Mature human eosinophils express functional Notch ligands mediating eosinophil autocrine regulation. *Blood* **113:**3092–3101.

61. **Robinson DS, Kariyawasam HH.** 2015. Mepolizumab for eosinophilic severe asthma: recent studies. *Expert Opin Biol Ther* **15:**909–914.

62. **Ortega HG, Liu MC, Pavord ID, Brusselle GG, FitzGerald JM, Chetta A, Humbert M, Katz LE, Keene ON, Yancey SW, Chanez P, MENSA Investigators.** 2014. Mepolizumab treatment in patients with severe eosinophilic asthma. *N Engl J Med* **371:**1198–1207.

63. **Pavord ID, Korn S, Howarth P, Bleecker ER, Buhl R, Keene ON, Ortega H, Chanez P.** 2012. Mepolizumab for severe eosinophilic asthma (DREAM): a multicentre, double-blind, placebo-controlled trial. *Lancet* **380:** 651–659.

64. **Castro M, Zangrilli J, Wechsler ME, Bateman ED, Brusselle GG, Bardin P, Murphy K, Maspero JF, O'Brien C, Korn S.** 2015. Reslizumab for inadequately controlled asthma with elevated blood eosinophil counts: results from two multicentre, parallel, double-blind, randomised, placebo-controlled, phase 3 trials. *Lancet Respir Med* **3:**355–366.

65. **Reichert JM.** 2015. Antibodies to watch in 2015. *MAbs* **7:**1–8.

66. **Rothenberg ME.** 2016. Humanized anti-IL-5 antibody therapy. *Cell* **165:**509. doi:10.1016/j.cell.2016.04.020.

67. **Persson T, Monsef N, Andersson P, Bjartell A, Malm J, Calafat J, Egesten A.** 2003. Expression of the neutrophil-activating CXC chemokine ENA-78/CXCL5 by human eosinophils. *Clin Exp Allergy* **33:**531–537.

68. **Page SM, Gleich GJ, Roebuck KA, Thomas LL.** 1999. Stimulation of neutrophil interleukin-8 production by eosinophil granule major basic protein. *Am J Respir Cell Mol Biol* **21:**230–237.

69. **Simpson JL, Grissell TV, Douwes J, Scott RJ, Boyle MJ, Gibson PG.** 2007. Innate immune activation in neutrophilic asthma and bronchiectasis. *Thorax* **62:**211–218.

70. **Singh N, Phillips RA, Iscove NN, Egan SE.** 2000. Expression of Notch receptors, Notch ligands, and Fringe genes in hematopoiesis. *Exp Hematol* **28:**527–534.

71. **Wang YC, He F, Feng F, Liu XW, Dong GY, Qin HY, Hu XB, Zheng MH, Liang L, Feng L, Liang YM, Han H.** 2010. Notch signaling determines the M1 versus M2 polarization of macrophages in antitumor immune responses. *Cancer Res* **70:**4840–4849.

72. **Bailis W, Yashiro-Ohtani Y, Fang TC, Hatton RD, Weaver CT, Artis D, Pear WS.** 2013. Notch simultaneously orchestrates multiple helper T cell programs independently of cytokine signals. *Immunity* **39:**148–159.

73. **Haldar P, Brightling CE, Hargadon B, Gupta S, Monteiro W, Sousa A, Marshall RP, Bradding P, Green RH, Wardlaw AJ, Pavord ID.** 2009. Mepolizumab and exacerbations of refractory eosinophilic asthma. *N Engl J Med* **360:**973–984.

74. **Al-Muhsen S, Johnson JR, Hamid Q.** 2011. Remodeling in asthma. *J Allergy Clin Immunol* **128:**451–462; quiz 463–464.

75. **Venge P.** 2010. The eosinophil and airway remodelling in asthma. *Clin Respir J* **4**(Suppl 1)**:**15–19.

76. **Zagai U, Lundahl J, Klominek J, Venge P, Sköld CM.** 2009. Eosinophil cationic protein stimulates migration of human lung fibroblasts *in vitro*. *Scand J Immunol* **69:** 381–386.

77. **Zagai U, Dadfar E, Lundahl J, Venge P, Sköld CM.** 2007. Eosinophil cationic protein stimulates TGF-β_1 release by human lung fibroblasts *in vitro*. *Inflammation* **30:**153–160.

78. **Abonia JP, Blanchard C, Butz BB, Rainey HF, Collins MH, Stringer K, Putnam PE, Rothenberg ME.** 2010. Involvement of mast cells in eosinophilic esophagitis. *J Allergy Clin Immunol* **126:**140–149.

79. **O'Sullivan S.** 1999. On the role of PGD_2 metabolites as markers of mast cell activation in asthma. *Acta Physiol Scand Suppl* **644:**1–74.

80. **Dahlén SE, Kumlin M.** 2004. Monitoring mast cell activation by prostaglandin D_2 in vivo. *Thorax* **59:**453–455.

81. **Kataoka N, Satoh T, Hirai A, Saeki K, Yokozeki H.** 2013. Indomethacin inhibits eosinophil migration to prostaglandin D_2: therapeutic potential of CRTH2 desensitization for eosinophilic pustular folliculitis. *Immunology* **140:**78–86.

82. **Kagawa S, Fukunaga K, Oguma T, Suzuki Y, Shiomi T, Sayama K, Kimura T, Hirai H, Nagata K, Nakamura M, Asano K.** 2011. Role of prostaglandin D_2 receptor CRTH2 in sustained eosinophil accumulation in the airways of mice with chronic asthma. *Int Arch Allergy Immunol* **155**(Suppl 1)**:**6–11.

83. **Cosmi L, Annunziato F, Galli MI, Maggi RM, Nagata K, Romagnani S.** 2000. CRTH2 is the most reliable marker for the detection of circulating human type 2 Th and type 2 T cytotoxic cells in health and disease. *Eur J Immunol* **30:**2972–2979.

84. **Otani IM, Anilkumar AA, Newbury RO, Bhagat M, Beppu LY, Dohil R, Broide DH, Aceves SS.** 2013. Anti-IL-5 therapy reduces mast cell and IL-9 cell numbers in pediatric patients with eosinophilic esophagitis. *J Allergy Clin Immunol* **131:**1576–1582.

85. **Osterfeld H, Ahrens R, Strait R, Finkelman FD, Renauld JC, Hogan SP.** 2010. Differential roles for the IL-9/IL-9 receptor α-chain pathway in systemic and oral antigen-induced anaphylaxis. *J Allergy Clin Immunol* **125:**469–476.e462. doi:10.1016/j.jaci.2009.09.054.

86. **Minai-Fleminger Y, Elishmereni M, Vita F, Soranzo MR, Mankuta D, Zabucchi G, Levi-Schaffer F.** 2010. Ultrastructural evidence for human mast cell-eosinophil interactions in vitro. *Cell Tissue Res* **341:**405–415.

87. **Numata Y, Terui T, Okuyama R, Hirasawa N, Sugiura Y, Miyoshi I, Watanabe T, Kuramasu A, Tagami H, Ohtsu H.** 2006. The accelerating effect of histamine on the cutaneous wound-healing process through the action of basic fibroblast growth factor. *J Invest Dermatol* **126:**1403–1409.

88. **Reher TM, Neumann D, Buschauer A, Seifert R.** 2012. Incomplete activation of human eosinophils via the histamine H_4-receptor: evidence for ligand-specific receptor conformations. *Biochem Pharmacol* **84:**192–203.

89. **Stenfeldt AL, Wenneras C.** 2004. Danger signals derived from stressed and necrotic epithelial cells activate human eosinophils. *Immunology* **112:**605–614.

90. **Todd R, Donoff BR, Chiang T, Chou MY, Elovic A, Gallagher GT, Wong DT.** 1991. The eosinophil as a cellular source of transforming growth factor alpha in healing cutaneous wounds. *Am J Pathol* **138:**1307–1313.

91. **Ohno I, Nitta Y, Yamauchi K, Hoshi H, Honma M, Woolley K, O'Byrne P, Dolovich J, Jordana M, Tamura G.** 1995. Eosinophils as a potential source of platelet-derived growth factor B-chain (PDGF-B) in nasal polyposis and bronchial asthma. *Am J Respir Cell Mol Biol* **13:**639–647.

92. **Horiuchi T, Weller PF.** 1997. Expression of vascular endothelial growth factor by human eosinophils: upregulation by granulocyte macrophage colony-stimulating factor and interleukin-5. *Am J Respir Cell Mol Biol* **17:** 70–77.

93. **Goh YP, Henderson NC, Heredia JE, Red Eagle A, Odegaard JI, Lehwald N, Nguyen KD, Sheppard D, Mukundan L, Locksley RM, Chawla A.** 2013.

Eosinophils secrete IL-4 to facilitate liver regeneration. *Proc Natl Acad Sci U S A* **110**:9914–9919.

94. **Elovic AE, Gallagher GT, Kabani S, Galli SJ, Weller PF, Wong DT.** 1996. Lack of TGF-alpha and TGF-beta 1 synthesis by human eosinophils in chronic oral ulcers. *Oral Surg Oral Med Oral Pathol Oral Radiol Endod* **81**:672–681.

95. **Yang J, Torio A, Donoff RB, Gallagher GT, Egan R, Weller PF, Wong DT.** 1997. Depletion of eosinophil infiltration by anti-IL-5 monoclonal antibody (TRFK-5) accelerates open skin wound epithelial closure. *Am J Pathol* **151**:813–819.

96. **Leitch VD, Strudwick XL, Matthaei KI, Dent LA, Cowin AJ.** 2009. IL-5-overexpressing mice exhibit eosinophilia and altered wound healing through mechanisms involving prolonged inflammation. *Immunol Cell Biol* **87**:131–140.

97. **Mills CD.** 2012. M1 and M2 macrophages: oracles of health and disease. *Crit Rev Immunol* **32**:463–488.

98. **Sica A, Mantovani A.** 2012. Macrophage plasticity and polarization: in vivo veritas. *J Clin Invest* **122**:787–795.

99. **Fujisaka S, Usui I, Bukhari A, Ikutani M, Oya T, Kanatani Y, Tsuneyama K, Nagai Y, Takatsu K, Urakaze M, Kobayashi M, Tobe K.** 2009. Regulatory mechanisms for adipose tissue M1 and M2 macrophages in diet-induced obese mice. *Diabetes* **58**:2574–2582.

100. **Wu D, Molofsky AB, Liang HE, Ricardo-Gonzalez RR, Jouihan HA, Bando JK, Chawla A, Locksley RM.** 2011. Eosinophils sustain adipose alternatively activated macrophages associated with glucose homeostasis. *Science* **332**:243–247.

101. **Foster EL, Simpson EL, Fredrikson LJ, Lee JJ, Lee NA, Fryer AD, Jacoby DB.** 2011. Eosinophils increase neuron branching in human and murine skin and *in vitro*. *PLoS One* **6**:e22029. doi:10.1371/journal.pone.0022029.

102. **Fryer AD, Stein LH, Nie Z, Curtis DE, Evans CM, Hodgson ST, Jose PJ, Belmonte KE, Fitch E, Jacoby DB.** 2006. Neuronal eotaxin and the effects of CCR3 antagonist on airway hyperreactivity and M2 receptor dysfunction. *J Clin Invest* **116**:228–236.

103. **Kobayashi H, Gleich GJ, Butterfield JH, Kita H.** 2002. Human eosinophils produce neurotrophins and secrete nerve growth factor on immunologic stimuli. *Blood* **99**: 2214–2220.

104. **Wen T, Stucke EM, Grotjan TM, Kemme KA, Abonia JP, Putnam PE, Franciosi JP, Garza JM, Kaul A, King EC, Collins MH, Kushner JP, Rothenberg ME.** 2013. Molecular diagnosis of eosinophilic esophagitis by gene expression profiling. *Gastroenterology* **145**:1289–1299.

105. **Noel RJ, Putnam PE, Rothenberg ME.** 2004. Eosinophilic esophagitis. *N Engl J Med* **351**:940–941.

106. **Hahn C, Islamian AP, Renz H, Nockher WA.** 2006. Airway epithelial cells produce neurotrophins and promote the survival of eosinophils during allergic airway inflammation. *J Allergy Clin Immunol* **117**:787–794.

107. **Sanico AM, Koliatsos VE, Stanisz AM, Bienenstock J, Togias A.** 1999. Neural hyperresponsiveness and nerve growth factor in allergic rhinitis. *Int Arch Allergy Immunol* **118**:154–158.

108. **Peters EM, Liezmann C, Spatz K, Daniltchenko M, Joachim R, Gimenez-Rivera A, Hendrix S, Botchkarev VA, Brandner JM, Klapp BF.** 2011. Nerve growth factor partially recovers inflamed skin from stress-induced worsening in allergic inflammation. *J Invest Dermatol* **131**:735–743.

109. **Furuta GT, Forbes D, Boey C, Dupont C, Putnam P, Roy S, Sabra A, Salvatierra A, Yamashiro Y, Husby S, Eosinophilic Gastrointestinal Diseases Working Group.** 2008. Eosinophilic gastrointestinal diseases (EGIDs). *J Pediatr Gastroenterol Nutr* **47**:234–238.

110. **Blanchard C, Rothenberg ME.** 2008. Basic pathogenesis of eosinophilic esophagitis. *Gastrointest Endosc Clin N Am* **18**:133–143; x.

111. **Jacoby DB, Costello RM, Fryer AD.** 2001. Eosinophil recruitment to the airway nerves. *J Allergy Clin Immunol* **107**:211–218.

112. **Yost BL, Gleich GJ, Fryer AD.** 1999. Ozone-induced hyperresponsiveness and blockade of M2 muscarinic receptors by eosinophil major basic protein. *J Appl Physiol* **87**:1272–1278.

113. **Kita H.** 2011. Eosinophils: multifaceted biological properties and roles in health and disease. *Immunol Rev* **242**:161–177.

114. **Carlens J, Wahl B, Ballmaier M, Bulfone-Paus S, Forster R, Pabst O.** 2009. Common γ-chain-dependent signals confer selective survival of eosinophils in the murine small intestine. *J Immunol* **183**:5600–5607.

115. **Matsuoka K, Shitara H, Taya C, Kohno K, Kikkawa Y, Yonekawa H.** 2013. Novel basophil- or eosinophil-depleted mouse models for functional analyses of allergic inflammation. *PLoS One* **8**:e60958. doi:10.1371/journal.pone.0060958.

116. **Moro K, Yamada T, Tanabe M, Takeuchi T, Ikawa T, Kawamoto H, Furusawa J, Ohtani M, Fujii H, Koyasu S.** 2010. Innate production of T_H2 cytokines by adipose tissue-associated c-Kit$^+$Sca-1$^+$ lymphoid cells. *Nature* **463**:540–544.

117. **Doyle AD, Jacobsen EA, Ochkur SI, Willetts L, Shim K, Neely J, Kloeber J, Lesuer WE, Pero RS, Lacy P, Moqbel R, Lee NA, Lee JJ.** 2013. Homologous recombination into the eosinophil peroxidase locus generates a strain of mice expressing Cre recombinase exclusively in eosinophils. *J Leukoc Biol* **94**:17–24.

Recruitment, Inflammation, and Repair

IV

Myeloid Cells in Health and Disease: A Synthesis
Edited by Siamon Gordon

doi:10.1128/microbiolspec.MCHD-0042-2016

Justin F. Deniset[1,3]
Paul Kubes[1,2,3]

15 Intravital Imaging of Myeloid Cells: Inflammatory Migration and Resident Patrolling

The first documented experiments using intravital microscopy were performed in the 19th century, in which very thin translucent tissues were used so that light could penetrate through the tissue and leukocyte trafficking could be observed (1). Neither human tissues nor solid organs in animal models could be used at the time. As such, tissues like the rodent mesentery, cremaster muscle, and ear and the bat wing were the preparations of choice for the next century. This type of imaging unveiled the very dynamic interaction of immune cells with vessel walls. The experimentalists tried to keep the conditions as close to the natural environment as was feasible. The bat wing and ear vasculatures required no surgery, making them likely the least perturbed approach. The mesentery and cremaster, which required only minor surgery, likely did induce a nonphysiologic baseline of leukocyte-vessel wall interactions. However, this came with the benefit of being able to examine cellular functions and behaviors under shear forces associated with blood flow as well as the surrounding architecture of capillaries and venules that was impossible to replicate *in vitro*. Indeed, as diligent as experimentalists were, *in vitro* settings could not completely replicate the behavior of immune cells as they interacted with each other, red blood cells and platelets in capillaries and postcapillary venules surrounded by pericytes and with macrophages, mast cells, and the myriad of other resident immune and parenchymal cells that constitute a living organ. Moreover, interorgan and neural communications were also not possible *in vitro*. However, it is always critical to remember that rodents, bats, and fish are not humans, and so all interpretations must be made with this in mind. It is also worth mentioning that many of the *in vivo* discoveries were made hand in hand with key *in vitro* experiments that allowed simplification of the complex model to elucidate cellular and molecular events.

In the last 3 decades, bright-field microscopy has helped to elucidate the molecular and biophysical mechanisms of leukocyte adhesion to blood vessels (2, 3). However, this very basic technique, which was used for 100 years, became somewhat obsolete as investigators tried to look (i) into less-translucent organs, (ii) at specific immune cell types, and (iii) into compartments found deep in organs such as the parenchyma of

[1]Department of Physiology and Pharmacology; [2]Department of Microbiology and Infectious Diseases; [3]Calvin, Phoebe, and Joan Snyder Institute for Chronic Diseases, University of Calgary, Calgary, AB T2N 4N1, Canada.

the lymph node. Fluorescence intravital imaging allowed tagging of specific immune cells with fluorescence-labeled antibodies to visualize their behavior in real time in live animals. Two additions that greatly improved fluorescence-based intravital microscopy are the different varieties of confocal microscopes and the transgenic mice that report on specific cell types that have no specific surface markers and require the expression of fluorescent proteins (e.g., green [GFP] and red [RFP] fluorescent protein) that are tagged to promoters of characteristic intracellular molecules. In the last decade, better subcellular resolution has been achieved by excluding out-of-focus light via point illumination and pinhole apertures (4, 5). For example, spinning-disk confocal intravital imaging systems provide rapid image acquisition and are extremely competent for dynamic observations of myeloid cell recruitment within the vasculature (6–8). These systems lack the ability to image deep tissue. Multiphoton microscope systems, with a pulsed infrared laser excitation to generate fluorescence, have the capacity to image deep into tissue, often up to 500-μm depth (9, 10). The limitation can be that very rapid interactions such as tethering of cells are more difficult to visualize and any motion artifact can disrupt image acquisition. The latest models of resonant scanners combine speed and depth for optimal imaging (11–13).

These improvements in microscopy have coincided and prompted the development of new tissue preparations allowing comparisons between the behavior of myeloid cells in different tissue settings. Limitations associated with tissue accessibility and motion artifacts have been overcome with strategies that carefully isolate (exteriorization) (7, 14, 15) and/or physically restrain (sutures, tissue adhesive, and custom mounts) (16–19) organs of interest. More recently, small imaging windows or endoscopes paired with gentle suction have been developed to increase stability in particularly difficult tissues such as the lung and the heart (20–22). Integration of computer software allowing for acquisition on a particular phase of respiration or cardiac rhythm combined with postprocessing has also been shown to enhance image quality and resolution (23, 24). To complement the numerous acute intravital preparations, chronic window approaches have been devised to image long term in multiple organs (25, 26).

LEUKOCYTE RECRUITMENT FROM VASCULATURE

The multistep leukocyte recruitment cascade data were originally derived from places like cremaster muscle and rat mesentery. This identified tethering, rolling, adhesion, subsequent crawling to junctions, and transmigration into tissues as key steps of the recruitment pathway (2, 3) (Fig. 1). The initial tethering and rolling can occur very rapidly due to histamine, oxidants, cysteinyl leukotrienes, and thrombin, all capable of mobilizing P-selectin from Weibel-Palade bodies within minutes (27). Cytokines can cause protein synthesis of E-selectin to further enhance leukocyte tethering and rolling after a few hours. P-selectin and E-selectin bind to their glycosylated ligands, including P-selectin glycoprotein ligand-1 (PSGL-1) (2, 28), to initiate tethering and rolling. Most immune cells found in blood can use these molecules in various combinations in peripheral tissues (29, 30). Molecules like histamine, oxidants, and thrombin generated by tissue sentinel cells (e.g., macrophages and mast cells) and blood factors stimulate endothelial cells to synthesize and secrete molecules, including platelet-activating factor (PAF), that activate the β_2 integrins on leukocytes to induce rapid adhesion (31). PAF can also be released during inflammatory conditions by activated cells including mast cells, neutrophils, monocytes, and platelets (32, 33). It is these molecules that contribute to efficient tethering, activation, and recruitment of neutrophils at sites of infection and sterile injury. Cytokines such as tumor necrosis factor α (TNF-α) and interleukin-1β (IL-1β) rapidly produced by innate immune cells (e.g., tissue-resident macrophages, dendritic cells [DCs], and neutrophils) can cause the release of chemokines by the endothelium and interstitial inflammatory cells that can immobilize on endothelial glycocalyx by binding to negatively charged heparan sulfates (34–36). Neutrophils can also amplify the recruitment of other immune cells through the local release of chemokines or granule proteins (22, 37, 38). These factors engage G protein-coupled chemokine receptors on the myeloid cell surface, which induces inside-out signaling, resulting in changes in the conformation of cell surface-expressed integrins (2). This leads to very rapid higher affinity for immunoglobulin-like cell adhesion molecules (intercellular [ICAMs] and vascular cell [VCAMs] adhesion molecules), and interactions between these endothelial cell surface receptors and leukocyte-bound integrins play an important role in leukocyte adhesion (2, 39).

Herein, different myeloid cells use different chemokines and different integrins, which likely accounts for selective recruitment. For example, IL-8 activates neutrophils specifically. Neutrophils express constitutively high levels of the integrins lymphocyte function-associated antigen-1 (LFA-1; $\alpha_1\beta_2$; CD11a/CD18) and Mac1 ($\alpha_M\beta_2$; CD11b/CD18), and IL-8 changes their

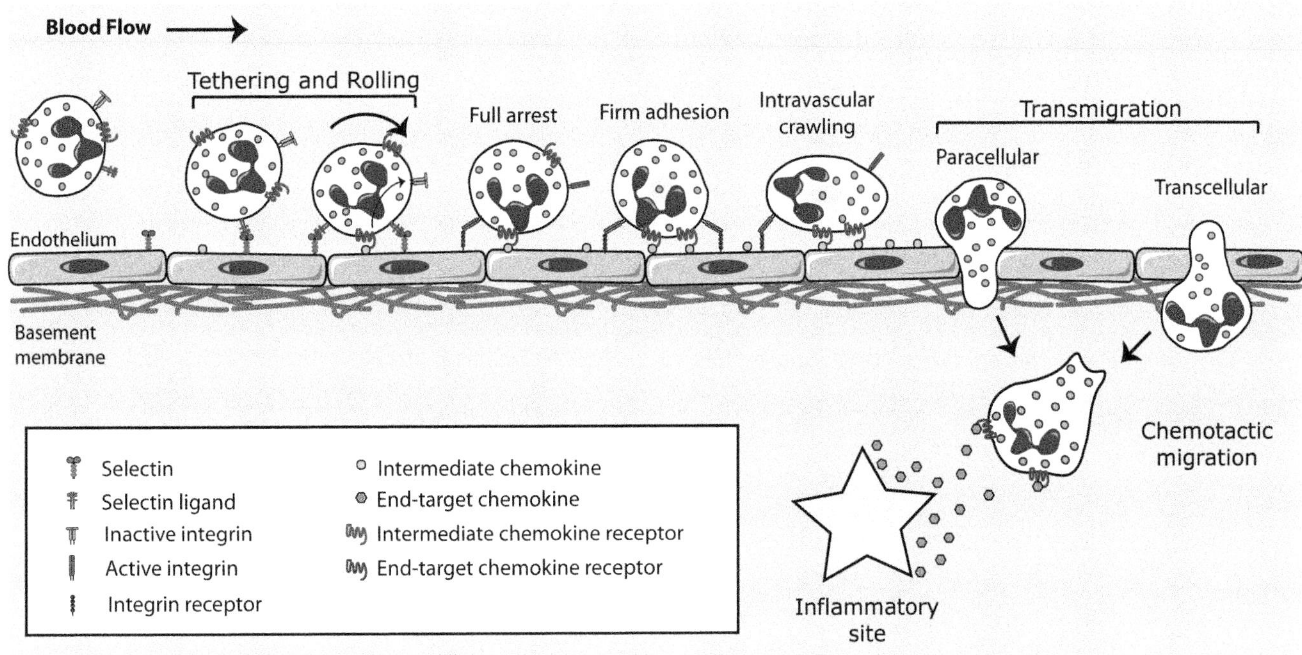

Figure 1 Classical leukocyte recruitment cascade. Depicted are the sequential steps of leukocyte recruitment from the vasculature into the tissue. Selectins and their ligands mediate initial tethering and rolling along the vascular wall. Engagement of intermediate chemokine receptors with their ligands lining the endothelium stimulates activation of integrins on the leukocyte cell surface, enabling their interaction with their respective receptors to facilitate arrest, adhesion, and subsequent transmigration by paracellular or transcellular routes. Chemotactic gradients of intermediate and end-target chemokines guide leukocytes to sites of transmigration and promote directed migration to the site of injury or infection within the tissue.

conformation, allowing binding to ICAM-1 and ICAM-2 presented on the endothelium. LFA-1 binding to ICAM-1 is essential for firm adhesion (40). By contrast, other chemokines, like monocyte chemoattractant protein-1 (MCP-1), recruit inflammatory monocytes through activation of both the CD18 integrin but also CD29 (β_1 integrin), and more specifically $\alpha_4\beta_1$ integrin, which binds VCAM-1 (41). These neutrophil- and monocyte-specific chemokines are thought to be produced by the endothelium, perivascular cells, and immune cells in response to cytokines such as TNF and IL-1 produced locally by innate immune cells (42–49). By contrast, IL-4 produced during parasite and allergic conditions induces the production of eotaxins for eosinophil recruitment (50). Eosinophils can also use CD18 and α_4 integrin just like the monocytes, making the type of chemokine potentially the key to selective recruitment (50).

While LFA-1 mediates the firm adhesion, Mac-1 (CD11b) appears to mediate subsequent crawling, although in cells that do not express the latter, α_4 integrin can also support crawling (40). For directional crawling to occur, generally gradients of chemotactic molecules are needed. Chemokines, which are positively charged molecules, are immobilized on endothelium by binding to negatively charged heparan sulfates, which serve as anchors to prevent the shear forces from washing these molecules away (35). This allows the formation of intravascular chemotactic gradients. Eventually, apically sequestered chemokines are removed by endothelial endocytosis, and this may limit neutrophil recruitment (51). The activation of G protein-coupled chemokine receptors on myeloid cells presumably allows the cells to crawl to junctions for subsequent emigration. If the chemokine gradient is primarily on the surface of endothelium, the myeloid cell should crawl to the highest concentration and remain at this site without emigrating. However, this is not the case as the cells begin to transmigrate across endothelium. A number of possibilities exist to explain this behavior. The initial release of chemokines by endothelium or tissue-resident leukocytes could occur subluminally so that the point source of chemokine emanates from outside the vasculature and leaks across junctions into the lumen of vessels. This would cause myeloid cells to follow the chemokine gradient out of the vasculature. It is also well known that myeloid cells also have the ability to differentiate between different chemoattractants and that a hierarchy of chemoattractants exists (7, 39, 52–55). This contributes to the sequential use of intermediate (e.g., IL-8) and end-target chemokines (e.g., formylated peptides and C5a) to reach the inflammatory site. Numerous mechanisms likely contribute to this phenomenon. Chemokine receptors for intermediate molecules (e.g., IL-8 and PAF) are constitutively localized at the plasma, whereas the majority of receptors for end-stage molecules (e.g., formylated peptide) are stored in intracellular vesicles ready to mobilize to the cell surface upon activation, leading to delayed and possibly more robust response for end-target chemokines (56, 57). This is compounded by the differential desensitization mechanisms described for both chemokine groups. Upon IL-8 engagement of its receptor, homologous desensitization through receptor internalization ensues following activation to prevent subsequent activation with the same chemokine without impacting signaling through formylated peptide receptor (57). In contrast, receptor engagement by formylated peptides leads to the heterologous desensitization that blocks subsequent activation by both end-target and intermediate chemokines (57). Finally, distinct signaling molecules inside the myeloid cells allow for this discrimination among different chemoattractants for migration (53).

The emigration *per se* remains a very complicated event, with many different molecules involved. It is the least well understood step in the recruitment cascade, perhaps in part because different subsets of immune cells behave differently. The molecular mechanisms of transmigration include platelet endothelial cell adhesion molecule-1, junctional adhesion molecules (JAMs), and CD99, to name a few, and are the topic of entire reviews (58, 59). The other problem with this field is the fact that emigration is a challenging process to observe. First and foremost, one requires delayed time-lapse imaging (longer intervals between frames), which is not optimal for seeing other, more rapid steps of the recruitment cascade such as rolling. Secondly, the exact site of emigration has been quite controversial, in part because definite proof that a myeloid cell is emigrating at a junction versus "near" a junction can be very difficult to dissect (40). Even serial sections using electron microscopy leave room for interpretation (60). Finally, while *in vitro* experiments can provide more-definitive answers about the emigration process, it is possible that the growth of endothelium *in vitro* in monolayers does not necessarily reflect the way endothelia overlap each other in blood vessels *in vivo*, and entirely different transmigrations may occur in these two significantly different settings.

Emigration through endothelial cells may require very different adhesion molecules compared to emigration between the endothelia (2). No one would dispute the existence of emigration at junctional sites, as many studies have shown almost exclusive paracellular emigration and those that have shown transcellular migration of myeloid cells still report that this is a minor route of migration (3). Nevertheless, migrations in a transcellular and paracellular fashion likely coexist (3). In very detailed experiments, neutrophils have been reported to migrate at tricellular corners where three endothelial cells meet (61) and in regions with low expression of extracellular matrix molecules (62). By contrast, monocytes migrate between two endothelial cells (63). Clearly, each cell type may use its own route of exit, and it is worth noting here that trying to find a "universal paradigm" of emigration for all immune cells is impossible. Indeed, one *in vitro* study showed that in the same monolayer, human T lymphocytes used the transcellular pathway whereas human neutrophils used the paracellular pathway, highlighting that differences cannot be attributed to different experimental conditions including type of endothelium and type of stimulus (64).

Exceptions to the Rule, or Is the Rule the Exception?

Our knowledge regarding the myeloid cell recruitment cascade is derived from translucent tissues such as cremaster muscle or mesentery (2, 3). Tethering, rolling, adhesion, and emigration occur in reasonably large postcapillary venules upon which either a chemokine or cytokine is applied (40, 65–70). A similar recruitment paradigm has been noted in skin and brain (3, 39). However, it is now well appreciated that in lung, liver, kidney, and perhaps heart, recruitment of myeloid cells occurs in capillaries as well as postcapillary venules (71–77). With the improvement of intravital microscopy and exploration of diverse tissues, it appears that myeloid cell recruitment may not always require selectins and/or integrins (7, 40, 71, 73, 78, 79). Indeed, neutrophil recruitment into the liver occurs primarily in capillaries, with only a minor component occurring in the larger vessels (portal and central venules) of the liver (76). In the latter, rolling and subsequent firm adhesion definitely occurs and is dependent on the classical molecules already described; however, in the sinusoidal capillaries of the liver, there is absolutely no rolling and as such no role for selectins (76). Adhesion can occur via integrins in models of sterile injury, but during infection, integrin-independent interactions were noted (7, 73, 80). Different adhesion mechanisms have also been noted between local and systemic bacterial stimuli (80). Sinusoids are coated with the carbohydrate hyaluronan, and under some infectious conditions, the hyaluronan molecule appears to bind proteins including serum hyaluronan-associated protein, which induces CD44-dependent adhesion (73). The CD44-dependent adhesion appears to be relevant for neutrophils, monocytes, stem cells, and cancer cells, suggesting that this may be a liver-specific recruitment mechanism.

Another organ where myeloid cells adhere primarily in capillaries is the lung. The pulmonary alveoli are surrounded by capillaries, and it is at these sites that most neutrophils exit the vasculature and enter the interstitium and ultimately the lumen of the lung. Early observations of selectin- and β_2-integrin-independent neutrophil recruitment *in vivo* were provided by Doerschuk and colleagues in certain models of lung infections (71). The lung microvasculature is a network of thin capillaries. As such, it has been postulated that neutrophils must significantly alter their normal cellular shape in order to pass through these conduits (71). This raises the possibility that neutrophil sequestration in the lung is due to mechanical trapping that bypasses the need for selectins and integrins. Neutrophils lacking selectin ligands and integrins still adhere to blood vessels in the lungs (P. Kubes and B. Yipp, unpublished observations), but it remains to be seen whether this is due to other adhesion molecules or physical trapping. Identification of specific adhesive mechanisms will shed light on how the lung recruits myeloid cells. Our own "opinion" is that there is selective recruitment of specific myeloid cells, making it quite difficult to invoke nonspecific physical trapping for all forms of lung inflammation. New imaging approaches, including a suction window (21) that allows visualization of moving organs like lung and heart, will almost certainly help resolve the adhesive mechanisms in these organs.

Finally, myeloid cell recruitment into tissues like kidney and brain may also harbor a modified version of the universal recruitment paradigm, in which platelets appear to be critical. Imaging the kidney microcirculation has been very challenging because of the location of the key vascular component, the glomerulus, deep within this solid organ. In fact, even the most superficial glomeruli are located at least 100 μm deep. The glomerulus is a highly specialized structure made up of afferent arterioles located at the beginning and efferent arterioles at the end of the capillary plexus. Glomerular capillaries are lined by specialized endothelial cells that do not express P-selectin but have been shown to express ICAM-1 constitutively and VCAM-1 and E-selectin under inflammatory conditions (81, 82). Intravital imaging revealed a very important role for platelet adhesion to endothelium, which formed a bridge to neutrophils to permit them to adhere in the unique vessels of the kidney glomerulus. Platelet depletion and P-selectin-deficient platelet transfer studies demonstrated that recruited platelets promoted glomerular neutrophil recruitment via their ability to express P-selectin within the inflamed glomerulus (83). In addition, integrins and ICAM and VCAM were involved, demonstrating that glomerular myeloid cell recruitment involved interactions with platelets and the inflamed endothelium.

The inflamed brain also has an increased propensity for platelet recruitment in this vasculature during inflammation. Intravital microscopy revealed an important role for platelets in the leukocyte recruitment into inflamed brain microvessels (84, 85). Platelets first adhere to the brain endothelium, and depletion of platelets reduced myeloid cell recruitment. The working hypothesis was that the platelets functioned as a bridge to tether and adhere myeloid cells. In this particular case, the brain endothelium could produce P-selectin, but the amount of expression was minuscule compared to the adhering platelets, which expressed large amounts of this adhesion molecule that can avidly recruit neutrophils (84). However, it is worth mentioning that platelets bound to activated brain endothelium expressed P-selectin in some (85) but not other models (84), suggesting that P-selectin expression on adhering platelets is context dependent. Nevertheless, P-selectin has been the key molecule for leukocyte recruitment in a number of models of brain inflammation including traumatic brain injury (86), permanent middle cerebral artery occlusion (87), and other inflammatory conditions (84). Furthermore, *in vitro* studies showed that platelets expressed the LFA-1 ligand ICAM-2 as well as the chemokine RANTES (CCL5) immobilized to their surface (88–90), and this can recruit various myeloid cells via PSGL-1, CCR1/CCR5, and LFA-1 to brain endothelium. Whether platelets adhering to brain endothelium can selectively affect neutrophil versus monocyte recruitment remains unclear and requires further study. However, this contention seems intriguing considering that platelets can express neutrophil and monocyte chemokines. Interestingly, the latest studies have argued that monocytes may be at least as important as neutrophils in inducing brain inflammation (91, 92).

During the course of acute inflammation, recruited neutrophils are thought to die in the tissue and are cleared by macrophages locally (93). However, the *in vivo* evidence for this process is less than robust, and evidence of alternative fates for neutrophils has started to emerge. Using a photoconversion system, Hampton et al. demonstrated that recruited neutrophils in response to skin infection could subsequently migrate to the draining lymph node and modulate adaptive immune responses (94). Recent work primarily in zebra fish embryos has suggested that neutrophils could enter sites of sterile injury and then return to the circulation (termed reverse transmigration) (95–97). Although this has not been shown in mammalian cells, one group has shown that neutrophils could at least extend a pseudopod or even their whole body out of the vasculature before returning to the circulation (98). This process in the mouse is noted to depend on the downregulation or cleavage of a junctional adhesion protein, JAM-C, which normally prevents reverse transmigration (98, 99). Further, small populations of neutrophils with a reverse-migrated phenotype have been described in patients with rheumatoid arthritis (100). Future studies will help to determine the functional consequences of this mechanism and the ability of reverse-migrated neutrophils to participate in subsequent inflammatory responses.

MYELOID CELL PATROLLING

With the advent of multiphoton and spinning-disk microscopy, it became clear that in addition to the intravascular immune cells that were primarily found in the mainstream of blood, there are also populations of cells that appear to patrol the vasculature. The best example of this type of situation was described by Geissmann and colleagues (101), who reported a CX3CR1^{hi}Ly6C^{lo} noninflammatory or alternative monocyte that lacked CCR2, the key molecule found on inflammatory monocytes. Using *in vivo* imaging in skin and mesentery, this group reported that these CX3CR1-GFP monocytes crawl on the endothelial surface under basal conditions throughout all areas of the vasculature. The brain has also been reported to have this population of cells (102). One caveat of imaging immunity is the significant effect that can occur when surgical interventions are required. Exteriorization could cause the monocytes to rapidly adhere to and crawl along vessel walls. Nevertheless, the fact that these cells can be seen in unperturbed or minimally perturbed preparations while neutrophils were not seen to be recruited does suggest that these are patrolling monocytes that serve an important role in immune surveillance. More recently, these cells have also been identified using multiphoton microscopy to crawl in glomerular capillaries, where they can detect endothelial infection via Toll-like receptor 7 and recruit neutrophils to eradicate infected cells (103). Although these monocytes were seen to patrol the vessels of various organs in the absence of inflammation, these cells were poised to rapidly enter sites of inflammation. Indeed, addition of irritants, aseptic wounding, or *Listeria monocytogenes* all induced rapid recruitment of these CX3CR1-positive monocytes. Interestingly, these cells were there even faster than neutrophils, although this has not been seen in other models of inflammation described below.

Many studies have shown that CCR2 is critical for monocyte recruitment in peritonitis (104), tuberculosis (105), atherosclerosis (106), and autoimmune encephalitis (107); however, many of these studies need to be reexamined inasmuch as CCR2 has been shown to be critical for monocyte recruitment out of the bone marrow and into the circulation (108, 109). As such, when Serbina and Pamer examined recruitment of wild-type and CCR2$^{-/-}$ monocytes via adoptive transfer (bypassing the efflux from bone marrow) in response to thioglycolate-induced peritonitis and *L. monocytogenes* infection into spleens, they found no migration defect (108). However, despite also reporting that CCR2$^{-/-}$ monocytes could not emigrate out of bone marrow when adoptively transferred into wild-type mice, Tsou et al. reported that these mice did have a recruitment defect into thioglycolate-induced peritonitis (109). These authors argued that CCR2 was important for early recruitment (24 h) but not the late recruitment (60 h) used by Serbina and Pamer, explaining the difference in these two studies. In addition, Tsou et al. went on to demonstrate that MCP-1 and, more dramatically, MCP-3 were the key molecules drawing monocytes out of the bone marrow. Therefore, while CCR2$^{-/-}$ mice do demonstrate defects in recruitment, the fact that these mice have very low circulating inflammatory monocyte counts complicates interpretation of results.

Recruitment of Ly6C^{hi} and Ly6C^{lo} monocytes is also somewhat controversial. Whether there are two separate waves of monocyte recruitment or whether the Ly6C^{hi} monocytes are recruited and then locally educated to become Ly6C^{lo} monocytes is an area of contention. Nahrendorf et al. reported Ly6C^{hi} monocyte influx into a myocardial infarction at day 3 and then a separate wave of Ly6C^{lo} monocyte influx that was delayed and peaked at day 7 (110). In the early phase, CCR2 was necessary for Ly6C^{hi} monocyte recruitment, whereas CX3CR1 was needed for Ly6C^{lo} monocyte recruitment in the later phase, and inhibiting one type of monocyte had no impact on recruitment of the other type of monocyte. Each monocyte appeared to have a different function: the Ly6C^{hi} monocyte had many proteases and inflammatory cytokines, whereas the Ly6C^{lo} monocyte harbored proangiogenic molecules. This sequential pathway may be disrupted in the context of chronic inflammation associated with atherosclerosis, leading to greater recruitment of Ly6C^{hi} and fewer Ly6C^{lo} monocytes. More recently, this group has shown a transition from Ly6C^{hi} monocytes to Ly6C^{lo}F4/80^{+} macrophages in myocardium (111). Interestingly, immature circulating monocytes that were Ly6C^{hi} were suggested to migrate to bone marrow and matured into Ly6C^{lo} monocytes before reaching sites of infection or inflammation, for the first time questioning the possibility of a single monocyte that matures with time (112).

This view was furthered subsequently by a number of groups. In skeletal muscle injury via a neurotoxin, Ly6C^{hi} monocytes infiltrated in small numbers and then converted to Ly6C^{lo}F4/80^{hi} macrophages (113). These macrophages, upon encountering muscle debris, rapidly converted to less-inflammatory repair macrophages producing transforming growth factor β (TGF-β) and IL-10 and stopped making inflammatory cytokines. No Ly6C^{lo}CX3CR1^{hi} monocytes could be seen to be recruited into the site using isolation, tagging, and reintroduction of cells into the bloodstream. Although these data are compelling evidence that there is conversion at the inflammatory site, one could argue that isolation and tagging of Ly6C^{lo} monocytes could modify their migration patterns. Moreover, it does not completely exclude the possibility that some Ly6C^{lo} monocytes infiltrated into the tissue and contributed to the formation of these macrophages.

Nevertheless, a very similar paradigm was reported in a model of sponge (foreign body) implantation below the skin (114). Ly6C^{hi} monocytes infiltrated into the tissue within the first day, with no evidence of infiltration by Ly6C^{lo} monocytes. These cells slowly transitioned into F4/80^{+} macrophages and helped in the healing process. Four different approaches were taken to demonstrate the transition from Ly6C^{hi} monocytes to repair macrophages. Fluorescent microparticles were injected and taken up by inflammatory monocytes, and these became macrophages at day 7. In this study, selective depletion of Ly6C^{lo} monocytes had no impact on the development of the tissue macrophages. By contrast, when Ly6C^{lo} monocytes were labeled with fluorescent microparticles, these cells did not enter into the wound and did not generate repair macrophages. Similarly, using congenic mice in which recipient mice were adoptively transferred with monocytes, F4/80^{+} macrophages were indeed derived from the donor monocytes. Similar cytokine profile switching was noted as in the aforementioned studies, inasmuch as the monocytes that entered tissue produced TNF-α and IL-1β at day 1 but the macrophages at day 14 were predominantly producing TGF-β and vascular endothelial growth factor.

Ideally, tracking the monocytes using imaging would provide further insight into the monocyte transition to macrophage. Indeed, one study used intravital imaging to track CCR2^{hi}Ly6C^{hi} monocytes and CX3CR1^{hi}Ly6C^{lo} monocytes (115). A liver sterile injury/repair model was used in which a few hundred liver cells were thermally killed. CCR2^{hi} monocytes surrounded the injury site in between 8 and 24 h, forming a concentric ring around the injury, with some of these monocytes entering the injury site. The CX3CR1^{hi} monocytes also made an appearance, however, about 24 h after the CCR2^{hi} monocytes (24 to 48 h after injury), but these cells did not enter the site through the blood vessels. Indeed, when both CCR2-RFP and CX3CR1-GFP reporters were incorporated into a single mouse, early CCR2^{hi} cells were seen infiltrating the injury site, but with time, more and more hybrid (CCR2^{+}CX3CR1^{+}) cells were seen in the ring surrounding the injury site. Over time, cells became more and more CX3CR1^{hi}Ly6C^{lo} until they became mostly CX3CR1^{hi}CCR2^{lo}Ly6C^{lo}. This study shows that local education of the monocytes allowed them to convert from a proinflammatory phenotype to a more alternative monocyte. The conversion could be delayed if IL-4 and IL-10 were blocked, and this led to delayed healing. While this study did not examine changes beyond 72 h, at this 3-day time point most of the cells had not yet begun expressing high levels of F4/80 or MER proto-oncogene tyrosine kinase (MerTK), two hallmark features of macrophages. Presumably, these markers would be expressed at later time points. A follow-up study from this group identified that there were indeed mature repair macrophages recruited to the injury site in as early as 1 h, entirely independent of monocyte recruitment (116), but the source of these cells was the peritoneum, which will be discussed later in this chapter.

RESIDENT NEUTROPHILS

There has always been the concept of a resident or marginated pool of neutrophils that could be rapidly mobilized during infection or inflammation. Presumably the largest pool of mature neutrophils have been identified in the bone marrow. These cells appear to be mobilized in response to various inflammatory molecules, including IL-8 (CXCR2 ligands), complement factors, and granulocyte colony-stimulating factor (G-CSF). In addition, blocking of CXCR4 or the ligand stromal cell-derived factor (CXCL12) also mobilizes neutrophils into the circulation, although some discrepancy around the latter issue exists. Numerous publications identify the bone marrow as one of the main sources of neutrophils during inflammation (117–120). However, this does exclude the lung as the alternative source. In fact, a recent study using imaging saw no evidence of emigration of neutrophils out of bone marrow into the vasculature and rather suggested that these neutrophils were localized in the lung vasculature and released from this compartment (121). This so-called marginated pool, found in animal models and humans, is estimated to contain 3-fold more neutrophils than the entire blood circulation (122–127). Recent imaging by Kreisel and colleagues (17) revealed that this population of neutrophils were attached within the alveolar capillaries, and we have recently imaged these neutrophils and find they constantly patrol this vasculature much like the aforementioned monocytes. There may be other localized pools of neutrophils at tissue sites. Unexpectedly, under resting conditions, neutrophils were also retained in glomerular capillaries. Some neutrophils remained static following arrest, while others migrated throughout the glomerular capillaries before detaching from the vessel wall and returning to the circulation. Similarly, there appears to be a population of neutrophils localized to the spleen, and perhaps even in lymph nodes (128, our unpublished observations). In skeletal muscle capillaries, by comparison, neutrophil arrest and migration was much less frequent and was shorter-lived, demonstrating that this phenomenon did not occur to a similar extent in all vascular beds. However, these observations raise the possibility that neutrophils as well as monocytes undertake a specialized form of immune surveillance that contributes to homeostasis of lung, kidney, lymph node, and other tissues. In fact, these were predicted to exist in all tissues and were termed pioneer neutrophils (129).

IMMUNE SURVEILLANCE BY TISSUE MACROPHAGE POPULATIONS

The mechanisms that regulate leukocyte recruitment to sites of infection or injury are dependent on the initial sensing and amplification of the inflammatory response by cells within that particular tissue. Tissue-resident macrophages are known to be key players in this network. Their extensive repertoire of pattern recognition receptors makes them well equipped to sense pathogens and damage locally and subsequently mediate the release of chemokines and cytokines (130). Although originally described to be uniquely bone marrow monocyte derived, it is now appreciated that many of these resident cells, such as microglia (brain), Kupffer cells (KCs) (liver), and Langerhans cells (LCs) (skin), are embryonically derived and are maintained by local proliferation (131). The local parenchymal and stromal cells provide both the maintenance signals (e.g., CSF-1, CSF-2, and IL-34) and tissue-specific cues that determine their functional phenotypes and tissue distribution (132, 133). This reliance on tissue microenvironment highlights why conventional *in vitro* approaches are inadequate for the functional assessment of tissue macrophages. As such, the development of intravital imaging has been instrumental in our understanding of macrophage function and behavior in many organs (Table 1).

Central Nervous System

Microglia are the tissue-resident macrophage population of the central nervous system (CNS) parenchyma. A pair of key studies using two-photon laser scanning microscopy provided the first description of the dynamic nature of these cells within an intact brain (18, 134). Under basal conditions, microglia have a distinct structure with a small soma and large ramifications that constantly extend and retract (18), dynamic movement dependent on Cl^- channels and actin polymerization (135). These dendrites continuously scan the parenchyma, contacting numerous cortical elements including astrocytes, neuronal bodies, and blood vessels, fulfilling an important housekeeping function by removing metabolic and tissue waste (18). In response to focal injury, ATP released locally binds purigenic receptors on microglia, causing a rapid, polarized movement of bulbous processes to the injury site (134). This activation response is typically accompanied by migration of microglia to the site, transition to an amoeboid shape, and expansion via local proliferation (136). Ischemic conditions in the brain severely reduce the dynamics and number of processes (137) and alter their interactive behavior with neuronal synapses (138). In experimental Alzheimer's disease, imaging studies have noted microglia being recruited to a plaque within the first 48 h, where they activate locally, contributing to neural loss in a CX3CR1-dependent fashion (139, 140). The sensitivity of the microglial system makes the tissue preparation an extremely important consideration, as tissue sections and even preparations in which the skull is removed can cause an activated phenotype (141). Many investigators now thin the skull and image

Table 1 Tissue-resident macrophage behavior and function *in vivo*

Tissue	Cell type	Detection markers (imaging modality)[a]	Behavior and function
CNS	Microglia	CX3CR1-GFP reporter mice (TPM)	Homeostatic surveillance Survival of neurons and synaptic remodeling Removal of metabolic and tissue debris
Liver	KC	F4/80 Ab (SDCM)	Clearance of microorganisms Promoting leukocyte recruitment Antigen presentation
Skin	LC	MHC-II–GFP (TPM), CD11c-YFP (TPM), and Langerin-EGFP (TPM) reporter mice	Homeostatic surveillance (dSEARCH) Skin antigen sampling and adaptive immune response modulation
	PVM	DPE-GFP reporter mice (TPM)	Modulating neutrophil recruitment and transmigration
Gastrointestinal tract	Intestinal macrophage	CX3CR1-GFP (TPM), CD11c-EGFP (TPM), and CX3CR1-GFP/CD11c-YFP (TPM) reporter mice	Homeostatic sampling of commensal bacteria Protective barrier to invasive pathogens
Lung	Alveolar macrophage	CD11c-YFP reporter mice (TPM)	Clearance of surfactant-opsonized particles and invading pathogens
Lymph node	SCS macrophage	CD169 Ab (TPM)	Antigen capture and transfer to lymphocytes
Serosal tissues	Peritoneal macrophage	LysM-GFP reporter mice (SDCM)	Homeostatic surveillance Modulating of B1-cell function Wound healing in the liver
Spleen	Red pulp macrophage	F4/80 Ab (SDCM)	Erythrocyte clearance Bacterial capture
	Marginal zone macrophage	CD209b Ab (TPM)	Encapsulated bacteria capture

[a]Abbreviations: Ab, antibody; SDCM, spinning-disk confocal microscopy; TPM, two-photon microscopy.

through bone deep into brain parenchyma using two-photon microscopy.

Liver

KCs residing in the liver make up roughly 80 to 90% of tissue macrophages in the body and play an important filtering role for blood contents for both the portal and arterial circulation. These sessile macrophages are strategically positioned within the liver sinusoids, with pseudopods that constantly sample the circulation. Intravital imaging of the liver by spinning-disk confocal microscopy has highlighted their efficient ability to rapidly bind and phagocytose bacteria under flow conditions, resulting in preferential sequestration of circulating Gram-positive and Gram-negative bacteria to the liver (6, 142–144) (Fig. 2). The arrest of pathogens and other opsonized material on their surface under flow conditions is predominantly dependent on their expression of the complement receptor of the immunoglobulin receptor family (CRIg) (Fig. 2). This receptor directly recognizes C3b and iC3b complement fragments bound to the surface of dying cells and pathogens (145) and has recently been shown to directly capture Gram-positive bacteria through lipoteichoic acid binding (144). Fcγ receptors also contribute to removal of antibody-opsonized cells. In fact, therapeutics antibodies have been used for the clearance of tumor cells via Fcγ receptor binding on KCs (146, 147).

The initial binding step of microbes or microbial components by KCs is crucial for subsequent innate and adaptive immune processes. KCs are involved in the recruitment and adhesion of neutrophils to the liver in response to lipopolysaccharide stimulation (148). Under steady-state conditions, circulating platelets engage in "touch and go" interactions with KCs (143). Following infection with methicillin-resistant *Staphylococcus aureus* and *Bacillus cereus*, this interaction quickly transitions to nucleation of platelets to surround the pathogen on the surface of the KC, a mechanism that is integral for bacterial clearance (143). KCs can contribute in an antigen-presenting role. Following uptake of *Borrelia burgdorferi*, KCs mediate invariant NKT cell recruitment via chemokine release and subsequent invariant NKT cell arrest and activation in a CD1d-dependent manner (6). Similar KC-dependent antigen presentation on major histocompatibility complex class II (MHC-II) and MHC-I molecules leading to antigen-specific activation of $CD4^+$ regulatory and $CD8^+$ T cells locally has also been visualized (149, 150). The adaptive immune response is likely to depend

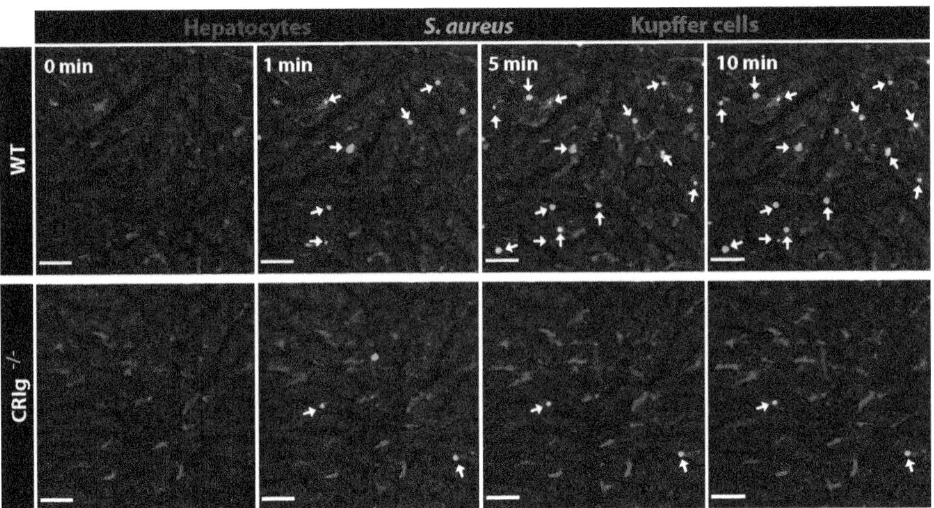

Figure 2 KC capture of circulating bacteria *in vivo*. Time-lapse spinning-disk confocal microscopy images of *S. aureus* (*S. aureus*-GFP; green) catching by liver KCs (F4/80; red) in wild-type (WT) (top) or CRIg$^{-/-}$ (bottom) animals. White arrows, KC-bound bacteria. Bars, 50 μm.

on both the antigen and the inflammatory state within the liver. Under conditions of chronic liver disease, KCs display an impaired ability to generate homeostatic regulatory T cells (150), again highlighting the importance of the local microenvironment for the function of tissue-resident macrophages.

Skin

The skin functions as an important barrier to limit entry of pathogens. A network of tissue-resident myeloid cells populates the layers of this organ to fulfill an important immune surveillance role. LCs, the best-described population, are long-lived resident cells within the epidermis that share characteristics of both DCs and tissue-resident macrophages (130, 151–153). Direct imaging of the epidermal layer of the skin combined with the use of a reporter mouse strain (MHC-II–GFP, CD11c-YFP [yellow fluorescent protein], or Langerin-EGFP [enhanced GFP]) allows for direct visualization of the cells *in vivo*. Under steady-state conditions, LCs are predominantly sessile with long dendritic processes (154, 155). In a small subset of these LCs, repetitive extension and retraction of the processes reminiscent of microglial sampling in the brain can be observed and has been termed dendrite surveillance extension and retraction cycling habitude (dSEARCH) (155). Contact hypersensitivity reactions and trauma result in increased dSEARCH activity and amoeboid-like lateral displacement of the cell body in an IL-1- and TNF-α-dependent manner (154, 156, 157). Similar to the brain, the use of skin tissue sections for imaging studies can result in an increase of the dSEARCH index of LCs, indicative of cell activation (155).

LCs are able to migrate from epidermis to the draining lymph nodes and mediate adaptive immune responses. An initiating step in this process under inflammatory conditions involves LC-dependent sampling of antigen near the surface of the skin by extension of their dendrites between keratinocytes without disrupting tight junctions (158). This allows for continuous detection of external stimuli and migration to the draining lymph nodes while maintaining the physical barrier. This sampling mechanism mediates early antibody production and protection in an experimental staphylococcal scalded skin syndrome model (159). The cancer-associated epithelial cell adhesion molecule (EpCAM) expressed on LCs is crucial for regulating activation, sampling, and subsequent migration to the lymph nodes. In the contact hypersensitivity model, EpCAM-deficient mice demonstrate reduced dSEARCH activity and decreased migration of hapten-bearing LCs to the draining lymph nodes (160). Studies in EpCAM- or LC-deficient mice support a tolerogenic role for LCs within the lymph nodes (160, 161).

Under conditions in which the skin barrier is breached, innate immune mechanisms including neutrophil recruitment are initiated in an effort to prevent dissemination of the pathogen into the bloodstream. Perivascular macrophages (PVMs) that surround post-capillary venules in the skin have been identified as important players in this response (162). Using a DPE-GFP transgenic mouse, Abtin and colleagues describe that these nonmigratory macrophages with motile dendrites cover nearly half of the outer surface of the capillaries (162). Following skin infection, these PVMs express high levels of neutrophil chemokines (e.g., keratinocyte-derived chemokine and macrophage inflammatory protein 2) and promote neutrophil transmigration predominantly across the vascular area they occupy and subsequent infiltration into the skin parenchyma. The importance of these PVMs in innate immune response in the skin is highlighted by the fact that *S. aureus* has developed an immune evasion mechanism targeting the lysis of these cells by one of its major virulence factors, α-toxin (162).

Similar to the skin, mucosal surfaces in the intestines and lungs represent common entry points for invading microorganisms. In both organ systems, tissue-resident macrophages reside along the barrier to perform homeostatic and surveillance functions.

Intestine

Intestinal macrophages reside in the lamina propria on the basolateral side of the epithelial layer. In contrast to many tissue-resident macrophage populations in other organs, intestinal macrophages are constantly replenished by proinflammatory monocytes (163–165). Analogous to LCs in the skin, imaging studies employing CX3CR1-GFP or CD11c-EGFP reporter mice note the ability of intestinal macrophages to sample contents of the lumen via the extension of transepithelial dendrites (166, 167). Transepithelial dendrite formation not only contributes to sampling of commensal bacteria but also serves as an important protective mechanism for clearance of enteroinvasive pathogens such as *Salmonella* (167). These macrophages can relocalize to mesenteric lymph nodes following infection (167); however, CD103$^+$ DCs, which also sample luminal contents through an alternative intraepithelial mechanism, likely play a more important role in mediating adaptive immunity in the gut (168, 169). The overlapping functions and common markers shared by both CX3CR1$^+$ macrophages and CD103$^+$ DCs highlights the importance of using multilabeling strategies (168) for future functional assessment of intestinal myeloid populations *in vivo*.

Lungs

Alveolar macrophages are long-lived embryonically derived cells that are localized on the alveolar side of the lung epithelium, where they are in direct contact with inhaled microbes, pollutants, and dust. Under steady-state conditions, they are maintained in a noninflammatory state through CD200- and TGF-β-dependent interactions with the epithelium (170, 171). They play an important phagocytic function within the airways that ensures clearance of surfactant-opsonized particles and invading pathogens (172–174). The characterization of their *in vivo* behavior has been limited due to the technical issues associated with respiration. As such, conventional approaches of imaging in the lung have relied on the use of perfused *ex vivo* lung sections or imaging following alteration or cessation of mechanical ventilation (175–178). Using perfused lung sections from CD11c-YFP reporter mice, Westphalen and colleagues characterized alveolar macrophages as sessile, phagocytic, and in close communication with the underlying epithelium via connexin-43 hemichannels, which allows for rapid propagation of Ca^{2+} waves across neighboring sections of the lung (178). Currently, it is unknown whether these characteristics observed are a true representation of the *in vivo* situation. Tissue section imaging approaches in the brain and skin have resulted in activated phenotypes for microglia and LCs. More-refined *in vivo* lung preparations that stabilize the lung while preserving ventilation and perfusion have been used for studying the behavior and migration of neutrophils, monocytes, and T cells within the pulmonary capillaries (17, 21, 179). Preliminary imaging of alveolar macrophages using this platform has revealed that these cells may in fact have a patrolling phenotype, moving along the epithelial layer to survey the environment (A. Neupane and P. Kubes, unpublished observations). However, the depth and liquid-air interface makes imaging very challenging in this compartment.

Lymph Nodes

A vast network of myeloid cells including DCs and macrophages populates lymphoid organs. CD169-expressing subcapsular sinus (SCS) macrophages that reside below the capsule in lymph nodes serve as an important filter for antigen. The positioning and dynamic movement of their dendrites as revealed by intravital imaging facilitate the rapid capture of viruses (180–183), immune complexes (184, 185), and tumor-derived antigens (186) entering the lymph node via afferent lymphatics. Depletion of SCS macrophages leads to viremia (182), highlighting the central role for these macrophages in mediating capture and triggering downstream events that promote viral clearance. Viral uptake can stimulate inflammasome activation in SCS macrophages *in vivo* via ASC speck formation, leading to subsequent cell death and recruitment of innate (neutrophils, monocytes, and NK cells) and T cells from the periphery (183). SCS macrophages can also mediate shuttling of antigen from the subcapsular space to follicular zone by migration and transfer of particles to follicular B cells (182, 184, 185). Disruption of this capture and shuttling pathway impairs B-cell activation and affinity maturation (182, 184). This mechanism is not limited to SCS macrophages, as medullary sinus macrophages and interfollicular macrophages also bind antigen (181, 185, 187) and participate in subsequent transfer to follicular B cells (187). The distribution of antigen within the lymph node is likely dependent on the repertoire of scavenging receptors (e.g., CD169, SIGN-R1 [specific ICAM-3-grabbing non-integrin related 1], and MARCO [macrophage receptor with collagenous structure]) expressed on these macrophage populations. The role of these macrophage populations in adaptive T-cell immunity is less clear. Dichotomous functions for lymph node DCs and macrophages have been noted in $CD8^+$ T-cell activation following vaccinia virus infection (181). Viruses containing medullary sinus macrophages can interact with $CD8^+$ T cells in a similar manner to DCs, but they lack the adequate priming signals for full T-cell activation and proliferation.

Cavity Macrophages

Our visceral organs, lungs, and heart are surrounded by the peritoneal, pleural, and pericardial cavities, respectively. Numerous types of immune cells, including resident macrophage populations, populate these serous cavities in humans and mice (188–192). In the mouse peritoneal cavity, two phenotypically, functionally, and developmentally distinct macrophage populations have been identified: large peritoneal macrophages (LPMs) and small peritoneal macrophages (SPMs) (190). The embryonically derived $F4/80^{hi}CD11b^{hi}MHC\text{-}II^{-}$ LPMs are the predominant population over $F4/80^{int}CD11b^{int}$ $MHC\text{-}II^{hi}$ SPMs during the steady state, representing roughly 90% of the macrophage compartment within the cavity (190). Retinoic acid secreted by the omentum dictates the LPM functional phenotype by stimulating the expression of transcription factor GATA-6 (193). GATA-6 is known to drive TGF-β expression, which supports IgA production by B1 cells within the cavity (193). During peritoneal cavity inflammation, the LPM population disappears and the monocyte-derived SPMs take over as the predominant macrophage population.

A novel role for LPMs in tissue injury of neighboring organs has been established (116). $GATA\text{-}6^+$ LPMs are able to migrate and localize to a sterile injury site in the liver via ATP- and CD44-dependent mechanisms (116). Imaging of these cells within the injury revealed their ability to break down necrotic tissue and promote wound healing (116). This represents the first description of an alternative source and mechanism of myeloid cell recruitment to sites of injury. Further studies are needed to determine the infiltration potential of LPMs and/or SPMs in other visceral organs and under different inflammatory settings (e.g., infection and cancer). Similar macrophage populations within the pleural cavity have been characterized (194) and noted to have functions in mediating inflammation within the pleural space (195, 196). Whether these pleural macrophages and lesser-known pericardial macrophages can migrate into the lung and heart tissue to mediate inflammation in these organs remains unknown. Intravital imaging provides an optimal platform to explore the function of these cavity macrophage at neighboring sites of injury and further characterize this new recruitment paradigm.

FUTURE DIRECTIONS

In addition to tissues discussed above, intravital imaging has been used to visualize immune cells in other organs such as the spleen (15), kidney (197), pancreas (14), adipose tissue (198), bone marrow (121), heart (72), and arteries (24). The evolving characterization of resident macrophage populations within these organs will drive the need to evaluate their function *in situ*. In the spleen, for example, distinct tissue-resident macrophage populations, including embryonically derived red pulp macrophages, as well as bone marrow-derived marginal zone macrophages and metallophilic macrophages, have been identified (174). These macrophage populations collectively are thought to play an important filtering role for circulating microbes, erythrocytes, and senescent cells (174). The spleen is particularly important for protection against encapsulated bacteria such as *Streptococcus pneumoniae*; however, the mechanisms involved remain unclear. Preliminary imaging of the spleen during *S. pneumoniae* infection highlights the importance of both red pulp macrophages and marginal zone macrophages in mediating initial binding of the bacteria and facilitating downstream protective mechanisms to ensure pneumococcal clearance (our unpublished observations).

Recent lineage-tracing and parabiosis approaches have contributed to the identification of tissue-resident macrophages in the heart (199) and arteries (200). In addition to homeostatic functions of these cells, there is a major interest in understanding their contribution to inflammatory responses following ischemic events (e.g., myocardial infarction) or chronically during atherosclerosis and hypertension. Imaging strategies have been developed to overcome major issues associated with mechanical pumping of the heart. Li and colleagues provided the first example of live imaging of the heart *in vivo* by heterotopically transplanting a heart to the neck to provide better stability (72). A more physiologically relevant imaging approach combining stabilization techniques with cardiac cycle gated acquisition and image processing algorithms has since been developed (23). A similar imaging strategy was adopted for imaging myeloid populations within an atherosclerotic carotid artery (24). These approaches will be key in determining the behavior of these novel macrophages in the cardiovascular system.

Beyond the direct study of specific tissue macrophage behavior, intravital imaging represents a unique strategy that could be equally useful in determining the fate of tissue macrophages over the course of an inflammatory response. The macrophage disappearance reaction was initially described in the peritoneal cavity, where LPMs vanish in response to local inflammation (201). This has subsequently been demonstrated to occur with resident macrophages in the lungs and the heart following acute inflammation (202, 203). Tissue emigration through draining lymphatics, tissue adherence, and cell death have all been proposed as potential mechanisms (201). The functional importance of this process to the inflammatory response remains unclear. The combination of tissue intravital imaging with photoconversion/activation approaches previously used to track migration of neutrophils in the skin (94) could provide valuable insight into the fate of tissue-resident macrophages following acute injury.

Cell plasticity is a fundamental concept in macrophage biology that describes the ability of a cell to change its functional phenotype in response to local cues. *In vitro* model systems have been predominantly used to evaluate this process, leading to identification of prototypical phenotypes associated with specific cytokines (e.g. TNF-α, IL-4, and TGF-β). *In vivo* studies have supported the idea that recruited monocyte/macrophage populations can polarize at the inflammatory site during the course of the immune response (111, 114, 204). However, the polarization capacity of tissue-resident macrophage populations is less well understood *in situ*. This is likely more complicated due to the influence of a preestablished tissue microenvironment transcriptional programming within these cells. Recently, the ability to visualize cell polarization *in vivo* using intravital imaging was demonstrated. Dal-Secco and colleagues described, with the use of $CCR2^{RFP/+}CX3CR1^{GFP/+}$ dual reporter mice, the phenoconversion of proinflammatory monocytes to reparative monocytes at the site of liver injury (115). Adopting a similar approach to evaluate the *in vivo* plasticity of tissue-resident macrophages during both acute and chronic inflammatory states could improve our understanding of the role of macrophage polarization in pathological settings.

The fate of both embryonic tissue-resident and monocyte-derived macrophages following the inflammatory response has become an important topic within the field. Following mild inflammation, tissue-resident macrophages are thought to repopulate the tissue via local proliferation (205). However, under chronic inflammation (e.g., atherosclerosis) or following severe injury/infection, monocyte-derived macrophages can contribute to the tissue-resident macrophage pool (205). Over the course of years, this may result in substantial replacement of the original embryonic-derived cells by macrophages originating from bone marrow progenitors. Macrophage populations of different origin have been noted in organs (e.g., liver and heart); however, it is unclear whether the source of the macrophages can influence their function. Depletion or ablation approaches in the liver have noted that bone marrow-derived KCs display similar transcription profiles (133, 206) and equivalent bacterial-capturing ability *in vivo* (G. B. Menezes and P. Kubes, unpublished observations) to that of embryonically derived KCs. However, dichotomous killing capacity among KC populations in the liver has also been observed (142), suggesting that certain functions may differ. Further *in vivo* imaging employing differential labeling strategies should facilitate direct functional comparisons between tissue-resident cells of different origin.

Citation. Deniset JF, Kubes P. 2016. Intravital imaging of myeloid cells: inflammatory migration and resident patrolling. Microbiol Spectrum 4(6):MCHD-0042-2016.

References

1. **Wagner R.** 1839. *Erlauterungstaflen zur Physiologie und Entwicklungsgeschichte*. Leopold Voss, Leipzig, Germany.
2. **Ley K, Laudanna C, Cybulsky MI, Nourshargh S.** 2007. Getting to the site of inflammation: the leukocyte adhesion cascade updated. *Nat Rev Immunol* **7**:678–689.
3. **Petri B, Phillipson M, Kubes P.** 2008. The physiology of leukocyte recruitment: an in vivo perspective. *J Immunol* **180**:6439–6446.
4. **Ntziachristos V.** 2010. Going deeper than microscopy: the optical imaging frontier in biology. *Nat Methods* **7**: 603–614.
5. **Pittet MJ, Weissleder R.** 2011. Intravital imaging. *Cell* **147**:983–991.
6. **Lee WY, Moriarty TJ, Wong CH, Zhou H, Strieter RM, van Rooijen N, Chaconas G, Kubes P.** 2010. An intravascular immune response to *Borrelia burgdorferi* involves Kupffer cells and iNKT cells. *Nat Immunol* **11**: 295–302.

7. **McDonald B, Pittman K, Menezes GB, Hirota SA, Slaba I, Waterhouse CC, Beck PL, Muruve DA, Kubes P.** 2010. Intravascular danger signals guide neutrophils to sites of sterile inflammation. *Science* **330:**362–366.
8. **Wong CH, Jenne CN, Lee WY, Léger C, Kubes P.** 2011. Functional innervation of hepatic iNKT cells is immunosuppressive following stroke. *Science* **334:**101–105.
9. **Helmchen F, Denk W.** 2005. Deep tissue two-photon microscopy. *Nat Methods* **2:**932–940.
10. **Ishii T, Ishii M.** 2011. Intravital two-photon imaging: a versatile tool for dissecting the immune system. *Ann Rheum Dis* **70**(Suppl 1)**:**i113–i115.
11. **Chodaczek G, Papanna V, Zal MA, Zal T.** 2012. Body-barrier surveillance by epidermal γδ TCRs. *Nat Immunol* **13:**272–282.
12. **Roth TL, Nayak D, Atanasijevic T, Koretsky AP, Latour LL, McGavern DB.** 2014. Transcranial amelioration of inflammation and cell death after brain injury. *Nature* **505:**223–228.
13. **Shaked I, Hanna RN, Shaked H, Chodaczek G, Nowyhed HN, Tweet G, Tacke R, Basat AB, Mikulski Z, Togher S, Miller J, Blatchley A, Salek-Ardakani S, Darvas M, Kaikkonen MU, Thomas GD, Lai-Wing-Sun S, Rezk A, Bar-Or A, Glass CK, Bandukwala H, Hedrick CC.** 2015. Transcription factor Nr4a1 couples sympathetic and inflammatory cues in CNS-recruited macrophages to limit neuroinflammation. *Nat Immunol* **16:**1228–1234.
14. **Coppieters K, Amirian N, von Herrath M.** 2012. Intravital imaging of CTLs killing islet cells in diabetic mice. *J Clin Invest* **122:**119–131.
15. **Swirski FK, Nahrendorf M, Etzrodt M, Wildgruber M, Cortez-Retamozo V, Panizzi P, Figueiredo JL, Kohler RH, Chudnovskiy A, Waterman P, Aikawa E, Mempel TR, Libby P, Weissleder R, Pittet MJ.** 2009. Identification of splenic reservoir monocytes and their deployment to inflammatory sites. *Science* **325:**612–616.
16. **Harding MG, Zhang K, Conly J, Kubes P.** 2014. Neutrophil crawling in capillaries; a novel immune response to *Staphylococcus aureus*. *PLoS Pathog* **10:**e1004379. doi:10.1371/journal.ppat.1004379.
17. **Kreisel D, Nava RG, Li W, Zinselmeyer BH, Wang B, Lai J, Pless R, Gelman AE, Krupnick AS, Miller MJ.** 2010. In vivo two-photon imaging reveals monocyte-dependent neutrophil extravasation during pulmonary inflammation. *Proc Natl Acad Sci U S A* **107:**18073–18078.
18. **Nimmerjahn A, Kirchhoff F, Helmchen F.** 2005. Resting microglial cells are highly dynamic surveillants of brain parenchyma in vivo. *Science* **308:**1314–1318.
19. **Chèvre R, González-Granado JM, Megens RT, Sreeramkumar V, Silvestre-Roig C, Molina-Sánchez P, Weber C, Soehnlein O, Hidalgo A, Andrés V.** 2014. High-resolution imaging of intravascular atherogenic inflammation in live mice. *Circ Res* **114:**770–779.
20. **Jung K, Kim P, Leuschner F, Gorbatov R, Kim JK, Ueno T, Nahrendorf M, Yun SH.** 2013. Endoscopic time-lapse imaging of immune cells in infarcted mouse hearts. *Circ Res* **112:**891–899.
21. **Looney MR, Thornton EE, Sen D, Lamm WJ, Glenny RW, Krummel MF.** 2011. Stabilized imaging of immune surveillance in the mouse lung. *Nat Methods* **8:**91–96.
22. **Thanabalasuriar A, Neupane AS, Wang J, Krummel MF, Kubes P.** 2016. iNKT cell emigration out of the lung vasculature requires neutrophils and monocyte-derived dendritic cells in inflammation. *Cell Rep* **16:**3260–3272.
23. **Aguirre AD, Vinegoni C, Sebas M, Weissleder R.** 2014. Intravital imaging of cardiac function at the single-cell level. *Proc Natl Acad Sci U S A* **111:**11257–11262.
24. **McArdle S, Chodaczek G, Ray N, Ley K.** 2015. Intravital live cell triggered imaging system reveals monocyte patrolling and macrophage migration in atherosclerotic arteries. *J Biomed Opt* **20:**26005. doi:10.1117/1.JBO.20.2.026005.
25. **Heo C, Park H, Kim YT, Baeg E, Kim YH, Kim SG, Suh M.** 2016. A soft, transparent, freely accessible cranial window for chronic imaging and electrophysiology. *Sci Rep* **6:**27818. doi:10.1038/srep27818.
26. **Ritsma L, Steller EJ, Ellenbroek SI, Kranenburg O, Borel Rinkes IH, van Rheenen J.** 2013. Surgical implantation of an abdominal imaging window for intravital microscopy. *Nat Protoc* **8:**583–594.
27. **Kansas GS.** 1996. Selectins and their ligands: current concepts and controversies. *Blood* **88:**3259–3287.
28. **Zarbock A, Ley K, McEver RP, Hidalgo A.** 2011. Leukocyte ligands for endothelial selectins: specialized glycoconjugates that mediate rolling and signaling under flow. *Blood* **118:**6743–6751.
29. **Ala A, Dhillon AP, Hodgson HJ.** 2003. Role of cell adhesion molecules in leukocyte recruitment in the liver and gut. *Int J Exp Pathol* **84:**1–16.
30. **Patel KD, Cuvelier SL, Wiehler S.** 2002. Selectins: critical mediators of leukocyte recruitment. *Semin Immunol* **14:**73–81.
31. **Lorant DE, Patel KD, McIntyre TM, McEver RP, Prescott SM, Zimmerman GA.** 1991. Coexpression of GMP-140 and PAF by endothelium stimulated by histamine or thrombin: a juxtacrine system for adhesion and activation of neutrophils. *J Cell Biol* **115:**223–234.
32. **Chung KF.** 1992. Platelet-activating factor in inflammation and pulmonary disorders. *Clin Sci (Lond)* **83:**127–138.
33. **Theoharides TC, Alysandratos KD, Angelidou A, Delivanis DA, Sismanopoulos N, Zhang B, Asadi S, Vasiadi M, Weng Z, Miniati A, Kalogeromitros D.** 2012. Mast cells and inflammation. *Biochim Biophys Acta* **1822:**21–33.
34. **Muller WA.** 2013. Getting leukocytes to the site of inflammation. *Vet Pathol* **50:**7–22.
35. **Massena S, Christoffersson G, Hjertström E, Zcharia E, Vlodavsky I, Ausmees N, Rolny C, Li JP, Phillipson M.** 2010. A chemotactic gradient sequestered on endothelial heparan sulfate induces directional intraluminal crawling of neutrophils. *Blood* **116:**1924–1931.
36. **Williams MR, Azcutia V, Newton G, Alcaide P, Luscinskas FW.** 2011. Emerging mechanisms of neutrophil recruitment across endothelium. *Trends Immunol* **32:**461–469.

37. **Charmoy M, Brunner-Agten S, Aebischer D, Auderset F, Launois P, Milon G, Proudfoot AE, Tacchini-Cottier F.** 2010. Neutrophil-derived CCL3 is essential for the rapid recruitment of dendritic cells to the site of *Leishmania major* inoculation in resistant mice. *PLoS Pathog* **6:**e1000755. doi:10.1371/journal.ppat.1000755.

38. **Döring Y, Drechsler M, Wantha S, Kemmerich K, Lievens D, Vijayan S, Gallo RL, Weber C, Soehnlein O.** 2012. Lack of neutrophil-derived CRAMP reduces atherosclerosis in mice. *Circ Res* **110:**1052–1056.

39. **Kolaczkowska E, Kubes P.** 2013. Neutrophil recruitment and function in health and inflammation. *Nat Rev Immunol* **13:**159–175.

40. **Phillipson M, Heit B, Colarusso P, Liu L, Ballantyne CM, Kubes P.** 2006. Intraluminal crawling of neutrophils to emigration sites: a molecularly distinct process from adhesion in the recruitment cascade. *J Exp Med* **203:**2569–2575.

41. **Randolph GJ, Furie MB.** 1995. A soluble gradient of endogenous monocyte chemoattractant protein-1 promotes the transendothelial migration of monocytes in vitro. *J Immunol* **155:**3610–3618.

42. **Colotta F, Borré A, Wang JM, Tattanelli M, Maddalena F, Polentarutti N, Peri G, Mantovani A.** 1992. Expression of a monocyte chemotactic cytokine by human mononuclear phagocytes. *J Immunol* **148:**760–765.

43. **Kasahara T, Mukaida N, Yamashita K, Yagisawa H, Akahoshi T, Matsushima K.** 1991. IL-1 and TNF-α induction of IL-8 and monocyte chemotactic and activating factor (MCAF) mRNA expression in a human astrocytoma cell line. *Immunology* **74:**60–67.

44. **Sica A, Wang JM, Colotta F, Dejana E, Mantovani A, Oppenheim JJ, Larsen CG, Zachariae CO, Matsushima K.** 1990. Monocyte chemotactic and activating factor gene expression induced in endothelial cells by IL-1 and tumor necrosis factor. *J Immunol* **144:**3034–3038.

45. **Strieter RM, Kunkel SL, Showell HJ, Remick DG, Phan SH, Ward PA, Marks RM.** 1989. Endothelial cell gene expression of a neutrophil chemotactic factor by TNF-α, LPS, and IL-1β. *Science* **243:**1467–1469.

46. **Strieter RM, Phan SH, Showell HJ, Remick DG, Lynch JP, Genord M, Raiford C, Eskandari M, Marks RM, Kunkel SL.** 1989. Monokine-induced neutrophil chemotactic factor gene expression in human fibroblasts. *J Biol Chem* **264:**10621–10626.

47. **Wang JM, Sica A, Peri G, Walter S, Padura IM, Libby P, Ceska M, Lindley I, Colotta F, Mantovani A.** 1991. Expression of monocyte chemotactic protein and interleukin-8 by cytokine-activated human vascular smooth muscle cells. *Arterioscler Thromb* **11:**1166–1174.

48. **Dinarello CA.** 1992. The biology of interleukin-1. *Chem Immunol* **51:**1–32.

49. **Vassalli P.** 1992. The pathophysiology of tumor necrosis factors. *Annu Rev Immunol* **10:**411–452.

50. **Cuvelier SL, Patel KD.** 2001. Shear-dependent eosinophil transmigration on interleukin 4-stimulated endothelial cells: a role for endothelium-associated eotaxin-3. *J Exp Med* **194:**1699–1709.

51. **Hillyer P, Male D.** 2005. Expression of chemokines on the surface of different human endothelia. *Immunol Cell Biol* **83:**375–382.

52. **Heit B, Robbins SM, Downey CM, Guan Z, Colarusso P, Miller BJ, Jirik FR, Kubes P.** 2008. PTEN functions to 'prioritize' chemotactic cues and prevent 'distraction' in migrating neutrophils. *Nat Immunol* **9:**743–752.

53. **Heit B, Tavener S, Raharjo E, Kubes P.** 2002. An intracellular signaling hierarchy determines direction of migration in opposing chemotactic gradients. *J Cell Biol* **159:**91–102.

54. **Chou RC, Kim ND, Sadik CD, Seung E, Lan Y, Byrne MH, Haribabu B, Iwakura Y, Luster AD.** 2010. Lipid-cytokine-chemokine cascade drives neutrophil recruitment in a murine model of inflammatory arthritis. *Immunity* **33:**266–278.

55. **Kim ND, Chou RC, Seung E, Tager AM, Luster AD.** 2006. A unique requirement for the leukotriene B_4 receptor BLT1 for neutrophil recruitment in inflammatory arthritis. *J Exp Med* **203:**829–835.

56. **Andréasson E, Önnheim K, Forsman H.** 2013. The subcellular localization of the receptor for platelet-activating factor in neutrophils affects signaling and activation characteristics. *Clin Dev Immunol* **2013:**456407.

57. **Fu H, Bylund J, Karlsson A, Pellmé S, Dahlgren C.** 2004. The mechanism for activation of the neutrophil NADPH-oxidase by the peptides formyl-Met-Leu-Phe and Trp-Lys-Tyr-Met-Val-Met differs from that for interleukin-8. *Immunology* **112:**201–210.

58. **Nourshargh S, Alon R.** 2014. Leukocyte migration into inflamed tissues. *Immunity* **41:**694–707.

59. **Petri B, Bixel MG.** 2006. Molecular events during leukocyte diapedesis. *FEBS J* **273:**4399–4407.

60. **Feng D, Nagy JA, Pyne K, Dvorak HF, Dvorak AM.** 1998. Neutrophils emigrate from venules by a transendothelial cell pathway in response to FMLP. *J Exp Med* **187:**903–915.

61. **Burns AR, Bowden RA, MacDonell SD, Walker DC, Odebunmi TO, Donnachie EM, Simon SI, Entman ML, Smith CW.** 2000. Analysis of tight junctions during neutrophil transendothelial migration. *J Cell Sci* **113:**45–57.

62. **Wang S, Voisin MB, Larbi KY, Dangerfield J, Scheiermann C, Tran M, Maxwell PH, Sorokin L, Nourshargh S.** 2006. Venular basement membranes contain specific matrix protein low expression regions that act as exit points for emigrating neutrophils. *J Exp Med* **203:**1519–1532.

63. **Carman CV, Springer TA.** 2004. A transmigratory cup in leukocyte diapedesis both through individual vascular endothelial cells and between them. *J Cell Biol* **167:**377–388.

64. **Nieminen M, Henttinen T, Merinen M, Marttila-Ichihara F, Eriksson JE, Jalkanen S.** 2006. Vimentin function in lymphocyte adhesion and transcellular migration. *Nat Cell Biol* **8:**156–162.

65. **Cara DC, Kaur J, Forster M, McCafferty DM, Kubes P.** 2001. Role of p38 mitogen-activated protein kinase in chemokine-induced emigration and chemotaxis in vivo. *J Immunol* **167:**6552–6558.

66. **Hickey MJ, Forster M, Mitchell D, Kaur J, De Caigny C, Kubes P.** 2000. L-selectin facilitates emigration and extravascular locomotion of leukocytes during acute inflammatory responses in vivo. *J Immunol* **165:**7164–7170.

67. **Johnston B, Burns AR, Suematsu M, Issekutz TB, Woodman RC, Kubes P.** 1999. Chronic inflammation upregulates chemokine receptors and induces neutrophil migration to monocyte chemoattractant protein-1. *J Clin Invest* **103:**1269–1276.

68. **Johnston B, Burns AR, Suematsu M, Watanabe K, Issekutz TB, Kubes P.** 2000. Increased sensitivity to the C-X-C chemokine CINC/gro in a model of chronic inflammation. *Microcirculation* **7:**109–118.

69. **Johnston B, Chee A, Issekutz TB, Ugarova T, Fox-Robichaud A, Hickey MJ, Kubes P.** 2000. α_4 Integrin-dependent leukocyte recruitment does not require VCAM-1 in a chronic model of inflammation. *J Immunol* **164:**3337–3344.

70. **Liu L, Cara DC, Kaur J, Raharjo E, Mullaly SC, Jongstra-Bilen J, Jongstra J, Kubes P.** 2005. LSP1 is an endothelial gatekeeper of leukocyte transendothelial migration. *J Exp Med* **201:**409–418.

71. **Doerschuk CM.** 2001. Mechanisms of leukocyte sequestration in inflamed lungs. *Microcirculation* **8:**71–88.

72. **Li W, Nava RG, Bribriesco AC, Zinselmeyer BH, Spahn JH, Gelman AE, Krupnick AS, Miller MJ, Kreisel D.** 2012. Intravital 2-photon imaging of leukocyte trafficking in beating heart. *J Clin Invest* **122:**2499–2508.

73. **McDonald B, McAvoy EF, Lam F, Gill V, de la Motte C, Savani RC, Kubes P.** 2008. Interaction of CD44 and hyaluronan is the dominant mechanism for neutrophil sequestration in inflamed liver sinusoids. *J Exp Med* **205:**915–927.

74. **Singbartl K, Forlow SB, Ley K.** 2001. Platelet, but not endothelial, P-selectin is critical for neutrophil-mediated acute postischemic renal failure. *FASEB J* **15:**2337–2344.

75. **Wang Q, Teder P, Judd NP, Noble PW, Doerschuk CM.** 2002. CD44 deficiency leads to enhanced neutrophil migration and lung injury in *Escherichia coli* pneumonia in mice. *Am J Pathol* **161:**2219–2228.

76. **Wong J, Johnston B, Lee SS, Bullard DC, Smith CW, Beaudet AL, Kubes P.** 1997. A minimal role for selectins in the recruitment of leukocytes into the inflamed liver microvasculature. *J Clin Invest* **99:**2782–2790.

77. **Block H, Herter JM, Rossaint J, Stadtmann A, Kliche S, Lowell CA, Zarbock A.** 2012. Crucial role of SLP-76 and ADAP for neutrophil recruitment in mouse kidney ischemia-reperfusion injury. *J Exp Med* **209:**407–421.

78. **Megens RT, Kemmerich K, Pyta J, Weber C, Soehnlein O.** 2011. Intravital imaging of phagocyte recruitment. *Thromb Haemost* **105:**802–810.

79. **Phillipson M, Heit B, Parsons SA, Petri B, Mullaly SC, Colarusso P, Gower RM, Neely G, Simon SI, Kubes P.** 2009. Vav1 is essential for mechanotactic crawling and migration of neutrophils out of the inflamed microvasculature. *J Immunol* **182:**6870–6878.

80. **Menezes GB, Lee WY, Zhou H, Waterhouse CC, Cara DC, Kubes P.** 2009. Selective down-regulation of neutrophil Mac-1 in endotoxemic hepatic microcirculation via IL-10. *J Immunol* **183:**7557–7568.

81. **Nagao T, Matsumura M, Mabuchi A, Ishida-Okawara A, Koshio O, Nakayama T, Minamitani H, Suzuki K.** 2007. Up-regulation of adhesion molecule expression in glomerular endothelial cells by anti-myeloperoxidase antibody. *Nephrol Dial Transplant* **22:**77–87.

82. **Ogawa T, Yorioka N, Ito T, Ogata S, Kumagai J, Kawanishi H, Yamakido M.** 1997. Precise ultrastructural localization of endothelial leukocyte adhesion molecule-1, vascular cell adhesion molecule-1, and intercellular adhesion molecule-1 in patients with IgA nephropathy. *Nephron* **75:**54–64.

83. **Kuligowski MP, Kitching AR, Hickey MJ.** 2006. Leukocyte recruitment to the inflamed glomerulus: a critical role for platelet-derived P-selectin in the absence of rolling. *J Immunol* **176:**6991–6999.

84. **Carvalho-Tavares J, Hickey MJ, Hutchison J, Michaud J, Sutcliffe IT, Kubes P.** 2000. A role for platelets and endothelial selectins in tumor necrosis factor-α-induced leukocyte recruitment in the brain microvasculature. *Circ Res* **87:**1141–1148.

85. **Ishikawa M, Cooper D, Russell J, Salter JW, Zhang JH, Nanda A, Granger DN.** 2003. Molecular determinants of the prothrombogenic and inflammatory phenotype assumed by the postischemic cerebral microcirculation. *Stroke* **34:**1777–1782.

86. **Grady MS, Cody RF Jr, Maris DO, McCall TD, Seckin H, Sharar SR, Winn HR.** 1999. P-selectin blockade following fluid-percussion injury: behavioral and immunochemical sequelae. *J Neurotrauma* **16:**13–25.

87. **Suzuki H, Hayashi T, Tojo SJ, Kitagawa H, Kimura K, Mizugaki M, Itoyama Y, Abe K.** 1999. Anti-P-selectin antibody attenuates rat brain ischemic injury. *Neurosci Lett* **265:**163–166.

88. **Kuijper PH, Gallardo Tores HI, Lammers JW, Sixma JJ, Koenderman L, Zwaginga JJ.** 1998. Platelet associated fibrinogen and ICAM-2 induce firm adhesion of neutrophils under flow conditions. *Thromb Haemost* **80:**443–448.

89. **von Hundelshausen P, Petersen F, Brandt E.** 2007. Platelet-derived chemokines in vascular biology. *Thromb Haemost* **97:**704–713.

90. **Weber C, Springer TA.** 1997. Neutrophil accumulation on activated, surface-adherent platelets in flow is mediated by interaction of Mac-1 with fibrinogen bound to $\alpha IIb\beta 3$ and stimulated by platelet-activating factor. *J Clin Invest* **100:**2085–2093.

91. **Hammond MD, Taylor RA, Mullen MT, Ai Y, Aguila HL, Mack M, Kasner SE, McCullough LD, Sansing LH.** 2014. $CCR2^{+}Ly6C^{hi}$ inflammatory monocyte recruitment exacerbates acute disability following intracerebral hemorrhage. *J Neurosci* **34:**3901–3909.

92. **Morganti JM, Jopson TD, Liu S, Riparip LK, Guandique CK, Gupta N, Ferguson AR, Rosi S.** 2015. CCR2 antagonism alters brain macrophage polarization and ameliorates cognitive dysfunction induced by traumatic brain injury. *J Neurosci* **35:**748–760.

93. **Buckley CD, Gilroy DW, Serhan CN, Stockinger B, Tak PP.** 2013. The resolution of inflammation. *Nat Rev Immunol* **13:**59–66.

94. **Hampton HR, Bailey J, Tomura M, Brink R, Chtanova T.** 2015. Microbe-dependent lymphatic migration of neutrophils modulates lymphocyte proliferation in lymph nodes. *Nat Commun* **6:**7139. doi:10.1038/ncomms8139.

95. **Hall C, Flores MV, Chien A, Davidson A, Crosier K, Crosier P.** 2009. Transgenic zebrafish reporter lines reveal conserved Toll-like receptor signaling potential in embryonic myeloid leukocytes and adult immune cell lineages. *J Leukoc Biol* **85:**751–765.

96. **Mathias JR, Perrin BJ, Liu TX, Kanki J, Look AT, Huttenlocher A.** 2006. Resolution of inflammation by retrograde chemotaxis of neutrophils in transgenic zebrafish. *J Leukoc Biol* **80:**1281–1288.

97. **Yoo SK, Huttenlocher A.** 2011. Spatiotemporal photolabeling of neutrophil trafficking during inflammation in live zebrafish. *J Leukoc Biol* **89:**661–667.

98. **Woodfin A, Voisin MB, Beyrau M, Colom B, Caille D, Diapouli FM, Nash GB, Chavakis T, Albelda SM, Rainger GE, Meda P, Imhof BA, Nourshargh S.** 2011. The junctional adhesion molecule JAM-C regulates polarized transendothelial migration of neutrophils *in vivo*. *Nat Immunol* **12:**761–769.

99. **Colom B, Bodkin JV, Beyrau M, Woodfin A, Ody C, Rourke C, Chavakis T, Brohi K, Imhof BA, Nourshargh S.** 2015. Leukotriene B_4-neutrophil elastase axis drives neutrophil reverse transendothelial cell migration in vivo. *Immunity* **42:**1075–1086.

100. **Buckley CD, Ross EA, McGettrick HM, Osborne CE, Haworth O, Schmutz C, Stone PC, Salmon M, Matharu NM, Vohra RK, Nash GB, Rainger GE.** 2006. Identification of a phenotypically and functionally distinct population of long-lived neutrophils in a model of reverse endothelial migration. *J Leukoc Biol* **79:**303–311.

101. **Auffray C, Fogg D, Garfa M, Elain G, Join-Lambert O, Kayal S, Sarnacki S, Cumano A, Lauvau G, Geissmann F.** 2007. Monitoring of blood vessels and tissues by a population of monocytes with patrolling behavior. *Science* **317:**666–670.

102. **Bellavance MA, Gosselin D, Yong VW, Stys PK, Rivest S.** 2015. Patrolling monocytes play a critical role in CX3CR1-mediated neuroprotection during excitotoxicity. *Brain Struct Funct* **220:**1759–1776.

103. **Carlin LM, Stamatiades EG, Auffray C, Hanna RN, Glover L, Vizcay-Barrena G, Hedrick CC, Cook HT, Diebold S, Geissmann F.** 2013. Nr4a1-dependent Ly6C^{low} monocytes monitor endothelial cells and orchestrate their disposal. *Cell* **153:**362–375.

104. **Boring L, Gosling J, Chensue SW, Kunkel SL, Farese RV Jr, Broxmeyer HE, Charo IF.** 1997. Impaired monocyte migration and reduced type 1 (Th1) cytokine responses in C-C chemokine receptor 2 knockout mice. *J Clin Invest* **100:**2552–2561.

105. **Peters W, Scott HM, Chambers HF, Flynn JL, Charo IF, Ernst JD.** 2001. Chemokine receptor 2 serves an early and essential role in resistance to *Mycobacterium tuberculosis*. *Proc Natl Acad Sci U S A* **98:**7958–7963.

106. **Boring L, Gosling J, Cleary M, Charo IF.** 1998. Decreased lesion formation in CCR2$^{-/-}$ mice reveals a role for chemokines in the initiation of atherosclerosis. *Nature* **394:**894–897.

107. **Izikson L, Klein RS, Charo IF, Weiner HL, Luster AD.** 2000. Resistance to experimental autoimmune encephalomyelitis in mice lacking the CC chemokine receptor (CCR)2. *J Exp Med* **192:**1075–1080.

108. **Serbina NV, Pamer EG.** 2006. Monocyte emigration from bone marrow during bacterial infection requires signals mediated by chemokine receptor CCR2. *Nat Immunol* **7:**311–317.

109. **Tsou CL, Peters W, Si Y, Slaymaker S, Aslanian AM, Weisberg SP, Mack M, Charo IF.** 2007. Critical roles for CCR2 and MCP-3 in monocyte mobilization from bone marrow and recruitment to inflammatory sites. *J Clin Invest* **117:**902–909.

110. **Nahrendorf M, Swirski FK, Aikawa E, Stangenberg L, Wurdinger T, Figueiredo JL, Libby P, Weissleder R, Pittet MJ.** 2007. The healing myocardium sequentially mobilizes two monocyte subsets with divergent and complementary functions. *J Exp Med* **204:**3037–3047.

111. **Hilgendorf I, Gerhardt LM, Tan TC, Winter C, Holderried TA, Chousterman BG, Iwamoto Y, Liao R, Zirlik A, Scherer-Crosbie M, Hedrick CC, Libby P, Nahrendorf M, Weissleder R, Swirski FK.** 2014. Ly-6C^{high} monocytes depend on Nr4a1 to balance both inflammatory and reparative phases in the infarcted myocardium. *Circ Res* **114:**1611–1622.

112. **Sunderkötter C, Nikolic T, Dillon MJ, Van Rooijen N, Stehling M, Drevets DA, Leenen PJ.** 2004. Subpopulations of mouse blood monocytes differ in maturation stage and inflammatory response. *J Immunol* **172:** 4410–4417.

113. **Arnold L, Henry A, Poron F, Baba-Amer Y, van Rooijen N, Plonquet A, Gherardi RK, Chazaud B.** 2007. Inflammatory monocytes recruited after skeletal muscle injury switch into antiinflammatory macrophages to support myogenesis. *J Exp Med* **204:** 1057–1069.

114. **Crane MJ, Daley JM, van Houtte O, Brancato SK, Henry WL Jr, Albina JE.** 2014. The monocyte to macrophage transition in the murine sterile wound. *PLoS One* **9:**e86660. doi:10.1371/journal.pone.0086660.

115. **Dal-Secco D, Wang J, Zeng Z, Kolaczkowska E, Wong CH, Petri B, Ransohoff RM, Charo IF, Jenne CN, Kubes P.** 2015. A dynamic spectrum of monocytes arising from the in situ reprogramming of CCR2^{+} monocytes at a site of sterile injury. *J Exp Med* **212:**447–456.

116. **Wang J, Kubes P.** 2016. A reservoir of mature cavity macrophages that can rapidly invade visceral organs to affect tissue repair. *Cell* **165:**668–678.

117. **Delano MJ, Kelly-Scumpia KM, Thayer TC, Winfield RD, Scumpia PO, Cuenca AG, Harrington PB, O'Malley KA, Warner E, Gabrilovich S, Mathews CE, Laface D, Heyworth PG, Ramphal R, Strieter RM, Moldawer LL, Efron PA.** 2011. Neutrophil mobilization from the bone marrow during polymicrobial sepsis is dependent on CXCL12 signaling. *J Immunol* **187:** 911–918.

118. Eash KJ, Greenbaum AM, Gopalan PK, Link DC. 2010. CXCR2 and CXCR4 antagonistically regulate neutrophil trafficking from murine bone marrow. *J Clin Invest* 120:2423–2431.

119. Eash KJ, Means JM, White DW, Link DC. 2009. CXCR4 is a key regulator of neutrophil release from the bone marrow under basal and stress granulopoiesis conditions. *Blood* 113:4711–4719.

120. Martin C, Burdon PC, Bridger G, Gutierrez-Ramos JC, Williams TJ, Rankin SM. 2003. Chemokines acting via CXCR2 and CXCR4 control the release of neutrophils from the bone marrow and their return following senescence. *Immunity* 19:583–593.

121. Devi S, Wang Y, Chew WK, Lima R, A-González N, Mattar CN, Chong SZ, Schlitzer A, Bakocevic N, Chew S, Keeble JL, Goh CC, Li JL, Evrard M, Malleret B, Larbi A, Renia L, Haniffa M, Tan SM, Chan JK, Balabanian K, Nagasawa T, Bachelerie F, Hidalgo A, Ginhoux F, Kubes P, Ng LG. 2013. Neutrophil mobilization via plerixafor-mediated CXCR4 inhibition arises from lung demargination and blockade of neutrophil homing to the bone marrow. *J Exp Med* 210:2321–2336.

122. Doerschuk CM, Beyers N, Coxson HO, Wiggs B, Hogg JC. 1993. Comparison of neutrophil and capillary diameters and their relation to neutrophil sequestration in the lung. *J Appl Physiol 1985* 74:3040–3045.

123. Martin BA, Wiggs BR, Lee S, Hogg JC. 1987. Regional differences in neutrophil margination in dog lungs. *J Appl Physiol 1985* 63:1253–1261.

124. Hogg JC, Doerschuk CM. 1995. Leukocyte traffic in the lung. *Annu Rev Physiol* 57:97–114.

125. Kuebler WM, Goetz AE. 2002. The marginated pool. *Eur Surg Res* 34:92–100.

126. Kuebler WM, Kuhnle GE, Goetz AE. 1999. Leukocyte margination in alveolar capillaries: interrelationship with functional capillary geometry and microhemodynamics. *J Vasc Res* 36:282–288.

127. Kuebler WM, Kuhnle GE, Groh J, Goetz AE. 1994. Leukocyte kinetics in pulmonary microcirculation: intravital fluorescence microscopic study. *J Appl Physiol (1985)* 76:65–71.

128. Puga I, Cols M, Barra CM, He B, Cassis L, Gentile M, Comerma L, Chorny A, Shan M, Xu W, Magri G, Knowles DM, Tam W, Chiu A, Bussel JB, Serrano S, Lorente JA, Bellosillo B, Lloreta J, Juanpere N, Alameda F, Baro T, de Heredia CD, Toran N, Catala A, Torrebadell M, Fortuny C, Cusi V, Carreras C, Diaz GA, Blander JM, Farber C, Silvestri G, Cunningham-Rundles C, Calvillo M, Dufour C, Notarangelo LD, Lougaris V, Plebani A, Casanova J, Ganal SC, Diefenbach A, Arostegui JI, Juan M, Yague J, Mahlaoui N, Donadieu J, Chen K, Cerutti A. 2011. B cell-helper neutrophils stimulate the diversification and production of immunoglobulin in the marginal zone of the spleen. *Nat Immunol* 13:170–180.

129. Ng LG, Qin JS, Roediger B, Wang Y, Jain R, Cavanagh LL, Smith AL, Jones CA, de Veer M, Grimbaldeston MA, Meeusen EN, Weninger W. 2011. Visualizing the neutrophil response to sterile tissue injury in mouse dermis reveals a three-phase cascade of events. *J Invest Dermatol* 131:2058–2068.

130. Davies LC, Jenkins SJ, Allen JE, Taylor PR. 2013. Tissue-resident macrophages. *Nat Immunol* 14:986–995.

131. Ginhoux F, Guilliams M. 2016. Tissue-resident macrophage ontogeny and homeostasis. *Immunity* 44:439–449.

132. Lavin Y, Mortha A, Rahman A, Merad M. 2015. Regulation of macrophage development and function in peripheral tissues. *Nat Rev Immunol* 15:731–744.

133. Lavin Y, Winter D, Blecher-Gonen R, David E, Keren-Shaul H, Merad M, Jung S, Amit I. 2014. Tissue-resident macrophage enhancer landscapes are shaped by the local microenvironment. *Cell* 159:1312–1326.

134. Davalos D, Grutzendler J, Yang G, Kim JV, Zuo Y, Jung S, Littman DR, Dustin ML, Gan WB. 2005. ATP mediates rapid microglial response to local brain injury *in vivo*. *Nat Neurosci* 8:752–758.

135. Hines DJ, Hines RM, Mulligan SJ, Macvicar BA. 2009. Microglia processes block the spread of damage in the brain and require functional chloride channels. *Glia* 57: 1610–1618.

136. Saijo K, Glass CK. 2011. Microglial cell origin and phenotypes in health and disease. *Nat Rev Immunol* 11: 775–787.

137. Masuda T, Croom D, Hida H, Kirov SA. 2011. Capillary blood flow around microglial somata determines dynamics of microglial processes in ischemic conditions. *Glia* 59:1744–1753.

138. Wake H, Moorhouse AJ, Jinno S, Kohsaka S, Nabekura J. 2009. Resting microglia directly monitor the functional state of synapses *in vivo* and determine the fate of ischemic terminals. *J Neurosci* 29:3974–3980.

139. Meyer-Luehmann M, Spires-Jones TL, Prada C, Garcia-Alloza M, de Calignon A, Rozkalne A, Koenigsknecht-Talboo J, Holtzman DM, Bacskai BJ, Hyman BT. 2008. Rapid appearance and local toxicity of amyloid-β plaques in a mouse model of Alzheimer's disease. *Nature* 451:720–724.

140. Fuhrmann M, Bittner T, Jung CK, Burgold S, Page RM, Mitteregger G, Haass C, LaFerla FM, Kretzschmar H, Herms J. 2010. Microglial *Cx3cr1* knockout prevents neuron loss in a mouse model of Alzheimer's disease. *Nat Neurosci* 13:411–413.

141. Nayak D, Zinselmeyer BH, Corps KN, McGavern DB. 2012. In vivo dynamics of innate immune sentinels in the CNS. *Intravital* 1:95–106.

142. Surewaard BGD, Deniset JF, Zemp FJ, Amrein M, Otto M, Conly J, Omri A, Yates RM, Kubes P. 2016. Identification and treatment of the *Staphylococcus aureus* reservoir in vivo. *J Exp Med* 213:1141–1151.

143. Wong CH, Jenne CN, Petri B, Chrobok NL, Kubes P. 2013. Nucleation of platelets with blood-borne pathogens on Kupffer cells precedes other innate immunity and contributes to bacterial clearance. *Nat Immunol* 14:785–792.

144. Zeng Z, Surewaard BG, Wong CH, Geoghegan JA, Jenne CN, Kubes P. 2016. CRIg functions as a macro-

phage pattern recognition receptor to directly bind and capture blood-borne Gram-positive bacteria. *Cell Host Microbe* 20:99–106.

145. Helmy KY, Katschke KJ Jr, Gorgani NN, Kljavin NM, Elliott JM, Diehl L, Scales SJ, Ghilardi N, van Lookeren Campagne M. 2006. CRIg: a macrophage complement receptor required for phagocytosis of circulating pathogens. *Cell* 124:915–927.

146. Gül N, Babes L, Siegmund K, Korthouwer R, Bögels M, Braster R, Vidarsson G, ten Hagen TL, Kubes P, van Egmond M. 2014. Macrophages eliminate circulating tumor cells after monoclonal antibody therapy. *J Clin Invest* 124:812–823.

147. Montalvao F, Garcia Z, Celli S, Breart B, Deguine J, Van Rooijen N, Bousso P. 2013. The mechanism of anti-CD20-mediated B cell depletion revealed by intravital imaging. *J Clin Invest* 123:5098–5103.

148. McDonald B, Jenne CN, Zhuo L, Kimata K, Kubes P. 2013. Kupffer cells and activation of endothelial TLR4 coordinate neutrophil adhesion within liver sinusoids during endotoxemia. *Am J Physiol Gastrointest Liver Physiol* 305:G797–G806.

149. Beattie L, Peltan A, Maroof A, Kirby A, Brown N, Coles M, Smith DF, Kaye PM. 2010. Dynamic imaging of experimental *Leishmania donovani*-induced hepatic granulomas detects Kupffer cell-restricted antigen presentation to antigen-specific CD8 T cells. *PLoS Pathog* 6:e1000805. doi:10.1371/journal.ppat.1000805.

150. Heymann F, Peusquens J, Ludwig-Portugall I, Kohlhepp M, Ergen C, Niemietz P, Martin C, van Rooijen N, Ochando JC, Randolph GJ, Luedde T, Ginhoux F, Kurts C, Trautwein C, Tacke F. 2015. Liver inflammation abrogates immunological tolerance induced by Kupffer cells. *Hepatology* 62:279–291.

151. Greter M, Lelios I, Pelczar P, Hoeffel G, Price J, Leboeuf M, Kündig TM, Frei K, Ginhoux F, Merad M, Becher B. 2012. Stroma-derived interleukin-34 controls the development and maintenance of Langerhans cells and the maintenance of microglia. *Immunity* 37: 1050–1060.

152. Satpathy AT, Wu X, Albring JC, Murphy KM. 2012. Re(de)fining the dendritic cell lineage. *Nat Immunol* 13: 1145–1154.

153. Wang Y, Szretter KJ, Vermi W, Gilfillan S, Rossini C, Cella M, Barrow AD, Diamond MS, Colonna M. 2012. IL-34 is a tissue-restricted ligand of CSF1R required for the development of Langerhans cells and microglia. *Nat Immunol* 13:753–760.

154. Kissenpfennig A, Henri S, Dubois B, Laplace-Builhé C, Perrin P, Romani N, Tripp CH, Douillard P, Leserman L, Kaiserlian D, Saeland S, Davoust J, Malissen B. 2005. Dynamics and function of Langerhans cells in vivo: dermal dendritic cells colonize lymph node areas distinct from slower migrating Langerhans cells. *Immunity* 22:643–654.

155. Vishwanath M, Nishibu A, Saeland S, Ward BR, Mizumoto N, Ploegh HL, Boes M, Takashima A. 2006. Development of intravital intermittent confocal imaging system for studying Langerhans cell turnover. *J Invest Dermatol* 126:2452–2457.

156. Nishibu A, Ward BR, Jester JV, Ploegh HL, Boes M, Takashima A. 2006. Behavioral responses of epidermal Langerhans cells *in situ* to local pathological stimuli. *J Invest Dermatol* 126:787–796.

157. Nishibu A, Ward BR, Boes M, Takashima A. 2007. Roles for IL-1 and TNFα in dynamic behavioral responses of Langerhans cells to topical hapten application. *J Dermatol Sci* 45:23–30.

158. Kubo A, Nagao K, Yokouchi M, Sasaki H, Amagai M. 2009. External antigen uptake by Langerhans cells with reorganization of epidermal tight junction barriers. *J Exp Med* 206:2937–2946.

159. Ouchi T, Kubo A, Yokouchi M, Adachi T, Kobayashi T, Kitashima DY, Fujii H, Clausen BE, Koyasu S, Amagai M, Nagao K. 2011. Langerhans cell antigen capture through tight junctions confers preemptive immunity in experimental staphylococcal scalded skin syndrome. *J Exp Med* 208:2607–2613.

160. Gaiser MR, Lämmermann T, Feng X, Igyarto BZ, Kaplan DH, Tessarollo L, Germain RN, Udey MC. 2012. Cancer-associated epithelial cell adhesion molecule (EpCAM; CD326) enables epidermal Langerhans cell motility and migration in vivo. *Proc Natl Acad Sci U S A* 109:E889–E897.

161. Kaplan DH, Jenison MC, Saeland S, Shlomchik WD, Shlomchik MJ. 2005. Epidermal Langerhans cell-deficient mice develop enhanced contact hypersensitivity. *Immunity* 23:611–620.

162. Abtin A, Jain R, Mitchell AJ, Roediger B, Brzoska AJ, Tikoo S, Cheng Q, Ng LG, Cavanagh LL, von Andrian UH, Hickey MJ, Firth N, Weninger W. 2014. Perivascular macrophages mediate neutrophil recruitment during bacterial skin infection. *Nat Immunol* 15:45–53.

163. Bain CC, Scott CL, Uronen-Hansson H, Gudjonsson S, Jansson O, Grip O, Guilliams M, Malissen B, Agace WW, Mowat AM. 2013. Resident and pro-inflammatory macrophages in the colon represent alternative context-dependent fates of the same Ly6C^{hi} monocyte precursors. *Mucosal Immunol* 6:498–510.

164. Rivollier A, He J, Kole A, Valatas V, Kelsall BL. 2012. Inflammation switches the differentiation program of Ly6C^{hi} monocytes from antiinflammatory macrophages to inflammatory dendritic cells in the colon. *J Exp Med* 209:139–155.

165. Tamoutounour S, Henri S, Lelouard H, de Bovis B, de Haar C, van der Woude CJ, Woltman AM, Reyal Y, Bonnet D, Sichien D, Bain CC, Mowat AM, Reis e Sousa C, Poulin LF, Malissen B, Guilliams M. 2012. CD64 distinguishes macrophages from dendritic cells in the gut and reveals the Th1-inducing role of mesenteric lymph node macrophages during colitis. *Eur J Immunol* 42:3150–3166.

166. Chieppa M, Rescigno M, Huang AY, Germain RN. 2006. Dynamic imaging of dendritic cell extension into the small bowel lumen in response to epithelial cell TLR engagement. *J Exp Med* 203:2841–2852.

167. Niess JH, Brand S, Gu X, Landsman L, Jung S, McCormick BA, Vyas JM, Boes M, Ploegh HL, Fox JG, Littman DR, Reinecker HC. 2005. CX3CR1-mediated dendritic cell access to the intestinal lumen and bacterial clearance. *Science* 307:254–258.

168. Farache J, Koren I, Milo I, Gurevich I, Kim KW, Zigmond E, Furtado GC, Lira SA, Shakhar G. 2013. Luminal bacteria recruit CD103+ dendritic cells into the intestinal epithelium to sample bacterial antigens for presentation. *Immunity* **38:**581–595.

169. Schulz O, Jaensson E, Persson EK, Liu X, Worbs T, Agace WW, Pabst O. 2009. Intestinal CD103+, but not CX3CR1+, antigen sampling cells migrate in lymph and serve classical dendritic cell functions. *J Exp Med* **206:** 3101–3114.

170. Morris DG, Huang X, Kaminski N, Wang Y, Shapiro SD, Dolganov G, Glick A, Sheppard D. 2003. Loss of integrin αvβ6-mediated TGF-β activation causes Mmp12-dependent emphysema. *Nature* **422:**169–173.

171. Snelgrove RJ, Goulding J, Didierlaurent AM, Lyonga D, Vekaria S, Edwards L, Gwyer E, Sedgwick JD, Barclay AN, Hussell T. 2008. A critical function for CD200 in lung immune homeostasis and the severity of influenza infection. *Nat Immunol* **9:**1074–1083.

172. Archambaud C, Salcedo SP, Lelouard H, Devilard E, de Bovis B, Van Rooijen N, Gorvel JP, Malissen B. 2010. Contrasting roles of macrophages and dendritic cells in controlling initial pulmonary *Brucella* infection. *Eur J Immunol* **40:**3458–3471.

173. Tate MD, Pickett DL, van Rooijen N, Brooks AG, Reading PC. 2010. Critical role of airway macrophages in modulating disease severity during influenza virus infection of mice. *J Virol* **84:**7569–7580.

174. Varol C, Mildner A, Jung S. 2015. Macrophages: development and tissue specialization. *Annu Rev Immunol* **33:**643–675.

175. Hasegawa A, Hayashi K, Kishimoto H, Yang M, Tofukuji S, Suzuki K, Nakajima H, Hoffman RM, Shirai M, Nakayama T. 2010. Color-coded real-time cellular imaging of lung T-lymphocyte accumulation and focus formation in a mouse asthma model. *J Allergy Clin Immunol* **125:**461–468.e6. doi:10.1016/j.jaci.2009.09.016.

176. Kiefmann R, Rifkind JM, Nagababu E, Bhattacharya J. 2008. Red blood cells induce hypoxic lung inflammation. *Blood* **111:**5205–5214.

177. Tabuchi A, Mertens M, Kuppe H, Pries AR, Kuebler WM. 2008. Intravital microscopy of the murine pulmonary microcirculation. *J Appl Physiol 1985* **104:**338–346.

178. Westphalen K, Gusarova GA, Islam MN, Subramanian M, Cohen TS, Prince AS, Bhattacharya J. 2014. Sessile alveolar macrophages communicate with alveolar epithelium to modulate immunity. *Nature* **506:**503–506.

179. Hanna RN, Cekic C, Sag D, Tacke R, Thomas GD, Nowyhed H, Herrley E, Rasquinha N, McArdle S, Wu R, Peluso E, Metzger D, Ichinose H, Shaked I, Chodaczek G, Biswas SK, Hedrick CC. 2015. Patrolling monocytes control tumor metastasis to the lung. *Science* **350:**985–990.

180. Carrasco YR, Batista FD. 2007. B cells acquire particulate antigen in a macrophage-rich area at the boundary between the follicle and the subcapsular sinus of the lymph node. *Immunity* **27:**160–171.

181. Hickman HD, Li L, Reynoso GV, Rubin EJ, Skon CN, Mays JW, Gibbs J, Schwartz O, Bennink JR, Yewdell JW. 2011. Chemokines control naive CD8+ T cell selection of optimal lymph node antigen presenting cells. *J Exp Med* **208:**2511–2524.

182. Junt T, Moseman EA, Iannacone M, Massberg S, Lang PA, Boes M, Fink K, Henrickson SE, Shayakhmetov DM, Di Paolo NC, van Rooijen N, Mempel TR, Whelan SP, von Andrian UH. 2007. Subcapsular sinus macrophages in lymph nodes clear lymph-borne viruses and present them to antiviral B cells. *Nature* **450:** 110–114.

183. Sagoo P, Garcia Z, Breart B, Lemaître F, Michonneau D, Albert ML, Levy Y, Bousso P. 2016. *In vivo* imaging of inflammasome activation reveals a subcapsular macrophage burst response that mobilizes innate and adaptive immunity. *Nat Med* **22:**64–71.

184. Phan TG, Green JA, Gray EE, Xu Y, Cyster JG. 2009. Immune complex relay by subcapsular sinus macrophages and noncognate B cells drives antibody affinity maturation. *Nat Immunol* **10:**786–793.

185. Phan TG, Grigorova I, Okada T, Cyster JG. 2007. Subcapsular encounter and complement-dependent transport of immune complexes by lymph node B cells. *Nat Immunol* **8:**992–1000.

186. Moalli F, Proulx ST, Schwendener R, Detmar M, Schlapbach C, Stein JV. 2015. Intravital and whole-organ imaging reveals capture of melanoma-derived antigen by lymph node subcapsular macrophages leading to widespread deposition on follicular dendritic cells. *Front Immunol* **6:**114. doi:10.3389/fimmu.2015.00114.

187. Park C, Arthos J, Cicala C, Kehrl JH. 2015. The HIV-1 envelope protein gp120 is captured and displayed for B cell recognition by SIGN-R1+ lymph node macrophages. *eLife* **4:**e06467. doi:10.7554/eLife.06467.

188. Benhaiem-Sigaux N, Mina E, Sigaux F, Lambré CR, Valensi F, Allégret C, Bernaudin JF. 1985. Characterization of human pericardial macrophages. *J Leukoc Biol* **38:**709–721.

189. Frankenberger M, Passlick B, Hofer T, Siebeck M, Maier KL, Ziegler-Heitbrock LH. 2000. Immunologic characterization of normal human pleural macrophages. *Am J Respir Cell Mol Biol* **23:**419–426.

190. Ghosn EE, Cassado AA, Govoni GR, Fukuhara T, Yang Y, Monack DM, Bortoluci KR, Almeida SR, Herzenberg LA, Herzenberg LA. 2010. Two physically, functionally, and developmentally distinct peritoneal macrophage subsets. *Proc Natl Acad Sci U S A* **107:** 2568–2573.

191. Kubicka U, Olszewski WL, Maldyk J, Wierzbicki Z, Orkiszewska A. 1989. Normal human immune peritoneal cells: phenotypic characteristics. *Immunobiology* **180:**80–92.

192. Weber GF. 2015. Immune targeting of the pleural space by intercostal approach. *BMC Pulm Med* **15:**14. doi: 10.1186/s12890-015-0010-6.

193. Okabe Y, Medzhitov R. 2014. Tissue-specific signals control reversible program of localization and functional polarization of macrophages. *Cell* **157:**832–844.

194. Kim KW, Williams JW, Wang YT, Ivanov S, Gilfillan S, Colonna M, Virgin HW, Gautier EL, Randolph GJ.

2016. MHC II+ resident peritoneal and pleural macrophages rely on IRF4 for development from circulating monocytes. *J Exp Med* **213**:1951–1959.

195. **Cailhier JF, Sawatzky DA, Kipari T, Houlberg K, Walbaum D, Watson S, Lang RA, Clay S, Kluth D, Savill J, Hughes J.** 2006. Resident pleural macrophages are key orchestrators of neutrophil recruitment in pleural inflammation. *Am J Respir Crit Care Med* **173**:540–547.
196. **Pace E, Profita M, Melis M, Bonanno A, Paternò A, Mody CH, Spatafora M, Ferraro M, Siena L, Vignola AM, Bonsignore G, Gjomarkaj M.** 2004. LTB4 is present in exudative pleural effusions and contributes actively to neutrophil recruitment in the inflamed pleural space. *Clin Exp Immunol* **135**:519–527.
197. **Devi S, Li A, Westhorpe CL, Lo CY, Abeynaike LD, Snelgrove SL, Hall P, Ooi JD, Sobey CG, Kitching AR, Hickey MJ.** 2013. Multiphoton imaging reveals a new leukocyte recruitment paradigm in the glomerulus. *Nat Med* **19**:107–112.
198. **Nishimura S, Manabe I, Nagasaki M, Seo K, Yamashita H, Hosoya Y, Ohsugi M, Tobe K, Kadowaki T, Nagai R, Sugiura S.** 2008. In vivo imaging in mice reveals local cell dynamics and inflammation in obese adipose tissue. *J Clin Invest* **118**:710–721.
199. **Epelman S, Lavine KJ, Beaudin AE, Sojka DK, Carrero JA, Calderon B, Brija T, Gautier EL, Ivanov S, Satpathy AT, Schilling JD, Schwendener R, Sergin I, Razani B, Forsberg EC, Yokoyama WM, Unanue ER, Colonna M, Randolph GJ, Mann DL.** 2014. Embryonic and adult-derived resident cardiac macrophages are maintained through distinct mechanisms at steady state and during inflammation. *Immunity* **40**:91–104.
200. **Ensan S, Li A, Besla R, Degousee N, Cosme J, Roufaiel M, Shikatani EA, El-Maklizi M, Williams JW, Robins L, Li C, Lewis B, Yun TJ, Lee JS, Wieghofer P, Khattar R, Farrokhi K, Byrne J, Ouzounian M, Zavitz CC, Levy GA, Bauer CM, Libby P, Husain M, Swirski FK, Cheong C, Prinz M, Hilgendorf I, Randolph GJ, Epelman S, Gramolini AO, Cybulsky MI, Rubin BB, Robbins CS.** 2016. Self-renewing resident arterial macrophages arise from embryonic CX3CR1+ precursors and circulating monocytes immediately after birth. *Nat Immunol* **17**:159–168.
201. **Barth MW, Hendrzak JA, Melnicoff MJ, Morahan PS.** 1995. Review of the macrophage disappearance reaction. *J Leukoc Biol* **57**:361–367.
202. **Lauder SN, Taylor PR, Clark SR, Evans RL, Hindley JP, Smart K, Leach H, Kidd EJ, Broadley KJ, Jones SA, Wise MP, Godkin AJ, O'Donnell V, Gallimore AM.** 2011. Paracetamol reduces influenza-induced immunopathology in a mouse model of infection without compromising virus clearance or the generation of protective immunity. *Thorax* **66**:368–374.
203. **Heidt T, Courties G, Dutta P, Sager HB, Sebas M, Iwamoto Y, Sun Y, Da Silva N, Panizzi P, van der Laan AM, Swirski FK, Weissleder R, Nahrendorf M.** 2014. Differential contribution of monocytes to heart macrophages in steady-state and after myocardial infarction. *Circ Res* **115**:284–295.
204. **Wang H, Melton DW, Porter L, Sarwar ZU, McManus LM, Shireman PK.** 2014. Altered macrophage phenotype transition impairs skeletal muscle regeneration. *Am J Pathol* **184**:1167–1184.
205. **Ginhoux F, Jung S.** 2014. Monocytes and macrophages: developmental pathways and tissue homeostasis. *Nat Rev Immunol* **14**:392–404.
206. **Scott CL, Zheng F, De Baetselier P, Martens L, Saeys Y, De Prijck S, Lippens S, Abels C, Schoonooghe S, Raes G, Devoogdt N, Lambrecht BN, Beschin A, Guilliams M.** 2016. Bone marrow-derived monocytes give rise to self-renewing and fully differentiated Kupffer cells. *Nat Commun* **7**:10321. doi:10.1038/ncomms10321.

Myeloid Cells in Health and Disease: A Synthesis
Edited by Siamon Gordon

doi:10.1128/microbiolspec.MCHD-0021-2015

Thomas A. Kufer[1]
Giulia Nigro[2,3]
Philippe J. Sansonetti[2,3,4]

16

Multifaceted Functions of NOD-Like Receptor Proteins in Myeloid Cells at the Intersection of Innate and Adaptive Immunity

HISTORY OF DISCOVERY OF NOD MOLECULES

The existence of extracellular pattern recognition was appreciated early on after formation of the pattern recognition receptor (PRR) theory (1) by the identification of Toll in *Drosophila* and cloning of the lipopolysaccharide sensor Toll-like receptor 4 (TLR4) in mammals (see reference 2 for a historical overview). However, it was evident that intracellular bacterial pathogens also induce inflammatory responses in infected cells, although the dedicated receptors for sensing of such threats remained elusive for some time. Analysis of invasive pathogenic bacteria and viruses showed that indeed such receptors exist. In recent decades, these cytosolic PRRs and their cognate microbe-associated molecular patterns (MAMPs) were identified.

Our current knowledge suggests that there are two broad families of such PRRs. On the one hand, intracellular viral sensors of the retinoic acid-inducible gene 1 protein (RIG-I)-like helicase family and the absent in melanoma 2 (AIM-2) family respond to foreign nucleic acid structures and cellular signaling intermediates including cyclic GMP-AMP synthase (cGAS) and subsequently activate the signaling adaptors STING (stimulator of interferon [IFN] genes) and mitochondrial antiviral signaling protein (MAVS) to drive type I IFN responses (3). These receptors show particularly high expression in myeloid cells, most prominently in plasmacytoid dendritic cells (DCs), which are prominent producers of type I IFNs upon viral infection. On the other hand, cytosolic NOD-like receptor (NLR) proteins ensure reactivity to a broad variety of bacterial and viral

[1]Institute of Nutritional Medicine, Department of Immunology, University of Hohenheim, 70593 Stuttgart, Germany; [2]Institut Pasteur, Unité de Pathogénie Microbienne Moléculaire, 75724 Paris, France; [3]INSERM, U1202, 75015 Paris, France; [4]Collège de France, 75005 Paris, France.

MAMPs as well as endogenous danger-associated molecular patterns (DAMPs) (4). The concept of "danger sensing" proposes that endogenous elicitors of immunity (DAMPs) are compartmentalized away from their cognate receptors in physiological conditions, gaining access to their sensors only upon tissue or cell damage (5). Hence intense research in the past 10 years has revealed, besides the TLRs, the existence of a vast and complex intracellular network of sensing and alert dedicated to early and adapted response to both microbial and sterile stresses. In this context, epithelial cells and immune and phagocytic cells that are directly confronted with microbial threats strongly rely on these systems, particularly the NLR proteins, to adapt their functions.

The first examples of intracellular MAMP-sensing NLR proteins were nucleotide-binding oligomerization domain-containing 1 (NOD1) and NOD2. Numerous independent studies meanwhile established that these NLR proteins confer inflammatory responses against invasive bacterial pathogens by reacting to peptidoglycan subunits (6–9). Initial evidence for the function of NLR proteins as bona fide PRRs was obtained in epithelial cells. These contributions were driven by the hypothesis that most cells of the mammalian organisms are able to respond to pathogenic stimuli independently of myeloid or lymphoid assessor cell types, a process also termed cell-autonomous immunity. Now it is well accepted that almost all cells, but most prominently cells of barrier tissues, can mount such responses and produce peptides with bactericidal activity as well as cytokines and IFNs to activate bystander cells and ultimately recruit granulocytes and activate other cells, particularly antigen-presenting cells, thereby establishing a direct link between innate responses to bacterial threat and the nature and intensity of the adaptive immune response.

In mammals, hematopoietic cells are the prime responders, upon activation with MAMPs, for systemic release of cytokines. Cells of the myeloid lineage are best known for their function as professional antigen-presenting cells that activate lymphocytes and their ability to rapidly engulf and destroy invading pathogens. The first evidence for the existence of PRRs actually was obtained in these cell types by the demonstration that activation of PRRs, such as TLRs and C-type lectin receptors (CLRs), is needed for the expression of costimulatory molecules on DCs and for macrophage activation. Starting with the seminal works on the identification of the mechanism of lipopolysaccharide-induced proinflammatory responses, it became evident that the members of the TLR family, the CLRs, and other PRRs including the NLR proteins were important drivers for proinflammatory responses, such as release of the cytokines interleukin-6 (IL-6) and tumor necrosis factor (TNF) and activation of myeloid cells. Myeloid cells have a high capacity to degrade endosomal material.

However, many bacterial pathogens have evolved to subvert these mechanisms, some of them by escape from the endocytic compartment into the cytosol. Autophagy, a process whereby cellular material is engulfed by a specialized double membrane, has evolved as a defense pathway used to degrade such invaders. A general mechanism of how cytosolic bacteria are recognized is by ubiquitination of their surface proteins, which are recognized by host proteins such as p62, nuclear domain 10 protein 52 (NDP52), optineurin (OPTN), and neighbor of BRCA1 (NBR1), which in turn trigger autophagy (10). NLR proteins such as NOD1 and NOD2 can directly trigger autophagy in a process that involves ATG16L1 (11–13). This activation of autophagy is mainly induced at membranes, where NOD1 and NOD2 are localized, suggesting that NOD1/2 serve as sentinels for membrane integrity. NOD1/2-induced autophagy thereby seems to correlate with maturation of myeloid cells from monocytes to macrophages and is only functional in the latter (14). NOD1 can localize to early endosome antigen 1 (EEA1)-positive endosomes to detect peptidoglycan and induce autophagy right at the first site of bacterial invasion (15). Notably, this mechanism is also activated by bacterial outer membrane vesicles (OMVs) taken up into cells. In DCs, the peptide transporters SLC15A3 and SLC15A4 are induced after PRR activation on endolysosomes and form a complex with NOD1 and Rip2 that is capable of reacting to endocytic-residing bacteria even before their escape to the cytosol (16). Endosomes are thus emerging as platforms for the integration of NOD1/2 signaling. Besides the role in activation of autophagy in a NOD-dependent manner, ATG16L1, but not the disease-associated forms, also negatively regulates NOD2-mediated inflammatory signals (17). Both the autophagy protein ATG16L1 and NOD2 are linked to Crohn's disease (CD), a severe inflammatory bowel disease; this suggests that negative regulation of bacterial sensing by ATG16L1 plays an important role in immune homeostasis in the mucosa. Here the expression of NOD1 and NOD2 in non-myeloid cells also plays an important role. NOD2, for example, is highly expressed in intestinal stem cells, providing a cytoprotective effect when these cells are submitted to a stress, such as drugs or irradiation (18). NOD2 is also expressed in Paneth cells, where its activation leads to the release of antimicrobial peptides in response to microbiota-derived stimuli (19). Accordingly, mice deficient in *Nod2* show alterations in the gut microbiota composition (20). Moreover, activation of NOD1 and NOD2 contributes to the development of secondary lymphoid tissue in the intestine (21, 22). Notably, different biological roles for NOD2 in myeloid versus nonmyeloid cells are emerging, whereas in nonhematopoietic cells NOD2 seems to contribute more toward the regulation of the microbiota, whereas its expression in myeloid cells might rather affect epithelial integrity and immune cell homeostasis (23). Some prevalent CD-associated polymorphisms in NOD2 show perturbation of the PRR function of NOD2, suggesting that both aberrant control of antibacterial peptide production and impaired immune homeostasis contribute to the development of CD.

Support for the contribution of NOD2 to the immune environment comes from the analysis of epithelioid cell granulomas, which are typical for patients with CD and Blau syndrome. Analysis of CD and Blau syndrome patients with defined NOD2 polymorphisms suggests that granulomas found in these patients are morphologically different and that an increased function of polymorphisms found in NOD2 of Blau syndrome patients might be related to a Th17 response (24). Analysis of the pathology of these diseases in more detail will help to better define the contribution and role of NOD2 in their etiology.

IL-1 CYTOKINES AND THE INFLAMMASOME

A particularly interesting cytokine produced in large amounts by myeloid cells upon activation by pathogenic stimuli is IL-1β. This cytokine, together with IL-6 and TNF, is a potent driver of the acute-phase response and induces fever. Seminal work led by Jörg Tschopp resulted in the identification that pyrin domain-containing NLR proteins can act as platforms for the activation of the pro-IL-1β processing enzyme (pro) caspase-1, which he called "inflammasomes" (25). The identification of the NLR family pyrin domain-containing 3 (NLRP3) inflammasome as the first described complex for formation of IL-1β and IL-18 inspired research in this field. Today, NLRP3 is by far the most-studied NLR protein, particularly in disease conditions. A vast amount of work has demonstrated that the NLRP3 inflammasome consists of the NLR protein NLRP3 and the adaptor ASC (apoptosis-associated speck-like protein containing a CARD domain), which forms prion-like aggregates upon activation, resulting in a single large aggregate in activated cells that can be visualized microscopically. This aggregate recruits procaspase-1 via homodimeric interaction of the CARD domains of ASC and procaspase-1 (26), resulting in procaspase-1 cleavage, processing of pro-IL-1β and pro-IL-18, and subsequent release of these cytokines by a still enigmatic secretion event that does not involve endoplasmic reticulum-mediated secretion. Triggering of this cascade likely is not reversible and comes at the price of cell death induced by caspase-1, named pyroptosis. Notably, NLRP3 and its adaptor are expressed at the highest levels in myeloid cells but need to be primed for inflammasome activity by transcriptional induction of the expression of the inflammasome components and its substrate pro-IL-1β. This is mediated by the activation of NF-κB responses induced upon engagement of PRRs such as TLR4 (27). In view of the potentially detrimental outcome of inflammasome activation, this two-step activation process warrants robustness of the response and ensures that "IL-1β commitment" is only induced when at least two PRRs are activated. Synergy in the activation of NLR and TLR receptors on myeloid cells and epithelial cells thereby is a common mechanism that is also observed for NOD1 and NOD2 in conjunction with TLR stimulation (for a summary, see reference 28).

Inflammasome complexes do not necessarily contain only one type of NLR. Functional pairing of different NLR proteins has been described, and currently the best-understood example is NLR family CARD domain containing 4 (NLRC4) and NLR family apoptosis inhibitory proteins (NAIPs). Whereas in most mammals NLRC4 is encoded by a single gene, the NAIPs are more heterogeneous among species, with multiple paralogs in mice but only one allele in humans. Alleles of the NAIP family can confer resistance to *Legionella pneumophila* in macrophages of susceptible A/J mice, and it was shown that NAIP5 (BIRC1e) is responsible for *L. pneumophila* resistance observed in C57BL/6 mice (29, 30). Subsequent contributions showed that this NLR senses bacterial flagellin from *Salmonella* and other bacteria (31, 32) and also that the single human NAIP gene mediates restriction of *Legionella* infection (33). Insight on how this is mediated was obtained by molecular analyses that revealed that NAIPs form a complex with NLRC4. NAIP proteins thereby physically interact with the rod components of the bacterial type II secretion system and induce inflammasome formation via interaction with NLRC4 (reviewed in reference 34). By contrast, the molecular mechanisms that trigger the activation of NLRP3 remain largely elusive. Many MAMPs and DAMPs are known to act as activators of the NLRP3 inflammasome; however, how these structurally extremely diverse molecules converge to activate NLRP3 is not clear. Induction of reactive oxygen species, changes in ion influx and the redox status, as well as posttranslational modification and/or cleavage of NLRs have been proposed as initiation mechanisms, although our current knowledge is not sufficient to pinpoint one of these mechanisms as being universal nor to clearly assign these to structural groups of elicitors (reviewed in references 25 and 35). Interestingly, the commonly used clinical adjuvant alum triggers the activation of the NLRP3 inflammasome (36–38). Although this would explain part of Janeway's "dirty little secret" (1), it is under debate if this NLRP3-mediated response results in adaptive immune instruction. Still, there is good evidence that activation of other NLR proteins might be suited to induce adjuvanticity (39). Sensing of mycobacterial muramyl dipeptide (MDP) by NOD2, for example, is largely responsible for the activity of complete Freund's adjuvant, and activation of NOD1 by a synthetic ligand is sufficient to induce T-cell-mediated adaptive immunity (40).

In any case, IL-1β is a physiologically very important cytokine that is involved in a variety of disorders. Identification of the molecular details of IL-1β activation inspired research that revealed that the inflammasome is triggered specifically by crystalline structures. Although the exact mechanism by which crystals lead to activation of NLRP3 remains elusive, the concept of frustrated phagocytosis that results in membrane rupture upon uptake of crystals is becoming widely accepted. Release of cathepsins from lysosomes that might act on NLRP3 or a yet to be identified activator thus has been proposed as a possible mechanism beside changes in ion fluxes induced by membrane damage, changes in the redox status, and release of mitochondrial DNA (35). Under disease conditions, crystalline substances can form in the human body, such as uric acid crystal in gout, cholesterol crystals in atherosclerosis, and amyloids in type 1 diabetes or Parkinson disease (summarized in reference 41). Common to all these cases is the absence of pathogen-derived MAMPs; hence the term "sterile inflammation" is emerging to describe these autoinflammatory processes. The physiological relevance of NLRP3 has further been appreciated with the identification of several polymorphisms in the encoding genes that are linked to monogenic hereditary periodic fever syndromes, a group of severe inflammatory disorders with the hallmark of elevated IL-1β levels. Clinical interventions with available IL-1β antagonists such as anakinra and canakinumab are emerging as powerful novel therapeutic options for these diseases as well as for pathologies with a clear NLRP3 link, such as gout (reviewed in reference 42).

NLRP proteins do not act solely via the formation of inflammasomes that drive IL-1 family cytokine responses but also contribute to immunity more directly. These effects, however, were so far described only in nonmyeloid cells. NLRP3, for example, is expressed in $CD4^+$ lymphocytes upon T-cell receptor activation and seems to be required for a Th2 decision by acting as a transcription factor in the nucleus and influencing IFN regulatory factor 4 (IRF4) activity (43). NLRP12, an NLR protein highly expressed in DCs, was shown to positively contribute to migration of DCs in the context of a delayed-type hypersensitivity model (44). Overall, a concept emerges according to which this NLR acts as a negative regulator of innate immunity-mediated proinflammatory responses (45) and T-cell responses. In a model of experimental autoimmune encephalomyelitis (EAE), it was shown that the lack of NLRP12 is associated with neuroinflammatory symptoms that were accompanied by increased IL-4 production (46).

It is now well established that activation of myeloid cells contributes to the metabolic syndrome and insulin resistance. Several studies highlight a contribution of NLRs to the adverse inflammatory effects observed in obese patients and identify IL-1β as an important contributor to the low-grade inflammatory state generally observed in the metabolic syndrome (47). However, other NLR proteins, such as NOD1 and NOD2, also contribute to metabolic syndrome (48, 49). Most interestingly, these effects might be mediated by altered gut microbiota (i.e., dysbiosis) appearing in animals lacking NLR proteins. Data in support of this hypothesis are provided by a study showing that high-fat diet-induced inflammation in NOD2 knockout mice is due to altered composition of the microbiota (50). Whether NLR triggering is causal of metabolic disorders—i.e., if increased fatty acid, glucose, or metabolite levels can directly activate NLR proteins—is less clear. By contrast, β-hydroxybutyrate, a substance increased upon caloric restriction, inhibits the NLRP3 inflammasome (51), and the same has been described for unsaturated fatty acids (52, 53). Taken together, these data show that, particularly in myeloid cells, the NLRP3 inflammasome and likely other NLRs are regulated by the host metabolic status (54).

TRANSITION TO ADAPTIVE IMMUNITY (INDUCTION ON MHC)

For the initiation of adaptive immunity, DCs play a pivotal role in presenting antigens to cognate naive

T lymphocytes. NLR proteins play important roles in this process and contribute to both activation of DCs and regulation of the expression of major histocompatibility complex (MHC) molecules (see below) (55). NOD1 and NOD2 activation in stromal cells thereby contributes also to Th2 lineage decision of lymphocytes (40, 56–58). DC intrinsic expression of NOD1/2 contributes to this process and might be involved in cross-priming of $CD8^+$ T cells (58, 59). Furthermore, NOD2-mediated induction of autophagy in DCs is linked to antigen presentation via MHC class II (MHC-II) and induction of $CD4^+$ T-cell responses and contributes to the etiology of CD (11).

NLR proteins, however, also contribute to adaptive immunity directly by regulating MHC gene expression. Probably the best-known and -understood NLR protein with a critical function in this process is the class II transcriptional activator protein (CIITA). Instead of acting as a PRR, CIITA serves as a transcriptional regulator of MHC-II genes. It was initially cloned by classical complementation for the bare lymphocyte syndrome (60), and a wealth of work has shown that CIITA can translocate to the nucleus and dock to MHC-II promoter sites. The N terminus of CIITA has transcriptional activity; however, CIITA does not function as a classical transcription factor as it lacks DNA-binding capacity. Instead, it is recruited to the X1 and X2 motifs present in MHC-II promoters via a multiprotein complex named the "MHC enhanceosome" that consists of the RFX proteins RFX, RFX-ANK, and RFX-AP and further auxiliary factors (reviewed in reference 61). CIITA subsequently acts as a scaffold for recruiting histone-modifying enzymes and transcriptional regulators to initiate and facilitate transcription of the targeted MHC-II genes. CIITA expression is driven mainly by IFN-γ, and it seems that this transcriptional regulation is both necessary and sufficient to drive MHC-II transcription, i.e., that MHC-II expression is regulated mainly by the amount of cellular CIITA protein.

Professional antigen-presenting cells extensively regulate MHC-II expression upon activation; however, MHC-I genes are also well known to be regulated upon stimulation and are differentially expressed in many tissues. Differences in the core promoter of MHC-I genes, which contains additional internal ribosome entry site and NF-κB elements in comparison to MHC-II core promoters, were long thought to be responsible for this. *In vitro* data furthermore suggested that CIITA also contributes to MHC-I expression, as these promoters also contain the X1, X2 boxes. However, genetic evidence using transgenic mouse models recently challenged this view and revealed that CIITA has no significant effect on MHC-I expression *in vivo* (reviewed in reference 62). Seminal work by the Kobayashi lab identified the NLR protein NLRC5 to show high homology to CIITA and to act as a major regulator of MHC-I genes (63). Several studies using human cell lines and different NLRC5 knockout mouse lines confirmed these findings and showed that NLRC5 acts with remarkable specificity for MHC-I promoters (summarized in reference 64). This comes as a surprise as both CIITA and NLRC5 use the MHC enhanceosome and the same DNA-binding motif for promoter binding (63, 65). Genome-wide analysis of promoters occupied by NLRC5 recently provided a partial answer to this paradox as NLRC5 prefers particular MHC-I promoter-specific binding motifs (66) and also the Y element of the MHC-I promoter contributes to MHC-I specificity (63). The N-terminal domain of NLRC5, which shows transcriptional activity, contributes to the specificity for MHC-I genes, and grafting this domain to CIITA results in a chimeric protein with dual activity for MHC-I and -II promoters (67). Future research will identify the factors that generate the specificity of NLRC5 and CIITA for MHC-I and -II genes, respectively. It is noteworthy that besides this well-established function in MHC gene regulation, NLRC5 might both positively and negatively contribute to innate immune responses (64, 68–73). It would make sense that NLRC5 positively contributes to type I IFN responses (64, 71, 74) to induce antiviral immunity both by the induction of MHC-I to present viral peptides to CD8 cells and production of IFNs to alert bystander cells.

HEMATOPOIETIC STEM CELLS

In addition to playing a major role in regulating the effector arm of the innate and adaptive immune response, NLR proteins also participate in maintaining and regulating the pool of cells constituting the immune compartment by directly controlling the hematopoietic stem cells (HSCs), from which the myeloid cells originate.

HSCs, located mainly in specialized niches in the bone marrow, are the reservoir of multipotent stem cells that provide a continuous supply of cells circulating in the peripheral blood. HSCs have the capability of self-renewal and multilineage differentiation. Their maturation into a terminally differentiated progeny involves the loss of self-renewal potential followed by progression through downstream subsets of increasingly restricted lineage potential, from oligopotent (lymphoid or myeloid restricted) and unipotent hematopoietic progenitor cells (HPCs) (75).

In steady-state conditions, the niche in which the HSCs are located provides signals to keep them in a dormant state. In response to stress-induced loss of hematopoietic cells, such as upon chemotherapy, irradiation, or heavy bleeding, dormant HSCs become activated and rapidly replenish the hematopoietic system (76).

Recent evidence suggests that HSCs and HPCs could respond to inflammatory cytokines through cytokine receptors and to MAMPs through PRRs (77). Nagai et al. have shown that TLR2 and TLR4 are expressed and are functionally active in murine HSCs and in some HPCs (78). *In vitro* stimulation with TLR ligands makes the cells proliferate and differentiate to monocytes/macrophages and DCs in the absence of exogenous growth and differentiation factors.

Several studies showed that other TLRs are expressed in these cells—not only in mice but also in human HSCs and HPCs (77). Instead, few investigations have been performed on the role of NLRs in HSC control and functional regulation. NOD2 has been shown to be expressed in freshly isolated human bone marrow HSCs. Stimulation of these cells with MDP led to expression of TNF-α, granulocyte-macrophage colony-stimulating factor, and IL-1β and increased expression of the transcription factor PU.1, which is important for myeloid differentiation (monocytes/macrophages and DCs). Moreover, MDP stimulation enhanced the responsiveness of HSCs to TLR2 ligands increasing intracellular α-defensin levels (79).

As most HSCs are located in specialized niches in the bone marrow, a key question is whether they can actually interact with bacteria or more likely bacterial products. Indeed, a small proportion of them circulate in the bloodstream, especially HPCs that have been recovered from extramedullary sites like the liver, spleen, and muscle (80). Moreover, Clarke et al. demonstrated that products from the intestinal microbiota, particularly peptidoglycan, can be found in the serum and bone marrow. Soluble peptidoglycan released by the intestinal microbiota may translocate through the epithelium, reach the bloodstream, and serve as a molecular mediator responsible for the remote systemic

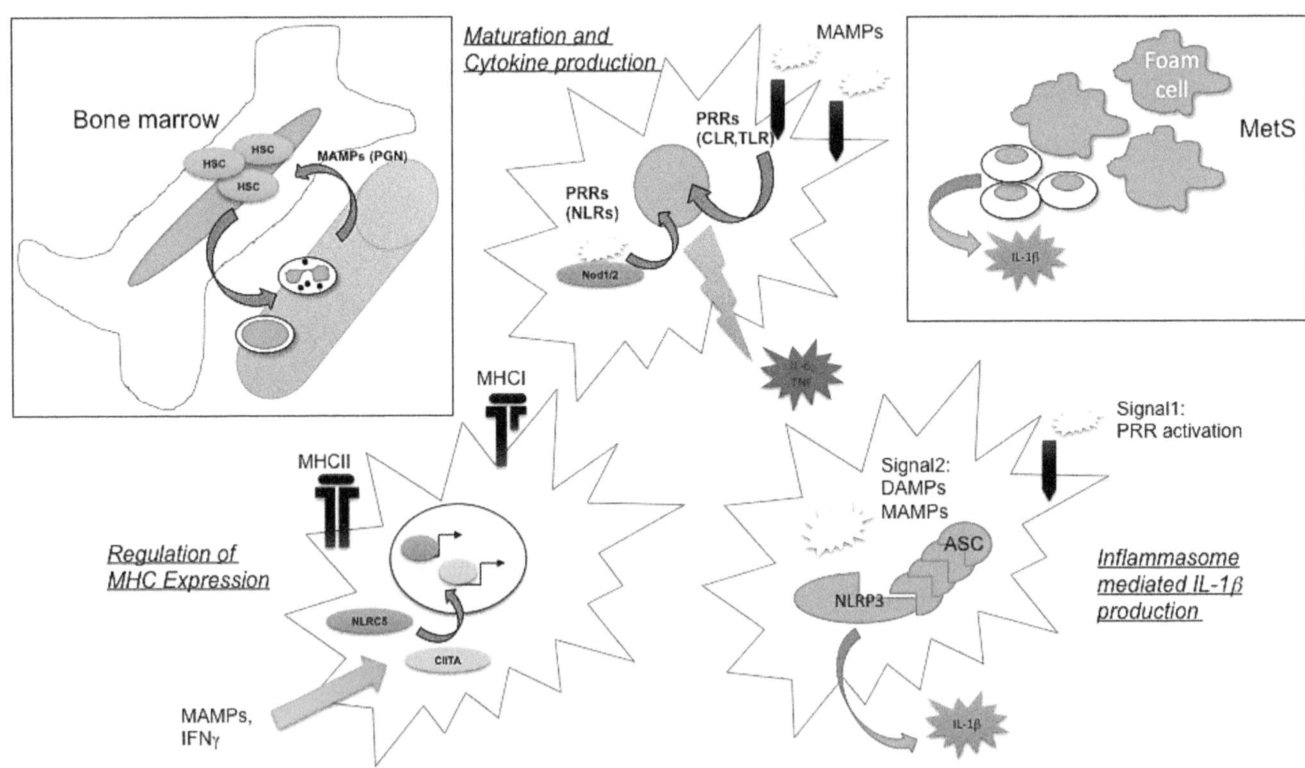

Figure 1 Different functions of NLRs in myeloid cells and progenitors. The circulating MAMPs, particularly peptidoglycan (PGN), contribute to activation and differentiation of HSCs. Stimulation by MAMPs, particularly PGN, activates innate immunity through recognition by PRR receptors, inducing their maturation and production of cytokines, such as IL-6 and TNF. NLRs also stimulate the adaptive immunity, particularly in regulating MHC expression levels and in stimulating the inflammasome.

priming of neutrophils in the bone marrow in a NOD1-dependent manner (81). More recently, NOD1 has been found to cooperate with TLRs in the synergistic production of granulocyte colony-stimulating factor, required for the mobilization of HSCs to the spleen, especially during an infection process. Mobilized HSCs and HPCs gave rise to neutrophils and monocytes, contributing to eradication of the infection (82).

CONCLUSION

One can draw a certain number of firm conclusions about the functions of NLR proteins in immunity in myeloid cells (Fig. 1). We still have limited knowledge about the expression and function of many of the mammalian NLR proteins in the myeloid lineage; however, the function of the pyrin domain-containing NLR proteins and NLRC4/NAIP as inflammasomes that drive IL-1β and IL-18 secretion upon pathogen stimulation and under sterile inflammation is well established. NOD1, NOD2, NLRP3, and NLRC4/NAIP act as bona fide PRRs that sense, in most cases, bacterial-derived MAMPs but also react to endogenous DAMPs. Ultimately, activation of these receptors ensures macrophage activation and maturation of DCs to drive antigen-specific adaptive immune responses. Under physiological conditions upon an infection, sensing of invading pathogens and likely of DAMPs is not a process that involves only one PRR. We know that bacterial pathogens trigger multiple PRRs in both myeloid and epithelial cells and that these are interconnected to coin tailored pathogen-adapted immune responses by induction of different cytokine profiles (28). These give rise to appropriate lymphocyte polarization. However, it should also be noted that the balance of proinflammatory mediators and type I IFNs is pivotal for the effective control of pathogens in cases such as *Listeria* and *Salmonella*.

More studies are warranted to understand the role of PRRs, especially NLRs, in the control and functional regulation of HSCs. The data available up to now indicate that HSCs are able to respond directly to MAMPs and that the recognition of these molecules can directly impact hematopoiesis.

The data reviewed in this chapter, taken together, indicate that a complex network exists in the myeloid compartment, from HSCs to progenitors and derived lineages, that, based on NLR sensing of MAMPs (largely peptidoglycan) and DAMPs that permanently signal local/systemic danger, maintains a state of active surveillance of the environment by the immune system. This concept of "armed peace" can be summarized as "physiological inflammation." Upon any aggression, massive production and diffusion of MAMPs and DAMPs mobilizes the NLR-based network into a state of "pathological inflammation" aimed at pathogen eradication and development of a properly oriented and gauged adaptive immune response that will prevent subsequent infection by the same pathogen in the immunized host.

Acknowledgments. *P.J.S. is a Howard Hughes Medical Institute foreign senior scholar.*

Citation. Kufer TA, Nigro G, Sansonetti PJ. 2016. Multifaceted functions of NOD-like receptor proteins in myeloid cells at the intersection of innate and adaptive immunity. Microbiol Spectrum 4(4):MCHD-0021-2015.

References

1. **Janeway CA Jr.** 1989. Approaching the asymptote? Evolution and revolution in immunology. *Cold Spring Harb Symp Quant Biol* **54:**1–13.
2. **Wagner H.** 2012. Innate immunity's path to the Nobel Prize 2011 and beyond. *Eur J Immunol* **42:**1089–1092.
3. **Unterholzner L.** 2013. The interferon response to intracellular DNA: why so many receptors? *Immunobiology* **218:**1312–1321.
4. **Fritz JH, Ferrero RL, Philpott DJ, Girardin SE.** 2006. Nod-like proteins in immunity, inflammation and disease. *Nat Immunol* **7:**1250–1257.
5. **Matzinger P.** 1994. Tolerance, danger, and the extended family. *Annu Rev Immunol* **12:**991–1045.
6. **Chamaillard M, Hashimoto M, Horie Y, Masumoto J, Qiu S, Saab L, Ogura Y, Kawasaki A, Fukase K, Kusumoto S, Valvano MA, Foster SJ, Mak TW, Nuñez G, Inohara N.** 2003. An essential role for NOD1 in host recognition of bacterial peptidoglycan containing diaminopimelic acid. *Nat Immunol* **4:**702–707.
7. **Girardin SE, Boneca IG, Carneiro LAM, Antignac A, Jéhanno M, Viala J, Tedin K, Taha MK, Labigne A, Zähringer U, Coyle AJ, DiStefano PS, Bertin J, Sansonetti PJ, Philpott DJ.** 2003. Nod1 detects a unique muropeptide from Gram-negative bacterial peptidoglycan. *Science* **300:**1584–1587.
8. **Girardin SE, Boneca IG, Viala J, Chamaillard M, Labigne A, Thomas G, Philpott DJ, Sansonetti PJ.** 2003. Nod2 is a general sensor of peptidoglycan through muramyl dipeptide (MDP) detection. *J Biol Chem* **278:**8869–8872.
9. **Girardin SE, Travassos LH, Hervé M, Blanot D, Boneca IG, Philpott DJ, Sansonetti PJ, Mengin-Lecreulx D.** 2003. Peptidoglycan molecular requirements allowing detection by Nod1 and Nod2. *J Biol Chem* **278:**41702–41708.
10. **Gomes LC, Dikic I.** 2014. Autophagy in antimicrobial immunity. *Mol Cell* **54:**224–233.
11. **Cooney R, Baker J, Brain O, Danis B, Pichulik T, Allan P, Ferguson DJ, Campbell BJ, Jewell D, Simmons A.** 2010. NOD2 stimulation induces autophagy in dendritic cells influencing bacterial handling and antigen presentation. *Nat Med* **16:**90–97.

12. Homer CR, Richmond AL, Rebert NA, Achkar JP, McDonald C. 2010. *ATG16L1* and *NOD2* interact in an autophagy-dependent antibacterial pathway implicated in Crohn's disease pathogenesis. *Gastroenterology* **139:** 1630–1641.

13. Travassos LH, Carneiro LA, Ramjeet M, Hussey S, Kim YG, Magalhães JG, Yuan L, Soares F, Chea E, Le Bourhis L, Boneca IG, Allaoui A, Jones NL, Nuñez G, Girardin SE, Philpott DJ. 2010. Nod1 and Nod2 direct autophagy by recruiting ATG16L1 to the plasma membrane at the site of bacterial entry. *Nat Immunol* **11:** 55–62.

14. Juárez E, Carranza C, Hernández-Sánchez F, Loyola E, Escobedo D, León-Contreras JC, Hernández-Pando R, Torres M, Sada E. 2014. Nucleotide-oligomerizing domain-1 (NOD1) receptor activation induces pro-inflammatory responses and autophagy in human alveolar macrophages. *BMC Pulm Med* **14:**152. doi:10.1186/1471-2466-14-152.

15. Irving AT, Mimuro H, Kufer TA, Lo C, Wheeler R, Turner LJ, Thomas BJ, Malosse C, Gantier MP, Casillas LN, Votta BJ, Bertin J, Boneca IG, Sasakawa C, Philpott DJ, Ferrero RL, Kaparakis-Liaskos M. 2014. The immune receptor NOD1 and kinase RIP2 interact with bacterial peptidoglycan on early endosomes to promote autophagy and inflammatory signaling. *Cell Host Microbe* **15:**623–635.

16. Nakamura N, Lill JR, Phung Q, Jiang Z, Bakalarski C, de Mazière A, Klumperman J, Schlatter M, Delamarre L, Mellman I. 2014. Endosomes are specialized platforms for bacterial sensing and NOD2 signalling. *Nature* **509:** 240–244.

17. Sorbara MT, Ellison LK, Ramjeet M, Travassos LH, Jones NL, Girardin SE, Philpott DJ. 2013. The protein ATG16L1 suppresses inflammatory cytokines induced by the intracellular sensors Nod1 and Nod2 in an autophagy-independent manner. *Immunity* **39:**858–873.

18. Nigro G, Rossi R, Commere P-H, Jay P, Sansonetti PJ. 2014. The cytosolic bacterial peptidoglycan sensor Nod2 affords stem cell protection and links microbes to gut epithelial regeneration. *Cell Host Microbe* **15:** 792–798.

19. Lala S, Ogura Y, Osborne C, Hor SY, Bromfield A, Davies S, Ogunbiyi O, Nuñez G, Keshav S. 2003. Crohn's disease and the NOD2 gene: a role for Paneth cells. *Gastroenterology* **125:**47–57.

20. Mondot S, Barreau F, Al Nabhani Z, Dussaillant M, Le Roux K, Doré J, Leclerc M, Hugot JP, Lepage P. 2012. Altered gut microbiota composition in immune-impaired *Nod2*$^{-/-}$ mice. *Gut* **61:**634–635.

21. Bouskra D, Brézillon C, Bérard M, Werts C, Varona R, Boneca IG, Eberl G. 2008. Lymphoid tissue genesis induced by commensals through NOD1 regulates intestinal homeostasis. *Nature* **456:**507–510.

22. Barreau F, Meinzer U, Chareyre F, Berrebi D, Niwa-Kawakita M, Dussaillant M, Foligne B, Ollendorff V, Heyman M, Bonacorsi S, Lesuffleur T, Sterkers G, Giovannini M, Hugot JP. 2007. CARD15/NOD2 is required for Peyer's patches homeostasis in mice. *PLoS One* **2:**e523–e523. doi:10.1371/journal.pone.0000523.

23. Alnabhani Z, Hugot JP, Montcuquet N, Le Roux K, Dussaillant M, Roy M, Leclerc M, Cerf-Bensussan N, Lepage P, Barreau F. 2016. Respective roles of hematopoietic and nonhematopoietic Nod2 on the gut microbiota and mucosal homeostasis. *Inflamm Bowel Dis* **22:** 763–773.

24. Janssen CE, Rose CD, De Hertogh G, Martin TM, Bader Meunier B, Cimaz R, Harjacek M, Quartier P, Ten Cate R, Thomee C, Desmet VJ, Fischer A, Roskams T, Wouters CH. 2012. Morphologic and immunohistochemical characterization of granulomas in the nucleotide oligomerization domain 2-related disorders Blau syndrome and Crohn disease. *J Allergy Clin Immunol* **129:**1076–1084.

25. Schroder K, Tschopp J. 2010. The inflammasomes. *Cell* **140:**821–832.

26. Elliott EI, Sutterwala FS. 2015. Initiation and perpetuation of NLRP3 inflammasome activation and assembly. *Immunol Rev* **265:**35–52.

27. Bauernfeind FG, Horvath G, Stutz A, Alnemri ES, MacDonald K, Speert D, Fernandes-Alnemri T, Wu J, Monks BG, Fitzgerald KA, Hornung V, Latz E. 2009. Cutting edge: NF-κB activating pattern recognition and cytokine receptors license NLRP3 inflammasome activation by regulating NLRP3 expression. *J Immunol* **183:** 787–791.

28. Kufer TA, Sansonetti PJ. 2007. Sensing of bacteria: NOD a lonely job. *Curr Opin Microbiol* **10:**62–69.

29. Diez E, Lee SH, Gauthier S, Yaraghi Z, Tremblay M, Vidal S, Gros P. 2003. *Birc1e* is the gene within the *Lgn1* locus associated with resistance to *Legionella pneumophila*. *Nat Genet* **33:**55–60.

30. Wright EK Jr, Goodart SA, Growney JD, Hadinoto V, Endrizzi MG, Long EM, Sadigh K, Abney AL, Bernstein-Hanley I, Dietrich WF. 2003. *Naip5* affects host susceptibility to the intracellular pathogen *Legionella pneumophila*. *Curr Biol* **13:**27–36.

31. Franchi L, Amer A, Body-Malapel M, Kanneganti TD, Ozören N, Jagirdar R, Inohara N, Vandenabeele P, Bertin J, Coyle A, Grant EP, Núñez G. 2006. Cytosolic flagellin requires Ipaf for activation of caspase-1 and interleukin 1β in salmonella-infected macrophages. *Nat Immunol* **7:**576–582.

32. Miao EA, Alpuche-Aranda CM, Dors M, Clark AE, Bader MW, Miller SI, Aderem A. 2006. Cytoplasmic flagellin activates caspase-1 and secretion of interleukin 1β via Ipaf. *Nat Immunol* **7:**569–575.

33. Vinzing M, Eitel J, Lippmann J, Hocke AC, Zahlten J, Slevogt H, N'guessan PD, Günther S, Schmeck B, Hippenstiel S, Flieger A, Suttorp N, Opitz B. 2008. NAIP and Ipaf control *Legionella pneumophila* replication in human cells. *J Immunol* **180:**6808–6815.

34. Vance RE. 2015. The NAIP/NLRC4 inflammasomes. *Curr Opin Immunol* **32:**84–89.

35. Sutterwala FS, Haasken S, Cassel SL. 2014. Mechanism of NLRP3 inflammasome activation. *Ann N Y Acad Sci* **1319:**82–95.

36. Eisenbarth SC, Colegio OR, O'Connor W, Sutterwala FS, Flavell RA. 2008. Crucial role for the Nalp3

inflammasome in the immunostimulatory properties of aluminium adjuvants. *Nature* **453**:1122–1126.

37. **Franchi L, Núñez G.** 2008. The Nlrp3 inflammasome is critical for aluminium hydroxide-mediated IL-1β secretion but dispensable for adjuvant activity. *Eur J Immunol* **38**:2085–2089.

38. **Li H, Willingham SB, Ting JP, Re F.** 2008. Cutting edge: inflammasome activation by alum and alum's adjuvant effect are mediated by NLRP3. *J Immunol* **181**:17–21.

39. **Maisonneuve C, Bertholet S, Philpott DJ, De Gregorio E.** 2014. Unleashing the potential of NOD- and Toll-like agonists as vaccine adjuvants. *Proc Natl Acad Sci U S A* **111**:12294–12299.

40. **Fritz JH, Le Bourhis L, Sellge G, Magalhaes JG, Fsihi H, Kufer TA, Collins C, Viala J, Ferrero RL, Girardin SE, Philpott DJ.** 2007. Nod1-mediated innate immune recognition of peptidoglycan contributes to the onset of adaptive immunity. *Immunity* **26**:445–459.

41. **Broderick L, De Nardo D, Franklin BS, Hoffman HM, Latz E.** 2015. The inflammasomes and autoinflammatory syndromes. *Annu Rev Pathol* **10**:395–424.

42. **Jesus AA, Goldbach-Mansky R.** 2014. IL-1 blockade in autoinflammatory syndromes. *Annu Rev Med* **65**:223–244.

43. **Bruchard M, Rebé C, Derangère V, Togbé D, Ryffel B, Boidot R, Humblin E, Hamman A, Chalmin F, Berger H, Chevriaux A, Limagne E, Apetoh L, Végran F, Ghiringhelli F.** 2015. The receptor NLRP3 is a transcriptional regulator of T_H2 differentiation. *Nat Immunol* **16**:859–870.

44. **Arthur JC, Lich JD, Ye Z, Allen IC, Gris D, Wilson JE, Schneider M, Roney KE, O'Connor BP, Moore CB, Morrison A, Sutterwala FS, Bertin J, Koller BH, Liu Z, Ting JP.** 2010. Cutting edge: NLRP12 controls dendritic and myeloid cell migration to affect contact hypersensitivity. *J Immunol* **185**:4515–4519.

45. **Tuncer S, Fiorillo MT, Sorrentino R.** 2014. The multifaceted nature of NLRP12. *J Leukoc Biol* **96**:991–1000.

46. **Lukens JR, Gurung P, Shaw PJ, Barr MJ, Zaki MH, Brown SA, Vogel P, Chi H, Kanneganti TD.** 2015. The NLRP12 sensor negatively regulates autoinflammatory disease by modulating interleukin-4 production in T cells. *Immunity* **42**:654–664.

47. **Gregor MF, Hotamisligil GS.** 2011. Inflammatory mechanisms in obesity. *Annu Rev Immunol* **29**:415–445.

48. **Schertzer JD, Tamrakar AK, Magalhães JG, Pereira S, Bilan PJ, Fullerton MD, Liu Z, Steinberg GR, Giacca A, Philpott DJ, Klip A.** 2011. NOD1 activators link innate immunity to insulin resistance. *Diabetes* **60**:2206–2215.

49. **Zhou YJ, Liu C, Li CL, Song YL, Tang YS, Zhou H, Li A, Li Y, Weng Y, Zheng FP.** 2015. Increased NOD1, but not NOD2, activity in subcutaneous adipose tissue from patients with metabolic syndrome. *Obesity (Silver Spring)* **23**:1394–1400.

50. **Denou E, Lolmède K, Garidou L, Pomie C, Chabo C, Lau TC, Fullerton MD, Nigro G, Zakaroff-Girard A, Luche E, Garret C, Serino M, Amar J, Courtney M, Cavallari JF, Henriksbo BD, Barra NG, Foley KP, McPhee JB, Duggan BM, O'Neill HM, Lee AJ, Sansonetti P, Ashkar AA, Khan WI, Surette MG, Bouloumié A, Steinberg GR, Burcelin R, Schertzer JD.** 2015. Defective NOD2 peptidoglycan sensing promotes diet-induced inflammation, dysbiosis, and insulin resistance. *EMBO Mol Med* **7**:259–274.

51. **Youm YH, Nguyen KY, Grant RW, Goldberg EL, Bodogai M, Kim D, D'Agostino D, Planavsky N, Lupfer C, Kanneganti TD, Kang S, Horvath TL, Fahmy TM, Crawford PA, Biragyn A, Alnemri E, Dixit VD.** 2015. The ketone metabolite β-hydroxybutyrate blocks NLRP3 inflammasome-mediated inflammatory disease. *Nat Med* **21**:263–269.

52. **L'homme L, Esser N, Riva L, Scheen A, Paquot N, Piette J, Legrand-Poels S.** 2013. Unsaturated fatty acids prevent activation of NLRP3 inflammasome in human monocytes/macrophages. *J Lipid Res* **54**:2998–3008.

53. **Yan Y, Jiang W, Spinetti T, Tardivel A, Castillo R, Bourquin C, Guarda G, Tian Z, Tschopp J, Zhou R.** 2013. Omega-3 fatty acids prevent inflammation and metabolic disorder through inhibition of NLRP3 inflammasome activation. *Immunity* **38**:1154–1163.

54. **McNelis JC, Olefsky JM.** 2014. Macrophages, immunity, and metabolic disease. *Immunity* **41**:36–48.

55. **Krishnaswamy JK, Chu T, Eisenbarth SC.** 2013. Beyond pattern recognition: NOD-like receptors in dendritic cells. *Trends Immunol* **34**:224–233.

56. **Duan W, Mehta AK, Magalhaes JG, Ziegler SF, Dong C, Philpott DJ, Croft M.** 2010. Innate signals from Nod2 block respiratory tolerance and program T_H2-driven allergic inflammation. *J Allergy Clin Immunol* **126**:1284–1293.e10. doi:10.1016/j.jaci.2010.09.021.

57. **Magalhaes JG, Fritz JH, Le Bourhis L, Sellge G, Travassos LH, Selvanantham T, Girardin SE, Gommerman JL, Philpott DJ.** 2008. Nod2-dependent Th2 polarization of antigen-specific immunity. *J Immunol* **181**:7925–7935.

58. **Magalhaes JG, Rubino SJ, Travassos LH, Le Bourhis L, Duan W, Sellge G, Geddes K, Reardon C, Lechmann M, Carneiro LA, Selvanantham T, Fritz JH, Taylor BC, Artis D, Mak TW, Comeau MR, Croft M, Girardin SE, Philpott DJ, Philpott DJ.** 2011. Nucleotide oligomerization domain-containing proteins instruct T cell helper type 2 immunity through stromal activation. *Proc Natl Acad Sci U S A* **108**:14896–14901.

59. **Asano J, Tada H, Onai N, Sato T, Horie Y, Fujimoto Y, Fukase K, Suzuki A, Mak TW, Ohteki T.** 2010. Nucleotide oligomerization binding domain-like receptor signaling enhances dendritic cell-mediated cross-priming in vivo. *J Immunol* **184**:736–745.

60. **Steimle V, Otten LA, Zufferey M, Mach B.** 1993. Complementation cloning of an MHC class II transactivator mutated in hereditary MHC class II deficiency (or bare lymphocyte syndrome). *Cell* **75**:135–146.

61. **Devaiah BN, Singer DS.** 2013. CIITA and its dual roles in MHC gene transcription. *Front Immunol* **4**:476. doi:10.3389/fimmu.2013.00476.

62. **Kobayashi KS, van den Elsen PJ.** 2012. NLRC5: a key regulator of MHC class I-dependent immune responses. *Nat Rev Immunol* **12**:813–820.

63. **Meissner TB, Li A, Biswas A, Lee KH, Liu YJ, Bayir E, Iliopoulos D, van den Elsen PJ, Kobayashi KS.** 2010.

NLR family member NLRC5 is a transcriptional regulator of MHC class I genes. *Proc Natl Acad Sci U S A* **107:** 13794–13799.

64. **Neerincx A, Lautz K, Menning M, Kremmer E, Zigrino P, Hösel M, Büning H, Schwarzenbacher R, Kufer TA.** 2010. A role for the human nucleotide-binding domain, leucine-rich repeat-containing family member NLRC5 in antiviral responses. *J Biol Chem* **285:**26223–26232.
65. **Neerincx A, Rodriguez GM, Steimle V, Kufer TA.** 2012. NLRC5 controls basal MHC class I gene expression in an MHC enhanceosome-dependent manner. *J Immunol* **188:**4940–4950.
66. **Ludigs K, Seguín-Estévez Q, Lemeille S, Ferrero I, Rota G, Chelbi S, Mattmann C, MacDonald HR, Reith W, Guarda G.** 2015. NLRC5 exclusively transactivates MHC class I and related genes through a distinctive SXY module. *PLoS Genet* **11:**e1005088. doi:10.1371/journal.pgen.1005088.
67. **Neerincx A, Jakobshagen K, Utermöhlen O, Büning H, Steimle V, Kufer TA.** 2014. The N-terminal domain of NLRC5 confers transcriptional activity for MHC class I and II gene expression. *J Immunol* **193:**3090–3100.
68. **Benko S, Magalhaes JG, Philpott DJ, Girardin SE.** 2010. NLRC5 limits the activation of inflammatory pathways. *J Immunol* **185:**1681–1691.
69. **Cui J, Zhu L, Xia X, Wang HY, Legras X, Hong J, Ji J, Shen P, Zheng S, Chen ZJ, Wang RF.** 2010. NLRC5 negatively regulates the NF-κB and type I interferon signaling pathways. *Cell* **141:**483–496.
70. **Davis BK, Roberts RA, Huang MT, Willingham SB, Conti BJ, Brickey WJ, Barker BR, Kwan M, Taxman DJ, Accavitti-Loper MA, Duncan JA, Ting JP.** 2011. Cutting edge: NLRC5-dependent activation of the inflammasome. *J Immunol* **186:**1333–1337.
71. **Kuenzel S, Till A, Winkler M, Häsler R, Lipinski S, Jung S, Grötzinger J, Fickenscher H, Schreiber S, Rosenstiel P.** 2010. The nucleotide-binding oligomerization domain-like receptor NLRC5 is involved in IFN-dependent antiviral immune responses. *J Immunol* **184:** 1990–2000.
72. **Lian L, Ciraci C, Chang G, Hu J, Lamont SJ.** 2012. NLRC5 knockdown in chicken macrophages alters response to LPS and poly (I:C) stimulation. *BMC Vet Res* **8:**23. doi:10.1186/1746-6148-8-23.
73. **Tong Y, Cui J, Li Q, Zou J, Wang HY, Wang RF.** 2012. Enhanced TLR-induced NF-κB signaling and type I interferon responses in *NLRC5* deficient mice. *Cell Res* **22:** 822–835.
74. **Ranjan P, Singh N, Kumar A, Neerincx A, Kremmer E, Cao W, Davis WG, Katz JM, Gangappa S, Lin R, Kufer TA, Sambhara S.** 2015. NLRC5 interacts with RIG-I to induce a robust antiviral response against influenza virus infection. *Eur J Immunol* **45:**758–772.
75. **Boiko JR, Borghesi L.** 2012. Hematopoiesis sculpted by pathogens: Toll-like receptors and inflammatory mediators directly activate stem cells. *Cytokine* **57:**1–8.
76. **Trumpp A, Essers M, Wilson A.** 2010. Awakening dormant haematopoietic stem cells. *Nat Rev Immunol* **10:** 201–209.
77. **King KY, Goodell MA.** 2011. Inflammatory modulation of HSCs: viewing the HSC as a foundation for the immune response. *Nat Rev Immunol* **11:**685–692.
78. **Nagai Y, Garrett KP, Ohta S, Bahrun U, Kouro T, Akira S, Takatsu K, Kincade PW.** 2006. Toll-like receptors on hematopoietic progenitor cells stimulate innate immune system replenishment. *Immunity* **24:**801–812.
79. **Sioud M, Fløisand Y.** 2009. NOD2/CARD15 on bone marrow $CD34^+$ hematopoietic cells mediates induction of cytokines and cell differentiation. *J Leukoc Biol* **85:** 939–946.
80. **Massberg S, Schaerli P, Knezevic-Maramica I, Köllnberger M, Tubo N, Moseman EA, Huff IV, Junt T, Wagers AJ, Mazo IB, von Andrian UH.** 2007. Immunosurveillance by hematopoietic progenitor cells trafficking through blood, lymph, and peripheral tissues. *Cell* **131:**994–1008.
81. **Clarke TB, Davis KM, Lysenko ES, Zhou AY, Yu Y, Weiser JN.** 2010. Recognition of peptidoglycan from the microbiota by Nod1 enhances systemic innate immunity. *Nat Med* **16:**228–231.
82. **Burberry A, Zeng MY, Ding L, Wicks I, Inohara N, Morrison SJ, Núñez G.** 2014. Infection mobilizes hematopoietic stem cells through cooperative NOD-like receptor and Toll-like receptor signaling. *Cell Host Microbe* **15:**779–791.

Myeloid Cells in Health and Disease: A Synthesis
Edited by Siamon Gordon

doi:10.1128/microbiolspec.MCHD-0049-2016

Sushmita Jha[1]
W. June Brickey[2]
Jenny Pan-Yun Ting[2,3]

17 Inflammasomes in Myeloid Cells: Warriors Within

INTRODUCTION

Inflammation is the body's response to injury, pathogen exposure, and irritants. Pattern recognition receptors allow our body to recognize a diverse array of patterns generated during exposure to these insults. In 2002, the nucleotide-binding domain leucine-rich repeat-containing (NLR, also known as NOD-like receptor) gene family of pattern recognition receptors was discovered (1–3). While several members were already recognized at that point, reports of the entire NLR family provided a global view. In the past 15 years of research, the physiological relevance of these genes has been revealed to include a diverse variety of functions. Gene mutations in some of the family members have been linked to autoinflammatory diseases in humans (Fig. 1). This association of mutations in NLR genes to autoinflammatory diseases indicates critical functions in the regulation of immunity and inflammation.

There are 22 NLR genes in humans and 34 identified in mice, with each gene encoding a protein with a characteristic tripartite structure of a central nucleotide-binding domain (NBD), an N-terminal effector domain, and a variable number of C-terminal leucine-rich repeats (LRRs) (Fig. 2) (4). Proteins with domain architecture similar to that of human NLRs exist in plants and in invertebrates such as sea urchins. These proteins are absent in nematodes and *Drosophila*, suggesting either convergent evolution between mammalian and plant NLRs or loss in invertebrates (5, 6). The effector domains of NLRs can be combinations of the following domain types: acidic transactivation domain (AD), baculoviral inhibitory repeat (BIR)-like domain, caspase-recruitment domain (CARD), pyrin domain, or domain of unknown function (X) (Fig. 2). The length of the LRR domains is highly variable. For example, an NLR may contain up to 30+ LRR domains (7). While each NLR has a unique capability to sense a variety of pathogen-associated molecular patterns (PAMPs) and danger-associated molecular patterns (DAMPs), the exact mechanisms of NLR-ligand binding have only recently emerged for NOD2 and NLRC4/NAIP proteins. Initially, NLRs were presumed to be expressed only in innate immune cells of monocyte/macrophage lineage. However, their ubiquitous expression throughout the human body is now widely accepted. Interestingly, different NLRs show distinct tissue, cellular, and intracellular distributions, suggesting variable roles in different cell types (8). This review will focus on the role of

[1]Department of Bio-Sciences and Bio-Engineering, Indian Institute of Technology Jodhpur, Rajasthan, 342011, India; [2]Lineberger Comprehensive Cancer Center; [3]Department of Genetics, The University of North Carolina, Chapel Hill, NC 27599.

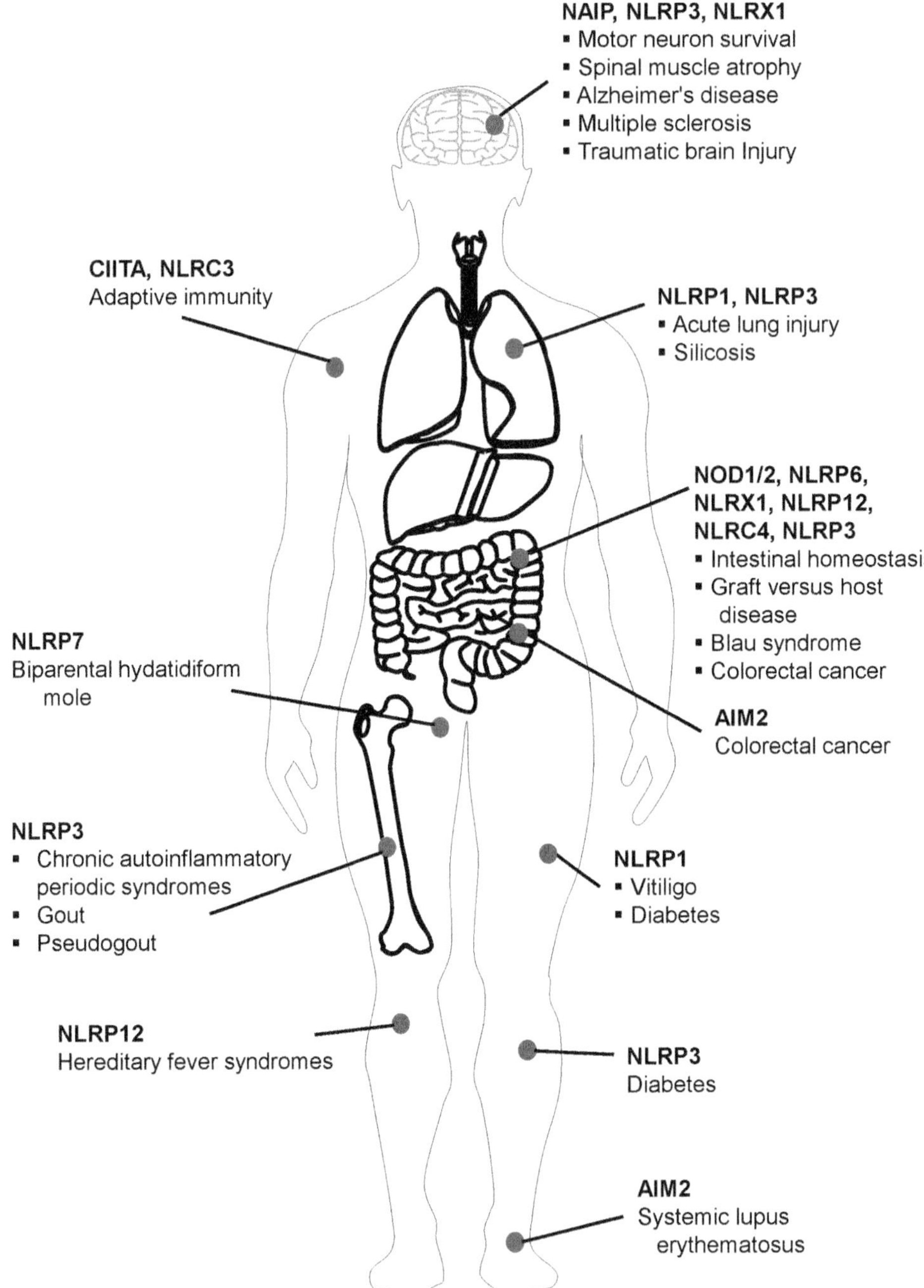

Figure 1 NLRs function in healthy and dysregulated disease states in the human body.

NLRs in myeloid cells in the normal host as well as in dysregulated immune states of disease.

ACTIVATION OF INFLAMMASOME NLRs

Upon activation, several NLRs form multiprotein complexes called "inflammasomes." These complexes consist of an NLR, an adaptor molecule known as ASC (apoptosis-associated speck-like protein containing a CARD), and the inflammatory protease procaspase-1 (4, 5). Even though the inflammasome-mediated roles of NLRs have been extensively studied, there are several non-inflammasome-mediated functions of NLRs, including NF-κB regulation, mitogen-activated protein kinase (MAPK) activation, cytokine and chemokine production, interferon (IFN) production, ribonuclease

L activation, and antimicrobial reactive oxygen species (ROS) production. The inflammasome-forming NLRs will be discussed in detail here, while readers are referred to other reviews for descriptions of the non-inflammasome-forming NLRs (9–11).

A subset of NLRs (NLRP1, NLRP3, NLRP6, NLRP7, NLRC4, NLRC5, and NAIP2/5/6) has been reported to form inflammasomes. Each NLR with its activating signal, inflammasome components, and disease association will be described here in some detail.

NLRP1

NLR family, pyrin domain-containing 1, or NLRP1 (formerly CARD7, DEFCAP, and NALP1) was first characterized as a member of the CED-4 family of apoptotic proteins that are required to initiate programmed cell death (12–14). The first caspase-1-activating inflammasome to be identified consisted of NLRP1, ASC, caspase-1, and caspase-5 (15). Overexpression of NLRP1 in mammalian cells led to apoptosis (12, 13). NLRP1 is a cytoplasmic protein that is highly expressed in peripheral blood lymphocytes (12). Initial studies on NLRP1 suggested that the NLRP1 inflammasome in humans consists of NLRP1, procaspase-1, caspase-5, and the adaptor ASC (12, 13). It was later revealed that even though the presence of ASC may not be required for processing of procaspase-1 by the NLRP1 inflammasome, ASC does augment processing of procaspase-1 (16).

There is one *NLRP1* gene in humans, in contrast to three paralogs in mice: *Nlrp1a*, *Nlrp1b*, and *Nlrp1c* (17). Interestingly, not all strains of mice express all isoforms. For example, some strains of inbred mice express different splice variants of *Nlrp1b*, while *Nlrp1a* is highly conserved (18). The NLRP1 protein in humans consists of an N-terminal pyrin domain, a central NBD-associated domain (NAD), LRRs, a function to find (FIIND) domain, and a C-terminal CARD domain. Polymorphisms in the *NLRP1* gene, in both the noncoding and coding sequences, have been associated with the dermatologic autoimmune disease vitiligo (19–21). The *NLRP1* haplotype associated with vitiligo and other autoimmune disorders leads to increased interleukin-1β (IL-1β) processing. Several coding polymorphisms have also been associated with heightened risk for other autoimmune diseases such as Addison's disease and type 1 diabetes (22).

The mouse Nlrp1 paralogs vary in structure from the human protein such that Nlrp1a lacks the N-terminal pyrin domain, Nlrp1b lacks both the pyrin and NAD domains, and Nlrp1c lacks all but the NBD and LRR domains. Due to these differences, mouse and human NLRP1 appear to exhibit functional differences. Specifically, susceptible and resistant mouse *Nlrp1b* loci were genetically associated with *Bacillus anthracis* susceptibility (23, 24). Additionally, anthrax lethal toxin was found to activate the mouse Nlrp1b and rat Nlrp1 inflammasome (25), resulting in caspase-1-dependent pyroptosis. Lethal toxin is composed of two proteins: protective antigen (PA) and lethal factor (LF), with PA binding to anthrax toxin receptors on host cells and subsequently translocating LF into the cytosol (26). LF was found to cause the proteolytic cleavage of rat Nlrp1 at the N terminus, presumably by cleaving an inhibitory domain. Mutation of this cleavage site transformed a responsive allele to a nonresponsive allele and resulted in the abrogation of caspase-1 activation (27). In the mouse system, an engineered Nlrp1b that contained an artificial TEV protease cleavage site activated inflammasome in the presence of TEV (26). Interestingly, this cleavage site coincided with the cleavage site found in rat NLRP1, although dissimilar in sequence. Overall, this association of LF cleavage with Nlrp1b in the intact animal is less straightforward, in that Nlrp1b proteins from both LF-responsive and -nonresponsive mouse strains were cleaved by LF (28).

In addition to pathogens, the human NLRP1 inflammasome is activated by the peptidoglycan component muramyl dipeptide (MDP) (16). MDP stimulation of a macrophage cell line also leads to association of overexpressed NLRP1 with NOD2, leading to formation of a multiprotein complex consisting of NLRP1, NOD2, and caspase-1 (25). These results suggest either the existence of an inflammasome containing NLRP1 plus NOD2 that is activated by MDP or that MDP activates both NLRP1 and NOD2 inflammasomes. However, a mouse cell line deficient in *Nlrp1b* and lacking inflammasome activation by anthrax lethal toxin shows no defect in the assembly of NLRP1 inflammasome by MDP (29). While investigating the crystal structure of the LRR domain of NLRP1, Reubold et al. concluded that the LRR domain is not likely to contain the MDP-binding domain (30). Thus, MDP may represent a species-specific activator of NLRP1, considering the structural difference between human and rodent NLRP1.

NLRP1 has also been shown to mediate inflammasome activation in response to *Toxoplasma gondii* in a human monocytic cell line (31). This protein was later found to mediate host response to *Toxoplasma* in mice (32), but the process was not dependent on the cleavage site of Nlrp1b found in anthrax (23). Both Nlrp1b and Nlrp3 form inflammasomes that restrict *T. gondii* infection via the production of IL-18 (33). Finally, additional studies have demonstrated that mouse *Nlrp1a*

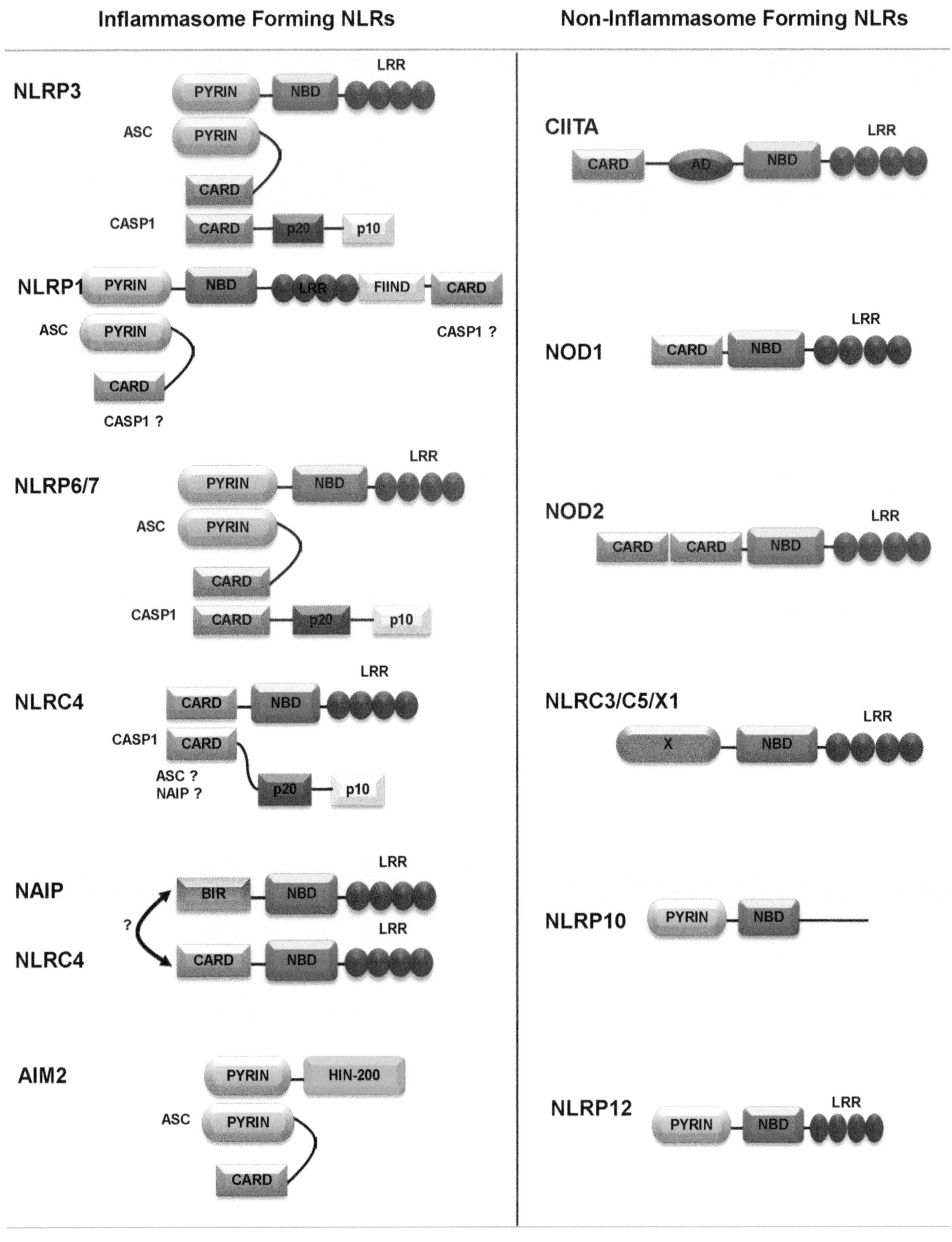
Inflammasome Forming NLRs
Non-Inflammasome Forming NLRs
NLRP3
PYRIN
NBD
LRR
ASC
PYRIN
CARD
CASP1
CARD
p20
p10
NLRP1
PYRIN
NBD
LRR
FIIND
CARD
ASC
PYRIN
CARD
CASP1 ?
CASP1 ?
NLRP6/7
PYRIN
NBD
LRR
ASC
PYRIN
CARD
CASP1
CARD
p20
p10
NLRC4
CARD
NBD
LRR
CASP1
CARD
ASC ?
NAIP ?
p20
p10
NAIP
BIR
NBD
LRR
?
NLRC4
CARD
NBD
LRR
AIM2
PYRIN
HIN-200
ASC
PYRIN
CARD
CIITA
CARD
AD
NBD
LRR
NOD1
CARD
NBD
LRR
NOD2
CARD
CARD
NBD
LRR
NLRC3/C5/X1
X
NBD
LRR
NLRP10
PYRIN
NBD
NLRP12
PYRIN
NBD
LRR

mediates inflammasome and pyroptosis function during lymphocytic choriomeningitis virus infection and upon chemotherapy treatment (34).

Aside from response to microbial pathogens, NLRP1 is associated with several disease pathologies, including acute glaucoma, traumatic brain injury, acute lung injury, colitis, and colitis-associated tumorigenesis (29, 35–37). Interestingly, a recent paper showed that Nlrp1 is also involved in metabolic disease, where it prevents obesity by the production of IL-18, which is known to prevent overeating (38). An inflammasome consisting of NLRP1, ASC, caspase-1, caspase-11 (the rodent ortholog of human caspase-5), and the X-linked inhibitor of apoptosis protein (XIAP) was shown to be present in rat spinal cord motor neurons in protein coimmunoprecipitation and immunofluorescence experiments (39). Remarkably, therapeutic neutralization of ASC with an antibody was shown to improve histopathology after traumatic brain injury via reduction of immune responses (40). This finding is consistent with a report that ASC specks accumulate in the extracellular space after cells undergo pyroptosis to promote IL-1β maturation. Although phagocytosis of these specks was shown to induce lysosomal damage, stimulate soluble ASC nucleation, and increase inflammation (41), "frustrated" phagocytosis has been noted as a consequence to injury (and inflammasome activation) found in various organs.

NLRP3

NLR family, pyrin domain-containing 3 (NLRP3; also known as cryopyrin, Nalp3, PYPAF1, and CIAS1) was discovered in 2001, in a seminal report that mapped a causative *NLRP3* mutation to rheumatologic autoinflammatory disorders, namely familial cold autoinflammatory syndrome (FCAS) and Muckle-Wells syndrome (MWS) (42). In 2002, with the discovery of NOMID/CINCA (neonatal-onset multisystem inflammatory disease and chronic infantile neurologic cutaneous and articular syndrome), FCAS and MWS were classified along with NOMID to form the cold-associated periodic syndromes (CAPS) (43, 44). To date, the primary focus of inflammasome research has been anchored by NLRP3, which is a cytoplasmic protein that is primarily expressed in monocytes, macrophages, granulocytes, dendritic cells, epithelial cells, and osteoblasts (45, 46). NLRP3 expression in myeloid cells is highly inducible (47). The protein is composed of three distinct domains: the N-terminal pyrin domain, the central NBD, and the C-terminal LRRs.

NLRP3 responds to a wide range of DAMPs and PAMPs, including bacterial and viral nucleic acids (48, 49), intracellular pathogens; ATP (50), uric acid (51, 52), β-amyloid (53), hyaluronan and heparan sulfate (54); silica (55–57), asbestos (56), cholesterol (58, 59) and alum crystals (57, 60); metabolites associated with type 2 diabetes such as ceramide, saturated fatty acids, islet amyloid peptides (61–63); hemozoin (64) by-product from blood-feeding parasites that cause malaria; and cyclic dinucleotides (65). Activation of the NLRP3 inflammasome requires two signals and is controlled at transcriptional and posttranslational levels (Fig. 3). The first signal, also referred to as the priming signal, is the induction of the Toll-like receptor (TLR)/NF-κB pathway to upregulate the expression of NLRP3 (50) and pro-IL-1β (66). Signal 2 is transduced by various PAMPs and DAMPs to activate the functional NLRP3 inflammasome by initiating assembly of a multiprotein complex consisting of NLRP3, the adaptor protein ASC, and procaspase-1 (67). Association of NLRP3 with ASC is required for recruitment of procaspase-1 (68). ASC utilizes its CARD domain to recruit procaspase-1 via homotypic CARD-CARD interactions. In the inflammasome complex, the inactive procaspase-1 undergoes autocatalytic cleavage to form active caspase-1. Caspase-1 in turn can cleave and activate multiple substrates ranging from chaperones, cytoskeletal and translation machinery and glycolysis, and immune proteins such as the proinflammatory cytokines IL-1β and IL-18 (69–71). While NLRP3 is known to respond to several PAMPs and DAMPs, evidence for direct binding of any ligand to NLRP3 remains indeterminate. One model is that NLRP3 activation is mediated via secondary intermediates such as potassium efflux (50), change in cell volume (72), calcium mobilization via the calcium channels TRPM2 or CASR (73–75), osmolarity changes (76), ROS (77), or mitochondrial DNA release (78, 79).

Figure 2 NLRs have a conserved tripartite structure with an N-terminal effector domain, a central NBD, and C-terminal LRRs. The effector domains of NLRs may include acidic transactivation domain (AD), baculoviral inhibitory repeat (BIR)-like domain, caspase-recruitment domain (CARD), pyrin domain, or domain of unknown function (X). In general, NLRP1, NLRP3, NLRP6, NLRP7, NLRC4, NAIP, and AIM2 are known to form inflammasomes, while CIITA, NOD1, NOD2, NLRC3, NLRC5, NLRX1, NLRP10, and NLRP12 do not.

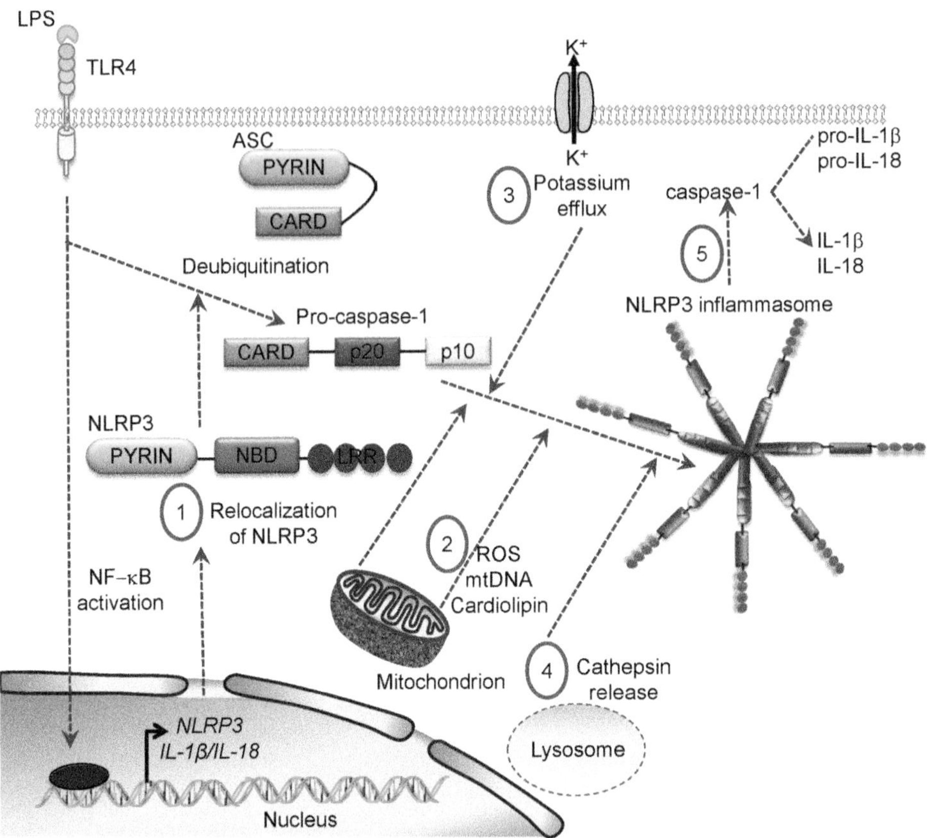

Figure 3 The NLRP3 inflammasome is activated in response to several PAMPs and DAMPs, including but not limited to nucleic acids, LPS, lipooligosaccharide (LOS), MDP, ATP, uric acid crystals, hyaluronan sulfate, heparan sulfate, β-amyloid, asbestos, and silica. NLRP3 inflammasome formation is a two-signal process. The first signal involves priming: LPS engagement of TLR4 leads to NF-κB activation, causing increased expression of NLRP3 and IL-1β (step 1). NLRP3 forms a multiprotein inflammasome complex with the adaptor ASC and procaspase-1. NLRP3 and ASC undergo deubiquitination prior to inflammasome assembly. After priming, canonical inflammasome activation requires a second signal. The second signal may be the release into the cytoplasm of mitochondrial factors such as ROS, mitochondrial DNA (mtDNA), or cardiolipin (step 2), potassium efflux (step 3), or lysosomal cathepsin release (step 4). After receiving the second signal, NLRP3 recruits ASC via pyrin-pyrin interactions. ASC utilizes its CARD domain to recruit procaspase-1 by CARD-CARD interactions, thus leading to processing of procaspase-1 to active caspase-1 (step 5). In turn, caspase-1 is critical for the processing and release of IL-1β and IL-18.

NLRP3 is associated with autoinflammatory, metabolic, and autoimmune diseases (80–84). Autoinflammatory diseases will be discussed here. Autosomal dominant mutations in NLRP3 lead to three CAPS autoinflammatory syndromes in humans, ranging from the mild FCAS to the intermediate MWS and the more severe NOMID/CINCA syndromes. Fever, urticaria-like rash, and varying degrees of arthropathy and neurological manifestations are present in all three syndromes (43, 85–87). FCAS consists of the mildest symptoms, including cold-induced urticaria and mild arthralgia. MWS is intermediate, with non-cold-induced spontaneous urticaria, sensorineural hearing loss, arthralgia, and in some cases renal amyloidosis. CINCA is the most severe, with spontaneous urticaria, deforming arthropathy, sensorineural hearing loss, and chronic aseptic meningitis. All CAPS are characterized by increased levels of IL-1β in the absence of infection and can be successfully treated with inhibitors of IL-1β (88–90). Gain-of-function mutations of *NLRP3* enhance IL-1β secretion even in the absence of a stimulus *in vitro* (15).

Major advances regarding NLRP3 inflammasome formation have been gleaned from elegant biochemical and cryo-electron microscopy (cryo-EM) studies. Biochemical studies showed that the pyrin domain is an evolutionarily conserved structure that can cause ASC to form a prion-like filament, which then activates downstream effector caspases. Cryo-EM (91) results similarly support a model in which activated NLRP3 forms an oligomeric platform where the pyrin domain nucleates ASC via the latter's pyrin domain to form a filamentous structure. The CARD domain of ASC then interacts with the CARD domain of caspase-1, causing proximal caspase-1 to undergo autocleavage (92). As is noted below, in addition to NLRP3, AIM2 inflammasome activation by DNA binding also undergoes the same process.

NLRC4 and *NAIP*

NLR family, CARD-containing 4 (NLRC4, also IPAF) is coupled to NAIP proteins (see below) that act as cytosolic receptors for PAMPs produced by flagellated pathogens, such as *Salmonella enterica* serovar Typhimurium (68, 93) and *Legionella pneumophila* (94); and nonflagellated pathogens, such as *Shigella flexneri* and *Pseudomonas aeruginosa* (95). NLRC4 forms a homooligomeric inflammasome with caspase-1 (68). The crystal structure of NLRC4 suggested that it remains in an autoinhibited state when unstimulated with ADP bound to its central NBD. Disruption of this ADP-NBD interaction leads to constitutive activation of NLRC4 (96).

NLRC4 is highly expressed in human brain, bone marrow, and the THP-1 human monocytic cell line (97). Initial characterization of NLRC4 in human tissues and cell lines demonstrated its direct association with the CARD domain of procaspase-1 through CARD-CARD interactions (97, 98). This interaction can cause autocatalytic processing of procaspase-1 to caspase-1 (97). A constitutively active NLRC4 causes autocatalytic processing of procaspase-1, generating caspase-1-dependent apoptosis in transfected cells (97). In macrophages, activation of the NLRC4 inflammasome by cytoplasmic flagellin leads to caspase-1 activation and IL-1β release (68, 93, 99). It is expected that NLRC4 interacts directly with procaspase-1 through CARD-CARD interactions. Although direct interaction of ASC with NLRC4 has not yet been demonstrated, *Asc*-deficient macrophages show defective caspase-1 activation and IL-1β release in response to *Salmonella*, *Shigella*, and *Pseudomonas* infections, indicating that ASC can enhance the function of NLRC4 (68, 95, 100).

NLR apoptosis-inhibitory proteins (NAIP, formerly called BIRC1 and NLRB1) represent prime examples of NLRs that recognize their cognate ligands, which promotes NLRC4 recruitment to form a multimeric inflammasome complex (101–103). NAIP is expressed in peripheral blood mononuclear cells and macrophages. While the human genome has one *NAIP* gene that is functionally similar to murine *Naip1* (101, 104), there are seven paralogs of NAIP in mice (Naip1 to -7), presumably to provide specificity of binding to a number of bacterial ligands.

Based on coimmunoprecipitation studies utilizing overexpressed NAIP and NLRC4, these two proteins were shown to associate, suggesting the potential for coengagement in the same caspase-1-activating inflammasome (105). NAIP5 inflammasome activation has been reported in response to the C terminus of flagellin after *L. pneumophila* infection (106). Transduction of macrophages with the C-terminal 35-amino-acid fragment of flagellin leads to NAIP5-dependent cell death, while full-length flagellin induces NAIP5-independent, NLRC4-dependent cell death and IL-1β release. Since NAIP5 does not have a caspase domain, it requires NLRC4 to activate procaspase-1. This suggests a mechanism for differential sensing of bacterial components whereby NAIP5 appears to possess NLRC4-dependent and -independent functions (106).

A recent study by Tenthorey et al. utilized a panel of chimeric NAIP molecules and identified the central NBD domain, rather than the expected LRR domain, to be associated with bacterial ligand binding (107). Moreover, ligand binding is essential for oligomerization of NAIP monomers into an inflammasome. In addition to NAIP5, NAIP6 can also recognize flagellin (102). In contrast, NAIP1 and -2 recognize bacterial type 3 secretory system needle protein and rod protein, respectively, but do not recognize flagellin. As a genetic test, mice with specific deletions in *Naip1* and *Naip2* provided biologic evidence for this specificity in mediating bacterial clearance (102, 108, 109).

Similar to NLRP3 and AIM2 (see below), cryo-EM has shed light on the assembly of the NLRC4/NAIP inflammasome. A contrasting model has emerged in which ligand (PrgJ)-bound NAIP is the initiating effector, but only one activated NAIP2 is necessary to cause the activation of NLRC4 protein, which undergoes a dramatic conformational change in the exposure of its catalytic surface to activate another NLRC4, eventually forming a wheel-like configuration. This wheel-like platform associates with caspase-1 to cause the autocatalytic cleavage of caspase-1 (110–112). In this model, ASC does not play a role since the CARD-

CARD domain interaction necessary for the final activation step is mediated through NLRC4 and caspase-1.

NAIP and NLRC4 also have roles that are distinct from the inflammasome and myeloid cells. For example, epithelium-intrinsic functions of this protein pair are associated with restriction of *Salmonella* proliferation in the gut epithelium, independent of inflammasome products (113). NAIP/NLRC4 is also known to inhibit caspase-3- and caspase-7-mediated pathways (114). NAIP interacts with procaspase-9 via its BIR3 domain. This association prevents the autoproteolysis of caspase-9 in the apoptosome complex, preventing caspase-9-mediated cell death (115). The NBD and BIR domains of NAIP are required for inhibition of procaspase-9 autoproteolysis. Human NAIP is involved in bacterial sensing and inducing pyroptosis in human macrophages and epithelial cells (116).

In 1995, two groups reported deletion mutants of the *Naip* gene in patients with spinal muscular atrophy (117, 118). Since then, multiple reports in a number of countries have shown that deletions in *Naip* represent one of the most frequent and consistent genomic changes associated with spinal muscular dystrophy and positively correlate with the clinical severity of this disease (119). The molecular basis for this disease association remains to be elucidated. More recently, a mutation in the NBD domain in NLRC4 was described that causes a gain-of-function phenotype that increases inflammasome activation and recurrent macrophage activation syndrome (120).

NLRP6

NLRP6 (formerly PYPAF5) plays a role in impeding clearance of both Gram-positive and -negative bacterial infections. Overexpression of ASC with NLRP6 leads to enhanced caspase-1 activation (121), suggesting that it serves an inflammasome function. Structurally, NLRP6 resembles NLRP3, with an N-terminal pyrin domain, a central NBD domain, and C-terminal LRRs. NLRP6 is expressed in myeloid cells such as granulocytes, dendritic cells, and macrophages; is found in CD4 and CD8 T cells (122–126); and is activated during development by peroxisome proliferator-activated receptor γ in intestinal epithelium (127).

The majority of studies have found that NLRP6 protects against experimental colitis and colitis-associated tumorigenesis (125, 128) and its function in monocytes contributes to this protective outcome (129). However, divergent mechanisms have been proposed to account for this activity. An analysis of the association between NLRP6 and the microbiome suggests that *Nlrp6* deletion in mice causes dysbiosis attributed to reduced IL-18, which then causes expanded pathobiont bacteria *Bacteroides* (*Prevotellaceae*) and TM7 microbiota (128). Reciprocally, the presence of the microbiota enhances caspase-1 maturation that is dependent on NLRP6 control (130). However, other reports indicate alternative roles for NLRP6. It has been implicated in the control of intestinal epithelium renewal (126) to preserve the epithelial barrier as a checkpoint regulator of NF-κB/MAPK pathways (124) or as a positive regulator of IFN-induced responses (i.e., IFN-stimulated gene factors) during viral infection (131). In addition, *NLRP6* deficiency has been found to cause defective autophagy in goblet cells and reduced mucus secretion, thus impeding pathogen clearance (132). A separate report also showed a similar dependency of goblet cell mucus secretion on NLRP6 and caspase-1/11, but not on IL-1/IL-18 (133). Precisely how caspase-1/11 are involved is unclear.

NLRC5, *NLRP7*, and *NLRP12*

NLRC5, NLRP7, and NLRP12 represent three NLRs that have other functions, but also have been reported to mediate inflammasome function. While the reported function of NLRC5 is to regulate class I major histocompatibility complex gene transcription, studies in human macrophage lines or primary monocytic cells indicated that it also mediates inflammasome activation in a similar fashion as NLRP3 and associates with NLRP3 (7). Similarly, rhinovirus induces NLRP3- and NLRC5-dependent inflammasome activation in bronchial cells (134). In *Nlrc5*-deficient mice, Nlrp3 inflammasome is partially impaired, hinting at the intersection of these two factors (135).

NLRP7 consists of an N-terminal pyrin domain, followed by an NBD domain and C-terminal LRRs. Structural analysis of the pyrin domain from NLRP7 indicated that it possesses a six-α-helix bundle death domain fold and forms a strong hydrophobic cluster upon pyrin-pyrin interactions (136). Khare et al. demonstrated the formation of an NLRP7-containing inflammasome in response to microbial lipopeptides in human macrophages (137). Activation of NLRP7 promoted ASC-dependent caspase-1 activation, IL-1β and IL-18 maturation, and restriction of intracellular bacterial replication, but not caspase-1-independent secretion of the proinflammatory cytokines IL-6 and tumor necrosis factor α. Radian et al. utilized the THP-1 monocytic cell line expressing a mutated Walker A motif to show defective NLRP7 inflammasome activation, thus suggesting that the NBD of NLRP7 is responsible for ATP binding and ATPase activity (138). This mutant cell line also showed defective IL-1β release and pyroptosis in response to acylated lipopeptides and *S. aureus* infection.

NLRP7 also has alternate functions outside of myeloid cells and is highly expressed in metaphase I and II oocytes. Mutations in the maternal gene *Nlrp7* are associated with biparental hydatidiform mole (HYDM1) in a number of patient cohorts, which is characterized by abnormal growth of the placenta and lack of proper embryonic development (139). To identify the molecular mechanism associated with HYDM1, Singer et al. utilized a yeast two-hybrid screen against an ovarian library with NLRP7 as the bait. This approach led to the identification of the transcriptional repressor ZBTB16 as an interacting protein of NLRP7 (140). This interaction was further verified in mammalian cells by immunoprecipitation and confocal microscopy; however, a clear mechanism for the molecular events leading to HYDM1 remains unknown.

NLRP12 (formerly called Monarch, PYPAF7, and CLR19.3) was one of the first NLRs reported to be a negative regulator of inflammation via suppression of NF-κB signaling (141, 142). NLRP12 protein consists of an N-terminal pyrin domain, a central NBD, and a C-terminal domain composed of at least 12 LRRs (143). Initial studies of NLRP12 utilizing overexpression systems suggested that it forms an inflammasome with ASC (144, 145). Additional research points to selective activation of the NLRP12 inflammasome by malaria and *Yersinia* (146, 147). However, aside from inflammasome activation, NLRP12 has prominent functions associated with inhibition of cytokine and inflammatory responses. At one level, it induces proteasome-mediated degradation of NF-κB-inducing kinase, leading to the suppression of the noncanonical NF-κB pathway and reduced expression of p52-dependent genes *Ccr4*, *Cxcl12*, and *Cxcl13* (141). Two reports confirmed these data by using *Nlrp12*$^{-/-}$ mice in the azoxymethane chemically-induced colorectal cancer model with the inflammatory agent dextran sodium sulfate (AOM-DSS). These reports congruently found that ablation of NLRP12 increases NF-κB canonical and noncanonical pathways, increases extracellular signal-related kinase phosphorylation in innate immune cells as well as in nonhematopoietic cells in the tumor model, and enhances proinflammatory cytokines and chemokines typically known to promote tumorigenesis (148, 149). NLRP12 also attenuates host response to *Salmonella* (150); however, this activity may be pathogen specific, in that *Nlrp12*$^{-/-}$ mice exhibited normal host response to other bacteria (151).

Mutations in *Nlrp12* lead to Guadeloupe variant periodic fever syndrome. In this syndrome, the following alterations were identified: two missense mutations within the *Nlrp12* gene, a nonsense mutation causing truncation within the NBD domain of the protein, and a deletion mutation leading to loss of the C-terminal LRRs. Both missense mutations caused reduced activity in the suppression of NF-κB signaling by NLRP12, while the NBD mutation caused a more significant impact on normal NLRP12-induced NF-κB signaling as compared to the LRR mutation. Since the symptoms are similar to FCAS, this syndrome is referred to as FCAS2. Individuals with this syndrome present with cold-induced heterogeneous symptoms including fever, arthralgia, myalgia, sensorineural hearing loss, aphthous ulcers, and lymphadenopathy (152).

AIM2

In addition to inflammasome NLRs, AIM2 (absent in melanoma 2) is a DNA sensor that activates the inflammasome (153–156). AIM2, an interferon-inducible gene also known as PYHIN4, was identified while screening tumor suppressor genes associated with melanoma (157). The AIM2 protein consists of an N-terminal pyrin domain, mediating homotypic interactions with ASC, and a C-terminal HIN-200 domain for DNA binding. AIM2 can associate into ASC specks to form a novel inflammasome platform inducing activation of ASC-mediated apoptotic and pyroptotic cell death pathways during host response to bacterial DNA such as from *Francisella tularensis* (153, 155, 157–159).

The crystal structure characterization of AIM2 has provided insight into interactions important for AIM2 autoinhibition and inflammasome assembly (160). AIM2 inflammasome-mediated and nonclassical IL-1β secretion induced by LC-3 autophagy are linked via the microtubule-associated protein EB-1 (161). AIM2 provides host defense against both cytosolic bacterial and viral pathogens, such as *F. tularensis*, *L. monocytogenes*, and *Mycobacterium tuberculosis* (162–164). AIM2 also contributes to inflammation in response to bacterial infection in the brain (165). Alternatively, investigation has also shown AIM2 to impede cell survival pathways that promote tumor growth (166–168). In the case of colorectal cancer, reduction or lack of expression of AIM2 is positively correlated with poor outcome (166). Two recent papers indicate that this role of AIM2 is inflammasome independent and is due to the negative regulation by AIM2 of proliferative signals such as Akt and c-myc signaling (166, 168). AIM2 is also protective in the case of breast cancer, where it prevents MCF-7 breast cancer cell growth *in vitro* and tumor growth *in vivo* (169).

NONCANONICAL INFLAMMASOMES

Aside from the activation of caspase-1, which is referred to as the canonical inflammasome pathway, a noncanonical pathway leading to caspase-11 maturation was first described by Kayagaki et al., who showed that this process is dependent on NLRP3 and ASC (170). Later, caspase-11 was shown to be activated by cytosolic LPS derived from Gram-negative bacteria that reside in the cytosol, thereby engaging in the protection of mice against infection by other LPS-producing bacteria (171, 172). Thus, while TLR4 mediates host response to extracellular LPS, the NLRP3-dependent caspase-11 pathway mediates host response to cytosolic LPS. The cytosolic presence of LPS is a crucial step, as *Salmonella* that reside in a vacuole do not elicit a caspase-11 response. In turn, the expression of caspase-11 is activated by STAT1 downstream of type I or type II IFN. Others showed that caspase-11 is an intracellular receptor of LPS (173). However, caspase-11 also binds to endogenous ligand-oxidized phospholipids to elicit inflammasome-dependent activities (174). More recently, a caspase-11 substrate, gasdermin D, was identified by differential genetic screening strategies to be important as an effector of pyroptosis and NLRP3-dependent inflammasome activation (175, 176).

In addition to the abovementioned noncanonical pathway, an unconventional one-step pathway of inflammasome activation exists in human monocytes in response to LPS alone. This pathway requires Syk activity and Ca^{2+} flux mediated by internalization of the CD14/TLR4 complex. Moreover, caspase-4 and caspase-5 have been shown to mediate IL-1α and IL-1β release from human monocytes after LPS stimulation (177).

INFLAMMASOME NLRs IN CANCERS

In addition to the disease associations described above, the association of chronic inflammation with cancer is well established, with chronic inflammation contributing to a tumor-promoting microenvironment. The following is not intended to be an exhaustive review of the field, but rather is presented to highlight the studies of NLRs and their roles in cancer. For example, the NLRP3 inflammasome remains the most investigated inflammasome with regard to cancer. Several groups have demonstrated the susceptibility of *Nlrp3*- and *Casp1*-deficient mice to DSS-induced colitis in a model of human ulcerative colitis. Defective inflammasome activation leads to loss of epithelial integrity, enhances leukocyte infiltration, and increases chemokine expression in *Nlrp3*$^{-/-}$ and *Casp1*$^{-/-}$ mice, leading to increased mortality (178). These results were supported by Zaki et al., who showed that NLRP3 inflammasome functions as a negative regulator of tumorigenesis during colitis-associated cancer, with NLRP3 inflammasome-dependent IL-18 production protecting against colorectal tumorigenesis (179). Another group showed that the suppressive impact of NLRP3 on colorectal cancer could be attributed to its enhancement of IL-18, which then activates natural killer cells (180). However, the protective or aggravating function of NLRP3 may be dependent on the severity of disease or factors in the local environment. The NLRP3 inflammasome also appears to play a central role in the pathology of melanomas, gastric cancer, and hepatocellular carcinoma (181). The NLRP3 inflammasome is constitutively expressed in human melanoma cells (182, 183). In the case of gastric cancer, *Mycoplasma hyorhinis* was shown to promote tumor development via NLRP3 inflammasome activation (184). Interestingly, IL-1β, but not IL-18, released from macrophages treated with *M. hyorhinis* promotes cell migration and invasion to exacerbate gastric cancer. The expression of NLRP3 inflammasome is downregulated in hepatic parenchymal cells in hepatocellular carcinoma (185). Loss of NLRP3 inflammasome activation positively correlates with a higher pathological grade in hepatocellular carcinogenesis.

NLRP3 inflammasome activation has also been implicated in adaptive immune responses to cancer vaccines (186). The NLRP3 inflammasome is activated during chemotherapy. Dying tumor cells release ATP, which is sensed by the $P2X_7$ receptors of dendritic cells, leading to NLRP3 inflammasome activation. However, when antitumor responses elicited by dendritic cell vaccination were tested, NLRP3 expression was found to be upregulated in tumor-associated myeloid-derived suppressor cells, thus suppressing antitumor response (187). *Nlrp3*$^{-/-}$ mice have fewer myeloid-derived suppressor cells accumulating at the tumor site and increased survival upon dendritic cell vaccination. Since this research focused on different adaptive immune cell populations, the differences may be attributable to differences in tumor cell types, in vaccine formulation, or in stimuli for NLRP3 inflammasome activation. Increased IL-1β secretion in the tumor microenvironment has also been linked to promotion of inflammation, early angiogenic response, as well as tumor induction and progression (188, 189).

Additional evidence has linked other inflammasomes to cancer. For instance, Hu et al. showed that regulation of inflammation-induced tumorigenesis is mediated by NLRC4 and caspase-1 (190). In the AOM-DSS inflammation-induced colorectal cancer model, *Casp1*$^{-/-}$ and *Nlrc4*$^{-/-}$ mice exhibited increased tumor load and number per mice. Caspase-1 and NLRC4 are relatively highly expressed in both colonic epithelial cells and CD45^{+} hematopoietic cells in the colon. In contrast, NLRP3 expression is primarily restricted to the hematopoietic compartment. These results concluded that an intrinsic epithelial cell effect exacerbates tumorigenesis in the absence of caspase-1 or NLRC4 activity.

Multiple associations of NAIP with cancers have been noted. NAIP expression is significantly elevated in malignantly transformed oral squamous cell carcinomas (191). The *Naip* allele is methylated in normal oral mucosa tissues. NAIP expression is increased in breast cancer (192) and is associated with an unfavorable prognosis. In the case of prostate cancer, several inhibitor of apoptosis protein members, including NAIP, are increased (193).

Linkage of NLRP6 to cancer has been derived from the study of mouse models of colon cancer. For example, Chen et al. established an association of NLRP6 with a colon cancer model, when they showed that *Nlrp6*-deficient mice are more susceptible to DSS-induced colitis and colitis-associated colon tumorigenesis as compared to wild-type controls (125). NLRP6 controls epithelial self-renewal and colorectal carcinogenesis upon injury due to DSS.

Lastly, studies of the adaptor ASC have demonstrated a role in various types of cancers. ASC is overexpressed in several tumors, triggering apoptosis and formation of ASC specks. Studies have shown methylation-associated silencing of ASC across many cancer types (194). However, the mechanisms underlying regulation of ASC silencing or overexpression remain largely undetermined. ASC is inactivated in almost 40% of breast cancers (195). Yokoyama et al. supported these findings by showing ASC methylation present in colorectal cancer tissues (196). Histone deacetylation of the ASC gene is also seen in ovarian cancer. Additionally, aberrant methylation and inactivation of ASC has been seen in glioblastoma, prostate cancer, lung cancer, hepatocellular carcinoma, and melanoma (197–199). Liu et al. identified a dual role of ASC in human melanoma tumorigenesis, with ASC expression in metastatic melanoma downregulated as compared to levels in primary melanoma (200). This reveals a complex role played by ASC in regulating cell proliferation. ASC may act as a potential modulator of inflammatory responses by coordinating the activity of NLRs and cytokine-activating caspases in mammalian cells.

INFLAMMASOME NLRs IN OTHER DISEASES

Above, we discussed the genetic and expression correlations of specific inflammasome genes to various diseases. Here, we intend to highlight where the inflammasome is implicated in other autoimmune and inflammatory diseases. In the case of Alzheimer's disease, caspase-1 expression was elevated in brain samples from Alzheimer's patients as well as from mice carrying mutations associated with familial Alzheimer's disease. Interestingly, mice lacking *Nlrp3* or *Casp1* showed less inflammasome activation and more protection from poor clinical outcomes associated with neuroinflammatory disease (201). It was found that phagocytosis of the β-amyloid protein by human microglia can activate the NLRP3 inflammasome and cause IL-1β release (53). This NLRP3 activation appears to be stimulated with lysosomal destabilization and subsequent release of cathepsin B caused by β-amyloid phagocytosis. The NLRP3 inflammasome also plays a significant role in the autoimmune demyelinating disease model of multiple sclerosis, in which experimental *Nlrp3*$^{-/-}$, *Casp1*$^{-/-}$, and *Il18*$^{-/-}$ mice displayed delayed demyelination (202, 203). Moreover, the efficacy of IFN-β in an experimental autoimmune encephalomyelitis model of multiple sclerosis was dependent on NLRP3 activity (204). Lastly, observations point to a link between the inflammasome and the autoimmune disease systemic lupus erythematosus (SLE). Leukocytes from SLE patients have increased AIM2 expression, even though there is no direct correlation between AIM2 expression and SLE disease activity (205). Similarly, in a mouse model of lupus, both IL-1β and IL-18 are important for disease progression, suggesting a possible inflammasome link (206).

CONCLUSIONS

The inflammasome has been a robust field of intensive investigation, uncovering significant revelations about an important family of regulators of health and disease. However, multiple questions remain unaddressed: the identity of ligands for several NLRs, the mechanism(s) of ligand binding, the specific signaling pathways for regulation, and cell-specific regulation of function in normal as well as in diseased hosts. Moreover, functions of NLRs beyond their roles in immunity remain largely unexplored. Discoveries emerging from investigating NLR biology promise to provide key insights into key pathways regulating immunity, inflammation, and homeostasis.

Acknowledgments. *We gratefully acknowledge the support of National Institutes of Health funding (U19-AI109965 and U19-AI067798) to J.P.-Y.T. and W.J.B. S.J.'s laboratory is* *funded by grants from the Department of Science and Technology (Young scientist scheme, SB/YS/LS-282/2013) and Board of Research in Nuclear Sciences (2013/36/72-BRNS/2415), Government of India. The software application Science Slides (VisiScience) was used to generate parts of figures.*

Citation. Jha S, Brickey WJ, Ting JP-Y. 2017. Inflammasomes in myeloid cells: warriors within. Microbiol Spectrum 5(1): MCHD-0049-2016.

References

1. **Harton JA, Linhoff MW, Zhang J, Ting JP.** 2002. Cutting edge: CATERPILLER: a large family of mammalian genes containing CARD, pyrin, nucleotide-binding, and leucine-rich repeat domains. *J Immunol* **169:**4088–4093.
2. **Inohara N, Nuñez G.** 2003. NODs: intracellular proteins involved in inflammation and apoptosis. *Nat Rev Immunol* **3:**371–382.
3. **Tschopp J, Martinon F, Burns K.** 2003. NALPs: a novel protein family involved in inflammation. *Nat Rev Mol Cell Biol* **4:**95–104.
4. **Davis BK, Wen H, Ting JP.** 2010. The inflammasome NLRs in immunity, inflammation, and associated diseases. *Annu Rev Immunol* **29:**707–735.
5. **Ting JP, Davis BK.** 2005. CATERPILLER: a novel gene family important in immunity, cell death, and diseases. *Annu Rev Immunol* **23:**387–414.
6. **Ausubel FM.** 2005. Are innate immune signaling pathways in plants and animals conserved? *Nat Immunol* **6:** 973–979.
7. **Davis BK, Roberts RA, Huang MT, Willingham SB, Conti BJ, Brickey WJ, Barker BR, Kwan M, Taxman DJ, Accavitti-Loper MA, Duncan JA, Ting JP.** 2011. Cutting edge: NLRC5-dependent activation of the inflammasome. *J Immunol* **186:**1333–1337.
8. **Kummer JA, Broekhuizen R, Everett H, Agostini L, Kuijk L, Martinon F, van Bruggen R, Tschopp J.** 2007. Inflammasome components NALP 1 and 3 show distinct but separate expression profiles in human tissues suggesting a site-specific role in the inflammatory response. *J Histochem Cytochem* **55:**443–452.
9. **Allen IC.** 2014. Non-inflammasome forming NLRs in inflammation and tumorigenesis. *Front Immunol* **5:**169. doi:10.3389/fimmu.2014.00169.
10. **Claes AK, Zhou JY, Philpott DJ.** 2015. NOD-like receptors: guardians of intestinal mucosal barriers. *Physiology (Bethesda)* **30:**241–250.
11. **Clay GM, Sutterwala FS, Wilson ME.** 2014. NLR proteins and parasitic disease. *Immunol Res* **59:**142–152.
12. **Hlaing T, Guo RF, Dilley KA, Loussia JM, Morrish TA, Shi MM, Vincenz C, Ward PA.** 2001. Molecular cloning and characterization of DEFCAP-L and -S, two isoforms of a novel member of the mammalian Ced-4 family of apoptosis proteins. *J Biol Chem* **276:**9230–9238.
13. **Chu ZL, Pio F, Xie Z, Welsh K, Krajewska M, Krajewski S, Godzik A, Reed JC.** 2001. A novel enhancer of the Apaf1 apoptosome involved in cytochrome *c*-dependent caspase activation and apoptosis. *J Biol Chem* **276:** 9239–9245.
14. **Yuan JY, Horvitz HR.** 1990. The *Caenorhabditis elegans* genes *ced-3* and *ced-4* act cell autonomously to cause programmed cell death. *Dev Biol* **138:**33–41.
15. **Martinon F, Burns K, Tschopp J.** 2002. The inflammasome: a molecular platform triggering activation of inflammatory caspases and processing of proIL-β. *Mol Cell* **10:**417–426.
16. **Faustin B, Lartigue L, Bruey JM, Luciano F, Sergienko E, Bailly-Maitre B, Volkmann N, Hanein D, Rouiller I, Reed JC.** 2007. Reconstituted NALP1 inflammasome reveals two-step mechanism of caspase-1 activation. *Mol Cell* **25:**713–724.
17. **Ting JP, Lovering RC, Alnemri ES, Bertin J, Boss JM, Davis BK, Flavell RA, Girardin SE, Godzik A, Harton JA, Hoffman HM, Hugot JP, Inohara N, Mackenzie A, Maltais LJ, Nunez G, Ogura Y, Otten LA, Philpott D, Reed JC, Reith W, Schreiber S, Steimle V, Ward PA.** 2008. The NLR gene family: a standard nomenclature. *Immunity* **28:**285–287.
18. **Sastalla I, Crown D, Masters SL, McKenzie A, Leppla SH, Moayeri M.** 2013. Transcriptional analysis of the three Nlrp1 paralogs in mice. *BMC Genomics* **14:**188. doi:10.1186/1471-2164-14-188.
19. **Dwivedi M, Laddha NC, Mansuri MS, Marfatia YS, Begum R.** 2013. Association of *NLRP1* genetic variants and mRNA overexpression with generalized vitiligo and disease activity in a Gujarat population. *Br J Dermatol* **169:**1114–1125.
20. **Jin Y, Mailloux CM, Gowan K, Riccardi SL, LaBerge G, Bennett DC, Fain PR, Spritz RA.** 2007. *NALP1* in vitiligo-associated multiple autoimmune disease. *N Engl J Med* **356:**1216–1225.
21. **Levandowski CB, Mailloux CM, Ferrara TM, Gowan K, Ben S, Jin Y, McFann KK, Holland PJ, Fain PR, Dinarello CA, Spritz RA.** 2013. *NLRP1* haplotypes associated with vitiligo and autoimmunity increase interleukin-1β processing via the NLRP1 inflammasome. *Proc Natl Acad Sci U S A* **110:**2952–2956.
22. **Magitta NF, Bøe Wolff AS, Johansson S, Skinningsrud B, Lie BA, Myhr KM, Undlien DE, Joner G, Njølstad PR, Kvien TK, Førre Ø, Knappskog PM, Husebye ES.** 2009. A coding polymorphism in NALP1 confers risk for autoimmune Addison's disease and type 1 diabetes. *Genes Immun* **10:**120–124.
23. **Newman ZL, Printz MP, Liu S, Crown D, Breen L, Miller-Randolph S, Flodman P, Leppla SH, Moayeri M.** 2010. Susceptibility to anthrax lethal toxin-induced rat death is controlled by a single chromosome 10 locus that includes *rNlrp1*. *PLoS Pathog* **6:**e1000906. doi: 10.1371/journal.ppat.1000906.
24. **Terra JK, Cote CK, France B, Jenkins AL, Bozue JA, Welkos SL, LeVine SM, Bradley KA.** 2010. Cutting edge: resistance to *Bacillus anthracis* infection mediated by a lethal toxin sensitive allele of *Nalp1b/Nlrp1b*. *J Immunol* **184:**17–20.
25. **Hsu LC, Ali SR, McGillivray S, Tseng PH, Mariathasan S, Humke EW, Eckmann L, Powell JJ, Nizet V, Dixit VM, Karin M.** 2008. A NOD2-NALP1 complex mediates caspase-1-dependent IL-1β secretion in response

to *Bacillus anthracis* infection and muramyl dipeptide. *Proc Natl Acad Sci U S A* **105**:7803–7808.

26. **Chavarría-Smith J, Vance RE.** 2013. Direct proteolytic cleavage of NLRP1B is necessary and sufficient for inflammasome activation by anthrax lethal factor. *PLoS Pathog* **9**:e1003452. doi:10.1371/journal.ppat.1003452.
27. **Levinsohn JL, Newman ZL, Hellmich KA, Fattah R, Getz MA, Liu S, Sastalla I, Leppla SH, Moayeri M.** 2012. Anthrax lethal factor cleavage of Nlrp1 is required for activation of the inflammasome. *PLoS Pathog* **8**: e1002638. doi:10.1371/journal.ppat.1002638.
28. **Hellmich KA, Levinsohn JL, Fattah R, Newman ZL, Maier N, Sastalla I, Liu S, Leppla SH, Moayeri M.** 2012. Anthrax lethal factor cleaves mouse Nlrp1b in both toxin-sensitive and toxin-resistant macrophages. *PLoS One* **7**:e49741. doi:10.1371/journal.pone.0049741.
29. **Kovarova M, Hesker PR, Jania L, Nguyen M, Snouwaert JN, Xiang Z, Lommatzsch SE, Huang MT, Ting JP, Koller BH.** 2012. NLRP1-dependent pyroptosis leads to acute lung injury and morbidity in mice. *J Immunol* **189**:2006–2016.
30. **Reubold TF, Hahne G, Wohlgemuth S, Eschenburg S.** 2014. Crystal structure of the leucine-rich repeat domain of the NOD-like receptor NLRP1: implications for binding of muramyl dipeptide. *FEBS Lett* **588**:3327–3332.
31. **Witola WH, Mui E, Hargrave A, Liu S, Hypolite M, Montpetit A, Cavailles P, Bisanz C, Cesbron-Delauw MF, Fournié GJ, McLeod R.** 2011. NALP1 influences susceptibility to human congenital toxoplasmosis, proinflammatory cytokine response, and fate of *Toxoplasma gondii*-infected monocytic cells. *Infect Immun* **79**: 756–766.
32. **Ewald SE, Chavarria-Smith J, Boothroyd JC.** 2014. NLRP1 is an inflammasome sensor for *Toxoplasma gondii*. *Infect Immun* **82**:460–468.
33. **Gorfu G, Cirelli KM, Melo MB, Mayer-Barber K, Crown D, Koller BH, Masters S, Sher A, Leppla SH, Moayeri M, Saeij JP, Grigg ME.** 2014. Dual role for inflammasome sensors NLRP1 and NLRP3 in murine resistance to *Toxoplasma gondii*. *mBio* **5**:e01117-13. doi:10.1128/mBio.01117-13.
34. **Masters SL, Gerlic M, Metcalf D, Preston S, Pellegrini M, O'Donnell JA, McArthur K, Baldwin TM, Chevrier S, Nowell CJ, Cengia LH, Henley KJ, Collinge JE, Kastner DL, Feigenbaum L, Hilton DJ, Alexander WS, Kile BT, Croker BA.** 2012. NLRP1 inflammasome activation induces pyroptosis of hematopoietic progenitor cells. *Immunity* **37**:1009–1023.
35. **Chi W, Li F, Chen H, Wang Y, Zhu Y, Yang X, Zhu J, Wu F, Ouyang H, Ge J, Weinreb RN, Zhang K, Zhuo Y.** 2014. Caspase-8 promotes NLRP1/NLRP3 inflammasome activation and IL-1β production in acute glaucoma. *Proc Natl Acad Sci U S A* **111**:11181–11186.
36. **de Rivero Vaccari JP, Lotocki G, Alonso OF, Bramlett HM, Dietrich WD, Keane RW.** 2009. Therapeutic neutralization of the NLRP1 inflammasome reduces the innate immune response and improves histopathology after traumatic brain injury. *J Cereb Blood Flow Metab* **29**:1251–1261.
37. **Williams TM, Leeth RA, Rothschild DE, Coutermarsh-Ott SL, McDaniel DK, Simmons AE, Heid B, Cecere TE, Allen IC.** 2015. The NLRP1 inflammasome attenuates colitis and colitis-associated tumorigenesis. *J Immunol* **194**:3369–3380.
38. **Murphy AJ, Kraakman MJ, Kammoun HL, Dragoljevic D, Lee MK, Lawlor KE, Wentworth JM, Vasanthakumar A, Gerlic M, Whitehead LW, DiRago L, Cengia L, Lane RM, Metcalf D, Vince JE, Harrison LC, Kallies A, Kile BT, Croker BA, Febbraio MA, Masters SL.** 2016. IL-18 production from the NLRP1 inflammasome prevents obesity and metabolic syndrome. *Cell Metab* **23**:155–164.
39. **de Rivero Vaccari JP, Lotocki G, Marcillo AE, Dietrich WD, Keane RW.** 2008. A molecular platform in neurons regulates inflammation after spinal cord injury. *J Neurosci* **28**:3404–3414.
40. **de Rivero Vaccari JP, Lotocki G, Alonso OF, Bramlett HM, Dietrich WD, Keane RW.** 2009. Therapeutic neutralization of the NLRP1 inflammasome reduces the innate immune response and improves histopathology after traumatic brain injury. *J Cereb Blood Flow Metab* **29**:1251–1261.
41. **Franklin BS, Bossaller L, De Nardo D, Ratter JM, Stutz A, Engels G, Brenker C, Nordhoff M, Mirandola SR, Al-Amoudi A, Mangan MS, Zimmer S, Monks BG, Fricke M, Schmidt RE, Espevik T, Jones B, Jarnicki AG, Hansbro PM, Busto P, Marshak-Rothstein A, Hornemann S, Aguzzi A, Kastenmüller W, Latz E.** 2014. The adaptor ASC has extracellular and 'prionoid' activities that propagate inflammation. *Nat Immunol* **15**:727–737.
42. **Hoffman HM, Mueller JL, Broide DH, Wanderer AA, Kolodner RD.** 2001. Mutation of a new gene encoding a putative pyrin-like protein causes familial cold autoinflammatory syndrome and Muckle-Wells syndrome. *Nat Genet* **29**:301–305.
43. **Aksentijevich I, Nowak M, Mallah M, Chae JJ, Watford WT, Hofmann SR, Stein L, Russo R, Goldsmith D, Dent P, Rosenberg HF, Austin F, Remmers EF, Balow JE Jr, Rosenzweig S, Komarow H, Shoham NG, Wood G, Jones J, Mangra N, Carrero H, Adams BS, Moore TL, Schikler K, Hoffman H, Lovell DJ, Lipnick R, Barron K, O'Shea JJ, Kastner DL, Goldbach-Mansky R.** 2002. De novo *CIAS1* mutations, cytokine activation, and evidence for genetic heterogeneity in patients with neonatal-onset multisystem inflammatory disease (NOMID): a new member of the expanding family of pyrin-associated autoinflammatory diseases. *Arthritis Rheum* **46**:3340–3348.
44. **Aganna E, Martinon F, Hawkins PN, Ross JB, Swan DC, Booth DR, Lachmann HJ, Bybee A, Gaudet R, Woo P, Feighery C, Cotter FE, Thome M, Hitman GA, Tschopp J, McDermott MF.** 2002. Association of mutations in the *NALP3/CIAS1/PYPAF1* gene with a broad phenotype including recurrent fever, cold sensitivity, sensorineural deafness, and AA amyloidosis. *Arthritis Rheum* **46**:2445–2452.
45. **Feldmann J, Prieur AM, Quartier P, Berquin P, Certain S, Cortis E, Teillac-Hamel D, Fischer A, de Saint Basile G.** 2002. Chronic infantile neurological cutaneous and articular syndrome is caused by mutations in *CIAS1*, a

gene highly expressed in polymorphonuclear cells and chondrocytes. *Am J Hum Genet* 71:198–203.

46. **Manji GA, Wang L, Geddes BJ, Brown M, Merriam S, Al-Garawi A, Mak S, Lora JM, Briskin M, Jurman M, Cao J, DiStefano PS, Bertin J.** 2002. PYPAF1, a PYRIN-containing Apaf1-like protein that assembles with ASC and regulates activation of NF-κB. *J Biol Chem* **277:** 11570–11575.

47. **Guarda G, Zenger M, Yazdi AS, Schroder K, Ferrero I, Menu P, Tardivel A, Mattmann C, Tschopp J.** 2011. Differential expression of NLRP3 among hematopoietic cells. *J Immunol* **186:**2529–2534.

48. **Kanneganti TD, Body-Malapel M, Amer A, Park JH, Whitfield J, Franchi L, Taraporewala ZF, Miller D, Patton JT, Inohara N, Núñez G.** 2006. Critical role for Cryopyrin/Nalp3 in activation of caspase-1 in response to viral infection and double-stranded RNA. *J Biol Chem* **281:**36560–36568.

49. **Kanneganti TD, Ozören N, Body-Malapel M, Amer A, Park JH, Franchi L, Whitfield J, Barchet W, Colonna M, Vandenabeele P, Bertin J, Coyle A, Grant EP, Akira S, Núñez G.** 2006. Bacterial RNA and small antiviral compounds activate caspase-1 through cryopyrin/Nalp3. *Nature* **440:**233–236.

50. **Mariathasan S, Weiss DS, Newton K, McBride J, O'Rourke K, Roose-Girma M, Lee WP, Weinrauch Y, Monack DM, Dixit VM.** 2006. Cryopyrin activates the inflammasome in response to toxins and ATP. *Nature* **440:**228–232.

51. **Martinon F, Pétrilli V, Mayor A, Tardivel A, Tschopp J.** 2006. Gout-associated uric acid crystals activate the NALP3 inflammasome. *Nature* **440:**237–241.

52. **Shi Y, Evans JE, Rock KL.** 2003. Molecular identification of a danger signal that alerts the immune system to dying cells. *Nature* **425:**516–521.

53. **Halle A, Hornung V, Petzold GC, Stewart CR, Monks BG, Reinheckel T, Fitzgerald KA, Latz E, Moore KJ, Golenbock DT.** 2008. The NALP3 inflammasome is involved in the innate immune response to amyloid-β. *Nat Immunol* **9:**857–865.

54. **Scheibner KA, Lutz MA, Boodoo S, Fenton MJ, Powell JD, Horton MR.** 2006. Hyaluronan fragments act as an endogenous danger signal by engaging TLR2. *J Immunol* **177:**1272–1281.

55. **Cassel SL, Eisenbarth SC, Iyer SS, Sadler JJ, Colegio OR, Tephly LA, Carter AB, Rothman PB, Flavell RA, Sutterwala FS.** 2008. The Nalp3 inflammasome is essential for the development of silicosis. *Proc Natl Acad Sci U S A* **105:**9035–9040.

56. **Dostert C, Pétrilli V, Van Bruggen R, Steele C, Mossman BT, Tschopp J.** 2008. Innate immune activation through Nalp3 inflammasome sensing of asbestos and silica. *Science* **320:**674–677.

57. **Hornung V, Bauernfeind F, Halle A, Samstad EO, Kono H, Rock KL, Fitzgerald KA, Latz E.** 2008. Silica crystals and aluminum salts activate the NALP3 inflammasome through phagosomal destabilization. *Nat Immunol* **9:** 847–856.

58. **Duewell P, Kono H, Rayner KJ, Sirois CM, Vladimer G, Bauernfeind FG, Abela GS, Franchi L, Nuñez G, Schnurr M, Espevik T, Lien E, Fitzgerald KA, Rock KL, Moore KJ, Wright SD, Hornung V, Latz E.** 2010. NLRP3 inflammasomes are required for atherogenesis and activated by cholesterol crystals. *Nature* **464:**1357–1361.

59. **Rajamäki K, Lappalainen J, Oörni K, Välimäki E, Matikainen S, Kovanen PT, Eklund KK.** 2010. Cholesterol crystals activate the NLRP3 inflammasome in human macrophages: a novel link between cholesterol metabolism and inflammation. *PLoS One* **5:**e11765. doi:10.1371/journal.pone.0011765.

60. **Li H, Willingham SB, Ting JP, Re F.** 2008. Cutting edge: inflammasome activation by alum and alum's adjuvant effect are mediated by NLRP3. *J Immunol* **181:**17–21.

61. **Masters SL, Dunne A, Subramanian SL, Hull RL, Tannahill GM, Sharp FA, Becker C, Franchi L, Yoshihara E, Chen Z, Mullooly N, Mielke LA, Harris J, Coll RC, Mills KH, Mok KH, Newsholme P, Nuñez G, Yodoi J, Kahn SE, Lavelle EC, O'Neill LA.** 2010. Activation of the NLRP3 inflammasome by islet amyloid polypeptide provides a mechanism for enhanced IL-1β in type 2 diabetes. *Nat Immunol* **11:**897–904.

62. **Vandanmagsar B, Youm YH, Ravussin A, Galgani JE, Stadler K, Mynatt RL, Ravussin E, Stephens JM, Dixit VD.** 2011. The NLRP3 inflammasome instigates obesity-induced inflammation and insulin resistance. *Nat Med* **17:**179–188.

63. **Wen H, Gris D, Lei Y, Jha S, Zhang L, Huang MT, Brickey WJ, Ting JP.** 2011. Fatty acid-induced NLRP3-ASC inflammasome activation interferes with insulin signaling. *Nat Immunol* **12:**408–415.

64. **Shio MT, Eisenbarth SC, Savaria M, Vinet AF, Bellemare MJ, Harder KW, Sutterwala FS, Bohle DS, Descoteaux A, Flavell RA, Olivier M.** 2009. Malarial hemozoin activates the NLRP3 inflammasome through Lyn and Syk kinases. *PLoS Pathog* **5:**e1000559. doi: 10.1371/journal.ppat.1000559.

65. **Abdul-Sater AA, Tattoli I, Jin L, Grajkowski A, Levi A, Koller BH, Allen IC, Beaucage SL, Fitzgerald KA, Ting JP, Cambier JC, Girardin SE, Schindler C.** 2013. Cyclic-di-GMP and cyclic-di-AMP activate the NLRP3 inflammasome. *EMBO Rep* **14:**900–906.

66. **Bauernfeind FG, Horvath G, Stutz A, Alnemri ES, MacDonald K, Speert D, Fernandes-Alnemri T, Wu J, Monks BG, Fitzgerald KA, Hornung V, Latz E.** 2009. Cutting edge: NF-κB activating pattern recognition and cytokine receptors license NLRP3 inflammasome activation by regulating NLRP3 expression. *J Immunol* **183:**787–791.

67. **Sutterwala FS, Ogura Y, Szczepanik M, Lara-Tejero M, Lichtenberger GS, Grant EP, Bertin J, Coyle AJ, Galán JE, Askenase PW, Flavell RA.** 2006. Critical role for NALP3/CIAS1/Cryopyrin in innate and adaptive immunity through its regulation of caspase-1. *Immunity* **24:** 317–327.

68. **Mariathasan S, Newton K, Monack DM, Vucic D, French DM, Lee WP, Roose-Girma M, Erickson S, Dixit VM.** 2004. Differential activation of the inflammasome by caspase-1 adaptors ASC and Ipaf. *Nature* **430:**213–218.

69. **Shao W, Yeretssian G, Doiron K, Hussain SN, Saleh M.** 2007. The caspase-1 digestome identifies the glycolysis pathway as a target during infection and septic shock. *J Biol Chem* **282:**36321–36329.

70. **Keller M, Rüegg A, Werner S, Beer HD.** 2008. Active caspase-1 is a regulator of unconventional protein secretion. *Cell* **132:**818–831.

71. **Li P, Allen H, Banerjee S, Seshadri T.** 1997. Characterization of mice deficient in interleukin-1β converting enzyme. *J Cell Biochem* **64:**27–32.

72. **Compan V, Baroja-Mazo A, López-Castejón G, Gomez AI, Martínez CM, Angosto D, Montero MT, Herranz AS, Bazán E, Reimers D, Mulero V, Pelegrín P.** 2012. Cell volume regulation modulates NLRP3 inflammasome activation. *Immunity* **37:**487–500.

73. **Lee GS, Subramanian N, Kim AI, Aksentijevich I, Goldbach-Mansky R, Sacks DB, Germain RN, Kastner DL, Chae JJ.** 2012. The calcium-sensing receptor regulates the NLRP3 inflammasome through $Ca2^+$ and cAMP. *Nature* **492:**123–127.

74. **Murakami T, Ockinger J, Yu J, Byles V, McColl A, Hofer AM, Horng T.** 2012. Critical role for calcium mobilization in activation of the NLRP3 inflammasome. *Proc Natl Acad Sci U S A* **109:**11282–11287.

75. **Zhong Z, Zhai Y, Liang S, Mori Y, Han R, Sutterwala FS, Qiao L.** 2013. TRPM2 links oxidative stress to NLRP3 inflammasome activation. *Nat Commun* **4:** 1611. doi:10.1038/ncomms2608.

76. **Schorn C, Frey B, Lauber K, Janko C, Strysio M, Keppeler H, Gaipl US, Voll RE, Springer E, Munoz LE, Schett G, Herrmann M.** 2011. Sodium overload and water influx activate the NALP3 inflammasome. *J Biol Chem* **286:**35–41.

77. **Cruz CM, Rinna A, Forman HJ, Ventura AL, Persechini PM, Ojcius DM.** 2007. ATP activates a reactive oxygen species-dependent oxidative stress response and secretion of proinflammatory cytokines in macrophages. *J Biol Chem* **282:**2871–2879.

78. **Shimada K, Crother TR, Karlin J, Dagvadorj J, Chiba N, Chen S, Ramanujan VK, Wolf AJ, Vergnes L, Ojcius DM, Rentsendorj A, Vargas M, Guerrero C, Wang Y, Fitzgerald KA, Underhill DM, Town T, Arditi M.** 2012. Oxidized mitochondrial DNA activates the NLRP3 inflammasome during apoptosis. *Immunity* **36:**401–414.

79. **Nakahira K, Haspel JA, Rathinam VAK, Lee S-J, Dolinay T, Lam HC, Englert JA, Rabinovitch M, Cernadas M, Kim HP, Fitzgerald KA, Ryter SW, Choi AMK.** 2011. Autophagy proteins regulate innate immune responses by inhibiting the release of mitochondrial DNA mediated by the NALP3 inflammasome. *Nat Immunol* **12:**222–230.

80. **Boschan C, Witt O, Lohse P, Foeldvari I, Zappel H, Schweigerer L.** 2006. Neonatal-onset multisystem inflammatory disease (NOMID) due to a novel S331R mutation of the *CIAS1* gene and response to interleukin-1 receptor antagonist treatment. *Am J Med Genet A* **140:**883–886.

81. **Koné-Paut I, Sanchez E, Le Quellec A, Manna R, Touitou I.** 2007. Autoinflammatory gene mutations in Behçet's disease. *Ann Rheum Dis* **66:**832–834.

82. **McDermott MF, Aksentijevich I.** 2002. The autoinflammatory syndromes. *Curr Opin Allergy Clin Immunol* **2:** 511–516.

83. **Robbins GR, Wen H, Ting JP.** 2014. Inflammasomes and metabolic disorders: old genes in modern diseases. *Mol Cell* **54:**297–308.

84. **Stojanov S, Kastner DL.** 2005. Familial autoinflammatory diseases: genetics, pathogenesis and treatment. *Curr Opin Rheumatol* **17:**586–599.

85. **Dodé C, Le Dû N, Cuisset L, Letourneur F, Berthelot JM, Vaudour G, Meyrier A, Watts RA, Scott DG, Nicholls A, Granel B, Frances C, Garcier F, Edery P, Boulinguez S, Domergues JP, Delpech M, Grateau G.** 2002. New mutations of *CIAS1* that are responsible for Muckle-Wells syndrome and familial cold urticaria: a novel mutation underlies both syndromes. *Am J Hum Genet* **70:**1498–1506.

86. **Hoffman HM, Gregory SG, Mueller JL, Tresierras M, Broide DH, Wanderer AA, Kolodner RD.** 2003. Fine structure mapping of *CIAS1*: identification of an ancestral haplotype and a common FCAS mutation, L353P. *Hum Genet* **112:**209–216.

87. **Ting JP, Kastner DL, Hoffman HM.** 2006. CATERPILLERs, pyrin and hereditary immunological disorders. *Nat Rev Immunol* **6:**183–195.

88. **Dinarello CA, van der Meer JW.** 2013. Treating inflammation by blocking interleukin-1 in humans. *Semin Immunol* **25:**469–484.

89. **Dinarello CA, Simon A, van der Meer JW.** 2012. Treating inflammation by blocking interleukin-1 in a broad spectrum of diseases. *Nat Rev Drug Discov* **11:** 633–652.

90. **Goldbach-Mansky R, Dailey NJ, Canna SW, Gelabert A, Jones J, Rubin BI, Kim HJ, Brewer C, Zalewski C, Wiggs E, Hill S, Turner ML, Karp BI, Aksentijevich I, Pucino F, Penzak SR, Haverkamp MH, Stein L, Adams BS, Moore TL, Fuhlbrigge RC, Shaham B, Jarvis JN, O'Neil K, Vehe RK, Beitz LO, Gardner G, Hannan WP, Warren RW, Horn W, Cole JL, Paul SM, Hawkins PN, Pham TH, Snyder C, Wesley RA, Hoffmann SC, Holland SM, Butman JA, Kastner DL.** 2006. Neonatal-onset multisystem inflammatory disease responsive to interleukin-1β inhibition. *N Engl J Med* 355:581–592.

91. **Cai X, Chen J, Xu H, Liu S, Jiang QX, Halfmann R, Chen ZJ.** 2014. Prion-like polymerization underlies signal transduction in antiviral immune defense and inflammasome activation. *Cell* **156:**1207–1222.

92. **Lu A, Magupalli VG, Ruan J, Yin Q, Atianand MK, Vos MR, Schröder GF, Fitzgerald KA, Wu H, Egelman EH.** 2014. Unified polymerization mechanism for the assembly of ASC-dependent inflammasomes. *Cell* **156:** 1193–1206.

93. **Miao EA, Alpuche-Aranda CM, Dors M, Clark AE, Bader MW, Miller SI, Aderem A.** 2006. Cytoplasmic flagellin activates caspase-1 and secretion of interleukin 1β via Ipaf. *Nat Immunol* **7:**569–575.

94. **Amer A, Franchi L, Kanneganti TD, Body-Malapel M, Özören N, Brady G, Meshinchi S, Jagirdar R, Gewirtz A, Akira S, Núñez G.** 2006. Regulation of *Legionella*

phagosome maturation and infection through flagellin and host Ipaf. *J Biol Chem* **281**:35217–35223.

95. **Suzuki T, Franchi L, Toma C, Ashida H, Ogawa M, Yoshikawa Y, Mimuro H, Inohara N, Sasakawa C, Nuñez G.** 2007. Differential regulation of caspase-1 activation, pyroptosis, and autophagy via Ipaf and ASC in *Shigella*-infected macrophages. *PLoS Pathog* **3**:e111. doi:10.1371/journal.ppat.0030111.
96. **Hu Z, Yan C, Liu P, Huang Z, Ma R, Zhang C, Wang R, Zhang Y, Martinon F, Miao D, Deng H, Wang J, Chang J, Chai J.** 2013. Crystal structure of NLRC4 reveals its autoinhibition mechanism. *Science* **341**:172–175.
97. **Poyet JL, Srinivasula SM, Tnani M, Razmara M, Fernandes-Alnemri T, Alnemri ES.** 2001. Identification of Ipaf, a human caspase-1-activating protein related to Apaf-1. *J Biol Chem* **276**:28309–28313.
98. **Geddes BJ, Wang L, Huang WJ, Lavellee M, Manji GA, Brown M, Jurman M, Cao J, Morgenstern J, Merriam S, Glucksmann MA, DiStefano PS, Bertin J.** 2001. Human CARD12 is a novel CED4/Apaf-1 family member that induces apoptosis. *Biochem Biophys Res Commun* **284**:77–82.
99. **Franchi L, Amer A, Body-Malapel M, Kanneganti TD, Ozören N, Jagirdar R, Inohara N, Vandenabeele P, Bertin J, Coyle A, Grant EP, Núñez G.** 2006. Cytosolic flagellin requires Ipaf for activation of caspase-1 and interleukin 1β in salmonella-infected macrophages. *Nat Immunol* **7**:576–582.
100. **Sutterwala FS, Mijares LA, Li L, Ogura Y, Kazmierczak BI, Flavell RA.** 2007. Immune recognition of *Pseudomonas aeruginosa* mediated by the IPAF/NLRC4 inflammasome. *J Exp Med* **204**:3235–3245.
101. **Yang J, Zhao Y, Shi J, Shao F.** 2013. Human NAIP and mouse NAIP1 recognize bacterial type III secretion needle protein for inflammasome activation. *Proc Natl Acad Sci U S A* **110**:14408–14413.
102. **Kofoed EM, Vance RE.** 2011. Innate immune recognition of bacterial ligands by NAIPs determines inflammasome specificity. *Nature* **477**:592–595.
103. **Zhao Y, Yang J, Shi J, Gong YN, Lu Q, Xu H, Liu L, Shao F.** 2011. The NLRC4 inflammasome receptors for bacterial flagellin and type III secretion apparatus. *Nature* **477**:596–600.
104. **Rayamajhi M, Zak DE, Chavarria-Smith J, Vance RE, Miao EA.** 2013. Cutting edge: mouse NAIP1 detects the type III secretion system needle protein. *J Immunol* **191**: 3986–3989.
105. **Zamboni DS, Kobayashi KS, Kohlsdorf T, Ogura Y, Long EM, Vance RE, Kuida K, Mariathasan S, Dixit VM, Flavell RA, Dietrich WF, Roy CR.** 2006. The Birc1e cytosolic pattern-recognition receptor contributes to the detection and control of *Legionella pneumophila* infection. *Nat Immunol* **7**:318–325.
106. **Lightfield KL, Persson J, Brubaker SW, Witte CE, von Moltke J, Dunipace EA, Henry T, Sun YH, Cado D, Dietrich WF, Monack DM, Tsolis RM, Vance RE.** 2008. Critical function for Naip5 in inflammasome activation by a conserved carboxy-terminal domain of flagellin. *Nat Immunol* **9**:1171–1178.
107. **Tenthorey JL, Kofoed EM, Daugherty MD, Malik HS, Vance RE.** 2014. Molecular basis for specific recognition of bacterial ligands by NAIP/NLRC4 inflammasomes. *Mol Cell* **54**:17–29.
108. **Rauch I, Tenthorey JL, Nichols RD, Al Moussawi K, Kang JJ, Kang C, Kazmierczak BI, Vance RE.** 2016. NAIP proteins are required for cytosolic detection of specific bacterial ligands in vivo. *J Exp Med* **213**: 657–665.
109. **Zhao Y, Shi J, Shi X, Wang Y, Wang F, Shao F.** 2016. Genetic functions of the NAIP family of inflammasome receptors for bacterial ligands in mice. *J Exp Med* **213**: 647–656.
110. **Diebolder CA, Halff EF, Koster AJ, Huizinga EG, Koning RI.** 2015. Cryoelectron tomography of the NAIP5/NLRC4 inflammasome: implications for NLR activation. *Structure* **23**:2349–2357.
111. **Hu Z, Zhou Q, Zhang C, Fan S, Cheng W, Zhao Y, Shao F, Wang HW, Sui SF, Chai J.** 2015. Structural and biochemical basis for induced self-propagation of NLRC4. *Science* **350**:399–404.
112. **Zhang L, Chen S, Ruan J, Wu J, Tong AB, Yin Q, Li Y, David L, Lu A, Wang WL, Marks C, Ouyang Q, Zhang X, Mao Y, Wu H.** 2015. Cryo-EM structure of the activated NAIP2-NLRC4 inflammasome reveals nucleated polymerization. *Science* **350**:404–409.
113. **Sellin ME, Müller AA, Felmy B, Dolowschiak T, Diard M, Tardivel A, Maslowski KM, Hardt WD.** 2014. Epithelium-intrinsic NAIP/NLRC4 inflammasome drives infected enterocyte expulsion to restrict *Salmonella* replication in the intestinal mucosa. *Cell Host Microbe* **16**: 237–248.
114. **Maier JK, Lahoua Z, Gendron NH, Fetni R, Johnston A, Davoodi J, Rasper D, Roy S, Slack RS, Nicholson DW, MacKenzie AE.** 2002. The neuronal apoptosis inhibitory protein is a direct inhibitor of caspases 3 and 7. *J Neurosci* **22**:2035–2043.
115. **Davoodi J, Ghahremani MH, Es-Haghi A, Mohammad-Gholi A, Mackenzie A.** 2010. Neuronal apoptosis inhibitory protein, NAIP, is an inhibitor of procaspase-9. *Int J Biochem Cell Biol* **42**:958–964.
116. **Vinzing M, Eitel J, Lippmann J, Hocke AC, Zahlten J, Slevogt H, N'guessan PD, Günther S, Schmeck B, Hippenstiel S, Flieger A, Suttorp N, Opitz B.** 2008. NAIP and Ipaf control *Legionella pneumophila* replication in human cells. *J Immunol* **180**:6808–6815.
117. **Roy N, Mahadevan MS, McLean M, Shutter G, Yaraghi Z, Farahani R, Baird S, Besner-Johnston A, Lefebvre C, Kang X, Salih M, Aubry H, Tamai K, Guan X, Ioannou P, Crawford TO, de Jong PJ, Surh L, Ikeda JE, Korneluk RG, MacKenzie A.** 1995. The gene for neuronal apoptosis inhibitory protein is partially deleted in individuals with spinal muscular atrophy. *Cell* **80**:167–178.
118. **Wirth B, Hahnen E, Morgan K, DiDonato CJ, Dadze A, Rudnik-Schöneborn S, Simard LR, Zerres K, Burghes AH.** 1995. Allelic association and deletions in autosomal recessive proximal spinal muscular atrophy: association of marker genotype with disease severity and candidate cDNAs. *Hum Mol Genet* **4**:1273–1284.

119. Theodorou L, Nicolaou P, Koutsou P, Georghiou A, Anastasiadou V, Tanteles G, Kyriakides T, Zamba-Papanicolaou E, Christodoulou K. 2015. Genetic findings of Cypriot spinal muscular atrophy patients. *Neurol Sci* **36:**1829–1834.

120. Canna SW, de Jesus AA, Gouni S, Brooks SR, Marrero B, Liu Y, DiMattia MA, Zaal KJ, Sanchez GA, Kim H, Chapelle D, Plass N, Huang Y, Villarino AV, Biancotto A, Fleisher TA, Duncan JA, O'Shea JJ, Benseler S, Grom A, Deng Z, Laxer RM, Goldbach-Mansky R. 2014. An activating *NLRC4* inflammasome mutation causes autoinflammation with recurrent macrophage activation syndrome. *Nat Genet* **46:**1140–1146.

121. Grenier JM, Wang L, Manji GA, Huang WJ, Al-Garawi A, Kelly R, Carlson A, Merriam S, Lora JM, Briskin M, DiStefano PS, Bertin J. 2002. Functional screening of five PYPAF family members identifies PYPAF5 as a novel regulator of NF-κB and caspase-1. *FEBS Lett* **530:**73–78.

122. Anand PK, Kanneganti TD. 2012. Targeting NLRP6 to enhance immunity against bacterial infections. *Future Microbiol* **7:**1239–1242.

123. Anand PK, Kanneganti TD. 2013. NLRP6 in infection and inflammation. *Microbes Infect* **15:**661–668.

124. Anand PK, Malireddi RK, Lukens JR, Vogel P, Bertin J, Lamkanfi M, Kanneganti TD. 2012. NLRP6 negatively regulates innate immunity and host defence against bacterial pathogens. *Nature* **488:**389–393.

125. Chen GY, Liu M, Wang F, Bertin J, Núñez G. 2011. A functional role for Nlrp6 in intestinal inflammation and tumorigenesis. *J Immunol* **186:**7187–7194.

126. Normand S, Delanoye-Crespin A, Bressenot A, Huot L, Grandjean T, Peyrin-Biroulet L, Lemoine Y, Hot D, Chamaillard M. 2011. Nod-like receptor pyrin domain-containing protein 6 (NLRP6) controls epithelial self-renewal and colorectal carcinogenesis upon injury. *Proc Natl Acad Sci U S A* **108:**9601–9606.

127. Kempster SL, Belteki G, Forhead AJ, Fowden AL, Catalano RD, Lam BY, McFarlane I, Charnock-Jones DS, Smith GC. 2011. Developmental control of the Nlrp6 inflammasome and a substrate, IL-18, in mammalian intestine. *Am J Physiol Gastrointest Liver Physiol* **300:**G253–G263.

128. Elinav E, Strowig T, Kau AL, Henao-Mejia J, Thaiss CA, Booth CJ, Peaper DR, Bertin J, Eisenbarth SC, Gordon JI, Flavell RA. 2011. NLRP6 inflammasome regulates colonic microbial ecology and risk for colitis. *Cell* **145:**745–757.

129. Seregin SS, Golovchenko N, Schaf B, Chen J, Eaton KA, Chen GY. 2016. NLRP6 function in inflammatory monocytes reduces susceptibility to chemically induced intestinal injury. *Mucosal Immunol* doi:10.1038/mi.2016.55.

130. Levy M, Thaiss CA, Zeevi D, Dohnalová L, Zilberman-Schapira G, Mahdi JA, David E, Savidor A, Korem T, Herzig Y, Pevsner-Fischer M, Shapiro H, Christ A, Harmelin A, Halpern Z, Latz E, Flavell RA, Amit I, Segal E, Elinav E. 2015. Microbiota-modulated metabolites shape the intestinal microenvironment by regulating NLRP6 inflammasome signaling. *Cell* **163:**1428–1443.

131. Wang P, Zhu S, Yang L, Cui S, Pan W, Jackson R, Zheng Y, Rongvaux A, Sun Q, Yang G, Gao S, Lin R, You F, Flavell R, Fikrig E. 2015. Nlrp6 regulates intestinal antiviral innate immunity. *Science* **350:**826–830.

132. Wlodarska M, Thaiss CA, Nowarski R, Henao-Mejia J, Zhang JP, Brown EM, Frankel G, Levy M, Katz MN, Philbrick WM, Elinav E, Finlay BB, Flavell RA. 2014. NLRP6 inflammasome orchestrates the colonic host-microbial interface by regulating goblet cell mucus secretion. *Cell* **156:**1045–1059.

133. Birchenough GM, Nyström EE, Johansson ME, Hansson GC. 2016. A sentinel goblet cell guards the colonic crypt by triggering Nlrp6-dependent Muc2 secretion. *Science* **352:**1535–1542.

134. Triantafilou K, Kar S, van Kuppeveld FJ, Triantafilou M. 2013. Rhinovirus-induced calcium flux triggers NLRP3 and NLRC5 activation in bronchial cells. *Am J Respir Cell Mol Biol* **49:**923–934.

135. Yao Y, Wang Y, Chen F, Huang Y, Zhu S, Leng Q, Wang H, Shi Y, Qian Y. 2012. NLRC5 regulates MHC class I antigen presentation in host defense against intracellular pathogens. *Cell Res* **22:**836–847.

136. Pinheiro AS, Proell M, Eibl C, Page R, Schwarzenbacher R, Peti W. 2010. Three-dimensional structure of the NLRP7 pyrin domain: insight into pyrin-pyrin-mediated effector domain signaling in innate immunity. *J Biol Chem* **285:**27402–27410.

137. Khare S, Dorfleutner A, Bryan NB, Yun C, Radian AD, de Almeida L, Rojanasakul Y, Stehlik C. 2012. An NLRP7-containing inflammasome mediates recognition of microbial lipopeptides in human macrophages. *Immunity* **36:**464–476.

138. Radian AD, Khare S, Chu LH, Dorfleutner A, Stehlik C. 2015. ATP binding by NLRP7 is required for inflammasome activation in response to bacterial lipopeptides. *Mol Immunol* **67**(2 Pt B)**:**294–302.

139. Slim R, Wallace EP. 2013. *NLRP7* and the genetics of hydatidiform moles: recent advances and new challenges. *Front Immunol* **4:**242. doi:10.3389/fimmu.2013.00242.

140. Singer H, Biswas A, Nuesgen N, Oldenburg J, El-Maarri O. 2015. NLRP7, involved in hydatidiform molar pregnancy (HYDM1), interacts with the transcriptional repressor ZBTB16. *PLoS One* **10:**e0130416. doi:10.1371/journal.pone.0130416.

141. Lich JD, Williams KL, Moore CB, Arthur JC, Davis BK, Taxman DJ, Ting JP. 2007. Monarch-1 suppresses non-canonical NF-κB activation and p52-dependent chemokine expression in monocytes. *J Immunol* **178:**1256–1260.

142. Williams KL, Lich JD, Duncan JA, Reed W, Rallabhandi P, Moore C, Kurtz S, Coffield VM, Accavitti-Loper MA, Su L, Vogel SN, Braunstein M, Ting JP. 2005. The CATERPILLER protein Monarch-1 is an antagonist of Toll-like receptor-, tumor necrosis factor α-, and *Mycobacterium tuberculosis*-induced pro-inflammatory signals. *J Biol Chem* **280:**39914–39924.

143. Pinheiro AS, Eibl C, Ekman-Vural Z, Schwarzenbacher R, Peti W. 2011. The NLRP12 pyrin domain: structure, dynamics, and functional insights. *J Mol Biol* **413:**790–803.

144. **Williams KL, Taxman DJ, Linhoff MW, Reed W, Ting JP.** 2003. Cutting edge: Monarch-1: a pyrin/nucleotide-binding domain/leucine-rich repeat protein that controls classical and nonclassical MHC class I genes. *J Immunol* **170:**5354–5358.

145. **Wang L, Manji GA, Grenier JM, Al-Garawi A, Merriam S, Lora JM, Geddes BJ, Briskin M, DiStefano PS, Bertin J.** 2002. PYPAF7, a novel PYRIN-containing Apaf1-like protein that regulates activation of NF-κB and caspase-1-dependent cytokine processing. *J Biol Chem* **277:**29874–29880.

146. **Ataide MA, Andrade WA, Zamboni DS, Wang D, Souza MC, Franklin BS, Elian S, Martins FS, Pereira D, Reed G, Fitzgerald KA, Golenbock DT, Gazzinelli RT.** 2014. Malaria-induced NLRP12/NLRP3-dependent caspase-1 activation mediates inflammation and hypersensitivity to bacterial superinfection. *PLoS Pathog* **10:** e1003885. doi:10.1371/journal.ppat.1003885.

147. **Vladimer GI, Weng D, Paquette SW, Vanaja SK, Rathinam VA, Aune MH, Conlon JE, Burbage JJ, Proulx MK, Liu Q, Reed G, Mecsas JC, Iwakura Y, Bertin J, Goguen JD, Fitzgerald KA, Lien E.** 2012. The NLRP12 inflammasome recognizes *Yersinia pestis*. *Immunity* **37:**96–107.

148. **Zaki MH, Vogel P, Malireddi RK, Body-Malapel M, Anand PK, Bertin J, Green DR, Lamkanfi M, Kanneganti TD.** 2011. The NOD-like receptor NLRP12 attenuates colon inflammation and tumorigenesis. *Cancer Cell* **20:** 649–660.

149. **Allen IC, Wilson JE, Schneider M, Lich JD, Roberts RA, Arthur JC, Woodford RM, Davis BK, Uronis JM, Herfarth HH, Jobin C, Rogers AB, Ting JP.** 2012. NLRP12 suppresses colon inflammation and tumorigenesis through the negative regulation of noncanonical NF-κB signaling. *Immunity* **36:**742–754.

150. **Zaki MH, Man SM, Vogel P, Lamkanfi M, Kanneganti TD.** 2014. *Salmonella* exploits NLRP12-dependent innate immune signaling to suppress host defenses during infection. *Proc Natl Acad Sci U S A* **111:**385–390.

151. **Allen IC, McElvania-TeKippe E, Wilson JE, Lich JD, Arthur JC, Sullivan JT, Braunstein M, Ting JP.** 2013. Characterization of NLRP12 during the *in vivo* host immune response to *Klebsiella pneumoniae* and Mycobacterium *tuberculosis*. *PLoS One* **8:**e60842. doi: 10.1371/journal.pone.0060842.

152. **Jéru I, Duquesnoy P, Fernandes-Alnemri T, Cochet E, Yu JW, Lackmy-Port-Lis M, Grimprel E, Landman-Parker J, Hentgen V, Marlin S, McElreavey K, Sarkisian T, Grateau G, Alnemri ES, Amselem S.** 2008. Mutations in *NALP12* cause hereditary periodic fever syndromes. *Proc Natl Acad Sci U S A* **105:**1614–1619.

153. **Hornung V, Ablasser A, Charrel-Dennis M, Bauernfeind F, Horvath G, Caffrey DR, Latz E, Fitzgerald KA.** 2009. AIM2 recognizes cytosolic dsDNA and forms a caspase-1-activating inflammasome with ASC. *Nature* **458:**514–518.

154. **Jones JW, Kayagaki N, Broz P, Henry T, Newton K, O'Rourke K, Chan S, Dong J, Qu Y, Roose-Girma M, Dixit VM, Monack DM.** 2010. Absent in melanoma 2 is required for innate immune recognition of *Francisella tularensis*. *Proc Natl Acad Sci U S A* **107:**9771–9776.

155. **Sagulenko V, Thygesen SJ, Sester DP, Idris A, Cridland JA, Vajjhala PR, Roberts TL, Schroder K, Vince JE, Hill JM, Silke J, Stacey KJ.** 2013. AIM2 and NLRP3 inflammasomes activate both apoptotic and pyroptotic death pathways via ASC. *Cell Death Differ* **20:**1149–1160.

156. **Warren SE, Armstrong A, Hamilton MK, Mao DP, Leaf IA, Miao EA, Aderem A.** 2010. Cutting edge: cytosolic bacterial DNA activates the inflammasome via Aim2. *J Immunol* **185:**818–821.

157. **DeYoung KL, Ray ME, Su YA, Anzick SL, Johnstone RW, Trapani JA, Meltzer PS, Trent JM.** 1997. Cloning a novel member of the human interferon-inducible gene family associated with control of tumorigenicity in a model of human melanoma. *Oncogene* **15:**453–457.

158. **Bürckstümmer T, Baumann C, Blüml S, Dixit E, Dürnberger G, Jahn H, Planyavsky M, Bilban M, Colinge J, Bennett KL, Superti-Furga G.** 2009. An orthogonal proteomic-genomic screen identifies AIM2 as a cytoplasmic DNA sensor for the inflammasome. *Nat Immunol* **10:**266–272.

159. **Fernandes-Alnemri T, Yu JW, Juliana C, Solorzano L, Kang S, Wu J, Datta P, McCormick M, Huang L, McDermott E, Eisenlohr L, Landel CP, Alnemri ES.** 2010. The AIM2 inflammasome is critical for innate immunity to *Francisella tularensis*. *Nat Immunol* **11:** 385–393.

160. **Jin T, Perry A, Smith P, Jiang J, Xiao TS.** 2013. Structure of the absent in melanoma 2 (AIM2) pyrin domain provides insights into the mechanisms of AIM2 autoinhibition and inflammasome assembly. *J Biol Chem* **288:**13225–13235.

161. **Wang LJ, Huang HY, Huang MP, Liou W, Chang YT, Wu CC, Ojcius DM, Chang YS.** 2014. The microtubule-associated protein EB1 links AIM2 inflammasomes with autophagy-dependent secretion. *J Biol Chem* **289:** 29322–29333.

162. **Rathinam VA, Jiang Z, Waggoner SN, Sharma S, Cole LE, Waggoner L, Vanaja SK, Monks BG, Ganesan S, Latz E, Hornung V, Vogel SN, Szomolanyi-Tsuda E, Fitzgerald KA.** 2010. The AIM2 inflammasome is essential for host defense against cytosolic bacteria and DNA viruses. *Nat Immunol* **11:**395–402.

163. **Saiga H, Kitada S, Shimada Y, Kamiyama N, Okuyama M, Makino M, Yamamoto M, Takeda K.** 2012. Critical role of AIM2 in *Mycobacterium tuberculosis* infection. *Int Immunol* **24:**637–644.

164. **Sauer JD, Witte CE, Zemansky J, Hanson B, Lauer P, Portnoy DA.** 2010. *Listeria monocytogenes* that lyse in the macrophage cytosol trigger AIM2-mediated pyroptosis. *Cell Host Microbe* **7:**412–419.

165. **Hanamsagar R, Aldrich A, Kielian T.** 2014. Critical role for the AIM2 inflammasome during acute central nervous system bacterial infection. *J Neurochem* **129:**704–711.

166. **Man SM, Zhu Q, Zhu L, Liu Z, Karki R, Malik A, Sharma D, Li L, Malireddi RK, Gurung P, Neale G, Olsen SR, Carter RA, McGoldrick DJ, Wu G, Finkelstein D, Vogel P, Gilbertson RJ, Kanneganti TD.** 2015. Critical role for the DNA sensor AIM2 in stem cell proliferation and cancer. *Cell* **162:**45–58.

167. **Patsos G, Germann A, Gebert J, Dihlmann S.** 2010. Restoration of absent in melanoma 2 (AIM2) induces G2/M cell cycle arrest and promotes invasion of colorectal cancer cells. *Int J Cancer* **126:**1838–1849.

168. **Wilson JE, Petrucelli AS, Chen L, Koblansky AA, Truax AD, Oyama Y, Rogers AB, Brickey WJ, Wang Y, Schneider M, Mühlbauer M, Chou WC, Barker BR, Jobin C, Allbritton NL, Ramsden DA, Davis BK, Ting JP.** 2015. Inflammasome-independent role of AIM2 in suppressing colon tumorigenesis via DNA-PK and Akt. *Nat Med* **21:**906–913.

169. **Chen IF, Ou-Yang F, Hung JY, Liu JC, Wang H, Wang SC, Hou MF, Hortobagyi GN, Hung MC.** 2006. AIM2 suppresses human breast cancer cell proliferation *in vitro* and mammary tumor growth in a mouse model. *Mol Cancer Ther* **5:**1–7.

170. **Kayagaki N, Warming S, Lamkanfi M, Vande Walle L, Louie S, Dong J, Newton K, Qu Y, Liu J, Heldens S, Zhang J, Lee WP, Roose-Girma M, Dixit VM.** 2011. Non-canonical inflammasome activation targets caspase-11. *Nature* **479:**117–121.

171. **Hagar JA, Powell DA, Aachoui Y, Ernst RK, Miao EA.** 2013. Cytoplasmic LPS activates caspase-11: implications in TLR4-independent endotoxic shock. *Science* **341:**1250–1253.

172. **Kayagaki N, Wong MT, Stowe IB, Ramani SR, Gonzalez LC, Akashi-Takamura S, Miyake K, Zhang J, Lee WP, Muszyński A, Forsberg LS, Carlson RW, Dixit VM.** 2013. Noncanonical inflammasome activation by intracellular LPS independent of TLR4. *Science* **341:** 1246–1249.

173. **Shi J, Zhao Y, Wang Y, Gao W, Ding J, Li P, Hu L, Shao F.** 2014. Inflammatory caspases are innate immune receptors for intracellular LPS. *Nature* **514:** 187–192.

174. **Zanoni I, Tan Y, Di Gioia M, Broggi A, Ruan J, Shi J, Donado CA, Shao F, Wu H, Springstead JR, Kagan JC.** 2016. An endogenous caspase-11 ligand elicits interleukin-1 release from living dendritic cells. *Science* **352:** 1232–1236.

175. **Kayagaki N, Stowe IB, Lee BL, O'Rourke K, Anderson K, Warming S, Cuellar T, Haley B, Roose-Girma M, Phung QT, Liu PS, Lill JR, Li H, Wu J, Kummerfeld S, Zhang J, Lee WP, Snipas SJ, Salvesen GS, Morris LX, Fitzgerald L, Zhang Y, Bertram EM, Goodnow CC, Dixit VM.** 2015. Caspase-11 cleaves gasdermin D for non-canonical inflammasome signalling. *Nature* **526:** 666–671.

176. **Shi J, Zhao Y, Wang K, Shi X, Wang Y, Huang H, Zhuang Y, Cai T, Wang F, Shao F.** 2015. Cleavage of GSDMD by inflammatory caspases determines pyroptotic cell death. *Nature* **526:**660–665.

177. **Viganò E, Diamond CE, Spreafico R, Balachander A, Sobota RM, Mortellaro A.** 2015. Human caspase-4 and caspase-5 regulate the one-step non-canonical inflammasome activation in monocytes. *Nat Commun* **6:**8761. doi:10.1038/ncomms9761.

178. **Allen IC, TeKippe EM, Woodford RM, Uronis JM, Holl EK, Rogers AB, Herfarth HH, Jobin C, Ting JP.** 2010. The NLRP3 inflammasome functions as a negative regulator of tumorigenesis during colitis-associated cancer. *J Exp Med* **207:**1045–1056.

179. **Zaki MH, Boyd KL, Vogel P, Kastan MB, Lamkanfi M, Kanneganti TD.** 2010. The NLRP3 inflammasome protects against loss of epithelial integrity and mortality during experimental colitis. *Immunity* **32:**379–391.

180. **Dupaul-Chicoine J, Arabzadeh A, Dagenais M, Douglas T, Champagne C, Morizot A, Rodrigue-Gervais IG, Breton V, Colpitts SL, Beauchemin N, Saleh M.** 2015. The Nlrp3 inflammasome suppresses colorectal cancer metastatic growth in the liver by promoting natural killer cell tumoricidal activity. *Immunity* **43:**751–763.

181. **Bauer C, Duewell P, Mayer C, Lehr HA, Fitzgerald KA, Dauer M, Tschopp J, Endres S, Latz E, Schnurr M.** 2010. Colitis induced in mice with dextran sulfate sodium (DSS) is mediated by the NLRP3 inflammasome. *Gut* **59:**1192–1199.

182. **Okamoto M, Liu W, Luo Y, Tanaka A, Cai X, Norris DA, Dinarello CA, Fujita M.** 2010. Constitutively active inflammasome in human melanoma cells mediating autoinflammation via caspase-1 processing and secretion of interleukin-1β. *J Biol Chem* **285:**6477–6488.

183. **Verma D, Bivik C, Farahani E, Synnerstad I, Fredrikson M, Enerbäck C, Rosdahl I, Söderkvist P.** 2012. Inflammasome polymorphisms confer susceptibility to sporadic malignant melanoma. *Pigment Cell Melanoma Res* **25:** 506–513.

184. **Xu Y, Li H, Chen W, Yao X, Xing Y, Wang X, Zhong J, Meng G.** 2013. *Mycoplasma hyorhinis* activates the NLRP3 inflammasome and promotes migration and invasion of gastric cancer cells. *PLoS One* **8:**e77955. doi:10.1371/journal.pone.0077955.

185. **Wei Q, Mu K, Li T, Zhang Y, Yang Z, Jia X, Zhao W, Huai W, Guo P, Han L.** 2014. Deregulation of the NLRP3 inflammasome in hepatic parenchymal cells during liver cancer progression. *Lab Invest* **94:**52–62.

186. **Ghiringhelli F, Apetoh L, Tesniere A, Aymeric L, Ma Y, Ortiz C, Vermaelen K, Panaretakis T, Mignot G, Ullrich E, Perfettini JL, Schlemmer F, Tasdemir E, Uhl M, Génin P, Civas A, Ryffel B, Kanellopoulos J, Tschopp J, André F, Lidereau R, McLaughlin NM, Haynes NM, Smyth MJ, Kroemer G, Zitvogel L.** 2009. Activation of the NLRP3 inflammasome in dendritic cells induces IL-1β-dependent adaptive immunity against tumors. *Nat Med* **15:**1170–1178.

187. **van Deventer HW, Burgents JE, Wu QP, Woodford RM, Brickey WJ, Allen IC, McElvania-Tekippe E, Serody JS, Ting JP.** 2010. The inflammasome component NLRP3 impairs antitumor vaccine by enhancing the accumulation of tumor-associated myeloid-derived suppressor cells. *Cancer Res* **70:**10161–10169.

188. **Carmi Y, Dotan S, Rider P, Kaplanov I, White MR, Baron R, Abutbul S, Huszar M, Dinarello CA, Apte RN, Voronov E.** 2013. The role of IL-1β in the early tumor cell-induced angiogenic response. *J Immunol* **190:** 3500–3509.

189. **Tarassishin L, Lim J, Weatherly DB, Angeletti RH, Lee SC.** 2014. Interleukin-1-induced changes in the glioblastoma secretome suggest its role in tumor progression. *J Proteomics* **99:**152–168.

190. **Hu B, Elinav E, Huber S, Booth CJ, Strowig T, Jin C, Eisenbarth SC, Flavell RA.** 2010. Inflammation-induced tumorigenesis in the colon is regulated by caspase-1 and NLRC4. *Proc Natl Acad Sci U S A* **107:**21635–21640.

191. **Chen YK, Huse SS, Lin LM.** 2011. Expression of inhibitor of apoptosis family proteins in human oral squamous cell carcinogenesis. *Head Neck* **33:**985–998.

192. **Choi J, Hwang YK, Choi YJ, Yoo KE, Kim JH, Nam SJ, Yang JH, Lee SJ, Yoo KH, Sung KW, Koo HH, Im YH.** 2007. Neuronal apoptosis inhibitory protein is overexpressed in patients with unfavorable prognostic factors in breast cancer. *J Korean Med Sci* **22**(Suppl)**:**S17–S23.

193. **Krajewska M, Krajewski S, Banares S, Huang X, Turner B, Bubendorf L, Kallioniemi OP, Shabaik A, Vitiello A, Peehl D, Gao GJ, Reed JC.** 2003. Elevated expression of inhibitor of apoptosis proteins in prostate cancer. *Clin Cancer Res* **9:**4914–4925.

194. **McConnell BB, Vertino PM.** 2004. TMS1/ASC: the cancer connection. *Apoptosis* **9:**5–18.

195. **Levine JJ, Stimson-Crider KM, Vertino PM.** 2003. Effects of methylation on expression of TMS1/ASC in human breast cancer cells. *Oncogene* **22:**3475–3488.

196. **Yokoyama T, Sagara J, Guan X, Masumoto J, Takeoka M, Komiyama Y, Miyata K, Higuchi K, Taniguchi S.** 2003. Methylation of *ASC/TMS1*, a proapoptotic gene responsible for activating procaspase-1, in human colorectal cancer. *Cancer Lett* **202:**101–108.

197. **Das PM, Ramachandran K, Vanwert J, Ferdinand L, Gopisetty G, Reis IM, Singal R.** 2006. Methylation mediated silencing of *TMS1/ASC* gene in prostate cancer. *Mol Cancer* **5:**28. doi:10.1186/1476-4598-5-28.

198. **Machida EO, Brock MV, Hooker CM, Nakayama J, Ishida A, Amano J, Picchi MA, Belinsky SA, Herman JG, Taniguchi S, Baylin SB.** 2006. Hypermethylation of *ASC/TMS1* is a sputum marker for late-stage lung cancer. *Cancer Res* **66:**6210–6218.

199. **Stone AR, Bobo W, Brat DJ, Devi NS, Van Meir EG, Vertino PM.** 2004. Aberrant methylation and down-regulation of TMS1/ASC in human glioblastoma. *Am J Pathol* **165:**1151–1161.

200. **Liu W, Luo Y, Dunn JH, Norris DA, Dinarello CA, Fujita M.** 2013. Dual role of apoptosis-associated speck-like protein containing a CARD (ASC) in tumorigenesis of human melanoma. *J Invest Dermatol* **133:** 518–527.

201. **Heneka MT, Kummer MP, Stutz A, Delekate A, Schwartz S, Vieira-Saecker A, Griep A, Axt D, Remus A, Tzeng TC, Gelpi E, Halle A, Korte M, Latz E, Golenbock DT.** 2013. NLRP3 is activated in Alzheimer's disease and contributes to pathology in APP/PS1 mice. *Nature* **493:**674–678.

202. **Gris D, Ye Z, Iocca HA, Wen H, Craven RR, Gris P, Huang M, Schneider M, Miller SD, Ting JP.** 2010. NLRP3 plays a critical role in the development of experimental autoimmune encephalomyelitis by mediating Th1 and Th17 responses. *J Immunol* **185:**974–981.

203. **Jha S, Ting JP.** 2009. Inflammasome-associated nucleotide-binding domain, leucine-rich repeat proteins and inflammatory diseases. *J Immunol* **183:**7623–7629.

204. **Inoue M, Williams KL, Oliver T, Vandenabeele P, Rajan JV, Miao EA, Shinohara ML.** 2012. Interferon-β therapy against EAE is effective only when development of the disease depends on the NLRP3 inflammasome. *Sci Signal* **5:**ra38. doi:10.1126/scisignal.2002767.

205. **Kimkong I, Avihingsanon Y, Hirankarn N.** 2009. Expression profile of HIN200 in leukocytes and renal biopsy of SLE patients by real-time RT-PCR. *Lupus* **18:** 1066–1072.

206. **Wozniacka A, Lesiak A, Narbutt J, McCauliffe DP, Sysa-Jedrzejowska A.** 2006. Chloroquine treatment influences proinflammatory cytokine levels in systemic lupus erythematosus patients. *Lupus* **15:**268–275.

Myeloid Cells in Health and Disease: A Synthesis
Edited by Siamon Gordon

doi:10.1128/microbiolspec.MCHD-0027-2016

Asif J. Iqbal[1]
Edward A. Fisher[2]
David R. Greaves[1]

18 Inflammation—a Critical Appreciation of the Role of Myeloid Cells

> The receptor concept is to pharmacology as homeostasis is to physiology, or metabolism to biochemistry. They provide the basic framework, and are the "Big Ideas" without which it is impossible to understand what the subjects are about.
>
> H. P. Rang. The receptor concept: pharmacology's big idea. 2008. *Br J Pharmacol* **147**(Suppl 1):S9–S16.

INTRODUCTION: A HISTORICAL AND EVOLUTIONARY SYNTHESIS

Multicellular organisms have had to develop a rapid response to infection and tissue injury. In all animals on our planet, this involves mobilization of specialized cells to the focus of the infection or injury. This important insight was beautifully illustrated by Elie Metchnikoff, whose detailed drawings of cells being recruited to the site of injury caused by a rose thorn in a starfish embryo gave us the first glimpse of cells he termed "macrophages," and neutrophils, which he termed "microphages." If not the first person to observe phagocytosis and leukocyte diapedesis, Metchnikoff was probably the first person to fully appreciate the role of these two important cellular processes in "natural" or innate immunity (1).

An important question arises from Metchnikoff's observations in model organisms and histological images of human tissues infected by pyogenic bacteria: What are the locally produced molecular signals that mediate innate immune cell mobilization? *A priori*, we would expect these signals and the receptors that recognize them to have arisen early in the evolution of multicellular organisms and for their function to be maintained by selection pressure exerted by everyday exposure to infectious disease and physical injury. A glimpse of how local signals to both pathogens and physical injury might have arisen comes from the Atlantic horseshoe crab, *Limulus polyphemus*. The horseshoe crab is often described as a "living fossil" due to its near identical form to species present in the Triassic period 230 million years ago. The blood (or hemolymph) of *Limulus* species coagulates in response to intact

[1]Sir William Dunn School of Pathology, University of Oxford, Oxford OX1 3RE, United Kingdom; [2]Departments of Medicine (Cardiology) and Cell Biology, New York University School of Medicine, New York, NY 10016.

bacteria and bacterial endotoxin, effectively walling off invading pathogens. The *Limulus* protein that recognizes lipopolysaccharide (LPS) and lipid A present in the outer membrane of Gram-negative bacteria is a 132-kDa protein called factor C. LPS binding to the LPS/lipid A recognition domain of factor C activates a serine protease domain within the same protein. Activated factor C activates hemolymph protein, factor B, which initiates the clotting cascade to cause local hemolymph coagulation. This observation formed the basis of the *Limulus* amoebocyte lysate (LAL) assay for detecting low-level endotoxin contamination of tissue culture reagents, biologicals, and medical devices (2). *Limulus* hemolymph also contains two conserved oligomeric serum proteins, the "short" pentraxins C-reactive protein (CRP) and serum amyloid P component (SAP) (3). These two pentraxins, together with the evolutionarily conserved "long" pentraxin PTX3, play important roles in mammalian host defense (4). A recent analysis of CRP knockout mice showed a marked sensitivity to *Streptococcus pneumoniae* infection in animals lacking endogenous CRP production that could be rescued by infusion of purified human plasma CRP or generation of anti-*S. pneumoniae* antibodies (5). These experiments strongly suggest that CRP has evolved to protect neonatal mammals from specific virulent bacterial pathogens. Specific roles for SAP in mammalian host defense have been harder to identify (possibly due to functional redundancy) but are likely to center around recognition of bacterial peptidoglycan, damaged host membranes, and complement activation (reviewed in reference 6). PTX3 was originally described as a nonredundant mammalian pattern recognition receptor essential for defense against the fungal pathogen *Aspergillus fumigatus* and later recognized to bind the bacterial pathogens *Pseudomonas aeruginosa* and uropathic *Escherichia coli*, as well as influenza virus (7, 8).

Mammalian hepatocytes synthesize a range of other host defense proteins as part of the acute-phase response, most notably ~30 proteins of the complement cascade. Complement is another evolutionary ancient defense against pathogens that shares with the coagulation cascade local activation and local amplification via serine protease cleavage of inactive enzymes (zymogens). It is clear that the complement system has evolved to be much more than a simple plasma pathogen recognition system that can kill microbes through deposition of a membrane attack complex (C5b and C6 to C9) (9). Proteolytic cleavage of the plasma protein C3 leads to deposition of the C3b protein fragment on target cells, greatly enhancing phagocytosis by professional phagocytes of the innate immune system (neutrophils and macrophages). Cleavage of the C5 complement protein by the C3 convertase complex generates a high local concentration of C5a, a potent chemoattractant for innate immune cells via the G protein-coupled receptor C5aR1 (10).

The unrelenting evolutionary pressure exerted by the twin drivers of infectious disease and tissue injury means that any germ line-encoded signaling molecule or cellular response system that enhances tissue defense can be rapidly fixed, duplicated, and mutated within eukaryote genomes. Multiple examples are provided from comparative genomics. One striking example is the *Drosophila* dorsoventral regulatory gene network, *spätzle/Toll/cactus*, which has been "reengineered" and "repurposed" during vertebrate evolution to give the cytokine-activated NF-κB signaling pathway (11). Gene duplication has generated a family of Toll-like receptors (TLRs), which act as cellular pattern recognition receptors for highly conserved molecules on microbial pathogens termed pathogen-associated molecular patterns (PAMPs) by Charles Janeway and Ruslan Medzhitov (12, 13).

Another striking example of duplication and diversification of immune defense genes comes from consideration of chemokines and their receptors. Comparative genomics reveals that the chemokine-chemokine receptor system has proven a useful module for directing cell-type-specific chemotaxis and activation. In addition to mediating T-cell chemotaxis, the CXCR4-CXCL12 interaction is used to keep hematopoietic stem cells within a specific bone marrow niche (14). Indeed, a small-molecule CXCR4 antagonist (AMD3100) has found clinical application in mobilizing donor hematopoietic stem cells from bone marrow to peripheral blood for more efficient and less painful harvesting. The chemokine-chemokine receptor system has been exploited by the adaptive immune system for dendritic cell migration to lymph nodes, homeostatic leukocyte trafficking, lymphocyte homing to different tissues (e.g., the gut), and recruitment of specific lymphocyte subsets to sites of inflammation (15). The original description of the chemokine system as an inflammatory cell recruitment system, and the impressive results obtained using chemokine receptor gene knockout animals in models of chronic inflammation (e.g., Boring et al., 1998 [16]), suggested that small-molecule drugs that inhibit chemokine receptor signaling would make potent, cell-type-specific anti-inflammatory drugs. To date, this initial optimism has not (yet) been converted into clinically useful drugs (17, 18).

No critical discussion of the role of myeloid cells in inflammation would be complete without consideration of inflammasome activation. The seminal contribution of Jürg Tschopp (1951-2011) to immunology and inflammation biology was the recognition that the secretion of active interleukin-1 (IL-1) (and IL-18) is critically dependent on the formation of a large (≥700-kDa), cytoplasmic, multisubunit caspase-activating complex (19). Since Tschopp's first description of this macromolecular structure, it has been shown that inflammasome activation can be triggered by a wide range of bacterial, viral, fungal, and even helminth PAMPs, as well as by a range of host damage-associated molecules. Inflammasome activation leads to high local concentrations of IL-1 and inflammatory cell death by pyroptosis and pyronecrosis. Recent experiments in murine models have revealed tantalizing glimpses of a link between inflammasome activation and the inflammatory component of metabolic diseases such as obesity and type 2 diabetes.

Finally, it is important to remember that the prototypic acute inflammatory response can be triggered by nonmicrobial stimuli. A good example of such "sterile inflammation" is provided by administration of substances such as monosodium urate crystals and calcium pyrophosphate crystals, the pathogenic drivers of gout and pseudogout, respectively. Sensing of these and other crystalline insults is absolutely dependent on the presence of a functional *NALP3* gene and subsequent caspase activation, showing the central role of inflammasome activation and IL-1β in this neutrophil-dominated response to tissue injury (20). Necrotic cell death releases specific intracellular molecules that can induce activation of innate immune cells *in vitro* and a prototypic acute inflammatory response *in vivo* (21). Following the intellectual lead provided by Charles Janeway, signaling molecules released by necrotic tissue have been termed damage-associated molecular patterns (DAMPs) by some, and alarmins by others (22). Such molecules include extracellular ATP, mitochondrial DNA, uric acid, and chromatin-associated proteins, including the chromatin high mobility group-1 protein (HMG-1). In a recent study, Kataoka et al. (23) directly compared the role of multiple signaling pathways in two mouse models of leukocyte recruitment in response to intraperitoneal injection of necrotic cells or induction of hepatocyte necrosis with paracetamol. The authors showed that neutrophil mobilization at early time points was significantly decreased in the absence of complement C3, natural antibodies, and the protease-activated receptor PAR2. Local depletion of ATP or deletion of the P2X7 receptor gene had no effect on leukocyte mobilization in these two models of necrotic cell-induced inflammation, in contrast to findings in cultured cells or other models of sterile injury.

Another potential "danger signal" in the context of tissue injury is damage and modification of the extracellular matrix (ECM) (24). Hyaluronan (HA) is an abundant nonsulfated glycosaminoglycan component of the ECM found in many tissues. Multiple biological functions have been ascribed to HA, and proinflammatory functions have been assigned to low-molecular-weight forms of HA (LMW HA) generated by the action of a range of hyaluronidase enzymes. An interesting preclinical study by Huang et al. (25) demonstrated that commercially available LMW HA and hyaluronidase enzyme preparations are contaminated with endotoxin and other proteins. This contamination of key reagents had led to the erroneous conclusion that LMW HA is a ligand of the TLR2 and TLR4 receptors, leading to proinflammatory cytokine production. By using an endotoxin-free pure preparation of the human hyaluronidase enzyme PH20 (rHuPH20) in an LPS dorsal air pouch model of acute inflammation, Huang et al. demonstrated a marked anti-inflammatory effect of hyaluronidase, characterized by no change in proinflammatory cytokine production but marked reduction of neutrophils. The proinflammatory role of ECM damage merits further study in the context of both acute and chronic inflammation as well as in the context of tissue repair and fibrosis.

EXPERIMENTAL MODELS OF INFLAMMATION

Animal models of inflammation are always open to criticism of how well they mimic clinical events in human disease. For instance, mouse models of atherosclerosis do not display the classical clinical sequelae of human atherosclerosis, i.e., myocardial infarction and ischemic stroke. Another obvious concern is that drugs that work well in preclinical models of human disease do not always translate into successful clinical trials. An early example was the success of anti-tumor necrosis factor (TNF) therapy in murine and primate models of endotoxic shock and the subsequent failure of anti-TNF monoclonal antibodies to impact on morbidity and mortality in human septic shock and sepsis in randomized clinical trials. Despite these obvious limitations, our current knowledge of inflammatory mediators, inflammatory cell biology, and the resolution of inflammation owes much to a wide range of well-characterized preclinical models, some of which are outlined below.

Animal Models of Acute Inflammation

Peritonitis is the inflammation of the peritoneum, a thin membrane lining the abdominal cavity. Experimentally it can be triggered by infectious stimuli, which if not treated immediately can spread to the blood and lead to septic shock. Peritonitis can also be induced by injection of sterile inflammagens or implantation of necrotic cells or tissues. Rodent models of peritonitis have provided insight into the generation of local mediators that aid the resolution and return to tissue homeostasis, such as annexin A1, lipoxins, resolvins, protectins, and maresins. A variety of inflammatory stimuli have been used to induce peritonitis, including zymosan, IL-1β, and Brewers thioglycolate. Zymosan-induced peritonitis is a simple and reproducible model of self-resolving inflammation. It has become a "go to" model to study not only the kinetics of leukocyte recruitment and proinflammatory mediator production but also proresolving actions on processes such as macrophage efferocytosis. Zymosan is a yeast cell wall extract of *Saccharomyces cerevisiae* and is recognized by TLR2 and Dectin-1(dendritic cell-associated C-type lectin 1). Intraperitoneal injection with low-dose zymosan (0.1 to 1 mg/ml) leads to an initial wave of polymorphonuclear leukocyte (PMN) recruitment into the peritoneal cavity that peaks in 6 to 8 h, followed by a second wave of mononuclear cells (>16 h) during the resolution phase (26).

Carrageenan (CG)-induced paw edema is a well-established model of acute inflammation used to test the effects of a variety of anti-inflammatory drugs/compounds and to understand the role of mediators during an acute inflammatory response. CG is a gelling agent consisting of sulfated galactans, and three main forms have been identified: ι-CG, κ-CG, and λ-CG. The λ species is most widely used in mice and rats and is given as a subplantar injection in one paw. A similar volume of saline is injected into the contralateral paw as a control, and edema is usually measured by plethysmometry (27). Upon subplantar injection with 1 to 3% CG, a biphasic inflammatory response ensues, with pain, increased vascular permeability, edema, and PMN influx observed as a result of the release of range of mediators being generated locally, including substance P, histamine, bradykinin, prostaglandins, complement, and reactive oxygen species. Peak edema levels in the first phase of the response are observed in between 4 and 6 h, followed by a second, more intense phase developing between 48 and 72 h. A study carried out by D'Agostino et al. (28) examined the role of peroxisome proliferator-activated receptor (PPAR)-α agonists in modulating CG-induced paw edema in mice. The authors found that intracerebroventricular administration of an endogenous PPAR-α agonist, palmitoylethanolamide, 30 min before CG administration reduced edema formation. This reduction was linked to a decrease in cyclooxygenase-2, inducible nitric oxide (NO) synthase expression, and IκB degradation. Mice lacking PPAR-α showed no reduction in edema formation following pretreatment with palmitoylethanolamide. This study elegantly demonstrated for the first time that activation of PPAR-α in the central nervous system could control peripheral inflammation.

The air pouch model is another simple model widely used to study inflammation *in vivo*. Two subcutaneous injections of air (days 0 and 3) into the dorsal intrascapular region leads to the formation of a discrete pouch. Injecting inflammatory stimuli such as zymosan, monosodium urate, or cytokines into the air pouch results in rapid influx of PMNs and the local generation of mediators, including IL-8 and C5a. Depending on the dose and the type of inflammagen used, a second wave of mononuclear cells can follow. The air pouch offers many advantages over the peritonitis model, including ease of multiple dosing and application of compounds with poor solubility profiles. It also offers users the ability to recover relatively clean exudate samples, which can be used to measure mediators with low levels of abundance (29).

Animal Models of Chronic Inflammation

The collagen-induced arthritis (CIA) model has been an invaluable tool in furthering our understanding of some of the key mechanisms involved in the pathogenesis of human rheumatoid arthritis (RA). CIA is an inflammatory polyarthritis that shares many of the clinical and histological manifestations associated with human RA, i.e., disease being centered on the joints and the destruction of cartilage and bone. CIA is induced following sensitization with heterologous type II collagen (CII) in complete Freund's adjuvant followed by a second immunization (day 21) with CII in incomplete Freund's adjuvant. This results in the initiation of cell-mediated and humoral adaptive immune responses, which are characterized by increased levels of anti-CII IgG, complement fixation in joints, and leukocyte infiltration and activation into the joint space and synovium, which leads to the generation of proinflammatory mediators such as TNF-α and IL-17/23 that continue to drive the disease. A study carried out by Notley et al. in 2008 (30) utilized the CIA model to assess the effects of TNF blockade on IL-17 production. They found that mice treated with TNFR-Fc fusion protein or anti-TNF monoclonal antibody had reduced arthritis severity and showed reduced accumulation of Th1 and Th17 cells in the joint but expanded Th1 and Th17 cell numbers in lymph nodes. Collectively, these findings suggested two opposing roles for TNF blockade in CIA: (i) blocking of the accumulation of Th1 and Th17 cells in joints and (ii) sequestration of pathogenic T-cell numbers in peripheral lymphoid organs.

In recent years, inducing RA in mice using serum transfer has become a more widely used model of RA, in part because the CIA model only works well in selected inbred mouse strains (a common feature of several mouse models of inflammation). The K/BxN serum transfer arthritis model was first described in the mid-1990s as an inadvertent by-product of crossing KRN T-cell receptor transgenic and nonobese diabetic (NOD) mice that led to the development of spontaneous arthritis (31). Transfer of pathogenic anti-glucose-6-phosphate isomerase antibodies from these mice to normal recipients resulted in the development of a severe arthritis as a result of alternative complement pathway activation and continual recruitment of neutrophils and mast cells into joints. A study from the laboratory of Mathis and Benoist (32) employed this model to delineate the role of chemokines in leukocyte recruitment during the initiation of autoantibody-mediated arthritis. The authors induced arthritis by transferring serum into several chemokine- and chemokine receptor-deficient mice (CCR1 to -7, CCR9, CXCR2, CXCR3, CXCR5, CX3CR1, CCL2, or CCL3) and found that only the absence of CXCR2, a classical neutrophil chemokine receptor, was critical for the development of autoantibody-mediated joint inflammation and arthritis in C57BL/6 mice.

The ability of $ApoE^{-/-}$ mice on a C57BL/6J background to develop atherosclerotic lesions was first reported in 1992, and a similar phenotype in $Ldlr^{-/-}$ mice fed a high-fat/high-cholesterol diet was reported a few years later. These two mouse strains have been used extensively to study the cell and molecular biology of atherosclerosis *in vivo* (33). Mouse models of atherosclerosis involve many of the key features of human atherosclerosis including trapping of apoB-containing lipoproteins in the subendothelial space of major arteries, monocyte recruitment, and macrophage differentiation.

ACUTE INFLAMMATION IS A PROCESS

Generally ascribed to the Roman physician Celsus, the cardinal signs of acute inflammation in response to infection or tissue injury have been recognized for 2,000 years, i.e., redness (rubor), heat (calor), pain (dolor), and swelling (edema). To paraphrase Billy Crystal's character Ben Sobel in the 1999 film Analyze This, acute inflammation, like grieving, is a process. The coordinated recruitment of plasma proteins, lipid mediators, and myeloid cells in the inflammatory exudate can be defined by careful examination of sterile inflammation in the experimental model systems outlined above and can be organized under three headings.

Initiation: The earliest event in acute inflammation is sensing of pathogens and tissue damage, typically by tissue-resident macrophages and mast cells. The most important role of these sentinel cells, which are relatively few in number, is to generate signaling molecules that lead to *local* endothelial cell activation in postcapillary venules and autocrine and paracrine activation of macrophage effector functions within tissues.

Amplification: Locally generated signaling molecules cause important changes in the properties of nearby endothelium, including the release of the contents of Weibel-Palade bodies, endothelial cell contraction, upregulation of cell adhesion molecules (e.g., intercellular adhesion molecule-1), and synthesis and presentation of chemokines. Acting in concert, all these changes in the endothelium allow for the local elaboration of an inflammatory exudate of plasma proteins and myeloid cells, especially neutrophils. Changes in the properties of endothelial cells cause the observed clinical signs of inflammation: redness and heat via increased local blood flow, edema through elevated oncotic pressure caused by elevated albumin in tissues, and localized pain through the action of locally produced mediators including bradykinin and prostaglandin E_2.

It is the recruitment of plasma proteins and their subsequent proteolytic cleavage at sites of tissue damage or suspected pathogen invasion that leads to the massive *amplification* phase of inflammation that is essential for the localized recruitment of myeloid cells from the blood and their subsequent activation. Local activation of serine protease cascades in response to DAMPs and PAMPs (using the same molecular mechanism adopted by the horseshoe crab 220 million years ago) generates large quantities of potent protein and peptide mediators such as complement C3a, C5a, and bradykinin.

Resolution: This is probably the least-understood aspect of the acute inflammatory process. Termination of further leukocyte recruitment requires catabolism of inflammatory mediators, neutrophil apoptosis, macrophage efferocytosis of apoptotic cells, lymphatic drainage, and the initiation of tissue repair processes. It is described by many as an "active" rather than a "passive" process. Definitive experiments addressing the cell biology of inflammation resolution using *in vivo* model systems are at a premium. An underresearched area of inflammation biology is the role of the lymphatic system in clearance of the inflammatory exudate during the resolution phase. One important anatomical arena where future research seems merited is the infarcted myocardium, following recent studies showing changes to the cardiac lymphatic system following injury (34).

Animal models of sterile peritonitis provide some idea of the potential timescale of the key events in acute inflammation. In the widely used mouse zymosan peritonitis model, resident F4/80^{hi} macrophages leave the serosal cavity to local lymph nodes within 60 min of inflammagen injection and neutrophil recruitment peaks at ~4 h. PMN recruitment is followed by a wave of inflammatory Ly6C^{hi} monocytes, which peaks in between 16 and 24 h. In this model, neutrophil and monocyte numbers in the cavity return to baseline within 96 h, but these timings and the absolute number of myeloid cells depend on the dose and nature of the inflammagen. A recent paper from the group of Derek Gilroy (35) extended analysis of the classic zymosan peritonitis model out past this 96-h window and showed that a true return to tissue homeostasis following clearance of zymosan particles took more than 3 weeks. After disappearance of PMNs from the peritoneum, Newson et al. (35) observed changes in monocyte-derived macrophage subsets and a significant increase in the number of B and T lymphocytes in the peritoneum. The recruitment and retention of this collection of lymphoid and myeloid cells long after the disappearance of the inciting sterile stimulus is intriguing. It will be important to see exactly how the mixture of cell types present postinflammation contribute to tissue repair and defense against infectious disease. An important myeloid cell type that we will not consider in this brief review is the dendritic cell. An important question for immunologists and pathologists is: Do dendritic cells play a unique role in the initiation, amplification, or resolution of inflammation that cannot be provided by monocyte-derived macrophages recruited to the site of inflammation? Perhaps the unique, nonredundant role of dendritic cells in inflamed tissues is to engage the anti-inflammatory/prorepair arm of the adaptive immune response, most notably regulatory T (Treg) cells and perhaps some lesser-studied B-lymphocyte subsets such as B1 and Breg cells. Published studies on the role of dendritic cells and B-cell subsets in animal models of atherosclerosis may lead the way in this regard.

A typical time course of an acute inflammatory response is shown in Fig. 1. Representing the intensity of the inflammatory response on the y axis (measured as leukocyte recruitment, leukocyte activation, or local inflammatory cytokine concentrations), the figure clearly demarcates the initiation, amplification, and resolution phases of a stereotypic, "healthy" acute inflammatory response. Failure to clear the initial inflammatory insult or failure of inflammation resolution leads to chronic inflammation, represented by the horizontal arrow. Hyperacute inflammation is represented by a continuing escalation of leukocyte recruitment, leukocyte activation (locally and systemically), and unrestrained inflammatory cytokine production.

A couple of important discussion points arise from consideration of this simplistic representation of the "classical" acute inflammatory response. The first is whether there is ever a complete absence of inflammation in host tissues, i.e., should the y axis be set to zero at $t = 0$? A paper published in 2010 by Jeffrey Weiser's laboratory showed that that low levels of systemic peptidoglycan derived from the gut microbiota prime the host innate immune system via the Nod1 receptor, leading to more-efficient neutrophil killing of *S. pneumoniae* and *Staphylococcus aureus* (36). A second important discussion point arising from consideration of Fig. 1 is: What regulates the magnitude of the acute inflammation response? And more specifically: Does the magnitude of the acute inflammation to the same inciting stimulus differ between tissues in the same individual and between individuals in the same population? A brief consideration of the mechanisms that might drive hyperacute inflammation as presented in Fig. 1 is also warranted. One simple explanation for the continuing amplification of acute inflammation seen in say, septic shock, could simply be the continuing proliferation of the initial bacterial infection that acted as the stimulus for the initiation of inflammation at $t = 0$. Circumstantial evidence supporting such an explanation comes from a retrospective analysis of mortality data for patients admitted to hospital with suspected sepsis. In a cohort of 17,990 patients with severe sepsis or septic shock given antibiotics upon admission to 165 intensive care units, the probability of in-hospital mortality increased steadily with time to antibiotic administration (37). An alternate explanation for hyperacute inflammation could be failure of endogenous pathways that act to limit myeloid cell responses to PAMPs, a phenomenon that has been termed "endotoxin tolerance"

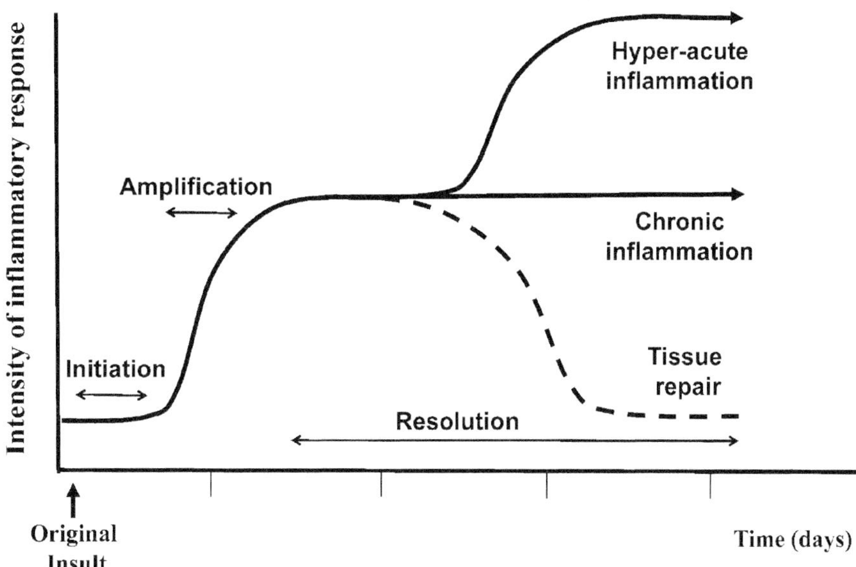

Figure 1 Time course of a typical acute inflammatory response. A schematic representation of the ideal outcome of an acute inflammatory response, i.e., resolution, is shown as a dashed line. Two potential outcomes leading to significant clinical sequelae are shown: hyperacute inflammation, e.g., septic shock; and nonresolving, chronic inflammation, e.g., RA. Adapted from an original figure by Christopher Buckley, University of Birmingham, United Kingdom.

(38), or the more recent appreciation of altered myeloid cell metabolism induced by sepsis (39).

In Fig. 2, we represent this question in a simple diagram. We propose that the *magnitude* of the host response to an inflammatory insult in any given anatomical location is the net result of the action of local proinflammatory mediators (upward arrows) and the opposing action of endogenous anti-inflammatory mediators (downward arrows). We have termed our model the "inflammatory set point hypothesis," and it stands apart from a previous theoretical consideration of the acute inflammatory response, which placed more weight on the *kinetics* of inflammation resolution by deriving a "resolution interval" for comparing the effects of different therapeutic interventions in experimental animal models, typically zymosan peritonitis (40). The inflammatory set point model as presented in Fig. 2 immediately suggests that pharmacological interventions that lower the activity of specific proinflammatory mediators (e.g., anti-TNF antibodies or chemokine receptor antagonists) will reduce the maximal intensity of the acute inflammatory response. Another approach to reduce the peak level of inflammation would be to augment the activity of relevant endogenous anti-inflammatory mediators. This therapeutic rationale could be particularly efficacious in the treatment of human diseases in which the magnitude of the initial inflammatory response overwhelms the inflammation resolution machinery, i.e., chronic ("nonresolving") inflammation and hyperacute inflammation (see Fig. 1).

Consideration of experimental evidence for the importance of the downward arrows in Fig. 2 comes from the enhanced inflammatory response seen in mice carrying gene deletions for the anti-inflammatory mediator annexin A1 and its receptor *Fpr2* (41). Finding evidence in support of the model presented in Fig. 2 using human rather than murine models will be challenging, but use of the beetle blister model could be instructive. Gilroy and colleagues performed an interesting experiment in which they gave human volunteers aspirin (75 mg orally once a day for 10 days) or placebo and then used a fixed dose of beetle cantharidin toxin to induce an acute inflammatory response in the skin characterized by dermal edema and localized leukocyte recruitment. Volunteers taking low-dose aspirin showed no change in blister edema volume at 24 h compared to volunteers taking placebo but did show reduced neutrophil and monocyte numbers in the inflammatory exudate (42). In terms of our set point model of Fig. 2, the effect of low-dose aspirin is likely 2-fold, reducing proinflammatory drive by reducing local prostaglandin E_2 production and simultaneously enhancing endogenous anti-inflammatory 15-epi-lipoxin A_4 production.

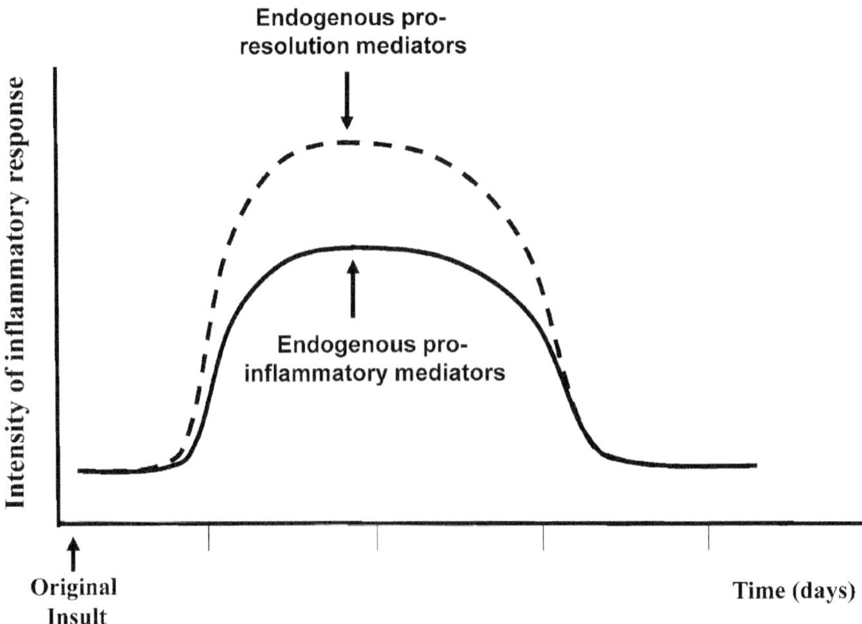

Figure 2 The inflammatory set point hypothesis. This schematic representation highlights the balance between locally produced proinflammatory and endogenous anti-inflammatory/ proresolution mediators in determining the magnitude of the inflammatory response to a given stimulus.

MONOCYTES AND INFLAMMATION

Macrophages have been described as the "mature forms of circulating monocytes that have left the blood and taken up residence in the tissues" (43). We now know that the origin of tissue macrophages is not exclusively from the circulating pool of monocytes. Under homeostatic conditions, many tissue-resident macrophage populations are probably of embryonic origin and capable of self-renewal (reviewed in Sieweke and Allen [44]). Most relevant to this chapter, however, is that an inflammatory stimulus will significantly increase the contribution of circulating monocytes to the tissue macrophage pool. In mice, the pool of circulating monocytes has been broadly divided into Ly6C^{hi} and Ly6C^{lo} subsets, which are also referred to as classical and nonclassical monocytes, respectively (reviewed in reference 45). In humans, the corresponding subsets are CD14^{+}CD16^{-} and CD14^{lo}CD16^{+}, but the proportions of classical to nonclassical monocytes vary by species (1:1 in mice and 9:1 in humans). Nonclassical monocytes are thought to be derived from classical monocytes, but there is also some evidence for independent origin (46).

The functions of circulating monocyte subsets in inflammation have been extensively studied, particularly in mice. CCR2-mediated mobilization and chemotaxis are major drivers of classical monocyte recruitment from bone marrow to blood and from blood to inflammatory sites. This has been borne out in multiple studies of CCR2$^{-/-}$ mice. CCR2-deficient mice have fewer monocyte-macrophages in models of both acute inflammation, e.g., thioglycolate-induced peritonitis (47), and chronic inflammation, e.g., atherosclerosis (16). Likely this reflects both the impaired release of classical monocytes from the bone marrow as well as the reduced chemokine-mediated recruitment of circulating monocytes to the site of inflammation.

Though nonclassical monocytes are low expressers of CCR2, in contrast to the classical cells they express high levels of CX3CR1. This chemokine receptor is thought to not only contribute to the migratory behavior of the cells but also to enhance the survival of both the nonclassical monocytes in the blood and the tissue macrophages derived from them (48). Like classical monocytes, nonclassical cells are recruited to sites of inflammation, but less abundantly so in both acute and chronic models (49). Also in contrast to the classical subset, they perform "patrolling" functions in the vasculature and can be recruited to noninflamed tissues (50). Another contrast between the two subsets is the polarization of the tissue macrophages derived from each type. It is thought to be toward the activated M1 state for those of classical, and toward the anti-inflammatory, tissue-repair M2 state for those of nonclassical origin (51). There are a number of exceptions, however, to this "rule" (52), suggesting that the phenotype of the macrophages derived from each subset is likely to be context dependent.

Though the bone marrow is typically the major source of circulating monocytes, in certain circumstances there can be acute and substantial contributions from extramedullary sources. One example is reported by Swirski, Nahrendorf, and their colleagues in a series of elegant papers (reviewed in reference 53). In a mouse myocardial infarction model, in which the myocardium experiences ischemia-reperfusion damage, they observed, not surprisingly, that the inflammatory process followed that in wound healing, namely, that there was biphasic entry of monocytes into the injured tissue, the first wave being classical ($Ly6C^{hi}$) cells responding to a burst of locally produced CCL2, the ligand of CCR2. These cells became activated, M1-like macrophages. The two surprises were that the second wave also consisted of $Ly6C^{hi}$ cells. These converted to $Ly6C^{lo}$ cells in the injured tissue, where they became tissue-repair, M2-like macrophages (54) in a variation to the "rule" that tissue M2 macrophages are derived from circulating nonclassical monocytes. The other surprise was that the source of the circulating monocytes was not the bone marrow directly, but rather the spleen, where monocyte progenitors (hematopoietic stem and progenitor cells) that traveled from bone marrow to specialized niches were stimulated to proliferate and to give rise to monocytes that entered the circulation and were recruited to the injured tissue (55). These observations leave open the question of whether drugs that inhibit CCR2 activity will ever find therapeutic utility in post-myocardial infarction patients. The Swirski-Nahrendorf results raise the possibility that blocking $CCR2^+$ cell recruitment post-myocardial infarction will interfere with tissue healing and remodeling, ultimately resulting in reduced cardiac output (56).

INFLAMMATORY MEDIATORS

Our brief consideration of acute inflammation and its resolution has highlighted the importance of sequential mobilization and differentiation of innate immune cells. Could sequential recruitment and differentiation of leukocyte subsets fit the bill for Humphrey Rang's "big idea" for inflammation biology? If so, it is clear that we will need to understand how different classes of inflammatory mediators are generated, how they change leukocyte migration and activation, and ultimately how these mediators work together to effect tissue repair programs.

The different classes of mediators are discussed briefly below.

Vasoactive amines: Activation of mast cells leads to rapid degranulation and the release of their potent proinflammatory granule contents, e.g., vasoactive amines histamine and serotonin. These mediators act via specific G protein-coupled receptors that can lead to vasodilatation and very rapid changes in cellular behavior, e.g., endothelial cell contraction. Systemic mast cell degranulation can be fatal, for instance, binding of food allergens to specific IgEs bound to mast cells via Fcε receptors.

Cytokines and chemokines: Cytokine genes are transcribed within minutes of exposure of macrophages to DAMPs and PAMPs, and these cytokines can act in an autocrine and paracrine manner to further amplify the transcription of proinflammatory cytokines. Typically, cytokine production by macrophages is measured in response to a single PAMP using primary cells (often bone marrow-derived macrophages) cultured *ex vivo*. The advent of single-cell transcriptomics will allow us to follow changes in the transcription pattern of the whole genome in sentinel cells responding to pathogens in infected tissues *in vivo*. To take full advantage of this surge in information, we will need to stop thinking about cytokines in isolation and start thinking in terms of cytokine networks.

Lipid mediators: In macrophages, TLR signaling increases prostaglandin synthesis by activating cytosolic phospholipase A_2, which releases arachidonic acid from membrane phospholipids and upregulates cyclooxygenase-2 and microsomal prostaglandin E synthase-1 expression. These changes in intracellular lipid pools and lipid-metabolizing enzymes set the scene for generation of multiple proinflammatory prostaglandins and leukotrienes. Later in the acute inflammatory response there is a marked "eicosanoid class switching," which is marked by a switch to production of anti-inflammatory, proresolution lipid mediators such as lipoxin A_4 (57, 58).

Gases as mediators: Since the Nobel Prize-winning discovery of NO as a signaling molecule, we now better appreciate two other gaseous signaling molecules that can act as anti-inflammatory mediators, carbon monoxide (CO) and hydrogen sulfide (H_2S) (59). Vascular endothelial cell production of NO from L-arginine is catalyzed by endothelial NO synthase (NOS3), and this almost ephemeral, very short-lived signaling molecule strongly influences vascular tone via cGMP signaling. CO is generated by heme catabolism by the enzyme heme oxidase-1, and CO exerts its anti-inflammatory effects via the mitogen-activated protein kinase pathway (60). The use of multiple H_2S donors and selective inhibitors of H_2S synthesis has helped to define the cellular actions of this gaseous signaling molecule. The therapeutic effects of H_2S-releasing drugs seen in animal models of inflammation has led to these compounds being taken forward into clinical trials (61).

A BRIEF NOTE ON INFLAMMASOMES AND AUTOINFLAMMATION

As alluded to above, the field of inflammation biology owes much to the seminal papers of Jürg Tschopp, who first identified the cytoplasmic molecular machinery for secretion of active IL-1β, a structure that he termed the "inflammasome" (62). The majority of inflammasomes are formed with one or two Nod-like receptor proteins (NLRs). Other non-NLR proteins, including absent in melanoma 2 (AIM-2) and pyrin, can also form inflammasomes (reviewed in reference 63). The N-terminal pyrin domain (PYD) within NLRs associates with apoptosis-associated speck-like protein containing a CARD domain (ASC), and this permits the recruitment of procaspase-1 to the inflammasome. The NACHT, LRR and PYD domains-containing protein 3 (NLRP3) inflammasome is the most extensively studied inflammasome to date. NLRP3 activation can occur in response to a wide range of stimuli, including intracellular bacterial products, extracellular ATP, monosodium urate, or cholesterol crystals, as well as changes in osmolarity or pH (64, 65). MCC950 is a highly selective inhibitor of NLRP3 that blocks canonical (ATP, monosodium urate) and noncanonical (cytosolic LPS) NLRP3 inflammasome activation at nanomolar concentration. Administration of MCC950 to mice with experimental autoimmune encephalomyelitis, a murine preclinical model of human multiple sclerosis, has been shown to improve clinical symptoms and attenuate IL-1β production (66). The ability to pharmacologically inhibit NLRP3 inflammasome activation in preclinical models will greatly aid investigation of the role of this signaling complex in the pathogenesis of inflammation and could be the starting point for development of novel small-molecule anti-inflammatory drugs.

Over the past 25 years, rheumatologists have come to recognize that autoinflammation is a distinct disease pathology from autoimmunity. TNF receptor-associated periodic syndrome (TRAPS) is a rare genetic disease that causes recurrent episodes of fever that are associated with chills and muscle pain. In 1999, Kastner's group showed that TRAPS did not involve T- or B-lymphocyte activation but rather inappropriate innate immune activation caused by germ line mutations in the 55-kDa TNF receptor 1 (67). The group coined the term "autoinflammation" for the observed defects in the regulation of systemic inflammation. Molecular analysis of other rare genetic diseases characterized by systemic inflammation with no obvious infectious disease or autoimmune component identified a group of rare monogenic diseases with defects in IL-1β production and the NALP3 inflammasome, including cryopyrin-associated periodic syndromes, Muckle-Wells syndrome, and familial cold autoinflammatory syndromes (68). Detailed analysis of other rare monogenic disorders continues to provide further examples of how loss of endogenous regulatory pathways can lead to inappropriate inflammatory and innate immune responses (69).

HYPERACUTE INFLAMMATION: BACTERIAL SEPTIC SHOCK AND VIRAL CYTOKINE STORM

Local activation of macrophages, mast cells, and endothelial cells is essential to mobilize an acute inflammatory response to sites of pathogen invasion and tissue injury. However, systemic activation of these cells by bacterial PAMPs can lead to life-threatening septic shock. An excessive systemic host reaction to viral pathogens such as influenza, frequently termed a "cytokine storm," can also have life-threatening consequences, often in young people with a robust immune system (shown diagrammatically in Fig. 1). The clinical sequelae of septic shock and cytokine storm show the importance of regulating the magnitude of the initial inflammatory response (Fig. 2). The recent appreciation of the need to distinguish sepsis from septic shock serves only to emphasize the importance of return to tissue homeostasis following the initial triggering of a hyperacute inflammatory response (70). There is a substantial unmet clinical need to develop better treatments for septic shock and sepsis, but a better understanding of the disease process is being held back by the lack of good experimental models and the continuing failure to translate basic science findings into effective new treatments (71). One fruitful area for future research might be to identify and augment endogenous pathways that can rapidly decrease the maximal host inflammatory response without "paralyzing" the innate immune system altogether.

CHRONIC INFLAMMATION: PATHOGENESIS AND CURRENT TREATMENTS

In an excellent 2010 review article, Carl Nathan and Aihhao Ding wrote, "The problem with inflammation is not how often it starts, but how often it fails to subside. Non-resolving inflammation is not a primary cause of atherosclerosis, obesity, cancer, chronic obstructive pulmonary disease, asthma, inflammatory bowel disease, multiple sclerosis, or RA, but it contributes significantly to their pathogenesis" (24). In their review, the authors lay out multiple overlapping and competing models for how chronic inflammation can arise *in vivo*.

1. Failure to clear a pathogen that was the original trigger for inciting acute inflammation. Classic examples would be failure to clear mycobacterial infection or chronic virus persistence in hepatitis.
2. Response to continuing tissue injury, for instance, necrotic cell damage and the release of DAMPs caused by ischemia-reperfusion injury.
3. Continuing presence of antigen, e.g., antigens recognized by autoantibodies in RA.
4. Nonresolving inflammation, for instance, the failure to clear macrophages and macrophage-derived foam cells from atherosclerotic lesions in major arteries, leading to the buildup of stable and unstable atherosclerotic plaques, a form of chronic inflammation that persists for decades.

Consideration of the multiplicity of pathogenic mechanisms in human diseases caused by chronic inflammation reminds one of the opening lines of Leo Tolstoy's novel *Anna Karenina*: "All happy families resemble one another, each unhappy family is unhappy in its own way." All successfully resolved bouts of inflammation resemble one another in showing a return to "happy" tissue architecture and essentially normal tissue function. In contrast, each chronic inflammatory disease shows "unhappy" tissue architecture caused by multiple defects in the return to homeostasis.

Current treatments for chronic inflammation include steroids, nonsteroidal anti-inflammatory drugs, disease-modifying antirheumatic drugs, and biologicals, i.e., recombinant monoclonal antibodies or decoy receptors that block inflammatory cytokine function. An important mode of action shared by these drugs is that they reduce inflammatory leukocyte recruitment and dampen adaptive immune responses. Although current anti-inflammatory drugs have different targets and widely different modes of action, they all share a common side effect, namely, a potentially life-threatening reduction in the host immune response to infectious disease. In a recent review, Ira Tabas and Chris Glass (72) give a good discussion of the challenges of developing new anti-inflammatory drugs and how we might achieve selective dampening of myeloid cell recruitment without compromise of "first responders" to infectious disease and wound repair. One promising approach might be to target proresolving agents specifically to sites of chronic inflammation; a proof of concept of this approach using nanoparticles was recently published by Tabas and coworkers (73).

UNANSWERED QUESTIONS AND FUTURE DIRECTIONS IN INFLAMMATION BIOLOGY

Following our brief overview of the role of myeloid cells in inflammation, we identify seven areas where further research is needed and offer seven questions for those working in the field.

Macrophage Differentiation and Plasticity *In Vivo*

For an excellent historical overview of the macrophage M0-M1-M2 concept and a critical assessment of the current literature, we thoroughly recommend a recent review by Fernando Martinez and Siamon Gordon (74). It is tempting to think about macrophage subsets in the same way we think about T-cell subsets in disease. However, advances in flow cytometry, including single-cell detection of cytokine production, have thrown up many more T-cell classifications than the simplistic Th1-Th2-Th17 classifications that we previously used to explain the role of T cells in disease pathogenesis.

Key questions for the field include the following. Do M1 macrophages, defined by surface expression of a handful of different markers, really do something *substantially* different from "regular" F4/80^{hi} tissue-resident macrophages in the context of an ongoing inflammatory response?' Do M2 (alternatively activated) macrophages, defined *in vitro* by a cluster of markers, really enhance tissue repair in the context of inflammation resolution or wound repair? Put another way: How plastic is macrophage differentiation *within tissues*?

The issue of macrophage plasticity *in vivo* has been placed center stage by two important papers published in 2014 (75, 76). Lavin et al. (76) elegantly demonstrated the plasticity of tissue-resident macrophage differentiation by taking F4/80^{hi} peritoneal macrophages from CD45.1 donor mice and instilling them into the lungs of CD45.2 recipient mice (without irradiation). In parallel they transferred F4/80^{hi} alveolar macrophages from CD45.1 donor mice and injected them into the peritoneum of CD45.2 recipient mice. CD45.1 donor cells were recovered from their new tissue microenvironments 3 weeks later and their transcriptomes and chromatin marks were analyzed and compared to those of resident peritoneal and alveolar macrophages in the same tissue. Strikingly, Lavin et al. showed a complete reprogramming of chromatin marks and gene expression patterns in donor macrophages so as to match the macrophages already resident in the recipient tissue. Combined with the intellectual framework provided by Chris Glass and coworkers from their studies of enhancer transcripts and transcription factor binding in myeloid cells, we now have a much better understanding of how macrophage subset gene expression patterns are established and reprogrammed (77).

Building on these seminal studies, we need to develop robust methodologies for switching specific macrophage M1 and M2 functions on or off *in situ*. Such an experimental approach will allow us to move from observing changes in gene expression to actually changing macrophage cellular behaviors within a site of chronic inflammation. The application of optogenetic technologies, until now largely confined to the CNS, might be one way to achieve this ambitious goal.

In the future, could changing macrophage behavior in chronic inflammation be used as a novel therapeutic modality, e.g., enhancing macrophage emigration from atherosclerotic plaques, or changing the behavior of tumor-associated macrophages?

Functional Assays for Proresolution Macrophages

Over the last 30 years, intense study of the molecular biology of inflammatory mediators, be they pro- or anti-inflammatory, has occurred at the expense of studying the cell biology of tissue repair and healing. Inflammation research would benefit greatly from developing useful *ex vivo* and *in vivo* models of tissue repair and fibrosis. Ideally such models should incorporate features of (patho)physiological ECM and tissue architecture, e.g., three-dimensional cultures rather than two-dimensional tissue cultures on plastic. In recent papers, Dalli, Serhan, and colleagues have shown that sulfido-conjugates of the proresolution lipid mediator maresin hasten the resolution of acute inflammatory responses. To test the ability of these mediators to promote tissue regeneration, they have turned to tissue regeneration models using a planaria flatworm model (78, 79). Tissue repair in mammals exhibits significant differences from the tissue regeneration seen in flatworms and urodeles following amputation. This important caveat emphasizes the need for developing more model systems to study the important process of tissue repair and healing.

Will a better understanding of successful tissue repair processes in mammals allow us to develop treatments to prevent the irreversible effects of chronic inflammation: tissue destruction and fibrosis?

Time of Day and Time of Life: Circadian Rhythms and Inflammaging

Nearly all aspects of mammalian physiology have been shown to be strongly influenced by the time of day. A series of experiments from the laboratory of Ajay Chawla elegantly demonstrated that even something as critical to the host as response to *Listeria monocytogenes* infection is susceptible to circadian rhythm. Having demonstrated a 2- to 3-fold circadian oscillation in Ly6C^{hi} monocyte numbers in blood and spleen, Nguyen et al. (80) infected two groups of C57BL/6J mice being kept on a 12-h light/dark cycle with the same intraperitoneal dose of *L. monocytogenes* 8 h apart in terms of their Zeitgeber time (ZT). Mice infected at ZT 8 had reduced bacterial counts in blood and tissues 2 days postinfection compared to mice infected at ZT 0, and this was correlated with improved recruitment of Ly6C^{hi} monocytes to the site of infection. Mice with myeloid-specific deletion of the clock gene *Bmal1* showed no rhythmic alterations in Ly6C^{hi} monocyte numbers in blood, spleen, or bone marrow and none of the circadian variation in chemokine gene expression seen in wild-type animals (80). The impact of circadian oscillation on innate immune cell numbers and myeloid cell gene expression patterns may explain the link between shift work; altered sleep patterns; and increased risk of obesity, diabetes, certain cancers, and cardiovascular disease (81).

Aging populations, including our own, are characterized by chronic, low-grade inflammation often accompanied by elevated cortisol levels. This phenomenon merits further investigation because with few exceptions all age-related diseases have a strong inflammatory component. The term "inflammaging" was coined by Franceschi and colleagues in 2000 and has become somewhat of a "poster child" for systems biology ever since (82). In Fig. 3, we have tried to represent key features of inflammaging in an idealized time course of inflammatory responses to a single inflammatory stimulus. The first point to note is that the inflammatory response in older subjects starts from an elevated baseline at $t = 0$, but an important and key question is whether older experimental subjects show an elevated maximal response to the same inflammatory stimulus and whether the resolution of inflammation occurs more slowly in comparison with younger test subjects. In contrast to studying how the inflammatory response varies in inbred mouse strains over a 24-h period, experimental investigation of how inflammatory responses are altered in older cohorts of experimental animals will be challenging but could bring novel mechanistic insights. Studies on exactly how inflammatory responses in humans change with age will be purely comparative (for instance, genome-wide association studies of people living to 100 years of age), but studying the efficacy of immune responses to, say, influenza vaccines in different human cohorts might be a promising place to start teasing apart changes in the immune response with age.

Future clinical and preclinical studies should take more account of the age of participants and the time of day when studies are performed. For instance, does the efficacy of anti-inflammatory drugs vary with the time of delivery and/or age of the patient? Another important challenge will be to decide whether anti-inflammatory/proresolution agents should be given in extended- versus short-release formulation. The optimal drug formulation may well depend on the "target" and the specific disease.

MicroRNAs: Nature's Own Cytokine Network Regulators?

MicroRNAs (miRNAs) are estimated to modulate the expression of ~30% of all protein-encoding genes in the mammalian genome, so it is not surprising that multiple miRNAs have been shown to modulate myeloid cell inflammatory responses in a number of different settings. Silencing miRNA expression *in vivo* through the use of chemically modified oligonucleotides called "antagomirs" has emerged as a new therapeutic modality for the development of novel anti-inflammatory drugs. The path to randomized clinical trials faces obvious problems, not least the mode of delivery and targeting of antagomirs or miRNAs to the site of inflammation. However, the recent FDA approval of Kynamro (mipomersen sodium), of a closely related therapeutic class of molecules—antisense oligonucleotides—gives rise to optimism for this approach. In a recent report, Wang et al. (83) compared the efficacy of systemic delivery of an anti-miR21 compound versus the same reagent delivered via a drug-eluting stent in an experimental animal model of in-stent restenosis. The authors showed that an anti-miR21 oligonucleotide could prevent in-stent restenosis in a balloon-injured human internal mammary artery transplanted into nude rats equally well with either

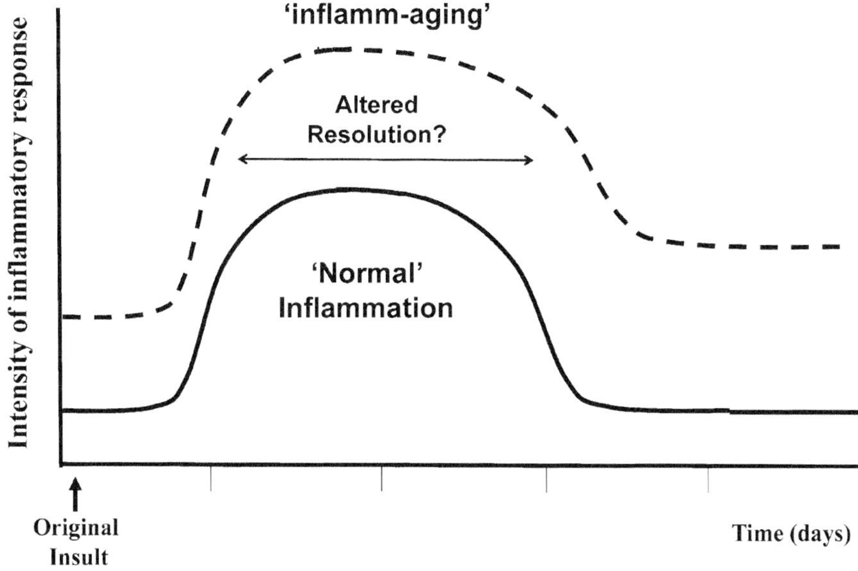

Figure 3 Inflammaging: do inflammatory responses change with age? This schematic representation compares the "normal" inflammatory response (solid line) with the inflammatory response seen in aged populations (dashed line). Aged populations show increased basal levels of systemic inflammation and may show differences in the magnitude of the response to inflammatory stimuli and/or altered resolution.

delivery method, but an anti-miR21-coated stent gave rise to fewer side effects.

Will a better understanding of the roles miRNAs play in modulating inflammatory responses lead to the development of new drugs for regulating cytokine networks?

Metabolic Inflammation and Obesity: Mediators and Microbiota

Increasing levels of obesity in the developed world have led to an alarming increase in the incidence of type 2 diabetes and cardiovascular disease. These conditions are associated with increased systemic inflammation and leukocyte infiltration into white adipose tissue (WAT). Healthy WAT in lean subjects is characterized by a population of M2 macrophages with the type 2 cytokines IL-4 and IL-13 for M2 polarization being provided by eosinophils, which are present in reduced numbers in obese WAT. Furthermore, proinflammatory monocytes recruited into WAT differentiate into M1 macrophages that contribute to insulin resistance and dyslipidemia (84). It is not yet clear whether targeting leukocyte recruitment to and leukocyte activation within adipose tissue will be a viable strategy to prevent the conversion of obesity and metabolic syndrome into type 2 diabetes.

Obesity is one of the more obvious manifestations of a "Western lifestyle," but some immunologists have postulated a link between the increased incidence of allergies, asthma, and autoimmunity and changes in diet and host microbiota-generated metabolites (85). Studies in preclinical models of inflammation have revealed the importance of a series of metabolite receptors, including GPR41, GPR43, GPR109, and GPR120, expressed by innate immune cells that can have potent anti-inflammatory actions. One of the first papers to reveal a link between diet, the gut microbiota, short-chain fatty acids, and inflammation was provided by the laboratory of Charles Mackay in 2009. Maslowski et al. (86) showed that $Gpr43^{-/-}$ mice had excessive inflammation in models of colitis, arthritis, and asthma. Germ-free mice showed a similar exacerbation of inflammatory responses consistent with bacterial fermentation of complex carbohydrate-generating short-chain fatty acids for stimulation of GPR43's anti-inflammatory actions.

In the future, will we be able to exploit our increasing knowledge of cross talk between the microbiota, microbial metabolites, and host metabolite receptors to modulate host innate and adaptive immune responses for therapeutic benefit?

Toward a More Molecular Definition of Inflammation

In all pathology textbooks, inflammation is defined by timing, being either short-term (acute) or long-standing (chronic). An alternative classification of inflammation could be envisaged that takes account of the inciting stimulus, e.g., metabolic inflammation, or the inflammatory cytokines that drive a specific chronic inflammatory disease process, e.g., TNF- versus IL-6-driven disease (87). Rapid advances in combined liquid chromatography-mass spectroscopy techniques for inflammatory exudates may soon allow us to draw up alternative classifications based on tissue responses to infectious diseases. A striking example is provided by two recent papers that followed the time course of lipid mediator appearance in the lungs of humans hospitalized with influenza infection and mice infected with the same titer of different strains of influenza that differed in their virulence. Morita et al. (88) and Tam et al. (89) undertook a detailed analysis of host lipid mediators and their regulation during the course of influenza infection and correlated these changes with viral replication, host immune responses including transcriptomics, and measurement of cytokine and chemokine profiles. One important result from these two extensive lipidomics studies in clinical cohorts and preclinical infection models was the identification of the endogenous lipid mediator protectin D1 as a potent inhibitor of influenza infection. In the longer term, advances in liquid chromatography-mass spectroscopy technology and our ability to interpret large data sets may allow us to better assess the severity of respiratory distress and virus pathology through analysis of nasal swabs of patients hospitalized with influenza (88, 89).

Ultimately, will better identification of the molecules and receptors that drive acute and chronic inflammation lead to better treatments and better patient outcomes?

Future Anti-Inflammatory Drug Targets and Clinical Trials

In 20 years' time, when the current front-line biologics du jour, such as anti-TNF-α, anti-IL-1β, anti-IL-6, anti-IL-17, etc., have gone generic, we may well view these pioneering monoclonal antibodies and decoy receptors as excellent "test reagents" used to identify the specific cytokine networks or specific cell types that cause chronic inflammatory responses. Currently a subgroup of RA and multiple sclerosis patients seem to respond better to B-lymphocyte-depleting antibodies, such as the anti-CD20 monoclonal rituximab, than they do to anti-TNF-α biologics. In the future, we would like to identify patients whose disease will respond better to anti-B-lymphocyte therapy as soon as possible to avoid "hit or miss" dosing with powerful and expensive drugs. Careful analysis of current clinical outcomes combined with biomarker analysis and pharmacogenomics studies will have the twin benefit of improved outcomes for patients and new insights into the pathogenesis of chronic inflammatory disease in human cohorts (87).

Rheumatologists have long sought biomarkers, haplotypes, environmental factors, and genetic markers that can identify patients at increased risk of developing RA. Recently biomarker panels have been expanded to include serum titers of anti-citrullinated peptide antibodies (ACPAs). Early reports have given rise to the idea that this class of autoantibodies may be driving chronic inflammation in tissues other than the joints, most notably within the lungs of patients before they present with joint inflammation and are diagnosed with RA (87). If ACPA screening could identify people who will go on to develop debilitating RA up to a decade later, what will we do with this knowledge? Should we treat this pre-RA lung inflammation aggressively with systemic anticytokine biologics or should we use inhaled glucocorticoids? Alternatively, should we call back people with high ACPA titers for radiological assessment every year and initiate aggressive anti-inflammatory treatment as soon as we see any sign of joint disease? An analogy can perhaps be drawn with the link between elevated plasma low-density lipoprotein, accelerated atherosclerosis, and the increased risk of cardiovascular disease, in which primary prevention emphasizes lifestyle changes and prescription of statins.

In the future, will we start prescribing anti-inflammatory drugs for people with prediabetes, pre-RA, or even predementia?

Acknowledgments. *D.R.G. thanks Siamon Gordon for introducing him to macrophage biology and Thomas Schall for introducing him to chemokines and their receptors. We thank Lewis Taylor and Sophia Valaris for constructive criticisms. Work in the laboratory of D.R.G. is supported by the British Heart Foundation (RG/10/15/28578, RG/15/10/23915). Work in the laboratory of E.A.F. is supported by the U.S. National Institutes of Health (HL098055 and DK095684). We are grateful to the Royal Society for an International Exchange Grant (IE120747).*

Citation. Iqbal AJ, Fisher EA, Greaves DR. 2016. Inflammation—a critical appreciation of the role of myeloid cells. Microbiol Spectrum 4(5):MCHD-0027-2016.

References

1. **Gordon S.** 2008. Elie Metchnikoff: father of natural immunity. *Eur J Immunol* **38:**3257–3264.
2. **Shrive AK, Metcalfe AM, Cartwright JR, Greenhough TJ.** 1999. C-reactive protein and SAP-like pentraxin are both present in *Limulus polyphemus* haemolymph: crystal structure of *Limulus* SAP. *J Mol Biol* **290:**997–1008.
3. **Tharia HA, Shrive AK, Mills JD, Arme C, Williams GT, Greenhough TJ.** 2002. Complete cDNA sequence of SAP-like pentraxin from *Limulus polyphemus*: implications for pentraxin evolution. *J Mol Biol* **316:**583–597.
4. **Mantovani A, Valentino S, Gentile S, Inforzato A, Bottazzi B, Garlanda C.** 2013. The long pentraxin PTX3: a paradigm for humoral pattern recognition molecules. *Ann N Y Acad Sci* **1285:**1–14.
5. **Simons JP, Loeffler JM, Al-Shawi R, Ellmerich S, Hutchinson WL, Tennent GA, Petrie A, Raynes JG, de Souza JB, Lawrence RA, Read KD, Pepys MB.** 2014. C-reactive protein is essential for innate resistance to pneumococcal infection. *Immunology* **142:**414–420.
6. **Du Clos TW.** 2013. Pentraxins: structure, function, and role in inflammation. *ISRN Inflamm* **379040:** doi:10.1155/2013/379040.
7. **Garlanda C, Hirsch E, Bozza S, Salustri A, De Acetis M, Nota R, Maccagno A, Riva F, Bottazzi B, Peri G, Doni A, Vago L, Botto M, De Santis R, Carminati P, Siracusa G, Altruda F, Vecchi A, Romani L, Mantovani A.** 2002. Non-redundant role of the long pentraxin PTX3 in anti-fungal innate immune response. *Nature* **420:**182–186.
8. **Reading PC, Bozza S, Gilbertson B, Tate M, Moretti S, Job ER, Crouch EC, Brooks AG, Brown LE, Bottazzi B, Romani L, Mantovani A.** 2008. Antiviral activity of the long chain pentraxin PTX3 against influenza viruses. *J Immunol* **180:**3391–3398.
9. **Oikonomopoulou K, Ricklin D, Ward PA, Lambris JD.** 2012. Interactions between coagulation and complement—their role in inflammation. *Semin Immunopathol* **34:**151–165.
10. **Gerard NP, Gerard C.** 1991. The chemotactic receptor for human C5a anaphylatoxin. *Nature* **349:**614–617.
11. **Lemaitre B, Nicolas E, Michaut L, Reichhart JM, Hoffmann JA.** 1996. The dorsoventral regulatory gene cassette spätzle/Toll/cactus controls the potent antifungal response in *Drosophila* adults. *Cell* **86:**973–983.
12. **Janeway CA Jr.** 1989. Approaching the asymptote? Evolution and revolution in immunology. *Cold Spring Harb Symp Quant Biol* **54:**1–13.
13. **Janeway CA Jr, Medzhitov R.** 2002. Innate immune recognition. *Annu Rev Immunol* **20:**197–216.
14. **Broxmeyer HE, Orschell CM, Clapp DW, Hangoc G, Cooper S, Plett PA, Liles WC, Li X, Graham-Evans B, Campbell TB, Calandra G, Bridger G, Dale DC, Srour EF.** 2005. Rapid mobilization of murine and human hematopoietic stem and progenitor cells with AMD3100, a CXCR4 antagonist. *J Exp Med* **201:**1307–1318.
15. **Zabel BA, Rott A, Butcher EC.** 2015. Leukocyte chemoattractant receptors in human disease pathogenesis. *Annu Rev Pathol* **10:**51–81.
16. **Boring L, Gosling J, Cleary M, Charo IF.** 1998. Decreased lesion formation in CCR2$^{-/-}$ mice reveals a role for chemokines in the initiation of atherosclerosis. *Nature* **394:**894–897.

17. **Mackay CR.** 2008. Moving targets: cell migration inhibitors as new anti-inflammatory therapies. *Nat Immunol* **9:** 988–998.
18. **White GE, Iqbal AJ, Greaves DR.** 2013. CC chemokine receptors and chronic inflammation—therapeutic opportunities and pharmacological challenges. *Pharmacol Rev* **65:**47–89.
19. **Martinon F, Burns K, Tschopp J.** 2002. The inflammasome: a molecular platform triggering activation of inflammatory caspases and processing of proIL-β. *Mol Cell* **10:**417–426.
20. **Martinon F, Pétrilli V, Mayor A, Tardivel A, Tschopp J.** 2006. Gout-associated uric acid crystals activate the NALP3 inflammasome. *Nature* **440:**237–241.
21. **Chen CJ, Kono H, Golenbock D, Reed G, Akira S, Rock KL.** 2007. Identification of a key pathway required for the sterile inflammatory response triggered by dying cells. *Nat Med* **13:**851–856.
22. **Chan JK, Roth J, Oppenheim JJ, Tracey KJ, Vogl T, Feldmann M, Horwood N, Nanchahal J.** 2012. Alarmins: awaiting a clinical response. *J Clin Invest* **122:**2711–2719.
23. **Kataoka H, Kono H, Patel Z, Kimura Y, Rock KL.** 2014. Evaluation of the contribution of multiple DAMPs and DAMP receptors in cell death-induced sterile inflammatory responses. *PLoS One* **9:**e104741 doi:10.1371/journal.pone.0104741.
24. **Nathan C, Ding A.** 2010. Nonresolving inflammation. *Cell* **140:**871–882.
25. **Huang Z, Zhao C, Chen Y, Cowell JA, Wei G, Kultti A, Huang L, Thompson CB, Rosengren S, Frost GI, Shepard HM.** 2014. Recombinant human hyaluronidase PH20 does not stimulate an acute inflammatory response and inhibits lipopolysaccharide-induced neutrophil recruitment in the air pouch model of inflammation. *J Immunol* **192:**5285–5295.
26. **Cash JL, White GE, Greaves DR.** 2009. Chapter 17. Zymosan-induced peritonitis as a simple experimental system for the study of inflammation. *Methods Enzymol* **461:**379–396.
27. **Di Rosa M.** 1972. Biological properties of carrageenan. *J Pharm Pharmacol* **24:**89–102.
28. **D'Agostino G, La Rana G, Russo R, Sasso O, Iacono A, Esposito E, Raso GM, Cuzzocrea S, Lo Verme J, Piomelli D, Meli R, Calignano A.** 2007. Acute intracerebroventricular administration of palmitoylethanolamide, an endogenous peroxisome proliferator-activated receptor-α agonist, modulates carrageenan-induced paw edema in mice. *J Pharmacol Exp Ther* **322:**1137–1143.
29. **Sedgwick AD, Sin YM, Edwards JC, Willoughby DA.** 1983. Increased inflammatory reactivity in newly formed lining tissue. *J Pathol* **141:**483–495.
30. **Notley CA, Inglis JJ, Alzabin S, McCann FE, McNamee KE, Williams RO.** 2008. Blockade of tumor necrosis factor in collagen-induced arthritis reveals a novel immunoregulatory pathway for Th1 and Th17 cells. *J Exp Med* **205:**2491–2497.
31. **Kouskoff V, Korganow AS, Duchatelle V, Degott C, Benoist C, Mathis D.** 1996. Organ-specific disease provoked by systemic autoimmunity. *Cell* **87:**811–822.
32. **Jacobs JP, Ortiz-Lopez A, Campbell JJ, Gerard CJ, Mathis D, Benoist C.** 2010. Deficiency of CXCR2, but not other chemokine receptors, attenuates autoantibody-mediated arthritis in a murine model. *Arthritis Rheum* **62:**1921–1932.
33. **Moore KJ, Tabas I.** 2011. Macrophages in the pathogenesis of atherosclerosis. *Cell* **145:**341–355.
34. **Klotz L, Norman S, Vieira JM, Masters M, Rohling M, Dubé KN, Bollini S, Matsuzaki F, Carr CA, Riley PR.** 2015. Cardiac lymphatics are heterogeneous in origin and respond to injury. *Nature* **522:**62–67.
35. **Newson J, Stables M, Karra E, Arce-Vargas F, Quezada S, Motwani M, Mack M, Yona S, Audzevich T, Gilroy DW.** 2014. Resolution of acute inflammation bridges the gap between innate and adaptive immunity. *Blood* **124:** 1748–1764.
36. **Clarke TB, Davis KM, Lysenko ES, Zhou AY, Yu Y, Weiser JN.** 2010. Recognition of peptidoglycan from the microbiota by Nod1 enhances systemic innate immunity. *Nat Med* **16:**228–231.
37. **Ferrer R, Martin-Loeches I, Phillips G, Osborn TM, Townsend S, Dellinger RP, Artigas A, Schorr C, Levy MM.** 2014. Empiric antibiotic treatment reduces mortality in severe sepsis and septic shock from the first hour: results from a guideline-based performance improvement program. *Crit Care Med* **42:**1749–1755.
38. **Collins PE, Carmody RJ.** 2015. The regulation of endotoxin tolerance and its impact on macrophage activation. *Crit Rev Immunol* **35:**293–323.
39. **Arts RJ, Gresnigt MS, Joosten LA, Netea MG.** 2016. Cellular metabolism of myeloid cells in sepsis. *J Leukoc Biol* doi:10.1189/jlb.4MR0216-066R.
40. **Bannenberg GL, Chiang N, Ariel A, Arita M, Tjonahen E, Gotlinger KH, Hong S, Serhan CN.** 2005. Molecular circuits of resolution: formation and actions of resolvins and protectins. *J Immunol* **174:**4345–4355.
41. **Yazid S, Norling LV, Flower RJ.** 2012. Anti-inflammatory drugs, eicosanoids and the annexin A1/FPR2 anti-inflammatory system. *Prostaglandins Other Lipid Mediat* **98:**94–100.
42. **Morris T, Stables M, Hobbs A, de Souza P, Colville-Nash P, Warner T, Newson J, Bellingan G, Gilroy DW.** 2009. Effects of low-dose aspirin on acute inflammatory responses in humans. *J Immunol* **183:**2089–2096.
43. **Parham P.** 2009. *The Immune System*, 3rd edition. Garland Science, New York, NY and London.
44. **Sieweke MH, Allen JE.** 2013. Beyond stem cells: self-renewal of differentiated macrophages. *Science* **342:** 1242974 doi:10.1126/science.1242974.
45. **Gautier EL, Jakubzick C, Randolph GJ.** 2009. Regulation of the migration and survival of monocyte subsets by chemokine receptors and its relevance to atherosclerosis. *Arterioscler Thromb Vasc Biol* **29:**1412–1418.
46. **Geissmann F, Manz MG, Jung S, Sieweke MH, Merad M, Ley K.** 2010. Development of monocytes, macrophages, and dendritic cells. *Science* **327:**656–661.
47. **Si Y, Tsou CL, Croft K, Charo IF.** 2010. CCR2 mediates hematopoietic stem and progenitor cell trafficking to sites of inflammation in mice. *J Clin Invest* **120:**1192–1203.

48. **White GE, McNeill E, Channon KM, Greaves DR.** 2014. Fractalkine promotes human monocyte survival via a reduction in oxidative stress. *Arterioscler Thromb Vasc Biol* **34:**2554–2562.

49. **Mildner A, Yona S, Jung S.** 2013. A close encounter of the third kind: monocyte-derived cells. *Adv Immunol* **120:**69–103.

50. **Geissmann F, Jung S, Littman DR.** 2003. Blood monocytes consist of two principal subsets with distinct migratory properties. *Immunity* **19:**71–82.

51. **Woollard KJ, Geissmann F.** 2010. Monocytes in atherosclerosis: subsets and functions. *Nat Rev Cardiol* **7:**77–86.

52. **Lu H, Huang D, Saederup N, Charo IF, Ransohoff RM, Zhou L.** 2011. Macrophages recruited via CCR2 produce insulin-like growth factor-1 to repair acute skeletal muscle injury. *FASEB J* **25:**358–369.

53. **Swirski FK, Nahrendorf M.** 2013. Leukocyte behavior in atherosclerosis, myocardial infarction, and heart failure. *Science* **339:**161–166.

54. **Hilgendorf I, Gerhardt LM, Tan TC, Winter C, Holderried TA, Chousterman BG, Iwamoto Y, Liao R, Zirlik A, Scherer-Crosbie M, Hedrick CC, Libby P, Nahrendorf M, Weissleder R, Swirski FK.** 2014. Ly-6C^{high} monocytes depend on Nr4a1 to balance both inflammatory and reparative phases in the infarcted myocardium. *Circ Res* **114:**1611–1622.

55. **Leuschner F, Rauch PJ, Ueno T, Gorbatov R, Marinelli B, Lee WW, Dutta P, Wei Y, Robbins C, Iwamoto Y, Sena B, Chudnovskiy A, Panizzi P, Keliher E, Higgins JM, Libby P, Moskowitz MA, Pittet MJ, Swirski FK, Weissleder R, Nahrendorf M.** 2012. Rapid monocyte kinetics in acute myocardial infarction are sustained by extramedullary monocytopoiesis. *J Exp Med* **209:**123–137.

56. **Dutta P, Sager HB, Stengel KR, Naxerova K, Courties G, Saez B, Silberstein L, Heidt T, Sebas M, Sun Y, Wojtkiewicz G, Feruglio PF, King K, Baker JN, van der Laan AM, Borodovsky A, Fitzgerald K, Hulsmans M, Hoyer F, Iwamoto Y, Vinegoni C, Brown D, Di Carli M, Libby P, Hiebert SW, Scadden DT, Swirski FK, Weissleder R, Nahrendorf M.** 2015. Myocardial infarction activates CCR2^{+} hematopoietic stem and progenitor cells. *Cell Stem Cell* **16:**477–487.

57. **Dennis EA, Norris PC.** 2015. Eicosanoid storm in infection and inflammation. *Nat Rev Immunol* **15:**511–523.

58. **Norris PC, Gosselin D, Reichart D, Glass CK, Dennis EA.** 2014. Phospholipase A_2 regulates eicosanoid class switching during inflammasome activation. *Proc Natl Acad Sci USA* **111:**12746–12751.

59. **Wallace JL, Ianaro A, Flannigan KL, Cirino G.** 2015. Gaseous mediators in resolution of inflammation. *Semin Immunol* **27:**227–233.

60. **Otterbein LE, Bach FH, Alam J, Soares M, Tao Lu H, Wysk M, Davis RJ, Flavell RA, Choi AM.** 2000. Carbon monoxide has anti-inflammatory effects involving the mitogen-activated protein kinase pathway. *Nat Med* **6:** 422–428.

61. **Wallace JL, Wang R.** 2015. Hydrogen sulfide-based therapeutics: exploiting a unique but ubiquitous gasotransmitter. *Nat Rev Drug Discov* **14:**329–345.

62. **O'Neill LA.** 2011. Retrospective. Jürg Tschopp (1951–2011). *Science* **332:**679 doi:10.1126/science.1207046.

63. **Guo H, Callaway JB, Ting JP.** 2015. Inflammasomes: mechanism of action, role in disease, and therapeutics. *Nat Med* **21:**677–687.

64. **Ip WK, Medzhitov R.** 2015. Macrophages monitor tissue osmolarity and induce inflammatory response through NLRP3 and NLRC4 inflammasome activation. *Nat Commun* **6:**6931 doi:10.1038/ncomms7931.

65. **Vanaja SK, Rathinam VA, Fitzgerald KA.** 2015. Mechanisms of inflammasome activation: recent advances and novel insights. *Trends Cell Biol* **25:**308–315.

66. **Coll RC, Robertson AA, Chae JJ, Higgins SC, Muñoz-Planillo R, Inserra MC, Vetter I, Dungan LS, Monks BG, Stutz A, Croker DE, Butler MS, Haneklaus M, Sutton CE, Núñez G, Latz E, Kastner DL, Mills KH, Masters SL, Schroder K, Cooper MA, O'Neill LA.** 2015. A small-molecule inhibitor of the NLRP3 inflammasome for the treatment of inflammatory diseases. *Nat Med* **21:**248–255.

67. **McDermott MF, Aksentijevich I, Galon J, McDermott EM, Ogunkolade BW, Centola M, Mansfield E, Gadina M, Karenko L, Pettersson T, McCarthy J, Frucht DM, Aringer M, Torosyan Y, Teppo AM, Wilson M, Karaarslan HM, Wan Y, Todd I, Wood G, Schlimgen R, Kumarajeewa TR, Cooper SM, Vella JP, Amos CI, Mulley J, Quane KA, Molloy MG, Ranki A, Powell RJ, Hitman GA, O'Shea JJ, Kastner DL.** 1999. Germline mutations in the extracellular domains of the 55 kDa TNF receptor, TNFR1, define a family of dominantly inherited autoinflammatory syndromes. *Cell* **97:**133–144.

68. **Henderson C, Goldbach-Mansky R.** 2010. Monogenic autoinflammatory diseases: new insights into clinical aspects and pathogenesis. *Curr Opin Rheumatol* **22:**567–578.

69. **de Jesus AA, Canna SW, Liu Y, Goldbach-Mansky R.** 2015. Molecular mechanisms in genetically defined autoinflammatory diseases: disorders of amplified danger signaling. *Annu Rev Immunol* **33:**823–874.

70. **Deutschman CS, Tracey KJ.** 2014. Sepsis: current dogma and new perspectives. *Immunity* **40:**463–475.

71. **Ward PA.** 2012. New approaches to the study of sepsis. *EMBO Mol Med* **4:**1234–1243.

72. **Tabas I, Glass CK.** 2013. Anti-inflammatory therapy in chronic disease: challenges and opportunities. *Science* **339:**166–172.

73. **Fredman G, Kamaly N, Spolitu S, Milton J, Ghorpade D, Chiasson R, Kuriakose G, Perretti M, Farokhzad O, Tabas I.** 2015. Targeted nanoparticles containing the proresolving peptide Ac2-26 protect against advanced atherosclerosis in hypercholesterolemic mice. *Sci Transl Med* **7:**275ra20 doi:10.1126/scitranslmed.aaa1065.

74. **Martinez FO, Gordon S.** 2014. The M1 and M2 paradigm of macrophage activation: time for reassessment. *F1000Prime Rep* **6:**13 doi:10.12703/P6-13.

75. **Gosselin D, Link VM, Romanoski CE, Fonseca GJ, Eichenfield DZ, Spann NJ, Stender JD, Chun HB, Garner H, Geissmann F, Glass CK.** 2014. Environment drives selection and function of enhancers controlling tissue-specific macrophage identities. *Cell* **159:**1327–1340.

76. **Lavin Y, Winter D, Blecher-Gonen R, David E, Keren-Shaul H, Merad M, Jung S, Amit I.** 2014. Tissue-resident macrophage enhancer landscapes are shaped by the local microenvironment. *Cell* **159:**1312–1326.

77. **Glass CK.** 2015. Genetic and genomic approaches to understanding macrophage identity and function. *Arterioscler Thromb Vasc Biol* **35:**755–762.

78. **Dalli J, Ramon S, Norris PC, Colas RA, Serhan CN.** 2015. Novel proresolving and tissue-regenerative resolvin and protectin sulfido-conjugated pathways. *FASEB J* **29:** 2120–2136.

79. **Serhan CN, Dalli J, Karamnov S, Choi A, Park CK, Xu ZZ, Ji RR, Zhu M, Petasis NA.** 2012. Macrophage proresolving mediator maresin 1 stimulates tissue regeneration and controls pain. *FASEB J* **26:**1755–1765.

80. **Nguyen KD, Fentress SJ, Qiu Y, Yun K, Cox JS, Chawla A.** 2013. Circadian gene *Bmal1* regulates diurnal oscillations of Ly6C^{hi} inflammatory monocytes. *Science* **341:** 1483–1488.

81. **Castanon-Cervantes O, Wu M, Ehlen JC, Paul K, Gamble KL, Johnson RL, Besing RC, Menaker M, Gewirtz AT, Davidson AJ.** 2010. Dysregulation of inflammatory responses by chronic circadian disruption. *J Immunol* **185:**5796–5805.

82. **Castellani GC, Menichetti G, Garagnani P, Giulia Bacalini M, Pirazzini C, Franceschi C, Collino S, Sala C, Remondini D, Giampieri E, Mosca E, Bersanelli M, Vitali S, Valle IF, Liò P, Milanesi L.** 2016. Systems medicine of inflammaging. *Brief Bioinform* **17:**527–540.

83. **Wang D, Deuse T, Stubbendorff M, Chernogubova E, Erben RG, Eken SM, Jin H, Li Y, Busch A, Heeger CH, Behnisch B, Reichenspurner H, Robbins RC, Spin JM, Tsao PS, Schrepfer S, Maegdefessel L.** 2015. Local microRNA modulation using a novel anti-miR-21-eluting stent effectively prevents experimental in-stent restenosis. *Arterioscler Thromb Vasc Biol* **35:**1945–1953.

84. **Carvalheira JB, Qiu Y, Chawla A.** 2013. Blood spotlight on leukocytes and obesity. *Blood* **122:**3263–3267.

85. **Thorburn AN, Macia L, Mackay CR.** 2014. Diet, metabolites, and "western-lifestyle" inflammatory diseases. *Immunity* **40:**833–842.

86. **Maslowski KM, Vieira AT, Ng A, Kranich J, Sierro F, Yu D, Schilter HC, Rolph MS, Mackay F, Artis D, Xavier RJ, Teixeira MM, Mackay CR.** 2009. Regulation of inflammatory responses by gut microbiota and chemoattractant receptor GPR43. *Nature* **461:**1282–1286.

87. **Schett G, Elewaut D, McInnes IB, Dayer JM, Neurath MF.** 2013. How cytokine networks fuel inflammation: toward a cytokine-based disease taxonomy. *Nat Med* **19:** 822–824.

88. **Morita M, Kuba K, Ichikawa A, Nakayama M, Katahira J, Iwamoto R, Watanebe T, Sakabe S, Daidoji T, Nakamura S, Kadowaki A, Ohto T, Nakanishi H, Taguchi R, Nakaya T, Murakami M, Yoneda Y, Arai H, Kawaoka Y, Penninger JM, Arita M, Imai Y.** 2013. The lipid mediator protectin D1 inhibits influenza virus replication and improves severe influenza. *Cell* **153:**112–125.

89. **Tam VC, Quehenberger O, Oshansky CM, Suen R, Armando AM, Treuting PM, Thomas PG, Dennis EA, Aderem A.** 2013. Lipidomic profiling of influenza infection identifies mediators that induce and resolve inflammation. *Cell* **154:**213–227.

Myeloid Cells in Health and Disease: A Synthesis
Edited by Siamon Gordon

doi:10.1128/microbiolspec.MCHD-0035-2016

Melanie Bennett[1]
Derek W. Gilroy[2]

19 Lipid Mediators in Inflammation

INFLAMMATION AND ITS ONSET

Before we discuss lipids and their role in homeostasis and host defense, we will recount the essence of the inflammatory response. Inflammation is a reaction of the microcirculation; it's a protective response initiated after infection or injury. While both local and systemic responses can be activated, inflammation is an essential biological process with the objective of eliminating the inciting stimulus, promoting tissue repair/wound healing, and, in the case of infection, establishing memory such that the host mounts a faster and more specific response upon a future encounter. The acute inflammatory response is a complex yet highly coordinated sequence of events involving a large number of molecular, cellular, and physiological changes. It begins with the production of soluble mediators (complement, chemokines, cytokines, eicosanoids—including prostaglandins [PGs], free radicals, vasoactive amines, etc.) by resident cells in the injured/infected tissue (i.e., tissue macrophages, dendritic cells, lymphocytes, endothelial cells, fibroblasts, and mast cells), concomitant with the upregulation of cell adhesion molecules on both leukocytes and endothelial cells that promote the exudation of proteins and influx of granulocytes from blood (1). Upon arrival, these leukocytes, typically polymorphonuclear leukocytes (PMNs) in the case of nonspecific inflammation or eosinophils in response to allergens, function primarily to phagocytose and eliminate foreign microorganisms via distinct intracellular (superoxide, myeloperoxidase, proteases, and lactoferrins) and/or extracellular (neutrophil extracellular traps) killing mechanisms (2). It is likely that the magnitude of the infectious load and its eventual neutralization signal the next phase of active anti-inflammation and proresolution (3).

RESOLUTION OF INFLAMMATION

It is important to distinguish between inflammatory resolution and inflammatory onset. At onset, local release/activation of soluble mediators (e.g., complement, vasoactive amines, cytokines, and lipids) from histiocytes and stromal cells and upregulation of cell adhesion molecules on the microvascular endothelium collectively facilitate extravascular leukocyte accumulation, manifesting in Celsus' cardinal signs of inflammation: heat, redness, swelling, and pain (Rudolf Virchow added loss of function in the 19th century) (4). This well-characterized phase of the inflammatory response is routinely targeted using drugs including nonsteroidal anti-inflammatory

[1]Roche Products Limited, Shire Park, Welwyn Garden City AL7 1TW, United Kingdom; [2]Centre for Clinical Pharmacology and Therapeutics, Division of Medicine, University College London, London WC1 E6JJ, United Kingdom.

drugs (NSAIDs) and anti-tumor necrosis factor α (TNF-α) agents that inhibit or antagonize the action of these inflammatory drivers, forming the mainstay for treating chronic inflammatory disease. Resolution, however, switches inflammation off. Inasmuch as onset is orchestrated by a host of sequentially released mediators, resolution is an active process that is no longer considered a passive event in which the response was hitherto thought to simply fizzle out (5, 6). For instance, a critical requirement for the inflammatory response to switch off is the elimination of the injurious agents that initiated it in the first place. Failure to achieve this first step will lead to chronic inflammation, as exemplified by chronic granulomatous disease, which results from a failure of the phagocytic NADPH oxidase enzyme system to produce superoxide and kill invading infections, leading to a predisposition to recurrent bacterial and fungal infections and the development of inflammatory granulomas (7). Successfully dispensing with the inciting stimulus will signal a cessation of proinflammatory mediator synthesis and lead to their catabolism. This will halt further leukocyte recruitment and edema formation. These are probably the very earliest determinants for the resolution of acute inflammation, the outcome of which signals the next stage of cell clearance. The clearance phase of resolution, be it PMN or eosinophil driven or adaptive (lymphocyte mediated) in nature, also has a number of mutually dependent steps. The clearance routes available to inflammatory leukocytes include systemic recirculation or local death by apoptosis/necrosis of influxed PMNs, eosinophils, or lymphocytes, followed by their phagocytosis or efferocytosis by recruited monocyte-derived macrophages. Once phagocytosis is complete, macrophages can leave the inflamed site by lymphatic drainage, with evidence that a small population may die locally by apoptosis (8).

Eliminating the injurious agent leads to the next phase of proinflammatory mediator catabolism, in which levels of cytokines, chemokines, eicosanoids, cell adhesion molecules, etc., must revert back to that expressed during the preinflamed state. In terms of chemokines, the atypical chemokine receptors such as D6 are unable to initiate classical signaling pathways after ligand binding, thereby acting as a type of scavenging system for proinflammatory signals, such that in tetradecanoyl phorbol acetate-induced skin inflammation, D6-deficient mice exhibit an excess concentration of chemokines, resulting in a notable inflammatory pathology with similarities to human psoriasis (for review, see reference 9). In addition, the work of Ariel and colleagues showed that CCL3 and CCL5 were increased in peritoneal exudates of $Ccr5^{-/-}$ mice during the resolution of acute peritonitis. Transfer of apoptotic PMNs resulted in CCR5-dependent scavenging of CCL3, CCL4, and CCL5. It transpires that CCR5 surface expression on apoptotic PMNs was reduced by proinflammatory cytokines and was increased by proresolution lipid mediators including lipoxin A_4 (LXA_4) (10), which will be discussed in detail later. Thus, endogenous systems exist to facilitate proinflammatory mediator clearance and whose function, when it becomes dysregulated, may lead to chronic inflammation. If all of these pathways of stimulus removal, inhibition of granulocyte trafficking, proinflammatory mediator catabolism, appropriate cell death/efferocytosis (phagocytosis of apoptotic cells), etc., are followed, then acute inflammation will resolve without causing excessive tissue damage and give little opportunity for the development of chronic, nonresolving inflammation.

Each stage of the resolution cascade represents an opportunity to be harnessed to drive ongoing inflammatory diseases down a proresolution pathway. Yet we caution that this will not be a panacea for all diseases driven by ongoing inflammation. We suspect that resolution processes may vary from tissue to tissue and be dependent on the nature of the injurious stimulus. Thus, designing proresolution drugs will have to be organ and disease specific. With that comes the need for more-appropriate animal models of ongoing inflammation that best reflect the intended human condition. In addition, more studies must be focused on examining resolution pathways in healthy and diseased humans.

RESOLUTION: A DYNAMIC PROCESS WITH CHECKS AND BALANCES

At this stage, it must be emphasized that inflammation leading to resolution is not a sequence of separate events that occur in isolation, but is a dynamic continuum of overlapping events where pro- and anti-inflammation blend seamlessly into proresolution. For instance, proinflammatory signals are activated in an immediate and early manner concomitant with anti-inflammatory signals that serve to temper the magnitude of the early-onset phase of acute inflammation, with PMN influx being a good barometer of inflammation severity. Over the course of hours and only after tissues have sensed that the injurious agent has been neutralized, is it safe to catabolize proinflammatory soluble mediators and switch off proinflammatory signaling pathways. Alongside this is the synthesis of factors that terminate further PMN trafficking and prepare the injured tissue for resolution.

In other words, while it is recognized that proinflammatory mediators generated in the inflamed tissue drive acute inflammation, there is also the systemic and local production of endogenous mediators that counterbalance these proinflammatory events. These internal checks and balances have evolved to avert development of pathologies such as those highlighted above. Lipid mediators derived from polyunsaturated fatty acids (PUFAs), such as arachidonic acid (AA) and the omega-3 PUFAs eicosapentaenoic acid (EPA) and docosahexaenoic acid (DHA), are synthesized during normal cell hemostasis or, more often, after cell activation and in conditions of stress, functioning as activators of counterregulatory, anti-inflammatory, and proresolution mechanisms. Interestingly, these immunomodulatory effects are also found with a family of lipids, called prostanoids, which help to drive some of the cardinal signs of inflammation (heat, redness, swelling, pain, and loss of function). As the role of lipids in inflammation is diverse, this review aims to provide an update of AA/DHA/EPA-derived signaling molecules that not only drive acute inflammation but also counterregulate its severity and bring about its timely resolution.

AA METABOLISM AND THE INFLAMMATORY RESPONSE

AA is a 20-carbon fatty acid and the main eicosanoid precursor and is a constituent of all cells. Although it is not freely available, stimulation by various cellular agonists, including receptor-mediated agonists (e.g., N-formyl-methionyl-leucyl-phenylalanine [fMLP], interleukin-8 [IL-8], and platelet-activating factor), microorganisms, phagocytic particles, and nonspecific stimuli such as damage or injury (11), activates several phospholipase enzymes (predominantly PLA_2), which releases AA from membrane phospholipid stores. Once in the cytosol, AA can be metabolized via three principal pathways to form an important family of oxygenated products, collectively termed eicosanoids, that are released from the source cell and act at nanomolar concentrations in an autocrine/paracrine manner on target cells. PGs and thromboxane (collectively termed prostanoids), formed by cyclooxygenase (COX); leukotrienes (LTs) and lipoxins (LXs), by lipoxygenases (LOXs) (12, 13); and epoxyeicosatrienoic acids (EETs), by cytochrome P450 (CYP) enzymes (14), are members of the eicosanoid family.

CYCLOOXYGENASE

COX is a bifunctional enzyme that acts successively as a bis-dioxygenase and peroxidase to carry out a complex free radical reaction. It begins by catalyzing the bisoxygenation and cyclization of AA to form the hydroperoxy arachidonate metabolite PGG_2 (15), after which the peroxidase element of the enzyme reduces the carbon 15-position hydroperoxide to its corresponding alcohol to form PGH_2 (16, 17). There are two main isoforms involved in the conversion of AA, COX-1 and COX-2. While COX-1 is constitutively expressed in most cells and tissues, COX-2 is rapidly induced when cells are challenged with inflammatory stimuli (18). Although not exclusive, it is generally accepted that COX-1 is involved in cellular housekeeping functions necessary for normal physiological activity, whereas COX-2 acts primarily at sites of inflammation. Formation of biologically active prostanoids from PGH_2 occurs through the actions of a set of synthases that are expressed in a tissue- and cell type-selective fashion. These synthases include prostaglandin D synthase (PGDS) (19), prostaglandin E synthase (PGES) (20), prostaglandin F synthase (PGFS) (21), prostaglandin I synthase (PGIS) (22), and thromboxane A synthase (TXAS) (23), which form PGD_2, PGE_2, $PGF_{2\alpha}$, PGI_2 (also known as prostacyclin), and TXA_2, respectively. It is the differential expression of these enzymes within cells that determines the profile of prostanoid production. For example, mast cells predominantly produce PGD_2 while macrophages produce PGE_2 and TXA_2. Moreover, alterations in the profile of prostanoid synthesis can occur upon cell activation such that resting macrophages produce TXA_2 in excess of PGE_2, but upon cell activation this ratio changes to favor PGE_2 (24). Several biochemical mechanisms have been proposed to explain this altered synthetic profile. First, it has been suggested that physical compartmentalization of COX-1 and COX-2 with specific terminal synthases could link the activity of these enzymes with the synthesis of specific prostanoid end products (25). Second, some of the synthases are inducible, and their expression may be regulated by environmental signals. For example, expression of the glutathione-dependent isoform of PGES is enhanced by IL-1β (26). Finally, it has been proposed that differences in substrate affinity and kinetics of PGES and TXAS account for different production profiles of resting and activated monocytes (27). There is also evidence that the two COX isoforms may preferentially contribute to the synthesis of distinct prostanoids. For instance, in primary peritoneal macrophages, expressing all terminal synthases, COX-1 yields a balance of prostanoids (i.e., PGE_2, PGD_2, PGI_2, and TXA_2) while COX-2 preferentially generates only PGE_2 and PGI_2 (28).

The biological effect of prostanoids is initiated by binding to specific cell surface receptors. Currently there are nine known prostanoid receptors in mice and man: the PGD receptors, DP1 and DP2; the PGE_2 receptors, EP1, EP2, EP3, and EP4; the PGF receptor, FP; the PGI receptor, IP; and the TXA receptor, TP. In addition, there are splice variants of the EP3, FP, and TP receptors differentiated only in their C-terminal tails. All belong to the G protein-coupled receptor (GPCR) superfamily of seven-transmembrane-spanning proteins, with the exception of DP2 (also known as CRTH2), which is a member of the chemoattractant receptor family (29–31). The IP, DP1, EP2, and EP4 receptors signal through G_s, resulting in increased intracellular cyclic AMP (cAMP), whereas the EP3 receptor couples to G_i to reduce cAMP. EP1, FP, and TP receptors signal through G_q to induce calcium mobilization.

PROSTANOIDS

In the mid-1930s, potent bioactive compounds in human semen were identified as prostanoids (32). Today it is appreciated that prostanoids are generated in most tissues and cells, modulating a wide range of biological processes such as smooth muscle tone (33–35), vascular permeability (36, 37), hyperalgesia (38), fever (39–41), and platelet aggregation (42). Indeed, the clinical importance of prostanoids is emphasized by the fact that prostanoid biosynthesis is the target of NSAIDs, one of the most widely used classes of pharmacotherapeutic agents for the treatment of chronic inflammatory diseases.

The more widely studied prostanoids, PGE_2 and PGI_2, both enhance vasodilation (43), edema formation, and vascular permeability, particularly in the presence of histamine, bradykinin, and 5-HT (44–49). Genetic depletion of their respective receptors (IP, EP2, and EP3) in mice significantly reduced pleural exudation after insult with carrageenin or zymosan (50, 51). PGE_2 is also one of the most potent pyretic agents known, with elevated concentrations found in cerebrospinal fluid taken from patients with bacterial or viral infections (52). Indeed, a number of lines of evidence from EP-deficient mice have shown that the febrile response to PGE_2 occurs through the action of PGE_2 on the EP3 receptor present on sensory neurons in the periphery and brain (53–56). This has been postulated to cause an increase in thermogenesis through activation of brown adipose tissue and reduced passive heat loss through the skin by tail artery vasoconstriction (57–61). Although none of the COX metabolites overtly cause pain, PGI_2 and PGE_2 cause peripheral and central hyperalgesia when bound to IP, EP1, EP3, and EP4 receptors by reducing the threshold of nociceptor sensory neurons to stimulation (34, 38, 62–70).

In addition, prostanoids play an important role in protecting against oxidative injury in cardiac tissue (71) and in maintaining cardiovascular (CV) homeostasis. Indeed, the protective effect has been demonstrated in clinical studies undertaken with NSAIDs, which found that COX-2-specific inhibitors increase the risk of stroke, myocardial infarction (MI), thrombosis, systemic and pulmonary hypertension, congestive heart failure, and sudden cardiac death (72, 73). Furthermore, deleting specific prostanoid synthases and receptors results in an augmentation of ischemia/reperfusion injury (74) as well as exacerbating the decline in cardiac function after MI (75, 76). The maintenance of CV health is dependent on a very fine balance between vasodilatory PGI_2 and prothrombotic TXA_2 (77, 78), where PGI_2 functions to counterbalance the actions of TXA_2 (73). Indeed, PGI_2 released from endothelial cells and in synergy with nitric oxide prevents TXA_2-induced platelet aggregation and thrombosis (42, 79, 80). TXA_2 is derived from platelet COX-1, causing platelet aggregation and vascular smooth muscle contraction (81–83). Clinical CV diseases, such as unstable angina, MI, and stroke, can be a result of overproduction of TXA_2. Importantly, the cardioprotective properties of aspirin can be attributed to the covalent inhibition of COX-1 (84).

As well as having "proinflammatory" properties, many prostanoids also exert immunosuppressive effects through upregulation of intracellular cAMP (85–87). For example, PGE_2 and PGI_2 reduce the ability of inflammatory leukocytes to phagocytose and kill microorganisms (88–93), as well as inhibit the production of downstream proinflammatory mediators (94–100) while, in contrast, enhancing the production of IL-10 and IL-6 (101, 102). Indeed, in a number of conditions associated with increased susceptibility to infection, including cancer (103), aging (104), and cystic fibrosis (105, 106), overexpression of PGE_2 has been reported. Interestingly, during the onset phase of inflammation, PGE_2 indirectly results in proresolution effects by switching on the transcription of enzymes required for the generation of LXs (107), resolvins (Rvs), and protectins (PDs) (108–111), other classes of bioactive lipids that are potent proresolution mediators.

As well as eliciting immunomodulatory and anti-inflammatory effects in the same manner as described for PGE_2 and PGI_2 via ligation to DP1, PGD_2 can also act independently of DP1 and DP2 receptor activation when nonenzymatically dehydrated into biologically active prostaglandins of the J_2 series (e.g., PGJ_2, $\Delta^{12,14}$-PGJ_2, and 15-deoxy-$\Delta^{12,14}$-PGJ_2 [15d-PGJ_2]) (112–116). These so called cyclopentenone PGs form covalent attachments with reactive sulfhydryl groups on intracellular regulatory proteins, which enables modulation of their function (117–119). For instance, 15d-PGJ_2, upon ligation to the nuclear receptor peroxisome proliferator-activated receptor (PPAR)-γ (120), decreases proinflammatory cytokine release and modifies gene expression (121, 122), as well as directly inhibiting the actions of IκB kinase (IKK), which is responsible for the activation of NF-κB (123–125). 15d-PGJ_2, independently of PPAR-γ, can preferentially inhibit monocyte rather than neutrophil trafficking through differential regulation of cell adhesion molecule and chemokine expression (8, 126–128); regulate macrophage activation and proinflammatory gene expression (129); and induce leukocyte apoptosis through a caspase-dependent mechanism (8, 115, 130–133). Moreover, it has been shown that PGD_2-derived compounds function as endogenous breaking signals for lymphocytes to stimulate resolution (134).

LIPOXYGENASE

LOX enzymes, including 5-, 12-, and 15-LOX in leukocytes, platelets, and endothelial cells, respectively, metabolize AA. The generation of the slow-reacting substances of anaphylaxis (LTC_4, LTD_4, and LTE_4—potent mediators of the allergic response) (135) and LTB_4, a powerful PMN (i.e., neutrophils and eosinophils) chemoattractant (136, 137), is elicited by leukocyte 5-LOX. Due to its involvement in LT synthesis, 5-LOX has received the most attention in inflammation research. Therefore, the remainder of this section will concentrate specifically on this pathway.

Once activated, 5-LOX converts AA into a hydroperoxide by inserting molecular oxygen into AA at position 5, aided by the 5-LOX activating protein (FLAP). Termed 5-hydroperoxyeicosatetraenoic acid (5-HPETE), this intermediate is then rapidly reduced to 5-hydroxyeicosatetraenoic acid (5-HETE). 5-HPETE can also be converted by removal of water to an unstable 5,6-epoxide containing a conjugated triene structure called LTA_4, which is then converted to either LTB_4 by insertion of a hydroxyl group at carbon-12 (C-12) through the action of LTA_4 hydrolase (138, 139) or LTC_4 by addition of the glutathionyl group at C-6 by γ-glutamyl-*S*-transferase (140). In most cases, LTB_4 and 5-HETE are subsequently secreted from the cell by an unidentified protein carrier (141). LTC_4 is also exported, but by the ATP-dependent multidrug resistance proteins (142), including MRP1 and MRP2. After export, LTC_4 is metabolized by the cleavage of glutamic acid by γ-glutamyl transpeptidase to form LTD_4, which can be further modified by removal of a glycine by cysteinyl glycinase to produce LTE_4. Unlike COX, 5-LOX is inactive in quiescent cells but becomes enzymatically functional after cell activation by increases in intracellular calcium (143), enhanced by ATP (144), or by phosphorylation, which can occur without an increase in calcium (145).

Heptahelical receptors of the rhodopsin class located on the outer leaflet of the plasma membrane of structural and inflammatory cells mediate the effects of LTs (146, 147). To date, four subtypes have been described, B LT receptors 1 and 2 (BLT1 and BLT2) and cysteinyl LT receptors 1 and 2 (CysLT1 and CysLT2). Once LTs have bound, a signal is sent via a G protein in the cytoplasm to increase intracellular calcium and block formation of cAMP, which then alters various cellular activities, ranging from motility to transcriptional activation. While CysLT1 mediates bronchoconstriction, mucus secretion, and edema accumulation in airways (148), CysLT2 contributes to inflammation, vascular permeability, and tissue fibrosis in lungs (149, 150). Indeed, overexpression of CysLT1 is seen in patients with asthma or chronic rhinosinusitis who have aspirin sensitivity (151). In contrast, BLT1 is a high-affinity receptor for LTB_4, mediating all of its chemoattractant and proinflammatory properties (147). Although BLT2 acts in a similar fashion to BLT1, LTB_4 affinity toward BLT1 is much higher. Interestingly, studies employing both an *in vitro* and a murine model of inflammation demonstrate that LTB_4 ligates and activates the anti-inflammatory nuclear receptor PPAR-α (152–155).

LTs IN INFLAMMATION

LTs are generated at sites of infection/inflammation primarily by inflammatory cells, including PMNs, macrophages, and mast cells, and play a critical role in the inflammatory response by acting as proinflammatory lipid mediators. Physiologically, each of the 5-LOX-derived compounds has a distinct role in driving different phases of inflammation. For example, LTB_4 attracts and activates neutrophils, monocytes, and lymphocytes, a hallmark of tissue inflammation (147, 156, 157), whereas LTD_4 is a potent chemoattractant for eosinophils (158). The cysteinyl LTs (LTC_4, LTD_4, and LTE_4), on the other hand, increase vascular permeability and plasma leakage, leading to edema that is characteristic of inflammation (159–163). Pathologically, LTs contribute to a variety of inflammatory and allergic diseases, such as rheumatoid arthritis, inflammatory bowel disease, psoriasis, allergic rhinitis, bronchial asthma, cancer, atherosclerosis, and osteoarthritis (164). This can be seen in asthmatic patients, in whom antileukotriene therapy (i.e., 5-LOX inhibition by zileuton and CysLT1 blockage by montelukast or zafirlukast) resulted in improved pulmonary function, symptoms, and overall quality of life (165–167).

The role of LTs in CV disease has been the subject of intense investigation. In atherosclerotic lesions, for example, 5-LOX activity and levels are associated with the severity of the lesion (168) and plaque instability (169). Furthermore, both LTB_4 and CysLTs participate in the development of atherosclerotic lesions in animals and *in vitro*. LTB_4 increases recruitment of monocytes and their differentiation to foam cells (170), as well as intimal hyperplasia (171). CysLTs, on the other hand, enhance the recruitment of leukocytes into the arterial wall and contribute to thrombosis and vascular remodeling (172, 173). Interestingly, in humans, the incidences of strokes and MI in certain populations have been linked to variants of the genes that encode FLAP and LTA_4 hydrolase, which cause an overproduction of LTs (174–176). Indeed, upon treatment with a FLAP inhibitor (veliflapon), a potent biomarker of inflammation, C-reactive protein, was reduced in one population of patients with a history of MI and one of the variants mentioned above (177). Despite their pathophysiologic role, it has now become apparent that LTs are important participants in the host response against infection (178). For instance, 5-LOX-deficient mice or pharmacological inhibition of LT synthesis caused increased mortality and reduced microbial clearance after challenge with a variety of microbes (e.g., bacteria, mycobacteria, fungi, and parasites) (179–184). Similarly, LT-deficient alveolar macrophages also displayed impaired phagocytosis and intracellular killing of bacteria, an effect that could be overcome with the exogenous introduction of LTB_4 or CysLTs (180, 185). Interestingly, LT deficiency is also a feature of a number of clinical conditions that are associated with impaired microbial clearance (HIV infection, malnutrition, cigarette smoking, vitamin D deficiency, and post-bone marrow transplantation) (186–191). It is believed that LTs enhance microbicidal activities in leukocytes by upregulating production of nitric oxide (192, 193) and the secretion of microbial peptides (194), as well as activating NADPH oxidase to generate reactive oxygen intermediates (185). Recently it has been demonstrated that LTB_4 may also possess anti-inflammatory properties through ligation to PPAR-α in its parent cell (155). It has been suggested that this activation in turn leads to its own catabolism, thus facilitating resolution of the inflammatory process. This hypothesis/theory/stipulation is conceivable and further demonstrates how inflammation is such a finely balanced process that is invoked when required, yet limited and resolved when it is no longer needed.

5-OXO-6,8,11,14-EICOSATETRAENOIC ACID

5-LOX activity also results in the generation of 5-oxo-6,8,11,14-eicosatetraenoic acid (5-oxo-ETE), which has potent biological activities that have only recently become appreciated, including eosinophil activation and chemoattraction. It is formed by the oxidation of 5*S*-HETE by 5-hydroxyeicosanoid dehydrogenase (5-HEDH), a microsomal enzyme widely distributed in both inflammatory and structural cells, including leukocytes and platelets (195). 5-HEDH, however, cannot generate 5-oxo-ETE without $NADP^+$, which is available in large quantities during the respiratory burst, neutrophil apoptosis, and oxidative stress (196, 197). Other endogenously occurring PUFAs (sebaleic acid, mead acid, and EPA) can also be converted to analogous 5-oxo-fatty acids following oxidation by 5-LOX-forming products that are also granulocyte chemoattractants (198–200). Furthermore, both enzymatic and nonenzymatic pathways can further modify 5-oxo-ETE to produce several additional eicosanoids (201).

5-oxo-ETE acts via the OXE receptor (OXE-R), a distinct orphan GPCR (202, 203) that is most highly expressed in human peripheral leukocytes, lungs, kidney, liver, and spleen (204, 205). The relative expression of OXE-R in eosinophils, neutrophils, and macrophages is 200:6:1 (205). OXE-R, once coupled to a $G_{i/o}$ protein (197, 206, 207), activates a number of distinct intracellular signaling pathways including PLC-β (208), phosphatidylinositol 3-kinase, and Akt (206, 208, 209); protein kinase C-δ/ζ (210); as well as extracellular signal-regulated kinase-1/2 and cytosolic PLA_2 (210, 211), which, in turn, could lead to further production of AA-derived metabolites. OXE-R may also inhibit the cascade mediated by adenylyl cyclase and cAMP (204).

5-OXO-ETE AND INFLAMMATION

5-oxo-ETE is produced by eosinophils, neutrophils, basophils, and monocytes, and like other inflammatory lipids, it acts in an autocrine manner. In addition to its most potent property as a chemoattractant for eosinophils (200), 5-oxo-ETE also induces calcium mobilization, actin polymerization, CD11b expression, and L-selectin shedding (201). Furthermore, 5-oxo-ETE induces degranulation and superoxide production in leukocytes primed with cytokines such as granulocyte-macrophage colony-stimulating factor (GM-CSF) and TNF-α, an effect not mirrored in naive cells (207, 211).

In addition, 5-oxo-ETE stimulates human monocytes to secrete GM-CSF (212), which is a potent survival factor for eosinophils. In prostate tumor cells, this lipid prevents apoptosis/proliferation (213, 214).

LXs—BIOSYNTHESIS AND RECEPTORS

LXs are a series of trihydroxytetraene-containing bioactive eicosanoids that were first isolated from human leukocytes in the mid-1980s (13). However, in contrast to LTs and 5-oxo-ETEs, which are manufactured by intracellular biosynthesis, LXs are generated through cell-cell interactions by a process known as transcellular biosynthesis. In different human cell types, during the first biosynthetic step of LX biosynthesis, LOX inserts molecular oxygen into AA. This can be achieved by two major routes—the first pathway involves the oxygenation of AA at C-15 by 15-LOX in eosinophils, monocytes, or epithelial cells (found in the respiratory tract, gastrointestinal tract, and oral cavity), yielding 15*S*-HPETE. Following secretion, 15*S*-HPETE is taken up by either PMNs or monocytes and rapidly converted into 5,6-epoxytetraene by 5-LOX, which is hydrolyzed within these recipient cells by either LXA_4 or LXB_4 hydrolase to bioactive LXA_4 or LXB_4. Interestingly, this process also markedly reduces the formation of LTs, which requires 5-LOX to convert AA into LTA_4 (215–217). Moreover, it has been found that the 15*S*-HETE synthesized via this pathway can also be esterified and stored within the membranes of neutrophils, specifically inositol-containing phospholipids. Upon cell stimulation, 15*S*-HETE is rapidly released and transformed to a second signal, such as LXA_4, to regulate the function of the neutrophil (218). The second major route of LX biosynthesis occurs in an LTA_4-dependent manner, involving peripheral blood platelet-leukocyte interactions. Leukocyte 5-LOX converts AA into LTA_4, which is released, taken up by adherent platelets, and subsequently transformed to LXA_4 and LXB_4 via the LX synthase activity of human 12-LOX (219). A third unorthodox route of LX generation occurs after the exogenous administration of aspirin (but not other conventional NSAIDs), which irreversibly acetylates COX-2 in endothelial cells and other cell types. Rather than COX-2 converting AA into PGG_2, acetylation causes the transformation of AA into 15*R*-HETE (C-15 alcohol carried in the *R*-configuration). This is then rapidly metabolized in a transcellular manner by adherent leukocyte, vascular endothelial or epithelial 5-LOX to form 15-epimeric-LXs (15-epi-LXs) or aspirin-triggered LXs (ATLs) that carry their C-15 alcohol in the *R*-configuration rather than 15*S* native LX. ATLs share many of the anti-inflammatory/proresolution characteristics of the native LXs.

LXA_4 and 15-epi-LXs elicit their multicellular responses via ALX (formyl peptide receptor-like-1 [FPRL1] receptor), a specific GPCR isolated and cloned in human, mouse, and rat tissues (220–222). Human ALX was subsequently identified and cloned in several types of leukocytes, including monocytes (223) and T cells (224), as well as resident cells such as macrophages, synovial fibroblasts (225), and intestinal epithelial cells (226). One of the functions attributed to ALX is in mediating the multicellular responses of LXA_4 and 15-epi-LXs. Studies in transgenic models have shown its selectivity toward LXA_4 and 15-epi-LPA_4 (not for LXB_4, LTB_4, LTD_4, or PGE_2) with high affinity (K_d [dissociation constant] = 1.7 nM) (231). ALX also has the ability to interact with other small peptides/proteins such as Ac2-26 and glucocorticoid-derived annexin-1, which carry out similar anti-inflammatory effects as LXs and 15-epi-LXs. Evidence that the protective effects of LXs and 15-epi-LXs are both ligand and receptor dependent arose from studies in transgenic mice overexpressing human ALX (227–229). In a zymosan-induced peritonitis model, infiltration of neutrophils was also substantially diminished in transgenic mice compared to their wild-type equivalents (227), with the site of lipoxin action being the leukocyte/endothelial interface mediated by the generation of nitric oxide's anti-adhesive properties (230).

ALX activation inhibits NADPH oxidase assembly, which, in turn, reduces superoxide anion generation by neutrophils through accumulation of polyisoprenoid presqualene diphosphate (231). Indeed, it has been demonstrated that inhibition of proinflammatory genes such as neutrophil chemoattractant IL-8 occurs via an ALX-dependent peroxynitrite-mediated signaling pathway (232). Moreover, peroxynitrite induction of IL-8 in response to lipopolysaccharide, TNF-α, or IL-β in human leukocytes occurs via an NF-κB- and activator protein 1 (AP1)-dependent pathway (233, 234). 15-epi-LX analogs also regulate an ALX-dependent p38/mitogen-activated protein kinase cascade known to promote chemotaxis by inhibiting leukocyte-specific AP1 phosphorylation and activation (235). In addition to ALX, LXs also function as partial agonists to a subclass of rhodopsin receptors (CysLT1) more commonly activated by LTs, mediating bioactions in several tissues and cell types other than leukocytes (221, 236). At nanomolar concentrations, LXA_4 has been shown to compete for binding with LTD_4 on mesangial cells (236) and human umbilical vein endothelial cells (222, 237) as well as opposing the proinflammatory effects of

LTD_4. There is also evidence that another intracellular receptor, the Ah receptor, mediates the bioactions of LXs. This receptor is a ligand-activated transcription factor that controls several of the biologic actions of LXs, such as increasing the expression of suppressor of cytokine signaling 2 (SOCS2) (238–240).

LXs IN INFLAMMATION

LXs are anti-inflammatory at nanomolar concentrations, controlling both granulocyte (neutrophil and eosinophil) and monocyte entry to sites of inflammation. Yet while they inhibit the transmigration of neutrophils and eosinophils down a chemokine gradient into inflamed sites (241–244), they promote noninflammatory infiltration of monocytes required for resolution and wound healing (245), without inducing neutrophil degranulation or release of other reactive oxygen species (232). Indeed, the ability of LXs to diminish neutrophil trafficking was corroborated when an analog of 15-epi-LX was intravenously administered to BLT1 knockout mice, which have dramatically elevated neutrophils in the lungs after high limb ischemia/reperfusion (246). Furthermore, research in our laboratory has uncovered in humans that 15-epi-LXs regulate PMN influx in forearm blisters, accounting for low-dose aspirin's anti-inflammatory properties (247). Our additional work on resolving inflammation has revealed that humans fall into two categories, those who resolve their acute inflammatory responses in an immediate manner and those who show a more delayed or prolonged healing process, with the severity and duration controlled by endogenous epi-LX/ALX expression (248).

At sites of inflammation, macrophages are stimulated by LXs to ingest and clear apoptotic neutrophils (249), which appears to be coupled to changes in the actin cytoskeleton (250). Furthermore, LXs elevate the levels of the anti-inflammatory cytokine transforming growth factor β1 (TGF-β1), which, in turn, downregulates a number of proinflammatory pathways (251–253). It is believed that these lipid mediators are generated *in situ* when neutrophils express 5-LOX at the onset of resolution as they begin to apoptose (107). LXs may also counteract the fibrotic response and thus improve tissue remodeling by reducing the proliferation of fibroblasts and mesangial cells induced by a number of factors, including connective tissue growth factor, platelet-derived growth factor, TNF-α, LTD_4, and TGF-β (254–257). 15-Epi-LXs exert the same biological effects as endogenously produced LXs, but with additional benefits that increase vasorelaxation (258), and induce endothelial cell production of anti-inflammatory nitric oxide synthesis (230, 259). Moreover, 15-epi-LPA_4 has been found to inhibit TNF-α-induced IL-1β in periodontitis *in vivo* (260, 261), dampen SOCS2 signaling (262), and inhibit TNF-α-induced IL-8 gene expression (226). Not surprisingly, both LXs and 15-epi-LXs have been identified and proven to exert beneficial effects in various experimental models of inflammation and human diseases, such as glomerulonephritis (263, 264), ischemia/reperfusion injury (246, 254), cystic fibrosis (265), periodontitis (266), acute pleuritis (230), asthma (267), wound healing processes in the eye (268), colitis, inflammation-induced hyperalgesia in rats, various cutaneous inflammation models (269), and microbial infection in mice (238, 270, 271).

OMEGA-3 PUFA PATHWAY

Omega-3 PUFAs have long been known to be important not only in maintaining organ function and health but also in reducing the incidence of infection and inflammation (110, 111, 272–275). A clinical trial (GISSI-Prevenzione) assessing the benefits of aspirin with and without omega-3 PUFA supplementation in patients recovering from MIs revealed a significant decrease in mortality in the group taking the supplement (276). More-recent evaluations have confirmed the importance of omega-3 PUFAs in reducing CV disease and inflammation associated with it (277, 278). It was initially hypothesized that fish oils demonstrate their antithrombotic, immunoregulatory, and anti-inflammatory bioactions by inhibiting PG and LT synthesis (279). However, current opinion is that it is likely that a series of novel compounds derived from EPA and DHA are responsible for eliciting these immunomodulatory effects. First identified in the resolving exudate of a mouse dorsal air pouch or peritonitis model using lipidomic and bioinformatic analysis (110, 111, 280, 281), these naturally occurring bioactive lipid mediators are termed resolvins (Rvs; derived from "resolution phase interaction products"), protectins (PDs), and maresins (MaRs; derived from "macrophage mediator in resolving inflammation"). All these omega-3 PUFA-derived products possess a plethora of stereospecific and potent anti-inflammatory and immunoregulatory actions that are protective *in vitro* and *in vivo* (282, 283).

Rvs AND PDs

Rvs can be generated from either EPA or DHA and are therefore categorized as either members of the E-series (from EPA) or D-series (from DHA). Rvs of both series were first isolated *in vivo* from murine dorsal air pouches treated with aspirin and EPA or DHA. Transcellular formation of E-series Rvs can occur with the conversion of EPA to 18*R*-hydroxyeicosapentanoic acid (18*R*-HEPE) by endothelial cells expressing COX-2 treated with aspirin. As with 15*R*-HETE in 15-epi-LX formation, 18*R*-HEPE can be released from endothelial cells to neighboring leukocytes for subsequent conversion by 5-LOX to either RvE1 or RvE2, via a 5(6)-epoxide-containing intermediate (110, 284). This interaction is blocked by selective COX-2 inhibition but not by indomethacin or paracetamol (110). RvE1 is spontaneously produced in healthy subjects, with levels increasing after treatment with either aspirin or EPA (285). D-series Rvs, aspirin-triggered RvD1 (AT-RvD1), and RvD1 are synthesized via a pathway involving sequential oxygenations, initiated by 15-LOX or aspirin-acetylated COX-2 in the microvasculature respectively, followed by 5-LOX in human neutrophils with an epoxide-containing intermediate. For AT-RvD1s, DHA is initially converted to epimeric 17*R*-hydroxydocosahexaenoic acid (17*R*-HDHA). In the absence of aspirin, however, DHA is enzymatically converted to 17*S*-HDHA (108). Interestingly, generation of E-series Rvs can also be mediated by microbial and mammalian CYP enzymes, which convert EPA into 18-HEPE. 18-HEPE can then be transformed by human neutrophils into either RvE1 or RvE2 (110). Hence, it is possible that microbes at sites of infection may contribute to the production of Rvs in a similar pathway.

DHA also serves as a precursor for the biosynthesis of PDs enzymatically converted by 15-LOX to a 17*S*-hydroperoxide-containing intermediate. Subsequently, this intermediate is rapidly converted by human leukocytes into a 16(17)-epoxide that is enzymatically converted in these cells to a 10,17-dihydroxy-containing compound (108, 286). PDs are distinguished by the presence of a conjugated triene double bond and by their potent bioactivity. One specific DHA-derived lipid mediator, 10,17*S*-docosatriene, was termed protectin D1 (PD1). When generated in neural tissue, however, this compound is called neuroprotectin D1 (NPD1). Moreover, PD1 exhibits tissue-specific bioactivity, as in humans this lipid is synthesized by peripheral blood mononuclear cells and Th2 $CD4^+$ T cells, while in mice it has been isolated from exudates and brain cells, human microglial cells (111), and in peripheral blood (108).

Rvs AND PDs IN INFLAMMATION

One of the broader immunomodulatory properties of RvE1 is its ability to inhibit neutrophil and dendritic cell accumulation at sites of inflammation by blocking transendothelial migration as well as enhancing their clearance from mucosal epithelial cells (110, 285, 287). Other bioactions of RvE1 include inhibition of neutrophil reactive oxygen intermediates in response to TNF-α and the bacterial peptide fMLP (288); abrogation of LTB_4-BLT1 signaling via NF-κB, and thus the production of proinflammatory cytokines and chemokines (251, 289, 290); stimulation of macrophages to ingest apoptotic neutrophils (291); enhancement of the percentage of phagocytes present in the lymph nodes (292); upregulation of CCR5 on late apoptotic neutrophils (10), which terminates chemokine signaling; and inhibition of dendritic cell migration. More recently, RvE1 has been demonstrated to regulate the leukocyte proinflammatory cell surface markers, such as L-selectin, while selectively disrupting TX-mediated platelet aggregation (293), adding further mechanistic insight into its anti-inflammatory/proresolution properties. In disease states, RvE1 suppresses *Porphyromonas gingivalis*-induced oral inflammation and alveolar bone loss during periodontitis (294), demonstrates protective actions in trinitrobenzene-sulfonic acid-induced colitis in mice (272), as well as causing reepithelization of mouse cornea after thermal injury (268). Over all, RvE1 initiates resolution of inflammation and causes decreased numbers of PMNs at sites of inflammation early during the response (reviewed in reference 283).

Structure-activity assays have elucidated that RvE1 binds to an orphan GPCR belonging to the same cluster as ALX (ChemR23), with a high affinity (K_d = 48.3 nm). This coupling downregulates the activity of NF-κB and hence TNF-α synthesis, as well as initiating signaling pathways involved in initiating mitogen-activated protein kinase (285). Indeed, ChemR23 activation has been demonstrated to inhibit one of the most prominent RvE1 actions, dendritic cell migration (285). Although it has been found in myeloid, gastrointestinal, kidney, brain, and CV tissue, the percentage of ChemR23 expression is highly variable. For example, it has been demonstrated that ChemR23 is markedly increased on the surface of human monocytes but less so on neutrophils by anti-inflammatory mediators such as TGF-β (295). Like ALX, ChemR23 acts as a receptor toward peptide ligands, including chemerin, that also act as anti-inflammatory mediators (296). RvE1 also appears to interact with the LTB_4 receptor, BLT1, and acts as a partial antagonist preventing neutrophil activation (289). Therefore, it can be concluded that RvE1 couples to two distinct receptors to both suppress proinflammatory mechanisms and enhance resolution pathways.

RvE2 is a second member of the EPA-derived family of E-series resolvins but is structurally distinct from RvE1. In human PMNs, it is generated at higher concentrations than RvE1, but is equipotent when given intravenously and additive when administered alongside RvE1 (292). As with RvE1, RvE2 suppresses PMN migration into the peritoneum after zymosan (292). Although it is still unclear what receptor RvE2 couples to, its identification is the subject of ongoing research.

D-series Rvs are derived from DHA and comprise four bioactive compounds, RvD1, RvD2, RvD3, and RvD4 (108). Like RvE1, RvD1/D2 exerts both anti-inflammatory and proresolution properties by blocking neutrophil infiltration, while in contrast enhancing macrophage phagocytosis of apoptotic PMNs (297–299). The latter occurs via the binding of RvD1 to either ALX or an orphan receptor, GPR32, present on the surface of both PMNs and monocytes, the expression of which is upregulated by inflammatory agonists, such as zymosan and GM-CSF (297). Interestingly, a member of the D-series Rvs has also been shown to contain microbicidal properties in septic mice initiated by cecal ligation and puncture. RvD2, whose receptor is GPR18 (300), in addition to blocking peritoneal PMN accumulation, markedly reduced bacterial load and proinflammatory cytokines, which subsequently led to increased survival and improved health (298).

As mentioned above, besides D-series Rvs, DHA also acts as a precursor for the biosynthesis of PDs. One member, PD1, has been demonstrated to be synthesized by microglial (111), peripheral blood mononuclear cells, and Th2 $CD4^+$ T cells (108, 286). Similarly to Rvs, PD1 exerts potent immunoregulatory effects that include inhibiting neutrophil migration and Toll-like receptor-mediated activation (301), suppression of Th2 inflammatory cytokines and proinflammatory lipid mediators (302), as well as the upregulation of CCR5 on PMNs (10). PD1 also blocks T-cell migration *in vivo* and promotes T-cell apoptosis (303). In disease states, PD1 has been proven to be protective in experimental models of ischemic stroke (109), oxidative stress (304–306), asthma (302), ischemia/reperfusion renal injury (301), and Alzheimer's disease (307). Indeed, Alzheimer's patients given DHA-rich dietary supplements have reduced production of IL-1β, IL-6, and granulocyte CSF in peripheral blood mononuclear cells (308). As with RvE2, a receptor has yet to be identified. It is likely, however, that it couples to a distinct receptor to RvE1, as its anti-inflammatory effects are additive with those of RvE1 *in vivo*.

MARESINS

MaRs were identified in 2008 after 17*S*-D-series Rvs, PDs, as well as 14*S*-HDHA were isolated from the resolution phase of mouse peritonitis and were added to stimulated resident peritoneal macrophages (281). Macrophages then convert these intermediates to novel dihydroxy-containing products, which possess potent anti-inflammatory and proresolving properties. Although the exact biosynthetic pathway has yet to be elucidated, a hypothetical scheme was proposed. It is thought that DHA is converted to 14*S*-hydroperoxy-docosahexaenoic acid (14*S*-HPDHA; maresin, MaR1) via 12- or 15-LOX, followed by reduction to 14*S*-HDHA and/or via double dioxygenation (e.g., sequential 12-LOX–5-LOX) to generate a metabolome of MaR1, 7*S*,14*S*-dihydroxydocasahexaenoic acid (7*S*,14*S*-diHDHA). Though MaRs have only been recently identified, it has been reported that, as with Rvs and PD1, MaR1 blocks the infiltration of PMNs, while stimulating macrophage phagocytosis of apoptotic PMNs/zymosan (281). Its metabolome 7*S*,14*S*-diHDHA was active but less potent.

CYTOCHROME P450

In the last decade, interest in a third, less well-characterized pathway of AA metabolism, CYP, has been rekindled. CYPs are families of membrane-bound, heme-containing enzymes found in the liver, brain, kidneys, lung, heart, and CV system, thought initially to be involved in catalyzing NADPH-dependent oxidation of drugs, chemicals, and carcinogens (309, 310). It is now well appreciated that CYPs also catalyze the conversion of fatty acids including AA into products that have been denoted EETs, HETEs, and dihydroxy-eicosatrienoic acids (DHETs) (311). For instance, AA is metabolized in the vascular endothelium by CYP epoxygenase to EETs (312), which can then be converted by epoxide hydrolase to the respective regio-isomer of DHETs (311). In the vascular smooth muscle, AA is catalyzed by CYP hydroxylases to 20-HETE (313). Indeed, one particular member, CYP4F3, is highly expressed in PMNs catalyzing the ω-hydroxylation of LTs (314). However, it is unknown whether CYP4F3 is the source of 20-HETE produced by PMNs (315). These metabolites play a large and complex role in maintaining renal, cardiac, and pulmonary homeostasis by regulating aspects such as vascular tone and reactivity, renal and pulmonary functions, ion transport, and growth responses (316–318). Interestingly, they have also been demonstrated to exert potent anti-inflammatory actions (319–321), detailed below.

CYP-DERIVED PRODUCTS AND INFLAMMATION

EETs catalyzed by CYPs 2C8, 2C9, and 2J2 prevent the adhesion of PMNs to the vascular wall by suppressing the expression of cell adhesion molecules, including intercellular adhesion molecule-1, vascular cell adhesion molecule-1, and E-selectin, on the surface of endothelial cells in response to cytokines (TNF-α and IL-1α) and lipopolysaccharide (316, 321). Mechanistically, this is associated with inhibiting the activation of the transcription factor NF-κB via the inhibitor of κB kinase (IKK) (321). As a consequence, EETs may therefore have the propensity to downregulate various cytokine-induced proinflammatory signaling pathways downstream of NF-κB activation. Indeed, it was recently reported that EETs display hyperalgesic bioactions during experimental inflammatory pain (319, 320). It was also shown that EETs could directly activate PPAR-γ in endothelial cells (322), with EET-mediated anti-inflammatory effects demonstrated to be blocked by PPAR-γ antagonists (322). EETs released from platelets have been shown to exert antithrombotic properties by inhibiting platelet aggregation induced by AA and vascular injury (323–325). It was also demonstrated that EETs could act in a profibrinolytic manner by increasing the expression of tissue plasminogen activator in a cAMP-dependent mechanism, thus suggesting that they could play an important role in controlling the fibrinolytic balance in the vessel wall (326). It was suggested that the anti-inflammatory properties of EETs occurred through their ligation to a cell surface receptor. It was reported that EETs bind with high affinity to an "EET receptor" on the surface of a monocytic cell line, belonging to a specific class of GPCRs (327). The identity of this receptor and its role, if any, in initiating the immunomodulatory actions of EETs have yet to be determined.

CYP hydroxylase metabolites also exhibit anti-inflammatory properties. Similarly to EETs, 16-HETE can also block the adhesion of leukocytes to the endothelium (315). In fact, it also suppresses the synthesis of LTs as well as inhibiting rises in cerebrospinal fluid pressure (index of tissue damage and swelling) in a thromboembolic model of stroke in rabbits (315). Furthermore, 20-HETE and 16-HETE released from PMNs in response to factors that activate phospholipase (platelet-activating factor, calcium, and thrombin) also inhibit TX-induced platelet aggregation (328). Therefore, it can be surmised that not only do metabolites of CYPs maintain renal and CV health, but they also regulate other multiple signaling pathways including inflammation, fibrinolysis, platelet aggregation, and cellular injury.

SUMMARY

Studies on inflammation and its resolution have advanced our understanding of leukocyte trafficking, efferocytosis, and proinflammatory leukocyte clearance, as well as immune-suppressive eicosanoids, specialized immune-regulatory cells, and cytokine catabolism. These pathways converge on the termination of acute inflammatory responses and contribute to the notion that chronic inflammation is avoided and wounds healed in an appropriate manner (329, 330). Implicit therein is that tolerance is not compromised, making the host susceptible to autoimmunity. AA metabolites were once considered proinflammatory due to the effective usage of NSAIDs in the treatment of chronic inflammatory diseases. While NSAIDs have been a valuable treatment in terms of anti-inflammation and pain relief, they have recently unmasked beneficial properties of some LOX and COX products. Thus, our understanding of eicosanoids in physiology and pathology has come a long way since the earliest observations of Kurzrok and Lieb (331). Hence, PGs may drive edema but prevent leukocyte trafficking, while at the same time elevating cAMP and impairing bacterial phagocytosis and killing. However, LTB_4/D_4 oppose the immune-suppressive actions of PGE_2, with 5-LOX metabolites thus enhancing macrophage antimicrobial functions and roles, including the phagocytosis of IgG-opsonized targets via the FcγR. COX/LOX-derived LXs, Rvs, and PCs attenuate innate immune responses, ameliorate or promote resolution, and are proving beneficial in experimental sepsis. Thus, the role of eicosanoids in inflammation is most likely dependent on the phase of the response during which they are synthesized, the tissues affected, and the nature of the inciting stimulus, with some AA metabolites counteracting the bioaction of others but also triggering the synthesis of other families of eicosanoids that terminate inflammation. And while eicosanoids act diversely in acute inflammation, their role in chronic, nonresolving inflammation may be far more complex. That notwithstanding, it now appears that not all eicosanoids are bad, as some attenuate innate immune-mediated functions and accelerate or facilitate their timely resolution. This offers a more accurate strategy in treating diseases driven by overexuberant inflammation.

Citation. Bennett M, Gilroy DW. 2016. Lipid mediators in inflammation. Microbiol Spectrum 4(6):MCHD-0035-2016.

References

1. **Serhan C, Ward P, Gilroy D.** 2010. *Fundamentals of Inflammation.* Cambridge University Press, Cambridge, United Kingdom.

2. **Segal AW.** 2005. How neutrophils kill microbes. *Annu Rev Immunol* **23:**197–223.

3. **Serhan CN, Savill J.** 2005. Resolution of inflammation: the beginning programs the end. *Nat Immunol* **6:**1191–1197.

4. **Majno G, Joris I.** 2004. *Cells, Tissues and Disease: Principles of General Pathology*, 2nd ed. Oxford University Press, New York, NY.

5. **Buckley CD, Gilroy DW, Serhan CN.** 2014. Proresolving lipid mediators and mechanisms in the resolution of acute inflammation. *Immunity* **40:**315–327.

6. **Buckley CD, Gilroy DW, Serhan CN, Stockinger B, Tak PP.** 2013. The resolution of inflammation. *Nat Rev Immunol* **13:**59–66.

7. **Segal AW, Geisow M, Garcia R, Harper A, Miller R.** 1981. The respiratory burst of phagocytic cells is associated with a rise in vacuolar pH. *Nature* **290:**406–409.

8. **Gilroy DW, Colville-Nash PR, McMaster S, Sawatzky DA, Willoughby DA, Lawrence T.** 2003. Inducible cyclooxygenase-derived 15deoxy$\Delta^{12-14}PGJ_2$ brings about acute inflammatory resolution in rat pleurisy by inducing neutrophil and macrophage apoptosis. *FASEB J* **17:**2269–2271.

9. **Nibbs RJ, Graham GJ.** 2013. Immune regulation by atypical chemokine receptors. *Nat Rev Immunol* **13:**815–829.

10. **Ariel A, Fredman G, Sun YP, Kantarci A, Van Dyke TE, Luster AD, Serhan CN.** 2006. Apoptotic neutrophils and T cells sequester chemokines during immune response resolution through modulation of CCR5 expression. *Nat Immunol* **7:**1209–1216.

11. **Piper P, Vane J.** 1971. The release of prostaglandins from lung and other tissues. *Ann N Y Acad Sci* **180:**363–385.

12. **Samuelsson B, Dahlén SE, Lindgren JA, Rouzer CA, Serhan CN.** 1987. Leukotrienes and lipoxins: structures, biosynthesis, and biological effects. *Science* **237:**1171–1176.

13. **Serhan CN, Hamberg M, Samuelsson B.** 1984. Trihydroxytetraenes: a novel series of compounds formed from arachidonic acid in human leukocytes. *Biochem Biophys Res Commun* **118:**943–949.

14. **Capdevila JH, Falck JR, Dishman E, Karara A.** 1990. Cytochrome P-450 arachidonate oxygenase. *Methods Enzymol* **187:**385–394.

15. **Pagels WR, Sachs RJ, Marnett LJ, Dewitt DL, Day JS, Smith WL.** 1983. Immunochemical evidence for the involvement of prostaglandin H synthase in hydroperoxide-dependent oxidations by ram seminal vesicle microsomes. *J Biol Chem* **258:**6517–6523.

16. **Hamberg M, Samuelsson B.** 1973. Detection and isolation of an endoperoxide intermediate in prostaglandin biosynthesis. *Proc Natl Acad Sci U S A* **70:**899–903.

17. **Nugteren DH, Hazelhof E.** 1973. Isolation and properties of intermediates in prostaglandin biosynthesis. *Biochim Biophys Acta* **326:**448–461.

18. **Dubois RN, Abramson SB, Crofford L, Gupta RA, Simon LS, Van De Putte LB, Lipsky PE.** 1998. Cyclooxygenase in biology and disease. *FASEB J* **12:**1063–1073.

19. **Shimizu T, Yamamoto S, Hayaishi O.** 1982. Purification of PGH-PGD isomerase from rat brain. *Methods Enzymol* **86:**73–77.

20. **Tanaka Y, Ward SL, Smith WL.** 1987. Immunochemical and kinetic evidence for two different prostaglandin H-prostaglandin E isomerases in sheep vesicular gland microsomes. *J Biol Chem* **262:**1374–1381.

21. **Hayashi H, Fujii Y, Watanabe K, Urade Y, Hayaishi O.** 1989. Enzymatic conversion of prostaglandin H_2 to prostaglandin $F_{2\alpha}$ by aldehyde reductase from human liver: comparison to the prostaglandin F synthetase from bovine lung. *J Biol Chem* **264:**1036–1040.

22. **DeWitt DL, Smith WL.** 1983. Purification of prostacyclin synthase from bovine aorta by immunoaffinity chromatography. Evidence that the enzyme is a hemoprotein. *J Biol Chem* **258:**3285–3293.

23. **Ullrich V, Haurand M.** 1983. Thromboxane synthase as a cytochrome P450 enzyme. *Adv Prostaglandin Thromboxane Leukot Res* **11:**105–110.

24. **Bezugla Y, Kolada A, Kamionka S, Bernard B, Scheibe R, Dieter P.** 2006. COX-1 and COX-2 contribute differentially to the LPS-induced release of PGE_2 and TxA_2 in liver macrophages. *Prostaglandins Other Lipid Mediat* **79:**93–100.

25. **Naraba H, Murakami M, Matsumoto H, Shimbara S, Ueno A, Kudo I, Oh-ishi S.** 1998. Segregated coupling of phospholipases A_2, cyclooxygenases, and terminal prostanoid synthases in different phases of prostanoid biosynthesis in rat peritoneal macrophages. *J Immunol* **160:**2974–2982.

26. **Jakobsson PJ, Thorén S, Morgenstern R, Samuelsson B.** 1999. Identification of human prostaglandin E synthase: a microsomal, glutathione-dependent, inducible enzyme, constituting a potential novel drug target. *Proc Natl Acad Sci U S A* **96:**7220–7225.

27. **Penglis PS, Cleland LG, Demasi M, Caughey GE, James MJ.** 2000. Differential regulation of prostaglandin E_2 and thromboxane A_2 production in human monocytes: implications for the use of cyclooxygenase inhibitors. *J Immunol* **165:**1605–1611.

28. **Brock TG, McNish RW, Peters-Golden M.** 1999. Arachidonic acid is preferentially metabolized by cyclooxygenase-2 to prostacyclin and prostaglandin E_2. *J Biol Chem* **274:**11660–11666.

29. **Hirai H, Tanaka K, Yoshie O, Ogawa K, Kenmotsu K, Takamori Y, Ichimasa M, Sugamura K, Nakamura M, Takano S, Nagata K.** 2001. Prostaglandin D_2 selectively induces chemotaxis in T helper type 2 cells, eosinophils, and basophils via seven-transmembrane receptor CRTH2. *J Exp Med* **193:**255–261.

30. **Monneret G, Gravel S, Diamond M, Rokach J, Powell WS.** 2001. Prostaglandin D_2 is a potent chemoattractant for human eosinophils that acts via a novel DP receptor. *Blood* **98:**1942–1948.

31. **Xue L, Gyles SL, Wettey FR, Gazi L, Townsend E, Hunter MG, Pettipher R.** 2005. Prostaglandin D_2 causes preferential induction of proinflammatory Th2 cytokine production through an action on chemoattractant receptor-like molecule expressed on Th2 cells. *J Immunol* **175:**6531–6536.

32. **von Euler US.** 1936. On the specific vaso-dilating and plain muscle stimulating substances from accessory genital glands in man and certain animals (prostaglandin and vesiglandin). *J Physiol* **88:**213–234.

33. **Eckenfels A, Vane JR.** 1972. Prostaglandins, oxygen tension and smooth muscle tone. *Br J Pharmacol* **45:** 451–462.

34. **Ferreira SH, Herman A, Vane JR.** 1972. Proceedings: prostaglandin generation maintains the smooth muscle tone of the rabbit isolated jejunum. *Br J Pharmacol* **44:** 328P–329P.

35. **Main IH.** 1964. The inhibitory actions of prostaglandins on respiratory smooth muscle. *Br Pharmacol Chemother* **22:**511–519.

36. **Williams TJ.** 1979. Prostaglandin E_2, prostaglandin I_2 and the vascular changes of inflammation. *Br J Pharmacol* **65:**517–524.

37. **Williams TJ, Jose PJ.** 1981. Mediation of increased vascular permeability after complement activation. Histamine-independent action of rabbit C5a. *J Exp Med* **153:**136–153.

38. **Ferreira SH, Nakamura M, de Abreu Castro MS.** 1978. The hyperalgesic effects of prostacyclin and prostaglandin E_2. *Prostaglandins* **16:**31–37.

39. **Feldberg W, Gupta KP.** 1973. Pyrogen fever and prostaglandin-like activity in cerebrospinal fluid. *J Physiol* **228:**41–53.

40. **Feldberg W, Saxena PN.** 1971. Fever produced by prostaglandin E_1. *J Physiol* **217:**547–556.

41. **Milton AS, Wendlandt S.** 1971. Effects on body temperature of prostaglandins of the A, E and F series on injection into the third ventricle of unanaesthetized cats and rabbits. *J Physiol* **218:**325–336.

42. **Moncada S, Gryglewski R, Bunting S, Vane JR.** 1976. An enzyme isolated from arteries transforms prostaglandin endoperoxides to an unstable substance that inhibits platelet aggregation. *Nature* **263:**663–665.

43. **Kaley G, Hintze TH, Panzenbeck M, Messina EJ.** 1985. Role of prostaglandins in microcirculatory function. *Adv Prostaglandin Thromboxane Leukot Res* **13:** 27–35.

44. **Hata AN, Breyer RM.** 2004. Pharmacology and signaling of prostaglandin receptors: multiple roles in inflammation and immune modulation. *Pharmacol Ther* **103:** 147–166.

45. **Higgs EA, Moncada S, Vane JR.** 1978. Inflammatory effects of prostacyclin (PGI_2) and 6-oxo-$PGF_{1\alpha}$ in the rat paw. *Prostaglandins* **16:**153–162.

46. **Komoriya K, Ohmori H, Azuma A, Kurozumi S, Hashimoto Y, Nicolaou KC, Barnette WE, Magolda RL.** 1978. Prostaglandin I_2 as a potentiator of acute inflammation in rats. *Prostaglandins* **15:**557–564.

47. **Lewis AJ, Nelson DJ, Sugrue MF.** 1975. On the ability of prostaglandin E_1, and arachidonic acid to modulate experimentally induced oedema in the rat paw. *Br J Pharmacol* **55:**51–56.

48. **Moncada S, Ferreira SH, Vane JR.** 1973. Prostaglandins, aspirin-like drugs and the oedema of inflammation. *Nature* **246:**217–219.

49. **Williams TJ, Morley J.** 1973. Prostaglandins as potentiators of increased vascular permeability in inflammation. *Nature* **246:**215–217.

50. **Yuhki K, Ueno A, Naraba H, Kojima F, Ushikubi F, Narumiya S, Oh-ishi S.** 2004. Prostaglandin receptors EP_2, EP_3, and IP mediate exudate formation in carrageenin-induced mouse pleurisy. *J Pharmacol Exp Ther* **311:**1218–1224.

51. **Yuhki K, Ushikubi F, Naraba H, Ueno A, Kato H, Kojima F, Narumiya S, Sugimoto Y, Matsushita M, Oh-Ishi S.** 2008. Prostaglandin I_2 plays a key role in zymosan-induced mouse pleurisy. *J Pharmacol Exp Ther* **325:**601–609.

52. **Saxena PN, Beg MM, Singhal KC, Ahmad M.** 1979. Prostaglandin-like activity in the cerebrospinal fluid of febrile patients. *Indian J Med Res* **70:**495–498.

53. **Dantzer R, Konsman JP, Bluthé RM, Kelley KW.** 2000. Neural and humoral pathways of communication from the immune system to the brain: parallel or convergent? *Auton Neurosci* **85:**60–65.

54. **Ek M, Kurosawa M, Lundeberg T, Ericsson A.** 1998. Activation of vagal afferents after intravenous injection of interleukin-1beta: role of endogenous prostaglandins. *J Neurosci* **18:**9471–9479.

55. **Lazarus M, Yoshida K, Coppari R, Bass CE, Mochizuki T, Lowell BB, Saper CB.** 2007. EP3 prostaglandin receptors in the median preoptic nucleus are critical for fever responses. *Nat Neurosci* **10:**1131–1133.

56. **Ushikubi F, Segi E, Sugimoto Y, Murata T, Matsuoka T, Kobayashi T, Hizaki H, Tuboi K, Katsuyama M, Ichikawa A, Tanaka T, Yoshida N, Narumiya S.** 1998. Impaired febrile response in mice lacking the prostaglandin E receptor subtype EP_3. *Nature* **395:**281–284.

57. **Madden CJ, Morrison SF.** 2003. Excitatory amino acid receptor activation in the raphe pallidus area mediates prostaglandin-evoked thermogenesis. *Neuroscience* **122:** 5–15.

58. **Madden CJ, Morrison SF.** 2004. Excitatory amino acid receptors in the dorsomedial hypothalamus mediate prostaglandin-evoked thermogenesis in brown adipose tissue. *Am J Physiol Regul Integr Comp Physiol* **286:** R320–R325.

59. **Morrison SF.** 2001. Differential regulation of sympathetic outflows to vasoconstrictor and thermoregulatory effectors. *Ann N Y Acad Sci* **940:**286–298.

60. **Morrison SF.** 2003. Raphe pallidus neurons mediate prostaglandin E_2-evoked increases in brown adipose tissue thermogenesis. *Neuroscience* **121:**17–24.

61. **Morrison SF.** 2004. Central pathways controlling brown adipose tissue thermogenesis. *News Physiol Sci* **19:**67–74.

62. **Ahmadi S, Lippross S, Neuhuber WL, Zeilhofer HU.** 2002. PGE_2 selectively blocks inhibitory glycinergic neurotransmission onto rat superficial dorsal horn neurons. *Nat Neurosci* **5:**34–40.

63. **Juhlin L, Michaëlsson G.** 1969. Cutaneous vascular reactions to prostaglandins in healthy subjects and in patients with urticaria and atopic dermatitis. *Acta Derm Venereol* **49:**251–261.

64. **Lin CR, Amaya F, Barrett L, Wang H, Takada J, Samad TA, Woolf CJ.** 2006. Prostaglandin E_2 receptor EP4 contributes to inflammatory pain hypersensitivity. *J Pharmacol Exp Ther* **319:**1096–1103.

65. **McAdam BF, Mardini IA, Habib A, Burke A, Lawson JA, Kapoor S, FitzGerald GA.** 2000. Effect of regulated expression of human cyclooxygenase isoforms on eicosanoid and isoeicosanoid production in inflammation. *J Clin Invest* **105:**1473–1482.

66. **Moriyama T, Higashi T, Togashi K, Iida T, Segi E, Sugimoto Y, Tominaga T, Narumiya S, Tominaga M.** 2005. Sensitization of TRPV1 by EP_1 and IP reveals peripheral nociceptive mechanism of prostaglandins. *Mol Pain* **1:**3. doi:10.1186/1744-8069-1-3.

67. **Murata T, Ushikubi F, Matsuoka T, Hirata M, Yamasaki A, Sugimoto Y, Ichikawa A, Aze Y, Tanaka T, Yoshida N, Ueno A, Oh-ishi S, Narumiya S.** 1997. Altered pain perception and inflammatory response in mice lacking prostacyclin receptor. *Nature* **388:**678–682.

68. **Reinold H, Ahmadi S, Depner UB, Layh B, Heindl C, Hamza M, Pahl A, Brune K, Narumiya S, Müller U, Zeilhofer HU.** 2005. Spinal inflammatory hyperalgesia is mediated by prostaglandin E receptors of the EP2 subtype. *J Clin Invest* **115:**673–679.

69. **Solomon LM, Juhlin L, Kirschenbaum MB.** 1968. Prostaglandin on cutaneous vasculature. *J Invest Dermatol* **51:**280–282.

70. **Ueno A, Matsumoto H, Naraba H, Ikeda Y, Ushikubi F, Matsuoka T, Narumiya S, Sugimoto Y, Ichikawa A, Oh-ishi S.** 2001. Major roles of prostanoid receptors IP and EP_3 in endotoxin-induced enhancement of pain perception. *Biochem Pharmacol* **62:**157–160.

71. **Smyth EM, Grosser T, Wang M, Yu Y, FitzGerald GA.** 2009. Prostanoids in health and disease. *J Lipid Res* **50** (Suppl)**:**S423–S428.

72. **García Rodríguez LA, Tacconelli S, Patrignani P.** 2008. Role of dose potency in the prediction of risk of myocardial infarction associated with nonsteroidal anti-inflammatory drugs in the general population. *J Am Coll Cardiol* **52:**1628–1636.

73. **Grosser T, Fries S, FitzGerald GA.** 2006. Biological basis for the cardiovascular consequences of COX-2 inhibition: therapeutic challenges and opportunities. *J Clin Invest* **116:**4–15.

74. **Xiao CY, Hara A, Yuhki K, Fujino T, Ma H, Okada Y, Takahata O, Yamada T, Murata T, Narumiya S, Ushikubi F.** 2001. Roles of prostaglandin I_2 and thromboxane A_2 in cardiac ischemia-reperfusion injury: a study using mice lacking their respective receptors. *Circulation* **104:**2210–2215.

75. **Degousee N, Fazel S, Angoulvant D, Stefanski E, Pawelzik SC, Korotkova M, Arab S, Liu P, Lindsay TF, Zhuo S, Butany J, Li RK, Audoly L, Schmidt R, Angioni C, Geisslinger G, Jakobsson PJ, Rubin BB.** 2008. Microsomal prostaglandin E_2 synthase-1 deletion leads to adverse left ventricular remodeling after myocardial infarction. *Circulation* **117:**1701–1710.

76. **Qian JY, Harding P, Liu Y, Shesely E, Yang XP, LaPointe MC.** 2008. Reduced cardiac remodeling and function in cardiac-specific EP_4 receptor knockout mice with myocardial infarction. *Hypertension* **51:**560–566.

77. **Bunting S, Moncada S, Vane JR.** 1983. The prostacyclin-thromboxane A_2 balance: pathophysiological and therapeutic implications. *Br Med Bull* **39:**271–276.

78. **FitzGerald GA.** 2003. COX-2 and beyond: approaches to prostaglandin inhibition in human disease. *Nat Rev Drug Discov* **2:**879–890.

79. **de Nucci G, Gryglewski RJ, Warner TD, Vane JR.** 1988. Receptor-mediated release of endothelium-derived relaxing factor and prostacyclin from bovine aortic endothelial cells is coupled. *Proc Natl Acad Sci U S A* **85:**2334–2338.

80. **Palmer RM, Ferrige AG, Moncada S.** 1987. Nitric oxide release accounts for the biological activity of endothelium-derived relaxing factor. *Nature* **327:**524–526.

81. **Ellis EF, Oelz O, Roberts LJ II, Payne NA, Sweetman BJ, Nies AS, Oates JA.** 1976. Coronary arterial smooth muscle contraction by a substance released from platelets: evidence that it is thromboxane A_2. *Science* **193:**1135–1137.

82. **Hamberg M, Svensson J, Samuelsson B.** 1975. Thromboxanes: a new group of biologically active compounds derived from prostaglandin endoperoxides. *Proc Natl Acad Sci U S A* **72:**2994–2998.

83. **Salzman PM, Salmon JA, Moncada S.** 1980. Prostacyclin and thromboxane A_2 synthesis by rabbit pulmonary artery. *J Pharmacol Exp Ther* **215:**240–247.

84. **Rocca B, Secchiero P, Ciabattoni G, Ranelletti FO, Catani L, Guidotti L, Melloni E, Maggiano N, Zauli G, Patrono C.** 2002. Cyclooxygenase-2 expression is induced during human megakaryopoiesis and characterizes newly formed platelets. *Proc Natl Acad Sci U S A* **99:**7634–7639.

85. **Aronoff DM, Carstens JK, Chen GH, Toews GB, Peters-Golden M.** 2006. Short communication: differences between macrophages and dendritic cells in the cyclic AMP-dependent regulation of lipopolysaccharide-induced cytokine and chemokine synthesis. *J Interferon Cytokine Res* **26:**827–833.

86. **Luo M, Jones SM, Phare SM, Coffey MJ, Peters-Golden M, Brock TG.** 2004. Protein kinase A inhibits leukotriene synthesis by phosphorylation of 5-lipoxygenase on serine 523. *J Biol Chem* **279:**41512–41520.

87. **van der Pouw Kraan TC, van Lier RA, Aarden LA.** 1995. PGE_2 and the immune response. A central role for prostaglandin E_2 in downregulating the inflammatory immune response. *Mol Med Today* **1:**61.

88. **Aronoff DM, Canetti C, Peters-Golden M.** 2004. Prostaglandin E_2 inhibits alveolar macrophage phagocytosis through an E-prostanoid 2 receptor-mediated increase in intracellular cyclic AMP. *J Immunol* **173:**559–565.

89. **Rossi AG, McCutcheon JC, Roy N, Chilvers ER, Haslett C, Dransfield I.** 1998. Regulation of macrophage phagocytosis of apoptotic cells by cAMP. *J Immunol* **160:**3562–3568.

90. **Serezani CH, Chung J, Ballinger MN, Moore BB, Aronoff DM, Peters-Golden M.** 2007. Prostaglandin E_2

suppresses bacterial killing in alveolar macrophages by inhibiting NADPH oxidase. *Am J Respir Cell Mol Biol* 37:562–570.

91. **Soares AC, Souza DG, Pinho V, Vieira AT, Barsante MM, Nicoli JR, Teixeira M.** 2003. Impaired host defense to *Klebsiella pneumoniae* infection in mice treated with the PDE4 inhibitor rolipram. *Br J Pharmacol* **140:** 855–862.

92. **Weinberg DA, Weston LK, Kaplan JE.** 1985. Influence of prostaglandin I_2 on fibronectin-mediated phagocytosis in vivo and in vitro. *J Leukoc Biol* **37:**151–159.

93. **Ydrenius L, Majeed M, Rasmusson BJ, Stendahl O, Särndahl E.** 2000. Activation of cAMP-dependent protein kinase is necessary for actin rearrangements in human neutrophils during phagocytosis. *J Leukoc Biol* **67:** 520–528.

94. **Aronoff DM, Peres CM, Serezani CH, Ballinger MN, Carstens JK, Coleman N, Moore BB, Peebles RS, Faccioli LH, Peters-Golden M.** 2007. Synthetic prostacyclin analogs differentially regulate macrophage function via distinct analog-receptor binding specificities. *J Immunol* **178:**1628–1634.

95. **Brandwein SR.** 1986. Regulation of interleukin 1 production by mouse peritoneal macrophages. Effects of arachidonic acid metabolites, cyclic nucleotides, and interferons. *J Biol Chem* **261:**8624–8632.

96. **Kunkel SL, Spengler M, May MA, Spengler R, Larrick J, Remick D.** 1988. Prostaglandin E_2 regulates macrophage-derived tumor necrosis factor gene expression. *J Biol Chem* **263:**5380–5384.

97. **Kunkel SL, Wiggins RC, Chensue SW, Larrick J.** 1986. Regulation of macrophage tumor necrosis factor production by prostaglandin E_2. *Biochem Biophys Res Commun* **137:**404–410.

98. **Takayama K, García-Cardena G, Sukhova GK, Comander J, Gimbrone MA Jr, Libby P.** 2002. Prostaglandin E_2 suppresses chemokine production in human macrophages through the EP4 receptor. *J Biol Chem* **277:**44147–44154.

99. **van der Pouw Kraan TC, Boeije LC, Snijders A, Smeenk RJ, Wijdenes J, Aarden LA.** 1996. Regulation of IL-12 production by human monocytes and the influence of prostaglandin E_2. *Ann N Y Acad Sci* **795:**147–157.

100. **Xu XJ, Reichner JS, Mastrofrancesco B, Henry WL Jr, Albina JE.** 2008. Prostaglandin E_2 suppresses lipopolysaccharide-stimulated IFN-β production. *J Immunol* **180:**2125–2131.

101. **Harizi H, Juzan M, Pitard V, Moreau JF, Gualde N.** 2002. Cyclooxygenase-2-issued prostaglandin E_2 enhances the production of endogenous IL-10, which down-regulates dendritic cell functions. *J Immunol* **168:** 2255–2263.

102. **Hinson RM, Williams JA, Shacter E.** 1996. Elevated interleukin 6 is induced by prostaglandin E_2 in a murine model of inflammation: possible role of cyclooxygenase-2. *Proc Natl Acad Sci U S A* **93:**4885–4890.

103. **Starczewski M, Voigtmann R, Peskar BA, Peskar BM.** 1984. Plasma levels of 15-keto-13,14-dihydro-prostaglandin E_2 in patients with bronchogenic carcinoma. *Prostaglandins Leukot Med* **13:**249–258.

104. **Hayek MG, Mura C, Wu D, Beharka AA, Han SN, Paulson KE, Hwang D, Meydani SN.** 1997. Enhanced expression of inducible cyclooxygenase with age in murine macrophages. *J Immunol* **159:**2445–2451.

105. **Medjane S, Raymond B, Wu Y, Touqui L.** 2005. Impact of CFTR ΔF508 mutation on prostaglandin E_2 production and type IIA phospholipase A_2 expression by pulmonary epithelial cells. *Am J Physiol Lung Cell Mol Physiol* **289:**L816–L824.

106. **Strandvik B, Svensson E, Seyberth HW.** 1996. Prostanoid biosynthesis in patients with cystic fibrosis. *Prostaglandins Leukot Essent Fatty Acids* **55:**419–425.

107. **Levy BD, Clish CB, Schmidt B, Gronert K, Serhan CN.** 2001. Lipid mediator class switching during acute inflammation: signals in resolution. *Nat Immunol* **2:**612–619.

108. **Hong S, Gronert K, Devchand PR, Moussignac RL, Serhan CN.** 2003. Novel docosatrienes and 17*S*-resolvins generated from docosahexaenoic acid in murine brain, human blood, and glial cells. Autacoids in anti-inflammation. *J Biol Chem* **278:**14677–14687.

109. **Marcheselli VL, Hong S, Lukiw WJ, Tian XH, Gronert K, Musto A, Hardy M, Gimenez JM, Chiang N, Serhan CN, Bazan NG.** 2003. Novel docosanoids inhibit brain ischemia-reperfusion-mediated leukocyte infiltration and pro-inflammatory gene expression. *J Biol Chem* **278:** 43807–43817.

110. **Serhan CN, Clish CB, Brannon J, Colgan SP, Chiang N, Gronert K.** 2000. Novel functional sets of lipid-derived mediators with antiinflammatory actions generated from omega-3 fatty acids via cyclooxygenase 2-nonsteroidal antiinflammatory drugs and transcellular processing. *J Exp Med* **192:**1197–1204.

111. **Serhan CN, Hong S, Gronert K, Colgan SP, Devchand PR, Mirick G, Moussignac RL.** 2002. Resolvins: a family of bioactive products of omega-3 fatty acid transformation circuits initiated by aspirin treatment that counter proinflammation signals. *J Exp Med* **196:** 1025–1037.

112. **Clark RB, Bishop-Bailey D, Estrada-Hernandez T, Hla T, Puddington L, Padula SJ.** 2000. The nuclear receptor PPARγ and immunoregulation: PPARγ mediates inhibition of helper T cell responses. *J Immunol* **164:**1364–1371.

113. **Combs CK, Johnson DE, Karlo JC, Cannady SB, Landreth GE.** 2000. Inflammatory mechanisms in Alzheimer's disease: inhibition of β-amyloid-stimulated proinflammatory responses and neurotoxicity by PPARγ agonists. *J Neurosci* **20:**558–567.

114. **Diab A, Deng C, Smith JD, Hussain RZ, Phanavanh B, Lovett-Racke AE, Drew PD, Racke MK.** 2002. Peroxisome proliferator-activated receptor-γ agonist 15-deoxy-$\Delta^{12,14}$-prostaglandin J_2 ameliorates experimental autoimmune encephalomyelitis. *J Immunol* **168:**2508–2515.

115. **Kawahito Y, Kondo M, Tsubouchi Y, Hashiramoto A, Bishop-Bailey D, Inoue K, Kohno M, Yamada R, Hla T, Sano H.** 2000. 15-Deoxy-$\Delta^{12,14}$-PGJ_2 induces synoviocyte apoptosis and suppresses adjuvant-induced arthritis in rats. *J Clin Invest* **106:**189–197.

116. **Reilly CM, Oates JC, Cook JA, Morrow JD, Halushka PV, Gilkeson GS.** 2000. Inhibition of mesangial cell nitric oxide in MRL/*lpr* mice by prostaglandin J_2 and proliferator activation receptor-γ agonists. *J Immunol* **164:**1498–1504.

117. **Kim EH, Na HK, Surh YJ.** 2006. Upregulation of VEGF by 15-deoxy-$\Delta^{12,14}$-prostaglandin J_2 via heme oxygenase-1 and ERK1/2 signaling in MCF-7 cells. *Ann N Y Acad Sci* **1090:**375–384.

118. **Oliva JL, Pérez-Sala D, Castrillo A, Martínez N, Cañada FJ, Boscá L, Rojas JM.** 2003. The cyclopentenone 15-deoxy-$\Delta^{12,14}$-prostaglandin J_2 binds to and activates H-Ras. *Proc Natl Acad Sci USA* **100:**4772–4777.

119. **Renedo M, Gayarre J, García-Domínguez CA, Pérez-Rodríguez A, Prieto A, Cañada FJ, Rojas JM, Pérez-Sala D.** 2007. Modification and activation of Ras proteins by electrophilic prostanoids with different structure are site-selective. *Biochemistry* **46:**6607–6616.

120. **Khan MM.** 1995. Regulation of IL-4 and IL-5 secretion by histamine and PGE_2. *Adv Exp Med Biol* **383:**35–42.

121. **Jiang C, Ting AT, Seed B.** 1998. PPAR-γ agonists inhibit production of monocyte inflammatory cytokines. *Nature* **391:**82–86.

122. **Ricote M, Li AC, Willson TM, Kelly CJ, Glass CK.** 1998. The peroxisome proliferator-activated receptor-γ is a negative regulator of macrophage activation. *Nature* **391:**79–82.

123. **Cernuda-Morollón E, Pineda-Molina E, Cañada FJ, Pérez-Sala D.** 2001. 15-Deoxy-$\Delta^{12,14}$-prostaglandin J_2 inhibition of NF-κB-DNA binding through covalent modification of the p50 subunit. *J Biol Chem* **276:** 35530–35536.

124. **Rossi A, Kapahi P, Natoli G, Takahashi T, Chen Y, Karin M, Santoro MG.** 2000. Anti-inflammatory cyclopentenone prostaglandins are direct inhibitors of IκB kinase. *Nature* **403:**103–108.

125. **Straus DS, Pascual G, Li M, Welch JS, Ricote M, Hsiang CH, Sengchanthalangsy LL, Ghosh G, Glass CK.** 2000. 15-Deoxy-$\Delta^{12,14}$-prostaglandin J_2 inhibits multiple steps in the NF-κB signaling pathway. *Proc Natl Acad Sci U S A* **97:**4844–4849.

126. **Jackson SM, Parhami F, Xi XP, Berliner JA, Hsueh WA, Law RE, Demer LL.** 1999. Peroxisome proliferator-activated receptor activators target human endothelial cells to inhibit leukocyte-endothelial cell interaction. *Arterioscler Thromb Vasc Biol* **19:**2094–2104.

127. **Pasceri V, Wu HD, Willerson JT, Yeh ET.** 2000. Modulation of vascular inflammation in vitro and in vivo by peroxisome proliferator-activated receptor-γ activators. *Circulation* **101:**235–238.

128. **Zhang X, Wang JM, Gong WH, Mukaida N, Young HA.** 2001. Differential regulation of chemokine gene expression by 15-deoxy-$\Delta^{12,14}$ prostaglandin J_2. *J Immunol* **166:**7104–7111.

129. **Lawrence T.** 2002. Modulation of inflammation in vivo through induction of the heat shock response, effects on NF-κB activation. *Inflamm Res* **51:**108–109.

130. **Bishop-Bailey D, Hla T.** 1999. Endothelial cell apoptosis induced by the peroxisome proliferator-activated receptor (PPAR) ligand 15-deoxy-$\Delta^{12,14}$-prostaglandin J_2. *J Biol Chem* **274:**17042–17048.

131. **Khoshnan A, Tindell C, Laux I, Bae D, Bennett B, Nel AE.** 2000. The NF-κB cascade is important in Bcl-x_L expression and for the anti-apoptotic effects of the CD28 receptor in primary human $CD4^+$ lymphocytes. *J Immunol* **165:**1743–1754.

132. **Lawrence T, Gilroy DW, Colville-Nash PR, Willoughby DA.** 2001. Possible new role for NF-κB in the resolution of inflammation. *Nat Med* **7:**1291–1297.

133. **Ward C, Dransfield I, Murray J, Farrow SN, Haslett C, Rossi AG.** 2002. Prostaglandin D_2 and its metabolites induce caspase-dependent granulocyte apoptosis that is mediated via inhibition of IκBα degradation using a peroxisome proliferator-activated receptor-γ-independent mechanism. *J Immunol* **168:**6232–6243.

134. **Trivedi SG, Newson J, Rajakariar R, Jacques TS, Hannon R, Kanaoka Y, Eguchi N, Colville-Nash P, Gilroy DW.** 2006. Essential role for hematopoietic prostaglandin D_2 synthase in the control of delayed type hypersensitivity. *Proc Natl Acad Sci U S A* **103:**5179–5184.

135. **Lewis RA, Austen KF, Drazen JM, Clark DA, Marfat A, Corey EJ.** 1980. Slow reacting substances of anaphylaxis: identification of leukotrienes C-1 and D from human and rat sources. *Proc Natl Acad Sci U S A* **77:** 3710–3714.

136. **Borgeat P, Samuelsson B.** 1979. Arachidonic acid metabolism in polymorphonuclear leukocytes: effects of ionophore A23187. *Proc Natl Acad Sci U S A* **76:**2148–2152.

137. **Smith MJ.** 1979. Prostaglandins and the polymorphonuclear leucocyte. *Agents Actions Suppl* **1979**(6):91–103.

138. **Funk CD.** 2001. Prostaglandins and leukotrienes: advances in eicosanoid biology. *Science* **294:**1871–1875.

139. **Minami M, Ohno S, Kawasaki H, Rådmark O, Samuelsson B, Jörnvall H, Shimizu T, Seyama Y, Suzuki K.** 1987. Molecular cloning of a cDNA coding for human leukotriene A_4 hydrolase. Complete primary structure of an enzyme involved in eicosanoid synthesis. *J Biol Chem* **262:**13873–13876.

140. **Hammarström S, Orning L, Bernström K.** 1985. Metabolism of leukotrienes. *Mol Cell Biochem* **69:**7–16.

141. **Lam BK, Gagnon L, Austen KF, Soberman RJ.** 1990. The mechanism of leukotriene B_4 export from human polymorphonuclear leukocytes. *J Biol Chem* **265:**13438–13441.

142. **Leier I, Jedlitschky G, Buchholz U, Keppler D.** 1994. Characterization of the ATP-dependent leukotriene C_4 export carrier in mastocytoma cells. *Eur J Biochem* **220:**599–606.

143. **Rouzer CA, Samuelsson B.** 1985. On the nature of the 5-lipoxygenase reaction in human leukocytes: enzyme purification and requirement for multiple stimulatory factors. *Proc Natl Acad Sci U S A* **82:**6040–6044.

144. **Ochi K, Yoshimoto T, Yamamoto S, Taniguchi K, Miyamoto T.** 1983. Arachidonate 5-lipoxygenase of guinea pig peritoneal polymorphonuclear leukocytes.

Activation by adenosine 5′-triphosphate. *J Biol Chem* 258:5754–5758.

145. **Werz O, Szellas D, Steinhilber D, Rådmark O.** 2002. Arachidonic acid promotes phosphorylation of 5-lipoxygenase at Ser-271 by MAPK-activated protein kinase 2 (MK2). *J Biol Chem* **277**:14793–14800.

146. **Kanaoka Y, Boyce JA.** 2004. Cysteinyl leukotrienes and their receptors: cellular distribution and function in immune and inflammatory responses. *J Immunol* **173**: 1503–1510.

147. **Tager AM, Luster AD.** 2003. BLT1 and BLT2: the leukotriene B_4 receptors. *Prostaglandins Leukot Essent Fatty Acids* **69**:123–134.

148. **Lynch KR, O'Neill GP, Liu Q, Im DS, Sawyer N, Metters KM, Coulombe N, Abramovitz M, Figueroa DJ, Zeng Z, Connolly BM, Bai C, Austin CP, Chateauneuf A, Stocco R, Greig GM, Kargman S, Hooks SB, Hosfield E, Williams DL Jr, Ford-Hutchinson AW, Caskey CT, Evans JF.** 1999. Characterization of the human cysteinyl leukotriene CysLT1 receptor. *Nature* **399**:789–793.

149. **Beller TC, Friend DS, Maekawa A, Lam BK, Austen KF, Kanaoka Y.** 2004. Cysteinyl leukotriene 1 receptor controls the severity of chronic pulmonary inflammation and fibrosis. *Proc Natl Acad Sci U S A* **101**:3047–3052.

150. **Hui Y, Cheng Y, Smalera I, Jian W, Goldhahn L, Fitzgerald GA, Funk CD.** 2004. Directed vascular expression of human cysteinyl leukotriene 2 receptor modulates endothelial permeability and systemic blood pressure. *Circulation* **110**:3360–3366.

151. **Sousa AR, Parikh A, Scadding G, Corrigan CJ, Lee TH.** 2002. Leukotriene-receptor expression on nasal mucosal inflammatory cells in aspirin-sensitive rhinosinusitis. *N Engl J Med* **347**:1493–1499.

152. **Devchand PR, Keller H, Peters JM, Vazquez M, Gonzalez FJ, Wahli W.** 1996. The PPARα-leukotriene B_4 pathway to inflammation control. *Nature* **384**:39–43.

153. **Krey G, Braissant O, L'Horset F, Kalkhoven E, Perroud M, Parker MG, Wahli W.** 1997. Fatty acids, eicosanoids, and hypolipidemic agents identified as ligands of peroxisome proliferator-activated receptors by coactivator-dependent receptor ligand assay. *Mol Endocrinol* **11**: 779–791.

154. **Lin Q, Ruuska SE, Shaw NS, Dong D, Noy N.** 1999. Ligand selectivity of the peroxisome proliferator-activated receptor α. *Biochemistry* **38**:185–190.

155. **Narala VR, Adapala RK, Suresh MV, Brock TG, Peters-Golden M, Reddy RC.** 2010. Leukotriene B_4 is a physiologically relevant endogenous peroxisome proliferator-activated receptor-α agonist. *J Biol Chem* **285**:22067–22074.

156. **Borgeat P, Naccache PH.** 1990. Biosynthesis and biological activity of leukotriene B_4. *Clin Biochem* **23**:459–468.

157. **Ott VL, Cambier JC, Kappler J, Marrack P, Swanson BJ.** 2003. Mast cell-dependent migration of effector $CD8^+$ T cells through production of leukotriene B_4. *Nat Immunol* **4**:974–981.

158. **Woodward DF, Krauss AH, Nieves AL, Spada CS.** 1991. Studies on leukotriene D_4 as an eosinophil chemoattractant. *Drugs Exp Clin Res* **17**:543–548.

159. **Björk J, Hedqvist P, Arfors KE.** 1982. Increase in vascular permeability induced by leukotriene B_4 and the role of polymorphonuclear leukocytes. *Inflammation* **6**: 189–200.

160. **Dahlén SE, Björk J, Hedqvist P, Arfors KE, Hammarström S, Lindgren JA, Samuelsson B.** 1981. Leukotrienes promote plasma leakage and leukocyte adhesion in postcapillary venules: *in vivo* effects with relevance to the acute inflammatory response. *Proc Natl Acad Sci U S A* **78**:3887–3891.

161. **Hedqvist P, Dahlén SE.** 1983. Pulmonary and vascular effects of leukotrienes imply involvement in asthma and inflammation. *Adv Prostaglandin Thromboxane Leukot Res* **11**:27–32.

162. **Orange RP, Stechschulte DJ, Austen KF.** 1969. Cellular mechanisms involved in the release of slow reacting substance of anaphylaxis. *Fed Proc* **28**:1710–1715.

163. **Svensjö E.** 1978. Bradykinin and prostaglandin E_1, E_2 and $F_{2\alpha}$-induced macromolecular leakage in the hamster cheek pouch. *Prostaglandins Med* **1**:397–410.

164. **Werz O, Steinhilber D.** 2005. Development of 5-lipoxygenase inhibitors—lessons from cellular enzyme regulation. *Biochem Pharmacol* **70**:327–333.

165. **Israel E, Rubin P, Kemp JP, Grossman J, Pierson W, Siegel SC, Tinkelman D, Murray JJ, Busse W, Segal AT, Fish J, Kaiser HB, Ledford D, Wenzel S, Rosenthal R, Cohn J, Lanni C, Pearlman H, Karahalios P, Drazen JM.** 1993. The effect of inhibition of 5-lipoxygenase by zileuton in mild-to-moderate asthma. *Ann Intern Med* **119**:1059–1066.

166. **Knorr B, Matz J, Bernstein JA, Nguyen H, Seidenberg BC, Reiss TF, Becker A, Pediatric Montelukast Study Group.** 1998. Montelukast for chronic asthma in 6- to 14-year-old children: a randomized, double-blind trial. *JAMA* **279**:1181–1186.

167. **Suissa S, Dennis R, Ernst P, Sheehy O, Wood-Dauphinee S.** 1997. Effectiveness of the leukotriene receptor antagonist zafirlukast for mild-to-moderate asthma. A randomized, double-blind, placebo-controlled trial. *Ann Intern Med* **126**:177–183.

168. **Spanbroek R, Grabner R, Lotzer K, Hildner M, Urbach A, Ruhling K, Moos MP, Kaiser B, Cohnert TU, Wahlers T, Zieske A, Plenz G, Robenek H, Salbach P, Kuhn H, Radmark O, Samuelsson B, Habenicht AJ.** 2003. Expanding expression of the 5-lipoxygenase pathway within the arterial wall during human atherogenesis. *Proc Natl Acad Sci U S A* **100**:1238–1243.

169. **Qiu H, Gabrielsen A, Agardh HE, Wan M, Wetterholm A, Wong CH, Hedin U, Swedenborg J, Hansson GK, Samuelsson B, Paulsson-Berne G, Haeggström JZ.** 2006. Expression of 5-lipoxygenase and leukotriene A_4 hydrolase in human atherosclerotic lesions correlates with symptoms of plaque instability. *Proc Natl Acad Sci U S A* **103**:8161–8166.

170. **Aiello RJ, Bourassa PA, Lindsey S, Weng W, Freeman A, Showell HJ.** 2002. Leukotriene B_4 receptor antago-

nism reduces monocytic foam cells in mice. *Arterioscler Thromb Vasc Biol* 22:443–449.

171. **Bäck M, Bu DX, Bränström R, Sheikine Y, Yan ZQ, Hansson GK.** 2005. Leukotriene B_4 signaling through NF-κB-dependent BLT1 receptors on vascular smooth muscle cells in atherosclerosis and intimal hyperplasia. *Proc Natl Acad Sci U S A* **102:**17501–17506.

172. **Uzonyi B, Lötzer K, Jahn S, Kramer C, Hildner M, Bretschneider E, Radke D, Beer M, Vollandt R, Evans JF, Funk CD, Habenicht AJ.** 2006. Cysteinyl leukotriene 2 receptor and protease-activated receptor 1 activate strongly correlated early genes in human endothelial cells. *Proc Natl Acad Sci U S A* **103:**6326–6331.

173. **Zhao L, Moos MP, Gräbner R, Pédrono F, Fan J, Kaiser B, John N, Schmidt S, Spanbroek R, Lötzer K, Huang L, Cui J, Rader DJ, Evans JF, Habenicht AJ, Funk CD.** 2004. The 5-lipoxygenase pathway promotes pathogenesis of hyperlipidemia-dependent aortic aneurysm. *Nat Med* **10:**966–973.

174. **Helgadottir A, Manolescu A, Helgason A, Thorleifsson G, Thorsteinsdottir U, Gudbjartsson DF, Gretarsdottir S, Magnusson KP, Gudmundsson G, Hicks A, Jonsson T, Grant SF, Sainz J, O'Brien SJ, Sveinbjornsdottir S, Valdimarsson EM, Matthiasson SE, Levey AI, Abramson JL, Reilly MP, Vaccarino V, Wolfe ML, Gudnason V, Quyyumi AA, Topol EJ, Rader DJ, Thorgeirsson G, Gulcher JR, Hakonarson H, Kong A, Stefansson K.** 2006. A variant of the gene encoding leukotriene A_4 hydrolase confers ethnicity-specific risk of myocardial infarction. *Nat Genet* **38:**68–74.

175. **Helgadottir A, Manolescu A, Thorleifsson G, Gretarsdottir S, Jonsdottir H, Thorsteinsdottir U, Samani NJ, Gudmundsson G, Grant SF, Thorgeirsson G, Sveinbjornsdottir S, Valdimarsson EM, Matthiasson SE, Johannsson H, Gudmundsdottir O, Gurney ME, Sainz J, Thorhallsdottir M, Andresdottir M, Frigge ML, Topol EJ, Kong A, Gudnason V, Hakonarson H, Gulcher JR, Stefansson K.** 2004. The gene encoding 5-lipoxygenase activating protein confers risk of myocardial infarction and stroke. *Nat Genet* **36:**233–239.

176. **Kajimoto K, Shioji K, Ishida C, Iwanaga Y, Kokubo Y, Tomoike H, Miyazaki S, Nonogi H, Goto Y, Iwai N.** 2005. Validation of the association between the gene encoding 5-lipoxygenase-activating protein and myocardial infarction in a Japanese population. *Circ J* **69:** 1029–1034.

177. **Hakonarson H, Thorvaldsson S, Helgadottir A, Gudbjartsson D, Zink F, Andresdottir M, Manolescu A, Arnar DO, Andersen K, Sigurdsson A, Thorgeirsson G, Jonsson A, Agnarsson U, Bjornsdottir H, Gottskalksson G, Einarsson A, Gudmundsdottir H, Adalsteinsdottir AE, Gudmundsson K, Kristjansson K, Hardarson T, Kristinsson A, Topol EJ, Gulcher J, Kong A, Gurney M, Thorgeirsson G, Stefansson K.** 2005. Effects of a 5-lipoxygenase-activating protein inhibitor on biomarkers associated with risk of myocardial infarction: a randomized trial. *JAMA* **293:**2245–2256.

178. **Peters-Golden M, Canetti C, Mancuso P, Coffey MJ.** 2005. Leukotrienes: underappreciated mediators of innate immune responses. *J Immunol* **174:**589–594.

179. **Bailie MB, Standiford TJ, Laichalk LL, Coffey MJ, Strieter R, Peters-Golden M.** 1996. Leukotriene-deficient mice manifest enhanced lethality from *Klebsiella* pneumonia in association with decreased alveolar macrophage phagocytic and bactericidal activities. *J Immunol* **157:**5221–5224.

180. **Mancuso P, Lewis C, Serezani CH, Goel D, Peters-Golden M.** 2010. Intrapulmonary administration of leukotriene B_4 enhances pulmonary host defense against pneumococcal pneumonia. *Infect Immun* **78:** 2264–2271.

181. **Medeiros AI, Sá-Nunes A, Turato WM, Secatto A, Frantz FG, Sorgi CA, Serezani CH, Deepe GS Jr, Faccioli LH.** 2008. Leukotrienes are potent adjuvant during fungal infection: effects on memory T cells. *J Immunol* **181:**8544–8551.

182. **Peres CM, de Paula L, Medeiros AI, Sorgi CA, Soares EG, Carlos D, Peters-Golden M, Silva CL, Faccioli LH.** 2007. Inhibition of leukotriene biosynthesis abrogates the host control of *Mycobacterium tuberculosis*. *Microbes Infect* **9:**483–489.

183. **Schultz MJ, Wijnholds J, Peppelenbosch MP, Vervoordeldonk MJ, Speelman P, van Deventer SJ, Borst P, van der Poll T.** 2001. Mice lacking the multidrug resistance protein 1 are resistant to *Streptococcus pneumoniae*-induced pneumonia. *J Immunol* **166:**4059–4064.

184. **Serezani CH, Perrela JH, Russo M, Peters-Golden M, Jancar S.** 2006. Leukotrienes are essential for the control of *Leishmania amazonensis* infection and contribute to strain variation in susceptibility. *J Immunol* **177:** 3201–3208.

185. **Serezani CH, Aronoff DM, Jancar S, Mancuso P, Peters-Golden M.** 2005. Leukotrienes enhance the bactericidal activity of alveolar macrophages against *Klebsiella pneumoniae* through the activation of NADPH oxidase. *Blood* **106:**1067–1075.

186. **Ballinger MN, Hubbard LL, McMillan TR, Toews GB, Peters-Golden M, Paine R III, Moore BB.** 2008. Paradoxical role of alveolar macrophage-derived granulocyte-macrophage colony-stimulating factor in pulmonary host defense post-bone marrow transplantation. *Am J Physiol Lung Cell Mol Physiol* **295:** L114–L122.

187. **Balter MS, Toews GB, Peters-Golden M.** 1989. Multiple defects in arachidonate metabolism in alveolar macrophages from young asymptomatic smokers. *J Lab Clin Med* **114:**662–673.

188. **Cederholm T, Lindgren JA, Palmblad J.** 2000. Impaired leukotriene C_4 generation in granulocytes from protein-energy malnourished chronically ill elderly. *J Intern Med* **247:**715–722.

189. **Coffey MJ, Phare SM, Kazanjian PH, Peters-Golden M.** 1996. 5-Lipoxygenase metabolism in alveolar macrophages from subjects infected with the human immunodeficiency virus. *J Immunol* **157:**393–399.

190. **Coffey MJ, Wilcoxen SE, Phare SM, Simpson RU, Gyetko MR, Peters-Golden M.** 1994. Reduced 5-lipoxygenase metabolism of arachidonic acid in macrophages from

1,25-dihydroxyvitamin D3-deficient rats. *Prostaglandins* **48:**313–329.

191. **Jubiz W, Draper RE, Gale J, Nolan G.** 1984. Decreased leukotriene B_4 synthesis by polymorphonuclear leukocytes from male patients with diabetes mellitus. *Prostaglandins Leukot Med* **14:**305–311.

192. **Lärfars G, Lantoine F, Devynck MA, Palmblad J, Gyllenhammar H.** 1999. Activation of nitric oxide release and oxidative metabolism by leukotrienes B_4, C_4, and D_4 in human polymorphonuclear leukocytes. *Blood* **93:**1399–1405.

193. **Talvani A, Machado FS, Santana GC, Klein A, Barcelos L, Silva JS, Teixeira MM.** 2002. Leukotriene B_4 induces nitric oxide synthesis in *Trypanosoma cruzi*-infected murine macrophages and mediates resistance to infection. *Infect Immun* **70:**4247–4253.

194. **Flamand L, Tremblay MJ, Borgeat P.** 2007. Leukotriene B_4 triggers the in vitro and in vivo release of potent antimicrobial agents. *J Immunol* **178:**8036–8045.

195. **Powell WS, Gravelle F, Gravel S.** 1992. Metabolism of 5(*S*)-hydroxy-6,8,11,14-eicosatetraenoic acid and other 5(*S*)-hydroxyeicosanoids by a specific dehydrogenase in human polymorphonuclear leukocytes. *J Biol Chem* **267:**19233–19241.

196. **Graham FD, Erlemann KR, Gravel S, Rokach J, Powell WS.** 2009. Oxidative stress-induced changes in pyridine nucleotides and chemoattractant 5-lipoxygenase products in aging neutrophils. *Free Radic Biol Med* **47:** 62–71.

197. **Powell WS, Gravelle F, Gravel S.** 1994. Phorbol myristate acetate stimulates the formation of 5-oxo-6, 8,11,14-eicosatetraenoic acid by human neutrophils by activating NADPH oxidase. *J Biol Chem* **269:** 25373–25380.

198. **Cossette C, Patel P, Anumolu JR, Sivendran S, Lee GJ, Gravel S, Graham FD, Lesimple A, Mamer OA, Rokach J, Powell WS.** 2008. Human neutrophils convert the sebum-derived polyunsaturated fatty acid sebaleic acid to a potent granulocyte chemoattractant. *J Biol Chem* **283:**11234–11243.

199. **Patel P, Cossette C, Anumolu JR, Gravel S, Lesimple A, Mamer OA, Rokach J, Powell WS.** 2008. Structural requirements for activation of the 5-oxo-6*E*,8*Z*, 11*Z*, 14*Z*-eicosatetraenoic acid (5-oxo-ETE) receptor: identification of a mead acid metabolite with potent agonist activity. *J Pharmacol Exp Ther* **325:**698–707.

200. **Powell WS, Chung D, Gravel S.** 1995. 5-Oxo-6,8,11, 14-eicosatetraenoic acid is a potent stimulator of human eosinophil migration. *J Immunol* **154:**4123–4132.

201. **Powell WS, Rokach J.** 2005. Biochemistry, biology and chemistry of the 5-lipoxygenase product 5-oxo-ETE. *Prog Lipid Res* **44:**154–183.

202. **Brink CB, Harvey BH, Bodenstein J, Venter DP, Oliver DW.** 2004. Recent advances in drug action and therapeutics: relevance of novel concepts in G-protein-coupled receptor and signal transduction pharmacology. *Br J Clin Pharmacol* **57:**373–387.

203. **Takeda S, Kadowaki S, Haga T, Takaesu H, Mitaku S.** 2002. Identification of G protein-coupled receptor genes from the human genome sequence. *FEBS Lett* **520:**97–101.

204. **Hosoi T, Koguchi Y, Sugikawa E, Chikada A, Ogawa K, Tsuda N, Suto N, Tsunoda S, Taniguchi T, Ohnuki T.** 2002. Identification of a novel human eicosanoid receptor coupled to $G_{i/o}$. *J Biol Chem* **277:**31459–31465.

205. **Jones CE, Holden S, Tenaillon L, Bhatia U, Seuwen K, Tranter P, Turner J, Kettle R, Bouhelal R, Charlton S, Nirmala NR, Jarai G, Finan P.** 2003. Expression and characterization of a 5-oxo-6*E*,8*Z*,11*Z*,14*Z*-eicosatetraenoic acid receptor highly expressed on human eosinophils and neutrophils. *Mol Pharmacol* **63:** 471–477.

206. **Norgauer J, Barbisch M, Czech W, Pareigis J, Schwenk U, Schröder JM.** 1996. Chemotactic 5-oxo-icosatetraenoic acids activate a unique pattern of neutrophil responses. Analysis of phospholipid metabolism, intracellular Ca^{2+} transients, actin reorganization, superoxide-anion production and receptor up-regulation. *Eur J Biochem* **236:** 1003–1009.

207. **O'Flaherty JT, Kuroki M, Nixon AB, Wijkander J, Yee E, Lee SL, Smitherman PK, Wykle RL, Daniel LW.** 1996. 5-Oxo-eicosanoids and hematopoietic cytokines cooperate in stimulating neutrophil function and the mitogen-activated protein kinase pathway. *J Biol Chem* **271:**17821–17828.

208. **Hosoi T, Sugikawa E, Chikada A, Koguchi Y, Ohnuki T.** 2005. TG1019/OXE, a $G\alpha_{i/o}$-protein-coupled receptor, mediates 5-oxo-eicosatetraenoic acid-induced chemotaxis. *Biochem Biophys Res Commun* **334:**987–995.

209. **O'Flaherty JT, Rogers LC, Chadwell BA, Owen JS, Rao A, Cramer SD, Daniel LW.** 2002. 5(*S*)-Hydroxy-6, 8,11,14-*E*,*Z*,*Z*,*Z*-eicosatetraenoate stimulates PC3 cell signaling and growth by a receptor-dependent mechanism. *Cancer Res* **62:**6817–6819.

210. **Langlois A, Chouinard F, Flamand N, Ferland C, Rola-Pleszczynski M, Laviolette M.** 2009. Crucial implication of protein kinase C (PKC)-δ, PKC-ζ, ERK-1/2, and p38 MAPK in migration of human asthmatic eosinophils. *J Leukoc Biol* **85:**656–663.

211. **O'Flaherty JT, Kuroki M, Nixon AB, Wijkander J, Yee E, Lee SL, Smitherman PK, Wykle RL, Daniel LW.** 1996. 5-Oxo-eicosatetraenoate is a broadly active, eosinophil-selective stimulus for human granulocytes. *J Immunol* **157:**336–342.

212. **Stamatiou PB, Chan CC, Monneret G, Ethier D, Rokach J, Powell WS.** 2004. 5-Oxo-6,8,11,14-eicosatetraenoic acid stimulates the release of the eosinophil survival factor granulocyte/macrophage colony-stimulating factor from monocytes. *J Biol Chem* **279:**28159–28164.

213. **Ghosh S, Karin M.** 2002. Missing pieces in the NF-κB puzzle. *Cell* **109**(Suppl)**:**S81–S96.

214. **Sundaram S, Ghosh J.** 2006. Expression of 5-oxoETE receptor in prostate cancer cells: critical role in survival. *Biochem Biophys Res Commun* **339:**93–98.

215. **Clària J, Serhan CN.** 1995. Aspirin triggers previously undescribed bioactive eicosanoids by human endothelial cell-leukocyte interactions. *Proc Natl Acad Sci U S A* **92:**9475–9479.

216. **Serhan CN.** 1989. On the relationship between leukotriene and lipoxin production by human neutrophils: evidence for differential metabolism of 15-HETE and 5-HETE. *Biochim Biophys Acta* **1004:**158–168.

217. **Serhan CN.** 1995. Leukocyte transmigration, chemotaxis, and oxygenated derivatives of arachidonic acid: when is chirality important? *Am J Respir Cell Mol Biol* **12:**251–253.

218. **Brezinski ME, Serhan CN.** 1990. Selective incorporation of (15*S*)-hydroxyeicosatetraenoic acid in phosphatidylinositol of human neutrophils: agonist-induced deacylation and transformation of stored hydroxyeicosanoids. *Proc Natl Acad Sci U S A* **87:**6248–6252.

219. **Romano M, Serhan CN.** 1992. Lipoxin generation by permeabilized human platelets. *Biochemistry* **31:**8269–8277.

220. **Chiang N, Takano T, Arita M, Watanabe S, Serhan CN.** 2003. A novel rat lipoxin A_4 receptor that is conserved in structure and function. *Br J Pharmacol* **139:** 89–98.

221. **Fiore S, Maddox JF, Perez HD, Serhan CN.** 1994. Identification of a human cDNA encoding a functional high affinity lipoxin A_4 receptor. *J Exp Med* **180:**253–260.

222. **Takano T, Fiore S, Maddox JF, Brady HR, Petasis NA, Serhan CN.** 1997. Aspirin-triggered 15-epi-lipoxin A_4 (LXA_4) and LXA_4 stable analogues are potent inhibitors of acute inflammation: evidence for anti-inflammatory receptors. *J Exp Med* **185:**1693–1704.

223. **Maddox JF, Hachicha M, Takano T, Petasis NA, Fokin VV, Serhan CN.** 1997. Lipoxin A_4 stable analogs are potent mimetics that stimulate human monocytes and THP-1 cells via a G-protein-linked lipoxin A_4 receptor. *J Biol Chem* **272:**6972–6978.

224. **Ariel A, Chiang N, Arita M, Petasis NA, Serhan CN.** 2003. Aspirin-triggered lipoxin A_4 and B_4 analogs block extracellular signal-regulated kinase-dependent TNF-α secretion from human T cells. *J Immunol* **170:**6266–6272.

225. **Sodin-Semrl S, Taddeo B, Tseng D, Varga J, Fiore S.** 2000. Lipoxin A_4 inhibits IL-1β-induced IL-6, IL-8, and matrix metalloproteinase-3 production in human synovial fibroblasts and enhances synthesis of tissue inhibitors of metalloproteinases. *J Immunol* **164:**2660–2666.

226. **Gronert K, Gewirtz A, Madara JL, Serhan CN.** 1998. Identification of a human enterocyte lipoxin A_4 receptor that is regulated by interleukin (IL)-13 and interferon γ and inhibits tumor necrosis factor α-induced IL-8 release. *J Exp Med* **187:**1285–1294.

227. **Devchand PR, Arita M, Hong S, Bannenberg G, Moussignac RL, Gronert K, Serhan CN.** 2003. Human ALX receptor regulates neutrophil recruitment in transgenic mice: roles in inflammation and host defense. *FASEB J* **17:**652–659.

228. **Fukunaga K, Kohli P, Bonnans C, Fredenburgh LE, Levy BD.** 2005. Cyclooxygenase 2 plays a pivotal role in the resolution of acute lung injury. *J Immunol* **174:** 5033–5039.

229. **Levy BD, De Sanctis GT, Devchand PR, Kim E, Ackerman K, Schmidt BA, Szczeklik W, Drazen JM, Serhan CN.** 2002. Multi-pronged inhibition of airway hyper-responsiveness and inflammation by lipoxin A_4. *Nat Med* **8:**1018–1023.

230. **Paul-Clark MJ, Van Cao T, Moradi-Bidhendi N, Cooper D, Gilroy DW.** 2004. 15-Epi-lipoxin A_4-mediated induction of nitric oxide explains how aspirin inhibits acute inflammation. *J Exp Med* **200:**69–78.

231. **Levy BD, Petasis NA, Serhan CN.** 1997. Polyisoprenyl phosphates in intracellular signalling. *Nature* **389:** 985–990.

232. **József L, Zouki C, Petasis NA, Serhan CN, Filep JG.** 2002. Lipoxin A_4 and aspirin-triggered 15-epi-lipoxin A_4 inhibit peroxynitrite formation, NF-κB and AP-1 activation, and IL-8 gene expression in human leukocytes. *Proc Natl Acad Sci U S A* **99:**13266–13271.

233. **Filep JG, Beauchamp M, Baron C, Paquette Y.** 1998. Peroxynitrite mediates IL-8 gene expression and production in lipopolysaccharide-stimulated human whole blood. *J Immunol* **161:**5656–5662.

234. **Zouki C, József L, Ouellet S, Paquette Y, Filep JG.** 2001. Peroxynitrite mediates cytokine-induced IL-8 gene expression and production by human leukocytes. *J Leukoc Biol* **69:**815–824.

235. **Ohira T, Bannenberg G, Arita M, Takahashi M, Ge Q, Van Dyke TE, Stahl GL, Serhan CN, Badwey JA.** 2004. A stable aspirin-triggered lipoxin A_4 analog blocks phosphorylation of leukocyte-specific protein 1 in human neutrophils. *J Immunol* **173:**2091–2098.

236. **Badr KF, DeBoer DK, Schwartzberg M, Serhan CN.** 1989. Lipoxin A_4 antagonizes cellular and *in vivo* actions of leukotriene D_4 in rat glomerular mesangial cells: evidence for competition at a common receptor. *Proc Natl Acad Sci U S A* **86:**3438–3442.

237. **Fiore S, Romano M, Reardon EM, Serhan CN.** 1993. Induction of functional lipoxin A_4 receptors in HL-60 cells. *Blood* **81:**3395–3403.

238. **Aliberti J, Serhan C, Sher A.** 2002. Parasite-induced lipoxin A_4 is an endogenous regulator of IL-12 production and immunopathology in *Toxoplasma gondii* infection. *J Exp Med* **196:**1253–1262.

239. **Machado FS, Johndrow JE, Esper L, Dias A, Bafica A, Serhan CN, Aliberti J.** 2006. Anti-inflammatory actions of lipoxin A_4 and aspirin-triggered lipoxin are SOCS-2 dependent. *Nat Med* **12:**330–334.

240. **Schaldach CM, Riby J, Bjeldanes LF.** 1999. Lipoxin A_4: a new class of ligand for the Ah receptor. *Biochemistry* **38:**7594–7600.

241. **Maddox JF, Colgan SP, Clish CB, Petasis NA, Fokin VV, Serhan CN.** 1998. Lipoxin B_4 regulates human monocyte/neutrophil adherence and motility: design of stable lipoxin B_4 analogs with increased biologic activity. *FASEB J* **12:**487–494.

242. **Patcha V, Wigren J, Winberg ME, Rasmusson B, Li J, Särndahl E.** 2004. Differential inside-out activation of β_2-integrins by leukotriene B_4 and fMLP in human neutrophils. *Exp Cell Res* **300:**308–319.

243. **Serhan CN, Takano T, Gronert K, Chiang N, Clish CB.** 1999. Lipoxin and aspirin-triggered 15-epi-lipoxin cellular interactions anti-inflammatory lipid mediators. *Clin Chem Lab Med* **37:**299–309.

244. **Soyombo O, Spur BW, Lee TH.** 1994. Effects of lipoxin A_4 on chemotaxis and degranulation of human eosinophils stimulated by platelet-activating factor and *N*-formyl-l-methionyl-l-leucyl-l-phenylalanine. *Allergy* **49:**230–234.

245. **Maddox JF, Serhan CN.** 1996. Lipoxin A_4 and B_4 are potent stimuli for human monocyte migration and adhesion: selective inactivation by dehydrogenation and reduction. *J Exp Med* **183:**137–146.

246. **Chiang N, Gronert K, Clish CB, O'Brien JA, Freeman MW, Serhan CN.** 1999. Leukotriene B_4 receptor transgenic mice reveal novel protective roles for lipoxins and aspirin-triggered lipoxins in reperfusion. *J Clin Invest* **104:**309–316.

247. **Morris T, Stables M, Hobbs A, de Souza P, Colville-Nash P, Warner T, Newson J, Bellingan G, Gilroy DW.** 2009. Effects of low-dose aspirin on acute inflammatory responses in humans. *J Immunol* **183:**2089–2096.

248. **Morris T, Stables M, Colville-Nash P, Newson J, Bellingan G, de Souza PM, Gilroy DW.** 2010. Dichotomy in duration and severity of acute inflammatory responses in humans arising from differentially expressed proresolution pathways. *Proc Natl Acad Sci U S A* **107:** 8842–8847.

249. **Godson C, Mitchell S, Harvey K, Petasis NA, Hogg N, Brady HR.** 2000. Cutting edge: lipoxins rapidly stimulate nonphlogistic phagocytosis of apoptotic neutrophils by monocyte-derived macrophages. *J Immunol* **164:** 1663–1667.

250. **Maderna P, Cottell DC, Berlasconi G, Petasis NA, Brady HR, Godson C.** 2002. Lipoxins induce actin reorganization in monocytes and macrophages but not in neutrophils: differential involvement of Rho GTPases. *Am J Pathol* **160:**2275–2283.

251. **Bannenberg GL, Chiang N, Ariel A, Arita M, Tjonahen E, Gotlinger KH, Hong S, Serhan CN.** 2005. Molecular circuits of resolution: formation and actions of resolvins and protectins. *J Immunol* **174:**4345–4355.

252. **Freire-de-Lima CG, Xiao YQ, Gardai SJ, Bratton DL, Schiemann WP, Henson PM.** 2006. Apoptotic cells, through transforming growth factor-β, coordinately induce anti-inflammatory and suppress pro-inflammatory eicosanoid and NO synthesis in murine macrophages. *J Biol Chem* **281:**38376–38384.

253. **Mitchell S, Thomas G, Harvey K, Cottell D, Reville K, Berlasconi G, Petasis NA, Erwig L, Rees AJ, Savill J, Brady HR, Godson C.** 2002. Lipoxins, aspirin-triggered epi-lipoxins, lipoxin stable analogues, and the resolution of inflammation: stimulation of macrophage phagocytosis of apoptotic neutrophils *in vivo*. *J Am Soc Nephrol* **13:**2497–2507.

254. **Leonard MO, Hannan K, Burne MJ, Lappin DW, Doran P, Coleman P, Stenson C, Taylor CT, Daniels F, Godson C, Petasis NA, Rabb H, Brady HR.** 2002. 15-Epi-16-(para-fluorophenoxy)-lipoxin A_4-methyl ester, a synthetic analogue of 15-epi-lipoxin A_4, is protective in experimental ischemic acute renal failure. *J Am Soc Nephrol* **13:**1657–1662.

255. **McMahon B, Mitchell D, Shattock R, Martin F, Brady HR, Godson C.** 2002. Lipoxin, leukotriene, and PDGF receptors cross-talk to regulate mesangial cell proliferation. *FASEB J* **16:**1817–1819.

256. **Sato Y, Kitasato H, Murakami Y, Hashimoto A, Endo H, Kondo H, Inoue M, Hayashi I.** 2004. Down-regulation of lipoxin A_4 receptor by thromboxane A_2 signaling in RAW246.7 cells in vitro and bleomycin-induced lung fibrosis in vivo. *Biomed Pharmacother* **58:** 381–387.

257. **Wu SH, Wu XH, Lu C, Dong L, Chen ZQ.** 2006. Lipoxin A_4 inhibits proliferation of human lung fibroblasts induced by connective tissue growth factor. *Am J Respir Cell Mol Biol* **34:**65–72.

258. **Serhan CN.** 1994. Lipoxin biosynthesis and its impact in inflammatory and vascular events. *Biochim Biophys Acta* **1212:**1–25.

259. **Tamaoki J, Tagaya E, Yamawaki I, Konno K.** 1995. Lipoxin A_4 inhibits cholinergic neurotransmission through nitric oxide generation in the rabbit trachea. *Eur J Pharmacol* **287:**233–238.

260. **Hachicha M, Pouliot M, Petasis NA, Serhan CN.** 1999. Lipoxin (LX)A_4 and aspirin-triggered 15-epi-LXA_4 inhibit tumor necrosis factor 1α-initiated neutrophil responses and trafficking: regulators of a cytokine-chemokine axis. *J Exp Med* **189:**1923–1930.

261. **Pouliot M, Serhan CN.** 1999. Lipoxin A_4 and aspirin-triggered 15-epi-LXA_4 inhibit tumor necrosis factor-α-initiated neutrophil responses and trafficking: novel regulators of a cytokine-chemokine axis relevant to periodontal diseases. *J Periodontal Res* **34:**370–373.

262. **Machado FS, Johndrow JE, Esper L, Dias A, Bafica A, Serhan CN, Aliberti J.** 2006. Anti-inflammatory actions of lipoxin A_4 and aspirin-triggered lipoxin are SOCS-2 dependent. *Nat Med* **12:**330–334.

263. **Munger KA, Montero A, Fukunaga M, Uda S, Yura T, Imai E, Kaneda Y, Valdivielso JM, Badr KF.** 1999. Transfection of rat kidney with human 15-lipoxygenase suppresses inflammation and preserves function in experimental glomerulonephritis. *Proc Natl Acad Sci U S A* **96:**13375–13380.

264. **O'Meara YM, Brady HR.** 1997. Lipoxins, leukocyte recruitment and the resolution phase of acute glomerulonephritis. *Kidney Int Suppl* **58:**S56–S61.

265. **Karp CL, Flick LM, Yang R, Uddin J, Petasis NA.** 2005. Cystic fibrosis and lipoxins. *Prostaglandins Leukot Essent Fatty Acids* **73:**263–270.

266. **Pouliot M, Clish CB, Petasis NA, Van Dyke TE, Serhan CN.** 2000. Lipoxin A_4 analogues inhibit leukocyte recruitment to *Porphyromonas gingivalis*: a role for cyclooxygenase-2 and lipoxins in periodontal disease. *Biochemistry* **39:**4761–4768.

267. **Levy BD, Bonnans C, Silverman ES, Palmer LJ, Marigowda G, Israel E, Severe Asthma Research Program, National Heart, Lung, and Blood Institute.** 2005. Diminished lipoxin biosynthesis in severe asthma. *Am J Respir Crit Care Med* **172:**824–830.

268. **Gronert K, Maheshwari N, Khan N, Hassan IR, Dunn M, Laniado Schwartzman M.** 2005. A role for the mouse 12/15-lipoxygenase pathway in promoting epithelial wound healing and host defense. *J Biol Chem* **280:**15267–15278.

269. **Schottelius AJ, Giesen C, Asadullah K, Fierro IM, Colgan SP, Bauman J, Guilford W, Perez HD, Parkinson JF.** 2002. An aspirin-triggered lipoxin A_4 stable analog displays a unique topical anti-inflammatory profile. *J Immunol* **169:**7063–7070.

270. **Aliberti J, Hieny S, Reis e Sousa C, Serhan CN, Sher A.** 2002. Lipoxin-mediated inhibition of IL-12 production by DCs: a mechanism for regulation of microbial immunity. *Nat Immunol* **3:**76–82.

271. **Bafica A, Scanga CA, Serhan C, Machado F, White S, Sher A, Aliberti J.** 2005. Host control of *Mycobacterium tuberculosis* is regulated by 5-lipoxygenase-dependent lipoxin production. *J Clin Invest* **115:** 1601–1606.

272. **Arita M, Yoshida M, Hong S, Tjonahen E, Glickman JN, Petasis NA, Blumberg RS, Serhan CN.** 2005. Resolvin E1, an endogenous lipid mediator derived from omega-3 eicosapentaenoic acid, protects against 2,4,6-trinitrobenzene sulfonic acid-induced colitis. *Proc Natl Acad Sci USA* **102:**7671–7676.

273. **Burr GO, Burr MM.** 1973. Nutrition classics from The Journal of Biological Chemistry 82:345-67, 1929. A new deficiency disease produced by the rigid exclusion of fat from the diet. *Nutr Rev* **31:**248–249.

274. **Galli C, Risé P.** 2009. Fish consumption, omega 3 fatty acids and cardiovascular disease. The science and the clinical trials. *Nutr Health* **20:**11–20.

275. **Riediger ND, Othman RA, Suh M, Moghadasian MH.** 2009. A systemic review of the roles of n-3 fatty acids in health and disease. *J Am Diet Assoc* **109:**668–679.

276. **GISSI-Prevenzione Investigators (Gruppo Italiano per lo Studio della Sopravvivenza nell'Infarto miocardico).** 1999. Dietary supplementation with n-3 polyunsaturated fatty acids and vitamin E after myocardial infarction: results of the GISSI-Prevenzione trial. *Lancet* **354:**447–455.

277. **León H, Shibata MC, Sivakumaran S, Dorgan M, Chatterley T, Tsuyuki RT.** 2008. Effect of fish oil on arrhythmias and mortality: systematic review. *BMJ* **337:** a2931. doi:10.1136/bmj.a2931.

278. **Ridker PM.** 2009. The JUPITER trial: results, controversies, and implications for prevention. *Circ Cardiovasc Qual Outcomes* **2:**279–285.

279. **De Caterina R, Caprioli R, Giannessi D, Sicari R, Galli C, Lazzerini G, Bernini W, Carr L, Rindi P.** 1993. n-3 fatty acids reduce proteinuria in patients with chronic glomerular disease. *Kidney Int* **44:**843–850.

280. **Lu Y, Hong S, Tjonahen E, Serhan CN.** 2005. Mediator-lipidomics: databases and search algorithms for PUFA-derived mediators. *J Lipid Res* **46:**790–802.

281. **Serhan CN, Yang R, Martinod K, Kasuga K, Pillai PS, Porter TF, Oh SF, Spite M.** 2009. Maresins: novel macrophage mediators with potent antiinflammatory and proresolving actions. *J Exp Med* **206:**15–23.

282. **Serhan CN.** 2008. Controlling the resolution of acute inflammation: a new genus of dual anti-inflammatory and proresolving mediators. *J Periodontol* **79**(Suppl): 1520–1526.

283. **Serhan CN.** 2008. Systems approach with inflammatory exudates uncovers novel anti-inflammatory and pro-resolving mediators. *Prostaglandins Leukot Essent Fatty Acids* **79:**157–163.

284. **Arita M, Clish CB, Serhan CN.** 2005. The contributions of aspirin and microbial oxygenase to the biosynthesis of anti-inflammatory resolvins: novel oxygenase products from omega-3 polyunsaturated fatty acids. *Biochem Biophys Res Commun* **338:**149–157.

285. **Arita M, Bianchini F, Aliberti J, Sher A, Chiang N, Hong S, Yang R, Petasis NA, Serhan CN.** 2005. Stereochemical assignment, antiinflammatory properties, and receptor for the omega-3 lipid mediator resolvin E1. *J Exp Med* **201:**713–722.

286. **Serhan CN.** 2006. Novel chemical mediators in the resolution of inflammation: resolvins and protectins. *Anesthesiol Clin* **24:**341–364.

287. **Campbell EL, Louis NA, Tomassetti SE, Canny GO, Arita M, Serhan CN, Colgan SP.** 2007. Resolvin E1 promotes mucosal surface clearance of neutrophils: a new paradigm for inflammatory resolution. *FASEB J* **21:**3162–3170.

288. **Gronert K, Kantarci A, Levy BD, Clish CB, Odparlik S, Hasturk H, Badwey JA, Colgan SP, Van Dyke TE, Serhan CN.** 2004. A molecular defect in intracellular lipid signaling in human neutrophils in localized aggressive periodontal tissue damage. *J Immunol* **172:**1856–1861.

289. **Arita M, Ohira T, Sun YP, Elangovan S, Chiang N, Serhan CN.** 2007. Resolvin E1 selectively interacts with leukotriene B_4 receptor BLT1 and ChemR23 to regulate inflammation. *J Immunol* **178:**3912–3917.

290. **Haworth O, Cernadas M, Yang R, Serhan CN, Levy BD.** 2008. Resolvin E1 regulates interleukin 23, interferon-γ and lipoxin A_4 to promote the resolution of allergic airway inflammation. *Nat Immunol* **9:**873–879.

291. **Schwab JM, Chiang N, Arita M, Serhan CN.** 2007. Resolvin E1 and protectin D1 activate inflammation-resolution programmes. *Nature* **447:**869–874.

292. **Tjonahen E, Oh SF, Siegelman J, Elangovan S, Percarpio KB, Hong S, Arita M, Serhan CN.** 2006. Resolvin E2: identification and anti-inflammatory actions: pivotal role of human 5-lipoxygenase in resolvin E series biosynthesis. *Chem Biol* **13:**1193–1202.

293. **Dona M, Fredman G, Schwab JM, Chiang N, Arita M, Goodarzi A, Cheng G, von Andrian UH, Serhan CN.** 2008. Resolvin E1, an EPA-derived mediator in whole blood, selectively counterregulates leukocytes and platelets. *Blood* **112:**848–855.

294. **Hasturk H, Kantarci A, Ohira T, Arita M, Ebrahimi N, Chiang N, Petasis NA, Levy BD, Serhan CN, Van Dyke TE.** 2006. RvE1 protects from local inflammation and osteoclast- mediated bone destruction in periodontitis. *FASEB J* **20:**401–403.

295. **Zabel BA, Ohyama T, Zuniga L, Kim JY, Johnston B, Allen SJ, Guido DG, Handel TM, Butcher EC.** 2006. Chemokine-like receptor 1 expression by macrophages in vivo: regulation by TGF-β and TLR ligands. *Exp Hematol* **34:**1106–1114.

296. **Cash JL, Hart R, Russ A, Dixon JP, Colledge WH, Doran J, Hendrick AG, Carlton MB, Greaves DR.**

2008. Synthetic chemerin-derived peptides suppress inflammation through ChemR23. *J Exp Med* **205:**767–775.

297. **Krishnamoorthy S, Recchiuti A, Chiang N, Yacoubian S, Lee CH, Yang R, Petasis NA, Serhan CN.** 2010. Resolvin D1 binds human phagocytes with evidence for proresolving receptors. *Proc Natl Acad Sci U S A* **107:**1660–1665.

298. **Spite M, Norling LV, Summers L, Yang R, Cooper D, Petasis NA, Flower RJ, Perretti M, Serhan CN.** 2009. Resolvin D2 is a potent regulator of leukocytes and controls microbial sepsis. *Nature* **461:**1287–1291.

299. **Sun YP, Oh SF, Uddin J, Yang R, Gotlinger K, Campbell E, Colgan SP, Petasis NA, Serhan CN.** 2007. Resolvin D1 and its aspirin-triggered 17*R* epimer. Stereochemical assignments, anti-inflammatory properties, and enzymatic inactivation. *J Biol Chem* **282:**9323–9334.

300. **Chiang N, Dalli J, Colas RA, Serhan CN.** 2015. Identification of resolvin D2 receptor mediating resolution of infections and organ protection. *J Exp Med* **212:**1203–1217.

301. **Duffield JS, Hong S, Vaidya VS, Lu Y, Fredman G, Serhan CN, Bonventre JV.** 2006. Resolvin D series and protectin D1 mitigate acute kidney injury. *J Immunol* **177:**5902–5911.

302. **Levy BD, Kohli P, Gotlinger K, Haworth O, Hong S, Kazani S, Israel E, Haley KJ, Serhan CN.** 2007. Protectin D1 is generated in asthma and dampens airway inflammation and hyperresponsiveness. *J Immunol* **178:**496–502.

303. **Ariel A, Li PL, Wang W, Tang WX, Fredman G, Hong S, Gotlinger KH, Serhan CN.** 2005. The docosatriene protectin D1 is produced by T_H2 skewing and promotes human T cell apoptosis via lipid raft clustering. *J Biol Chem* **280:**43079–43086.

304. **Mukherjee PK, Marcheselli VL, Barreiro S, Hu J, Bok D, Bazan NG.** 2007. Neurotrophins enhance retinal pigment epithelial cell survival through neuroprotectin D1 signaling. *Proc Natl Acad Sci U S A* **104:**13152–13157.

305. **Mukherjee PK, Marcheselli VL, de Rivero Vaccari JC, Gordon WC, Jackson FE, Bazan NG.** 2007. Photoreceptor outer segment phagocytosis attenuates oxidative stress-induced apoptosis with concomitant neuroprotectin D1 synthesis. *Proc Natl Acad Sci U S A* **104:**13158–13163.

306. **Mukherjee PK, Marcheselli VL, Serhan CN, Bazan NG.** 2004. Neuroprotectin D1: a docosahexaenoic acid-derived docosatriene protects human retinal pigment epithelial cells from oxidative stress. *Proc Natl Acad Sci U S A* **101:**8491–8496.

307. **Lukiw WJ, Cui JG, Marcheselli VL, Bodker M, Botkjaer A, Gotlinger K, Serhan CN, Bazan NG.** 2005. A role for docosahexaenoic acid-derived neuroprotectin D1 in neural cell survival and Alzheimer disease. *J Clin Invest* **115:**2774–2783.

308. **Vedin I, Cederholm T, Freund Levi Y, Basun H, Garlind A, Faxén Irving G, Jönhagen ME, Vessby B, Wahlund LO, Palmblad J.** 2008. Effects of docosahexaenoic acid-rich n-3 fatty acid supplementation on cytokine release from blood mononuclear leukocytes: the OmegAD study. *Am J Clin Nutr* **87:**1616–1622.

309. **Nelson DR, Koymans L, Kamataki T, Stegeman JJ, Feyereisen R, Waxman DJ, Waterman MR, Gotoh O, Coon MJ, Estabrook RW, Gunsalus IC, Nebert DW.** 1996. P450 superfamily: update on new sequences, gene mapping, accession numbers and nomenclature. *Pharmacogenetics* **6:**1–42.

310. **Scarborough PE, Ma J, Qu W, Zeldin DC.** 1999. P450 subfamily CYP2J and their role in the bioactivation of arachidonic acid in extrahepatic tissues. *Drug Metab Rev* **31:**205–234.

311. **Roman RJ.** 2002. P-450 metabolites of arachidonic acid in the control of cardiovascular function. *Physiol Rev* **82:**131–185.

312. **Zhang Y, Oltman CL, Lu T, Lee HC, Dellsperger KC, VanRollins M.** 2001. EET homologs potently dilate coronary microvessels and activate BK_{Ca} channels. *Am J Physiol Heart Circ Physiol* **280:**H2430–H2440.

313. **Wang MH, Guan H, Nguyen X, Zand BA, Nasjletti A, Laniado-Schwartzman M.** 1999. Contribution of cytochrome P-450 4A1 and 4A2 to vascular 20-hydroxyeicosatetraenoic acid synthesis in rat kidneys. *Am J Physiol* **276:**F246–F253.

314. **Kikuchi Y, Miyauchi M, Oomori K, Kita T, Kizawa I, Kato K.** 1986. Inhibition of human ovarian cancer cell growth *in vitro* and in nude mice by prostaglandin D_2. *Cancer Res* **46:**3364–3366.

315. **Bednar MM, Gross CE, Balazy MK, Belosludtsev Y, Colella DT, Falck JR, Balazy M.** 2000. 16(*R*)-hydroxy-5,8,11,14-eicosatetraenoic acid, a new arachidonate metabolite in human polymorphonuclear leukocytes. *Biochem Pharmacol* **60:**447–455.

316. **Fleming I.** 2007. DiscrEET regulators of homeostasis: epoxyeicosatrienoic acids, cytochrome P450 epoxygenases and vascular inflammation. *Trends Pharmacol Sci* **28:**448–452.

317. **Moreno JJ.** 2009. New aspects of the role of hydroxyeicosatetraenoic acids in cell growth and cancer development. *Biochem Pharmacol* **77:**1–10.

318. **Spector AA, Norris AW.** 2007. Action of epoxyeicosatrienoic acids on cellular function. *Am J Physiol Cell Physiol* **292:**C996–C1012.

319. **Inceoglu B, Jinks SL, Ulu A, Hegedus CM, Georgi K, Schmelzer KR, Wagner K, Jones PD, Morisseau C, Hammock BD.** 2008. Soluble epoxide hydrolase and epoxyeicosatrienoic acids modulate two distinct analgesic pathways. *Proc Natl Acad Sci U S A* **105:**18901–18906.

320. **Inceoglu B, Schmelzer KR, Morisseau C, Jinks SL, Hammock BD.** 2007. Soluble epoxide hydrolase inhibition reveals novel biological functions of epoxyeicosatrienoic acids (EETs). *Prostaglandins Other Lipid Mediat* **82:**42–49.

321. **Node K, Huo Y, Ruan X, Yang B, Spiecker M, Ley K, Zeldin DC, Liao JK.** 1999. Anti-inflammatory properties of cytochrome P450 epoxygenase-derived eicosanoids. *Science* **285:**1276–1279.

322. **Liu Y, Zhang Y, Schmelzer K, Lee TS, Fang X, Zhu Y, Spector AA, Gill S, Morisseau C, Hammock BD, Shyy JY.** 2005. The antiinflammatory effect of laminar flow: the role of PPARγ, epoxyeicosatrienoic acids, and soluble epoxide hydrolase. *Proc Natl Acad Sci U S A* **102:** 16747–16752.

323. **Briggs WH, Xiao H, Parkin KL, Shen C, Goldman IL.** 2000. Differential inhibition of human platelet aggregation by selected *Allium* thiosulfinates. *J Agric Food Chem* **48:**5731–5735.

324. **Fitzpatrick FA, Ennis MD, Baze ME, Wynalda MA, McGee JE, Liggett WF.** 1986. Inhibition of cyclooxygenase activity and platelet aggregation by epoxyeicosatrienoic acids. Influence of stereochemistry. *J Biol Chem* **261:**15334–15338.

325. **Heizer ML, McKinney JS, Ellis EF.** 1991. 14,15-Epoxyeicosatrienoic acid inhibits platelet aggregation in mouse cerebral arterioles. *Stroke* **22:**1389–1393.

326. **Node K, Ruan XL, Dai J, Yang SX, Graham L, Zeldin DC, Liao JK.** 2001. Activation of Gα_s mediates induction of tissue-type plasminogen activator gene transcription by epoxyeicosatrienoic acids. *J Biol Chem* **276:** 15983–15989.

327. **Behm DJ, Ogbonna A, Wu C, Burns-Kurtis CL, Douglas SA.** 2009. Epoxyeicosatrienoic acids function as selective, endogenous antagonists of native thromboxane receptors: identification of a novel mechanism of vasodilation. *J Pharmacol Exp Ther* **328:**231–239.

328. **Hill E, Fitzpatrick F, Murphy RC.** 1992. Biological activity and metabolism of 20-hydroxyeicosatetraenoic acid in the human platelet. *Br J Pharmacol* **106:**267–274.

329. **Buckley CD, Gilroy DW, Serhan CN, Stockinger B, Tak PP.** 2013. The resolution of inflammation. *Nat Rev Immunol* **13:**59–66.

330. **Serhan CN, Brain SD, Buckley CD, Gilroy DW, Haslett C, O'Neill LA, Perretti M, Rossi AG, Wallace JL.** 2007. Resolution of inflammation: state of the art, definitions and terms. *FASEB J* **21:**325–332.

331. **Kurzrok R, Lieb CC.** 1930. Biochemical studies of human semen. II. The action of semen on the human uterus. *Proc Soc Exp Biol Med* **28:**268–272.

Myeloid Cells in Health and Disease: A Synthesis
Edited by Siamon Gordon

doi:10.1128/microbiolspec.MCHD-0001-2014

Jesmond Dalli[1]
Charles Serhan[1]

20 Macrophage Proresolving Mediators—the When and Where

INTRODUCTION

Inflammation is the organism's response to local injury in vascularized tissues programmed to traffic leukocytes and plasma delivery to an injured site or point of bacterial invasion (1); this protective response, when uncontrolled in humans, is associated with many widely occurring diseases. These include cardiovascular, metabolic, and the classic inflammatory diseases, e.g., arthritis and periodontal disease, along with cancers (reviewed in reference 2). Nonresolving inflammation is now widely acknowledged as a major driver in most of these diseases (for a review, see reference 3). The classical view of the resolution phase of the acute inflammatory response as understood and presented in pathology textbooks (1, 4) as well as in medical dictionaries (5) was that local inflammatory chemical messengers and cells were *passively* diluted at the site (dilution of chemotactic gradient), hence halting further leukocyte recruitment and resolving the exudate or battlefield of inflammation (6, 7). The historical perspective on the origins and concepts in the medical community regarding the resolution of inflammation apparently trace back as early as 11th-century Europe, and interested readers can refer to a recent review (8).

Regarding the origins that led pathologists to consider the resolution of inflammation as a passive process, the reader is directed to reference 9. In considering the outcomes of acute inflammation, pathologists considered four major paths: (i) complete resolution, (ii) abscess formation, (iii) healing by connective tissue replacement (fibrosis), and (iv) progression of the tissue on to chronic inflammation. Specifically, in considering complete resolution, it was noted that in an ideal situation all inflammatory responses, once they have succeeded in neutralizing injurious stimuli, must come to a termination or end with the restoration of the site of acute inflammation and its return to normality. This process, termed resolution, is the usual outcome when the injury is limited or short-lived or when there has been little tissue destruction and the damaged parenchymal cells can regenerate (9). Resolution thus involves "the neutralization or spontaneous decay of the local chemical mediators with the return of normal vascular permeability, cessation of leukocytic infiltration"

[1]Center for Experimental Therapeutics and Reperfusion Injury, Harvard Institutes of Medicine, Brigham and Women's Hospital and Harvard Medical School, Boston, MA 02115.

(9) and cell death (largely by cellular apoptosis of neutrophils) and their removal, as well as the removal of edema, foreign agents, and necrotic debris from the inflammatory site. These events were held to take place primarily via phagocytic processes of phagocytes and the lymphatics (4, 9). Thus, while the events leading to resolution were thought to be seemingly chaotic, the process is highly organized as observable via microscopy of histological tissue sections, leading pathologists to the original ideas regarding resolution as a passive and spontaneous process.

The study of chemical mediators initially focused on those signals, local-acting autacoids, that play critical roles in the initiation process. Local mediators such as histamine, complement split products (C5a and C3b), and chemokines serve as chemoattractants to bring neutrophils, which are the first responders in many instances, via diapedesis from postcapillary venules into tissues to combat injury and invading organisms. In this process of initiation of inflammation intended to neutralize foreign invaders, arachidonic acid (AA)- or n-6-derived mediators play a critical role. In this regard, prostaglandins are known to play pivotal roles in inflammation, and leukotrienes, in particular leukotriene B_4 (LTB_4), play a pivotal role as the key chemoattractants involved in the hierarchy of chemoattractants (10, 11). The gradient of chemoattractants such as LTB_4 is critical in enabling neutrophils to smell the chemical signal *per se* and move directedly along the gradient. Prostaglandin E_2 (PGE_2) plays a critical role in the hemodynamic changes and responses of endothelial cells, which permit neutrophils to migrate past them during diapedesis. All of these events are critical in the initiation process, and for the better part of the last century investigators focused their efforts on defining the mediators and mechanisms involved in initiating the acute inflammatory response, since this has led to therapeutic agents that can limit the dynamics and extent of the inflammatory response (12–14).

It became evident that there are lipid mediators (LMs) that are also involved in endogenous anti-inflammation. In this regard, both PGE_2 and PGD_2 play special roles. There are multiple receptors for these eicosanoids, and their cell-specific signaling is dictated by specific G protein coupling, which is cell type specific. PGE_2 has anti-inflammatory actions such as stopping neutrophil migration, as does PGD_2, but these mediators have a distinction in that they can induce the apparatus responsible for lipoxin production (15). The distinction between anti-inflammation and resolution is defined below, but at this juncture it is important to point out that anti-inflammation and resolution are distinct processes. Proresolving processes are active events stimulating the uptake of, for example, apoptotic neutrophils and/or bacteria and their debris, while anti-inflammation is generally an inhibitory process, stopping leukocytic infiltration, for example. The distinction between these processes is now clear. One is an agonist, namely a proresolving mediator, while anti-inflammation can be initiated by a pharmacologic inhibitor or antagonist (16, 17).

Results from this laboratory demonstrated that resolution of self-limited inflammatory exudates is a biochemically active process that involves the local and temporal biosynthesis of a new genus of specialized proresolving mediators (SPMs) with their novel functions mapped employing resolution indices (7, 18–20). SPMs encompass several families of structurally and chemically distinct mediators. These chemical mediator families include lipoxins biosynthesized from AA; E-series resolvins from eicosapentaenoic acid (EPA); and docosahexaenoic acid (DHA)-derived D-series resolvins, protectins, and maresins. Each potent bioactive member of these families shares a defining action in resolving local inflammation. By definition, they each limit further polymorphonuclear neutrophil (PMN) recruitment to the site of injury and/or microbial invasion and enhance macrophage uptake of cellular debris and apoptotic PMNs to bring about tissue homeostasis (2, 21). The placement of these mediators within the resolution of inflammation as well as the state-of-the art definitions are reviewed elsewhere (16, 21).

The process of macrophage clearance is a critical process conserved throughout evolution. Elie Metchnikoff first described it in the late 19th century in starfish larvae. Since then, the role of this process has been extensively studied in mammalian systems, where it is appreciated to be critical in both the maintenance of homeostasis and the clearance of cellular debris, bacteria (22, 23), and apoptotic cells, a process termed efferocytosis (24). Along with these defining properties, specific members of the SPM genus carry out more-specialized tasks within programmed resolution (25), and hence the scope of their individual actions is non-overlapping and evoked via specific cell surface receptors that are G protein-coupled receptors (GPCRs) (2). The role for specific SPMs in mediating the biological actions of SPMs is established using both mice overexpressing these receptors that give enhanced host protective actions as well as mice lacking the receptors that demonstrate excessive inflammation and loss of response to their specific SPM agonist(s). For example, in mice overexpressing the human lipoxin receptor (ALX), which also mediates the biological actions of resolvin

D1 (RvD1), we found a reduction in the amplitude of the inflammatory response, accelerated resolution, and enhanced responses to lipoxin A_4 (LXA_4) (26) and RvD1 (27). Similar results were also observed with transgenic mice overexpressing the receptor for RvE1, ChemR23/ERV (28), whereas in mice lacking the murine homolog of the ALX receptor there is excessive inflammation and decreased responses to these proresolving mediators (27, 29, 30). For a detailed recent review, the reader is directed to reference 31.

From the recent literature, it is clear that specific LMs interact with GPCRs that can evoke cell type-specific responses. The specific responses of a given cell type could be mediated by different intracellular signals, such as cyclic AMP in the case of RvD2 (32) or other intracellular signals such as phosphorylation and receptor cross-desensitization (33). Recent studies have also shown that proresolving actions of GPCRs, as in the case of ALX receptor, can differentially signal if they are in the heterodimer versus homodimer state (34). These recent studies from Perretti and colleagues clearly demonstrate that heterodimer receptors with ALX and LXA_4 as the ligand can differentially signal. Further research is needed along these lines to decode each of the SPM receptors and their target cell type and tissue responses and intracellular signal transduction.

A systems approach led to the identification of novel bioactive structures coined resolvins and protectins in murine inflammatory exudates and isolated human cells based on liquid chromatography-tandem mass spectrometry (LC-MS/MS)-based LM lipidomics and tandem assessment of their functions in anti-inflammation and proresolution (18, 35). The complete stereochemistry and total organic synthesis of several key resolvins, protectins, as well as their aspirin-triggered forms have been established (recently reviewed in references 21 and 36). These include RvD1 (7*S*,8*R*,17*S*-trihydroxy-4*Z*,9*E*,11*E*,13*Z*,15*E*,19*Z*-DHA), RvD2 (7*S*,16*R*,17*S*-trihydroxy-4*Z*,8*E*,10*Z*,12*E*,14*E*,19*Z*-DHA), 17*R*-HDHA (17*R*-hydroxy-4*Z*,7*Z*,10*Z*,13*Z*,15*E*,19*Z*-DHA), neuroprotectin D1 (NPD1) (10*R*,17*S*-dihydroxy-4*Z*,7*Z*,11*E*,13*E*,15*Z*,19*Z*-DHA), RvE1 (5*S*,12*R*,18*R*-trihydroxy-6*Z*,8*E*,10*E*,14*Z*,16*E*-EPA), and most recently maresin 1 (MaR1) (7*R*,14*S*-dihydroxy-4*Z*,8*E*,10*E*,12*Z*,16*Z*,19*Z*-DHA) (37). In addition to confirming the original structural assignments and potent anti-inflammatory and proresolving actions of resolvins, lipoxins, and maresins *in vivo* (36), new studies from others demonstrate their potent actions in experimental colitis (38), arthritis (39), arthritic pain (40), ocular diseases (41), and resolving obesity (42) and diabetes (43). Importantly, synthetic SPMs permitted their identification in other biological sources, including, for example, human inflammatory responses (44, 45), human serum (46), fish (47) and other marine organisms (48), and the invertebrate phyla (37). These mediators also counterregulate the production of proinflammatory signals including LMs (e.g., leukotrienes and prostaglandins) and cytokines (e.g., tumor necrosis factor α and interleukin-6 [IL-6]) (21) and upregulate levels of host protective and anti-inflammatory mediators including proteins such as annexin A1 (reviewed in reference 49), gases including endogenous heme oxygenase-1-derived carbon monoxide (50) as well as endogenous hydrogen sulfide (51), and cytokines including transforming growth factor β and IL-10 (reviewed in reference 52). For a detailed review of the role of cytokines in regulating steps during the resolution of inflammation, readers are directed to reference 52.

The discovery of potent bioactive resolvins and protectins in resolving exudates indicated that multiple cell types are involved in their biosynthesis, given the dynamic process ongoing within evolving inflammatory exudates (7, 18). With authenticated SPMs and proinflammatory LMs, we can now carry out targeted LM metabololipidomics via profiling with monitoring >50 individual mediators and pathway markers from these autacoid pathways in initiation and resolution of inflammation (53).

Microparticles in human synovium can influence the course of inflammation (54). Microparticles are small, membrane-bound vesicles first identified by Peter Wolf in 1967, and were thought to be by-products of platelet activation carrying no significant biological activity (55). It is now appreciated that microparticles are produced by a large number of cells including leukocytes, muscle cells, and endothelial cells (54). These microstructures are now appreciated to carry distinct functions in both the initiation and resolution of acute inflammation (reviewed in reference 56). Microparticles can act as vehicles carrying a wide range of molecular cargo, including microRNAs that are implicated in the regulation of hematopoiesis (57) as well as in protection during ischemia/reperfusion-mediated kidney injury (58). The proinflammatory cytokine IL-1β is also carried by microparticles, and its loading into these microvesicles is thought to be one of the mechanisms by which this cytokine, which does not possess a secretion motif, is released from cells (59). Morphogens, such as Sonic hedgehog, are also part of the cargo carried by subsets of these microparticles, whereby microparticles enriched in this morphogen were found to promote angiogenesis and thus may play a role in tumor growth (60). Microparticles also express receptors that are incorporated from the parent cell plasma membrane during formation. These receptors may be functionally transferred to recipient cells, thereby expanding the receptor repertoire of the recipient cell, providing either a nongenomic gain of function or being involved in the pathogenesis of a number of infectious diseases, including HIV infection. In this context, monocyte microparticles carrying the C-C chemokine receptor type 5 (CCR5), the principal coreceptor for macrophage-tropic HIV-1, transfer this receptor to endothelial cells during monocyte transendothelial migration, making the endothelial cells susceptible to infection by HIV-1 (61).

Recently, we found that leukocyte microparticles also display anti-inflammatory actions (62), carry precursors for the biosynthesis of proresolving mediators (63, 64), and display potent proresolving and host protective actions (63, 65). Since evolving self-limited inflammatory exudates produce functional microparticles that also signal to stimulate resolution of inflammation in mice (63), we systematically profiled LMs and SPMs produced by individual human cell types and microparticles involved in initiation and gauged their contributions to resolution of inflammatory responses. Employing a targeted LM metabololipidomic approach, we found that apoptotic PMNs possess a proresolving LM profile and their uptake stimulated SPM biosynthesis in macrophages, a process that is also regulated by PMN microparticles and transcellular biosynthesis (64) that can also impact aging (66), pain (67), as well as vagus nerve control of resolution (68).

APOPTOTIC PMN AND MICROPARTICLES STIMULATE MACROPHAGE SPM PRODUCTION

We assessed endogenous pro- and anti-inflammatory LM biosynthesis by human macrophages during uptake of apoptotic PMNs using targeted LC-MS/MS metabololipidomics emphasizing functional lipidomics. We identified mediators from both lipoxygenase and cyclooxygenase pathways, including RvD1, MaR1, PGE_2, and $PGF_{2\alpha}$, (Fig. 1A). All were identified in accordance with published criteria (53) that included matching retention times, fragmentation patterns, and at least six characteristic and diagnostic ions for each as illustrated with results obtained for RvD1 (e.g., 375 = M-H; 357 = M-H-H_2O; 339 = M-H-2H_2O; 331 = M-H-CO_2; 313 = M-H-H_2O-CO_2; 295 = M-H-2H_2O-CO_2; 287 = 303-H_2O; and 241 = 277-H_2O), RvD2, LXB_4, and 5S,15S-dihydroxyeicosatetraenoic acid (diHETE) (Fig. 1B and insets for diagnostic ions).

LM quantification is achieved via multiple reaction monitoring (MRM) of signature ion pairs Q1 (parent ion) and Q3 (characteristic daughter ion) (Fig. 1C). Macrophage efferocytosis of apoptotic PMNs increased SPM biosynthesis, primarily RvD1 (18 ± 10 versus 127 ± 14 pg/2.5 × 10^5 cells), RvD2 (28 ± 8 versus 124 ± 12 pg/2.5 × 10^5 cells), and LXB_4 (19 ± 2 versus 44 ± 6 pg/2.5 × 10^5 cells). This was accompanied by increases in prostanoids including PGE_2 (7 ± 1 versus 44 ± 12 pg/2.5 × 10^5 cells; Fig. 1C). Having found that microparticles stimulate efferocytosis, we assessed their actions on SPM biosynthesis during apoptotic PMN uptake. Microparticles stimulated macrophage biosynthesis of RvD2 (219 ± 44 versus 124 ± 12 pg/2.5 × 10^5 cells), LXB_4 (84 ± 15 versus 44 ± 6 pg/2.5 × 10^5 cells), and RvE2 (478 ± 57 versus 199 ± 26 pg/2.5 × 10^5 cells) to a greater extent than apoptotic PMNs, while reducing $PGF_{2\alpha}$ (259 ± 29 versus 459 ± 54 pg/2.5 × 10^5 cells) and thromboxane B_2 (TXB_2) (328 ± 21 versus 456 ± 62 pg/2.5 × 10^5 cells) (Fig. 1C).

Temporal SPM biosynthesis contributes to regulating the onset of resolution programs (2). Therefore, we assessed the cumulative levels for each LM family identified (64). During human macrophage efferocytosis there was an increase in DHA- (Fig. 2A) and AA-derived SPM (Fig. 2B) biosynthesis, along with an augmentation in prostanoid production (Fig. 2C). Microparticles further enhanced apoptotic PMN-stimulated SPM biosynthesis (Fig. 2A and B) while reducing prostanoid production (Fig. 2C). To rule out the possibility that microparticles carried mediators in addition to their known ability to carry SPM precursors (63), we profiled the microparticles. Microparticles did not carry mature SPMs. Hence, we profiled the amounts of LMs associated with apoptotic PMNs and microparticles in the absence of efferocytosis. Although specific SPMs and LMs were demonstrable, their values were baseline compared to levels produced via macrophages during efferocytosis (64).

Whether PMN microparticles are able to stimulate macrophage LM biosynthesis was also investigated. Incubation of PMN microparticles with macrophages led to the biosynthesis of lipoxygenase- and cyclooxygenase-derived LMs (64). These included RvD5, protectin D1 (PD1), and PGE_2 in accordance with criteria (53) that included matching retention times and at least six diagnostic ions as illustrated for MaR1, PD1, RvE2, and RvE1. These results are the first to demonstrate each of these proresolving mediators in human macrophage incubations and their interrelationships.

In these experiments we observed, for example, that PMN microparticles led to an increase in MaR1

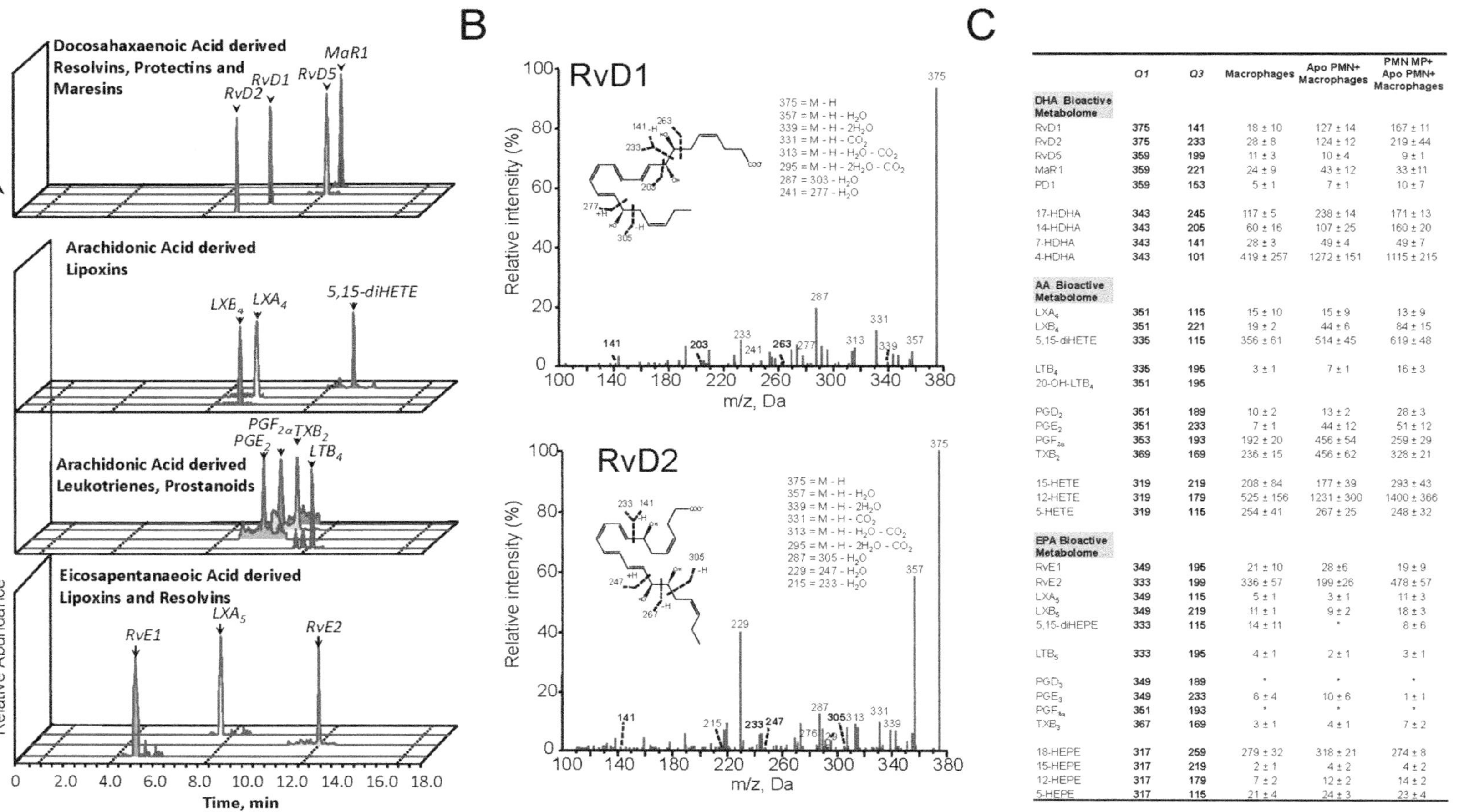

	Q1	Q3	Macrophages	Apo PMN+ Macrophages	PMN MP+ Apo PMN+ Macrophages
DHA Bioactive Metabolome					
RvD1	375	141	18 ± 10	127 ± 14	167 ± 11
RvD2	375	233	28 ± 8	124 ± 12	219 ± 44
RvD5	359	199	11 ± 3	10 ± 4	9 ± 1
MaR1	359	221	24 ± 9	43 ± 12	33 ±11
PD1	359	153	5 ± 1	7 ± 1	10 ± 7
17-HDHA	343	245	117 ± 5	238 ± 14	171 ± 13
14-HDHA	343	205	60 ± 16	107 ± 25	160 ± 20
7-HDHA	343	141	28 ± 3	49 ± 4	49 ± 7
4-HDHA	343	101	419 ± 257	1272 ± 151	1115 ± 215
AA Bioactive Metabolome					
LXA_4	351	115	15 ± 10	15 ± 9	13 ± 9
LXB_4	351	221	19 ± 2	44 ± 6	84 ± 15
5,15-diHETE	335	115	356 ± 61	514 ± 45	619 ± 48
LTB_4	335	195	3 ± 1	7 ± 1	16 ± 3
20-OH-LTB_4	351	195			
PGD_2	351	189	10 ± 2	13 ± 2	28 ± 3
PGE_2	351	233	7 ± 1	44 ± 12	51 ± 12
$PGF_{2\alpha}$	353	193	192 ± 20	456 ± 54	259 ± 29
TXB_2	369	169	236 ± 15	456 ± 62	328 ± 21
15-HETE	319	219	208 ± 84	177 ± 39	293 ± 43
12-HETE	319	179	525 ± 156	1231 ± 300	1400 ± 366
5-HETE	319	115	254 ± 41	267 ± 25	248 ± 32
EPA Bioactive Metabolome					
RvE1	349	195	21 ± 10	28 ±6	19 ± 9
RvE2	333	199	336 ± 57	199 ±26	478 ± 57
LXA_5	349	115	5 ± 1	3 ± 1	11 ± 3
LXB_5	349	219	11 ± 1	9 ± 2	18 ± 3
5,15-diHEPE	333	115	14 ± 11	*	8 ± 6
LTB_5	333	195	4 ± 1	2 ± 1	3 ± 1
PGD_3	349	189	*	*	*
PGE_3	349	233	6 ± 4	10 ± 6	1 ± 1
$PGF_{3\alpha}$	351	193	*	*	*
TXB_3	367	169	3 ± 1	4 ± 1	7 ± 2
18-HEPE	317	259	279 ± 32	318 ± 21	274 ± 8
15-HEPE	317	219	2 ± 1	4 ± 2	4 ± 2
12-HEPE	317	179	7 ± 2	12 ± 2	14 ± 2
5-HEPE	317	115	21 ± 4	24 ± 3	23 ± 4

Figure 1 SPM identification in human macrophages: regulation during efferocytosis and by neutrophil microparticles. Macrophages were incubated with PMN microparticles before addition of apoptotic (Apo) PMNs. (**A**) Representative MRM traces for identified LMs. (**B**) Accompanying MS/MS spectra used for SPM identification. (**C**) Specific bioactive LM and precursor/pathway markers where Q1, M-H (parent ion); and Q3, diagnostic ion in the MS/MS (daughter ion), along with mean ± standard error of the mean (SEM) values for each of the mediators are identified. Quantification and values obtained after PMN (3×10^5 PMNs) and microparticle (2×10^5 microparticles) incubations. The detection limit was ~1 pg. *Below limits. $n = 4$ distinct cell preparations. See reference 64 and the text for further details.

(13 ± 4 versus 21 ± 2 pg/2.5 × 10^5 macrophages), PD1 (7 ± 1 versus 28 ± 5 pg/2.5 × 10^5 macrophages), LXB_4 (78 ± 4 versus 248 ± 26 pg/2.5 × 10^5 macrophages), and RvE1 (14 ± 8 versus 72 ± 9 pg/2.5 × 10^5 macrophages) production. Also, we found that PMN microparticles stimulated macrophage prostanoids, in particular PGE_2 (18 ± 2 versus 40 ± 6 pg/2.5 × 10^5 macrophages) and TXB_2 (28 ± 7 versus 63 ± 10 pg/2.5 × 10^5 macrophages). Cumulative LM metabololipidomics for each LM family identified demonstrated that PMN microparticles stimulated biosynthesis of DHA- (Fig. 2A), AA- (Fig. 2B), and EPA-derived SPMs along with prostanoids (Fig. 2C). Of interest, incubation of macrophages with G protein inhibitors (pertussis toxin and cholera toxin) reduced SPM biosynthesis without altering prostanoid levels. Together, these results demonstrate that microparticles selectively stimulate macrophage SPM production in a GPCR-dependent manner (64, 66).

MICROPARTICLES STIMULATE EFFEROCYTOSIS OF APOPTOTIC PMNs

Since PMN microparticles exert anti-inflammatory and proresolving actions *in vivo* (63), we assessed their ability to regulate macrophage uptake of apoptotic cells (efferocytosis) (17). Fluorescently labeled PMN microparticles were obtained from activated PMNs and incubated with macrophages prior to addition of fluorescent apoptotic PMNs that dose dependently increased macrophage-associated fluorescence (Fig. 2D). These results indicate that microparticles regulate efferocytosis

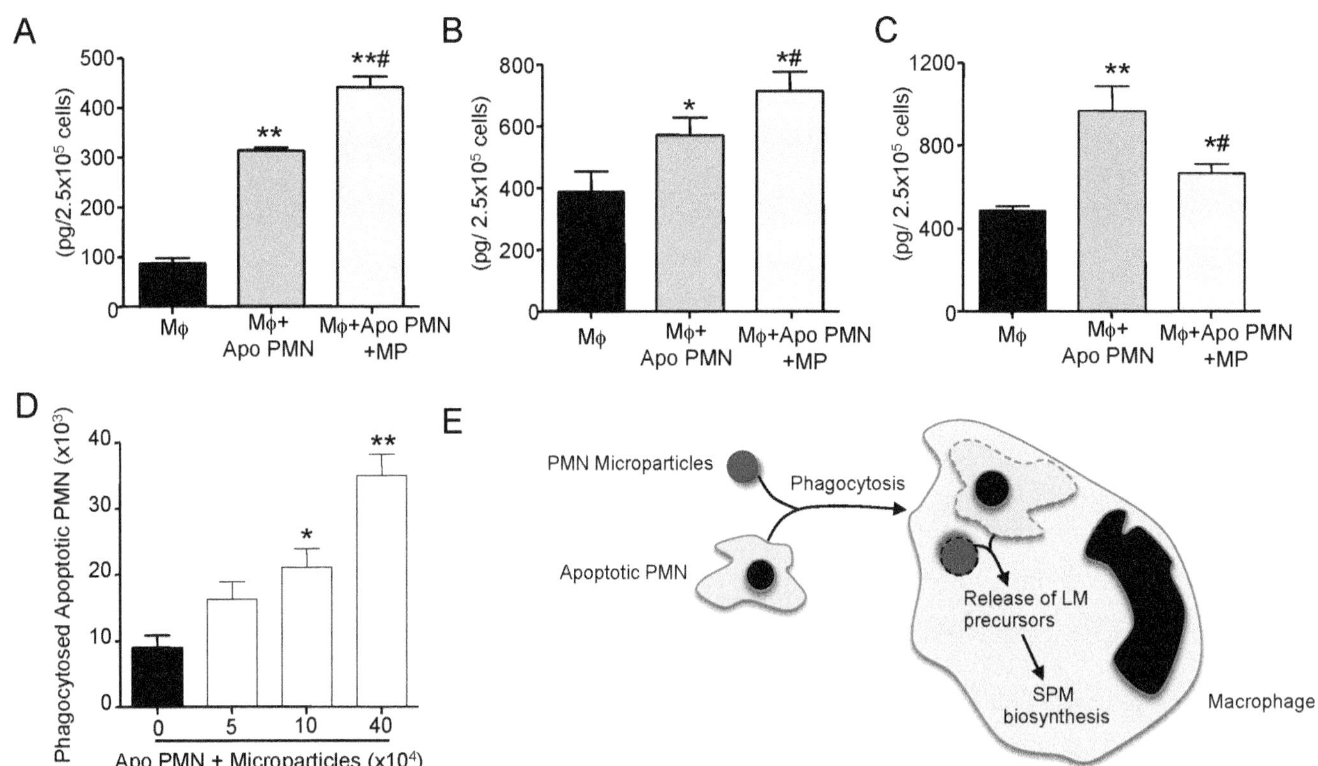

Figure 2 Upregulation of SPMs in macrophages (Mϕ) by apoptotic PMNs and PMN microparticles (MPs). Incubations were conducted as in Fig. 1; LMs were identified and quantified by LC-MS/MS. (**A**) D-series resolvins, protectins, and maresins. (**B**) Lipoxins. (**C**) Prostaglandins and thromboxanes. Results are expressed as mean ± SEM; n = 4 distinct cell preparations. $^*P < 0.05$ versus macrophage group; $^{**}P < 0.01$ versus macrophage group; $^{\#}P < 0.05$ versus macrophage plus Apo PMN group. (**D**) Uptake of the carboxyfluorescein diacetate succinimidyl ester (CFDA)-labeled Apo PMNs was monitored after incubation with the indicated MP concentrations. Results are expressed as mean ± SEM (n = 4 distinct cell preparations). $^*P < 0.05$ versus macrophage plus PMN group; $^{**}P < 0.01$ versus macrophage plus Apo PMN group. (**E**) Schematic highlighting the contribution of apoptotic PMNs and MPs to SPM biosynthesis in macrophages. See reference 64 and the text for further details.

of apoptotic cells and stimulate the biosynthesis of SPMs (Fig. 2E).

APOPTOTIC PMNs POSSESS A PRORESOLVING LM PROFILE

Because apoptotic PMNs stimulate macrophage SPMs during efferocytosis (64), we investigated mediator profiles of apoptotic PMNs in comparison to zymosan-stimulated PMNs. Using targeted LM metabololipidomics, we identified a profile signature of LMs that included RvD1, LXB_4, PGD_2, and PGE_2. Identification was achieved (*vide supra*) as illustrated for RvE2, PD1 (also known as NPD1/PD1 when produced in neural systems [19]), RvD2, and LTB_4. MRM LM quantification of signature ion pairs demonstrated that AA-derived LMs represents >~85% of the apoptotic PMN AA, EPA, and DHA identified LM metabolome. In these cells, the DHA-derived mediators represented ~6% while EPA-derived mediators amounted to ~11% of the targeted LMs. On the other hand, with zymosan-stimulated PMNs, AA-derived LMs constituted ~81% of the targeted AA, EPA, and DHA LM metabolomes.

Comparison of endogenous LMs identified in apoptotic versus zymosan-activated PMNs revealed that RvD1 (42 ± 12 versus 5 ± 2 pg/5 × 10^6 cells) and LXB_4 (142 ± 37 versus 14 ± 7 pg/5 × 10^6 cells) biosynthesis was reduced in zymosan-stimulated cells. Similarly, there was a downregulation in a number of cyclooxygenase-derived LMs in zymosan-stimulated PMNs, including PGD_2 (123 ± 29 versus 82 ± 15 pg/5 × 10^6 cells), PGE_2 (390 ± 75 versus 82 ± 16 pg/5 × 10^6 cells), and $PGF_{2\alpha}$ (127 ± 40 versus 52 ± 18 pg/5 × 10^6 cells). Production of the chemoattractant LTB_4 (including its 20-OH P450 metabolite) was elevated in zymosan-stimulated cells as opposed to apoptotic PMNs. For intact PMNs without exposure to agonists, the levels for these mediators were close to or below limits. In order to gauge the effector functions endowed upon PMNs by their individual LM profiles, we assessed the cumulative LM amounts for the distinct LM families identified using LM metabololipidomics. We found that apoptotic PMNs produced significantly higher SPMs and prostanoids than zymosan or intact PMNs. Assessment of AA-, EPA-, and DHA-derived monohydroxy products did not reveal significant differences between apoptotic, zymosan-stimulated, or intact PMNs (64).

To appreciate the sequence of events involved in LM production during human PMN apoptosis, we also identified LMs present at distinct stages of PMN progression to apoptosis determined by annexin V staining. Since, during inflammation, edema formation supplies the inflammatory response with extracellular substrate (69), we determined the time course of LM biosynthesis by apoptotic cells in the presence of extracellular substrate. MRM quantification of identified LMs demonstrated a significant increase in SPM biosynthesis by apoptotic cells, including D- and E-series resolvins, that reached maximum within 30 min (64). Prostanoids were also elevated, while LTB_4 (including its 20-OH metabolite) occurred during the initial phase and subsequently declined, both in the presence and absence of substrate. Although we were not able to rule out that a small proportion of potentially nonapoptotic PMNs that might have been present could have contributed to LM profiles in incubations with apoptotic PMNs, it is more than likely that at the 18-h interval LM profiles reflect the capacity of apoptotic PMNs to biosynthesize specific LMs.

APOPTOTIC PMNs AND MICROPARTICLES, A NIDUS FOR MACROPHAGE SPM BIOSYNTHESIS

The potential contribution of transcellular biosynthesis was assessed, in particular for SPMs, during macrophage efferocytosis (64). For this purpose we utilized deuterium-labeled (*d*) precursors, which allowed for the identification of mediators from substrate/precursors derived from either apoptotic PMNs or microparticles. LMs derived from *d*-substrate had a higher mass-to-charge ratio (*m/z*) than those biosynthesized from endogenous substrate. Specifically, LMs produced from d_8-AA were 8 atomic mass units (amu) and those from d_5-EPA and d_5-DHA were 5 amu greater than LMs biosynthesized from endogenous precursors. We first fortified microparticles and apoptotic PMNs with *d*-labeled precursors and assessed levels of *d*-labeled LMs, essential fatty acids (AA, EPA, and DHA), and biosynthetic precursors (including 18-hydroxy-eicosapentaenoic acid (HEPE), 15-HEPE, 15-hydroxy-eicosatetraenoic acid (HETE), 17-HDHA, and 14-HDHA). We identified LM precursors including d_5-DHA, d_5-17-HDHA, and d_8-AA along with a select number of mediators, including d_5-RvD5 and d_5-RvE2, in accordance with published criteria (53). MRM quantitation of the identified LMs demonstrated that the levels of these *d*-labeled mediators within microparticles and apoptotic PMNs were near limits of detection (<5 pg LM/6 × 10^5 microparticles and <1 pg LM/1 × 10^6 PMNs). We also identified *d*-labeled biosynthetic precursors that included 16 ± 3 pg d_5-17-HDHA/5 × 10^5 microparticles and 18 ± 1 pg d_5-17-HDHA/7.5 × 10^5 apoptotic PMNs and *d*-labeled essential fatty acids that included 423 ± 35 pg

d_5-DHA/5 × 10^5 microparticles and 459 ± 133 pg d_5-DHA/7.5 × 10^5 apoptotic PMNs. These results suggested that while there was a significant enrichment of *d*-labeled precursors in both microparticles and apoptotic PMNs, only a relatively minor proportion was further converted to bioactive LMs in the absence of macrophages.

In these studies, we added either *d*-labeled microparticles or *d*-labeled apoptotic PMNs to macrophages, assessing LM transcellular biosynthesis during macrophage efferocytosis (64). Apoptotic PMN uptake by macrophages led to the biosynthesis of d_5-RvD2 and d_5-RvD5 from the D-series resolvins, d_5-PD1 from the protectins, d_8-LXB_4 from the lipoxins, and d_5-RvE2 from the E-series resolvins, while d_8-PGD_2 and d_8-$PGF_{2\alpha}$ were the only prostanoids identified, and d_5-RvD5, d_5-RvE2, and d_8-LXB_4 where the atomic mass of the parent molecules was increased by either 5 or 8 amu. MRM quantification of *d*-LMs biosynthesized during macrophage efferocytosis suggested that microparticles acted as a nexus for d_5-RvD2, d_5-RvD5, d_8-LXA_4, d_8-PGE_2, and d_8-$PGF_{2\alpha}$ production. d_5-MaR1 and d_8-LXB_4 biosynthesis were primarily sustained by precursors derived from within apoptotic PMNs. d_5-PD1 and d_5-RvE2 biosynthesis were equally reliant on precursors from apoptotic cells and microparticles. These results demonstrate that transcellular biosynthesis contributes to LM production during macrophage efferocytosis in which both apoptotic PMNs and microparticles contribute to and sustain a nidus for LM biosynthesis.

HUMAN MACROPHAGE SUBTYPES YIELD DISTINCT LM AND SPM PROFILES

Following LM metabolomics of M1 and M2 macrophages that were prepared from human peripheral blood monocytes, we identified 35 distinct LMs, which included RvD1, PD1, MaR1, LXA_5, LXB_5, PGD_2, and PGE_2. These mediators were identified using their diagnostic fragmentation patterns, matching a minimum of six characteristic fragments, as shown for RvD2, PD1, RvE2, and LXA_5. Using LM metabololipidomics with M1 macrophages, we found that AA-derived LMs amounted to ~48% of LMs identified. On the other hand, LM metabololipidomics of M2 macrophages demonstrated that SPMs represented ~50% of the identified AA, EPA, and DHA metabolome, consisting of D-series resolvins, protectins, and maresins (~11%); lipoxins (~15%); and E-series resolvins and lipoxins (~24%).

Using MRM, we compared the endogenous biosynthesis of individual LMs between the two macrophage subtypes. In these experiments, we found that biosynthesis of distinct SPMs was elevated in M2 in comparison to M1 macrophages; these included RvD5 (196 ± 26 versus 43 ± 5 pg/2.5 × 10^5 cells), MaR1 (299 ± 8 versus 45 ± 8 pg/2.5 × 10^5 cells), PD1 (3,442 ± 206 versus 1,339 ± 206 pg/2.5 × 10^5 cells), LXA_4 (977 ± 173 versus 675 ± 167 pg/2.5 × 10^5 cells), LXB_4 (11,750 ± 724 versus 7,560 ± 659 pg/2.5 × 10^5 cells), LXB_5 (4,370 ± 397 versus 2,551 ± 859 pg/2.5 × 10^5 cells), and RvE2 (20,680 ± 4,910 versus 10,910 ± 4,232 pg/2.5 × 10^5 cells). Conversely, production of cyclooxygenase-derived LMs was elevated in M1 macrophages; these included PGE_2 (2,285 ± 435 versus 1,573 ± 490 pg/2.5 × 10^5 cells), $PGF_{2\alpha}$ (11,325 ± 428 versus 2,108 ± 256 pg/2.5 × 10^5 cells), TXB_2 (550 ± 135 versus 222 ± 73 pg/2.5 × 10^5 cells), and the EPA-derived prostanoids.

Because different human macrophage subtypes possess distinct LM profiles, we compared the cumulative amounts of different LM families identified in M1 and M2 macrophages in order to assess effector functions that these endow for the individual macrophage subtypes. SPM biosynthesis was higher in M2 compared to M1 macrophages, while production of prostanoids and leukotrienes LTB_4 and LTB_5 was elevated in M1 macrophages compared to M2 (Fig. 3). Assessment of LM biosynthesis following apoptotic cell efferocytosis demonstrated that apoptotic PMN uptake gave SPMs in M1 macrophages while reducing prostanoids and leukotrienes. Apoptotic PMNs with M2 macrophages reduced overall LM production, including SPMs. These results indicate that human M1 and M2 macrophages, as defined (70), possess distinct endogenous LM signature profiles.

MaR1 IS PRORESOLVING AND TISSUE REGENERATIVE

Self-resolving inflammatory exudates and LM metabolomics recently uncovered a new family of potent anti-inflammatory and proresolving mediators biosynthesized by macrophages, coined maresins (71). It was essential to establish the stereochemistry of this pathway. We determined that MaR1 produced by human macrophages from endogenous DHA matched synthetic 7*R*,14*S*-dihydroxydocosa-4*Z*,8*E*,10*E*,12*Z*,16*Z*,19*Z*-hexaenoic acid. The MaR1 alcohol groups and *Z*/*E* geometry of conjugated double bonds were matched using isomers prepared by total organic synthesis. MaR1's potent defining actions were confirmed with synthetic MaR1, i.e., limiting neutrophil (PMN) infiltration in murine peritonitis in the ng per mouse dose range as well as

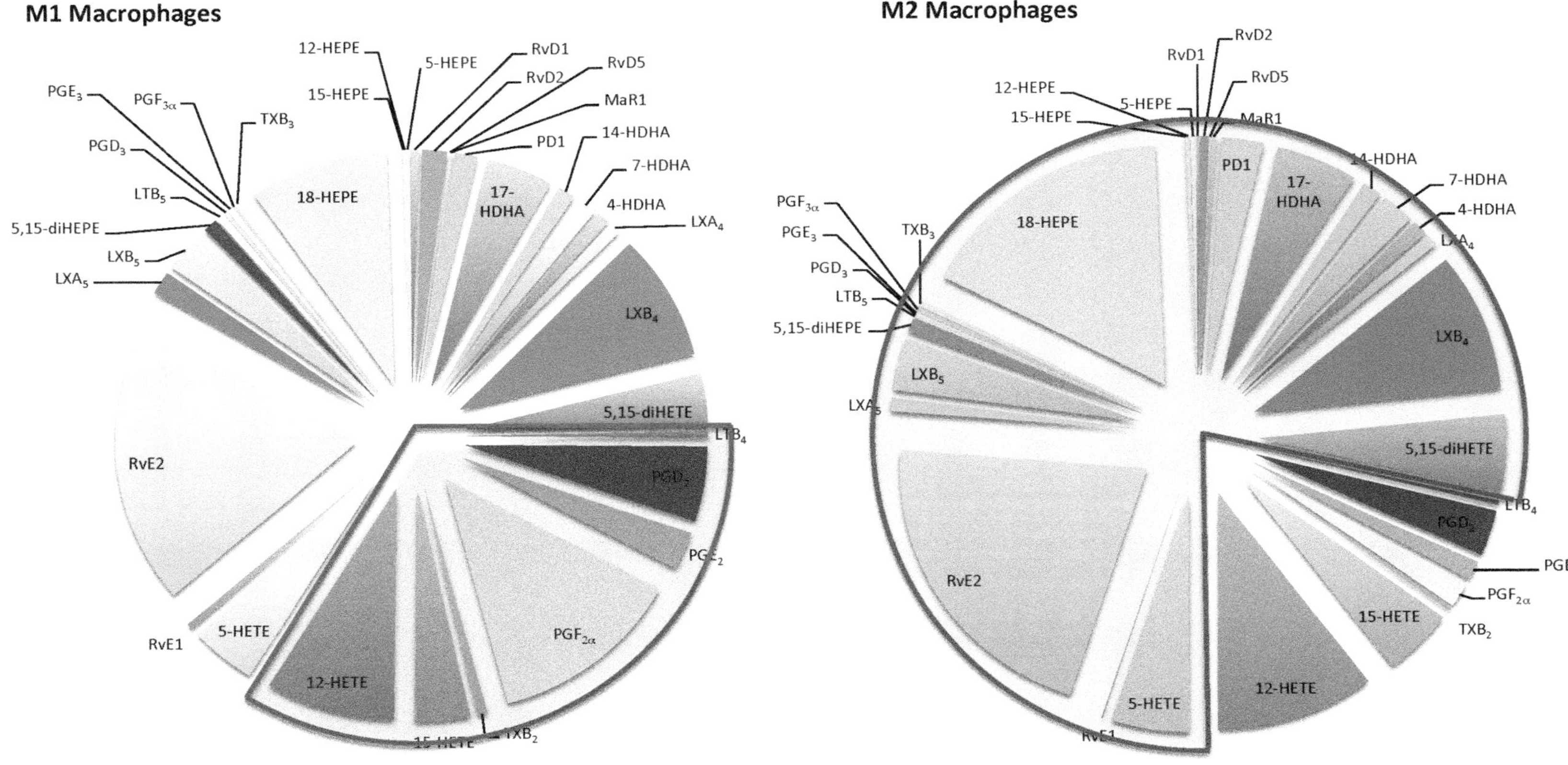

Figure 3 Distinct macrophage subtypes display characteristic LM profiles. Human M1 and M2 macrophages were obtained by incubating peripheral blood monocytes with granulocyte-macrophage colony-stimulating factor, gamma interferon, and lipopolysaccharide or macrophage colony-stimulating factor and IL-4, respectively, and LM profiles assessed by LC-MS/MS. Results are representative on $n = 6$ to 8 separate cell preparations. The original results were obtained in reference 64 and are replotted here.

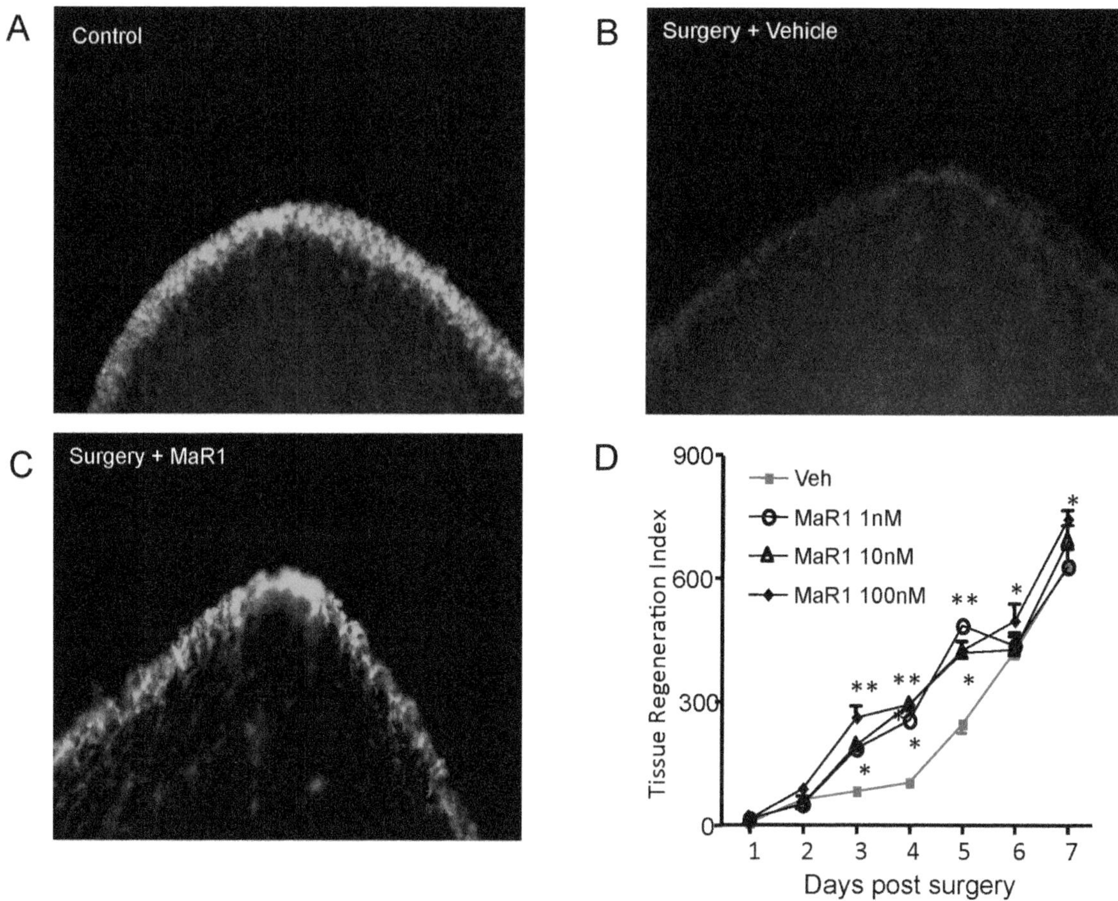

Figure 4 The macrophage-derived proresolving mediator MaR1 stimulates tissue regeneration. (**A to C**) Anterior portions of brown planaria were surgically removed, planaria were exposed to vehicle or MaR1, and regeneration was assessed 4 days after surgery using fluorescently conjugated *Erythrina cristagalli* lectin to visualize secretory cells. (**D**) After surgery, planaria were kept in water-containing vehicle (Veh) or the indicated concentrations of MaR1. We direct interested readers to reference 37 for detailed experimental conditions. Results are mean ± SEM (n = 5 planaria/group). *$P < 0.05$; **$P < 0.01$; ***$P < 0.001$ versus vehicle alone.

enhancing human macrophage uptake of apoptotic PMNs. At 1 nM, MaR1 was slightly more potent than RvD1 in stimulating human macrophage efferocytosis, an action not shared by LTB_4.

MaR1 also accelerated surgical regeneration in planaria, increasing the rate of head reappearance (37). Upon injury of planaria, MaR1 was biosynthesized from deuterium-labeled (d_5)-DHA that was blocked with lipoxygenase inhibitor. Planaria are simple organisms capable of rapid regeneration, the process wherein mammalian tissue macrophages play distinct roles. Hence, we questioned whether MaR1 demonstrated properties in tissue regeneration. To this end, planaria were grown and their anterior portions were surgically removed and exposed to proresolving mediators (Fig. 4). Planaria are increasingly recognized as a useful model system for tissue regeneration (72). Factors are known to be involved in planaria tissue regeneration (72). However, they remained to be identified. MaR1 and RvE1 at 100 nM each enhanced the rate of tissue regeneration, with the appearance of the anterior portion, i.e., head regeneration, evident as early as 3 days postsurgery (Fig. 4). We examined MaR1 dose dependency (Fig. 4). MaR1 at doses as low as 1, 10, and 100 nM enhanced tissue regeneration in the anterior head, which was statistically significant by days 3 and 4. By days 6 and 7, the actions were less prominent. MaR1-enhanced tissue regeneration was concentration dependent (Fig. 4) (37).

MaR1 also possesses potent antinociceptive actions, dose-dependently inhibiting TRPV1 currents in neurons,

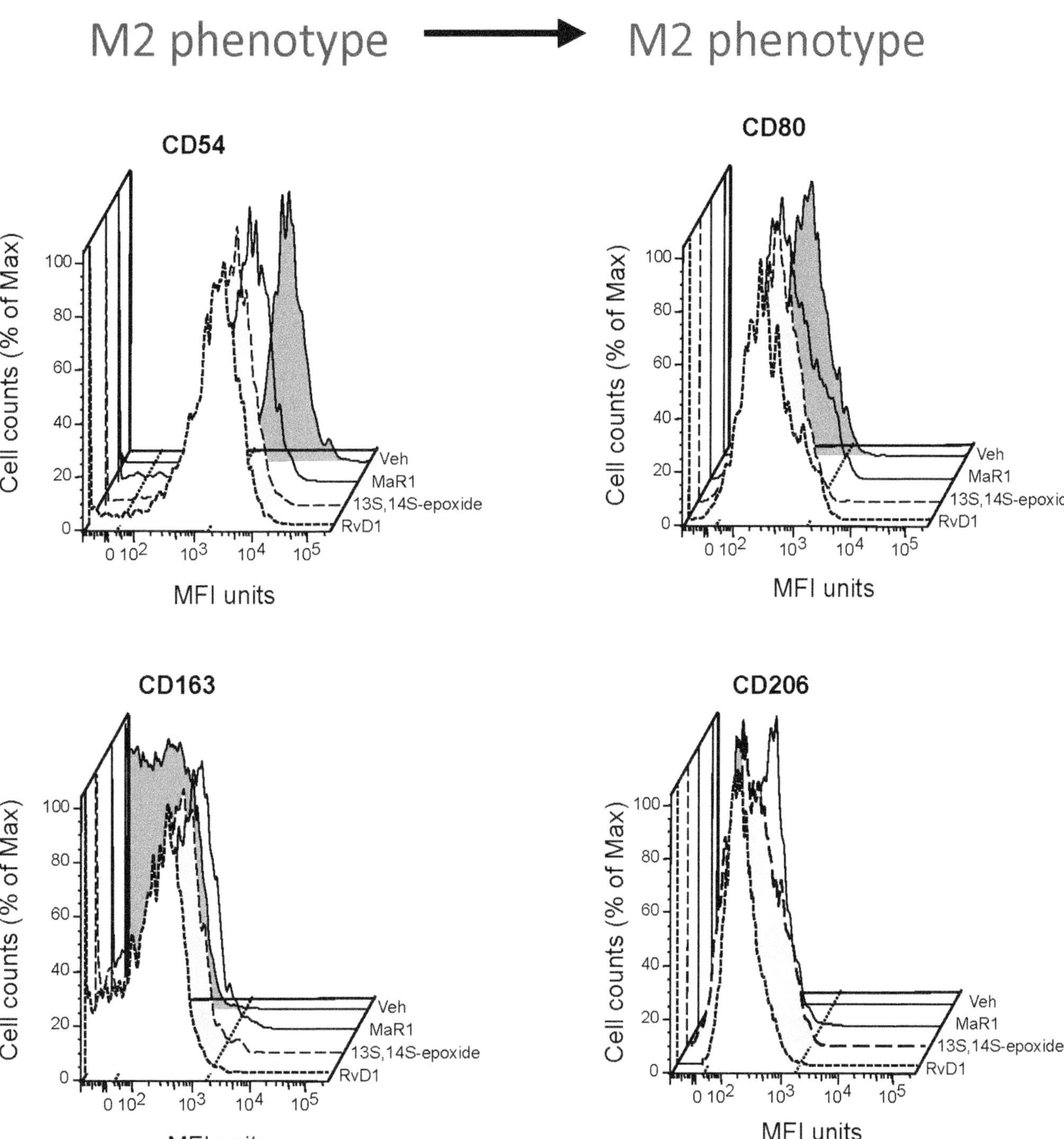

Figure 5 Maresins and RvD1 promote macrophage phenotype switch. M1 macrophages were incubated with vehicle, 13*S*,14*S*-epoxy-maresin, MaR1, or RvD1 for 6 h. Expression of M1 (CD54, CD80) and M2 (CD163, CD206) was assessed by flow cytometry using fluorescently conjugated antibodies. Results are representative of $n = 3$ for each incubation condition. MFI, mean fluorescence intensity. See the text for additional details and reference 64 for the original data set.

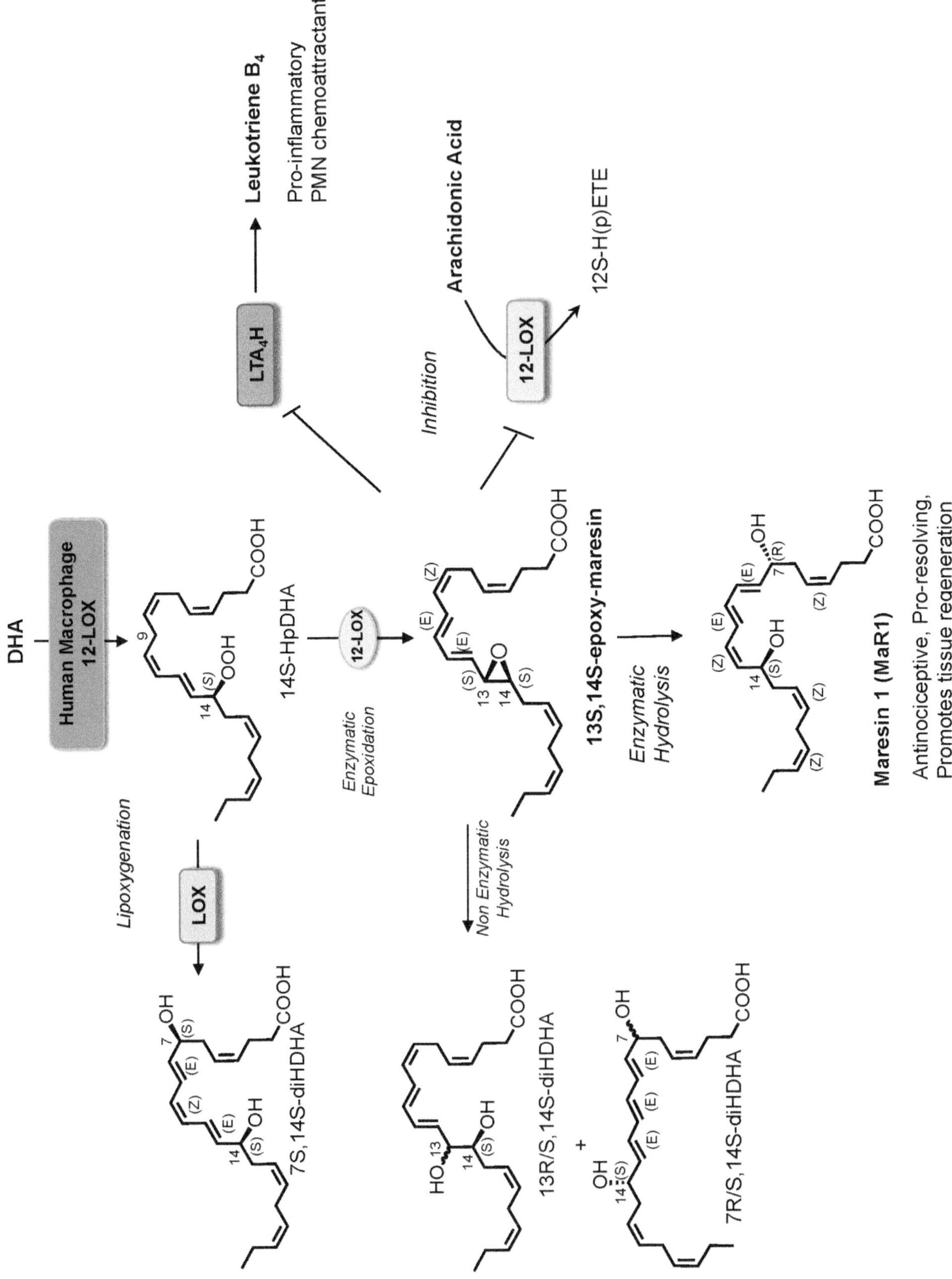

DHA
Human Macrophage 12-LOX
14S-HpDHA
Lipoxygenation
LOX
7S,14S-diHDHA
Enzymatic Epoxidation
12-LOX
13S,14S-epoxy-maresin
Non Enzymatic Hydrolysis
13R/S,14S-diHDHA
+
7R/S,14S-diHDHA
LTA$_4$H
Leukotriene B$_4$
Pro-inflammatory
PMN chemoattractant
Inhibition
Arachidonic Acid
12-LOX
12S-H(p)ETE
Enzymatic Hydrolysis
Maresin 1 (MaR1)
Antinociceptive, Pro-resolving,
Promotes tissue regeneration

blocking capsaicin (100 nM)-induced inward currents (IC_{50} = 0.49 ± 0.02 ng/ml), and reducing both inflammatory and chemotherapy-induced neuropathic pain in mice. These results demonstrate the potent actions of MaR1 in regulating inflammation resolution and tissue regeneration and resolving pain (8, 37, 67). These findings suggest that chemical signals are shared in resolutive cellular trafficking important in tissue regeneration. Moreover, immunoresolvents of the innate immune response (8), such as MaR1, offer new opportunities for assessing macrophages and their local DHA metabolome in the return to tissue homeostasis.

THE MARESIN BIOSYNTHETIC PATHWAY

In the biosynthesis of maresins, the 14-lipoxygenation of DHA by human macrophage 12-lipoxygenase (12-LOX) gave 14-hydro(peroxy)-docosahexaenoic acid (14-HpDHA) as well as several dihydroxy-docosahexaenoic acids, implicating an epoxide intermediate formation by this enzyme. Using a stereocontrolled synthesis, enantiomerically pure 13*S*,14*S*-epoxy-docosa-4*Z*,7*Z*,9*E*,11*E*,16*Z*,19*Z*-hexaenoic acid (13*S*,14*S*-epoxy-DHA) was prepared and its stereochemistry was confirmed by nuclear magnetic resonance spectroscopy. When this 13*S*,14*S*-epoxide was incubated with human macrophages, it was converted to MaR1. The synthetic 13*S*,14*S*-epoxide inhibited LTB_4 formation by human LTA_4 hydrolase by ~40% ($P < 0.05$), to a similar extent as LTA_4 (~50%; $P < 0.05$), but was not converted to MaR1 by this enzyme. 13*S*,14*S*-Epoxy-DHA also reduced (~60%; $P < 0.05$) AA conversion by human 12-LOX. Incubation of 13*S*,14*S*-epoxy-DHA with M1 macrophages led to a change in macrophage phenotype toward an M2 profile (73). Macrophage phenotype was determined using flow cytometry in which M1 macrophages expressed higher CD80 and CD54 while M2 macrophages displayed higher CD163 and CD206 levels. Incubation of M1 macrophages with either 13*S*,14*S*-epoxy-DHA (10 nM) or MaR1 (10 nM) led to significant reductions in CD54 and CD80 expression and a concomitant upregulation of CD163 and CD206 (Fig. 5). For comparison, this regulation was also found with RvD1 (10 nM). We also investigated the conversion of 13*S*,14*S*-epoxy-DHA to MaR1 by different human macrophage subtypes. Incubation of 13*S*,14*S*-epoxy-DHA (2 µM) with M2 macrophages gave higher MaR1 levels than when the 13*S*,14*S*-epoxy-DHA was incubated with M1 macrophages as determined by LM metabololipidomics.

These results establish the biosynthesis of the 13*S*,14*S*-epoxide, its absolute stereochemistry, its precursor role in MaR1 biosynthesis, and its own intrinsic bioactivity (73). Given its actions and role in MaR1 biosynthesis, this epoxide is now termed 13,14-epoxy-maresin (13,14-eMaR) and exhibits new mechanisms in resolution of inflammation in its ability to inhibit proinflammatory mediator production by LTA_4 hydrolase and to block arachidonate conversion by human 12-LOX rather than merely terminating phagocyte involvement. Thus, these findings identify novel regulatory actions of DHA-derived mediators within the eicosanoid cascade and in macrophage phenotypes (Fig. 6) that can play a role in vascular integrity (74) as well as in aging (66), tissue regeneration (37, 71), pain (67), and host defense mechanisms (75), all critical life processes that are regulated by neutrophils and ultimately by specific macrophage subtypes in which resolution phase signals such as the proresolving mediators (resolvins, protectins, and maresins) actively turn on homeostatic mechanisms.

Therefore, SPMs emerge as potent regulators of macrophage responses of interest during the resolution phase of acute inflammatory responses. Since the means and methodologies to identify these mediators have only recently become widely available, a number of aspects of their biology in humans have only just started to be explored, such as the production of these mediators in human tissues (76–78). In addition, a number of questions remain to be addressed, such as when and where these mediators are produced in human tissues in health and disease and the relevance

Figure 6 Maresin biosynthesis and actions. DHA is converted by human 12-LOX through a lipoxygenase reaction involving the abstraction of hydrogen and the antarafacial addition of molecular oxygen at carbon 14 to produce 14*S*-HpDHA, which is also converted to the 13,14-epoxide by the same enzyme (88), denoted 13*S*,14*S*-epoxide-maresin, which is further converted to the bioactive MaR1, which promotes resolution, reduces pain signaling, and promotes tissue regeneration. See the text for further details, as well as references 37, 71, and 73. The epoxide 13*S*,14*S*-epoxy-maresin is also bioactive and inhibits LTA_4 hydrolase, limiting the production of LTB_4 as well as the conversion of AA to 12*S*-HpETE. These actions of the maresin epoxide indicate that activation of the pathway leads to downregulation of proinflammatory mediators during resolution.

of resolution processes that may fail, giving rise to human disease. For those interested in the future directions and questions to be addressed, we direct readers to reference 21.

LOOKING AHEAD

In the near future, given the ability now to profile both proinflammatory and anti-inflammatory proresolving mediators such as the SPMs in human tissues (78), and in other human organs (79–81) including human placenta (82), plasma (78, 81), and milk (83), it is possible that LM and SPM signature profiles will be very useful in both precision and personalized medicine. In this regard, the profile of SPMs can serve as a barometer of diagnostic value for ranking the nutritional status of individuals and whether or not they can be timely or slow resolvers of a local acute inflammatory challenge (44). For example, a randomized controlled trial in human chronic kidney disease showed that n-3 fatty acid supplementation increases SPMs, which may help improve inflammation in this disease (84). Most importantly, we will be able to establish whether specific SPMs (resolvins, protectins, or maresins) act locally to evoke their biological responses or if they have a distal site of action. In this regard, we recently identified resolvins, protectins, and maresins in human milk, which suggests that they could have a local site of action but also could be active in the newborn (85). Most importantly, such methods of profiling initiators of inflammation and proresolving mediators in human tissues will help to assess the nutritional status of an individual. For example, what are the optimal levels of omega-3 essential fatty acids such as EPA and DHA as well as n-6 AA needed at a local tissue site to ensure timely neutralization of invading foreign injury as well as timely resolution and homeostasis? There are clearly age- and gender-related components in the resolution of inflammation, as we have found that the ability to resolve acute inflammatory responses in a timely fashion diminishes with age in mouse models (66). Perhaps it will be possible to personalize our nutrition to impact resolution in both an age- and gender-dependent manner to improve health and reduce disease burden by enhancing local resolution of inflammation that could diminish chronic inflammation and the impact of chronic inflammatory diseases on organs throughout the body. In addition to tailoring designer medical foods to personalized nutrition, it is also possible to use each of the proresolving mediators as templates for analogs that can be used as therapeutics to stimulate resolution of inflammation as a new approach to resolution pharmacology (86). This has already been extended to ongoing human clinical trials in ocular indications (http://www.auventx.com/auven/products/rx10045php) and for periodontal disease (87).

Given that resolution of inflammation is a fundamental process in all human tissues and that phagocytes and specifically macrophages play a central role in orchestrating this response, as well as tissue repair and regeneration, immunoresolvents (such as resolvins, protectins and maresins) may provide a novel therapeutic approach for diseases characterized by uncontrolled inflammation and failed resolution.

Acknowledgments. *The authors thank Mary Halm Small for expert assistance in manuscript preparation. Studies in the authors' lab were supported by the National Institutes of Health (grant numbers R01 GM38765 and P01 GM095467).*

Citation. Dalli J, Serhan C. 2016. Macrophage proresolving mediators—the when and where. Microbiol Spectrum 4(3): MCHD-0001-2014.

References

1. **Majno G, Joris I.** 2004. *Cells, Tissues, and Disease: Principles of General Pathology*, 2nd ed. Oxford University Press, New York, NY.
2. **Serhan CN.** 2010. Novel resolution mechanisms in acute inflammation: to resolve or not? *Am J Pathol* **177:**1576–1591.
3. **Nathan C, Ding A.** 2010. Nonresolving inflammation. *Cell* **140:**871–882.
4. **Robbins SL, Cotran R.** 1979. *Pathologic Basis of Disease*, 2nd ed. W.B. Saunders Co, Philadelphia, PA.
5. **Taber CW.** 1970. *Taber's Cyclopedic Medical Dictionary*, 11th ed. F. A. Davis Co, Philadelphia, PA.
6. **Willoughby DA, Moore AR, Colville-Nash PR, Gilroy D.** 2000. Resolution of inflammation. *Int J Immunopharmacol* **22:**1131–1135.
7. **Serhan CN, Clish CB, Brannon J, Colgan SP, Chiang N, Gronert K.** 2000. Novel functional sets of lipid-derived mediators with antiinflammatory actions generated from omega-3 fatty acids via cyclooxygenase 2-nonsteroidal antiinflammatory drugs and transcellular processing. *J Exp Med* **192:**1197–1204.
8. **Serhan CN.** 2011. The resolution of inflammation: the devil in the flask and in the details. *FASEB J* **25:**1441–1448.
9. **Cotran RS, Kumar V, Collins T.** 1999. *Robbins Pathologic Basis of Disease*, 6th ed, p. 78, W.B. Saunders Co, Philadelphia, PA.
10. **Lammermann T, Afonso PV, Angermann BR, Wang JM, Kastenmüller W, Parent CA, Germain RN.** 2013. Neutrophil swarms require LTB4 and integrins at sites of cell death *in vivo*. *Nature* **498:**371–375.
11. **Malawista SE, de Boisfleury Chevance A, van Damme J, Serhan CN.** 2008. Tonic inhibition of chemotaxis in human plasma. *Proc Natl Acad Sci U S A* **105:**17949–17954.

12. **Samuelsson B, Dahlén SE, Lindgren JÅ, Rouzer CA, Serhan CN.** 1987. Leukotrienes and lipoxins: structures, biosynthesis, and biological effects. *Science* **237:**1171–1176.

13. **Samuelsson B.** 2012. Role of basic science in the development of new medicines: examples from the eicosanoid field. *J Biol Chem* **287:**10070–10080.

14. **Vane JR.** 1982. Adventures and excursions in bioassay: the stepping stones to prostacyclin, p 181–206. *In* Nobel Foundation (ed), *Les Prix Nobel: Nobel Prizes, Presentations, Biographies and Lectures.* Almqvist & Wiksell, Stockholm, Sweden.

15. **Levy BD, Clish CB, Schmidt B, Gronert K, Serhan CN.** 2001. Lipid mediator class switching during acute inflammation: signals in resolution. *Nat Immunol* **2:**612–619.

16. **Serhan CN, Brain SD, Buckley CD, Gilroy DW, Haslett C, O'Neill LA, Perretti M, Rossi AG, Wallace JL.** 2007. Resolution of inflammation: state of the art, definitions and terms. *FASEB J* **21:**325–332.

17. **Serhan CN, Savill J.** 2005. Resolution of inflammation: the beginning programs the end. *Nat Immunol* **6:**1191–1197.

18. **Serhan CN, Hong S, Gronert K, Colgan SP, Devchand PR, Mirick G, Moussignac RL.** 2002. Resolvins: a family of bioactive products of omega-fatty acid transformation circuits initiated by aspirin treatment that counter pro-inflammation signals. *J Exp Med* **196:**1025–1037.

19. **Bannenberg GL, Chiang N, Ariel A, Arita M, Tjonahen E, Gotlinger KH, Hong S, Serhan CN.** 2005. Molecular circuits of resolution: formation and actions of resolvins and protectins. *J Immunol* **174:**4345–4355.

20. **Schwab JM, Chiang N, Arita M, Serhan CN.** 2007. Resolvin E1 and protectin D1 activate inflammation-resolution programmes. *Nature* **447:**869–874.

21. **Serhan CN.** 2014. Pro-resolving lipid mediators are leads for resolution physiology. *Nature* **510:**92–101.

22. **Tauber AI, Chernyak L.** 1991. *Metchnikoff and the Origins of Immunology: from Metaphor to Theory.* Oxford University Press, New York, NY.

23. **Gordon S.** 2007. The macrophage: past, present and future. *Eur J Immunol* **37**(Suppl 1):S9–S17.

24. **Savill JS, Wyllie AH, Henson JE, Walport MJ, Henson PM, Haslett C.** 1989. Macrophage phagocytosis of aging neutrophils in inflammation. Programmed cell death in the neutrophil leads to its recognition by macrophages. *J Clin Invest* **83:**865–875.

25. **Buckley CD, Gilroy DW, Serhan CN.** 2014. Proresolving lipid mediators and mechanisms in the resolution of acute inflammation. *Immunity* **40:**315–327.

26. **Devchand PR, Arita M, Hong S, Bannenberg G, Moussignac RL, Gronert K, Serhan CN.** 2003. Human ALX receptor regulates neutrophil recruitment in transgenic mice: roles in inflammation and host-defense. *FASEB J* **17:**652–659.

27. **Krishnamoorthy S, Recchiuti A, Chiang N, Fredman G, Serhan CN.** 2012. Resolvin D1 receptor stereoselectivity and regulation of inflammation and proresolving microRNAs. *Am J Pathol* **180:**2018–2027.

28. **Gao L, Faibish D, Fredman G, Herrera BS, Chiang N, Serhan CN, Van Dyke TE, Gyurko R.** 2013. Resolvin E1 and chemokine-like receptor 1 mediate bone preservation. *J Immunol* **190:**689–694.

29. **Norling LV, Dalli J, Flower RJ, Serhan CN, Perretti M.** 2012. Resolvin D1 limits polymorphonuclear leukocytes recruitment to inflammatory loci: receptor dependent actions. *Arterioscler Thromb Vasc Biol* **32:**1970–1978.

30. **Brancaleone V, Gobbetti T, Cenac N, le Faouder P, Colom B, Flower RJ, Vergnolle N, Nourshargh S, Perretti M.** 2013. A vasculo-protective circuit centered on lipoxin A_4 and aspirin-triggered 15-epi-lipoxin A_4 operative in murine microcirculation. *Blood* **122:**608–617.

31. **Serhan CN, Chiang N.** 2013. Resolution phase lipid mediators of inflammation: agonists of resolution. *Curr Opin Pharmacol* **13:**632–640.

32. **Chiang N, Dalli J, Colas RA, Serhan CN.** 2015. Identification of resolvin D2 receptor mediating resolution of infections and organ protection. *J Exp Med* **212:**1203–1217.

33. **Li D, Hodges RR, Jiao J, Carozza RB, Shatos MA, Chiang N, Serhan CN, Dartt DA.** 2013. Resolvin D1 and aspirin-triggered resolvin D1 regulate histamine-stimulated conjunctival goblet cell secretion. *Mucosal Immunol* **6:** 1119–1130.

34. **Cooray SN, Gobbetti T, Montero-Melendez T, McArthur S, Thompson D, Clark AJ, Flower RJ, Perretti M.** 2013. Ligand-specific conformational change of the G-protein-coupled receptor ALX/FPR2 determines proresolving functional responses. *Proc Natl Acad Sci U S A* **110:**18232–18237.

35. **Lu Y, Hong S, Tjonahen E, Serhan CN.** 2005. Mediator-lipidomics: databases and search algorithms for PUFA-derived mediators. *J Lipid Res* **46:**790–802.

36. **Serhan CN, Petasis NA.** 2011. Resolvins and protectins in inflammation-resolution. *Chem Rev* **111:**5922–5943.

37. **Serhan CN, Dalli J, Karamnov S, Choi A, Park CK, Xu ZZ, Ji RR, Zhu M, Petasis NA.** 2012. Macrophage pro-resolving mediator maresin 1 stimulates tissue regeneration and controls pain. *FASEB J* **26:**1755–1765.

38. **Bento AF, Claudino RF, Dutra RC, Marcon R, Calixto JB.** 2011. Omega-3 fatty acid-derived mediators 17(*R*)-hydroxy docosahexaenoic acid, aspirin-triggered resolvin D1 and resolvin D2 prevent experimental colitis in mice. *J Immunol* **187:**1957–1969.

39. **Chan MM, Moore AR.** 2010. Resolution of inflammation in murine autoimmune arthritis is disrupted by cyclooxygenase-2 inhibition and restored by prostaglandin E_2-mediated lipoxin A_4 production. *J Immunol* **184:** 6418–6426.

40. **Bang S, Yoo S, Yang TJ, Cho H, Hwang SW.** 2012. 17(R)-Resolvin D1 specifically inhibits transient receptor potential ion channel vanilloid 3 leading to peripheral antinociception. *Br J Pharmacol* **165:**683–692.

41. **Rajasagi NK, Reddy PB, Suryawanshi A, Mulik S, Gjorstrup P, Rouse BT.** 2011. Controlling herpes simplex virus-induced ocular inflammatory lesions with the lipid-derived mediator resolvin E1. *J Immunol* **186:** 1735–1746.

42. **Titos E, Rius B, González-Périz A, López-Vicario C, Morán-Salvador E, Martínez-Clemente M, Arroyo V, Clària J.** 2011. Resolvin D1 and its precursor docosahexaenoic acid promote resolution of adipose tissue inflammation by eliciting macrophage polarization toward an M2-like phenotype. *J Immunol* **187:**5408–5418.
43. **White PJ, Arita M, Taguchi R, Kang JX, Marette A.** 2010. Transgenic restoration of long-chain n-3 fatty acids in insulin target tissues improves resolution capacity and alleviates obesity-linked inflammation and insulin resistance in high-fat-fed mice. *Diabetes* **59:**3066–3073.
44. **Morris T, Stables M, Colville-Nash P, Newson J, Bellingan G, de Souza PM, Gilroy DW.** 2010. Dichotomy in duration and severity of acute inflammatory responses in humans arising from differentially expressed proresolution pathways. *Proc Natl Acad Sci U S A* **107:** 8842–8847.
45. **Morris T, Stables M, Hobbs A, de Souza P, Colville-Nash P, Warner T, Newson J, Bellingan G, Gilroy DW.** 2009. Effects of low-dose aspirin on acute inflammatory responses in humans. *J Immunol* **183:**2089–2096.
46. **Psychogios N, Hau DD, Peng J, Guo AC, Mandal R, Bouatra S, Sinelnikov I, Krishnamurthy R, Eisner R, Gautam B, Young N, Xia J, Knox C, Dong E, Huang P, Hollander Z, Pedersen TL, Smith SR, Bamforth F, Greiner R, McManus B, Newman JW, Goodfriend T, Wishart DS.** 2011. The human serum metabolome. *PLoS One* **6:**e16957. doi:10.1371/journal.pone.0016957.
47. **Raatz SK, Golovko MY, Brose SA, Rosenberger TA, Burr GS, Wolters WR, Picklo MJ Sr.** 2011. Baking reduces prostaglandin, resolvin, and hydroxy-fatty acid content of farm-raised Atlantic salmon (*Salmo salar*). *J Agric Food Chem* **59:**11278–11286.
48. **Tobin DM, Roca FJ, Oh SF, McFarland R, Vickery TW, Ray JP, Ko DC, Zou Y, Bang ND, Chau TT, Vary JC, Hawn TR, Dunstan SJ, Farrar JJ, Thwaites GE, King MC, Serhan CN, Ramakrishnan L.** 2012. Host genotype-specific therapies can optimize the inflammatory response to mycobacterial infections. *Cell* **148:**434–446.
49. **Perretti M, Dalli J.** 2009. Exploiting the Annexin A1 pathway for the development of novel anti-inflammatory therapeutics. *Br J Pharmacol* **158:**936–946.
50. **Chiang N, Shinohara M, Dalli J, Mirakaj V, Kibi M, Choi AMK, Serhan CN.** 2013. Inhaled carbon monoxide accelerates resolution of inflammation via unique proresolving mediator–heme oxygenase-1 circuits. *J Immunol* **190:**6378–6388.
51. **Flannigan KL, Agbor TA, Blackler RW, Kim JJ, Khan WI, Verdu EF, Ferraz JG, Wallace JL.** 2014. Impaired hydrogen sulfide synthesis and IL-10 signaling underlie hyperhomocysteinemia-associated exacerbation of colitis. *Proc Natl Acad Sci U S A* **111:**13559–13564.
52. **Dinarello CA.** 2010. Anti-inflammatory agents: present and future. *Cell* **140:**935–950.
53. **Yang R, Chiang N, Oh SF, Serhan CN.** 2011. Metabolomics-lipidomics of eicosanoids and docosanoids generated by phagocytes. *Curr Protoc Immunol* **Chapter 14:**Unit 14.26. doi:10.1002/0471142735.im1426s95.
54. **Boilard E, Nigrovic PA, Larabee K, Watts GF, Coblyn JS, Weinblatt ME, Massarotti EM, Remold-O'Donnell E, Farndale RW, Ware J, Lee DM.** 2010. Platelets amplify inflammation in arthritis via collagen-dependent microparticle production. *Science* **327:**580–583.
55. **Wolf P.** 1967. The nature and significance of platelet products in human plasma. *Br J Haematol* **13:**269–288.
56. **Norling LV, Dalli J.** 2013. Microparticles are novel effectors of immunity. *Curr Opin Pharmacol* **13:**570–575.
57. **Hunter MP, Ismail N, Zhang X, Aguda BD, Lee EJ, Yu L, Xiao T, Schafer J, Lee ML, Schmittgen TD, Nana-Sinkam SP, Jarjoura D, Marsh CB.** 2008. Detection of microRNA expression in human peripheral blood microvesicles. *PLoS One* **3:**e3694. doi:10.1371/journal.pone.0003694.
58. **Cantaluppi V, Gatti S, Medica D, Figliolini F, Bruno S, Deregibus MC, Sordi A, Biancone L, Tetta C, Camussi G.** 2012. Microvesicles derived from endothelial progenitor cells protect the kidney from ischemia-reperfusion injury by microRNA-dependent reprogramming of resident renal cells. *Kidney Int* **82:**412–427.
59. **Brown GT, McIntyre TM.** 2011. Lipopolysaccharide signaling without a nucleus: kinase cascades stimulate platelet shedding of proinflammatory IL-1β-rich microparticles. *J Immunol* **186:**5489–5496.
60. **Agouni A, Mostefai HA, Porro C, Carusio N, Favre J, Richard V, Henrion D, Martínez MC, Andriantsitohaina R.** 2007. Sonic hedgehog carried by microparticles corrects endothelial injury through nitric oxide release. *FASEB J* **21:**2735–2741.
61. **Mack M, Kleinschmidt A, Bruhl H, Klier C, Nelson PJ, Cihak J, Plachy J, Stangassinger M, Erfle V, Schlondorff D.** 2000. Transfer of the chemokine receptor CCR5 between cells by membrane-derived microparticles: a mechanism for cellular human immunodeficiency virus 1 infection. *Nat Med* **6:**769–775.
62. **Dalli J, Norling LV, Renshaw D, Cooper D, Leung KY, Perretti M.** 2008. Annexin 1 mediates the rapid anti-inflammatory effects of neutrophil-derived microparticles. *Blood* **112:**2512–2519.
63. **Norling LV, Spite M, Yang R, Flower RJ, Perretti M, Serhan CN.** 2011. Cutting edge: humanized nano-proresolving medicines mimic inflammation-resolution and enhance wound healing. *J Immunol* **186:**5543–5547.
64. **Dalli J, Serhan CN.** 2012. Specific lipid mediator signatures of human phagocytes: microparticles stimulate macrophage efferocytosis and pro-resolving mediators. *Blood* **120:**e60–e72. doi:10.1182/blood-2012-04-423525.
65. **Dalli J, Norling LV, Montero-Melendez T, Federici Canova D, Lashin H, Pavlov AM, Sukhorukov GB, Hinds CJ, Perretti M.** 2014. Microparticle alpha-2-macroglobulin enhances pro-resolving responses and promotes survival in sepsis. *EMBO Mol Med* **6:**27–42.
66. **Arnardottir HH, Dalli J, Colas RA, Shinohara M, Serhan CN.** 2014. Aging delays resolution of acute inflammation in mice: reprogramming the host response with novel nano-proresolving medicines. *J Immunol* **193:** 4235–4244.

67. **Ji RR, Xu ZZ, Gao YJ.** 2014. Emerging targets in neuroinflammation-driven chronic pain. *Nat Rev Drug Discov* **13:**533–548.

68. **Mirakaj V, Dalli J, Granja T, Rosenberger P, Serhan CN.** 2014. Vagus nerve controls resolution and pro-resolving mediators of inflammation. *J Exp Med* **211:**1037–1048.

69. **Kasuga K, Yang R, Porter TF, Agrawal N, Petasis NA, Irimia D, Toner M, Serhan CN.** 2008. Rapid appearance of resolvin precursors in inflammatory exudates: novel mechanisms in resolution. *J Immunol* **181:**8677–8687.

70. **Bellora F, Castriconi R, Dondero A, Reggiardo G, Moretta L, Mantovani A, Moretta A, Bottino C.** 2010. The interaction of human natural killer cells with either unpolarized or polarized macrophages results in different functional outcomes. *Proc Natl Acad Sci U S A* **107:** 21659–21664.

71. **Serhan CN, Yang R, Martinod K, Kasuga K, Pillai PS, Porter TF, Oh SF, Spite M.** 2009. Maresins: novel macrophage mediators with potent anti-inflammatory and pro-resolving actions. *J Exp Med* **206:**15–23.

72. **Pellettieri J, Fitzgerald P, Watanabe S, Mancuso J, Green DR, Alvarado AS.** 2010. Cell death and tissue remodeling in planarian regeneration. *Dev Biol* **338:**76–85.

73. **Dalli J, Zhu M, Vlasenko NA, Deng B, Haeggström JZ, Petasis NA, Serhan CN.** 2013. The novel 13*S*,14*S*-epoxy-maresin is converted by human macrophages to maresin1 (MaR1), inhibits leukotriene A_4 hydrolase (LTA_4H), and shifts macrophage phenotype. *FASEB J* **27:**2573–2583.

74. **Libby P, Tabas I, Fredman G, Fisher EA.** 2014. Inflammation and its resolution as determinants of acute coronary syndromes. *Circ Res* **114:**1867–1879.

75. **Legg K.** 2012. Infectious disease: pro-resolving lipids offer a helping hand to antibiotics. *Nat Rev Drug Discov* **11:**441.

76. **Markworth JF, Vella LD, Lingard BS, Tull DL, Rupasinghe TW, Sinclair AJ, Maddipati KR, Cameron-Smith D.** 2013. Human inflammatory and resolving lipid mediator responses to resistance exercise and ibuprofen treatment. *Am J Physiol Regul Integr Comp Physiol* **305:** R1281–R1296.

77. **Sasaki A, Fukuda H, Shiida N, Tanaka N, Furugen A, Ogura J, Shuto S, Mano N, Yamaguchi H.** 2015. Determination of ω-6 and ω-3 PUFA metabolites in human urine samples using UPLC/MS/MS. *Anal Bioanal Chem* **407:**1625–1639.

78. **Colas RA, Shinohara M, Dalli J, Chiang N, Serhan CN.** 2014. Identification and signature profiles for pro-resolving and inflammatory lipid mediators in human tissue. *Am J Physiol Cell Physiol* **307:**C39–C54.

79. **Barden A, Mas E, Croft KD, Phillips M, Mori TA.** 2014. Short-term n-3 fatty acid supplementation but not aspirin increases plasma proresolving mediators of inflammation. *J Lipid Res* **55:**2401–2407.

80. **Jones ML, Mark PJ, Keelan JA, Barden A, Mas E, Mori TA, Waddell BJ.** 2013. Maternal dietary omega-3 fatty acid intake increases resolvin and protectin levels in the rat placenta. *J Lipid Res* **54:**2247–2254.

81. **Mas E, Croft KD, Zahra P, Barden A, Mori TA.** 2012. Resolvins D1, D2, and other mediators of self-limited resolution of inflammation in human blood following n-3 fatty acid supplementation. *Clin Chem* **58:**1476–1484.

82. **Keelan JA, Mas E, D'Vaz N, Dunstan JA, Li S, Barden AE, Mark PJ, Waddell BJ, Prescott SL, Mori TA.** 2015. Effects of maternal n-3 fatty acid supplementation on placental cytokines, pro-resolving lipid mediators and their precursors. *Reproduction* **149:**171–178.

83. **Weiss GA, Troxler H, Klinke G, Rogler D, Braegger C, Hersberger M.** 2013. High levels of anti-inflammatory and pro-resolving lipid mediators lipoxins and resolvins and declining docosahexaenoic acid levels in human milk during the first month of lactation. *Lipids Health Dis* **12:** 89. doi:10.1186/1476-511X-12-89.

84. **Mas E, Barden A, Burke V, Beilin LJ, Watts GF, Huang RC, Puddey IB, Irish AB, Mori TA.** 2015. A randomized controlled trial of the effects of n-3 fatty acids on resolvins in chronic kidney disease. *Clin Nutr* **35:**331–336.

85. **Arnardottir H, Orr SK, Dalli J, Serhan CN.** 2015. Human milk proresolving mediators stimulate resolution of acute inflammation. *Mucosal Immunol* doi:10.1038/mi.2015.99.

86. **Arita M, Oh S, Chonan T, Hong S, Elangovan S, Sun YP, Uddin J, Petasis NA, Serhan CN.** 2006. Metabolic inactivation of resolvin E1 and stabilization of its anti-inflammatory actions. *J Biol Chem* **281:**22847–22854.

87. **Hasturk H, Kantarci A, Ohira T, Arita M, Ebrahimi N, Chiang N, Petasis NA, Levy BD, Serhan CN, Van Dyke TE.** 2006. RvE1 protects from local inflammation and osteoclast mediated bone destruction in periodontitis. *FASEB J* **20:**401–403.

88. **Deng B, Wang CW, Arnardottir HH, Li Y, Cheng CY, Dalli J, Serhan CN.** 2014. Maresin biosynthesis and identification of maresin 2, a new anti-inflammatory and pro-resolving mediator from human macrophages. *PLoS One* **9:**e102362. doi:10.1371/journal.pone.0102362.

Myeloid Cells in Health and Disease: A Synthesis
Edited by Siamon Gordon

doi:10.1128/microbiolspec.MCHD-0017-2015

Jenna L. Cash[1]
Paul Martin[2]

21 Myeloid Cells in Cutaneous Wound Repair

INTRODUCTION

Wound repair is a complex and dynamic process that aims to restore cellular structures and tissue layers to damaged organs. Damage to the anatomical barriers against the environment, including the skin and gastrointestinal and respiratory systems, opens the organism to microbial invasion, and so the first function of the repair process is to temporarily seal this breach with a platelet plug and to counter infection as rapidly as possible. In skin wounds in healthy adults, barrier function is efficiently restored; however, repair of deeper dermal structures culminates in scar formation with loss of the original tissue structure and function. Tissue damage can be inflicted to a variety of organs by diverse stimuli, including ischemia (heart attack), burns (chemical, heat, or electrical), trauma, surgery, or infection. Although the anatomical sites injured may be distinct, for example, the immune-privileged cornea versus the gut with its complex microbiome, or the heart with its propensity for fibrosis, it is generally acknowledged that repair of all tissues shares basic commonalities. The wound-healing response can thus be typically divided into four overlapping phases: hemostasis, inflammation, cell migration/proliferation, and remodeling (Fig. 1) (1).

Tissue injury generally results in rupture or severing of blood vessels and thus platelet activation and aggregation to form a platelet plug. Activated platelets also degranulate to release various chemoattractants and growth factors as well as initiating the clotting cascade, which converts fibrinogen found in serum into long strands of insoluble fibrin at the wound site to form a clot (2). Before hemostasis is complete, circulating leukocytes passively leak into the wound from damaged blood vessels, while tissue-resident immune cells, e.g., Langerhans cells and dermal macrophages (Mϕs) in the skin, become activated by damage-associated molecular patterns and/or pathogen-associated molecular patterns to release a plethora of growth factors, inflammatory cytokines, and chemoattractants (3). Injured or inflamed keratinocytes in the wound-adjacent epidermis produce several signaling molecules, which also aid leukocyte recruitment to the wound (4).

A period of transendothelial recruitment of neutrophils follows, with these cells often exhibiting a swarming behavior, such that acute wounds are typically associated with a neutrophil-rich cellular infiltrate within hours of tissue damage (5–8). At the wound site, and once activated, neutrophils degranulate, releasing

[1]MRC Centre for Inflammation Research, The University of Edinburgh, The Queen's Medical Research Institute, Edinburgh, EH16 4TJ; [2]School of Biochemistry, Medical Sciences, University Walk, Bristol University, Bristol BS8 1TD, United Kingdom.

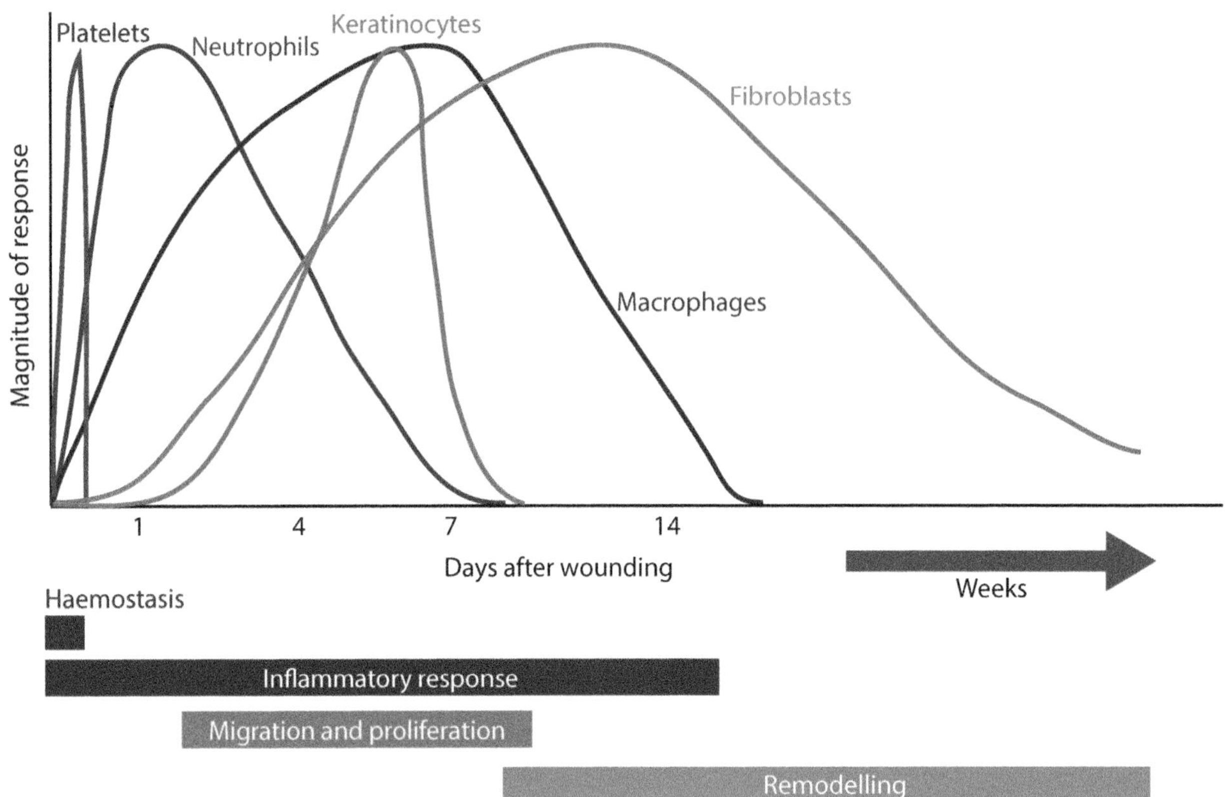

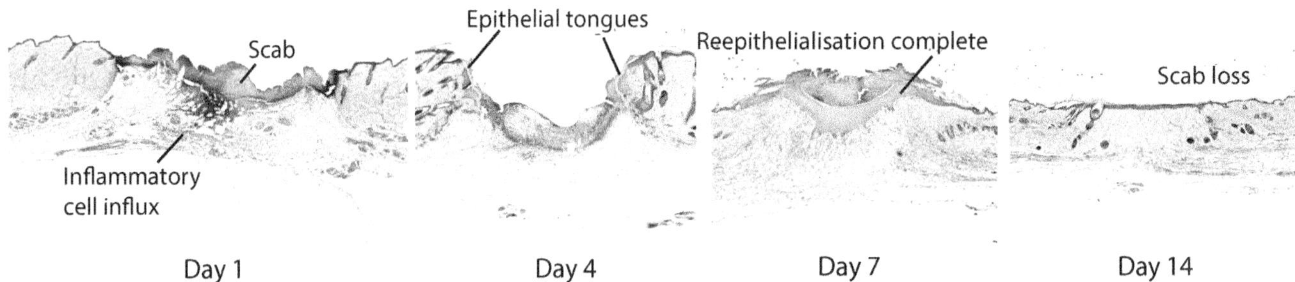

Figure 1 Time course of the cutaneous wound repair response. (Top) The time relationship between different wound repair processes and the cells involved. Wound repair is often thought of as occurring in four phases: hemostasis (platelet-mediated blood coagulation and immediate damage-signaling events), inflammation (leukocyte recruitment to the site of injury), migration and proliferation (keratinocyte proliferation and migration to reepithelialize the wound, fibroblast migration, contraction, and collagen deposition leading to scar formation), and remodeling (resolution of wound vessels and remodeling of the scar tissue). (Bottom) Representative hematoxylin and eosin-stained wound midsections from days 1, 4, 7, and 14 after excisional wounding are shown. These depict important features of each stage of repair, including scab formation and loss, inflammatory cell influx, and reepithelialization.

antimicrobial cytotoxic molecules, including reactive oxygen species (ROS) and enzymes (e.g., elastase). An overlapping wave of monocyte recruitment follows, with monocyte-derived Mϕs functioning as the major phagocytes at the wound site, engulfing microorganisms, apoptotic cells (especially spent neutrophils), and other cell and extracellular matrix (ECM) debris. Mϕs also dramatically influence the wound microenvironment through expression of growth factors, cytokines, and chemokines (4, 9–12). As such, the Mϕ, perhaps more than any other myeloid cell, has the capacity to regulate many aspects of wound repair. Besides their phagocytic roles, Mϕs are thought to also regulate angiogenesis (including sprouting and vessel permeability) and lymphangiogenesis (lymphatic regrowth) and to direct fibroblast-mediated collagen deposition to form the scar. Myeloid cells recruited to wounds during the inflammatory phase are thought to play both positive and negative roles during the repair process (13–15).

The proliferative phase of wound repair overlaps the inflammatory phase and involves reepithelialization, angiogenesis, collagen deposition, and wound contraction. Reepithelialization of skin wounds is achieved by the proliferation and migration of keratinocytes from the wound edge at the interface between the scab and the newly forming granulation tissue and is thought to be, in part, regulated by Mϕ-derived keratinocyte growth factors (11, 12). Unwounded epidermis is stratified with a superficial cornified dead layer, but the advancing edge of the epithelial tongue that "crawls" forward between the scab and the forming granulation tissue is generally only one or two cells thick. Many genes, including proteases for cutting a pathway and enzymes to sequester ROS and protect the migrating keratinocytes, are upregulated to enable reepithelialization (16, 17). A subset of these genes are first epigenetically unsilenced by removal of polycomb-mediated histone marks (18). Beneath the epithelial layer, wound contraction is mediated by contractile myofibroblasts that differentiate from fibroblasts under the influence of mechanical cues and growth factors, including transforming growth factor β (TGF-β) and shrink the wound by tugging on collagen fibers (19, 20). Fibroblasts also synthesize and lay down collagen fibers in addition to other ECM components, such as proteoglycans, to generate granulation tissue. Wound neovascularization gives granulation tissue its name and is thought to be supported by Mϕs and required to provide sufficient nutrients and oxygen to the metabolically demanding tissue (21–23). Remodeling of the wound granulation tissue is the final phase of repair and refers to multiple events, such as maturation of the newly formed blood vessel network, including trimming of superfluous branches and remodeling of the collagen-rich scar, which may continue for weeks to years after the original injury and never completely regenerates the unwounded dermal structure (24).

Ultimately, the repair process needs to be swift to eliminate microorganisms and to reestablish barrier function. It avoids excessive collateral damage to host tissues by instigating a controlled, coordinated inflammatory response, which results in an as complete as possible return to tissue homeostasis. However, with the exception of early embryos, which have a much higher regenerative capacity and a considerably reduced wound inflammatory response, most mammalian tissues heal with a robust inflammatory response leading to formation and retention of a collagen-rich scar (14, 25, 26). The majority of wounds, regardless of organ, heal acutely with scar formation, but deviations from the "normal" wound repair processes can result in pathological conditions. These include extreme fibroproliferative responses, as observed in keloid scars, and, at the other end of the spectrum, nonhealing chronic wounds that can persist for many months or even decades. Chronic wounds, including diabetic foot ulcers, venous leg ulcers, and pressure ulcers, are estimated to cost $25 billion per year in the United States alone; they are debilitating and are the most common cause of limb amputation. Chronic wounds are thought to become "stuck" in the inflammatory phase of repair rather than effectively progressing to the proliferative and remodeling phases; thus the influence of myeloid cells on chronic wound formation is likely to be significant (27, 28).

MYELOID CELLS IN WOUNDS

Myeloid cells, especially neutrophils and Mϕs, are indispensable to host defense against infection, as without the inflammatory response all breaches of barrier layers would ultimately lead to septicemia and death. However, tissue repair *per se* is not absolutely dependent on recruitment of myeloid cells to wounds. At early embryonic stages, prior to development of myeloid lineages or soon after, when only a small inflammatory response is triggered at the wound site, tissues can heal rapidly and effectively (29, 30). Indeed, wound repair at these early embryonic stages in mouse (up to two-thirds through gestation), and following surgical procedures carried out on pre-24-week human fetuses, occurs in the absence of a robust inflammatory response and results in restoration of normal dermal architecture in a process resembling regeneration (31, 32). This has led to the idea that perhaps inflammation might be, at least in part, causal of fibrosis and scarring. Various genetic and pharmacological approaches for knocking down the wound inflammatory response have also suggested that myeloid cells may deliver signals that lead to scarring (14, 33, 34). For example, studies of the neonatal PU.1 mouse, which lacks neutrophils, Mϕ, and mast cells and so cannot raise an innate immune response, exhibit almost perfect healing without a scar (1, 2, 31). These studies suggest that a "conversation," whether through physical interactions or paracrine factors, occurs between innate immune cells and fibroblasts, which leads to fibroblasts laying down collagen in an aberrant way to form scar tissue.

A variety of strategies and models have been used to determine the roles of myeloid cell lineages in the wound repair process. The most fully characterized model organism is the mouse (*Mus musculus*), in which transgenic and knockout studies have helped refine our understanding of the function of each of these lineages in the repair process (6, 35, 36). Intravital microscopy enables live imaging of immune cell recruitment to mouse wounds, but the translucent zebrafish (*Danio rerio*) also offers opportunities here with ease of *in vivo* imaging and genetic advantages over the mouse. In addition, the zebrafish may provide key clues from its regenerative capacity on how to optimize mammalian repair (37, 38). For some basic studies of innate immune cell function and migration, the fruit fly (*Drosophila melanogaster*) offers the best genetic tractability of all (39). Despite the specific advantages of conventional model organisms (e.g., *M. musculus*, *D. rerio*, and *D. melanogaster*), it is nevertheless important for the scientific community to adopt a species-inclusive approach, as underscored by the discovery of mammalian skin regeneration in the African spiny mouse and the MRL mouse (40, 41).

Among the myeloid cells, the roles of neutrophils and Mϕ in wound repair have been most extensively characterized; while eosinophils, mast cells, and dendritic cells (DCs) are also present, their influence on repair is somewhat less clearly defined. We will now provide an overview of our current understanding of the roles of myeloid cells in healing, with a focus on cutaneous repair (Fig. 2).

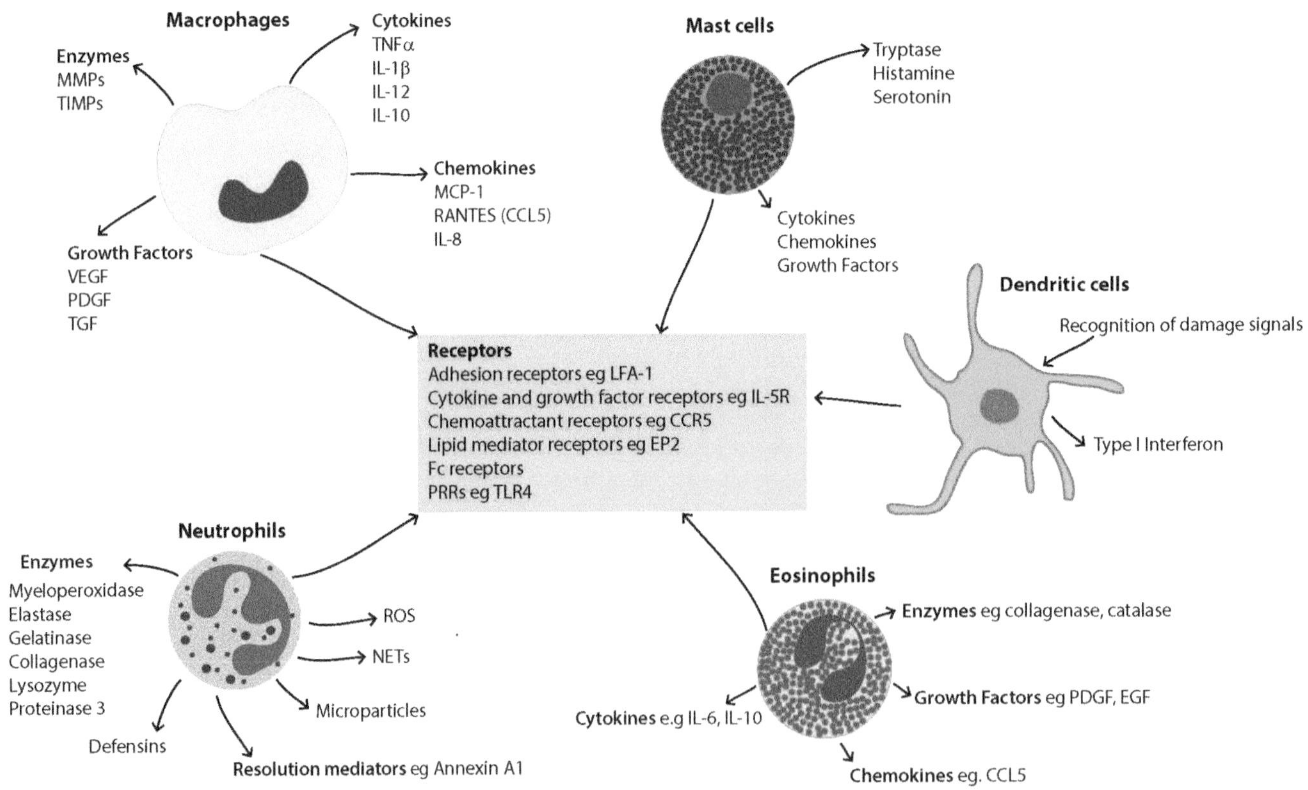

Figure 2 Myeloid cells in wounds. Diagram depicting myeloid cells involved in cutaneous repair along with some of the key receptors with which they sense wound signals, and signaling molecules and enzymes released in response to the specific signals that these cells process. Key receptors shared by these cells are noted in the central green box, while neutrophil, mast cell, and eosinophil granules are shown as purple or pink filled circles. Cells are not drawn to scale. Abbreviations: LFA, lymphocyte function-associated antigen; MCP, monocyte chemoattractant protein; PRR, pattern recognition receptor.

Neutrophils

In adult wounds, neutrophils are recruited to wound sites within minutes to hours of damage and play crucial roles in killing invading microorganisms through degranulation of azurophil and gelatinase granules and specific vesicles and production of locally high levels of ROS and other cytotoxic materials (42, 43). Model organism studies have begun to reveal some of the wound leukocyte damage attractants. Using transgenic zebrafish larvae with a genetically encoded H_2O_2 sensor, Niethammer et al. (44) showed that rapid wound detection by neutrophils is driven by an H_2O_2 gradient synthesized by a dual oxidase, while Yoo et al. (45) demonstrated that Lyn acts as a redox sensor to detect H_2O_2 at wound sites. *Drosophila* studies have indicated that the same signal draws fly Mϕ (hemocytes) to wounds and that epidermal calcium flashes in turn drive dual oxidase activation (46, 47). Murine intravital microscopy studies have shown that sterile burn injuries trigger ATP release from necrotic cells to generate an inflammatory microenvironment, in part consequent to NLRP3 inflammasome activation, which promotes neutrophil intravascular adhesion to the endothelium to enable neutrophil intravascular crawling to the wound site. However, within the necrotic part of a wound, it appears that neutrophil migration is dependent on a macrophage inflammatory protein 2 (MIP-2) gradient and formyl peptide receptor 1 (FPR1)-dependent chemotaxis (6). Amplification of the neutrophil recruitment response is, in part, a consequence of death of the first neutrophils to arrive at the wound site, which drives neutrophil swarming to the damaged tissue, with neutrophils being retained at the wound through integrin signaling (8). These and other elegant live imaging studies have defined the cascade of events involved in wound leukocyte signaling, however, we have a startling lack of knowledge of the functions of neutrophils after they have reached damaged tissues, along with how, and if, they resolve from the wound site once their job is done. Some insight has again been achieved using zebrafish, wherein recent studies have shown that wound-recruited neutrophils can disperse via the vasculature back into the body (37).

Nevertheless, the positive actions of neutrophils in repair, beyond wound disinfection, are commonly thought to be negligible. Indeed, studies have suggested that the weapons used by neutrophils to kill pathogens (e.g., ROS) can cause significant collateral damage to host tissues, in particular through damage of cellular macromolecules such as DNA (43). Furthermore, the oral mucosa, which heals rapidly with minimal inflammation and scar formation, displays significantly reduced neutrophil recruitment (49, 50), while mice null for secretory leukocyte protease inhibitor, a potent inhibitor of serine proteases, including neutrophil elastase and cathepsin G, show impaired cutaneous healing with increased inflammation and elastase activity (51).

One significant way in which neutrophils may influence the later wound inflammatory response is through their apoptotic death and subsequent engulfment by Mϕ, which acts to reprogram the Mϕ toward an anti-inflammatory phenotype (52, 53). With good reason, this is the generally accepted dogma, but there is also some support for apoptotic neutrophil phagocytosis driving proinflammatory Mϕ programs, depending on the stimulus that first triggered neutrophil apoptosis. Pathogen-induced apoptotic neutrophils, for example, express heat shock proteins Hsp60 and Hsp70 and drive Mϕ activation to a proinflammatory state with increased production of tumor necrosis factor α (TNF-α) (54). Efferocytosis in the context of wound repair has, to our knowledge, not been studied. However, since neutrophil-depleted mice are capable of healing wounds, with the only significant phenotype being accelerated reepithelialization, current data strongly suggest that neutrophil debridement of wounds, their apoptosis, and their subsequent phagocytosis are not essential for healing competence and that whatever roles they may play can be compensated for by other cell lineages (55).

Monocytes and Macrophages

Mϕs, the "big eaters" of the myeloid lineage, are recruited to damaged tissue. One source of wound Mϕs is tissue-resident pools, but they can also differentiate from inflammatory monocytes that spill into the damaged tissue through injured blood vessels or circulating monocytes that extravasate from the locally inflamed microvasculature and migrate to the wound site. It has recently been shown by Dal-Secco et al. that monocytes recruited to focal necrotic injury of the liver rapidly undergo phenotypic conversion from $CCR2^{hi}CX3CR1^{lo}$ to $CCR2^{lo}CX3CR1^{hi}$ cells on reaching a sterile wound and this conversion is essential for optimal clearance of dead cells (56). However, others have shown, using a different set of markers, that $Ly6C^{hi}$ wound monocytes do not undergo rapid maturation but rather persist in the wound as $Ly6C^{hi}F4/80^{+}CD64^{+}MerTK^{-}$ cells expressing high levels of TNF-α and interleukin-1β (IL-1β). These $Ly6C^{hi}$ cells later mature to $Ly6C^{lo}F4/80^{+}CD64^{+}MerTK^{+}$ Mϕ in the wound, which gradually upregulate markers including CD206, major histocompatibility complex class II (MHC-II), vascular endothelial growth factor (VEGF), and TGF-β during the course of repair (57).

Having differentiated from monocytes, the recruited wound Mϕ display extraordinary phenotypic plasticity as a result of complex reciprocal interactions with neighboring cells, including other immune cells or stromal cells, as well as microorganisms, soluble mediators, and ECM components. However, their eventual phenotype also appears to partially reflect that of the precursor circulating monocytes (3, 56). Early attempts to describe Mϕ phenotypic states classified Mϕs into two distinct phenotypes, classical (M1-type) and alternative (M2-type) Mϕs (58, 59). Classical Mϕ activation describes enhanced microbicidal activity in response to gamma interferon with or without lipopolysaccharide, which was later linked with Th1 responses and extended to cytotoxic and antitumoral properties. Alternative Mϕ, proposed by Stein et al., upregulate mannose receptor in response to Th2 stimuli, IL-4, and IL-13 and exhibit high clearance of mannosylated ligands, increased MHC-II expression, and reduced proinflammatory cytokine secretion (58). However, many other factors, including IL-10, TGF-β, glucocorticoids, and, more recently, proresolving mediators, do not fit clearly into the Th1/Th2 context of M1/M2 Mϕ, but appear to elicit similar alternative Mϕ activation, including downregulation of inflammatory cytokines and reduced use of reactive nitrogen species and ROS killing mechanisms (58). Albina et al. demonstrated that classically activated (M1) and alternatively activated (M2) Mϕs also differ in terms of arginine metabolism. While M1 Mϕs express more nitric oxide synthase (NOS) to catalyze arginine to cytotoxic nitric oxide, in M2 Mϕs arginase dominates and converts arginine into ornithine, which is a polyamine and proline precursor required ultimately for cell proliferation and collagen synthesis (60). Thus, alternatively activated Mϕ are often purported to drive wound repair, yet the phenotype of Mϕs throughout the course of cutaneous repair has yet to be studied by direct and detailed examination of these cells. A study by Daley et al. goes some way to address this deficiency in our knowledge, by showing that wound Mϕs exhibit complex and dynamic phenotypes that change as the cutaneous wound matures, including simultaneous expression of traits typically associated with both alternative and classical Mϕ activation. For example, simultaneous production of TNF-α and expression of mannose receptor can be observed in wound Mϕs (61).

The role of Mϕ subsets in the inflammatory and fibrotic components of liver and lung repair is somewhat clearer (reviewed in reference 62), with Ly6C^{hi}F4/80^{int}CD11b^{hi} macrophages identified as the restorative macrophages responsible for remodeling hepatic scars, through elevated matrix metalloproteinase (MMP) expression (63). In schistosomiasis, a parasitic infection causing chronic injury and inflammation and resulting in irreversible fibrosis, alternatively activated macrophages control inflammation and scarring, which slows the development of lethal fibrosis. These alternatively activated macrophages are derived from LysM^{lo}F4/80^{hi}CD11b^{hi}IL-4Ra^{+} macrophages, which in response to IL-4 and IL-13 upregulate arginase-1 (64). Others have demonstrated that resistin-like molecule-α (RELMα/FIZZ1), an alternatively activated macrophage marker, restrains *Schistosoma* parasite-induced lung inflammation and fibrosis (65). Cutaneous wound repair research could potentially benefit from the knowledge gleaned from other organs, even when the injury stimuli are contrasting, as it is likely that some common pathways will be preserved across organs.

It is clear that Mϕs sense an extraordinary array of stimuli in their microenvironment that combine to determine complex phenotypes that cannot be conveniently partitioned into bipolar groups. It is also almost certainly the case that Mϕ phenotypes vary throughout the duration of the repair process in response to exposure to changing signals, and this capacity enables them to fulfill different functions during different phases of repair. Mosser and Edwards proposed that Mϕ phenotype/activation represents a continuum in which Mϕs can share features of classically and alternatively activated cells (66). In the context of wound repair, current data suggest that Mϕs exhibit a more proinflammatory/cytotoxic phenotype during the early inflammatory phase (e.g., with elevated inducible NOS), whereas in the later parts of the repair response a switch toward a growth factor-producing/anti-inflammatory Mϕ phenotype exists, corresponding to elevated arginase-1 expression (61, 66, 67). What remains unclear is precisely when and how this transition takes place, in addition to detailed information on what Mϕ subpopulations are present throughout repair and their corresponding roles (61, 67, 68). To date, we also lack compelling evidence that alternatively activated (M2-type) Mϕs do drive repair, as they are purported to. Importantly, despite wound Mϕs expressing IL-4Rα, the corresponding ligands IL-4 and IL-13 are not known to be robustly expressed in wounds, suggesting that wound healing and wound Mϕ phenotype may occur independently of these prototypical M2 stimuli. In agreement with this, wound Mϕs from IL-4Rα$^{-/-}$ mice show no substantial change in phenotype, and adoptive transfer of "M2" *in vitro*-cultured Mϕ to mouse wounds has no detectable effect on repair (61, 69). These data align with observations from Campbell et al. suggesting that "M2" alternatively activated Mϕs are more similar in phenotype to resting Mϕs, whereas regulatory Mϕs can play important roles in dampening inflammatory immune responses (68, 70). Interestingly, inhibition of local arginase-1 activity or inducible Mϕ-specific arginase-1 ablation results in increased inflammation, including increased neutrophil and Mϕ influx, impaired reepithelialization, and poor wound contraction (68). Likewise, in *Schistosoma* infection, Mϕ-specific arginase-1 deletion results in enhanced fibrosis and inflammation (71).

Specific and conditional ablation of Mϕs using transgenic mice has enabled the mechanistic testing of the function of Mϕs during different phases of cutaneous wound repair, but because of issues of redundancy, all roles are still not entirely clear. As already described, neonatal PU.1 knockout mice that have no leukocyte lineages at all, including Mϕs, heal wounds rapidly and efficiently without scarring. However, in adult animals the situation becomes much more complex. The earliest studies of the role of Mϕs in repair used antimacrophage serum to deplete these cells and demonstrated impaired wound repair (72). More recent studies have confirmed and extended these findings, showing that specific Mϕ ablation throughout the time course of repair results in delayed reepithelialization and wound contraction, reduced collagen deposition, impaired angiogenesis, and decreased cell proliferation in healing wounds (73). Conditional Mϕ depletion at specific time points during repair, using mice expressing the diphtheria toxin receptor under control of a Mϕ-specific promoter, has revealed that Mϕs clearly fulfill temporally specific roles in wound healing. During early repair, Mϕ depletion impairs granulation tissue formation, wound contraction, and reepithelialization, whereas depletion specifically during the midstage of repair impairs tissue maturation, with neutrophil persistence within the wound bed and severe hemorrhaging, presumably through improper vessel development. These adverse changes are associated with elevated TNF-α and reduced TGF-β and VEGF levels in the wound and atypical endothelial cell apoptosis (36, 73). In corneal wound healing models, Mϕ depletion using clodronate liposomes or using a neutralizing antibody to the Mϕ chemoattractant MIP-1α resulted in impaired healing, associated with disrupted ECM deposition and reduced angiogenesis (74, 75). Mϕ and neutrophil-derived MMPs play important beneficial roles in acute wound repair, with MMPs removing damaged ECM from the wound site, which enables new ECM components to integrate correctly with the ECM at the wound edges. MMPs also loosen attachments between biofilms and the wound bed and degrade the capillary basement membrane to allow vascular endothelial cells to migrate to establish new blood vessels in the wound bed (76).

Wound Mϕs play important and temporally specific roles in wound repair; future studies should strive to improve our understanding of what cues drive Mϕ phenotypic changes in repairing wounds and when, along with how, Mϕ behavior changes in aberrant repair scenarios.

Mast Cells

The role of mast cells in wound repair has long been debated, with mouse studies yielding confusing, contradictory results. Wound treatment with disodium cromoglycate, a drug that prevents mast cell degranulation, results in thicker mature collagen fibers than in control mice, which still have immature fibers 7 days after wounding, accelerated wound closure, and reduced proinflammatory cytokine expression and neutrophil recruitment. Interestingly, fibroblast treatment with tryptase, the main mast cell enzyme released on degranulation, stimulates fibroblast α-smooth muscle actin and collagen expression in *in vitro* studies (77). Previous work using the mast cell-deficient $Kit^{-/-}$ mice have demonstrated both accelerated (78) and delayed (79) wound closure. However, although $Kit^{-/-}$ mice are mast cell deficient, they also have various other immune system alterations including anemia, neutropenia, and deficient γδ T cells, which may have an impact on healing. Two more recent research studies using Mcpt5-Cre mice, in which mast cells can be specifically and inducibly ablated, revealed almost no change in fibrosis, inflammatory cell recruitment, or rate of dermal wound repair (80, 81). These results suggest that connective tissue mast cells do not play a substantial role in wound repair, but leave open the question of whether circulating basophils, the mast cell precursor cell, may be recruited to the wound following mast cell ablation.

Eosinophils

The role of eosinophils in wound repair has largely been neglected. However, eosinophil depletion versus enrichment using IL-5-deficient versus -overexpressing mice has demonstrated a role for these cells in driving keratinocyte migration to achieve timely reepithelialization. IL-5-overexpressing mice exhibit elevated wound eosinophil infiltration with delayed reepithelialization and scab loss (82), whereas eosinophil depletion accelerates skin wound epithelial closure (83). During repair, eosinophils are a source of TGF-α (84), a growth factor known to promote keratinocyte migration to close *in vitro* scratch wounds through the epidermal growth factor (EGF) receptor (85). Thus, eosinophils might regulate reepithelialization of cutaneous wounds by releasing TGF-α. In the context of bleomycin-induced lung fibrosis, eosinophils appear to play a role in driving fibrosis by stimulating lung fibroblasts to upregulate α-smooth muscle actin and collagen expression (86). It remains to be seen whether eosinophils play a role in the pathophysiology of aberrant healing, including chronic wounds and keloids.

Dendritic Cells

Langerhans cells in the epidermis are morphologically distinct, F4/80$^+$ DCs that reside in close contact with keratinocytes and reciprocally exchange survival and trophic factors. Langerhans cells can be divided into plasmacytoid DCs (pDCs) and dermal (interstitial) DCs. pDCs rapidly infiltrate wounds, whereas dermal DCs are induced by stimuli such as antigens or TNF-α to migrate to draining lymph nodes, where they can activate or inhibit lymphocytes (87, 88). Cathelicidin antimicrobial peptides are rapidly induced in skin wounds and facilitate pDC recognition of host-derived nucleic acids released during dermal injury through Toll-like receptor 7 (TLR7) and TLR9, leading to type I interferon production. pDC depletion using a neutralizing antibody inhibits expression of the inflammatory cytokines IL-17 and IL-22 in wounds and retards wound reepithelialization, suggesting that early wound pDC infiltration might play an important role in timely reepithelialization (89). Indeed, human acute wounds and healing diabetic foot ulcers exhibit increased epidermal Langerhans cells in comparison to nonhealing diabetic foot ulcers, suggesting that Langerhans cells may play an important role in ulcer healing outcome (90).

In recent years, lineage-specific knockdown studies have begun to reveal the functions of myeloid cells in repair. Here we discuss the likely roles that myeloid cells may play in each of the key wound repair processes: clearance of microorganisms and apoptotic cells, wound neovascularization and leukocyte-stromal cell conversations that may direct keratinocyte reepithelialization, and fibroblast wound contraction and collagen deposition.

APOPTOSIS, NECROSIS, AND EFFEROCYTOSIS

Cell death by apoptosis is an active, highly regulated, and noninflammatory process and a key event in the resolution of inflammation for several reasons. First, apoptosis acts as a nonphlogistic strategy to remove inflammatory cells from the wound in order to allow the healing process to progress. Uptake of the apoptotic cell prior to its lysis prevents the release of histotoxic, proinflammatory, and immunogenic material and actively suppresses inflammation (91–93). Within the context of an acute cutaneous wound, neutrophil numbers in the granulation tissue decline during the latter stages of healing, though there is debate as to whether the neutrophils undergo apoptosis or reenter the circulation to traffic to other organs (37, 94). Clearance of cellular corpses, including neutrophils, is performed largely by Mϕ and acts as the terminating event of apoptosis *in vivo*. Apoptotic cells express phosphatidylserine externally, which is recognized by Mϕ through a plethora of evolutionarily conserved receptors, including scavenger receptor A (95–106). Second, efferocytosis (apoptotic cell phagocytosis) occurs quickly and efficiently and triggers a phenotype change in Mϕ, as evidenced by their reduced ability to secrete inflammatory cytokines and increased production of anti-inflammatory mediators IL-10, TGF-β, and prostaglandin E_2 (97, 104, 105, 107). Proresolving mediators, a particular class of anti-inflammatory molecules that drive the termination of the inflammatory response, can also be released by the apoptotic cells themselves, including annexin A1 protein and annexin peptides (108–110). Inefficient clearance of apoptotic cells, for example, when the number of apoptotic cells exceeds the capacity of local phagocytes, can result in these cells undergoing secondary necrosis and exacerbating inflammation by releasing cytotoxic cellular contents that activate myeloid cells (53, 111–114). Indeed, human and mouse diabetic wounds have a higher apoptotic cell burden with dysfunctional Mϕ efferocytosis (115). Thus, abnormalities in neutrophil apoptosis or clearance may contribute to the pathogenesis of chronic wounds or other abnormal healing processes.

In contrast, necrosis is a passive catabolic process resulting from traumatic insult to the cell. Primary necrotic cell debris is generally loaded with endogenous danger signals (damage-associated molecular patterns), including heat shock proteins, nuclear proteins (high-mobility group box-1 [HMGB1]), histones, and DNA, which stimulate Mϕ activation to produce proinflammatory mediators (116, 117). Keratinocytes at the margins of E18 (embryonic day 18) mouse dermal wounds release higher levels of HMGB1 in comparison to E15 wounds, resulting in increased Mϕ infiltration and scarring. Indeed, injection of HMGB1 into E15 wounds, which normally heal scarlessly, results in scar formation associated with Mϕ recruitment and increased wound fibroblasts (118). Recently, Chen et al. demonstrated that Mϕ-specific deficiency of peroxisome proliferator-activated receptor γ results in elevated TNF-α levels, leading to impaired apoptotic cell clearance, which in turn resulted in poor wound repair with reduced collagen deposition and angiogenesis (35).

We do not currently have a good understanding of what role neutrophil apoptosis plays within the context of healing, as descriptions are generally derived from acute inflammation models that lack the level of tissue damage observed in typical repair models. Neutrophil apoptosis and subsequent efferocytosis during the repair of wounds of different etiologies (e.g., burn versus trauma) or at distinct anatomical sites (e.g., liver versus skin) is clearly an area of tissue repair research that requires more attention. Furthermore, it is unclear whether neutrophil apoptosis and subsequent efferocytosis varies between wounds of different etiologies (e.g., burn versus trauma). Most descriptions of the role of apoptosis in wound repair are thus derived from speculation from such acute inflammation models.

WHAT ROLE DO MACROPHAGES PLAY IN THE REVASCULARIZATION OF REPAIRING TISSUE?

Mϕs mediate the inflammatory phase of tissue repair but are also implicated in guiding the angiogenic response, which is a key part of the proliferative phase of tissue repair. Angiogenesis in the healing wound is required to reestablish normal circulation and thus tissue oxygenation. At the cellular level, angiogenesis involves the migration and proliferation of not only endothelial cells but also supporting cells such as pericytes and vascular smooth muscle cells. After blood vessel injury following wounding, growth factors such as VEGF guide angiogenic sprouting by directing endothelial tip cell migration at specific sites on the blood vessel. As the new sprout proliferates, it elongates, branches, and connects with other sprouts to form an expanding network of endothelial tip and stalk cells (119, 120). This immature vascular network is subsequently remodeled to remove redundant blood vessels, a process called pruning, and matures through recruitment of pericytes and smooth muscle cells. Thus, when discussing the role of Mϕs in angiogenesis, we need to consider that Mϕs might regulate specific phases of angiogenesis, the entire process, or even exert opposing effects on particular angiogenic events. We can glean insight from studies of developmental and tumor angiogenesis that have been interrogated in much greater detail (24, 121, 122). Mϕs are known to secrete a plethora of proangiogenic factors, especially VEGF and platelet-derived growth factor (PDGF), with the former having been shown to be crucial for angiogenic sprouting. Neutralizing VEGF antibodies reduce neovascularization and granulation tissue formation in porcine wounds and angiogenic activity of human wound fluids *in vitro*, which demonstrates the clear importance of VEGF in wound angiogenesis but does not define the source of this growth factor (123, 124).

As described in "Monocytes and Macrophages" above, Mϕ ablation studies have demonstrated impaired angiogenesis, severe hemorrhaging, and poor transition into later phases of repair (36, 73), whereas neonatal PU.1 mice that lack Mϕs and other innate immune cells throughout development exhibit faster wound revascularization (13). These observations in PU.1 mice and also the fact that endothelial cells in a three-dimensional *in vitro* environment can self-assemble branching tubes, suggest that angiogenesis *per se* is not absolutely dependent on Mϕs. Nevertheless, it does seem likely that Mϕs are involved in regulating wound angiogenesis, but their precise role is still far from clear (125, 126). A relatively novel hypothesis supported by Stefater et al. is that Mϕ may actually restrain angiogenesis during early repair using a Wnt-Calcineurin-Flt1 axis, to ensure that vascular networks are reestablished at the appropriate stage of repair (122). During the proliferative phase of repair, Mϕ may also provide physical bridging support for vascular anastomosis, though this has only been shown so far in development of the retinal vasculature (121, 170).

Future work utilizing live imaging techniques in mice and other model organisms will characterize the precise role of Mϕ in all stages of angiogenesis during repair, from initial sprouting to anastomosis through to subsequent pruning, and help define the molecular mechanisms and cellular communications that regulate these processes.

EVIDENCE FOR MACROPHAGE-STROMAL CELL SIGNALS THAT REGULATE SCARRING AND HYPERPIGMENTATION

Fibroblasts and Scarring

As discussed in the previous sections, inflammation is a prerequisite for healing with a scar, with abundant, convincing evidence linking wound Mϕ, TGF-β production, and scar formation. Briefly, embryonic cutaneous wounds do not elicit an inflammatory response, exhibit reduced levels of TGF-β1 and -β2 (produced

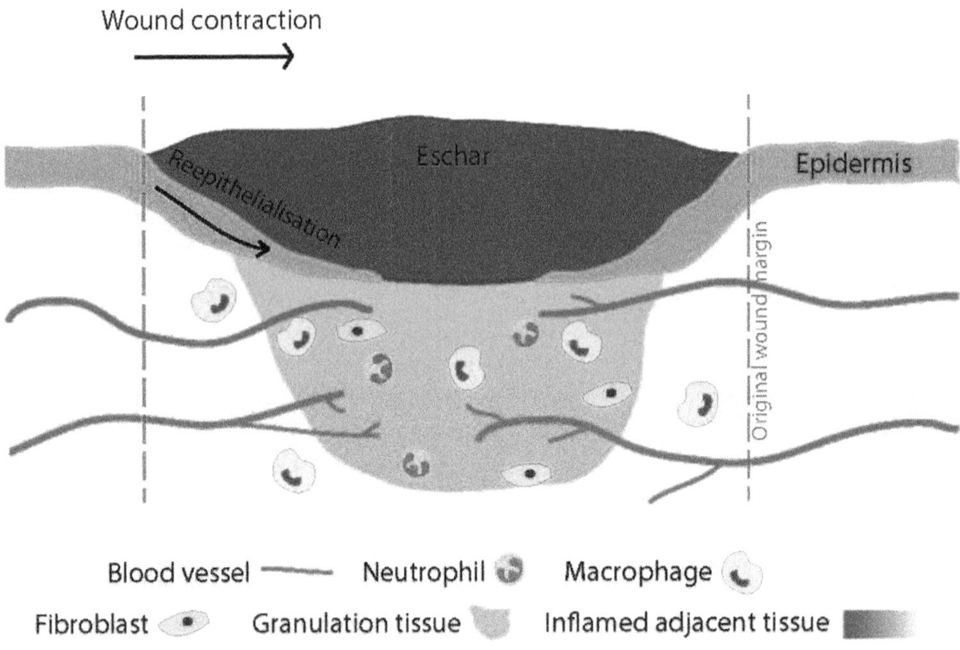

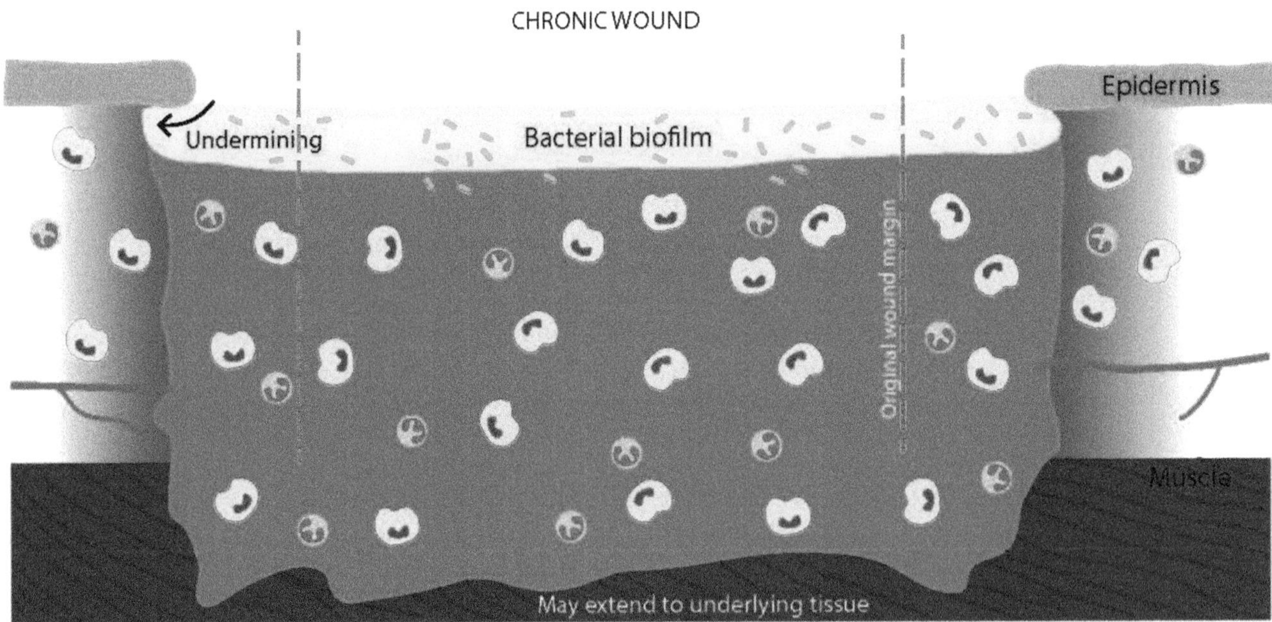

Figure 3 Acute versus chronic wound healing. A healthy repairing acute wound is protected by a scab throughout much of the healing response. During this period, the various missing tissue layers are replaced by cell migration and proliferation, and this is supported by an influx of myeloid cells, which subsequently resolve after the wound has healed. In a chronic wound, a scab may not be present but a bacterial biofilm invariably is, and certain cells migrate poorly. There is a prolonged and elevated influx of myeloid cells, with the inflammatory response overflowing into the adjacent tissue and often extending into the underlying muscle or bone.

during embryonic healing, but very rapidly cleared), and heal scarlessly (127). Interestingly, Smad3 knockout mice, which are TGF-β signaling deficient, exhibit reduced wound Mϕ recruitment, have reduced myofibroblasts, and also heal without scarring (128). We again lack detail of the molecular, or indeed physical, Mϕ-fibroblast "conversations," the signaling mechanisms triggering these episodes, and how these conversations ultimately regulate collagen deposition.

Just as Mϕ appear to be required for the initiation of scarring, they are also involved in matrix remodeling and in the resolution of fibrosis. Duffield et al. have demonstrated, using a liver fibrosis model, that Mϕ ablation during fibrosis development results in decreased myofibroblast accumulation and scarring. However, depleting Mϕ during the resolution of fibrosis impairs scar clearance through matrix degradation (129). This study was the first to demonstrate that distinct Mϕ populations exist in the same tissue, playing temporally distinct roles in the repair process. The Mϕ population which orchestrates liver fibrosis regression was shown to be $CD11b^{hi}$, $F4/80^{int}$, $Ly6C^{lo}$ cells expressing MMP9 and MMP12 (171). The molecular mechanism underpinning the role of macrophage scar remodeling may also involve a growth factor, Mfge8 (milk fat globule epidermal growth factor 8), expressed by Mϕ, which directly binds collagen and mediates its uptake. $Mfge8^{-/-}$ mice exhibit a defect in collagen turnover at the wound site that results in poor resolution of fibrosis. This phenotype can be rescued by administering recombinant Mfge8, thus demonstrating the important role of Mϕ Mfge8 in facilitating the removal of accumulated collagen to enable fibrosis resolution (130).

Hyperpigmentation

Dermal melanocytes protect the skin from damage by UV light and determine our skin color. Cutaneous wounding triggers a repair response that frequently leads to alterations in skin pigmentation, in particular wound hyperpigmentation (darkening) (131, 132). Melanocytes and their undifferentiated precursors (melanoblasts) migrate to the wound site after neutrophil and Mϕ recruitment. Myeloid cell depletion in zebrafish results in reduced wound pigmentation through reduced melanocyte and melanoblast recruitment, suggesting that wound hyperpigmentation is driven by the inflammatory response (133). Future studies will hopefully study the signals produced by wound myeloid cells and how these regulate melanocyte recruitment. Identification of the particular signaling mechanisms involved, for example, could enable the development of therapeutics to specifically inhibit myeloid cell-driven melanocyte recruitment to prevent wound hyperpigmentation.

EVIDENCE FOR THE INVOLVEMENT OF MYELOID CELLS IN ABERRANT WOUND REPAIR

The previous sections have touched on the functions of myeloid cells in aberrant healing. We will now look in more detail at potential roles of myeloid cells in chronic wounds and other examples of aberrant healing, including keloids and age-related impaired repair (Fig. 3).

A chronic wound is defined as a cutaneous barrier defect that has remained unhealed for 3 months or more; these wounds are becoming increasingly common in the Western world as the incidence of predisposing comorbidities, including diabetes, aging, obesity, and vascular diseases, rises. These predisposing conditions contribute to the highly heterogeneous nature of chronic wounds, which are clinically categorized into three main groups: diabetic foot ulcers, pressure ulcers, and leg ulcers. Chronic wounds are thought to fail to progress through the typical four phases of wound repair, but instead become stuck in the inflammatory phase and engage in a vicious cycle of inflammation and tissue destruction. These wounds can be extremely painful and debilitating and are the most common cause of limb amputation. Indeed, diabetics have a 25% chance of developing a foot ulcer, and if this ulcer results in amputation, the patient then has only a 50% 5-year survival probability. Chronic wounds, despite their extreme heterogeneity, tend to share certain hallmark cellular and molecular features, including oxidative stress, protease imbalance, biofilm formation or wound infection, edema, and prolonged, persistent leukocyte recruitment (27, 28, 134). Our understanding of chronic wounds lags well behind that of acute wounds, in part due to the lack of an animal model that adequately recapitulates the human pathology. However, it is entirely possible, indeed likely, that myeloid cells play an important detrimental role in chronic wound pathophysiology.

We discussed in previous sections how myeloid cells can have profound effects on the wound microenvironment, in part through the large number of cytokines, growth factors, and enzymes they secrete. In the context of chronic wounds, there is often an imbalance in MMPs and their neutralizing counterparts the tissue inhibitors of metalloproteinases (TIMPs), such that elevated levels of wound MMPs (up to 60 times that found in acute wounds), which are produced for too long or in the wrong places and often coupled with lower TIMP levels, result in off-target destruction of proteins including growth factors, receptors, and ECM components. Mϕ-derived cytokines, including IL-1β and TNF-α, for example, have been shown to promote uncoordinated MMP production. This could impair the repair process by degrading growth factors such as VEGF and PDGF, resulting in impaired wound reepithelialization and contraction by reducing or eliminating the stimulus for stromal cell proliferation. Indeed, chronic wound exudate contains copious proteases, which degrade recombinant growth factors including VEGF and PDGF in *in vitro* assays (135–139). In addition, Mϕ isolated from diabetic human and mouse wounds exhibit sustained NLRP3 inflammasome activity, with corresponding increases in IL-1β expression. Inhibiting inflammasome activity in diabetic mouse wounds using topical pharmacological inhibitors rescues the impaired healing, suggesting that wound Mϕ contribute to poor healing through their elevated inflammasome activity (140, 141).

All wounds are inevitably invaded by a population of microorganisms, but in the majority of acute wounds, the immune system efficiently sterilizes the wound site. Biofilms, however, are present in 60% of chronic wounds and are complex multispecies microbial communities contained within a slimy barrier composed of sugars, proteins, and bacterial DNA (142). Biofilms currently provide an intractable challenge for wound management. Multispecies bacterial biofilms, commonly containing *Pseudomonas aeruginosa*, are antibiotic resistant and virtually inaccessible to immune cells and thus provide a persistent inflammatory stimulus that is likely to be, in part, the reason for the continued presence of immune cells in the wound. In acute wounds, neutrophils decontaminate the wound bed by releasing ROS and nitric oxide to kill pathogens and prevent wound infection. Since biofilms stimulate a chronic inflammatory response, Mϕ and neutrophils are thought to be continually recruited to the wound site in high numbers, releasing ROS and MMPs in an attempt to kill the bacteria and help break attachments between the biofilm and the wound bed. Ultimately this repetitive cycle overloads endogenous antioxidant machinery, resulting in elevated oxidative stress and a profoundly proinflammatory and tissue-destructive state, with damage inflicted to host cell proteins, lipids, and DNA (25, 143–149). Indeed, the first truly chronic wound model in the mouse hinges on locally elevating oxidative stress at the wound site, demonstrating the devastating potential of dysregulation of oxidative balance (150, 151).

The notion that a chronic neutrophil response in wounds may be detrimental to healing is corroborated by neutrophil depletion studies in mice, which have been shown to result in accelerated cutaneous wound reepithelialization. In diabetic mice, which exhibit impaired healing, delayed wound reepithelialization is associated with excessive neutrophil infiltration, which can be rescued by neutrophil depletion approaches (50, 152, 153). Indeed, human chronic wounds exhibit a protracted inflammatory response with elevated and prolonged neutrophil infiltration and neutrophil-derived proteases (154, 155). The neutrophil may also contribute to impaired healing via another of its antimicrobial defense mechanisms: neutrophil extracellular trap (NET) formation (NETosis). Diabetes in mice and in human patients primes neutrophils to undergo NETosis, which can also induce tissue damage. In the context of mouse diabetic wounds, administration of DNase 1, which digests the NETs, has been shown to accelerate wound repair, strongly suggesting that NETs may contribute to impaired healing in diabetics (156).

HARNESSING PRORESOLVING PATHWAYS TO DRIVE REPAIR

Finding the balance between sufficient leukocyte recruitment to eliminate any invading microorganisms versus excessive pathological inflammation is clearly of critical importance in any tissue repair response (14, 34). Endogenous resolution pathways exert anti-inflammatory effects and direct the termination of the inflammatory response by driving proresolving events such as efferocytosis. Genetic deletion of immune cells is evidently not a practical option to control scarring and chronic wounds in the clinic, and until recently it was thought that resolution of the acute inflammatory response was a passive process that simply fizzled out. It is now evident that endogenous anti-inflammatory and proresolving pathways have evolved to actively promote inflammatory resolution without compromising host defense (157–159). These pathways offer enormous potential to therapeutically modulate the resolution phase of inflammation and thus impact on the repair process (158, 160). Indeed, we have previously shown that topical application of the peptide resolution mediator chemerin15 dampens the wound inflammatory response and skews the Mϕ phenotype to one typically associated with repair (reduced inducible NOS and TNF-α; increased arginase-1), resulting in accelerated healing with reduced scarring (161).

Possibly the most extensively characterized resolution mediator, adenosine, is an astonishingly multifunctional molecule exhibiting anti-inflammatory and angiogenic properties (162). Adenosine signals through the A_2 receptor to inhibit neutrophil recruitment to sites of inflammation and repair, impairs ROS generation by neutrophils, and promotes a Mϕ phenotypic switch toward one that is typically associated with repair (163–168). A clinical trial has recently provided the first evidence in humans that administration of proresolving mediators to chronic wounds, in this case the A_{2A} agonist polydeoxyribonucleotide, can promote diabetic foot ulcer healing (169). Resolution mediators thus offer an extremely promising avenue for therapeutic developments to treat aberrant healing.

CONCLUDING DISCUSSION

The two extreme pathologies of tissue repair—chronic wounds that fail to heal, and excessive scarring/fibrosis after repair—represent a major clinical unmet need. The incidence of chronic wounds is rising due to an aging population and the increasing prevalence of diabetes, obesity, and autoimmune diseases, which, along with the continued lack of effective treatments, further contributes to the scope of the problem (10, 11). An improved understanding of the acute-to-chronic wound transition, including the role of myeloid cells in the process, is vital to provide improved treatment options. In addition, there remains the need for therapeutic interventions that promote healing with reduced scar formation in skin and other tissues so that the injured tissue can continue to function. A better understanding of the repair inflammatory response and how it becomes dysregulated will guide us toward therapeutic strategies that could be based on harnessing endogenous resolution pathways.

Citation. Cash JL, Martin P. 2016. Myeloid cells in cutaneous wound repair. Microbiol Spectrum 4(3):MCHD-0017-2015.

References

1. **Eming SA, Martin P, Tomic-Canic M.** 2014. Wound repair and regeneration: mechanisms, signaling, and translation. *Sci Transl Med* **6:**265sr6. doi:10.1126/scitranslmed.3009337.
2. **Jenne CN, Kubes P.** 2015. Platelets in inflammation and infection. *Platelets* **26:**286–292.
3. **Rodero MP, Licata F, Poupel L, Hamon P, Khosrotehrani K, Combadiere C, Boissonnas A.** 2014. In vivo imaging reveals a pioneer wave of monocyte recruitment into mouse skin wounds. *PLoS One* **9:**e108212. doi:10.1371/journal.pone.0108212.
4. **Rappolee DA, Mark D, Banda MJ, Werb Z.** 1988. Wound macrophages express TGF-alpha and other growth factors in vivo: analysis by mRNA phenotyping. *Science* **241:**708–712.
5. **McDonald B, Kubes P.** 2011. Cellular and molecular choreography of neutrophil recruitment to sites of sterile inflammation. *J Mol Med (Berl)* **89:**1079–1088.
6. **McDonald B, Pittman K, Menezes GB, Hirota SA, Slaba I, Waterhouse CC, Beck PL, Muruve DA, Kubes P.** 2010. Intravascular danger signals guide neutrophils to sites of sterile inflammation. *Science* **330:**362–366.
7. **Lämmermann T, Bader BL, Monkley SJ, Worbs T, Wedlich-Söldner R, Hirsch K, Keller M, Förster R, Critchley DR, Fässler R, Sixt M.** 2008. Rapid leukocyte migration by integrin-independent flowing and squeezing. *Nature* **453:**51–55.
8. **Lämmermann T, Afonso PV, Angermann BR, Wang JM, Kastenmüller W, Parent CA, Germain RN.** 2013. Neutrophil swarms require LTB4 and integrins at sites of cell death *in vivo*. *Nature* **498:**371–375.
9. **Murray PJ, Allen JE, Biswas SK, Fisher EA, Gilroy DW, Goerdt S, Gordon S, Hamilton JA, Ivashkiv LB, Lawrence T, Locati M, Mantovani A, Martinez FO, Mege JL, Mosser DM, Natoli G, Saeij JP, Schultze JL, Shirey KA, Sica A, Suttles J, Udalova I, van Ginderachter JA, Vogel SN, Wynn TA.** 2014. Macrophage activation and polarization: nomenclature and experimental guidelines. *Immunity* **41:**14–20.
10. **Martinez FO, Gordon S.** 2014. The M1 and M2 paradigm of macrophage activation: time for reassessment. *F1000Prime Rep* **6:**13. doi:10.12703/P6-13.
11. **Hancock GE, Kaplan G, Cohn ZA.** 1988. Keratinocyte growth regulation by the products of immune cells. *J Exp Med* **168:**1395–1402.
12. **Edwards JP, Zhang X, Mosser DM.** 2009. The expression of heparin-binding epidermal growth factor-like growth factor by regulatory macrophages. *J Immunol* **182:**1929–1939.
13. **Martin P, D'Souza D, Martin J, Grose R, Cooper L, Maki R, McKercher SR.** 2003. Wound healing in the PU.1 null mouse—tissue repair is not dependent on inflammatory cells. *Curr Biol* **13:**1122–1128.
14. **Stramer BM, Mori R, Martin P.** 2007. The inflammation-fibrosis link? A Jekyll and Hyde role for blood cells during wound repair. *J Invest Dermatol* **127:**1009–1017.
15. **Silva MT.** 2010. When two is better than one: macrophages and neutrophils work in concert in innate immunity as complementary and cooperative partners of a myeloid phagocyte system. *J Leukoc Biol* **87:**93–106.
16. **Shaw TJ, Martin P.** 2009. Wound repair at a glance. *J Cell Sci* **122:**3209–3213.
17. **Werner S, Grose R.** 2003. Regulation of wound healing by growth factors and cytokines. *Physiol Rev* **83:**835–870.
18. **Shaw T, Martin P.** 2009. Epigenetic reprogramming during wound healing: loss of polycomb-mediated silencing may enable upregulation of repair genes. *EMBO Rep* **10:**881–886.
19. **Hinz B.** 2007. Formation and function of the myofibroblast during tissue repair. *J Invest Dermatol* **127:**526–537.

20. **Hinz B, Phan SH, Thannickal VJ, Galli A, Bochaton-Piallat ML, Gabbiani G.** 2007. The myofibroblast: one function, multiple origins. *Am J Pathol* **170:**1807–1816.
21. **Fantin A, Vieira JM, Plein A, Denti L, Fruttiger M, Pollard JW, Ruhrberg C.** 2013. NRP1 acts cell autonomously in endothelium to promote tip cell function during sprouting angiogenesis. *Blood* **121:**2352–2362.
22. **Driskell RR, Lichtenberger BM, Hoste E, Kretzschmar K, Simons BD, Charalambous M, Ferron SR, Herault Y, Pavlovic G, Ferguson-Smith AC, Watt FM.** 2013. Distinct fibroblast lineages determine dermal architecture in skin development and repair. *Nature* **504:** 277–281.
23. **Driskell RR, Watt FM.** 2015. Understanding fibroblast heterogeneity in the skin. *Trends Cell Biol* **25:**92–99.
24. **Adams RH, Alitalo K.** 2007. Molecular regulation of angiogenesis and lymphangiogenesis. *Nat Rev Mol Cell Biol* **8:**464–478.
25. **Martin P, Leibovich SJ.** 2005. Inflammatory cells during wound repair: the good, the bad and the ugly. *Trends Cell Biol* **15:**599–607.
26. **Gurtner GC, Werner S, Barrandon Y, Longaker MT.** 2008. Wound repair and regeneration. *Nature* **453:** 314–321.
27. **Sen CK, Gordillo GM, Roy S, Kirsner R, Lambert L, Hunt TK, Gottrup F, Gurtner GC, Longaker MT.** 2009. Human skin wounds: a major and snowballing threat to public health and the economy. *Wound Repair Regen* **17:**763–771.
28. **Nunan R, Harding KG, Martin P.** 2014. Clinical challenges of chronic wounds: searching for an optimal animal model to recapitulate their complexity. *Dis Model Mech* **7:**1205–1213.
29. **Hopkinson-Woolley J, Hughes D, Gordon S, Martin P.** 1994. Macrophage recruitment during limb development and wound healing in the embryonic and foetal mouse. *J Cell Sci* **107:**1159–1167.
30. **Ferguson MW, O'Kane S.** 2004. Scar-free healing: from embryonic mechanisms to adult therapeutic intervention. *Philos Trans R Soc Lond B Biol Sci* **359:**839–850.
31. **Adzick NS, Harrison MR, Glick PL, Beckstead JH, Villa RL, Scheuenstuhl H, Goodson WH III.** 1985. Comparison of fetal, newborn, and adult wound healing by histologic, enzyme-histochemical, and hydroxyproline determinations. *J Pediatr Surg* **20:**315–319.
32. **Lorenz HP, Adzick NS.** 1993. Scarless skin wound repair in the fetus. *West J Med* **159:**350–355.
33. **Eming SA, Hammerschmidt M, Krieg T, Roers A.** 2009. Interrelation of immunity and tissue repair or regeneration. *Semin Cell Dev Biol* **20:**517–527.
34. **Eming SA, Krieg T, Davidson JM.** 2007. Inflammation in wound repair: molecular and cellular mechanisms. *J Invest Dermatol* **127:**514–525.
35. **Chen H, Shi R, Luo B, Yang X, Qiu L, Xiong J, Jiang M, Liu Y, Zhang Z, Wu Y.** 2015. Macrophage peroxisome proliferator-activated receptor γ deficiency delays skin wound healing through impairing apoptotic cell clearance in mice. *Cell Death Dis* **6:**e1597. doi:10.1038/cddis.2014.544.
36. **Lucas T, Waisman A, Ranjan R, Roes J, Krieg T, Müller W, Roers A, Eming SA.** 2010. Differential roles of macrophages in diverse phases of skin repair. *J Immunol* **184:**3964–3977.
37. **Ellett F, Elks PM, Robertson AL, Ogryzko NV, Renshaw SA.** 2015. Defining the phenotype of neutrophils following reverse migration in zebrafish. *J Leukoc Biol* **98:**975–981.
38. **Renshaw SA, Trede NS.** 2012. A model 450 million years in the making: zebrafish and vertebrate immunity. *Dis Model Mech* **5:**38–47.
39. **Evans IR, Rodrigues FS, Armitage EL, Wood W.** 2015. Draper/CED-1 mediates an ancient damage response to control inflammatory blood cell migration in vivo. *Curr Biol* **25:**1606–1612.
40. **McBrearty BA, Clark LD, Zhang XM, Blankenhorn EP, Heber-Katz E.** 1998. Genetic analysis of a mammalian wound-healing trait. *Proc Natl Acad Sci U S A* **95:** 11792–11797.
41. **Seifert AW, Kiama SG, Seifert MG, Goheen JR, Palmer TM, Maden M.** 2012. Skin shedding and tissue regeneration in African spiny mice (*Acomys*). *Nature* **489:** 561–565.
42. **Nathan C.** 2006. Neutrophils and immunity: challenges and opportunities. *Nat Rev Immunol* **6:**173–182.
43. **Lekstrom-Himes JA, Gallin JI.** 2000. Immunodeficiency diseases caused by defects in phagocytes. *N Engl J Med* **343:**1703–1714.
44. **Niethammer P, Grabher C, Look AT, Mitchison TJ.** 2009. A tissue-scale gradient of hydrogen peroxide mediates rapid wound detection in zebrafish. *Nature* **459:**996–999.
45. **Yoo SK, Starnes TW, Deng Q, Huttenlocher A.** 2011. Lyn is a redox sensor that mediates leukocyte wound attraction in vivo. *Nature* **480:**109–112.
46. **Razzell W, Evans IR, Martin P, Wood W.** 2013. Calcium flashes orchestrate the wound inflammatory response through DUOX activation and hydrogen peroxide release. *Curr Biol* **23:**424–429.
47. **Moreira S, Stramer B, Evans I, Wood W, Martin P.** 2010. Prioritization of competing damage and developmental signals by migrating macrophages in the *Drosophila* embryo. *Curr Biol* **20:**464–470.
48. **Beyer TA, Auf dem Keller U, Braun S, Schäfer M, Werner S.** 2007. Roles and mechanisms of action of the Nrf2 transcription factor in skin morphogenesis, wound repair and skin cancer. *Cell Death Differ* **14:**1250–1254.
49. **Szpaderska AM, Zuckerman JD, DiPietro LA.** 2003. Differential injury responses in oral mucosal and cutaneous wounds. *J Dent Res* **82:**621–626.
50. **Dovi JV, Szpaderska AM, DiPietro LA.** 2004. Neutrophil function in the healing wound: adding insult to injury? *Thromb Haemost* **92:**275–280.
51. **Ashcroft GS, Lei K, Jin W, Longenecker G, Kulkarni AB, Greenwell-Wild T, Hale-Donze H, McGrady G, Song XY, Wahl SM.** 2000. Secretory leukocyte protease inhibitor mediates non-redundant functions necessary for normal wound healing. *Nat Med* **6:**1147–1153.

52. **Lucas M, Stuart LM, Savill J, Lacy-Hulbert A.** 2003. Apoptotic cells and innate immune stimuli combine to regulate macrophage cytokine secretion. *J Immunol* **171:**2610–2615.

53. **Savill J, Dransfield I, Gregory C, Haslett C.** 2002. A blast from the past: clearance of apoptotic cells regulates immune responses. *Nat Rev Immunol* **2:**965–975.

54. **Zheng L, He M, Long M, Blomgran R, Stendahl O.** 2004. Pathogen-induced apoptotic neutrophils express heat shock proteins and elicit activation of human macrophages. *J Immunol* **173:**6319–6326.

55. **Simpson DM, Ross R.** 1972. The neutrophilic leukocyte in wound repair a study with antineutrophil serum. *J Clin Invest* **51:**2009–2023.

56. **Dal-Secco D, Wang J, Zeng Z, Kolaczkowska E, Wong CH, Petri B, Ransohoff RM, Charo IF, Jenne CN, Kubes P.** 2015. A dynamic spectrum of monocytes arising from the in situ reprogramming of $CCR2^+$ monocytes at a site of sterile injury. *J Exp Med* **12:**447–456.

57. **Crane MJ, Daley JM, van Houtte O, Brancato SK, Henry WL Jr, Albina JE.** 2014. The monocyte to macrophage transition in the murine sterile wound. *PLoS One* **9:**e86660. doi:10.1371/journal.pone.0086660.

58. **Stein M, Keshav S, Harris N, Gordon S.** 1992. Interleukin 4 potently enhances murine macrophage mannose receptor activity: a marker of alternative immunologic macrophage activation. *J Exp Med* **176:**287–292.

59. **Mills CD, Kincaid K, Alt JM, Heilman MJ, Hill AM.** 2000. M-1/M-2 macrophages and the Th1/Th2 paradigm. *J Immunol* **164:**6166–6173.

60. **Albina JE, Mills CD, Henry WL Jr, Caldwell MD.** 1990. Temporal expression of different pathways of L-arginine metabolism in healing wounds. *J Immunol* **144:**3877–3880.

61. **Daley JM, Brancato SK, Thomay AA, Reichner JS, Albina JE.** 2010. The phenotype of murine wound macrophages. *J Leukoc Biol* **87:**59–67.

62. **Wynn TA, Ramalingam TR.** 2012. Mechanisms of fibrosis: therapeutic translation for fibrotic disease. *Nat Med* **18:**1028–1040.

63. **Ramachandran P, Pellicoro A, Vernon MA, Boulter L, Aucott RL, Ali A, Hartland SN, Snowdon VK, Cappon A, Gordon-Walker TT, Williams MJ, Dunbar DR, Manning JR, van Rooijen N, Fallowfield JA, Forbes SJ, Iredale JP.** 2012. Differential Ly-6C expression identifies the recruited macrophage phenotype, which orchestrates the regression of murine liver fibrosis. *Proc Natl Acad Sci U S A* **109:**E3186–E3195.

64. **Vannella KM, Barron L, Borthwick LA, Kindrachuk KN, Narasimhan PB, Hart KM, Thompson RW, White S, Cheever AW, Ramalingam TR, Wynn TA.** 2014. Incomplete deletion of IL-4Rα by $LysM^{Cre}$ reveals distinct subsets of M2 macrophages controlling inflammation and fibrosis in chronic schistosomiasis. *PLoS Pathog* **10:**e1004372. doi:10.1371/journal.ppat.1004372.

65. **Nair MG, Du Y, Perrigoue JG, Zaph C, Taylor JJ, Goldschmidt M, Swain GP, Yancopoulos GD, Valenzuela DM, Murphy A, Karow M, Stevens S, Pearce EJ, Artis D.** 2009. Alternatively activated macrophage-derived RELM-α is a negative regulator of type 2 inflammation in the lung. *J Exp Med* **206:**937–952.

66. **Mosser DM, Edwards JP.** 2008. Exploring the full spectrum of macrophage activation. *Nat Rev Immunol* **8:** 958–969.

67. **Sindrilaru A, Peters T, Wieschalka S, Baican C, Baican A, Peter H, Hainzl A, Schatz S, Qi Y, Schlecht A, Weiss JM, Wlaschek M, Sunderkotter C, Scharffetter-Kochanek K.** 2011. An unrestrained proinflammatory M1 macrophage population induced by iron impairs wound healing in humans and mice. *J Clin Invest* **121:** 985–997.

68. **Campbell L, Saville CR, Murray PJ, Cruickshank SM, Hardman MJ.** 2013. Local arginase 1 activity is required for cutaneous wound healing. *J Invest Dermatol* **133:**2461–2470.

69. **Stout RD.** 2010. Editorial: macrophage functional phenotypes: no alternatives in dermal wound healing? *J Leukoc Biol* **87:**19–21.

70. **Fleming BD, Mosser DM.** 2011. Regulatory macrophages: setting the threshold for therapy. *Eur J Immunol* **41:**2498–2502.

71. **Pesce JT, Ramalingam TR, Mentink-Kane MM, Wilson MS, El Kasmi KC, Smith AM, Thompson RW, Cheever AW, Murray PJ, Wynn TA.** 2009. Arginase-1-expressing macrophages suppress Th2 cytokine-driven inflammation and fibrosis. *PLoS Pathog* **5:**e1000371. doi: 10.1371/journal.ppat.1000371.

72. **Leibovich SJ, Ross R.** 1975. The role of the macrophage in wound repair. A study with hydrocortisone and antimacrophage serum. *Am J Pathol* **78:**71–100.

73. **Mirza R, DiPietro LA, Koh TJ.** 2009. Selective and specific macrophage ablation is detrimental to wound healing in mice. *Am J Pathol* **175:**2454–2462.

74. **Li S, Li B, Jiang H, Wang Y, Qu M, Duan H, Zhou Q, Shi W.** 2013. Macrophage depletion impairs corneal wound healing after autologous transplantation in mice. *PLoS One* **8:**e61799. doi:10.1371/journal.pone.0061799.

75. **DiPietro LA, Burdick M, Low QE, Kunkel SL, Strieter RM.** 1998. MIP-1α as a critical macrophage chemoattractant in murine wound repair. *J Clin Invest* **101:** 1693–1698.

76. **Moldovan NI, Goldschmidt-Clermont PJ, Parker-Thornburg J, Shapiro SD, Kolattukudy PE.** 2000. Contribution of monocytes/macrophages to compensatory neovascularization: the drilling of metalloelastase-positive tunnels in ischemic myocardium. *Circ Res* **87:** 378–384.

77. **Chen L, Schrementi ME, Ranzer MJ, Wilgus TA, DiPietro LA.** 2014. Blockade of mast cell activation reduces cutaneous scar formation. *PLoS One* **9:**e85226. doi:10.1371/journal.pone.0085226.

78. **Wulff BC, Parent AE, Meleski MA, DiPietro LA, Schrementi ME, Wilgus TA.** 2012. Mast cells contribute to scar formation during fetal wound healing. *J Invest Dermatol* **132:**458–465.

79. **Weller K, Foitzik K, Paus R, Syska W, Maurer M.** 2006. Mast cells are required for normal healing of skin wounds in mice. *FASEB J* **20:**2366–2368.

80. **Willenborg S, Eckes B, Brinckmann J, Krieg T, Waisman A, Hartmann K, Roers A, Eming SA.** 2014. Genetic ablation of mast cells redefines the role of mast cells in skin wound healing and bleomycin-induced fibrosis. *J Invest Dermatol* **134:**2005–2015.

81. **Antsiferova M, Martin C, Huber M, Feyerabend TB, Forster A, Hartmann K, Rodewald HR, Hohl D, Werner S.** 2013. Mast cells are dispensable for normal and activin-promoted wound healing and skin carcinogenesis. *J Immunol* **191:**6147–6155.

82. **Leitch VD, Strudwick XL, Matthaei KI, Dent LA, Cowin AJ.** 2009. IL-5-overexpressing mice exhibit eosinophilia and altered wound healing through mechanisms involving prolonged inflammation. *Immunol Cell Biol* **87:**131–140.

83. **Yang J, Torio A, Donoff RB, Gallagher GT, Egan R, Weller PF, Wong DT.** 1997. Depletion of eosinophil infiltration by anti-IL-5 monoclonal antibody (TRFK 5) accelerates open skin wound epithelial closure. *Am J Pathol* **151:**813–819.

84. **Todd R, Donoff BR, Chiang T, Chou MY, Elovic A, Gallagher GT, Wong DT.** 1991. The eosinophil as a cellular source of transforming growth factor alpha in healing cutaneous wounds. *Am J Pathol* **138:**1307–1313.

85. **Cha D, O'Brien P, O'Toole EA, Woodley DT, Hudson LG.** 1996. Enhanced modulation of keratinocyte motility by transforming growth factor-α (TGF-α) relative to epidermal growth factor (EGF). *J Invest Dermatol* **106:** 590–597.

86. **Huaux F, Liu T, McGarry B, Ullenbruch M, Xing Z, Phan SH.** 2003. Eosinophils and T lymphocytes possess distinct roles in bleomycin-induced lung injury and fibrosis. *J Immunol* **171:**5470–5481.

87. **Hieronymus T, Zenke M, Baek JH, Sere K.** 2015. The clash of Langerhans cell homeostasis in skin: should I stay or should I go? *Semin Cell Dev Biol* **41:**30–38.

88. **Malissen B, Tamoutounour S, Henri S.** 2014. The origins and functions of dendritic cells and macrophages in the skin. *Nat Rev Immunol* **14:**417–428.

89. **Gregorio J, Meller S, Conrad C, Di Nardo A, Homey B, Lauerma A, Arai N, Gallo RL, Digiovanni J, Gilliet M.** 2010. Plasmacytoid dendritic cells sense skin injury and promote wound healing through type I interferons. *J Exp Med* **207:**2921–2930.

90. **Stojadinovic O, Yin N, Lehmann J, Pastar I, Kirsner RS, Tomic-Canic M.** 2013. Increased number of Langerhans cells in the epidermis of diabetic foot ulcers correlates with healing outcome. *Immunol Res* **57:**222–228.

91. **Vandivier RW, Henson PM, Douglas IS.** 2006. Burying the dead: the impact of failed apoptotic cell removal (efferocytosis) on chronic inflammatory lung disease. *Chest* **129:**1673–1682.

92. **Serhan CN, Savill J.** 2005. Resolution of inflammation: the beginning programs the end. *Nat Immunol* **6:**1191–1197.

93. **Haslett C.** 1992. Resolution of acute inflammation and the role of apoptosis in the tissue fate of granulocytes. *Clin Sci (Lond)* **83:**639–648.

94. **Woodfin A, Voisin MB, Beyrau M, Colom B, Caille D, Diapouli FM, Nash GB, Chavakis T, Albelda SM, Rainger GE, Meda P, Imhof BA, Nourshargh S.** 2011. The junctional adhesion molecule JAM-C regulates polarized transendothelial migration of neutrophils *in vivo*. *Nat Immunol* **12:**761–769.

95. **Fadok VA, Chimini G.** 2001. The phagocytosis of apoptotic cells. *Semin Immunol* **13:**365–372.

96. **Fadok VA, Bratton DL, Guthrie L, Henson PM.** 2001. Differential effects of apoptotic versus lysed cells on macrophage production of cytokines: role of proteases. *J Immunol* **166:**6847–6854.

97. **Fadok VA, Bratton DL, Henson PM.** 2001. Phagocyte receptors for apoptotic cells: recognition, uptake, and consequences. *J Clin Invest* **108:**957–962.

98. **Savill J, Fadok V, Henson P, Haslett C.** 1993. Phagocyte recognition of cells undergoing apoptosis. *Immunol Today* **14:**131–136.

99. **Savill JS, Wyllie AH, Henson JE, Walport MJ, Henson PM, Haslett C.** 1989. Macrophage phagocytosis of aging neutrophils in inflammation. Programmed cell death in the neutrophil leads to its recognition by macrophages. *J Clin Invest* **83:**865–875.

100. **Ren Y, Savill J.** 1998. Apoptosis: the importance of being eaten. *Cell Death Differ* **5:**563–568.

101. **Fadok VA, Savill JS, Haslett C, Bratton DL, Doherty DE, Campbell PA, Henson PM.** 1992. Different populations of macrophages use either the vitronectin receptor or the phosphatidylserine receptor to recognize and remove apoptotic cells. *J Immunol* **149:**4029–4035.

102. **Metchnikoff I.** 1908. On the present state of the question of immunity in infectious diseases. *Nobel lecture*, December 11, 1908.

103. **Gordon S.** 2007. The macrophage: past, present and future. *Eur J Immunol* **37**(Suppl 1)**:**S9–S17.

104. **Fadok VA, Bratton DL, Konowal A, Freed PW, Westcott JY, Henson PM.** 1998. Macrophages that have ingested apoptotic cells in vitro inhibit proinflammatory cytokine production through autocrine/paracrine mechanisms involving TGF-β, PGE2, and PAF. *J Clin Invest* **101:**890–898.

105. **Huynh ML, Fadok VA, Henson PM.** 2002. Phosphatidylserine-dependent ingestion of apoptotic cells promotes TGF-β1 secretion and the resolution of inflammation. *J Clin Invest* **109:**41–50.

106. **Neumann J, Sauerzweig S, Rönicke R, Gunzer F, Dinkel K, Ullrich O, Gunzer M, Reymann KG.** 2008. Microglia cells protect neurons by direct engulfment of invading neutrophil granulocytes: a new mechanism of CNS immune privilege. *J Neurosci* **28:**5965–5975.

107. **Fadok VA, McDonald PP, Bratton DL, Henson PM.** 1998. Regulation of macrophage cytokine production by phagocytosis of apoptotic and post-apoptotic cells. *Biochem Soc Trans* **26:**653–656.

108. **Kurosaka K, Watanabe N, Kobayashi Y.** 2002. Potentiation by human serum of anti-inflammatory cytokine production by human macrophages in response to apoptotic cells. *J Leukoc Biol* **71:**950–956.

109. **Scannell M, Flanagan MB, deStefani A, Wynne KJ, Cagney G, Godson C, Maderna P.** 2007. Annexin-1 and peptide derivatives are released by apoptotic cells and stimulate phagocytosis of apoptotic neutrophils by macrophages. *J Immunol* **178:**4595–4605.

110. **Maderna P, Yona S, Perretti M, Godson C.** 2005. Modulation of phagocytosis of apoptotic neutrophils by supernatant from dexamethasone-treated macrophages and annexin-derived peptide Ac2-26. *J Immunol* **174:** 3727–3733.

111. **Baumann I, Kolowos W, Voll RE, Manger B, Gaipl U, Neuhuber WL, Kirchner T, Kalden JR, Herrmann M.** 2002. Impaired uptake of apoptotic cells into tingible body macrophages in germinal centers of patients with systemic lupus erythematosus. *Arthritis Rheum* **46:** 191–201.

112. **Ren Y, Tang J, Mok MY, Chan AW, Wu A, Lau CS.** 2003. Increased apoptotic neutrophils and macrophages and impaired macrophage phagocytic clearance of apoptotic neutrophils in systemic lupus erythematosus. *Arthritis Rheum* **48:**2888–2897.

113. **Wu X, Molinaro C, Johnson N, Casiano CA.** 2001. Secondary necrosis is a source of proteolytically modified forms of specific intracellular autoantigens: implications for systemic autoimmunity. *Arthritis Rheum* **44:**2642–2652.

114. **Schrijvers DM, De Meyer GR, Kockx MM, Herman AG, Martinet W.** 2005. Phagocytosis of apoptotic cells by macrophages is impaired in atherosclerosis. *Arterioscler Thromb Vasc Biol* **25:**1256–1261.

115. **Khanna S, Biswas S, Shang Y, Collard E, Azad A, Kauh C, Bhasker V, Gordillo GM, Sen CK, Roy S.** 2010. Macrophage dysfunction impairs resolution of inflammation in the wounds of diabetic mice. *PLoS One* **5:** e9539. doi:10.1371/journal.pone.0009539.

116. **Kitanaka C, Kuchino Y.** 1999. Caspase-independent programmed cell death with necrotic morphology. *Cell Death Differ* **6:**508–515.

117. **Silva MT, do Vale A, dos Santos NM.** 2008. Secondary necrosis in multicellular animals: an outcome of apoptosis with pathogenic implications. *Apoptosis* **13:**463–482.

118. **Dardenne AD, Wulff BC, Wilgus TA.** 2013. The alarmin HMGB-1 influences healing outcomes in fetal skin wounds. *Wound Repair Regen* **21:**282–291.

119. **Gerhardt H, Golding M, Fruttiger M, Ruhrberg C, Lundkvist A, Abramsson A, Jeltsch M, Mitchell C, Alitalo K, Shima D, Betsholtz C.** 2003. VEGF guides angiogenic sprouting utilizing endothelial tip cell filopodia. *J Cell Biol* **161:**1163–1177.

120. **Ruhrberg C, Gerhardt H, Golding M, Watson R, Ioannidou S, Fujisawa H, Betsholtz C, Shima DT.** 2002. Spatially restricted patterning cues provided by heparin-binding VEGF-A control blood vessel branching morphogenesis. *Genes Dev* **16:**2684–2698.

121. **Fantin A, Vieira JM, Gestri G, Denti L, Schwarz Q, Prykhozhij S, Peri F, Wilson SW, Ruhrberg C.** 2010. Tissue macrophages act as cellular chaperones for vascular anastomosis downstream of VEGF-mediated endothelial tip cell induction. *Blood* **116:**829–840.

122. **Stefater JA III, Rao S, Bezold K, Aplin AC, Nicosia RF, Pollard JW, Ferrara N, Lang RA.** 2013. Macrophage Wnt-Calcineurin-Flt1 signaling regulates mouse wound angiogenesis and repair. *Blood* **121:**2574–2578.

123. **Nissen NN, Polverini PJ, Koch AE, Volin MV, Gamelli RL, DiPietro LA.** 1998. Vascular endothelial growth factor mediates angiogenic activity during the proliferative phase of wound healing. *Am J Pathol* **152:**1445–1452.

124. **Howdieshell TR, McGuire L, Maestas J, McGuire PG.** 2011. Pattern recognition receptor gene expression in ischemia-induced flap revascularization. *Surgery* **150:** 418–428.

125. **Montesano R, Vassalli JD, Baird A, Guillemin R, Orci L.** 1986. Basic fibroblast growth factor induces angiogenesis *in vitro*. *Proc Natl Acad Sci U S A* **83:**7297–7301.

126. **Arnaoutova I, Kleinman HK.** 2010. *In vitro* angiogenesis: endothelial cell tube formation on gelled basement membrane extract. *Nat Protoc* **5:**628–635.

127. **Martin P, Dickson MC, Millan FA, Akhurst RJ.** 1993. Rapid induction and clearance of TGF beta 1 is an early response to wounding in the mouse embryo. *Dev Genet* **14:**225–238.

128. **Ashcroft GS, Yang X, Glick AB, Weinstein M, Letterio JL, Mizel DE, Anzano M, Greenwell-Wild T, Wahl SM, Deng C, Roberts AB.** 1999. Mice lacking Smad3 show accelerated wound healing and an impaired local inflammatory response. *Nat Cell Biol* **1:**260–266.

129. **Duffield JS, Forbes SJ, Constandinou CM, Clay S, Partolina M, Vuthoori S, Wu S, Lang R, Iredale JP.** 2005. Selective depletion of macrophages reveals distinct, opposing roles during liver injury and repair. *J Clin Invest* **115:**56–65.

130. **Atabai K, Jame S, Azhar N, Kuo A, Lam M, McKleroy W, Dehart G, Rahman S, Xia DD, Melton AC, Wolters P, Emson CL, Turner SM, Werb Z, Sheppard D.** 2009. Mfge8 diminishes the severity of tissue fibrosis in mice by binding and targeting collagen for uptake by macrophages. *J Clin Invest* **119:**3713–3722.

131. **Snell RS.** 1963. A study of the melanocytes and melanin in a healing deep wound. *J Anat* **97:**243–253.

132. **Chadwick SL, Yip C, Ferguson MW, Shah M.** 2013. Repigmentation of cutaneous scars depends on original wound type. *J Anat* **223:**74–82.

133. **Levesque M, Feng Y, Jones RA, Martin P.** 2013. Inflammation drives wound hyperpigmentation in zebrafish by recruiting pigment cells to sites of tissue damage. *Dis Model Mech* **6:**508–515.

134. **Reiber GE, Lipsky BA, Gibbons GW.** 1998. The burden of diabetic foot ulcers. *Am J Surg* **176:**5S–10S.

135. **Beidler SK, Douillet CD, Berndt DF, Keagy BA, Rich PB, Marston WA.** 2009. Inflammatory cytokine levels in chronic venous insufficiency ulcer tissue before and after compression therapy. *J Vasc Surg* **49:**1013–1020.

136. **Barrientos S, Stojadinovic O, Golinko MS, Brem H, Tomic-Canic M.** 2008. Growth factors and cytokines in wound healing. *Wound Repair Regen* **16:**585–601.

137. **McCarty SM, Percival SL.** 2013. Proteases and delayed wound healing. *Adv Wound Care (New Rochelle)* **2:** 438–447.

138. **McCarty SM, Cochrane CA, Clegg PD, Percival SL.** 2012. The role of endogenous and exogenous enzymes in chronic wounds: a focus on the implications of aberrant levels of both host and bacterial proteases in wound healing. *Wound Repair Regen* **20:**125–136.

139. **Tarnuzzer RW, Schultz GS.** 1996. Biochemical analysis of acute and chronic wound environments. *Wound Repair Regen* **4:**321–325.

140. **Mirza RE, Fang MM, Ennis WJ, Koh TJ.** 2013. Blocking interleukin-1β induces a healing-associated wound macrophage phenotype and improves healing in type 2 diabetes. *Diabetes* **62:**2579–2587.

141. **Mirza RE, Fang MM, Weinheimer-Haus EM, Ennis WJ, Koh TJ.** 2014. Sustained inflammasome activity in macrophages impairs wound healing in type 2 diabetic humans and mice. *Diabetes* **63:**1103–1114.

142. **Percival SL, Hill KE, Williams DW, Hooper SJ, Thomas DW, Costerton JW.** 2012. A review of the scientific evidence for biofilms in wounds. *Wound Repair Regen* **20:** 647–657.

143. **Schafer M, Werner S.** 2008. Oxidative stress in normal and impaired wound repair. *Pharmacol Res* **58:**165–171.

144. **Pastar I, Nusbaum AG, Gil J, Patel SB, Chen J, Valdes J, Stojadinovic O, Plano LR, Tomic-Canic M, Davis SC.** 2013. Interactions of methicillin resistant *Staphylococcus aureus* USA300 and *Pseudomonas aeruginosa* in polymicrobial wound infection. *PLoS One* **8:**e56846. doi:10.1371/journal.pone.0056846.

145. **Frank DN, Wysocki A, Specht-Glick DD, Rooney A, Feldman RA, St Amand AL, Pace NR, Trent JD.** 2009. Microbial diversity in chronic open wounds. *Wound Repair Regen* **17:**163–172.

146. **Roche ED, Renick PJ, Tetens SP, Ramsay SJ, Daniels EQ, Carson DL.** 2012. Increasing the presence of biofilm and healing delay in a porcine model of MRSA-infected wounds. *Wound Repair Regen* **20:**537–543.

147. **Kirketerp-Moller K, Jensen PO, Fazli M, Madsen KG, Pedersen J, Moser C, Tolker-Nielsen T, Hoiby N, Givskov M, Bjarnsholt T.** 2008. Distribution, organization, and ecology of bacteria in chronic wounds. *J Clin Microbiol* **46:**2717–2722.

148. **Bjarnsholt T, Kirketerp-Moller K, Jensen PO, Madsen KG, Phipps R, Krogfelt K, Hoiby N, Givskov M.** 2008. Why chronic wounds will not heal: a novel hypothesis. *Wound Repair Regen* **16:**2–10.

149. **Mittal M, Siddiqui MR, Tran K, Reddy SP, Malik AB.** 2014. Reactive oxygen species in inflammation and tissue injury. *Antioxid Redox Signal* **20:**1126–1167.

150. **Dhall S, Do D, Garcia M, Wijesinghe DS, Brandon A, Kim J, Sanchez A, Lyubovitsky J, Gallagher S, Nothnagel EA, Chalfant CE, Patel RP, Schiller N, Martins-Green M.** 2014. A novel model of chronic wounds: importance of redox imbalance and biofilm-forming bacteria for establishment of chronicity. *PLoS One* **9:**e109848. doi:10.1371/journal.pone.0109848.

151. **Dhall S, Do DC, Garcia M, Kim J, Mirebrahim SH, Lyubovitsky J, Lonardi S, Nothnagel EA, Schiller N, Martins-Green M.** 2014. Generating and reversing chronic wounds in diabetic mice by manipulating wound redox parameters. *J Diabetes Res* **2014:**562625. doi:10.1155/2014/562625.

152. **Dovi JV, He LK, DiPietro LA.** 2003. Accelerated wound closure in neutrophil-depleted mice. *J Leukoc Biol* **73:**448–455.

153. **Lan CC, Wu CS, Huang SM, Wu IH, Chen GS.** 2013. High-glucose environment enhanced oxidative stress and increased interleukin-8 secretion from keratinocytes: new insights into impaired diabetic wound healing. *Diabetes* **62:**2530–2538.

154. **Yager DR, Nwomeh BC.** 1999. The proteolytic environment of chronic wounds. *Wound Repair Regen* **7:** 433–441.

155. **Rosner K, Ross C, Karlsmark T, Skovgaard GL.** 2001. Role of LFA-1/ICAM-1, CLA/E-selectin and VLA-4/VCAM-1 pathways in recruiting leukocytes to the various regions of the chronic leg ulcer. *Acta Derm Venereol* **81:**334–339.

156. **Wong SL, Demers M, Martinod K, Gallant M, Wang Y, Goldfine AB, Kahn CR, Wagner DD.** 2015. Diabetes primes neutrophils to undergo NETosis, which impairs wound healing. *Nat Med* **21:**815–819.

157. **Serhan CN, Brain SD, Buckley CD, Gilroy DW, Haslett C, O'Neill LA, Perretti M, Rossi AG, Wallace JL.** 2007. Resolution of inflammation: state of the art, definitions and terms. *FASEB J* **21:**325–332.

158. **Serhan CN, Chiang N, Van Dyke TE.** 2008. Resolving inflammation: dual anti-inflammatory and pro-resolution lipid mediators. *Nat Rev Immunol* **8:**349–361.

159. **Serhan CN, Brain SD, Buckley CD, Gilroy DW, Haslett C, O'Neill LA, Perretti M, Rossi AG, Wallace JL.** 2007. Resolution of inflammation: state of the art, definitions and terms. *FASEB J* **21:**325–332.

160. **Lawrence T, Willoughby DA, Gilroy DW.** 2002. Anti-inflammatory lipid mediators and insights into the resolution of inflammation. *Nat Rev Immunol* **2:**787–795.

161. **Cash JL, Bass MD, Campbell J, Barnes M, Kubes P, Martin P.** 2014. Resolution mediator chemerin15 reprograms the wound microenvironment to promote repair and reduce scarring. *Curr Biol* **24:**1406–1414.

162. **Cronstein BN.** 2011. Adenosine receptors and fibrosis: a translational review. *F1000 Biol Rep* **3:**21. doi:10.3410/B3-21.

163. **Leibovich SJ, Chen JF, Pinhal-Enfield G, Belem PC, Elson G, Rosania A, Ramanathan M, Montesinos C, Jacobson M, Schwarzschild MA, Fink JS, Cronstein B.** 2002. Synergistic up-regulation of vascular endothelial growth factor expression in murine macrophages by adenosine A_{2A} receptor agonists and endotoxin. *Am J Pathol* **160:**2231–2244.

164. **Cronstein BN, Kramer SB, Weissmann G, Hirschhorn R.** 1983. Adenosine: a physiological modulator of superoxide anion generation by human neutrophils. *J Exp Med* **158:**1160–1177.

165. **Cronstein BN, Levin RI, Philips M, Hirschhorn R, Abramson SB, Weissmann G.** 1992. Neutrophil adherence to endothelium is enhanced via adenosine A1 receptors and inhibited via adenosine A2 receptors. *J Immunol* **148:**2201–2206.

166. **Cronstein BN, Rosenstein ED, Kramer SB, Weissmann G, Hirschhorn R.** 1985. Adenosine; a physiologic modulator of superoxide anion generation by human neutrophils. Adenosine acts via an A2 receptor on human neutrophils. *J Immunol* **135:**1366–1371.

167. **Macedo L, Pinhal-Enfield G, Alshits V, Elson G, Cronstein BN, Leibovich SJ.** 2007. Wound healing is impaired in MyD88-deficient mice: a role for MyD88 in the regulation of wound healing by adenosine A_{2A} receptors. *Am J Pathol* **171:**1774–1788.

168. **Montesinos MC, Gadangi P, Longaker M, Sung J, Levine J, Nilsen D, Reibman J, Li M, Jiang CK, Hirschhorn R, Recht PA, Ostad E, Levin RI, Cronstein BN.** 1997. Wound healing is accelerated by agonists of adenosine A_2 ($G_{\alpha s}$-linked) receptors. *J Exp Med* **186:**1615–1620.

169. **Squadrito F, Bitto A, Altavilla D, Arcoraci V, De Caridi G, De Feo ME, Corrao S, Pallio G, Sterrantino C, Minutoli L, Saitta A, Vaccaro M, Cucinotta D.** 2014. The effect of PDRN, an adenosine receptor A_{2A} agonist, on the healing of chronic diabetic foot ulcers: results of a clinical trial. *J Clin Endocrinol Metab* **99:**E746–753.

170. **Liu C, Wu C, Yang Q, Gao J, Li L, Yang D, Luo L.** 2016. Macrophages mediate the repair of brain vascular rupture through direct physical adhesion and mechanical traction. *Immunity* **44:**1162–1176.

171. **Ramachandran P, Pellicoro A, Vernon MA, Boulter L, Aucott RL, Ali A, Hartland SN, Snowdon VK, Cappon A, Gordon-Walker TT, Williams MJ, Dunbar DR, Manning JR, van Rooijen N, Fallowfield JA, Forbes SJ, Iredale JP.** 2012. Differential Ly-6C expression identifies the recruited macrophage phenotype, which orchestrates the regression of murine liver fibrosis. *Proc Natl Acad Sci U S A* **109:**E3186–3195.

Recognition, Signaling, and Gene Expression

Myeloid Cells in Health and Disease: A Synthesis
Edited by Siamon Gordon

doi:10.1128/microbiolspec.MCHD-0045-2016

Stylianos Bournazos[1]
Taia T. Wang[1]
Jeffrey V. Ravetch[1]

22 The Role and Function of Fcγ Receptors on Myeloid Cells

A key determinant for the survival of an organism is the ability to recognize and respond to invading pathogens without damaging host tissues. This is accomplished largely by the concerted activity of the innate and adaptive branches of the immune system, which efficiently eliminate invading pathogens and restore tissue homeostasis. An initial step in the generation of robust immune responses is the recognition of pathogens by host cells, triggering subsequent immune cell activation and induction of proinflammatory responses. This initial recognition is facilitated via pathogen-associated molecular patterns (PAMPs) that represent highly conserved molecular structures uniquely found in bacterial, viral, and fungal pathogens but not in host tissues. Such structures include peptidoglycans, zymosan, lipopolysaccharides, flagellin, double-stranded and single-stranded RNA, and CpG-containing DNA (1). So far, an ever increasing number of receptors with the capacity to sense and respond to these PAMPs have been identified and are broadly categorized into distinct receptor families: Toll-like receptors (TLRs), RIG-I (retinoic acid-inducible gene I)-like receptors, NOD-like receptors, and C-type lectin receptors (2–4). Following engagement, these pattern recognition receptors trigger the activation of several inflammatory pathways essential to mediate robust antimicrobial activity and induce sustained immune responses. This central role in immunity for pathogen sensing by innate immune receptors is also reflected by the emergence of pattern recognition receptors early in evolutionary history, as evidenced by the presence of highly conserved gene orthologs in invertebrate species. Additionally, genetic analysis of human *TLR* and *NOD* genes provided evidence for strong positive selection pressure in human populations, and several nonsynonymous polymorphisms influencing receptor activity have been associated with disease susceptibility (1).

Despite the key role for innate immune receptors in mediating pathogen recognition and responses by effector leukocytes, they present limited capacity to recognize pathogens of infinite diversity. Another key mechanism for recognition and clearance of nonself material is accomplished by IgG antibodies, which provide specific recognition of antigens of almost unlimited diversity. Indeed, through the diversification of their variable domains (V_H and V_L), antibodies have the capacity to specifically recognize diverse antigens, providing effective host protection during an immune

[1]Laboratory of Molecular Genetics and Immunology, The Rockefeller University, New York, NY 10065.

response. Contrary to the antigen-binding Fab domain that exhibits astonishing variability, antibodies also comprise a relatively constant domain, the Fc domain. Recognition and binding of antibodies to the surface of the leukocytes is mediated through interactions of their Fc domains with specialized receptors, Fc receptors, expressed by several types of circulating and tissue-resident leukocytes (5). By directly linking molecules of the adaptive immunity with innate leukocytes, Fc receptors represent an important component that links both branches of immunity, enabling innate immune cells to specifically recognize and respond to antigens of unlimited diversity.

While traditionally termed the "constant domain/region" of the antibody molecule, the Fc domain is, in fact, heterogeneous both in primary amino acid sequence (IgG subclass) and in the composition of the Fc-associated glycan (5–7). These two determinants regulate the structure and conformational flexibility of the IgG Fc domain and, in turn, determine interactions with various type I and type II Fc receptors (FcγRs). Indeed, recent crystallographic studies support the existence of two main conformational states for the Fc domain: an "open" and a "closed" state that are determined by the Fc-associated glycan structure; a highly conserved glycan site present in all human IgG subclasses and among many mammalian species (8, 9). In view of the two conformational states of the Fc domain, FcγRs can be categorized into type I and type II receptors, based on their capacity to interact with the "open" or the "closed" Fc domain conformation, respectively (8, 9). Engagement of type I and type II FcγRs by the Fc domain is a tightly regulated process that is determined primarily by the conformational flexibility of the Fc domain and results in the induction of pleiotropic activities by effector leukocytes (5, 10).

IgG Fc DOMAIN HETEROGENEITY AND STRUCTURAL FLEXIBILITY

The highly flexible structure of the Fc region is indicative of the unique structural organization of its different domains. In particular, the Fc region is composed of the two constant domains (C_H2 and C_H3) of the two heavy chains that form homodimers through tight association of the two C_H3 domains at the C-terminal-proximal region of the IgG as well as the presence of disulfide bonds in the C_H2-proximal hinge region (11). This results in a characteristic horseshoe-like conformation, with the two C_H2 domains forming a hydrophobic cleft, where the central *N*-linked glycan structure is localized. This Fc-associated glycan is conjugated to the amino acid backbone of the C_H2 domain at position Asn297 and is required to maintain the Fc domain in a structural conformation permissive for interactions with FcγRs (12, 13) (Fig. 1A). Indeed, genetic deletion of the Asn297 site or enzymatic removal of the Fc-associated glycan abrogates the capacity of the Fc domain to interact with FcγRs (14–17). In addition, an ever increasing body of evidence strongly suggests a crucial role for the Fc-associated glycan in the regulation of Fc domain interactions with type I and type II FcγRs through modulation of the Fc domain flexibility (15, 16).

The Fc-associated glycan structure consists of a core heptasaccharide structure that is attached to the C_H2 domains of each of the two heavy chains that comprise the Fc domain (12, 15) (Fig. 1A). Previous analyses of the Fc glycan composition revealed substantial heterogeneity at steady-state conditions, due to the selective addition of fucose, galactose, *N*-acetylglucosamine, and sialic acid residues to the core glycan structure (5, 6, 18). Such heterogeneity in the glycan structure and composition finely tunes the affinity of the Fc domain for binding to the different classes of type I and type II FcγRs (6, 14, 19) (Fig. 1B). For example, the branching fucose residue regulates the Fc domain affinity specifically for the activating type I FcγR, FcγRIIIa, without any impact on the capacity of the Fc domain to interact with other classes of type I FcγRs (20, 21). Based on the selectively enhanced binding affinity for FcγRIIIa, Fc glycoengineering has been successfully used as a strategy to generate antibodies with improved *in vivo* efficacy, as afucosylated IgG glycovariants exhibited enhanced Fc effector activity compared to their fucosylated counterparts (22–26).

In contrast, the presence of terminal sialic acid residues is associated with reduced binding to type I FcγRs and preferential engagement of type II FcγRs (18, 27, 28), an effect attributed to the induction of a conformational change of the C_H2 domains upon sialylation that affects type I and type II FcγR binding (5, 8, 9). Indeed, sialylation of the Fc-associated glycan exposes a region at the C_H2-C_H3 interface that serves as the binding site for type II FcγRs (8, 9). This conformational change also results in the obstruction of the type I FcγR-binding site at the hinge-proximal region of the C_H2 domain, suggesting that Fc domain glycosylation modulates the capacity of the Fc domain to adopt two mutually exclusive conformations: an "open" that enables for type I but not type II FcγR binding, and a "closed" that is induced upon sialylation and preferentially engages type II but not type I FcγRs (9).

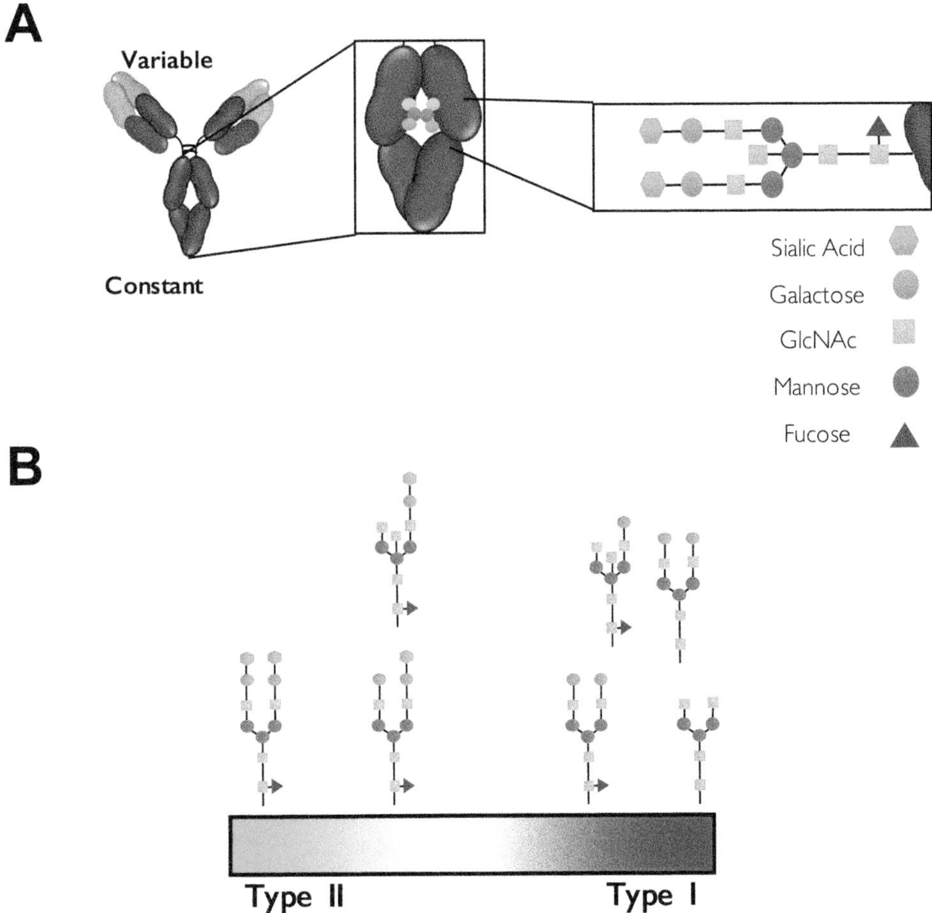

Figure 1 Structure and composition of the Fc-associated N-linked glycan regulates binding to type I and type II FcγRs. (**A**) An N-linked glycan structure is attached at the C_H2 domain of each of the two heavy chains of IgG and consists of a core heptasaccharide structure that is composed of fucose, galactose, *N*-acetylglucosamine, and sialic acid residues. (**B**) The structure and composition of this Fc-associated glycan determine binding specificity and affinity of the IgG Fc domain for different types of FcγRs. Several glycoform combinations exist with distinct binding capacity to interact with type I and type II FcγRs.

FcγRs: TYPES, FUNCTION, AND DOWNSTREAM SIGNALING

The conformational state of the Fc domain regulates the interactions with two distinct types of FcγRs—type I and type II—that differ in terms of their structural domain organization as well as the stoichiometry by which they interact with the Fc domain (Fig. 2A). Type I FcγRs belong to the immunoglobulin (Ig) receptor superfamily, and their extracellular, IgG-binding region consists of two or three Ig-like domains (29, 30). Type I FcγR-Fc interactions are mediated through 1:1 binding of the FcγR loop region between the Ig-like domains with the hinge-proximal C_H2 domain of the IgG Fc (31–33). In contrast, all type II FcγRs are members of the C-type lectin receptors and recognize the Fc domain only at the closed conformation at a 1:2 (Fc:FcγR) stoichiometry through interactions with the C_H2-C_H3 domain interface that is exposed following acquisition of the closed conformation state induced upon sialylation (5, 8, 9).

In humans, type I FcγRs are encoded by eight different genes, each with multiple transcriptional isoforms, located at a locus on the long arm of chromosome 1 (1q21-23). Genetic analysis suggested that the majority of the FcγR genes (with the exception of the genes encoding FcγRI) emerged throughout evolutionary history through sequential gene duplication events of the ancestral FcγR gene locus, giving rise to the unique organization of the human IgG locus, which comprises several FcγR genes that share a high degree of

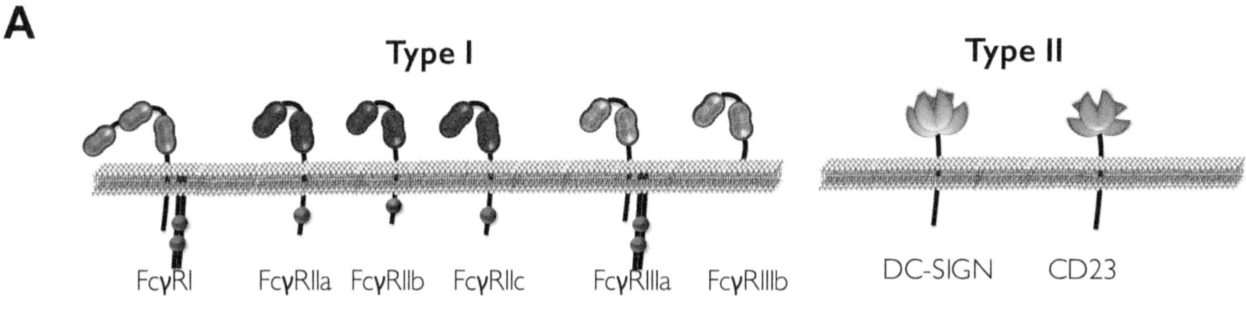

B

	Type I						Type II	
	FcγRI	FcγRIIa	FcγRIIb	FcγRIIc	FcγRIIIa	FcγRIIIb	DC-SIGN	CD23
Neutrophils	#	+	+	-	-	+	-	#
Eosinophils	#	+	+	-	-	#	-	#
Basophils	#	+	+	-	-	+/-	-	#
Monocytes	+	+	+	-	+/-	-	-	#
Macrophages	+/-	+	+	-	+/-	-	+/-	#
Dendritic cells	-/#	+	+	-	-/#	-	+	-
Platelets	-	+	-	-	-	-	-	-

\+ Constitutive expression

\- No expression

\# Inducible expression

Figure 2 Overview of type I and type II FcγR structure and expression pattern in myeloid cell populations. (**A**) Structural characteristics of type I and type II FcγRs. Type I FcγRs belong to the Ig receptor superfamily and are composed of two or three extracellular Ig-like domains that interact with the IgG Fc domain at the hinge-proximal region of the C_H2 domain. Type II FcγRs comprise DC-SIGN and CD23, both C-type lectin receptors, which bind sialylated Fc IgGs. (**B**) Expression pattern of type I and type II FcγRs in myeloid leukocyte populations. The expression of several FcγRs is determined by cell differentiation status and is regulated by cytokines like IL-4 and IFN-γ.

homology (34, 35). Type I FcγRs are further subdivided into high- and low-affinity receptors, based on their capacity to interact with monomeric or multimeric IgG, respectively. FcγRI represents the high-affinity FcγR and is capable of interacting with monomeric IgG with relatively high affinity (K_d [dissociation constant], 10^{-9} to 10^{-10} M). Three genes coding for FcγRI have been described: *FCGR1A*, *FCGR1B*, and *FCGR1C*; however, only *FCGR1A* encodes the prototypic high-affinity receptor, FcγRI, which is capable of high-affinity IgG binding (36). *FCGR1B* and *FCGR1C* possibly represent pseudogenes that arose from duplication of the *FCGR1A* gene and express truncated or soluble forms of FcγRI, with unknown or poorly characterized function (36, 37). In particular, the FcγRIb1 and FcγRIc transcripts contain premature stop codons within the extracellular domain of FcγRI, possibly representing soluble forms of the FcγRI receptor (36, 37). In contrast, the FcγRIb2 isoform is expressed as an intracellular protein, retained predominantly within the endoplasmic reticulum; however, its function remains unknown (37, 38). The increased affinity of FcγRI for IgG is attributed to its unique structure, as contrary to the low-affinity FcγRs (like FcγRII and FcγRIII), the α ligand-binding chain of FcγRI comprises three V-type, Ig-like domains that greatly stabilize Fc-FcγR interactions (39). The FcγRI α chain associates with a disulfide-bonded dimer of the Fc receptor γ chain (encoded by *FCER1G*), a signal-transducing polypeptide initially described as a component of the high-affinity IgE receptor FcεRI (40–42). Association of the FcγRI α chain with the FcR γ chain is necessary for receptor expression and signaling following receptor engagement (41, 43). Indeed, the FcR γ chain carries activating signaling motifs (immunoreceptor tyrosine-based activation motifs [ITAMs]), which serve as the docking site for Syk family kinases and mediate signal transduction following IgG-FcγRI interactions (40, 44). FcγRI is constitutively expressed by monocytes and macrophages, as well as by many myeloid progenitor cells. FcγRI expression can also be induced in other myeloid cell types, like neutrophils and eosinophils, following stimulation with primarily gamma interferon (IFN-γ) and to a lesser extent by granulocyte colony-stimulating factor, IFN-α, and interleukin-12 (IL-12) (45, 46) (Fig. 2B).

Apart from the high-affinity FcγRI, FcγRs (like FcγRII and FcγRIII) display low affinity for monomeric IgG. Instead of engaging monomeric IgG, low-affinity FcγRs are only capable of binding to aggregated IgG through multiple multivalent, low-affinity, high-avidity interactions (33). This is particularly important, as low-affinity FcγRs can be engaged only by antibody-antigen complexes during an immune response but not at steady-state conditions by monomeric IgG in the absence of antigen. This ensures that engagement and cross-linking of low-affinity FcγR occurs only during an immune response, thereby preventing inappropriate or excessive FcγR-mediated signaling at physiological conditions. All genes encoding the low-affinity FcγRs are clustered into a single FcγR locus, located at chromosome 1, and represent a series of multiple gene duplication and recombination events, followed by gain-of-function mutations of the ancestral FcγR locus, which can be traced back early in mammalian evolution as well as in marsupials (35). In humans, five low-affinity FcγRs have been described, along with several genetic variants (either in the form of nonsynonymous substitutions or copy number variants) that impact FcγR expression and ligand binding affinity, thereby influencing disease susceptibility and severity for a number of chronic inflammatory and autoimmune pathologies (29). A common feature of all low-affinity FcγRs is the presence of two V-type, Ig-like domains that comprise the extracellular, ligand-binding region (Fig. 2A). A series of crystallographic, molecular modeling and mutational analyses have previously identified the ligand-binding interface for each low-affinity FcγR, indicating that the IgG-binding region is localized in the second, membrane-proximal Ig-like domain, which is normally accessible for IgG interactions due to the horizontal orientation of the extracellular domain of the receptor (31–33, 47–50).

Despite the similarities between the low-affinity FcγRs in relation to their extracellular domains and binding interface, each FcγR exhibits unique characteristics in terms of their intracellular structure and signaling activities. FcγRIIa, FcγRIIb, and FcγRIIc (encoded by *FCGR2A*, *FCGR2B*, and *FCGR2C*, respectively) share a characteristic structure, which is the presence of functional signaling motifs in their intracellular domains. FcγRIIa, -b, and -c are expressed as single-chain (α chain) membrane receptors that comprise both the extracellular, IgG-binding region as well as an intracellular signaling domain, without requiring association with the FcR γ chain for signaling or expression (30). The intracellular motifs of FcγRII include either an ITAM for FcγRIIa and FcγRIIc or an ITIM (immunoreceptor tyrosine-based inhibitory motif) for FcγRIIb (10). FcγRIIa is expressed by diverse cell types, including predominantly cells of myeloid origin, like polymorphonuclear leukocytes, macrophages, dendritic cells (DCs), monocytes, and platelets (10). Additionally, FcγRIIa expression has been reported in certain types of endothelial and epithelial cells, as well as in subsets of memory T cells (memory αβ T cells). FcγRIIb, the sole FcγR capable of transducing inhibitory signals upon cross-linking, exhibits two main splicing isoforms: b1 and b2. FcγRIIb1 represents the full-length transcript, which encodes a signal sequence that inhibits receptor internalization (51–54). In contrast, FcγRIIb2 is generated by skipping the exon that codes for this signal sequence, resulting in a variant that is capable of receptor internalization following IgG cross-linking (55). In addition to the differential internalization capacity, FcγRIIb1 and FcγRIIb2 also exhibit characteristic expression patterns. More specifically, FcγRIIb1, the full-length variant that is incapable of internalization, is predominantly expressed by cells of the lymphoid lineage, like B cells, whereas FcγRIIb2 expression is restricted to phagocytic myeloid cells, like monocytes, neutrophils, macrophages, DCs, and eosinophils (52) (Fig. 2B). FcγRIIc is encoded by the *FCGR2C* gene and shares a high degree of similarity with both FcγRIIa and FcγRIIb, probably reflecting a gene recombination event between *FCGR2A* and *FCGR2B* late in human evolution (35). Indeed, FcγRIIc represents a chimeric receptor comprising the extracellular, ligand-binding region of FcγRIIb and the intracellular, ITAM-containing domain of FcγRIIa. Its expression is mainly restricted to NK cells (56); however, for the majority of human populations (70 to 90%), FcγRIIc expression is absent due to the presence of a premature stop codon at exon 3 (56, 57).

Similar to FcγRII receptors, FcγRIIIa and FcγRIIIb are characterized by relatively high sequence similarity, indicating a common origin for *FCGR3A* and *FCGR3B* genes, which encode FcγRIIIa and FcγRIIIb, respectively. Indeed, previous genetic analyses and comparison of the FcγR locus among the various primate species revealed that the *FCGR3B* gene emerged relatively late in evolutionary history as a result of gene duplication of *FCGR3A* followed by a point mutation at the extracellular, membrane-proximal region of the receptor (35, 58). This key point mutation created a glycophosphatidylinositol (GPI) anchor signal sequence, resulting in the processing of FcγRIIIb as a GPI-anchored protein (59). This difference accounts for the distinct structural differences between FcγRIIIa and FcγRIIIb (Fig. 2A). Whereas FcγRIIIa is a transmembrane protein that requires the association with the FcR γ chain for expression and signaling (44, 60), FcγRIIIb is posttranslationally processed as a GPI-anchored protein lacking intracellular signaling domains (61). This difference also influences the receptor immunostimulatory activity following engagement. In contrast to FcγRIIIa, which is capable of transducing potent activating signals following receptor cross-linking through the FcR γ chain, FcγRIIIb lacks robust signaling capacity (59, 62). FcγRIIIb often relies on other receptors (like FcγRIIa) or accessory chains (like the ζ chain or FcR γ chain) for signaling activity (42). The differences between FcγRIIIa and FcγRIIIb are not limited to their structure and signaling activity, but also extend to their expression pattern and distribution. FcγRIIIb expression is restricted to neutrophils, which constitutively express high levels of FcγRIIIb on their surface (61, 63). Other granulocyte subsets, like eosinophils, also express FcγRIIIb, but only following induction with IFN-γ (5, 64). In contrast, FcγRIIIa is widely expressed by several leukocyte cell types, including macrophages, NK cells, and a subset of monocytes ($CD14^{int}CD16^{hi}$; patrolling monocytes) (5, 10, 65) (Fig. 2B). FcγRIIIa expression levels are also greatly variable among the different tissue-resident macrophage and DC subsets (10).

Since almost all FcγRs are capable of transducing intracellular signals following cross-linking by IgG complexes, FcγRs are often divided into activating or inhibitory, based on their ability to transduce immunostimulatory or inhibitory signals (16). Common to all myeloid cell types is the concurrent expression of both activating and inhibitory FcγRs on their surface with competing signaling activity. Differential engagement of activating or inhibitory FcγRs could therefore influence the outcome of IgG-mediated inflammation, a concept that has been experimentally evaluated in several *in vivo* models of antibody-mediated cellular cytotoxicity (16, 64, 66, 67). Activating FcγRs have the capacity to initiate a range of cellular activation processes through their activating signaling motifs (ITAMs), in processes resembling those described for other antigen receptors, like the T- and B-cell receptors (68). Most Fc receptors consist of heterodimers of the α subunit, which constitutes the ligand-binding domain, along with the FcR γ chain (either one or two), which contains the ITAM motifs necessary for signaling (40, 61, 69) (Fig. 2A). The FcR γ chain (encoded by *FCER1G*) is structurally and functionally related to the ζ chain of the T-cell receptor, and both are located at the same locus on chromosome 1. The FcR γ chain normally forms γ-γ homodimers that associate with the α subunit through transmembrane domain interactions of an aspartic acid residue of the γ chain with basic residues of the α subunit (51, 70). Apart from γ-γ homodimers, heterodimers between the FcR γ chain and ζ chain have been reported to associate with the FcR α subunit *in vitro*; however, it is unclear whether this association occurs naturally (71).

In contrast to the FcγRs that require the association with the FcR γ chain for expression and signaling, FcγRIIa and FcγRIIc comprise only a single α subunit that not only has the capacity to interact with the Fc domain of IgG but also possesses an ITAM signaling motif within its intracellular domain. These receptors, which are uniquely found in humans, have the capacity to transduce activating signals following receptor cross-linking, in a process similar to that in FcγRs that require association with the FcR γ chain, like FcγRIIIa. Several biochemical and biophysical studies have elucidated the precise signaling pathways that are activated following binding of IgG complexes to activating FcγRs (10, 72). Since signaling of activating FcγRs is mediated through motifs (ITAMs) or accessory signaling proteins (like the FcR γ chain) that are also found in the high-affinity IgE receptor, FcεRI, both Fcγ and Fcε receptors share a number of common features related to their signaling activity following receptor cross-linking. Indeed, initial studies on the high-affinity FcεRI receptor have provided useful insights into the mechanisms by which activating FcγRs transduce intracellular signals (73–75). Apart from the difference in their capacity to interact with monomeric IgGs, high- and low-affinity FcγRs also differ in relation to the sequence of events that are required for the initiation of downstream signaling. For example, high-affinity receptors require binding of monomeric antibodies that are subsequently aggregated through the recognition and binding of multivalent antigens to their Fab domains (73, 75). In contrast, low-affinity FcγRs can only engage antibody-antigen complexes, but not monomeric antibodies. Binding of such complexes is accomplished through multiple low-affinity, high-avidity interactions that facilitate receptor clustering and subsequent initiation of intrinsic signals (76, 77). Despite the differences between high- and low-affinity FcγRs in the events preceding receptor cross-linking, all activating FcγRs share a common pattern of signal cascade activation following cross-linking. In particular, upon receptor aggregation, two or more ITAM domains present in the FcR γ chain or in the intracellular region of the α subunit of FcγRIIa and FcγRIIc become cross-phosphorylated by Src and Syk family kinases, including Lyn, Lck, Hck, and Fgr (62, 78–83). In addition to these kinases, phosphorylation of additional proteins including FAK, ZAP-70, and ζ-chain subunits have been reported following receptor aggregation (81, 83, 84). Initial activation of Syk and Src family kinases is accompanied by subsequent activation of downstream signaling pathways, including the activation of phospholipase C-γ (PLC-γ), which in turn generates metabolites that lead to the activation of protein kinase C (PKC) and inositol triphosphate, triggering a rapid increase in intracellular Ca^{2+} levels through mobilization of the endoplasmic reticulum Ca^{2+} pool (85, 86). Elevated intracellular Ca^{2+} levels also activate a number of Ca^{2+}-regulated signaling proteins, including PKC (87, 88). Finally, a number of late signaling pathways become activated following FcγR cross-linking, including the Ras pathways, kinases of the MEK and mitogen-activated protein kinase family, as well as transcriptional induction of several cytokine and chemokine genes along with genes involved in cell survival, differentiation, and motility through the activation of transcription factors, such as NF-κB and NFAT (nuclear factor of activated T cells) (85, 89, 90).

Contrary to the activating, ITAM-bearing FcγRs, two FcγRs, FcγRIIIb and FcγRIIb, have been described that lack intrinsic signaling capacity and initiate immunosuppressive signals, respectively. As mentioned above, FcγRIIIb is uniquely found in humans and exhibits structural features characterized by the absence of intracellular domains necessary for signaling. Although FcγRIIIb is incapable of transducing intracellular signals following cross-linking by IgG complexes, FcγRIIIb-Fc interactions have been previously shown to result in the induction of cellular activation, mainly through the synergistic activity of FcγRIIIb with other receptors, like FcγRIIa and complement receptors, expressed on the surface of neutrophils (63, 91).

FcγRIIb represents the sole FcγR with inhibitory activity that has the capacity to transduce intracellular signals that directly antagonize the immunostimulatory signals of activating FcγRs. In contrast to the other FcγRII receptors, FcγRIIb comprises an ITIM domain at its intracellular region that mediates inhibitory cellular signaling. Several studies, mainly on B cells, have previously dissected the exact molecular mechanisms of FcγRIIb-mediated cellular inhibition. In particular, coaggregation of FcγRIIb with activating receptors, like the B-cell receptor or activating FcγRs, results in phosphorylation of FcγRIIb ITIM domains, which in turn recruits SHIP (Src homology 2 domain-containing inositol 5′-phosphatase) and SHP-2 (Src homology-2 domain-containing phosphatase) phosphatases (68, 92, 93). In the case of FcγRIIb-activating FcγR cross-linking, recruitment of these phosphatases to the FcγRIIb ITIM domains prevents phosphorylation of Syk and Src family kinases, thereby counterbalancing any activating signals originating from activating FcγRs (94, 95). Similarly, on B cells, recruited SHIP and SHP-2 phosphatases interfere with B-cell receptor signaling by hydrolyzing phosphoinositide intermediates, such as phosphatidylinositol 3,4,5-triphosphate, thereby preventing the recruitment of PH domain-containing kinases, like PLC-γ and Bruton's tyrosine kinase, and the activation of downstream signaling pathways (68, 93). Since FcγRIIb is the only FcγR with inhibitory activity, it plays a central role in regulating FcγR-mediated inflammation and B-cell receptor activity. Indeed, FcγRIIb regulates key processes in B cells related to cell activation, selection, and antibody production (95, 96), as well as in myeloid cells, like macrophages, influencing the outcome of IgG-mediated inflammation. Several examples from *Fcgr2b* knockout mice exist that highlight the importance of FcγRIIb in regulating B-cell activation, antibody affinity selection, and antibody serum levels (96, 97). Likewise, genetic deletion of *Fcgr2b* results in enhanced proinflammatory activity in macrophages in murine models of immune complex-mediated alveolitis and collagen-induced arthritis (66, 98–100). Additionally, studies in humans revealed that polymorphisms in the promoter region or transmembrane domain of FcγRIIb that influence receptor expression or activity, respectively, are associated with susceptibility to autoimmune disorders, further highlighting the significance of FcγRIIb in maintaining peripheral tolerance and regulating IgG-mediated inflammatory processes (29, 101–104).

As mentioned above, type II FcγRs belong to the C-type lectin family of receptors that exhibit ligand binding specificity exclusively for the "closed" conformation of the IgG Fc domain (5, 8, 9). Two main type II FcγRs have been identified so far: DC-SIGN and CD23, with distinct biological activity *in vivo* (8, 9, 28, 105) (Fig. 2A). Although sialylated IgG Fc binding specificity has been recently shown for other C-type lectin receptors, such as CD22 and DCIR (DC inhibitory receptor), their precise biological significance has not been fully investigated (106–108). DC-SIGN is encoded by the *CD209* gene, which is mapped at the same locus with the gene that codes for CD23 (*FCER2*), indicative of a common functional and structural origin. Indeed, both CD23 and DC-SIGN are heavily glycosylated membrane proteins that belong to the C-type lectin receptor superfamily. Since the capacity of type II FcγRs to engage IgG has been only recently described, the precise signaling pathways have not been defined as extensively as for type I receptors. However, a number of recent studies have elucidated the mechanisms that regulate IgG binding to type II FcγRs, as well as the downstream biological consequences following receptor engagement (9, 27, 105, 109, 110).

DC-SIGN has the capacity to engage diverse carbohydrate ligands and heavily glycosylated glycoproteins, including mannose-rich glycan structures; and pathogen surface glycoproteins, such as gp160, the HIV-1 envelope glycoprotein (28, 111). Binding of these ligands to DC-SIGN is accomplished predominantly through carbohydrate-mediated interactions. In contrast, sialylated IgG Fc–DC-SIGN binding is mediated through protein interactions with the amino acid backbone of the Fc domain at the C_H2-C_H3 interface, a region that becomes exposed following conformational alterations induced by the Fc-associated glycan structure (8, 9, 28). Since DC-SIGN has the capacity to interact with diverse ligands, downstream signaling cascades following receptor engagement depend largely on the nature of the ligand. In the case of DC-SIGN engagement by sialylated IgG Fc, expression of IL-33 is induced following receptor engagement on regulatory macrophages. This effect triggers a robust Th2 polarizing response that induces potent regulatory T-cell responses, effectively suppressing Th1- and Th17-mediated inflammation (109, 110). Likewise, IL-33 triggers the production and release of IL-4 by basophils that results in FcγRIIb upregulation on effector myeloid cells, including monocytes and macrophages, at sites of inflammation (109).

CD23 was initially described as a low-affinity receptor for IgE; however, recent studies showed that CD23 exhibits ligand-binding activity for sialylated IgG in addition to IgE (9, 105). Molecular modeling of the CD23-IgE interaction and the potential CD23-sialylated IgG Fc complex first suggested that CD23 might have the capacity for binding to both ligands (9). IgE interactions with CD23 are attributed largely to the intrinsic flexibility of the Cε3 domain (functionally related to the C_H2 domain of IgG). In an analogous model, sialylation of the Fc-associated glycan increases the flexibility of the Cγ2 domain, thereby conferring flexibility to the Fc and enabling Fc interactions with CD23 (9, 112, 113). In terms of expression, CD23 exhibits two splice variants, CD23a and CD23b, which differ both in relation to their expression pattern and responsiveness to IL-4 (114). CD23a is constitutively expressed by B cells, whereas CD23b expression is induced only following IL-4 treatment and is limited to myeloid cells, such as monocytes, macrophages, and granulocytes, as well as in certain T-cell subsets (114). Apart from sialylated IgG, CD23 can also interact with many other ligands, including CD21, CD11c, and CD11b; however, the functional consequences of these interactions have not been characterized (115).

CELLULAR EFFECTS OF FcγR SIGNALING ON MYELOID CELLS

FcγR engagement and the associated signaling pathways that are activated upon receptor cross-linking modulate the function of diverse immunoregulatory pathways that shape immune responses, regulate cellular activation, and modulate IgG-mediated inflammation (10) (Fig. 3). B cells, a lymphoid cell type expressing only one class of each FcγR type (the type I FcγRIIb and the type II CD23), represent a bona fide example of complex FcγR-mediated signaling cellular regulation. In particular, CD23-mediated signaling on B cells following engagement with sialylated IgG complexes modulates the expression of FcγRIIb in these cells (105). Increased FcγRIIb expression on B cells drives the generation of high-affinity antibody responses, through the regulation of B-cell receptor signaling and modulation of B-cell selection (105). Likewise, FcγRIIb on antibody-producing plasma cells regulates cell survival and antibody expression, a key process during the resolution phase of the humoral immune response (93, 97).

In contrast to lymphoid cells, cells of the myeloid lineage, including granulocytes, macrophages, and DCs, coexpress several classes of FcγRs on their surface during all stages of their differentiation. It is therefore anticipated that FcγR engagement in these cells could have diverse immunomodulatory consequences that regulate key processes during an immune response, affecting the outcome of IgG-mediated signaling. Indeed, Fc-FcγR interactions on myeloid cells mediate diverse effector functions including cellular activation, cytotoxicity, and phagocytosis of IgG-opsonized targets (5, 10). Additionally, FcγR-mediated uptake of IgG complexes by phagocytes influences antigen processing and presentation and regulates the differentiation and maturation of antigen-presenting cells, thereby modulating T-cell responses (10, 67). Since downstream FcγR

Granulocytes Cell activation
ROS production
Degranulation
Phagocytosis
Chemokine & cytokine expression

Macrophages Macrophage polarization
Cell survival & differentiation
Phagocytosis
Antigen processing & presentation
Chemokine & cytokine expression

Platelets Degranulation
Aggregation
Thrombogenesis

Dendritic Cells Dendritic cell maturation
Upregulation of co-stimulatory molecules
Antigen uptake & presentation
Enhanced T cell activation
Cytokine & chemokine responses

Monocytes Cell activation
Phagocytosis
Cell survival
Chemokine & cytokine expression

Figure 3 Effector functions and processes that are regulated by Fc-FcγR interactions. Engagement of type I and type II FcγRs by the Fc domain of IgG initiates signaling cascades with diverse proinflammatory, anti-inflammatory, and immunomodulatory consequences on myeloid cells, including granulocytes, monocytes, macrophages, DCs, and platelets. ROS, reactive oxygen species.

signaling leads to the transcriptional activation of several cytokine and chemokine genes, diverse effects from such responses impact cell mobilization, migration, differentiation, and survival. In the next sections, the cellular and biological effects of FcγR signaling initiated upon interactions with the Fc domain of IgG are discussed in detail, focusing on the significance of FcγR pathways in the modulation of myeloid cell functional activity (Fig. 3).

Cell Activation

Engagement of ITAM-bearing type I FcγRs by IgG complexes initiates a number of signaling cascades that lead to cellular activation and subsequent induction of effector functions. Cellular responses to Fc-FcγR interactions vary between myeloid cell types; however, FcγR aggregation typically leads to rapid internalization of FcγRs and activation of different signaling pathways that influence cell activation (61, 68, 69, 116). Among myeloid cell types, activation induced upon FcγR engagement is most profound in granulocytes and platelets, as these cell types rapidly mediate biological effects within a few minutes following stimulation with IgG complexes (61, 63, 76, 117, 118). Granulocytes represent the most abundant leukocyte cell type in circulation and mediate pleiotropic effector functions during an inflammatory response to invading pathogens (119). Indeed, granulocytes, which include neutrophils, eosinophils, and basophils, are the first leukocyte cell type recruited to sites of inflammation in response to infection or injury and play a central role in immunity against bacterial, viral, and fungal pathogens (120). Several studies have examined the role of FcγR-mediated cellular activation and determined the downstream functional consequences upon FcγR engagement of granulocytes. Granulocyte effector responses initiated following cellular activation in response to Fc-FcγR interactions aim at the destruction of invading pathogens and involve the release of an array of microbicidal molecules. These molecules are either preformed and stored in specialized granules in the cytoplasm or are rapidly *de novo* generated (120). Activation of Syk and Src family kinases upon FcγR cross-linking triggers the generation of reactive oxygen intermediates through the formation and activation of the NADPH-dependent oxidase complex, a multicomponent protein complex that consists of $p40^{phox}$, $p47^{phox}$, $p67^{phox}$, $p22^{phox}$, and $gp91^{phox}$ (121–124). Phosphorylation of $p47^{phox}$ promotes its physical association with cytochrome b_{558}, triggering the rapid generation of superoxide anions. Superoxide anions can have direct cytotoxic activity or can lead to the generation of reactive oxygen or nitrogen intermediates, including hydroxyl radical (OHO), hypochlorous acid (HOCl), and peroxynitrite ($ONOO^-$) (122, 125, 126). Additionally, superoxide anions can be dismutated by superoxide dismutase, leading to generation of hydrogen peroxide (127). The release of antimicrobial molecules by granulocytes in response to FcγR cross-linking represents one of the most powerful immune mechanisms in the human body. Apart from these chemical compounds, activation of the PKC pathway and increase in the intracellular Ca^{2+} concentration following FcγR-mediated cellular activation trigger the mobilization and release of preformed molecules stored in specialized granules (128). The content of these cytoplasmic granules varies among the different granulocyte subtypes; however, it typically comprises proteases (elastase, cathepsins, and collagenases), antimicrobial peptides and proteins (lysozyme, defensins, and lactoferrin), enzymes (peroxidase and alkaline phosphatase), lipid mediators (leukotrienes), as well as cell surface receptors (CD11b) (129–135). Release of these mediators at inflammatory sites represents a hallmark for granulocyte activation and constitutes an important effector mechanism by which granulocytes mediate *in vivo* activity against invading pathogens.

In an analogy to granulocyte activation and degranulation, FcγR-mediated cross-linking also has the capacity to induce platelet activation. Human platelets express FcγRIIa, an activating type I FcγR, which, upon cross-linking, mediates potent signaling activity (136–138). Additionally, they express FcR γ chain, an important accessory signaling subunit for platelet function that associates with several platelet surface receptors, including GPVI (139, 140). Activation of downstream signaling pathways triggers platelet activation and degranulation in a process similar to that observed for granulocytes (141). Elevation of the intracellular Ca^{2+} concentration following FcγR cross-linking triggers rapid degranulation and release of platelet granule content (139, 142). This includes cell surface receptors that participate in cell-cell adhesion and platelet aggregation, as well as molecules involved in the activation of fibrin cascade, including fibronectin, fibrinogen, and coagulation factors V and XIII (139). Additionally, platelet degranulation is associated with the release of proinflammatory cytokines and chemokines, including IL-1β and IL-8, as well as growth factors and prosurvival factors that influence leukocyte cell survival, differentiation, and effector activity (143–145). These events highlight the potential of FcγR-mediated signaling to trigger platelet activation and thrombogenesis, as well as to influence leukocyte function.

Phagocytosis and Antigen Presentation

A common function of all type I FcγRs is their capacity to mediate efficient phagocytosis of IgG-opsonized particles, which can range from small antigens (toxins, infectious pathogens, etc.) to whole cells. Although the phagocytic capacity varies greatly among different myeloid phagocytes, the downstream signaling events and mechanisms that characterize FcγR-mediated phagocytosis follow the general pattern of FcγR cross-linking and cellular activation. In particular, FcγR cross-linking by immune complexes triggers receptor internalization and activation of downstream signaling pathways that facilitate actin remodeling and endosomal uptake and sorting (10, 63, 77, 83, 146). Although in myeloid cells both activating and inhibitory FcγRs are capable of internalization and phagocytosis, uptake through activating FcγRs mediates more-potent effector responses associated with the induction of proinflammatory signaling pathways (69, 83, 147). Activation of Syk family kinases upon receptor cross-linking mediates cellular activation, thereby influencing leukocyte effector function. Indeed, activating FcγR-mediated phagocytosis is associated with enhanced endosomal maturation and lysosomal fusion, facilitating antigen processing and presentation on major histocompatibility complex class II molecules (69, 148–150). These effects result in the induction of more robust T-cell responses, further augmenting the *in vivo* protective activity of antibodies (66, 98, 151, 152).

Several studies have demonstrated the key role of Fc-FcγR interactions in inducing clearance of IgG-coated particles *in vivo*. For example, seminal experiments using IgG-opsonized erythrocytes demonstrated that their uptake is mediated exclusively through FcγR-mediated pathways by splenic macrophages (153, 154). Likewise, the *in vivo* activity of antibodies against toxins and bacterial pathogens depends on interactions with FcγRs expressed by effector leukocytes; experimentally, this can be shown using mice that are genetically modified to lack expression of specific FcγRs required for IgG-mediated protection or through modification of the Fc domain on protective antibodies to abrogate FcγR interactions, thus greatly reducing their *in vivo* activity (58, 155–162). Similarly, clearance of tumors or infected cells targeted by antibodies against surface antigens requires interactions with activating FcγRs (16, 66, 67, 154, 156, 163–167).

DC Maturation

Under homeostatic conditions, human DCs express two type I FcγRs, the inhibitory FcγRIIb and the activating FcγRIIa (5). These receptors are coexpressed on the surface of DCs, and activating signals from FcγRIIa engagement are counterbalanced by the inhibitory activity of FcγRIIb, which prevents undesired DC maturation and differentiation. The balance of activating and inhibitory signaling is a key regulatory process controlling DC activity; therefore, the level of expression of these two FcγRs must be tightly regulated. Inflammatory microenvironments can trigger expression of additional activating type I FcγRs, such as FcγRI and FcγRIIIa, as well as influence the expression of FcγRIIb on DCs. For example, IL-4 has been previously shown to induce FcγRIIb upregulation in DCs, whereas IFN-γ decreases FcγRIIb expression and stimulates FcγRI expression (46, 109, 151, 168, 169). This finely tuned balance between activating and inhibitory FcγR expression and activity determines the threshold for immune complex-mediated DC activation and responsiveness to PAMPs. Indeed, stimulation of DCs with IgG immune complexes often does not induce robust cell maturation, but requires costimulatory signals, such as TLR signaling, to overcome the inhibitory activity of FcγRIIb (67, 98, 116, 151, 152, 168, 170). Skewing the balance of the contrasting signaling activity of DC FcγRIIa and FcγRIIb has profound consequences for cell maturation and the development of subsequent T-cell responses. Indeed, genetic deletion or antibody-mediated block of FcγRIIb ligand-binding activity on DCs greatly augments immune complex-mediated cell maturation, resulting in the upregulation of major histocompatibility complex and costimulatory molecules, as well as in enhanced antigen presentation and T-cell activation (116, 151, 152, 168, 171). Likewise, in a model of $CD20^+$ lymphoma, preferential engagement of DC FcγRIIa through Fc domain engineering of anti-CD20 antibodies resulted in improved antigen-specific T-cell responses, an effect attributed to an increase in the threshold for FcγRIIb-mediated inhibition of DC maturation (67). These studies highlight the importance of the balancing activity of activating and inhibitory FcγRs in regulating DC activation, thereby influencing adaptive immune responses.

Macrophage Polarization

Human macrophage populations represent a continuum of diverse activation states with distinct functional and phenotypic characteristics (172). Macrophage polarization was originally divided into two broad phenotypes—M1 and M2, induced by the contrasting activity of Th1 cytokines, like IFN-γ, and Th2 cytokines, like IL-4, respectively (173). However, it was soon appreciated that immune complex-mediated effector pathways represent an additional determinant for macrophage polarization. In monocytes and polarized macrophages, signaling through the activating FcγRs is associated with the upregulation of several proinflammatory cytokines and chemokines (10, 58). However, when activating FcγR signaling is coupled with stimulation through TLRs, like TLR4 in nonpolarized macrophages, this synergistic signaling activity triggers induction of a specific polarization state that resembles the M2 phenotype (174–176). This phenotype, which is generally termed as M2b or "regulatory," is characterized by increased IL-10, IL-1, and IL-6 expression as well as by increased migratory and phagocytic capacity (174–176). A number of *in vivo* studies utilizing mouse strains with genetic deletion of *Fcgr2b* have provided useful insights into the contribution of FcγR-mediating signaling in regulating macrophage polarization and functional activity. For example, FcγRIIb-deficient mice exhibit lower macrophage activation threshold upon challenge with immune complexes and present a more severe phenotype in models of immune complex-induced shock, arthritis, and alveolitis (100, 153, 174). Also, FcγRIIb expression can determine susceptibility to infection; for example, $Fcgr2b^{-/-}$ mice exhibit improved bacterial clearance in models of pneumococcal peritonitis (177, 178), whereas overexpression of FcγRIIb is associated with increased mortality upon challenge with *Streptococcus pneumoniae* (177, 178). These findings highlight the importance of FcγR-mediated signaling in macrophage polarization and in the regulation of macrophage effector function *in vivo*.

Fc RECEPTORS IN DISEASE

Balanced signaling through Fc receptors is required for proper immune activity and health; indeed, FcγR polymorphisms that affect signaling are associated with a variety of human diseases. In particular, specific autoimmune diseases are more commonly found in association with FcγR polymorphisms that confer lower-affinity Fc-FcγR interactions. For example, the R131 variant of FcγRIIa has lower affinity for the IgG2 Fc domain when compared with the H131 form and is more commonly found in people with systemic lupus erythematosus (179), antiphospholipid syndrome (180), myasthenia gravis (181), and severe Guillain-Barré syndrome (182). The frequency of individuals homozygous for either the R131 or H131 variant of FcγRIIa varies between ~25 and 35% depending on ethnicity (183). In addition, polymorphisms that reduce transcription or disrupt membrane localization of the inhibitory type I FcR, FcγRIIb, are associated with systemic lupus erythematosus in humans (102–104, 184). The role of FcγRIIb in maintenance of immune tolerance and production of high-affinity antibody responses has been studied in some detail: FcγRIIb signaling on B cells increases the threshold for the affinity of B-cell receptor that is required for cell survival; thus low FcγRIIb expression or signaling can result in autoantibody and low-affinity antibody production (96, 97, 105). Many FcγR variants have been described and are reviewed elsewhere (29, 185). Studies that provide clear mechanistic evidence for the role of specific FcγRs in health and disease are often performed in animal models. Here, we briefly review works demonstrating the role of specific FcγRs in infection, immunosuppression, and cytotoxicity along with clinical studies on FcγR polymorphisms that may support the basic findings.

Immunity to Infectious Pathogens

Type I FcγRs contribute to protection against a variety of infectious pathogens *in vivo*. This is exemplified by experiments demonstrating that antibodies specific for influenza virus proteins, which do not exhibit neutralizing activity *in vitro*, can have protective activity *in vivo* that depends on activating type I FcγRs (157, 158). Similarly, suppression of simian-human immunodeficiency virus viremia by anti-HIV monoclonal antibodies (MAbs) in macaques was dependent on Fc interactions with activating FcγRs (156). Numerous studies have implicated FcγRs in antibody-mediated protection against bacterial, viral, and fungal pathogens (58). Clinical studies have found a correlation between the low-affinity variant of FcγRIIa (R131) and increased susceptibility to severe bacterial infections and sepsis (186–188), suggesting a protective role for myeloid cell FcγRIIa-mediated effector functions in protection from infection in humans.

Disease-Enhancing IgG

A long-appreciated yet not fully understood immunomodulatory property of type I FcγRs is their occasional ability to mediate enhanced infectious disease. An example of this phenomenon can be observed in secondary infection with dengue virus; while primary infection is often asymptomatic or mild in presentation, subsequent infection with a distinct dengue serotype can be associated with enhanced viral replication and disease (189, 190). This enhanced secondary disease is, at least in part, thought to be mediated by cross-reactive, nonneutralizing IgGs generated during the primary virus exposure that enhance infection of type I FcγR-bearing cells, including monocytes, macrophages, and DCs. Antibody-dependent enhancement in dengue virus infection has been demonstrated in a variety of *in vitro* and *in vivo* dengue virus infection models, but the mechanisms underlying enhancement are not well understood (191–194). Studies have shown that FcγR-mediated internalization of dengue viruses can result in more infected cells (195, 196), enhanced viral fusion activity (197), and suppression of innate immune signaling (198, 199). Severe dengue virus disease has been associated with specific combinations of virus serotypes and preexisting serotype immunity (200), viral genetic factors (201–204), and several host factors (205–209).

Antibody-dependent enhancement of disease that is not secondary to increased microbial replication has been observed during respiratory syncytial virus outbreaks in patients previously vaccinated with formalin-inactivated viral proteins and in some severe influenza virus infections. In these circumstances, immune complexes formed from nonneutralizing IgGs with viral proteins are thought to have mediated cytotoxicity and/or complement deposition and inflammation (210, 211). Why some microbes can cause increased infectivity and/or clinical disease through antibody-mediated mechanisms is not well understood. Adaptation to productive replication in monocytes or macrophages may be one determinant of cytokine-associated disease enhancement. A second determinant might be antigenic variability of a pathogen, which often results in production of cross-reactive, nonneutralizing antibodies that can enhance pathogen uptake by FcγR-expressing cells or, in rare circumstances, may form insoluble immune complexes that cause disease due to type I FcγR-mediated inflammation or direct cytotoxicity.

Immunosuppressive IgG

A prime example of immunosuppressive activity through FcγRs is the CD209/DC-SIGN-mediated anti-inflammatory activity achieved clinically through administration of high-dose intravenous immunoglobulin (28). This activity is dependent on the presence of IgGs modified by glycans with α2,6-linked terminal sialic acid, which induce conformational flexibility of the Fc C_H2 domain, enabling binding to type II receptors. A mechanism underlying the anti-inflammatory signaling induced by intravenous immunoglobulin has been described whereby sialylated Fc domains interact with DC-SIGN on regulatory myeloid cells, triggering IL-33 production, which in turn induces expansion of IL-4-producing basophils that promote increased expression of the inhibitory Fc receptor FcγRIIb on effector macrophages (109).

Cytotoxic IgG and Immunity to Tumors

Passively administered antitumor MAbs are key therapeutics in a number of cancers and often mediate their cytotoxic effects through mechanisms that depend on FcγR expression on effector myeloid cells (66, 212). FcγRIIIa expression on macrophages has been shown to mediate cytotoxicity of an anti-CD20 MAb in mice that were humanized for Fc receptor expression (67).

Several clinical studies support the dependence of cytotoxic MAb activity on FcγRIIIa. Most prominently, the high-binding variant of FcγRIIIa, V158, which confers up to 10-fold-higher binding to IgG1 and significantly enhanced antibody-dependent cell-mediated cytotoxicity, has been associated with improved survival in several studies of cancer patients administered antibody therapeutics. Significantly increased response rates and survival are documented in lymphoma patients treated with rituximab who were heterozygous or homozygous for the FcγRIIIa V158 polymorphism, for example (213–215). In addition to direct cytotoxicity, long-term antitumor immunity has been demonstrated in some patients after administration of rituximab (216). In mice, induction of antitumor immunity by an IgG1 anti-CD20 MAb could be generated by a mechanism dependent on FcγRIIa, expressed on DCs in humans (67).

CONCLUSIONS

IgG antibodies recruit effector cells through engagement of type I and type II FcγRs. In this way, Fc-FcγR interactions represent a central tie between the humoral and cellular immune divisions. As outlined briefly here, these interactions are essential for immune mechanisms that can protect against infectious organisms and tumors and can, on occasion, mediate enhanced disease. Antibody therapeutics for the treatment of inflammatory or infectious diseases or cancers require specific effector functions, and their development must consider not only target specificity but also downstream effector functions that will be required for optimal therapeutic efficacy.

Acknowledgments. *We thank The Rockefeller University and community for its continued support. Research conducted in the Laboratory of Molecular Genetics and Immunology was funded in part by various sponsors including the National Institutes of Health, Bill and Melinda Gates Foundation, Defense Advanced Research Projects Agency, Cancer Research Institute, and American Foundation for AIDS Research. Stylianos Bournazos and Taia T. Wang contributed equally to this work.*

Citation. Bournazos S, Wang TT, Ravetch JV. 2016. The role and function of Fcγ receptors on myeloid cells. Microbiol Spectrum 4(6):MCHD-0045-2016.

References

1. **Netea MG, Wijmenga C, O'Neill LA.** 2012. Genetic variation in Toll-like receptors and disease susceptibility. *Nat Immunol* **13:**535–542.
2. **Gack MU.** 2014. Mechanisms of RIG-I-like receptor activation and manipulation by viral pathogens. *J Virol* **88:**5213–5216.
3. **Yoo JS, Kato H, Fujita T.** 2014. Sensing viral invasion by RIG-I like receptors. *Curr Opin Microbiol* **20:**131–138.
4. **Sparrer KM, Gack MU.** 2015. Intracellular detection of viral nucleic acids. *Curr Opin Microbiol* **26:**1–9.
5. **Pincetic A, Bournazos S, DiLillo DJ, Maamary J, Wang TT, Dahan R, Fiebiger BM, Ravetch JV.** 2014. Type I and type II Fc receptors regulate innate and adaptive immunity. *Nat Immunol* **15:**707–716.
6. **Anthony RM, Wermeling F, Ravetch JV.** 2012. Novel roles for the IgG Fc glycan. *Ann N Y Acad Sci* **1253:**170–180.
7. **Borrok MJ, Jung ST, Kang TH, Monzingo AF, Georgiou G.** 2012. Revisiting the role of glycosylation in the structure of human IgG Fc. *ACS Chem Biol* **7:**1596–1602.
8. **Ahmed AA, Giddens J, Pincetic A, Lomino JV, Ravetch JV, Wang LX, Bjorkman PJ.** 2014. Structural characterization of anti-inflammatory immunoglobulin G Fc proteins. *J Mol Biol* **426:**3166–3179.
9. **Sondermann P, Pincetic A, Maamary J, Lammens K, Ravetch JV.** 2013. General mechanism for modulating immunoglobulin effector function. *Proc Natl Acad Sci U S A* **110:**9868–9872.
10. **Bournazos S, Ravetch JV.** 2015. Fcγ receptor pathways during active and passive immunization. *Immunol Rev* **268:**88–103.
11. **Narciso JE, Uy ID, Cabang AB, Chavez JF, Pablo JL, Padilla-Concepcion GP, Padlan EA.** 2011. Analysis of the antibody structure based on high-resolution crystallographic studies. *N Biotechnol* **28:**435–447.
12. **Krapp S, Mimura Y, Jefferis R, Huber R, Sondermann P.** 2003. Structural analysis of human IgG-Fc glycoforms reveals a correlation between glycosylation and structural integrity. *J Mol Biol* **325:**979–989.
13. **Teplyakov A, Zhao Y, Malia TJ, Obmolova G, Gilliland GL.** 2013. IgG2 Fc structure and the dynamic features of the IgG CH_2-CH_3 interface. *Mol Immunol* **56:**131–139.
14. **Albert H, Collin M, Dudziak D, Ravetch JV, Nimmerjahn F.** 2008. *In vivo* enzymatic modulation of IgG glycosylation inhibits autoimmune disease in an IgG subclass-dependent manner. *Proc Natl Acad Sci U S A* **105:**15005–15009.
15. **Lux A, Nimmerjahn F.** 2011. Impact of differential glycosylation on IgG activity. *Adv Exp Med Biol* **780:**113–124.
16. **Nimmerjahn F, Ravetch JV.** 2005. Divergent immunoglobulin G subclass activity through selective Fc receptor binding. *Science* **310:**1510–1512.
17. **Nimmerjahn F, Anthony RM, Ravetch JV.** 2007. Agalactosylated IgG antibodies depend on cellular Fc receptors for *in vivo* activity. *Proc Natl Acad Sci U S A* **104:**8433–8437.
18. **Anthony RM, Nimmerjahn F, Ashline DJ, Reinhold VN, Paulson JC, Ravetch JV.** 2008. Recapitulation of IVIG anti-inflammatory activity with a recombinant IgG Fc. *Science* **320:**373–376.
19. **Baudino L, Shinohara Y, Nimmerjahn F, Furukawa J, Nakata M, Martínez-Soria E, Petry F, Ravetch JV,**

Nishimura S, Izui S. 2008. Crucial role of aspartic acid at position 265 in the CH2 domain for murine IgG2a and IgG2b Fc-associated effector functions. *J Immunol* **181:**6664–6669.

20. **Ferrara C, Grau S, Jäger C, Sondermann P, Brünker P, Waldhauer I, Hennig M, Ruf A, Rufer AC, Stihle M, Umaña P, Benz J.** 2011. Unique carbohydrate-carbohydrate interactions are required for high affinity binding between FcγRIII and antibodies lacking core fucose. *Proc Natl Acad Sci U S A* **108:**12669–12674.
21. **Shinkawa T, Nakamura K, Yamane N, Shoji-Hosaka E, Kanda Y, Sakurada M, Uchida K, Anazawa H, Satoh M, Yamasaki M, Hanai N, Shitara K.** 2003. The absence of fucose but not the presence of galactose or bisecting *N*-acetylglucosamine of human IgG1 complex-type oligosaccharides shows the critical role of enhancing antibody-dependent cellular cytotoxicity. *J Biol Chem* **278:**3466–3473.
22. **Cramer P, Hallek M, Eichhorst B.** 2016. State-of-the-art treatment and novel agents in chronic lymphocytic leukemia. *Oncol Res Treat* **39:**25–32.
23. **Goede V, Fischer K, Busch R, Engelke A, Eichhorst B, Wendtner CM, Chagorova T, de la Serna J, Dilhuydy MS, Illmer T, Opat S, Owen CJ, Samoylova O, Kreuzer KA, Stilgenbauer S, Döhner H, Langerak AW, Ritgen M, Kneba M, Asikanius E, Humphrey K, Wenger M, Hallek M.** 2014. Obinutuzumab plus chlorambucil in patients with CLL and coexisting conditions. *N Engl J Med* **370:**1101–1110.
24. **Natsume A, Niwa R, Satoh M.** 2009. Improving effector functions of antibodies for cancer treatment: enhancing ADCC and CDC. *Drug Des Devel Ther* **3:** 7–16.
25. **Hiatt A, Bohorova N, Bohorov O, Goodman C, Kim D, Pauly MH, Velasco J, Whaley KJ, Piedra PA, Gilbert BE, Zeitlin L.** 2014. Glycan variants of a respiratory syncytial virus antibody with enhanced effector function and in vivo efficacy. *Proc Natl Acad Sci U S A* **111:** 5992–5997.
26. **Shields RL, Lai J, Keck R, O'Connell LY, Hong K, Meng YG, Weikert SH, Presta LG.** 2002. Lack of fucose on human IgG1 *N*-linked oligosaccharide improves binding to human FcγRIII and antibody-dependent cellular toxicity. *J Biol Chem* **277:**26733–26740.
27. **Kaneko Y, Nimmerjahn F, Ravetch JV.** 2006. Anti-inflammatory activity of immunoglobulin G resulting from Fc sialylation. *Science* **313:**670–673.
28. **Anthony RM, Wermeling F, Karlsson MC, Ravetch JV.** 2008. Identification of a receptor required for the anti-inflammatory activity of IVIG. *Proc Natl Acad Sci U S A* **105:**19571–19578.
29. **Bournazos S, Woof JM, Hart SP, Dransfield I.** 2009. Functional and clinical consequences of Fc receptor polymorphic and copy number variants. *Clin Exp Immunol* **157:**244–254.
30. **Nimmerjahn F, Ravetch JV.** 2006. Fcγ receptors: old friends and new family members. *Immunity* **24:**19–28.
31. **Shields RL, Namenuk AK, Hong K, Meng YG, Rae J, Briggs J, Xie D, Lai J, Stadlen A, Li B, Fox JA, Presta LG.** 2001. High resolution mapping of the binding site on human IgG1 for FcγRI, FcγRII, FcγRIII, and FcRn and design of IgG1 variants with improved binding to the FcγR. *J Biol Chem* **276:**6591–6604.
32. **Sondermann P, Huber R, Oosthuizen V, Jacob U.** 2000. The 3.2-Å crystal structure of the human IgG1 Fc fragment-FcγRIII complex. *Nature* **406:**267–273.
33. **Sondermann P, Kaiser J, Jacob U.** 2001. Molecular basis for immune complex recognition: a comparison of Fc-receptor structures. *J Mol Biol* **309:**737–749.
34. **Qiu WQ, de Bruin D, Brownstein BH, Pearse R, Ravetch JV.** 1990. Organization of the human and mouse low-affinity FcγR genes: duplication and recombination. *Science* **248:**732–735.
35. **Su K, Wu J, Edberg JC, McKenzie SE, Kimberly RP.** 2002. Genomic organization of classical human low-affinity Fcγ receptor genes. *Genes Immun* **3**(Suppl 1): S51–S56.
36. **Ernst LK, van de Winkel JG, Chiu IM, Anderson CL.** 1992. Three genes for the human high affinity Fc receptor for IgG (FcγRI) encode four distinct transcription products. *J Biol Chem* **267:**15692–15700.
37. **Ernst LK, Duchemin AM, Miller KL, Anderson CL.** 1998. Molecular characterization of six variant Fcγ receptor class I (CD64) transcripts. *Mol Immunol* **35:** 943–954.
38. **van Vugt MJ, Reefman E, Zeelenberg I, Boonen G, Leusen JH, van de Winkel JG.** 1999. The alternatively spliced CD64 transcript FcγRIb2 does not specify a surface-expressed isoform. *Eur J Immunol* **29:**143–149.
39. **Kiyoshi M, Caaveiro JM, Kawai T, Tashiro S, Ide T, Asaoka Y, Hatayama K, Tsumoto K.** 2015. Structural basis for binding of human IgG1 to its high-affinity human receptor FcγRI. *Nat Commun* **6:**6866. doi: 10.1038/ncomms7866.
40. **Duchemin AM, Ernst LK, Anderson CL.** 1994. Clustering of the high affinity Fc receptor for immunoglobulin G (FcγRI) results in phosphorylation of its associated γ-chain. *J Biol Chem* **269:**12111–12117.
41. **Ernst LK, Duchemin AM, Anderson CL.** 1993. Association of the high-affinity receptor for IgG (FcγRI) with the γ subunit of the IgE receptor. *Proc Natl Acad Sci U S A* **90:**6023–6027.
42. **Indik ZK, Hunter S, Huang MM, Pan XQ, Chien P, Kelly C, Levinson AI, Kimberly RP, Schreiber AD.** 1994. The high affinity Fcγ receptor (CD64) induces phagocytosis in the absence of its cytoplasmic domain: the γ subunit of FcγRIIIA imparts phagocytic function to FcγRI. *Exp Hematol* **22:**599–606.
43. **van Vugt MJ, Heijnen AF, Capel PJ, Park SY, Ra C, Saito T, Verbeek JS, van de Winkel JG.** 1996. FcR γ-chain is essential for both surface expression and function of human FcγRI (CD64) in vivo. *Blood* **87:**3593–3599.
44. **Masuda M, Roos D.** 1993. Association of all three types of FcγR (CD64, CD32, and CD16) with a γ-chain homodimer in cultured human monocytes. *J Immunol* **151:**7188–7195.
45. **Li Y, Lee PY, Sobel ES, Narain S, Satoh M, Segal MS, Reeves WH, Richards HB.** 2009. Increased expression

of FcγRI/CD64 on circulating monocytes parallels ongoing inflammation and nephritis in lupus. *Arthritis Res Ther* **11**:R6. doi:10.1186/ar2590.

46. **Uciechowski P, Schwarz M, Gessner JE, Schmidt RE, Resch K, Radeke HH.** 1998. IFN-gamma induces the high-affinity Fc receptor I for IgG (CD64) on human glomerular mesangial cells. *Eur J Immunol* **28**:2928–2935.
47. **Maxwell KF, Powell MS, Hulett MD, Barton PA, McKenzie IF, Garrett TP, Hogarth PM.** 1999. Crystal structure of the human leukocyte Fc receptor, FcγRIIa. *Nat Struct Biol* **6**:437–442.
48. **Radaev S, Motyka S, Fridman WH, Sautes-Fridman C, Sun PD.** 2001. The structure of a human type III Fcγ receptor in complex with Fc. *J Biol Chem* **276**:16469–16477.
49. **Radaev S, Sun P.** 2002. Recognition of immunoglobulins by Fcγ receptors. *Mol Immunol* **38**:1073–1083.
50. **Maenaka K, van der Merwe PA, Stuart DI, Jones EY, Sondermann P.** 2001. The human low affinity Fcγ receptors IIa, IIb, and III bind IgG with fast kinetics and distinct thermodynamic properties. *J Biol Chem* **276**: 44898–44904.
51. **Ravetch JV, Bolland S.** 2001. IgG Fc receptors. *Annu Rev Immunol* **19**:275–290.
52. **Brooks DG, Qiu WQ, Luster AD, Ravetch JV.** 1989. Structure and expression of human IgG FcRII(CD32). Functional heterogeneity is encoded by the alternatively spliced products of multiple genes. *J Exp Med* **170**: 1369–1385.
53. **Lewis VA, Koch T, Plutner H, Mellman I.** 1986. A complementary DNA clone for a macrophage-lymphocyte Fc receptor. *Nature* **324**:372–375.
54. **Latour S, Fridman WH, Daëron M.** 1996. Identification, molecular cloning, biologic properties, and tissue distribution of a novel isoform of murine low-affinity IgG receptor homologous to human Fc gamma RIIB1. *J Immunol* **157**:189–197.
55. **Hibbs ML, Bonadonna L, Scott BM, McKenzie IF, Hogarth PM.** 1988. Molecular cloning of a human immunoglobulin G Fc receptor. *Proc Natl Acad Sci U S A* **85**:2240–2244.
56. **Metes D, Ernst LK, Chambers WH, Sulica A, Herberman RB, Morel PA.** 1998. Expression of functional CD32 molecules on human NK cells is determined by an allelic polymorphism of the FcγRIIC gene. *Blood* **91**:2369–2380.
57. **Ernst LK, Metes D, Herberman RB, Morel PA.** 2002. Allelic polymorphisms in the FcγRIIC gene can influence its function on normal human natural killer cells. *J Mol Med (Berl)* **80**:248–257.
58. **Bournazos S, DiLillo DJ, Ravetch JV.** 2015. The role of Fc-FcγR interactions in IgG-mediated microbial neutralization. *J Exp Med* **212**:1361–1369.
59. **Hibbs ML, Selvaraj P, Carpén O, Springer TA, Kuster H, Jouvin MH, Kinet JP.** 1989. Mechanisms for regulating expression of membrane isoforms of Fc gamma RIII (CD16). *Science* **246**:1608–1611.
60. **Masuda M, Verhoeven AJ, Roos D.** 1993. Tyrosine phosphorylation of a gamma-chain homodimer associated with Fc gamma RIII (CD16) in cultured human monocytes. *J Immunol* **151**:6382–6388.
61. **Unkeless JC, Shen Z, Lin CW, DeBeus E.** 1995. Function of human Fc gamma RIIA and Fc gamma RIIIB. *Semin Immunol* **7**:37–44.
62. **Selvaraj P, Carpén O, Hibbs ML, Springer TA.** 1989. Natural killer cell and granulocyte Fc gamma receptor III (CD16) differ in membrane anchor and signal transduction. *J Immunol* **143**:3283–3288.
63. **Edberg JC, Kimberly RP.** 1994. Modulation of Fc gamma and complement receptor function by the glycosylphosphatidylinositol-anchored form of Fc gamma RIII. *J Immunol* **152**:5826–5835.
64. **Smith P, DiLillo DJ, Bournazos S, Li F, Ravetch JV.** 2012. Mouse model recapitulating human Fcγ receptor structural and functional diversity. *Proc Natl Acad Sci U S A* **109**:6181–6186.
65. **Passlick B, Flieger D, Ziegler-Heitbrock HW.** 1989. Identification and characterization of a novel monocyte subpopulation in human peripheral blood. *Blood* **74**: 2527–2534.
66. **Clynes RA, Towers TL, Presta LG, Ravetch JV.** 2000. Inhibitory Fc receptors modulate *in vivo* cytotoxicity against tumor targets. *Nat Med* **6**:443–446.
67. **DiLillo DJ, Ravetch JV.** 2015. Differential Fc-receptor engagement drives an anti-tumor vaccinal effect. *Cell* **161**:1035–1045.
68. **Amigorena S, Bonnerot C, Drake JR, Choquet D, Hunziker W, Guillet JG, Webster P, Sautes C, Mellman I, Fridman WH.** 1992. Cytoplasmic domain heterogeneity and functions of IgG Fc receptors in B lymphocytes. *Science* **256**:1808–1812.
69. **Amigorena S, Salamero J, Davoust J, Fridman WH, Bonnerot C.** 1992. Tyrosine-containing motif that transduces cell activation signals also determines internalization and antigen presentation via type III receptors for IgG. *Nature* **358**:337–341.
70. **Cosson P, Lankford SP, Bonifacino JS, Klausner RD.** 1991. Membrane protein association by potential intramembrane charge pairs. *Nature* **351**:414–416.
71. **Orloff DG, Ra CS, Frank SJ, Klausner RD, Kinet JP.** 1990. Family of disulphide-linked dimers containing the ζ and η chains of the T-cell receptor and the γ chain of Fc receptors. *Nature* **347**:189–191.
72. **Nimmerjahn F, Ravetch JV.** 2008. Fcγ receptors as regulators of immune responses. *Nat Rev Immunol* **8**: 34–47.
73. **Segal DM, Taurog JD, Metzger H.** 1977. Dimeric immunoglobulin E serves as a unit signal for mast cell degranulation. *Proc Natl Acad Sci U S A* **74**:2993–2997.
74. **Kulczycki A Jr, Metzger H.** 1974. The interaction of IgE with rat basophilic leukemia cells. II. Quantitative aspects of the binding reaction. *J Exp Med* **140**:1676–1695.
75. **Ishizaka T, Ishizaka K.** 1978. Triggering of histamine release from rat mast cells by divalent antibodies against IgE-receptors. *J Immunol* **120**:800–805.
76. **Salmon JE, Millard SS, Brogle NL, Kimberly RP.** 1995. Fcγ receptor IIIb enhances Fcγ receptor IIa function in

an oxidant-dependent and allele-sensitive manner. *J Clin Invest* **95:**2877–2885.

77. **Sobota A, Strzelecka-Kiliszek A, Gładkowska E, Yoshida K, Mrozińska K, Kwiatkowska K.** 2005. Binding of IgG-opsonized particles to FcγR is an active stage of phagocytosis that involves receptor clustering and phosphorylation. *J Immunol* **175:**4450–4457.
78. **Durden DL, Liu YB.** 1994. Protein-tyrosine kinase p72[syk] in FcγRI receptor signaling. *Blood* **84:**2102–2108.
79. **Durden DL, Kim HM, Calore B, Liu Y.** 1995. The Fc gamma RI receptor signals through the activation of hck and MAP kinase. *J Immunol* **154:**4039–4047.
80. **Eiseman E, Bolen JB.** 1992. Engagement of the high-affinity IgE receptor activates *src* protein-related tyrosine kinases. *Nature* **355:**78–80.
81. **Jouvin MH, Adamczewski M, Numerof R, Letourneur O, Vallé A, Kinet JP.** 1994. Differential control of the tyrosine kinases Lyn and Syk by the two signaling chains of the high affinity immunoglobulin E receptor. *J Biol Chem* **269:**5918–5925.
82. **Pignata C, Prasad KV, Robertson MJ, Levine H, Rudd CE, Ritz J.** 1993. Fc gamma RIIIA-mediated signaling involves src-family lck in human natural killer cells. *J Immunol* **151:**6794–6800.
83. **Swanson JA, Hoppe AD.** 2004. The coordination of signaling during Fc receptor-mediated phagocytosis. *J Leukoc Biol* **76:**1093–1103.
84. **Hamawy MM, Minoguchi K, Swaim WD, Mergenhagen SE, Siraganian RP.** 1995. A 77-kDa protein associates with pp125[FAK] in mast cells and becomes tyrosine-phosphorylated by high affinity IgE receptor aggregation. *J Biol Chem* **270:**12305–12309.
85. **García-García E, Sánchez-Mejorada G, Rosales C.** 2001. Phosphatidylinositol 3-kinase and ERK are required for NF-κB activation but not for phagocytosis. *J Leukoc Biol* **70:**649–658.
86. **Kanakaraj P, Duckworth B, Azzoni L, Kamoun M, Cantley LC, Perussia B.** 1994. Phosphatidylinositol-3 kinase activation induced upon FcγRIIIA-ligand interaction. *J Exp Med* **179:**551–558.
87. **Ninomiya N, Hazeki K, Fukui Y, Seya T, Okada T, Hazeki O, Ui M.** 1994. Involvement of phosphatidylinositol 3-kinase in Fcγ receptor signaling. *J Biol Chem* **269:**22732–22737.
88. **Sánchez-Mejorada G, Rosales C.** 1998. Fcγ receptor-mediated mitogen-activated protein kinase activation in monocytes is independent of Ras. *J Biol Chem* **273:**27610–27619.
89. **Aramburu J, Azzoni L, Rao A, Perussia B.** 1995. Activation and expression of the nuclear factors of activated T cells, NFATp and NFATc, in human natural killer cells: regulation upon CD16 ligand binding. *J Exp Med* **182:**801–810.
90. **Bracke M, Coffer PJ, Lammers JW, Koenderman L.** 1998. Analysis of signal transduction pathways regulating cytokine-mediated Fc receptor activation on human eosinophils. *J Immunol* **161:**6768–6774.
91. **Zhou MJ, Brown EJ.** 1994. CR3 (Mac-1, $\alpha_M\beta_2$, CD11b/CD18) and FcγRIII cooperate in generation of a neutrophil respiratory burst: requirement for FcγRIII and tyrosine phosphorylation. *J Cell Biol* **125:**1407–1416.
92. **Muta T, Kurosaki T, Misulovin Z, Sanchez M, Nussenzweig MC, Ravetch JV.** 1994. A 13-amino-acid motif in the cytoplasmic domain of FcγRIIB modulates B-cell receptor signalling. *Nature* **368:**70–73.
93. **Pearse RN, Kawabe T, Bolland S, Guinamard R, Kurosaki T, Ravetch JV.** 1999. SHIP recruitment attenuates FcγRIIB-induced B cell apoptosis. *Immunity* **10:**753–760.
94. **Ono M, Bolland S, Tempst P, Ravetch JV.** 1996. Role of the inositol phosphatase SHIP in negative regulation of the immune system by the receptor FcγRIIB. *Nature* **383:**263–266.
95. **Ono M, Okada H, Bolland S, Yanagi S, Kurosaki T, Ravetch JV.** 1997. Deletion of SHIP or SHP-1 reveals two distinct pathways for inhibitory signaling. *Cell* **90:**293–301.
96. **Bolland S, Yim YS, Tus K, Wakeland EK, Ravetch JV.** 2002. Genetic modifiers of systemic lupus erythematosus in FcγRIIB$^{-/-}$ mice. *J Exp Med* **195:**1167–1174.
97. **Bolland S, Ravetch JV.** 2000. Spontaneous autoimmune disease in FcγRIIB-deficient mice results from strain-specific epistasis. *Immunity* **13:**277–285.
98. **Desai DD, Harbers SO, Flores M, Colonna L, Downie MP, Bergtold A, Jung S, Clynes R.** 2007. Fcγ receptor IIB on dendritic cells enforces peripheral tolerance by inhibiting effector T cell responses. *J Immunol* **178:**6217–6226.
99. **Takai T, Ono M, Hikida M, Ohmori H, Ravetch JV.** 1996. Augmented humoral and anaphylactic responses in FcγRII-deficient mice. *Nature* **379:**346–349.
100. **Yuasa T, Kubo S, Yoshino T, Ujike A, Matsumura K, Ono M, Ravetch JV, Takai T.** 1999. Deletion of Fcγ receptor IIB renders H-2^b mice susceptible to collagen-induced arthritis. *J Exp Med* **189:**187–194.
101. **Floto RA, Clatworthy MR, Heilbronn KR, Rosner DR, MacAry PA, Rankin A, Lehner PJ, Ouwehand WH, Allen JM, Watkins NA, Smith KG.** 2005. Loss of function of a lupus-associated FcγRIIb polymorphism through exclusion from lipid rafts. *Nat Med* **11:**1056–1058.
102. **Kono H, Kyogoku C, Suzuki T, Tsuchiya N, Honda H, Yamamoto K, Tokunaga K, Honda Z.** 2005. FcγRIIB Ile232Thr transmembrane polymorphism associated with human systemic lupus erythematosus decreases affinity to lipid rafts and attenuates inhibitory effects on B cell receptor signaling. *Hum Mol Genet* **14:**2881–2892.
103. **Blank MC, Stefanescu RN, Masuda E, Marti F, King PD, Redecha PB, Wurzburger RJ, Peterson MG, Tanaka S, Pricop L.** 2005. Decreased transcription of the human *FCGR2B* gene mediated by the -343 G/C promoter polymorphism and association with systemic lupus erythematosus. *Hum Genet* **117:**220–227.
104. **Su K, Wu J, Edberg JC, Li X, Ferguson P, Cooper GS, Langefeld CD, Kimberly RP.** 2004. A promoter haplotype of the immunoreceptor tyrosine-based inhibitory motif-bearing FcγRIIb alters receptor expression and

associates with autoimmunity. I. Regulatory *FCGR2B* polymorphisms and their association with systemic lupus erythematosus. *J Immunol* **172**:7186–7191.

105. **Wang TT, Maamary J, Tan GS, Bournazos S, Davis CW, Krammer F, Schlesinger SJ, Palese P, Ahmed R, Ravetch JV.** 2015. Anti-HA glycoforms drive B cell affinity selection and determine influenza vaccine efficacy. *Cell* **162**:160–169.

106. **Jellusova J, Nitschke L.** 2012. Regulation of B cell functions by the sialic acid-binding receptors Siglec-G and CD22. *Front Immunol* **2**:96. doi:10.3389/fimmu.2011.00096.

107. **Schwab I, Seeling M, Biburger M, Aschermann S, Nitschke L, Nimmerjahn F.** 2012. B cells and CD22 are dispensable for the immediate anti-inflammatory activity of intravenous immunoglobulins in vivo. *Eur J Immunol* **42**:3302–3309.

108. **Böhm S, Kao D, Nimmerjahn F.** 2014. Sweet and sour: the role of glycosylation for the anti-inflammatory activity of immunoglobulin G. *Curr Top Microbiol Immunol* **382**:393–417.

109. **Anthony RM, Kobayashi T, Wermeling F, Ravetch JV.** 2011. Intravenous gammaglobulin suppresses inflammation through a novel T_H2 pathway. *Nature* **475**:110–113.

110. **Fiebiger BM, Maamary J, Pincetic A, Ravetch JV.** 2015. Protection in antibody- and T cell-mediated autoimmune diseases by anti-inflammatory IgG Fcs requires type II FcRs. *Proc Natl Acad Sci U S A* **112**:E2385–E2394.

111. **Soilleux EJ.** 2003. DC-SIGN (dendritic cell-specific ICAM-grabbing non-integrin) and DC-SIGN-related (DC-SIGNR): friend or foe? *Clin Sci (Lond)* **104**:437–446.

112. **Borthakur S, Andrejeva G, McDonnell JM.** 2011. Basis of the intrinsic flexibility of the Cε3 domain of IgE. *Biochemistry* **50**:4608–4614.

113. **Dhaliwal B, Yuan D, Pang MO, Henry AJ, Cain K, Oxbrow A, Fabiane SM, Beavil AJ, McDonnell JM, Gould HJ, Sutton BJ.** 2012. Crystal structure of IgE bound to its B-cell receptor CD23 reveals a mechanism of reciprocal allosteric inhibition with high affinity receptor FcεRI. *Proc Natl Acad Sci U S A* **109**:12686–12691.

114. **Yokota A, Kikutani H, Tanaka T, Sato R, Barsumian EL, Suemura M, Kishimoto T.** 1988. Two species of human Fc epsilon receptor II (FcεRII/CD23): tissue-specific and IL-4-specific regulation of gene expression. *Cell* **55**:611–618.

115. **Weskamp G, Ford JW, Sturgill J, Martin S, Docherty AJ, Swendeman S, Broadway N, Hartmann D, Saftig P, Umland S, Sehara-Fujisawa A, Black RA, Ludwig A, Becherer JD, Conrad DH, Blobel CP.** 2006. ADAM10 is a principal 'sheddase' of the low-affinity immunoglobulin E receptor CD23. *Nat Immunol* **7**:1293–1298.

116. **Regnault A, Lankar D, Lacabanne V, Rodriguez A, Théry C, Rescigno M, Saito T, Verbeek S, Bonnerot C, Ricciardi-Castagnoli P, Amigorena S.** 1999. Fcγ receptor-mediated induction of dendritic cell maturation and major histocompatibility complex class I-restricted antigen presentation after immune complex internalization. *J Exp Med* **189**:371–380.

117. **Jakus Z, Berton G, Ligeti E, Lowell CA, Mócsai A.** 2004. Responses of neutrophils to anti-integrin antibodies depends on costimulation through low affinity FcγRs: full activation requires both integrin and nonintegrin signals. *J Immunol* **173**:2068–2077.

118. **Mócsai A, Abram CL, Jakus Z, Hu Y, Lanier LL, Lowell CA.** 2006. Integrin signaling in neutrophils and macrophages uses adaptors containing immunoreceptor tyrosine-based activation motifs. *Nat Immunol* **7**:1326–1333.

119. **Dale DC, Boxer L, Liles WC.** 2008. The phagocytes: neutrophils and monocytes. *Blood* **112**:935–945.

120. **Nathan C.** 2006. Neutrophils and immunity: challenges and opportunities. *Nat Rev Immunol* **6**:173–182.

121. **Hampton MB, Kettle AJ, Winterbourn CC.** 1998. Inside the neutrophil phagosome: oxidants, myeloperoxidase, and bacterial killing. *Blood* **92**:3007–3017.

122. **Martyn KD, Kim MJ, Quinn MT, Dinauer MC, Knaus UG.** 2005. p21-activated kinase (Pak) regulates NADPH oxidase activation in human neutrophils. *Blood* **106**:3962–3969.

123. **Suh CI, Stull ND, Li XJ, Tian W, Price MO, Grinstein S, Yaffe MB, Atkinson S, Dinauer MC.** 2006. The phosphoinositide-binding protein $p40^{phox}$ activates the NADPH oxidase during FcγIIA receptor-induced phagocytosis. *J Exp Med* **203**:1915–1925.

124. **Yamauchi A, Kim C, Li S, Marchal CC, Towe J, Atkinson SJ, Dinauer MC.** 2004. Rac2-deficient murine macrophages have selective defects in superoxide production and phagocytosis of opsonized particles. *J Immunol* **173**:5971–5979.

125. **Nathan CF, Brukner LH, Silverstein SC, Cohn ZA.** 1979. Extracellular cytolysis by activated macrophages and granulocytes. I. Pharmacologic triggering of effector cells and the release of hydrogen peroxide. *J Exp Med* **149**:84–99.

126. **Nathan CF, Silverstein SC, Brukner LH, Cohn ZA.** 1979. Extracellular cytolysis by activated macrophages and granulocytes. II. Hydrogen peroxide as a mediator of cytotoxicity. *J Exp Med* **149**:100–113.

127. **Nathan C, Cunningham-Bussel A.** 2013. Beyond oxidative stress: an immunologist's guide to reactive oxygen species. *Nat Rev Immunol* **13**:349–361.

128. **Jönsson F, Mancardi DA, Albanesi M, Bruhns P.** 2013. Neutrophils in local and systemic antibody-dependent inflammatory and anaphylactic reactions. *J Leukoc Biol* **94**:643–656.

129. **Sørensen O, Arnljots K, Cowland JB, Bainton DF, Borregaard N.** 1997. The human antibacterial cathelicidin, hCAP-18, is synthesized in myelocytes and metamyelocytes and localized to specific granules in neutrophils. *Blood* **90**:2796–2803.

130. **Cowland JB, Johnsen AH, Borregaard N.** 1995. hCAP-18, a cathelin/pro-bactenecin-like protein of human neutrophil specific granules. *FEBS Lett* **368**:173–176.

131. **Egesten A, Breton-Gorius J, Guichard J, Gullberg U, Olsson I.** 1994. The heterogeneity of azurophil granules

in neutrophil promyelocytes: immunogold localization of myeloperoxidase, cathepsin G, elastase, proteinase 3, and bactericidal/permeability increasing protein. *Blood* 83:2985–2994.

132. **Fouret P, du Bois RM, Bernaudin JF, Takahashi H, Ferrans VJ, Crystal RG.** 1989. Expression of the neutrophil elastase gene during human bone marrow cell differentiation. *J Exp Med* **169:**833–845.

133. **Owen CA, Campbell MA, Boukedes SS, Campbell EJ.** 1995. Inducible binding of bioactive cathepsin G to the cell surface of neutrophils. A novel mechanism for mediating extracellular catalytic activity of cathepsin G. *J Immunol* **155:**5803–5810.

134. **Panyutich AV, Hiemstra PS, van Wetering S, Ganz T.** 1995. Human neutrophil defensin and serpins form complexes and inactivate each other. *Am J Respir Cell Mol Biol* **12:**351–357.

135. **Gabay JE, Almeida RP.** 1993. Antibiotic peptides and serine protease homologs in human polymorphonuclear leukocytes: defensins and azurocidin. *Curr Opin Immunol* **5:**97–102.

136. **Ankersmit HJ, Roth GA, Moser B, Zuckermann A, Brunner M, Rosin C, Buchta C, Bielek E, Schmid W, Jensen-Jarolim E, Wolner E, Boltz-Nitulescu G, Volf I.** 2003. CD32-mediated platelet aggregation *in vitro* by anti-thymocyte globulin: implication of therapy-induced *in vivo* thrombocytopenia. *Am J Transplant* **3:** 754–759.

137. **Pedicord DL, Dicker I, O'Neil K, Breth L, Wynn R, Hollis GF, Billheimer JT, Stern AM, Seiffert D.** 2003. CD32-dependent platelet activation by a drug-dependent antibody to glycoprotein IIb/IIIa antagonists. *Thromb Haemost* **89:**513–521.

138. **Poole A, Gibbins JM, Turner M, van Vugt MJ, van de Winkel JG, Saito T, Tybulewicz VL, Watson SP.** 1997. The Fc receptor γ-chain and the tyrosine kinase Syk are essential for activation of mouse platelets by collagen. *EMBO J* **16:**2333–2341.

139. **Cerletti C, Tamburrelli C, Izzi B, Gianfagna F, de Gaetano G.** 2012. Platelet-leukocyte interactions in thrombosis. *Thromb Res* **129:**263–266.

140. **Gibbins JM, Okuma M, Farndale R, Barnes M, Watson SP.** 1997. Glycoprotein VI is the collagen receptor in platelets which underlies tyrosine phosphorylation of the Fc receptor γ-chain. *FEBS Lett* **413:**255–259.

141. **Nieswandt B, Bergmeier W, Schulte V, Rackebrandt K, Gessner JE, Zirngibl H.** 2000. Expression and function of the mouse collagen receptor glycoprotein VI is strictly dependent on its association with the FcRγ chain. *J Biol Chem* **275:**23998–24002.

142. **Martini F, Riondino S, Pignatelli P, Gazzaniga PP, Ferroni P, Lenti L.** 2002. Involvement of GD3 in platelet activation. A novel association with Fcγ receptor. *Biochim Biophys Acta* **1583:**297–304.

143. **Hansson GK, Hermansson A.** 2011. The immune system in atherosclerosis. *Nat Immunol* **12:**204–212.

144. **Steinhubl SR, Moliterno DJ.** 2005. The role of the platelet in the pathogenesis of atherothrombosis. *Am J Cardiovasc Drugs* **5:**399–408.

145. **Bournazos S, Rennie J, Hart SP, Fox KA, Dransfield I.** 2008. Monocyte functional responsiveness after PSGL-1-mediated platelet adhesion is dependent on platelet activation status. *Arterioscler Thromb Vasc Biol* **28:** 1491–1498.

146. **Odin JA, Edberg JC, Painter CJ, Kimberly RP, Unkeless JC.** 1991. Regulation of phagocytosis and $[Ca2+]_i$ flux by distinct regions of an Fc receptor. *Science* **254:** 1785–1788.

147. **Miettinen HM, Rose JK, Mellman I.** 1989. Fc receptor isoforms exhibit distinct abilities for coated pit localization as a result of cytoplasmic domain heterogeneity. *Cell* **58:**317–327.

148. **Bergtold A, Desai DD, Gavhane A, Clynes R.** 2005. Cell surface recycling of internalized antigen permits dendritic cell priming of B cells. *Immunity* **23:**503–514.

149. **Hoffmann E, Kotsias F, Visentin G, Bruhns P, Savina A, Amigorena S.** 2012. Autonomous phagosomal degradation and antigen presentation in dendritic cells. *Proc Natl Acad Sci U S A* **109:**14556–14561.

150. **Bonnerot C, Briken V, Brachet V, Lankar D, Cassard S, Jabri B, Amigorena S.** 1998. syk protein tyrosine kinase regulates Fc receptor γ-chain-mediated transport to lysosomes. *EMBO J* **17:**4606–4616.

151. **Dhodapkar KM, Kaufman JL, Ehlers M, Banerjee DK, Bonvini E, Koenig S, Steinman RM, Ravetch JV, Dhodapkar MV.** 2005. Selective blockade of inhibitory Fcγ receptor enables human dendritic cell maturation with IL-12p70 production and immunity to antibody-coated tumor cells. *Proc Natl Acad Sci U S A* **102:** 2910–2915.

152. **Kalergis AM, Ravetch JV.** 2002. Inducing tumor immunity through the selective engagement of activating Fcγ receptors on dendritic cells. *J Exp Med* **195:**1653–1659.

153. **Takai T, Li M, Sylvestre D, Clynes R, Ravetch JV.** 1994. FcR gamma chain deletion results in pleiotrophic effector cell defects. *Cell* **76:**519–529.

154. **Clynes R, Ravetch JV.** 1995. Cytotoxic antibodies trigger inflammation through Fc receptors. *Immunity* **3:** 21–26.

155. **Abboud N, Chow SK, Saylor C, Janda A, Ravetch JV, Scharff MD, Casadevall A.** 2010. A requirement for FcγR in antibody-mediated bacterial toxin neutralization. *J Exp Med* **207:**2395–2405.

156. **Bournazos S, Klein F, Pietzsch J, Seaman MS, Nussenzweig MC, Ravetch JV.** 2014. Broadly neutralizing anti-HIV-1 antibodies require Fc effector functions for *in vivo* activity. *Cell* **158:**1243–1253.

157. **DiLillo DJ, Palese P, Wilson PC, Ravetch JV.** 2016. Broadly neutralizing anti-influenza antibodies require Fc receptor engagement for in vivo protection. *J Clin Invest* **126:**605–610.

158. **DiLillo DJ, Tan GS, Palese P, Ravetch JV.** 2014. Broadly neutralizing hemagglutinin stalk-specific antibodies require FcγR interactions for protection against influenza virus *in vivo*. *Nat Med* **20:**143–151.

159. **Varshney AK, Wang X, Aguilar JL, Scharff MD, Fries BC.** 2014. Isotype switching increases efficacy of antibody protection against staphylococcal enterotoxin

B-induced lethal shock and *Staphylococcus aureus* sepsis in mice. *MBio* 5:e01007-14. doi:10.1128/mBio.01007-14.

160. **Weber S, Tian H, van Rooijen N, Pirofski LA.** 2012. A serotype 3 pneumococcal capsular polysaccharide-specific monoclonal antibody requires Fcγ receptor III and macrophages to mediate protection against pneumococcal pneumonia in mice. *Infect Immun* **80:**1314–1322.
161. **Sanford JE, Lupan DM, Schlageter AM, Kozel TR.** 1990. Passive immunization against *Cryptococcus neoformans* with an isotype-switch family of monoclonal antibodies reactive with cryptococcal polysaccharide. *Infect Immun* **58:**1919–1923.
162. **Schlageter AM, Kozel TR.** 1990. Opsonization of *Cryptococcus neoformans* by a family of isotype-switch variant antibodies specific for the capsular polysaccharide. *Infect Immun* **58:**1914–1918.
163. **Nimmerjahn F, Lux A, Albert H, Woigk M, Lehmann C, Dudziak D, Smith P, Ravetch JV.** 2010. FcγRIV deletion reveals its central role for IgG2a and IgG2b activity in vivo. *Proc Natl Acad Sci U S A* **107:**19396–19401.
164. **Lu CL, Murakowski DK, Bournazos S, Schoofs T, Sarkar D, Halper-Stromberg A, Horwitz JA, Nogueira L, Golijanin J, Gazumyan A, Ravetch JV, Caskey M, Chakraborty AK, Nussenzweig MC.** 2016. Enhanced clearance of HIV-1-infected cells by broadly neutralizing antibodies against HIV-1 in vivo. *Science* **352:** 1001–1004.
165. **Cartron G, Dacheux L, Salles G, Solal-Celigny P, Bardos P, Colombat P, Watier H.** 2002. Therapeutic activity of humanized anti-CD20 monoclonal antibody and polymorphism in IgG Fc receptor FcγRIIIa gene. *Blood* **99:**754–758.
166. **Uchida J, Hamaguchi Y, Oliver JA, Ravetch JV, Poe JC, Haas KM, Tedder TF.** 2004. The innate mononuclear phagocyte network depletes B lymphocytes through Fc receptor-dependent mechanisms during anti-CD20 antibody immunotherapy. *J Exp Med* **199:**1659–1669.
167. **Hamaguchi Y, Xiu Y, Komura K, Nimmerjahn F, Tedder TF.** 2006. Antibody isotype-specific engagement of Fcγ receptors regulates B lymphocyte depletion during CD20 immunotherapy. *J Exp Med* **203:**743–753.
168. **Boruchov AM, Heller G, Veri MC, Bonvini E, Ravetch JV, Young JW.** 2005. Activating and inhibitory IgG Fc receptors on human DCs mediate opposing functions. *J Clin Invest* **115:**2914–2923.
169. **te Velde AA, de Waal Malefijt R, Huijbens RJ, de Vries JE, Figdor CG.** 1992. IL-10 stimulates monocyte Fc gamma R surface expression and cytotoxic activity. Distinct regulation of antibody-dependent cellular cytotoxicity by IFN-gamma, IL-4, and IL-10. *J Immunol* **149:**4048–4052.
170. **Guilliams M, Bruhns P, Saeys Y, Hammad H, Lambrecht BN.** 2014. The function of Fcγ receptors in dendritic cells and macrophages. *Nat Rev Immunol* **14:** 94–108.
171. **Diaz de Ståhl T, Heyman B.** 2001. IgG2a-mediated enhancement of antibody responses is dependent on FcRγ+ bone marrow-derived cells. *Scand J Immunol* **54:** 495–500.
172. **Gordon S, Plüddemann A, Martinez Estrada F.** 2014. Macrophage heterogeneity in tissues: phenotypic diversity and functions. *Immunol Rev* **262:**36–55.
173. **Mosser DM, Edwards JP.** 2008. Exploring the full spectrum of macrophage activation. *Nat Rev Immunol* **8:** 958–969.
174. **Clynes R, Maizes JS, Guinamard R, Ono M, Takai T, Ravetch JV.** 1999. Modulation of immune complex-induced inflammation in vivo by the coordinate expression of activation and inhibitory Fc receptors. *J Exp Med* **189:**179–185.
175. **Sutterwala FS, Noel GJ, Clynes R, Mosser DM.** 1997. Selective suppression of interleukin-12 induction after macrophage receptor ligation. *J Exp Med* **185:**1977–1985.
176. **Dhodapkar KM, Banerjee D, Connolly J, Kukreja A, Matayeva E, Veri MC, Ravetch JV, Steinman RM, Dhodapkar MV.** 2007. Selective blockade of the inhibitory Fcγ receptor (FcγRIIB) in human dendritic cells and monocytes induces a type I interferon response program. *J Exp Med* **204:**1359–1369.
177. **Brownlie RJ, Lawlor KE, Niederer HA, Cutler AJ, Xiang Z, Clatworthy MR, Floto RA, Greaves DR, Lyons PA, Smith KG.** 2008. Distinct cell-specific control of autoimmunity and infection by FcγRIIb. *J Exp Med* **205:**883–895.
178. **Clatworthy MR, Smith KG.** 2004. FcγRIIb balances efficient pathogen clearance and the cytokine-mediated consequences of sepsis. *J Exp Med* **199:**717–723.
179. **Karassa FB, Trikalinos TA, Ioannidis JP, FcgammaRIIa-SLE Meta-Analysis Investigators.** 2002. Role of the Fcγ receptor IIa polymorphism in susceptibility to systemic lupus erythematosus and lupus nephritis: a meta-analysis. *Arthritis Rheum* **46:**1563–1571.
180. **Karassa FB, Trikalinos TA, Ioannidis JP.** 2004. The role of FcγRIIA and IIIA polymorphisms in autoimmune diseases. *Biomed Pharmacother* **58:**286–291.
181. **van der Pol WL, Jansen MD, Kuks JB, de Baets M, Leppers-van de Straat FG, Wokke JH, van de Winkel JG, van den Berg LH.** 2003. Association of the Fc gamma receptor IIA-R/R131 genotype with myasthenia gravis in Dutch patients. *J Neuroimmunol* **144:**143–147.
182. **van der Pol WL, van den Berg LH, Scheepers RH, van der Bom JG, van Doorn PA, van Koningsveld R, van den Broek MC, Wokke JH, van de Winkel JG.** 2000. IgG receptor IIa alleles determine susceptibility and severity of Guillain-Barré syndrome. *Neurology* **54:** 1661–1665.
183. **Lehrnbecher T, Foster CB, Zhu S, Leitman SF, Goldin LR, Huppi K, Chanock SJ.** 1999. Variant genotypes of the low-affinity Fcγ receptors in two control populations and a review of low-affinity Fcγ receptor polymorphisms in control and disease populations. *Blood* **94:**4220–4232.
184. **Tackenberg B, Jelcic I, Baerenwaldt A, Oertel WH, Sommer N, Nimmerjahn F, Lünemann JD.** 2009. Impaired inhibitory Fcγ receptor IIB expression on B cells

in chronic inflammatory demyelinating polyneuropathy. *Proc Natl Acad Sci U S A* **106**:4788–4792.

185. **Li X, Gibson AW, Kimberly RP.** 2014. Human FcR polymorphism and disease. *Curr Top Microbiol Immunol* **382**:275–302.

186. **Beppler J, Koehler-Santos P, Pasqualim G, Matte U, Alho CS, Dias FS, Kowalski TW, Velasco IT, Monteiro RC, Pinheiro da Silva F.** 2016. Fc gamma receptor IIA (CD32A) R131 polymorphism as a marker of genetic susceptibility to sepsis. *Inflammation* **39**:518–525.

187. **Endeman H, Cornips MC, Grutters JC, van den Bosch JM, Ruven HJ, van Velzen-Blad H, Rijkers GT, Biesma DH.** 2009. The Fcγ receptor IIA-R/R131 genotype is associated with severe sepsis in community-acquired pneumonia. *Clin Vaccine Immunol* **16**:1087–1090.

188. **Salmon JE, Edberg JC, Brogle NL, Kimberly RP.** 1992. Allelic polymorphisms of human Fcγ receptor IIA and Fcγ receptor IIIB. Independent mechanisms for differences in human phagocyte function. *J Clin Invest* **89**: 1274–1281.

189. **González D, Castro OE, Kourí G, Perez J, Martinez E, Vazquez S, Rosario D, Cancio R, Guzman MG.** 2005. Classical dengue hemorrhagic fever resulting from two dengue infections spaced 20 years or more apart: Havana, Dengue 3 epidemic, 2001-2002. *Int J Infect Dis* **9**:280–285.

190. **Kliks SC, Nimmanitya S, Nisalak A, Burke DS.** 1988. Evidence that maternal dengue antibodies are important in the development of dengue hemorrhagic fever in infants. *Am J Trop Med Hyg* **38**:411–419.

191. **Beltramello M, Williams KL, Simmons CP, Macagno A, Simonelli L, Quyen NT, Sukupolvi-Petty S, Navarro-Sanchez E, Young PR, de Silva AM, Rey FA, Varani L, Whitehead SS, Diamond MS, Harris E, Lanzavecchia A, Sallusto F.** 2010. The human immune response to Dengue virus is dominated by highly cross-reactive antibodies endowed with neutralizing and enhancing activity. *Cell Host Microbe* **8**:271–283.

192. **Vaughn DW, Green S, Kalayanarooj S, Innis BL, Nimmannitya S, Suntayakorn S, Endy TP, Raengsakulrach B, Rothman AL, Ennis FA, Nisalak A.** 2000. Dengue viremia titer, antibody response pattern, and virus serotype correlate with disease severity. *J Infect Dis* **181**: 2–9.

193. **Halstead SB, O'Rourke EJ.** 1977. Dengue viruses and mononuclear phagocytes. I. Infection enhancement by non-neutralizing antibody. *J Exp Med* **146**:201–217.

194. **Moi ML, Takasaki T, Saijo M, Kurane I.** 2013. Dengue virus infection-enhancing activity of undiluted sera obtained from patients with secondary dengue virus infection. *Trans R Soc Trop Med Hyg* **107**:51–58.

195. **Blackley S, Kou Z, Chen H, Quinn M, Rose RC, Schlesinger JJ, Coppage M, Jin X.** 2007. Primary human splenic macrophages, but not T or B cells, are the principal target cells for dengue virus infection in vitro. *J Virol* **81**:13325–13334.

196. **Kou Z, Lim JY, Beltramello M, Quinn M, Chen H, Liu S, Martinez-Sobrido L, Diamond MS, Schlesinger JJ, de Silva A, Sallusto F, Jin X.** 2011. Human antibodies against dengue enhance dengue viral infectivity without suppressing type I interferon secretion in primary human monocytes. *Virology* **410**:240–247.

197. **Flipse J, Wilschut J, Smit JM.** 2013. Molecular mechanisms involved in antibody-dependent enhancement of dengue virus infection in humans. *Traffic* **14**:25–35.

198. **Modhiran N, Kalayanarooj S, Ubol S.** 2010. Subversion of innate defenses by the interplay between DENV and pre-existing enhancing antibodies: TLRs signaling collapse. *PLoS Negl Trop Dis* **4**:e924. doi:10.1371/journal.pntd.0000924.

199. **Ubol S, Phuklia W, Kalayanarooj S, Modhiran N.** 2010. Mechanisms of immune evasion induced by a complex of dengue virus and preexisting enhancing antibodies. *J Infect Dis* **201**:923–935.

200. **OhAinle M, Balmaseda A, Macalalad AR, Tellez Y, Zody MC, Saborío S, Nuñez A, Lennon NJ, Birren BW, Gordon A, Henn MR, Harris E.** 2011. Dynamics of dengue disease severity determined by the interplay between viral genetics and serotype-specific immunity. *Sci Transl Med* **3**:114ra128. doi:10.1126/scitranslmed.3003084.

201. **Leitmeyer KC, Vaughn DW, Watts DM, Salas R, Villalobos I, de Chacon, Ramos C, Rico-Hesse R.** 1999. Dengue virus structural differences that correlate with pathogenesis. *J Virol* **73**:4738–4747.

202. **Balmaseda A, Hammond SN, Pérez L, Tellez Y, Saborío SI, Mercado JC, Cuadra R, Rocha J, Pérez MA, Silva S, Rocha C, Harris E.** 2006. Serotype-specific differences in clinical manifestations of dengue. *Am J Trop Med Hyg* **74**:449–456.

203. **Rico-Hesse R, Harrison LM, Salas RA, Tovar D, Nisalak A, Ramos C, Boshell J, de Mesa MT, Nogueira RM, da Rosa AT.** 1997. Origins of dengue type 2 viruses associated with increased pathogenicity in the Americas. *Virology* **230**:244–251.

204. **Morrison J, Laurent-Rolle M, Maestre AM, Rajsbaum R, Pisanelli G, Simon V, Mulder LC, Fernandez-Sesma A, García-Sastre A.** 2013. Dengue virus co-opts UBR4 to degrade STAT2 and antagonize type I interferon signaling. *PLoS Pathog* **9**:e1003265. doi:10.1371/journal.ppat.1003265.

205. **Stephens HA, Klaythong R, Sirikong M, Vaughn DW, Green S, Kalayanarooj S, Endy TP, Libraty DH, Nisalak A, Innis BL, Rothman AL, Ennis FA, Chandanayingyong D.** 2002. HLA-A and -B allele associations with secondary dengue virus infections correlate with disease severity and the infecting viral serotype in ethnic Thais. *Tissue Antigens* **60**:309–318.

206. **Ryan EJ, Dring M, Ryan CM, McNulty C, Stevenson NJ, Lawless MW, Crowe J, Nolan N, Hegarty JE, O'Farrelly C.** 2010. Variant in CD209 promoter is associated with severity of liver disease in chronic hepatitis C virus infection. *Hum Immunol* **71**:829–832.

207. **Loke H, Bethell D, Phuong CX, Day N, White N, Farrar J, Hill A.** 2002. Susceptibility to dengue hemorrhagic fever in Vietnam: evidence of an association with variation in the vitamin D receptor and Fcγ receptor IIa genes. *Am J Trop Med Hyg* **67**:102–106.

208. **Mohsin SN, Mahmood S, Amar A, Ghafoor F, Raza SM, Saleem M.** 2015. Association of FcγRIIa polymorphism with clinical outcome of dengue infection: first insight from Pakistan. *Am J Trop Med Hyg* **93:** 691–696.

209. **García G, Sierra B, Pérez AB, Aguirre E, Rosado I, Gonzalez N, Izquierdo A, Pupo M, Danay Díaz DR, Sánchez L, Marcheco B, Hirayama K, Guzmán MG.** 2010. Asymptomatic dengue infection in a Cuban population confirms the protective role of the RR variant of the FcγRIIa polymorphism. *Am J Trop Med Hyg* **82:** 1153–1156.

210. **Monsalvo AC, Batalle JP, Lopez MF, Krause JC, Klemenc J, Hernandez JZ, Maskin B, Bugna J, Rubinstein C, Aguilar L, Dalurzo L, Libster R, Savy V, Baumeister E, Aguilar L, Cabral G, Font J, Solari L, Weller KP, Johnson J, Echavarria M, Edwards KM, Chappell JD, Crowe JE Jr, Williams JV, Melendi GA, Polack FP.** 2011. Severe pandemic 2009 H1N1 influenza disease due to pathogenic immune complexes. *Nat Med* **17:**195–199.

211. **Guihot A, Luyt CE, Parrot A, Rousset D, Cavaillon JM, Boutolleau D, Fitting C, Pajanirassa P, Mallet A, Fartoukh M, Agut H, Musset L, Zoorob R, Kirilovksy A, Combadière B, van der Werf S, Autran B, Carcelain G, Flu BALSG, FluBAL Study Group.** 2014. Low titers of serum antibodies inhibiting hemagglutination predict fatal fulminant influenza A(H1N1) 2009 infection. *Am J Respir Crit Care Med* **189:**1240–1249.

212. **Stavenhagen JB, Gorlatov S, Tuaillon N, Rankin CT, Li H, Burke S, Huang L, Vijh S, Johnson S, Bonvini E, Koenig S.** 2007. Fc optimization of therapeutic antibodies enhances their ability to kill tumor cells *in vitro* and controls tumor expansion *in vivo* via low-affinity activating Fcgamma receptors. *Cancer Res* **67:**8882–8890.

213. **Weng WK, Levy R.** 2003. Two immunoglobulin G fragment C receptor polymorphisms independently predict response to rituximab in patients with follicular lymphoma. *J Clin Oncol* **21:**3940–3947.

214. **Persky DO, Dornan D, Goldman BH, Braziel RM, Fisher RI, Leblanc M, Maloney DG, Press OW, Miller TP, Rimsza LM.** 2012. Fc gamma receptor 3a genotype predicts overall survival in follicular lymphoma patients treated on SWOG trials with combined monoclonal antibody plus chemotherapy but not chemotherapy alone. *Haematologica* **97:**937–942.

215. **Kim DH, Jung HD, Kim JG, Lee JJ, Yang DH, Park YH, Do YR, Shin HJ, Kim MK, Hyun MS, Sohn SK.** 2006. *FCGR3A* gene polymorphisms may correlate with response to frontline R-CHOP therapy for diffuse large B-cell lymphoma. *Blood* **108:**2720–2725.

216. **Hilchey SP, Hyrien O, Mosmann TR, Livingstone AM, Friedberg JW, Young F, Fisher RI, Kelleher RJ Jr, Bankert RB, Bernstein SH.** 2009. Rituximab immunotherapy results in the induction of a lymphoma idiotype-specific T-cell response in patients with follicular lymphoma: support for a "vaccinal effect" of rituximab. *Blood* **113:**3809–3812.

Myeloid Cells in Health and Disease: A Synthesis
Edited by Siamon Gordon

doi:10.1128/microbiolspec.MCHD-0034-2016

Michael L. Dustin[1]

23 Complement Receptors in Myeloid Cell Adhesion and Phagocytosis

INTRODUCTION

Complement is a system of blood plasma proteins that play critical roles in host defense through attracting leukocytes to sites of inflammation, mediating myeloid cell uptake and destruction of microbes, and guiding B- and T-cell activation (1, 2). Regardless of the activation mechanism, the complement cascade converges on generation of third component of complement (C3) convertases that cleave C3 to C3a and C3b. The N-terminus of the C3α subunit is the anaphylatoxin (ANA) domain that becomes C3a after cleavage. C3b consists of two subunits containing eight macroglobulin-like domains (MG1 to -8). The β subunit consists of MG1 to MG5 plus the N-terminal half of MG6. The α subunit starts with the C-terminal half of MG6; a C1r/C1s, Uegf, and bone morphogenetic protein-1 (CUB) domain and a thioester domain (TED) inserted between MG7 and MG8; followed by the "anchor" and C345C domain (the trapezoid in Fig. 1).

Classical antibody-mediated, lectin-mediated, and the alternative thioester hydrolysis-mediated pathways for complement activation operate through self-amplifying zymogen cascades focused around proteolytic processing of C3 to generate C3 convertases that include C3b. The activating step is cleavage of C3 to C3a and C3b. This results in a large conformational change in the TED domain that exposes the thioester bond, and this reacts covalently to immune complexes and microbial surfaces (3). The proteases that mediate this cleavage include components of C1q in antibody-mediated complement activation (classical pathway), mannose-binding protein by surfaces such as yeast cell walls (lectin pathway), and hydrolysis of the thioester bond in soluble C3 (alternative pathway). C3b then partners with C4b or C5b to form complexes that convert more C3 to C3a and C3b to amplify the response. The C3 convertases also activate C5 to C5a and C5b, with C5b leading to assembly of the pore-forming membrane attack complex by C6 to C9. The small, soluble C3a and C5a fragments are referred to as "anaphylatoxins" and play an important role as chemoattractants for myeloid cells.

Host cells express a number of regulatory factors that inhibit the complement cascade at different steps, by blocking formation of the C3 convertases, accelerating their inactivation, or blocking effector mechanisms such as membrane attack complex assembly.

[1]Kennedy Institute of Rheumatology, Nuffield Department of Orthopedics, Rheumatology and Musculoskeletal Sciences, The University of Oxford, Headington, OX3 7FY, United Kingdom.

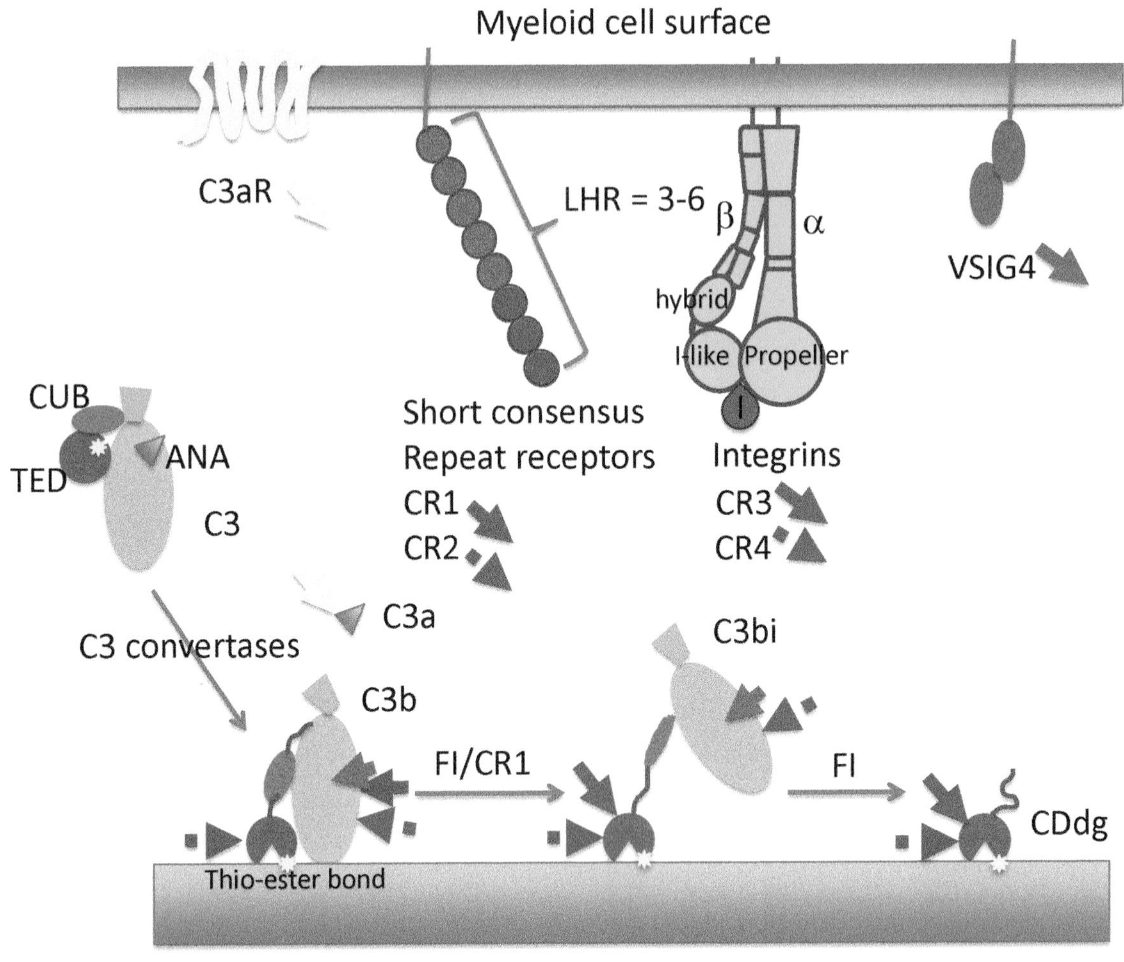

Figure 1 Receptors for products of C3 expressed on human macrophages. Human macrophages differentiated from CD14⁺ monocytes with granulocyte-monocyte colony-stimulating factor express all the major complement receptors, including C3aR, C5aR, CR1, CR3, and VSIG4. The arrows with the receptor names indicate approximate binding-site location within the schematic of C3 breakdown products that are released in the production of C3a, C3b, and its covalently attached products iC3b and C3d. The upper part of the schematic is the macrophage surface and the lower part is a microbial surface bearing the complement components.

This protects host cells to some extent from background complement activation and means that the alternative pathway activation can target microbial surfaces lacking these regulatory components by a "missing self" process. Complement deposition mediated by the classical pathway can overcome this regulation and damage host cells in the context of autoimmunity or excessive immune complex formation. Some of the complement receptors we will discuss also act as regulatory components (Table 1).

Soluble C3a, C3b clusters, and the covalently attached breakdown products of C3b, anchored to microbial surfaces, serve as ligands for three groups of type I transmembrane complement receptors (2, 4). Complement receptors 1 and 2 (CR1 and CR2) are members of the SCR family; complement receptors 3 and 4 (CR3 and CR4) are members of the integrin family; and complement receptor immunoglobulin-like (CRIg), which we will refer to using its gene name as variable-set immunoglobulin-like domain 4 (VSIG4), is a member of the Ig superfamily. Human CR1 is highly expressed on myeloid cells in addition to erythrocytes, whereas VSIG4 is expressed on selected macrophage subpopulations including Kupffer cells in the liver and

Table 1 Complement receptors involved in adhesion, migration, and phagocytosis[a]

Protein	Gene(s)	Ligand(s)	Cells	Function(s)
C3aR	*C3aR*	C3a	MF, DC, hN, hMC, hEO	Chemotaxis, activation
C5aR	*C5aR*	C5a	MF, MO, N, T	Chemotaxis, activation
CR1	*hCr1/mCr2*	C3b, C4b	N, EO, BASO, MF	IC clearance, C3b/C4b regulation, phagocytosis
CR2	*hCr2/mCr2*	C3d	B, FDC	IC capture
CR3	*ITGAM* *ITGB2*	iC3b, C3d, ICAM-1, fibrinogen	MF, N, EO, BASO, DC, actT	Phagocytosis, migration
CR4	*ITAGAX* *ITGB2*	iC3b, ICAM-1, ICAM-2, VCAM, denatured proteins	DC, N, MF, actT	Phagocytosis, migration
CRIg/VSIG4	*VSIG4*	C3b, iC3b	MF, MC, Kupffer	IC clearance, phagocytosis, regulation

[a]Abbreviations: B, B cell; Baso, basophil; EO, eosinophil; FDC, follicular dendritic cell; h, human; IC, immune complex; m, mouse; MC, mast cell; MF, macrophage; MO, monocyte; N, neutrophil; T, T cell.

peritoneal macrophages in the mouse and macrophages and mast cells in humans (http://www.immgen.org). Further processing of C3b by factor I to iC3b (attached) and C3f (released) and then to C3d (attached) and C3c (released) through a second cleavage. C3b, iC3b, and C3d tend to cluster on microbial surfaces due to the focal nature of the amplification process, and this clustered configuration is optimal for CR3 function (4). CR3, CR4, and VSIG4 all bind to iC3b. C3d is a ligand for CR2 and CR3. Human CR2 is expressed on B cells and follicular dendritic cells (DCs) and interacts with C3d on soluble immune complexes, whereas CR3 and CR4 interact with clustered iC3b on opsonized particles in different ways (5–7). CR3 retains binding to C3d, which is essentially the TED domain of C3, whereas VSIG4 and CR4 bind to C3c, which is released as a soluble fragment when iC3b is cleaved a second time by factor I (8). Due to the low affinity of these interactions, it is unlikely that soluble C3c will interfere significantly with binding to clustered iC3b on surfaces.

While acting as complement receptors in adhesion and phagocytosis, CR1 and VSIG4 both have regulatory activity. CR1 is a cofactor for factor I-mediated cleavage of C3b to iC3b, and VSIG4 blocks the interaction of C3b in the alternative pathway C3 convertase with C5, which is necessary for the alternative pathway of complement activation to mediate C5a release and formation of membrane attack complexes (9). Thus, VSIG4 is a negative regulator of complement activation by the alternative pathway (10). Interestingly, VSIG4 has also been studied as a checkpoint regulator for T-cell activation through an unknown mechanism, although it is assumed to be a checkpoint receptor like programmed death 1 or cytotoxic-T-lymphocyte-associated antigen 4 (11). We will discuss this effect of VSIG4 in the context of proposed roles for complement proteins in T-cell activation. This chapter will attempt to reconcile what we know about the cellular distribution of these myeloid cell complement receptors, their biophysical capabilities, and data on their contributions to immune responses *in vivo*.

G PROTEIN-COUPLED RECEPTOR FAMILY

C3a and C5a are small, basic, four-helix bundle proteins that bind to the synonymous G protein-coupled receptors (GPCRs) C3aR and C5aR (12). GPCRs are a large family of multispanning transmembrane receptors with an extracellular N terminus lacking a signal peptide, seven transmembrane helices, and a C-terminal cytoplasmic domain. The extracellular N terminus and loops form a ligand-binding surface that mediates a conformational change in the transmembrane helices that is directly transmitted to the cytoplasmic loops and C terminus, where heterotrimeric G proteins are docked. Ligand binding to the GPCR liberates the α subunit of the heterotrimeric G protein, leaving the active $\beta\gamma$ complex, which can activate phosphatidylinositol 3-kinase and phospholipase C-β (PLC-β), associated with the receptor. C3aR and C5aR are also coupled to signal transduction pathways through pertussis toxin-sensitive Gα_i (13). It has been shown that C5aR also activates p38 mitogen-activated kinase and this inhibits chemotaxis to leukotriene B_4 and interleukin-8. It is thought that this p38-mediated signaling prioritizes the microbe-targeting C5a gradient over host cell-derived attractants that provide more general guidance into the tissue (14). C3aR and C5aR are both highly expressed on basophils and more moderately expressed on mast cells and smooth muscle cells. C3aR is expressed strongly on thioglycolate-elicited macrophages in mice and at lower levels on DCs and other macrophage populations. C5aR is highly expressed on neutrophils and a wide range of tissue macrophages, and more moderately on monocytes. C3aR and C5aR are also coupled to contraction in smooth muscle cells through a cascade involving release of the nucleotide UDP, which binds to a GPCR coupled to $G\alpha_q$ and PLC-β, that triggers Ca^{2+} elevation (15). Due to its expression pattern in neutrophils and monocytes, C5aR is particularly important in recruitment of leukocytes into sites of inflammation. C3aR and C5aR are reported to play a role in T-cell activation, as will be discussed below (16), but microarray and RNA sequencing data in mouse suggest very low expression compared to macrophages (C3aR) or macrophages and neutrophils (C5aR). Both C3a and C5a can be further processed by carboxypeptidases to yield des-Arg forms with modified biological activity (17). There is a second C5aR, referred to as C5L2 (encoded by gene *C5AR2*), which is either a decoy receptor that removes C5a and may reinforce gradients or may also directly modulate the activity of C5aR by recruiting β-arrestin (18). In other systems, decoy receptors can be involved in setting up gradients (19), but this is not established for C5L2.

Genetic studies demonstrate nonredundant roles of C3aR and C5aR in various pathological settings. C3aR has a pronounced role in allergic inflammation in mouse models, and C3a is particularly abundant in lavage fluid of humans exposed to allergens (20). C5aR has a nonredundant role in protection of the host against bacterial pathogens at mucosal surfaces, which goes beyond neutrophil recruitment (21). Either C3aR or C5aR deficiency in mice reduced the severity of collagen-induced arthritis in mice, with C5aR appearing to play a greater role (22). Consistent with this, antibody-mediated blockade of C5aR eliminates mouse collagen-induced arthritis and reduces levels of multiple cytokines (23). However, attempts to treat human rheumatoid arthritis with a small-molecule C5aR antagonist were not successful (24).

SHORT CONSENSUS REPEAT SUPERFAMILY

Many complement regulatory proteins, scavenger receptors, and a family of low-density lipoprotein-like receptors share a repeated domain referred to as a short consensus repeat (SCR). The SCR domain is ~60 amino acids long and has six β-strands in two β-sheets stabilized by two highly conserved disulfide bonds. CR1 (CD35) and CR2 (CD21) have extracellular domains that are entirely composed of long homologous repeats (LHRs) containing six or seven SCRs that reflect functional complement protein-binding units. The genomic structures of the loci encoding CR1 and CR2 are different in primates (including humans) and nonprimates (including mice), making mouse models only partially applicable to the human situation (25) (Fig. 2).

In mice and other nonprimates, CR2 consists of four LHRs and binds C3d on immune complexes. The expression of mouse CR2 protein is largely restricted to B cells (26). In contrast, follicular DCs, a stromal cell type that define the B-cell follicle and retain immune complexes on their surface for capture by germinal-center B cells, express the CR1 protein based on alternative splicing that adds an additional LHR that binds C3b and C4b (26). This N-terminal LHR is highly related to a distinct and more widely expressed C4b- and

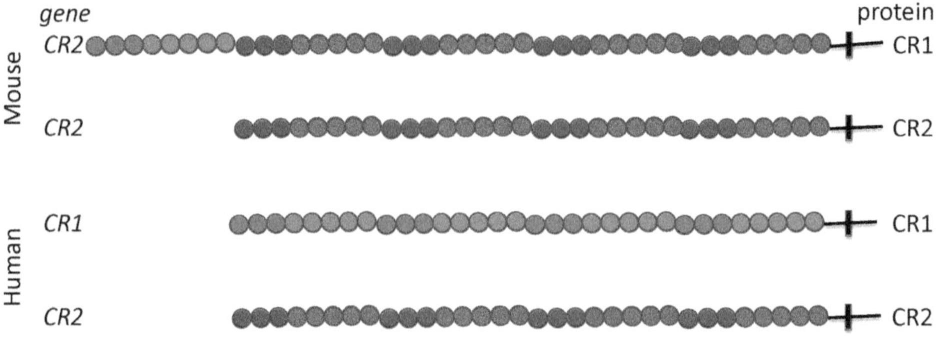

Figure 2 Schematic of the mouse and human complement receptor gene products. In the mouse, CR1 and CR2 proteins are derived from alternative splicing of the *Cr2* gene. Mouse CR1 and CR2 are not present on myeloid cells. In humans, the CR1 and CR2 proteins are products of different genes. Human CR1 is expressed on myeloid cells and functions in phagocytosis in addition to clearance of immune complexes bearding C3b and/or C4b. CR1 also acts as a cofactor for factor I in conversion of C3b to iC3b and, further, to C3d. Each ball is an SCR, and each group of seven repeats is referred to as an LHR.

C3b-binding protein called Crry (also referred to as CR1-like, or CR1L). Both the N-terminal LHR of mouse CR1 and the single LHR of Crry have cofactor activity for factor I, and thus have an important role in control of C3 convertase activity by converting C3b into iC3b and C3d (25), although a number of soluble and surface-bound complement regulatory factors also inhibit or facilitate inactivation of C3 convertases. Neither mouse CR1 nor CR2 is highly expressed in myeloid cells. Crry is expressed on myeloid cells, but at moderate levels, and has not been clearly associated with effector functions like adhesion or phagocytosis.

In humans, a distinct gene encodes the CR1 mRNA and protein, with three to six C4b- and/or C3b-binding LHRs in different allelic variants, with four LHRs in the most common alleles (27). Human CR1 is widely expressed, including on myeloid cells, erythrocytes, and some epithelial cells. For example, on the surface of kidney podocytes, CR1 facilitates the conversion of C3b to C3d with a half time of ~30 min *in vitro*, potentially contributing to protection of the kidney filtration apparatus from immune complex-mediated damage, along with other regulators such as decay-accelerating factor (28). Monomeric C3b binds to CR1 LHRs with a K_d (dissociation constant) in the 10^{-6} M range in physiological salt conditions (29). Caution is needed in reading the literature, as measurements are often performed with oligomeric forms of C3b and/or at low salt, where the apparent K_d is in the 10^{-9} M range (29). Thus, CR1 is a high-avidity receptor for polyvalent binding to clustered ligands on immune complexes or surfaces. The low affinity and fast off rate are typical of surface receptors that bind multivalent or particulate ligands.

Given the differences between mouse and human CR1 genes, it is difficult to predict the function of human CR1 based on mouse models and knockout mice. Mice deficient in all variants of CR1 and CR2, but expressing Crry, have impaired B-cell responses, including loss of IgG3, and are more susceptible to *Pseudomonas*-mediated pneumonia than are wild-type mice (30). Mice lacking only CR1 on follicular DCs have milder defects and are not as susceptible to *Pseudomonas* pneumonia as are mice lacking both CR1 and CR2 (26). The role of CR1 in myeloid cells' effector functions, such as phagocytosis, cannot be studied in mice, although studies in mice expressing human CR1 on erythrocytes demonstrate a role in immune complex clearance (31).

In vitro studies with human neutrophils and macrophages demonstrate that CR1 plays a role in phagocytosis (6, 32), in parallel with CR3 (see below). There is a general concept that phagocytosis mediated by CR1 requires activation of the myeloid cell through either Fc receptors or other immunoreceptors or through prior activation/differentiation of the myeloid cell (33). CR1 is more clustered on erythrocytes compared to neutrophils (34). Erythrocytes shuttle immune complexes from the blood to phagocytes in the liver and spleen. The greater basal clustering of CR1 on erythrocytes may enable greater binding of soluble immune complexes, whereas neutrophils with less basal clustering bound few $C3b^+$ immune complexes despite higher CR1 expression. Activation of neutrophils with the chemotactic tripeptide formyl-methionyl-leucyl-phenylalanine increased CR1 expression but didn't increase its clustering. The potential importance of clustering in ligand discrimination is consistent with the relatively low affinity and transient interaction of monomeric C3b with single CR1 LHR. CR1 has a short cytoplasmic domain that has received surprisingly little detailed attention, although studies suggest that CR1 participates in PLC activation (6). The C terminus of CR1 interacts with the PDZ domains of Fas-associated phosphatase-1 (FAP-1) in erythrocytes, which may account for CR1 clustering (35). The biophysical implications for the large size of CR1 in the context of immune receptor signaling will be discussed further below.

INTEGRIN FAMILY COMPLEMENT RECEPTORS

Integrins are a large family of heterodimeric transmembrane glycoproteins expressed in vertebrates and some invertebrates (36). They are specialized for cell-extracellular matrix and cell-cell adhesion and particularly the integration of adhesion with the cytoskeleton, transduction of mechanical signals, and transmission of force to the cell's environment (37). Ligand binding by integrins is unique among complement receptors in being metal ion dependent. In the immune system, physical processes such as leukocyte extravasation, some modes of interstitial migration, and phagocytosis of particles by myeloid cells are dependent on members of the integrin family. Subfamilies of integrin family adhesion molecules are organized around shared subunits, with both subunits contributing to ligand specificity and specific cytoskeletal interactions that define functions. The subunits are named in the traditional biochemical convention with the larger subunit referred to as α and the smaller as β, and then subscripts are used to designate the specific gene products that make up the receptor (38). For example, pairing of the α_M (gene *ITGAM*) and β_2 (gene *ITGB2*) subunits forms CR3, the major receptor for iC3b and the further breakdown product C3d, whereas pairing of the α_X (*ITGAX*) and β_2 (*ITGB2*) subunits forms CR4, a receptor with affinity for iC3b but not C3d. The key role of integrins in myeloid cell function is dramatically illustrated by the primary immunodeficiency leukocyte adhesion deficiency type 1 (LAD1), in which four leukocyte integrins, including CR3 and CR4, are deficient due to mutation or deletion of the shared β_2 subunit (39). These patients suffer recurrent bacterial infections and die in childhood unless treated by allogeneic bone marrow transplantation. The β_2 integrins play key roles in the extravasation of blood-borne leukocytes through binding intercellular adhesion molecule-1 (ICAM-1) and other ligands on endothelial cells. Thus, interpreting the phenotype of LAD1 patients required consideration of both the complement type and noncomplement ligands. Myeloid cells also express a number of other integrin family members that function through interactions with extracellular matrix proteins, cell surface adhesion ligands, and transforming growth factor β activation (40), as will be discussed below.

INTEGRIN STRUCTURE

Integrin α subunits are type I transmembrane proteins with an N-terminal β-propeller domain, a series of globular domains forming a large, segmented leg connected to the single transmembrane domain and short cytoplasmic domain (41). Some of the α subunits, including CR3 and CR4, have an insertion into the β-propeller domain of a 200-amino-acid inserted, or I, domain with a "dinucleotide binding fold"—essentially a β-sheet flanked by seven α-helices (7, 36). Divalent metal ion-binding helix-loop-helix motifs similar to the EF hand of parvalbumin are present in the β-propeller domain, and a single divalent metal ion-binding metal-dependent adhesion site (MIDAS) is present at the apex of the I domain. The typically smaller β subunit has a conserved dinucleotide binding fold (I-like) domain inserted into the "hybrid" domain that is connected to both the N and C termini of the I-like domain. Changes in the angle of the hybrid domain with respect to the I-like domain shift the C-terminal α-helix of the I-like domain and change the conformation of the MIDAS and affinity for divalent cation-dependent ligand binding (41). This hybrid domain links to a segmented leg domain characterized by an abundance of disulfide bonds, a single transmembrane domain, and a short cytoplasmic domain. When the α and β subunits are paired, the α subunit's β-propeller domain and the β subunit's I-like domain come together to form a single globular headpiece, from which the hybrid domain projects with a variable angle. The segmented leg domains can either be extended or bend at an intersegment genu. When the legs are straight, the integrin headpiece is projected ~20 nm from the cell surface, making it a large structure on the cell surface, but when the genu is bent, the headpiece is held a few nanometers from the cell surface. The transmembrane domains are together in the genu bent conformation, whereas they are separated in the extended form such that cytoplasmic interactions that cause the integrin to extend are thought to separate the transmembrane domains by disrupting weak interactions between the transmembrane domains and the membrane-proximal segments of the cytoplasmic domains (the "clasp"). The ligand-binding subunit of integrins lacking I domains binds ligands with the RGD core sequence (or similar with a conserved D or E) that are extended from globular protein domains on loops. The α-subunit I domain is the ligand-binding site for integrins with this insertion. The conformational change in the headpiece apparently operates like a "bell rope" at the C-terminal end of the I domain to alter its conformation and ligand-binding properties by moving the C-terminal α-helix to control the conformation of the ligand-binding site. The MIDAS motifs bind Mg^{2+} in either CR3 or CR4. The divalent cation coordinates with a key acidic residue in the ligand to complete the metal coordination. The affinity for ligand is low when the hybrid domain is close to the α-subunit thigh domain (closed = bell rope slack) or high when the hybrid domain is pulled away from the thigh domain (open = bell rope tight) (42). The range of affinities that can be achieved varies with the particular integrin-ligand pair and functional context. Cryoelectron microscopy of integrins has revealed additional possible structures, one of which appears to crouch, rather than genuflect, and may be capable of binding ligand in close membrane contacts (43).

The cytoplasmic tails of integrins bind to a number of proteins, but the best defined is talin, which has an N-terminal trefoil FERM domain and a long tail that may set the optimal spacing for functional integrin clusters (44, 45). Talin binds to a membrane-proximal motif in the β_2 cytoplasmic domain that is associated with integrin extension (44). Talin forms a bridge between the engaged integrin and F-actin (46). It was first described in focal adhesion but is also prominently accumulated with the immunological synapse between T cells and B cells (47). Talin is required for CR3-mediated phagocytosis (48). Application of forces to talin through the link between the extracellular matrix and actin-myosin contractile machinery exposes binding sites for vinculin, which translates force exerted on talin into a biochemical binding event and signal (49). The force required to induce vinculin binding is 5 pN. Thus, CR3 and CR4 may impart mechanical sensitivity to the phagocytosis process. Force sensing is also important for bacterial adhesion through fimbria protein FimH (50), which interacts with the Ig family ligand CD48 on phagocytes and diverted internalized bacteria into nonacidified organelles in which they can survive within the phagocyte (51).

LEUKOCYTE INTEGRINS

Integrin subfamilies are defined by three subunits that engage in the most functional heterodimers. These are the α_V, β_1, and β_2. Members of each of these families are expressed on leukocytes, but the β_2 subfamily is restricted to leukocytes (with the exception of some transformed stromal cells) and is thus referred to as the leukocyte integrin subfamily. We can discuss all three subfamilies in terms of functional relevance to myeloid cells, but the complement-binding integrins are members of the β_2 subfamily, as discussed already.

The β_2 integrins were first identified by monoclonal antibodies to myeloid differentiation antigens and lymphocyte function-associated antigens (LFAs), which were selected by inhibition of cytotoxic-T-lymphocyte-mediated killing (38). Springer was the first to note that these distinct heterodimers were related by a shared β subunit, and subsequently the cloning of the cDNA for the β_2 subunit defined the integrin family after the β_1 (CD29) subunit was cloned by Hynes (52). There are four members of the β_2 subfamily: αL (LFA-1; CD11a/CD18), αM (Mac-1; CR3; CD11b/CD18), αX (p150/95; CR4; CD11c/CD18), and αD (CD11d/CD18). For simplicity we will use LFA-1, CR3, CR4, and CD11d to refer to these heterodimers, as these are the most relevant for this discussion around myeloid complement receptors. There are function-blocking antibodies to CD18, which will act on all heterodimers, and blocking antibodies to the α subunits that are specific for individual heterodimers.

The same screens that generated function-blocking antibodies to LFA-1 also identified another set of heterodimers that were identified based on expression on the activated lymphocytes, but not on most lymphocytes freshly isolated from blood. These antigens were expressed "late" in activation of lymphocytes after the cells had undergone clonal expansion and became postmitotic effector cells and were thus called very late activation antigens (VLAs) (53). These turned out to be β_1 integrins, and the literature still contains many references to VLA-4 ($\alpha_4\beta_1$), which is the target for an important therapeutic antibody used in treatment of multiple sclerosis (54).

There are a number of useful antibodies to β_2 integrins in addition to the function-blocking antibodies (55). For example, KIM127 is an epitope on the β_2-subunit calf domain that only binds to the extended conformation due to masking of the epitope in the genuflected conformation (56). Thus, it can be used to study the temperature and activation-dependent extension of the β_2 integrins. Monoclonal antibody 24 (MAb 24) binds to the hybrid domain of the β_2 subunit (8). MAb 24 binding stabilizes the high-affinity open conformation and is also induced by ligand (57). Thus, it has a net activating effect on integrin-mediated adhesion, but the prolongation of the high-affinity state may impair some functions. The conformation-specific antibodies have been useful tools in analyzing the location of different conformations on cells (58). Caution is needed as both KIM127 and Mab 24 are synergistic with ligand in stabilizing extended or high-affinity forms of β_2 integrins.

LAD1, which was mentioned earlier, is the result of a variety of mutations in the β_2 subunit that either eliminate the protein completely or reduce expression (59). The α and β subunits assemble in the endoplasmic reticulum, and when the β subunit is absent, the α subunit is not expressed on the surface. The consequence of this deficiency is the inability to recruit myeloid cells into sites of inflammation in the gingiva and the skin, resulting in lack of pus formation and life-threatening bacterial infections. The condition can be effectively treated by bone marrow transplantation, and functional defects in lymphocyte may account for a higher success rate for bone marrow transplantation despite the history of the patients with infections (60). The major defect driving this disease appears to be the localization of polymorphonuclear leukocytes to some, but not all, inflamed tissues.

A third variant of LAD to be described, LAD3, is based on the deficiency of the β_2 cytoplasmic domain-binding protein kindlin-3 and results in a disease that combines features of LAD1 and platelet function deficiency, such as Glanzmann thrombasthenia, due to the role of kindlin-3 in the function of leukocyte and platelet integrins (61). Kindlin-3, encoded by gene *FERMT3*, is related to talin but binds to a distinct motif that is more distal from the membrane. While binding of talin alone is sufficient for integrin extension, binding of both talin and kindlin-3 is required to generate the high-affinity form of leukocyte and platelet integrins. Kindlin-3, like talin, is also required for phagocytosis and signaling mediated by CR3 (62).

The most widely expressed leukocyte integrin is LFA-1. It is expressed on essentially all leukocytes and is important for anchorage of some cells in the vasculature, extravasation, and cell-cell interactions. Its ligands are referred to as ICAMs, of which there are five. The major ligands with demonstrated functions are ICAM-1 (CD54), ICAM-2 (CD105), and ICAM-3 (CD50). ICAM-1 is also a ligand for CR3, although LFA-1 and CR3 bind to different regions within ICAM-1 (63, 64). All of the ICAMs are members of the Ig superfamily, which share variable numbers of ~100-amino-acid β-sandwich domains. ICAM-1 and ICAM-2 are highly expressed on vascular endothelial cells, whereas ICAM-3 is mostly expressed on leukocytes in humans, but a similar gene has not been identified in rodents. ICAM-1 expression is regulated by inflammatory cytokines, lymphocytes, and by leukocyte activation (65). ICAM-2, in contrast, is constitutively expressed on endothelial cells and some leukocytes (66). LFA-1 is only known to participate in cell-cell adhesion through binding to ICAMs. It is important for efficient entry of lymphocytes into sites of inflammation and lymphoid tissues. LFA-1 is also required for the retention of marginal-zone B cells in the splenic marginal zone (67). LFA-1 is also a defining component of the immunological synapse formed by T cells with antigen-presenting cells (68, 69). Interestingly, LFA-1 is readily activated for binding to ICAM-1 in monocytes and immature DCs but becomes inactive and refractory to activation in mature DCs (70). The reason for this is not clear, but it may prevent DCs from forming homotypic aggregates that could exclude T cells when ICAM-1 is upregulated during maturation. The low-affinity form of LFA-1 binds ICAM-1 with a K_d of ~1.5 mM, whereas the intermediate- and high-affinity forms bind with K_ds of ~15 M and ~150 nM, respectively, an up-to-10,000-fold increase in affinity (42). The LFA-1–ICAM-1 interaction is highly dynamic *in vivo*, suggesting that leukocytes likely utilize a mixture of intermediate- and high-affinity conformations with a continuum of interaction time frames modulated by force-dependent effects. The structure of the LFA-1 I domain in a complex with the first Ig-like domain of ICAM-1 has been solved and shows the coordination of the Mg^{2+} ion with the E34 residue, which is critical for binding (42).

CR3 is expressed on myeloid cells and activated T-cell subsets and plays an important role in leukocyte interactions with endothelial cells by binding to ICAM-1, although at a distinct site from that bound by LFA-1 (64), and it plays a critical role in phagocytosis of particles that are coated with fragments of the third component of complement, including iC3b (71, 72). Recently the crystal structure of the CR3 I-domain bound to a further breakdown product of iC3b, called C3d, has been solved, revealing the central interaction of the bound Mg^{2+} with an aspartic acid in C3d (7). It is also apparent from this structure that CR3 cannot bind C3b due to a steric conflict with the CUB domain, which is partly unfolded in iC3b (7). In addition to high-affinity ligands like ICAM-1 and iC3b/C3d, CR3 also appears to have multiple low-affinity interactions with large recombinant ligands like iC3b, and it is likely that this is based on recognition of acidic side chains in denatured/unstructured peptides (7, 73).

CR3-deficient macrophages have defects in uptake of *Mycobacterium tuberculosis*, and a significant component of this defect is directly attributable to complement receptor function (74, 75). In mouse models of Alzheimer's disease, C1q, C3, and CR3 expressed by microglial cells are required for destruction of synapses early in the disease process (76). This was a particularly interesting study because unlike others that viewed microglia as reacting to plaques, this study suggests a more direct pathogen role. The role of inflammation and complement in neurodegeneration is complicated, with both protective and pathological roles of complement in different models (77).

CR4 is expressed on DCs, tissue macrophages, and effector T cells. High expression of CR4 is used as a marker of DCs for isolation, and the CR4 (CD11c) promoter has been used to drive expression of fluorescent proteins (78) and the diphtheria toxin receptor for visualization and deletion of DCs, respectively (79). Despite the progress enabled by these tools, there are caveats to exclusive use of CR4 as a marker for DCs. Recently Nussenzweig and colleagues generated an alternative set of tools based on Zbtb46 as being a cleaner promoter for marking or deleting the classical DC lineage (80). CR4 has lower affinity for iC3b than CR3 and does not interact with C3d, instead binding to the released C3c fragment (8). CR4 also has a functionally relevant interaction with ICAM-1, ICAM-2, and vascular cell adhesion molecule-1 (VCAM-1) and thus can mediate attachment of cells to inflamed endothelial cells. As discussed for CR3, CR4 also has the capacity to bind many ligands with exposed acidic side chains, and it has been proposed that an important function of CR4 is to sense the denatured proteins as a danger signal associated with tissue injury (73). The CR4 (CD11c) knockout mouse has a phenotype of reduced atherosclerosis and autoimmunity (81, 82). Interestingly,

T cells elicited by immunization of CR4-deficient mice with myelin oligodendrocyte glycoprotein are not encephalogenic in wild-type mice. It is not clear if this phenotype is T cell intrinsic or if it is imparted to the T cells by defective antigen presentation or microenvironment-related defects in myeloid cells. Effector T cells express significant levels of CR3 and CR4, and complement activation, particularly in relation to CD46, has been implicated in T-cell activation (83–85). It seems likely that some of these effects are dependent on myeloid cells (16), and it will be interesting to see how our understanding of T-cell versus myeloid functions of these receptors progresses.

The integrin CD11d is the least studied of the β_2 integrins and has no known complement receptor function. It is highly expressed in red pulp macrophages and some $\gamma\delta$ T-cell subsets. It has also been shown that T-cell responses to superantigens are reduced in the absence of CD11d (86), similar to results with CR3, although expression of CD11d in T cells seems more restricted than either CR3 or CR4. Antibodies to CD11d have shown promise in enhancing recovery of spinal cord and cerebral cortex following traumatic injury (87).

Myeloid cells express a number of other integrins, including receptors for fibronectin ($\alpha_5\beta_1$), vitronectin ($\alpha_V\beta_3$), laminin ($\alpha_6\beta_1$), and collagen (α_1 and $\alpha_2\beta_1$). Myeloid cells generally lack expression of the VCAM-1 receptor ($\alpha_4\beta_1$), which renders them more dependent on β_2 integrins for extravasation at sites of inflammation. The E-cadherin-binding integrin $\alpha_E\beta_7$ is a marker for a subset of intestinal lamina propria DCs that play important roles in homeostasis with gut microbes (88). DCs also utilize the integrin $\alpha_V\beta_8$ for activation of transforming growth factor β, which is critical for maintaining regulatory T cells (40). Surprisingly, DCs lacking all of the major integrins including α_V, β_1, and β_3 were able to migrate from the skin to the lymph node and localize in the T-cell zones (89). This finding resonates with earlier studies suggesting that amoeboid locomotion can operate by mechanical coupling between cell shape and confining features of the three-dimensional matrix and cellular environment with very low or no adhesion (90).

Ig SUPERFAMILY COMPLEMENT RECEPTORS

Most of the complement regulatory proteins are in the SCR structural family, except VSIG4 (also called CRIg), which competes for binding to C3b with C5, which connects the alternative C3 convertase to terminal complement components. VSIG4 is part of a subfamily of the Ig superfamily with two Ig-like domains and a short cytoplasmic domain (10). It is expressed on Kupffer cells and peritoneal macrophages. It is critical for the ability of Kupffer cells to clear iC3b-opsonized bacteria from the blood and thus contributes to controlling infection (9, 91). VSIG4 also appears to direct phagocytosed cargo to autophagic compartments that also improve control of intracellular bacteria (92). VSIG4 suppresses complement-mediated injury in arthritis models when administered as an Ig-fusion protein (93). Recently VSIG4 has been shown to have direct pattern recognition function in clearance of Gram-positive bacteria from the blood (94).

INTEGRATIVE FUNCTIONS

Myeloid cells move from the bone marrow to tissue sites via the blood and need to cross endothelial barriers to gain access to tissues. Complement proteins, particularly the anaphylatoxins, play an important role but punctuate processes that depend on noncomplement adhesion and guidance systems. At sites of inflammation the endothelium becomes modified to increase adhesiveness in a process that can be acute or chronic. There are three key steps that are classically served by three distinct molecular families: initial adhesion mediated by selectins, rapid activation mediated by chemokine receptors, and firm adhesion and motility mediated by integrins (95, 96).

The rolling adhesion is mediated by the brief and low-valence interaction of selectins to specific glycans that are present in high abundance. The selectins are single-chain type I transmembrane glycoproteins that have N-terminal lectin domains. CD62L is critical for leukocyte rolling in secondary lymphoid tissues, where it binds sialyl Lewis X-based carbohydrate moieties carried on multiple core proteins of the high endothelial cells. Note that the second described form of LAD, LAD2, is based on deficiency in the ability to add fucose sugar chains that are required to make selectin ligands (97). Inflamed endothelial cells rapidly upregulate CD62P on their surface in response to inflammation due to stored CD62P in Weibel-Palade bodies in the endothelial cells. CD62P binds sialyl Lewis X determinants in the context of the core protein P-selectin glycoprotein ligand-1 (PSGL-1), which is also sulfated on tyrosines to form the optimal ligand. In some contexts, PSGL-1 can be expressed without the appropriate carbohydrate modifications, in which case it does not serve as a functional ligand for CD62P. CD62E is upregulated on endothelial cells by inflammation over a period of hours to further increase adhesion of leukocyte-bearing sialyl Lewis X glycans (98).

Once the leukocyte is in transient contact with the endothelial wall, an activating signal is needed to induce firm, integrin-mediated adhesion, or else the leukocyte will detach and resume flowing with the blood until it reaches another capillary bed. The activating signal is generally provided by a "chemoattractant" that is acting in an acute activation mode. Chemokines can be presented by the endothelial cells because they bind to heparan sulfate proteoglycans, but can also be provided by other types of ligands of pertussis toxin-sensitive GPCRs with the requirement that activation is rapid (99). Nonchemokine attractants that can induce arrest of rolling leukocytes include platelet-activating factor and C5a (100).

Chemokine receptor signals rapidly activate integrins such as LFA-1 and CR3 and induce spreading of the leukocyte on the endothelial surface. Leukocytes then extravasate using either junctional or transcellular routes (101). The junctional route requires disengaging or rearranging multiple junctional adhesion mechanisms of the endothelial cells and can be achieved without or with significant vascular leakage depending on other signals (102). An alternative mode of interaction between leukocytes and endothelial cells is intravascular scanning, which was first observed in the liver for natural killer T cells (103) but has now been described more widely for the CX3CR1$^+$ subpopulation of monocytes (104). It is assumed that this is a form of surveillance of the endothelium and endothelial wall, as it is also observed in arterioles and large arteries in which the endothelium is not accessible from the tissue.

The behavior of myeloid cells in tissues is characterized by neutrophil swarming in acute responses to sterile tissue injury or microbe-driven inflammation. The process can be coordinated by cytokines like interleukin-1 in the case of staphylococcal infection or neutrophil apoptosis with release of leukotriene B_4 (105, 106). Neutrophil movement can be highly directed but is not integrin dependent. The convergence of neutrophils in foci above a critical cell number leads to formation of integrin-dependent cell aggregates. These aggregates are sufficiently forceful to rapidly exclude extracellular matrix fibrils without proteolysis. *In vitro* studies suggest a hierarchy of signals, with endogenous signals taking a back seat to microbial signals, like guidance to formylated peptides, that provide direct targeting to the microbe, but this has not been clearly demonstrated *in vivo* (107). As complement activation takes places on microbial surfaces, which will then be sources of anaphylatoxins C3a and C5a, it has been found that C5a is dominant over leukotriene B_4 in attracting neutrophils, as mentioned above (14). In the context of bacterial infections that are not cleared by neutrophils, CCR2$^+$ monocytes are recruited along with NK cells and eventually effector T cells (108). T-cell production of cytokines may guide a response down inappropriate pathways to resolution, but in some situations can lead to inappropriate responses that lead to fatal immunopathology or chronicity (102, 109). Intravascular complement activation in lipopolysaccharide-primed mice results in fatal shock that depends on C5aR (110), but it is not clear if complement deposition on the endothelial cells plays a role in leukocyte interactions or if these interactions are mediated by classical endothelial adhesion molecules like ICAM-1, ICAM-2, and VCAM.

PHAGOCYTIC SYNAPSES

Phagocytosis of foreign bodies and microbes is a critical function of myeloid cells. Foreign objects and microbes are likely to become opsonized with complement and natural antibodies; and may also have evolutionarily conserved patterns that are directly recognized by pattern recognition receptors on the myeloid cells. Activating FcR and pattern recognition receptors like Dectin-1, which recognizes mannans found in fungal cell walls, signal through tyrosine kinase cascades driven by Src family and Syk family kinases, leading to activation of PLC-γ (111). Activating FcRs have noncovalently associated subunits with immunoreceptor tyrosine-based activation motifs (ITAMs), in which a pair of precisely spaced tyrosines are phosphorylated by a Src family kinase and then recruits a Syk family kinase through its two SH2 domains that in turn phosphorylates an adapter to allow PLC-γ recruitment. In the case of Dectin-1, each receptor has a single tyrosine motif and dimerization of the receptors by mannans creates a functional ITAM. As such, each receptor has a half or hemi-TAM (112). This tyrosine kinase cascade is the same mode of signaling utilized by the T-cell antigen receptors, for which exclusion of large phosphatases CD45 and CD148 is a common feature of the immunological synapse—a specialized junction that facilitates signaling and effector functions (113, 114). Dectin-1-mediated phagocytosis of yeast particles has been shown to induce CD45 exclusion as part of the formation of the phagocytic cup, suggesting that a phagocytic synapse may use similar topological strategies (115). This is consistent with the relatively small size of Dectin-1 and its interaction with a relatively flat surface like a yeast cell wall or, experimentally, a latex bead, which together drive CD45 exclusion.

Small receptors like FcR and Dectin-1 generate close contacts that may exclude CD45 and allow local propagation of tyrosine kinase cascades, but it has not been thought that large receptors like integrins or CR1 would be able to mediate such close contacts. It is not clear if CR3-mediated interactions with iC3b or C3d on opsonized particles or in combination with FcR would generate similar phagocytic synapses as part of the activation process. A recent study from Grinstein and colleagues suggests that integrins are capable of orchestrating CD45 exclusion in an F-actin-dependent manner (116). In this study, a ligand for CR3, fibrinogen, was interspersed on a surface with a pattern of IgG to locally engage Fc receptors, which provided activating signals. Rather than simply excluding CD45 from small Fc receptor clusters, a CR3-mediated Arp2/3 complex-dependent mechanism excluded CD45 from large areas that extended 1 to 2 m from the site of FcR activation (116). The Fc receptor signaling activates WASP (Wiskott-Aldrich syndrome protein), which in turn activates the Arp2/3 complex to nucleate branched F-actin networks that can generate protrusive forces, apparently capable of CD45 exclusion. It is not clear how the integrins in this setting are linked to the F-actin, but it is likely that talin and kindlin-3 are involved. It is also of interest to consider if the very large CR1 molecules could similarly become involved in an F-actin-dependent close contact process (Fig. 3). Each LHR of CR1 could extend 25 nm based on the electron microscopy structure of C4-binding protein (117). Thus, CR1 could have a total length approaching 100 nm. Even if it is assumed that CR1 would be hinged and flexible, it would likely generate too large an intermembrane separation when binding to its ligands to exclude CD45 when bound to C3b, which is itself a relatively compact globular protein with the receptor-binding sites very close to the membrane. Evaluating the potential of CR1-mediated interaction with C3b on a surface to exclude CD45 in support of local tyrosine kinase cascades will require direct measurements. Such experiments could provide a complementary data set to the one generated by Grinstein's group for integrins. If CR1 is capable of generating similar F-actin-dependent close contacts following triggering by a smaller Fc receptor, then this would place additional constraints on models to explain this phenomenon. Since the function of CR1 and CR3 overlap in many contexts *in vitro*, it seems plausible that CR1 may in some way access a similar active mechanism for promoting close contacts despite its apparently long reach. The function of this extended CD45 exclusion was not entirely clear, but it was proposed that it allows extension of the phagocytic cup past sparse sites of opsonization. Such a mechanism may also allow CR3 and CR4 to mediate phagocytosis of iC3b-coated particles without an FcR signal.

COMPLEMENT IN T-CELL ACTIVATION

Complement activation has been proposed to play a role in T-cell activation. C3aR and C5aR mediate one mode of complement dependent activation in T-cell priming (16). Kubes and colleagues have also reported that CD4 T cells use C5aR to mediate delayed-type hypersensitivity reactions *in vivo* (118). The complement regulatory protein CD46 mediates a second mechanism (83, 84, 119). CD46 is an SCR family member with four SCR domains, a transmembrane domain, and a cytoplasmic domain that links to polarity proteins (120). CD46 shares activities with CR1 in that it binds C3b and C4b and protects host cells from complement-mediated damage. It is intriguing that VSIG4, which protects the vascular system from alternative pathway activation by clearing hydrolyzed C3 from the blood and preventing its association with C5 (10), is also a negative regulator of T-cell function (11, 92, 121, 122). The high expression of VSIG4 in the liver and peritoneal cavity may help establish a level of immune privilege in these sites (123), although the role of VSIG4 has not been investigated. The receptor for VSIG4 on T cells is not known, but it is intriguing to consider that VSIG4-mediated inhibition of complement activation could be related to its ability to inhibit T-cell activation.

FUTURE DIRECTIONS

The SCR superfamily and integrin family receptors play diverse and indispensable roles in innate immunity, including the regulation of complement cascades, leukocyte localization, and myeloid cell effector function. The relationship between extension and headpiece conformation is still a matter of debate and is challenging to address. While integrins have been studied extensively with electron microscopy and modern data-processing methods (124), most of the electron microscopy studies on complement regulators were carried out earlier. Zhu and colleagues have suggested that LFA-1 may undergo extension and genuflection while bound to ligand with high/intermediate affinity (125). This is significant because the genuflected interaction will bring membranes closer together and may form a better seal. Can large receptors like CR1 participate in such close contact or are their roles more in initial capture of particles, which are then "reeled in" by integrins in an F-actin-dependent

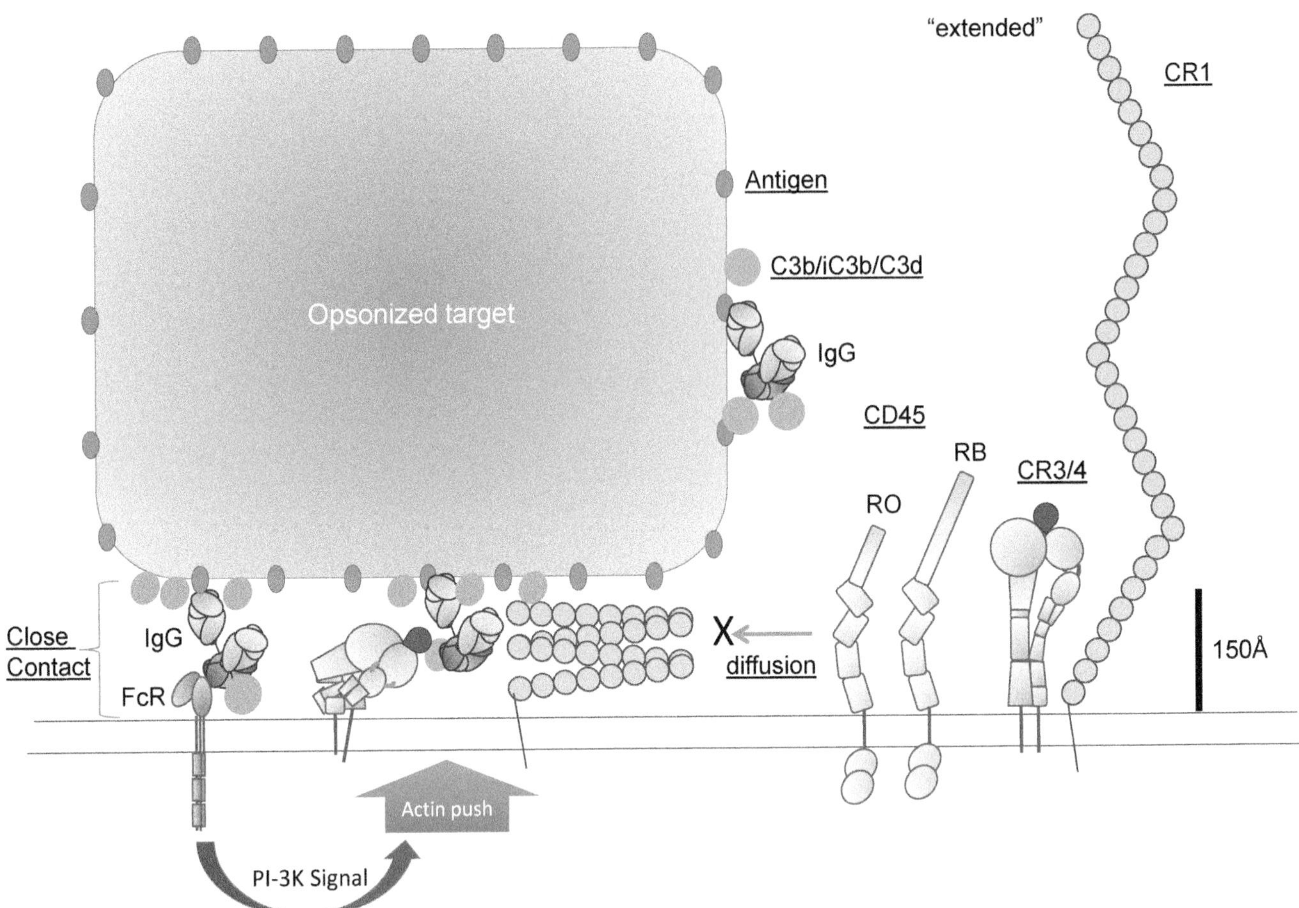

Figure 3 Fitting integrins and complement receptors into a diffusion barrier model for the phagocytic synapse. Close contacts are inherent to immunological synapses. Fc receptors and T-cell receptors naturally fit into a 15-nm gap that generates a diffusion barrier for entry of the RO and RB splice variants of CD45 and thus tips the local kinase/phosphatase balance in favor of the kinases. Large receptors like CR3 and CR4 are too large to fit into the <15-nm space when fully extended. Active F-actin-mediated processes induced by phosphatidylinositol-3 kinase (PI-3K) signaling can work with integrins to expand close contacts in phagocytic immune synapses and increase the area from which CD45 is excluded. The relevant integrin conformations that mediate this close contact formation are not known, but may include alternative crouching conformations recently described by electron microscopy or tilted extended conformations generated by forces tangential to the membrane. CR1 function overlaps extensively with CR3/4, and thus it is possible that CR1 can also adopt conformations that facilitate close contact in an F-actin-dependent manner, despite its apparent large size. Further study is needed to understand whether CR1 also participates in close contact formation and how CR1's structure is adapted to this task.

manner? There are also a number of questions about differential regulation of integrins on myeloid cells; for example, how on DCs LFA-1 is turned off while integrins like $\alpha_V\beta_8$ remain active in regulation of transforming growth factor β activation. What are the mechanisms for differential regulation of integrins? Recent studies on the regional diversity of macrophages may add further opportunities for tissue-specific regulation of such processes (126).

Citation. Dustin ML. 2016. Complement receptors in myeloid cell adhesion and phagocytosis. Microbiol Spectrum 4(6):MCHD-0034-2016.

References

1. **Holers VM.** 2014. Complement and its receptors: new insights into human disease. *Annu Rev Immunol* **32:** 433–459.
2. **Merle NS, Church SE, Fremeaux-Bacchi V, Roumenina LT.** 2015. Complement system part I—molecular mech-

anisms of activation and regulation. *Front Immunol* **6**: 262. doi:10.3389/fimmu.2015.00262.

3. **Janssen BJ, Huizinga EG, Raaijmakers HC, Roos A, Daha MR, Nilsson-Ekdahl K, Nilsson B, Gros P.** 2005. Structures of complement component C3 provide insights into the function and evolution of immunity. *Nature* **437**:505–511.
4. **Hermanowski-Vosatka A, Detmers PA, Götze O, Silverstein SC, Wright SD.** 1988. Clustering of ligand on the surface of a particle enhances adhesion to receptor-bearing cells. *J Biol Chem* **263**:17822–17827.
5. **Rothlein R, Springer TA.** 1985. Complement receptor type three-dependent degradation of opsonized erythrocytes by mouse macrophages. *J Immunol* **135**:2668–2672.
6. **Fällman M, Andersson R, Andersson T.** 1993. Signaling properties of CR3 (CD11b/CD18) and CR1 (CD35) in relation to phagocytosis of complement-opsonized particles. *J Immunol* **151**:330–338.
7. **Bajic G, Yatime L, Sim RB, Vorup-Jensen T, Andersen GR.** 2013. Structural insight on the recognition of surface-bound opsonins by the integrin I domain of complement receptor 3. *Proc Natl Acad Sci U S A* **110**: 16426–16431.
8. **Chen X, Yu Y, Mi LZ, Walz T, Springer TA.** 2012. Molecular basis for complement recognition by integrin $\alpha_X\beta_2$. *Proc Natl Acad Sci U S A* **109**:4586–4591.
9. **Gorgani NN, He JQ, Katschke KJ Jr, Helmy KY, Xi H, Steffek M, Hass PE, van Lookeren Campagne M.** 2008. Complement receptor of the Ig superfamily enhances complement-mediated phagocytosis in a subpopulation of tissue resident macrophages. *J Immunol* **181**:7902–7908.
10. **Wiesmann C, Katschke KJ, Yin J, Helmy KY, Steffek M, Fairbrother WJ, McCallum SA, Embuscado L, DeForge L, Hass PE, van Lookeren Campagne M.** 2006. Structure of C3b in complex with CRIg gives insights into regulation of complement activation. *Nature* **444**:217–220.
11. **Vogt L, Schmitz N, Kurrer MO, Bauer M, Hinton HI, Behnke S, Gatto D, Sebbel P, Beerli RR, Sonderegger I, Kopf M, Saudan P, Bachmann MF.** 2006. VSIG4, a B7 family-related protein, is a negative regulator of T cell activation. *J Clin Invest* **116**:2817–2826.
12. **Bajic G, Yatime L, Klos A, Andersen GR.** 2013. Human C3a and C3a desArg anaphylatoxins have conserved structures, in contrast to C5a and C5a desArg. *Protein Sci* **22**:204–212.
13. **Skokowa J, Ali SR, Felda O, Kumar V, Konrad S, Shushakova N, Schmidt RE, Piekorz RP, Nürnberg B, Spicher K, Birnbaumer L, Zwirner J, Claassens JW, Verbeek JS, van Rooijen N, Köhl J, Gessner JE.** 2005. Macrophages induce the inflammatory response in the pulmonary Arthus reaction through $G\alpha_{i2}$ activation that controls C5aR and Fc receptor cooperation. *J Immunol* **174**:3041–3050.
14. **Heit B, Tavener S, Raharjo E, Kubes P.** 2002. An intracellular signaling hierarchy determines direction of migration in opposing chemotactic gradients. *J Cell Biol* **159**:91–102.
15. **Flaherty P, Radhakrishnan ML, Dinh T, Rebres RA, Roach TI, Jordan MI, Arkin AP.** 2008. A dual receptor crosstalk model of G-protein-coupled signal transduction. *PLOS Comput Biol* **4**:e1000185. doi:10.1371/journal.pcbi.1000185.
16. **Strainic MG, Liu J, Huang D, An F, Lalli PN, Muqim N, Shapiro VS, Dubyak GR, Heeger PS, Medof ME.** 2008. Locally produced complement fragments C5a and C3a provide both costimulatory and survival signals to naive $CD4^+$ T cells. *Immunity* **28**:425–435.
17. **Senior RM, Griffin GL, Perez HD, Webster RO.** 1988. Human C5a and C5a des Arg exhibit chemotactic activity for fibroblasts. *J Immunol* **141**:3570–3574.
18. **Bamberg CE, Mackay CR, Lee H, Zahra D, Jackson J, Lim YS, Whitfeld PL, Craig S, Corsini E, Lu B, Gerard C, Gerard NP.** 2010. The C5a receptor (C5aR) C5L2 is a modulator of C5aR-mediated signal transduction. *J Biol Chem* **285**:7633–7644.
19. **Venkiteswaran G, Lewellis SW, Wang J, Reynolds E, Nicholson C, Knaut H.** 2013. Generation and dynamics of an endogenous, self-generated signaling gradient across a migrating tissue. *Cell* **155**:674–687.
20. **Humbles AA, Lu B, Nilsson CA, Lilly C, Israel E, Fujiwara Y, Gerard NP, Gerard C.** 2000. A role for the C3a anaphylatoxin receptor in the effector phase of asthma. *Nature* **406**:998–1001.
21. **Höpken UE, Lu B, Gerard NP, Gerard C.** 1996. The C5a chemoattractant receptor mediates mucosal defence to infection. *Nature* **383**:86–89.
22. **Banda NK, Hyatt S, Antonioli AH, White JT, Glogowska M, Takahashi K, Merkel TJ, Stahl GL, Mueller-Ortiz S, Wetsel R, Arend WP, Holers VM.** 2012. Role of C3a receptors, C5a receptors, and complement protein C6 deficiency in collagen antibody-induced arthritis in mice. *J Immunol* **188**:1469–1478.
23. **Andersson C, Wenander CS, Usher PA, Hebsgaard JB, Sondergaard BC, Rønø B, Mackay C, Friedrichsen B, Chang C, Tang R, Hornum L.** 2014. Rapid-onset clinical and mechanistic effects of anti-C5aR treatment in the mouse collagen-induced arthritis model. *Clin Exp Immunol* **177**:219–233.
24. **Vergunst CE, Gerlag DM, Dinant H, Schulz L, Vinkenoog M, Smeets TJ, Sanders ME, Reedquist KA, Tak PP.** 2007. Blocking the receptor for C5a in patients with rheumatoid arthritis does not reduce synovial inflammation. *Rheumatology (Oxford)* **46**:1773–1778.
25. **Jacobson AC, Weis JH.** 2008. Comparative functional evolution of human and mouse CR1 and CR2. *J Immunol* **181**:2953–2959.
26. **Donius LR, Handy JM, Weis JJ, Weis JH.** 2013. Optimal germinal center B cell activation and T-dependent antibody responses require expression of the mouse complement receptor Cr1. *J Immunol* **191**:434–447.
27. **Hourcade D, Miesner DR, Bee C, Zeldes W, Atkinson JP.** 1990. Duplication and divergence of the amino-terminal coding region of the complement receptor 1 (CR1) gene. An example of concerted (horizontal) evolution within a gene. *J Biol Chem* **265**:974–980.

28. **Java A, Liszewski MK, Hourcade DE, Zhang F, Atkinson JP.** 2015. Role of complement receptor 1 (CR1; CD35) on epithelial cells: a model for understanding complement-mediated damage in the kidney. *Mol Immunol* **67**(2 Pt B):584–595.
29. **Schramm EC, Roumenina LT, Rybkine T, Chauvet S, Vieira-Martins P, Hue C, Maga T, Valoti E, Wilson V, Jokiranta S, Smith RJ, Noris M, Goodship T, Atkinson JP, Fremeaux-Bacchi V.** 2015. Mapping interactions between complement C3 and regulators using mutations in atypical hemolytic uremic syndrome. *Blood* **125:** 2359–2369.
30. **Molina H, Holers VM, Li B, Fung Y, Mariathasan S, Goellner J, Strauss-Schoenberger J, Karr RW, Chaplin DD.** 1996. Markedly impaired humoral immune response in mice deficient in complement receptors 1 and 2. *Proc Natl Acad Sci U S A* **93:**3357–3361.
31. **Repik A, Pincus SE, Ghiran I, Nicholson-Weller A, Asher DR, Cerny AM, Casey LS, Jones SM, Jones SN, Mohamed N, Klickstein LB, Spitalny G, Finberg RW.** 2005. A transgenic mouse model for studying the clearance of blood-borne pathogens via human complement receptor 1 (CR1). *Clin Exp Immunol* **140:**230–240.
32. **Newman SL, Becker S, Halme J.** 1985. Phagocytosis by receptors for C3b (CR1), iC3b (CR3), and IgG (Fc) on human peritoneal macrophages. *J Leukoc Biol* **38:** 267–278.
33. **Holers VM, Kinoshita T, Molina H.** 1992. The evolution of mouse and human complement C3-binding proteins: divergence of form but conservation of function. *Immunol Today* **13:**231–236.
34. **Paccaud JP, Carpentier JL, Schifferli JA.** 1990. Difference in the clustering of complement receptor type 1 (CR1) on polymorphonuclear leukocytes and erythrocytes: effect on immune adherence. *Eur J Immunol* **20:** 283–289.
35. **Ghiran I, Glodek AM, Weaver G, Klickstein LB, Nicholson-Weller A.** 2008. Ligation of erythrocyte CR1 induces its clustering in complex with scaffolding protein FAP-1. *Blood* **112:**3465–3473.
36. **Springer TA, Dustin ML.** 2012. Integrin inside-out signaling and the immunological synapse. *Curr Opin Cell Biol* **24:**107–115.
37. **Vogel V, Sheetz MP.** 2009. Cell fate regulation by coupling mechanical cycles to biochemical signaling pathways. *Curr Opin Cell Biol* **21:**38–46.
38. **Sanchez-Madrid F, Nagy JA, Robbins E, Simon P, Springer TA.** 1983. A human leukocyte differentiation antigen family with distinct α-subunits and a common β-subunit: the lymphocyte function-associated antigen (LFA-1), the C3bi complement receptor (OKM1/Mac-1), and the p150,95 molecule. *J Exp Med* **158:**1785–1803.
39. **Anderson DC, Springer TA.** 1987. Leukocyte adhesion deficiency: an inherited defect in the Mac-1, LFA-1, and p150,95 glycoproteins. *Annu Rev Med* **38:**175–194.
40. **Worthington JJ, Kelly A, Smedley C, Bauché D, Campbell S, Marie JC, Travis MA.** 2015. Integrin αvβ8-mediated TGF-β activation by effector regulatory T cells is essential for suppression of T-cell-mediated inflammation. *Immunity* **42:**903–915.
41. **Xiong JP, Stehle T, Zhang R, Joachimiak A, Frech M, Goodman SL, Arnaout MA.** 2002. Crystal structure of the extracellular segment of integrin αVβ3 in complex with an Arg-Gly-Asp ligand. *Science* **296:**151–155.
42. **Shimaoka M, Xiao T, Liu JH, Yang Y, Dong Y, Jun CD, McCormack A, Zhang R, Joachimiak A, Takagi J, Wang JH, Springer TA.** 2003. Structures of the αL I domain and its complex with ICAM-1 reveal a shape-shifting pathway for integrin regulation. *Cell* **112:** 99–111.
43. **Choi WS, Rice WJ, Stokes DL, Coller BS.** 2013. Three-dimensional reconstruction of intact human integrin αIIbβ3: new implications for activation-dependent ligand binding. *Blood* **122:**4165–4171.
44. **García-Alvarez B, de Pereda JM, Calderwood DA, Ulmer TS, Critchley D, Campbell ID, Ginsberg MH, Liddington RC.** 2003. Structural determinants of integrin recognition by talin. *Mol Cell* **11:**49–58.
45. **Cavalcanti-Adam EA, Volberg T, Micoulet A, Kessler H, Geiger B, Spatz JP.** 2007. Cell spreading and focal adhesion dynamics are regulated by spacing of integrin ligands. *Biophys J* **92:**2964–2974.
46. **Kanchanawong P, Shtengel G, Pasapera AM, Ramko EB, Davidson MW, Hess HF, Waterman CM.** 2010. Nanoscale architecture of integrin-based cell adhesions. *Nature* **468:**580–584.
47. **Kupfer A, Singer SJ.** 1989. The specific interaction of helper T cells and antigen-presenting B cells. IV. Membrane and cytoskeletal reorganizations in the bound T cell as a function of antigen dose. *J Exp Med* **170:** 1697–1713.
48. **Lim J, Wiedemann A, Tzircotis G, Monkley SJ, Critchley DR, Caron E.** 2007. An essential role for talin during $\alpha_M\beta_2$-mediated phagocytosis. *Mol Biol Cell* **18:** 976–985.
49. **Yao M, Goult BT, Chen H, Cong P, Sheetz MP, Yan J.** 2014. Mechanical activation of vinculin binding to talin locks talin in an unfolded conformation. *Sci Rep* **4:** 4610. doi:10.1038/srep04610.
50. **Le Trong I, Aprikian P, Kidd BA, Forero-Shelton M, Tchesnokova V, Rajagopal P, Rodriguez V, Interlandi G, Klevit R, Vogel V, Stenkamp RE, Sokurenko EV, Thomas WE.** 2010. Structural basis for mechanical force regulation of the adhesin FimH via finger trap-like β sheet twisting. *Cell* **141:**645–655.
51. **Baorto DM, Gao Z, Malaviya R, Dustin ML, van der Merwe A, Lublin DM, Abraham SN.** 1997. Survival of FimH-expressing enterobacteria in macrophages relies on glycolipid traffic. *Nature* **389:**636–639.
52. **Hynes RO.** 1987. Integrins: a family of cell surface receptors. *Cell* **48:**549–554.
53. **Hemler ME, Sanchez-Madrid F, Flotte TJ, Krensky AM, Burakoff SJ, Bhan AK, Springer TA, Strominger JL.** 1984. Glycoproteins of 210,000 and 130,000 m.w. on activated T cells: cell distribution and antigenic relation to components on resting cells and T cell lines. *J Immunol* **132:**3011–3018.

54. **Elices MJ, Osborn L, Takada Y, Crouse C, Luhowskyj S, Hemler ME, Lobb RR.** 1990. VCAM-1 on activated endothelium interacts with the leukocyte integrin VLA-4 at a site distinct from the VLA-4/fibronectin binding site. *Cell* **60:**577–584.
55. **Springer TA, Dustin ML.** 2012. Integrin inside-out signaling and the immunological synapse. *Curr Opin Cell Biol* **24:**107–115.
56. **Nishida N, Xie C, Shimaoka M, Cheng Y, Walz T, Springer TA.** 2006. Activation of leukocyte β_2 integrins by conversion from bent to extended conformations. *Immunity* **25:**583–594.
57. **Stewart MP, Cabanas C, Hogg N.** 1996. T cell adhesion to intercellular adhesion molecule-1 (ICAM-1) is controlled by cell spreading and the activation of integrin LFA-1. *J Immunol* **156:**1810–1817.
58. **Comrie WA, Babich A, Burkhardt JK.** 2015. F-actin flow drives affinity maturation and spatial organization of LFA-1 at the immunological synapse. *J Cell Biol* **208:** 475–491.
59. **Kishimoto TK, Hollander N, Roberts TM, Anderson DC, Springer TA.** 1987. Heterogeneous mutations in the β subunit common to the LFA-1, Mac-1, and p150,95 glycoproteins cause leukocyte adhesion deficiency. *Cell* **50:**193–202.
60. **Thomas C, Le Deist F, Cavazzana-Calvo M, Benkerrou M, Haddad E, Blanche S, Hartmann W, Friedrich W, Fischer A.** 1995. Results of allogeneic bone marrow transplantation in patients with leukocyte adhesion deficiency. *Blood* **86:**1629–1635.
61. **Manevich-Mendelson E, Feigelson SW, Pasvolsky R, Aker M, Grabovsky V, Shulman Z, Kilic SS, Rosenthal-Allieri MA, Ben-Dor S, Mory A, Bernard A, Moser M, Etzioni A, Alon R.** 2009. Loss of Kindlin-3 in LAD-III eliminates LFA-1 but not VLA-4 adhesiveness developed under shear flow conditions. *Blood* **114:**2344–2353.
62. **Xue ZH, Feng C, Liu WL, Tan SM.** 2013. A role of kindlin-3 in integrin αMβ2 outside-in signaling and the Syk-Vav1-Rac1/Cdc42 signaling axis. *PLoS One* **8:** e56911. doi:10.1371/journal.pone.0056911.
63. **Diamond MS, Staunton DE, de Fougerolles AR, Stacker SA, Garcia-Aguilar J, Hibbs ML, Springer TA.** 1990. ICAM-1 (CD54): a counter-receptor for Mac-1 (CD11b/CD18). *J Cell Biol* **111:**3129–3139.
64. **Diamond MS, Staunton DE, Marlin SD, Springer TA.** 1991. Binding of the integrin Mac-1 (CD11b/CD18) to the third immunoglobulin-like domain of ICAM-1 (CD54) and its regulation by glycosylation. *Cell* **65:**961–971.
65. **Dustin ML, Rothlein R, Bhan AK, Dinarello CA, Springer TA.** 1986. Induction by IL 1 and interferon-gamma: tissue distribution, biochemistry, and function of a natural adherence molecule (ICAM-1). *J Immunol* **137:**245–254.
66. **Staunton DE, Dustin ML, Springer TA.** 1989. Functional cloning of ICAM-2, a cell adhesion ligand for LFA-1 homologous to ICAM-1. *Nature* **339:**61–64.
67. **Lu TT, Cyster JG.** 2002. Integrin-mediated long-term B cell retention in the splenic marginal zone. *Science* **297:** 409–412.
68. **Grakoui A, Bromley SK, Sumen C, Davis MM, Shaw AS, Allen PM, Dustin ML.** 1999. The immunological synapse: a molecular machine controlling T cell activation. *Science* **285:**221–227.
69. **Scholer A, Hugues S, Boissonnas A, Fetler L, Amigorena S.** 2008. Intercellular adhesion molecule-1-dependent stable interactions between T cells and dendritic cells determine CD8$^+$ T cell memory. *Immunity* **28:**258–270.
70. **Cambi A, Joosten B, Koopman M, de Lange F, Beeren I, Torensma R, Fransen JA, Garcia-Parajó M, van Leeuwen FN, Figdor CG.** 2006. Organization of the integrin LFA-1 in nanoclusters regulates its activity. *Mol Biol Cell* **17:**4270–4281.
71. **Wright SD, Rao PE, Van Voorhis WC, Craigmyle LS, Iida K, Talle MA, Westberg EF, Goldstein G, Silverstein SC.** 1983. Identification of the C3bi receptor of human monocytes and macrophages by using monoclonal antibodies. *Proc Natl Acad Sci U S A* **80:**5699–5703.
72. **Wright SD, Silverstein SC.** 1983. Receptors for C3b and C3bi promote phagocytosis but not the release of toxic oxygen from human phagocytes. *J Exp Med* **158:** 2016–2023.
73. **Vorup-Jensen T, Carman CV, Shimaoka M, Schuck P, Svitel J, Springer TA.** 2005. Exposure of acidic residues as a danger signal for recognition of fibrinogen and other macromolecules by integrin $\alpha_X\beta_2$. *Proc Natl Acad Sci U S A* **102:**1614–1619.
74. **Melo MD, Catchpole IR, Haggar G, Stokes RW.** 2000. Utilization of CD11b knockout mice to characterize the role of complement receptor 3 (CR3, CD11b/CD18) in the growth of *Mycobacterium tuberculosis* in macrophages. *Cell Immunol* **205:**13–23.
75. **Rooyakkers AW, Stokes RW.** 2005. Absence of complement receptor 3 results in reduced binding and ingestion of *Mycobacterium tuberculosis* but has no significant effect on the induction of reactive oxygen and nitrogen intermediates or on the survival of the bacteria in resident and interferon-gamma activated macrophages. *Microb Pathog* **39:**57–67.
76. **Hong S, Beja-Glasser VF, Nfonoyim BM, Frouin A, Li S, Ramakrishnan S, Merry KM, Shi Q, Rosenthal A, Barres BA, Lemere CA, Selkoe DJ, Stevens B.** 2016. Complement and microglia mediate early synapse loss in Alzheimer mouse models. *Science* **352:**712–716.
77. **Wyss-Coray T, Rogers J.** 2012. Inflammation in Alzheimer disease—a brief review of the basic science and clinical literature. *Cold Spring Harb Perspect Med* **2:**a006346. doi:10.1101/cshperspect.a006346.
78. **Lindquist RL, Shakhar G, Dudziak D, Wardemann H, Eisenreich T, Dustin ML, Nussenzweig MC.** 2004. Visualizing dendritic cell networks *in vivo*. *Nat Immunol* **5:**1243–1250.
79. **Jung S, Unutmaz D, Wong P, Sano G, De los Santos K, Sparwasser T, Wu S, Vuthoori S, Ko K, Zavala F, Pamer EG, Littman DR, Lang RA.** 2002. In vivo depletion of CD11c$^+$ dendritic cells abrogates priming of CD8$^+$ T cells by exogenous cell-associated antigens. *Immunity* **17:**211–220.

80. **Meredith MM, Liu K, Darrasse-Jeze G, Kamphorst AO, Schreiber HA, Guermonprez P, Idoyaga J, Cheong C, Yao KH, Niec RE, Nussenzweig MC.** 2012. Expression of the zinc finger transcription factor zDC (Zbtb46, Btbd4) defines the classical dendritic cell lineage. *J Exp Med* **209:**1153–1165.

81. **Bullard DC, Hu X, Adams JE, Schoeb TR, Barnum SR.** 2007. p150/95 (CD11c/CD18) expression is required for the development of experimental autoimmune encephalomyelitis. *Am J Pathol* **170:**2001–2008.

82. **Wu H, Gower RM, Wang H, Perrard XY, Ma R, Bullard DC, Burns AR, Paul A, Smith CW, Simon SI, Ballantyne CM.** 2009. Functional role of CD11c$^+$ monocytes in atherogenesis associated with hypercholesterolemia. *Circulation* **119:**2708–2717.

83. **Astier A, Trescol-Biémont MC, Azocar O, Lamouille B, Rabourdin-Combe C.** 2000. Cutting edge: CD46, a new costimulatory molecule for T cells, that induces p120CBL and LAT phosphorylation. *J Immunol* **164:** 6091–6095.

84. **Kemper C, Chan AC, Green JM, Brett KA, Murphy KM, Atkinson JP.** 2003. Activation of human CD4$^+$ cells with CD3 and CD46 induces a T-regulatory cell 1 phenotype. *Nature* **421:**388–392.

85. **Ghannam A, Fauquert JL, Thomas C, Kemper C, Drouet C.** 2014. Human complement C3 deficiency: Th1 induction requires T cell-derived complement C3a and CD46 activation. *Mol Immunol* **58:**98–107.

86. **Wu H, Rodgers JR, Perrard XY, Perrard JL, Prince JE, Abe Y, Davis BK, Dietsch G, Smith CW, Ballantyne CM.** 2004. Deficiency of CD11b or CD11d results in reduced staphylococcal enterotoxin-induced T cell response and T cell phenotypic changes. *J Immunol* **173:**297–306.

87. **Geremia NM, Bao F, Rosenzweig TE, Hryciw T, Weaver L, Dekaban GA, Brown A.** 2012. CD11d antibody treatment improves recovery in spinal cord-injured mice. *J Neurotrauma* **29:**539–550.

88. **Arnold IC, Mathisen S, Schulthess J, Danne C, Hegazy AN, Powrie F.** 2016. CD11c$^+$ monocyte/macrophages promote chronic *Helicobacter hepaticus*-induced intestinal inflammation through the production of IL-23. *Mucosal Immunol* **9:**352–363.

89. **Lämmermann T, Bader BL, Monkley SJ, Worbs T, Wedlich-Söldner R, Hirsch K, Keller M, Förster R, Critchley DR, Fässler R, Sixt M.** 2008. Rapid leukocyte migration by integrin-independent flowing and squeezing. *Nature* **453:**51–55.

90. **Haston WS, Shields JM.** 1984. Contraction waves in lymphocyte locomotion. *J Cell Sci* **68:**227–241.

91. **Helmy KY, Katschke KJ Jr, Gorgani NN, Kljavin NM, Elliott JM, Diehl L, Scales SJ, Ghilardi N, van Lookeren Campagne M.** 2006. CRIg: a macrophage complement receptor required for phagocytosis of circulating pathogens. *Cell* **124:**915–927.

92. **Kim KH, Choi BK, Kim YH, Han C, Oh HS, Lee DG, Kwon BS.** 2016. Extracellular stimulation of VSIG4/complement receptor Ig suppresses intracellular bacterial infection by inducing autophagy. *Autophagy* **12:** 1647–1659.

93. **Katschke KJ Jr, Helmy KY, Steffek M, Xi H, Yin J, Lee WP, Gribling P, Barck KH, Carano RA, Taylor RE, Rangell L, Diehl L, Hass PE, Wiesmann C, van Lookeren Campagne M.** 2007. A novel inhibitor of the alternative pathway of complement reverses inflammation and bone destruction in experimental arthritis. *J Exp Med* **204:**1319–1325.

94. **Zeng Z, Surewaard BG, Wong CH, Geoghegan JA, Jenne CN, Kubes P.** 2016. CRIg functions as a macrophage pattern recognition receptor to directly bind and capture blood-borne Gram-positive bacteria. *Cell Host Microbe* **20:**99–106.

95. **von Andrian UH, Chambers JD, McEvoy LM, Bargatze RF, Arfors KE, Butcher EC.** 1991. Two-step model of leukocyte-endothelial cell interaction in inflammation: distinct roles for LECAM-1 and the leukocyte β_2 integrins *in vivo*. *Proc Natl Acad Sci U S A* **88:**7538–7542.

96. **Lawrence MB, Springer TA.** 1991. Leukocytes roll on a selectin at physiologic flow rates: distinction from and prerequisite for adhesion through integrins. *Cell* **65:** 859–873.

97. **Wild MK, Lühn K, Marquardt T, Vestweber D.** 2002. Leukocyte adhesion deficiency II: therapy and genetic defect. *Cells Tissues Organs* **172:**161–173.

98. **Pober JS, Bevilacqua MP, Mendrick DL, Lapierre LA, Fiers W, Gimbrone MA Jr.** 1986. Two distinct monokines, interleukin 1 and tumor necrosis factor, each independently induce biosynthesis and transient expression of the same antigen on the surface of cultured human vascular endothelial cells. *J Immunol* **136:**1680–1687.

99. **Shamri R, Grabovsky V, Gauguet JM, Feigelson S, Manevich E, Kolanus W, Robinson MK, Staunton DE, von Andrian UH, Alon R.** 2005. Lymphocyte arrest requires instantaneous induction of an extended LFA-1 conformation mediated by endothelium-bound chemokines. *Nat Immunol* **6:**497–506.

100. **Chan JR, Hyduk SJ, Cybulsky MI.** 2001. Chemoattractants induce a rapid and transient upregulation of monocyte $\alpha 4$ integrin affinity for vascular cell adhesion molecule 1 which mediates arrest: an early step in the process of emigration. *J Exp Med* **193:**1149–1158.

101. **Carman CV, Sage PT, Sciuto TE, de la Fuente MA, Geha RS, Ochs HD, Dvorak HF, Dvorak AM, Springer TA.** 2007. Transcellular diapedesis is initiated by invasive podosomes. *Immunity* **26:**784–797.

102. **Kim JV, Kang SS, Dustin ML, McGavern DB.** 2009. Myelomonocytic cell recruitment causes fatal CNS vascular injury during acute viral meningitis. *Nature* **457:** 191–195.

103. **Geissmann F, Cameron TO, Sidobre S, Manlongat N, Kronenberg M, Briskin MJ, Dustin ML, Littman DR.** 2005. Intravascular immune surveillance by CXCR6$^+$ NKT cells patrolling liver sinusoids. *PLoS Biol* **3:**e113. doi:10.1371/journal.pbio.0030113.

104. **Carlin LM, Stamatiades EG, Auffray C, Hanna RN, Glover L, Vizcay-Barrena G, Hedrick CC, Cook HT, Diebold S, Geissmann F.** 2013. *Nr4a1*-dependent

Ly6C^{low} monocytes monitor endothelial cells and orchestrate their disposal. *Cell* **153:**362–375.

105. **Liese J, Rooijakkers SH, van Strijp JA, Novick RP, Dustin ML.** 2013. Intravital two-photon microscopy of host-pathogen interactions in a mouse model of *Staphylococcus aureus* skin abscess formation. *Cell Microbiol* **15:**891–909.
106. **Lämmermann T, Afonso PV, Angermann BR, Wang JM, Kastenmüller W, Parent CA, Germain RN.** 2013. Neutrophil swarms require LTB4 and integrins at sites of cell death *in vivo*. *Nature* **498:**371–375.
107. **Foxman EF, Campbell JJ, Butcher EC.** 1997. Multistep navigation and the combinatorial control of leukocyte chemotaxis. *J Cell Biol* **139:**1349–1360.
108. **Waite JC, Leiner I, Lauer P, Rae CS, Barbet G, Zheng H, Portnoy DA, Pamer EG, Dustin ML.** 2011. Dynamic imaging of the effector immune response to listeria infection *in vivo*. *PLoS Pathog* **7:**e1001326. doi:10.1371/journal.ppat.1001326.
109. **Zinselmeyer BH, Heydari S, Sacristán C, Nayak D, Cammer M, Herz J, Cheng X, Davis SJ, Dustin ML, McGavern DB.** 2013. PD-1 promotes immune exhaustion by inducing antiviral T cell motility paralysis. *J Exp Med* **210:**757–774.
110. **Mizuno M, Nishikawa K, Okada N, Matsuo S, Ito K, Okada H.** 1999. Inhibition of a membrane complement regulatory protein by a monoclonal antibody induces acute lethal shock in rats primed with lipopolysaccharide. *J Immunol* **162:**5477–5482.
111. **Tohyama Y, Yamamura H.** 2009. Protein tyrosine kinase, Syk: a key player in phagocytic cells. *J Biochem* **145:**267–273.
112. **Rogers NC, Slack EC, Edwards AD, Nolte MA, Schulz O, Schweighoffer E, Williams DL, Gordon S, Tybulewicz VL, Brown GD, Reis e Sousa C.** 2005. Syk-dependent cytokine induction by Dectin-1 reveals a novel pattern recognition pathway for C type lectins. *Immunity* **22:**507–517.
113. **Varma R, Campi G, Yokosuka T, Saito T, Dustin ML.** 2006. T cell receptor-proximal signals are sustained in peripheral microclusters and terminated in the central supramolecular activation cluster. *Immunity* **25:**117–127.
114. **Chang VT, Fernandes RA, Ganzinger KA, Lee SF, Siebold C, McColl J, Jönsson P, Palayret M, Harlos K, Coles CH, Jones EY, Lui Y, Huang E, Gilbert RJ, Klenerman D, Aricescu AR, Davis SJ.** 2016. Initiation of T cell signaling by CD45 segregation at 'close contacts.' *Nat Immunol* **17:**574–582.
115. **Goodridge HS, Simmons RM, Underhill DM.** 2007. Dectin-1 stimulation by *Candida albicans* yeast or zymosan triggers NFAT activation in macrophages and dendritic cells. *J Immunol* **178:**3107–3115.
116. **Freeman SA, Goyette J, Furuya W, Woods EC, Bertozzi CR, Bergmeier W, Hinz B, van der Merwe PA, Das R, Grinstein S.** 2016. Integrins form an expanding diffusional barrier that coordinates phagocytosis. *Cell* **164:**128–140.
117. **Dahlbäck B, Smith CA, Müller-Eberhard HJ.** 1983. Visualization of human C4b-binding protein and its complexes with vitamin K-dependent protein S and complement protein C4b. *Proc Natl Acad Sci U S A* **80:**3461–3465.
118. **Norman MU, Hulliger S, Colarusso P, Kubes P.** 2008. Multichannel fluorescence spinning disk microscopy reveals early endogenous CD4 T cell recruitment in contact sensitivity via complement. *J Immunol* **180:**510–521.
119. **Kolev M, Dimeloe S, Le Friec G, Navarini A, Arbore G, Povoleri GA, Fischer M, Belle R, Loeliger J, Develioglu L, Bantug GR, Watson J, Couzi L, Afzali B, Lavender P, Hess C, Kemper C.** 2015. Complement regulates nutrient influx and metabolic reprogramming during Th1 cell responses. *Immunity* **42:**1033–1047.
120. **Oliaro J, Pasam A, Waterhouse NJ, Browne KA, Ludford-Menting MJ, Trapani JA, Russell SM.** 2006. Ligation of the cell surface receptor, CD46, alters T cell polarity and response to antigen presentation. *Proc Natl Acad Sci U S A* **103:**18685–18690.
121. **Jung K, Seo SK, Choi I.** 2015. Endogenous VSIG4 negatively regulates the helper T cell-mediated antibody response. *Immunol Lett* **165:**78–83.
122. **Li Y, Wang YQ, Wang DH, Hou WP, Zhang Y, Li M, Li FR, Mu J, Du X, Pang F, Yuan FH.** 2014. Costimulatory molecule VSIG4 exclusively expressed on macrophages alleviates renal tubulointerstitial injury in VSIG4 KO mice. *J Nephrol* **27:**29–36.
123. **Fuller MJ, Callendret B, Zhu B, Freeman GJ, Hasselschwert DL, Satterfield W, Sharpe AH, Dustin LB, Rice CM, Grakoui A, Ahmed R, Walker CM.** 2013. Immunotherapy of chronic hepatitis C virus infection with antibodies against programmed cell death-1 (PD-1). *Proc Natl Acad Sci U S A* **110:**15001–15006.
124. **Takagi J, Petre BM, Walz T, Springer TA.** 2002. Global conformational rearrangements in integrin extracellular domains in outside-in and inside-out signaling. *Cell* **110:**599–611.
125. **Chen W, Lou J, Evans EA, Zhu C.** 2012. Observing force-regulated conformational changes and ligand dissociation from a single integrin on cells. *J Cell Biol* **199:**497–512.
126. **Lavin Y, Winter D, Blecher-Gonen R, David E, Keren-Shaul H, Merad M, Jung S, Amit I.** 2014. Tissue-resident macrophage enhancer landscapes are shaped by the local microenvironment. *Cell* **159:**1312–1326.

Myeloid Cells in Health and Disease: A Synthesis
Edited by Siamon Gordon

doi:10.1128/microbiolspec.MCHD-0040-2016

Takashi Satoh[1,2]
Shizuo Akira[1,2]

24 Toll-Like Receptor Signaling and Its Inducible Proteins

TLR SIGNALING PATHWAYS

Various cell types, including macrophages, dendritic cells (DCs), neutrophils, natural killer cells, and fibroblasts, express Toll-like receptors (TLRs) that activate the immune system (1, 2). TLRs are type I transmembrane proteins with ectodomains containing leucine-rich repeats that mediate the recognition of pathogen-associated molecular patterns (PAMPs) derived from pathogens, and damage-associated molecular patterns (DAMPs) from dying or injured cells. TLRs harbor transmembrane domains and intracellular Toll–interleukin-1 (IL-1) receptor (TIR) domains in addition to leucine-rich repeat domains, and some adaptor molecules bind to them to activate the downstream signaling pathways. Various organisms express the TLR family, especially mammals, and 13 types of TLRs have been reported. TLR1 to -9 are conserved in both the mouse and human. However, in mice, a retroviral insertion has rendered the TLR10 molecule nonfunctional. TLR11, -12, and -13 do not occur in humans. The active TLRs localize differently. TLR1, -2, -4, -5, -6, and -10 are expressed on the cell surface, whereas TLR3, -7, -8, -9, -11, -12, and -13 are expressed in the endosome (3, 4). Studies of mice deficient in each TLR have shown that each TLR has a distinct function in terms of PAMP recognition and the immune responses.

Cell surface TLRs mainly recognize microbial membrane components, such as lipids, lipoproteins, and proteins. TLR4 recognizes lipopolysaccharide (LPS) derived from Gram-negative bacteria and the envelope proteins of syncytial viruses, glycoinositol phospholipids from trypanosomes, and heat shock proteins 60 and 70. Recent studies have clearly shown that several transmembrane accessory molecules contribute to the activation of TLR signaling. CD14 is a typical glycosylphosphatidylinositol-anchored protein, and this molecule is associated with the recognition of LPS. CD14 acts cooperatively with TLR4 and myeloid differentiation protein 2 (MD-2). LPS recognition by these molecules results in the internalization of TLR4 into endosomes to activate the TLR4 signaling pathway through the action of immunoreceptor tyrosine-based activation motif-mediated spleen tyrosine kinase (Syk) and phospholipase C-γ2 (5). In addition to LPS and other bacterial ligands, TLR4 also recognizes S100a8, a ligand derived from dying cells (6, 7). Furthermore, the macrophages in adipose tissues produce inflammatory cytokines in response to free fatty acids via TLR4

[1]Department of Host Defense, Research Institute for Microbial Diseases, Osaka University, Osaka 565-0871, Japan; [2]Laboratory of Host Defense, WPI Immunology Frontier Research Center, Osaka University, Osaka 565-0871, Japan.

(8, 9). From these data, it is clear that a single TLR can recognize a variety of ligands to trigger an immune response.

Once PAMPs and DAMPs bind to the corresponding TLR, adaptor molecules, such as MyD88 (myeloid differentiation primary-response protein 88), TRIF (TIR-domain-containing adaptor protein inducing beta interferon [IFN-β]), TIRAP (TIR-domain-containing adaptor protein), or TRAM (TIR-domain-containing adaptor molecule), are recruited to the receptor. The signaling pathways downstream from these TLR family molecules are activated to produce inflammatory cytokines, IFNs, chemokines, and a variety of inducible proteins. Ligand recognition by TLR4 causes one of the TIR-domain-containing adaptor molecules, such as MyD88, to form a complex with an IRAK (IL-1 receptor-associated kinase) family molecule via its cytoplasmic region of adaptors (10). After the formation of the complex, IRAK4 phosphorylates IRAK1 (11) and the modified IRAK1 is released from MyD88 (12). IRAK1 then associates with TRAF6 (tumor necrosis factor [TNF] receptor-associated factor 6), a member of the RING-domain E3 ubiquitin ligase family. Then TRAF6, together with the ubiquitin E2-conjugating enzyme complex (UBC13 and UEV1A), promotes the K63-linked polyubiquitination of both TRAF6 and the TAK1 (transforming growth factor β-activated kinase 1) protein kinase complex. A member of the MAPKKK (mitogen-activated protein kinase kinase kinase) family, TAK1 forms a complex with the regulatory subunits TAB1, TAB2, and TAB3, and this complex then binds to TRAF6 (13, 14). TAK1 then activates two different signaling pathways, both the IKK (IκB kinase) complex–NF-κB and IKK complex-MAPK pathways. TAK1 binds to the IKK complex, which consists of IKKα, IKKβ, and NEMO (NF-κB essential modulator), through ubiquitin chains. The IKK complex phosphorylates the NF-κB inhibitory protein IκBα, which is then degraded, causing NF-κB to translocate to the nucleus. Activated TAK1 also activates the MAPK signaling pathway, which initiates the inflammatory response (1, 4).

In addition to MyD88, other TIR-domain-containing adaptors, such as TRIF, act as adaptor proteins for TLR4. After LPS engages with TLR4, this molecule is recruited to the cytosolic region of TLR4 and triggers another signaling pathway that activates IFN regulatory factor 3 (IRF3) and NF-κB, which induce the expression of type I IFNs and inflammatory cytokines. To bind both TLR4 and TRIF, a third TIR-domain-containing molecule, TRAM, is recruited between these two molecules. Thus, the downstream signaling from TLR4 is divided into two pathways, the MyD88-dependent and MyD88-independent (TRIF-dependent) pathways. Although MyD88 utilizes IRAK family molecules, TRIF uses TRAF6 and TRAF3 to activate downstream signaling. TRAF6 binds to RIP-1 (receptor-interacting protein kinase 1), and this complex then cooperatively activates TAK1, leading to the activation of NF-κB, which induces the expression of inflammatory cytokines. In contrast to TRAF6, TRAF3 activates the IKK-related kinases TBK1 (TANK-binding kinase 1) and IKKi, and these molecules phosphorylate IRF3. Phosphorylated IRF3 then translocates to the nucleus from the cytoplasm, leading to the expression of type I IFNs (1, 4).

Pellino-1 is a member of the RING-like domain family that contains the E3 ubiquitin ligases, which are implicated in TLR signaling (15). The loss of Pellino-1 in macrophages impairs the activation of TRIF-dependent NF-κB and cytokine production because RIP-1 remains unmodified (16). The kinase downstream from TRIF, TBK1/IKKi, phosphorylates Pellino-1, and this modification results in the ubiquitination of RIP-1. TRIF also forms a signaling complex and acts cooperatively with Pellino-1, as well as TRAF6, RIP-1, and TAK1, to activate NF-κB. Pellino-1 also regulates IRF3 activation by binding to DEAF-1 (deformed epidermal autoregulatory factor 1) to promote the expression of IFN-β (15). A recent study clearly showed that an inositol lipid is involved in the production of type I IFNs. The inositol lipid phosphatidylinositol 5-phosphate associates with both IRF3 and TBK1, and this association leads to the formation of a complex between TBK1 and IRF3. PIKFYVE also acts as a kinase in the production of phosphatidylinositol 5-phosphate as part of the antiviral innate immune response (17).

Another TLR family molecule, such as TLR2, acts coordinately with TLR1 or TLR6 to recognize a wide variety of PAMPs, including lipoproteins, peptidoglycans, lipoteichoic acids, zymosan, and mannan, derived from Gram-positive bacteria. Whereas the TLR2-TLR1 heterodimer recognizes triacylated lipoproteins, the TLR2-TLR6 heterodimer recognizes diacylated lipopeptides. The lipid capture structure of both TLR1 and TLR6 is critical for the recognition of the corresponding ligands. TLR5 also recognizes bacterial flagellin. After the formation of the TLR1-TLR2 heterodimer, the TLR2-TLR6 heterodimer or TLR5 engages each ligand, and the MyD88-dependent pathway is activated to produce inflammatory cytokines and chemokines (1, 4).

So far, we have discussed the extracellular TLRs and their downstream signaling pathways. In contrast, TLRs expressed in intracellular components such as endosomes recognize PAMPs, such as nucleic acids, derived from viruses and bacteria, and also DAMPs, such as "self" nucleic acids, in disease conditions such as autoimmunity (18). TLR3 recognizes synthesized and various virus-derived double-stranded RNAs and self RNAs derived from damaged cells via its ectodomain (19–21). Another intracellular TLR family molecule, TLR7, is predominantly expressed in plasmacytoid DCs and recognizes single-stranded RNA from various viruses and imidazoquinoline derivatives, such as imiquimod. On conventional DCs, this receptor recognizes the RNA from group B *Streptococcus* bacteria and is involved in the production of type I IFNs (22). Another intracellular TLR family molecule, TLR9, recognizes bacterial and viral unmethylated CpG-DNA motifs and the insoluble crystal hemozoin from *Plasmodium falciparum* during the process of detoxification (23). TLR3, but not TLR4, binds only to TRIF, and thus activates the downstream signaling pathway. In contrast, TLR7 and TLR9 bind MyD88, activating downstream signaling to trigger an antiviral immune response through the production of type I IFNs and inflammatory cytokines (1, 4). This intracellular TLR subfamily, including TLR3, -7, and -9, is trafficked in the endoplasmic reticulum to endosomes after they are bound by their ligands. This recruitment is regulated by the action of the endoplasmic reticulum-located protein UNC93B1 (24). Thus, the trafficking of some TLR family members is critical for the activation of downstream signaling pathways.

TLR13 recognizes only 23S rRNA (25–27). TLR11 is preferentially expressed in the kidney and bladder, and recognizes flagellin (28) or uropathogenic *Escherichia coli*, as well as a profilin-like molecule derived from *Toxoplasma gondii* (29) in mice. TLR12 is predominantly expressed in myeloid cells and is very similar to TLR11, recognizing profilin from *T. gondii* (30). TLR12 forms with TLR11 and recognizes the corresponding ligands (31, 32).

ROLES OF TLR SIGNAL-INDUCIBLE PROTEINS

The TLRs recognize various ligands, including PAMPs and DAMPs, and their downstream signaling pathways must be activated for the proper function of the innate immune system. Macrophages expressing TLRs, in particular, cause the expression of various TLR-inducible proteins, as well as cytokines and chemokines. Although the roles of cytokines and chemokines are well studied, those of TLR-inducible proteins remain unclear. It is possible that novel TLR-inducible genes are critical in regulating downstream immune responses. A bioinformatic analysis using comprehensive gene expression data from macrophages stimulated with LPS in the early phase of infection clearly showed that >200 genes were upregulated (33). Hierarchical clustering of these LPS-inducible genes from wild-type and MyD88$^{-/-}$ mice produced three clusters, and several LPS-inducible genes expressed in the early phase were included in one cluster. *NFKBIZ* and *ATF3*, which have already been reported as LPS-induced immediate-early genes, were also included in this cluster. For example, NFKBIZ, which is an IκB family member, positively regulates the expression of several genes, such as *IL6*, by interacting with NF-κBp50 and inducing histone-based epigenetic modifications (34, 35). ATF3 (activating transcription factor 3) negatively regulates the gene expression induced in the late phase (36). Moreover, LPS-inducible early genes, such as *NFKBID* and *BCL6*, are reported to regulate the induction of several genes in the late phase (37–40). Therefore, some TLR-inducible proteins expressed in the early phase are critical for the proper regulation of the immune response in the late phase.

Other TLR-inducible proteins and their physiological roles have been reported. We discuss some of them here. Recent studies have clearly shown that Regnase-1 is also found in the hierarchical cluster that includes *Iκbζ* and other inducible genes. The regulatory molecule Regnase-1, which contains a CCCH-type zinc-finger domain and an RNase domain, is also rapidly induced in response to TLR ligand stimulation. Regnase-1 degrades target mRNAs in macrophages, such IL-6 mRNA and IL-12B mRNA, but not that of TNF-α, via their 3′ untranslated regions (3′ UTRs) using its RNase activity. Therefore, large amounts of IL-6 and IL-12p40, but not TNF, are produced in Regnase-1-deficient macrophages in response to TLR ligand stimulation. Conventional Regnase-1 knockout mice also show elevated serum levels of immunoglobulin and increased autoantibody production (33).

After the activation of TLR4 signaling, the IKKs phosphorylate the Regnase-1 protein and Regnase-1 mRNA is synthesized. Phosphorylated Regnase-1 is rapidly degraded via the ubiquitin-proteasome machinery (41). With the loss of Regnase-1, IL-6 mRNA accumulates. Overall, Regnase-1 functions as a brake on the inflammatory response by controlling specific mRNAs via their 3′ UTRs. However, Regnase-1 degrades different target mRNAs in other cell types. For example, it degrades a set of C-Rel, OX40, and IL-2 mRNAs in CD4$^+$ T cells to maintain the normal effector CD4$^+$ T-cell status (42). Conditional CD4$^+$ T-cell-specific Regnase-1 knockout mice display autoimmunity arising from the aberrant proliferation of effector T cells. Therefore, this protein controls mRNA stability via the mRNA 3′ UTR, and the dysregulation of Regnase-1 causes the development of immune disorders. In addition to the control of mRNAs by Regnase-1, other molecules associate with the characteristic *cis*-element-containing AU-rich regions located in their 3′ UTRs. Recent studies have shown that HuR stabilizes its target mRNAs (43), and other regulatory molecules, such as tristetraprolin (44, 45), Roquin (46, 47), and KSRP (KH-type splicing regulatory protein) (44, 45, 48, 49), degrade various target mRNAs. All these molecules are involved in the activation of the immune system or the maintenance of homeostasis.

JMJD3 is another TLR-inducible gene, included in the same hierarchical cluster as *Iκbζ* and Regnase-1. JMJD3 functions as a histone H3 Lys27 (H3K27)-specific demethylase (50–52). Although UTX and UTY also act as H3K27 demethylases, only JMJD3 is rapidly induced by the activation of TLR signaling (51). Although JMJD3$^{-/-}$ mice show normal M1 macrophage activation in response to TLR ligands and bacterial infection, these mice show no M2 macrophage polarization against helminth infections or allergens (53). Therefore, although *JMJD3* is one of the TLR-inducible genes in macrophages, it specifically affects M2 macrophage polarization without affecting M1 macrophage activation or inflammation. Consequently, TLR-inducible genes modulate various types of immune responses.

CONCLUSION

As stated above, most TLR signaling pathways have been identified and the critical molecules reported in recent studies. In addition to TLR signaling, the various molecules involved in signaling by retinoic acid-inducible gene I (RIG-I)-like receptors (RLRs) and nucleotide-binding oligomerization domain (NOD)-like receptors have also been studied. For example, RNA virus infection is initially recognized by some endogenous receptors, including RIG-I, melanoma differentiation-associated gene 5 (MDA5), and laboratory of genetics and physiology 2 (LGP2), in addition to TLRs, and these induce antiviral responses that include the production of type I IFNs and proinflammatory cytokines (4, 54). After an RLR recognizes an RNA virus, an adaptor protein, IPS-1 (IFN-β promoter stimulator 1), associates with the two receptors, RIG-I and MDA5 (55–57). IPS-1 then activates TBK1/IKKi, which phosphorylate IRF3 and IRF7 to produce type I IFNs. Whereas extracellular PAMPs from viruses are recognized by TLRs, intracellular components derived from viruses are detected by RLRs. Interestingly, the endogenous receptor RIG-I is also strongly and rapidly induced in response to the activation of TLR signaling. Therefore, the innate immune system contains multistep PAMP-sensing mechanisms to eliminate pathogens.

The TLR-inducible genes are critical to the immune response and are involved in the development of various disorders. The TLR-inducible proteins Regnase-1 and Roquin are involved in the regulation of mRNA stability to ensure a proper immune response, and studies of these proteins are critical in understanding inflammatory disorders. The system that maintains mRNA stability has also attracted attention because it is an important mechanism regulating TLR-dependent inflammation.

Another TLR-inducible gene, *JMJD3*, is associated with the differentiation of macrophage subtypes. In recent immunological studies, various macrophage/monocyte subtypes and their differentiation mechanisms have been investigated. JMJD3 controls the differentiation of M2 macrophages in response to allergic stimuli and helminth infection. TRIB1 regulates tissue-resident M2-like macrophages, which maintain the homeostasis of adipose tissue (58). As well as the M1 and M2 macrophages (59–62), various functional macrophage/monocyte subtypes have been reported in recent studies that are involved in the development of specific disorders. The regulatory molecules in each cell type have also been reported (63–69). For example, DC-SIGN^{+} CD209 monocyte-derived DCs act like classical DCs in response to LPS, whereas SpiC-regulated red pulp macrophages trap and phagocytose red blood cells to maintain iron homeostasis. Peritoneal cavity-resident macrophages require GATA6 for their proliferation and are involved in immunoglobulin production in the gut. Reservoir monocytes participate in wound healing in response to myocardial injury. Each macrophage/monocyte subtype is a different cell type and is regulated by distinct genes. It is likely that the macrophage/monocyte subtypes are even more complex *in vivo* and that disorder-specific macrophages are present in our bodies.

In summary, TLR family molecules are critical for initiation of immune response. Further clarification of the TLR signal-inducible genes should eventually allow us to manipulate them to treat various diseases that are intimately associated with a dysregulated innate immune system.

Citation. Satoh T, Akira S. 2016. Toll-like receptor signaling and its inducible proteins. Microbiol Spectrum 4(6):MCHD-0040-2016.

References

1. **Akira S, Uematsu S, Takeuchi O.** 2006. Pathogen recognition and innate immunity. *Cell* **124:**783–801.
2. **Janeway CA Jr, Medzhitov R.** 2002. Innate immune recognition. *Annu Rev Immunol* **20:**197–216.
3. **Celhar T, Magalhães R, Fairhurst AM.** 2012. TLR7 and TLR9 in SLE: when sensing self goes wrong. *Immunol Res* **53:**58–77.
4. **Kawai T, Akira S.** 2010. The role of pattern-recognition receptors in innate immunity: update on Toll-like receptors. *Nat Immunol* **11:**373–384.
5. **Zanoni I, Ostuni R, Marek LR, Barresi S, Barbalat R, Barton GM, Granucci F, Kagan JC.** 2011. CD14 controls the LPS-induced endocytosis of Toll-like receptor 4. *Cell* **147:**868–880.
6. **Vogl T, Tenbrock K, Ludwig S, Leukert N, Ehrhardt C, van Zoelen MA, Nacken W, Foell D, van der Poll T, Sorg C, Roth J.** 2007. Mrp8 and Mrp14 are endogenous activators of Toll-like receptor 4, promoting lethal, endotoxin-induced shock. *Nat Med* **13:**1042–1049.
7. **Pouliot P, Plante I, Raquil MA, Tessier PA, Olivier M.** 2008. Myeloid-related proteins rapidly modulate macrophage nitric oxide production during innate immune response. *J Immunol* **181:**3595–3601.
8. **Suganami T, Mieda T, Itoh M, Shimoda Y, Kamei Y, Ogawa Y.** 2007. Attenuation of obesity-induced adipose tissue inflammation in C3H/HeJ mice carrying a Toll-like receptor 4 mutation. *Biochem Biophys Res Commun* **354:**45–49.
9. **Suganami T, Tanimoto-Koyama K, Nishida J, Itoh M, Yuan X, Mizuarai S, Kotani H, Yamaoka S, Miyake K, Aoe S, Kamei Y, Ogawa Y.** 2007. Role of the Toll-like receptor 4/NF-κB pathway in saturated fatty acid-induced inflammatory changes in the interaction between adipocytes and macrophages. *Arterioscler Thromb Vasc Biol* **27:**84–91.
10. **Lin SC, Lo YC, Wu H.** 2010. Helical assembly in the MyD88-IRAK4-IRAK2 complex in TLR/IL-1R signalling. *Nature* **465:**885–890.
11. **Kollewe C, Mackensen AC, Neumann D, Knop J, Cao P, Li S, Wesche H, Martin MU.** 2004. Sequential autophosphorylation steps in the interleukin-1 receptor-associated kinase-1 regulate its availability as an adapter in interleukin-1 signaling. *J Biol Chem* **279:**5227–5236.
12. **Jiang Z, Ninomiya-Tsuji J, Qian Y, Matsumoto K, Li X.** 2002. Interleukin-1 (IL-1) receptor-associated kinase-dependent IL-1-induced signaling complexes phosphorylate TAK1 and TAB2 at the plasma membrane and activate TAK1 in the cytosol. *Mol Cell Biol* **22:**7158–7167.
13. **Ajibade AA, Wang HY, Wang RF.** 2013. Cell type-specific function of TAK1 in innate immune signaling. *Trends Immunol* **34:**307–316.
14. **Chen ZJ.** 2012. Ubiquitination in signaling to and activation of IKK. *Immunol Rev* **246:**95–106.
15. **Jiang X, Chen ZJ.** 2011. The role of ubiquitylation in immune defence and pathogen evasion. *Nat Rev Immunol* **12:**35–48.
16. **Chang M, Jin W, Sun SC.** 2009. Peli1 facilitates TRIF-dependent Toll-like receptor signaling and proinflammatory cytokine production. *Nat Immunol* **10:**1089–1095.
17. **Kawasaki T, Takemura N, Standley DM, Akira S, Kawai T.** 2013. The second messenger phosphatidylinositol-5-phosphate facilitates antiviral innate immune signaling. *Cell Host Microbe* **14:**148–158.
18. **Blasius AL, Beutler B.** 2010. Intracellular Toll-like receptors. *Immunity* **32:**305–315.
19. **Bernard JJ, Cowing-Zitron C, Nakatsuji T, Muehleisen B, Muto J, Borkowski AW, Martinez L, Greidinger EL, Yu BD, Gallo RL.** 2012. Ultraviolet radiation damages self noncoding RNA and is detected by TLR3. *Nat Med* **18:**1286–1290.
20. **Takemura N, Kawasaki T, Kunisawa J, Sato S, Lamichhane A, Kobiyama K, Aoshi T, Ito J, Mizuguchi K, Karuppuchamy T, Matsunaga K, Miyatake S, Mori N, Tsujimura T, Satoh T, Kumagai Y, Kawai T, Standley DM, Ishii KJ, Kiyono H, Akira S, Uematsu S.** 2014. Blockade of TLR3 protects mice from lethal radiation-induced gastrointestinal syndrome. *Nat Commun* **5:** 3492. doi:10.1038/ncomms4492.
21. **Zhang SY, Jouanguy E, Ugolini S, Smahi A, Elain G, Romero P, Segal D, Sancho-Shimizu V, Lorenzo L, Puel A, Picard C, Chapgier A, Plancoulaine S, Titeux M, Cognet C, von Bernuth H, Ku CL, Casrouge A, Zhang XX, Barreiro L, Leonard J, Hamilton C, Lebon P, Héron B, Vallée L, Quintana-Murci L, Hovnanian A, Rozenberg F, Vivier E, Geissmann F, Tardieu M, Abel L, Casanova JL.** 2007. TLR3 deficiency in patients with herpes simplex encephalitis. *Science* **317:**1522–1527.
22. **Mancuso G, Gambuzza M, Midiri A, Biondo C, Papasergi S, Akira S, Teti G, Beninati C.** 2009. Bacterial recognition by TLR7 in the lysosomes of conventional dendritic cells. *Nat Immunol* **10:**587–594.
23. **Coban C, Igari Y, Yagi M, Reimer T, Koyama S, Aoshi T, Ohata K, Tsukui T, Takeshita F, Sakurai K, Ikegami T, Nakagawa A, Horii T, Nuñez G, Ishii KJ, Akira S.** 2010. Immunogenicity of whole-parasite vaccines against *Plasmodium falciparum* involves malarial hemozoin and host TLR9. *Cell Host Microbe* **7:**50–61.
24. **Tabeta K, Hoebe K, Janssen EM, Du X, Georgel P, Crozat K, Mudd S, Mann N, Sovath S, Goode J, Shamel L, Herskovits AA, Portnoy DA, Cooke M, Tarantino LM, Wiltshire T, Steinberg BE, Grinstein S, Beutler B.** 2006. The *Unc93b1* mutation 3d disrupts exogenous antigen presentation and signaling via Toll-like receptors 3, 7 and 9. *Nat Immunol* **7:**156–164.
25. **Hidmark A, von Saint Paul A, Dalpke AH.** 2012. Cutting edge: TLR13 is a receptor for bacterial RNA. *J Immunol* **189:**2717–2721.
26. **Li XD, Chen ZJ.** 2012. Sequence specific detection of bacterial 23S ribosomal RNA by TLR13. *eLife* **1:**e00102. doi:10.7554/eLife.00102.
27. **Oldenburg M, Krüger A, Ferstl R, Kaufmann A, Nees G, Sigmund A, Bathke B, Lauterbach H, Suter M, Dreher S, Koedel U, Akira S, Kawai T, Buer J, Wagner H, Bauer S, Hochrein H, Kirschning CJ.** 2012. TLR13 recognizes bacterial 23S rRNA devoid of erythromycin resistance-forming modification. *Science* **337:**1111–1115.
28. **Mathur R, Oh H, Zhang D, Park SG, Seo J, Koblansky A, Hayden MS, Ghosh S.** 2012. A mouse model of *Salmonella* Typhi infection. *Cell* **151:**590–602.

29. **Yarovinsky F, Zhang D, Andersen JF, Bannenberg GL, Serhan CN, Hayden MS, Hieny S, Sutterwala FS, Flavell RA, Ghosh S, Sher A.** 2005. TLR11 activation of dendritic cells by a protozoan profilin-like protein. *Science* **308:**1626–1629.
30. **Koblansky AA, Jankovic D, Oh H, Hieny S, Sungnak W, Mathur R, Hayden MS, Akira S, Sher A, Ghosh S.** 2013. Recognition of profilin by Toll-like receptor 12 is critical for host resistance to *Toxoplasma gondii*. *Immunity* **38:** 119–130.
31. **Andrade WA, Souza MC, Ramos-Martinez E, Nagpal K, Dutra MS, Melo MB, Bartholomeu DC, Ghosh S, Golenbock DT, Gazzinelli RT.** 2013. Combined action of nucleic acid-sensing Toll-like receptors and TLR11/TLR12 heterodimers imparts resistance to *Toxoplasma gondii* in mice. *Cell Host Microbe* **13:**42–53.
32. **Broz P, Monack DM.** 2013. Newly described pattern recognition receptors team up against intracellular pathogens. *Nat Rev Immunol* **13:**551–565.
33. **Matsushita K, Takeuchi O, Standley DM, Kumagai Y, Kawagoe T, Miyake T, Satoh T, Kato H, Tsujimura T, Nakamura H, Akira S.** 2009. Zc3h12a is an RNase essential for controlling immune responses by regulating mRNA decay. *Nature* **458:**1185–1190.
34. **Kayama H, Ramirez-Carrozzi VR, Yamamoto M, Mizutani T, Kuwata H, Iba H, Matsumoto M, Honda K, Smale ST, Takeda K.** 2015. Class-specific regulation of pro-inflammatory genes by MyD88 pathways and IκBζ. *J Biol Chem* **290:**22446.
35. **Yamamoto M, Yamazaki S, Uematsu S, Sato S, Hemmi H, Hoshino K, Kaisho T, Kuwata H, Takeuchi O, Takeshige K, Saitoh T, Yamaoka S, Yamamoto N, Yamamoto S, Muta T, Takeda K, Akira S.** 2004. Regulation of Toll/IL-1-receptor-mediated gene expression by the inducible nuclear protein IκBζ. *Nature* **430:**218–222.
36. **Gilchrist M, Thorsson V, Li B, Rust AG, Korb M, Roach JC, Kennedy K, Hai T, Bolouri H, Aderem A.** 2006. Systems biology approaches identify ATF3 as a negative regulator of Toll-like receptor 4. *Nature* **441:**173–178.
37. **Bours V, Franzoso G, Azarenko V, Park S, Kanno T, Brown K, Siebenlist U.** 1993. The oncoprotein Bcl-3 directly transactivates through κB motifs via association with DNA-binding p50B homodimers. *Cell* **72:**729–739.
38. **Kuwata H, Matsumoto M, Atarashi K, Morishita H, Hirotani T, Koga R, Takeda K.** 2006. IκBNS inhibits induction of a subset of Toll-like receptor-dependent genes and limits inflammation. *Immunity* **24:**41–51.
39. **Wessells J, Baer M, Young HA, Claudio E, Brown K, Siebenlist U, Johnson PF.** 2004. BCL-3 and NF-κB p50 attenuate lipopolysaccharide-induced inflammatory responses in macrophages. *J Biol Chem* **279:**49995–50003.
40. **Hirotani T, Lee PY, Kuwata H, Yamamoto M, Matsumoto M, Kawase I, Akira S, Takeda K.** 2005. The nuclear IκB protein IκBNS selectively inhibits lipopolysaccharide-induced IL-6 production in macrophages of the colonic lamina propria. *J Immunol* **174:**3650–3657.
41. **Iwasaki H, Takeuchi O, Teraguchi S, Matsushita K, Uehata T, Kuniyoshi K, Satoh T, Saitoh T, Matsushita M, Standley DM, Akira S.** 2011. The IκB kinase complex regulates the stability of cytokine-encoding mRNA induced by TLR-IL-1R by controlling degradation of regnase-1. *Nat Immunol* **12:**1167–1175.
42. **Uehata T, Iwasaki H, Vandenbon A, Matsushita K, Hernandez-Cuellar E, Kuniyoshi K, Satoh T, Mino T, Suzuki Y, Standley DM, Tsujimura T, Rakugi H, Isaka Y, Takeuchi O, Akira S.** 2013. Malt1-induced cleavage of regnase-1 in CD4[+] helper T cells regulates immune activation. *Cell* **153:**1036–1049.
43. **Wagner BJ, DeMaria CT, Sun Y, Wilson GM, Brewer G.** 1998. Structure and genomic organization of the human AUF1 gene: alternative pre-mRNA splicing generates four protein isoforms. *Genomics* **48:**195–202.
44. **Carballo E, Lai WS, Blackshear PJ.** 1998. Feedback inhibition of macrophage tumor necrosis factor-α production by tristetraprolin. *Science* **281:**1001–1005.
45. **Taylor GA, Carballo E, Lee DM, Lai WS, Thompson MJ, Patel DD, Schenkman DI, Gilkeson GS, Broxmeyer HE, Haynes BF, Blackshear PJ.** 1996. A pathogenetic role for TNFα in the syndrome of cachexia, arthritis, and autoimmunity resulting from tristetraprolin (TTP) deficiency. *Immunity* **4:**445–454.
46. **Glasmacher E, Hoefig KP, Vogel KU, Rath N, Du L, Wolf C, Kremmer E, Wang X, Heissmeyer V.** 2010. Roquin binds inducible costimulator mRNA and effectors of mRNA decay to induce microRNA-independent post-transcriptional repression. *Nat Immunol* **11:**725–733.
47. **Yu D, Tan AH, Hu X, Athanasopoulos V, Simpson N, Silva DG, Hutloff A, Giles KM, Leedman PJ, Lam KP, Goodnow CC, Vinuesa CG.** 2007. Roquin represses autoimmunity by limiting inducible T-cell co-stimulator messenger RNA. *Nature* **450:**299–303.
48. **Chen CY, Gherzi R, Ong SE, Chan EL, Raijmakers R, Pruijn GJ, Stoecklin G, Moroni C, Mann M, Karin M.** 2001. AU binding proteins recruit the exosome to degrade ARE-containing mRNAs. *Cell* **107:**451–464.
49. **Gherzi R, Lee KY, Briata P, Wegmüller D, Moroni C, Karin M, Chen CY.** 2004. A KH domain RNA binding protein, KSRP, promotes ARE-directed mRNA turnover by recruiting the degradation machinery. *Mol Cell* **14:** 571–583.
50. **Agger K, Cloos PA, Christensen J, Pasini D, Rose S, Rappsilber J, Issaeva I, Canaani E, Salcini AE, Helin K.** 2007. UTX and JMJD3 are histone H3K27 demethylases involved in *HOX* gene regulation and development. *Nature* **449:**731–734.
51. **De Santa F, Totaro MG, Prosperini E, Notarbartolo S, Testa G, Natoli G.** 2007. The histone H3 lysine-27 demethylase Jmjd3 links inflammation to inhibition of polycomb-mediated gene silencing. *Cell* **130:**1083–1094.
52. **Lan F, Bayliss PE, Rinn JL, Whetstine JR, Wang JK, Chen S, Iwase S, Alpatov R, Issaeva I, Canaani E, Roberts TM, Chang HY, Shi Y.** 2007. A histone H3 lysine 27 demethylase regulates animal posterior development. *Nature* **449:**689–694.
53. **Satoh T, Takeuchi O, Vandenbon A, Yasuda K, Tanaka Y, Kumagai Y, Miyake T, Matsushita K, Okazaki T, Saitoh T, Honma K, Matsuyama T, Yui K, Tsujimura T,**

Standley DM, Nakanishi K, Nakai K, Akira S. 2010. The Jmjd3-Irf4 axis regulates M2 macrophage polarization and host responses against helminth infection. *Nat Immunol* **11:**936–944.

54. **Takeuchi O, Akira S.** 2010. Pattern recognition receptors and inflammation. *Cell* **140:**805–820.

55. **Seth RB, Sun L, Ea CK, Chen ZJ.** 2005. Identification and characterization of MAVS, a mitochondrial antiviral signaling protein that activates NF-κB and IRF 3. *Cell* **122:**669–682.

56. **Xu LG, Wang YY, Han KJ, Li LY, Zhai Z, Shu HB.** 2005. VISA is an adapter protein required for virus-triggered IFN-β signaling. *Mol Cell* **19:**727–740.

57. **Kawai T, Takahashi K, Sato S, Coban C, Kumar H, Kato H, Ishii KJ, Takeuchi O, Akira S.** 2005. IPS-1, an adaptor triggering RIG-I- and Mda5-mediated type I interferon induction. *Nat Immunol* **6:**981–988.

58. **Satoh T, Kidoya H, Naito H, Yamamoto M, Takemura N, Nakagawa K, Yoshioka Y, Morii E, Takakura N, Takeuchi O, Akira S.** 2013. Critical role of Trib1 in differentiation of tissue-resident M2-like macrophages. *Nature* **495:**524–528.

59. **Biswas SK, Mantovani A.** 2010. Macrophage plasticity and interaction with lymphocyte subsets: cancer as a paradigm. *Nat Immunol* **11:**889–896.

60. **Gordon S, Plüddemann A, Martinez Estrada F.** 2014. Macrophage heterogeneity in tissues: phenotypic diversity and functions. *Immunol Rev* **262:**36–55.

61. **Sica A, Invernizzi P, Mantovani A.** 2014. Macrophage plasticity and polarization in liver homeostasis and pathology. *Hepatology* **59:**2034–2042.

62. **Wynn TA, Chawla A, Pollard JW.** 2013. Macrophage biology in development, homeostasis and disease. *Nature* **496:**445–455.

63. **Asano K, Nabeyama A, Miyake Y, Qiu CH, Kurita A, Tomura M, Kanagawa O, Fujii S, Tanaka M.** 2011. CD169-positive macrophages dominate antitumor immunity by crosspresenting dead cell-associated antigens. *Immunity* **34:**85–95.

64. **Auffray C, Fogg D, Garfa M, Elain G, Join-Lambert O, Kayal S, Sarnacki S, Cumano A, Lauvau G, Geissmann F.** 2007. Monitoring of blood vessels and tissues by a population of monocytes with patrolling behavior. *Science* **317:**666–670.

65. **Carlin LM, Stamatiades EG, Auffray C, Hanna RN, Glover L, Vizcay-Barrena G, Hedrick CC, Cook HT, Diebold S, Geissmann F.** 2013. *Nr4a1*-dependent $Ly6C^{low}$ monocytes monitor endothelial cells and orchestrate their disposal. *Cell* **153:**362–375.

66. **Cheong C, Matos I, Choi JH, Dandamudi DB, Shrestha E, Longhi MP, Jeffrey KL, Anthony RM, Kluger C, Nchinda G, Koh H, Rodriguez A, Idoyaga J, Pack M, Velinzon K, Park CG, Steinman RM.** 2010. Microbial stimulation fully differentiates monocytes to DC-SIGN/$CD209^+$ dendritic cells for immune T cell areas. *Cell* **143:**416–429.

67. **Kohyama M, Ise W, Edelson BT, Wilker PR, Hildner K, Mejia C, Frazier WA, Murphy TL, Murphy KM.** 2009. Role for Spi-C in the development of red pulp macrophages and splenic iron homeostasis. *Nature* **457:**318–321.

68. **Kumamoto Y, Linehan M, Weinstein JS, Laidlaw BJ, Craft JE, Iwasaki A.** 2013. $CD301b^+$ dermal dendritic cells drive T helper 2 cell-mediated immunity. *Immunity* **39:**733–743.

69. **Rosas M, Davies LC, Giles PJ, Liao CT, Kharfan B, Stone TC, O'Donnell VB, Fraser DJ, Jones SA, Taylor PR.** 2014. The transcription factor Gata6 links tissue macrophage phenotype and proliferative renewal. *Science* **344:**645–648.

Myeloid Cells in Health and Disease: A Synthesis
Edited by Siamon Gordon

doi:10.1128/microbiolspec.MCHD-0036-2016

Gordon D. Brown[1]
Paul R. Crocker[2]

25 Lectin Receptors Expressed on Myeloid Cells

INTRODUCTION

Lectins, defined as proteins that recognize carbohydrates, perform numerous essential biological functions. Recognizing a diverse array of carbohydrate structures, vertebrate lectins have been subdivided into several structurally distinct families which can be located intracellularly (such as the intracellular M-type family of lectins, which function primarily in the glycoprotein secretory pathway), in the plasma membrane (such as some members of the C-type lectin and Siglec [sialic acid-binding immunoglobulin-type lectin] families, which are involved in pathogen recognition and immune regulation), or are secreted into the extracellular milieu (such as some members of the galectin family, which serve several homeostatic and immune functions) (Table 1). We will restrict our discussion here to selected myeloid- and plasma membrane-expressed members of only two families, the C-type lectins and Siglecs. We will provide a brief overview of each family and then focus on selected illustrative and detailed examples that highlight how these lectins influence myeloid cell functioning in health and disease. For an overview on the other lectin families, the reader is referred to an excellent website (http://www.imperial.ac.uk/research/animallectins/ctld/lectins.html).

SIGLECS

Siglecs are a distinct subgroup of the immunoglobulin (Ig) superfamily that have evolved to use sialylated glycans as their predominant ligands (1). Siglecs have been mainly defined in mammalian species, but clear orthologs are also present in amphibia and fish (2). Siglec-like molecules have also been identified in streptococcal bacteria (3) and in an adenovirus capsid protein (4). In mammals, there are two subgroups of Siglecs: (i) sialoadhesin (Sn; Siglec-1), CD22 (Siglec-2), MAG (Siglec-4), and Siglec-15, which are present in all species; and (ii) CD33-related Siglecs that vary considerably in composition between species and appear to be undergoing rapid evolution (Fig. 1). Due to uncertainties in gene ontologies, the human CD33-related Siglecs have been assigned numerical suffixes, whereas the mouse CD33-related Siglecs have been assigned alphabetical suffixes. All Siglecs are type 1 membrane proteins containing a homologous N-terminal V-set Ig-like domain that recognizes sialylated glycans, followed by variable numbers of C2 set domains. Recognition of sialic acid depends on a conserved structural template involving both hydrogen bonding networks, ionic and hydrophobic interactions, together with variable inter-strand loops that make contact with additional glycan

[1]MRC Centre for Medical Mycology, University of Aberdeen, Foresterhill, Aberdeen AB25 2ZD, United Kingdom; [2]Division of Cell Signalling and Immunology, School of Life Sciences, University of Dundee, Dundee DD1 5EH, United Kingdom.

Table 1 Lectin families

Family name	Selected ligands
Calnexin	Glc_1Man_9
Chitinase-like lectins	GlcN, GalN, chitin, chito-oligosaccharides
C-type lectins	Various (e.g., mannose, fucose, GalNAc, β-glucan)
F-box lectins	High-mannose and sulfated glycoproteins
Ficolins	GlcNAc, GalNAc, fucose
F-type lectins	Fucose and others (e.g., 3-O-methyl-D-galactose)
Galectins	β-Galactosides (e.g., *N*-acetyllactosamine)
Intelectins	Galactofuranose, pentoses
L-type lectins	Various (e.g., oligomannose)
M-type lectins	High-mannose glycans (e.g., $Man_8GlcNAc_2$)
P-type lectins	Mannose 6-phosphate
R-type lectins	Various (e.g., GalNAc, sialic acid, sulfated glycans)
Siglecs (I-type lectins)	Sialic acid

residues and confer extended specificity to Siglecs (5). A Siglec-like Ig fold was recently seen in the regulatory myeloid receptors PILR (paired immunoglobulin-like type 2 receptor)-α and -β, which mediate high-affinity binding to mucin-like sialylated ligands via both protein-protein and protein-sialic acid interactions (6, 7). Siglecs are expressed broadly across the hematopoietic and immune systems, except for MAG, which is restricted to the myelin-forming cells of the nervous system, oligodendrocytes, and Schwann cells (Fig. 1). As discussed below, some Siglecs are highly restricted to particular cell types, whereas others are more broadly expressed. In humans and mice, the major subgroup of cells lacking Siglecs are CD4 and CD8 T cells, although in other species such as chimpanzees, Siglec-5 is reported to be expressed on all circulating T cells (8).

When expressed naturally at the cell surface, Siglecs interact with sialic acid-containing ligands both in *cis* (on the same plasma membrane) and in *trans* (on the plasma membrane of different cells). The degree to which each occurs likely depends on the relative affinity, density, and display of sialylated ligands as well as other poorly understood topographical constraints. Both types of interaction have been shown to play important roles in immune modulation (reviewed in references 9 and 10). The cytoplasmic tails of most Siglecs contain two or more tyrosine-based motifs that can be phosphorylated and recruit SH2 domain-containing effector molecules. The most common motif is the immunoreceptor tyrosine-based inhibitory motif (ITIM) that has been identified in hundreds of receptors of the immune system (11). Phosphorylated ITIMs exhibit high affinity for the tandem SH2 domain-containing protein tyrosine phosphatases SHP-1 and SHP-2, which are activated on binding to ITIMs and thereby potentially capable of modulating signaling functions of Siglec-expressing cells (9). Many Siglecs also possess ITIM-like motifs that appear to synergize with the ITIMs for efficient recruitment of SHP-1 and SHP-2 (12, 13). The same motifs are also important for the endocytic functions of many Siglecs. Some Siglecs possess a basic residue in the transmembrane region (Fig. 1) that leads to formation of a membrane complex with DNAX activation protein of 12 kDa (DAP12), an adaptor with an immunoreceptor tyrosine-based activation motif (ITAM). As a consequence, these Siglecs have the potential to directly trigger signal transduction via Syk recruitment and activation. Links between glycan recognition by Siglecs and subsequent intracellular signaling have been a major focus for many laboratories since their discovery, but the specific downstream targets and biochemical pathways remain elusive for the most part. The focus of this section will be Siglecs that are expressed predominantly on human myeloid cells, namely Sn (Siglec-1), CD33 (Siglec-3), and Siglec-5, -7, -8, -9, -10, -11, -14, and -15. Given the importance of mouse models in determining the biological functions of Siglecs, the murine CD33-related Siglec-E, -F, and -G will be grouped with their most closely related human counterparts for comparative purposes.

Sialoadhesin (Siglec-1; CD169)

Sn is a prototypic Siglec that was identified as a sialic acid-dependent erythrocyte receptor expressed by subsets of mouse tissue-resident macrophages (14). Sn has an unusually large number of 17 Ig domains that appear conserved in mammals and reptiles (15). These are important for extending the sialic acid-binding site away from the plasma membrane to promote intercellular interactions. Sn prefers α2,3-linked sialic acids over α2,6- and α2,8-linked sialic acids and does not bind sialic acids modified by hydroxylation (Neu5Gc) or 9-O-acetylation (e.g., Neu5,9Ac2) (16, 17).

In humans and mice, Sn expression appears specific for tissue macrophage subsets described as "$CD169^+$ macrophages" (18–20). These cells are abundant in lymphoid tissues, notably subcapsular sinus macrophages in lymph nodes and marginal metallophilic macrophages in the spleens of rodents or perifollicular capillary sheaths in spleens of humans (21). These macrophage populations are strategically positioned to capture viruses and immune complexes from the afferent lymphatics and splenic sinuses, respectively, and may therefore share similar biological functions (22). Emerging evidence clearly shows that $CD169^+$ macro-

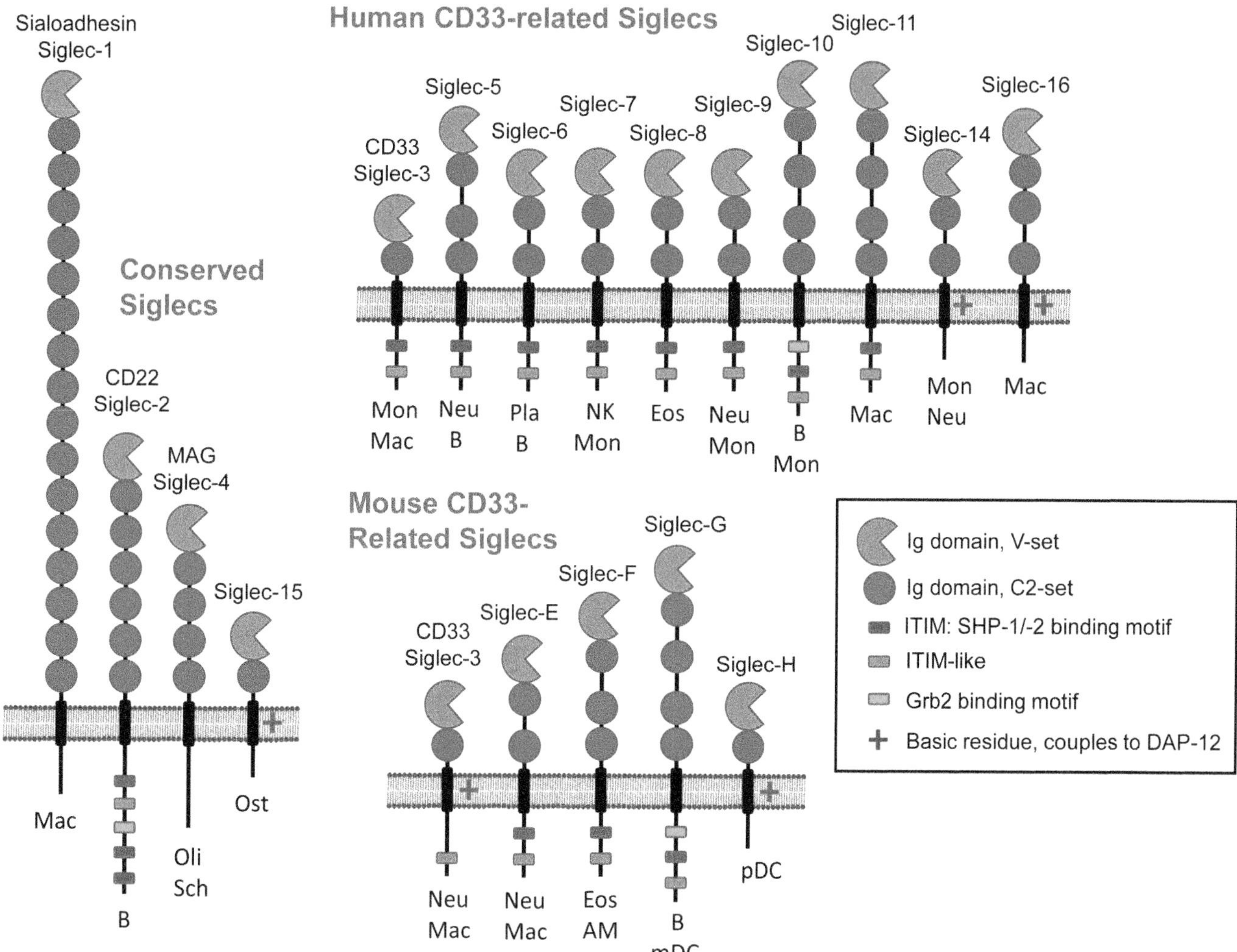

Figure 1 Siglecs in humans and mice. There are two subgroups of Siglecs: One group contains Siglecs that are conserved in all mammalian species and the other group contains CD33-related Siglecs that appear to be undergoing rapid evolution in primates. The cell types expressing highest levels of each Siglec are indicated. B, B cell; Eos, eosinophil; Mac, macrophage; mDC, myeloid dendritic cell; Mon, monocyte; Neu, neutrophil; NK, NK cell; Oli, oligodendrocyte; Ost, osteoclast; pDC, plasmacytoid dendritic cell; Pla, placental syncytiotrophoblast; Sch, Schwann cell.

phages play a key role in the capture of a broad range of viruses, including arteriviruses (23), retroviruses (24), herpesviruses (25, 26), and adenoviruses (27), and for the first two, this directly involves viral recognition by Sn. CD169^{+} macrophage capture of viruses is important for restraining viral spread to distal sites (28), but it can also promote viral transfer to neighboring cells such as T cells (29, 30) and B cells (31) and directly prime CD8^{+} cytotoxic T cells for responses to viral antigens via cross-presentation (27). Interestingly, CD169^{+} macrophages in the spleen have been shown to act as "Trojan horses" for vesicular stomatitis viruses, permitting high viral replication that is important for stimulation of protective adaptive immune responses (32).

In addition to viruses, CD169^{+} macrophages can capture apoptotic tumor cells and cross-present tumor antigens to drive antitumor cytotoxic CD8 T-cell responses (33). Conversely, uptake of apoptotic cells by CD169^{+} macrophages can drive tolerance of self-reactive T cells via induction of the chemokine CCL22 (34). CD169^{+} macrophages can also transfer exogenous antigens to dendritic cells (DCs) and promote cross-presentation to CD8 T cells (35) and transfer antigens to B cells (36) and NKT cells (37) to promote cellular activation. Besides its constitutive high expression on tissue macrophage subsets, Sn can also be induced strongly on monocytes, macrophages, and monocyte-derived DCs *in vitro* by type I interferons or agents such as viruses and Toll-like receptor (TLR) ligands that induce interferon production (24). Accordingly, Sn is upregulated on circulating monocytes in HIV-infected individuals and on macrophages in rheumatoid arthritis (RA) (38), primary biliary cirrhosis (39), systemic sclerosis (40), and systemic lupus erythematosus (41). Sn expression on inflammatory macrophages has been associated with favorable prognosis in colorectal cancer (42) and in endometrial carcinoma (43), but with a more severe disease in proliferative glomerulonephritis (44). Many of the above disease associations may reflect exposure of macrophages to interferons rather than being causally related. Indeed, in the BWF1 murine model of spontaneous systemic lupus erythematosus, there was no influence of Sn deficiency on disease severity (45). However, in mouse models of inherited neuropathy (46–48), autoimmune uveoretinitis (49), and experimental allergic encephalomyelitis (50), Sn-deficient mice exhibited reduced inflammation accompanied by reduced levels of T-cell and macrophage activation. In the experimental allergic encephalomyelitis model, this appears to be due to an Sn-dependent suppression of $CD4^{+}FoxP3^{+}$ regulatory T-cell (Treg) expansion, thereby promoting inflammation (50), whereas in the other central nervous system models, Sn-dependent regulation of CD8 T cells is important (46). The upregulation of sialylated ligands for Sn on activated T-cell populations is likely to be an important determinant in mediating the Sn-dependent suppression of T-cell subsets and function (51). Sn can also efficiently mediate the capture and uptake of exosomes released from B lymphocytes following apoptosis and therefore play a role in antigen presentation to T cells (52, 53).

A role for Sn in phagocytic interactions of macrophages with various sialylated bacterial and protozoal pathogens was initially demonstrated, including *Neisseria meningitidis* (54), *Campylobacter jejuni* (55), and *Trypanosoma cruzi* (56). Sn-dependent targeting of heat-killed *C. jejuni* to splenic red pulp macrophages led to a rapid induction of type I interferon and proinflammatory cytokines, in a MyD88-dependent manner, suggesting a host protective role for Sn against sialylated bacteria (57). This was also supported in an infection model using a sialylated strain of group B streptococcus (GBS), where Sn-deficient mice exhibited reduced bacterial spread (58). However, this protective role for Sn was only seen in neutrophil-depleted mice, suggesting that Sn-dependent macrophage bacterial uptake can provide a backup defense in the event of neutrophils failing to clear the bacteria. Conversely, Sn expression on macrophages and monocyte-derived DCs can be exploited by enveloped viruses displaying host-derived sialic acids, leading to their capture, uptake, and dissemination. This was first seen with the porcine reproductive and respiratory syndrome virus, which targets lung alveolar macrophages of pigs (23), and more recently with HIV (24) and other retroviruses (31, 59). On HIV, Sn can recognize both gp120, a sialylated glycoprotein, and GM3, a monosialylated ganglioside terminating in NeuAcα2-3Gal (30, 60–62). GM3 is packaged into the HIV envelope during the budding from infected cells that occurs in lipid rafts (63). On monocyte-derived DCs, Sn interactions with HIV lead to membrane invaginations containing viral particles that are very efficiently transferred to T cells in a process known as *trans*-infection (30). *In vivo* evidence that Sn promotes retroviral *trans*-infection was obtained following infection of mice using murine leukemia virus, where *trans*-infection of B cells depended on the expression of Sn on lymph node sinus-lining macrophages (31).

Although Sn is unusual among Siglecs in not having well-defined signaling motifs in its cytoplasmic tail, recent reports have suggested it can associate with the ITAM adaptor DAP12 and either suppress type I interferon production or stimulate transforming growth factor β production (64, 65). These studies were both done using RNA knockdown approaches to suppress Sn expression, and the physiological significance of the findings should be confirmed using primary macrophages from Sn-deficient mice.

CD33 (Siglec-3)

CD33 is a marker of early human myeloid progenitors and leukemic cells, and is also expressed on monocytes, tissue macrophages, NK-cell subsets, and weakly on neutrophils. It has two Ig domains and was the first of the CD33-related Siglecs to be characterized as an inhibitory receptor, suppressing activation of FcγRI and recruiting SHP-1 and SHP-2 (66). CD33 has some preference for α2,6- over α2,3-sialylated glycans and binds strongly to sialylated ligands on myeloid leukemic cell lines (67). The restricted expression of CD33 has been exploited in the treatment of acute myeloid leukemia using gemtuzumab, a humanized anti-CD33 monoclonal antibody (mAb) coupled to the toxic antibiotic calicheamicin. Binding of anti-CD33 mAbs to CD33 triggers endocytosis of the bound antibody. This depends on ITIM phosphorylation, recruitment of the E3 ligase Cbl, and ubiquitylation of the CD33 cytoplasmic tail (68–70). Selective expression of CD33 on leukemic progenitor cells also makes it an attractive target for therapy using chimeric antigen receptors expressed on cytotoxic T cells.

Recently, two coinherited single-nucleotide polymorphisms (SNPs) have been associated with protection of humans against late-onset Alzheimer's disease in genome-wide association studies. These SNPs result in increased exon 2 skipping, leading to raised levels of CD33 lacking the V-set domain and reduced levels of full-length CD33 (71). Since full-length CD33 can inhibit microglial cell uptake of amyloid-β protein in a sialic acid-dependent manner (72, 73), it is thought that individuals lacking the protective SNPs may accumulate more toxic amyloid-β proteins, thus driving pathology. Targeting CD33 using antibodies that either inhibit function or promote internalization and degradation may be a useful approach to treating Alzheimer's disease.

The murine ortholog of CD33 exists as two spliced forms that differ in the cytoplasmic tail, neither containing the typical ITIM found in most other CD33-related Siglecs (74). Furthermore, mCD33 has a lysine residue in the transmembrane sequence and may therefore couple to the DAP12 transmembrane adaptor, as shown for mouse Siglec-H (75) and human Siglec-14 (76), -15 (77), and -16 (78). In contrast to hCD33, mCD33 in the blood is expressed mainly on neutrophils rather than monocytes, which also suggests a nonconserved function of this receptor (79).

Siglec-5 (CD170) and Siglec-14

The *SIGLEC5* and *SIGLEC14* genes are adjacent to each other on chromosome 19 and encode proteins containing four and three Ig-like domains, respectively. The first two Ig domains of Siglec-5 and -14 share >99% sequence identity but then diverge. Siglec-5 is an inhibitory receptor with typical ITIMs, whereas Siglec-14 is complexed with DAP12 and mediates activatory signaling. Both Siglec-5 and -14 bind similar ligands, with a preference for the sialyl-Tn structure (Neu5Acα2-6GalNAcα) (76). Although many antibodies to Siglec-5 cross-react with Siglec-14, specific antibodies have shown that while Siglec-5 is expressed on neutrophils and B cells, Siglec-14 is found at low levels on neutrophils and monocytes. A *SIGLEC14* null allele is frequently present in Asian populations but is less common in Europeans (80). This is due to a recombination event between the 5′ region of the *SIGLEC14* gene and the 3′ region of the *SIGLEC5* gene, resulting in a fusion protein that is identical to Siglec-5 but expressed in a Siglec-14-like manner. Individuals with chronic obstructive pulmonary disease who are *SIGLEC14* null exhibited reduced exacerbation attacks compared with individuals expressing Siglec-14 (80). Siglec-5 can bind sialylated strains of *N. meningitidis*, and both Siglec-5 and -14 can bind sialylated strains of *Haemophilus influenzae* implicated in chronic obstructive pulmonary disease exacerbations and trigger inhibitory and activatory responses, respectively (80). Thus, the absence of Siglec-14 on neutrophils would lead to reduced inflammatory responses in *SIGLEC14*-null individuals. Besides expression on leukocytes, both Siglec-5 and -14 are found on human amniotic epithelium and may influence responses to GBS infection and the frequency of preterm births in infected mothers (81). Besides mediating sialic acid-dependent interactions with host cells and pathogens, Siglec-5 and -14 can mediate sugar-independent interactions with some strains of GBS via recognition of the β protein (81). A recent study also demonstrated that the nonglycosylated danger-associated molecular pattern (DAMP) protein heat shock protein 70 (Hsp70) can bind to Siglec-5 and -14 and modulate cellular responses (82). There are no obvious equivalents of Siglec-5 or -14 in mice, making it difficult to study this interesting pair of receptors in animal models.

Siglec-7, -9, and -E

Siglec-7 and -9 share a high degree of sequence similarity and appear to have evolved by gene duplication from an ancestral gene encoding a three-Ig-domain inhibitory Siglec, represented in mice by Siglec-E. Siglec-7 is the major Siglec on human NK cells and is also seen at lower levels on monocytes, macrophages, DCs, and a minor subset of CD8 T cells (83–85). Siglec-7 has also been detected in platelets, basophils, and mast cells, where it may modulate survival and activation (85). Siglec-9 is prominently expressed on neutrophils, monocytes, macrophages, and DCs; ~30% of NK cells; and minor subsets of CD4 and CD8 T cells (86, 87). Despite high sequence similarity, Siglec-7 binds strongly to α2,8-linked sialic acids present in "b-series" gangliosides (and some glycoproteins, whereas Siglec-9 prefers α2,3-linked sialic acids) (88). Sulfation of the sialyl Lewis X (sLeX) structure can strongly influence recognition by both Siglecs, with Siglec-9 preferring 6-sulfo-sLeX and Siglec-7 binding well to both 6-sulfo-sLeX and 6′-sulfo-sLeX (89). It has recently been shown that Siglec-9 can bind strongly to high-molecular-weight hyaluronan, and that its ligation on neutrophils leads to suppression of cellular activation (90). Siglec-E in mice exhibits a combination of some features of Siglec-7 and Siglec-9, being mainly expressed on neutrophils, monocytes, and macrophages, with sialic acid-binding preferences that span those of both Siglec-7 and -9 (91). Similar to T cells, NK cells in mice appear to lack expression of inhibitory Siglecs. Siglec-E is an important inhibitory receptor of neutrophils, as initially demonstrated in a lipopolysaccharide (LPS)-induced lung inflammation model in which Siglec-E-deficient mice exhibited exaggerated CD11b-dependent neutrophil influx (92). This was found to be linked to Siglec-E-dependent production of reactive oxygen species (ROS) by neutrophils triggered on the CD11b ligand fibrinogen, which suppressed neutrophil recruitment to the lung (93). Siglec-E-dependent inhibition of neutrophil function has also been proposed to be a mechanism underlying an exaggerated aging phenotype observed in one strain of Siglec-E-deficient mice (94). Several studies have also demonstrated inhibitory functions of Siglec-E on macrophages and DCs, including suppression of proinflammatory cytokine production in response to TLR ligands and promotion of Tregs in response to sialylated antigens (95–99). Furthermore, targeting Siglec-E on macrophages with sialylated nanoparticles was shown to block inflammatory responses *in vitro* and *in vivo* (100).

Tumor cells often upregulate cell surface sialylated glycans, and it appears that these may be important in Siglec-dependent dampening of antitumor immunity. Siglec-7 and -9 can both suppress NK-cell cytotoxicity against tumor cells expressing relevant glycan ligands (101–103). Siglec-9 and Siglec-E can also dampen neutrophil activation and tumor-cell killing, while ligation of Siglec-9 or Siglec-E on macrophages by tumor glycans seems to suppress formation of tumor-promoting M2 macrophages (98). Studies with GBS have also demonstrated that sialylated bacteria can subvert innate immune responses by targeting Siglec-9 and Siglec-E on neutrophils and macrophages, resulting in attenuation of phagocytosis, killing, and proinflammatory cytokine production (104, 105).

Siglec-8 and -F

Siglec-8 has three Ig domains and is expressed on eosinophils and mast cells, with weaker expression on basophils (106, 107). It binds strongly to 6′-sulfo-sLeX and to mucins isolated from bronchial tissues (108, 109), but endogenous mucin ligands do not seem to require sulfation for strong binding (110). In mast cells, antibodies to Siglec-8 can inhibit FcεRI-triggered degranulation responses, in line with its role as an inhibitory receptor (111). In eosinophils, much attention has focused on the role of Siglec-8 in triggering apoptosis, which can occur following cross-linking with anti-Siglec-8 antibodies or sialoglycan polymers (112, 113). Apoptosis depends on generation of ROS and caspase activation and is paradoxically enhanced in the presence of cytokine "survival" factors such as granulocyte-macrophage colony-stimulating factor and interleukin-5 (IL-5) (113). A role for Siglec-8 in the pathogenesis of asthma has been suggested by upregulation of Siglec-8 ligands in inflamed lung tissue (109) and by associations of Siglec-8 polymorphisms with asthma (114).

Although there is no ortholog of Siglec-8 in mice, the four-Ig-domain mouse Siglec-F is expressed in a similar way to Siglec-8 on eosinophils, has a similar glycan-binding preference for to 6′-sulfo-sLeX, and appears to have acquired similar functions through convergent evolution (115–117). There are some important differences, however. Siglec-F can recognize a broader range of α2,3-linked sialic acids; it is also expressed on alveolar macrophages and triggers weaker apoptosis using different signaling pathways (118). Siglec-F-null mice show exaggerated eosinophilic responses in certain lung allergy models, suggesting that Siglec-F negatively regulates eosinophil production and/or survival following immunological challenge (119, 120). Interestingly, Siglec-F ligands in the airways and lung parenchyma were also upregulated during allergic inflammation, but these did not appear to require sulfation to mediate strong binding to Siglec-F (110).

Siglec-10 and -G

Siglec-10 has five Ig-like domains and, in addition to the ITIM and ITIM-like motifs, displays an additional tyrosine-based motif in its cytoplasmic tail (121–123). It is expressed at relatively low levels on several cells of the immune system, including B cells, monocytes, and eosinophils (122). It can also be strongly upregulated on tumor-infiltrating NK cells in hepatocellular carcinoma, where its expression was negatively associated with patient survival (124). It is the only CD33-related human Siglec that has a clear-cut ortholog in mice, designated Siglec-G (9). Both Siglec-10 and Siglec-G prefer Neu5Gc over Neu5Ac in both α2,3 and α2,6 linkages (125). Similar to Siglec-10 in humans and pigs (126), Siglec-G is expressed mainly on B cells and subsets of DCs and weakly on eosinophils (127, 128). Mice deficient in Siglec-G show a 10-fold increase in numbers of a specialized subset of B lymphocytes, the B1a cells, which make natural antibodies (129). These Siglec-G-deficient B1a cells also show exaggerated Ca fluxing following B-cell receptor cross-linking. Studies using knock-in mice carrying an inactivating mutation in the sialic acid-binding site of Siglec-G show a similar phenotype (127). This appears to be due to a requirement of sialic acid-dependent *cis*-interactions between Siglec-G and the B-cell receptor. On DCs, Siglec-G has been proposed to regulate cytokine responses to DAMPs released by necrotic cells in sterile inflammation. This is thought to be due to a dampening effect of *cis*-interactions between Siglec-G and the heavily sialylated DAMP receptor CD24 (130). Disruption of this interaction through sialidases released by bacteria such as *Streptococcus pneumoniae* may be important in triggering inflammatory responses in sepsis (131). A recent study has also shown that pseudaminic acid expressed on the flagella of *C. jejuni* can be recognized by Siglec-10 and trigger IL-10 production in DCs (132). This suggests a novel form of glycan recognition by Siglec-10 that is exploited by some pathogens.

Siglec-11 and -16

Siglec-11 and -16 are paired inhibitory and activatory receptors, with five and four Ig domains, respectively (78, 133). In most humans, the *SIGLEC16* gene has a 4-bp deletion and only ~35% of humans express one or two functional alleles. The extracellular regions of these proteins are >99% identical due to gene conversion events, and anti-Siglec-11 mAb 4C4 cross-reacts with Siglec-16. Siglec-11 binds weakly to α2,8-linked sialic acids *in vitro*. Siglec-11 appears to be absent from circulating leukocytes but is expressed widely on populations of tissue macrophages, including resident microglia in the brain, where high levels of α2,8-linked sialic acids are present on gangliosides. Expression of Siglec-11 on microglia can impair their phagocytosis of apoptotic cells and neurotoxicity (134). Polysialic acid presented by neural cell adhesion molecule is also α2,8 linked and was shown to be recognized by Siglec-11 on macrophages and suppress LPS-dependent tumor necrosis factor α (TNF-α) production and phagocytosis triggered by LPS exposure (134, 135). Interestingly, microglial expression appears to be unique to humans (136). In mice, its function may be mediated by Siglec-E, which is similarly expressed on microglia and able to mediate neuroprotective effects in response to inflammatory signals (96). The activating receptor Siglec-16 is also present on macrophages, including those in the brain, but functional studies have not been reported (78).

Siglec-15

Siglec-15 was first described in 2007 as a highly conserved and ancient Siglec found in vertebrates (77). It lacks the typical arrangement of cysteines seen in the V-set Ig domain of other Siglecs and has an unusual intron-exon arrangement. Nevertheless, it can bind the sialyl-Tn structure (Neu5Acα2-6GalNAcα), with weaker binding to 3′-sialyllactose. It is associated with DAP12 and also has a tyrosine-based motif in the cytoplasmic tail. On macrophages, interactions with sialyl-Tn antigens expressed by tumor cells were shown to trigger transforming growth factor β production, which could be important in immunosuppression and promoting tumor growth (137).

Although it was first reported as being expressed on macrophages and DCs in human lymphoid tissues, subsequent work has established that Siglec-15 is most strongly expressed in osteoclasts and their precursors, where it plays an important role together with receptor activator of NF-κB ligand (RANKL) in triggering osteoclast differentiation (138–141). Osteoclasts are key cells involved in bone degradation and share a common hematopoietic progenitor with macrophages. Mice lacking Siglec-15 show a mild osteopetrosis and impaired osteoclast differentiation (140, 141). Specific antibodies directed to Siglec-15 are able to phenocopy this due to antibody-induced internalization and degradation of Siglec-15 (142). Siglec-15 therefore provides a novel target for diseases involving excessive osteoclast activation and bone loss, such as menopause-related osteoporosis.

C-TYPE LECTIN RECEPTORS

C-type lectin receptors (CLRs) are a diverse collection of >1,000 proteins and are the largest lectin family (143). All of these receptors possess at least one C-type lectin-like domain (CTLD), a characteristic fold formed by disulfide linkages between highly conserved cysteine residues (143). Based on their phylogeny and structure, CLRs have been divided into 17 groups that are either membrane bound or secreted (143). The term "C-type lectin" originated from initial observations that these receptors required Ca^{2+} for carbohydrate recognition. However, we now know that not all CLRs require Ca^{2+} for ligand recognition and that these receptors can recognize a much more diverse range of ligands, such as lipids and proteins, for example. Many of these receptors have also been shown to bind to different classes of ligands (i.e., they are multivalent) and can recognize both endogenous and exogenous ligands.

A great many CLRs have essential roles in immunity. A key example is the endothelial-expressed selectins that function as adhesion molecules by binding cell surface glycoproteins on leukocytes and play a critical part in leukocyte migration during inflammation (144). Other examples include the secreted collectins, such as the surfactant proteins, which function in both pulmonary physiology and immunity, and serum mannose-binding protein (MBL), which has an essential role in triggering complement activation through MBL-associated serine proteases in response to microbial infection (145). The focus of the rest of this section, however, will be on selected transmembrane receptors that are widely expressed by myeloid cells and have been extensively characterized in murine models, including Dectin-1, MICL (myeloid inhibitory C-type lectin), Dectin-2, Mincle (macrophage inducible C-type lectin receptor)/MCL (macrophage C-type lectin), and the macrophage mannose receptor (MR). Detailed descriptions of these molecules will serve as illustrative examples of the varied nature of C-type lectins and their importance in myeloid cell function in health and disease. The functions and properties of other myeloid-expressed CLRs, including well-characterized receptors such as dendritic cell-specific intercellular adhesion molecule-3-grabbing non-integrin (DC-SIGN), can be found in several excellent reviews (146–148).

Dectin-1 (CLEC-7A)

Dectin-1 is one of the best-characterized myeloid-expressed CLRs, and this type II transmembrane receptor belongs to group V within the CLR family. Dectin-1 contains a single extracellular CTLD, a stalk region, a single-pass transmembrane domain, and a cytoplasmic tail containing signaling motifs, including an ITAM-like (or hem-ITAM) motif and a triacidic motif (Fig. 2). Dectin-1 is alternatively spliced into two major isoforms, differing by the presence or absence of the stalk region, which are expressed differentially in different cell types and mouse strains and which have slightly different functionalities (149). The receptor is *N*-glycosylated, which can affect its expression and function (150), and is predominantly expressed by myeloid cells, including monocytes, macrophages, DCs, and neutrophils (151). There is also evidence for expression of this receptor on B cells and subsets of T cells, and it may be unregulated on epithelial cells during inflammation (151–154).

Through mechanisms that are not yet completely understood, the CTLD of Dectin-1 is able to recognize β-1,3-glucan-containing carbohydrates (155). These carbohydrates are found predominantly in fungal cell walls, and consequently there has been considerable focus on the role of Dectin-1 in antifungal immunity. Indeed, Dectin-1 recognizes many fungal species, including major human pathogens such as *Aspergillus*, *Candida*, *Coccidioides*, and *Pneumocystis* (156). There is now substantial evidence that Dectin-1 plays an essential role in antifungal immunity: several polymorphisms of this receptor in humans (including a Y238X polymorphism that essentially renders homozygous individuals Dectin-1 deficient) have been linked to increased susceptibility to mucocutaneous fungal infections or fungal-induced inflammation in the gut (157–159). Moreover, Dectin-1 knockout mice are more susceptible to systemic and mucocutaneous infections with several pathogens (160–162). However, the requirement for Dectin-1 for controlling *Candida albicans in vivo* is dependent on the fungal strains, which undergo differential changes in their cell wall during infection (163).

In addition to fungi, Dectin-1 can recognize mycobacteria. How Dectin-1 recognizes these pathogens is unknown, and although shown to promote IL-12 responses *in vitro*, the receptor does not appear to play an essential role in antimycobacterial immunity *in vivo* (164, 165). Dectin-1 has also been implicated in the recognition of other pathogens, including *Leishmania* (166).

Dectin-1 was originally identified as acting as a T-cell costimulatory molecule through recognition of an endogenous ligand (167), but the nature of this ligand remains elusive. Several other endogenous ligands have been described, including vimentin, through which Dectin-1 was thought to be involved in driving lipid oxidation in atherosclerosis (168). However, Dectin-1 deficiency was subsequently found not to affect atherosclerosis development in mouse models (169). Dectin-1 has also been implicated in the reverse transcytosis of secretory IgA-antigen complexes by intestinal M cells and induction of subsequent mucosal and systemic antibody responses (170). Moreover, in the presence of galactosylated IgG1, Dectin-1 associates with FcγRIIB, resulting in the inhibition of complement-mediated inflammation (171). In response to intestinal mucus, FcγRIIB, along with another lectin, galectin-3, complexes with Dectin-1 to promote the anti-inflammatory properties of DCs, enhancing homeostasis and oral tolerance (172). Most recently, a protective role for Dectin-1 in antitumor immunity has been demonstrated. Mechanistically, Dectin-1-mediated recognition of *N*-glycan structures on tumor cells was shown to augment NK-mediated killing and, in a model of hepatocarcinogenesis, act protectively by suppressing TLR4 signaling (173, 174).

Upon recognition of β-glucans, Dectin-1 can activate Syk-dependent and Syk-independent intracellular signaling cascades. Surprisingly, the activation of Syk was shown to require the tyrosine phosphatase SHP-2, which acted as a scaffold and facilitated the recruitment of Syk to Dectin-1 (175). The ability of Dectin-1 to induce Syk-dependent signaling pathways is mediated by a single phosphorylated tyrosine residue in the

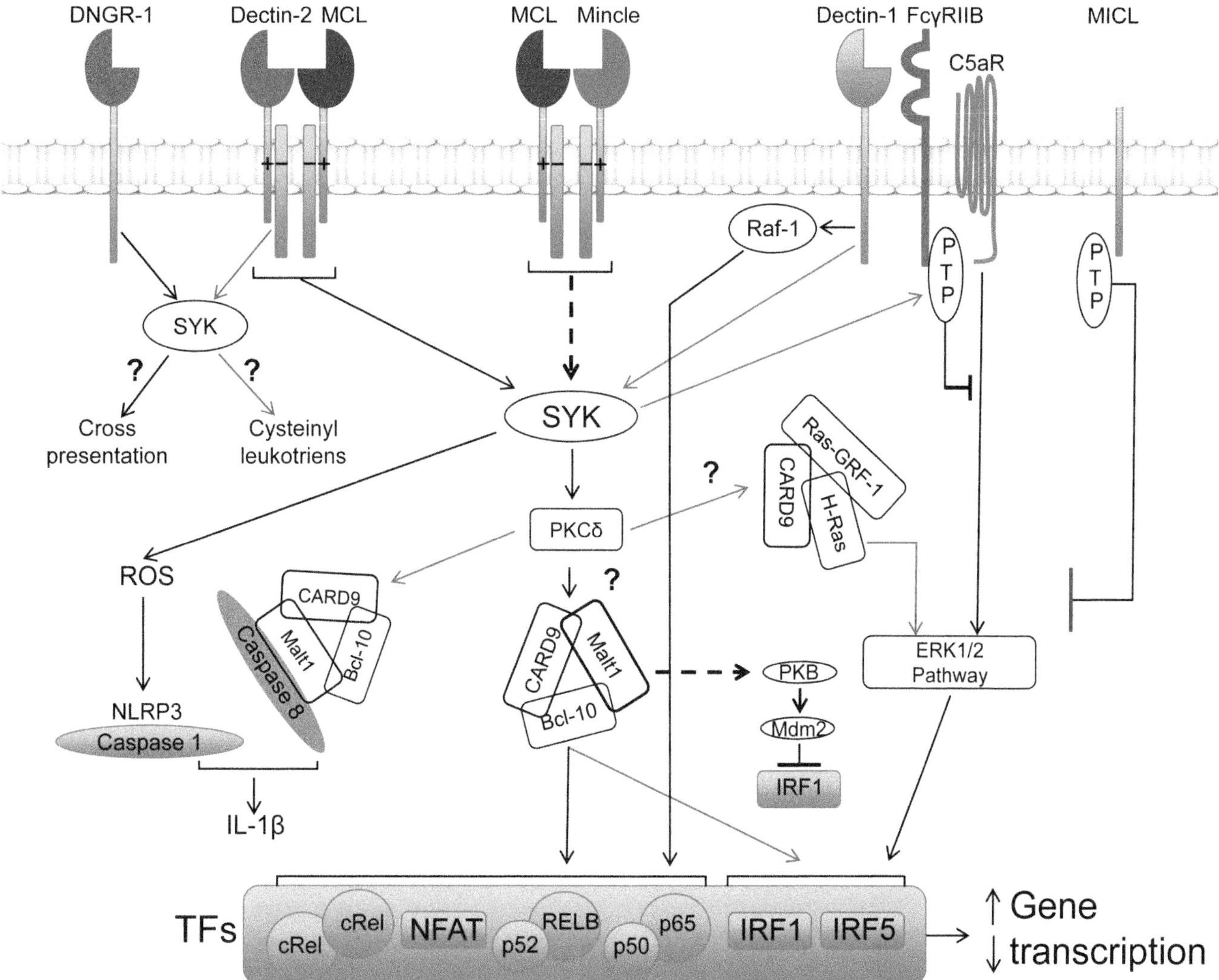

Figure 2 Selected signal transduction cascades induced by C-type lectin receptors. Activation receptors, such as Dectin-1, Dectin-2, Mincle, and MCL, induce cellular responses primarily through Syk kinase, although other pathways can be involved, such as those induced by Raf-1. Inhibitory receptors, such as MICL, activate protein tyrosine phosphatases (PTPs, such as SHP-1), which attenuate activation pathways. DNGR-1 (CLEC9A), not discussed in the text, is an actin-binding receptor expressed by CD8[+] DCs and involved in antigen cross-presentation. Reprinted from reference 329, with permission.

ITAM-like motif within the cytoplasmic tail and is likely to require receptor dimerization (176). Signaling through this pathway involves protein kinase C-δ (PKC-δ) and the caspase recruitment domain family member 9 (CARD9)-Bcl10-Malt1 complex and leads to the induction of canonical and noncanonical NF-κB subunits and interferon regulatory factor 1 (IRF1), resulting in gene transcription (177, 178). Recently, CARD9 was found to be dispensable for NF-κB activation, but regulated extracellular signal-regulated kinase (ERK) activation by linking Ras-GRF1 to H-Ras (179). The CARD9 pathway is utilized by several other receptors (see also below) and is essential for protective antimicrobial immunity, particularly against fungi (180–182). Syk activation by Dectin-1 induces IRF5 and nuclear factor of activated T cells (NFAT), through phospholipase C-γ (PLC-γ) and calcineurin (183, 184), a pathway inhibited by immunosuppressive drugs, such as cyclosporine, and linked to the increased susceptibility to fungal infection that occurs following administration of these compounds (185). The Syk-independent pathway from Dectin-1 involves activation of Raf-1, which integrates with the Syk-dependent pathway at the point of NF-κB activation (186). Other pathways also exist. For example, the induction of phagocytosis by Dectin-1 in macrophages is Syk independent, requiring Bruton's tyrosine kinase (Btk) and Vav-1 (187, 188). The ability of Dectin-1 to induce productive intracellular signaling (i.e., leading to cellular responses) requires receptor clustering into a "phagocytic synapse" and exclusion of regulatory tyrosine phosphatases (189). Moreover, the ability of Dectin-1 to induce productive responses to purified agonists can be cell type specific, an effect linked to differential utilization of CARD9 (190).

Activation of Dectin-1 signaling pathways can induce multiple cellular responses, including actin-mediated phagocytosis (Fig. 3), phagosome maturation, activation of the respiratory burst, regulation of neutrophil extracellular trap (NET) formation in neutrophils, DC maturation, and antigen presentation, in part through the use of autophagy machinery (191–194). Dectin-1 can activate inflammasomes, facilitating the production of IL-1β. Indeed, this receptor has been implicated in activation of the Nlrp3 (NLR family, pyrin domain-containing 3) inflammasome, although the pathways involved are unclear, and can directly induce the noncanonical caspase-8 inflammasome, through CARD9 and Malt1 (195–197). Assembly and activation of the caspase-8 inflammasome was recently shown to require the nonreceptor tyrosine kinase Tec (198). Dectin-1 also induces the production of eicosanoids and several cytokines and chemokines (including TNF, IL-10, IL-6, IL-23, CCL2, and CCL3), and can modulate cytokine production and cellular functions induced by other pattern recognition receptors. For example, costimulation of Dectin-1 and MyD88-coupled TLRs leads to the synergistic production of cytokines, such as TNF and IL-23, while simultaneously repressing the induction of others, such as IL-12 (199, 200). Another example is the ability of Dectin-1 to activate complement receptor 3 (CR3; alternatively Mac-1), through activation of Vav1, Vav3, and PLC-γ, which results in enhanced neutrophil phagocytosis and ROS production (201). These two receptors also act collaboratively in macrophages, through association in lipid rafts and activation of the Syk-Jun N-terminal kinase–activator protein 1 (AP1) pathway, to enhance inflammatory cytokine responses (202).

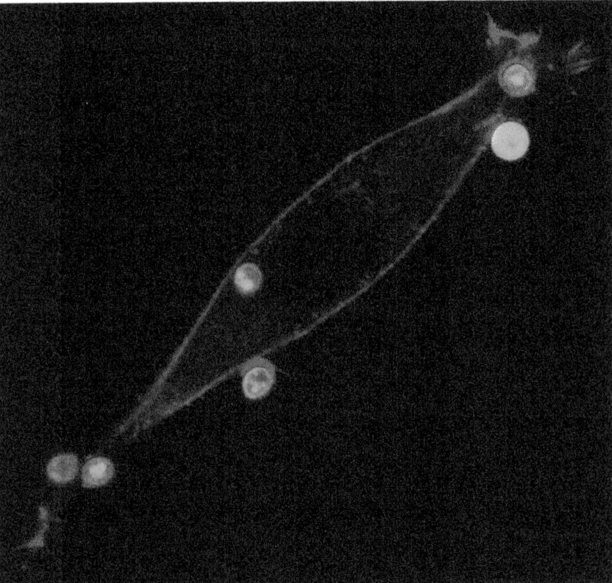

Figure 3 Dectin-1 can mediate the nonopsonic phagocytosis of fluorescently labeled fungal particles (green) via actin (red)-based phagocytic cups. Reprinted from reference 155, with permission.

Like the TLRs, Dectin-1 is capable of instructing the development of adaptive immune responses, particularly Th1 and Th17 immunity (203). Interestingly, Dectin-1-activated DCs can also instruct Tregs ($CD25^+Foxp3^+$) to express IL-17 (204). While Th1 responses are important for the control of systemic infections, Th17 responses are critical for controlling fungal infections at the mucosa. Indeed, several human diseases associated with chronic mucocutaneous candidiasis, including CARD9 deficiency, have been linked to alterations in components of the Th17 response (205). How Dectin-1 promotes Th17 responses is incompletely understood, but involves Malt1-dependent activation of the NF-κB subunit c-Rel, which is required for the induction of polarizing cytokines such as IL-1β and IL-23p19 (206). Dectin-1 can also induce humoral responses (207), stimulate cytotoxic T-cell responses (208), and induce myeloid-derived suppressor cells, which can suppress T- and NK-cell responses (209). In addition to classic adaptive immunity, activation of Dectin-1 has been shown to induce innate immune memory (or trained immunity), through the epigenetic reprogramming of monocytes that occurs following aerobic glycolysis induced through an Akt–mammalian target of rapamycin (mTOR)–hypoxia-inducible factor-1α (HIF1α) pathway (210, 211).

The role of Dectin-1 in driving adaptive immunity during infection is still not completely understood, but there has been some recent progress. For example, Dectin-1 was found not to be essential for IL-17 production in mice systemically infected with *C. albicans* (203), yet was required to drive Th17 polarization during pulmonary infection with *Aspergillus fumigatus*

(162, 212). The ability of Dectin-1 to induce T helper cell differentiation during a skin infection model with *C. albicans* was recently shown to be dependent on fungal morphology (correlating with β-glucan exposure) and the DC subset involved (213). In the gastrointestinal tract, Dectin-1 was found to be essential for driving fungal-specific $CD4^+$ T-cell responses and for the maintenance of the cellularity of gastrointestinal-associated lymphoid tissues (214). Dectin-1 can also regulate intestinal Treg-cell differentiation through modification of the microbiota, following exposure to dietary β-glucans (215).

MICL (CLEC-12A)

MICL (also called DCAL-2, CLL-1, and KLRL-1) is structurally similar to Dectin-1 and located in the same genomic region (192). Unlike Dectin-1, MICL is one of the few myeloid-expressed CLRs that contains an ITIM in its cytoplasmic tail, and, like some of the Siglecs described above, can induce inhibitory intracellular signaling through SHP-1 and SHP-2 phosphates (216). Human MICL is alternatively spliced into at least three isoforms (α, β, and γ), and the receptor is expressed as a monomer and heavily glycosylated (216). These latter features differ in the murine ortholog, which is expressed as a dimer and is only moderately glycosylated (217). In both species, MICL is expressed primarily by myeloid cells, including macrophages, monocytes, DCs, and granulocytes, although the receptor is also expressed on B cells, $CD8^+$ T cells, and bone marrow NK cells in the mouse (217, 218). Expression levels of MICL are substantially regulated during inflammatory processes both *in vitro* and *in vivo* (217, 218). Interestingly, MICL is highly expressed on acute myeloid leukemia cells, and the receptor has been put forward as a marker of this disease as well as for developing antibody-directed immunotherapies (219–222). In addition, murine MICL has been proposed to be a marker for a distinct subset of $CD8\alpha^-$ DCs (223). In mouse, targeting of antigens to MICL was found to induce CD4 and CD8 T-cell proliferation and enhance antibody responses (224).

MICL functions as an inhibitory receptor, and experiments with receptor chimeras have directly demonstrated that MICL can inhibit the activation signals induced through other pattern recognition receptors (216). Moreover, antibody cross-linking experiments have shown that MICL can inhibit NK-cell cytotoxicity (225) and differentially modulate DC responses, such as IL-12 production, depending on the mode of activation (226). Recently, MICL was shown to be regulated in an ATG16L1 (autophagy-related protein 16-like 1)-dependent manner and play a key role in antibacterial autophagy through a functional interaction with an E3-ubiquitin ligase complex (227).

MICL recognizes an endogenous ligand in many tissues and was recently identified as a receptor for dead cells and uric acid (217, 228). MICL was shown to be required to suppress the inflammatory responses induced by these ligands (228). Similar observations have been made with human leukocytes (229). Thus, MICL appears to have an important role in controlling damage-induced inflammation and may be involved in autoimmune diseases. Indeed, MICL was recently found to play an essential role in regulating myeloid cell-mediated inflammation in a murine model of RA (230). Although polymorphisms of *CLEC12A* do not associate with RA, autoantibodies to MICL were identified in a subset of RA patients, which, in mouse models, could exacerbate the disease (230). These findings suggest that the threshold of myeloid cell activation can be modulated by autoantibodies that bind to these types of inhibitory receptors. Downregulation of this receptor has also been proposed to underlie hyperinflammatory responses observed in Behçet's syndrome and gout (231).

Dectin-2 (CLEC4n)

Dectin-2 has a structure similar to that of Dectin-1, except that it possesses a short cytoplasmic tail lacking recognizable signaling motifs (232). To mediate intracellular signaling, this receptor associates with the ITAM-containing FcRγ adaptor molecule (232). Dectin-2 is unusual in this respect in that its interaction with the adaptor is mediated by a membrane-proximal region within its intracellular tail rather than through a transmembrane arginine residue as occurs with other similarly structured receptors (233). As with Dectin-1, signaling from the ITAM motif following Dectin-2 ligation occurs through the Syk, PKC-δ, and CARD9-Bcl10-Malt1 pathway, but also involves PLC-γ2 (233–238). Dectin-2 is expressed primarily by myeloid cells, including macrophages, subsets of DCs, and neutrophils, as well as monocytes, where its expression can be markedly upregulated during inflammation (206, 239–241).

Dectin-2 recognizes high-mannose-based structures through its "classical" carbohydrate-binding CTLD, which possesses a conserved EPN motif (242). This ligand specificity enables recognition of a variety of pathogens (including bacteria and nematodes, for example) and pathogen-derived molecules (including house dust mite allergens) (232). Recently, Dectin-2 was shown to recognize mannose-capped lipoarabinomannan of mycobacteria and play a role in antimycobacterial immunity (243). However, most attention has focused on the role of Dectin-2 in antifungal immunity, where it is required for protection against infection with selected fungal species, including *C. albicans* (through recognition of α-mannans on specific morphological forms) and *Candida glabrata* (233–236, 244). Dectin-2 can also recognize species of *Malassezia*, through an O-linked mannobiose-rich glycoprotein; *Blastomyces dermatitidis*; *Cryptococcus neoformans*; *Fonsecaea pedrosoi*; and *A. fumigatus* (245–249). In addition to pathogens, Dectin-2 may recognize an endogenous ligand and be involved in modulating UV-induced immunosuppression (250).

Like Dectin-1, Dectin-2 induces several cellular responses in response to microbial stimuli and can influence the development of adaptive immunity. In response to *C. albicans*, for example, Dectin-2 was shown to drive inflammatory host cytokine responses, including TNF, IL-6, and IL-12, and the development of Th17 and Th1 immunity (233, 235, 236, 248). Notably, Dectin-2 was found to selectively induce Th17-polarizing cytokines, including IL-23 and IL-1β, by activating the NF-κB subunit, c-Rel, via Malt1 (206). More recently, Dectin-2 was found to regulate a key neutrophil IL-17 autocrine loop during fungal infection (247). This receptor also plays a role in the physical recognition of fungi, and signaling from Dectin-2 can induce Nlrp3 inflammasome activation, extracellular trap formation, the respiratory burst, and production of cysteinyl leukotrienes (238, 241, 251–254). Dectin-2 may also form heterodimeric complexes with MCL (Dectin-3), although this is still controversial (255, 256).

The induction of cysteinyl leukotrienes, in particular, has led to a great deal of interest in the role of Dectin-2 in airway inflammation induced by house dust mite. This CLR can recognize a glycan component of house dust mite, inducing the production of cysteinyl leukotrienes by DCs and stimulating the development of Th2 responses (252, 257). In mouse models of house dust mite-mediated pulmonary inflammation, Dectin-2 drove eosinophilic and neutrophilic responses by promoting both Th17 and Th2 immunity (257–260). Despite a clear role for Dectin-2 in allergy and host defense in mouse models, there is only one report demonstrating a link between polymorphism in this receptor and human disease (pulmonary cryptococcosis) (261).

Mincle/MCL (CLEC4d/CLEC4e)

Mincle and MCL (also known as CLECSF8 or Dectin-3) are similar in structure to Dectin-2, but are discussed here together because they form a heterotrimeric complex (along with the ITAM-containing FcRγ adaptor) that is required for expression at the cell surface (256, 262–265). The FcRγ adaptor appears to associate primarily with Mincle, through a positively charged arginine residue in the transmembrane domain, and can induce signaling through the Syk, PKC-δ, CARD9-Bcl10-Malt1, and mitogen-activated protein kinase pathways, leading to activation of transcription factors, including NF-κB (237, 262, 266). This adaptor can also associate with MCL, but this occurs in an unusual fashion independently of any charged amino acid residue in the transmembrane or cytoplasmic domain (264, 267). The association of MCL with Mincle is mediated by the stalk region, and expression of these receptors is coordinately regulated under naive and inflammatory conditions (256, 265). The CTLD of MCL has also been shown to be involved in regulating surface expression (267).

Unsurprisingly, given that they function as a complex, there is significant overlap in the reports describing the expression and function of Mincle and MCL. Both of these receptors have been described as being predominantly expressed on myeloid cells, including macrophages, neutrophils, monocytes, and DCs, although there is also evidence of expression on other leukocytes, including some subsets of B cells (256, 262, 267–273). Expression of these receptors can be upregulated following exposure to inflammatory stimuli, including microbial components such as LPS, and for Mincle this has been shown to occur in a MyD88- and CCAAT/enhancer-binding protein β (C/EBPβ)-dependent manner (256, 274, 275). Mincle has also been reported to be reciprocally expressed on neutrophils and monocytes within individuals, which has functional implications (269).

Mincle and MCL have both been shown to induce and/or regulate numerous cellular responses including endocytosis, phagocytosis, the respiratory burst, activation of the Nlrp3 inflammasome, NET formation, and the production of proinflammatory cytokines and chemokines (TNF, macrophage inflammatory protein 2 [MIP-2], IL-1β, MIP-1α, IL-6, keratinocyte-derived chemokine, and granulocyte colony-stimulating factor, for example) (237, 255, 262–264, 267, 272, 276–279). Moreover, both receptors can modulate the development of adaptive immunity, promoting Th1 and Th17 responses (264, 277, 278). Mincle has also been shown to promote Th2 development by suppressing Dectin-1-mediated IL-12 production (178).

The CTLD of Mincle contains a classical mannose-recognition EPN motif, but the receptor appears to primarily recognize microbial glycolipids (280–283).

Specific microbial ligands have been identified, including mycobacterial cord factor, trehalose dimycolate (TDM), its synthetic analog trehalose dibehenate (TDB), glycerol monomycolate from mycobacteria, and glyceroglycolipid and mannitol-linked mannosyl fatty acids from fungi (245, 277, 284). Structural analysis suggests that Mincle's CTLD has binding sites for both the sugar and fatty acid moieties of these ligands (282, 283). In contrast, the CTLD of MCL is unable to directly recognize carbohydrates (267), but this receptor can recognize TDM (264). Structural analysis has suggested that, like Mincle, the CTLD may interact with both the sugar and fatty acid moieties of this glycolipid (282).

Given the ability of these CLRs to recognize mycobacterial components, it is not surprising that both receptors have been implicated in antimycobacterial immunity. In response to mycobacterial ligands, for example, Mincle induces the production of inflammatory cytokines and nitric oxide, granuloma formation, and Th1 and Th17 responses (271, 272, 277, 278, 285). These activities of Mincle contribute to the adjuvant activities of complete Freund's adjuvant (286). MCL was similarly found to be required for the adjuvant activity of TDM, and loss of this receptor impaired both innate (inflammation and granuloma formation) and adaptive responses (T-cell function) induced by this glycolipid (264). Mincle has also been shown to recognize intact mycobacteria *in vitro*, but its actual role *in vivo* during infection is still unclear (277, 278, 287). One group has reported no effect of Mincle deficiency on infections with *Mycobacterium tuberculosis* H37Rv, whereas other groups have described some alterations in inflammation and bacterial burdens following infection with *Mycobacterium bovis* BCG or *M. tuberculosis* Erdman (271, 287, 288). In contrast, MCL was recently discovered to have an essential role in the nonopsonic recognition of mycobacteria by myeloid cells, and loss of this receptor resulted in higher extracellular mycobacterial burdens that drove neutrophilic inflammation and increased mortality in mouse models (289). Importantly, a polymorphism of MCL was also shown to be associated with susceptibility to tuberculosis in humans (289).

Both Mincle and MCL can also recognize other bacteria, including *Klebsiella pneumoniae*. During *K. pneumoniae* infection, for example, $MCL^{-/-}$ mice showed increased susceptibility and presented with increased bacterial burdens, inflammatory neutrophilic responses, and severe lung pathology (290). Mincle has similarly been found to be required for the control of *K. pneumoniae* infection in mouse models (279).

Mincle was first characterized as a receptor for *C. albicans* (291), and in response to this fungal pathogen, Mincle can induce protective immune responses including phagocytosis, fungal killing, and inflammatory cytokine production (269, 291). As mentioned above, Mincle was found to be reciprocally expressed on leukocytes within the same individuals, and in monocytes expression correlated with reduced fungal uptake and killing, but enhanced inflammatory cytokine production. In contrast, expression of Mincle on neutrophils correlated with enhanced fungal uptake and killing (269). $MCL^{-/-}$ mice were also shown to display increased susceptibility to *C. albicans*, with higher fungal burdens, and defective inflammatory responses (255). However, these observations are controversial, as other groups have not found any evidence for a role of MCL in anti-*Candida* immunity (263, 267, 289).

Mincle has been implicated in immunity to *Malassezia* and was found to be required for cytokine induction and inflammation during *in vivo* infection (280). In addition, Mincle recognizes *F. pedrosoi* and *Fonsecaea monophora*, causative agents of chromoblastomycosis (178, 292). In contrast to other fungal pathogens, this recognition was found to be inefficient, due to a lack of TLR costimulation, and contributed to chronicity of the infection (292). Moreover, in response to *F. monophora*, Mincle can suppress Dectin-1-mediated IL-12 production, promoting Th2 responses (178). Similarly, Mincle was shown to suppress Th17 cell differentiation induced by Dectin-2 (248).

Mincle also recognizes endogenous ligands. Spliceosome-associated protein (SAP) 130, released from necrotic cells, was shown to be a ligand for Mincle, although recognition occurred through a different binding site on its CTLD (262). Mincle recognition of SAP130 induces inflammatory cytokine production (MIP-2 and TNF, for example) and neutrophil accumulation (262). Recently, human Mincle was also shown to recognize cholesterol crystals (293). This recognition of endogenous ligands suggests a role for Mincle in homeostasis, although our understanding of this function is still poor. There is emerging evidence, however, suggesting that Mincle may be involved in RA, pathogenesis of ischemic stroke and early brain injury after subarachnoid hemorrhage, and obesity-induced adipose tissue inflammation and fibrosis (294–298).

The Macrophage MR

The MR (CD206) is a type I transmembrane protein that contains a heavily glycosylated extracellular region consisting of a cysteine-rich domain, a fibronectin type II domain, and eight CTLDs (Fig. 4) (299, 300). Two conformations of the MR have been proposed: an extended form and a more compact "bent" form that is influenced by pH (301). The MR is expressed predominantly intracellularly, as part of the endocytic pathway, in subsets of macrophages and DCs, as well as some other nonmyeloid cell types including endothelial cells (300, 302). The expression of this receptor can be influenced by several cytokines, including IL-4, which causes marked upregulation of the MR (303). In fact, this upregulation has led to the MR being used as a marker for alternatively activated macrophages (304). Within the MR gene is a coregulated microRNA (miR-511-3p) that modulates cellular activation in tumor-associated and other macrophages and was recently shown to contribute to intestinal inflammation (305, 306). The extracellular domain of the MR can also be cleaved by metalloproteinases following cellular activation, through Dectin-1 signaling, for example, releasing a functional soluble form (sMR) (307).

The extracellular domains of the MR each recognize different structures. The cysteine-rich domain binds sulfated carbohydrates, the fibronectin domain binds collagen, while the CTLDs (specifically CTLDs 4 to 8) bind terminal mannose- and fucose-based structures as well as *N*-acetylglucosamine in a Ca^{2+}-dependent manner (300). The MR can also recognize CpG motif-containing oligodeoxynucleotides (308). The recognition of such a broad range of structures has led to substantial literature implicating the MR in both homeostasis and antimicrobial immunity. Indeed, the MR has been shown to recognize multiple types of pathogens, including viruses, helminths, trypanosomes, fungi,

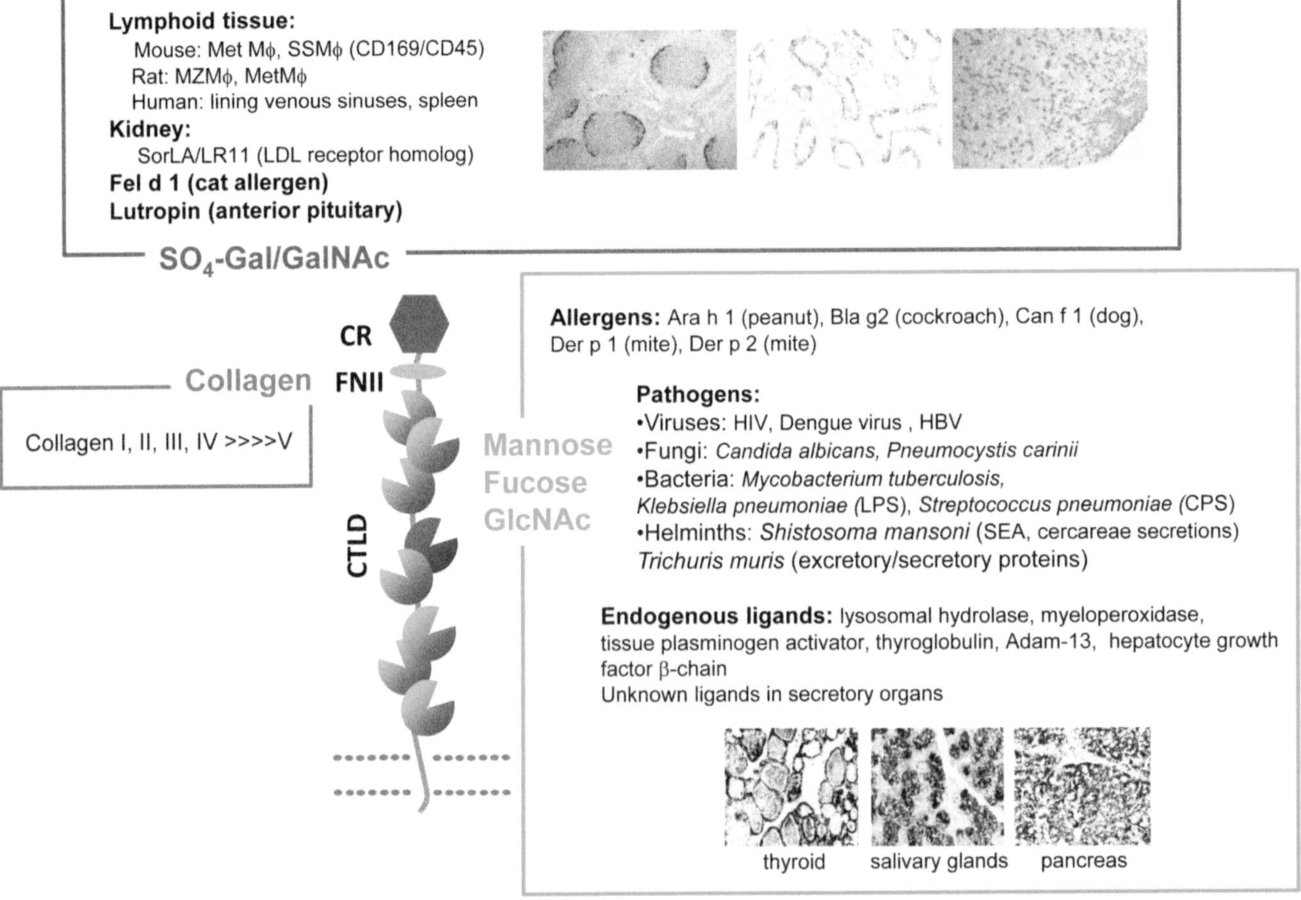

Figure 4 The macrophage mannose receptor. Structure of the MR indicating its exogenous and endogenous ligands (including those in tissues). Mϕ, macrophage; LDL, low-density lipoprotein; HBV, hepatitis B virus; CPS, capsular polysaccharide; SEA, secreted egg antigen; Adam-13, a disintegrin and metalloprotease 13. Reprinted from reference 300, with permission.

and bacteria (166, 300). Recognition by the MR has been proposed to induce several cellular responses, including endocytosis, phagocytosis, antigen cross-presentation, and cytokine production, and modulate the development of adaptive immunity (300, 309, 310). How the MR actually mediates many of these responses is unclear, as the receptor lacks known signaling motifs in its cytoplasmic tail, although its ability to mediate antigen cross-presentation was shown to involve ubiquitination (311). In fact, its role in some of these responses is now controversial. For example, the MR was initially described as a phagocytic receptor, but was subsequently shown not to be directly capable of mediating this activity (312, 313).

Several lines of evidence suggest that the effects ascribed to the MR may stem from collaboration with other receptors. For example, this receptor has been proposed to collaborate with Dectin-1 and the TLRs in the response to fungi such as *C. albicans* and *Paracoccidioides brasiliensis*, inducing the production of IL-17, Th17, and Tc17 cells (314–316). The differential responses to various MR ligands also support a notion for collaboration with other receptors from intracellular signaling. For example, mannan had no effect on DC cytokine production, yet other MR ligands, including mannose-capped lipoarabinomannan (a mycobacterial cell envelope molecule) and biglycan (an extracellular matrix proteoglycan), were found to influence cytokine responses in these cells (317). However, many of the ligands of the MR are recognized by other receptors, such as Dectin-2 and Mincle (discussed above), and this overlapping specificity casts doubt on much of the early work.

Despite the considerable literature implicating the MR in immunity, studies of MR-deficient mice have suggested that the functions of this receptor are largely redundant. These mice are viable and do not show significantly increased susceptibility to most infectious agents. For example, loss of the MR did not alter the susceptibility of mice to infection with *M. tuberculosis*, *C. albicans*, or *Pneumocystis murina* (318–320). On the other hand, the MR knockout mice were found to have slightly increased susceptibility to infections with *C. neoformans*, due to alterations in development of protective CD4 T-cell responses (321). Deficiency of the MR has also been shown to lead to alterations in the regulation of serum glycoprotein homeostasis and the development of crescentic glomerulonephritis, as well as allergic responses to cat allergens such as Fel D1 (322–324). In humans, polymorphisms in MR have been linked to susceptibility to asthma, sarcoidosis, and tuberculosis (325–327).

CONCLUSION

Research over the last few decades has provided exciting new insights into the wide and varied functions of lectins. Through their ability to recognize carbohydrates and other ligands, we now appreciate that these molecules are an essential component of multicellular existence. As our understanding of the physiological roles of these receptors increases, opportunities for novel therapeutic approaches are emerging, such as the targeting of these receptors to drive vaccine responses (328). Yet there is still much we need to learn. For example, we tend to study these molecules in isolation, but it is clear that these receptors function in a coordinated and cooperative fashion. Indeed, the recognition of intact pathogens involves numerous receptors that trigger multiple intracellular signaling pathways, producing an integrated cellular response. Despite the importance of such receptor cross talk, we still understand very little about how such signaling is integrated and how this directs the final immunological response. We also know relatively little about the regulation and influence of glycosylation on homeostasis and immune function or the recognition mechanisms that are involved. Tackling these important problems is a priority for future research.

Acknowledgments. *The authors thank the Wellcome Trust, Medical Research Council, and Arthritis Research UK for funding.*

Citation. Brown GD, Crocker PR. 2016. Lectin receptors expressed on myeloid cells. Microbiol Spectrum 4(5):MCHD-0036-2016.

References

1. **Crocker PR, Clark EA, Filbin M, Gordon S, Jones Y, Kehrl JH, Kelm S, Le Douarin N, Powell L, Roder J, Schnaar RL, Sgroi DC, Stamenkovic K, Schauer R, Schachner M, van den Berg TK, van der Merwe PA, Watt SM, Varki A.** 1998. Siglecs: a family of sialic-acid binding lectins. *Glycobiology* **8:**v.
2. **Cao H, Crocker PR.** 2011. Evolution of CD33-related siglecs: regulating host immune functions and escaping pathogen exploitation? *Immunology* **132:**18–26.
3. **Bensing BA, Khedri Z, Deng L, Yu H, Prakobphol A, Fisher SJ, Chen X, Iverson TM, Varki A, Sullam PM.** 2016. Novel aspects of sialoglycan recognition by the Siglec-like domains of streptococcal SRR glycoproteins. *Glycobiology* doi:10.1093/glycob/cww042.
4. **Rademacher C, Bru T, McBride R, Robison E, Nycholat CM, Kremer EJ, Paulson JC.** 2012. A Siglec-like sialic-acid-binding motif revealed in an adenovirus capsid protein. *Glycobiology* **22:**1086–1091.
5. **Attrill H, Imamura A, Sharma RS, Kiso M, Crocker PR, van Aalten DM.** 2006. Siglec-7 undergoes a major conformational change when complexed with the α(2,8)-

disialylganglioside GT1b. *J Biol Chem* **281:**32774–32783.

6. **Kuroki K, Wang J, Ose T, Yamaguchi M, Tabata S, Maita N, Nakamura S, Kajikawa M, Kogure A, Satoh T, Arase H, Maenaka K.** 2014. Structural basis for simultaneous recognition of an O-glycan and its attached peptide of mucin family by immune receptor PILRα. *Proc Natl Acad Sci U S A* **111:**8877–8882.
7. **Lu Q, Lu G, Qi J, Wang H, Xuan Y, Wang Q, Li Y, Zhang Y, Zheng C, Fan Z, Yan J, Gao GF.** 2014. PILRα and PILRβ have a siglec fold and provide the basis of binding to sialic acid. *Proc Natl Acad Sci U S A* **111:** 8221–8226.
8. **Soto PC, Stein LL, Hurtado-Ziola N, Hedrick SM, Varki A.** 2010. Relative over-reactivity of human versus chimpanzee lymphocytes: implications for the human diseases associated with immune activation. *J Immunol* **184:**4185–4195.
9. **Crocker PR, Paulson JC, Varki A.** 2007. Siglecs and their roles in the immune system. *Nat Rev Immunol* **7:** 255–266.
10. **Macauley MS, Crocker PR, Paulson JC.** 2014. Siglec-mediated regulation of immune cell function in disease. *Nat Rev Immunol* **14:**653–666.
11. **Daëron M, Jaeger S, Du Pasquier L, Vivier E.** 2008. Immunoreceptor tyrosine-based inhibition motifs: a quest in the past and future. *Immunol Rev* **224:**11–43.
12. **Avril T, Floyd H, Lopez F, Vivier E, Crocker PR.** 2004. The membrane-proximal immunoreceptor tyrosine-based inhibitory motif is critical for the inhibitory signaling mediated by Siglecs-7 and -9, CD33-related Siglecs expressed on human monocytes and NK cells. *J Immunol* **173:**6841–6849.
13. **Avril T, Freeman SD, Attrill H, Clarke RG, Crocker PR.** 2005. Siglec-5 (CD170) can mediate inhibitory signaling in the absence of immunoreceptor tyrosine-based inhibitory motif phosphorylation. *J Biol Chem* **280:**19843–19851.
14. **Crocker PR, Gordon S.** 1986. Properties and distribution of a lectin-like hemagglutinin differentially expressed by murine stromal tissue macrophages. *J Exp Med* **164:**1862–1875.
15. **Klaas M, Crocker PR.** 2012. Sialoadhesin in recognition of self and non-self. *Semin Immunopathol* **34:**353–364.
16. **Crocker PR, Kelm S, Dubois C, Martin B, McWilliam AS, Shotton DM, Paulson JC, Gordon S.** 1991. Purification and properties of sialoadhesin, a sialic acid-binding receptor of murine tissue macrophages. *EMBO J* **10:** 1661–1669.
17. **Kelm S, Brossmer R, Isecke R, Gross HJ, Strenge K, Schauer R.** 1998. Functional groups of sialic acids involved in binding to siglecs (sialoadhesins) deduced from interactions with synthetic analogues. *Eur J Biochem* **255:**663–672.
18. **Chávez-Galán L, Olleros ML, Vesin D, Garcia I.** 2015. Much more than M1 and M2 macrophages, there are also CD169+ and TCR+ macrophages. *Front Immunol* **6:**263. doi:10.3389/fimmu.2015.00263.
19. **Crocker PR, Gordon S.** 1989. Mouse macrophage hemagglutinin (sheep erythrocyte receptor) with specificity for sialylated glycoconjugates characterized by a monoclonal antibody. *J Exp Med* **169:**1333–1346.
20. **Hartnell A, Steel J, Turley H, Jones M, Jackson DG, Crocker PR.** 2001. Characterization of human sialoadhesin, a sialic acid binding receptor expressed by resident and inflammatory macrophage populations. *Blood* **97:**288–296.
21. **Steiniger B, Barth P, Herbst B, Hartnell A, Crocker PR.** 1997. The species-specific structure of microanatomical compartments in the human spleen: strongly sialoadhesin-positive macrophages occur in the perifollicular zone, but not in the marginal zone. *Immunology* **92:**307–316.
22. **Martinez-Pomares L, Gordon S.** 2012. CD169+ macrophages at the crossroads of antigen presentation. *Trends Immunol* **33:**66–70.
23. **Vanderheijden N, Delputte PL, Favoreel HW, Vandekerckhove J, Van Damme J, van Woensel PA, Nauwynck HJ.** 2003. Involvement of sialoadhesin in entry of porcine reproductive and respiratory syndrome virus into porcine alveolar macrophages. *J Virol* **77:** 8207–8215.
24. **Rempel H, Calosing C, Sun B, Pulliam L.** 2008. Sialoadhesin expressed on IFN-induced monocytes binds HIV-1 and enhances infectivity. *PLoS One* **3:**e1967. doi: 10.1371/journal.pone.0001967.
25. **Farrell HE, Davis-Poynter N, Bruce K, Lawler C, Dolken L, Mach M, Stevenson PG.** 2015. Lymph node macrophages restrict murine cytomegalovirus dissemination. *J Virol* **89:**7147–7158.
26. **Frederico B, Chao B, Lawler C, May JS, Stevenson PG.** 2015. Subcapsular sinus macrophages limit acute gammaherpesvirus dissemination. *J Gen Virol* **96:**2314–2327.
27. **Bernhard CA, Ried C, Kochanek S, Brocker T.** 2015. CD169+ macrophages are sufficient for priming of CTLs with specificities left out by cross-priming dendritic cells. *Proc Natl Acad Sci U S A* **112:**5461–5466.
28. **Iannacone M, Moseman EA, Tonti E, Bosurgi L, Junt T, Henrickson SE, Whelan SP, Guidotti LG, von Andrian UH.** 2010. Subcapsular sinus macrophages prevent CNS invasion on peripheral infection with a neurotropic virus. *Nature* **465:**1079–1083.
29. **Akiyama H, Miller C, Patel HV, Hatch SC, Archer J, Ramirez NG, Gummuluru S.** 2014. Virus particle release from glycosphingolipid-enriched microdomains is essential for dendritic cell-mediated capture and transfer of HIV-1 and henipavirus. *J Virol* **88:**8813–8825.
30. **Izquierdo-Useros N, Lorizate M, Puertas MC, Rodriguez-Plata MT, Zangger N, Erikson E, Pino M, Erkizia I, Glass B, Clotet B, Keppler OT, Telenti A, Kräusslich HG, Martinez-Picado J.** 2012. Siglec-1 is a novel dendritic cell receptor that mediates HIV-1 trans-infection through recognition of viral membrane gangliosides. *PLoS Biol* **10:**e1001448. doi:10.1371/journal.pbio.1001448.
31. **Sewald X, Ladinsky MS, Uchil PD, Beloor J, Pi R, Herrmann C, Motamedi N, Murooka TT, Brehm MA,**

Greiner DL, Shultz LD, Mempel TR, Bjorkman PJ, Kumar P, Mothes W. 2015. Retroviruses use CD169-mediated trans-infection of permissive lymphocytes to establish infection. *Science* **350:**563–567.

32. **Honke N, Shaabani N, Cadeddu G, Sorg UR, Zhang DE, Trilling M, Klingel K, Sauter M, Kandolf R, Gailus N, van Rooijen N, Burkart C, Baldus SE, Grusdat M, Löhning M, Hengel H, Pfeffer K, Tanaka M, Häussinger D, Recher M, Lang PA, Lang KS.** 2011. Enforced viral replication activates adaptive immunity and is essential for the control of a cytopathic virus. *Nat Immunol* **13:**51–57.
33. **Asano K, Nabeyama A, Miyake Y, Qiu CH, Kurita A, Tomura M, Kanagawa O, Fujii S, Tanaka M.** 2011. CD169-positive macrophages dominate antitumor immunity by crosspresenting dead cell-associated antigens. *Immunity* **34:**85–95.
34. **Ravishankar B, Shinde R, Liu H, Chaudhary K, Bradley J, Lemos HP, Chandler P, Tanaka M, Munn DH, Mellor AL, McGaha TL.** 2014. Marginal zone CD169^{+} macrophages coordinate apoptotic cell-driven cellular recruitment and tolerance. *Proc Natl Acad Sci U S A* **111:**4215–4220.
35. **Backer R, Schwandt T, Greuter M, Oosting M, Jüngerkes F, Tüting T, Boon L, O'Toole T, Kraal G, Limmer A, den Haan JM.** 2010. Effective collaboration between marginal metallophilic macrophages and CD8^{+} dendritic cells in the generation of cytotoxic T cells. *Proc Natl Acad Sci U S A* **107:**216–221.
36. **Veninga H, Borg EG, Vreeman K, Taylor PR, Kalay H, van Kooyk Y, Kraal G, Martinez-Pomares L, den Haan JM.** 2015. Antigen targeting reveals splenic CD169^{+} macrophages as promoters of germinal center B-cell responses. *Eur J Immunol* **45:**747–757.
37. **Kawasaki N, Vela JL, Nycholat CM, Rademacher C, Khurana A, van Rooijen N, Crocker PR, Kronenberg M, Paulson JC.** 2013. Targeted delivery of lipid antigen to macrophages via the CD169/sialoadhesin endocytic pathway induces robust invariant natural killer T cell activation. *Proc Natl Acad Sci U S A* **110:**7826–7831.
38. **Xiong YS, Cheng Y, Lin QS, Wu AL, Yu J, Li C, Sun Y, Zhong RQ, Wu LJ.** 2014. Increased expression of Siglec-1 on peripheral blood monocytes and its role in mononuclear cell reactivity to autoantigen in rheumatoid arthritis. *Rheumatology (Oxford)* **53:**250–259.
39. **Bao G, Han Z, Yan Z, Wang Q, Zhou Y, Yao D, Gu M, Chen B, Chen S, Deng A, Zhong R.** 2010. Increased Siglec-1 expression in monocytes of patients with primary biliary cirrhosis. *Immunol Invest* **39:**645–660.
40. **York MR, Nagai T, Mangini AJ, Lemaire R, van Seventer JM, Lafyatis R.** 2007. A macrophage marker, Siglec-1, is increased on circulating monocytes in patients with systemic sclerosis and induced by type I interferons and Toll-like receptor agonists. *Arthritis Rheum* **56:**1010–1020.
41. **Biesen R, Demir C, Barkhudarova F, Grün JR, Steinbrich-Zöllner M, Backhaus M, Häupl T, Rudwaleit M, Riemekasten G, Radbruch A, Hiepe F, Burmester GR, Grützkau A.** 2008. Sialic acid-binding Ig-like lectin 1 expression in inflammatory and resident monocytes is a potential biomarker for monitoring disease activity and success of therapy in systemic lupus erythematosus. *Arthritis Rheum* **58:**1136–1145.
42. **Ohnishi K, Komohara Y, Saito Y, Miyamoto Y, Watanabe M, Baba H, Takeya M.** 2013. CD169-positive macrophages in regional lymph nodes are associated with a favorable prognosis in patients with colorectal carcinoma. *Cancer Sci* **104:**1237–1244.
43. **Ohnishi K, Yamaguchi M, Erdenebaatar C, Saito F, Tashiro H, Katabuchi H, Takeya M, Komohara Y.** 2016. Prognostic significance of CD169-positive lymph node sinus macrophages in patients with endometrial carcinoma. *Cancer Sci* **107:**846–852.
44. **Ikezumi Y, Suzuki T, Hayafuji S, Okubo S, Nikolic-Paterson DJ, Kawachi H, Shimizu F, Uchiyama M.** 2005. The sialoadhesin (CD169) expressing a macrophage subset in human proliferative glomerulonephritis. *Nephrol Dial Transplant* **20:**2704–2713.
45. **Kidder D, Richards HE, Lyons PA, Crocker PR.** 2013. Sialoadhesin deficiency does not influence the severity of lupus nephritis in New Zealand black x New Zealand white F1 mice. *Arthritis Res Ther* **15:**R175. doi:10.1186/ar4364.
46. **Groh J, Ribechini E, Stadler D, Schilling T, Lutz MB, Martini R.** 2016. Sialoadhesin promotes neuroinflammation-related disease progression in two mouse models of CLN disease. *Glia* **64:**792–809.
47. **Ip CW, Kroner A, Crocker PR, Nave KA, Martini R.** 2007. Sialoadhesin deficiency ameliorates myelin degeneration and axonopathic changes in the CNS of PLP overexpressing mice. *Neurobiol Dis* **25:**105–111.
48. **Kobsar I, Oetke C, Kroner A, Wessig C, Crocker P, Martini R.** 2006. Attenuated demyelination in the absence of the macrophage-restricted adhesion molecule sialoadhesin (Siglec-1) in mice heterozygously deficient in P0. *Mol Cell Neurosci* **31:**685–691.
49. **Jiang HR, Hwenda L, Makinen K, Oetke C, Crocker PR, Forrester JV.** 2006. Sialoadhesin promotes the inflammatory response in experimental autoimmune uveoretinitis. *J Immunol* **177:**2258–2264.
50. **Wu C, Rauch U, Korpos E, Song J, Loser K, Crocker PR, Sorokin LM.** 2009. Sialoadhesin-positive macrophages bind regulatory T cells, negatively controlling their expansion and autoimmune disease progression. *J Immunol* **182:**6508–6516.
51. **Kidder D, Richards HE, Ziltener HJ, Garden OA, Crocker PR.** 2013. Sialoadhesin ligand expression identifies a subset of CD4^{+}Foxp3^{-} T cells with a distinct activation and glycosylation profile. *J Immunol* **190:**2593–2602.
52. **Black LV, Saunderson SC, Coutinho FP, Muhsin-Sharafaldine MR, Damani TT, Dunn AC, McLellan AD.** 2015. The CD169 sialoadhesin molecule mediates cytotoxic T-cell responses to tumour apoptotic vesicles. *Immunol Cell Biol* **94:**430–438.
53. **Saunderson SC, Dunn AC, Crocker PR, McLellan AD.** 2014. CD169 mediates the capture of exosomes in spleen and lymph node. *Blood* **123:**208–216.

54. **Jones C, Virji M, Crocker PR.** 2003. Recognition of sialylated meningococcal lipopolysaccharide by siglecs expressed on myeloid cells leads to enhanced bacterial uptake. *Mol Microbiol* **49:**1213–1225.
55. **Heikema AP, Bergman MP, Richards H, Crocker PR, Gilbert M, Samsom JN, van Wamel WJ, Endtz HP, van Belkum A.** 2010. Characterization of the specific interaction between sialoadhesin and sialylated *Campylobacter jejuni* lipooligosaccharides. *Infect Immun* **78:** 3237–3246.
56. **Monteiro VG, Lobato CS, Silva AR, Medina DV, de Oliveira MA, Seabra SH, de Souza W, DaMatta RA.** 2005. Increased association of *Trypanosoma cruzi* with sialoadhesin positive mice macrophages. *Parasitol Res* **97:**380–385.
57. **Klaas M, Oetke C, Lewis LE, Erwig LP, Heikema AP, Easton A, Willison HJ, Crocker PR.** 2012. Sialoadhesin promotes rapid proinflammatory and type I IFN responses to a sialylated pathogen, *Campylobacter jejuni*. *J Immunol* **189:**2414–2422.
58. **Chang YC, Olson J, Louie A, Crocker PR, Varki A, Nizet V.** 2014. Role of macrophage sialoadhesin in host defense against the sialylated pathogen group B *Streptococcus*. *J Mol Med (Berl)* **92:**951–959.
59. **Erikson E, Wratil PR, Frank M, Ambiel I, Pahnke K, Pino M, Azadi P, Izquierdo-Useros N, Martinez-Picado J, Meier C, Schnaar RL, Crocker PR, Reutter W, Keppler OT.** 2015. Mouse Siglec-1 mediates *trans*-infection of surface-bound murine leukemia virus in a sialic acid *N*-acyl side chain-dependent manner. *J Biol Chem* **290:**27345–27359.
60. **Akiyama H, Ramirez NG, Gudheti MV, Gummuluru S.** 2015. CD169-mediated trafficking of HIV to plasma membrane invaginations in dendritic cells attenuates efficacy of anti-gp120 broadly neutralizing antibodies. *PLoS Pathog* **11:**e1004751. doi:10.1371/journal.ppat.1004751.
61. **Puryear WB, Akiyama H, Geer SD, Ramirez NP, Yu X, Reinhard BM, Gummuluru S.** 2013. Interferon-inducible mechanism of dendritic cell-mediated HIV-1 dissemination is dependent on Siglec-1/CD169. *PLoS Pathog* **9:**e1003291. doi:10.1371/journal.ppat.1003291.
62. **Zou Z, Chastain A, Moir S, Ford J, Trandem K, Martinelli E, Cicala C, Crocker P, Arthos J, Sun PD.** 2011. Siglecs facilitate HIV-1 infection of macrophages through adhesion with viral sialic acids. *PLoS One* **6:** e24559. doi:10.1371/journal.pone.0024559.
63. **Izquierdo-Useros N, Lorizate M, McLaren PJ, Telenti A, Kräusslich HG, Martinez-Picado J.** 2014. HIV-1 capture and transmission by dendritic cells: the role of viral glycolipids and the cellular receptor Siglec-1. *PLoS Pathog* **10:**e1004146. doi:10.1371/journal.ppat.1004146.
64. **Zheng Q, Hou J, Zhou Y, Yang Y, Xie B, Cao X.** 2015. Siglec1 suppresses antiviral innate immune response by inducing TBK1 degradation via the ubiquitin ligase TRIM27. *Cell Res* **25:**1121–1136.
65. **Wu Y, Lan C, Ren D, Chen GY.** 2016. Induction of Siglec-1 by endotoxin tolerance suppresses the innate immune response by promoting TGF-β1 production. *J Biol Chem* **291:**12370–12382.
66. **Taylor VC, Buckley CD, Douglas M, Cody AJ, Simmons DL, Freeman SD.** 1999. The myeloid-specific sialic acid-binding receptor, CD33, associates with the protein-tyrosine phosphatases, SHP-1 and SHP-2. *J Biol Chem* **274:**11505–11512.
67. **Freeman SD, Kelm S, Barber EK, Crocker PR.** 1995. Characterization of CD33 as a new member of the sialoadhesin family of cellular interaction molecules. *Blood* **85:**2005–2012.
68. **Walter RB, Häusermann P, Raden BW, Teckchandani AM, Kamikura DM, Bernstein ID, Cooper JA.** 2008. Phosphorylated ITIMs enable ubiquitylation of an inhibitory cell surface receptor. *Traffic* **9:**267–279.
69. **Walter RB, Raden BW, Kamikura DM, Cooper JA, Bernstein ID.** 2005. Influence of CD33 expression levels and ITIM-dependent internalization on gemtuzumab ozogamicin-induced cytotoxicity. *Blood* **105:**1295–1302.
70. **Walter RB, Raden BW, Zeng R, Häusermann P, Bernstein ID, Cooper JA.** 2008. ITIM-dependent endocytosis of CD33-related Siglecs: role of intracellular domain, tyrosine phosphorylation, and the tyrosine phosphatases, Shp1 and Shp2. *J Leukoc Biol* **83:**200–211.
71. **Raj T, Ryan KJ, Replogle JM, Chibnik LB, Rosenkrantz L, Tang A, Rothamel K, Stranger BE, Bennett DA, Evans DA, De Jager PL, Bradshaw EM.** 2014. CD33: increased inclusion of exon 2 implicates the Ig V-set domain in Alzheimer's disease susceptibility. *Hum Mol Genet* **23:**2729–2736.
72. **Bradshaw EM, Chibnik LB, Keenan BT, Ottoboni L, Raj T, Tang A, Rosenkrantz LL, Imboywa S, Lee M, Von Korff A, Morris MC, Evans DA, Johnson K, Sperling RA, Schneider JA, Bennett DA, De Jager PL, Alzheimer Disease Neuroimaging Initiative.** 2013. CD33 Alzheimer's disease locus: altered monocyte function and amyloid biology. *Nat Neurosci* **16:**848–850.
73. **Griciuc A, Serrano-Pozo A, Parrado AR, Lesinski AN, Asselin CN, Mullin K, Hooli B, Choi SH, Hyman BT, Tanzi RE.** 2013. Alzheimer's disease risk gene CD33 inhibits microglial uptake of amyloid beta. *Neuron* **78:** 631–643.
74. **Tchilian EZ, Beverley PC, Young BD, Watt SM.** 1994. Molecular cloning of two isoforms of the murine homolog of the myeloid CD33 antigen. *Blood* **83:**3188–3198.
75. **Blasius AL, Cella M, Maldonado J, Takai T, Colonna M.** 2006. Siglec-H is an IPC-specific receptor that modulates type I IFN secretion through DAP12. *Blood* **107:**2474–2476.
76. **Angata T, Hayakawa T, Yamanaka M, Varki A, Nakamura M.** 2006. Discovery of Siglec-14, a novel sialic acid receptor undergoing concerted evolution with Siglec-5 in primates. *FASEB J* **20:**1964–1973.
77. **Angata T, Tabuchi Y, Nakamura K, Nakamura M.** 2007. Siglec-15: an immune system Siglec conserved throughout vertebrate evolution. *Glycobiology* **17:** 838–846.
78. **Cao H, Lakner U, de Bono B, Traherne JA, Trowsdale J, Barrow AD.** 2008. SIGLEC16 encodes a DAP12-associated receptor expressed in macrophages that evolved from its inhibitory counterpart SIGLEC11 and

has functional and non-functional alleles in humans. *Eur J Immunol* 38:2303–2315.

79. **Brinkman-Van der Linden EC, Angata T, Reynolds SA, Powell LD, Hedrick SM, Varki A.** 2003. CD33/Siglec-3 binding specificity, expression pattern, and consequences of gene deletion in mice. *Mol Cell Biol* **23:** 4199–4206.

80. **Angata T, Ishii T, Motegi T, Oka R, Taylor RE, Soto PC, Chang YC, Secundino I, Gao CX, Ohtsubo K, Kitazume S, Nizet V, Varki A, Gemma A, Kida K, Taniguchi N.** 2013. Loss of Siglec-14 reduces the risk of chronic obstructive pulmonary disease exacerbation. *Cell Mol Life Sci* **70:**3199–3210.

81. **Ali SR, Fong JJ, Carlin AF, Busch TD, Linden R, Angata T, Areschoug T, Parast M, Varki N, Murray J, Nizet V, Varki A.** 2014. Siglec-5 and Siglec-14 are polymorphic paired receptors that modulate neutrophil and amnion signaling responses to group B *Streptococcus*. *J Exp Med* **211:**1231–1242.

82. **Fong JJ, Sreedhara K, Deng L, Varki NM, Angata T, Liu Q, Nizet V, Varki A.** 2015. Immunomodulatory activity of extracellular Hsp70 mediated via paired receptors Siglec-5 and Siglec-14. *EMBO J* **34:**2775–2788.

83. **Falco M, Biassoni R, Bottino C, Vitale M, Sivori S, Augugliaro R, Moretta L, Moretta A.** 1999. Identification and molecular cloning of p75/AIRM1, a novel member of the sialoadhesin family that functions as an inhibitory receptor in human natural killer cells. *J Exp Med* **190:**793–802.

84. **Nicoll G, Ni J, Liu D, Klenerman P, Munday J, Dubock S, Mattei MG, Crocker PR.** 1999. Identification and characterization of a novel siglec, siglec-7, expressed by human natural killer cells and monocytes. *J Biol Chem* **274:**34089–34095.

85. **Mizrahi S, Gibbs BF, Karra L, Ben-Zimra M, Levi-Schaffer F.** 2014. Siglec-7 is an inhibitory receptor on human mast cells and basophils. *J Allergy Clin Immunol* **134:**230–233.

86. **Angata T, Varki A.** 2000. Cloning, characterization, and phylogenetic analysis of siglec-9, a new member of the CD33-related group of siglecs. Evidence for co-evolution with sialic acid synthesis pathways. *J Biol Chem* **275:**22127–22135.

87. **Zhang JQ, Nicoll G, Jones C, Crocker PR.** 2000. Siglec-9, a novel sialic acid binding member of the immunoglobulin superfamily expressed broadly on human blood leukocytes. *J Biol Chem* **275:**22121–22126.

88. **Yamaji T, Teranishi T, Alphey MS, Crocker PR, Hashimoto Y.** 2002. A small region of the natural killer cell receptor, Siglec-7, is responsible for its preferred binding to α2,8-disialyl and branched α2,6-sialyl residues: a comparison with Siglec-9. *J Biol Chem* **277:** 6324–6332.

89. **Campanero-Rhodes MA, Childs RA, Kiso M, Komba S, Le Narvor C, Warren J, Otto D, Crocker PR, Feizi T.** 2006. Carbohydrate microarrays reveal sulphation as a modulator of siglec binding. *Biochem Biophys Res Commun* **344:**1141–1146.

90. **Secundino I, Lizcano A, Roupé KM, Wang X, Cole JN, Olson J, Ali SR, Dahesh S, Amayreh LK, Henningham A, Varki A, Nizet V.** 2016. Host and pathogen hyaluronan signal through human Siglec-9 to suppress neutrophil activation. *J Mol Med (Berl)* **94:**219–233.

91. **Redelinghuys P, Antonopoulos A, Liu Y, Campanero-Rhodes MA, McKenzie E, Haslam SM, Dell A, Feizi T, Crocker PR.** 2011. Early murine T-lymphocyte activation is accompanied by a switch from *N*-glycolyl- to *N*-acetyl-neuraminic acid and generation of ligands for siglec-E. *J Biol Chem* **286:**34522–34532.

92. **McMillan SJ, Sharma RS, McKenzie EJ, Richards HE, Zhang J, Prescott A, Crocker PR.** 2013. Siglec-E is a negative regulator of acute pulmonary neutrophil inflammation and suppresses CD11b β2-integrin-dependent signaling. *Blood* **121:**2084–2094.

93. **McMillan SJ, Sharma RS, Richards HE, Hegde V, Crocker PR.** 2014. Siglec-E promotes β2-integrin-dependent NADPH oxidase activation to suppress neutrophil recruitment to the lung. *J Biol Chem* **289:**20370–20376.

94. **Schwarz F, Pearce OM, Wang X, Samraj AN, Läubli H, Garcia JO, Lin H, Fu X, Garcia-Bingman A, Secrest P, Romanoski CE, Heyser C, Glass CK, Hazen SL, Varki N, Varki A, Gagneux P.** 2015. Siglec receptors impact mammalian lifespan by modulating oxidative stress. *eLife* **4:**4. doi:10.7554/eLife.06184.

95. **Boyd CR, Orr SJ, Spence S, Burrows JF, Elliott J, Carroll HP, Brennan K, Ní Gabhann J, Coulter WA, Jones C, Crocker PR, Johnston JA, Jefferies CA.** 2009. Siglec-E is up-regulated and phosphorylated following lipopolysaccharide stimulation in order to limit TLR-driven cytokine production. *J Immunol* **183:**7703–7709.

96. **Claude J, Linnartz-Gerlach B, Kudin AP, Kunz WS, Neumann H.** 2013. Microglial CD33-related Siglec-E inhibits neurotoxicity by preventing the phagocytosis-associated oxidative burst. *J Neurosci* **33:**18270–18276.

97. **Chen GY, Brown NK, Wu W, Khedri Z, Yu H, Chen X, van de Vlekkert D, D'Azzo A, Zheng P, Liu Y.** 2014. Broad and direct interaction between TLR and Siglec families of pattern recognition receptors and its regulation by Neu1. *eLife* **3:**e04066. doi:10.7554/eLife.04066.

98. **Läubli H, Pearce OM, Schwarz F, Siddiqui SS, Deng L, Stanczak MA, Deng L, Verhagen A, Secrest P, Lusk C, Schwartz AG, Varki NM, Bui JD, Varki A.** 2014. Engagement of myelomonocytic Siglecs by tumor-associated ligands modulates the innate immune response to cancer. *Proc Natl Acad Sci U S A* **111:**14211–14216.

99. **Perdicchio M, Ilarregui JM, Verstege MI, Cornelissen LA, Schetters ST, Engels S, Ambrosini M, Kalay H, Veninga H, den Haan JM, van Berkel LA, Samsom JN, Crocker PR, Sparwasser T, Berod L, Garcia-Vallejo JJ, van Kooyk Y, Unger WW.** 2016. Sialic acid-modified antigens impose tolerance via inhibition of T-cell proliferation and de novo induction of regulatory T cells. *Proc Natl Acad Sci U S A* **113:**3329–3334.

100. **Spence S, Greene MK, Fay F, Hams E, Saunders SP, Hamid U, Fitzgerald M, Beck J, Bains BK, Smyth P, Themistou E, Small DM, Schmid D, O'Kane CM, Fitzgerald DC, Abdelghany SM, Johnston JA, Fallon PG, Burrows JF, McAuley DF, Kissenpfennig A, Scott

CJ. 2015. Targeting Siglecs with a sialic acid-decorated nanoparticle abrogates inflammation. *Sci Transl Med* **7:** 303ra140. doi:10.1126/scitranslmed.aab3459.

101. **Hudak JE, Canham SM, Bertozzi CR.** 2014. Glycocalyx engineering reveals a Siglec-based mechanism for NK cell immunoevasion. *Nat Chem Biol* **10:**69–75.

102. **Jandus C, Boligan KF, Chijioke O, Liu H, Dahlhaus M, Démoulins T, Schneider C, Wehrli M, Hunger RE, Baerlocher GM, Simon HU, Romero P, Münz C, von Gunten S.** 2014. Interactions between Siglec-7/9 receptors and ligands influence NK cell-dependent tumor immunosurveillance. *J Clin Invest* **124:**1810–1820.

103. **Nicoll G, Avril T, Lock K, Furukawa K, Bovin N, Crocker PR.** 2003. Ganglioside GD3 expression on target cells can modulate NK cell cytotoxicity via siglec-7-dependent and -independent mechanisms. *Eur J Immunol* **33:**1642–1648.

104. **Carlin AF, Uchiyama S, Chang YC, Lewis AL, Nizet V, Varki A.** 2009. Molecular mimicry of host sialylated glycans allows a bacterial pathogen to engage neutrophil Siglec-9 and dampen the innate immune response. *Blood* **113:**3333–3336.

105. **Chang YC, Olson J, Beasley FC, Tung C, Zhang J, Crocker PR, Varki A, Nizet V.** 2014. Group B *Streptococcus* engages an inhibitory Siglec through sialic acid mimicry to blunt innate immune and inflammatory responses *in vivo*. *PLoS Pathog* **10:**e1003846. doi:10.1371/journal.ppat.1003846.

106. **Floyd H, Ni J, Cornish AL, Zeng Z, Liu D, Carter KC, Steel J, Crocker PR.** 2000. Siglec-8. A novel eosinophil-specific member of the immunoglobulin superfamily. *J Biol Chem* **275:**861–866.

107. **Kikly KK, Bochner BS, Freeman SD, Tan KB, Gallagher KT, D'alessio KJ, Holmes SD, Abrahamson JA, Erickson-Miller CL, Murdock PR, Tachimoto H, Schleimer RP, White JR.** 2000. Identification of SAF-2, a novel siglec expressed on eosinophils, mast cells, and basophils. *J Allergy Clin Immunol* **105:**1093–1100.

108. **Bochner BS, Alvarez RA, Mehta P, Bovin NV, Blixt O, White JR, Schnaar RL.** 2005. Glycan array screening reveals a candidate ligand for Siglec-8. *J Biol Chem* **280:**4307–4312.

109. **Jia Y, Yu H, Fernandes SM, Wei Y, Gonzalez-Gil A, Motari MG, Vajn K, Stevens WW, Peters AT, Bochner BS, Kern RC, Schleimer RP, Schnaar RL.** 2015. Expression of ligands for Siglec-8 and Siglec-9 in human airways and airway cells. *J Allergy Clin Immunol* **135:**799–810.e7. doi:10.1016/j.jaci.2015.01.004.

110. **Patnode ML, Cheng CW, Chou CC, Singer MS, Elin MS, Uchimura K, Crocker PR, Khoo KH, Rosen SD.** 2013. Galactose 6-O-sulfotransferases are not required for the generation of Siglec-F ligands in leukocytes or lung tissue. *J Biol Chem* **288:**26533–26545.

111. **Yokoi H, Choi OH, Hubbard W, Lee HS, Canning BJ, Lee HH, Ryu SD, von Gunten S, Bickel CA, Hudson SA, Macglashan DW Jr, Bochner BS.** 2008. Inhibition of FcεRI-dependent mediator release and calcium flux from human mast cells by sialic acid-binding immunoglobulin-like lectin 8 engagement. *J Allergy Clin Immunol* **121:**499–505.

112. **Hudson SA, Bovin NV, Schnaar RL, Crocker PR, Bochner BS.** 2009. Eosinophil-selective binding and proapoptotic effect in vitro of a synthetic Siglec-8 ligand, polymeric 6′-sulfated sialyl Lewis X. *J Pharmacol Exp Ther* **330:**608–612.

113. **Nutku E, Aizawa H, Hudson SA, Bochner BS.** 2003. Ligation of Siglec-8: a selective mechanism for induction of human eosinophil apoptosis. *Blood* **101:**5014–5020.

114. **Gao PS, Shimizu K, Grant AV, Rafaels N, Zhou LF, Hudson SA, Konno S, Zimmermann N, Araujo MI, Ponte EV, Cruz AA, Nishimura M, Su SN, Hizawa N, Beaty TH, Mathias RA, Rothenberg ME, Barnes KC, Bochner BS.** 2010. Polymorphisms in the sialic acid-binding immunoglobulin-like lectin-8 (Siglec-8) gene are associated with susceptibility to asthma. *Eur J Hum Genet* **18:**713–719.

115. **Angata T, Hingorani R, Varki NM, Varki A.** 2001. Cloning and characterization of a novel mouse Siglec, mSiglec-F: differential evolution of the mouse and human (CD33) Siglec-3-related gene clusters. *J Biol Chem* **276:**45128–45136.

116. **Tateno H, Crocker PR, Paulson JC.** 2005. Mouse Siglec-F and human Siglec-8 are functionally convergent paralogs that are selectively expressed on eosinophils and recognize 6′-sulfo-sialyl Lewis X as a preferred glycan ligand. *Glycobiology* **15:**1125–1135.

117. **Zhang JQ, Biedermann B, Nitschke L, Crocker PR.** 2004. The murine inhibitory receptor mSiglec-E is expressed broadly on cells of the innate immune system whereas mSiglec-F is restricted to eosinophils. *Eur J Immunol* **34:**1175–1184.

118. **Mao H, Kano G, Hudson SA, Brummet M, Zimmermann N, Zhu Z, Bochner BS.** 2013. Mechanisms of Siglec-F-induced eosinophil apoptosis: a role for caspases but not for SHP-1, Src kinases, NADPH oxidase or reactive oxygen. *PLoS One* **8:**e68143. doi:10.1371/journal.pone.0068143.

119. **McMillan SJ, Richards HE, Crocker PR.** 2014. Siglec-F-dependent negative regulation of allergen-induced eosinophilia depends critically on the experimental model. *Immunol Lett* **160:**11–16.

120. **Zhang M, Angata T, Cho JY, Miller M, Broide DH, Varki A.** 2007. Defining the in vivo function of Siglec-F, a CD33-related Siglec expressed on mouse eosinophils. *Blood* **109:**4280–4287.

121. **Li N, Zhang W, Wan T, Zhang J, Chen T, Yu Y, Wang J, Cao X.** 2001. Cloning and characterization of Siglec-10, a novel sialic acid binding member of the Ig superfamily, from human dendritic cells. *J Biol Chem* **276:**28106–28112.

122. **Munday J, Kerr S, Ni J, Cornish AL, Zhang JQ, Nicoll G, Floyd H, Mattei MG, Moore P, Liu D, Crocker PR.** 2001. Identification, characterization and leucocyte expression of Siglec-10, a novel human sialic acid-binding receptor. *Biochem J* **355:**489–497.

123. **Whitney G, Wang S, Chang H, Cheng KY, Lu P, Zhou XD, Yang WP, McKinnon M, Longphre M.** 2001. A new siglec family member, siglec-10, is expressed in cells of the immune system and has signaling properties similar to CD33. *Eur J Biochem* **268:**6083–6096.

124. **Zhang P, Lu X, Tao K, Shi L, Li W, Wang G, Wu K.** 2015. Siglec-10 is associated with survival and natural killer cell dysfunction in hepatocellular carcinoma. *J Surg Res* **194:**107–113.

125. **Duong BH, Tian H, Ota T, Completo G, Han S, Vela JL, Ota M, Kubitz M, Bovin N, Paulson JC, Nemazee D.** 2010. Decoration of T-independent antigen with ligands for CD22 and Siglec-G can suppress immunity and induce B cell tolerance in vivo. *J Exp Med* **207:** 173–187.

126. **Escalona Z, Álvarez B, Uenishi H, Toki D, Yuste M, Revilla C, del Moral MG, Alonso F, Ezquerra A, Domínguez J.** 2015. Molecular characterization of porcine Siglec-10 and analysis of its expression in blood and tissues. *Dev Comp Immunol* **48:**116–123.

127. **Hutzler S, Özgör L, Naito-Matsui Y, Kläsener K, Winkler TH, Reth M, Nitschke L.** 2014. The ligand-binding domain of Siglec-G is crucial for its selective inhibitory function on B1 cells. *J Immunol* **192:**5406–5414.

128. **Pfrengle F, Macauley MS, Kawasaki N, Paulson JC.** 2013. Copresentation of antigen and ligands of Siglec-G induces B cell tolerance independent of CD22. *J Immunol* **191:**1724–1731.

129. **Hoffmann A, Kerr S, Jellusova J, Zhang J, Weisel F, Wellmann U, Winkler TH, Kneitz B, Crocker PR, Nitschke L.** 2007. Siglec-G is a B1 cell-inhibitory receptor that controls expansion and calcium signaling of the B1 cell population. *Nat Immunol* **8:**695–704.

130. **Chen GY, Tang J, Zheng P, Liu Y.** 2009. CD24 and Siglec-10 selectively repress tissue damage-induced immune responses. *Science* **323:**1722–1725.

131. **Chen GY, Brown NK, Zheng P, Liu Y.** 2014. Siglec-G/10 in self-nonself discrimination of innate and adaptive immunity. *Glycobiology* **24:**800–806.

132. **Stephenson HN, Mills DC, Jones H, Milioris E, Copland A, Dorrell N, Wren BW, Crocker PR, Escors D, Bajaj-Elliott M.** 2014. Pseudaminic acid on *Campylobacter jejuni* flagella modulates dendritic cell IL-10 expression via Siglec-10 receptor: a novel flagellin-host interaction. *J Infect Dis* **210:**1487–1498.

133. **Angata T, Kerr SC, Greaves DR, Varki NM, Crocker PR, Varki A.** 2002. Cloning and characterization of human Siglec-11. A recently evolved signaling molecule that can interact with SHP-1 and SHP-2 and is expressed by tissue macrophages, including brain microglia. *J Biol Chem* **277:**24466–24474.

134. **Wang Y, Neumann H.** 2010. Alleviation of neurotoxicity by microglial human Siglec-11. *J Neurosci* **30:** 3482–3488.

135. **Shahraz A, Kopatz J, Mathy R, Kappler J, Winter D, Kapoor S, Schütza V, Scheper T, Gieselmann V, Neumann H.** 2015. Anti-inflammatory activity of low molecular weight polysialic acid on human macrophages. *Sci Rep* **5:**16800. doi:10.1038/srep16800.

136. **Wang X, Mitra N, Cruz P, Deng L, NISC Comparative Sequencing Program, Varki N, Angata T, Green ED, Mullikin J, Hayakawa T, Varki A, Varki A.** 2012. Evolution of Siglec-11 and Siglec-16 genes in hominins. *Mol Biol Evol* **29:**2073–2086.

137. **Takamiya R, Ohtsubo K, Takamatsu S, Taniguchi N, Angata T.** 2013. The interaction between Siglec-15 and tumor-associated sialyl-Tn antigen enhances TGF-β secretion from monocytes/macrophages through the DAP12-Syk pathway. *Glycobiology* **23:**178–187.

138. **Hiruma Y, Hirai T, Tsuda E.** 2011. Siglec-15, a member of the sialic acid-binding lectin, is a novel regulator for osteoclast differentiation. *Biochem Biophys Res Commun* **409:**424–429.

139. **Ishida-Kitagawa N, Tanaka K, Bao X, Kimura T, Miura T, Kitaoka Y, Hayashi K, Sato M, Maruoka M, Ogawa T, Miyoshi J, Takeya T.** 2012. Siglec-15 protein regulates formation of functional osteoclasts in concert with DNAX-activating protein of 12 kDa (DAP12). *J Biol Chem* **287:**17493–17502.

140. **Hiruma Y, Tsuda E, Maeda N, Okada A, Kabasawa N, Miyamoto M, Hattori H, Fukuda C.** 2013. Impaired osteoclast differentiation and function and mild osteopetrosis development in Siglec-15-deficient mice. *Bone* **53:**87–93.

141. **Kameda Y, Takahata M, Komatsu M, Mikuni S, Hatakeyama S, Shimizu T, Angata T, Kinjo M, Minami A, Iwasaki N.** 2013. Siglec-15 regulates osteoclast differentiation by modulating RANKL-induced phosphatidylinositol 3-kinase/Akt and Erk pathways in association with signaling Adaptor DAP12. *J Bone Miner Res* **28:**2463–2475.

142. **Stuible M, Moraitis A, Fortin A, Saragosa S, Kalbakji A, Filion M, Tremblay GB.** 2014. Mechanism and function of monoclonal antibodies targeting Siglec-15 for therapeutic inhibition of osteoclastic bone resorption. *J Biol Chem* **289:**6498–6512.

143. **Zelensky AN, Gready JE.** 2005. The C-type lectin-like domain superfamily. *FEBS J* **272:**6179–6217.

144. **McEver RP.** 2015. Selectins: initiators of leucocyte adhesion and signalling at the vascular wall. *Cardiovasc Res* **107:**331–339.

145. **Willment JA, Brown GD.** 2008. C-type lectin receptors in antifungal immunity. *Trends Microbiol* **16:**27–32.

146. **Sancho D, Reis e Sousa C.** 2012. Signaling by myeloid C-type lectin receptors in immunity and homeostasis. *Annu Rev Immunol* **30:**491–529.

147. **Geijtenbeek TB, Gringhuis SI.** 2009. Signalling through C-type lectin receptors: shaping immune responses. *Nat Rev Immunol* **9:**465–479.

148. **Garcia-Vallejo JJ, van Kooyk Y.** 2013. The physiological role of DC-SIGN: a tale of mice and men. *Trends Immunol* **34:**482–486.

149. **Willment JA, Gordon S, Brown GD.** 2001. Characterization of the human β-glucan receptor and its alternatively spliced isoforms. *J Biol Chem* **276:**43818–43823.

150. **Kato Y, Adachi Y, Ohno N.** 2006. Contribution of *N*-linked oligosaccharides to the expression and functions of β-glucan receptor, Dectin-1. *Biol Pharm Bull* **29:**1580–1586.

151. **Willment JA, Marshall AS, Reid DM, Williams DL, Wong SY, Gordon S, Brown GD.** 2005. The human β-glucan receptor is widely expressed and functionally

equivalent to murine Dectin-1 on primary cells. *Eur J Immunol* **35**:1539–1547.

152. **Martin B, Hirota K, Cua DJ, Stockinger B, Veldhoen M.** 2009. Interleukin-17-producing γδ T cells selectively expand in response to pathogen products and environmental signals. *Immunity* **31**:321–330.

153. **Sun WK, Lu X, Li X, Sun QY, Su X, Song Y, Sun HM, Shi Y.** 2012. Dectin-1 is inducible and plays a crucial role in *Aspergillus*-induced innate immune responses in human bronchial epithelial cells. *Eur J Clin Microbiol Infect Dis* **31**:2755–2764.

154. **Bertuzzi M, Schrettl M, Alcazar-Fuoli L, Cairns TC, Muñoz A, Walker LA, Herbst S, Safari M, Cheverton AM, Chen D, Liu H, Saijo S, Fedorova ND, Armstrong-James D, Munro CA, Read ND, Filler SG, Espeso EA, Nierman WC, Haas H, Bignell EM.** 2014. The pH-responsive PacC transcription factor of *Aspergillus fumigatus* governs epithelial entry and tissue invasion during pulmonary aspergillosis. *PLoS Pathog* **10**: e1004413. doi:10.1371/journal.ppat.1004413.

155. **Brown GD, Gordon S.** 2001. Immune recognition. A new receptor for β-glucans. *Nature* **413**:36–37.

156. **Drummond RA, Brown GD.** 2011. The role of Dectin-1 in the host defence against fungal infections. *Curr Opin Microbiol* **14**:392–399.

157. **Ferwerda B, Ferwerda G, Plantinga TS, Willment JA, van Spriel AB, Venselaar H, Elbers CC, Johnson MD, Cambi A, Huysamen C, Jacobs L, Jansen T, Verheijen K, Masthoff L, Morré SA, Vriend G, Williams DL, Perfect JR, Joosten LA, Wijmenga C, van der Meer JW, Adema GJ, Kullberg BJ, Brown GD, Netea MG.** 2009. Human dectin-1 deficiency and mucocutaneous fungal infections. *N Engl J Med* **361**:1760–1767.

158. **Iliev ID, Funari VA, Taylor KD, Nguyen Q, Reyes CN, Strom SP, Brown J, Becker CA, Fleshner PR, Dubinsky M, Rotter JI, Wang HL, McGovern DP, Brown GD, Underhill DM.** 2012. Interactions between commensal fungi and the C-type lectin receptor Dectin-1 influence colitis. *Science* **336**:1314–1317.

159. **Sainz J, Lupiáñez CB, Segura-Catena J, Vazquez L, Ríos R, Oyonarte S, Hemminki K, Försti A, Jurado M.** 2012. Dectin-1 and DC-SIGN polymorphisms associated with invasive pulmonary *Aspergillosis* infection. *PLoS One* **7**:e32273. doi:10.1371/journal.pone.0032273.

160. **Taylor PR, Tsoni SV, Willment JA, Dennehy KM, Rosas M, Findon H, Haynes K, Steele C, Botto M, Gordon S, Brown GD.** 2007. Dectin-1 is required for β-glucan recognition and control of fungal infection. *Nat Immunol* **8**:31–38.

161. **Saijo S, Fujikado N, Furuta T, Chung SH, Kotaki H, Seki K, Sudo K, Akira S, Adachi Y, Ohno N, Kinjo T, Nakamura K, Kawakami K, Iwakura Y.** 2007. Dectin-1 is required for host defense against *Pneumocystis carinii* but not against *Candida albicans*. *Nat Immunol* **8**: 39–46.

162. **Werner JL, Metz AE, Horn D, Schoeb TR, Hewitt MM, Schwiebert LM, Faro-Trindade I, Brown GD, Steele C.** 2009. Requisite role for the Dectin-1 β-glucan receptor in pulmonary defense against *Aspergillus fumigatus*. *J Immunol* **182**:4938–4946.

163. **Marakalala MJ, Vautier S, Potrykus J, Walker LA, Shepardson KM, Hopke A, Mora-Montes HM, Kerrigan A, Netea MG, Murray GI, Maccallum DM, Wheeler R, Munro CA, Gow NA, Cramer RA, Brown AJ, Brown GD.** 2013. Differential adaptation of *Candida albicans* in vivo modulates immune recognition by Dectin-1. *PLoS Pathog* **9**:e1003315. doi:10.1371/journal.ppat.1003315.

164. **Rothfuchs AG, Bafica A, Feng CG, Egen JG, Williams DL, Brown GD, Sher A.** 2007. Dectin-1 interaction with *Mycobacterium tuberculosis* leads to enhanced IL-12p40 production by splenic dendritic cells. *J Immunol* **179**:3463–3471.

165. **Marakalala MJ, Guler R, Matika L, Murray G, Jacobs M, Brombacher F, Rothfuchs AG, Sher A, Brown GD.** 2011. The Syk/CARD9-coupled receptor Dectin-1 is not required for host resistance to *Mycobacterium tuberculosis* in mice. *Microbes Infect* **13**:198–201.

166. **Lefèvre L, Lugo-Villarino G, Meunier E, Valentin A, Olagnier D, Authier H, Duval C, Dardenne C, Bernad J, Lemesre JL, Auwerx J, Neyrolles O, Pipy B, Coste A.** 2013. The C-type lectin receptors Dectin-1, MR, and SIGNR3 contribute both positively and negatively to the macrophage response to *Leishmania infantum*. *Immunity* **38**:1038–1049.

167. **Ariizumi K, Shen GL, Shikano S, Xu S, Ritter R III, Kumamoto T, Edelbaum D, Morita A, Bergstresser PR, Takashima A.** 2000. Identification of a novel, dendritic cell-associated molecule, dectin-1, by subtractive cDNA cloning. *J Biol Chem* **275**:20157–20167.

168. **Thiagarajan PS, Yakubenko VP, Elsori DH, Yadav SP, Willard B, Tan CD, Rodriguez ER, Febbraio M, Cathcart MK.** 2013. Vimentin is an endogenous ligand for the pattern recognition receptor Dectin-1. *Cardiovasc Res* **99**:494–504.

169. **Szilagyi K, Gijbels MJ, van der Velden S, Heinsbroek SE, Kraal G, de Winther MP, van den Berg TK.** 2015. Dectin-1 deficiency does not affect atherosclerosis development in mice. *Atherosclerosis* **239**:318–321.

170. **Rochereau N, Drocourt D, Perouzel E, Pavot V, Redelinghuys P, Brown GD, Tiraby G, Roblin X, Verrier B, Genin C, Corthésy B, Paul S.** 2013. Dectin-1 is essential for reverse transcytosis of glycosylated SIgA-antigen complexes by intestinal M cells. *PLoS Biol* **11**:e1001658. doi:10.1371/journal.pbio.1001658.

171. **Karsten CM, Pandey MK, Figge J, Kilchenstein R, Taylor PR, Rosas M, McDonald JU, Orr SJ, Berger M, Petzold D, Blanchard V, Winkler A, Hess C, Reid DM, Majoul IV, Strait RT, Harris NL, Köhl G, Wex E, Ludwig R, Zillikens D, Nimmerjahn F, Finkelman FD, Brown GD, Ehlers M, Köhl J.** 2012. Anti-inflammatory activity of IgG1 mediated by Fc galactosylation and association of FcγRIIB and dectin-1. *Nat Med* **18**:1401–1406.

172. **Shan M, Gentile M, Yeiser JR, Walland AC, Bornstein VU, Chen K, He B, Cassis L, Bigas A, Cols M, Comerma L, Huang B, Blander JM, Xiong H, Mayer L, Berin C, Augenlicht LH, Velcich A, Cerutti A.** 2013. Mucus enhances gut homeostasis and oral tolerance by delivering immunoregulatory signals. *Science* **342**:447–453.

173. **Seifert L, Deutsch M, Alothman S, Alqunaibit D, Werba G, Pansari M, Pergamo M, Ochi A, Torres-Hernandez A, Levie E, Tippens D, Greco SH, Tiwari S, Ly NN, Eisenthal A, van Heerden E, Avanzi A, Barilla R, Zambirinis CP, Rendon M, Daley D, Pachter HL, Hajdu C, Miller G.** 2015. Dectin-1 regulates hepatic fibrosis and hepatocarcinogenesis by suppressing TLR4 signaling pathways. *Cell Rep* **13:**1909–1921.

174. **Chiba S, Ikushima H, Ueki H, Yanai H, Kimura Y, Hangai S, Nishio J, Negishi H, Tamura T, Saijo S, Iwakura Y, Taniguchi T.** 2014. Recognition of tumor cells by Dectin-1 orchestrates innate immune cells for anti-tumor responses. *eLife* **3:**e04177. doi:10.7554/eLife.04177.

175. **Deng Z, Ma S, Zhou H, Zang A, Fang Y, Li T, Shi H, Liu M, Du M, Taylor PR, Zhu HH, Chen J, Meng G, Li F, Chen C, Zhang Y, Jia XM, Lin X, Zhang X, Pearlman E, Li X, Feng GS, Xiao H.** 2015. Tyrosine phosphatase SHP-2 mediates C-type lectin receptor-induced activation of the kinase Syk and anti-fungal TH17 responses. *Nat Immunol* **16:**642–652.

176. **Rogers NC, Slack EC, Edwards AD, Nolte MA, Schulz O, Schweighoffer E, Williams DL, Gordon S, Tybulewicz VL, Brown GD, Reis e Sousa C.** 2005. Syk-dependent cytokine induction by Dectin-1 reveals a novel pattern recognition pathway for C type lectins. *Immunity* **22:** 507–517.

177. **Drummond RA, Brown GD.** 2013. Signalling C-type lectins in antimicrobial immunity. *PLoS Pathog* **9:** e1003417. doi:10.1371/journal.ppat.1003417.

178. **Wevers BA, Kaptein TM, Zijlstra-Willems EM, Theelen B, Boekhout T, Geijtenbeek TB, Gringhuis SI.** 2014. Fungal engagement of the C-type lectin mincle suppresses dectin-1-induced antifungal immunity. *Cell Host Microbe* **15:**494–505.

179. **Jia XM, Tang B, Zhu LL, Liu YH, Zhao XQ, Gorjestani S, Hsu YM, Yang L, Guan JH, Xu GT, Lin X.** 2014. CARD9 mediates Dectin-1-induced ERK activation by linking Ras-GRF1 to H-Ras for antifungal immunity. *J Exp Med* **211:**2307–2321.

180. **Glocker EO, Hennigs A, Nabavi M, Schäffer AA, Woellner C, Salzer U, Pfeifer D, Veelken H, Warnatz K, Tahami F, Jamal S, Manguiat A, Rezaei N, Amirzargar AA, Plebani A, Hannesschläger N, Gross O, Ruland J, Grimbacher B.** 2009. A homozygous *CARD9* mutation in a family with susceptibility to fungal infections. *N Engl J Med* **361:**1727–1735.

181. **Gross O, Gewies A, Finger K, Schäfer M, Sparwasser T, Peschel C, Förster I, Ruland J.** 2006. Card9 controls a non-TLR signalling pathway for innate anti-fungal immunity. *Nature* **442:**651–656.

182. **Dorhoi A, Desel C, Yeremeev V, Pradl L, Brinkmann V, Mollenkopf HJ, Hanke K, Gross O, Ruland J, Kaufmann SH.** 2010. The adaptor molecule CARD9 is essential for tuberculosis control. *J Exp Med* **207:**777–792.

183. **del Fresno C, Soulat D, Roth S, Blazek K, Udalova I, Sancho D, Ruland J, Ardavín C.** 2013. Interferon-β production via Dectin-1-Syk-IRF5 signaling in dendritic cells is crucial for immunity to *C. albicans*. *Immunity* **38:**1176–1186.

184. **Goodridge HS, Simmons RM, Underhill DM.** 2007. Dectin-1 stimulation by *Candida albicans* yeast or zymosan triggers NFAT activation in macrophages and dendritic cells. *J Immunol* **178:**3107–3115.

185. **Greenblatt MB, Aliprantis A, Hu B, Glimcher LH.** 2010. Calcineurin regulates innate antifungal immunity in neutrophils. *J Exp Med* **207:**923–931.

186. **Gringhuis SI, den Dunnen J, Litjens M, van der Vlist M, Wevers B, Bruijns SC, Geijtenbeek TB.** 2009. Dectin-1 directs T helper cell differentiation by controlling noncanonical NF-κB activation through Raf-1 and Syk. *Nat Immunol* **10:**203–213.

187. **Herre J, Marshall AS, Caron E, Edwards AD, Williams DL, Schweighoffer E, Tybulewicz V, Reis e Sousa C, Gordon S, Brown GD.** 2004. Dectin-1 uses novel mechanisms for yeast phagocytosis in macrophages. *Blood* **104:**4038–4045.

188. **Strijbis K, Tafesse FG, Fairn GD, Witte MD, Dougan SK, Watson N, Spooner E, Esteban A, Vyas VK, Fink GR, Grinstein S, Ploegh HL.** 2013. Bruton's tyrosine kinase (BTK) and Vav1 contribute to Dectin1-dependent phagocytosis of *Candida albicans* in macrophages. *PLoS Pathog* **9:**e1003446. doi:10.1371/journal.ppat.1003446.

189. **Goodridge HS, Reyes CN, Becker CA, Katsumoto TR, Ma J, Wolf AJ, Bose N, Chan AS, Magee AS, Danielson ME, Weiss A, Vasilakos JP, Underhill DM.** 2011. Activation of the innate immune receptor Dectin-1 upon formation of a 'phagocytic synapse.' *Nature* **472:**471–475.

190. **Goodridge HS, Shimada T, Wolf AJ, Hsu YM, Becker CA, Lin X, Underhill DM.** 2009. Differential use of CARD9 by Dectin-1 in macrophages and dendritic cells. *J Immunol* **182:**1146–1154.

191. **Ma J, Becker C, Lowell CA, Underhill DM.** 2012. Dectin-1-triggered recruitment of light chain 3 protein to phagosomes facilitates major histocompatibility complex class II presentation of fungal-derived antigens. *J Biol Chem* **287:**34149–34156.

192. **Plato A, Willment JA, Brown GD.** 2013. C-type lectin-like receptors of the Dectin-1 cluster: ligands and signaling pathways. *Int Rev Immunol* **32:**134–156.

193. **Branzk N, Lubojemska A, Hardison SE, Wang Q, Gutierrez MG, Brown GD, Papayannopoulos V.** 2014. Neutrophils sense microbe size and selectively release neutrophil extracellular traps in response to large pathogens. *Nat Immunol* **15:**1017–1025.

194. **Mansour MK, Tam JM, Khan NS, Seward M, Davids PJ, Puranam S, Sokolovska A, Sykes DB, Dagher Z, Becker C, Tanne A, Reedy JL, Stuart LM, Vyas JM.** 2013. Dectin-1 activation controls maturation of β-1,3-glucan-containing phagosomes. *J Biol Chem* **288:** 16043–16054.

195. **Gringhuis SI, Kaptein TM, Wevers BA, Theelen B, van der Vlist M, Boekhout T, Geijtenbeek TB.** 2012. Dectin-1 is an extracellular pathogen sensor for the induction and processing of IL-1β via a noncanonical caspase-8 inflammasome. *Nat Immunol* **13:**246–254.

196. **Gross O, Poeck H, Bscheider M, Dostert C, Hannesschläger N, Endres S, Hartmann G, Tardivel A, Schweighoffer E,**

Tybulewicz V, Mocsai A, Tschopp J, Ruland J. 2009. Syk kinase signalling couples to the Nlrp3 inflammasome for anti-fungal host defence. *Nature* **459:**433–436.

197. **Hise AG, Tomalka J, Ganesan S, Patel K, Hall BA, Brown GD, Fitzgerald KA.** 2009. An essential role for the NLRP3 inflammasome in host defense against the human fungal pathogen *Candida albicans*. *Cell Host Microbe* **5:**487–497.

198. **Zwolanek F, Riedelberger M, Stolz V, Jenull S, Istel F, Köprülü AD, Ellmeier W, Kuchler K.** 2014. The non-receptor tyrosine kinase Tec controls assembly and activity of the noncanonical caspase-8 inflammasome. *PLoS Pathog* **10:**e1004525. doi:10.1371/journal.ppat.1004525.

199. **Dennehy KM, Ferwerda G, Faro-Trindade I, Pyz E, Willment JA, Taylor PR, Kerrigan A, Tsoni SV, Gordon S, Meyer-Wentrup F, Adema GJ, Kullberg BJ, Schweighoffer E, Tybulewicz V, Mora-Montes HM, Gow NA, Williams DL, Netea MG, Brown GD.** 2008. Syk kinase is required for collaborative cytokine production induced through Dectin-1 and Toll-like receptors. *Eur J Immunol* **38:**500–506.

200. **Dennehy KM, Willment JA, Williams DL, Brown GD.** 2009. Reciprocal regulation of IL-23 and IL-12 following co-activation of Dectin-1 and TLR signaling pathways. *Eur J Immunol* **39:**1379–1386.

201. **Li X, Utomo A, Cullere X, Choi MM, Milner DA Jr, Venkatesh D, Yun SH, Mayadas TN.** 2011. The β-glucan receptor Dectin-1 activates the integrin Mac-1 in neutrophils via Vav protein signaling to promote *Candida albicans* clearance. *Cell Host Microbe* **10:**603–615.

202. **Huang JH, Lin CY, Wu SY, Chen WY, Chu CL, Brown GD, Chuu CP, Wu-Hsieh BA.** 2015. CR3 and Dectin-1 collaborate in macrophage cytokine response through association on lipid rafts and activation of Syk-JNK-AP-1 pathway. *PLoS Pathog* **11:**e1004985. doi:10.1371/journal.ppat.1004985.

203. **LeibundGut-Landmann S, Gross O, Robinson MJ, Osorio F, Slack EC, Tsoni SV, Schweighoffer E, Tybulewicz V, Brown GD, Ruland J, Reis e Sousa C.** 2007. Syk- and CARD9-dependent coupling of innate immunity to the induction of T helper cells that produce interleukin 17. *Nat Immunol* **8:**630–638.

204. **Osorio F, LeibundGut-Landmann S, Lochner M, Lahl K, Sparwasser T, Eberl G, Reis e Sousa C.** 2008. DC activated via dectin-1 convert Treg into IL-17 producers. *Eur J Immunol* **38:**3274–3281.

205. **Hernández-Santos N, Gaffen SL.** 2012. Th17 cells in immunity to *Candida albicans*. *Cell Host Microbe* **11:**425–435.

206. **Gringhuis SI, Wevers BA, Kaptein TM, van Capel TM, Theelen B, Boekhout T, de Jong EC, Geijtenbeek TB.** 2011. Selective C-Rel activation via Malt1 controls anti-fungal T_H-17 immunity by dectin-1 and dectin-2. *PLoS Pathog* **7:**e1001259. doi:10.1371/journal.ppat.1001259.

207. **Carter RW, Thompson C, Reid DM, Wong SY, Tough DF.** 2006. Preferential induction of $CD4^+$ T cell responses through in vivo targeting of antigen to dendritic cell-associated C-type lectin-1. *J Immunol* **177:**2276–2284.

208. **Leibundgut-Landmann S, Osorio F, Brown GD, Reis e Sousa C.** 2008. Stimulation of dendritic cells via the dectin-1/Syk pathway allows priming of cytotoxic T-cell responses. *Blood* **12:**4971–4980.

209. **Rieber N, Singh A, Öz H, Carevic M, Bouzani M, Amich J, Ost M, Ye Z, Ballbach M, Schäfer I, Mezger M, Klimosch SN, Weber AN, Handgretinger R, Krappmann S, Liese J, Engeholm M, Schüle R, Salih HR, Marodi L, Speckmann C, Grimbacher B, Ruland J, Brown GD, Beilhack A, Loeffler J, Hartl D.** 2015. Pathogenic fungi regulate immunity by inducing neutrophilic myeloid-derived suppressor cells. *Cell Host Microbe* **17:**507–514.

210. **Quintin J, Saeed S, Martens JH, Giamarellos-Bourboulis EJ, Ifrim DC, Logie C, Jacobs L, Jansen T, Kullberg BJ, Wijmenga C, Joosten LA, Xavier RJ, van der Meer JW, Stunnenberg HG, Netea MG.** 2012. *Candida albicans* infection affords protection against reinfection via functional reprogramming of monocytes. *Cell Host Microbe* **12:**223–232.

211. **Cheng SC, Quintin J, Cramer RA, Shepardson KM, Saeed S, Kumar V, Giamarellos-Bourboulis EJ, Martens JH, Rao NA, Aghajanirefah A, Manjeri GR, Li Y, Ifrim DC, Arts RJ, van der Veer BM, Deen PM, Logie C, O'Neill LA, Willems P, van de Veerdonk FL, van der Meer JW, Ng A, Joosten LA, Wijmenga C, Stunnenberg HG, Xavier RJ, Netea MG.** 2014. mTOR- and HIF-1α-mediated aerobic glycolysis as metabolic basis for trained immunity. *Science* **345:**1250684. doi:10.1126/science.1250684.

212. **Rivera A, Hohl TM, Collins N, Leiner I, Gallegos A, Saijo S, Coward JW, Iwakura Y, Pamer EG.** 2011. Dectin-1 diversifies *Aspergillus fumigatus*-specific T cell responses by inhibiting T helper type 1 CD4 T cell differentiation. *J Exp Med* **208:**369–381.

213. **Kashem SW, Igyártó BZ, Gerami-Nejad M, Kumamoto Y, Mohammed J, Jarrett E, Drummond RA, Zurawski SM, Zurawski G, Berman J, Iwasaki A, Brown GD, Kaplan DH.** 2015. *Candida albicans* morphology and dendritic cell subsets determine T helper cell differentiation. *Immunity* **42:**356–366.

214. **Drummond RA, Dambuza IM, Vautier S, Taylor JA, Reid DM, Bain CC, Underhill DM, Masopust D, Kaplan DH, Brown GD.** 2016. $CD4^+$ T-cell survival in the GI tract requires Dectin-1 during fungal infection. *Mucosal Immunol* **9:**492–502.

215. **Tang C, Kamiya T, Liu Y, Kadoki M, Kakuta S, Oshima K, Hattori M, Takeshita K, Kanai T, Saijo S, Ohno N, Iwakura Y.** 2015. Inhibition of Dectin-1 signaling ameliorates colitis by inducing *Lactobacillus*-mediated regulatory T cell expansion in the intestine. *Cell Host Microbe* **18:**183–197.

216. **Marshall AS, Willment JA, Lin HH, Williams DL, Gordon S, Brown GD.** 2004. Identification and characterization of a novel human myeloid inhibitory C-type lectin-like receptor (MICL) that is predominantly expressed on granulocytes and monocytes. *J Biol Chem* **279:**14792–14802.

217. Pyz E, Huysamen C, Marshall AS, Gordon S, Taylor PR, Brown GD. 2008. Characterisation of murine MICL (CLEC12A) and evidence for an endogenous ligand. *Eur J Immunol* 38:1157–1163.

218. Marshall AS, Willment JA, Pyz E, Dennehy KM, Reid DM, Dri P, Gordon S, Wong SY, Brown GD. 2006. Human MICL (CLEC12A) is differentially glycosylated and is down-regulated following cellular activation. *Eur J Immunol* 36:2159–2169.

219. van Rhenen A, van Dongen GA, Kelder A, Rombouts EJ, Feller N, Moshaver B, Stigter-van Walsum M, Zweegman S, Ossenkoppele GJ, Jan Schuurhuis G. 2007. The novel AML stem cell associated antigen CLL-1 aids in discrimination between normal and leukemic stem cells. *Blood* 110:2659–2666.

220. Larsen HO, Roug AS, Just T, Brown GD, Hokland P. 2012. Expression of the hMICL in acute myeloid leukemia—a highly reliable disease marker at diagnosis and during follow-up. *Cytometry B Clin Cytom* 82:3–8.

221. Zhao X, Singh S, Pardoux C, Zhao J, Hsi ED, Abo A, Korver W. 2010. Targeting C-type lectin-like molecule-1 for antibody-mediated immunotherapy in acute myeloid leukemia. *Haematologica* 95:71–78.

222. Roug AS, Larsen HO, Nederby L, Just T, Brown G, Nyvold CG, Ommen HB, Hokland P. 2014. hMICL and CD123 in combination with a CD45/CD34/CD117 backbone—a universal marker combination for the detection of minimal residual disease in acute myeloid leukaemia. *Br J Haematol* 164:212–222.

223. Kasahara S, Clark EA. 2012. Dendritic cell-associated lectin 2 (DCAL2) defines a distinct CD8α^- dendritic cell subset. *J Leukoc Biol* 91:437–448.

224. Lahoud MH, Proietto AI, Ahmet F, Kitsoulis S, Eidsmo L, Wu L, Sathe P, Pietersz S, Chang HW, Walker ID, Maraskovsky E, Braley H, Lew AM, Wright MD, Heath WR, Shortman K, Caminschi I. 2009. The C-type lectin Clec12A present on mouse and human dendritic cells can serve as a target for antigen delivery and enhancement of antibody responses. *J Immunol* 182: 7587–7594.

225. Han Y, Zhang M, Li N, Chen T, Zhang Y, Wan T, Cao X. 2004. KLRL1, a novel killer cell lectinlike receptor, inhibits natural killer cell cytotoxicity. *Blood* 104: 2858–2866.

226. Chen CH, Floyd H, Olson NE, Magaletti D, Li C, Draves K, Clark EA. 2006. Dendritic-cell-associated C-type lectin 2 (DCAL-2) alters dendritic-cell maturation and cytokine production. *Blood* 107:1459–1467.

227. Begun J, Lassen KG, Jijon HB, Baxt LA, Goel G, Heath RJ, Ng A, Tam JM, Kuo SY, Villablanca EJ, Fagbami L, Oosting M, Kumar V, Schenone M, Carr SA, Joosten LA, Vyas JM, Daly MJ, Netea MG, Brown GD, Wijmenga C, Xavier RJ. 2015. Integrated genomics of Crohn's disease risk variant identifies a role for CLEC12A in antibacterial autophagy. *Cell Rep* 11: 1905–1918.

228. Neumann K, Castiñeiras-Vilariño M, Höckendorf U, Hannesschläger N, Lemeer S, Kupka D, Meyermann S, Lech M, Anders HJ, Kuster B, Busch DH, Gewies A, Naumann R, Groß O, Ruland J. 2014. Clec12a is an inhibitory receptor for uric acid crystals that regulates inflammation in response to cell death. *Immunity* 40: 389–399.

229. Gagné V, Marois L, Levesque JM, Galarneau H, Lahoud MH, Caminschi I, Naccache PH, Tessier P, Fernandes MJ. 2013. Modulation of monosodium urate crystal-induced responses in neutrophils by the myeloid inhibitory C-type lectin-like receptor: potential therapeutic implications. *Arthritis Res Ther* 15:R73. doi: 10.1186/ar4250.

230. Redelinghuys P, Whitehead L, Augello A, Drummond RA, Levesque JM, Vautier S, Reid DM, Kerscher B, Taylor JA, Nigrovic PA, Wright J, Murray GI, Willment JA, Hocking LJ, Fernandes MJ, De Bari C, McInnes IB, Brown GD. 2016. MICL controls inflammation in rheumatoid arthritis. *Ann Rheum Dis* 75:1386–1391.

231. Oğuz AK, Yılmaz S, Akar N, Özdağ H, Gürler A, Ateş A, Oygür CS, Kılıçoğlu SS, Demirtaş S. 2015. C-type lectin domain family 12, member A: a common denominator in Behçet's syndrome and acute gouty arthritis. *Med Hypotheses* 85:186–191.

232. Kerscher B, Willment JA, Brown GD. 2013. The Dectin-2 family of C-type lectin-like receptors: an update. *Int Immunol* 25:271–277.

233. Sato K, Yang XL, Yudate T, Chung JS, Wu J, Luby-Phelps K, Kimberly RP, Underhill D, Cruz PD Jr, Ariizumi K. 2006. Dectin-2 is a pattern recognition receptor for fungi that couples with the Fc receptor γ chain to induce innate immune responses. *J Biol Chem* 281:38854–38866.

234. Bi L, Gojestani S, Wu W, Hsu YM, Zhu J, Ariizumi K, Lin X. 2010. CARD9 mediates Dectin-2-induced IκBα kinase ubiquitination leading to activation of NF-κB in response to stimulation by the hyphal form of *Candida albicans*. *J Biol Chem* 285:25969–25977.

235. Robinson MJ, Osorio F, Rosas M, Freitas RP, Schweighoffer E, Gross O, Verbeek JS, Ruland J, Tybulewicz V, Brown GD, Moita LF, Taylor PR, Reis e Sousa C. 2009. Dectin-2 is a Syk-coupled pattern recognition receptor crucial for Th17 responses to fungal infection. *J Exp Med* 206:2037–2051.

236. Saijo S, Ikeda S, Yamabe K, Kakuta S, Ishigame H, Akitsu A, Fujikado N, Kusaka T, Kubo S, Chung SH, Komatsu R, Miura N, Adachi Y, Ohno N, Shibuya K, Yamamoto N, Kawakami K, Yamasaki S, Saito T, Akira S, Iwakura Y. 2010. Dectin-2 recognition of α-mannans and induction of Th17 cell differentiation is essential for host defense against *Candida albicans*. *Immunity* 32:681–691.

237. Strasser D, Neumann K, Bergmann H, Marakalala MJ, Guler R, Rojowska A, Hopfner KP, Brombacher F, Urlaub H, Baier G, Brown GD, Leitges M, Ruland J. 2012. Syk kinase-coupled C-type lectin receptors engage protein kinase C-σ to elicit Card9 adaptor-mediated innate immunity. *Immunity* 36:32–42.

238. Gorjestani S, Yu M, Tang B, Zhang D, Wang D, Lin X. 2011. Phospholipase Cγ2 (PLCγ2) is key component in Dectin-2 signaling pathway, mediating anti-fungal innate immune responses. *J Biol Chem* 286:43651–43659.

239. Ariizumi K, Shen GL, Shikano S, Ritter R III, Zukas P, Edelbaum D, Morita A, Takashima A. 2000. Cloning of a second dendritic cell-associated C-type lectin (dectin-2) and its alternatively spliced isoforms. *J Biol Chem* **275**:11957–11963.

240. Taylor PR, Reid DM, Heinsbroek SE, Brown GD, Gordon S, Wong SY. 2005. Dectin-2 is predominantly myeloid restricted and exhibits unique activation-dependent expression on maturing inflammatory monocytes elicited *in vivo*. *Eur J Immunol* **35**:2163–2174.

241. Loures FV, Röhm M, Lee CK, Santos E, Wang JP, Specht CA, Calich VL, Urban CF, Levitz SM. 2015. Recognition of *Aspergillus fumigatus* hyphae by human plasmacytoid dendritic cells is mediated by Dectin-2 and results in formation of extracellular traps. *PLoS Pathog* **11**:e1004643. doi:10.1371/journal.ppat.1004643.

242. McGreal EP, Rosas M, Brown GD, Zamze S, Wong SY, Gordon S, Martinez-Pomares L, Taylor PR. 2006. The carbohydrate-recognition domain of Dectin-2 is a C-type lectin with specificity for high mannose. *Glycobiology* **16**:422–430.

243. Yonekawa A, Saijo S, Hoshino Y, Miyake Y, Ishikawa E, Suzukawa M, Inoue H, Tanaka M, Yoneyama M, Oh-Hora M, Akashi K, Yamasaki S. 2014. Dectin-2 is a direct receptor for mannose-capped lipoarabinomannan of mycobacteria. *Immunity* **41**:402–413.

244. Ifrim DC, Bain JM, Reid DM, Oosting M, Verschueren I, Gow NA, van Krieken JH, Brown GD, Kullberg BJ, Joosten LA, van der Meer JW, Koentgen F, Erwig LP, Quintin J, Netea MG. 2014. Role of Dectin-2 for host defense against systemic infection with *Candida glabrata*. *Infect Immun* **82**:1064–1073.

245. Ishikawa T, Itoh F, Yoshida S, Saijo S, Matsuzawa T, Gonoi T, Saito T, Okawa Y, Shibata N, Miyamoto T, Yamasaki S. 2013. Identification of distinct ligands for the C-type lectin receptors Mincle and Dectin-2 in the pathogenic fungus *Malassezia*. *Cell Host Microbe* **13**: 477–488.

246. Wang H, LeBert V, Hung CY, Galles K, Saijo S, Lin X, Cole GT, Klein BS, Wüthrich M. 2014. C-type lectin receptors differentially induce Th17 cells and vaccine immunity to the endemic mycosis of North America. *J Immunol* **192**:1107–1119.

247. Taylor PR, Roy S, Leal SM Jr, Sun Y, Howell SJ, Cobb BA, Li X, Pearlman E. 2014. Activation of neutrophils by autocrine IL-17A-IL-17RC interactions during fungal infection is regulated by IL-6, IL-23, RORγt and dectin-2. *Nat Immunol* **15**:143–151.

248. Wüthrich M, Wang H, Li M, Lerksuthirat T, Hardison SE, Brown GD, Klein B. 2015. *Fonsecaea pedrosoi*-induced Th17-cell differentiation in mice is fostered by Dectin-2 and suppressed by Mincle recognition. *Eur J Immunol* **45**:2542–2552.

249. Nakamura Y, Sato K, Yamamoto H, Matsumura K, Matsumoto I, Nomura T, Miyasaka T, Ishii K, Kanno E, Tachi M, Yamasaki S, Saijo S, Iwakura Y, Kawakami K. 2015. Dectin-2 deficiency promotes Th2 response and mucin production in the lungs after pulmonary infection with *Cryptococcus neoformans*. *Infect Immun* **83**:671–681.

250. Aragane Y, Maeda A, Schwarz A, Tezuka T, Ariizumi K, Schwarz T. 2003. Involvement of dectin-2 in ultraviolet radiation-induced tolerance. *J Immunol* **171**: 3801–3807.

251. Ritter M, Gross O, Kays S, Ruland J, Nimmerjahn F, Saijo S, Tschopp J, Layland LE, Prazeres da Costa C. 2010. *Schistosoma mansoni* triggers Dectin-2, which activates the Nlrp3 inflammasome and alters adaptive immune responses. *Proc Natl Acad Sci U S A* **107**: 20459–20464.

252. Barrett NA, Maekawa A, Rahman OM, Austen KF, Kanaoka Y. 2009. Dectin-2 recognition of house dust mite triggers cysteinyl leukotriene generation by dendritic cells. *J Immunol* **182**:1119–1128.

253. Suram S, Gangelhoff TA, Taylor PR, Rosas M, Brown GD, Bonventre JV, Akira S, Uematsu S, Williams DL, Murphy RC, Leslie CC. 2010. Pathways regulating cytosolic phospholipase A_2 activation and eicosanoid production in macrophages by *Candida albicans*. *J Biol Chem* **285**:30676–30685.

254. McDonald JU, Rosas M, Brown GD, Jones SA, Taylor PR. 2012. Differential dependencies of monocytes and neutrophils on dectin-1, dectin-2 and complement for the recognition of fungal particles in inflammation. *PLoS One* **7**:e45781. doi:10.1371/journal.pone.0045781.

255. Zhu LL, Zhao XQ, Jiang C, You Y, Chen XP, Jiang YY, Jia XM, Lin X. 2013. C-type lectin receptors Dectin-3 and Dectin-2 form a heterodimeric pattern-recognition receptor for host defense against fungal infection. *Immunity* **39**:324–334.

256. Kerscher B, Wilson GJ, Reid DM, Mori D, Taylor JA, Besra GS, Yamasaki S, Willment JA, Brown GD. 2015. The mycobacterial receptor, Clec4d (CLECSF8, MCL) is co-regulated with Mincle and upregulated on mouse myeloid cells following microbial challenge. *Eur J Immunol* **46**:381–389.

257. Barrett NA, Rahman OM, Fernandez JM, Parsons MW, Xing W, Austen KF, Kanaoka Y. 2011. Dectin-2 mediates Th2 immunity through the generation of cysteinyl leukotrienes. *J Exp Med* **208**:593–604.

258. Parsons MW, Li L, Wallace AM, Lee MJ, Katz HR, Fernandez JM, Saijo S, Iwakura Y, Austen KF, Kanaoka Y, Barrett NA. 2014. Dectin-2 regulates the effector phase of house dust mite-elicited pulmonary inflammation independently from its role in sensitization. *J Immunol* **192**:1361–1371.

259. Norimoto A, Hirose K, Iwata A, Tamachi T, Yokota M, Takahashi K, Saijo S, Iwakura Y, Nakajima H. 2014. Dectin-2 promotes house dust mite-induced Th2 and Th17 cell differentiation and allergic airway inflammation in mice. *Am J Respir Cell Mol Biol* **51**: 201–209.

260. Clarke DL, Davis NH, Campion CL, Foster ML, Heasman SC, Lewis AR, Anderson IK, Corkill DJ, Sleeman MA, May RD, Robinson MJ. 2014. Dectin-2 sensing of house dust mite is critical for the initiation of airway inflammation. *Mucosal Immunol* **7**: 558–567.

261. Hu XP, Wang RY, Wang X, Cao YH, Chen YQ, Zhao HZ, Wu JQ, Weng XH, Gao XH, Sun RH, Zhu LP.

2015. Dectin-2 polymorphism associated with pulmonary cryptococcosis in HIV-uninfected Chinese patients. *Med Mycol* **53**:810–816.

262. **Yamasaki S, Ishikawa E, Sakuma M, Hara H, Ogata K, Saito T.** 2008. Mincle is an ITAM-coupled activating receptor that senses damaged cells. *Nat Immunol* **9**: 1179–1188.

263. **Lobato-Pascual A, Saether PC, Fossum S, Dissen E, Daws MR.** 2013. Mincle, the receptor for mycobacterial cord factor, forms a functional receptor complex with MCL and FcεRI-γ. *Eur J Immunol* **43**:3167–3174.

264. **Miyake Y, Toyonaga K, Mori D, Kakuta S, Hoshino Y, Oyamada A, Yamada H, Ono K, Suyama M, Iwakura Y, Yoshikai Y, Yamasaki S.** 2013. C-type lectin MCL is an FcRγ-coupled receptor that mediates the adjuvanticity of mycobacterial cord factor. *Immunity* **38**: 1050–1062.

265. **Miyake Y, Masatsugu OH, Yamasaki S.** 2015. C-type lectin receptor MCL facilitates Mincle expression and signaling through complex formation. *J Immunol* **194**: 5366–5374.

266. **Zhao XQ, Zhu LL, Chang Q, Jiang C, You Y, Luo T, Jia XM, Lin X.** 2014. C-type lectin receptor Dectin-3 mediates trehalose 6,6′-dimycolate (TDM)-induced Mincle expression through CARD9/Bcl10/MALT1-dependent nuclear factor (NF)-κB activation. *J Biol Chem* **289**:30052–30062.

267. **Graham LM, Gupta V, Schafer G, Reid DM, Kimberg M, Dennehy KM, Hornsell WG, Guler R, Campanero-Rhodes MA, Palma AS, Feizi T, Kim SK, Sobieszczuk P, Willment JA, Brown GD.** 2012. The C-type lectin receptor CLECSF8 (CLEC4D) is expressed by myeloid cells and triggers cellular activation through Syk kinase. *J Biol Chem* **287**:25964–25974.

268. **Flornes LM, Bryceson YT, Spurkland A, Lorentzen JC, Dissen E, Fossum S.** 2004. Identification of lectin-like receptors expressed by antigen presenting cells and neutrophils and their mapping to a novel gene complex. *Immunogenetics* **56**:506–517.

269. **Vijayan D, Radford KJ, Beckhouse AG, Ashman RB, Wells CA.** 2012. Mincle polarizes human monocyte and neutrophil responses to *Candida albicans*. *Immunol Cell Biol* **90**:889–895.

270. **Kawata K, Illarionov P, Yang GX, Kenny TP, Zhang W, Tsuda M, Ando Y, Leung PS, Ansari AA, Eric Gershwin M.** 2012. Mincle and human B cell function. *J Autoimmun* **39**:315–322.

271. **Lee WB, Kang JS, Yan JJ, Lee MS, Jeon BY, Cho SN, Kim YJ.** 2012. Neutrophils promote mycobacterial trehalose dimycolate-induced lung inflammation via the Mincle pathway. *PLoS Pathog* **8**:e1002614. doi: 10.1371/journal.ppat.1002614.

272. **Schweneker K, Gorka O, Schweneker M, Poeck H, Tschopp J, Peschel C, Ruland J, Gross O.** 2012. The mycobacterial cord factor adjuvant analogue trehalose-6,6′-dibehenate (TDB) activates the Nlrp3 inflammasome. *Immunobiology* **218**:664–673.

273. **Balch SG, McKnight AJ, Seldin MF, Gordon S.** 1998. Cloning of a novel C-type lectin expressed by murine macrophages. *J Biol Chem* **273**:18656–18664.

274. **Schoenen H, Huber A, Sonda N, Zimmermann S, Jantsch J, Lepenies B, Bronte V, Lang R.** 2014. Differential control of Mincle-dependent cord factor recognition and macrophage responses by the transcription factors C/EBPβ and HIF1α. *J Immunol* **193**:3664–3675.

275. **Matsumoto M, Tanaka T, Kaisho T, Sanjo H, Copeland NG, Gilbert DJ, Jenkins NA, Akira S.** 1999. A novel LPS-inducible C-type lectin is a transcriptional target of NF-IL6 in macrophages. *J Immunol* **163**: 5039–5048.

276. **Arce I, Martínez-Muñoz L, Roda-Navarro P, Fernández-Ruiz E.** 2004. The human C-type lectin CLECSF8 is a novel monocyte/macrophage endocytic receptor. *Eur J Immunol* **34**:210–220.

277. **Ishikawa E, Ishikawa T, Morita YS, Toyonaga K, Yamada H, Takeuchi O, Kinoshita T, Akira S, Yoshikai Y, Yamasaki S.** 2009. Direct recognition of the mycobacterial glycolipid, trehalose dimycolate, by C-type lectin Mincle. *J Exp Med* **206**:2879–2888.

278. **Schoenen H, Bodendorfer B, Hitchens K, Manzanero S, Werninghaus K, Nimmerjahn F, Agger EM, Stenger S, Andersen P, Ruland J, Brown GD, Wells C, Lang R.** 2010. Cutting edge: mincle is essential for recognition and adjuvanticity of the mycobacterial cord factor and its synthetic analog trehalose-dibehenate. *J Immunol* **184**:2756–2760.

279. **Sharma A, Steichen AL, Jondle CN, Mishra BB, Sharma J.** 2014. Protective role of Mincle in bacterial pneumonia by regulation of neutrophil mediated phagocytosis and extracellular trap formation. *J Infect Dis* **209**: 1837–1846.

280. **Yamasaki S, Matsumoto M, Takeuchi O, Matsuzawa T, Ishikawa E, Sakuma M, Tateno H, Uno J, Hirabayashi J, Mikami Y, Takeda K, Akira S, Saito T.** 2009. C-type lectin Mincle is an activating receptor for pathogenic fungus, *Malassezia*. *Proc Natl Acad Sci U S A* **106**: 1897–1902.

281. **Lee RT, Hsu TL, Huang SK, Hsieh SL, Wong CH, Lee YC.** 2011. Survey of immune-related, mannose/fucose-binding C-type lectin receptors reveals widely divergent sugar-binding specificities. *Glycobiology* **21**:512–520.

282. **Furukawa A, Kamishikiryo J, Mori D, Toyonaga K, Okabe Y, Toji A, Kanda R, Miyake Y, Ose T, Yamasaki S, Maenaka K.** 2013. Structural analysis for glycolipid recognition by the C-type lectins Mincle and MCL. *Proc Natl Acad Sci U S A* **110**:17438–17443.

283. **Feinberg H, Jégouzo SA, Rowntree TJ, Guan Y, Brash MA, Taylor ME, Weis WI, Drickamer K.** 2013. Mechanism for recognition of an unusual mycobacterial glycolipid by the macrophage receptor mincle. *J Biol Chem* **288**:28457–28465.

284. **Hattori Y, Morita D, Fujiwara N, Mori D, Nakamura T, Harashima H, Yamasaki S, Sugita M.** 2014. Glycerol monomycolate is a novel ligand for the human, but not mouse macrophage inducible C-type lectin, Mincle. *J Biol Chem* **289**:15405–15412.

285. **Werninghaus K, Babiak A, Gross O, Hölscher C, Dietrich H, Agger EM, Mages J, Mocsai A, Schoenen H, Finger K, Nimmerjahn F, Brown GD, Kirschning C,**

Heit A, Andersen P, Wagner H, Ruland J, Lang R. 2009. Adjuvanticity of a synthetic cord factor analogue for subunit *Mycobacterium tuberculosis* vaccination requires FcRγ-Syk-Card9-dependent innate immune activation. *J Exp Med* **206:**89–97.

286. Shenderov K, Barber DL, Mayer-Barber KD, Gurcha SS, Jankovic D, Feng CG, Oland S, Hieny S, Caspar P, Yamasaki S, Lin X, Ting JP, Trinchieri G, Besra GS, Cerundolo V, Sher A. 2013. Cord factor and peptidoglycan recapitulate the Th17-promoting adjuvant activity of mycobacteria through mincle/CARD9 signaling and the inflammasome. *J Immunol* **190:**5722–5730.

287. Heitmann L, Schoenen H, Ehlers S, Lang R, Hölscher C. 2013. Mincle is not essential for controlling *Mycobacterium tuberculosis* infection. *Immunobiology* **218:** 506–516.

288. Behler F, Steinwede K, Balboa L, Ueberberg B, Maus R, Kirchhof G, Yamasaki S, Welte T, Maus UA. 2012. Role of Mincle in alveolar macrophage-dependent innate immunity against mycobacterial infections in mice. *J Immunol* **189:**3121–3129.

289. Wilson GJ, Marakalala MJ, Hoving JC, van Laarhoven A, Drummond RA, Kerscher B, Keeton R, van de Vosse E, Ottenhoff TH, Plantinga TS, Alisjahbana B, Govender D, Besra GS, Netea MG, Reid DM, Willment JA, Jacobs M, Yamasaki S, van Crevel R, Brown GD. 2015. The C-type lectin receptor CLECSF8/CLEC4D is a key component of anti-mycobacterial immunity. *Cell Host Microbe* **17:**252–259.

290. Steichen AL, Binstock BJ, Mishra BB, Sharma J. 2013. C-type lectin receptor Clec4d plays a protective role in resolution of Gram-negative pneumonia. *J Leukoc Biol* **94:**393–398.

291. Wells CA, Salvage-Jones JA, Li X, Hitchens K, Butcher S, Murray RZ, Beckhouse AG, Lo YL, Manzanero S, Cobbold C, Schroder K, Ma B, Orr S, Stewart L, Lebus D, Sobieszczuk P, Hume DA, Stow J, Blanchard H, Ashman RB. 2008. The macrophage-inducible C-type lectin, Mincle, is an essential component of the innate immune response to *Candida albicans*. *J Immunol* **180:** 7404–7413.

292. Sousa MG, Reid DM, Schweighoffer E, Tybulewicz V, Ruland J, Langhorne J, Yamasaki S, Taylor PR, Almeida SR, Brown GD. 2011. Restoration of pattern recognition receptor costimulation to treat chromoblastomycosis, a chronic fungal infection of the skin. *Cell Host Microbe* **9:**436–443.

293. Kiyotake R, Oh-Hora M, Ishikawa E, Miyamoto T, Ishibashi T, Yamasaki S. 2015. Human Mincle binds to cholesterol crystals and triggers innate immune responses. *J Biol Chem* **290:**25322–25332.

294. Tanaka M, Ikeda K, Suganami T, Komiya C, Ochi K, Shirakawa I, Hamaguchi M, Nishimura S, Manabe I, Matsuda T, Kimura K, Inoue H, Inagaki Y, Aoe S, Yamasaki S, Ogawa Y. 2014. Macrophage-inducible C-type lectin underlies obesity-induced adipose tissue fibrosis. *Nat Commun* **5:**4982. doi:10.1038/ncomms5982.

295. Wu XY, Guo JP, Yin FR, Lu XL, Li R, He J, Liu X, Li ZG. 2012. Macrophage-inducible C-type lectin is associated with anti-cyclic citrullinated peptide antibodies-positive rheumatoid arthritis in men. *Chin Med J (Engl)* **125:**3115–3119.

296. Suzuki Y, Nakano Y, Mishiro K, Takagi T, Tsuruma K, Nakamura M, Yoshimura S, Shimazawa M, Hara H. 2013. Involvement of Mincle and Syk in the changes to innate immunity after ischemic stroke. *Sci Rep* **3:**3177. doi:10.1038/srep03177.

297. Ichioka M, Suganami T, Tsuda N, Shirakawa I, Hirata Y, Satoh-Asahara N, Shimoda Y, Tanaka M, Kim-Saijo M, Miyamoto Y, Kamei Y, Sata M, Ogawa Y. 2011. Increased expression of macrophage-inducible C-type lectin in adipose tissue of obese mice and humans. *Diabetes* **60:**819–826.

298. He Y, Xu L, Li B, Guo ZN, Hu Q, Guo Z, Tang J, Chen Y, Zhang Y, Tang J, Zhang JH. 2015. Macrophage-inducible C-type lectin/spleen tyrosine kinase signaling pathway contributes to neuroinflammation after subarachnoid hemorrhage in rats. *Stroke* **46:**2277–2286.

299. Taylor ME, Conary JT, Lennartz MR, Stahl PD, Drickamer K. 1990. Primary structure of the mannose receptor contains multiple motifs resembling carbohydrate-recognition domains. *J Biol Chem* **265:**12156–12162.

300. Martinez-Pomares L. 2012. The mannose receptor. *J Leukoc Biol* **92:**1177–1186.

301. Boskovic J, Arnold JN, Stilion R, Gordon S, Sim RB, Rivera-Calzada A, Wienke D, Isacke CM, Martinez-Pomares L, Llorca O. 2006. Structural model for the mannose receptor family uncovered by electron microscopy of Endo180 and the mannose receptor. *J Biol Chem* **281:**8780–8787.

302. McKenzie EJ, Taylor PR, Stillion RJ, Lucas AD, Harris J, Gordon S, Martinez-Pomares L. 2007. Mannose receptor expression and function define a new population of murine dendritic cells. *J Immunol* **178:**4975–4983.

303. Stein M, Keshav S, Harris N, Gordon S. 1992. Interleukin 4 potently enhances murine macrophage mannose receptor activity: a marker of alternative immunologic macrophage activation. *J Exp Med* **176:**287–292.

304. Gordon S. 2003. Alternative macrophage activation. *Nat Rev Immunol* **3:**23–35.

305. Heinsbroek SE, Squadrito ML, Schilderink R, Hilbers FW, Verseijden C, Hofmann M, Helmke A, Boon L, Wildenberg ME, Roelofs JJ, Ponsioen CY, Peters CP, Te Velde AA, Gordon S, De Palma M, de Jonge WJ. 2015. miR-511-3p, embedded in the macrophage mannose receptor gene, contributes to intestinal inflammation. *Mucosal Immunol* **9:**960–973.

306. Squadrito ML, Pucci F, Magri L, Moi D, Gilfillan GD, Ranghetti A, Casazza A, Mazzone M, Lyle R, Naldini L, De Palma M. 2012. miR-511-3p modulates genetic programs of tumor-associated macrophages. *Cell Rep* **1:** 141–154.

307. Martínez-Pomares L, Mahoney JA, Káposzta R, Linehan SA, Stahl PD, Gordon S. 1998. A functional soluble form of the murine mannose receptor is produced by macrophages *in vitro* and is present in mouse serum. *J Biol Chem* **273:**23376–23380.

308. Moseman AP, Moseman EA, Schworer S, Smirnova I, Volkova T, von Andrian U, Poltorak A. 2013. Mannose

receptor 1 mediates cellular uptake and endosomal delivery of CpG-motif containing oligodeoxynucleotides. *J Immunol* **191:**5615–5624.

309. **Royer PJ, Emara M, Yang C, Al-Ghouleh A, Tighe P, Jones N, Sewell HF, Shakib F, Martinez-Pomares L, Ghaemmaghami AM.** 2010. The mannose receptor mediates the uptake of diverse native allergens by dendritic cells and determines allergen-induced T cell polarization through modulation of IDO activity. *J Immunol* **185:**1522–1531.

310. **Everts B, Hussaarts L, Driessen NN, Meevissen MH, Schramm G, van der Ham AJ, van der Hoeven B, Scholzen T, Burgdorf S, Mohrs M, Pearce EJ, Hokke CH, Haas H, Smits HH, Yazdanbakhsh M.** 2012. Schistosome-derived omega-1 drives Th2 polarization by suppressing protein synthesis following internalization by the mannose receptor. *J Exp Med* **209:**1753–1767.

311. **Zehner M, Chasan AI, Schuette V, Embgenbroich M, Quast T, Kolanus W, Burgdorf S.** 2011. Mannose receptor polyubiquitination regulates endosomal recruitment of p97 and cytosolic antigen translocation for cross-presentation. *Proc Natl Acad Sci U S A* **108:**9933–9938.

312. **Ezekowitz RA, Sastry K, Bailly P, Warner A.** 1990. Molecular characterization of the human macrophage mannose receptor: demonstration of multiple carbohydrate recognition-like domains and phagocytosis of yeasts in Cos-1 cells. *J Exp Med* **172:**1785–1794.

313. **Le Cabec V, Emorine LJ, Toesca I, Cougoule C, Maridonneau-Parini I.** 2005. The human macrophage mannose receptor is not a professional phagocytic receptor. *J Leukoc Biol* **77:**934–943.

314. **Galès A, Conduché A, Bernad J, Lefevre L, Olagnier D, Béraud M, Martin-Blondel G, Linas MD, Auwerx J, Coste A, Pipy B.** 2010. PPARγ controls Dectin-1 expression required for host antifungal defense against *Candida albicans*. *PLoS Pathog* **6:**e1000714. doi:10.1371/journal.ppat.1000714.

315. **van de Veerdonk FL, Marijnissen RJ, Kullberg BJ, Koenen HJ, Cheng SC, Joosten I, van den Berg WB, Williams DL, van der Meer JW, Joosten LA, Netea MG.** 2009. The macrophage mannose receptor induces IL-17 in response to *Candida albicans*. *Cell Host Microbe* **5:**329–340.

316. **Loures FV, Araújo EF, Feriotti C, Bazan SB, Calich VL.** 2015. TLR-4 cooperates with Dectin-1 and mannose receptor to expand Th17 and Tc17 cells induced by *Paracoccidioides brasiliensis* stimulated dendritic cells. *Front Microbiol* **6:**261. doi:10.3389/fmicb.2015.00261.

317. **Chieppa M, Bianchi G, Doni A, Del Prete A, Sironi M, Laskarin G, Monti P, Piemonti L, Biondi A, Mantovani A, Introna M, Allavena P.** 2003. Cross-linking of the mannose receptor on monocyte-derived dendritic cells activates an anti-inflammatory immunosuppressive program. *J Immunol* **171:**4552–4560.

318. **Lee SJ, Zheng NY, Clavijo M, Nussenzweig MC.** 2003. Normal host defense during systemic candidiasis in mannose receptor-deficient mice. *Infect Immun* **71:**437–445.

319. **Swain SD, Lee SJ, Nussenzweig MC, Harmsen AG.** 2003. Absence of the macrophage mannose receptor in mice does not increase susceptibility to *Pneumocystis carinii* infection in vivo. *Infect Immun* **71:**6213–6221.

320. **Court N, Vasseur V, Vacher R, Frémond C, Shebzukhov Y, Yeremeev VV, Maillet I, Nedospasov SA, Gordon S, Fallon PG, Suzuki H, Ryffel B, Quesniaux VF.** 2010. Partial redundancy of the pattern recognition receptors, scavenger receptors, and C-type lectins for the long-term control of *Mycobacterium tuberculosis* infection. *J Immunol* **184:**7057–7070.

321. **Dan JM, Kelly RM, Lee CK, Levitz SM.** 2008. Role of the mannose receptor in a murine model of *Cryptococcus neoformans* infection. *Infect Immun* **76:**2362–2367.

322. **Lee SJ, Evers S, Roeder D, Parlow AF, Risteli J, Risteli L, Lee YC, Feizi T, Langen H, Nussenzweig MC.** 2002. Mannose receptor-mediated regulation of serum glycoprotein homeostasis. *Science* **295:**1898–1901.

323. **Emara M, Royer PJ, Abbas Z, Sewell HF, Mohamed GG, Singh S, Peel S, Fox J, Shakib F, Martinez-Pomares L, Ghaemmaghami AM.** 2011. Recognition of the major cat allergen Fel d 1 through the cysteine-rich domain of the mannose receptor determines its allergenicity. *J Biol Chem* **286:**13033–13040.

324. **Chavele KM, Martinez-Pomares L, Domin J, Pemberton S, Haslam SM, Dell A, Cook HT, Pusey CD, Gordon S, Salama AD.** 2010. Mannose receptor interacts with Fc receptors and is critical for the development of crescentic glomerulonephritis in mice. *J Clin Invest* **120:**1469–1478.

325. **Hattori T, Konno S, Hizawa N, Isada A, Takahashi A, Shimizu K, Shimizu K, Gao P, Beaty TH, Barnes KC, Huang SK, Nishimura M.** 2009. Genetic variants in the mannose receptor gene (*MRC1*) are associated with asthma in two independent populations. *Immunogenetics* **61:**731–738.

326. **Hattori T, Konno S, Takahashi A, Isada A, Shimizu K, Shimizu K, Taniguchi N, Gao P, Yamaguchi E, Hizawa N, Huang SK, Nishimura M.** 2010. Genetic variants in mannose receptor gene (*MRC1*) confer susceptibility to increased risk of sarcoidosis. *BMC Med Genet* **11:**151. doi:10.1186/1471-2350-11-151.

327. **Zhang X, Li X, Zhang W, Wei L, Jiang T, Chen Z, Meng C, Liu J, Wu F, Wang C, Li F, Sun X, Li Z, Li JC.** 2013. The novel human *MRC1* gene polymorphisms are associated with susceptibility to pulmonary tuberculosis in Chinese Uygur and Kazak populations. *Mol Biol Rep* **40:**5073–5083.

328. **Park CG.** 2014. Vaccine strategies utilizing C-type lectin receptors on dendritic cells in vivo. *Clin Exp Vaccine Res* **3:**149–154.

329. **Dambuza IM, Brown GD.** 2015. C-type lectins in immunity: recent developments. *Curr Opin Immunol* **32:**21–27.

Myeloid Cells in Health and Disease: A Synthesis
Edited by Siamon Gordon

doi:10.1128/microbiolspec.MCHD-0028-2016

Hsi-Hsien Lin[1,2]
Martin Stacey[3]

G Protein-Coupled Receptors in Macrophages

26

GPCR CLASSIFICATION AND SIGNALING

G protein-coupled receptors (GPCRs) have been reviewed in depth elsewhere (1); however, the following section summarizes our present knowledge of GPCR classification and mechanisms involved in GPCR-mediated signaling.

Based on the proposed GRAFS classification system (glutamate, rhodopsin, adhesion, frizzled/taste 2, secretin), GPCRs are grouped into five distinct phylogenetic subfamilies (2). The rhodopsin family, the largest group, includes members of the chemokine, purinergic, odorant, and lipid receptors, among many others. The second-largest family is the adhesion group, containing 33 diverse receptors that have been proposed to have a dual role in cellular adhesion and signal transduction. The secretin, glutamate, and frizzled/taste 2 groups have fewer members; however, their receptors still play major roles in multiple physiological processes, including macrophage biology.

Despite of the diversity of GPCRs, the majority are believed to signal though a similar mechanism, that is, the transduction of external stimuli into intracellular secondary messengers via associated heterotrimeric G protein complexes (Fig. 1). These complexes consist of α, β, and γ subunits that prior to ligand binding exist in an inactive, GDP-bound state. However, upon ligand binding, conformational changes through the seven-transmembrane helices of the receptor facilitate the GDP-GTP exchange within the α subunit, resulting in dissociation from the βγ complex. The α and βγ subunits are then able to regulate various effectors, leading to the modulation of downstream secondary messengers, which can lead to rapid changes in cellular phenotype, or the activation of various transcription factors and concomitant alterations in gene expression. Specificity of signaling is mediated in part by the regulated expression of numerous types of α and βγ subunits that engage different effectors and through the regulation of their activity. Their activity is controlled by numerous protein families, including guanine nucleotide exchange factors (GEFs) and regulators of G protein signaling (RGSs), which enhance α-subunit GDP-GTP exchange and GTPase activity, activating and deactivating G protein signaling, respectively. Further regulation of signaling is mediated via receptor desensitization by GPCR kinases (GRKs) and arrestin proteins, which again possess regulated and restricted expression profiles.

[1]Department of Microbiology and Immunology, College of Medicine, Chang Gung University, Tao-Yuan, Taiwan; [2]Chang Gung Immunology Consortium and Department of Anatomic Pathology, Chang Gung Memorial Hospital-Linkou, 333 Tao-Yuan, Taiwan; [3]School of Molecular and Cellular Biology, University of Leeds, Leeds, LS2 9JT, United Kingdom.

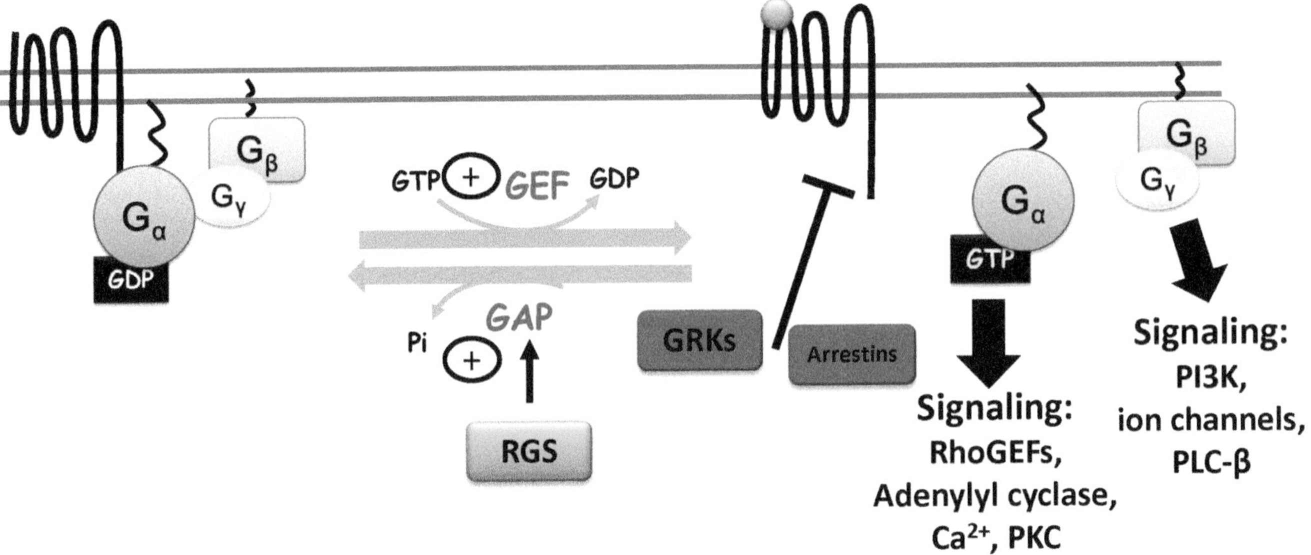

Figure 1 GPCR signaling and the activation/deactivation of heterotrimeric G proteins. Green circle, ligand; GAP, GTPase-activating protein; PI3K, phosphatidylinositol 3-kinase; PKC, protein kinase C; PLC, phospholipase C.

MACROPHAGE-SPECIFIC GPCRs

Through the use of RNA-seq data and other transcriptional profiling techniques, it is now evident that macrophages and other myeloid cells express a large repertoire of GPCRs (3). Those that are highly enriched or tightly restricted, however, have the greatest potential to shed light on the unique traits of these cells. The GPCRs discussed in this review are summarized in Table 1.

The example *par excellence* of a macrophage-restricted marker subsequently discovered to be a GPCR is the murine glycoprotein F4/80 (Emr1). Since its description in the 1980s, F4/80 has been the best surface marker for defining mouse tissue macrophage subpopulations (4–6).

Many resident tissue macrophages, including liver Kupffer cells, spleen red pulp macrophages, brain microglia cells, and the resident macrophages in kidney, lymph nodes, bone marrow stroma, gut lamina propria, testis, and peritoneum, all express high levels of the F4/80 antigen. On the contrary, macrophages within the T-cell areas of spleen (white pulp), lymph nodes (paracortex), and Peyer's patches are usually F4/80 negative. Weak F4/80 expression is found in other tissue macrophages such as alveolar macrophages in the lung and marginal zone and subcapsular sinus macrophages in the spleen and lymph nodes (6, 7). Apart from macrophages, eosinophils and Langerhans cells, the resident immature dendritic cells (iDCs) in the epidermis, are also F4/80 positive. Interestingly, F4/80 is downregulated during DC migration and maturation in the draining lymph nodes (6, 8). In tumor development, F4/80-expressing tumor-associated macrophages (TAMs) have been identified in many types of tumors (9–12). In addition, F4/80 has also been detected in tumor-infiltrating myeloid-derived suppressor cells (12–15). When cloned, the F4/80 gene (*Emr1*) was found to encode an unusually large, 904-amino-acid GPCR that was to become one of the founding members of the adhesion GPCR class of receptors (discussed in greater detail below). The F4/80 receptor comprises seven epidermal growth factor (EGF)-like modules coupled to a seven-span GPCR moiety via a highly glycosylated domain. Although the receptor's function remained elusive for many years, data now show F4/80 to be involved in the generation of antigen-specific efferent $CD8^+$ regulatory T cells associated with peripheral immune tolerance (16). It is believed that F4/80 is required for the cellular interactions among antigen-presenting cells and other immune effector cells during the induction of peripheral immune tolerance. In addition, F4/80 has been shown to be involved in macrophage-NK cell interaction in a *Listeria monocytogenes* infection model (17). At present, little is known of the molecular ligand of F4/80. Based on these findings, F4/80 is believed to play an immune regulatory role through the interaction with an unidentified cellular ligand expressed on immune effector cells, although the reason behind its highly restricted nature remains a mystery (16, 18). The facts that gene targeting of *Emr1*

Table 1 Physiological and pathological roles of macrophage GPCRs

Receptor	Function (Reference[s])	Disease association (Reference[s])
Complement receptors (C3aR, C5aR)	Inflammatory activation Chemotaxis, phagocytosis Proinflammatory cytokine production (45, 46, 51, 52) Negative modulator of anaphylatoxins (47)	Sepsis, rheumatoid arthritis, allergic responses (83–88) Pathogen evasion (89)
Formyl peptide receptors (FPR1–3)	Recognition of bacterial peptide and host protein chemotaxis M1 polarization (36, 53–56)	Pathogen evasion (89)
ChemR23	Anti-inflammatory responses (78)	
Chemokine receptors (CCR2, CCR5, etc.)	M1 skewing (37) CCR2l, CCR1, CCR2, and CX3CR1 cell migration	Atherosclerosis (79) Tumorigenesis (81) Viral subversion (82)
Purinergic receptors	P2RY6 inflammatory and antiviral immune responses (25–27) GPR86, GPR105/P2Y14, P2Y11, and P2Y12 chemotaxis and phagocytosis (43) Adenosine A_{2a} receptor attenuation of inflammation and macrophage polarization (29–32) P2Y-related P2RY6/12/13, GPR34, and ADORA3 microglia enriched (44)	
Fatty acid receptors	GPR84 migration, myeloid activation (22, 23) GPR84, GPR43, GPR109a, and GPR120 anti-inflammatory responses (24)	
Eicosanoid/arachidonic acid receptors (PGD_2Rs, PGE_2Rs, ChemR23, ALX/FPR2)	M2 skewing (38–40) Proresolution (76–78)	Arthritis, atherosclerosis (91) TAM polarization (92)
Leukotriene B_4 receptors (LTB_4R1, LTB_4R2)	Chemotaxis, bacterial phagocytosis and killing (58)	Cystic fibrosis, chronic obstructive pulmonary disease, inflammatory disease (56, 58, 59, 93) Pathogen subversion (98, 99)
Lipid receptors (PAFR, S1PR, LPARs, etc.)	M2 skewing (42)	Cancer, arthritis, atherosclerosis, diabetes, osteoporosis (102) Pathogen subversion (100)
PAR receptors (PAR1–4)	Inflammatory activation (70–74) Macrophage differentiation (70–75)	Arthritis (103) Allergy (108, 109) HIV-induced encephalitis (112)
F4/80	Cell adhesion (17, 48) Peripheral tolerance (16)	
EMR2	Inflammatory activation (124, 125)	Systemic inflammatory response syndrome (124)
EMR4	Dendritic cell activation (132)	
CD97	Host defense, leukocyte trafficking (140–143) T-cell proliferation/cytokine production (144, 145)	Autoinflammatory disease
BAI1	Clearance of apoptotic cells, phagocytosis (150, 151, 154)	
GPR116	Regulation of macrophage-induced inflammation (160)	

does not grossly affect macrophage distribution/function in mice and that successive research has shown the protein of the human homolog (EMR1) to be specific to eosinophils (19, 20) suggest that its expression does not *per se* define macrophage function.

Although F4/80 is still universally used as a defining marker of mouse macrophages, global expression profiling techniques including RNA-seq have now shown that other GPCRs are highly enriched in or restricted to macrophages. In one microarray study, 33 GPCRs including F4/80 were found to be enriched in bone marrow-derived or thioglycolate-elicited peritoneal macrophages (>10-fold higher than median of the 91 cell/tissue types tested) (21). The list of enriched genes included a number of known complement and chemokine receptors such as *C3aR*, *C5aR*, *Ccr2l*, *Ccr1*, *Ccr2*, and *Cx3cr1*. GPCRs were further enriched/restricted within these two macrophage populations, illustrating the potential importance of GPCRs in generating the heterogeneous phenotypes characteristic of macrophages. In this study, the predominantly macrophage-restricted GPCRs were found to be F4/80, GPR84, and the purinergic receptor P2ry6. The physiological role of GPR84 is still unknown; however, medium-chain fatty acids have been shown to act as ligands enhancing chemotaxis and the secretion of tumor necrosis factor α (TNF-α) and interleukin (IL)-12p40 by myeloid cells (22, 23). It is tempting to speculate that the GPR84 receptor could sense free fatty acid metabolites in response to cellular stress or differentiation state, as has been shown for other macrophage GPCRs such as the succinate GPCR receptor (SUCNR1) and the purinergic receptors. GPR84 signaling could also support the link between macrophage function and metabolic syndromes by sensing increased free fatty acid levels and the inducing proinflammatory mediators within adipose tissues. If this was the case, GPR84 antagonists would provide powerful therapeutics and potentially complement the beneficial anti-inflammatory effects of gut microbe-derived short-chain fatty acids, which are agonists for the macrophage GPCRs GPR43 and GPR109a and fish-derived omega-3 fatty acids that bind GPR120 (24).

P2RY6 has not yet been extensively studied with regard to its importance in macrophage function; however, reports indicate that P2RY6 on monocytes is required for Toll-like receptor (TLR) 1/2-induced IL-8 secretion and the induction of CCL2 on microglia (25, 26). UDP/P2Y6 signaling in the RAW 264.7 macrophage cell line as well as bone marrow-derived macrophages also enhances resistance to vesicular stomatitis virus, demonstrating a role in the innate antiviral response. UDP released by viral-infected cells via pannexin-1 channels results in P2RY6-dependent signaling and blocking of viral infectivity through the increased secretion of beta interferon (IFN-β) (27). Much remains to be elucidated with regard to the roles of macrophage-restricted GPCRs and whether these receptors alone are able to determine macrophage-specific function or define cellular subtypes. It is, however, highly likely that cross talk and integration of macrophage GPCR signaling with other receptors, such as the pattern recognition receptor, along with macrophage-specific transcriptional networks and tissue-specific niches will ultimately determine the differential fate and function of these cells.

MONOCYTE/MACROPHAGE DIFFERENTIATION

Cells of the monocyte/macrophage lineage are highly diverse and may be differentiated, activated, and polarized to generate a continuum of phenotypes. It is therefore not surprising considering their functional heterogeneity that the GPCRs' expression profiles in blood-borne monocytes and tissue-resident macrophages can differ significantly. An initial transcriptional profiling study using microarray analysis showed that monocytes had enriched expression of 11 GPCRs when compared with *in vitro* differentiation macrophages (28). These included the chemoattractant receptors *CCR2*, *CCR5*, *CCR7*, *CX3CR1*, *FPR1*, and another myeloid GPCR, adenosine A_{2a} receptor, which has been reported to attenuate immune response in inflammatory and tumor environments and facilitate macrophage polarization (29–32). Subsequent quantitative PCR arrays of primary human monocytes have since confirmed these results and examined the whole repertoire of GPCRs during monocyte-to-macrophage differentiation (33). The majority of GPCRs were observed to be downregulated upon maturation; however, in these studies a set of ~10 genes were upregulated (>5-fold) in all macrophages irrespective of differentiation stimuli (granulocyte-macrophage colony-stimulating factor [GM-CSF] or M-CSF), suggesting their critical nature in generating the mature macrophage phenotype. These included *MRGPRF*, *SUCNR1*, *LGR4*, and the adhesion GPCRs *CELSR1* and *GPR125* (33). Although little is known of these upregulated GPCRs in myeloid biology, emerging data suggest the metabolite succinate, the ligand of SUCNR1, may play a role in directing the inflammatory response. The increased metabolic strain imposed upon activated myeloid cells is now known to shift their metabolism from oxidation phosphorylation to glycolysis and pentose pathways, akin to the Warburg effect observed in tumors. The accumulation of metabolic intermediates, including succinate, enhances inflammation through stabilization of hypoxia-inducible factor-1α and concomitant IL-1β expression. Direct G protein-dependent signaling via SUCNR1 also synergizes with TLR signaling and mediates myeloid migration. Moreover, SUCNR1-deficient mice were less likely to reject allographs, identifying SUCNR1 antagonists as potential immune-suppressive therapeutics (34). LGR4 has been shown to attenuate lipopolysaccharide (LPS) responses via the classical cyclic AMP-protein kinase A signaling pathway by downregulating cell surface CD14 (35). Of the GPCRs downregulated in the quantitative PCR array study, a set including CX3CR1, GPR97, VIPR1 (vasoactive intestinal peptide receptor 1), CNR2 (cannabinoid receptor 2), P2RY10, and CCR7 were dramatically altered (>40-fold) upon macrophage differentiation. Of these, CX3CR1 and CCR7 are well-characterized myeloid chemokine receptors, with roles in monocyte patrolling in the lumen of the blood vessels and emigration into the lymphatic vessels/lymphoid nodes, whereas much remains to be investigated in the biology and function of GPR97, VIPR1, and CNR2.

MACROPHAGE POLARIZATION

The maturation of circulating monocytes into mature tissue macrophages can result in a continuum of heterogeneous phenotypes depending on their tissue environment and stimulation, the extremes of which, at least *in vitro*, have been designated as M1 and M2 macrophages, analogous to the functionally distinct Th1/Th2 T-cell subsets. The M1, or classically activated, macrophages are polarized via the engagement of IFN-γ and TLR ligands, whereas M2, or alternatively activated, phenotypes are generated via IL-4 and IL-13. A myriad of signaling pathways, epigenetic changes, and transcriptional networks actually determine macrophage polarization; however, the predominance of NF-κB and STAT1 activation leads to the M1 phenotype, whereas STAT3 and STAT6 skew toward an M2 phenotype. Ultimately, M1 skewing leads to the expression of proinflammatory cytokines including IL-12, TNF, and IL-23, which promotes Th1 and Th17 responses and the formation of reactive oxygen and nitrogen intermediates that enhance tumoricidal and microbicidal macrophage activity. M2-polarized macrophages are characterized by the expression of YM1, arginase-1, CCL24, and CCL17. These cells facilitate parasite clearance, reduce inflammation, and enhance tissue remodeling and repair through their increased activity in arginine metabolism, which is shifted to ornithine and polyamines. Whether GPCRs play a direct role in M1/M2 polarization has not been extensively studied; however, a limited number of studies are emerging in which signaling via GPCRs can facilitate macrophage polarization. Formyl peptide receptor 2 (FPR2)-deficient mice display a decreased survival rate following Lewis lung carcinoma implantation, and their macrophages have decreased expression of M1 markers when incubated with supernatant from tumor cells. It is therefore believed that the FPR2 receptor promotes antitumor activity by limiting M2 polarization of macrophages (36). Similarly, CXCR3 has been shown to skew macrophages toward the M1 antitumor phenotype, as deficient mice exhibit enhanced M2 polarization and tumor progression (37).

Ligands to GPCRs can also skew toward M2 polarization. Prostaglandin E_2 (PGE_2) derived from the tumor environment is known to skew TAMs toward to the M2 phenotype (38). PGE_2 in combination with LPS is also able to promote an anti-inflammatory phenotype characterized by high expression of IL-10 and the regulatory markers SPHK1 and LIGHT (39). Other studies show that PGE_2 induces the polarization to the proresolving M2b subtype through its GPCR eicosanoid receptors (40). Interestingly, this effect appears to be exploited by the more virulent strains of the fungus *Paracoccidioides brasiliensis*, which induce PGE_2, thereby reducing the microbicidal M1 polarization and evading killing (41). Another study shows that macrophages undergo M2 polarization in an IL-4-dependent manner in response to sphingosine 1-phosphate (S1P) binding to its cognate GPCR receptor. As S1P is associated with lipoproteins, especially with high-density lipoprotein, it has been proposed that the antiatherogenic properties of high-density lipoprotein are through the S1P receptor (S1PR) signaling and resultant M2 polarization of macrophages (42).

One transcriptional profiling study of monocytes and M1/M2 skewing surprisingly did not show a number of differentially regulated GPCRs (28). In this study, 53 receptors were shown to be altered during monocyte differentiation and macrophage polarization. The combination of IFN-γ and LPS had a broad effect, with no clear GPCR family being overrepresented. However, M2 polarization showed a dramatic upregulation of a group of purinergic receptors (28). Independent studies have shown that M1 polarization results in the increased use of differential transcriptional start sites of the *ChemR23* gene, leading to increased expression of functionally active receptor in M1 macrophages. The ChemR23 receptor is expressed on monocytes, macrophages, and DCs and binds to the peptide chemerin, resulting in the recruitment of ChemR23-positive leukocytes to sites of inflammation. Interestingly, use of the alternative ChemR23 ligand, resolvin E1, partially repolarized M1 macrophages to an intermediate proresolving phenotype, resulting in increased IL-10 transcription and zymosan phagocytosis.

The functional consequences of the M2-upregulated purinergic receptors (GPR86, GPR105/P2Y14, P2Y8, P2Y11, and P2Y12) remain to be fully explored. However, chemotaxis and phagocytosis by microglia in response to extracellular ATP, and damaged neurons releasing UDP, respectively, have been shown to be dependent on PY2 receptors. A number of studies have also shown that P2Y receptors play crucial chemotactic roles in locating apoptotic cells releasing the "find me" ATP signals (43). Further dissection of these receptors in macrophage biology is clearly required. More recently, RNA-seq data from a study examining the expression landscape of seven types of tissue macrophages and monocytes have further highlighted the heterogeneity of macrophages (44). Approximately 60 GPCRs and associated genes were shown to be differentially expressed between at least two of the macrophage subsets or monocytes tested. Multiple GPCRs were enriched in just one of the tissue macrophage subtypes (microglia; Kupffer cells; or splenic, lung, peritoneal, ileal, or colonic macrophages and monocytes), perhaps explaining the functional differences in these populations. The greatest number of differentially expressed genes was observed in microglia; these included multiple purinergic-related receptors (P2RY6/12/13, GPR34, and ADORA3) among others, indicating larger differences between microglia and other macrophage subtypes.

In the same study, the importance of environmental cues and the plasticity of macrophage differentiation was highlighted by the fact that peritoneal macrophages adopted a largely lung macrophage expression profile when transferred into the lung. The recruitment of macrophage precursors into these different environmental niches and hence their ultimate phenotype is probably determined largely by monocyte GPCRs including chemokine receptors and other GPCRs involved in migration.

INFLAMMATION AND PATHOGEN ELIMINATION

As important innate effector cells, monocytes/macrophages are pivotal in the recognition and elimination of pathogens. GPCR signaling induced by both host- and pathogen-derived ligands is critical for the initiation and maintenance of the inflammatory response, promoting cellular recruitment and mediating macrophage activation and pathogen elimination. GPCRs known to activate macrophages are listed below.

The Complement C3a and C5a Receptors

The complement system is essential in the innate as well as adaptive immune responses. Soluble products of the complement activation pathways, especially C3a and C5a (also called anaphylatoxins), mediate their biological activities through two specific GPCRs, designated the C3aR and C5aR (CD88), respectively (45, 46). C5a also binds to another seven-transmembrane receptor called C5L2 (GPR77), which, interestingly, is not coupled to G proteins. Instead, C5L2 is thought to act as a C5a decoy receptor and a negative modulator of C5aR-mediated signal transduction pathways (47). The majority of myeloid immune cells, including monocytes/macrophages, DCs, granulocytes, and mast cells, express C3aR and C5aR at variable levels (45, 46). Indeed, anaphylatoxins (C3a, C4a, and C5a) function mainly by triggering the degranulation of these immune cells as well as endothelial cells, hence producing a strong inflammatory response. Under inflammatory conditions, expression of C3aR and C5aR can be further upregulated upon cell activation.

While both C3aR and C5aR are classical GPCRs, C3aR contains a uniquely large second extracellular loop that is richly modified by tyrosine sulfation, important for high-affinity binding of C3a. Multiple phosphorylation sites and a potential S-palmitoylation site are found in the intracellular regions of C3aR. Similar to C3aR, C5aR also contains intracellular phosphorylation sites but no S-acylation site. C5aR is similarly modified by extracellular tyrosine sulfation. Both C3aR and C5aR are also decorated by N-link glycosylation and phosphorylated upon ligand binding (45, 46).

Signaling mediated by C3aR and C5aR is predominantly transmitted by pertussis toxin-sensitive $G\alpha_i$ subunits, though coupling to $G\alpha_{12/13}$ and $G\alpha_{16}$ has been reported (45, 46). In addition, G protein-independent β-arrestin binding was also noted. Regulation of signaling by GRKs and Na^+/H^+ exchange regulatory factor (NHERF1 and NHERF2) have also been implicated (48–50). These differential signaling pathways of C3aR and C5aR are employed in different cell types or by different stimuli. Activation of C3aR and C5aR is usually accompanied by changes of intracellular cAMP and Ca^{2+} levels as well as phosphorylation of signaling molecules such as extracellular signal-related kinase 1/2 and Akt (45, 46).

Initially identified as anaphylatoxins that induced histamine degranulation from mast cells, C3a and C5a are now known as pleiotropic molecules with functional roles inside and outside the immune system. Within the immune system, C3a and C5a mainly act as inflammatory mediators that promote chemotaxis, phagocytosis, antimicrobial response, and production of proinflammatory cytokines (45, 46, 51, 52).

The FPR Family

Identified initially as the receptor for the bacteria-derived formyl-methionyl-leucyl-phenylalanine (fMLF) peptide, FPR1 and the later-discovered homologous receptors, FPR2 (FPR like-1 [FPRL1]) and FPR3 (FPRL2), form the FPR family of GPCRs (53, 54). The genes encoding these three receptors are clustered in the human chromosome next to that of the human C5a receptor, suggesting a possible evolutionary and functional link (46). The human FPRs are highly expressed in myeloid cells such as neutrophils, monocytes/macrophages, and DCs with some differential patterns for the individual members. Blood monocytes express all three FPRs, and the expression remains relatively stable when differentiated to macrophages. In contrast, the expression of FPR1 and FPR2 declines gradually following monocyte differentiation to DCs, while FPR3 expression stays unchanged. Recent murine studies have shown FPR1 and FPR2 to be enriched in lung macrophages when compared to multiple other tissue macrophage subpopulations (44). Although bacteria-derived fMLF is the widely used prototypic *N*-formyl peptide ligand of the FPRs, other microbe-derived *N*-formyl and nonformyl peptides are known to potently activate the FPRs. For example, FPR2 responds to certain α-helical, amphipathic peptides released by pathogenic bacteria such as HP(2-20) peptide from *Helicobacter pylori* and phenol-soluble modulin peptides that are secreted by *Staphylococcus aureus* and coagulase-negative staphylococci (55). In addition, host cell-derived peptide and nonpeptide agonists are identified for individual FPRs. Of particular interest among the host-derived FPR agonists are peptides associated with aseptic inflammatory and antimicrobial responses, including serum amyloid A, β-amyloid peptide, prion protein fragment, annexin 1, and LL-37, an enzymatic cleavage fragment of cathelicidin, as well as mitochondrial-derived formyl peptides. In most cases, activation of FPRs stimulates cellular migration and activation of myeloid cells, inducing the production of reactive oxygen and nitrogen species and various proinflammatory cytokines such as IL-1β and IL-6. Nevertheless, due to the high ligand promiscuity of FPRs, differing types or concentrations of ligands can promote anti-inflammatory or inhibitory immune reactions (53, 54, 56).

The LTB_4 Receptors

Leukotriene B_4 (LTB_4) is an oxidized fatty acid derivative catalyzed from arachidonic acid by 5-lipoxygenase (5-LOX) and acts as a chemoattractant for various phagocytes (57). Due to the high expression levels of 5-LOX in innate immune cells, granulocytes, macrophages, and mast cells are the main producers of LTB_4 (56, 58). LTB_4 was initially discovered by its chemotactic activity on neutrophils and results in increased reactive oxygen species production, enhanced degranulation, and bacterial phagocytosis and killing (56, 58, 59). LTB_4 is also able to induce integrin activation, leading to tight adhesion of leukocytes (60, 61). Upon LTB_4 stimulation, human monocytes enhance IL-6 production, while mouse macrophages promote bacterial phagocytosis and killing (59). LTB_4 activates and mediates chemotaxis and recruitment of murine and human mast cells (62). In osteoclasts, LTB_4 treatment increases the production of tartrate-resistant acid phosphatase and enhances the resorption of calcified matrices (63). Apart from cell migration and activation, LTB_4 promotes neutrophil cell survival by preventing apoptosis. In summary, as a highly potent lipid mediator, LTB_4 plays an important role in inflammatory responses and immune regulation.

LTB_4-induced cellular responses are mediated by two related GPCRs, LTB_4R1 (BLT1) and LTB_4R2 (BLT2) (58, 59). BLT1 is a high-affinity receptor for LTB_4, expressed restrictedly in myeloid leukocytes, while BLT2 is a low-affinity LTB_4 receptor with ubiquitous expression patterns. Both GPCRs are coupled to the $G\alpha_i$ protein predominantly and activate cells via the adenylate cyclase pathway (64–66). LTB_4-induced BLT1 signaling was shown to regulate NF-κB activation via STAT1-dependent regulation of MyD88 expression in macrophages (67). Similarly, LTB_4-BLT1 signaling was found to regulate the transcription of the fungal pattern recognition receptor Dectin-1 (DC-associated C-type lectin 1) via GM-CSF and PU.1, thus controlling Dectin-1-dependent phagocytosis and cytokine production (68). The role of LTB_4-BLT1 signaling in macrophage phagocytosis was further revealed by the attenuated FcγR-mediated opsonized phagocytosis of IgG-coated particles in BLT1 knockout macrophages. Indeed, the cross talk between LTB_4-BLT1 and IgG-FcγR signaling pathways was shown by the clustering of BLT1 and FcγR complex and signaling molecules in the lipid raft microdomain (69).

Taken together, these findings suggest that the LTB4-induced cellular activities mediated by BLT1 play a critical role in inflammatory responses as well as immune regulatory functions.

The PAR Family

The protease-activated receptors (PARs) are activated upon proteolytic cleavage by serine proteases such as thrombin and trypsin that unmask a cryptic peptide sequence at the new amino terminus, which acts as a tethered ligand for the activation of the exposed cryptic peptide's own receptors (70). There are four PARs (PAR1 to -4) that display similar receptor structures and activation modes but possess distinct protease cleavage sites and tethered ligand pharmacology (71). Although highly expressed in platelets and critically involved in hemostasis and thrombosis, PARs are also expressed in various immune cells (70, 71). For example, PARs show differential expression patterns during human monocyte differentiation and are most likely regulated by cytokines. Peripheral blood monocytes express

PAR1 and PAR3 at the protein level, whereas monocyte-derived macrophages additionally express PAR2, while only mRNAs have been found in monocyte-derived DCs (70, 71). Human alveolar macrophages show positive staining of PAR1 and PAR2, while vascular macrophage-derived foam cells express only PAR2. Human liver Kupffer cells express PAR4, and murine microglia cells express all PAR RNA transcripts at different levels (70, 71). Activation of mouse microglial PAR1 induces intracellular Ca^{2+} increase and transient mitogen-activated protein kinase activation (72). Interestingly, activation of PAR1 and PAR2 in monocytes/macrophages triggers similar cytosolic Ca^{2+} responses but activates distinct expression patterns of genes involved in inflammation (73). As such, activation of PAR1 but not PAR2 in human monocytes or macrophages induces the expression of the proinflammatory mediator monocyte chemoattractant protein-1. On the other hand, stimulation of PAR2 in human monocytes increased the production of IL-6, IL-8, and IL-1β (74). Recent studies show that activation of PAR2 during M1 and M2 macrophage maturation modulates the differentiation phenotypes and effector functions (75).

Similar differential expression patterns and functions of PARs are identified in other myeloid cells. Human neutrophils express significant levels of PAR2 but very low, if any, PAR1. PAR2 activation in neutrophils increases intracellular Ca^{2+} and primes the cells for further activation (70, 71). In addition, neutrophils stimulated by PAR2 agonists express higher levels of surface Mac-1 (macrophage-1 antigen, also called CR3) and VLA-4 (very late antigen-4); stimulate IL-1, IL-8, and IL-6 secretion; and display enhanced motility in collagen gels. Similar to neutrophils, human eosinophils express PAR2 variably but very little PAR1. PAR2 activation in eosinophils induces reactive oxygen species production and degranulation. Human basophils do not express any PARs. Human tissue mast cells in tonsil, skin, and colon express PAR2 protein, while PAR1, -3, and -4 RNAs are detected in human skin mast cells. Activation of human tissue mast cells or mast cell lines by PAR agonists enhances secretion of histamine and TNF-α (70, 71, 73).

In addition to the classical PAR-activating serine proteases and activating peptides, PARs can also be alternatively activated or inactivated by various cell- and pathogen-derived proteases such as cathepsin G, granzyme A, leukocyte elastase, and gingipains RgpB and HRgpA, based on the location of the proteolytic site in the N-terminal region of PARs (70, 71, 73). This is of particular interest when the role of PARs in immune regulation during infection is considered.

In summary, PARs are variably expressed in immune cells, where their activities are able to modulate innate and adaptive immune responses positively or negatively under physiological and pathological (see later section) situations.

RESOLUTION OF INFLAMMATION

Although inflammation is critical for the elimination of pathogens and the initiation of wound repair, the resolution phase of inflammation is equally important in promoting cell clearance, preventing collateral tissue damage, and instigating tissue remodeling. Though it was previously believed to be a passive process in which inflammation simply dwindled away, it is now increasingly clear that the resolution is an active process involving numerous cellular and soluble mediators culminating in tissue homeostasis. It is now appreciated that many proresolving mediators and mechanisms act through GPCRs.

One such proresolving mediator is the glucocorticoid-regulated protein annexin A1. This protein possesses potent anti-inflammatory and proresolving functions, including the reduction of neutrophil-endothelial interactions, stimulation of macrophage efferocytosis, and acceleration of neutrophil apoptosis. These proresolving activities are dependent on the receptor ALX/FPR2, which from a therapeutic perspective is encouragingly agonized by a protease-resistant 25-amino-acid N-terminal fragment of annexin A1 (76). As well as annexin A1, the ALX/FPR2 GPCR is also able to bind the lipid lipoxin A_4. Lipoxin A_4 is an arachidonic acid metabolite generated by platelets through their interactions with neutrophils. Unlike chemokine receptors, ALX signaling lacks increases in Ca^{2+} but results in phosphorylation of cytoskeletal proteins resulting in cell arrest. Multiple *in vitro* and *in vivo* models have shown that signaling affects leukocyte trafficking by preventing inflammatory neutrophil recruitment and enhancing monocyte migration. Resolution is further mediated through blockade of superoxide anion generation by neutrophils and increases in nonphlogistic phagocytosis by macrophages. Multiple other complex disease models ranging from colitis and periodontitis to asthma and cystic fibrosis have also illustrated the potent anti-inflammatory/proresolving effects of lipoxin A_4 or aspirin-triggered lipoxin A_4 (77). In a number of these reports, however, the effects of nonmyeloid cells have not been ruled out. Other proresolving lipid mediators include the resolvins and protectins derived from eicosapentaenoic and docosahexaenoic acid. Resolvin E1 binds the myeloid receptor ChemR23, attenuating

TNF-α-mediated NF-κB signaling, and directly antagonizes the proinflammatory effects of LTB_4 binding to the monocyte GPCR BLT1 (78). The receptors of lipid mediator protectins and maresins remain elusive, but structural and activity assays suggest they are likely to be $G\alpha_i$-dependent GPCRs (76).

DETRIMENTAL ROLES OF MYELOID GPCRs

In addition to the beneficial roles in host defense and tissue homeostasis, the inappropriate activities of macrophages and monocytes can also have detrimental pathological effects. Macrophages are known to exacerbate multiple inflammatory disorders such as rheumatoid arthritis and colitis, and their inappropriate polarization within the tumor microenvironment is known to facilitate the maintenance, progression, and metastasis of many cancers. Increasingly, macrophage involvement in a collection of metabolic disorders including atherosclerosis, insulin resistance, and dyslipidemia is also being demonstrated. This is not surprising: given the abundance and repertoire of GPCRs in myeloid cells, they often play key roles in these pathologies.

CHEMOKINE RECEPTORS IN DISEASE

The chemokine GPCR receptors, in particular CCR2, are direct mediators of monocyte recruitment in many disease processes. Ccr2-null mice have markedly reduced plaques when crossed with atherosclerosis-prone $apoE^{-/-}$ mice (79), revealing a role in the initiation of atherosclerosis. In addition, CCR2-mediated monocyte recruitment also contributes to the development of multiple sclerosis, scleroderma, rheumatoid arthritis, and ischemia-reperfusion injury (80).

CCR2 also recruits monocytes to primary tumors and metastases in the PYMT tumor model and enhances extravasation and metastasis of tumor cells in part by expression of vascular endothelial growth factor. Ablation of the CCR1 ligand, CCL9, resulted in loss of monocytes and/or TAMs and a resultant inhibition of malignancy (81). The chemokine receptors on macrophages are also exploited by pathogens, especially by viruses as entry receptors and as strategies of immune subversion. A well-characterized example is CCR5, which acts as the cell fusion coreceptor for macrophage-tropic strains of HIV-1. Moreover, virally encoded chemokines from a number of herpesviruses have also been shown to interact with monocyte/macrophage GPCRs. The viral homologs of macrophage inflammatory protein I (vMIP-I) and vCCL4 from Kaposi's sarcoma associated herpesvirus and human herpesvirus 6, respectively, are thought to facilitate the spread of virus via the recruitment of macrophages through the binding of their cognate chemokine receptors CCR8 and CCR5. Conversely, the antagonistic Kaposi's sarcoma-associated herpesvirus chemokine homolog vMIP-II has also been shown to prevent the recruitment of leukocytes during viral infection (82).

fMLF/COMPLEMENT RECEPTORS

Due to their role in physiological inflammation, it is not surprising that C3aR and C5aR GPCRs have detrimental roles in conditions including rheumatoid arthritis, sepsis, tissue injury, infection, and asthma/allergy. In a neutrophil-dependent mouse model of intestinal ischemia-reperfusion injury, C3aR was shown to play a role in constraining neutrophil mobilization in acute tissue injuries (83). In the collagen-induced arthritis (CIA) and collagen antibody-induced arthritis (CAIA) animal models, C5aR deficiency reduced the recruitment of neutrophils, macrophages, and T cells to joints and led to the secretion of lower levels of inflammatory mediators including IL-1β, TNF, MIP-1α, and MIP-2α (84). Similar findings were observed in the C3aR knockout mice in the CAIA model and in wild-type mice receiving systemic administration of anti-C5a monoclonal antibody (mAb) in the CIA model (85). In a pulmonary allergy model in mice, C3aR ablation attenuated airway hyperresponsiveness with a significant decrease in infiltrating eosinophils and IL-4-producing cells in the airway, and diminished levels of the Th2 cytokines (IL-5 and IL-13) in bronchoalveolar lavage (86). In influenza type A virus-infected mice, C5aR antagonist treatment greatly reduced the frequency and absolute numbers of flu-specific $CD8^+$ T cells with attenuated antiviral cytolytic activity in the lungs (87). Likewise, C5aR and neutrophils were identified as key mediators of fetal injury in a miscarriage model of antiphospholipid syndrome (88). The generation of C5a within the tumor microenvironment is also believed to attract myeloid suppressor cells, whose generation of reactive oxygen species and nitric oxide synthase impairs the antitumor activity of T cells. As with many receptors in the immune system, the GPCRs FPR1 and C5aR are targeted by pathogens to evade the immune response. For instance, *Haemophilus influenzae* releases a small unidentified factor that inhibits fMLF-mediated migration, and pertussis toxin from *Bordetella pertussis* acts as a potent inhibitor of chemotaxis through the inactivation of G_i subunits involved in G protein-coupled signaling. Interestingly, the *Staphylococcus aureus* chemotaxis inhibitory protein (CHIP) is known to antagonize FPR1 and C5aR, inhibiting the recruitment of both macrophages and neutrophils (89). The related *S. aureus* FLIPr protein is also known to be a potent antagonist of both FLP1 and FLPR2.

LIPID RECEPTORS IN DISEASE

Under most physiological settings, monocyte/macrophage GPCRs of lipid intermediates are crucial contributors to inflammation, initiating pathogen elimination and tissue repair. However, inappropriate generation of lipid mediators leads to a number of pathologies. Prostaglandins, small lipid mediators derived from arachidonic acid, are known to have potent inflammatory activities that exacerbate numerous pathologies. Indeed, many of the beneficial effects of nonsteroidal anti-inflammatory drugs are due to their blockade of prostaglandin synthesis through the inhibition of cyclooxygenase-1 (COX-1) and COX-2. PGE_2, for example, can signal through four canonical GPCRs of the rhodopsin class, EP1 to -4. EP2 to -4 are abundantly expressed by macrophages; $EP4^{-/-}$ mice show an impressive resistance to development of experimental arthritis (90). Similarly, $LDLR^{-/-}$ mice chimeric for $EP4^{-/-}$ in hematopoietic cells show suppressed early atherosclerosis due to the increased apoptosis of macrophages. Moreover, PGE_2-mediated signaling via EP2/EP4 is known to induce matrix metalloproteinase-1 (MMP-1) in monocytes, which could conceivably lead to increased joint damage in arthritis or instability in atherosclerotic plaques (91). In addition, tumor-derived PGE_2 is believed to regulate TAM polarization through EP2 and EP4 receptors, thereby leading to the progression of tumors and explaining the potential antitumor effects of COX-2 inhibitors (92).

Other arachidonic acid derivatives also play detrimental roles in inflammation-driven disease. For example, high levels of the LTB_4 are detected in patients with cystic fibrosis, chronic obstructive pulmonary disease, asthma, and the acute respiratory distress syndrome (ARDS). Similarly, patients with rheumatoid arthritis, multiple sclerosis, inflammatory bowel disease, and psoriasis all produce elevated amounts of LTB_4 (56, 58). Its importance in many immunological disorders is clearly revealed in studies using animals deficient in its receptor (BLT1) and specific LTB_4R antagonists. In a CIA model in mice, BLT1 is shown to be required for the infiltration of inflammatory cells into joints for disease development (93). Similarly, studies in BLT1 knockout mice reveal a role for BLT1 in regulating neutrophil recruitment in inflammatory arthritis, eosinophil migration in drug-induced peritonitis, and *in vivo* homing of Langerhans cells to draining lymph nodes. As such, BLT1 knockout mice showed much reduced severity in many disease models, including autoimmune uveitis, asthma, atherogenesis, experimental autoimmune encephalomyelitis, experimental spinal cord injury, and lung transplantation (58).

Likewise, utilization of specific receptor antagonists has implicated BLT1 in monocyte recruitment and foam cell formation, the migration of eosinophils into the spinal cord in an animal model of multiple sclerosis, and neutrophil migration into the airways and the development of airway hyperresponsiveness in an asthma disease model (58, 59, 94). In a porcine model of shock and ARDS, LTB_4R antagonist significantly reduced syndromes associated with endotoxin shock such as pulmonary arterial hypertension, pulmonary edema, and arterial hypoxemia (95). Moreover, LTB_4R antagonists have also been shown to be effective in prolonging allograft survival in tissue transplantation experiments (96, 97). Interestingly, pathogens are able to exploit lipid signaling via arachidonic acid-derived mediators.

Studies of *Pseudomonas aeruginosa* and *Toxoplasma gondii* show that they encode their own lipoxygenases, causing increasing local levels of the proresolving lipoxin A_4, thereby potentially skewing the immune response to a more conducive proresolving microenvironment (98, 99).

Platelet-activating factor (PAF) is as phospholipid activator produced by a number of cells including platelets, monocytes, macrophages, endothelial cells, and neutrophils in response to pathogens and inflammatory stimuli. Through binding of its receptor, PAFR, PAF stimulates phagocyte chemotaxis and oxidative metabolism and amplifies and propagates the early stages of inflammatory responses, potentially mediating tissue destruction. Indeed, PAF inhibitors or receptor antagonists reduce tissue injury in a number of mouse models challenged with toxins including LPS, paracetamol, carbon tetrachloride, and bleomycin, implying that PAF participates in the pathogenesis of these toxins. Furthermore, the PAFR has been proposed to act as an adhesion receptor for the opportunistic pathogen *Streptococcus pneumoniae* (100); however, whether bacterial adhesion and invasion is mediated solely through airway epithelial PAFR or whether alveolar macrophages are also targets remains to be determined. Other GPCRs involved in transducing signals of lipids are the LPAR (1–6) and S1PR (1–5) family of GPCRs, many of which are expressed by macrophages (101). They bind the lysophosphatidic acids (LPAs) and S1P, which are now recognized as critical regulators of many physiological processes, ranging from cancer and arthritis to atherosclerosis, diabetes, and osteoporosis (102).

PAR RECEPTORS IN DISEASE

To date, many studies have implicated PARs as key mediators/regulators of inflammatory diseases. Both PAR1 and PAR2 expression are prevalent in the synovial lining of rheumatoid arthritis and osteoarthritis patients (103). In a mouse model of antigen-induced arthritis, significantly reduced arthritis severity was observed in PAR1 knockout mice, which secreted reduced levels of proinflammatory mediators including IL-1, IL-6, and MMP-13 (104). Likewise, PAR2 knockout mice were resistant to adjuvant-induced arthritis and exhibited reduced contact hypersensitivity responses (105, 106). On the other hand, the intra-articular injection of PAR2-specific agonists resulted in joint swelling and synovial hyperemia (107). A role for PAR1 and PAR2 in the development of skin and airway inflammation disease models has also been reported, suggesting their involvement in the occurrence of allergy (108, 109). In the progression of gastrointestinal inflammatory disorders such as Crohn's disease, ulcerative colitis, and infectious colitis, both anti-inflammatory and proinflammatory effects were associated with PARs, which showed distinct specific effects on different lymphocyte populations (110, 111). PAR1 and PAR2 are also implicated in the central nervous system inflammation and the progression of neurodegenerative diseases such as experimentally induced autoimmune encephalomyelitis and multiple sclerosis. Interestingly, PARs did not seem to play a role in LPS-induced endotoxemia (112).

ADHESION GPCRs IN MYELOID CELLS

The adhesion GPCRs make up the second-largest group of the human GPCR superfamily and are believed to be evolutionarily ancient molecules conserved from tetrahymena to mammals (113). There are 33 members found in mammals, and they have multiple features that make them distinct from the other classes of GPCRs.

Common Characteristics of the Adhesion GPCRs

The most unique feature of the adhesion GPCRs is the structural composition and formation of the receptors. In general, three distinct domains can be found in the adhesion GPCRs. The long N termini usually contain protein modules with potential cell adhesion functions. This is followed immediately by an evolutionarily conserved autoproteolytic protein fold named the GPCR autoproteolysis-inducing (GAIN) domain. Finally, a typical GPCR-signature seven-transmembrane segment with various lengths of intracellular sequences is located at the C terminus (113, 114).

The adhesion GPCRs are usually cleaved by an autocatalytic proteolytic process at the conserved GPCR proteolysis site (GPS) motif embedded in the GAIN domain (115, 116). Hence, the mature receptor is a noncovalent dimeric complex of the N-terminal fragment (NTF), containing the cell adhesion extracellular domain and the GAIN domains and the seven-transmembrane C-terminal fragment (CTF). Apart from a few exceptions, the GPS autoproteolysis is exclusively found in the adhesion GPCRs and is shown to be essential for the stability, maturation, intracellular trafficking, and function of the receptors (115, 116). In addition to the GPS cleavage, other posttranslational modifications such as glycosylation and further proteolytic processing by sheddases are also noted. In general, the adhesion GPCRs are believed to perform their cellular functions by cell-cell and/or cell-matrix interaction via the NTF and cell signaling through the CTF. As a result, a role in innate as well as adaptive immune responses, with an emphasis in cell-cell/matrix and cell-pathogen interaction, has been predicted for the myeloid adhesion GPCRs.

Interestingly, the majority of the EGF-TM7 subclass of adhesion GPCRs, namely EMR1/F4/80 (discussed in the previous section), EMR2, EMR3, EMR4, and CD97 (8, 117, 118), are restricted to cells of myeloid lineage, including monocytes/macrophages, DCs, and granulocytes. This suggests that the EGF-TM7 receptors are conserved not only genetically but also functionally in myeloid cell biology. Similarly, some other adhesion GPCRs have recently been found to play a role in macrophage function. In this section, the adhesion GPCRs of myeloid cells are described and their functions discussed.

EMR2

Interestingly, EMR2 is found only in humans, and no rodent EMR2 ortholog exists (119). In fact, sequence comparison among the EGF-TM7 receptors showed that EMR2 represents a hybrid between CD97 and EMR3 (119, 120), and as such these receptors are believed to be derived from gene duplication and conversion events during evolution. EMR2 expression is detected in monocytes, macrophages, neutrophils, and DCs. Specifically, $CD16^+$ blood monocytes, macrophages, and $BDCA3^+$ myeloid DCs display the strongest EMR2 signal. Upregulation of EMR2 expression is found during the differentiation and maturation of macrophages *in vitro*, but its expression is reduced during DC maturation (121). In normal tissues, EMR2 is detected in macrophage subpopulations of the skin, spleen, lung, placenta, and tonsil. Macrophages of liver and kidney do not seem to express EMR2. In inflamed tissues, EMR2 is found in subpopulations of macrophages and neutrophils (121). More recent studies showed that foamy macrophages in atherosclerotic vessels and Gaucher cells in spleen strongly express EMR2. In contrast, foam cells in multiple sclerosis brain expressed little if any EMR2, but strong CD97 (122). Certain tumor-infiltrating macrophages in carcinomas expressed strong surface EMR2. Interestingly, by staining breast cancer tissue sections, we recently identified strong EMR2 reactivity within tumor epithelial cells (123). Ligation of EMR2 receptor by a specific mAb has been shown to strongly enhance the inflammatory responses of both neutrophils and macrophage rafts (124, 125). These reports, in conjunction with the identification of the extracellular matrix ligand dermatan sulfate, are highly suggestive of a role in myeloid migration and activation (126).

EMR3 and EMR4

Both EMR3 and EMR4 contain two EGF-like motifs at their N terminus (127, 128). Similar to EMR2, the EMR3 gene is present only in the human but not the mouse genome. The expression of EMR3 is restricted in monocytes/macrophages, myeloid DCs, and mature granulocytes (129). In a survey of the expression profile of the GPCR repertoire during primary human monocyte/macrophage differentiation, EMR3 was found to be highly upregulated in GM-CSF-stimulated macrophages compared to CSF-1-activated macrophages (33). A potential cellular ligand of EMR3 is found on human macrophages and activated neutrophils (127). In contrast to EMR2 and EMR3, the human EMR4 gene is inactivated because of a single nucleotide deletion, hence a pseudogene (128, 130). Mouse EMR4 (also called FIRE) is expressed in monocytes/macrophages and $CD8^-$ DC subsets (131). Specifically, its expression was detected in blood monocytes and some tissue macrophages including peritoneum, spleen (red pulp and marginal zone), and lymph node (subcapsular sinus and paracortex). $CD8^-$ DCs but not $CD8^+$ DCs specifically express mEMR4, which is downregulated upon DC activation (131). Targeting mEMR4 on the surface of immature $CD8^-$ DCs by specific mAb was found to enhance humoral immunity in mice, resulting in an ~100- to 1000-fold increase in antibody production (132). A potential ligand of mEMR4 was found on the B-lymphoma cell line A20 (128).

CD97

CD97 has a highly similar structure to EMR2, containing a total of five EGF-like motifs. As with EMR2, numerous CD97 isoforms are expressed from alternatively spliced transcripts, the largest of which also interacts with dermatan sulfate (126, 133–135). The smallest isoform of CD97 interacts with the complement control protein CD55 (decay-accelerating factor). CD97 has also been shown to interact with several other cellular ligands, $\alpha_5\beta_1/\alpha_v\beta_3$ integrins (136) and CD90 (Thy-1) (137). CD97 was first identified as an early activation marker for T and B lymphocytes, which weakly express CD97 in resting conditions (133, 134). Later studies showed that CD97 is widely expressed on hematopoietic stem and progenitor cells, many types of leukocytes, epithelial cells, and muscle cells, as well as their malignant counterparts (113, 138). Immunohistochemical analyses of normal human tissues revealed abundant CD97 expression in most resident tissue macrophages (139). These include Kupffer cells and periportal histiocytes (liver); alveolar macrophages (lung); epidermal macrophages and Langerhans cells (skin); perivascular macrophages but not microglia (brain); mesangial cells of the glomeruli (kidney); and those of secondary lymphoid organs such as lymph nodes, spleen, tonsil, and mucosa-associated lymphoid tissues. DCs in most of the lymphoid tissues also express CD97, whereas CD97-expressing lymphocytes are mostly restricted to intraepithelial and subepithelial locations. Lymphocytes in paracortical areas, follicles, and germinal centers are mostly CD97 weak or negative.

CD97 is known to function as an immune effector molecule involved in host defense (140, 141). CD97 is upregulated in activated leukocytes to promote adhesion and migration to sites of inflammation (142). In addition, CD97 is shown to regulate granulocyte homeostasis *in vivo*, and enhanced granulopoiesis was observed in mice lacking CD97 or its ligand CD55 (141). Administration of CD97-specific antibodies in animals depleted granulocytes and diminished various inflammatory disorders (143). The interaction of CD97 and CD55 modulates T-cell activation, increasing cell proliferation and cytokine production (144, 145). Thus, it is believed that CD97 on myeloid cells plays a role in certain autoinflammatory diseases, such as rheumatoid arthritis and multiple sclerosis.

BAI1

The brain-specific angiogenesis inhibitor-1 (BAI1) was identified originally in a search of p53-regulated genes (146). Distinct from the EGF-TM7 receptors, BAI1 and two closely related receptors, BAI2 and BAI3, form the subfamily VII of the adhesion GPCRs (113). The BAI1 molecule possesses an Arg-Gly-Asp (RGD) motif, five thrombospondin type-1 repeats (TSRs), a hormone-binding domain, and a GAIN domain within the extracellular domain. The CTF contains a considerably long cytoplasmic tail with a QTEV motif able to interact with PDZ domain-containing proteins. BAP1, a PDZ domain-containing protein, was identified as a binding partner of BAI1 in a yeast two-hybrid screen using BAI1 as bait (147).

Earlier studies showed that BAI1 expression was brain specific and downregulated in glioblastoma cell lines through epigenetic mechanisms (146). However, more-recent analyses have detected BAI1 expression in different cell types and tissues, including many phagocytic cells such as monocytes and macrophages, microglia, gastric phagocytes, and osteoclasts (148–151).

As suggested by its name, many antiangiogenic and antiproliferative activities have been ascribed to BAI1 (146, 152, 153). However, since these initial findings BAI1 has been shown to act as an engulfment receptor on macrophages, binding phosphatidylserine (PS) exposed on apoptotic cells via its TSRs (154). BAI1 promoted the internalization of apoptotic cell corpses by coupling its cytoplasmic tail with the ELMO (engulfment and cell motility)/Dock180/Rac proteins, a conserved signaling apparatus for the efficient disposal of apoptotic cells. Later studies indicated that BAI1 facilitates the clearance of human apoptotic gastric epithelial cells by gastric phagocytes during *H. pylori* infection (148) and that during the engulfment of dying neurons by microglia BAI1 was found to regulate phagosome formation and cargo transport (151). Intriguingly, recognition of PS on apoptotic myoblasts by BAI1 normally expressed on healthy myoblasts provides a contact-dependent signal to promote fusion of healthy myoblasts (155). Whether BAI1 is involved in the formation of multinucleated giant cells found in granulomas, foreign body giant cells, or osteoclasts remains to be dissected. Finally, BAI1 was demonstrated recently to be a pattern recognition receptor for LPS and mediate the binding and phagocytosis of various Gram-negative bacteria by macrophages (150). As such, BAI1 is the only phagocytic adhesion GPCR identified to date.

GPR116 (Ig-Hepta)

GPR116 (Ig-Hepta) is an adhesion GPCR containing two EGF-like motifs, a SEA module, two immunoglobulin-like repeats, and a GAIN domain within its extracellular domain (113). Interestingly, GPR116 undergoes multiple proteolytic processing events during its biosynthesis and maturation process, including two autoproteolytic cleavages at the SEA and GPS motifs and one at the N-terminal part of the EGF-like motif by furin-like protease (156). As a result, the mature receptor contains three protein fragments associated noncovalently on the cell surface.

GPR116 expression is detected mostly in alveolar walls of lung and intercalated cells in the collecting duct of kidney (157–159). GPR116 is believed to be involved in the regulation of lung surfactant homeostasis; however, more recently GPR116-deficient alveolar macrophages were shown to display activated phenotypes with enhanced NF-κB activation and express high levels of inflammatory mediators, lipid hydroperoxides, reactive oxygen species, and MMPs. The GPR116-deficient animals exhibited emphysema-like symptoms with foamy alveolar macrophages accumulating in the lung (160).

REGULATION OF GPCR SIGNALING

Due to the diverse and critical nature of GPCR signaling in macrophages, described above, it is hardly surprisingly that their activity and expression require tight regulation. They are controlled at numerous levels ranging from mRNA degradation via microRNA (miRNA) and receptor desensitization via GRKs and arrestins to the modulation of G protein activity via RGSs.

Although much remains to be discovered regarding the regulation of GPCR expression, a number of studies now show the emerging role of targeting via miRNAs. Examples include miR-181b, which is upregulated in human macrophages upon zymosan challenge. Targeting of the 3′ untranslated region of the ALX/FPR2 gene by miR-181b and the resultant degradation of mRNA suppresses the proresolving effects of its ligands lipoxin A_4 and resolvin D1, thereby reducing phagocytosis and increasing zymozan-induced TNF-α generation (161). The expression of CCR2 is also regulated by miR-146a, altering both proliferation and trafficking of monocytes (162). As well as its expression in macrophages, mi126b can be derived from apoptotic bodies and can also target the adhesion GPCR CD97 (163), suggesting an important potential role in regulating macrophage function.

In addition to the downregulation of GPCR expression via miRNA, receptor activity and signal integration are modulated via the differential expression and regulation of receptor desensitization and downstream signaling components. Dissection of mouse mRNA (www.immgen.org) reveals differential expression of numerous GPCR-associated proteins not only between macrophages and other leukocyte types including DCs but also between peritoneal, bone marrow, and splenic macrophages and Ly6C$^{+/-}$ subsets (164). For example, a number of desensitizing GRKs, including *GRK2*,

GRK5, and *GRK6*, are abundantly expressed in monocytes and macrophages. Mice bearing a GRK2 deletion in myeloid cells exhibit exaggerated inflammatory cytokine/chemokine production and organ damage in response to LPS (165), whereas mouse studies have shown GRK5 to specifically desensitize a number of migratory receptors, including the monocyte chemokine CCR2 toward atherogenic stimuli. Interestingly, in other studies GRK5 did not reduce CXCR4-dependent migration, reinforcing the fact that GPCR-mediated signaling is a highly specific and integrated process. Studies of GRK6 have shown that it facilitates the engulfment of senescent red blood cells by splenic red pulp macrophages and that deficiency results in increased levels of anti–double-stranded DNA antibodies and glomerular deposition of immune complexes in a phenotype resembling that of human systemic lupus erythematosus (166). Whether GRK6 is able to regulate phagocytosis via the macrophage engulfment adhesion GPCR BAI1, and to what extent GRK6 activity is GPCR dependent, remains to be fully investigated. One of the major consequences of phosphorylation of agonist-bound GPCRs by GRKs is the recruitment of an important class of cytosolic proteins, known as arrestins. β-Arrestins classically desensitize GPCRs by blocking G protein interactions and also serve as adaptors that link the receptors to elements of the endocytotic machinery, such as clathrins, which mediate the internalization of numerous macrophage GPCRs (167, 168). Arrestins also mediate both G protein-dependent and -independent signaling pathways and act as critical regulators of inflammation (168). *In vivo* experiments show that β-arrestin-2 negatively regulates inflammatory responses of macrophages recruited to the areas of myocardial infarct (169) and negatively regulates LPS-induced signaling via its production of IL-10 and interaction with p38 mitogen-activated protein kinase. Interestingly, the anti-inflammatory effects of the antidepressant drug fluoxetine on microglial cells have been shown to be dependent on β-arrestin-2 (170).

The type and extent of secondary messenger generation by GPCR signaling is obviously determined by the cell's complement of G proteins. The human genome contains numerous G proteins, and as with other signaling components, mRNA analysis shows that these are differentially expressed by macrophages. For example, macrophages lack the G protein $G\alpha_{i1}$ but express abundant amounts of $G\alpha_{i2}$ and $G\alpha_{i3}$. $G\alpha_{i2}$ deficiency results in a significant defect in leukocyte migration, whereas $G\alpha_{i3}$ has little effect, demonstrating the importance of specific G proteins and the apparent redundancy of others (171). Abundant levels of $G\alpha_{12/13}$ mRNA are also found in macrophage subsets including red pulp splenic macrophages, which may facilitate signaling of receptors, including BAI1, EMR2, CD97, and S1PR, known to couple to these G proteins. Other downstream components of GPCR signaling are also differentially expressed in monocytes and macrophage lineages. One such family is the RGSs, a group of ~30 members that accelerate the intrinsic GTPase activity of G protein, thereby attenuating prolonged GPCR-mediated signaling. Although much remains unknown with regard to the RGS proteins, RGS1, -2 (very high in monocytes) -10, -14, -18, and -19 are enriched in macrophages. Indeed, RGS1 has been found to be upregulated with monocyte-macrophage activation and with atherosclerotic plaque progression and therefore may be important in preventing sustained chemokine signaling in macrophage foam cells within atherosclerotic plaques (172). $Rgs10^{-/-}$ macrophages display dysregulated M1 responses along with blunted M2 alternative activation responses, suggesting that RGS10 plays an important role in determining macrophage activation responses and differentiation (173). Again, as with GPCRs, the GRKs are differentially expressed between mouse tissue macrophage types (44). Most notably, RGS2, -10, and -19 are all highly enriched in microglia, and RGS3 is enriched in splenic macrophages.

CONCLUSION

Historical antibody studies and more recent transcriptome analysis have demonstrated beyond doubt that macrophages express an impressive array of GPCRs. Functional research has shown these receptors to be involved in all aspects of macrophage biology from cellular maturation, polarization, and migration to their physiological roles in the initiation, maintenance, and resolution of inflammation, and engulfment of pathogens and apoptotic cells. GPCRs have also been implicated in the more detrimental aspects of macrophage biology, namely, diseases ranging from cancer and atherosclerosis to autoinflammatory conditions and viral entry. However, the continued deorphanization and molecular dissection of cell signaling mechanisms, along with the regulatory and functional insights of next-generation sequencing and recent advances in gene targeting technologies, will hopefully provide logical therapeutic strategies for the future.

Acknowledgments. *H.-H. Lin's research is supported by grants from the Ministry of Science and Technology, Taiwan (MOST101-2320-B-182-029-MY3), and the Chang Gung Memorial Hospital (CMRPD1C0632-3, CMRPD1A0181-3,* *and CMRPD1D0391-2). M. Stacey's research is supported by grants from the U.K. Research Council (BBSRC-GSK STU100036918).*

Citation. Lin HH, Stacey M. 2016. G protein-coupled receptors in macrophages. Microbiol Spectrum 4(4):MCHD-0028-2016.

References

1. **Venkatakrishnan AJ, Deupi X, Lebon G, Tate CG, Schertler GF, Babu MM.** 2013. Molecular signatures of G-protein-coupled receptors. *Nature* **494:**185–194.
2. **Fredriksson R, Lagerström MC, Lundin LG, Schiöth HB.** 2003. The G-protein-coupled receptors in the human genome form five main families. Phylogenetic analysis, paralogon groups, and fingerprints. *Mol Pharmacol* **63:**1256–1272.
3. **Groot-Kormelink PJ, Fawcett L, Wright PD, Gosling M, Kent TC.** 2012. Quantitative GPCR and ion channel transcriptomics in primary alveolar macrophages and macrophage surrogates. *BMC Immunol* **13:**57. doi: 10.1186/1471-2172-13-57.
4. **Austyn JM, Gordon S.** 1981. F4/80, a monoclonal antibody directed specifically against the mouse macrophage. *Eur J Immunol* **11:**805–815.
5. **Gordon S, Taylor PR.** 2005. Monocyte and macrophage heterogeneity. *Nat Rev Immunol* **5:**953–964.
6. **Taylor PR, Martinez-Pomares L, Stacey M, Lin HH, Brown GD, Gordon S.** 2005. Macrophage receptors and immune recognition. *Annu Rev Immunol* **23:**901–944.
7. **Gordon S, Hamann J, Lin HH, Stacey M.** 2011. F4/80 and the related adhesion-GPCRs. *Eur J Immunol* **41:** 2472–2476.
8. **McKnight AJ, Gordon S.** 1998. The EGF-TM7 family: unusual structures at the leukocyte surface. *J Leukoc Biol* **63:**271–280.
9. **Mantovani A, Sozzani S, Locati M, Allavena P, Sica A.** 2002. Macrophage polarization: tumor-associated macrophages as a paradigm for polarized M2 mononuclear phagocytes. *Trends Immunol* **23:**549–555.
10. **Qian BZ, Pollard JW.** 2010. Macrophage diversity enhances tumor progression and metastasis. *Cell* **141:** 39–51.
11. **Rabinovich GA, Gabrilovich D, Sotomayor EM.** 2007. Immunosuppressive strategies that are mediated by tumor cells. *Annu Rev Immunol* **25:**267–296.
12. **Umemura N, Saio M, Suwa T, Kitoh Y, Bai J, Nonaka K, Ouyang GF, Okada M, Balazs M, Adany R, Shibata T, Takami T.** 2008. Tumor-infiltrating myeloid-derived suppressor cells are pleiotropic-inflamed monocytes/macrophages that bear M1- and M2-type characteristics. *J Leukoc Biol* **83:**1136–1144.
13. **Gordon S, Martinez FO.** 2010. Alternative activation of macrophages: mechanism and functions. *Immunity* 32:593–604.
14. **Martinez FO, Helming L, Gordon S.** 2009. Alternative activation of macrophages: an immunologic functional perspective. *Annu Rev Immunol* **27:**451–483.
15. **Sinha P, Clements VK, Ostrand-Rosenberg S.** 2005. Reduction of myeloid-derived suppressor cells and induction of M1 macrophages facilitate the rejection of established metastatic disease. *J Immunol* **174:**636–645.
16. **Lin HH, Faunce DE, Stacey M, Terajewicz A, Nakamura T, Zhang-Hoover J, Kerley M, Mucenski ML, Gordon S, Stein-Streilein J.** 2005. The macrophage F4/80 receptor is required for the induction of antigen-specific efferent regulatory T cells in peripheral tolerance. *J Exp Med* **201:**1615–1625.
17. **Warschkau H, Kiderlen AF.** 1999. A monoclonal antibody directed against the murine macrophage surface molecule F4/80 modulates natural immune response to *Listeria monocytogenes*. *J Immunol* **163:**3409–3416.
18. **van den Berg TK, Kraal G.** 2005. A function for the macrophage F4/80 molecule in tolerance induction. *Trends Immunol* **26:**506–509.
19. **Legrand F, Tomasevic N, Simakova O, Lee CC, Wang Z, Raffeld M, Makiya MA, Palath V, Leung J, Baer M, Yarranton G, Maric I, Bebbington C, Klion AD.** 2014. The eosinophil surface receptor epidermal growth factor-like module containing mucin-like hormone receptor 1 (EMR1): a novel therapeutic target for eosinophilic disorders. *J Allergy Clin Immunol* **133:**1439–1447.
20. **Hamann J, Koning N, Pouwels W, Ulfman LH, van Eijk M, Stacey M, Lin HH, Gordon S, Kwakkenbos MJ.** 2007. EMR1, the human homolog of F4/80, is an eosinophil-specific receptor. *Eur J Immunol* **37:**2797–2802.
21. **Lattin JE, Schroder K, Su AI, Walker JR, Zhang J, Wiltshire T, Saijo K, Glass CK, Hume DA, Kellie S, Sweet MJ.** 2008. Expression analysis of G protein-coupled receptors in mouse macrophages. *Immunome Res* **4:**5. doi:10.1186/1745-7580-4-5.
22. **Suzuki M, Takaishi S, Nagasaki M, Onozawa Y, Iino I, Maeda H, Komai T, Oda T.** 2013. Medium-chain fatty acid-sensing receptor, GPR84, is a proinflammatory receptor. *J Biol Chem* **288:**10684–10691.
23. **Wang J, Wu X, Simonavicius N, Tian H, Ling L.** 2006. Medium-chain fatty acids as ligands for orphan G protein-coupled receptor GPR84. *J Biol Chem* **281:** 34457–34464.
24. **Maslowski KM, Mackay CR.** 2011. Diet, gut microbiota and immune responses. *Nat Immunol* **12:**5–9.
25. **Ben Yebdri F, Kukulski F, Tremblay A, Sévigny J.** 2009. Concomitant activation of $P2Y_2$ and $P2Y_6$ receptors on monocytes is required for TLR1/2-induced neutrophil migration by regulating IL-8 secretion. *Eur J Immunol* **39:**2885–2894.
26. **Kim B, Jeong HK, Kim JH, Lee SY, Jou I, Joe EH.** 2011. Uridine 5′-diphosphate induces chemokine expression in microglia and astrocytes through activation of the P2Y6 receptor. *J Immunol* **186:**3701–3709.
27. **Li R, Tan B, Yan Y, Ma X, Zhang N, Zhang Z, Liu M, Qian M, Du B.** 2014. Extracellular UDP and $P2Y_6$ function as a danger signal to protect mice from vesicular stomatitis virus infection through an increase in IFN-β production. *J Immunol* **193:**4515–4526.

28. **Martinez FO, Gordon S, Locati M, Mantovani A.** 2006. Transcriptional profiling of the human monocyte-to-macrophage differentiation and polarization: new molecules and patterns of gene expression. *J Immunol* **177:** 7303–7311.

29. **Welihinda AA, Amento EP.** 2014. Positive allosteric modulation of the adenosine A_{2a} receptor attenuates inflammation. *J Inflamm (Lond)* **11:**37. doi:10.1186/s12950-014-0037-0.

30. **Dubois-Colas N, Petit-Jentreau L, Barreiro LB, Durand S, Soubigou G, Lecointe C, Klibi J, Rezaï K, Lokiec F, Coppée JY, Gicquel B, Tailleux L.** 2014. Extracellular adenosine triphosphate affects the response of human macrophages infected with *Mycobacterium tuberculosis*. *J Infect Dis* **210:**824–833.

31. **Cekic C, Day YJ, Sag D, Linden J.** 2014. Myeloid expression of adenosine A_{2A} receptor suppresses T and NK cell responses in the solid tumor microenvironment. *Cancer Res* **74:**7250–7259.

32. **Csóka B, Selmeczy Z, Koscsó B, Németh ZH, Pacher P, Murray PJ, Kepka-Lenhart D, Morris SM Jr, Gause WC, Leibovich SJ, Haskó G.** 2012. Adenosine promotes alternative macrophage activation *via* A_{2A} and A_{2B} receptors. *FASEB J* **26:**376–386.

33. **Hohenhaus DM, Schaale K, Le Cao KA, Seow V, Iyer A, Fairlie DP, Sweet MJ.** 2013. An mRNA atlas of G protein-coupled receptor expression during primary human monocyte/macrophage differentiation and lipopolysaccharide-mediated activation identifies targetable candidate regulators of inflammation. *Immunobiology* **218:**1345–1353.

34. **Mills E, O'Neill LA.** 2014. Succinate: a metabolic signal in inflammation. *Trends Cell Biol* **24:**313–320.

35. **Du B, Luo W, Li R, Tan B, Han H, Lu X, Li D, Qian M, Zhang D, Zhao Y, Liu M.** 2013. Lgr4/Gpr48 negatively regulates TLR2/4-associated pattern recognition and innate immunity by targeting CD14 expression. *J Biol Chem* **288:**15131–15141.

36. **Liu Y, Chen K, Wang C, Gong W, Yoshimura T, Liu M, Wang JM.** 2013. Cell surface receptor FPR2 promotes antitumor host defense by limiting M2 polarization of macrophages. *Cancer Res* **73:**550–560.

37. **Oghumu S, Varikuti S, Terrazas C, Kotov D, Nasser MW, Powell CA, Ganju RK, Satoskar AR.** 2014. CXCR3 deficiency enhances tumor progression by promoting macrophage M2 polarization in a murine breast cancer model. *Immunology* **143:**109–119.

38. **Eruslanov E, Daurkin I, Ortiz J, Vieweg J, Kusmartsev S.** 2010. Pivotal advance: tumor-mediated induction of myeloid-derived suppressor cells and M2-polarized macrophages by altering intracellular PGE_2 catabolism in myeloid cells. *J Leukoc Biol* **88:**839–848.

39. **Heusinkveld M, de Vos van Steenwijk PJ, Goedemans R, Ramwadhdoebe TH, Gorter A, Welters MJ, van Hall T, van der Burg SH.** 2011. M2 macrophages induced by prostaglandin E_2 and IL-6 from cervical carcinoma are switched to activated M1 macrophages by $CD4^+$ Th1 cells. *J Immunol* **187:**1157–1165.

40. **MacKenzie KF, Clark K, Naqvi S, McGuire VA, Nöehren G, Kristariyanto Y, van den Bosch M, Mudaliar M, McCarthy PC, Pattison MJ, Pedrioli PG, Barton GJ, Toth R, Prescott A, Arthur JS.** 2013. PGE_2 induces macrophage IL-10 production and a regulatory-like phenotype via a protein kinase A-SIK-CRTC3 pathway. *J Immunol* **190:**565–577.

41. **Bordon-Graciani AP, Dias-Melicio LA, Acorci-Valério MJ, Araujo JP Jr, de Campos Soares AM.** 2012. Inhibitory effect of PGE_2 on the killing of *Paracoccidioides brasiliensis* by human monocytes can be reversed by cellular activation with cytokines. *Med Mycol* **50:**726–734.

42. **Park SJ, Lee KP, Kang S, Lee J, Sato K, Chung HY, Okajima F, Im DS.** 2014. Sphingosine 1-phosphate induced anti-atherogenic and atheroprotective M2 macrophage polarization through IL-4. *Cell Signal* **26:** 2249–2258.

43. **Koizumi S, Ohsawa K, Inoue K, Kohsaka S.** 2013. Purinergic receptors in microglia: functional modal shifts of microglia mediated by P2 and P1 receptors. *Glia* **61:**47–54.

44. **Lavin Y, Winter D, Blecher-Gonen R, David E, Keren-Shaul H, Merad M, Jung S, Amit I.** 2014. Tissue-resident macrophage enhancer landscapes are shaped by the local microenvironment. *Cell* **159:**1312–1326.

45. **Klos A, Wende E, Wareham KJ, Monk PN.** 2013. International Union of Basic and Clinical Pharmacology. [corrected]. LXXXVII. Complement peptide C5a, C4a, and C3a receptors. *Pharmacol Rev* **65:**500–543.

46. **Rabiet MJ, Huet E, Boulay F.** 2007. The *N*-formyl peptide receptors and the anaphylatoxin C5a receptors: an overview. *Biochimie* **89:**1089–1106.

47. **Li R, Coulthard LG, Wu MC, Taylor SM, Woodruff TM.** 2013. C5L2: a controversial receptor of complement anaphylatoxin, C5a. *FASEB J* **27:**855–864.

48. **Guo Q, Subramanian H, Gupta K, Ali H.** 2011. Regulation of C3a receptor signaling in human mast cells by G protein coupled receptor kinases. *PLoS One* **6:**e22559. doi:10.1371/journal.pone.0022559.

49. **Vibhuti A, Gupta K, Subramanian H, Guo Q, Ali H.** 2011. Distinct and shared roles of β-arrestin-1 and β-arrestin-2 on the regulation of C3a receptor signaling in human mast cells. *PLoS One* **6:**e19585. doi:10.1371/journal.pone.0019585.

50. **Subramanian H, Gupta K, Ali H.** 2012. Roles for NHERF1 and NHERF2 on the regulation of C3a receptor signaling in human mast cells. *PLoS One* **7:**e51355. doi:10.1371/journal.pone.0051355.

51. **Karsten CM, Köhl J.** 2012. The immunoglobulin, IgG Fc receptor and complement triangle in autoimmune diseases. *Immunobiology* **217:**1067–1079.

52. **Monk PN, Scola AM, Madala P, Fairlie DP.** 2007. Function, structure and therapeutic potential of complement C5a receptors. *Br J Pharmacol* **152:**429–448.

53. **Ye RD, Boulay F, Wang JM, Dahlgren C, Gerard C, Parmentier M, Serhan CN, Murphy PM.** 2009. International Union of Basic and Clinical Pharmacology. LXXIII. Nomenclature for the formyl peptide receptor (FPR) family. *Pharmacol Rev* **61:**119–161.

54. **Migeotte I, Communi D, Parmentier M.** 2006. Formyl peptide receptors: a promiscuous subfamily of G protein-

coupled receptors controlling immune responses. *Cytokine Growth Factor Rev* **17:**501–519.

55. **Bloes DA, Kretschmer D, Peschel A.** 2015. Enemy attraction: bacterial agonists for leukocyte chemotaxis receptors. *Nat Rev Microbiol* **13:**95–104.
56. **Lattin J, Zidar DA, Schroder K, Kellie S, Hume DA, Sweet MJ.** 2007. G-protein-coupled receptor expression, function, and signaling in macrophages. *J Leukoc Biol* **82:**16–32.
57. **Samuelsson B.** 1983. Leukotrienes: mediators of immediate hypersensitivity reactions and inflammation. *Science* **220:**568–575.
58. **Yokomizo T.** 2011. Leukotriene B_4 receptors: novel roles in immunological regulations. *Adv Enzyme Regul* **51:**59–64.
59. **Tager AM, Luster AD.** 2003. BLT1 and BLT2: the leukotriene B_4 receptors. *Prostaglandins Leukot Essent Fatty Acids* **69:**123–134.
60. **van Pelt JP, de Jong EM, van Erp PE, Mitchell MI, Marder P, Spaethe SM, van Hooijdonk CA, Kuijpers AL, van de Kerkhof PC.** 1997. The regulation of CD11b integrin levels on human blood leukocytes and leukotriene B_4-stimulated skin by a specific leukotriene B_4 receptor antagonist (LY293111). *Biochem Pharmacol* **53:** 1005–1012.
61. **Dahlén SE, Björk J, Hedqvist P, Arfors KE, Hammarström S, Lindgren JA, Samuelsson B.** 1981. Leukotrienes promote plasma leakage and leukocyte adhesion in postcapillary venules: *in vivo* effects with relevance to the acute inflammatory response. *Proc Natl Acad Sci U S A* **78:**3887–3891.
62. **Lundeen KA, Sun B, Karlsson L, Fourie AM.** 2006. Leukotriene B_4 receptors BLT1 and BLT2: expression and function in human and murine mast cells. *J Immunol* **177:**3439–3447.
63. **Flynn MA, Qiao M, Garcia C, Dallas M, Bonewald LF.** 1999. Avian osteoclast cells are stimulated to resorb calcified matrices by and possess receptors for leukotriene B_4. *Calcif Tissue Int* **64:**154–159.
64. **Yokomizo T, Izumi T, Chang K, Takuwa Y, Shimizu T.** 1997. A G-protein-coupled receptor for leukotriene B_4 that mediates chemotaxis. *Nature* **387:**620–624.
65. **Yokomizo T, Kato K, Terawaki K, Izumi T, Shimizu T.** 2000. A second leukotriene B_4 receptor, BLT2. A new therapeutic target in inflammation and immunological disorders. *J Exp Med* **192:**421–432.
66. **Gaudreau R, Le Gouill C, Métaoui S, Lemire S, Stankovà J, Rola-Pleszczynski M.** 1998. Signalling through the leukotriene B_4 receptor involves both α_i and α_{16}, but not α_q or α_{11} G-protein subunits. *Biochem J* **335:**15–18.
67. **Serezani CH, Lewis C, Jancar S, Peters-Golden M.** 2011. Leukotriene B_4 amplifies NF-κB activation in mouse macrophages by reducing SOCS1 inhibition of MyD88 expression. *J Clin Invest* **121:**671–682.
68. **Serezani CH, Kane S, Collins L, Morato-Marques M, Osterholzer JJ, Peters-Golden M.** 2012. Macrophage dectin-1 expression is controlled by leukotriene B_4 via a GM-CSF/PU.1 axis. *J Immunol* **189:**906–915.
69. **Serezani CH, Aronoff DM, Sitrin RG, Peters-Golden M.** 2009. FcγRI ligation leads to a complex with BLT1 in lipid rafts that enhances rat lung macrophage antimicrobial functions. *Blood* **114:**3316–3324.
70. **Shpacovitch V, Feld M, Hollenberg MD, Luger TA, Steinhoff M.** 2008. Role of protease-activated receptors in inflammatory responses, innate and adaptive immunity. *J Leukoc Biol* **83:**1309–1322.
71. **Macfarlane SR, Seatter MJ, Kanke T, Hunter GD, Plevin R.** 2001. Proteinase-activated receptors. *Pharmacol Rev* **53:**245–282.
72. **Suo Z, Wu M, Ameenuddin S, Anderson HE, Zoloty JE, Citron BA, Andrade-Gordon P, Festoff BW.** 2002. Participation of protease-activated receptor-1 in thrombin-induced microglial activation. *J Neurochem* **80:**655–666.
73. **Colognato R, Slupsky JR, Jendrach M, Burysek L, Syrovets T, Simmet T.** 2003. Differential expression and regulation of protease-activated receptors in human peripheral monocytes and monocyte-derived antigen-presenting cells. *Blood* **102:**2645–2652.
74. **Johansson U, Lawson C, Dabare M, Syndercombe-Court D, Newland AC, Howells GL, Macey MG.** 2005. Human peripheral blood monocytes express protease receptor-2 and respond to receptor activation by production of IL-6, IL-8, and IL-1β. *J Leukoc Biol* **78:** 967–975.
75. **Steven R, Crilly A, Lockhart JC, Ferrell WR, McInnes IB.** 2013. Proteinase-activated receptor-2 modulates human macrophage differentiation and effector function. *Innate Immun* **19:**663–672.
76. **Headland SE, Norling LV.** 2015. The resolution of inflammation: principles and challenges. *Semin Immunol* **27:**149–160.
77. **Serhan CN, Chiang N, Van Dyke TE.** 2008. Resolving inflammation: dual anti-inflammatory and pro-resolution lipid mediators. *Nat Rev Immunol* **8:**349–361.
78. **Serhan CN.** 2014. Pro-resolving lipid mediators are leads for resolution physiology. *Nature* **510:**92–101.
79. **Boring L, Gosling J, Cleary M, Charo IF.** 1998. Decreased lesion formation in CCR2$^{-/-}$ mice reveals a role for chemokines in the initiation of atherosclerosis. *Nature* **394:**894–897.
80. **Yan Q, Sun L, Zhu Z, Wang L, Li S, Ye RD.** 2014. Jmjd3-mediated epigenetic regulation of inflammatory cytokine gene expression in serum amyloid A-stimulated macrophages. *Cell Signal* **26:**1783–1791.
81. **Noy R, Pollard JW.** 2014. Tumor-associated macrophages: from mechanisms to therapy. *Immunity* **41:** 49–61.
82. **Sodhi A, Montaner S, Gutkind JS.** 2004. Viral hijacking of G-protein-coupled-receptor signalling networks. *Nat Rev Mol Cell Biol* **5:**998–1012.
83. **Wu MC, Brennan FH, Lynch JP, Mantovani S, Phipps S, Wetsel RA, Ruitenberg MJ, Taylor SM, Woodruff TM.** 2013. The receptor for complement component C3a mediates protection from intestinal ischemia-reperfusion injuries by inhibiting neutrophil mobilization. *Proc Natl Acad Sci U S A* **110:**9439–9444.

84. **Grant EP, Picarella D, Burwell T, Delaney T, Croci A, Avitahl N, Humbles AA, Gutierrez-Ramos JC, Briskin M, Gerard C, Coyle AJ.** 2002. Essential role for the C5a receptor in regulating the effector phase of synovial infiltration and joint destruction in experimental arthritis. *J Exp Med* **196:**1461–1471.

85. **Banda NK, Hyatt S, Antonioli AH, White JT, Glogowska M, Takahashi K, Merkel TJ, Stahl GL, Mueller-Ortiz S, Wetsel R, Arend WP, Holers VM.** 2012. Role of C3a receptors, C5a receptors, and complement protein C6 deficiency in collagen antibody-induced arthritis in mice. *J Immunol* **188:**1469–1478.

86. **Drouin SM, Corry DB, Hollman TJ, Kildsgaard J, Wetsel RA.** 2002. Absence of the complement anaphylatoxin C3a receptor suppresses Th2 effector functions in a murine model of pulmonary allergy. *J Immunol* **169:** 5926–5933.

87. **Kim AH, Dimitriou ID, Holland MC, Mastellos D, Mueller YM, Altman JD, Lambris JD, Katsikis PD.** 2004. Complement C5a receptor is essential for the optimal generation of antiviral $CD8^+$ T cell responses. *J Immunol* **173:**2524–2529.

88. **Girardi G, Berman J, Redecha P, Spruce L, Thurman JM, Kraus D, Hollmann TJ, Casali P, Caroll MC, Wetsel RA, Lambris JD, Holers VM, Salmon JE.** 2003. Complement C5a receptors and neutrophils mediate fetal injury in the antiphospholipid syndrome. *J Clin Invest* **112:** 1644–1654.

89. **de Haas CJ, Veldkamp KE, Peschel A, Weerkamp F, Van Wamel WJ, Heezius EC, Poppelier MJ, Van Kessel KP, van Strijp JA.** 2004. Chemotaxis inhibitory protein of *Staphylococcus aureus*, a bacterial antiinflammatory agent. *J Exp Med* **199:**687–695.

90. **McCoy JM, Wicks JR, Audoly LP.** 2002. The role of prostaglandin E2 receptors in the pathogenesis of rheumatoid arthritis. *J Clin Invest* **110:**651–658.

91. **Foudi N, Gomez I, Benyahia C, Longrois D, Norel X.** 2012. Prostaglandin E_2 receptor subtypes in human blood and vascular cells. *Eur J Pharmacol* **695:**1–6.

92. **Kambayashi T, Alexander HR, Fong M, Strassmann G.** 1995. Potential involvement of IL-10 in suppressing tumor-associated macrophages. Colon-26-derived prostaglandin E_2 inhibits TNF-α release via a mechanism involving IL-10. *J Immunol* **154:**3383–3390.

93. **Shao WH, Del Prete A, Bock CB, Haribabu B.** 2006. Targeted disruption of leukotriene B_4 receptors BLT1 and BLT2: a critical role for BLT1 in collagen-induced arthritis in mice. *J Immunol* **176:**6254–6261.

94. **Saiwai H, Ohkawa Y, Yamada H, Kumamaru H, Harada A, Okano H, Yokomizo T, Iwamoto Y, Okada S.** 2010. The LTB4-BLT1 axis mediates neutrophil infiltration and secondary injury in experimental spinal cord injury. *Am J Pathol* **176:**2352–2366.

95. **Fink MP, O'Sullivan BP, Menconi MJ, Wollert PS, Wang H, Youssef ME, Fleisch JH.** 1993. A novel leukotriene B4-receptor antagonist in endotoxin shock: a prospective, controlled trial in a porcine model. *Crit Care Med* **21:**1825–1837.

96. **Weringer EJ, Perry BD, Sawyer PS, Gilman SC, Showell HJ.** 1999. Antagonizing leukotriene B4 receptors delays cardiac allograft rejection in mice. *Transplantation* **67:** 808–815.

97. **Ii T, Izumi R, Shimizu K.** 1996. The immunosuppressive effects of a leukotriene B4 receptor antagonist on liver allotransplantation in rats. *Surg Today* **26:**419–426.

98. **Vance RE, Hong S, Gronert K, Serhan CN, Mekalanos JJ.** 2004. The opportunistic pathogen *Pseudomonas aeruginosa* carries a secretable arachidonate 15-lipoxygenase. *Proc Natl Acad Sci U S A* **101:**2135–2139.

99. **Bannenberg GL, Aliberti J, Hong S, Sher A, Serhan C.** 2004. Exogenous pathogen and plant 15-lipoxygenase initiate endogenous lipoxin A_4 biosynthesis. *J Exp Med* **199:**515–523.

100. **Iovino F, Brouwer MC, van de Beek D, Molema G, Bijlsma JJ.** 2013. Signalling or binding: the role of the platelet-activating factor receptor in invasive pneumococcal disease. *Cell Microbiol* **15:**870–881.

101. **Hornuss C, Hammermann R, Fuhrmann M, Juergens UR, Racké K.** 2001. Human and rat alveolar macrophages express multiple EDG receptors. *Eur J Pharmacol* **429:**303–308.

102. **Blaho VA, Hla T.** 2011. Regulation of mammalian physiology, development, and disease by the sphingosine 1-phosphate and lysophosphatidic acid receptors. *Chem Rev* **111:**6299–6320.

103. **Cirino G, Napoli C, Bucci M, Cicala C.** 2000. Inflammation-coagulation network: are serine protease receptors the knot? *Trends Pharmacol Sci* **21:**170–172.

104. **Yang YH, Hall P, Little CB, Fosang AJ, Milenkovski G, Santos L, Xue J, Tipping P, Morand EF.** 2005. Reduction of arthritis severity in protease-activated receptor-deficient mice. *Arthritis Rheum* **52:**1325–1332.

105. **Lindner JR, Kahn ML, Coughlin SR, Sambrano GR, Schauble E, Bernstein D, Foy D, Hafezi-Moghadam A, Ley K.** 2000. Delayed onset of inflammation in protease-activated receptor-2-deficient mice. *J Immunol* **165:** 6504–6510.

106. **Ferrell WR, Lockhart JC, Kelso EB, Dunning L, Plevin R, Meek SE, Smith AJ, Hunter GD, McLean JS, McGarry F, Ramage R, Jiang L, Kanke T, Kawagoe J.** 2003. Essential role for proteinase-activated receptor-2 in arthritis. *J Clin Invest* **111:**35–41.

107. **Seeliger S, Derian CK, Vergnolle N, Bunnett NW, Nawroth R, Schmelz M, Von Der Weid PY, Buddenkotte J, Sunderkötter C, Metze D, Andrade-Gordon P, Harms E, Vestweber D, Luger TA, Steinhoff M.** 2003. Proinflammatory role of proteinase-activated receptor-2 in humans and mice during cutaneous inflammation *in vivo*. *FASEB J* **17:**1871–1885.

108. **Reed CE, Kita H.** 2004. The role of protease activation of inflammation in allergic respiratory diseases. *J Allergy Clin Immunol* **114:**997–1008, quiz 1009.

109. **Lan RS, Stewart GA, Henry PJ.** 2002. Role of protease-activated receptors in airway function: a target for therapeutic intervention? *Pharmacol Ther* **95:**239–257.

110. **Hansen KK, Sherman PM, Cellars L, Andrade-Gordon P, Pan Z, Baruch A, Wallace JL, Hollenberg MD,**

Vergnolle N. 2005. A major role for proteolytic activity and proteinase-activated receptor-2 in the pathogenesis of infectious colitis. *Proc Natl Acad Sci U S A* **102:** 8363–8368.

111. **Vergnolle N, Cellars L, Mencarelli A, Rizzo G, Swaminathan S, Beck P, Steinhoff M, Andrade-Gordon P, Bunnett NW, Hollenberg MD, Wallace JL, Cirino G, Fiorucci S.** 2004. A role for proteinase-activated receptor-1 in inflammatory bowel diseases. *J Clin Invest* **114:** 1444–1456.

112. **Pawlinski R, Pedersen B, Schabbauer G, Tencati M, Holscher T, Boisvert W, Andrade-Gordon P, Frank RD, Mackman N.** 2004. Role of tissue factor and protease-activated receptors in a mouse model of endotoxemia. *Blood* **103:**1342–1347.

113. **Hamann J, Aust G, Araç D, Engel FB, Formstone C, Fredriksson R, Hall RA, Harty BL, Kirchhoff C, Knapp B, Krishnan A, Liebscher I, Lin HH, Martinelli DC, Monk KR, Peeters MC, Piao X, Prömel S, Schöneberg T, Schwartz TW, Singer K, Stacey M, Ushkaryov YA, Vallon M, Wolfrum U, Wright MW, Xu L, Langenhan T, Schiöth HB.** 2015. International Union of Basic and Clinical Pharmacology. XCIV. Adhesion G protein-coupled receptors. *Pharmacol Rev* **67:**338–367.

114. **Yona S, Lin HH, Siu WO, Gordon S, Stacey M.** 2008. Adhesion-GPCRs: emerging roles for novel receptors. *Trends Biochem Sci* **33:**491–500.

115. **Lin HH, Stacey M, Yona S, Chang GW.** 2010. GPS proteolytic cleavage of adhesion-GPCRs. *Adv Exp Med Biol* **706:**49–58.

116. **Prömel S, Langenhan T, Araç D.** 2013. Matching structure with function: the GAIN domain of adhesion-GPCR and PKD1-like proteins. *Trends Pharmacol Sci* **34:**470–478.

117. **McKnight AJ, Gordon S.** 1996. EGF-TM7: a novel subfamily of seven-transmembrane-region leukocyte cell-surface molecules. *Immunol Today* **17:**283–287.

118. **Stacey M, Lin HH, Gordon S, McKnight AJ.** 2000. LNB-TM7, a group of seven-transmembrane proteins related to family-B G-protein-coupled receptors. *Trends Biochem Sci* **25:**284–289.

119. **Kwakkenbos MJ, Matmati M, Madsen O, Pouwels W, Wang Y, Bontrop RE, Heidt PJ, Hoek RM, Hamann J.** 2006. An unusual mode of concerted evolution of the EGF-TM7 receptor chimera EMR2. *FASEB J* **20:**2582–2584.

120. **Kwakkenbos MJ, Kop EN, Stacey M, Matmati M, Gordon S, Lin HH, Hamann J.** 2004. The EGF-TM7 family: a postgenomic view. *Immunogenetics* **55:**655–666.

121. **Chang GW, Davies JQ, Stacey M, Yona S, Bowdish DM, Hamann J, Chen TC, Lin CY, Gordon S, Lin HH.** 2007. CD312, the human adhesion-GPCR EMR2, is differentially expressed during differentiation, maturation, and activation of myeloid cells. *Biochem Biophys Res Commun* **353:**133–138.

122. **van Eijk M, Aust G, Brouwer MS, van Meurs M, Voerman JS, Dijke IE, Pouwels W, Sändig I, Wandel E, Aerts JM, Boot RG, Laman JD, Hamann J.** 2010. Differential expression of the EGF-TM7 family members CD97 and EMR2 in lipid-laden macrophages in atherosclerosis, multiple sclerosis and Gaucher disease. *Immunol Lett* **129:**64–71.

123. **Davies JQ, Lin HH, Stacey M, Yona S, Chang GW, Gordon S, Hamann J, Campo L, Han C, Chan P, Fox SB.** 2011. Leukocyte adhesion-GPCR EMR2 is aberrantly expressed in human breast carcinomas and is associated with patient survival. *Oncol Rep* **25:**619–627.

124. **Yona S, Lin HH, Dri P, Davies JQ, Hayhoe RP, Lewis SM, Heinsbroek SE, Brown KA, Perretti M, Hamann J, Treacher DF, Gordon S, Stacey M.** 2008. Ligation of the adhesion-GPCR EMR2 regulates human neutrophil function. *FASEB J* **22:**741–751.

125. **Huang YS, Chiang NY, Hu CH, Hsiao CC, Cheng KF, Tsai WP, Yona S, Stacey M, Gordon S, Chang GW, Lin HH.** 2012. Activation of myeloid cell-specific adhesion class G protein-coupled receptor EMR2 via ligation-induced translocation and interaction of receptor subunits in lipid raft microdomains. *Mol Cell Biol* **32:** 1408–1420.

126. **Stacey M, Chang GW, Davies JQ, Kwakkenbos MJ, Sanderson RD, Hamann J, Gordon S, Lin HH.** 2003. The epidermal growth factor-like domains of the human EMR2 receptor mediate cell attachment through chondroitin sulfate glycosaminoglycans. *Blood* **102:**2916–2924.

127. **Stacey M, Lin HH, Hilyard KL, Gordon S, McKnight AJ.** 2001. Human epidermal growth factor (EGF) module-containing mucin-like hormone receptor 3 is a new member of the EGF-TM7 family that recognizes a ligand on human macrophages and activated neutrophils. *J Biol Chem* **276:**18863–18870.

128. **Stacey M, Chang GW, Sanos SL, Chittenden LR, Stubbs L, Gordon S, Lin HH.** 2002. EMR4, a novel epidermal growth factor (EGF)-TM7 molecule up-regulated in activated mouse macrophages, binds to a putative cellular ligand on B lymphoma cell line A20. *J Biol Chem* **277:** 29283–29293.

129. **Matmati M, Pouwels W, van Bruggen R, Jansen M, Hoek RM, Verhoeven AJ, Hamann J.** 2007. The human EGF-TM7 receptor EMR3 is a marker for mature granulocytes. *J Leukoc Biol* **81:**440–448.

130. **Hamann J, Kwakkenbos MJ, de Jong EC, Heus H, Olsen AS, van Lier RA.** 2003. Inactivation of the EGF-TM7 receptor EMR4 after the Pan-Homo divergence. *Eur J Immunol* **33:**1365–1371.

131. **Caminschi I, Lucas KM, O'Keeffe MA, Hochrein H, Laâbi Y, Köntgen F, Lew AM, Shortman K, Wright MD.** 2001. Molecular cloning of F4/80-like-receptor, a seven-span membrane protein expressed differentially by dendritic cell and monocyte-macrophage subpopulations. *J Immunol* **167:**3570–3576.

132. **Corbett AJ, Caminschi I, McKenzie BS, Brady JL, Wright MD, Mottram PL, Hogarth PM, Hodder AN, Zhan Y, Tarlinton DM, Shortman K, Lew AM.** 2005. Antigen delivery via two molecules on the $CD8^-$ dendritic cell subset induces humoral immunity in the absence of conventional "danger." *Eur J Immunol* **35:** 2815–2825.

133. Gray JX, Haino M, Roth MJ, Maguire JE, Jensen PN, Yarme A, Stetler-Stevenson MA, Siebenlist U, Kelly K. 1996. CD97 is a processed, seven-transmembrane, heterodimeric receptor associated with inflammation. *J Immunol* **157**:5438–5447.

134. Hamann J, Eichler W, Hamann D, Kerstens HM, Poddighe PJ, Hoovers JM, Hartmann E, Strauss M, van Lier RA. 1995. Expression cloning and chromosomal mapping of the leukocyte activation antigen CD97, a new seven-span transmembrane molecule of the secretion receptor superfamily with an unusual extracellular domain. *J Immunol* **155**:1942–1950.

135. Kwakkenbos MJ, Pouwels W, Matmati M, Stacey M, Lin HH, Gordon S, van Lier RA, Hamann J. 2005. Expression of the largest CD97 and EMR2 isoforms on leukocytes facilitates a specific interaction with chondroitin sulfate on B cells. *J Leukoc Biol* **77**:112–119.

136. Wang T, Ward Y, Tian L, Lake R, Guedez L, Stetler-Stevenson WG, Kelly K. 2005. CD97, an adhesion receptor on inflammatory cells, stimulates angiogenesis through binding integrin counterreceptors on endothelial cells. *Blood* **105**:2836–2844.

137. Wandel E, Saalbach A, Sittig D, Gebhardt C, Aust G. 2012. Thy-1 (CD90) is an interacting partner for CD97 on activated endothelial cells. *J Immunol* **188**:1442–1450.

138. Aust G, Steinert M, Schütz A, Boltze C, Wahlbuhl M, Hamann J, Wobus M. 2002. CD97, but not its closely related EGF-TM7 family member EMR2, is expressed on gastric, pancreatic, and esophageal carcinomas. *Am J Clin Pathol* **118**:699–707.

139. Jaspars LH, Vos W, Aust G, Van Lier RA, Hamann J. 2001. Tissue distribution of the human CD97 EGF-TM7 receptor. *Tissue Antigens* **57**:325–331.

140. Veninga H, Becker S, Hoek RM, Wobus M, Wandel E, van der Kaa J, van der Valk M, de Vos AF, Haase H, Owens B, van der Poll T, van Lier RA, Verbeek JS, Aust G, Hamann J. 2008. Analysis of CD97 expression and manipulation: antibody treatment but not gene targeting curtails granulocyte migration. *J Immunol* **181**:6574–6583.

141. Veninga H, de Groot DM, McCloskey N, Owens BM, Dessing MC, Verbeek JS, Nourshargh S, van Eenennaam H, Boots AM, Hamann J. 2011. CD97 antibody depletes granulocytes in mice under conditions of acute inflammation via a Fc receptor-dependent mechanism. *J Leukoc Biol* **89**:413–421.

142. Leemans JC, te Velde AA, Florquin S, Bennink RJ, de Bruin K, van Lier RA, van der Poll T, Hamann J. 2004. The epidermal growth factor-seven transmembrane (EGF-TM7) receptor CD97 is required for neutrophil migration and host defense. *J Immunol* **172**:1125–1131.

143. Hamann J, Veninga H, de Groot DM, Visser L, Hofstra CL, Tak PP, Laman JD, Boots AM, van Eenennaam H. 2010. CD97 in leukocyte trafficking. *Adv Exp Med Biol* **706**:128–137.

144. Abbott RJ, Spendlove I, Roversi P, Fitzgibbon H, Knott V, Teriete P, McDonnell JM, Handford PA, Lea SM. 2007. Structural and functional characterization of a novel T cell receptor co-regulatory protein complex, CD97-CD55. *J Biol Chem* **282**:22023–22032.

145. Capasso M, Durrant LG, Stacey M, Gordon S, Ramage J, Spendlove I. 2006. Costimulation via CD55 on human CD4+ T cells mediated by CD97. *J Immunol* **177**:1070–1077.

146. Nishimori H, Shiratsuchi T, Urano T, Kimura Y, Kiyono K, Tatsumi K, Yoshida S, Ono M, Kuwano M, Nakamura Y, Tokino T. 1997. A novel brain-specific p53-target gene, BAI1, containing thrombospondin type 1 repeats inhibits experimental angiogenesis. *Oncogene* **15**:2145–2150.

147. Shiratsuchi T, Futamura M, Oda K, Nishimori H, Nakamura Y, Tokino T. 1998. Cloning and characterization of BAI-associated protein 1: a PDZ domain-containing protein that interacts with BAI1. *Biochem Biophys Res Commun* **247**:597–604.

148. Das S, Sarkar A, Ryan KA, Fox S, Berger AH, Juncadella IJ, Bimczok D, Smythies LE, Harris PR, Ravichandran KS, Crowe SE, Smith PD, Ernst PB. 2014. Brain angiogenesis inhibitor 1 is expressed by gastric phagocytes during infection with *Helicobacter pylori* and mediates the recognition and engulfment of human apoptotic gastric epithelial cells. *FASEB J* **28**:2214–2224.

149. Harre U, Keppeler H, Ipseiz N, Derer A, Poller K, Aigner M, Schett G, Herrmann M, Lauber K. 2012. Moonlighting osteoclasts as undertakers of apoptotic cells. *Autoimmunity* **45**:612–619.

150. Das S, Owen KA, Ly KT, Park D, Black SG, Wilson JM, Sifri CD, Ravichandran KS, Ernst PB, Casanova JE. 2011. Brain angiogenesis inhibitor 1 (BAI1) is a pattern recognition receptor that mediates macrophage binding and engulfment of Gram-negative bacteria. *Proc Natl Acad Sci U S A* **108**:2136–2141.

151. Mazaheri F, Breus O, Durdu S, Haas P, Wittbrodt J, Gilmour D, Peri F. 2014. Distinct roles for BAI1 and TIM-4 in the engulfment of dying neurons by microglia. *Nat Commun* **5**:4046. doi:10.1038/ncomms5046.

152. Koh JT, Kook H, Kee HJ, Seo YW, Jeong BC, Lee JH, Kim MY, Yoon KC, Jung S, Kim KK. 2004. Extracellular fragment of brain-specific angiogenesis inhibitor 1 suppresses endothelial cell proliferation by blocking alphavbeta5 integrin. *Exp Cell Res* **294**:172–184.

153. Kaur B, Brat DJ, Devi NS, Van Meir EG. 2005. Vasculostatin, a proteolytic fragment of brain angiogenesis inhibitor 1, is an antiangiogenic and antitumorigenic factor. *Oncogene* **24**:3632–3642.

154. Park D, Tosello-Trampont AC, Elliott MR, Lu M, Haney LB, Ma Z, Klibanov AL, Mandell JW, Ravichandran KS. 2007. BAI1 is an engulfment receptor for apoptotic cells upstream of the ELMO/Dock180/Rac module. *Nature* **450**:430–434.

155. Hochreiter-Hufford AE, Lee CS, Kinchen JM, Sokolowski JD, Arandjelovic S, Call JA, Klibanov AL, Yan Z, Mandell JW, Ravichandran KS. 2013. Phosphatidylserine receptor BAI1 and apoptotic cells as new promoters of myoblast fusion. *Nature* **497**:263–267.

156. Fukuzawa T, Hirose S. 2006. Multiple processing of Ig-Hepta/GPR116, a G protein-coupled receptor with immunoglobulin (Ig)-like repeats, and generation of EGF2-like fragment. *J Biochem* **140**:445–452.

157. **Bridges JP, Ludwig MG, Mueller M, Kinzel B, Sato A, Xu Y, Whitsett JA, Ikegami M.** 2013. Orphan G protein-coupled receptor GPR116 regulates pulmonary surfactant pool size. *Am J Respir Cell Mol Biol* **49:** 348–357.

158. **Yang MY, Hilton MB, Seaman S, Haines DC, Nagashima K, Burks CM, Tessarollo L, Ivanova PT, Brown HA, Umstead TM, Floros J, Chroneos ZC, St Croix B.** 2013. Essential regulation of lung surfactant homeostasis by the orphan G protein-coupled receptor GPR116. *Cell Reports* **3:**1457–1464.

159. **Fukuzawa T, Ishida J, Kato A, Ichinose T, Ariestanti DM, Takahashi T, Ito K, Abe J, Suzuki T, Wakana S, Fukamizu A, Nakamura N, Hirose S.** 2013. Lung surfactant levels are regulated by Ig-Hepta/GPR116 by monitoring surfactant protein D. *PLoS One* **8:**e69451. doi:10.1371/journal.pone.0069451.

160. **Ariestanti DM, Ando H, Hirose S, Nakamura N.** 2015. Targeted disruption of Ig-Hepta/Gpr116 causes emphysema-like symptoms that are associated with alveolar macrophage activation. *J Biol Chem* **290:**11032–11040.

161. **Pierdomenico AM, Recchiuti A, Simiele F, Codagnone M, Mari VC, Davì G, Romano M.** 2015. MicroRNA-181b regulates ALX/FPR2 receptor expression and proresolution signaling in human macrophages. *J Biol Chem* **290:**3592–3600.

162. **Etzrodt M, Cortez-Retamozo V, Newton A, Zhao J, Ng A, Wildgruber M, Romero P, Wurdinger T, Xavier R, Geissmann F, Meylan E, Nahrendorf M, Swirski FK, Baltimore D, Weissleder R, Pittet MJ.** 2012. Regulation of monocyte functional heterogeneity by miR-146a and Relb. *Cell Rep* **19:**317–324.

163. **Lu YY, Sweredoski MJ, Huss D, Lansford R, Hess S, Tirrell DA.** 2014. Prometastatic GPCR CD97 is a direct target of tumor suppressor microRNA-126. *ACS Chem Biol* **9:**334–338.

164. **Boularan C, Kehrl JH.** 2014. Implications of non-canonical G-protein signaling for the immune system. *Cell Signal* **26:**1269–1282.

165. **Patial S, Saini Y, Parvataneni S, Appledorn DM, Dorn GW II, Lapres JJ, Amalfitano A, Senagore P, Parameswaran N.** 2011. Myeloid-specific GPCR kinase-2 negatively regulates NF-κB1p105-ERK pathway and limits endotoxemic shock in mice. *J Cell Physiol* **226:**627–637.

166. **Nakaya M, Tajima M, Kosako H, Nakaya T, Hashimoto A, Watari K, Nishihara H, Ohba M, Komiya S, Tani N, Nishida M, Taniguchi H, Sato Y, Matsumoto M, Tsuda M, Kuroda M, Inoue K, Kurose H.** 2013. GRK6 deficiency in mice causes autoimmune disease due to impaired apoptotic cell clearance. *Nat Commun* **4:**1532. doi:10.1038/ncomms2540.

167. **Ferguson SS.** 2001. Evolving concepts in G protein-coupled receptor endocytosis: the role in receptor desensitization and signaling. *Pharmacol Rev* **53:**1–24.

168. **Fan H.** 2014. β-Arrestins 1 and 2 are critical regulators of inflammation. *Innate Immun* **20:**451–460.

169. **Watari K, Nakaya M, Nishida M, Kim KM, Kurose H.** 2013. β-Arrestin2 in infiltrated macrophages inhibits excessive inflammation after myocardial infarction. *PLoS One* **8:**e68351. doi:10.1371/journal.pone.0068351.

170. **Du RW, Du RH, Bu WG.** 2014. β-Arrestin 2 mediates the anti-inflammatory effects of fluoxetine in lipopolysaccharide-stimulated microglial cells. *J Neuroimmune Pharmacol* **9:**582–590.

171. **Wiege K, Le DD, Syed SN, Ali SR, Novakovic A, Beer-Hammer S, Piekorz RP, Schmidt RE, Nürnberg B, Gessner JE.** 2012. Defective macrophage migration in $G\alpha_{i2}$- but not $G\alpha_{i3}$-deficient mice. *J Immunol* **189:**980–987.

172. **Patel J, McNeill E, Douglas G, de Bono JP, Greaves DR, Channon KM.** 2014. A new role for the regulator of G-protein signalling-1 in inflammatory cell function in angiotensin II-induced aortic aneurysm formation. *Atherosclerosis* **232:**E7–E7.

173. **Patel J, McNeill E, Douglas G, Hale AB, de Bono J, Lee R, Iqbal AJ, Regan-Komito D, Stylianou E, Greaves DR, Channon KM.** 2015. RGS1 regulates myeloid cell accumulation in atherosclerosis and aortic aneurysm rupture through altered chemokine signalling. *Nat Commun* **6:**6614.

Myeloid Cells in Health and Disease: A Synthesis
Edited by Siamon Gordon

doi:10.1128/microbiolspec.MCHD-0013-2015

Valentin Jaumouillé[1]
Sergio Grinstein[1]

27 Molecular Mechanisms of Phagosome Formation

INTRODUCTION

Phagocytosis culminates with the entrapment of the target particles within large vacuoles called phagosomes. Because of the multiplicity of phagocytic receptors, it is becoming apparent that a variety of different signaling cascades can be activated during the process. However, several aspects of phagocytosis appear to be conserved, distinguishing it from other mechanisms of cellular uptake such as endocytosis and macropinocytosis. First, phagocytosis can accommodate a wide variety of particle sizes, from hundreds of nanometers to tens of micrometers (1, 2), as well as complex particle morphologies (3, 4). Second, phagocytosis requires the progressive engagement of phagocyte surface receptors around the entire particle (5). This ratchet mechanism has been described as the "zipper" model, which contrasts with the limited number of independent receptors that need to be activated by soluble ligands to trigger macropinocytosis (6). Third, phagocytosis is an active mechanism that involves local remodeling of the actin cytoskeleton, which drives the deformation of the plasma membrane and the progression of the receptor/ligand "ratchet" around the particle (7–10). In addition, as the actin cytoskeleton is tightly associated with the plasma membrane, signaling mediated by phospholipids appears to be a common feature of phagocytosis. Phosphoinositides in particular play a critical role, as phosphatidylinositol 3-kinase (PI3K) is seemingly involved in virtually all known phagocytic systems (11–14). These different features impose a temporal progression of the phagosome formation, which can be described by the following sequence of events: (i) binding of the ligand to surface receptors; (ii) activation of receptor-mediated signaling cascades; (iii) remodeling of the actin cytoskeleton; (iv) progressive engagement of additional receptors around the particle; and (v) membrane fusion, leading to the closure of the phagosome (Fig. 1). Yet despite these conserved traits, one cannot fully appreciate the molecular mechanisms involved in phagosome formation without taking into account the diversity of phagocytic receptors and the variety of signaling cascades they induce individually and cooperatively. Thus, here, we chose to focus on some of the best-characterized receptors and signaling pathways in order to give an overview of the many roads that lead to phagosome formation, whereas phagosome maturation and subsequent responses will be described elsewhere.

[1]Cell Biology Program, The Hospital for Sick Children, Toronto, Ontario M5G 1X8, Canada.

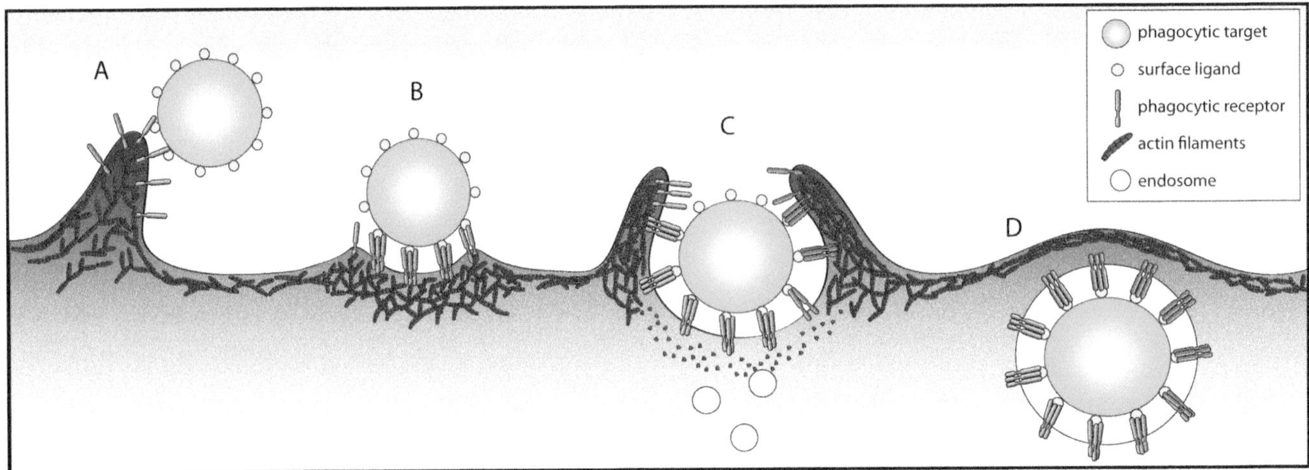

Figure 1 Temporal sequence of particle uptake by phagocytosis. (**A**) Particle surface molecules are engaged by phagocyte receptors. Actin-driven membrane dynamics facilitate the detection of surrounding particles. (**B**) Engagement and activation of the receptor lead to the induction of signaling cascades that elicit actin reorganization. (**C**) Actin polymerization progresses around the particle, accompanied by further engagement of receptors. Actin clearance and focal exocytosis at the base of the cup facilitate particle engulfment. (**D**) Once the particle is fully surrounded, membrane fusion at the rims of the cup seals the phagosome and separates it from the plasma membrane.

TARGET RECOGNITION: RECEPTOR DIVERSITY AND CELLULAR DYNAMICS

As our body needs to dispose of a large variety of particulate material, it is not surprising that a plethora of different phagocytic receptors exist (Table 1). These receptors have various degrees of ligand specificity, and individual targets—which often display a diverse array of ligands on their surface—can engage multiple receptor types. Importantly, a number of the receptors that can play a role in phagocytosis are not *bona fide* phagocytic receptors, as they are incapable on their own of inducing the particle uptake. For instance, cytokine receptors and Toll-like receptors (TLRs) can activate phagocytic integrins (15, 16) or modulate the expression level of Fc receptors (FcRs) (17, 18), but are not themselves sufficient to elicit phagocytosis. The demonstration that a given receptor has intrinsic phagocytic properties is challenging. Heterologous expression of some receptors, such as FcγR and Dectin-1 (dendritic cell-associated C-type lectin 1), can confer phagocytic properties to otherwise nonphagocytic cells (19, 20), implying that they are *bona fide* phagocytic receptors. By contrast, other receptors can act as tethers of the target but require the contribution of coreceptors to elicit particle engulfment, making the definition of phagocytic receptors more ambiguous.

Different types of phagocytic receptors are described below. These can be classified in accordance with the type of ligands they recognize: foreign molecules identifiable by unique molecular patterns, opsonins, or apoptotic bodies.

Pattern Recognition Receptors and Scavenger Receptors

To clear very diverse foreign materials, professional phagocytes have the ability to detect molecular cues that are not present in endogenous (self) cells. Interestingly, because of their early evolutionary divergence, bacteria, fungi, and parasites display numerous molecules that are not expressed by higher organisms. Phagocytes display an extended array of pattern recognition receptors (PRRs) that bind specifically some of these pathogen-associated molecular patterns (PAMPs). In this way, PRRs constitute a primitive means of detection of foreign bodies within the tissue. Whereas many of the PRRs, such as the TLRs and the NOD-like receptors, are primarily involved in inducing proinflammatory pathways that may indirectly activate phagocytic receptors (21), several PRRs are themselves phagocytic. For instance, the mannose receptor and Dectin-1 induce the phagocytosis of fungi displaying particular polysaccharides on their surface (12, 22). In

Table 1 Phagocytic receptors and their specific ligands

Receptors	Ligands	Reference(s)
PRRs		
Mannose receptor (CD206)	Mannan	22
Dectin-1 (CLEC7A)	β-1,3-Glucan	12
CLECSF8 (CLEC4D)	β-Glucan	166
CD14	Lipopolysaccharide-binding protein	30
SR-A (CD204)	Lipopolysaccharide, lipoteichoic acid	24, 25
CD36	*Plasmodium falciparum*-infected erythrocytes	167
MARCO	Bacteria	23
Opsonic receptors		
FcγRI (CD64)	IgG1 = IgG3 > IgG4	27
FcγRIIa (CD32a)	IgG3 ≥ IgG1 = IgG2	27
FcγRIIc (CD32c)	IgG	27
FcγRIIIa (CD16a)	IgG	27
FcαRI (CD89)	IgA1, IgA2	168
FcεRI	IgE	169
CR1 (CD35)	Mannan-binding lectin, C1q, C4b, C3b	170
CR3 ($\alpha_M\beta_2$, CD11b/CD18, Mac1)	iC3b, β-Glucan, lipopolysccharide	171
CR4 ($\alpha_X\beta_2$, CD11c/CD18, gp150/95)	iC3b	171
Integrin $\alpha_5\beta_1$	Fibronectin, vitronectin	29
Receptors of apoptotic corpses		
TIM-1	PS	35
TIM-3	PS	36
TIM-4	PS	35
BAI1	PS	37
RAGE	PS	34
Stabilin-2	PS	38
KIM-1[a]	PS	172
CD300f	PS	173
TAM family	Gas6, protein S	40, 174
LRP1 (CD91)	C1q/calreticulin	43
Integrin $\alpha_v\beta_3$	MFG-E8	39
Integrin $\alpha_v\beta_5$	Apoptotic cells	175
CD36	Oxidized lipids	176

[a]KIM-1, kidney injury molecule-1.

addition, several scavenger receptors initiate phagocytosis upon PAMP recognition; these include the scavenger receptor A (SR-A) and the macrophage receptor with collagenous structure (MARCO), which bind surface molecules of Gram-negative and -positive bacteria (23–25). Interestingly, while these two receptors detect molecules exposed by the same bacteria, the outcome of their engagement is quite different, since MARCO induces the production of the proinflammatory cytokine interleukin-12 whereas SR-A inhibits it (26). The array of phagocytic PRRs known to date seems to be insufficient to detect all the possible pathogens. For this reason and because of their decisive role in the early detection of pathogen invasion, PRRs are currently of great interest. It is safe to predict that the number of identified phagocytic PRRs will expand considerably in the coming years.

Opsonic Receptors

The fact that serum molecules can enhance phagocytosis was recognized by Almroth Wright as early as 1903. Indeed, several soluble molecules, called opsonins, can be deposited onto foreign surfaces and serve as adaptors that bind and activate potent phagocytic receptors. For instance, IgG specifically bound to microbial surface antigens associates with phagocyte FcγRs that recognize their fragment crystallizable (Fc) region (27, 28). C3b and iC3b molecules of the complement system can also decorate foreign particles. Their deposition can proceed via the classical pathway (which requires prior binding of IgMs or IgGs onto the particle), the lectin pathway (which involves binding of mannan-binding lectin onto microbial carbohydrates), or the less efficient constitutively activated alternative pathway. Consequently, whereas Fc-mediated phagocytosis is linked to the adaptive immune response and immunological memory, complement-mediated phagocytosis can also take place in the sole context of innate immunity. In addition to these canonical opsonins, deposition of several other soluble proteins could be considered as opsonization, as they can enhance phagocytosis. Binding of the lipopolysaccharide-binding protein onto Gram-negative bacteria or deposition of fibronectin or vitronectin onto particles could stimulate their phagocytosis (29, 30), whereas milk fat globule-EGF factor 8 (MFG-E8), growth arrest-specific 6 (Gas6), and protein S participate in phagocytosis of apoptotic corpses (see next paragraph). These molecules serve as an alternative type of opsonins and could be viewed as soluble pattern recognition molecules that link molecular patterns to phagocyte surface receptors. As such, they function in a manner related to yet distinct from PRRs, which bind molecular patterns directly.

Receptors of Apoptotic Corpses

In addition to the clearance of foreign particles, phagocytosis is also critical for cell turnover within the organism. Indeed, billions of cells die by apoptosis every day and must be disposed of. Phagocytosis of apoptotic cells, however, presents an interesting problem, since instead of detecting foreign material, phagocytes must distinguish between two types of self: dead versus alive. This distinction relies on the balance between defined "eat me" signals that are landmarks of apoptotic bodies and "don't eat me" signals displayed by healthy cells.

Perhaps the best-characterized signature of apoptotic cells is the increased surface exposure of the lipid phosphatidylserine (PS) (31). In healthy cells, PS is localized almost exclusively on the inner leaflet of the plasma membrane. However, once the apoptosis pathway is triggered, this asymmetrical distribution is lost, leading to a 300-fold increase in the concentration of PS on the outside leaflet of the plasmalemma (32). The mechanism underlying PS relocation has not been fully elucidated, but evidence indicates that it involves the Cl^- channel-related protein TMEM16F, which mediates phospholipid scrambling in a Ca^{2+}-dependent manner (33). Several phagocytic receptors can detect PS directly: the brain-specific angiogenesis inhibitor 1 (BAI1), stabilin-2, the receptor for advanced glycation end-products (RAGE), and members of the T-cell immunoglobulin and mucin domain-containing family (TIM) (34–38). In addition, soluble proteins, such as MFG-E8, Gas6, and protein S, bind to exposed PS and act as linkers, akin to opsonins. MFG-E8 (also called lactadherin) binds phagocyte $\alpha_v\beta_3$ integrins via their RGD motif (39). Protein S and Gas6 induce phagocytosis by binding to the Tyro 3, Axl, Mer (TAM) family of receptors (40). Interestingly, forced exposure of PS at the surface of viable cells does not lead to their engulfment (41), suggesting that phagocytes need multiple converging signals to initiate phagocytosis of apoptotic cells. Indeed, surface changes during apoptosis go far beyond PS exposure. Additional "eat me" signals include changes in glycosylation of surface protein (which can bind lectins); surface accumulation of serum proteins like thrombospondin or C1q (which bind CD36 and lipoprotein receptor-related protein-1 [LRP1], respectively); expression of intercellular adhesion molecule-3 (ICAM-3, which binds CD14 and lymphocyte function-associated antigen-1 [LFA-1]); as well as exposure of the intracellular proteins annexin I and calreticulin, a ligand of LRP1 (42, 43).

Exposure of "eat me" signals, however, is not restricted to apoptotic cells. For instance, activated lymphocytes display elevated amounts of PS on their surface, and all leukocytes express ICAM-3. Yet this does not lead to their engulfment by phagocytes, likely because living cells also display "don't eat me" signals that protect them against phagocytosis. CD47 (also known as integrin-associated protein) is a ubiquitously expressed transmembrane protein that binds SIRPα (signal regulatory protein α), expressed on the phagocyte surface (44). This interaction stimulates tyrosine phosphatases, leading to the inhibition of myosin II and of cell engulfment. Importantly, CD47 expression is downregulated during apoptosis, relieving the inhibition. In addition, the platelet endothelial cell adhesion molecule-1 (CD31), expressed by endothelial and hematopoietic cells, inhibits phagocytosis when engaged in homotypic interactions (45).

Cellular Factors

Productive interactions between targets and phagocytes depend on multiple parameters. In a passive system, engagement of the phagocytic receptors by their cognate ligands is dictated by their respective surface density and lateral mobility—which determines the frequency of their interactions—as well as their mutual affinity (46, 47). Expression levels of phagocytic receptors can vary greatly depending on the cell type and their activation and polarization state. For instance, macrophage activation by alpha interferon leads to increased surface expression of FcγRI and MARCO, whereas alternative activation by interleukin-4 stimulates surface expression of SR-A and mannose receptor (48). In addition, the lateral mobility of membrane proteins, such as the FcγR, is restricted by the actin-based compartmentalization of the plasma membrane in dynamic microdomains (49). The structure and dynamics of the actin-based microdomains is governed by the kinase Syk, which is regulated by environmental cues, such as integrin engagement and TLR signaling. Finally, receptor affinity is not always a constant. For instance, the affinity of integrins is increased to an intermediate level by extension of their conformation upon association with talin, which links the integrins to the actin cytoskeleton. This regulation is controlled by a Rap-1 GTPase-dependent "inside-out" pathway that can be elicited by various stimuli, including TLR signaling and FcγR engagement (15, 50). Moreover, the tensile force generated upon engagement of ligands can induce another conformational change that increases the integrin-ligand affinity, a phenomenon called "catch bond" (51).

Professional phagocytes are very dynamic cells that can form large membrane protrusions, constantly extending and retracting from the cell body. The formation of these protrusions, also called ruffles, varies with the state of activation of the cells. Converging studies demonstrated that membrane ruffles can capture particles at a distance, resulting in increased binding and engulfment of targets (52–55). Extension of these protrusions is driven by the actin cytoskeleton and can be elicited acutely by various stimuli, such as growth factors and TLR ligands. Surprisingly, whereas PI3K and Rac activities are critical for acute induction of ruffling and fluid uptake by macropinocytosis, they appear to be dispensable for the constitutive ruffling of resting macrophages and immature dendritic cells (56–59). The molecular mechanisms underlying constitutive ruffling remain largely unexplored; however, a recent study showed that it involves the accumulation of phosphatidic acid (PA) in the plasma membrane, generated upon phosphorylation of diacylglycerol (DAG) by DAG kinases, under the control of an unidentified G protein-coupled receptor (60). Having bound a target particle, macrophage protrusions undergo stepwise retraction, developing forces greater than 15 pN, suggesting the involvement of molecular motors that have yet to be identified (53).

MULTIPLE PATHWAYS TO FORM A PHAGOSOME

As described previously, phagocytic receptors belong to various membrane protein families: immunoreceptors, integrins, scavenger receptors, etc. Physiological targets often engage several of these receptors at the same time, simultaneously activating numerous signaling cascades that will act in parallel or in a cooperative fashion to induce particle uptake. Analyzing these events in a systematic manner remains an arduous endeavor. However, the implementation of experimental models that restrict phagocytosis to the engagement of a single receptor—or at least a single class of receptors—has enabled investigators to dissect individual signaling cascades. Summing up this information allows a greater appreciation of how different pathways function and intersect, and highlights the key factors involved in the physiological context.

FcγR-Mediated Phagocytosis

Coating with IgG is perhaps the most efficient way to induce or accelerate particle engulfment by professional phagocytes. This model is definitively the most studied and best understood and has therefore become a paradigm of phagocytosis. Phagocytosis mediated by the association of particle-bound IgG to FcγR is an iterative process involving (i) activation of engaged receptors; (ii) local induction of a signaling cascade; and (iii) actin-mediated membrane deformation and juxtaposition along the particle, leading to new receptor engagement. Iteration of these steps leads to complete "zippering" of the phagocyte membrane around the target particle. However, internalization requires a membrane fusion event that disconnects the particle-containing vesicle from the plasma membrane.

FcγR activation

Soluble IgGs are extremely abundant proteins, typically reaching concentrations of >5 mg/ml in interstitial fluids (61). It is therefore critical that phagocytes distinguish IgG-associated particles and immune complexes from unbound IgGs. Compelling structural work demonstrated that FcRs do not undergo significant conformational changes upon binding to soluble immunoglobulins (62). Instead, activation occurs when receptors are clustered laterally by bivalent or multivalent ligands (63, 64). Similar observations have been made for other immunoreceptors (FcεR, T-cell receptors [TCRs], and B-cell receptors) that are also activated upon formation of small clusters, mediated by the engagement of multivalent ligands (65). Activating immunoreceptors display a cytosolic signaling domain, called the immunoreceptor tyrosine-based activation motif (ITAM), characterized by a double YxxI/L sequence. FcγR clustering leads to the phosphorylation of the two tyrosines of the ITAM by the Src family kinases

(SFKs) Lyn, Hck, and Fgr (66–68). FcγR-mediated phagocytosis is largely attenuated and occurs only slowly in bone marrow-derived macrophages from *hck*$^{-/-}$*fgr*$^{-/-}$*lyn*$^{-/-}$ triple-knockout mice, suggesting that, although less important, other SFKs can also phosphorylate the ITAM (69). Doubly phosphorylated ITAMs bind the cytosolic spleen tyrosine kinase (Syk), leading to its recruitment and activation by autophosphorylation (70, 71). Syk can participate as well in receptor activation by phosphorylating nearby ITAMs (72, 73). Interestingly, in *syk*$^{-/-}$ macrophages, FcγR-mediated phagocytosis is initiated, as apparent by the accumulation of actin filaments beneath the nascent phagocytic cup. However, the pseudopods fail to extend and particle engulfment is completely abolished (72, 74).

How receptor clustering initiates ITAM phosphorylation is still a matter of debate. It has been proposed that immunoreceptor clusters recruit SFKs by generating cholesterol-enriched lipid rafts. This model is supported by membrane fractionation experiments showing that FcγRIIA associates with detergent-resistant membranes upon cross-linking and ITAM phosphorylation is inhibited by treatment with methyl-β-cyclodextrin, which removes cholesterol (75, 76). However, the evidence has been questioned because methyl-β-cyclodextrin treatment drastically alters the cellular architecture and large lipid domains can form *in vitro* during fractionation experiments. Moreover, while FcγRIIA association with detergent-resistant membranes requires its palmitoylation, other FcγRs are not lipidated (77). Furthermore, multiple studies on other immunoreceptors indicate that SFK recruitment and assembly of signaling clusters occur independently of lipid rafts (78–80).

SFK activation is mediated by the transphosphorylation of the tyrosine located in the catalytic domain. When dephosphorylated by various phosphatases, SFKs can be switched to an inactive conformation by Csk, which phosphorylates a conserved tyrosine in the amino-terminal domain. Interestingly, the transmembrane SFK phosphatases CD45 and CD148 (protein tyrosine phosphatase receptor type C and J, respectively) are specifically excluded from immunological synapses and also from phagocytic cups (81, 82). This exclusion is attributed to their very large ectodomains, which are seemingly pushed aside by the very close apposition of the immunoreceptor-bearing membrane and its target (83). Indeed, interactions between phagocytes and IgG-coated surfaces are so tight that the contact area is impermeable to molecules as small as 50 kDa; by comparison, the space beneath adherent cells is accessible to molecules as large as 200 kDa (84). These observations led to the proposition of the kinetic segregation model, which attributes receptor activation to the local exclusion of SFK phosphatases. Consistently, CD45 CD148 double-knockout macrophages (*Ptprc*$^{-/-}$*Ptprj*$^{-/-}$) are largely, though not entirely, unable to perform FcγR-mediated phagocytosis (85). However, whether CD45 and CD148 are excluded from FcγR clusters induced by the association with immune complexes remains to be demonstrated. In addition, the kinetic segregation model is not sufficient to explain the initial activation, because CD45 activity is important for the dephosphorylation of the SFK amino-terminal tyrosine. Consistently, Csk inhibition in T cells leads to an increased TCR signaling that requires CD45 (86).

How FcγRs form signaling clusters during phagocytosis is also poorly understood. Converging evidence points to the actin cytoskeleton as a key factor in this process. Multiple observations indicate that pharmacologically induced depolymerization of actin filaments increases immunoreceptor-mediated signaling (65, 87, 88). Furthermore, when Csk is inhibited, the initiation of full TCR signaling requires actin remodeling or depolymerization (89). Like that of many membrane-associated proteins, FcγR mobility is restricted by plasma membrane microdomains formed by interactions between the lipid bilayer, transmembrane proteins, and the subjacent cortical actin network (49). Interestingly, *in silico* simulations and tracking experiments demonstrated that, at equal density, FcγRs undergoing free motion coalesce with a greater frequency than FcγRs confined in actin-based corrals. Consequently, increased actin dynamics or filament disassembly facilitates FcγR cluster formation. During FcγR-mediated phagocytosis, FcγR mobility is increased at the leading edge of the cup, favoring the formation of new clusters, whereas the dense actin filament assembly behind the leading edge blocks FcγR lateral diffusion. By preventing further FcγR clustering, actin polymerization creates a negative feedback loop that facilitates forward progression of the cup. Furthermore, the structure of the cortical actin network appears to be regulated by the distal activity of Syk and SFKs (49). Thus, signaling pathways that activate these kinases, such as those triggered by integrins and TLRs, can prime FcγR responsiveness by increasing its lateral mobility. Whether actin filament accumulation around the clusters also maintains the integrity of these signaling platforms, as proposed for the B-cell receptor, remains to be demonstrated (88).

The FcγR signaling platform

FcγR activation leads to the recruitment and activation of many adaptor and signaling proteins, as well as multiple lipid modifications, which creates a complex signaling platform (Fig. 2). Syk recruits and phosphorylates the transmembrane adaptor linker of activated T cells (LAT), leading to the recruitment of Grb2 and the Grb2-associated binder Gab2 (90, 91). Gab2 and Syk cooperate to recruit and activate subunits of the class I PI3K (92, 93), whereas LAT and Syk mediate the recruitment and activation of the phospholipase C-γ (PLC-γ) (94, 95). The recruitment of adaptor SH2-domain-containing leukocyte protein of 76 kDa (SLP-76) and the Fyn-binding/SLP-76-associated protein (Fyb/SLAP) leads to the accumulation of the actin regulatory proteins Ena/vasodilator-stimulated phosphoprotein (VASP) (96). The adaptor CrkII is also recruited via its SH2 domain and associates with the Dock180-ELMO (engulfment and cell motility) heterodimer, which acts as a guanine nucleotide exchange factor (GEF) for the Rho GTPase Rac (97). Dock180-ELMO recruitment is also associated with the GTPase RhoG, which plays a potentially important, yet poorly characterized, role in FcγR-mediated phagocytosis (98, 99). The adaptor Nck is also recruited and phosphorylated upon FcγR activation, promoting the recruitment of the nucleation-promoting factor WASP (Wiskott-Aldrich syndrome protein) and the GTPase Cdc42 (96, 100, 101).

Multiple lipid modifications occur at the level of the forming phagocytic cup. The concentration of phosphatidylinositol 4,5-bisphosphate [$PI(4,5)P_2$], present

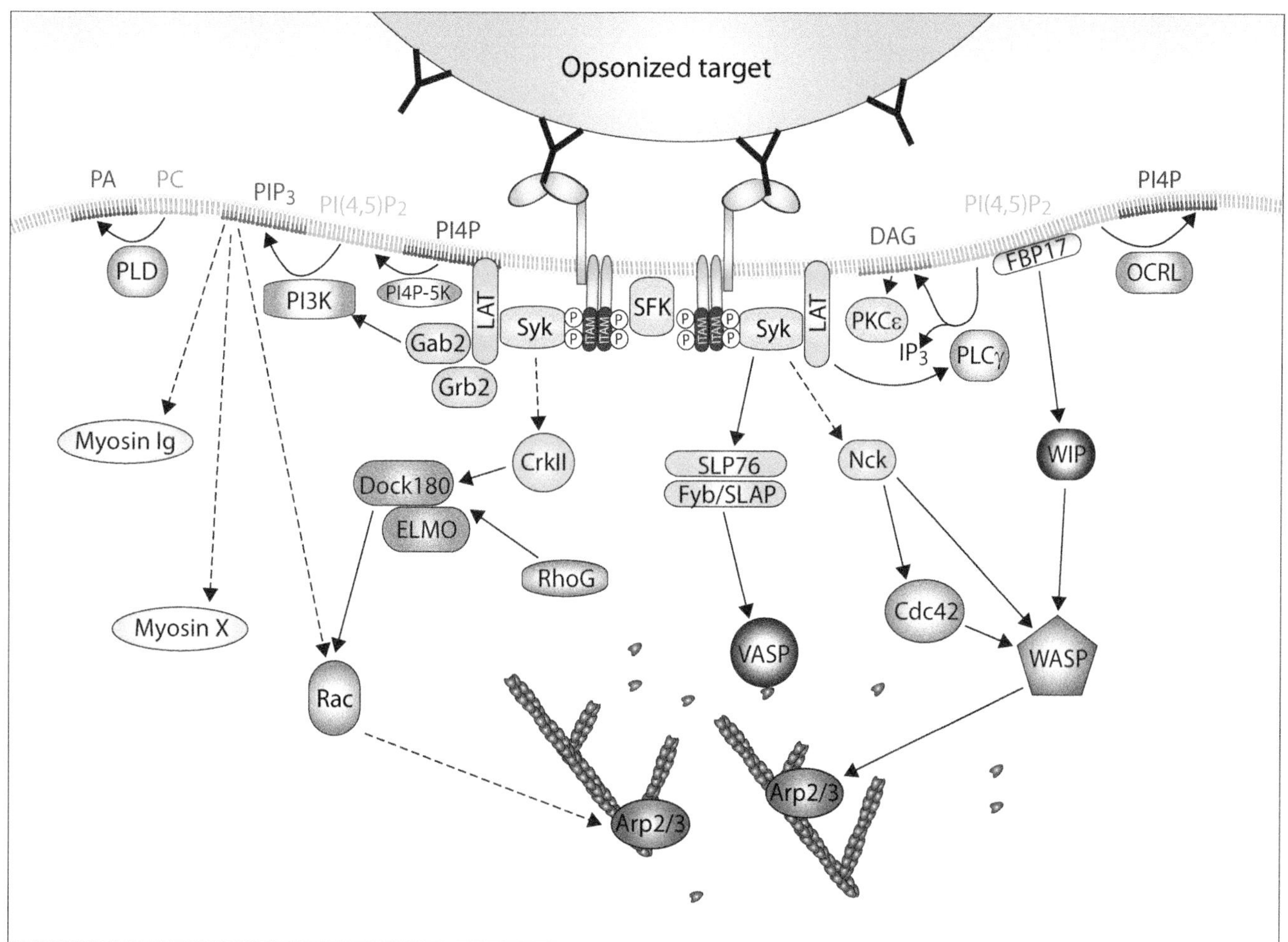

Figure 2 Signaling cascades leading to actin reorganization during FcγR-mediated phagocytosis. Engagement and aggregation of FcγRs activate tyrosine kinases (yellow), which recruit multiple adaptor proteins (green). The FcγR signaling complex activates lipid modification enzymes (orange), GEFs (pink), actin modulators (navy blue), and Rho GTPases (purple). By activating nucleation-promoting factors (brown), they stimulate the activity of the Arp2/3 complex (red), which nucleates actin polymerization into a branched network. Abbreviations: PIP_3, phosphatidylinositol 3,4,5-trisphosphate; PLD, phospholipase D.

exclusively in the inner leaflet of the plasma membrane, transiently increases by the phosphorylation of phosphatidylinositol 4-phosphate (PI4P) by phosphatidylinositol 4-phosphate 5-kinases (102). Shortly after its modest accumulation, the concentration of PI(4,5)P_2 decreases drastically, facilitating actin disassembly, which allows particle penetration into the cell (103). The depletion of PI(4,5)P_2 at the phagocytic cup is mediated by multiple enzymes. First, class I PI3K phosphorylates PI(4,5)P_2 into phosphatidylinositol 3,4,5-trisphosphate [PI(3,4,5)P_3], which appears at the base of the phagocytic cup (14, 104). Second, PLC-γ hydrolyzes PI(4,5)P_2 into inositol 3,4,5-trisphosphate and DAG (94), which in turn induces the recruitment of the protein kinase C-ε (PKC-ε) (105). Pharmacological evidence suggests that PLC-γ recruitment and/or activation at the forming phagosome requires PI3K activity (103). Interestingly, PI3K has a much more important role in the phagocytosis of large particles (>3 μm) than of small ones (56, 106). Finally PI(4,5)P_2 is dephosphorylated to produce PI4P by the phosphatases OCRL (oculocerebrorenal syndrome protein) and Inpp5B (107, 108). These phosphatases are delivered to the forming phagosome by Rab5-positive early endosomes through a mechanism involving the protein B-cell lymphoma/leukemia-10 (BCL-10), the clathrin adaptors AP1 (activator protein 1) and EpsinR, and the adaptor protein APPL1 (adaptor protein, phosphotyrosin interaction, PH domain and leucine zipper containing 1). These observations suggest that, in addition to providing additional membrane to increase the surface area available to form the phagocytic cup, focal exocytosis of endomembrane vesicles also contributes to the signaling required for phagosome sealing.

During phagocytosis, the fusion of recycling endosomes and late endosomes is observed at the base of the phagocytic cup (109–111). Inhibition of this so-called focal exocytosis decreases the uptake of large particles (110–112). In addition, the depletion of PI(4,5)P_2 at the base of the phagocytic cup leads to the deactivation of Cdc42, which is required for large particle engulfment (113). Besides phosphoinositides, hydrolysis of phosphatidylcholine into PA by phospholipase D also facilitates phagosome formation by a mechanism that remains to be clarified (114).

Actin reorganization during FcγR-mediated phagocytosis

Actin, the driving force of phagocytosis, is an ATPase that polymerizes into dynamic microfilaments. These microfilaments assemble and disassemble to form a constantly changing network that underpins the dynamic cellular morphology. Formation of new filaments requires the enzymatic activity of nucleation-promoting factors, such as the Arp2/3 complex, formins or proteins that display tandem WASP homology 2 (WH2) domains. Whereas the role of many of these nucleators has not been thoroughly explored, the Arp2/3 complex undoubtedly plays a critical role in actin polymerization during FcγR-mediated phagocytosis (115). Activation of the Arp2/3 heteroheptamer requires the association with VCA (verprolin homology, cofilin homology, acidic) domain-bearing proteins, also called nucleation-promoting factors. Typically, the WASP and WAVE family proteins are the nucleation-promoting factors found associated with the plasma membrane, whereas WASH, WHAMM, and JMY associate with intracellular compartments. Whereas WAVE inhibition has little effect on FcγR-mediated phagocytosis (116), multiple lines of evidence indicate that WASP is required (117–119). The protein adaptor Nck and the BAR domain-containing formin-binding protein 17 (FBP17), which recognizes inward-bent membranes, are important for the recruitment of WASP and the WASP-interacting protein (WIP) to the phagocytic cup (100, 120). WASP activation requires binding to PI(4,5)P_2 and active Cdc42, which are both localized at the leading edge of the phagocytic cup (94, 121). Accordingly, expression of dominant-negative mutants and gene silencing indicate that Cdc42 is required for FcγR-mediated phagocytosis (99, 118, 122). Cdc42 recruitment to the cup is facilitated by Nck, but the GEF responsible for its activation remains to be identified (100).

In addition to Cdc42, other small GTPases play a role in FcγR-mediated phagocytosis, yet their exact contribution is not fully understood. Expression of dominant-negative mutants indicates that GEFs for Rac are required (122). Many of these GEFs, however, can activate several GTPases. Despite being activated at the forming phagocytic cup (121), gene silencing and knockout experiments demonstrate that Rac1 is dispensable for FcγR-mediated phagocytosis, whereas the related GTPases Rac2 and RhoG are required (99, 123, 124). RhoG is known to activate the bipartite GEF Dock180/ELMO, which can activate Rac1 and Rac2. However, since WAVE, the nucleation-promoting factor activated by the Rac GTPases, is dispensable for phagocytosis, the role of Rac2 and RhoG might not be related to the induction of actin polymerization. This model is supported by the distinct localizations of Rac1 and Rac2 activities and most of the actin filaments around the forming phagosome (121). In a similar fashion, inhibition of the small GTPase Arf6 (ADP ribosylation factor 6) leads to a marked reduction of FcγR-mediated phagocytosis (125). Arf6 inhibition has no discernible effect on actin reorganization but appears to promote focal exocytosis of endocytic vesicles. The contribution of the GTPases RhoA/B/C in FcγR-mediated phagocytosis has been a matter of debate, since their simultaneous inhibition by the *Clostridium botulinum* toxin C3 led to inconsistent results (115, 122, 126). However, expression of a RhoA dominant-negative mutant and individual gene silencing of each GTPase suggest that they are dispensable (99, 127).

In addition to the nucleation of its polymerization, actin treadmilling is regulated by numerous regulatory proteins. In particular, *in vitro* experiments using reconstituted systems demonstrated that availability of actin monomers, which are released at the filament pointed end, limits the rate of treadmilling (128). During FcγR-mediated phagocytosis, Ena/VASP is recruited to the cup and, through the interaction with G-actin-associated profilin, favors the recruitment and recycling of actin monomers that fuel the polymerization (96). Capping proteins, which bind the barbed (polymerizing) end of the filament and prevent further polymerization, are a required component of many actin-driven mechanisms. When capping proteins are present, actin monomers released at the pointed ends of capped and uncapped filaments are "funneled" to feed the uncapped barbed end, resulting in an increased filament growth rate. In macrophages, CapG is a very abundant capping protein and deletion of the corresponding gene leads to marked reduction of the rate of FcγR-mediated phagocytosis (129). Severing proteins regulate actin treadmilling by chopping off actin filaments, generating new barbed ends and new pointed ends. The severing activity itself does not affect treadmilling but, in conjunction with other activities, such as barbed end capping, leads to filament depolymerization. In neutrophils, in which FcγR-mediated phagocytosis is Ca^{2+} dependent, gelsolin, a severing protein activated by Ca^{2+}, is necessary for the engulfment of IgG-opsonized particles, whereas it is dispensable in macrophages (130).

Finally, myosins are molecular motors that shape the actin network architecture, provide contractile forces, and facilitate actin depolymerization. Myosin Ic, Ig, II, IXb, and X have been observed lining the forming phagosome; however, their respective roles remain unclear (131–134). Pharmacological inhibition of myosin II prevents phagocytosis and is associated with defective exclusion of the phosphatase CD45 from the phagocytic cup and with a reduced level of FcγR activation (82, 135, 136). Myosin Ig and X are downstream effectors of PI3K and are important for the uptake of large particles (131, 132). Finally, observations of simultaneous phagocytosis of one erythrocyte by two macrophages showed that the common prey is squeezed by the two forming phagosomes. Myosin Ic was localized at the site where constriction occurred, suggesting a possible role in closure of the phagosome (134).

Nonopsonic Phagocytosis Mediated by Dectin-1

The C-type lectin Dectin-1 is currently the best-characterized nonopsonic phagocytic receptor and plays an important role in the clearance of fungi. Since the cytosolic signaling domain of Dectin-1 resembles an ITAM, it is not surprising that the Dectin-1- and FcγR-mediated signaling cascades share many common features. Interestingly, however, only the membrane-proximal tyrosine of the ITAM-like motif of Dectin-1 is required for phagocytosis (Fig. 3), in contrast to FcγRs, where both tyrosines need to be phosphorylated to engage both SH2 domains of Syk. This suggests that two adjacent hemi-ITAMs of two Dectin molecules may cooperate to bind one Syk molecule simultaneously (12, 137). Strikingly, little is known about Dectin-1 clustering during phagocytosis. Clustering in this instance might be uniquely effective, because β-1,3-glucans bind Dectin-1 with high affinity and form multivalent polymers, which can cluster multiple receptors (138). Nevertheless, Dectin-1 signaling is induced by particulate β-glucans but not by soluble polymers (139). This is at least partly attributable to the tyrosine phosphatases CD45 and CD148, which play a critical role in regulating the Dectin-1 activation threshold. They remain in the proximity of receptors attached to soluble β-glucans, while the close interaction of Dectin-1 with particulate ligands promotes their physical exclusion, thereby allowing increased tyrosine phosphorylation and thus productive downstream signaling (139). In contrast to CD45/CD148, Dectin-1 becomes highly enriched in the area where the phagocyte makes contact with particulate ligands, forming a "bull's-eye" pattern reminiscent of immunological synapses (137, 139). This increased local concentration of Dectin-1 is expected to augment the avidity for its ligands, which would strengthen the association and possibly accelerate zippering around the particle.

Heterologous expression of Dectin-1 in fibroblasts (Fig. 3) indicates that, as in the case of FcγR stimulation, phagocytosis requires SFKs, Syk, PKC, Cdc42, and Rac (12). However, major discrepancies were observed between different cell types: SFKs, Syk, PI3K, and the GTPases play only a moderate role in macrophage cell lines (12); Syk is required for zymosan phagocytosis in monocytes and dendritic cells, but not

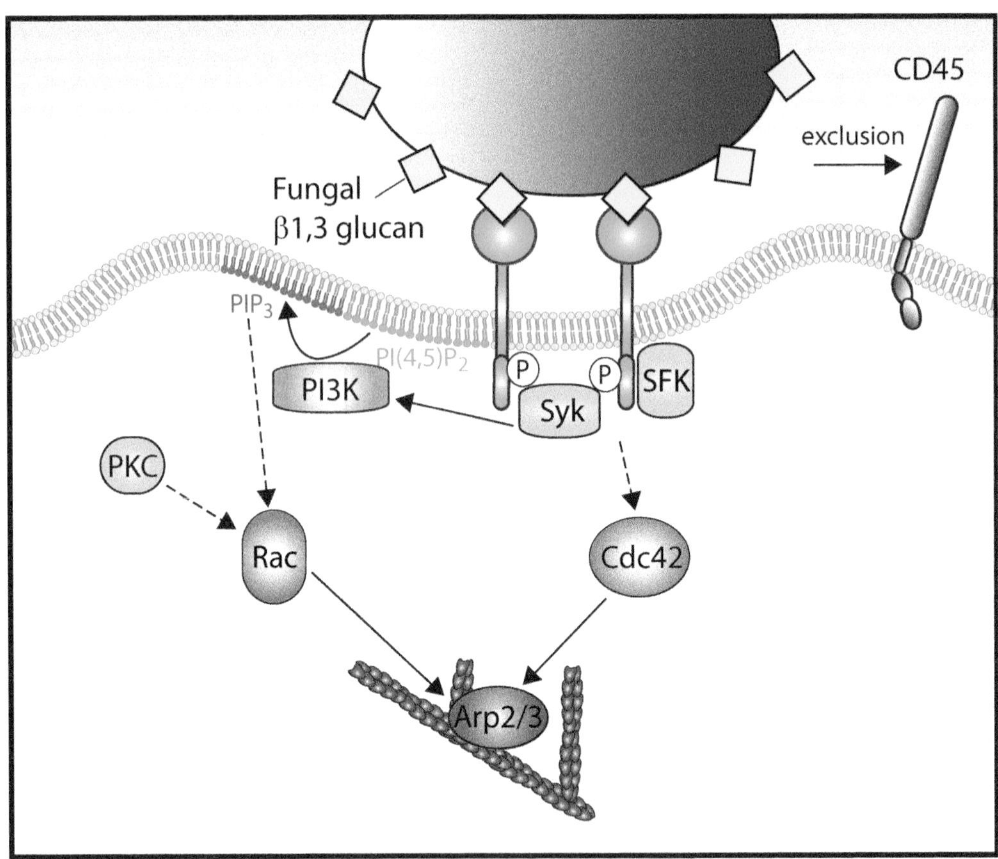

Figure 3 Molecular mechanism of fungi phagocytosis by Dectin-1. Engagement of Dectin-1 leads to the phosphorylation of its hemi-ITAM by SFKs and the recruitment of Syk. Activation of these kinases is facilitated by the exclusion of the tyrosine phosphatases CD45 and CD148 from the phagocytic cup. The combined action of SFKs, Syk, PI3K, and PKC lead to the activation of the small GTPases Rac and Cdc42, which activate Arp2/3-driven actin polymerization.

in bone marrow-derived macrophages (137, 140, 141); and SFKs are seemingly dispensable in activated bone marrow-derived macrophages (141). Finally, PKC-δ is required for zymosan phagocytosis in human monocytes but not in murine bone marrow dendritic cells (140, 142). Whether these contradictory observations are related to distinct experimental procedures or reveal actual cell type specificities remains to be determined.

Integrin-Mediated Phagocytosis: the CR3 Paradigm

Integrins form a large family of heterodimeric adhesion molecules expressed by most cell types. They are involved in cellular migration, cell-cell interactions, as well as in the phagocytosis of multiple targets, whether serving as primary phagocytic receptors or as coreceptors (Table 1). In the context of phagocytosis, the integrin $\alpha_M\beta_2$, also called complement receptor 3 (CR3), Mac-1, and CD11b/CD18, has been the most extensively studied. It is a specific receptor for the processed complement molecule iC3b but can also bind a plethora of ligands, including extracellular matrix proteins (fibronectin, laminin, collagen, and vitronectin), surface receptors (ICAM-1 and -2), blood coagulation proteins (fibrinogen, factor X, and kininogen), and microbial surface molecules (lipopolysaccharide and β-glucans). In contrast to immunoreceptors, integrin signaling is associated with conformational changes of the receptor (143). At rest, integrins keep a bent conformation. Activation by the "inside-out" signaling pathway leads to extension of the heterodimer and an increase in its affinity for ligands. Ligand engagement to activated receptors induces a separation of the legs of the α and β subunits, which adopt a triangular conformation, and elicits a downstream "outside-in" signaling cascade. As reported for other integrins (Fig. 4),

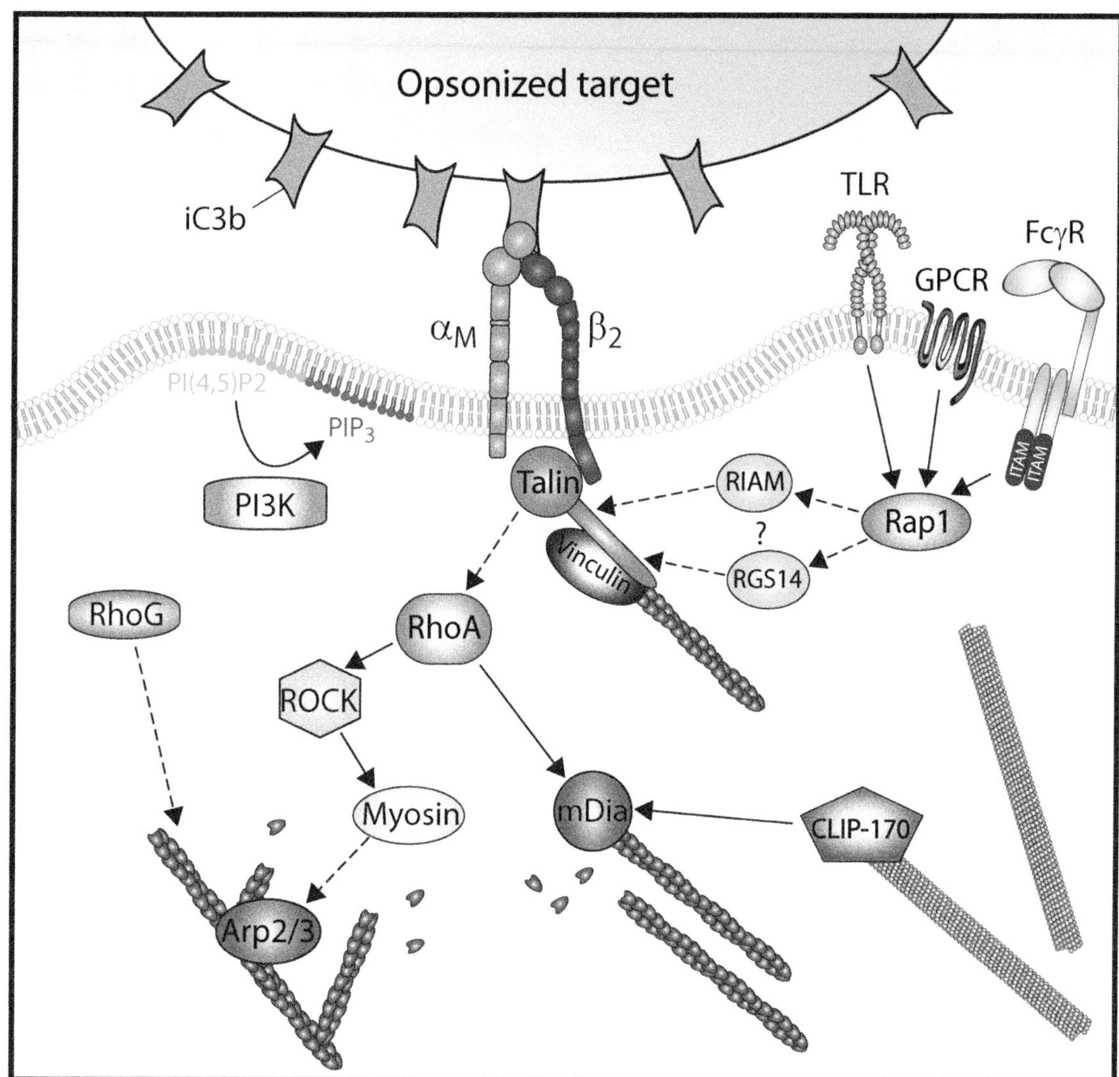

Figure 4 Actin reorganization during complement-mediated phagocytosis by the integrin $\alpha_M\beta_2$ (CR3). Rap-1-mediated inside-out activation of CR3 via its association with talin can be induced by various receptors, including TLRs, G protein-coupled receptors (GPCRs), and Fc receptors. Engagement of CR3 leads to the activation of PI3K and the small GTPases RhoA and RhoG. RhoA activates the actin nucleator of the formin family mDia1, which stimulates actin polymerization into a linear network, whereas the serine/threonine kinase ROCK activates myosin II, which favors the recruitment of the Arp2/3 complex, leading to the polymerization of a branched actin network.

CR3 activation by inside-out signaling involves the small GTPase Rap1 and talin-1 (15, 144). The head domain of talin-1, which binds the β_2 subunit, is sufficient to stabilize the active form of CR3, whereas the rod domain of talin-1, which associates with actin filaments, is required for phagocytosis (144). Typically, Rap1-mediated talin-integrin association requires an additional effector. The role of Rap1-GTP-interacting adaptor molecule (RIAM), which binds both Rap1 and talin and is implicated in $\alpha_{IIb}\beta_3$ activation, has been a matter of debate (145, 146). On the other hand, G-protein signaling-14 (RGS14) appears to play a role in CR3 activation, by an unknown mechanism (147).

The outside-in signaling cascade triggered by engagement of CR3 during phagocytosis appears to be quite distinct compared to the mechanisms we have described so far. In contrast to FcγR-mediated phagocytosis, tyrosine kinase inhibitors do not prevent complement-mediated phagocytosis (148, 149). In addition, macrophages derived from the bone marrow of *hck*$^{-/-}$*fgr*$^{-/-}$

lyn$^{-/-}$ triple knockouts or from *syk*$^{-/-}$ mice have impaired phagocytosis of IgG-opsonized, but not iC3b-opsonized, erythrocytes (69, 72). Yet Syk was reported to be involved in phagocytosis of iC3b-opsonized zymosan particle (150). This discrepancy might be explained by zymosan engagement of other receptors, particularly Dectin-1. In contrast to tyrosine kinases, PI3K activity is required for phagocytosis of iC3b-opsonized erythrocytes (11). However, PI3K is dispensable for actin accumulation around the phagosome, and its exact role in particle internalization remains to be clarified.

Expression of dominant-negative mutants, treatment with specific inhibitors, and gene silencing show that complement-mediated particle uptake requires the small GTPases RhoA and RhoG but not Cdc42 and Rac (99, 115, 122, 127). In addition, a marked activation of RhoA is observed during the phagocytosis of iC3b-opsonized erythrocytes, whereas Rac activity remains unchanged (151). Surprisingly, the fraction of macrophages that ingest iC3b-opsonized erythrocytes has been found to be drastically decreased by knocking out the *rac1* and *rac2* genes (126). Furthermore, knockout of the genes encoding the Vav proteins, which are GEFs for Rac1/2/3, RhoA/B/C, and RhoG but not Cdc42, also impaired CR3-mediated phagocytosis (126). Expression of constitutively active Rac complements the defect of the *vav1/3* knockout, suggesting that RhoA and RhoG activation might be mediated by other GEFs. However, knocking out *rac1/2* and *vav1/3* had no effect on CR3-mediated phagocytosis in neutrophils (152). It is conceivable that Rac plays a role mainly in the ruffling elicited by priming agents (like phorbol esters), which may be essential to make contact with target particles rather than for their internalization (54) (Fig. 4).

CR3-mediated phagocytosis involves both the actin and microtubule cytoskeletons (9). RhoA activates actin polymerization by recruiting the formin mDia1 to the phagocytic cup (127). Formins nucleate actin polymerization and associate with the barbed end of the filament, increasing the rate of actin monomer incorporation and preventing binding of capping proteins. The microtubule-associated protein CLIP-170 (cytoplasmic linker protein 170) is also involved in mDia1 recruitment at the phagocytic cup, providing a connection between the microtubule cytoskeleton and actin reorganization (153). RhoA also orchestrates the actin cytoskeleton reorganization by activation of the serine/threonine kinase ROCK (Rho kinase), which in turn phosphorylates and activates the regulatory subunit of myosin II. ROCK and myosin II are required for CR3-mediated phagocytosis and appear to promote the recruitment of the actin nucleation complex Arp2/3 (136). Heterologous overexpression of the VCA domain of SCAR inhibits Arp2/3 recruitment and prevents phagocytosis of iC3b-opsonized beads (115). How RhoG participates in particle uptake remains unclear. However, studies in fibroblasts showed that RhoG induces membrane ruffling and cellular migration by both Rac1-dependent and -independent processes, suggesting a possible role in Arp2/3 activation during CR3-mediated phagocytosis (154).

Phagocytosis of Apoptotic Cells Mediated by TIM-4 in Cooperation with Integrins

As introduced earlier, multiple different phagocytic receptors are capable of binding apoptotic corpses (Table 1); their simultaneous engagement makes studies of the specific downstream signaling pathways associated with each receptor rather challenging. Nevertheless, recent progress has been made in the understanding of TIM-4, which appears to be a major receptor for phagocytosis of effete cells by murine macrophages (35, 155). Surprisingly, whereas heterologous expression of TIM-4 in nonphagocytic cells is sufficient to induce apoptotic cell uptake, its cytosolic tail and transmembrane domains are dispensable, suggesting that TIM-4 does not signal by itself (35, 156). In fact, TIM-4-mediated phagocytosis is promoted by the expression of β_1, β_3, and β_5 integrins (157, 158). Since the close association of a particle is not sufficient to induce its phagocytosis, and artificial clustering of TIM-4 increases its colocalization with β_1 integrins, it has been proposed that TIM-4 interacts directly with integrins. Consistent with this, TIM-4-mediated phagocytosis involves SFKs, the focal adhesion kinase (FAK), and PI3K, as well as the small GTPases Rac1, Rac2, and Rho and their GEF, Vav3 (157) (Fig. 5). Whereas this molecular pathway is clearly distinct from CR3-mediated phagocytosis, the same signaling cascade is involved in integrin-mediated leukocyte adhesion and migration (159). It is also noteworthy that the crystal structure of TIM-4 reveals an accessible, conserved RGD motif, which might allow a direct association with integrins (160).

CONCLUSION

Exploration of the molecular mechanisms underlying particle uptake led to the identification of numerous phagocytic receptors and several distinct signaling pathways that lead to the reorganization of the actin cytoskeleton and ultimately to phagosome formation. The study of these signaling pathways individually enabled the discovery of key mechanisms. Yet to date

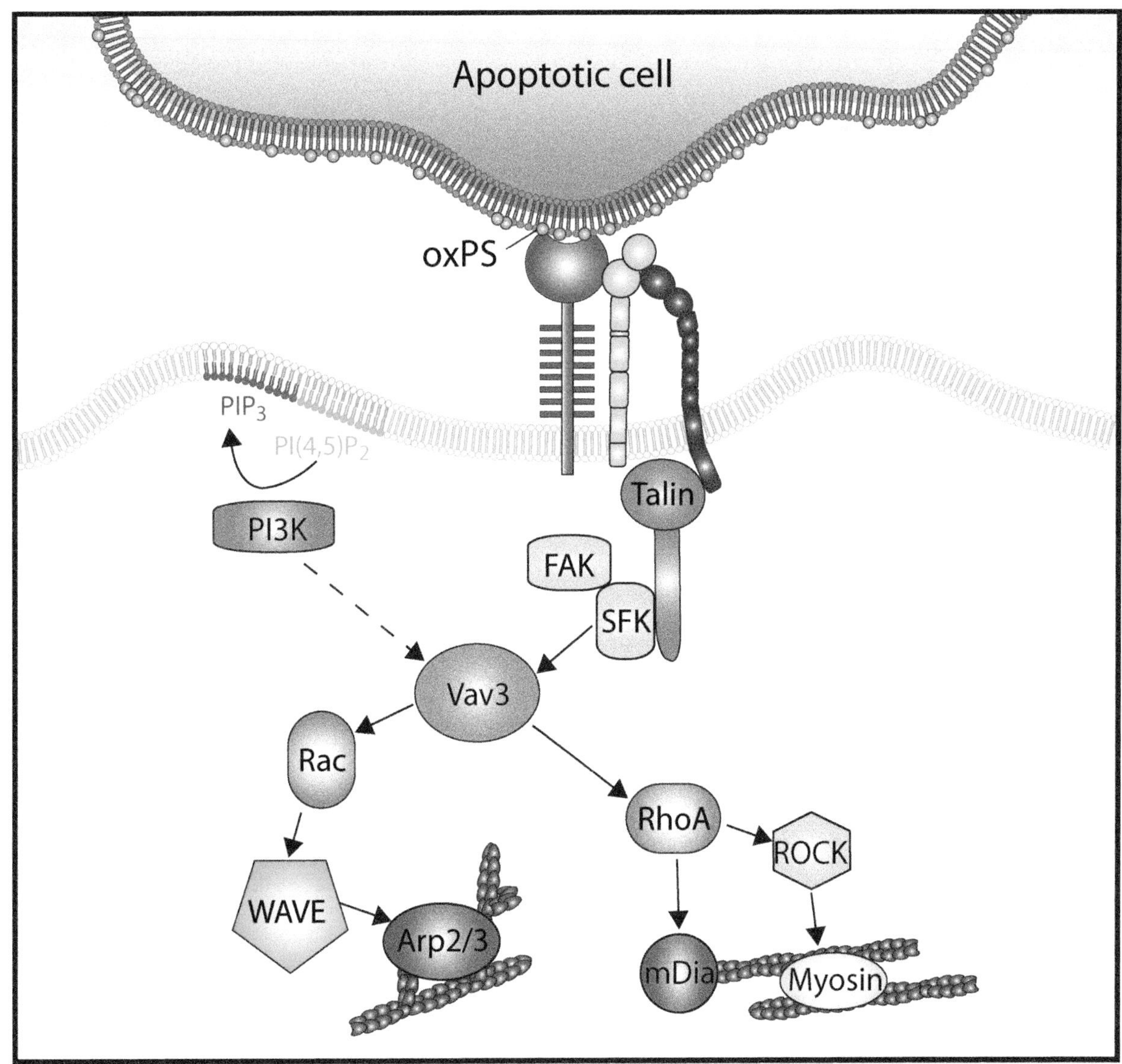

Figure 5 Signaling events in phagocytosis of apoptotic bodies mediated by TIM-4. TIM-4 and integrins cooperate to take up apoptotic bodies. SFK, FAK, and PI3K activities lead to the stimulation of Vav3, a GEF for RhoA and Rac, which activate the actin nucleators mDia and Arp2/3, respectively. oxPS; oxidized phosphatidylserine.

none of them have been fully elucidated, and several important questions are still pending. How do the membranes of the extending pseudopods fuse to seal the phagosome? How is actin polymerization terminated and reversed? How can phagocytic integrins sometimes elicit anti-inflammatory responses, yet produce inflammation in other cases? What specific receptors and signaling pathways are implicated in physiological scenarios, when phagocytes are confronted with particles bearing multiple PAMPs as well as opsonins? As observed for FcγR and CR3, it is very likely that multiple receptors cooperate to efficiently clear unwanted materials (161, 162). The example of TIM-4 highlights the role of coreceptors in particle uptake. In a similar fashion, the mechanisms involved in phagocytosis by CD36 or Mer are very reminiscent of integrin-mediated signaling, suggesting that integrins might be broadly used as coreceptors for phagocytosis (163, 164).

Clearly, despite the wealth of accumulated knowledge, much remains to be learned about phagocytosis and the manner in which phagocytic receptors cooperate with other immunological signaling pathways in order to stage an appropriate response to both exogenous and endogenous threats (165).

Acknowledgments. *Original work in the authors' laboratory is supported by the Canadian Institutes of Health Research. V.J. is supported by the Fondation Bettencourt Schueller. We thank Spencer Freeman for his technical support in preparing the figures.*

Citation. Jaumouillé V, Grinstein S. 2016. Molecular mechanisms of phagosome formation. Microbiol Spectrum 4(3): MCHD-0013-2015.

References

1. **Cannon GJ, Swanson JA.** 1992. The macrophage capacity for phagocytosis. *J Cell Sci* **101:**907–913.
2. **Roberts J, Quastel JH.** 1963. Particle uptake by polymorphonuclear leucocytes and Ehrlich ascites-carcinoma cells. *Biochem J* **89:**150–156.
3. **Champion JA, Mitragotri S.** 2006. Role of target geometry in phagocytosis. *Proc Natl Acad Sci U S A* **103:** 4930–4934.
4. **Paul D, Achouri S, Yoon YZ, Herre J, Bryant CE, Cicuta P.** 2013. Phagocytosis dynamics depends on target shape. *Biophys J* **105:**1143–1150.
5. **Griffin FM Jr, Griffin JA, Leider JE, Silverstein SC.** 1975. Studies on the mechanism of phagocytosis. I. Requirements for circumferential attachment of particle-bound ligands to specific receptors on the macrophage plasma membrane. *J Exp Med* **142:**1263–1282.
6. **Swanson JA, Baer SC.** 1995. Phagocytosis by zippers and triggers. *Trends Cell Biol* **5:**89–93.
7. **Axline SG, Reaven EP.** 1974. Inhibition of phagocytosis and plasma membrane mobility of the cultivated macrophage by cytochalasin B. Role of subplasmalemmal microfilaments. *J Cell Biol* **62:**647–659.
8. **Malawista SE, Gee JB, Bensch KG.** 1971. Cytochalasin B reversibly inhibits phagocytosis: functional, metabolic, and ultrastructural effects in human blood leukocytes and rabbit alveolar macrophages. *Yale J Biol Med* **44:** 286–300.
9. **Newman SL, Mikus LK, Tucci MA.** 1991. Differential requirements for cellular cytoskeleton in human macrophage complement receptor- and Fc receptor-mediated phagocytosis. *J Immunol* **146:**967–974.
10. **Tollis S, Dart AE, Tzircotis G, Endres RG.** 2010. The zipper mechanism in phagocytosis: energetic requirements and variability in phagocytic cup shape. *BMC Syst Biol* **4:**149. doi:10.1186/1752-0509-4-149.
11. **Cox D, Dale BM, Kashiwada M, Helgason CD, Greenberg S.** 2001. A regulatory role for Src homology 2 domain-containing inositol 5′-phosphatase (SHIP) in phagocytosis mediated by Fcγ receptors and complement receptor 3 ($\alpha_M\beta_2$; CD11b/CD18). *J Exp Med* **193:**61–71.
12. **Herre J, Marshall AS, Caron E, Edwards AD, Williams DL, Schweighoffer E, Tybulewicz V, Reis e Sousa C, Gordon S, Brown GD.** 2004. Dectin-1 uses novel mechanisms for yeast phagocytosis in macrophages. *Blood* **104:**4038–4045.
13. **Leverrier Y, Ridley AJ.** 2001. Requirement for Rho GTPases and PI 3-kinases during apoptotic cell phagocytosis by macrophages. *Curr Biol* **11:**195–199.
14. **Ninomiya N, Hazeki K, Fukui Y, Seya T, Okada T, Hazeki O, Ui M.** 1994. Involvement of phosphatidylinositol 3-kinase in Fcγ receptor signaling. *J Biol Chem* **269:**22732–22737.
15. **Caron E, Self AJ, Hall A.** 2000. The GTPase Rap1 controls functional activation of macrophage integrin αMβ2 by LPS and other inflammatory mediators. *Curr Biol* **10:**974–978.
16. **Griffin FM, Griffin JA.** 1980. Augmentation of macrophage complement receptor function in vitro. II. Characterization of the effects of a unique lymphokine upon the phagocytic capabilities of macrophages. *J Immunol* **125:**844–849.
17. **Fridman WH, Gresser I, Bandu MT, Aguet M, Neauport-Sautes C.** 1980. Interferon enhances the expression of Fc gamma receptors. *J Immunol* **124:**2436–2441.
18. **Itoh K, Inoue M, Kataoka S, Kumagai K.** 1980. Differential effect of interferon expression of IgG- and IgM-Fc receptors on human lymphocytes. *J Immunol* **124:** 2589–2595.
19. **Allen JM, Seed B.** 1989. Isolation and expression of functional high-affinity Fc receptor complementary DNAs. *Science* **243:**378–381.
20. **Brown GD, Gordon S.** 2001. Immune recognition. A new receptor for β-glucans. *Nature* **413:**36–37.
21. **Doyle SE, O'Connell RM, Miranda GA, Vaidya SA, Chow EK, Liu PT, Suzuki S, Suzuki N, Modlin RL, Yeh WC, Lane TF, Cheng G.** 2004. Toll-like receptors induce a phagocytic gene program through p38. *J Exp Med* **199:**81–90.
22. **Ezekowitz RA, Sastry K, Bailly P, Warner A.** 1990. Molecular characterization of the human macrophage mannose receptor: demonstration of multiple carbohydrate recognition-like domains and phagocytosis of yeasts in Cos-1 cells. *J Exp Med* **172:**1785–1794.
23. **van der Laan LJ, Döpp EA, Haworth R, Pikkarainen T, Kangas M, Elomaa O, Dijkstra CD, Gordon S, Tryggvason K, Kraal G.** 1999. Regulation and functional involvement of macrophage scavenger receptor MARCO in clearance of bacteria in vivo. *J Immunol* **162:**939–947.
24. **Peiser L, Gough PJ, Kodama T, Gordon S.** 2000. Macrophage class A scavenger receptor-mediated phagocytosis of *Escherichia coli*: role of cell heterogeneity, microbial strain, and culture conditions in vitro. *Infect Immun* **68:**1953–1963.
25. **Thomas CA, Li Y, Kodama T, Suzuki H, Silverstein SC, El Khoury J.** 2000. Protection from lethal gram-positive infection by macrophage scavenger receptor-dependent phagocytosis. *J Exp Med* **191:**147–156.
26. **Józefowski S, Arredouani M, Sulahian T, Kobzik L.** 2005. Disparate regulation and function of the class A scavenger receptors SR-AI/II and MARCO. *J Immunol* **175:**8032–8041.
27. **Anderson CL, Shen L, Eicher DM, Wewers MD, Gill JK.** 1990. Phagocytosis mediated by three distinct Fcγ receptor classes on human leukocytes. *J Exp Med* **171:**1333–1345.
28. **Nimmerjahn F, Ravetch JV.** 2008. Fcγ receptors as regulators of immune responses. *Nat Rev Immunol* **8:** 34–47.

29. **Blystone SD, Graham IL, Lindberg FP, Brown EJ.** 1994. Integrin $\alpha_v\beta_3$ differentially regulates adhesive and phagocytic functions of the fibronectin receptor $\alpha_5\beta_1$. *J Cell Biol* **127:**1129–1137.

30. **Schiff DE, Kline L, Soldau K, Lee JD, Pugin J, Tobias PS, Ulevitch RJ.** 1997. Phagocytosis of Gram-negative bacteria by a unique CD14-dependent mechanism. *J Leukoc Biol* **62:**786–794.

31. **Fadok VA, Voelker DR, Campbell PA, Cohen JJ, Bratton DL, Henson PM.** 1992. Exposure of phosphatidylserine on the surface of apoptotic lymphocytes triggers specific recognition and removal by macrophages. *J Immunol* **148:**2207–2216.

32. **Borisenko GG, Matsura T, Liu SX, Tyurin VA, Jianfei J, Serinkan FB, Kagan VE.** 2003. Macrophage recognition of externalized phosphatidylserine and phagocytosis of apoptotic Jurkat cells—existence of a threshold. *Arch Biochem Biophys* **413:**41–52.

33. **Suzuki J, Umeda M, Sims PJ, Nagata S.** 2010. Calcium-dependent phospholipid scrambling by TMEM16F. *Nature* **468:**834–838.

34. **He M, Kubo H, Morimoto K, Fujino N, Suzuki T, Takahasi T, Yamada M, Yamaya M, Maekawa T, Yamamoto Y, Yamamoto H.** 2011. Receptor for advanced glycation end products binds to phosphatidylserine and assists in the clearance of apoptotic cells. *EMBO Rep* **12:**358–364.

35. **Kobayashi N, Karisola P, Peña-Cruz V, Dorfman DM, Jinushi M, Umetsu SE, Butte MJ, Nagumo H, Chernova I, Zhu B, Sharpe AH, Ito S, Dranoff G, Kaplan GG, Casasnovas JM, Umetsu DT, Dekruyff RH, Freeman GJ.** 2007. TIM-1 and TIM-4 glycoproteins bind phosphatidylserine and mediate uptake of apoptotic cells. *Immunity* **27:**927–940.

36. **Nakayama M, Akiba H, Takeda K, Kojima Y, Hashiguchi M, Azuma M, Yagita H, Okumura K.** 2009. Tim-3 mediates phagocytosis of apoptotic cells and cross-presentation. *Blood* **113:**3821–3830.

37. **Park D, Tosello-Trampont AC, Elliott MR, Lu M, Haney LB, Ma Z, Klibanov AL, Mandell JW, Ravichandran KS.** 2007. BAI1 is an engulfment receptor for apoptotic cells upstream of the ELMO/Dock180/Rac module. *Nature* **450:**430–434.

38. **Park SY, Jung MY, Kim HJ, Lee SJ, Kim SY, Lee BH, Kwon TH, Park RW, Kim IS.** 2008. Rapid cell corpse clearance by stabilin-2, a membrane phosphatidylserine receptor. *Cell Death Differ* **15:**192–201.

39. **Hanayama R, Tanaka M, Miwa K, Shinohara A, Iwamatsu A, Nagata S.** 2002. Identification of a factor that links apoptotic cells to phagocytes. *Nature* **417:** 182–187.

40. **Anderson HA, Maylock CA, Williams JA, Paweletz CP, Shu H, Shacter E.** 2003. Serum-derived protein S binds to phosphatidylserine and stimulates the phagocytosis of apoptotic cells. *Nat Immunol* **4:**87–91.

41. **Segawa K, Suzuki J, Nagata S.** 2011. Constitutive exposure of phosphatidylserine on viable cells. *Proc Natl Acad Sci U S A* **108:**19246–19251.

42. **Arur S, Uche UE, Rezaul K, Fong M, Scranton V, Cowan AE, Mohler W, Han DK.** 2003. Annexin I is an endogenous ligand that mediates apoptotic cell engulfment. *Dev Cell* **4:**587–598.

43. **Gardai SJ, McPhillips KA, Frasch SC, Janssen WJ, Starefeldt A, Murphy-Ullrich JE, Bratton DL, Oldenborg PA, Michalak M, Henson PM.** 2005. Cell-surface calreticulin initiates clearance of viable or apoptotic cells through *trans*-activation of LRP on the phagocyte. *Cell* **123:**321–334.

44. **Tsai RK, Discher DE.** 2008. Inhibition of "self" engulfment through deactivation of myosin-II at the phagocytic synapse between human cells. *J Cell Biol* **180:** 989–1003.

45. **Brown S, Heinisch I, Ross E, Shaw K, Buckley CD, Savill J.** 2002. Apoptosis disables CD31-mediated cell detachment from phagocytes promoting binding and engulfment. *Nature* **418:**200–203.

46. **Gibbons MM, Chou T, D'Orsogna MR.** 2010. Diffusion-dependent mechanisms of receptor engagement and viral entry. *J Phys Chem B* **114:**15403–15412.

47. **Jaumouillé V, Grinstein S.** 2011. Receptor mobility, the cytoskeleton, and particle binding during phagocytosis. *Curr Opin Cell Biol* **23:**22–29.

48. **Sica A, Mantovani A.** 2012. Macrophage plasticity and polarization: in vivo veritas. *J Clin Invest* **122:**787–795.

49. **Jaumouillé V, Farkash Y, Jaqaman K, Das R, Lowell CA, Grinstein S.** 2014. Actin cytoskeleton reorganization by Syk regulates Fcγ receptor responsiveness by increasing its lateral mobility and clustering. *Dev Cell* **29:** 534–546.

50. **Botelho RJ, Harrison RE, Stone JC, Hancock JF, Philips MR, Jongstra-Bilen J, Mason D, Plumb J, Gold MR, Grinstein S.** 2009. Localized diacylglycerol-dependent stimulation of Ras and Rap1 during phagocytosis. *J Biol Chem* **284:**28522–28532.

51. **Kong F, García AJ, Mould AP, Humphries MJ, Zhu C.** 2009. Demonstration of catch bonds between an integrin and its ligand. *J Cell Biol* **185:**1275–1284.

52. **Flannagan RS, Harrison RE, Yip CM, Jaqaman K, Grinstein S.** 2010. Dynamic macrophage "probing" is required for the efficient capture of phagocytic targets. *J Cell Biol* **191:**1205–1218.

53. **Kress H, Stelzer EH, Holzer D, Buss F, Griffiths G, Rohrbach A.** 2007. Filopodia act as phagocytic tentacles and pull with discrete steps and a load-dependent velocity. *Proc Natl Acad Sci U S A* **104:**11633–11638.

54. **Patel PC, Harrison RE.** 2008. Membrane ruffles capture C3bi-opsonized particles in activated macrophages. *Mol Biol Cell* **19:**4628–4639.

55. **Vonna L, Wiedemann A, Aepfelbacher M, Sackmann E.** 2007. Micromechanics of filopodia mediated capture of pathogens by macrophages. *Eur Biophys J* **36:**145–151.

56. **Araki N, Johnson MT, Swanson JA.** 1996. A role for phosphoinositide 3-kinase in the completion of macropinocytosis and phagocytosis by macrophages. *J Cell Biol* **135:**1249–1260.

57. **Papakonstanti EA, Zwaenepoel O, Bilancio A, Burns E, Nock GE, Houseman B, Shokat K, Ridley AJ, Vanhaesebroeck B.** 2008. Distinct roles of class IA PI3K isoforms in primary and immortalised macrophages. *J Cell Sci* **121**:4124–4133.

58. **West MA, Prescott AR, Eskelinen EL, Ridley AJ, Watts C.** 2000. Rac is required for constitutive macropinocytosis by dendritic cells but does not control its downregulation. *Curr Biol* **10**:839–848.

59. **Wheeler AP, Wells CM, Smith SD, Vega FM, Henderson RB, Tybulewicz VL, Ridley AJ.** 2006. Rac1 and Rac2 regulate macrophage morphology but are not essential for migration. *J Cell Sci* **119**:2749–2757.

60. **Bohdanowicz M, Schlam D, Hermansson M, Rizzuti D, Fairn GD, Ueyama T, Somerharju P, Du G, Grinstein S.** 2013. Phosphatidic acid is required for the constitutive ruffling and macropinocytosis of phagocytes. *Mol Biol Cell* **24**:1700–1712, S1–7.

61. **Poulsen HL.** 1974. Interstitial fluid concentrations of albumin and immunoglobulin G in normal men. *Scand J Clin Lab Invest* **34**:119–122.

62. **Woof JM, Burton DR.** 2004. Human antibody-Fc receptor interactions illuminated by crystal structures. *Nat Rev Immunol* **4**:89–99.

63. **Jones DH, Nusbacher J, Anderson CL.** 1985. Fc receptor-mediated binding and endocytosis by human mononuclear phagocytes: monomeric IgG is not endocytosed by U937 cells and monocytes. *J Cell Biol* **100**: 558–564.

64. **Odin JA, Edberg JC, Painter CJ, Kimberly RP, Unkeless JC.** 1991. Regulation of phagocytosis and $[Ca^{2+}]_i$ flux by distinct regions of an Fc receptor. *Science* **254**:1785–1788.

65. **Holowka D, Sil D, Torigoe C, Baird B.** 2007. Insights into immunoglobulin E receptor signaling from structurally defined ligands. *Immunol Rev* **217**:269–279.

66. **Ghazizadeh S, Bolen JB, Fleit HB.** 1994. Physical and functional association of Src-related protein tyrosine kinases with FcγRII in monocytic THP-1 cells. *J Biol Chem* **269**:8878–8884.

67. **Hamada F, Aoki M, Akiyama T, Toyoshima K.** 1993. Association of immunoglobulin G Fc receptor II with Src-like protein-tyrosine kinase Fgr in neutrophils. *Proc Natl Acad Sci U S A* **90**:6305–6309.

68. **Wang AV, Scholl PR, Geha RS.** 1994. Physical and functional association of the high affinity immunoglobulin G receptor (FcγRI) with the kinases Hck and Lyn. *J Exp Med* **180**:1165–1170.

69. **Fitzer-Attas CJ, Lowry M, Crowley MT, Finn AJ, Meng F, DeFranco AL, Lowell CA.** 2000. Fcγ receptor-mediated phagocytosis in macrophages lacking the Src family tyrosine kinases Hck, Fgr, and Lyn. *J Exp Med* **191**:669–682.

70. **Ghazizadeh S, Bolen JB, Fleit HB.** 1995. Tyrosine phosphorylation and association of Syk with FcγRII in monocytic THP-1 cells. *Biochem J* **305**:669–674.

71. **Johnson SA, Pleiman CM, Pao L, Schneringer J, Hippen K, Cambier JC.** 1995. Phosphorylated immunoreceptor signaling motifs (ITAMs) exhibit unique abilities to bind and activate Lyn and Syk tyrosine kinases. *J Immunol* **155**:4596–4603.

72. **Kiefer F, Brumell J, Al-Alawi N, Latour S, Cheng A, Veillette A, Grinstein S, Pawson T.** 1998. The Syk protein tyrosine kinase is essential for Fcγ receptor signaling in macrophages and neutrophils. *Mol Cell Biol* **18**: 4209–4220.

73. **Mukherjee S, Zhu J, Zikherman J, Parameswaran R, Kadlecek TA, Wang Q, Au-Yeung B, Ploegh H, Kuriyan J, Das J, Weiss A.** 2013. Monovalent and multivalent ligation of the B cell receptor exhibit differential dependence upon Syk and Src family kinases. *Sci Signal* **6**:ra1. doi:10.1126/scisignal.2003220.

74. **Crowley MT, Costello PS, Fitzer-Attas CJ, Turner M, Meng F, Lowell C, Tybulewicz VL, DeFranco AL.** 1997. A critical role for Syk in signal transduction and phagocytosis mediated by Fcγ receptors on macrophages. *J Exp Med* **186**:1027–1039.

75. **Kwiatkowska K, Sobota A.** 2001. The clustered Fcγ receptor II is recruited to Lyn-containing membrane domains and undergoes phosphorylation in a cholesterol-dependent manner. *Eur J Immunol* **31**:989–998.

76. **Rollet-Labelle E, Marois S, Barbeau K, Malawista SE, Naccache PH.** 2004. Recruitment of the cross-linked opsonic receptor CD32A (FcγRIIA) to high-density detergent-resistant membrane domains in human neutrophils. *Biochem J* **381**:919–928.

77. **García-García E, Brown EJ, Rosales C.** 2007. Transmembrane mutations to FcγRIIA alter its association with lipid rafts: implications for receptor signaling. *J Immunol* **178**:3048–3058.

78. **Douglass AD, Vale RD.** 2005. Single-molecule microscopy reveals plasma membrane microdomains created by protein-protein networks that exclude or trap signaling molecules in T cells. *Cell* **121**:937–950.

79. **Hashimoto-Tane A, Yokosuka T, Ishihara C, Sakuma M, Kobayashi W, Saito T.** 2010. T-cell receptor microclusters critical for T-cell activation are formed independently of lipid raft clustering. *Mol Cell Biol* **30**: 3421–3429.

80. **Kovárová M, Tolar P, Arudchandran R, Dráberová L, Rivera J, Dráber P.** 2001. Structure-function analysis of Lyn kinase association with lipid rafts and initiation of early signaling events after Fcε receptor I aggregation. *Mol Cell Biol* **21**:8318–8328.

81. **Johnson KG, Bromley SK, Dustin ML, Thomas ML.** 2000. A supramolecular basis for CD45 tyrosine phosphatase regulation in sustained T cell activation. *Proc Natl Acad Sci U S A* **97**:10138–10143.

82. **Yamauchi S, Kawauchi K, Sawada Y.** 2012. Myosin II-dependent exclusion of CD45 from the site of Fcγ receptor activation during phagocytosis. *FEBS Lett* **586**: 3229–3235.

83. **Cordoba S-P, Choudhuri K, Zhang H, Bridge M, Basat AB, Dustin ML, van der Merwe PA.** 2013. The large ectodomains of CD45 and CD148 regulate their segregation from and inhibition of ligated T-cell receptor. *Blood* **121**:4295–4302.

84. **Wright SD, Silverstein SC.** 1984. Phagocytosing macrophages exclude proteins from the zones of contact with opsonized targets. *Nature* **309:**359–361.

85. **Zhu JW, Brdicka T, Katsumoto TR, Lin J, Weiss A.** 2008. Structurally distinct phosphatases CD45 and CD148 both regulate B cell and macrophage immunoreceptor signaling. *Immunity* **28:**183–196.

86. **Schoenborn JR, Tan YX, Zhang C, Shokat KM, Weiss A.** 2011. Feedback circuits monitor and adjust basal Lck-dependent events in T cell receptor signaling. *Sci Signal* **4:**ra59. doi:10.1126/scisignal.2001893.

87. **Coxon PY, Rane MJ, Powell DW, Klein JB, McLeish KR.** 2000. Differential mitogen-activated protein kinase stimulation by Fcγ receptor IIa and Fcγ receptor IIIb determines the activation phenotype of human neutrophils. *J Immunol* **164:**6530–6537.

88. **Treanor B, Depoil D, Gonzalez-Granja A, Barral P, Weber M, Dushek O, Bruckbauer A, Batista FD.** 2010. The membrane skeleton controls diffusion dynamics and signaling through the B cell receptor. *Immunity* **32:** 187–199.

89. **Tan YX, Manz BN, Freedman TS, Zhang C, Shokat KM, Weiss A.** 2014. Inhibition of the kinase Csk in thymocytes reveals a requirement for actin remodeling in the initiation of full TCR signaling. *Nat Immunol* **15:** 186–194.

90. **Tridandapani S, Lyden TW, Smith JL, Carter JE, Coggeshall KM, Anderson CL.** 2000. The adapter protein LAT enhances Fcγ receptor-mediated signal transduction in myeloid cells. *J Biol Chem* **275:**20480–20487.

91. **Zhang W, Sloan-Lancaster J, Kitchen J, Trible RP, Samelson LE.** 1998. LAT: the ZAP-70 tyrosine kinase substrate that links T cell receptor to cellular activation. *Cell* **92:**83–92.

92. **Gu H, Botelho RJ, Yu M, Grinstein S, Neel BG.** 2003. Critical role for scaffolding adapter Gab2 in FcγR-mediated phagocytosis. *J Cell Biol* **161:**1151–1161.

93. **Moon KD, Post CB, Durden DL, Zhou Q, De P, Harrison ML, Geahlen RL.** 2005. Molecular basis for a direct interaction between the Syk protein-tyrosine kinase and phosphoinositide 3-kinase. *J Biol Chem* **280:** 1543–1551.

94. **Botelho RJ, Teruel M, Dierckman R, Anderson R, Wells A, York JD, Meyer T, Grinstein S.** 2000. Localized biphasic changes in phosphatidylinositol-4,5-bisphosphate at sites of phagocytosis. *J Cell Biol* **151:**1353–1368.

95. **Liao F, Shin HS, Rhee SG.** 1992. Tyrosine phosphorylation of phospholipase C-γ1 induced by cross-linking of the high-affinity or low-affinity Fc receptor for IgG in U937 cells. *Proc Natl Acad Sci U S A* **89:**3659–3663.

96. **Coppolino MG, Krause M, Hagendorff P, Monner DA, Trimble W, Grinstein S, Wehland J, Sechi AS.** 2001. Evidence for a molecular complex consisting of Fyb/SLAP, SLP-76, Nck, VASP and WASP that links the actin cytoskeleton to Fcγ receptor signalling during phagocytosis. *J Cell Sci* **114:**4307–4318.

97. **Lee WL, Cosio G, Ireton K, Grinstein S.** 2007. Role of CrkII in Fcγ receptor-mediated phagocytosis. *J Biol Chem* **282:**11135–11143.

98. **Jankowski A, Zhu P, Marshall JG.** 2008. Capture of an activated receptor complex from the surface of live cells by affinity receptor chromatography. *Anal Biochem* **380:**235–248.

99. **Tzircotis G, Braga VMM, Caron E.** 2011. RhoG is required for both FcγR- and CR3-mediated phagocytosis. *J Cell Sci* **124:**2897–2902.

100. **Dart AE, Donnelly SK, Holden DW, Way M, Caron E.** 2012. Nck and Cdc42 co-operate to recruit N-WASP to promote FcγR-mediated phagocytosis. *J Cell Sci* **125:** 2825–2830.

101. **Izadi KD, Erdreich-Epstein A, Liu Y, Durden DL.** 1998. Characterization of Cbl-Nck and Nck-Pak1 interactions in myeloid FcγRII signaling. *Exp Cell Res* **245:**330–342.

102. **Coppolino MG, Dierckman R, Loijens J, Collins RF, Pouladi M, Jongstra-Bilen J, Schreiber AD, Trimble WS, Anderson R, Grinstein S.** 2002. Inhibition of phosphatidylinositol-4-phosphate 5-kinase Iα impairs localized actin remodeling and suppresses phagocytosis. *J Biol Chem* **277:**43849–43857.

103. **Scott CC, Dobson W, Botelho RJ, Coady-Osberg N, Chavrier P, Knecht DA, Heath C, Stahl P, Grinstein S.** 2005. Phosphatidylinositol-4,5-bisphosphate hydrolysis directs actin remodeling during phagocytosis. *J Cell Biol* **169:**139–149.

104. **Marshall JG, Booth JW, Stambolic V, Mak T, Balla T, Schreiber AD, Meyer T, Grinstein S.** 2001. Restricted accumulation of phosphatidylinositol 3-kinase products in a plasmalemmal subdomain during Fcγ receptor-mediated phagocytosis. *J Cell Biol* **153:**1369–1380.

105. **Larsen EC, Ueyama T, Brannock PM, Shirai Y, Saito N, Larsson C, Loegering D, Weber PB, Lennartz MR.** 2002. A role for PKC-ε in FcγR-mediated phagocytosis by RAW 264.7 cells. *J Cell Biol* **159:**939–944.

106. **Cox D, Tseng CC, Bjekic G, Greenberg S.** 1999. A requirement for phosphatidylinositol 3-kinase in pseudopod extension. *J Biol Chem* **274:**1240–1247.

107. **Bohdanowicz M, Balkin DM, De Camilli P, Grinstein S.** 2012. Recruitment of OCRL and Inpp5B to phagosomes by Rab5 and APPL1 depletes phosphoinositides and attenuates Akt signaling. *Mol Biol Cell* **23:**176–187.

108. **Marion S, Mazzolini J, Herit F, Bourdoncle P, Kambou-Pene N, Hailfinger S, Sachse M, Ruland J, Benmerah A, Echard A, Thome M, Niedergang F.** 2012. The NF-κB signaling protein Bcl10 regulates actin dynamics by controlling AP1 and OCRL-bearing vesicles. *Dev Cell* **23:** 954–967.

109. **Bajno L, Peng XR, Schreiber AD, Moore HP, Trimble WS, Grinstein S.** 2000. Focal exocytosis of VAMP3-containing vesicles at sites of phagosome formation. *J Cell Biol* **149:**697–706.

110. **Braun V, Fraisier V, Raposo G, Hurbain I, Sibarita J-B, Chavrier P, Galli T, Niedergang F.** 2004. TI-VAMP/VAMP7 is required for optimal phagocytosis of opsonised particles in macrophages. *EMBO J* **23:**4166–4176.

111. **Cox D, Lee DJ, Dale BM, Calafat J, Greenberg S.** 2000. A Rab11-containing rapidly recycling compartment in macrophages that promotes phagocytosis. *Proc Natl Acad Sci U S A* **97:**680–685.

112. **Lee WL, Mason D, Schreiber AD, Grinstein S.** 2007. Quantitative analysis of membrane remodeling at the phagocytic cup. *Mol Biol Cell* **18:**2883–2892.

113. **Beemiller P, Zhang Y, Mohan S, Levinsohn E, Gaeta I, Hoppe AD, Swanson JA.** 2010. A Cdc42 activation cycle coordinated by PI 3-kinase during Fc receptor-mediated phagocytosis. *Mol Biol Cell* **21:**470–480.

114. **Iyer SS, Barton JA, Bourgoin S, Kusner DJ.** 2004. Phospholipases D1 and D2 coordinately regulate macrophage phagocytosis. *J Immunol* **173:**2615–2623.

115. **May RC, Caron E, Hall A, Machesky LM.** 2000. Involvement of the Arp2/3 complex in phagocytosis mediated by FcγR or CR3. *Nat Cell Biol* **2:**246–248.

116. **Kheir WA, Gevrey J-C, Yamaguchi H, Isaac B, Cox D.** 2005. A WAVE2-Abi1 complex mediates CSF-1-induced F-actin-rich membrane protrusions and migration in macrophages. *J Cell Sci* **118:**5369–5379.

117. **Lorenzi R, Brickell PM, Katz DR, Kinnon C, Thrasher AJ.** 2000. Wiskott-Aldrich syndrome protein is necessary for efficient IgG-mediated phagocytosis. *Blood* **95:** 2943–2946.

118. **Park H, Cox D.** 2009. Cdc42 regulates Fcγ receptor-mediated phagocytosis through the activation and phosphorylation of Wiskott-Aldrich syndrome protein (WASP) and neural-WASP. *Mol Biol Cell* **20:**4500–4508.

119. **Tsuboi S, Meerloo J.** 2007. Wiskott-Aldrich syndrome protein is a key regulator of the phagocytic cup formation in macrophages. *J Biol Chem* **282:**34194–34203.

120. **Tsuboi S, Takada H, Hara T, Mochizuki N, Funyu T, Saitoh H, Terayama Y, Yamaya K, Ohyama C, Nonoyama S, Ochs HD.** 2009. FBP17 mediates a common molecular step in the formation of podosomes and phagocytic cups in macrophages. *J Biol Chem* **284:**8548–8556.

121. **Hoppe AD, Swanson JA.** 2004. Cdc42, Rac1, and Rac2 display distinct patterns of activation during phagocytosis. *Mol Biol Cell* **15:**3509–3519.

122. **Caron E, Hall A.** 1998. Identification of two distinct mechanisms of phagocytosis controlled by different Rho GTPases. *Science* **282:**1717–1721.

123. **Koh ALY, Sun CX, Zhu F, Glogauer M.** 2005. The role of Rac1 and Rac2 in bacterial killing. *Cell Immunol* **235:**92–97.

124. **Utomo A, Cullere X, Glogauer M, Swat W, Mayadas TN.** 2006. Vav proteins in neutrophils are required for FcγR-mediated signaling to Rac GTPases and nicotinamide adenine dinucleotide phosphate oxidase component p40(phox). *J Immunol* **177:**6388–6397.

125. **Niedergang F, Colucci-Guyon E, Dubois T, Raposo G, Chavrier P.** 2003. ADP ribosylation factor 6 is activated and controls membrane delivery during phagocytosis in macrophages. *J Cell Biol* **161:**1143–1150.

126. **Hall AB, Gakidis MA, Glogauer M, Wilsbacher JL, Gao S, Swat W, Brugge JS.** 2006. Requirements for Vav guanine nucleotide exchange factors and Rho GTPases in FcγR- and complement-mediated phagocytosis. *Immunity* **24:**305–316.

127. **Colucci-Guyon E, Niedergang F, Wallar BJ, Peng J, Alberts AS, Chavrier P.** 2005. A role for mammalian diaphanous-related formins in complement receptor (CR3)-mediated phagocytosis in macrophages. *Curr Biol* **15:**2007–2012.

128. **Carlier MF, Laurent V, Santolini J, Melki R, Didry D, Xia GX, Hong Y, Chua NH, Pantaloni D.** 1997. Actin depolymerizing factor (ADF/cofilin) enhances the rate of filament turnover: implication in actin-based motility. *J Cell Biol* **136:**1307–1322.

129. **Witke W, Li W, Kwiatkowski DJ, Southwick FS.** 2001. Comparisons of CapG and gelsolin-null macrophages: demonstration of a unique role for CapG in receptor-mediated ruffling, phagocytosis, and vesicle rocketing. *J Cell Biol* **154:**775–784.

130. **Serrander L, Skarman P, Rasmussen B, Witke W, Lew DP, Krause KH, Stendahl O, Nüsse O.** 2000. Selective inhibition of IgG-mediated phagocytosis in gelsolin-deficient murine neutrophils. *J Immunol* **165:**2451–2457.

131. **Cox D, Berg JS, Cammer M, Chinegwundoh JO, Dale BM, Cheney RE, Greenberg S.** 2002. Myosin X is a downstream effector of PI(3)K during phagocytosis. *Nat Cell Biol* **4:**469–477.

132. **Dart AE, Tollis S, Bright MD, Frankel G, Endres RG.** 2012. The motor protein myosin 1G functions in FcγR-mediated phagocytosis. *J Cell Sci* **125:**6020–6029.

133. **Diakonova M, Bokoch G, Swanson JA.** 2002. Dynamics of cytoskeletal proteins during Fcγ receptor-mediated phagocytosis in macrophages. *Mol Biol Cell* **13:**402–411.

134. **Swanson JA, Johnson MT, Beningo K, Post P, Mooseker M, Araki N.** 1999. A contractile activity that closes phagosomes in macrophages. *J Cell Sci* **112:** 307–316.

135. **Mansfield PJ, Shayman JA, Boxer LA.** 2000. Regulation of polymorphonuclear leukocyte phagocytosis by myosin light chain kinase after activation of mitogen-activated protein kinase. *Blood* **95:**2407–2412.

136. **Olazabal IM, Caron E, May RC, Schilling K, Knecht DA, Machesky LM.** 2002. Rho-kinase and myosin-II control phagocytic cup formation during CR, but not FcγR, phagocytosis. *Curr Biol* **12:**1413–1418.

137. **Rogers NC, Slack EC, Edwards AD, Nolte MA, Schulz O, Schweighoffer E, Williams DL, Gordon S, Tybulewicz VL, Brown GD, Reis e Sousa C.** 2005. Syk-dependent cytokine induction by Dectin-1 reveals a novel pattern recognition pathway for C type lectins. *Immunity* **22:** 507–517.

138. **Adams EL, Rice PJ, Graves B, Ensley HE, Yu H, Brown GD, Gordon S, Monteiro MA, Papp-Szabo E, Lowman DW, Power TD, Wempe MF, Williams DL.** 2008. Differential high-affinity interaction of dectin-1 with natural or synthetic glucans is dependent upon primary structure and is influenced by polymer chain length and side-chain branching. *J Pharmacol Exp Ther* **325:**115–123.

139. **Goodridge HS, Reyes CN, Becker CA, Katsumoto TR, Ma J, Wolf AJ, Bose N, Chan AS, Magee AS, Danielson ME, Weiss A, Vasilakos JP, Underhill DM.** 2011. Activation of the innate immune receptor Dectin-1 upon formation of a 'phagocytic synapse'. *Nature* **472:**471–475.

140. **Elsori DH, Yakubenko VP, Roome T, Thiagarajan PS, Bhattacharjee A, Yadav SP, Cathcart MK.** 2011. Protein kinase Cδ is a critical component of Dectin-1 signaling in primary human monocytes. *J Leukoc Biol* **90:**599–611.

141. **Underhill DM, Rossnagle E, Lowell CA, Simmons RM.** 2005. Dectin-1 activates Syk tyrosine kinase in a dynamic subset of macrophages for reactive oxygen production. *Blood* **106:**2543–2550.

142. **Strasser D, Neumann K, Bergmann H, Marakalala MJ, Guler R, Rojowska A, Hopfner KP, Brombacher F, Urlaub H, Baier G, Brown GD, Leitges M, Ruland J.** 2012. Syk kinase-coupled C-type lectin receptors engage protein kinase C-σ to elicit Card9 adaptor-mediated innate immunity. *Immunity* **36:**32–42.

143. **Zhu J, Luo BH, Xiao T, Zhang C, Nishida N, Springer TA.** 2008. Structure of a complete integrin ectodomain in a physiologic resting state and activation and deactivation by applied forces. *Mol Cell* **32:**849–861.

144. **Lim J, Wiedemann A, Tzircotis G, Monkley SJ, Critchley DR, Caron E.** 2007. An essential role for talin during $\alpha_M\beta_2$-mediated phagocytosis. *Mol Biol Cell* **18:** 976–985.

145. **Lim J, Dupuy AG, Critchley DR, Caron E.** 2010. Rap1 controls activation of the $\alpha_M\beta_2$ integrin in a talin-dependent manner. *J Cell Biochem* **111:**999–1009.

146. **Medraño-Fernandez I, Reyes R, Olazabal I, Rodriguez E, Sanchez-Madrid F, Boussiotis VA, Reche PA, Cabañas C, Lafuente EM.** 2013. RIAM (Rap1-interacting adaptor molecule) regulates complement-dependent phagocytosis. *Cell Mol Life Sci* **70:**2395–2410.

147. **Lim J, Thompson J, May RC, Hotchin NA, Caron E.** 2013. Regulator of G-protein signalling-14 (RGS14) regulates the activation of αMβ2 integrin during phagocytosis. *PLoS One* **8:**e69163. doi:10.1371/journal.pone.0069163.

148. **Allen LA, Aderem A.** 1996. Molecular definition of distinct cytoskeletal structures involved in complement- and Fc receptor-mediated phagocytosis in macrophages. *J Exp Med* **184:**627–637.

149. **Lutz MA, Correll PH.** 2003. Activation of CR3-mediated phagocytosis by MSP requires the RON receptor, tyrosine kinase activity, phosphatidylinositol 3-kinase, and protein kinase C ζ. *J Leukoc Biol* **73:**802–814.

150. **Shi Y, Tohyama Y, Kadono T, He J, Miah SM, Hazama R, Tanaka C, Tohyama K, Yamamura H.** 2006. Protein-tyrosine kinase Syk is required for pathogen engulfment in complement-mediated phagocytosis. *Blood* **107:**4554–4562.

151. **Wiedemann A, Patel JC, Lim J, Tsun A, van Kooyk Y, Caron E.** 2006. Two distinct cytoplasmic regions of the β_2 integrin chain regulate RhoA function during phagocytosis. *J Cell Biol* **172:**1069–1079.

152. **Utomo A, Hirahashi J, Mekala D, Asano K, Glogauer M, Cullere X, Mayadas TN.** 2008. Requirement for Vav proteins in post-recruitment neutrophil cytotoxicity in IgG but not complement C3-dependent injury. *J Immunol* **180:**6279–6287.

153. **Lewkowicz E, Herit F, Le Clainche C, Bourdoncle P, Perez F, Niedergang F.** 2008. The microtubule-binding protein CLIP-170 coordinates mDia1 and actin reorganization during CR3-mediated phagocytosis. *J Cell Biol* **183:**1287–1298.

154. **Meller J, Vidali L, Schwartz MA.** 2008. Endogenous RhoG is dispensable for integrin-mediated cell spreading but contributes to Rac-independent migration. *J Cell Sci* **121:**1981–1989.

155. **Martinez J, Almendinger J, Oberst A, Ness R, Dillon CP, Fitzgerald P, Hengartner MO, Green DR.** 2011. Microtubule-associated protein 1 light chain 3 alpha (LC3)-associated phagocytosis is required for the efficient clearance of dead cells. *Proc Natl Acad Sci U S A* **108:**17396–17401.

156. **Park D, Hochreiter-Hufford A, Ravichandran KS.** 2009. The phosphatidylserine receptor TIM-4 does not mediate direct signaling. *Curr Biol* **19:**346–351.

157. **Flannagan RS, Canton J, Furuya W, Glogauer M, Grinstein S.** 2014. The phosphatidylserine receptor TIM4 utilizes integrins as coreceptors to effect phagocytosis. *Mol Biol Cell* **25:**1511–1522.

158. **Toda S, Hanayama R, Nagata S.** 2012. Two-step engulfment of apoptotic cells. *Mol Cell Biol* **32:**118–125.

159. **Huveneers S, Danen EHJ.** 2009. Adhesion signaling—crosstalk between integrins, Src and Rho. *J Cell Sci* **122:** 1059–1069.

160. **Santiago C, Ballesteros A, Martínez-Muñoz L, Mellado M, Kaplan GG, Freeman GJ, Casasnovas JM.** 2007. Structures of T cell immunoglobulin mucin protein 4 show a metal-ion-dependent ligand binding site where phosphatidylserine binds. *Immunity* **27:**941–951.

161. **Huang ZY, Hunter S, Chien P, Kim MK, Han-Kim TH, Indik ZK, Schreiber AD.** 2011. Interaction of two phagocytic host defense systems: Fcγ receptors and complement receptor 3. *J Biol Chem* **286:**160–168.

162. **Jongstra-Bilen J, Harrison R, Grinstein S.** 2003. Fcγ-receptors induce Mac-1 (CD11b/CD18) mobilization and accumulation in the phagocytic cup for optimal phagocytosis. *J Biol Chem* **278:**45720–45729.

163. **Heit B, Kim H, Cosío G, Castaño D, Collins R, Lowell CA, Kain KC, Trimble WS, Grinstein S.** 2013. Multimolecular signaling complexes enable Syk-mediated signaling of CD36 internalization. *Dev Cell* **24:**372–383.

164. **Wu Y, Singh S, Georgescu MM, Birge RB.** 2005. A role for Mer tyrosine kinase in αvβ5 integrin-mediated phagocytosis of apoptotic cells. *J Cell Sci* **118:**539–553.

165. **Bezbradica JS, Rosenstein RK, DeMarco RA, Brodsky I, Medzhitov R.** 2014. A role for the ITAM signaling module in specifying cytokine-receptor functions. *Nat Immunol* **15:**333–342.

166. **Graham LM, Gupta V, Schafer G, Reid DM, Kimberg M, Dennehy KM, Hornsell WG, Guler R, Campanero-Rhodes MA, Palma AS, Feizi T, Kim SK, Sobieszczuk P, Willment JA, Brown GD.** 2012. The C-type lectin receptor CLECSF8 (CLEC4D) is expressed by myeloid cells and triggers cellular activation through Syk kinase. *J Biol Chem* **287:**25964–25974.

167. **Patel SN, Serghides L, Smith TG, Febbraio M, Silverstein RL, Kurtz TW, Pravenec M, Kain KC.** 2004. CD36 mediates the phagocytosis of *Plasmodium*

falciparum-infected erythrocytes by rodent macrophages. *J Infect Dis* **189**:204–213.

168. **van Spriel AB, van den Herik-Oudijk IE, van Sorge NM, Vilé HA, van Strijp JA, van de Winkel JG.** 1999. Effective phagocytosis and killing of *Candida albicans* via targeting FcγRI (CD64) or FcαRI (CD89) on neutrophils. *J Infect Dis* **179**:661–669.

169. **Daëron M, Malbec O, Bonnerot C, Latour S, Segal DM, Fridman WH.** 1994. Tyrosine-containing activation motif-dependent phagocytosis in mast cells. *J Immunol* **152**:783–792.

170. **Ghiran I, Barbashov SF, Klickstein LB, Tas SW, Jensenius JC, Nicholson-Weller A.** 2000. Complement receptor 1/CD35 is a receptor for mannan-binding lectin. *J Exp Med* **192**:1797–1808.

171. **Ross GD, Reed W, Dalzell JG, Becker SE, Hogg N.** 1992. Macrophage cytoskeleton association with CR3 and CR4 regulates receptor mobility and phagocytosis of iC3b-opsonized erythrocytes. *J Leukoc Biol* **51**:109–117.

172. **Ichimura T, Asseldonk EJ, Humphreys BD, Gunaratnam L, Duffield JS, Bonventre JV.** 2008. Kidney injury molecule-1 is a phosphatidylserine receptor that confers a phagocytic phenotype on epithelial cells. *J Clin Invest* **118**:1657–1668.

173. **Choi SC, Simhadri VR, Tian L, Gil-Krzewska A, Krzewski K, Borrego F, Coligan JE.** 2011. Cutting edge: mouse CD300f (CMRF-35-like molecule-1) recognizes outer membrane-exposed phosphatidylserine and can promote phagocytosis. *J Immunol* **187**:3483–3487.

174. **Nagata K, Ohashi K, Nakano T, Arita H, Zong C, Hanafusa H, Mizuno K.** 1996. Identification of the product of growth arrest-specific gene 6 as a common ligand for Axl, Sky, and Mer receptor tyrosine kinases. *J Biol Chem* **271**:30022–30027.

175. **Albert ML, Kim JI, Birge RB.** 2000. αvβ5 integrin recruits the CrkII-Dock180-rac1 complex for phagocytosis of apoptotic cells. *Nat Cell Biol* **2**:899–905.

176. **Greenberg ME, Sun M, Zhang R, Febbraio M, Silverstein R, Hazen SL.** 2006. Oxidized phosphatidylserine-CD36 interactions play an essential role in macrophage-dependent phagocytosis of apoptotic cells. *J Exp Med* **203**:2613–2625.

Myeloid Cells in Health and Disease: A Synthesis
Edited by Siamon Gordon

doi:10.1128/microbiolspec.MCHD-0029-2016

Noah Fine,[1] Samira Khaliq,[1]
Siavash Hassanpour,[1] and Michael Glogauer[1]

28 Role of the Cytoskeleton in Myeloid Cell Function

INTRODUCTION

The intracellular cytoskeleton, consisting of filamentous actin (F-actin), microtubules (MTs), and intermediate filaments, makes up a network of dynamic polymeric structures that regulate cell shape and function (1). Chemotaxis and phagocytosis, two essential myeloid cell functions that enable them to defend the host against harmful opportunistic microorganisms, rely heavily on a highly dynamic and multifunctional cytoskeletal network (2–4). The importance of the cytoskeleton in myeloid cells is emphasized by human innate immune dysfunction syndromes that arise from defects in cytoskeletal proteins or the proteins that regulate cytoskeletal function (5–7). Although most of our knowledge of cytoskeletal structure and function comes from fibroblasts and other cell types, myeloid cells are valuable model systems that have been exploited to study the diverse roles and functions of the cytoskeleton. In this chapter, we will review the current knowledge with respect to cytoskeletal regulation of myeloid cell function, with an emphasis on neutrophils and macrophages.

ACTIN

The actin cytoskeleton is intricately involved in many fundamental aspects of eukaryotic cell biology. Myeloid cells, as important protective cells of the innate immune system, have many unique and diverse applications of common cellular functions, which require distinct regulation of the actin cytoskeleton compared to other cell types, such as fibroblasts. Myeloid cell functions that require unique regulation of the actin cytoskeleton include phagocytosis of pathogens, the ability to squeeze into tight spaces, and regulation of cell morphology and adhesion to facilitate fast cell migration. The dynamic regulation of the actin cytoskeleton is an essential component of membrane receptor-initiated myeloid cell responses including changes in cell shape (8, 9), rapid cell motility (10), phagocytosis (11, 12), transepithelial migration (9), and signal transduction and transcriptional regulation (13, 14). The actin cytoskeleton is composed of a network of actin filaments (F-actin), which arise from the polymerization of actin monomers. Six highly homologous actin monomer isoforms are expressed in humans. Four of the six actin isoforms ($\alpha_{skeletal}$-actin, $\alpha_{cardiac}$-actin, α_{smooth}-actin, and γ_{smooth}-actin) are specific to muscle tissues, while B_{cyto}-actin and γ_{cyto}-actin are ubiquitously expressed in all cells, including those of myeloid origin (13). Due to the intrinsic polarity of actin monomers, actin polymers are characterized by a distinct barbed (+) end and a pointed (–) end. Actin monomers actively assemble at

[1]Department of Dentistry, Matrix Dynamics Group, University of Toronto, Toronto, ON M5S 3E2, Canada.

barbed ends and disassemble at pointed ends in a process known as F-actin treadmilling. Although this linear model of polymerization and depolymerization suitably describes actin dynamics in a cell-free system, actin regulation is much more complex *in vivo* (15). Living cells organize their actin filaments into complex three-dimensional networks, bundles, and gels via interaction with a variety of actin-binding proteins. These include proteins that promote *de novo* nucleation of actin filaments (e.g., Arp2/3 complex, formins, Cobl, Spire, and leiomodin), proteins that sever actin filaments and generate new barbed ends and thus promote F-actin remodeling (e.g., actin-depolymerizing factor [ADF]/cofilins, gelsolin, and adseverin), proteins that sequester actin monomers and retard actin filament polymerization (e.g., profilin, twinfilin, and β-thymosins), proteins that limit F-actin polymerization by capping the barbed end of F-actin strands (e.g., CapG, Eps8, and Ena/vasodilator-stimulated phosphoprotein [VASP]), and finally actin-binding proteins that bundle and cross-link preexisting actin filaments (e.g., α-actinin and fascin) (reviewed in 15–17). Together, these actin-binding proteins mediate the length, flexibility, and viscosity of the actin network, altering its emergent properties, to orchestrate changes in cell morphology and function.

Rho GTPases in Actin Assembly

Leukocytes recognize and respond to bacterial byproducts and endogenous proinflammatory mediators, aggregate at sites of inflammation, and eliminate invading pathogens (18). The coordination of these events requires the recognition of an extracellular stimulus, the transduction of the signal across the plasma membrane, and an appropriate change in cell state. Rho family small GTPases are critical signaling nodes that interpret extracellular stimuli and orchestrate a wide range of cellular responses (19–21), including dynamic regulation of the actin cytoskeleton (22, 23). In leukocytes, they are important regulators of chemotaxis (24), phagocytosis (2), degranulation (25), and production of reactive oxygen species (ROS) (26, 27). Their activity is regulated by cycling between an inactive, GDP-bound form and an active, GTP-bound form. Loading of the small GTPases with GTP by guanine nucleotide exchange factors (GEFs) induces a conformational change that promotes the activation of downstream effector molecules. The prototypical Rho family GTPases are RhoA, Rac1, and Cdc42. In crawling fibroblasts, Rac1 and Cdc42 are responsible for protrusive F-actin-based structures at the leading edge during migration (28), and RhoA is required for formation of F-actin stress fibers and integrin-based focal adhesions (29–31). These factors play somewhat analogous roles in myeloid cells (32); however, the cytoskeletal structures formed are different. For example, in macrophages the small GTPases generate distinct focal complexes and fine actin cables (33). Important downstream effector molecules of the Rho GTPases include serine/threonine kinases, such as p21-activated kinase (PAK) (34) and Rho-associated protein kinase (ROCK) (35); actin nucleating proteins such as the formin mDia (mammalian homolog of *Drosophila* diaphanous) (36); and members of the WASP (Wiskott–Aldrich syndrome protein)/WAVE (WASP-family verprolin homologous protein) family that promote nucleation through Arp2/3 (37). Mutations in WASP cause Wiskott-Aldrich syndrome, a severe human immunodeficiency syndrome characterized by myeloid cell defects (38–40).

During neutrophil chemotaxis, Rac GTPases promote F-actin formation at the leading edge and coordinate directional migration (41–43). Rac2 is the predominant Rac isoform in neutrophils, making up 80 to 95% of total Rac protein (44). Although Rac1 and Rac2 have >90% homology, they have nonredundant roles in neutrophil function (45). Rac1-deficient neutrophils fail to properly orient in a chemotactic gradient, accompanied by the formation of multiple randomly oriented lamellipodia (45, 46), while Rac2-deficient cells show defective F-actin assembly and reduced cell migration speed; however, they are still able to orient in a chemotactic gradient (47). Rac1 controls the uncapping of existing F-actin barbed ends, whereas Rac2 plays a major role in the extension of actin filaments via cofilin and Arp2/3-dependent mechanisms (48). Distinct activation kinetics and different roles for Rac1 and Rac2 have been demonstrated in response to low versus high concentrations of the formylated tripeptide chemoattractant N-formyl-methionyl-leucyl-phenylalanine (fMLP) (42). In the low range of stimulatory fMLP concentrations, Rac1 is activated and initiates cell spreading. At high stimulatory fMLP concentrations, Rac1 is first activated to initiate the formation of a lamellipod at the leading edge, but Rac2 is further required for continuous expansion of the lamellipod, which drives effective migration. In addition, Rac1 and Rac2, through their roles as part of the NADPH oxidase complex, are both required to produce ROS at the neutrophil leading edge, which is essential for directional migration through a redox-mediated feedback loop to phosphatidylinositol 3,4,5-trisphosphate (PIP_3) (49).

Several Rac GEFs have been implicated in stimulation of F-actin at the leading edge in neutrophils. The Rac GEF P-Rex1 is activated downstream of G protein-coupled receptors (GPCRs) and promotes neutrophil polarization and leading edge F-actin formation (50). This occurs through stimulation of RhoG, which activates another Rac GEF, DOCK2, to stimulate Rac1 activity at the leading edge. DOCK2 was shown to localize at the leading edge and trigger neutrophil migration and polarity in a phosphatidylinositol 3-kinase (PI3K)-dependent manner (41). Also, the adaptor protein 3BP2 was shown to be required for localization of P-Rex1 and another Rac GEF, Vav1, at the leading edge to stimulate Rac2 activity, neutrophil activation, and F-actin polymerization (51).

RhoA promotes cell contractility through ROCK and stimulates actin polymerization through mDia. ROCK phosphorylates LIM kinase, MLC phosphatase, and the myosin light chain (MLC), leading to the stabilization of actin filaments and formation of actomyosin bundles, whereas mDia nucleates actin polymerization (52). Stimulation of cell surface cytokine and chemokine GPCRs stimulates Rho-dependent actin polymerization and actomyosin contractility to orchestrate cell polarization, adhesion, and protrusion during migration (53). In neutrophils stimulated with fMLP, active RhoA localizes at the sides and back (54), where it promotes uropod contractility through actomyosin. Inhibition of RhoA causes tail retraction defects in monocytes (55), neutrophils (46, 56, 57), eosinophils (58), and lymphocytes (59). In addition to deadhesion at the rear, RhoA-dependent contractility also strengthens integrin-based adhesions in neutrophils (60), lymphocytes (61), and fibroblasts (62). In resting neutrophils, RhoA deficiency caused random migration, F-actin cap formation, elongated uropods, increased granule exocytosis, and elevated ROS production (63). The authors show that in the absence of stimulation RhoA acts to limit neutrophil priming and hyperresponsiveness through formins.

Inhibition of the RhoA pathway may represent a novel target for immunosuppression and antirejection therapies through effects on neutrophil migration. Modulation of the cytoskeleton through specific activation of RhoA and Rac1 is thought to be a critical mechanism of action of the small-molecule anti-inflammatory drug pomalidomide (64). The Rho kinase inhibitor fasudil has therapeutic efficacy in a number of human conditions (65, 66) and may also have immunosuppressive effects.

THE ROLE OF ACTIN IN MYELOID CELL RECRUITMENT

The recruitment of leukocytes from the bloodstream to a site of infection, inflammation, or injury entails multiple bidirectional signaling and adhesive interactions between the leukocytes and vascular endothelial cells, resulting in attachment of leukocytes to the endothelium near the affected area, rolling and arrest, crawling, and transmigration of the leukocyte across the endothelium (2, 67). Each of these processes relies on intricate regulation of the actin cytoskeleton, as discussed below.

Rolling

Nonpolarized neutrophils in the circulation typically have a robust cortical-actin cytoskeleton. The ability of rolling neutrophils to be captured by the endothelial surface depends on deformability of the cell. Computer simulations and experimental evidence suggest that loss of cell deformability limits rolling interactions and cell capture under shear stress (68–70). Treatment with cytochalasin, an inhibitor of actin polymerization, increases neutrophil deformability (71), resulting in a larger contact area with the endothelium and consequently slower and more stable rolling. This suggests that changes in the polymerization state of leukocyte cortical actin can raise or lower the threshold at which stable rolling can occur. Neutrophils isolated from the wounds and peripheral blood of burn victims display increased polymerized actin content, increased actin stability, and reduced random motility and chemotaxis (72). The increase in actin stability in neutrophils of burn patients results in increased neutrophil stiffness, decreased deformation, and hindered passage through distal capillaries (73).

Shear Stress

Leukocytes are well adapted to mechanical shear forces that they experience in the circulation. Leukocytes need to adhere and migrate on the endothelial surface only when the appropriate stimuli are detected, and these responses have evolved in the presence of, and are closely regulated by, shear stress due to blood flow. Fluid shear stress in the range of 1 to 10 dyn/cm^2 promotes leukocyte crawling and diapedesis under inflammatory conditions (74–76). The mechanisms and pathways that contribute to the mechanosensory response of leukocytes to shear stress are likely to be initiated by physical perturbation of load-bearing subcellular structures including integrin-based adhesions and the cytoskeleton (77). Human leukocytes undergo myosin- and F-actin-dependent cell stiffening within seconds of exposure to shear stress (78) and show pseudopod projection and spreading 5 min after exposure to shear stress (79). In contrast, promyelocytic HL-60 cells, differentiated into neutrophils, and mouse blood neutrophils had a cell rounding response to shear stress (80). The shear stress-induced cell retraction response of HL-60 cells was accompanied by a transient increase in RhoA activity, while active Rac1 and Rac2 levels were reduced. When HL-60s were prestimulated with fMLP, cell rounding and reduced Rac1 and Rac2 activity were reversed. This is suggestive of a mechanism whereby Rac activity, and thereby neutrophil pseudopod projection, is inhibited by neutrophil exposure to shear stress in the absence of stimulation, limiting unsolicited recruitment (81), while shear stress enhances transmigratory responses in the presence of proinflammatory signals.

Neutrophils rolling on the endothelial surface are captured through an integrin catch-bond mechanism, and adhesions are reinforced by F-actin. RhoA promotes neutrophil adhesion through F-actin-dependent regulation of integrin function (60, 82–84), facilitating spreading and the formation of F-actin-rich podosomes. In one study, monocytes plated on physiological ligands formed F-actin-based structures at $\alpha_4\beta_1$ integrin adhesion sites in response to shear stress (85). These F-actin-based anchors formed as a result of a signaling cascade that depended on Rap1, PI3Kγ, and Rac, but not Cdc42, and are thought to help leukocytes brace themselves against the force of flowing blood. Also, *in vivo* studies of the inflamed mouse microvasculature showed that the Rac GEF Vav1 was required to induce neutrophil crawling in the direction perpendicular to the flow of blood (86). Since endothelial cells elongate parallel to the direction of flow, this directional crawling ensures the shortest distance of travel to the closest endothelial junction when leukocytes are exposed to shear stress on the vascular surface.

Polarization

In order to crawl on a two-dimensional surface or within the tissue matrix, leukocytes undergo polarization by creating a morphologically distinct leading edge as well as a trailing edge. The leading edge is characterized by F-actin-rich membrane extensions, while the trailing uropod region is rich in contractile actomyosin

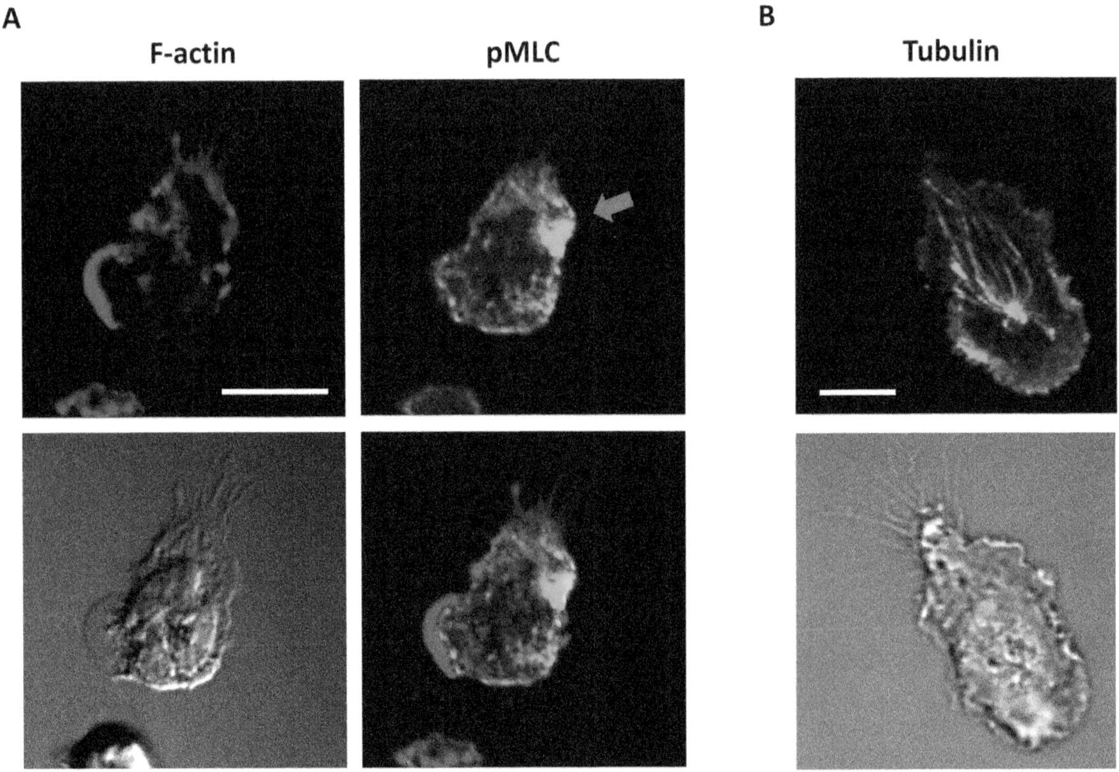

Figure 1 The neutrophil F-actin and MT cytoskeletons. (**A**) Mouse bone marrow neutrophils were plated on ICAM-1 (Intercellular Adhesion Molecule-1)-coated glass and treated with 1 M fMLP for 10 min. Cells were fixed with paraformaldehyde and labeled for F-actin (red) and phosphorylated MLC (Thr18/Ser19) (green). An arrow indicates the position of the uropod. (**B**) Cells were treated as above, fixed with glutaraldehyde as in reference 120 and labeled for MTs (green). Bars, 10 m.

fibers (Fig. 1A). Unique signaling modules have been identified that designate cell "frontness" and "backness," dictating an axis of symmetry that can be oriented in a chemotactic gradient to facilitate directional homing. The neutrophil axis of polarity is stable, as demonstrated by their ability to reorient toward a chemotactic stimulus that is repositioned (87).

GPCRs, membrane phosphoinositides, and small GTPases are important regulators of cell polarity and the chemotactic compass. PIP_3 stabilizes polarity by locally enhancing Rac activity at the leading edge and by stimulating the activation of Cdc42, which promotes RhoA-dependent backness at the trailing edge. Inhibition of PI3K results in the formation of multiple weak and transient pseudopods culminating in reduced directionality of migration (88). Furthermore, the internal cellular gradient of PIP_3 was impaired when cells were treated with the F-actin inhibitors latrunculin or jasplakinolide, indicating that F-actin plays a reciprocal role in stabilizing membrane phosphoinositide distribution and the chemotactic compass (88). Inhibitor studies suggest that initial stimulation of neutrophil polarity with fMLP, characterized by protrusive F-actin, occurs through two distinct pathways (89): a PI3Kγ-dependent pathway, which also involves protein kinase B (Akt/PKB) and protein kinase C-ζ (PKC-ζ); and a PI3Kγ-independent pathway, which depends on RhoA, ROCK, Src family kinases, and NADPH. Once established, a positive feedback loop linking PIP_3, Rac, and F-actin maintains a robust pseudopod at the leading edge (88, 90, 91), while RhoA and ROCK maintain backness, associated with actomyosin-based contractility in the uropod (87, 92). Cross talk between the Rho GTPases helps to establish and maintain polarity. Leading edge signals locally inhibit myosin activity at the leading edge, which effectively limits the domain of Rho-induced contractile activity to the uropod (93). Cdc42 activation at the leading edge induces RhoA-mediated contractility in the uropod at a distance (94) and maintains polar alignment of the MT cytoskeleton (95). Contractility in the uropod in turn generates membrane tension that inhibits frontness. Establishing and maintaining these mutually exclusive domains serves to coordinate and promote a stable axis of polarity during neutrophil chemotaxis (46, 54).

PAK kinases are important downstream effectors of Rac and Cdc42 that facilitate cross talk between Rho family GTPases to establish frontness and limit Rho-dependent signals to the posterior end of the neutrophil. PAK1 phosphorylates numerous cytoskeletal regulators and promotes the formation of protrusive F-actin and inhibition of contractile actomyosin at the leading edge (96). Activated PAK2 also accumulates at the neutrophil leading edge in response to fMLP to support Rac/Cdc42-mediated actin dynamics (97). PAK inhibition alters the subcellular localization of active RhoA and induces aberrant formation of vinculin-rich complexes, increased spreading, decreased migration speed, and loss of directionality.

Crawling

Resting leukocytes possess an actin-myosin-rich cortex that runs parallel to the plasma membrane. This actin-myosin cortex is important in determining cell shape and regulating cortical tension and undergoes dynamic reorganization to allow for adhesion and migration (98). The kinetics of cortical F-actin polymerization/depolymerization determines changes in cell morphology and is the driving force for cell migration. Unstimulated neutrophils lack motility, and treatment with cytochalasin, a toxin that binds to barbed ends and inhibits actin polymerization, of resting neutrophils does not decrease basal F-actin levels (99, 100), suggesting limited dynamics of the F-actin cytoskeleton under these conditions. The polymerization state of actin filaments is subject to regulation by extracellular signals and in this way coupled to cellular responses to environmental stimuli. In neutrophils, stimulation with a variety of chemoattractants and cytokines triggers actin polymerization at the leading edge (101, 102). The number of actin filaments per cell doubles 90 s after stimulation of resting neutrophils with fMLP, which can be reversed by addition of cytochalasin (103). The force generated by polymerization of actin is sufficient to produce a membrane extension at the leading edge. F-actin-based structures in the lamellipod, in association with integrin-based adhesions, generate force and traction at the leading edge to facilitate crawling (104). Rapid cell migration is thought to depend on a dynamic balance between actomyosin-based contraction and integrin-based adhesions (105). Furthermore, integrin-based adhesions are themselves subject to regulation by actomyosin contractility. Remodeling of the F-actin cytoskeleton is necessary to promote stable integrin-based adhesions in monocytes, and the RhoA effector ROCK limits adhesion formation by phosphorylating and inactivating cofilin through LIM kinase (92). Monocytes that were treated with the ROCK inhibitor Y-27632 were strongly adherent and could not undergo productive migration (92).

In addition to pulling forces at the leading edge, crawling neutrophils are propelled by actomyosin-based pushing forces at the rear. Contractile forces that concentrate in the uropod under the control of RhoA facilitate deadhesion and tail retraction of crawling leukocytes (55, 58). The neutrophil uropod is formed by segregation of detergent-insoluble membrane rafts into functional domains that are enriched for flotilins and phosphorylated ezrin/radixin/moesin (pERM) proteins (106–108) and phosphorylated MLC. Uropods are functional repositories of adhesion molecules where rearward tractional stresses are generated during migration (109). Uropods are also thought to be important sites for immunological interactions due to the abundance of receptors and coreceptors in these domains (110). Contractile forces in the uropod are important for an amoeboid mode of migration in a three-dimensional matrix that has been demonstrated in neutrophils and dendritic cells (111).

Transmigration

Prior to actively participating in the elimination of invasive pathogens, leukocytes must exit the bloodstream and enter infected tissues through transendothelial migration (112). Actin polymerization at the leading edge initiates neutrophil transmigration through the endothelial cell-cell junctions. Neutrophils failed to undergo transendothelial migration when actin polymerization or depolymerization (113) was inhibited by latrunculin-A or jasplakinolide, respectively. The lack of transendothelial migration by latrunculin-A-treated neutrophils also indicates that neutrophil transmigration is an active process that requires more than a decrease in cellular stiffness (114) and "oozing" of cells through endothelial cell junctions.

THE ROLE OF MTs IN MYELOID CELL MIGRATION

The MT cytoskeleton is a dynamic intracellular structure that regulates diverse cellular phenomena. MTs enable vesicle trafficking; regulate cell polarity and migration; are the basic structural elements of cilia and flagella; and form the mitotic spindle, essential for cellular division. MTs act as a platform for intracellular transport through interaction with the molecular motors dynein and kinesin. The MT array is formed by polymerization of α- and β-tubulin dimers and has inherent polarity, growing through polymerization at the plus end, with the minus end generally nucleated at a microtubule organizing center (MTOC). MTs display dynamic instability with steady-state levels depending on the balance between polymerization and depolymerization. MTs undergo posttranslational modifications including acetylation and detyrosination (115) that can influence cytoskeletal function (116). Detyrosination is a marker of a distinct subset of stable MTs that are thought to be important for generating asymmetries associated with directional cell migration (117, 118).

Myeloid cells and other leukocytes are among the fastest-moving cells in the body and are quickly recruited from the circulation to sites of inflammation. As discussed previously, myeloid cells must undergo complex morphological changes as they transition from rolling to crawling cells, cross the endothelial barrier, and navigate the three-dimensional tissue matrix. Irrespective of the role of MTs in cell polarity during migration, myeloid cells require a highly labile and dynamic cytoskeleton in order to undergo the morphological contortions necessary during extravasation and to navigate the three-dimensional tissue matrix. Given this, it is not surprising that the MT array is much more dynamic in monocytes (119) and neutrophils (120) relative to other cell types. When human neutrophils and monocytes are treated with nocodazole, an inhibitor of MT polymerization, the MT array is lost with a half-life of ~30 s (120), 10 to 40 times faster than for most mammalian cells. Regrowth of the MTs begins as early as 1 min after washout of nocodazole and is completed in between 5 and 10 min in neutrophils and monocytes (119, 120).

There are some striking differences in the organization of the MT array in neutrophils compared to other cell types. In contrast to most cell types, including macrophages (Fig. 2), where the MTs grow into the leading lamellipod, the MT array extends backwards from the MTOC into the uropod in polarized neutrophils (121) (Fig. 1B). Upon stimulation with chemotactic peptides, the MTs, which initially have a radial appearance, reorient toward the uropod of the newly polarized cell in a Cdc42/WASP-dependent manner (95). Cdc42 promotes polarity and orients the MTOC in front of the nucleus in fibroblasts and astrocytes (122, 123). In neutrophils, Cdc42 also promotes polarity; however, it does so through a unique "at a distance" mechanism of action, through signaling to the uropod (124). When neutrophils from $Cdc42^{-/-}$ mice were exposed to a gradient of fMLP, they lost uropod sequestration of the MT array, developed multiple F-actin-rich protrusions, and turned frequently (95). It was demonstrated that the Cdc42 effector protein, WASP, promotes sequestration of CD11b in uropods, which in turn recruits the MT end-binding protein EB1 to capture and stabilize the MT array. Cdc42, WASP, and CD11b were all necessary for proper orientation of the MT array toward the uropod in polarized neutrophils. It is unclear whether MT polarization actively promotes contractility and membrane tension at the rear of the cell,

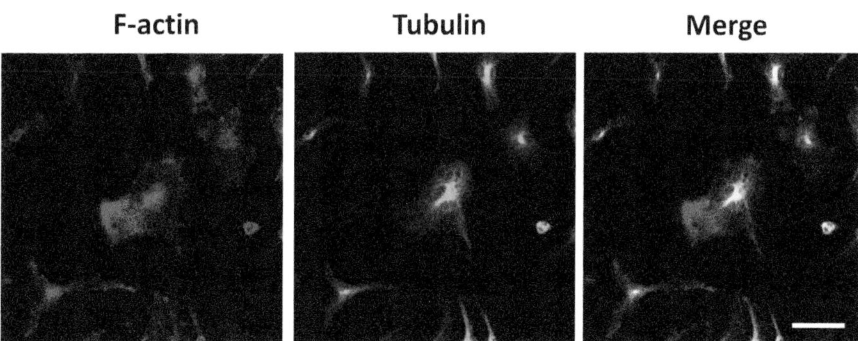

Figure 2 The macrophage F-actin and MT cytoskeletons. Colony-stimulating factor-1-cultured mouse bone marrow macrophages were fixed and labeled for F-actin (red) and MTs (green). Bar, 50 m.

through cross talk to the Rho pathway, for example, or is simply a by-product of cell polarization.

In addition to the unique orientation of the MT array toward the uropod, neutrophils also respond differently from other cells to depolymerization of the MT array. While MT depolymerization promotes Rho-dependent stress fiber formation and cellular contractility in fibroblasts and macrophages, in neutrophils it induces cell polarity and random migration (107, 125–127). Therefore, if sequestration of the MT array toward the uropod is necessary to maintain neutrophil polarity, the complete loss of polymerized MTs also produces neutrophil polarity. Induction of neutrophil polarity by MT-depolymerizing agents is characterized by formation of an F-actin-rich leading edge and a uropod at the posterior of the cell (128); however, the mechanism of action is different from that of chemoattractant-mediated polarization. Based on the observed migration of neutrophils upon MT depolymerization, early studies concluded that the MT array serves to limit cell migration in the absence of chemokine (129). Indeed, when neutrophils with a depleted MT array were stimulated with chemotactic peptide, a subpopulation of cells became multipolar (121), suggesting that the MTs serve to maintain a single axis of polarization. However, more-recent results have demonstrated that MT depolymerization increased the speed of differentiated HL-60 cell migration toward a gradient of fMLP, while simultaneously impairing purposeful and directional migration due to excessive activation of Rho, which inhibits fMLP-induced frontness (130). Although it has been assumed that the MT array maintains polarity in migrating neutrophils, as it does in many other cell types (123, 131, 132), these MT inhibitor studies demonstrate that the MT array may act simply to limit excessive Rho activity and therefore to allow directional migration. Consistent with inhibition of directional migration *in vitro*, *in vivo* imaging using zebrafish showed that recruitment of neutrophils was inhibited by nocodazole (133). Although nocodazole induced random migration, increased Rho and Rac activity, and increased polar F-actin levels in live zebrafish, leading edge PI3K and purposeful migration was defective. Similar to neutrophils, macrophages treated with MT-depolymerizing agents form multiple opposing lamellipodia with a consequent loss of random migration (134), and nocodazole inhibited recruitment of macrophages to a wound in live zebrafish (135). Interestingly, normal *in vivo* wound recruitment was restored by the ROCK inhibitor Y-27632, suggesting that macrophages can chemotax normally in the complete absence of MTs, as long as excessive Rho activation is blocked. It would be interesting to see if Y-27632 has a similar effect on neutrophils *in vivo*.

The asymmetrical distribution of the MT array toward the uropod is thought to reinforce neutrophil polarity during migration by buffering front-to-back signaling, so that the uropod is stable despite fluctuations at the leading edge (136). However, since the leading edge orientation of the MT array accomplishes a similar task in other cells, it is not clear what additional benefits are afforded to neutrophils by this innovation. One possibility is that neutrophils rely more heavily on a Rho-dependent amoeboid form of migration that puts unique emphasis on uropod regulation, and that this is somehow accomplished by sequestering the MTs, and associated factors, in the uropod. Also, uropod sequestration of MTs could be related to the unique ability of neutrophils to establish stable self-organizing polarity that is not dependent on the presence of a gradient of chemoattractant (54, 87). An alternative interpretation is that the MT array acts to sequester factors that might otherwise elicit excessive Rho activity and tightly regulates their activation in a localized manner. One potential activator of Rho signaling in response to MT depolymerization is the MT-associated GEF GEF-H1, which has been shown to trigger RhoA-induced contractility in response to MT depolymerization in other cell types (137, 138).

In addition to their role in cell migration, MTs play an important role in intracellular trafficking. MTs are necessary for macrophage secretion of MMP-9 (139), a matrix metalloproteinase important for degradation of extracellular matrix by migrating cells. In one study, lipopolysaccharide was shown to induce MT acetylation in macrophages and increased secretion of the anti-inflammatory cytokine interleukin-10. Interleukin-10 secretion could also be induced by treatment of macrophages with the MT-stabilizing agent paclitaxel (140).

CROSS TALK

Although MT and actin cytoskeletons are distinct polymeric structures that independently influence cell shape, polarity, and movement, they also collaborate. Evidence indicates that there is direct interaction and signaling between these two systems (141). Formins, which nucleate actin polymerization, can also bind to and stabilize MTs (142, 143). This promotes alignment of MTs along actin filaments and the reorientation of a subset of MTs into the leading edge of migrating cells.

In fibroblasts, depolymerization of the MT network triggers polymerization of the actin cytoskeleton and increased actomyosin contractility through stimulation of RhoA (144, 145), indicating that these two systems are fundamentally coupled. This phenomenon is mediated by activation of the RhoA-specific MT-associated GEF GEF-H1 (137, 138). As mentioned previously, neutrophils also undergo actomyosin-mediated contractility and random migration in response to MT depolymerization (126, 127); however, it is not yet known whether GEF-H1 is responsible for this effect or what is the function of this biochemical feedback pathway under normal physiological conditions.

PHAGOCYTOSIS

Phagocytosis is a specialized form of endocytosis in which large particles such as pathogenic organisms, dead cells, mineral particles, and neutrophil extracellular traps (NETs) are recognized via specific tags, internalized, and destroyed after fusion with lysosomes (146). Phagocytic cells recognize their prey through the surface chemistry of their targets, either through pattern recognition receptor interaction with pathogen-associated molecular patterns or through FcR or complement receptor (CR) interaction with host molecules (opsonins) on the surface of pathogens. Although FcR- and CR-mediated phagocytosis occur through distinct mechanisms (24), they both require dynamic regulation of the actin cytoskeleton and actomyosin contractility (4).

The Role of F-Actin in Phagocytosis

The fundamentals of actin polymerization/depolymerization during phagocytosis remain constant, but the function of F-actin, the protein interactions, and the signaling pathways differ depending on the type of phagocytosis. Recognition of a particle by Fc receptors triggers a signaling cascade that leads to activation of Src and Syk family serine/threonine kinases and Rho GTPases. Actin assembly at this site is dually regulated: (i) phosphatidylinositol 4,5-bisphosphate and Cdc42 signaling activates WASP, which in turn binds and activates the Arp2/3 complex (147, 148); and (ii) Fc-mediated Syk signaling stimulates Vav1, which in turn actives Cdc42 and Rac to stimulate actin polymerization through PAK1 (148–152). The Arp2/3 complex induces branching of actin filaments, leading to an increase in the number of uncapped ends and to isotropic growth of the actin network, which forms a phagocytic cup around the target particle (153). The developing phagocyte is enriched in gelsolin (154, 155) and cofilin (156). Further, gene targeting approaches in macrophages have demonstrated the importance of the F-actin capping protein CapG in regulating phagocytosis (157).

Phagocytes stained with fluorescein isothiocyanate (FITC)-phalloidin show a dense band of F-actin progressing from the bottom toward the top of the phagocytic cup during pseudopod extension (Fig. 3). Experiments based on fluorescent speckle microscopy show that actin polymerization does not directly push the membrane outwards but prevents the membrane from moving backwards by acting as a ratchet and making the ligand-receptor bonds effectively irreversible (158, 159). In FcR-mediated phagocytosis, advancing membrane-particle tethering occurs through a zipper model, with F-actin contributing structural stability.

In CR-mediated phagocytosis, particles "sink" into phagocytes without formation of a protrusive structure. Internalization occurs through F-actin and contractility-dependent pinching of particles. The Rho effector mDia is recruited early on during CR-mediated phagocytosis and colocalizes with polymerized actin in the phagocytic cup (160). Inhibiting mDia activity has a negative effect on CR-mediated phagocytosis but not FcR-mediated phagocytosis. Although myosin-mediated contractility is necessary for particle internalization in both

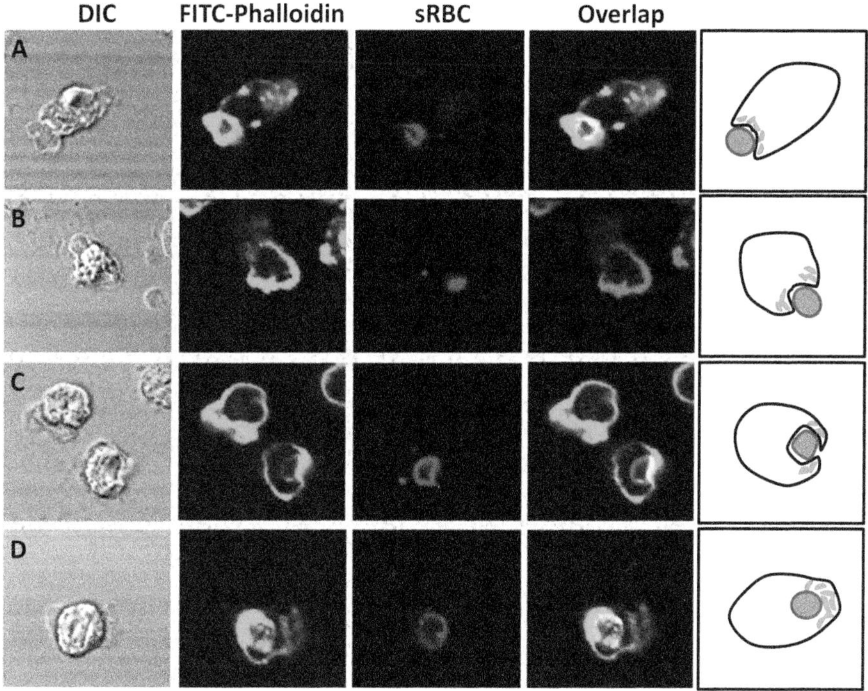

Figure 3 F-actin localization during Fc-mediated phagocytosis in neutrophils. Mouse bone marrow-derived neutrophils were incubated with IgG-tagged sheep red blood cells (sRBCs) at 37°C. Cells were fixed during phagocytosis, labeled, and visualized by confocal microscopy. F-actin was labeled with FITC-phalloidin and RBCs were labeled with anti-IgG Alexa-568. (**A**) F-actin concentrates on the site of recognition of the IgG-labeled RBC. (**B**) It then extends toward the tips of the nascent phagosome, creating protrusions at the site. (**C**) Consequently, cell membrane is pushed forward and receptor-ligand binding ensures enwrapping of the particle completely. (**D**) Lastly, the tips of the protrusions fuse due to being pushed into close proximity and the particle is internalized (S. Khaliq, unpublished data).

types of phagocytosis, the MLC kinase inhibitor ML-7 blocks actin filament assembly and phagocytic cup formation in CR-mediated phagocytosis but not in FcγR-mediated phagocytosis (161).

Maturation of phagosomes requires Rac in both Fc- and CR-mediated phagocytosis. Macrophages lacking Rac1 and -2 exhibited defective phagosome maturation (162), and macrophages that lacked β_1 integrin expression had reduced F-actin accumulation in the periphagosomal region, reduced maturation of the phagosome, and impaired bactericidal activity. Defective phagosome maturation in β_1 integrin-deficient macrophages could be rescued by ectopic expression of Rac1 but not Cdc42 (162). The oral pathogen *Treponema denticola* evades phagocytosis by inhibiting neutrophil Rac1 (163).

Localized PIP_3 production at the phagosome triggers downstream Rac and PKC signaling, which regulates actin assembly. It has been shown that PKC-α localizes to nascent phagosomes in macrophages (164). Myristoylated, alanine-rich C-kinase substrate (MARCKS), a major substrate of PKC-α, is rapidly phosphorylated during particle uptake and recruited to the nascent phagosome (164). MARCKS cross-links F-actin, and this activity is prevented by PKC-dependent phosphorylation (165). PKC inhibitors block recruitment of MARCKS and F-actin to the site of bound particles and prevent phagocytosis (164). Another PIP_3 effector molecule, phospholipase D (PLD), is also important for actin dynamics during phagocytosis. Lack of PLD1 or PLD2 in macrophages causes isoform-specific actin cytoskeleton abnormalities leading to decreased phagocytosis (166).

In addition to the roles of actin in cell crawling and phagocytosis, recent results have highlighted its importance in other effector functions of neutrophils, including NET formation. NET formation is a highly regulated terminal process in which neutrophils entrap and kill microorganisms through the release of relaxed chromatin strands, coated with antimicrobial compounds, into the extracellular space (167). Following signal transduction of appropriate cues from the microenvironment, neutrophils initiate DNA breakdown followed by the dissolution of the nuclear membrane and plasma membrane and release of NETs. Neutrophil elastase released from granules prior to NET formation degrades actin polymers, which triggers the release of chromatin from neutrophils (168). These results indicate that F-actin may act as a barrier to NET formation and release. Furthermore, it has been demonstrated that the MT cytoskeleton as well as active polymerization of actin are both critical for lipopolysaccharide-induced NET formation (169), since depolymerization of MTs with nocodazole or inhibition of actin polymerization with cytochalasin D limited nuclear breakdown and NET release. Furthermore, resolution of inflammation depends on active ingestion of NETs by macrophages in an F-actin-dependent endocytic process (146).

The Role of MTs in Myeloid Effector Function

The MT array has a specific role in CR-mediated phagocytosis, since disruption of MTs inhibits CR-mediated but not Fc receptor-mediated phagocytosis (170, 171). This is thought to be due to the ability of MTs to regulate β_2 integrin mobility and clustering rather than through direct effects on phagosome closure (172).

The MT array is also required for maturation of phagosomes due to the vesicle trafficking function of MTs, which facilitates vesicle fusion. This is confirmed by the observation that nocodazole-treated macrophages show defects in phagosome maturation (173, 174). Early-stage phagosomes have a strong preference for MT plus ends, whereas late-stage phagosomes do not. This plus-end affinity has been attributed to the presence of MT-associated proteins (175). Most phagosomes show centripetal movement toward the MTOC, where phagosome-lysosome fusion frequently occurs. Both dynein and dynactin are required for this minus-end-directed movement along the MTs. Although some phagosomes move to the plus end along the MTs using kinesin, the significance of this centrifugal movement is unknown (176).

In neutrophils, primary (azurophil) granules, which are lysosomal in nature, are exocytosed near the site of phagosome formation by focal exocytosis (177, 178). MTs contribute to the vectorial nature of the response, since the localized delivery of primary granules can be disrupted by colchicine (177). However, since the exocytosis of secondary granules is less polarized than that of primary granules, the transport of secondary granules may not be MT dependent. Similarly, macrophages treated with colchicine fail to localize antigens to the trans-Golgi network, indicating a requirement for MT in antigen presentation in these cells (179).

CONCLUSION

Myeloid cells perform unique cell biological functions related to pathogen killing, processes that require distinctive cytoskeletal responses. The importance of the cytoskeleton in myeloid function is underlined by mutations in cytoskeletal components that lead to myeloid dysfunction and human disease. Defects in actin polymerization and actin-dependent neutrophil function including chemotaxis, phagocytosis, and degranulation have been noted in a number of pathological conditions including chronic myeloid leukemia (180), localized aggressive periodontitis (181), neutrophil actin dysfunction disorder (182), and as a result of burn trauma (72). Mutations in WASP (183, 184) and Rac2 (6, 185, 186) cause defects in myeloid cell chemotaxis that lead to severe infections, and a mutation in the human β-actin gene is associated with recurrent infections due to dysfunctional neutrophil migration and ROS production (5, 187, 188). Furthermore, several strains of mice predisposed to spontaneous autoimmune disease all show reduced RhoA activation and dysregulation of F-actin in serum-treated macrophages (189, 190). These macrophages also display elevated adhesion and an abnormal morphology due to altered regulation of integrin dynamics resulting from the Rho and F-actin defects. In addition to primary defects of the cytoskeleton and cytoskeletal regulation, a number of pathogenic organisms have learned to subvert the cytoskeleton in order to overcome destruction by phagocytes (191–194). A thorough understanding of cytoskeleton function in myeloid cells could lead to development of novel antiphlogistic approaches and other therapeutic interventions.

Citation. Fine N, Khaliq S, Hassanpour S, Glogauer M. 2016. Role of the cytoskeleton in myeloid cell function. Microbiol Spectrum 4(4):MCHD-0029-2016.

References

1. **Wang YL.** 1991. Dynamics of the cytoskeleton in live cells. *Curr Opin Cell Biol* **3:**27–32.
2. **Fenteany G, Glogauer M.** 2004. Cytoskeletal remodeling in leukocyte function. *Curr Opin Hematol* **11:** 15–24.
3. **May RC, Machesky LM.** 2001. Phagocytosis and the actin cytoskeleton. *J Cell Sci* **114:**1061–1077.
4. **Freeman SA, Grinstein S.** 2014. Phagocytosis: receptors, signal integration, and the cytoskeleton. *Immunol Rev* **262:**193–215.
5. **Nunoi H, Yamazaki T, Kanegasaki S.** 2001. Neutrophil cytoskeletal disease. *Int J Hematol* **74:**119–124.
6. **Gu Y, Williams DA.** 2002. RAC2 GTPase deficiency and myeloid cell dysfunction in human and mouse. *J Pediatr Hematol Oncol* **24:**791–794.

7. **Dinauer MC.** 2014. Disorders of neutrophil function: an overview. *Methods Mol Biol* **1124:**501–515.
8. **Watts RG, Crispens MA, Howard TH.** 1991. A quantitative study of the role of F-actin in producing neutrophil shape. *Cell Motil Cytoskeleton* **19:**159–168.
9. **Watts RG, Howard TH.** 1993. Mechanisms for actin reorganization in chemotactic factor-activated polymorphonuclear leukocytes. *Blood* **81:**2750–2757.
10. **Carlier MF, Pantaloni D.** 1997. Control of actin dynamics in cell motility. *J Mol Biol* **269:**459–467.
11. **Jones GE.** 2000. Cellular signaling in macrophage migration and chemotaxis. *J Leukoc Biol* **68:**593–602.
12. **Cicchetti G, Allen PG, Glogauer M.** 2002. Chemotactic signaling pathways in neutrophils: from receptor to actin assembly. *Crit Rev Oral Biol Med* **13:** 220–228.
13. **Perrin BJ, Ervasti JM.** 2010. The actin gene family: function follows isoform. *Cytoskeleton (Hoboken)* **67:** 630–634.
14. **Xu YZ, Thuraisingam T, Morais DA, Rola-Pleszczynski M, Radzioch D.** 2010. Nuclear translocation of β-actin is involved in transcriptional regulation during macrophage differentiation of HL-60 cells. *Mol Biol Cell* **21:** 811–820.
15. **Welch MD, Mullins RD.** 2002. Cellular control of actin nucleation. *Annu Rev Cell Dev Biol* **18:**247–288.
16. **Schafer DA, Cooper JA.** 1995. Control of actin assembly at filament ends. *Annu Rev Cell Dev Biol* **11:** 497–518.
17. **Silacci P, Mazzolai L, Gauci C, Stergiopulos N, Yin HL, Hayoz D.** 2004. Gelsolin superfamily proteins: key regulators of cellular functions. *Cell Mol Life Sci* **61:** 2614–2623.
18. **Nauseef WM, Borregaard N.** 2014. Neutrophils at work. *Nat Immunol* **15:**602–611.
19. **Olson MF, Ashworth A, Hall A.** 1995. An essential role for Rho, Rac, and Cdc42 GTPases in cell cycle progression through G1. *Science* **269:**1270–1272.
20. **Aznar S, Lacal JC.** 2001. Rho signals to cell growth and apoptosis. *Cancer Lett* **165:**1–10.
21. **Hill CS, Wynne J, Treisman R.** 1995. The Rho family GTPases RhoA, Rac1, and CDC42Hs regulate transcriptional activation by SRF. *Cell* **81:**1159–1170.
22. **Etienne-Manneville S, Hall A.** 2002. Rho GTPases in cell biology. *Nature* **420:**629–635.
23. **Jaffe AB, Hall A.** 2005. Rho GTPases: biochemistry and biology. *Annu Rev Cell Dev Biol* **21:**247–269.
24. **Caron E, Hall A.** 1998. Identification of two distinct mechanisms of phagocytosis controlled by different Rho GTPases. *Science* **282:**1717–1721.
25. **Lacy P.** 2006. Mechanisms of degranulation in neutrophils. *Allergy Asthma Clin Immunol* **2:**98–108.
26. **Knaus UG, Heyworth PG, Evans T, Curnutte JT, Bokoch GM.** 1991. Regulation of phagocyte oxygen radical production by the GTP-binding protein Rac 2. *Science* **254:**1512–1515.
27. **Bokoch GM.** 2005. Regulation of innate immunity by Rho GTPases. *Trends Cell Biol* **15:**163–171.
28. **Nobes CD, Hall A.** 1999. Rho GTPases control polarity, protrusion, and adhesion during cell movement. *J Cell Biol* **144:**1235–1244.
29. **Paterson HF, Self AJ, Garrett MD, Just I, Aktories K, Hall A.** 1990. Microinjection of recombinant $p21^{rho}$ induces rapid changes in cell morphology. *J Cell Biol* **111:**1001–1007.
30. **Ridley AJ, Hall A.** 1992. The small GTP-binding protein rho regulates the assembly of focal adhesions and actin stress fibers in response to growth factors. *Cell* **70:** 389–399.
31. **Hotchin NA, Hall A.** 1995. The assembly of integrin adhesion complexes requires both extracellular matrix and intracellular rho/rac GTPases. *J Cell Biol* **131:** 1857–1865.
32. **Jones GE, Allen WE, Ridley AJ.** 1998. The Rho GTPases in macrophage motility and chemotaxis. *Cell Adhes Commun* **6:**237–245.
33. **Allen WE, Jones GE, Pollard JW, Ridley AJ.** 1997. Rho, Rac and Cdc42 regulate actin organization and cell adhesion in macrophages. *J Cell Sci* **110:**707–720.
34. **Manser E, Leung T, Salihuddin H, Zhao ZS, Lim L.** 1994. A brain serine/threonine protein kinase activated by Cdc42 and Rac1. *Nature* **367:**40–46.
35. **Fujisawa K, Fujita A, Ishizaki T, Saito Y, Narumiya S.** 1996. Identification of the Rho-binding domain of $p160^{ROCK}$, a Rho-associated coiled-coil containing protein kinase. *J Biol Chem* **271:**23022–23028.
36. **Watanabe N, Kato T, Fujita A, Ishizaki T, Narumiya S.** 1999. Cooperation between mDia1 and ROCK in Rho-induced actin reorganization. *Nat Cell Biol* **1:**136–143.
37. **Lane J, Martin T, Weeks HP, Jiang WG.** 2014. Structure and role of WASP and WAVE in Rho GTPase signalling in cancer. *Cancer Genomics Proteomics* **11:** 155–165.
38. **Thrasher AJ, Burns SO.** 2010. WASP: a key immunological multitasker. *Nat Rev Immunol* **10:**182–192.
39. **Ishihara D, Dovas A, Park H, Isaac BM, Cox D.** 2012. The chemotactic defect in Wiskott-Aldrich syndrome macrophages is due to the reduced persistence of directional protrusions. *PLoS One* **7:**e30033. doi:10.1371/journal.pone.0030033.
40. **Jones RA, Feng Y, Worth AJ, Thrasher AJ, Burns SO, Martin P.** 2013. Modelling of human Wiskott-Aldrich syndrome protein mutants in zebrafish larvae using *in vivo* live imaging. *J Cell Sci* **126:**4077–4084.
41. **Kunisaki Y, Nishikimi A, Tanaka Y, Takii R, Noda M, Inayoshi A, Watanabe K, Sanematsu F, Sasazuki T, Sasaki T, Fukui Y.** 2006. DOCK2 is a Rac activator that regulates motility and polarity during neutrophil chemotaxis. *J Cell Biol* **174:**647–652.
42. **Zhang H, Sun C, Glogauer M, Bokoch GM.** 2009. Human neutrophils coordinate chemotaxis by differential activation of Rac1 and Rac2. *J Immunol* **183:**2718–2728.
43. **Yoo SK, Deng Q, Cavnar PJ, Wu YI, Hahn KM, Huttenlocher A.** 2010. Differential regulation of protrusion and polarity by PI3K during neutrophil motility in live zebrafish. *Dev Cell* **18:**226–236.

44. **Heyworth PG, Bohl BP, Bokoch GM, Curnutte JT.** 1994. Rac translocates independently of the neutrophil NADPH oxidase components $p47^{phox}$ and $p67^{phox}$. Evidence for its interaction with flavocytochrome b_{558}. *J Biol Chem* **269:**30749–30752.

45. **Koh AL, Sun CX, Zhu F, Glogauer M.** 2005. The role of Rac1 and Rac2 in bacterial killing. *Cell Immunol* **235:**92–97.

46. **Pestonjamasp KN, Forster C, Sun C, Gardiner EM, Bohl B, Weiner O, Bokoch GM, Glogauer M.** 2006. Rac1 links leading edge and uropod events through Rho and myosin activation during chemotaxis. *Blood* **108:** 2814–2820.

47. **Glogauer M, Marchal CC, Zhu F, Worku A, Clausen BE, Foerster I, Marks P, Downey GP, Dinauer M, Kwiatkowski DJ.** 2003. Rac1 deletion in mouse neutrophils has selective effects on neutrophil functions. *J Immunol* **170:**5652–5657.

48. **Sun CX, Magalhães MA, Glogauer M.** 2007. Rac1 and Rac2 differentially regulate actin free barbed end formation downstream of the fMLP receptor. *J Cell Biol* **179:** 239–245.

49. **Kuiper JW, Sun C, Magalhães MA, Glogauer M.** 2011. Rac regulates $PtdInsP_3$ signaling and the chemotactic compass through a redox-mediated feedback loop. *Blood* **118:**6164–6171.

50. **Damoulakis G, Gambardella L, Rossman KL, Lawson CD, Anderson KE, Fukui Y, Welch HC, Der CJ, Stephens LR, Hawkins PT.** 2014. P-Rex1 directly activates RhoG to regulate GPCR-driven Rac signalling and actin polarity in neutrophils. *J Cell Sci* **127:**2589–2600.

51. **Chen G, Dimitriou I, Milne L, Lang KS, Lang PA, Fine N, Ohashi PS, Kubes P, Rottapel R.** 2012. The 3BP2 adapter protein is required for chemoattractant-mediated neutrophil activation. *J Immunol* **189:**2138–2150.

52. **Thumkeo D, Watanabe S, Narumiya S.** 2013. Physiological roles of Rho and Rho effectors in mammals. *Eur J Cell Biol* **92:**303–315.

53. **Ridley AJ, Schwartz MA, Burridge K, Firtel RA, Ginsberg MH, Borisy G, Parsons JT, Horwitz AR.** 2003. Cell migration: integrating signals from front to back. *Science* **302:**1704–1709.

54. **Wong K, Pertz O, Hahn K, Bourne H.** 2006. Neutrophil polarization: spatiotemporal dynamics of RhoA activity support a self-organizing mechanism. *Proc Natl Acad Sci U S A* **103:**3639–3644.

55. **Worthylake RA, Lemoine S, Watson JM, Burridge K.** 2001. RhoA is required for monocyte tail retraction during transendothelial migration. *J Cell Biol* **154:** 147–160.

56. **Yoshinaga-Ohara N, Takahashi A, Uchiyama T, Sasada M.** 2002. Spatiotemporal regulation of moesin phosphorylation and rear release by Rho and serine/threonine phosphatase during neutrophil migration. *Exp Cell Res* **278:**112–122.

57. **Niggli V.** 2003. Signaling to migration in neutrophils: importance of localized pathways. *Int J Biochem Cell Biol* **35:**1619–1638.

58. **Alblas J, Ulfman L, Hordijk P, Koenderman L.** 2001. Activation of RhoA and ROCK are essential for detachment of migrating leukocytes. *Mol Biol Cell* **12:**2137–2145.

59. **Liu L, Schwartz BR, Lin N, Winn RK, Harlan JM.** 2002. Requirement for RhoA kinase activation in leukocyte de-adhesion. *J Immunol* **169:**2330–2336.

60. **Laudanna C, Campbell JJ, Butcher EC.** 1996. Role of Rho in chemoattractant-activated leukocyte adhesion through integrins. *Science* **271:**981–983.

61. **Giagulli C, Scarpini E, Ottoboni L, Narumiya S, Butcher EC, Constantin G, Laudanna C.** 2004. RhoA and ζ PKC control distinct modalities of LFA-1 activation by chemokines: critical role of LFA-1 affinity triggering in lymphocyte in vivo homing. *Immunity* **20:** 25–35.

62. **Chrzanowska-Wodnicka M, Burridge K.** 1996. Rho-stimulated contractility drives the formation of stress fibers and focal adhesions. *J Cell Biol* **133:**1403–1415.

63. **Jennings RT, Strengert M, Hayes P, El-Benna J, Brakebusch C, Kubica M, Knaus UG.** 2014. RhoA determines disease progression by controlling neutrophil motility and restricting hyperresponsiveness. *Blood* **123:**3635–3645.

64. **Xu Y, Li J, Ferguson GD, Mercurio F, Khambatta G, Morrison L, Lopez-Girona A, Corral LG, Webb DR, Bennett BL, Xie W.** 2009. Immunomodulatory drugs reorganize cytoskeleton by modulating Rho GTPases. *Blood* **114:**338–345.

65. **Lu Q, Longo FM, Zhou H, Massa SM, Chen YH.** 2009. Signaling through Rho GTPase pathway as viable drug target. *Curr Med Chem* **16:**1355–1365.

66. **Surma M, Wei L, Shi J.** 2011. Rho kinase as a therapeutic target in cardiovascular disease. *Future Cardiol* **7:** 657–671.

67. **Springer TA.** 1994. Traffic signals for lymphocyte recirculation and leukocyte emigration: the multistep paradigm. *Cell* **76:**301–314.

68. **Lei X, Lawrence MB, Dong C.** 1999. Influence of cell deformation on leukocyte rolling adhesion in shear flow. *J Biomech Eng* **121:**636–643.

69. **Jadhav S, Eggleton CD, Konstantopoulos K.** 2005. A 3-D computational model predicts that cell deformation affects selectin-mediated leukocyte rolling. *Biophys J* **88:**96–104.

70. **Bose S, Das SK, Karp JM, Karnik R.** 2010. A semi-analytical model to study the effect of cortical tension on cell rolling. *Biophys J* **99:**3870–3879.

71. **Sheikh S, Nash GB.** 1998. Treatment of neutrophils with cytochalasins converts rolling to stationary adhesion on P-selectin. *J Cell Physiol* **174:**206–216.

72. **Hasslen SR, Ahrenholz DH, Solem LD, Nelson RD.** 1992. Actin polymerization contributes to neutrophil chemotactic dysfunction following thermal injury. *J Leukoc Biol* **52:**495–500.

73. **Worthen GS, Schwab B III, Elson EL, Downey GP.** 1989. Mechanics of stimulated neutrophils: cell stiffening induces retention in capillaries. *Science* **245:**183–186.

74. **Kitayama J, Hidemura A, Saito H, Nagawa H.** 2000. Shear stress affects migration behavior of polymorphonuclear cells arrested on endothelium. *Cell Immunol* **203:**39–46.

75. **Alon R, Dustin ML.** 2007. Force as a facilitator of integrin conformational changes during leukocyte arrest on blood vessels and antigen-presenting cells. *Immunity* **26:**17–27.

76. **Zarbock A, Ley K.** 2009. Neutrophil adhesion and activation under flow. *Microcirculation* **16:**31–42.

77. **Hoffman BD, Grashoff C, Schwartz MA.** 2011. Dynamic molecular processes mediate cellular mechanotransduction. *Nature* **475:**316–323.

78. **Coughlin MF, Sohn DD, Schmid-Schönbein GW.** 2008. Recoil and stiffening by adherent leukocytes in response to fluid shear. *Biophys J* **94:**1046–1051.

79. **Coughlin MF, Schmid-Schönbein GW.** 2004. Pseudopod projection and cell spreading of passive leukocytes in response to fluid shear stress. *Biophys J* **87:**2035–2042.

80. **Makino A, Glogauer M, Bokoch GM, Chien S, Schmid-Schönbein GW.** 2005. Control of neutrophil pseudopods by fluid shear: role of Rho family GTPases. *Am J Physiol Cell Physiol* **288:**C863–C871.

81. **Makino A, Shin HY, Komai Y, Fukuda S, Coughlin M, Sugihara-Seki M, Schmid-Schönbein GW.** 2007. Mechanotransduction in leukocyte activation: a review. *Biorheology* **44:**221–249.

82. **Sheikh S, Gratzer WB, Pinder JC, Nash GB.** 1997. Actin polymerisation regulates integrin-mediated adhesion as well as rigidity of neutrophils. *Biochem Biophys Res Commun* **238:**910–915.

83. **Anderson SI, Hotchin NA, Nash GB.** 2000. Role of the cytoskeleton in rapid activation of CD11b/CD18 function and its subsequent downregulation in neutrophils. *J Cell Sci* **113:**2737–2745.

84. **Laudanna C, Mochly-Rosen D, Liron T, Constantin G, Butcher EC.** 1998. Evidence of ζ protein kinase C involvement in polymorphonuclear neutrophil integrin-dependent adhesion and chemotaxis. *J Biol Chem* **273:** 30306–30315.

85. **Rullo J, Becker H, Hyduk SJ, Wong JC, Digby G, Arora PD, Cano AP, Hartwig J, McCulloch CA, Cybulsky MI.** 2012. Actin polymerization stabilizes α4β1 integrin anchors that mediate monocyte adhesion. *J Cell Biol* **197:** 115–129.

86. **Phillipson M, Heit B, Parsons SA, Petri B, Mullaly SC, Colarusso P, Gower RM, Neely G, Simon SI, Kubes P.** 2009. Vav1 is essential for mechanotactic crawling and migration of neutrophils out of the inflamed microvasculature. *J Immunol* **182:**6870–6878.

87. **Xu J, Wang F, Van Keymeulen A, Herzmark P, Straight A, Kelly K, Takuwa Y, Sugimoto N, Mitchison T, Bourne HR.** 2003. Divergent signals and cytoskeletal assemblies regulate self-organizing polarity in neutrophils. *Cell* **114:** 201–214.

88. **Wang F, Herzmark P, Weiner OD, Srinivasan S, Servant G, Bourne HR.** 2002. Lipid products of PI(3)Ks maintain persistent cell polarity and directed motility in neutrophils. *Nat Cell Biol* **4:**513–518.

89. **Chodniewicz D, Zhelev DV.** 2003. Chemoattractant receptor-stimulated F-actin polymerization in the human neutrophil is signaled by 2 distinct pathways. *Blood* **101:**1181–1184.

90. **Niggli V.** 2000. A membrane-permeant ester of phosphatidylinositol 3,4,5-trisphosphate (PIP_3) is an activator of human neutrophil migration. *FEBS Lett* **473:** 217–221.

91. **Weiner OD, Neilsen PO, Prestwich GD, Kirschner MW, Cantley LC, Bourne HR.** 2002. A $PtdInsP_3$- and Rho GTPase-mediated positive feedback loop regulates neutrophil polarity. *Nat Cell Biol* **4:**509–513.

92. **Worthylake RA, Burridge K.** 2003. RhoA and ROCK promote migration by limiting membrane protrusions. *J Biol Chem* **278:**13578–13584.

93. **Weiner OD, Rentel MC, Ott A, Brown GE, Jedrychowski M, Yaffe MB, Gygi SP, Cantley LC, Bourne HR, Kirschner MW.** 2006. Hem-1 complexes are essential for Rac activation, actin polymerization, and myosin regulation during neutrophil chemotaxis. *PLoS Biol* **4:**e38. doi: 10.1371/journal.pbio.0040038.

94. **Van Keymeulen A, Wong K, Knight ZA, Govaerts C, Hahn KM, Shokat KM, Bourne HR.** 2006. To stabilize neutrophil polarity, PIP3 and Cdc42 augment RhoA activity at the back as well as signals at the front. *J Cell Biol* **174:**437–445.

95. **Kumar S, Xu J, Perkins C, Guo F, Snapper S, Finkelman FD, Zheng Y, Filippi MD.** 2012. Cdc42 regulates neutrophil migration via crosstalk between WASp, CD11b, and microtubules. *Blood* **120:**3563–3574.

96. **Bokoch GM.** 2003. Biology of the p21-activated kinases. *Annu Rev Biochem* **72:**743–781.

97. **Itakura A, Aslan JE, Kusanto BT, Phillips KG, Porter JE, Newton PK, Nan X, Insall RH, Chernoff J, McCarty OJ.** 2013. p21-activated kinase (PAK) regulates cytoskeletal reorganization and directional migration in human neutrophils. *PLoS One* **8:**e73063. doi:10.1371/journal.pone.0073063.

98. **Svetina S, Bozic B, Derganc J, Zeks B.** 2001. Mechanical and functional aspects of membrane skeletons. *Cell Mol Biol Lett* **6:**677–690.

99. **White JR, Naccache PH, Sha'afi RI.** 1983. Stimulation by chemotactic factor of actin association with the cytoskeleton in rabbit neutrophils. Effects of calcium and cytochalasin B. *J Biol Chem* **258:**14041–14047.

100. **Cassimeris L, McNeill H, Zigmond SH.** 1990. Chemoattractant-stimulated polymorphonuclear leukocytes contain two populations of actin filaments that differ in their spatial distributions and relative stabilities. *J Cell Biol* **110:**1067–1075.

101. **Gomez-Cambronero J, Horn J, Paul CC, Baumann MA.** 2003. Granulocyte-macrophage colony-stimulating factor is a chemoattractant cytokine for human neutrophils: involvement of the ribosomal p70 S6 kinase signaling pathway. *J Immunol* **171:**6846–6855.

102. **Kamata N, Kutsuna H, Hato F, Kato T, Oshitani N, Arakawa T, Kitagawa S.** 2004. Activation of human neutrophils by granulocyte colony-stimulating factor, granulocyte-macrophage colony-stimulating factor, and

tumor necrosis factor alpha: role of phosphatidylinositol 3-kinase. *Int J Hematol* **80**:421–427.

103. **Cano ML, Lauffenburger DA, Zigmond SH.** 1991. Kinetic analysis of F-actin depolymerization in polymorphonuclear leukocyte lysates indicates that chemoattractant stimulation increases actin filament number without altering the filament length distribution. *J Cell Biol* **115**:677–687.

104. **Rougerie P, Miskolci V, Cox D.** 2013. Generation of membrane structures during phagocytosis and chemotaxis of macrophages: role and regulation of the actin cytoskeleton. *Immunol Rev* **256**:222–239.

105. **Gupton SL, Waterman-Storer CM.** 2006. Spatiotemporal feedback between actomyosin and focal-adhesion systems optimizes rapid cell migration. *Cell* **125**:1361–1374.

106. **Seveau S, Eddy RJ, Maxfield FR, Pierini LM.** 2001. Cytoskeleton-dependent membrane domain segregation during neutrophil polarization. *Mol Biol Cell* **12**:3550–3562.

107. **Rossy J, Schlicht D, Engelhardt B, Niggli V.** 2009. Flotillins interact with PSGL-1 in neutrophils and, upon stimulation, rapidly organize into membrane domains subsequently accumulating in the uropod. *PLoS One* **4**: e5403. doi:10.1371/journal.pone.0005403.

108. **Ludwig A, Otto GP, Riento K, Hams E, Fallon PG, Nichols BJ.** 2010. Flotillin microdomains interact with the cortical cytoskeleton to control uropod formation and neutrophil recruitment. *J Cell Biol* **191**:771–781.

109. **Smith LA, Aranda-Espinoza H, Haun JB, Dembo M, Hammer DA.** 2007. Neutrophil traction stresses are concentrated in the uropod during migration. *Biophys J* **92**:L58–L60.

110. **Fais S, Malorni W.** 2003. Leukocyte uropod formation and membrane/cytoskeleton linkage in immune interactions. *J Leukoc Biol* **73**:556–563.

111. **Lämmermann T, Bader BL, Monkley SJ, Worbs T, Wedlich-Söldner R, Hirsch K, Keller M, Förster R, Critchley DR, Fässler R, Sixt M.** 2008. Rapid leukocyte migration by integrin-independent flowing and squeezing. *Nature* **453**:51–55.

112. **Muller WA.** 2011. Mechanisms of leukocyte transendothelial migration. *Annu Rev Pathol* **6**:323–344.

113. **Stroka KM, Hayenga HN, Aranda-Espinoza H.** 2013. Human neutrophil cytoskeletal dynamics and contractility actively contribute to trans-endothelial migration. *PLoS One* **8**:e61377. doi:10.1371/journal.pone.0061377.

114. **Wang N.** 1998. Mechanical interactions among cytoskeletal filaments. *Hypertension* **32**:162–165.

115. **Janke C.** 2014. The tubulin code: molecular components, readout mechanisms, and functions. *J Cell Biol* **206**:461–472.

116. **Reed NA, Cai D, Blasius TL, Jih GT, Meyhofer E, Gaertig J, Verhey KJ.** 2006. Microtubule acetylation promotes kinesin-1 binding and transport. *Curr Biol* **16**: 2166–2172.

117. **Gundersen GG, Kalnoski MH, Bulinski JC.** 1984. Distinct populations of microtubules: tyrosinated and nontyrosinated alpha tubulin are distributed differently in vivo. *Cell* **38**:779–789.

118. **Bulinski JC, Gundersen GG.** 1991. Stabilization of post-translational modification of microtubules during cellular morphogenesis. *BioEssays* **13**:285–293.

119. **Cassimeris LU, Wadsworth P, Salmon ED.** 1986. Dynamics of microtubule depolymerization in monocytes. *J Cell Biol* **102**:2023–2032.

120. **Ding M, Robinson JM, Behrens BC, Vandré DD.** 1995. The microtubule cytoskeleton in human phagocytic leukocytes is a highly dynamic structure. *Eur J Cell Biol* **66**:234–245.

121. **Eddy RJ, Pierini LM, Maxfield FR.** 2002. Microtubule asymmetry during neutrophil polarization and migration. *Mol Biol Cell* **13**:4470–4483.

122. **Palazzo AF, Joseph HL, Chen YJ, Dujardin DL, Alberts AS, Pfister KK, Vallee RB, Gundersen GG.** 2001b. Cdc42, dynein, and dynactin regulate MTOC reorientation independent of Rho-regulated microtubule stabilization. *Curr Biol* **11**:1536–1541.

123. **Etienne-Manneville S, Hall A.** 2003. Cdc42 regulates GSK-3β and adenomatous polyposis coli to control cell polarity. *Nature* **421**:753–756.

124. **Szczur K, Zheng Y, Filippi MD.** 2009. The small Rho GTPase Cdc42 regulates neutrophil polarity via CD11b integrin signaling. *Blood* **114**:4527–4537.

125. **Dziezanowski MA, DeStefano MJ, Rabinovitch M.** 1980. Effect of antitubulins on spontaneous and chemotactic migration of neutrophils under agarose. *J Cell Sci* **42**:379–388.

126. **Keller HU, Naef A, Zimmermann A.** 1984. Effects of colchicine, vinblastine and nocodazole on polarity, motility, chemotaxis and cAMP levels of human polymorphonuclear leukocytes. *Exp Cell Res* **153**:173–185.

127. **Niggli V.** 2003. Microtubule-disruption-induced and chemotactic-peptide-induced migration of human neutrophils: implications for differential sets of signalling pathways. *J Cell Sci* **116**:813–822.

128. **Keller HU, Niggli V.** 1993. Colchicine-induced stimulation of PMN motility related to cytoskeletal changes in actin, alpha-actinin, and myosin. *Cell Motil Cytoskeleton* **25**:10–18.

129. **Rich AM, Hoffstein ST.** 1981. Inverse correlation between neutrophil microtubule numbers and enhanced random migration. *J Cell Sci* **48**:181–191.

130. **Xu J, Wang F, Van Keymeulen A, Rentel M, Bourne HR.** 2005. Neutrophil microtubules suppress polarity and enhance directional migration. *Proc Natl Acad Sci U S A* **102**:6884–6889.

131. **Wittmann T, Waterman-Storer CM.** 2001. Cell motility: can Rho GTPases and microtubules point the way? *J Cell Sci* **114**:3795–3803.

132. **Kodama A, Lechler T, Fuchs E.** 2004. Coordinating cytoskeletal tracks to polarize cellular movements. *J Cell Biol* **167**:203–207.

133. **Yoo SK, Lam PY, Eichelberg MR, Zasadil L, Bement WM, Huttenlocher A.** 2012. The role of microtubules in neutrophil polarity and migration in live zebrafish. *J Cell Sci* **125**:5702–5710.

134. **Glasgow JE, Daniele RP.** 1994. Role of microtubules in random cell migration: stabilization of cell polarity. *Cell Motil Cytoskeleton* **27**:88–96.

135. **Redd MJ, Kelly G, Dunn G, Way M, Martin P.** 2006. Imaging macrophage chemotaxis in vivo: studies of microtubule function in zebrafish wound inflammation. *Cell Motil Cytoskeleton* **63:**415–422.

136. **Wang Y, Ku CJ, Zhang ER, Artyukhin AB, Weiner OD, Wu LF, Altschuler SJ.** 2013. Identifying network motifs that buffer front-to-back signaling in polarized neutrophils. *Cell Rep* **3:**1607–1616.

137. **Krendel M, Zenke FT, Bokoch GM.** 2002. Nucleotide exchange factor GEF-H1 mediates cross-talk between microtubules and the actin cytoskeleton. *Nat Cell Biol* **4:**294–301.

138. **Chang YC, Nalbant P, Birkenfeld J, Chang ZF, Bokoch GM.** 2008. GEF-H1 couples nocodazole-induced microtubule disassembly to cell contractility via RhoA. *Mol Biol Cell* **19:**2147–2153.

139. **Hanania R, Sun HS, Xu K, Pustylnik S, Jeganathan S, Harrison RE.** 2012. Classically activated macrophages use stable microtubules for matrix metalloproteinase-9 (MMP-9) secretion. *J Biol Chem* **287:**8468–8483.

140. **Wang B, Rao YH, Inoue M, Hao R, Lai CH, Chen D, McDonald SL, Choi MC, Wang Q, Shinohara ML, Yao TP.** 2014. Microtubule acetylation amplifies p38 kinase signalling and anti-inflammatory IL-10 production. *Nat Commun* **5:**3479. doi:10.1038/ncomms4479.

141. **Rodriguez OC, Schaefer AW, Mandato CA, Forscher P, Bement WM, Waterman-Storer CM.** 2003. Conserved microtubule-actin interactions in cell movement and morphogenesis. *Nat Cell Biol* **5:**599–609.

142. **Palazzo AF, Cook TA, Alberts AS, Gundersen GG.** 2001a. mDia mediates Rho-regulated formation and orientation of stable microtubules. *Nat Cell Biol* **3:**723–729.

143. **Bartolini F, Moseley JB, Schmoranzer J, Cassimeris L, Goode BL, Gundersen GG.** 2008. The formin mDia2 stabilizes microtubules independently of its actin nucleation activity. *J Cell Biol* **181:**523–536.

144. **Danowski BA.** 1989. Fibroblast contractility and actin organization are stimulated by microtubule inhibitors. *J Cell Sci* **93:**255–266.

145. **Enomoto T.** 1996. Microtubule disruption induces the formation of actin stress fibers and focal adhesions in cultured cells: possible involvement of the rho signal cascade. *Cell Struct Funct* **21:**317–326.

146. **Farrera C, Fadeel B.** 2013. Macrophage clearance of neutrophil extracellular traps is a silent process. *J Immunol* **191:**2647–2656.

147. **Swanson JA, Baer SC.** 1995. Phagocytosis by zippers and triggers. *Trends Cell Biol* **5:**89–93.

148. **Greenberg S.** 1995. Signal transduction of phagocytosis. *Trends Cell Biol* **5:**93–99.

149. **Cox D, Chang P, Zhang Q, Reddy PG, Bokoch GM, Greenberg S.** 1997. Requirements for both Rac1 and Cdc42 in membrane ruffling and phagocytosis in leukocytes. *J Exp Med* **186:**1487–1494.

150. **Greenberg S, Chang P, Silverstein SC.** 1993. Tyrosine phosphorylation is required for Fc receptor-mediated phagocytosis in mouse macrophages. *J Exp Med* **177:**529–534.

151. **Dharmawardhane S, Brownson D, Lennartz M, Bokoch GM.** 1999. Localization of p21-activated kinase 1 (PAK1) to pseudopodia, membrane ruffles, and phagocytic cups in activated human neutrophils. *J Leukoc Biol* **66:**521–527.

152. **Hoppe AD, Swanson JA.** 2004. Cdc42, Rac1, and Rac2 display distinct patterns of activation during phagocytosis. *Mol Biol Cell* **15:**3509–3519.

153. **Groves E, Dart AE, Covarelli V, Caron E.** 2008. Molecular mechanisms of phagocytic uptake in mammalian cells. *Cell Mol Life Sci* **65:**1957–1976.

154. **Yin HL, Albrecht JH, Fattoum A.** 1981. Identification of gelsolin, a $Ca2^{+}$-dependent regulatory protein of actin gel-sol transformation, and its intracellular distribution in a variety of cells and tissues. *J Cell Biol* **91:**901–906.

155. **Serrander L, Skarman P, Rasmussen B, Witke W, Lew DP, Krause KH, Stendahl O, Nüsse O.** 2000. Selective inhibition of IgG-mediated phagocytosis in gelsolin-deficient murine neutrophils. *J Immunol* **165:**2451–2457.

156. **Robinson JM, Badwey JA.** 2002. Rapid association of cytoskeletal remodeling proteins with the developing phagosomes of human neutrophils. *Histochem Cell Biol* **118:**117–125.

157. **Witke W, Li W, Kwiatkowski DJ, Southwick FS.** 2001. Comparisons of CapG and gelsolin-null macrophages: demonstration of a unique role for CapG in receptor-mediated ruffling, phagocytosis, and vesicle rocketing. *J Cell Biol* **154:**775–784.

158. **Herant M, Heinrich V, Dembo M.** 2006. Mechanics of neutrophil phagocytosis: experiments and quantitative models. *J Cell Sci* **119:**1903–1913.

159. **Tollis S, Dart AE, Tzircotis G, Endres RG.** 2010. The zipper mechanism in phagocytosis: energetic requirements and variability in phagocytic cup shape. *BMC Syst Biol* **4:**149. doi:10.1186/1752-0509-4-149.

160. **Colucci-Guyon E, Niedergang F, Wallar BJ, Peng J, Alberts AS, Chavrier P.** 2005. A role for mammalian diaphanous-related formins in complement receptor (CR3)-mediated phagocytosis in macrophages. *Curr Biol* **15:**2007–2012.

161. **Olazabal IM, Caron E, May RC, Schilling K, Knecht DA, Machesky LM.** 2002. Rho-kinase and myosin-II control phagocytic cup formation during CR, but not FcγR, phagocytosis. *Curr Biol* **12:**1413–1418.

162. **Wang QQ, Li H, Oliver T, Glogauer M, Guo J, He YW.** 2008. Integrin β_1 regulates phagosome maturation in macrophages through Rac expression. *J Immunol* **180:**2419–2428.

163. **Magalhães MA, Sun CX, Glogauer M, Ellen RP.** 2008. The major outer sheath protein of *Treponema denticola* selectively inhibits Rac1 activation in murine neutrophils. *Cell Microbiol* **10:**344–354.

164. **Allen LH, Aderem A.** 1995. A role for MARCKS, the α isozyme of protein kinase C and myosin I in zymosan phagocytosis by macrophages. *J Exp Med* **182:**829–840.

165. **Hartwig JH, Thelen M, Rosen A, Janmey PA, Nairn AC, Aderem A.** 1992. MARCKS is an actin filament crosslinking protein regulated by protein kinase C and calcium-calmodulin. *Nature* **356:**618–622.

166. **Ali WH, Chen Q, Delgiorno KE, Su W, Hall JC, Hongu T, Tian H, Kanaho Y, Di Paolo G, Crawford HC, Frohman MA.** 2013. Deficiencies of the lipid-signaling enzymes phospholipase D1 and D2 alter cytoskeletal

organization, macrophage phagocytosis, and cytokine-stimulated neutrophil recruitment. *PLoS One* **8**:e55325. doi:10.1371/journal.pone.0055325.

167. **Brinkmann V, Zychlinsky A.** 2007. Beneficial suicide: why neutrophils die to make NETs. *Nat Rev Microbiol* **5**:577–582.

168. **Metzler KD, Goosmann C, Lubojemska A, Zychlinsky A, Papayannopoulos V.** 2014. A myeloperoxidase-containing complex regulates neutrophil elastase release and actin dynamics during NETosis. *Cell Reports* **8**:883–896.

169. **Neeli I, Dwivedi N, Khan S, Radic M.** 2009. Regulation of extracellular chromatin release from neutrophils. *J Innate Immun* **1**:194–201.

170. **Newman SL, Mikus LK, Tucci MA.** 1991. Differential requirements for cellular cytoskeleton in human macrophage complement receptor- and Fc receptor-mediated phagocytosis. *J Immunol* **146**:967–974.

171. **Allen LA, Aderem A.** 1996. Molecular definition of distinct cytoskeletal structures involved in complement- and Fc receptor-mediated phagocytosis in macrophages. *J Exp Med* **184**:627–637.

172. **Zhou X, Li J, Kucik DF.** 2001. The microtubule cytoskeleton participates in control of β_2 integrin avidity. *J Biol Chem* **276**:44762–44769.

173. **Desjardins M, Huber LA, Parton RG, Griffiths G.** 1994. Biogenesis of phagolysosomes proceeds through a sequential series of interactions with the endocytic apparatus. *J Cell Biol* **124**:677–688.

174. **Damiani MT, Colombo MI.** 2003. Microfilaments and microtubules regulate recycling from phagosomes. *Exp Cell Res* **289**:152–161.

175. **Blocker A, Griffiths G, Olivo JC, Hyman AA, Severin FF.** 1998. A role for microtubule dynamics in phagosome movement. *J Cell Sci* **111**:303–312.

176. **Blocker A, Severin FF, Burkhardt JK, Bingham JB, Yu H, Olivo JC, Schroer TA, Hyman AA, Griffiths G.** 1997. Molecular requirements for bi-directional movement of phagosomes along microtubules. *J Cell Biol* **137**:113–129.

177. **Tapper H, Grinstein S.** 1997. Fc receptor-triggered insertion of secretory granules into the plasma membrane of human neutrophils: selective retrieval during phagocytosis. *J Immunol* **159**:409–418.

178. **Botelho RJ, Tapper H, Furuya W, Mojdami D, Grinstein S.** 2002. FcγR-mediated phagocytosis stimulates localized pinocytosis in human neutrophils. *J Immunol* **169**:4423–4429.

179. **Peachman KK, Rao M, Palmer DR, Zidanic M, Sun W, Alving CR, Rothwell SW.** 2004. Functional microtubules are required for antigen processing by macrophages and dendritic cells. *Immunol Lett* **95**:13–24.

180. **Radhika V, Naik NR, Advani SH, Bhisey AN.** 2000. Actin polymerization in response to different chemoattractants is reduced in granulocytes from chronic myeloid leukemia patients. *Cytometry* **42**:379–386.

181. **Bhansali RS, Yeltiwar RK, Bhat KG.** 2013. Assessment of peripheral neutrophil functions in patients with localized aggressive periodontitis in the Indian population. *J Indian Soc Periodontol* **17**:731–736.

182. **Southwick FS, Dabiri GA, Stossel TP.** 1988. Neutrophil actin dysfunction is a genetic disorder associated with partial impairment of neutrophil actin assembly in three family members. *J Clin Invest* **82**:1525–1531.

183. **Derry JM, Ochs HD, Francke U.** 1994. Isolation of a novel gene mutated in Wiskott-Aldrich syndrome. *Cell* **78**:635–644.

184. **Binks M, Jones GE, Brickell PM, Kinnon C, Katz DR, Thrasher AJ.** 1998. Intrinsic dendritic cell abnormalities in Wiskott-Aldrich syndrome. *Eur J Immunol* **28**:3259–3267.

185. **Williams DA, Tao W, Yang F, Kim C, Gu Y, Mansfield P, Levine JE, Petryniak B, Derrow CW, Harris C, Jia B, Zheng Y, Ambruso DR, Lowe JB, Atkinson SJ, Dinauer MC, Boxer L.** 2000. Dominant negative mutation of the hematopoietic-specific Rho GTPase, Rac2, is associated with a human phagocyte immunodeficiency. *Blood* **96**:1646–1654.

186. **Ambruso DR, Knall C, Abell AN, Panepinto J, Kurkchubasche A, Thurman G, Gonzalez-Aller C, Hiester A, deBoer M, Harbeck RJ, Oyer R, Johnson GL, Roos D.** 2000. Human neutrophil immunodeficiency syndrome is associated with an inhibitory Rac2 mutation. *Proc Natl Acad Sci U S A* **97**:4654–4659.

187. **Nunoi H, Yamazaki T, Tsuchiya H, Kato S, Malech HL, Matsuda I, Kanegasaki S.** 1999. A heterozygous mutation of β-actin associated with neutrophil dysfunction and recurrent infection. *Proc Natl Acad Sci U S A* **96**:8693–8698.

188. **Hundt N, Preller M, Swolski O, Ang AM, Mannherz HG, Manstein DJ, Müller M.** 2014. Molecular mechanisms of disease-related human β-actin mutations p.R183W and p.E364K. *FEBS J* **281**:5279–5291.

189. **Longacre A, Koh JS, Hsiao KK, Gilligan H, Fan H, Patel VA, Levine JS.** 2004. Macrophages from lupus-prone MRL mice are characterized by abnormalities in Rho activity, cytoskeletal organization, and adhesiveness to extracellular matrix proteins. *J Leukoc Biol* **76**:971–984.

190. **Fan H, Patel VA, Longacre A, Levine JS.** 2006. Abnormal regulation of the cytoskeletal regulator Rho typifies macrophages of the major murine models of spontaneous autoimmunity. *J Leukoc Biol* **79**:155–165.

191. **Radtke K, Döhner K, Sodeik B.** 2006. Viral interactions with the cytoskeleton: a hitchhiker's guide to the cell. *Cell Microbiol* **8**:387–400.

192. **Stamm LM, Morisaki JH, Gao LY, Jeng RL, McDonald KL, Roth R, Takeshita S, Heuser J, Welch MD, Brown EJ.** 2003. *Mycobacterium marinum* escapes from phagosomes and is propelled by actin-based motility. *J Exp Med* **198**:1361–1368.

193. **Alpuche-Aranda CM, Racoosin EL, Swanson JA, Miller SI.** 1994. *Salmonella* stimulate macrophage macropinocytosis and persist within spacious phagosomes. *J Exp Med* **179**:601–608.

194. **Alpuche-Aranda CM, Berthiaume EP, Mock B, Swanson JA, Miller SI.** 1995. Spacious phagosome formation within mouse macrophages correlates with *Salmonella* serotype pathogenicity and host susceptibility. *Infect Immun* **63**:4456–4462.

Myeloid Cells in Health and Disease: A Synthesis
Edited by Siamon Gordon

doi:10.1128/microbiolspec.MCHD-0025-2015

Toby Lawrence[1]

29 Coordinated Regulation of Signaling Pathways during Macrophage Activation

NF-κB SIGNALING

NF-κB is the prototypical proinflammatory transcription factor and plays an important role in production of cytokines, chemokines, and other proinflammatory mediators during the inflammatory response (1). NF-κB activation in macrophages is best characterized in response to proinflammatory cytokines, such as tumor necrosis factor α (TNF-α) and interleukin-1β (IL-1β), or the recognition of microbial products by Toll-like receptors (TLRs) or NOD-like receptors (NLRs). Using various upstream signaling pathways, these receptors converge on activation of the IKK (IκB kinase) complex, leading to activation of NF-κB. The IKK complex consists of two kinase subunits, IKKα (CHUK) and IKKβ (IKBKB), and the ubiquitin-binding protein IKKγ (IKBKG) (2). Despite its well-established role as a proinflammatory transcription factor in many cells, the NF-κB pathway in macrophages also has important anti-inflammatory roles (3). This was revealed by studies using the targeted deletion of IKKβ in macrophages, which was shown to have proinflammatory effects in several models of inflammation (4–7).

Several mechanisms for the anti-inflammatory action of NF-κB in macrophages have been proposed. As with many proinflammatory signaling pathways, NF-κB induces the expression of negative feedback regulators to limit activation of the pathway; the most obvious example is the induction of expression of IκBα (NFKBIA), the endogenous inhibitor of NF-κB activation. But NF-κB also induces expression of negative regulators of other signaling pathways, such as the deubiquitinase A20 (TNFIAP3), which targets upstream regulators of both NF-κB and mitogen-activated protein kinase (MAPK) activation downstream of TNF, IL-1, and TLR signaling (8). Deletion of A20 expression in macrophages leads to spontaneous inflammatory disease (9), illustrating its important role in limiting macrophage activation. NF-κB also induces expression of caspase-1 inhibitors in macrophages that block the production of mature IL-1β, despite the fact that NF-κB is required for induction of pro-IL-β mRNA (4). In other studies, IKKβ deletion in macrophages was shown to increase activation of STAT1 in response to autocrine beta interferon (IFN-β) or exogenous IFN-γ (5). This was associated with increased expression of major histocompatibility complex class II (MHC-II), nitric oxide synthase 2 (NOS2), and IL-12 and conferred enhanced resistance to infection. Collectively, these studies show

[1]Centre d'Immunologie de Marseille-Luminy, Aix Marseille Université, Inserm, CNRS, Marseille, France.

that activation of NF-κB signaling in macrophages through IKKβ incorporates important feedback control mechanisms to limit the inflammatory response.

NFKB1 (p105/p50) and Macrophage Activation

After TLR-induced activation, macrophages can undergo a prolonged period of cell-autonomous hyporesponsiveness—often referred to as macrophage tolerance. Homodimers of the p50 subunit of NF-κB, which lacks a transactivation domain, are thought to play an important role in repression of proinflammatory genes in tolerant macrophages (10–12), while promoting the expression of anti-inflammatory genes such as IL-10 through interaction with coactivators such as BCL-3 (13, 14). Mature p50 is generated by the proteolytic processing of p105 (the gene product of *NFKB1*). p105 also participates in the regulation of parallel signaling pathways in macrophages; p105 stabilizes tumor progression locus 2 (TPL2; COT), which activates extracellular signal-regulated kinase (ERK) in response to TLR stimulation (15), which in turn has been shown to inhibit IFN-β expression in macrophages (16). Therefore, in *Nfkb1*$^{-/-}$ macrophages, IFN-β expression is elevated due to impaired TPL2 activity (17, 18). TPL2-mediated ERK activation is also associated with increased c-Fos and IL-10 expression in macrophages (16, 18); c-Fos has been shown to inhibit NF-κB-mediated proinflammatory gene expression in macrophages (19), and IL-10 has well-established inhibitory effects on macrophage activation. Therefore, p105/p50 expression in macrophages mediates several mechanisms of negative feedback to control proinflammatory macrophage activation. Interestingly, IKKβ-mediated phosphorylation of p105 has also been shown to be required for TPL2 activation (15); this may provide an additional anti-inflammatory mechanism for IKKβ in macrophages through TPL2-ERK-dependent activation of c-Fos and IL-10 expression.

Noncanonical NF-κB Signaling in Macrophages

The IKK complex consists of two kinase subunits (IKKα/IKKβ), but only IKKβ is required for canonical NF-κB activation (20). An "alternative" pathway for NF-κB activation was described through the IKKα-dependent phosphorylation of NFKB2 (p100) and the specific activation of p52:RelB heterodimers. This pathway was independent of IKKγ and IKKβ, but required the upstream kinase NIK (NF-κB-inducing kinase; MAP3K14) (21). The alternative pathway is triggered by TNF family member ligands—including BAFF (B-cell-activating factor), lymphotoxin β, and CD40L—but the significance of this pathway in macrophages is unclear. However, IKKα has been shown to have an important role as a negative regulator of canonical NF-κB activation in macrophages. We showed that IKKα-mediated phosphorylation of RelA and c-Rel led to their proteasomal degradation and limited NF-κB activation in macrophages exposed to invasive bacteria; this was required to promote pathogen-induced apoptosis of infected macrophages, therefore limiting their activation and persistence (20). Another study showed that IKKα regulated the activity of the SUMO (small ubiquitin-like modifier) ligase PIAS1 (protein inhibitor of activated STAT1)—which subsequently mediated the degradation of NF-κB and STAT1 transcription factors to attenuate proinflammatory gene expression (22). Finally, IKKα was also shown to participate in A2-mediated inhibition of NF-κB signaling through the phosphorylation of TAX1BP1, a component of the A20 deubiquitinase complex (23). These studies suggest that IKKα is an important negative regulator of canonical NF-κB signaling required to limit the activation and persistence of inflammatory macrophages.

MAPK SIGNALING

Along with the NF-κB pathway, activation of MAPKs, including ERK, Jun N-terminal protein kinase (JNK), and p38, is critical for proinflammatory cytokine expression during macrophage activation. However, in addition to their well-described functions in regulating proinflammatory cytokine production and promoting inflammation, triggering of MAPKs by TLR ligands or cytokines also leads to activation of feedback control mechanisms to limit macrophage activation. The TPL2-ERK-dependent regulation of IL-10 expression in macrophages is mentioned above (16, 18). Certain TLRs, for example, TLR4 and TLR9, coordinately induce IFN-β and IL-10 expression in macrophages. However, in the absence of TPL2 signaling, ERK-mediated IL-10 induction is inhibited and there is a reciprocal increase in IFN-β expression (16). Interestingly, ERK-dependent negative regulation of IFN-β was independent of IL-10 induction but was instead mediated by upregulation of the AP-1 transcription factor c-Fos. Thus, ERK activation is required for two independent feedback control mechanisms of TLR-induced macrophage activation: induction of the anti-inflammatory cytokine IL-10 and c-Fos-dependent repression of IFN-β expression.

The p38 MAPK was originally identified for its critical role in lipopolysaccharide (LPS) and IL-1 signaling. Activation of p38 leads to phosphorylation of multiple substrates, including downstream kinases, sequence-specific transcription factors, chromatin-binding proteins, and factors regulating mRNA stability and translation, that all are important regulators of proinflammatory gene expression. As an example of the coordinated regulation of gene expression by p38 and NF-κB activation in macrophages, p38-dependent phosphorylation of histone H3 on selected NF-κB target genes during LPS-induced macrophage activation marks these genes for increased NF-κB recruitment and enhanced expression (24). Activation of p38 has also been shown to be an important negative regulator of proinflammatory signaling in macrophages. The p38-mediated phosphorylation of TAB1 (TAK1-binding protein 1)—a regulatory subunit of TAK1 (transforming growth factor-β-activated kinase 1)—which is an essential upstream activator of both NF-κB and MAPK in response to TLR or IL-1 signaling, leads to inhibition of TAK1 activity and downstream activation of NF-κB and MAPK (25). The p38- and ERK-dependent induction of dual-specificity phosphatase (DUSP1; MKP1) is also an important mechanism to limit MAPK activation, including p38, JNK, and ERK, in a classic feedback control mechanism. The lineage-specific targeting of the p38 (*Mapk14*) gene has shown that the anti-inflammatory role of p38 is particularly important in macrophages in response to TLR signaling. The targeted deletion of *Mapk14* in epithelial cells significantly abrogated inflammation, while deletion of *Mapk14* in macrophages blocked anti-inflammatory gene expression and exacerbated the inflammatory response (26). The inhibition of macrophage activation by p38 was mediated through activation of the downstream kinases MSK1 (mitogen- and stress-activated kinase 1) and MSK2, which activated CREB/ATF1 (activating transcription factor 1)-dependent expression of IL-10 and DUSP1 (26, 27). This is reminiscent of the effects on cell-specific targeting of IKKβ/NF-κB activation: targeted deletion of IKKβ in lung epithelial cells reduced lung inflammation in the context of infection, but the deletion of IKKβ in macrophages enhanced proinflammatory signaling and macrophage activation (5). These studies suggest that proinflammatory signaling pathways are tightly regulated in macrophages by IKKβ/NF-κB- and MAPK-mediated feedback control mechanisms to prevent protracted macrophage activation and tissue injury or chronic inflammation, and these anti-inflammatory mechanisms can in fact be dominant in the context of inflammation *in vivo*.

JAK/STAT SIGNALING

Cytokine signaling is a major driver of macrophage polarization during the inflammatory response, particularly type I and type II cytokines signaling through JAK/STAT pathways.

IFN Signaling

IFN-γ produced by innate lymphocytes, such as natural killer (NK) cells, or antigen-specific T cells dramatically enhances the microbicidal activity and antigen-presenting cell function of macrophages, associated with increased NOS2, MHC-II, and IL-12 expression. These characteristics are promoted by IFN-γ-mediated JAK-STAT signaling. The IFN-γ receptor (IFNGR) triggers JAK-mediated tyrosine phosphorylation and dimerization of STAT1, the promoters for NOS2, the MHC-II transactivator (CIITA) and IL-12 contain *cis* elements called gamma-activated sequence (GAS) that bind STAT1 homodimers (28). Studies with macrophages from STAT1-deficient mice clearly show that NOS2, MHC-II, and IL-12 are all STAT1 dependent in this context.

TLR signaling also upregulates MHC-II, NOS2, and IL-12 expression on macrophages; however, expression of these genes is dramatically enhanced by the autocrine/paracrine production of type I IFN (IFN-α/β), which signals through a distinct receptor, IFNAR (29). JAK activation downstream of IFNAR triggers both STAT1 homodimers and activation of STAT1/STAT2 heterodimers, which subsequently recruit IFN regulatory factor 9 (IRF9) to form the IFN-stimulated gene factor 3 (ISGF3) complex; ISGF3 binds distinct *cis* elements called IFN-stimulated regulatory element (ISRE).

NF-κB and STAT1 can also coordinately regulate expression of specific genes in macrophages. For example, NF-κB binding to the NOS2 gene promoter was shown to be required for recruitment of a component of the basal transcriptional machinery, TFIIA; the subsequent activation of STAT1 recruited RNA Pol II and TFIIA-mediated phosphorylation of Pol II was required for full induction of NOS2 expression (30). This mechanism of coincidence detection could help dictate the functional outcome of macrophage activation in specific contexts, in this case infection with an invasive intracellular bacterium that induces both NF-κB activation through TLR signaling and type I IFN expression after invasion into the macrophage cytosol. The high expression of NOS2 in this context is critical to restrict the intracellular replication of the bacteria. Another coincidence detection mechanism in macrophages for enhancing STAT1 activation is the engagement of phagocytic antibody Fcγ receptors. These receptors have immunoreceptor tyrosine-based activation motifs (ITAMs) leading to activation of tyrosine kinase and phosphatidylinositol 3-kinase (PI3K). Triggering of Fcγ-mediated PI3K activation during phagocytosis was shown to promote cell-autonomous antimicrobial activity through enhanced IFN-γ-mediated STAT1 activation (31). Previous studies had shown that ITAM-associated activation of Syk kinase (spleen tyrosine kinase) enhanced STAT1 activation in IFN-stimulated macrophages through the increased expression of STAT1 (32). The roles of ITAM signaling in macrophage activation are further discussed later in this chapter.

It is clear that IFN-γ/STAT1 signaling in macrophages is critical for resistance to infection by intracellular pathogens including viruses, *Listeria monocytogenes*, and *Mycobacterium tuberculosis* (33–36). However, the role of IFNAR signaling and ISGF3 in macrophage activation *in vivo* is less clear. IFN-α/β can have profound anti-inflammatory activity in certain contexts (37); furthermore, IFNAR signaling in macrophages inhibits resistance to *L. monocytogenes* infection by promoting macrophage apoptosis and thus limiting pathogen replication (38). This suggests quite contrasting roles for IFN-γ and IFN-α/β signaling in macrophage activation *in vivo* that may diverge at the level of ISGF3 activation.

IL-4/IL-13 Signaling

The role for IL-4/IL-13 signaling in the development of alternatively activated macrophages (also called M2 macrophages) has been well established both *in vitro* and *in vivo* (39). This has been facilitated by the generation of myeloid cell-specific IL-4Rα (*Il4ra*) knockout mice that are resistant to M2 macrophage development in several mouse models of infection and Th2-mediated inflammation where IL-4/IL-13 plays a major role (40, 41). IL-4 signals through JAK-STAT6 (42), and many of the genes associated with M2 macrophages are regulated by STAT6, including arginase-1 (*Arg1*), mannose receptor (*Mrc1*), Fizz1 (*Retnla*), and Ym-1 (*Chil3*) (39). However, IL-4 also triggers other signaling pathways such as PI3K, which is also critical for cell responses to IL-4 (43). In fact, studies with mice deficient in SHIP (Src homology 2 domain-containing inositol 5′-phosphatase)—an important negative regulator of PI3K signaling—suggest that PI3K activity is an important factor in M2 macrophage activation (44).

It is generally well appreciated that classical macrophage activation, typified by IFN-γ/STAT1 signaling, and the M2 macrophage phenotype are mutually exclusive (45); however, the molecular mechanisms for this mutual antagonism are still not clearly defined.

Suppressors of Cytokine Signaling

In the synonymous context of Th1/Th2 cell polarization, suppressors of cytokine signaling (SOCS) proteins are thought to play a pivotal role in the polarization of cellular responses to IL-4 and IFN-γ (46). SOCS proteins are induced by JAK/STAT signaling and act as negative feedback regulators; all eight SOCS family proteins (SOCS1 to SOCS7 and CIS) have an SH2 domain that mediates recruitment to activated JAK and a highly conserved SOCS box domain that recruits an E3 ubiquitin ligase the cytokine receptor complex. Recruitment of the E3 ligase leads to ubiquitination and targeted degradation of the receptor complex including JAK. SOCS1, SOCS3, and CIS expression can also be induced by macrophages in response to TLR signaling and thus regulate cytokine signaling in the context of infection (47).

SOCS1 is well established as a critical negative regulator of IFN-γ-induced STAT1 activation. There have been reports that SOCS1 can negatively regulate TLR signaling in macrophages (48); however, subsequent studies suggest that this is likely due to indirect inhibition of JAK/STAT activation by autocrine type I IFN (49). On the other hand, Ryo et al. showed that SOCS1 inhibited cytokine-induced NF-κB activation by targeting RelA for proteasomal degradation (50). The targeted deletion of SOCS3 in macrophages leads to deregulation of IL-6 signaling and increased activation of both STAT3 and STAT1—leading to IL-6-mediated induction of genes normally restricted to IFN-stimulated macrophages (51). However, IL-10-induced STAT3 activation has anti-inflammatory effects in macrophages, at least in part through antagonism of NF-κB-mediated transcription (52). SOCS3 specifically targets gp130-mediated signaling by IL-6 and therefore doesn't directly target IL-10-mediated activation of STAT3, because IL-10 doesn't use gp130 (51). However, a parallel study showed that SOCS3 deletion led to anti-inflammatory IL-6 signaling in TLR-stimulated macrophages; this was attributed to increased STAT3 activation by IL-6 in the absence of SOCS3, mimicking IL-10-like signaling (53). TLR-induced SOCS3 induction also inhibited IFN-induced STAT1 activation in macrophages—and therefore classical macrophage activation (54). Thus, SOCS3 expression can affect both proinflammatory and anti-inflammatory signaling in macrophages and skew macrophage activation depending on the cytokine milieu.

Several SOCS proteins have been shown to regulate IL-4 signaling in T cells, including SOCS2 and SOCS5, which coordinately inhibit IL-4-induced gene expression (46). However, the role of SOCS proteins in IL-4/IL-13-induced macrophage activation and cross-regulation of IFN-γ signaling is not clear. IL-4 can induce expression of both SOCS1 and SOCS2 in macrophages, whereas IFN-γ induces SOCS1 and SOCS3 but not SOCS2 (55). SOCS1 expression in macrophages was shown to inhibit STAT6 activation, but the role of IL-4-induced SOCS1 and SOCS2 in cross-regulation of IFN-γ signaling and polarization of M2 macrophages has not been adequately addressed.

ITAM SIGNALING

Macrophages express a number of ITAM-associated receptors, including phagocytic Fc receptors, C-type lectins, and β_2 integrins (56). These receptors are often coactivated with TLR or cytokine signaling during macrophage activation *in vivo*. Although these receptors have specific functions, such as phagocytosis and adhesion, recent research has highlighted the important roles for ITAM signaling in cross-regulation of heterologous receptors and inflammatory signaling pathways during macrophage activation (56, 57).

Triggering of ITAM receptors leads to activation of Syk kinase and the downstream activation of phospholipase C-γ (PLC-γ)-protein kinase C (PKC) and caspase recruitment domain family member 9 (CARD9)-coupled MAPK and NF-κB activation (56). Several studies have shown that ITAM signaling and activation of Syk can amplify proinflammatory macrophage activation in response to microbial stimuli and certain cytokines (31, 32, 58, 59). ITAM-mediated Syk activation was shown to enhance IFN-α signaling in IFN-γ-primed macrophages by upregulating expression of STAT1 (32). Integrin-associated ITAM signaling was also shown to increase IFN-α-mediated JAK/STAT1 activation through calcium-activated kinases CaMK (Ca^{2+}/calmodulin-dependent protein kinase) and Pyk2 (58). A recent study showed that Fcγ-associated ITAM signaling and activation of the IFN-γ receptor cooperatively regulate STAT1-dependent gene expression and antimicrobial functions of macrophages (31). Activation of ITAM-associated C-type lectins on macrophages during fungal infection was shown to be required for both CARD9–NF-κB-mediated induction of proinflammatory cytokine expression and activation of the Nlrp3-dependent inflammasome, required for mature IL-1β production (59). C-type lectins have also been shown to collaborate with MyD88-dependent TLR signaling during macrophage activation, including Dectin-1 (dendritic cell-associated C-type lectin 1; CLEC7A) and TLR2 during recognition of fungal β-glucan (60, 61).

Despite many examples of cooperative macrophage activation through ITAM-coupled receptors, ITAM signaling has also been shown to inhibit proinflammatory macrophage activation in certain contexts (56). DAP12 (DNAX activation protein of 12 kDa) is a signaling adaptor recruited to ITAM-associated receptors; macrophages from mice deficient in DAP12 or the downstream kinase Syk were shown to have enhanced responses to a number of TLR ligands and increased resistance to *L. monocytogenes* infection *in vivo* (62). Other studies demonstrated that coactivation of ITAM-associated β_2 integrin or Fcγ receptors during TLR-mediated macrophage activation inhibited proinflammatory cytokine expression and increased IL-10 production (63). This was associated with induction of several negative regulators of TLR signaling, including A20 and SOCS3, through ITAM-mediated activation of Syk. These studies demonstrate that ITAM signaling in macrophages has an important role in "fine-tuning" macrophage activation to either enhance proinflammatory functions and pathogen clearance or downregulate macrophage activation and promote the resolution of inflammation.

CONCLUSIONS AND PERSPECTIVES

We have reviewed the canonical pathways that regulate macrophage activation after engagement of different receptors by microbes or cytokines, and how cross talk between these pathways can regulate macrophage activation and effector functions. However, there are also other intrinsic factors that can affect innate immune signaling in macrophages, including metabolic pathways. The influence of cell metabolism on immune cell function is a rapidly developing field, and macrophages are no exception. Glycolysis, the Krebs cycle, and fatty acid metabolism have specific effects on macrophage activation during encounters with pathogens that profoundly influence their effector functions (64). TLR signaling has a profound influence on metabolic pathways in macrophages; LPS-activated macrophages metabolically shift from the Krebs cycle and fatty acid oxidation to glycolytic metabolism driven by increased glucose uptake by upregulation of expression of GLUT1—the glucose transporter (65). This metabolic shift increases proinflammatory cytokine production and macrophage activation (66). For example, the oxygen-sensing transcription factor hypoxia-inducible factor-1α (HIF1α) has been shown to drive IL-1β expression directly during LPS-induced glycolysis in macrophages (67). Other studies have shown that NF-κB activation downstream of TLR signaling can drive expression of HIF1α itself, suggesting a feed-forward loop in this pathway through TLR-mediated NF-κB activation (68). This metabolic shift in LPS-activated macrophages is quite distinct from the effects of IL-4- and IL-13-induced JAK/STAT signaling on macrophage metabolism. Alternatively activated macrophages shift toward oxidative phosphorylation, driving the Krebs cycle and catabolic pathways rather than glycolysis (69); this shift away from the more proinflammatory glycolytic phenotype of classically activated macrophages may contribute to their anti-inflammatory phenotype by suppressing proinflammatory signaling pathways. However, the exact mechanisms of the metabolic control of macrophage polarization are only beginning to be unraveled.

Another emerging theme in the regulation of macrophage activation is the concept of trained immunity or innate immunological memory (70). This stems from the observation that primary challenge with attenuated pathogens as vaccines, for example, *Mycobacterium bovis* BCG, leads to nonspecific protection against subsequent heterologous infections (71). For example, BCG vaccination can protect against systemic candidiasis, and vice versa, attenuated *Candida albicans* vaccination provides protection against not only subsequent *Candida* infection but also bacterial infections (72, 73), which was unexpectedly independent of T lymphocytes but dependent on macrophages. In fact, mechanistically it was determined that β-glucan receptor (CLEC7A)-mediated Raf-1 signaling triggered an epigenetic reprogramming of proinflammatory genes, namely, increased histone H3 (H3K4) trimethylation, which held the chromatin around these genes in an open conformation, increasing their transcription upon a secondary stimulation (74). An alternative mechanism was unraveled for trained immunity in macrophages induced by BCG, but again involving the stable epigenetic reprogramming of proinflammatory genes; in this case NOD2-dependent activation of receptor-interacting protein kinase 2 (RIPK2) was the upstream signal required for changes in H3K4 trimethylation during primary infection with BCG (75). These studies illustrate that signaling pathways downstream of innate immune receptors can profoundly influence the epigenetic reprogramming of macrophage activation and function, which can have profound implications for host defense.

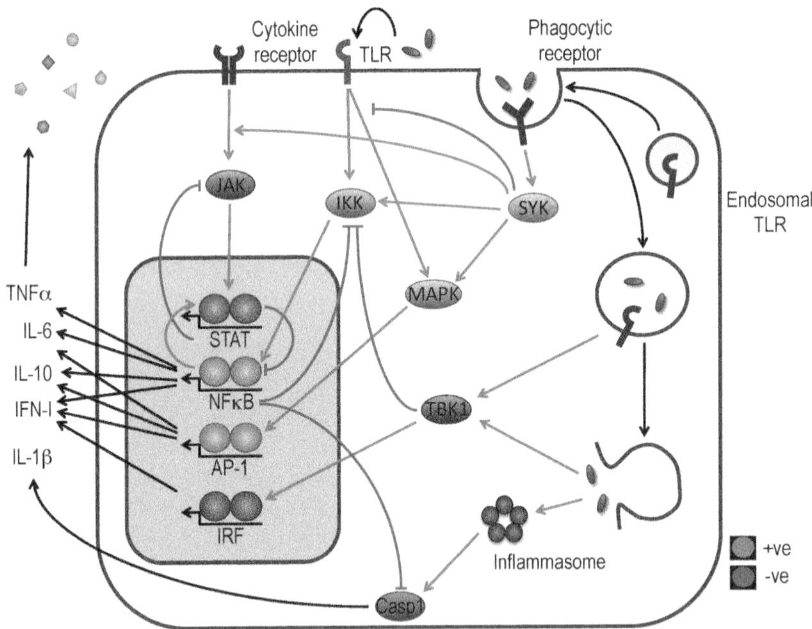

Figure 1 Cross talk between signaling pathways in macrophages during encounters with microbial stimuli. Macrophages express a number of receptors that can recognize microbial patterns, including TLRs and phagocytic receptors. In addition, cytosolic detectors such as inflammasomes can respond to microbes that invade the cytoplasm of cells or escape the phagocytic compartment. TLRs directly trigger activation of NF-κB, through the IKK complex and MAPK pathways, to drive proinflammatory and anti-inflammatory cytokine expression. Endosomal TLR signaling also activates the TBK1 (TANK-binding kinase 1)-dependent expression of type I IFN (IFN-I) through triggering of IRF activation. Cytokines such as IFN-I, IL-6, and IL-10 can also trigger autocrine signaling through JAK/STAT activation, whereas other proinflammatory cytokines such as TNF-α and IL-1β activate IKK and MAPK pathways. There are a number of mechanisms for negative (red) and positive (green) cross talk between these signaling pathways in macrophages to fine-tune the magnitude and duration of their activation.

The interplay between signaling pathways in macrophages during encounters with microbial stimuli is extremely complicated. The coordinated triggering of phagocytic receptors, TLRs-NLRs-RLRs, inflammasomes, and autocrine/paracrine cytokine signaling generates an extensive network of positive and negative feedback to orchestrate macrophage activation and effector functions (Fig. 1). Despite the mind-boggling complexity of this scenario, this is the true context in which we need to understand macrophage activation. The challenges we face in understanding the coordination of the signaling pathways that regulate macrophage activation are further complicated by the variation in responses at the single-cell level. Only recently, with the development of powerful single-cell analysis tools, have we come to appreciate the quantitative and kinetic differences in the response of individual cells, including macrophages, to microbial stimuli of even intact pathogens (76–78); in the latter case this is further complicated by the individual variation in pathogen responses within host cells (76). To date these studies have focused mainly on high-content gene expression analysis at fixed time points; however, the extension of this single-cell approach to the temporal and spatial analysis of signaling pathways is an exciting new dimension. One can envisage this approach being combined with the use of reporters and fluorescently tagged signaling molecules and the live imaging of individual cells to reveal the dynamic interactions of these pathways during the host cell response combined with the quantitative analysis of signaling outputs.

Citation. Lawrence T. 2016. Coordinated regulation of signaling pathways during macrophage activation. Microbiol Spectrum 4(5):MCHD-0025-2015.

References

1. **Hayden MS, Ghosh S.** 2012. NF-κB, the first quarter-century: remarkable progress and outstanding questions. *Genes Dev* **26:**203–234.
2. **Vallabhapurapu S, Karin M.** 2009. Regulation and function of NF-κB transcription factors in the immune system. *Annu Rev Immunol* **27:**693–733.
3. **Lawrence T.** 2009. The nuclear factor NF-κB pathway in inflammation. *Cold Spring Harb Perspect Biol* **1:** a001651. doi:10.1101/cshperspect.a001651.
4. **Greten FR, Arkan MC, Bollrath J, Hsu LC, Goode J, Miething C, Göktuna SI, Neuenhahn M, Fierer J, Paxian S, Van Rooijen N, Xu Y, O'Cain T, Jaffee BB, Busch DH, Duyster J, Schmid RM, Eckmann L, Karin M.** 2007. NF-κB is a negative regulator of IL-1β secretion as revealed by genetic and pharmacological inhibition of IKKβ. *Cell* **130:**918–931.
5. **Fong CH, Bebien M, Didierlaurent A, Nebauer R, Hussell T, Broide D, Karin M, Lawrence T.** 2008. An antiinflammatory role for IKKβ through the inhibition of "classical" macrophage activation. *J Exp Med* **205:** 1269–1276.
6. **Mankan AK, Canli O, Schwitalla S, Ziegler P, Tschopp J, Korn T, Greten FR.** 2011. TNF-α-dependent loss of IKKβ-deficient myeloid progenitors triggers a cytokine loop culminating in granulocytosis. *Proc Natl Acad Sci U S A* **108:**6567–6572.
7. **Kanters E, Pasparakis M, Gijbels MJ, Vergouwe MN, Partouns-Hendriks I, Fijneman RJ, Clausen BE, Förster I, Kockx MM, Rajewsky K, Kraal G, Hofker MH, de Winther MP.** 2003. Inhibition of NF-κB activation in macrophages increases atherosclerosis in LDL receptor-deficient mice. *J Clin Invest* **112:**1176–1185.
8. **Ma A, Malynn BA.** 2012. A20: linking a complex regulator of ubiquitylation to immunity and human disease. *Nat Rev Immunol* **12:**774–785.
9. **Matmati M, Jacques P, Maelfait J, Verheugen E, Kool M, Sze M, Geboes L, Louagie E, Mc Guire C, Vereecke L, Chu Y, Boon L, Staelens S, Matthys P, Lambrecht BN, Schmidt-Supprian M, Pasparakis M, Elewaut D, Beyaert R, van Loo G.** 2011. A20 (TNFAIP3) deficiency in myeloid cells triggers erosive polyarthritis resembling rheumatoid arthritis. *Nat Genet* **43:**908–912.
10. **Ziegler-Heitbrock L.** 2001. The p50-homodimer mechanism in tolerance to LPS. *J Endotoxin Res* **7:**219–222.
11. **Bohuslav J, Kravchenko VV, Parry GC, Erlich JH, Gerondakis S, Mackman N, Ulevitch RJ.** 1998. Regulation of an essential innate immune response by the p50 subunit of NF-κB. *J Clin Invest* **102:**1645–1652.
12. **Porta C, Rimoldi M, Raes G, Brys L, Ghezzi P, Di Liberto D, Dieli F, Ghisletti S, Natoli G, De Baetselier P, Mantovani A, Sica A.** 2009. Tolerance and M2 (alternative) macrophage polarization are related processes orchestrated by p50 nuclear factor κB. *Proc Natl Acad Sci U S A* **106:**14978–14983.
13. **Tomczak MF, Erdman SE, Davidson A, Wang YY, Nambiar PR, Rogers AB, Rickman B, Luchetti D, Fox JG, Horwitz BH.** 2006. Inhibition of *Helicobacter hepaticus*-induced colitis by IL-10 requires the p50/p105 subunit of NF-κB. *J Immunol* **177:**7332–7339.
14. **Wessells J, Baer M, Young HA, Claudio E, Brown K, Siebenlist U, Johnson PF.** 2004. BCL-3 and NF-κB p50 attenuate lipopolysaccharide-induced inflammatory responses in macrophages. *J Biol Chem* **279:**49995–50003.
15. **Gantke T, Sriskantharajah S, Sadowski M, Ley SC.** 2012. IκB kinase regulation of the TPL-2/ERK MAPK pathway. *Immunol Rev* **246:**168–182.
16. **Kaiser F, Cook D, Papoutsopoulou S, Rajsbaum R, Wu X, Yang HT, Grant S, Ricciardi-Castagnoli P, Tsichlis PN, Ley SC, O'Garra A.** 2009. TPL-2 negatively regulates interferon-β production in macrophages and myeloid dendritic cells. *J Exp Med* **206:**1863–1871.
17. **Yang HT, Wang Y, Zhao X, Demissie E, Papoutsopoulou S, Mambole A, O'Garra A, Tomczak MF, Erdman SE, Fox JG, Ley SC, Horwitz BH.** 2011. NF-κB1 inhibits TLR-induced IFN-β production in macrophages through TPL-2-dependent ERK activation. *J Immunol* **186:**1989–1996.

18. **Tomczak MF, Gadjeva M, Wang YY, Brown K, Maroulakou I, Tsichlis PN, Erdman SE, Fox JG, Horwitz BH.** 2006. Defective activation of ERK in macrophages lacking the p50/p105 subunit of NF-κB is responsible for elevated expression of IL-12 p40 observed after challenge with *Helicobacter hepaticus*. *J Immunol* **176:**1244–1251.
19. **Koga K, Takaesu G, Yoshida R, Nakaya M, Kobayashi T, Kinjyo I, Yoshimura A.** 2009. Cyclic adenosine monophosphate suppresses the transcription of proinflammatory cytokines via the phosphorylated c-Fos protein. *Immunity* **30:**372–383.
20. **Lawrence T, Bebien M, Liu GY, Nizet V, Karin M.** 2005. IKKα limits macrophage NF-κB activation and contributes to the resolution of inflammation. *Nature* **434:**1138–1143.
21. **Senftleben U, Cao Y, Xiao G, Greten FR, Krähn G, Bonizzi G, Chen Y, Hu Y, Fong A, Sun SC, Karin M.** 2001. Activation by IKKα of a second, evolutionary conserved, NF-κB signaling pathway. *Science* **293:**1495–1499.
22. **Liu B, Yang Y, Chernishof V, Loo RR, Jang H, Tahk S, Yang R, Mink S, Shultz D, Bellone CJ, Loo JA, Shuai K.** 2007. Proinflammatory stimuli induce IKKα-mediated phosphorylation of PIAS1 to restrict inflammation and immunity. *Cell* **129:**903–914.
23. **Shembade N, Pujari R, Harhaj NS, Abbott DW, Harhaj EW.** 2011. The kinase IKKα inhibits activation of the transcription factor NF-κB by phosphorylating the regulatory molecule TAX1BP1. *Nat Immunol* **12:**834–843.
24. **Saccani S, Pantano S, Natoli G.** 2002. p38-dependent marking of inflammatory genes for increased NF-κB recruitment. *Nat Immunol* **3:**69–75.
25. **Cheung PC, Campbell DG, Nebreda AR, Cohen P.** 2003. Feedback control of the protein kinase TAK1 by SAPK2a/p38α. *EMBO J* **22:**5793–5805.
26. **Kim C, Sano Y, Todorova K, Carlson BA, Arpa L, Celada A, Lawrence T, Otsu K, Brissette JL, Arthur JS, Park JM.** 2008. The kinase p38α serves cell type-specific inflammatory functions in skin injury and coordinates pro- and anti-inflammatory gene expression. *Nat Immunol* **9:**1019–1027.
27. **Ananieva O, Darragh J, Johansen C, Carr JM, McIlrath J, Park JM, Wingate A, Monk CE, Toth R, Santos SG, Iversen L, Arthur JS.** 2008. The kinases MSK1 and MSK2 act as negative regulators of Toll-like receptor signaling. *Nat Immunol* **9:**1028–1036.
28. **Darnell JE Jr, Kerr IM, Stark GR.** 1994. Jak-STAT pathways and transcriptional activation in response to IFNs and other extracellular signaling proteins. *Science* **264:**1415–1421.
29. **Park C, Li S, Cha E, Schindler C.** 2000. Immune response in Stat2 knockout mice. *Immunity* **13:**795–804.
30. **Farlik M, Reutterer B, Schindler C, Greten F, Vogl C, Müller M, Decker T.** 2010. Nonconventional initiation complex assembly by STAT and NF-κB transcription factors regulates nitric oxide synthase expression. *Immunity* **33:**25–34.
31. **Bezbradica JS, Rosenstein RK, DeMarco RA, Brodsky I, Medzhitov R.** 2014. A role for the ITAM signaling module in specifying cytokine-receptor functions. *Nat Immunol* **15:**333–342.
32. **Tassiulas I, Hu X, Ho H, Kashyap Y, Paik P, Hu Y, Lowell CA, Ivashkiv LB.** 2004. Amplification of IFN-α-induced STAT1 activation and inflammatory function by Syk and ITAM-containing adaptors. *Nat Immunol* **5:**1181–1189.
33. **Meraz MA, White JM, Sheehan KC, Bach EA, Rodig SJ, Dighe AS, Kaplan DH, Riley JK, Greenlund AC, Campbell D, Carver-Moore K, DuBois RN, Clark R, Aguet M, Schreiber RD.** 1996. Targeted disruption of the *Stat1* gene in mice reveals unexpected physiologic specificity in the JAK-STAT signaling pathway. *Cell* **84:**431–442.
34. **Durbin JE, Hackenmiller R, Simon MC, Levy DE.** 1996. Targeted disruption of the mouse *Stat1* gene results in compromised innate immunity to viral disease. *Cell* **84:**443–450.
35. **Varinou L, Ramsauer K, Karaghiosoff M, Kolbe T, Pfeffer K, Müller M, Decker T.** 2003. Phosphorylation of the Stat1 transactivation domain is required for full-fledged IFN-γ-dependent innate immunity. *Immunity* **19:**793–802.
36. **Soudja SM, Chandrabos C, Yakob E, Veenstra M, Palliser D, Lauvau G.** 2014. Memory-T-cell-derived interferon-γ instructs potent innate cell activation for protective immunity. *Immunity* **40:**974–988.
37. **Kovarik P, Sauer I, Schaljo B.** 2007. Molecular mechanisms of the anti-inflammatory functions of interferons. *Immunobiology* **212:**895–901.
38. **Stockinger S, Kastner R, Kernbauer E, Pilz A, Westermayer S, Reutterer B, Soulat D, Stengl G, Vogl C, Frenz T, Waibler Z, Taniguchi T, Rülicke T, Kalinke U, Müller M, Decker T.** 2009. Characterization of the interferon-producing cell in mice infected with *Listeria monocytogenes*. *PLoS Pathog* **5:**e1000355. doi:10.1371/journal.ppat.1000355.
39. **Martinez FO, Helming L, Gordon S.** 2009. Alternative activation of macrophages: an immunologic functional perspective. *Annu Rev Immunol* **27:**451–483.
40. **Herbert DR, Hölscher C, Mohrs M, Arendse B, Schwegmann A, Radwanska M, Leeto M, Kirsch R, Hall P, Mossmann H, Claussen B, Förster I, Brombacher F.** 2004. Alternative macrophage activation is essential for survival during schistosomiasis and downmodulates T helper 1 responses and immunopathology. *Immunity* **20:**623–635.
41. **Brombacher F, Arendse B, Peterson R, Hölscher A, Hölscher C.** 2009. Analyzing classical and alternative macrophage activation in macrophage/neutrophil-specific IL-4 receptor-alpha-deficient mice. *Methods Mol Biol* **531:**225–252.
42. **Takeda K, Tanaka T, Shi W, Matsumoto M, Minami M, Kashiwamura S, Nakanishi K, Yoshida N, Kishimoto T, Akira S.** 1996. Essential role of Stat6 in IL-4 signalling. *Nature* **380:**627–630.
43. **Fruman DA, Snapper SB, Yballe CM, Davidson L, Yu JY, Alt FW, Cantley LC.** 1999. Impaired B cell development and proliferation in absence of phosphoinositide 3-kinase p85α. *Science* **283:**393–397.

44. **Sly LM, Ho V, Antignano F, Ruschmann J, Hamilton M, Lam V, Rauh MJ, Krystal G.** 2007. The role of SHIP in macrophages. *Front Biosci* **12:**2836–2848.

45. **Lawrence T, Natoli G.** 2011. Transcriptional regulation of macrophage polarization: enabling diversity with identity. *Nat Rev Immunol* **11:**750–761.

46. **Knosp CA, Johnston JA.** 2012. Regulation of $CD4^{+}$ T-cell polarization by suppressor of cytokine signaling proteins. *Immunology* **135:**101–111.

47. **Strebovsky J, Walker P, Dalpke AH.** 2012. Suppressor of cytokine signaling proteins as regulators of innate immune signaling. *Front Biosci (Landmark Ed)* **17:**1627–1639.

48. **Kinjyo I, Hanada T, Inagaki-Ohara K, Mori H, Aki D, Ohishi M, Yoshida H, Kubo M, Yoshimura A.** 2002. SOCS1/JAB is a negative regulator of LPS-induced macrophage activation. *Immunity* **17:**583–591.

49. **Gingras S, Parganas E, de Pauw A, Ihle JN, Murray PJ.** 2004. Re-examination of the role of suppressor of cytokine signaling 1 (SOCS1) in the regulation of Toll-like receptor signaling. *J Biol Chem* **279:**54702–54707.

50. **Ryo A, Suizu F, Yoshida Y, Perrem K, Liou YC, Wulf G, Rottapel R, Yamaoka S, Lu KP.** 2003. Regulation of NF-κB signaling by Pin1-dependent prolyl isomerization and ubiquitin-mediated proteolysis of p65/RelA. *Mol Cell* **12:**1413–1426.

51. **Lang R, Pauleau AL, Parganas E, Takahashi Y, Mages J, Ihle JN, Rutschman R, Murray PJ.** 2003. SOCS3 regulates the plasticity of gp130 signaling. *Nat Immunol* **4:** 546–550.

52. **Smallie T, Ricchetti G, Horwood NJ, Feldmann M, Clark AR, Williams LM.** 2010. IL-10 inhibits transcription elongation of the human *TNF* gene in primary macrophages. *J Exp Med* **207:**2081–2088.

53. **Yasukawa H, Ohishi M, Mori H, Murakami M, Chinen T, Aki D, Hanada T, Takeda K, Akira S, Hoshijima M, Hirano T, Chien KR, Yoshimura A.** 2003. IL-6 induces an anti-inflammatory response in the absence of SOCS3 in macrophages. *Nat Immunol* **4:**551–556.

54. **Qin H, Holdbrooks AT, Liu Y, Reynolds SL, Yanagisawa LL, Benveniste EN.** 2012. SOCS3 deficiency promotes M1 macrophage polarization and inflammation. *J Immunol* **189:**3439–3448.

55. **Dickensheets H, Vazquez N, Sheikh F, Gingras S, Murray PJ, Ryan JJ, Donnelly RP.** 2007. Suppressor of cytokine signaling-1 is an IL-4-inducible gene in macrophages and feedback inhibits IL-4 signaling. *Genes Immun* **8:** 21–27.

56. **Ivashkiv LB.** 2009. Cross-regulation of signaling by ITAM-associated receptors. *Nat Immunol* **10:**340–347.

57. **Bezbradica JS, Medzhitov R.** 2012. Role of ITAM signaling module in signal integration. *Curr Opin Immunol* **24:** 58–66.

58. **Wang L, Tassiulas I, Park-Min KH, Reid AC, Gil-Henn H, Schlessinger J, Baron R, Zhang JJ, Ivashkiv LB.** 2008. 'Tuning' of type I interferon-induced Jak-STAT1 signaling by calcium-dependent kinases in macrophages. *Nat Immunol* **9:**186–193.

59. **Gross O, Poeck H, Bscheider M, Dostert C, Hannesschläger N, Endres S, Hartmann G, Tardivel A, Schweighoffer E, Tybulewicz V, Mocsai A, Tschopp J, Ruland J.** 2009. Syk kinase signaling couples to the Nlrp3 inflammasome for anti-fungal host defence. *Nature* **459:**433–436.

60. **Rogers NC, Slack EC, Edwards AD, Nolte MA, Schulz O, Schweighoffer E, Williams DL, Gordon S, Tybulewicz VL, Brown GD, Reis e Sousa C.** 2005. Syk-dependent cytokine induction by Dectin-1 reveals a novel pattern recognition pathway for C type lectins. *Immunity* **22:** 507–517.

61. **Gantner BN, Simmons RM, Canavera SJ, Akira S, Underhill DM.** 2003. Collaborative induction of inflammatory responses by dectin-1 and Toll-like receptor 2. *J Exp Med* **197:**1107–1117.

62. **Hamerman JA, Tchao NK, Lowell CA, Lanier LL.** 2005. Enhanced Toll-like receptor responses in the absence of signaling adaptor DAP12. *Nat Immunol* **6:**579–586.

63. **Wang L, Gordon RA, Huynh L, Su X, Park Min KH, Han J, Arthur JS, Kalliolias GD, Ivashkiv LB.** 2010. Indirect inhibition of Toll-like receptor and type I interferon responses by ITAM-coupled receptors and integrins. *Immunity* **32:**518–530.

64. **O'Neill LA, Pearce EJ.** 2016. Immunometabolism governs dendritic cell and macrophage function. *J Exp Med* **213:** 15–23.

65. **Fukuzumi M, Shinomiya H, Shimizu Y, Ohishi K, Utsumi S.** 1996. Endotoxin-induced enhancement of glucose influx into murine peritoneal macrophages via GLUT1. *Infect Immun* **64:**108–112.

66. **O'Neill LA, Hardie DG.** 2013. Metabolism of inflammation limited by AMPK and pseudo-starvation. *Nature* **493:**346–355.

67. **Tannahill GM, Curtis AM, Adamik J, Palsson-McDermott EM, McGettrick AF, Goel G, Frezza C, Bernard NJ, Kelly B, Foley NH, Zheng L, Gardet A, Tong Z, Jany SS, Corr SC, Haneklaus M, Caffrey BE, Pierce K, Walmsley S, Beasley FC, Cummins E, Nizet V, Whyte M, Taylor CT, Lin H, Masters SL, Gottlieb E, Kelly VP, Clish C, Auron PE, Xavier RJ, O'Neill LA.** 2013. Succinate is an inflammatory signal that induces IL-1β through HIF-1α. *Nature* **496:**238–242.

68. **Rius J, Guma M, Schachtrup C, Akassoglou K, Zinkernagel AS, Nizet V, Johnson RS, Haddad GG, Karin M.** 2008. NF-κB links innate immunity to the hypoxic response through transcriptional regulation of HIF-1α. *Nature* **453:** 807–811.

69. **Odegaard JI, Chawla A.** 2011. Alternative macrophage activation and metabolism. *Annu Rev Pathol* **6:** 275–297.

70. **Netea MG, Joosten LA, Latz E, Mills KH, Natoli G, Stunnenberg HG, O'Neill LA, Xavier RJ.** 2016. Trained immunity: a program of innate immune memory in health and disease. *Science* **352:**aaf1098. doi:10.1126/science.aaf1098.

71. **van der Meer JW, Joosten LA, Riksen N, Netea MG.** 2015. Trained immunity: a smart way to enhance innate immune defence. *Mol Immunol* **68:**40–44.

72. **Bistoni F, Vecchiarelli A, Cenci E, Puccetti P, Marconi P, Cassone A.** 1986. Evidence for macrophage-mediated protection against lethal *Candida albicans* infection. *Infect Immun* **51:**668–674.

73. **Bistoni F, Verducci G, Perito S, Vecchiarelli A, Puccetti P, Marconi P, Cassone A.** 1988. Immunomodulation by a low-virulence, agerminative variant of *Candida albicans*. Further evidence for macrophage activation as one of the effector mechanisms of nonspecific anti-infectious protection. *J Med Vet Mycol* **26:**285–299.

74. **Saeed S, Quintin J, Kerstens HH, Rao NA, Aghajanirefah A, Matarese F, Cheng SC, Ratter J, Berentsen K, van der Ent MA, Sharifi N, Janssen-Megens EM, Ter Huurne M, Mandoli A, van Schaik T, Ng A, Burden F, Downes K, Frontini M, Kumar V, Giamarellos-Bourboulis EJ, Ouwehand WH, van der Meer JW, Joosten LA, Wijmenga C, Martens JH, Xavier RJ, Logie C, Netea MG, Stunnenberg HG.** 2014. Epigenetic programming of monocyte-to-macrophage differentiation and trained innate immunity. *Science* **345:**1251086. doi: 10.1126/science.1251086.

75. **Kleinnijenhuis J, Quintin J, Preijers F, Joosten LA, Ifrim DC, Saeed S, Jacobs C, van Loenhout J, de Jong D, Stunnenberg HG, Xavier RJ, van der Meer JW, van Crevel R, Netea MG.** 2012. Bacille Calmette-Guerin induces NOD2-dependent nonspecific protection from reinfection via epigenetic reprogramming of monocytes. *Proc Natl Acad Sci U S A* **109:**17537–17542.

76. **Avraham R, Haseley N, Brown D, Penaranda C, Jijon HB, Trombetta JJ, Satija R, Shalek AK, Xavier RJ, Regev A, Hung DT.** 2015. Pathogen cell-to-cell variability drives heterogeneity in host immune responses. *Cell* **162:**1309–1321.

77. **Tay S, Hughey JJ, Lee TK, Lipniacki T, Quake SR, Covert MW.** 2010. Single-cell NF-κB dynamics reveal digital activation and analogue information processing. *Nature* **466:**267–271.

78. **Snijder B, Sacher R, Rämö P, Damm EM, Liberali P, Pelkmans L.** 2009. Population context determines cell-to-cell variability in endocytosis and virus infection. *Nature* **461:**520–523.

Myeloid Cells in Health and Disease: A Synthesis
Edited by Siamon Gordon

doi:10.1128/microbiolspec.MCHD-0039-2016

Gregory J. Fonseca[1]
Jason S. Seidman[1,2]
Christopher K. Glass[1,3]

30

Genome-Wide Approaches to Defining Macrophage Identity and Function

DIVERSE MACROPHAGE PHENOTYPES IN HEALTH AND DISEASE

Macrophages are among the most phenotypically diverse cell types of mammalian organisms (1, 2). They inhabit all or nearly all tissues under healthy conditions, where they play important roles as sentinels of infection and injury. These functions are enabled by the expression of a multitude of cell surface and internal receptors that recognize microbial-associated molecular patterns and/or damage-associated molecular patterns, exemplified by Toll-like receptors (TLRs) (3). Engagement of these receptors by microbial components, such as bacterial lipopolysaccharide (LPS), initiates signaling cascades that lead to the activation of latent transcription factors, including NF-κB, interferon regulatory factors (IRFs), and members of the activator protein 1 (AP1) family (4, 5). These factors, in turn, function to activate hundreds of genes that play key roles in the orchestration of the innate immune response and that influence the development of adaptive immunity (6, 7). In addition to this sentinel function, macrophages are professional phagocytes, serving to clear bacteria, apoptotic cells, and a diverse range of host-derived and environmental debris, thereby contributing to an additional layer of immunity and tissue homeostasis (2).

While the sentinel and phagocytic functions comprise central and shared macrophage characteristics, the various populations of tissue-resident macrophages also exhibit a striking range of phenotypic diversity (1). Consider, for example, the diverse morphologies and functions of microglia, Kupffer cells, alveolar macrophages, Langerhans cells, peritoneal macrophages, and splenic red pulp macrophages. All of these cell types retain phagocytic and sentinel functions, but have also acquired distinct patterns of gene expression that are linked to their tissue-specific functional roles, e.g., synaptic pruning in the case of microglia (8), clearance of surfactant by alveolar macrophages (9), and removal of senescent red blood cells by splenic macrophages (10).

Although macrophages normally play adaptive roles in immunity, tissue repair, and homeostasis, they are also implicated in a broad spectrum of human diseases

[1]Department of Cellular and Molecular Medicine; [2]Biomedical Sciences Graduate Program; [3]Department of Medicine, University of California, San Diego, La Jolla, CA 92093.

(2). For example, macrophages contribute to all phases of the development of atherosclerosis, from formation of the initial fatty streaks to the rupture of complex lesions that result in myocardial infarction (11). Adipose tissue macrophages and Kupffer cells are implicated in metabolic diseases that include insulin resistance and nonalcoholic steatohepatitis (12–14). Microglia have been linked to numerous neurodegenerative diseases, including Alzheimer's disease, Parkinson's disease, and Huntington's disease (15). As an example, genome-wide association studies provide strong evidence that genomic variants in or near genes expressed in microglia, such as *TREM2*, are associated with increased risk of Alzheimer's disease (16). In cancer, macrophages have been shown to play complex roles in tumor initiation, growth, metastasis, and immune evasion (17, 18).

These observations raise a number of questions, including the following. (i) How are distinct macrophage identities achieved? (ii) To what extent does their developmental origin specify their functional properties in relation to tissue-specific environmental signals? (iii) If tissue environment is important, what are the signals driving macrophage specialization? (iv) What are the mechanisms that lead to pathogenic roles of macrophages in diseases such as atherosclerosis, metabolic disease, neurodegenerative diseases, and cancer? (v) Do the phenotypes of resident macrophages change in response to a primary disease process and/or are pathogenic activities the result of infiltration by monocyte-derived macrophages? (vi) Is it possible to alter macrophage phenotypes for therapeutic purposes?

GENOME-WIDE APPROACHES TO DEFINE MACROPHAGE IDENTITY AND FUNCTION

Among the most widely used and successful approaches to define cellular identity is the use of antibodies to mark specific cell surface or internal proteins and to quantify these proteins by flow cytometry. This approach can be considered to define what a cell is "wearing." In much the same way that a uniform enables one to deduce roles of a person dressed as a policeman or fireman, specific combinations of markers can be used define cell types with different functional properties, e.g., macrophages, T cells, and B cells. An alternative approach to define cell identity, made possible by the development of chromatin immunoprecipitation linked to massively parallel DNA sequencing (ChIP-seq), is to define the total repertoire of transcriptional regulatory elements in that cell, i.e., its enhancers and promoters (19). These elements are selected from the genome in a cell-specific fashion and provide the cell with its transcriptional identity and regulatory potential. Knowledge of such elements not only enables an understanding of the program of gene expression observed in that cell but in principle also enables predictions of how the cell will respond to new internal and external signals. Thus, in contrast to flow cytometry, genomic approaches can be considered to provide insights into what the cell is "thinking." From this perspective, answers to some of the questions posed above may be attained by systematic evaluation of the enhancer and promoter elements of diverse tissue-resident macrophage populations under normal and disease conditions. In the following sections, we briefly review general principles of enhancer selection and function and recent findings relevant to this process in various macrophage model systems and in tissue-resident macrophages.

ENHANCERS AND SUPERENHANCERS REGULATE CELLULAR IDENTITY AND FUNCTION

Gene regulation at the level of transcription is achieved through the coordinated functions of enhancers and promoters. Promoters activated by RNA polymerase II represent the sites of initiation of mRNAs and long noncoding RNAs. While promoters are often key points of regulation by signal-dependent transcription factors, they are often not sufficient to direct the highly divergent levels of transcription of primary transcripts observed across cell types or to fully capture cell-specific responses to developmental and homeostatic signals (20, 21). These functions are frequently mediated by enhancers, which are defined as DNA sequences that "enhance" transcription from their target promoters (22). Enhancers are generally thought to act by looping to their target promoters and increasing the recruitment of RNA polymerase II and/or its transition from a paused to elongating form (23). Thus, a major effort in molecular and developmental biology is to define the mechanisms by which different cell types select their specific complement of enhancers, and how they, in turn, act to regulate target genes.

General features of enhancers are indicated in Fig. 1. Eukaryotic DNA is packaged in the nucleus through its association with histone octamers to form nucleosomes, the basic units of chromatin (23). Each nucleosome consists of ~147 nucleotides of DNA wrapped around a heterodimeric octamer core of two copies each of histone proteins H2A, H2B, H3, and H4. Similar to promoters, enhancers are characterized by nucleosome-free regions that are occupied by sequence-specific transcription factors. These factors in turn recruit various coactivator proteins that facilitate nucleosome remodeling, histone tail modifications, and recruitment of core transcription factors, including RNA polymerase II (24, 25). Enhancers are defined by a distinct array of chromatin modifications as compared to promoters. Whereas promoters are characterized by a high trimethylated histone H3 lysine 4 (H3K4me3) to low H3K4me1 ratio, enhancers show a contrary makeup of low H3K4me3 to high H3K4me1 (26). However, active enhancers and promoters are both associated with histone tail acetylation, for example, of histone H3 lysine 27 (27). The use of ChIP-seq for these histone modifications and global DNase I hypersensitivity assays to define open regions of chromatin in the genomes of many different cell types and tissues revealed the presence of hundreds of thousands of these putative enhancer elements in the human genome (28). Each cell type selects a subset of ~20,000 to 30,000 of this vast repertoire of potential regulatory elements, which are presumed to play essential roles in establishing cell-specific gene expression.

An important caveat to these global approaches is that while genomic regions that have been genetically established to function as enhancers almost always exhibit features shown in Fig. 1, genomic regions that have these characteristics are not necessarily active enhancers (29). To define enhancers, combinations of marks improve predictive power. For example, regions of the genome exhibiting open chromatin, transcription factor binding, and H3K4me1, but not H3K27 acetylation (H3K27ac), are often considered "poised," while the further addition of H3K27ac designates "active" enhancer regions (27). However, the most rigorous test

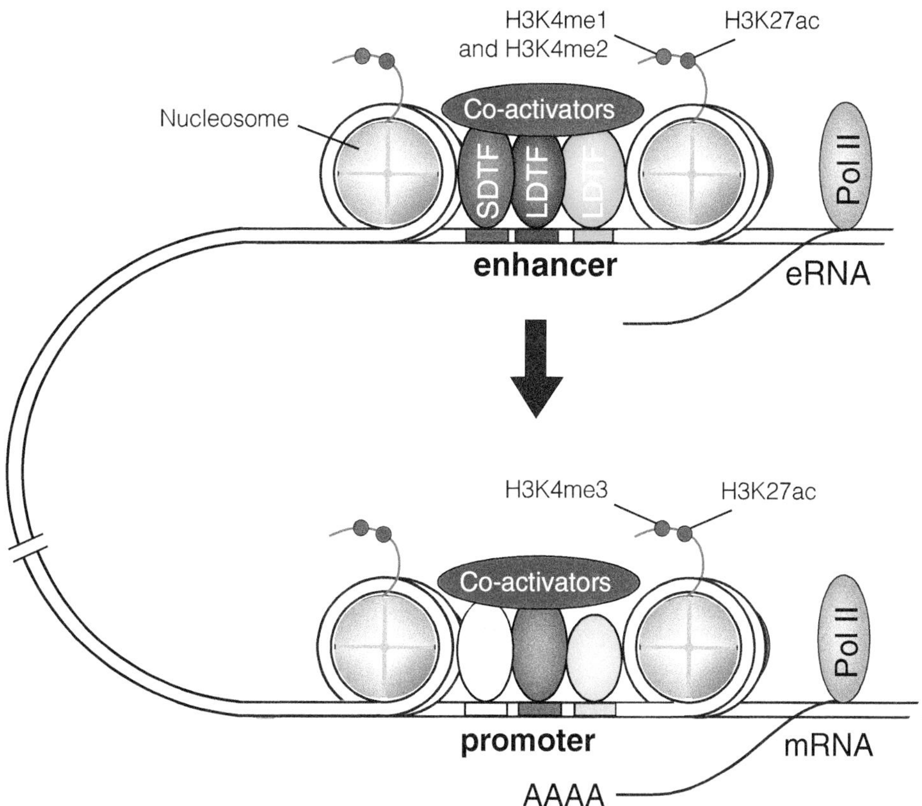

Figure 1 General organization of enhancers and promoters. DNA is packaged into nucleosomes that are displaced by sequence-specific transcription factors and coactivators. Promoters are primarily occupied by broadly expressed transcription factors, whereas enhancers are enriched for the binding of LDTFs. SDTFs can bind to enhancers or promoters (here shown only at the enhancer). Promoters are distinguished by high levels of H3K4me3 compared to H3K4me1 and H3K4me2. Enhancers are characterized by high levels of H3K4me1 relative to H3K4me3. Active enhancers and promoters are associated with transcriptional coactivators and acetylated histones, such as H3K27ac. Active enhancers are frequently associated with RNA polymerase II (Pol II) enzymes that generate eRNAs.

for enhancer function is to delete or otherwise mutate key elements of the DNA sequence of a putative enhancer and demonstrate reduced expression of the target promoter. While advances in genomic engineering greatly facilitate these types of experiments (30), the vast number of enhancer-like regions in the genome makes even these approaches insufficient for global analysis, and improved methods for prediction of enhancer function are needed.

In addition to being occupied by sequence-specific transcription factors and transcriptional coregulators, a subset of enhancers recruit core transcription factors and RNA polymerase II and generate noncoding RNAs referred to as enhancer RNAs (eRNAs) (31, 32) (Fig. 1). These are typically in the range of 200 to 1,000 nucleotides long and are thus classified as long noncoding RNAs. The majority of these RNAs are capped at the 5′ end but are not spliced or polyadenylated and are rapidly degraded in an exosome-dependent manner (32). The gain or loss of eRNA expression in response to signals, such as TLR4 ligation or ligands for nuclear receptors, is highly correlated with gain or loss in nearby gene expression (31). Thus, eRNA production may be an additional feature of enhancers that is predictive of enhancer function. The functional importance of enhancer transcription has not been fully established. Studies of enhancers that are newly selected in response to TLR4 ligation suggested that the process of enhancer transcription itself is important by serving to recruit histone methyltransferases that write the histone H3 lysine 4 mono- and dimethylation marks that are characteristic of enhancers (33). This was proposed to be due to recruitment of the Mll2 and Mll4 histone methyltransferases to the C-terminal domain of RNA polymerase II.

In addition, several lines of evidence suggest that at least some eRNAs contribute to enhancer function. In macrophages, the Rev-erb nuclear receptors were shown to repress gene expression by repressing enhancer transcription. This activity could be reproduced by using antisense oligonucleotides to reduce the expression of specific eRNAs in the nucleus (34). These and other studies suggest that if active, eRNAs primarily function in a local manner to regulate the expression of the particular target of the enhancer that they are generated from (34, 35). Several mechanisms have been suggested to account for eRNAs' function, including promoting enhancer/promoter looping (35), release of the negative elongation factor NELF (36), and trapping of transcription factors such as YY1 (37). Disruption of the integrator complex, which is necessary for the 3′-end processing of nonpolyadenylated, RNA polymerase II-dependent, uridylate-rich, small nuclear RNA genes, which includes enhancers, results in a reduction of eRNA production and subsequently abrogates enhancer-promoter looping (38). Conversely, WDR82 and SET1 function to control termination of eRNAs, such that their loss of function results in unusually long eRNAs (39).

Examination of the distribution of genomic features associated with enhancers in various cell types, such as components of the Mediator complex and histone acetylation, revealed that they were clustered at very high density at a few hundred regions in each cell type. Such regions have been referred to as superenhancers or stretch enhancers (40–42). Generally, superenhancers are composed of clusters of ordinary enhancers that together extend more than an order of magnitude larger than individual ordinary enhancers (kilobases versus hundreds of bases) (40, 41). Interestingly, the separate enhancers within superenhancer regions have been shown to not work in an additive manner. Instead, there is a much more complex interplay where some regions are more important for activation and some regions negatively affect enhancer activity and may play a role in regulating enhancer strength (43). Furthermore, the eRNA transcription of the separate regions within superenhancers is uniformly regulated rather than individually regulated (44).

Superenhancer regions have been seen to overlap with and are thought to activate the transcriptional activators that are vital in the establishment of cell fate and identity (41). In macrophages, superenhancers have been reported to overlap genes that are essential for macrophage development and function, including *Spi1* (encoding PU.1), *Cebpa*, members of the *Irf* family, *Csf1r*, *Fcgr2b*, and *Ctsb* (45) (Fig. 2). Similarly, in mouse embryonic stem cells (mESCs), genes required for pluripotency such as *Oct4*, *Sox2*, and *Nanog* are associated with superenhancers (41). As well as having high levels of Mediator, superenhancers are much more reliant on Mediator than are regular enhancers. This is evidenced by knockdown experiments in mESCs against Med12, an integral component of the Mediator complex. Loss of Med12 resulted in a disproportionate downregulation of superenhancer-adjacent genes such as *Oct4*, *Sox2*, and *Nanog* (41). Further, a loss of Mediator in mESCs affected pluripotency (46). To date, little work has been done to establish the role of superenhancers in macrophages. However, data in other cell types suggest that superenhancers play a disproportionately large part in controlling the expression of genes that establish cell fate, making them an ideal target for future studies of macrophage enhancer biology.

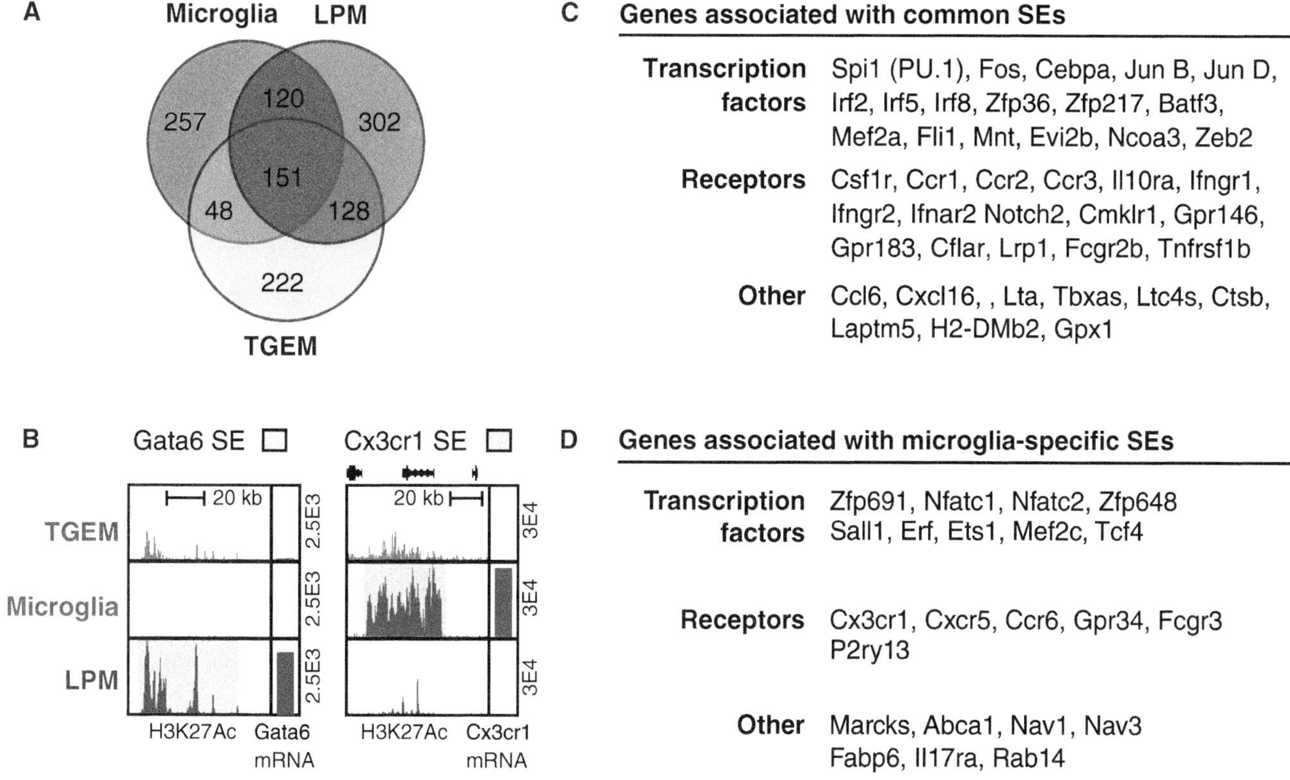

Figure 2 Superenhancers (SEs) in macrophages. (A) Venn diagram of shared and subset-specific superenhancers in thioglycolate-elicited macrophages (TGEMs), large peritoneal macrophages (LPMs), and microglia. (B) Examples of subset-specific superenhancers near the *Gata6* and *Cx3cr1* genes in TGEMs, LPMs, and microglia. (C) Partial listing of the 151 genes associated with superenhancers found in all three macrophage subsets. (D) Partial listing of genes associated with the 257 superenhancers selectively found in microglia.

SELECTION AND ACTIVATION OF MACROPHAGE-SPECIFIC ENHANCERS

The development of multicellular organisms involves hierarchically organized progenitor cells that ultimately give rise to terminally differentiated cell types with specialized functions. Cell fates are specified by lineage-determining transcription factors (LDTFs) whose expression is often not limited to a single cell type (47). Most transcription factors recognize short DNA motifs of ~8 to 12 bp in length, and considerable degeneracy can be tolerated for sequence-specific binding *in vivo*. As a consequence, there are potentially tens of millions of binding sites for most transcription factors within the mammalian genome. Initial use of ChIP-seq to study the genome-wide binding patterns of transcription factors in a variety of species and cell types demonstrated that only a small fraction of these sites are actually occupied. Furthermore, different factors in the same cell type were often found to colocalize (48, 49), and the same factor in different cell types or at different stages of development exhibited different genome-wide binding patterns (50–52). However, mechanisms accounting for these binding patterns were unknown.

PU.1 is an LDTF required for the normal development of macrophages and B cells (53, 54), where it drives divergent programs of gene expression in each cell type. An initial application of ChIP-seq approaches to investigate the genome-wide binding patterns of the Ets domain transcription factor PU.1 in thioglycolate-elicited macrophages and splenic B cells provided evidence for a collaborative/hierarchical model (55) (Fig. 3). Here, relatively simple combinations of LDTFs were suggested to play dominant roles in the selection of a large fraction of each cell type's enhancers. While the binding of PU.1 to promoter regions was found to be similar in the two cell types, binding to regions of the genome distal to promoters, which include enhancers, was highly cell type specific. The recognition motif for PU.1 was identical at these sites in both cell types, indicating that the motif itself did not contribute to cell-specific binding

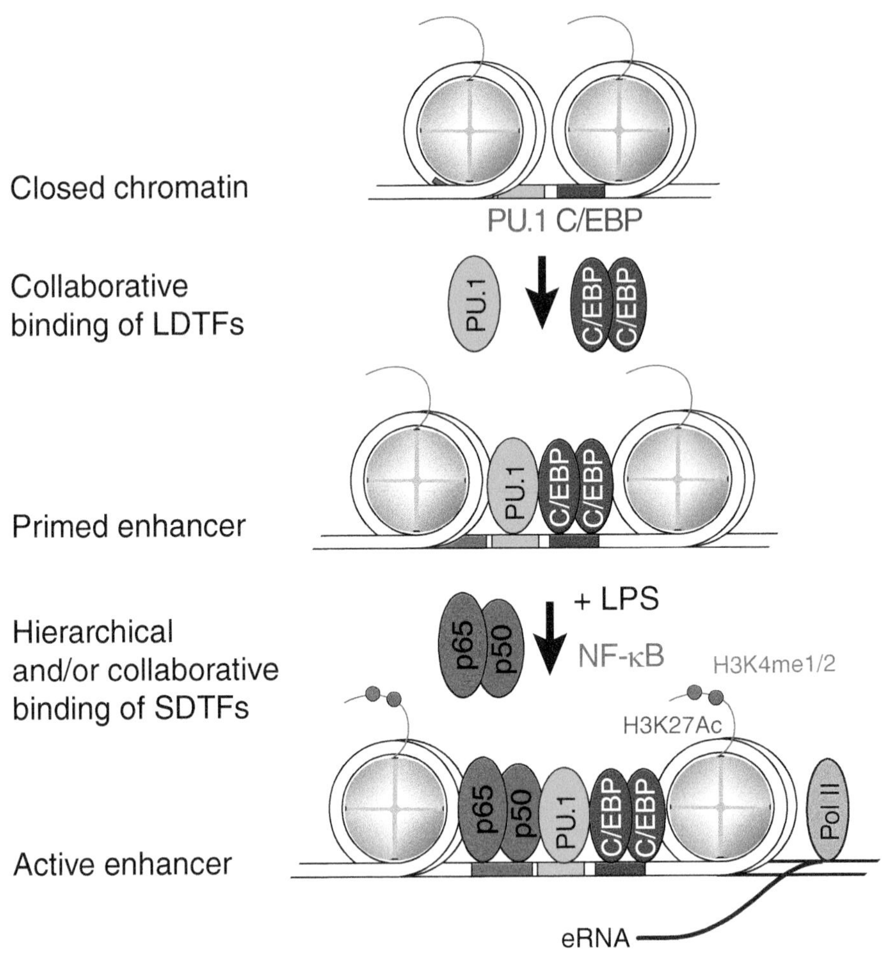

Figure 3 A collaborative/hierarchical model for selection and activation of macrophage enhancers. Macrophage LDTFs, exemplified by PU.1 and C/EBPs, collaborate with each other to bind to genomic regions containing closely spaced PU.1 and C/EBP recognition motifs to establish a primed enhancer. Signal-dependent activation of NF-κB (here shown as p50 and p65) leads to its binding to primed enhancers and enhancer activation, resulting in histone acetylation and production of eRNAs.

patterns. In contrast, genomic regions within 100 bp of PU.1 binding sites in macrophages were highly enriched for motifs for other macrophage LDTFs, particularly CCAAT/enhancer-binding protein (C/EBP) and AP1 motifs. Alternatively, genomic regions within 100 bp of PU.1 binding sites in B cells were highly enriched for motifs for B-cell lineage-determining factors, including Oct, E2A, EBF, and κB motifs. Gain-of-function experiments indicated that at regions of the genome exhibiting closely spaced binding sites for PU.1 and C/EBP, binding of C/EBP was dependent on PU.1. Conversely, loss-of-function experiments in B cells indicated that in regions of the genome exhibiting closely spaced binding sites for PU.1 and E2A, binding of PU.1 was dependent on E2A (55). These observations suggested a "collaborative" model of enhancer selection, in which PU.1 and alternative lineage-determining factors such as C/EBPs and E2A are alone unable to bind to specific regions of the genome, but do so at genomic locations containing the appropriate combination of recognition motifs when they are coexpressed at high levels.

Many regions of the genome that are bound by PU.1 and C/EBPs in macrophages exhibit H3K4me1, but not H3K27ac, consistent with features of poised enhancers. Notably, the majority of binding of signal-dependent transcription factors (SDTFs) in response to activating ligands in thioglycolate-elicited macrophages was found to occur at preexisting poised or active enhancer-like regions. Initially observed for the nuclear receptors liver X receptor (LXR)α and LXRβ (55), this pattern was subsequently established for NF-κB (33), the glucocorticoid receptor (56), and Rev-erbs (34). These results are of interest because they provide a potential explanation for how broadly expressed SDTFs can exert cell-specific effects on gene expression. In essence, these findings suggest that SDTFs are "instructed" as to where to localize in each cell type based on the open regions of chromatin established by that cell type's LDTFs and, consequently, how the cell will respond to signals based on the previous binding of LDTFs. For this class of interactions, the binding of LDTFs and SDTFs is "hierarchical," in that SDTF binding is proposed to be dependent on prior binding of LDTFs, whereas the binding of LDTFs is independent of the SDTFs (Fig. 3). This relationship was established by gain- and loss-of-function experiments for PU.1 and LXRs in the thioglycolate-elicited macrophage model system: at sites of colocalization, LXR binding depended on the preexisting binding of PU.1, but PU.1 binding was not dependent on LXRs (55).

The natural genetic variation existing between inbred strains of mice has been leveraged to provide a genetic test of the collaborative/hierarchical model of LDTF and SDTF binding in macrophages (57). These studies focused on regions of the genome at which PU.1 and C/EBP bound to their respective motifs in close proximity and were subsequently occupied by the p65 component of NF-κB. At these locations, selective mutations in the binding sites for PU.1 resulted not only in loss of PU.1 binding but also in the loss of C/EBP and the signal-induced binding of p65. Similarly, selective mutations in binding sites for C/EBPs resulted in loss of C/EBP, PU.1, and p65. In contrast, mutations in NF-κB binding sites abolished binding of p65 but rarely affected the preexisting binding of PU.1 and C/EBP (57). These findings supported both the collaborative model of LDTF binding and the hierarchical relationship of SDTFs and LDTFs. Furthermore, mutation of sites that abolished PU.1 and C/EBP binding also generally resulted in a loss of other features of enhancers, such as histone methylation and acetylation, indicating that these posttranslational modifications are dependent on the binding of the LDTFs.

DE NOVO OR LATENT ENHANCERS

Although most of the genome-wide binding of p65 occurs at preexisting enhancer-like regions as described above, two independent studies demonstrated that a subset of binding events occurs at regions of the genome that do not have features of enhancers in resting macrophages but acquire them after TLR4 ligation (33, 58). Macrophage activation by the TLR4 agonist Kdo2 lipid A (KLA) resulted in the selection of a few thousand enhancers, called "*de novo*" or "latent" enhancers. These enhancers did not possess detectable levels of enhancer-associated chromatin modifications, nor LDTF binding before KLA treatment. As well, unlike the vast majority of enhancers found after KLA stimulation, where LDTFs primed the environment and recruited SDTFs to further activate enhancer transcription, these enhancers required collaboration with LDTFs and SDTFs for the initial selection of the enhancer and subsequent indicative chromatin marks. A requirement for NF-κB was demonstrated through inhibition of NF-κB via an IκB kinase (IKK) inhibitor, where downregulation of NF-κB prevented the establishment of these *de novo* enhancers (33). These data suggest that there are state-dependent enhancer landscapes that are only activated via external signals acting through SDTFs. Notably, histone modifications associated with latent enhancers were shown to persist following removal of the initiating signal. Genes associated with these regions exhibited more rapid and robust responses when cells were restimulated, suggesting that such modifications may be associated with an epigenetic "memory" (58). These findings support the broader concept of a state of "trained immunity" in which the epigenetic state of innate immune cells can be modulated to influence subsequent responses to a challenge (59).

EVIDENCE FOR TISSUE ENVIRONMENT AS A DETERMINANT OF MACROPHAGE PHENOTYPE

Transcriptional profiling of diverse macrophage populations has revealed striking differences in patterns of mRNA expression according to tissue bed (60). For example, a comparison between microglia and large peritoneal macrophages revealed nearly 1,000 mRNAs that are more than 16-fold differentially expressed in each direction (45). While the functional roles of most differentially regulated genes remain poorly understood, these observations suggest a molecular basis for the phenotypic diversity of these cells. At the other end of the spectrum, at least two distinct populations of macrophages can be isolated from the mouse peritoneal cavity, referred to as large and small peritoneal macrophages and further distinguished by relatively high levels of major histocompatibility complex class II expression in the small peritoneal macrophages (61). In contrast to the substantial differences in gene expression observed between large peritoneal macrophages and microglia, only ~100 genes show >16-fold-higher expression in comparing large and small peritoneal macrophages. Most of these are more highly expressed in the small peritoneal macrophage population and are enriched for functional annotations related to antigen presentation (45). Regardless of whether these differences are due to different developmental origins, different times of residence within the peritoneal cavity, or other mechanisms, these experiments indicate that macrophages can exhibit distinct phenotypes within a common tissue environment.

However, there is also emerging evidence that tissue environment is a strong determinant of macrophage phenotype, and that these phenotypes exhibit remarkable plasticity. One line of evidence is based on bone marrow transplantation, in which macrophages derived from adult hematopoietic stem cells replace tissue macrophages of embryonic origin. For example, lung and peritoneal macrophages from transplanted adult bone marrow acquire gene expression signatures similar to those of embryonically derived macrophages (62). Second, peritoneal macrophages adoptively transferred to the lung adopt a lung macrophage-like pattern of gene expression (62). Third, transfer of peritoneal macrophages and microglia to tissue culture environments led to significant changes in the expression of hundreds of genes. Significantly, genes exhibiting downregulation in peritoneal macrophages were enriched for those that made them most different from microglia, and vice versa (45). Collectively, these experiments indicate that environmental factors are significant drivers of distinct macrophage phenotypes.

ENHANCER LANDSCAPES OF TISSUE-RESIDENT MACROPHAGES

Improvements in the sensitivity of genomic assays have enabled the recent analyses of enhancer landscapes in various tissue-resident macrophage populations (45, 62, 63). These studies documented the presence of tens of thousands of enhancer-like regions in each population, the majority of which are shared. Putting these regions into the context of the collaborative/hierarchical model of enhancer selection and function, a large subset exhibit features of both "priming" (presence of only H3K4 methylation) and "activation" (presence of H3K4 methylation and H3K27 acetylation). For example, multiple enhancer-like regions exhibiting high levels of both H3K4me2 and H3K27ac reside in the vicinity of the *Spi1* gene, encoding PU.1, in all macrophage populations examined. This is consistent with the requirement for PU.1 in all macrophage subsets. In contrast, the *Rarb* gene, encoding the retinoic acid receptor β, exhibits H3K4 methylation in both microglia and peritoneal macrophages, but the H3K27ac mark of activation is only observed in the peritoneal macrophage population (45). This observation of the enhancer region being poised in microglia but active in peritoneal macrophages is consistent with the recent finding that the development of peritoneal macrophages is under the control of locally produced retinoic acid and long-standing evidence that the *Rarb* gene is itself retinoic acid inducible (64). These findings suggest that the enhancers driving Rarb expression are selected in both microglia and peritoneal macrophages by a common set of macrophage lineage-determining factors, but that these enhancers only become active in the peritoneal cavity due to the selective presence of sufficient concentrations of retinoic acid in that tissue environment.

In addition to shared enhancers, each macrophage population also exhibits enhancer-like regions, corresponding to ~15 to 20% of the total, that are either specific to that cell type or are restricted to a subset of the various macrophage populations examined (45, 62, 63). The existence of such subset-specific enhancers thus suggests that there are additional, context-specific transcription factor interactions necessary for their selection. Motif enrichment analysis of subset-specific enhancers returns binding sites for PU.1 as among the most highly enriched motifs. Furthermore, ChIP-seq experiments confirmed binding of PU.1 to both common and subset-specific enhancer-like regions, consistent with a requirement of PU.1 for the development of nearly all tissue-resident macrophages (45). However, within subset-specific enhancers, different transcription factor recognition motifs were identified in the vicinity of PU.1 binding. For example, a motif for Gata6, established to be essential for development of peritoneal macrophages, is significantly enriched near PU.1 in enhancers that are specific for peritoneal macrophages (45).

Using H3K27ac, superenhancers were defined in large peritoneal macrophages, microglia, and thioglycolate-elicited macrophages (45). This analysis revealed a core set of ~150 superenhancers that were present in all three subsets (Fig. 2). Genes associated with common superenhancers included genes encoding transcription factors and receptors essential for development and survival of all macrophages, such as *Spi1*, which encodes PU.1, and *Csf1R*, which encodes the receptor for macrophage colony-stimulating factor. Intriguingly, each macrophage subset exhibited ~200 superenhancers that were specific for that subset and were highly correlated with subset-specific gene expression. Genes associated with superenhancers specific to large peritoneal macrophages, including the *Rarb* and *Gata6* genes, were required for specification of the large peritoneal macrophage phenotype. Conversely, superenhancers observed selectively in microglia were associated with numerous receptors and cell surface proteins associated with functions in the brain, including *Cx3cr1*, *Nav1*, and *Bin1* (45). Notably, very little is known with respect to functions of many of the genes associated with superenhancers in each macrophage subset. Given the strong enrichment of genes with essential roles in regulation of macrophage development and function in regions of the genome marked by superenhancers, this designation may be a useful means for prioritization of analysis of genes with unknown functions. Further, the establishment of unique superenhancers during differentiation may play a vital role in biasing the cell-specific responses to extracellular differentiation signals (41, 65–67). Not surprisingly, superenhancers have been shown to play an important role in determining cell function, acting as a fast switch to aid in cell-state transitions. For example, treatment of endothelial cells with tumor necrosis factor α was found to result in drastic changes to cell superenhancer selection through the activation of NF-κB, inducing a proinflammatory gene expression program (65).

In keeping with the dramatic changes in gene expression observed following transfer of peritoneal macrophages from an *in vivo* to an *in vitro* tissue culture environment, a corresponding change was observed in the enhancer landscape of these cells as measured by H3K4me2 and H3K27ac (45). Nearly half of the enhancer-like regions defined immediately after recovery from the peritoneal cavity exhibited a >50% reduction in signal for one or both of these marks by day 7 of tissue culture. This was observed at both regular enhancers and superenhancers. Lost enhancers were generally associated with genes exhibiting reduced expression *in vitro*, providing correlative evidence for a functional relationship (45). These data support the concept that enhancers and superenhancers are transcriptional modules that integrate multiple signals to regulate responsiveness of target genes (43).

EXTENDING THE COLLABORATIVE/ HIERARCHICAL MODEL

While supported by gain- and loss-of-function experiments and the effects of natural genetic variation, the collaborative/hierarchical model depicted in Fig. 3 is vastly oversimplified. For example, the model does not account for the functions of the vast majority of transcription factors that are expressed in macrophages. Furthermore, since it is derived from responses of tissue culture macrophages to selective signals, its relevance to the mechanisms leading to selection and function of enhancers in tissue-resident macrophages is unclear. We next consider several different approaches to define transcription factors required for macrophage development, and place findings derived from these methods into the context of enhancer selection and function.

GENE DELETION STUDIES

One of the most powerful approaches to delineating gene function is through targeted generation of null alleles, either systemically or conditionally. Here, we briefly survey a representative subset of studies examining the consequences of loss of function of specific transcription factors on macrophage development and tissue-specific phenotypes. As noted previously, PU.1 is a key LDTF for macrophages, neutrophils, and B cells, such that its deletion results in neonatal death and a general lack of these cell types (53, 54). Indeed, blocks in differentiation resulting from loss of PU.1 result in leukemic transformation of myeloid cells (68, 69). Genomic studies of PU.1 binding during distinct stages of B-cell differentiation indicated progressive remodeling of its genomic locations in concert with sequential expression of additional LDTFs for B-cell development (55).

Specifications of branch points in cellular differentiation are achieved in part through expression of additional LDTFs. For example, IRF8 has been shown to be necessary for dendritic cell differentiation in knockout mouse models. In the absence of IRF8, myeloid progenitors undergo a dendritic cell-to-neutrophil reprogramming, indicating a requirement for IRF8 in dendritic cell commitment (70, 71). In B cells, loss-of-function studies indicate that IRF8 and IRF4 compete to establish the differentiation of B cells to plasma cells (72). Whereas expression of IRF4 promotes differentiation of plasma cells, expression of IRF8 inhibits this differentiation (72). Ultimately, this fate decision is determined by the relative expression of these two LDTFs. LDTFs can also function to inhibit each other to specify differentiation, as seen in the case of IRF8 inhibiting C/EBP chromatin association through a direct inhibitory interaction. The result of this interaction is a block in the differentiation into neutrophils from mononuclear phagocyte progenitors (73). Studies in macrophages characterizing the genome-wide binding patterns of IRF8 in resting and LPS-stimulated macrophages indicate that it contributes to the selection of both basal enhancer elements as well as latent enhancers (74).

Several transcription factors have been identified that are required for tissue-specific macrophage subsets. The PU.1-related family member Spi-C was recently shown to be necessary for red pulp macrophage development. The Spi-C knockout mouse has a cell-autonomous defect in splenic iron homeostasis, which led to the discovery of a specific lack of red pulp macrophages (75). One of the best-characterized examples of a subset-specific LDTF is c-Fos. Mice lacking c-Fos were shown to have severe growth retardation and osteopetrosis resulting from a loss of osteoclasts (76–79). In the lung, differentiation of fetal monocytes into alveolar macrophages requires the expression of peroxisome proliferator-activated receptor γ. Knockout of peroxisome proliferator-activated receptor γ results in diminished lipid catabolism-associated gene expression and enhanced cholesterol esterification (80). Nr4a1-deficient mice result in an absence of $Ly6C^-$ monocytes, which are patrolling monocytes (81). Nr4a1-deficient mice exhibit an increase in atherosclerosis in hypercholesterolemic low-density lipoprotein receptor knockout mice, which suggests that $Ly6C^-$ monocytes are necessary for controlling the inflammatory phenotype that leads to atherosclerotic plaque development (81). In the brain, transforming growth factor β (TGF-β)-induced SMAD activity is necessary for maintenance of the microglia phenotype (82).

While these gene deletion experiments provide strong evidence for functional roles of the corresponding transcription factors, they do not establish the mechanisms by which they exert their effects on macrophage development and function. Of the factors discussed above, only PU.1 and IRF8 have been studied thus far at a genome-wide level in macrophages. The subset-selective activities of c-Fos in osteoclasts or Spi-C in splenic macrophages are as yet not understood. Given the biological insights derived from these studies, it will be of considerable interest to evaluate the genome-wide binding and functions of these and other transcription factors required for the acquisition of tissue-specific phenotypes. A further limitation of gene deletion studies is that phenotypes may be less pronounced for factors that are members of gene families and where functional redundancies may be present.

EXPLOITING SUPERENHANCERS

As previously discussed, superenhancers have been shown to overlap with factors necessary for cell fate. As such, a logical progression of superenhancer discovery is the concomitant discovery of the cell fate proteins these superenhancers regulate. Indeed, several groups have successfully used this approach to demonstrate that superenhancers are associated with transcription factors necessary for cell fate specification (41, 83, 84). For example, in embryonic stem cells, superenhancers covered genes that define cell identity such as *Oct4*, *Sox2*, and *Nanog* as well as several novel factors (41). Additionally, in a study of somatic copy number alterations in cancer pathogenesis, focally amplified lineage-specific superenhancers in human epithelial cancers were targeted to known oncogenes such as *Klf5*, *Usp12*, *Pard6B*, and *Myc*, and this resulted in overexpression (84). An alternative method used gene ontology analysis on the top-ranked superenhancers. Several transcriptional regulators came up as regulators of these programs, including Sox9, which was then confirmed as a crucial chromatin rheostat of hair follicle stem cell regulation and identity (83). Thus, transcription factors identified to be associated with superenhancers in macrophage subsets are candidates for prioritization. For example, numerous known and putative transcription factors are associated with superenhancers in microglia that have not been studied in this cell type, including *Zfp691*, *Sall1*, and *Nfat1c* (45).

COEXPRESSION ANALYSIS

A common starting point for identifying potential cell fate-specific transcription factors is to perform RNA expression analyses that allow comparisons of gene expression across cell types and/or during cell differentiation. Temporal changes in expression of transcription factors during developmental transitions can suggest contributing transcription factors. For example, analysis of RNA expression during neutrophil differentiation from hematopoietic stem cells from human patients revealed changes in transcriptional activators that included GATA2, AML1/RUNX1, SCL/TAL1, C/EBPα, and PU.1 (85). Various software tools have been developed to enable placing transcription factor expression into the context of networks and to predict cause-effect relationships, such as Cytoscape, Ingenuity Pathway Analysis, and Pathway Studio (86). For example, Ingenuity Pathway tools were used to define physical and regulatory interactions that predict distinct roles of C/EBPs, Maf, NFE2, and nuclear receptor family members in directing the expression of 14 macrophage-associated modules (60).

DISCOVERY OF TRANSCRIPTION FACTOR MOTIFS USING ATAC-seq

A recently reported method for identifying important transcription factors is through analysis of open chromatin identified using the transposase-accessible chromatin sequencing (ATAC-seq) method (87). ATAC-seq is analogous to DNase I hypersensitivity in defining open chromatin, but uses a transposase-assisted integration of DNA primer sequences. The practical importance of this difference is that open regions of chromatin can be identified in relatively small populations of cells (reportedly as few as 500) (87). It is thus amenable to analysis of the often limited numbers of cells that can be obtained from mouse tissues or clinical samples. Several groups have used this method to define open chromatin and then have leveraged this information to define the protein-specific DNA-binding motifs. For example, ATAC-seq was used in the identification of putative enhancers in tissue-resident macrophages (62). These analyses identified Mef2c and Gata6 as candidate LDTFs in microglia and peritoneal macrophages, respectively. This method has also been used to define several LDTFs, including BAF and p63 mutual recruitment to almost 15,000 open chromatin regions in adherent human keratinocytes (88); PDX1 and NKX6.1 in β cells; and NKX2.2, MAFB, and FOXA2 in islet cells (89). Interestingly, this approach has also been used to define the LDTFs responsible for driving *in vivo* tumor development, specifically in Ras-dependent oncogenesis in mice (90). Here, open chromatin regions in Ras-driven tumors were found to contain both AP1 and Stat92E. Further, the introduction of a loss-of-function Stat92E mutant rescued the tumor phenotype, confirming the applicability of using ATAC-seq in the novel discovery of LDTFs necessary for tumor development. A limitation of ATAC-seq experiments is that open regions of chromatin are not necessarily active regulatory regions. In addition, it is not yet clear what the "false-negative" rate is for this method. Despite these limitations, it appears that ATAC-seq will be a powerful method for defining the most important transcription factor motifs within specific cell types.

NATURAL GENETIC VARIATION AS A "MUTAGENESIS" STRATEGY

An emerging approach for novel LDTF and SDTF discovery uses the natural genetic variation provided by different strains of inbred mice. Natural genetic variation was initially used as a way to test the collaborative and hierarchical model of enhancer selection involving PU.1, C/EBP, and AP1, as described above (57). The demonstration that mutations in PU.1 motifs led to loss of not only PU.1 binding but also nearby C/EBP binding suggested that natural genetic variation could be used to discover unknown collaborative binding partners. To test this approach, ChIP-seq was used to define the binding sites of PU.1 in resident peritoneal macrophages and microglia in three inbred mouse strains providing 5 million to 40 million single nucleotide polymorphisms (SNPs) (45). Genomic regions exhibiting differential binding of PU.1 were then analyzed for SNPs. Motifs altered by SNPs that were statistically correlated with strain-specific PU.1 binding were considered to represent collaborative binding partners of PU.1. This analysis led to the identification of several dozen motifs that were highly correlated with nearby binding of PU.1. Importantly, these included motifs for C/EBP factors, which were independently established as important collaborative binding partners for PU.1. In addition, GATA6 motifs were identified near PU.1 binding sites in large peritoneal macrophages and SMAD motifs in microglia, providing proof of principle for the utility of this approach as a discovery strategy. Many of the additional motifs are recognized by transcription factors that are expressed in macrophages but that have not as yet been studied in this context.

These varying approaches to uncovering putative LDTFs and SDTFs exploit different angles and offer different advantages and disadvantages. For example, defining superenhancers and performing RNA expression analysis provide candidates at a relatively low cost; however, these methods do not explain binding patterns and assume protein expression and activity. Using measures of open chromatin provides likely motifs, but these are correlative and cannot distinguish specific proteins from families of factors that bind similar motifs. Using SNPs in natural genetic variation provides genetic evidence for the importance of transcription factors recognizing a particular motif in driving collaborative binding interactions, but also does not specify the particular factor among a transcription factor family. Thus, an essential step is to validate roles of specific transcription factors through loss-of-function studies. These studies in themselves can be challenging to interpret due to transcription factor redundancy, requiring simultaneous loss-of-function strategies. In addition, loss-of-function studies may lead to early embryonic lethality, effects in multiple tissues, or loss of a progenitor cell, precluding analysis in specific macrophage populations. Thus, conditional methods for gene deletion are often required. Optimally, this can be achieved by Cre recombinases that are directed to the particular macrophage population of interest. However, a relatively limited set of macrophage-specific Cre drivers are available at present, and new and highly specific drivers that target gene deletion in particular macrophage subsets would be valuable additions to this field.

CONCLUSIONS AND FUTURE DIRECTIONS

A central aspect of macrophage biology is the cells' ability to sense the environment through their expression of a multitude of cell surface receptors that control the expression and/or activity of downstream transcription factors. The emergent picture is that specific phenotypes arise from the combinatorial actions of broadly expressed signal-dependent factors and more restricted lineage-determining factors at macrophage-specific enhancers. Thus, the complement of enhancers within each macrophage is proposed to correspond to tens of thousands of distinct analog genomic sensors that integrate diverse signaling inputs to regulate the expression of target genes. Although these principles appear to be general for the many cell types examined thus far, macrophages may be particularly specialized as sensors and responders to environmental signals.

Integration of findings from gain- and loss-of-function experiments and epigenetic profiling suggest a revised model for selection and activation of tissue-specific macrophage enhancers (Fig. 4). In this model, a core set of LDTFs, exemplified by PU.1 and C/EBP factors, select primed enhancers in most or all macrophage populations. These enhancers can be acted upon by environment-specific signals to induce direct target genes of those signals. For example, retinoic acid is present in high concentrations in the peritoneal cavity compared to the adult brain, and thus retinoic acid-responsive enhancers and their target genes are preferentially activated in peritoneal macrophages. Conversely, TGF-β is more abundant in the adult brain than within the peritoneal cavity, resulting in preferential activation of SMAD-dependent enhancers and genes in microglia. Among the spectrum of genes induced by these enhancers are genes encoding transcription factors that have the potential to collaborate with PU.1, and presumably other macrophage LDTFs, to select new enhancers that are environment specific. For example, GATA6 expression is positively regulated by retinoic acid and functions as a collaborative partner of PU.1 to select peritoneal macrophage-specific enhancers. Thus, this model proposes that the full complement of tissue-

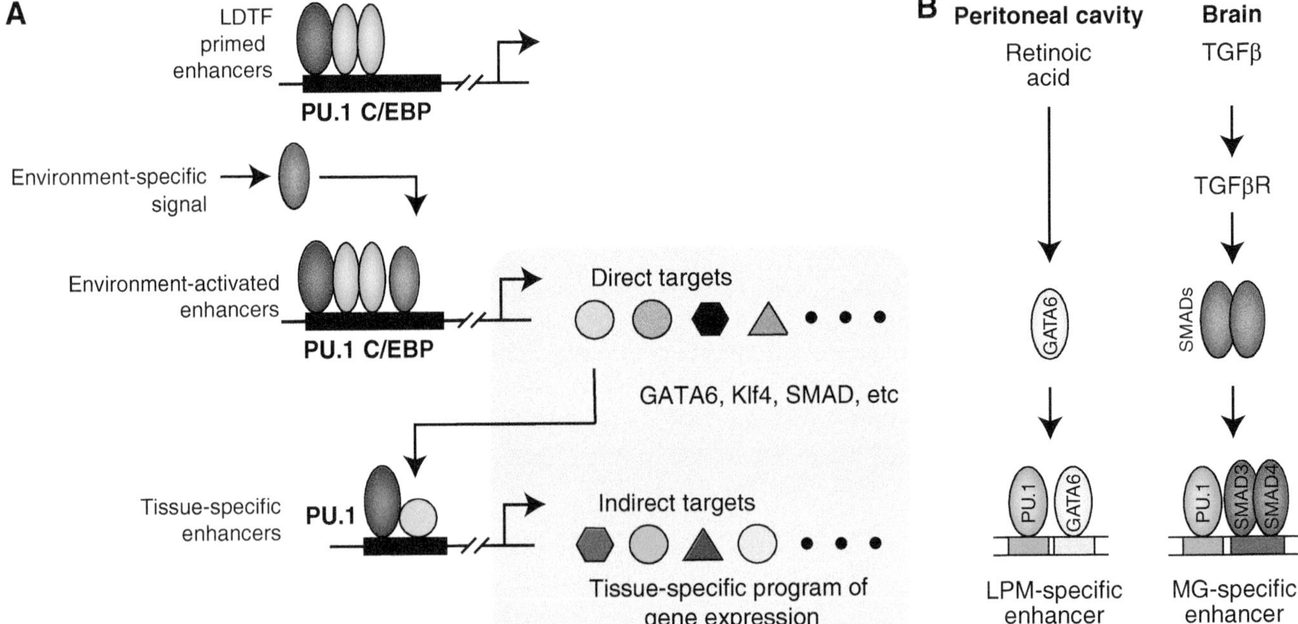

Figure 4 Selection and activation of tissue-specific macrophage enhancers. (A) Generic model. A core set of macrophage LDTFs, exemplified by PU.1 and C/EBP factors, prime a common set of enhancers in many or all macrophage subsets. These enhancers can be acted upon by environment-specific signals to drive the expression of direct target genes. A subset of these genes includes transcription factors that can collaborate with macrophage LDTFs, such as PU.1, to select a secondary, tissue-specific set of enhancers that drive expression of additional target genes. The tissue-specific gene expression program thus results from both direct and indirect environmental effects. (B) Examples of signals preferential for the peritoneal cavity (retinoic acid) or brain (TGF-β), resulting in expression of collaborative factors Gata6 or SMADs, respectively. MG, microglia.

specific gene expression is driven by the combination of direct effects of environmental factors on common enhancers and induced expression of transcription factors that drive the selection and function of tissue-specific enhancers.

CRISPR-Cas9 (clustered regularly interspaced short palindromic repeats–CRISPR-associated protein 9) technology could be used to untangle many of the genetic and biochemical questions that still surround enhancer selection and activation. CRISPR-Cas9 is a bacterial-based system that has been adapted as a tool for genome editing. It contains a single-guide RNA (sgRNA) sequence that has a target sequence that binds to complementary host DNA and forms a complex with the DNA endonuclease protein Cas9. Once targeted to a genomic DNA sequence complementary to the sgRNA, Cas9 causes a double-stranded break in the target DNA, which may result in a deletion if the break is repaired by nonhomologous end joining (30, 91). Thus, the CRISPR-Cas9 system could be used to delete enhancers in order to ascertain which gene or genes a particular enhancer regulates (92, 93). Alternatively, a mutant form of Cas9, denoted dCas9, that is unable to induce a break in the DNA could be used to modify enhancer activity. This can be accomplished by adding an activation (VP16) or repression (Krüppel-associated box domain of KOX1) domain to the dCas9 (94, 95). This technique allows an activation or repression domain to be targeted to any enhancer in the genome. More-specific modifications could be made using a combination of CRISPR-Cas9 and homologous recombination. Here, a homologous DNA sequence with small mutations would be cointroduced with CRISPR-Cas9. Recombination of this sequence would allow for subtle changes of specific DNA sequences to either alter transcription factor binding or remove eRNA start sites (96, 97). CRISPR-Cas9 could also be used for biochemical analysis of specific enhancers. Using enChIP, a tagged dCas9-based system, Cas9 can be specifically targeted to a specific enhancer, and ChIP coupled with mass spectrometry protein sequencing would be used to determine the complexes bound at the target enhancer (98). Using CRISPR-Cas9 technologies should allow for a more thorough understanding of enhancer activity, specifically relating to enhancer targeting and redundancy through direct knockout or targeted activation/repression, transcription factor cooperation through base-pair modification, and eRNA requirements and function, among many others.

A remaining limitation of this model is that it does not take into account the potential origin of macrophages as a determinant of enhancer landscapes or the question of whether there are transcriptional circuits that are "hard-wired." The analysis of large and small peritoneal macrophages indicates that small peritoneal macrophages exhibit a program of gene expression that is distinct from the large peritoneal macrophages that live in the same environment. The basis for this difference is as yet unexplained. In addition, while many genes that are specific for large peritoneal macrophages and microglia fall dramatically when transferred from the *in vivo* environment to an *in vitro* environment, many other genes that are specific to these subsets do not change. The basis for this retention of tissue-specific programs of expression is also not understood. Going forward, genomic studies, including studies of DNA methylation, which could provide a more long-lasting epigenetic mark, need to be combined with lineage tracing to help establish the potential importance of origin in defining tissue-specific phenotypes.

In addition to providing insights into mechanisms that specify macrophage identity and function, the delineation of tissue-specific LDTFs and SDTFs has direct relevance to better understanding of roles of macrophages in disease. First, knowledge of these factors is likely to facilitate efforts to reprogram patient-derived stem cells to specific macrophage phenotypes *in vitro*. Such reprogrammed cells are useful for studying cell-autonomous disease mechanisms and responses to drugs. Second, the ascertainment of the genome-wide binding patterns of these factors and the effects of natural genetic variation will help inform interpretation of risk alleles identified by genome-wide association studies. Third, the observation that macrophage enhancer landscapes are dependent on constant input from their environment implies that these landscapes will change in the context of disease. Because these changes can now be measured in relatively small populations of macrophages isolated from tissues, it should be possible to determine the corresponding transcription factors that are gained or lost using motif analysis. This information might then be used to "reverse engineer" the potential signaling pathways that are gained or lost in the particular disease state and thereby gain insights into mechanisms driving macrophage phenotypic conversion.

Lastly, there is emerging interest in the potential to alter cellular phenotypes by targeting enhancers. The observation that at least some eRNAs contribute to enhancer function has suggested the possibility of using antisense oligonucleotides to knock down expression of cell-specific eRNAs as a means of altering gene expression in a cell-specific manner (34). In addition, a relatively new class of pharmaceuticals is targeted at proteins that are "readers, writers, and erasers" of the epigenetic code associated with regulation of gene expression. Such molecules include inhibitors of histone deacetylases, histone methyltransferases, and histone tail mimetics. In addition to acting at promoters, these molecules also exert effects at enhancer elements. A striking demonstration of the potential of this general class of molecules to have translational potential is provided by histone tail mimetics that prevent the interaction of the bromodomain and extra-terminal domain (BET) family of proteins with acetylated histones (99). These compounds disrupt chromatin complexes responsible for the expression of key inflammatory genes in activated macrophages and confer protection against LPS-induced endotoxic shock and bacteria-induced sepsis. Subsequent studies demonstrated that super-enhancers are particularly susceptible to this class of compounds, further raising the possibility of targeting the enhancer landscape for therapeutic purposes (100). Intriguingly, this class of histone tail mimetics alters inflammation-induced enhancer selection in endothelial cells and inhibits the development of atherosclerosis in a mouse model (65).

In conclusion, the expanding appreciation of the tissue-specific homeostatic functions of macrophages and their various roles in human disease reinforces the importance of efforts to understand the mechanisms by which they achieve their distinct phenotypes. Genome-wide approaches to defining macrophage enhancer selection and function are likely to provide a fruitful avenue of investigation toward this goal for the foreseeable future.

Citation. Fonseca GJ, Seidman JS, Glass CK. 2016. Genome-wide approaches to defining macrophage identity and function. Microbiol Spectrum 4(5):MCHD-0039-2016.

References

1. **Gordon S, Plüddemann A, Martinez Estrada F.** 2014. Macrophage heterogeneity in tissues: phenotypic diversity and functions. *Immunol Rev* **262:**36–55.
2. **Wynn TA, Chawla A, Pollard JW.** 2013. Macrophage biology in development, homeostasis and disease. *Nature* **496:**445–455.
3. **Barton GM, Medzhitov R.** 2002. Toll-like receptors and their ligands. *Curr Top Microbiol Immunol* **270:**81–92.
4. **Medzhitov R, Horng T.** 2009. Transcriptional control of the inflammatory response. *Nat Rev Immunol* **9:**692–703.
5. **Smale ST.** 2012. Transcriptional regulation in the innate immune system. *Curr Opin Immunol* **24:**51–57.
6. **Pasare C, Medzhitov R.** 2005. Toll-like receptors: linking innate and adaptive immunity. *Adv Exp Med Biol* **560:** 11–18.
7. **Glass CK, Natoli G.** 2016. Molecular control of activation and priming in macrophages. *Nat Immunol* **17:**26–33.
8. **Hong S, Dissing-Olesen L, Stevens B.** 2016. New insights on the role of microglia in synaptic pruning in health and disease. *Curr Opin Neurobiol* **36:**128–134.
9. **Lumeng CN.** 2016. Lung macrophage diversity and asthma. *Ann Am Thorac Soc* **13**(Suppl 1):S31–S34. doi: 10.1513/AnnalsATS.201506-384MG.
10. **Kurotaki D, Uede T, Tamura T.** 2015. Functions and development of red pulp macrophages. *Microbiol Immunol* **59:**55–62.
11. **Tabas I, Bornfeldt KE.** 2016. Macrophage phenotype and function in different stages of atherosclerosis. *Circ Res* **118:**653–667.
12. **Olefsky JM, Glass CK.** 2010. Macrophages, inflammation, and insulin resistance. *Annu Rev Physiol* **72:**219–246.
13. **Ju C, Mandrekar P.** 2015. Macrophages and alcohol-related liver inflammation. *Alcohol Res* **37:**251–262.
14. **Huang W, Metlakunta A, Dedousis N, Zhang P, Sipula I, Dube JJ, Scott DK, O'Doherty RM.** 2010. Depletion of liver Kupffer cells prevents the development of diet-induced hepatic steatosis and insulin resistance. *Diabetes* **59:**347–357.
15. **Ransohoff RM, El Khoury J.** 2015. Microglia in health and disease. *Cold Spring Harb Perspect Biol* **8:**a020560. doi:10.1101/cshperspect.a020560.
16. **Villegas-Llerena C, Phillips A, Garcia-Reitboeck P, Hardy J, Pocock JM.** 2016. Microglial genes regulating neuroinflammation in the progression of Alzheimer's disease. *Curr Opin Neurobiol* **36:**74–81.
17. **Noy R, Pollard JW.** 2014. Tumor-associated macrophages: from mechanisms to therapy. *Immunity* **41:**49–61.
18. **De Vlaeminck Y, González-Rascón A, Goyvaerts C, Breckpot K.** 2016. Cancer-associated myeloid regulatory cells. *Front Immunol* **7:**113.
19. **Winter DR, Jung S, Amit I.** 2015. Making the case for chromatin profiling: a new tool to investigate the immune-regulatory landscape. *Nat Rev Immunol* **15:**585–594.
20. **Pennacchio LA, Ahituv N, Moses AM, Prabhakar S, Nobrega MA, Shoukry M, Minovitsky S, Dubchak I, Holt A, Lewis KD, Plajzer-Frick I, Akiyama J, De Val S, Afzal V, Black BL, Couronne O, Eisen MB, Visel A, Rubin EM.** 2006. *In vivo* enhancer analysis of human conserved non-coding sequences. *Nature* **444:**499–502.
21. **Woolfe A, Goodson M, Goode DK, Snell P, McEwen GK, Vavouri T, Smith SF, North P, Callaway H, Kelly K, Walter K, Abnizova I, Gilks W, Edwards YJ, Cooke JE, Elgar G.** 2005. Highly conserved non-coding sequences are associated with vertebrate development. *PLoS Biol* **3:** e7. doi:10.1371/journal.pbio.0030007.
22. **Banerji J, Rusconi S, Schaffner W.** 1981. Expression of a β-globin gene is enhanced by remote SV40 DNA sequences. *Cell* **27:**299–308.
23. **Heinz S, Romanoski CE, Benner C, Glass CK.** 2015. The selection and function of cell type-specific enhancers. *Nat Rev Mol Cell Biol* **16:**144–154.
24. **Liu Z, Merkurjev D, Yang F, Li W, Oh S, Friedman MJ, Song X, Zhang F, Ma Q, Ohgi KA, Krones A, Rosenfeld MG.** 2014. Enhancer activation requires *trans*-recruitment of a mega transcription factor complex. *Cell* **159:**358–373.

25. **De Santa F, Barozzi I, Mietton F, Ghisletti S, Polletti S, Tusi BK, Muller H, Ragoussis J, Wei CL, Natoli G.** 2010. A large fraction of extragenic RNA Pol II transcription sites overlap enhancers. *PLoS Biol* **8:**e1000384. doi:10.1371/journal.pbio.1000384.

26. **Heintzman ND, Stuart RK, Hon G, Fu Y, Ching CW, Hawkins RD, Barrera LO, Van Calcar S, Qu C, Ching KA, Wang W, Weng Z, Green RD, Crawford GE, Ren B.** 2007. Distinct and predictive chromatin signatures of transcriptional promoters and enhancers in the human genome. *Nat Genet* **39:**311–318.

27. **Creyghton MP, Cheng AW, Welstead GG, Kooistra T, Carey BW, Steine EJ, Hanna J, Lodato MA, Frampton GM, Sharp PA, Boyer LA, Young RA, Jaenisch R.** 2010. Histone H3K27ac separates active from poised enhancers and predicts developmental state. *Proc Natl Acad Sci U S A* **107:**21931–21936.

28. **ENCODE Project Consortium.** 2012. An integrated encyclopedia of DNA elements in the human genome. *Nature* **489:**57–74.

29. **Huang J, Liu X, Li D, Shao Z, Cao H, Zhang Y, Trompouki E, Bowman TV, Zon LI, Yuan GC, Orkin SH, Xu J.** 2016. Dynamic control of enhancer repertoires drives lineage and stage-specific transcription during hematopoiesis. *Dev Cell* **36:**9–23.

30. **Wright AV, Nuñez JK, Doudna JA.** 2016. Biology and applications of CRISPR systems: harnessing nature's toolbox for genome engineering. *Cell* **164:**29–44.

31. **Lam MTY, Li W, Rosenfeld MG, Glass CK.** 2014. Enhancer RNAs and regulated transcriptional programs. *Trends Biochem Sci* **39:**170–182.

32. **Li W, Notani D, Rosenfeld MG.** 2016. Enhancers as non-coding RNA transcription units: recent insights and future perspectives. *Nat Rev Genet* **17:**207–223.

33. **Kaikkonen MU, Spann NJ, Heinz S, Romanoski CE, Allison KA, Stender JD, Chun HB, Tough DF, Prinjha RK, Benner C, Glass CK.** 2013. Remodeling of the enhancer landscape during macrophage activation is coupled to enhancer transcription. *Mol Cell* **51:**310–325.

34. **Lam MT, Cho H, Lesch HP, Gosselin D, Heinz S, Tanaka-Oishi Y, Benner C, Kaikkonen MU, Kim AS, Kosaka M, Lee CY, Watt A, Grossman TR, Rosenfeld MG, Evans RM, Glass CK.** 2013. Rev-Erbs repress macrophage gene expression by inhibiting enhancer-directed transcription. *Nature* **498:**511–515.

35. **Li W, Notani D, Ma Q, Tanasa B, Nunez E, Chen AY, Merkurjev D, Zhang J, Ohgi K, Song X, Oh S, Kim HS, Glass CK, Rosenfeld MG.** 2013. Functional roles of enhancer RNAs for oestrogen-dependent transcriptional activation. *Nature* **498:**516–520.

36. **Schaukowitch K, Joo JY, Liu X, Watts JK, Martinez C, Kim TK.** 2014. Enhancer RNA facilitates NELF release from immediate early genes. *Mol Cell* **56:**29–42.

37. **Sigova AA, Abraham BJ, Ji X, Molinie B, Hannett NM, Guo YE, Jangi M, Giallourakis CC, Sharp PA, Young RA.** 2015. Transcription factor trapping by RNA in gene regulatory elements. *Science* **350:**978–981.

38. **Lai F, Gardini A, Zhang A, Shiekhattar R.** 2015. Integrator mediates the biogenesis of enhancer RNAs. *Nature* **525:**399–403.

39. **Austenaa LM, Barozzi I, Simonatto M, Masella S, Della Chiara G, Ghisletti S, Curina A, de Wit E, Bouwman BA, de Pretis S, Piccolo V, Termanini A, Prosperini E, Pelizzola M, de Laat W, Natoli G.** 2015. Transcription of mammalian **cis**-regulatory elements is restrained by actively enforced early termination. *Mol Cell* **60:**460–474.

40. **Hnisz D, Abraham BJ, Lee TI, Lau A, Saint-André V, Sigova AA, Hoke HA, Young RA.** 2013. Super-enhancers in the control of cell identity and disease. *Cell* **155:**934–947.

41. **Whyte WA, Orlando DA, Hnisz D, Abraham BJ, Lin CY, Kagey MH, Rahl PB, Lee TI, Young RA.** 2013. Master transcription factors and mediator establish super-enhancers at key cell identity genes. *Cell* **153:**307–319.

42. **Parker SC, Stitzel ML, Taylor DL, Orozco JM, Erdos MR, Akiyama JA, van Bueren KL, Chines PS, Narisu N, NISC Comparative Sequencing Program, Black BL, Visel A, Pennacchio LA, Collins FS, National Institutes of Health Intramural Sequencing Center Comparative Sequencing Program Authors, NISC Comparative Sequencing Program Authors.** 2013. Chromatin stretch enhancer states drive cell-specific gene regulation and harbor human disease risk variants. *Proc Natl Acad Sci U S A* **110:**17921–17926.

43. **Hnisz D, Schuijers J, Lin CY, Weintraub AS, Abraham BJ, Lee TI, Bradner JE, Young RA.** 2015. Convergence of developmental and oncogenic signaling pathways at transcriptional super-enhancers. *Mol Cell* **58:**362–370.

44. **Hah N, Benner C, Chong LW, Yu RT, Downes M, Evans RM.** 2015. Inflammation-sensitive super enhancers form domains of coordinately regulated enhancer RNAs. *Proc Natl Acad Sci U S A* **112:**E297–E302.

45. **Gosselin D, Link VM, Romanoski CE, Fonseca GJ, Eichenfield DZ, Spann NJ, Stender JD, Chun HB, Garner H, Geissmann F, Glass CK.** 2014. Environment drives selection and function of enhancers controlling tissue-specific macrophage identities. *Cell* **159:**1327–1340.

46. **Kagey MH, Newman JJ, Bilodeau S, Zhan Y, Orlando DA, van Berkum NL, Ebmeier CC, Goossens J, Rahl PB, Levine SS, Taatjes DJ, Dekker J, Young RA.** 2010. Mediator and cohesin connect gene expression and chromatin architecture. *Nature* **467:**430–435.

47. **Tronche F, Yaniv M.** 1992. HNF1, a homeoprotein member of the hepatic transcription regulatory network. *BioEssays* **14:**579–587.

48. **Chen X, Xu H, Yuan P, Fang F, Huss M, Vega VB, Wong E, Orlov YL, Zhang W, Jiang J, Loh YH, Yeo HC, Yeo ZX, Narang V, Govindarajan KR, Leong B, Shahab A, Ruan Y, Bourque G, Sung WK, Clarke ND, Wei CL, Ng HH.** 2008. Integration of external signaling pathways with the core transcriptional network in embryonic stem cells. *Cell* **133:**1106–1117.

49. **MacArthur S, Li XY, Li J, Brown JB, Chu HC, Zeng L, Grondona BP, Hechmer A, Simirenko L, Keränen SV, Knowles DW, Stapleton M, Bickel P, Biggin MD, Eisen MB.** 2009. Developmental roles of 21 *Drosophila* transcription factors are determined by quantitative differences in binding to an overlapping set of thousands of genomic regions. *Genome Biol* **10:**R80. doi:10.1186/gb-2009-10-7-r80.

50. **Lupien M, Eeckhoute J, Meyer CA, Wang Q, Zhang Y, Li W, Carroll JS, Liu XS, Brown M.** 2008. FoxA1 translates epigenetic signatures into enhancer-driven lineage-specific transcription. *Cell* **132:**958–970.

51. **Odom DT, Zizlsperger N, Gordon DB, Bell GW, Rinaldi NJ, Murray HL, Volkert TL, Schreiber J, Rolfe PA, Gifford DK, Fraenkel E, Bell GI, Young RA.** 2004. Control of pancreas and liver gene expression by HNF transcription factors. *Science* **303:**1378–1381.

52. **Sandmann T, Jensen LJ, Jakobsen JS, Karzynski MM, Eichenlaub MP, Bork P, Furlong EE.** 2006. A temporal map of transcription factor activity: Mef2 directly regulates target genes at all stages of muscle development. *Dev Cell* **10:**797–807.

53. **Scott EW, Simon MC, Anastasi J, Singh H.** 1994. Requirement of transcription factor PU.1 in the development of multiple hematopoietic lineages. *Science* **265:**1573–1577.

54. **McKercher SR, Torbett BE, Anderson KL, Henkel GW, Vestal DJ, Baribault H, Klemsz M, Feeney AJ, Wu GE, Paige CJ, Maki RA.** 1996. Targeted disruption of the *PU.1* gene results in multiple hematopoietic abnormalities. *EMBO J* **15:**5647–5658.

55. **Heinz S, Benner C, Spann N, Bertolino E, Lin YC, Laslo P, Cheng JX, Murre C, Singh H, Glass CK.** 2010. Simple combinations of lineage-determining transcription factors prime *cis*-regulatory elements required for macrophage and B cell identities. *Mol Cell* **38:**576–589.

56. **Uhlenhaut NH, Barish GD, Yu RT, Downes M, Karunasiri M, Liddle C, Schwalie P, Hübner N, Evans RM.** 2013. Insights into negative regulation by the glucocorticoid receptor from genome-wide profiling of inflammatory cistromes. *Mol Cell* **49:**158–171.

57. **Heinz S, Romanoski CE, Benner C, Allison KA, Kaikkonen MU, Orozco LD, Glass CK.** 2013. Effect of natural genetic variation on enhancer selection and function. *Nature* **503:**487–492.

58. **Ostuni R, Piccolo V, Barozzi I, Polletti S, Termanini A, Bonifacio S, Curina A, Prosperini E, Ghisletti S, Natoli G.** 2013. Latent enhancers activated by stimulation in differentiated cells. *Cell* **152:**157–171.

59. **Netea MG, Joosten LA, Latz E, Mills KH, Natoli G, Stunnenberg HG, O'Neill LA, Xavier RJ.** 2016. Trained immunity: a program of innate immune memory in health and disease. *Science* **352:**aaf1098. doi:10.1126/science.aaf1098.

60. **Gautier EL, Shay T, Miller J, Greter M, Jakubzick C, Ivanov S, Helft J, Chow A, Elpek KG, Gordonov S, Mazloom AR, Ma'ayan A, Chua WJ, Hansen TH, Turley SJ, Merad M, Randolph GJ, Immunological Genome Consortium.** 2012. Gene-expression profiles and transcriptional regulatory pathways that underlie the identity and diversity of mouse tissue macrophages. *Nat Immunol* **13:**1118–1128.

61. **Ghosn EE, Cassado AA, Govoni GR, Fukuhara T, Yang Y, Monack DM, Bortoluci KR, Almeida SR, Herzenberg LA, Herzenberg LA.** 2010. Two physically, functionally, and developmentally distinct peritoneal macrophage subsets. *Proc Natl Acad Sci U S A* **107:**2568–2573.

62. **Lavin Y, Winter D, Blecher-Gonen R, David E, Keren-Shaul H, Merad M, Jung S, Amit I.** 2014. Tissue-resident macrophage enhancer landscapes are shaped by the local microenvironment. *Cell* **159:**1312–1326.

63. **Lara-Astiaso D, Weiner A, Lorenzo-Vivas E, Zaretsky I, Jaitin DA, David E, Keren-Shaul H, Mildner A, Winter D, Jung S, Friedman N, Amit I.** 2014. Immunogenetics. Chromatin state dynamics during blood formation. *Science* **345:**943–949.

64. **Okabe Y, Medzhitov R.** 2014. Tissue-specific signals control reversible program of localization and functional polarization of macrophages. *Cell* **157:**832–844.

65. **Brown JD, Lin CY, Duan Q, Griffin G, Federation AJ, Paranal RM, Bair S, Newton G, Lichtman AH, Kung AL, Yang T, Wang H, Luscinskas FW, Croce KJ, Bradner JE, Plutzky J.** 2014. NF-κB directs dynamic super enhancer formation in inflammation and atherogenesis. *Mol Cell* **56:**219–231.

66. **Di Micco R, Fontanals-Cirera B, Low V, Ntziachristos P, Yuen SK, Lovell CD, Dolgalev I, Yonekubo Y, Zhang G, Rusinova E, Gerona-Navarro G, Cañamero M, Ohlmeyer M, Aifantis I, Zhou M-M, Tsirigos A, Hernando E.** 2014. Control of embryonic stem cell identity by BRD4-dependent transcriptional elongation of super-enhancer-associated pluripotency genes. *Cell Rep* **9:**234–247.

67. **Herranz D, Ambesi-Impiombato A, Palomero T, Schnell SA, Belver L, Wendorff AA, Xu L, Castillo-Martin M, Llobet-Navás D, Cordon-Cardo C, Clappier E, Soulier J, Ferrando AA.** 2014. A NOTCH1-driven *MYC* enhancer promotes T cell development, transformation and acute lymphoblastic leukemia. *Nat Med* **20:**1130–1137.

68. **Moreau-Gachelin F, Wendling F, Molina T, Denis N, Titeux M, Grimber G, Briand P, Vainchenker W, Tavitian A.** 1996. Spi-1/PU.1 transgenic mice develop multistep erythroleukemias. *Mol Cell Biol* **16:**2453–2463.

69. **Rosenbauer F, Wagner K, Kutok JL, Iwasaki H, Le Beau MM, Okuno Y, Akashi K, Fiering S, Tenen DG.** 2004. Acute myeloid leukemia induced by graded reduction of a lineage-specific transcription factor, PU.1. *Nat Genet* **36:**624–630.

70. **Schönheit J, Kuhl C, Gebhardt ML, Klett FF, Riemke P, Scheller M, Huang G, Naumann R, Leutz A, Stocking C, Priller J, Andrade-Navarro MA, Rosenbauer F.** 2013. PU.1 level-directed chromatin structure remodeling at the *Irf8* gene drives dendritic cell commitment. *Cell Rep* **3:**1617–1628.

71. **Terry RL, Miller SD.** 2014. Molecular control of monocyte development. *Cell Immunol* **291:**16–21.

72. **Carotta S, Willis SN, Hasbold J, Inouye M, Pang SHM, Emslie D, Light A, Chopin M, Shi W, Wang H, Morse HC III, Tarlinton DM, Corcoran LM, Hodgkin PD, Nutt SL.** 2014. The transcription factors IRF8 and PU.1 negatively regulate plasma cell differentiation. *J Exp Med* **211:**2169–2181.

73. **Kurotaki D, Yamamoto M, Nishiyama A, Uno K, Ban T, Ichino M, Sasaki H, Matsunaga S, Yoshinari M, Ryo A, Nakazawa M, Ozato K, Tamura T.** 2014. IRF8 inhibits C/EBPα activity to restrain mononuclear phagocyte

progenitors from differentiating into neutrophils. *Nat Commun* 5:4978. doi:10.1038/ncomms5978.

74. **Mancino A, Termanini A, Barozzi I, Ghisletti S, Ostuni R, Prosperini E, Ozato K, Natoli G.** 2015. A dual *cis*-regulatory code links IRF8 to constitutive and inducible gene expression in macrophages. *Genes Dev* **29**:394–408.
75. **Kohyama M, Ise W, Edelson BT, Wilker PR, Hildner K, Mejia C, Frazier WA, Murphy TL, Murphy KM.** 2009. Role for Spi-C in the development of red pulp macrophages and splenic iron homeostasis. *Nature* **457**:318–321.
76. **Alfaqeeh S, Oralova V, Foxworthy M, Matalova E, Grigoriadis AE, Tucker AS.** 2015. Root and eruption defects in *c-Fos* mice are driven by loss of osteoclasts. *J Dent Res* **94**:1724–1731.
77. **Grigoriadis AE, Wang ZQ, Cecchini MG, Hofstetter W, Felix R, Fleisch HA, Wagner EF.** 1994. c-Fos: a key regulator of osteoclast-macrophage lineage determination and bone remodeling. *Science* **266**:443–448.
78. **Johnson RS, Spiegelman BM, Papaioannou V.** 1992. Pleiotropic effects of a null mutation in the *c-fos* proto-oncogene. *Cell* **71**:577–586.
79. **Wang ZQ, Ovitt C, Grigoriadis AE, Möhle-Steinlein U, Rüther U, Wagner EF.** 1992. Bone and haematopoietic defects in mice lacking *c-fos*. *Nature* **360**:741–745.
80. **Schneider C, Nobs SP, Kurrer M, Rehrauer H, Thiele C, Kopf M.** 2014. Induction of the nuclear receptor PPAR-γ by the cytokine GM-CSF is critical for the differentiation of fetal monocytes into alveolar macrophages. *Nat Immunol* **15**:1026–1037.
81. **Hanna RN, Shaked I, Hubbeling HG, Punt JA, Wu R, Herrley E, Zaugg C, Pei H, Geissmann F, Ley K, Hedrick CC.** 2012. NR4A1 (Nur77) deletion polarizes macrophages toward an inflammatory phenotype and increases atherosclerosis. *Circ Res* **110**:416–427.
82. **Butovsky O, Jedrychowski MP, Moore CS, Cialic R, Lanser AJ, Gabriely G, Koeglsperger T, Dake B, Wu PM, Doykan CE, Fanek Z, Liu L, Chen Z, Rothstein JD, Ransohoff RM, Gygi SP, Antel JP, Weiner HL.** 2014. Identification of a unique TGF-β-dependent molecular and functional signature in microglia. *Nat Neurosci* **17**:131–143.
83. **Adam RC, Yang H, Rockowitz S, Larsen SB, Nikolova M, Oristian DS, Polak L, Kadaja M, Asare A, Zheng D, Fuchs E.** 2015. Pioneer factors govern super-enhancer dynamics in stem cell plasticity and lineage choice. *Nature* **521**:366–370.
84. **Zhang X, Choi PS, Francis JM, Imielinski M, Watanabe H, Cherniack AD, Meyerson M.** 2016. Identification of focally amplified lineage-specific super-enhancers in human epithelial cancers. *Nat Genet* **48**:176–182.
85. **Sweeney CL, Teng R, Wang H, Merling RK, Lee J, Choi U, Koontz S, Wright DG, Malech HL.** 2016. Molecular analysis of neutrophil differentiation from human iPSCs delineates the kinetics of key regulators of hematopoiesis. *Stem Cells* **34**:1513–1526.
86. **Thomas S, Bonchev D.** 2010. A survey of current software for network analysis in molecular biology. *Hum Genomics* **4**:353–360.
87. **Buenrostro JD, Giresi PG, Zaba LC, Chang HY, Greenleaf WJ.** 2013. Transposition of native chromatin for fast and sensitive epigenomic profiling of open chromatin, DNA-binding proteins and nucleosome position. *Nat Methods* **10**:1213–1218.
88. **Bao X, Rubin AJ, Qu K, Zhang J, Giresi PG, Chang HY, Khavari PA.** 2015. A novel ATAC-seq approach reveals lineage-specific reinforcement of the open chromatin landscape via cooperation between BAF and p63. *Genome Biol* **16**:284. doi:10.1186/s13059-015-0840-9.
89. **Ackermann AM, Wang Z, Schug J, Naji A, Kaestner KH.** 2016. Integration of ATAC-seq and RNA-seq identifies human alpha cell and beta cell signature genes. *Mol Metab* **5**:233–244.
90. **Davie K, Jacobs J, Atkins M, Potier D, Christiaens V, Halder G, Aerts S.** 2015. Discovery of transcription factors and regulatory regions driving in vivo tumor development by ATAC-seq and FAIRE-seq open chromatin profiling. *PLoS Genet* **11**:e1004994. doi:10.1371/journal.pgen.1004994.
91. **Jinek M, Chylinski K, Fonfara I, Hauer M, Doudna JA, Charpentier E.** 2012. A programmable dual-RNA-guided DNA endonuclease in adaptive bacterial immunity. *Science* **337**:816–821.
92. **Jinek M, East A, Cheng A, Lin S, Ma E, Doudna J.** 2013. RNA-programmed genome editing in human cells. *eLife* **2**:e00471. doi:10.7554/eLife.00471.
93. **Cong L, Ran FA, Cox D, Lin S, Barretto R, Habib N, Hsu PD, Wu X, Jiang W, Marraffini LA, Zhang F.** 2013. Multiplex genome engineering using CRISPR/Cas systems. *Science* **339**:819–823.
94. **Perez-Pinera P, Kocak DD, Vockley CM, Adler AF, Kabadi AM, Polstein LR, Thakore PI, Glass KA, Ousterout DG, Leong KW, Guilak F, Crawford GE, Reddy TE, Gersbach CA.** 2013. RNA-guided gene activation by CRISPR-Cas9-based transcription factors. *Nat Methods* **10**:973–976.
95. **Gilbert LA, Larson MH, Morsut L, Liu Z, Brar GA, Torres SE, Stern-Ginossar N, Brandman O, Whitehead EH, Doudna JA, Lim WA, Weissman JS, Qi LS.** 2013. CRISPR-mediated modular RNA-guided regulation of transcription in eukaryotes. *Cell* **154**:442–451.
96. **Lewis WR, Malarkey EB, Tritschler D, Bower R, Pasek RC, Porath JD, Birket SE, Saunier S, Antignac C, Knowles MR, Leigh MW, Zariwala MA, Challa AK, Kesterson RA, Rowe SM, Drummond IA, Parant JM, Hildebrandt F, Porter ME, Yoder BK, Berbari NF.** 2016. Mutation of growth arrest specific 8 reveals a role in motile cilia function and human disease. *PLoS Genet* **12**:e1006220. doi:10.1371/journal.pgen.1006220.
97. **Lee JS, Grav LM, Pedersen LE, Lee GM, Kildegaard HF.** 2016. Accelerated homology-directed targeted integration of transgenes in Chinese hamster ovary cells via CRISPR/Cas9 and fluorescent enrichment. *Biotechnol Bioeng* doi:10.1002/bit.26002.
98. **Fujita T, Fujii H.** 2013. Efficient isolation of specific genomic regions and identification of associated proteins by engineered DNA-binding molecule-mediated chromatin immunoprecipitation (enChIP) using CRISPR. *Biochem Biophys Res Commun* **439**:132–136.

99. **Nicodeme E, Jeffrey KL, Schaefer U, Beinke S, Dewell S, Chung CW, Chandwani R, Marazzi I, Wilson P, Coste H, White J, Kirilovsky J, Rice CM, Lora JM, Prinjha RK, Lee K, Tarakhovsky A.** 2010. Suppression of inflammation by a synthetic histone mimic. *Nature* **468:**1119–1123.

100. **Lovén J, Hoke HA, Lin CY, Lau A, Orlando DA, Vakoc CR, Bradner JE, Lee TI, Young RA.** 2013. Selective inhibition of tumor oncogenes by disruption of super-enhancers. *Cell* **153:**320–334.

Myeloid Cells in Health and Disease: A Synthesis
Edited by Siamon Gordon

doi:10.1128/microbiolspec.MCHD-0010-2015

Lionel B. Ivashkiv[1,2]
Sung Ho Park[1]

31 Epigenetic Regulation of Myeloid Cells

INTRODUCTION

Epigenetics in the context of cell differentiation and activation refers to mechanisms that regulate and potentially stabilize gene expression in response to developmental and environmental cues. Epigenetic regulation is mediated by posttranslational modification of DNA or chromatin and by noncoding RNAs. In myeloid cells, the predominant focus of research on epigenetic mechanisms has been chromatin-mediated regulation of macrophages, which will be the focus of this review (1–3). Differentiation of macrophages from myeloid precursors is regulated by developmental signals and pioneer transcription factors that impart an "epigenetic landscape" that helps determine macrophage phenotype and how cells respond to environmental challenges. Macrophages protect the host from pathogenic microorganisms and other environmental insults, providing a rapid response as an initial line of defense. Accordingly, recognition of pathogen-associated molecular patterns through germ line-encoded receptors initiates and subsequently amplifies the adaptive immune response through cytokine production and antigen presentation. Importantly, macrophage phenotype is plastic, and macrophages carry out their distinct roles while maintaining the ability to adapt to local or systemic environmental changes (4–7).

Macrophages generate a transcriptional response that is both cell type and stimulus specific, helping the host develop specific innate and adaptive responses to successfully control various infections. In addition, macrophages must coordinate responses to, and have the ability to "remember," many stimuli, including signaling cues from other cells, the extracellular matrix, hormones, and active components of bacteria or viruses. Based on their repertoire of pattern recognition receptors, receptor-mediated signaling events, or their underlying differentiation phenotype, macrophages exhibit diverse responses to numerous stimuli and also drive rapid and appropriate immune responses. For example, a number of functionally distinct macrophage subsets has been described that can be broadly categorized in terms of ontogeny, homeostatic role, or maturation status (5, 8). Recent studies have highlighted the complex ontogeny of macrophages and dendritic cells and the unexpected complexity of the myeloid system, with highly specialized function and distribution in various tissues (6, 7, 9). Furthermore, during inflammatory processes, macrophages exhibit extensive plasticity of their

[1]Arthritis and Tissue Degeneration Program and David Z. Rosensweig Genomics Center, Hospital for Special Surgery, New York, NY 10021;
[2]Department of Medicine and Immunology and Microbial Pathogenesis Program, Weill Cornell Medical College, New York, NY 10021.

phenotypes, including polarization and activation, and such characteristics are controlled by both changes at the transcriptional level and an unanticipated degree of epigenetic control (1, 2, 9). Epigenetic regulation is not only coupled with transcription factor-mediated regulation but is also linked with upstream signaling pathways that connect external signals to gene function to shape the identity and function of macrophages during differentiation and activation. Chromatin-mediated epigenetic mechanisms also participate in innate memory-like phenomena that can either promote tolerance to a stimulus or prime cells for a more robust response.

In this article, we first discuss myeloid lineage decisions during development, with an emphasis on the emerging role of pioneer transcription factors and distal regulatory elements (enhancers) in regulating cell lineage and cell-type-specific responses in a chromatin-regulated manner. We then discuss how epigenetic regulation affects the response of macrophages to activation and classify stimulus response genes in concert with signaling pathways and transcription factors. Furthermore, we illustrate how chromatin can provide a memory of prior stimuli, and discuss the basis for extensive plasticity in heterogeneous macrophage populations present in different tissues.

EPIGENETIC PRINCIPLES IN MYELOID CELLS

The traditional definition of the term "epigenetics" refers to stably inherited changes in gene expression and phenotype that do not involve changes to the underlying DNA sequence. More recently, within the context of cell differentiation and activation, the term epigenetics is often used to connote mechanisms that stabilize gene expression even after environmental signals that regulate gene expression have resolved. These changes are not necessarily heritable or propagated across cell mitosis, and mechanistically are mediated by histone modifications and other chromatin changes and modification of DNA such as (hydroxyl)methylation (1–3). In this review, epigenetic changes typically refer to stable, chromatin-mediated alterations in the transcriptional potential of a cell.

In myeloid cells, as in all mammalian cells, nuclear DNA is wrapped around histones to form nucleosomes, which are compacted into chromatin. These interactions not only enable the marked compaction of DNA required for packaging in the nucleus but also impose a barrier to transcription. This organization generates a default state of inaccessibility, and hence, the most fundamental issue of epigenetic regulation is how to ensure access of transcription machinery to DNA despite the compact and protective chromatin organization (10). Epigenetic mechanisms are typically mediated by posttranslational modifications of histones and DNA methylation, ATP-dependent chromatin-remodeling complexes, and noncoding RNA, and function at gene regulatory regions such as promoters and enhancers (11–14). The sum total and pattern of epigenetic modifications determine accessibility of DNA, linking epigenetic regulation with transcription.

Another issue is that epigenetic marks have traditionally been considered to be stable and are a suitable mechanism to explain stable phenotypes in various differentiated cell types that express different patterns of gene expression (15). At the cellular level, signaling cascades initiated at the cell surface relay messages via effector proteins, ultimately culminating in the nucleus, where transcription factors are targeted to induce or shut down a particular gene expression signature. This event may be transient in nature, for example, when a cell needs to respond acutely to an event and then return to its previous steady state. However, the development, health, and/or adaptation of an organism often require an environmental signal to be converted into a long-lived phenotypic change. Stabilization of cell phenotypes by epigenetic mechanisms includes changes in chromatin landscapes in response to external stimuli that control gene expression (10, 14, 15). Epigenetic modifications that persist after the original stimulus could provide a mechanism for extending transient, short-lived signals into a more stable and sustained cellular response lasting several hours or days. This epigenetic landscape, in turn, can be reprogrammed in response to subsequent stimulation for activation and polarization. Such reprogramming of the epigenetic landscape helps integrate signaling over time and maintains distinct cell fates to respond rapidly and properly to subsequent stimuli (10, 14, 15).

Investigation of the epigenetics of macrophages to date has focused primarily on posttranslational modification of histones. More than 60 different residues on histone tails are known to be posttranslationally modified by various covalent modifications such as acetylation, methylation, phosphorylation, ubiquitination, sumoylation, ADP-ribosylation, and others, and there are more than 100 possible histone modifications identified to date (16). These histone marks are "written" and "erased" by enzymes that exhibit specificity for particular histone amino acid residues and marks. Some histone marks, such as H4 lysine 16 acetylation (H4K16ac) and H2B ubiquitination (H2Bub), can directly alter nucleosome structure, and these marks generally increase nucleosome mobility along the DNA strand and accessibility of DNA to transcription factors, which correlate with gene expression (17, 18). Other histone marks, such as H3 lysine 4 methylation (H3K4me), contribute to a sort of "code" that is "read" and interpreted by additional chromatin regulators that recognize and bind to these marks and serve as "signaling platforms" for recruitment of additional regulatory proteins (14). DNA sequence-specific transcription factors and interacting coactivators/corepressors function in concert with chromatin regulators to determine the rates of transcription initiation and elongation, and orchestrate changes in the transcriptional program. However, it is often not clear whether these histone modifications play a causal role in regulating transcription or are passive by-products or markers of transcription, although it has been shown that H3 lysine 4 trimethylation (H3K4me3) or H3 lysine 27 acetylation (H3K27ac) at promoters directly facilitates RNA polymerase (Pol) II recruitment and polymerase initiation complex formation *in vitro* in cell-free systems (19, 20).

In addition to histone modifications, chromatin remodelers, such as the SWI/SNF family of proteins in humans, can "open" chromatin through nucleosome sliding, eviction, or breathing where the associated DNA is unwrapped to expose regulatory sites (12). In this way, remodeling of nucleosomes can regulate the accessibility of DNA to transcription factors to control gene expression. Whereas analysis of histone modifications reveals overwhelming complexity and leaves mechanistic questions unaddressed, DNA accessibility provides a simple testable paradigm for understanding the role of nucleosomes in gene regulatory processes (13). Many high-throughput assays have been developed to take advantage of next-generation sequencing technologies in assessing the genome-wide chromatin state of mammalian cells. These include DNase I digestion followed by high-throughput sequencing (DNase-seq), formaldehyde-assisted isolation of regulatory elements (FAIRE-seq), and assay for transposase-accessible chromatin using sequencing (ATAC-seq), which reveal nucleosome-depleted regions known as open chromatin. Chromatin immunoprecipitation using specific antibodies for histone marks or transcription factors followed by sequencing (ChIP-seq) reveals genome-wide patterns of histone marks and transcription factor occupancy, known as the "cistrome" (21). On a finer scale, DNase-seq data can be used to identify binding sites of transcription factors by analyzing the distribution of the signal or "footprint" within a DNase-hypersensitive site (21, 22). ChIP experiments can be used for unbiased profiling of chromatin remodelers, transcription factors, and other proteins that are associated with DNA as well as histone modifications through enriching for bound sequences by means of a specific antibody (21). Taken together, the positioning and modifications of nucleosomes throughout the genome define the chromatin state of a cell, and genome-wide data in macrophages have rapidly accumulated. We discuss the current understanding of epigenetic regulation in macrophages and how these epigenetic principles can be applied to define the identity and function of macrophages during differentiation and activation while the cells remember previous experiences.

EPIGENETICS IN DEVELOPMENT OF MACROPHAGES

Macrophages can be classified by cell surface marker expression, ontogeny, or differential dependence on lineage-defining transcription factors and growth factor signaling, which sharply distinguish specific populations in different tissues. During cellular differentiation, common myeloid precursors undergo a series of progressive choices that eventually change their transcriptional program and chromatin, which play a critical part in the selective activation and repression of genes important in various macrophage homeostatic and stress-related functions (2, 5, 6, 23).

It is thought that expression of transcription factors called pioneer factors or master regulators determines myeloid cell lineage commitment, and the composition and function of those are likely to vary between cell types (24–26). High levels of the lineage-determining transcription factor PU.1 are important for differentiation of macrophages and neutrophils (27, 28). PU.1 instructs progenitors to upregulate myeloid-specific cell surface antigens and to downregulate other cell-specific markers and transcription factors (29). By definition, pioneer factors can bind target DNA sequences even in a silent or native chromatin environment, differing from other transcription factors, which readily bind only to already accessible DNA (30). Indeed, recent genome-wide studies in macrophages and dendritic cells (24, 31–33) indicate that a small set of myeloid lineage factors such as PU.1 and members of the CCAAT/enhancer-binding protein (C/EBP) and activator protein 1 (AP1) families function in a collaborative manner to shape the genome and establish a large fraction of the enhancer-like elements. Their binding to tens of thousands of chromatin sites is established during differentiation, and PU.1 genomic distribution in differentiated macrophages and B cells shows highly distinct patterns

(31, 34–36). Accordingly, PU.1 is associated with nearly all genomic enhancers marked by H3K4me1 (31, 35, 36). Moreover, reexpression of PU.1 in PU.1-deficient myeloid precursors, or even in fibroblasts, leads to the local deposition of H3K4me1 and increased accessibility of the underlying DNA at PU.1-bound genomic regions (31, 36). Another example of pioneer factors in myeloid cells is the C/EBP family. C/EBPα can switch B cells into macrophages by increasing accessibility of relevant chromatin (37), and depending on the order of expression, C/EBPα and GATA-2 regulate the differentiation of different types of myeloid cells such as neutrophils and mast cells (26). C/EBPα-mediated conversion of primary pre-B cells into macrophages also results in the downregulation of histone deacetylase 7 (HDAC7), which inhibits the induction of key genes for macrophage function, such as inflammatory response and phagocytosis (38). In addition, natural genetic variation between inbred strains of mice in the binding sites of the PU.1 and C/EBPα impairs not only the binding of pioneer factors but also that of closely bound NF-κB in response to signals, which is consistent with a model in which pioneer factors form enhancers and set up the epigenetic landscape, which then determines the binding profiles of signaling-dependent transcription factors (35).

The enhancer landscape, established by pioneer factors, is unique to a particular cell lineage. The defining characteristic of enhancers is their ability to drive gene expression at a distance (39), and modularity of enhancer elements permits a given gene to be transcribed in multiple tissues, at varied levels, in response to different signaling cascades, and at different states during development. Recent studies of the ENCODE consortium revealed that the great majority of binding sites for transcription factors are in enhancers rather than in promoters, suggesting that enhancers are likely to be specific to the lineage and represent future regulatory potential for the lineage (40–42). Many different enhancer states can be defined by enrichment for specific combinations of histone modifications, low nucleosomal occupancy (open chromatin), co-occupancy by multiple transcription factors, binding of coactivators and RNA Pol II, and production of enhancer RNAs (eRNAs) (43–46). In contrast to promoters, which exhibit relatively high levels of H3K4me3 as compared to H3K4me1 or H3K4me2, enhancers exhibit relatively low H3K4me3 and high H3K4me1 (43–45). H3K27ac in H3K4me1-associated regions separates active from poised enhancers (47), and many transcription coregulators, such as myeloid/lymphoid or mixed-lineage leukemia (MLL) proteins, p300 and CREB-binding protein, the mammalian SWI/SNF complexes, and the Mediator complex, can promote enhancer activity (48–51). Importantly, enhancer landscapes vary between cell types and predict developmental states (52). For example, recent genome-wide studies in macrophages and dendritic cells found that enhancer landscape, defined by mapping histone modifications and a comprehensive panel of myeloid transcription factors, is unique to the cell lineage and is correlated with lineage-specific gene expression (24). In addition, tissue-resident macrophages from mouse have a large repertoire of tissue-specific enhancers, which drive tissue-specific transcriptional and functional responses by inducing the expression of divergent secondary transcription factors that collaborate with PU.1 (53, 54). In peritoneal macrophages, it has been shown that local tissue-derived retinoic acid promotes the peritoneal macrophage phenotype through the reversible induction of GATA-6 (55), implying that the local microenvironment may reprogram the tissue-dependent characteristics of enhancers and transcription factors.

Although many studies identify and/or predict enhancer position and activity using a combination of histone marks, coactivator and Pol II occupancy, and eRNA expression, major challenges are to connect enhancers with their target genes and to validate the physiological function of enhancers. Techniques that measure physical interactions between DNA loci, using variations of chromosome conformation capture (3C), allow inference of DNA looping and have been helpful in linking enhancers with their target genes. A newly developed variant of 3C, termed Hi-C, and chromatin interaction analysis by paired-end tag sequencing (ChIA-PET) allows for genome-wide, unbiased mapping of long-range interactions (56, 57). Most recent resolution of the Hi-C technology is ~0.1 to 1 Mb for mammalian genomes (58), making it possible to map enhancer-promoter interactions genome-wide. It has been shown that chromatin compartments called topologically associated domains (TADs) are largely conserved in their position between cell types and developmental states (59, 60). However, TADs can be further substratified into smaller domains, and inside TADs the chromatin structure of regions <100 kb does differ in a cell-type-specific manner (56), which implies that different enhancer-promoter interactions within TADs can be dynamically regulated in a context-dependent manner. Accordingly, a recent ChIA-PET study in embryonic stem cells revealed that large genomic domains with a high density of enhancer-associated marks, called super-enhancers, exist within TADs and interact with promoters of adjacent genes within the TAD (61). Furthermore, genome editing of putative enhancers mediated by CRISPR (clustered regularly interspaced short palindromic repeats) or TALEN (transcription activator-like effector nuclease) provides insights into the functional importance of these elements (57, 61). Finally, eRNAs have been shown to modulate the levels of nearby gene expression, potentially by facilitating enhancer-promoter interactions through chromatin looping, recruitment of cofactors, the release of the negative elongation factor complex (44, 62), and transcription-associated changes in histone marks. RNA interference-mediated inhibition of eRNAs results in reduced expression of nearby mRNAs, suggesting a direct role of these eRNAs in enhancer function and providing potential tools for validating the physiological function of enhancers (63–65).

Finally, DNA methylation has been shown to play a central role in development and function of myeloid cells. Methylation of cytosines generally silences certain genomic regions by modulating the binding of particular transcription factors to their binding sites or the binding of other factors that display different affinity for methylated cytosines. Terminally differentiated human myeloid cells show a marked hypomethylation pattern compared to lymphoid cells (66). DNA methylation levels decrease sharply as myeloid differentiation progresses, while they increase upon lymphoid commitment (66, 67). For instance, the *in vitro* differentiation of monocytes into macrophages or dendritic cells involves active DNA demethylation. Genes encoding transcription factors that are important for myeloid cell differentiation, such as C/EBPα, display low levels of DNA methylation and are enriched in both activating H3K4me3 and repressive H3K27me3 marks, known as bivalent domains (68, 69). Recent studies revealed that ten-eleven translocation (TET) enzymes catalyze the oxidation of 5-methylcytosine (5mC) to 5-hydroxymethylcytosine (5hmC) and to further oxidation products, and these oxidized nucleotides can lead to active/passive demethylation in many different ways (70, 71). Enrichment of 5hmC at the transcription start site (TSS) and gene bodies indicates the potential role of such modification in transcriptional control, and TET proteins contribute to histone modification by interacting with chromatin-modifying enzymes. TET proteins can contribute to both activating and repressive effects on gene expression, probably depending on the context and location of binding (72–74). For example, PU.1 binds TET2 and recruits it to the promoters of genes that become demethylated. PU.1 is also able to recruit DNA methyltransferase DNMT3B to target the deposition of *de novo* DNA methylation (75). Although genome-wide analysis of 5hmC has not yet been described for myeloid cells, it seems that dynamic regulation of the methylation and hydroxymethylation of cytosine at gene-specific loci during differentiation and activation of myeloid cells may contribute to modulation of the immune responses.

EPIGENETICS IN MACROPHAGE ACTIVATION

Macrophages respond to external stimulation by activating hundreds of response genes. Different transcriptional responses of cytokine and inflammatory/antiviral genes define unique macrophage biological functions. In a classic model of stimulus-dependent activation, each stimulus elicits a characteristic transcriptional response solely by activating a unique combination of signaling pathways (76, 77). Although signal transduction pathways and transcription factors are clearly central mediators of transcriptional activation, the possible roles of chromatin must also be considered. For example, nucleosome remodeling, as detected by DNase I hypersensitivity and restriction enzyme accessibility, was initially observed at the promoters of the *IL12B* and *IFNB1* genes (78, 79). Inducible histone modifications relevant to an immune response were first observed by ChIP at the human *IFNB1* gene (80). Like most plastic cells of the myeloid system, macrophages are responsive to a wide spectrum of regulatory molecules and provide a robust model system for investigation of chromatin regulation for transcriptional responses at a genome-wide level. Most data in this area have been generated by analyses of the macrophage in response to prototypical inflammatory stimuli such as Toll-like receptor (TLR) agonists.

Epigenetics in Resting States

The resting state of a macrophage is associated with its development history. An epigenetic landscape is established during differentiation, and pioneer factors bind to and open the regulatory regions that are the future binding sites of transcription factors in response to stimuli. For example, PU.1 is associated not only with macrophage-specific enhancers but also with most lipopolysaccharide (LPS)-induced enhancers in both unstimulated and stimulated macrophages (Fig. 1) (36). This finding suggests that epigenetic landscapes for inducible genes are directly targeted by PU.1, rendering the genes they control susceptible to transcriptional induction when the mature cell encounters a stimulus (31). In addition to pioneer factors, H3K4me1 enrichment as well as binding of p300 to inducible gene enhancers was usually observed in both unstimulated

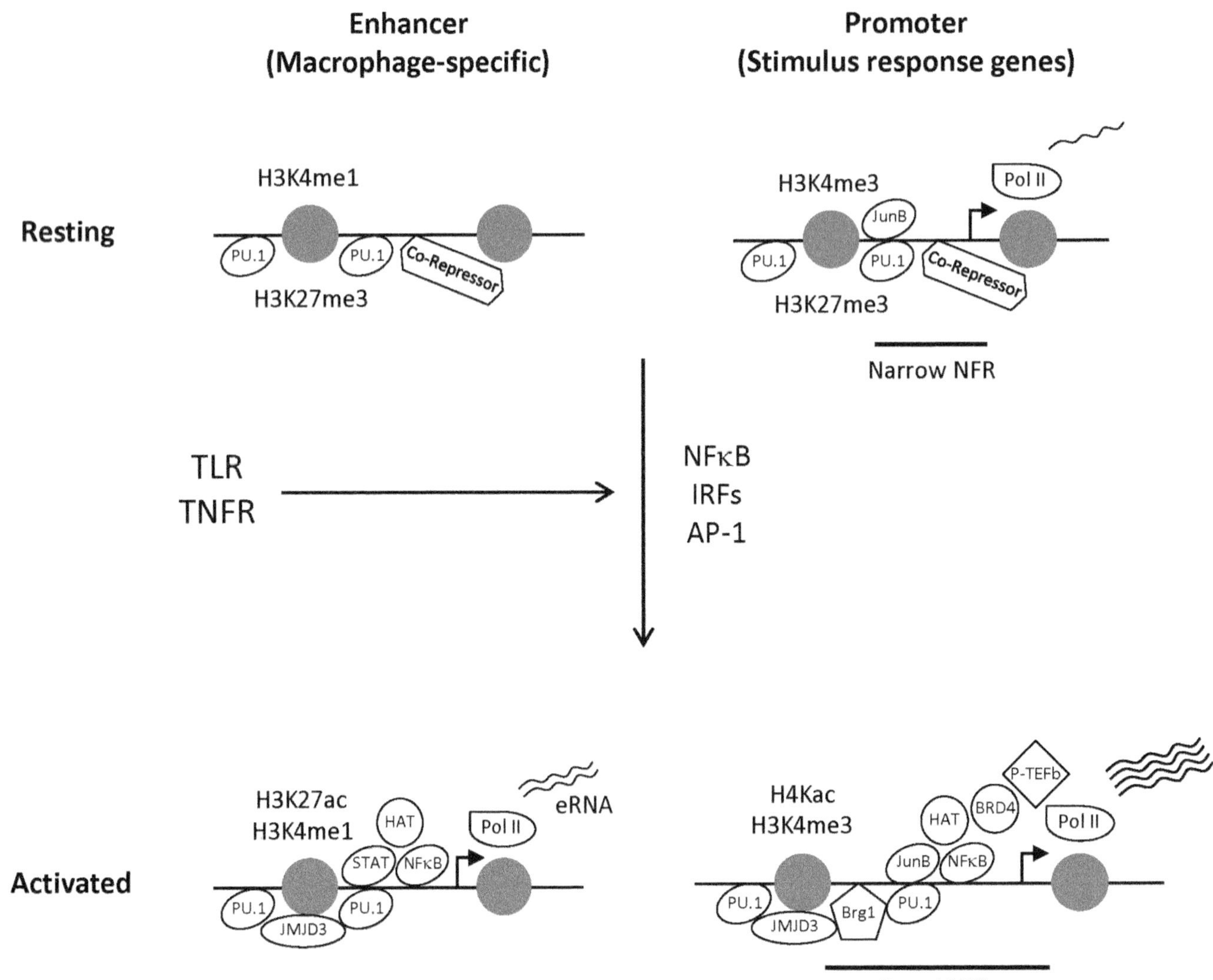

Figure 1 Epigenetic features of promoters and enhancers of inflammatory genes under basal and acute activated conditions. During macrophage differentiation, PU.1 binds to promoters and enhancers and facilitates the opening of chromatin and an increase in H3K4me3 at promoters and H3K4me1 at enhancers. The negative histone marks/corepressors also are distributed at both promoters and enhancers. There is often low or nonproductive basal transcription. Inflammatory stimulation of macrophages leads to release and loss of corepressors and negative histone marks, increased histone acetylation, additional nucleosome remodeling by BRG1, and recruitment of signaling transcription factors such as NF-κB. The subsequent recruitment of histone acetyltransferases (HATs) results in histone acetylation, which is subsequently bound by Brd4/P-TEFb complex. This results in successive rounds of Pol II elongation and active transcription. Enhancers of active genes are characterized by occupancy by HATs, H3K27ac, and low levels of transcription of eRNA. Interaction between enhancers and promoters involves structural connections that include the Mediator complex and cohesions to promote formation of the preinitiation complex, to initiate transcription. Substantial data support that many interactions between enhancers and promoters ("looping") are constitutive, although it is possible that such interactions may be enhanced by stimulation. NFR, nucleosome free region.

and stimulated macrophages (31, 36). Promoters also are marked by basal permissive histone marks such as H3K4me3 and a nucleosome-depleted region upstream of the TSS. There is often basal low-level transcription of TLR-inducible genes, and the promoters of primary response genes are occupied at baseline by Pol II that is paused in the vicinity of the TSS (81, 82). Basal transcription rates can be set by prebound primer factors, including JunB, activating transcription factor 3 (ATF3), and interferon regulatory factor 4 (IRF4), which may also recruit additional factors after cell stimulation (24). The negative histone marks, such as H3K9me3 (83), H3K27me3 (84), and H4K20me3 (85), also are distributed at promoters of inflammatory genes. Histone H3K9me, which generally correlates with transcriptional repression, is readily apparent at the promoters of some but not all inducible genes in unstimulated human dendritic cells (86). In turn, the activation of cells was associated with the rapid loss of H3K9me at these promoters. However, macrophages and dendritic cells display relatively very low levels of H3K9me2 at antiviral and antimicrobial genes as compared to fibroblasts, myocytes, or neurons (83), suggesting a potential explanation of their professional role in immune defense. H3K27me, another repressive histone modification, which acts by recruiting polycomb complexes, has also been identified at a selective subset of inducible genes prior to their activation (84, 87). H3K27me3 is downregulated at immune response genes after stimulation by the action of Jmjd3, a JmjC family histone demethylase, which is required for activation of a subset of TLR4-dependent genes (84). Finally, many corepressors that act prior and/or subsequent to stimulus responses restrain inflammatory cytokines in the basal state. These include occupancy of gene loci by repressors such as B-cell lymphoma 6 (BCL-6) and nuclear receptors that recruit corepressor complexes that contain HDACs and histone demethylases that limit the amount of positive histone marks (88). BCL-6 has been shown to corepress many LPS-inducible genes, such that its loss results in hypersensitivity to proinflammatory signals (88). The Nuclear receptor corepressor (NCoR1)-HDAC3 corepressor contributes to maintaining low H3K9/14 acetylation levels under resting conditions (89) and contains the histone methyltransferase SET and MYND domain-containing protein 5 (SMYD5), which contributes to repression by catalyzing H4K20me3, inhibiting the expression of TLR4 target genes (85). The binding of these corepressors to distinct subsets of target genes in resting cells has the potential to selectively regulate their activation.

Epigenetics in Acute Activation

In macrophages, TLR stimulation results in the release of the aforementioned epigenetic brakes, such as dismissal of BCL-6 and corepressors from gene loci and concomitant induction or activation of demethylases that erase the negative histone marks H3K27me3, H3K9me3, and H4K20me3 (Fig. 1) (84, 85, 87, 88). In addition to histone modifications, induction of a subset of genes that includes *IL6* and *IL12B* requires nucleosome remodeling by the SWI/SNF complexes (82, 90). This epigenetic remodeling facilitates recruitment of signaling transcription factors such as NF-κB, an increase in positive histone marks such H4ac and H3K4me3, and release of paused Pol II (81) to promote transcription elongation. Although low levels of precursor transcripts are constitutively produced at these genes, NF-κB and possibly other inducible factors are needed to enhance the efficiency of transcription elongation and pre-mRNA processing, in addition to enhancing the frequency of transcription initiation (81, 91). These inducible factors promote H4 acetylation, which is then recognized by the bromodomain-containing adaptor protein Brd4. Brd4 recruits the positive transcription elongation factor (P-TEFb), which promotes elongation and pre-mRNA processing through its ability to phosphorylate the C-terminal domain of RNA Pol II (81).

Enhancers also are activated, as demonstrated by recruitment of p300, increased histone acetylation, binding of signaling transcription factors, and transcription of eRNA (Fig. 1) (31, 36, 92, 93). Although some enhancers can be activated solely by pioneer factors and/or lineage-dependent transcription factors, signal-dependent transcription factors will be required for other enhancers to be fully activated (35, 94, 95). Signal-dependent transcription factors frequently activate common sets of genes in different cell types but can also regulate gene expression in a cell-type-specific manner. For example, a large set of enhancers appears to be specifically activated upon LPS stimulation, which is commonly associated with p300 and its homolog CBP (36). They promote transcriptional activation by acetylating histones and transcription factors and possibly also by mechanisms unrelated to acetylation that have not yet been clarified. Interestingly, ~ 90% of the binding of the p65 subunit of NF-κB occurs at enhancers that are already poised (where chromatin is in an accessible state), whereas the remainder is associated with the *de novo* induction of latent enhancers in collaboration with PU.1 and C/EBPα (35, 95). In addition, modulation of the enhancer repertoire and transcriptional induction in response to gamma interferon

(IFN-γ) or interleukin-4 (IL-4) was almost completely abolished in STAT1 or STAT6 knockout macrophages (94), implicating signal-activated transcription factors such as STATs in enhancer remodeling. However, little is known about the mechanisms by which the enhancer-bound transcription factors and the chromatin state of enhancers influence the recruitment of the general transcription machinery and RNA Pol II to the promoter.

Furthermore, early loss-of-function studies have revealed variable requirements of nucleosome remodeling by the SWI/SNF complexes at distinct LPS-induced genes in mouse macrophages (90). A majority of primary response genes were induced in a SWI/SNF-independent manner, whereas other primary response genes and most secondary response genes required signal-induced SWI/SNF complexes for their transcriptional induction (82). Interestingly, a substantial subset of SWI/SNF-dependent, LPS-induced primary response genes was found to require IRF3 for activation, and the promoters for these genes usually contain consensus IRF3-binding sites (82). Inducible nucleosome remodeling at these promoters, analyzed by restriction enzyme accessibility, was absent in LPS-stimulated macrophages from *IRF3* knockout mice, demonstrating that IRF3 promotes nucleosome remodeling, either directly or indirectly, at this select subset of LPS-induced primary response genes. Taken together, the above results suggest that the nucleosomes found at the promoters of selectively regulated genes serve as a barrier to transcriptional activation, such that these genes will be activated only by stimuli capable of inducing factors that bind nucleosome-embedded sites and recruit SWI/SNF complexes to their promoters. Although it seems clear that enhancers determine gene selectivity, little is known about the role of promoters in determining gene selectivity. Interestingly, recent genome-wide studies showed that broad H3K4me3 domains (the top 5%) preferentially mark cell identity and function genes in a given cell type. The broadest H3K4me3 domains also have more paused Pol II at their promoters and enhanced transcriptional consistency rather than increased transcriptional levels (96). On promoters of active genes, H3K4me1, a well-established feature of enhancers, was associated with transcriptional silencing in macrophages, embryonic fibroblasts, and embryonic stem cells together with H3K27me3 and H4K20me1, suggesting an unexpected role of H3K4me1 in restricting the recruitment of chromatin modifiers to defined regions within promoters (97). Cap analysis of gene expression (CAGE) across a large collection of primary cell types has revealed that many mammalian promoters are composite entities composed of multiple closely separated TSSs, with independent cell-type-specific expression profiles (98). In addition, ChIA-PET results show that some promoters were actually acting like enhancers (99), suggesting a more complex layer of promoter-dependent regulation.

EPIGENETICS IN MAINTAINING DISTINCT FUNCTIONAL STATES UPON ACTIVATION

With respect to distinct transcriptional responses in the innate immune system, the chromatin-dependent mechanisms described above that contribute to macrophage activation and development may reflect an ability to adapt to different environments and maintain functional states in order to properly respond to subsequent challenges. Activation of macrophages plays a key role in tissue homeostasis, disease pathogenesis, and resolving and nonresolving inflammation and is involved in the outcome of many diseases, including autoimmune diseases, metabolic diseases, and bacterial and viral infections (4, 5, 8, 100). It has been well established that macrophages change their activation states in response to growth factors (e.g., colony-stimulating factor-1 [CSF-1] and granulocyte-macrophage CSF [GM-CSF]) and external cues, such as cytokines, microbes, microbial products, and other modulators. Activated macrophages polarize toward various functional phenotypes depending on the pathogen and cytokines expressed in the microenvironment, with a spectrum of activation states commonly observed (4, 5, 8, 100). Recently it has become clear that the signaling pathways and transcription factors important for distinct macrophage activation states induces epigenetic changes, as exemplified by alterations in histone modifications and chromatin accessibility (101). Epigenetic regulation could provide a suitable mechanism to explain how transient signals are transformed into more sustained patterns of functional distinct gene expression, and how the epigenetic landscape of the macrophage "remembers" its history of differentiation and previous environmental stimulation. Epigenetically conferred transcriptional memory provides the molecular basis for integration of various polarizing signals into functional distinct states and for reprogramming of macrophages for altered responses to subsequent environmental challenges.

Epigenetics in Macrophage Polarization

Macrophage polarization states are defined by the inducing stimulus and by the ensuing patterns of gene expression, which determine function. *In vivo*, macrophage phenotype is heterogeneous, and multiple polarization states have been described. These states exist on a spectrum of overlapping phenotypes and gene expression patterns related to the original classification of M1 and M2 macrophages (4, 5, 10). Although it is unclear how the epigenetic landscape differently regulates macrophage polarization to explain spectrum of activation, recent studies have revealed that epigenetic changes contribute to macrophage polarization in classical M1 and M2 models, two extreme ends of a functional spectrum.

Notably, macrophage polarization has been shown to be associated with histone modifiers such as writers, erasers, and readers of histone marks (Fig. 2). The histone methyltransferase MLL was recently shown to be a regulator of M1 activation (102). MLL expression increases upon stimulation with IFN-γ/LPS, M1 polarizing factors, and inhibition of MLL reduces the induction of the M1 gene *CXCL10*. In addition, H3K4 methylation at *de novo* enhancers is primarily dependent on the histone methyltransferases MLL1, MLL2/4, and MLL3, and H4K20me3 by SMYD5 restricts the expression of TLR4 target genes in macrophages (95). Two histone mark erasers, Jmjd3 and HDAC3, also have been shown to regulate M1/M2 polarization. Expression of a subset of LPS-induced genes is strongly decreased in Jmjd3-deficient macrophages (87), and Jmjd3/UTX inhibition reduces LPS-induced inflammatory cytokine production in human macrophages (103). Interestingly, Jmjd3 is also induced by IL-4 (104) and was shown to be crucial for M2 polarization and immune responses against helminth infections through IRF4 (105). In addition, H3K27me3 demethylation in the *Nfatc1* locus by Jmjd3 promotes receptor activator of NF-κB ligand (RANKL)-induced osteoclast differentiation (106), suggesting that Jmjd3 allows various responses to different external stimuli and regulates macrophage polarization in a context-dependent manner. Another histone eraser, HDAC3, promotes M1 responses and at the same time acts as a brake for M2 polarization. Macrophages lacking HDAC3 show an M2-like phenotype in the absence of external stimuli and are hyperresponsive to IL-4 (107). HDAC3 is required for the activation of hundreds of mainly STAT1-dependent, inflammatory genes in M1 macrophages, and this was shown to be dependent on defective IFN-β responses in macrophages (108). In contrast to histone deacetylation by HDAC3, bromodomain and extraterminal domain family (BET) proteins, such as the Brd2 to -4 readers of histone acetylation, are crucial for inducing inflammatory genes (81, 109), and the BET protein inhibitors I-BET and JQ1 block macrophage inflammatory responses (109, 110). A more recent study showed that I-BET also suppresses RANKL-induced

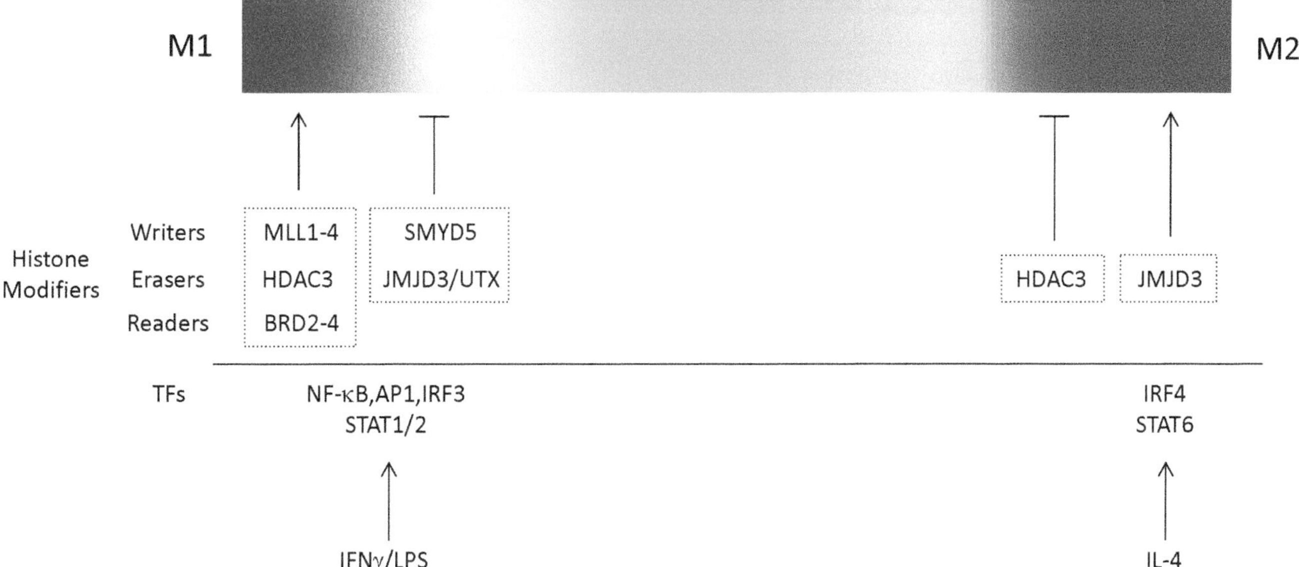

Figure 2 Epigenetic regulators of macrophage polarization. A color spectrum illustrates different polarized macrophage populations, showing the linear scale of two macrophage designations, M1 and M2. Histone modifiers implicated in M1 and M2 macrophage polarization are indicated. TF, transcription factor.

osteoclastogenesis by inhibiting the MYC-NFAT axis (111), suggesting that epigenetic modulation broadly regulates macrophage polarization and differentiation.

TLR-induced expression of core M1 inflammatory cytokines is transient, and gene expression is rapidly repressed to near baseline levels (112). In contrast to activation, less is known about mechanisms of gene repression. Nuclear receptors, TLR-induced transcriptional repressors ATF3 and hairy and enhancer of split 1 (Hes1), feedback inhibitors induced by IL-10, and the p50 NF-κB subunit can recruit corepressor complexes that contain HDACs and histone demethylases and decrease gene expression (113, 114). However, the precise mechanisms of action of these repressors and how chromatin states are regulated during deactivation of M1 inflammatory genes are not known. Histone marks can be long-lived, and the extent to which positive marks are removed, and whether negative marks are installed, during gene repression is not clear. In addition, there is evidence for a repressive role for the nucleosome remodeling and deacetylase (NuRD), which presumably shifts nucleosomes to a configuration that limits access of gene loci to transcription factors and general transcriptional machinery (90, 115, 116). Interestingly, deactivation of cytokine gene expression is delayed by the M1-promoting cytokines IFN-γ and GM-CSF, which work in part by suppressing expression of transcriptional repressors such as Hes1 (117).

EPIGENETICS IN PRIMING MACROPHAGE ACTIVATION STATES

Of note, priming by IFNs enables macrophages to be fully activated with augmented and sustained expression of inflammatory cytokine genes in response to pathogen-associated molecular patterns and inflammatory cytokines. Basal IFN-β production under homeostatic conditions can systemically calibrate type I IFN response and maintain macrophages in a primed state of increased readiness to respond rapidly and strongly to infectious challenges (118). Recent reports show that in mice the microbiome, comprising commensal microorganisms that colonize body surfaces, promotes a partial and low-grade M1-like phenotype in macrophages throughout the body, including those in lymphoid organs (119, 120). This M1-like priming of macrophages induces chromatin remodeling with increased H3K4me3 marks at *Ifnb*, *Il6*, and *Tnf* promoters, which is associated with increased binding of NF-κB p65, IRF3, and Pol II upon cell stimulation (119, 120). Interestingly, poising of IFN response genes is targeted by the influenza A protein NS1, which mimics H3K4-containing peptides and thereby suppresses the positive functions of H3K4me3 by blocking interactions with the human Pol II-associated factor 1 (PAF1) transcription elongation complex (hPAF1C), the reader of this epigenetic mark (121).

IFN-γ is the most potent M1-activating cytokine and promotes a TLR-induced classical inflammatory activation state. IFN-γ and STAT1 have been implicated in nucleosome remodeling and opening of chromatin (122). Recent genome-wide studies of human macrophages revealed that IFN-γ induced stable and coordinated recruitment of STAT1, IRF1, and associated histone acetylation to enhancers and promoters of genes that are synergistically activated by IFN-γ and LPS, such as *TNF*, *IL6*, and *IL12B* (123, 124). Importantly, CRISPR-mediated deletion established the functional importance of IFN-γ-primed enhancer at the *TNF* locus. In T cells, p300 recruitment to a large percentage of enhancers in T-bet-deficient cells was dependent on the STAT1/4 proteins, indicating the potential role of STAT1 in shaping enhancer landscape independently of pioneer factors (125). Overall, these results suggest that the IFN-mediated STAT pathway may create a primed chromatin environment to augment TLR-induced gene transcription to explain a synergy mechanism in concert with signaling pathways.

EPIGENETICS IN MEMORY-LIKE BEHAVIOR OF MONOCYTES/MACROPHAGES

It has long been known that monocytes exposed to endotoxins can enter a refractory functional state, characterized by incapacity to produce proinflammatory cytokines (126). More recently, another state termed "trained immunity" was introduced to describe how an initial infection or vaccinations could increase the responsiveness of monocytes to a secondary infection (Fig. 3) (127). Both tolerance and training represent clinically relevant functional states, such as the immune paralysis encountered during bacterial sepsis or endotoxic shock or the nonspecific protective effects of live microorganism vaccination that strongly influence susceptibility to secondary infections (126–129). Importantly, tolerance and training states are persistent, but fully reversible, after the initial challenge. They affect responses to subsequent stimulation with similar or different stimuli by "remembering" prior environmental stimulation. Thus, the essential features of such phenomena are their persistence beyond the initial stimulation and their decay over time, and in this way they can be considered short-term memories of environmental exposure. Tolerance and training are likely initial examples of a more pervasive phenomenon of the conditioning of biological systems by exposure to environmental changes. This conditioning reflects the need to adapt to the external milieu and to maintain such an adapted state after stimulus termination to avoid the adverse consequences of exaggerated, chronic, or repeated stimulation.

EPIGENETICS IN TOLERANCE

Exposure to a potent inflammatory stimulus can lead to acquired refractoriness to induction of genes encoding inflammatory molecules after subsequent stimulation (Fig. 3). This observation was first made with LPS as the stimulus and is referred to as LPS tolerance or endotoxin tolerance (126). However, tumor necrosis factor (TNF) has similarly been shown to induce tolerance (130). Tolerant macrophages exhibit a selective defect in the induction of a subset of genes, including inflammatory cytokine genes, called tolerized genes. By contrast, nontolerized genes, for example, those encoding antimicrobial products and some chemokines, are expressed (131). Many molecular mechanisms seem to contribute to tolerance, ranging from mechanisms for suppressing the transduction of inflammatory signals, to expression of microRNAs, to active repression of genes encoding inflammatory molecules through chromatin regulation (126).

Earlier studies of endotoxin-tolerant human monocytic cell lines reported RelB to direct deposition of histone H3K9me2, resulting in impaired p65 NF-κB transactivation of the *IL1B* promoter (132). Similarly, binding of high-mobility group box-1 protein (HMGB1) and histone H1 linker at the promoters of *TNF* and *IL1B* genes also suppresses chromatin remodeling (133). Recent works have clarified that epigenetic mechanisms

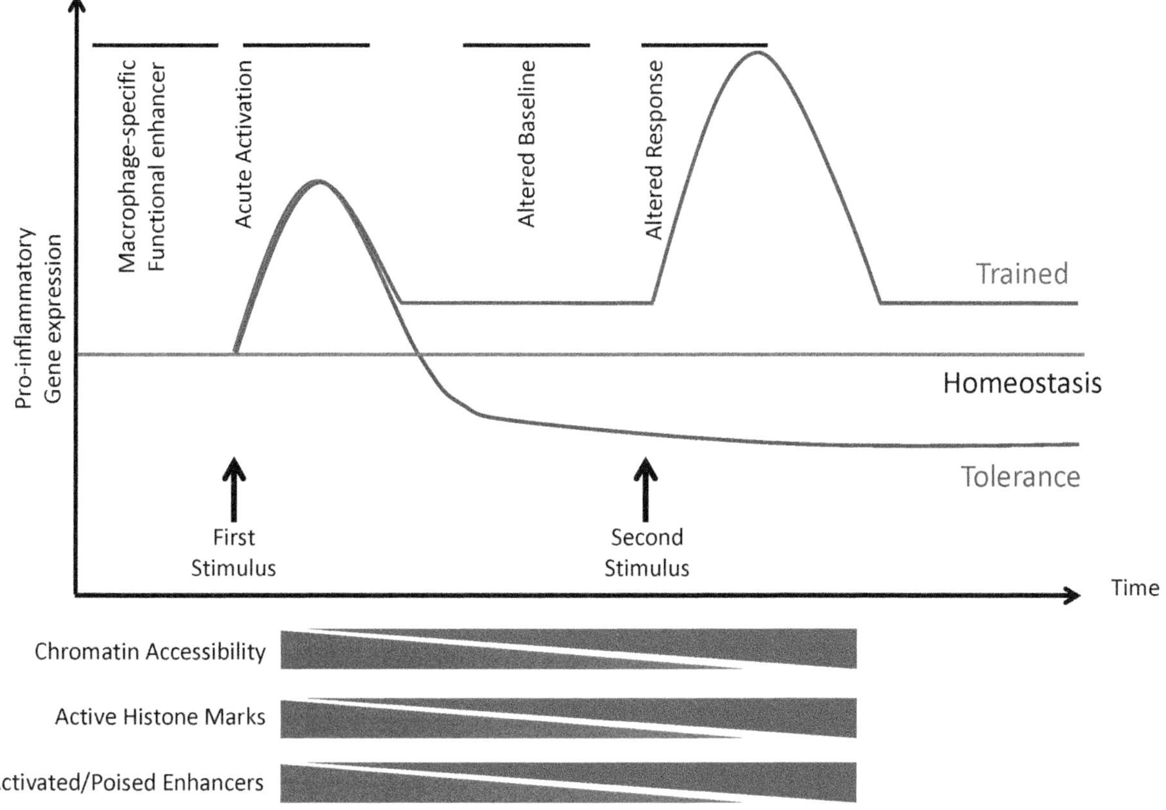

Figure 3 Epigenetic regulation of endotoxin tolerance and trained immunity in macrophages. A proposed model of endotoxin tolerance and trained immunity is depicted in the graph. The red line indicates tolerized genes that remain refractory to a second stimulus, and the blue line represents trained genes that show enhanced expression in response to subsequent stimulation. Innate immune responses during and after first stimuli can lead to epigenetic reprogramming, which translates into decreased or increased immune response to subsequent stimulation.

likely explain the gene-specific nature of tolerance. LPS-induced, TNF-induced, and sepsis-induced tolerance are governed by changes in chromatin accessibility and histone modification, more specifically at the level of H3K4me3, H3K27me2, and histone methyltransferase complexes (122, 130, 131, 134, 135). Tolerized genes exhibit decreased chromatin accessibility, as assessed by restriction enzyme accessibility assays, and diminished recruitment of transcription factors such as p65. The molecular explanation for diminished accessibility involves decreased TLR-induced recruitment of Brahma-related gene 1 (BRG1)-containing nucleosome-remodeling complexes and complex changes in histone acetylation and methylation. The signals to explain gene-specific regulation in tolerance are not clear, although newly transcribed gene products arc important to establish tolerance (131), and glycogen synthase kinase 3 (GSK3) plays a key role in TNF-induced tolerance (130). Recent motif-based analyses of TLR4-induced genes revealed that NF-κB motifs were the key regulatory elements significantly enriched in tolerizable genes, by facilitating the formation of an NCoR-HDAC3-p50 repressosome (136). Interestingly, IFN-γ prevents tolerance by preserving expression of the receptor-interacting protein 140 (RIP140) coactivator and promoting TLR-induced chromatin accessibility upon secondary TLR challenge (122, 137). By contrast, nontolerized genes maintain an open chromatin state. Nontolerized genes exhibit more H4 acetylation and maintain H3K4me3 after restimulation (131). Like the genes in naive macrophages, these genes are capable of recruiting the BRG1 and chromodomain helicase DNA-binding protein 4 (CHD4) chromatin-remodeling complexes to their promoters. Interestingly, nontolerized genes are induced with enhanced kinetics and magnitude in tolerant macrophages. Differences in their sensitivity to inhibitors of protein synthesis, such as cycloheximide, showed that a subset of nontolerized genes that were secondary response genes in naive macrophages was converted into primary response genes in tolerant macrophages (131). IFN-γ also altered the transcriptional requirements for secondary tolerized *IL6* gene expression, and converted *IL6* into a primary response gene in tolerant monocytes by increasing *IL6* promoter accessibility in the absence of new protein synthesis (122). These results suggest that removal of a requirement for new protein synthesis for chromatin remodeling represents one mechanism by which a secondary tolerized gene can become converted into a primary nontolerized gene by IFN-γ. Epigenetic mechanisms that regulate polarization of macrophages to a tolerized state need to be further clarified by genome-wide analysis of chromatin landscape and by identification of key transcription factors and chromatin-remodeling complexes that regulate the tolerization process. It will be interesting to determine whether defects in establishing tolerance states contribute to chronic inflammation and disease.

EPIGENETICS IN TRAINED IMMUNITY

Conversely, trained macrophages have been described as cells responsible for nonspecific protection against reinfection independently of adaptive immunity, and increased production of proinflammatory cytokines is characteristic of trained macrophages (Fig. 3) (127). It has been shown that infections with bacteria, yeasts, or viruses (e.g., *Candida albicans*, BCG, or herpesvirus) are capable of conferring a beneficial protection against reinfection with a second, unrelated pathogen, due to the upregulation of the basal level of innate immunity such as macrophage activation, as well as heightened TNF and IFN-γ production (138, 139). Interestingly, two recent studies have shown that helminth coinfection resulted in impaired antiviral immunity and was associated with STAT6-dependent, helminth-induced alternative activation of macrophages (140, 141), suggesting that infection with one pathogen predisposes the host to respond to subsequent infection, and macrophage phenotypes and the local cytokine environment may have a role in immunomodulation for protective immune responses.

It has been shown that *C. albicans*- and β-glucan-dependent functional reprogramming of monocytes requires the C-type lectin receptor Dectin-1 and the noncanonical Raf-1 pathway (142), whereas protection associated with BCG vaccination/training is induced through NOD2 (nucleotide-binding oligomerization domain-containing protein 2) recognition of peptidoglycans (143). *In vivo*, training of both human and murine monocytes/macrophages was associated with enhanced H3K4me3 levels but not changes in H3K27me3. These genome-wide modifications in H3K4me3 correlate with changes in gene expression. Of interest, the H3K4me3 signature is already apparent after 24 h of β-glucan incubation and H3K4me3 patterns do not diminish after 7 days. More recently, it has been demonstrated that β-glucan priming specifically induced about 3,000 distal regulatory elements with distinct motif profiles for transcription factors in DNase I-hypersensitive sites at dynamic epigenomic regions (144). Notably, β-glucan-mediated training of monocytes also leads to epigenetic remodeling at genes involved in the mTOR pathway and glycolysis. Accordingly, trained immunity could not be induced in monocytes from patients deficient in Dectin-1 or in monocytes treated with inhibitors of AKT, mTOR, or hypoxia-inducible factor-1α (HIF1α) (145). β-Glucan-trained monocytes displayed reduced oxygen consumption and increased glucose consumption, which is consistent with a switch from oxidative metabolism to glycolysis.

THE ROLE OF LATENT ENHANCERS FOR MEMORY-LIKE BEHAVIOR

In line with memory-like behavior for selective gene expression, another mechanistic insight recently emerged from an analysis of enhancers for LPS-induced genes in macrophages that do not exhibit an H3K4me1 mark in the basal state (94). Although most LPS-induced enhancers possess the properties described above, with PU.1 binding, nucleosome remodeling and open chromatin, and H3K4me1 deposition, a significant subset of enhancers appears to remain unmarked prior to LPS exposure. LPS stimulation induces the acquisition of H3K4me1/H3K4me3 and H3K27ac at several thousand regulatory regions that are completely unmarked in naive cells (94). Importantly, when the stimulus has ceased, H3K27ac and H3K4me3 disappear, whereas H3K4me1 persists at a new class of enhancers, termed latent enhancers, despite the release of PU.1 and the partner transcription factors that contributed to initial activation. The maintenance of this modification correlates with more rapid induction of enhancer acetylation in response to a subsequent stimulus. Stimulation of macrophages by a variety of stimuli such as IFN-γ and IL-4 also induces the acquisition of different patterns of latent enhancers, suggesting that stimulus-specific latent enhancers may modulate gene expression in response to a subsequent stimulus. Similar to tolerance and training, these chromatin features may represent a chromatin-mediated memory mechanism, whereby the H3K4me1 mark is maintained in the absence of the transcription factors responsible for its initial deposition.

CONCLUDING REMARKS

Macrophages continuously sample the environment and quickly react to it. Thus, the differentiation and functional specialization of macrophages must be tightly regulated to ensure that these cells execute their proper function. In this review, we have attempted to emphasize the dynamic role of epigenetic regulation in macrophages with respect to differentiation, activation, and memory-like behaviors on many levels and over broad time frames. Recent genome-wide studies provide chromatin dynamics at extraordinary resolution and testify that chromatin is far more dynamic than was appreciated previously. We described how individual or combinations of pioneer factors establish the resting enhancer landscape throughout differentiation and determine various responses to activation. Genome-wide mapping of diverse enhancer features, their functional relationship with promoters, and their ultimate transcriptional outputs have provided a number of striking discoveries, ranging from the identification of the large number of enhancers to the widespread production of eRNAs. We discussed that macrophage function, upon activation, is coordinated by regulation of the chromatin landscape in the context of enhancers, promoters, and rapid response genes. Preexisting chromatin marks deposited during differentiation interpret, calibrate, and transmit environmental signals to determine the magnitude and specificity of gene expression. In addition, epigenetic control mechanisms may have a role in providing a flexible mechanism for the stable preactivation of cells that ensures macrophages mount the correct level of response under inflammatory conditions. Finally, we presented the epigenetic mechanisms governing the polarization of macrophages and the alternative perspectives of chromatin state mediating memory in trained or tolerant states. Epigenetic regulation provides stable, yet reversible, marks that bestow flexibility on these processes. Evidence reviewed here suggests that epigenetic changes fundamentally reprogram macrophages to exhibit altered gene expression programs in response to environmental stimuli. Such reprogramming would allow transcriptional memory to shape macrophage phenotype in response to environmental changes and may contribute to the complex macrophage phenotypes. Thus, the epigenetic mechanisms that are supposed to control the memory of the environmental impact may also contribute to the persistence of disease-associated phenotypes, particularly in sustaining chronic inflammation and in mediating the interactions of genes and environment that lead to disease. Of note, many single nucleotide polymorphisms associated with autoimmune/inflammatory diseases, affecting chromatin states, are concentrated in regulatory regions that are subjected to epigenetic regulation (45, 146–148). In this context, it would be interesting to consider the possibility of treating chronic inflammatory states by targeting specific epigenetic modifications. Investigation of epigenetic regulation in myeloid cells is still at an early stage, and there are many exciting areas for future research. As advances in technology rapidly increase the complexity of interacting transcription factors, *cis*-regulatory elements, chromatin regulators, and posttranslational modifications of histones and DNA methylation, it is a major challenging issue to decipher global regulatory processes on a mechanistic level and integrate new findings with known genomic features. Finally, despite many features conserved between the human and mouse systems (149, 150), there are important known species differences in myeloid cells such as macrophages and dendritic cells (149–153). Systematically determining the similarities and differences between mouse and human regulatory networks in myeloid cells will help to interpret biomedical insights derived from research performed on mouse models.

Citation. Ivashkiv LB, Park SH. 2016. Epigenetic regulation of myeloid cells. Microbiol Spectrum 4(3):MCHD-0010-2015.

References

1. **Smale ST, Tarakhovsky A, Natoli G.** 2014. Chromatin contributions to the regulation of innate immunity. *Annu Rev Immunol* **32:**489–511.
2. **Winter DR, Amit I.** 2014. The role of chromatin dynamics in immune cell development. *Immunol Rev* **261:** 9–22.
3. **Heinz S, Romanoski CE, Benner C, Glass CK.** 2015. The selection and function of cell type-specific enhancers. *Nat Rev Mol Cell Biol* **16:**144–154.
4. **Gordon S, Martinez FO.** 2010. Alternative activation of macrophages: mechanism and functions. *Immunity* **32:**593–604.
5. **Murray PJ, Allen JE, Biswas SK, Fisher EA, Gilroy DW, Goerdt S, Gordon S, Hamilton JA, Ivashkiv LB, Lawrence T, Locati M, Mantovani A, Martinez FO, Mege JL, Mosser DM, Natoli G, Saeij JP, Schultze JL, Shirey KA, Sica A, Suttles J, Udalova I, van Ginderachter JA, Vogel SN, Wynn TA.** 2014. Macrophage activation and polarization: nomenclature and experimental guidelines. *Immunity* **41:**14–20.
6. **Merad M, Sathe P, Helft J, Miller J, Mortha A.** 2013. The dendritic cell lineage: ontogeny and function of dendritic cells and their subsets in the steady state and the inflamed setting. *Annu Rev Immunol* **31:**563–604.
7. **Ginhoux F, Jung S.** 2014. Monocytes and macrophages: developmental pathways and tissue homeostasis. *Nat Rev Immunol* **14:**392–404.
8. **Sica A, Mantovani A.** 2012. Macrophage plasticity and polarization: in vivo veritas. *J Clin Invest* **122:**787–795.
9. **Xue J, Schmidt SV, Sander J, Draffehn A, Krebs W, Quester I, De Nardo D, Gohel TD, Emde M, Schmidleithner L, Ganesan H, Nino-Castro A, Mallmann MR, Labzin L, Theis H, Kraut M, Beyer M, Latz E, Freeman TC, Ulas T, Schultze JL.** 2014. Transcriptome-based network analysis reveals a spectrum model of human macrophage activation. *Immunity* **40:**274–288.
10. **Li B, Carey M, Workman JL.** 2007. The role of chromatin during transcription. *Cell* **128:**707–719.
11. **Sadeh R, Allis CD.** 2011. Genome-wide "re"-modeling of nucleosome positions. *Cell* **147:**263–266.
12. **Becker PB, Workman JL.** 2013. Nucleosome remodeling and epigenetics. *Cold Spring Harb Perspect Biol* **5:** a017905. doi:10.1101/cshperspect.a017905.
13. **Henikoff S, Shilatifard A.** 2011. Histone modification: cause or cog? *Trends Genet* **27:**389–396.
14. **Badeaux AI, Shi Y.** 2013. Emerging roles for chromatin as a signal integration and storage platform. *Nat Rev Mol Cell Biol* **14:**211–224.
15. **Monticelli S, Natoli G.** 2013. Short-term memory of danger signals and environmental stimuli in immune cells. *Nat Immunol* **14:**777–784.
16. **Kouzarides T.** 2007. Chromatin modifications and their function. *Cell* **128:**693–705.
17. **Shogren-Knaak M, Ishii H, Sun JM, Pazin MJ, Davie JR, Peterson CL.** 2006. Histone H4-K16 acetylation controls chromatin structure and protein interactions. *Science* **311:**844–847.
18. **Fierz B, Chatterjee C, McGinty RK, Bar-Dagan M, Raleigh DP, Muir TW.** 2011. Histone H2B ubiquitylation disrupts local and higher-order chromatin compaction. *Nat Chem Biol* **7:**113–119.
19. **Lauberth SM, Nakayama T, Wu X, Ferris AL, Tang Z, Hughes SH, Roeder RG.** 2013. H3K4me3 interactions with TAF3 regulate preinitiation complex assembly and selective gene activation. *Cell* **152:**1021–1036.
20. **Stasevich TJ, Hayashi-Takanaka Y, Sato Y, Maehara K, Ohkawa Y, Sakata-Sogawa K, Tokunaga M, Nagase T, Nozaki N, McNally JG, Kimura H.** 2014. Regulation of RNA polymerase II activation by histone acetylation in single living cells. *Nature* **516:**272–275.
21. **Meyer CA, Liu XS.** 2014. Identifying and mitigating bias in next-generation sequencing methods for chromatin biology. *Nat Rev Genet* **15:**709–721.
22. **Thurman RE, Rynes E, Humbert R, Vierstra J, Maurano MT, Haugen E, Sheffield NC, Stergachis AB, Wang H, Vernot B, Garg K, John S, Sandstrom R, Bates D, Boatman L, Canfield TK, Diegel M, Dunn D, Ebersol AK, Frum T, Giste E, Johnson AK, Johnson EM, Kutyavin T, Lajoie B, Lee BK, Lee K, London D, Lotakis D, Neph S, Neri F, Nguyen ED, Qu H, Reynolds AP, Roach V, Safi A, Sanchez ME, Sanyal A, Shafer A, Simon JM, Song L, Vong S, Weaver M, Yan Y, Zhang Z, Zhang Z, Lenhard B, Tewari M, Dorschner MO, Hansen RS, Navas PA, Stamatoyannopoulos G, Iyer VR, Lieb JD, Sunyaev SR, Akey JM, Sabo PJ, Kaul R, Furey TS, Dekker J, Crawford GE, Stamatoyannopoulos JA.** 2012. The accessible chromatin landscape of the human genome. *Nature* **489:**75–82.
23. **Geissmann F, Manz MG, Jung S, Sieweke MH, Merad M, Ley K.** 2010. Development of monocytes, macrophages, and dendritic cells. *Science* **327:**656–661.
24. **Garber M, Yosef N, Goren A, Raychowdhury R, Thielke A, Guttman M, Robinson J, Minie B, Chevrier N, Itzhaki Z, Blecher-Gonen R, Bornstein C, Amann-Zalcenstein D, Weiner A, Friedrich D, Meldrim J, Ram O, Cheng C, Gnirke A, Fisher S, Friedman N, Wong B, Bernstein BE, Nusbaum C, Hacohen N, Regev A, Amit I.** 2012. A high-throughput chromatin immunoprecipitation approach reveals principles of dynamic gene regulation in mammals. *Mol Cell* **47:**810–822.

25. Mullen AC, Orlando DA, Newman JJ, Lovén J, Kumar RM, Bilodeau S, Reddy J, Guenther MG, DeKoter RP, Young RA. 2011. Master transcription factors determine cell-type-specific responses to TGF-β signaling. *Cell* 147:565–576.

26. Graf T, Enver T. 2009. Forcing cells to change lineages. *Nature* 462:587–594.

27. Scott EW, Simon MC, Anastasi J, Singh H. 1994. Requirement of transcription factor PU.1 in the development of multiple hematopoietic lineages. *Science* 265: 1573–1577.

28. Scott EW, Fisher RC, Olson MC, Kehrli EW, Simon MC, Singh H. 1997. PU.1 functions in a cell-autonomous manner to control the differentiation of multipotential lymphoid-myeloid progenitors. *Immunity* 6:437–447.

29. Nerlov C, Graf T. 1998. PU.1 induces myeloid lineage commitment in multipotent hematopoietic progenitors. *Genes Dev* 12:2403–2412.

30. Zaret KS, Carroll JS. 2011. Pioneer transcription factors: establishing competence for gene expression. *Genes Dev* 25:2227–2241.

31. Heinz S, Benner C, Spann N, Bertolino E, Lin YC, Laslo P, Cheng JX, Murre C, Singh H, Glass CK. 2010. Simple combinations of lineage-determining transcription factors prime *cis*-regulatory elements required for macrophage and B cell identities. *Mol Cell* 38:576–589.

32. Pham TH, Minderjahn J, Schmidl C, Hoffmeister H, Schmidhofer S, Chen W, Längst G, Benner C, Rehli M. 2013. Mechanisms of *in vivo* binding site selection of the hematopoietic master transcription factor PU.1. *Nucleic Acids Res* 41:6391–6402.

33. Schonheit J, Kuhl C, Gebhardt ML, Klett FF, Riemke P, Scheller M, Huang G, Naumann R, Leutz A, Stocking C, Priller J, Andrade-Navarro MA, Rosenbauer F. 2013. PU.1 level-directed chromatin structure remodeling at the *8* gene drives dendritic cell commitment. *Cell Rep* 3:1617–1628.

34. Escoubet-Lozach L, Benner C, Kaikkonen MU, Lozach J, Heinz S, Spann NJ, Crotti A, Stender J, Ghisletti S, Reichart D, Cheng CS, Luna R, Ludka C, Sasik R, Garcia-Bassets I, Hoffmann A, Subramaniam S, Hardiman G, Rosenfeld MG, Glass CK. 2011. Mechanisms establishing TLR4-responsive activation states of inflammatory response genes. *PLoS Genet* 7:e1002401. doi: 10.1371/journal.pgen.1002401.

35. Heinz S, Romanoski CE, Benner C, Allison KA, Kaikkonen MU, Orozco LD, Glass CK. 2013. Effect of natural genetic variation on enhancer selection and function. *Nature* 503:487–492.

36. Ghisletti S, Barozzi I, Mietton F, Polletti S, De Santa F, Venturini E, Gregory L, Lonie L, Chew A, Wei CL, Ragoussis J, Natoli G. 2010. Identification and characterization of enhancers controlling the inflammatory gene expression program in macrophages. *Immunity* 32: 317–328.

37. Di Stefano B, Sardina JL, van Oevelen C, Collombet S, Kallin EM, Vicent GP, Lu J, Thieffry D, Beato M, Graf T. 2014. C/EBPα poises B cells for rapid reprogramming into induced pluripotent stem cells. *Nature* 506:235–239.

38. Barneda-Zahonero B, Román-González L, Collazo O, Rafati H, Islam AB, Bussmann LH, di Tullio A, De Andres L, Graf T, López-Bigas N, Mahmoudi T, Parra M. 2013. HDAC7 is a repressor of myeloid genes whose downregulation is required for transdifferentiation of pre-B cells into macrophages. *PLoS Genet* 9: e1003503. doi:10.1371/journal.pgen.1003503.

39. Calo E, Wysocka J. 2013. Modification of enhancer chromatin: what, how, and why? *Mol Cell* 49:825–837.

40. ENCODE Project Consortium. 2012. An integrated encyclopedia of DNA elements in the human genome. *Nature* 489:57–74.

41. Gerstein MB, Kundaje A, Hariharan M, Landt SG, Yan KK, Cheng C, Mu XJ, Khurana E, Rozowsky J, Alexander R, Min R, Alves P, Abyzov A, Addleman N, Bhardwaj N, Boyle AP, Cayting P, Charos A, Chen DZ, Cheng Y, Clarke D, Eastman C, Euskirchen G, Frietze S, Fu Y, Gertz J, Grubert F, Harmanci A, Jain P, Kasowski M, Lacroute P, Leng J, Lian J, Monahan H, O'Geen H, Ouyang Z, Partridge EC, Patacsil D, Pauli F, Raha D, Ramirez L, Reddy TE, Reed B, Shi M, Slifer T, Wang J, Wu L, Yang X, Yip KY, Zilberman-Schapira G, Batzoglou S, Sidow A, Farnham PJ, Myers RM, Weissman SM, Snyder M. 2012. Architecture of the human regulatory network derived from ENCODE data. *Nature* 489:91–100.

42. Neph S, Vierstra J, Stergachis AB, Reynolds AP, Haugen E, Vernot B, Thurman RE, John S, Sandstrom R, Johnson AK, Maurano MT, Humbert R, Rynes E, Wang H, Vong S, Lee K, Bates D, Diegel M, Roach V, Dunn D, Neri J, Schafer A, Hansen RS, Kutyavin T, Giste E, Weaver M, Canfield T, Sabo P, Zhang M, Balasundaram G, Byron R, MacCoss MJ, Akey JM, Bender MA, Groudine M, Kaul R, Stamatoyannopoulos JA. 2012. An expansive human regulatory lexicon encoded in transcription factor footprints. *Nature* 489: 83–90.

43. Ernst J, Kheradpour P, Mikkelsen TS, Shoresh N, Ward LD, Epstein CB, Zhang X, Wang L, Issner R, Coyne M, Ku M, Durham T, Kellis M, Bernstein BE. 2011. Mapping and analysis of chromatin state dynamics in nine human cell types. *Nature* 473:43–49.

44. Andersson R, Gebhard C, Miguel-Escalada I, Hoof I, Bornholdt J, Boyd M, Chen Y, Zhao X, Schmidl C, Suzuki T, Ntini E, Arner E, Valen E, Li K, Schwarzfischer L, Glatz D, Raithel J, Lilje B, Rapin N, Bagger FO, Jørgensen M, Andersen PR, Bertin N, Rackham O, Burroughs AM, Baillie JK, Ishizu Y, Shimizu Y, Furuhata E, Maeda S, Negishi Y, Mungall CJ, Meehan TF, Lassmann T, Itoh M, Kawaji H, Kondo N, Kawai J, Lennartsson A, Daub CO, Heutink P, Hume DA, Jensen TH, Suzuki H, Hayashizaki Y, Müller F, FANTOM Consortium, Forrest AR, Carninci P, Rehli M, Sandelin A. 2014. An atlas of active enhancers across human cell types and tissues. *Nature* 507:455–461.

45. Roadmap Epigenomics Consortium, Kundaje A, Meuleman W, Ernst J, Bilenky M, Yen A, Heravi-Moussavi A, Kheradpour P, Zhang Z, Wang J, Ziller MJ, Amin V, Whitaker JW, Schultz MD, Ward LD, Sarkar A, Quon G, Sandstrom RS, Eaton ML, Wu YC, Pfenning AR, Wang X, Claussnitzer M, Liu Y, Coarfa C, Harris RA, Shoresh N,

Epstein CB, Gjoneska E, Leung D, Xie W, Hawkins RD, Lister R, Hong C, Gascard P, Mungall AJ, Moore R, Chuah E, Tam A, Canfield TK, Hansen RS, Kaul R, Sabo PJ, Bansal MS, Carles A, Dixon JR, Farh KH, Feizi S, Karlic R, Kim AR, et al. 2015. Integrative analysis of 111 reference human epigenomes. *Nature* **518**:317–330.

46. **Wang P, Xue Y, Han Y, Lin L, Wu C, Xu S, Jiang Z, Xu J, Liu Q, Cao X.** 2014. The STAT3-binding long noncoding RNA lnc-DC controls human dendritic cell differentiation. *Science* **344**:310–313.

47. **Creyghton MP, Cheng AW, Welstead GG, Kooistra T, Carey BW, Steine EJ, Hanna J, Lodato MA, Frampton GM, Sharp PA, Boyer LA, Young RA, Jaenisch R.** 2010. Histone H3K27ac separates active from poised enhancers and predicts developmental state. *Proc Natl Acad Sci U S A* **107**:21931–21936.

48. **Herz HM, Hu D, Shilatifard A.** 2014. Enhancer malfunction in cancer. *Mol Cell* **53**:859–866.

49. **Wang Z, Zang C, Cui K, Schones DE, Barski A, Peng W, Zhao K.** 2009. Genome-wide mapping of HATs and HDACs reveals distinct functions in active and inactive genes. *Cell* **138**:1019–1031.

50. **Morris SA, Baek S, Sung MH, John S, Wiench M, Johnson TA, Schiltz RL, Hager GL.** 2014. Overlapping chromatin-remodeling systems collaborate genome wide at dynamic chromatin transitions. *Nat Struct Mol Biol* **21**:73–81.

51. **Kagey MH, Newman JJ, Bilodeau S, Zhan Y, Orlando DA, van Berkum NL, Ebmeier CC, Goossens J, Rahl PB, Levine SS, Taatjes DJ, Dekker J, Young RA.** 2010. Mediator and cohesin connect gene expression and chromatin architecture. *Nature* **467**:430–435.

52. **Lara-Astiaso D, Weiner A, Lorenzo-Vivas E, Zaretsky I, Jaitin DA, David E, Keren-Shaul H, Mildner A, Winter D, Jung S, Friedman N, Amit I.** 2014. Immunogenetics. Chromatin state dynamics during blood formation. *Science* **345**:943–949.

53. **Lavin Y, Winter D, Blecher-Gonen R, David E, Keren-Shaul H, Merad M, Jung S, Amit I.** 2014. Tissue-resident macrophage enhancer landscapes are shaped by the local microenvironment. *Cell* **159**:1312–1326.

54. **Gosselin D, Link VM, Romanoski CE, Fonseca GJ, Eichenfield DZ, Spann NJ, Stender JD, Chun HB, Garner H, Geissmann F, Glass CK.** 2014. Environment drives selection and function of enhancers controlling tissue-specific macrophage identities. *Cell* **159**:1327–1340.

55. **Okabe Y, Medzhitov R.** 2014. Tissue-specific signals control reversible program of localization and functional polarization of macrophages. *Cell* **157**:832–844.

56. **Jin F, Li Y, Dixon JR, Selvaraj S, Ye Z, Lee AY, Yen CA, Schmitt AD, Espinoza CA, Ren B.** 2013. A high-resolution map of the three-dimensional chromatin interactome in human cells. *Nature* **503**:290–294.

57. **Kieffer-Kwon KR, Tang Z, Mathe E, Qian J, Sung MH, Li G, Resch W, Baek S, Pruett N, Grøntved L, Vian L, Nelson S, Zare H, Hakim O, Reyon D, Yamane A, Nakahashi H, Kovalchuk AL, Zou J, Joung JK, Sartorelli V, Wei CL, Ruan X, Hager GL, Ruan Y, Casellas R.** 2013. Interactome maps of mouse gene regulatory domains reveal basic principles of transcriptional regulation. *Cell* **155**:1507–1520.

58. **Rao SS, Huntley MH, Durand NC, Stamenova EK, Bochkov ID, Robinson JT, Sanborn AL, Machol I, Omer AD, Lander ES, Aiden EL.** 2014. A 3D map of the human genome at kilobase resolution reveals principles of chromatin looping. *Cell* **159**:1665–1680.

59. **Dixon JR, Selvaraj S, Yue F, Kim A, Li Y, Shen Y, Hu M, Liu JS, Ren B.** 2012. Topological domains in mammalian genomes identified by analysis of chromatin interactions. *Nature* **485**:376–380.

60. **Ghavi-Helm Y, Klein FA, Pakozdi T, Ciglar L, Noordermeer D, Huber W, Furlong EE.** 2014. Enhancer loops appear stable during development and are associated with paused polymerase. *Nature* **512**:96–100.

61. **Dowen JM, Fan ZP, Hnisz D, Ren G, Abraham BJ, Zhang LN, Weintraub AS, Schuijers J, Lee TI, Zhao K, Young RA.** 2014. Control of cell identity genes occurs in insulated neighborhoods in mammalian chromosomes. *Cell* **159**:374–387.

62. **Schaukowitch K, Joo JY, Liu X, Watts JK, Martinez C, Kim TK.** 2014. Enhancer RNA facilitates NELF release from immediate early genes. *Mol Cell* **56**:29–42.

63. **Lam MT, Cho H, Lesch HP, Gosselin D, Heinz S, Tanaka-Oishi Y, Benner C, Kaikkonen MU, Kim AS, Kosaka M, Lee CY, Watt A, Grossman TR, Rosenfeld MG, Evans RM, Glass CK.** 2013. Rev-Erbs repress macrophage gene expression by inhibiting enhancer-directed transcription. *Nature* **498**:511–515.

64. **Li W, Notani D, Ma Q, Tanasa B, Nunez E, Chen AY, Merkurjev D, Zhang J, Ohgi K, Song X, Oh S, Kim HS, Glass CK, Rosenfeld MG.** 2013. Functional roles of enhancer RNAs for oestrogen-dependent transcriptional activation. *Nature* **498**:516–520.

65. **Ilott NE, Heward JA, Roux B, Tsitsiou E, Fenwick PS, Lenzi L, Goodhead I, Hertz-Fowler C, Heger A, Hall N, Donnelly LE, Sims D, Lindsay MA.** 2014. Long non-coding RNAs and enhancer RNAs regulate the lipopolysaccharide-induced inflammatory response in human monocytes. *Nat Commun* **5**:3979. doi:10.1038/ncomms4979.

66. **Ji H, Ehrlich LI, Seita J, Murakami P, Doi A, Lindau P, Lee H, Aryee MJ, Irizarry RA, Kim K, Rossi DJ, Inlay MA, Serwold T, Karsunky H, Ho L, Daley GQ, Weissman IL, Feinberg AP.** 2010. Comprehensive methylome map of lineage commitment from haematopoietic progenitors. *Nature* **467**:338–342.

67. **Zilbauer M, Rayner TF, Clark C, Coffey AJ, Joyce CJ, Palta P, Palotie A, Lyons PA, Smith KG.** 2013. Genome-wide methylation analyses of primary human leukocyte subsets identifies functionally important cell-type-specific hypomethylated regions. *Blood* **122**:e52–e60. doi:10.1182/blood-2013-05-503201.

68. **Klug M, Schmidhofer S, Gebhard C, Andreesen R, Rehli M.** 2013. 5-Hydroxymethylcytosine is an essential intermediate of active DNA demethylation processes in primary human monocytes. *Genome Biol* **14**:R46. doi:10.1186/gb-2013-14-5-r46.

69. **Zhang X, Ulm A, Somineni HK, Oh S, Weirauch MT, Zhang HX, Chen X, Lehn MA, Janssen EM, Ji H.**

2014. DNA methylation dynamics during *ex vivo* differentiation and maturation of human dendritic cells. *Epigenetics Chromatin* 7:21. doi:10.1186/1756-8935-7-21.

70. Tahiliani M, Koh KP, Shen Y, Pastor WA, Bandukwala H, Brudno Y, Agarwal S, Iyer LM, Liu DR, Aravind L, Rao A. 2009. Conversion of 5-methylcytosine to 5-hydroxymethylcytosine in mammalian DNA by MLL partner TET1. *Science* **324**:930–935.

71. Ito S, Shen L, Dai Q, Wu SC, Collins LB, Swenberg JA, He C, Zhang Y. 2011. Tet proteins can convert 5-methylcytosine to 5-formylcytosine and 5-carboxylcytosine. *Science* 333:1300–1303.

72. Ficz G, Branco MR, Seisenberger S, Santos F, Krueger F, Hore TA, Marques CJ, Andrews S, Reik W. 2011. Dynamic regulation of 5-hydroxymethylcytosine in mouse ES cells and during differentiation. *Nature* **473**: 398–402.

73. Williams K, Christensen J, Pedersen MT, Johansen JV, Cloos PA, Rappsilber J, Helin K. 2011. TET1 and hydroxymethylcytosine in transcription and DNA methylation fidelity. *Nature* **473**:343–348.

74. Wu H, D'Alessio AC, Ito S, Wang Z, Cui K, Zhao K, Sun YE, Zhang Y. 2011. Genome-wide analysis of 5-hydroxymethylcytosine distribution reveals its dual function in transcriptional regulation in mouse embryonic stem cells. *Genes Dev* **25**:679–684.

75. de la Rica L, Rodríguez-Ubreva J, García M, Islam AB, Urquiza JM, Hernando H, Christensen J, Helin K, Gómez-Vaquero C, Ballestar E. 2013. PU.1 target genes undergo Tet2-coupled demethylation and DNMT3b-mediated methylation in monocyte-to-osteoclast differentiation. *Genome Biol* **14**:R99. doi:10.1186/gb-2013-14-9-r99.

76. Medzhitov R, Horng T. 2009. Transcriptional control of the inflammatory response. *Nat Rev Immunol* **9**: 692–703.

77. Smale ST, Natoli G. 2014. Transcriptional control of inflammatory responses. *Cold Spring Harb Perspect Biol* **6**:a016261. doi:10.1101/cshperspect.a016261.

78. Weinmann AS, Plevy SE, Smale ST. 1999. Rapid and selective remodeling of a positioned nucleosome during the induction of IL-12 p40 transcription. *Immunity* **11**: 665–675.

79. Agalioti T, Lomvardas S, Parekh B, Yie J, Maniatis T, Thanos D. 2000. Ordered recruitment of chromatin modifying and general transcription factors to the IFN-β promoter. *Cell* **103**:667–678.

80. Parekh BS, Maniatis T. 1999. Virus infection leads to localized hyperacetylation of histones H3 and H4 at the IFN-β promoter. *Mol Cell* **3**:125–129.

81. Hargreaves DC, Horng T, Medzhitov R. 2009. Control of inducible gene expression by signal-dependent transcriptional elongation. *Cell* **138**:129–145.

82. Ramirez-Carrozzi VR, Braas D, Bhatt DM, Cheng CS, Hong C, Doty KR, Black JC, Hoffmann A, Carey M, Smale ST. 2009. A unifying model for the selective regulation of inducible transcription by CpG islands and nucleosome remodeling. *Cell* **138**:114–128.

83. Fang TC, Schaefer U, Mecklenbrauker I, Stienen A, Dewell S, Chen MS, Rioja I, Parravicini V, Prinjha RK, Chandwani R, MacDonald MR, Lee K, Rice CM, Tarakhovsky A. 2012. Histone H3 lysine 9 di-methylation as an epigenetic signature of the interferon response. *J Exp Med* **209**:661–669.

84. De Santa F, Totaro MG, Prosperini E, Notarbartolo S, Testa G, Natoli G. 2007. The histone H3 lysine-27 demethylase Jmjd3 links inflammation to inhibition of polycomb-mediated gene silencing. *Cell* **130**:1083–1094.

85. Stender JD, Pascual G, Liu W, Kaikkonen MU, Do K, Spann NJ, Boutros M, Perrimon N, Rosenfeld MG, Glass CK. 2012. Control of proinflammatory gene programs by regulated trimethylation and demethylation of histone H4K20. *Mol Cell* **48**:28–38.

86. Saccani S, Natoli G. 2002. Dynamic changes in histone H3 Lys 9 methylation occurring at tightly regulated inducible inflammatory genes. *Genes Dev* **16**:2219–2224.

87. De Santa F, Narang V, Yap ZH, Tusi BK, Burgold T, Austenaa L, Bucci G, Caganova M, Notarbartolo S, Casola S, Testa G, Sung WK, Wei CL, Natoli G. 2009. Jmjd3 contributes to the control of gene expression in LPS-activated macrophages. *EMBO J* **28**:3341–3352.

88. Barish GD, Yu RT, Karunasiri M, Ocampo CB, Dixon J, Benner C, Dent AL, Tangirala RK, Evans RM. 2010. Bcl-6 and NF-κB cistromes mediate opposing regulation of the innate immune response. *Genes Dev* **24**:2760–2765.

89. Ogawa S, Lozach J, Jepsen K, Sawka-Verhelle D, Perissi V, Sasik R, Rose DW, Johnson RS, Rosenfeld MG, Glass CK. 2004. A nuclear receptor corepressor transcriptional checkpoint controlling activator protein 1-dependent gene networks required for macrophage activation. *Proc Natl Acad Sci U S A* **101**:14461–14466.

90. Ramirez-Carrozzi VR, Nazarian AA, Li CC, Gore SL, Sridharan R, Imbalzano AN, Smale ST. 2006. Selective and antagonistic functions of SWI/SNF and Mi-2β nucleosome remodeling complexes during an inflammatory response. *Genes Dev* **20**:282–296.

91. Amir-Zilberstein L, Ainbinder E, Toube L, Yamaguchi Y, Handa H, Dikstein R. 2007. Differential regulation of NF-κB by elongation factors is determined by core promoter type. *Mol Cell Biol* **27**:5246–5259.

92. Jin F, Li Y, Ren B, Natarajan R. 2011. PU.1 and C/EBPα synergistically program distinct response to NF-κB activation through establishing monocyte specific enhancers. *Proc Natl Acad Sci U S A* **108**:5290–5295.

93. De Santa F, Barozzi I, Mietton F, Ghisletti S, Polletti S, Tusi BK, Muller H, Ragoussis J, Wei CL, Natoli G. 2010. A large fraction of extragenic RNA Pol II transcription sites overlap enhancers. *PLoS Biol* **8**: e1000384. doi:10.1371/journal.pbio.1000384.

94. Ostuni R, Piccolo V, Barozzi I, Polletti S, Termanini A, Bonifacio S, Curina A, Prosperini E, Ghisletti S, Natoli G. 2013. Latent enhancers activated by stimulation in differentiated cells. *Cell* **152**:157–171.

95. Kaikkonen MU, Spann NJ, Heinz S, Romanoski CE, Allison KA, Stender JD, Chun HB, Tough DF, Prinjha

RK, Benner C, Glass CK. 2013. Remodeling of the enhancer landscape during macrophage activation is coupled to enhancer transcription. *Mol Cell* **51**:310–325.

96. **Benayoun BA, Pollina EA, Ucar D, Mahmoudi S, Karra K, Wong ED, Devarajan K, Daugherty AC, Kundaje AB, Mancini E, Hitz BC, Gupta R, Rando TA, Baker JC, Snyder MP, Cherry JM, Brunet A.** 2014. H3K4me3 breadth is linked to cell identity and transcriptional consistency. *Cell* **158**:673–688.

97. **Cheng J, Blum R, Bowman C, Hu D, Shilatifard A, Shen S, Dynlacht BD.** 2014. A role for H3K4 monomethylation in gene repression and partitioning of chromatin readers. *Mol Cell* **53**:979–992.

98. **FANTOM Consortium and the RIKEN PMI and CLST (DGT), Forrest AR, Kawaji H, Rehli M, Baillie JK, de Hoon MJ, Haberle V, Lassman T, Kulakovskiy IV, Lizio M, Itoh M, Andersson R, Mungall CJ, Meehan TF, Schmeier S, Bertin N, Jorgensen M, Dimont E, Arner E, Schmidl C, Schaefer U, Medvedeva YA, Plessy C, Vitezic M, Severin J, Semple C, Ishizu Y, Young RS, Francescatto M, Alam I, Albanese D, Altschuler GM, Arakawa T, Archer JA, Arner P, Babina M, Rennie S, Balwierz PJ, Beckhouse AG, Pradhan-Bhatt S, Blake JA, Blumenthal A, Bodega B, Bonetti A, Briggs J, Brombacher F, Burroughs AM, Califano A, et al.** 2014. A promoter-level mammalian expression atlas. *Nature* **507**:462–470.

99. **Li G, Ruan X, Auerbach RK, Sandhu KS, Zheng M, Wang P, Poh HM, Goh Y, Lim J, Zhang J, Sim HS, Peh SQ, Mulawadi FH, Ong CT, Orlov YL, Hong S, Zhang Z, Landt S, Raha D, Euskirchen G, Wei CL, Ge W, Wang H, Davis C, Fisher-Aylor KI, Mortazavi A, Gerstein M, Gingeras T, Wold B, Sun Y, Fullwood MJ, Cheung E, Liu E, Sung WK, Snyder M, Ruan Y.** 2012. Extensive promoter-centered chromatin interactions provide a topological basis for transcription regulation. *Cell* **148**:84–98.

100. **Mosser DM, Edwards JP.** 2008. Exploring the full spectrum of macrophage activation. *Nat Rev Immunol* **8**: 958–969.

101. **Lawrence T, Natoli G.** 2011. Transcriptional regulation of macrophage polarization: enabling diversity with identity. *Nat Rev Immunol* **11**:750–761.

102. **Kittan NA, Allen RM, Dhaliwal A, Cavassani KA, Schaller M, Gallagher KA, Carson WF IV, Mukherjee S, Grembecka J, Cierpicki T, Jarai G, Westwick J, Kunkel SL, Hogaboam CM.** 2013. Cytokine induced phenotypic and epigenetic signatures are key to establishing specific macrophage phenotypes. *PLoS One* **8**:e78045. doi:10.1371/journal.pone.0078045.

103. **Kruidenier L, Chung CW, Cheng Z, Liddle J, Che K, Joberty G, Bantscheff M, Bountra C, Bridges A, Diallo H, Eberhard D, Hutchinson S, Jones E, Katso R, Leveridge M, Mander PK, Mosley J, Ramirez-Molina C, Rowland P, Schofield CJ, Sheppard RJ, Smith JE, Swales C, Tanner R, Thomas P, Tumber A, Drewes G, Oppermann U, Patel DJ, Lee K, Wilson DM.** 2012. A selective jumonji H3K27 demethylase inhibitor modulates the proinflammatory macrophage response. *Nature* **488**:404–408.

104. **Ishii M, Wen H, Corsa CA, Liu T, Coelho AL, Allen RM, Carson WF IV, Cavassani KA, Li X, Lukacs NW, Hogaboam CM, Dou Y, Kunkel SL.** 2009. Epigenetic regulation of the alternatively activated macrophage phenotype. *Blood* **114**:3244–3254.

105. **Satoh T, Takeuchi O, Vandenbon A, Yasuda K, Tanaka Y, Kumagai Y, Miyake T, Matsushita K, Okazaki T, Saitoh T, Honma K, Matsuyama T, Yui K, Tsujimura T, Standley DM, Nakanishi K, Nakai K, Akira S.** 2010. The Jmjd3-Irf4 axis regulates M2 macrophage polarization and host responses against helminth infection. *Nat Immunol* **11**:936–944.

106. **Yasui T, Hirose J, Tsutsumi S, Nakamura K, Aburatani H, Tanaka S.** 2011. Epigenetic regulation of osteoclast differentiation: possible involvement of Jmjd3 in the histone demethylation of Nfatc1. *J Bone Miner Res* **26**: 2665–2671.

107. **Mullican SE, Gaddis CA, Alenghat T, Nair MG, Giacomin PR, Everett LJ, Feng D, Steger DJ, Schug J, Artis D, Lazar MA.** 2011. Histone deacetylase 3 is an epigenomic brake in macrophage alternative activation. *Genes Dev* **25**:2480–2488.

108. **Chen X, Barozzi I, Termanini A, Prosperini E, Recchiuti A, Dalli J, Mietton F, Matteoli G, Hiebert S, Natoli G.** 2012. Requirement for the histone deacetylase Hdac3 for the inflammatory gene expression program in macrophages. *Proc Natl Acad Sci U S A* **109**:E2865–E2874. doi:10.1073/pnas.1121131109.

109. **Nicodeme E, Jeffrey KL, Schaefer U, Beinke S, Dewell S, Chung CW, Chandwani R, Marazzi I, Wilson P, Coste H, White J, Kirilovsky J, Rice CM, Lora JM, Prinjha RK, Lee K, Tarakhovsky A.** 2010. Suppression of inflammation by a synthetic histone mimic. *Nature* **468**: 1119–1123.

110. **Belkina AC, Nikolajczyk BS, Denis GV.** 2013. BET protein function is required for inflammation: Brd2 genetic disruption and BET inhibitor JQ1 impair mouse macrophage inflammatory responses. *J Immunol* **190**: 3670–3678.

111. **Park-Min KH, Lim E, Lee MJ, Park SH, Giannopoulou E, Yarilina A, van der Meulen M, Zhao B, Smithers N, Witherington J, Lee K, Tak PP, Prinjha RK, Ivashkiv LB.** 2014. Inhibition of osteoclastogenesis and inflammatory bone resorption by targeting BET proteins and epigenetic regulation. *Nat Commun* **5**:5418. doi: 10.1038/ncomms6418.

112. **Ivashkiv LB.** 2011. Inflammatory signaling in macrophages: transitions from acute to tolerant and alternative activation states. *Eur J Immunol* **41**:2477–2481.

113. **Ivashkiv LB.** 2013. Epigenetic regulation of macrophage polarization and function. *Trends Immunol* **34**:216–223.

114. **Murray PJ, Smale ST.** 2012. Restraint of inflammatory signaling by interdependent strata of negative regulatory pathways. *Nat Immunol* **13**:916–924.

115. **Roger T, Lugrin J, Le Roy D, Goy G, Mombelli M, Koessler T, Ding XC, Chanson AL, Reymond MK, Miconnet I, Schrenzel J, François P, Calandra T.** 2011. Histone deacetylase inhibitors impair innate immune responses to Toll-like receptor agonists and to infection. *Blood* **117**:1205–1217.

116. **Shimizu-Hirota R, Xiong W, Baxter BT, Kunkel SL, Maillard I, Chen XW, Sabeh F, Liu R, Li XY, Weiss SJ.**

2012. MT1-MMP regulates the PI3Kδ·Mi-2/NuRD-dependent control of macrophage immune function. *Genes Dev* **26**:395–413.

117. **Hu X, Ivashkiv LB.** 2009. Cross-regulation of signaling pathways by interferon-γ: implications for immune responses and autoimmune diseases. *Immunity* **31**:539–550.

118. **Ivashkiv LB, Donlin LT.** 2014. Regulation of type I interferon responses. *Nat Rev Immunol* **14**:36–49.

119. **Abt MC, Osborne LC, Monticelli LA, Doering TA, Alenghat T, Sonnenberg GF, Paley MA, Antenus M, Williams KL, Erikson J, Wherry EJ, Artis D.** 2012. Commensal bacteria calibrate the activation threshold of innate antiviral immunity. *Immunity* **37**:158–170.

120. **Ganal SC, Sanos SL, Kallfass C, Oberle K, Johner C, Kirschning C, Lienenklaus S, Weiss S, Staeheli P, Aichele P, Diefenbach A.** 2012. Priming of natural killer cells by nonmucosal mononuclear phagocytes requires instructive signals from commensal microbiota. *Immunity* **37**:171–186.

121. **Marazzi I, Ho JS, Kim J, Manicassamy B, Dewell S, Albrecht RA, Seibert CW, Schaefer U, Jeffrey KL, Prinjha RK, Lee K, García-Sastre A, Roeder RG, Tarakhovsky A.** 2012. Suppression of the antiviral response by an influenza histone mimic. *Nature* **483**:428–433.

122. **Chen J, Ivashkiv LB.** 2010. IFN-γ abrogates endotoxin tolerance by facilitating Toll-like receptor-induced chromatin remodeling. *Proc Natl Acad Sci U S A* **107**: 19438–19443.

123. **Qiao Y, Giannopoulou EG, Chan CH, Park SH, Gong S, Chen J, Hu X, Elemento O, Ivashkiv LB.** 2013. Synergistic activation of inflammatory cytokine genes by interferon-γ-induced chromatin remodeling and Toll-like receptor signaling. *Immunity* **39**:454–469.

124. **Chow NA, Jasenosky LD, Goldfeld AE.** 2014. A distal locus element mediates IFN-γ priming of lipopolysaccharide-stimulated *TNF* gene expression. *Cell Rep* **9**:1718–1728.

125. **Vahedi G, Takahashi H, Nakayamada S, Sun HW, Sartorelli V, Kanno Y, O'Shea JJ.** 2012. STATs shape the active enhancer landscape of T cell populations. *Cell* **151**:981–993.

126. **Biswas SK, Lopez-Collazo E.** 2009. Endotoxin tolerance: new mechanisms, molecules and clinical significance. *Trends Immunol* **30**:475–487.

127. **Netea MG, Quintin J, van der Meer JW.** 2011. Trained immunity: a memory for innate host defense. *Cell Host Microbe* **9**:355–361.

128. **Shalova IN, Lim JY, Chittezhath M, Zinkernagel AS, Beasley F, Hernández-Jiménez E, Toledano V, Cubillos-Zapata C, Rapisarda A, Chen J, Duan K, Yang H, Poidinger M, Melillo G, Nizet V, Arnalich F, López-Collazo E, Biswas SK.** 2015. Human monocytes undergo functional re-programming during sepsis mediated by hypoxia-inducible factor-1α. *Immunity* **42**:484–498.

129. **Benn CS, Netea MG, Selin LK, Aaby P.** 2013. A small jab—a big effect: nonspecific immunomodulation by vaccines. *Trends Immunol* **34**:431–439.

130. **Park SH, Park-Min KH, Chen J, Hu X, Ivashkiv LB.** 2011. Tumor necrosis factor induces GSK3 kinase-mediated cross-tolerance to endotoxin in macrophages. *Nat Immunol* **12**:607–615.

131. **Foster SL, Hargreaves DC, Medzhitov R.** 2007. Gene-specific control of inflammation by TLR-induced chromatin modifications. *Nature* **447**:972–978.

132. **Chen X, El Gazzar M, Yoza BK, McCall CE.** 2009. The NF-κB factor RelB and histone H3 lysine methyltransferase G9a directly interact to generate epigenetic silencing in endotoxin tolerance. *J Biol Chem* **284**: 27857–27865.

133. **El Gazzar M, Yoza BK, Chen X, Garcia BA, Young NL, McCall CE.** 2009. Chromatin-specific remodeling by HMGB1 and linker histone H1 silences proinflammatory genes during endotoxin tolerance. *Mol Cell Biol* **29**:1959–1971.

134. **Carson WF, Cavassani KA, Dou Y, Kunkel SL.** 2011. Epigenetic regulation of immune cell functions during post-septic immunosuppression. *Epigenetics* **6**:273–283.

135. **Wen H, Dou Y, Hogaboam CM, Kunkel SL.** 2008. Epigenetic regulation of dendritic cell-derived interleukin-12 facilitates immunosuppression after a severe innate immune response. *Blood* **111**:1797–1804.

136. **Yan Q, Carmody RJ, Qu Z, Ruan Q, Jager J, Mullican SE, Lazar MA, Chen YH.** 2012. Nuclear factor-κB binding motifs specify Toll-like receptor-induced gene repression through an inducible repressosome. *Proc Natl Acad Sci U S A* **109**:14140–14145.

137. **Ho PC, Tsui YC, Feng X, Greaves DR, Wei LN.** 2012. NF-κB-mediated degradation of the coactivator RIP140 regulates inflammatory responses and contributes to endotoxin tolerance. *Nat Immunol* **13**:379–386.

138. **Barton ES, White DW, Cathelyn JS, Brett-McClellan KA, Engle M, Diamond MS, Miller VL, Virgin HW IV.** 2007. Herpesvirus latency confers symbiotic protection from bacterial infection. *Nature* **447**:326–329.

139. **Bistoni F, Vecchiarelli A, Cenci E, Puccetti P, Marconi P, Cassone A.** 1986. Evidence for macrophage-mediated protection against lethal *Candida albicans* infection. *Infect Immun* **51**:668–674.

140. **Osborne LC, Monticelli LA, Nice TJ, Sutherland TE, Siracusa MC, Hepworth MR, Tomov VT, Kobuley D, Tran SV, Bittinger K, Bailey AG, Laughlin AL, Boucher JL, Wherry EJ, Bushman FD, Allen JE, Virgin HW, Artis D.** 2014. Coinfection. Virus-helminth coinfection reveals a microbiota-independent mechanism of immunomodulation. *Science* **345**:578–582.

141. **Reese TA, Wakeman BS, Choi HS, Hufford MM, Huang SC, Zhang X, Buck MD, Jezewski A, Kambal A, Liu CY, Goel G, Murray PJ, Xavier RJ, Kaplan MH, Renne R, Speck SH, Artyomov MN, Pearce EJ, Virgin HW.** 2014. Coinfection. Helminth infection reactivates latent gamma-herpesvirus via cytokine competition at a viral promoter. *Science* **345**:573–577.

142. **Quintin J, Saeed S, Martens JH, Giamarellos-Bourboulis EJ, Ifrim DC, Logie C, Jacobs L, Jansen T, Kullberg BJ, Wijmenga C, Joosten LA, Xavier RJ, van der Meer JW, Stunnenberg HG, Netea MG.** 2012. *Candida albicans* infection affords protection against reinfection via functional reprogramming of monocytes. *Cell Host Microbe* **12**:223–232.

143. Kleinnijenhuis J, Quintin J, Preijers F, Joosten LA, Ifrim DC, Saeed S, Jacobs C, van Loenhout J, de Jong D, Stunnenberg HG, Xavier RJ, van der Meer JW, van Crevel R, Netea MG. 2012. Bacille Calmette-Guerin induces NOD2-dependent nonspecific protection from reinfection via epigenetic reprogramming of monocytes. *Proc Natl Acad Sci U S A* **109**:17537–17542.

144. Saeed S, Quintin J, Kerstens HH, Rao NA, Aghajanirefah A, Matarese F, Cheng SC, Ratter J, Berentsen K, van der Ent MA, Sharifi N, Janssen-Megens EM, Ter Huurne M, Mandoli A, van Schaik T, Ng A, Burden F, Downes K, Frontini M, Kumar V, Giamarellos-Bourboulis EJ, Ouwehand WH, van der Meer JW, Joosten LA, Wijmenga C, Martens JH, Xavier RJ, Logie C, Netea MG, Stunnenberg HG. 2014. Epigenetic programming of monocyte-to-macrophage differentiation and trained innate immunity. *Science* **345**:1251086. doi:10.1126/science.1251086.

145. Cheng SC, Quintin J, Cramer RA, Shepardson KM, Saeed S, Kumar V, Giamarellos-Bourboulis EJ, Martens JH, Rao NA, Aghajanirefah A, Manjeri GR, Li Y, Ifrim DC, Arts RJ, van der Veer BM, Deen PM, Logie C, O'Neill LA, Willems P, van de Veerdonk FL, van der Meer JW, Ng A, Joosten LA, Wijmenga C, Stunnenberg HG, Xavier RJ, Netea MG. 2014. mTOR- and HIF-1α-mediated aerobic glycolysis as metabolic basis for trained immunity. *Science* **345**:1250684.

146. Farh KK, Marson A, Zhu J, Kleinewietfeld M, Housley WJ, Beik S, Shoresh N, Whitton H, Ryan RJ, Shishkin AA, Hatan M, Carrasco-Alfonso MJ, Mayer D, Luckey CJ, Patsopoulos NA, De Jager PL, Kuchroo VK, Epstein CB, Daly MJ, Hafler DA, Bernstein BE. 2015. Genetic and epigenetic fine mapping of causal autoimmune disease variants. *Nature* **518**:337–343.

147. 1000 Genomes Project Consortium, Abecasis GR, Auton A, Brooks LD, DePristo MA, Durbin RM, Handsaker RE, Kang HM, Marth GT, McVean GA. 2012. An integrated map of genetic variation from 1,092 human genomes. *Nature* **491**:56–65.

148. McVicker G, van de Geijn B, Degner JF, Cain CE, Banovich NE, Raj A, Lewellen N, Myrthil M, Gilad Y, Pritchard JK. 2013. Identification of genetic variants that affect histone modifications in human cells. *Science* **342**:747–749.

149. Cheng Y, Ma Z, Kim BH, Wu W, Cayting P, Boyle AP, Sundaram V, Xing X, Dogan N, Li J, Euskirchen G, Lin S, Lin Y, Visel A, Kawli T, Yang X, Patacsil D, Keller CA, Giardine B, Mouse EC, Kundaje A, Wang T, Pennacchio LA, Weng Z, Hardison RC, Snyder MP. 2014. Principles of regulatory information conservation between mouse and human. *Nature* **515**:371–375.

150. Lin S, Lin Y, Nery JR, Urich MA, Breschi A, Davis CA, Dobin A, Zaleski C, Beer MA, Chapman WC, Gingeras TR, Ecker JR, Snyder MP. 2014. Comparison of the transcriptional landscapes between human and mouse tissues. *Proc Natl Acad Sci U S A* **111**:17224–17229.

151. Shay T, Jojic V, Zuk O, Rothamel K, Puyraimond-Zemmour D, Feng T, Wakamatsu E, Benoist C, Koller D, Regev A, ImmGen Consortium. 2013. Conservation and divergence in the transcriptional programs of the human and mouse immune systems. *Proc Natl Acad Sci U S A* **110**:2946–2951.

152. Schmidt D, Wilson MD, Ballester B, Schwalie PC, Brown GD, Marshall A, Kutter C, Watt S, Martinez-Jimenez CP, Mackay S, Talianidis I, Flicek P, Odom DT. 2010. Five-vertebrate ChIP-seq reveals the evolutionary dynamics of transcription factor binding. *Science* **328**:1036–1040.

153. Seok J, Warren HS, Cuenca AG, Mindrinos MN, Baker HV, Xu W, Richards DR, McDonald-Smith GP, Gao H, Hennessy L, Finnerty CC, López CM, Honari S, Moore EE, Minei JP, Cuschieri J, Bankey PE, Johnson JL, Sperry J, Nathens AB, Billiar TR, West MA, Jeschke MG, Klein MB, Gamelli RL, Gibran NS, Brownstein BH, Miller-Graziano C, Calvano SE, Mason PH, Cobb JP, Rahme LG, Lowry SF, Maier RV, Moldawer LL, Herndon DN, Davis RW, Xiao W, Tompkins RG; Inflammation and Host Response to Injury, Large Scale Collaborative Research Program. 2013. Genomic responses in mouse models poorly mimic human inflammatory diseases. *Proc Natl Acad Sci U S A* **110**:3507–3512.

Secretion and Defense

VI

Myeloid Cells in Health and Disease: A Synthesis
Edited by Siamon Gordon

doi:10.1128/microbiolspec.MCHD-0030-2016

Gillian M. Griffiths[1]

32 Secretion from Myeloid Cells: Secretory Lysosomes

Over the last 20 years, a great deal of progress has been made in understanding the mechanisms that control the biogenesis and secretion of the modified lysosomes found in immune cells. The picture that emerges is that of a series of very successful "variations on a theme," with different combinations of related proteins interacting to provide different mechanisms for secretion in each specialized cell type. This allows cytotoxic T lymphocytes (CTLs) to provide a very focused secretion toward a single point, while mast cells and platelets give a generalized release all around the plasma membrane.

In most cell types, the lysosomes are not thought of as secretory organelles. Roles in plasma membrane repair have been identified, although this seems to involve a very small population of lysosomes, possibly those closest to the plasma membrane, as only small amounts of lysosomal hydrolases are released during this process. In contrast, both CTL and mast cell secretion can be monitored by release of lysosomal hydrolases or by appearance of lysosomal membrane proteins on the plasma membrane, revealing release of a more significant proportion of the lysosomal population. This enhanced ability to use lysosomes as a regulated secretory pathway in immune cells appears to be due to the expression of a number of key proteins that control both the delivery to and release at the plasma membrane (1, 2).

Studies of human genetic disease, and in particular, immunodeficiencies, have led to the identification of proteins required for lysosomal secretion in immune cells. The rationale for this approach emerged from a seminal observation that identified a mutation in perforin as giving rise to a familial form of hemophagocytic lymphohistiocytosis (FHL) (3). Perforin is a critical mediator of target cell destruction, able to form pores in the membranes of targets, and so it made sense that loss of perforin might give rise to a severe immunodeficiency like FHL. However, although some FHL patients were deficient in perforin, others were not, yet had the same profound immunodeficiency. This led to the idea that defects in secretion of the perforin-containing granules (the secretory lysosomes) might lead to the same disease, as perforin would no longer be delivered. This proved to be the case, and it soon emerged that mutations in Rab27a, Munc13-4, Syntaxin11, and Munc18-2 all gave rise to FHL as a result of their roles in secretory lysosome exocytosis (reviewed in reference 4). Intriguingly, Rab27a not only gave rise to FHL but was also associated with reduced pigmentation, revealing a role for Rab27a in secretion of both the secretory lysosomes of immune cells and melanosomes, a lysosome-related organelle. A similar role has also been identified for LYST, defects in which give rise

[1]Cambridge Institute for Medical Research, Cambridge Biomedical Campus, Cambridge CB2 0XY, United Kingdom.

to Chediak-Higashi syndrome, an FHL immunodeficiency with profound pigmentation defects and greatly enlarged lysosomes found in all cells.

Rab27a provided clues as to how flexible this system could be at producing variations on a secretory theme. This is because Rab27a has been found to have many different possible effector proteins (5). While in CTLs Rab27a interacts with Munc13-4 and loss of either of these proteins affects CTL secretion, in melanocytes Rab27a interacts with melanophilin and loss of either of these proteins gives rise to loss of pigmentation. Intriguingly, loss of melanophilin does not give rise to immunodeficiency. This illustrates the concept that by interacting with Munc13-4 (which is expressed in CTLs but not melanocytes) and melanophilin, which interacts with myosin Va and actin in melanocytes, Rab27a can control secretion differentially in two different cell types.

A series of very elegant studies from the Hammer and Seabra labs revealed a mechanism whereby Rab27a on melanosomes mediated a "handover" from microtubules to actin as melanosomes approached the plasma membrane (see, e.g., references 6 and 7). These studies had important implications in demonstrating that melanosomes move along microtubules in a plus-end direction toward the periphery of the cell, where they are tethered by actin prior to secretion. This, in turn, raised the question as to how secretory lysosomes were secreted from CTLs, as the laboratories of Kupfer and Berke had both shown that the minus ends of microtubules (the microtubule-organizing center or centrosome) focus toward the point of secretion between CTL and target (8, 9), an area known as the immunological synapse (10). Additional studies showed that the centrosome contacted the plasma membrane at the synapse and that minus-end microtubule movement alone was sufficient to deliver secretion at the immunological synapse (11). Furthermore, it emerged that actin was depleted across the area where secretion occurred. Taken together, these observations illustrated that Rab27a was mediating two very different routes for secretion of the modified lysosomes found in CTLs and melanocytes: one dependent on actin tethering and the other not (Fig. 1).

At the same time, Rab27a was found to have many different potential interacting proteins, including not only Munc13-4 and melanophilin but also a series of synaptotagmin-like proteins (Slps) (5), with differential expression and roles in immune cells. While Slp1 was required for granule secretion from platelets, there was redundancy between Slp1 to -3 in CTLs (12, 13). These multiple potential interactions highlighted the pleiotropic nature of Rab27a in controlling different mechanisms of secretory lysosome release in immune cells.

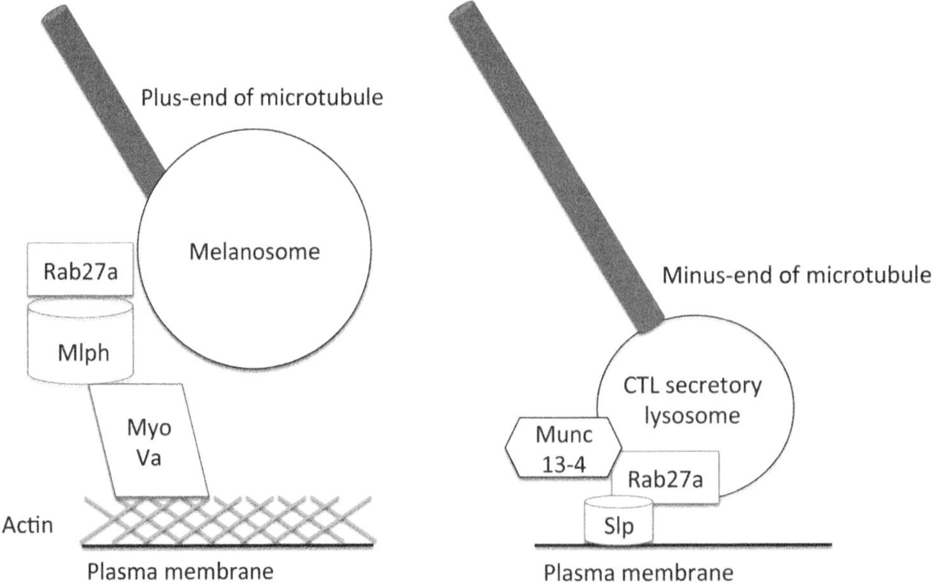

Figure 1 Rab27a interacts with different effector proteins to provide different modes of secretion. (Left) In melanosomes, Rab27a interacts with the Slp melanophilin (Mlph), which in turn interacts with myosin (Myo) Va and captures melanosomes onto cortical actin at the plus ends of microtubules. (Right) In CTLs, Rab27a interacts with Munc13-4 on secretory lysosomes and also interacts with Slp1 to -3, which are localized to the plasma membrane.

The observation that the centrosome contacted the plasma membrane at the immunological synapse revealed some remarkable parallels with other biological systems that were completely unexpected. The images of the centrosome at the synapse bore striking similarities to images of cilia formation (11, 14, 15). In most cells, the centrosome nestles next to the nuclear envelope near the center of the cell and does not approach the plasma membrane. However, during cilia formation, the centrosome contacts the plasma membrane via the distal appendages of the "mother" centriole and the inner microtubules then extend to form the axoneme of the cilium. The close apposition of the centrosome to the plasma membrane of the immunological synapse was reminiscent of cilia formation, although neither an axoneme nor a cilium form at the synapse.

There are two forms of cilia: motile cilia that contain a 9+2 inner core of microtubules required for ciliary-driven movement, and nonmotile or primary cilia that lack the central pair of microtubules (9+0) and play a role in signaling (16, 17). In particular, the Hedgehog (Hh) signaling pathway, important in development, is focused at the primary cilium. This central role in signaling for both the synapse and the primary cilium suggested that there might be more than simply morphological parallels between these seemingly different structures. Further intrigue was added to the question of how similar these structures were, by the long-held observation from electron microscopy studies that "all cells make primary cilia, except lymphocytes"! This led us to suggest that the immunological synapse might be a "frustrated cilium" (15).

Another functional similarity that was evident between synapse and cilium was that both are focal points for secretion, with the cytolytic granules of CTLs being delivered to the point of centrosome docking at the immunological synapse, and the ciliary pocket well characterized as the site of secretion (perhaps best in the specialized flagellar pocket of trypanosomes). Taken together with the finding that the centrosome comes to the point of abscission during cytokinesis, a point where secretion and endocytosis is also focused, this suggested that the fundamental role of centrosome docking at the plasma membrane might be to define an area of membrane specialized for endo- and exocytosis (15).

More similarities emerged, with the intraflagellar transport proteins, which had been assumed to be specific for ciliated cells, found to be expressed in T cells and implicated in T-cell receptor (TCR) recycling at the synapse (18). Furthermore, it was discovered that Hh signaling, which is important during early lymphocyte development, was initiated by TCR signaling in CD8 T cells (19). Inhibition of this pathway, either chemically or genetically, reduced the ability of CTLs to kill target cells. While the levels of TCR signaling and granule proteins were unaffected, two important steps required for CTL secretion were disrupted when the Hh component of signaling was selectively perturbed. One of the first events that occurs when a CTL synapse forms is depletion of cortical actin across the synapse (11). Both centrosome and granules are then delivered to this area of depleted actin (Fig. 2). The depletion of actin from the synapse appears to be triggered for TCR signaling. When TCR-associated kinases Lck or Zap70 are impaired, or when TCR signaling is inhibited by SLAM (signaling lymphocyte activation molecule) family receptors, actin depletion from the interface fails to occur. Under these conditions, TCR is unable to cluster into the center of the synapse, and neither centrosome nor granules polarize to the interface between CTL and target (20–22).

What happens when the TCR-triggered Hh signaling is selectively depleted? Both actin clearance and centrosome polarization were disrupted, suggesting that it is the Hh component of TCR signaling that somehow controls the ability of CTLs to reorganize their actin and microtubule cytoskeletons to deliver the secretory granules to the synapse. As Hh is a transcription factor, controlling gene expression, this suggests that Hh target genes might play an important role in priming CTLs for polarized secretion. What emerged is that one of the Hh target genes in CTLs is Rac1. Naive CD8 T cells express very low levels of Rac1, but upon TCR activation Rac1 is dramatically upregulated so that

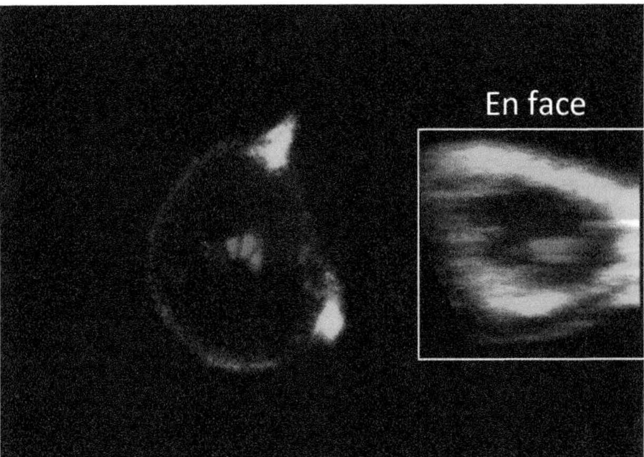

Figure 2 Secretory lysosomes are delivered to an area of plasma membrane depleted in cortical actin. CTLs expressing actin-green fluorescent protein (green), fixed and stained with CD63 (red), forming a synapse viewed from the side and en face. Reprinted from reference (11), with permission.

there are high levels in CTLs. However, selective inhibition of Hh signaling reduced CTL levels to those of naive cells, and these cells were unable to deliver their granules effectively to the synapse as both actin reorganization and centrosome polarization were perturbed. As Rac1 plays a role in both actin reorganization and is also implicated in centrosome polarization, then this provided a model for the role of Hh signaling in CTL secretion (19) (Fig. 3).

How frequently centrosome polarization (and formation of a "frustrated cilium") is involved in secretion from myeloid cells remains to be fully discovered. Dendritic cells, B cells, NK cells, and both CD4 and NKT cells appear to polarize the centrosome to the synapse (23–25). In contrast, mast cells that have been found to undergo polarized secretion toward synapses formed with B cells do not appear to polarize the centrosome or Golgi apparatus toward the synapse. Nevertheless, the depletion of actin observed at secretory synapses is also seen when mast cells form an "antibody-dependent degranulatory synapse" with B cells (26, 27). This suggests that secretion occurs in a relatively actin-depleted area, although exactly how the granules polarize toward this site is not yet known.

Another important secretory cell type from the myeloid lineage is neutrophils, which kill microorganisms by delivering the contents of secretory granules into the phagosome. This secretory step has been shown to precede closure of the phagosomal space, with azurophilic granules, containing the antimicrobial proteins, presumably moving along microtubules toward the centrosome that polarizes to the site of phagocytosis (28).

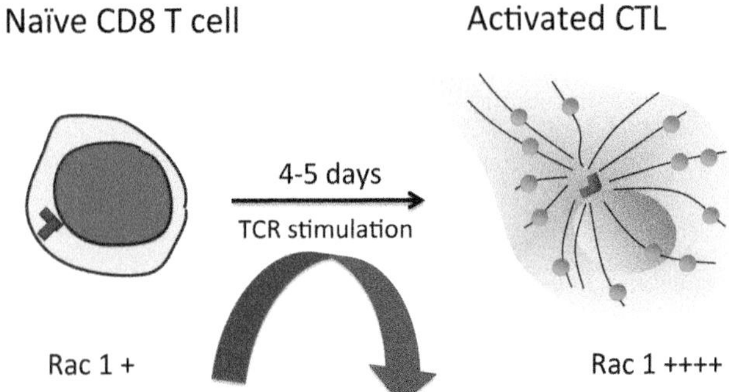

Figure 3 Naive CD8 T cells express only low levels of Rac1 and no secretory lysosomes, Upon activation, TCR signaling triggers secretory lysosome biogenesis and prearms the CTL to kill. TCR signaling also activates Hh signaling, increasing levels of Rac1, required for actin reorganization and microtubule organization and thereby secretory lysosome polarization and release.

Intriguingly, the phagosome, which is initially enriched in actin, rapidly depletes in actin upon internalization (29). This raises the possibility that azurophilic granules might be delivered to a membrane depleted in actin, as in CTLs and mast cells.

Although both Rab27a and its partner Munc13-4 play roles in neutrophil degranulation, only Munc13-4 is thought to be important for azurophilic granule secretion into the phagosome, and Rab27a-deficient mice show normal phagocytosis in neutrophils. Secretion of neutrophil granules at the plasma membrane is dependent on Rab27a, but whether centrosome polarization is involved is not yet clear. What emerges is that even within a single cell, Rab27a can interact with different effector proteins to control the release of different secretory granules by different mechanisms (as reviewed in reference 30).

Identifying the proteins that control the final secretory events with the plasma membrane has been challenging, as it is always difficult to determine where in the secretory pathway a fusion event is occurring. In CTLs, the fact that mutations in Munc18-2 and Stx11 give rise to FHL (reviewed in reference 4) indicates that these proteins are important for perforin secretion. It seems likely that Stx11 forms part of the final SNARE (soluble *N*-ethylmaleimide-sensitive factor attachment protein receptor) complex during granule release, as Stx11 localizes to the plasma membrane. Intriguingly, this localization is lost when Munc18-2 is absent (31). These same proteins have also been found to play a role in mast cell degranulation (32, 33).

Our current state of knowledge supports the idea that myeloid cells have developed to make use of slightly different ways to secrete their modified lysosomes, both by packaging specific effector proteins into the secretory compartments and also by using different mechanisms to deliver and release their secretory lysosomes.

Acknowledgments. *The author is funded by the Wellcome Trust.*

Citation. Griffiths GM. 2016. Secretion from myeloid cells: secretory lysosomes. Microbiol Spectrum 4(4):MCHD-0030-2016.

References

1. **Blott EJ, Griffiths GM.** 2002. Secretory lysosomes. *Nat Rev Mol Cell Biol* **3:**122–131.
2. **Griffiths G.** 2002. What's special about secretory lysosomes? *Semin Cell Dev Biol* **13:**279–284.
3. **Stepp SE, Dufourcq-Lagelouse R, Le Deist F, Bhawan S, Certain S, Mathew PA, Henter JI, Bennett M, Fischer A, de Saint Basile G, Kumar V.** 1999. Perforin gene defects in familial hemophagocytic lymphohistiocytosis. *Science* **286:**1957–1959.

4. **Luzio JP, Hackmann Y, Dieckmann NM, Griffiths GM.** 2014. The biogenesis of lysosomes and lysosome-related organelles. *Cold Spring Harb Perspect Biol* **6:**a016840. doi:10.1101/cshperspect.a016840.
5. **Fukuda M.** 2013. Rab27 effectors, pleiotropic regulators in secretory pathways. *Traffic* **14:**949–963.
6. **Hume AN, Collinson LM, Rapak A, Gomes AQ, Hopkins CR, Seabra MC.** 2001. Rab27a regulates the peripheral distribution of melanosomes in melanocytes. *J Cell Biol* **152:**795–808.
7. **Wu X, Rao K, Bowers MB, Copeland NG, Jenkins NA, Hammer JA III.** 2001. Rab27a enables myosin Va-dependent melanosome capture by recruiting the myosin to the organelle. *J Cell Sci* **114:**1091–1100.
8. **Kupfer A, Louvard D, Singer SJ.** 1982. Polarization of the Golgi apparatus and the microtubule-organizing center in cultured fibroblasts at the edge of an experimental wound. *Proc Natl Acad Sci U S A* **79:**2603–2607.
9. **Geiger B, Rosen D, Berke G.** 1982. Spatial relationships of microtubule-organizing centers and the contact area of cytotoxic T lymphocytes and target cells. *J Cell Biol* **95:**137–143.
10. **Kupfer A, Kupfer H.** 2003. Imaging immune cell interactions and functions: SMACs and the immunological synapse. *Semin Immunol* **15:**295–300.
11. **Stinchcombe JC, Majorovits E, Bossi G, Fuller S, Griffiths GM.** 2006. Centrosome polarization delivers secretory granules to the immunological synapse. *Nature* **443:**462–465.
12. **Holt O, Kanno E, Bossi G, Booth S, Daniele T, Santoro A, Arico M, Saegusa C, Fukuda M, Griffiths GM.** 2008. Slp1 and Slp2-a localize to the plasma membrane of CTL and contribute to secretion from the immunological synapse. *Traffic* **9:**446–457.
13. **Kurowska M, Goudin N, Nehme NT, Court M, Garin J, Fischer A, de Saint Basile G, Ménasché G.** 2012. Terminal transport of lytic granules to the immune synapse is mediated by the kinesin-1/Slp3/Rab27a complex. *Blood* **119:**3879–3889.
14. **Stinchcombe JC, Griffiths GM.** 2007. Secretory mechanisms in cell-mediated cytotoxicity. *Annu Rev Cell Dev Biol* **23:**495–517.
15. **Griffiths GM, Tsun A, Stinchcombe JC.** 2010. The immunological synapse: a focal point for endocytosis and exocytosis. *J Cell Biol* **189:**399–406.
16. **Satir P, Pedersen LB, Christensen ST.** 2010. The primary cilium at a glance. *J Cell Sci* **123:**499–503.
17. **Singla V, Reiter JF.** 2006. The primary cilium as the cell's antenna: signaling at a sensory organelle. *Science* **313:** 629–633.
18. **Finetti F, Paccani SR, Riparbelli MG, Giacomello E, Perinetti G, Pazour GJ, Rosenbaum JL, Baldari CT.** 2009. Intraflagellar transport is required for polarized recycling of the TCR/CD3 complex to the immune synapse. *Nat Cell Biol* **11:**1332–1339.
19. **de la Roche M, Ritter AT, Angus KL, Dinsmore C, Earnshaw CH, Reiter JF, Griffiths GM.** 2013. Hedgehog signaling controls T cell killing at the immunological synapse. *Science* **342:**1247–1250.
20. **Tsun A, Qureshi I, Stinchcombe JC, Jenkins MR, de la Roche M, Kleczkowska J, Zamoyska R, Griffiths GM.** 2011. Centrosome docking at the immunological synapse is controlled by Lck signaling. *J Cell Biol* **192:**663–674.
21. **Jenkins MR, Tsun A, Stinchcombe JC, Griffiths GM.** 2009. The strength of T cell receptor signal controls the polarization of cytotoxic machinery to the immunological synapse. *Immunity* **31:**621–631.
22. **Zhao F, Cannons JL, Dutta M, Griffiths GM, Schwartzberg PL.** 2012. Positive and negative signaling through SLAM receptors regulate synapse organization and thresholds of cytolysis. *Immunity* **36:**1003–1016.
23. **Pulecio J, Petrovic J, Prete F, Chiaruttini G, Lennon-Dumenil AM, Desdouets C, Gasman S, Burrone OR, Benvenuti F.** 2010. Cdc42-mediated MTOC polarization in dendritic cells controls targeted delivery of cytokines at the immune synapse. *J Exp Med* **207:**2719–2732.
24. **Yuseff MI, Reversat A, Lankar D, Diaz J, Fanget I, Pierobon P, Randrian V, Larochette N, Vascotto F, Desdouets C, Jauffred B, Bellaiche Y, Gasman S, Darchen F, Desnos C, Lennon-Duménil AM.** 2011. Polarized secretion of lysosomes at the B cell synapse couples antigen extraction to processing and presentation. *Immunity* **35:**361–374.
25. **Stinchcombe JC, Salio M, Cerundolo V, Pende D, Arico M, Griffiths GM.** 2011. Centriole polarisation to the immunological synapse directs secretion from cytolytic cells of both the innate and adaptive immune systems. *BMC Biol* **9:**45. doi:10.1186/1741-7007-9-45.
26. **Joulia R, Gaudenzio N, Rodrigues M, Lopez J, Blanchard N, Valitutti S, Espinosa E.** 2015. Mast cells form antibody-dependent degranulatory synapse for dedicated secretion and defence. *Nat Commun* **6:**6174. doi:10.1038/ncomms7174.
27. **Gaudenzio N, Espagnolle N, Mars LT, Liblau R, Valitutti S, Espinosa E.** 2009. Cell-cell cooperation at the T helper cell/mast cell immunological synapse. *Blood* **114:**4979–4988.
28. **Tapper H, Furuya W, Grinstein S.** 2002. Localized exocytosis of primary (lysosomal) granules during phagocytosis: role of Ca^{2+}-dependent tyrosine phosphorylation and microtubules. *J Immunol* **168:**5287–5296.
29. **Scott CC, Dobson W, Botelho RJ, Coady-Osberg N, Chavrier P, Knecht DA, Heath C, Stahl P, Grinstein S.** 2005. Phosphatidylinositol-4,5-bisphosphate hydrolysis directs actin remodeling during phagocytosis. *J Cell Biol* **169:**139–149.
30. **Catz SD.** 2014. The role of Rab27a in the regulation of neutrophil function. *Cell Microbiol* **16:**1301–1310.
31. **Hackmann Y, Graham SC, Ehl S, Höning S, Lehmberg K, Aricò M, Owen DJ, Griffiths GM.** 2013. Syntaxin binding mechanism and disease-causing mutations in Munc18-2. *Proc Natl Acad Sci U S A* **110:**E4482–E4491.
32. **Bin NR, Jung CH, Piggott C, Sugita S.** 2013. Crucial role of the hydrophobic pocket region of Munc18 protein in mast cell degranulation. *Proc Natl Acad Sci U S A* **110:** 4610–4615.
33. **D'Orlando O, Zhao F, Kasper B, Orinska Z, Müller J, Hermans-Borgmeyer I, Griffiths GM, Zur Stadt U, Bulfone-Paus S.** 2013. Syntaxin 11 is required for NK and $CD8^+$ T-cell cytotoxicity and neutrophil degranulation. *Eur J Immunol* **43:**194–208.

Myeloid Cells in Health and Disease: A Synthesis
Edited by Siamon Gordon

doi:10.1128/microbiolspec.MCHD-0018-2015

Adam P. Levine[1]
Anthony W. Segal[1]

33

The NADPH Oxidase and Microbial Killing by Neutrophils, With a Particular Emphasis on the Proposed Antimicrobial Role of Myeloperoxidase within the Phagocytic Vacuole

THE RESPIRATORY BURST OF PROFESSIONAL PHAGOCYTIC CELLS IS ACCOMPLISHED THROUGH THE ACTION OF THE NADPH OXIDASE NOX2

In 1933, Baldridge and Gerard observed the "extra respiration of phagocytosis" when dog leukocytes were mixed with Gram-positive bacteria and assumed that it was associated with the production of energy required for engulfment of the organisms. It was later shown that this "respiratory burst" was not inhibited by the mitochondrial poisons cyanide (1) or azide (2), which indicated that this is a nonmitochondrial process. The hunt for the neutrophil oxidase was then on because oxidative phosphorylation is far and away the major mechanism by which oxygen is consumed in mammalian cells, and another system that could consume oxygen at a similar rate was of considerable interest. Although many oxidases use NADH or NADPH as substrate, this oxidase was specifically called the NADPH oxidase because in the early days there was quite a controversy as to whether the substrate was NADH (3) or NADPH (4), and the matter was finally settled in favor of NADPH.

In 1957, a new disease called "fatal granulomatosus of childhood" was discovered (5, 6). Neutrophils from subjects with this "chronic granulomatous disease," or

[1]Division of Medicine, University College London, London WC1E 6JF, United Kingdom.

CGD, demonstrated an impaired ability to kill *Staphylococcus aureus*, as well as an absent respiratory burst, thereby linking the NADPH oxidase to the bacterial-killing process.

The next big development was the discovery of an enzyme, superoxide dismutase (SOD), which rapidly converted superoxide O_2^- (7). The inclusion of SOD allowed the detection of superoxide O_2^- in biological systems because it specifically inhibited superoxide-induced reactions. The use of this enzyme enabled Babior and colleagues to demonstrate that activated neutrophils generate O_2^-, which was in agreement with the previous observation that these cells generate hydrogen peroxide (H_2O_2) (8), which is produced by the dismutation of O_2^-.

The search was then on for an "enzyme" that would transport an electron from NADPH to oxygen to form O_2^-. This turned out to be a demanding task because the activity disappeared with simple purification techniques, for reasons that will become obvious later. In 1978, we discovered a very low-potential cytochrome *b* that was absent from neutrophils of patients with CGD, which appeared to fulfill the requirements of the oxidase (9, 10). Borregaard and colleagues (11) found the cytochrome *b* in patients with the autosomal pattern of inheritance, and surprisingly also in another such patient with the X-linked pattern of inheritance, which led them to question the relationship between the cytochrome *b* and the oxidase. We went on to confirm that patients with autosomal recessive CGD did in fact have the cytochrome *b* in their neutrophils, but were unable to transfer electrons onto it when the cells were activated (12), indicating the absence of an upstream electron-transporting molecule or of an activation mechanism. The latter case turned out to be the correct one. Over time it was demonstrated that the cytochrome comprised gp91^{phox}, the electron-transporting component, together with another membrane component, p22^{phox} (13, 14), in stoichiometric equivalence (15). In addition, five cytosolic proteins were required, p47^{phox} (also called NOX organizer 1) (16), p67^{phox} (NOX activator 1) (16–18), p40^{phox} (19), and the small GTP-binding protein Rac, which dissociates from its binding partner GDI (GDP dissociation inhibitor) (20). All of these components, apart from GDI, move to the membranes and activate electron transport through the cytochrome. The requirement for the coordinated integration of these multiple components for an active oxidase system explains why it proved impossible to obtain the active oxidase by classical biochemical purification techniques, because these resulted in dissociation of the complex with loss of enzymic activity.

It proved possible to recombine most of the components of the oxidase in a "cell-free" assay (21, 22), in which solubilized membranes or purified cytochrome *b* could be mixed with cytosol or the purified cytosolic protein components, and NADPH. Electron transport was then induced by the addition of arachidonic acid or a detergent, which must have changed the conformation of the cytochrome, giving it access to substrate and the accessory binding proteins. This "cell-free" assay system was also useful in characterizing some cases of autosomal recessive CGD, where the missing or abnormal cytosolic components of the oxidase were identified by complementation of the deficient cytosol with semipure proteins (23).

Once the cytochrome *b* had been identified, it could be isolated by following its spectrographic signature (24) through the different stages of purification. We showed that it was a flavocytochrome and characterized it in terms of its NADPH, flavin adenine dinucleotide (FAD) (25), heme (26), and carbohydrate (27) binding sites. We purified and obtained amino acid sequence from gp91^{phox} (28) and demonstrated that it was coded for by the gene, abnormalities of which cause X-CGD, that had been cloned based on its chromosomal localization (29).

THE RELATIONSHIP BETWEEN THE PHAGOCYTE OXIDASE AND MICROBIAL KILLING

It was known that stimulated neutrophils generated H_2O_2 and that they contained a very high concentration of a peroxidase, myeloperoxidase (MPO) (30), in their granules (31). Klebanoff demonstrated the fixation of iodine by neutrophils phagocytosing bacteria, and that concentrated MPO can kill *Escherichia coli* in combination with iodide and H_2O_2 *in vitro*, suggesting that this may contribute to bacterial killing by neutrophils *in vivo* (32). The discovery that absence of the oxidase in CGD was accompanied by greatly increased incidence of clinical infections, together with a clear impairment of bacterial killing by neutrophils from these patients, established the requirement of the activity of this oxidase for efficient microbial killing. Defective iodination by their neutrophils reinforced the feasibility of a physiological role for the MPO-H_2O_2-iodide antimicrobial system (33).

The demonstration that activated neutrophils can generate O_2^- (34) produced enormous excitement. Superoxide, H_2O_2, and their reaction products were proposed as effector molecules that could themselves cause the microbicidal lethal reactions. In addition, because of the predicted toxicity of these molecules (35) that could be generated by neutrophils, and because neutrophils were found in abundance in inflammatory sites where tissue damage occurred, it was a natural progression to suggest that the tissue damage was being produced by the neutrophil-generated radicals (36). This implied that antioxidant molecules that reacted with and consumed these reactive oxygen molecules might provide valuable therapeutic anti-inflammatory agents (37).

THE CASE FOR AND AGAINST A PRIMARY ROLE FOR MPO IN THE KILLING OF INGESTED MICROBES WITHIN THE PHAGOCYTIC VACUOLE

The MPO-Halide-Hydrogen Peroxide Antibacterial System

Much evidence has been generated to support the concept that MPO oxidizes chloride and other halides to generate their hypohalous acids and that these compounds kill the organisms within the phagocytic vacuole. It is important, however, to be aware of two important factors. The first is that these theories originated when it was believed that the pH within the phagocytic vacuole of the neutrophil is acidic, with a pH of 6.5, falling to 4.0 (38). The *in vitro* MPO-halide-hydrogen peroxide antibacterial system is critically dependent on the ambient pH, being highly efficient at pH 5.0 and ineffective at pHs of 7.0 and above (32, 39). Newly obtained data demonstrate that in normal human neutrophils the respiratory burst elevates the pH within the phagocytic vacuole to ~9.0, at which pH the *in vitro* peroxidatic effect of MPO is almost nonexistent (40). The second issue of consequence is that in the test tube it is exceedingly difficult to accurately distinguish between organisms that have been fully engulfed into a closed vacuole and those that remain in suspension in the medium or attached to the neutrophils but not fully taken up. The latter situation is particularly relevant when dealing with bacteria that have a tendency to form microcolonies and biofilms, such as staphylococci (41), *Burkholderia cenocepacia* (42), *Pseudomonas* (42), and *Serratia* (43); and fungi, like *Aspergillus* (44) and *Candida* (45).

The Consequences of Generating H_2O_2 in the Phagocytic Vacuole

The NADPH oxidase generates O_2^- in the vacuole, which naturally dismutates to H_2O_2, and this process is accelerated by MPO, which has SOD activity (40, 46). There is very strong evidence for the generation of H_2O_2 in the vacuole (47) as initially proposed on the basis of the oxidation of formate to CO_2 (8). It is not clear what then happens to the H_2O_2. The dogma is that it is used as substrate for MPO to generate HOCl, but this is very unlikely at the pH of 9.0 recently demonstrated in the neutrophil vacuole, at which the peroxidatic and chlorinating activities of MPO were shown to be very low (40). An alternative scenario is that MPO acts first as a SOD and then as a catalase (40, 48) to safely remove the H_2O_2 and regenerate its SOD activity, thereby preventing the damaging oxidation of the important granule proteins. It seems highly improbable that the neutrophil would synthesize, store, and release highly specialized enzymes into the phagocytic vacuole, only to have them degraded by HOCl (49, 50). One piece of evidence used to support the MPO-halide-hydrogen peroxide antibacterial system has been the misconception that patients with CGD are less susceptible to infections with catalase-negative than -positive organisms, because it was supposed that the former produced H_2O_2 that was then utilized by the MPO system for their own destruction. This was in fact shown not to be the case, as catalase-negative *S. aureus* (51) and *Aspergillus nidulans* (52) were found to be just as virulent as the catalase-positive strains in CGD mouse models.

Evidence for the Generation of HOCl in the Phagocytic Vacuole

Several attempts have been made to measure HOCl in the vacuole (53). Hurst's group used fluorescein coupled to polyacrylamide microspheres to generate fluorescent reporter groups that could then be uncoupled from the particles by reduction of the cystamide disulfide bond. The mono-, di-, and trichlorofluorescein compounds were separated and measured by liquid chromatography. They demonstrated that the generation of these chlorinated species by the MPO-H_2O_2-Cl^- system had a pH optimum of 6 to 7.5, with virtually no activity at physiological pH of >8.5 (54). They assessed the efficiency of conversion of O_2 to HOCl at about 11%, which they judged to all be within closed vacuoles because the fluorescence was not quenched by extracellular methyl viologen, which effectively quenched the fluorescence of the beads alone. However, some of these quenching experiments were performed in 100% serum, which we have found to largely abolish the quenching of the fluorescence of fluorescein isothiocyanate-labeled bacteria by methyl viologen (our unpublished data).

Various other attempts have been made to develop HOCl-specific probes, for example, that by Koide and colleagues (55), but once again specificity was not demonstrated under vacuolar conditions. They demonstrated fluorescence of their intravacuolar probe, but this did not occur by 90 s after phagocytosis of opsonized zymosan, by which time the respiratory burst would have been well advanced if not completed (40).

Painter and colleagues metabolically labeled *Pseudomonas* with [$^{13}C_9$]-l-tyrosine, which was opsonized with serum and mixed with neutrophils in the presence of catalase and taurine, which were included to suppress extracellular but not intracellular chlorination. The largely ingested bacteria were subjected to protease hydrolysis, and levels of $^{13}C_9$-derived chlorotyrosine and tyrosine were determined by gas chromatography and mass spectrometry (GC-MS) using the isotope dilution technique. Because neutrophil-derived tyrosine contains no [$^{13}C_9$]-l-tyrosine or its derivatives, this is a direct measure of the chlorination of intracellular *Pseudomonas*-derived tyrosine. They did find these products, but only a very small amount, ~0.15%, of the [$^{13}C_9$]-l-tyrosine was chlorinated (56, 57).

Chapman and colleagues (58) also used GC-MS to measure chlorinated tyrosine compounds. They found that ~0.2% of the tyrosines became chlorinated and that 94% of chlorination was of neutrophil rather than bacterial tyrosine residues even though tyrosine residues in bacteria were chlorinated ~2.5 times as efficiently as those in neutrophils. An interesting, and unexplained, observation in this study was that when azide was added as an inhibitor of MPO, it had exactly the opposite effect to that expected, resulting in a 3-fold increase in chlorination (Fig. 8 in reference 58). This group subsequently performed similar experiments with purified phagocytic vacuoles containing magnetic beads and achieved a chlorination efficiency of 1.1% of the vacuolar tyrosines (59).

We examined the targets of iodination after bacteria were phagocytosed (60). Less than 0.3% of the oxygen consumed was utilized for iodination and almost all the iodine was incorporated into neutrophil proteins rather than those of the engulfed bacteria (61).

So how can we explain these results where some evidence of chlorination is observed while at the same time we know that conditions within the vacuole are most unfavorable for such chlorination reactions? We know that under certain experimental conditions up to one half of all phagocytic vacuoles can fail to close (62, 63), and in these vacuoles that communicate with the extracellular medium the MPO would be interacting with H_2O_2 and Cl^- at the pH of this extracellular medium, which is ~7.4. The failure of closure of only a small proportion of the vacuoles could give rise to the observed chlorination results. It is also possible, as will be discussed below, that MPO has a dual function. It may be that in the closed vacuole it acts as a catalase and SOD and that when the vacuole is unable to close it acts as a peroxidase.

Superoxide, H_2O_2, and HOCl Are Not as Microbicidal as Previously Thought

The discovery of the production of O_2^- by neutrophils was associated with the expectation that this free radical might itself be microbicidal (34), but it is now generally accepted that neither O_2^- nor its dismutation product, H_2O_2, is damaging enough to be directly microbicidal (64–66), and as a result the emphasis has been placed on the toxicity of HOCl generated by MPO.

We conducted experiments to directly determine the antimicrobial actions of O_2^-, H_2O_2, and HOCl on *S. aureus* and *E. coli* (61) on their own and in the presence of granule proteins. It is important to assess their effects in the presence of the granule proteins because these are the circumstances under which these products of the oxidase would be generated. HOCl would not be produced in the absence of MPO, indicating that degranulation would have occurred into the vacuole where the bulk of the chlorination is observed (59) and where the granule proteins have been shown to be chlorinated (58).

In the presence of 25 mg/ml of granule proteins, which while being a practical experimental concentration is almost certainly an underestimate of that pertaining in the vacuole, neither 100 mM O_2^- or H_2O_2 nor 1 mM HOCl killed either organism at neutral pH.

MPO Deficiency

If the main function of the oxidase is to produce H_2O_2 as substrate for MPO-mediated generation of HOCl, and if HOCl is the major microbicidal species, then MPO deficiency would be expected to predispose to serious infection. This does not turn out to be the case. In large studies involving between 57,000 and 70,000 individuals in the U.S., Japan, and Germany, the incidence of MPO deficiency was found to be between 0.05 and 0.15% (67–69). The incidence of infection in these subjects was very low. In the largest, German, study, the distribution of diseases in the deficient patients did not differ from that of the general hospital population and only ~6% of these subjects had infectious disease. The MPO knockout mouse demonstrated variable sensitivity to infection with a range of different organisms (70). MPO-deficient and control mice were infected intranasally with various fungi and bacteria, and the number of residual microorganisms in the lungs was compared 48 h later. MPO-deficient mice showed severely reduced cytotoxicity to *Candida albicans*, *Candida tropicalis*, *Trichosporon asahii*, and *Pseudomonas aeruginosa*. However, the mutant mice showed a slight but significantly delayed clearance of *Aspergillus fumigatus* and *Klebsiella pneumoniae* and had comparable levels of resistance to the wild type against *Candida glabrata*, *Cryptococcus neoformans*, *S. aureus*, and *Streptococcus pneumoniae*. It is of note that infection with these agents generally occurs by multiplication from a relatively small inoculum rather than the instillation in massive numbers as a bolus, and that those organisms that were eliminated more slowly from the MPO-deficient lungs tend to grow in clumps or mycelia that would be difficult for neutrophils to fully engulf (*vide infra*).

Whereas few, if any, MPO-deficient subjects develop serious pyogenic bacterial or fungal infections, the penetrance of the molecular lesions of CGD is almost complete. Even though the rare individual with CGD has been reported as presenting in later life (71), >90% of patients present by the age of 20 years (72), and we are unaware of an individual with the molecular lesion of CGD, i.e., the relative of an affected patient, who did not develop clinically serious infections. The knockout mice mirror the discrepancy observed in humans. CGD mice demonstrate a major vulnerability to infections with a wide variety of organisms while the propensity is only slightly increased in the MPO-deficient mouse (73, 74). This argues that the function of the NADPH oxidase cannot simply be to generate H_2O_2 as substrate for MPO-dependent generation of HOCl. The MPO-deficient mice appear to have a particular problem resisting infection with organisms like *Candida*, *Aspergillus*, and *Pseudomonas* that form clumps, hyphae, or biofilms. It is possible that these organisms are not phagocytosed and that they produce a state of frustrated phagocytosis in which the MPO acts at the pH of the extracellular medium, which is acid at sites of infection (75), under which circumstances HOCl could be produced and the extracellular organisms killed.

Approaching the issue from the aspect of a diagnostic laboratory investigating leukocyte function in patients with pyogenic infections (76), of 81 patients with a diagnosis of a primary phagocytic disorder, 48 had CGD and only 2 had MPO deficiency, of whom 1 had other neutrophil abnormalities as evidenced by delayed separation of the umbilical cord and severely impaired chemotaxis, and the other the relatively mild infection of furunculosis. The incidence of CGD is at most about 1 in 250,000 and that of MPO deficiency about 125 times as common; thus the predisposition to pyogenic infection in CGD is about 6,000 times as great as that of MPO deficiency. It thus seems totally implausible that MPO acts as the effector system for H_2O_2 produced by the NADPH oxidase that is defective in CGD.

GRANULE PROTEINS

The unique structural characteristic of neutrophils is their content of cytoplasmic granules, which comprise ~10% of the total cellular protein. The antimicrobial proteins and peptides are largely contained within the azurophilic granules, where they are almost all strongly cationic and are bound to a negatively charged sulfated proteoglycan matrix (77). The granule contents comprise a diverse complement of enzymes and peptides that have evolved to kill a wide variety of microorganisms and to digest diverse organic material (78, 79). The most abundant of these enzymes are the neutral proteases, cathepsin G, elastase and proteinase 3, lysozyme, MPO, and the defensins, but there are a host of other enzymes and proteins (80).

In the absence of clear evidence of microbial killing by products of the oxidase, we investigated the possibility that the neutral proteases might be involved (81), which was particularly pertinent in view of the observation that we had made that the phagocytic vacuole was alkalinized by the oxidase (82) (as will be discussed later). We knocked out the genes for cathepsin G and for elastase and examined the susceptibility of the single- and double-knockout mice to infection and the competence of their neutrophils to kill bacteria and fungi *in vitro*. Despite normal neutrophil development and recruitment, the mice lacking either of these proteins were more susceptible to infection with *A. fumigatus*, and they were even more susceptible when deficient in both enzymes (83). Differential responses were seen when these animals were challenged with *S. aureus* or *C. albicans* (84). The elastase-deficient mice were normally resistant to *S. aureus* but unduly sensitive to *C. albicans*, and the opposite was true for cathepsin G deficiency. The double-knockout mice were susceptible to both organisms. *In vitro*, the purified neutrophils from susceptible animals exhibited deficient microbial killing. These findings provided a clear example of the selectivity of the microbicidal activity of the granule enzymes for different organisms. The same mice were used by others to demonstrate the requirement for cathepsin G, and to a greater extent both cathepsin G and elastase, for resistance to infection with *S. pneumoniae* (85) or with mycobacteria (86).

The susceptibility of the mice made deficient in the neutral proteases was observed despite the presence of a normal respiratory burst and of iodination, further evidence against a microbicidal action of the products of the oxidase or MPO (84).

Further evidence for the role of the neutral proteases was that selective inhibitors impaired the killing of *S. pneumoniae* by human neutrophils, and purified elastase and cathepsin G were independently able to kill these organisms in *in vitro* assays (87).

Another set of protease-deficient mice were made. These elastase-deficient mice were found to be abnormally vulnerable to infection with *K. pneumoniae* and *E. coli* but resistant to *S. aureus* (88), while the cathepsin G-deficient mice were resistant to all three organisms (89). Mice lacking both cathepsin G and elastase demonstrated normal resistance to *A. fumigatus* and *Burkholderia cepacia*, while p47$^{phox-/-}$ NADPH oxidase-deficient mice were very susceptible to both these organisms (90). It was suggested that these results cast doubt upon the initial studies describing the importance of these neutral proteases in antimicrobial resistance. However, the mice we constructed and studied were on a 129/SvJ background (84), whereas the background of those studied by Brahm Segal and the Shapiro group was C57black6 (90). This is important because

> "Inbred laboratory mouse strains are highly divergent in their immune response patterns as a result of genetic mutations and polymorphisms. ... Although common inbred mice are considered 'immune competent,' many have variations in their immune system—some of which have been described—that may affect the phenotype. Recognition of these immune variations among commonly used inbred mouse strains is essential for the accurate interpretation of expected phenotypes, or those that may arise unexpectedly" (91).

Papillon-Lèfevre Syndrome

Papillon-Lèfevre syndrome (PLS) is an interesting and informative syndrome that is characterized by symmetrical palmoplantar hyperkeratosis and periodontal inflammation, causing the loss of both primary and permanent teeth, and pyogenic infection. It is very rare, and only about 300 to 400 cases have been described. It is caused by mutations in the *CTSC* gene, which codes for cathepsin C (92) (also known as dipeptidyl peptidase-1; EC 3.4.14.1), which is a cysteine dipeptidyl aminopeptidase expressed in the cytoplasmic granules of several tissues, with levels highest in lung, macrophages, neutrophils, CD8$^+$ T cells, and mast cells. The enzyme activates cathepsin G; proteinase 3; neutrophil elastase; granzymes A, B, and C; and mast cell chymase and tryptase by removing inhibitory N-terminal dipeptides (93). It is not clear where and when this dipeptidase acts. It is composed of a dimer of disulfide-linked heavy and light chains, both produced from a single protein precursor, and a residual portion of the propeptide acts as an intramolecular chaperone for the folding and stabilization of the mature enzyme. The quaternary structure of this protein confers this strict dipeptidase activity (94).

Knocking out this gene in mice (95) led to failure of the activation of granzymes A and B, serine proteases most commonly found in the granules of cytotoxic T lymphocytes, natural killer cells, and cytotoxic T cells (96). In neutrophils this led to an accumulation of neutrophil elastase and a disappearance of cathepsin G, with the loss of activity of both. Evidence against the antimicrobial role of neutral proteases has been proffered by the description of a case of PLS in which neutrophils appeared to lack cathepsin G and proteinase 3 but appeared to kill *S. aureus* and *K. pneumoniae* normally (97). In a similar study of three such cases, concentrations of cathepsin G, neutrophil elastase, and proteinase 3 were severely reduced and killing of *E. coli* and *S. aureus* was impaired in two of the three (98), as had been previously described (99). Killing of *C. albicans* was found to be impaired in all 15 Egyptian cases tested (100). It is possible that the lack of detection of a killing defect by Sørensen et al. (97) was caused by the method they used to measure bacterial killing. They initially incubated the neutrophils and opsonized bacteria at 37°C for 10 min, centrifuged them, and reincubated them at 37°C for a further 10 and 30 min, taking the numbers of bacteria at the start of the second incubation as time 0 and expressing killing as a percentage of the viable bacteria at that time. It is very possible that much of the killing had already occurred in the first 10 min and that differences in the function of the patients' and control cells had been missed.

In addition to the characteristic very aggressive periodontitis, serious infections are common in PLS. Seventeen percent of patients have serious cutaneous sepsis, and 15 cases of liver abscess, 1 renal abscess, 1 case of multiple brain abscesses, and 1 case of pyelonephritis have been described. *S. aureus* and *E. coli* were the main organisms isolated, and where documented the histology has been that of a granulomatous inflammation.

A discordant feature is that the periodontitis that characterizes PLS is much more severe than in any of the other primary immunodeficiency diseases. It leads to bone resorption around the affected teeth, resulting in mobility and loss of most of the teeth. Several organisms have been isolated from the crevicular space of these patients (101), but the bulk of evidence suggests that *Actinobacillus actinomycetemcomitans* are largely responsible for these infections. Neutrophil serine proteases are important for resistance to these organisms (102), which are killed *in vitro* by neutrophil elastase and cathepsin G (103). The cathepsin C knockout mice have no dental problems (104).

It is also of interest that although cathepsin C is thought to be required for the activation of granzymes (95), which are important for T-cell and natural killer cell activity, these patients do not exhibit evidence of T-cell deficiencies such as chronic viral infections.

In addition, in the patient described by Sørensen et al. (97), the neutrophils do produce some cathepsin G and elastase (Supplemental Table 1 in reference 97), and the lighter, more primitive bone marrow cells produce cathepsin C, which disappears in older cells, which in this case had been incubated *in vitro* at 37°C for 4 h before analysis. These young myeloid cells might also contain cathepsin G and elastase, which is important because in serious infections there is a shift to the left of the myeloid series of cells, and these primitive cells enter the circulation (105) and could play an important antimicrobial role.

The interesting feature of this case, which is probably generally applicable to PLS, is why is there a major reduction in the content of these neutral proteases in azurophil granules? The tertiary and quaternary structure of cathepsin C is designed to restrict its activity to that of an exodipeptidase. If this structure were disrupted by activating mutations that lifted this restriction, producing endopeptidase activity, the enzyme, or another protease activated by it, would be able to degrade proteins within the granule, the interior of which has a pH close to the optimum of about 6.0 for cathepsin C (106). The proteomics data presented by Sørensen et al. (97) indicated that a number of granule proteins other than the three major neutral proteases were degraded in their patient's cells, indicating that digestion of the neutral proteases is not simply due to failure to remove their terminal dipeptide. The release of an active endopeptidase from neutrophils attracted to the periodontal space by infection could digest the periodontal ligament and surrounding bone because the pH is also acidic in this compartment (107).

Based on the scanty, incomplete, and conflicting data from investigations of the very rare PLS patients, the conclusion that "proteases in human neutrophils are dispensable for protection against bacterial infection" (108) appears premature.

THE ROLE OF THE NADPH OXIDASE IN ELEVATING AND REGULATING THE pH OF THE PHAGOCYTIC VACUOLE

Initially the pH of the neutrophil phagocytic vacuole was thought to be acidic (109). We showed that through the action of the NADPH oxidase the pH actually rises to what was originally thought to be ~7.5 to 8 (82), and these findings have been replicated several times (62, 110–112). In those studies, fluorescein, coupled to phagocytosed organisms, was employed as pH indicator. This is not an ideal indicator for this purpose because it cannot measure changes in pH above ~8 (110) and is not generally used in a ratiometric manner. In addition, it has been thought by some to become bleached by the action of MPO in the phagocytic vacuole (54, 113).

In recent studies we have used seminaphtorhodafluor (SNARF) (114) to make these measurements (40). SNARF has a dynamic range of between 6 and 11, is ratiometric, and different preparations can be used to simultaneously measure the pH in the vacuole and cytoplasm. We have shown in several ways that its fluorescent properties are not altered by the conditions in the phagocytic vacuole. In addition, we only measured the pH of *Candida* inside vacuoles, so that the fluorescence signal was not influenced by that of the nonphagocytosed organisms that are maintained at the pH of the buffered extracellular medium. Synchronizing the changes in fluorescence to the time of particle uptake gave a more accurate indication of the temporal pH changes.

Almost immediately following engulfment of the SNARF-labeled *Candida* by human neutrophils, the vacuole underwent a significant alkalinization (40). The mean maximum pH reached postphagocytosis was 9.0 (standard deviation, 8.3 to 10.2), and this elevated pH was maintained for 20 to 30 min (Fig. 1). When the NADPH oxidase was inhibited by DPI, the vacuole acidified to 6.3 (standard deviation, 6.1 to 6.6), and when DPI was added to cells that had phagocytosed *Candida*, where the vacuolar pH was very alkaline, the vacuolar pH rapidly dropped, which indicates that continued activity of the oxidase is required to maintain this alkaline pH. The addition of 5 mM azide had a dramatic effect by acidifying the vacuole by about 2 pH units. This effect was not due to the inhibition of the activity of MPO because it was not observed with two other more specific MPO inhibitors, 4-aminobenzoic acid hydrazide and KCN (40).

The observation that the addition of azide to phagocytosing neutrophils produces an acidification of the phagocytic vacuole is important. Previous studies

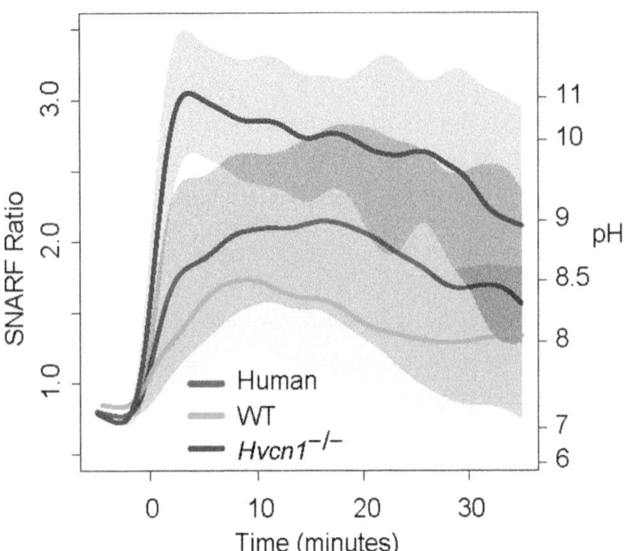

Figure 1 Time courses of changes in vacuolar pH in human and *Hvcn1*$^{-/-}$ and wild-type (WT) mouse neutrophils phagocytosing SNARF-labeled *Candida*. Reprinted from reference 40, with permission.

that were unable to detect an elevation of vacuolar pH above neutral in normal neutrophils (115, 116) included azide in all the solutions to counteract bleaching of the fluorescein by MPO. In addition, azide has been included in microbicidal assays to inhibit the action of MPO in order to demonstrate its importance in the killing process (117, 118). We now know that this inhibition by azide of the killing of microbes by neutrophils could be produced either by blocking the peroxidatic function of MPO or by impeding the proteolytic activity of the neutral proteases, which are much less efficient in the more acidic environment induced by the presence of this agent.

HOW IS THE VACUOLAR pH ELEVATED?

The transport of electrons into the phagocytic vacuole is electrogenic, causing a large, rapid membrane depolarization that will itself curtail further electron transport unless there is compensatory ion movement (119) by the passage of cations into the vacuole and/or anions in the opposite direction (Fig. 2). The nature of the ions that compensate the charge will have a direct effect on the pH within the vacuole and the cytosol. The cytoplasmic granules are very acidic, with a pH of about 5.5, and they download their acid contents into the vacuole. The electrons that pass into the vacuole produce O_2^-, which dismutates to form peroxide (O_2^{2-}) that is then protonated to form H_2O_2. The source of these protons will govern the alterations in the vacuolar pH. If all the charge is compensated by protons passing into the vacuole, then none of the protons released into the vacuole from the granules will be consumed and the pH will remain acidic. In fact, most of the charge is compensated by protons passing from the cytoplasm into the vacuole through the HVCN1 channel (120), because if this channel is knocked out in mice the vacuolar pH becomes grossly elevated to about 11 (Fig. 1) (40). Under normal physiological conditions, about 5 to 10% of the compensating charge is contributed by nonproton ions, some of which are K^+, passing into the vacuole (84); the residual ion flux could be the egress of chloride (40). These nonproton ion channels remain to be identified.

INFLUENCE OF pH ON THE ACTIVITIES OF CATHEPSIN G, ELASTASE, AND MPO

The accurate measurement of vacuolar pH in normal human neutrophils is absolutely central to understanding the mechanisms by which these cells kill bacteria and fungi. A strong body of opinion supports the concept that MPO is "a front line defender against phagocytosed microorganisms" (53) and that it kills microbes within this compartment through the generation of HOCl. We measured the effect of pH on peroxidase and chlorinating activities of MPO and found these to be maximal at acidic pH, with both activities falling off substantially as the pH was elevated, until both activities were almost completely abolished at the pH of ~9 pertaining in the vacuole (40). However, as shown here, the elevated pH provides an optimal milieu for the microbicidal and digestive functions of the major granule proteases, elastase, cathepsin G, and proteinase 3 (Fig. 3), which are activated by this elevated pH and the influx of K^+ into the vacuole (84).

The alkaline vacuolar pH will assist the dissociation of the cationic proteins from the negatively charged sulfated proteoglycan matrix to which they are bound. The pKs of cathepsin G, elastase, proteinase 3, and MPO are all ~10 (121).

In CGD, digestion of ingested microbes is inefficient because the vacuolar pH is very acidic and the granule contents do not disperse within the vacuole (82). This retained undigested material results in the observed granulomatous tissue response and hyperinflammatory tissue reactions observed in patients (122) and in experimental animals (123, 124).

These results demonstrate that at the physiological pH pertaining in the phagocytic vacuole, cathepsin G and elastase will be active but that MPO has virtually

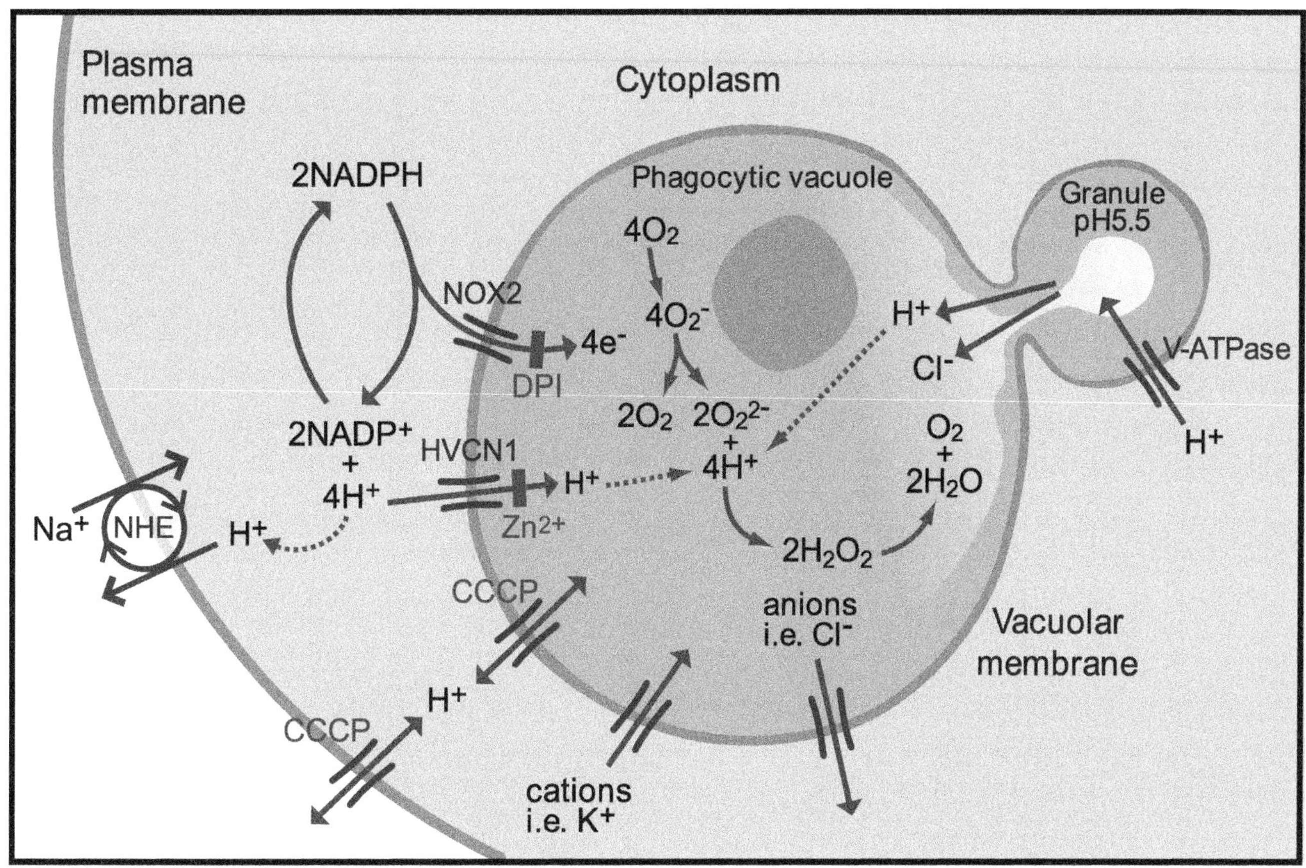

Figure 2 Schematic representation of the neutrophil phagocytic vacuole showing the consequences of electron transport by NOX2 onto oxygen. The proposed ion fluxes that might be required to compensate the movement of charge across the phagocytic membrane together with modulators of ion fluxes are shown. CCCP, carbonyl cyanide m-chlorophenyl hydrazone; NHE, sodium proton exchanger. Reprinted from reference 40, with permission.

no activity as a chlorinating peroxidase, although it retains SOD and catalase activities. They also explain the anomalous results produced when azide has been used as an inhibitor of MPO to demonstrate the functional relevance of this enzyme for microbial killing. Azide not only inhibits MPO, but the vacuolar acidification it produces impairs the function of the neutral proteases, so that the effects of azide ascribed to the inhibition of MPO activity might just as well have been, and probably were, due to acidification of the phagocytic vacuole and the negating effect this had on the digestive enzyme activity within this compartment.

It is possible that MPO has evolved to serve two different functions *in vivo*, depending on the local environment. The pH of ~9 in the vacuole favors SOD and catalase (48, 125) activity rather that of a peroxidase, whereas when the neutrophil encounters an organism that it is unable to fully engulf, such as a fungal mycelium, the situation is different (126). The concentration of MPO and H_2O_2 will be much lower; the pH, which is low in inflammatory foci at ~6 (75), is optimal for peroxidatic activity; and the supply of chloride is limitless.

THE FAMILY OF NOXs AND DUOXs

With the development of high-throughput DNA sequencing and efficient computer-based analysis and comparison of sequences, it emerged that a large family of NADPH oxidases, termed NOXs, exist throughout the plant and animal kingdoms (127–129). All these NOXs conformed to the structural organization of the prototype gp91^{phox} (25, 27, 130), which inexplicably was called NOX2 (127) even though it was discovered two decades before the other NOXs. Variants of the NOXs are the dual oxidases (DUOXs), which have a very similar structure but have additional domains with homology to peroxidases, but which act as SODs,

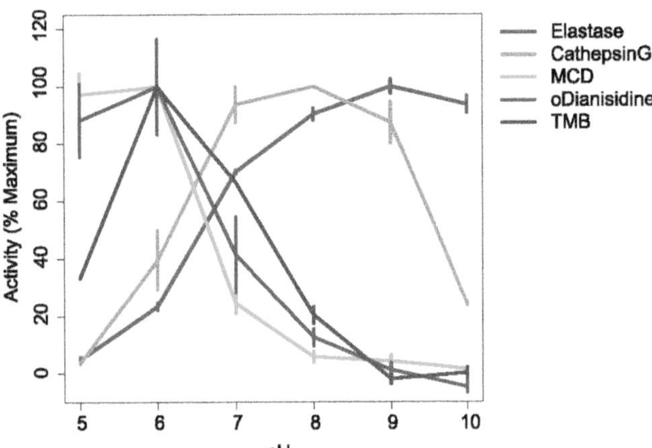

Figure 3 The effect of variations in pH on peroxidatic (TMB and o-Dianisidine) and chlorinating monochlorodimedone (MCD) activities of MPO and on the protease activities of cathepsin G and of elastase are shown. TMB, tetramethylbenzidine. From Levine et al. (40).

causing the DUOXs to generate H_2O_2 rather than O_2^- (131).

NOX3 and Inner Ear Function

The NOX3 knockout mouse gives important clues as to the general function of NOXs. These mice lack normal spatial awareness because of abnormalities of their otoconia (132). These small bodies are structures in the saccule and utricle of the inner ear and give information as to position and changes in movement. Mass is added to these proteinaceous bodies by the deposition of calcium carbonate, and it is the failure to calcify their otoconia that produces the malfunction in the NOX3 mouse. The question is why this failure of otoconial calcification occurs. The ionic composition of the endolymph of the inner ear has similarities with that of the vacuole, having a high concentration of K^+ and an alkaline pH (133, 134), the latter being required for the deposition of $CaCO_3$, the solubility of which is very pH dependent. It is probable that NOX3 plays an important part in the establishment or maintenance of these ionic conditions.

NOXs in Plants

The NOXs in plants are called RBOHs (respiratory burst oxidase homologs) and are important for the plant defense response (135). They also have more-specific functions in specialized organelles of the plant, including the pollen tubes, lateral roots, and stomatal guard cells as examples. These cells all contain large vacuoles surrounded by membranes containing a variety of ion channels. In general, K^+ enters these vacuoles, increasing the osmotic pressure and causing growth of the pollen tube (136, 137), extension of lateral roots (138), and stomatal opening (139).

NOXs in Fungi

There is growing evidence that NOXs are important for many aspects of fungal life including vegetative hyphal growth, differentiation of conidial anastomosis tubes, fruiting body and infection structure formation, and the induction of apoptosis (140).

CONCLUSION

With the discovery of the production of oxygen free radicals by neutrophils, free radical chemistry was transported from the aegis of the radiation chemist to that of the cell biologist and clinical investigator. The initial concept was that these oxygen radicals were very toxic and that they themselves could be held responsible for microbial killing in professional phagocytes and for tissue damage at sites of inflammation.

After a lot of work and a clearer understanding of free radical reactivity and toxicity, together with clinical trials of radical scavengers and antioxidants that proved therapeutically ineffective (141, 142), a consensus has emerged that these molecules are not as toxic as was originally envisaged. In the case of the neutrophil, the mantle of a toxic mechanism reverted to that of MPO-mediated halogenation. However, it was subsequently discovered that these NOXs are widely distributed throughout the biological world, and a function or functions had to be ascribed to them. With the observation that the production of these reduced oxygen products could be induced by a stimulus, and that another cellular response was also evoked, and in the absence of NOX activity this additional response was abrogated, several functions, including that of signaling molecules, have been given to these reduced oxygen products.

Careful investigation has demonstrated that the neutrophil oxidase has a dramatic effect upon the pH and ionic composition of the phagocytic vacuole, which in turn influences the functional efficiency of the myriad of enzymes released into this compartment from the cytoplasmic granules. This has a direct effect of the efficiency with which these enzymes kill and digest the phagocytosed microbes.

With the phagocytic vacuole as an example, the function of the NOXs housed in different biological niches can be examined in a different light. Rather than simply assuming that knockout or knockdown of NOXs at different sites in plants exerts its effect by the removal of a signaling mechanism, examination of the effect of the removal of an electromotive force that produces ion fluxes with secondary osmotic and concomitant mechanical forces needs to be considered, together with the consequent pH changes.

In summary, the NOXs provide a simple and highly efficient electrochemical generator. The pathways to the generation of the substrate, NADPH, are metabolically efficient, particularly in plants, where it is a primary product of photosynthesis. The NOX enzymes contain one molecule of FAD and two hemes, one close to the inner and the other to the outer surface of the membrane, and the acceptor is oxygen, which is largely ubiquitous. This simple system generates an electromotive force that can be used to drive cations in the same direction as the electrons, or anions in the opposite direction. The separation of electrons and protons alters the pH on both sides of the membrane. These physicochemical changes can be adapted to a variety of biological applications.

Acknowledgments. *We thank the Wellcome Trust, Medical Research Council, Charles Wolfson Charitable Trust, and Irwin Joffe Memorial Trust for financial support.*

Citation. Levine AP, Segal AW. 2016. The NADPH oxidase and microbial killing by neutrophils, with a particular emphasis on the proposed antimicrobial role of myeloperoxidase within the phagocytic vacuole. Microbiol Spectrum 4(4): MCHD-0018-2015.

References

1. **Sbarra AJ, Karnovsky ML.** 1959. The biochemical basis of phagocytosis. I. Metabolic changes during the ingestion of particles by polymorphonuclear leukocytes. *J Biol Chem* **234:**1355–1362.
2. **Klebanoff SJ, Hamon CB.** 1972. Role of myeloperoxidase-mediated antimicrobial systems in intact leukocytes. *J Reticuloendothel Soc* **12:**170–196.
3. **Baehner RL, Gilman N, Karnovsky ML.** 1970. Respiration and glucose oxidation in human and guinea pig leukocytes: comparative studies. *J Clin Invest* **49:**692–700.
4. **Zatti M, Rossi F.** 1965. Early changes of hexose monophosphate pathway activity and of NADPH oxidation in phagocytizing leucocytes. *Biochim Biophys Acta* **99:**557–561.
5. **Berendes H, Bridges RA, Good RA.** 1957. A fatal granulomatosus of childhood: the clinical study of a new syndrome. *Minn Med* **40:**309–312.
6. **Bridges RA, Berendes H, Good RA.** 1959. A fatal granulomatous disease of childhood; the clinical, pathological, and laboratory features of a new syndrome. *AMA J Dis Child* **97:**387–408.
7. **McCord JM, Fridovich I.** 1969. Superoxide dismutase. An enzymic function for erythrocuprein (hemocuprein). *J Biol Chem* **244:**6049–6055.
8. **Iyer GY, Islam DM, Quastel JH.** 1961. Biochemical aspects of phagocytosis. *Nature* **192:**535–541.
9. **Segal AW, Jones OT, Webster D, Allison AC.** 1978. Absence of a newly described cytochrome *b* from neutrophils of patients with chronic granulomatous disease. *Lancet* **2:**446–449.
10. **Segal AW, Jones OT.** 1978. Novel cytochrome *b* system in phagocytic vacuoles of human granulocytes. *Nature* **276:**515–517.
11. **Borregaard N, Johansen KS, Taudorff E, Wandall JH.** 1979. Cytochrome *b* is present in neutrophils from patients with chronic granulomatous disease. *Lancet* **1:** 949–951.
12. **Segal AW, Jones OT.** 1980. Absence of cytochrome *b* reduction in stimulated neutrophils from both female and male patients with chronic granulomatous disease. *FEBS Lett* **110:**111–114.
13. **Segal AW.** 1987. Absence of both cytochrome b_{-245} subunits from neutrophils in X-linked chronic granulomatous disease. *Nature* **326:**88–91.
14. **Parkos CA, Allen RA, Cochrane CG, Jesaitis AJ.** 1987. Purified cytochrome *b* from human granulocyte plasma membrane is comprised of two polypeptides with relative molecular weights of 91,000 and 22,000. *J Clin Invest* **80:**732–742.
15. **Wallach TM, Segal AW.** 1996. Stoichiometry of the subunits of flavocytochrome b_{558} of the NADPH oxidase of phagocytes. *Biochem J* **320**(Pt 1)**:**33–38.
16. **Takeya R, Ueno N, Kami K, Taura M, Kohjima M, Izaki T, Nunoi H, Sumimoto H.** 2003. Novel human homologues of $p47^{phox}$ and $p67^{phox}$ participate in activation of superoxide-producing NADPH oxidases. *J Biol Chem* **278:**25234–25246.
17. **Clark RA, Volpp BD, Leidal KG, Nauseef WM.** 1990. Two cytosolic components of the human neutrophil respiratory burst oxidase translocate to the plasma membrane during cell activation. *J Clin Invest* **85:**714–721.
18. **Segal AW, Heyworth PG, Cockcroft S, Barrowman MM.** 1985. Stimulated neutrophils from patients with autosomal recessive chronic granulomatous disease fail to phosphorylate a Mr-44,000 protein. *Nature* **316:** 547–549.
19. **Wientjes FB, Hsuan JJ, Totty NF, Segal AW.** 1993. $p40^{phox}$, a third cytosolic component of the activation complex of the NADPH oxidase to contain *src* homology 3 domains. *Biochem J* **296:**557–561.
20. **Abo A, Pick E, Hall A, Totty N, Teahan CG, Segal AW.** 1991. Activation of the NADPH oxidase involves the small GTP-binding protein $p21^{rac1}$. *Nature* **353:**668–670.
21. **Bromberg Y, Pick E.** 1985. Activation of NADPH-dependent superoxide production in a cell-free system by sodium dodecyl sulfate. *J Biol Chem* **260:**13539–13545.
22. **Curnutte JT.** 1985. Activation of human neutrophil nicotinamide adenine dinucleotide phosphate, reduced (triphosphopyridine nucleotide, reduced) oxidase by arachidonic acid in a cell-free system. *J Clin Invest* **75:** 1740–1743.

23. **Curnutte JT, Scott PJ, Mayo LA.** 1989. Cytosolic components of the respiratory burst oxidase: resolution of four components, two of which are missing in complementing types of chronic granulomatous disease. *Proc Natl Acad Sci U S A* **86:**825–829.

24. **Harper AM, Dunne MJ, Segal AW.** 1984. Purification of cytochrome b_{-245} from human neutrophils. *Biochem J* **219:**519–527.

25. **Segal AW, West I, Wientjes F, Nugent JH, Chavan AJ, Haley B, Garcia RC, Rosen H, Scrace G.** 1992. Cytochrome b_{-245} is a flavocytochrome containing FAD and the NADPH-binding site of the microbicidal oxidase of phagocytes. *Biochem J* **284:**781–788.

26. **Finegold AA, Shatwell KP, Segal AW, Klausner RD, Dancis A.** 1996. Intramembrane bis-heme motif for transmembrane electron transport conserved in a yeast iron reductase and the human NADPH oxidase. *J Biol Chem* **271:**31021–31024.

27. **Wallach TM, Segal AW.** 1997. Analysis of glycosylation sites on gp91*phox*, the flavocytochrome of the NADPH oxidase, by site-directed mutagenesis and translation *in vitro*. *Biochem J* **321:**583–585.

28. **Teahan C, Rowe P, Parker P, Totty N, Segal AW.** 1987. The X-linked chronic granulomatous disease gene codes for the β-chain of cytochrome b_{-245}. *Nature* **327:** 720–721.

29. **Royer-Pokora B, Kunkel LM, Monaco AP, Goff SC, Newburger PE, Baehner RL, Cole FS, Curnutte JT, Orkin SH.** 1986. Cloning the gene for the inherited disorder chronic granulomatous disease on the basis of its chromosomal location. *Cold Spring Harbor Symp Quant Biol* **51**(Pt 1):177–183.

30. **Agner K.** 1947. Detoxicating effect of verdoperoxidase on toxins. *Nature* **159:**271–272.

31. **Segal AW, Dorling J, Coade S.** 1980. Kinetics of fusion of the cytoplasmic granules with phagocytic vacuoles in human polymorphonuclear leukocytes. Biochemical and morphological studies. *J Cell Biol* **85:**42–59.

32. **Klebanoff SJ.** 1967. Iodination of bacteria: a bactericidal mechanism. *J Exp Med* **126:**1063–1078.

33. **Klebanoff SJ, White LR.** 1969. Iodination defect in the leukocytes of a patient with chronic granulomatous disease of childhood. *N Engl J Med* **280:**460–466.

34. **Babior BM, Kipnes RS, Curnutte JT.** 1973. Biological defense mechanisms. The production by leukocytes of superoxide, a potential bactericidal agent. *J Clin Invest* **52:**741–744.

35. **McCord JM, Wong K.** 1979. Phagocyte-produced free radicals: roles in cytotoxicity and inflammation, p 343–360. *In* Fitzsimons DW (ed), *Oxygen Free Radicals and Tissue Damage*. Elsevier, North-Holland, Amsterdam, The Netherlands.

36. **Ward PA.** 1983. Role of toxic oxygen products from phagocytic cells in tissue injury. *Adv Shock Res* **10:** 27–34.

37. **Greenwald RA.** 1991. Oxygen radicals, inflammation, and arthritis: pathophysiological considerations and implications for treatment. *Semin Arthritis Rheum* **20:** 219–240.

38. **Jacques YV, Bainton DF.** 1978. Changes in pH within the phagocytic vacuoles of human neutrophils and monocytes. *Lab Invest* **39:**179–185.

39. **Klebanoff SJ.** 1968. Myeloperoxidase-halide-hydrogen peroxide antibacterial system. *J Bacteriol* **95:**2131–2138.

40. **Levine AP, Duchen MR, de Villiers S, Rich PR, Segal AW.** 2015. Alkalinity of neutrophil phagocytic vacuoles is modulated by HVCN1 and has consequences for myeloperoxidase activity. *PLoS One* **10:**e0125906. doi: 10.1371/journal.pone.0125906.

41. **Götz F.** 2002. *Staphylococcus* and biofilms. *Mol Microbiol* **43:**1367–1378.

42. **Loutet SA, Valvano MA.** 2010. A decade of *Burkholderia cenocepacia* virulence determinant research. *Infect Immun* **78:**4088–4100.

43. **Wei JR, Lai HC.** 2006. *N*-Acylhomoserine lactone-dependent cell-to-cell communication and social behavior in the genus *Serratia*. *Int J Med Microbiol* **296:** 117–124.

44. **Mowat E, Williams C, Jones B, McChlery S, Ramage G.** 2009. The characteristics of *Aspergillus fumigatus* mycetoma development: is this a biofilm? *Med Mycol* **47** (Suppl 1):S120–S126.

45. **Mathé L, Van Dijck P.** 2013. Recent insights into *Candida albicans* biofilm resistance mechanisms. *Curr Genet* **59:**251–264.

46. **Cuperus RA, Muijsers AO, Wever R.** 1986. The superoxide dismutase activity of myeloperoxidase; formation of compound III. *Biochim Biophys Acta* **871:**78–84.

47. **Briggs RT, Karnovsky ML, Karnovsky MJ.** 1977. Hydrogen peroxide production in chronic granulomatous disease. A cytochemical study of reduced pyridine nucleotide oxidases. *J Clin Invest* **59:**1088–1098.

48. **Kettle AJ, Winterbourn CC.** 2001. A kinetic analysis of the catalase activity of myeloperoxidase. *Biochemistry* **40:**10204–10212.

49. **Clark RA, Borregaard N.** 1985. Neutrophils autoinactivate secretory products by myeloperoxidase-catalyzed oxidation. *Blood* **65:**375–381.

50. **Vissers MC, Winterbourn CC.** 1988. Oxidative inactivation of neutrophil granule proteinases: implications for neutrophil-mediated proteolysis. *Basic Life Sci* **49:** 845–848.

51. **Messina CG, Reeves EP, Roes J, Segal AW.** 2002. Catalase negative *Staphylococcus aureus* retain virulence in mouse model of chronic granulomatous disease. *FEBS Lett* **518:**107–110.

52. **Chang YC, Segal BH, Holland SM, Miller GF, Kwon-Chung KJ.** 1998. Virulence of catalase-deficient *Aspergillus nidulans* in p47$^{phox-/-}$ mice. Implications for fungal pathogenicity and host defense in chronic granulomatous disease. *J Clin Invest* **101:**1843–1850.

53. **Klebanoff SJ, Kettle AJ, Rosen H, Winterbourn CC, Nauseef WM.** 2013. Myeloperoxidase: a front-line defender against phagocytosed microorganisms. *J Leukoc Biol* **93:**185–198.

54. **Jiang Q, Griffin DA, Barofsky DF, Hurst JK.** 1997. Intraphagosomal chlorination dynamics and yields

determined using unique fluorescent bacterial mimics. *Chem Res Toxicol* **10**:1080–1089.

55. **Koide Y, Urano Y, Hanaoka K, Terai T, Nagano T.** 2011. Development of an Si-rhodamine-based far-red to near-infrared fluorescence probe selective for hypochlorous acid and its applications for biological imaging. *J Am Chem Soc* **133**:5680–5682.

56. **Painter RG, Valentine VG, Lanson NA Jr, Leidal K, Zhang Q, Lombard G, Thompson C, Viswanathan A, Nauseef WM, Wang G, Wang G.** 2006. CFTR expression in human neutrophils and the phagolysosomal chlorination defect in cystic fibrosis. *Biochemistry* **45**: 10260–10269.

57. **Painter RG, Bonvillain RW, Valentine VG, Lombard GA, LaPlace SG, Nauseef WM, Wang G.** 2008. The role of chloride anion and CFTR in killing of *Pseudomonas aeruginosa* by normal and CF neutrophils. *J Leukoc Biol* **83**:1345–1353.

58. **Chapman AL, Hampton MB, Senthilmohan R, Winterbourn CC, Kettle AJ.** 2002. Chlorination of bacterial and neutrophil proteins during phagocytosis and killing of *Staphylococcus aureus*. *J Biol Chem* **277**: 9757–9762.

59. **Green JN, Kettle AJ, Winterbourn CC.** 2014. Protein chlorination in neutrophil phagosomes and correlation with bacterial killing. *Free Radic Biol Med* **77**:49–56.

60. **Segal AW, Garcia RC, Harper AM, Banga JP.** 1983. Iodination by stimulated human neutrophils. Studies on its stoichiometry, subcellular localization and relevance to microbial killing. *Biochem J* **210**:215–225.

61. **Reeves EP, Nagl M, Godovac-Zimmermann J, Segal AW.** 2003. Reassessment of the microbicidal activity of reactive oxygen species and hypochlorous acid with reference to the phagocytic vacuole of the neutrophil granulocyte. *J Med Microbiol* **52**:643–651.

62. **Cech P, Lehrer RI.** 1984. Phagolysosomal pH of human neutrophils. *Blood* **63**:88–95.

63. **Cech P, Lehrer RI.** 1984. Heterogeneity of human neutrophil phagolysosomes: functional consequences for candidacidal activity. *Blood* **64**:147–151.

64. **Paiva CN, Bozza MT.** 2014. Are reactive oxygen species always detrimental to pathogens? *Antioxid Redox Signal* **20**:1000–1037.

65. **Winterbourn CC, Kettle AJ.** 2013. Redox reactions and microbial killing in the neutrophil phagosome. *Antioxid Redox Signal* **18**:642–660.

66. **Hurst JK, Barrette WC Jr.** 1989. Leukocytic oxygen activation and microbicidal oxidative toxins. *Crit Rev Biochem Mol Biol* **24**:271–328.

67. **Parry MF, Root RK, Metcalf JA, Delaney KK, Kaplow LS, Richar WJ.** 1981. Myeloperoxidase deficiency: prevalence and clinical significance. *Ann Intern Med* **95**:293–301.

68. **Becker R, Pflüger KH.** 1994. Myeloperoxidase deficiency: an epidemiological study and flow-cytometric detection of other granular enzymes in myeloperoxidase-deficient subjects. *Ann Hematol* **69**:199–203.

69. **Nunoi H, Kohi F, Kajiwara H, Suzuki K.** 2003. Prevalence of inherited myeloperoxidase deficiency in Japan. *Microbiol Immunol* **47**:527–531.

70. **Aratani Y, Kura F, Watanabe H, Akagawa H, Takano Y, Suzuki K, Maeda N, Koyama H.** 2000. Differential host susceptibility to pulmonary infections with bacteria and fungi in mice deficient in myeloperoxidase. *J Infect Dis* **182**:1276–1279.

71. **Schapiro BL, Newburger PE, Klempner MS, Dinauer MC.** 1991. Chronic granulomatous disease presenting in a 69-year-old man. *N Engl J Med* **325**:1786–1790.

72. **Winkelstein JA, Marino MC, Johnston RB Jr, Boyle J, Curnutte J, Gallin JI, Malech HL, Holland SM, Ochs H, Quie P, Buckley RH, Foster CB, Chanock SJ, Dickler H.** 2000. Chronic granulomatous disease. Report on a national registry of 368 patients. *Medicine (Baltimore)* **79**:155–169.

73. **Aratani Y, Kura F, Watanabe H, Akagawa H, Takano Y, Suzuki K, Dinauer MC, Maeda N, Koyama H.** 2002. Relative contributions of myeloperoxidase and NADPH-oxidase to the early host defense against pulmonary infections with *Candida albicans* and *Aspergillus fumigatus*. *Med Mycol* **40**:557–563.

74. **Aratani Y, Kura F, Watanabe H, Akagawa H, Takano Y, Suzuki K, Dinauer MC, Maeda N, Koyama H.** 2002. Critical role of myeloperoxidase and nicotinamide adenine dinucleotide phosphate-oxidase in high-burden systemic infection of mice with *Candida albicans*. *J Infect Dis* **185**:1833–1837.

75. **Lardner A.** 2001. The effects of extracellular pH on immune function. *J Leukoc Biol* **69**:522–530.

76. **Wolach B, Gavrieli R, Roos D, Berger-Achituv S.** 2012. Lessons learned from phagocytic function studies in a large cohort of patients with recurrent infections. *J Clin Immunol* **32**:454–466.

77. **Kolset SO, Gallagher JT.** 1990. Proteoglycans in haemopoietic cells. *Biochim Biophys Acta* **1032**:191–211.

78. **Borregaard N, Cowland JB.** 1997. Granules of the human neutrophilic polymorphonuclear leukocyte. *Blood* **89**:3503–3521.

79. **Pham CT.** 2006. Neutrophil serine proteases: specific regulators of inflammation. *Nat Rev Immunol* **6**:541–550.

80. **Cohn ZA, Hirsch JG.** 1960. The isolation and properties of the specific cytoplasmic granules of rabbit polymorphonuclear leucocytes. *J Exp Med* **112**:983–1004.

81. **Odeberg H, Olsson I.** 1976. Mechanisms for the microbicidal activity of cationic proteins of human granulocytes. *Infect Immun* **14**:1269–1275.

82. **Segal AW, Geisow M, Garcia R, Harper A, Miller R.** 1981. The respiratory burst of phagocytic cells is associated with a rise in vacuolar pH. *Nature* **290**:406–409.

83. **Tkalcevic J, Novelli M, Phylactides M, Iredale JP, Segal AW, Roes J.** 2000. Impaired immunity and enhanced resistance to endotoxin in the absence of neutrophil elastase and cathepsin G. *Immunity* **12**:201–210.

84. **Reeves EP, Lu H, Jacobs HL, Messina CGM, Bolsover S, Gabella G, Potma EO, Warley A, Roes J, Segal AW.** 2002. Killing activity of neutrophils is mediated through activation of proteases by K^+ flux. *Nature* **416**:291–297.

85. **Hahn I, Klaus A, Janze AK, Steinwede K, Ding N, Bohling J, Brumshagen C, Serrano H, Gauthier F, Paton**

JC, Welte T, Maus UA. 2011. Cathepsin G and neutrophil elastase play critical and nonredundant roles in lung-protective immunity against *Streptococcus pneumoniae* in mice. *Infect Immun* **79:**4893–4901.

86. **Steinwede K, Maus R, Bohling J, Voedisch S, Braun A, Ochs M, Schmiedl A, Länger F, Gauthier F, Roes J, Welte T, Bange FC, Niederweis M, Bühling F, Maus UA.** 2012. Cathepsin G and neutrophil elastase contribute to lung-protective immunity against mycobacterial infections in mice. *J Immunol* **188:**4476–4487.
87. **Standish AJ, Weiser JN.** 2009. Human neutrophils kill *Streptococcus pneumoniae* via serine proteases. *J Immunol* **183:**2602–2609.
88. **Belaaouaj A, McCarthy R, Baumann M, Gao Z, Ley TJ, Abraham SN, Shapiro SD.** 1998. Mice lacking neutrophil elastase reveal impaired host defense against Gram negative bacterial sepsis. *Nat Med* **4:**615–618.
89. **MacIvor DM, Shapiro SD, Pham CT, Belaaouaj A, Abraham SN, Ley TJ.** 1999. Normal neutrophil function in cathepsin G-deficient mice. *Blood* **94:**4282–4293.
90. **Vethanayagam RR, Almyroudis NG, Grimm MJ, Lewandowski DC, Pham CT, Blackwell TS, Petraitiene R, Petraitis V, Walsh TJ, Urban CF, Segal BH.** 2011. Role of NADPH oxidase versus neutrophil proteases in antimicrobial host defense. *PLoS One* **6:**e28149. doi:10.1371/journal.pone.0028149.
91. **Sellers RS, Clifford CB, Treuting PM, Brayton C.** 2012. Immunological variation between inbred laboratory mouse strains: points to consider in phenotyping genetically immunomodified mice. *Vet Pathol* **49:**32–43.
92. **Hewitt C, McCormick D, Linden G, Turk D, Stern I, Wallace I, Southern L, Zhang L, Howard R, Bullon P, Wong M, Widmer R, Gaffar KA, Awawdeh L, Briggs J, Yaghmai R, Jabs EW, Hoeger P, Bleck O, Rüdiger SG, Petersilka G, Battino M, Brett P, Hattab F, Al-Hamed M, Sloan P, Toomes C, Dixon M, James J, Read AP, Thakker N.** 2004. The role of cathepsin C in Papillon-Lefèvre syndrome, prepubertal periodontitis, and aggressive periodontitis. *Hum Mutat* **23:**222–228.
93. **Adkison AM, Raptis SZ, Kelley DG, Pham CTN.** 2002. Dipeptidyl peptidase I activates neutrophil-derived serine proteases and regulates the development of acute experimental arthritis. *J Clin Invest* **109:**363–371.
94. **Turk D, Janjić V, Stern I, Podobnik M, Lamba D, Dahl SW, Lauritzen C, Pedersen J, Turk V, Turk B.** 2001. Structure of human dipeptidyl peptidase I (cathepsin C): exclusion domain added to an endopeptidase framework creates the machine for activation of granular serine proteases. *EMBO J* **20:**6570–6582.
95. **Pham CT, Ley TJ.** 1999. Dipeptidyl peptidase I is required for the processing and activation of granzymes A and B *in vivo*. *Proc Natl Acad Sci U S A* **96:**8627–8632.
96. **Adkison AM, Raptis SZ, Kelley DG, Pham CT.** 2002. Dipeptidyl peptidase I activates neutrophil-derived serine proteases and regulates the development of acute experimental arthritis. *J Clin Invest* **109:**363–371.
97. **Sørensen OE, Clemmensen SN, Dahl SL, Østergaard O, Heegaard NH, Glenthøj A, Nielsen FC, Borregaard N.** 2014. Papillon-Lefèvre syndrome patient reveals species-dependent requirements for neutrophil defenses. *J Clin Invest* **124:**4539–4548.
98. **Pham CT, Ivanovich JL, Raptis SZ, Zehnbauer B, Ley TJ.** 2004. Papillon-Lefèvre syndrome: correlating the molecular, cellular, and clinical consequences of cathepsin C/dipeptidyl peptidase I deficiency in humans. *J Immunol* **173:**7277–7281.
99. **Djawari D.** 1978. Deficient phagocytic function in Papillon-Lefèvre syndrome. *Dermatologica* **156:**189–192.
100. **Ghaffer KA, Zahran FM, Fahmy HM, Brown RS.** 1999. Papillon-Lefèvre syndrome: neutrophil function in 15 cases from 4 families in Egypt. *Oral Surg Oral Med Oral Pathol Oral Radiol Endod* **88:**320–325.
101. **Clerehugh V, Drucker DB, Seymour GJ, Bird PS.** 1996. Microbiological and serological investigations of oral lesions in Papillon-Lefèvre syndrome. *J Clin Pathol* **49:**255–257.
102. **de Haar SF, Hiemstra PS, van Steenbergen MT, Everts V, Beertsen W.** 2006. Role of polymorphonuclear leukocyte-derived serine proteinases in defense against *Actinobacillus actinomycetemcomitans*. *Infect Immun* **74:**5284–5291.
103. **Miyasaki KT, Bodeau AL.** 1991. In vitro killing of oral *Capnocytophaga* by granule fractions of human neutrophils is associated with cathepsin G activity. *J Clin Invest* **87:**1585–1593.
104. **de Haar SF, Tigchelaar-Gutter W, Everts V, Beertsen W.** 2006. Structure of the periodontium in cathepsin C-deficient mice. *Eur J Oral Sci* **114:**171–173.
105. **Ansari-Lari MA, Kickler TS, Borowitz MJ.** 2003. Immature granulocyte measurement using the Sysmex XE-2100. Relationship to infection and sepsis. *Am J Clin Pathol* **120:**795–799.
106. **Nauland U, Rijken DC.** 1994. Activation of thrombin-inactivated single-chain urokinase-type plasminogen activator by dipeptidyl peptidase I (cathepsin C). *Eur J Biochem* **223:**497–501.
107. **Sissons CH, Wong L, Shu M.** 1998. Factors affecting the resting pH of *in vitro* human microcosm dental plaque and *Streptococcus mutans* biofilms. *Arch Oral Biol* **43:**93–102.
108. **Nauseef WM.** 2014. Proteases, neutrophils, and periodontitis: the NET effect. *J Clin Invest* **124:**4237–4239.
109. **Jensen MS, Bainton DF.** 1973. Temporal changes in pH within the phagocytic vacuole of the polymorphonuclear neutrophilic leukocyte. *J Cell Biol* **56:**379–388.
110. **Dri P, Presani G, Perticarari S, Albèri L, Prodan M, Decleva E.** 2002. Measurement of phagosomal pH of normal and CGD-like human neutrophils by dual fluorescence flow cytometry. *Cytometry* **48:**159–166.
111. **Bernardo J, Long HJ, Simons ER.** 2010. Initial cytoplasmic and phagosomal consequences of human neutrophil exposure to *Staphylococcus epidermidis*. *Cytometry A* **77:**243–252.
112. **Mayer SJ, Keen PM, Craven N, Bourne FJ.** 1989. Regulation of phagolysosome pH in bovine and human neutrophils: the role of NADPH oxidase activity and an Na^+/H^+ antiporter. *J Leukoc Biol* **45:**239–248.

113. **Hurst JK, Albrich JM, Green TR, Rosen H, Klebanoff S.** 1984. Myeloperoxidase-dependent fluorescein chlorination by stimulated neutrophils. *J Biol Chem* **259:** 4812–4821.

114. **Bassnett S, Reinisch L, Beebe DC.** 1990. Intracellular pH measurement using single excitation-dual emission fluorescence ratios. *Am J Physiol* **258:**C171–C178.

115. **El Chemaly A, Nunes P, Jimaja W, Castelbou C, Demaurex N.** 2014. Hv1 proton channels differentially regulate the pH of neutrophil and macrophage phagosomes by sustaining the production of phagosomal ROS that inhibit the delivery of vacuolar ATPases. *J Leukoc Biol* **95:**1–13.

116. **Jankowski A, Scott CC, Grinstein S.** 2002. Determinants of the phagosomal pH in neutrophils. *J Biol Chem* **277:**6059–6066.

117. **Klebanoff SJ.** 1970. Myeloperoxidase: contribution to the microbicidal activity of intact leukocytes. *Science* **169:**1095–1097.

118. **Hampton MB, Kettle AJ, Winterbourn CC.** 1996. Involvement of superoxide and myeloperoxidase in oxygen-dependent killing of *Staphylococcus aureus* by neutrophils. *Infect Immun* **64:**3512–3517.

119. **Henderson LM, Chappell JB, Jones OT.** 1988. Superoxide generation by the electrogenic NADPH oxidase of human neutrophils is limited by the movement of a compensating charge. *Biochem J* **255:**285–290.

120. **Ramsey IS, Ruchti E, Kaczmarek JS, Clapham DE.** 2009. Hv1 proton channels are required for high-level NADPH oxidase-dependent superoxide production during the phagocyte respiratory burst. *Proc Natl Acad Sci U S A* **106:**7642–7647.

121. **Korkmaz B, Horwitz MS, Jenne DE, Gauthier F.** 2010. Neutrophil elastase, proteinase 3, and cathepsin G as therapeutic targets in human diseases. *Pharmacol Rev* **62:**726–759.

122. **Magnani A, Brosselin P, Beauté J, de Vergnes N, Mouy R, Debré M, Suarez F, Hermine O, Lortholary O, Blanche S, Fischer A, Mahlaoui N.** 2014. Inflammatory manifestations in a single-center cohort of patients with chronic granulomatous disease. *J Allergy Clin Immunol* **134:**655–662.e8. doi:10.1016/j.jaci.2014.04.014.

123. **Rieber N, Hector A, Kuijpers T, Roos D, Hartl D.** 2012. Current concepts of hyperinflammation in chronic granulomatous disease. *Clin Dev Immunol* **2012:**252460. doi:10.1155/2012/252460.

124. **Schäppi MG, Jaquet V, Belli DC, Krause KH.** 2008. Hyperinflammation in chronic granulomatous disease and anti-inflammatory role of the phagocyte NADPH oxidase. *Semin Immunopathol* **30:**255–271.

125. **Kettle AJ, Anderson RF, Hampton MB, Winterbourn CC.** 2007. Reactions of superoxide with myeloperoxidase. *Biochemistry* **46:**4888–4897.

126. **Feldmesser M.** 2006. Role of neutrophils in invasive aspergillosis. *Infect Immun* **74:**6514–6516.

127. **Lassègue B, Sorescu D, Szöcs K, Yin Q, Akers M, Zhang Y, Grant SL, Lambeth JD, Griendling KK.** 2001. Novel gp91^{phox} homologues in vascular smooth muscle cells: nox1 mediates angiotensin II-induced superoxide formation and redox-sensitive signaling pathways. *Circ Res* **88:**888–894.

128. **Lambeth JD.** 2002. Nox/Duox family of nicotinamide adenine dinucleotide (phosphate) oxidases. *Curr Opin Hematol* **9:**11–17.

129. **Bedard K, Lardy B, Krause KH.** 2007. NOX family NADPH oxidases: not just in mammals. *Biochimie* **89:** 1107–1112.

130. **Shatwell KP, Dancis A, Cross AR, Klausner RD, Segal AW.** 1996. The FRE1 ferric reductase of *Saccharomyces cerevisiae* is a cytochrome *b* similar to that of NADPH oxidase. *J Biol Chem* **271:**14240–14244.

131. **Meitzler JL, Ortiz de Montellano PR.** 2009. *Caenorhabditis elegans* and human dual oxidase 1 (DUOX1) "peroxidase" domains: insights into heme binding and catalytic activity. *J Biol Chem* **284:**18634–18643.

132. **Paffenholz R, Bergstrom RA, Pasutto F, Wabnitz P, Munroe RJ, Jagla W, Heinzmann U, Marquardt A, Bareiss A, Laufs J, Russ A, Stumm G, Schimenti JC, Bergstrom DE.** 2004. Vestibular defects in head-tilt mice result from mutations in Nox3, encoding an NADPH oxidase. *Genes Dev* **18:**486–491.

133. **Trune DR.** 2010. Ion homeostasis in the ear: mechanisms, maladies, and management. *Curr Opin Otolaryngol Head Neck Surg* **18:**413–419.

134. **Hibino H, Nin F, Tsuzuki C, Kurachi Y.** 2010. How is the highly positive endocochlear potential formed? The specific architecture of the stria vascularis and the roles of the ion-transport apparatus. *Pflugers Arch* **459:** 521–533.

135. **Torres MA, Dangl JL, Jones JD.** 2002. *Arabidopsis* gp91^{phox} homologues *AtrbohD* and *AtrbohF* are required for accumulation of reactive oxygen intermediates in the plant defense response. *Proc Natl Acad Sci U S A* **99:**517–522.

136. **Potocký M, Jones MA, Bezvoda R, Smirnoff N, Zárský V.** 2007. Reactive oxygen species produced by NADPH oxidase are involved in pollen tube growth. *New Phytol* **174:**742–751.

137. **Tavares B, Domingos P, Dias PN, Feijó JA, Bicho A.** 2011. The essential role of anionic transport in plant cells: the pollen tube as a case study. *J Exp Bot* **62:**2273–2298.

138. **Montiel J, Arthikala MK, Quinto C.** 2013. *Phaseolus vulgaris* RbohB functions in lateral root development. *Plant Signal Behav* **8:**e22694. doi:10.4161/psb.22694.

139. **Zhang H, Fang Q, Zhang Z, Wang Y, Zheng X.** 2009. The role of respiratory burst oxidase homologues in elicitor-induced stomatal closure and hypersensitive response in *Nicotiana benthamiana*. *J Exp Bot* **60:** 3109–3122.

140. **Tudzynski P, Heller J, Siegmund U.** 2012. Reactive oxygen species generation in fungal development and pathogenesis. *Curr Opin Microbiol* **15:**653–659.

141. **Steinhubl SR.** 2008. Why have antioxidants failed in clinical trials? *Am J Cardiol* **101:**14D–19D.

142. **Polidori MC, Nelles G.** 2014. Antioxidant clinical trials in mild cognitive impairment and Alzheimer's disease—challenges and perspectives. *Curr Pharm Des* **20:**3083–3092.

Myeloid Cells in Health and Disease: A Synthesis
Edited by Siamon Gordon

doi:10.1128/microbiolspec.MCHD-0050-2016

Ryan G. Gaudet[1,2,3]
Clinton J. Bradfield[1,2,3]
John D. MacMicking[1,2,3]

34 Evolution of Cell-Autonomous Effector Mechanisms in Macrophages versus Non-Immune Cells

INTRODUCTION

Microorganisms maintain an evolutionary advantage over their slowly replicating eukaryotic hosts. High mutation rates, rapid doubling times, and free genetic exchange between microbial species often place a considerable burden on the infected host to counter virulence escape mechanisms. This selective pressure has driven the acquisition of numerous eukaryotic defense strategies to protect host genome integrity and promote survival at the level of the individual cell (1). These cell-autonomous effector mechanisms, often considered unique to the immune cells of advanced metazoans, have in fact been largely inherited and repurposed from our eukaryotic ancestors (Fig. 1). For example, phagocytosis developed as a trophic mechanism in unicellular amoebae long before its adaptation as a tool for immunity in the specialized "immune-like" cells of early invertebrates (2, 3). Amebocytes, hemocytes, and coelomocytes present in lower organisms likewise predate professional phagocytes in animals with their ability to bind, engulf, and kill foreign microorganisms (4).

The appearance of multicellular organization ~600 million years ago (Mya), coupled with the evolutionary arms race between host and microbe, saw the eventual emergence of a dedicated immune system (4). With this came a remarkable degree of specialization to counter the temporal and genetic advantage held by pathogens. In animals, hematopoietic cells developed extensive machinery to detect and respond to microbial and cellular host signatures through unique immunoreceptors. Such receptors endow immune cells with a capacity to survey, sequester, and ultimately destroy microbial pathogens as well as produce paracrine and autocrine signaling molecules that invoke drastic changes in local tissue microenvironments (5).

With the advent of multicellularity, however, came a new challenge: pathogen cell tropism. Here, the increased number and diversity of host cell types offered

[1]Howard Hughes Medical Institute, Chevy Chase, MD 20815; [2]Yale Systems Biology Institute; [3]Departments of Microbial Pathogenesis and Immunobiology, Yale University School of Medicine, New Haven, CT 06520.

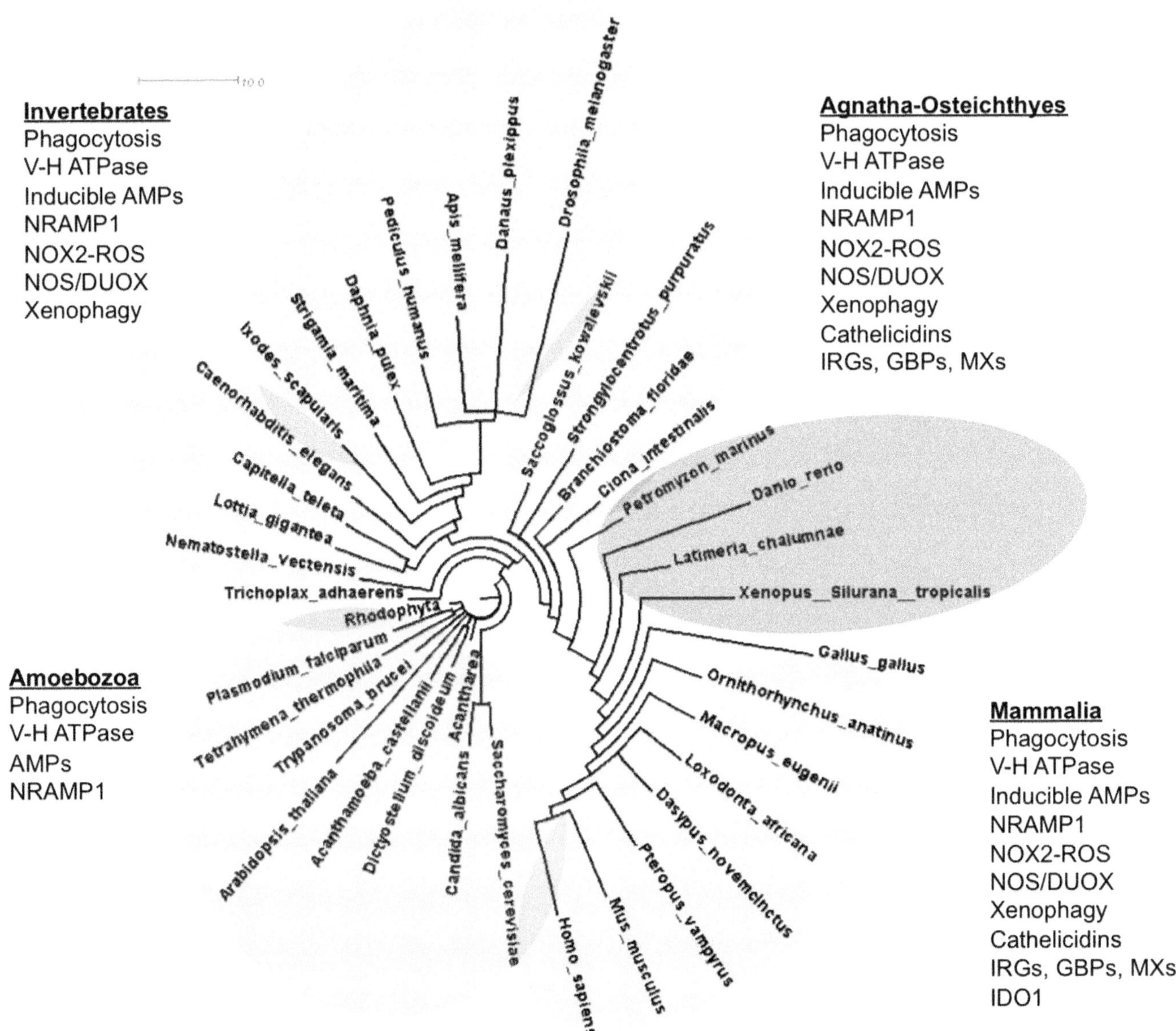

Figure 1 Evolution of antimicrobial effector mechanisms. Depicted is a phylogenetic tree of the unikonts (*Amoebozoa* and *Opisthokonta*) and a summary of accompanying cell-autonomous effector mechanisms common to each major group. Scale indicates divergent nodal distance across NCBI taxa. Phylogram generated in Dendroscope 3.

potential refuge for taxonomically distinct microbes that target selected lineages for replication. As a consequence, intrinsic defense mechanisms also emerged in non-immune cells as well (1, 6). Many of these restriction factors evolved from the defense arsenal of lower organisms like that seen for the classical immune system, and are thus operative in most nucleated cells. For example, nitric oxide synthases (NOSs) serve an antimicrobial function not just in mammalian macrophages but also in hepatocytes, neurons, fibroblasts, and smooth muscle (6, 7). These mammalian NO-mediated killing mechanisms were, in turn, presaged by those found in flies (8), crustaceans (9), and even Gram-positive bacteria (10).

In this chapter, we probe the evolutionary record for clues about the ancient and diverse phylogenetic origins of macrophage killing mechanisms. We demonstrate how some of their properties are shared with cell lineages outside the traditional bounds of immunity in higher vertebrates such as mammals and speculate on their historical legacy for cell-autonomous defense.

PHYLOGENETIC ORIGINS OF MACROPHAGE KILLING MECHANISMS

Amoeboid Defenses

The phylum *Amoebozoa* emerged soon after the divergence of plants, forming a sister group to animals and fungi that provides a glimpse into the effector mechanisms operating before the divergence of metazoans (11–13). These unicellular organisms are highly phagocytic cells in perpetual contact with bacteria in the environment. Much of the basic machinery and signal transduction pathways of phagocytosis are evolutionarily conserved between amoeba and vertebrate macrophages, reflecting the ancient origins of this process (14). Model amoebae like *Dictyostelium discoideum* can offer invaluable insights into specific bacterial killing mechanisms that have been retained in the common ancestor of plants and animals. The production of superoxide radicals (O_2^-) by the NADPH complex that becomes targeted to the phagosome during phagocytosis is one such example (15). In mammalian macrophages, this machinery is a potent antimicrobial pathway, and there is some evidence for respiratory burst activity in *Acanthamoeba* (16). *Dictyostelium* strains lacking NADPH oxidases, however, still display normal phagocytosis and bacterial killing profiles (17). Thus, *Dictyostelium* reveals the existence of other cell-intrinsic killing mechanisms besides oxidant defense. For these amoebae, the bulk of bacterial restriction appears reliant on vacuole acidification, when acidic vacuoles containing the V-H^+-ATPase and lysosomal hydrolases fuse with the contractile vacuole to generate the phagolysosome (18–20). Mammalian macrophages also enlist phagolysosomal killing in certain settings as well (6).

A survey of the *Dictyostelium* genome reveals an extensive repertoire of pore-forming peptides that resemble human defensins (21), and the amoebapore used by *Entamoeba histolytica* is a homolog of the NK-lysin produced by cytotoxic T lymphocytes (22). Indeed, the amoebapore shares an unusual pattern of cysteines with other saposin-like proteins found across phylogeny from nematodes to mammals (23). Like mammalian antimicrobial defensins, these peptides use highly charged residues to permeabilize the bacterial membrane and cause bacteriolysis. In *Amoebozoa*, they are delivered to the contractile vacuole to aid in bacterial killing before degradation by the lysosomal hydrolases. Among these hydrolases is lysozyme, which plays a key role in host defense in animals. Several lysozyme-encoding genes are present within the *D. discoideum* genome, belonging to either the bacteriophage T4 or conventional (C-type) families that exhibit preferential activity against Gram-positive bacteria (14).

Whereas amoebae engulf, kill, and digest bacteria primarily as a food source, they can occasionally serve as hosts for parasitic bacteria like *Legionella* and *Mycobacterium* that enter during phagocytosis to survive intracellularly. Accordingly, amoebae have evolved effector mechanisms that attempt to restrict the growth of these facultative bacteria. Natural resistance-associated membrane protein 1 (NRAMP1), a divalent metal transporter that is widely distributed from bacteria to humans and found on prelysosomal vesicles prior to vacuole acidification (24), is deployed against both *Mycobacterium* and *Legionella* spp. to combat infection. Eukaryotic NRAMPs transport Fe^{2+} and Mn^{2+} (25, 26), whereas bacterial homologs solely transport Mn^{2+} (27). In mammals, two NRAMP isoforms exist—NRAMP1 (SLC11A1) and NRAMP2 (SLC11A2)—with the former expressed exclusively within the endocytic pathway of macrophages and the latter exhibiting plasma membrane expression across many tissue types (25).

The *Dictyostelium* genome likewise encodes two *NRAMP* genes, with NRAMP1 orthologous to mammalian NRAMP1, as it contains a single N-terminal intron conserved in all eukaryotes (28) and is localized to phagosomal membranes. This transporter is essential for amoeboid defense against *Mycobacterium* and *Legionella* since genetic disruption of *NRAMP1* renders *Dictyostelium* more susceptible to these pathogenic bacilli (29, 30). Interestingly, phagocytosis and digestion of nonpathogenic bacteria are unimpaired in the *NRAMP1* mutant, suggesting that it is positioned solely for defense against virulent organisms (29). *Dictyostelium* NRAMP1 arguably represents the first appearance of a so-called nutriprive mechanism during evolution, using the proton gradient in the phagosome to electrogenically pump Fe^{2+} out of the bacteria-containing vacuole (31). This depletion restricts the intracellular growth of *Legionella* and *Mycobacterium*, which normally replicate within the phagosome by interfering with recruitment of the V-H^+-ATPase and concomitant vacuole acidification.

In addition to the free-living form that feed on soil bacteria, *Dictyostelium* is a social amoeba, insomuch that during starvation conditions, solitary amoebae aggregate into large groups to form a multicellular slug. This slug can migrate for several days before forming a fruiting body complete with a cellular stock and environment-resistant spores (32). During the multicellular slug stage, the selective pressure to avoid parasitic exploitation is increased (33). Remarkably, this pressure appears to have driven a subgroup of these amoebae, termed sentinel cells (S-cells), to develop immune-like functions complete with Toll/interleukin-1 receptor domain signaling pathways (34) and the ability to deploy extracellular DNA nets (or traps) that can sequester and kill bacteria (35). Thus, amoebae enlist both offensive and defensive effector mechanisms that are conserved throughout the evolution of mammals, implying that the ability to kill bacteria, either as a food source or for self-defense, emerged well before the appearance of metazoans.

Invertebrate Defenses

The rise of multicellularity during the Paleozoic era (541 Mya to 252 Mya) brought about both an expansion of amoeboid effector mechanisms and the evolution of a new group of effector molecules that protect larger multicellular animals from infection. Contrary to popular belief, both invertebrates and vertebrates employ specialized cell types that harness the phagocytic and antimicrobial abilities of amoebae to function as specialized "immune-like" progenitors (36). The most primitive groups, the sponges and cnidarians, carry single specialized cell types, amebocytes or interstitial cells, respectively, that perform a phagocytic function in their respective organ (3). The evolutionary importance of these cell types is reflected by their rapid diversification in more-advanced invertebrates (echinoderms and urochordates), where both granulocyte- and macrophage-like cells become apparent (37–39). However, even in invertebrate animals, these defense mechanisms are not exclusive to immune-like cells. Antimicrobial effector mechanisms are found in organs that are not traditionally considered to have evolved immune-like functions, such as the epithelia of the reproductive, respiratory, and digestive tracts in *Drosophila* (40) or the midgut epithelium of the *Anopheles gambiae* mosquito (41).

Together with phagocytosis, antimicrobial peptides (AMPs) represent the most fundamental defense mechanism across all phylogeny and comprise the main effector arm of the invertebrate immune system. In contrast with *Amoebozoa*, in which AMP production is constitutive (42), pathogens or their products selectively induce AMPs in complex invertebrates such as *Caenorhabditis elegans* (43) and *Drosophila melanogaster* (44). AMPs can be produced by phagocytic cells, such as in arthropods, annelids, and urochordates to aid bacterial destruction (36, 45), or by somatic cells of the fat body and epithelia of renal tubules and the digestive tract in *Drosophila* (8, 40). Here, flies infected with entomopathogenic fungi trigger signaling via the Toll and Imd pathways to elicit AMPs with specific antifungal activities (46). In addition to their role in cell-autonomous defense, AMPs serve as the systemic arm of the invertebrate immune response, some circulating in the hemolymph for several days.

Invertebrate phagocytes readily produce reactive oxygen (ROS) and nitrogen species (RNS) during phagocytosis to aid in microbial killing. Nitric oxide (·NO) is a highly reactive labile gas made by NOSs from the guanidino nitrogen of L-arginine and molecular oxygen. It serves numerous biological roles related to metamorphosis, feeding, and nervous system and vascular signaling, along with its fundamental role in host defense (7, 9). Genomic analysis of marine invertebrate NOSs indicate that the cofactor binding sites for flavin adenine dinucleotide, NADPH, heme, tetrahydrobiopterin, and calmodulin found in mammalian NOSs are all well conserved in these lower organisms (9). Moreover, some marine species, such as the sea urchin *Arbacia punctulata*, encode multiple *NOS* genes that display striking homology to both constitutive (NOS1, NOS3) and inducible (NOS2) NOSs in mammals (47), reaffirming the long ancestry of this system. The *Anopheles* mosquito NOS displays the highest homology with mammalian inducible NOSs, complete with a lipopolysaccharide (LPS) and inflammatory cytokine response elements in its promoter (48). *Anopheles* NOS is induced by the JAK/STAT pathway in the midgut during infection with *Plasmodium falciparum* and acts to inhibit oocyst development and promote immunity (41).

Perhaps the most primordial ·NO defense network in eukaryotes appeared with the emergence of the sponges some 600 Mya, in which dendritic-like cells in the parenchyma generate ·NO under a variety of stress conditions (49). As a reflection of the system's antiquity, hemocytes from the living fossil horseshoe crab *Limulus polyphemus*, an ancestor to both mammalian platelets and macrophages, deploy ·NO to regulate their aggregation (platelet function) and antimicrobial properties (macrophage function) (9). Similarly, hemocytes of the molluscs *Mytilus edulis* and *Viviparus ater* produce ·NO during the later stages of phagocytosis and confer cytotoxicity by inducing bacterial clumping (50, 51). Again, a role for NOS in antibacterial defense is not limited to immune-like cells, as peripheral tissue in renal tubules in *Drosophila* and *Anopheles* can autonomously respond to microbial challenge by inducing NOS (8). In addition to being directly antimicrobial, ·NO in turn increases systemic production of the AMP diptericin, which is required for whole-organism survival following bacterial challenge (52).

Besides RNS, ROS are also generated from reduction of molecular oxygen to superoxide ($O_2\cdot^-$) by the NADPH oxidase (NOX2) complex in a reaction known as the respiratory burst. A variety of highly reactive oxygen species are then formed that are directly injurious to microbes (53). In contrast to NOSs, which can be produced by immune and non-immune cells, NOX2 generation of ROS is tightly linked with phagocytosis and is produced by immune-like cells across phylogeny; these span hemocytes from the cockroach *Blaberus discoidalis* (54) or the mussel *Mytilus galloprovincialis* (55) to the dendritic-like cells of the marine sponge *Sycon* sp. (56). In mammals, NOX2 is a multicomponent complex consisting of membrane-bound subunits ($p22^{phox}$ and $gp91^{phox}$ [cytochrome b_{558}]), cytosolic subunits ($p40^{phox}$, $p47^{phox}$, and $p67^{phox}$), and members of the Ras superfamily of GTP-binding proteins (Rac1 or Rac2). During phagocytosis or cell activation, the cytosolic components are phosphorylated and relocate to the phagocytic cup, assembling into an active holoenzyme that uses molecular oxygen on the luminal side of the phagosome to generate ROS for killing engulfed prey (36, 57).

Homologous enzymes can be found in most eukaryotes, but to date no evidence of their utilization is found among prokaryotes. Even so, the presence of a *bis*-heme binding motif in $gp91^{phox}$ suggests possible evolutionary links with prokaryotic *b*-type cytochromes (58). $gp91^{phox}$ appears to be the most ancient subunit, having been present before the divergence of choanoflagellates, the closest living relatives to animals. However, $gp91^{phox}$ seems to have been lost in some exuviating invertebrates (*Ecdysozoa*) during evolution of the protostomes, and is missing from the hemocytes of *Drosophila* and *C. elegans* (36). Similarly, the other membrane-associated subunit, $p22^{phox}$, can be found in *Choanoflagellata*, *Mollusca*, and *Cnidaria* but not in *Annelida*, *Ecdysozoa*, and *Echinodermata*, suggesting that ROS generation by *Drosophila* and *C. elegans* differs from the mechanism used by other metazoans (59). Surprisingly, the intracellular killing mechanisms used by the hemocytes of these invertebrate model organisms remain largely undescribed.

Other sources of ROS besides NOX2 are evident in non-immune cells across phylogeny. In *Drosophila*, dual oxidases (DUOXs) harbor an NADPH domain and an N-terminal peroxidase domain to generate $O_2\cdot^-$ and hydrogen peroxide (H_2O_2). DUOXs are located on the plasma membranes of epithelial cells lining the digestive tract and produce extracellular H_2O_2 in response to bacteria (60) or their metabolites such as uracil (61). DUOXs serve a critical function in maintaining gut-microbe homeostasis, as flies silenced for DUOX succumb to foodborne bacterial infections (62). This mucosal expression pattern is conserved among mammals with DUOX1 and DUOX2 in the human gastrointestinal and respiratory tracts, respectively (63, 64). However, unlike their noted impact on *Drosophila* and *C. elegans* immunity, the importance of DUOX in mammalian immunity remains largely uncharacterized. Invertebrates thus use ROS, NOS, phagocytosis, and AMPs like many other taxa to combat infection. In addition, they have evolved their own defense modules, namely, lamellocyte-driven encapsulation and melanization, that are unique to the invertebrate group (65). For the purpose of comparison with mammalian macrophages and other myeloid cells, however, these divergent mechanisms are not discussed here.

Jawless and Jawed Fishes

As the oldest vertebrates, fish are best known for the traceable evolution of the adaptive immune system. The development of lymphoid follicles, immunoglobulin and major histocompatibility complex (MHC) molecules, and the B- and T-cell subsets can each be ascribed in a linear fashion to the appearance of the jawless (agnathans), cartilaginous, and bony fish, respectively. This new adaptive arm complemented the more ancient innate effector mechanisms described above. Indeed, many of the primitive phagocytic and antimicrobial responses of invertebrates and amoebae are conserved in fish (and mammals) as well. In addition, however, an inducible defense circuitry specific to vertebrates has also emerged, possibly to limit self-injury by regulating the production of harmful effector molecules mobilized during repeated microbial encounters over a longer life span.

The family of AMPs termed cathelicidins can be traced to hagfish, which diverged from the gnathostomes 500 Mya (66). Cathelicidins differ from the more ancient defensin-like AMPs, as they are synthesized as precursor proteins containing a conserved signal peptide and cathelin-like spacer domain followed by a highly variable antimicrobial domain (67). In hagfish, three cathelicidin proteins are synthesized by groups of myeloid cells within the loose connective tissue of the gut wall, perhaps a precursor to gut-associated lymphoid tissue, and defend against intestinal bacteria that accompany ingestion of dead and decaying food sources (66). Similarities emerge with human cathelicidin LL-37, which is found not only within granules of macrophages and neutrophils but also in gut epithelia (68), implying a broader role in host defense than previously thought. In a manner similar to that of fish cathelicidins, LL-37 is cleaved into its active form in response to inflammatory stimuli and released by exocytosis. Electrostatic interactions between the positively charged region of these cathelicidins and negatively charged bacterial membrane promote bacteriolysis (68). Humans lacking a functional LL-37 develop Chediak-Higashi syndrome and are highly predisposed to bacterial infections (69).

Gene duplication events after the agnathan-gnathostome split also gave rise to two of the three NOS isoforms currently found in mammals, namely neuronal NOS (nNOS; NOS1) and inducible NOS (iNOS; NOS2) (59). In contrast to invertebrate NOS, which bears closest homology to vertebrate NOS1 and can be constitutive or inducible depending on the species (70), vertebrate NOS2 is robustly induced in immune cells by microbial components like LPS or immunoregulatory cytokines like gamma interferon (IFN-γ). This generates copious amounts of ·NO during pathogen challenge, and fish NOS2 has been described as essential for resisting challenge to the trout pathogen *Yersinia ruckeri* (71) and the salmonid pathogens *Aeromonas salmonicida* and *Renibacterium salmoninarum* (72, 73).

Nutriprive defense mechanisms are also subject to inducible control. Teleost NRAMP1, which is constitutively expressed in amoebae, is responsive to LPS and IFN-γ transcriptional regulation, akin to mammalian NRAMP1 (74). Like the mammalian isoform expressed in macrophages, fish NRAMP1 is involved in regulating chemokine secretion, iNOS activation, MHC class II upregulation, and the respiratory burst (75). These novel roles supplement its antecedent function as a cation antiporter that depletes Fe^{2+} from the pathogen-containing vacuole (75, 76). Other teleost proteins regulate Fe^{2+} homeostasis for immunity. In mammals, the immune-inducible hormone hepcidin binds to the macrophage iron exporter ferroportin-1 and drastically decreases both extracellular and intracellular iron levels (77). Teleosts express several hepcidin orthologs depending on species, and consistent with their antimicrobial role, they are robustly induced during pathogen challenge (78). Interestingly, whereas mammalian hepcidin is produced mainly by the liver, macrophages themselves are the primary source of teleost hepcidin (79), hinting at a divergence in the cell-autonomous defense mechanisms of mammalian and fish hepcidin orthologs.

Marsupial Mammals

The marsupial and placental mammals diverged 130 Mya and thus offer clues to the phylogenetic origins of mammal-specific effector mechanisms. One such mechanism is intracellular starvation by IFN-γ-induced degradation of the amino acid tryptophan through the kynurenine pathway. IFN-γ induces robust expression of the enzyme indoleamine 2,3-dioxygenase-1 (IDO1), which depletes cytosolic tryptophan stores and restricts the growth of tryptophan auxotrophic microbes like *Toxoplasma gondii* and *Chlamydia trachomatis* (80). IDO has also been implicated in preventing the spread of viral pathogens, namely, herpes simplex virus, human cytomegalovirus, and measles virus, in both immune and non-immune cells (81–83). Moreover, in myeloid cells, the kynurenine catabolites generated during tryptophan degradation can exert direct antibacterial activity, as evidenced by their toxicity toward intracellular *Listeria monocytogenes* (84).

IDO orthologs exist in molluscs, amphibians, and bony fish; however, biochemical characterization revealed that the K_m value for these enzymes was 500- to 1,000-fold higher than for mammalian IDOs, suggesting that L-tryptophan is not the cognate substrate for these ancient IDO orthologs (74). In fact, these genes have been renamed proto-IDOs, and their biochemical and physiological roles are unknown. Interestingly, tandem copies of IDO are present on the chromosomes of both the marsupial opossum and humans, whereas only single copies of the proto-IDOs are found in the genomes of lower vertebrates and molluscs (85). This implies that the mammalian IDOs arose from a proto-IDO after a gene duplication event that occurred before the divergence of marsupial and placental mammals. In addition to IFN-γ, IDO is robustly induced by a variety of activating stimuli in macrophages, including LPS, tumor necrosis factor, and IFN-α/β (80, 86). The responsiveness of IDO to these stimuli varies considerably in immune and non-immune tissues, and in some organs, like the lung, IDO can even be constitutively expressed (80).

SPECIALIZED ADAPTATIONS OF MAMMALIAN MACROPHAGES: COMPARISON WITH ANTIMICROBIAL MACHINERIES IN NONMYELOID CELLS

Cell-autonomous immunity is a process embodied by the classical macrophage. These cells harbor a vast armamentarium that restricts and in many cases eliminates the microbial pathogens they encounter (Fig. 2). Robust production of cytotoxic gases (e.g., $O_2\cdot^-$, H_2O_2, and ·NO), AMPs, lysosomal hydrolases, and nutriprive immunity all ensure that classical macrophages play key roles in pathogen restriction. The evolutionary

A - Enzymatically Employed Metabolites

Iron Starvation
(NRAMP1) Fe^{2+}
Acidification
(v-ATPase)
Nitric Oxide
NOS/iNOS
NO
H^+
Tryptophan
Degradation
(IDO1)
Trp → Kynurenine
O_2^-
H_2O_2
Zn^{2+}
Cu^{2+}
Superoxide
(NOX2/DUOX1-2)
Copper/Zinc Overload
(ATP7/SLC30A/SLC39A)

C - Autophagic Sequestration

Ubiquitin:LC3 adapter
Glycan: Lectin
Antibody: Trim21

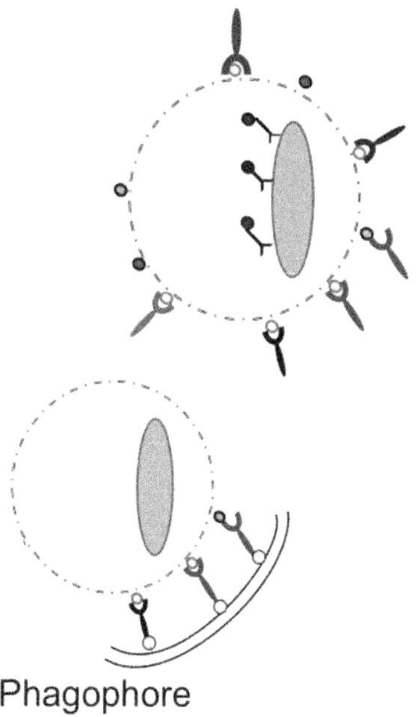

B - Trafficked Zymogen Antimicrobials

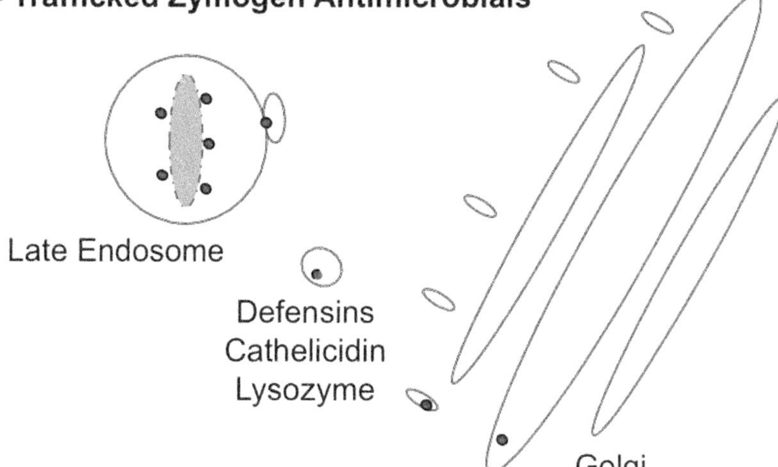

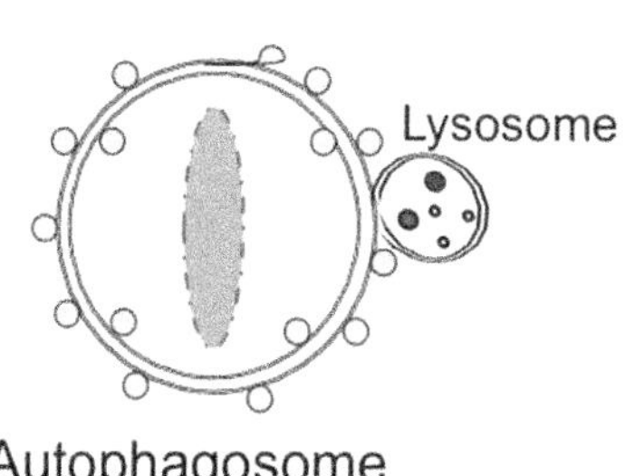

Figure 2 Canonical effector mechanisms in myeloid cells. (A) Enzymatically derived metabolites are potent antimicrobials. Macrophages deploy nutriprive mechanisms to reduce accessible nutrients and inhibit microbial growth, including NRAMP1 (SLC11A1), which pumps iron out of the phagosome; and IFN-γ-induced IDO1, which degrades intracellular tryptophan into kynurenine metabolites, which in themselves can be toxic to microbes. Superoxide generation by NOX2 and possibly DUOX1/2 generate ROS on the luminal side of the phagosome, which is directly antimicrobial. NO generation by NOS is also cytotoxic. Acidification by lysosomal V-ATPases or import of copper/zinc ions also functions to create a bactericidal environment within the phagosome. (B) AMPs are generated within the Golgi and trafficked to the pathogen-containing endosome. Here, they are processed into their active form to induce bacterial lysis. (C) Intracellular pathogens, either vacuolar or cytosolic, can become targeted by the autophagic machinery via host adaptor proteins that recognize signals of microbial infection. Proposed mechanisms include adaptors like p62/SQSTM1, which bind to ubiquitinated microbes or vacuoles; lectins like galectin 8 (LGALS8), which bind glycans exposed on damaged vacuoles; and TRIM21, which binds antibodies on the surface of incoming bacteria and viruses. These adaptors promote phagophore formation around their cargo by binding to autophagosome adaptor LC3. Microbes are then sequestered within the autophagosome, a process that requires the ATG proteins, and fuse with the lysosome for degradation.

record suggests that these pathways have largely been inherited from the defense repertoires of early vertebrates, invertebrates, amoebae, and, in some cases, prokaryotic bacterial or archaeal systems. A common ancestry implies that cell types other than macrophages also benefit from genetic inheritance of their own. Indeed, antimicrobial properties are already apparent in the somatic cells of lower organisms. Many of these defense pathways, however, are placed under inducible control in mammals by inflammatory stimuli such as IFN-γ, a powerful transactivating signal that induces hundreds of host defense genes (6). Herein, we focus on major effector mechanisms employed by mammalian macrophages against intracellular pathogens and contrast this evidence with what is known about these mechanisms in non-immune cells.

NOXs and NOSs: Radical Gases

Diatomic radical gases like $O_2\cdot^-$ and ·NO arose very early during eukaryotic evolution as effector mechanisms and are still an essential component of mammalian host defense. In this respect, humans who lack superoxide-generating activity suffer from recurrent life-threatening infections (87, 88) and mice with combined $gp91^{phox}$ and Nos2 deficiencies are highly susceptible to infection by their own commensal flora (89), underscoring the importance of gas-based protection for microbial control. In mammals, $O_2\cdot^-$ is generated by mammalian NOX complexes 1 to 5 as well as by DUOX1/2. Macrophages prototypically employ NOX2, the phagocyte oxidase. Upon assembling together with active GTP-bound Rac1, this complex pumps an electron through two sequential cytochrome units and onto a terminal oxygen acceptor. The resultant superoxide molecule is able to covalently link free amines, cysteines, and DNA, especially when converted into other antimicrobial metabolites (e.g., ·OH and H_2O_2) (90). Interestingly, $O_2\cdot^-$ and ·OH inefficiently permeate the bacterial membrane, suggesting that membrane-permeable peroxide may be a more efficient antimicrobial. Lastly, H_2O_2 can be coopted by myeloperoxidase in neutrophils or lactoperoxidase in goblet cells to produce antimicrobial hypochlorite and thiocyanate, respectively (57, 90).

DUOX1 and DUOX2 are large enzymes generating $O_2\cdot^-$, which is rapidly converted to H_2O_2 on the apical side of epithelial cells. A favored hypothesis is that the combination of DUOX and lactoperoxidase enzymes generate thiocyanate in the extracellular space; it is unclear, however, if these enzymes play intracellular roles in bacterial restriction. In mammalian epithelial cells, NOX1 and DUOX systems appear sufficient to help combat infection despite the absence of a NOX2 machinery. NOX1 and DUOX systems are each regulated by distinct organizer and activator protein subunits, with certain non-immune cells (e.g., HeLa cells) that lack DUOX1 and the activator component of DUOX2 (DUOXA2) useful for reconstitution assays (91, 92). Hence, cell type specificity may dictate whether oxidant or nitrosative mechanisms predominate.

In mammals, iNOS (NOS2) is robustly elicited by IFNs and contributes to host defense by reduction of the terminal amine of L-arginine. A 5-electron oxidation is employed in which electrons are shuttled from NADPH through two tetrabiopterin (BH_4) groups before heme transfer, coupling oxygen to arginine, yielding NO gas and L-citrulline (7, 93). Nitrosamines and nitrosothiols are other derivatives arising from ·NO production by iNOS to target intraphagosomal bacteria (74, 94). ·NO can likewise react with ROS to yield the more potent congener, peroxynitrite [$ONOO^-$] (95). RNS covalently link free amines and thiols and *S*-nitrosylate DNA to cripple their microbial targets. Interestingly, species divergence for NOS2 is seen in mammals; rodent phagocytes produce large amounts of ·NO during pathogenic challenge, but in human macrophages, ·NO release is greatly diminished, perhaps as a consequence of limited BH_4 cofactor availability (96).

Lysosomal Repertoire

In 1955, Christian de Duve isolated a highly degradative organelle now known as the lysosome (97). These acidic organelles are the major bacterial killing machines of single-celled amoebae, harboring many acid hydrolases to digest macromolecules and deliver nutrients to the cytosol. Echoing the early studies of Elie Metchnikoff, who fed litmus paper to macrophages, later studies using acid-sensitive dyes revealed that lysosomes actively maintain an acidic pH of ~4.5 through the electrogenic action of $V\text{-}H^+$-ATPases (98). Interestingly, not all lysosomal compartments equilibrate to the same pH, and in some myeloid cells, this is coupled to an ability for specialized functions. Dendritic cells, for example, maintain a more alkaline pH (~pH 5.5) to limit proteolytic digestion and aid MHC-II loading (99, 100), whereas macrophages can acidify lysosomes to pH ~4.5 for microbial killing (101, 102). Not surprisingly, macrophages use this acidic environment to activate hydrolases capable of disassembling most naturally occurring macromolecules; at least 50 hydrolytic enzymes are deployed for this purpose (103, 104). Fusion of the lysosome with the pathogen-containing phagosomes is a critical step for microbial killing, and many pathogenic bacteria, like virulent *Mycobacterium* spp., employ secreted effectors to prevent this step (105). IFN-γ activation helps overcome this blockage to induce phagosome-lysosomal fusion and concomitant control of *Mycobacterium*, *Salmonella*, and *Legionella*, which preferentially survive within immature phagosomes (101, 106, 107). Proteomic studies indicate that IFN-γ drives extensive remodeling of the phagosome, upregulating lipidases, maturation factors, and the MHC-I peptide loading complex (108). Besides phagocytosed bacteria, the lysosome also acts as a major killing device for invasive cytosolic bacteria. Here bacteria are flagged for lysosomal delivery by the autophagic pathway, as discussed below.

Autophagy

"Autophagy" was first coined in 1966 to describe cellular digestion of endogenous cytosolic components (97). Our current mechanistic understanding of autophagy stems from yeast genetic studies performed by Tsukada and Ohsumi, which uncovered 15 of the 31 ATG proteins required for this process (109, 110). Similarly, Klionsky et al. uncovered a cytoplasm-to-vesicle protein transport (Cvt) pathway (111). Although functionally distinct, the robust molecular characterization of these two systems largely shaped our view of autophagic sequestration of cytosolic materials. Autophagy is the process of cellular "self-eating," and is aptly named given that the autophagic process physically sequesters cytosolic contents. Currently, three independent forms of autophagy have been elucidated: chaperone-mediated autophagy, microautophagy, and macroautophagy, which rely on distinct molecular machineries and are phenotypically separable. Macroautophagy is characterized by sequestration of cytosolic contents, often by a double-bilayer isolation membrane (autophagosome), followed by fusion with lysosomes for degradation of their contents. The molecular details of macroautophagy have been extensively reviewed elsewhere (112–114), so they will not be covered in detail here.

Cues that drive autophagosome formation are intimately tied to recognition of cellular homeostatic imbalances. These sensory components are coupled to diverse families of ubiquitin (Ub) ligases that generally respond to lack of nutrients, protein aggregates, or foreign macromolecules in the cytosol (115). In each case, particular subsets of E3 Ub ligases conjugate Ub-like modifiers harboring C-terminal LRLRGG motifs to lysine residues in target proteins. These modifications can be either mono- or polyubiquitination (generally K11, K48, or K63). Recent studies indicate that specific ubiquitination machinery tags intracellular organelles, or in some cases intracellular bacteria (known as xenophagy), for lysosomal delivery and degradation. These modes of selective autophagy are triggered by exposure of "hidden" glycan chains to the cytosol, phosphorylation of mitochondria-associated Ub molecules, or ubiquitination of the peroxisomal protein PEX5 (116–119). Upon Ub tagging, one of five Ub-light chain 3 (LC3) bridging proteins (TAX1BP1, NBR1, p62/SQSTM1, NDP52, or OPTN) mobilizes autophagosomal isolation membranes to the intended cargo (120–123). Surprisingly, adaptor proteins are mobilized independently of one another and typically associate with specific forms of autophagy via distinct interactions with ATG8 homologs (124–126). Currently it is unknown whether their predominant role is to simply facilitate acquisition of ATG8 homolog-positive isolation membranes or help complete autophagosome closure (127). In the case of bacteria, it is also unclear which of several triggers—nutrient deprivation, vacuolar damage, or bacterial products—predominates to mobilize this autophagic cascade (128–130). Whatever the signal, both ubiquitination and Ub-LC3 bridging proteins appear to be critical for targeting intracellular bacteria to the phagophore and subsequent delivery of bacteria-containing $LC3^+$ autophagosomes to the lysosome for degradation (129, 131).

Autophagy-related homologs exist in all eukaryotic organisms examined to date (113) and confer immunity in lower as well as higher organisms. Autophagy genes are needed, for example, to restrict *Salmonella enterica* serovar Typhimurium in the amoeba *D. discoideum* and nematode *C. elegans* (132), as well as *L. monocytogenes* in *Drosophila* (133). In mammals, control of *Salmonella* Typhimurium and *L. monocytogenes* similarly enlists autophagy proteins to control bacterial proliferation, suggesting preadaptation from earlier species (116, 120, 122, 125, 134). In addition, activating signals like IFN-γ and CD40L engagement stimulate autophagic protection of macrophages against *Mycobacterium tuberculosis* (135) and non-immune cells against the protozoan parasite *T. gondii*, respectively (136). Dorsal root ganglionic neurons also enlist autophagy in an IFN-independent pathway to control neurotropic viral infection (137). Thus, autophagic defense is a widely coopted strategy to prevent colonization and, in some cases, dissemination of invasive pathogens among different cell types and species (138). A wider list of targeted pathogens appears in an accompanying autophagy-related chapter by Grinstein and colleagues of this volume.

Nutritional Immunity

All homeostatic events occur at physiological salt and ion concentrations. It is therefore not surprising that shifts in ionic balance result in permissive or restrictive niches for intracellular pathogens. Both Cu^{2+} and Zn^{2+} overload and Fe^{2+}/Fe^{3+} sequestration are now accepted as important macrophage defense mechanisms (6). Cu^{2+} and Zn^{2+} accumulate at high concentrations in macrophage lysosomes during infection (139–142). While no specific mechanism has been identified, the P-type ATPase ATP7A can import Cu^{2+} into the phagosome, where it has the potential to form toxic hydroxyl radicals (140). This together with the sequestration of Fe^{2+}/Fe^{3+} by NRAMP1 indicate that the availability of specific trace metals within the phagosome is a powerful antibacterial strategy. These trace elements can also have indirect microbicidal effects, as Cu^{2+}, Zn^{2+}, and Fe^{2+}/Fe^{3+} likely incorporate into the semiselective binding pockets of iron-binding proteins and activate metalloproteases within the lysosome (143). Additionally, Fe^{2+} cations convert H_2O_2 into the more powerful ·OH radical via Fenton chemistry, thereby augmenting ROS generation.

Whereas trace metal manipulation appears most evident in immune phagocytes, tryptophan depletion by IDO1 is more broadly deployed, in part reflecting its emergence with marsupials early during mammalian evolution. In human epithelial cells and fibroblasts, IFN-γ-induced IDO1 expression has been shown to deplete intracellular tryptophan stores, inhibiting the growth of tryptophan auxotrophic organisms such as *C. trachomatis* (144) and *T. gondii* (145) as well as auxotrophic mutants of *M. tuberculosis* (146). In mice, combined IDO1 and IDO2 deficiency renders animals highly susceptible to *Toxoplasma* challenge, although more resistant to infection with *Leishmania* (147). Hence, pathogen-specific resistance of IDOs coincides with tryptophan dependence of the infecting parasite.

Recent evidence points to the importance of microbial carbon (C2) catabolism as a target for cell-autonomous immunity. Virulent bacteria and fungi resort to a carbon-salvaging process termed anaplerosis to survive nutrient shortage in the face of the host immune pressure. LPS and IFNs induce the mitochondrial protein immunoresponsive gene 1 (IRG1; also termed aconitate decarboxylase 1), which contains an MmgE/PrpD-like methylcitrate dehydratase domain and decarboxylates the tricarboxylic acid intermediate *cis*-aconitase, needed for isocitrate synthesis (148, 149). This diverts carbons away from the anaplerotic glyoxylate shunt that is essential for the persistence of several pathogens including *M. tuberculosis*, *S.* Typhimurium, *Leptospira*, and fungi such as *Candida* and *Magnaporthe* (149–151). In addition to a described effector function, IRG1 also appears to regulate mitochondrial respiration and inflammatory cytokine production by inhibiting succinate dehydrogenase-mediated oxidation of succinate (152). Thus, carbon and amino acids are heavily contested between host and microbe to influence infection.

Antimicrobial Peptides

Mammalian AMPs arose from primitive gene duplication events resulting in the following structurally distinct protein families: α-defensins, β-defensins, θ-defensins, cathelicidins, histatins, and anionic AMPs (153). In humans, α- and β-defensins predominate in phagocytes and epithelial secretions, whereas θ-defensins and cathelicidins are genomically sparse. Of these protein families, human β-defensins are constitutively expressed in the epithelium, whereas α-defensin expression predominates in lymphoid cells (153). In other species, there have been cell-specific adaptations, such as mouse neutrophils that lack all α-defensins (154) or marsupials that possess a greatly expanded cathelicidin repertoire, the likely consequence of giving birth to underdeveloped and immunologically naive young (155). Further cell-specific adaptations are seen in macrophages where bacterial killing is enhanced by the mobilization of preformed or acid-responsive AMPs to the phagosome during phagocytosis (156). This achieves a higher local concentration of AMPs than is observed during extracellular secretion by nonimmune epithelial cells.

Different families of mammalian AMPs adopt unique structures. Nonetheless, the molecular mechanisms that bestow their bacterial restriction are thought to be mechanistically similar (157). AMPs are small ($<$100-amino-acid) peptides containing high surface density and use positively charged residues as a means to target anionic, glycosylated components in microbial outer membranes. Ultimately this promotes permeabilization of the bacterial membrane or viral envelope. Different structural aspects of AMPs suggest three models of action: the carpet, toroidal-pore, and barrel-stave models. The carpet model posits that accumulation of AMPs on negatively charged membranes facilitates bending and loss of membrane integrity (158). This is the likely mechanism of action for AMPs that lack significant amounts of hydrophobic regions. In contrast, the toroidal-pore model proposes that AMPs insert themselves into lipid bilayers to biophysically cluster lipid tails (159). This ultimately results in membrane involution and pore formation. Lastly, the barrel-stave model consists of peptides bridging the lipid bilayer to form a proteinaceous pore that is reminiscent of the action of bacterial cholesterol-dependent cytolysins (160). Even though the biological significance and evolutionary expansion suggests importance for cationic AMPs, it is curious that charge dependency necessitates their functionality. On this point, many bacteria have found ways to modify their outer membrane to be less anionic, thereby avoiding actions of cationic AMPs (161, 162).

In addition to AMPs, other enzymes capable of dismantling bacterial structural components like peptidoglycan have long been utilized as a means to lyse bacteria *in vitro* and may also promote bacterial restriction *in vivo*. The most famous of these enzymes is lysozyme, which can be traced to its digestive function in the amoeba contractile vacuole. The term "lysozyme" was first coined by Alexander Fleming in 1922 to describe a substance that exhibits remarkable bacteriolytic effects in tears, nasal secretions, and hen egg whites (163). Today the term "lysozyme" is more refined to describe enzymes that catalytically break down bacterial peptidoglycan via the hydrolysis of the glycosidic bond linking *N*-acetylmuramic acid and *N*-acetylglucosamine polymers. Interestingly, the lysozyme family is heavily expanded in simple eukaryotes, but has been limited to four members in humans (LYZC, LYZL1, LYZL2, and LYZL4), many of which are predominantly expressed in the testes. Specific lysozyme family members may help generate hydrolyzed peptides reported to subsume antimicrobial roles in mammalian macrophage lysosomes, such as cleaved Ub that can restrict *L. monocytogenes*, *S.* Typhimurium, and *M. tuberculosis in vitro* to confer cellular defense (164, 165).

Immune GTPases

Among the most potent cell-autonomous defense proteins operating in vertebrates are an emerging superfamily of immune GTPases (166, 167). These IFN-inducible proteins orchestrate antimicrobial activities against bacteria, viruses, and protozoa. In many cases, they do so by directly targeting the pathogen-containing vacuole (PCV), as first seen for *M. tuberculosis* and *T. gondii* in IFN-γ-activated macrophages and fibroblasts, respectively (101, 168). IFN-inducible GTPases fall into four subgroups based on paralogy and molecular mass: 21- to 47-kDa immunity-related GTPases (IRGs), 65- to 73-kDa guanylate-binding proteins (GBPs), 72- to 82-kDa myxoma (MX) resistance proteins, and ~200- to 285-kDa GVINs (GTPase, very large inducible) (166). The last group has not been assigned host defense functions; however, the other subfamilies exhibit pathogen-specific activities. Their evolutionary origins and distinct antimicrobial profiles are briefly discussed below.

IRGs

The IRG proteins have been identified in early protochordates such as amphioxi or lancelets (*Branchiostoma floridae* and *Branchiostoma japonicum*), suggesting origins before the radiation of chordate subphyla (169). A mammalian IRG ortholog (IRGD/IRG47) was first identified in pre-B-cell lines treated with IFN-γ (170), and five additional IRGs (IRGM1 to -3, IRGB6, and IRGA6) were discovered within a decade (171–174). Bioinformatic approaches then uncovered 14 to 16 additional IRGs in the euchromatic mouse genome, bringing the complete family to 21 to 23 genes (169, 175, 176). This expanded *Irg* repertoire in mice appears atypical among vertebrates, however, because humans possess only 4 *IRG* genes, with rats, dogs, zebrafish, and pufferfish harboring 7, 9, 11, and 2 members, respectively (169, 175, 176). Thus, gene loss and acquisition show that IRGs are a fast-evolving family of defense proteins at the interface of host-pathogen interactions. Indeed, recent evidence in outbred wild mice suggests that an expanded polymorphic IRG repertoire may arise due to parasite-host-prey relationships in which IRGs help mice combat asexual *Toxoplasma* transmission as part of the parasite life cycle completed within a natural predator, cats (177).

The first functional evidence revealed that IRGs confer some of their cell-autonomous defense activities by targeting the PCV. Studies in IFN-γ-activated macrophages infected with *M. tuberculosis* found that IRGM1 (formerly lipopolysaccharide response gene of 47 kDa [Lrg-47]) targeted mycobacterial phagosomes to facilitate fusion of this compartment with microbicidal lysosomes (101, 178). The molecular basis of this targeting arose from recognizing host-derived lipid species, including membrane phosphoinositides and diphosphatidylglycerol, on the phagocytic cup and PCVs shortly after bacterial uptake (178). This fit the hypothesis that IRGs and other IFN-induced GTPases may recognize "altered self" signals rather than pathogen-associated molecular patterns (179). A similar outcome was subsequently observed with the human IRGM ortholog in IFN-γ-activated monocytes, wherein it promoted *Mycobacterium bovis* BCG phagosomal maturation via the autophagic pathway following recognition of diphosphatidylglycerol (180). IRGs are likewise recruited to PCVs harboring *T. gondii* and *C. trachomatis* in murine fibroblasts, which may occur through "missing self" recognition (168, 181). In the case of *T. gondii*, multiple IRGs hierarchically associate with PCVs, directly or indirectly helping to disrupt this organelle (168). To counter such activity, virulent *Toxoplasma* strains have evolved mechanisms to inactivate IRG effectors (e.g., IRGB6) through ROP18-mediated phosphorylation (182).

GBPs

A second family of immune GTPases, the GBPs, have recently emerged as potent cell-intrinsic defense factors against intracellular bacteria, viruses, and protozoa (167, 183). Comprehensive hidden Markov modeling screens across 91 taxa revealed that GBPs have ancestral orthologs as far back as deuterostomes, plants, amoebae, and algae (184). Close inspection of the fossil record found that teleost GBPs harbored caspase activation and recruitment domains (CARDs) like those in human NLRP1 and ASC, suggesting that mammalian GBPs may also regulate inflammasome responses, except here the relevant domains reside on separate GBP and inflammasome proteins (183, 184). This proved to be the case: IFN-induced GBP5 and GBP2 promote NLRP3 and AIM-2 (absent in melanoma 2) inflammasome responses to microbial triggers, while other GBP family members also direct inflammasome responses to Gram-negative bacteria via the caspase-11 pathway (185–189). In *Danio rerio*, zGBP4 is responsible for inflammasome-dependent prostaglandin E_2 synthesis by neutrophils for protection against *Salmonella* (190), reinforcing this relationship across taxa (183).

Prior to the connections with inflammasome machinery emerging, a number of studies had reported the ability of GBPs to restrict bacterial (*Mycobacterium*, *Listeria*, *Salmonella*, and *Chlamydia*), viral (hepatitis C virus and influenza), or protozoan (*Toxoplasma*) replication in IFN-activated macrophages, hepatocytes, fibroblasts, and epithelia (167, 191–195). Ascription of these new roles benefited from the identification of complete *GBP* gene clusters in both human and mouse genomes (167, 175, 196, 197). It also enabled family-wide loss-of-function screens to uncover member-specific activities along with susceptibility profiles in the first Gbp1-deficient mice (167). This has now been extended to bacterial and protozoan vulnerability in Gbp2-, Gbp5-, and $Gbp^{chr.3}$-deficient animals (184, 189, 195, 196, 198, 199).

At the protein level, human GBPs share 40.0 to 98.4% amino acid identity and possess a bidomain architecture composed of similarly sized N-terminal GD and CTHD, the latter of which harbors 12 or 13 amphipathic α-helices involved in tail-to-tail dimer contacts (184, 200). Both the GD and CTHD contribute to homotypic self-assembly akin to the ~90-kDa dynamin-like GTPases, with which they share structural similarities (184, 201–204). GBPs bind GTP, GDP, or GMP with equimolar affinity and exhibit high intrinsic rates of GTPase and GDPase activity (k_{cat}, ~100 to 150 min $^{-1}$) (201, 205) due to an internal GTPase-activating protein (GAP) domain, obviating the need for external GAPs to accelerate GTP or GDP hydrolysis and oligomerization (206). Other accessory proteins, however, may be required; a Rab GDP dissociation inhibitor, RabGDIα, was recently implicated in turning off Gbp2 activity during *T. gondii* infection (207). Whether RabGDIα-mediated suppression impacts inflammasome kinetics or selectivity awaits testing.

Macromolecular assembly of the GBPs facilitates interactions with native protein partners and, for some members, association with lipid membranes. Gbp7 engages NADPH oxidase subunits ($gp91^{phox}$ and $p67^{phox}$) to help assemble Nox2 on PCVs (167). Genetic GBP1, GBP2, or GBP5 mutants unable to undergo nucleotide-dependent tetramerization also fail to engage antimicrobial effectors and localize to the Golgi, autolysosomes, PCVs, or target escaped bacteria or HIV-1 inside host cells (167, 198, 208–211). Conformational changes induced by transition-state tetramerization are also needed for human GBP1 to bind liposomes in cell-free assays (212). This membrane targeting may be further facilitated by isoprenylation at a C-terminal CaaX motif in GBPs 1, 2, and 5; here, CVIL or CTIS mutants likewise abolish relocation to organelles and PVCs (167, 198, 210, 213). Thus, higher-order assembly and/or lipid anchorage appear to be important for GBPs to target membrane-bound as well as cytosolically escaped pathogens.

Interestingly, many of these biochemical properties are retained across large evolutionary distances (169, 214). This is due to high levels of intrafamilial protein conservation in the G and middle domains that facilitate GTP hydrolysis and mechanotransduction, respectively (215). It ensures that "alternative mechanical coupling" of antimicrobial machinery to membrane surfaces is a conserved feature of many GBPs within most vertebrate species. Where the functional differences emerge are in the C-terminal domain; variation in this region often helps dictate the location at which GBPs interfere with infection inside host cells (215, 216).

MXs

A third group of IFN-induced dynamin-like GTPases, the Mx proteins, are potent antiviral restriction factors (214). Originally identified as an inherited trait in influenza A-susceptible mouse strains harboring deletions or nonsense mutations in the murine *Mx1* (myxoma resistance gene 1) locus, the MX proteins are major antiviral proteins operating against flu and influenza-like togaviruses, plus bunyaviruses, rhabdoviruses, and Thogoto, coxsackie, and hepatitis B viruses (6, 217). Their importance is underscored by the ability of the human MxA (MX1) transgene to rescue mice lacking IFN type I (IFN-α/β) receptors from influenza infection (217). This protein has undergone strong purifying selection in higher species such as primates, where a surface-exposed L4 loop is critical for species-specific activity against influenza A and Thogoto viruses (218). MxA assembles into large oligomeric complexes that adopt ring-like structures to block viral nucleoprotein formation (219).

In contrast, MxB (MX2) appears to be more specific for lentiviruses such as HIV-1, in which most of the structural determinants for activity are in the C-terminal tail. Here, MX2 interacts with nuclear pores to prevent HIV from integrating into the host genome (220, 221). Multiple functional losses have occurred across evolution for MxB (MX2), either by pseudogenization or via gene conversion from *MxA* genes. Notably, both *MX* genes are nonfunctional in Odontoceti cetaceans (toothed whales, including dolphins and orcas), and this loss occurred soon after the divergence of odontocetes and mysticetes (baleen whales) ~33 Mya to 37 Mya (222). Presumably, compensation occurs elsewhere in the genome, since examinations of 56 other mammalian genomes have found both *MX* genes to be intact. Hence, this potent family of IFN-induced antiviral proteins is a prominent feature of most mammalian hosts.

RETROSPECT AND PROSPECT

This chapter has focused on the evolution of the cell-autonomous effector mechanisms that not only bestow potent antimicrobial functions on specialized immune cells such as macrophages but also guard the sanctity of non-immune lineages. This ancient system exhibits many tissue-specific phenotypes, and could be viewed as offensive when deployed by phagocytic cells to detect and destroy microbes within the endolysosomal system (ROS, NO, AMPs, and xenophagy) or defensive when mobilized in non-immune cells to restrict invasive and cytosolic intracellular microbes (e.g., IDOs, IRGs, and GBPs).

Gene duplication has been a driving force in the evolution of these new effector mechanisms, allowing one copy to accumulate mutations subject to Darwinian selection while the other maintains its original function. Gene conversion has been another route, especially within the MX family. Several of these effector genes, such as the GBPs, MXs, and IDO1, have evolved rapidly, leading to a remarkable degree of functional plasticity even within closely related species. Hence, selective pressure provided by pathogens has been a dominant factor in the acquisition of new effector mechanisms across species. It has also extended the boundaries of cell-autonomous defense beyond macrophages and other myeloid derivatives to encompass cell types outside the classical immune system.

Acknowledgments. *Support for some of the work described herein has come from the following sources: Howard Hughes Medical Institute, National Institutes of Health National Institute of Allergy and Infectious Diseases (grants R01 AI068041-09 and R01 AI108834-01A1), Burroughs Wellcome Fund Investigators in the Pathogenesis of Infectious Disease award (1007845), and Canadian Institutes of Health Research Postdoctoral Fellowship.*

Citation. Gaudet RG, Bradfield CJ, MacMicking JD. 2016. Evolution of cell-autonomous effector mechanisms in macrophages versus non-immunecells. Microbiol Spectrum 4(6): MCHD-0050-2016.

References

1. **Randow F, MacMicking JD, James LC.** 2013. Cellular self-defense: how cell-autonomous immunity protects against pathogens. *Science* **340:**701–706.
2. **Tauber AI.** 2003. Metchnikoff and the phagocytosis theory. *Nat Rev Mol Cell Biol* **4:**897–901.
3. **Stuart LM, Ezekowitz RA.** 2005. Phagocytosis: elegant complexity. *Immunity* **22:**539–550.
4. **Buchmann K.** 2014. Evolution of innate immunity: clues from invertebrates via fish to mammals. *Front Immunol* **5:**459. doi:10.3389/fimmu.2014.00459.
5. **MacMicking JD.** 2009. Recognizing macrophage activation and host defense. *Cell Host Microbe* **5:**405–407.
6. **MacMicking JD.** 2012. Interferon-inducible effector mechanisms in cell-autonomous immunity. *Nat Rev Immunol* **12:**367–382.
7. **MacMicking J, Xie QW, Nathan C.** 1997. Nitric oxide and macrophage function. *Annu Rev Immunol* **15:** 323–350.
8. **McGettigan J, McLennan RK, Broderick KE, Kean L, Allan AK, Cabrero P, Regulski MR, Pollock VP, Gould GW, Davies SA, Dow JA.** 2005. Insect renal tubules constitute a cell-autonomous immune system that protects the organism against bacterial infection. *Insect Biochem Mol Biol* **35:**741–754.
9. **Palumbo A.** 2005. Nitric oxide in marine invertebrates: a comparative perspective. *Comp Biochem Physiol A Mol Integr Physiol* **142:**241–248.
10. **Gusarov I, Shatalin K, Starodubtseva M, Nudler E.** 2009. Endogenous nitric oxide protects bacteria against a wide spectrum of antibiotics. *Science* **325:**1380–1384.
11. **Baldauf SL, Doolittle WF.** 1997. Origin and evolution of the slime molds (Mycetozoa). *Proc Natl Acad Sci U S A* **94:**12007–12012.

12. **Bapteste E, Brinkmann H, Lee JA, Moore DV, Sensen CW, Gordon P, Duruflé L, Gaasterland T, Lopez P, Müller M, Philippe H.** 2002. The analysis of 100 genes supports the grouping of three highly divergent amoebae: *Dictyostelium*, *Entamoeba*, and *Mastigamoeba*. *Proc Natl Acad Sci U S A* **99:**1414–1419.

13. **Song J, Xu Q, Olsen R, Loomis WF, Shaulsky G, Kuspa A, Sucgang R.** 2005. Comparing the *Dictyostelium* and *Entamoeba* genomes reveals an ancient split in the Conosa lineage. *PLoS Comput Biol* **1:**e71. doi:10.1371/journal.pcbi.0010071.

14. **Bozzaro S, Bucci C, Steinert M.** 2008. Phagocytosis and host-pathogen interactions in *Dictyostelium* with a look at macrophages. *Int Rev Cell Mol Biol* **271:**253–300.

15. **Minakami R, Sumimotoa H.** 2006. Phagocytosis-coupled activation of the superoxide-producing phagocyte oxidase, a member of the NADPH oxidase (Nox) family. *Int J Hematol* **84:**193–198.

16. **Yan L, Cerny RL, Cirillo JD.** 2004. Evidence that hsp90 is involved in the altered interactions of *Acanthamoeba castellanii* variants with bacteria. *Eukaryot Cell* **3:** 567–578.

17. **Lardy B, Bof M, Aubry L, Paclet MH, Morel F, Satre M, Klein G.** 2005. NADPH oxidase homologs are required for normal cell differentiation and morphogenesis in *Dictyostelium discoideum*. *Biochim Biophys Acta* **1744:**199–212.

18. **Rodriguez-Paris JM, Nolta KV, Steck TL.** 1993. Characterization of lysosomes isolated from *Dictyostelium discoideum* by magnetic fractionation. *J Biol Chem* **268:** 9110–9116.

19. **Souza GM, Mehta DP, Lammertz M, Rodriguez-Paris J, Wu R, Cardelli JA, Freeze HH.** 1997. *Dictyostelium* lysosomal proteins with different sugar modifications sort to functionally distinct compartments. *J Cell Sci* **110:**2239–2248.

20. **Gotthardt D, Warnatz HJ, Henschel O, Brückert F, Schleicher M, Soldati T.** 2002. High-resolution dissection of phagosome maturation reveals distinct membrane trafficking phases. *Mol Biol Cell* **13:**3508–3520.

21. **Leippe M, Bruhn H, Hecht O, Grötzinger J.** 2005. Ancient weapons: the three-dimensional structure of amoebapore A. *Trends Parasitol* **21:**5–7.

22. **Leippe M.** 1995. Ancient weapons: NK-lysin, is a mammalian homolog to pore-forming peptides of a protozoan parasite. *Cell* **83:**17–18.

23. **Leippe M.** 1999. Antimicrobial and cytolytic polypeptides of amoeboid protozoa—effector molecules of primitive phagocytes. *Dev Comp Immunol* **23:**267–279.

24. **Courville P, Chaloupka R, Cellier MFM.** 2006. Recent progress in structure-function analyses of Nramp proton-dependent metal-ion transporters. *Biochem Cell Biol* **84:**960–978.

25. **Forbes JR, Gros P.** 2001. Divalent-metal transport by NRAMP proteins at the interface of host-pathogen interactions. *Trends Microbiol* **9:**397–403.

26. **Nevo Y, Nelson N.** 2006. The NRAMP family of metal-ion transporters. *Biochim Biophys Acta* **1763:** 609–620.

27. **Cellier MF.** 2012. Nutritional immunity: homology modeling of Nramp metal import. *Adv Exp Med Biol* **946:**335–351.

28. **Richer E, Courville P, Bergevin I, Cellier MFM.** 2003. Horizontal gene transfer of "prototype" Nramp in bacteria. *J Mol Evol* **57:**363–376.

29. **Peracino B, Wagner C, Balest A, Balbo A, Pergolizzi B, Noegel AA, Steinert M, Bozzaro S.** 2006. Function and mechanism of action of *Dictyostelium* Nramp1 (Slc11a1) in bacterial infection. *Traffic* **7:**22–38.

30. **Peracino B, Buracco S, Bozzaro S.** 2013. The Nramp (Slc11) proteins regulate development, resistance to pathogenic bacteria and iron homeostasis in *Dictyostelium discoideum*. *J Cell Sci* **126:**301–311.

31. **Appelberg R.** 2006. Macrophage nutriprive antimicrobial mechanisms. *J Leukoc Biol* **79:**1117–1128.

32. **Schaap P.** 2007. Evolution of size and pattern in the social amoebas. *BioEssays* **29:**635–644.

33. **Janeway CA Jr, Medzhitov R.** 2002. Innate immune recognition. *Annu Rev Immunol* **20:**197–216.

34. **Chen G, Zhuchenko O, Kuspa A.** 2007. Immune-like phagocyte activity in the social amoeba. *Science* **317:** 678–681.

35. **Zhang X, Zhuchenko O, Kuspa A, Soldati T.** 2016. Social amoebae trap and kill bacteria by casting DNA nets. *Nat Commun* **7:**10938. doi:10.1038/ncomms10938.

36. **Dzik JM.** 2010. The ancestry and cumulative evolution of immune reactions. *Acta Biochim Pol* **57:**443–466.

37. **Salazar-Jaramillo L, Paspati A, van de Zande L, Vermeulen CJ, Schwander T, Wertheim B.** 2014. Evolution of a cellular immune response in *Drosophila*: a phenotypic and genomic comparative analysis. *Genome Biol Evol* **6:**273–289.

38. **Rhodes CP, Ratcliffe NA, Rowley AF.** 1982. Presence of coelomocytes in the primitive chordate amphioxus (*Branchiostoma lanceolatum*). *Science* **217:**263–265.

39. **Ribatti D, Crivellato E.** 2014. Mast cell ontogeny: an historical overview. *Immunol Lett* **159:**11–14.

40. **Lemaitre B, Hoffmann J.** 2007. The host defense of *Drosophila melanogaster*. *Annu Rev Immunol* **25:**697–743.

41. **Clayton AM, Dong Y, Dimopoulos G.** 2014. The *Anopheles* innate immune system in the defense against malaria infection. *J Innate Immun* **6:**169–181.

42. **Andrä J, Herbst R, Leippe M.** 2003. Amoebapores, archaic effector peptides of protozoan origin, are discharged into phagosomes and kill bacteria by permeabilizing their membranes. *Dev Comp Immunol* **27:** 291–304.

43. **Mallo GV, Kurz CL, Couillault C, Pujol N, Granjeaud S, Kohara Y, Ewbank JJ.** 2002. Inducible antibacterial defense system in *C. elegans*. *Curr Biol* **12:**1209–1214.

44. **Lemaitre B, Reichhart JM, Hoffmann JA.** 1997. *Drosophila* host defense: differential induction of antimicrobial peptide genes after infection by various classes of microorganisms. *Proc Natl Acad Sci U S A* **94:**14614–14619.

45. **Salzet M, Tasiemski A, Cooper E.** 2006. Innate immunity in lophotrochozoans: the annelids. *Curr Pharm Des* **12:**3043–3050.

46. **Hoffmann J.** 2007. Antifungal defense in *Drosophila. Nat Immunol* **8:**543–545.

47. **Cox RL, Mariano T, Heck DE, Laskin JD, Stegeman JJ.** 2001. Nitric oxide synthase sequences in the marine fish *Stenotomus chrysops* and the sea urchin *Arbacia punctulata*, and phylogenetic analysis of nitric oxide synthase calmodulin-binding domains. *Comp Biochem Physiol B Biochem Mol Biol* **130:**479–491.

48. **Luckhart S, Rosenberg R.** 1999. Gene structure and polymorphism of an invertebrate nitric oxide synthase gene. *Gene* **232:**25–34.

49. **Giovine M, Pozzolini M, Favre A, Bavestrello G, Cerrano C, Ottaviani F, Chiarantini L, Cerasi A, Cangiotti M, Zocchi E, Scarfì S, Sarà M, Benatti U.** 2001. Heat stress-activated, calcium-dependent nitric oxide synthase in sponges. *Nitric Oxide* **5:**427–431.

50. **Ottaviani E, Paeman LR, Cadet P, Stefano GB.** 1993. Evidence for nitric oxide production and utilization as a bacteriocidal agent by invertebrate immunocytes. *Eur J Pharmacol* **248:**319–324.

51. **Franchini A, Conte A, Ottaviani E.** 1995. Nitric oxide: an ancestral immunocyte effector molecule. *Adv Neuroimmunol* **5:**463–478.

52. **Foley E, O'Farrell PH.** 2003. Nitric oxide contributes to induction of innate immune responses to gram-negative bacteria in *Drosophila. Genes Dev* **17:**115–125.

53. **Babior BM, Kipnes RS, Curnutte JT.** 1973. Biological defense mechanisms. The production by leukocytes of superoxide, a potential bactericidal agent. *J Clin Invest* **52:**741–744.

54. **Whitten MM, Ratcliffe NA.** 1999. In vitro superoxide activity in the haemolymph of the West Indian leaf cockroach, *Blaberus discoidalis. J Insect Physiol* **45:** 667–675.

55. **García-García E, Prado-Alvarez M, Novoa B, Figueras A, Rosales C.** 2008. Immune responses of mussel hemocyte subpopulations are differentially regulated by enzymes of the PI 3-K, PKC, and ERK kinase families. *Dev Comp Immunol* **32:**637–653.

56. **Peskin AV, Labas YA, Tikhonov AN.** 1998. Superoxide radical production by sponges *Sycon* sp. *FEBS Lett* **434:** 201–204.

57. **Babior BM.** 1999. NADPH oxidase: an update. *Blood* **93:**1464–1476.

58. **Sumimoto H.** 2008. Structure, regulation and evolution of Nox-family NADPH oxidases that produce reactive oxygen species. *FEBS J* **275:**3249–3277.

59. **Andreakis N, D'Aniello S, Albalat R, Patti FP, Garcia-Fernàndez J, Procaccini G, Sordino P, Palumbo A.** 2011. Evolution of the nitric oxide synthase family in metazoans. *Mol Biol Evol* **28:**163–179.

60. **Ritsick DR, Edens WA, McCoy JW, Lambeth JD.** 2004. The use of model systems to study biological functions of Nox/Duox enzymes. *Biochem Soc Symp* **71:**85–96.

61. **Lee KA, Kim SH, Kim EK, Ha EM, You H, Kim B, Kim MJ, Kwon Y, Ryu JH, Lee WJ.** 2013. Bacterial-derived uracil as a modulator of mucosal immunity and gut-microbe homeostasis in *Drosophila. Cell* **153:**797–811.

62. **Ha EM, Oh CT, Bae YS, Lee WJ.** 2005. A direct role for dual oxidase in *Drosophila* gut immunity. *Science* **310:**847–850.

63. **Geiszt M, Witta J, Baffi J, Lekstrom K, Leto TL.** 2003. Dual oxidases represent novel hydrogen peroxide sources supporting mucosal surface host defense. *FASEB J* **17:** 1502–1504.

64. **El Hassani RA, Benfares N, Caillou B, Talbot M, Sabourin J-C, Belotte V, Morand S, Gnidehou S, Agnandji D, Ohayon R, Kaniewski J, Noël-Hudson MS, Bidart JM, Schlumberger M, Virion A, Dupuy C.** 2005. Dual oxidase2 is expressed all along the digestive tract. *Am J Physiol Gastrointest Liver Physiol* **288:**G933–G942.

65. **Kounatidis I, Ligoxygakis P.** 2012. *Drosophila* as a model system to unravel the layers of innate immunity to infection. *Open Biol* **2:**120075. doi:10.1098/rsob.120075.

66. **Uzzell T, Stolzenberg ED, Shinnar AE, Zasloff M.** 2003. Hagfish intestinal antimicrobial peptides are ancient cathelicidins. *Peptides* **24:**1655–1667.

67. **Gennaro R, Zanetti M.** 2000. Structural features and biological activities of the cathelicidin-derived antimicrobial peptides. *Biopolymers* **55:**31–49.

68. **Bals R, Wilson JM.** 2003. Cathelicidins—a family of multifunctional antimicrobial peptides. *Cell Mol Life Sci* **60:**711–720.

69. **Cole AM, Shi J, Ceccarelli A, Kim YH, Park A, Ganz T.** 2001. Inhibition of neutrophil elastase prevents cathelicidin activation and impairs clearance of bacteria from wounds. *Blood* **97:**297–304.

70. **Toni M, De Angelis F, di Patti MC, Cioni C.** 2015. Nitric oxide synthase in the central nervous system and peripheral organs of *Stramonita haemastoma*: protein distribution and gene expression in response to thermal stress. *Mar Drugs* **13:**6636–6664.

71. **Chettri JK, Raida MK, Kania PW, Buchmann K.** 2012. Differential immune response of rainbow trout (*Oncorhynchus mykiss*) at early developmental stages (larvae and fry) against the bacterial pathogen *Yersinia ruckeri. Dev Comp Immunol* **36:**463–474.

72. **Grayson TH, Cooper LF, Wrathmell AB, Roper J, Evenden AJ, Gilpin ML.** 2002. Host responses to *Renibacterium salmoninarum* and specific components of the pathogen reveal the mechanisms of immune suppression and activation. *Immunology* **106:**273–283.

73. **Fast MD, Tse B, Boyd JM, Johnson SC.** 2009. Mutations in the *Aeromonas salmonicida* subsp. *salmonicida* type III secretion system affect Atlantic salmon leucocyte activation and downstream immune responses. *Fish Shellfish Immunol* **27:**721–728.

74. **Grayfer L, Hodgkinson JW, Belosevic M.** 2014. Antimicrobial responses of teleost phagocytes and innate immune evasion strategies of intracellular bacteria. *Dev Comp Immunol* **43:**223–242.

75. **Blackwell JM, Goswami T, Evans CA, Sibthorpe D, Papo N, White JK, Searle S, Miller EN, Peacock CS, Mohammed H, Ibrahim M.** 2001. SLC11A1 (formerly NRAMP1) and disease resistance. *Cell Microbiol* **3:** 773–784.

76. **Alter-Koltunoff M, Goren S, Nousbeck J, Feng CG, Sher A, Ozato K, Azriel A, Levi BZ.** 2008. Innate immunity to intraphagosomal pathogens is mediated by interferon regulatory factor 8 (IRF-8) that stimulates the expression of macrophage-specific Nramp1 through antagonizing repression by c-Myc. *J Biol Chem* **283:** 2724–2733.

77. **Nemeth E, Tuttle MS, Powelson J, Vaughn MB, Donovan A, Ward DM, Ganz T, Kaplan J.** 2004. Hepcidin regulates cellular iron efflux by binding to ferroportin and inducing its internalization. *Science* **306:**2090–2093.

78. **Douglas SE, Gallant JW, Liebscher RS, Dacanay A, Tsoi SC.** 2003. Identification and expression analysis of hepcidin-like antimicrobial peptides in bony fish. *Dev Comp Immunol* **27:**589–601.

79. **Costa MM, Maehr T, Diaz-Rosales P, Secombes CJ, Wang T.** 2011. Bioactivity studies of rainbow trout (*Oncorhynchus mykiss*) interleukin-6: effects on macrophage growth and antimicrobial peptide gene expression. *Mol Immunol* **48:**1903–1916.

80. **Taylor MW, Feng GS.** 1991. Relationship between interferon-gamma, indoleamine 2,3-dioxygenase, and tryptophan catabolism. *FASEB J* **5:**2516–2522.

81. **Adams O, Besken K, Oberdörfer C, MacKenzie CR, Rüssing D, Däubener W.** 2004. Inhibition of human herpes simplex virus type 2 by interferon gamma and tumor necrosis factor alpha is mediated by indoleamine 2,3-dioxygenase. *Microbes Infect* **6:**806–812.

82. **Obojes K, Andres O, Kim KS, Däubener W, Schneider-Schaulies J.** 2005. Indoleamine 2,3-dioxygenase mediates cell type-specific anti-measles virus activity of gamma interferon. *J Virol* **79:**7768–7776.

83. **Bodaghi B, Goureau O, Zipeto D, Laurent L, Virelizier JL, Michelson S.** 1999. Role of IFN-γ-induced indoleamine 2,3 dioxygenase and inducible nitric oxide synthase in the replication of human cytomegalovirus in retinal pigment epithelial cells. *J Immunol* **162:**957–964.

84. **Niño-Castro A, Abdullah Z, Popov A, Thabet Y, Beyer M, Knolle P, Domann E, Chakraborty T, Schmidt SV, Schultze JL.** 2014. The IDO1-induced kynurenines play a major role in the antimicrobial effect of human myeloid cells against *Listeria monocytogenes*. *Innate Immun* **20:**401–411.

85. **Yuasa HJ, Takubo M, Takahashi A, Hasegawa T, Noma H, Suzuki T.** 2007. Evolution of vertebrate indoleamine 2,3-dioxygenases. *J Mol Evol* **65:**705–714.

86. **Ball HJ, Jusof FF, Bakmiwewa SM, Hunt NH, Yuasa HJ.** 2014. Tryptophan-catabolizing enzymes—party of three. *Front Immunol* **5:**485. doi:10.3389/fimmu.2014.00485.

87. **Roos D, de Boer M, Kuribayashi F, Meischl C, Weening RS, Segal AW, Ahlin A, Nemet K, Hossle JP, Bernatowska-Matuszkiewicz E, Middleton-Price H.** 1996. Mutations in the X-linked and autosomal recessive forms of chronic granulomatous disease. *Blood* **87:**1663–1681.

88. **Heyworth PG, Cross AR, Curnutte JT.** 2003. Chronic granulomatous disease. *Curr Opin Immunol* **15:**578–584.

89. **Shiloh MU, MacMicking JD, Nicholson S, Brause JE, Potter S, Marino M, Fang F, Dinauer M, Nathan C.** 1999. Phenotype of mice and macrophages deficient in both phagocyte oxidase and inducible nitric oxide synthase. *Immunity* **10:**29–38.

90. **Bedard K, Krause K-H.** 2007. The NOX family of ROS-generating NADPH oxidases: physiology and pathophysiology. *Physiol Rev* **87:**245–313.

91. **Pacquelet S, Lehmann M, Luxen S, Regazzoni K, Frausto M, Noack D, Knaus UG.** 2008. Inhibitory action of NoxA1 on dual oxidase activity in airway cells. *J Biol Chem* **283:**24649–24658.

92. **Grasberger H, Refetoff S.** 2006. Identification of the maturation factor for dual oxidase. Evolution of an eukaryotic operon equivalent. *J Biol Chem* **281:**18269–18272.

93. **Wink DA, Hines HB, Cheng RYS, Switzer CH, Flores-Santana W, Vitek MP, Ridnour LA, Colton CA.** 2011. Nitric oxide and redox mechanisms in the immune response. *J Leukoc Biol* **89:**873–891.

94. **Nathan C, Xie QW.** 1994. Nitric oxide synthases: roles, tolls, and controls. *Cell* **78:**915–918.

95. **Denicola A, Rubbo H, Rodríguez D, Radi R.** 1993. Peroxynitrite-mediated cytotoxicity to *Trypanosoma cruzi*. *Arch Biochem Biophys* **304:**279–286.

96. **Schneemann M, Schoedon G, Hofer S, Blau N, Guerrero L, Schaffner A.** 1993. Nitric oxide synthase is not a constituent of the antimicrobial armature of human mononuclear phagocytes. *J Infect Dis* **167:**1358–1363.

97. **De Duve C, Wattiaux R.** 1966. Functions of lysosomes. *Annu Rev Physiol* **28:**435–492.

98. **Ohkuma S, Poole B.** 1978. Fluorescence probe measurement of the intralysosomal pH in living cells and the perturbation of pH by various agents. *Proc Natl Acad Sci U S A* **75:**3327–3331.

99. **Trombetta ES, Ebersold M, Garrett W, Pypaert M, Mellman I.** 2003. Activation of lysosomal function during dendritic cell maturation. *Science* **299:**1400–1403.

100. **Jancic C, Savina A, Wasmeier C, Tolmachova T, El-Benna J, Dang PM, Pascolo S, Gougerot-Pocidalo MA, Raposo G, Seabra MC, Amigorena S.** 2007. Rab27a regulates phagosomal pH and NADPH oxidase recruitment to dendritic cell phagosomes. *Nat Cell Biol* **9:**367–378.

101. **MacMicking JD, Taylor GA, McKinney JD.** 2003. Immune control of tuberculosis by IFN-γ-inducible LRG-47. *Science* **302:**654–659.

102. **Vandal OH, Pierini LM, Schnappinger D, Nathan CF, Ehrt S.** 2008. A membrane protein preserves intrabacterial pH in intraphagosomal *Mycobacterium tuberculosis*. *Nat Med* **14:**849–854.

103. **Lübke T, Lobel P, Sleat DE.** 2009. Proteomics of the lysosome. *Biochim Biophys Acta* **1793:**625–635.

104. **Schröder BA, Wrocklage C, Hasilik A, Saftig P.** 2010. The proteome of lysosomes. *Proteomics* **10:**4053–4076.

105. **Weiss G, Schaible UE.** 2015. Macrophage defense mechanisms against intracellular bacteria. *Immunol Rev* **264:**182–203.

106. **Santic M, Molmeret M, Abu Kwaik Y.** 2005. Maturation of the *Legionella pneumophila*-containing phagosome into a phagolysosome within gamma interferon-activated macrophages. *Infect Immun* **73:** 3166–3171.

107. **Ishibashi Y, Arai T.** 1990. Effect of γ-interferon on phagosome-lysosome fusion in *Salmonella typhimurium*-infected murine macrophages. *FEMS Microbiol Immunol* **2:**75–82.

108. **Jutras I, Houde M, Currier N, Boulais J, Duclos S, LaBoissière S, Bonneil E, Kearney P, Thibault P, Paramithiotis E, Hugo P, Desjardins M.** 2008. Modulation of the phagosome proteome by interferon-γ. *Mol Cell Proteomics* **7:**697–715.

109. **Tsukada M, Ohsumi Y.** 1993. Isolation and characterization of autophagy-defective mutants of *Saccharomyces cerevisiae*. *FEBS Lett* **333:**169–174.

110. **Klionsky DJ, Cregg JM, Dunn WA Jr, Emr SD, Sakai Y, Sandoval IV, Sibirny A, Subramani S, Thumm M, Veenhuis M, Ohsumi Y.** 2003. A unified nomenclature for yeast autophagy-related genes. *Dev Cell* **5:**539–545.

111. **Klionsky DJ, Cueva R, Yaver DS.** 1992. Aminopeptidase I of *Saccharomyces cerevisiae* is localized to the vacuole independent of the secretory pathway. *J Cell Biol* **119:**287–299.

112. **Deretic V.** 2011. Autophagy in immunity and cell-autonomous defense against intracellular microbes. *Immunol Rev* **240:**92–104.

113. **Levine B, Klionsky DJ.** 2004. Development by self-digestion: molecular mechanisms and biological functions of autophagy. *Dev Cell* **6:**463–477.

114. **Levine B.** 2005. Eating oneself and uninvited guests: autophagy-related pathways in cellular defense. *Cell* **120:**159–162.

115. **Deshaies RJ, Joazeiro CAP.** 2009. RING domain E3 ubiquitin ligases. *Annu Rev Biochem* **78:**399–434.

116. **Thurston TLM, Wandel MP, von Muhlinen N, Foeglein A, Randow F.** 2012. Galectin 8 targets damaged vesicles for autophagy to defend cells against bacterial invasion. *Nature* **482:**414–418.

117. **Zhang J, Tripathi DN, Jing J, Alexander A, Kim J, Powell RT, Dere R, Tait-Mulder J, Lee JH, Paull TT, Pandita RK, Charaka VK, Pandita TK, Kastan MB, Walker CL.** 2015. ATM functions at the peroxisome to induce pexophagy in response to ROS. *Nat Cell Biol* **17:**1259–1269.

118. **Narendra D, Tanaka A, Suen D-F, Youle RJ.** 2009. Parkin-induced mitophagy in the pathogenesis of Parkinson disease. *Autophagy* **5:**706–708.

119. **Koyano F, Okatsu K, Kosako H, Tamura Y, Go E, Kimura M, Kimura Y, Tsuchiya H, Yoshihara H, Hirokawa T, Endo T, Fon EA, Trempe JF, Saeki Y, Tanaka K, Matsuda N.** 2014. Ubiquitin is phosphorylated by PINK1 to activate parkin. *Nature* **510:**162–166.

120. **Wild P, Farhan H, McEwan DG, Wagner S, Rogov VV, Brady NR, Richter B, Korac J, Waidmann O, Choudhary C, Dötsch V, Bumann D, Dikic I.** 2011. Phosphorylation of the autophagy receptor optineurin restricts *Salmonella* growth. *Science* **333:**228–233.

121. **Pankiv S, Clausen TH, Lamark T, Brech A, Bruun JA, Outzen H, Øvervatn A, Bjørkøy G, Johansen T.** 2007. p62/SQSTM1 binds directly to Atg8/LC3 to facilitate degradation of ubiquitinated protein aggregates by autophagy. *J Biol Chem* **282:**24131–24145.

122. **Thurston TL, Ryzhakov G, Bloor S, von Muhlinen N, Randow F.** 2009. The TBK1 adaptor and autophagy receptor NDP52 restricts the proliferation of ubiquitin-coated bacteria. *Nat Immunol* **10:**1215–1221.

123. **Kirkin V, Lamark T, Sou YS, Bjørkøy G, Nunn JL, Bruun JA, Shvets E, McEwan DG, Clausen TH, Wild P, Bilusic I, Theurillat JP, Øvervatn A, Ishii T, Elazar Z, Komatsu M, Dikic I, Johansen T.** 2009. A role for NBR1 in autophagosomal degradation of ubiquitinated substrates. *Mol Cell* **33:**505–516.

124. **Matsumoto G, Wada K, Okuno M, Kurosawa M, Nukina N.** 2011. Serine 403 phosphorylation of p62/SQSTM1 regulates selective autophagic clearance of ubiquitinated proteins. *Mol Cell* **44:**279–289.

125. **Mostowy S, Sancho-Shimizu V, Hamon MA, Simeone R, Brosch R, Johansen T, Cossart P.** 2011. p62 and NDP52 proteins target intracytosolic *Shigella* and *Listeria* to different autophagy pathways. *J Biol Chem* **286:**26987–26995.

126. **Cemma M, Kim PK, Brumell JH.** 2011. The ubiquitin-binding adaptor proteins p62/SQSTM1 and NDP52 are recruited independently to bacteria-associated microdomains to target *Salmonella* to the autophagy pathway. *Autophagy* **7:**341–345.

127. **Lazarou M, Sliter DA, Kane LA, Sarraf SA, Wang C, Burman JL, Sideris DP, Fogel AI, Youle RJ.** 2015. The ubiquitin kinase PINK1 recruits autophagy receptors to induce mitophagy. *Nature* **524:**309–314.

128. **He C, Klionsky DJ.** 2009. Regulation mechanisms and signaling pathways of autophagy. *Annu Rev Genet* **43:** 67–93.

129. **Huang J, Brumell JH.** 2014. Bacteria-autophagy interplay: a battle for survival. *Nat Rev Microbiol* **12:**101–114.

130. **Randow F, Youle RJ.** 2014. Self and nonself: how autophagy targets mitochondria and bacteria. *Cell Host Microbe* **15:**403–411.

131. **Tattoli I, Sorbara MT, Philpott DJ, Girardin SE.** 2012. Bacterial autophagy: the trigger, the target and the timing. *Autophagy* **8:**1848–1850.

132. **Jia K, Thomas C, Akbar M, Sun Q, Adams-Huet B, Gilpin C, Levine B.** 2009. Autophagy genes protect against *Salmonella typhimurium* infection and mediate insulin signaling-regulated pathogen resistance. *Proc Natl Acad Sci U S A* **106:**14564–14569.

133. **Yano T, Mita S, Ohmori H, Oshima Y, Fujimoto Y, Ueda R, Takada H, Goldman WE, Fukase K, Silverman N, Yoshimori T, Kurata S.** 2008. Autophagic control of *Listeria* through intracellular innate immune recognition in *Drosophila*. *Nat Immunol* **9:**908–916.

134. **Campoy E, Colombo MI.** 2009. Autophagy in intracellular bacterial infection. *Biochim Biophys Acta* **1793:** 1465–1477.

135. **Gutierrez MG, Master SS, Singh SB, Taylor GA, Colombo MI, Deretic V.** 2004. Autophagy is a defense

mechanism inhibiting BCG and *Mycobacterium tuberculosis* survival in infected macrophages. *Cell* **119:** 753–766.

136. **Van Grol J, Muniz-Feliciano L, Portillo JA, Bonilha VL, Subauste CS.** 2013. CD40 induces anti-*Toxoplasma gondii* activity in nonhematopoietic cells dependent on autophagy proteins. *Infect Immun* **81:**2002–2011.

137. **Yordy B, Iijima N, Huttner A, Leib D, Iwasaki A.** 2012. A neuron-specific role for autophagy in antiviral defense against herpes simplex virus. *Cell Host Microbe* **12:**334–345.

138. **Benjamin JL, Sumpter R Jr, Levine B, Hooper LV.** 2013. Intestinal epithelial autophagy is essential for host defense against invasive bacteria. *Cell Host Microbe* **13:**723–734.

139. **Wagner D, Maser J, Lai B, Cai Z, Barry CE III, Höner Zu Bentrup K, Russell DG, Bermudez LE.** 2005. Elemental analysis of *Mycobacterium avium-*, *Mycobacterium tuberculosis-*, and *Mycobacterium smegmatis-*containing phagosomes indicates pathogen-induced microenvironments within the host cell's endosomal system. *J Immunol* **174:**1491–1500.

140. **White C, Lee J, Kambe T, Fritsche K, Petris MJ.** 2009. A role for the ATP7A copper-transporting ATPase in macrophage bactericidal activity. *J Biol Chem* **284:** 33949–33956.

141. **Nairz M, Schleicher U, Schroll A, Sonnweber T, Theurl I, Ludwiczek S, Talasz H, Brandacher G, Moser PL, Muckenthaler MU, Fang FC, Bogdan C, Weiss G.** 2013. Nitric oxide-mediated regulation of ferroportin-1 controls macrophage iron homeostasis and immune function in *Salmonella* infection. *J Exp Med* **210:** 855–873.

142. **Botella H, Peyron P, Levillain F, Poincloux R, Poquet Y, Brandli I, Wang C, Tailleux L, Tilleul S, Charrière GM, Waddell SJ, Foti M, Lugo-Villarino G, Gao Q, Maridonneau-Parini I, Butcher PD, Castagnoli PR, Gicquel B, de Chastellier C, Neyrolles O.** 2011. Mycobacterial P_1-type ATPases mediate resistance to zinc poisoning in human macrophages. *Cell Host Microbe* **10:**248–259.

143. **Nagase H, Visse R, Murphy G.** 2006. Structure and function of matrix metalloproteinases and TIMPs. *Cardiovasc Res* **69:**562–573.

144. **Ibana JA, Belland RJ, Zea AH, Schust DJ, Nagamatsu T, AbdelRahman YM, Tate DJ, Beatty WL, Aiyar AA, Quayle AJ.** 2011. Inhibition of indoleamine 2,3-dioxygenase activity by levo-1-methyl tryptophan blocks gamma interferon-induced *Chlamydia trachomatis* persistence in human epithelial cells. *Infect Immun* **79:**4425–4437.

145. **Pfefferkorn ER.** 1984. Interferon γ blocks the growth of *Toxoplasma gondii* in human fibroblasts by inducing the host cells to degrade tryptophan. *Proc Natl Acad Sci U S A* **81:**908–912.

146. **Zhang YJ, Reddy MC, Ioerger TR, Rothchild AC, Dartois V, Schuster BM, Trauner A, Wallis D, Galaviz S, Huttenhower C, Sacchettini JC, Behar SM, Rubin EJ.** 2013. Tryptophan biosynthesis protects mycobacteria from CD4 T-cell-mediated killing. *Cell* **155:**1296–1308.

147. **Divanovic S, Sawtell NM, Trompette A, Warning JI, Dias A, Cooper AM, Yap GS, Arditi M, Shimada K, Duhadaway JB, Prendergast GC, Basaraba RJ, Mellor AL, Munn DH, Aliberti J, Karp CL.** 2012. Opposing biological functions of tryptophan catabolizing enzymes during intracellular infection. *J Infect Dis* **205:**152–161.

148. **Degrandi D, Hoffmann R, Beuter-Gunia C, Pfeffer K.** 2009. The proinflammatory cytokine-induced IRG1 protein associates with mitochondria. *J Interferon Cytokine Res* **29:**55–67.

149. **Michelucci A, Cordes T, Ghelfi J, Pailot A, Reiling N, Goldmann O, Binz T, Wegner A, Tallam A, Rausell A, Buttini M, Linster CL, Medina E, Balling R, Hiller K.** 2013. Immune-responsive gene 1 protein links metabolism to immunity by catalyzing itaconic acid production. *Proc Natl Acad Sci U S A* **110:**7820–7825.

150. **Muñoz-Elías EJ, McKinney JD.** 2006. Carbon metabolism of intracellular bacteria. *Cell Microbiol* **8:**10–22.

151. **MacMicking JD.** 2014. Cell-autonomous effector mechanisms against *Mycobacterium tuberculosis*. *Cold Spring Harb Perspect Med* **4:**a018507. doi:10.1101/cshperspect.a018507.

152. **Lampropoulou V, Sergushichev A, Bambouskova M, Nair S, Vincent EE, Loginicheva E, Cervantes-Barragan L, Ma X, Huang SC, Griss T, Weinheimer CJ, Khader S, Randolph GJ, Pearce EJ, Jones RG, Diwan A, Diamond MS, Artyomov MN.** 2016. Itaconate links inhibition of succinate dehydrogenase with macrophage metabolic remodeling and regulation of inflammation. *Cell Metab* **24:**158–166.

153. **Ganz T.** 2003. Defensins: antimicrobial peptides of innate immunity. *Nat Rev Immunol* **3:**710–720.

154. **Eisenhauer PB, Lehrer RI.** 1992. Mouse neutrophils lack defensins. *Infect Immun* **60:**3446–3447.

155. **Wang J, Wong ES, Whitley JC, Li J, Stringer JM, Short KR, Renfree MB, Belov K, Cocks BG.** 2011. Ancient antimicrobial peptides kill antibiotic-resistant pathogens: Australian mammals provide new options. *PLoS One* **6:**e24030. doi:10.1371/journal.pone.0024030.

156. **Diamond G, Beckloff N, Weinberg A, Kisich KO.** 2009. The roles of antimicrobial peptides in innate host defense. *Curr Pharm Des* **15:**2377–2392.

157. **Brogden KA.** 2005. Antimicrobial peptides: pore formers or metabolic inhibitors in bacteria? *Nat Rev Microbiol* **3:**238–250.

158. **Dean RE, O'Brien LM, Thwaite JE, Fox MA, Atkins H, Ulaeto DO.** 2010. A carpet-based mechanism for direct antimicrobial peptide activity against vaccinia virus membranes. *Peptides* **31:**1966–1972.

159. **Sengupta D, Leontiadou H, Mark AE, Marrink SJ.** 2008. Toroidal pores formed by antimicrobial peptides show significant disorder. *Biochim Biophys Acta* **1778:** 2308–2317.

160. **Yeaman MR, Yount NY.** 2003. Mechanisms of antimicrobial peptide action and resistance. *Pharmacol Rev* **55:**27–55.

161. **Gunn JS.** 2008. The *Salmonella* PmrAB regulon: lipopolysaccharide modifications, antimicrobial peptide resistance and more. *Trends Microbiol* **16:**284–290.

162. **Beceiro A, Tomás M, Bou G.** 2013. Antimicrobial resistance and virulence: a successful or deleterious association in the bacterial world? *Clin Microbiol Rev* **26:** 185–230.

163. **Nakatsuji T, Gallo RL.** 2012. Antimicrobial peptides: old molecules with new ideas. *J Invest Dermatol* **132:** 887–895.

164. **Alonso S, Pethe K, Russell DG, Purdy GE.** 2007. Lysosomal killing of *Mycobacterium* mediated by ubiquitin-derived peptides is enhanced by autophagy. *Proc Natl Acad Sci U S A* **104:**6031–6036.

165. **Hiemstra PS, van den Barselaar MT, Roest M, Nibbering PH, van Furth R.** 1999. Ubiquicidin, a novel murine microbicidal protein present in the cytosolic fraction of macrophages. *J Leukoc Biol* **66:**423–428.

166. **Kim BH, Shenoy AR, Kumar P, Bradfield CJ, MacMicking JD.** 2012. IFN-inducible GTPases in host cell defense. *Cell Host Microbe* **12:**432–444.

167. **Kim BH, Shenoy AR, Kumar P, Das R, Tiwari S, MacMicking JD.** 2011. A family of IFN-γ-inducible 65-kD GTPases protects against bacterial infection. *Science* **332:**717–721.

168. **Martens S, Parvanova I, Zerrahn J, Griffiths G, Schell G, Reichmann G, Howard JC.** 2005. Disruption of *Toxoplasma gondii* parasitophorous vacuoles by the mouse p47-resistance GTPases. *PLoS Pathog* **1:**e24. doi:10.1371/journal.ppat.0010024.

169. **Li G, Zhang J, Sun Y, Wang H, Wang Y.** 2009. The evolutionarily dynamic IFN-inducible GTPase proteins play conserved immune functions in vertebrates and cephalochordates. *Mol Biol Evol* **26:**1619–1630.

170. **Gilly M, Wall R.** 1992. The IRG-47 gene is IFN-gamma induced in B cells and encodes a protein with GTP-binding motifs. *J Immunol* **148:**3275–3281.

171. **Taylor GA, Jeffers M, Largaespada DA, Jenkins NA, Copeland NG, Vande Woude GF.** 1996. Identification of a novel GTPase, the inducibly expressed GTPase, that accumulates in response to interferon γ. *J Biol Chem* **271:**20399–20405.

172. **Carlow DA, Marth J, Clark-Lewis I, Teh HS.** 1995. Isolation of a gene encoding a developmentally regulated T cell-specific protein with a guanine nucleotide triphosphate-binding motif. *J Immunol* **154:**1724–1734.

173. **Sorace JM, Johnson RJ, Howard DL, Drysdale BE.** 1995. Identification of an endotoxin and IFN-inducible cDNA: possible identification of a novel protein family. *J Leukoc Biol* **58:**477–484.

174. **Boehm U, Guethlein L, Klamp T, Ozbek K, Schaub A, Fütterer A, Pfeffer K, Howard JC.** 1998. Two families of GTPases dominate the complex cellular response to IFN-γ. *J Immunol* **161:**6715–6723.

175. **Shenoy AR, Kim B-H, Choi H-P, Matsuzawa T, Tiwari S, MacMicking JD.** 2007. Emerging themes in IFN-γ-induced macrophage immunity by the p47 and p65 GTPase families. *Immunobiology* **212:**771–784.

176. **Bekpen C, Hunn JP, Rohde C, Parvanova I, Guethlein L, Dunn DM, Glowalla E, Leptin M, Howard JC.** 2005. The interferon-inducible p47 (IRG) GTPases in vertebrates: loss of the cell autonomous resistance mechanism in the human lineage. *Genome Biol* **6:**R92. doi: 10.1186/gb-2005-6-11-r92.

177. **Lilue J, Müller UB, Steinfeldt T, Howard JC.** 2013. Reciprocal virulence and resistance polymorphism in the relationship between *Toxoplasma gondii* and the house mouse. *eLife* **2:**e01298. doi:10.7554/eLife.01298.

178. **Tiwari S, Choi HP, Matsuzawa T, Pypaert M, MacMicking JD.** 2009. Targeting of the GTPase Irgm1 to the phagosomal membrane via PtdIns(3,4)P_2 and PtdIns(3,4,5)P_3 promotes immunity to mycobacteria. *Nat Immunol* **10:**907–917.

179. **MacMicking JD.** 2004. IFN-inducible GTPases and immunity to intracellular pathogens. *Trends Immunol* **25:** 601–609.

180. **Singh SB, Ornatowski W, Vergne I, Naylor J, Delgado M, Roberts E, Ponpuak M, Master S, Pilli M, White E, Komatsu M, Deretic V.** 2010. Human IRGM regulates autophagy and cell-autonomous immunity functions through mitochondria. *Nat Cell Biol* **12:**1154–1165.

181. **Haldar AK, Saka HA, Piro AS, Dunn JD, Henry SC, Taylor GA, Frickel EM, Valdivia RH, Coers J.** 2013. IRG and GBP host resistance factors target aberrant, "non-self" vacuoles characterized by the missing of "self" IRGM proteins. *PLoS Pathog* **9:**e1003414. doi: 10.1371/journal.ppat.1003414.

182. **Steinfeldt T, Könen-Waisman S, Tong L, Pawlowski N, Lamkemeyer T, Sibley LD, Hunn JP, Howard JC.** 2010. Phosphorylation of mouse immunity-related GTPase (IRG) resistance proteins is an evasion strategy for virulent *Toxoplasma gondii*. *PLoS Biol* **8:**e1000576. doi: 10.1371/journal.pbio.1000576.

183. **Kim BH, Chee JD, Bradfield CJ, Park ES, Kumar P, MacMicking JD.** 2016. Interferon-induced guanylate-binding proteins in inflammasome activation and host defense. *Nat Immunol* **17:**481–489.

184. **Shenoy AR, Wellington DA, Kumar P, Kassa H, Booth CJ, Cresswell P, MacMicking JD.** 2012. GBP5 promotes NLRP3 inflammasome assembly and immunity in mammals. *Science* **336:**481–485.

185. **Meunier E, Dick MS, Dreier RF, Schürmann N, Kenzelmann Broz D, Warming S, Roose-Girma M, Bumann D, Kayagaki N, Takeda K, Yamamoto M, Broz P.** 2014. Caspase-11 activation requires lysis of pathogen-containing vacuoles by IFN-induced GTPases. *Nature* **509:**366–370.

186. **Finethy R, Jorgensen I, Haldar AK, de Zoete MR, Strowig T, Flavell RA, Yamamoto M, Nagarajan UM, Miao EA, Coers J.** 2015. Guanylate binding proteins enable rapid activation of canonical and noncanonical inflammasomes in *Chlamydia*-infected macrophages. *Infect Immun* **83:**4740–4749.

187. **Pilla DM, Hagar JA, Haldar AK, Mason AK, Degrandi D, Pfeffer K, Ernst RK, Yamamoto M, Miao EA, Coers J.** 2014. Guanylate binding proteins promote caspase-11-dependent pyroptosis in response to cytoplasmic LPS. *Proc Natl Acad Sci U S A* **111:**6046–6051.

188. **Man SM, Karki R, Malireddi RK, Neale G, Vogel P, Yamamoto M, Lamkanfi M, Kanneganti TD.** 2015. The transcription factor IRF1 and guanylate-binding proteins target activation of the AIM2 inflammasome by *Francisella* infection. *Nat Immunol* **16:**467–475.

189. Meunier E, Wallet P, Dreier RF, Costanzo S, Anton L, Rühl S, Dussurgey S, Dick MS, Kistner A, Rigard M, Degrandi D, Pfeffer K, Yamamoto M, Henry T, Broz P. 2015. Guanylate-binding proteins promote activation of the AIM2 inflammasome during infection with *Francisella novicida*. *Nat Immunol* 16:476–484.

190. Tyrkalska SD, Candel S, Angosto D, Gómez-Abellán V, Martín-Sánchez F, García-Moreno D, Zapata-Pérez R, Sánchez-Ferrer Á, Sepulcre MP, Pelegrín P, Mulero V. 2016. Neutrophils mediate *Salmonella* Typhimurium clearance through the GBP4 inflammasome-dependent production of prostaglandins. *Nat Commun* 7:12077. doi:10.1038/ncomms12077.

191. Itsui Y, Sakamoto N, Kakinuma S, Nakagawa M, Sekine-Osajima Y, Tasaka-Fujita M, Nishimura-Sakurai Y, Suda G, Karakama Y, Mishima K, Yamamoto M, Watanabe T, Ueyama M, Funaoka Y, Azuma S, Watanabe M. 2009. Antiviral effects of the interferon-induced protein guanylate binding protein 1 and its interaction with the hepatitis C virus NS5B protein. *Hepatology* 50:1727–1737.

192. Nordmann A, Wixler L, Boergeling Y, Wixler V, Ludwig S. 2012. A new splice variant of the human guanylate-binding protein 3 mediates anti-influenza activity through inhibition of viral transcription and replication. *FASEB J* 26:1290–1300.

193. Rupper AC, Cardelli JA. 2008. Induction of guanylate binding protein 5 by gamma interferon increases susceptibility to *Salmonella enterica* serovar Typhimurium-induced pyroptosis in RAW 264.7 cells. *Infect Immun* 76:2304–2315.

194. Tietzel I, El-Haibi C, Carabeo RA. 2009. Human guanylate binding proteins potentiate the anti-chlamydia effects of interferon-γ. *PLoS One* 4:e6499. doi:10.1371/journal.pone.0006499.

195. Yamamoto M, Okuyama M, Ma JS, Kimura T, Kamiyama N, Saiga H, Ohshima J, Sasai M, Kayama H, Okamoto T, Huang DCS, Soldati-Favre D, Horie K, Takeda J, Takeda K. 2012. A cluster of interferon-γ-inducible p65 GTPases plays a critical role in host defense against *Toxoplasma gondii*. *Immunity* 37:302–313.

196. Degrandi D, Konermann C, Beuter-Gunia C, Kresse A, Würthner J, Kurig S, Beer S, Pfeffer K. 2007. Extensive characterization of IFN-induced GTPases mGBP1 to mGBP10 involved in host defense. *J Immunol* 179:7729–7740.

197. Olszewski MA, Gray J, Vestal DJ. 2006. *In silico* genomic analysis of the human and murine guanylate-binding protein (GBP) gene clusters. *J Interferon Cytokine Res* 26:328–352.

198. Degrandi D, Kravets E, Konermann C, Beuter-Gunia C, Klümpers V, Lahme S, Wischmann E, Mausberg AK, Beer-Hammer S, Pfeffer K. 2013. Murine guanylate binding protein 2 (mGBP2) controls *Toxoplasma gondii* replication. *Proc Natl Acad Sci U S A* 110:294–299.

199. Selleck EM, Fentress SJ, Beatty WL, Degrandi D, Pfeffer K, Virgin HW IV, Macmicking JD, Sibley LD. 2013. Guanylate-binding protein 1 (Gbp1) contributes to cell-autonomous immunity against *Toxoplasma gondii*. *PLoS Pathog* 9:e1003320. doi:10.1371/journal.ppat.1003320.

200. Wehner M, Kunzelmann S, Herrmann C. 2012. The guanine cap of human guanylate-binding protein 1 is responsible for dimerization and self-activation of GTP hydrolysis. *FEBS J* 279:203–210.

201. Prakash B, Praefcke GJ, Renault L, Wittinghofer A, Herrmann C. 2000. Structure of human guanylate-binding protein 1 representing a unique class of GTP-binding proteins. *Nature* 403:567–571.

202. Kunzelmann S, Praefcke GJK, Herrmann C. 2006. Transient kinetic investigation of GTP hydrolysis catalyzed by interferon-γ-induced hGBP1 (human guanylate binding protein 1). *J Biol Chem* 281:28627–28635.

203. Syguda A, Bauer M, Benscheid U, Ostler N, Naschberger E, Ince S, Stürzl M, Herrmann C. 2012. Tetramerization of human guanylate-binding protein 1 is mediated by coiled-coil formation of the C-terminal α-helices. *FEBS J* 279:2544–2554.

204. Wehner M, Herrmann C. 2010. Biochemical properties of the human guanylate binding protein 5 and a tumor-specific truncated splice variant. *FEBS J* 277:1597–1605.

205. Ghosh A, Praefcke GJK, Renault L, Wittinghofer A, Herrmann C. 2006. How guanylate-binding proteins achieve assembly-stimulated processive cleavage of GTP to GMP. *Nature* 440:101–104.

206. Abdullah N, Srinivasan B, Modiano N, Cresswell P, Sau AK. 2009. Role of individual domains and identification of internal gap in human guanylate binding protein-1. *J Mol Biol* 386:690–703.

207. Ohshima J, Sasai M, Liu J, Yamashita K, Ma JS, Lee Y, Bando H, Howard JC, Ebisu S, Hayashi M, Takeda K, Standley DM, Frickel EM, Yamamoto M. 2015. RabGDIα is a negative regulator of interferon-γ-inducible GTPase-dependent cell-autonomous immunity to *Toxoplasma gondii*. *Proc Natl Acad Sci U S A* 112:E4581–E4590.

208. Modiano N, Lu YE, Cresswell P. 2005. Golgi targeting of human guanylate-binding protein-1 requires nucleotide binding, isoprenylation, and an IFN-γ-inducible cofactor. *Proc Natl Acad Sci U S A* 102:8680–8685.

209. Britzen-Laurent N, Bauer M, Berton V, Fischer N, Syguda A, Reipschläger S, Naschberger E, Herrmann C, Stürzl M. 2010. Intracellular trafficking of guanylate-binding proteins is regulated by heterodimerization in a hierarchical manner. *PLoS One* 5:e14246. doi:10.1371/journal.pone.0014246.

210. Kravets E, Degrandi D, Weidtkamp-Peters S, Ries B, Konermann C, Felekyan S, Dargazanli JM, Praefcke GJ, Seidel CA, Schmitt L, Smits SH, Pfeffer K. 2012. The GTPase activity of murine guanylate-binding protein 2 (mGBP2) controls the intracellular localization and recruitment to the parasitophorous vacuole of *Toxoplasma gondii*. *J Biol Chem* 287:27452–27466.

211. Krapp C, Hotter D, Gawanbacht A, McLaren PJ, Kluge SF, Stürzel CM, Mack K, Reith E, Engelhart S, Ciuffi A, Hornung V, Sauter D, Telenti A, Kirchhoff F. 2016. Guanylate binding protein (GBP) 5 is an interferon-inducible inhibitor of HIV-1 infectivity. *Cell Host Microbe* 19:504–514.

212. **Fres JM, Müller S, Praefcke GJ.** 2010. Purification of the CaaX-modified, dynamin-related large GTPase hGBP1 by coexpression with farnesyltransferase. *J Lipid Res* **51:**2454–2459.

213. **Virreira Winter S, Niedelman W, Jensen KD, Rosowski EE, Julien L, Spooner E, Caradonna K, Burleigh BA, Saeij JPJ, Ploegh HL, Frickel EM.** 2011. Determinants of GBP recruitment to *Toxoplasma gondii* vacuoles and the parasitic factors that control it. *PLoS One* **6:** e24434. doi:10.1371/journal.pone.0024434.

214. **Horisberger MA, Staeheli P, Haller O.** 1983. Interferon induces a unique protein in mouse cells bearing a gene for resistance to influenza virus. *Proc Natl Acad Sci U S A* **80:**1910–1914.

215. **Pendin D, Tosetto J, Moss TJ, Andreazza C, Moro S, McNew JA, Daga A.** 2011. GTP-dependent packing of a three-helix bundle is required for atlastin-mediated fusion. *Proc Natl Acad Sci U S A* **108:**16283–16288.

216. **Schulte K, Pawlowski N, Faelber K, Fröhlich C, Howard J, Daumke O.** 2016. The immunity-related GTPase Irga6 dimerizes in a parallel head-to-head fashion. *BMC Biol* **14:**14. doi:10.1186/s12915-016-0236-7.

217. **Haller O, Staeheli P, Schwemmle M, Kochs G.** 2015. Mx GTPases: dynamin-like antiviral machines of innate immunity. *Trends Microbiol* **23:**154–163.

218. **Mitchell PS, Young JM, Emerman M, Malik HS.** 2015. Evolutionary analyses suggest a function of MxB immunity proteins beyond lentivirus restriction. *PLoS Pathog* **11:**e1005304. doi:10.1371/journal.ppat.1005304.

219. **Haller O, Staeheli P, Kochs G.** 2007. Interferon-induced Mx proteins in antiviral host defense. *Biochimie* **89:** 812–818.

220. **Kane M, Yadav SS, Bitzegeio J, Kutluay SB, Zang T, Wilson SJ, Schoggins JW, Rice CM, Yamashita M, Hatziioannou T, Bieniasz PD.** 2013. MX2 is an interferon-induced inhibitor of HIV-1 infection. *Nature* **502:**563–566.

221. **Goujon C, Moncorgé O, Bauby H, Doyle T, Ward CC, Schaller T, Hué S, Barclay WS, Schulz R, Malim MH.** 2013. Human MX2 is an interferon-induced post-entry inhibitor of HIV-1 infection. *Nature* **502:**559–562.

222. **Braun BA, Marcovitz A, Camp JG, Jia R, Bejerano G.** 2015. Mx1 and Mx2 key antiviral proteins are surprisingly lost in toothed whales. *Proc Natl Acad Sci U S A* **112:**8036–8040.

Myeloid Cells in Health and Disease: A Synthesis
Edited by Siamon Gordon

doi:10.1128/microbiolspec.MCHD-0022-2015

Irina Udalova,[1] Claudia Monaco,[1]
Jagdeep Nanchahal,[1] and Marc Feldmann[1]

35 Anti-TNF Therapy

INTRODUCTION

Tumor necrosis factor (TNF) is a member of the large family of cytokines; not hormones, but important local signaling molecules that transmit information from one cell to another (1). Different cytokines convey different messages, but cytokines are key players in every important biological process, including immunity, inflammation, cell growth, migration, fibrosis, vascularization, etc. So it is not surprising that abnormalities of these key mediators, molecules that are important enough to be conserved through evolution, may be involved in disease processes. What might not have been predicted was that removal of a single upregulated cytokine can make a clinical difference. This is best documented for anti-TNF (2) but is also true for blockade of interleukin-6 (IL-6), granulocyte-macrophage colony-stimulating factor (GM-CSF), and IL-1. In this review, we will summarize the current state of knowledge about cytokine expression and dysregulation in rheumatoid arthritis (RA) and other diseases and the role of TNF, the great majority of which is produced by macrophages. The knowledge gained has impacted our understanding of and therapy for other diseases also, and by focusing on cytokines, major rate-limiting steps, and hence therapeutic targets, there are opportunities for planning of therapy for many unmet needs.

RA: MORE THAN JUST A CYTOKINE DYSREGULATION

Knowledge about the pathogenesis of RA has been augmented in recent years. RA was first defined as an autoimmune disease due to the presence of rheumatoid factor, an autoantibody to the Ig hinge regions, and then by its linkage to HLA and to HLA-DR4 especially (3). More recently our knowledge of its autoimmune nature has been greatly amplified by the revelation that the RA autoantigens are posttranslational modifications of several abundant proteins, with loss of an amino (NH_2) group from arginine to form citrullinated proteins. The ones that appear to be most important are citrullinated enolase and vimentin, but also fibrinogen and collagen type II (4). Some of the autoantibodies are pathogenic, for example, activating osteoclasts (5). There is clear heterogeneity in RA, with the patients expressing HLA-DR4 also producing these antibodies, but not RA patients with other HLA-DRs, who are thus "seronegative." Both types of RA patients, however, respond equally well to anti-TNF therapy (3).

Our work on the role of cytokines in RA was initially triggered by two events. First was the exploration of an old conundrum of the 1970s and 1980s: was there an immune response to cancer in humans that might be therapeutically useful? In mice it was clear at

[1]Kennedy Institute of Rheumatology, NDORMS, University of Oxford, Botnar Research Centre, Headington, Oxford OX3 7LD, United Kingdom.

the time that there was an immune response to cancer, but the murine cancers were "foreign," as they expressed viral antigens. In contrast, immune responses were not detectable in human cancers. One of the authors (Marc Feldmann) had trained scientifically at the Walter and Eliza Hall Institute of Medical Research (WEHI), where the legacy of Sir Frank McFarlane Burnet, whose clonal selection theory underpinned research in the field of immunity and autoimmunity, was still notable. The concept of autoimmunity suggested an approach. What was needed for immunotherapy of cancer was to induce an autoimmune response to it. But this was not so simple, as it was not known how an autoimmune response was induced.

However, in the early 1980s there was an increasing amount of data genetically associating different autoimmune diseases with various HLA-DR specificities (6). Subsequently the histological demonstration of upregulated expression of HLA-DR in local sites of autoimmune diseases (e.g., type 1 diabetes, thyroiditis, and RA) was linked to the role of HLA-DR in antigen presentation (7, 8) by Feldmann and colleagues (Bottazzo, Pujol-Borrell, and Hanafusa), who published a new hypothesis in 1983 proposing that locally upregulated HLA-DR augmented antigen presentation of local autoantigens. As self-reactive T cells are present in normal humans (and mice), this might be sufficient to initiate autoimmunity (8) if regulatory pathways were ineffective. Since at the time the only known regulators of HLA-DR expression were cytokines (9), this hypothesis focused attention on the role of cytokines in autoimmune disease. A "cottage industry" was engendered of expressing cytokines, e.g., gamma interferon (IFN-γ), locally in transgenic mice, which as predicted triggered autoimmunity in these mice (10).

HOW WAS TNF IDENTIFIED AS A THERAPEUTIC TARGET?

The cellular basis of the hypothesis of upregulated antigen presentation and presence of autoantigen-reactive T cells was established using human thyroid autoimmune disease tissue with the able assistance of Marco Londei (11, 12). But determining which cytokines were of importance was not possible in thyroid tissue as it was rendered quiescent before surgery. However, in RA, a much more important unmet need with poor prognosis, it was possible to sample active disease tissue and explore cytokine production (13).

How could cytokines be evaluated in disease tissue? Fortunately, the advent of the molecular biology revolution with the cloning of cDNAs for cytokines provided the necessary research tools for this work in RA (14). Our approach was to evaluate which cytokines were produced at site of the disease, and RA is one of the most accessible autoimmune diseases. Glenn Buchan was the postdoctoral fellow who first succeeded in modifying the conventional techniques to be useful for small human disease tissue samples, and we chose to measure mRNA as a close reflection of local cytokine synthesis. We were surprised to find that all the RA synovial samples produced all the cytokines that we could measure (15, 16). At the same time, other groups were also investigating cytokine production in joints. Gordon Duff's group detected IL-1 and TNF (17); Jean-Michel Dayer (18) and Firestein and Zvaifler (19) were also active in this field. Since in health, cytokine production, at least for immune/inflammatory cytokines, is transient, detecting them in all samples was abnormal and was considered to reflect long-term production. This was not due to long half-life of mRNA, so it was due to continued stimulation of production. This result was exciting as it validated that in RA there was cytokine dysregulation, but from the therapeutic perspective, the important question remained: which cytokine might be a therapeutic target (20)? This was relevant as the biotech industry had developed the capacity to generate specific cytokine inhibitors, monoclonal antibodies or antibody-like receptor-Ig fusion proteins. Which single cytokine to target when there are many present with similar actions was a major dilemma, and led to many of our competitors in this field deciding, on the basis of "cytokine redundancy," that cytokines were not good therapeutic targets. Fionula Brennan was our postdoctoral fellow in 1988-89 who resolved this dilemma by analyzing the regulation of cytokine expression in a dissociated synovial cell culture model of arthritis that we had developed.

Prior to Brennan's publications, culture of RA synovial cells was simple and serial passage kept only the synovial fibroblasts alive. The inflammatory/immune cells, which are the great majority of an RA sample, died or were discarded. Feldmann's Ph.D. at WEHI began in 1969 with study of how to optimize the *in vitro* immune response of mouse spleen cells (21). So the challenge of maintaining the great majority of immune/inflammatory cells from synovium alive to study cytokine dysregulation and this attempt to define a therapeutic target was not beyond our skill set. By modifying conditions (serum type and amount, oxygenation, cell density), cultures sustaining mixed synovial cell function for 5 to 6 days were established and used to analyze synovial cytokine dysregulation. Brennan obtained the paradoxical result that blocking a single cytokine,

TNF, totally abrogated IL-1 production within 2 to 3 days in seven consecutive RA operative samples (16).

This led to a new concept of the "TNF-dependent cytokine cascade." As is often the case, work from other directions led to a similar conclusion. Tony Cerami and his colleagues were studying bacterial sepsis in mice and showed that the first cytokine detectable in mouse blood was TNF, within 30 min, and then IL-1 and later still IL-6 (22). As had been found in synovial cultures, neutralization of TNF by anti-TNF inhibited the production of IL-1 (22). This suggested that the TNF-dependent cytokine cascade could also operate *in vivo* in mice.

The unexpected TNF-dependent cytokine cascade concept was tested in synovial cultures for all the proinflammatory cytokines produced in RA synovial cultures that could be evaluated. TNF blockade of synovial cultures rapidly downregulated GM-CSF (23), IL-6, IL-8 (24), and others; later it was found to also apply to the anti-inflammatory mediators IL-10, IL-1 receptor antagonist, and soluble TNF receptor. The TNF-dependent cascade clearly showed that TNF blockade was different from other cytokine blockade, as IL-1 blockade did not diminish TNF. This suggested that TNF might be the elusive therapeutic target for RA.

At the time (1989), the predictive capacity of animal models of disease was considered poor; for example, collagen-induced arthritis had been used to predict that killing CD4 T cells would be therapeutic for RA, which it wasn't, and little has changed in that context. So testing anti-TNF in animal models of arthritis only was not a high priority and only happened after the key human synovial experiments had been completed; once suitable anti-murine TNF antibodies had been generously donated to us by Robert Schrieber, Richard Williams was able to show the ameliorative but not curative effects of anti-TNF given after disease onset in mice with collagen-induced arthritis (25).

HOW DID ANTI-TNF BECOME STATE-OF-THE-ART THERAPY FOR RA?

With a presumed rationale for anti-TNF therapy in RA, the next challenge was to test this in patients. It was fortunate that specific TNF inhibitors had been generated by several groups in the biotech industry, based on Tony Cerami's influential concept that TNF was the driver of bacterial septic shock, which if blocked could save hundreds of thousands of lives (26). But our ideas of the role of TNF in RA at the time (1989 onwards) were not widely accepted, and the response of several companies that had produced suitable inhibitors that we talked to at length was eventually negative. But when James Woody, a former colleague, joined Centocor, a monoclonal antibody-focused company, as their chief scientist, we found an ally. He was convinced that Centocor should work with us, and as his clinical colleagues were at the time preoccupied with testing anti-CD4 antibody in RA, he and a few helpers worked with Ravinder Maini and Feldmann to test our concept clinically. They initially provided a modest grant ($100,000) and the necessary amount of chimeric anti-TNF antibody, which was then in short supply, then known as cA2, now infliximab (Remicade), to test in 10 patients. Not a lot for an academic proof of principle, but enough. No dose response could be done, so we chose the highest dose that had been safely evaluated for sepsis, 10 mg/kg, coincidently the same dose as another anti-TNF antibody that we had used in mouse experiments.

The results were dramatic, and the first patients reported relief from fatigue, as if a load had lifted, even while the slow (3-h) infusion was still ongoing (27). The enthusiasm of the patients in an open-label trial and their hope for a good outcome argues for caution in interpreting results, but the unprecedented degree of rapid clinical change, coupled with marked changes in blood tests within days in many cells (monocytes and granulocytes) and inflammatory proteins (IL-6 and C-reactive protein), suggested real clinical benefit. Despite the initial great benefit, all patients relapsed in 12 to 18 weeks, and as the first patients desperately wanted more such treatment, approval was sought from the hospital ethical review board to retreat them as they relapsed, which was granted. This was a key experiment, as it began to address an important question, whether if TNF is blocked another "redundant" proinflammatory cytokine pathway would take over. Retreatment reinduced the same degree of benefit in seven patients, but the duration of benefit "tended" to get shorter (28). The latter was not a scientific conclusion, given the limited number of patients involved (8) and also the halving of the therapeutic dose of cA2 administered.

From the successful proof of principle described above, a formal double-blind, randomized, placebo-controlled trial was performed comparing 1 and 10 mg/ml cA2 and a placebo (human serum albumin, with the same appearance), as a single infusion, assessed at 4 weeks. The short treatment duration was because a major benefit had been noted within that period with 10 mg/kg, and since patients had been taken off all disease-modifying drugs, too long a duration would have led to dropouts, especially in the placebo controls, with loss of statistical power and thus a major risk of a failed clinical trial (29).

Both doses were effective as compared to placebo, and this together with the open retreatment study opened the path to longer-term studies. The key trial was to define the optimal "unmet need" indication for anti-TNF therapy. As an important unmet need was treatment for methotrexate (MTX) failures, this was the population treated. Unusually, we chose to continue MTX therapy despite its inadequate benefit, and so this was a form of adjunctive or combination therapy. To permit easier evaluation and promote safety, the MTX dose was standardized at the very low end, 7.5 mg/week, and three doses of anti-TNF were used: 1, 3, and 10 mg/kg at monthly intervals. Again the results were very interesting. MTX even at the very low dose in the MTX failures augmented the response after administration of anti-TNF, notably, dramatically at 1 mg/kg cA2, but also at the higher doses (3 and 10 mg/kg) (30).

From this trial has sprung the major use of all the anti-TNFs in combination with MTX, and also successful use patents held by the Kennedy Trust for Rheumatology Research, which have gained royalties, permitting among other things the relocation and rebuilding of the Kennedy Institute of Rheumatology to the University of Oxford.

Based on the trials of cA2 (now termed infliximab) with MTX, all the other anti-TNF agents were tested in combination with MTX and the positive results in each case have led to its routine use in combination therapy (31, 32).

MECHANISM-OF-ACTION STUDIES

While some pharmaceutical companies think that if a medicine works clinically the mechanism of action is irrelevant, academics are very curious and keen to understand why a medicine actually works. This is because successful therapy is a very powerful probe of human biology, one of few available. Hence, there have been many studies probing anti-TNF therapy during its early academic-led development, which regrettably were curtailed in phase 3 as it became clear that anti-TNF was going to be going on sale.

From the first proof-of-principle studies, changes in inflammatory proteins and blood cellular counts were noted longitudinally, using the same patient's prior samples as a control. More informative was the first randomized study, as it had both placebo dose response and longitudinal comparisons. Much of what we know mechanistically was first elucidated then. Very painfully, the samples from a subsequent longer-term study defrosted in a freezer disaster.

Rapid changes in inflammatory proteins, within a few hours in IL-6, for example, confirm that TNF has direct effects on production of other cytokines, including chemokines, and that the TNF-dependent cytokine cascade defined in synovial culture experiments operates in RA *in vivo* (29). Other inflammatory protein levels, including CRP and serum amyloid A protein, fall in a few days. Inflammatory cell numbers, granulocytes, and monocytes, elevated in active disease, rapidly decrease, as do platelets, while T lymphocytes, relatively low in active RA, increase rapidly, so rapidly (hours) that the only possible plausible mechanism is egress from joints, probably due to reduced adhesion molecule expression (33).

Immunohistology has certain problems including potential sampling errors due to very small samples, problems in the timing of biopsy, and lack of quantitation of changes by the techniques of the 1990s. But certain changes are so dramatic as to be unequivocal, such as the reduced cellularity of both mononuclear and lymphoid cells. The mechanism of reduced cellularity is not fully understood; reduced ingress and augmented exit are documented in other studies, and increase in apoptosis remains controversial. Other important findings were normalization of hematologic abnormalities (see above), including reduction in fibrinogen and platelets, presumably reducing the risk of thrombosis and of cardiovascular disease. Abnormal immunity is normalized; the low response to exogenous antigen of blood cells and by skin test is augmented within a week. Regulatory T cells become normalized. Markers of probable tissue damage are reduced, e.g., levels of matrix metalloproteinases and of cleavage products of connective tissue. Vascular endothelial growth factor is partly reduced, suggesting that angiogenesis may be partly reduced, which was confirmed in histology studies (reviewed in references 34 and 35).

From the retreatment studies that were performed on the first cohort of patients, when they had relapsed, it was clear that the reinduced benefit, while similar in magnitude, had shorter duration. This led to investigation of the immunogenicity of the anti-TNF antibody.

There is an extensive literature indicating that all proteins, even autologous ones, are immunogenic, from the time of Jacques Oudin and Rodkey that anti-idiotype antibodies to reinjected self-immunoglobulin in rabbits (36, 37).

IFNs, which upregulate antigen presentation, are immunogenic in most patients, limiting the duration of their usefulness. In contrast, there is extensive literature from the 1960s and 1970s showing that while aggregated Ig administered subcutaneously is immunogenic, deaggregated Ig injected intravenously is nonimmunogenic, and Weigle, Dresser, Mitchison, Basten, and others described the induction of immunological tolerance by deaggregated Ig at either very low or very high doses (38, 39), termed "low-zone" or "high-zone" tolerance. Therapeutic antibodies are rigorously deaggregated, or else there would be toxic side effects.

It was found that intravenous infliximab was tolerogenic at high doses. At 1 mg/kg, anti-infliximab antibody was detected in about half of patients, but only about an eighth at 10 mg/kg, in keeping with that concept. In the presence of low-dose MTX, anti-infliximab antibodies were further reduced (40).

While the presence of anti-therapeutic antibodies does not interfere with response short term, long term there is evidence of better outcomes in patients who do not form antibodies (41). Currently there is little evidence that utility and sales are restricted by immunogenicity; currently antibodies form half of the world's top 10 best-selling medicines, with anti-TNF the best-selling drug class.

LIMITATIONS OF ANTI-TNF THERAPY

While there is excitement and a sense of achievement in initiating a new therapy, upon closer examination, with time, there are always limitations that emerge. And while patients are initially mostly greatly relieved, with time there is realization that more is needed. The patients appreciate progress, but they would like to get closer to a cure. What are the residual problems after anti-TNF therapy? The most important symptom is residual pain, varying greatly between patients; many have persistent fatigue and tenderness. A major problem, joint damage, is markedly reduced, cartilage damage less so than bone damage (34).

As might be expected from blocking TNF, a major host defense mediator, there are increases in some infections. First noted is infection with intracellular organisms, such as tuberculosis (TB), less often with *Listeria*, etc. Regrettably, with reduced inflammation the symptoms of TB are different, and the first patient with recrudescence of TB died as it was diagnosed late. But with effective screening the risk of recrudescence of TB falls from about 1/2,000 to 1/20,000 (42) and does not limit its use in the Western world. Other infections are reduced; there is more skin infection. But in the big postmarketing registries such as in the United Kingdom, the overall risk of infection does not change compared to that in severe RA treated by other means (43, 44). This is probably due to a combination of two reasons: (i) RA patients' immune function is compromised by the disease itself, and (ii) anti-TNF does not block all TNF signaling. It is very unlikely that most TNF, a local mediator, can be effectively blocked by an antibody from signaling to an adjacent cell in direct contact. This limited inhibition probably reduces TNF levels from elevated to normal but is not a total blockade, which intracellular TNF synthesis inhibitors could do, as do knockouts.

With time, the percentage of patients responding, initially 70%, is reduced, and the degree of response can also diminish. This has opened up great debate and interest as to what is the best treatment for anti-TNF low responders (45).

WHAT MIGHT BE DONE TO GET CLOSER TO A CURE? COMBINATIONS

Sales of the anti-TNF medicines (five on the market) dwarf those of other new drugs for RA; more than 10-fold higher. Thus it appears that the clinicians prescribing new treatment for RA consider its efficacy and safety superior to its competitors, and their familiarity with how to use anti-TNFs is greater. This may be partly helped by regulatory and purchaser priorities. But it is worth mentioning that in the 20 years since anti-TNF clinical trials first reported success, a multitude of new medicines have been tested and those that have completed trials and come to market have roughly comparable efficacy. It does not seem, in a complex heterogeneous disease with multiple pathways deregulated, that a single drug could yield a cure (46). That prediction may be wrong, but it is reasonable on the basis of evidence available today.

So what could be the future? Looking sideways, i.e., comparison, has a long tradition in science. In two fields, HIV and hematologic malignancies, it is clear that multiple drugs, i.e., combination therapy, has dramatically changed the prognosis for HIV from certain death (47) to life with some niggling side effects of the drugs.

So what combinations might get many patients closer to a cure? There have been attempts to augment the effect of the anti-TNF agent etanercept (TNFR2-Ig fusion protein) by combination with anakinra (IL-1ra) (48) or abatacept (CTLA4-Ig) (49). These trials were smallish, but there was no increase in efficacy at all, but a marked increase in infections. What might be learned from this? Probably that depressing inflammatory function (e.g., with etanercept and anakinra) too much is not helpful. It should be noted that the redundancy, the similarity between the effects of TNF and IL-1 on their targets, is probably about 90%. Combination of etanercept, with some inhibition of host defense, with abatacept, which is designed to block the initial antigen presentation between T lymphocytes and antigen-presenting cells at the beginning of an immune response (e.g., to infectious organisms), is also risky, with a degree of hindsight. Of course, these conclusions might depend on dose or spacing between administration, but it is now unlikely that these combinations might be revisited.

But more promising might be the combination of anti-TNF therapy with medicines that target mechanisms that maintain disease chronicity but do not interfere with host defense against infection. The mechanisms that comes to mind are blocking angiogenesis and blocking the fibroblast-like synoviocytes that erode into cartilage. There is evidence to support these concepts that blocking angiogenesis (34, 35) and fibroblast-like synoviocytes (50) might be effective, including some combinations in an animal model, collagen-induced arthritis (51). But it needs a concerted effort to define the best combination to add to existing therapy. The most likely existing therapy would be anti-TNF, as the biggest unmet need is how to enhance low responsiveness to anti-TNF. This is discussed in detail elsewhere (46).

FUTURE ANTI-TNF AGENTS

In parallel to finding ways to enhance low responsiveness to anti-TNF therapy, further research into the molecular mechanisms that govern inflammatory response may lead to identification of new targets for specific and ideally orally administrated therapeutic intervention.

Since macrophages are the main producers of TNF (52), one approach to new target identification could consist of mapping the molecular pathways and signaling events that lead to TNF production in macrophages. Macrophages are considered to be of central importance in the pathogenesis of RA. The increase in the number of sublining macrophages is an early hallmark of active rheumatic disease (53), with high numbers of macrophages being a prominent feature of inflammatory lesions (54). The degree of synovial macrophage infiltration correlates with the degree of joint erosion (55), and their depletion from inflamed tissue has a profound therapeutic benefit (56). Due to the major role of macrophages in RA, the effect of antirheumatic treatment on macrophage numbers, activation, and function is considered to be an objective readout of their efficacy (57).

Macrophages are heterogeneous, with many subpopulations now known. They also demonstrate remarkable plasticity that allows them to efficiently respond to environmental signals and change their phenotype and their physiology in response to cytokines and microbial signals. For example, CSFs play a key role in macrophage polarization. Macrophage CSF is constitutively produced by host tissue cells, such as fibroblasts, stromal cells, and osteoblasts, even in the absence of inflammation and has a largely homeostatic and resolving role, whereas GM-CSF is produced by the same cells during inflammation and has a clear proinflammatory effect on macrophages (58). These changes can give rise to populations of macrophages with diverse functions, which are phenotypically characterized by production of proinflammatory and anti-inflammatory mediators (59).

Although the extent of heterogeneity of macrophages in RA has not been fully uncovered, there is believed to be a mix of proinflammatory infiltrating macrophages and less inflammatory tissue-resident ones. Efforts to understand the polarization of macrophages and how it influences disease progression have led to identification of novel signaling pathways and strategies that target components of these pathways with varied specificity and selectivity. For example, blockade of GM-CSF inhibits the development of arthritis (60). Since depletion of GM-CSF is effective, several therapies are in phase 1 or phase 2 clinical trials or preclinical studies using GM-CSF-specific antibodies, and GM-CSF receptor-specific antibodies.

Our recent work has indicated that GM-CSF (along with IFN-γ) is a major inducer of interferon regulatory factor 5 (IRF5), a transcription factor defining proinflammatory macrophage polarization (61). IRF5 is involved in the positive regulation of Th1/Th17-associated mediators, such as IL-1, IL-12, IL-23, and tumor necrosis factor α (TNF-α) (61). We have reported that IRF5 forms a protein complex with NF-κB RelA to drive a sustained induction of the human TNF gene (62), and lately extended this observation to demonstrating that interactions of IRF5 with RelA are a common mechanism of proinflammatory gene regulation by IRF5 (63).

IRF5 is a member of a family of transcription factors originally implicated in antiviral response and IFN production (64). Subsequent studies revealed their multifaceted role in regulation of antimicrobial responses and cell differentiation (64). In a murine model of antigen-induced arthritis, synovial macrophages are characterized by high levels of IRF5 (65). IRF5 is also a genetic risk factor for many autoimmune diseases, including RA, in which it was identified as a new risk locus by a genome-wide association study (66) and contributed to the modulation of the erosive phenotype (67).

Considering the key role of macrophages in synovial inflammation, it is tempting to speculate that IRF5 expression is a finely tuned balance between macrophage adaptive versus pathological responses and thus IRF5 represents a new post anti-TNF target for therapeutic intervention. In fact, a recent study reported that silencing IRF5 in infarct macrophages resulted in reprogramming of macrophage phenotype, resolution of inflammation, and improved infarct healing (68). How blockade of the IRF5-RelA interaction would compare with TNF blockade in terms of efficacy and safety is not currently understood.

OTHER INDICATIONS

Cardiovascular

One of the first pieces of indirect evidence of cytokine production in vascular disease was the upregulation of HLA-DR in human atherosclerotic lesions by smooth muscle cells (69). TNF-α was the first cytokine to be identified in human atherosclerotic plaque (70) shortly thereafter. TNF-α has a significant role in the activation of the endothelium (71) and in the upregulation of adhesion molecules, a key event in the first steps of atherosclerotic disease.

TNF-α is involved in multiple actions on different cell types in the lesion, such as adhesion molecule expression and foam cell formation. Blocking TNF-α by TNF binding protein or IL-1 by IL-1 receptor antagonist partially protected apoE knockout mice from atherosclerosis (72). Myeloid cell production of TNF-α was the most relevant to atherogenesis. In mice on a high-fat diet, the plaque area in apoE$^{-/-}$ TNF$^{-/-}$ mice was 50% smaller than in apoE$^{-/-}$ mice (73). Brånén et al. showed that transplantation of bone marrow from apoE$^{-/-}$ TNF$^{-/-}$ into apoE$^{-/-}$ mice resulted in an impressive 83% reduction in lesion size after 25 weeks. In the same study, treatment of apoE$^{-/-}$ mice with the TNF blocker recombinant soluble p55 also led to a reduction in lesion size, indicating that TNF-α plays an important role in atherosclerosis (73). Similarly, Ohta et al. showed a slightly smaller decrease in lesion size in apoE$^{-/-}$ TNF$^{-/-}$ mice on a normal chow diet (74). In this model, expression levels of intercellular adhesion molecule-1, vascular cell adhesion molecule-1, and chemokine monocyte chemoattractant protein-1 (CCL-2) were also significantly decreased, along with reduced scavenger receptor expression and uptake of oxidized low-density lipoprotein in double-knockout mice.

Peripheral blood levels of TNF-α also predict the development of myocardial infarction (75) in patients with known cardiovascular disease. NF-κB is an important pathway leading to the production of TNF-α in human atherosclerotic plaques (76). However, TNF blockade has never been tested properly in cardiovascular disease after the failure of TNF blockade in heart failure patients (77–79), leading to failed opportunities for the use of biologics in cardiovascular disease.

Acute Injuries and Fibrosis

Macrophages undoubtedly have an important role in wound healing. They have been shown to be key in orchestrating fracture healing (80). They also play a crucial role in cutaneous wound healing: their early deletion leads to impaired granulation tissue formation, while in the mid-phase of healing, depletion results in severe wound hemorrhage (81). Macrophage subtypes are in part defined by the types of cytokines they produce. For example, the M2 phenotype is characterized by the expression of transforming growth factor β (TGF-β), while the classically activated M1 cells secrete proinflammatory cytokines. It has become accepted that during the early phases of healing classically activated M1 macrophages predominate, while the repair phase is dominated by cells with the alternatively activated M2 phenotype (82). Subsets within this grouping have been ascribed specific functions, for example, M2a profibrotic macrophages in cutaneous wound healing (82). However, it is not clear whether classifications based on *in vitro* phenotypes are representative of what happens *in vivo*. In particular, M2a macrophages may not necessarily predominate *in vivo* during tissue repair (83).

The emphasis on the accepted roles of certain cytokines produced by subsets of immune cells has formed the basis of numerous experimental studies (84). An example of this is that inflammation driven by TNF tends to be detrimental, and the archetypal profibrotic cytokine is TGF-β. Fibrosis is especially difficult to study as primary human tissues, especially at early stages of the disease, are difficult to access and animal models are of necessity based on toxic injuries that are rarely encountered in human disease. While data from these models have highlighted the importance of TGF-β, all late-phase clinical trials to date of TGF-β1 inhibition have failed, with some reporting significant adverse effects (85, 86). The cell responsible for both the matrix deposition and contraction in all fibrotic diseases is the myofibroblast. Our studies of Dupuytren's disease, a common fibrotic condition of the palm of the hand that affects ~4% of the general U.K. and U.S. populations, were based exclusively on human samples from patients with relatively early-stage disease. In addition to myofibroblasts, we found significant numbers of immune cells, including classically and alternatively activated macrophages. Utilizing a system similar to the one previously used to identify TNF as a therapeutic target in RA, we determined the cytokines secreted by freshly disaggregated cells from Dupuytren's nodules. We then examined the effects of these on myofibroblast precursor cells from patients with Dupuytren's disease compared to normal fibroblasts from the same individuals and from healthy donors. Exogenous addition of TNF, but not other cytokines, including IL-6 and IL-1β, promoted differentiation specifically of palmar dermal fibroblasts from Dupuytren's patients into myofibroblasts. A previous genome-wide association study identified the role of Wnt in Dupuytren's disease (87). We demonstrated that TNF acts via the Wnt signaling pathway to drive contraction and profibrotic signaling in these cells. Neutralizing antibodies to TNF inhibited the contractile activity of myofibroblasts derived from Dupuytren's patients, reduced their expression of α-smooth muscle actin, and mediated disassembly of the contractile apparatus (88). These data form the basis of ongoing clinical trials funded by the Wellcome Trust and the U.K. Department of Health to assess the efficacy of local administration of anti-TNF directly into the nodules of patients with early Dupuytren's disease.

The importance of studying tissues from the early stages of the disease is emphasized by accumulating evidence that inflammation precedes almost all fibrotic processes (89). We found that freshly disaggregated cells from Dupuytren's disease secreted TNF at the levels that were optimal for promoting the differentiation of palmar fibroblasts from affected patients into myofibroblasts. At passage 2, TNF levels were negligible while levels of TGF-β1 increased 3-fold through autocrine production by myofibroblasts (88). Importantly, the levels of TNF secreted *in vitro* were unaffected by the total number of primary disaggregated cells over almost a log range, indicating considerable autoregulation. Our findings emphasize the importance of studying primary human diseased tissues (90) and comparison with appropriate controls and may go some way toward explaining the failure of late-phase clinical trials targeting TGF-β1.

The early macrophage response following injury is typically described as infiltration by the M1 subset and related to clearance of microbes and tissue debris, while the secondary M2 response is usually considered to orchestrate tissue repair (84). However, this is probably somewhat simplistic, and emerging evidence suggests that while persistent or high levels of inflammation are detrimental to tissue healing, low levels of proinflammatory cytokines can initiate downstream healing responses. Again using primary human tissues, we found that supernatants derived from fractured bone fragments but not from surgically cut bone promoted the chemotaxis and osteogenic differentiation of human mesenchymal stromal cells (MSCs). The cytokine primarily responsible was TNF, which promoted MSC chemotaxis at 1 pg/ml and osteogenic differentiation of MSCs at 1 ng/ml. Preincubation of MSCs with TNF also enhanced the effects of other chemokines (91). Having elucidated a target utilizing representative primary human samples, we went on to investigate the mechanism using a murine model. Local administration of TNF only within the first 24 h postinjury enhanced fracture healing, while anti-TNF and IL-10 impaired healing. Addition of exogenous TNF enhanced recruitment of neutrophils and monocytes through CCL2 production, while depletion of neutrophils or inhibition of CCR2 resulted in significantly impaired fracture healing (92). Local administration of TNF at the fracture site in osteoporotic mice improved healing during the early phase of fracture repair.

Our data would suggest that TNF can modulate healing in a variety of ways, ranging from initiating the events culminating in fracture healing through to driving fibrosis. The levels of this cytokine as well as the duration of secretion appear to be important in determining tissue response.

Citation. Udalova I, Monaco C, Nanchahal J, Feldmann M. 2016. Anti-TNF therapy. Microbiol Spectrum 4(4):MCHD-0022-2015.

References

1. **Thomson AW, Lotze MT (ed).** 2003. *The Cytokine Handbook*, 4th ed. Academic Press, San Diego, CA.
2. **Feldmann M, Maini RN.** 2003. Lasker Clinical Medical Research Award. TNF defined as a therapeutic target for rheumatoid arthritis and other autoimmune diseases. *Nat Med* **9:**1245–1250.
3. **Klareskog L, Catrina AI, Paget S.** 2009. Rheumatoid arthritis. *Lancet* **373:**659–672.
4. **Wegner N, Lundberg K, Kinloch A, Fisher B, Malmström V, Feldmann M, Venables PJ.** 2010. Autoimmunity to specific citrullinated proteins gives the first clues to the etiology of rheumatoid arthritis. *Immunol Rev* **233:**34–54.
5. **Harre U, Georgess D, Bang H, Bozec A, Axmann R, Ossipova E, Jakobsson PJ, Baum W, Nimmerjahn F, Szarka E, Sarmay G, Krumbholz G, Neumann E, Toes R, Scherer HU, Catrina AI, Klareskog L, Jurdic P, Schett G.** 2012. Induction of osteoclastogenesis and bone loss by human autoantibodies against citrullinated vimentin. *J Clin Invest* **122:**1791–1802.
6. **Bell JI, Todd JA, McDevitt HO.** 1989. The molecular basis of HLA-disease association. *Adv Hum Genet* **18:** 1–41.

7. **Klareskog L, Forsum U, Scheynius A, Kabelitz D, Wigzell H.** 1982. Evidence in support of a self-perpetuating HLA-DR-dependent delayed-type cell reaction in rheumatoid arthritis. *Proc Natl Acad Sci U S A* **79:**3632–3636.

8. **Bottazzo GF, Pujol-Borrell R, Hanafusa T, Feldmann M.** 1983. Role of aberrant HLA-DR expression and antigen presentation in induction of endocrine autoimmunity. *Lancet* **2:**1115–1119.

9. **Sztein MB, Steeg PS, Johnson HM, Oppenheim JJ.** 1984. Regulation of human peripheral blood monocyte DR antigen expression *in vitro* by lymphokines and recombinant interferons. *J Clin Invest* **73:**556–565.

10. **Sarvetnick N, Shizuru J, Liggitt D, Martin L, McIntyre B, Gregory A, Parslow T, Stewart T.** 1990. Loss of pancreatic islet tolerance induced by β-cell expression of interferon-γ. *Nature* **346:**844–847.

11. **Londei M, Lamb JR, Bottazzo GF, Feldmann M.** 1984. Epithelial cells expressing aberrant MHC class II determinants can present antigen to cloned human T cells. *Nature* **312:**639–641.

12. **Londei M, Bottazzo GF, Feldmann M.** 1985. Human T-cell clones from autoimmune thyroid glands: specific recognition of autologous thyroid cells. *Science* **228:**85–89.

13. **Erhardt CC, Mumford PA, Venables PJ, Maini RN.** 1989. Factors predicting a poor life prognosis in rheumatoid arthritis: an eight year prospective study. *Ann Rheum Dis* **48:**7–13.

14. **Goeddel DV, Aggarwal BB, Gray PW, Leung DW, Nedwin GE, Palladino MA, Patton JS, Pennica D, Shepard HM, Sugarman BJ, Wong GH.** 1986. Tumor necrosis factors: gene structure and biological activities. *Cold Spring Harb Symp Quant Biol* **51:**597–609.

15. **Buchan G, Barrett K, Turner M, Chantry D, Maini RN, Feldmann M.** 1988. Interleukin-1 and tumour necrosis factor mRNA expression in rheumatoid arthritis: prolonged production of IL-1α. *Clin Exp Immunol* **73:**449–455.

16. **Brennan FM, Chantry D, Jackson A, Maini R, Feldmann M.** 1989. Inhibitory effect of TNFα antibodies on synovial cell interleukin-1 production in rheumatoid arthritis. *Lancet* **2:**244–247.

17. **Wood NC, Dickens E, Symons JA, Duff GW.** 1992. In situ hybridization of interleukin-1 in CD14-positive cells in rheumatoid arthritis. *Clin Immunol Immunopathol* **62:**295–300.

18. **Poubelle P, Damon M, Blotman F, Dayer JM.** 1985. Production of mononuclear cell factor by mononuclear phagocytes from rheumatoid synovial fluid. *J Rheumatol* **12:**412–417.

19. **Xu WD, Firestein GS, Taetle R, Kaushansky K, Zvaifler NJ.** 1989. Cytokines in chronic inflammatory arthritis. II. Granulocyte-macrophage colony-stimulating factor in rheumatoid synovial effusions. *J Clin Invest* **83:**876–882.

20. **Feldmann M, Brennan FM, Maini RN.** 1996. Role of cytokines in rheumatoid arthritis. *Annu Rev Immunol* **14:**397–440.

21. **Feldmann M, Easten A.** 1971. The relationship between antigenic structure and the requirement for thymus-derived cells in the immune response. *J Exp Med* **134:**103–119.

22. **Fong Y, Tracey KJ, Moldawer LL, Hesse DG, Manogue KB, Kenney JS, Lee AT, Kuo GC, Allison AC, Lowry SF, Cerami A.** 1989. Antibodies to cachectin/tumor necrosis factor reduce interleukin 1β and interleukin 6 appearance during lethal bacteremia. *J Exp Med* **170:**1627–1633.

23. **Haworth C, Brennan FM, Chantry D, Turner M, Maini RN, Feldmann M.** 1991. Expression of granulocyte-macrophage colony-stimulating factor in rheumatoid arthritis: regulation by tumor necrosis factor-α. *Eur J Immunol* **21:**2575–2579.

24. **Butler DM, Maini RN, Feldmann M, Brennan FM.** 1995. Modulation of proinflammatory cytokine release in rheumatoid synovial membrane cell cultures. Comparison of monoclonal anti TNF-α antibody with the interleukin-1 receptor antagonist. *Eur Cytokine Netw* **6:**225–230.

25. **Williams RO, Feldmann M, Maini RN.** 1992. Anti-tumor necrosis factor ameliorates joint disease in murine collagen-induced arthritis. *Proc Natl Acad Sci U S A* **89:**9784–9788.

26. **Beutler B, Cerami A.** 1988. Cachectin, cachexia, and shock. *Annu Rev Med* **39:**75–83.

27. **Elliott MJ, Maini RN, Feldmann M, Long-Fox A, Charles P, Katsikis P, Brennan FM, Walker J, Bijl H, Ghrayeb J, Woody J.** 1993. Treatment of rheumatoid arthritis with chimeric monoclonal antibodies to tumor necrosis factor α. *Arthritis Rheum* **36:**1681–1690.

28. **Elliott MJ, Maini RN, Feldmann M, Long-Fox A, Charles P, Bijl H, Woody JN.** 1994. Repeated therapy with monoclonal antibody to tumour necrosis factor α (cA2) in patients with rheumatoid arthritis. *Lancet* **344:**1125–1127.

29. **Elliott MJ, Maini RN, Feldmann M, Kalden JR, Antoni C, Smolen JS, Leeb B, Breedveld FC, Macfarlane JD, Bijl H, Woody JN.** 1994. Randomised double-blind comparison of chimeric monoclonal antibody to tumour necrosis factor α (cA2) versus placebo in rheumatoid arthritis. *Lancet* **344:**1105–1110.

30. **Maini RN, Breedveld FC, Kalden JR, Smolen JS, Davis D, Macfarlane JD, Antoni C, Leeb B, Elliott MJ, Woody JN, Schaible TF, Feldmann M.** 1998. Therapeutic efficacy of multiple intravenous infusions of anti-tumor necrosis factor α monoclonal antibody combined with low-dose weekly methotrexate in rheumatoid arthritis. *Arthritis Rheum* **41:**1552–1563.

31. **Weinblatt ME, Kremer JM, Bankhurst AD, Bulpitt KJ, Fleischmann RM, Fox RI, Jackson CG, Lange M, Burge DJ.** 1999. A trial of etanercept, a recombinant tumor necrosis factor receptor:Fc fusion protein, in patients with rheumatoid arthritis receiving methotrexate. *N Engl J Med* **340:**253–259.

32. **Weisman MH, Moreland LW, Furst DE, Weinblatt ME, Keystone EC, Paulus HE, Teoh LS, Velagapudi RB, Noertersheuser PA, Granneman GR, Fischkoff SA, Chartash EK.** 2003. Efficacy, pharmacokinetic, and safety assessment of adalimumab, a fully human anti-tumor necrosis factor-alpha monoclonal antibody, in adults with rheumatoid arthritis receiving concomitant methotrexate: a pilot study. *Clin Ther* **25:**1700–1721.

33. **Charles P, Elliott MJ, Davis D, Potter A, Kalden JR, Antoni C, Breedveld FC, Smolen JS, Eberl G, Woody JN, Feldmann M, Maini RN.** 1999. Regulation of cytokines, cytokine inhibitors, and acute phase-proteins following anti-TNF-α therapy in rheumatoid arthritis. *J Immunol* **163:**1521–1528.

34. **Feldmann M, Maini RN.** 2001. Anti-TNFα therapy of rheumatoid arthritis: what have we learned? *Annu Rev Immunol* **19:**163–196.

35. **Paleolog EM.** 2002. Angiogenesis in rheumatoid arthritis. *Arthritis Res* **4**(Suppl 3):S81–S90.

36. **Rodkey LS.** 1974. Studies of idiotypic antibodies. Production and characterization of autoantiidiotypic antisera. *J Exp Med* **139:**712–720.

37. **Oudin J, Michel M.** 1969. Idiotypy of rabbit antibodies. II. Comparison of idiotypy of various kinds of antibodies formed in the same rabbits against *Salmonella* Typhi. *J Exp Med* **130:**619–642.

38. **Mitchison NA.** 1964. Induction of immunological paralysis in two zones of dosage. *Proc R Soc Lond B Biol Sci* **161:**275–292.

39. **Golub ES, Weigle WO.** 1969. Studies on the induction of immunologic unresponsiveness. 3. Antigen form and mouse strain variation. *J Immunol* **102:**389–396.

40. **Feldmann M, Maini RN.** 2001. Anti-TNFα therapy of rheumatoid arthritis: what have we learned? *Annu Rev Immunol* **19:**163–196.

41. **Pascual-Salcedo D, Plasencia C, Ramiro S, Nuño L, Bonilla G, Nagore D, Ruiz Del Agua A, Martínez A, Aarden L, Martín-Mola E, Balsa A.** 2011. Influence of immunogenicity on the efficacy of long-term treatment with infliximab in rheumatoid arthritis. *Rheumatology (Oxford)* **50:**1445–1452.

42. **Askling J, Fored CM, Brandt L, Baecklund E, Bertilsson L, Cöster L, Geborek P, Jacobsson LT, Lindblad S, Lysholm J, Rantapää-Dahlqvist S, Saxne T, Romanus V, Klareskog L, Feltelius N.** 2005. Risk and case characteristics of tuberculosis in rheumatoid arthritis associated with tumor necrosis factor antagonists in Sweden. *Arthritis Rheum* **52:**1986–1992.

43. **Hashimoto A, Matsui T.** 2015. Risk of serious infection in patients with rheumatoid arthritis. *Nihon Rinsho Meneki Gakkai Kaishi* **38:**109–115. (In Japanese.)

44. **Listing J, Gerhold K, Zink A.** 2013. The risk of infections associated with rheumatoid arthritis, with its comorbidity and treatment. *Rheumatology (Oxford)* **52:** 53–61.

45. **Smolen JS, Aletaha D.** 2013. Forget personalised medicine and focus on abating disease activity. *Ann Rheum Dis* **72:**3–6.

46. **Feldmann M, Maini RN.** 2015. Can we get closer to a cure for rheumatoid arthritis? *Arthritis Rheumatol* **67:** 2283–2291.

47. **Sabin CA.** 2013. Do people with HIV infection have a normal life expectancy in the era of combination antiretroviral therapy? *BMC Med* **11:**251–257.

48. **Genovese MC, Cohen S, Moreland L, Lium D, Robbins S, Newmark R, Bekker P, 20000223 Study Group.** 2004. Combination therapy with etanercept and anakinra in the treatment of patients with rheumatoid arthritis who have been treated unsuccessfully with methotrexate. *Arthritis Rheum* **50:**1412–1419.

49. **Weinblatt M, Schiff M, Goldman A, Kremer J, Luggen M, Li T, Chen D, Becker JC.** 2007. Selective costimulation modulation using abatacept in patients with active rheumatoid arthritis while receiving etanercept: a randomised clinical trial. *Ann Rheum Dis* **66:**228–234.

50. **Aletaha D, Funovits J, Smolen JS.** 2011. Physical disability in rheumatoid arthritis is associated with cartilage damage rather than bone destruction. *Ann Rheum Dis* **70:**733–739.

51. **Kaneko K, Williams RO, Dransfield DT, Nixon AE, Sandison A, Itoh Y.** 2016. Selective inhibition of membrane type 1 matrix metalloproteinase abrogates progression of experimental inflammatory arthritis: synergy with tumor necrosis factor blockade. *Arthritis Rheumatol* **68:** 521–531.

52. **Grivennikov SI, Tumanov AV, Liepinsh DJ, Kruglov AA, Marakusha BI, Shakhov AN, Murakami T, Drutskaya LN, Förster I, Clausen BE, Tessarollo L, Ryffel B, Kuprash DV, Nedospasov SA.** 2005. Distinct and nonredundant in vivo functions of TNF produced by T cells and macrophages/neutrophils: protective and deleterious effects. *Immunity* **22:**93–104.

53. **Tak PP, Bresnihan B.** 2000. The pathogenesis and prevention of joint damage in rheumatoid arthritis: advances from synovial biopsy and tissue analysis. *Arthritis Rheum* **43:**2619–2633.

54. **Smeets TJ, Kraan MC, Galjaard S, Youssef PP, Smith MD, Tak PP.** 2001. Analysis of the cell infiltrate and expression of matrix metalloproteinases and granzyme B in paired synovial biopsy specimens from the cartilage-pannus junction in patients with RA. *Ann Rheum Dis* **60:** 561–565.

55. **Mulherin D, Fitzgerald O, Bresnihan B.** 1996. Synovial tissue macrophage populations and articular damage in rheumatoid arthritis. *Arthritis Rheum* **39:**115–124.

56. **Barrera P, Blom A, van Lent PL, van Bloois L, Beijnen JH, van Rooijen N, de Waal Malefijt MC, van de Putte LB, Storm G, van den Berg WB.** 2000. Synovial macrophage depletion with clodronate-containing liposomes in rheumatoid arthritis. *Arthritis Rheum* **43:**1951–1959.

57. **Haringman JJ, Gerlag DM, Zwinderman AH, Smeets TJ, Kraan MC, Baeten D, McInnes IB, Bresnihan B, Tak PP.** 2005. Synovial tissue macrophages: a sensitive biomarker for response to treatment in patients with rheumatoid arthritis. *Ann Rheum Dis* **64:**834–838.

58. **Hamilton JA.** 2008. Colony-stimulating factors in inflammation and autoimmunity. *Nat Rev Immunol* **8:** 533–544.

59. **Gordon S, Taylor PR.** 2005. Monocyte and macrophage heterogeneity. *Nat Rev Immunol* **5:**953–964.

60. **Cook AD, Braine EL, Campbell IK, Rich MJ, Hamilton JA.** 2001. Blockade of collagen-induced arthritis postonset by antibody to granulocyte-macrophage colony-stimulating factor (GM-CSF): requirement for GM-CSF in the effector phase of disease. *Arthritis Res* **3:**293–298.

61. **Krausgruber T, Blazek K, Smallie T, Alzabin S, Lockstone H, Sahgal N, Hussell T, Feldmann M, Udalova IA.** 2011.

IRF5 promotes inflammatory macrophage polarization and T_H1-T_H17 responses. *Nat Immunol* **12:**231–238.

62. **Krausgruber T, Saliba D, Ryzhakov G, Lanfrancotti A, Blazek K, Udalova IA.** 2010. IRF5 is required for late-phase TNF secretion by human dendritic cells. *Blood* **115:**4421–4430.
63. **Saliba DG, Heger A, Eames HL, Oikonomopoulos S, Teixeira A, Blazek K, Androulidaki A, Wong D, Goh FG, Weiss M, Byrne A, Pasparakis M, Ragoussis J, Udalova IA.** 2014. IRF5:RelA interaction targets inflammatory genes in macrophages. *Cell Rep* **8:**1308–1317.
64. **Tamura T, Yanai H, Savitsky D, Taniguchi T.** 2008. The IRF family transcription factors in immunity and oncogenesis. *Annu Rev Immunol* **26:**535–584.
65. **Weiss M, Blazek K, Byrne AJ, Perocheau DP, Udalova IA.** 2013. IRF5 is a specific marker of inflammatory macrophages *in vivo*. *Mediators Inflamm* **2013:**245804. doi: 10.1155/2013/245804.
66. **Stahl EA, Raychaudhuri S, Remmers EF, Xie G, Eyre S, Thomson BP, Li Y, Kurreeman FA, Zhernakova A, Hinks A, Guiducci C, Chen R, Alfredsson L, Amos CI, Ardlie KG; BIRAC Consortium, Barton A, Bowes J, Brouwer E, Burtt NP, Catanese JJ, Coblyn J, Coenen MJ, Costenbader KH, Criswell LA, Crusius JB, Cui J, de Bakker PI, De Jager PL, Ding B, Emery P, Flynn E, Harrison P, Hocking LJ, Huizinga TW, Kastner DL, Ke X, Lee AT, Liu X, Martin P, Morgan AW, Padyukov L, Posthumus MD, Radstake TR, Reid DM, Seielstad M, Seldin MF, Shadick NA, Steer S, Tak PP, et al.** 2010. Genome-wide association study meta-analysis identifies seven new rheumatoid arthritis risk loci. *Nat Genet* **42:** 508–514.
67. **Dawidowicz K, Allanore Y, Guedj M, Pierlot C, Bombardieri S, Balsa A, Westhovens R, Barrera P, Alves H, Teixeira VH, Petit-Teixeira E, van de Putte L, van Riel P, Prum B, Bardin T, Meyer O, Cornélis F, Dieudé P, ECRAF.** 2011. The interferon regulatory factor 5 gene confers susceptibility to rheumatoid arthritis and influences its erosive phenotype. *Ann Rheum Dis* **70:**117–121.
68. **Courties G, Heidt T, Sebas M, Iwamoto Y, Jeon D, Truelove J, Tricot B, Wojtkiewicz G, Dutta P, Sager HB, Borodovsky A, Novobrantseva T, Klebanov B, Fitzgerald K, Anderson DG, Libby P, Swirski FK, Weissleder R, Nahrendorf M.** 2014. *In vivo* silencing of the transcription factor IRF5 reprograms the macrophage phenotype and improves infarct healing. *J Am Coll Cardiol* **63:** 1556–1566.
69. **Hansson GK, Jonasson L, Holm J, Claesson-Welsh L.** 1986. Class II MHC antigen expression in the atherosclerotic plaque: smooth muscle cells express HLA-DR, HLA-DQ and the invariant gamma chain. *Clin Exp Immunol* **64:**261–268.
70. **Barath P, Fishbein MC, Cao J, Berenson J, Helfant RH, Forrester JS.** 1990. Detection and localization of tumor necrosis factor in human atheroma. *Am J Cardiol* **65:** 297–302.
71. **Bevilacqua MP, Pober JS, Majeau GR, Fiers W, Cotran RS, Gimbrone MA Jr.** 1986. Recombinant tumor necrosis factor induces procoagulant activity in cultured human vascular endothelium: characterization and comparison with the actions of interleukin 1. *Proc Natl Acad Sci U S A* **83:**4533–4537.
72. **Elhage R, Maret A, Pieraggi MT, Thiers JC, Arnal JF, Bayard F.** 1998. Differential effects of interleukin-1 receptor antagonist and tumor necrosis factor binding protein on fatty-streak formation in apolipoprotein E-deficient mice. *Circulation* **97:**242–244.
73. **Brånén L, Hovgaard L, Nitulescu M, Bengtsson E, Nilsson J, Jovinge S.** 2004. Inhibition of tumor necrosis factor-α reduces atherosclerosis in apolipoprotein E knockout mice. *Arterioscler Thromb Vasc Biol* **24:**2137–2142.
74. **Ohta H, Wada H, Niwa T, Kirii H, Iwamoto N, Fujii H, Saito K, Sekikawa K, Seishima M.** 2005. Disruption of tumor necrosis factor-alpha gene diminishes the development of atherosclerosis in ApoE-deficient mice. *Atherosclerosis* **180:**11–17.
75. **Ridker PM, Rifai N, Pfeffer M, Sacks F, Lepage S, Braunwald E.** 2000. Elevation of tumor necrosis factor-alpha and increased risk of recurrent coronary events after myocardial infarction. *Circulation* **101:**2149–2153.
76. **Monaco C, Andreakos E, Kiriakidis S, Mauri C, Bicknell C, Foxwell B, Cheshire N, Paleolog E, Feldmann M.** 2004. Canonical pathway of nuclear factor kappa B activation selectively regulates proinflammatory and prothrombotic responses in human atherosclerosis. *Proc Natl Acad Sci U S A* **101:**5634–5639.
77. **Anker SD, Coats AJ.** 2002. How to RECOVER from RENAISSANCE? The significance of the results of RECOVER, RENAISSANCE, RENEWAL and ATTACH. *Int J Cardiol* **86:**123–130.
78. **Coletta AP, Clark AL, Banarjee P, Cleland JG.** 2002. Clinical trials update: RENEWAL (RENAISSANCE and RECOVER) and ATTACH. *Eur J Heart Fail* **4:**559–561.
79. **Chung ES, Packer M, Lo KH, Fasanmade AA, Willerson JT, Anti-TNF Therapy Against Congestive Heart Failure Investigators.** 2003. Randomized, double-blind, placebo-controlled, pilot trial of infliximab, a chimeric monoclonal antibody to tumor necrosis factor-alpha, in patients with moderate-to-severe heart failure: results of the anti-TNF Therapy Against Congestive Heart Failure (ATTACH) trial. *Circulation* **107:**3133–3140.
80. **Raggatt LJ, Wullschleger ME, Alexander KA, Wu AC, Millard SM, Kaur S, Maugham ML, Gregory LS, Steck R, Pettit AR.** 2014. Fracture healing via periosteal callus formation requires macrophages for both initiation and progression of early endochondral ossification. *Am J Pathol* **184:**3192–3204.
81. **Lucas T, Waisman A, Ranjan R, Roes J, Krieg T, Müller W, Roers A, Eming SA.** 2010. Differential roles of macrophages in diverse phases of skin repair. *J Immunol* **184:** 3964–3977.
82. **Sindrilaru A, Scharffetter-Kochanek K.** 2013. Disclosure of the Culprits: Macrophages-Versatile Regulators of Wound Healing. *Adv Wound Care (New Rochelle)* **2:**357–368.
83. **Novak ML, Koh TJ.** 2013. Macrophage phenotypes during tissue repair. *J Leukoc Biol* **93:**875–881.
84. **Murray PJ, Wynn TA.** 2011. Protective and pathogenic functions of macrophage subsets. *Nat Rev Immunol* **11:** 723–737.

85. **Varga J, Pasche B.** 2009. Transforming growth factor beta as a therapeutic target in systemic sclerosis. *Nat Rev Rheumatol* **5**:200–206.

86. **Hawinkels LJ, Ten Dijke P.** 2011. Exploring anti-TGF-β therapies in cancer and fibrosis. *Growth Factors* **29**: 140–152.

87. **Dolmans GH, Werker PM, Hennies HC, Furniss D, Festen EA, Franke L, Becker K, van der Vlies P, Wolffenbuttel BH, Tinschert S, Toliat MR, Nothnagel M, Franke A, Klopp N, Wichmann HE, Nürnberg P, Giele H, Ophoff RA, Wijmenga C, Dutch Dupuytren Study Group, German Dupuytren Study Group, LifeLines Cohort Study, BSSH-GODD Consortium.** 2011. Wnt signaling and Dupuytren's disease. *N Engl J Med* **365**:307–317.

88. **Verjee LS, Verhoekx JS, Chan JK, Krausgruber T, Nicolaidou V, Izadi D, Davidson D, Feldmann M, Midwood KS, Nanchahal J.** 2013. Unraveling the signaling pathways promoting fibrosis in Dupuytren's disease reveals TNF as a therapeutic target. *Proc Natl Acad Sci U S A* **110**:E928–E937.

89. **Wick G, Grundtman C, Mayerl C, Wimpissinger TF, Feichtinger J, Zelger B, Sgonc R, Wolfram D.** 2013. The immunology of fibrosis. *Annu Rev Immunol* **31**:107–135.

90. **Edwards AM, et al, SGC Open Source Target-Discovery Partnership.** 2015. Preclinical target validation using patient-derived cells. *Nat Rev Drug Discov* **14**: 149–150.

91. **Glass GE, Chan JK, Freidin A, Feldmann M, Horwood NJ, Nanchahal J.** 2011. TNF-alpha promotes fracture repair by augmenting the recruitment and differentiation of muscle-derived stromal cells. *Proc Natl Acad Sci U S A* **108**:1585–1590.

92. **Chan JK, Glass GE, Ersek A, Freidin A, Williams GA, Gowers K, Espirito Santo AI, Jeffery R, Otto WR, Poulsom R, Feldmann M, Rankin SM, Horwood NJ, Nanchahal J.** 2015. Low-dose TNF augments fracture healing in normal and osteoporotic bone by up-regulating the innate immune response. *EMBO Mol Med* **7**:547–561.

Myeloid Cells in Health and Disease: A Synthesis
Edited by Siamon Gordon

doi:10.1128/microbiolspec.MCHD-0002-2015

Jonathan Chou[1,2]
Matilda F. Chan[3]
Zena Werb[1]

36 Metalloproteinases: a Functional Pathway for Myeloid Cells

INTRODUCTION

Myeloid cells play significant roles in tissue remodeling, including the turnover of extracellular matrix (ECM), regulation of inflammation, and progression of cancer. Proteinases are important mediators of these processes and myeloid cells are major sources of proteinases, including the matrix metalloproteinases (MMPs). Indeed, the first cellular sources of MMPs, collagenase (MMP8) and gelatinase B (MMP9), were discovered in neutrophils (1, 2). Papers published in the 1970s and 1980s demonstrated that macrophages also secreted collagenolytic (MMP8), gelatinolytic (MMP9), and elastinolytic (MMP12) metalloproteinases that degrade components of the ECM in the extracellular, pericellular, and lysosomal compartments (3–7). Taken together, these early discoveries suggested that metalloproteinases produced by myeloid cells have a major role in the remodeling of the microenvironment.

The metalloproteinase family includes the MMPs; the adamalysins, which include ADAM (a disintegrin and metalloproteinase) and ADAMTS (a disintegrin and metalloproteinase with thrombospondin motifs) enzymes; as well as the astacins. These multimodular enzymes share a common, highly conserved motif (HEXXHXXGXXH) containing three histidine residues that coordinate a zinc ion at the catalytic site, which is critical for carrying out hydrolysis of protein substrates. Together, metalloproteinases can cleave nearly all components of ECM (8) and are widely expressed during normal development as well as in disease states such as fibrosis, regeneration following tissue damage, and cancer (reviewed in reference 9). We focus this report on the MMPs, because they have been more actively studied, and to a lesser extent discuss the role of adamalysins in myeloid cells.

MEMBERS OF THE METALLOPROTEINASE FAMILY

The MMP family of endopeptidases includes 23 human and 24 mouse members. The soluble and cell surface-bound MMPs process substrates through proteolysis

[1]Department of Anatomy; [2]Department of Medicine; [3]Department of Ophthalmology, University of California, San Francisco, CA 94143.

and clip short fragments from cytokines, chemokines, and growth factors to alter their bioactive properties. In addition, they actively participate in ectodomain shedding of cell surface receptors, which affects the protein composition of the plasma membrane and the cellular interactions with growth factors and the microenvironment. After cleavage of cell surface receptors by MMPs, the intracellular and intramembrane portions of these receptors can undergo further cleavage to release proteins that have additional intracellular functions, allowing MMPs to serve as crucial regulators of cell signaling.

Structurally, MMPs consist of an amino-terminal signal peptide required for extracellular secretion, a propeptide domain required for activation, a catalytic domain responsible for proteolysis, a linker peptide (hinge region), and a carboxy-terminal hemopexin (HPX) domain required for recognizing protein partners and determining substrate specificity (Fig. 1). MMPs are initially expressed and secreted as zymogens, which are enzymati-

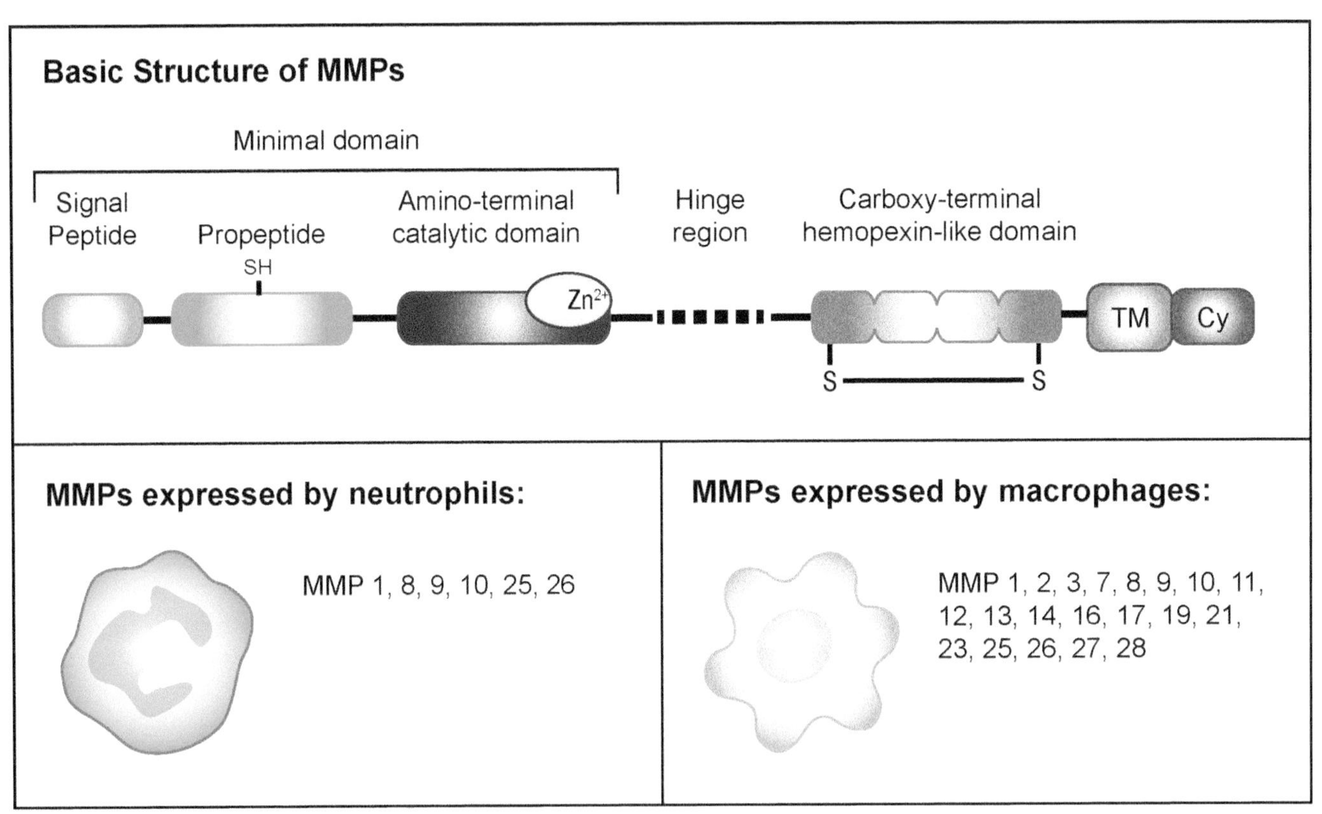

Figure 1 MMP structure and expression in myeloid cells. (Top) MMPs have a basic structure composed of functional subdomains. All MMPs have a "minimal domain" comprising an amino-terminal signal sequence that directs them to the endoplasmic reticulum, a propeptide domain with a cysteine that provides a zinc-interacting thiol (SH) group to maintain them as inactive zymogens, and a catalytic domain with three histidines that form a zinc-binding site (Zn). Most MMPs contain additional domains, the most common of which is the carboxy-terminal HPX-like domain, which mediates interactions with TIMPs, cell surface molecules, and proteolytic substrates. This domain is composed of a four-β-propeller structure and contains a disulfide bond (S-S) between the first and the last subdomains. MT-MMPs have an additional single-span transmembrane domain (TM) and a very short cytoplasmic domain (Cy). (Bottom) Expression of MMPs in myeloid cells. Neutrophils and macrophages release a variety of proteinases into the extracellular space during diverse biological processes including infection, tumorigenesis, and tissue repair. While neutrophils and macrophages are able to express several MMPs, the specific MMPs expressed by each cell type depend on the tissue microenvironment. In addition, both neutrophils and macrophages express a number of ADAM and ADAMTS proteins (not depicted here) that are important for their function and for regulating inflammation and signaling.

cally inactive (also called pro-MMPs). This is achieved by the interaction of a cysteine residue of the propeptide domain with the zinc-binding site within the catalytic domain (termed the cysteine switch), which characterizes proteinases as *metallo*proteinases. Disrupting the cysteine-zinc interaction, either chemically (by reactive oxygen species or oxidized glutathione) or proteolytically (by removing the propeptide domain), activates the MMP (10). *In vivo*, this activation is mediated by membrane-tethered MMPs (MT-MMPs) and other serine proteinases such as plasmin and neutrophil elastase. Once activated, MMPs bind and recognize substrates via the HPX domain and enzymatically cleave these substrates, causing degradation and/or generation of new biologically active fragments.

The diversity of MMPs within the genome allows this family of proteinases to have a broad range of substrates. MMPs have traditionally been classified based on their ECM substrate profile as well as structural similarities. These include collagenases (MMP1, MMP8, MMP13, and MMP19), gelatinases (MMP2 and MMP9), matrilysins (MMP7 and MMP26), and stromelysins (MMP3, MMP10, and MMP11). In addition, there are four transmembrane (MT)-MMPs—MMP14 (MT1-MMP), MMP15 (MT2-MMP), MMP16 (MT3-MMP), and MMP24 (MT5-MMP)—and two glycosylphosphatidylinositol-anchored MMPs—MMP17 (MT4-MMP) and MMP25 (MT6-MMP) (11)—that have critical roles in myeloid cell migration and processing chemokines (reviewed in reference 12). However, these traditional MMP subgroups may not reflect the metalloproteinase's dominant role *in vivo* or its most important set of substrates since many MMPs have multiple and context-dependent substrates. For example, MMP14 is an MT-MMP that is a collagenase but also activates other MMPs such as MMP2; MMP9 cleaves gelatin as well as other ECM proteins such as type IV collagen and non-ECM proteins such as vascular endothelial growth factor (VEGF) sequestered in the ECM.

The adamalysins include ADAM and ADAMTS proteins (13). To date, 21 ADAM genes have been identified but only 12 encode proteolytic proteins, indicating that many ADAMs have nonproteolytic functions (14). Those with proteolytic activities are called sheddases because of their major role in cleaving transmembrane protein ectodomains (15). In lieu of the HPX domain, ADAMs instead contain a cysteine-rich region, a series of epidermal growth factor-like repeats, and a disintegrin domain, which mediate cell-ECM interactions by binding integrins that allow ADAMs to bind and interact with substrates on neighboring cells. Most ADAMs are membrane anchored and function in the pericellular space. In contrast, ADAMTS proteins are secreted proteinases with thrombospondin type I-like repeats and cleave many ECM proteins, including aggrecan (16).

Given their potential to have broad effects on the structural and biochemical milieu of the microenvironment, tight regulation of metalloproteinase activity must be achieved. Not surprisingly, metalloproteinases are regulated at multiple levels, including RNA transcription; protein synthesis, secretion, intracellular trafficking, and localization; zymogen activation; and inhibition by extracellular inhibitors (11). For example, metalloproteinases can be compartmentalized in intracellular or extracellular locations, such as in granules, or sequestered by ECM components such as glycosaminoglycans. In addition, MMPs are tightly regulated in the extracellular space (17). MT-MMPs, for example, are internalized by clathrin-dependent and -independent modes of endocytosis and transported to lysosomes or recycled back to the cell surface (18, 19). MMPs are cleared from extracellular fluids by binding to low-density lipoprotein receptor-related protein-1 (LRP-1), leading to internalization and degradation (20, 21). Thus, extracellular and pericellular control are important mechanisms by which MMPs are locally regulated.

Importantly, MMPs and adamalysins are also regulated by tissue inhibitors of metalloproteinases (TIMPs), which are the most important endogenous inhibitors of metalloproteinase activity (22). The TIMP family consists of four members (TIMP1 to −4), which can reversibly bind and inhibit the activity of all MMPs and ADAMs. TIMPs contain an amino-terminal domain that specifically folds within itself and wedges into the active site of MMPs to inhibit these proteins (23). TIMP3 is sequestered in the ECM, whereas the other TIMPs are soluble *in vivo*. Not surprisingly, TIMP levels are also regulated by endocytosis by binding LRP-1, demonstrating that endocytic mechanisms play an important role in regulating metalloproteinase activity (24, 25).

Relative concentrations of metalloproteinase to TIMP levels determine overall proteolytic activity, with higher MMP/ADAM levels favoring proteolysis and higher TIMP levels favoring inhibition. In neutrophils, MMP9 is stored in tertiary granules for immediate release following stimulation with interleukin-8 or tumor necrosis factor (TNF) (26). Interestingly, neutrophils lack TIMP1, which allows these cells to release MMP9 rapidly in an active form upon stimulation, consistent with their roles as first responders to infection or tissue damage and control processes such as angiogenesis (27). The regulation of metalloproteinase expression and activity is therefore critical in determining myeloid cell function.

FUNCTIONS OF METALLOPROTEINASES IN MYELOID CELLS

Metalloproteinases Are Important in Myeloid Cell Migration

Because metalloproteinases degrade and remodel the ECM, they play a significant role in facilitating cell migration through basement membranes and the interstitium. Collectively, MMPs are able to degrade components of the ECM including collagens, laminin, fibronectin, and fibrin, as well as aggrecan, perlecan, and nidogen. In addition, MMPs also regulate the turnover of ECM proteins (such as fibronectin) through endocytosis, as a separate mechanism to remodel the ECM (28). Expression of both secreted and membrane-bound MMPs is a common mechanism by which cells loosen the ECM in order to move through the dense deposits of ECM proteins. Mast cells, for example, utilize MMP9 to migrate through tissue, which can be induced by proinflammatory cytokines such as TNF-α (29). MMP2 and MMP9 synergistically promote neutrophil infiltration, as knockout mice deficient in *Mmp2* and *Mmp9* have impaired neutrophil influx in multiple infection and inflammation models, including influenza virus infection and acute pancreatitis (30, 31).

In addition to cleaving ECM components, metalloproteinases also regulate chemokine-induced migration. For example, macrophage-derived MMP12 cleaves and inactivates CXC-chemokine ligand 2 (CXCL2) and CXCL3, which reduces the influx of neutrophils and other macrophages, thereby attenuating the acute immune response (32). Furthermore, metalloproteinases such as MMP14 can cleave cell surface adhesion molecules such as CD44 and syndecan to promote migration. Finally, the adamalysins also regulate migration, for example, in acute lung inflammation; deletion of *Adam10* in myeloid cells results in reduced lipopolysaccharide-induced pulmonary inflammation and decreased pulmonary edema (33). Taken together, these examples illustrate that metalloproteinases can regulate myeloid cell migration by cleaving ECM components, as well as chemokines and cell surface proteins.

Metalloproteinases Are Involved in Regulating Inflammation and Fibrosis

ADAM17 was discovered in 1997 as an enzyme that releases the membrane-bound TNF-α precursor into its soluble form (34, 35). This discovery provided the first evidence for the physiologic activity of ADAMs. ADAM17, also known as the TNF-α-converting enzyme (TACE), is ubiquitously expressed. Increased ADAM17 catalytic activity has been associated with a variety of inflammatory diseases characterized by elevated levels of tissue TNF-α, including rheumatoid arthritis, inflammatory bowel disease, and osteoarthritis (14). ADAM17 is the principal enzyme responsible for the release of soluble TNF-α from myeloid cells during endotoxin-mediated shock (36). ADAM17 also cleaves the TNF receptor 1 and interleukin-6 receptor to control downstream inflammatory signaling, as well as adhesion proteins such as L-selectin (reviewed in reference 37). In addition, MMPs contribute to pathologic inflammatory conditions, including arthritis, in which local macrophages express higher levels of MMP2, MMP9, MMP13, and MMP14, ultimately leading to destruction of cartilage and bone. Alveolar macrophages also express greater amounts of MMP1 and MMP12 in emphysema, and their expression levels correlate with the severity of pulmonary disease. MMP12 is responsible for generating elastin fragments that promote inflammatory cell recruitment (38). These studies demonstrate that myeloid cell-derived metalloproteinases are important components that initiate and sustain inflammation.

Conversely, metalloproteinases are also involved in dampening the inflammatory response. For example, ADAM17 downregulates macrophage colony-stimulating factor receptor in macrophages undergoing activation (39). In addition, ADAM8, which is expressed in lung bronchial epithelial cells and myeloid cells such as eosinophils, monocytes, macrophages, and dendritic cells, protects against allergic airway inflammation. This is mediated by the ability of ADAM8 to activate the apoptotic pathway (40). MMPs like MMP12 also regulate the immune response by cleaving and inactivating chemokines. Following corneal injury, for example, MMP12 alters levels of CXCL1 and CCL2, which facilitates wound repair by regulating leukocyte infiltration and angiogenesis (41). Taken together, these studies highlight the role of metalloproteinases in regulating both physiologic and pathologic systemic immune responses.

Metalloproteinases in Myeloid Cells Have Nonproteolytic Functions

Although the majority of research on MMPs has primarily focused on their catalytic activities, recent studies suggest that other domains also play critical roles. For example, MMP12 in macrophages exhibits antibacterial properties important in clearing Gram-positive and Gram-negative bacterial infections, and this activity lies within the carboxy-terminal and not the catalytic domain (42). Remarkably, MMP12 also has antiviral properties by functioning as a transcription factor intracellularly to promote alpha interferon

(IFN-α) expression. Interestingly, its protease function is also important in the extracellular space to dampen IFN-α signaling, which provides an autofeedback mechanism (Fig. 2) (43).

MMP14 also regulates macrophage gene expression by localizing in the nucleus and having transcription factor-like activity. MMP14 triggers phosphatidylinositol 3-kinase δ signaling to regulate the Mi-2/NuRD nucleosome remodeling complex, which controls the expression of immune response genes (44). In addition, MMP14 regulates myeloid cell migration by controlling Rac1 activity via its cytoplasmic tail, which is necessary for macrophage fusion during osteoclast and giant-cell formation. Again, this function appears to be independent of its catalytic activity (45). These unexpected findings illustrate that MMPs have multiple activities independent from their catalytic domains. Future work aimed at identifying MMPs with alternative functions and characterizing additional MMPs with transcription factor-like activities will yield insights into how these metalloproteinases regulate myeloid cell function.

Metalloproteinases Supplied by Myeloid Cells Promote and Restrict Cancer

Metalloproteinases are aberrantly expressed during all stages of tumorigenesis (reviewed in reference 46), whereby altered proteolysis leads to unregulated cell growth, tissue remodeling, invasion, and ultimately, metastasis. MMPs are prominently expressed in myeloid cells responding to progressing tumors (47, 48).

In addition to their role in ECM turnover and cancer cell migration, MMPs regulate signaling pathways that control tumor growth, inflammation, and angiogenesis (Fig. 2). For example, MMP9 is upregulated in invasive, human papillomavirus-induced skin cancer. Mice lacking MMP9 have decreased keratinocyte proliferation and incidence of invasive tumors. Interestingly, MMP9 is predominantly expressed in neutrophils, macrophages, and mast cells rather than in the tumor cells. Transplanting *Mmp9*-deficient mice with MMP9-sufficient bone marrow cells restores skin cancer progression in the mouse model (49). In models of pancreatic islet carcinoma, MMP2 and MMP9 from infiltrating myeloid cells stimulate tumor cell growth and release VEGF, one of the main drivers of angiogenesis, from the ECM (50). These findings suggest that myeloid cell-derived MMPs promote tumor growth, not only by structurally reorganizing the ECM but also by releasing growth factors from the ECM that directly stimulate cell growth or enhance angiogenesis. Myeloid cells can also affect the vascular permeability of tumors and tumor responsiveness to chemotherapeutics like doxorubicin in an MMP-dependent manner (51), and myeloid cell-derived MMPs are involved in preparing the metastatic niche to facilitate colonization of new organs by circulating tumor cells (52, 53). Whether MMPs might have nonproteolytic roles in promoting cancer remains to be determined. One possibility is that MMPs might regulate self-renewal pathways (54) or enhance invasion (55), which may be important in maintaining the cancer stem cell niche or promoting metastasis.

Finally, MMPs can also restrict tumor progression. Loss-of-function mutations in MMP8, which is highly expressed in neutrophils, increases melanoma progression, while inactivating *ADAMTS15* mutations are found in human colorectal cancer samples; expression of the wild-type metalloproteinase restricts tumor growth and progression (56, 57). MMP12 from macrophages also suppresses growth of lung metastases and is associated with decreased tumor-associated blood vessel density *in vivo* (58). In addition, conditional inactivation of ADAM10, a major proteinase that regulates Notch cleavage and activation in hematopoietic stem cells, results in a myeloproliferative disorder characterized by splenomegaly and increased numbers of myeloid cells (59). Together, these studies highlight the remarkable contributions of MMPs and adamalysins to both promoting and restricting cancer progression and metastasis.

CONCLUSIONS AND FUTURE RESEARCH DIRECTIONS

As we have discussed, metalloproteinases produced by myeloid cells have broad roles in development and disease, and their perturbation may be therapeutically beneficial (Fig. 2). Most MMP inhibitors, however, have been nonspecific and have focused only on targeting the MMP catalytic domain. Given their expanding nonproteolytic roles, it will be important to develop MMP inhibitors that target different domains and functions. It is also important to consider that some metalloproteinases restrict disease progression and so their pan-inhibition may result in worse outcomes. These issues may help explain why initial clinical trials with MMP inhibitors were disappointing in cancer patients (60). In addition, the HPX domain is an important determinant of stem cell function and cell invasion (54, 55). Inhibitors that selectively target specific metalloproteinases and domains, therefore, need to be developed and investigated for therapeutic potential (61). Noncatalytic inhibitors may need to be used in combi-

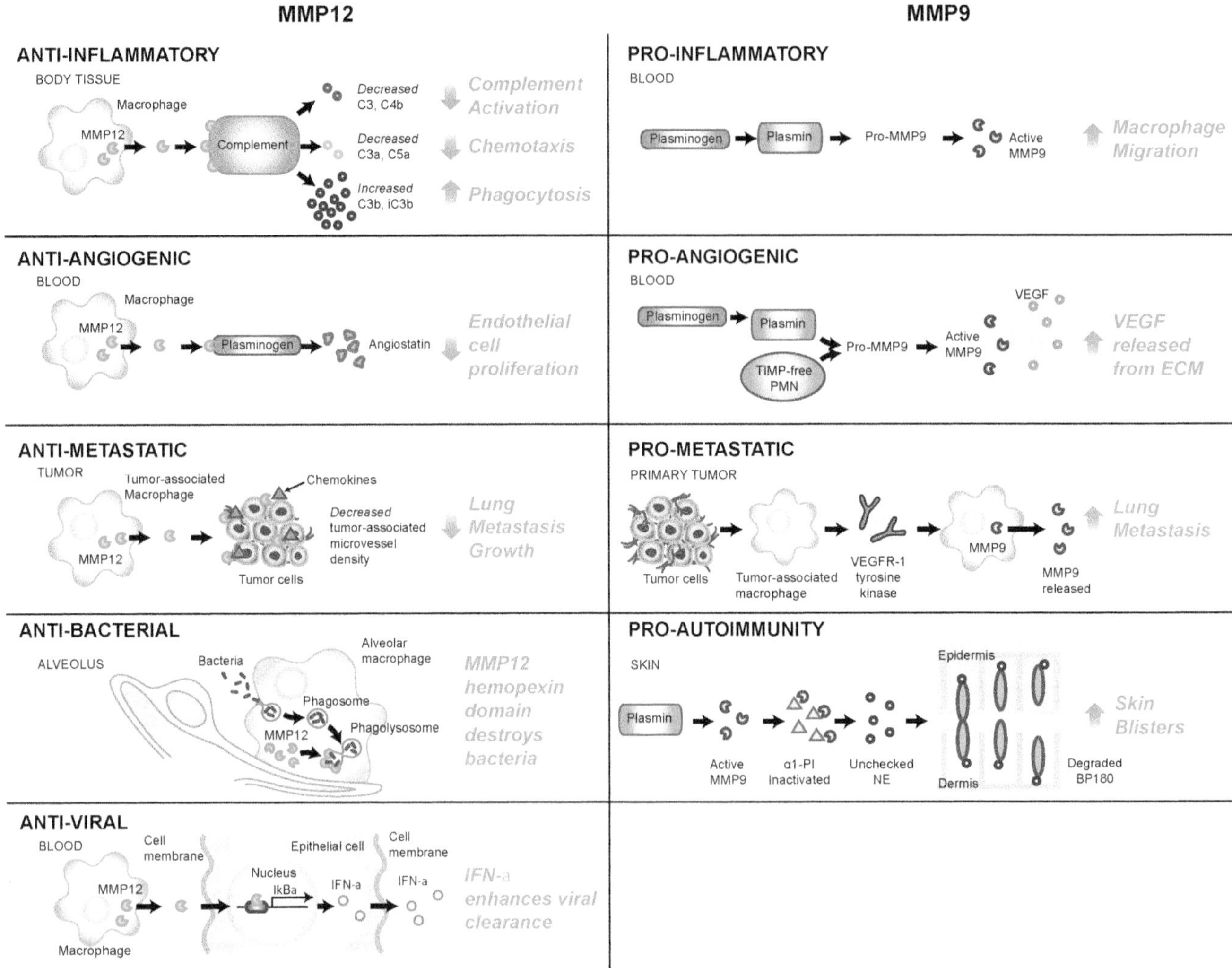

Figure 2 Functions of MMP12 and MMP9 in biological processes. MMP12 and MMP9 have contrasting roles in the processes of inflammation, angiogenesis, and metastasis, with MMP12 inhibiting these processes and MMP9 promoting them. MMP12 protects against inflammation by cleaving complements C3 and C4b and reducing complement activation, cleaving complements C3a and C5a and reducing neutrophil recruitment, and creating cleaved forms of C3b and iC3b that are potent phagocytosis enhancers (65). In contrast, MMP9 promotes inflammation by stimulating macrophage migration and infiltration upon being activated by plasminogen (66). MMPs also promote autoimmune disease. For example, in the skin disease bullous pemphigoid, MMP9 activated by plasmin proteolytically inactivates α1-proteinase inhibitor (α1-PI), the physiological inhibitor of neutrophil elastase (NE), which allows unrestrained activity of NE (67). NE degrades BP180, which results in dermal-epidermal separation (68, 69). MMP12 inhibits angiogenesis through its cleavage of plasminogen to generate angiostatin, which results in decreased endothelial cell proliferation (70). MMP9, however, promotes angiogenesis through the release of VEGF into the extracellular matrix following activation by plasmin or upon secretion from TIMP-free neutrophils (27, 50). Lung metastatic growth is reduced by MMP12 produced by tumor-associated macrophages, which interact with chemokines to decrease tumor-associated microvessel density (58). Lung metastasis is increased by MMP9 produced by tumor-associated macrophages via VEGFR-1/Flt-1 tyrosine kinase, and MMP9 levels in the lungs of patients with distant tumors are significantly elevated compared with the lungs of control patients (71). Finally, in addition to the above functions, MMP12 has direct antimicrobial activity through its HPX domain by disrupting bacterial membranes in phagolysosomes (42). MMP12 also enhances antiviral clearance by binding to the promoter of the gene encoding IκBα (*NFKBIA*) and enhancing the production of IκBα, which promotes IFN-α secretion from the cell (43). Together, these examples illustrate the diverse functions of MMPs in myeloid cells.

nation with catalytic inhibitors for complete MMP inhibition. Since MMP HPX domains are unique, this strategy may also increase specificity. In addition, antibodies that specifically recognize certain MMP conformations or domains will also help elucidate MMP function. This would allow targeting of specific pathogenic configurations without affecting its other roles.

A more comprehensive understanding of the various functions of myeloid-derived MMPs and adamalysins will also be important, as well as of the ability to control metalloproteinase activation, localization, and expression. Given MMP12 and MMP14's surprising transcription factor activity, it will be important to characterize their gene targets, which will likely be cell type specific, and to determine whether other MMPs or adamalysins might also have transcriptional activity. It will also be crucial to elucidate how the transcriptional complex is assembled and what other proteins might be within that complex in order to better understand what novel functions are conferred. In addition, the complete substrate repertoire of myeloid cell-derived metalloproteinases in specific physiologic and pathologic conditions will be important to determine through proteomic approaches (62). Because MMPs have unique functions depending on their location, identifying drugs that inhibit the extracellular but not the intracellular activity (or vice versa) may be necessary. Finally, utilizing microRNAs that target genes for MMPs, ADAMs, and ADAMTSs may increase specificity and be a useful strategy to modulate their expression.

Myeloid cells regulate the local microenvironment via MMPs and adamalysins. Depending on what substrates or protein-binding partners are in the vicinity, MMPs can facilitate different, sometimes opposite, biological processes (Fig. 2). Thus, appreciating the spatial context of myeloid cells and their secreted metalloproteinases and understanding the pericellular activity of these metalloproteinases are critical. Better methods, especially imaging tools, to assess MMP and adamalysin function *in vivo* and in real time are needed to fully appreciate their complexity. Recent advancements in MMP imaging have come forward (47, 63). For example, the development of dynamic high-resolution multimodal microscopy, which uses a green fluorescent protein-tagged proteinase in combination with a fluorophore-labeled ECM polymer (63), can reveal proteinase localization and activity in live cells. Finally, it will be critical to understand the timing with which different myeloid metalloproteinases act during physiologic and pathologic processes. Because the same metalloproteinase may play opposite roles during the early versus late stages, it is important to consider the timing of using MMP and adamalysin inhibitors (64). These exciting concepts in the biology of metalloproteinases underscore their complex roles in physiology, immunity, and diseases mediated by myeloid cells and offer critical insights into strategies and considerations for the design of more-effective therapeutics.

Acknowledgments. *We thank Suling Wang for assistance with preparing and illustrating the figures. This study was supported by funds from the National Institutes of Health (R01 CA057621 and U01 ES019458 to Z.W., R01 EY022739 to M.F.C., and P30 EY002162).*

Citation. Chou J, Chan MF, Werb Z. 2016. Metalloproteinases: a functional pathway for myeloid cells. Microbiol Spectrum 4(2):MCHD-0002-2015.

References

1. **Lazarus GS, Brown RS, Daniels JR, Fullmer HM.** 1968. Human granulocyte collagenase. *Science* **159:**1483–1485.
2. **Sopata I, Dancewicz AM.** 1974. Presence of a gelatin-specific proteinase and its latent form in human leucocytes. *Biochim Biophys Acta* **370:**510–523.
3. **Gordon S, Werb Z.** 1976. Secretion of macrophage neutral proteinase is enhanced by colchicine. *Proc Natl Acad Sci U S A* **73:**872–876.
4. **Werb Z, Bainton DF, Jones PA.** 1980. Degradation of connective tissue matrices by macrophages. III. Morphological and biochemical studies on extracellular, pericellular, and intracellular events in matrix proteolysis by macrophages in culture. *J Exp Med* **152:**1537–1553.
5. **Werb Z, Banda MJ, Jones PA.** 1980. Degradation of connective tissue matrices by macrophages. I. Proteolysis of elastin, glycoproteins, and collagen by proteinases isolated from macrophages. *J Exp Med* **152:**1340–1357.
6. **Werb Z, Gordon S.** 1975. Elastase secretion by stimulated macrophages. Characterization and regulation. *J Exp Med* **142:**361–377.
7. **Werb Z, Gordon S.** 1975. Secretion of a specific collagenase by stimulated macrophages. *J Exp Med* **142:**346–360.
8. **Lu P, Takai K, Weaver VM, Werb Z.** 2011. Extracellular matrix degradation and remodeling in development and disease. *Cold Spring Harb Perspect Biol* **3:**a005058. doi: 10.1101/cshperspect.a005058.
9. **Bonnans C, Chou J, Werb Z.** 2014. Remodelling the extracellular matrix in development and disease. *Nat Rev Mol Cell Biol* **15:**786–801.
10. **Van Wart HE, Birkedal-Hansen H.** 1990. The cysteine switch: a principle of regulation of metalloproteinase activity with potential applicability to the entire matrix metalloproteinase gene family. *Proc Natl Acad Sci U S A* **87:**5578–5582.
11. **Page-McCaw A, Ewald AJ, Werb Z.** 2007. Matrix metalloproteinases and the regulation of tissue remodelling. *Nat Rev Mol Cell Biol* **8:**221–233.
12. **Marco M, Fortin C, Fulop T.** 2013. Membrane-type matrix metalloproteinases: key mediators of leukocyte function. *J Leukoc Biol* **94:**237–246.

13. **Kuno K, Kanada N, Nakashima E, Fujiki F, Ichimura F, Matsushima K.** 1997. Molecular cloning of a gene encoding a new type of metalloproteinase-disintegrin family protein with thrombospondin motifs as an inflammation associated gene. *J Biol Chem* **272:**556–562.
14. **Lisi S, D'Amore M, Sisto M.** 2014. ADAM17 at the interface between inflammation and autoimmunity. *Immunol Lett* **162:**159–169.
15. **Murphy G.** 2008. The ADAMs: signalling scissors in the tumour microenvironment. *Nat Rev Cancer* **8:**929–941.
16. **Apte SS.** 2009. A disintegrin-like and metalloprotease (reprolysin-type) with thrombospondin type 1 motif (ADAMTS) superfamily: functions and mechanisms. *J Biol Chem* **284:**31493–31497.
17. **Yamamoto K, Murphy G, Troeberg L.** 2015. Extracellular regulation of metalloproteinases. *Matrix Biol* **44–46:** 255–263.
18. **Uekita T, Itoh Y, Yana I, Ohno H, Seiki M.** 2001. Cytoplasmic tail-dependent internalization of membrane-type 1 matrix metalloproteinase is important for its invasion-promoting activity. *J Cell Biol* **155:**1345–1356.
19. **Piccard H, Van den Steen PE, Opdenakker G.** 2007. Hemopexin domains as multifunctional liganding modules in matrix metalloproteinases and other proteins. *J Leukoc Biol* **81:**870–892.
20. **Barmina OY, Walling HW, Fiacco GJ, Freije JM, López-Otín C, Jeffrey JJ, Partridge NC.** 1999. Collagenase-3 binds to a specific receptor and requires the low density lipoprotein receptor-related protein for internalization. *J Biol Chem* **274:**30087–30093.
21. **Hahn-Dantona E, Ruiz JF, Bornstein P, Strickland DK.** 2001. The low density lipoprotein receptor-related protein modulates levels of matrix metalloproteinase 9 (MMP-9) by mediating its cellular catabolism. *J Biol Chem* **276:**15498–15503.
22. **Khokha R, Murthy A, Weiss A.** 2013. Metalloproteinases and their natural inhibitors in inflammation and immunity. *Nat Rev Immunol* **13:**649–665.
23. **Nagase H, Visse R, Murphy G.** 2006. Structure and function of matrix metalloproteinases and TIMPs. *Cardiovasc Res* **69:**562–573.
24. **Scilabra SD, Troeberg L, Yamamoto K, Emonard H, Thøgersen I, Enghild JJ, Strickland DK, Nagase H.** 2012. Differential regulation of extracellular tissue inhibitor of metalloproteinases-3 levels by cell membrane-bound and shed low density lipoprotein receptor-related protein 1. *J Biol Chem* **288:**332–342.
25. **Thevenard J, Verzeaux L, Devy J, Etique N, Jeanne A, Schneider C, Hachet C, Ferracci G, David M, Martiny L, Charpentier E, Khrestchatisky M, Rivera S, Dedieu S, Emonard H.** 2014. Low-density lipoprotein receptor-related protein-1 mediates endocytic clearance of tissue inhibitor of metalloproteinases-1 and promotes its cytokine-like activities. *PLoS One* **9:**e103839. doi:10.1371/journal.pone.0103839.
26. **Masure S, Proost P, Van Damme J, Opdenakker G.** 1991. Purification and identification of 91-kDa neutrophil gelatinase. Release by the activating peptide interleukin-8. *Eur J Biochem* **198:**391–398.
27. **Ardi VC, Kupriyanova TA, Deryugina EI, Quigley JP.** 2007. Human neutrophils uniquely release TIMP-free MMP-9 to provide a potent catalytic stimulator of angiogenesis. *Proc Natl Acad Sci U S A* **104:**20262–20267.
28. **Shi F, Sottile J.** 2011. MT1-MMP regulates the turnover and endocytosis of extracellular matrix fibronectin. *J Cell Sci* **124:**4039–4050.
29. **Di Girolamo N, Indoh I, Jackson N, Wakefield D, McNeil HP, Yan W, Geczy C, Arm JP, Tedla N.** 2006. Human mast cell-derived gelatinase B (matrix metalloproteinase-9) is regulated by inflammatory cytokines: role in cell migration. *J Immunol* **177:**2638–2650.
30. **Bradley LM, Douglass MF, Chatterjee D, Akira S, Baaten BJ.** 2012. Matrix metalloprotease 9 mediates neutrophil migration into the airways in response to influenza virus-induced toll-like receptor signaling. *PLoS Pathog* **8:**e1002641. doi:10.1371/journal.ppat.1002641.
31. **Awla D, Abdulla A, Syk I, Jeppsson B, Regner S, Thorlacius H.** 2012. Neutrophil-derived matrix metalloproteinase-9 is a potent activator of trypsinogen in acinar cells in acute pancreatitis. *J Leukoc Biol* **91:**711–719.
32. **Dean RA, Cox JH, Bellac CL, Doucet A, Starr AE, Overall CM.** 2008. Macrophage-specific metalloelastase (MMP-12) truncates and inactivates ELR^+ CXC chemokines and generates CCL2, -7, -8, and -13 antagonists: potential role of the macrophage in terminating polymorphonuclear leukocyte influx. *Blood* **112:**3455–3464.
33. **Pruessmeyer J, Hess FM, Alert H, Groth E, Pasqualon T, Schwarz N, Nyamoya S, Kollert J, van der Vorst E, Donners M, Martin C, Uhlig S, Saftig P, Dreymueller D, Ludwig A.** 2014. Leukocytes require ADAM10 but not ADAM17 for their migration and inflammatory recruitment into the alveolar space. *Blood* **123:**4077–4088.
34. **Black RA, Rauch CT, Kozlosky CJ, Peschon JJ, Slack JL, Wolfson MF, Castner BJ, Stocking KL, Reddy P, Srinivasan S, Nelson N, Boiani N, Schooley KA, Gerhart M, Davis R, Fitzner JN, Johnson RS, Paxton RJ, March CJ, Cerretti DP.** 1997. A metalloproteinase disintegrin that releases tumour-necrosis factor-α from cells. *Nature* 385:729–733.
35. **Moss ML, Jin SL, Milla ME, Bickett DM, Burkhart W, Carter HL, Chen WJ, Clay WC, Didsbury JR, Hassler D, Hoffman CR, Kost TA, Lambert MH, Leesnitzer MA, McCauley P, McGeehan G, Mitchell J, Moyer M, Pahel G, Rocque W, Overton LK, Schoenen F, Seaton T, Su JL, Becherer JD.** 1997. Cloning of a disintegrin metalloproteinase that processes precursor tumour-necrosis factor-α. *Nature* 385:733–736.
36. **Horiuchi K, Kimura T, Miyamoto T, Takaishi H, Okada Y, Toyama Y, Blobel CP.** 2007. Cutting edge: TNF-α-converting enzyme (TACE/ADAM17) inactivation in mouse myeloid cells prevents lethality from endotoxin shock. *J Immunol* **179:**2686–2689.
37. **Scheller J, Chalaris A, Garbers C, Rose-John S.** 2011. ADAM17: a molecular switch to control inflammation and tissue regeneration. *Trends Immunol* **32:**380–387.
38. **Houghton AM, Quintero PA, Perkins DL, Kobayashi DK, Kelley DG, Marconcini LA, Mecham RP, Senior RM, Shapiro SD.** 2006. Elastin fragments drive disease

progression in a murine model of emphysema. *J Clin Invest* **116**:753–759.

39. **Rovida E, Paccagnini A, Del Rosso M, Peschon J, Dello Sbarba P.** 2001. TNF-α-converting enzyme cleaves the macrophage colony-stimulating factor receptor in macrophages undergoing activation. *J Immunol* **166**:1583–1589.

40. **Knolle MD, Nakajima T, Hergrueter A, Gupta K, Polverino F, Craig VJ, Fyfe SE, Zahid M, Permaul P, Cernadas M, Montano G, Tesfaigzi Y, Sholl L, Kobzik L, Israel E, Owen CA.** 2013. Adam8 limits the development of allergic airway inflammation in mice. *J Immunol* **190**: 6434–6449.

41. **Chan MF, Li J, Bertrand A, Casbon AJ, Lin JH, Maltseva I, Werb Z.** 2013. Protective effects of matrix metalloproteinase-12 following corneal injury. *J Cell Sci* **126**:3948–3960.

42. **Houghton AM, Hartzell WO, Robbins CS, Gomis-Ruth FX, Shapiro SD.** 2009. Macrophage elastase kills bacteria within murine macrophages. *Nature* **460**:637–641.

43. **Marchant DJ, Bellac CL, Moraes TJ, Wadsworth SJ, Dufour A, Butler GS, Bilawchuk LM, Hendry RG, Robertson AG, Cheung CT, Ng J, Ang L, Luo Z, Heilbron K, Norris MJ, Duan W, Bucyk T, Karpov A, Devel L, Georgiadis D, Hegele RG, Luo H, Granville DJ, Dive V, McManus BM, Overall CM.** 2014. A new transcriptional role for matrix metalloproteinase-12 in antiviral immunity. *Nat Med* **20**:493–502.

44. **Shimizu-Hirota R, Xiong W, Baxter BT, Kunkel SL, Maillard I, Chen XW, Sabeh F, Liu R, Li XY, Weiss SJ.** 2012. MT1-MMP regulates the PI3Kδ·Mi-2/NuRD-dependent control of macrophage immune function. *Genes Dev* **26**:395–413.

45. **Gonzalo P, Guadamillas MC, Hernández-Riquer MV, Pollán A, Grande-García A, Bartolomé RA, Vasanji A, Ambrogio C, Chiarle R, Teixidó J, Risteli J, Apte SS, del Pozo MA, Arroyo AG.** 2010. MT1-MMP is required for myeloid cell fusion via regulation of Rac1 signaling. *Dev Cell* **18**:77–89.

46. **Kessenbrock K, Plaks V, Werb Z.** 2010. Matrix metalloproteinases: regulators of the tumor microenvironment. *Cell* **141**:52–67.

47. **Lohela M, Casbon AJ, Olow A, Bonham L, Branstetter D, Weng N, Smith J, Werb Z.** 2014. Intravital imaging reveals distinct responses of depleting dynamic tumor-associated macrophage and dendritic cell subpopulations. *Proc Natl Acad Sci U S A* **111**:E5086–E5095.

48. **Casbon AJ, Reynaud D, Park C, Khuc E, Gan DD, Schepers K, Passegué E, Werb Z.** 2015. Invasive breast cancer reprograms early myeloid differentiation in the bone marrow to generate immunosuppressive neutrophils. *Proc Natl Acad Sci U S A* **112**:E566–E575.

49. **Coussens LM, Tinkle CL, Hanahan D, Werb Z.** 2000. MMP-9 supplied by bone marrow-derived cells contributes to skin carcinogenesis. *Cell* **103**:481–490.

50. **Bergers G, Brekken R, McMahon G, Vu TH, Itoh T, Tamaki K, Tanzawa K, Thorpe P, Itohara S, Werb Z, Hanahan D.** 2000. Matrix metalloproteinase-9 triggers the angiogenic switch during carcinogenesis. *Nat Cell Biol* **2**:737–744.

51. **Nakasone ES, Askautrud HA, Kees T, Park JH, Plaks V, Ewald AJ, Fein M, Rasch MG, Tan YX, Qiu J, Park J, Sinha P, Bissell MJ, Frengen E, Werb Z, Egeblad M.** 2012. Imaging tumor-stroma interactions during chemotherapy reveals contributions of the microenvironment to resistance. *Cancer Cell* **21**:488–503.

52. **Kaplan RN, Riba RD, Zacharoulis S, Bramley AH, Vincent L, Costa C, MacDonald DD, Jin DK, Shido K, Kerns SA, Zhu Z, Hicklin D, Wu Y, Port JL, Altorki N, Port ER, Ruggero D, Shmelkov SV, Jensen KK, Rafii S, Lyden D.** 2005. VEGFR1-positive haematopoietic bone marrow progenitors initiate the pre-metastatic niche. *Nature* **438**:820–827.

53. **Huang Y, Song N, Ding Y, Yuan S, Li X, Cai H, Shi H, Luo Y.** 2009. Pulmonary vascular destabilization in the premetastatic phase facilitates lung metastasis. *Cancer Res* **69**:7529–7537.

54. **Kessenbrock K, Dijkgraaf GJ, Lawson DA, Littlepage LE, Shahi P, Pieper U, Werb Z.** 2013. A role for matrix metalloproteinases in regulating mammary stem cell function via the Wnt signaling pathway. *Cell Stem Cell* **13**:300–313.

55. **Correia AL, Mori H, Chen EI, Schmitt FC, Bissell MJ.** 2013. The hemopexin domain of MMP3 is responsible for mammary epithelial invasion and morphogenesis through extracellular interaction with HSP90β. *Genes Dev* **27**:805–817.

56. **Balbin M, Fueyo A, Tester AM, Pendás AM, Pitiot AS, Astudillo A, Overall CM, Shapiro SD, López-Otín C.** 2003. Loss of collagenase-2 confers increased skin tumor susceptibility to male mice. *Nat Genet* **35**:252–257.

57. **Palavalli LH, Prickett TD, Wunderlich JR, Wei X, Burrell AS, Porter-Gill P, Davis S, Wang C, Cronin JC, Agrawal NS, Lin JC, Westbroek W, Hoogstraten-Miller S, Molinolo AA, Fetsch P, Filie AC, O'Connell MP, Banister CE, Howard JD, Buckhaults P, Weeraratna AT, Brody LC, Rosenberg SA, Samuels Y.** 2009. Analysis of the matrix metalloproteinase family reveals that *MMP8* is often mutated in melanoma. *Nat Genet* **41**:518–520.

58. **Houghton AM, Grisolano JL, Baumann ML, Kobayashi DK, Hautamaki RD, Nehring LC, Cornelius LA, Shapiro SD.** 2006. Macrophage elastase (matrix metalloproteinase-12) suppresses growth of lung metastases. *Cancer Res* **66**: 6149–6155.

59. **Yoda M, Kimura T, Tohmonda T, Uchikawa S, Koba T, Takito J, Morioka H, Matsumoto M, Link DC, Chiba K, Okada Y, Toyama Y, Horiuchi K.** 2011. Dual functions of cell-autonomous and non-cell-autonomous ADAM10 activity in granulopoiesis. *Blood* **118**:6939–6942.

60. **Coussens LM, Fingleton B, Matrisian LM.** 2002. Matrix metalloproteinase inhibitors and cancer: trials and tribulations. *Science* **295**:2387–2392.

61. **Remacle AG, Golubkov VS, Shiryaev SA, Dahl R, Stebbins JL, Chernov AV, Cheltsov AV, Pellecchia M, Strongin AY.** 2012. Novel MT1-MMP small-molecule inhibitors based on insights into hemopexin domain function in tumor growth. *Cancer Res* **72**:2339–2349.

62. **Tam EM, Morrison CJ, Wu YI, Stack MS, Overall CM.** 2004. Membrane protease proteomics: isotope-coded

affinity tag MS identification of undescribed MT1-matrix metalloproteinase substrates. *Proc Natl Acad Sci U S A* **101**:6917–6922.

63. **Wolf K, Wu YI, Liu Y, Geiger J, Tam E, Overall C, Stack MS, Friedl P.** 2007. Multi-step pericellular proteolysis controls the transition from individual to collective cancer cell invasion. *Nat Cell Biol* **9**:893–904.

64. **Zeisberg M, Khurana M, Rao VH, Cosgrove D, Rougier JP, Werner MC, Shield CF III, Werb Z, Kalluri R.** 2006. Stage-specific action of matrix metalloproteinases influences progressive hereditary kidney disease. *PLoS Med* **3**: e100. doi:10.1371/journal.pmed.0030100.

65. **Bellac CL, Dufour A, Krisinger MJ, Loonchanta A, Starr AE, Auf dem Keller U, Lange PF, Goebeler V, Kappelhoff R, Butler GS, Burtnick LD, Conway EM, Roberts CR, Overall CM.** 2014. Macrophage matrix metalloproteinase-12 dampens inflammation and neutrophil influx in arthritis. *Cell Rep* **9**:618–632.

66. **Gong Y, Hart E, Shchurin A, Hoover-Plow J.** 2008. Inflammatory macrophage migration requires MMP-9 activation by plasminogen in mice. *J Clin Invest* **118**: 3012–3024.

67. **Liu Z, Zhou X, Shapiro SD, Shipley JM, Twining SS, Diaz LA, Senior RM, Werb Z.** 2000. The serpin α1-proteinase inhibitor is a critical substrate for gelatinase B/MMP-9 in vivo. *Cell* **102**:647–655.

68. **Shimanovich I, Mihai S, Oostingh GJ, Ilenchuk TT, Brocker EB, Opdenakker G, Zillikens D, Sitaru C.** 2004. Granulocyte-derived elastase and gelatinase B are required for dermal-epidermal separation induced by autoantibodies from patients with epidermolysis bullosa acquisita and bullous pemphigoid. *J Pathol* **204**:519–527.

69. **Liu Z, Li N, Diaz LA, Shipley M, Senior RM, Werb Z.** 2005. Synergy between a plasminogen cascade and MMP-9 in autoimmune disease. *J Clin Invest* **115**:879–887.

70. **Cornelius LA, Nehring LC, Harding E, Bolanowski M, Welgus HG, Kobayashi DK, Pierce RA, Shapiro SD.** 1998. Matrix metalloproteinases generate angiostatin: effects on neovascularization. *J Immunol* **161**:6845–6852.

71. **Hiratsuka S, Nakamura K, Iwai S, Murakami M, Itoh T, Kijima H, Shipley JM, Senior RM, Shibuya M.** 2002. MMP9 induction by vascular endothelial growth factor receptor-1 is involved in lung-specific metastasis. *Cancer Cell* **2**:289–300.

Immunoregulation and Infection

VIII

Myeloid Cells in Health and Disease: A Synthesis
Edited by Siamon Gordon

doi:10.1128/microbiolspec.MCHD-0051-2016

Sébastien Jaillon,[1,2] Eduardo Bonavita,[1,*] Cecilia Garlanda,[1,2] and Alberto Mantovani[1,2]

37 Interplay between Myeloid Cells and Humoral Innate Immunity

INTRODUCTION

The immune system of mammalians is organized around two components: the innate immunity and the adaptive immunity. Older in terms of evolution, the innate immune system constitutes the first line of defense against microorganisms. This system is supplemented by the adaptive immunity, which is more recent in terms of evolution and provides the basis of immunological memory. Both the adaptive and innate immune systems are composed of a cellular and a humoral arm acting in a complementary and coordinated manner to regulate the innate response.

The induction of a protective immune response against pathogens resides in the capacity to identify them. Adaptive immunity uses specific receptors encoded by gene rearrangements. In contrast, innate immune molecules involved in the recognition of pathogens and in the initiation of the immune response are germ line-encoded receptors. These receptors are called pattern recognition molecules (PRMs) and recognize highly conserved motifs expressed by microorganisms, called pathogen-associated molecular patterns (PAMPs) (1).

Based on their localization, PRMs are divided into cell-associated receptors and soluble molecules. Cell-associated receptors include endocytic receptors (e.g., scavenger receptors), signaling receptors (e.g., Toll-like receptors [TLRs]), nucleotide-binding oligomerization domain (NOD)-like receptors (NLRs), and RNA helicases such as melanoma differentiation-associated gene 5 (MDA5) and retinoic acid-inducible gene I (RIG-I). Fluid-phase molecules belonging to the humoral arm of innate immunity represent the functional ancestor of antibodies (2).

Soluble PRMs form a heterogeneous group of molecules, comprising collectins, ficolins, and pentraxins, and share basic functions, including pathogen opsonization, recognition of modified self, and regulation of complement activation. These molecules are expressed by a variety of cells, including myeloid, epithelial, and endothelial cells (2). This diversification supports the production of key molecules over time and in specific sites (Fig. 1).

Here, we will review the key soluble PRMs, their production by phagocytes, and their roles in innate immunity and inflammation (Table 1). In particular, we will focus our attention on the prototypic long pentraxin 3 (PTX3), whose regulation and function have been conserved in evolution.

[1]Humanitas Clinical Research Center; [2]Humanitas University, 20089, Rozzano (Milano), Italy. [*]Present address: Cancer Research UK Manchester Institute, The University of Manchester, Manchester M20 4QL, United Kingdom.

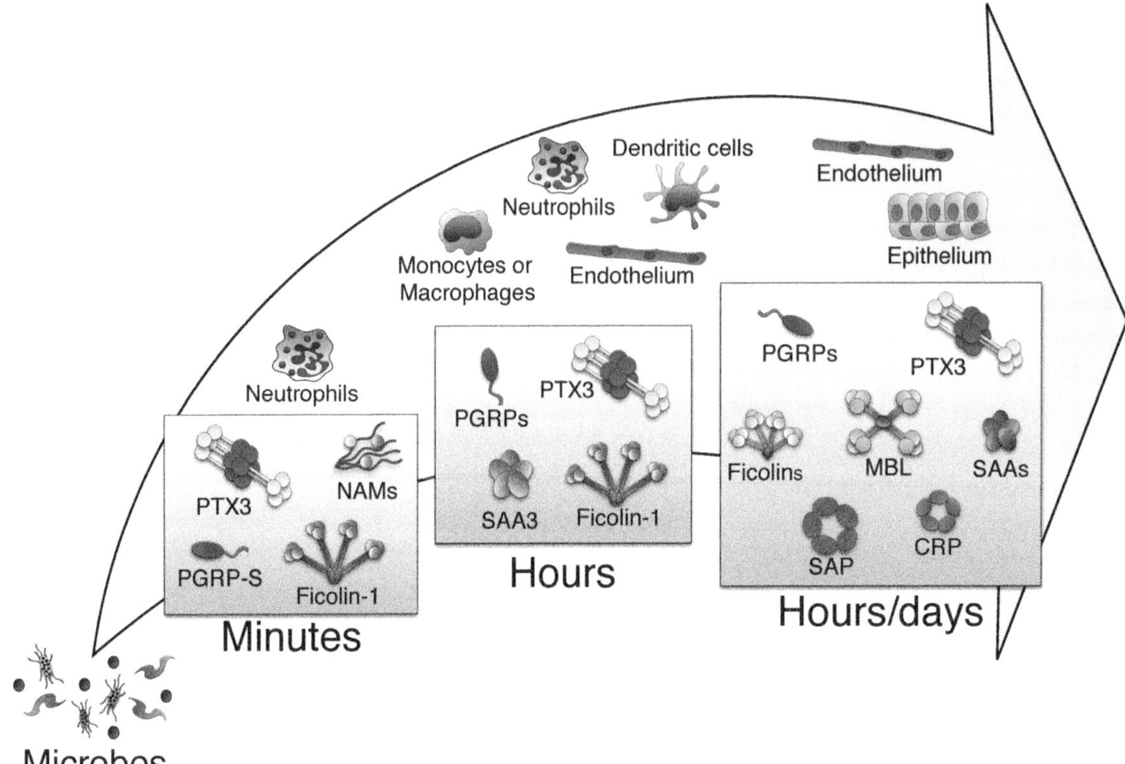

Figure 1 Humoral innate immunity. Soluble PRMs are expressed and secreted by a variety of cells, including in particular myeloid cells, allowing the production of PRMs over time. Some PRMs (e.g., PTX3, PGRP-S, and ficolin-1) are stored in neutrophil granules for rapid release in minutes. PRMs such as PTX3 and PGRP-S are also found among NET-associated molecules (NAMs). Production of PRMs by mononuclear phagocytes, dendritic cells, and endothelium in a gene expression-dependent manner sustains the presence of these molecules over time. Finally, epithelial tissues (e.g., liver) sustain systemic mass production.

COMPONENTS OF HUMORAL INNATE IMMUNITY

Collectins

Collectins are a family of multimeric proteins belonging to the superfamily of Ca^{2+}-dependent lectins (C-type lectins) (3). Nine members have been identified: mannose-binding lectin (MBL), conglutinin (only in bovidae), surfactant protein (SP)-A, SP-D, collectin (CL)-43 (only in bovidae), CL-46 (only in bovidae), CL-P1 (in placenta), CL-L1 (in liver), and CL-K1 (in kidney) (4). In addition, heteromeric complexes formed by CL-L1 and CL-K1 have been detected in the circulation and called CL-LK (5). This heterocomplex binds mannan-binding lectin-associated serine proteases (MASPs) and can mediate complement activation (5). Whereas all collectins are soluble and secreted, CL-P1 is a type II transmembrane protein (6, 7). However, shedding of this transmembrane molecule generates a soluble form of CL-P1 with effector activities (i.e., recognition of pathogens and activation of the complement cascade) (8). In addition to these molecules, the complement component C1q is related to this family, based on structure and function similarities. For instance, the classical and lectin pathways of complement activation are induced by interaction between C1q and MBL, respectively, with their ligands. Phylogenetic analysis revealed that components of the classical and lectin pathways of complement can be traced back to lamprey and ascidians, respectively (9, 10). For instance, an MBL-like lectin was isolated from the plasma of a urochordate, and an ortholog of mammalian C1q was found in lamprey (10, 11). Therefore, before the appearance of adaptive immunity and immunoglobulins in jawed vertebrates, C1q may have functioned as a PRM in innate immunity. Subsequently, the emergence of adaptive immunity has directed the specificity of C1q for the Fc region of immunoglobulins.

Collectins are formed by subunits composed of three identical polypeptide chains, with the exceptions

Table 1 Expression sites, ligands, and activities of soluble PRMs[a]

PRMs	Expression sites	Ligands	Activities
Collectins (MBL, SP-A, SP-D)	Liver (hepatocytes) Lung (type II alveolar cells)	Microorganisms (bacteria, fungi, viruses) and microbial moieties (LPS, LTA, LOS, PDG) Carbohydrates and lipids exposed on pathogens (gp55, viral glycoprotein envelopes)	Activation and regulation of the complement system Opsonic activity
Ficolins	Liver (hepatocytes) Lung (type II alveolar cells) Myeloid cells (neutrophils, monocytes, macrophages)	Microorganisms (bacteria, fungi, viruses) and microbial moieties [LPS, LTA, β-(1, 3)-D-glucan] Carbohydrates	Activation and regulation of the complement system Opsonic activity Inhibition of viral infectivity
Short pentraxins (CRP, SAP)	Liver (hepatocytes)	Microorganisms (bacteria, fungi, viruses) and microbial moieties (phosphorylcholine [CRP], LPS [SAP]) Complement components Apoptotic cells Phosphorylcholine, carbohydrates Extracellular matrix protein (fibronectin, collagen IV, laminin, proteoglycan) Amyloid fibrils	Activation and regulation of the complement system Opsonic activity (controversial data) Elimination of apoptotic cells
Long pentraxin PTX3	Myeloid cells (neutrophils, monocytes, macrophages, dendritic cells) Epithelial cells Endothelial cells Fibroblasts Adipocytes	Microorganisms (bacteria, fungi, viruses) and microbial moieties (OmpA) Complement components Extracellular matrix protein (IαI, TSG-6, fibrin) Plasminogen	Activation and regulation of the complement system Opsonic activity Inhibition of viral infectivity Elimination of apoptotic cells Matrix remodeling Fibrinolysis Regulation of P-selectin-dependent leukocyte recruitment
SAA	Liver (hepatocytes) Myeloid cells (monocytes, macrophages) Synovial cells Adipocytes	Microorganisms (bacteria, viruses) and microbial moieties (Omp)	Opsonic activity Inhibition of viral infectivity
PGLYRPs	Epithelial cells Liver Neutrophils	Bacteria PDG	Bactericidal activity

[a]Abbreviations: LOS, lipooligosaccharide; PDG, peptidoglycan; IαI, inter-α-trypsin inhibitor; TSG-6, TNF-α-induced protein 6.

of human SP-A, formed by two polypeptide chains, SP-A1 and SP-A2; and CL-LK (see above). It is thought that two SP-A1 molecules combine with one SP-A2 molecule to form a heterotrimer (4, 12, 13). The number of trimeric units required per molecule differs among collectins, and the degree of multimerization of collectins can affect their functions. Each chain of collectins is formed by a globular C-terminal carbohydrate recognition domain (CRD), which mediates the lectin activity; a short α-helical hydrophobic neck region composed of 24 to 28 amino acids; a collagen-like region consisting of *n* repetitions of the triplet Gly-Xaa-Yaa, where Xaa and Yaa are usually proline or hydroxyproline; and an N-terminal region composed of 7 to 28 amino acids (14). The collagen-like region is involved in the oligomerization and stability of the molecule, and the multimeric organization is stabilized by hydrophobic interactions and interchain disulfide bonds. SP-A is an octadecamer formed by six trimeric subunits, giving rise to a bouquet-like structure, whereas SP-D and conglutinin are dodecamers formed by four trimeric subunits, giving rise to a cruciform-like structure (15–17). A dodecameric structure was also proposed for CL-46 (18). Different oligomers of the homotrimeric structural unit of MBL were found in serum, ranging from dimers to octamers (19–21). The two major forms, termed MBL-I and MBL-II, are trimers and tetramers of the structural unit, respectively. Among them, MBL-II is the most prominent oligomer (20–22). The tertiary structure of C1q resembles that of collectins, but three different chains compose the basic building block of C1q.

Collectins interact with bacterial PAMPs, such as lipopolysaccharide (LPS) from Gram-negative bacteria, lipoteichoic acid (LTA) from *Bacillus subtilis*, and peptidoglycan from *Staphylococcus aureus* (4, 23). The CRD of collectins recognizes neutral sugar on the surface of microorganisms. In particular, collectins interact with 3- and 4-hydroxyl groups. All collectins have a mannose-type CRD with the exception of CL-P1, which has a galactose-type CRD (6). The recognition of mannose- and galactose-type ligands depends on the orientation of hydroxyl groups on sugars presented. Mannose-type CRDs interact with hydroxyls in horizontal position, while galactose-type CRDs interact with hydroxyls in vertical orientation. The selectivity may be changed from horizontal 4-OH (i.e., mannose) to vertical 4-OH (i.e., galactose) by site-directed mutagenesis introducing the residues found in galactose-type CRD (24). Glycosylated proteins found on surface of fungi, such as gp55 and gp45 from *Aspergillus fumigatus* or mannan and β-glucan from *Saccharomyces cerevisiae*, are recognized by collectins (4, 25, 26). Collectins also bind to viruses, such as HIV, influenza A virus (IAV), severe acute respiratory syndrome-associated coronavirus, and herpes simplex virus, via recognition of viral envelope glycoproteins; and to parasites, including *Leishmania major*, *Leishmania mexicana*, *Plasmodium falciparum*, and *Trypanosoma cruzi* (4).

Collectins can also bind to and regulate the activity of immune cells. For instance, SP-D can interact with T lymphocytes via its CRD, inducing the expression of cytotoxic-T-lymphocyte-associated antigen 4 (CTLA-4) in T lymphocytes (27). Therefore, administration of a fragment of SP-D containing the C-terminal lectin domain had the capacity to decrease lymphoproliferation and production of interleukin-2 (IL-2) from activated splenocytes *in vitro* as well as allergen-induced inflammation *in vivo* (27). In addition, SP-A was involved in protection against the development of gastrointestinal graft-versus-host disease (28). Therefore, mice deficient in SP-A that had received allogeneic bone marrow transplantation showed increased incidence of gastrointestinal graft-versus-host disease associated with increased Th17 cells and decreased regulatory T cells (28).

The recognition of pathogens by collectins was shown to be protective for the host and to be associated with the induction of an appropriate immune response. For instance, collectins have opsonic activity, enhancing phagocytosis of pathogens, and the capacity to activate the lectin pathway of the complement system, leading to the formation of the membrane attack complex on microbial surfaces (3). However, coating of protozoa with MBL was shown to increase infectivity of cells, and MBL levels were associated with increased risk of developing visceral leishmaniasis, suggesting that MBL can act as a "double-edged sword" (4). Collectins were shown to have antiviral activity. For instance, SP-D had a protective role against IAV, reducing hemagglutination activity and causing aggregation of the virus (29). In addition, the preincubation of IAV with dodecameric SP-D had an opsonic effect, increasing the uptake of IAV by neutrophils and the respiratory burst (29, 30). However, the preincubation of neutrophils with SP-D increased neutrophil uptake of IAV but reduced the respiratory burst activity (30, 31). The neutrophil oxidative response was also reduced in the presence of other innate immune proteins, such as salivary scavenger and agglutinin (SALSA; also known as gp340) (30). These innate immune proteins may regulate the neutrophil response to IAV by reducing an excessive and potentially harmful burst response (30, 31).

In humans, three polymorphisms were reported in the *MBL2* gene and were associated with decreased systemic concentrations of MBL and increased susceptibility to infections (32). MBL is primarily expressed in the liver (32). Hormones (e.g., growth hormone and thyroid hormones) and IL-6 can induce the expression of MBL in hepatocytes in a gene expression-dependent fashion, and early reports have suggested that MBL can act as an acute-phase reactant in different scenarios (e.g., malaria and patients with major surgery) (33–36). However, other studies showed discordant results, and in patients with sepsis and septic shock, the levels of MBL were variable and not associated with the acute-phase response (37, 38). In addition to the activity related to pathogen recognition, MBL can recognize endogenous oligosaccharides (e.g., fucosylated type 1 Lewis glycans) expressed by human colorectal carcinoma cells but not present in nonmalignant tissues (39). The expression of MBL ligands was detected more frequently in transverse, descending, and sigmoid colon and was associated with an infiltration of HLA-DR^+ cells and a favorable prognosis in patients with colorectal cancer (39).

Ficolins

Ficolins are lectin proteins with a general structure resembling that of collectins. Ficolins constitute a family of several members identified in vertebrates (40). Three members have been identified in humans: M-ficolin (also called ficolin-1), L-ficolin (also called ficolin-2), and H-ficolin (also called Hakata antigen or ficolin-3)

(40). L-ficolin and H-ficolin are serum proteins expressed mainly in the liver (L-ficolin) and in the liver and lung (H-ficolin). M-ficolin was initially identified on the cell surface of circulating monocytes and granulocytes (41). Subsequently, M-ficolin was identified in serum and characterized as a secreted protein expressed in neutrophils, monocytes, macrophages, and type II alveolar epithelial cells (42). Two ficolins have been identified in mouse (ficolin-A and ficolin-B) and pig (ficolin-A and ficolin-B), and one ficolin has been identified in the horseshoe crab *Tachypleus tridentatus* (42).

Ficolins are oligomeric proteins assembled from an N-terminal region with two functionally important cysteine residues, a collagen-like domain, and a C-terminal fibrinogen-like domain organized in a globular structure and involved in the recognition of pathogens (43). The proposed structure for L-ficolin is a dodecamer supported by the cross-linking of subunits via disulfide bonds (44, 45). H-ficolin and M-ficolin are oligomers consisting of octadecamers with similar structures (44, 45).

The fibrinogen-like domain of ficolins is functionally similar to the CRD of collectins and is responsible for the recognition of carbohydrate by ficolins. L-ficolin, H-ficolin, and M-ficolin can interact with GlcNAc, and M-ficolin also recognizes sialic acid. In addition, L-ficolin and H-ficolin were shown to interact with microbial moieties [e.g., LTA and β-(1,3)-D-glucan for L-ficolin and LPS for H-ficolin] (43, 46, 47).

Ficolins are involved in defense against invading pathogens via a mechanism of opsonophagocytosis and the initiation of the lectin pathway of the complement system (42, 43). For instance, L-ficolin has the capacity to interact with Gram-negative bacteria, such as the rough type of *Salmonella enterica* serovar Typhimurium TV119 and *Pseudomonas aeruginosa*, and Gram-positive bacteria, such as *S. aureus*, and binding of L-ficolin on pathogens enhanced their clearance by phagocytes (42). M-ficolin and H-ficolin can also interact with Gram-negative bacteria (e.g., *Salmonella* Minnesota and *Escherichia* spp.) and Gram-positive bacteria (e.g., *S. aureus* for M-ficolin and *Aerococcus viridans* for H-ficolin) (42). L-ficolin also has the capacity to bind the virulent strain of *Mycobacterium tuberculosis* H37Rv, reducing the infection of human lung epithelial cells. Importantly, serum levels of L-ficolin decreased in patients infected with pulmonary tuberculosis, and administration of L-ficolin had therapeutic activity in H37Rv-infected mice (48).

In addition to bacteria, ficolins interact with viruses and can have antiviral effects. For instance, L-ficolin interacts with hepatitis C virus, leading to inhibition of cell infection and activation of the complement lectin pathway, and binding of H-ficolin to IAV inhibits viral infectivity and hemagglutination (42, 49, 50).

Pentraxins

Pentraxins constitute a superfamily of multifunctional proteins that are evolutionarily conserved and are characterized by a multimeric structure (51). All pentraxins contain a "pentraxin domain" in their carboxy terminus, which is a conserved 8-amino-acid-long sequence (HxCxS/TWxS, where x is any amino acid). Based on the primary structure of the protomer, pentraxins are divided into the short pentraxins C-reactive protein (CRP) and serum amyloid P (SAP) and the long pentraxins (51).

Short pentraxins

During the 1930s, the short pentraxin CRP was the first PRM identified and purified from the serum of infected patients. Its name derives from its capacity to recognize the C-polysaccharide of *Streptococcus pneumoniae* (51). Subsequently, human SAP was identified as closely related to CRP. Indeed, the degree of amino acid sequence identity between human SAP and human CRP is 51% (51).

CRP and SAP are ~25-kDa proteins organized in five identical subunits arranged in a pentameric radial symmetry (2, 51). CRP and SAP are produced by hepatocytes and constitute the main acute-phase proteins in human and mouse, respectively (2). The plasma level of CRP is low in healthy adults (≤3 mg/liter) and increases as much as 1,000-fold during inflammation. In human serum, SAP is constitutively present at 30 to 50 mg/liter (2, 51).

Besides their structural similarity, CRP and SAP also share functional properties, such as activation of the complement system and recognition of pathogens. CRP and SAP have the capacity to interact with Fcγ receptors (FcγRs), resulting in phagocytosis of microorganisms and cytokine secretion (52). Pentraxins share the binding site on FcγRs with IgG and have the capacity to inhibit immune complex-mediated phagocytosis, suggesting antibody-like functions for pentraxins (52).

As mentioned above, C-polysaccharide of *S. pneumoniae* was the first reported ligand for CRP. Subsequently, CRP was shown to bind to numerous pathogens including fungi, yeasts, and bacteria, leading to pathogen phagocytosis and resistance to infection. For instance, CRP binds to the cell wall carbohydrate of *S. pneumoniae*, and transgenic mice overexpressing CRP are resistant to *S. pneumoniae* infection (53).

Similarly to CRP, SAP binds various bacteria, such as *Streptococcus pyogenes* and *Neisseria meningitidis* (2). SAP interacts with LPS, leading to inhibition of LPS-mediated complement activation and LPS toxicity (54). The relationship between recognition of pathogens and the functions of CRP and SAP is still a matter of debate (2). Indeed, the host defense function of CRP and SAP may also occur against pathogens that they do not recognize (2). In addition, and in contrast to its prophagocytic activity, SAP has been shown to enhance virulence of *S.* Typhimurium and rough strains of *Escherichia coli* by inhibiting the phagocytosis of bacteria (54).

The short pentraxins have the capacity to regulate the activation of the three complement system pathways, by interacting with C1q, ficolins, C4b-binding protein (C4BP), and factor H (55). Complement activation by short pentraxins has been suggested to favor removal of apoptotic cells and cell debris, preventing the onset of autoimmune diseases (56). For instance, the binding of CRP to apoptotic cells promotes the activation of the classical pathway of the complement system without assembly of the terminal complement components (57). Therefore, CRP amplifies the phagocytosis of apoptotic cells by phagocytes associated with the expression of transforming growth factor β (57). In addition, CRP can also recruit factor H, which is known to inhibit the alternative pathway of complement, and C4BP, a regulator of the classical and lectin pathways of complement (55).

The long pentraxin PTX3

During the early 1990s, PTX3, a new secreted protein containing a C-terminal pentraxin domain, was identified as the first long pentraxin (58, 59). Subsequently to PTX3, other molecules, including guinea pig apexin, neuronal pentraxin (NP) 1, NP2, and neuronal pentraxin receptor (NPR), a transmembrane molecule, have been identified. More recently, the long pentraxin PTX4 was identified, and transcript expression analysis showed that PTX4 has a unique pattern of mRNA expression, distinct from that of other members of the family (60).

The human and murine genes coding for PTX3 are localized on chromosome 3 and organized in three exons (Fig. 2). The first two exons code for the leader signal peptide and the N-terminal domain, and the third exon codes for the C-terminal pentraxin domain (51). Many transcription factors have potential binding sites in the promoter of both the human and murine *PTX3* gene, including Pu1, AP-1, NF-κB, Sp-1, and NF-IL-6 (61). NF-κB is involved in the response to proinflammatory cytokines, and AP-1 is involved in the control of basal transcription. The production of PTX3 has been associated with different pathways, including the phosphatidylinositol 3-kinase/Akt axis and Jun N-terminal protein kinase (55). Moreover, it has been shown that the IL-1 receptor type I pathway was responsible for the production of PTX3 in the heart during acute myocardial infarction in mice and in both local and systemic sites in the model of 3-methylcholanthrene-induced carcinogenesis (62, 63). In skin wound healing, induction of PTX3 was almost completely abolished in *MyD88*$^{-/-}$ mice and partially reduced in *Il1r*$^{-/-}$, *Tlr3*$^{-/-}$, *Irf3*$^{-/-}$, and *Ticam1*$^{-/-}$ mice, suggesting that expression of PTX3 is downstream of TLR sensing and IL-1 amplification (64). In addition, the TLR4/MyD88 pathway controlled the production of PTX3 in uroepithelial cells during urinary tract infections mediated by uropathogenic *E. coli* (UPEC) (61).

PTX3 has a protomer of 381 amino acids composed of a signal peptide (17 amino acids), an N-terminal domain unrelated to any known protein, and a C-terminal pentraxin domain homologous to the short pentraxins CRP and SAP. The primary structure of PTX3 is highly conserved among species, with 82% identical amino acids in human and mouse, suggesting that the functional role played by PTX3 was maintained by an evolutionary pressure (51).

PTX3 is a multimeric glycoprotein with a complex quaternary structure characterized by two tetramers linked together by interchain bridges to form an octamer of 340 kDa (65). Electron microscopy and small-angle X-ray scattering (SAXS) showed that PTX3 folds into an elongated structure with a large and a small domain interconnected by a stalk region. This quaternary structure was involved in the biological function of PTX3 (Fig. 2). For instance, the N-terminal region of PTX3 forms a tetramer, which supports the recognition of several PTX3 ligands (2). In addition to the quaternary structure, the N-linked glycosidic moiety localized at Asn220 was shown to play an essential role in the interaction between PTX3 and P-selectin (66).

Inflammatory cytokines (e.g., tumor necrosis factor α [TNF-α] and IL-1β), TLR agonists (e.g., LPS), microbial moieties (e.g., outer membrane protein A of *Klebsiella pneumoniae* [KpOmpA]), or pathogens (e.g., UPEC and *A. fumigatus*) induce the expression of PTX3 in various cell types, including dendritic cells, monocytes, macrophages, epithelial cells, endothelial cells, fibroblasts, and adipocytes (55). In addition, PTX3 is also stored in neutrophil-specific granules in a ready-to-use form and rapidly released in response to microorganisms or TLR agonists (67). Upon release, a part of the molecule was found in neutrophil extracellular traps

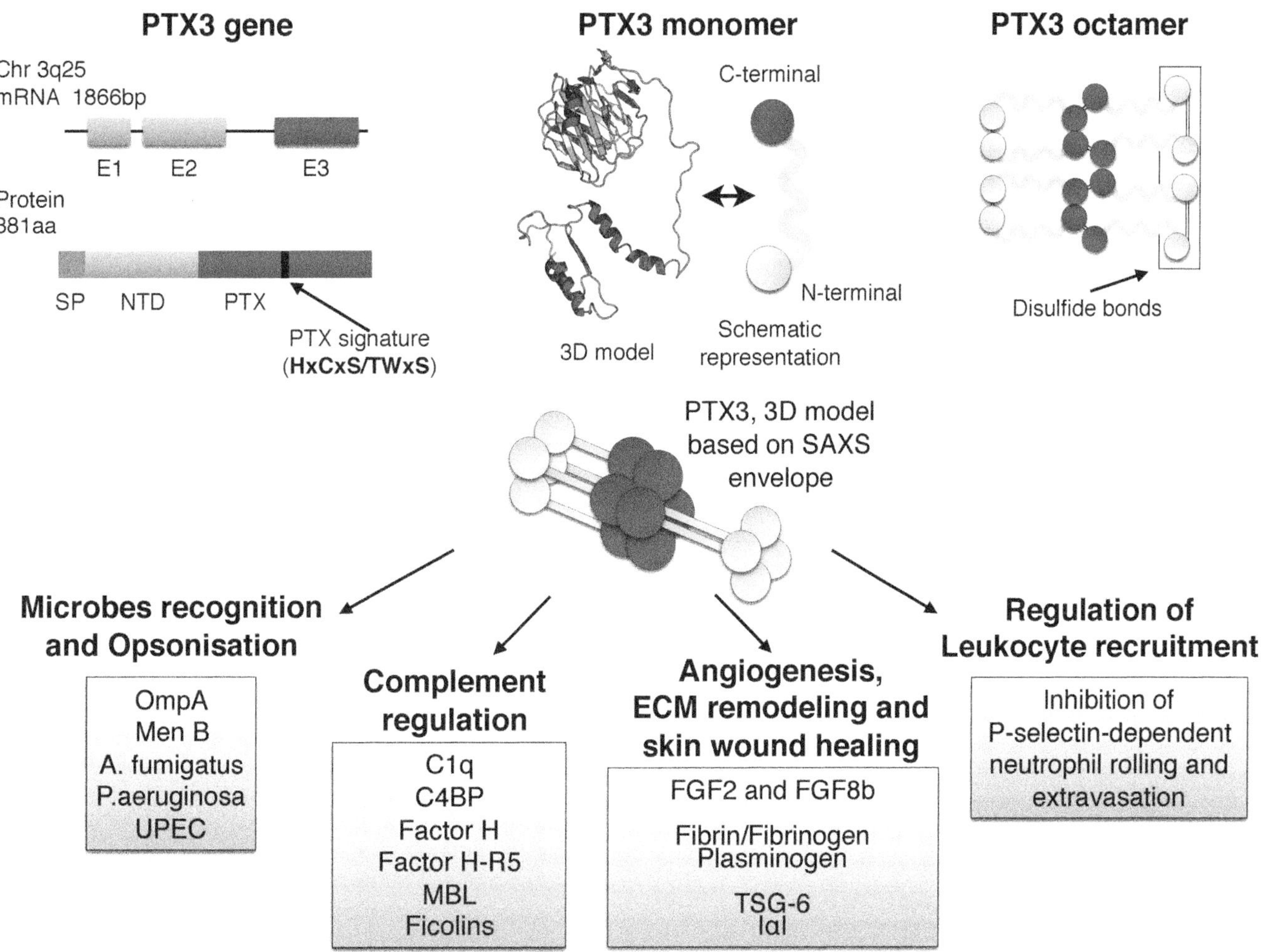

Figure 2 Gene organization, protein structures, and roles of PTX3. The PTX3 gene is organized in three exons. The first two exons code for the signal peptide (SP) and the N-terminal domain of the protein (NTD), respectively, and the third exon codes for the pentraxin domain (PTX). A three-dimensional model of the pentraxin domain has been generated based on the crystallographic structures of CRP and SAP, showing that the pentraxin domain of PTX3 adopts a β-jelly roll topology. PTX3 has a unique quaternary structure with eight subunits, associated together to form an octamer by disulfide bonds between cysteine residues present on both the N-terminal and C-terminal domains. Once released, PTX3 plays a role in pathogen opsonization and agglutination, complement activation, regulation of inflammation and leukocyte recruitment, angiogenesis, extracellular matrix (ECM) remodeling, and wound healing. Men B, meningococcus type B; FGF, fibroblast growth factor 2; TSG-6, tumor necrosis factor-inducible gene 6 protein; IαI, inter-alpha-trypsin inhibitor.

(NETs), which are known for their antimicrobial activity (67).

PTX3 binds to a wide range of microorganisms, including fungi (e.g., *A. fumigatus* conidia and *Paracoccidioides brasiliensis*), bacteria (e.g., *P. aeruginosa*, *K. pneumoniae*, UPEC, and *N. meningitidis*), and viruses (e.g., human and murine cytomegalovirus [CMV] and selected strains of influenza virus) (55). The role played by PTX3 in host defense and inflammation was revealed by the generation of PTX3-deficient mice. For instance, *Ptx3*$^{-/-}$ mice showed increased susceptibility to invasive pulmonary aspergillosis associated with defective phagocytosis and clearance of *A. fumigatus* conidia by PTX3-deficient phagocytes and a low protective Th1 response, which were restored by treatment with recombinant PTX3 (67–69). Neutrophil-associated PTX3 was essential for resistance against this pathogen (67–69). Similarly, PTX3 was involved in defense against *P. aeruginosa*, some viruses, and UPEC (55). These protective activities and the molecular mechanisms involved will be discussed below. More recently, it has been shown that PTX3 binds to fibrin/fibrinogen and plasminogen, increasing plasmin-mediated fibrinolysis (64). Therefore, PTX3 deficiency was associated with defective wound healing of skin (64).

PTX3 participates in the fine-tuning of the inflammatory response. For instance, PTX3 binds selectively to P-selectin and competes with the interaction of P-selectin with P-selectin glycoprotein ligand-1 (PSGL-1) (66). Therefore, PTX3 can prevent exacerbated P-selectin-dependent recruitment of neutrophils in mouse models of inflammation (66, 70, 71). In contrast, PTX3 can positively regulate the inflammatory response, as demonstrated upon the recognition of the KpOmpA by PTX3 (72). Despite the fact that PTX3 does not affect the recognition of KpOmpA by cellular receptors, PTX3 amplifies the inflammation induced by KpOmpA *in vivo*, through a complement-dependent mechanism (72, 73). Accordingly, PTX3-overexpressing mice infected by *K. pneumoniae* present an increased production of proinflammatory mediators, including nitric oxide and TNF-α (74).

As reported for CRP and SAP, PTX3 has the capacity to regulate the activation of the complement system. The complement component C1q, which is the main activator of the classical pathway of the complement system, was the first ligand identified for PTX3. This interaction occurs between the globular head of C1q and PTX3 and can activate or inhibit the complement cascade, depending on the way in which the binding occurs. Plastic-immobilized PTX3 recruits C1q and induces the activation of the classical cascade of complement, whereas the interaction between PTX3 and C1q in the fluid phase inhibits complement activation by blocking the interaction of C1q with immunoglobulins (55). In addition, PTX3 has the capacity to regulate the lectin pathway of the complement system via direct interaction with ficolin-1, ficolin-2, and MBL. For instance, the formation of heterocomplexes PTX3/ficolin-2 and PTX3/MBL can promote the deposition of complement, as observed on the surface of *A. fumigatus* and *Candida albicans*, respectively (75, 76). In addition, the deposition of ficolin-1 on late apoptotic cells is amplified in the presence of PTX3, leading to increased phagocytosis of apoptotic cells by macrophages (77).

PTX3 also interacts with factor H-related protein 5 and negative complement regulators, such as factor H and C4BP, favoring their deposition on PTX3-coated surfaces (78–80). For instance, PTX3 binds to apoptotic cells and induces the recruitment of C4BP, limiting complement activation and an exacerbated inflammatory response (78). In a murine model of myocardial infarction, PTX3-deficient mice showed higher myocardial damage associated with increased neutrophil infiltration and C3 deposition, likely due to the missing deposition of factor H (62, 79).

Recently, PTX3 deficiency was found to be associated with increased susceptibility to mesenchymal and epithelial carcinogenesis and increased cancer-related inflammation (63). Tumors developed in a PTX3-deficient context were characterized by exacerbated inflammation, amplification of complement activation, recruitment of tumor-promoting macrophages, higher frequency of *Trp53* mutations, increased DNA oxidative damage, and higher expression of DNA damage response markers (63). It has been shown that PTX3 acts as an oncosuppressor through recruitment of the negative regulator factor H and that exacerbated inflammation observed in PTX3-deficient mice contributes to cancer genetic instability (63).

Other Soluble PRMs

Peptidoglycan recognition proteins

Peptidoglycan recognition proteins (PGRPs) are innate immune molecules highly conserved from insects to humans and playing a key part in defense against bacteria (81, 82). PGRPs constitute a family of four proteins named PGRP-S, PGRP-L, PGRP-Iα, and PGRP-Iβ. The Human Genome Organization Gene Nomenclature Committee modified their names to PGLYRP (peptidoglycan recognition protein)-1, -2, -3, and -4, respectively (81).

The short peptidoglycan recognition protein, PGLYRP-1, is 196 and 182 amino acids long in humans and mice, respectively, and composed of a signal peptide and a type 2 amidase domain. This domain is called the PGRP domain and is homologous to bacteriophage and bacterial type 2 amidases (81). Long (i.e., PGLYRP-3 and PGLYRP-4) and intermediate-sized (i.e., PGLYRP-2) molecules present a C-terminal PGRP domain and a unique amino-terminal sequence not conserved and with variable length among PGRPs (81).

Mammalian PGRPs are expressed in various cell types and involved in amidase activity and antibacterial activity. PGLYRP-1 is expressed in bone marrow and stored in neutrophil granules, and PGLYRP-3 and PGLYRP-4 are expressed by epithelial cells from tissues that come in contact with the external environment (e.g., tongue, eyes, skin, salivary gland, throat, and esophagus) (83). PGLYRP-2 is an *N*-acetylmuramoyl-L-alanine amidase produced by the liver and secreted into the circulation. PGLYRP-2 hydrolyzes the lactyl bond between the MurNAc and the L-Ala in peptidoglycan, reducing the proinflammatory activity of polymeric peptidoglycan (83).

PGLYRP-1 (PGRP-S), PGLYRP-3 (PGRP-Iα), and PGLYRP-4 (PGRP-Iβ) bind bacterial peptidoglycan and have bactericidal activities against nonpathogenic and pathogenic bacteria (83). PGRPs kill bacteria by mechanisms similar to those employed by antibiotics. Briefly, PGRPs bind to the cell wall of Gram-positive bacteria and to the outer cell membrane of Gram-negative bacteria, inducing a bacterial stress response that leads to membrane depolarization, arrest of macromolecule synthesis, accumulation of toxic hydroxyl radicals, and bacterial death (82, 84).

Serum amyloid A

Serum amyloid A (SAA) is a family of α-helical proteins divided into acute-phase SAAs (A-SAAs) and constitutive SAAs (C-SAAs). Four mouse SAAs (SAA1, SAA2, SAA3, and SAA4) and three human SAAs (SAA1, SAA2, and SAA4) were identified. SAA1 and SAA2 are acute-phase proteins and SAA4 is expressed constitutively (85). SAA3 is a truncated protein expressed in mice by extrahepatic cells, including adipocytes and macrophages, but is a pseudogene in humans (85).

A-SAAs are produced mainly by the hepatocytes upon proinflammatory stimuli, including proinflammatory cytokines and bacterial moieties (2). In addition to the liver, atherosclerotic plaque cells (especially macrophages and smooth muscle cells), synovial cells, chondrocytes, and epithelial cell lines can produce SAAs, and adipocytes represent the major site of SAA expression in obese individuals (85).

SAA interacts with a large variety of Gram-negative bacteria (e.g., *E. coli*, *K. pneumoniae*, *Shigella flexneri*, and *P. aeruginosa*) through binding to OmpA family members (2). SAA has opsonic activity toward bacteria, increasing their phagocytosis and the production of TNF-α and IL-10 by phagocytes (86). In addition to bacteria, SAA also recognizes hepatitis C virus and blocks viral entry into cells (87).

SAA can also interfere with the inflammatory response. For instance, SAA3 has been shown to interact with myeloid differentiation protein 2 (MD-2) and activate the MyD88-dependent TLR4/MD-2 pathway (88). Accordingly, injection of SAA3 peptide in mice increased the recruitment of wild-type but not *MD-2*$^{-/-}$ CD11b^{+}Gr-1^{+} cells in the lungs, suggesting that SAA3 serves as an endogenous signal to induce myeloid cell mobilization from the bone marrow in an MD-2-dependent manner (88). In addition, recent investigations showed that SAA3 activates the NOD-like receptor pyrin domain containing 3 (NLRP3) inflammasome, leading to the promotion of Th17 allergic asthma (89). Most recently, SAA was involved in the expression of IL-33 by monocytes and macrophages (90). TLR2, known to be a receptor for SAA, is responsible for the induction of IL-33, and interferon (IFN) regulatory factor 7 (IRF7) is a critical transcription factor for SAA-induced IL-33 expression (90).

MACROPHAGES AS A SOURCE OF PRMs

Macrophages are major components of innate immune defense, expressing a large variety of PRMs (91). Macrophages express cell surface receptors, including TLRs, scavenger receptors (e.g., SR-A, MARCO [macrophage receptor with collagenous structure], SRCL-1 [scavenger receptor C-type lectin 1], and Lox-1), C-type lectin receptors (e.g., dendritic cell (DC)-specific intercellular adhesion molecule (ICAM)-3-grabbing non-integrin (DC-SIGN), mannose receptor, Dectin-1 [dendritic cell-associated C-type lectin 1], Dectin-2, Mincle, and CLEC5A), and intracellular receptors (e.g. NOD-1, NOD-2, RIG-I, and MDA5) (91, 92). These PRMs are involved in the recognition of self, nonself, and modified self and in the subsequent responses. For instance, scavenger receptors were shown to contribute to the recognition and phagocytosis of pathogens and apoptotic cells by macrophages and to regulate the inflammatory response (91, 93). Also, the receptors of the C-type lectin family were involved in processes of pathogen recognition, cellular interactions, and regulation of the inflammatory response. For instance, Dectin-1 was involved in the recognition and uptake of β-glucans and yeast, and the immunoreceptor tyrosine-based activation motif in its intracellular tail was involved in cell activation (91).

Macrophages are also a source of fluid-phase PRMs, including PTX3, ficolins (human M-ficolin and mouse ficolin-B), and SAA3 (2). Expression of PTX3 in macrophages was induced in response to microbial components (e.g., zymosan and LPS) and proinflammatory cytokines (e.g., TNF-α and IL-1β) (55). In turn, PTX3 has the capacity to interact with zymosan, inducing aggregation and phagocytosis of a high number of particles by macrophages through a Dectin-1-dependent mechanism (94). Accordingly, macrophages from PTX3 transgenic mice showed increased phagocytosis of zymosan particles as well as the yeast form of *P. brasiliensis* (54). Similarly, the production of PTX3 was enhanced upon stimulation by KpOmpA, and PTX3 in turn recognized this microbial moiety, enhancing the binding of this microbial moiety to the scavenger receptors SREC-1 and Lox-1 and inducing an amplification loop of the inflammatory response (72).

Macrophages also express M-ficolin and ficolin-B in humans and mice, respectively. Ficolin-B localizes in lysozymes of activated macrophages and has the capacity to interact with GalNAc and sialic acid. Mouse ficolin-B can recognize and aggregate bacteria, such as *S. aureus*, leading to increased phagocytosis of the pathogen by macrophages (43). In humans, macrophages express and secrete M-ficolin, and this expression was increased upon exposure to TLR2 or TLR4 ligands (95). Recent studies showed the formation of a ficolin-1 (M-ficolin)/PTX3 complex on the surface of apoptotic cells, facilitating the clearance of apoptotic cells and downregulating the release of IL-8 by macrophages (77).

NEUTROPHILS AS A SOURCE OF PRMs

Neutrophils express a vast repertoire of cellular-associated PRMs, including all TLRs with the exception of TLR3, C-type lectin receptors (e.g., Dectin-1, CLEC2, CLEC4E [also called Mincle], and CLEC4D [also called CLECSF8]), and cytoplasmic receptors (such as NOD-1, RIG-I, MDA5, and the DNA sensor IFN-inducible protein 16 [IFI16]) (96). These receptors are involved in the activation and regulation of neutrophil effector functions, such as phagocytosis; production of cytokines, antimicrobial peptides, and reactive oxygen species; and formation of NETs (96). For instance, phagocytosis and elimination of *C. albicans* and *A. fumigatus* by neutrophils was shown to occur through a Dectin-1-dependent mechanism (97). Interestingly, the blockage of Dectin-1-mediated phagocytosis of *C. albicans* in neutrophils can be compensated for by exogenous MBL, suggesting that a cooperation exists between cell-associated and fluid-phase PRMs in neutrophils for the development of an optimal antifungal response (98). Neutrophils also express the formyl peptide receptors (FPRs) FPR1 and FPR2, two seven-transmembrane G protein-coupled receptors with a high and low affinity, respectively, for the bacterial-derived peptide *N*-formyl-methionyl-leucyl-phenylalanine (99, 100). Activation of FPR1 or FPR2 leads to activation or inhibition, respectively, of neutrophil chemotaxis (99).

Neutrophils express a set of cytosolic DNA sensors, such as IFI16, MDA5, RIG-I, leucine-rich repeat flightless-interacting protein 1 (LRRFIP1), Asp-Glu-Ala-Asp (DEAD) box protein 41 (DDX41), and STING (stimulator of IFN genes), promoting the expression of IFN-β and CXCL10 in neutrophils upon stimulation by intracellular pathogens or transfection with plasmid DNA (101).

Neutrophils have emerged as a source of fluid-phase receptors, including PTX3, M-ficolin, and PGRP-S. These molecules are stored in neutrophil granules in a ready-to-use form. Therefore, neutrophils serve as a reservoir, ready for rapid release and covering a temporal window preceding gene expression-dependent production.

NETs are an extracellular fibrillary network formed by activated neutrophils during the process of "NETosis" (102). NETs are decorated by nuclear components and a set of proteins from primary, secondary, and tertiary granules, comprising the PRMs PTX3 and PGRP-S (96). NETs trap bacteria (e.g., *E. coli*, *S. flexneri*, and *S. aureus*) and fungi (e.g., *A. fumigatus* and *C. albicans*), promoting their interaction with effector molecules and their elimination (102). More recently, NETs were involved in the host defense against viruses, such as HIV-1 and myxoma virus (103, 104). In addition to neutrophil-associated molecules, SP-D can simultaneously recognize NETs and carbohydrate ligands *in vivo*. Therefore, agglutination of bacteria by SP-D allows an efficient bacterial trapping in NETs (105).

The expression of PTX3 mRNA is restricted to immature cells, and mature neutrophils stored the protein into the secondary granules (also known as specific granules) (67). Released PTX3 has opsonic activity against pathogens (e.g., *A. fumigatus*, *P. aeruginosa*, and UPEC), and part of the molecule is localized into NETs. Molecular mechanisms involved in the protective effects of PTX3 during bacterial, fungal, and viral infections are discussed below. PTX3 accumulates in blebs at the surface of late apoptotic neutrophils, resulting from its active translocation from granules to the membrane, and acts as a late "eat me" molecule involved in the recognition and elimination of apoptotic neutrophils by macrophages (106).

PGRP-S and M-ficolin are stored in secondary and tertiary granules (107, 108). PGRP-S was identified in bovine, murine, and human neutrophils, and similarly to PTX3, PGRP-S was found in NETs, where it can exert bacteriostatic and bactericidal activities against selected microorganisms (e.g., *Micrococcus luteus*, *S. aureus*, and *B. subtilis*) (84, 107). PGRP-S deficiency in mice was associated with increased susceptibility to intraperitoneal infection with *M. luteus* and *B. subtilis* (109). Neutrophils isolated from PGRP-S-deficient mice were able to recognize and phagocytose bacteria but failed to generate an oxidative burst (109). In addition, both PGRP-S and lysozyme recognize peptidoglycan and showed a synergistic antimicrobial effect against *E. coli* (containing mesodiaminopimelic acid-type peptidoglycan) (107).

M-ficolin released from neutrophil granules can bind to the neutrophil surface through a direct interaction with CD43, leading to cell adhesion and aggregation and activation of the lectin complement pathway on neutrophils (108, 110).

REGULATION OF PHAGOCYTE FUNCTION BY FLUID-PHASE PRMs: PTX3 AS A PARADIGM

PTX3 in Defense Against Viruses

PTX3 recognizes both human and murine CMV, coronavirus murine hepatitis virus strain 1, and specific strains of influenza virus (H3N2). PTX3-deficient mice showed increased susceptibility to infections by these viruses, and PTX3 had therapeutic activity (111–113).

PTX3 interacts with H3N2 via the recognition of the hemagglutinin glycoprotein found on the surface of viruses and the glycosidic moiety of PTX3 (113). This interaction was shown to have antiviral activities, inhibiting viral hemagglutination and neuraminidase activity and neutralizing the virus infectivity (113). In contrast, both seasonal and pandemic H1N1 influenza virus and other subtypes of H3N2 viruses were not recognized by PTX3 and were resistant to the antiviral activity of PTX3 (114, 115).

PTX3 also has the capacity to bind to human and murine CMV, inhibiting the entry of virus into dendritic cells (112). In addition, it has been proposed that PTX3 can protect mice from murine CMV primary infection and reactivation *in vivo*, as well as *Aspergillus* superinfection, through the activation of IRF3 in dendritic cells and the promotion of the IL-12/IFN-γ-dependent effector pathway (112).

PTX3 in Defense Against Fungi

Zymosan was shown to induce the expression of PTX3 in macrophages, and, in turn, PTX3 can bind to zymosan and the yeast form of *P. brasiliensis* (94). As previously mentioned, PTX3 has opsonic activity, and phagocytosis of *P. brasiliensis* was increased in PTX3-overexpressing macrophages. Mechanistically, opsonization of zymosan by PTX3 induced the aggregation of zymosan and the phagocytosis of a high number of particles by macrophages via a mechanism dependent on Dectin-1 (94).

As previously mentioned, PTX3 has the capacity to bind to *A. fumigatus* conidia, and PTX3$^{-/-}$ mice are highly susceptible to invasive pulmonary aspergillosis (69). Interestingly, PTX3-deficient neutrophils and macrophages have defective phagocytosis of conidia, and opsonization of spores by recombinant PTX3 or neutrophil-associated PTX3 reverses this phenotype (67, 69). Accordingly, PTX3 expressed by neutrophils is essential to control fungal growth *in vitro* and *in vivo* (67).

Investigations showed that the opsonic activity of PTX3 against fungi occurred through an FcγRII- and complement-dependent mechanism (68). Indeed, PTX3 can increase the phagocytosis of *A. fumigatus* conidia by neutrophils by interacting with FcγRII, which has been proposed as pentraxin receptor (52). Briefly, the binding of PTX3-opsonized conidia to FcγRII was shown to induce activation of the complement receptor 3 (CD11b/CD18), with subsequent increased phagocytosis of C3b-opsonized conidia by neutrophils (Fig. 3) (68). Recently, it has been proposed that PTX3 binds to MD-2, inducing protective antifungal activity through TLR4/MD-2-mediated signaling (116).

PTX3 can also interact with ficolin-2, and the binding of PTX3 on conidia was amplified by ficolin-2 and vice versa (75). Subsequently, ficolin-2-dependent complement deposition on the surface of *A. fumigatus* was enhanced by PTX3, suggesting that PTX3 and ficolin-2 can cooperate to amplify microbial recognition and effector functions (75). Similarly, PTX3 can interact with MBL on the surface of *C. albicans*, and the formation of this heterocomplex triggered complement deposition and phagocytosis of *C. albicans* by neutrophils (117) (Fig. 3).

In humans, single-nucleotide polymorphisms within the *PTX3* gene were associated with reduction of the intracellular stock of PTX3 in neutrophils and PTX3 plasma levels and increased susceptibility to *A. fumigatus* infection in patients undergoing bone marrow transplantation (118, 119).

PTX3 in Defense Against Bacteria

PTX3 interacts with selected bacteria, including *P. aeruginosa*, *N. meningitidis*, *K. pneumoniae*, and UPEC (55, 120).

PTX3 displays opsonic activity against *P. aeruginosa*, facilitating the phagocytosis of this pathogen by neutrophils. As observed for *A. fumigatus*, PTX3 amplifies the phagocytosis of *P. aeruginosa* through an interplay between complement and FcγRs (121). Importantly, PTX3 showed therapeutic activity in a mouse model of chronic *P. aeruginosa* lung infection,

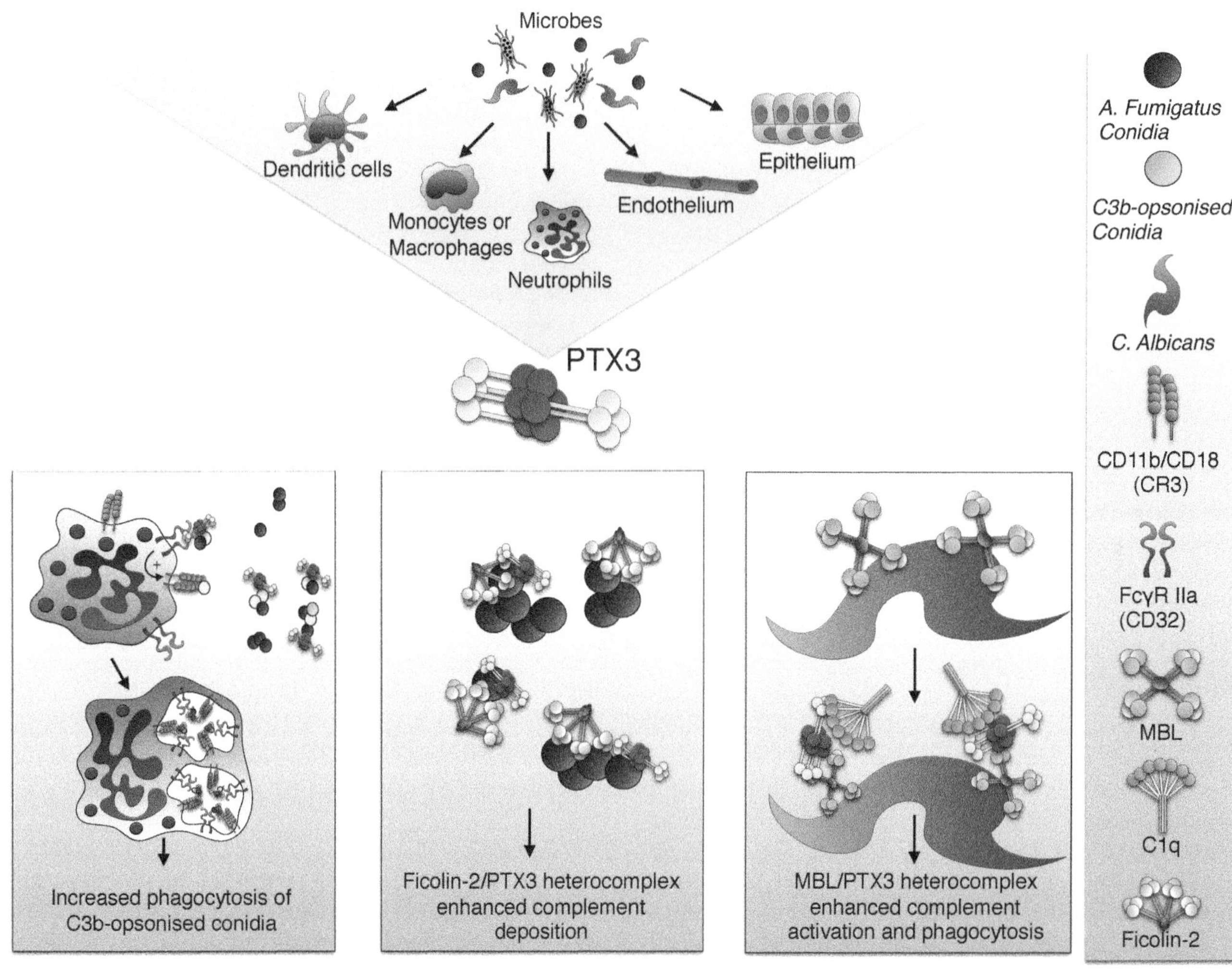

Figure 3 Role of PTX3 in defense against fungi. In the presence of PTX3-opsonized conidia, FcγRIIA induces inside-out CD11b/CD18 activation, recruitment to the phagocytic cup, and amplification of C3b-opsonized conidia phagocytosis (left panel). PTX3 interacts with ficolin-2 and MBL on the surface of conidia and *C. albicans*, respectively, triggering complement deposition and phagocytosis of pathogens.

reducing bacterial load and preventing excessive activation of the inflammatory response (121). Moreover, orally administered PTX3 in neonate mice can diffuse rapidly in tissues and provided protection against *P. aeruginosa* lung infection (122).

Recently, PTX3 was identified as the first soluble PRM essential in defense against UPEC-induced urinary tract infection (61). UPEC induced a rapid production and secretion of PTX3 by uroepithelial cells in a TLR4- and MyD88-dependent manner. In turn, PTX3 opsonized UPEC, enhancing phagocytosis and phagosome maturation in neutrophils (61). Therefore, PTX3 deficiency in mice was associated with exacerbated inflammation and tissue damage, demonstrating a fundamental role played by PTX3 in defense against urinary tract infections (61).

In humans, genetic studies showed that *PTX3* single-nucleotide polymorphisms were associated with increased susceptibility to pulmonary tuberculosis, acute pyelonephritis, cystitis, and *P. aeruginosa* infections (61, 123, 124).

PTX3 in Clearance of Apoptotic Cells

An efficient and rapid elimination of apoptotic cells is required to maintain tissue homeostasis (93). The short pentraxins CRP and SAP interact with apoptotic cells, promoting their elimination by phagocytes. In contrast, the presence of PTX3 in fluid phase was shown to inhibit the clearance of apoptotic cells by phagocytes, likely due to the sequestration of C1q by PTX3 (125). In contrast, when PTX3 was preincubated with apoptotic cells, the deposition of C1q and C3 on apoptotic cells was increased (126). Moreover, opsonization of apoptotic cells by PTX3 increased the deposition of factor H to their surface, suggesting that PTX3 may limit the complement-mediated lysis of apoptotic cells (79).

PTX3 was found accumulated in blebs at the surface of late apoptotic neutrophils, resulting from the translocation of PTX3 from granules to blebs mediated by activation of a caspase- and rho-associated protein kinase 1 (ROCK-1)-dependent mechanism (106). In addition to neutrophils, PTX3 was also observed at the surface of late apoptotic macrophages and membrane-associated PTX3 acted as a late "eat-me" molecule, increasing the phagocytosis of late apoptotic neutrophils and macrophages by phagocytes (106, 127). Therefore, the membrane-associated form of PTX3 may promote the elimination of apoptotic cells before loss of cell membrane integrity, whereas fluid-phase PTX3 released during inflammation can inhibit the capture of apoptotic cells.

PTX3 in Leukocyte Recruitment

The process of leukocyte extravasation is essential for the development of innate and adaptive immune responses. This process is divided in sequential cell migration events, including chemoattraction, tethering, rolling, adhesion, diapedesis, and transmigration, and is regulated by cellular adhesion molecules, including selectins (128). P-selectin is found on the surface of activated endothelial cells and interacts with PSGL-1 expressed on leukocytes, favoring tethering and rolling of leukocytes on activated endothelial cells (128).

Studies have reported that PTX3 was engaged in regulatory loops to finely tune the inflammatory response and leukocyte extravasation. Indeed, PTX3 interacted selectively with P-selectin, and not with E-selectin or L-selectin, via its N-linked glycosidic moiety and competed with the interaction of P-selectin with the leukocyte receptor PSGL-1 (66). Therefore, endogenous PTX3 from hematopoietic cells or administration of recombinant PTX3 acted as a negative feedback loop, preventing an excessive recruitment of neutrophils in inflamed tissues (66, 71).

CONCLUDING REMARKS

The humoral arm of innate immunity is composed of members of the complement cascade and soluble PRMs. These PRMs are diverse in term of structure but share fundamental mechanisms of the immune response, including activation and regulation of the classical, alternative, and lectin pathways of the complement system; opsonization of pathogens to facilitate their elimination by phagocytes; aggregation and neutralization of viral particles; and elimination of apoptotic cells.

Generation of genetically modified animals showed the importance of these PRMs in different pathological conditions, including infections, exacerbated inflammation, autoimmunity, and cancer. Importantly, genetic and epigenetic evidence supports these activities also in humans.

Therefore, PRMs are part of the immune response and participate in the fine-tuning of the inflammatory response. The expression and release of different PRMs by different cell types allow their presence at different tempos of the immune and inflammatory responses. Indeed, PRMs can be stored in neutrophil granules, ready for a rapid release, or *de novo* synthesized by mononuclear phagocytes and dendritic cells. Then epithelial tissues including the liver can sustain systemic production of soluble PRMs.

Citation. Jaillon S, Bonavita E, Garlanda C, Mantovani A. 2016. Interplay between myeloid cells and humoral innate immunity. Microbiol Spectrum 4(5):MCHD-0051-2016.

References

1. **Iwasaki A, Medzhitov R.** 2010. Regulation of adaptive immunity by the innate immune system. *Science* **327:** 291–295.
2. **Bottazzi B, Doni A, Garlanda C, Mantovani A.** 2010. An integrated view of humoral innate immunity: pentraxins as a paradigm. *Annu Rev Immunol* **28:**157–183.
3. **Holmskov U, Thiel S, Jensenius JC.** 2003. Collections and ficolins: humoral lectins of the innate immune defense. *Annu Rev Immunol* **21:**547–578.
4. **Gupta G, Surolia A.** 2007. Collectins: sentinels of innate immunity. *BioEssays* **29:**452–464.
5. **Henriksen ML, Brandt J, Andrieu JP, Nielsen C, Jensen PH, Holmskov U, Jorgensen TJ, Palarasah Y, Thielens NM, Hansen S.** 2013. Heteromeric complexes of native collectin kidney 1 and collectin liver 1 are found in the circulation with MASPs and activate the complement system. *J Immunol* **191:**6117–6127.
6. **Ohtani K, Suzuki Y, Eda S, Kawai T, Kase T, Keshi H, Sakai Y, Fukuoh A, Sakamoto T, Itabe H, Suzutani T, Ogasawara M, Yoshida I, Wakamiya N.** 2001. The membrane-type collectin CL-P1 is a scavenger receptor on vascular endothelial cells. *J Biol Chem* **276:**44222–44228.
7. **Nakamura K, Funakoshi H, Miyamoto K, Tokunaga F, Nakamura T.** 2001. Molecular cloning and functional characterization of a human scavenger receptor with

C-type lectin (SRCL), a novel member of a scavenger receptor family. *Biochem Biophys Res Commun* **280:** 1028–1035.

8. **Ma YJ, Hein E, Munthe-Fog L, Skjoedt MO, Bayarri-Olmos R, Romani L, Garred P.** 2015. Soluble collectin-12 (CL-12) is a pattern recognition molecule initiating complement activation via the alternative pathway. *J Immunol* **195:**3365–3373.
9. **Fujita T.** 2002. Evolution of the lectin-complement pathway and its role in innate immunity. *Nat Rev Immunol* **2:**346–353.
10. **Matsushita M, Matsushita A, Endo Y, Nakata M, Kojima N, Mizuochi T, Fujita T.** 2004. Origin of the classical complement pathway: lamprey orthologue of mammalian C1q acts as a lectin. *Proc Natl Acad Sci U S A* **101:**10127–10131.
11. **Sekine H, Kenjo A, Azumi K, Ohi G, Takahashi M, Kasukawa R, Ichikawa N, Nakata M, Mizuochi T, Matsushita M, Endo Y, Fujita T.** 2001. An ancient lectin-dependent complement system in an ascidian: novel lectin isolated from the plasma of the solitary ascidian, *Halocynthia roretzi*. *J Immunol* **167:**4504–4510.
12. **Katyal SL, Singh G, Locker J.** 1992. Characterization of a second human pulmonary surfactant-associated protein SP-A gene. *Am J Respir Cell Mol Biol* **6:**446–452.
13. **Floros J, Hoover RR.** 1998. Genetics of the hydrophilic surfactant proteins A and D. *Biochim Biophys Acta* **1408:**312–322.
14. **Uemura T, Sano H, Katoh T, Nishitani C, Mitsuzawa H, Shimizu T, Kuroki Y.** 2006. Surfactant protein A without the interruption of Gly-X-Y repeats loses a kink of oligomeric structure and exhibits impaired phospholipid liposome aggregation ability. *Biochemistry* **45:** 14543–14551.
15. **Crouch E, Persson A, Chang D, Heuser J.** 1994. Molecular structure of pulmonary surfactant protein D (SP-D). *J Biol Chem* **269:**17311–17319.
16. **Kishore U, Greenhough TJ, Waters P, Shrive AK, Ghai R, Kamran MF, Bernal AL, Reid KB, Madan T, Chakraborty T.** 2006. Surfactant proteins SP-A and SP-D: structure, function and receptors. *Mol Immunol* **43:**1293–1315.
17. **Spissinger T, Schäfer KP, Voss T.** 1991. Assembly of the surfactant protein SP-A. Deletions in the globular domain interfere with the correct folding of the molecule. *Eur J Biochem* **199:**65–71.
18. **Hansen S, Holm D, Moeller V, Vitved L, Bendixen C, Reid KB, Skjoedt K, Holmskov U.** 2002. CL-46, a novel collectin highly expressed in bovine thymus and liver. *J Immunol* **169:**5726–5734.
19. **Lipscombe RJ, Sumiya M, Summerfield JA, Turner MW.** 1995. Distinct physicochemical characteristics of human mannose binding protein expressed by individuals of differing genotype. *Immunology* **85:**660–667.
20. **Dahl MR, Thiel S, Matsushita M, Fujita T, Willis AC, Christensen T, Vorup-Jensen T, Jensenius JC.** 2001. MASP-3 and its association with distinct complexes of the mannan-binding lectin complement activation pathway. *Immunity* **15:**127–135.
21. **Teillet F, Dublet B, Andrieu JP, Gaboriaud C, Arlaud GJ, Thielens NM.** 2005. The two major oligomeric forms of human mannan-binding lectin: chemical characterization, carbohydrate-binding properties, and interaction with MBL-associated serine proteases. *J Immunol* **174:**2870–2877.
22. **Jensenius H, Klein DC, van Hecke M, Oosterkamp TH, Schmidt T, Jensenius JC.** 2009. Mannan-binding lectin: structure, oligomerization, and flexibility studied by atomic force microscopy. *J Mol Biol* **391:**246–259.
23. **van de Wetering JK, van Eijk M, van Golde LM, Hartung T, van Strijp JA, Batenburg JJ.** 2001. Characteristics of surfactant protein A and D binding to lipoteichoic acid and peptidoglycan, 2 major cell wall components of gram-positive bacteria. *J Infect Dis* **184:** 1143–1151.
24. **Drickamer K.** 1992. Engineering galactose-binding activity into a C-type mannose-binding protein. *Nature* **360:**183–186.
25. **Madan T, Kaur S, Saxena S, Singh M, Kishore U, Thiel S, Reid KB, Sarma PU.** 2005. Role of collectins in innate immunity against aspergillosis. *Med Mycol* **43** (Suppl 1):S155–S163.
26. **Allen MJ, Voelker DR, Mason RJ.** 2001. Interactions of surfactant proteins A and D with *Saccharomyces cerevisiae* and *Aspergillus fumigatus*. *Infect Immun* **69:** 2037–2044.
27. **Lin KW, Jen KY, Suarez CJ, Crouch EC, Perkins DL, Finn PW.** 2010. Surfactant protein D-mediated decrease of allergen-induced inflammation is dependent upon CTLA4. *J Immunol* **184:**6343–6349.
28. **Gowdy KM, Cardona DM, Nugent JL, Giamberardino C, Thomas JM, Mukherjee S, Martinu T, Foster WM, Plevy SE, Pastva AM, Wright JR, Palmer SM.** 2012. Novel role for surfactant protein A in gastrointestinal graft-versus-host disease. *J Immunol* **188:**4897–4905.
29. **Hartshorn KL, Crouch EC, White MR, Eggleton P, Tauber AI, Chang D, Sastry K.** 1994. Evidence for a protective role of pulmonary surfactant protein D (SP-D) against influenza A viruses. *J Clin Invest* **94:**311–319.
30. **White MR, Crouch E, Vesona J, Tacken PJ, Batenburg JJ, Leth-Larsen R, Holmskov U, Hartshorn KL.** 2005. Respiratory innate immune proteins differentially modulate the neutrophil respiratory burst response to influenza A virus. *Am J Physiol Lung Cell Mol Physiol* **289:** L606–L616.
31. **Reichhardt MP, Meri S.** 2016. SALSA: a regulator of the early steps of complement activation on mucosal surfaces. *Front Immunol* **7:**85. doi:10.3389/fimmu.2016.00085.
32. **Garred P, Honoré C, Ma YJ, Munthe-Fog L, Hummelshøj T.** 2009. *MBL2*, *FCN1*, *FCN2* and *FCN3*—the genes behind the initiation of the lectin pathway of complement. *Mol Immunol* **46:**2737–2744.
33. **Sørensen CM, Hansen TK, Steffensen R, Jensenius JC, Thiel S.** 2006. Hormonal regulation of mannan-binding lectin synthesis in hepatocytes. *Clin Exp Immunol* **145:** 173–182.
34. **Ezekowitz RA, Day LE, Herman GA.** 1988. A human mannose-binding protein is an acute-phase reactant that

shares sequence homology with other vertebrate lectins. *J Exp Med* **167**:1034–1046.

35. **Arai T, Tabona P, Summerfield JA.** 1993. Human mannose-binding protein gene is regulated by interleukins, dexamethasone and heat shock. *Q J Med* **86**: 575–582.

36. **Thiel S, Holmskov U, Hviid L, Laursen SB, Jensenius JC.** 1992. The concentration of the C-type lectin, mannan-binding protein, in human plasma increases during an acute phase response. *Clin Exp Immunol* **90**: 31–35.

37. **Perez-Castellano M, Peñaranda M, Payeras A, Milà J, Riera M, Vidal J, Pujalte F, Pareja A, Villalonga C, Matamoros N.** 2006. Mannose-binding lectin does not act as an acute-phase reactant in adults with community-acquired pneumococcal pneumonia. *Clin Exp Immunol* **145**:228–234.

38. **Dean MM, Minchinton RM, Heatley S, Eisen DP.** 2005. Mannose binding lectin acute phase activity in patients with severe infection. *J Clin Immunol* **25**: 346–352.

39. **Nonaka M, Imaeda H, Matsumoto S, Yong Ma B, Kawasaki N, Mekata E, Andoh A, Saito Y, Tani T, Fujiyama Y, Kawasaki T.** 2014. Mannan-binding protein, a C-type serum lectin, recognizes primary colorectal carcinomas through tumor-associated Lewis glycans. *J Immunol* **192**:1294–1301.

40. **Fujita T, Matsushita M, Endo Y.** 2004. The lectin-complement pathway—its role in innate immunity and evolution. *Immunol Rev* **198**:185–202.

41. **Thielens NM.** 2011. The double life of M-ficolin: what functions when circulating in serum and tethered to leukocyte surfaces? *J Leukoc Biol* **90**:410–412.

42. **Ren Y, Ding Q, Zhang X.** 2014. Ficolins and infectious diseases. *Virol Sin* **29**:25–32.

43. **Matsushita M.** 2010. Ficolins: complement-activating lectins involved in innate immunity. *J Innate Immun* **2**: 24–32.

44. **Lacroix M, Dumestre-Pérard C, Schoehn G, Houen G, Cesbron JY, Arlaud GJ, Thielens NM.** 2009. Residue Lys57 in the collagen-like region of human L-ficolin and its counterpart Lys47 in H-ficolin play a key role in the interaction with the mannan-binding lectin-associated serine proteases and the collectin receptor calreticulin. *J Immunol* **182**:456–465.

45. **Endo Y, Matsushita M, Fujita T.** 2007. Role of ficolin in innate immunity and its molecular basis. *Immunobiology* **212**:371–379.

46. **Ma YG, Cho MY, Zhao M, Park JW, Matsushita M, Fujita T, Lee BL.** 2004. Human mannose-binding lectin and L-ficolin function as specific pattern recognition proteins in the lectin activation pathway of complement. *J Biol Chem* **279**:25307–25312.

47. **Lynch NJ, Roscher S, Hartung T, Morath S, Matsushita M, Maennel DN, Kuraya M, Fujita T, Schwaeble WJ.** 2004. L-Ficolin specifically binds to lipoteichoic acid, a cell wall constituent of Gram-positive bacteria, and activates the lectin pathway of complement. *J Immunol* **172**:1198–1202.

48. **Luo F, Sun X, Wang Y, Wang Q, Wu Y, Pan Q, Fang C, Zhang XL.** 2013. Ficolin-2 defends against virulent *Mycobacteria tuberculosis* infection *in vivo*, and its insufficiency is associated with infection in humans. *PLoS One* **8**:e73859. doi:10.1371/journal.pone.0073859.

49. **Liu J, Ali MA, Shi Y, Zhao Y, Luo F, Yu J, Xiang T, Tang J, Li D, Hu Q, Ho W, Zhang X.** 2009. Specifically binding of L-ficolin to *N*-glycans of HCV envelope glycoproteins E1 and E2 leads to complement activation. *Cell Mol Immunol* **6**:235–244.

50. **Verma A, White M, Vathipadiekal V, Tripathi S, Mbianda J, Ieong M, Qi L, Taubenberger JK, Takahashi K, Jensenius JC, Thiel S, Hartshorn KL.** 2012. Human H-ficolin inhibits replication of seasonal and pandemic influenza A viruses. *J Immunol* **189**:2478–2487.

51. **Garlanda C, Bottazzi B, Bastone A, Mantovani A.** 2005. Pentraxins at the crossroads between innate immunity, inflammation, matrix deposition, and female fertility. *Annu Rev Immunol* **23**:337–366.

52. **Lu J, Marnell LL, Marjon KD, Mold C, Du Clos TW, Sun PD.** 2008. Structural recognition and functional activation of FcγR by innate pentraxins. *Nature* **456**:989–992.

53. **Szalai AJ.** 2002. The antimicrobial activity of C-reactive protein. *Microbes Infect* **4**:201–205.

54. **Noursadeghi M, Bickerstaff MC, Gallimore JR, Herbert J, Cohen J, Pepys MB.** 2000. Role of serum amyloid P component in bacterial infection: protection of the host or protection of the pathogen. *Proc Natl Acad Sci U S A* **97**:14584–14589.

55. **Jaillon S, Bonavita E, Gentile S, Rubino M, Laface I, Garlanda C, Mantovani A.** 2014. The long pentraxin PTX3 as a key component of humoral innate immunity and a candidate diagnostic for inflammatory diseases. *Int Arch Allergy Immunol* **165**:165–178.

56. **Nauta AJ, Daha MR, van Kooten C, Roos A.** 2003. Recognition and clearance of apoptotic cells: a role for complement and pentraxins. *Trends Immunol* **24**:148–154.

57. **Gershov D, Kim S, Brot N, Elkon KB.** 2000. C-reactive protein binds to apoptotic cells, protects the cells from assembly of the terminal complement components, and sustains an antiinflammatory innate immune response: implications for systemic autoimmunity. *J Exp Med* **192**:1353–1364.

58. **Breviario F, d'Aniello EM, Golay J, Peri G, Bottazzi B, Bairoch A, Saccone S, Marzella R, Predazzi V, Rocchi M, Della Valle G, Dejana E, Mantovani A, Introna M.** 1992. Interleukin-1-inducible genes in endothelial cells. Cloning of a new gene related to C-reactive protein and serum amyloid P component. *J Biol Chem* **267**: 22190–22197.

59. **Lee GW, Lee TH, Vilcek J.** 1993. TSG-14, a tumor necrosis factor- and IL-1-inducible protein, is a novel member of the pentaxin family of acute phase proteins. *J Immunol* **150**:1804–1812.

60. **Martinez de la Torre Y, Fabbri M, Jaillon S, Bastone A, Nebuloni M, Vecchi A, Mantovani A, Garlanda C.** 2010. Evolution of the pentraxin family: the new entry PTX4. *J Immunol* **184**:5055–5064.

61. **Jaillon S, Moalli F, Ragnarsdottir B, Bonavita E, Puthia M, Riva F, Barbati E, Nebuloni M, Cvetko Krajinovic L, Markotic A, Valentino S, Doni A, Tartari S, Graziani G, Montanelli A, Delneste Y, Svanborg C, Garlanda C, Mantovani A.** 2014. The humoral pattern recognition molecule PTX3 is a key component of innate immunity against urinary tract infection. *Immunity* **40:**621–632.
62. **Salio M, Chimenti S, De Angelis N, Molla F, Maina V, Nebuloni M, Pasqualini F, Latini R, Garlanda C, Mantovani A.** 2008. Cardioprotective function of the long pentraxin PTX3 in acute myocardial infarction. *Circulation* **117:**1055–1064.
63. **Bonavita E, Gentile S, Rubino M, Maina V, Papait R, Kunderfranco P, Greco C, Feruglio F, Molgora M, Laface I, Tartari S, Doni A, Pasqualini F, Barbati E, Basso G, Galdiero MR, Nebuloni M, Roncalli M, Colombo P, Laghi L, Lambris JD, Jaillon S, Garlanda C, Mantovani A.** 2015. PTX3 is an extrinsic oncosuppressor regulating complement-dependent inflammation in cancer. *Cell* **160:**700–714.
64. **Doni A, Musso T, Morone D, Bastone A, Zambelli V, Sironi M, Castagnoli C, Cambieri I, Stravalaci M, Pasqualini F, Laface I, Valentino S, Tartari S, Ponzetta A, Maina V, Barbieri SS, Tremoli E, Catapano AL, Norata GD, Bottazzi B, Garlanda C, Mantovani A.** 2015. An acidic microenvironment sets the humoral pattern recognition molecule PTX3 in a tissue repair mode. *J Exp Med* **212:**905–925.
65. **Inforzato A, Rivieccio V, Morreale AP, Bastone A, Salustri A, Scarchilli L, Verdoliva A, Vincenti S, Gallo G, Chiapparino C, Pacello L, Nucera E, Serlupi-Crescenzi O, Day AJ, Bottazzi B, Mantovani A, De Santis R, Salvatori G.** 2008. Structural characterization of PTX3 disulfide bond network and its multimeric status in cumulus matrix organization. *J Biol Chem* **283:**10147–10161.
66. **Deban L, Russo RC, Sironi M, Moalli F, Scanziani M, Zambelli V, Cuccovillo I, Bastone A, Gobbi M, Valentino S, Doni A, Garlanda C, Danese S, Salvatori G, Sassano M, Evangelista V, Rossi B, Zenaro E, Constantin G, Laudanna C, Bottazzi B, Mantovani A.** 2010. Regulation of leukocyte recruitment by the long pentraxin PTX3. *Nat Immunol* **11:**328–334.
67. **Jaillon S, Peri G, Delneste Y, Frémaux I, Doni A, Moalli F, Garlanda C, Romani L, Gascan H, Bellocchio S, Bozza S, Cassatella MA, Jeannin P, Mantovani A.** 2007. The humoral pattern recognition receptor PTX3 is stored in neutrophil granules and localizes in extracellular traps. *J Exp Med* **204:**793–804.
68. **Moalli F, Doni A, Deban L, Zelante T, Zagarella S, Bottazzi B, Romani L, Mantovani A, Garlanda C.** 2010. Role of complement and Fcγ receptors in the protective activity of the long pentraxin PTX3 against *Aspergillus fumigatus*. *Blood* **116:**5170–5180.
69. **Garlanda C, Hirsch E, Bozza S, Salustri A, De Acetis M, Nota R, Maccagno A, Riva F, Bottazzi B, Peri G, Doni A, Vago L, Botto M, De Santis R, Carminati P, Siracusa G, Altruda F, Vecchi A, Romani L, Mantovani A.** 2002. Non-redundant role of the long pentraxin PTX3 in antifungal innate immune response. *Nature* **420:**182–186.
70. **Han B, Haitsma JJ, Zhang Y, Bai X, Rubacha M, Keshavjee S, Zhang H, Liu M.** 2011. Long pentraxin PTX3 deficiency worsens LPS-induced acute lung injury. *Intensive Care Med* **37:**334–342.
71. **Lech M, Römmele C, Gröbmayr R, Eka Susanti H, Kulkarni OP, Wang S, Gröne HJ, Uhl B, Reichel C, Krombach F, Garlanda C, Mantovani A, Anders HJ.** 2013. Endogenous and exogenous pentraxin-3 limits postischemic acute and chronic kidney injury. *Kidney Int* **83:**647–661.
72. **Jeannin P, Bottazzi B, Sironi M, Doni A, Rusnati M, Presta M, Maina V, Magistrelli G, Haeuw JF, Hoeffel G, Thieblemont N, Corvaia N, Garlanda C, Delneste Y, Mantovani A.** 2005. Complexity and complementarity of outer membrane protein A recognition by cellular and humoral innate immunity receptors. *Immunity* **22:**551–560.
73. **Cotena A, Maina V, Sironi M, Bottazzi B, Jeannin P, Vecchi A, Corvaia N, Daha MR, Mantovani A, Garlanda C.** 2007. Complement dependent amplification of the innate response to a cognate microbial ligand by the long pentraxin PTX3. *J Immunol* **179:**6311–6317.
74. **Soares AC, Souza DG, Pinho V, Vieira AT, Nicoli JR, Cunha FQ, Mantovani A, Reis LF, Dias AA, Teixeira MM.** 2006. Dual function of the long pentraxin PTX3 in resistance against pulmonary infection with *Klebsiella pneumoniae* in transgenic mice. *Microbes Infect* **8:** 1321–1329.
75. **Ma YJ, Doni A, Hummelshøj T, Honoré C, Bastone A, Mantovani A, Thielens NM, Garred P.** 2009. Synergy between ficolin-2 and pentraxin 3 boosts innate immune recognition and complement deposition. *J Biol Chem* **284:**28263–28275.
76. **Ma YJ, Doni A, Skjoedt MO, Honoré C, Arendrup M, Mantovani A, Garred P.** 2011. Heterocomplexes of mannose-binding lectin and the pentraxins PTX3 or serum amyloid P component trigger cross-activation of the complement system. *J Biol Chem* **286:**3405–3417.
77. **Ma YJ, Doni A, Romani L, Jürgensen HJ, Behrendt N, Mantovani A, Garred P.** 2013. Ficolin-1-PTX3 complex formation promotes clearance of altered self-cells and modulates IL-8 production. *J Immunol* **191:**1324–1333.
78. **Braunschweig A, Józsi M.** 2011. Human pentraxin 3 binds to the complement regulator c4b-binding protein. *PLoS One* **6:**e23991. doi:10.1371/journal.pone.0023991.
79. **Deban L, Jarva H, Lehtinen MJ, Bottazzi B, Bastone A, Doni A, Jokiranta TS, Mantovani A, Meri S.** 2008. Binding of the long pentraxin PTX3 to factor H: interacting domains and function in the regulation of complement activation. *J Immunol* **181:**8433–8440.
80. **Csincsi AI, Kopp A, Zöldi M, Bánlaki Z, Uzonyi B, Hebecker M, Caesar JJ, Pickering MC, Daigo K, Hamakubo T, Lea SM, Goicoechea de Jorge E, Józsi M.** 2015. Factor H-related protein 5 interacts with pentraxin 3 and the extracellular matrix and modulates complement activation. *J Immunol* **194:**4963–4973.
81. **Dziarski R, Gupta D.** 2006. Mammalian PGRPs: novel antibacterial proteins. *Cell Microbiol* **8:**1059–1069.
82. **Kietzman C, Tuomanen E.** 2011. PGRPs kill with an ancient weapon. *Nat Med* **17:**665–666.

83. **Dziarski R, Gupta D.** 2010. Review: mammalian peptidoglycan recognition proteins (PGRPs) in innate immunity. *Innate Immun* **16:**168–174.

84. **Kashyap DR, Wang M, Liu LH, Boons GJ, Gupta D, Dziarski R.** 2011. Peptidoglycan recognition proteins kill bacteria by activating protein-sensing two-component systems. *Nat Med* **17:**676–683.

85. **O'Brien KD, Chait A.** 2006. Serum amyloid A: the "other" inflammatory protein. *Curr Atheroscler Rep* **8:** 62–68.

86. **Shah C, Hari-Dass R, Raynes JG.** 2006. Serum amyloid A is an innate immune opsonin for Gram-negative bacteria. *Blood* **108:**1751–1757.

87. **Cai Z, Cai L, Jiang J, Chang KS, van der Westhuyzen DR, Luo G.** 2007. Human serum amyloid A protein inhibits hepatitis C virus entry into cells. *J Virol* **81:** 6128–6133.

88. **Deguchi A, Tomita T, Omori T, Komatsu A, Ohto U, Takahashi S, Tanimura N, Akashi-Takamura S, Miyake K, Maru Y.** 2013. Serum amyloid A3 binds MD-2 to activate p38 and NF-κB pathways in a MyD88-dependent manner. *J Immunol* **191:**1856–1864.

89. **Ather JL, Ckless K, Martin R, Foley KL, Suratt BT, Boyson JE, Fitzgerald KA, Flavell RA, Eisenbarth SC, Poynter ME.** 2011. Serum amyloid A activates the NLRP3 inflammasome and promotes Th17 allergic asthma in mice. *J Immunol* **187:**64–73.

90. **Sun L, Zhu Z, Cheng N, Yan Q, Ye RD.** 2014. Serum amyloid A induces interleukin-33 expression through an IRF7-dependent pathway. *Eur J Immunol* **44:**2153–2164.

91. **Bowdish DM, Loffredo MS, Mukhopadhyay S, Mantovani A, Gordon S.** 2007. Macrophage receptors implicated in the "adaptive" form of innate immunity. *Microbes Infect* **9:**1680–1687.

92. **Takeuchi O, Akira S.** 2010. Pattern recognition receptors and inflammation. *Cell* **140:**805–820.

93. **Jeannin P, Jaillon S, Delneste Y.** 2008. Pattern recognition receptors in the immune response against dying cells. *Curr Opin Immunol* **20:**530–537.

94. **Diniz SN, Nomizo R, Cisalpino PS, Teixeira MM, Brown GD, Mantovani A, Gordon S, Reis LF, Dias AA.** 2004. PTX3 function as an opsonin for the dectin-1-dependent internalization of zymosan by macrophages. *J Leukoc Biol* **75:**649–656.

95. **Frankenberger M, Schwaeble W, Ziegler-Heitbrock L.** 2008. Expression of M-Ficolin in human monocytes and macrophages. *Mol Immunol* **45:**1424–1430.

96. **Jaillon S, Galdiero MR, Del Prete D, Cassatella MA, Garlanda C, Mantovani A.** 2013. Neutrophils in innate and adaptive immunity. *Semin Immunopathol* **35:**377–394.

97. **Greenblatt MB, Aliprantis A, Hu B, Glimcher LH.** 2010. Calcineurin regulates innate antifungal immunity in neutrophils. *J Exp Med* **207:**923–931.

98. **Li D, Dong B, Tong Z, Wang Q, Liu W, Wang Y, Liu W, Chen J, Xu L, Chen L, Duan Y.** 2012. MBL-mediated opsonophagocytosis of *Candida albicans* by human neutrophils is coupled with intracellular Dectin-1-triggered ROS production. *PLoS One* **7:**e50589. doi: 10.1371/journal.pone.0050589.

99. **Liu X, Ma B, Malik AB, Tang H, Yang T, Sun B, Wang G, Minshall RD, Li Y, Zhao Y, Ye RD, Xu J.** 2012. Bidirectional regulation of neutrophil migration by mitogen-activated protein kinases. *Nat Immunol* **13:**457–464.

100. **McDonald B, Pittman K, Menezes GB, Hirota SA, Slaba I, Waterhouse CC, Beck PL, Muruve DA, Kubes P.** 2010. Intravascular danger signals guide neutrophils to sites of sterile inflammation. *Science* **330:**362–366.

101. **Tamassia N, Bazzoni F, Le Moigne V, Calzetti F, Masala C, Grisendi G, Bussmeyer U, Scutera S, De Gironcoli M, Costantini C, Musso T, Cassatella MA.** 2012. IFN-β expression is directly activated in human neutrophils transfected with plasmid DNA and is further increased via TLR-4-mediated signaling. *J Immunol* **189:**1500–1509.

102. **Brinkmann V, Reichard U, Goosmann C, Fauler B, Uhlemann Y, Weiss DS, Weinrauch Y, Zychlinsky A.** 2004. Neutrophil extracellular traps kill bacteria. *Science* **303:**1532–1535.

103. **Saitoh T, Komano J, Saitoh Y, Misawa T, Takahama M, Kozaki T, Uehata T, Iwasaki H, Omori H, Yamaoka S, Yamamoto N, Akira S.** 2012. Neutrophil extracellular traps mediate a host defense response to human immunodeficiency virus-1. *Cell Host Microbe* **12:**109–116.

104. **Jenne CN, Wong CH, Zemp FJ, McDonald B, Rahman MM, Forsyth PA, McFadden G, Kubes P.** 2013. Neutrophils recruited to sites of infection protect from virus challenge by releasing neutrophil extracellular traps. *Cell Host Microbe* **13:**169–180.

105. **Douda DN, Jackson R, Grasemann H, Palaniyar N.** 2011. Innate immune collectin surfactant protein D simultaneously binds both neutrophil extracellular traps and carbohydrate ligands and promotes bacterial trapping. *J Immunol* **187:**1856–1865.

106. **Jaillon S, Jeannin P, Hamon Y, Frémaux I, Doni A, Bottazzi B, Blanchard S, Subra JF, Chevailler A, Mantovani A, Delneste Y.** 2009. Endogenous PTX3 translocates at the membrane of late apoptotic human neutrophils and is involved in their engulfment by macrophages. *Cell Death Differ* **16:**465–474.

107. **Cho JH, Fraser IP, Fukase K, Kusumoto S, Fujimoto Y, Stahl GL, Ezekowitz RA.** 2005. Human peptidoglycan recognition protein S is an effector of neutrophil-mediated innate immunity. *Blood* **106:**2551–2558.

108. **Rørvig S, Honore C, Larsson LI, Ohlsson S, Pedersen CC, Jacobsen LC, Cowland JB, Garred P, Borregaard N.** 2009. Ficolin-1 is present in a highly mobilizable subset of human neutrophil granules and associates with the cell surface after stimulation with fMLP. *J Leukoc Biol* **86:**1439–1449.

109. **Dziarski R, Platt KA, Gelius E, Steiner H, Gupta D.** 2003. Defect in neutrophil killing and increased susceptibility to infection with nonpathogenic gram-positive bacteria in peptidoglycan recognition protein-S (PGRP-S)-deficient mice. *Blood* **102:**689–697.

110. **Moreno-Amaral AN, Gout E, Danella-Polli C, Tabarin F, Lesavre P, Pereira-da-Silva G, Thielens NM, Halbwachs-**

Mecarelli L. 2012. M-Ficolin and leukosialin (CD43): new partners in neutrophil adhesion. *J Leukoc Biol* **91:** 469–474.

111. **Han B, Ma X, Zhang J, Zhang Y, Bai X, Hwang DM, Keshavjee S, Levy GA, McGilvray I, Liu M.** 2012. Protective effects of long pentraxin PTX3 on lung injury in a severe acute respiratory syndrome model in mice. *Lab Invest* **92:**1285–1296.

112. **Bozza S, Bistoni F, Gaziano R, Pitzurra L, Zelante T, Bonifazi P, Perruccio K, Bellocchio S, Neri M, Iorio AM, Salvatori G, De Santis R, Calvitti M, Doni A, Garlanda C, Mantovani A, Romani L.** 2006. Pentraxin 3 protects from MCMV infection and reactivation through TLR sensing pathways leading to IRF3 activation. *Blood* **108:**3387–3396.

113. **Reading PC, Bozza S, Gilbertson B, Tate M, Moretti S, Job ER, Crouch EC, Brooks AG, Brown LE, Bottazzi B, Romani L, Mantovani A.** 2008. Antiviral activity of the long chain pentraxin PTX3 against influenza viruses. *J Immunol* **180:**3391–3398.

114. **Job ER, Bottazzi B, Short KR, Deng YM, Mantovani A, Brooks AG, Reading PC.** 2014. A single amino acid substitution in the hemagglutinin of H3N2 subtype influenza A viruses is associated with resistance to the long pentraxin PTX3 and enhanced virulence in mice. *J Immunol* **192:**271–281.

115. **Job ER, Deng YM, Tate MD, Bottazzi B, Crouch EC, Dean MM, Mantovani A, Brooks AG, Reading PC.** 2010. Pandemic H1N1 influenza A viruses are resistant to the antiviral activities of innate immune proteins of the collectin and pentraxin superfamilies. *J Immunol* **185:**4284–4291.

116. **Bozza S, Campo S, Arseni B, Inforzato A, Ragnar L, Bottazzi B, Mantovani A, Moretti S, Oikonomous V, De Santis R, Carvalho A, Salvatori G, Romani L.** 2014. PTX3 binds MD-2 and promotes TRIF-dependent immune protection in aspergillosis. *J Immunol* **193:**2340–2348.

117. **Davey MS, Tamassia N, Rossato M, Bazzoni F, Calzetti F, Bruderek K, Sironi M, Zimmer L, Bottazzi B, Mantovani A, Brandau S, Moser B, Eberl M, Cassatella MA.** 2011. Failure to detect production of IL-10 by activated human neutrophils. *Nat Immunol* **12:**1017–1018, author reply 1018–1020.

118. **Barbati E, Specchia C, Villella M, Rossi ML, Barlera S, Bottazzi B, Crociati L, d'Arienzo C, Fanelli R, Garlanda C, Gori F, Mango R, Mantovani A, Merla G, Nicolis EB, Pietri S, Presbitero P, Sudo Y, Villella A, Franzosi MG.** 2012. Influence of pentraxin 3 (PTX3) genetic variants on myocardial infarction risk and PTX3 plasma levels. *PLoS One* **7:**e53030. doi:10.1371/journal.pone.0053030.

119. **Cunha C, Aversa F, Lacerda JF, Busca A, Kurzai O, Grube M, Löffler J, Maertens JA, Bell AS, Inforzato A, Barbati E, Almeida B, Santos e Sousa P, Barbui A, Potenza L, Caira M, Rodrigues F, Salvatori G, Pagano L, Luppi M, Mantovani A, Velardi A, Romani L, Carvalho A.** 2014. Genetic PTX3 deficiency and aspergillosis in stem-cell transplantation. *N Engl J Med* **370:**421–432.

120. **Bottazzi B, Santini L, Savino S, Giuliani MM, Dueñas Díez AI, Mancuso G, Beninati C, Sironi M, Valentino S, Deban L, Garlanda C, Teti G, Pizza M, Rappuoli R, Mantovani A.** 2015. Recognition of *Neisseria meningitidis* by the long pentraxin PTX3 and its role as an endogenous adjuvant. *PLoS One* **10:**e0120807. doi: 10.1371/journal.pone.0120807.

121. **Moalli F, Paroni M, Véliz Rodriguez T, Riva F, Polentarutti N, Bottazzi B, Valentino S, Mantero S, Nebuloni M, Mantovani A, Bragonzi A, Garlanda C.** 2011. The therapeutic potential of the humoral pattern recognition molecule PTX3 in chronic lung infection caused by *Pseudomonas aeruginosa*. *J Immunol* **186:** 5425–5434.

122. **Jaillon S, Mancuso G, Hamon Y, Beauvillain C, Cotici V, Midiri A, Bottazzi B, Nebuloni M, Garlanda C, Frémaux I, Gauchat JF, Descamps P, Beninati C, Mantovani A, Jeannin P, Delneste Y.** 2013. Prototypic long pentraxin PTX3 is present in breast milk, spreads in tissues, and protects neonate mice from *Pseudomonas aeruginosa* lung infection. *J Immunol* **191:**1873–1882.

123. **Chiarini M, Sabelli C, Melotti P, Garlanda C, Savoldi G, Mazza C, Padoan R, Plebani A, Mantovani A, Notarangelo LD, Assael BM, Badolato R.** 2010. *PTX3* genetic variations affect the risk of *Pseudomonas aeruginosa* airway colonization in cystic fibrosis patients. *Genes Immun* **11:**665–670.

124. **Olesen R, Wejse C, Velez DR, Bisseye C, Sodemann M, Aaby P, Rabna P, Worwui A, Chapman H, Diatta M, Adegbola RA, Hill PC, Østergaard L, Williams SM, Sirugo G.** 2007. DC-SIGN (CD209), pentraxin 3 and vitamin D receptor gene variants associate with pulmonary tuberculosis risk in West Africans. *Genes Immun* **8:**456–467.

125. **Baruah P, Dumitriu IE, Peri G, Russo V, Mantovani A, Manfredi AA, Rovere-Querini P.** 2006. The tissue pentraxin PTX3 limits C1q-mediated complement activation and phagocytosis of apoptotic cells by dendritic cells. *J Leukoc Biol* **80:**87–95.

126. **Nauta AJ, Bottazzi B, Mantovani A, Salvatori G, Kishore U, Schwaeble WJ, Gingras AR, Tzima S, Vivanco F, Egido J, Tijsma O, Hack EC, Daha MR, Roos A.** 2003. Biochemical and functional characterization of the interaction between pentraxin 3 and C1q. *Eur J Immunol* **33:**465–473.

127. **Guo T, Ke L, Qi B, Wan J, Ge J, Bai L, Cheng B.** 2012. PTX3 is located at the membrane of late apoptotic macrophages and mediates the phagocytosis of macrophages. *J Clin Immunol* **32:**330–339.

128. **Ley K, Reutershan J.** 2006. Leucocyte-endothelial interactions in health and disease. *Handbook Exp Pharmacol* **2006:**97–133.

Myeloid Cells in Health and Disease: A Synthesis
Edited by Siamon Gordon

doi:10.1128/microbiolspec.MCHD-0023-2015

Rob J. W. Arts[1]
Mihai G. Netea[1]

38 Adaptive Characteristics of Innate Immune Responses in Macrophages

MACROPHAGES ARE CRUCIAL FOR BOTH INNATE AND ADAPTIVE IMMUNE RESPONSES

Macrophages are a central component of antimicrobial host defense, described as being crucial for both innate immune mechanisms and adaptive immunity (1). The dichotomy between the immediate antimicrobial responses seen as nonspecific and the relatively late-onset specific T- and B-cell responses has driven our understanding of host defense for more than half a century. Innate immunity reacts instantly upon an encounter with a pathogen but has been viewed as nonspecific and incapable of building immunological memory. In contrast, adaptive immune responses can specifically recognize pathogenic microorganisms and build memory capable of protection against reinfection. Macrophages are involved in both of these responses: on the one hand, macrophages have the capacity to phagocytose and kill microorganisms in a nonspecific fashion, as well as to release proinflammatory mediators that drive inflammation; but on the other hand, they can also present antigens and initiate and modulate the specific T-cell responses through expression of costimulatory molecules and specific cytokines (2).

The last decade has dramatically changed the dogma of innate immunity being nonspecific, through the discovery of pattern recognition receptors (PRRs) such as the Toll-like receptors (TLRs), C-type lectin receptors, NOD-like receptors, and retinoic acid-inducible gene 1 protein (RIG-I)-like receptors. These receptors are expressed either on the surface of macrophages (and other immune cells) or intracellularly to allow specific recognition of conserved structures of different classes of microorganisms (3). PRRs thus allow semispecific recognition of different types of microorganisms, though not individual microbial species. The widespread concept that innate immunity in general, and macrophages in particular, cannot adapt to a previous encounter with a pathogen and is incapable of building the memory of a previous infection has, however, not been contested until recently.

MACROPHAGE PRIMING, ACTIVATION, AND TOLERANCE AS ADAPTIVE RESPONSES

Since the late 1960s, it has been known that following exposure to microbes or to microbial components,

[1]Department of Internal Medicine, Radboud University Medical Center, Nijmegen, The Netherlands.

mononuclear phagocytes show increased effector functions as revealed by their augmented microbicidal and tumoricidal activity (4–6) (Fig. 1). An unbiased survey of immunological literature reveals important clues that innate immunity is able to display adaptive features and that these mechanisms can offer protection against reinfection, independently of the classic adaptive immune memory (7, 8). The adaptive characteristics of innate immunity, also called "trained immunity" (8), are manifested as protection against reinfection by the same or different pathogens in organisms lacking adaptive immune responses, such as plants (9), invertebrates (10, 11), or mammals lacking functional T and B cells (12–14) (Table 1). Classic activation of macrophages requires hours, and the enhanced effector functions are transient. It was subsequently discovered that coexposure to selected cytokines (the prototype of which is gamma interferon [IFN-γ]) resulted in a dramatic increase in the responsiveness of macrophages to microbial moieties (5, 15) (Fig. 1). After exposure to IFN-γ, macrophages retain a primed state for several hours, with a dramatic augmentation of responsiveness, for instance, to lipopolysaccharide (LPS), in terms of toxic mediators (nitric oxide), cytokines (tumor necrosis factor α [TNF-α] and interleukin-12 [IL-12]), chemokines, or costimulatory molecules (5, 15, 16). Priming of human monocytes is in part sustained by increased expression of PRRs (TLR4 and myeloid differentiation factor 2 [MD2]) and transducers (myeloid differentiation primary-response protein 88 [MyD88]) (17). The discovery of alternative forms of macrophage activation, originally by IL-4 (18), has opened a new perspective on the plasticity of macrophages and their activation states (19). Moreover, transcriptional profiling

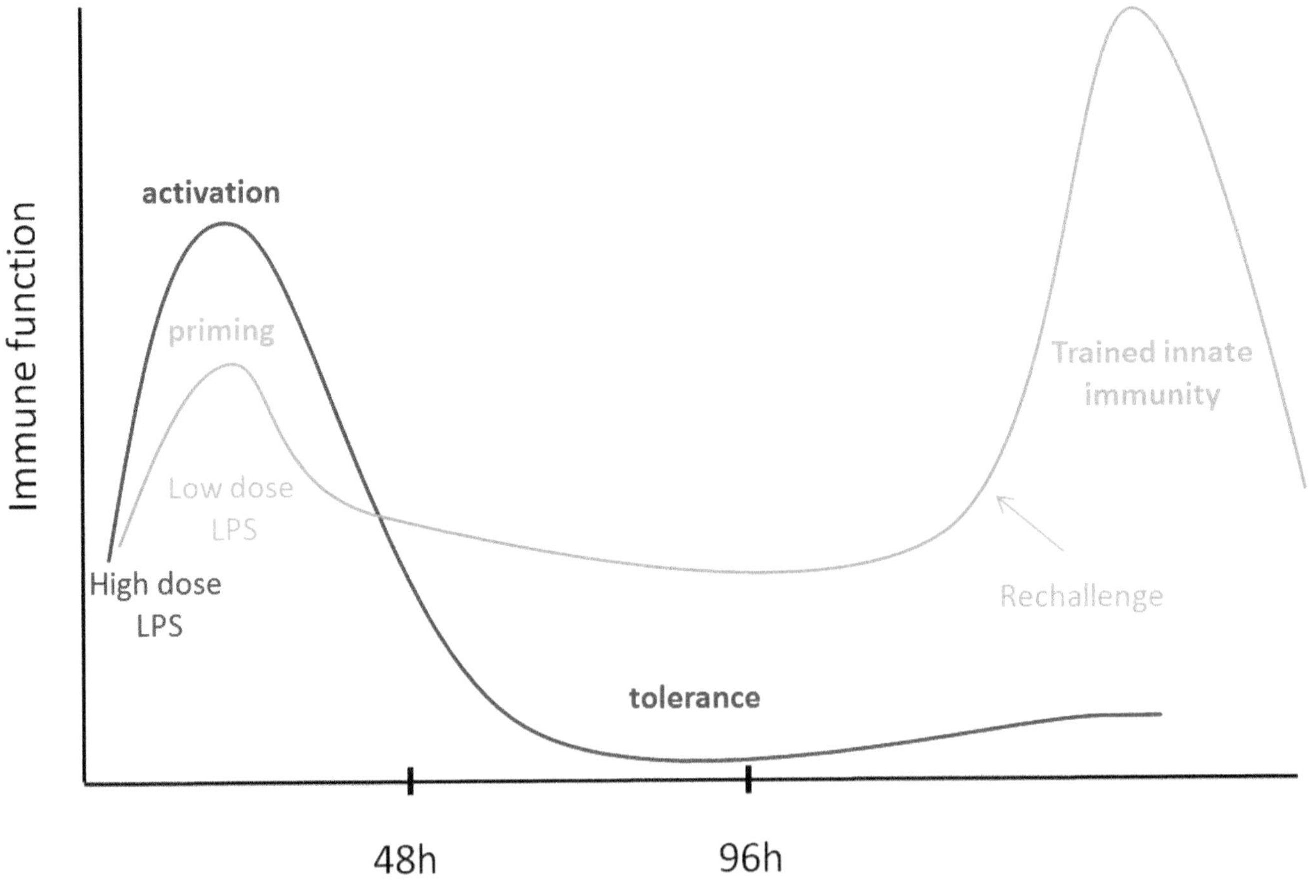

Figure 1 Priming, activation, and tolerance as adaptive macrophage responses. A schematic representation of the possible responses of macrophages to a novel stimulus (e.g., LPS). While LPS causes macrophage activation followed by tolerance, low doses of LPS prime macrophages for enhanced innate immune responses: trained innate immunity. Both tolerance and trained innate immunity are results of long-term epigenetic changes in macrophage function ("memory").

Table 1 Examples of memory of innate immunity in plants, invertebrates, and mammals[a]

Organism	Experimental model	Biological effect	Specificity	Mechanism
Plants—SAR				
Large variety of plants	Viruses, bacteria, and fungi	Protection against reinfection	Variable	Salicylic acid Epigenetic mechanisms
Adaptive immune responses in nonvertebrates				
Beetle	LPS or bacterial prechallenge	Protection against reinfection	–	Transgenerational priming
Drosophila	*Streptococcus pneumoniae* *Beauveria bassiana* *Serratia marcescens*	Protection	+	Serine protease CG33462
Anopheles gambiae	Midgut flora	Protection against *Plasmodium*	+	Toll-dependent hematocyte differentiation factor
Adaptive characteristics of innate immunity in mammals				
Mice	*C. albicans* BCG	Protection against candidiasis	–	Monocyte epigenetic reprogramming
Mice	Murine cytomegalovirus	NK-dependent hypersensitization	+	Ly49$^+$ NK cells Hepatic CXCR6$^+$ NK cells
Humans	BCG vaccination	Protection against nonrelated infections	–	Monocyte reprogramming

[a]+, specific protection; –, nonspecific protection; SAR, systemic acquired resistance.

(see, e.g., reference 20) has revealed the complexity of gene regulation during "activation" and paved the way to the dissection of underlying mechanisms (19, 21).

Deactivation can also occur as a consequence of microbial encounters (Fig. 1). Microbial signals (e.g., LPS) under appropriate conditions can result in innate immune tolerance, i.e., in hyporesponsiveness to a subsequent challenge at the macrophage and organism level. The immunosuppressive phenotype observed in late sepsis is likely a reflection of tolerance to LPS and/or other bacterial ligands. Tolerance is generally viewed as a defense strategy to limit inflammation-caused tissue damage (22). The definition of "tolerance" is actually as misleading as that of "activation." Analysis of the macrophage transcriptome has revealed that LPS tolerance is in fact a manifestation of reorientation of macrophage functions. The transcriptional profile of LPS-tolerant macrophages has some similarity to that expressed by alternatively activated M2-polarized macrophages (23, 24). Tolerant macrophages have an increased expression of IL-10, arginase-1, and the chemokines CCL17 and CCL22. Thus, what has long been viewed as endotoxin tolerance is not simple unresponsiveness: it represents an adaptive response of macrophages with reorientation to an immunoregulatory phenotype.

The interaction with microbial components not only affects, but also profoundly alters, the receptor repertoire expressed by cells of the monocyte-macrophage lineage (7, 25). It has been shown that microbes and cytokines change the levels of Dectin-1 (dendritic cell-associated C-type lectin 1) and of the scavenger receptor MARCO (macrophage receptor with collagenous structure) and hence modify subsequent responses to their ligands (7, 25).

In addition to changing the levels and repertoire of macrophage surface receptors, inflammatory signals also induce production of fluid-phase pattern recognition molecules (PRMs). These molecules constitute the humoral arm of innate immunity together with complement (26, 27), and they function as ancestors of antibodies ("ante-antibodies") (28). Myelomonocytic cells are the source of a vast range of fluid-phase PRMs. These include collectins (e.g., mannose-binding lectin), ficolins (e.g., L-, H-, and M-ficolin), and pentraxins (e.g., pentraxin 3 [PTX3]) (28). PTX3 has served as a paradigm for the mode of action of the interplay between the cellular and the humoral arms of innate immunity (29). This "long" pentraxin is produced in a gene expression-dependent way by mononuclear phagocytes. In addition, it is stored in granular compartments in neutrophils. PTX3 is an essential component of resistance to selected pathogens such as *Aspergillus fumigatus* and *Pseudomonas aeruginosa*. The effector mechanisms utilized by PTX3 include recognition and binding to microbial components, activation and regulation of the complement cascade, and opsonization-mediated destruction of pathogens (26, 28, 29). In addition, PTX3 binds P-selectin and reduces neutrophil recruitment to sites of inflammation, dampening inflammation (27, 30). Thus, PTX3 and other soluble

PRMs produced by phagocytes serve as an amplification and regulatory loop in the phagocyte-mediated resistance and tissue damage.

The regulation of macrophage function in the lungs provides a good example of long-term conditioning of macrophage responsiveness (31, 32). After resolution of viral infections (influenza, respiratory syncytial virus), long-lasting lung macrophage desensitization was observed (32). Interestingly, unresponsiveness was mediated by TLR4 desensitization, a finding reminiscent of previous *in vitro* and *in vivo* studies (17, 33). On the other hand, after exposure to allergic bronchial inflammation, alveolar macrophages show increased inflammatory responses to TLR ligands, with acquisition of the capacity to produce IFN-β (31). Thus, infectious or inflammatory conditions may imprint lung macrophages for attenuated or increased responsiveness. Defining the molecular basis of adaptive responses of lung macrophages may have implications for better understanding of the pathogenesis of diverse clinical conditions.

MOLECULAR MECHANISMS DRIVING THE ADAPTIVE CHARACTERISTICS OF INNATE IMMUNITY

The importance of immune adaptation as an evolutionarily driven property of complex defense systems is underlined by the discovery of an independently developed second type of adaptive immune system in jawless vertebrates, based on variable lymphocyte receptors (34). Complementary to these phylogenetic arguments, studies investigating the ontology of the immune system have demonstrated that TLR-induced maturation of innate immunity (as assessed by cytokine profiles) is an adaptive feature of mammalian host defense that is designed to reduce the impact and severity of subsequent infections (35, 36).

Building on these evolutionary arguments, an overview of the rich body of research documenting adaptive traits of innate immunity in plants and invertebrates provides a first set of strong arguments for innate immune memory and the molecular mechanisms that drive it (8, 11). Systemic acquired resistance (SAR) has been described as the central process providing protection against reinfection in plants. The study of the biochemical mechanisms mediating SAR has provided the first clues that epigenetic processes are crucial for innate immune memory (37, 38). Interestingly, epigenetic changes have been shown to be able to provide transgenerational transmission of resistance, with the acetylation of histone 3 lysine 9 (H3K9) being central in this process (39). The epigenetic programs induced in plants during SAR represent a first clue regarding the mechanisms that induce long-term changes in the functional phenotype of innate immune cells.

How the function of macrophages is modulated by microbial stimuli is an important aspect in determining host defense. Recent studies have reported that the dose of LPS determines whether tolerance (by high LPS concentrations) or priming (by ultra-low LPS concentrations) is attained when macrophages are stimulated by activation of different signaling pathways (40). This dual role of priming and tolerance depending on the concentration applies not only to LPS but also to other ligands (41). Epigenetic mechanisms were shown to be central in the process of LPS tolerance (42), and links with immunometabolism become more apparent. NAD^+ production is an important signal leading to activation of sirtuins (histone deacetylases), which are key regulators in the epigenetic regulation of metabolism of monocytes (43, 44). Important new insights shed more light on epigenetic regulation during cell stimulation: several types of epigenetic modifications involving both histone methylation and acetylation are induced by LPS in a cell type- and stimulus-specific manner (45), with the most pronounced role for H3K27 acetylation (ac) (46). However, while most of these markers fade in time, histone methylation at H3K4 monomethylation (me1) in so-called latent enhancers remains present as an epigenetic marker of this process, conferring long-term immunological memory (45).

The importance of histone methylation and acetylation for the epigenetic reprogramming of monocytes has been also demonstrated following exposure to β-glucans, leading to an increased response to a secondary stimulation (14, 46) (Fig. 2). Trained innate immunity induced *in vivo* through epigenetic reprogramming results in protection of T/B cell-deficient *Scid* and *Rag1*$^{-/-}$ mice against lethal systemic candidiasis, underlining the therapeutic potential of the adaptive characteristics induced in macrophages. An important aspect observed in these studies is the lack of specificity conferred by trained immunity, in which protection is given against not only the original microorganism but additional nonrelated microorganisms as well. This was mediated by an increased phosphorylation of the p38 mitogen-activated protein kinase pathway and epigenetic changes promoting transcription in several genes, among others, PRRs, signaling molecules, and cytokines (14). A whole-genome mapping of histone changes induced by β-glucan showed the importance of induction of H3K27ac and to a lesser extent H3K4 trimethylation (me3). Cluster analysis of genes that

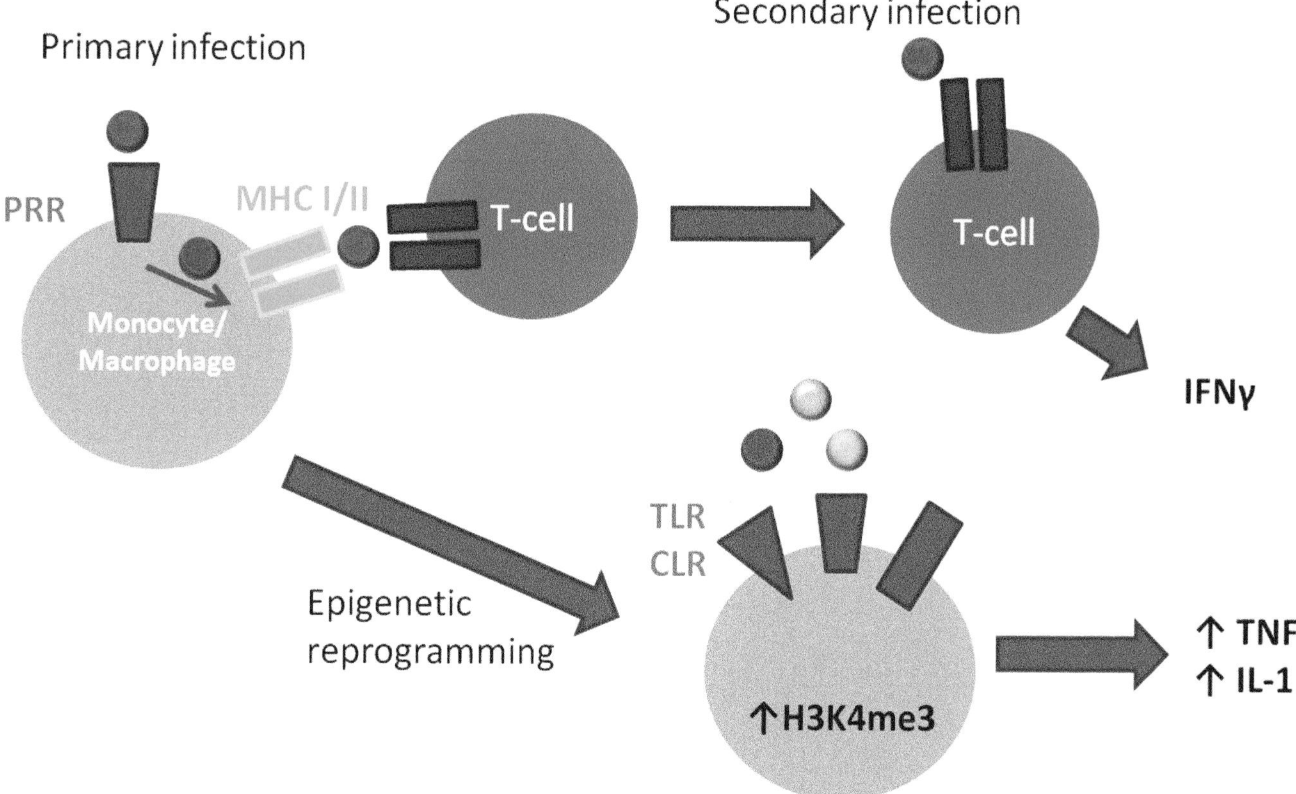

Figure 2 Macrophages play a double role in inducing immunological memory following an infectious insult: on the one hand, they initiate adaptive immune responses; and on the other hand, they undergo epigenetic reprogramming to respond with an increased array of PRR expression and inflammatory cytokine production to a secondary infection ("trained innate immunity"). MHC, major histocompatibility complex.

were epigenetically most modified revealed an upregulation of the mTOR (mammalian target of rapamycin) pathway and cell metabolism, especially glycolysis (46). Stimulation of monocytes with β-glucan was shown to induce the Akt/mTOR/hypoxia-inducible factor-1α (HIF1α) pathway. HIF1α is an important transcriptional regulator of metabolism, and in the β-glucan-trained monocytes it was shown to induce anaerobic glycolysis (which results in lactate production) and reduce activity of the tricarboxylic acid cycle. This shift from oxidative phosphorylation (OxPhos) to anaerobic glycolysis is known as the Warburg effect and results in fast production of ATP. It was first described in cancer cells, and later also in activated macrophages (47, 48). Apparently, trained monocytes rely on this shift to ensure their energy requirements, thus enhancing their effector functions (49).

To conclude, these data suggest a picture in which the innate immune system is characterized by adaptive features and can be trained to provide partial protection against infection independent of the classic T/B-cell adaptive immunity. Functional reprogramming, especially through epigenetically and metabolically mediated mechanisms, mediates these effects.

CONSEQUENCES FOR HUMAN DISEASES

What are the consequences of the capacity of innate immunity to build immunological memory? Despite the data demonstrating the role of innate immune memory in plants and invertebrates, one may speculate that in mammals the biological relevance of the adaptive characteristics of innate immunity was lost during evolution, because classic T/B-cell-dependent adaptive immunity conferred the specificity needed during reinfection. However, there are several arguments to support that trained immunity remains an important component of host defense. For example, the commensalism of certain microorganisms with the mammalian host may mirror their function in inducing nonspecific immune protection. It is tempting to hypothesize that this may be the case for *Candida albicans*, a very common colonizer of human skin and mucosa that strongly induces trained immunity (14), and this has also been suggested for herpesvirus latency (50).

In addition, inducing an adaptive response of innate immunity is likely to have important therapeutic potential. Vaccines such as *Mycobacterium bovis* BCG have been shown to nonspecifically protect animals against fungal and bacterial sepsis (51), and BCG can nonspecifically protect mice against influenza infection (52). If this effect was observed in humans, this would represent an unorthodox approach to vaccination in situations in which specific vaccines are not (yet) available, such as in a global pandemic with a novel influenza virus. Indeed, nonspecific protective effects against nonmycobacterial infections have been reported both to be very quick and to last for years when children are vaccinated with live vaccines such as BCG or measles vaccine (53, 54), and the number and strength of these epidemiologic data sets strongly advocate a sustained effort to understand trained immunity and its potential therapeutic effects. In addition, not only could children in developing countries benefit from this approach, but also children in developed countries such as Spain and Denmark seem to benefit from vaccination with live attenuated vaccines, with lower hospital admission rates and fewer cases of respiratory syncytial virus infections among vaccinated children (55–57). Furthermore, it was very recently shown that trained immunity with BCG is dependent on autophagy. Certain single nucleotide polymorphisms in autophagy genes reduced the training capacity of BCG. Subsequently, the recurrence-free survival of bladder carcinoma patients who were treated with BCG instillations was associated with these same single nucleotide polymorphisms (58).

The results and concepts discussed here call for monitoring of the impact of new adjuvants on innate immunity and resistance to unrelated microorganisms. The challenge for the coming years is thus to deconstruct the molecular mechanisms mediating the adaptive characteristics of macrophages and other innate immune cells, as well as to harness these protective effects in clinical practice. Inducing trained innate immunity may be advantageous in various clinical conditions, from large-scale vaccination programs in settings of high infectious pressure in developing countries, to vaccination of patients lacking a functional immune system (e.g., HIV patients), to the reversal of immunoparalysis in sepsis.

In conclusion, the adaptive characteristics of innate immunity in general, and of macrophages in particular, have emerged as an important new property of innate host defense mechanisms. Its study in the coming years promises to become an area of very active immunological research, with a direct impact on our understanding of immune responses and on the design of novel immunotherapies.

Citation. Arts RJW, Netea MG. 2016. Adaptive characteristics of innate immune responses in macrophages. Microbiol Spectrum4(4):MCHD-0023-2015.

References

1. **Gordon S, Mantovani A.** 2011. Diversity and plasticity of mononuclear phagocytes. *Eur J Immunol* **41:**2470–2472.
2. **Taylor PR, Martinez-Pomares L, Stacey M, Lin HH, Brown GD, Gordon S.** 2005. Macrophage receptors and immune recognition. *Annu Rev Immunol* **23:**901–944.
3. **Takeuchi O, Akira S.** 2010. Pattern recognition receptors and inflammation. *Cell* **140:**805–820.
4. **Mackaness GB.** 1969. The influence of immunologically committed lymphoid cells on macrophage activity in vivo. *J Exp Med* **129:**973–992.
5. **Adams DO, Hamilton TA.** 1984. The cell biology of macrophage activation. *Annu Rev Immunol* **2:**283–318.
6. **Evans R, Alexander P.** 1970. Cooperation of immune lymphoid cells with macrophages in tumour immunity. *Nature* **228:**620–622.
7. **Bowdish DM, Loffredo MS, Mukhopadhyay S, Mantovani A, Gordon S.** 2007. Macrophage receptors implicated in the "adaptive" form of innate immunity. *Microbes Infect* **9:**1680–1687.
8. **Netea MG, Quintin J, van der Meer JW.** 2011. Trained immunity: a memory for innate host defense. *Cell Host Microbe* **9:**355–361.
9. **Durrant WE, Dong X.** 2004. Systemic acquired resistance. *Annu Rev Phytopathol* **42:**185–209.
10. **Pham LN, Dionne MS, Shirasu-Hiza M, Schneider DS.** 2007. A specific primed immune response in *Drosophila* is dependent on phagocytes. *PLoS Pathog* **3:**e26. doi: 10.1371/journal.ppat.0030026.g001.
11. **Kurtz J.** 2005. Specific memory within innate immune systems. *Trends Immunol* **26:**186–192.
12. **Sun JC, Beilke JN, Lanier LL.** 2009. Adaptive immune features of natural killer cells. *Nature* **457:**557–561.
13. **Paust S, Gill HS, Wang BZ, Flynn MP, Moseman EA, Senman B, Szczepanik M, Telenti A, Askenase PW, Compans RW, von Andrian UH.** 2010. Critical role for the chemokine receptor CXCR6 in NK cell-mediated antigen-specific memory of haptens and viruses. *Nat Immunol* **11:**1127–1135.
14. **Quintin J, Saeed S, Martens JHA, Giamarellos-Bourboulis EJ, Ifrim DC, Logie C, Jacobs L, Jansen T, Kullberg BJ, Wijmenga C, Joosten LAB, Xavier RJ, van der Meer JWM, Stunnenberg HG, Netea MG.** 2012. *Candida albicans* infection affords protection against reinfection via functional reprogramming of monocytes. *Cell Host Microbe* **12:**223–232.
15. **Nathan CF, Prendergast TJ, Wiebe ME, Stanley ER, Platzer E, Remold HG, Welte K, Rubin BY, Murray HW.**

1984. Activation of human macrophages. Comparison of other cytokines with interferon-γ. *J Exp Med* **160:** 600–605.

16. **Meltzer MS, Occhionero M, Ruco LP.** 1982. Macrophage activation for tumor cytotoxicity: regulatory mechanisms for induction and control of cytotoxic activity. *Fed Proc* **41:**2198–2205.
17. **Bosisio D, Polentarutti N, Sironi M, Bernasconi S, Miyake K, Webb GR, Martin MU, Mantovani A, Muzio M.** 2002. Stimulation of toll-like receptor 4 expression in human mononuclear phagocytes by interferon-γ: a molecular basis for priming and synergism with bacterial lipopolysaccharide. *Blood* **99:**3427–3431.
18. **Stein M, Keshav S, Harris N, Gordon S.** 1992. Interleukin 4 potently enhances murine macrophage mannose receptor activity: a marker of alternative immunologic macrophage activation. *J Exp Med* **176:**287–292.
19. **Sica A, Mantovani A.** 2012. Macrophage plasticity and polarization: in vivo veritas. *J Clin Invest* **122:**787–795.
20. **Martinez FO, Gordon S, Locati M, Mantovani A.** 2006. Transcriptional profiling of the human monocyte-to-macrophage differentiation and polarization: new molecules and patterns of gene expression. *J Immunol* **177:** 7303–7311.
21. **Monticelli S, Natoli G.** 2013. Short-term memory of danger signals and environmental stimuli in immune cells. *Nat Immunol* **14:**777–784.
22. **Medzhitov R, Schneider DS, Soares MP.** 2012. Disease tolerance as a defense strategy. *Science* **335:**936–941.
23. **Akey JM.** 2009. Constructing genomic maps of positive selection in humans: where do we go from here? *Genome Res* **19:**711–722.
24. **Biswas SK, Lopez-Collazo E.** 2009. Endotoxin tolerance: new mechanisms, molecules and clinical significance. *Trends Immunol* **30:**475–487.
25. **Willment JA, Lin HH, Reid DM, Taylor PR, Williams DL, Wong SY, Gordon S, Brown GD.** 2003. Dectin-1 expression and function are enhanced on alternatively activated and GM-CSF-treated macrophages and are negatively regulated by IL-10, dexamethasone, and lipopolysaccharide. *J Immunol* **171:**4569–4573.
26. **Ricklin D, Lambris JD.** 2013. Complement in immune and inflammatory disorders: pathophysiological mechanisms. *J Immunol* **190:**3831–3838.
27. **Hajishengallis G, Chavakis T.** 2013. Endogenous modulators of inflammatory cell recruitment. *Trends Immunol* **34:**1–6.
28. **Bottazzi B, Doni A, Garlanda C, Mantovani A.** 2010. An integrated view of humoral innate immunity: pentraxins as a paradigm. *Annu Rev Immunol* **28:**157–183.
29. **Jeannin P, Bottazzi B, Sironi M, Doni A, Rusnati M, Presta M, Maina V, Magistrelli G, Haeuw JF, Hoeffel G, Thieblemont N, Corvaia N, Garlanda C, Delneste Y, Mantovani A.** 2005. Complexity and complementarity of outer membrane protein A recognition by cellular and humoral innate immunity receptors. *Immunity* **22:**551–560.
30. **Deban L, Russo RC, Sironi M, Moalli F, Scanziani M, Zambelli V, Cuccovillo I, Bastone A, Gobbi M, Valentino S, Doni A, Garlanda C, Danese S, Salvatori G, Sassano M, Evangelista V, Rossi B, Zenaro E, Constantin G, Laudanna C, Bottazzi B, Mantovani A.** 2010. Regulation of leukocyte recruitment by the long pentraxin PTX3. *Nat Immunol* **11:**328–334.
31. **Naessens T, Vander Beken S, Bogaert P, Van Rooijen N, Lienenklaus S, Weiss S, De Koker S, Grooten J.** 2012. Innate imprinting of murine resident alveolar macrophages by allergic bronchial inflammation causes a switch from hypoinflammatory to hyperinflammatory reactivity. *Am J Pathol* **181:**174–184.
32. **Didierlaurent A, Goulding J, Patel S, Snelgrove R, Low L, Bebien M, Lawrence T, van Rijt LS, Lambrecht BN, Sirard JC, Hussell T.** 2008. Sustained desensitization to bacterial Toll-like receptor ligands after resolution of respiratory influenza infection. *J Exp Med* **205:**323–329.
33. **Kleinnijenhuis J, Quintin J, Preijers F, Joosten LA, Ifrim DC, Saeed S, Jacobs C, van Loenhout J, de Jong D, Stunnenberg HG, Xavier RJ, van der Meer JW, van Crevel R, Netea MG.** 2012. Bacille Calmette-Guerin induces NOD2-dependent nonspecific protection from reinfection via epigenetic reprogramming of monocytes. *Proc Natl Acad Sci U S A* **109:**17537–17542.
34. **Boehm T, McCurley N, Sutoh Y, Schorpp M, Kasahara M, Cooper MD.** 2012. VLR-based adaptive immunity. *Annu Rev Immunol* **30:**203–220.
35. **Levy O.** 2007. Innate immunity of the newborn: basic mechanisms and clinical correlates. *Nat Rev Immunol* **7:** 379–390.
36. **Philbin VJ, Dowling DJ, Gallington LC, Cortes G, Tan Z, Suter EE, Chi KW, Shuckett A, Stoler-Barak L, Tomai M, Miller RL, Mansfield K, Levy O.** 2012. Imidazoquinoline Toll-like receptor 8 agonists activate human newborn monocytes and dendritic cells through adenosine-refractory and caspase-1-dependent pathways. *J Allergy Clin Immunol* **130:**195–204.e9. doi:10.1016/j.jaci.2012.02.042.
37. **van den Burg HA, Takken FL.** 2009. Does chromatin remodeling mark systemic acquired resistance? *Trends Plant Sci* **14:**286–294.
38. **Conrath U.** 2011. Molecular aspects of defence priming. *Trends Plant Sci* **16:**524–531.
39. **Slaughter A, Daniel X, Flors V, Luna E, Hohn B, Mauch-Mani B.** 2012. Descendants of primed *Arabidopsis* plants exhibit resistance to biotic stress. *Plant Physiol* **158:**835–843.
40. **Maitra U, Deng H, Glaros T, Baker B, Capelluto DG, Li Z, Li L.** 2012. Molecular mechanisms responsible for the selective and low-grade induction of proinflammatory mediators in murine macrophages by lipopolysaccharide. *J Immunol* **189:**1014–1023.
41. **Ifrim DC, Quintin J, Joosten LA, Jacobs C, Jansen T, Jacobs L, Gow NA, Williams DL, van der Meer JW, Netea MG.** 2014. Trained immunity or tolerance: opposing functional programs induced in human monocytes after engagement of various pattern recognition receptors. *Clin Vaccine Immunol* **21:**534–545.
42. **Foster SL, Hargreaves DC, Medzhitov R.** 2007. Gene-specific control of inflammation by TLR-induced chromatin modifications. *Nature* **447:**972–978.

43. **Liu TF, Yoza BK, El Gazzar M, Vachharajani VT, McCall CE.** 2011. NAD+-dependent SIRT1 deacetylase participates in epigenetic reprogramming during endotoxin tolerance. *J Biol Chem* **286:**9856–9864.

44. **Liu TF, Vachharajani VT, Yoza BK, McCall CE.** 2012. NAD+-dependent sirtuin 1 and 6 proteins coordinate a switch from glucose to fatty acid oxidation during the acute inflammatory response. *J Biol Chem* **287:**25758–25769.

45. **Ostuni R, Piccolo V, Barozzi I, Polletti S, Termanini A, Bonifacio S, Curina A, Prosperini E, Ghisletti S, Natoli G.** 2013. Latent enhancers activated by stimulation in differentiated cells. *Cell* **152:**157–171.

46. **Saeed S, Quintin J, Kerstens HH, Rao NA, Aghajanirefah A, Matarese F, Cheng SC, Ratter J, Berentsen K, van der Ent MA, Sharifi N, Janssen-Megens EM, Ter Huurne M, Mandoli A, van Schaik T, Ng A, Burden F, Downes K, Frontini M, Kumar V, Giamarellos-Bourboulis EJ, Ouwehand WH, van der Meer JW, Joosten LA, Wijmenga C, Martens JH, Xavier RJ, Logie C, Netea MG, Stunnenberg HG.** 2014. Epigenetic programming of monocyte-to-macrophage differentiation and trained innate immunity. *Science* **345:**1251086. doi:10.1126/science.1251086.

47. **Warburg O, Wind F, Negelein E.** 1927. The metabolism of tumors in the body. *J Gen Physiol* **8:**519–530.

48. **Rodríguez-Prados JC, Través PG, Cuenca J, Rico D, Aragonés J, Martín-Sanz P, Cascante M, Boscá L.** 2010. Substrate fate in activated macrophages: a comparison between innate, classic, and alternative activation. *J Immunol* **185:**605–614.

49. **Cheng SC, Quintin J, Cramer RA, Shepardson KM, Saeed S, Kumar V, Giamarellos-Bourboulis EJ, Martens JH, Rao NA, Aghajanirefah A, Manjeri GR, Li Y, Ifrim DC, Arts RJ, van der Veer BM, Deen PM, Logie C, O'Neill LA, Willems P, van de Veerdonk FL, van der Meer JW, Ng A, Joosten LA, Wijmenga C, Stunnenberg HG, Xavier RJ, Netea MG.** 2014. mTOR- and HIF-1α-mediated aerobic glycolysis as metabolic basis for trained immunity. *Science* **345:**1250684. doi:10.1126/science.1250684.

50. **Barton ES, White DW, Cathelyn JS, Brett-McClellan KA, Engle M, Diamond MS, Miller VL, Virgin HW IV.** 2007. Herpesvirus latency confers symbiotic protection from bacterial infection. *Nature* **447:**326–329.

51. **van 't Wout JW, Poell R, van Furth R.** 1992. The role of BCG/PPD-activated macrophages in resistance against systemic candidiasis in mice. *Scand J Immunol* **36:**713–719.

52. **Spencer JC, Ganguly R, Waldman RH.** 1977. Nonspecific protection of mice against influenza virus infection by local or systemic immunization with Bacille Calmette-Guérin. *J Infect Dis* **136:**171–175.

53. **Aaby P, Roth A, Ravn H, Napirna BM, Rodrigues A, Lisse IM, Stensballe L, Diness BR, Lausch KR, Lund N, Biering-Sørensen S, Whittle H, Benn CS.** 2011. Randomized trial of BCG vaccination at birth to low-birth-weight children: beneficial nonspecific effects in the neonatal period? *J Infect Dis* **204:**245–252.

54. **Roth AE, Stensballe LG, Garly ML, Aaby P.** 2006. Beneficial non-targeted effects of BCG—ethical implications for the coming introduction of new TB vaccines. *Tuberculosis (Edinb)* **86:**397–403.

55. **Lopez MJ, Pardo-Seco JJ, Martinon-Torres F.** 2015. Nonspecific (heterologous) protection of neonatal BCG vaccination against hospitalization due to respiratory infection and sepsis. *Clin Infect Dis* **60:**1611–1619.

56. **Sørup S, Benn CS, Poulsen A, Krause TG, Aaby P, Ravn H.** 2014. Live vaccine against measles, mumps, and rubella and the risk of hospital admissions for nontargeted infections. *JAMA* **311:**826–835.

57. **Sørup S, Benn CS, Stensballe LG, Aaby P, Ravn H.** 2015. Measles-mumps-rubella vaccination and respiratory syncytial virus-associated hospital contact. *Vaccine* **33:**237–245.

58. **Buffen K, Oosting M, Quintin J, Ng A, Kleinnijenhuis J, Kumar V, van de Vosse E, Wijmenga C, van Crevel R, Oosterwijk E, Grotenhuis AJ, Vermeulen SH, Kiemeney LA, van de Veerdonk FL, Chamilos G, Xavier RJ, van der Meer JW, Netea MG, Joosten LA.** 2014. Autophagy controls BCG-induced trained immunity and the response to intravesical BCG therapy for bladder cancer. *PLoS Pathog* **10:**e1004485. doi:10.1371/journal.ppat.1004485.

Myeloid Cells in Health and Disease: A Synthesis
Edited by Siamon Gordon

doi:10.1128/microbiolspec.MCHD-0047-2016

Theodore J. Sanders[1]
Ulf Yrlid[2]
Kevin J. Maloy[1]

39 Intestinal Mononuclear Phagocytes in Health and Disease

INTRODUCTION

The intestine constitutes a crucial interface for the acquisition of essential nutrients. In addition, the gut is colonized by a huge and diverse microbiota that confers several benefits to the host, including improved nutrient absorption and protection from pathogenic invasion. Thus, the intestine is the tissue of the body with the highest constitutive exposure to foreign antigen and is also a common entry portal for many local and systemic pathogens. Therefore, the local immune system, known as the gastrointestinal lymphoid tissue, has the unenviable task of balancing efficient responses to dangerous pathogens with tolerance toward beneficial microbiota and food antigens. As in most tissues, the decision between tolerance and immunity is critically governed by the activity of local myeloid cells. However, the unique challenges posed by the intestinal environment have necessitated the development of several specialized myeloid populations with distinct phenotypic and functional characteristics that have vital roles in maintaining barrier function and immune homeostasis in the intestine.

Intestinal mononuclear phagocyte populations, comprising dendritic cells (DCs) and macrophages (Mϕ), are crucial for raising appropriate active immune responses against ingested pathogens. These two subsets of phagocytic cells have distinct and complementary roles in intestinal homeostasis. Following microbial activation, DCs rapidly migrate from the intestinal wall to the draining lymph nodes to activate naive T cells that both aid B cells to generate high-affinity antibody responses and differentiate into effector T cells with intestinal-homing properties. In contrast, Mϕ do not enter afferent lymphatics and instead perform innate immune effector functions through phagocytosis and destruction of bacteria, as well as secretion of cytokines that promote leukocyte recruitment, thereby driving the local inflammatory response. Importantly, DCs and Mϕ also collaborate to maintain tolerance against innocuous food antigens and beneficial commensals. DCs continuously migrate from the intestine to the lymph node and induce T cells with regulatory functions (Tregs). Under steady-state conditions, Mϕ in the intestinal wall secrete cytokines that expand these Tregs

[1]Sir William Dunn School of Pathology, Oxford, OX1 3RE, United Kingdom; [2]Department of Microbiology and Immunology, Institute of Biomedicine, University of Gothenburg, S-405 30 Gothenburg, Sweden.

and also release factors that help maintain the integrity of the epithelial barrier.

Intestinal DCs and Mϕ share several surface receptors, many of which have been routinely used to distinguish these cells in other tissues. This has caused considerable confusion, but recent technical advances, including microsurgical approaches allowing collection of cells migrating in intestinal lymph, intravital microscopy, and novel gene-targeting approaches, have led to clearer distinctions between mononuclear phagocyte populations in intestinal tissue. In this review, we present an overview of the various subpopulations of intestinal mononuclear phagocytes and discuss their phenotypic and functional characteristics. We also outline their roles in host protection from infection and their regulatory functions in maintaining immune tolerance toward beneficial intestinal antigens.

INTESTINAL DCs

Subsets of Intestinal DCs in Mice and Humans

Murine DCs in lymphoid tissues have historically been defined as cells expressing the integrin CD11c and high surface levels of major histocompatibility complex class II (MHC-II). In the intestinal wall, this is not sufficient to discriminate DCs, as intestinal Mϕ may also express these markers (1). However, in contrast to Mϕ, intestinal DCs do not express high levels of the chemokine CX3C receptor 1 (CX3CR1) or the pan-Mϕ markers F4/80 or FcγRI (CD64) (2–4). Intestinal DCs constitute a heterogeneous population of cells with the majority expressing the αε integrin (CD103) (5). Moreover, murine intestinal CD103$^+$ DCs can be further subdivided based on coexpression of CD11b (with CD172a/SIRPα [signal regulatory protein α] being used similarly in humans) (5, 6) (Fig. 1). In the murine small intestinal lamina propria (SI-LP), the CD103$^+$CD11b$^+$ DC subset predominates, whereas in the colonic LP, CD103$^+$ CD11b$^-$ DCs are the most prevalent (7–9).

The CD103$^+$CD11b$^-$ LP DCs may also express CD8α (9), which initially led to speculation that these cells were derived from Peyer's patches and isolated lymphoid follicles, as CD8α$^+$ DCs were originally identified in lymphoid tissues. However, CD103$^+$CD11b$^-$ CD8α$^+$ DCs were also found to migrate in intestinal lymph, and the frequency of CD11b$^-$CD8α$^+$ LP DCs was not reduced in mice deficient in the transcription factor RAR-related orphan receptor γt (RORγt), which lack secondary lymphoid tissues (4). The CD103$^+$ CD11b$^-$CD8α$^+$ intestinal lymph DCs, like lymph node-resident CD8α$^+$ DCs and CD103$^+$CD11b$^-$ LP DCs, were dramatically reduced in BatF3-deficient mice (10). However, the expression of CD8α by CD103$^+$CD11b$^-$ lymph DCs was generally lower than that observed on lymph node-resident CD11b$^-$ DCs (10), which could explain why CD8α is not consistently detected on intestinal LP DCs. Furthermore, Toll-like receptor (TLR) 7/8 ligand administration led to a dramatic increase in all intestinal lymph migratory DCs, including the CD103$^+$CD11b$^-$(CD172a$^-$) subset, but this was not accompanied by any loss of DCs from Peyer's patches (10, 11). Hence, CD103$^+$CD11b$^-$ intestinal DCs expressing intermediate levels of CD8α appear to constitute a bona fide LP-resident DC population.

In addition to the two subsets of CD103$^+$ intestinal DCs, CD11c$^+$MHC-II$^+$ cells that express intermediate levels of CX3CR1 but not CD103 have recently been described (2, 4, 12). These CD11c$^+$CX3CR1^{int} cells are a heterogeneous population comprising mainly a differentiation continuum of monocytes to Mϕ, but a subpopulation of CD11c$^+$CX3CR1^{int} cells lacking F4/80 and CD64 expression have been identified in murine intestinal lymph (4, 12). These cells also express CCR7 and the DC-restricted transcription factor Zbt46, and expanded *in vivo* in response to Flt3L (Fms-like tyrosine kinase 3 ligand), indicating that a proportion of the intestinal CD11c$^+$CX3CR1^{int} cells are by definition DCs (4, 12, 13). This is further supported by observations that CD103$^-$CD11c$^+$CX3CR1^{int} cells are migratory and found in intestinal lymph under steady state (4, 12, 13). CD103$^-$ lymph DCs can be further subdivided based on CD11b expression, and the CD11b$^-$ subset might be derived from intestinal lymphoid tissues, as it is significantly reduced in RORγt-deficient mice (4).

The study of human intestinal DCs from unperturbed tissue is limited. Recently, however, DCs enriched from proximal human small intestine were obtained from patients undergoing bariatric surgery and compared with murine SI-LP DCs. Based on expression of CD103 and CD172a, this study identified three distinct subsets of human SI-LP DCs: CD103$^+$ CD172a$^+$, CD103$^+$CD172a$^-$, and CD103$^-$CD172a$^+$ (6). Human DC subsets in other tissues have been defined by CD141 and CD1c expression, and extended transcriptional analysis revealed that the CD103$^+$CD172a$^-$ human intestinal DCs were related to human blood CD141$^+$ DCs and to mouse intestinal CD103$^+$CD11b$^-$ DCs (6). Conversely, human intestinal CD103$^+$CD172a$^+$ DCs aligned with human blood CD1c$^+$ DCs and mouse intestinal CD103$^+$CD11b$^+$ DCs (6). This suggests a conserved program among the intestinal DC subsets in humans and mice. Interestingly, similar changes in

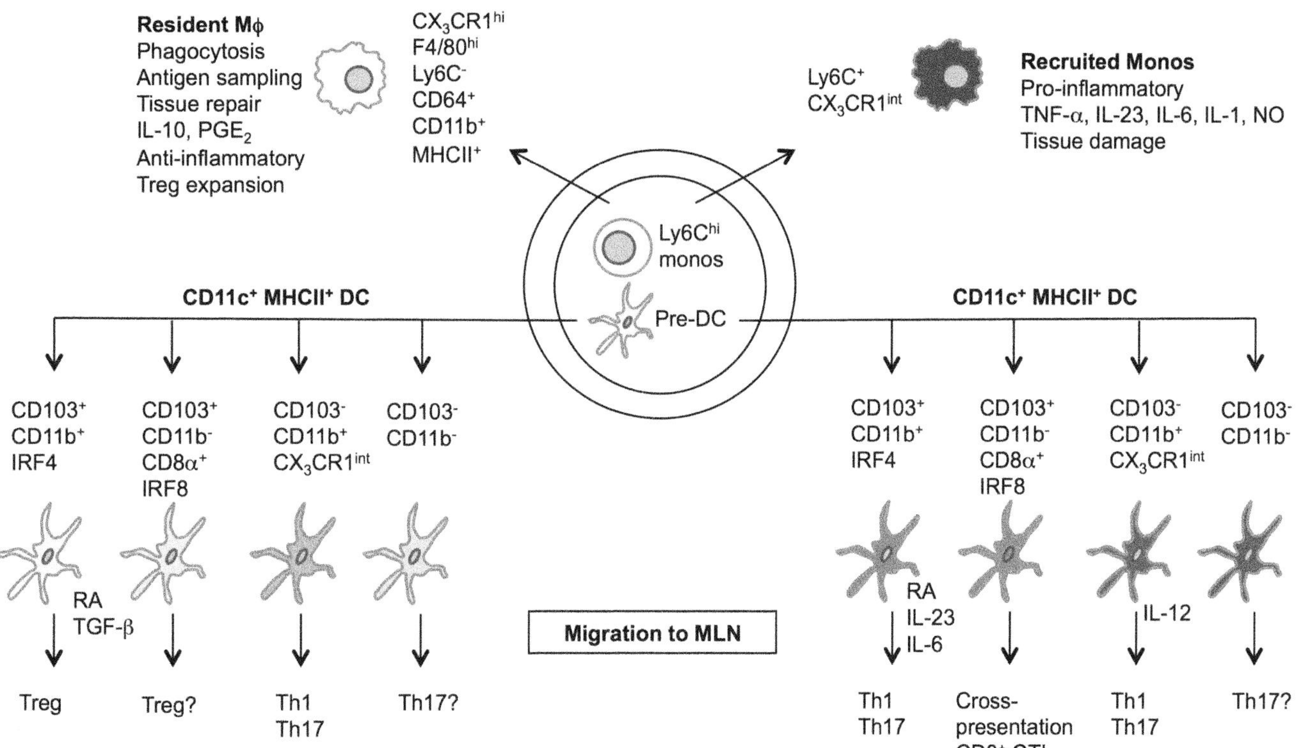

Figure 1 Major subsets of murine mononuclear phagocytes. In the steady-state intestine, resident Mϕ and migratory DCs act synergistically to maintain homeostasis and prevent inflammation. In response to infectious or inflammatory insults, recruited Ly6C$^+$ monocytes and migratory DCs exhibit a proinflammatory phenotype that coordinates protective innate and adaptive immune responses. However, sustained activation of mononuclear phagocytes can drive chronic intestinal inflammation, leading to tissue damage and impaired function. CTL, cytotoxic T lymphocyte; NO, nitric oxide.

relative proportions of DC subsets in human SI-LP and colonic LP were observed, with the CD103$^+$CD172a$^+$ predominating in the small intestine but present at reduced frequencies in the colon (6, 14).

Ontogeny of Intestinal DCs

Intestinal CD103$^+$ DC populations arise from a pre-classical DC (pre-cDC) progenitor and are dependent on Flt3L for their development, as CD103$^+$ LP DCs are absent in mice deficient in either Flt3 or Flt3L (15, 16), and *in vivo* administration of Flt3L dramatically increases the number of both CD103$^+$CD11b$^+$ DCs and CD103$^+$CD11b$^-$ DCs (17). However, the two intestinal CD103$^+$ DC subsets are dependent on different growth factors and transcription factors. For example, the development of CD103$^+$CD11b$^+$ DCs is selectively impaired in mice deficient in colony-stimulating factor-2 receptor (CSF-2R) or Notch-2 (18, 19). In addition, targeted deletion of interferon (IFN) regulatory factor 4 (IRF4) in CD11c-expressing cells results in a selective reduction of CD103$^+$CD11b$^+$ DCs in the intestine and mesenteric lymph nodes (MLNs), suggesting that this factor is important for both their development and survival (20, 21). In contrast, the intestinal CD103$^+$ CD11b$^-$ DCs are, like CD103$^+$CD11b$^-$ DCs in peripheral tissues, dependent on the transcription factors Batf3 and IRF8 for their development (22, 23). As noted above, the intestinal CD11c$^+$CX3CR1^{int} population is rather heterogeneous, and this has complicated analysis of their ontogeny. However, the CD11c$^+$ CX3CR1^{int} DC subset is at least partly derived from a common pre-cDC precursor and proliferates in response to Flt3L (4, 12, 24). Moreover, in a recent study in which intestinal CD11c$^+$CX3CR1^{int} DCs were further subdivided according to CCR2 expression, targeted ablation of IRF4 in all CD11c$^+$ cells resulted in a selective reduction in the CCR2$^+$ fraction of intestinal CD11c$^+$CX3CR1^{int} DCs (25). Taken together, these results suggest that although intestinal DC subsets exhibit some shared ontological requirements, such as Flt3 dependency, the development of distinct intestinal DC subpopulations is controlled through differential activation of transcription factors and growth receptors. Whether and how local environmental signals influence the development of the different populations of intestinal DCs remains to be determined.

Intestinal DCs Migrate at Steady State and Induce T-Cell Tolerance

Intestinal DCs continuously migrate to the draining lymph node, although how this constitutive migration is regulated is still poorly understood (5). The migration of intestinal CD103$^+$ DC subsets is dependent on CCR7, whereas the accumulation of CD11c$^+$ CX3CR1^{int} DCs in the draining MLN was not significantly impaired in CCR7-deficient mice (26). The gut microbiota has a large impact on many intestinal immune functions, and it was therefore somewhat surprising when initial studies suggested that intestinal lymph-borne DC migration to the MLN was not altered in mice lacking TLR signaling, or in germ-free mice (27). However, as described above, the composition of DC subsets migrating from the intestine is complex, and a recent study revealed that deficiency in the TLR signaling adapter molecule myeloid differentiation primary-response protein 88 (MyD88) was associated with a 50 to 60% reduction in migration of intestinal CD103$^+$ DCs at steady state (26). This effect was observed for both the CD103$^+$CD11b$^+$ DC and CD103$^+$ CD11b$^-$ DC subsets, and MyD88 signaling in CD11c$^+$ cells was shown to promote DC migration (26). In addition, by directly collecting intestinal lymph, the same study confirmed that the frequency of migratory intestinal DCs was not altered in germ-free mice (26). Thus, steady-state egress of DCs from the intestine is at least partially mediated by MyD88 signaling, but appears to be independent of the microbiota. This suggests that other stimuli, such as endogenous damage-associated molecular patterns released, for example, from dying intestinal epithelial cells, could trigger MyD88 signaling in intestinal DCs, leading to their exit from the intestine. The steady-state accumulation of DCs in skin-draining lymph nodes has recently been shown to be dependent on NF-κB signaling, but whether this also regulates intestinal DC migration remains to be determined (28).

The main function of migratory intestinal DCs is to induce adaptive immune responses or tolerance in the draining lymph node. In addition to migration via lymph, this also requires the uptake of antigens in the intestinal tissue. In organized intestinal lymphoid tissue, such as the Peyer's patches, the transfer of luminal antigens to underlying DCs is mediated by specialized M cells. In contrast, how DCs in the SI-LP acquire antigen under steady-state conditions is still not completely understood. Unlike CX3CR1$^+$ Mϕ, which have been proposed to extend transepithelial dendrites to access luminal antigens (29, 30), intestinal CD103$^+$ DCs do not utilize transepithelial dendrites to sample particulate luminal antigens at steady state—although they may do this in the presence of certain pathogens, such as *Salmonella* (31). Several mechanisms have recently been suggested to potentially supply CD103$^+$ DCs with luminal antigens under steady state. Goblet cells may act as transporting cells, forming conduits that deliver antigen to the underlying DCs (32). In addition, a recent study also suggested that the porous nature of the mucus layer in the small intestine facilitates access of bacteria to the epithelial surface, whereupon coating with the major mucus protein MUC2 promotes uptake by CD103$^+$ DCs and CX3CR1$^+$ cells through an unusual complex of innate receptors involving galectin-3, Dectin-1 (DC-associated C-type lectin 1), and FcγRIIB (33). Alternatively, it has also been proposed that CX3CR1$^+$ Mϕ that have acquired luminal antigen, possibly through extending transepithelial dendrites, may also transfer antigen to CD103$^+$ DCs via gap junctions (34).

Intestinal CD103$^+$ DCs have a strong capacity to drive the differentiation of immune-suppressive FoxP3$^+$ Tregs (35, 36). This is linked to their dual expression of retinaldehyde dehydrogenase (RALDH), an enzyme that enables the CD103$^+$ DCs to metabolize vitamin A into retinoic acid (RA) (2, 35, 37), together with the $\alpha_v\beta_8$ integrin that allows them to convert latent transforming growth factor β (TGF-β) into the active form (38). Generation of RA by CD103$^+$RALDH$^+$ DCs also explains their ability to induce the expression of gut-homing properties in T cells, such as CCR9 and $\alpha_4\beta_7$ integrin (38–40). In addition, intestinal CD103$^+$ DCs also express indoleamine 2,3-dioxygenase, an enzyme that catalyzes the conversion of tryptophan into immune-suppressive metabolites that inhibit effector T cells and favor Treg cell development (41). Finally, the MUC2-mediated pathway of antigen uptake described above was associated with an anti-inflammatory profile in intestinal DCs, mediated through activation of β-catenin, which inhibited NF-κB activation and proinflammatory cytokine release (33). Consistent with these activities, oral tolerance induction has been shown to be dependent on lymphatic drainage and CCR7-dependent DC migration (42), and ablation of CD103$^+$ DCs also results in lower numbers of intestinal FoxP3$^+$ Tregs (43). As CD103$^+$CD11b$^-$ DCs express the highest levels of RALDH and $\alpha_v\beta_8$, they might be expected to be critical inducers of Tregs (44). However, the capacity to induce FoxP3$^+$ Treg cells does not seem to be linked to a particular subset of CD103$^+$ DCs, as LP FoxP3$^+$ Treg cells were not reduced in mice deficient in either CD103$^+$CD11b$^+$ DCs or CD103$^+$CD11b$^-$ DCs (21, 22). Indeed, it is noteworthy that CD103$^-$ DCs from intestinal lymph also exhibit the capacity to metabolize vitamin A into RA and induce the expression of gut-homing molecules (4). Therefore, it is possible that this subset of migratory DCs may induce FoxP3$^+$ Treg cells, but this remains to be formally demonstrated. In contrast to the redundancy in FoxP3$^+$ Treg induction, targeted deletion of IRF4 in CD11c-expressing cells led to reduction in both CD103$^+$CD11b$^+$ DCs and in intestinal T cells that produce interleukin-17 (IL-17) (21). Nevertheless, intestinal lymph CD103$^-$ DCs collected under steady-state conditions also exhibited the capacity to induce Th17 cells, suggesting that this subset might also contribute to differentiation of Th17 cells in the gut (4).

The impact of intestinal CD103$^+$ DCs on intestinal T-cell homeostasis extends beyond effects on Tregs and Th17 cells. For example, CD103$^+$CD11b$^-$ DCs that migrate in intestinal lymph, like their counterparts in other tissues, have the capacity to cross-present soluble antigens to activate CD8$^+$ T cells, making them unique among intestinal DCs (4). Recently, studies using mice selectively deficient in IRF8 in CD11c$^+$ cells demonstrated that CD103$^+$CD11b$^-$ intestinal DCs play complex, nonredundant roles in intestinal T-cell homeostasis (23). These mice exhibited reduced numbers of small intestine T cells, with a paucity of CD8$^+$ and CD4$^+$CD8$^+$ T cells. In addition, they also showed impaired Th1 responses at steady state and following pathogen challenge (23).

Taken together, these studies indicate that the various subpopulations of intestinal DCs that migrate from the intestine at steady state have both overlapping and distinct functional capabilities (Fig. 1).

INTESTINAL Mϕ—PHENOTYPE AND FUNCTION

Mϕ in the intestinal LP act primarily as innate effector cells. Through engulfment and clearance of bacteria and apoptotic cellular debris, they are key agents in the maintenance of intestinal homeostasis. However, recent evidence indicates that they can also contribute to regulation of adaptive immune responses in the gut.

Murine Intestinal Mϕ in the Steady-State Gut

Mϕ resident in the murine intestine are MHC-II$^+$ and CD11b$^+$ with variable CD11c expression. They are CX3CR1^{hi}, and may be distinguished from CX3CR1^{int} DC populations by expression of F4/80 and the Mϕ-specific marker FcγRI (CD64) (5). They have cytoplasmic vacuoles, abundant cytoplasm, high phagocytic activity, and, unlike DCs, neither migrate in lymphatic vessels nor efficiently activate naive T cells (2, 12, 45).

The extent to which mucosal immune cells including Mϕ acquire antigen directly from the intestinal lumen remains controversial. As noted above, transepithelial dendrite formation by CX3CR1^{hi} cells was originally proposed as a mechanism of antigen acquisition by DCs (29, 46). While these cells are now identified as resident Mϕ (2), the importance of this pathway as key source of antigen uptake from the gut remains to be determined (5). However, strategically positioned beneath the intestinal epithelium, Mϕ are known to take up antigenic material, including particulates like bacteria, as it emerges from the basolateral epithelial surface (47). Indeed, there is some evidence that bacterial debris and other small particulate antigens may transit across the epithelium into the SI-LP, where they may be taken up by local mononuclear phagocytes (48).

Murine intestinal resident Mϕ generate remarkably low levels of proinflammatory cytokines such as tumor necrosis factor α (TNF-α) and IL-6 upon microbial stimulation (49). This is attributed to both low TLR expression and to increased expression of negative regulators of NF-κB signaling, such as A20 and IL-1 receptor-associated kinase M (49, 50). These Mϕ also constitutively produce IL-10, which is further increased upon microbial stimulation (12, 45). *In vivo*, the signals driving this IL-10 production may originate from the microbiota as Mϕ production of IL-10 is significantly reduced in germ-free mice (51). Over all, this suggests a system in which bacterial encroachment into the intestinal mucosa is efficiently cleared by resident Mϕ without generating a damaging inflammatory response (Fig. 1).

Production of IL-10 by intestinal Mϕ has been proposed as an important factor in supporting expansion of Foxp3$^+$ Tregs during oral tolerance induction (52) and in experimental T-cell transfer colitis (53). However, it has been proposed that it is the ability of Mϕ to sense IL-10 through IL-10R, and not the production of

IL-10 itself, which is crucial in the maintenance of intestinal homeostasis (50). Furthermore, abrogation of IL-10 responsiveness in intestinal Mϕ, through ablation of downstream STAT3 signaling, leads to strong inflammatory responses to TLR stimulation and spontaneous colitis (54, 55).

Unlike bona fide DCs, Mϕ do not migrate in the lymphatics to lymph nodes in the steady state, or upon stimulation with TLR ligands (2). Although it has been proposed that antibiotic-induced dysbiosis triggers migration of Mϕ upon *Salmonella* infection, the distinction between Mϕ and CX3CR1^{int} DCs in this work remains controversial (56). Interactions between Mϕ and CD4$^+$ T cells are therefore likely to be limited to the intestinal mucosa. Resident Mϕ have been reported to constitutively produce pro-IL-1β, which has been proposed to play a role in supporting the differentiation of Th17 cells within the mucosa itself (57). While this *Il1b* expression is microbiota and MyD88 dependent, how inflammasomes are activated to induce processing and release of mature IL-1β in the absence of pathogenic infection, and how this occurs in the context of their reduced TLR sensitivity, remain unclear. A recent study suggested that sensing of the intestinal microbiota at steady state led to IL-1β release by intestinal Mϕ, which activated type 3 innate lymphoid cells (ILC3) to secrete granulocyte-macrophage CSF (GM-CSF) (58). In turn, the ILC3-derived GM-CSF stimulated intestinal Mϕ and DCs to produce IL-10 and RA, which promote Treg cell responses (58). Therefore, cross talk between intestinal Mϕ and local ILC3 reinforces tolerogenic pathways in the healthy intestine.

Intestinal Mϕ Are Constantly Replenished from Ly6C^{hi} Blood Monocytes

Mϕ populations resident in most tissues derive from yolk sac and fetal liver precursors and self-replenish throughout adult life. This includes microglia of the central nervous system and liver Kupffer cells (59). Intestinal Mϕ are distinct, however, in that they require constant replenishment from the circulating pool of Ly6C^{hi} monocytes (60). While Ly6C^{lo} monocytes patrol and maintain the vasculature (61, 62), they appear to have no role in replenishing the intestinal Mϕ population (12, 17).

Upon entering the intestinal mucosa in the steady state, Ly6C^{hi} monocytes begin a process of differentiation in which Ly6C expression is lost alongside an increase in expression of MHC-II, F4/80, CD64, and CD11c. They also transition from CX3CR1^{int} to CX3CR1^{hi} during this process, which occurs over the course of 4 days (3, 12). In parallel to changes in surface phenotype, the cells gradually take on the features of resident intestinal Mϕ, such as increased phagocytic activity and reduced responsiveness to TLR stimulation (3, 12). The factors driving this differentiation process remain unclear, but may include TGF-β, CX3CL1, and peroxisome proliferator-activated receptor γ ligands, whereas a role for the microbiota in this process remains controversial (5).

The arrival of Ly6C^{hi} monocytes in the steady-state intestinal mucosa is regulated to a large extent by CCR2-CCL2 signaling. This is supported by depleted intestinal Mϕ numbers observed in *Ccr2*$^{-/-}$ mice (12, 63) and by the almost exclusive repopulation of intestinal Mϕ from wild-type (WT) bone marrow cells in WT:*Ccr2*$^{-/-}$ mixed bone marrow chimera experiments (3). It is important to note that CCR2 also regulates egress of monocytes from the bone marrow (64). Resident Mϕ are a source of CCL2 in the steady state and during inflammation (50), and any additional factors supporting monocyte recruitment remain to be determined. Macrophage CSF (M-CSF or CSF-1) is also required for the differentiation of intestinal Mϕ, as they derive almost exclusively from WT bone marrow in WT:*Csf1r*$^{-/-}$ mixed bone marrow chimera experiments (20), and the absolute number of Ly6C^{hi} monocytes is significantly reduced in *Csf1r*$^{-/-}$ mice (15). Moreover, intestinal Mϕ are depleted in mice treated with anti-CSF-1R antibodies (65). In addition, the transcription factor PU.1 must be expressed at high levels to induce and maintain Mϕ differentiation (66).

Human Intestinal Mϕ

Mϕ resident in the human intestine express MHC-II, CD64, CD68, the scavenger receptor CD163, and low to intermediate levels of the LPS coreceptor CD14 (67). Human intestinal Mϕ have high phagocytic and bactericidal activity and do not produce proinflammatory cytokines or reactive oxygen species (ROS) upon microbial stimulation. Unlike their murine counterparts, human intestinal Mϕ also do not constitutively produce IL-10 (68, 69). Lack of response to TLR stimulation in these cells is attributed to downregulation of key TLR signaling molecules such as MyD88 (70). The steady-state intestinal mucosa also contains a low frequency of CD14$^+$ Mϕ, which generate TNF-α and IL-6 in response to microbial stimulation (71). The conserved response to bacterial encroachment is therefore clearance by phagocytosis and intracellular killing without the induction of chronic inflammation.

Human monocytes cultured in TGF-β develop a hyporesponsive resident Mϕ-like phenotype (69), similar to the differentiation of Ly6C^{hi} monocytes in the steady-state murine intestine. Moreover, the inflamed human intestine contains a significantly increased frequency of $CD14^+$ Mϕ compared with healthy controls (3, 12, 71, 72). These cells share numerous phenotypic features of blood $CD14^+$ monocytes, including expression of CD64 and responsiveness to microbial stimulation (12, 71). Furthermore, direct recruitment of $CD14^+$ monocytes into the inflamed intestinal mucosa of inflammatory bowel disease (IBD) patients has been observed (73). This suggests that, as in the mouse, human $CD14^+$ blood monocytes that are recruited into the intestine may differentiate into a resident or inflammatory Mϕ phenotype depending on the context.

INTESTINAL MONONUCLEAR PHAGOCYTES DURING INFECTION AND INFLAMMATION

Although the precise roles of distinct subsets of mononuclear phagocytes in different immune and inflammatory reactions in the gut are incompletely defined, there has been significant progress in our understanding of the key contributions that these cells make to both protective and pathogenic responses in the intestine.

Mϕ in Intestinal Inflammation and Infection

Even during intestinal inflammation, resident $CX3CR1^{hi}$ intestinal Mϕ show remarkable retention of their steady-state phenotype, including expression of IL-10, and do not secrete inflammatory cytokines (12, 50, 53, 71, 74). Furthermore, depletion of resident Mϕ increases the severity of disease in dextran sulfate sodium-induced (DSS) colitis (75). In addition to their anti-inflammatory role, it has been proposed that resident Mϕ support the resolution phase of inflammation through provision of prostaglandin E_2 (PGE_2), which promotes proliferation of epithelial progenitors (76).

Nevertheless, many studies have reported that monocyte-derived Mϕ are increased in the intestinal mucosa during infection or inflammation (5). However, $Ly6C^{hi}$ monocytes that are drawn into the inflamed gut do not differentiate through to the $CX3CR1^{hi}$ hyporesponsive phenotype, but instead remain $Ly6C^+$ and $CX3CR1^{int}$, and are acutely sensitive to microbial stimulation. As a result, they generate large amounts of proinflammatory cytokines, including IL-6, IL-23, and TNF-α, as well as mediators such as nitric oxide, which promote the inflammatory response (3, 12, 50). Microbial stimulation of $Ly6C^{hi}$ monocytes during DSS colitis leads to the expression and assembly of inflammasome machinery, triggering release of mature IL-1β in response to pathobionts present in the mucosa (77). The key role of $Ly6C^{hi}$ monocytes in driving acute inflammatory responses in the gut is demonstrated by the attenuated disease severity of DSS colitis in *Ccr2*$^{-/-}$ mice, or following blockade of CCR2 in WT mice (49, 50). During intestinal infection, $Ly6C^{hi}$ monocyte-derived proinflammatory Mϕ perform crucial host-protective functions; they can efficiently ingest pathogens and are required for protection from invasive pathogens such as *Salmonella enterica* serovar Typhimurium and *Toxoplasma gondii* (47, 78, 79). $CX3CR1^+$ cells have also been reported to contribute to protection against the attaching/effacing intestinal pathogen *Citrobacter rodentium*, as the IL-23 and IL-1β that they produce activate local ILC3 to secrete IL-22, which plays a critical role during the acute stages of infection (80, 81). Intestinal Mϕ and recruited monocytes may also contribute to protective type 2 immune responses during parasitic infection, as mice lacking CCL2 (which plays a key role in recruitment of $CCR2^+$ monocytes) exhibited susceptibility to *Trichuris muris* infection (82) and liposome-mediated depletion of monocytes and Mϕ impaired clearance of the intestinal nematode *Heligmosomoides polygyrus* (83).

In addition to the induction of potent antimicrobial defenses, it was recently shown that $Ly6C^{hi}$-dependent populations may limit pathology in the terminal ileum during *T. gondii* infection by secreting PGE_2, which inhibits neutrophil inflammatory responses (84). These results suggest that recruited $Ly6C^{hi}$ cells may also mediate important anti-inflammatory activities in the gut to limit infection-associated pathology. During inflammatory conditions, $Ly6C^{hi}$ monocytes may also differentiate into "inflammatory DCs" in the intestine and thus could potentially shape T-cell priming and differentiation in the MLNs (45, 50). However, further studies are required to confirm that these cells act as key primers of pathogenic T-cell responses *in vivo*, because it is difficult to distinguish them from $CD103^-$ DCs (5). In addition, such cells appear in both the colon and MLN independently of CCR7 expression, suggesting that they originate from the bloodstream and do not migrate between sites in the conventional sense (5). Moreover, it should be noted that the precise distinction between "inflammatory monocytes" and "inflammatory DCs" remains to be clarified.

In humans, numerous studies have identified a potential role for $CD14^+$ Mϕ in the pathogenesis of IBD. Microbial stimulation of these cells from IBD patients induces production of TNF-α and IL-23, which is enhanced in Crohn's disease (CD) patients compared with those afflicted by ulcerative colitis (UC) (71). IL-23 produced by $CD14^+$ Mϕ in the inflamed intestinal mucosa of CD patients enhances IFN-γ production by a subset of NK cells (85), and IL-23 also synergizes with TNF-α to enhance production of IFN-γ by CD4$^+$ T cells in the intestinal mucosa (71). CD14$^+$ Mϕ in the inflamed intestinal mucosa of CD patients express high levels of TNF-like factor 1A (TL1A) (86). Synergy between IL-23 and TL1A enhances expression of both IFN-γ and IL-17 by intestinal mucosal T cells in CD patients (86). TL1A expression is further enhanced through stimulation of CD64 with immobilized IgG, triggering abundant production of TNF-α and IL-1β (87). Moreover, both CD14$^+$ monocytes and CD14$^+$ Mϕ in inflamed intestinal mucosa of CD and UC patients express TREM-1 (triggering receptor expressed on myeloid cells 1) (88, 89). While the physiological ligand of TREM-1 is unknown, stimulation of intestinal CD14$^+$ Mϕ from CD and UC patients with anti-TREM-1 monoclonal antibodies significantly increases production of a range of proinflammatory cytokines including TNF-α, IL-6, and IL-1β (89).

Mϕ from the inflamed large intestinal mucosa of CD and UC patients also display elevated production of ROS as compared with healthy controls (90), which may contribute to inflammation in the intestine by damaging the epithelium (91). As circulating human monocytes produce ROS, including hydrogen peroxide and superoxide (92), this further supports the concept that these Mϕ arise from CD14$^+$ monocytes that are recruited into the intestinal mucosa in IBD. CD14$^+$ Mϕ obtained from intestinal mucosa of healthy controls and CD patients are able to activate naive CD4$^+$ T cells *in vitro* (93). However, as human intestinal CD14$^+$ Mϕ do not express CCR7 (71), they are unlikely to migrate into draining lymph nodes and interact with naive T cells. In this manner, they are reminiscent of monocyte-derived "inflammatory DCs" in the murine intestine. Furthermore, depletion of CD14$^+$ cells from a mixed LP cell preparation derived from the inflamed intestinal mucosa of CD patients significantly reduces production of TNF-α following microbial stimulation (71). This suggests that human resident Mϕ conserve their non-responsive phenotype during inflammation, as reported with murine resident Mϕ.

It is important to emphasize that human IBD is a heterogeneous disease and there is evidence that defective bacterial killing by Mϕ and monocytes might contribute to disease susceptibility. Thus, Mϕ and monocytes isolated from patients with CD exhibited decreased ROS production and impaired bacterial phagocytosis and killing (94, 95). It is postulated that this deficiency results in prolonged bacterial persistence leading to excessive proinflammatory cytokine release that exacerbates tissue inflammation (95). The concept of IBD arising as a consequence of impaired bacterial control by phagocytes is supported by findings that many patients suffering from primary immune deficiencies, such as those afflicted by chronic granulomatous disease, frequently develop IBD-like intestinal inflammation (96).

In summary, although resident intestinal Mϕ largely maintain their anti-inflammatory and tissue-repair functions during infection and inflammation, recruited Ly6C^{hi} cells exhibit a proinflammatory phenotype that contributes to protection against pathogens, but also drives pathology and tissue damage during chronic inflammation.

Intestinal DCs in Immunity and Inflammation

Consistent with their role as immune sentinels in peripheral tissues, upon sensing of danger or infection, intestinal DCs can efficiently prime T-cell effector responses. For example, CD103$^+$CD11b$^+$ DCs express TLR5 and secrete IL-23 in response to bacterial flagellin, leading to enhanced priming of Th1 and Th17 responses (16, 97). The rapid release of IL-23 by CD103$^+$ CD11b$^+$ DCs also activates local innate immune defenses, by triggering ILCs to secrete IL-22, which reinforces the epithelial barrier and induces intestinal epithelial cells to produce antimicrobial peptides (16). Similarly, CD103$^+$ intestinal DCs mount strong responses to stimulation through other TLRs and can efficiently prime both CD4$^+$ and CD8$^+$ T-cell responses (4, 9, 98). These DCs also produce RA, which promotes the generation of IgA$^+$ plasma cells in the LP (97).

These types of responses would be expected to confer resistance against some enteric pathogens, and indeed IL-23 production by CD103$^+$CD11b$^+$ DCs has been reported to be required for protection from *C. rodentium* (13). This has been correlated with decreased intestinal Th17 cell responses in the absence of CD103$^+$CD11b$^+$ DCs (25). In addition, during *Salmonella* infection, CD103$^+$CD11b$^+$ DCs are recruited from the LP into the intestinal epithelium, where they may acquire bacterial antigens (31). Conversely, however, other investigators reported that total ablation of CD103$^+$ DCs, using transgenic mice in which the diphtheria toxin receptor was expressed under the control of the human Langerin promoter, did not result in enhanced susceptibility to *C. rodentium* or *Salmonella* infection (43). This finding suggests that other DC subsets are also capable of contributing to these protective responses, which is consistent with recent observations that CD103$^-$ DCs are able to prime effector Th cells in the MLN (4). Again, these proinflammatory activities can in some instances have deleterious consequences for the host, most notably in the context of chronic intestinal inflammation. Indeed, CD103$^+$ DCs that migrate to the MLN during experimental colitis exhibit a proinflammatory phenotype and contribute to pathogenesis by enhancing Th1 and Th17 effector T-cell responses (99). In addition, a recent study found that the mortality observed in IL-23-deficient mice infected with *C. rodentium* was driven by excess secretion of IL-12 by CD103$^+$CD11b$^-$ DCs (100). It was also shown that the IL-23 released by CX3CR1$^+$ Mϕ and/or CD103$^-$CD11b$^+$ DCs during this infection not only stimulates antimicrobial immunity in the gut but also represses IL-12 production by CD103$^+$CD11b$^-$ DCs and thus prevents fatal immunopathology (100). These findings indicate that cross talk between different populations of mononuclear phagocytes is important to balance protective immunity with host immunopathology.

Human intestinal DCs (identified as HLA-DR$^+$ CD11c$^+$ cells negative for markers of other cell lineages) in both CD and UC patients express increased levels of TLR2 and TLR4 compared with healthy controls, and produce increased TNF-α and IL-8 following stimulation with LPS (101, 102). This suggests an increased sensitivity to bacterial stimulation in IBD. However, the level of DC maturation as determined by CD80 and CD86 expression is not significantly different between healthy control and CD patients (103). MLN DCs from CD patients induce a significantly greater level of IFN-γ production by CD4$^+$ T cells compared with DCs from UC patients and noninflamed healthy controls (104). This is attributed to increased IL-23 and reduced IL-10 production by these cells in CD patients (104), and is consistent with the increased production of IFN-γ by LP T cells in CD (105). The studies outlined above analyzed intestinal DCs as an overall group and not at the subset level. More recently, it has been shown that human MLN CD103$^+$ DCs in CD patients retain the ability to induce $\alpha_4\beta_7$ and CCR9 expression on naive CD8$^+$ T cells (40), indicating that the production of RA by RALDH activity in human intestinal CD103$^+$ DCs is conserved between healthy controls and CD patients. Furthermore, RALDH activity of human intestinal DCs may actually be increased in CD while being reduced in UC patients relative to healthy controls (106, 107). Thus, similar to what has been reported in murine models of infection and inflammation, human DCs isolated from the inflamed colons of IBD patients show a more proinflammatory phenotype and prime gut-homing effector T cells that exacerbate tissue pathology.

CONCLUDING REMARKS

Recent findings have greatly advanced our understanding of the ontogeny and functions of the diverse subsets of mononuclear phagocytes found within the gastrointestinal tract. During steady state, intestinal DCs and Mϕ induce and reinforce immunological tolerance toward beneficial antigens present in the intestinal lumen, while remaining poised to respond to signs of infection or tissue damage. When necessary, they coordinate a variety of local innate immune activities and direct appropriate adaptive immune responses, which together usually result in pathogen clearance and restoration of tissue homeostasis. However, when aberrantly induced, or when directed against persistent stimuli, the protective immune circuits triggered by intestinal Mϕ and DCs can drive harmful immunopathology that characterizes chronic intestinal disorders. Over the last decade, there has been a move away from the simplistic binary view of pattern recognition receptors as the crucial determinants of responsiveness of mononuclear phagocytes, toward a more integrated view of these cells acting as sentinels of tissue homeostasis. In the intestine, this entails Mϕ and DCs operating in a complex immunological environment, where signals from several sources, including the microbiota, intestinal epithelial cells, and local innate and adaptive leukocytes, must be integrated to ensure balanced and appropriate immune responses are induced and regulated. Continued progress in our understanding of the complex cross-talk circuits in which intestinal Mϕ and DCs participate should identify new therapeutic avenues for the treatment of chronic intestinal inflammation and lead to improved treatments and vaccines against intestinal infection.

Citation. Sanders TJ, Yrlid U, Maloy KJ. 2017. Intestinal mononuclear phagocytes in health and disease. Microbiol Spectrum 5(1):MCHD-0047-2016.

References

1. **Bradford BM, Sester DP, Hume DA, Mabbott NA.** 2011. Defining the anatomical localisation of subsets of the murine mononuclear phagocyte system using integrin alpha X (Itgax, CD11c) and colony stimulating factor 1 receptor (Csf1r, CD115) expression fails to discriminate dendritic cells from macrophages. *Immunobiology* **216:**1228–1237.
2. **Schulz O, Jaensson E, Persson EK, Liu X, Worbs T, Agace WW, Pabst O.** 2009. Intestinal CD103$^+$, but not CX3CR1$^+$, antigen sampling cells migrate in lymph and serve classical dendritic cell functions. *J Exp Med* **206:** 3101–3114.
3. **Tamoutounour S, Henri S, Lelouard H, de Bovis B, de Haar C, van der Woude CJ, Woltman AM, Reyal Y, Bonnet D, Sichien D, Bain CC, Mowat AM, Reis e Sousa C, Poulin LF, Malissen B, Guilliams M.** 2012.

CD64 distinguishes macrophages from dendritic cells in the gut and reveals the Th1-inducing role of mesenteric lymph node macrophages during colitis. *Eur J Immunol* 42:3150–3166.

4. Cerovic V, Houston SA, Scott CL, Aumeunier A, Yrlid U, Mowat AM, Milling SW. 2013. Intestinal CD103^{-} dendritic cells migrate in lymph and prime effector T cells. *Mucosal Immunol* 6:104–113.
5. Cerovic V, Bain CC, Mowat AM, Milling SW. 2014. Intestinal macrophages and dendritic cells: what's the difference? *Trends Immunol* 35:270–277.
6. Watchmaker PB, Lahl K, Lee M, Baumjohann D, Morton J, Kim SJ, Zeng R, Dent A, Ansel KM, Diamond B, Hadeiba H, Butcher EC. 2014. Comparative transcriptional and functional profiling defines conserved programs of intestinal DC differentiation in humans and mice. *Nat Immunol* 15:98–108.
7. Jakubzick C, Bogunovic M, Bonito AJ, Kuan EL, Merad M, Randolph GJ. 2008. Lymph-migrating, tissue-derived dendritic cells are minor constituents within steady-state lymph nodes. *J Exp Med* 205:2839–2850.
8. Denning TL, Norris BA, Medina-Contreras O, Manicassamy S, Geem D, Madan R, Karp CL, Pulendran B. 2011. Functional specializations of intestinal dendritic cell and macrophage subsets that control Th17 and regulatory T cell responses are dependent on the T cell/APC ratio, source of mouse strain, and regional localization. *J Immunol* 187:733–747.
9. Fujimoto K, Karuppuchamy T, Takemura N, Shimohigoshi M, Machida T, Haseda Y, Aoshi T, Ishii KJ, Akira S, Uematsu S. 2011. A new subset of CD103^{+}CD8α^{+} dendritic cells in the small intestine expresses TLR3, TLR7, and TLR9 and induces Th1 response and CTL activity. *J Immunol* 186:6287–6295.
10. Cerovic V, Houston SA, Westlund J, Utriainen L, Davison ES, Scott CL, Bain CC, Joeris T, Agace WW, Kroczek RA, Mowat AM, Yrlid U, Milling SW. 2015. Lymph-borne CD8α^{+} dendritic cells are uniquely able to cross-prime CD8^{+} T cells with antigen acquired from intestinal epithelial cells. *Mucosal Immunol* 8:38–48.
11. Yrlid U, Cerovic V, Milling S, Jenkins CD, Klavinskis LS, MacPherson GG. 2006. A distinct subset of intestinal dendritic cells responds selectively to oral TLR7/8 stimulation. *Eur J Immunol* 36:2639–2648.
12. Bain CC, Scott CL, Uronen-Hansson H, Gudjonsson S, Jansson O, Grip O, Guilliams M, Malissen B, Agace WW, Mowat AM. 2013. Resident and pro-inflammatory macrophages in the colon represent alternative context-dependent fates of the same Ly6C^{hi} monocyte precursors. *Mucosal Immunol* 6:498–510.
13. Satpathy AT, Briseño CG, Lee JS, Ng D, Manieri NA, Kc W, Wu X, Thomas SR, Lee WL, Turkoz M, McDonald KG, Meredith MM, Song C, Guidos CJ, Newberry RD, Ouyang W, Murphy TL, Stappenbeck TS, Gommerman JL, Nussenzweig MC, Colonna M, Kopan R, Murphy KM. 2013. Notch2-dependent classical dendritic cells orchestrate intestinal immunity to attaching-and-effacing bacterial pathogens. *Nat Immunol* 14:937–948.
14. Mann ER, Bernardo D, English NR, Landy J, Al-Hassi HO, Peake ST, Man R, Elliott TR, Spranger H, Lee GH, Parian A, Brant SR, Lazarev M, Hart AL, Li X, Knight SC. 2016. Compartment-specific immunity in the human gut: properties and functions of dendritic cells in the colon versus the ileum. *Gut* 65:256–270.
15. Bogunovic M, Ginhoux F, Helft J, Shang L, Hashimoto D, Greter M, Liu K, Jakubzick C, Ingersoll MA, Leboeuf M, Stanley ER, Nussenzweig M, Lira SA, Randolph GJ, Merad M. 2009. Origin of the lamina propria dendritic cell network. *Immunity* 31:513–525.
16. Kinnebrew MA, Buffie CG, Diehl GE, Zenewicz LA, Leiner I, Hohl TM, Flavell RA, Littman DR, Pamer EG. 2012. Interleukin 23 production by intestinal CD103^{+}CD11b^{+} dendritic cells in response to bacterial flagellin enhances mucosal innate immune defense. *Immunity* 36:276–287.
17. Varol C, Vallon-Eberhard A, Elinav E, Aychek T, Shapira Y, Luche H, Fehling HJ, Hardt WD, Shakhar G, Jung S. 2009. Intestinal lamina propria dendritic cell subsets have different origin and functions. *Immunity* 31:502–512.
18. Greter M, Helft J, Chow A, Hashimoto D, Mortha A, Agudo-Cantero J, Bogunovic M, Gautier EL, Miller J, Leboeuf M, Lu G, Aloman C, Brown BD, Pollard JW, Xiong H, Randolph GJ, Chipuk JE, Frenette PS, Merad M. 2012. GM-CSF controls nonlymphoid tissue dendritic cell homeostasis but is dispensable for the differentiation of inflammatory dendritic cells. *Immunity* 36:1031–1046.
19. Lewis KL, Caton ML, Bogunovic M, Greter M, Grajkowska LT, Ng D, Klinakis A, Charo IF, Jung S, Gommerman JL, Ivanov II, Liu K, Merad M, Reizis B. 2011. Notch2 receptor signaling controls functional differentiation of dendritic cells in the spleen and intestine. *Immunity* 35:780–791.
20. Schlitzer A, McGovern N, Teo P, Zelante T, Atarashi K, Low D, Ho AW, See P, Shin A, Wasan PS, Hoeffel G, Malleret B, Heiseke A, Chew S, Jardine L, Purvis HA, Hilkens CM, Tam J, Poidinger M, Stanley ER, Krug AB, Renia L, Sivasankar B, Ng LG, Collin M, Ricciardi-Castagnoli P, Honda K, Haniffa M, Ginhoux F. 2013. IRF4 transcription factor-dependent CD11b^{+} dendritic cells in human and mouse control mucosal IL-17 cytokine responses. *Immunity* 38:970–983.
21. Persson EK, Uronen-Hansson H, Semmrich M, Rivollier A, Hägerbrand K, Marsal J, Gudjonsson S, Håkansson U, Reizis B, Kotarsky K, Agace WW. 2013. IRF4 transcription-factor-dependent CD103^{+}CD11b^{+} dendritic cells drive mucosal T helper 17 cell differentiation. *Immunity* 38:958–969.
22. Edelson BT, Kc W, Juang R, Kohyama M, Benoit LA, Klekotka PA, Moon C, Albring JC, Ise W, Michael DG, Bhattacharya D, Stappenbeck TS, Holtzman MJ, Sung SS, Murphy TL, Hildner K, Murphy KM. 2010. Peripheral CD103^{+} dendritic cells form a unified subset developmentally related to CD8α^{+} conventional dendritic cells. *J Exp Med* 207:823–836.
23. Luda KM, Joeris T, Persson EK, Rivollier A, Demiri M, Sitnik KM, Pool L, Holm JB, Melo-Gonzalez F, Richter L, Lambrecht BN, Kristiansen K, Travis MA, Svensson-Frej M, Kotarsky K, Agace WW. 2016. IRF8

transcription-factor-dependent classical dendritic cells are essential for intestinal T cell homeostasis. *Immunity* **44**:860–874.

24. **Schraml BU, van Blijswijk J, Zelenay S, Whitney PG, Filby A, Acton SE, Rogers NC, Moncaut N, Carvajal JJ, Reis e Sousa C.** 2013. Genetic tracing via DNGR-1 expression history defines dendritic cells as a hematopoietic lineage. *Cell* **154**:843–858.
25. **Scott CL, Bain CC, Wright PB, Sichien D, Kotarsky K, Persson EK, Luda K, Guilliams M, Lambrecht BN, Agace WW, Milling SW, Mowat AM.** 2015. $CCR2^+$ $CD103^-$ intestinal dendritic cells develop from DC-committed precursors and induce interleukin-17 production by T cells. *Mucosal Immunol* **8**:327–339.
26. **Hägerbrand K, Westlund J, Yrlid U, Agace W, Johansson-Lindbom B.** 2015. MyD88 signaling regulates steady-state migration of intestinal $CD103^+$ dendritic cells independently of TNF-α and the gut microbiota. *J Immunol* **195**:2888–2899.
27. **Wilson NS, Young LJ, Kupresanin F, Naik SH, Vremec D, Heath WR, Akira S, Shortman K, Boyle J, Maraskovsky E, Belz GT, Villadangos JA.** 2008. Normal proportion and expression of maturation markers in migratory dendritic cells in the absence of germs or Toll-like receptor signaling. *Immunol Cell Biol* **86**:200–205.
28. **Baratin M, Foray C, Demaria O, Habbeddine M, Pollet E, Maurizio J, Verthuy C, Davanture S, Azukizawa H, Flores-Langarica A, Dalod M, Lawrence T.** 2015. Homeostatic NF-κB signaling in steady-state migratory dendritic cells regulates immune homeostasis and tolerance. *Immunity* **42**:627–639.
29. **Niess JH, Brand S, Gu X, Landsman L, Jung S, McCormick BA, Vyas JM, Boes M, Ploegh HL, Fox JG, Littman DR, Reinecker HC.** 2005. CX3CR1-mediated dendritic cell access to the intestinal lumen and bacterial clearance. *Science* **307**:254–258.
30. **Chieppa M, Rescigno M, Huang AY, Germain RN.** 2006. Dynamic imaging of dendritic cell extension into the small bowel lumen in response to epithelial cell TLR engagement. *J Exp Med* **203**:2841–2852.
31. **Farache J, Koren I, Milo I, Gurevich I, Kim KW, Zigmond E, Furtado GC, Lira SA, Shakhar G.** 2013. Luminal bacteria recruit $CD103^+$ dendritic cells into the intestinal epithelium to sample bacterial antigens for presentation. *Immunity* **38**:581–595.
32. **McDole JR, Wheeler LW, McDonald KG, Wang B, Konjufca V, Knoop KA, Newberry RD, Miller MJ.** 2012. Goblet cells deliver luminal antigen to $CD103^+$ dendritic cells in the small intestine. *Nature* **483**:345–349.
33. **Shan M, Gentile M, Yeiser JR, Walland AC, Bornstein VU, Chen K, He B, Cassis L, Bigas A, Cols M, Comerma L, Huang B, Blander JM, Xiong H, Mayer L, Berin C, Augenlicht LH, Velcich A, Cerutti A.** 2013. Mucus enhances gut homeostasis and oral tolerance by delivering immunoregulatory signals. *Science* **342**: 447–453.
34. **Mazzini E, Massimiliano L, Penna G, Rescigno M.** 2014. Oral tolerance can be established via gap junction transfer of fed antigens from $CX3CR1^+$ macrophages to $CD103^+$ dendritic cells. *Immunity* **40**:248–261.
35. **Coombes JL, Siddiqui KR, Arancibia-Cárcamo CV, Hall J, Sun CM, Belkaid Y, Powrie F.** 2007. A functionally specialized population of mucosal $CD103^+$ DCs induces $Foxp3^+$ regulatory T cells via a TGF-β- and retinoic acid-dependent mechanism. *J Exp Med* **204**: 1757–1764.
36. **Sun CM, Hall JA, Blank RB, Bouladoux N, Oukka M, Mora JR, Belkaid Y.** 2007. Small intestine lamina propria dendritic cells promote de novo generation of Foxp3 T reg cells via retinoic acid. *J Exp Med* **204**: 1775–1785.
37. **Jaensson-Gyllenbäck E, Kotarsky K, Zapata F, Persson EK, Gundersen TE, Blomhoff R, Agace WW.** 2011. Bile retinoids imprint intestinal $CD103^+$ dendritic cells with the ability to generate gut-tropic T cells. *Mucosal Immunol* **4**:438–447.
38. **Worthington JJ, Czajkowska BI, Melton AC, Travis MA.** 2011. Intestinal dendritic cells specialize to activate transforming growth factor-β and induce $Foxp3^+$ regulatory T cells via integrin αvβ8. *Gastroenterology* **141**:1802–1812.
39. **Johansson-Lindbom B, Svensson M, Pabst O, Palmqvist C, Marquez G, Förster R, Agace WW.** 2005. Functional specialization of gut $CD103^+$ dendritic cells in the regulation of tissue-selective T cell homing. *J Exp Med* **202**: 1063–1073.
40. **Jaensson E, Uronen-Hansson H, Pabst O, Eksteen B, Tian J, Coombes JL, Berg PL, Davidsson T, Powrie F, Johansson-Lindbom B, Agace WW.** 2008. Small intestinal $CD103^+$ dendritic cells display unique functional properties that are conserved between mice and humans. *J Exp Med* **205**:2139–2149.
41. **Matteoli G, Mazzini E, Iliev ID, Mileti E, Fallarino F, Puccetti P, Chieppa M, Rescigno M.** 2010. Gut $CD103^+$ dendritic cells express indoleamine 2,3-dioxygenase which influences T regulatory/T effector cell balance and oral tolerance induction. *Gut* **59**:595–604.
42. **Worbs T, Bode U, Yan S, Hoffmann MW, Hintzen G, Bernhardt G, Förster R, Pabst O.** 2006. Oral tolerance originates in the intestinal immune system and relies on antigen carriage by dendritic cells. *J Exp Med* **203**: 519–527.
43. **Welty NE, Staley C, Ghilardi N, Sadowsky MJ, Igyártó BZ, Kaplan DH.** 2013. Intestinal lamina propria dendritic cells maintain T cell homeostasis but do not affect commensalism. *J Exp Med* **210**:2011–2024.
44. **Esterházy D, Loschko J, London M, Jove V, Oliveira TY, Mucida D.** 2016. Classical dendritic cells are required for dietary antigen-mediated induction of peripheral T_{reg} cells and tolerance. *Nat Immunol* **17**:545–555.
45. **Rivollier A, He J, Kole A, Valatas V, Kelsall BL.** 2012. Inflammation switches the differentiation program of $Ly6C^{hi}$ monocytes from antiinflammatory macrophages to inflammatory dendritic cells in the colon. *J Exp Med* **209**:139–155.
46. **Rescigno M, Urbano M, Valzasina B, Francolini M, Rotta G, Bonasio R, Granucci F, Kraehenbuhl JP, Ricciardi-Castagnoli P.** 2001. Dendritic cells express tight junction proteins and penetrate gut epithelial monolayers to sample bacteria. *Nat Immunol* **2**:361–367.

47. **Müller AJ, Kaiser P, Dittmar KE, Weber TC, Haueter S, Endt K, Songhet P, Zellweger C, Kremer M, Fehling HJ, Hardt WD.** 2012. *Salmonella* gut invasion involves TTSS-2-dependent epithelial traversal, basolateral exit, and uptake by epithelium-sampling lamina propria phagocytes. *Cell Host Microbe* **11:**19–32.

48. **Howe SE, Lickteig DJ, Plunkett KN, Ryerse JS, Konjufca V.** 2014. The uptake of soluble and particulate antigens by epithelial cells in the mouse small intestine. *PLoS One* **9:**e86656. doi:10.1371/journal.pone.0086656.

49. **Platt AM, Bain CC, Bordon Y, Sester DP, Mowat AM.** 2010. An independent subset of TLR expressing CCR2-dependent macrophages promotes colonic inflammation. *J Immunol* **184:**6843–6854.

50. **Zigmond E, Varol C, Farache J, Elmaliah E, Satpathy AT, Friedlander G, Mack M, Shpigel N, Boneca IG, Murphy KM, Shakhar G, Halpern Z, Jung S.** 2012. $Ly6C^{hi}$ monocytes in the inflamed colon give rise to proinflammatory effector cells and migratory antigen-presenting cells. *Immunity* **37:**1076–1090.

51. **Ueda Y, Kayama H, Jeon SG, Kusu T, Isaka Y, Rakugi H, Yamamoto M, Takeda K.** 2010. Commensal microbiota induce LPS hyporesponsiveness in colonic macrophages via the production of IL-10. *Int Immunol* **22:**953–962.

52. **Hadis U, Wahl B, Schulz O, Hardtke-Wolenski M, Schippers A, Wagner N, Müller W, Sparwasser T, Förster R, Pabst O.** 2011. Intestinal tolerance requires gut homing and expansion of $FoxP3^+$ regulatory T cells in the lamina propria. *Immunity* **34:**237–246.

53. **Murai M, Turovskaya O, Kim G, Madan R, Karp CL, Cheroutre H, Kronenberg M.** 2009. Interleukin 10 acts on regulatory T cells to maintain expression of the transcription factor Foxp3 and suppressive function in mice with colitis. *Nat Immunol* **10:**1178–1184.

54. **Takeda K, Clausen BE, Kaisho T, Tsujimura T, Terada N, Förster I, Akira S.** 1999. Enhanced Th1 activity and development of chronic enterocolitis in mice devoid of Stat3 in macrophages and neutrophils. *Immunity* **10:**39–49.

55. **Hirotani T, Lee PY, Kuwata H, Yamamoto M, Matsumoto M, Kawase I, Akira S, Takeda K.** 2005. The nuclear IκB protein IκBNS selectively inhibits lipopolysaccharide-induced IL-6 production in macrophages of the colonic lamina propria. *J Immunol* **174:**3650–3657.

56. **Diehl GE, Longman RS, Zhang JX, Breart B, Galan C, Cuesta A, Schwab SR, Littman DR.** 2013. Microbiota restricts trafficking of bacteria to mesenteric lymph nodes by CX_3CR1^{hi} cells. *Nature* **494:**116–120.

57. **Shaw MH, Kamada N, Kim YG, Núñez G.** 2012. Microbiota-induced IL-1β, but not IL-6, is critical for the development of steady-state T_H17 cells in the intestine. *J Exp Med* **209:**251–258.

58. **Mortha A, Chudnovskiy A, Hashimoto D, Bogunovic M, Spencer SP, Belkaid Y, Merad M.** 2014. Microbiota-dependent crosstalk between macrophages and ILC3 promotes intestinal homeostasis. *Science* **343:**1249288.

59. **Ginhoux F, Jung S.** 2014. Monocytes and macrophages: developmental pathways and tissue homeostasis. *Nat Rev Immunol* **14:**392–404.

60. **Bain CC, Bravo-Blas A, Scott CL, Gomez Perdiguero E, Geissmann F, Henri S, Malissen B, Osborne LC, Artis D, Mowat AM.** 2014. Constant replenishment from circulating monocytes maintains the macrophage pool in the intestine of adult mice. *Nat Immunol* **15:**929–937.

61. **Auffray C, Fogg DK, Narni-Mancinelli E, Senechal B, Trouillet C, Saederup N, Leemput J, Bigot K, Campisi L, Abitbol M, Molina T, Charo I, Hume DA, Cumano A, Lauvau G, Geissmann F.** 2009. $CX3CR1^+$ $CD115^+$ $CD135^+$ common macrophage/DC precursors and the role of CX_3CR1 in their response to inflammation. *J Exp Med* **206:**595–606.

62. **Carlin LM, Stamatiades EG, Auffray C, Hanna RN, Glover L, Vizcay-Barrena G, Hedrick CC, Cook HT, Diebold S, Geissmann F.** 2013. Nr4a1-dependent $Ly6C^{low}$ monocytes monitor endothelial cells and orchestrate their disposal. *Cell* **153:**362–375.

63. **Takada Y, Hisamatsu T, Kamada N, Kitazume MT, Honda H, Oshima Y, Saito R, Takayama T, Kobayashi T, Chinen H, Mikami Y, Kanai T, Okamoto S, Hibi T.** 2010. Monocyte chemoattractant protein-1 contributes to gut homeostasis and intestinal inflammation by composition of IL-10-producing regulatory macrophage subset. *J Immunol* **184:**2671–2676.

64. **Serbina NV, Pamer EG.** 2006. Monocyte emigration from bone marrow during bacterial infection requires signals mediated by chemokine receptor CCR2. *Nat Immunol* **7:**311–317.

65. **MacDonald KP, Palmer JS, Cronau S, Seppanen E, Olver S, Raffelt NC, Kuns R, Pettit AR, Clouston A, Wainwright B, Branstetter D, Smith J, Paxton RJ, Cerretti DP, Bonham L, Hill GR, Hume DA.** 2010. An antibody against the colony-stimulating factor 1 receptor depletes the resident subset of monocytes and tissue- and tumor-associated macrophages but does not inhibit inflammation. *Blood* **116:**3955–3963.

66. **Olson MC, Scott EW, Hack AA, Su GH, Tenen DG, Singh H, Simon MC.** 1995. PU.1 is not essential for early myeloid gene expression but is required for terminal myeloid differentiation. *Immunity* **3:**703–714.

67. **Bain CC, Mowat AM.** 2014. The monocyte-macrophage axis in the intestine. *Cell Immunol* **291:**41–48.

68. **Mahida YR, Wu KC, Jewell DP.** 1989. Respiratory burst activity of intestinal macrophages in normal and inflammatory bowel disease. *Gut* **30:**1362–1370.

69. **Smythies LE, Sellers M, Clements RH, Mosteller-Barnum M, Meng G, Benjamin WH, Orenstein JM, Smith PD.** 2005. Human intestinal macrophages display profound inflammatory anergy despite avid phagocytic and bacteriocidal activity. *J Clin Invest* **115:**66–75.

70. **Smythies LE, Shen R, Bimczok D, Novak L, Clements RH, Eckhoff DE, Bouchard P, George MD, Hu WK, Dandekar S, Smith PD.** 2010. Inflammation anergy in human intestinal macrophages is due to Smad-induced IκBα expression and NF-κB inactivation. *J Biol Chem* **285:**19593–19604.

71. **Kamada N, Hisamatsu T, Okamoto S, Chinen H, Kobayashi T, Sato T, Sakuraba A, Kitazume MT, Sugita A, Koganei K, Akagawa KS, Hibi T.** 2008. Unique $CD14^+$ intestinal macrophages contribute to the patho-

genesis of Crohn disease via IL-23/IFN-γ axis. *J Clin Invest* **118**:2269–2280.

72. **Grimm MC, Pavli P, Van de Pol E, Doe WF.** 1995. Evidence for a CD14⁺ population of monocytes in inflammatory bowel disease mucosa—implications for pathogenesis. *Clin Exp Immunol* **100**:291–297.

73. **Grimm MC, Pullman WE, Bennett GM, Sullivan PJ, Pavli P, Doe WF.** 1995. Direct evidence of monocyte recruitment to inflammatory bowel disease mucosa. *J Gastroenterol Hepatol* **10**:387–395.

74. **Weber B, Saurer L, Schenk M, Dickgreber N, Mueller C.** 2011. CX3CR1 defines functionally distinct intestinal mononuclear phagocyte subsets which maintain their respective functions during homeostatic and inflammatory conditions. *Eur J Immunol* **41**:773–779.

75. **Qualls JE, Kaplan AM, van Rooijen N, Cohen DA.** 2006. Suppression of experimental colitis by intestinal mononuclear phagocytes. *J Leukoc Biol* **80**:802–815.

76. **Pull SL, Doherty JM, Mills JC, Gordon JI, Stappenbeck TS.** 2005. Activated macrophages are an adaptive element of the colonic epithelial progenitor niche necessary for regenerative responses to injury. *Proc Natl Acad Sci U S A* **102**:99–104.

77. **Seo SU, Kamada N, Muñoz-Planillo R, Kim YG, Kim D, Koizumi Y, Hasegawa M, Himpsl SD, Browne HP, Lawley TD, Mobley HL, Inohara N, Núñez G.** 2015. Distinct commensals induce interleukin-1β via NLRP3 inflammasome in inflammatory monocytes to promote intestinal inflammation in response to injury. *Immunity* **42**:744–755.

78. **Dunay IR, Damatta RA, Fux B, Presti R, Greco S, Colonna M, Sibley LD.** 2008. Gr1⁺ inflammatory monocytes are required for mucosal resistance to the pathogen *Toxoplasma gondii*. *Immunity* **29**:306–317.

79. **Schulthess J, Meresse B, Ramiro-Puig E, Montcuquet N, Darche S, Bègue B, Ruemmele F, Combadière C, Di Santo JP, Buzoni-Gatel D, Cerf-Bensussan N.** 2012. Interleukin-15-dependent NKp46⁺ innate lymphoid cells control intestinal inflammation by recruiting inflammatory monocytes. *Immunity* **37**:108–121.

80. **Manta C, Heupel E, Radulovic K, Rossini V, Garbi N, Riedel CU, Niess JH.** 2013. CX_3CR1^+ macrophages support IL-22 production by innate lymphoid cells during infection with *Citrobacter rodentium*. *Mucosal Immunol* **6**:177–188.

81. **Longman RS, Diehl GE, Victorio DA, Huh JR, Galan C, Miraldi ER, Swaminath A, Bonneau R, Scherl EJ, Littman DR.** 2014. CX_3CR1^+ mononuclear phagocytes support colitis-associated innate lymphoid cell production of IL-22. *J Exp Med* **211**:1571–1583.

82. **deSchoolmeester ML, Little MC, Rollins BJ, Else KJ.** 2003. Absence of CC chemokine ligand 2 results in an altered Th1/Th2 cytokine balance and failure to expel *Trichuris muris* infection. *J Immunol* **170**:4693–4700.

83. **Anthony RM, Urban JF Jr, Alem F, Hamed HA, Rozo CT, Boucher JL, Van Rooijen N, Gause WC.** 2006. Memory T_H2 cells induce alternatively activated macrophages to mediate protection against nematode parasites. *Nat Med* **12**:955–960.

84. **Grainger JR, Wohlfert EA, Fuss IJ, Bouladoux N, Askenase MH, Legrand F, Koo LY, Brenchley JM, Fraser ID, Belkaid Y.** 2013. Inflammatory monocytes regulate pathologic responses to commensals during acute gastrointestinal infection. *Nat Med* **19**:713–721.

85. **Takayama T, Kamada N, Chinen H, Okamoto S, Kitazume MT, Chang J, Matuzaki Y, Suzuki S, Sugita A, Koganei K, Hisamatsu T, Kanai T, Hibi T.** 2010. Imbalance of NKp44⁺NKp46⁻ and NKp44⁻NKp46⁺ natural killer cells in the intestinal mucosa of patients with Crohn's disease. *Gastroenterology* **139**:882–892, 892.e1–892.e3. doi:10.1053/j.gastro.2010.05.040.

86. **Kamada N, Hisamatsu T, Honda H, Kobayashi T, Chinen H, Takayama T, Kitazume MT, Okamoto S, Koganei K, Sugita A, Kanai T, Hibi T.** 2010. TL1A produced by lamina propria macrophages induces Th1 and Th17 immune responses in cooperation with IL-23 in patients with Crohn's disease. *Inflamm Bowel Dis* **16**:568–575.

87. **Uo M, Hisamatsu T, Miyoshi J, Kaito D, Yoneno K, Kitazume MT, Mori M, Sugita A, Koganei K, Matsuoka K, Kanai T, Hibi T.** 2013. Mucosal CXCR4⁺ IgG plasma cells contribute to the pathogenesis of human ulcerative colitis through FcγR-mediated CD14 macrophage activation. *Gut* **62**:1734–1744.

88. **Schenk M, Bouchon A, Birrer S, Colonna M, Mueller C.** 2005. Macrophages expressing triggering receptor expressed on myeloid cells-1 are underrepresented in the human intestine. *J Immunol* **174**:517–524.

89. **Schenk M, Bouchon A, Seibold F, Mueller C.** 2007. TREM-1-expressing intestinal macrophages crucially amplify chronic inflammation in experimental colitis and inflammatory bowel diseases. *J Clin Invest* **117**:3097–3106.

90. **Rugtveit J, Haraldsen G, Høgåsen AK, Bakka A, Brandtzaeg P, Scott H.** 1995. Respiratory burst of intestinal macrophages in inflammatory bowel disease is mainly caused by CD14⁺L1⁺ monocyte derived cells. *Gut* **37**:367–373.

91. **Nathan C, Cunningham-Bussel A.** 2013. Beyond oxidative stress: an immunologist's guide to reactive oxygen species. *Nat Rev Immunol* **13**:349–361.

92. **Nakagawara A, Nathan CF, Cohn ZA.** 1981. Hydrogen peroxide metabolism in human monocytes during differentiation in vitro. *J Clin Invest* **68**:1243–1252.

93. **Kamada N, Hisamatsu T, Honda H, Kobayashi T, Chinen H, Kitazume MT, Takayama T, Okamoto S, Koganei K, Sugita A, Kanai T, Hibi T.** 2009. Human CD14⁺ macrophages in intestinal lamina propria exhibit potent antigen-presenting ability. *J Immunol* **183**:1724–1731.

94. **Caradonna L, Amati L, Lella P, Jirillo E, Caccavo D.** 2000. Phagocytosis, killing, lymphocyte-mediated antibacterial activity, serum autoantibodies, and plasma endotoxins in inflammatory bowel disease. *Am J Gastroenterol* **95**:1495–1502.

95. **Smith AM, Rahman FZ, Hayee B, Graham SJ, Marks DJ, Sewell GW, Palmer CD, Wilde J, Foxwell BM, Gloger IS, Sweeting T, Marsh M, Walker AP, Bloom SL, Segal AW.** 2009. Disordered macrophage cytokine

secretion underlies impaired acute inflammation and bacterial clearance in Crohn's disease. *J Exp Med* **206:** 1883–1897.

96. **Marks DJ, Miyagi K, Rahman FZ, Novelli M, Bloom SL, Segal AW.** 2009. Inflammatory bowel disease in CGD reproduces the clinicopathological features of Crohn's disease. *Am J Gastroenterol* **104:**117–124.

97. **Uematsu S, Fujimoto K, Jang MH, Yang BG, Jung YJ, Nishiyama M, Sato S, Tsujimura T, Yamamoto M, Yokota Y, Kiyono H, Miyasaka M, Ishii KJ, Akira S.** 2008. Regulation of humoral and cellular gut immunity by lamina propria dendritic cells expressing Toll-like receptor 5. *Nat Immunol* **9:**769–776.

98. **Cerovic V, Jenkins CD, Barnes AG, Milling SW, MacPherson GG, Klavinskis LS.** 2009. Hyporesponsiveness of intestinal dendritic cells to TLR stimulation is limited to TLR4. *J Immunol* **182:**2405–2415.

99. **Laffont S, Siddiqui KR, Powrie F.** 2010. Intestinal inflammation abrogates the tolerogenic properties of MLN CD103$^+$ dendritic cells. *Eur J Immunol* **40:**1877–1883.

100. **Aychek T, Mildner A, Yona S, Kim KW, Lampl N, Reich-Zeliger S, Boon L, Yogev N, Waisman A, Cua DJ, Jung S.** 2015. IL-23-mediated mononuclear phagocyte crosstalk protects mice from *Citrobacter rodentium*-induced colon immunopathology. *Nat Commun* **6:** 6525. doi:10.1038/ncomms7525.

101. **Hart AL, Al-Hassi HO, Rigby RJ, Bell SJ, Emmanuel AV, Knight SC, Kamm MA, Stagg AJ.** 2005. Characteristics of intestinal dendritic cells in inflammatory bowel diseases. *Gastroenterology* **129:**50–65.

102. **Baumgart DC, Thomas S, Przesdzing I, Metzke D, Bielecki C, Lehmann SM, Lehnardt S, Dörffel Y, Sturm A, Scheffold A, Schmitz J, Radbruch A.** 2009. Exaggerated inflammatory response of primary human myeloid dendritic cells to lipopolysaccharide in patients with inflammatory bowel disease. *Clin Exp Immunol* **157:** 423–436.

103. **Bell SJ, Rigby R, English N, Mann SD, Knight SC, Kamm MA, Stagg AJ.** 2001. Migration and maturation of human colonic dendritic cells. *J Immunol* **166:**4958–4967.

104. **Sakuraba A, Sato T, Kamada N, Kitazume M, Sugita A, Hibi T.** 2009. Th1/Th17 immune response is induced by mesenteric lymph node dendritic cells in Crohn's disease. *Gastroenterology* **137:**1736–1745.

105. **Fuss IJ, Neurath M, Boirivant M, Klein JS, de la Motte C, Strong SA, Fiocchi C, Strober W.** 1996. Disparate CD4$^+$ lamina propria (LP) lymphokine secretion profiles in inflammatory bowel disease. Crohn's disease LP cells manifest increased secretion of IFN-γ, whereas ulcerative colitis LP cells manifest increased secretion of IL-5. *J Immunol* **157:**1261–1270.

106. **Sanders TJ, McCarthy NE, Giles EM, Davidson KL, Haltalli ML, Hazell S, Lindsay JO, Stagg AJ.** 2014. Increased production of retinoic acid by intestinal macrophages contributes to their inflammatory phenotype in patients with Crohn's disease. *Gastroenterology* **146:** 1278–88.e2. doi:10.1053/j.gastro.2014.01.057.

107. **Magnusson MK, Brynjólfsson SF, Dige A, Uronen-Hansson H, Börjesson LG, Bengtsson JL, Gudjonsson S, Öhman L, Agnholt J, Sjövall H, Agace WW, Wick MJ.** 2016. Macrophage and dendritic cell subsets in IBD: ALDH$^+$ cells are reduced in colon tissue of patients with ulcerative colitis regardless of inflammation. *Mucosal Immunol* **9:**171–182.

Myeloid Cells in Health and Disease: A Synthesis
Edited by Siamon Gordon

doi:10.1128/microbiolspec.MCHD-0012-2015

Gabriel Mitchell[1,*]
Chen Chen[1,*]
Daniel A. Portnoy[1,2]

40 Strategies Used by Bacteria to Grow in Macrophages

INTRODUCTION

Intracellular bacterial pathogens cause a wide range of diseases and significantly contribute to the morbidity and mortality associated with infectious diseases worldwide (1–16) (Table 1). These bacteria use several different strategies to replicate in host cells and influence host processes such as membrane trafficking, signaling pathways, metabolism, cell death, and survival (17–19). Broadly, intracellular bacteria colonize two topologically distinct regions of the host cell and are divided into cytosolic and intravacuolar bacteria according to their intracellular lifestyle. However, most intracellular bacterial pathogens have unique intracellular life cycles with features strikingly different from one another (Fig. 1). It should also be noted that intravacuolar pathogens gain access to the host cytosol to some extent, and that cytosolic bacteria might spend an underestimated part of their intracellular life cycle within membrane-bound compartments (20–22).

Cytosolic bacteria escape from the endocytic pathway and replicate in the host cytosol. The host cytosol indeed constitutes an attractive replicative niche for intracellular bacteria because this subcellular compartment provides an environment rich in nutrients. The cytoplasm also offers the distinct advantage of being separated from the extracellular environment, and thereby may constitute an ideal hideout where pathogens can evade extracellular immune surveillance and killing.

Alternatively, intracellular bacterial pathogens reside and replicate within the host endomembrane system, which comprises an intricate network of membrane-bound organelles and vesicular trafficking intermediates. The replication of intracellular bacteria in these vesicular compartments is accompanied by concomitant vacuolar membrane expansion, which is driven by adaptive strategies from pathogens. Even though the vacuolar intracellular lifestyle requires complex host-pathogen interactions in order to maintain the unique membrane-bound replication niche, bacterial pathogens benefit from this lifestyle that provides protection from the host cytosolic innate immune defenses.

Most intracellular bacteria replicate in myeloid cells, especially in macrophages (2, 23). Macrophages are remarkably plastic cells characterized by their phenotypic diversity and are involved in pathogen detection,

[1]Department of Molecular and Cell Biology; [2]School of Public Health, University of California, Berkeley, Berkeley, CA 94720. [*]Contributed equally to this work.

Table 1 Characteristics and diseases associated with intracellular pathogens that infect human myeloid cells

Pathogen	Targeted myeloid cells	Secretion system[a]	Replication niche	Disease	Reference
Anaplasma phagocytophilum	Neutrophil	T4SS	Vacuole	Human granulocytic anaplasmosis	3
Brucella spp.[b]	Macrophage	T4SS	Vacuole	Brucellosis	4
Burkholderia pseudomallei	Macrophage	T3SS, T6SS	Cytosol	Melioidosis	5
Chlamydia pneumoniae	Macrophage	T3SS	Vacuole	Pneumonia, bronchitis	6
Citrobacter koseri	Macrophage	T3SS	Vacuole	Meningitis	7
Coxiella burnetii	Macrophage	T4SS	Vacuole	Q fever	8
Ehrlichia chaffeensis	Macrophage	T4SS	Vacuole	Ehrlichiosis	9
Francisella tularensis	Macrophage	T6SS	Cytosol	Tularemia	10
Legionella pneumophila	Macrophage	T4SS	Vacuole	Legionnaires' disease	11
Listeria monocytogenes	Macrophage	T2SS (Sec)	Cytosol	Listeriosis	12
Mycobacterium tuberculosis	Macrophage	T7SS	Vacuole	Tuberculosis	13
Rhodococcus equi	Macrophage	T2SS (Sec)	Vacuole	Pneumonia	14
Rickettsia rickettsii	Macrophage	T4SS (putative)	Cytosol	Rocky Mountain spotted fever	15
Salmonella enterica	Macrophage	T3SS, T6SS	Vacuole	Salmonellosis	16

[a]Major bacterial secretion systems involved in pathogenesis.
[b]*B. abortus*, *B. canis*, *B. suis*, and *B. melitensis*.

antigen presentation, cytokine production, tissue repair, and, more notoriously, microbial killing (24). These cells indeed possess an extensive antimicrobial arsenal and are endowed with the ability to ingest and destroy microorganisms (25). The observation that most pathogenic bacteria preferentially replicate in macrophages thus constitutes a paradox (2, 17).

This manuscript focuses on bacterial pathogens that have the ability to replicate in macrophages and aims to provide an overview of the strategies deployed by these bacteria to grow intracellularly. A brief description of the defense mechanisms used by macrophages against these intracellular bacteria is provided, and the current knowledge about the pathogenic strategies specifically used by cytosolic and intravacuolar bacteria is reviewed.

DEFENSE MECHANISMS AGAINST INTRACELLULAR BACTERIA

Detection of Intracellular Bacterial Infection by Macrophages

Macrophages express a wide range of receptors that trigger innate immune responses and antimicrobial defenses upon bacterial infection (23, 26). These sensors are referred to as pattern recognition receptors (PRRs). PRRs recognize conserved microbial molecules named pathogen-associated molecular patterns (PAMPs) as well as damage-associated molecular patterns (DAMPs) released in response to stress and tissue damage. There are two main classes of PRRs: the membrane-bound receptors (e.g., the Toll-like receptors [TLRs]) and the cytosolic receptors (e.g., the NOD-like receptors [NLRs]).

TLRs are localized on the plasma membrane (e.g., TLR4) or on endosomal membrane compartments (e.g., TLR9) and recognize PAMPs such as lipoproteins, lipopolysaccharide, flagellin, or nucleic acids (27). Upon ligand recognition, TLRs activate signaling pathways and regulate downstream cytokine expression by interacting with adaptor proteins such as MyD88 (myeloid differentiation primary-response protein 88) and TRIF (TIR-domain-containing adaptor protein inducing interferon-β [IFN-β]) (26).

In the cytosol, the NLR proteins NOD1 (nucleotide-binding oligomerization domain-containing protein 1) and NOD2 are triggered by the presence of peptidoglycan fragments, and activate NF-κB (28). Interestingly, it was shown that the activation of the NOD1 signaling pathway by peptidoglycan fragments is dependent on the small Rho GTPase Rac1 and, more broadly, that the manipulation of small Rho GTPases by pathogens is a process that can be detected by the host in a NOD1-dependent manner (29). Recognition of PAMPs or DAMPs by other cytosolic NLRs, such as NLRC4 (NLR family, CARD domain-containing 4), NLRP1 (NLR family, pyrin domain-containing 1), and NLRP3, leads to the assembly of cytosolic multiprotein oligomers termed inflammasomes. In turn, inflammasomes activate caspase-1, induce the extracellular release of interleukin-1β (IL-1β) and IL-18, and trigger a type of inflammatory cell death called pyroptosis (30, 31). The PYHIN member protein AIM2 (absent in melanoma 2) also activates an inflammasome in response to cytosolic

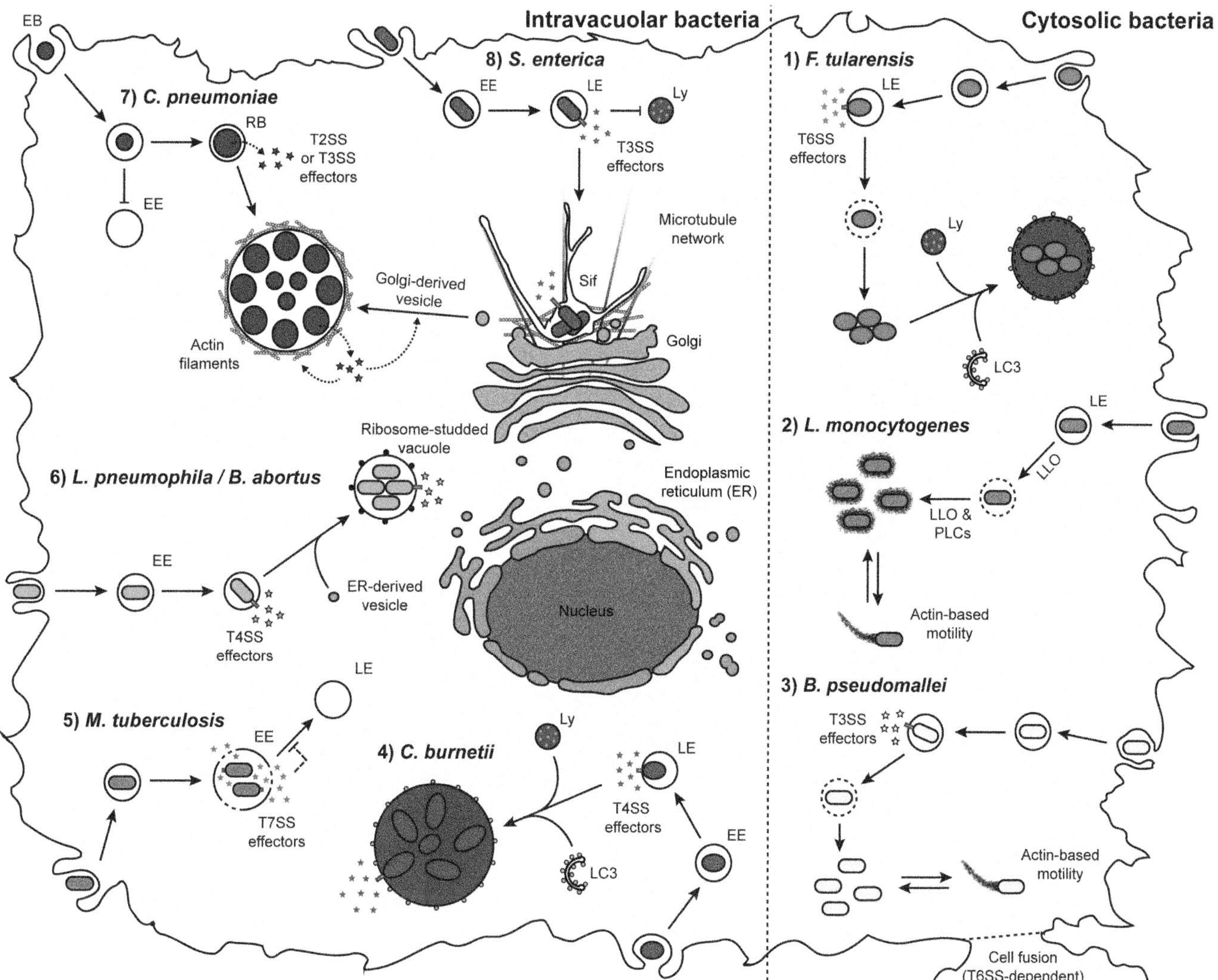

Figure 1 Lifestyles of intracellular bacterial pathogens. (1) *F. tularensis* escapes a late endosome (LE)-like vacuole in a T6SS-dependent manner. Following replication in the cytosol, *F. tularensis* may retranslocate to a membrane-bound compartment resembling an autolysosome. (2) *L. monocytogenes* escapes the phagolysosomal pathway using the T2SS (Sec) effectors LLO and PLCs. *L. monocytogenes* replicates rapidly in the cytosol and hijacks the host actin polymerization machinery to move within and between cells. (3) *B. pseudomallei* escapes into the cytosol in a T3SS-dependent manner. *B. pseudomallei* performs actin-based motility and promotes host cell fusion. (4) *C. burnetii* is adapted to the phagolysosomal pathway and resides in a spacious phagolysosomal-like compartment. The Dot/Icm system (T4SS) is required for recruiting the autophagosomal marker LC3 and for vacuole biogenesis. (5) *M. tuberculosis* arrests phagosome maturation at the early endosome (EE) stage in a T7SS-dependent manner. (6) *L. pneumophila* and *B. abortus* segregate from the endocytic route at the EE stage, recruit ER-derived vesicles, and form ribosome-studded specialized vacuoles in a T4SS-dependent manner. (7) *C. pneumoniae* segregates from the endocytic route and forms a unique inclusion vacuole by recruiting Golgi-derived vesicles. *C. pneumoniae* effectors promote Golgi fragmentation and generate actin filaments around the inclusion. *Chlamydia* is found in two different forms: the nonreplicating infectious elementary body (EB) and the intracytoplasmic replicative reticulate body (RB). T2SS and T3SS effectors are thought to be involved in the intracellular life cycle of *Chlamydia*. (8) *S. enterica* replicates in an LE-like compartment that excludes lysosomal degradation enzymes. The *S. enterica*-containing vacuole migrates to the microtubule-organizing center and forms *Salmonella*-induced filaments (Sif) along microtubules in a T3SS-dependent manner.

DNA (32, 33) released from intracellular bacteria (34–39).

Besides AIM2, there are other PRRs that detect the presence of foreign nucleic acids in the cytosol and trigger distinct immune responses (26). STING (stimulator of IFN genes) is important in the cytosolic response to nucleic acids (40, 41), such as DNA (42) and cyclic dinucleotides (43), and triggers a type I IFN response upon bacterial infection. STING binds directly to cyclic dinucleotides but not to DNA, serving as a direct sensor and an adaptor molecule (44). The host cytosolic DNA sensor cyclic GMP-AMP (cGAMP) synthase (cGAS) is able to synthesize cGAMP, which is an endogenous secondary messenger that binds and activates STING (45, 46). Accordingly, cGAS is involved in the secretion of IFN-β following the detection of bacterial DNA in the host cytosol (47–49). It is noteworthy that other cytosolic PRRs also contribute to nucleic acid sensing and induce a type I IFN response, including DEAD box polypeptide 41 (DDX41) (50, 51) and IFN-γ-inducible protein 16 (IFI16) (52).

Besides antiviral immune response, the role of type I IFNs during bacterial infection is enigmatic. Many intracellular bacteria induce type I IFNs during infection, but rather than being protective for the host, type I IFNs enhance the host susceptibility to intracellular pathogens such as *Listeria monocytogenes* (53, 54), *Salmonella enterica* serovar Typhimurium (55), *Chlamydia trachomatis* (56, 57), and *Mycobacterium tuberculosis* (58–61). The mechanism(s) by which type I IFN signaling increases host susceptibility to bacterial infection is an active area of study and is not well defined.

The Phagolysosomal Pathway Is the Front-Line Defense against Intracellular Bacterial Pathogens

Macrophages are proficient in the internalization and destruction of bacteria and in promoting innate immune responses (25). Following phagocytosis, a series of coordinated fusion and fission events with specific compartments of the endocytic pathway ultimately leads to the generation of a phagolysosome. The phagolysosome possesses potent microbicidal features and constitutes a highly acidic, oxidative, and degradative environment.

The acidification of the phagosome lumen is dependent on proton pumping by the V-ATPase, which is required for the optimal activity of lysosomal hydrolytic enzymes and interferes with bacterial growth by impairing the metabolism of some bacteria (25, 62). The generation of reactive oxygen species (ROS) and reactive nitrogen species by the NOX2 NADPH oxidase (63, 64) and the inducible NOS2 nitric oxide synthase (65) are also important antimicrobial mechanisms of the phagolysosomal pathway. Limitation in the availability of essential nutrients may also impact the growth of bacteria in the phagosome. For instance, NRAMP1 (natural resistance-associated macrophage protein 1) interferes with housekeeping and antioxidative functions of some bacteria by extruding Fe^{2+}, Zn^{2+}, and Mn^{2+} from the phagosomal lumen (66). Molecules that directly compromise the integrity of bacteria such as defensins, cathelicidins, lyzozyme, endopeptidases, exopeptidases, as well as hydrolases targeting carbohydrates and lipids are also delivered to the lumen of phagosomes (25). The above-mentioned antimicrobial mechanisms defines the harsh environment that culminates in the phagolysosome, an organelle that intracellular bacteria need to avoid or adapt to in order to survive and proliferate.

Bacterial Autophagy Is a Cell-Autonomous Defense Mechanism

Macroautophagy (hereafter autophagy) involves the formation of a double-membrane vesicle (i.e., the autophagosome) that encloses and targets intracellular components for lysosomal degradation. During starvation, autophagy nonselectively targets parts of the cytoplasm in an attempt to maintain cellular homeostasis. Autophagy also selectively targets organelles (e.g., damaged mitochondria) and intracellular microbes. As such, autophagy contributes to innate immunity by restricting the intracellular growth of pathogens (67). More than 30 autophagy-related (ATG) proteins are involved in orchestrating autophagosome formation (68). However, some of these proteins have functions outside of the autophagy pathway and may affect host-pathogen interactions in an autophagy-independent manner (67, 69).

Initiation of autophagy is controlled by the availability of cellular nutrients and energy (70, 71) or by recognition of specific cargo (72). During autophagy, the production of phosphatidylinositol 3-phosphate (PI3P) results in the recruitment of PI3P-binding proteins and downstream ATG proteins that catalyze covalent binding of LC3 proteins to the isolation membrane (67, 72, 73). This isolation membrane is then recruited to and engulfs the cytosolic cargo. The resulting autophagosome matures into an autolysosome, in which the intraluminal content is digested (73). In the case of selective autophagy, cargo-specific receptors (74–76) and adaptor proteins containing LC3-interacting region (LIR) recruit the LC3-decorated autophagosome to the cargo (77). Ubiquitylation and ubiquitin-binding protein adaptors such as NBR1 (neighbor of BRCA1), NDP52 (nuclear dot protein 52), optineurin, and p62 are involved in the specific targeting of intracellular bacteria by autophagy (78–83).

Several alternative or noncanonical autophagy pathways do not require some of the core autophagy machinery components (84–86). These alternative pathways also play a role in cell-autonomous defense against intracellular microorganisms, as exemplified for *Toxoplasma gondii* (87) and *Staphylococcus aureus* (88). In addition, under some circumstances, the autophagy machinery participates in a degradative process named LC3-associated phagocytosis (LAP) (89–91). In contrast to canonical autophagy, LC3 is conjugated directly to the single-membrane phagosome and the ULK1 complex is dispensable during LAP (90, 91). Besides the engagement of TLRs (89) and the NOX2-dependent production of ROS are involved in the formation of LC3-associated phagosomes (92). By promoting phagosome acidification and maturation, LAP restricts the intracellular growth of bacterial pathogens such as *S.* Typhimurium (92) and *Burkholderia pseudomallei* (93, 94).

Programmed Cell Death Is a Host Defense Mechanism That Destroys the Replication Niche of Intracellular Bacteria

Programmed cell death plays an important role in innate immunity and is an effective way to eliminate infected cells and control infection. There are three major programmed cell death pathways: apoptosis, necroptosis, and pyroptosis. These cell death pathways all contribute to the host defense against microbial infections (95, 96).

Apoptosis is a noninflammatory form of cell death triggered by extrinsic (receptor-mediated) and intrinsic (mitochondria-mediated) pathways. Apoptosis is characterized by membrane blebbing, cell shrinkage, DNA fragmentation, increased mitochondrial permeability, and the eventual cell breakdown and release of apoptotic bodies (95, 96). Both external and internal stimuli trigger apoptosis through the activation of cysteine proteases termed caspases, which target intracellular components to induce cell death. Mitochondria are central in modulating host cell death and survival pathways, especially apoptosis (97). As such, mitochondria-dependent cell death pathways are important in host-bacteria interactions (95, 97).

Although necrosis was traditionally regarded as an accidental and uncontrolled cell death, it is now appreciated that some forms of necrosis, referred to as necroptosis, are genetically controlled (96, 98). Necroptosis is a caspase-independent cell death pathway characterized by organelle damage, cell swelling, rupture of the plasma membrane, and release of the cellular content. Necrotic cell death is induced upon bacterial infection and physical damage and is triggered by ROS, lysosomal permeabilization, calpain activation, and depletion of ATP. More specifically, necroptosis is initiated by the activation of RIP1 (receptor-interacting protein 1) and RIP3 kinases, which activate downstream targets by phosphorylation and induce cell death, especially when caspase-8 is compromised (99, 100). Many proteins involved in the regulation of apoptosis also regulate necroptosis, highlighting the importance of cross talk between these cell death pathways (101).

Pyroptosis is a nonapoptotic cell death pathway induced following inflammasome-mediated caspase-1 activation. Pyroptosis is characterized by DNA fragmentation, loss of membrane integrity, and release of the cell content. In addition, pyroptosis is associated with the secretion of the proinflammatory cytokines IL-1β and IL-18 (30, 31). The induction of pyroptosis decreases the replication of intracellular bacteria within host cells and exposes them to the extracellular immune response (102).

Although some microbes block or delay host cell death to promote their intracellular replication at early times of infection, escape and dissemination may eventually require host cell lysis. In some cases, cell death pathways are coopted by pathogens as a strategy of pathogenesis (103, 104). Furthermore, by inducing host cell death, bacterial pathogens can eliminate key immune cells and, consequently, evade host defenses (95).

INFECTION OF MACROPHAGES BY INTRAVACUOLAR BACTERIA

Remodeling Host Pathways through Specialized Secretion Systems

During their intracellular life cycle, intravacuolar bacteria generally reside in a remodeled membrane-bound compartment. In order to successfully survive and replicate in this "sealed" environment, one of the main challenges encountered by intravacuolar pathogens is to exert actions beyond the vacuolar membrane. This is mainly accomplished using systems that secrete effectors that actively modify the host physiology to create an environment permissive to bacterial proliferation.

Protein secretion plays a central role in modulating the interactions of bacteria with their environment.

This process is more complex for intravacuolar bacteria because secretion requires translocation across both the bacterial surface (plasma membrane and cell wall) and the host (plasma/vacuole) membrane. Four accessory secretion systems are known to play a role in establishing intravacuolar bacterial replication niches by delivering effector proteins into the host cytosol and by modulating a large variety of host cell functions, including vesicular trafficking and the host immune responses (Table 1).

The type III secretion system (T3SS), also called an injectisome, is found in Gram-negative bacteria that interact with both plant and animal hosts. The T3SS appears to have a common evolutionary origin with the flagellum (105). It is composed of up to 25 proteins that form a series of rings that span the bacterial membranes and connect the bacterial and the host cytosol with a hollow filament (106). There are two T3SSs present in *S. enterica*, each encoded in distinct genomic islands known as *Salmonella* pathogenicity island 1 (SPI–1) and SPI-2. SPI-1 confers the ability to invade nonphagocytic cells, while SPI-2 allows *Salmonella* to survive within mammalian cells and spread to internal organs (105). In contrast, *Chlamydia* species possess only one T3SS, which is active at both early and late phases of infection (107). Several putative *Chlamydia* T3SS substrates were identified using *Salmonella* (108), *Shigella* (109, 110), or *Yersinia* (111) as genetically tractable heterologous hosts.

Type IV secretion systems (T4SSs) are extensively used by intravacuolar Gram-negative bacteria to colonize host cells. In comparison to other secretion systems, T4SS is unique in its ability to transport nucleic acids in addition to proteins into plant and animal cells, and is evolutionarily related to bacterial conjugation systems (112, 113). The T4SS can be divided into the canonical VirB (type IVA) and Dot/Icm (type IVB) secretion systems. Both systems are multiprotein complexes that can span the bacterial surface and the host membrane. Essential roles for T4SS in pathogenesis are established in several important intravacuolar bacterial pathogens, including *Brucella suis* (VirB) (114), *Legionella pneumophila* (Dot/Icm), (115, 116), and *Coxiella burnetii* (Dot/Icm) (117). In addition, other intravacuolar bacterial pathogens, including *Anaplasma phagocytophilum* and *Ehrlichia chaffeensis*, may also encode functional type IVA secretion systems that have a role in pathogenesis, as their putative T4SS genes were upregulated during infection and several secreted effectors have been identified (118–120).

Type VI secretion systems (T6SSs) are widely distributed in Gram-negative bacteria. The molecular mechanism by which T6SSs translocate effector proteins into target cells is similar to that of bacteriophage-like injection devices. Based on their function, T6SSs can be classified into two categories: eukaryotic cell-targeting T6SSs and competitor bacterial cell-targeting T6SSs (121, 122). For intravacuolar bacteria, the function of cell targeting is more relevant to their intracellular life cycle. *Salmonella* spp. harbor five phylogenetically distinct T6SSs, which are differentially distributed among serotypes (123). Two of them, SPI-6 and SPI-19, contribute to the pathogenesis of serotypes *S.* Typhimurium and *S.* Gallinarum in mouse and chicken macrophages, respectively (124, 125).

Mycobacterium spp. are unique among intracellular bacteria due to the composition of their cell wall, which is heavily modified by lipids and termed the mycomembrane. Perhaps as a consequence of this special structure that forms an outer membrane bilayer (126), these species use a family of specialized secretion system named the type VII secretion system (T7SS) (127). Putative T7SSs have been identified in some other Gram-positive organisms and are defined by the presence of two conserved elements, a membrane-bound ATPase, EccC, and a small secretion substrate, EsxB (128, 129). *M. tuberculosis* encodes up to five T7SSs (ESX-1 to ESX-5) that do not functionally complement each other. The importance of these secretion systems is highlighted by the fact that loss of the ESX-1 T7SS in *M. tuberculosis* is the most important genetic difference between virulent strains and the attenuated vaccine strain *Mycobacterium bovis* BCG (130, 131). The detailed structure and function of T7SSs are an active area of investigation (132). A current model suggests that the *M. tuberculosis*-containing vacuole is permeabilized by the T7SS effectors ESAT-6 (early secretory antigenic target-6) and CFP-10 (10-kDa culture filtrate protein), which allows bacterial products to access the cytosol (61, 127, 133).

Avoidance and Adaptation to the Phagolysosomal Pathway

The majority of intravacuolar bacterial pathogens have mechanisms to avoid being trafficked to the terminal phagolysosome and generate specialized remodeled membrane-bounded compartments. Although the common function of these vacuolar compartments is to support the intracellular replication of these bacteria, the specific features of these compartments are determined in a pathogen-specific fashion. As such, intravacuolar bacteria exploit several strategies to generate their specialized vacuoles, leading to hybrid organelles that are biochemically and morphologically distinct from typical compartments found in uninfected cells (Fig. 1).

Some intravacuolar bacteria actively evade the phagolysosomal pathway. Following entry, vacuoles containing brucellae, chlamydiae, and legionellae rapidly diverge from the phagolysosomal pathway and fail to acquire late phagosomal/lysosomal proteins (134). Both *Legionella*- and *Brucella*-containing vacuoles are endoplasmic reticulum (ER)-like membrane compartments and are associated with ribosomes (135, 136). Bacterial secretion systems are required for this divergent trafficking process (115, 116, 137). For example, *L. pneumophila* type IV effector proteins hijack the early secretory pathway and regulate vesicle traffic between the Golgi and the ER by targeting the host small GTPases Arf1 and Rab1, respectively (138–143). *C. trachomatis* also resides within a vacuole that is segregated from the phagolysosomal pathway. The *Chlamydia*-containing vacuole fuses with Golgi-derived vesicles and traffics to the Golgi area (144). The molecular mechanism of this process is still largely unknown, but the host GTPase dynamin and golgin-84 are involved in *Chlamydia*-mediated Golgi fragmentation, which is required for generating the *Chlamydia* replicative vacuoles (145, 146).

Mycobacteria and salmonellae exploit another strategy to avoid terminal phagolysosome formation and block phagosome maturation. Mycobacterial vacuoles are decorated with the early endosomal marker Rab5 and exclude the late endosomal GTPase Rab7, indicating an early arrest of the vacuole maturation (147). The type VII effector EsxH interacts with the host endosomal-sorting complex required for transport (ESCRT) and disrupts phagosome maturation (148). Moreover, the secreted *M. tuberculosis* protein tyrosine phosphatase (PtpA) prevents acidification of mycobacterial vacuoles by binding and excluding the V-ATPase machinery from the phagosome membrane (149). Interestingly, *Legionella* also targets and inhibits the host V-ATPase using the type IV effector SidK (150). These results suggest that targeting the vacuolar proton pump is a common strategy utilized by intravacuolar bacteria for the biogenesis of their specialized vacuoles. In addition, the active exclusion of PI3P from the *M. tuberculosis*-containing vacuole (151, 152) represents another strategy of maturation arrest by removing the docking site for the Rab5 effectors EEA1 (early endosome antigen 1) and Hrs (hepatocyte growth factor-regulated tyrosine kinase substrate) (153–155). Conversely, *S.* Typhimurium increases PI3P levels on the vacuole to stimulate biogenesis of a unique compartment with properties of late endosomes, and characterized by transient acquisition of EEA1, gradual lumen acidification, and acquisition of some lysosomal glycoproteins (156). However, unlike typical phagolysosomes, *Salmonella*-containing vacuoles are depleted of lysosomal degradation enzymes (157).

C. burnetii is a unique intravacuolar bacterium that does not follow this paradigm. *C. burnetii* resides in a vacuole that resembles a terminal phagolysosome and contains a variety of antimicrobial agents (158). However, instead of passively replicating in its vacuole, *C. burnetii* actively remodels it. For example, *Coxiella* is able to recruit the autophagy marker LC3 to its residing vacuole early after internalization and generate a specialized spacious vacuole in a T4SS-dependent manner (158–160).

Maintenance of Vacuole Integrity

The specialized vacuolar environment provides an ideal hideout from recognition by cytosolic innate immune sensors. Several studies suggest that damage to the bacterial vacuole membrane exposes the intravacuolar bacteria to the host cytosol and compromises the ability of these bacteria to colonize host cells (161–166). In macrophages and epithelial cells, cytosolic galectins are able to bind β-galactoside-associated membrane remnants derived from ruptured vacuoles. This leads to the recruitment of autophagy adaptor proteins and targeting by the autophagy machinery (161, 162). In addition, the presence of bacterial components in the host cytosol may activate immune responses and trigger cell death (167). Therefore, maintenance of the integrity of bacteria-containing vacuoles is essential for intravacuolar bacterial pathogens. Accordingly, recent studies suggest that intravacuolar bacteria target multiple host components and signaling pathways to maintain the stability of their residing vacuoles (164, 165).

The formation of specialized bacteria-containing vacuoles largely depends on vesicle trafficking, which relies on the host cytoskeleton network. Thus, it is not surprising that intravacuolar bacterial pathogens exploit effector proteins to remodel the cytoskeleton network. Interestingly, in addition to their role as a physical and structural support for trafficking, the cytoskeleton and cytoskeletal motors also contribute to vacuole stability, as exemplified by *Salmonella*- and *Chlamydia*-containing vacuoles (168, 169). Interestingly, loss of vacuole integrity is observed during infection with a *Chlamydia* mutant lacking the bacterial protease CPAF (chlamydial protease-like activity factor), which digests intermediate filament proteins to form a filamentous structure around its containing vacuoles (170). Similarly, during *Salmonella* infection, the type III effector

SifA regulates the interaction between cytoskeletal motors and the bacteria-containing vacuole (171, 172), and deletion of SifA leads to loss of vacuole integrity (165, 172, 173).

The formation of specialized bacteria-containing vacuoles is a highly ordered and tightly controlled process that requires the active modifications of the vacuole. The majority of these modifications involve the regulation of small GTPases by secreted bacterial effectors. Remarkably, in addition to their essential role in vacuole development, the timely acquisition of these small GTPases also contributes to vacuole stability (164). For example, one *Legionella* type IV effector, LidA, is involved in recruiting Rab1 and contributes to vacuole integrity (174). Comparatively, during *S.* Typhimurium infection, perturbation of GTPase activity by overexpressing dominant-positive Rab5 or dominant-negative Rab7 causes vacuole rupture and the release of bacteria into the cytosol (175).

A variety of bacterial pathogens affect the stability of the residing vacuoles by subverting vacuolar membrane lipid composition for their own benefit. For example, certain phosphatidylinositol phosphate species can be either enriched or excluded from bacteria-containing vacuoles in order to regulate the binding of host signaling molecules and bacterial effectors (176). Nevertheless, instead of just providing a docking site for the recruitment of molecules, some vacuolar membrane lipids may play additional roles in vacuole stability. *C. trachomatis* recruits host sphingomyelin to its vacuole by incorporating lipid droplets (111). Although the precise role of this molecule in vacuole stability is not yet clear, host sphingomyelin acquisition is absolutely required for vacuole expansion and prevents *Chlamydia* vacuole fragmentation (177, 178). *S.* Typhimurium and *L. pneumophila* are able to reduce cholesterol levels from their specialized vacuoles by secreting phospholipases SseJ and PlaA (179, 180), respectively. This process is thought to be beneficial to these intravacuolar bacteria by increasing membrane fluidity and inhibiting lipid raft formation. However, exclusion of cholesterol from the vacuolar surface adversely affects its stability. *S.* Typhimurium and *L. pneumophila* use additional secreted effectors, SifA and SdhA, to counter this defect (164, 181, 182).

Autophagy Is a Double-Edged Sword for Intravacuolar Bacteria

Although vacuole integrity is critical for intravacuolar bacteria, interacting with the host cytosol requires some level of vacuole permeabilization. However, membrane permeabilization leads to targeting by the selective autophagy pathway, as exemplified by *M. tuberculosis* (183). Exposure of *M. tuberculosis* DNA following vacuole permeabilization triggers ubiquitin-mediated autophagy targeting through the cGAS/STING/TANK-binding kinase 1 (TBK1) pathway (49, 183). Similarly, *S.* Typhimurium accidently exposed to the cytosol can also be targeted by autophagy, as shown by the recruitment of LC3 and other ATG proteins to the bacteria. Adaptor proteins, such as NDP52 and optineurin, as well as TBK1, have a role in recognizing ubiquitylated *Salmonella* in the cytosol (82, 184).

To survive and propagate in the face of host cell-autonomous defense mechanisms, intravacuolar bacteria such as *M. tuberculosis* and *L. pneumophila* actively antagonize the host autophagy machinery. The *M. tuberculosis* secreted redox regulator Eis inhibits autophagy by indirectly blocking Jun N-terminal kinase (JNK) activation (185). The suppression of the PI3-kinase VPS34 (vacuolar protein sorting 34) (151) and exclusion of PI3P from *M. tuberculosis*-containing vacuoles (152) could be another active strategy to avoid autophagy, as PI3P is required for autophagy initiation. In addition, a recent study suggests that *M. tuberculosis* inhibits autophagy flux in a T7SS-dependent manner, which suggests the existence of an uncharacterized autophagy modulator (186). Remarkably, the *L. pneumophila* effector RavZ directly and irreversibly uncouples ATG8 proteins attached to phosphatidylethanolamine on preautophagosomal structures (187).

In contrast, some bacteria, such as *C. burnetii* (159, 188, 189), *B. abortus* (190), and *A. phagocytophilum* (191, 192), subvert the autophagic pathway for the formation of their specialized vacuole. For example, pharmacological inhibition of autophagy flux arrests the biogenesis of *Coxiella*- and *Anaplasma*-containing vacuoles and impairs their growth in macrophages (189, 192). Both *C. burnetii* and *A. phagocytophilum* vacuoles are decorated with the autophagic protein beclin-1 (191, 193), a protein that forms a complex with VPS34 and is required for the initial steps of the autophagy pathway. Interestingly, *C. burnetii* infection also induces recruitment of the antiapoptotic protein B-cell lymphoma 2 (BCL-2) to its vacuole surface, and the interaction between beclin-1 and BCL-2 inhibits host apoptosis (193). Therefore, some intravacuolar bacteria use the autophagy pathway to promote intracellular replication and to block apoptosis (194). It is also possible that the autophagy pathway plays an important role in membrane biogenesis and in providing nutrients to the specialized vacuoles of these pathogens (67, 194). For example, one *C. burnetii* T4SS effector protein, Cig2, is involved in recruiting autophagosomes to *Coxiella* spacious vacuoles (159). The detailed mechanism of how autophagosomes are recruited and incorporated into vacuoles still needs to be elucidated, although bacterial secretion systems and effectors are likely to play a pivotal role in these processes (67).

Manipulation of Host Cell Death by Intravacuolar Bacteria

Even though intravacuolar bacteria reside in an environment that prevents host cytosolic sensing, the subversion of the host processes can pose stresses to eukaryotic cells and trigger host cell death. Hence, intravacuolar bacterial pathogens have developed a myriad of strategies to modulate host cell death pathways (96, 195).

Mitochondria play a central role in modulating host cell death and survival. As such, many intravacuolar bacteria subvert mitochondria-dependent cell death pathways. *L. pneumophila* and *C. burnetii* secrete the type IV effector proteins SidF and AnkG into the host cytosol and target the proapoptotic factors BNIP3 (BCL-2 and adenovirus E1B 19 kDa-interacting protein 3) and p32, respectively, thereby inhibiting mitochondria-mediated apoptotic signaling (196, 197). Similarly, *C. trachomatis* inhibits apoptosis by secreting the protease CPAF, which degrades proapoptotic BH3-only proteins, including BIM (BCL-2-interacting mediator of cell death), PUMA (p53 upregulated modulator of apoptosis), and BAD (BCL-2-associated death promoter) (198, 199).

In addition to dampening proapoptotic pathways, activation of prosurvival pathways is another mechanism utilized by intravacuolar bacteria to prevent host cell death. The master regulator of the innate immune response NF-κB promotes cell survival, and the subversion of the NF-κB pathway also represents a common bacterial strategy to modulate host cell death. For example, the *L. pneumophila* effector LegK1 phosphorylates and inactivates the NF-κB inhibitor IκBα and p100, thus enhancing the activation of the NF-κB prosurvival pathway (200). Some intravacuolar bacteria alter the activity of host kinases that are linked to NF-κB activation. One *S.* Typhimurium effector, AvrA, has acetyltransferase activity that targets mitogen-activated protein kinase kinases and strongly inhibits the JNK signaling pathway (201), which helps dampen inflammatory and cell death responses. Moreover, although the molecular mechanisms are still largely unknown, Akt activation and enhancement of cell survival have been observed during *C. trachomatis* (202), *C. burnetii* (203), *and S.* Typhimurium (204) infections.

Although intravacuolar bacteria are separated from most cytosolic immune sensing mechanisms, their specialized secretion systems poke holes in host membranes, which can be detected by the host cell. It has been shown that the basal body rod component of the T3SS SPI-1 apparatus of *S.* Typhimurium (PrgJ) is detected by NLRC4 and triggers inflammasome activation. However, SsaI, the equivalent basal body rod component of the T3SS SPI-2, is not detected by NLRC4. This constitutes an evasion strategy required for *S.* Typhimurium virulence (205).

INFECTION OF MACROPHAGES BY CYTOSOLIC BACTERIA

General Strategies Deployed by Cytosolic Bacteria To Infect Macrophages

During their intracellular life cycle, *B. pseudomallei*, *Francisella tularensis*, and *L. monocytogenes* escape from a phagosome, replicate in the macrophage cytosol, and manipulate host immune responses (18). In contrast to *F. tularensis*, both *B. pseudomallei* and *L. monocytogenes* exploit the host actin polymerization machinery to move within the cytosol and spread from cell to cell (206). The intracellular life cycles of *B. pseudomallei* and *F. tularensis* have unique features such as the formation of multinucleated giant cells (207, 208) and the translocation to a membrane-bound compartment subsequent to replication in the host cytosol (22), respectively. It is noteworthy that other cytosolic bacterial pathogens such as *Mycobacterium marinum*, *Rickettsia* spp., and *Shigella flexneri* are also known to infect macrophages (18). However, macrophages are not the primary cells targeted during rickettsial infections (209), and *Shigella* spp. induce macrophage cell death instead of replicating intracellularly as a strategy of pathogenesis (210–212).

Cytosolic bacteria secrete virulence factors within the host cells in order to establish infection. In *L. monocytogenes*, the virulence factors involved in phagosomal escape and cell-to-cell spread are well characterized and are mostly encoded on the *Listeria* pathogenicity island 1 (213, 214). *L. monocytogenes* relies mainly on the canonical Sec translocation system to secrete virulence factors (215, 216), while the translocation of effectors in *B. pseudomallei* and *F. tularensis* depends on specialized secretion systems. In *B. pseudomallei*, a genetic locus with similarity to the *Salmonella* and *Shigella* T3SS is required for replication in host cells (217, 218). A T6SS also contributes to the cell-to-cell spread ability and the virulence of *B. pseudomallei*, but the effectors secreted by this system are not well characterized (219, 220). Interestingly, it was recently shown that the T6SS effector VrgG5 mediates host cell fusion (221, 222), which is suggested to be required for intercellular spread (223). Several virulence determinants involved in the intracellular life cycle of *F. tularensis* are encoded on the *Francisella* pathogenicity island (224, 225). A number of the genes within this island share sequence homology to the T6SS gene clusters found in *Vibrio cholerae* and *Pseudomonas aeruginosa* (226, 227), and the products of at least two of these genes are translocated in the host cytosol during infection (228).

Mechanisms Used by Cytosolic Bacteria To Escape Phagosomes

The ability to escape from vacuoles is crucial to cytosolic pathogens. Following internalization, cytosolic bacteria escape from a phagosomal compartment in as early as 5 min, which may reflect a need to escape the phagocytic pathway before fusion with the lysosome (18). Cytosolic bacteria that exploit actin-based motility to spread from cell to cell also encounter a double-membrane vesicle referred to as the secondary vacuole. Escape from the secondary vacuole seems to process in a similar manner to the escape from the primary vacuole, at least for *L. monocytogenes* and *S. flexneri* (18).

L. monocytogenes avoids fusion with lysosomes and escapes an acidified compartment with features of a late endosome (229, 230). More specifically, *L. monocytogenes* escapes the phagosome using the cholesterol-dependent cytolysin listeriolysin O (LLO) (231), which forms pores in vacuoles, blocks the trafficking of the bacteria within the endosomal pathway (229, 230), and ultimately promotes membrane disruption (232–234). LLO activity is compartmentalized by its acidic pH optimum, which prevents damage to the host cell (235). In addition, the reduction of LLO by GILT (IFN-γ-inducible lysosomal thiol reductase) (236) and the CFTR (cystic fibrosis transmembrane conductance regulator)-dependent increase in chloride concentration (237) potentiate the activity of LLO within the phagosome. The escape of *L. monocytogenes* from vacuoles is also facilitated by a phosphatidylinositol-specific phospholipase C (PlcA) and a broad-range phospholipase C (PlcB) (238). Whereas Goldfine and colleagues showed that PlcA promotes phagosomal escape by activating the host protein kinase C-β (239, 240), the involvement of PlcB in this process is attributed to wide-spectrum activity against phospholipids and to an ability to mediate membrane fusion (241, 242). The interplay between LLO and the phospholipases C (PLCs) is not completely understood, but it is possible that these listerial factors synergistically destabilize the membrane to promote optimal escape from the phagosome. It has also been suggested that PLCs translocate into the host cytosol through LLO pores (243).

The detailed molecular mechanisms involved in phagosomal escape are mostly uncharacterized in most cytosolic bacterial pathogens. *B. pseudomallei* mutants lacking components of the T3SS and, more specifically, the T3SS effector BopA have a defect in the escape from the phagosome (93, 217). In *F. tularensis*, the secretion system encoded by the *Francisella* pathogenicity island is required for phagosomal escape, but the specific set of virulence factors involved is not well characterized (225, 228, 244–246). Interestingly, the *Shigella* effectors IpaB and IpaC, like LLO, form a pore complex that binds cholesterol and inserts into cell membranes (247, 248). Similarly, *Rickettsia* spp. produce a hemolysin and phospholipases that may play a role in escape from vacuoles (249–252).

Sensing of the Vacuole-to-Cytosol Transition by Cytosolic Bacteria

Cytosolic bacteria are likely to exploit distinct sets of virulence factors at different stages of their intracellular life cycle. The sensing of the environmental changes encountered throughout their passage within the host cell is likely to be critical for the tight regulation of their virulence factors. However, although it is established that cytosolic bacteria induce the expression of specific sets of genes during intracellular infections (253–256), less is known about their spatial and temporal regulation within the intracellular niche. The intricacies of virulence factor regulation are exemplified by the on/off expression of the *S. flexneri* T3SS apparatus during cell infection, which is specifically activated during bacterial entry (257).

The Crp family member transcription factor PrfA upregulates virulence factor expression once *L. monocytogenes* reaches the host cytosol (258, 259). The activity of PrfA integrates both environmental and bacterial signals and is regulated by temperature (260), a bacterial autorepressor (261), and the availability of specific nutrients (262–268). Reniere et al. (269) recently identified that bacterial- and host-derived glutathione allosterically binds and activates PrfA in the host cytosol. Given that enhanced PrfA activation is dispensable for vacuole escape (270), the authors suggested a two-step activation mechanism based on the oxidation-reduction states of PrfA thiols and glutathione in the vacuole (oxidizing/repressing environment) and in the cytosol (reducing/activating environment). Although it is suggested that *S. flexneri* also takes advantage of an intracellular signal to activate the MxiE transcription factor during infection (271), the environmental cues and the bacterial regulatory machinery used to regulate virulence factors expression are still largely unknown in other cytosolic bacteria.

Manipulation of Host Sensing Pathways by Cytosolic Bacteria

Although cytosolic bacteria are protected from the extracellular immune responses and escape the microbicidal lysosomal environment, their detection by cytosolic PRRs activates defense mechanisms including autophagy, host cell death pathways, and secretion of cytokines. However, in some instances, it should be noted that the detection by the host cell and the subsequent consequences might be advantageous for the pathogen. For example, as discussed above, there is increasing evidence that some intracellular bacteria benefit from IFN-β signaling (272).

The notion that cytosolic pathogens employ active mechanisms to delay or redirect the immune response is attractive but is supported by few studies. *L. monocytogenes* might dampen the host immune response by secreting the virulence factor InlC and by interfering with the activation of NF-κB during intracellular infection (273), but it is unclear whether this has an impact on virulence (274). It is also speculated that the active modulation of the immune response by *F. tularensis* contributes to pathogenesis, although the mechanisms remain poorly understood (224, 275). The ability of *L. monocytogenes* to grow in host cells without inducing inflammasome activation is quite striking, and might involve regulatory mechanisms that limit flagellin expression during infection in order to avoid detection by NAIP5/NLRC4 (30, 276, 277). It is also plausible that *L. monocytogenes* avoids autolysis in host cells to minimize detection by the AIM2 inflammasome. Sauer et al. (37) demonstrated that while *L. monocytogenes* mutants with cell wall defects lyse within host cells and induce the AIM2 inflammasome, wild-type bacteria showed very little intracellular bacteriolysis and induced low levels of pyroptosis. Conversely, the activation of the AIM2/ASC inflammasome plays a pivotal role in the innate immune response to *F. tularensis* and leads to the control of bacterial replication both in macrophages and *in vivo* (34–36, 278). However, the *F. tularensis* protein encoded by FTL-0325 delays inflammasome activation during infection (279), possibly by contributing to the structural integrity of the bacterial surface rather than by actively repressing host inflammasome activation (280). Interestingly, it was recently shown that the *S. flexneri* T3SS effector OspC3 inhibits the induction of a noncanonical inflammasome by directly targeting caspase-4, the human ortholog of mouse caspase-11 (281). Nevertheless, it is not known whether this mechanism is active in myeloid cells or whether other cytosolic pathogens use a similar strategy to inhibit inflammasome activation.

Interaction of Cytosolic Bacteria with the Autophagy Pathway

Infection of host cells by cytosolic pathogens such as *L. monocytogenes* and *S. flexneri* leads to induction of autophagy (282–284), and not surprisingly, mechanisms of autophagy evasion are described for most cytosolic bacterial pathogens. *L. monocytogenes* uses ActA and InlK to recruit host proteins (Arp2/3 and Ena/VASP, and the major vault protein, respectively) to the bacterial surface and interfere with autophagic recognition (78, 285). *L. monocytogenes* also causes stalling of preautophagosomal structures by reducing the autophagic flux and PI3P levels in a PLC-dependent manner (283). Accordingly, it was recently demonstrated that the inactivation of both ActA and PlcA drastically impairs the ability of *L. monocytogenes* to avoid autophagy and to grow within macrophages (286). *S. flexneri* interferes with autophagic recognition by secreting IcsB, which competes with ATG5 for binding to the bacterial protein VirG/IcsA (287). In addition, the *F. tularensis* polysaccharidic O antigen contributes to autophagy evasion by preventing ubiquitylation and the recruitment of the autophagy adaptor p62 (288). Overall, these studies clearly demonstrate that interference with autophagy is a common strategy utilized by cytosolic bacteria.

Instead of being targeted by macroautophagy, *B. pseudomallei* is targeted by LAP (93). *B. pseudomallei* mutants lacking the T3SS effector BopA showed a defect in escaping single-membrane vacuoles and an increased colocalization with LC3 (93). Although the *B. pseudomallei* proteins BimA and BopA have high homology with the *S. flexneri* proteins IcsA and IcsB, it was suggested that they interfere with autophagy through different mechanisms (289, 290). However, recent data suggest that IcsB not only interferes with autophagic recognition of the bacterial surface but also inhibits LAP or LC3 recruitment to vacuolar remnants early during *S. flexneri* infection (291). Interestingly, *L. monocytogenes* is also targeted by LAP early during macrophage infection (292). Overall, these studies suggest that the autophagy machinery can target cytosolic bacteria at the entry stage of their intracellular life cycle.

In most cases, it is assumed that the activity of the autophagy pathway is detrimental for cytosolic bacteria, although it might not always be the case. For example, a proportion of *F. tularensis* reenters a membrane-bound compartment with autolysosomal features subsequent to replication in the cytosol (22), but it is not clear whether this represents a mechanism of dissemination or a host cellular defense mechanism. Evidence also suggests that autophagy provides *F. tularensis* with the nutrients required for intracellular replication (293). It is thus plausible that cytosolic bacteria benefit from an increase in nutrient availability resulting from autophagy activation.

Manipulation of Host Cell Death Pathways by Cytosolic Bacteria

Although cytosolic bacteria may protect the intracellular replication niche by suppressing host cell death pathways, death of the infected host cell is inevitable because the host cell has multilayered mechanisms to induce cell death pathways in response to bacterial infections. For example, infection of macrophages with *B. pseudomallei* induces an early NLRC4-dependent pyroptosis and late caspase-1-dependent and -independent cell death pathways (294). More specifically, induction of apoptosis was observed in caspase-1/11-deficient macrophages infected with *B. pseudomallei* (294). Similarly, caspase-1-deficient macrophages infected with *F. tularensis* undergo AIM2/ASC-dependent caspase-8-mediated apoptosis (295). As already mentioned, *F. tularensis* delays inflammasome activation during the early stages of infection (279) but also uses TolC-secreted effectors to delay the activation of the intrinsic apoptotic pathway in order to preserve the host cell (296). This suggests that cytosolic bacteria have to interact with several host pathways in order to inhibit, or at least delay, cell death. Nevertheless, our understanding of the mechanisms used by cytosolic bacteria to inhibit host cell death pathways remains mostly superficial.

Another strategy used by cytosolic bacteria to maintain their replication niche is to induce host prosurvival pathways. For example, *S. flexneri* and *R. rickettsii* activate prosurvival pathways to dampen cell death signals in nonphagocytic cells (297–299), but it is unknown whether these pathways also promote macrophage survival during infection. Cytosolic bacteria such as *L. monocytogenes* damage host membranes during their intracellular life cycle. However, in the case of *L. monocytogenes*, LLO has several mechanisms to minimize damage to the host cell membranes (231). In addition, adaptive mechanisms are likely to be deployed by the host cell to withstand and repair the lesions and thus promote survival (300). The induction of autophagy is a common host response to several bacterial toxins and might constitute a host protection or detoxification mechanism (300, 301). Interestingly, a tight relationship exists between autophagy and apoptosis, and it is possible that bacterial pathogens target host components shared by both pathways to promote survival of the host cell (301).

Some cytosolic bacteria may benefit from inducing a specific host cell death pathway during infection. For example, it was suggested that the accumulation of apoptotic cell debris leads to alternative macrophage activation and decreased bacterial clearance during

Table 2 General strategies used by intravacuolar and cytosolic bacteria to deal with challenges encountered within host cells

	Strategies utilized by intracellular bacteria	
Challenge	Intravacuolar	Cytosolic
Phagolysosomal pathway	Avoidance, blockage, and adaptation	Escape into the cytosol
Access to nutrients	Hijack host vesicle trafficking to promote vacuole biogenesis and nutrient acquisition	Direct access to cytosolic nutrients
Microenvironment	Intravacuolar bacteria create and enlarge their replication space using a variety of effectors	The cytosol constitutes a spacious and favorable environment
Innate recognition	The vacuolar compartment provides a hideout from cytosolic innate sensing[a]	Cytosolic bacteria are directly exposed to the cytosolic sensing machinery and need to downregulate PAMP expression to delay recognition
Autophagy	Maintenance of vacuole integrity, inhibition of autophagy initiation, inhibition of autophagy flux	Inhibition of autophagy initiation, inhibition of autophagy flux, prevention of autophagic recognition
Host cell death	Dampening of proapoptotic pathways, activation of prosurvival pathways	Interference with inflammasome activation, interaction with proapoptotic pathways, activation of prosurvival pathways

[a]Some components of intravacuolar bacterial pathogens can be recognized by host cytosolic sensing mechanisms due to damage to the pathogen-containing vacuoles or effector translocation by specialized secretion systems. For example, DNA from *M. tuberculosis*, PrgJ from *S. enterica*, and the flagellum of *L. pneumophila* are recognized by the host cytosolic sensing machinery.

Table 3 Examples of factors used by intracellular bacteria to counteract host defense mechanisms

Subverted host defense mechanism	Strategies	Pathogen	Example of bacterial factors
Immune detection	Actin-based cell-to-cell spread	*B. pseudomallei*	BimA
		L. monocytogenes	ActA
	Modification of the immune response	*F. tularensis*	T6SS effectors
		L. monocytogenes	InlC
	Low flagellin expression	*L. monocytogenes*	MogR
	Maintenance of vacuole integrity	*C. pneumoniae*	CPAF
		L. pneumophila	LidA, PlaA, SdhA
		S. Typhimurium	SseJ, SifA
Phagolysosomal pathway	Blockage of phagosome maturation	*L. pneumophila*	SidK
		M. tuberculosis	EsxH, PtpA
		S. Typhimurium	Unknown
	Escape from the vacuole	*B. pseudomallei*	BopA
		F. tularensis	T6SS effector(s)
		L. monocytogenes	LLO, PLCs
	Evasion of the endocytic pathway	*Brucella* spp.	VirB(T4SS) effectors
		C. trachomatis	Unknown
		L. pneumophila	Dot/Icm(T4SS) effectors
	Phagosome permeabilization	*M. tuberculosis*	ESAT-6, CFP-10
	Survival in a lysosome-like structure	*C. burnetii*	Dot/Icm(T4SS) effectors
Autophagy	Escape from LAP	*B. pseudomallei*	BopA
	Formation of a replication niche	*A. phagocytophilum*	Ats-1
		Brucella spp.	VirB(T4SS) effectors
		C. burnetii	Dot/Icm(T4SS) effectors
	Interference with recognition	*F. tularensis*	O antigen
		L. monocytogenes	ActA, InlK
	Inhibition of autophagy initiation/flux	*L. monocytogenes*	PLCs
		L. pneumophila	RavZ
		M. tuberculosis	Eis
	Nutrient acquisition	*F. tularensis*	Unknown
Cell death	Inhibition/delay of apoptosis	*C. burnetii*	AnkG
		C. trachomatis	CPAF
		F. tularensis	TolC-dependent
		L. pneumophila	SidF
		L. pneumophila	LegK1
		S. Typhimurium	AvrA
	Induction of apoptosis	*B. pseudomallei*	CHBP
	Hijacking of efferocytosis	*L. monocytogenes*	LLO, ActA
	Evasion of inflammasome	*S.* Typhimurium	SsaI

F. tularensis infection (302). It thus seems conceivable that *F. tularensis* prevents early cell death induction to allow intracellular bacterial replication (279, 296), but then specifically induces apoptosis to promote an anti-inflammatory immune response (302, 303). Interestingly, *B. pseudomallei* delivers a cycle-inhibiting factor homolog (CHBP) that triggers macrophage-specific apoptosis (304), but whether this leads to an anti-inflammatory immune response remains to be determined. Although macrophages infected with *L. monocytogenes* do not undergo apoptotic cell death (305), *L. monocytogenes* exploits the process of apoptotic cell death clearance by macrophages (i.e., efferocytosis) to facilitate cell-to-cell spread and dissemination (306). Overall, these studies suggest that cytosolic bacteria manipulate the host apoptotic pathway to their own advantage during intracellular replication.

CONCLUDING REMARKS

Macrophages, as major immune sentinel cells, play an essential role in sensing and destroying invading microorganisms and release cytokines that shape the immune response of other cells. In comparison to extracellular bacterial pathogens, which are fully exposed to all immune cells and components, intracellular bacteria hide and replicate within host cells. Paradoxically, many intracellular bacterial pathogens preferentially replicate within macrophages. Following long-term evolutionary selection, these highly adapted intracellular bacteria have selected a specific replication niche inside these host cells and have evolved sophisticated mechanisms that facilitate their intracellular replication (Fig. 1). The study of these bacterial pathogens has led to numerous mechanistic insights on basic host processes including endocytosis, vesicle trafficking, and innate immune responses.

As reviewed here, intracellular bacteria have to deal with a variety of host defense mechanisms after being phagocytosed by macrophages (Table 2). To facilitate a comparative analysis of bacterial counterstrategies, we exemplify some pathogen-host interactions that illustrate how these organisms use their effector arsenal to face the challenges encountered during their intracellular life cycle (Table 3). A delicate balance exists between the ability of the host cell to control invading microbes and the ability of pathogens to hide from or subvert these defenses. Due to the length of this review, we are not able to cover all aspects of host-pathogen interaction, but we highlight some of the most important and current topics in the field. Importantly, several interesting questions are still not fully understood. For example, both *M. tuberculosis* and *M. marinum* encode ESX-1 T7SSs; however, they replicate in a vacuolar compartment and in the cytosol, respectively. It is puzzling why these genetically related bacteria that share common secretion systems replicate in distinct intracellular niches. Also, the function of type I IFN induction during intracellular bacterial infection is still elusive. Elucidating the molecular mechanisms underlying pathogen subversion strategies and how they specifically affect innate immune responses is critical to a comprehensive understanding of bacterial pathogenesis and the host response. In addition, the study of host-intracellular bacteria interactions may provide clues for advancing the development of novel therapeutics against pathogens and inflammatory diseases.

Acknowledgments. *We thank Thomas P. Burke for the critical reading of this manuscript. This work was supported by National Institutes of Health grants 1P01 AI63302 (D.A.P.) and 1R01 AI27655 (D.A.P.). G.M. was supported by fellowships from Fonds de recherche santé Québec and the Natural Sciences and Engineering Research Council of Canada. C.C. was supported by National Research Service Award Institutional Training Grants 1T32AI100829. G.M. and C.C. contributed equally to this work. D.A.P. has a consulting relationship with and a financial interest in Aduro Biotech, and both he and the company stand to benefit from the commercialization of the results of his research.*

Citation. Mitchell G, Chen C, Portnoy DA. 2016. Strategies used by bacteria to grow in macrophages. Microbiol Spectrum 4(3):MCHD-0012-2015.

References

1. **Fauci AS, Morens DM.** 2012. The perpetual challenge of infectious diseases. *N Engl J Med* **366:**454–461.
2. **Price JV, Vance RE.** 2014. The macrophage paradox. *Immunity* **41:**685–693.
3. **Dumler JS, Choi KS, Garcia-Garcia JC, Barat NS, Scorpio DG, Garyu JW, Grab DJ, Bakken JS.** 2005. Human granulocytic anaplasmosis and *Anaplasma phagocytophilum. Emerg Infect Dis* **11:**1828–1834.
4. **Franco MP, Mulder M, Gilman RH, Smits HL.** 2007. Human brucellosis. *Lancet Infect Dis* **7:**775–786.
5. **Piggott JA, Hochholzer L.** 1970. Human melioidosis. A histopathologic study of acute and chronic melioidosis. *Arch Pathol* **90:**101–111.
6. **Grayston JT, Aldous MB, Easton A, Wang SP, Kuo CC, Campbell LA, Altman J.** 1993. Evidence that *Chlamydia pneumoniae* causes pneumonia and bronchitis. *J Infect Dis* **168:**1231–1235.
7. **Gross RJ, Rowe B, Easton JA.** 1973. Neonatal meningitis caused by *Citrobacter koseri. J Clin Pathol* **26:** 138–139.
8. **Maurin M, Raoult D.** 1999. Q fever. *Clin Microbiol Rev* **12:**518–553.
9. **Dumler JS, Madigan JE, Pusterla N, Bakken JS.** 2007. Ehrlichioses in humans: epidemiology, clinical presentation, diagnosis, and treatment. *Clin Infect Dis* **45** (Suppl 1):S45–S51.
10. **Sjöstedt A.** 2007. Tularemia: history, epidemiology, pathogen physiology, and clinical manifestations. *Ann N Y Acad Sci* **1105:**1–29.
11. **Phin N, Parry-Ford F, Harrison T, Stagg HR, Zhang N, Kumar K, Lortholary O, Zumla A, Abubakar I.** 2014. Epidemiology and clinical management of Legionnaires' disease. *Lancet Infect Dis* **14:**1011–1021.
12. **Hof H.** 2003. History and epidemiology of listeriosis. *FEMS Immunol Med Microbiol* **35:**199–202.
13. **Daniel TM.** 2006. The history of tuberculosis. *Respir Med* **100:**1862–1870.
14. **Prescott JF.** 1991. *Rhodococcus equi*: an animal and human pathogen. *Clin Microbiol Rev* **4:**20–34.
15. **Harrell GT.** 1949. Rocky Mountain spotted fever. *Medicine (Baltimore)* **28:**333–370.
16. **Blaser MJ, Newman LS.** 1982. A review of human salmonellosis: I. Infective dose. *Rev Infect Dis* **4:**1096–1106.
17. **Thi EP, Lambertz U, Reiner NE.** 2012. Sleeping with the enemy: how intracellular pathogens cope with a macrophage lifestyle. *PLoS Pathog* **8:**e1002551. doi: 10.1371/journal.ppat.1002551.
18. **Ray K, Marteyn B, Sansonetti PJ, Tang CM.** 2009. Life on the inside: the intracellular lifestyle of cytosolic bacteria. *Nat Rev Microbiol* **7:**333–340.
19. **Alix E, Mukherjee S, Roy CR.** 2011. Subversion of membrane transport pathways by vacuolar pathogens. *J Cell Biol* **195:**943–952.

20. **Fredlund J, Enninga J.** 2014. Cytoplasmic access by intracellular bacterial pathogens. *Trends Microbiol* **22:** 128–137.

21. **Birmingham CL, Canadien V, Kaniuk NA, Steinberg BE, Higgins DE, Brumell JH.** 2008. Listeriolysin O allows *Listeria monocytogenes* replication in macrophage vacuoles. *Nature* **451:**350–354.

22. **Checroun C, Wehrly TD, Fischer ER, Hayes SF, Celli J.** 2006. Autophagy-mediated reentry of *Francisella tularensis* into the endocytic compartment after cytoplasmic replication. *Proc Natl Acad Sci U S A* **103:**14578–14583.

23. **Plüddemann A, Mukhopadhyay S, Gordon S.** 2011. Innate immunity to intracellular pathogens: macrophage receptors and responses to microbial entry. *Immunol Rev* **240:**11–24.

24. **Murray PJ, Wynn TA.** 2011. Protective and pathogenic functions of macrophage subsets. *Nat Rev Immunol* **11:** 723–737.

25. **Flannagan RS, Cosío G, Grinstein S.** 2009. Antimicrobial mechanisms of phagocytes and bacterial evasion strategies. *Nat Rev Microbiol* **7:**355–366.

26. **Broz P, Monack DM.** 2013. Newly described pattern recognition receptors team up against intracellular pathogens. *Nat Rev Immunol* **13:**551–565.

27. **Kawai T, Akira S.** 2006. TLR signaling. *Cell Death Differ* **13:**816–825.

28. **Strober W, Murray PJ, Kitani A, Watanabe T.** 2006. Signalling pathways and molecular interactions of NOD1 and NOD2. *Nat Rev Immunol* **6:**9–20.

29. **Keestra AM, Winter MG, Auburger JJ, Frässle SP, Xavier MN, Winter SE, Kim A, Poon V, Ravesloot MM, Waldenmaier JF, Tsolis RM, Eigenheer RA, Bäumler AJ.** 2013. Manipulation of small Rho GTPases is a pathogen-induced process detected by NOD1. *Nature* **496:**233–237.

30. **von Moltke J, Ayres JS, Kofoed EM, Chavarría-Smith J, Vance RE.** 2013. Recognition of bacteria by inflammasomes. *Annu Rev Immunol* **31:**73–106.

31. **Vanaja SK, Rathinam VA, Fitzgerald KA.** 2015. Mechanisms of inflammasome activation: recent advances and novel insights. *Trends Cell Biol* **25:**308–315.

32. **Hornung V, Ablasser A, Charrel-Dennis M, Bauernfeind F, Horvath G, Caffrey DR, Latz E, Fitzgerald KA.** 2009. AIM2 recognizes cytosolic dsDNA and forms a caspase-1-activating inflammasome with ASC. *Nature* **458:** 514–518.

33. **Fernandes-Alnemri T, Yu JW, Datta P, Wu J, Alnemri ES.** 2009. AIM2 activates the inflammasome and cell death in response to cytoplasmic DNA. *Nature* **458:** 509–513.

34. **Fernandes-Alnemri T, Yu JW, Juliana C, Solorzano L, Kang S, Wu J, Datta P, McCormick M, Huang L, McDermott E, Eisenlohr L, Landel CP, Alnemri ES.** 2010. The AIM2 inflammasome is critical for innate immunity to *Francisella tularensis*. *Nat Immunol* **11:** 385–393.

35. **Jones JW, Kayagaki N, Broz P, Henry T, Newton K, O'Rourke K, Chan S, Dong J, Qu Y, Roose-Girma M, Dixit VM, Monack DM.** 2010. Absent in melanoma 2 is required for innate immune recognition of *Francisella tularensis*. *Proc Natl Acad Sci U S A* **107:**9771–9776.

36. **Rathinam VA, Jiang Z, Waggoner SN, Sharma S, Cole LE, Waggoner L, Vanaja SK, Monks BG, Ganesan S, Latz E, Hornung V, Vogel SN, Szomolanyi-Tsuda E, Fitzgerald KA.** 2010. The AIM2 inflammasome is essential for host defense against cytosolic bacteria and DNA viruses. *Nat Immunol* **11:**395–402.

37. **Sauer JD, Witte CE, Zemansky J, Hanson B, Lauer P, Portnoy DA.** 2010. *Listeria monocytogenes* triggers AIM2-mediated pyroptosis upon infrequent bacteriolysis in the macrophage cytosol. *Cell Host Microbe* **7:** 412–419.

38. **Tsuchiya K, Hara H, Kawamura I, Nomura T, Yamamoto T, Daim S, Dewamitta SR, Shen Y, Fang R, Mitsuyama M.** 2010. Involvement of absent in melanoma 2 in inflammasome activation in macrophages infected with *Listeria monocytogenes*. *J Immunol* **185:**1186–1195.

39. **Warren SE, Armstrong A, Hamilton MK, Mao DP, Leaf IA, Miao EA, Aderem A.** 2010. Cutting edge: cytosolic bacterial DNA activates the inflammasome via Aim2. *J Immunol* **185:**818–821.

40. **Ishikawa H, Barber GN.** 2008. STING is an endoplasmic reticulum adaptor that facilitates innate immune signalling. *Nature* **455:**674–678.

41. **Zhong B, Yang Y, Li S, Wang YY, Li Y, Diao F, Lei C, He X, Zhang L, Tien P, Shu HB.** 2008. The adaptor protein MITA links virus-sensing receptors to IRF3 transcription factor activation. *Immunity* **29:**538–550.

42. **Ishikawa H, Ma Z, Barber GN.** 2009. STING regulates intracellular DNA-mediated, type I interferon-dependent innate immunity. *Nature* **461:**788–792.

43. **Sauer JD, Sotelo-Troha K, von Moltke J, Monroe KM, Rae CS, Brubaker SW, Hyodo M, Hayakawa Y, Woodward JJ, Portnoy DA, Vance RE.** 2011. The *N*-ethyl-*N*-nitrosourea-induced *Goldenticket* mouse mutant reveals an essential function of *Sting* in the *in vivo* interferon response to *Listeria monocytogenes* and cyclic dinucleotides. *Infect Immun* **79:**688–694.

44. **Burdette DL, Monroe KM, Sotelo-Troha K, Iwig JS, Eckert B, Hyodo M, Hayakawa Y, Vance RE.** 2011. STING is a direct innate immune sensor of cyclic di-GMP. *Nature* **478:**515–518.

45. **Sun L, Wu J, Du F, Chen X, Chen ZJ.** 2013. Cyclic GMP-AMP synthase is a cytosolic DNA sensor that activates the type I interferon pathway. *Science* **339:** 786–791.

46. **Wu J, Sun L, Chen X, Du F, Shi H, Chen C, Chen ZJ.** 2013. Cyclic GMP-AMP is an endogenous second messenger in innate immune signaling by cytosolic DNA. *Science* **339:**826–830.

47. **Zhang Y, Yeruva L, Marinov A, Prantner D, Wyrick PB, Lupashin V, Nagarajan UM.** 2014. The DNA sensor, cyclic GMP-AMP synthase, is essential for induction of IFN-β during *Chlamydia trachomatis* infection. *J Immunol* **193:**2394–2404.

48. **Hansen K, Prabakaran T, Laustsen A, Jørgensen SE, Rahbæk SH, Jensen SB, Nielsen R, Leber JH, Decker T,**

Horan KA, Jakobsen MR, Paludan SR. 2014. *Listeria monocytogenes* induces IFNβ expression through an IFI16-, cGAS- and STING-dependent pathway. *EMBO J* 33:1654–1666.

49. **Watson RO, Bell SL, MacDuff DA, Kimmey JM, Diner EJ, Olivas J, Vance RE, Stallings CL, Virgin HW, Cox JS.** 2015. The cytosolic sensor cGAS detects *Mycobacterium tuberculosis* DNA to induce type I interferons and activate autophagy. *Cell Host Microbe* **17:**811–819.
50. **Zhang Z, Yuan B, Bao M, Lu N, Kim T, Liu YJ.** 2011. The helicase DDX41 senses intracellular DNA mediated by the adaptor STING in dendritic cells. *Nat Immunol* **12:**959–965.
51. **Parvatiyar K, Zhang Z, Teles RM, Ouyang S, Jiang Y, Iyer SS, Zaver SA, Schenk M, Zeng S, Zhong W, Liu ZJ, Modlin RL, Liu YJ, Cheng G.** 2012. The helicase DDX41 recognizes the bacterial secondary messengers cyclic di-GMP and cyclic di-AMP to activate a type I interferon immune response. *Nat Immunol* **13:**1155–1161.
52. **Unterholzner L, Keating SE, Baran M, Horan KA, Jensen SB, Sharma S, Sirois CM, Jin T, Latz E, Xiao TS, Fitzgerald KA, Paludan SR, Bowie AG.** 2010. IFI16 is an innate immune sensor for intracellular DNA. *Nat Immunol* **11:**997–1004.
53. **Auerbuch V, Brockstedt DG, Meyer-Morse N, O'Riordan M, Portnoy DA.** 2004. Mice lacking the type I interferon receptor are resistant to *Listeria monocytogenes*. *J Exp Med* **200:**527–533.
54. **O'Connell RM, Saha SK, Vaidya SA, Bruhn KW, Miranda GA, Zarnegar B, Perry AK, Nguyen BO, Lane TF, Taniguchi T, Miller JF, Cheng G.** 2004. Type I interferon production enhances susceptibility to *Listeria monocytogenes* infection. *J Exp Med* **200:**437–445.
55. **Robinson N, McComb S, Mulligan R, Dudani R, Krishnan L, Sad S.** 2012. Type I interferon induces necroptosis in macrophages during infection with *Salmonella enterica* serovar Typhimurium. *Nat Immunol* **13:** 954–962.
56. **Nagarajan UM, Prantner D, Sikes JD, Andrews CW Jr, Goodwin AM, Nagarajan S, Darville T.** 2008. Type I interferon signaling exacerbates *Chlamydia muridarum* genital infection in a murine model. *Infect Immun* **76:** 4642–4648.
57. **Qiu H, Fan Y, Joyee AG, Wang S, Han X, Bai H, Jiao L, Van Rooijen N, Yang X.** 2008. Type I IFNs enhance susceptibility to *Chlamydia muridarum* lung infection by enhancing apoptosis of local macrophages. *J Immunol* **181:**2092–2102.
58. **Desvignes L, Wolf AJ, Ernst JD.** 2012. Dynamic roles of type I and type II IFNs in early infection with *Mycobacterium tuberculosis*. *J Immunol* **188:**6205–6215.
59. **Manca C, Tsenova L, Freeman S, Barczak AK, Tovey M, Murray PJ, Barry C III, Kaplan G.** 2005. Hypervirulent *M. tuberculosis* W/Beijing strains upregulate type I IFNs and increase expression of negative regulators of the Jak-Stat pathway. *J Interferon Cytokine Res* **25:** 694–701.
60. **Dorhoi A, Yeremeev V, Nouailles G, Weiner J III, Jörg S, Heinemann E, Oberbeck-Müller D, Knaul JK, Vogelzang A, Reece ST, Hahnke K, Mollenkopf HJ, Brinkmann V, Kaufmann SH.** 2014. Type I IFN signaling triggers immunopathology in tuberculosis-susceptible mice by modulating lung phagocyte dynamics. *Eur J Immunol* **44:**2380–2393.
61. **Manzanillo PS, Shiloh MU, Portnoy DA, Cox JS.** 2012. *Mycobacterium tuberculosis* activates the DNA-dependent cytosolic surveillance pathway within macrophages. *Cell Host Microbe* **11:**469–480.
62. **Huynh KK, Grinstein S.** 2007. Regulation of vacuolar pH and its modulation by some microbial species. *Microbiol Mol Biol Rev* **71:**452–462.
63. **Quinn MT, Gauss KA.** 2004. Structure and regulation of the neutrophil respiratory burst oxidase: comparison with nonphagocyte oxidases. *J Leukoc Biol* **76:**760–781.
64. **Minakami R, Sumimotoa H.** 2006. Phagocytosis-coupled activation of the superoxide-producing phagocyte oxidase, a member of the NADPH oxidase (Nox) family. *Int J Hematol* **84:**193–198.
65. **Fang FC.** 2004. Antimicrobial reactive oxygen and nitrogen species: concepts and controversies. *Nat Rev Microbiol* **2:**820–832.
66. **Cellier MF, Courville P, Campion C.** 2007. Nramp1 phagocyte intracellular metal withdrawal defense. *Microbes Infect* **9:**1662–1670.
67. **Huang J, Brumell JH.** 2014. Bacteria-autophagy interplay: a battle for survival. *Nat Rev Microbiol* **12:** 101–114.
68. **Mizushima N, Yoshimori T, Ohsumi Y.** 2011. The role of Atg proteins in autophagosome formation. *Annu Rev Cell Dev Biol* **27:**107–132.
69. **Bestebroer J, V'kovski P, Mauthe M, Reggiori F.** 2013. Hidden behind autophagy: the unconventional roles of ATG proteins. *Traffic* **14:**1029–1041.
70. **Kim DH, Sarbassov DD, Ali SM, King JE, Latek RR, Erdjument-Bromage H, Tempst P, Sabatini DM.** 2002. mTOR interacts with raptor to form a nutrient-sensitive complex that signals to the cell growth machinery. *Cell* **110:**163–175.
71. **Hosokawa N, Hara T, Kaizuka T, Kishi C, Takamura A, Miura Y, Iemura S, Natsume T, Takehana K, Yamada N, Guan JL, Oshiro N, Mizushima N.** 2009. Nutrient-dependent mTORC1 association with the ULK1-Atg13-FIP200 complex required for autophagy. *Mol Biol Cell* **20:**1981–1991.
72. **He C, Klionsky DJ.** 2009. Regulation mechanisms and signaling pathways of autophagy. *Annu Rev Genet* **43:** 67–93.
73. **Parzych KR, Klionsky DJ.** 2014. An overview of autophagy: morphology, mechanism, and regulation. *Antioxid Redox Signal* **20:**460–473.
74. **Hanna RA, Quinsay MN, Orogo AM, Giang K, Rikka S, Gustafsson AB.** 2012. Microtubule-associated protein 1 light chain 3 (LC3) interacts with Bnip3 protein to selectively remove endoplasmic reticulum and mitochondria via autophagy. *J Biol Chem* **287:**19094–19104.
75. **Liu L, Feng D, Chen G, Chen M, Zheng Q, Song P, Ma Q, Zhu C, Wang R, Qi W, Huang L, Xue P, Li B, Wang X, Jin H, Wang J, Yang F, Liu P, Zhu Y, Sui S, Chen Q.** 2012. Mitochondrial outer-membrane protein FUNDC1

mediates hypoxia-induced mitophagy in mammalian cells. *Nat Cell Biol* **14**:177–185.

76. **Novak I, Kirkin V, McEwan DG, Zhang J, Wild P, Rozenknop A, Rogov V, Löhr F, Popovic D, Occhipinti A, Reichert AS, Terzic J, Dötsch V, Ney PA, Dikic I.** 2010. Nix is a selective autophagy receptor for mitochondrial clearance. *EMBO Rep* **11**:45–51.

77. **Boyle KB, Randow F.** 2013. The role of 'eat-me' signals and autophagy cargo receptors in innate immunity. *Curr Opin Microbiol* **16**:339–348.

78. **Yoshikawa Y, Ogawa M, Hain T, Yoshida M, Fukumatsu M, Kim M, Mimuro H, Nakagawa I, Yanagawa T, Ishii T, Kakizuka A, Sztul E, Chakraborty T, Sasakawa C.** 2009. *Listeria monocytogenes* ActA-mediated escape from autophagic recognition. *Nat Cell Biol* **11**:1233–1240.

79. **Mostowy S, Sancho-Shimizu V, Hamon MA, Simeone R, Brosch R, Johansen T, Cossart P.** 2011. p62 and NDP52 proteins target intracytosolic *Shigella* and *Listeria* to different autophagy pathways. *J Biol Chem* **286**:26987–26995.

80. **Zheng YT, Shahnazari S, Brech A, Lamark T, Johansen T, Brumell JH.** 2009. The adaptor protein p62/SQSTM1 targets invading bacteria to the autophagy pathway. *J Immunol* **183**:5909–5916.

81. **von Muhlinen N, Thurston T, Ryzhakov G, Bloor S, Randow F.** 2010. NDP52, a novel autophagy receptor for ubiquitin-decorated cytosolic bacteria. *Autophagy* **6**: 288–289.

82. **Wild P, Farhan H, McEwan DG, Wagner S, Rogov VV, Brady NR, Richter B, Korac J, Waidmann O, Choudhary C, Dötsch V, Bumann D, Dikic I.** 2011. Phosphorylation of the autophagy receptor optineurin restricts *Salmonella* growth. *Science* **333**:228–233.

83. **Chong A, Wehrly TD, Child R, Hansen B, Hwang S, Virgin HW, Celli J.** 2012. Cytosolic clearance of replication-deficient mutants reveals *Francisella tularensis* interactions with the autophagic pathway. *Autophagy* **8**:1342–1356.

84. **Cheong H, Lindsten T, Wu J, Lu C, Thompson CB.** 2011. Ammonia-induced autophagy is independent of ULK1/ULK2 kinases. *Proc Natl Acad Sci U S A* **108**: 11121–11126.

85. **Nishida Y, Arakawa S, Fujitani K, Yamaguchi H, Mizuta T, Kanaseki T, Komatsu M, Otsu K, Tsujimoto Y, Shimizu S.** 2009. Discovery of Atg5/Atg7-independent alternative macroautophagy. *Nature* **461**:654–658.

86. **Grishchuk Y, Ginet V, Truttmann AC, Clarke PG, Puyal J.** 2011. Beclin 1-independent autophagy contributes to apoptosis in cortical neurons. *Autophagy* **7**: 1115–1131.

87. **Choi J, Park S, Biering SB, Selleck E, Liu CY, Zhang X, Fujita N, Saitoh T, Akira S, Yoshimori T, Sibley LD, Hwang S, Virgin HW.** 2014. The parasitophorous vacuole membrane of *Toxoplasma gondii* is targeted for disruption by ubiquitin-like conjugation systems of autophagy. *Immunity* **40**:924–935.

88. **Mestre MB, Fader CM, Sola C, Colombo MI.** 2010. Alpha-hemolysin is required for the activation of the autophagic pathway in *Staphylococcus aureus*-infected cells. *Autophagy* **6**:110–125.

89. **Sanjuan MA, Dillon CP, Tait SW, Moshiach S, Dorsey F, Connell S, Komatsu M, Tanaka K, Cleveland JL, Withoff S, Green DR.** 2007. Toll-like receptor signalling in macrophages links the autophagy pathway to phagocytosis. *Nature* **450**:1253–1257.

90. **Henault J, Martinez J, Riggs JM, Tian J, Mehta P, Clarke L, Sasai M, Latz E, Brinkmann MM, Iwasaki A, Coyle AJ, Kolbeck R, Green DR, Sanjuan MA.** 2012. Noncanonical autophagy is required for type I interferon secretion in response to DNA-immune complexes. *Immunity* **37**:986–997.

91. **Martinez J, Almendinger J, Oberst A, Ness R, Dillon CP, Fitzgerald P, Hengartner MO, Green DR.** 2011. Microtubule-associated protein 1 light chain 3 alpha (LC3)-associated phagocytosis is required for the efficient clearance of dead cells. *Proc Natl Acad Sci U S A* **108**:17396–17401.

92. **Huang J, Canadien V, Lam GY, Steinberg BE, Dinauer MC, Magalhaes MA, Glogauer M, Grinstein S, Brumell JH.** 2009. Activation of antibacterial autophagy by NADPH oxidases. *Proc Natl Acad Sci U S A* **106**: 6226–6231.

93. **Gong L, Cullinane M, Treerat P, Ramm G, Prescott M, Adler B, Boyce JD, Devenish RJ.** 2011. The *Burkholderia pseudomallei* type III secretion system and BopA are required for evasion of LC3-associated phagocytosis. *PLoS One* **6**:e17852. doi:10.1371/journal.pone.0017852.

94. **Cullinane M, Gong L, Li X, Lazar-Adler N, Tra T, Wolvetang E, Prescott M, Boyce JD, Devenish RJ, Adler B.** 2008. Stimulation of autophagy suppresses the intracellular survival of *Burkholderia pseudomallei* in mammalian cell lines. *Autophagy* **4**:744–753.

95. **Lamkanfi M, Dixit VM.** 2010. Manipulation of host cell death pathways during microbial infections. *Cell Host Microbe* **8**:44–54.

96. **Ashida H, Mimuro H, Ogawa M, Kobayashi T, Sanada T, Kim M, Sasakawa C.** 2011. Cell death and infection: a double-edged sword for host and pathogen survival. *J Cell Biol* **195**:931–942.

97. **Rudel T, Kepp O, Kozjak-Pavlovic V.** 2010. Interactions between bacterial pathogens and mitochondrial cell death pathways. *Nat Rev Microbiol* **8**:693–705.

98. **Sridharan H, Upton JW.** 2014. Programmed necrosis in microbial pathogenesis. *Trends Microbiol* **22**:199–207.

99. **Kaiser WJ, Upton JW, Long AB, Livingston-Rosanoff D, Daley-Bauer LP, Hakem R, Caspary T, Mocarski ES.** 2011. RIP3 mediates the embryonic lethality of caspase-8-deficient mice. *Nature* **471**:368–372.

100. **Holler N, Zaru R, Micheau O, Thome M, Attinger A, Valitutti S, Bodmer JL, Schneider P, Seed B, Tschopp J.** 2000. Fas triggers an alternative, caspase-8-independent cell death pathway using the kinase RIP as effector molecule. *Nat Immunol* **1**:489–495.

101. **Mocarski ES, Upton JW, Kaiser WJ.** 2011. Viral infection and the evolution of caspase 8-regulated apoptotic and necrotic death pathways. *Nat Rev Immunol* **12**:79–88.

102. **Miao EA, Leaf IA, Treuting PM, Mao DP, Dors M, Sarkar A, Warren SE, Wewers MD, Aderem A.** 2010. Caspase-1-induced pyroptosis is an innate immune effector mechanism against intracellular bacteria. *Nat Immunol* **11:**1136–1142.

103. **Ross TM.** 2001. Using death to one's advantage: HIV modulation of apoptosis. *Leukemia* **15:**332–341.

104. **Early J, Fischer K, Bermudez LE.** 2011. *Mycobacterium avium* uses apoptotic macrophages as tools for spreading. *Microb Pathog* **50:**132–139.

105. **Cornelis GR.** 2006. The type III secretion injectisome. *Nat Rev Microbiol* **4:**811–825.

106. **Blocker AJ, Deane JE, Veenendaal AK, Roversi P, Hodgkinson JL, Johnson S, Lea SM.** 2008. What's the point of the type III secretion system needle? *Proc Natl Acad Sci U S A* **105:**6507–6513.

107. **Valdivia RH.** 2008. *Chlamydia* effector proteins and new insights into chlamydial cellular microbiology. *Curr Opin Microbiol* **11:**53–59.

108. **Ho TD, Starnbach MN.** 2005. The *Salmonella enterica* serovar Typhimurium-encoded type III secretion systems can translocate *Chlamydia trachomatis* proteins into the cytosol of host cells. *Infect Immun* **73:**905–911.

109. **Subtil A, Delevoye C, Balañá ME, Tastevin L, Perrinet S, Dautry-Varsat A.** 2005. A directed screen for chlamydial proteins secreted by a type III mechanism identifies a translocated protein and numerous other new candidates. *Mol Microbiol* **56:**1636–1647.

110. **Pennini ME, Perrinet S, Dautry-Varsat A, Subtil A.** 2010. Histone methylation by NUE, a novel nuclear effector of the intracellular pathogen *Chlamydia trachomatis*. *PLoS Pathog* **6:**e1000995. doi:10.1371/journal.ppat.1000995.

111. **da Cunha M, Milho C, Almeida F, Pais SV, Borges V, Maurício R, Borrego MJ, Gomes JP, Mota LJ.** 2014. Identification of type III secretion substrates of *Chlamydia trachomatis* using *Yersinia enterocolitica* as a heterologous system. *BMC Microbiol* **14:**40. doi:10.1186/1471-2180-14-40.

112. **Christie PJ, Cascales E.** 2005. Structural and dynamic properties of bacterial type IV secretion systems (review). *Mol Membr Biol* **22:**51–61.

113. **Vogel JP, Andrews HL, Wong SK, Isberg RR.** 1998. Conjugative transfer by the virulence system of *Legionella pneumophila*. *Science* **279:**873–876.

114. **O'Callaghan D, Cazevieille C, Allardet-Servent A, Boschiroli ML, Bourg G, Foulongne V, Frutos P, Kulakov Y, Ramuz M.** 1999. A homologue of the *Agrobacterium tumefaciens* VirB and *Bordetella pertussis* Ptl type IV secretion systems is essential for intracellular survival of *Brucella suis*. *Mol Microbiol* **33:**1210–1220.

115. **Sadosky AB, Wiater LA, Shuman HA.** 1993. Identification of *Legionella pneumophila* genes required for growth within and killing of human macrophages. *Infect Immun* **61:**5361–5373.

116. **Berger KH, Merriam JJ, Isberg RR.** 1994. Altered intracellular targeting properties associated with mutations in the *Legionella pneumophila dotA* gene. *Mol Microbiol* **14:**809–822.

117. **Carey KL, Newton HJ, Lührmann A, Roy CR.** 2011. The *Coxiella burnetii* Dot/Icm system delivers a unique repertoire of type IV effectors into host cells and is required for intracellular replication. *PLoS Pathog* **7:**e1002056. doi:10.1371/journal.ppat.1002056.

118. **Liu H, Bao W, Lin M, Niu H, Rikihisa Y.** 2012. *Ehrlichia* type IV secretion effector ECH0825 is translocated to mitochondria and curbs ROS and apoptosis by upregulating host MnSOD. *Cell Microbiol* **14:**1037–1050.

119. **Lin M, den Dulk-Ras A, Hooykaas PJ, Rikihisa Y.** 2007. *Anaplasma phagocytophilum* AnkA secreted by type IV secretion system is tyrosine phosphorylated by Abl-1 to facilitate infection. *Cell Microbiol* **9:**2644–2657.

120. **Rikihisa Y, Lin M.** 2010. *Anaplasma phagocytophilum* and *Ehrlichia chaffeensis* type IV secretion and Ank proteins. *Curr Opin Microbiol* **13:**59–66.

121. **Kudryashev M, Wang RY, Brackmann M, Scherer S, Maier T, Baker D, DiMaio F, Stahlberg H, Egelman EH, Basler M.** 2015. Structure of the type VI secretion system contractile sheath. *Cell* **160:**952–962.

122. **Coulthurst SJ.** 2013. The Type VI secretion system—a widespread and versatile cell targeting system. *Res Microbiol* **164:**640–654.

123. **Pezoa D, Blondel CJ, Silva CA, Yang HJ, Andrews-Polymenis H, Santiviago CA, Contreras I.** 2014. Only one of the two type VI secretion systems encoded in the *Salmonella enterica* serotype Dublin genome is involved in colonization of the avian and murine hosts. *Vet Res* **45:**2. doi:10.1186/1297-9716-45-2.

124. **Mulder DT, Cooper CA, Coombes BK.** 2012. Type VI secretion system-associated gene clusters contribute to pathogenesis of *Salmonella enterica* serovar Typhimurium. *Infect Immun* **80:**1996–2007.

125. **Blondel CJ, Jiménez JC, Leiva LE, Alvarez SA, Pinto BI, Contreras F, Pezoa D, Santiviago CA, Contreras I.** 2013. The type VI secretion system encoded in *Salmonella* pathogenicity island 19 is required for *Salmonella enterica* serotype Gallinarum survival within infected macrophages. *Infect Immun* **81:**1207–1220.

126. **Bansal-Mutalik R, Nikaido H.** 2014. Mycobacterial outer membrane is a lipid bilayer and the inner membrane is unusually rich in diacyl phosphatidylinositol dimannosides. *Proc Natl Acad Sci U S A* **111:**4958–4963.

127. **Abdallah AM, Gey van Pittius NC, Champion PA, Cox J, Luirink J, Vandenbroucke-Grauls CM, Appelmelk BJ, Bitter W.** 2007. Type VII secretion—mycobacteria show the way. *Nat Rev Microbiol* **5:**883–891.

128. **Bitter W, Houben EN, Bottai D, Brodin P, Brown EJ, Cox JS, Derbyshire K, Fortune SM, Gao LY, Liu J, Gey van Pittius NC, Pym AS, Rubin EJ, Sherman DR, Cole ST, Brosch R.** 2009. Systematic genetic nomenclature for type VII secretion systems. *PLoS Pathog* **5:**e1000507. doi:10.1371/journal.ppat.1000507.

129. **Pallen MJ.** 2002. The ESAT-6/WXG100 superfamily—and a new Gram-positive secretion system? *Trends Microbiol* **10:**209–212.

130. **Brodin P, Majlessi L, Marsollier L, de Jonge MI, Bottai D, Demangel C, Hinds J, Neyrolles O, Butcher PD, Leclerc C, Cole ST, Brosch R.** 2006. Dissection of ESAT-6 system 1 of *Mycobacterium tuberculosis* and impact on immunogenicity and virulence. *Infect Immun* **74:**88–98.

131. **Pym AS, Brodin P, Majlessi L, Brosch R, Demangel C, Williams A, Griffiths KE, Marchal G, Leclerc C, Cole ST.** 2003. Recombinant BCG exporting ESAT-6 confers enhanced protection against tuberculosis. *Nat Med* **9:** 533–539.

132. **Ekiert DC, Cox JS.** 2014. Structure of a PE-PPE-EspG complex from *Mycobacterium tuberculosis* reveals molecular specificity of ESX protein secretion. *Proc Natl Acad Sci U S A* **111:**14758–14763.

133. **van der Wel N, Hava D, Houben D, Fluitsma D, van Zon M, Pierson J, Brenner M, Peters PJ.** 2007. *M. tuberculosis* and *M. leprae* translocate from the phagolysosome to the cytosol in myeloid cells. *Cell* **129:**1287–1298.

134. **Méresse S, Steele-Mortimer O, Moreno E, Desjardins M, Finlay B, Gorvel JP.** 1999. Controlling the maturation of pathogen-containing vacuoles: a matter of life and death. *Nat Cell Biol* **1:**E183–E188.

135. **Vogel JP, Isberg RR.** 1999. Cell biology of *Legionella pneumophila*. *Curr Opin Microbiol* **2:**30–34.

136. **Pizarro-Cerdá J, Méresse S, Parton RG, van der Goot G, Sola-Landa A, Lopez-Goñi I, Moreno E, Gorvel JP.** 1998. *Brucella abortus* transits through the autophagic pathway and replicates in the endoplasmic reticulum of nonprofessional phagocytes. *Infect Immun* **66:**5711–5724.

137. **Celli J, de Chastellier C, Franchini DM, Pizarro-Cerda J, Moreno E, Gorvel JP.** 2003. *Brucella* evades macrophage killing via VirB-dependent sustained interactions with the endoplasmic reticulum. *J Exp Med* **198:**545–556.

138. **Franco IS, Shuman HA, Charpentier X.** 2009. The perplexing functions and surprising origins of *Legionella pneumophila* type IV secretion effectors. *Cell Microbiol* **11:**1435–1443.

139. **Nagai H, Kagan JC, Zhu X, Kahn RA, Roy CR.** 2002. A bacterial guanine nucleotide exchange factor activates ARF on *Legionella* phagosomes. *Science* **295:** 679–682.

140. **Neunuebel MR, Chen Y, Gaspar AH, Backlund PS Jr, Yergey A, Machner MP.** 2011. De-AMPylation of the small GTPase Rab1 by the pathogen *Legionella pneumophila*. *Science* **333:**453–456.

141. **Neunuebel MR, Machner MP.** 2012. The taming of a Rab GTPase by *Legionella pneumophila*. *Small GTPases* **3:**28–33.

142. **Derré I, Isberg RR.** 2004. *Legionella pneumophila* replication vacuole formation involves rapid recruitment of proteins of the early secretory system. *Infect Immun* **72:** 3048–3053.

143. **Kagan JC, Roy CR.** 2002. *Legionella* phagosomes intercept vesicular traffic from endoplasmic reticulum exit sites. *Nat Cell Biol* **4:**945–954.

144. **Scidmore MA, Fischer ER, Hackstadt T.** 1996. Sphingolipids and glycoproteins are differentially trafficked to the *Chlamydia trachomatis* inclusion. *J Cell Biol* **134:** 363–374.

145. **Heuer D, Rejman Lipinski A, Machuy N, Karlas A, Wehrens A, Siedler F, Brinkmann V, Meyer TF.** 2009. *Chlamydia* causes fragmentation of the Golgi compartment to ensure reproduction. *Nature* **457:**731–735.

146. **Boleti H, Benmerah A, Ojcius DM, Cerf-Bensussan N, Dautry-Varsat A.** 1999. *Chlamydia* infection of epithelial cells expressing dynamin and Eps15 mutants: clathrin-independent entry into cells and dynamin-dependent productive growth. *J Cell Sci* **112:**1487–1496.

147. **Via LE, Deretic D, Ulmer RJ, Hibler NS, Huber LA, Deretic V.** 1997. Arrest of mycobacterial phagosome maturation is caused by a block in vesicle fusion between stages controlled by rab5 and rab7. *J Biol Chem* **272:**13326–13331.

148. **Mehra A, Zahra A, Thompson V, Sirisaengtaksin N, Wells A, Porto M, Köster S, Penberthy K, Kubota Y, Dricot A, Rogan D, Vidal M, Hill DE, Bean AJ, Philips JA.** 2013. *Mycobacterium tuberculosis* type VII secreted effector EsxH targets host ESCRT to impair trafficking. *PLoS Pathog* **9:**e1003734. doi:10.1371/journal.ppat.1003734.

149. **Wong D, Bach H, Sun J, Hmama Z, Av-Gay Y.** 2011. *Mycobacterium tuberculosis* protein tyrosine phosphatase (PtpA) excludes host vacuolar-H+-ATPase to inhibit phagosome acidification. *Proc Natl Acad Sci U S A* **108:**19371–19376.

150. **Xu L, Shen X, Bryan A, Banga S, Swanson MS, Luo ZQ.** 2010. Inhibition of host vacuolar H^+-ATPase activity by a *Legionella pneumophila* effector. *PLoS Pathog* **6:**e1000822. doi:10.1371/journal.ppat.1000822.

151. **Fratti RA, Chua J, Vergne I, Deretic V.** 2003. *Mycobacterium tuberculosis* glycosylated phosphatidylinositol causes phagosome maturation arrest. *Proc Natl Acad Sci U S A* **100:**5437–5442.

152. **Vergne I, Chua J, Lee HH, Lucas M, Belisle J, Deretic V.** 2005. Mechanism of phagolysosome biogenesis block by viable *Mycobacterium tuberculosis*. *Proc Natl Acad Sci U S A* **102:**4033–4038.

153. **Fratti RA, Backer JM, Gruenberg J, Corvera S, Deretic V.** 2001. Role of phosphatidylinositol 3-kinase and Rab5 effectors in phagosomal biogenesis and mycobacterial phagosome maturation arrest. *J Cell Biol* **154:**631–644.

154. **Philips JA.** 2008. Mycobacterial manipulation of vacuolar sorting. *Cell Microbiol* **10:**2408–2415.

155. **Alix E, Mukherjee S, Roy CR.** 2011. Subversion of membrane transport pathways by vacuolar pathogens. *J Cell Biol* **195:**943–952.

156. **Steele-Mortimer O, Méresse S, Gorvel JP, Toh BH, Finlay BB.** 1999. Biogenesis of *Salmonella typhimurium*-containing vacuoles in epithelial cells involves interactions with the early endocytic pathway. *Cell Microbiol* **1:**33–49.

157. **Steele-Mortimer O.** 2008. The *Salmonella*-containing vacuole: moving with the times. *Curr Opin Microbiol* **11**:38–45.
158. **van Schaik EJ, Chen C, Mertens K, Weber MM, Samuel JE.** 2013. Molecular pathogenesis of the obligate intracellular bacterium *Coxiella burnetii*. *Nat Rev Microbiol* **11**:561–573.
159. **Newton HJ, Kohler LJ, McDonough JA, Temoche-Diaz M, Crabill E, Hartland EL, Roy CR.** 2014. A screen of *Coxiella burnetii* mutants reveals important roles for Dot/Icm effectors and host autophagy in vacuole biogenesis. *PLoS Pathog* **10**:e1004286. doi:10.1371/journal.ppat.1004286.
160. **Larson CL, Beare PA, Voth DE, Howe D, Cockrell DC, Bastidas RJ, Valdivia RH, Heinzen RA.** 2015. *Coxiella burnetii* effector proteins that localize to the parasitophorous vacuole membrane promote intracellular replication. *Infect Immun* **83**:661–670.
161. **Thurston TL, Wandel MP, von Muhlinen N, Foeglein A, Randow F.** 2012. Galectin 8 targets damaged vesicles for autophagy to defend cells against bacterial invasion. *Nature* **482**:414–418.
162. **Paz I, Sachse M, Dupont N, Mounier J, Cederfur C, Enninga J, Leffler H, Poirier F, Prevost MC, Lafont F, Sansonetti P.** 2010. Galectin-3, a marker for vacuole lysis by invasive pathogens. *Cell Microbiol* **12**:530–544.
163. **Perrin AJ, Jiang X, Birmingham CL, So NS, Brumell JH.** 2004. Recognition of bacteria in the cytosol of mammalian cells by the ubiquitin system. *Curr Biol* **14**:806–811.
164. **Creasey EA, Isberg RR.** 2014. Maintenance of vacuole integrity by bacterial pathogens. *Curr Opin Microbiol* **17**:46–52.
165. **Kumar Y, Valdivia RH.** 2009. Leading a sheltered life: intracellular pathogens and maintenance of vacuolar compartments. *Cell Host Microbe* **5**:593–601.
166. **Radtke AL, O'Riordan MX.** 2008. Homeostatic maintenance of pathogen-containing vacuoles requires TBK1-dependent regulation of aquaporin-1. *Cell Microbiol* **10**:2197–2207.
167. **Meunier E, Dick MS, Dreier RF, Schürmann N, Kenzelmann Broz D, Warming S, Roose-Girma M, Bumann D, Kayagaki N, Takeda K, Yamamoto M, Broz P.** 2014. Caspase-11 activation requires lysis of pathogen-containing vacuoles by IFN-induced GTPases. *Nature* **509**:366–370.
168. **Kumar Y, Valdivia RH.** 2008. Actin and intermediate filaments stabilize the *Chlamydia trachomatis* vacuole by forming dynamic structural scaffolds. *Cell Host Microbe* **4**:159–169.
169. **Méresse S, Unsworth KE, Habermann A, Griffiths G, Fang F, Martínez-Lorenzo MJ, Waterman SR, Gorvel JP, Holden DW.** 2001. Remodelling of the actin cytoskeleton is essential for replication of intravacuolar *Salmonella*. *Cell Microbiol* **3**:567–577.
170. **Jorgensen I, Bednar MM, Amin V, Davis BK, Ting JP, McCafferty DG, Valdivia RH.** 2011. The *Chlamydia* protease CPAF regulates host and bacterial proteins to maintain pathogen vacuole integrity and promote virulence. *Cell Host Microbe* **10**:21–32.
171. **Wasylnka JA, Bakowski MA, Szeto J, Ohlson MB, Trimble WS, Miller SI, Brumell JH.** 2008. Role for myosin II in regulating positioning of *Salmonella*-containing vacuoles and intracellular replication. *Infect Immun* **76**:2722–2735.
172. **Dumont A, Boucrot E, Drevensek S, Daire V, Gorvel JP, Poüs C, Holden DW, Méresse S.** 2010. SKIP, the host target of the *Salmonella* virulence factor SifA, promotes kinesin-1-dependent vacuolar membrane exchanges. *Traffic* **11**:899–911.
173. **Beuzón CR, Méresse S, Unsworth KE, Ruíz-Albert J, Garvis S, Waterman SR, Ryder TA, Boucrot E, Holden DW.** 2000. *Salmonella* maintains the integrity of its intracellular vacuole through the action of SifA. *EMBO J* **19**:3235–3249.
174. **Cheng W, Yin K, Lu D, Li B, Zhu D, Chen Y, Zhang H, Xu S, Chai J, Gu L.** 2012. Structural insights into a unique *Legionella pneumophila* effector LidA recognizing both GDP and GTP bound Rab1 in their active state. *PLoS Pathog* **8**:e1002528. doi:10.1371/journal.ppat.1002528.
175. **Brumell JH, Tang P, Zaharik ML, Finlay BB.** 2002. Disruption of the *Salmonella*-containing vacuole leads to increased replication of *Salmonella enterica* serovar Typhimurium in the cytosol of epithelial cells. *Infect Immun* **70**:3264–3270.
176. **Hilbi H, Weber S, Finsel I.** 2011. Anchors for effectors: subversion of phosphoinositide lipids by *Legionella*. *Front Microbiol* **2**:91. doi:10.3389/fmicb.2011.00091.
177. **Robertson DK, Gu L, Rowe RK, Beatty WL.** 2009. Inclusion biogenesis and reactivation of persistent *Chlamydia trachomatis* requires host cell sphingolipid biosynthesis. *PLoS Pathog* **5**:e1000664. doi:10.1371/journal.ppat.1000664.
178. **Elwell CA, Jiang S, Kim JH, Lee A, Wittmann T, Hanada K, Melancon P, Engel JN.** 2011. *Chlamydia trachomatis* co-opts GBF1 and CERT to acquire host sphingomyelin for distinct roles during intracellular development. *PLoS Pathog* **7**:e1002198. doi:10.1371/journal.ppat.1002198.
179. **Nawabi P, Catron DM, Haldar K.** 2008. Esterification of cholesterol by a type III secretion effector during intracellular *Salmonella* infection. *Mol Microbiol* **68**:173–185.
180. **Flieger A, Neumeister B, Cianciotto NP.** 2002. Characterization of the gene encoding the major secreted lysophospholipase A of *Legionella pneumophila* and its role in detoxification of lysophosphatidylcholine. *Infect Immun* **70**:6094–6106.
181. **Creasey EA, Isberg RR.** 2012. The protein SdhA maintains the integrity of the *Legionella*-containing vacuole. *Proc Natl Acad Sci U S A* **109**:3481–3486.
182. **Ruiz-Albert J, Yu XJ, Beuzón CR, Blakey AN, Galyov EE, Holden DW.** 2002. Complementary activities of SseJ and SifA regulate dynamics of the *Salmonella typhimurium* vacuolar membrane. *Mol Microbiol* **44**:645–661.

183. **Watson RO, Manzanillo PS, Cox JS.** 2012. Extracellular *M. tuberculosis* DNA targets bacteria for autophagy by activating the host DNA-sensing pathway. *Cell* **150:** 803–815.

184. **Thurston TL, Ryzhakov G, Bloor S, von Muhlinen N, Randow F.** 2009. The TBK1 adaptor and autophagy receptor NDP52 restricts the proliferation of ubiquitin-coated bacteria. *Nat Immunol* **10:**1215–1221.

185. **Shin DM, Jeon BY, Lee HM, Jin HS, Yuk JM, Song CH, Lee SH, Lee ZW, Cho SN, Kim JM, Friedman RL, Jo EK.** 2010. *Mycobacterium tuberculosis* Eis regulates autophagy, inflammation, and cell death through redox-dependent signaling. *PLoS Pathog* **6:**e1001230. doi: 10.1371/journal.ppat.1001230.

186. **Romagnoli A, Etna MP, Giacomini E, Pardini M, Remoli ME, Corazzari M, Falasca L, Goletti D, Gafa V, Simeone R, Delogu G, Piacentini M, Brosch R, Fimia GM, Coccia EM.** 2012. ESX-1 dependent impairment of autophagic flux by *Mycobacterium tuberculosis* in human dendritic cells. *Autophagy* **8:**1357–1370.

187. **Choy A, Dancourt J, Mugo B, O'Connor TJ, Isberg RR, Melia TJ, Roy CR.** 2012. The *Legionella* effector RavZ inhibits host autophagy through irreversible Atg8 deconjugation. *Science* **338:**1072–1076.

188. **Gutierrez MG, Vázquez CL, Munafó DB, Zoppino FC, Berón W, Rabinovitch M, Colombo MI.** 2005. Autophagy induction favours the generation and maturation of the *Coxiella*-replicative vacuoles. *Cell Microbiol* **7:** 981–993.

189. **Berón W, Gutierrez MG, Rabinovitch M, Colombo MI.** 2002. *Coxiella burnetii* localizes in a Rab7-labeled compartment with autophagic characteristics. *Infect Immun* **70:**5816–5821.

190. **Starr T, Child R, Wehrly TD, Hansen B, Hwang S, López-Otin C, Virgin HW, Celli J.** 2012. Selective subversion of autophagy complexes facilitates completion of the *Brucella* intracellular cycle. *Cell Host Microbe* **11:**33–45.

191. **Niu H, Xiong Q, Yamamoto A, Hayashi-Nishino M, Rikihisa Y.** 2012. Autophagosomes induced by a bacterial Beclin 1 binding protein facilitate obligatory intracellular infection. *Proc Natl Acad Sci U S A* **109:** 20800–20807.

192. **Niu H, Yamaguchi M, Rikihisa Y.** 2008. Subversion of cellular autophagy by *Anaplasma phagocytophilum*. *Cell Microbiol* **10:**593–605.

193. **Vázquez CL, Colombo MI.** 2010. *Coxiella burnetii* modulates Beclin 1 and Bcl-2, preventing host cell apoptosis to generate a persistent bacterial infection. *Cell Death Differ* **17:**421–438.

194. **Campoy E, Colombo MI.** 2009. Autophagy in intracellular bacterial infection. *Biochim Biophys Acta* **1793:** 1465–1477.

195. **Luo ZQ.** 2011. Striking a balance: modulation of host cell death pathways by *Legionella pneumophila*. *Front Microbiol* **2:**36. doi:10.3389/fmicb.2011.00036.

196. **Banga S, Gao P, Shen X, Fiscus V, Zong WX, Chen L, Luo ZQ.** 2007. *Legionella pneumophila* inhibits macrophage apoptosis by targeting pro-death members of the Bcl2 protein family. *Proc Natl Acad Sci U S A* **104:** 5121–5126.

197. **Lührmann A, Nogueira CV, Carey KL, Roy CR.** 2010. Inhibition of pathogen-induced apoptosis by a *Coxiella burnetii* type IV effector protein. *Proc Natl Acad Sci U S A* **107:**18997–19001.

198. **Fischer SF, Vier J, Kirschnek S, Klos A, Hess S, Ying S, Häcker G.** 2004. *Chlamydia* inhibit host cell apoptosis by degradation of proapoptotic BH3-only proteins. *J Exp Med* **200:**905–916.

199. **Ying S, Seiffert BM, Häcker G, Fischer SF.** 2005. Broad degradation of proapoptotic proteins with the conserved Bcl-2 homology domain 3 during infection with *Chlamydia trachomatis*. *Infect Immun* **73:**1399–1403.

200. **Ge J, Xu H, Li T, Zhou Y, Zhang Z, Li S, Liu L, Shao F.** 2009. A *Legionella* type IV effector activates the NF-κB pathway by phosphorylating the IκB family of inhibitors. *Proc Natl Acad Sci USA* **106:**13725–13730.

201. **Jones RM, Wu H, Wentworth C, Luo L, Collier-Hyams L, Neish AS.** 2008. *Salmonella* AvrA coordinates suppression of host immune and apoptotic defenses via JNK pathway blockade. *Cell Host Microbe* **3:**233–244.

202. **Verbeke P, Welter-Stahl L, Ying S, Hansen J, Häcker G, Darville T, Ojcius DM.** 2006. Recruitment of BAD by the *Chlamydia trachomatis* vacuole correlates with host-cell survival. *PLoS Pathog* **2:**e45. doi:10.1371/journal.ppat.0020045.

203. **Voth DE, Heinzen RA.** 2009. Sustained activation of Akt and Erk1/2 is required for *Coxiella burnetii* antiapoptotic activity. *Infect Immun* **77:**205–213.

204. **Knodler LA, Finlay BB, Steele-Mortimer O.** 2005. The *Salmonella* effector protein SopB protects epithelial cells from apoptosis by sustained activation of Akt. *J Biol Chem* **280:**9058–9064.

205. **Miao EA, Mao DP, Yudkovsky N, Bonneau R, Lorang CG, Warren SE, Leaf IA, Aderem A.** 2010. Innate immune detection of the type III secretion apparatus through the NLRC4 inflammasome. *Proc Natl Acad Sci U S A* **107:**3076–3080.

206. **Ireton K.** 2013. Molecular mechanisms of cell-cell spread of intracellular bacterial pathogens. *Open Biol* **3:**130079. doi:10.1098/rsob.130079.

207. **Kespichayawattana W, Rattanachetkul S, Wanun T, Utaisincharoen P, Sirisinha S.** 2000. *Burkholderia pseudomallei* induces cell fusion and actin-associated membrane protrusion: a possible mechanism for cell-to-cell spreading. *Infect Immun* **68:**5377–5384.

208. **Harley VS, Dance DA, Drasar BS, Tovey G.** 1998. Effects of *Burkholderia pseudomallei* and other *Burkholderia* species on eukaryotic cells in tissue culture. *Microbios* **96:**71–93.

209. **Sahni SK, Narra HP, Sahni A, Walker DH.** 2013. Recent molecular insights into rickettsial pathogenesis and immunity. *Future Microbiol* **8:**1265–1288.

210. **Suzuki T, Nakanishi K, Tsutsui H, Iwai H, Akira S, Inohara N, Chamaillard M, Nuñez G, Sasakawa C.** 2005. A novel caspase-1/Toll-like receptor 4-independent pathway of cell death induced by cytosolic *Shigella* in infected macrophages. *J Biol Chem* **280:**14042–14050.

211. Schroeder GN, Hilbi H. 2008. Molecular pathogenesis of *Shigella* spp.: controlling host cell signaling, invasion, and death by type III secretion. *Clin Microbiol Rev* **21:** 134–156.

212. Cunha LD, Zamboni DS. 2013. Subversion of inflammasome activation and pyroptosis by pathogenic bacteria. *Front Cell Infect Microbiol* 3:76. doi:10.3389/fcimb.2013.00076.

213. Vázquez-Boland JA, Kuhn M, Berche P, Chakraborty T, Domínguez-Bernal G, Goebel W, González-Zorn B, Wehland J, Kreft J. 2001. *Listeria* pathogenesis and molecular virulence determinants. *Clin Microbiol Rev* **14:** 584–640.

214. Cossart P. 2011. Illuminating the landscape of host-pathogen interactions with the bacterium *Listeria monocytogenes*. *Proc Natl Acad Sci U S A* **108:**19484–19491.

215. Burg-Golani T, Pozniak Y, Rabinovich L, Sigal N, Nir Paz R, Herskovits AA. 2013. Membrane chaperone SecDF plays a role in the secretion of *Listeria monocytogenes* major virulence factors. *J Bacteriol* **195:** 5262–5272.

216. Desvaux M, Hébraud M. 2006. The protein secretion systems in *Listeria*: inside out bacterial virulence. *FEMS Microbiol Rev* **30:**774–805.

217. Stevens MP, Wood MW, Taylor LA, Monaghan P, Hawes P, Jones PW, Wallis TS, Galyov EE. 2002. An Inv/Mxi-Spa-like type III protein secretion system in *Burkholderia pseudomallei* modulates intracellular behaviour of the pathogen. *Mol Microbiol* **46:**649–659.

218. Wiersinga WJ, van der Poll T, White NJ, Day NP, Peacock SJ. 2006. Melioidosis: insights into the pathogenicity of *Burkholderia pseudomallei*. *Nat Rev Microbiol* **4:**272–282.

219. Pilatz S, Breitbach K, Hein N, Fehlhaber B, Schulze J, Brenneke B, Eberl L, Steinmetz I. 2006. Identification of *Burkholderia pseudomallei* genes required for the intracellular life cycle and *in vivo* virulence. *Infect Immun* **74:**3576–3586.

220. Shalom G, Shaw JG, Thomas MS. 2007. *In vivo* expression technology identifies a type VI secretion system locus in *Burkholderia pseudomallei* that is induced upon invasion of macrophages. *Microbiology* **153:**2689–2699.

221. Schwarz S, Singh P, Robertson JD, LeRoux M, Skerrett SJ, Goodlett DR, West TE, Mougous JD. 2014. VgrG-5 is a *Burkholderia* type VI secretion system-exported protein required for multinucleated giant cell formation and virulence. *Infect Immun* **82:**1445–1452.

222. Toesca IJ, French CT, Miller JF. 2014. The type VI secretion system spike protein VgrG5 mediates membrane fusion during intercellular spread by Pseudomallei group *Burkholderia* species. *Infect Immun* **82:** 1436–1444.

223. French CT, Toesca IJ, Wu TH, Teslaa T, Beaty SM, Wong W, Liu M, Schröder I, Chiou PY, Teitell MA, Miller JF. 2011. Dissection of the *Burkholderia* intracellular life cycle using a photothermal nanoblade. *Proc Natl Acad Sci U S A* **108:**12095–12100.

224. Steiner DJ, Furuya Y, Metzger DW. 2014. Host-pathogen interactions and immune evasion strategies in *Francisella tularensis* pathogenicity. *Infect Drug Resist* 7:239–251.

225. Nano FE, Schmerk C. 2007. The *Francisella* pathogenicity island. *Ann N Y Acad Sci* **1105:**122–137.

226. de Bruin OM, Duplantis BN, Ludu JS, Hare RF, Nix EB, Schmerk CL, Robb CS, Boraston AB, Hueffer K, Nano FE. 2011. The biochemical properties of the *Francisella* pathogenicity island (FPI)-encoded proteins IglA, IglB, IglC, PdpB and DotU suggest roles in type VI secretion. *Microbiology* **157:**3483–3491.

227. Bröms JE, Meyer L, Lavander M, Larsson P, Sjöstedt A. 2012. DotU and VgrG, core components of type VI secretion systems, are essential for *Francisella* LVS pathogenicity. *PLoS One* 7:e34639. doi:10.1371/journal.pone.0034639.

228. Barker JR, Chong A, Wehrly TD, Yu JJ, Rodriguez SA, Liu J, Celli J, Arulanandam BP, Klose KE. 2009. The *Francisella tularensis* pathogenicity island encodes a secretion system that is required for phagosome escape and virulence. *Mol Microbiol* **74:**1459–1470.

229. Henry R, Shaughnessy L, Loessner MJ, Alberti-Segui C, Higgins DE, Swanson JA. 2006. Cytolysin-dependent delay of vacuole maturation in macrophages infected with *Listeria monocytogenes*. *Cell Microbiol* **8:**107–119.

230. Shaughnessy LM, Hoppe AD, Christensen KA, Swanson JA. 2006. Membrane perforations inhibit lysosome fusion by altering pH and calcium in *Listeria monocytogenes* vacuoles. *Cell Microbiol* **8:**781–792.

231. Schnupf P, Portnoy DA. 2007. Listeriolysin O: a phagosome-specific lysin. *Microbes Infect* **9:**1176–1187.

232. Beauregard KE, Lee KD, Collier RJ, Swanson JA. 1997. pH-dependent perforation of macrophage phagosomes by listeriolysin O from *Listeria monocytogenes*. *J Exp Med* **186:**1159–1163.

233. Bielecki J, Youngman P, Connelly P, Portnoy DA. 1990. *Bacillus subtilis* expressing a haemolysin gene from *Listeria monocytogenes* can grow in mammalian cells. *Nature* **345:**175–176.

234. Portnoy DA, Tweten RK, Kehoe M, Bielecki J. 1992. Capacity of listeriolysin O, streptolysin O, and perfringolysin O to mediate growth of *Bacillus subtilis* within mammalian cells. *Infect Immun* **60:**2710–2717.

235. Glomski IJ, Gedde MM, Tsang AW, Swanson JA, Portnoy DA. 2002. The *Listeria monocytogenes* hemolysin has an acidic pH optimum to compartmentalize activity and prevent damage to infected host cells. *J Cell Biol* **156:**1029–1038.

236. Singh R, Jamieson A, Cresswell P. 2008. GILT is a critical host factor for *Listeria monocytogenes* infection. *Nature* **455:**1244–1247.

237. Radtke AL, Anderson KL, Davis MJ, DiMagno MJ, Swanson JA, O'Riordan MX. 2011. *Listeria monocytogenes* exploits cystic fibrosis transmembrane conductance regulator (CFTR) to escape the phagosome. *Proc Natl Acad Sci U S A* **108:**1633–1638.

238. **Smith GA, Marquis H, Jones S, Johnston NC, Portnoy DA, Goldfine H.** 1995. The two distinct phospholipases C of *Listeria monocytogenes* have overlapping roles in escape from a vacuole and cell-to-cell spread. *Infect Immun* **63:**4231–4237.

239. **Poussin MA, Leitges M, Goldfine H.** 2009. The ability of *Listeria monocytogenes* PI-PLC to facilitate escape from the macrophage phagosome is dependent on host PKCβ. *Microb Pathog* **46:**1–5.

240. **Poussin MA, Goldfine H.** 2005. Involvement of *Listeria monocytogenes* phosphatidylinositol-specific phospholipase C and host protein kinase C in permeabilization of the macrophage phagosome. *Infect Immun* **73:**4410–4413.

241. **Montes LR, Goñi FM, Johnston NC, Goldfine H, Alonso A.** 2004. Membrane fusion induced by the catalytic activity of a phospholipase C/sphingomyelinase from *Listeria monocytogenes. Biochemistry* **43:**3688–3695.

242. **Alberti-Segui C, Goeden KR, Higgins DE.** 2007. Differential function of *Listeria monocytogenes* listeriolysin O and phospholipases C in vacuolar dissolution following cell-to-cell spread. *Cell Microbiol* **9:**179–195.

243. **Sibelius U, Chakraborty T, Krögel B, Wolf J, Rose F, Schmidt R, Wehland J, Seeger W, Grimminger F.** 1996. The listerial exotoxins listeriolysin and phosphatidylinositol-specific phospholipase C synergize to elicit endothelial cell phosphoinositide metabolism. *J Immunol* **157:**4055–4060.

244. **Barker JR, Klose KE.** 2007. Molecular and genetic basis of pathogenesis in *Francisella tularensis. Ann N Y Acad Sci* **1105:**138–159.

245. **de Bruin OM, Ludu JS, Nano FE.** 2007. The *Francisella* pathogenicity island protein IglA localizes to the bacterial cytoplasm and is needed for intracellular growth. *BMC Microbiol* **7:**1. doi:10.1186/1471-2180-7-1.

246. **Buchan BW, McCaffrey RL, Lindemann SR, Allen LA, Jones BD.** 2009. Identification of *migR*, a regulatory element of the *Francisella tularensis* live vaccine strain *iglABCD* virulence operon required for normal replication and trafficking in macrophages. *Infect Immun* **77:** 2517–2529.

247. **Hayward RD, Cain RJ, McGhie EJ, Phillips N, Garner MJ, Koronakis V.** 2005. Cholesterol binding by the bacterial type III translocon is essential for virulence effector delivery into mammalian cells. *Mol Microbiol* **56:** 590–603.

248. **Picking WL, Nishioka H, Hearn PD, Baxter MA, Harrington AT, Blocker A, Picking WD.** 2005. IpaD of *Shigella flexneri* is independently required for regulation of Ipa protein secretion and efficient insertion of IpaB and IpaC into host membranes. *Infect Immun* **73:**1432–1440.

249. **Whitworth T, Popov VL, Yu XJ, Walker DH, Bouyer DH.** 2005. Expression of the *Rickettsia prowazekii pld* or *tlyC* gene in *Salmonella enterica* serovar Typhimurium mediates phagosomal escape. *Infect Immun* **73:** 6668–6673.

250. **Renesto P, Dehoux P, Gouin E, Touqui L, Cossart P, Raoult D.** 2003. Identification and characterization of a phospholipase D-superfamily gene in rickettsiae. *J Infect Dis* **188:**1276–1283.

251. **Silverman DJ, Santucci LA, Meyers N, Sekeyova Z.** 1992. Penetration of host cells by *Rickettsia rickettsii* appears to be mediated by a phospholipase of rickettsial origin. *Infect Immun* **60:**2733–2740.

252. **Winkler HH, Miller ET.** 1982. Phospholipase A and the interaction of *Rickettsia prowazekii* and mouse fibroblasts (L-929 cells). *Infect Immun* **38:**109–113.

253. **Lucchini S, Liu H, Jin Q, Hinton JC, Yu J.** 2005. Transcriptional adaptation of *Shigella flexneri* during infection of macrophages and epithelial cells: insights into the strategies of a cytosolic bacterial pathogen. *Infect Immun* **73:**88–102.

254. **Chieng S, Carreto L, Nathan S.** 2012. *Burkholderia pseudomallei* transcriptional adaptation in macrophages. *BMC Genomics* **13:**328. doi:10.1186/1471-2164-13-328.

255. **Wehrly TD, Chong A, Virtaneva K, Sturdevant DE, Child R, Edwards JA, Brouwer D, Nair V, Fischer ER, Wicke L, Curda AJ, Kupko JJ III, Martens C, Crane DD, Bosio CM, Porcella SF, Celli J.** 2009. Intracellular biology and virulence determinants of *Francisella tularensis* revealed by transcriptional profiling inside macrophages. *Cell Microbiol* **11:**1128–1150.

256. **Chatterjee SS, Hossain H, Otten S, Kuenne C, Kuchmina K, Machata S, Domann E, Chakraborty T, Hain T.** 2006. Intracellular gene expression profile of *Listeria monocytogenes. Infect Immun* **74:**1323–1338.

257. **Campbell-Valois FX, Schnupf P, Nigro G, Sachse M, Sansonetti PJ, Parsot C.** 2014. A fluorescent reporter reveals on/off regulation of the *Shigella* type III secretion apparatus during entry and cell-to-cell spread. *Cell Host Microbe* **15:**177–189.

258. **Scortti M, Monzó HJ, Lacharme-Lora L, Lewis DA, Vázquez-Boland JA.** 2007. The PrfA virulence regulon. *Microbes Infect* **9:**1196–1207.

259. **Chakraborty T, Leimeister-Wächter M, Domann E, Hartl M, Goebel W, Nichterlein T, Notermans S.** 1992. Coordinate regulation of virulence genes in *Listeria monocytogenes* requires the product of the *prfA* gene. *J Bacteriol* **174:**568–574.

260. **Johansson J, Mandin P, Renzoni A, Chiaruttini C, Springer M, Cossart P.** 2002. An RNA thermosensor controls expression of virulence genes in *Listeria monocytogenes. Cell* **110:**551–561.

261. **Ermolaeva S, Novella S, Vega Y, Ripio MT, Scortti M, Vázquez-Boland JA.** 2004. Negative control of *Listeria monocytogenes* virulence genes by a diffusible autorepressor. *Mol Microbiol* **52:**601–611.

262. **Milenbachs AA, Brown DP, Moors M, Youngman P.** 1997. Carbon-source regulation of virulence gene expression in *Listeria monocytogenes. Mol Microbiol* **23:** 1075–1085.

263. **Brehm K, Ripio MT, Kreft J, Vázquez-Boland JA.** 1999. The *bvr* locus of *Listeria monocytogenes* mediates virulence gene repression by β-glucosides. *J Bacteriol* **181:** 5024–5032.

264. **Ripio MT, Brehm K, Lara M, Suárez M, Vázquez-Boland JA.** 1997. Glucose-1-phosphate utilization by *Listeria monocytogenes* is PrfA dependent and coordinately expressed with virulence factors. *J Bacteriol* **179:** 7174–7180.

265. **Herro R, Poncet S, Cossart P, Buchrieser C, Gouin E, Glaser P, Deutscher J.** 2005. How seryl-phosphorylated HPr inhibits PrfA, a transcription activator of *Listeria monocytogenes* virulence genes. *J Mol Microbiol Biotechnol* **9:**224–234.

266. **Park SF, Kroll RG.** 1993. Expression of listeriolysin and phosphatidylinositol-specific phospholipase C is repressed by the plant-derived molecule cellobiose in *Listeria monocytogenes*. *Mol Microbiol* **8:**653–661.

267. **Lobel L, Sigal N, Borovok I, Belitsky BR, Sonenshein AL, Herskovits AA.** 2015. The metabolic regulator CodY links *Listeria monocytogenes* metabolism to virulence by directly activating the virulence regulatory gene *prfA*. *Mol Microbiol* **95:**624–644.

268. **Lobel L, Sigal N, Borovok I, Ruppin E, Herskovits AA.** 2012. Integrative genomic analysis identifies isoleucine and CodY as regulators of *Listeria monocytogenes* virulence. *PLoS Genet* **8:**e1002887. doi:10.1371/journal.pgen.1002887.

269. **Reniere ML, Whiteley AT, Hamilton KL, John SM, Lauer P, Brennan RG, Portnoy DA.** 2015. Glutathione activates virulence gene expression of an intracellular pathogen. *Nature* **517:**170–173.

270. **Deshayes C, Bielecka MK, Cain RJ, Scortti M, de las Heras A, Pietras Z, Luisi BF, Núñez Miguel R, Vázquez-Boland JA.** 2012. Allosteric mutants show that PrfA activation is dispensable for vacuole escape but required for efficient spread and *Listeria* survival *in vivo*. *Mol Microbiol* **85:**461–477.

271. **Kane CD, Schuch R, Day WA Jr, Maurelli AT.** 2002. MxiE regulates intracellular expression of factors secreted by the *Shigella flexneri* 2a type III secretion system. *J Bacteriol* **184:**4409–4419.

272. **Dhariwala MO, Anderson DM.** 2014. Bacterial programming of host responses: coordination between type I interferon and cell death. *Front Microbiol* **5:**545. doi:10.3389/fmicb.2014.00545.

273. **Gouin E, Adib-Conquy M, Balestrino D, Nahori MA, Villiers V, Colland F, Dramsi S, Dussurget O, Cossart P.** 2010. The *Listeria monocytogenes* InlC protein interferes with innate immune responses by targeting the IκB kinase subunit IKKα. *Proc Natl Acad Sci U S A* **107:** 17333–17338.

274. **Leung N, Gianfelice A, Gray-Owen SD, Ireton K.** 2013. Impact of the *Listeria monocytogenes* protein InlC on infection in mice. *Infect Immun* **81:**1334–1340.

275. **Jones BD, Faron M, Rasmussen JA, Fletcher JR.** 2014. Uncovering the components of the *Francisella tularensis* virulence stealth strategy. *Front Cell Infect Microbiol* **4:** 32. doi:10.3389/fcimb.2014.00032.

276. **Dons L, Rasmussen OF, Olsen JE.** 1992. Cloning and characterization of a gene encoding flagellin of *Listeria monocytogenes*. *Mol Microbiol* **6:**2919–2929.

277. **Gründling A, Burrack LS, Bouwer HG, Higgins DE.** 2004. *Listeria monocytogenes* regulates flagellar motility gene expression through MogR, a transcriptional repressor required for virulence. *Proc Natl Acad Sci U S A* **101:**12318–12323.

278. **Mariathasan S, Weiss DS, Dixit VM, Monack DM.** 2005. Innate immunity against *Francisella tularensis* is dependent on the ASC/caspase-1 axis. *J Exp Med* **202:** 1043–1049.

279. **Dotson RJ, Rabadi SM, Westcott EL, Bradley S, Catlett SV, Banik S, Harton JA, Bakshi CS, Malik M.** 2013. Repression of inflammasome by *Francisella tularensis* during early stages of infection. *J Biol Chem* **288:** 23844–23857.

280. **Robertson GT, Case ED, Dobbs N, Ingle C, Balaban M, Celli J, Norgard MV.** 2014. FTT0831c/FTL_0325 contributes to *Francisella tularensis* cell division, maintenance of cell shape, and structural integrity. *Infect Immun* **82:**2935–2948.

281. **Kobayashi T, Ogawa M, Sanada T, Mimuro H, Kim M, Ashida H, Akakura R, Yoshida M, Kawalec M, Reichhart JM, Mizushima T, Sasakawa C.** 2013. The *Shigella* OspC3 effector inhibits caspase-4, antagonizes inflammatory cell death, and promotes epithelial infection. *Cell Host Microbe* **13:**570–583.

282. **Tattoli I, Sorbara MT, Vuckovic D, Ling A, Soares F, Carneiro LA, Yang C, Emili A, Philpott DJ, Girardin SE.** 2012. Amino acid starvation induced by invasive bacterial pathogens triggers an innate host defense program. *Cell Host Microbe* **11:**563–575.

283. **Tattoli I, Sorbara MT, Yang C, Tooze SA, Philpott DJ, Girardin SE.** 2013. *Listeria* phospholipases subvert host autophagic defenses by stalling pre-autophagosomal structures. *EMBO J* **32:**3066–3078.

284. **Meyer-Morse N, Robbins JR, Rae CS, Mochegova SN, Swanson MS, Zhao Z, Virgin HW, Portnoy D.** 2010. Listeriolysin O is necessary and sufficient to induce autophagy during *Listeria monocytogenes* infection. *PLoS One* **5:**e8610. doi:10.1371/journal.pone.0008610.

285. **Dortet L, Mostowy S, Samba-Louaka A, Gouin E, Nahori MA, Wiemer EA, Dussurget O, Cossart P.** 2011. Recruitment of the major vault protein by InlK: a *Listeria monocytogenes* strategy to avoid autophagy. *PLoS Pathog* **7:**e1002168. doi:10.1371/journal.ppat.1002168.

286. **Mitchell G, Ge L, Huang Q, Chen C, Kianian S, Roberts MF, Schekman R, Portnoy DA.** 2015. Avoidance of autophagy mediated by PlcA or ActA is required for *Listeria monocytogenes* growth in macrophages. *Infect Immun* **83:**2175–2184.

287. **Ogawa M, Yoshimori T, Suzuki T, Sagara H, Mizushima N, Sasakawa C.** 2005. Escape of intracellular *Shigella* from autophagy. *Science* **307:**727–731.

288. **Case ED, Chong A, Wehrly TD, Hansen B, Child R, Hwang S, Virgin HW, Celli J.** 2014. The *Francisella* O-antigen mediates survival in the macrophage cytosol via autophagy avoidance. *Cell Microbiol* **16:**862–877.

289. **Devenish RJ, Lai SC.** 2015. Autophagy and *Burkholderia*. *Immunol Cell Biol* **93:**18–24.

290. **Galyov EE, Brett PJ, DeShazer D.** 2010. Molecular insights into *Burkholderia pseudomallei* and *Burkholderia mallei* pathogenesis. *Annu Rev Microbiol* **64:** 495–517.

291. **Baxt LA, Goldberg MB.** 2014. Host and bacterial proteins that repress recruitment of LC3 to *Shigella* early during infection. *PLoS One* **9:**e94653. doi:10.1371/journal.pone.0094653.

292. **Lam GY, Cemma M, Muise AM, Higgins DE, Brumell JH.** 2013. Host and bacterial factors that regulate LC3 recruitment to *Listeria monocytogenes* during the early stages of macrophage infection. *Autophagy* **9:** 985–995.

293. **Steele S, Brunton J, Ziehr B, Taft-Benz S, Moorman N, Kawula T.** 2013. *Francisella tularensis* harvests nutrients derived via ATG5-independent autophagy to support intracellular growth. *PLoS Pathog* **9:**e1003562. doi:10.1371/journal.ppat.1003562.

294. **Bast A, Krause K, Schmidt IH, Pudla M, Brakopp S, Hopf V, Breitbach K, Steinmetz I.** 2014. Caspase-1-dependent and -independent cell death pathways in *Burkholderia pseudomallei* infection of macrophages. *PLoS Pathog* **10:**e1003986. doi:10.1371/journal.ppat.1003986.

295. **Pierini R, Juruj C, Perret M, Jones CL, Mangeot P, Weiss DS, Henry T.** 2012. AIM2/ASC triggers caspase-8-dependent apoptosis in *Francisella*-infected caspase-1-deficient macrophages. *Cell Death Differ* **19:**1709–1721.

296. **Doyle CR, Pan JA, Mena P, Zong WX, Thanassi DG.** 2014. TolC-dependent modulation of host cell death by the *Francisella tularensis* live vaccine strain. *Infect Immun* **82:**2068–2078.

297. **Carneiro LA, Travassos LH, Soares F, Tattoli I, Magalhaes JG, Bozza MT, Plotkowski MC, Sansonetti PJ, Molkentin JD, Philpott DJ, Girardin SE.** 2009. *Shigella* induces mitochondrial dysfunction and cell death in nonmyleoid cells. *Cell Host Microbe* **5:**123–136.

298. **Joshi SG, Francis CW, Silverman DJ, Sahni SK.** 2004. NF-κB activation suppresses host cell apoptosis during *Rickettsia rickettsii* infection via regulatory effects on intracellular localization or levels of apoptogenic and anti-apoptotic proteins. *FEMS Microbiol Lett* **234:** 333–341.

299. **Joshi SG, Francis CW, Silverman DJ, Sahni SK.** 2003. Nuclear factor κB protects against host cell apoptosis during *Rickettsia rickettsii* infection by inhibiting activation of apical and effector caspases and maintaining mitochondrial integrity. *Infect Immun* **71:**4127–4136.

300. **Cassidy SK, O'Riordan MX.** 2013. More than a pore: the cellular response to cholesterol-dependent cytolysins. *Toxins (Basel)* **5:**618–636.

301. **Mestre MB, Colombo MI.** 2013. Autophagy and toxins: a matter of life or death. *Curr Mol Med* **13:**241–251.

302. **Mares CA, Sharma J, Li Q, Rangel EL, Morris EG, Enriquez MI, Teale JM.** 2011. Defect in efferocytosis leads to alternative activation of macrophages in *Francisella* infections. *Immunol Cell Biol* **89:**167–172.

303. **Lai XH, Golovliov I, Sjöstedt A.** 2001. *Francisella tularensis* induces cytopathogenicity and apoptosis in murine macrophages via a mechanism that requires intracellular bacterial multiplication. *Infect Immun* **69:** 4691–4694.

304. **Yao Q, Cui J, Wang J, Li T, Wan X, Luo T, Gong YN, Xu Y, Huang N, Shao F.** 2012. Structural mechanism of ubiquitin and NEDD8 deamidation catalyzed by bacterial effectors that induce macrophage-specific apoptosis. *Proc Natl Acad Sci U S A* **109:**20395–20400.

305. **Cassidy SK, Hagar JA, Kanneganti TD, Franchi L, Nuñez G, O'Riordan MX.** 2012. Membrane damage during *Listeria monocytogenes* infection triggers a caspase-7 dependent cytoprotective response. *PLoS Pathog* **8:**e1002628. doi:10.1371/journal.ppat.1002628.

306. **Czuczman MA, Fattouh R, van Rijn JM, Canadien V, Osborne S, Muise AM, Kuchroo VK, Higgins DE, Brumell JH.** 2014. *Listeria monocytogenes* exploits efferocytosis to promote cell-to-cell spread. *Nature* **509:** 230–234.

Myeloid Cells in Health and Disease: A Synthesis
Edited by Siamon Gordon

doi:10.1128/microbiolspec.MCHD-0006-2015

Barry R. Bloom[1]
Robert L. Modlin[2]

41 Mechanisms of Defense against Intracellular Pathogens Mediated by Human Macrophages

SOME NECESSARY CONDITIONS FOR PROTECTION

While animal models of human disease have provided great insights that would have been difficult to achieve in studies in humans, our initial experiments on protection against *Mycobacterium tuberculosis* were designed to take advantage of transgenic knockout mice. While animal models have contributed enormously to our understanding, it must be noted that none faithfully reproduces the pathology or course of disease of human tuberculosis (TB) or leprosy. With that caveat, we explored the question of immunologically necessary conditions for protection of mice against *M. tuberculosis* infection. We found that mice lacking the gene for gamma interferon (IFN-γ) died from *M. tuberculosis* challenge in a matter of 2 to 3 weeks after intravenous challenge and within a month after aerosol challenge (1). We hypothesized that the pathology observed in the lungs would likely be mediated by local production of tumor necrosis factor alpha (TNF-α) and were quite surprised to learn that TNF-α-depleted mice succumbed with precisely the same time to death as the IFN-γ knockouts (2). Additionally, we (3) and others (4, 5) found that mice whose major histocompatibility complex (MHC) class I presentation to cytotoxic T lymphocytes (CTLs) was deficient, e.g., β_2-microglobulin deficient or TAP (transporter associated with antigen processing) deficient, succumbed to TB infection far earlier than control mice, but many weeks later than the IFN-γ- and TNF-α-deficient mice (1) (Fig. 1). These results established that IFN-γ and TNF-α are necessary for initial protection in mice, likely mediated by innate immunity and cytokine-activated macrophages, and suggested that CTLs may play a role later in infection.

MOUSE MACROPHAGE MICROBICIDAL MECHANISMS

The experiments *in vivo* were followed by *in vitro* experiments on mouse macrophages to learn how the

[1]Harvard School of Public Health, Boston, MA 02115; [2]David Geffen School of Medicine, University of California at Los Angeles, Los Angeles, CA 90095.

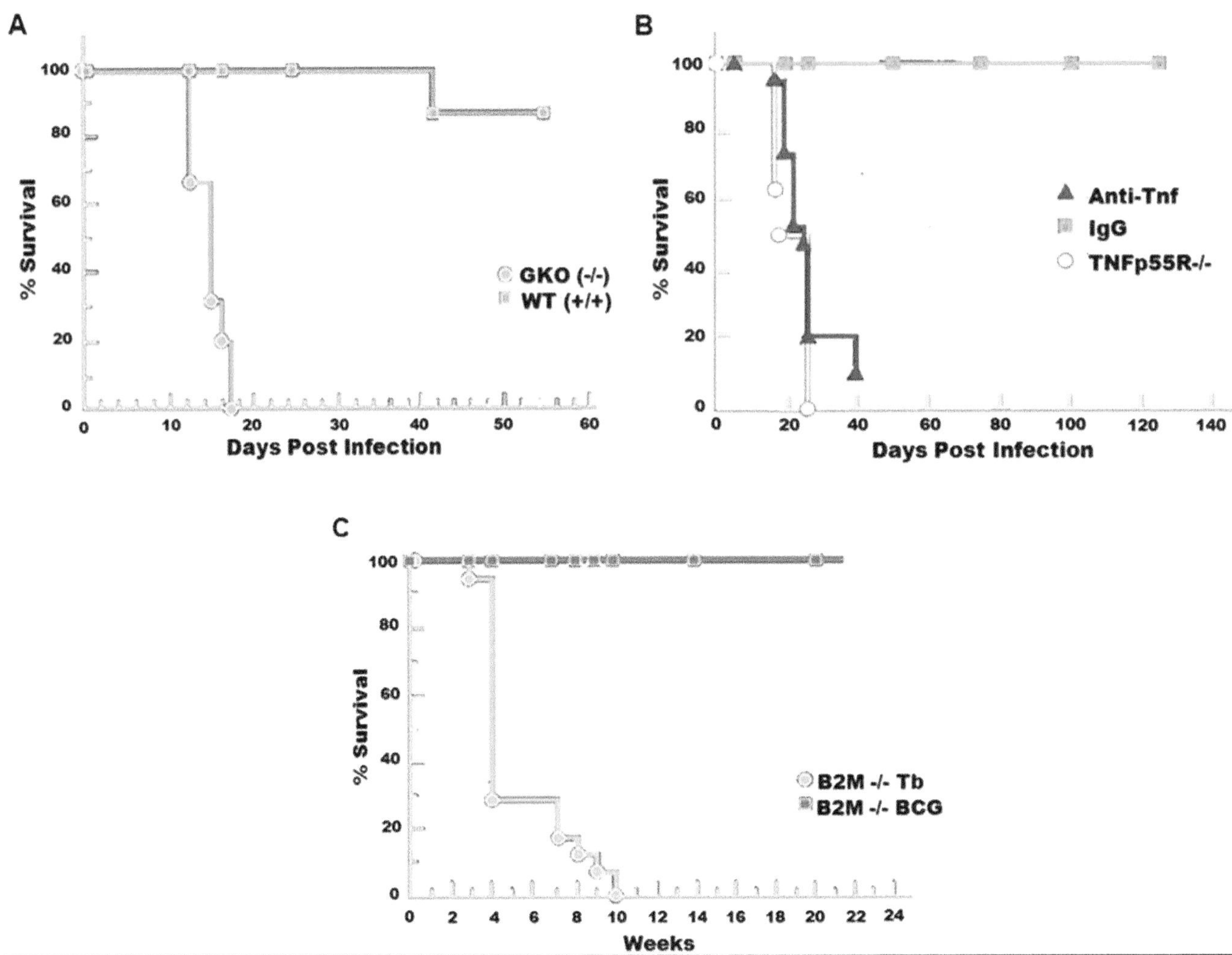

Figure 1 Necessary conditions for protection in mice. Depletion of IFN-γ (**A**), TNF-α (**B**), or MHC class I (**C**) results in increased susceptibility and death following *M. tuberculosis* challenge in C57BL/6 mice. B2M, beta-2 microglobulin; BCG, Bacillus Calmette–Guérin; GKO, IFN-gamma gene knock out; WT, wild type. Sources: references 1–3.

cytokines IFN-γ and TNF-α contributed to protection. Activation by IFN-γ and TNF-α revealed that mouse macrophages were able to kill *M. tuberculosis*, but the mechanism was mediated by the action of reactive oxygen intermediates, which were also effective against many other pathogens (6). In 1991, Chan et al. established that while the killing of *M. tuberculosis* by mouse macrophages was not affected by inhibitors of reactive oxygen intermediates, the killing was dramatically reduced by inhibitors of nitric oxide synthase (NOS2) and that *M. tuberculosis* was directly killed by NO itself (7). In elegant experiments, MacMicking et al. established the importance of macrophage killing by NO *in vivo* by demonstrating that NOS2 knockout mice died rapidly from *M. tuberculosis*. Of interest, the time of death was not quite as rapid as in the case of the IFN-γ knockouts (8). These results indicated that the major mediator of macrophage killing of *M. tuberculosis* in the mouse was reactive nitrogen intermediates. Mycobacteria have evolved multiple mechanisms to resist oxygen radicals (6, 9) and nitrogen radicals (10–12), including scavenging of radicals by surface carbohydrates, and enzymatic mechanisms.

LEARNING FROM HUMAN LESIONS

The battle between the tubercle or leprosy bacillus and the host macrophages is fought at the site of disease in the tissue lesions. Establishing what is happening in the lungs of TB patients at any stage in the disease has been extremely difficult. In contrast, leprosy is essentially a skin disease, accessible for study by biopsy used for diagnosis, staging, and assessing prognosis. Consequently, we sought to characterize the events occurring in lesions of human leprosy. One of the fascinations of leprosy is that it does not exist as a single clinical entity, but rather is a spectrum. At one pole of the spectrum, tuberculoid leprosy, there are well-organized granulomas and well-differentiated macrophages but almost no acid-fast bacilli. At the other pole, lepromatous leprosy, there are macrophages full of acid-fast *Mycobacterium leprae*, very few lymphocytes, and the granulomas are disorganized (Fig. 2). Most patients are classified as "borderline" and show lesions with mixed characteristics. Leprosy forms are not always static and patients can undergo reactional states moving toward either pole. Tuberculoid lesions have a preponderance of CD4 Th1 cells, while lepromatous lesions have few CD4 and CD8 Th2 cells (13). Analysis of gene expression in lesions revealed that in tuberculoid lesions IFN-γ, interleukin-2 (IL-2), and lymphotoxin predominate, while in polar lepromatous lesions IL-4, IL-5, and IL-10 predominate (14, 15), which was the first example of the Th1 and Th2 dichotomy, previously described in mice (16), existing in human disease lesions.

THE INNATE AND ACQUIRED VITAMIN D-DEPENDENT ANTIMICROBIAL MECHANISM OF HUMAN MACROPHAGES

If, as is generally assumed, the infectious dose of *M. tuberculosis* required to cause infection and disease in humans is of the order of 1 to 400 bacilli, our first approach was to explore the ability of cells of the innate immune system to respond to the pathogen. This is of particular interest in light of the finding of many in the field that there are people with high levels of exposure to *M. tuberculosis* who remain healthy and fail to convert their tuberculin skin test. That suggests that the innate immune system may have the capability of effectively killing the initial low level of transmitted bacilli so rapidly that the bacilli are unable to multiply to a level required to engage the acquired T-cell response. In that context, it seemed important to pursue innate mechanisms of killing of intracellular mycobacteria by human monocytes and monocyte-derived macrophages stimulated through the Toll-like receptors (TLRs) through *in vitro* studies. Human peripheral blood precursors can be differentiated *in vitro* into macrophages or dendritic cells (DCs) when cultured in different cytokine-containing growth media. Macrophages were derived from cultures with macrophage colony-stimulating factor (M-CSF) or granulocyte-macrophage CSF (GM-CSF), and DCs were differentiated with GM-CSF + IL-4, and each possessed characteristic surface markers. When these cells were cultured in appropriate human sera for 4 days followed by activation via the innate immune receptor TLR2, we found a striking reduction in *M. tuberculosis* CFU by macrophages but not by the DCs (17, 18). Surprisingly, while the NOS2 inhibitor L-NIL was able to almost completely block killing of *M. tuberculosis* by TLR2-activated mouse macrophages, it had almost no effect on killing by activated human macrophages (Fig. 3). This was the first evidence that human macrophages, at least *in vitro*, kill *M. tuberculosis* by a mechanism different from that of mouse macrophages.

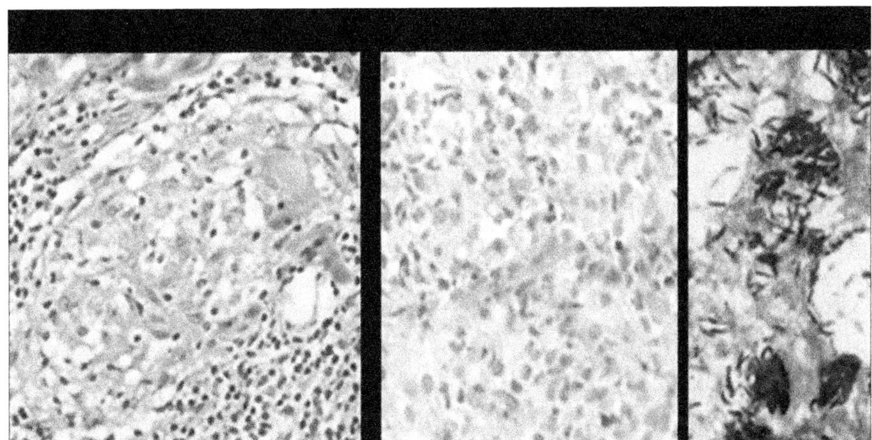

Figure 2 Macrophages in leprosy lesions. Cellular infiltrates in tuberculoid and lepromatous leprosy lesions. The third panel shows an acid-fast stain, indicating the abundance of bacilli in lepromatous lesions. Courtesy of Thomas H. Rea.

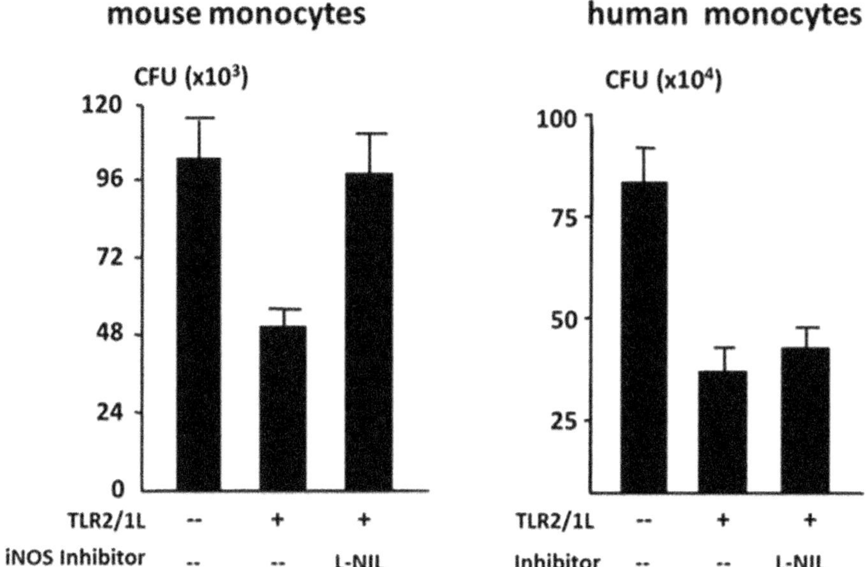

Figure 3 Inducible NOS2 (iNOS) is essential for killing of *M. tuberculosis* by activated mouse macrophages *in vitro* but not for human monocyte-derived macrophages. L-NIL is an inhibitor of NOS2. Source: reference 17.

Those findings led to experiments seeking to learn which genes were expressed by macrophages but not expressed by DCs following TLR2 stimulation (18). Using two-way analysis of variance, there were three genes predominantly expressed in the microbicidal macrophages and not in DCs: S100A12; the vitamin D receptor (VDR); and Cyp27B1, the enzyme that converts 25-hydroxyvitamin D to the active 1,25-dihydroxyvitamin D. That, for the first time, suggested a role for vitamin D in the microbicidal process. When cathelicidin and β-defensin 2 (DEFB4), known downstream products of Cyp27B1 related to antimicrobial activity, were screened for in macrophages activated by TLR2 ligand alone, neither of the mRNAs for these antimicrobial peptides was expressed. However, when the macrophages were stimulated by TLR2 ligand in the presence of 25-hydroxyvitamin D, mRNAs for cathelicidin and DEFB4 were stimulated 5- to 20-fold (19). We also demonstrated induction of the active form of cathelicidin, a 37-amino-acid amphipathic α-helical antimicrobial peptide cleaved from a large precursor, encoded by the CAMP gene (18). We were able to establish that the cathelicidin peptide itself was capable of killing *M. tuberculosis* in culture. Further, although it is well known that *M. tuberculosis* in macrophage phagocytic vacuoles is able to block fusion of lysosomes and is refractory to many cellular proteins, including the NADP oxidase, we were able to observe that the cathelicidin peptide in TLR2-activated cells was colocalized with green fluorescent protein-labeled mycobacteria within the phagocytic vacuole (18). The formal demonstration that cathelicidin and DEFB4 were critical for killing of *M. tuberculosis* in human macrophages *in vitro* was the demonstration that small interfering RNA to cathelicidin essentially eliminated the antimicrobial activity (20) (Fig. 4).

Multiple reports in the literature concluded that, in contrast to murine macrophages, IFN-γ, the best-studied T-cell mediator of macrophage activation, was not able to activate human monocyte-derived macrophages *in vitro* to kill *M. tuberculosis* (21–28). However, we were able to show that the vitamin D-dependent antimicrobial pathway, with expression of cathelicidin and DEFB4, is induced in human macrophages by the acquired immune response *in vitro* and mediated by IFN-γ (Fig. 5B) (29). Thus both innate and acquired immune-activated human macrophages, through TLR2 triggering of the MyD88 signaling pathway or through activation by the type II IFN IFN-γ of a STAT1-mediated pathway, were able to kill *M. tuberculosis*. It is perhaps significant that the same antimicrobial pathway is conserved through evolution of the innate and adaptive immune system, though induced by two different immune mechanisms of activation acting through two different signaling pathways (29).

In all these experiments, inhibition of the VDR by a chemical VDR antagonist, VAZ, blocked the induction of cathelicidin, DEFB4, and the antimicrobial activity.

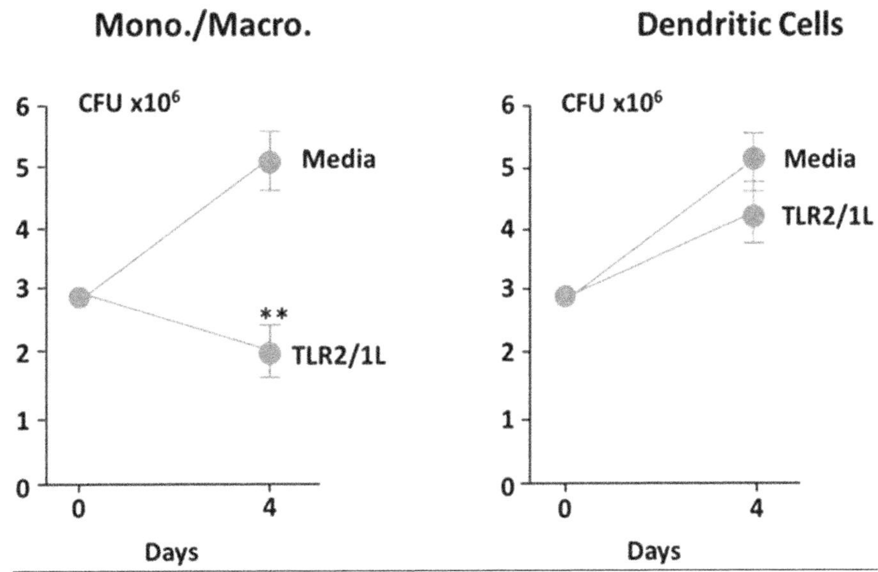

Figure 4 TLR2-activated human monocyte-derived macrophages are able to kill *M. tuberculosis*, whereas activated DCs (derived by culture of monocytes with GM-CSF and IL-4) were unable to kill under the same conditions. $^{**}P \leq 0.01$. Source: reference 18.

Thus the killing of *M. tuberculosis* in human macrophages *in vitro* was mediated by the vitamin D antimicrobial mechanism, not by oxygen or nitrogen radicals. The mechanism does not appear to be important in mouse macrophages, which have a CAMP homolog, but one which lacks any vitamin D responsive elements (30). This makes sense since mice are essentially nocturnal animals that have little exposure to sunlight. Only some primates and humans have the vitamin D responsive elements in their CAMP. The failure of previous

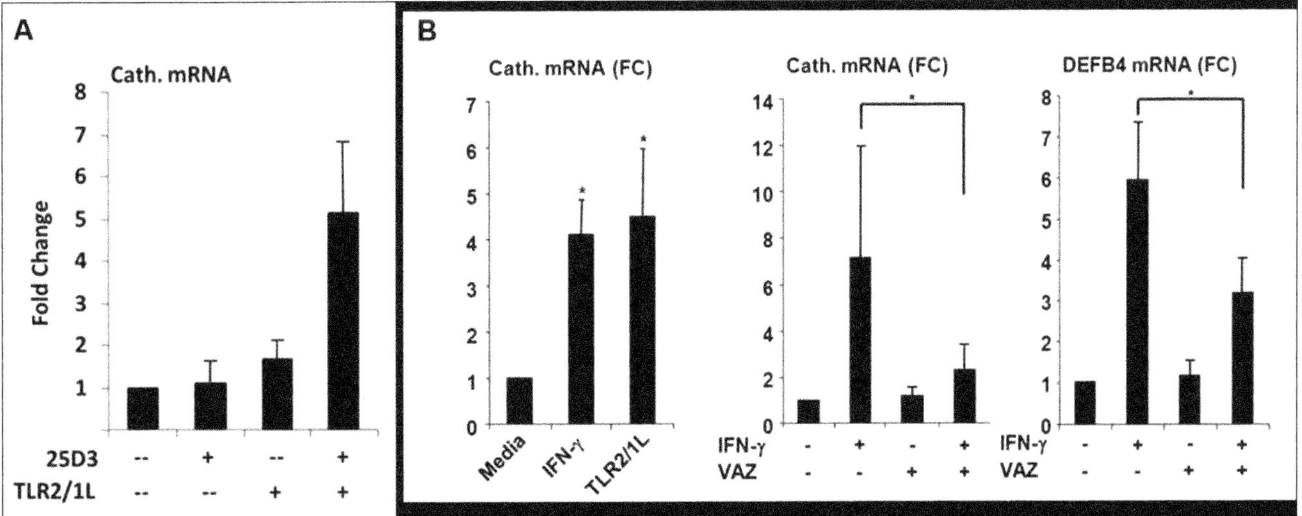

Figure 5 (A) Failure of TLR2-activated human macrophages to induce cathelicidin (Cath.) mRNA in culture with sera from African Americans can be reversed by addition of 25-hydroxyvitamin D. (B) IFN-γ activates human macrophages to produce cathelicidin to a comparable extent as the innate response to TLR2 agonists. Production of cathelicidin mRNA is dependent on vitamin D and inhibited by VAZ, a VDR antagonist. Vitamin D is essential for human macrophages activated by either TLR2 or IFN-γ to kill *M. tuberculosis in vitro*. FC, fold change. Sources: references 18, 29.

studies to observe killing of *M. tuberculosis* in human macrophages is most likely due to the culture of the cells either in fetal calf serum, which has negligible levels of 25-hydroxyvitamin D, or in human sera that lacked sufficient levels. While our experiments clearly demonstrate a novel vitamin D-dependent antimicrobial mechanism used by human macrophages in culture, we would caution that there may well be additional mechanisms for killing intracellular pathogens, including *M. tuberculosis*, by the macrophages of the human lung and other tissues.

THE QUESTION OF THE ROLE OF VITAMIN D IN HUMAN TB AND LEPROSY

Vitamin D is produced in human skin exposed to UV light. The circulating form is 25-hydroxyvitamin D, where the hydroxyl group is added by the liver. The biologically active form is 1,25-dihydroxyvitamin D, which we found to be produced by the action of Cyp27B1 in activated human macrophages (31). It had long been thought in the pre-antibiotic era that exposure of TB patients to sunshine and fresh air was helpful for recovery, and this was the rationale for the sanatorium movement in Europe and the United States. Clearly exposure to UV light would be expected to increase vitamin D levels, which might have been helpful in recovery. There is a fascinating historical study of cod liver oil, which is rich in vitamin D, in TB patients at Brompton Hospital in London in 1848 (32). The results indicated that while cod liver oil cured none of the patients, it arrested the disease in 18% of patients and reduced mortality from 33 to 19%.

It is well known that African Americans have a greater susceptibility to *M. tuberculosis* infection (33) and that in the pre-antibiotic era they developed more virulent forms of TB (34). Since melanin is known to absorb UV light, this history suggested to us that dark-skinned individuals might not have sufficient levels of vitamin D. In fact, a literature existed reporting that African Americans had vitamin D levels at 10 to 20% of what is considered normal. Consequently, we tested the ability of human sera from African Americans and white individuals to support the induction of cathelicidin in TLR2-stimulated macrophages. The result was striking: only 20% of the sera from African Americans had levels of 25-hydroxyvitamin D, in the normal range and were able to induce expression of cathelicidin mRNA. Addition of 25-hydroxyvitamin D, to the cultures enabled the induction of cathelicidin mRNA to levels comparable to those in the serum of white individuals and corrected the defect (Fig. 5).

These results suggest the possibility that the increased susceptibility of dark-skinned individuals from Africa or Asia to TB might be related to insufficient levels of vitamin D. Consistent with these findings are time-series epidemiological findings showing a parallel seasonal variation in vitamin D levels and notifications of TB in South Africa (35). There have been several *in vivo* studies of vitamin D supplementation, with sometimes conflicting results. A few contemporary clinical studies have described a potential benefit for vitamin D supplementation as an adjuvant to chemotherapy, measuring various endpoints including clinical and radiological improvement (36–38), sputum conversion (37, 39), and immune responses (38, 40, 41). However, these studies were generally inconclusive due to a number of study design confounders. While some studies showed shortening of the time to sputum clearance with chemotherapy, others found no beneficial effect. The studies were subject to many confounding factors, including being underpowered and being characterized by a low baseline prevalence of vitamin D deficiency; seasonality and increased levels of vitamin D in placebo controls; inadequate levels of supplementation due to dose and/or time; lack of sputum cultures in some studies; and the absence of agreed-upon, clearly defined clinical endpoints. However, a recent study in which patients were treated with high doses of vitamin D found clinical and radiological improvement, including a reduction in the number of cavities (38).

It is clearly challenging to ascertain a beneficial effect of vitamin D in patients who have active disease and are of necessity receiving antimycobacterial drugs. Vitamin D is not itself an antimicrobial drug, and it is asking a great deal to see dramatic effects in patients with active TB in which the tubercle bacillus has escaped protective mechanisms such that there are enormous numbers of bacilli in their tissues. If, as is believed, humans with disease are infected with between 1 and 400 CFU of tubercle bacilli, it is possible that the greatest effect of vitamin D would be in raising the threshold of innate resistance to infection, and the effects of vitamin D supplementation might be best found in healthy contacts or patients with latent infection.

THE PARADOXICAL ROLE OF IFNs

From a variety of animal and human studies it has long been known that type I IFNs (IFN-α and IFN-β) can engender protection against many viral infections but little protection against bacterial infections. As described above, in both humans and mice, IFN-γ is essential for protection against *M. tuberculosis*. Using analysis of gene expression arrays on peripheral blood of TB patients in Europe and in Africa, several laboratories have identified sets of genes that distinguish individuals with active TB disease from those with latent infection (42, 43). The most striking characteristic of the "signature" for active TB thus far is the increase in a set of genes regulated by IFNs, in particular the type I IFNs IFN-β and IFN-α (42, 44). The induction of the type I IFN gene program was associated with the extent of disease (42), and reversed completely by 2 months of effective treatment (45). At the same time, studies have shown that *M. tuberculosis* induces type I IFNs in macrophages, requiring the ESX-1 secretion pathway (46) for *M. tuberculosis* double-stranded DNA to gain access to the cytoplasm, where it activates DNA sensors including STING (47). In a mouse model, the hypervirulent *M. tuberculosis* strain HN878 induced high levels of type I IFN expression and low levels of IL-12p40 (48, 49), which augments Th1 responses. The spectrum of leprosy offered a unique opportunity to investigate the involvement of type I and type II IFNs in lesions of the human disease. The results were striking: in tuberculoid leprosy, type I IFN-associated mRNAs were hardly detectable, but there was evidence of type II IFN-induced genes. In contrast, in lepromatous leprosy lesions, there was marked elevation of type I IFN-induced genes, particularly IL-10 (50). In parallel, mRNAs for Cyp27B1 and the VDR were elevated in tuberculoid leprosy lesions.

When human macrophages were stimulated by IFN-γ *in vitro*, there was elevation of mRNAs for the microbicidal peptides cathelicidin and DEFB4. This stimulation of the antimicrobial response was found to be vitamin D dependent (29). In contrast, treatment of peripheral blood adherent cells with IFN-β induced IL-10. Simultaneous addition of IFN-γ and IFN-β resulted in suppression of cathelicidin and DEFB4 mRNAs. This inhibition of type II IFN activation was blocked by anti-IL-10 antibody, indicating that the suppressive effect of type I IFNs was mediated by IL-10 (50). When macrophages were activated by IFN-γ in serum sufficient for vitamin D, they were able to kill and reduce the viability of *M. leprae* and *M. tuberculosis*. The process required autophagy, and surprisingly, vitamin D was required for autophagy as well (29). When type I IFN or IL-10 was added, IFN-γ-induced killing was abrogated, and treatment with anti-IL-10 restored the antimicrobial effect (Fig. 6). These results demonstrate that type I and type II IFNs have opposing effects on the microbicidal activity of human macrophages and identify IL-10 as the mediator of type I IFN suppression of antimicrobial activity (Fig. 7).

OBSERVATIONS ON MAJOR DIFFERENCES BETWEEN HUMAN AND MURINE ANTIMICROBIAL RESPONSES

One general theme that has arisen in these experiments is that, despite many similarities, there are significant differences between human immune responses relating to antimicrobial activity and those of the common animal models. Some examples include the following observations. (i) Most animal models of TB fail to exhibit latency and human-like pathology. (ii) Mouse macrophages kill *M. tuberculosis* predominantly by NO, while human macrophages utilize microbicidal peptides, e.g.,

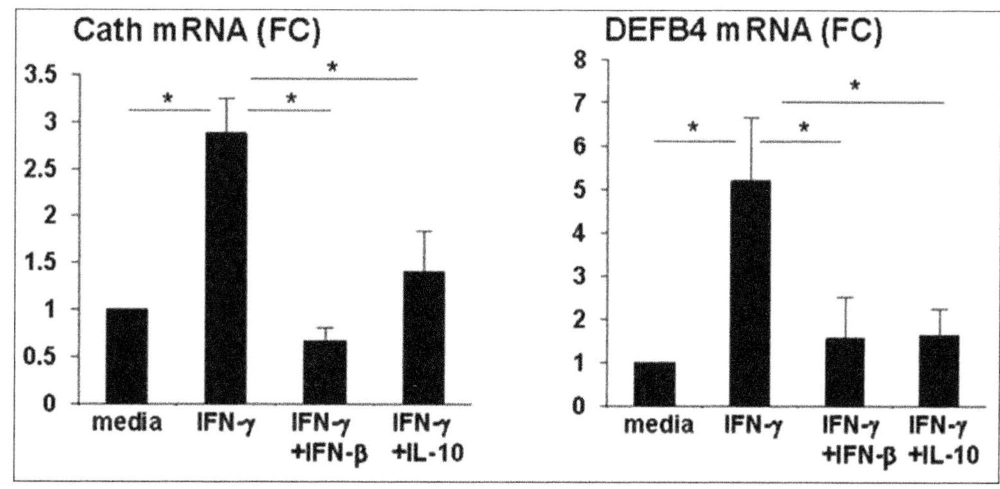

Figure 6 Inhibition of the effect of IFN-γ in stimulating induction of cathelicidin (Cath) mRNA by IFN-β is mediated by IL-10. FC, fold change. $*P \leq 0.05$. Source: reference 50.

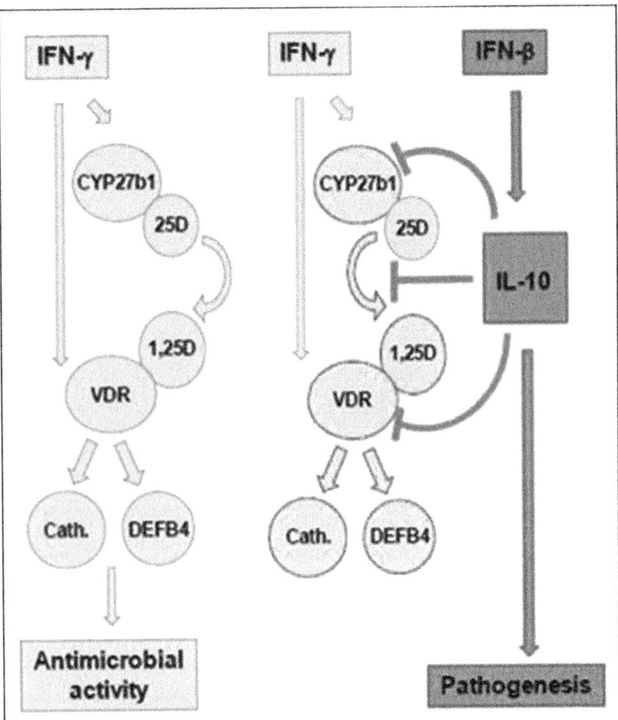

Figure 7 Common activation pathway for human macrophages stimulated through the innate receptor TLR2 or the acquired immune activator IFN-γ. IFN-β suppresses that activation through IL-10 by inhibiting induction of both CYP27B1 and the VDR. Source: reference 50.

cathelicidin and DEFB4, and likely other mechanisms. (iii) Human macrophage killing is vitamin D dependent, while that by mouse macrophages is vitamin D independent. (iv) While the human CAMP gene, the precursor of cathelicidin, has three vitamin D responsive elements, its homolog in the mouse, a nocturnal animal, has none. (v) DEFB4 has no mouse homolog; autophagy in human *M. tuberculosis*-infected macrophages is vitamin D dependent. (vi) CD1a, -b, and -c, which present nonpeptide antigens to human T cells, have no homologs in the mouse (which does have CD1d). (vii) Human CD8 CTLs release the antimicrobial peptide granulysin, which is lacking in mice. (viii) For IL-32, no homolog has yet been found in mice.

FINDING A MARKER OF PROTECTION AGAINST TB

One of the major problems in TB, particularly affecting evaluation of new vaccines, is that we have no molecular correlate of protection. Both the last major *Mycobacterium bovis* BCG trial in south India, with 360,000 subjects followed for 15 years (51), and a recent trial of a new candidate vaccine, MVA85A, in South African infants, regrettably showed no protection against *M. tuberculosis* infection or disease (52, 53). With about 40 vaccine candidates in preclinical trials, unless there is a way to prioritize them on some rational scientific basis, it is unlikely that hugely expensive large-scale efficacy trials will be undertaken. One exciting approach to developing such correlates is systems biology approaches, specifically the study of gene expression profiles in peripheral blood of TB patients. Recently, several laboratories have identified a set of genes that distinguish individuals with active TB disease from those with latent infection (42, 43, 45, 54). The most striking characteristic of the "signature" for active TB thus far is the increase in the type I IFN-regulated genes (42, 44). Because the type I IFN gene program was associated with extent of disease (42), and response to treatment (45), the type I IFN gene signature should be considered a "correlate of risk or pathogenesis" for TB.

In contrast, in BCG-vaccinated children in South Africa followed for 3 years for development of TB, studies of production of cytokines by peripheral blood cells, including IFN-γ, TNF-α, IL-12, and IL-17, were unable to distinguish those who developed disease from those who did not (55). We reasoned, however, that it might be possible to gain insight into candidate molecular "correlates of protection" by focusing on individuals with evidence of *M. tuberculosis* infection who do not progress to active disease. Consequently, we sought to identify, in currently available data sets, those genes that were more highly expressed in the blood of individuals with latent TB infection, as defined by a positive blood IFN-γ release assay, compared to individuals with active TB and healthy controls. Using a sophisticated bioinformatics approach known as weighted gene coexpression network analysis, which searches for pairwise gene expression, an IL-15-induced host defense module was identified in antimicrobial macrophages, which contained among other genes IL-32 (56). When these results were overlapped with the five existing data sets of gene expression in peripheral blood of patients with latent, active, or no TB, following informatics analysis, there were eight genes expressed to a greater extent in latent than in active patients, in healthy controls, or in sarcoidosis and other diseases in which TB was in the clinical differential diagnosis. Among those genes was IL-32, which was induced in human monocytes and macrophages by stimulation with IFN-γ (57) or activation of NOD2 by muramyl dipeptide (58) or of TLR4 by lipopolysaccharide (59). In the clinical data sets, IL-32 was elevated in latent TB, depressed in active

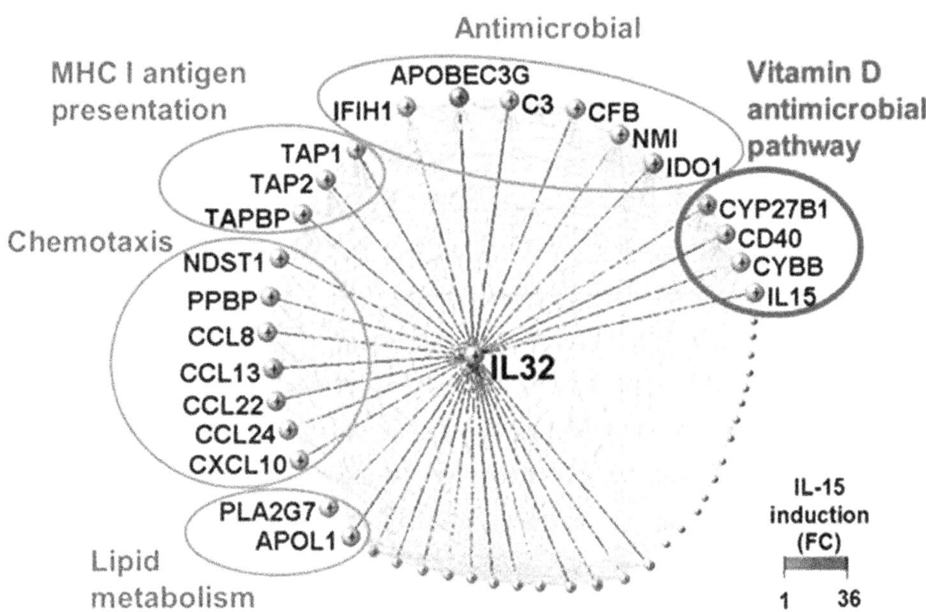

Figure 8 IL-15 defense response network links IL-32 to the vitamin D antimicrobial pathway. The IL-15-induced host defense network reveals IL-32 as a hub gene, connected to sets of genes involved in host defense, including the vitamin D antimicrobial pathway. The color of each node depicts fold change (FC) induction by IL-15 at 24 h. Source: reference 56.

disease, and returned to an intermediate level following successful treatment of the patients. IL-32 was found to be a central node in an IL-15-induced host defense network of 48 genes, 35 of which were expressed at baseline in the myeloid cell lineage in addition to IL-32 (Fig. 8). The IL-15-induced IL-32 gene network included vitamin D-related genes, e.g. cathelicidin, and genes related to MHC class I activity and chemotaxis, all of which could plausibly be related to antimicrobial activity.

When human monocytes or macrophages were treated with IL-32 *in vitro*, the vitamin D-dependent antimicrobial mechanism was activated to levels comparable to activation with IFN-γ or IL-15 and *M. tuberculosis* was killed by the IL-32-treated macrophages. This once again was a vitamin D-dependent process.

Since IL-32 expression correlates with the latent state of TB, which reflects a sufficient immune response to prevent the infection from progressing to active disease, we believe IL-32 should be considered as one "correlate of protection" against TB. Other biomarkers for protection are urgently needed to assess the likelihood of success of any of the many vaccine candidates before large-scale efficacy trials are undertaken. We would suggest that that may require a different scale of translational research on vaccines than has been considered previously, with multiple small-scale vaccination studies of human volunteers to develop a panel of molecular biomarkers that might be useful markers of protection. Such markers could markedly accelerate the development and testing of effective vaccines against TB. Clearly establishing any useful biomarkers for protection would require independent verification in other data sets and ideally in a longitudinal vaccine trial that provided protection to a significant number of recipients, where the correlation with protection against disease could be formally established. In the case of the vitamin D antimicrobial mechanism described here, while we do not know whether it is a necessary mechanism, we believe the evidence we have developed over the past several years clearly indicates that it is a sufficient mechanism by which human macrophages can kill intracellular bacterial pathogens such as *M. tuberculosis* and *M. leprae*.

Acknowledgments. *We wish to express our sincere appreciation for our many students and fellows who carried out the experiments summarized in this paper and to our clinical collaborators, Thomas H. Rea, Maria Teresa Ochoa, and Euzenir Sarno, and the many patients who generously volunteered for these studies. We also are grateful for research support from the National Institutes of Health over many years.*

Citation. Bloom BR, Modlin RL. 2016. Mechanisms of defense against intracellular pathogens mediated by human macrophages. Microbiol Spectrum 4(3):MCHD-0006-2015.

References

1. **Flynn JL, Chan J, Triebold KJ, Dalton DK, Stewart TA, Bloom BR.** 1993. An essential role for interferon γ in resistance to *Mycobacterium tuberculosis* infection. *J Exp Med* **178:**2249–2254.
2. **Flynn JL, Goldstein MM, Chan J, Triebold KJ, Pfeffer K, Lowenstein CJ, Schreiber R, Mak TW, Bloom BR.** 1995. Tumor necrosis factor-α is required in the protective immune response against *Mycobacterium tuberculosis* in mice. *Immunity* **2:**561–572.
3. **Flynn JL, Goldstein MM, Triebold KJ, Koller B, Bloom BR.** 1992. Major histocompatibility complex class I-restricted T cells are required for resistance to *Mycobacterium tuberculosis* infection. *Proc Natl Acad Sci U S A* **89:**12013–12017.
4. **Dunn PL, North RJ.** 1991. Resolution of primary murine listeriosis and acquired resistance to lethal secondary infection can be mediated predominantly by Thy-1$^+$ CD4$^-$ CD8$^-$ cells. *J Infect Dis* **164:**869–877.
5. **Turner J, D'Souza CD, Pearl JE, Marietta P, Noel M, Frank AA, Appelberg R, Orme IM, Cooper AM.** 2001. CD8- and CD95/95L-dependent mechanisms of resistance in mice with chronic pulmonary tuberculosis. *Am J Respir Cell Mol Biol* **24:**203–209.
6. **Chan J, Fan XD, Hunter SW, Brennan PJ, Bloom BR.** 1991. Lipoarabinomannan, a possible virulence factor involved in persistence of *Mycobacterium tuberculosis* within macrophages. *Infect Immun* **59:**1755–1761.
7. **Chan J, Tanaka K, Carroll D, Flynn J, Bloom BR.** 1995. Effects of nitric oxide synthase inhibitors on murine infection with *Mycobacterium tuberculosis*. *Infect Immun* **63:**736–740.
8. **MacMicking JD, North RJ, LaCourse R, Mudgett JS, Shah SK, Nathan CF.** 1997. Identification of nitric oxide synthase as a protective locus against tuberculosis. *Proc Natl Acad Sci U S A* **94:**5243–5248.
9. **Bryk R, Lima CD, Erdjument-Bromage H, Tempst P, Nathan C.** 2002. Metabolic enzymes of mycobacteria linked to antioxidant defense by a thioredoxin-like protein. *Science* **295:**1073–1077.
10. **Chan J, Xing Y, Magliozzo RS, Bloom BR.** 1992. Killing of virulent *Mycobacterium tuberculosis* by reactive nitrogen intermediates produced by activated murine macrophages. *J Exp Med* **175:**1111–1122.
11. **Chan J, Fujiwara T, Brennan P, McNeil M, Turco SJ, Sibille JC, Snapper M, Aisen P, Bloom BR.** 1989. Microbial glycolipids: possible virulence factors that scavenge oxygen radicals. *Proc Natl Acad Sci U S A* **86:**2453–2457.
12. **Darwin KH, Ehrt S, Gutierrez-Ramos JC, Weich N, Nathan CF.** 2003. The proteasome of *Mycobacterium tuberculosis* is required for resistance to nitric oxide. *Science* **302:**1963–1966.
13. **Modlin RL, Melancon-Kaplan J, Young SM, Pirmez C, Kino H, Convit J, Rea TH, Bloom BR.** 1988. Learning from lesions: patterns of tissue inflammation in leprosy. *Proc Natl Acad Sci U S A* **85:**1213–1217.
14. **Yamamura M, Uyemura K, Deans RJ, Weinberg K, Rea TH, Bloom BR, Modlin RL.** 1991. Defining protective responses to pathogens: cytokine profiles in leprosy lesions. *Science* **254:**277–279.
15. **Salgame P, Abrams JS, Clayberger C, Goldstein H, Convit J, Modlin RL, Bloom BR.** 1991. Differing lymphokine profiles of functional subsets of human CD4 and CD8 T cell clones. *Science* **254:**279–282.
16. **Mosmann TR, Cherwinski H, Bond MW, Giedlin MA, Coffman RL.** 1986. Two types of murine helper T cell clones. I. Definition according to profiles of lymphokine activities and secreted proteins. *J Immunol* **136:**2348–2357.
17. **Thoma-Uszynski S, Stenger S, Takeuchi O, Ochoa MT, Engele M, Sieling PA, Barnes PF, Rollinghoff M, Bolcskei PL, Wagner M, Akira S, Norgard MV, Belisle JT, Godowski PJ, Bloom BR, Modlin RL.** 2001. Induction of direct antimicrobial activity through mammalian Toll-like receptors. *Science* **291:**1544–1547.
18. **Liu PT, Stenger S, Li H, Wenzel L, Tan BH, Krutzik SR, Ochoa MT, Schauber J, Wu K, Meinken C, Kamen DL, Wagner M, Bals R, Steinmeyer A, Zugel U, Gallo RL, Eisenberg D, Hewison M, Hollis BW, Adams JS, Bloom BR, Modlin RL.** 2006. Toll-like receptor triggering of a vitamin D-mediated human antimicrobial response. *Science* **311:**1770–1773.
19. **Liu PT, Schenk M, Walker VP, Dempsey PW, Kanchanapoomi M, Wheelwright M, Vazirnia A, Zhang X, Steinmeyer A, Zügel U, Hollis BW, Cheng G, Modlin RL.** 2009. Convergence of IL-1β and VDR activation pathways in human TLR2/1-induced antimicrobial responses. *PLoS One* **4:**e5810. doi:10.1371/journal.pone.0005810.
20. **Liu PT, Stenger S, Tang DH, Modlin RL.** 2007. Cutting edge: vitamin D-mediated human antimicrobial activity against *Mycobacterium tuberculosis* is dependent on the induction of cathelicidin. *J Immunol* **179:**2060–2063.
21. **Douvas GS, Looker DL, Vatter AE, Crowle AJ.** 1985. Gamma interferon activates human macrophages to become tumoricidal and leishmanicidal but enhances replication of macrophage-associated mycobacteria. *Infect Immun* **50:**1–8.
22. **Rook GA, Steele J, Fraher L, Barker S, Karmali R, O'Riordan J, Stanford J.** 1986. Vitamin D_3, gamma interferon, and control of proliferation of *Mycobacterium tuberculosis* by human monocytes. *Immunology* **57:**159–163.
23. **Rook GA, Taverne J, Leveton C, Steele J.** 1987. The role of gamma-interferon, vitamin D_3 metabolites and tumour necrosis factor in the pathogenesis of tuberculosis. *Immunology* **62:**229–234.
24. **Denis M.** 1991. Killing of *Mycobacterium tuberculosis* within human monocytes: activation by cytokines and calcitriol. *Clin Exp Immunol* **84:**200–206.
25. **Robertson AK, Andrew PW.** 1991. Interferon gamma fails to activate human monocyte-derived macrophages to kill or inhibit the replication of a non-pathogenic mycobacterial species. *Microb Pathog* **11:**283–288.
26. **Olakanmi O, Britigan BE, Schlesinger LS.** 2000. Gallium disrupts iron metabolism of mycobacteria residing within human macrophages. *Infect Immun* **68:**5619–5627.

27. **Bonecini-Almeida MG, Chitale S, Boutsikakis I, Geng J, Doo H, He S, Ho JL.** 1998. Induction of in vitro human macrophage anti-*Mycobacterium tuberculosis* activity: requirement for IFN-γ and primed lymphocytes. *J Immunol* **160:**4490–4499.

28. **Kumar D, Nath L, Kamal MA, Varshney A, Jain A, Singh S, Rao KV.** 2010. Genome-wide analysis of the host intracellular network that regulates survival of *Mycobacterium tuberculosis*. *Cell* **140:**731–743.

29. **Fabri M, Stenger S, Shin DM, Yuk JM, Liu PT, Realegeno S, Lee HM, Krutzik SR, Schenk M, Sieling PA, Teles R, Montoya D, Iyer SS, Bruns H, Lewinsohn DM, Hollis BW, Hewison M, Adams JS, Steinmeyer A, Zügel U, Cheng GH, Jo EK, Bloom BR, Modlin RL.** 2011. Vitamin D is required for IFN-γ-mediated antimicrobial activity of human macrophages. *Sci Transl Med* **3:** 104ra102. doi:10.1126/scitranslmed.3003045.

30. **Gombart AF, Borregaard N, Koeffler HP.** 2005. Human cathelicidin antimicrobial peptide (CAMP) gene is a direct target of the vitamin D receptor and is strongly up-regulated in myeloid cells by 1,25-dihydroxyvitamin D_3. *FASEB J* **19:**1067–1077.

31. **Wang TT, Nestel FP, Bourdeau V, Nagai Y, Wang Q, Liao J, Tavera-Mendoza L, Lin R, Hanrahan JW, Mader S, White JH.** 2004. Cutting edge: 1,25-dihydroxyvitamin D_3 is a direct inducer of antimicrobial peptide gene expression. *J Immunol* **173:**2909–2912.

32. **Green M.** 2011. Cod liver oil and tuberculosis. *BMJ* **343:** d7505. doi:10.1136/bmj.d7505.

33. **Stead WW, Senner JW, Reddick WT, Lofgren JP.** 1990. Racial differences in susceptibility to infection by *Mycobacterium tuberculosis*. *N Engl J Med* **322:**422–427.

34. **Rich AR.** 1944. *Pathogenesis of Tuberculosis*. Charles C Thomas, Publisher, Springfield, IL.

35. **Martineau AR, Nhamoyebonde S, Oni T, Rangaka MX, Marais S, Bangani N, Tsekela R, Bashe L, de Azevedo V, Caldwell J, Venton TR, Timms PM, Wilkinson KA, Wilson RJ.** 2011. Reciprocal seasonal variation in vitamin D status and tuberculosis notifications in Cape Town, South Africa. *Proc Natl Acad Sci U S A* **108:** 19013–19017.

36. **Morcos MM, Gabr AA, Samuel S, Kamel M, el Baz M, el Beshry M, Michail RR.** 1998. Vitamin D administration to tuberculous children and its value. *Boll Chim Farm* **137:**157–164.

37. **Nursyam EW, Amin Z, Rumende CM.** 2006. The effect of vitamin D as supplementary treatment in patients with moderately advanced pulmonary tuberculous lesion. *Acta Med Indones* **38:**3–5.

38. **Salahuddin N, Ali F, Hasan Z, Rao N, Aqeel M, Mahmood F.** 2013. Vitamin D accelerates clinical recovery from tuberculosis: results of the SUCCINCT Study [Supplementary Cholecalciferol in recovery from tuberculosis]. A randomized, placebo-controlled, clinical trial of vitamin D supplementation in patients with pulmonary tuberculosis]. *BMC Infect Dis* **13:**22. doi:10.1186/1471-2334-13-22.

39. **Martineau AR, Timms PM, Bothamley GH, Hanifa Y, Islam K, Claxton AP, Packe GE, Moore-Gillon JC, Darmalingam M, Davidson RN, Milburn HJ, Baker LV, Barker RD, Woodward NJ, Venton TR, Barnes KE, Mullett CJ, Coussens AK, Rutterford CM, Mein CA, Davies GR, Wilkinson RJ, Nikolayevskyy V, Drobniewski FA, Eldridge SM, Griffiths CJ.** 2011. High-dose vitamin D_3 during intensive-phase antimicrobial treatment of pulmonary tuberculosis: a double-blind randomised controlled trial. *Lancet* **377:**242–250.

40. **Coussens AK, Wilkinson RJ, Hanifa Y, Nikolayevskyy V, Elkington PT, Islam K, Timms PM, Venton TR, Bothamley GH, Packe GE, Darmalingam M, Davidson RN, Milburn HJ, Baker LV, Barker RD, Mein CA, Bhaw-Rosun L, Nuamah R, Young DB, Drobniewski FA, Griffiths CJ, Martineau AR.** 2012. Vitamin D accelerates resolution of inflammatory responses during tuberculosis treatment. *Proc Natl Acad Sci U S A* **109:**15449–15454.

41. **Ganmaa D, Giovannucci E, Bloom BR, Fawzi W, Burr W, Batbaatar D, Sumberzul N, Holick MF, Willett WC.** 2012. Vitamin D, tuberculin skin test conversion, and latent tuberculosis in Mongolian school-age children: a randomized, double-blind, placebo-controlled feasibility trial. *Am J Clin Nutr* **96:**391–396.

42. **Berry MP, Graham CM, McNab FW, Xu Z, Bloch SA, Oni T, Wilkinson KA, Banchereau R, Skinner J, Wilkinson RJ, Quinn C, Blankenship D, Dhawan R, Cush JJ, Mejias A, Ramilo O, Kon OM, Pascual V, Banchereau J, Chaussabel D, O'Garra A.** 2010. An interferon-inducible neutrophil-driven blood transcriptional signature in human tuberculosis. *Nature* **466:**973–977.

43. **Maertzdorf J, Weiner J III, Mollenkopf HJ, TBornotTB Network, Bauer T, Prasse A, Müller-Quernheim J, Kaufmann SH.** 2012. Common patterns and disease-related signatures in tuberculosis and sarcoidosis. *Proc Natl Acad Sci U S A* **109:**7853–7858.

44. **Ottenhoff TH, Dass RH, Yang N, Zhang MM, Wong HE, Sahiratmadja E, Khor CC, Alisjahbana B, van Crevel R, Marzuki S, Seielstad M, van de Vosse E, Hibberd ML.** 2012. Genome-wide expression profiling identifies type 1 interferon response pathways in active tuberculosis. *PLoS One* **7:**e45839. doi:10.1371/journal.pone.0045839.

45. **Bloom CI, Graham CM, Berry MP, Wilkinson KA, Oni T, Rozakeas F, Xu Z, Rossello-Urgell J, Chaussabel D, Banchereau J, Pascual V, Lipman M, Wilkinson RJ, O'Garra A.** 2012. Detectable changes in the blood transcriptome are present after two weeks of antituberculosis therapy. *PLoS One* **7:**e46191. doi:10.1371/journal.pone.0046191.

46. **Stanley SA, Johndrow JE, Manzanillo P, Cox JS.** 2007. The type I IFN response to infection with *Mycobacterium tuberculosis* requires ESX-1-mediated secretion and contributes to pathogenesis. *J Immunol* **178:**3143–3152.

47. **Manzanillo PS, Shiloh MU, Portnoy DA, Cox JS.** 2012. *Mycobacterium tuberculosis* activates the DNA-dependent cytosolic surveillance pathway within macrophages. *Cell Host Microbe* **11:**469–480.

48. **Manca C, Tsenova L, Bergtold A, Freeman S, Tovey M, Musser JM, Barry CE III, Freedman VH, Kaplan G.** 2001. Virulence of a *Mycobacterium tuberculosis* clinical isolate in mice is determined by failure to induce Th1 type immunity and is associated with induction of IFN-α/β. *Proc Natl Acad Sci U S A* **98:**5752–5757.

49. **Manca C, Tsenova L, Freeman S, Barczak AK, Tovey M, Murray PJ, Barry C, Kaplan G.** 2005. Hypervirulent *M. tuberculosis* W/Beijing strains upregulate type I IFNs and increase expression of negative regulators of the Jak-Stat pathway. *J Interferon Cytokine Res* **25:**694–701.

50. **Teles RM, Graeber TG, Krutzik SR, Montoya D, Schenk M, Lee DJ, Komisopoulou E, Kelly-Scumpia K, Chun R, Iyer SS, Sarno EN, Rea TH, Hewison M, Adams JS, Popper SJ, Relman DA, Stenger S, Bloom BR, Cheng G, Modlin RL.** 2013. Type I interferon suppresses type II interferon-triggered human anti-mycobacterial responses. *Science* **339:**1448–1453.

51. **Anonymous.** 1999. Fifteen year follow up of trial of BCG vaccines in south India for tuberculosis prevention. Tuberculosis Research Centre (ICMR), Chennai. *Indian J Med Res* **110:**56–69.

52. **Scriba TJ, Tameris M, Mansoor N, Smit E, van der Merwe L, Isaacs F, Keyser A, Moyo S, Brittain N, Lawrie A, Gelderbloem S, Veldsman A, Hatherill M, Hawkridge A, Hill AV, Hussey GD, Mahomed H, McShane H, Hanekom WA.** 2010. Modified vaccinia Ankara-expressing Ag85A, a novel tuberculosis vaccine, is safe in adolescents and children, and induces polyfunctional $CD4^+$ T cells. *Eur J Immunol* **40:**279–290.

53. **Matsumiya M, Stylianou E, Griffiths K, Lang Z, Meyer J, Harris SA, Rowland R, Minassian AM, Pathan AA, Fletcher H, McShane H.** 2013. Roles for Treg expansion and HMGB1 signaling through the TLR1-2-6 axis in determining the magnitude of the antigen-specific immune response to MVA85A. *PLoS One* **8:**e67922. doi: 10.1371/journal.pone.0067922.

54. **Kaforou M, Wright VJ, Oni T, French N, Anderson ST, Bangani N, Banwell CM, Brent AJ, Crampin AC, Dockrell HM, Eley B, Heyderman RS, Hibberd ML, Kern F, Langford PR, Ling L, Mendelson M, Ottenhoff TH, Zgambo F, Wilkinson RJ, Coin LJ, Levin M.** 2013. Detection of tuberculosis in HIV-infected and -uninfected African adults using whole blood RNA expression signatures: a case-control study. *PLoS Med* **10:**e1001538. doi:10.1371/journal.pmed.1001538.

55. **Kagina BM, Abel B, Scriba TJ, Hughes EJ, Keyser A, Soares A, Gamieldien H, Sidibana M, Hatherill M, Gelderbloem S, Mahomed H, Hawkridge A, Hussey G, Kaplan G, Hanekom WA.** 2010. Specific T cell frequency and cytokine expression profile do not correlate with protection against tuberculosis after bacillus Calmette-Guerin vaccination of newborns. *Am J Respir Crit Care Med* **182:**1073–1079.

56. **Montoya D, Inkeles MS, Liu PT, Realegeno S, Teles RM, Vaidya P, Munoz MA, Schenk M, Swindell WR, Chun R, Zavala K, Hewison M, Adams JS, Horvath S, Pellegrini M, Bloom BR, Modlin RL.** 2014. IL-32 is a molecular marker of a host defense network in human tuberculosis. *Sci Transl Med* **6:**250ra114. doi:10.1126/scitranslmed.3009546.

57. **Netea MG, Azam T, Lewis EC, Joosten LA, Wang M, Langenberg D, Meng X, Chan ED, Yoon DY, Ottenhoff T, Kim SH, Dinarello CA.** 2006. *Mycobacterium tuberculosis* induces interleukin-32 production through a caspase-1/IL-18/interferon-γ-dependent mechanism. *PLoS Med* **3:**e277. doi:10.1371/journal.pmed.0030277.

58. **Schenk M, Krutzik SR, Sieling PA, Lee DJ, Teles RM, Ochoa MT, Komisopoulou E, Sarno EN, Rea TH, Graeber TG, Kim S, Cheng G, Modlin RL.** 2012. NOD2 triggers an interleukin-32-dependent human dendritic cell program in leprosy. *Nat Med* **18:**555–563.

59. **Barksby HE, Nile CJ, Jaedicke KM, Taylor JJ, Preshaw PM.** 2009. Differential expression of immunoregulatory genes in monocytes in response to *Porphyromonas gingivalis* and *Escherichia coli* lipopolysaccharide. *Clin Exp Immunol* **156:**479–487.

Myeloid Cells in Health and Disease: A Synthesis
Edited by Siamon Gordon

doi:10.1128/microbiolspec.MCHD-0053-2016

Bart N. Lambrecht[1,2]
Emma K. Persson[1]
Hamida Hammad[1,2]

Myeloid Cells in Asthma

42

ASTHMA IS A HETEROGENEOUS DISEASE, DEFINED BY ENDOTYPES

Asthma is clinically defined by variable airway obstruction that causes recurrent periods of shortness of breath, chest tightness, wheezing, and coughing. Patients also often have altered mucus production and have problems in expectorating sputa because of reduced viscosity of the mucus. One of the characteristic changes to lung physiology is the occurrence of bronchial hyperreactivity, which is defined as a tendency of the smooth muscle layer to contract to nonspecific stimuli like cold air or exercise, and measured in the lung function lab as increased bronchoconstriction to very low amounts of histamine or methacholine. We now realize that asthma is not one single disorder, but rather a syndrome or a spectrum of diseases, characterized by endotypes that rely on distinct pathomechanisms and controlled by various adaptive or innate immune cells (1–5). In early life, asthma is often allergic, driven by CD4 Th2 lymphocytes and associated with allergic comorbidity like atopic dermatitis and rhinitis. On histology, target organs often contain many eosinophils. In adult-onset asthma, almost half of the cases are not associated with allergy. Some of these patients have eosinophilic airway inflammation, whereas others have a neutrophil-predominant inflammation, a mixed neutrophil-eosinophil infiltration, or even pauci-immune disease. Important comorbidities are obesity, acid-reflux disease, and chronic rhinosinusitis (2, 5). Across all age groups, the presence of a more neutrophilic infiltrate is associated with more (therapy-resistant) severe disease, and it is possible that this disease variant relies more on interleukin-17 (IL-17)-producing Th17 cells (1, 6).

Inhaled steroids represent the cornerstone of anti-inflammatory asthma treatment. The presence of neutrophils is a marker for steroid resistance. The presence of eosinophils in sputum on first diagnosis is a good marker for Th2 cell-controlled and steroid-responsive asthma. However, when eosinophilia persists in the bloodstream or tissues despite steroid treatment, this is a sign of severe asthma and defines a specific endotype of asthma, often associated with nasal polyps or chronic rhinosinusitis. In this endotype, type 2 innate lymphoid cells (ILC2), rather than $CD4^+$ Th2 lymphocytes, might be the drivers of inflammation (7, 8).

ALLERGIC ASTHMA

Allergy is defined by the presence of serum IgE and/or a positive skin prick test to common antigens like house

[1]VIB Center for Inflammation Research, Ghent University, 9000 Gent, Belgium; [2]Department of Pulmonary Medicine, Ghent University Hospital, 9000 Gent, Belgium.

dust mite (HDM), cockroach, animal dander, grass and tree pollen, food constituents, or fungal spores. In allergic individuals, experimental or environmental exposure to a relevant allergen, e.g., via inhalation, will lead to an almost immediate allergic reaction, occurring within minutes after inhalation. This inhalation can lead to nasal itching and blockade, but also to bronchospasm, and in some cases even to fatal anaphylaxis. This *immediate* or *early allergic reaction* is caused by degranulation of mast cells and basophils that are armed with IgE on their surface. Cross-linking of IgE leads to the immediate release of preformed mediators like histamine and platelet-activating factor. In about 50% of cases, the immediate response is followed by a delayed or late allergic response that starts ~3 to 6 h after exposure and again leads to symptoms. This *delayed* or *late allergic response* can last for days and is caused by cellular influx into the target organ. In most cases, allergic asthma results from continuous inhalation of allergens, which leads to chronic eosinophil-rich inflammation, goblet cell hyper-/metaplasia, and bronchial hyperreactivity. As allergen exposure is often continuous (e.g., in the case of HDM allergen or other perennial allergens), it is hard to discriminate the early- from the late-phase response. The end result of these alterations is airway obstruction.

Typically, IgE synthesis, eosinophil expansion and activation, mucus overproduction, and hyperreactivity of smooth muscle cells are driven by the cytokines IL-4, IL-5, and IL-13, which are produced by Th2 lymphocytes (3, 9, 10). For most allergens, including spores of the fungus *Alternaria* or cockroach allergens, there might be a spectrum of innate immune cells (like basophils, ILC2, and mast cells) that come into play at various time points after allergen exposure. The relative contribution of these innate immune cells to induction of Th2 immunity in the lung depends on, among other factors, the dose of extract used, the presence of contaminating microbial products in the allergen extracts, and the timing of the experimental observations made mostly in mice. Nevertheless, studies in mouse models employing relevant allergens have taught us much about the immunology of this disease.

DENDRITIC CELLS IN ALLERGIC ASTHMA

A crucial contribution for antigen-presenting cells (APCs) like dendritic cells (DCs) in asthma is not so surprising, as allergic inflammation is often controlled by an adaptive CD4 Th2 lymphocyte response. DCs originate in the bone marrow from committed $Flt3^+$ progenitors that give rise to conventional DCs (cDCs) and plasmacytoid DCs (pDCs). In mice, all DCs express major histocompatibility complex class I (MHC-I) and MHC-II, the integrin CD11c, and the dipeptidyl peptidase CD26 (11). These markers can also be employed in humans. In mice and humans, the cDCs are now best separated into two subsets: Batf3- and interferon (IFN) regulatory factor 8 (IRF8)-dependent $XCR1^+CADM1^+$ cDC1s, whose main function is cross-presentation to CD8 T cells; and ZEB2- and IRF4-dependent $CD172^+$ cDC2s, whose function is to present antigens to CD4 T cells (12, 13). Also in human lungs, it is now possible to delineate the various DC subsets and discriminate them from monocytes and macrophages (14).

The contribution of DCs to the induction of Th2 immune responses has been studied in mouse and rat models of asthma and allergic rhinitis, where mainly ovalbumin (OVA) and HDM have been used as allergens (15), as well as in models employing the proteinase papain (16–18). Adoptive transfer of granulocyte-macrophage colony-stimulating factor (GM-CSF) generated cDCs and macrophages that were pulsed with OVA antigen *in vitro* and then injected in the lungs of mice, causing CD4 Th2 sensitization and subsequent eosinophilic inflammation and asthma features when the mice were challenged with OVA aerosol (19, 20). In addition to these effects of DCs in acute asthma models, repeated injection of DCs into the lungs induced irreversible airway remodeling involving sub-basement membrane deposition of components of the extracellular matrix, such as typically seen in chronic asthma (21). Likewise, when DCs from the lungs of mice exposed to allergen are transferred to naive recipients, they could cause features of asthma (22–24). Recent work demonstrated that induction of CD4 Th2 immunity, bronchial hyperresponsiveness, and mucus overproduction in response to HDM allergen or papain required endogenous $CD11c^+$ lung DCs (16–18, 24, 25), even in very young mice before weaning (26). The idea that $CD11b^+$ cDC2s are required for Th2 immunity was also demonstrated in *CD11cCre* × $Irf4^{fl/fl}$ mice, which selectively lack this subset of cDCs. These mice failed to mount Th2 immunity driven by OVA and the Th2 adjuvant alum and to HDM allergen (27; our unpublished observations). The precise role of pDCs in asthma is still controversial, but likely mainly involves a role in immunoregulation. Indeed, depletion of pDCs during sensitization or challenge to OVA or HDM allergen has a tendency to exacerbate inflammation, as immunoregulatory regulatory T cells fail to function properly in the absence of pDCs (28–30).

Activation of DCs by Epithelial Cytokines in Asthma

A precise account of the cytokines that activate lung DCs in allergic inflammation has recently been given (9). The general consensus is that epithelial cells release cytokines like IL-1a, GM-CSF, IL-33, thymic stromal lymphopoietin (TSLP), and IL-25 that have the potential to activate DCs and turn them into inducers of Th2 immune responses (31–34). At the same time, these epithelial cytokines can also activate local basophils, ILC2, and Th2 effector cells to become cytokine-producing effector cells, and thus the tissue response to allergens is an important checkpoint for development of allergic type 2 inflammation (35). In neonatal mice, lung epithelial cells overproduce IL-33, which induces the expression of OX40L on DCs and suppresses their potential to produce Th1-polarizing cytokine IL-12, explaining the bias for Th2 responses in early life (26). Environmental triggers of asthma like cigarette smoking, diesel particles, and viral infections have the potential to elicit these cytokines, thus driving DC activation indirectly. On the other hand, protective environments, such as living on a farm and farm dust or lipopolysaccharide (LPS) exposure, have the potential to suppress these cytokines and ensuing DC activation (36).

Monocytes and Monocyte-Derived DCs in Asthma

Monocytes are circulating cells that develop in the bone marrow under the influence of M-CSF, and leave the bone marrow in a CCR2-dependent manner, to be recruited into tissues in a CCL2/monocyte chemoattractant protein-1-dependent manner (37). In humans and mice, several subsets of $CD14^+$ monocytes have been described, some of which express high levels of CD16. In human severe asthmatics, there is an increase in the frequency of circulating $CD14^{hi}CD16^+$ monocytes, and these respond poorly to steroids (38). Whereas resident alveolar macrophages seem suppressive to asthma development, recruited monocytes can stimulate allergic reactions in mice, and they probably do so by differentiating into a cell type that closely resembles DCs (39). These monocyte-derived DCs (moDCs) are rapidly recruited to the inflammation site in a CCR2-dependent way, and Ly6C monocytic marker has commonly been used to differentiate $CD11b^+$ monocyte-derived DCs from cDC2s that also express CD11b. However, as Ly6C is rapidly downregulated, it is probably better to use expression of FcεRI (recognized by antibody MAR-1) or the macrophage markers CD64 and Mer tyrosine kinase (MerTK), to rightly identify moDCs (11–13, 23). In fact, when we sorted DCs based on these characteristics, and gene array expression data were analyzed on various APCs of the lung, moDCs tended to cluster with macrophages (our unpublished observations). Recently, similar conclusions about the resemblance of moDCs to macrophages were also obtained in a study in which macrophages were traced using MafB-Cre lineage tracing (40). It is therefore not surprising that moDCs are poorly migratory like macrophages and do not express high levels of CCR7. Studies in mice from which $CD11c^{hi}$ cells (cDCs, pDCs, moDCs, and alveolar macrophages) can be conditionally depleted have shown that DCs are also crucial during the allergen challenge period when primed mice repeatedly encounter inhaled allergens (21, 41, 42). In areas of eosinophil-rich type 2 inflammation, moDCs have an activated phenotype expressing higher levels of costimulatory molecules OX40L, CD80, CD86, programmed death ligand 1 (PDL-1), and PDL-2 (41). In this regard, it is again interesting to note that PDL-2 is classically seen as a marker of alternatively activated macrophages (43), again pointing out the many similarities between moDCs and macrophages. The exact role of PDL-2 expression on moDCs is currently unclear, but PDL-2 was identified as a marker for Th2-inducing DCs of the skin (18).

In a secondary immune response, DCs are found very close to effector T cells and T resident memory (Trm) cells in the regions of airways and large blood vessels (3, 9, 10, 44–46). At these sites, they might secrete chemokines to attract effector T cells or they might restimulate Trm cells by providing costimulatory molecules (23, 47). Allergen-specific IgE and IgG1, by stimulating FcεRI and FcεRIII, respectively, provide strong enhancement because they target inhaled allergens to DCs in primed mice; this further boosts Th2 immunity (48, 49). As they are poorly migratory, moDCs likely interact with effector Th2 cells that migrate back to the lung or with Trm cells (50). The observed attraction of effector Th2 cells by chemokines produced by moDCs provides support of this model (23, 47, 51). MoDCs are also the predominant APCs expressing FcεRI and FcεRIII.

ROLE OF MACROPHAGES IN ASTHMA

The precise role of tissue-resident macrophages in asthma has also been poorly studied, because many of the markers used to identify these cells overlap with moDCs (see above) or other myeloid cells. Alveolar macrophages are sessile, long-lived, and self-renewing cells that derive from fetal monocytes under the influence of GM-CSF (52–55). Studies performed by Holt

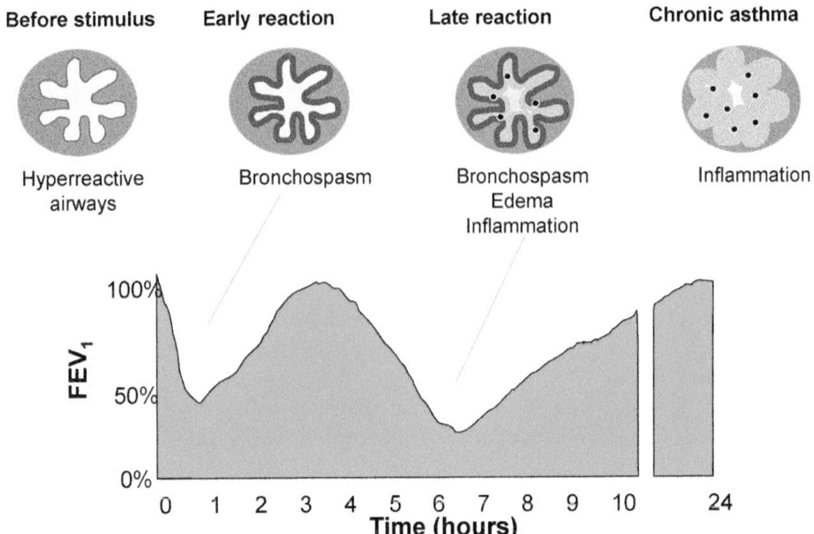

Figure 1 The early and late allergic response. This figure depicts the alterations in the airways, as well as the lung function alterations (measured as the amount of air being exhaled in 1 s in a forced expiratory maneuver, or FEV_1) in allergic patients before, during, and after an experimental allergen challenge. Before the challenge, the airways of asthmatics are hyperreactive (this does not always lead to reduced FEV_1). Minutes after allergen challenge, there is bronchoconstriction, and this is reflected by a drop in FEV_1 that can sometimes be as much as 50% in very severe asthma attacks. This is called the early or immediate allergic response. After some 3 to 6 h, there is a second drop in FEV_1, this time accompanied by edema of the airway wall, cellular influx with inflammatory cells, and bronchoconstriction. This phase is called the late or delayed allergic response.

et al. in rats have clearly demonstrated that resident alveolar macrophages tend to dampen immune responses to inhaled antigens (56–60). Recently, this work has been repeated in mice, where prostaglandin E_2 seems to play an important role in inducing suppressive phenotype in resident alveolar macrophages, through secretion of exosomes that suppress epithelial cell responses (39, 55, 61–65). Interstitial macrophages of the lung also seem to suppress T-cell activation by induction of regulatory T cells (66–68).

Macrophages have also been broadly characterized as either classically activated (M1) or alternatively activated (M2) based on phenotypes observed when macrophages are cultured *in vitro* in the presence of LPS and IFN-γ (M1) or IL-4 or IL-13 (M2). M2 macrophages generally express increased levels of receptors involved in phagocytosis, such as CD206, CD163, and macrophage galactose C-type lectin (CLEC10A/CD301), as well as important Th2 cell chemokines, including the CCR4 ligands CCL17 and CCL22. Studies in mice have nicely shown the presence of IL-4 receptor-dependent M2 macrophages in allergen-induced asthma, and these M2 macrophages express markers like Ym1 (Chil313), arginase, Fizz, and the macrophage mannose receptor, induced by various type 2 immune cytokines like IL-4, IL-13, TSLP, and IL-33 and suppressed by serum amyloid P protein (35, 69–75). The precise contribution of each of these cytokines, and the exact contribution of the M2 polarization to asthma pathogenesis, remain to be precisely identified, as some studies demonstrated unaltered asthma severity in the absence of M2 polarization, whereas others showed clear protection from asthma (71, 72, 76).

In humans, alveolar macrophages of asthmatics have an activated phenotype, expressing higher levels of costimulatory molecules, although this is not the case for every member of the CD80 family (62, 77). There is also evidence of altered phenotype in alveolar macrophages of human asthmatics, and reduced expression of the Th1-inducing cytokine IL-12 (61, 62, 78). In one study, these cells only expressed some M2 markers, yet did overexpress the chemokine CCL17 mRNA, which has the potential to recruit $CCR4^+$ Th2 cells (79). On the contrary, other studies suggested clear signs of M2 polarization in human asthmatic alveolar macrophages, including increased expression of MHC-II, E-cadherin, and histamine receptor H1 (37, 78, 80, 81). There might be important sex differences in the importance of M2 macrophages in human asthma pathogenesis (82).

Novel tools are urgently needed to deplete alveolar macrophages in a selective manner in mice or to alter the behavior of macrophages in humans to understand better how these cells contribute to asthma.

EOSINOPHILS

Eosinophils are the primary inflammatory myeloid cells in type 2 inflammation typically seen in sites of allergic inflammation, such as asthma and rhinitis (83, 84). Eosinophils are circulating effector cells that contribute to a range of inflammatory processes. In noninflammatory conditions, eosinophils travel at relatively low levels in the bloodstream and constitute only a minor fraction of the white blood cells, about 1 to 3% of the total pool (85). In the gastrointestinal tract and in the bone marrow, tissue eosinophils are relatively abundant also in the absence of inflammation. In the intestine, this may reflect important functions in immune responses to gastrointestinal parasitic helminths, by boosting the function of DCs through release of eosinophilic toxins like eosinophil-derived neurotoxin (86). In the bone marrow, eosinophils have been suggested to play a role in the regulation of antibody responses by providing survival factors to long-lived bone marrow plasma cells (87). Recently, it was also shown that eosinophils accumulate spontaneously in developing lungs before weaning, driven by ILC2 and IL-33 signals (26).

During inflammatory/allergic disease states, and in response to inflammatory signals, eosinophils are also recruited from the bloodstream into other tissues including the lungs, skin, and brain. Infiltration of eosinophils into the lungs is one of the hallmark features of allergic asthma, in patients as well as in animal models (88). Thus, asthmatic patients have an increase in eosinophils in peripheral blood, lung tissue, and bronchoalveolar lavage fluid, with degree of eosinophilia generally correlating with severity of disease (89). Moreover, presence of eosinophils in the airway lumen has shown to be predictive of loss of asthma control after discontinuation of inhaled corticosteroids (90, 91). Although the contribution of eosinophils to asthma and other allergic diseases remains to be fully understood, evidence suggests that eosinophils play a causal role in disease pathogenesis (3). The effector functions of eosinophils are largely attributed to the secretion of toxic proteins stored in intracellular granules and reactive oxygen species, which may contribute to tissue damage during allergic inflammation (92). It is also becoming increasingly clear that apart from end-stage cytotoxic functions, eosinophils also influence immune responses and contribute to innate and adaptive parts of immune responses through various other mechanisms, via the release of immunomodulatory mediators such as cytokines and lipid mediators.

Development and Trafficking of Eosinophils

Eosinophils develop in the bone marrow from hematopoietic stem cells. During hematopoiesis, eosinophil lineage-committed precursors develop that can be defined by the expression of CD34 and the IL-5α receptor (93–95). These eosinophilic precursors undergo maturation in the bone marrow upon exposure to IL-3, IL-5, and GM-CSF (96, 97). IL-5 signaling is unique to and critical for eosinophil expansion, differentiation, and activation, and expression of the IL-5α receptor is maintained throughout eosinophil differentiation (93). IL-5 also stimulates the release of eosinophils into the peripheral circulation. The importance of IL-5 in eosinophil development is evidenced in mice lacking expression of IL-5 or IL-5 receptor, which also lack eosinophils (98–100), and in mice overexpressing IL-5, which have largely increased numbers of eosinophils (101). IL-5 is thus a potentially attractive target for modulation of eosinophilic inflammation. The critical role of IL-5 in regulating human eosinophils has also been demonstrated in a number of clinical trials with a humanized anti-IL-5 antibody (mepolizumab), which lowers eosinophil counts in blood, peripheral tissues, and bone marrow, accompanied by a reduction in airway remodeling and reduced need for systemic steroids (102–111).

Eosinophil development is dependent on the expression and interplay of specific transcription factors, including GATA-binding protein 1 (GATA-1), CCAAT/enhancer-binding protein-α (C/EBPα), PU box-binding protein (PU1), and IFN consensus sequence-binding protein (ICSBP). The transcription factor GATA-1 is critical for eosinophil development, and mice in which a high-affinity palindromic GATA site in the Gata1 promoter is lacking (ΔdblGATA mice) have a selective loss of eosinophils (112). Mouse models in which eosinophils are selectively depleted allow studies of the role of eosinophils in allergic disease (see below). The number of eosinophilic precursors in the bone marrow is increased in several inflammatory conditions, including allergic responses. Increased numbers of eosinophil-lineage-committed progenitors have also been observed in the lungs of allergen-challenged mice (113). Recently, these studies have been extended to human asthmatics, in whom cells with eosinophil progenitor potential were found in induced sputum (114). These observations suggest that enhanced production of eosinophil progenitors may be a checkpoint in the regulation of eosinophilic inflammation.

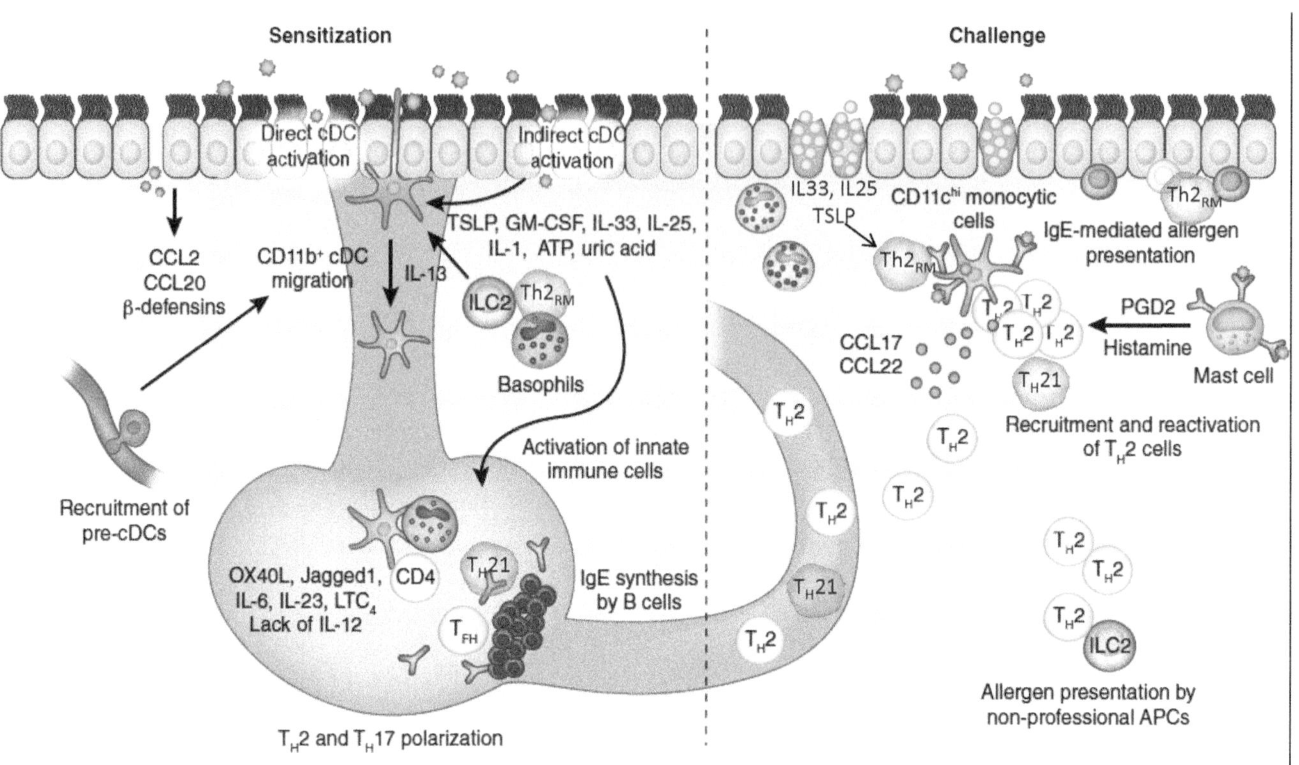

Figure 2 The cellular interplay of myeloid and lymphoid cells in allergic asthma. (Left) A first allergen exposure leads to activation of $CD11b^+$ cDC2s. This activation is direct or indirect, because lung epithelial cells make cytokines like IL-1, IL-33, IL-25, TSLP, and GM-CSF that mature the DCs. The same cytokines also activate basophils and ILC2, which control immediate innate eosinophilia. After a few hours, DCs will also arrive in the draining mediastinal nodes, where they will polarize adaptive immune cells to become Th2, Th21, and Th17 cells. B cells will be induced to secrete IgE. (Right) A recall response to allergens, which occurs continuously in patients allergic to perennial allergens. Upon repeated encounter with allergens, tissue mast cells and basophils are armed with IgE and release immediate mediators into the lung tissue, causing bronchoconstriction and local edema. Monocyte-derived cells (DCs and activated macrophages) will also take up allergens via IgE and this time present these allergens locally to T effector cells and resident memory Th2 cells. These effector cells reach the lungs because of chemokine production by monocyte-derived cells. The effector lymphocytes will also produce loss of IL-5, which boosts the production of eosinophils. These eosinophils migrate into the lungs and cause damage to the lung epithelium. In some cases, particularly when there is a Th17 response, neutrophils also will accumulate in the lungs. PGD_2, prostaglandin D_2. Modified from reference 3, with permission.

Eosinophils exit the bone marrow and are released into the bloodstream as terminally differentiated cells. Eosinophil trafficking and extravasation of eosinophils into tissues are dependent on a network of adhesion molecules, cytokines, chemokines, and their receptors. The extravasation is largely regulated by the chemokine receptor CCR3, which is abundantly and selectively expressed by eosinophils. CCR3 is the primary receptor for the eotaxin subfamily of chemokines, CCL11 (eotaxin-1), CCL24 (eotaxin-2), and CCL26 (eotaxin-3) (CCL26 is a pseudogene in mice but functional in humans), which are high-affinity agonists for CCR3 (84). Recruitment of eosinophils to lungs is mediated primarily by eotaxin-1 and eotaxin-2, which is produced by airway epithelial cells under the influence of cytokines such as IL-13 (115–130). Under homeostatic conditions, eosinophils in tissues are relatively short-lived. However, it has been demonstrated that, via various mediators, eosinophil cell death is delayed at inflammatory sites. Such mechanisms also contribute to the increased numbers of eosinophils within tissues (131, 132). At least in humans, IL-5 seems to be the most important and specific mediator promoting eosinophil survival (133, 134).

Eosinophils in Asthma

In the lungs, eosinophils can potentially contribute to asthmatic inflammation in a number of ways, and via the release of various mediators, including toxic granule proteins or reactive oxygen species, and secretion of cytokines, chemokines, and leukotrienes (84, 135–138). Although the functions of many of these mediators are known, they are also often produced by other immune cell types like basophils, mast cells, or neutrophils, and therefore the specific contribution of eosinophil-derived products often remains unclear. In general, eosinophil effector functions are largely attributed to the secretion of toxic proteins stored in intracellular granules (for example, major basic protein [MBP]), as well as reactive oxygen species. Upon eosinophil accumulation and degranulation in bronchial mucosa during asthma, these may contribute to tissue damage through cytotoxic effects on airway cells, including nerve endings. MBP is a highly cationic protein stored in crystalline form in the core of the crystalline granule (135–138). This protein can also trigger the degranulation of other effector cells, like mast cells and basophils, that may also be involved in disease pathogenesis. Moreover, it is thought to be indirectly involved in airway hyperreactivity through its ability to increase smooth muscle reactivity (139). MBP is also found closely associated within eosinophil extracellular traps that form upon eosinophil degranulation and contain large amounts of extracellular DNA (140–144).

Eosinophils are also believed to modulate immune responses through release of immunomodulatory mediators such as cytokines (e.g., transforming growth factor β), chemokines, and lipid mediators. For example, cysteinyl leukotrienes (LTC_4 and LTB_4) are proinflammatory mediators with several potential roles in lung inflammation, including increased vascular permeability, increased mucus secretion, and smooth muscle contraction (145, 146). Eosinophils may also influence the functioning of other immune cells via cytokines and/or chemokines, and promote Th2 polarization through production of the Th2-associated cytokines IL-4 and IL-13 and via release of T-cell chemoattractants (92, 147–151). A role for eosinophil-derived transforming growth factor β in tissue remodeling has also been suggested, explaining how treatment with mepolizumab (anti-IL-5) or blocking antibodies to IL-5 in mice can reduce airway wall remodeling (152–156), although this idea has also been refuted (157).

To understand the contribution of eosinophils to the development of asthma, models of asthma have been performed in specific mouse models that specifically lack eosinophils, either genetically or due to injection of depleting antibodies to Siglec-F. Lee and colleagues have assessed asthma development in a transgenic mouse model expressing diphtheria toxin A under the eosinophil-specific peroxidase promoter for selective depletion of eosinophils (TgPHIL mice) (148, 150, 158). In an OVA-induced model of acute airway inflammation, mice lacking eosinophils had significantly reduced signs of asthma, including airway hypertrophy, goblet cell metaplasia/mucus accumulation, and airway hyperresponsiveness, when compared with eosinophil-sufficient controls. The observation that mucus accumulation in eosinophil-deficient mice was reduced but still higher than in allergen-naive animals suggested that eosinophil-independent mechanisms contribute to this response (158). Eosinophil contribution to asthma has been tested in ΔdblGATA mice, lacking eosinophils. Airway physiology in ΔdblGATA mice was similar to that in control mice, suggesting that eosinophils are redundant for allergen-induced changes in airway physiology (156, 157). Th2 responses and mucus secretion were also normal, suggesting that allergen-driven Th2 responses develop in the absence of eosinophils. Matrix deposition was, however, reduced in mice lacking eosinophils, suggesting that eosinophils contribute to allergen-induced subepithelial collagen deposition. Results demonstrate that eosinophils are in part responsible for both collagen and smooth muscle changes in a chronic model of asthma, but are not obligatory for airway physiology changes associated with this disease. Discrepancies between studies may be an indication of mouse strain-specific differences in the requirement for eosinophils in asthma development. Studies in eosinophil-deficient mice have also indicated that eosinophils are required for the recruitment of Th2 cytokine-secreting T cells into the lung upon allergen provocation, likely due to release of an eosinophil-regulated chemokine that recruits Th2 cells (159).

Recently, it has been shown that eosinophils play a crucial role in a specific endotype of severe asthma characterized by steroid-resistant persistent blood and tissue eosinophilia and frequent exacerbations. In these patients, treatment with mepolizumab (anti-IL-5 monoclonal antibody) was able to reduce exacerbation frequency, reduce need of systemic steroids, and improve lung function in some cases (108, 110, 111, 160, 161).

NEUTROPHILS IN ASTHMA

Studies in humans have clearly demonstrated that bronchoalveolar lavage fluid of asthmatics contains neutrophil chemoattractants such as complement C5a, IL-8, and CXCL2, and that neutrophils extravasate from the bloodstream into the lung upon allergen challenge (162–164). Recruitment of neutrophils into asthmatic airways is suppressed by a dual CXCR1/CXCR2 antagonist (165) and by inhibiting the blood-clotting cascade with activated protein C (166, 167). It is only fairly recently that neutrophils have been implicated in causing asthmatic inflammation and determining response to standard therapies. Neutrophil-rich asthma is notoriously steroid resistant and runs a much more severe clinical course (1). In these patients, higher levels of IL-17, derived from Th17 cells or γδ TCR (T-cell receptor) T cells, are seen (1, 168–178). There is also evidence that these steroid-resistant asthmatics have higher levels of environmental LPS exposure, driving M1 macrophage polarization, and production of neutrophil chemokines like IL-8 and CXCL1 and -2, suggesting that coexposure of allergens with microbial compounds (e.g., derived from the environment or from the microbiome) or even viruses might determine the neutrophilic nature of asthma (179, 180). In mouse models of HDM-induced asthma, increased neutrophil influx and IL-17 production is seen when the allergen extracts contain more fungal-derived β-glucan components (181), suggesting that also in clinical asthma fungal sensitization might contribute to severe neutrophil asthma. If and how exactly neutrophilic inflammation should be targeted in asthma is a matter of intense study. If there is concomitant smoking, smoking cessation is paramount. Additional therapies might consist of chronic low-dose macrolides, also frequently used in non-cystic fibrosis bronchiectasis (182, 183). The precise potential of blocking IL-17A or IL-17RA in treating neutrophilic asthma remains uncertain at present.

MAST CELLS

Although mast cells and closely related basophils are traditionally seen as myeloid cells, relatively little is known about mast cell development. In mice, mast cells are generated in the bone marrow from an IRF8$^+$ granulocytic progenitor that gives rise to a GATA2$^+$ STAT5$^+$ progenitor that also leads to formation of basophils (184). Mast cell development subsequently diverges from basophil development through differential upregulation of transcription factors C/EBPα and microphthalmia-associated transcription factor, respectively (185, 186). Mast cell and basophil progenitors also express the IL-33 receptor T1/ST and produce type 2 cytokines like IL-13 already at the progenitor stage when stimulated with IL-33 (95). Mast cells are thought to reside in peripheral tissues like skin, gut, and lung for prolonged periods, surviving for years. The growth and proliferation of mast cells is stimulated by stem cell factor (the ligand for the c-Kit receptor), IL-3, and IL-9 (187, 188). The evolutionary benefit of having mast cells has long remained elusive. Recently, however, the group of Galli has shown that mast cells are mainly effective in neutralizing toxins and venoms of various life-threatening organisms such as Russell's viper, Gila monster, and bees, potentially offering an evolutionary benefit to allergic-type inflammation (189, 190). The way in which mast cells achieve this function is through IgE-mediated release of toxin-neutralizing enzymes (191, 192). The function of mast cells is indeed intimately associated with IgE effector functions, and cross-linking of IgE on the high-affinity IgE receptor (FcεRI) expressed on mast cells leads to release of preformed mediators (such as histamine, serotonin, and proteases) and *de novo* synthesis of novel proteins, lipid mediators, and enzymes. In allergic individuals, tissue mast cells are armed with IgE bound to FcεRI and exposure to allergens results in the cross-linking and aggregation of surface FcεRI and the immediate phase of the allergic reaction. Other mediators (cytokines, chemokines, and growth factors) are secreted several hours after mast cell activation (reviewed in reference 193). If allergen exposure occurred in the airways, the early-released mediators will induce vasodilation, edema formation, and/or bronchoconstriction. The production of chemotactic factors results in the infiltration of inflammatory cells to the tissue. Although mast cell activation is thought to be driven mostly by IgE, release of mediators by mast cells can also occur independently of IgE and specific allergens. For instance, mast cells express several pattern recognition receptors such as Toll-like receptors, and activation of mast cells with Toll-like receptor ligands was shown to induce cytokine and chemokine production by mast cells (194, 195). Many other stimuli like TSLP (196) or IL-33 (197), produced upon exposure to inhaled allergens (31, 32), are able to directly activate mast cells. The role of mast cells in allergic inflammation has been extensively studied in mice and in humans. Analyses of lung epithelial brushings, biopsies (198, 199), or tissues obtained at autopsy (200, 201) have shown increased numbers of mast cells in the airway smooth muscle layer in mild asthmatics, associated with bronchial hyperreactivity (202). Patients with more severe asthma showed a significant trend to higher numbers of chymase-positive mast cells in the proximal airway epithelium (200, 201). Patients with mild to moderate asthma with an IL-13-responsive gene signature in their epithelium brushings (Th2-high asthmatics) had higher levels of intraepithelial mast cells compared to healthy donors or to Th2-low patients, and those mast cells were tryptase positive but chymase negative (198). Autopsy specimens obtained from individuals with fatal asthma showed a higher degranulation of mast cells in airway smooth muscle compared to healthy individuals (201).

Studying the contribution of mast cells to the pathogenesis of asthma has been made possible using mast cell-deficient mice (WBB6F1-Kit$^{W/Wv}$ or C57BL/6-Kit$^{W\text{-}sh/W\text{-}sh}$) or mast cell-reconstituted mice (194). Using these mice, it has been shown that mast cells directly contributed several features of the disease, including goblet cell metaplasia or airway remodeling (203, 204). However, it seems that mast cells only play a role in models that do not use strong adjuvants and that use low doses of allergens (reviewed in reference 205). We have also studied the contribution of mast cells in HDM-driven asthma by studying mice in which mast cells were hyperactive due to a mast cell-specific deficiency of the ubiquitin-modifying enzyme Tnfaip3. In these mice, HDM-induced plasma extravasation was increased, and so were features of asthma (206).

However, many other effector cells, especially Th2 cells, also have the potential to contribute to late-phase response inflammatory features, and the relative contribution of mast cells and other effector cells to these features remains to be fully addressed, especially in humans. In addition, although asthma features seem to be reduced in mast cell-deficient mice, it is very unclear from these models in which phase of the immune response (primary or secondary response) mast cells are really important. It has been hypothesized that mast cells may be involved in the process of sensitization to allergens mostly because mast cells can regulate epithelial barrier integrity, and because they have been shown to express MHC-II and costimulatory molecules (207). Although there is no evidence of mast cells being able to directly activate naive CD4$^+$ T cells *in vivo*, it cannot be formally excluded that mast cells may contribute indirectly to Th2 sensitization. Indeed, in a mouse model of asthma, mast cells have been shown to promote the transport of inhaled antigen by DCs to the lung-draining lymph node (208), and in this way to contribute to the primary activation of T cells.

Mast cells might also have anti-inflammatory and tissue-healing effects in asthma, through their production of IL-10 (209). Mast cell-derived IL-10 reduces B-lymphocyte functions and antibody production (210). Moreover, mast cells express Siglec-8, a lectin that can trigger eosinophil apoptosis (211, 212). At present, it is unclear if the function of mast cells to suppress inflammation is model dependent or can be extended to all models and humans with asthma.

BASOPHILS

Basophils are closely related to mast cells, and develop from a shared progenitor in the bone marrow. Basophils also express the high-affinity IgE receptor FcεRI and upon FcεRI cross-linking produce similar mediators, including histamine, platelet-activating factor, Th2-associated cytokines, and lipid mediators. However, unlike tissue-resident mast cells, mature basophils have a much shorter life span (~60 h) under steady-state conditions (213). As with mast cells, the main cytokine driving basophil development is IL-3, produced by T cells and promoting basophil recruitment to draining lymph nodes during inflammatory reactions (214, 215). TSLP is a cytokine produced in the lungs of HDM-exposed animals and human severe asthmatics (32). TSLP induces basophil development and activation in mice (216). A role for basophils in Th2 cell polarization in mice has been proposed (217). However, whether this occurs *in vivo* is very controversial. The fact that basophils are recruited to the lymph nodes of mice administered with different agents inducing Th2 immunity (papain, allergens) is very clear (24, 217). Some studies have reported that basophils expressed high levels of MHC-II after papain administration in the skin (218). However, this was not observed when mice were administered HDM allergen in the airways (24). The depletion of basophils using antibodies to FcεRI resulted in very poor Th2 responses to the cysteine protease papain (218). Strikingly, in this same study, the depletion of DCs using CD11c-DTR had no effect. Of note, other groups have reported on the importance of DCs in the induction of Th2 immunity (19, 41, 219), and it has been shown that the development of Th2 responses required the cooperation between DCs and basophils (220). Our group has shown that the depletion of basophils with anti-FcεRI antibodies depleted a subset of inflammatory DCs involved in Th2 immunity to inhaled allergens (23, 24). In addition, a different group has recently created a basophil-deficient mouse strain (Mcpt8-Cre) and demonstrated that Th2 polarization induced by papain and asthma features induced by OVA-alum administration occurred independently of basophils (221). These data were also confirmed by others using a different strain of basophil-deficient mice (222).

Due to the controversy in mice, researchers have sought to investigate whether basophils could act as APCs in humans, and found that circulating basophils did not express HLA-DR or costimulatory molecules and did not promote Th2 cytokine production by cocultured CD4+ T cells (223, 224). It therefore seems that basophils are dispensable for the induction of Th2 immunity in humans and in mice. However, it is very likely that they might play a role in the activation of effector responses and a more prominent role in chronic asthma (221, 225). Indeed, their number is increased in the lungs of asthmatic patients (226), where they mostly localize in the basement membrane but not in smooth muscles. Moreover, basophils were also found in large numbers in postmortem biopsies of fatal asthma, and were a copious source of type 2 cytokines in chronic asthma, driven by a spliced form of the epithelial cytokine IL-33 (225, 227). Basophils were able to enhance DC-driven Th2 differentiation and exacerbate ongoing airway inflammation upon adoptive transfer into sensitized mice (228). This capacity to enhance Th2 cytokine production by effector CD4+ T cells was also noticed in human basophils (228).

CONCLUSION

Asthma is a very heterogeneous disease with many endotypes based on discrete pathophysiological mechanisms. Many myeloid cells play a crucial role in asthma pathogenesis. APCs, including DCs and freshly recruited monocytes, are the first to recognize the allergens, pollutants, and viruses that are implicated in asthma pathogenesis, and subsequently initiate the adaptive immune component. Eosinophils are the hallmark of type 2 inflammation, releasing toxic compounds in the airways and contributing to airway remodeling. Mast cells and basophils control both the early- and late-phase allergic response and contribute to alterations in smooth muscle reactivity. Finally, relatively little is known about neutrophils and macrophages in this disease, but tools to dissect these in mouse models of asthma are being developed. Although many of these myeloid cells respond well to therapy with inhaled steroids, there is now an increasing armamentarium of targeted biologicals that can specifically eliminate only one myeloid cell population, like eosinophils. It is only with those clinically available monoclonal antibodies that we will be able to fully understand the role of myeloid cells in chronic asthma in humans.

Citation. Lambrecht BN, Persson EK, Hammad H. 2017. Myeloid cells in asthma. Microbiol Spectrum 5(1):MCHD-0053-2016.

References

1. **Moore WC, Hastie AT, Li X, Li H, Busse WW, Jarjour NN, Wenzel SE, Peters SP, Meyers DA, Bleecker ER.** 2014. Sputum neutrophil counts are associated with more severe asthma phenotypes using cluster analysis. *J Allergy Clin Immunol* **133:**1557–1563.e5. doi:10.1016/j.jaci.2013.10.011.
2. **Wenzel SE.** 2012. Asthma phenotypes: the evolution from clinical to molecular approaches. *Nat Med* **18:** 716–725.
3. **Lambrecht BN, Hammad H.** 2015. The immunology of asthma. *Nat Immunol* **16:**45–56.
4. **Anderson GP.** 2008. Endotyping asthma: new insights into key pathogenic mechanisms in a complex, heterogeneous disease. *Lancet* **372:**1107–1119.
5. **Lotvall J, Akdis CA, Bacharier LB, Bjermer L, Casale TB, Custovic A, Lemanske RF Jr, Wardlaw AJ, Wenzel SE, Greenberger PA.** 2011. Asthma endotypes: a new approach to classification of disease entities within the asthma syndrome. *J Allergy Clin Immunol* **127:**355–360.
6. **Wu W, Bleecker E, Moore W, Busse WW, Castro M, Chung KF, Calhoun WJ, Erzurum S, Gaston B, Israel E, Curran-Everett D, Wenzel SE.** 2014. Unsupervised phenotyping of Severe Asthma Research Program participants using expanded lung data. *J Allergy Clin Immunol* **133:** 1280–1288.
7. **Mjösberg JM, Trifari S, Crellin NK, Peters CP, van Drunen CM, Piet B, Fokkens WJ, Cupedo T, Spits H.** 2011. Human IL-25- and IL-33-responsive type 2 innate lymphoid cells are defined by expression of CRTH2 and CD161. *Nat Immunol* **12:**1055–1062.
8. **Mjösberg J, Bernink J, Golebski K, Karrich JJ, Peters CP, Blom B, te Velde AA, Fokkens WJ, van Drunen CM, Spits H.** 2012. The transcription factor GATA3 is essential for the function of human type 2 innate lymphoid cells. *Immunity* **37:**649–659.
9. **Hammad H, Lambrecht BN.** 2015. Barrier epithelial cells and the control of type 2 immunity. *Immunity* **43:** 29–40.
10. **Lambrecht BN, Galli SJ.** 2015. SnapShot: Integrated Type 2 Immune Responses. *Immunity* **43:**408–408.e1. doi:10.1016/j.immuni.2015.07.019.
11. **Guilliams M, Dutertre CA, Scott CL, McGovern N, Sichien D, Chakarov S, Van Gassen S, Chen J, Poidinger M, De Prijck S, Tavernier SJ, Low I, Irac SE, Mattar CN, Sumatoh HR, Low GH, Chung TJ, Chan DK, Tan KK, Hon TL, Fossum E, Bogen B, Choolani M, Chan JK, Larbi A, Luche H, Henri S, Saeys Y, Newell EW, Lambrecht BN, Malissen B, Ginhoux F.** 2016. Unsupervised high-dimensional analysis aligns dendritic cells across tissues and species. *Immunity* **45:**669–684.
12. **Sichien D, Scott CL, Martens L, Vanderkerken M, Van Gassen S, Plantinga M, Joeris T, De Prijck S, Vanhoutte L, Vanheerswynghels M, Van Isterdael G, Toussaint W,**

Madeira FB, Vergote K, Agace WW, Clausen BE, Hammad H, Dalod M, Saeys Y, Lambrecht BN, Guilliams M. 2016. IRF8 transcription factor controls survival and function of terminally differentiated conventional and plasmacytoid dendritic cells, respectively. *Immunity* **45**:626–640.

13. **Scott CL, Soen B, Martens L, Skrypek N, Saelens W, Taminau J, Blancke G, Van Isterdael G, Huylebroeck D, Haigh J, Saeys Y, Guilliams M, Lambrecht BN, Berx G.** 2016. The transcription factor Zeb2 regulates development of conventional and plasmacytoid DCs by repressing Id2. *J Exp Med* **213**:897–911.
14. **Desch AN, Gibbings SL, Goyal R, Kolde R, Bednarek J, Bruno T, Slansky JE, Jacobelli J, Mason R, Ito Y, Messier E, Randolph GJ, Prabagar M, Atif SM, Segura E, Xavier RJ, Bratton DL, Janssen WJ, Henson PM, Jakubzick CV.** 2016. Flow cytometric analysis of mononuclear phagocytes in nondiseased human lung and lung-draining lymph nodes. *Am J Respir Crit Care Med* **193**:614–626.
15. **Lambrecht BN, Hammad H.** 2012. Lung dendritic cells in respiratory viral infection and asthma: from protection to immunopathology. *Annu Rev Immunol* **30**: 243–270.
16. **Tang H, Cao W, Kasturi SP, Ravindran R, Nakaya HI, Kundu K, Murthy N, Kepler TB, Malissen B, Pulendran B.** 2010. The T helper type 2 response to cysteine proteases requires dendritic cell-basophil cooperation via ROS-mediated signaling. *Nat Immunol* **11**:608–617.
17. **Kumamoto Y, Linehan M, Weinstein JS, Laidlaw BJ, Craft JE, Iwasaki A.** 2013. CD301b^{+} dermal dendritic cells drive T helper 2 cell-mediated immunity. *Immunity* **39**:733–743.
18. **Gao Y, Nish SA, Jiang R, Hou L, Licona-Limon P, Weinstein JS, Zhao H, Medzhitov R.** 2013. Control of T helper 2 responses by transcription factor IRF4-dependent dendritic cells. *Immunity* **39**:722–732.
19. **Lambrecht BN, De Veerman M, Coyle AJ, Gutierrez-Ramos JC, Thielemans K, Pauwels RA.** 2000. Myeloid dendritic cells induce Th2 responses to inhaled antigen, leading to eosinophilic airway inflammation. *J Clin Invest* **106**:551–559.
20. **Raymond M, Van VQ, Wakahara K, Rubio M, Sarfati M.** 2011. Lung dendritic cells induce T_H17 cells that produce T_H2 cytokines, express GATA-3, and promote airway inflammation. *J Allergy Clin Immunol* **128**:192–201.e.6. doi:10.1016/j.jaci.2011.04.029.
21. **van Rijt LS, Vos N, Willart M, Muskens F, Tak PP, van der Horst C, Hoogsteden HC, Lambrecht BN.** 2011. Persistent activation of dendritic cells after resolution of allergic airway inflammation breaks tolerance to inhaled allergens in mice. *Am J Respir Crit Care Med* **184**:303–311.
22. **Krishnamoorthy N, Oriss TB, Paglia M, Fei M, Yarlagadda M, Vanhaesebroeck B, Ray A, Ray P.** 2008. Activation of c-Kit in dendritic cells regulates T helper cell differentiation and allergic asthma. *Nat Med* **14**: 565–573.
23. **Plantinga M, Guilliams M, Vanheerswynghels M, Deswarte K, Branco-Madeira F, Toussaint W, Vanhoutte L, Neyt K, Killeen N, Malissen B, Hammad H, Lambrecht BN.** 2013. Conventional and monocyte-derived CD11b^{+} dendritic cells initiate and maintain T helper 2 cell-mediated immunity to house dust mite allergen. *Immunity* **38**:322–335.
24. **Hammad H, Plantinga M, Deswarte K, Pouliot P, Willart MA, Kool M, Muskens F, Lambrecht BN.** 2010. Inflammatory dendritic cells—not basophils—are necessary and sufficient for induction of Th2 immunity to inhaled house dust mite allergen. *J Exp Med* **207**:2097–2111.
25. **Phythian-Adams AT, Cook PC, Lundie RJ, Jones LH, Smith KA, Barr TA, Hochweller K, Anderton SM, Hammerling GJ, Maizels RM, MacDonald AS.** 2010. CD11c depletion severely disrupts Th2 induction and development in vivo. *J Exp Med* **207**:2089–2096.
26. **De Kleer I, Kool M, de Bruijn M, Willart M, Van Moorleghem J, Schuijs M, Plantinga M, Beyaert R, Hams E, Fallon GP, Hammad H, Hendriks R, Lambrecht BN.** 2016. Perinatal activation of the interleukin-33 pathway promotes type 2 immunity in the developing lung. *Immunity* doi:10.1016/j.immuni.2016.1010.1031.
27. **Williams JW, Tjota MY, Clay BS, Vander Lugt B, Bandukwala HS, Hrusch CL, Decker DC, Blaine KM, Fixsen BR, Singh H, Sciammas R, Sperling AI.** 2013. Transcription factor IRF4 drives dendritic cells to promote Th2 differentiation. *Nat Commun* **4**:2990. doi: 10.1038/ncomms3990.
28. **Kool M, van Nimwegen M, Willart MA, Muskens F, Boon L, Smit JJ, Coyle A, Clausen BE, Hoogsteden HC, Lambrecht BN, Hammad H.** 2009. An anti-inflammatory role for plasmacytoid dendritic cells in allergic airway inflammation. *J Immunol* **183**:1074–1082.
29. **de Heer HJ, Hammad H, Soullie T, Hijdra D, Vos N, Willart MA, Hoogsteden HC, Lambrecht BN.** 2004. Essential role of lung plasmacytoid dendritic cells in preventing asthmatic reactions to harmless inhaled antigen. *J Exp Med* **200**:89–98.
30. **Lombardi V, Speak AO, Kerzerho J, Szely N, Akbari O.** 2012. $CD8\alpha^{+}\beta^{-}$ and $CD8\alpha^{+}\beta^{+}$ plasmacytoid dendritic cells induce Foxp3^{+} regulatory T cells and prevent the induction of airway hyper-reactivity. *Mucosal Immunol* **5**:432–443.
31. **Hammad H, Chieppa M, Perros F, Willart MA, Germain RN, Lambrecht BN.** 2009. House dust mite allergen induces asthma via Toll-like receptor 4 triggering of airway structural cells. *Nat Med* **15**:410–416.
32. **Willart MA, Deswarte K, Pouliot P, Braun H, Beyaert R, Lambrecht BN, Hammad H.** 2012. Interleukin-1α controls allergic sensitization to inhaled house dust mite via the epithelial release of GM-CSF and IL-33. *J Exp Med* **209**:1505–1517.
33. **Lambrecht BN, Hammad H.** 2013. Asthma: the importance of dysregulated barrier immunity. *Eur J Immunol* **43**:3125–3137.
34. **Lambrecht BN, Hammad H.** 2014. Dendritic cell and epithelial cell interactions at the origin of murine asthma. *Ann Am Thorac Soc* **11**(Suppl 5):S236–S243.

35. **Van Dyken SJ, Nussbaum JC, Lee J, Molofsky AB, Liang HE, Pollack JL, Gate RE, Haliburton GE, Ye CJ, Marson A, Erle DJ, Locksley RM.** 2016. A tissue checkpoint regulates type 2 immunity. *Nat Immunol* **17:** 1381–1387.

36. **Schuijs MJ, Willart MA, Vergote K, Gras D, Deswarte K, Ege MJ, Madeira FB, Beyaert R, van Loo G, Bracher F, von Mutius E, Chanez P, Lambrecht BN, Hammad H.** 2015. Farm dust and endotoxin protect against allergy through A20 induction in lung epithelial cells. *Science* **349:**1106–1110.

37. **Lee YG, Jeong JJ, Nyenhuis S, Berdyshev E, Chung S, Ranjan R, Karpurapu M, Deng J, Qian F, Kelly EA, Jarjour NN, Ackerman SJ, Natarajan V, Christman JW, Park GY.** 2015. Recruited alveolar macrophages, in response to airway epithelial-derived monocyte chemoattractant protein 1/CCl2, regulate airway inflammation and remodeling in allergic asthma. *Am J Respir Cell Mol Biol* **52:**772–784.

38. **Moniuszko M, Bodzenta-Lukaszyk A, Kowal K, Lenczewska D, Dabrowska M.** 2009. Enhanced frequencies of CD14^{++}CD16^{+}, but not CD14^{+}CD16^{+}, peripheral blood monocytes in severe asthmatic patients. *Clin Immunol* **130:**338–346.

39. **Zaslona Z, Przybranowski S, Wilke C, van Rooijen N, Teitz-Tennenbaum S, Osterholzer JJ, Wilkinson JE, Moore BB, Peters-Golden M.** 2014. Resident alveolar macrophages suppress, whereas recruited monocytes promote, allergic lung inflammation in murine models of asthma. *J Immunol* **193:**4245–4253.

40. **Wu X, Briseno CG, Durai V, Albring JC, Haldar M, Bagadia P, Kim KW, Randolph GJ, Murphy TL, Murphy KM.** 2016. Mafb lineage tracing to distinguish macrophages from other immune lineages reveals dual identity of Langerhans cells. *J Exp Med* **213:**2553–2565.

41. **van Rijt LS, Jung S, Kleinjan A, Vos N, Willart M, Duez C, Hoogsteden HC, Lambrecht BN.** 2005. In vivo depletion of lung CD11c^{+} dendritic cells during allergen challenge abrogates the characteristic features of asthma. *J Exp Med* **201:**981–991.

42. **Lambrecht BN, Salomon B, Klatzmann D, Pauwels RA.** 1998. Dendritic cells are required for the development of chronic eosinophilic airway inflammation in response to inhaled antigen in sensitized mice. *J Immunol* **160:** 4090–4097.

43. **Huber S, Hoffmann R, Muskens F, Voehringer D.** 2010. Alternatively activated macrophages inhibit T-cell proliferation by Stat6-dependent expression of PD-L2. *Blood* **116:**3311–3320.

44. **Huh JC, Strickland DH, Jahnsen FL, Turner DJ, Thomas JA, Napoli S, Tobagus I, Stumbles PA, Sly PD, Holt PG.** 2003. Bidirectional interactions between antigen-bearing respiratory tract dendritic cells (DCs) and T cells precede the late phase reaction in experimental asthma: DC activation occurs in the airway mucosa but not in the lung parenchyma. *J Exp Med* **198:**19–30.

45. **Thornton EE, Looney MR, Bose O, Sen D, Sheppard D, Locksley R, Huang X, Krummel MF.** 2012. Spatiotemporally separated antigen uptake by alveolar dendritic cells and airway presentation to T cells in the lung. *J Exp Med* **209:**1183–1199.

46. **Turner DL, Bickham KL, Thome JJ, Kim CY, D'Ovidio F, Wherry EJ, Farber DL.** 2014. Lung niches for the generation and maintenance of tissue-resident memory T cells. *Mucosal Immunol* **7:**501–510.

47. **Medoff BD, Seung E, Hong S, Thomas SY, Sandall BP, Duffield JS, Kuperman DA, Erle DJ, Luster AD.** 2009. CD11b^{+} myeloid cells are the key mediators of Th2 cell homing into the airway in allergic inflammation. *J Immunol* **182:**623–635.

48. **Sallmann E, Reininger B, Brandt S, Duschek N, Hoflehner E, Garner-Spitzer E, Platzer B, Dehlink E, Hammer M, Holcmann M, Oettgen HC, Wiedermann U, Sibilia M, Fiebiger E, Rot A, Maurer D.** 2011. High-affinity IgE receptors on dendritic cells exacerbate Th2-dependent inflammation. *J Immunol* **187:** 164–171.

49. **Tjota MY, Williams JW, Lu T, Clay BS, Byrd T, Hrusch CL, Decker DC, de Araujo CA, Bryce PJ, Sperling AI.** 2013. IL-33-dependent induction of allergic lung inflammation by FcγRIII signaling. *J Clin Invest* **123:** 2287–2297.

50. **Nakano H, Burgents JE, Nakano K, Whitehead GS, Cheong C, Bortner CD, Cook DN.** 2013. Migratory properties of pulmonary dendritic cells are determined by their developmental lineage. *Mucosal Immunol* **6:** 678–691.

51. **Parsons MW, Li L, Wallace AM, Lee MJ, Katz HR, Fernandez JM, Saijo S, Iwakura Y, Austen KF, Kanaoka Y, Barrett NA.** 2014. Dectin-2 regulates the effector phase of house dust mite-elicited pulmonary inflammation independently from its role in sensitization. *J Immunol* **192:**1361–1371.

52. **Guilliams M, De Kleer I, Henri S, Post S, Vanhoutte L, De Prijck S, Deswarte K, Malissen B, Hammad H, Lambrecht BN.** 2013. Alveolar macrophages develop from fetal monocytes that differentiate into long-lived cells in the first week of life via GM-CSF. *J Exp Med* **210:**1977–1992.

53. **van de Laar L, Saelens W, De Prijck S, Martens L, Scott CL, Van Isterdael G, Hoffmann E, Beyaert R, Saeys Y, Lambrecht BN, Guilliams M.** 2016. Yolk sac macrophages, fetal liver, and adult monocytes can colonize an empty niche and develop into functional tissue-resident macrophages. *Immunity* **44:**755–768.

54. **Westphalen K, Gusarova GA, Islam MN, Subramanian M, Cohen TS, Prince AS, Bhattacharya J.** 2014. Sessile alveolar macrophages communicate with alveolar epithelium to modulate immunity. *Nature* **506:**503–506.

55. **Bhattacharya J, Westphalen K.** 2016. Macrophage-epithelial interactions in pulmonary alveoli. *Semin Immunopathol* **38:**461–469.

56. **Thepen T, McMenamin C, Oliver J, Kraal G, Holt PG.** 1991. Regulation of immune responses to inhaled antigen by alveolar macrophages (AM) : differential effects of AM elimination in vivo on the induction of tolerance versus immunity. *Eur J Immunol* **21:**2845–2850.

57. **Bilyk N, Holt PG.** 1993. Inhibition of the immunosuppressive activity of resident pulmonary alveolar macro-

phages by granulocyte/macrophage colony-stimulating factor. *J Exp Med* **177**:1773–1777.

58. **Holt PG, Oliver J, Bilyk N, McMenamin C, McMenamin PG, Kraal G, Thepen T.** 1993. Downregulation of the antigen presenting cell function(s) of pulmonary dendritic cells in vivo by resident alveolar macrophages. *J Exp Med* **177**:397–407.
59. **Bilyk N, Holt PG.** 1995. Cytokine modulation of the immunosuppressive phenotype of pulmonary alveolar macrophage populations. *Immunology* **86**:231–237.
60. **Upham JW, Strickland DH, Bilyk N, Robinson BW, Holt PG.** 1995. Alveolar macrophages from humans and rodents selectively inhibit T-cell proliferation but permit T-cell activation and cytokine secretion. *Immunology* **84**:142–147.
61. **Tang C, Inman MD, van Rooijen N, Yang P, Shen H, Matsumoto K, O'Byrne PM.** 2001. Th type 1-stimulating activity of lung macrophages inhibits Th2- mediated allergic airway inflammation by an IFN-γ-dependent mechanism. *J Immunol* **166**:1471–1481.
62. **Tang C, Ward C, Reid D, Bish R, O'Byrne PM, Walters EH.** 2001. Normally suppressing CD40 coregulatory signals delivered by airway macrophages to T_H2 lymphocytes are defective in patients with atopic asthma. *J Allergy Clin Immunol* **107**:863–870.
63. **Draijer C, Boorsma CE, Reker-Smit C, Post E, Poelstra K, Melgert BN.** 2016. PGE_2-treated macrophages inhibit development of allergic lung inflammation in mice. *J Leukoc Biol* **100**:95–102.
64. **Bourdonnay E, Zaslona Z, Penke LR, Speth JM, Schneider DJ, Przybranowski S, Swanson JA, Mancuso P, Freeman CM, Curtis JL, Peters-Golden M.** 2015. Transcellular delivery of vesicular SOCS proteins from macrophages to epithelial cells blunts inflammatory signaling. *J Exp Med* **212**:729–742.
65. **Speth JM, Bourdonnay E, Penke LR, Mancuso P, Moore BB, Weinberg JB, Peters-Golden M.** 2016. Alveolar epithelial cell-derived prostaglandin E_2 serves as a request signal for macrophage secretion of suppressor of cytokine signaling 3 during innate inflammation. *J Immunol* **196**:5112–5120.
66. **Bedoret D, Wallemacq H, Marichal T, Desmet C, Quesada Calvo F, Henry E, Closset R, Dewals B, Thielen C, Gustin P, de Leval L, Van Rooijen N, Le Moine A, Vanderplasschen A, Cataldo D, Drion PV, Moser M, Lekeux P, Bureau F.** 2009. Lung interstitial macrophages alter dendritic cell functions to prevent airway allergy in mice. *J Clin Invest* **119**:3723–3738.
67. **Albacker LA, Yu S, Bedoret D, Lee WL, Umetsu SE, Monahan S, Freeman GJ, Umetsu DT, Dekruyff RH.** 2013. TIM-4, expressed by medullary macrophages, regulates respiratory tolerance by mediating phagocytosis of antigen-specific T cells. *Mucosal Immunol* **6**: 580–590.
68. **Soroosh P, Doherty TA, Duan W, Mehta AK, Choi H, Adams YF, Mikulski Z, Khorram N, Rosenthal P, Broide DH, Croft M.** 2013. Lung-resident tissue macrophages generate Foxp3+ regulatory T cells and promote airway tolerance. *J Exp Med* **210**:775–788.
69. **Kurowska-Stolarska M, Stolarski B, Kewin P, Murphy G, Corrigan CJ, Ying S, Pitman N, Mirchandani A, Rana B, van Rooijen N, Shepherd M, McSharry C, McInnes IB, Xu D, Liew FY.** 2009. IL-33 amplifies the polarization of alternatively activated macrophages that contribute to airway inflammation. *J Immunol* **183**: 6469–6477.
70. **Byers DE, Holtzman MJ.** 2011. Alternatively activated macrophages and airway disease. *Chest* **140**:768–774.
71. **Ford AQ, Dasgupta P, Mikhailenko I, Smith EM, Noben-Trauth N, Keegan AD.** 2012. Adoptive transfer of IL-4Rα+ macrophages is sufficient to enhance eosinophilic inflammation in a mouse model of allergic lung inflammation. *BMC Immunol* **13**:6. doi:10.1186/1471-2172-13-6.
72. **Nieuwenhuizen NE, Kirstein F, Jayakumar J, Emedi B, Hurdayal R, Horsnell WG, Lopata AL, Brombacher F.** 2012. Allergic airway disease is unaffected by the absence of IL-4Rα-dependent alternatively activated macrophages. *J Allergy Clin Immunol* **130**:743–750 e748.
73. **Han H, Headley MB, Xu W, Comeau MR, Zhou B, Ziegler SF.** 2013. Thymic stromal lymphopoietin amplifies the differentiation of alternatively activated macrophages. *J Immunol* **190**:904–912.
74. **Robbe P, Draijer C, Borg TR, Luinge M, Timens W, Wouters IM, Melgert BN, Hylkema MN.** 2015. Distinct macrophage phenotypes in allergic and nonallergic lung inflammation. *Am J Physiol Lung Cell Mol Physiol* **308**: L358–L367.
75. **Moreira AP, Cavassani KA, Hullinger R, Rosada RS, Fong DJ, Murray L, Hesson DP, Hogaboam CM.** 2010. Serum amyloid P attenuates M2 macrophage activation and protects against fungal spore-induced allergic airway disease. *J Allergy Clin Immunol* **126**:712–721.e7. doi:10.1016/j.jaci.2010.06.010.
76. **Braza F, Dirou S, Forest V, Sauzeau V, Hassoun D, Chesne J, Cheminant-Muller MA, Sagan C, Magnan A, Lemarchand P.** 2016. Mesenchymal stem cells induce suppressive macrophages through phagocytosis in a mouse model of asthma. *Stem Cells* **34**:1836–1845.
77. **Jaffar ZH, Roberts K, Pandit A, Linsley P, Djukanovic R, Holgate ST.** 1999. B7 costimulation is required for IL-5 and IL-13 secretion by bronchial biopsy tissue of atopic asthmatic subjects in response to allergen stimulation. *Am J Respir Cell Mol Biol* **20**:153–162.
78. **Girodet PO, Nguyen D, Mancini JD, Hundal M, Zhou X, Israel E, Cernadas M.** 2016. Alternative macrophage activation is increased in asthma. *Am J Respir Cell Mol Biol* **55**:467–475.
79. **Staples KJ, Hinks TS, Ward JA, Gunn V, Smith C, Djukanović R.** 2012. Phenotypic characterization of lung macrophages in asthmatic patients: overexpression of CCL17. *J Allergy Clin Immunol* **130**:1404–1412.e7. doi:10.1016/j.jaci.2012.07.023.
80. **Melgert BN, ten Hacken NH, Rutgers B, Timens W, Postma DS, Hylkema MN.** 2011. More alternative activation of macrophages in lungs of asthmatic patients. *J Allergy Clin Immunol* **127**:831–833.
81. **Martinez FO, Helming L, Milde R, Varin A, Melgert BN, Draijer C, Thomas B, Fabbri M, Crawshaw A,**

Ho LP, Ten Hacken NH, Cobos Jiménez V, Kootstra NA, Hamann J, Greaves DR, Locati M, Mantovani A, Gordon S. 2013. Genetic programs expressed in resting and IL-4 alternatively activated mouse and human macrophages: similarities and differences. *Blood* **121**:e57–e69. doi:10.1182/blood-2012-06-436212.

82. **Melgert BN, Oriss TB, Qi Z, Dixon-McCarthy B, Geerlings M, Hylkema MN, Ray A.** 2010. Macrophages: regulators of sex differences in asthma? *Am J Respir Cell Mol Biol* **42**:595–603.

83. **Bousquet J, Chanez P, Lacoste JY, Barneon G, Ghavanian N, Enander P, Venge P, Ahlstedt S, Simony-Lafontaine J, Godard P.** 1990. Eosinophilic inflammation in asthma. *N Engl J Med* **323**:1033–1039.

84. **Travers J, Rothenberg ME.** 2015. Eosinophils in mucosal immune responses. *Mucosal Immunol* **8**:464–475.

85. **Possa SS, Leick EA, Prado CM, Martins MA, Tibério IF.** 2013. Eosinophilic inflammation in allergic asthma. *Front Pharmacol* **4**:46. doi:10.3389/fphar.2013.00046.

86. **Chu DK, Jimenez-Saiz R, Verschoor CP, Walker TD, Goncharova S, Llop-Guevara A, Shen P, Gordon ME, Barra NG, Bassett JD, Kong J, Fattouh R, McCoy KD, Bowdish DM, Erjefalt JS, Pabst O, Humbles AA, Kolbeck R, Waserman S, Jordana M.** 2014. Indigenous enteric eosinophils control DCs to initiate a primary Th2 immune response in vivo. *J Exp Med* **211**:1657–1672.

87. **Chu VT, Frohlich A, Steinhauser G, Scheel T, Roch T, Fillatreau S, Lee JJ, Lohning M, Berek C.** 2011. Eosinophils are required for the maintenance of plasma cells in the bone marrow. *Nat Immunol* **12**:151–159.

88. **Brusselle GG, Kips JC, Tavernier J, Van Der Heyden JG, Cuvelier CA, Pauwels RA, Bluethmann H.** 1994. Attenuation of allergic airway inflammation in IL-4 deficient mice. *Clin Exp Allergy* **24**:73–80.

89. **Bousquet J, Chanez P, Lacoste JY, Barnéon G, Ghavanian N, Enander I, Venge P, Ahlstedt S, Simony-Lafontaine J, Godard P, Michel FB.** 1990. Eosinophilic inflammation in asthma. *N Engl J Med* **323**:1033–1039.

90. **Jatakanon A, Lim S, Barnes PJ.** 2000. Changes in sputum eosinophils predict loss of asthma control. *Am J Respir Crit Care Med* **161**:64–72.

91. **Deykin A, Lazarus SC, Fahy JV, Wechsler ME, Boushey HA, Chinchilli VM, Craig TJ, Dimango E, Kraft M, Leone F, Lemanske RF, Martin RJ, Pesola GR, Peters SP, Sorkness CA, Szefler SJ, Israel E; Asthma Clinical Research Network, National Heart, Lung, and Blood Institute/NIH.** 2005. Sputum eosinophil counts predict asthma control after discontinuation of inhaled corticosteroids. *J Allergy Clin Immunol* **115**:720–727.

92. **Jacobsen EA, Lee NA, Lee JJ.** 2014. Re-defining the unique roles for eosinophils in allergic respiratory inflammation. *Clin Exp Allergy* **44**:1119–1136.

93. **Mori Y, Iwasaki H, Kohno K, Yoshimoto G, Kikushige Y, Okeda A, Uike N, Niiro H, Takenaka K, Nagafuji K, Miyamoto T, Harada M, Takatsu K, Akashi K.** 2009. Identification of the human eosinophil lineage-committed progenitor: revision of phenotypic definition of the human common myeloid progenitor. *J Exp Med* **206**:183–193.

94. **Iwasaki H, Mizuno S, Mayfield R, Shigematsu H, Arinobu Y, Seed B, Gurish MF, Takatsu K, Akashi K.** 2005. Identification of eosinophil lineage-committed progenitors in the murine bone marrow. *J Exp Med* **201**:1891–1897.

95. **Tsuzuki H, Arinobu Y, Miyawaki K, Takaki A, Ota SI, Ota Y, Mitoma H, Akahoshi M, Mori Y, Iwasaki H, Niiro H, Tsukamoto H, Akashi K.** 2017. Functional IL-33 receptors are expressed in early progenitor stages of allergy-related granulocytes. *Immunology* **150**:64–73.

96. **Lopez AF, Begley CG, Williamson DJ, Warren DJ, Vadas MA, Sanderson CJ.** 1986. Murine eosinophil differentiation factor. An eosinophil-specific colony-stimulating factor with activity for human cells. *J Exp Med* **163**:1085–1099.

97. **Asquith KL, Ramshaw HS, Hansbro PM, Beagley KW, Lopez AF, Foster PS.** 2008. The IL-3/IL-5/GM-CSF common receptor plays a pivotal role in the regulation of Th2 immunity and allergic airway inflammation. *J Immunol* **180**:1199–1206.

98. **Kopf M, Brombacher F, Hodgkin PD, Ramsay AJ, Milbourne EA, Dai WJ, Ovington KS, Behm CA, Kohler G, Young IG, Matthaei KI.** 1996. IL-5-deficient mice have a developmental defect in $CD5^+$ B-1 cells and lack eosinophilia but have normal antibody and cytotoxic T cell responses. *Immunity* **4**:15–24.

99. **Nakajima H, Iwamoto I, Tomoe S, Matsumura R, Tomioka H, Takatsu K, Yoshida S.** 1992. $CD4^+$ T-lymphocytes and interleukin-5 mediate antigen-induced eosinophil infiltration into the mouse trachea. *Am Rev Respir Dis* **146**:374–377.

100. **Yoshida T, Ikuta K, Sugaya H, Maki K, Takagi M, Kanazawa H, Sunaga S, Kinashi T, Yoshimura K, Miyazaki J, Takaki S, Takatsu K.** 1996. Defective B-1 cell development and impaired immunity against *Angiostrongylus cantonensis* in IL-5Rα-deficient mice. *Immunity* **4**:483–494.

101. **Dent LA, Strath M, Mellor AL, Sanderson CJ.** 1990. Eosinophilia in transgenic mice expressing interleukin 5. *J Exp Med* **172**:1425–1431.

102. **Leckie MJ, ten Brinke A, Khan J, Diamant Z, O'Connor BJ, Walls CM, Mathur AK, Cowley HC, Chung KF, Djukanovic R, Hansel TT, Holgate ST, Sterk PJ, Barnes PJ.** 2000. Effects of an interleukin-5 blocking monoclonal antibody on eosinophils, airway hyper-responsiveness, and the late asthmatic response. *Lancet* **356**:2144–2148.

103. **Flood-Page P, Menzies-Gow A, Phipps S, Ying S, Wangoo A, Ludwig MS, Barnes N, Robinson D, Kay AB.** 2003. Anti-IL-5 treatment reduces deposition of ECM proteins in the bronchial subepithelial basement membrane of mild atopic asthmatics. *J Clin Invest* **112**:1029–1036.

104. **Menzies-Gow A, Flood-Page P, Sehmi R, Burman J, Hamid Q, Robinson DS, Kay AB, Denburg J.** 2003. Anti-IL-5 (mepolizumab) therapy induces bone marrow eosinophil maturational arrest and decreases eosinophil progenitors in the bronchial mucosa of atopic asthmatics. *J Allergy Clin Immunol* **111**:714–719.

105. Garrett JK, Jameson SC, Thomson B, Collins MH, Wagoner LE, Freese DK, Beck LA, Boyce JA, Filipovich AH, Villanueva JM, Sutton SA, Assa'ad AH, Rothenberg ME. 2004. Anti-interleukin-5 (mepolizumab) therapy for hypereosinophilic syndromes. *J Allergy Clin Immunol* **113**:115–119.

106. Phipps S, Flood-Page P, Menzies-Gow A, Ong YE, Kay AB. 2004. Intravenous anti-IL-5 monoclonal antibody reduces eosinophils and tenascin deposition in allergen-challenged human atopic skin. *J Invest Dermatol* **122**: 1406–1412.

107. Stein ML, Villanueva JM, Buckmeier BK, Yamada Y, Filipovich AH, Assa'ad AH, Rothenberg ME. 2008. Anti-IL-5 (mepolizumab) therapy reduces eosinophil activation ex vivo and increases IL-5 and IL-5 receptor levels. *J Allergy Clin Immunol* **121**:1473–1483.

108. Haldar P, Brightling CE, Hargadon B, Gupta S, Monteiro W, Sousa A, Marshall RP, Bradding P, Green RH, Wardlaw AJ, Pavord ID. 2009. Mepolizumab and exacerbations of refractory eosinophilic asthma. *N Engl J Med* **360**:973–984.

109. Roufosse FE, Kahn JE, Gleich GJ, Schwartz LB, Singh AD, Rosenwasser LJ, Denburg JA, Ring J, Rothenberg ME, Sheikh J, Haig AE, Mallett SA, Templeton DN, Ortega HG, Klion AD. 2012. Long-term safety of mepolizumab for the treatment of hypereosinophilic syndromes. *J Allergy Clin Immunol* **131**:461–467.e5. doi:10.1016/j.jaci.2012.07.055.

110. Bel EH, Wenzel SE, Thompson PJ, Prazma CM, Keene ON, Yancey SW, Ortega HG, Pavord ID. 2014. Oral glucocorticoid-sparing effect of mepolizumab in eosinophilic asthma. *N Engl J Med* **371**:1189–1197.

111. Ortega HG, Liu MC, Pavord ID, Brusselle GG, FitzGerald JM, Chetta A, Humbert M, Katz LE, Keene ON, Yancey SW, Chanez P, MENSA Investigators. 2014. Mepolizumab treatment in patients with severe eosinophilic asthma. *N Engl J Med* **371**:1198–1207.

112. Yu C, Cantor AB, Yang H, Browne C, Wells RA, Fujiwara Y, Orkin SH. 2002. Targeted deletion of a high-affinity GATA-binding site in the GATA-1 promoter leads to selective loss of the eosinophil lineage in vivo. *J Exp Med* **195**:1387–1395.

113. Southam DS, Widmer N, Ellis R, Hirota JA, Inman MD, Sehmi R. 2005. Increased eosinophil-lineage committed progenitors in the lung of allergen-challenged mice. *J Allergy Clin Immunol* **115**:95–102.

114. Sehmi R, Smith SG, Kjarsgaard M, Radford K, Boulet LP, Lemiere C, Prazma CM, Ortega H, Martin JG, Nair P. 2016. Role of local eosinophilopoietic processes in the development of airway eosinophilia in prednisone-dependent severe asthma. *Clin Exp Allergy* **46**:793–802.

115. Foster PS, Mould AW, Yang M, Mackenzie J, Mattes J, Hogan SP, Mahalingam S, McKenzie AN, Rothenberg ME, Young IG, Matthaei KI, Webb DC. 2001. Elemental signals regulating eosinophil accumulation in the lung. *Immunol Rev* **179**:173–181.

116. MacKenzie JR, Mattes J, Dent LA, Foster PS. 2001. Eosinophils promote allergic disease of the lung by regulating $CD4^+$ Th2 lymphocyte function. *J Immunol* **167**:3146–3155.

117. Foster PS, Hogan SP, Yang M, Mattes J, Young IG, Matthaei KI, Kumar RK, Mahalingam S, Webb DC. 2002. Interleukin-5 and eosinophils as therapeutic targets for asthma. *Trends Mol Med* **8**:162–167.

118. Mattes J, Yang M, Mahalingam S, Kuehr J, Webb DC, Simson L, Hogan SP, Koskinen A, McKenzie AN, Dent LA, Rothenberg ME, Matthaei KI, Young IG, Foster PS. 2002. Intrinsic defect in T cell production of interleukin (IL)-13 in the absence of both IL-5 and eotaxin precludes the development of eosinophilia and airways hyperreactivity in experimental asthma. *J Exp Med* **195**: 1433–1444.

119. Rothenberg ME, Luster AD, Leder P. 1995. Murine eotaxin: an eosinophil chemoattractant inducible in endothelial cells and in interleukin 4-induced tumor suppression. *Proc Natl Acad Sci U S A* **92**:8960–8964.

120. Rothenberg ME, Luster AD, Lilly CM, Drazen JM, Leder P. 1995. Constitutive and allergen-induced expression of eotaxin mRNA in the guinea pig lung. *J Exp Med* **181**:1211–1216.

121. Garcia-Zepeda EA, Rothenberg ME, Ownbey RT, Celestin J, Leder P, Luster AD. 1996. Human eotaxin is a specific chemoattractant for eosinophil cells and provides a new mechanism to explain tissue eosinophilia. *Nat Med* **2**:449–456.

122. Rothenberg ME, MacLean JA, Pearlman E, Luster AD, Leder P. 1997. Targeted disruption of the chemokine eotaxin partially reduces antigen-induced tissue eosinophilia. *J Exp Med* **185**:785–790.

123. Hogan SP, Mould AW, Young JM, Rothenberg ME, Ramsay AJ, Matthaei K, Young IG, Foster PS. 1998. Cellular and molecular regulation of eosinophil trafficking to the lung. *Immunol Cell Biol* **76**:454–460.

124. Mould AW, Ramsay AJ, Matthaei KI, Young IG, Rothenberg ME, Foster PS. 2000. The effect of IL-5 and eotaxin expression in the lung on eosinophil trafficking and degranulation and the induction of bronchial hyperreactivity. *J Immunol* **164**:2142–2150.

125. Zimmermann N, Hershey GK, Foster PS, Rothenberg ME. 2003. Chemokines in asthma: cooperative interaction between chemokines and IL-13. *J Allergy Clin Immunol* **111**:227–242; quiz 243.

126. Fulkerson PC, Fischetti CA, Rothenberg ME. 2006. Eosinophils and CCR3 regulate interleukin-13 transgene-induced pulmonary remodeling. *Am J Pathol* **169**: 2117–2126.

127. Ravensberg AJ, Ricciardolo FL, van Schadewijk A, Rabe KF, Sterk PJ, Hiemstra PS, Mauad T. 2005. Eotaxin-2 and eotaxin-3 expression is associated with persistent eosinophilic bronchial inflammation in patients with asthma after allergen challenge. *J Allergy Clin Immunol* **115**:779–785.

128. Ying S, Robinson DS, Meng Q, Rottman J, Kennedy R, Ringler DJ, Mackay CR, Daugherty BL, Springer MS, Durham SR, Williams TJ, Kay AB. 1997. Enhanced expression of eotaxin and CCR3 mRNA and protein in atopic asthma. Association with airway hyperresponsiveness and predominant co-localization of eotaxin mRNA to bronchial epithelial and endothelial cells. *Eur J Immunol* **27**:3507–3516.

129. **Ganzalo JA, Jia GQ, Aguirre V, Friend D, Coyle AJ, Jenkins NA, Lin GS, Katz H, Lichtman A, Copeland N, Kopf M, Gutierrez-Ramos JC.** 1996. Mouse Eotaxin expression parallels eosinophil accumulation during lung allergic inflammation but it is not restricted to a Th2-type response. *Immunity* **4:**1–14.

130. **Gonzalo JA, Lloyd CM, Wen D, Albar JP, Wells TN, Proudfoot A, Martinez-A C, Dorf M, Bjerke T, Coyle AJ, Gutierrez-Ramos JC.** 1998. The coordinated action of CC chemokines in the lung orchestrates allergic inflammation and airway hyperresponsiveness. *J Exp Med* **188:**157–167.

131. **Tsuyuki S, Bertrand C, Erard F, Trifilieff A, Tsuyuki J, Wesp M, Anderson GP, Coyle AJ.** 1995. Activation of the Fas receptor on lung eosinophils leads to apoptosis and the resolution of eosinophilic inflammation of the airways. *J Clin Invest* **96:**2924–2931.

132. **Park YM, Bochner BS.** 2010. Eosinophil survival and apoptosis in health and disease. *Allergy Asthma Immunol Res* **2:**87–101.

133. **Kolbeck R, Kozhich A, Koike M, Peng L, Andersson CK, Damschroder MM, Reed JL, Woods R, Dall'acqua WW, Stephens GL, Erjefalt JS, Bjermer L, Humbles AA, Gossage D, Wu H, Kiener PA, Spitalny GL, Mackay CR, Molfino NA, Coyle AJ.** 2010. MEDI-563, a humanized anti-IL-5 receptor α mAb with enhanced antibody-dependent cell-mediated cytotoxicity function. *J Allergy Clin Immunol* **125:**1344–1353.e2. doi: 10.1016/j.jaci.2010.04.004.

134. **Bagley CJ, Lopez AF, Vadas MA.** 1997. New frontiers for IL-5. *J Allergy Clin Immunol* **99:**725–728.

135. **Coyle AJ, Ackerman SJ, Irvin CG.** 1993. Cationic proteins induce airway hyperresponsiveness dependent on charge interactions. *Am Rev Respir Dis* **147:**896–900.

136. **Coyle AJ, Perretti F, Manzini S, Irvin CG.** 1994. Cationic protein-induced sensory nerve activation: role of substance P in airway hyperresponsiveness and plasma protein extravasation. *J Clin Invest* **94:**2301–2306.

137. **Coyle AJ, Uchida D, Ackerman SJ, Mitzner W, Irvin CG.** 1994. Role of cationic proteins in the airway. Hyperresponsiveness due to airway inflammation. *Am J Respir Crit Care Med* **150:**S63–S71.

138. **Coyle AJ, Ackerman SJ, Burch R, Proud D, Irvin CG.** 1995. Human eosinophil-granule major basic protein and synthetic polycations induce airway hyperresponsiveness in vivo dependent on bradykinin generation. *J Clin Invest* **95:**1735–1740.

139. **Jacoby DB, Gleich GJ, Fryer AD.** 1993. Human eosinophil major basic protein is an endogenous allosteric antagonist at the inhibitory muscarinic M2 receptor. *J Clin Invest* **91:**1314–1318.

140. **Morshed M, Yousefi S, Stockle C, Simon HU, Simon D.** 2012. Thymic stromal lymphopoietin stimulates the formation of eosinophil extracellular traps. *Allergy* **67:** 1127–1137.

141. **Schorn C, Janko C, Latzko M, Chaurio R, Schett G, Herrmann M.** 2012. Monosodium urate crystals induce extracellular DNA traps in neutrophils, eosinophils, and basophils but not in mononuclear cells. *Front Immunol* 3:277. doi:10.3389/fimmu.2012.00277.

142. **Yousefi S, Simon D, Simon HU.** 2012. Eosinophil extracellular DNA traps: molecular mechanisms and potential roles in disease. *Curr Opin Immunol* **24:**736–739.

143. **Ueki S, Melo RC, Ghiran I, Spencer LA, Dvorak AM, Weller PF.** 2013. Eosinophil extracellular DNA trap cell death mediates lytic release of free secretion-competent eosinophil granules in humans. *Blood* **121:** 2074–2083.

144. **Simon D, Radonjic-Hosli S, Straumann A, Yousefi S, Simon HU.** 2015. Active eosinophilic esophagitis is characterized by epithelial barrier defects and eosinophil extracellular trap formation. *Allergy* **70:**443–452.

145. **Fulkerson PC, Rothenberg ME.** 2013. Targeting eosinophils in allergy, inflammation and beyond. *Nat Rev Drug Discov* **12:**117–129.

146. **Henderson WR Jr, Lewis DB, Albert RK, Zhang Y, Lamm WJ, Chiang GK, Jones F, Eriksen P, Tien YT, Jonas M, Chi EY.** 1996. The importance of leukotrienes in airway inflammation in a mouse model of asthma. *J Exp Med* **184:**1483–1494.

147. **Jacobsen EA, Ochkur SI, Pero RS, Taranova AG, Protheroe CA, Colbert DC, Lee NA, Lee JJ.** 2008. Allergic pulmonary inflammation in mice is dependent on eosinophil-induced recruitment of effector T cells. *J Exp Med* **205:**699–710.

148. **Song DJ, Cho JY, Lee SY, Miller M, Rosenthal P, Soroosh P, Croft M, Zhang M, Varki A, Broide DH.** 2009. Anti-Siglec-F antibody reduces allergen-induced eosinophilic inflammation and airway remodeling. *J Immunol* **183:**5333–5341.

149. **Doyle AD, Jacobsen EA, Ochkur SI, McGarry MP, Shim KG, Nguyen DT, Protheroe C, Colbert D, Kloeber J, Neely J, Shim KP, Dyer KD, Rosenberg HF, Lee JJ, Lee NA.** 2013. Expression of the secondary granule proteins major basic protein 1 (MBP-1) and eosinophil peroxidase (EPX) is required for eosinophilopoiesis in mice. *Blood* **122:**781–790.

150. **Doyle AD, Jacobsen EA, Ochkur SI, Willetts L, Shim K, Neely J, Kloeber J, Lesuer WE, Pero RS, Lacy P, Moqbel R, Lee NA, Lee JJ.** 2013. Homologous recombination into the eosinophil peroxidase locus generates a strain of mice expressing Cre recombinase exclusively in eosinophils. *J Leukoc Biol* **94:**17–24.

151. **Jacobsen EA, Doyle AD, Colbert DC, Zellner KR, Protheroe CA, LeSuer WE, Lee NA, Lee JJ.** 2015. Differential activation of airway eosinophils induces IL-13-mediated allergic Th2 pulmonary responses in mice. *Allergy* **70:**1148–1159.

152. **Trifilieff A, Fujitani Y, Coyle AJ, Kopf M, Bertrand C.** 2001. IL-5 deficiency abolishes aspects of airway remodelling in a murine model of lung inflammation. *Clin Exp Allergy* **31:**934–942.

153. **Torrego A, Hew M, Oates T, Sukkar M, Fan Chung K.** 2007. Expression and activation of TGF-β isoforms in acute allergen-induced remodelling in asthma. *Thorax* **62:**307–313.

154. Al Heialy S, McGovern TK, Martin JG. 2011. Insights into asthmatic airway remodelling through murine models. *Respirology* 16:589–597.

155. Saglani S, Lloyd CM. 2014. Eosinophils in the pathogenesis of paediatric severe asthma. *Curr Opin Allergy Clin Immunol* 14:143–148.

156. Humbles AA, Lloyd CM, McMillan SJ, Friend DS, Xanthou G, McKenna EE, Ghiran S, Gerard NP, Yu C, Orkin SH, Gerard C. 2004. A critical role for eosinophils in allergic airways remodeling. *Science* 305:1776–1779.

157. Fattouh R, Al-Garawi A, Fattouh M, Arias K, Walker TD, Goncharova S, Coyle AJ, Humbles AA, Jordana M. 2011. Eosinophils are dispensable for allergic remodeling and immunity in a model of house dust mite-induced airway disease. *Am J Respir Crit Care Med* 183:179–188.

158. Lee JJ, Dimina D, Macias MP, Ochkur SI, McGarry MP, O'Neill KR, Protheroe C, Pero R, Nguyen T, Cormier SA, Lenkiewicz E, Colbert D, Rinaldi L, Ackerman SJ, Irvin CG, Lee NA. 2004. Defining a link with asthma in mice congenitally deficient in eosinophils. *Science* 305:1773–1776.

159. Walsh ER, Sahu N, Kearley J, Benjamin E, Kang BH, Humbles A, August A. 2008. Strain-specific requirement for eosinophils in the recruitment of T cells to the lung during the development of allergic asthma. *J Exp Med* 205:1285–1292.

160. Nair P, Pizzichini MM, Kjarsgaard M, Inman MD, Efthimiadis A, Pizzichini E, Hargreave FE, O'Byrne PM. 2009. Mepolizumab for prednisone-dependent asthma with sputum eosinophilia. *N Engl J Med* 360:985–993.

161. Prazma CM, Wenzel S, Barnes N, Douglass JA, Hartley BF, Ortega H. 2014. Characterisation of an OCS-dependent severe asthma population treated with mepolizumab. *Thorax* 69:1141–1142.

162. Lukawska JJ, Livieratos L, Sawyer BM, Lee T, O'Doherty M, Blower PJ, Kofi M, Ballinger JR, Corrigan CJ, Gnanasegaran G, Sharif-Paghaleh E, Mullen GE. 2014. Real-time differential tracking of human neutrophil and eosinophil migration *in vivo*. *J Allergy Clin Immunol* 133:233–239e231.

163. Teran LM, Campos MG, Begishvilli BT, Schroder JM, Djukanovic R, Shute JK, Church MK, Holgate ST, Davies DE. 1997. Identification of neutrophil chemotactic factors in bronchoalveolar lavage fluid of asthmatic patients. *Clin Exp Allergy* 27:396–405.

164. Lukawska JJ, Livieratos L, Sawyer BM, Lee T, O'Doherty M, Blower PJ, Kofi M, Ballinger JR, Corrigan CJ, Gnanasegaran G, Sharif-Paghaleh E, Mullen GE. 2014. Imaging inflammation in asthma: real time, differential tracking of human neutrophil and eosinophil migration in allergen challenged, atopic asthmatics in vivo. *EBioMedicine* 1:173–180.

165. Todd CM, Salter BM, Murphy DM, Watson RM, Howie KJ, Milot J, Sadeh J, Boulet LP, O'Byrne PM, Gauvreau GM. 2016. The effects of a CXCR1/CXCR2 antagonist on neutrophil migration in mild atopic asthmatic subjects. *Pulm Pharmacol Ther* 41:34–39.

166. de Boer JD, Berger M, Majoor CJ, Kager LM, Meijers JC, Terpstra S, Nieuwland R, Boing AN, Lutter R, Wouters D, van Mierlo GJ, Zeerleder SS, Bel EH, van't Veer C, de Vos AF, van der Zee JS, van der Poll T. 2015. Activated protein C inhibits neutrophil migration in allergic asthma: a randomised trial. *Eur Respir J* 46:1636–1644.

167. Lambrecht BN, Hammad H. 2013. Asthma and coagulation. *N Engl J Med* 369:1964–1966.

168. Tonnel AB, Gosset P, Tillie-Leblond I. 2001. Characteristics of the inflammatory response in bronchial lavage fluids from patients with status asthmaticus. *Int Arch Allergy Immunol* 124:267–271.

169. Wark PA, Johnston SL, Moric I, Simpson JL, Hensley MJ, Gibson PG. 2002. Neutrophil degranulation and cell lysis is associated with clinical severity in virus-induced asthma. *Eur Respir J* 19:68–75.

170. Hellings PW, Kasran A, Liu Z, Vandekerckhove P, Wuyts A, Overbergh L, Mathieu C, Ceuppens JL. 2003. Interleukin-17 orchestrates the granulocyte influx into airways after allergen inhalation in a mouse model of allergic asthma. *Am J Respir Cell Mol Biol* 28:42–50.

171. Schleimer RP. 2004. Glucocorticoids suppress inflammation but spare innate immune responses in airway epithelium. *Proc Am Thorac Soc* 1:222–230.

172. Bhakta NR, Woodruff PG. 2011. Human asthma phenotypes: from the clinic, to cytokines, and back again. *Immunol Rev* 242:220–232.

173. Dworski R, Simon HU, Hoskins A, Yousefi S. 2011. Eosinophil and neutrophil extracellular DNA traps in human allergic asthmatic airways. *J Allergy Clin Immunol* 127:1260–1266.

174. Fei M, Bhatia S, Oriss TB, Yarlagadda M, Khare A, Akira S, Saijo S, Iwakura Y, Fallert Junecko BA, Reinhart TA, Foreman O, Ray P, Kolls J, Ray A. 2011. TNF-α from inflammatory dendritic cells (DCs) regulates lung IL-17A/IL-5 levels and neutrophilia versus eosinophilia during persistent fungal infection. *Proc Natl Acad Sci U S A* 108:5360–5365.

175. Nakagome K, Matsushita S, Nagata M. 2012. Neutrophilic inflammation in severe asthma. *Int Arch Allergy Immunol* 158(Suppl 1):96–102.

176. Kupczyk M, Dahlen B, Sterk PJ, Nizankowska-Mogilnicka E, Papi A, Bel EH, Chanez P, Howarth PH, Holgate ST, Brusselle G, Siafakas NM, Gjomarkaj M, Dahlen SE. 2014. Stability of phenotypes defined by physiological variables and biomarkers in adults with asthma. *Allergy* 69:1198–1204.

177. Manni ML, Trudeau JB, Scheller EV, Mandalapu S, Elloso MM, Kolls JK, Wenzel SE, Alcorn JF. 2014. The complex relationship between inflammation and lung function in severe asthma. *Mucosal Immunol* 7:1186–1198.

178. Zhao J, Lloyd CM, Noble A. 2013. Th17 responses in chronic allergic airway inflammation abrogate regulatory T-cell-mediated tolerance and contribute to airway remodeling. *Mucosal Immunol* 6:335–346.

179. Goleva E, Hauk PJ, Hall CF, Liu AH, Riches DW, Martin RJ, Leung DY. 2008. Corticosteroid-resistant

asthma is associated with classical antimicrobial activation of airway macrophages. *J Allergy Clin Immunol* **122:**550–559.e3. doi:10.1016/j.jaci.2008.07.007.

180. **Goleva E, Jackson LP, Harris JK, Robertson CE, Sutherland ER, Hall CF, Good JT Jr, Gelfand EW, Martin RJ, Leung DY.** 2013. The effects of airway microbiome on corticosteroid responsiveness in asthma. *Am J Respir Crit Care Med* **188:**1193–1201.

181. **Hadebe S, Kirstein F, Fierens K, Chen K, Drummond RA, Vautier S, Sajaniemi S, Murray G, Williams DL, Redelinghuys P, Reinhart TA, Fallert Junecko BA, Kolls JK, Lambrecht BN, Brombacher F, Brown GD.** 2015. Microbial ligand costimulation drives neutrophilic steroid-refractory asthma. *PLoS One* **10:**e0134219. doi:10.1371/journal.pone.0134219.

182. **Brusselle GG, Vanderstichele C, Jordens P, Deman R, Slabbynck H, Ringoet V, Verleden G, Demedts IK, Verhamme K, Delporte A, Demeyere B, Claeys G, Boelens J, Padalko E, Verschakelen J, Van Maele G, Deschepper E, Joos GF.** 2013. Azithromycin for prevention of exacerbations in severe asthma (AZISAST): a multicentre randomised double-blind placebo-controlled trial. *Thorax* **68:**322–329.

183. **Brusselle GG.** 2015. Are the antimicrobial properties of macrolides required for their therapeutic efficacy in chronic neutrophilic airway diseases? *Thorax* **70:**401–403.

184. **Sasaki H, Kurotaki D, Osato N, Sato H, Sasaki I, Koizumi S, Wang H, Kaneda C, Nishiyama A, Kaisho T, Aburatani H, Morse HC III, Ozato K, Tamura T.** 2015. Transcription factor IRF8 plays a critical role in the development of murine basophils and mast cells. *Blood* **125:**358–369.

185. **Li Y, Qi X, Liu B, Huang H.** 2015. The STAT5-GATA2 pathway is critical in basophil and mast cell differentiation and maintenance. *J Immunol* **194:**4328–4338.

186. **Qi X, Hong J, Chaves L, Zhuang Y, Chen Y, Wang D, Chabon J, Graham B, Ohmori K, Li Y, Huang H.** 2013. Antagonistic regulation by the transcription factors C/EBPα and MITF specifies basophil and mast cell fates. *Immunity* **39:**97–110.

187. **Ohmori K, Luo Y, Jia Y, Nishida J, Wang Z, Bunting KD, Wang D, Huang H.** 2009. IL-3 induces basophil expansion in vivo by directing granulocyte-monocyte progenitors to differentiate into basophil lineage-restricted progenitors in the bone marrow and by increasing the number of basophil/mast cell progenitors in the spleen. *J Immunol* **182:**2835–2841.

188. **Matsuzawa S, Sakashita K, Kinoshita T, Ito S, Yamashita T, Koike K.** 2003. IL-9 enhances the growth of human mast cell progenitors under stimulation with stem cell factor. *J Immunol* **170:**3461–3467.

189. **Starkl P, Marichal T, Gaudenzio N, Reber LL, Sibilano R, Tsai M, Galli SJ.** 2016. IgE antibodies, FcεRIα, and IgE-mediated local anaphylaxis can limit snake venom toxicity. *J Allergy Clin Immunol* **137:**246–257.e11. doi:10.1016/j.jaci.2015.08.005.

190. **Mukai K, Tsai M, Starkl P, Marichal T, Galli SJ.** 2016. IgE and mast cells in host defense against parasites and venoms. *Semin Immunopathol* **38:**581–603.

191. **Marichal T, Starkl P, Reber LL, Kalesnikoff J, Oettgen HC, Tsai M, Metz M, Galli SJ.** 2013. A beneficial role for immunoglobulin E in host defense against honeybee venom. *Immunity* **39:**963–975.

192. **Akahoshi M, Song CH, Piliponsky AM, Metz M, Guzzetta A, Abrink M, Schlenner SM, Feyerabend TB, Rodewald HR, Pejler G, Tsai M, Galli SJ.** 2011. Mast cell chymase reduces the toxicity of Gila monster venom, scorpion venom, and vasoactive intestinal polypeptide in mice. *J Clin Invest* **121:**4180–4191.

193. **Galli SJ, Tsai M.** 2012. IgE and mast cells in allergic disease. *Nat Med* **18:**693–704.

194. **Galli SJ, Tsai M.** 2010. Mast cells in allergy and infection: versatile effector and regulatory cells in innate and adaptive immunity. *Eur J Immunol* **40:**1843–1851.

195. **Moon TC, St Laurent CD, Morris KE, Marcet C, Yoshimura T, Sekar Y, Befus AD.** Advances in mast cell biology: new understanding of heterogeneity and function. *Mucosal Immunol* **3:**111–128.

196. **Olivera A, Rivera J.** 2011. An emerging role for the lipid mediator sphingosine-1-phosphate in mast cell effector function and allergic disease. *Adv Exp Med Biol* **716:**123–142.

197. **Allakhverdi Z, Smith DE, Comeau MR, Delespesse G.** 2007. Cutting edge: The ST2 ligand IL-33 potently activates and drives maturation of human mast cells. *J Immunol* **179:**2051–2054.

198. **Dougherty RH, Sidhu SS, Raman K, Solon M, Solberg OD, Caughey GH, Woodruff PG, Fahy JV.** 2010. Accumulation of intraepithelial mast cells with a unique protease phenotype in T_H2-high asthma. *J Allergy Clin Immunol* **125:**1046–1053.e8. doi:10.1016/j.jaci.2010.03.003.

199. **Balzar S, Fajt ML, Comhair SA, Erzurum SC, Bleecker E, Busse WW, Castro M, Gaston B, Israel E, Schwartz LB, Curran-Everett D, Moore CG, Wenzel SE.** 2011. Mast cell phenotype, location, and activation in severe asthma. Data from the Severe Asthma Research Program. *Am J Respir Crit Care Med* **183:**299–309.

200. **Carroll NG, Mutavdzic S, James AL.** 2002. Distribution and degranulation of airway mast cells in normal and asthmatic subjects. *Eur Respir J* **19:**879–885.

201. **Carroll NG, Mutavdzic S, James AL.** 2002. Increased mast cells and neutrophils in submucosal mucous glands and mucus plugging in patients with asthma. *Thorax* **57:**677–682.

202. **Brightling CE, Bradding P, Symon FA, Holgate ST, Wardlaw AJ, Pavord ID.** 2002. Mast-cell infiltration of airway smooth muscle in asthma. *N Engl J Med* **346:** 1699–1705.

203. **Yu M, Tsai M, Tam SY, Jones C, Zehnder J, Galli SJ.** 2006. Mast cells can promote the development of multiple features of chronic asthma in mice. *J Clin Invest* **116:**1633–1641.

204. **Yu M, Eckart MR, Morgan AA, Mukai K, Butte AJ, Tsai M, Galli SJ.** 2011. Identification of an IFN-γ/mast cell axis in a mouse model of chronic asthma. *J Clin Invest* **121:**3133–3143.

205. **Galli SJ, Kalesnikoff J, Grimbaldeston MA, Piliponsky AM, Williams CM, Tsai M.** 2005. Mast cells as "tunable" effector and immunoregulatory cells: recent advances. *Annu Rev Immunol* **23:**749–786.

206. **Heger K, Fierens K, Vahl JC, Aszodi A, Peschke K, Schenten D, Hammad H, Beyaert R, Saur D, van Loo G, Roers A, Lambrecht BN, Kool M, Schmidt-Supprian M.** 2014. A20-deficient mast cells exacerbate inflammatory responses *in vivo*. *PLoS Biol* **12:**e1001762. doi: 10.1371/journal.pbio.1001762.

207. **Suurmond J, van Heemst J, van Heiningen J, Dorjée AL, Schilham MW, van der Beek FB, Huizinga TW, Schuerwegh AJ, Toes RE.** 2013. Communication between human mast cells and $CD4^+$ T cells through antigen-dependent interactions. *Eur J Immunol* **43:** 1758–1768.

208. **Reuter S, Dehzad N, Martin H, Heinz A, Castor T, Sudowe S, Reske-Kunz AB, Stassen M, Buhl R, Taube C.** 2010. Mast cells induce migration of dendritic cells in a murine model of acute allergic airway disease. *Int Arch Allergy Immunol* **151:**214–222.

209. **Grimbaldeston MA, Nakae S, Kalesnikoff J, Tsai M, Galli SJ.** 2007. Mast cell-derived interleukin 10 limits skin pathology in contact dermatitis and chronic irradiation with ultraviolet B. *Nat Immunol* **8:**1095–1104.

210. **Chacon-Salinas R, Limon-Flores AY, Chavez-Blanco AD, Gonzalez-Estrada A, Ullrich SE.** 2011. Mast cell-derived IL-10 suppresses germinal center formation by affecting T follicular helper cell function. *J Immunol* **186:**25–31.

211. **Na HJ, Hudson SA, Bochner BS.** 2011. IL-33 enhances Siglec-8 mediated apoptosis of human eosinophils. *Cytokine* **57:**169–174.

212. **Yokoi H, Choi OH, Hubbard W, Lee HS, Canning BJ, Lee HH, Ryu SD, von Gunten S, Bickel CA, Hudson SA, MacGlashan DW Jr, Bochner BS.** 2008. Inhibition of FcεRI-dependent mediator release and calcium flux from human mast cells by sialic acid-binding immunoglobulin-like lectin 8 engagement. *J Allergy Clin Immunol* **121:**499–505.e1.doi:10.1016/j.jaci.2007.10.004

213. **Ohnmacht C, Voehringer D.** 2009. Basophil effector function and homeostasis during helminth infection. *Blood* **113:**2816–2825.

214. **Kim S, Prout M, Ramshaw H, Lopez AF, LeGros G, Min B.** 2010. Cutting edge: basophils are transiently recruited into the draining lymph nodes during helminth infection via IL-3, but infection-induced Th2 immunity can develop without basophil lymph node recruitment or IL-3. *J Immunol* **184:**1143–1147.

215. **Shen T, Kim S, Do JS, Wang L, Lantz C, Urban JF, Le Gros G, Min B.** 2008. T cell-derived IL-3 plays key role in parasite infection-induced basophil production but is dispensable for *in vivo* basophil survival. *Int Immunol* **20:**1201–1209.

216. **Siracusa MC, Saenz SA, Hill DA, Kim BS, Headley MB, Doering TA, Wherry EJ, Jessup HK, Siegel LA, Kambayashi T, Dudek EC, Kubo M, Cianferoni A, Spergel JM, Ziegler SF, Comeau MR, Artis D.** 2011. TSLP promotes interleukin-3-independent basophil haematopoiesis and type 2 inflammation. *Nature* **477:** 229–233.

217. **Sokol CL, Barton GM, Farr AG, Medzhitov R.** 2008. A mechanism for the initiation of allergen-induced T helper type 2 responses. *Nat Immunol* **9:**310–318.

218. **Sokol CL, Chu NQ, Yu S, Nish SA, Laufer TM, Medzhitov R.** 2009. Basophils function as antigen-presenting cells for an allergen-induced T helper type 2 response. *Nat Immunol* **10:**713–720.

219. **Marichal T, Bedoret D, Mesnil C, Pichavant M, Goriely S, Trottein F, Cataldo D, Goldman M, Lekeux P, Bureau F, Desmet CJ.** 2010. Interferon response factor 3 is essential for house dust mite-induced airway allergy. *J Allergy Clin Immunol* **126:**836–844.e13. doi: 10.1016/j.jaci.2010.06.009.

220. **Tang H, Cao W, Kasturi SP, Ravindran R, Nakaya HI, Kundu K, Murthy N, Kepler TB, Malissen B, Pulendran B.** 2010. The T helper type 2 response to cysteine proteases requires dendritic cell-basophil cooperation via RAS-mediated signaling. *Nat Immunol* **11:** 608–617.

221. **Ohnmacht C, Schwartz C, Panzer M, Schiedewitz I, Naumann R, Voehringer D.** 2010. Basophils orchestrate chronic allergic dermatitis and protective immunity against helminths. *Immunity* **33:**364–374.

222. **Sullivan BM, Liang HE, Bando JK, Wu D, Cheng LE, McKerrow JK, Allen CD, Locksley RM.** 2011. Genetic analysis of basophil function *in vivo*. *Nat Immunol* **12:** 527–535.

223. **Kitzmuller C, Nagl B, Deifl S, Walterskirchen C, Jahn-Schmid B, Zlabinger GJ, Bohle B.** 2012. Human blood basophils do not act as antigen-presenting cells for the major birch pollen allergen Bet v 1. *Allergy* **67:**593–600.

224. **Sharma M, Hegde P, Aimanianda V, Beau R, Sénéchal H, Poncet P, Latgé JP, Kaveri SV, Bayry J.** 2013. Circulating human basophils lack the features of professional antigen presenting cells. *Sci Rep* **3:**1188. doi:10.1038/srep01188.

225. **Gordon ED, Simpson LJ, Rios CL, Ringel L, Lachowicz-Scroggins ME, Peters MC, Wesolowska-Andersen A, Gonzalez JR, MacLeod HJ, Christian LS, Yuan S, Barry L, Woodruff PG, Ansel KM, Nocka K, Seibold MA, Fahy JV.** 2016. Alternative splicing of interleukin-33 and type 2 inflammation in asthma. *Proc Natl Acad Sci U S A* **113:**8765–8770.

226. **Macfarlane AJ, Kon OM, Smith SJ, Zeibecoglou K, Khan LN, Barata LT, McEuen AR, Buckley MG, Walls AF, Meng Q, Humbert M, Barnes NC, Robinson DS, Ying S, Kay AB.** 2000. Basophils, eosinophils, and mast cells in atopic and nonatopic asthma and in late-phase allergic reactions in the lung and skin. *J Allergy Clin Immunol* **105:**99–107.

227. **Kepley CL, McFeeley PJ, Oliver JM, Lipscomb MF.** 2001. Immunohistochemical detection of human basophils in postmortem cases of fatal asthma. *Am J Respir Crit Care Med* **164:**1053–1058.

228. **Wakahara K, Van VQ, Baba N, Begin P, Rubio M, Delespesse G, Sarfati M.** 2013. Basophils are recruited to inflamed lungs and exacerbate memory Th2 responses in mice and humans. *Allergy* **68:**180–189.

Myeloid Cells in Health and Disease: A Synthesis
Edited by Siamon Gordon

doi:10.1128/microbiolspec.MCHD-0043-2016

Rick M. Maizels[1]
James P. Hewitson[2]

43

Myeloid Cell Phenotypes in Susceptibility and Resistance to Helminth Parasite Infections

INTRODUCTION

The immune system is fundamentally divided into the innate and adaptive arms, predominantly represented by the myeloid and lymphoid lineages, respectively, and largely derived from bone marrow progenitors. This simplistic classification belies an intricate circuitry in which the innate and adaptive cells communicate, stimulate, and regulate each other throughout the course of every immune response. Hence, in every respect myeloid cell populations are instrumental to successful defense against parasitic infections.

Myeloid cells include the heterogeneous monocyte-macrophage lineage, which permeates all tissues of the body, and first emerge as self-renewing progeny of embryonic yolk sac progenitors (1). Subsequent populations of macrophages are derived from the bone marrow (2), as are the closely related dendritic cells (DCs), crucial to initiating immune responses (3); the neutrophils, which are most populous in the circulation; and several other granulocyte subsets (eosinophils, basophils, and mast cells), which expand rapidly in either the bloodstream or tissues during particular parasite infections. In addition, the myeloid cell family includes megakaryocytes, which give rise to platelets in the blood. Each of these cell types is known to play critical roles in one or more parasite infections.

Not surprisingly, parasitic organisms target myeloid cells to divert or block the immune response; some parasitic protozoa, such as *Leishmania* species, *Toxoplasma gondii*, and *Trypanosoma cruzi*, even invade myeloid cells such as neutrophils and macrophages, to survive and propagate in an intracellular lifestyle.

In addition, extracellular parasites such as African trypanosome protozoa and multicellular (metazoan) helminth worms manipulate myeloid cell populations to ensure their survival. The interactions of intracellular parasites with myeloid cells has been dissected and described in fascinating detail (4–7), and hence this review will primarily focus on recent findings implicating the different subsets of myeloid cells in resistance to the metazoan helminth parasites.

Extracellular parasites cause dramatic alterations in host myeloid cell populations (8, 9). Perhaps the first such observation was of >60% peripheral blood

[1]Wellcome Trust Centre for Molecular Parasitology, Institute of Infection, Immunity and Inflammation, University of Glasgow, Glasgow G12 8TA, United Kingdom; [2]Department of Biology, University of York, York YO10 5DD, United Kingdom.

eosinophilia in a patient infected with the nematode *Trichinella spiralis* (10). Eosinophilia is now recognized as an enduring hallmark of helminth infection, although uncertainty remains over the cells' role in eliminating parasites (11, 12). In addition, basophilia is also commonly observed in these infections (13), as is mucosal mast cell hyperplasia in the gut epithelium, where parasites infest the gastrointestinal tract (14).

A more qualitative analysis through molecular markers and gene expression also reveals that each of these myeloid cell types adopts a different phenotype in infections with extracellular parasites, contrasting with the pattern conventionally associated with microbial and intracellular infections. In the case of basophils and eosinophils, this may involve the production of type 2 cytokines such as interleukin-4 (IL-4) (15, 16), while within the macrophage compartment, a distinct profile designated as the alternatively activated macrophage emerges, driven by IL-4 and the related cytokine IL-13 (17). In addition, DCs and neutrophils influenced by the helminth-driven type 2 environment can express gene sets similar to the pattern of alternative activation (18, 19). In this fashion, innate myeloid cells can both set the tone of the adaptive immune response and be instructed by cytokine-producing adaptive cells in the phenotype they adopt. These features will be discussed below for each lineage in turn.

INITIATION OF IMMUNITY: DCs

Helminth infections are the archetypal inducers of the type 2 response, and indeed may have been the selective force that drove the evolution of this mode of immunity (17, 20). The type 2 response begins on a local scale with innate cells (such as innate lymphoid cells) responding to epithelial alarmins (21), but requires the adaptive arm of immunity to gather sufficient strength and attain systemic effects through the differentiation of Th2 cells.

Th2 induction is highly dependent on DCs; for instance, the *in vivo* transfer of bone marrow-derived DCs pulsed with helminth products such as schistosome egg antigen (SEA) (22) or *Nippostrongylus brasiliensis* excretory-secretory antigens (23) is sufficient to stimulate subsequent Th2 differentiation.

Conversely, depletion of $CD11c^+$ DCs *in vivo* greatly impairs Th2 induction in *Schistosoma mansoni* and gut nematode infection (24–26). Despite this, other innate aspects of type 2 immunity, for instance, eosinophilia and alternative macrophage activation, are evoked as normal in a DC-independent manner (26), confirming the unique importance of DCs in recruiting and activating the adaptive immune compartment (27).

DCs represent a heterogeneous set of cells of differing origin and phenotype, suggesting that specialized subsets may be responsible for recognizing and responding to helminth infection. For example, in the dermis, DCs expressing the macrophage galactose-type C-type lectin 2, CD301b, are primarily responding to infection with skin-penetrating larvae of *N. brasiliensis* (28). Conversely, Th2 immunity is elevated in the absence of Batf3-dependent conventional DC (cDC) populations (primarily lymphoid-resident $CD8\alpha^+$ DCs and migratory $CD103^+$ cells), owing to the constitutive production of Th1-promoting IL-12 by these cells (29).

Use of mice with specific defects in particular DC subsets has revealed the importance of interferon regulatory factor 4 (IRF4)-dependent cDC populations, as in animals in which this factor is deleted from the $CD11c^+$ subset, Th2 responses to *N. brasiliensis* are greatly impaired (30). In addition, the Krüppel-like factor KLF4 is also required for normal Th2 responsiveness, and mice lacking this protein within the DCs show poor survival when infected with *S. mansoni* (31). On the other side of the coin, DCs express a surface receptor kinase, Tyro3, that conveys signals that inhibit Th2 induction by the cell; Tyro3-deficient mice mount stronger Th2 responses, clear *N. brasiliensis* more rapidly, and harbor DCs that, when pulsed and transferred into wild-type mice, induce higher levels of type 2 cytokines (32).

In some instances, helminth products can also modulate DCs to drive a stronger regulatory cell component, as *in vitro*, for example, in DC–T-cell cocultures incubated with SEA (33). Similarly, more tolerogenic DCs are induced by coincubation with molecules released by the liver fluke *Fasciola hepatica* (34, 35) and the nematode *T. spiralis* (36, 37). *In vivo*, the immunoregulatory parasite *Heligmosomoides polygyrus* changes the composition of intestinal DCs toward a predominance of $CD11c^{lo}$ cells, which preferentially induce regulatory T cells (Tregs) (25). Interestingly, intestinal DCs from *H. polygyrus*-infected mice could, on transfer into RAG-deficient mice, protect recipients from T-cell-mediated colitis (38). In addition, DCs pulsed with products of the tapeworm *Hymenolepis diminuta* protected recipient mice from pathology in a dinitrobenzene sulfonic acid-induced colitis model (39), and those exposed to *T. spiralis* larval secretions protected from experimental autoimmune encephalitis (40).

Helminth infection also favors DCs adopting an "alternate activation" phenotype (18) akin to that commonly observed in macrophages, and also dependent on IL-4Rα-mediated signaling. In such DCs, there is significant upregulation of Ym1 and resistin-like molecule-α (RELMα) expression, the latter being found to be essential for DC-driven IL-10 production by *in vitro*-polarized Th2 cells.

A major question in the field is how DCs detect the presence of helminth products and discriminate them from microbial organisms to adopt a Th2- (or Treg-) driving program (41). Generally, immune sensing of helminths does not depend on Toll-like receptor (TLR)-mediated interactions and differs from TLR stimulation in key respects. Recognition of SEA by DCs does not upregulate the same pathways of costimulatory surface proteins (e.g., CD40, CD80, and CD86) and inflammatory cytokines (IL-6, IL-12, and tumor necrosis factor) observed when cells encounter a strong TLR ligand such as lipopolysaccharide (LPS) (42). Moreover, some helminth molecules can directly interfere with the response to LPS and other TLR ligands (23, 43–46), raising the question of whether the inability of DCs to fully activate in response to helminths is a host adaptation to this class of parasite or a parasite strategy to dampen host reactivity.

A key component of SEA from schistosome eggs that promotes DC Th2 induction has been identified as a ribonuclease, omega-1, which in native or recombinant form can reproduce the Th2-driving effects of SEA itself (47, 48). Omega-1 is internalized via the mannose receptor, and subsequently degrades RNA within DCs (49), accompanied by cytoskeletal changes within the DC that impair interactions with antigen-specific $CD4^+$ T cells (48). Such low-level DC–T-cell conjugate formation may favor Th2 responses through suboptimal signal delivery. Exposure of DCs to SEA also leads to epigenetic modification crucial for their Th2-polarizing ability, as DCs deficient in methyl-binding protein-2 have altered (predominantly downregulated) gene expression and impaired ability to prime *in vivo* Th2 responses (50).

THE ALTERNATIVELY ACTIVATED MACROPHAGE

Alternatively activated macrophages (AAMs) are those driven through the IL-4/IL-13 type 2 STAT6-dependent pathway, in contrast to cells activated in the classical gamma interferon-dependent manner (51, 52). AAMs are also termed M2 macrophages, in distinction to the classically activated (M1) cells; although inarguably an oversimplification (53), these designations remain useful especially when analyzing *in vivo* macrophage populations in the complex setting of helminth infections.

The AAM phenotype is particularly prominent in parasite infections, having been identified in mice infected with the filarial nematode *Brugia malayi* (54) and subsequently in many other helminth infections (55), as well as in animals infected with the extracellular protozoan parasite *Trypanosoma brucei* (56, 57). In these infections, macrophages present a characteristic pattern of gene expression producing high levels of arginase-1 (Arg-1), RELMα, and the chitinase-like molecule Ym1 (Chi3L3) (58, 59). Macrophage expression of Arg-1 is, for example, essential to inhibit both Th2-mediated liver fibrosis (60) and IL-12/IL-23-dependent gut inflammation in murine schistosomiasis (61). In addition, the metabolism of AAM cells uses oxidative phosphorylation, markedly different from classically activated (M1) macrophages in which the Krebs cycle is interrupted and glycolysis predominates (62).

As discussed above for DCs, helminths and their products are frequently associated with inhibition of the TLR response of macrophages, to the extent that mice infected with the filarial parasite *Litomosoides sigmodontis* show a switch in macrophage phenotype that protects against sepsis during acute bacterial exposure (63).

AAMs may differ from inflammatory M1 macrophages not only in function but also in provenance. Analysis of macrophage populations expanding in the pleural cavity following migration of *L. sigmodontis* showed that stimulation of resident cell division, through IL-4, was the major response to infection (64), in contrast to the M1 inflammatory setting in which circulating monocytes infiltrate into tissue suffering microbial infection. However, this distinction is not absolute and may be either parasite or tissue site specific, since CCR2-dependent monocytes preferentially contribute to the expanded liver AAM population observed in schistosome infection (65, 66). Moreover, while both resident and monocyte-derived macrophages acquired the alternative activation profile in response to IL-4, they differed substantially in transcriptomic profile, and only the blood-derived subset was able to induce FoxP3 expression in T cells (67). Nevertheless, there is ample evidence that macrophages are highly adaptable, acquiring tissue-specific epigenetic marks in response to their environment (68), and are able to adopt similar phenotypes in the tissues irrespective of their anatomical origin (69).

AAMs are of increasing interest also for their physiological roles in homeostasis, repair, and metabolism. These macrophages are required for wound repair in an acute model of helminth parasite tissue damage caused by migrating larvae of *N. brasiliensis* transiting the lung, which is rapidly resolved in wild-type mice but not in immune-deficient SCID mice (70), or IL-4R-deficient animals unable to generate AAMs (70, 71). In addition, hemorrhage and erythrocyte egress into the bronchoalveolar spaces is controlled by macrophages, as depletion with anti-F4/80 antibody caused blood loss in mice that would otherwise be protected by prior immunization (72).

The combination of anti-inflammatory and repair-promoting functions of AAMs and the ability of helminths to induce this cell type has generated much interest in the potential therapeutic use of macrophages conditioned by helminths or by helminth products (73). So far, investigations have been limited to mouse models, but with promising results including inhibition of colitis with macrophages transferred from schistosome-infected mice (74). Most strikingly, *in vitro* treatment of macrophages with a cysteine protease inhibitor, AvCystatin, induced a strongly regulatory population that was able, on transfer to recipient mice, to suppress both airway allergic inflammation and intestinal colitis (75).

Metabolic dysfunction reflected by insulin resistance and obesity has also been linked to the phenotype of macrophages under the influence of helminth parasites. In *N. brasiliensis* infection, activated eosinophils produced IL-4 that in turn induced AAMs in the adipose tissue, which counteracted obesity and maintained glucose tolerance (16). In another study, SEA, which drives a strong AAM differentiation, was found to reduce atherosclerotic plaque formation in hyperlipidemic mice, with increased IL-10 levels from macrophages (76). Hence, helminth modulation of macrophages can also give rise to beneficial physiological consequences for the host.

The AAM phenotype may become imprinted through epigenetic changes; demethylation at the H3K27 residue of histones associated with the AAM-associated genes Arg-1, RELMα, and Ym1 (Chi3L3) is mediated by the Jmjd3 demethylase enzyme, induced by the IL-4/STAT6 pathway (77). Furthermore, *ex vivo* macrophages recovered from mice exposed to schistosome eggs were found to be demethylated at these loci, providing a physiological backdrop to the findings.

MACROPHAGES AS EFFECTOR CELLS

In recent years, strong evidence has emerged that macrophages are key effectors in the antiparasite response. In *H. polygyrus* infection, depletion of phagocytes through clodronate-loaded liposomes compromised both primary (78) and secondary (79) immunity, while transfer of macrophages (activated by *in vitro* IL-33 treatment) induced clearance of parasites (80). In an *in vivo* chamber implantation model, activated AAMs, but not conventionally activated macrophages, could kill larvae of the nematode *Strongyloides stercoralis* (81), while in the lung, *N. brasiliensis* larvae killing is attenuated in mice depleted of interstitial macrophages with anti-F4/80 antibody (72). Moreover, clearance of adult *N. brasiliensis* is also macrophage dependent, as it is ablated in mice treated with clodronate liposomes (82).

Mechanistically, macrophages may directly trap and attack the helminths (83), release key mediators such as Arg-1 (79), or simply produce necessary cytokines at the site of infection (80). Different parasite species are undoubtedly susceptible or resistant to different pathways of attack, perhaps driving the diversity of mechanisms in play. Some parasites even show a contrary profile, with immunity to the cestode tapeworm *Taenia crassiceps* actually enhanced by AAM depletion (84), reflecting that in this relatively unusual case a type 1 response is protective and is inhibited by the immunosuppressive properties of AAMs.

BASOPHILS—RARE OR WELL DONE?

Basophils are FcεR1$^+$ granulocytes that are scarce in uninfected peripheral blood but expand rapidly following helminth infection through IL-3 and thymic stromal lymphopoietin stimulation (85, 86), and populate tissues such as the liver and lung (87), as well as the skin if ectoparasites such as ticks attempt to feed. Recently several basophil-deficient animal models have been reported, ranging from antibody depletion to lineage ablation, which demonstrate, for example, that immunity to ticks is dependent on IgE-armed basophils (88), probably acting through release of granule contents such as the basophil-specific granzyme mast cell serine protease-8 (MCP-8) (89). Basophil-deficient mice, however, retain the ability to expel primary infections with *N. brasiliensis* but lack the rapid expulsion of secondary challenge infections that occurs in wild-type mice (90). Interestingly, in the case of *H. polygyrus*, basophil-deficient mice are fully competent to clear parasites when immunity is induced by vaccination (91) but in the setting of repeated live infections show impaired clearance of challenge parasites (92).

Controversy has surrounded the role of basophils in induction of the Th2 response. While they are among the first cell types to respond to infection through the production of IL-4 and were reported to present antigen to naive T cells (93, 94), basophil depletion or ablation does not compromise the generation of Th2 responses *in vivo* to either schistosomes (24) or intestinal nematodes (26, 95, 96). Together with similar data from the house dust mite allergy model (27), a role for basophils in inducing the antigen-specific Th2 response is now effectively excluded. Nevertheless, basophil-derived IL-4 plays an essential role in the skin to induce alternatively activated macrophages (97), and activation of basophils to release IL-4 is itself sufficient to drive a Th2 response (98). Hence the basophil has evolved a critical role in cutaneous defense against ectoparasites while also being an important contributor to the fully developed type 2 response at the systemic level (95).

MAST CELLS

Mast cells are long-lived tissue-resident cells with a characteristic highly granulated morphology associated with both allergic and antiparasite responses (86, 99); like the basophils to which they are closely related, they are promoted by IL-3 but also IL-9 and stem cell factor, for which they carry the c-Kit receptor. Thus, IL-3 administration can accelerate expulsion of the nematode parasite *Strongyloides* (100), although recently this cytokine has also been linked to alternative activation of macrophages (101), clouding interpretation of the data. Likewise, IL-9 promotes both mastocytosis and expulsion of the *T. spiralis* (102), yet is now also known to expand innate lymphoid cells in helminth infection (103).

Historically, many studies were performed with mutants of c-Kit (such as the *W, Wv,* or *Sash* alleles) that lack mast cells, although again more recently it has emerged that innate lymphoid cells also express this receptor. Nevertheless, c-Kit-deficient mice are more susceptible to most helminth parasites that have been reported (reviewed in reference 104), and in the case of *H. polygyrus*, worm burdens are reduced if these mice receive exogenous mast cells (105), arguing that this cell type is a significant component of antiparasite immunity.

EOSINOPHILS

Eosinophilia is the classic corollary of helminth infection, sufficiently so to be an indicative diagnostic feature. While their close association with helminthiases reflects a common pathway for eosinophil activation (through, for example, IL-5 and eotaxin), the part they play is highly dependent on the parasite in question (12). For some helminths, eosinophils fulfill important protective functions, particularly where they intercept tissue-migrating larvae (106). In another example, eosinophils are required to clear the blood-borne first-stage larval microfilariae of *Brugia malayi* (107). However, in schistosomiasis, despite strong evidence for protective effects *in vitro* and in the semipermissive rat model (108), eosinophil-deficient mice show no difference in the course of *S. mansoni* infection compared to their eosinophil-replete counterparts (109).

Studies with IL-5-overexpressing transgenic mice have also indicated that, in sufficient number, eosinophils can kill migrating *N. brasiliensis* larvae (110); notably, larvae from another species, *Toxocara canis*, are unscathed in these mice, perhaps reflecting a long-standing observation that they slough off adhering eosinophils by shedding their surface coat (111). Hence, a picture emerges of this cell type playing very different roles according to the precise nature of the infective parasite.

A further twist is provided by the case of *T. spiralis* infection, in which eosinophils in fact promote infection, with greater killing of parasite larvae in eosinophil-deficient mice, which can be reversed by eosinophil transfer, and which is attributed to their production of IL-10 to block larvicidal nitric oxide production by other innate myeloid cells (12, 112). This instance reiterates the importance of eosinophils as cytokine-producing cells, including IL-4, which, as mentioned above, is key to the activation of AAMs in adipose tissue for glucose homeostasis (16).

NEUTROPHILS

Classically activated neutrophils are the primary defense against bacteria, which they can engulf and degrade through reactive oxygen intermediates; their role in antiparasite responses is much less well defined. Classical studies with neutrophil-depleting antibodies showed impaired immunity to *H. polygyrus*, while parasite burdens were reduced in mice receiving neutrophils from immune mice (113, 114). In an immunization model, neutrophil depletion had no impact on immunity of vaccinated mice, but worm loads in controls undergoing primary infection were significantly higher in the absence of neutrophils (91). Moreover, recently it was found that neutrophil extracellular traps (NETs) form around larvae of *S. stercoralis* in a mouse model system (115), while antibody-mediated neutrophil depletion reduced the ability of immune mice to intercept skin-penetrating larvae of *N. brasiliensis* (72); notably, these effects are partial rather than complete ablation of protection.

In helminth infections, neutrophils may amplify the type 2 response without being the active agents of worm killing. Thus, in *N. brasiliensis* infection, macrophages from parasite-primed animals were able to transfer protection to naive mice, but only if the donor mice had an intact neutrophil population; depletion of neutrophils negated effective priming of macrophages, which was dependent on neutrophil IL-13 production (19).

A further key role for neutrophils was recently elucidated in the *N. brasiliensis* model, in the context of tissue damage in the lung: the chitinase-like product Ym1 stimulated γδ T cells to produce IL-17, which in turn recruited neutrophils; in the lung setting, neutrophils were able to degrade the parasite larvae, compromising their ability to migrate and mature in the gut. At the same time, neutrophils aggravated the injury to the lung, illustrating a complex balance between immunity and pathology with this cell type at the nexus (116).

MYELOID-DERIVED SUPPRESSOR CELLS

An intriguing parallel exists between tumor-associated macrophages as well as the overlapping populations of myeloid-derived suppressor cells (MDSCs; which may present with either a monocytic or a granulocytic phenotype) (117). Such cells inhibit the protective T-cell response to tumors and are largely promoted by STAT3 and STAT6 signals, including IL-4, IL-10, and IL-13; they also characteristically express Arg-1 in a similar manner to AAMs.

In a novel recent study, it was shown that transfer of granulocytic, but not monocytic, MDSCs induced early expulsion of *N. brasiliensis* (118), in a manner that also depended on recipient expression of wild-type c-Kit alleles, while depletion of MDSCs with gemcitabine resulted in greater worm loads in both *N. brasiliensis* and *T. spiralis* infections (119). In contrast, worm burdens in *H. polygyrus*-infected mice actually increased following adoptive transfer of MDSCs due to greater suppression of the Th2 response (120).

TRAINED IMMUNITY AND INNATE "MEMORY"

A consistent and surprising feature of macrophage activation in the lung of *N. brasiliensis*-infected mice is the longevity of the AAM state; although parasites transit the lung for not much more than 24 h, macrophages at the site appear to make a long-term commitment to the AAM phenotype very evident 1 month postinfection (121), which has detrimental consequences as emphysema develops in the lung up to 300 days following the single episode of helminth disruption (122). These prolonged effects in type 2 conditions may be akin to the new concepts of imprinting activation phenotypes of innate myeloid cells following exposure to inflammatory stimuli such as bacterial LPS (123). The parallel is even more striking in that both type 1 "trained immunity" following microbial exposure and type 2 alternative activation are associated with major epigenetic changes to key genomic loci (77, 124).

HOST-PARASITE COEVOLUTION AND THE INNATE IMMUNE SYSTEM

In conclusion, it is interesting to consider how the dialogue between parasites and the myeloid populations may have evolved. Parasites induce major phenotypic changes in host myeloid populations, but the degree to which this is directed by specific parasite products or results from host response mechanisms remains poorly defined. Some parasite mediators, however, have been identified, for example, the cystatins (cysteine protease inhibitors), which block key antigen-processing enzymes (125) as well as cytokine production in macrophages (126) and DCs (127).

More broadly, the diversity of myeloid cell types has clearly evolved to counter the evolution of many classes of pathogens, including protozoa and helminths; with several specialized cells appearing to target helminths in particular, this may reflect the selective pressure to accommodate, regulate, and survive helminth infections that has so strongly shaped the innate immune system that exists today.

Acknowledgments. *R.M.M. gratefully acknowledges support through a Wellcome Trust Senior Investigator Award (Ref 106122).*

Citation. Maizels RM, Hewitson JP. 2016. Myeloid cell phenotypes in susceptibility and resistance to helminth parasite infections. Microbiol Spectrum 4(5):MCHD-0043-2016.

References

1. **Gomez Perdiguero E, Klapproth K, Schulz C, Busch K, Azzoni E, Crozet L, Garner H, Trouillet C, de Bruijn MF, Geissmann F, Rodewald HR.** 2015. Tissue-resident macrophages originate from yolk-sac-derived erythro-myeloid progenitors. *Nature* **518:**547–551.
2. **Bain CC, Hawley CA, Garner H, Scott CL, Schridde A, Steers NJ, Mack M, Joshi A, Guilliams M, Mowat AM, Geissmann F, Jenkins SJ.** 2016. Long-lived self-renewing bone marrow-derived macrophages displace embryo-derived cells to inhabit adult serous cavities. *Nat Commun* **7:**ncomms11852. doi:10.1038/ncomms11852.
3. **Merad M, Sathe P, Helft J, Miller J, Mortha A.** 2013. The dendritic cell lineage: ontogeny and function of dendritic cells and their subsets in the steady state and the inflamed setting. *Annu Rev Immunol* **31:**563–604.
4. **Bogdan C.** 2008. Mechanisms and consequences of persistence of intracellular pathogens: leishmaniasis as an example. *Cell Microbiol* **10:**1221–1234.

5. **Mashayekhi M, Sandau MM, Dunay IR, Frickel EM, Khan A, Goldszmid RS, Sher A, Ploegh HL, Murphy TL, Sibley LD, Murphy KM.** 2011. CD8α[+] dendritic cells are the critical source of interleukin-12 that controls acute infection by *Toxoplasma gondii* tachyzoites. *Immunity* **35:**249–259.

6. **Ribeiro-Gomes FL, Sacks D.** 2012. The influence of early neutrophil-*Leishmania* interactions on the host immune response to infection. *Front Cell Infect Microbiol* **2:**59. doi:10.3389/fcimb.2012.00059.

7. **Beattie L, d'El-Rei Hermida M, Moore JW, Maroof A, Brown N, Lagos D, Kaye PM.** 2013. A transcriptomic network identified in uninfected macrophages responding to inflammation controls intracellular pathogen survival. *Cell Host Microbe* **14:**357–368.

8. **Cadman ET, Lawrence RA.** 2010. Granulocytes: effector cells or immunomodulators in the immune response to helminth infection? *Parasite Immunol* **32:**1–19.

9. **Barron L, Wynn TA.** 2011. Macrophage activation governs schistosomiasis-induced inflammation and fibrosis. *Eur J Immunol* **41:**2509–2514.

10. **Brown TR.** 1898. Studies on trichinosis, with especial reference to the increase of the eosinophilic cells in the blood and the muscle, the origin of these cells and their diagnostic importance. *J Exp Med* **3:**315–347.

11. **Klion AD, Nutman TB.** 2004. The role of eosinophils in host defense against helminth parasites. *J Allergy Clin Immunol* **113:**30–37.

12. **Huang L, Appleton JA.** 2016. Eosinophils in helminth infection: defenders and dupes. *Trends Parasitol* **32:** 798–807.

13. **Ohnmacht C, Voehringer D.** 2009. Basophil effector function and homeostasis during helminth infection. *Blood* **113:**2816–2825.

14. **Miller HR.** 1996. Mucosal mast cells and the allergic response against nematode parasites. *Vet Immunol Immunopathol* **54:**331–336.

15. **van Panhuys N, Prout M, Forbes E, Min B, Paul WE, Le Gros G.** 2011. Basophils are the major producers of IL-4 during primary helminth infection. *J Immunol* **186:** 2719–2728.

16. **Wu D, Molofsky AB, Liang HE, Ricardo-Gonzalez RR, Jouihan HA, Bando JK, Chawla A, Locksley RM.** 2011. Eosinophils sustain adipose alternatively activated macrophages associated with glucose homeostasis. *Science* **332:**243–247.

17. **Allen JE, Maizels RM.** 2011. Diversity and dialogue in immunity to helminths. *Nat Rev Immunol* **11:**375–388.

18. **Cook PC, Jones LH, Jenkins SJ, Wynn TA, Allen JE, MacDonald AS.** 2012. Alternatively activated dendritic cells regulate CD4[+] T-cell polarization in vitro and in vivo. *Proc Natl Acad Sci U S A* **109:**9977–9982.

19. **Chen F, Wu W, Millman A, Craft JF, Chen E, Patel N, Boucher JL, Urban JF Jr, Kim CC, Gause WC.** 2014. Neutrophils prime a long-lived effector macrophage phenotype that mediates accelerated helminth expulsion. *Nat Immunol* **15:**938–946.

20. **Allen JE, Wynn TA.** 2011. Evolution of Th2 immunity: a rapid repair response to tissue destructive pathogens. *PLoS Pathog* **7:**e1002003. doi:10.1371/journal.ppat.1002003.

21. **Saenz SA, Taylor BC, Artis D.** 2008. Welcome to the neighborhood: epithelial cell-derived cytokines license innate and adaptive immune responses at mucosal sites. *Immunol Rev* **226:**172–190.

22. **MacDonald AS, Straw AD, Bauman B, Pearce EJ.** 2001. CD8[−] dendritic cell activation status plays an integral role in influencing Th2 response development. *J Immunol* **167:**1982–1988.

23. **Balic A, Harcus Y, Holland MJ, Maizels RM.** 2004. Selective maturation of dendritic cells by *Nippostrongylus brasiliensis*-secreted proteins drives Th2 immune responses. *Eur J Immunol* **34:**3047–3059.

24. **Phythian-Adams AT, Cook PC, Lundie RJ, Jones LH, Smith KA, Barr TA, Hochweller K, Anderton SM, Hämmerling GJ, Maizels RM, MacDonald AS.** 2010. CD11c depletion severely disrupts Th2 induction and development *in vivo*. *J Exp Med* **207:**2089–2096.

25. **Smith KA, Hochweller K, Hämmerling GJ, Boon L, Macdonald AS, Maizels RM.** 2011. Chronic helminth infection mediates tolerance in vivo through dominance of CD11c[lo] CD103[−] DC population. *J Immunol* **186:** 7098–7109.

26. **Smith KA, Harcus Y, Garbi N, Hämmerling GJ, MacDonald AS, Maizels RM.** 2012. Type 2 innate immunity in helminth infection is induced redundantly and acts autonomously following CD11c[+] cell depletion. *Infect Immun* 80:3481–3489.

27. **Hammad H, Plantinga M, Deswarte K, Pouliot P, Willart MA, Kool M, Muskens F, Lambrecht BN.** 2010. Inflammatory dendritic cells—not basophils—are necessary and sufficient for induction of Th2 immunity to inhaled house dust mite allergen. *J Exp Med* **207:**2097–2111.

28. **Kumamoto Y, Linehan M, Weinstein JS, Laidlaw BJ, Craft JE, Iwasaki A.** 2013. CD301b[+] dermal dendritic cells drive T helper 2 cell-mediated immunity. *Immunity* **39:**733–743.

29. **Everts B, Tussiwand R, Dreesen L, Fairfax KC, Huang SC, Smith AM, O'Neill CM, Lam WY, Edelson BT, Urban JF Jr, Murphy KM, Pearce EJ.** 2016. Migratory CD103[+] dendritic cells suppress helminth-driven type 2 immunity through constitutive expression of IL-12. *J Exp Med* **213:**35–51.

30. **Gao Y, Nish SA, Jiang R, Hou L, Licona-Limón P, Weinstein JS, Zhao H, Medzhitov R.** 2013. Control of T helper 2 responses by transcription factor IRF4-dependent dendritic cells. *Immunity* **39:**722–732.

31. **Tussiwand R, Everts B, Grajales-Reyes GE, Kretzer NM, Iwata A, Bagaitkar J, Wu X, Wong R, Anderson DA, Murphy TL, Pearce EJ, Murphy KM.** 2015. *Klf4* expression in conventional dendritic cells is required for T helper 2 cell responses. *Immunity* **42:**916–928.

32. **Chan PY, Carrera Silva EA, De Kouchkovsky D, Joannas LD, Hao L, Hu D, Huntsman S, Eng C, Licona-Limón P, Weinstein JS, Herbert DR, Craft JE, Flavell RA, Repetto S, Correale J, Burchard EG, Torgerson DG, Ghosh S, Rothlin CV.** 2016. The TAM family receptor tyrosine kinase TYRO3 is a negative regulator of type 2 immunity. *Science* **352:**99–103.

33. **Zaccone P, Burton O, Miller N, Jones FM, Dunne DW, Cooke A.** 2009. *Schistosoma mansoni* egg antigens induce Treg that participate in diabetes prevention in NOD mice. *Eur J Immunol* **39:**1098–1107.
34. **Dowling DJ, Hamilton CM, Donnelly S, La Course J, Brophy PM, Dalton J, O'Neill SM.** 2010. Major secretory antigens of the helminth *Fasciola hepatica* activate a suppressive dendritic cell phenotype that attenuates Th17 cells but fails to activate Th2 immune responses. *Infect Immun* **78:**793–801.
35. **Falcón C, Carranza F, Martínez FF, Knubel CP, Masih DT, Motrán CC, Cervi L.** 2010. Excretory-secretory products (ESP) from *Fasciola hepatica* induce tolerogenic properties in myeloid dendritic cells. *Vet Immunol Immunopathol* **137:**36–46.
36. **Gruden-Movsesijan A, Ilic N, Colic M, Majstorovic I, Vasilev S, Radovic I, Sofronic-Milosavljevic L.** 2011. The impact of *Trichinella spiralis* excretory-secretory products on dendritic cells. *Comp Immunol Microbiol Infect Dis* **34:**429–439.
37. **Aranzamendi C, Fransen F, Langelaar M, Franssen F, van der Ley P, van Putten JP, Rutten V, Pinelli E.** 2012. *Trichinella spiralis*-secreted products modulate DC functionality and expand regulatory T cells *in vitro*. *Parasite Immunol* **34:**210–223.
38. **Blum AM, Hang L, Setiawan T, Urban JP Jr, Stoyanoff KM, Leung J, Weinstock JV.** 2012. *Heligmosomoides polygyrus bakeri* induces tolerogenic dendritic cells that block colitis and prevent antigen-specific gut T cell responses. *J Immunol* **189:**2512–2520.
39. **Matisz CE, Leung G, Reyes JL, Wang A, Sharkey KA, McKay DM.** 2015. Adoptive transfer of helminth antigen-pulsed dendritic cells protects against the development of experimental colitis in mice. *Eur J Immunol* **45:**3126–3139.
40. **Sofronic-Milosavljevic LJ, Radovic I, Ilic N, Majstorovic I, Cvetkovic J, Gruden-Movsesijan A.** 2013. Application of dendritic cells stimulated with *Trichinella spiralis* excretory-secretory antigens alleviates experimental autoimmune encephalomyelitis. *Med Microbiol Immunol (Berl)* **202:**239–249.
41. **Everts B, Smits HH, Hokke CH, Yazdankbakhsh M.** 2010. Sensing of helminth infections by dendritic cells via pattern recognition receptors and beyond: consequences for T helper 2 and regulatory T cell polarization. *Eur J Immunol* **40:**1525–1537.
42. **Marshall FA, Pearce EJ.** 2008. Uncoupling of induced protein processing from maturation in dendritic cells exposed to a highly antigenic preparation from a helminth parasite. *J Immunol* **181:**7562–7570.
43. **Cervi L, MacDonald AS, Kane C, Dzierszinski F, Pearce EJ.** 2004. Cutting edge: dendritic cells copulsed with microbial and helminth antigens undergo modified maturation, segregate the antigens to distinct intracellular compartments, and concurrently induce microbe-specific Th1 and helminth-specific Th2 responses. *J Immunol* **172:**2016–2020.
44. **Segura M, Su Z, Piccirillo C, Stevenson MM.** 2007. Impairment of dendritic cell function by excretory-secretory products: a potential mechanism for nematode-induced immunosuppression. *Eur J Immunol* **37:**1887–1904.
45. **Langelaar M, Aranzamendi C, Franssen F, Van Der Giessen J, Rutten V, van der Ley P, Pinelli E.** 2009. Suppression of dendritic cell maturation by *Trichinella spiralis* excretory/secretory products. *Parasite Immunol* **31:**641–645.
46. **Terrazas CA, Alcántara-Hernández M, Bonifaz L, Terrazas LI, Satoskar AR.** 2013. Helminth-excreted/secreted products are recognized by multiple receptors on DCs to block the TLR response and bias Th2 polarization in a cRAF dependent pathway. *FASEB J* **27:**4547–4560.
47. **Everts B, Perona-Wright G, Smits HH, Hokke CH, van der Ham AJ, Fitzsimmons CM, Doenhoff MJ, van der Bosch J, Mohrs K, Haas H, Mohrs M, Yazdanbakhsh M, Schramm G.** 2009. Omega-1, a glycoprotein secreted by *Schistosoma mansoni* eggs, drives Th2 responses. *J Exp Med* **206:**1673–1680.
48. **Steinfelder S, Andersen JF, Cannons JL, Feng CG, Joshi M, Dwyer D, Caspar P, Schwartzberg PL, Sher A, Jankovic D.** 2009. The major component in schistosome eggs responsible for conditioning dendritic cells for Th2 polarization is a T2 ribonuclease (omega-1). *J Exp Med* **206:**1681–1690.
49. **Everts B, Hussaarts L, Driessen NN, Meevissen MH, Schramm G, van der Ham AJ, van der Hoeven B, Scholzen T, Burgdorf S, Mohrs M, Pearce EJ, Hokke CH, Haas H, Smits HH, Yazdanbakhsh M.** 2012. Schistosome-derived omega-1 drives Th2 polarization by suppressing protein synthesis following internalization by the mannose receptor. *J Exp Med* **209:**1753–1767, S1.
50. **Cook PC, Owen H, Deaton AM, Borger JG, Brown SL, Clouaire T, Jones GR, Jones LH, Lundie RJ, Marley AK, Morrison VL, Phythian-Adams AT, Wachter E, Webb LM, Sutherland TE, Thomas GD, Grainger JR, Selfridge J, McKenzie AN, Allen JE, Fagerholm SC, Maizels RM, Ivens AC, Bird A, MacDonald AS.** 2015. A dominant role for the methyl-CpG-binding protein Mbd2 in controlling Th2 induction by dendritic cells. *Nat Commun* **6:**6920. doi:10.1038/ncomms7920.
51. **Gordon S.** 2003. Alternative activation of macrophages. *Nat Rev Immunol* **3:**23–35.
52. **Martinez FO, Helming L, Gordon S.** 2009. Alternative activation of macrophages: an immunologic functional perspective. *Annu Rev Immunol* **27:**451–483.
53. **Murray PJ, Allen JE, Biswas SK, Fisher EA, Gilroy DW, Goerdt S, Gordon S, Hamilton JA, Ivashkiv LB, Lawrence T, Locati M, Mantovani A, Martinez FO, Mege JL, Mosser DM, Natoli G, Saeij JP, Schultze JL, Shirey KA, Sica A, Suttles J, Udalova I, van Ginderachter JA, Vogel SN, Wynn TA.** 2014. Macrophage activation and polarization: nomenclature and experimental guidelines. *Immunity* **41:**14–20.
54. **Allen JE, Lawrence RA, Maizels RM.** 1996. Antigen presenting cells from mice harboring the filarial nematode, *Brugia malayi*, prevent cellular proliferation but not cytokine production. *Int Immunol* **8:**143–151.

55. **Kreider T, Anthony RM, Urban JF Jr, Gause WC.** 2007. Alternatively activated macrophages in helminth infections. *Curr Opin Immunol* **19:**448–453.

56. **Raes G, De Baetselier P, Noël W, Beschin A, Brombacher F, Hassanzadeh Gh G.** 2002. Differential expression of FIZZ1 and Ym1 in alternatively versus classically activated macrophages. *J Leukoc Biol* **71:**597–602.

57. **Raes G, Beschin A, Ghassabeh GH, De Baetselier P.** 2007. Alternatively activated macrophages in protozoan infections. *Curr Opin Immunol* **19:**454–459.

58. **Nair MG, Gallagher IJ, Taylor MD, Loke P, Coulson PS, Wilson RA, Maizels RM, Allen JE.** 2005. Chitinase and Fizz family members are a generalized feature of nematode infection with selective upregulation of Ym1 and Fizz1 by antigen-presenting cells. *Infect Immun* **73:** 385–394.

59. **Sutherland TE, Maizels RM, Allen JE.** 2009. Chitinases and chitinase-like proteins: potential therapeutic targets for the treatment of T-helper type 2 allergies. *Clin Exp Allergy* **39:**943–955.

60. **Pesce JT, Ramalingam TR, Mentink-Kane MM, Wilson MS, El Kasmi KC, Smith AM, Thompson RW, Cheever AW, Murray PJ, Wynn TA.** 2009. Arginase-1-expressing macrophages suppress Th2 cytokine-driven inflammation and fibrosis. *PLoS Pathog* **5:**e1000371. doi: 10.1371/journal.ppat.1000371.

61. **Herbert DR, Orekov T, Roloson A, Ilies M, Perkins C, O'Brien W, Cederbaum S, Christianson DW, Zimmermann N, Rothenberg ME, Finkelman FD.** 2010. Arginase I suppresses IL-12/IL-23p40-driven intestinal inflammation during acute schistosomiasis. *J Immunol* **184:** 6438–6446.

62. **O'Neill LA, Pearce EJ.** 2016. Immunometabolism governs dendritic cell and macrophage function. *J Exp Med* **213:**15–23.

63. **Gondorf F, Berbudi A, Buerfent BC, Ajendra J, Bloemker D, Specht S, Schmidt D, Neumann AL, Layland LE, Hoerauf A, Hübner MP.** 2015. Chronic filarial infection provides protection against bacterial sepsis by functionally reprogramming macrophages. *PLoS Pathog* **11:** e1004616. doi:10.1371/journal.ppat.1004616.

64. **Jenkins SJ, Ruckerl D, Cook PC, Jones LH, Finkelman FD, van Rooijen N, MacDonald AS, Allen JE.** 2011. Local macrophage proliferation, rather than recruitment from the blood, is a signature of T_H2 inflammation. *Science* **332:**1284–1288.

65. **Girgis NM, Gundra UM, Ward LN, Cabrera M, Frevert U, Loke P.** 2014. Ly6C^{high} monocytes become alternatively activated macrophages in schistosome granulomas with help from CD4$^+$ cells. *PLoS Pathog* **10:**e1004080. doi:10.1371/journal.ppat.1004080.

66. **Nascimento M, Huang SC, Smith A, Everts B, Lam W, Bassity E, Gautier EL, Randolph GJ, Pearce EJ.** 2014. Ly6C^{hi} monocyte recruitment is responsible for Th2 associated host-protective macrophage accumulation in liver inflammation due to schistosomiasis. *PLoS Pathog* **10:**e1004282. doi:10.1371/journal.ppat.1004282.

67. **Gundra UM, Girgis NM, Ruckerl D, Jenkins S, Ward LN, Kurtz ZD, Wiens KE, Tang MS, Basu-Roy U, Mansukhani A, Allen JE, Loke P.** 2014. Alternatively activated macrophages derived from monocytes and tissue macrophages are phenotypically and functionally distinct. *Blood* **123:**e110–e122. doi:10.1182/blood-2013-08-520619.

68. **Lavin Y, Winter D, Blecher-Gonen R, David E, Keren-Shaul H, Merad M, Jung S, Amit I.** 2014. Tissue-resident macrophage enhancer landscapes are shaped by the local microenvironment. *Cell* **159:**1312–1326.

69. **van de Laar L, Saelens W, De Prijck S, Martens L, Scott CL, Van Isterdael G, Hoffmann E, Beyaert R, Saeys Y, Lambrecht BN, Guilliams M.** 2016. Yolk sac macrophages, fetal liver, and adult monocytes can colonize an empty niche and develop into functional tissue-resident macrophages. *Immunity* **44:**755–768.

70. **Reece JJ, Siracusa MC, Scott AL.** 2006. Innate immune responses to lung-stage helminth infection induce alternatively activated alveolar macrophages. *Infect Immun* **74:**4970–4981.

71. **Chen F, Liu Z, Wu W, Rozo C, Bowdridge S, Millman A, Van Rooijen N, Urban JF Jr, Wynn TA, Gause WC.** 2012. An essential role for T_H2-type responses in limiting acute tissue damage during experimental helminth infection. *Nat Med* **18:**260–266.

72. **Bouchery T, Kyle R, Camberis M, Shepherd A, Filbey K, Smith A, Harvie M, Painter G, Johnston K, Ferguson P, Jain R, Roediger B, Delahunt B, Weninger W, Forbes-Blom E, Le Gros G.** 2015. ILC2s and T cells cooperate to ensure maintenance of M2 macrophages for lung immunity against hookworms. *Nat Commun* **6:** 6970. doi:10.1038/ncomms7970.

73. **Steinfelder S, O'Regan NL, Hartmann S.** 2016. Diplomatic assistance: can helminth-modulated macrophages act as treatment for inflammatory disease? *PLoS Pathog* **12:**e1005480. doi:10.1371/journal.ppat.1005480.

74. **Smith P, Mangan NE, Walsh CM, Fallon RE, McKenzie AN, van Rooijen N, Fallon PG.** 2007. Infection with a helminth parasite prevents experimental colitis via a macrophage-mediated mechanism. *J Immunol* **178:** 4557–4566.

75. **Ziegler T, Rausch S, Steinfelder S, Klotz C, Hepworth MR, Kühl AA, Burda PC, Lucius R, Hartmann S.** 2015. A novel regulatory macrophage induced by a helminth molecule instructs IL-10 in CD4$^+$ T cells and protects against mucosal inflammation. *J Immunol* **194:**1555–1564.

76. **Wolfs IM, Stöger JL, Goossens P, Pöttgens C, Gijbels MJ, Wijnands E, van der Vorst EP, van Gorp P, Beckers L, Engel D, Biessen EA, Kraal G, van Die I, Donners MM, de Winther MP.** 2014. Reprogramming macrophages to an anti-inflammatory phenotype by helminth antigens reduces murine atherosclerosis. *FASEB J* **28:** 288–299.

77. **Ishii M, Wen H, Corsa CA, Liu T, Coelho AL, Allen RM, Carson WF IV, Cavassani KA, Li X, Lukacs NW, Hogaboam CM, Dou Y, Kunkel SL.** 2009. Epigenetic regulation of the alternatively activated macrophage phenotype. *Blood* **114:**3244–3254.

78. **Filbey KJ, Grainger JR, Smith KA, Boon L, van Rooijen N, Harcus Y, Jenkins S, Hewitson JP, Maizels RM.**

2014. Innate and adaptive type 2 immune cell responses in genetically controlled resistance to intestinal helminth infection. *Immunol Cell Biol* 92:436–448.

79. **Anthony RM, Urban JF Jr, Alem F, Hamed HA, Rozo CT, Boucher JL, Van Rooijen N, Gause WC.** 2006. Memory T_H2 cells induce alternatively activated macrophages to mediate protection against nematode parasites. *Nat Med* **12:**955–960.

80. **Yang Z, Grinchuk V, Urban JF Jr, Bohl J, Sun R, Notari L, Yan S, Ramalingam T, Keegan AD, Wynn TA, Shea-Donohue T, Zhao A.** 2013. Macrophages as IL-25/IL-33-responsive cells play an important role in the induction of type 2 immunity. *PLoS One* **8:**e59441. doi:10.1371/journal.pone.0059441.

81. **Bonne-Année S, Kerepesi LA, Hess JA, O'Connell AE, Lok JB, Nolan TJ, Abraham D.** 2013. Human and mouse macrophages collaborate with neutrophils to kill larval *Strongyloides stercoralis*. *Infect Immun* **81:**3346–3355.

82. **Zhao A, Urban JF Jr, Anthony RM, Sun R, Stiltz J, van Rooijen N, Wynn TA, Gause WC, Shea-Donohue T.** 2008. Th2 cytokine-induced alterations in intestinal smooth muscle function depend on alternatively activated macrophages. *Gastroenterology* **135:**217–225.e1. doi:10.1053/j.gastro.2008.03.077.

83. **Esser-von Bieren J, Volpe B, Kulagin M, Sutherland DB, Guiet R, Seitz A, Marsland BJ, Verbeek JS, Harris NL.** 2015. Antibody-mediated trapping of helminth larvae requires CD11b and Fcγ receptor I. *J Immunol* **194:**1154–1163.

84. **Reyes JL, Terrazas CA, Alonso-Trujillo J, van Rooijen N, Satoskar AR, Terrazas LI.** 2010. Early removal of alternatively activated macrophages leads to *Taenia crassiceps* cysticercosis clearance in vivo. *Int J Parasitol* **40:**731–742.

85. **Sullivan BM, Locksley RM.** 2009. Basophils: a nonredundant contributor to host immunity. *Immunity* **30:**12–20.

86. **Voehringer D.** 2013. Protective and pathological roles of mast cells and basophils. *Nat Rev Immunol* **13:**362–375.

87. **Min B, Prout M, Hu-Li J, Zhu J, Jankovic D, Morgan ES, Urban JF Jr, Dvorak AM, Finkelman FD, LeGros G, Paul WE.** 2004. Basophils produce IL-4 and accumulate in tissues after infection with a Th2-inducing parasite. *J Exp Med* **200:**507–517.

88. **Wada T, Ishiwata K, Koseki H, Ishikura T, Ugajin T, Ohnuma N, Obata K, Ishikawa R, Yoshikawa S, Mukai K, Kawano Y, Minegishi Y, Yokozeki H, Watanabe N, Karasuyama H.** 2010. Selective ablation of basophils in mice reveals their nonredundant role in acquired immunity against ticks. *J Clin Invest* **120:**2867–2875.

89. **Karasuyama H, Mukai K, Obata K, Tsujimura Y, Wada T.** 2011. Nonredundant roles of basophils in immunity. *Annu Rev Immunol* **29:**45–69.

90. **Ohnmacht C, Schwartz C, Panzer M, Schiedewitz I, Naumann R, Voehringer D.** 2010. Basophils orchestrate chronic allergic dermatitis and protective immunity against helminths. *Immunity* **33:**364–374.

91. **Hewitson JP, Filbey KJ, Esser-von Bieren J, Camberis M, Schwartz C, Murray J, Reynolds LA, Blair N, Robertson E, Harcus Y, Boon L, Huang SC, Yang L, Tu Y, Miller MJ, Voehringer D, Le Gros G, Harris N, Maizels RM.** 2015. Concerted activity of IgG1 antibodies and IL-4/IL-25-dependent effector cells trap helminth larvae in the tissues following vaccination with defined secreted antigens, providing sterile immunity to challenge infection. *PLoS Pathog* **11:**e1004676. doi:10.1371/journal.ppat.1004676.

92. **Schwartz C, Turqueti-Neves A, Hartmann S, Yu P, Nimmerjahn F, Voehringer D.** 2014. Basophil-mediated protection against gastrointestinal helminths requires IgE-induced cytokine secretion. *Proc Natl Acad Sci U S A* **111:**E5169–E5177. doi:10.1073/pnas.1412663111.

93. **Perrigoue JG, Saenz SA, Siracusa MC, Allenspach EJ, Taylor BC, Giacomin PR, Nair MG, Du Y, Zaph C, van Rooijen N, Comeau MR, Pearce EJ, Laufer TM, Artis D.** 2009. MHC class II-dependent basophil-$CD4^+$ T cell interactions promote T_H2 cytokine-dependent immunity. *Nat Immunol* **10:**697–705.

94. **Sokol CL, Chu NQ, Yu S, Nish SA, Laufer TM, Medzhitov R.** 2009. Basophils function as antigen-presenting cells for an allergen-induced T helper type 2 response. *Nat Immunol* **10:**713–720.

95. **Sullivan BM, Liang HE, Bando JK, Wu D, Cheng LE, McKerrow JK, Allen CD, Locksley RM.** 2011. Genetic analysis of basophil function *in vivo*. *Nat Immunol* **12:**527–535.

96. **Kim S, Prout M, Ramshaw H, Lopez AF, LeGros G, Min B.** 2010. Cutting edge: basophils are transiently recruited into the draining lymph nodes during helminth infection via IL-3, but infection-induced Th2 immunity can develop without basophil lymph node recruitment or IL-3. *J Immunol* **184:**1143–1147.

97. **Egawa M, Mukai K, Yoshikawa S, Iki M, Mukaida N, Kawano Y, Minegishi Y, Karasuyama H.** 2013. Inflammatory monocytes recruited to allergic skin acquire an anti-inflammatory M2 phenotype via basophil-derived interleukin-4. *Immunity* **38:**570–580.

98. **Khodoun MV, Orekhova T, Potter C, Morris S, Finkelman FD.** 2004. Basophils initiate IL-4 production during a memory T-dependent response. *J Exp Med* **200:**857–870.

99. **Abraham SN, St John AL.** 2010. Mast cell-orchestrated immunity to pathogens. *Nat Rev Immunol* **10:**440–452.

100. **Abe T, Nawa Y.** 1988. Worm expulsion and mucosal mast cell response induced by repetitive IL-3 administration in *Strongyloides ratti*-infected nude mice. *Immunology* **63:**181–185.

101. **Borriello F, Longo M, Spinelli R, Pecoraro A, Granata F, Staiano RI, Loffredo S, Spadaro G, Beguinot F, Schroeder J, Marone G.** 2015. IL-3 synergises with basophil-derived IL-4 and IL-13 to promote the alternative activation of human monocytes. *Eur J Immunol* **45:**2042–2051.

102. **Faulkner H, Humphreys N, Renauld J-C, Van Snick J, Grencis R.** 1997. Interleukin-9 is involved in host protective immunity to intestinal nematode infection. *Eur J Immunol* **27:**2536–2540.

103. **Turner J-E, Morrison PJ, Wilhelm C, Wilson M, Ahlfors H, Renauld J-C, Panzer U, Helmby H, Stockinger B.**

2013. IL-9-mediated survival of type 2 innate lymphoid cells promotes damage control in helminth-induced lung inflammation. *J Exp Med* **210**:2951–2965.

104. **Reber LL, Sibilano R, Mukai K, Galli SJ.** 2015. Potential effector and immunoregulatory functions of mast cells in mucosal immunity. *Mucosal Immunol* **8**:444–463.

105. **Hepworth MR, Daniłowicz-Luebert E, Rausch S, Metz M, Klotz C, Maurer M, Hartmann S.** 2012. Mast cells orchestrate type 2 immunity to helminths through regulation of tissue-derived cytokines. *Proc Natl Acad Sci U S A* **109**:6644–6649.

106. **Huang L, Gebreselassie NG, Gagliardo LF, Ruyechan MC, Luber KL, Lee NA, Lee JJ, Appleton JA.** 2015. Eosinophils mediate protective immunity against secondary nematode infection. *J Immunol* **194**:283–290.

107. **Cadman ET, Thysse KA, Bearder S, Cheung AY, Johnston AC, Lee JJ, Lawrence RA.** 2014. Eosinophils are important for protection, immunoregulation and pathology during infection with nematode microfilariae. *PLoS Pathog* **10**:e1003988. doi:10.1371/journal.ppat.1003988.

108. **Capron M, Capron A.** 1992. Effector functions of eosinophils in schistosomiasis. *Mem Inst Oswaldo Cruz* **87**(Suppl 4):167–170.

109. **Swartz JM, Dyer KD, Cheever AW, Ramalingam T, Pesnicak L, Domachowske JB, Lee JJ, Lee NA, Foster PS, Wynn TA, Rosenberg HF.** 2006. *Schistosoma mansoni* infection in eosinophil lineage-ablated mice. *Blood* **108**:2420–2427.

110. **Dent LA, Daly CM, Mayrhofer G, Zimmerman T, Hallett A, Bignold LP, Creaney J, Parsons JC.** 1999. Interleukin-5 transgenic mice show enhanced resistance to primary infections with *Nippostrongylus brasiliensis* but not primary infections with *Toxocara canis*. *Infect Immun* **67**:989–993.

111. **Fattah DI, Maizels RM, McLaren DJ, Spry CJ.** 1986. *Toxocara canis*: interaction of human blood eosinophils with the infective larvae. *Exp Parasitol* **61**:421–431.

112. **Gebreselassie NG, Moorhead AR, Fabre V, Gagliardo LF, Lee NA, Lee JJ, Appleton JA.** 2012. Eosinophils preserve parasitic nematode larvae by regulating local immunity. *J Immunol* **188**:417–425.

113. **Penttila IA, Ey PL, Jenkin CR.** 1984. Infection of mice with *Nematospiroides dubius*: demonstration of neutrophil-mediated immunity *in vivo* in the presence of antibodies. *Immunology* **53**:147–154.

114. **Penttila IA, Ey PL, Jenkin CR.** 1984. Reduced infectivity of *Nematospiroides dubius* larvae after incubation *in vitro* with neutrophils or eosinophils from infected mice and a lack of effect by neutrophils from normal mice. *Parasite Immunol* **6**:295–308.

115. **Bonne-Année S, Kerepesi LA, Hess JA, Wesolowski J, Paumet F, Lok JB, Nolan TJ, Abraham D.** 2014. Extracellular traps are associated with human and mouse neutrophil and macrophage mediated killing of larval *Strongyloides stercoralis*. *Microbes Infect* **16**:502–511.

116. **Sutherland TE, Logan N, Rückerl D, Humbles AA, Allan SM, Papayannopoulos V, Stockinger B, Maizels RM, Allen JE.** 2014. Chitinase-like proteins promote IL-17-mediated neutrophilia in a tradeoff between nematode killing and host damage. *Nat Immunol* **15**: 1116–1125.

117. **Van Ginderachter JA, Beschin A, De Baetselier P, Raes G.** 2010. Myeloid-derived suppressor cells in parasitic infections. *Eur J Immunol* **40**:2976–2985.

118. **Saleem SJ, Martin RK, Morales JK, Sturgill JL, Gibb DR, Graham L, Bear HD, Manjili MH, Ryan JJ, Conrad DH.** 2012. Cutting edge: mast cells critically augment myeloid-derived suppressor cell activity. *J Immunol* **189**:511–515.

119. **Morales JK, Saleem SJ, Martin RK, Saunders BL, Barnstein BO, Faber TW, Pullen NA, Kolawole EM, Brooks KB, Norton SK, Sturgill J, Graham L, Bear HD, Urban JF Jr, Lantz CS, Conrad DH, Ryan JJ.** 2014. Myeloid-derived suppressor cells enhance IgE-mediated mast cell responses. *J Leukoc Biol* **95**:643–650.

120. **Valanparambil RM, Tam M, Jardim A, Geary TG, Stevenson MM.** 2016. Primary *Heligmosomoides polygyrus bakeri* infection induces myeloid-derived suppressor cells that suppress $CD4^+$ Th2 responses and promote chronic infection. *Mucosal Immunol* doi: 10.1038/mi.2016.36.

121. **Reece JJ, Siracusa MC, Southard TL, Brayton CF, Urban JF Jr, Scott AL.** 2008. Hookworm-induced persistent changes to the immunological environment of the lung. *Infect Immun* **76**:3511–3524.

122. **Marsland BJ, Kurrer M, Reissmann R, Harris NL, Kopf M.** 2008. *Nippostrongylus brasiliensis* infection leads to the development of emphysema associated with the induction of alternatively activated macrophages. *Eur J Immunol* **38**:479–488.

123. **Netea MG, Quintin J, van der Meer JW.** 2011. Trained immunity: a memory for innate host defense. *Cell Host Microbe* **9**:355–361.

124. **Saeed S, Quintin J, Kerstens HH, Rao NA, Aghajanirefah A, Matarese F, Cheng SC, Ratter J, Berentsen K, van der Ent MA, Sharifi N, Janssen-Megens EM, Ter Huurne M, Mandoli A, van Schaik T, Ng A, Burden F, Downes K, Frontini M, Kumar V, Giamarellos-Bourboulis EJ, Ouwehand WH, van der Meer JW, Joosten LA, Wijmenga C, Martens JH, Xavier RJ, Logie C, Netea MG, Stunnenberg HG.** 2014. Epigenetic programming of monocyte-to-macrophage differentiation and trained innate immunity. *Science* **345**:1251086. doi:10.1126/science.1251086.

125. **Manoury B, Gregory WF, Maizels RM, Watts C.** 2001. *Bm*-CPI-2, a cystatin homolog secreted by the filarial parasite *Brugia malayi*, inhibits class II MHC-restricted antigen processing. *Curr Biol* **11**:447–451.

126. **Klotz C, Ziegler T, Figueiredo AS, Rausch S, Hepworth MR, Obsivac N, Sers C, Lang R, Hammerstein P, Lucius R, Hartmann S.** 2011. A helminth immunomodulator exploits host signaling events to regulate cytokine production in macrophages. *PLoS Pathog* **7**:e1001248. doi:10.1371/journal.ppat.1001248.

127. **Sun Y, Liu G, Li Z, Chen Y, Liu Y, Liu B, Su Z.** 2012. Modulation of dendritic cell function and immune response by cysteine protease inhibitor from murine nematode parasite *Heligmosomoides polygyrus*. *Immunology* **138**:370–381.

Myeloid Cells in Health and Disease: A Synthesis
Edited by Siamon Gordon

doi:10.1128/microbiolspec.MCHD-0019-2015

Herman Waldmann[1]
Duncan Howie[1]
Stephen Cobbold[1]

44 Induction of Immunological Tolerance as a Therapeutic Procedure

INTRODUCTION

A major goal of immunosuppressive therapy in management of chronic inflammatory diseases and allogeneic transplants has been to harness long-term tolerance processes from short-term treatments. This should limit morbidity from long-term undermining of immune mechanisms, which is the hallmark of current immunosuppression.

Historically, dendritic cells (DCs) have represented one arm of the innate immune system; these cells need to interact with the adaptive system to provide immunity from microbial infection. More recently, myeloid cells have also been seen as partners for acquisition of immunological tolerance in adaptive lymphocytes. Not only can they ensure appropriate antigen presentation to induce tolerance, but evidence is accumulating that they can also act as active participants to regulate or suppress the adaptive system. In this article, we examine the interplay between lymphocytes and DCs as the basis for therapeutic reprogramming of the immune system toward tolerance.

A long-standing dogma of immunology was that self-tolerance was mediated by clonal inactivation and deletion of antigen-specific lymphocytes. Much of this was thought to occur in the primary lymphoid organs, with the job completed in the so-called peripheral immune system. The identification of patients suffering from autoimmune diseases associated with the APECED (autoimmune polyendocrinopathy-candidiasis-ectodermal dystrophy) syndrome, due to mutations in the *AIRE* gene, provided confirmation of the importance of antigen presentation in the thymus as a key primary lymphoid organ engaged in tolerogenesis (1). Much work in the early 1990s pointed to additional regulatory mechanisms that operated to prevent autoimmune disease and gut immunopathology, as well as prevention of graft rejection. The clinical relevance of these early studies was established clearly with the identification of the IPEX (immune dysregulation, polyendocrinopathy, enteropathy, X-linked) syndrome (2, 3), in which pathology resulted from defects in expression of the *FOXP3* gene, which determines in large

[1]Sir William Dunn School of Pathology, Oxford OX1 3RE, United Kingdom.

part the phenotype of a subset of CD4 T cells that have come to be referred to as regulatory T cells (Tregs).

The advent of monoclonal antibodies (MAbs) directed to many surface molecules of T cells not only provided magic bullets to ablate lymphocytes and their subsets but also allowed for the identification of key molecules involved in translating immunogenic signals from DCs to T lymphocytes. Nonlytic antibodies were developed to block these signals (4). Using these probes, it has been possible, in preclinical models, to induce tolerance to transplants and to restore tolerance in autoimmune disease models.

In this chapter, we will describe what we know about mechanisms that underlie such therapeutic tolerance. The harnessing of Foxp3$^+$ regulatory T cells emerges as a major feature of one form of therapeutic tolerance in these experimental systems, and orchestration of protective immunosuppressive processes in the protected tissues has a key role for innate cells. Parallels can likely be drawn between the changed microenvironment of tolerized tissues and those of tumors. We will focus largely on studies using coreceptor and costimulation blockade with CD4, CD8, and CD154 (anti-CD40L) antibodies in mouse models of allogeneic transplantation to draw mechanistic conclusions. The authors underwent major conceptual conversions in the course of this work, and we thought it would be of value to the readers of this chapter for us to give a historical overview of how that conversion took place. In particular, we wish to highlight the importance of regulatory activity operating within tolerated tissues.

Until the late 1980s immunologists favored deletional/inactivation (passive) models of tolerance and had come to regard the idea of T-cell-mediated suppression (active tolerance) as "unsafe." Against this background we were attempting to induce therapeutic tolerance to foreign proteins and to transplanted tissues, and became confronted with data that seemed incompatible with the passive interpretation. The findings, starting some 30 years ago, and summarized below, have led us to the conclusion that therapeutic tolerance can be achieved when regulatory T cells are empowered to dominate the immune response to antigen.

TOLERANCE TO FOREIGN PROTEINS CAN BE ACHIEVED WITH A SHORT PULSE OF MAbs

Our work began with the observation that rat MAbs to mouse CD4 were not, unlike other antibodies targeting lymphocytes, immunogenic (5, 6). They could, furthermore, induce tolerance to other therapeutic antibodies and to aggregated human IgG, normally a strong immunogen in mice. Particular nonlytic anti-CD4 antibodies could even induce tolerance without depleting CD4 T helper cells (7–9).

The tolerance achieved was, however, not permanent and was lost over time unless, paradoxically, animals were rechallenged with the immunogen (9). In other words, tolerance involved "memory" for the tolerizing antigen. Also, and quite surprising to us, tolerance could not be broken by transfusions of normal lymphocytes—a phenomenon for which we coined the term "resistance." At that time, it was hard to think of this form of tolerance as anything other than active, but given the then-current skepticism over suppression, we compromised on an explanation that we called the "civil service model" (10, 11). We proposed that T cells, inactivated or anergized by antigen, competed with competent T cells for antigen or activation molecules in the local microenvironment of individual "myeloid" DCs. Strong competition would result in strong suppression.

INDUCTION OF TOLERANCE TO TRANSPLANTED TISSUES

The observation that memory of tolerance could be sustained by booster doses of immunogen led us to speculate that coreceptor blockade might also enable tolerance to transplants, where the engrafted tissue would provide continuous supplies of the "booster" antigens. That indeed proved to be the case for skin grafts mismatched for multiple minor antigens (minors), as well as for major histocompatibility complex (MHC)-mismatched grafts (12).

Tolerance could even be induced this way in mice previously primed to minors (13). Tolerance across MHC barriers was somewhat more difficult to achieve and could be enhanced by costimulation blockade through the addition of an antibody to the costimulatory ligand CD40L (CD154) (14). Once again we observed "resistance," as tolerance could not be overcome by infusion of normal lymphocytes (12). Strikingly, though, we could also demonstrate that the tolerance could spread to third-party antigens when these were presented together with the tolerated antigens in the challenge grafts (linked suppression) (15, 16) (Fig. 1). Functional studies on blood T cells and splenocytes indicated that "passive" tolerance was unlikely as mixed lymphocyte reaction (MLR), cytotoxic-T-lymphocyte generation, and interleukin-2 (IL-2), IL-4, and gamma interferon cytokine release were comparable between tolerized hosts and control mice (17, 18). This left us having to consider that a significant component of

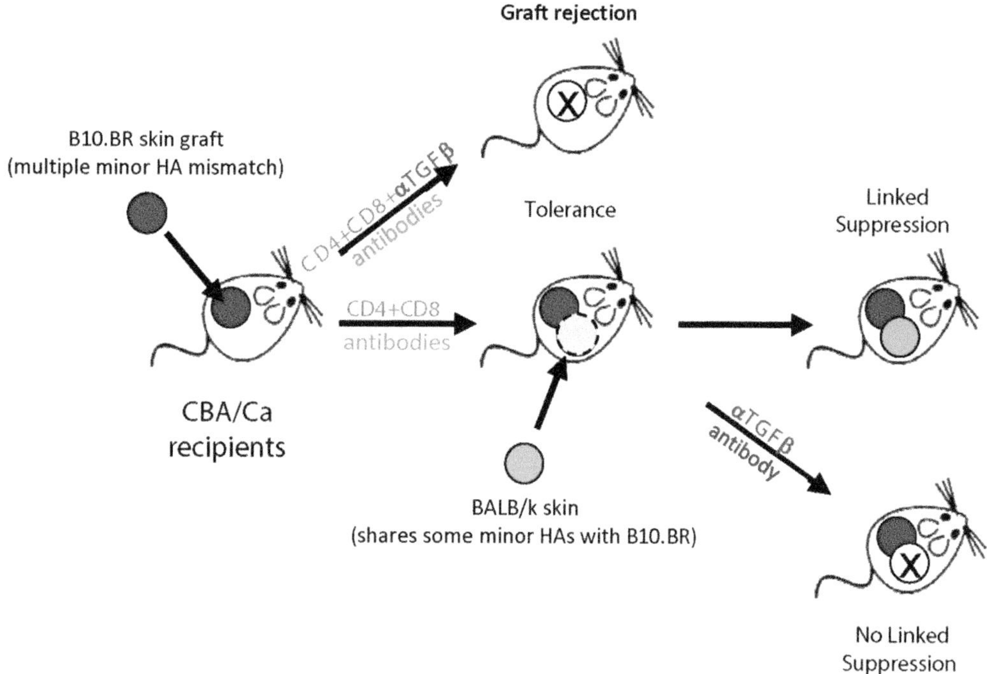

Figure 1 Coreceptor blockade induces transplantation tolerance and linked suppression in adult mice. CBA/Ca mice would normally rapidly reject skin grafts from either B10.BR or BALB/k mice, as they differ in minor histocompatibility antigens. If the recipients are given a brief course of nondepleting MAbs that block the CD4 and CD8 coreceptors at the time of grafting, B10.BR skin grafts are, however, permanently accepted. Remarkably, such mice made tolerant of B10.BR skin can accept third-party grafts, even from "multiple minor" different BALB/k donors, if they share sufficient antigens in common with the tolerated graft, in a process known as linked suppression. Both the induction of tolerance and the process of linked suppression are blocked by antibodies that neutralize active TGF-β.

therapeutic tolerance might operate in the tolerated tissue or its draining lymph nodes. How could that be?

TRANSFERABLE SUPPRESSION BY CD4 T CELLS

Despite the universal skepticism about suppressor T cells that was rife at the time, we performed adoptive transfer studies to ask if tolerant T cells might suppress rejection by naive T cells. Not only did this turn out to be the case, but the active suppressors turned out to be $CD4^+$ T cells (13, 19, 20) and not $CD8^+$ T cells, as the discredited old literature had claimed.

We asked whether a tolerized immune system removed from its tolerated antigens would stay tolerant. The answer proved unequivocal. Tolerance was lost once lymphocytes were removed from a source of persisting antigen (12, 13, 17, 21). This definitively established that tolerance could not have arisen just through passive deletion of all antigen-specific T cells. The alternatives were that tolerance involved a reversible inactivation of T cells or dominance of an active suppressive or regulatory population, that persisting dominance being dependent on a continuous supply of antigen.

If the latter were the case, would the need for antigen be to provide boosts of the first cohort of $CD4^+$ Tregs, or might it be required to recruit further cohorts of Tregs into graft protection, or perhaps both? By using marked populations of CD4 T cells, we were able to show that new cohorts of $CD4^+$ Tregs were being continuously recruited within tolerant animals, under the influence of previous cohorts (19, 22). We dubbed this form of regulation "infectious tolerance." T cells mediating infectious tolerance were also responsible for linked suppression, suggesting that these two phenomena were all manifestations of the same $CD4^+$ Tregs (23) (Fig. 2).

HOW MIGHT ANTIBODY THERAPY BE FAVORING THE EMERGENCE OF Tregs?

We asked whether the tolerizing effects of antibody blockade were immediate or whether they required

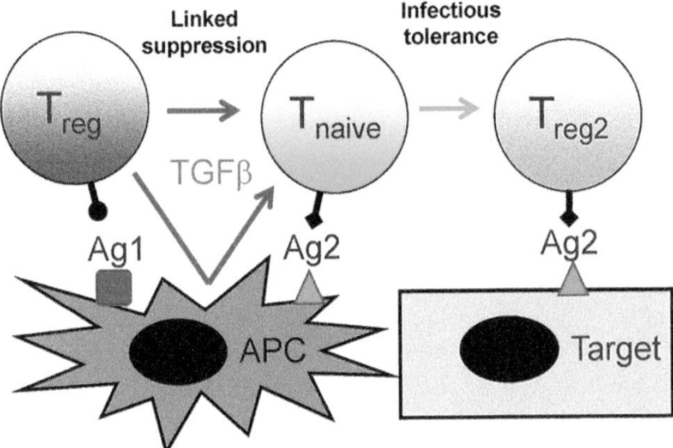

Figure 2 Linked suppression depends on an interaction between regulatory T cells and antigen-presenting cells (APCs). The copresentation of tolerated (Ag1) and third-party (Ag2) antigen by the same APCs promotes an interaction between the Tregs maintaining tolerance and naive T cells that would otherwise have the potential to develop into effector cells against tissues expressing the target antigen. This "linking" of the two antigens by the APC allows the Treg to suppress the naive T cell (i.e., linked suppression) and, through the action of TGF-β (and other additional mechanisms, not shown), to also guide the naive T cell to differentiate into a second cohort of regulatory T cells, thereby further enforcing tolerance to the target antigen (i.e., infectious tolerance).

time to develop. Splenocytes from mice given the tolerizing antibody protocol were transferred into lymphopenic hosts (away from the therapeutic agent) at different times during the treatment regimen. In the case of tolerance to multiple minor antigens, we observed that it took some 3 weeks of antibody administration before sampled splenocytes behaved as if tolerant (21). Before then, they remained fully capable of rejecting donor grafts. In other words, tolerance could not be induced immediately but required some time to reach completion.

This led us to conclude that the therapeutic antibodies likely operated by fulfilling two requirements. First, they fully controlled T-cell rejection events by blindfolding T cells for as long as it took the immune system to (secondly) establish a numerical or functional advantage for Tregs. The issue of how and where regulation operated required some way of distinguishing Tregs from other T cells, and this will be discussed later.

HOW IS ANTIGEN PRESENTED TO MAINTAIN TOLERANCE AND REGULATION?

We were able to implicate host DCs as key elements in ensuring continued regulation by demonstrating that graft antigens could be processed and presented by DCs to induce tolerance and to amplify Tregs (16, 24–26) (Fig. 3). We were accustomed to the notion that DCs present antigen for immunity, but were here having to postulate that in the course of "infectious tolerance" host DCs were also presenting for tolerance and regulation.

To reconcile how this could be, we envisaged that host DCs were processing antigen from well-healed grafts that would be free of inflammatory signals (danger-free), either from modulation of DC function or simply lack of activation signals. In such an environment where inflammatory quiescence was also enforced by Tregs, the DCs might only be able to present antigen in a way that would encourage tolerance and Tregs but not provoke damaging immunity. This tolerogenic presentation of antigen by myeloid cells could then be at the root of infectious tolerance. In short, continuous antigen presentation by DCs that are compromised for activation enables them to maintain the active tolerant state (16).

DCs "DECOMMISSIONED" BY IL-10 AND TGF-β CAN PRESENT FOR TOLERANCE AND INDUCE Tregs IN THE PERIPHERY

The notion that resting DCs, or DCs decommissioned from activation, could present for tolerance was tested by asking whether IL-10- or transforming growth factor β (TGF-β)-conditioned DCs could present Dby (male) peptide for tolerance rather than for immunity in the same TCR (T-cell receptor) transgenic mice as discussed above (25). Presentation by decommissioned DCs did indeed allow tolerance, and this was associated with induction of peripheral Tregs in one TCR transgenic strain of mice but not in another. This finding in the one strain was consistent with earlier studies showing that DCs can constitutively present antigen for tolerance and regulation (27) but (on the basis of the differing result in the other strain) that outcome would depend on properties of the responding T cell.

Foxp3 AS A MARKER OF REGULATORY T CELLS

Until some 15 years ago, it had been difficult to pursue the biology of regulatory T cells because of the lack of a surface marker by which they could be manipulated. The expression of $CD25^{hi}$ (28) or $CD45RB^{lo}$ (29) offered some opportunities, but it was only when the forkhead transcription factor Foxp3 was identified as a unique marker for Tregs (2, 30–32) that research into the mechanisms underlying transplantation tolerance could progress further.

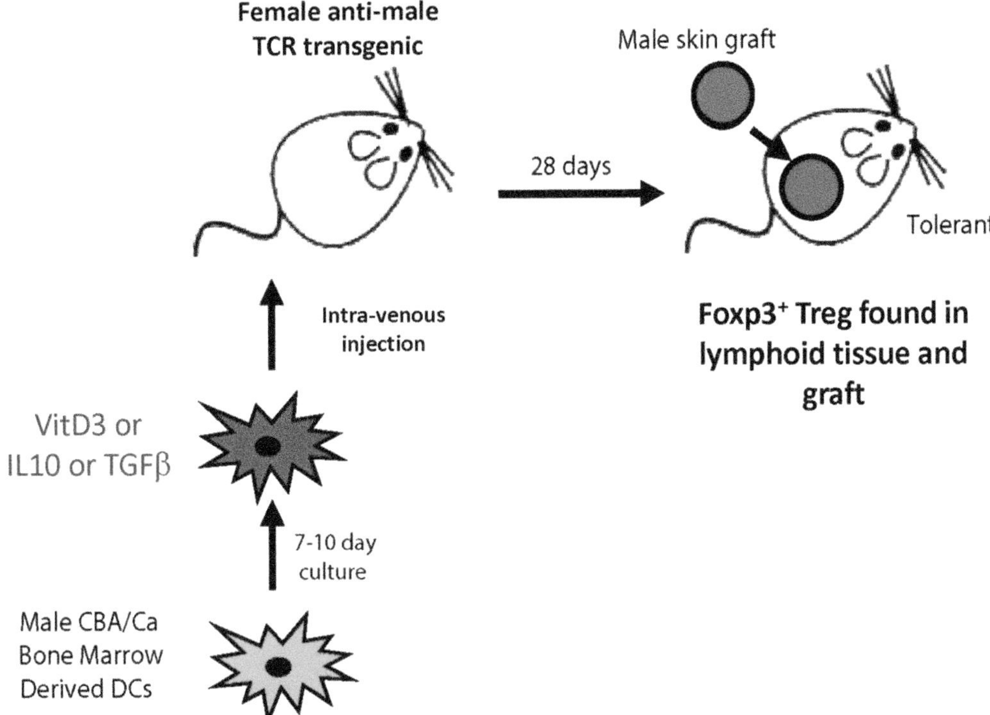

Figure 3 Tolerance can be induced by pharmacological modulation of DCs. Female anti-male TCR transgenic mice would normally reject male skin grafts rapidly. If such mice are administered mature (lipopolysaccharide-activated), bone marrow-derived DCs from male mice, they are further primed for graft rejection. In contrast, pretreatment of the bone marrow-derived DCs by any one of a number of modulatory agents, such as VitD3, IL-10, or TGF-β (which suppress DC maturation), induces a state of tolerance to the male antigen. Such tolerant mice now accept male skin grafts indefinitely, and while this is not associated with deletion of male-specific T cells, it does lead to the systemic induction of Foxp3$^+$ anti-male Tregs.

The need for reductionist systems to study therapeutic transplantation tolerance led us to use TCR transgenic mice whose only adaptive lymphocytes were CD4 T cells carrying a TCR directed to a single minor transplantation antigen (the male antigen Dby) (33–35). Female mice carried no T cells expressing Foxp3, as they could not generate the natural Tregs that conventional mice produce in their thymus. Females could be tolerized to male skin grafts by a short pulse of anti-CD4 treatment (18). Despite exhibiting tolerance, many TCR$^+$ CD4 T cells could still be detected in their lymphoid organs. Foxp3$^+$ T cells could be seen to accumulate over time in the periphery of these mice, with a particularly high frequency within the tolerated tissue itself. By definition, these must have been peripheral Tregs, induced outside of the thymus.

A precedent for extrathymic Tregs had been postulated by Wanjun Chen, who observed that CD4 T cells whose TCR was cross-linked *in vitro* in the presence of TGF-β would also convert to expression of Foxp3 (36). Consistent with our Foxp3$^+$ cells in tolerant mice, cells being programmed by TGF-β, we showed that tolerance and linked suppression were not inducible if TGF-β was neutralized by an appropriate antibody (Fig. 1), and that TCR transgenic T cells would convert to Foxp3 expression and suppressive function when exposed to antigen and TGF-β *in vitro* (14, 18, 37). In contrast, therapeutic tolerance could not be induced in TCR transgenic mice genetically defective in TGF-β-mediated signaling to T cells (14, 37). To establish that it was not just TGF-β but the Foxp3 induced by TGF-β that mattered, we went on to show that TCR transgenic mice lacking a functional Foxp3 gene failed to be tolerized by CD4 blockade (37).

FUNCTIONING Tregs CAN BE FOUND WITHIN THE TOLERATED GRAFT

We asked whether functional Tregs could be found within a tolerated graft. To do this we retransplanted tolerated grafts onto lymphopenic mice, waited sufficient time for their immune system to reconstitute from the graft, and then tested these mice for "resistance" to breakdown of tolerance following infusion of naive lymphocytes (38). "Resistance" is what we observed. This experiment was the first to show Treg localization to tolerated grafts and was a vindication of our early thinking that active suppression must have a strong local tissue element to it.

The importance of Foxp3 expression in tolerance induction was further established when we were able to exploit a cell surface marker (human CD2; hCD2) in a novel transgenic strain in which a cDNA encoding the hCD2 was knocked in downstream of the Foxp3 locus (39). After exposure to antigen in the presence of TGF-β *in vitro*, only the T cells expressing Foxp3 were able to suppress transplant rejection *in vivo*, while their Foxp3-negative counterparts could not (37). Use of an anti-hCD2 antibody to ablate Tregs *in vivo* showed us that intragraft Tregs were active in ensuring graft protection from potential aggressors also located in the graft (40) (Fig. 4).

Finally, using the hCD2 marker and an adoptive transfer protocol, we showed that some TCR transgenic CD4 T cells that had been "suppressed" had converted to Foxp3 expression (40, 41). In other words, "infectious tolerance" could be explained, at least in part, by the first cohort of Tregs enabling naive T cells to convert to peripheral Tregs under the influence of TGF-β.

THERAPEUTIC TOLERANCE REQUIRES CONSTANT VIGILANCE FROM Tregs

With this information from TCR transgenic mice, we went back to conventional mice that had been tolerized to skin grafts and, using ablative antibodies, could demonstrate that the suppression we had seen with adoptive transfer of splenic T cells some 20 years earlier was

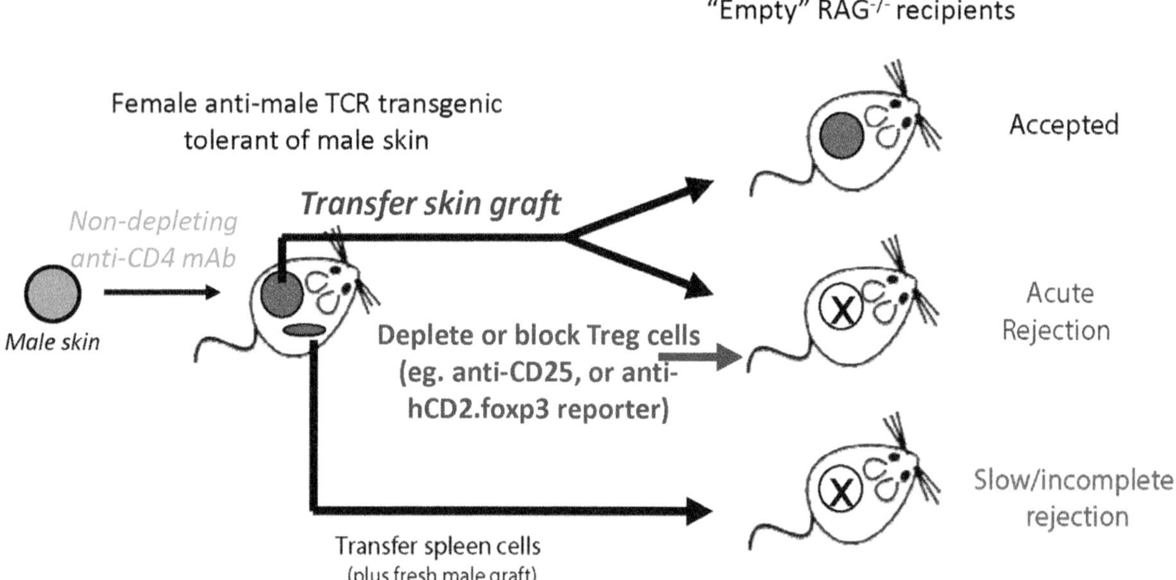

Figure 4 Regulatory T cells actively maintain tolerance within accepted skin grafts. Female anti-male TCR transgenic mice can be made tolerant of male skin grafts by a single injection of nondepleting anti-CD4 MAb. This is associated with the induction of Foxp3$^+$ Tregs, which are particularly concentrated in the tolerated skin but are often only at a low frequency systemically. This low frequency may not be sufficient to transfer tolerance with spleen cells (unless boosted by systemic antigen), but if the tolerated graft itself is adoptively transferred to empty mice, the grafts are accepted. This is not just passive acceptance by the mice that have no adaptive immune system of their own, because the administration of antibodies that block or deplete Tregs (e.g., antibodies to CD25, the hCD2. Foxp3 reporter when hCD2. Foxp3 reporter mice were the original graft recipients, or other functional blocking antibodies such as anti-CTLA4 or anti-TGF-β) all lead to rapid graft rejection. This demonstrates that Tregs are required to actively and continuously block the action of effector T cells that are also present in the grafted and tolerated skin.

due to Foxp3+ cells (40). This established clearly that suppression was a continually active process in tolerant animals maintaining constant vigilance over residual effector T cells capable of rejection.

AIMING FOR AN OVERALL SCHEME TO EXPLAIN THERAPEUTIC TOLERANCE

In studies of gene expression between DCs interacting with Tregs, and between tolerated and rejecting skin grafts, we observed an upregulation of enzymes that can catabolize essential amino acids (EAAs) (42). We wondered whether local depletion of EAAs by these enzymes might control the immune responses through a nutrient-sensing mechanism within the mTOR (mammalian target of rapamycin) pathway. A precedent for such thinking comes from the pioneering work of Mellor on the role of indoleamine 2,3-dioxygenase (IDO)-mediated tryptophan catabolism in the control of maternal alloreactivity to the fetus (43, 44). The suggestion had been that tryptophan depletion was sensed by general control nonrepressed 2 (GCN2) through the integrated stress response. We had noticed, however, that the induction of Foxp3 in naive CD4+ T cells in the presence of low doses of TGF-β *in vitro* was not affected by activating the GCN2 pathway, whereas inhibition of the mTOR pathway using rapamycin enhanced Foxp3 expression. This observation offered us one attractive route by which tolerance and regulation might be maintained: through depletion of essential nutrients at sites of antigen encounter (Fig. 5).

DEPLETION OF EAAs PROVIDES ONE ROUTE TO MAINTAINING A TOLEROGENIC MICROENVIRONMENT IN LYMPHOID ORGANS AND WITHIN TOLERATED TISSUES

The identification of IDO as a player in acquired tolerance has offered the clearest example of immune regulation due to amino acid catabolism, perhaps because tryptophan is normally at the lowest concentration in body fluids when compared to other EAAs. This might explain why various routes to catabolizing tryptophan have been associated with graft protection in models of transplantation tolerance. Arginase (ARG1) expression has also been implicated in regulating maternal alloreactivity during pregnancy (45, 46), and is also described as upregulated in type 2 macrophages (47) within tissues (48). When arginine is limiting, arginase and also inducible nitric oxide synthase can reduce levels of this EAA further, to the point of causing mTOR inhibition and reduced T-cell effector function (49). IL4-induced 1 (IL4i1), known to be induced in myeloid cells under Th2 conditions and able to deplete EAAs such as phenylalanine, is also upregulated in DCs cocultured with Tregs (49).

We reported that many of these EAA-consuming enzymes could be induced in DCs *in vitro* by a cognate interaction with Tregs, through cytokines such as gamma interferon, IL-4, or TGF-β or via coinhibitory molecules such as cytotoxic-T-lymphocyte-associated antigen 4 (CTLA4) (49). In addition, among these induced enzymes, threonine-catabolizing enzymes such as threonine dehydrogenase and Bcat1 (branched-chain amino acid aminotransferase) appeared upregulated soon after skin grafting, even in grafts placed onto recipients with no adaptive immune system. It may be then that many tissues carry innate nutrient-sensing mechanisms to protect them against inflammatory damage, with that mechanism contributing to the induction and maintenance of tolerance.

One such cell source might be mast cells, which seem to be needed for induction of therapeutic tolerance (34, 50, 51), and also to allow transplantation of syngeneic tumor cell lines (52). Tryptophan hydroxylase within murine mast cells may be responsible for this "innate" tolerance (53).

We propose, therefore, that tolerance is maintained, at least in part, by Tregs that perpetuate tolerogenic microenvironments within lymphoid organs and tissues, where induction of diverse enzymes deplete local EAAs. The resulting nutrient deficiency is then being sensed by T cells through the mTOR pathway (49, 54). This would result in the inhibition of effector T-cell priming and function, while Foxp3+ Tregs would be selectively amplified. This pathway based on "infectious" nutrient deficiency may offer one route to "infectious tolerance" within the host.

If the above were an avenue by which tolerance could be induced and maintained, how might mTOR inhibition bring about the necessary controlling effects on the immune system?

mTOR INTEGRATES NUTRIENT SENSING AND ACTIVATION OF T CELLS

The mTOR pathway (Fig. 6) coordinates cell growth and metabolism. mTOR comprises two different signaling complexes (TORC1 and TORC2) (55, 56). The former contributes the nutrient-sensing complex and comprises the serine/threonine kinase mTOR itself, raptor, FKB12, deptor, mLST8, and the regulatory subunit PRAS40, a target of AKT downstream of phosphatidylinositol

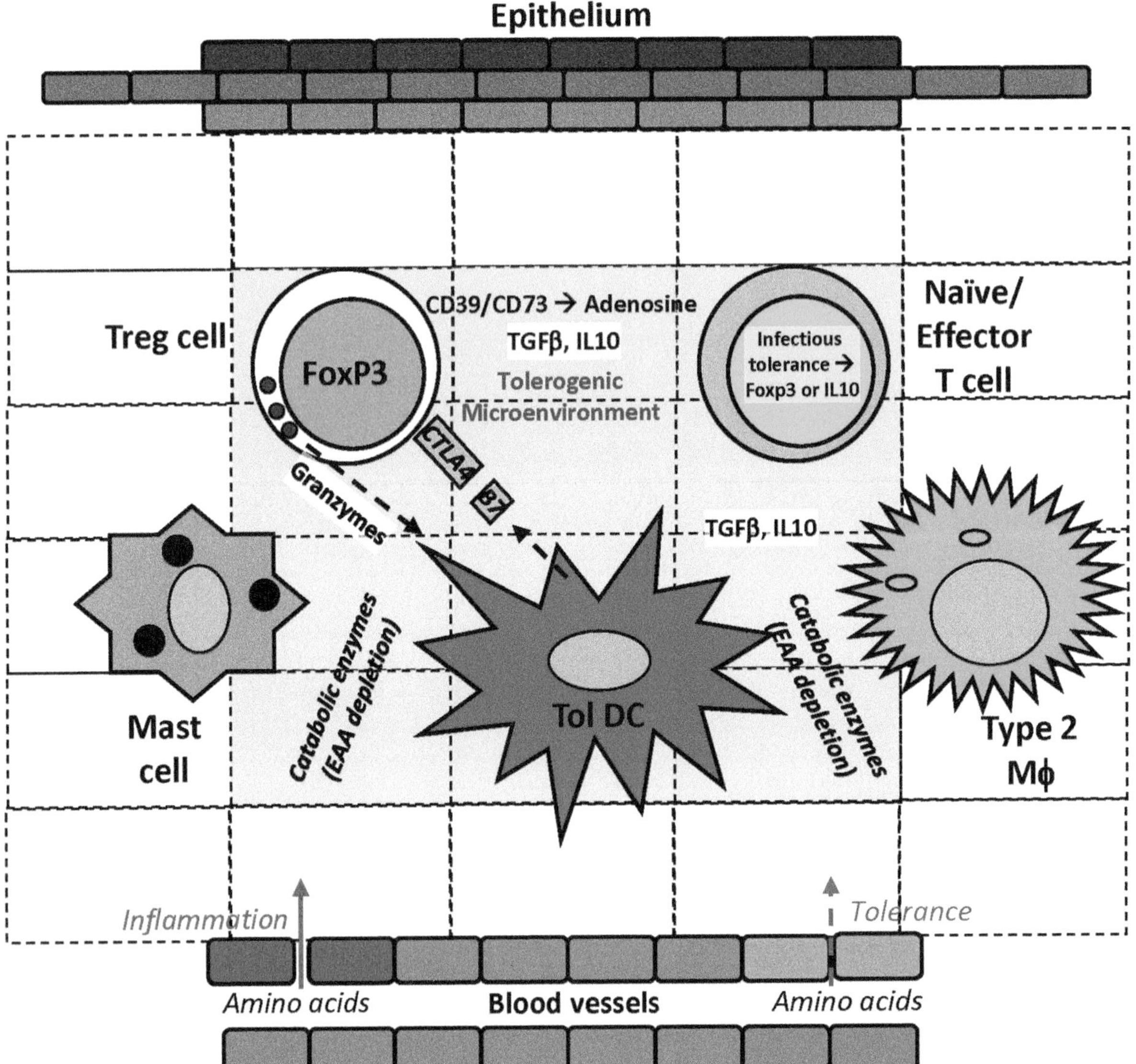

Figure 5 Interactions between regulatory T cells and myeloid cells maintain a tolerogenic microenvironment within tolerated tissues. Myeloid cells, including certain dendritic cells (Tol DC), type II macrophages (Mϕ), and mast cells can all express a range of enzymes, either intracellularly or secreted, that catabolize or utilize EAAs. In the context of an intact vasculature, this leads to the local depletion of amino acids, which represents one component of an immune-privileged or tolerogenic microenvironment. In addition, CD39 and CD73 coexpression on Tregs and other cell types, when enhanced by the local secretion of TGF-β, generates the anti-inflammatory mediator adenosine. Some Tregs also secrete IL-10, which further inhibits the ability of dendritic cells and Mϕ to present antigens for effector cell differentiation. Costimulatory ligands are also depleted from myeloid APCs by transendocytosis after capture by CTLA4 at the surface of Tregs. All these components of the tolerogenic microenvironment cooperate to limit the activity of naive and effector T cells while promoting Foxp3 expression and infectious tolerance. Graft rejection, on the other hand, is associated with leaking blood vessels and edema, which disrupts the tolerogenic niche, overwhelms the catabolic enzymes, and provides essential nutrients for T-cell activation and effector function.

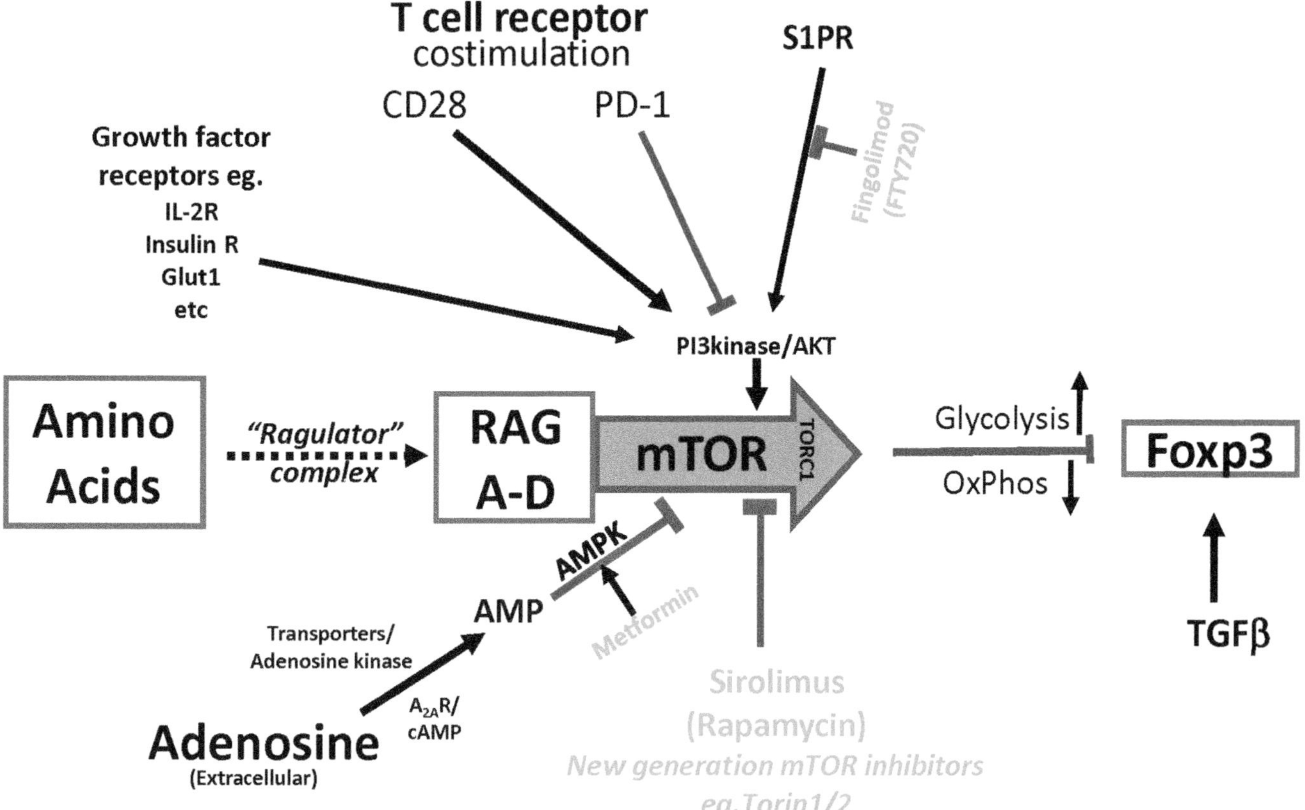

Figure 6 Nutrient and environmental sensing via mTOR regulates metabolism and Foxp3 expression. mTOR acts as an integrator of signals that arrive from a range of cell surface receptors and nutrient-sensing pathways and is critical in regulating cell metabolism and Foxp3 expression. The majority of these signals, including the TCR, growth factors, positive and negative costimulation (CD28 and PD-1), and the sphingosine 1-phosphate receptor (S1PR), converge on mTOR via the PI3K/AKT pathway, and are all dependent on recruiting phosphorylated Rheb to form the active TORC1 complex. The TORC1 complex can only be formed if there are sufficient EAAs to activate the regulator complex and the RAG proteins A to D. This means that a lack of amino acids effectively trumps all other signals via mTOR, inhibiting TORC1 activation, leading to enhanced oxidative phosphorylation (OxPhos), and allowing the Foxp3 gene to respond to TGF-β-mediated induction. Adenosine may also act indirectly on mTOR via cell surface receptors and/or adenosine transporters via AMP kinase (AMPK) signaling. It is interesting to note that at least three different classes of licensed drugs with immunomodulatory or metabolic activity (rapamycin, fingolimod, and metformin) target components of the mTOR signaling network.

3-kinase (PI3K) signaling. Most of the signals that lead to TORC1 activation do so via PI3K signaling. This results in phosphorylation of mTORC1 via the tuberous sclerosis (TSC) 1/2 complex and the Ras homolog expressed in brain (Rheb). Rheb is contained within a lysosomal compartment, and its interaction with TORC1 depends on sensing of amino acids. Downstream signaling requires the four Ras-related GTP-binding (or RAG GTPase-RRAG) proteins (A to D) in conjunction with the ragulator complex (57, 58). Through this route, amino acid deprivation acts to inhibit TORC1 activity. Rapamycin, an immunosuppressive drug when bound to FKB12, also inhibits formation and function of mTORC1 (59).

TORC1 is important for the initiation of mRNA translation and the upregulation of amino acid transporters at the cell surface. It also activates lipid oxidation and cell proliferation and inhibits the expression of *FOXP3* and Treg differentiation while favoring Th1 and Th17 cells (60). The TORC2 complex can sense reactive oxygen and also glucose availability via a cyclic AMP/protein kinase A pathway. TORC2 controls spatial aspects of cell growth, such as cell polarity and responses to chemotactic signals (61).

mTOR SIGNALING INHIBITS Foxp3 EXPRESSION

mTOR inhibition might enhance Foxp3 expression through a number of different pathways. These could be in part through effects on Foxp3 translation via inhibition of S6K1 and reduced phosphorylation of the ribosomal protein S6. In addition, mTOR could act indirectly through suppressor of cytokine signaling 3 (SOCS3) or directly on signal transducer and activator of transcription 3 (STAT3). In addition, FOXO3a and SMAD3, which promote FOXP3 expression, are themselves inhibited by AKT, which is downstream of TORC2 (62–64). Evidence from knockout mice with T-cell-targeted deficiencies in TORC1 or TORC2 suggests that TORC1 activation promotes Th1 differentiation (65), while TORC2 may have a bias toward Th2 (66). Inhibition of both complexes seems to be required for the optimal induction of Foxp3$^+$ Tregs.

Foxp3 EXPRESSION REQUIRES mTOR INHIBITION, WHILE mTOR ACTIVATION IS REQUIRED FOR REGULATORY FUNCTION

Mice with T-cell-specific mTOR inactivation show an enhanced capacity to generate Foxp3$^+$ Treg cells over other effector cells (65). However, the genetic inactivation of TORC1 in Foxp3$^+$ Treg cells resulted in a scurfy/IPEX-like syndrome (67). Inactivation of TORC1 activity in all T cells did not produce disease, suggesting that effector T cells were also compromised. This raises the possibility that the optimal induction and expansion of Foxp3$^+$ Tregs takes place in mTOR-inhibited microenvironments but that Tregs can exert their suppressive function only when there is a trigger of inflammation involving mTOR activation. It has been suggested that optimal induction of Treg cells requires alternate cycles of mTOR inhibition to promote induction and mTOR activation to promote proliferation (68).

Tregs, TGF-β, AND ADENOSINE GENERATION

TGF-β is able to induce the coexpression of two surface membrane ectoenzymes, CD39 and CD73, in many cell types (69, 70). These two enzymes are constitutively expressed at high levels in murine Tregs (71, 72). They act to convert extracellular sources of ATP into the anti-inflammatory adenosine. Extracellular adenosine can, in turn, act on specific G protein-coupled receptors on diverse immune cells to exert these inhibitory effects (73). It may be that generation of adenosine offers an additional element to the creation of a tolerogenic microenvironment by Tregs (Fig. 7). The other well-defined contributors would be mTOR inhibition (as discussed above) and inhibitory ligands such as CTLA4 interacting with costimulatory receptors on DCs. Not only may CTLA4 on Tregs suppress immune function by competing with costimulatory molecules for binding to their receptors on DCs (74), but it may also induce IDO induction by a subset of DCs (75) and effectively strip costimulatory receptors off the DC membrane (76), three routes to decommissioning the immune effects of DC.

CLINICAL RELEVANCE

For the past 25 years we have endeavored to take infectious tolerance to the clinic, searching for long-term benefit derived from a short-term treatment. We accumulated substantive preclinical data that tolerance can be induced not only in mouse models of transplantation and autoimmune disease, but also in prevention and correction of unwanted immunity to therapeutic proteins. To this end we humanized antibodies to human CD4 and CD8 as potential therapeutic candidates. Tolerance induced to foreign proteins with anti-CD4 therapy has already been demonstrated in nonhuman primates (77). Sadly, coreceptor blockade has not yet attracted pharmaceutical partners committed to short-term therapy, and this is probably due to a number of logistical and commercial reasons. A humanized antibody to the costimulatory molecule CD40L did enter the clinic, but development was curtailed due to risks of thromboembolism (78).

We have, however, observed long-term benefit from short-term therapy following lymphocyte depletion with our humanized anti-CD52 antibody CAMPATH-1H (alemtuzumab) (79). In light of our ideas on therapeutic tolerance requiring the harnessing of regulatory T cells, we sought evidence of whether long-term tolerance could be harnessed following T-cell depletion in our rodent models of transplantation.

LYMPHOCYTE DEPLETION, HOMEOSTATIC EXPANSION, AND "PHYSICIAN-AIDED RECONSTITUTION OF THE IMMUNE SYSTEM"

Using such a triple short-term combination, we were able to generate tolerance in CBA/Ca mice to very immunogenic MHC-incompatible skin grafts. To model the clinical situation, we created mice transgenic for human CD52 in their T cells and then gave short pulses of the depleting anti-CD52 antibody CAMPATH-1 to produce an ~2-log reduction in their peripheral T cells

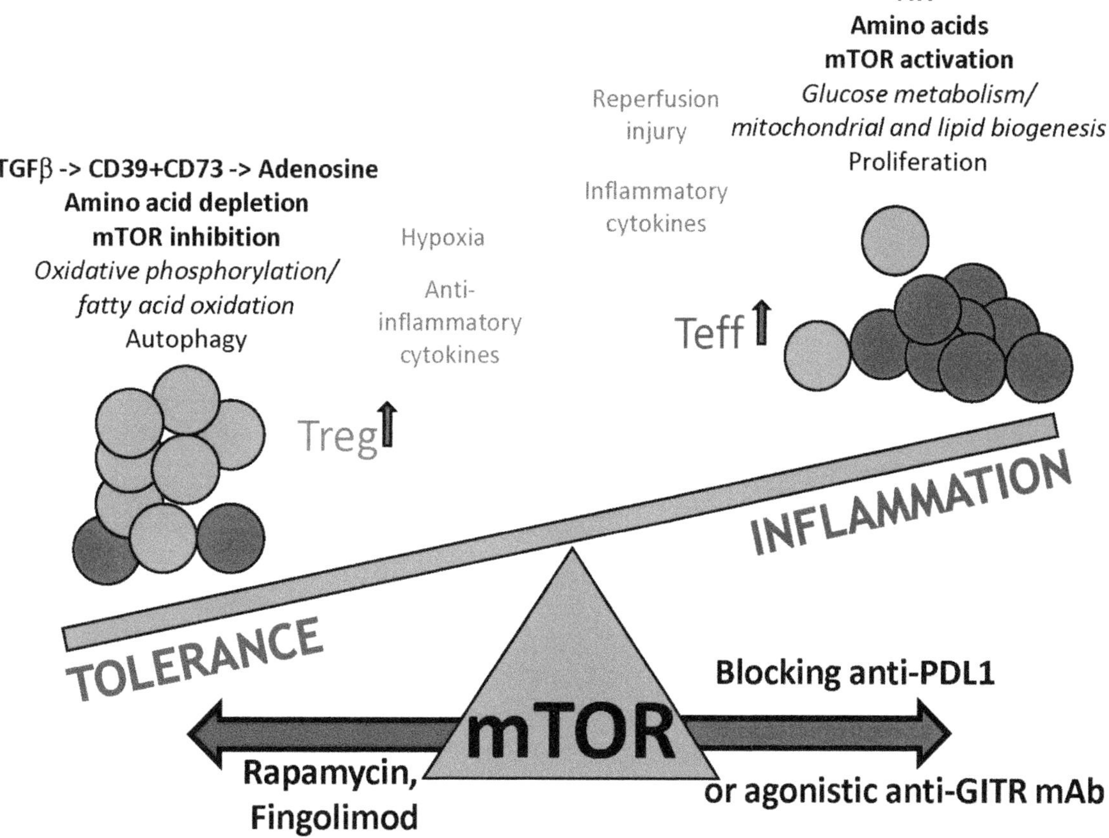

Figure 7 Tolerance or inflammation depends on the balance of Treg to T effector cell (Teff) and the response of mTOR to a range of microenvironmental factors. Whether the outcome of an immunological response is one of tolerance or inflammation is increasingly being considered as a balance between the number of Tregs and Teffs. This needs to be tempered by the influence of a number of other factors, particularly within the local microenvironment of the tolerated tissue, including the balance of adenosine to ATP, the effectiveness of amino acid depletion, and probably many other factors that may be associated with particular tissues and their state of health, such as inflammatory or anti-inflammatory cytokines, hypoxia, and reperfusion injury. Drugs that inhibit or activate mTOR might be able to adjust the balance point in favor of tolerance (for autoimmune disease and transplantation) or inflammation (for treatment of cancer) and provide a long-lasting therapeutic outcome from a short course of treatment.

(80). Despite such depletion, all mice rejected MHC-mismatched skin grafts, albeit at a reduced tempo when compared to controls. Rejection could only have been mediated by T cells spared by the therapy or those that had derived by expansion from those residual cells. Based on our concepts emerging from coreceptor blockade, we speculated that the reconstitution of T cells after depletion (homeostatic expansion) may not have provided Tregs with a numerical advantage over T effectors. To test that idea, we attempted to guide the reconstitution phase by addition of short-term therapy with the mTOR inhibitor rapamycin, and also MAb blockade of the IL-7 receptor (80). The latter was based on the observation that resting Tregs carry very little IL-7 receptor compared to other lymphocytes, and that the cytokine IL-7 is important in maintaining the normal balance of B cells and T cells in the immune system.

One week after the initiation of treatment we observed a strong numerical bias of detectable regulatory T cells in the spleens, that bias disappearing after 1 month (80). Tolerant mice also exhibited linked suppression whereby they could not reject grafts carrying a mix of the tolerated antigens with additional third-party antigens. This is strong evidence of active, dominant tolerance mediated by regulatory T cells, just like that operating after coreceptor blockade.

These data suggest that enhancing Treg numbers and function early in the process of reconstitution may give Tregs the protracted advantage needed to establish a long-term tolerant state.

We also speculate that some of the unwanted autoimmune phenomena that follow lymphocyte depletion and homeostatic expansion in the lymphocyte reconstitution phase (80, 81) can be prevented by such guided antilymphocyte therapy early in the inductive process.

CONCLUSION

On the basis of these studies, we feel that there are opportunities to therapeutically reprogram the immune system based on dominant tolerance by influencing initial lymphocyte perceptions of antigen so as to favor regulation. Thus far, MAbs have figured strongly in our protocols, but as we achieve a greater understanding of the metabolic processes underlying infectious tolerance, and the key elements contributing to tolerogenic microenvironments, hopefully new combinations of "rational" drugs will emerge to enable reprogramming to be safe and more predictable. Understanding of tolerance mechanisms that operate may also permit the development of "smart" biomarker sets that guide the physician to better enhance particular regulatory systems. Importantly, we should also recognize that this is not just a question of controlling adaptive immune responses, but that innate tolerance mechanisms are essential elements in ensuring the creation of protective tissue microenvironments.

Citation. Waldmann H, Howie D, Cobbold S. 2016. Induction of immunological tolerance as a therapeutic procedure. Microbiol Spectrum 4(4):MCHD-0019-2015.

References

1. **Aaltonen J, Björses P, Sandkuijl L, Perheentupa J, Peltonen L.** 1994. An autosomal locus causing autoimmune disease: autoimmune polyglandular disease type I assigned to chromosome 21. *Nat Genet* **8:**83–87.
2. **Bennett CL, Christie J, Ramsdell F, Brunkow ME, Ferguson PJ, Whitesell L, Kelly TE, Saulsbury FT, Chance PF, Ochs HD.** 2001. The immune dysregulation, polyendocrinopathy, enteropathy, X-linked syndrome (IPEX) is caused by mutations of *FOXP3*. *Nat Genet* **27:**20–21.
3. **Wildin RS, Ramsdell F, Peake J, Faravelli F, Casanova JL, Buist N, Levy-Lahad E, Mazzella M, Goulet O, Perroni L, Bricarelli FD, Byrne G, McEuen M, Proll S, Appleby M, Brunkow ME.** 2001. X-linked neonatal diabetes mellitus, enteropathy and endocrinopathy syndrome is the human equivalent of mouse scurfy. *Nat Genet* **27:**18–20.
4. **Cobbold SP, Jayasuriya A, Nash A, Prospero TD, Waldmann H.** 1984. Therapy with monoclonal antibodies by elimination of T-cell subsets in vivo. *Nature* **312:**548–551.
5. **Benjamin RJ, Cobbold SP, Clark MR, Waldmann H.** 1986. Tolerance to rat monoclonal antibodies. Implications for serotherapy. *J Exp Med* **163:**1539–1552.
6. **Benjamin RJ, Waldmann H.** 1986. Induction of tolerance by monoclonal antibody therapy. *Nature* **320:**449–451.
7. **Benjamin RJ, Qin SX, Wise MP, Cobbold SP, Waldmann H.** 1988. Mechanisms of monoclonal antibody-facilitated tolerance induction: a possible role for the CD4 (L3T4) and CD11a (LFA-1) molecules in self-non-self discrimination. *Eur J Immunol* **18:**1079–1088.
8. **Qin S, Cobbold S, Tighe H, Benjamin R, Waldmann H.** 1987. CD4 monoclonal antibody pairs for immunosuppression and tolerance induction. *Eur J Immunol* **17:** 1159–1165.
9. **Qin SX, Wise M, Cobbold SP, Leong L, Kong YC, Parnes JR, Waldmann H.** 1990. Induction of tolerance in peripheral T cells with monoclonal antibodies. *Eur J Immunol* **20:**2737–2745.
10. **Waldmann H, Qin S, Cobbold S.** 1992. Monoclonal antibodies as agents to reinduce tolerance in autoimmunity. *J Autoimmun* **5**(Suppl A):93–102.
11. **Waldmann H, Cobbold S.** 1998. How do monoclonal antibodies induce tolerance? A role for infectious tolerance? *Annu Rev Immunol* **16:**619–644.
12. **Qin SX, Cobbold S, Benjamin R, Waldmann H.** 1989. Induction of classical transplantation tolerance in the adult. *J Exp Med* **169:**779–794.
13. **Marshall SE, Cobbold SP, Davies JD, Martin GM, Phillips JM, Waldmann H.** 1996. Tolerance and suppression in a primed immune system. *Transplantation* **62:** 1614–1621.
14. **Daley SR, Ma J, Adams E, Cobbold SP, Waldmann H.** 2007. A key role for TGF-β signaling to T cells in the long-term acceptance of allografts. *J Immunol* **179:**3648–3654.
15. **Davies JD, Leong LY, Mellor A, Cobbold SP, Waldmann H.** 1996. T cell suppression in transplantation tolerance through linked recognition. *J Immunol* **156:**3602–3607.
16. **Wise MP, Bemelman F, Cobbold SP, Waldmann H.** 1998. Linked suppression of skin graft rejection can operate through indirect recognition. *J Immunol* **161:** 5813–5816.
17. **Cobbold SP, Qin S, Leong LY, Martin G, Waldmann H.** 1992. Reprogramming the immune system for peripheral tolerance with CD4 and CD8 monoclonal antibodies. *Immunol Rev* **129:**165–201.
18. **Cobbold SP, Castejon R, Adams E, Zelenika D, Graca L, Humm S, Waldmann H.** 2004. Induction of *foxP3*[+] regulatory T cells in the periphery of T cell receptor transgenic mice tolerized to transplants. *J Immunol* **172:** 6003–6010.
19. **Qin S, Cobbold SP, Pope H, Elliott J, Kioussis D, Davies J, Waldmann H.** 1993. "Infectious" transplantation tolerance. *Science* **259:**974–977.
20. **Davies JD, Martin G, Phillips J, Marshall SE, Cobbold SP, Waldmann H.** 1996. T cell regulation in adult transplantation tolerance. *J Immunol* **157:**529–533.
21. **Scully R, Qin S, Cobbold S, Waldmann H.** 1994. Mechanisms in CD4 antibody-mediated transplantation

tolerance: kinetics of induction, antigen dependency and role of regulatory T cells. *Eur J Immunol* **24:**2383–2392.

22. **Chen Z, Cobbold S, Metcalfe S, Waldmann H.** 1992. Tolerance in the mouse to major histocompatibility complex-mismatched heart allografts, and to rat heart xenografts, using monoclonal antibodies to CD4 and CD8. *Eur J Immunol* **22:**805–810.
23. **Waldmann H, Cobbold S.** 2001. Regulating the immune response to transplants: a role for CD4$^+$ regulatory cells? *Immunity* **14:**399–406.
24. **Yates SF, Paterson AM, Nolan KF, Cobbold SP, Saunders NJ, Waldmann H, Fairchild PJ.** 2007. Induction of regulatory T cells and dominant tolerance by dendritic cells incapable of full activation. *J Immunol* **179:**967–976.
25. **Farquhar CA, Paterson AM, Cobbold SP, Garcia Rueda H, Fairchild PJ, Yates SF, Adams E, Saunders NJ, Waldmann H, Nolan KF.** 2010. Tolerogenicity is not an absolute property of a dendritic cell. *Eur J Immunol* **40:** 1728–1737.
26. **Nolan KF, Strong V, Soler D, Fairchild PJ, Cobbold SP, Croxton R, Gonzalo JA, Rubio A, Wells M, Waldmann H.** 2004. IL-10-conditioned dendritic cells, decommissioned for recruitment of adaptive immunity, elicit innate inflammatory gene products in response to danger signals. *J Immunol* **172:**2201–2209.
27. **Hawiger D, Inaba K, Dorsett Y, Guo M, Mahnke K, Rivera M, Ravetch JV, Steinman RM, Nussenzweig MC.** 2001. Dendritic cells induce peripheral T cell unresponsiveness under steady state conditions in vivo. *J Exp Med* **194:**769–779.
28. **Sakaguchi S, Sakaguchi N, Asano M, Itoh M, Toda M.** 1995. Immunologic self-tolerance maintained by activated T cells expressing IL-2 receptor alpha-chains (CD25). Breakdown of a single mechanism of self-tolerance causes various autoimmune diseases. *J Immunol* **155:**1151–1164.
29. **Powrie F, Mason D.** 1990. OX-22^{high} CD4$^+$ T cells induce wasting disease with multiple organ pathology: prevention by the OX-22^{low} subset. *J Exp Med* **172:**1701–1708.
30. **Hori S, Nomura T, Sakaguchi S.** 2003. Control of regulatory T cell development by the transcription factor Foxp3. *Science* **299:**1057–1061.
31. **Fontenot JD, Gavin MA, Rudensky AY.** 2003. Foxp3 programs the development and function of CD4$^+$CD25$^+$ regulatory T cells. *Nat Immunol* **4:**330–336.
32. **Khattri R, Cox T, Yasayko SA, Ramsdell F.** 2003. An essential role for Scurfin in CD4$^+$CD25$^+$ T regulatory cells. *Nat Immunol* **4:**337–342.
33. **Zelenika D, Adams E, Mellor A, Simpson E, Chandler P, Stockinger B, Waldmann H, Cobbold SP.** 1998. Rejection of H-Y disparate skin grafts by monospecific CD4$^+$ Th1 and Th2 cells: no requirement for CD8$^+$ T cells or B cells. *J Immunol* **161:**1868–1874.
34. **Zelenika D, Adams E, Humm S, Lin CY, Waldmann H, Cobbold SP.** 2001. The role of CD4$^+$ T-cell subsets in determining transplantation rejection or tolerance. *Immunol Rev* **182:**164–179.
35. **Zelenika D, Adams E, Humm S, Graca L, Thompson S, Cobbold SP, Waldmann H.** 2002. Regulatory T cells overexpress a subset of Th2 gene transcripts. *J Immunol* **168:**1069–1079.
36. **Chen W, Jin W, Hardegen N, Lei KJ, Li L, Marinos N, McGrady G, Wahl SM.** 2003. Conversion of peripheral CD4$^+$CD25$^-$ naive T cells to CD4$^+$CD25$^+$ regulatory T cells by TGF-β induction of transcription factor *Foxp3*. *J Exp Med* **198:**1875–1886.
37. **Regateiro FS, Chen Y, Kendal AR, Hilbrands R, Adams E, Cobbold SP, Ma J, Andersen KG, Betz AG, Zhang M, Madhiwalla S, Roberts B, Waldmann H, Nolan KF, Howie D.** 2012. Foxp3 expression is required for the induction of therapeutic tissue tolerance. *J Immunol* **189:** 3947–3956.
38. **Graca L, Cobbold SP, Waldmann H.** 2002. Identification of regulatory T cells in tolerated allografts. *J Exp Med* **195:**1641–1646.
39. **Komatsu N, Mariotti-Ferrandiz ME, Wang Y, Malissen B, Waldmann H, Hori S.** 2009. Heterogeneity of natural Foxp3$^+$ T cells: a committed regulatory T-cell lineage and an uncommitted minor population retaining plasticity. *Proc Natl Acad Sci USA* **106:**1903–1908.
40. **Kendal AR, Chen Y, Regateiro FS, Ma J, Adams E, Cobbold SP, Hori S, Waldmann H.** 2011. Sustained suppression by Foxp3$^+$ regulatory T cells is vital for infectious transplantation tolerance. *J Exp Med* **208:**2043–2053.
41. **Cobbold SP, Adams E, Graca L, Daley S, Yates S, Paterson A, Robertson NJ, Nolan KF, Fairchild PJ, Waldmann H.** 2006. Immune privilege induced by regulatory T cells in transplantation tolerance. *Immunol Rev* **213:**239–255.
42. **Cobbold SP, Adams E, Farquhar CA, Nolan KF, Howie D, Lui KO, Fairchild PJ, Mellor AL, Ron D, Waldmann H.** 2009. Infectious tolerance via the consumption of essential amino acids and mTOR signaling. *Proc Natl Acad Sci USA* **106:**12055–12060.
43. **Munn DH, Zhou M, Attwood JT, Bondarev I, Conway SJ, Marshall B, Brown C, Mellor AL.** 1998. Prevention of allogeneic fetal rejection by tryptophan catabolism. *Science* **281:**1191–1193.
44. **Munn DH, Sharma MD, Baban B, Harding HP, Zhang Y, Ron D, Mellor AL.** 2005. GCN2 kinase in T cells mediates proliferative arrest and anergy induction in response to indoleamine 2,3-dioxygenase. *Immunity* **22:** 633–642.
45. **Kropf P, Baud D, Marshall SE, Munder M, Mosley A, Fuentes JM, Bangham CR, Taylor GP, Herath S, Choi BS, Soler G, Teoh T, Modolell M, Müller I.** 2007. Arginase activity mediates reversible T cell hyporesponsiveness in human pregnancy. *Eur J Immunol* **37:**935–945.
46. **Chabtini L, Mfarrej B, Mounayar M, Zhu B, Batal I, Dakle PJ, Smith BD, Boenisch O, Najafian N, Akiba H, Yagita H, Guleria I.** 2013. TIM-3 regulates innate immune cells to induce fetomaternal tolerance. *J Immunol* **190:**88–96.
47. **Stein M, Keshav S, Harris N, Gordon S.** 1992. Interleukin 4 potently enhances murine macrophage mannose receptor activity: a marker of alternative immunologic macrophage activation. *J Exp Med* **176:**287–292.
48. **Lumeng CN, Bodzin JL, Saltiel AR.** 2007. Obesity induces a phenotypic switch in adipose tissue macrophage polarization. *J Clin Invest* **117:**175–184.

49. **Cobbold SP, Adams E, Graca L, Waldmann H.** 2003. Serial analysis of gene expression provides new insights into regulatory T cells. *Semin Immunol* **15:**209–214.

50. **Lu LF, Lind EF, Gondek DC, Bennett KA, Gleeson MW, Pino-Lagos K, Scott ZA, Coyle AJ, Reed JL, Van Snick J, Strom TB, Zheng XX, Noelle RJ.** 2006. Mast cells are essential intermediaries in regulatory T-cell tolerance. *Nature* **442:**997–1002.

51. **Waldmann H.** 2006. Immunology: protection and privilege. *Nature* **442:**987–988.

52. **Dalton DK, Noelle RJ.** 2012. The roles of mast cells in anticancer immunity. *Cancer Immunol Immunother* **61:** 1511–1520.

53. **Nowak EC, de Vries VC, Wasiuk A, Ahonen C, Bennett KA, Le Mercier I, Ha DG, Noelle RJ.** 2012. Tryptophan hydroxylase-1 regulates immune tolerance and inflammation. *J Exp Med* **209:**2127–2135.

54. **Cobbold SP, Adams E, Nolan KF, Regateiro FS, Waldmann H.** 2010. Connecting the mechanisms of T-cell regulation: dendritic cells as the missing link. *Immunol Rev* **236:**203–218.

55. **Loewith R, Jacinto E, Wullschleger S, Lorberg A, Crespo JL, Bonenfant D, Oppliger W, Jenoe P, Hall MN.** 2002. Two TOR complexes, only one of which is rapamycin sensitive, have distinct roles in cell growth control. *Mol Cell* **10:**457–468.

56. **Howie D, Waldmann H, Cobbold S.** 2014. Nutrient sensing via mTOR in T cells maintains a tolerogenic microenvironment. *Front Immunol* **5:**409 10.3389/fimmu.2014.00409.

57. **Sancak Y, Peterson TR, Shaul YD, Lindquist RA, Thoreen CC, Bar-Peled L, Sabatini DM.** 2008. The Rag GTPases bind raptor and mediate amino acid signaling to mTORC1. *Science* **320:**1496–1501.

58. **Sancak Y, Bar-Peled L, Zoncu R, Markhard AL, Nada S, Sabatini DM.** 2010. Ragulator-Rag complex targets mTORC1 to the lysosomal surface and is necessary for its activation by amino acids. *Cell* **141:**290–303.

59. **Sabatini DM, Erdjument-Bromage H, Lui M, Tempst P, Snyder SH.** 1994. RAFT1: a mammalian protein that binds to FKBP12 in a rapamycin-dependent fashion and is homologous to yeast TORs. *Cell* **78:**35–43.

60. **Kopf H, de la Rosa GM, Howard OM, Chen X.** 2007. Rapamycin inhibits differentiation of Th17 cells and promotes generation of FoxP3$^+$ T regulatory cells. *Int Immunopharmacol* **7:**1819–1824.

61. **Charest PG, Shen Z, Lakoduk A, Sasaki AT, Briggs SP, Firtel RA.** 2010. A Ras signaling complex controls the RasC-TORC2 pathway and directed cell migration. *Dev Cell* **18:**737–749.

62. **Harada Y, Harada Y, Elly C, Ying G, Paik JH, DePinho RA, Liu YC.** 2010. Transcription factors Foxo3a and Foxo1 couple the E3 ligase Cbl-b to the induction of Foxp3 expression in induced regulatory T cells. *J Exp Med* **207:**1381–1391.

63. **Ouyang W, Beckett O, Ma Q, Paik JH, DePinho RA, Li MO.** 2010. Foxo proteins cooperatively control the differentiation of Foxp3$^+$ regulatory T cells. *Nat Immunol* **11:**618–627.

64. **Zhang Q, Cui F, Fang L, Hong J, Zheng B, Zhang JZ.** 2013. TNF-α impairs differentiation and function of TGF-β-induced Treg cells in autoimmune diseases through Akt and Smad3 signaling pathway. *J Mol Cell Biol* **5:** 85–98.

65. **Delgoffe GM, Kole TP, Zheng Y, Zarek PE, Matthews KL, Xiao B, Worley PF, Kozma SC, Powell JD.** 2009. The mTOR kinase differentially regulates effector and regulatory T cell lineage commitment. *Immunity* **30:**832–844.

66. **Lee K, Gudapati P, Dragovic S, Spencer C, Joyce S, Killeen N, Magnuson MA, Boothby M.** 2010. Mammalian target of rapamycin protein complex 2 regulates differentiation of Th1 and Th2 cell subsets via distinct signaling pathways. *Immunity* **32:**743–753.

67. **Zeng H, Yang K, Cloer C, Neale G, Vogel P, Chi H.** 2013. mTORC1 couples immune signals and metabolic programming to establish T_{reg}-cell function. *Nature* **499:** 485–490.

68. **Procaccini C, De Rosa V, Galgani M, Abanni L, Calì G, Porcellini A, Carbone F, Fontana S, Horvath TL, La Cava A, Matarese G.** 2010. An oscillatory switch in mTOR kinase activity sets regulatory T cell responsiveness. *Immunity* **33:**929–941.

69. **Regateiro FS, Howie D, Nolan KF, Agorogiannis EI, Greaves DR, Cobbold SP, Waldmann H.** 2011. Generation of anti-inflammatory adenosine by leukocytes is regulated by TGF-β. *Eur J Immunol* **41:**2955–2965.

70. **Regateiro FS, Cobbold SP, Waldmann H.** 2013. CD73 and adenosine generation in the creation of regulatory microenvironments. *Clin Exp Immunol* **171:**1–7.

71. **Kobie JJ, Shah PR, Yang L, Rebhahn JA, Fowell DJ, Mosmann TR.** 2006. T regulatory and primed uncommitted CD4 T cells express CD73, which suppresses effector CD4 T cells by converting 5′-adenosine monophosphate to adenosine. *J Immunol* **177:**6780–6786.

72. **Deaglio S, Dwyer KM, Gao W, Friedman D, Usheva A, Erat A, Chen JF, Enjyoji K, Linden J, Oukka M, Kuchroo VK, Strom TB, Robson SC.** 2007. Adenosine generation catalyzed by CD39 and CD73 expressed on regulatory T cells mediates immune suppression. *J Exp Med* **204:** 1257–1265.

73. **Ohta A, Sitkovsky M.** 2014. Extracellular adenosine-mediated modulation of regulatory T cells. *Front Immunol* **5:**304. doi:10.3389/fimmu.2014.00304.

74. **Wing K, Onishi Y, Prieto-Martin P, Yamaguchi T, Miyara M, Fehervari Z, Nomura T, Sakaguchi S.** 2008. CTLA-4 control over Foxp3$^+$ regulatory T cell function. *Science* **322:**271–275.

75. **Mellor AL, Chandler P, Baban B, Hansen AM, Marshall B, Pihkala J, Waldmann H, Cobbold S, Adams E, Munn DH.** 2004. Specific subsets of murine dendritic cells acquire potent T cell regulatory functions following CTLA4-mediated induction of indoleamine 2,3 dioxygenase. *Int Immunol* **16:**1391–1401.

76. **Qureshi OS, Zheng Y, Nakamura K, Attridge K, Manzotti C, Schmidt EM, Baker J, Jeffery LE, Kaur S, Briggs Z, Hou TZ, Futter CE, Anderson G, Walker LS, Sansom DM.** 2011. Trans-endocytosis of CD80 and CD86: a molecular basis for the cell-extrinsic function of CTLA-4. *Science* **332:**600–603.

77. **Winsor-Hines D, Merrill C, O'Mahony M, Rao PE, Cobbold SP, Waldmann H, Ringler DJ, Ponath PD.** 2004. Induction of immunological tolerance/hyporesponsiveness in baboons with a nondepleting CD4 antibody. *J Immunol* **173:**4715–4723.

78. **Kawai T, Andrews D, Colvin RB, Sachs DH, Cosimi AB.** 2000. Thromboembolic complications after treatment with monoclonal antibody against CD40 ligand. *Nat Med* **6:**114.

79. **Coles AJ, Cox A, Le Page E, Jones J, Trip SA, Deans J, Seaman S, Miller DH, Hale G, Waldmann H, Compston DA.** 2006. The window of therapeutic opportunity in multiple sclerosis: evidence from monoclonal antibody therapy. *J Neurol* **253:**98–108.

80. **Piotti G, Ma J, Adams E, Cobbold S, Waldmann H.** 2014. Guiding postablative lymphocyte reconstitution as a route toward transplantation tolerance. *Am J Transplant* **14:**1678–1689.

81. **Cox AL, Thompson SA, Jones JL, Robertson VH, Hale G, Waldmann H, Compston DA, Coles AJ.** 2005. Lymphocyte homeostasis following therapeutic lymphocyte depletion in multiple sclerosis. *Eur J Immunol* **35:** 3332–3342.

Metabolic and Malignant Disease

VIII

Myeloid Cells in Health and Disease: A Synthesis
Edited by Siamon Gordon

doi:10.1128/microbiolspec.MCHD-0044-2016

Mark M. Hughes[1]
Anne F. McGettrick[1]
Luke A. J. O'Neill[1]

45

Glutathione and Glutathione Transferase Omega 1 as Key Posttranslational Regulators in Macrophages

INTRODUCTION

The cytoplasm of an activated macrophage is a dangerous place due to the accumulation of reactive oxygen species (ROS), which can perturb cytoplasmic oxidative balance. Macrophage activation through monocyte recruitment in tissues occurs in a highly regulated manner upon detection of microbial pathogen-associated molecular patterns (PAMPs), such as the Gram-negative bacterial cell wall component lipopolysaccharide (LPS), stimulating ROS production. ROS can be generated from many cellular processes, either directly or as a result of incomplete reduction of free radicals (1). One of the main sources of ROS is the NADPH oxidase (NOX), a specialized transmembrane protein complex that generates superoxide (O_2^-), which has been implicated in several inflammatory diseases (2). The generation of ROS can also stem from the mitochondrial electron transport chain, due to insufficient reduction of superoxide anions. Another source is the inducible form of nitric oxide synthase 2 (iNOS). iNOS produces nitric oxide from L-arginine and is known to play key roles in macrophage function (3, 4).

While ROS have benefits, acting as endogenous signaling molecules (5), activated macrophages produce high levels of ROS, which can go unregulated under inflammatory conditions, indiscriminately targeting redox-sensitive protein machinery. Glutathione (γ-L-glutamyl-L-cysteinyl-glycine; GSH) is the fundamental nonprotein tripeptide redox agent that detoxifies ROS. Composed of the amino acids glutamate, cysteine, and glycine, GSH is utilized within the cell to maintain redox homeostasis (6). Overproduction of ROS can target the thiol group (-SH) of cysteine amino acids on proteins. GSH can prevent oxidation of proteins via formation of mixed disulfides (-SSG) to redox-sensitive cysteine amino acids in proteins (7). Cells contain a variety of redox enzymes, such as glutathione transferases and glutaredoxins, which utilize GSH to detoxify ROS

[1]School of Biochemistry and Immunology, Trinity Biomedical Science Institute, Trinity College Dublin, Dublin 2, Ireland.

and maintain protein integrity. GSH is predominantly found in the thiol-reduced form, reaching intracellular concentrations of up to 10 mM (8), and the ratio of reduced GSH to oxidized GSH (termed GSSG) can be used as a measure of cellular oxidative imbalance (9). GSH may, however, not be limited to antioxidant defense, with posttranslational modifications placing GSH as both a positive and negative regulator of protein function. Here we will discuss emerging themes in the field of GSH biology, notably the enzymatic addition or removal of GSH to proteins, which modifies their function. Glutathionylation is emerging as a key process in macrophage activation, targeting proteins such as caspase-1, hypoxia-inducible factor-1α (HIF1α), and STAT3. Here we will discuss this process and its regulation during macrophage activation.

GSH AS AN ANTIOXIDANT

GSH is ubiquitously expressed in all cell types and tissues. Aside from cellular detoxification, GSH is also a source of readily accessible cysteine, a rate-limiting amino acid for many cellular processes, including GSH synthesis. The amino acid cysteine is also crucial for maintaining protein tertiary structure via disulfide bond formation (10, 11) and is highly redox sensitive. Due to the presence of a thiol group, cysteine is readily oxidized to cystine, a cysteine dimer, and can subsequently generate ROS. Excessive ROS can promote the formation of thiol intermediates on cysteines, termed sulfenic acid, sulfinic acid, and sulfonic acid in proteins (12). Overoxidation of cysteine residues to sulfonic acid is irreversible and promotes proteasomal degradation. GSH can reverse the formation of sulfenic and sulfinic acid in conjunction with reductase (sulfenic) or sulfiredoxin (sulfinic) enzymes, resulting in regeneration of native protein structure and generation of nontoxic intermediates as a by-product, such as H_2O. High levels of GSH may thereby have a dual role in acting as a cysteine storage pool for physiological use as well as oxidative defense.

GSH is synthesized in a two-step enzymatic process requiring ATP investment at each step to generate GSH from the amino acids glutamate, cysteine, and glycine (Fig. 1). Upon exposure to oxidative stress inducers such as O_2^- or hydrogen peroxide (H_2O_2), GSH scavenges ROS and forms GSSG, which can be recycled in the cytoplasm back to reduced GSH by glutathione reductase (13). One of the main sources of endogenous ROS is mitochondria. Although the sizes of mitochondria are modest (ranging from 0.5 to 3 μm), they contain a concentrated level of GSH, comprising roughly 10 to 15% of cellular GSH (14), that can buffer ROS production. Mitochondria are metabolic powerhouses that not only promote ATP production through oxidative phosphorylation (OXPHOS), but also have been recently reported to regulate the proinflammatory status of macrophages when challenged with Toll-like receptor (TLR) agonists, such as LPS, via the tricarboxylic acid cycle component succinate (15, 16). Due to the critical nature of OXPHOS, the pathway is tightly regulated, and GSH has been shown to play a role in OXPHOS regulation. Complex V, otherwise known as ATP synthase, and succinyl-coenzyme A transferase are both negatively regulated by glutathionylation (17). The main purpose of OXPHOS is to produce ATP via cofactors such as NADH, which donate their energy-rich electrons into the complex assembly chain. This process, however, is not perfect, with some electrons donated directly to O_2, resulting in superoxide anion generation that can lead to pathophysiological conditions if unregulated (18). GSH conjugates to these superoxide anions to help protect mitochondrial integrity. It is likely that GSH controls ROS generated by succinate, which drives ROS production by complex I in LPS-activated macrophages (16). Utilization of molecular oxygen to generate ATP can also produce free radicals, leading to lipid peroxidation and generation of a harmful by-product, 4-hydroxynonenal (4-HNE) (19). GSH can therefore prevent formation 4-HNE by preventing ROS accumulation (Fig. 2). While lipid peroxidation is not uncommon, GSH in this context is crucial for redox homeostasis, as high levels of 4-HNE (up to 20 μM) have been shown to inhibit NF-κB signaling (20) and promote caspase-3 activation via Akt inhibition to promote apoptosis (21). Damaged mitochondria are known to undergo autophagy (22), or deliberate breakdown of cellular components and organelles to generate basic components for new synthesis pathway intermediates. Recent evidence has identified that the immunosuppressive mycotoxin patulin induces depletion of GSH, inhibiting TLRs in macrophages and enhancing mitophagy in a p62-dependent manner (23), further emphasizing the critical role of GSH in maintaining mitochondrial integrity.

GSH REGULATES PROTEIN ACTIVITY IN MACROPHAGES

Within the past decade, studies have revealed a role for GSH in posttranslational modification of proteins. This reversible modification regulates some of the best-characterized enzymes and pathways. The ability of GSH to bind to a cysteine residue can be influenced by

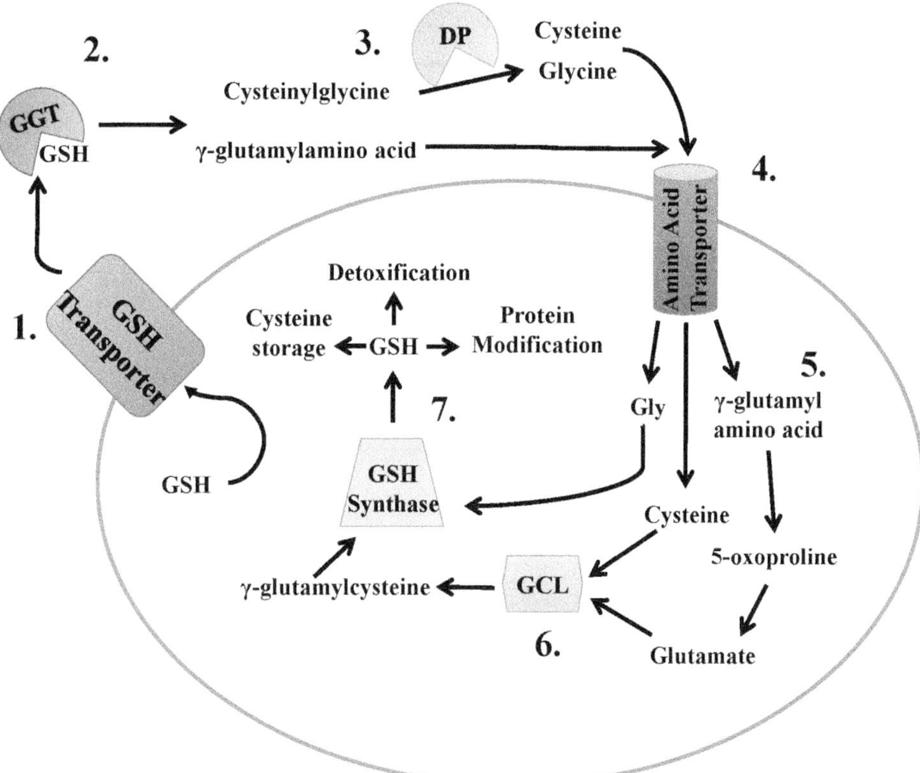

Figure 1 GSH synthesis pathway in mammalian cells. GSH is exported from the cell to the extracellular environment via a glutathione transporter (1). Extracellular GSH is targeted by γ-glutamylpeptidase (GGT), transferring the γ-glutamyl component of GSH to an amino acid, forming γ-glutamylamino acid and cysteinylglycine (2). Cysteinylglycine is further cleaved by dipeptidase (DP) into the amino acids cysteine and glycine (3). γ-glutamylamino acid, cysteine, and glycine reenter the cell via an amino acid transporter (4). Release of the amino acid component of γ-glutamylamino acid forms 5-oxoproline, which is further converted to glutamate in an ATP-dependent process (5). Newly formed glutamate and cysteine are ligated into γ-glutamylcysteine via glutamate cysteine ligase (GCL) in an ATP-dependent process (6). γ-Glutamylcysteine and glycine are ligated via GSH synthase to re-form GSH (7).

the neighboring amino acids (24). Positively charged amino acids, such as histidine, lysine, and arginine, can directly influence the probability of GSH binding. Cysteines are more susceptible to oxidation by ROS due to positively charged amino acids lowering their thiol pK_a, which deprotonates cysteines, promoting glutathionylation (25). A recent study has used a Web-based software program called GSHSite to rank potentially glutathionylated proteins in mouse macrophages based on conserved cysteine amino acids, positively charged flanking amino acids, and the ability of GSH to access cysteines in folded proteins (26). Using this approach, two known glutathionylated proteins, thioredoxin (TRX) and protein tyrosine phosphatase 1b, were correctly identified as glutathionylated proteins. Proteomic approaches to quantify cellular glutathionylation can also be employed using mass spectrometry (MS) (27). Su et al. were able to quantify 690 mouse macrophage proteins that were susceptible to glutathionylation when Raw 264.7 cells were pretreated with the oxidant diamide, and 290 proteins that were glutathionylated in response to hydrogen peroxide treatment. Similarly, Ullevig et al. identified >130 glutathionylated proteins in THP-1 monocytic cells treated with hydrogen peroxide to induce metabolic stress (28). While MS can identify glutathionylated proteins, methodologies differ in sample preparation, which may account for the differences in reported glutathionylation targets. As a result, glutathionylation can also be confirmed by Western blotting conditions, using nonreducing SDS-PAGE and an anti-GSH antibody. These studies have highlighted a number of proteins in mac-

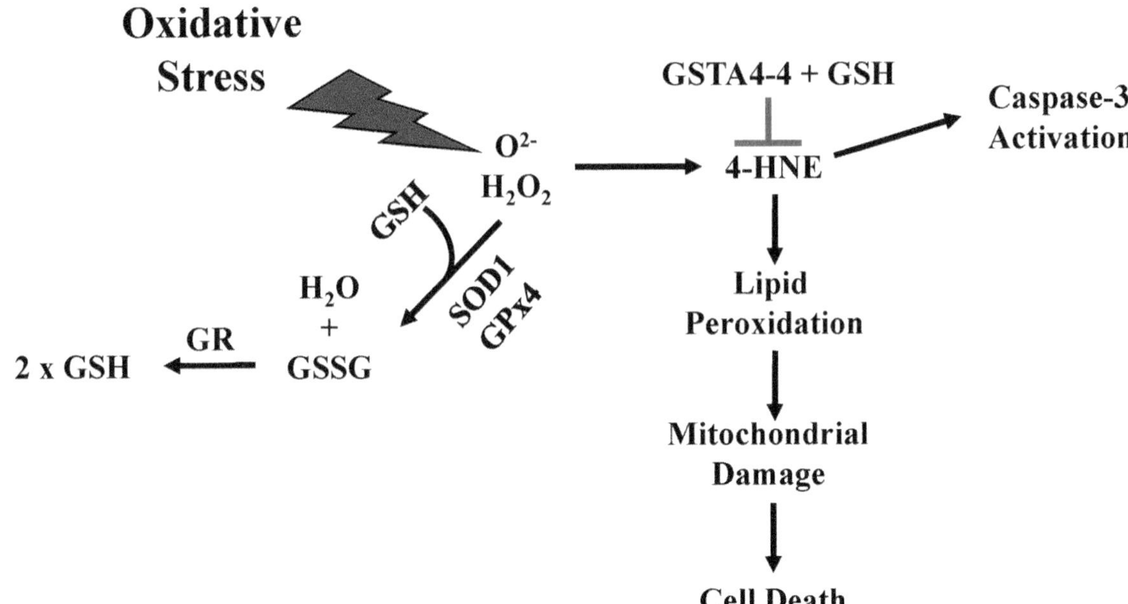

Figure 2 Cellular roles of GSH. Oxidative stress can be induced by many exogenous factors, including LPS, or endogenously by the mitochondrial electron transport chain. ROS can build up within the cell, in the form of O_2^- and H_2O_2, which damages organelles and results in 4-HNE production. 4-HNE can promote lipid peroxidation and caspase-3 activation. The redox enzyme GSTA4-4 can target 4-HNE, utilizing GSH to prevent lipid peroxidation and maintain homeostasis. O_2^- and H_2O_2 can be reduced via redox enzymes SOD1 and glutathione peroxidase 4 (GPx4), respectively. Both enzymes utilize GSH to remove ROS, forming glutathione disulfide (GSSG). GSSG can then be reduced to GSH by glutathione reductase (GR).

rophages as potential GSH targets, which may alter their function; however, GSH concentration can also impact macrophage function.

Research by Yang et al. has discovered that GSH regulates CD36 activity (29). CD36 is critical for oxidized low-density lipoprotein (oxLDL) clearance, a process that can go awry in atherosclerosis and promote foam cell generation. Macrophages treated with buthionine sulfoximine (BSO), a compound that inhibits GSH synthesis, resulting in decreased intracellular GSH levels, had increased ROS production and CD36 levels. This increased oxLDL uptake, driving foam cell generation. Importantly, the addition of antioxidants such as *N*-acetylcysteine and antioxidant enzymes catalase or superoxide dismutase 1 (SOD1) to peritoneal macrophages inhibited the effects of BSO on CD36 protein levels, preventing oxLDL uptake. GSH therefore displays antiatherogenic properties and regulates the levels of CD36 within macrophages to limit oxLDL uptake.

In related research, Vasamsetti et al. found that monocyte-to-macrophage differentiation in atherosclerotic plaque formation was dependent on GSH levels (30). Treatment of monocytes with the antioxidant resveratrol dose-dependently inhibited phorbol myristate acetate (PMA)-induced monocyte-to-macrophage differentiation. BSO-mediated GSH depletion was found to induce monocyte differentiation and promote inflammation, while resveratrol and BSO cotreatment prevented monocyte-to-macrophage transition, restoring GSH levels. Maintaining high GSH levels within monocytes is therefore protective in preventing monocyte differentiation in atherosclerotic plaques and maintaining lower CD36 levels, thereby preventing foam cell generation. Examination of the redox status of the cell has yielded a number of proteins in macrophages that are regulated by glutathionylation, which will now be discussed.

GLUTATHIONYLATED PRDX2 IS SECRETED FROM MACROPHAGES AND DRIVES TNF-α RELEASE

Using a redox proteomics MS approach, LPS-treated macrophages were found to secrete glutathionylated proteins from the cytoplasm. LPS treatment promoted oxidation of proteins due to increased ROS production, which were subsequently glutathionylated (31). Of particular note, the enzyme peroxiredoxin-2 (PRDX2), involved in hydrogen peroxide regulation, was secreted in the glutathionylated form and was identified to act as an inflammatory mediator, promoting tumor necrosis factor α (TNF-α) release. LPS did not induce PRDX2 expression; rather, it promoted oxidation of PRDX2 from the mainly reduced cytosolic form, promoting glutathionylation of PRDX2, which could drive TNF-α. Similarly, recombinant PRDX2 was found to drive TNF-α production. The identification of a redox enzyme with cytokine-like properties has resulted in the coining of the term "redoxkine." Furthermore, the redox enzyme TRX1, a substrate for PRDX2, was also found to be glutathionylated and secreted. PRDX2 uses reduced TRX1 as an electron donor for hydrogen peroxide reduction. TRX1 is an oxidoreductase enzyme and can target proteins due to its thiol-disulfide exchange properties and has been found to have immunomodulatory effects. Indeed, TRX1 has been shown to regulate lymphocyte response to HIV-1 entry into lymphocytes by controlling CD4 membrane localization (32). Many cytokine receptors can be activated by disulfide bond-mediated dimerization, including the interleukin-3 (IL-3), IL-5, and granulocyte-macrophage colony-stimulating factor receptors (33). The enzymatic activity of both PRDX2 and TRX1 could therefore initiate the dimerization of cytokine receptors and promote inflammation. The glutathionylation and secretion of PRDX2 and TRX1 by activated macrophages thereby identifies a new danger signal whereby redoxkines could alter extracellular receptor complexes, further driving inflammatory responses.

GSH NEGATIVELY REGULATES CASPASE-1 ACTIVATION

The cytokine IL-1β is a central proinflammatory cytokine (34), whose overproduction can give rise to a range of diseases, including autoinflammatory diseases such as Muckle-Wells syndrome, as well as common diseases such as gout. Upon exposure to PAMPs, such as LPS, IL-1β transcription is induced by the transcription factor NF-κB to produce pro-IL-1β, also known as "signal 1." Pro-IL-1β must be cleaved to elicit an inflammatory phenotype. PAMP or danger-associated molecular pattern recognition stimulates the formation of a multicomponent complex termed the "inflammasome," which processes pro-IL-1β and pro-IL-18 into bioactive IL-1β and IL-18 via proteolytic cleavage by caspase-1, termed "signal 2" (35, 36). Several inflammasomes have been identified. Nucleotide-binding domain leucine-rich repeat-containing receptors (NLRs) are critical components. A feature common to NLRs is a central nucleotide-binding (NACHT) domain and C-terminal leucine-rich repeat (LRR) domain. The NLRs can further contain an N-terminal pyrin domain (PYD) and a caspase recruitment domain (CARD). NLR family members, such as NLRP3, form inflammasomes with apoptosis-associated speck-like protein containing a CARD (ASC) and procaspase-1 (37). The NLRP3 inflammasome is the most-studied inflammasome and has been shown to be activated by mitochondrial ROS (38, 39). GSH has been shown to play a regulatory role in this process. Upstream of caspase-1, GSH has been found to limit inflammasome activation. In a model of dextran sulfate sodium-induced murine colitis, which mimics inflammatory bowel disease, pretreatment of mice with dimethyl fumarate (DMF) dose-dependently reduced inflammatory bowel disease progression and weight loss. The mechanism of DMF protection was attributed to increased GSH levels and activity of the antioxidant enzyme NRF2, which transcribes a host of antioxidant genes, subsequently dampening mitochondrial ROS levels. NRF2 has also been shown to upregulate GCLc, the catalytic subunit of the rate-limiting enzyme in GSH synthesis, further driving GSH levels (40). DMF was found to decrease activation of the NLRP3 inflammasome, subsequently limiting caspase-1 activity (41). Aside from ROS disruption, research by Meissner et al. has reported that glutathionylation of caspase-1 on cysteine 362 and 397 directly inhibits its activity (Fig. 3). They have shown that SOD1, an enzyme that converts O_2^- into O_2 and H_2O_2, regulates the activity of caspase-1 and subsequent endotoxic shock response, with no effect on mitogen-activated protein kinase or NF-κB signaling (42). SOD1-deficient macrophages had higher levels of O_2^-, affecting the cellular redox environment, causing oxidation and glutathionylation of caspase-1 on Cys362 and Cys397. Overexpression of caspase-1-containing cysteine-to-serine mutants (C362S, C397S) yielded a higher production of IL-1β compared to wild-type caspase-1, due to GSH being unable to glutathionylate and inhibit caspase-1. When caspase-1-containing C362S or C397S was transfected into caspase-1-deficient macrophages, both mutants were overactive under normoxic or oxygen-rich conditions promoting IL-1β production. Conversely, when subjected to hypoxia, neither mutant of caspase-1 could increase IL-1β processing, indicating that Cys362 and Cys397 are redox sensitive and potential glutathionylation targets when oxidized to limit caspase-1 activation. Finally, SOD1-deficient mice were shown to be more resistant to LPS-induced sepsis when compared to both caspase-1-deficient and wild-type mice.

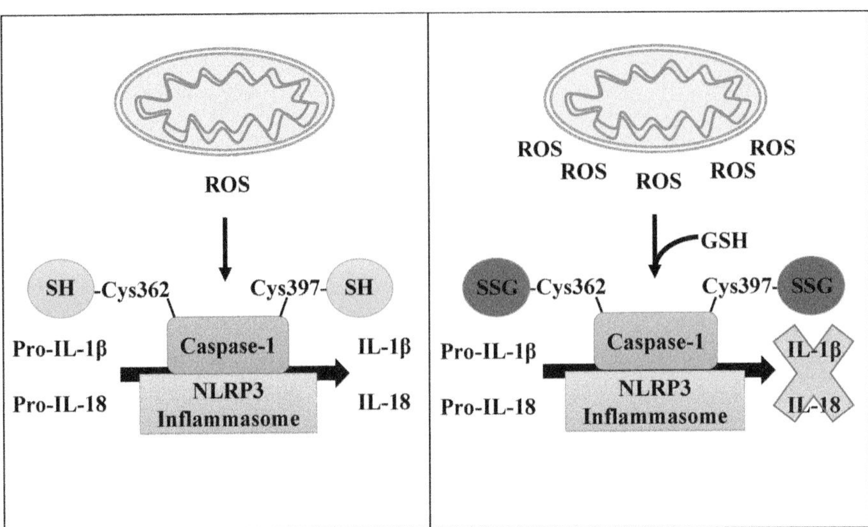

Figure 3 GSH-dependent inhibition of caspase-1 activity. (Left) Production of ROS from mitochondria drives NLRP3 inflammasome activation, producing mature IL-1β and IL-18 via caspase-1 activity. (Right) Over time, ROS accumulation leads to GSH binding to Cys362 and Cys397 on caspase-1, inhibiting caspase-1 catalytic activity. Oxidizing agents, such as superoxide and hydrogen peroxide, can induce glutathionylation of caspase-1, preventing cleavage of pro-IL-1β and pro-IL-18.

Serum cytokine analyses identified that SOD1-deficient mice had decreased IL-1β and IL-18 compared to wild-type mice. The ratio of GSH/GSSG is commonly used as a measure of oxidative imbalance (43). SOD1-deficient mice challenged with 3-h *Escherichia coli* infection and 30-min ATP, which drives inflammasome activation, were found to have >10-fold-higher GSSG levels compared to wild-type mice. SOD1-deficient mice, therefore, have a higher oxidative burden, which would result in oxidation and glutathionylation of caspase-1. GSH is therefore a negative regulator of caspase-1 activity both *in vitro* and *in vivo*. This presents something of a paradox. ROS are a positive signal for NLRP3 and yet inhibit caspase-1 by glutathionylation. It is possible that this is a matter of timing, whereby ROS will initially activate NLRP3 but then later limit caspase-1 by glutathionylation. This implicates GSH in the regulation of inflammation by ROS.

GSH STABILIZES HIF1α PROTEIN

Watanabe et al. have found that the transcription factor HIF1α is redox regulated by GSH in a mouse model of ischemia (44). GSH was shown to promote stability of HIF1α protein. Furthermore, the cell-permeable oxidizing agent oxidized GSH (GSSG-ethyl ester) was shown to increase HIF1α levels. Deletion of the deglutathionylating enzyme glutaredoxin-1 (Glrx) improved ischemic revascularization. *In vivo* analyses of Glrx deletion discovered that blood flow recovery after ischemic hind limb revascularization was significantly improved, concomitant with increased levels of vascular endothelial growth factor A (VEGF-A) and HIF1α. The deletion of Glrx is thereby highly advantageous during ischemic tissue recovery, as its absence will promote GSH binding, subsequently stabilizing HIF1α and thereby increasing angiogenesis and revascularization. To further implicate GSH as a modulator of HIF1α, MS identified that cysteine 520 (or cysteine 533 in mice) was bound to a GSH adduct on HIF1α. Functional mutation of cysteine 520 to serine (C520S) yielded decreased HIF1α stability (Fig. 4). These data clearly identify GSH as a positive regulator of HIF1α activity that may play a vital role in angiogenesis in response to ischemic tissue damage. Since HIF1α is also required for a number of macrophage responses (45, 46), GSH may positively regulate the expression of these genes via stabilization of HIF1α.

GSH IMPAIRS STAT3 PHOSPHORYLATION

Since its discovery as a potential oncogene, STAT3 has been the subject of intense research. STAT3 is activated by multiple macrophage-activating cytokines, including IL-6. Glutathionylation was shown to negatively regulate STAT3. Initial studies by Xie et al. identified that STAT3 could be glutathionylated when treated with diamide, which induces oxidative stress (47). Such

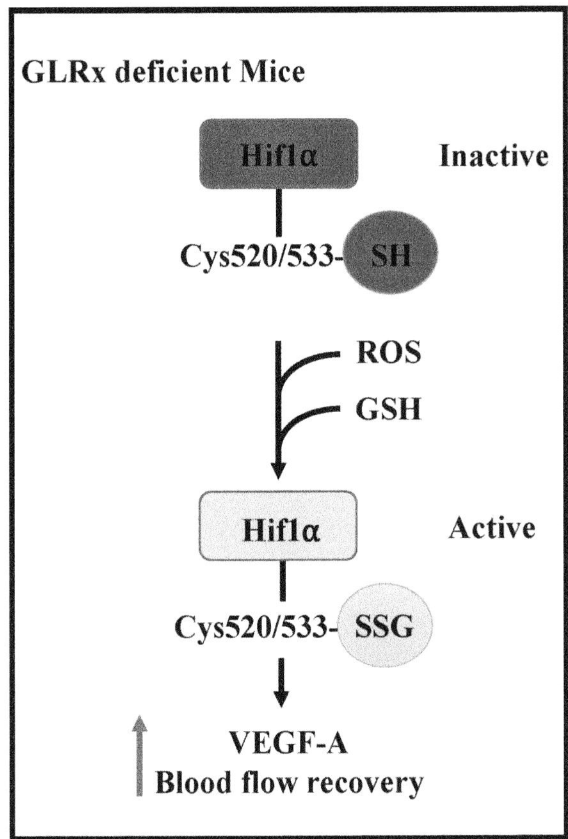

Figure 4 HIF1α protein is stabilized by glutathionylation. MS revealed two redox-sensitive cysteines that undergo glutathionylation, Cys520 in humans or Cys533 in mice. Ischemia-reperfusion of Glrx-deficient mice showed improved hind limb revascularization and increased HIF1α levels. HIF1α stability was increased due to deficiency of Glrx, increasing expression of VEGF-A and blood flow recovery.

treatment was found to decrease IL-6-dependent STAT3 signaling, along with decreased tyrosine phosphorylation and nuclear accumulation (Fig. 5). Further research by Butturini et al. identified the redox-sensitive cysteine residues in STAT3, cysteines 328 and 542, which directly impair phosphorylation when glutathionylated (48). Cotreatment with diamide and reduced GSH was found to induce glutathionylation of STAT3. Furthermore, STAT3 phosphorylation by JAK2 was abrogated with diamide and GSH treatment, indicating that GSH inhibits phosphorylation of STAT3. Functional mutagenesis of the key cysteine amino acids to serine (C328S, C542S) prevented STAT3 from being glutathionylated when pretreated with diamide. Further work is required to verify the downstream consequences of this negative regulatory event on STAT3, considering it plays a vital role in many essential pathways, from autophagy to oncogenesis.

GSH PROMOTES VIRULENCE GENE EXPRESSION DURING MACROPHAGE INFECTION WITH *LISTERIA MONOCYTOGENES*

GSH has been found to aid the activation of the master transcription factor of *Listeria monocytogenes*, PrfA, via allosteric regulation. *L. monocytogenes* is a Gram-positive intracellular pathogen that infects macrophages, and the success of its survival is entirely dependent on PrfA activation, as PrfA-deficient strains are avirulent (49). *L. monocytogenes* is one of the few Gram-positive strains of bacteria to produce GSH, as most other strains utilize other low-molecular-weight thiols, such as bacillithiol and mycothiol (50, 51). Reniere et al. have shown that the bifunctional *L. monocytogenes* glutathione synthase (gshF), a multidomain protein capable of synthesizing GSH without GCL, is critical for the survival of *L. monocytogenes* in macrophages (52). GSH was shown to react with PrfA. Mutation of key cysteines in PrfA (C38A, C144A, C205A, and C229A) decreased its ability to induce ActA, which is responsible for actin-based motility, limiting pathogenicity. GSH allosterically regulates PrfA, acting as a cofactor, as gshF-deficient strains are avirulent. In this instance, GSH acts as a cofactor to upregulate virulence genes in *L. monocytogenes* during macrophage infection (Fig. 6).

CONTROLLING GLUTATHIONYLATION IN MACROPHAGES: GSTO1-1

Our understanding of how glutathionylation is controlled in macrophages has also improved recently. A particular focus has been the enzyme glutathione transferase omega 1 (GSTO1-1). Glutathione transferases (previously termed glutathione *S*-transferases; GSTs) have come to the forefront in recent years as phase II detoxification enzymes that play a critical role in the regulation of cellular responses. Seven classes of GSTs have been identified, each class containing up to several isoforms that are involved in GSH conjugation reactions (53). As a broad family of GSH-conjugating enzymes, GST diversity can be viewed as evolutionarily advantageous. GSTO1-1 stands out from this family due to an atypical cysteine amino acid active site, which allows GSTO1-1 to carry out different enzymatic reactions, including thioltransferase activity. GSTO1-1 appears to play a part in multiple processes in macrophages, including antioxidant defense, enhanced glycolysis, xenobiotic detoxification, ROS production, and GSH cycling and cell survival crosstalk, which will be discussed later. Staining for GSTO1-1 by

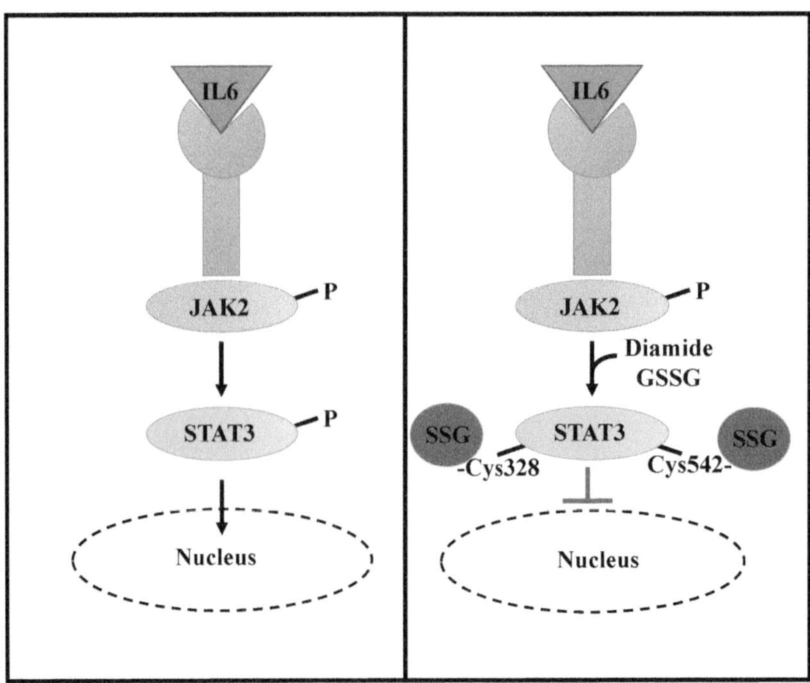

Figure 5 Glutathionylation of STAT3 prevents STAT3 phosphorylation and nuclear translocation. Activation of the IL-6 receptor by the cytokine IL-6 drives phosphorylation of JAK2. JAK2 subsequently phosphorylates STAT3, promoting nuclear translocation for upregulation of STAT3-dependent gene transcription. The oxidant diamide induces glutathionylation of two redox-sensitive cysteines on STAT3, Cys328 and Cys542, preventing JAK2-dependent STAT3 phosphorylation and nuclear translocation.

immunohistochemistry has revealed that the expression of GSTO1-1 is widespread in different tissues, but notably in macrophages (54). GSTO1-1 was detected in macrophages in the skin and lung but not in endothelial cells, lymphocytes, or smooth muscle cells in these tissues.

In response to LPS and interferon gamma (IFN-γ), macrophages can differentiate into a proinflammatory, M1 or M(LPS IFN-γ) macrophage, or alternatively they can be activated by the cytokine IL-4 into a resolving anti-inflammatory macrophage, termed M2 or M(IL–4) (55). The need to clarify specific subsets of genes upregulated by these stimuli can thereby aid as potential biomarkers for macrophage populations in disease states. A recent screen comparing undifferentiated human and mouse macrophages with IL-4-differentiated macrophages both *ex vivo* and *in vitro* revealed novel IL-4-controlled gene subsets. The authors identified transglutaminase 2 as a novel M2 marker that is consistently induced in both human and mouse macrophages with IL-4 treatment (56). Interestingly, the authors found that GSTO1-1 was also consistently detected within their analyses as a signature gene in both human and mouse IL-4-treated macrophages. GSTO1-1 could therefore become a potential biomarker for M2 macrophage populations. This represents somewhat of a paradox, as GSTO1-1 has been found to play roles in response to proinflammatory stimuli. GSTO1-1 has, however, been reported to play a role in the glutathionylation cycle (57), and endogenous IL-4 has been shown to drive GSH synthesis in a mouse model of liver injury through glutamate cysteine ligase enzymatic activity (58). In this manner, IL-4 signaling may increase GSTO1-1 levels; however, further examination into the role of GSTO1-1 in IL-4 signaling is required.

GSTO1-1 AND NF-κB TRANSLOCATION

NF-κB, one of the most widely characterized transcription factors that upregulates a plethora of inflammatory genes, has been shown to require GSTO1-1 for nuclear translocation in macrophages. NF-κB has previously been reported as a redox-sensitive protein. Oxidation of Cys38 in the p65 subunit of NF-κB and Cys62 in the p50 subunit will prevent DNA binding (59). Menon et al., examining the effect of GSTO1-1 deficiency via knockdown approaches in macrophages, discovered that the absence of GSTO1-1 prevented

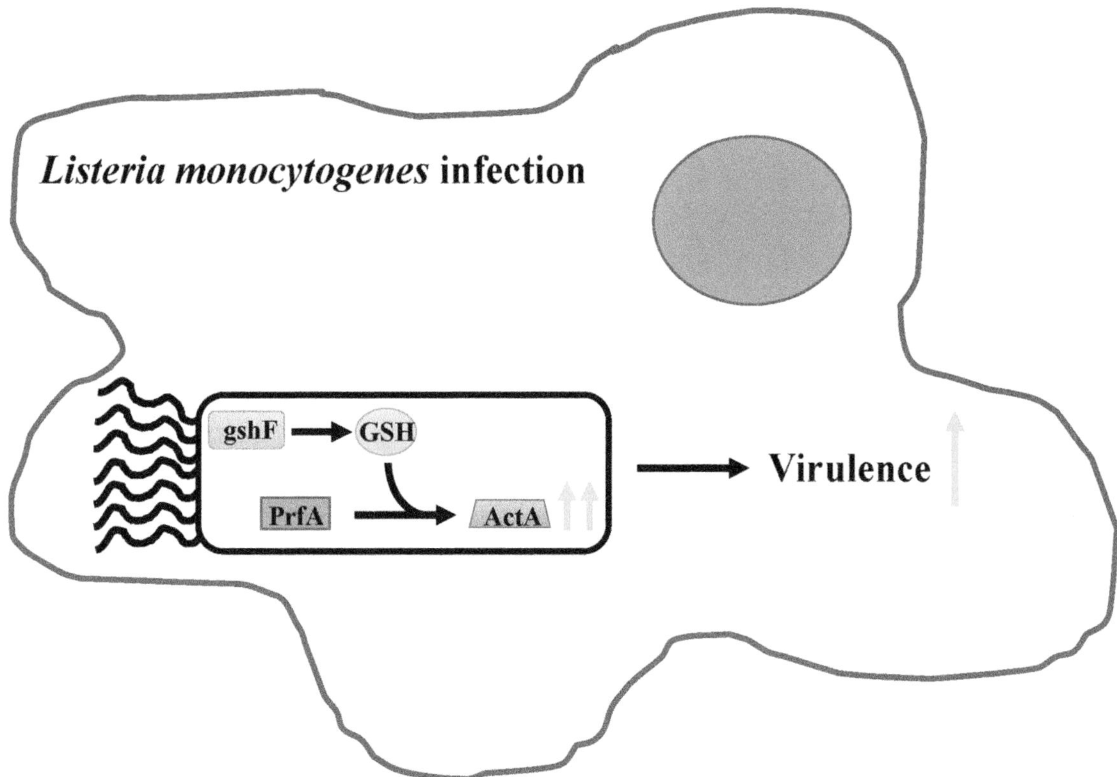

Figure 6 GSH acts as a cofactor to promote virulence gene expression by *L. monocytogenes*. Macrophage infection by *L. monocytogenes* is dependent on the master virulence gene transcription factor PrfA. PrfA transcribes ActA, an actin-mobility protein that promotes virulence success of *L. monocytogenes*. *L. monocytogenes* synthesizes its own GSH with the enzyme gshF. PrfA is only active when GSH binds as a cofactor, an event essential for virulence.

nuclear translocation of NF-κB upon LPS stimulation (60). This indicates that NF-κB may require GSTO1-1 for dissociation from IκBα, possibly by deglutathionylation. This would place GSTO1-1 as a potential downstream effector of TLR4 signaling. Components of the TLR4 signaling pathway are possibly deglutathionylated by GSTO1-1, allowing them to become active, which would promote NF-κB activation.

GSTO1-1 AS AN AUTOPHAGY GATEKEEPER

GSTO1-1 has recently been reported to play a key role in cell survival following treatment of macrophages with aflatoxin B_1 (AFB_1), a mycotoxin considered to be the most potent carcinogen found in contaminated food, with powerful immunosuppressive effects (61). GSTO1-1 has previously been linked to apoptosis regulation, preventing apoptosis via the activity of Jun N-terminal protein kinase 1 (JNK1) (62). Interestingly, Paul et al. have identified a cytoprotective role for GSTO1-1 in AFB_1-treated macrophages. A 6-h treatment of AFB_1 was reported to induce an elevation in ROS levels from mitochondria, resulting in loss of mitochondrial membrane potential and subsequent JNK-mediated caspase-dependent cell death (63). GSTO1-1 was identified as the causative link between autophagy and apoptosis, as small interfering RNA-mediated knockdown of GSTO1-1 mimicked 6-h AFB_1 treatment results, promoting apoptosis. Taken together, these data indicate that GSTO1-1 plays a role in cell survival via autophagy induction by agents such as AFB_1. Increased ROS and NF-κB activity was found to promote mitochondrial fission in hepatocellular carcinoma, enhancing autophagy and preventing apoptosis (64). GSTO1-1-mediated NF-κB translocation may thereby promote autophagy and could therefore be exploited by cancer cells. Further studies will be required to verify the role of GSTO1-1 in autophagy induction and if other xenobiotics could induce similar autophagy-related cell responses.

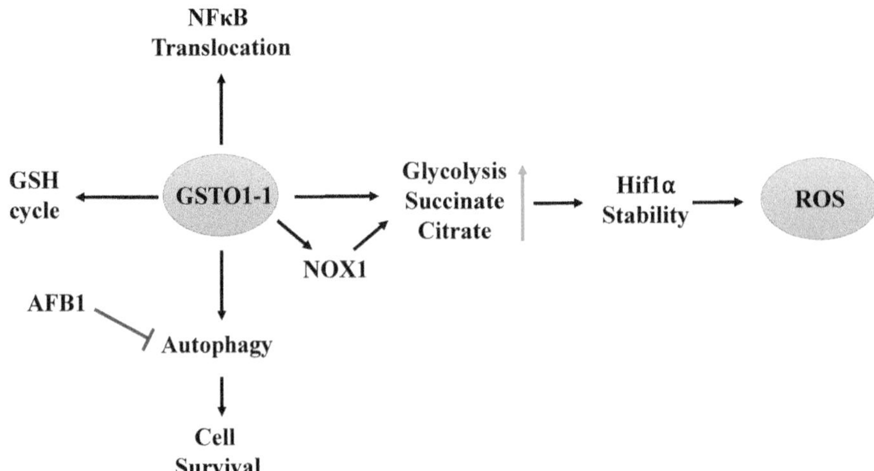

Figure 7 GSTO1-1 regulates macrophage responses to cellular processes. GSTO1-1 protects macrophages from AFB_1-induced apoptosis. GSTO1-1 is also essential for LPS action in macrophages, including translocation of NF-κB to the nucleus, increased glycolysis, elevated TCA cycle intermediate succinate levels, NOX-1 activation, and subsequent ROS induction. GSTO1-1 is therefore likely to regulate components in the TLR4 signaling pathway.

GSTO1-1 AND METABOLIC REPROGRAMMING IN MACROPHAGES

GSTO1-1 has been implicated in multiple responses induced by LPS in macrophages, notably induction of the ROS generator NOX1, enhanced mitochondrial ROS production by succinate, and enhanced glycolysis. The knockdown of GSTO1-1 in macrophages decreased expression of NOX1 after LPS stimulation. NOX1 is one of the main producers of ROS, along with mitochondria, and similar results were obtained utilizing the GSTO1-1 inhibitor ML175. The authors suggested that GSTO1-1 acts upstream of NOX1 on the TLR4 pathway (65), and further studies published by the same authors identified a critical role for GSTO1-1 in the modulation of macrophage metabolism via LPS. Indeed, LPS activation will skew the phenotype of macrophages toward a proglycolytic environment (66); however, this switch is significantly attenuated in GSTO1-1-deficient cells. In addition to the effect on NOX1, GSTO1-1-deficient cells did not increase mitochondrial ROS generation with LPS treatment, compared to a significant ROS enhancement in control cells, implicating GSTO1-1 in mitochondrial ROS production. Succinate accumulates in LPS-treated macrophages, which stabilizes HIF1α to promote IL-1β levels (15). The buildup of succinate and citrate was abolished in GSTO1-1-deficient cells, concomitant with decreased HIF1α production (67). These data clearly place GSTO1-1 as a key modulator of LPS responsiveness within macrophages. It is possible that GSTO1-1 regulates a key component of TLR4 signaling, which might impact deglutathionylation of adaptor proteins. This identifies GSTO1-1 as a key regulator of LPS responsiveness in macrophage activation (Fig. 7).

CONCLUDING REMARKS

Glutathionylation of proteins is emerging as a key posttranslational regulator in macrophages, with target proteins including HIF1α, STAT3, and caspase-1. A unifying aspect here is ROS, which promote glutathionylation, with HIF1α being stabilized and STAT3 and caspase-1 being inhibited. Additional proteins are likely to be identified in the so-called glutathionylome, with further complexities being revealed as to how ROS affect macrophage function. The GSH flux has also been found to affect monocyte-to-macrophage differentiation, being detrimental in atherosclerosis. GSTO1-1 has emerged as a potential regulator of these events, being highly expressed in macrophages and acting on LPS signaling. Several studies have linked GSTO1-1 to neurological disorders such as Alzheimer's disease and Parkinson's disease (68, 69). Further work on glutathionylation and GSTO1-1 in macrophages may lead to a greater understanding of these and possibly other inflammatory diseases.

Acknowledgments. *This work was funded by Science Foundation Ireland. The authors declare no financial or commercial conflict of interest.*

Citation. Hughes MM, McGettrick AF, O'Neill LAJ. 2017. Glutathione and glutathione transferase omega 1 as key posttranslational regulators in macrophages. Microbiol Spectrum 5(1):MCHD-0044-2016.

References

1. **Panth N, Paudel KR, Parajuli K.** 2016. Reactive oxygen species: a key hallmark of cardiovascular disease. *Adv Med* **2016:**9152732. doi:10.1155/2016/9152732.
2. **Gimenez M, Schickling BM, Lopes LR, Miller FJ Jr.** 2016. Nox1 in cardiovascular diseases: regulation and pathophysiology. *Clin Sci (Lond)* **130:**151–165.
3. **Zhang G, Li X, Sheng C, Chen X, Chen Y, Zhu D, Gao P.** 2016. Macrophages activate iNOS signaling in adventitial fibroblasts and contribute to adventitia fibrosis. *Nitric Oxide* **61:**20–28.
4. **Dey P, Panga V, Raghunathan S.** 2016. A cytokine signalling network for the regulation of inducible nitric oxide synthase expression in rheumatoid arthritis. *PLoS One* **11:**e0161306. doi:10.1371/journal.pone.0161306.
5. **Sandalio LM, Rodríguez-Serrano M, Romero-Puertas MC, del Río LA.** 2013. Role of peroxisomes as a source of reactive oxygen species (ROS) signaling molecules. *Subcell Biochem* **69:**231–255.
6. **Morris D, Khurasany M, Nguyen T, Kim J, Guilford F, Mehta R, Gray D, Saviola B, Venketaraman V.** 2013. Glutathione and infection. *Biochim Biophys Acta* **1830:** 3329–3349.
7. **Alanazi AM, Mostafa GA, Al-Badr AA.** 2015. Glutathione. *Profiles Drug Subst Excip Relat Methodol* **40:** 43–158.
8. **Montero D, Tachibana C, Rahr Winther J, Appenzeller-Herzog C.** 2013. Intracellular glutathione pools are heterogeneously concentrated. *Redox Biol* **1:**508–513.
9. **Zhou Y, Harrison DE, Love-Myers K, Chen Y, Grider A, Wickwire K, Burgess JR, Stochelski MA, Pazdro R.** 2014. Genetic analysis of tissue glutathione concentrations and redox balance. *Free Radic Biol Med* **71:**157–164.
10. **Perlman JH, Wang W, Nussenzveig DR, Gershengorn MC.** 1995. A disulfide bond between conserved extracellular cysteines in the thyrotropin-releasing hormone receptor is critical for binding. *J Biol Chem* **270:**24682–24685.
11. **Patil NA, Tailhades J, Hughes RA, Separovic F, Wade JD, Hossain MA.** 2015. Cellular disulfide bond formation in bioactive peptides and proteins. *Int J Mol Sci* **16:** 1791–1805.
12. **Paulech J, Liddy KA, Engholm-Keller K, White MY, Cordwell SJ.** 2015. Global analysis of myocardial peptides containing cysteines with irreversible sulfinic and sulfonic acid post-translational modifications. *Mol Cell Proteomics* **14:**609–620.
13. **Lu SC.** 2013. Glutathione synthesis. *Biochim Biophys Acta* **1830:**3143–3153.
14. **Marí M, Morales A, Colell A, García-Ruiz C, Fernández-Checa JC.** 2009. Mitochondrial glutathione, a key survival antioxidant. *Antioxid Redox Signal* **11:**2685–2700.
15. **Tannahill GM, Curtis AM, Adamik J, Palsson-McDermott EM, McGettrick AF, Goel G, Frezza C, Bernard NJ, Kelly B, Foley NH, Zheng L, Gardet A, Tong Z, Jany SS, Corr SC, Haneklaus M, Caffrey BE, Pierce K, Walmsley S, Beasley FC, Cummins E, Nizet V, Whyte M, Taylor CT, Lin H, Masters SL, Gottlieb E, Kelly VP, Clish C, Auron PE, Xavier RJ, O'Neill LA.** 2013. Succinate is an inflammatory signal that induces IL-1β through HIF-1α. *Nature* **496:**238–242.
16. **Mills EL, Kelly B, Logan A, Costa AS, Varma M, Bryant CE, Tourlomousis P, Däbritz JH, Gottlieb E, Latorre I, Corr SC, McManus G, Ryan D, Jacobs HT, Szibor M, Xavier RJ, Braun T, Frezza C, Murphy MP, O'Neill LA.** 2016. Succinate dehydrogenase supports metabolic repurposing of mitochondria to drive inflammatory macrophages. *Cell* **167:**457–470.e13. doi:10.1016/j.cell.2016.08.064.
17. **Garcia J, Han D, Sancheti H, Yap LP, Kaplowitz N, Cadenas E.** 2010. Regulation of mitochondrial glutathione redox status and protein glutathionylation by respiratory substrates. *J Biol Chem* **285:**39646–39654.
18. **Gao L, Laude K, Cai H.** 2008. Mitochondrial pathophysiology, reactive oxygen species, and cardiovascular diseases. *Vet Clin North Am Small Anim Pract* **38:**137–155, vi.
19. **Zhong H, Yin H.** 2015. Role of lipid peroxidation derived 4-hydroxynonenal (4-HNE) in cancer: focusing on mitochondria. *Redox Biol* **4:**193–199.
20. **Dou X, Li S, Wang Z, Gu D, Shen C, Yao T, Song Z.** 2012. Inhibition of NF-κB activation by 4-hydroxynonenal contributes to liver injury in a mouse model of alcoholic liver disease. *Am J Pathol* **181:**1702–1710.
21. **Ji GR, Yu NC, Xue X, Li ZG.** 2014. 4-Hydroxy-2-nonenal induces apoptosis by inhibiting AKT signaling in human osteosarcoma cells. *ScientificWorldJournal* **2014:** 873525. doi:10.1155/2014/873525.
22. **Youle RJ, Narendra DP.** 2011. Mechanisms of mitophagy. *Nat Rev Mol Cell Biol* **12:**9–14.
23. **Tsai WT, Lo YC, Wu MS, Li CY, Kuo YP, Lai YH, Tsai Y, Chen KC, Chuang TH, Yao CH, Lee JC, Hsu LC, Hsu JT, Yu GY.** 2016. Mycotoxin patulin suppresses innate immune responses by mitochondrial dysfunction and p62/sequestosome-1-dependent mitophagy. *J Biol Chem* **291:**19299–19311.
24. **Grek CL, Zhang J, Manevich Y, Townsend DM, Tew KD.** 2013. Causes and consequences of cysteine *S*-glutathionylation. *J Biol Chem* **288:**26497–26504.
25. **Forman HJ, Fukuto JM, Torres M.** 2004. Redox signaling: thiol chemistry defines which reactive oxygen and nitrogen species can act as second messengers. *Am J Physiol Cell Physiol* **287:**C246–C256.
26. **Chen YJ, Lu CT, Huang KY, Wu HY, Chen YJ, Lee TY.** 2015. GSHSite: exploiting an iteratively statistical method to identify S-glutathionylation sites with substrate specificity. *PLoS One* **10:**e0118752. doi:10.1371/journal.pone.0118752.
27. **Su D, Gaffrey MJ, Guo J, Hatchell KE, Chu RK, Clauss TR, Aldrich JT, Wu S, Purvine S, Camp DG, Smith RD, Thrall BD, Qian WJ.** 2014. Proteomic identification and quantification of S-glutathionylation in mouse macrophages using resin-assisted enrichment and isobaric labeling. *Free Radic Biol Med* **67:**460–470.
28. **Ullevig SL, Kim HS, Short JD, Tavakoli S, Weintraub ST, Downs K, Asmis R.** 2016. Protein *S*-glutathionylation mediates macrophage responses to metabolic cues from the extracellular environment. *Antioxid Redox Signal* **25:** 836–851.

29. **Yang X, Yao H, Chen Y, Sun L, Li Y, Ma X, Duan S, Li X, Xiang R, Han J, Duan Y.** 2015. Inhibition of glutathione production induces macrophage CD36 expression and enhances cellular-oxidized low density lipoprotein (oxLDL) uptake. *J Biol Chem* **290:**21788–21799.

30. **Vasamsetti SB, Karnewar S, Gopoju R, Gollavilli PN, Narra SR, Kumar JM, Kotamraju S.** 2016. Resveratrol attenuates monocyte-to-macrophage differentiation and associated inflammation via modulation of intracellular GSH homeostasis: relevance in atherosclerosis. *Free Radic Biol Med* **96:**392–405.

31. **Salzano S, Checconi P, Hanschmann EM, Lillig CH, Bowler LD, Chan P, Vaudry D, Mengozzi M, Coppo L, Sacre S, Atkuri KR, Sahaf B, Herzenberg LA, Herzenberg LA, Mullen L, Ghezzi P.** 2014. Linkage of inflammation and oxidative stress via release of glutathionylated peroxiredoxin-2, which acts as a danger signal. *Proc Natl Acad Sci U S A* **111:**12157–12162.

32. **Moolla N, Killick M, Papathanasopoulos M, Capovilla A.** 2016. Thioredoxin (Trx1) regulates CD4 membrane domain localization and is required for efficient CD4-dependent HIV-1 entry. *Biochim Biophys Acta* **1860:** 1854–1863.

33. **Stomski FC, Woodcock JM, Zacharakis B, Bagley CJ, Sun Q, Lopez AF.** 1998. Identification of a Cys motif in the common β chain of the interleukin 3, granulocyte-macrophage colony-stimulating factor, and interleukin 5 receptors essential for disulfide-linked receptor heterodimerization and activation of all three receptors. *J Biol Chem* **273:**1192–1199.

34. **McGettrick AF, O'Neill LA.** 2013. NLRP3 and IL-1β in macrophages as critical regulators of metabolic diseases. *Diabetes Obes Metab* **15**(Suppl 3)**:**19–25.

35. **Li P, Allen H, Banerjee S, Franklin S, Herzog L, Johnston C, McDowell J, Paskind M, Rodman L, Salfeld J, Towne E, Tracey D, Wardwell S, Wei FY, Wong W, Kamen R, Seshadri T.** 1995. Mice deficient in IL-1β-converting enzyme are defective in production of mature IL-1β and resistant to endotoxic shock. *Cell* **80:** 401–411.

36. **Kuida K, Lippke JA, Ku G, Harding MW, Livingston DJ, Su MS, Flavell RA.** 1995. Altered cytokine export and apoptosis in mice deficient in interleukin-1 beta converting enzyme. *Science* **267:**2000–2003.

37. **Haneklaus M, O'Neill LA, Coll RC.** 2013. Modulatory mechanisms controlling the NLRP3 inflammasome in inflammation: recent developments. *Curr Opin Immunol* **25:**40–45.

38. **Kim SR, Kim DI, Kim SH, Lee H, Lee KS, Cho SH, Lee YC.** 2014. NLRP3 inflammasome activation by mitochondrial ROS in bronchial epithelial cells is required for allergic inflammation. *Cell Death Dis* **5:**e1498. doi: 10.1038/cddis.2014.460.

39. **Ren JD, Wu XB, Jiang R, Hao DP, Liu Y.** 2016. Molecular hydrogen inhibits lipopolysaccharide-triggered NLRP3 inflammasome activation in macrophages by targeting the mitochondrial reactive oxygen species. *Biochim Biophys Acta* **1863:**50–55.

40. **Kim AD, Zhang R, Kang KA, You HJ, Hyun JW.** 2011. Increased glutathione synthesis following Nrf2 activation by vanadyl sulfate in human Chang liver cells. *Int J Mol Sci* **12:**8878–8894.

41. **Liu X, Zhou W, Zhang X, Lu P, Du Q, Tao L, Ding Y, Wang Y, Hu R.** 2016. Dimethyl fumarate ameliorates dextran sulfate sodium-induced murine experimental colitis by activating Nrf2 and suppressing NLRP3 inflammasome activation. *Biochem Pharmacol* **112:**37–49.

42. **Meissner F, Molawi K, Zychlinsky A.** 2008. Superoxide dismutase 1 regulates caspase-1 and endotoxic shock. *Nat Immunol* **9:**866–872.

43. **Schafer FQ, Buettner GR.** 2001. Redox environment of the cell as viewed through the redox state of the glutathione disulfide/glutathione couple. *Free Radic Biol Med* **30:**1191–1212.

44. **Watanabe Y, Murdoch CE, Sano S, Ido Y, Bachschmid MM, Cohen RA, Matsui R.** 2016. Glutathione adducts induced by ischemia and deletion of glutaredoxin-1 stabilize HIF-1α and improve limb revascularization. *Proc Natl Acad Sci U S A* **113:**6011–6016.

45. **Cramer T, Yamanishi Y, Clausen BE, Förster I, Pawlinski R, Mackman N, Haase VH, Jaenisch R, Corr M, Nizet V, Firestein GS, Gerber HP, Ferrara N, Johnson RS.** 2003. HIF-1α is essential for myeloid cell-mediated inflammation. *Cell* **112:**645–657.

46. **Cheng SC, Quintin J, Cramer RA, Shepardson KM, Saeed S, Kumar V, Giamarellos-Bourboulis EJ, Martens JH, Rao NA, Aghajanirefah A, Manjeri GR, Li Y, Ifrim DC, Arts RJ, van der Veer BM, Deen PM, Logie C, O'Neill LA, Willems P, van de Veerdonk FL, van der Meer JW, Ng A, Joosten LA, Wijmenga C, Stunnenberg HG, Xavier RJ, Netea MG.** 2014. mTOR- and HIF-1α-mediated aerobic glycolysis as metabolic basis for trained immunity. *Science* **345:**1250684. doi:10.1126/science.1250684.

47. **Xie Y, Kole S, Precht P, Pazin MJ, Bernier M.** 2009. *S*-Glutathionylation impairs signal transducer and activator of transcription 3 activation and signaling. *Endocrinology* **150:**1122–1131.

48. **Butturini E, Darra E, Chiavegato G, Cellini B, Cozzolino F, Monti M, Pucci P, Dell'Orco D, Mariotto S.** 2014. S-Glutathionylation at Cys328 and Cys542 impairs STAT3 phosphorylation. *ACS Chem Biol* **9:**1885–1893.

49. **Chakraborty T, Leimeister-Wächter M, Domann E, Hartl M, Goebel W, Nichterlein T, Notermans S.** 1992. Coordinate regulation of virulence genes in *Listeria monocytogenes* requires the product of the *prfA* gene. *J Bacteriol* **174:**568–574.

50. **Perera VR, Newton GL, Parnell JM, Komives EA, Pogliano K.** 2014. Purification and characterization of the *Staphylococcus aureus* bacillithiol transferase BstA. *Biochim Biophys Acta* **1840:**2851–2861.

51. **Liu Y, Yang X, Yin Y, Lin J, Chen C, Pan J, Si M, Shen X.** 2016. Mycothiol protects *Corynebacterium glutamicum* against acid stress via maintaining intracellular pH homeostasis, scavenging ROS, and *S*-mycothiolating MetE. *J Gen Appl Microbiol* **62:**144–153.

52. **Reniere ML, Whiteley AT, Hamilton KL, John SM, Lauer P, Brennan RG, Portnoy DA.** 2015. Glutathione activates virulence gene expression of an intracellular pathogen. *Nature* **517:**170–173.

53. **Board PG, Menon D.** 2013. Glutathione transferases, regulators of cellular metabolism and physiology. *Biochim Biophys Acta* **1830:**3267–3288.

54. **Yin ZL, Dahlstrom JE, Le Couteur DG, Board PG.** 2001. Immunohistochemistry of omega class glutathione S-transferase in human tissues. *J Histochem Cytochem* **49:**983–987.

55. **Murray PJ, Allen JE, Biswas SK, Fisher EA, Gilroy DW, Goerdt S, Gordon S, Hamilton JA, Ivashkiv LB, Lawrence T, Locati M, Mantovani A, Martinez FO, Mege JL, Mosser DM, Natoli G, Saeij JP, Schultze JL, Shirey KA, Sica A, Suttles J, Udalova I, van Ginderachter JA, Vogel SN, Wynn TA.** 2014. Macrophage activation and polarization: nomenclature and experimental guidelines. *Immunity* **41:**14–20.

56. **Martinez FO, Helming L, Milde R, Varin A, Melgert BN, Draijer C, Thomas B, Fabbri M, Crawshaw A, Ho LP, Ten Hacken NH, Cobos Jiménez V, Kootstra NA, Hamann J, Greaves DR, Locati M, Mantovani A, Gordon S.** 2013. Genetic programs expressed in resting and IL-4 alternatively activated mouse and human macrophages: similarities and differences. *Blood* **121:**e57–e69. doi:10.1182/blood-2012-06-436212.

57. **Menon D, Board PG.** 2013. A role for glutathione transferase Omega 1 (GSTO1-1) in the glutathionylation cycle. *J Biol Chem* **288:**25769–25779.

58. **Ryan PM, Bourdi M, Korrapati MC, Proctor WR, Vasquez RA, Yee SB, Quinn TD, Chakraborty M, Pohl LR.** 2012. Endogenous interleukin-4 regulates glutathione synthesis following acetaminophen-induced liver injury in mice. *Chem Res Toxicol* **25:**83–93.

59. **Toledano MB, Leonard WJ.** 1991. Modulation of transcription factor NF-κB binding activity by oxidation-reduction *in vitro*. *Proc Natl Acad Sci U S A* **88:**4328–4332.

60. **Menon D, Coll R, O'Neill LA, Board PG.** 2014. Glutathione transferase omega 1 is required for the lipopolysaccharide-stimulated induction of NADPH oxidase 1 and the production of reactive oxygen species in macrophages. *Free Radic Biol Med* **73:**318–327.

61. **Meissonnier GM, Pinton P, Laffitte J, Cossalter AM, Gong YY, Wild CP, Bertin G, Galtier P, Oswald IP.** 2008. Immunotoxicity of aflatoxin B1: impairment of the cell-mediated response to vaccine antigen and modulation of cytokine expression. *Toxicol Appl Pharmacol* **231:**142–149.

62. **Piaggi S, Raggi C, Corti A, Pitzalis E, Mascherpa MC, Saviozzi M, Pompella A, Casini AF.** 2010. Glutathione transferase omega 1-1 (GSTO1-1) plays an anti-apoptotic role in cell resistance to cisplatin toxicity. *Carcinogenesis* **31:**804–811.

63. **Paul S, Jakhar R, Bhardwaj M, Kang SC.** 2015. Glutathione-S-transferase omega 1 (GSTO1-1) acts as mediator of signaling pathways involved in aflatoxin B1-induced apoptosis-autophagy crosstalk in macrophages. *Free Radic Biol Med* **89:**1218–1230.

64. **Huang Q, Zhan L, Cao H, Li J, Lyu Y, Guo X, Zhang J, Ji L, Ren T, An J, Liu B, Nie Y, Xing J.** 2016. Increased mitochondrial fission promotes autophagy and hepatocellular carcinoma cell survival through the ROS-modulated coordinated regulation of the NFKB and TP53 pathways. *Autophagy* **12:**999–1014.

65. **Menon D, Coll R, O'Neill LA, Board PG.** 2014. Glutathione transferase omega 1 is required for the lipopolysaccharide-stimulated induction of NADPH oxidase 1 and the production of reactive oxygen species in macrophages. *Free Radic Biol Med* **73:**318–327.

66. **Palsson-McDermott EM, Curtis AM, Goel G, Lauterbach MA, Sheedy FJ, Gleeson LE, van den Bosch MW, Quinn SR, Domingo-Fernandez R, Johnston DG, Jiang JK, Israelsen WJ, Keane J, Thomas C, Clish C, Vander Heiden M, Xavier RJ, O'Neill LA.** 2015. Pyruvate kinase M2 regulates Hif-1α activity and IL-1β induction and is a critical determinant of the Warburg effect in LPS-activated macrophages. *Cell Metab* **21:**65–80.

67. **Menon D, Coll R, O'Neill LA, Board PG.** 2015. GSTO1-1 modulates metabolism in macrophages activated through the LPS and TLR4 pathway. *J Cell Sci* **128:** 1982–1990.

68. **Li YJ, Oliveira SA, Xu P, Martin ER, Stenger JE, Scherzer CR, Hauser MA, Scott WK, Small GW, Nance MA, Watts RL, Hubble JP, Koller WC, Pahwa R, Stern MB, Hiner BC, Jankovic J, Goetz CG, Mastaglia F, Middleton LT, Roses AD, Saunders AM, Schmechel DE, Gullans SR, Haines JL, Gilbert JR, Vance JM, Pericak-Vance MA, Hulette C, Welsh-Bohmer KA.** 2003. Glutathione *S*-transferase omega-1 modifies age-at-onset of Alzheimer disease and Parkinson disease. *Hum Mol Genet* **12:** 3259–3267.

69. **Li YJ, Scott WK, Zhang L, Lin PI, Oliveira SA, Skelly T, Doraiswamy MP, Welsh-Bohmer KA, Martin ER, Haines JL, Pericak-Vance MA, Vance JM.** 2006. Revealing the role of glutathione *S*-transferase omega in age-at-onset of Alzheimer and Parkinson diseases. *Neurobiol Aging* **27:** 1087–1093.

Myeloid Cells in Health and Disease: A Synthesis
Edited by Siamon Gordon

doi:10.1128/microbiolspec.MCHD-0037-2016

Tomas Ganz[1]

46 Macrophages and Iron Metabolism

INTRODUCTION

Macrophages perform a vast range of biological functions, including regulation of embryonic development; scavenging and recycling of redundant, aged, or injured cells; modulation of tissue repair; and coordination and effector activity in host defense. By recycling and storing iron from senescent erythrocytes and other damaged cells, the macrophage controls iron homeostasis, supplying most of the iron needed for hemoglobin synthesis in erythrocyte precursors, and for the much smaller but important iron requirements of other cell populations. The macrophages serve a crucial regulatory role by functioning as a regulated storage compartment for iron. In response to systemic iron requirements, the release of iron from macrophages into plasma is negatively regulated by the interaction of the hepatic hormone hepcidin with its receptor/iron exporter ferroportin. In humans, macrophages contribute most of the iron entering the plasma compartment, with the rest of the iron influx into plasma made up from duodenal iron absorption and release of stored iron from hepatocytes. During infection and inflammation, interleukin-6, and to a lesser extent other cytokines, increases hepcidin synthesis. Hepcidin binds to macrophage ferroportin, induces its endocytosis and proteolysis, and thereby causes iron sequestration in macrophages. The resulting decrease in iron availability in other tissues can limit the growth and pathogenicity of invading extracellular microbes and is as an important means of host defense. Finally, bone marrow macrophages also have an important role in supporting efficient and rapid production of erythrocytes. The involvement of macrophages in iron metabolism thus serves both trophic and host defense functions. This review addresses the role of macrophages in iron metabolism in all these contexts, and represents an update and expansion of my previous discussion of the same subject (1).

ORGANISMAL IRON FLOWS

The average human male adult contains ~4 g of iron, of which ~2.5 g is in hemoglobin of erythrocytes. Women have a lower total body iron and fewer erythrocytes, even when adjustments are made for body weight. In the adult male, ~1 g of iron is stored predominantly in the hepatocytes and macrophages of the liver and to a lesser extent in splenic macrophages; women of reproductive age store much less iron and are prone to iron deficiency, as are children, who also require iron for body growth and the corresponding expansion of erythrocyte volume. Natural diets contain relatively little bioavailable iron, so that humans and most other

[1]Departments of Medicine and Pathology, David Geffen School of Medicine, University of California, Los Angeles, CA 90095.

vertebrates evolved to strictly conserve and internally recycle iron. In men and nonmenstruating women, the daily loss of iron from the body by desquamation and minor bleeding is only ~1 to 2 mg (or <0.05% of total body iron content) and must be replaced by intestinal iron absorption from the diet. Much more iron (~25 mg/day) is obtained from recycling of senescent erythrocytes and other damaged cells and reused to supply the iron content of the replacement cells. Because erythrocytes contain much more iron than any other cell type, nearly all the major flow of iron in the body is in the hemoglobin cycle, where iron is recovered by macrophages from the hemoglobin of old erythrocytes and supplied to erythrocyte precursors in the bone marrow for hemoglobin synthesis. Since the life span of erythrocytes is ~120 days, 20 to 25 mg of iron/day flows through the hemoglobin cycle, and 20 to 25 ml (or 2×10^{11}) of packed erythrocytes/day are removed and replaced by new ones. In plasma, iron is bound to transferrin, which is typically saturated to about one-third of its iron-carrying capacity, containing ~2 mg of iron. Total extracellular fluid volume is ~5 times larger, containing ~10 mg of iron. This means that the transferrin iron pool turns over several times a day.

SYSTEMIC IRON REGULATION BY HEPCIDIN AND FERROPORTIN

The concentration of iron in plasma and extracellular fluid is determined by the balance between the supply of iron from dietary iron absorption and release from stores versus the consumption of iron, predominantly for hemoglobin synthesis in erythrocyte precursors. The absorption of dietary iron and its release from stores are regulated by hepcidin, a 25-amino-acid peptide hormone produced predominantly by hepatocytes. Hepcidin controls the transfer of iron from cells involved in dietary absorption and storage to plasma and extracellular fluid. The membrane protein ferroportin is the sole known conduit through which iron can cross the cellular membrane (2), and it is also the receptor for the hepatic peptide hormone hepcidin (3). Export of iron through ferroportin is therefore the most important control point for the regulation of iron concentration in plasma. Hepcidin causes the endocytosis and eventual proteolysis of ferroportin and thereby controls iron export from macrophages and other professional iron-transporting cells, including hepatocytes, duodenal enterocytes, and placental syncytiotrophoblasts. The production of hepcidin in hepatocytes is feedback regulated by plasma and tissue iron concentrations as well as by the erythroid demand for iron. All known regulation of hepcidin is transcriptional, with the bone morphogenetic protein (BMP) receptor and its canonical SMAD transduction pathway as the principal regulator of hepcidin (4). It appears that several iron sensors and adaptors modulate the sensitivity of the receptor to its iron-related ligand, BMP-6, and also affect the production of BMP-6 (5). Although hepcidin mRNA is also detected in small amounts in macrophages, the fate and biological role of macrophage-derived hepcidin is not well understood.

IRON SEQUESTRATION AND HOST DEFENSE

Iron is an essential trace element for nearly all living organisms. The tenuous supply of bioavailable iron is the Achilles' heel of many pathogenic microbes, and it is not surprising that defense mechanisms evolved to target this susceptibility. However, some microbes have evolved multiple mechanisms that allow them to obtain iron even under very restrictive conditions. Although such mechanisms promote microbial survival, they have energetic and metabolic costs that impair microbial fitness in the competition with other microbes for niches within specific hosts and varied environments.

Multiple host mechanisms restrict iron availability during infections (6). Upon contact with mucous epithelia, microbes encounter secretions that contain lactoferrin, an avid iron binder related to the carrier protein transferrin. When neutrophils are recruited to the site of infection, they also carry and secrete iron-sequestering lactoferrin from their secondary granules, so that iron becomes unavailable to microbes that lack countermeasures for freeing it up. Siderophores, small organic iron chelators used by microbes to secure iron from the environment, allow microbes to wrest iron from many ferroproteins. However, siderophores can also be "stolen" and utilized by other microbes that compete for the same niche, compromising the potential advantages of this iron acquisition strategy. The host also evolved ways of decreasing siderophore concentration as both neutrophils and epithelia synthesize siderocalin (lipocalin-2; neutrophil gelatinase-associated lipocalin), a protein that avidly binds certain siderophores. Highly host-adapted pathogens, such as *Neisseria*, can take up host ferroproteins by dedicated receptors, exemplified by a host-specific mechanism for the uptake of transferrin (7).

Hypoferremia of inflammation is a common response to infection in vertebrates. Within hours after infection, the concentration of iron bound to transferrin is greatly decreased by inflammatory signals that override the normal homeostatic regulation of iron concentrations and fluxes. Although it is taken for granted that this response somehow contributes to host defense, how this works is far from obvious. In particular, transferrin binds iron very tightly and transferrin-bound iron is in principle not available to most microbes regardless of its concentration. Alternatively, the function of the hypoferremic response may be to prevent the generation of much more accessible non-transferrin-bound iron, produced when transferrin is (nearly) saturated with iron. Without this mechanism, plasma iron concentration and the saturation of transferrin may greatly increase (8) when the efflux of iron from macrophages is augmented by scavenging of infection/inflammation-damaged erythrocytes and other injured tissues, and the consumption of iron is decreased by inhibitory effects of cytokines on erythropoiesis.

ALTERED IRON HOMEOSTASIS DURING INFECTION AND INFLAMMATION

During infection and inflammation, increased production of inflammatory cytokines drives the reprogramming of the synthetic repertoire of the liver to serve host defense, the so-called acute-phase response. Interleukin-6, and perhaps other cytokines, increases the transcription of the iron regulatory hormone hepcidin (9, 10), as well as the synthesis of a number of other proteins involved in the scavenging and sequestration of iron in tissues, including ceruloplasmin, siderocalin, haptoglobin, hemopexin, and ferritin. Increased hepcidin binds to macrophage ferroportin and causes its endocytosis and degradation, thereby trapping iron in cytoplasmic ferritin of macrophages that recycle aged erythrocytes. The decreased delivery of iron from macrophages to blood plasma coupled with continuing consumption of iron for erythropoiesis cause the characteristic hypoferremia, and eventually anemia of inflammation. Paradoxically, transferrin synthesis is decreased, leading eventually to lower transferrin concentrations as the protein turns over. The lower concentration of this protein leads to increased utilization of both of its iron-binding sites and preserves the supply of diferric transferrin for erythropoiesis in the face of prolonged inflammation-induced hypoferremia. The importance of hepcidin for host defense is documented by the deleterious effects of hepcidin deficiency: increased susceptibility of patients with hepcidin deficiency and iron overload to infections with highly iron-dependent microbes, such as *Yersinia* and *Vibrio* species (11–14). Detailed experiments in mouse models of these infections indicate that the critical action of hepcidin in this context may be its ability to decrease the generation of non-transferrin-bound iron, which acts as a potent stimulus for the rapid growth of some bacteria (15).

THE ROLE OF MACROPHAGES IN IRON RECYCLING

The macrophage is a key agent in iron homeostasis as well as in inflammatory hypoferremia. Unlike other cell types that take up iron predominantly in the form of diferric transferrin, macrophages in the spleen, in the liver (Kupffer cells), and perhaps elsewhere recognize damaged or senescent erythrocytes, phagocytize them, and digest them to extract heme and eventually iron (16). It has been estimated that in laboratory rats each of these macrophages ingests about one erythrocyte per day, without any apparent harmful effect on the macrophage. However, higher rates of erythrophagocytosis, as may occur during hemolytic diseases, may cause injury to macrophages (17). Heme is degraded through the action of heme oxygenase (predominantly heme oxygenase-1 [HO-1]) to release iron into the cytoplasm for eventual export by plasma membrane ferroportin. In humans, iron flux from macrophages greatly exceeds inflows from dietary iron absorption and from iron stored in hepatocytes. In laboratory mice, which consume much more food relative to their body mass, the iron flow from enterocytes is comparable to that from macrophages.

ERYTHROPHAGOCYTOSIS

As erythrocytes age, they progressively change, becoming stiffer and displaying membrane markers of aging, but in humans they are not normally ingested by macrophages until they are close to their terminal age of ~120 days or undergo eryptosis (18), a suicidal process distantly resembling apoptosis but occurring in the absence of nuclei or mitochondria, and triggered by chemical or osmotic stress or energy depletion. Based on the fraction of blood flow to the spleen, each human erythrocyte transits the splenic circulation every 20 min, and there is subjected to several types of quality control (19). Erythrocytes come into contact with macrophage receptors in the red pulp probing for the display of opsonic antibodies or pathological surface markers, and are subjected to deformability challenge as erythrocytes return to the circulation through narrow endothelial slits. Aging changes must reach a specific threshold that is recognized by macrophages, initiating erythrophagocytosis. Alterations that may be recognized by macrophages as markers of erythrocyte aging (20–23) include (i) modifications of the most abundant erythrocyte membrane protein, Band 3, inducing its membrane clustering (probably the most important factor); (ii) the appearance of phosphatidylserine on the outer leaflet of the plasma membrane, where it can be recognized by specific macrophage or endothelial receptors, or opsonins that bind to phosphatidylserine (21, 24); (iii) increased membrane rigidity; and possibly (iv) the loss of sialic acid and the CD47 antigen. Clustering of Band 3 protein is caused by the binding of denatured hemoglobin to its cytoplasmic domain, and may be promoted also by age-related proteolytic cleavage or covalent modifications. The modified and clustered Band 3 protein is recognized by opsonic natural antibodies and complement, triggering conventional antibody- and complement-mediated phagocytosis. Eryptosis (25) is initiated in many pathological states that injure erythrocytes, including systemic infections, malaria, hemoglobinopathies, and genetic defects of erythrocyte metabolism. Eryptotic cells are shrunken, with blebbed membranes and phosphatidylserine exposed in the outer leaflet of the plasma membrane. Phosphatidylserine receptors and phosphatidylserine-specific opsonins are particularly important for the phagocytosis of eryptotic erythrocytes. The erythrophagocytosis of stressed or damaged erythrocytes by macrophages also has a host defense function because it removes from circulation cells that have been parasitized by malarial or other parasites of erythrocytes. Mild genetic erythrocyte abnormalities (sickle cell trait, glucose-6-phosphate dehydrogenase deficiency, and others) protect the carriers from lethal malaria in part through the enhanced clearance of parasitized erythrocytes in these conditions (26).

RECOVERY OF IRON FROM ERYTHROCYTES

In the phagocytic vacuole, the ingested erythrocyte is digested by exposure to reactive oxygen species and to hydrolytic enzymes, leading to the release of hemoglobin and eventually heme into vacuolar fluid. Subsequently, the heme-inducible enzyme HO-1 uses molecular oxygen and NADPH to cleave heme into equimolar amounts of iron, carbon monoxide, and biliverdin (27–29). Biliverdin is then reduced to bilirubin by biliverdin reductase. HO-1 is a membrane-anchored protein highly expressed in the liver and the spleen, and primarily but not exclusively located in the endoplasmic reticulum, from where it could be delivered when this compartment contributes to the formation of the phagosomal membrane. Although there is evidence that the extraction of iron from heme occurs in the phagosomal lumen, with consequent iron transport across the membrane (30), recent studies indicate that heme is transported across the membrane by the heme transporter HRG-1 (heme-responsive gene-1) and converted to iron on the cytoplasmic side (31; reviewed in 29). HO-1 is essential for the recovery of iron from heme, as indicated by the consequences of HO-1 deficiency for iron homeostasis in mouse models (27, 28). Ablation of HO-1 resulted in an iron-restricted anemia with low serum iron, low transferrin saturation, and low mean corpuscular volume but iron deposition in the liver and other tissues. This pattern is indicative of a defect in iron recycling accompanied by maldistribution of stored iron. The many other manifestations of HO-1 deficiency in mice and two human patients (32, 33) are presumed to result from the toxicity of free heme normally neutralized by HO-1, or the role of HO-1 and its enzymatic products in the regulation of inflammation and apoptosis. In HO-1-deficient patients and mice, macrophages are among the cell types primarily affected by heme toxicity, and the destruction of erythrophagocytosing macrophages in the spleen and the liver then further exacerbates heme-mediated injury to endothelia and other tissues (28). A second heme oxygenase, HO-2, is constitutively expressed in many tissues but at a lower level, apparently insufficient to compensate for HO-1 deficiency.

ALTERNATIVE RECYCLING PATHWAYS

Although most senescent or damaged erythrocytes are removed from the circulation by macrophages, intravascular hemolysis may reach comparable levels under pathological or stress conditions, including such common situations as vigorous exercise. Ruptured erythrocytes release hemoglobin into plasma, where hemoglobin binds to its dedicated carrier, haptoglobin. Hemoglobin-haptoglobin complexes are endocytosed by hepatocytes and macrophages via the CD163 receptor (34, 35), and the iron recovered by processes similar to those that follow erythrophagocytosis. Oxidation of hemoglobin that results in the oxidation of its heme-associated ferrous iron to ferric iron causes the release of the resulting hemin (heme containing ferric instead of ferrous iron) into circulation, where it binds to hemopexin. Hemopexin-heme complexes are cleared by hepatocytes and macrophages via the CD91 receptor (34, 36), and the iron is extracted by heme oxygenase. In HO-1-deficient mice and humans, the loss of macrophages engaged in iron recycling (28) causes intravascular hemolysis and the release of hemoglobin, leading to the generation of hemoglobin-haptoglobin complexes. At the same time, increased release of heme into the circulation generates heme-hemopexin complexes. The complexes are taken up by hepatocytes and other tissues expressing the CD163 and CD91 receptors (34), where they cause abnormal iron deposition, presumably facilitated by remaining constitutive HO-2 activity.

CONSTITUTIVE ERYTHROPHAGOCYTOSIS IS CARRIED OUT BY SPECIALIZED MACROPHAGES

Splenic red pulp macrophages are a specialized, highly erythrophagocytic cell type, characterized in mice as F4/80^{hi}CD68^{+}CD11b$^{lo/-}$ splenic cells with intense autofluorescence (37). Their development is dependent on the expression of the transcription factor Spi-C, selectively expressed in red pulp macrophages. Mice null for Spi-C lack this macrophage subset but not other macrophage subtypes in the spleen or other tissues. Spi-C-deficient mice accumulate iron in the spleen, consistent with their low expression of splenic macrophage ferroportin, resulting in a restricted capacity for exporting recycled iron. Spi-C is induced by macrophage colony-stimulating factor but much less by granulocyte-macrophage colony-stimulating factor, consistent with the proposed role of these growth factors in the differentiation of macrophages for erythrophagocytosis and other scavenging tasks versus inflammation/host defense. Compared to other macrophage subtypes, red pulp macrophages show increased expression of genes required for erythrophagocytosis and iron recycling (CD163, ferroportin, and heme oxygenase) and the adhesion molecule vascular cell adhesion molecule-1. Vascular cell adhesion molecule-1 appears to be directly under Spi-C control, but the other molecules characteristic of erythrophagocytosis, HO-1 and ferroportin, may be induced by exposure of macrophages to heme and iron rather than directly by Spi-C (38).

REGULATION OF MACROPHAGE PHENOTYPE BY IRON AND HEME

Iron and heme release in the macrophage induce the expression of molecules necessary for iron recycling. HO-1 transcription is increased by the heme-dependent dissociation of the repressor Bach1 from multiple sites in the HO-1 promoter ("Maf recognition elements," or MARE sites) allowing the Maf activators to bind to these sites to activate HO-1 transcription (39). Both activators and repressors heterodimerize with Nrf transcription factors. Ferroportin and ferritin transcription is also induced by heme by similar mechanisms (38, 40, 41). In addition, iron increases the synthesis of ferritin and ferroportin by relieving the translational block mediated by the attachment of iron regulatory proteins (IRP-1 and IRP-2) to the 5′ iron regulatory elements in their mRNAs (42). More recently, the microRNA MiR-485-3p induced during iron deficiency has also been found to repress ferroportin translation, so it could function in parallel with IRPs (43).

MACROPHAGE ACTIVATION BY EXCESS HEME

In some hemolytic disorders, especially malaria and sickle cell anemia, a prominent inflammatory component is seen. This has been attributed to the ability of heme to bind weakly to Toll-like receptor 4 and to catalyze the production of reactive oxygen species (44, 45). Injections of heme into mice elicit hepatic inflammatory damage and fibrosis and cause phenotypic changes in macrophages indicative of inflammation. These changes were seen to be exacerbated in hemopexin-deficient mice and partially prevented by the administration of hemopexin (46). The proinflammatory effect of hemolysis is not prominent in all hemolytic disorders, arguing for likely synergistic effects with other inflammatory stimuli, such as the presence of malaria parasites and sickling-induced ischemia and leukocyte and platelet activation.

SUBCELLULAR TRAFFICKING OF IRON IN MACROPHAGES

Iron recovered within the phagolysosome must be moved across the phagosomal membrane, undergo chaperoned transport through the cytoplasm, and be delivered as ferrous iron to plasma membrane ferroportin, for insertion into iron-containing proteins or storage within ferritin. Ferric reductases and ferroxidases participate in transport by facilitating the interconversion of ferrous and ferric iron to match the chemical requirements of iron transporters and storage mechanisms. Cytoplasmic ferritin serves to store macrophage iron when systemic demand for iron is low or inflammation leads to iron sequestration in macrophages. Two related ferrous iron transporters, divalent metal transporter 1 (DMT-1; also called Nramp2, or natural resistance-associated macrophage protein 2) and Nramp1, are required for normal export of iron from phagosomes (30, 47), and appear to function in a partially redundant manner. The subsequent journey of iron through the cytoplasm must be facilitated by chaperones that negate the chemical reactivity and toxicity of iron, but the identity of these molecules is unknown. Two related chaperones, human poly(rC)-binding proteins (PCBP1 and -2), have been implicated in the delivery of iron to ferritin and to certain other nonheme ferroproteins (48, 49), but it is not yet clear if the same (50) or different proteins deliver cytoplasmic iron for iron export from macrophages.

IRON EXPORT FROM MACROPHAGES

Although the details of iron transport through ferroportin are not yet understood, the structure of a bacterial homolog was recently resolved in two conformations (51). The structure confirmed that ferroportin is a member of the major facilitator superfamily of transporters and that it likely functions by binding ferrous iron to its cytoplasmic opening and then undergoing a conformational change that releases the iron atom to extracellular fluid. It is likely that an anionic species is cotransported, but the details are not yet understood. Optimal function of ferroportin requires a ferroxidase to convert the exported ferrous iron to the ferric form and to load it onto its plasma carrier transferrin. The copper-containing ferroxidase ceruloplasmin catalyzes the oxidation reaction and thereby facilitates iron efflux (52, 53), as indicated by the impaired export of iron from macrophages and hepatocytes in ceruloplasmin-deficient mice and humans. In addition to the abundant soluble ceruloplasmin originating from hepatocytes and circulating in plasma, a membrane-anchored glycosylphosphatidylinositol-containing form of ceruloplasmin is expressed in macrophages (52). Both forms may stabilize ferroportin by converting ferrous to ferric iron and thus facilitating the flow of ferrous iron through the molecule. Plasma transferrin with its two iron-binding sites is the ultimate carrier of ferric iron, with most of the cargo destined for erythropoiesis, which is wholly dependent on transferrin-bound iron. When the carrying capacity of transferrin is exceeded in iron overload or in transferrin deficiency, non-transferrin-bound iron is generated (54, 55), consisting of iron associated with albumin or with small organic molecules such as citrate or acetate. This form of iron is avidly taken up by hepatocytes but also by cardiac myocytes and by pancreas and endocrine tissues not normally involved in iron storage, and accounts for the development of liver disease and other organ damage in hereditary hemochromatosis and other iron overload diseases.

THE SUBCELLULAR DISTRIBUTION OF IRON TRANSPORTERS IN MACROPHAGES

Macrophages express iron transporters that are capable of moving iron across the membrane in both directions, DMT-1 into the cytoplasm and ferroportin out of the cytoplasm. In other cell types, e.g., the duodenal enterocyte, the importer DMT-1 is on the apical membrane and the exporter ferroportin on the basolateral membrane, effecting the flow of iron from the intestinal lumen to the blood plasma. In macrophages, there must be equivalent segregation of the importers and exporters, presumably to the phagosomal membrane, as was observed for the DMT-1 homolog Nramp1, and the plasma membrane, as for ferroportin (56). However, the phagosomal membrane originates in the plasma membrane, so the process requires either subcellular storage of the importers and exporters, with selective delivery to different membranes, or some other mechanism for their segregation. Cytoplasmic vesicles containing ferroportin have been observed in resting macrophages (57) and apparently translocate to the cell membrane when macrophages are iron loaded or subjected to erythrophagocytosis. It is not yet clear how iron importers traffic to their subcellular sites and how the trafficking of iron transporters in macrophages is regulated in response to iron loading or other stimuli.

MACROPHAGE IRON AND HOST DEFENSE AGAINST INTRACELLULAR MICROBES

The iron transporter Nramp1 (SLC11a1 in systematic nomenclature) expressed in murine macrophages mediates their resistance to intracellular microbes, including *Mycobacterium bovis* BCG, *Leishmania donovani*, and *Salmonella enterica* serovar Typhimurium (58). Mice or humans deficient in Nramp1 have increased susceptibility to infections with these or closely related pathogens. In agreement with its role in host defense, Nramp1 synthesis is increased in macrophages activated by gamma interferon (59). In further support of its proposed role, Nramp1 is not only found in macrophages but is also abundant in human neutrophils in the membranes of gelatinase-containing granules and is delivered to phagosomes during phagocytosis. There is controversy (reviewed in reference 60) over whether Nramp1 functions similarly to DMT-1 (Nramp2) to move iron and other divalent metals along a proton gradient, i.e., out of acidified phagosomes and into the cytoplasm (61), presumably depriving phagocytosed microbes of essential trace metals, or in the opposite direction, into the phagosome (62, 63), to elicit antimicrobial activity via the Fenton-Haber-Weiss reactions that generate the toxic hydroxyl radical.

THE EFFECT OF IRON ON THE IMMUNE AND INFLAMMATORY FUNCTIONS OF MACROPHAGES

Like erythrophagocytic macrophages in the spleen and liver, macrophages in hemorrhagic tissues (e.g., from physical injury, hemorrhagic stroke, or bleeding into an atheroma) must also scavenge damaged or lysed erythrocytes. Remarkably, iron-loaded macrophages in chronic venous leg ulcers exhibit a phenotype that is reported to promote inflammation by increased release of tumor necrosis factor α and reactive oxygen species and retard wound healing by causing DNA damage to fibroblasts and activating their senescence program (64). In experimental models of skin injury in mice, these effects could be reproduced by coadministering iron-dextran and ameliorated by desferrioxamine, suggesting that they were iron dependent. However, proteomic analysis of human monocyte-macrophages cultured *in vitro* with hemoglobin-haptoglobin complexes (65) showed the expected increase in proteins involved in hemoglobin and heme clearance and of antioxidant proteins, with concomitant suppression of HLA class II proteins, indicating that the *in vivo* observations of the iron-dependent proinflammatory macrophage phenotype are not simply a result of exposure to extravascular hemoglobin-haptoglobin complexes but most likely require an as yet unidentified cofactor. Favoring this interpretation, the resolution of simple hematomas involves only low-grade inflammation. Another observation that favors the idea that erythrophagocytosis is not generally an inflammatory stimulus for macrophages is the tendency of *Salmonella* Typhimurium to select such macrophages as its niche (66), thereby perhaps accessing a source of iron as well.

INTERACTIONS BETWEEN MACROPHAGES AND THEIR RESIDENT PATHOGENS

Sequestration of iron in macrophages during infections would be expected to impair the access of extracellular microbes to iron but could have the opposite effect on intracellular microbes resident in macrophages. Some experiments, mainly with macrophages and macrophage-like cell lines in cell culture systems, support this hypothesis (reviewed in reference 67). However, during *in vivo* infections, most iron is sequestered in erythrophagocytic macrophages while other macrophage types are relatively unaffected by hepcidin because they do not contain much ferroportin and do not engage in erythrophagocytosis, which is quantitatively the largest source of iron for macrophages. Moreover, different microbes may have varying extracellular residence times, enter macrophages by diverse routes, reside in different subcellular compartments, elicit a variably intense systemic inflammatory response, and utilize diverse mechanisms to manipulate their intracellular environment. These important nuances are not readily modeled in experiments with macrophages in cell culture. Even mouse models of infection with intracellular microbes have yielded contradictory results on whether the iron sequestration response favors macrophage-resident pathogens or not (reviewed in reference 68). With the notable exception of Nramp1, no strong genetic links have been identified to support a major role for iron-dependent susceptibility to intracellular infections. The availability of extreme mouse genetic models that accumulate iron in macrophages owing to dysregulated high hepcidin (69) or loss-of-function mutations in ferroportin (70), or have iron-depleted macrophages because of the absence of hepcidin (71) or ferroportin insensitivity to hepcidin (72), may clarify the role of macrophage iron in host defense against intracellular pathogens. Although corresponding human genetic diseases also exist, their rarity precludes the detection of all but the strongest associations between macrophage iron and susceptibility to infections with intracellular microbes.

THE TROPHIC ROLE OF BONE MARROW MACROPHAGES IN ERYTHROPOIESIS

Erythropoiesis in humans takes place in the bone marrow, where hematopoietic stem cells progressively differentiate through distinct identifiable stages into hemoglobin-synthesizing erythroblasts and then enucleated mature erythrocytes. Although it is possible to recapitulate this sequence *in vitro* by exposing purified hematopoietic stem cells to a mixture of hematopoietic growth factors, such erythropoiesis is slow and inefficient in comparison to erythropoiesis *in vivo* (73). In the marrow, each cluster of differentiating erythroid cells is associated with a central macrophage in a structure called an erythroblastic island (reviewed in reference 74). These islands form through multiple receptor-mediated adhesive interactions between macrophages and erythroid precursors, which promote erythroid precursor proliferation and maturation. How this occurs is not well understood. One well-documented function of the central macrophage is to phagocytize the erythrocyte fragment (pyrenocyte) that contains the ejected nucleus of the erythroblast. Another proposed but not sufficiently documented function of the central macrophage is to supply iron (perhaps as ferritin) to the developing erythroblast.

SUMMARY AND FUTURE DIRECTIONS

Iron scavenging, recycling, and storage by macrophages serves important trophic and host defense functions. Although much has been learned about the mechanisms involved, further work is needed to provide detailed understanding of iron movement in the macrophage and the subcellular location and regulation of molecules that mediate it. The contribution of extracellular or phagosomal iron deprivation to host defense against specific pathogens is also of great interest, and if understood, could be manipulated for therapeutic purposes. Finally, the effect of macrophage iron and other erythrocyte breakdown products on inflammatory processes in various pathological settings should be a fruitful area for future studies.

Citation. Ganz T. 2016. Macrophages and iron metabolism. Microbiol Spectrum 4(5):MCHD-0037-2016.

References

1. **Ganz T.** 2012. Macrophages and systemic iron homeostasis. *J Innate Immun* **4:**446–453.
2. **Donovan A, Lima CA, Pinkus JL, Pinkus GS, Zon LI, Robine S, Andrews NC.** 2005. The iron exporter ferroportin/Slc40a1 is essential for iron homeostasis. *Cell Metab* **1:**191–200.
3. **Nemeth E, Tuttle MS, Powelson J, Vaughn MB, Donovan A, Ward DM, Ganz T, Kaplan J.** 2004. Hepcidin regulates cellular iron efflux by binding to ferroportin and inducing its internalization. *Science* **306:**2090–2093.
4. **Babitt JL, Huang FW, Wrighting DM, Xia Y, Sidis Y, Samad TA, Campagna JA, Chung RT, Schneyer AL, Woolf CJ, Andrews NC, Lin HY.** 2006. Bone morphogenetic protein signaling by hemojuvelin regulates hepcidin expression. *Nat Genet* **38:**531–539.
5. **Corradini E, Rozier M, Meynard D, Odhiambo A, Lin HY, Feng Q, Migas MC, Britton RS, Babitt JL, Fleming RE.** 2011. Iron regulation of hepcidin despite attenuated Smad1,5,8 signaling in mice without transferrin receptor 2 or Hfe. *Gastroenterology* **141:**1907–1914.
6. **Johnson EE, Wessling-Resnick M.** 2012. Iron metabolism and the innate immune response to infection. *Microbes Infect* **14:**207–216.
7. **Barber MF, Elde NC.** 2014. Nutritional immunity. Escape from bacterial iron piracy through rapid evolution of transferrin. *Science* **346:**1362–1366.
8. **Kim A, Fung E, Parikh SG, Valore EV, Gabayan V, Nemeth E, Ganz T.** 2014. A mouse model of anemia of inflammation: complex pathogenesis with partial dependence on hepcidin. *Blood* **123:**1129–1136.
9. **Rodriguez R, Jung CL, Gabayan V, Deng JC, Ganz T, Nemeth E, Bulut Y, Roy CR.** 2014. Hepcidin induction by pathogens and pathogen-derived molecules is strongly dependent on interleukin-6. *Infect Immun* **82:**745–752.
10. **Nemeth E, Rivera S, Gabayan V, Keller C, Taudorf S, Pedersen BK, Ganz T.** 2004. IL-6 mediates hypoferremia of inflammation by inducing the synthesis of the iron regulatory hormone hepcidin. *J Clin Invest* **113:**1271–1276.
11. **Barton JC, Acton RT.** 2009. Hemochromatosis and *Vibrio vulnificus* wound infections. *J Clin Gastroenterol* **43:** 890–893.
12. **Frank KM, Schneewind O, Shieh WJ.** 2011. Investigation of a researcher's death due to septicemic plague. *N Engl J Med* **364:**2563–2564.
13. **Bergmann TK, Vinding K, Hey H.** 2001. Multiple hepatic abscesses due to *Yersinia enterocolitica* infection secondary to primary haemochromatosis. *Scand J Gastroenterol* **36:**891–895.
14. **Höpfner M, Nitsche R, Rohr A, Harms D, Schubert S, Fölsch UR.** 2001. *Yersinia enterocolitica* infection with multiple liver abscesses uncovering a primary hemochromatosis. *Scand J Gastroenterol* **36:**220–224.
15. **Arezes J, Jung G, Gabayan V, Valore E, Ruchala P, Gulig PA, Ganz T, Nemeth E, Bulut Y.** 2015. Hepcidin-induced hypoferremia is a critical host defense mechanism against the siderophilic bacterium *Vibrio vulnificus*. *Cell Host Microbe* **17:**47–57.
16. **Beaumont C, Delaby C.** 2009. Recycling iron in normal and pathological states. *Semin Hematol* **46:**328–338.
17. **Kondo H, Saito K, Grasso JP, Aisen P.** 1988. Iron metabolism in the erythrophagocytosing Kupffer cell. *Hepatology* **8:**32–38.
18. **Lang F, Qadri SM.** 2012. Mechanisms and significance of eryptosis, the suicidal death of erythrocytes. *Blood Purif* **33:**125–130.
19. **Buffet PA, Safeukui I, Deplaine G, Brousse V, Prendki V, Thellier M, Turner GD, Mercereau-Puijalon O.** 2011. The pathogenesis of *Plasmodium falciparum* malaria in humans: insights from splenic physiology. *Blood* **117:** 381–392.
20. **Low PS, Waugh SM, Zinke K, Drenckhahn D.** 1985. The role of hemoglobin denaturation and band 3 clustering in red blood cell aging. *Science* **227:**531–533.
21. **Lee SJ, Park SY, Jung MY, Bae SM, Kim IS.** 2011. Mechanism for phosphatidylserine-dependent erythrophagocytosis in mouse liver. *Blood* **117:**5215–5223.
22. **Pantaleo A, Giribaldi G, Mannu F, Arese P, Turrini F.** 2008. Naturally occurring anti-band 3 antibodies and red blood cell removal under physiological and pathological conditions. *Autoimmun Rev* **7:**457–462.
23. **Bosman GJCG, Werre JM, Willekens FLA, Novotný VM.** 2008. Erythrocyte ageing *in vivo* and *in vitro*: structural aspects and implications for transfusion. *Transfus Med* **18:**335–347.
24. **Ravichandran KS.** 2010. Find-me and eat-me signals in apoptotic cell clearance: progress and conundrums. *J Exp Med* **207:**1807–1817.
25. **Föller M, Huber SM, Lang F.** 2008. Erythrocyte programmed cell death. *IUBMB Life* **60:**661–668.
26. **Mohandas N, An X.** 2012. Malaria and human red blood cells. *Med Microbiol Immunol (Berl)* **201:**593–598.
27. **Poss KD, Tonegawa S.** 1997. Heme oxygenase 1 is required for mammalian iron reutilization. *Proc Natl Acad Sci U S A* **94:**10919–10924.

28. **Kovtunovych G, Eckhaus MA, Ghosh MC, Ollivierre-Wilson H, Rouault TA.** 2010. Dysfunction of the heme recycling system in heme oxygenase 1-deficient mice: effects on macrophage viability and tissue iron distribution. *Blood* **116:**6054–6062.

29. **Korolnek T, Hamza I.** 2015. Macrophages and iron trafficking at the birth and death of red cells. *Blood* **125:** 2893–2897.

30. **Soe-Lin S, Apte SS, Mikhael MR, Kayembe LK, Nie G, Ponka P.** 2010. Both Nramp1 and DMT1 are necessary for efficient macrophage iron recycling. *Exp Hematol* **38:** 609–617.

31. **White C, Yuan X, Schmidt PJ, Bresciani E, Samuel TK, Campagna D, Hall C, Bishop K, Calicchio ML, Lapierre A, Ward DM, Liu P, Fleming MD, Hamza I.** 2013. HRG1 is essential for heme transport from the phagolysosome of macrophages during erythrophagocytosis. *Cell Metab* **17:**261–270.

32. **Radhakrishnan N, Yadav SP, Sachdeva A, Pruthi PK, Sawhney S, Piplani T, Wada T, Yachie A.** 2011. Human heme oxygenase-1 deficiency presenting with hemolysis, nephritis, and asplenia. *J Pediatr Hematol Oncol* **33:** 74–78.

33. **Yachie A, Niida Y, Wada T, Igarashi N, Kaneda H, Toma T, Ohta K, Kasahara Y, Koizumi S.** 1999. Oxidative stress causes enhanced endothelial cell injury in human heme oxygenase-1 deficiency. *J Clin Invest* **103:** 129–135.

34. **Nielsen MJ, Møller HJ, Moestrup SK.** 2010. Hemoglobin and heme scavenger receptors. *Antioxid Redox Signal* **12:**261–273.

35. **Kristiansen M, Graversen JH, Jacobsen C, Sonne O, Hoffman HJ, Law SK, Moestrup SK.** 2001. Identification of the haemoglobin scavenger receptor. *Nature* **409:** 198–201.

36. **Hvidberg V, Maniecki MB, Jacobsen C, Højrup P, Møller HJ, Moestrup SK.** 2005. Identification of the receptor scavenging hemopexin-heme complexes. *Blood* **106:**2572–2579.

37. **Kohyama M, Ise W, Edelson BT, Wilker PR, Hildner K, Mejia C, Frazier WA, Murphy TL, Murphy KM.** 2009. Role for Spi-C in the development of red pulp macrophages and splenic iron homeostasis. *Nature* **457:**318–321.

38. **Delaby C, Pilard N, Puy H, Canonne-Hergaux F.** 2008. Sequential regulation of ferroportin expression after erythrophagocytosis in murine macrophages: early mRNA induction by haem, followed by iron-dependent protein expression. *Biochem J* **411:**123–131.

39. **Igarashi K, Sun J.** 2006. The heme-Bach1 pathway in the regulation of oxidative stress response and erythroid differentiation. *Antioxid Redox Signal* **8:**107–118.

40. **Hintze KJ, Katoh Y, Igarashi K, Theil EC.** 2007. Bach1 repression of ferritin and thioredoxin reductase1 is heme-sensitive in cells and *in vitro* and coordinates expression with heme oxygenase1, β-globin, and NADP(H) quinone (oxido) reductase1. *J Biol Chem* **282:**34365–34371.

41. **Marro S, Chiabrando D, Messana E, Stolte J, Turco E, Tolosano E, Muckenthaler MU.** 2010. Heme controls ferroportin1 (FPN1) transcription involving Bach1, Nrf2 and a MARE/ARE sequence motif at position -7007 of the FPN1 promoter. *Haematologica* **95:**1261–1268.

42. **Muckenthaler MU, Galy B, Hentze MW.** 2008. Systemic iron homeostasis and the iron-responsive element/iron-regulatory protein (IRE/IRP) regulatory network. *Annu Rev Nutr* **28:**197–213.

43. **Sangokoya C, Doss JF, Chi JT.** 2013. Iron-responsive miR-485-3p regulates cellular iron homeostasis by targeting ferroportin. *PLoS Genet* **9:**e1003408. doi:10.1371/journal.pgen.1003408.

44. **Belcher JD, Chen C, Nguyen J, Milbauer L, Abdulla F, Alayash AI, Smith A, Nath KA, Hebbel RP, Vercellotti GM.** 2014. Heme triggers TLR4 signaling leading to endothelial cell activation and vaso-occlusion in murine sickle cell disease. *Blood* **123:**377–390.

45. **Chiabrando D, Vinchi F, Fiorito V, Mercurio S, Tolosano E.** 2014. Heme in pathophysiology: a matter of scavenging, metabolism and trafficking across cell membranes. *Front Pharmacol* **5:**61. doi:10.3389/fphar.2014.00061.

46. **Vinchi F, Costa da Silva M, Ingoglia G, Petrillo S, Brinkman N, Zuercher A, Cerwenka A, Tolosano E, Muckenthaler MU.** 2016. Hemopexin therapy reverts heme-induced proinflammatory phenotypic switching of macrophages in a mouse model of sickle cell disease. *Blood* **127:**473–486.

47. **Soe-Lin S, Sheftel AD, Wasyluk B, Ponka P.** 2008. Nramp1 equips macrophages for efficient iron recycling. *Exp Hematol* **36:**929–937.

48. **Shi H, Bencze KZ, Stemmler TL, Philpott CC.** 2008. A cytosolic iron chaperone that delivers iron to ferritin. *Science* **320:**1207–1210.

49. **Nandal A, Ruiz JC, Subramanian P, Ghimire-Rijal S, Sinnamon RA, Stemmler TL, Bruick RK, Philpott CC.** 2011. Activation of the HIF prolyl hydroxylase by the iron chaperones PCBP1 and PCBP2. *Cell Metab* **14:** 647–657.

50. **Yanatori I, Yasui Y, Tabuchi M, Kishi F.** 2014. Chaperone protein involved in transmembrane transport of iron. *Biochem J* **462:**25–37.

51. **Taniguchi R, Kato HE, Font J, Deshpande CN, Wada M, Ito K, Ishitani R, Jormakka M, Nureki O.** 2015. Outward- and inward-facing structures of a putative bacterial transition-metal transporter with homology to ferroportin. *Nat Commun* **6:**8545. doi:10.1038/ncomms9545.

52. **Marques L, Auriac A, Willemetz A, Banha J, Silva B, Canonne-Hergaux F, Costa L.** 2012. Immune cells and hepatocytes express glycosylphosphatidylinositol-anchored ceruloplasmin at their cell surface. *Blood Cells Mol Dis* **48:**110–120.

53. **Cherukuri S, Tripoulas NA, Nurko S, Fox PL.** 2004. Anemia and impaired stress-induced erythropoiesis in aceruloplasminemic mice. *Blood Cells Mol Dis* **33:** 346–355.

54. **Brissot P, Ropert M, Le Lan C, Loréal O.** 2012. Non-transferrin bound iron: a key role in iron overload and iron toxicity. *Biochim Biophys Acta* **1820:**403–410.

55. **Breuer W, Ronson A, Slotki IN, Abramov A, Hershko C, Cabantchik ZI.** 2000. The assessment of serum

nontransferrin-bound iron in chelation therapy and iron supplementation. *Blood* 95:2975–2982.

56. **Delaby C, Rondeau C, Pouzet C, Willemetz A, Pilard N, Desjardins M, Canonne-Hergaux F.** 2012. Subcellular localization of iron and heme metabolism related proteins at early stages of erythrophagocytosis. *PLoS One* 7: e42199. doi:10.1371/journal.pone.0042199.

57. **Canonne-Hergaux F, Donovan A, Delaby C, Wang HJ, Gros P.** 2006. Comparative studies of duodenal and macrophage ferroportin proteins. *Am J Physiol Gastrointest Liver Physiol* 290:G156–G163.

58. **Vidal S, Gros P, Skamene E.** 1995. Natural resistance to infection with intracellular parasites: molecular genetics identifies Nramp1 as the Bcg/Ity/Lsh locus. *J Leukoc Biol* 58:382–390.

59. **Alter-Koltunoff M, Goren S, Nousbeck J, Feng CG, Sher A, Ozato K, Azriel A, Levi BZ.** 2008. Innate immunity to intraphagosomal pathogens is mediated by interferon regulatory factor 8 (IRF-8) that stimulates the expression of macrophage-specific Nramp1 through antagonizing repression by c-Myc. *J Biol Chem* 283:2724–2733.

60. **Wessling-Resnick M.** 2015. Nramp1 and other transporters involved in metal withholding during infection. *J Biol Chem* 290:18984–18990.

61. **Forbes JR, Gros P.** 2003. Iron, manganese, and cobalt transport by Nramp1 (Slc11a1) and Nramp2 (Slc11a2) expressed at the plasma membrane. *Blood* 102:1884–1892.

62. **Techau ME, Valdez-Taubas J, Popoff JF, Francis R, Seaman M, Blackwell JM.** 2007. Evolution of differences in transport function in Slc11a family members. *J Biol Chem* 282:35646–35656.

63. **Goswami T, Bhattacharjee A, Babal P, Searle S, Moore E, Li M, Blackwell JM.** 2001. Natural-resistance-associated macrophage protein 1 is an H^+/bivalent cation antiporter. *Biochem J* 354:511–519.

64. **Sindrilaru A, Peters T, Wieschalka S, Baican C, Baican A, Peter H, Hainzl A, Schatz S, Qi Y, Schlecht A, Weiss JM, Wlaschek M, Sunderkötter C, Scharffetter-Kochanek K.** 2011. An unrestrained proinflammatory M1 macrophage population induced by iron impairs wound healing in humans and mice. *J Clin Invest* 121:985–997.

65. **Kaempfer T, Duerst E, Gehrig P, Roschitzki B, Rutishauser D, Grossmann J, Schoedon G, Vallelian F, Schaer DJ.** 2011. Extracellular hemoglobin polarizes the macrophage proteome toward Hb-clearance, enhanced antioxidant capacity and suppressed HLA class 2 expression. *J Proteome Res* 10:2397–2408.

66. **Nix RN, Altschuler SE, Henson PM, Detweiler CS.** 2007. Hemophagocytic macrophages harbor *Salmonella enterica* during persistent infection. *PLoS Pathog* 3:e193. doi:10.1371/journal.ppat.0030193.

67. **Soares MP, Weiss G.** 2015. The Iron age of host-microbe interactions. *EMBO Rep* 16:1482–1500.

68. **Michels K, Nemeth E, Ganz T, Mehrad B.** 2015. Hepcidin and host defense against infectious diseases. *PLoS Pathog* 11:e1004998. doi:10.1371/journal.ppat.1004998.

69. **Du X, She E, Gelbart T, Truksa J, Lee P, Xia Y, Khovananth K, Mudd S, Mann N, Moresco EM, Beutler E, Beutler B.** 2008. The serine protease TMPRSS6 is required to sense iron deficiency. *Science* 320:1088–1092.

70. **Zohn IE, De Domenico I, Pollock A, Ward DM, Goodman JF, Liang X, Sanchez AJ, Niswander L, Kaplan J.** 2007. The flatiron mutation in mouse ferroportin acts as a dominant negative to cause ferroportin disease. *Blood* 109:4174–4180.

71. **Lesbordes-Brion JC, Viatte L, Bennoun M, Lou DQ, Ramey G, Houbron C, Hamard G, Kahn A, Vaulont S.** 2006. Targeted disruption of the hepcidin 1 gene results in severe hemochromatosis. *Blood* 108:1402–1405.

72. **Altamura S, Kessler R, Gröne HJ, Gretz N, Hentze MW, Galy B, Muckenthaler MU.** 2014. Resistance of ferroportin to hepcidin binding causes exocrine pancreatic failure and fatal iron overload. *Cell Metab* 20:359–367.

73. **Rhodes MM, Kopsombut P, Bondurant MC, Price JO, Koury MJ.** 2008. Adherence to macrophages in erythroblastic islands enhances erythroblast proliferation and increases erythrocyte production by a different mechanism than erythropoietin. *Blood* 111:1700–1708.

74. **An X, Mohandas N.** 2011. Erythroblastic islands, terminal erythroid differentiation and reticulocyte maturation. *Int J Hematol* 93:139–143.

Myeloid Cells in Health and Disease: A Synthesis
Edited by Siamon Gordon

doi:10.1128/microbiolspec.MCHD-0026-2015

Filip K. Swirski[1]
Matthias Nahrendorf[1]
Peter Libby[2]

47 Mechanisms of Myeloid Cell Modulation of Atherosclerosis

INTRODUCTION

Concepts of the pathogenesis of atherosclerosis have evolved substantially through the decades. Viewing it as an inevitable degenerative process, Sir William Osler attributed atherosclerosis to the stress and strain of modern life at the dawn of the 20th century (1). Indeed, the pathogenesis of atherosclerosis had given rise to great controversy in the middle portion of the 19th century, particularly among German pathologists. Von Rokitansky postulated a role of incorporated thrombus into the artery wall as the primary event in atherosclerosis (2). Rudolf Virchow posited a role for proliferation of medial cells, now recognized as arterial smooth muscle cells (SMCs), in the pathogenesis of atherosclerosis (3). Virchow also recognized cell death as a component of atherogenesis and observed bone formation in atherosclerotic plaques. While von Rokitansky's notion of the incorporation of thrombus lost popularity, the concept of atherosclerosis as a proliferative disorder of SMCs received considerable attention by pathologists and cell biologists in the latter part of the 20th century. Earl Benditt provided evidence for monotypic accumulation of SMCs in atherosclerotic plaques (4). Russell Ross focused on the role of platelet products, notably platelet-derived growth factor, as a causal stimulus for SMC growth in atherosclerotic plaques (5, 6).

The participation of lipid in atherosclerotic plaques and the recognition of cholesterol as a key component of the atheroma came into sharp focus with the experiments of Anitschkow and Chalatow at the dawn of the 20th century (7, 8). These investigators demonstrated that diets enriched in cholesterol could promote the formation of atherosclerotic plaques in animals that ordinarily resist this disease, initially rabbits. The excellent illustrations of the microscopic observations of this Russian group showed collections of lipid droplets in cells we now recognize as mononuclear phagocytes. Thus, the Russian investigators were able to reproduce experimentally the Schaumzellen or foam cells described in human lesions by the 19th-century pathologists. The simplistic notion that atheromata represented passive accumulations of lipid, comprising a lipid storage disease, grew out of the observations regarding atheroma formation in response to excessive dietary cholesterol. The description of the inflammatory aspect of atherosclerosis that emerged from the careful observations of 19th-century pathologists fell

[1]Center for Systems Biology, Massachusetts General Hospital, Boston, MA 02114; [2]Division of Cardiovascular Medicine, Department of Medicine, Brigham and Women's Hospital, Harvard Medical School, Boston, MA 02115.

from favor as different schools of atherosclerosis researchers focused on passive lipid accumulation or SMC proliferation (9).

Interest in inflammation reemerged from microscopic studies. Poole and Florey, working at Oxford University, called attention to the association of mononuclear leukocytes on the surface of endothelial cells as an early event in experimental atherogenesis produced by dietary manipulations in rabbits (10). Joris and Majno showed leukocyte accumulation in rats subjected to experimental atherogenic manipulations (11). Gerrity provided similar observations in swine (12). The group of Russell Ross, which had emphasized so heavily SMC proliferation, refocused on inflammation in response to their striking morphologic studies of experimentally produced atherosclerotic lesions in nonhuman primates by electron microscopy (13, 14). They showed captivating images of blood monocytes adhering to intact endothelium in early atherogenesis, preceding the accumulation of mononuclear cells in the arterial intima and the formation of lipid-laden macrophages at later stages of atherosclerosis. These persuasive and well-illustrated observations controverted previous assumptions of endothelial denudation as an initial step in atherosclerosis that would permit the local recruitment and activation of platelets and release of their growth factors. By the end of the 20th century, the primordial role of inflammation and inflammatory cells in atherosclerosis became firmly rooted (15–17).

As laboratory researchers focused on competing schemata for the pathogenesis of atherosclerotic plaques, clinical investigators and epidemiologists defined a number of risk factors for atherosclerosis and atherosclerotic events (18). Indeed, high plasma levels of total cholesterol and ultimately the fraction borne by low-density lipoprotein (LDL) particles emerged as a prime precursor of atherosclerotic disease (19–21).

We have promoted the concept that inflammation, while not necessarily an instigator of atherosclerosis *per se*, provides a series of pathogenic pathways that couple the risk factors for atherosclerosis defined by human observation with altered behavior of the intrinsic cells of the arterial wall, endothelium, and smooth muscle in a manner that gives rise to the disease and its complications (17). The current synoptic view accords a central role to inflammatory mechanisms and mediators as transducers of the traditional risk factors for atherosclerosis. Thus, the implication of inflammation in atherosclerosis by no means challenges the role of traditional risk factors, but provides mechanisms by which they promote the disease.

The study of inflammation in atherosclerosis has progressed beyond morphologic observations over the last quarter century. Both arms of immunity likely participate in atherogenesis (22). A number of authoritative recent reviews explicate the evidence supporting involvement of adaptive immunity in atherogenesis (23–25). In keeping with the focus on myeloid cells in this volume, the balance of this chapter will focus on innate immunity and contemporary studies, many derived from studies in mice, which support a pivotal role for myeloid cells in atherosclerosis. We will divide our discussion into phases of the initiation, progression, and complications of atherosclerosis. We will further consider myeloid cells as possible targets for treatment of atherosclerosis.

MECHANISMS OF MYELOID CELL PARTICIPATION IN THE INITIATION OF ATHEROSCLEROSIS

The introductory comments above alluded to the morphologic observations of Florey, Gerrity, Ross, Majno, and others showing early recruitment of leukocytes to experimental atherosclerosis lesions. The pathways of leukocyte recruitment to the nascent atherosclerotic plaque have undergone considerable investigation. Although knowledge of the mechanisms of leukocyte recruitment has matured, the precise identity of instigating stimuli, particularly components of putative triggers such as oxidized lipoproteins, remain incompletely validated (26). The lack of certitude regarding inciting factors resembles the situation in other inflammatory diseases such as rheumatoid arthritis, lupus erythematosus, and multiple sclerosis.

The expression of leukocyte adhesion molecules by endothelial cells in the region of inflammation initiates local leukocyte recruitment (Fig. 1). Among myeloid cells, monocytes arrive first on the scene during atherogenesis. While granulocytes localize in experimental atheromata in mice (27), their primary participation in human atherogenesis remains uncertain. In general, selectins expressed by the endothelial cells promote rolling of leukocytes and members of the IgG superfamily of adhesion molecules mediate the adherence of leukocytes, which tarry in their transit through the circulation due to local selectin expression (28). In the case of atherosclerosis, evidence in mouse experiments supports the participation of P-selectin and of vascular cell adhesion molecule-1 (VCAM-1) in leukocyte recruitment to early atherosclerotic lesions. Despite these definitive results in mice with experimental atherosclerosis, the observations in humans do not substantiate

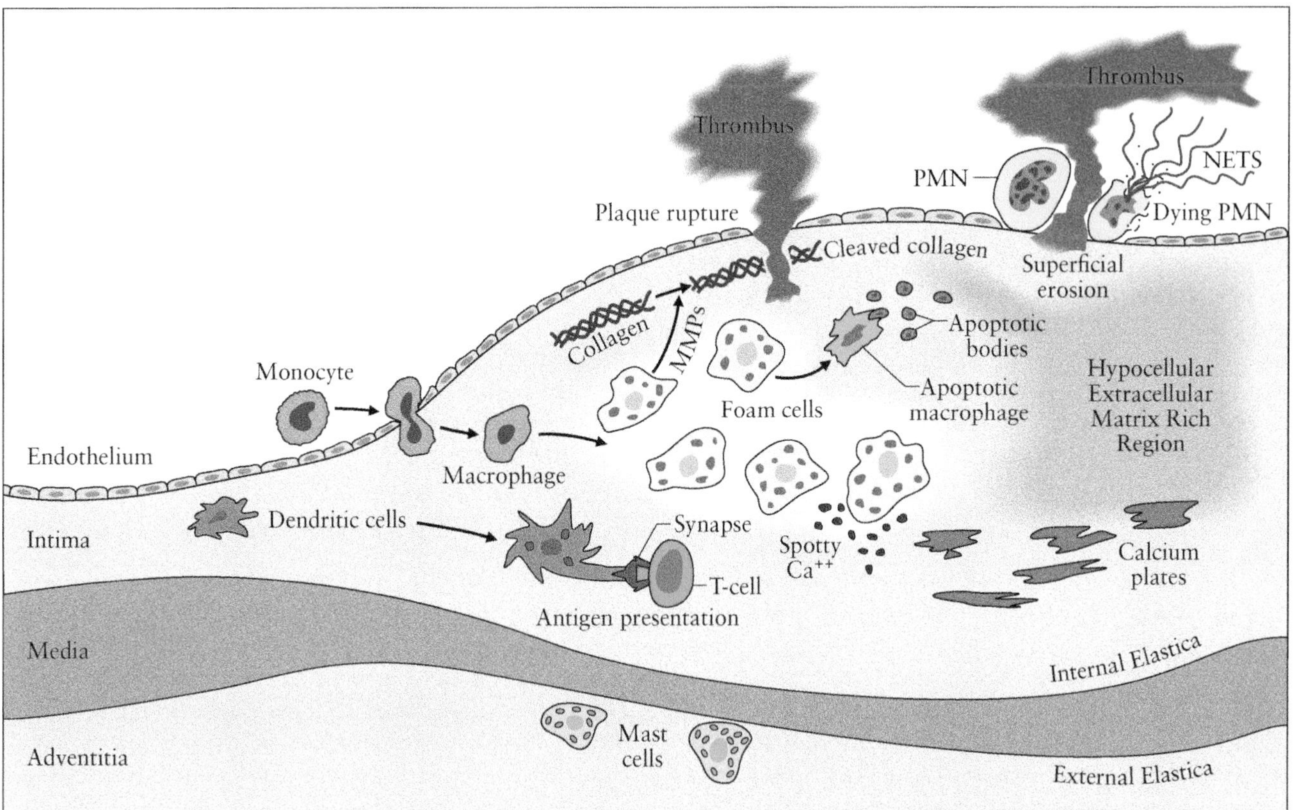

Figure 1 Roles of myeloid cells in the evolution of the atherosclerotic plaque. This figure depicts the participation of myeloid cells in the evolution of the atheroma from the inception (left) through the progression and complication of the plaque (middle to right). Dendritic cells may populate the normal arterial intima. Early in the atherogenesis, monocytes adhere to endothelial cells (left). The adherent monocytes can diapedese into the intima and mature into macrophages. These cells imbibe lipid and become foam cells, a hallmark of the atheromatous plaque. These cells populate the lipid core of the evolving atheroma (yellow central potion, middle.) These foamy macrophages can elaborate many mediators that amplify and sustain the atherogenic process. In particular, they can secrete interstitial collagenases, members of the MMP family, that can degrade the collagen fibrils that lend strength to the fibrous cap that overlies the lipid core of the established atheroma. Cleavage of collagen in the plaque's fibrous cap allows contact of the blood coagulation components with the procoagulant tissue factor produced by the plaque macrophages. Thus, disruption of the plaque by a fracture of the fibrous cap triggers thrombosis that leads to the most dreaded clinical complications of atherosclerosis such as the acute coronary syndromes and many ischemic strokes. Macrophages can die within plaques, as shown by the cell with the pyknotic nucleus casting off apoptotic bodies that can bear tissue factor. Macrophages can also release microparticles that can provide a nidus for spotty calcification associated with plaque instability. Calcium mineral can coalesce into plates that complicate the advanced atherosclerotic plaque. As plaques mature, they can become less cellular and accumulate more extracellular matrix. In regions of plaques rich in proteoglycan and glycosaminoglycans, endothelial cells can detach, exposing blood to underlying collagen and other thrombogenic mediators that can instigate clot formation. This process, denoted superficial erosion, can lead to recruitment of polymorphonuclear leukocytes (PMNs). Dying granulocytes release DNA that can associate with the pro-oxidant enzyme myeloperoxidase, the procoagulant tissue factor, and other enzymes associated with further endothelial damage and thrombosis. These NETs can further entrap platelets, promoting propagation of thrombi. The dendritic cells in plaques can present antigen to T cells, providing a link between innate and adaptive immunity. Mast cells in the adventitia of arteries can elaborate numerous mediators including histamine, heparin, serine proteinases, and cytokines that can amplify atherogenesis and lesion complication. Thus, myeloid cells participate in all phases of atherogenesis, from lesion initiation (left) through thrombotic complications (middle and right).

such a clear role for P-selectin or VCAM-1 expression by macrovascular endothelium as a mechanism of leukocyte recruitment. While in humans microvessels in established atherosclerotic plaques express VCAM-1, even analysis of fatty streaks (early precursors of atherosclerosis in humans) have not disclosed consistent expression of these adhesion molecules on macrovascular luminal endothelial cells (29–31). Thus, while VCAM-1 may promote leukocyte recruitment to established lesions through the microvessels that form in plaques, the precise nature of the adhesion molecules that operate during atherogenesis in humans remains uncertain.

Much of the recent expansion in knowledge of the role of myeloid cells in atherosclerosis revolves around functional subsets of monocytes (32–35). Mouse experiments suggest that proinflammatory monocytes that express high levels of Ly6C or Gr-1 preferentially accumulate in atherosclerotic plaques and adhere to cytokine-stimulated endothelial cells *in vitro* (Fig. 2).

Once bound to the endothelial surface, chemoattractant molecules direct the migration of the adherent cells into the intima, penetrating the endothelial lining between intercellular junctions or possibly by transcytosis (Fig. 1). Experimental evidence supports the involvement of a number of chemokines in intimal leukocyte accumulation. The iconic chemoattractant monocyte chemoattractant protein-1 (MCP-1/CCL2) appears to promote particularly accumulation of the proinflammatory subset of monocytes in mice (33, 34, 36). Fractalkine binding to its receptor, CX3CR1, may mediate the recruitment of Ly6C^{lo} cells (37, 38). As gauged by the relative surface expression and by kinetic studies, monocyte recruitment to mouse atheromata does not occur just during the initial phase of atherogenesis but persists during lesion progression. Indeed, monocyte recruitment appears proportional to the extent of disease (39).

Beyond adhesion and chemoattraction, retention of mononuclear phagocytes within the arterial intima contributes to their accumulation. The neural guidance molecule netrin-1 inhibits leukocyte migration. Macrophages within mouse atheromata express high levels of netrin-1; its receptor, UNC5B; as well as semaphorin E3 (40, 41). *In vitro* studies in lipid-loaded monocytoid cells show that netrin-1 limits the directed migration of macrophages. Mice with selective lack of netrin-1 in bone marrow-derived cells show less macrophage accumulation, smaller lesions, and evidence for retarded efflux of macrophages from lesions. Thus, accumulation of mononuclear phagocytes in the atherosclerotic plaque depends on the balance between adhesion, chemoattraction, retention, and efflux (42).

Some observations highlight a potential very early role for dendritic cells in the nascent atheroma, even preceding the appearance of many macrophages (43). Dendritic cells provide a potential link between innate and adaptive immunity in view of their antigen-presenting function. The dendritic cells may patrol arteries and present potential antigens to adaptive immune cells, including those that reside in lymphoid-like structures in the adventitia of arteries. Even when engorged with lipid, dendritic cells retain capacity to present antigens, and thus retain the ability to provide a bridge between myeloid cells that mediate innate immunity and the adaptive immune response (44).

Atheromata localize in regions of the arterial tree particularly prone to their formation. In contrast, risk factors for lesion formation such as dyslipidemia bathe the circulatory tree homogeneously. Sites predilected to atheroma formation characteristically localize in regions of low endothelial shear stress or disturbed flow (45, 46). Such hydrodynamic disturbances characterize zones prone to atheroma formation, including those around flow dividers or branch points in the arterial tree. Sites of predilection to form atheromata display early signs of inflammatory activation, for example, nuclear translocation of NF-κB, a master transcription factor that regulates the expression of many inflammatory genes (47).

MECHANISMS OF MYELOID CELL MODULATION OF THE PROGRESSION OF ATHEROSCLEROTIC PLAQUES

Once resident in the intima, blood monocytes mature into macrophages (Fig. 1). The formation of foam cells, lipid-laden macrophages that constitute the hallmark of atheromata, requires the expression of scavenger receptors that evade suppression by cellular cholesterol accumulation and have high capacity for internalization of lipoprotein particles. Among the stimuli for monocyte maturation into macrophages and augmented expression of scavenger receptors, macrophage colony-stimulating factor (M-CSF or CSF-1), acting through its cognate receptor c-Fms, participates in this process (48). Early observations in humans showed mitotic figures in cells identified as mononuclear phagocytes in plaques (49). Rabbit studies also showed evidence for myeloid cell proliferation in atherosclerotic lesions (50). Studies in mice have addressed quantitatively the relative contributions of monocyte recruitment, as discussed above, and proliferation *in situ* to leukocyte accumulation in various phases of atherosclerosis (51). Early in the life history of a mouse atheroma,

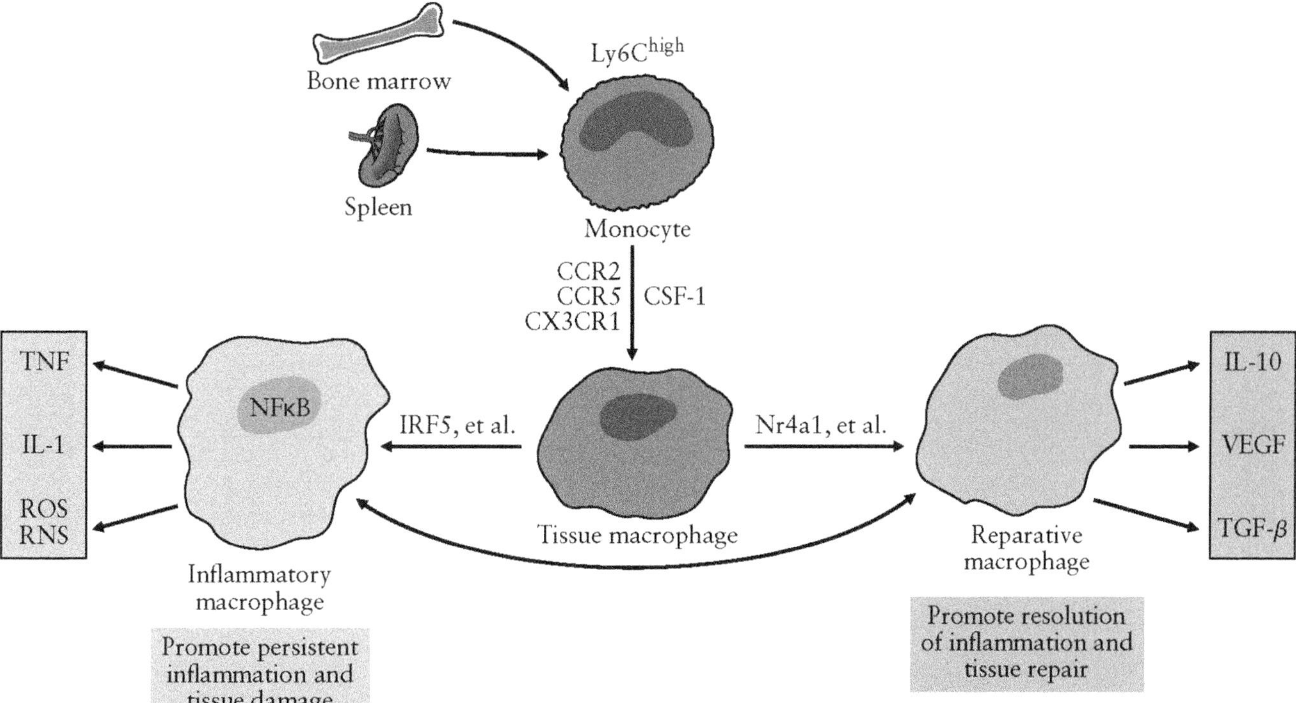

Figure 2 Monocyte and macrophage diversity in relation to cardiovascular homeostasis. Blood monocytes derive from hematopoiesis in the bone marrow or from extramedullary hematopoiesis in the spleen that harbors a preformed pool of proinflammatory monocytes in mice. Monocytes enter tissues in response to chemoattractants that engage the chemokine receptors CCR2, CCR5, and CX3CR1. Hematopoietic growth factors, notably CSF-1 (M-CSF), promote the maturation of monocytes into tissue macrophages. Once recruited to and resident in tissues, macrophages can polarize toward a proinflammatory set of functions including production of cytokines such as TNF or IL-1 isoforms. IRF5 provides an example of a transcription factor that promotes the polarization of macrophages toward the proinflammatory palette of functions. The transcription factor NF-κB regulates the genes that encode a number of proinflammatory mediators elaborated by the proinflammatory subset of macrophages (pink box). Monocytes/macrophages can modulate toward a reparative and resolution-promoting set of functions under the influence of various transcription factors including nuclear receptor subfamily 4 group a member 1 (NR4a1), also known as Nur77. Mediators elaborated by the reparative subset of macrophages include IL-10; vascular endothelial growth factor (VEGF), a stimulator of angiogenesis; and transforming growth factor β (TGF-β), an anti-inflammatory and fibrogenic protein. The balance between the proinflammatory versus anti-inflammatory, reparative, and resolution functions of macrophages determines critical aspects of atherogenesis and aspects of myocardial disease related to repair of ischemic injury and the responses to hemodynamic overload conditions, features of myocardial remodeling critical in determining clinical course including the development of heart failure. ROS, reactive oxygen species (e.g., superoxide anion and hypochlorous acid); RNS, reactive nitrogen species (e.g., nitric oxide).

macrophage accumulation depends principally on recruitment. In the later stages of atherosclerosis, proliferation predominates as a mechanism of accumulation of mononuclear phagocytes.

Mononuclear phagocytes in plaques also die (Fig. 1). These cells can undergo programmed cell death by apoptosis or succumb by oncosis (52). The formation of the central necrotic core of the mature atherosclerotic plaque reflects the accumulation of extracellular lipid that may arise from foam cells that have died (53). In general, clearance mechanisms rapidly remove cells that have undergone apoptosis. A failure to rapidly clear dead mononuclear phagocytes may prevail in the atherosclerotic plaque (52, 54). This process, known as efferocytosis, impaired in atherosclerotic plaques, likely contributes to formation of the lipid core of atheromata, sometimes referred to as the "necrotic core," indicating the paucity of viable cells in this locale. The net accumulation of mononuclear phagocytes thus depends not only on recruitment and proliferation but on the rate of cell death. The clearance of dead cells appears to influence prominently the progression of atheromata by contributing to formation of the lipid core (52, 55, 56).

During the phase of progression of atherosclerosis, neovessels can form, usually in the base of the plaque arising as extensions of adventitial microvessels. The neovessels in the plaque, like those in the diabetic retina, appear fragile and prone to developing local hemorrhage and thrombosis *in situ* within the plaque. Regions of neovascularization in plaques colocalize with macrophages, providing a link between inflammatory cells and plaque angiogenesis (57, 58). Episodes of local thrombin generation due to plaque hemorrhage and thrombosis may not manifest clinically, but through stimulating bouts of proliferation of SMCs, their production of extracellular matrix, and recruiting further SMCs from the tunica media may contribute to the transition from fatty, lipid-rich lesions to more-fibrous plaques characteristic of the advancing disease.

As noted above, Virchow recognized that plaque formation can recapitulate osteogenesis and lead to accumulation of calcium mineral (Fig. 1). Rather than a passive process, or mere "degeneration" leading to dystrophic calcification, plaque calcification appears highly regulated. In animals that lack M-CSF, impaired macrophage maturation likely limits their acquisition of osteoclastic activity, yielding calcium accumulation as shown by mouse experiments (59). Thus calcium accumulation depends not only on accretion but also on ongoing breakdown by mononuclear phagocytes that exert osteoclastic functions regulated by M-CSF. The genesis of calcium accumulation within the plaque may depend on "seeding" by microparticles or microvesicles derived from mononuclear phagocytes (60). Hence, myeloid cells appear critical both in the accretion of calcium mineral and in its catabolism.

As atheromata take root in the arterial intima, they first grow outward, in an ablumenal direction, preserving the caliber of the arterial lumen until the later stages of the disease. This outward remodeling or compensatory enlargement likely depends on multiple mechanisms including matrix-degrading proteinase expression by SMCs, but also potentially by myeloid cells. The outward growth of the plaque likely necessitates degradation of the external elastic lamina that forms the border of the tunica media and the arterial adventitia. Activated mononuclear phagocytes as well as SMCs can elaborate several elastolytic enzymes that may permit this outward remodeling of atherosclerotic plaques during the progression phase (61).

THE ROLES OF MYELOID CELLS IN COMPLICATIONS OF ATHEROSCLEROSIS PLAQUES

The dreaded complications of atherosclerosis result from thrombosis. Myocardial infarction, ischemic stroke, and abrupt limb ischemia all arise primarily from thrombotic complications of atherosclerotic plaques. Myeloid cells play pivotal roles in these thrombotic complications. One important mechanism of plaque thrombosis involves a physical disruption, fissure, or rupture of the fibrous cap of the established atherosclerotic lesion that overlies the lipid core (62, 63). The fibrous cap owes its integrity by and large to interstitial forms of collagen, type I and type III. Considerable evidence supports a key contribution of myeloid cells to the catabolism of collagen that can deplete the fibrous cap of this critical extracellular matrix macromolecule, weaken it, and render it susceptible to rupture (Fig. 1). Human atherosclerotic plaques contain many macrophages that express matrix metalloproteinases (MMPs) specialized in breakdown of interstitial collagen. The interstitial collagenases expressed by macrophages in human atherosclerotic plaques include MMPs 1, 8, and 13. Extensive experiments in mice demonstrate a causal role for interstitial collagenases in collagen accumulation in atherosclerotic plaques, by both loss-of-function and gain-of-function experimental manipulations (62, 64). *In situ* studies of human atherosclerotic plaques indicate the primacy of MMP-13 as a collagenolytic enzyme expressed by myeloid cells (65).

When plaques fracture, they permit coagulation factors in blood to gain access to the innards of plaque. Myeloid cells within the plaque overexpress the potent procoagulant tissue factor. Contact with tissue factor accelerates the activation of coagulation factors VII and X manyfold, triggering thrombus formation (Fig. 1). Myeloid cells therefore contribute not only to the fragility of atheromata but also to their thrombogenicity, which promotes clot formation following fracture of the plaque's fibrous cap.

Superficial erosion of the endothelial monolayer without a frank fracture of the plaque's fibrous cap also can serve as a nidus for thrombus formation, complicating atherosclerotic plaques (63) (Fig. 1). In contrast to thin-capped, lipid-rich plaques associated with rupture, those that typically cause thrombosis due to superficial erosion lack prominent populations of mononuclear phagocytes. Lesions that provoke superficial erosion do accumulate extracellular matrix macromolecules, including proteoglycan and glycosaminoglycans. Thus, the morphology of plaques associated with superficial erosion contrasts starkly with those that cause thrombosis due to plaque rupture (Fig. 3). While these lesions associated with superficial erosion lack mononuclear phagocytes, once a discontinuity occurs in the endothelial monolayer that lines the intima, polymorphonuclear cells can join the forming thrombus and contribute to the ensuing catastrophe (66). When they die, neutrophils can elaborate strands of DNA, forming neutrophil extracellular traps (NETs) (67) (Fig. 1). These structures can entrap platelets and weave a nefarious network with forming fibrin multimers that promotes the growth and persistence of thrombi (68). NETs can further stimulate thrombosis by localizing tissue factor procoagulant (69). NETs can also amplify inflammation by triggering the inflammasome, boosting the local production of active interleukin-1β (IL-1β) and IL-18 (70).

Some observations in mice with atherosclerosis suggest the presence of neutrophils in undisrupted plaques (27). As mentioned above, evidence for neutrophil involvement in undisrupted human plaques remains scant. Yet, once an initial intimal rent appears, the subsequent recruitment of granulocytes may potentiate thrombosis and through the production of reactive oxygen species and elaboration of neutrophil elastase and gelatinases that attack the basement membrane to which endothelial cells adhere can extend the desquamation of these cells and potentiate the pathogenic process (Fig. 4). Mechanisms of superficial erosion remain poorly explored compared to the plaque rupture mechanism. While fissure of the fibrous cap depends heavily on mononuclear phagocytes, the complications of superficial erosion appear to involve granulocytes to a greater extent. Thus, cells of both the monocytic and the polymorphonuclear series can contribute to the thrombotic complications of atherosclerosis.

Atherosclerotic plaques also harbor mast cells. In particular, the adventitial layer of arteries contains mast cells. In experimental atherosclerosis in mice, IL-6 and gamma interferon derived from mast cells drive atherogenesis (71–75). In addition, mast cells produce the serine proteinases chymase and tryptase, which can process angiotensin and metalloproteinase precursors to their active forms. These cells can elaborate heparin, which can facilitate angiogenesis, and thus contribute to neovascularization of the plaque from adventitial microvessels. Histamine released by activated mast cells can increase local vascular permeability as well. Although mast cells can augment atherosclerosis in mice, their role in human lesions remains less well established.

MYELOID CELLS AS TARGETS FOR THERAPY OF ATHEROSCLEROSIS

Given the primordial importance of leukocyte adhesion in the early recruitment of these cells and in their ongoing accumulation, a number of therapeutic strategies

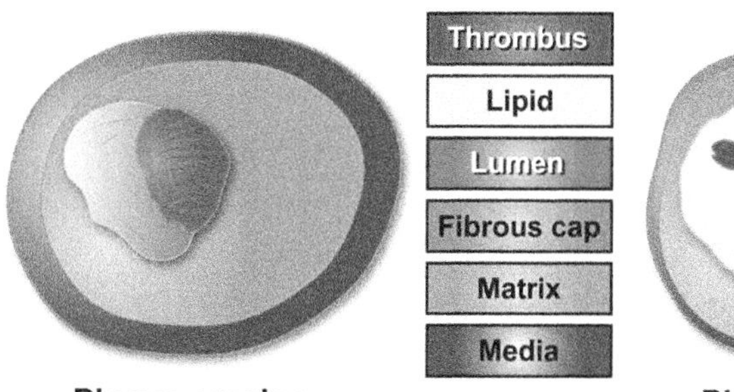

Figure 3 Distinctions between superficial erosion and fibrous cap rupture as causes of arterial thrombosis. Source: reference 84, with permission.

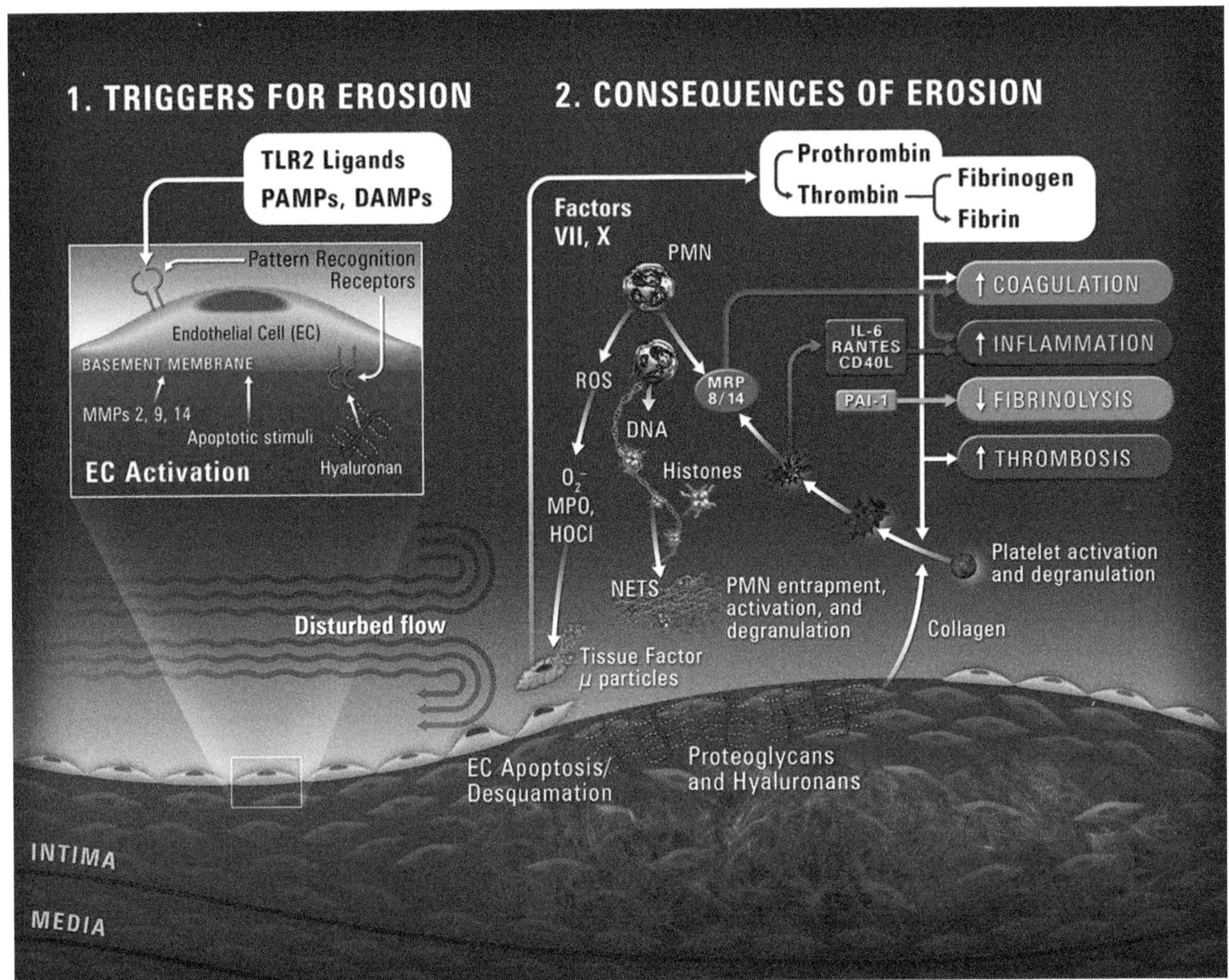

Figure 4 A depiction of potential pathophysiologic pathways that yield superficial erosion and thrombosis on atherosclerotic plaques. The bottom of the diagram depicts a longitudinal section of an artery that harbors an extracellular matrix-rich atheroma. The darker brown indicates accumulation of the proteoglycans such as versican and biglycan and of the glycosaminoglycan hyaluronic acid. (1) Some possible triggers for endothelial damage that causes superficial erosion. Inciting stimuli could include pathogen-associated molecular patterns (PAMPs), danger-associated molecular patterns (DAMPs), or other ligands for innate immune receptors, including Toll-like receptor 2 (TLR2). These ligands bind to pattern recognition receptors on the surface of the endothelial cell. Hyaluronan, a common constituent of plaques that have undergone superficial erosion, can activate TLR2. Various apoptotic stimuli elaborated by myeloid cells in plaques, as well as oxidized lipoproteins, can unleash endothelial apoptosis. Matrix-degrading enzymes including MMPs can catabolize constituents of the basement membrane that comprises a substrate for endothelial cell adherence via integrins or other adhesion molecules. The nonfibrillar collagenases MMP-2 and MMP-9 and the activator of MMP-2, MMP-14, enzymes that localize in atheromata, may cleave the tethers of the endothelial cell to the basement membrane. (2) Some of the downstream effects of erosion. Once an endothelial cell has sloughed (as portrayed by the endothelial cell with nuclear pycnosis), the moribund endothelial cell can spew forth microparticles rich in tissue factor, a potent procoagulant. Uncovering the subendothelial matrix stimulates sticking of granulocytes and their activation and degranulation. Granulocytes elaborate reactive oxygen species (ROS) such as hypochlorous acid, HOCl, produced by myeloperoxidase (MPO), as well as superoxide anion (O_2^-). This scheme postulates arrival of granulocytes secondarily, only after the initial disturbance of the endothelial monolayer. Granulocytes can also release myeloid-related protein (MRP) 8/14, a calgranulin family member implicated in inflammation and other aspects of atherothrombosis. Agonal granulocytes release DNA and citrullinated histones that contribute to NETosis, the formation of NETs. NETs can promote propagation of thrombosis and entrap further myeloid cells and platelets that amplify regional inflammation. Exposure of the subendothelial extracellular matrix macromolecules can also activate platelets, triggering their degranulation and elaboration of proinflammatory mediators including IL-6 and RANTES. In addition, activated platelets release plasminogen activator inhibitor-1 (PAI-1), which inhibits endogenous fibrinolysis, stabilizing clots. PMN, polymorphonuclear leukocyte. Source: reference 66, with permission.

have envisioned inhibiting this process. Approaches that target leukocyte adhesion have undergone evaluation in acute coronary syndromes in small studies and in myocardial ischemia/reperfusion. Despite a large preclinical database supporting the potential efficacy of such strategies, we lack convincing data in humans that inhibiting this key step in host defenses can modulate atherosclerosis or its complications. Selective targeting of adhesion molecules at sites of lesions by nanoparticle theranostics might obviate some of the impairment of host defenses that might result from global inhibition of leukocyte adhesion. Examples include targeting CCR2 and interferon regulatory factor 5 (IRF5) (76, 77).

The use of the statin class of drugs has revolutionized preventive cardiology. These drugs prevent first ever and recurrent events due to the complications of atherosclerosis in broad categories of patients at various levels of risk. While initially developed because of their ability to lower LDL concentrations, statins possess anti-inflammatory effects independent of LDL lowering (78). Through augmented activity of transcription factors such as Krüppel-like factor 2 (KLF2) and the prenylation of small G proteins, the statins may interfere with the function of myeloid cells, accounting in part for their protective effects on atherosclerotic lesion complication.

Various strategies for altering innate immunity involving myeloid cells have attracted interest. Preliminary results suggest that colchicine administration may limit complications of atherosclerosis (79). Two large-scale trials are currently evaluating anti-inflammatory therapies that do not appreciably affect cholesterol levels. These studies involve administration of low doses of methotrexate weekly, mimicking the strategy that has proven successful in treatment of rheumatoid arthritis (80). Biological therapeutics exist that target various proinflammatory cytokines. The use of such agents (e.g., anti-tumor necrosis factor [TNF] treatment) has transformed the practice of rheumatology and provided novel tools for treating certain inflammatory dermatologic and gastrointestinal diseases. The Canakinumab Anti-inflammatory Thrombosis Outcomes Study (CANTOS) will evaluate the effect of a monoclonal antibody that targets IL-1β as an innate immune-directed intervention that could influence cardiovascular events (81). Attempts to harness adaptive immunity, upstream of myeloid cells, involve vaccination strategies to elicit the formation of natural antibodies that may alleviate atherosclerosis, among other approaches that have emerged from animal experiments and that require evaluation in large-scale human studies to assess their efficacy (82, 83).

CONCLUSION

Myeloid cells participate critically in all phases of atherosclerosis from initiation through progression and ultimately the thrombotic complications of this disease. While most attention is focused on mononuclear phagocytes, polymorphonuclear leukocytes likely participate in the consequences of plaque disruption and may potentiate and amplify disruptions to plaque and their thrombotic consequences. Existing therapies such as statins may exert some of their protective effect by altering the functions of myeloid cells, as supported by a large body of experimental evidence. The pathways of innate immunity that involve myeloid cells provide a myriad of potential targets for modifying atherosclerosis and its complications, providing a fertile field for future attempts to address the residual burden of this disease whose global prevalence is on the rise.

Citation. Swirski FK, Nahrendorf M, Libby P. 2016. Mechanisms of myeloid cell modulation of atherosclerosis. Microbiol Spectrum 4(4):MCHD-0026-2015.

References

1. **Osler W.** 1892. *The Principles and Practice of Medicine.* D. Appleton and Company, New York, NY.
2. **Rokitansky K.** 1855. *A Manual of Pathological Anatomy*, vol IV, p 201–208. Blanchard and Lea, Philadelphia, PA.
3. **Virchow R.** 1858. *Cellular Pathology.* John Churchill, London, United Kingdom.
4. **Benditt EP, Benditt JM.** 1973. Evidence for a monoclonal origin of human atherosclerotic plaques. *Proc Natl Acad Sci U S A* **70:**1753–1756.
5. **Ross R, Glomset JA.** 1973. Atherosclerosis and the arterial smooth muscle cells. *Science* **180:**1332–1339.
6. **Ross R, Glomset J, Kariya B, Harker L.** 1974. A platelet-dependent serum factor that stimulates the proliferation of arterial smooth muscle cells *in vitro*. *Proc Natl Acad Sci U S A* **71:**1207–1210.
7. **Anitschkow N, Chalatow S.** 1983. On experimental cholesterin steatosis and its significance in the origin of some pathological processes (1913). *Arteriosclerosis* **3:** 178–182.
8. **Steinberg D.** 2013. In celebration of the 100th anniversary of the lipid hypothesis of atherosclerosis. *J Lipid Res* **54:**2946–2949.
9. **Ross R, Harker L.** 1976. Hyperlipidemia and atherosclerosis. *Science* **193:**1094–1100.
10. **Poole JC, Florey HW.** 1958. Changes in the endothelium of the aorta and the behaviour of macrophages in experimental atheroma of rabbits. *J Pathol Bacteriol* **75:** 245–251.
11. **Joris I, Zand T, Nunnari JJ, Krolikowski FJ, Majno G.** 1983. Studies on the pathogenesis of atherosclerosis. I. Adhesion and emigration of mononuclear cells in the aorta of hypercholesterolemic rats. *Am J Pathol* **113:** 341–358.

12. **Gerrity RG.** 1981. The role of the monocyte in atherogenesis: I. Transition of blood-borne monocytes into foam cells in fatty lesions. *Am J Pathol* **103:**181–190.

13. **Faggiotto A, Ross R.** 1984. Studies of hypercholesterolemia in the nonhuman primate. II. Fatty streak conversion to fibrous plaque. *Arteriosclerosis* **4:**341–356.

14. **Faggiotto A, Ross R, Harker L.** 1984. Studies of hypercholesterolemia in the nonhuman primate. I. Changes that lead to fatty streak formation. *Arteriosclerosis* **4:** 323–340.

15. **Libby P.** 1990. Inflammatory and immune mechanisms in atherogenesis, p 79–89. *In* Leaf A, Weber P (ed), *Atheroclerosis Reviews*, **vol 21**. Raven Press, New York, NY.

16. **Hansson GK, Jonasson L.** 2009. The discovery of cellular immunity in the atherosclerotic plaque. *Arterioscler Thromb Vasc Biol* **29:**1714–1717.

17. **Libby P.** 2012. Inflammation in atherosclerosis. *Arterioscler Thromb Vasc Biol* **32:**2045–2051.

18. **Kannel WB, Dawber TR, Kagan A, Revotskie N, Stokes J III.** 1961. Factors of risk in the development of coronary heart disease—six year follow-up experience. The Framingham Study. *Ann Intern Med* **55:**33–50.

19. **Brown MS, Goldstein JL.** 1986. A receptor-mediated pathway for cholesterol homeostasis. *Science* **232:**34–47.

20. **Steinberg D.** 2005. Thematic review series: the pathogenesis of atherosclerosis. An interpretive history of the cholesterol controversy: part II: the early evidence linking hypercholesterolemia to coronary disease in humans. *J Lipid Res* **46:**179–190.

21. **Steinberg D.** 2004. Thematic review series: the pathogenesis of atherosclerosis. An interpretive history of the cholesterol controversy: part I. *J Lipid Res* **45:**1583–1593.

22. **Hansson GK, Libby P, Schönbeck U, Yan ZQ.** 2002. Innate and adaptive immunity in the pathogenesis of atherosclerosis. *Circ Res* **91:**281–291.

23. **Libby P, Hansson GK.** 2015. Inflammation and immunity in diseases of the arterial tree: players and layers. *Circ Res* **116:**307–311.

24. **Libby P.** 2013. Collagenases and cracks in the plaque. *J Clin Invest* **123:**3201–3203.

25. **Lahoute C, Herbin O, Mallat Z, Tedgui A.** 2011. Adaptive immunity in atherosclerosis: mechanisms and future therapeutic targets. *Nat Rev Cardiol* **8:**348–358.

26. **Libby P, Ridker PM, Hansson GK.** 2011. Progress and challenges in translating the biology of atherosclerosis. *Nature* **473:**317–325.

27. **Soehnlein O.** 2012. Multiple roles for neutrophils in atherosclerosis. *Circ Res* **110:**875–888.

28. **Mestas J, Ley K.** 2008. Monocyte-endothelial cell interactions in the development of atherosclerosis. *Trends Cardiovasc Med* **18:**228–232.

29. **Munro JM, Cotran RS.** 1988. The pathogenesis of atherosclerosis: atherogenesis and inflammation. *Lab Invest* **58:**249–261.

30. **O'Brien KD, Allen MD, McDonald TO, Chait A, Harlan JM, Fishbein D, McCarty J, Ferguson M, Hudkins K, Benjamin CD, Lobb R, Alpers C.** 1993. Vascular cell adhesion molecule-1 is expressed in human coronary atherosclerotic plaques. Implications for the mode of progression of advanced coronary atherosclerosis. *J Clin Invest* **92:**945–951.

31. **Libby P, Li H.** 1993. Vascular cell adhesion molecule-1 and smooth muscle cell activation during atherogenesis. *J Clin Invest* **92:**538–539.

32. **Murray PJ, Allen JE, Biswas SK, Fisher EA, Gilroy DW, Goerdt S, Gordon S, Hamilton JA, Ivashkiv LB, Lawrence T, Locati M, Mantovani A, Martinez FO, Mege JL, Mosser DM, Natoli G, Saeij JP, Schultze JL, Shirey KA, Sica A, Suttles J, Udalova I, van Ginderachter JA, Vogel SN, Wynn TA.** 2014. Macrophage activation and polarization: nomenclature and experimental guidelines. *Immunity* **41:**14–20.

33. **Tacke F, Alvarez D, Kaplan TJ, Jakubzick C, Spanbroek R, Llodra J, Garin A, Liu J, Mack M, van Rooijen N, Lira SA, Habenicht AJ, Randolph GJ.** 2007. Monocyte subsets differentially employ CCR2, CCR5, and CX3CR1 to accumulate within atherosclerotic plaques. *J Clin Invest* **117:**185–194.

34. **Swirski FK, Libby P, Aikawa E, Alcaide P, Luscinskas FW, Weissleder R, Pittet MJ.** 2007. Ly-6C^{hi} monocytes dominate hypercholesterolemia-associated monocytosis and give rise to macrophages in atheromata. *J Clin Invest* **117:**195–205.

35. **Woollard KJ, Geissmann F.** 2010. Monocytes in atherosclerosis: subsets and functions. *Nat Rev Cardiol* **7:**77–86.

36. **Weber C, Zernecke A, Libby P.** 2008. The multifaceted contributions of leukocyte subsets to atherosclerosis: lessons from mouse models. *Nat Rev Immunol* **8:**802–815.

37. **Saederup N, Chan L, Lira SA, Charo IF.** 2008. Fractalkine deficiency markedly reduces macrophage accumulation and atherosclerotic lesion formation in CCR2$^{-/-}$ mice: evidence for independent chemokine functions in atherogenesis. *Circulation* **117:**1642–1648.

38. **Soehnlein O, Drechsler M, Döring Y, Lievens D, Hartwig H, Kemmerich K, Ortega-Gómez A, Mandl M, Vijayan S, Projahn D, Garlichs CD, Koenen RR, Hristov M, Lutgens E, Zernecke A, Weber C.** 2013. Distinct functions of chemokine receptor axes in the atherogenic mobilization and recruitment of classical monocytes. *EMBO Mol Med* **5:**471–481.

39. **Swirski FK, Pittet MJ, Kircher MF, Aikawa E, Jaffer FA, Libby P, Weissleder R.** 2006. Monocyte accumulation in mouse atherogenesis is progressive and proportional to extent of disease. *Proc Natl Acad Sci U S A* **103:** 10340–10345.

40. **van Gils JM, Derby MC, Fernandes LR, Ramkhelawon B, Ray TD, Rayner KJ, Parathath S, Distel E, Feig JL, Alvarez-Leite JI, Rayner AJ, McDonald TO, O'Brien KD, Stuart LM, Fisher EA, Lacy-Hulbert A, Moore KJ.** 2012. The neuroimmune guidance cue netrin-1 promotes atherosclerosis by inhibiting the emigration of macrophages from plaques. *Nat Immunol* **13:**136–143.

41. **Wanschel A, Seibert T, Hewing B, Ramkhelawon B, Ray TD, van Gils JM, Rayner KJ, Feig JE, O'Brien ER, Fisher EA, Moore KJ.** 2013. Neuroimmune guidance cue

Semaphorin 3E is expressed in atherosclerotic plaques and regulates macrophage retention. *Arterioscler Thromb Vasc Biol* 33:886–893.

42. **Swirski FK, Nahrendorf M, Libby P.** 2012. The ins and outs of inflammatory cells in atheromata. *Cell Metab* **15:** 135–136.

43. **Cybulsky MI, Jongstra-Bilen J.** 2010. Resident intimal dendritic cells and the initiation of atherosclerosis. *Curr Opin Lipidol* **21:**397–403.

44. **Packard RR, Maganto-García E, Gotsman I, Tabas I, Libby P, Lichtman AH.** 2008. CD11c+ dendritic cells maintain antigen processing, presentation capabilities, and CD4+ T-cell priming efficacy under hypercholesterolemic conditions associated with atherosclerosis. *Circ Res* **103:**965–973.

45. **Gimbrone MA Jr, García-Cardeña G.** 2013. Vascular endothelium, hemodynamics, and the pathobiology of atherosclerosis. *Cardiovasc Pathol* **22:**9–15.

46. **Chatzizisis YS, Blankstein R, Libby P.** 2014. Inflammation goes with the flow: implications for non-invasive identification of high-risk plaque. *Atherosclerosis* **234:** 476–478.

47. **Jongstra-Bilen J, Haidari M, Zhu SN, Chen M, Guha D, Cybulsky MI.** 2006. Low-grade chronic inflammation in regions of the normal mouse arterial intima predisposed to atherosclerosis. *J Exp Med* **203:**2073–2083.

48. **Clinton SK, Underwood R, Hayes L, Sherman ML, Kufe DW, Libby P.** 1992. Macrophage colony-stimulating factor gene expression in vascular cells and in experimental and human atherosclerosis. *Am J Pathol* **140:**301–316.

49. **Stary HC.** 1989. Evolution and progression of atherosclerotic lesions in coronary arteries of children and young adults. *Arteriosclerosis* **9**(Suppl):I19–I32.

50. **Rosenfeld ME, Ross R.** 1990. Macrophage and smooth muscle cell proliferation in atherosclerotic lesions of WHHL and comparably hypercholesterolemic fat-fed rabbits. *Arteriosclerosis* **10:**680–687.

51. **Robbins CS, Hilgendorf I, Weber GF, Theurl I, Iwamoto Y, Figueiredo JL, Gorbatov R, Sukhova GK, Gerhardt LM, Smyth D, Zavitz CC, Shikatani EA, Parsons M, van Rooijen N, Lin HY, Husain M, Libby P, Nahrendorf M, Weissleder R, Swirski FK.** 2013. Local proliferation dominates lesional macrophage accumulation in atherosclerosis. *Nat Med* **19:**1166–1172.

52. **Geng YJ, Libby P.** 1995. Evidence for apoptosis in advanced human atheroma. Colocalization with interleukin-1β-converting enzyme. *Am J Pathol* **147:**251–266.

53. **Geng YJ, Libby P.** 2002. Progression of atheroma: a struggle between death and procreation. *Arterioscler Thromb Vasc Biol* **22:**1370–1380.

54. **Li S, Sun Y, Liang CP, Thorp EB, Han S, Jehle AW, Saraswathi V, Pridgen B, Kanter JE, Li R, Welch CL, Hasty AH, Bornfeldt KE, Breslow JL, Tabas I, Tall AR.** 2009. Defective phagocytosis of apoptotic cells by macrophages in atherosclerotic lesions of ob/ob mice and reversal by a fish oil diet. *Circ Res* **105:**1072–1082.

55. **Thorp E, Tabas I.** 2009. Mechanisms and consequences of efferocytosis in advanced atherosclerosis. *J Leukoc Biol* **86:**1089–1095.

56. **Libby P, Tabas I, Fredman G, Fisher EA.** 2014. Inflammation and its resolution as determinants of acute coronary syndromes. *Circ Res* **114:**1867–1879.

57. **Brogi E, Winkles JA, Underwood R, Clinton SK, Alberts GF, Libby P.** 1993. Distinct patterns of expression of fibroblast growth factors and their receptors in human atheroma and nonatherosclerotic arteries. Association of acidic FGF with plaque microvessels and macrophages. *J Clin Invest* **92:**2408–2418.

58. **Sluimer JC, Daemen MJ.** 2009. Novel concepts in atherogenesis: angiogenesis and hypoxia in atherosclerosis. *J Pathol* **218:**7–29.

59. **Rajavashisth T, Qiao JH, Tripathi S, Tripathi J, Mishra N, Hua M, Wang XP, Loussararian A, Clinton S, Libby P, Lusis A.** 1998. Heterozygous osteopetrotic (*op*) mutation reduces atherosclerosis in LDL receptor-deficient mice. *J Clin Invest* **101:**2702–2710.

60. **New SE, Goettsch C, Aikawa M, Marchini JF, Shibasaki M, Yabusaki K, Libby P, Shanahan CM, Croce K, Aikawa E.** 2013. Macrophage-derived matrix vesicles: an alternative novel mechanism for microcalcification in atherosclerotic plaques. *Circ Res* **113:**72–77.

61. **Liu J, Sukhova GK, Sun JS, Xu WH, Libby P, Shi GP.** 2004. Lysosomal cysteine proteases in atherosclerosis. *Arterioscler Thromb Vasc Biol* **24:**1359–1366.

62. **Libby P.** 2013. Mechanisms of acute coronary syndromes and their implications for therapy. *N Engl J Med* **368:** 2004–2013.

63. **Bentzon JF, Otsuka F, Virmani R, Falk E.** 2014. Mechanisms of plaque formation and rupture. *Circ Res* **114:** 1852–1866.

64. **Ueno T, Dutta P, Keliher E, Leuschner F, Majmudar M, Marinelli B, Iwamoto Y, Figueiredo JL, Christen T, Swirski FK, Libby P, Weissleder R, Nahrendorf M.** 2013. Nanoparticle PET-CT detects rejection and immunomodulation in cardiac allografts. *Circ Cardiovasc Imaging* **6:**568–573.

65. **Quillard T, Araújo HA, Franck G, Tesmenitsky Y, Libby P.** 2014. Matrix metalloproteinase-13 predominates over matrix metalloproteinase-8 as the functional interstitial collagenase in mouse atheromata. *Arterioscler Thromb Vasc Biol* **34:**1179–1186.

66. **Quillard T, Araújo HA, Franck G, Shvartz E, Sukhova G, Libby P.** 2015. TLR2 and neutrophils potentiate endothelial stress, apoptosis and detachment: implications for superficial erosion. *Eur Heart J* **36:**1394–1404.

67. **Megens RT, Vijayan S, Lievens D, Döring Y, van Zandvoort MA, Grommes J, Weber C, Soehnlein O.** 2012. Presence of luminal neutrophil extracellular traps in atherosclerosis. *Thromb Haemost* **107:**597–598.

68. **Borissoff JI, Joosen IA, Versteylen MO, Brill A, Fuchs TA, Savchenko AS, Gallant M, Martinod K, Ten Cate H, Hofstra L, Crijns HJ, Wagner DD, Kietselaer BL.** 2013. Elevated levels of circulating DNA and chromatin are independently associated with severe coronary atherosclerosis and a prothrombotic state. *Arterioscler Thromb Vasc Biol* **33:**2032–2040.

69. **Stakos DA, Kambas K, Konstantinidis T, Mitroulis I, Apostolidou E, Arelaki S, Tsironidou V, Giatromanolaki**

A, Skendros P, Konstantinides S, Ritis K. 2015. Expression of functional tissue factor by neutrophil extracellular traps in culprit artery of acute myocardial infarction. *Eur Heart J* **36**:1405–1414.

70. **Warnatsch A, Ioannou M, Wang Q, Papayannopoulos V.** 2015. Inflammation. Neutrophil extracellular traps license macrophages for cytokine production in atherosclerosis. *Science* **349**:316–320.

71. **Sun J, Sukhova GK, Wolters PJ, Yang M, Kitamoto S, Libby P, MacFarlane LA, Mallen-St Clair J, Shi GP.** 2007. Mast cells promote atherosclerosis by releasing proinflammatory cytokines. *Nat Med* **13**:719–724.

72. **Libby P, Shi GP.** 2007. Mast cells as mediators and modulators of atherogenesis. *Circulation* **115**:2471–2473.

73. **Wang J, Lindholt JS, Sukhova GK, Shi MA, Xia M, Chen H, Xiang M, He A, Wang Y, Xiong N, Libby P, Wang JA, Shi GP.** 2014. IgE actions on CD4$^+$ T cells, mast cells, and macrophages participate in the pathogenesis of experimental abdominal aortic aneurysms. *EMBO Mol Med* **6**:952–969.

74. **Bot I, de Jager SC, Zernecke A, Lindstedt KA, van Berkel TJ, Weber C, Biessen EA.** 2007. Perivascular mast cells promote atherogenesis and induce plaque destabilization in apolipoprotein E-deficient mice. *Circulation* **115**: 2516–2525.

75. **Willems S, Vink A, Bot I, Quax PH, de Borst GJ, de Vries JP, van de Weg SM, Moll FL, Kuiper J, Kovanen PT, de Kleijn DP, Hoefer IE, Pasterkamp G.** 2013. Mast cells in human carotid atherosclerotic plaques are associated with intraplaque microvessel density and the occurrence of future cardiovascular events. *Eur Heart J* **34**: 3699–3706.

76. **Sager HB, Dutta P, Dahlman JE, Hulsmans M, Courties G, Sun Y, Heidt T, Vinegoni C, Borodovsky A, Fitzgerald K, Wojtkiewicz GR, Iwamoto Y, Tricot B, Khan OF, Kauffman KJ, Xing Y, Shaw TE, Libby P, Langer R, Weissleder R, Swirski FK, Anderson DG, Nahrendorf M.** 2016. RNAi targeting multiple cell adhesion molecules reduces immune cell recruitment and vascular inflammation after myocardial infarction. *Sci Transl Med* **8**: 342ra80.

77. **Courties G, Heidt T, Sebas M, Iwamoto Y, Jeon D, Truelove J, Tricot B, Wojtkiewicz G, Dutta P, Sager HB, Borodovsky A, Novobrantseva T, Klebanov B, Fitzgerald K, Anderson DG, Libby P, Swirski FK, Weissleder R, Nahrendorf M.** 2014. In vivo silencing of the transcription factor IRF5 reprograms the macrophage phenotype and improves infarct healing. *J Am Coll Cardiol* **63**: 1556–1566.

78. **Jain MK, Ridker PM.** 2005. Anti-inflammatory effects of statins: clinical evidence and basic mechanisms. *Nat Rev Drug Discov* **4**:977–987.

79. **Nidorf SM, Eikelboom JW, Budgeon CA, Thompson PL.** 2013. Low-dose colchicine for secondary prevention of cardiovascular disease. *J Am Coll Cardiol* **61**:404–410.

80. **Everett BM, Pradhan AD, Solomon DH, Paynter N, Macfadyen J, Zaharris E, Gupta M, Clearfield M, Libby P, Hasan AA, Glynn RJ, Ridker PM.** 2013. Rationale and design of the Cardiovascular Inflammation Reduction Trial: a test of the inflammatory hypothesis of atherothrombosis. *Am Heart J* **166**:199–207.e15. doi: 10.1016/j.ahj.2013.03.018.

81. **Ridker PM, Thuren T, Zalewski A, Libby P.** 2011. Interleukin-1β inhibition and the prevention of recurrent cardiovascular events: rationale and design of the Canakinumab Anti-inflammatory Thrombosis Outcomes Study (CANTOS). *Am Heart J* **162**:597–605.

82. **Shah PK, Chyu KY, Dimayuga PC, Nilsson J.** 2014. Vaccine for atherosclerosis. *J Am Coll Cardiol* **64**:2779–2791.

83. **Nilsson J, Lichtman A, Tedgui A.** 2015. Atheroprotective immunity and cardiovascular disease: therapeutic opportunities and challenges. *J Intern Med* **278**:507–519.

84. **Libby P, Pasterkamp G.** 2015. Requiem for the 'vulnerable plaque'. *Eur Heart J* **36**:2984–2987.

Myeloid Cells in Health and Disease: A Synthesis
Edited by Siamon Gordon

doi:10.1128/microbiolspec.MCHD-0048-2016

Emil R. Unanue[1]

48 Macrophages in Endocrine Glands, with Emphasis on Pancreatic Islets

That macrophages normally inhabit endocrine organs became evident from the early studies in mice made in the laboratories of Siamon Gordon and David Hume (1). They used the macrophage surface marker F4/80 to examine tissue sections by immunohistochemistry and found macrophages in all endocrine organs. In most organs, the macrophages were found associated with the vessels. To note are several important examples. In the adrenal gland, the zona glomerulosa contained abundant macrophages that "wrapped around capillaries or line vascular sinuses, but membrane processes extend into the surrounding tissue." In the pituitary, they found them distributed differentially in the various areas. In the thyroid, many surrounded the follicles. In the testes, the macrophages were next to the Leydig cells and not inside the seminiferous tubules. In the ovaries, macrophages were abundant around follicles and usually surrounding the vessels.

In sum, they found three distinctive features of endocrine-resident macrophages: (i) most were associated with blood vessels; (ii) some surrounded the endocrine cells or glands, extending filopodia among them; and (iii) in some organs such as the pituitary, there were differences in the distribution depending on the anatomy.

An understanding of the important role of macrophages in the normal homeostasis of endocrine organs has come from many studies, but particularly from Jeffrey Pollard's group, many carried out in the *Csfm*$^{op/op}$ [op/op] mouse strain (2–4). ("Homeostasis" is the term used by Walter B. Cannon in discussing the *milieu interieur* of Claude Bernard: at the cellular level, we refer to those cell functions needed to maintain a tissue under steady state [5]). The op/op mouse strain has a natural null mutation in the *Csf1* gene and lacks functional colony-stimulating factor-1 (CSF-1), resulting in marked absence of macrophages in most tissues (6). CSF-1 is a protein made in several tissues that regulates the differentiation of the macrophage lineage (7). The op/op mice were first described for their osteopetrosis resulting from a lack of osteoclasts, a differentiated cell of the macrophage lineage required to remove bone matrix. The production of CSF-1 was absent from all tissues examined (7). In the initial evaluation of this strain carried out by Stanley, Pollard, and associates, a partial correction was obtained by infusing mice with CSF-1. The op/op mice also showed a number of abnormalities, particularly in endocrine glands (4). The strain is an eloquent example of the role that macrophages play beyond the classical

[1]Department of Pathology and Immunology, Washington University School of Medicine, St. Louis, MO 63103.

scavenger and phagocytic function accentuated since their classical description by Metchnikoff that resulted in the "phagocytic theory" of immunity.

THE ISLETS OF LANGERHANS

Pancreatic islets in all species contain a resident set of macrophages. These were first considered as "passenger leukocytes," referring to blood leukocytes thought to be trapped in isolated islets. In a series of studies carried out using isolated islets for transplantation to diabetic patients, it was found that a preculture of the islets resulted in the escape of leukocytes from them; hence the conclusion that these cells derived from blood cells. These islets transplanted in certain allogeneic combinations were not rejected. The implication was that the macrophages were the target of the transplantation reaction since they expressed the proteins encoded in the major histocompatibility gene complex (MHC). These findings are extensively reviewed with appropriate references in reference 8. Studying the phagocytes in islets is important particularly because this mini-organ is the target of diabetic autoimmunity (9). Macrophages have an antigen-presenting function that is relevant in the autoimmune process.

Recent analyses have been carried out examining islets obtained from the pancreas using combinations of cell surface markers and gene expression analysis. Islets are harvested by techniques that separate them from the acinar tissue. The islets are then disrupted so individual cells can be examined. These recent studies clearly established that islets under steady-state conditions, that is, in noninflammatory situations, harbor a normal set of resident macrophages (10). These macrophages are represented by a single set identified by their display of membrane proteins typical of macrophages as well as by their expression of macrophage-specific genes.

Most of the detailed analyses of islet macrophages have been carried out in the mouse. Islets isolated from nondiabetic mice contain a small number of macrophages, usually from 5 to 10. There is an interesting correlation between the size of the islets and their number of macrophages: for example, about 5 to 10% of islets are of large size and can harbor as many as 30 macrophages. The macrophages are located next to blood vessels and in very close contact with β cells; in live image analysis of islets, the macrophages are seen extending long filopodia in between β cells, some reaching up to the edge of the islet. Moreover, filopodia also extend along the vessel wall and into the blood vessel lumen.

Of great interest are the findings showing that every macrophage takes up secretory granules from β cells (11). This uptake was substantiated by electron microscopy of islets that showed typical insulin-containing dense core granules inside the phagocytic vesicles of the macrophages. Examination of the macrophages using antibodies to insulin or insulin products confirmed the presence of insulin and insulin peptides inside the macrophage phagocytic vesicles (12). Such uptake is a normal feature of the islet macrophage found in all strains examined. The uptake requires live β cells and is unrelated to β-cell death.

Pertinent to this finding are the studies made in the NOD mouse, an inbred strain that spontaneously develops autoimmune diabetes. In most mouse colonies, the majority of NOD mice become diabetic within the 18th to 24th week of life, with islets heavily infiltrated by lymphocytes and inflammatory cells. Diabetic autoimmunity is an autoimmune disease caused by autoreactive T cells that recognize peptides derived from the intracellular processing of β-cell proteins. These peptides associate with the MHC class II (MHC-II) molecules to form the molecular complex recognized by the T cells. The MHC-II genes that confer the propensity for diabetes are very similar structurally between the NOD mouse and humans with type 1 diabetes (13, 14). In both situations, their MHC-II molecules lack a critical amino acid, an aspartic acid at the β-chain residue 57. The MHC-II peptidomes of the diabetes-propensity molecules from humans and mice are very similar. Analysis by mass spectrometry of peptides eluted from the MHC-II molecules showed them to be rich in peptides with acidic residues at their carboxy end (15). In the NOD mouse, which develops spontaneous diabetic autoimmunity, the uptake of β-cell granules has biological significance: macrophages degrade the content of the granules, incorporating some of the peptides to their MHC-II proteins to be presented to autoreactive T cells (10, 11). β Cells do not express MHC-II proteins but only MHC-I at low levels. Importantly, besides macrophages, the islets also contain dendritic cells that infiltrate at a very early stage in the process and that are essential for the development of autoimmunity, presumably in cooperativity with the resident macrophages (16). One autoantigen that is prominent both in NOD and in patients with type 1 diabetes is insulin (9, 17, 18).

The physiological role of the islet macrophages is to maintain islet homeostasis (2, 10). There is a truly symbiotic relationship between β cells and macrophages, with both influencing each other. Examination of the osteopetrotic op/op mouse is to the point: their islets are half the size of normal and hypoinsulinemic (2, 10). Most islets lack macrophages; at the most, a few may have one or two macrophages per islet. Many op/op mice have an impaired glucose tolerance test. On the macrophage side, and in contrast to macrophages from other tissues, the islet macrophages in wild-type mice have a gene signature of activation, with expression of tumor necrosis factor and interleukin-1β and very high levels of MHC-II molecules (10). Their gene expression pattern is compatible with an M1-like activated macrophage. There is a range of genes expressed in macrophages under conditions of activation, and in broad terms they can be grouped into an M1 or M2 pattern (19). The M2 gene activation signature was defined after interleukin-4 treatment (20) and is believed to be involved in tissue repair.

The recent report from Calderon et al. made a critical analysis of pancreatic macrophages in nondiabetic mice. Islet macrophages were found during embryological development from definitive hematopoiesis (10) (Fig. 1). (For a review on recent developments in macrophage differentiation, see reference 21.) Normally in noninflammatory conditions islet macrophages show a low level of replication and are not supplied by blood monocytes. Parabiosis experiments were carried out between members of the pair having a different CD45 allele: there was no exchange between the two partners, while monocytes and lymphocytes in the spleen exchanged evenly between the two mice (10). These resident macrophages were maintained by a low level of proliferation.

Macrophages are also found in the interacinar stroma, about 10 times more than in the islets. But these stromal macrophages are very different from those in the islets (Table 1). The stromal macrophages are composed of two distinct sets based on their embryological development and by their expression of the mannose receptor, CD206, and the protein CLEC10A (CD301). One set express low levels of CD206/CD301, high levels of MHC-II molecules, and are constantly maintained from blood monocytes. This set of macrophages are dispersed throughout the stroma. A second set express high levels of CD206/CD301 and derive from yolk sac hematopoiesis. This set is enriched around the pancreatic ducts. Thus, both sets of stromal macrophages have a different anatomical localization. Both sets of stromal macrophages express an M2-like pattern of gene transcripts, in contrast to the islet macrophages, which express an M1-like pattern.

The gene expression signatures in islets and stromal macrophages are maintained with time.

An important experimental manipulation made in the report of Calderon et al. was to heavily irradiate the mice in order to deplete the islet macrophages, and to then replace them with bone marrow stem cells. After a period of time, the pancreas was examined. The

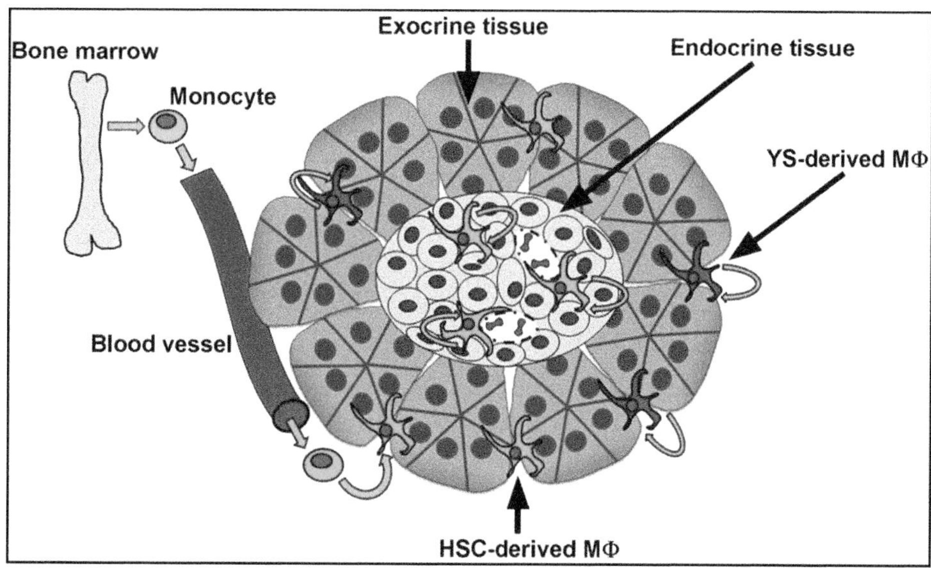

Figure 1 Pancreatic macrophages (MΦ). The pancreas contains three sets of macrophages. One set reside in the islets, labeled endocrine tissue, and have an M1-like transcriptional signature. Two sets are in the stroma in between the acinar gland, labeled as exocrine tissue. Both have an M2-like transcriptional signature. One set derives from blood monocytes; the second derives from yolk sac (YS) hematopoiesis. Reprinted from reference 10, with permission.

Table 1 Features of pancreatic macrophages

Feature	Islet	Stromal set 1	Stromal set 2
Derivation	From embryo Definite hematopoiesis	Yolk sac Hematopoiesis	Blood monocytes
Anatomy	Next to blood vessels	Enriched in peritubular areas	Dispersed
Turnover	Resident	Resident	Supplied by blood monocytes
Gene signature	M1	M2	M2
Gene signature after irradiation	M1	M2	M2
MHC-II	High	Low	High
Role	Homeostatic capture granules	Not known	Not known
Pathology	In autoimmune diabetes	Not known Role in pancreatic cancer?	Not known

new macrophages in islets and stroma had the same gene expression pattern as before, M1-like for those of the islets and M2-like for the stroma ones. The differences most likely are explained by the anatomy of the pancreas, its microenvironment. The Calderon et al. study concluded that the "pancreas anatomy conditions the origin and properties of the resident macrophages" (10).

REPRODUCTIVE ORGANS: TESTES AND OVARIES

Macrophages are found in the testes, where they play a major physiological role (4, 22–25). As in the pancreas, macrophages are located in two distinct anatomical sites in the mammalian testes (26). One site is in the interstitium between seminiferous tubules: the macrophages are closely associated with Leydig cells, the testosterone-producing cells. The second site is around the border of the seminiferous tubules, forming part of the peritubular capsule (26; see Comment in reference 27) (Fig. 2). At both sites, macrophages profoundly influence testicular function (4, 22–24). In the interstitium, the macrophages form tight clusters with Leydig cell (28, 29). There is a striking close contact between both cells, with cytoplasmic extensions of Leydig cells contacting the macrophages in close membrane-to-membrane contacts (29).

The macrophages are found during embryonal differentiation of the testis and have a major role in its development and early vascularization. Lineage-tracing studies indicate that the macrophages derive from early hematopoietic stem cells (HSCs) in the yolk sac (30). Such macrophages have a gene signature compatible with M2 features and express proteins typical of macrophages, like Fc receptors. When isolated, they can be shown to release cytokines (22). Studies in the op/op mouse showed markedly reduced numbers of macrophages. Accompanying this reduction was a parallel reduction in the number of Leydig cells and a low level of testosterone (4, 25). Leydig cells in the op/op mouse have a dilated endoplasmic reticulum, together with abnormal vesicles. The macrophages produce 25-hydroxycholesterol, an intermediate in the biosynthesis of testosterone that may be a substrate for its production by the Leydig cell (28).

As mentioned, macrophages are also found tightly associated with the seminiferous tubules (26). The cells are situated on the outside surface of tubules close to the stem cell niche that gives rise to spermatogonia (26). These macrophages show long filopodia extending along the surface of the tubule. They produce CSF-1, which contributes to the differentiation of the sperm cells. Macrophages from the two sites, i.e., the interstitium next to Leydig cells and on the seminiferous tubule surface, can be differentiated by their expression of the receptor for CSF-1 and expression of MHC-II molecules. Those in the interstitium have the receptor for CSF-1 but are MHC-II negative, and the reverse was shown for the macrophages of the seminiferous tubules (26). The relationship between these two sets of macrophages is not known, but the testes, like the pancreas, represent another example of the tissue conditioning its macrophage resident population while being profoundly influenced by it: a true symbiosis.

The testis has been considered an immune-privileged tissue, and the macrophages may be part of the barrier for the protection of the sperm cells (reviewed in references 31 and 32). Whether the macrophages have an antigen-presenting function in autoimmune orchitis or autoimmune aspermatogenesis has not been established. In experimental autoimmune orchitis, in which the animal is immunized with testes antigen, an initial reaction is at the surface of the seminiferous tubules (33). The thinking is that the macrophages at that site contain antigens derived from spermatogonia and may have a presenting function for autoreactive T cells.

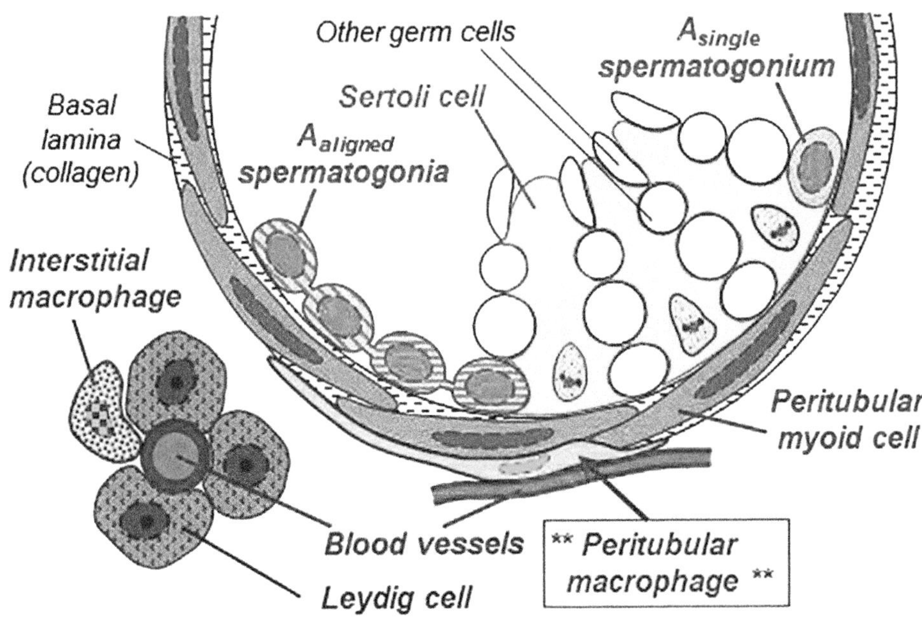

Figure 2 Testicular macrophages. The macrophages in the testes are found in two locations. One set are in the interstitium, forming a cluster with Leydig cells. The second set surround the seminiferous tubules—the peritubular macrophages. Reproduced from reference 27, with permission.

Macrophages are also found in the ovaries (4, 34). They are situated at different sites within the interstitial tissue that surrounds the ovarian follicles but never inside them. CSF-1 is made by cells in the granulosa layer of the follicle and has been thought to have a role in maintaining the macrophage population (4). During ovulation, the number of macrophages increases in the theca layer outside of the follicle. After the ovum is released, macrophages are prominent in the corpus luteum (35–37). The op/op mouse has defective ovarian function, ascribed in part to the profound reduction in macrophage numbers (4, 38). Estrous cycles are much longer and the degree of ovulation is reduced. In brief, as in the other endocrine organs, the macrophage modulates the function of the organ (3, 4, 39).

OTHER ENDOCRINE ORGANS

As reported by Gordon and Hume and collaborators (1), macrophages also are found in adrenal glands, the thyroid, and the pituitary. Analyses of these tissue-resident macrophages in terms of their role and embryonic derivation have yet to be done. In the adrenal, the macrophages are found in both cortex and medulla, although the number in each layer varies. Macrophages are reported to extend long projections between the endocrine cells and localizing close to blood vessels (1). Addison's disease is a rare autoimmune disease that results in the destruction of the adrenal cortex. Most patients have autoantibodies to steroid 21-hydroxylase, an intracellular endoplasmic reticulum protein enzyme involved in steroid biosynthesis (40). CD4 and CD8 T cells directed to peptides of the enzyme have been reported (41). Although not examined, it is likely that the macrophages are responsible for the uptake of the intracellular material and participate in the initiation of the autoimmune process.

Concerning the thyroid, few studies have examined macrophages under steady-state conditions. Macrophages have been detected in the pituitary gland. Their distribution varies, with the greatest number in the anterior and posterior lobes, and they are practically absent in the intermediate lobe (1, 34). The hypothalamus-pituitary-gonadal axis is markedly defective in the op/op mouse (4), but the nature of the macrophage involvement has not been examined.

FINAL COMMENT

Of the various endocrine organs, the pancreatic islets and the testes are the two in which macrophages have been examined most extensively. Three important results have come out of their analysis. The first concerns the different gene signatures among the macrophages.

The studies in the pancreas, where those in the interacinar stroma differ from those in the islets, are particularly informative and point to the microenvironment controlling the biological response of the macrophage. A second important conclusion is that the endocrine cells require the trophic function of the macrophage. This requirement was shown best through the examination of the op/op mouse. These mice show defective islets and testes function. It is also apparent that macrophages also influence ovarian function and the pituitary-gonadal axis. However, no very detailed examinations of the macrophages in the latter two sites have been carried out. The conclusion is that there is a striking symbiosis between the endocrine cells and the macrophages, each deeply influencing each other. Finally, macrophages take up products of the surrounding endocrine cells. It is likely that these are the molecules that influence the behavior of the macrophage. As we discussed for pancreatic islets, the macrophages take up whole dense core granules that contain not only insulin, about 3×10^5 molecules per granule, but other bioactive molecules such as ATP and granins.

Finally, it is important to note that endocrine organs are among the most frequent targets for autoimmunity, and this needs to be explained. Having macrophages expressing high levels of MHC-II plus containing the molecules responsible for the autoimmune reaction places them as a central player.

Citation. Unanue ER. 2016. Macrophages in endocrine glands, with emphasis on pancreatic islets. Microbiol Spectrum 4(6):MCHD-0048-2016.

References

1. **Hume DA, Halpin D, Charlton H, Gordon S.** 1984. The mononuclear phagocyte system of the mouse defined by immunohistochemical localization of antigen F4/80: macrophages of endocrine organs. *Proc Natl Acad Sci U S A* **81**:4174–4177.
2. **Banaei-Bouchareb L, Gouon-Evans V, Samara-Boustani D, Castellotti MC, Czernichow P, Pollard JW, Polak M.** 2004. Insulin cell mass is altered in *Csf1^op/Csf1^op* macrophage-deficient mice. *J Leukoc Biol* **76**:359–367.
3. **Pollard JW.** 2009. Trophic macrophages in development and disease. *Nat Rev Immunol* **9**:259–270.
4. **Cohen PE, Nishimura K, Zhu L, Pollard JW.** 1999. Macrophages: important accessory cells for reproductive function. *J Leukoc Biol* **66**:765–772.
5. **Cannon WB.** 1929. Organization for physiological homeostasis. *Physiol Rev* **9**:399–431.
6. **Yoshida H, Hayashi S, Kunisada T, Ogawa M, Nishikawa S, Okamura H, Sudo T, Shultz LD, Nishikawa S.** 1990. The murine mutation osteopetrosis is in the coding region of the macrophage colony stimulating factor gene. *Nature* **345**:442–444.
7. **Wiktor-Jedrzejczak W, Bartocci A, Ferrante AW Jr, Ahmed-Ansari A, Sell KW, Pollard JW, Stanley ER.** 1990. Total absence of colony-stimulating factor 1 in the macrophage-deficient osteopetrotic (*op/op*) mouse. *Proc Natl Acad Sci U S A* **87**:4828–4832.
8. **Calderon B, Unanue ER.** 2012. Antigen presentation events in autoimmune diabetes. *Curr Opin Immunol* **24**:119–128.
9. **Unanue ER.** 2014. Antigen presentation in the autoimmune diabetes of the NOD mouse. *Annu Rev Immunol* **32**:579–608.
10. **Calderon B, Carrero JA, Ferris ST, Sojka DK, Moore L, Epelman S, Murphy KM, Yokoyama WM, Randolph GJ, Unanue ER.** 2015. The pancreas anatomy conditions the origin and properties of resident macrophages. *J Exp Med* **212**:1497–1512.
11. **Vomund AN, Zinselmeyer BH, Hughes J, Calderon B, Valderrama C, Ferris ST, Wan X, Kanekura K, Carrero JA, Urano F, Unanue ER.** 2015. Beta cells transfer vesicles containing insulin to phagocytes for presentation to T cells. *Proc Natl Acad Sci U S A* **112**:E5496–E5502. doi:10.1073/pnas.1515954112.
12. **Mohan JF, Levisetti MG, Calderon B, Herzog JW, Petzold SJ, Unanue ER.** 2010. Unique autoreactive T cells recognize insulin peptides generated within the islets of Langerhans in autoimmune diabetes. *Nat Immunol* **11**:350–354.
13. **Todd JA, Bell JI, McDevitt HO.** 1987. HLA-DQ_{β} gene contributes to susceptibility and resistance to insulin-dependent diabetes mellitus. *Nature* **329**:599–604.
14. **Acha-Orbea H, McDevitt HO.** 1987. The first external domain of the nonobese diabetic mouse class II I-A β chain is unique. *Proc Natl Acad Sci USA* **84**:2435–2439.
15. **Suri A, Walters JJ, Gross ML, Unanue ER.** 2005. Natural peptides selected by diabetogenic DQ8 and murine I-A^{g7} molecules show common sequence specificity. *J Clin Invest* **115**:2268–2276.
16. **Ferris ST, Carrero JA, Mohan JF, Calderon B, Murphy KM, Unanue ER.** 2014. A minor subset of Batf3-dependent antigen-presenting cells in islets of Langerhans is essential for the development of autoimmune diabetes. *Immunity* **41**:657–669.
17. **Zhang L, Nakayama M, Eisenbarth GS.** 2008. Insulin as an autoantigen in NOD/human diabetes. *Curr Opin Immunol* **20**:111–118.
18. **Brezar V, Carel JC, Boitard C, Mallone R.** 2011. Beyond the hormone: insulin as an autoimmune target in type 1 diabetes. *Endocr Rev* **32**:623–669.
19. **Murray PJ, Allen JE, Biswas SK, Fisher EA, Gilroy DW, Goerdt S, Gordon S, Hamilton JA, Ivashkiv LB, Lawrence T, Locati M, Mantovani A, Martinez FO, Mege JL, Mosser DM, Natoli G, Saeij JP, Schultze JL, Shirey KA, Sica A, Suttles J, Udalova I, van Ginderachter JA, Vogel SN, Wynn TA.** 2014. Macrophage activation and polarization: nomenclature and experimental guidelines. *Immunity* **41**:14–20.
20. **Gordon S.** 2003. Alternative activation of macrophages. *Nat Rev Immunol* **3**:23–35.
21. **Perdiguero EG, Geissmann F.** 2016. The development and maintenance of resident macrophages. *Nat Immunol* **17**:2–8.

22. **Hutson JC.** 2006. Physiologic interactions between macrophages and Leydig cells. *Exp Biol Med (Maywood)* **231:**1–7.

23. **Hutson JC.** 1994. Testicular macrophages. *Int Rev Cytol* **149:**99–143.

24. **Bhushan S, Meinhardt A.** 2016. The macrophages in testes function. *J Reprod Immunol* doi:10.1016/j.jri.2016.06.008.

25. **Pollard JW, Dominguez MG, Mocci S, Cohen PE, Stanley ER.** 1997. Effect of the colony-stimulating factor-1 null mutation, osteopetrotic ($csfm^{op}$), on the distribution of macrophages in the male mouse reproductive tract. *Biol Reprod* **56:**1290–1300.

26. **DeFalco T, Potter SJ, Williams AV, Waller B, Kan MJ, Capel B.** 2015. Macrophages contribute to the spermatogonial niche in the adult testis. *Cell Rep* **12:**1107–1119.

27. **Meistrich ML, Shetty G.** 2015. The new director of "the spermatogonial niche": introducing the peritubular macrophage. *Cell Rep* **12:**1069–1070.

28. **Lukyanenko YO, Chen JJ, Hutson JC.** 2001. Production of 25-hydroxycholesterol by testicular macrophages and its effects on Leydig cells. *Biol Reprod* **64:**790–796.

29. **Hutson JC.** 1992. Development of cytoplasmic digitations between Leydig cells and testicular macrophages of the rat. *Cell Tissue Res* **267:**385–389.

30. **DeFalco T, Bhattacharya I, Williams AV, Sams DM, Capel B.** 2014. Yolk-sac-derived macrophages regulate fetal testis vascularization and morphogenesis. *Proc Natl Acad Sci U S A* **111:**E2384–E2393. doi:10.1073/pnas.1400057111.

31. **Stanton PG.** 2016. Regulation of the blood-testis barrier. *Semin Cell Dev Biol* doi:10.1016/j.semcdb.2016.06.018.

32. **Mruk DD, Cheng CY.** 2015. The mammalian blood-testis barrier: its biology and regulation. *Endocr Rev* **36:**564–591.

33. **Tung KS, Yule TD, Mahi-Brown CA, Listrom MB.** 1987. Distribution of histopathology and Ia positive cells in actively induced and passively transferred experimental autoimmune orchitis. *J Immunol* **138:**752–759.

34. **Hoek A, Allaerts W, Leenen PJ, Schoemaker J, Drexhage HA.** 1997. Dendritic cells and macrophages in the pituitary and the gonads. Evidence for their role in the fine regulation of the reproductive endocrine response. *Eur J Endocrinol* **136:**8–24.

35. **Kirsch TM, Friedman AC, Vogel RL, Flickinger GL.** 1981. Macrophages in corpora lutea of mice: characterization and effects on steroid secretion. *Biol Reprod* **25:**629–638.

36. **Brännström M, Giesecke L, Moore IC, van den Heuvel CJ, Robertson SA.** 1994. Leukocyte subpopulations in the rat corpus luteum during pregnancy and pseudopregnancy. *Biol Reprod* **50:**1161–1167.

37. **Care AS, Diener KR, Jasper MJ, Brown HM, Ingman WV, Robertson SA.** 2013. Macrophages regulate corpus luteum development during embryo implantation in mice. *J Clin Invest* **123:**3472–3487.

38. **Cohen PE, Zhu L, Pollard JW.** 1997. Absence of colony stimulating factor-1 in osteopetrotic ($csfm^{op}/csfm^{op}$) mice disrupts estrous cycles and ovulation. *Biol Reprod* **56:**110–118.

39. **Chua AC, Hodson LJ, Moldenhauer LM, Robertson SA, Ingman WV.** 2010. Dual roles for macrophages in ovarian cycle-associated development and remodelling of the mammary gland epithelium. *Development* **137:**4229–4238.

40. **Winqvist O, Karlsson FA, Kämpe O.** 1992. 21-Hydroxylase, a major autoantigen in idiopathic Addison's disease. *Lancet* **339:**1559–1562.

41. **Mitchell AL, Pearce SH.** 2012. Autoimmune Addison disease: pathophysiology and genetic complexity. *Nat Rev Endocrinol* **8:**306–316.

Myeloid Cells in Health and Disease: A Synthesis
Edited by Siamon Gordon

doi:10.1128/microbiolspec.MCHD-0016-2015

Francesco De Sanctis[1]
Vincenzo Bronte[1]
Stefano Ugel[1]

49

Tumor-Induced Myeloid-Derived Suppressor Cells

MDSCs AS A HALLMARK OF CANCER PROGRESSION

Tumors are composed of heterogeneous, transformed cell populations with different morphologies and phenotypes, which are organized in a pyramidal architecture determined by self-renewal ability, differentiation grade, and tumorigenic and clonogenic potential (1). During tumor progression, cancer cells secrete tumor-derived factors (TDFs), like cytokines, chemokines, and metabolites, which promote the development of a flexible microenvironment inducing both the generation of new vessels and the modification of the immune responses (2). Tumors can escape the immune system by three main mechanisms: (i) cancer cells can veil their identity to escape recognition by immune effectors, (ii) they can directly modify antitumor immunity, or (iii) they can recruit other immune regulatory cells whose normal function is to inhibit immune reactions and prevent the unfavorable effects of uncontrolled immune stimulation (3). Probably the most pervasive and efficient strategy of "tumor escape" relies on the tumor's ability to create a tolerant microenvironment by modification of normal hematopoiesis. In fact, cancers can induce the proliferation and differentiation of myeloid precursors into myeloid cells with immunosuppressive functions, in both the bone marrow and other hematopoietic organs such as the spleen, at the expense of additional myeloid cell subsets, such as dendritic cells (DCs) (4). Additionally, the persisting imbalance in the number and type of myeloid cells can deeply influence myeloid cell recruitment and function at the tumor site and secondary lymphoid organs. In the bone marrow, hematopoietic progenitor cells give rise to immature DCs (iDCs). To reach complete maturation, iDCs require inflammation-related stimuli because, although able to take up, process, and present antigens, they express few or none of the costimulatory molecules, such as CD80, CD86, and CD40, necessary to exert their functions (5). The higher number of iDCs found at the tumor site stems from defects in myelopoiesis rather than simply from the lack of appropriate activation signals at the tumor site. *In vitro* treatment of tumor-infiltrating DCs with appropriate stimuli—such as granulocyte-macrophage colony-stimulating factor (GM-CSF), tumor necrosis factor α (TNF-α), or CD40L—was not sufficient to induce DC maturation; this evidence supports the concept that the reduced functionality of DCs is most likely due to defects in

[1]Immunology Section, Department of Pathology and Diagnostics, University of Verona, 37135, Verona, Italy.

differentiation from their iDC progenitors (6). However, iDCs are not the only myeloid cell populations modified in cancer. In postnatal life, hematopoietic stem cells present in hemopoietic compartments give rise to lymphoid and myeloid multipotent precursor cells. Other pluripotent cell types originate from the myeloid precursors: the common DCs and the immature myeloid cell precursors (IMCs). The first originates iDCs and plasmacytoid DCs, and the second is the common progenitor for macrophages, granulocytes, and monocyte-derived DCs (7). In healthy mice, IMCs rapidly differentiate into their descendant lineages; consequently, they represent a relatively low percentage of circulating myeloid cells. However, under pathological conditions, including cancer, there is a partial block in IMC differentiation, leading to the accumulation of CD11b$^+$/Gr-1$^+$ myeloid cells with immunosuppressive function, named myeloid-derived suppressor cells (MDSCs) (8).

About 30 years ago, Strober (9) highlighted, for the first time, the immunosuppressive properties of myeloid cells. These cells were originally defined as natural suppressor (NS) cells; they lacked common markers for lymphocytes, natural killer (NK) cells, and macrophages and showed the peculiar property of suppressing T-cell functions, affecting alloreactive immune responses in allogeneic bone marrow chimera experiments (9). NS cells were deemed to appear only transiently in some life phases, such as in the placenta during pregnancy, in fetal newborn tissues, and in the neonatal maturation of lymphoid tissues. However, they could be induced by immune system manipulation, for example, through total body irradiation or chemotherapy; during chronic inflammatory pathologies, such as graft-versus-host disease; and in cancer. Technical limitations in identification, purification, and *in vitro* culture conditions delayed the definition of biological properties and phenotype of NS cells. The first clear involvement of myeloid cells in lowering immune surveillance and in promoting tumor growth was provided in 1995. The administration of an antibody directed against the antigen Gr-1 (recognizing the cross-reacting molecules of lymphocyte antigen 6 complex locus C and G, i.e., Ly6C and Ly6G) to immunocompetent mice reduced the growth of a UV light-induced tumor (10). The effect of the *in vivo* anti-Gr-1 administration was originally attributed to the elimination of granulocytes, but successive reports described that the Gr-1$^+$ cells were mostly CD11b$^+$ and comprised both polymorphonuclear and mononuclear cells, including elements at different maturation stages along the myelomonocytic differentiation lineage (11, 12).

CD11b$^+$/Gr-1$^+$ cells have been identified in noncancerous disease settings such as sepsis (13), toxoplasmosis (14), candidiasis (15), and leishmaniasis (16), as well as during autoimmune diseases (17), stress (18), and aging (19, 20). The heterogeneity of the CD11b$^+$/Gr-1$^+$ cells has for many years generated some misunderstanding, somehow amplified by the use of different acronyms to define the same cell population (i.e., NS cells, immature myeloid cells, or myeloid suppressor cells) (21). In 2007, a board of investigators agreed to use the common term "MDSC," which highlights the frequent (but not absolute) finding of immune regulatory properties coupled with enhanced myelopoiesis in tumor-bearing hosts; *de facto*, the term "MDSC" acknowledges the incomplete understanding of the relationship between the cell subsets originated by the tumor-driven, enhanced myelopoiesis (22). Even though MDSCs are not a uniform cell population—rather, they include several subgroups distinct in their morphology and expression of surface markers—the recent identification of specialized molecular programs that orchestrate MDSC differentiation, joined with innovative new technologies, has provided insights into understanding the complex and unique myeloid deviation leading to MDSC generation.

DEFINITION OF MOUSE AND HUMAN MDSCs

Mouse MDSC Phenotype and Differentiation

MDSC composition is flexible and peculiar for each disease scenario and often changes, following the kinetics and development of the disease. In healthy mice, CD11b$^+$/Gr-1$^+$ cells can be detected in sufficient numbers only in the bone marrow (about 30 to 40%), but they do not show a relevant suppressive activity *ex vivo*. A similar scenario is observed even when CD11b$^+$/Gr-1$^+$ cells are isolated from bone marrow of tumor-bearing mice (23). Suppression of T-cell function can be observed only when supraphysiologic numbers of cells are used in *in vitro* assays (24) or when bone marrow cells are previously cultured for a few days in the presence of GM-CSF, granulocyte CSF (G-CSF), and interleukin-6 (IL-6) cytokines (23). Indeed, bone marrow CD11b$^+$/Gr-1$^+$ cells contain pluripotent cells that can differentiate (25), depending on the cytokine/chemokine context, into cells able to either enhance (e.g., myeloid DCs) or restrain (MDSCs) the immune response (25, 26). In many models, indeed, the dysfunctional immune responses of T lymphocytes in tumor-bearing mice depended almost entirely on the accumulation of MDSCs in the blood and secondary lymphoid organs. Primary tumor resection, Gr-1$^+$ depletion, pharmacological inhibition, or genetic MDSC inactivation often resulted, in fact, in a complete correction of T-cell dysfunctions; as further endorsement of their dominant role, MDSC adoptive cell transfer into vaccinated, tumor-bearing mice dramatically contracted the effectiveness of immunotherapy (27, 28). Moreover, while CD11b$^+$/Gr-1$^+$ cells from naïve mice adoptively transferred into congenic mice differentiated into mature CD11c-positive, major histocompatibility complex class II (MHC-II)-positive DCs and Gr-1$^-$F4/80$^+$ macrophages within 5 days, MDSCs from tumor-bearing mice preserved their immature phenotype longer (CD11b$^+$/Gr-1$^+$), and the differentiation to mature macrophages was significantly impaired (29). Despite this panoply of differentiation options, for practical reasons, mouse MDSCs have been divided into two main subsets: monocytic (MO-MDSCs) and polymorphonuclear/granulocytic (PMN-MDSCs) (24).

In tumor-bearing mice, MO-MDSCs (Gr-1$^{lo/int}$ CD11b$^+$Ly6C^{hi}Ly6G$^-$) are highly immunosuppressive and exert their effect largely in an antigen-nonspecific manner, whereas PMN-MDSCs (Gr-1^{hi}CD11b$^+$Ly6C^{lo} Ly6G$^+$) are moderately immunosuppressive and promote T-cell tolerance via antigen-specific mechanisms. MO-MDSCs are side scatter low (SSC^{lo}), while PMN-MDSCs are SSC^{hi}. The same phenotypes in tumor-free mice define inflammatory monocytes and polymorphonuclear neutrophils, respectively, both lacking the immunosuppressive activity (30). MO-MDSCs usually express higher levels of F4/80 (macrophage marker), CD115 (c-Fms, the receptor for M-CSF/CSF-1), CD124 (receptor α of IL-4), and CCR2 (receptor for monocyte chemoattractant protein, also known as CCL2), although these markers are not uniformly expressed by MDSCs induced by all tumors. Moreover, MO-MDSCs but not PMN-MDSCs mature *in vitro* and acquire F4/80 and CD11c expression when cultured with GM-CSF (24, 31, 32). In addition, the expression of CD49d marker was associated with MO-MDSC immune-suppressive function: the CD49d$^+$ MDSC cell subset strongly inhibited antigen-specific T-cell proliferation in a nitric oxide (NO)-dependent fashion (33).

In many tumor models, as well as in cancer patients, PMN-MDSCs are the predominant subset, representing 70 to 80% of the tumor-induced MDSCs, compared to 20 to 30% of the cells reflecting the monocytic lineage (34). The two subpopulations also differ in the effector pathways used to suppress T-cell activation. MO-MDSCs suppress CD8$^+$ T-lymphocyte proliferation mainly through activation of inducible NO synthase (iNOS) and arginase 1 (ARG1) enzymes and through the production of reactive nitrogen species (RNS) (24, 32). PMN-MDSCs, instead, can express some levels of ARG1 but suppress CD8$^+$ T cells mainly through the release of reactive oxygen species (ROS) (32). A schematic summary of the main mouse and human MDSC markers is presented in Table 1. The main MDSC subsets are not two completely distinct and fully differentiated myeloid populations but rather differentiation states along a common lineage (35). A relevant fraction of MO-MDSCs, in tumor-bearing but not tumor-free mice, acquires phenotypic, morphological, and functional features of PMN-MDSCs by a mechanism that involves the epigenetic downregulation of the retinoblastoma protein (Rb1) by histone deacetylases (35). Thus, MO-MDSCs not only have the capacity to strongly downmodulate antitumor immunity but also serve as "precursors" that maintain the PMN-MDSC pool. Indeed, mouse MO-MDSCs proliferate faster than either PMN-MDSCs or the normal monocytic counterpart, can form colonies in agar, and generate a wide range of myeloid cells when either adoptively transferred to tumor-bearing hosts or exposed to TDFs, as well as the GM-CSF and IL-6 cytokines *in vitro* (28). The interplay between PMN-MDSCs and MO-MDSCs might also require reciprocal influence among the cytokines that each subset secretes: PMN-MDSCs produce high levels of gamma interferon (IFN-γ) and discrete levels of IL-13, whereas MO-MDSCs secrete low levels of both cytokines. When primed by interaction with

Table 1 Surface and molecular markers of mouse and human MDSCs

Subset	Markers	Subset	Markers
Mouse MO-MDSCs	CD11b$^+$	Mouse PMN-MDSCs	CD11b$^+$
	Gr-1^{int}		Gr-1^{hi}
	Ly6C^{hi}		Ly6C^{int}
	Ly6G$^-$		Ly6G$^+$
	F4/80$^+$		F4/80$^-$
	CD115$^+$		CD115^{lo}
	CCR2$^+$		CCR2$^-$
	CD124$^+$		CD124^{+-}
	ARG1$^+$		ARG1$^+$
	iNOS$^+$		iNOS$^-$
	ROS$^+$		ROS$^+$
Human MO-MDSCs	CD11b$^+$	Human PMN-MDSCs	CD11b$^+$
	CD33$^+$		CD33$^+$
	CD14$^+$		CD14$^-$
	CD15$^-$		CD15$^+$
	HLA-DR$^-$		HLA-DR$^-$
	CD124$^+$		CD124$^+$

activated T lymphocytes, MDSCs produce both IL-13 and IFN-γ, which are utilized in an autocrine manner to enhance the production and activity of both ARG1 and iNOS enzymes; in this loop, IFN-γ is required for the upregulation of IL-4Rα, which mediates IL-13 signaling and promotes the survival of MO-MDSCs (36). Thus, the elevated production of IFN-γ by PMN-MDSCs may serve to maintain MO-MDSCs' suppressive activity and prevent their apoptotic death (37).

The plasticity of MDSCs depends on the ability of myeloid cells to lose lineage identity in response to specific microenvironmental signals (38). Furthermore, TDFs induce and promote tumor-infiltrating MDSC differentiation into other myeloid cell subsets, such as tumor-associated macrophages (TAMs) and tumor-associated neutrophils (TANs). However, reports on this topic are still scattered and often controversial, and some of the discrepancies could be clarified by the fact that often the expression of markers for the myeloid lineage does not univocally define these cells (39). The current view of macrophages depicts two extremes: M1 "classically" activated macrophages (MHC-II^{hi}CD80/CD86^{+}NOS2^{+}IL-12^{hi}IL-6^{+}IL-1^{+}TNF-α^{+}IL-10^{lo}) respond to IFN-γ by releasing proinflammatory cytokines such as IL-12 and IL-23 and are involved in Th1 cell-mediated responses; M2 "alternatively" activated macrophages (MHC-II^{lo}CD11c^{-}CD206^{+}CD163^{+}IL-10^{hi}TGF-β^{hi}) react to IL-4/IL-13 and are involved in Th2-type responses, fibrosis, and tissue repair (40). This clear-cut dichotomy was recently revised toward a continuum of activation states with partially overlapping phenotypes and functions, ranging over host defense, wound healing, and immune regulation by taking into consideration three main factors: the source of macrophages, the activating molecules, and the specific set of markers that define macrophage activation (41). Tumor-infiltrating MDSCs (defined as CD11b^{+}Gr-1^{lo}F4/80^{+}IL-4Rα^{+}CCR2^{+}CX3CR1^{+}) expressed macrophage markers of both M1 and M2 polarization (42). The expression of CCR2 receptor can highlight a key role for CCL2 chemokine in MDSC recruitment from bone marrow and secondary hematopoietic organs (like the spleen), to tumor mass, where they can differentiate into macrophages. Ly6C^{hi}Ly6G^{-}F4/80^{+}CX3CR1^{lo} CCR2^{hi}CD62L^{+} monocyte precursors, which are phenotypically related to MO-MDSCs, are able to replenish nonproliferating TAM populations (43). The MDSC conversion to immunosuppressive TAMs is mediated mainly by CSF-1 (44), but also by molecular pathways associated with the hypoxia inducible factor-1α (HIF1α), as discussed below, or through a hypoxia-independent stabilization of HIF1α by the lactic acid produced via the Warburg effect in cancer metabolic state (45). Neutrophils, like all other leukocytes, are also able to migrate to tissues from the blood under the influence of specific chemokines (i.e., Chemokine C-X-C motif ligand (CXCL)-1, CXCL-2, and CXCL-6), cytokines (i.e., TNF-α), and adhesion molecules located on their own surface (i.e., CD11b) and on the surface of endothelial cells (i.e., selectins, intercellular adhesion molecule-1, and platelet endothelial cell adhesion molecule-1) (46). When they traffic to tumors, they are often indicated as TANs. TANs sustain not only tumor growth but also metastatic tumor cell spread through the secretion of cytokines, such as transforming growth factor β (TGF-β) (47). PMN-MDSCs and naïve neutrophils (NNs) are functionally and phenotypically different. PMN-MDSCs, but not neutrophils, are immunosuppressive (48) and they express higher levels of CD115 and CD244 and lower levels of CXC-chemokine receptor 1 (CXCR1) and CXCR2 (49). Moreover, transcriptomic analysis comparing TANs with either PMN-MDSCs from mesothelioma-bearing mice or NNs revealed that the gene profiles of NNs and PMN-MDSCs are more closely related to each other than to TANs, suggesting that TANs are not simply "tissue-based MDSCs" but are a distinct population of neutrophils (50). However, a simple analysis of surface marker expression may not be sufficient to distinguish NNs from PMN-MDSCs and inflammatory monocytes from MO-MDSCs. Substantial help on this topic can be provided by the functional and molecular characterization of these myeloid subsets. For example, PMN-MDSCs produce higher levels of ROS, ARG1, and myeloperoxidase than NNs (32, 48), whereas MO-MDSCs can be distinguished from inflammatory monocytes because they upregulate iNOS, ARG1, IL-10, and TGF-β (36).

Myeloid cell plasticity creates particular situations in which unexpected cell development can take place. For example, mitogen-activated protein kinase kinase 6 (MKK6)–p38 mitogen-activated protein kinase (MAPK) activation in human peripheral blood- or CD34-derived neutrophils starts a molecular program that culminates in the transdifferentiation to monocytes. This process requires proteasomal degradation of CCAAT/enhancer-binding protein α (C/EBPα) and is, at least in part, mediated by c-Jun induction and phosphorylation (51). Accordingly, Ly6G^{+}F4/80^{-} neutrophils adoptively transferred to mice 4 h post-induction of thioglycollate peritonitis progressively acquired a monocyte phenotype characterized by progressive upregulation of F4/80 marker and downregulation of Ly6G (51). Moreover, a peculiar MDSC fraction (CD33$^{lo/-}$CD15^{+}CD66b^{+}

IL-4Rα$^+$CD11b$^+$HLA-DR$^-$) that displayed classic fibrocyte markers, such as α smooth muscle actin and collagen I/V, possessed a potent tolerogenic activity by suppressing T-cell proliferation and promoting regulatory T cell (Treg) expansion primarily via indoleamine deoxygenase (52, 53). Moreover, CCR2$^+$Ly6C^{int}CD11b$^+$ MDSCs are able to differentiate into fibrocytes by the activation of the transcriptional factor Kruppel-like factor 4 in tumor metastasis (54). As an example of the "ectopic" differentiation in cancer, GM-CSF secreted by tumor cells can induce the conversion of CD11b^{hi}CD27^{hi} NK cells into Ly6C^{hi}Ly6G^{hi} MDSCs (55).

Definition of Human MDSCs and Use as Novel Biomarkers in Oncology

Human MDSCs can be found in the blood of patients with solid tumors; many if not all patients have elevated levels of MDSCs that correlate with clinical cancer stage and metastatic tumor burden. There is a long list of solid tumors in which MDSCs have been identified: breast cancer; non-small-cell lung cancer; colon and colorectal carcinoma; pancreatic adenocarcinoma; prostate cancer; sarcoma; carcinoid; gall bladder, adrenocortical, thyroid, and hepatocellular carcinoma; and melanoma (56); head and neck squamous carcinoma (57); renal cell carcinoma (RCC) (58); gastrointestinal cancer and esophageal cancer (59); bladder cancer (60); and urothelial tract cancer (49). MDSCs have also been detected in the blood of patients with hematologic malignancies, including multiple myeloma and non-Hodgkin's lymphoma (61). However, a substantial heterogeneity of expression and levels of cell surface markers in human MDSCs was found within different studies and tumor types. Initially, human MDSCs were defined as tumor-infiltrating and blood-circulating CD34$^+$ cells that were amplified by tumor-released GM-CSF and able to suppress immune functions (62). Nowadays, CD34 is no longer considered a universal marker of immunosuppressive myeloid cells in humans. The most immature MDSCs are positive for the common myeloid marker CD33 and negative for HLA-DR and other lineage-specific markers of differentiated lymphocytes and NK cells and hence defined as Lin1$^{-/lo}$ (Lin$^-$ cocktail usually contains antibodies to CD3, CD14, CD16, CD19, CD20, and CD56). Lin$^-$HLA-DR$^-$CD33$^+$CD11b$^+$ MDSCs share features and granule content in common with promyelocytes (even though normal promyelocytes are not immune suppressive) (63) and are increased in the blood of patients with different tumors including glioblastoma and breast, colon, lung, and kidney cancer (61, 64). This immature fraction might be indicative of the overall tumor burden, and the increased circulating levels correlate with worse prognosis and radiographic progression in breast and colorectal cancer patients (56).

Although there are disputes in unanimously defining human MDSC subsets and also for the lack of a human ortholog for the mouse Gr-1 marker, it is possible to divide the more differentiated MDSCs into two main subsets: PMN-MDSCs (CD11b$^+$CD14$^-$CD15$^+$HLA-DR$^-$) and MO-MDSCs (CD11b$^+$CD14$^+$IL-4Rα$^+$CD15$^-$HLA-DR$^-$) (65). PMN-MDSCs are negative for co-stimulatory molecules (CD80, CD86, and CD83) and very cryosensitive (66); thus, functional studies can be performed only on fresh samples. PBMCs collected from RCC patients were contaminated with PMN-MDSCs able to suppress the proliferation of T cells stimulated with CD3/CD28 antibodies by ARG1-dependent, CD3 ζ-chain downregulation (67). Circulating MO-MDSCs (CD11b$^+$Lin$^-$CD33$^+$HLA-DR$^-$CD14$^+$) and Tregs are increased in metastatic prostate cancer patients compared to healthy donors and negatively correlate with patient survival. Moreover, MDSCs isolated from prostate cancer patients' blood possessed immune-suppressive capabilities on activated T cells, probably through upregulation of iNOS (68). Circulating MDSCs might thus represent a predictive biomarker, and there is an ongoing effort to relate MDSC subsets in blood with tumor progression, serological concentration of immune regulatory molecules, and clinical outcome. In this context, a positive association between higher MDSC (HLA-DR$^-$CD14$^-$CD11b$^+$CD33$^+$) and IL-13 and IL-6 cytokine levels in blood and increased CD163$^+$ TAM infiltration can be detected in esophageal cancer; accordingly, IFN-γ and IL-12 serum concentrations were reduced in this cancer patient group (69). A similar correlation was described in pancreatic adenocarcinoma cases. Indeed, patients with stable disease showed significantly lower MDSC levels in blood than those undergoing progression. Accordingly, MDSC systemic accumulation correlated with higher serum concentrations of mediators associated with MDSC proliferation and acquisition of an immune-suppressive phenotype, such as angiogenic factors (fibroblast growth factor-2 and vascular endothelial growth factor [VEGF]), interleukins (IL-4, IL-8, IL-17), chemokines (CCL5), and platelet-derived growth factor (70). Similarly, CD11b$^+$Lin$^-$CD33$^+$HLA-DR$^-$CD14$^+$ MO-MDSCs endowed with T-cell-suppressive properties accumulate early in the blood of melanoma patients and are associated with higher serum IL-8 concentrations (71). MO-MDSC levels are inversely correlated with the presence of either NY-ESO-1- or Melan-A-specific T cells and with clinical outcome (72). Recently, phase I/II clinical trials showed that vaccines based on tumor-associated peptides could prolong survival in patients with RCC and colorectal cancer who showed signs of a multipeptide-specific immunization. Moreover, positive and negative predictors of clinical responses could be found in the blood among leukocyte subsets (Tregs and MDSCs) and serum proteins (chemokines and apolipoproteins) (28, 64). In these studies, a panel of antibodies was developed to identify six MDSC phenotypes in a single multicolor staining. Levels of all MDSC subsets except one were significantly increased in the blood of patients with RCC, suggesting a global myelopoiesis alteration, as seen in mice. However, in a restrospective analysis, only two MDSC phenotypes were negatively associated with survival: $CD14^{+}HLA\text{-}DR^{-/lo}$ and $CD11b^{+}CD14^{-}CD15^{+}$. Interestingly, in RCC and colorectal cancer patients, the prevaccination serum levels of the chemokine CCL2 inversely correlated with the clinical response to the cancer vaccine in subjects responding to multiple peptides present in the vaccine formulation. Indeed, in addition to its known chemotactic activity, CCL2, together with IL-6, can have a double function to induce antiapoptotic effects on $CD11b^{+}$ cells isolated from human peripheral blood by the enforced expression of c-$FLIP_L$ (cellular FLICE [FADD-like IL-1β-converting enzyme]-inhibitory protein large), and drive *in vitro* differentiation to M2-polarized macrophages, as assessed by the overexpression of CD206 (mannose receptor) marker (73). Mouse models also allowed unveiling of a tolerogenic role for this chemokine. CCL2 released by mesenchymal and myeloid cells in the spleen of tumor-bearing mice, in fact, was essential for the generation of a tolerogenic environment in the marginal zone of spleen. In this specific compartment of the white pulp, peculiar $Ly6C^{+}$ monocytic cells (i.e., MO-MDSCs) attracted by the chemokine cross-presented tumor antigen to $CD8^{+}$ T cells, inhibiting their activity, and likely formed a pool of precursors for other MDSC subsets (28). Taken together, the existing data on human MDSCs indicate that these cells share many of the functional properties found in mice.

MDSCs AND IMMUNE DYSFUNCTIONS

MDSCs' ability to suppress T-cell activity is considered a hallmark of tumor progression (74). As discussed, MDSCs exert their immunosuppressive effect in both an antigen-specific and -nonspecific manner, depending on their localization and the specific characteristics of the tumor. MDSCs can restrain the immune response through different mechanisms including essential metabolite (L-arginine, L-cysteine, and L-tryptophan) consumption and/or their metabolic conversion to active by-products, ROS and RNS production, as well as display of inhibitory surface molecules that alter T-cell trafficking and viability. All these mechanisms can operate singularly or in combination. Moreover, these mechanisms can be either direct or indirect, in this latter case involving the generation or the expansion of other regulatory populations such as $CD4^{+}CD25^{+}$ Tregs (Fig. 1).

MDSC Immunosuppressive Mechanisms Related to the Depletion of Essential Metabolites

T-cell proliferation and fitness rely on the availability of L-cysteine. Cells can import cystine, the oxidized form of cysteine, from exogenous sources either through the x_c^- transporter or from the conversion of methionine by cystathionase. However, T cells lack both enzymes, so the only way to collect cysteine is during the immunologic synapse occurring between antigen-presenting cell (APC) and T cell. During this process, APCs release in the extracellular space, through the ASC transporter, reduced cysteine, which can be taken up by T cells. For this reason, media used for *in vitro* T-cell culture are supplied with a reducing agent (β_2-mercaptoethanol) (75). MDSCs lack ASC transporter and cystathionase and do not export cysteine, thus reducing the availability of both oxidized and reduced amino acid. Cystine/cysteine extracellular deprivation limits as well the supplying of cysteine to T cells because APCs can obtain it only from the modification of methionine (75). Another amino acid involved in T-cell function and immune regulation is tryptophan. The two isoenzymes responsible for its catabolism, indoleamine 2,3-dioxygenase 1 (IDO1) and IDO2, catalyze the degradation of the amino acid along the kynurenine pathway. L-Tryptophan starvation activates GCN2 kinase, which in turn inhibits $CD8^{+}$ T-cell proliferation by causing cell cycle arrest and inducing anergy, and directs $CD4^{+}$ T-cell differentiation toward the Treg phenotype by Foxp3 upregulation (76). The kynurenines produced by the reaction have similar immune-modulatory properties on $CD4^{+}$ T cells and can reprogram DCs toward a tolerogenic function by binding the aryl hydrocarbon receptor (77). IDO1 is an enzyme expressed by tumor cells and specific leukocyte subsets, such as TAMs, plasmacytoid DCs, and MDSCs (78), involved in regulation of local inflammation that contribute to tumor-acquired immune tolerance. Accordingly, its expression inversely correlates with T-cell trafficking and clinical outcome in many tumors and is associated with higher Treg tumor presence and disease progression (79). L-Arginine

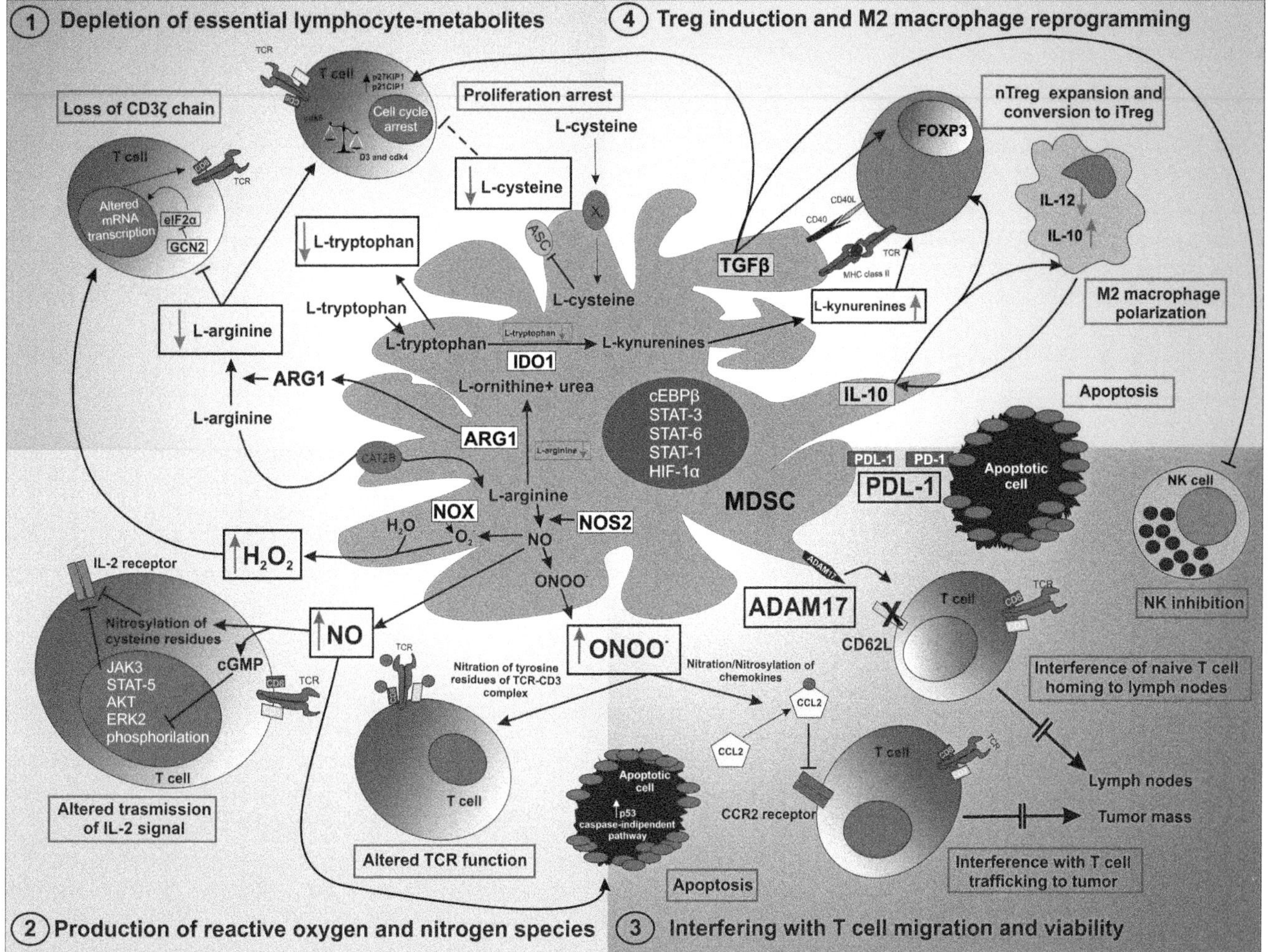

Figure 1 MDSCs suppress the immune response by four main mechanisms. (1) MDSCs deplete essential metabolites for T lymphocyte fitness, such as L-cysteine, L-tryptophan (by the activation of IDO1), and L-arginine (by the activation of both ARG1 and NOS2), inducing the T-cell proliferation arrest. T-cell proliferation block is exacerbated by MDSC-released TGF-β. L-Arginine depletion by ARG1 activity also induces the translational repression of the CD3 ζ chain, which prevents T cells from responding to various stimuli. NO production inhibits T cells by interfering with the signaling cascade downstream of the IL-2 receptor. (2) High arginase activity in combination with increased NO production by the MDSCs not only results in more pronounced T-cell apoptosis but also leads to an increased production of ROS and RNS, such as the free radical peroxynitrite ($ONOO^-$), by the MDSCs. This process requires collaboration with NOX2 enzyme, which contributes to large amounts of ROS, such as H_2O_2, which then affect T-cell fitness by downregulating CD3 ζ-chain expression and reducing cytokine secretion. RNS can act on α and β TCR chains, preventing TCR signaling and promoting dissociation of CD3 ζ chain from the complex. (3) MDSCs interfere with T-cell migration and viability. MDSCs express the metalloproteinase ADAM17, able to cut the integrin CD62L on the T-cell membrane. RNS also modify leukocyte trafficking, promoting homing of immune-suppressive subsets other than T cells by tyrosine nitration of selective chemokines (like CCL2) or their receptors. MDSCs expressing PD-L1 can induce T-cell apoptosis by engaging PD-1. Moreover, NO produced by MDSCs has a direct pro-apoptotic role mediated by the accumulation of p53 and signaling by Fas, TNF receptor family members, and caspase-independent pathways. Finally, the MDSC-derived TGF-β can promote NK-cell inhibition. (4) MDSCs drive the differentiation of specific subsets into regulatory cells: by TGF-β release, MDSCs promote the clonal expansion of antigen-specific natural (n) Treg cells and drive the conversion of naive $CD4^+$ T cells into induced (i) Treg cells. MDSCs skew macrophages toward an M2 phenotype by release of IL-10. For abbreviations and more details, see the text.

represents the common substrate for two enzymes: iNOS, which generates NO; and ARG1, which converts L-arginine to urea and L-ornithine. MDSCs express high levels of both ARG1 and iNOS, and a direct role for both of these enzymes in the inhibition of T-cell function is well established (80). The production of NO could inhibit, in an antigen-independent way, the T-cell signaling cascade downstream of the IL-2 receptor by different mechanisms, involving the blockade of phosphorylation and activation of JAK3 and STAT5 transcription factor, inhibition of MHC-II gene expression, and induction of T-cell apoptosis (81). Moreover, a direct proapoptotic effect has also been observed in T cells exposed to high concentrations of NO, likely mediated by the accumulation of the tumor suppressor protein p53, signaling by CD95 (also known as Fas) and TNF receptor family members, or signaling through caspase-independent pathways (82). ARG1 activity causes the depletion of L-arginine and translational blockade of the ζ chain of CD3, which prevents T cells from responding to various stimuli (83). Moreover, L-arginine starvation blocks protein translation through the accumulation of empty aminoacyl-tRNAs, which in turn activate GCN2 kinase and phosphorylate the translation initiation factor eIF2α. The phosphorylation increases the affinity of eIF2α for eIF2β, which can no longer exchange GDP with GTP, thus interfering with protein synthesis (84). Finally, L-arginine consumption blocks the lymphocyte cell cycle in the G_0-G_1 phase because of the imbalance between cdk6 (increased) and the reduced levels of cyclin D3 and cdk4 (85). ARG1 is a player in T-cell immune dysfunction since its inhibition with the selective antagonist N^{ω}-hydroxy-nor-L-arginine affects, in a dose-dependent manner, tumor growth of a mouse lung carcinoma (86).

MDSC Immunosuppressive Mechanisms Related to ROS and RNS Production

High arginase activity in combination with increased NO production by MDSCs also leads to increased production of ROS and RNS, such as the free radical peroxynitrite ($ONOO^-$), by the MDSCs. These species have multiple inhibitory effects on T cells. ROS and RNS are radical compounds with high reactivity for macromolecules such as DNA, lipids, and proteins. By modifying their tertiary and quaternary structure, oxygen/nitrogen-derived posttranslational protein modification does not just induce inactivation/activation of the target but can modify signaling cascades and finely modulate biological processes, including immune responses. ROS production comprises superoxide anion (O_2) and peroxide hydrogen (H_2O_2) and relies on the activity of proteins of the NADPH oxidase (NOX) family, which includes NOX1 to -5, DUOX1, and DUOX2 (87). NOX2, the main actor for ROS production in leukocytes (such as granulocytes, macrophages, and MDSCs) and endothelial cells, is finely tuned at a transcriptional and posttranslational level: cytokines (i.e., TNF-α and IFN-γ), growth factors (VEGF), and transcription factors (STAT3) upregulate NOX2 levels, whereas specific phosphorylation patterns and subunit availability modulate its activity (88). As outlined above, ROS can also result from iNOS uncoupling reaction during deprivation of L-arginine. Tumor-infiltrating MDSCs produce large amounts of ROS (H_2O_2), which affect T-cell fitness by downregulating CD3 ζ-chain expression and reducing cytokine secretion, as observed in pancreatic cancer and melanoma (89, 90). In a physiological context, cells can limit the oxidative stress by an enzymatic detoxifying system such as superoxide dismutase and glutathione *S*-transferase. However, at high ROS concentration, radicals can directly react with macromolecules or combine with NO to generate more dangerous RNS, such as peroxynitrite and dinitrogen trioxide, which can nitrate/nitrosylate tyrosine, cysteine, methionine, and tryptophan in different proteins and enzymes, thus changing their biological functions (91). Under pathological conditions (i.e., tumor), RNS by inducing apoptosis and autophagy direct tumor evolution, and more importantly suppress T-cell trafficking and cytotoxic functions, contributing to shaping an immune-privileged environment that promotes tumor outgrowth. RNS can indeed alter the formation of a correct peptide-MHC complex by MHC-I or induce modification in the immune-dominant peptides. This results in unsuccessful peptide loading on MHC-I of target cells or failure in binding by CD8-TCR (T-cell receptor) lymphocyte antigen recognition complex (92). Moreover, RNS can act on α and β chains of the lymphocyte TCR, preventing signaling and promoting dissociation of the CD3 ζ chain from the complex (93). Finally, RNS also modify trafficking of leukocytes, promoting homing of immune-suppressive subsets other than T cells (further addressed below). This is in part mediated by tyrosine nitration of either chemokines (CCL2, CCL5, CCL21, and CXCL12) or receptors (CXCR4) (94).

MDSC Immunosuppressive Mechanisms Related to T-Cell Migration Interference

T-cell activation and effector functions require trafficking to lymph nodes and tumor sites. Both of these processes can be modified by MDSCs. L-Selectin (CD62L) is a homing receptor for T cells and is critical for directing naïve T cells to lymph nodes. Peripheral blood

T cells in tumor-bearing mice present a cleaved CD62L marker on their surface due to the activity of the metalloproteinase ADAM17 (disintegrin and metalloproteinase domain-containing protein 17) expressed on the MDSC surface (95). MDSC and cytotoxic T lymphocyte (CTL) trafficking to tumor is regulated by the CCL2/CCR2 axis. Intratumoral production of RNS induced nitration/nitrosylation of CCL2 in different human and mouse cancers. As a result, modified CCL2 could no longer attract tumor-specific CTLs but could still recruit myeloid cells to the tumor. Notably, this novel mechanism of tumor escape could be pharmacologically targeted. *In vivo* administration of a novel drug ([3-(aminocarbonyl)furoxan-4-yl]methyl salicylate; AT38) that blocks intratumoral RNS production induced a robust T-cell infiltration within the tumor and enabled transferred CTLs to reject solid tumors (96). In addition to regulating T-cell trafficking, MDSCs also decrease the number and inhibit the function of mouse and human NK cells, mostly through membrane contact-dependent mechanisms (97).

MDSC Immunosuppressive Mechanisms Related to Treg Induction and M2 Macrophage Reprogramming

MDSCs produce large amounts of TGF-β and IL-10 in the tumor microenvironment. TGF-β exerts direct antiproliferative effects on T cells, arresting their cell cycle typically in the G_1 phase by inducing the expression of the cell cycle inhibitors $p27^{Kip1}$ and $p21^{Cip1}$ (98) or by inhibiting IL-2 secretion (99). Importantly, TGF-β was shown to inhibit the differentiation of $CD4^+$ T cells into Th1 or Th2 cells by suppressing the expression of T-bet and GATA-3 master regulators of Th1 and Th2 conversion, respectively (100). TGF-β-producing MDSCs also promote the clonal expansion of antigen-specific natural Treg cells and induce the conversion of naive $CD4^+$ T cells into induced Treg cells. The mechanisms are not completely understood, but may involve cell-to-cell contact (including CD40-CD40L interactions) and the production of soluble factors in combination with released TGF-β, such as IFN-γ and IL-10 (101). Human $CD14^+HLA\text{-}DR^{lo/-}$ MDSCs promote the transdifferentiation of Th17 cells into $Foxp3^+$-induced Treg cells by producing TGF-β and retinoic acid (102). Moreover, through an IL-10- and cell contact-dependent mechanism, MDSCs skew macrophages toward an M2 phenotype by decreasing macrophage production of IL-12 (103). The downregulation of IL-12 is further exacerbated by the macrophages themselves, since macrophages promote the production of IL-10 by MDSCs, creating a self-maintaining, negative loop.

MDSC-INDUCED MECHANISMS OF TUMOR PROMOTION

MDSC activity is not simply directed to building up an immune-suppressive environment that keeps T cells at bay and protects tumors from the effector arm of the immune system, but includes mechanisms that sustain and promote tumor growth as well as metastatic spreading. These actions can be classified as direct tumor-promoting activities and include the control of cancer stemness, angiogenesis and stroma deposition, epithelial-to-mesenchymal transition (EMT), and metastasis formation.

Cancer Stemness

Normal cells undergo a limited number of divisions before reaching a proliferative block characterized by absence of response to growth factors, modification in morphology, and metabolism. This process, known as senescence, protects cells from immortalization and is triggered by telomere shortening, DNA damage, and upregulation of the *CDKN2* locus (involved in cell cycle regulation). Molecularly, senescence follows oncogene-sustained proliferation, depletion of suppressor genes (*Pten* and *Rb*), or activation of the p53 tumor suppressor (104). Senescence is a characteristic of normal tissues and premalignant tumor cells; for example, while observed in pancreatic intraductal neoplasia and lung adenomas, it is absent in their advanced, malignant stages (105, 106). Myeloid cells finely tune tumor senescence by promoting cellular stemness. In two different spontaneous, senescence-inducing tumor models (conditional oncosuppressor $PTEN^{-/-}$ prostate adenocarcinoma and oncogene-mediated $Ki\text{-}ras^{G12/V}$ lung adenocarcinoma), at tumor onset, neoplastic cells showed senescence phenotype, a condition reversed by MDSC action (107). Interestingly, MDSC interferes with the senescence-associated secretory phenotype by releasing IL-1RA (IL-1 receptor antagonist), which interrupts the IL-1α–IL-1R axis and activates a reprogramming activity in tumor cells. Accordingly, impairment in MDSC tumor trafficking (i.e., by CXCL1 and -2/CXCR2 targeting) enhanced chemotherapy-induced cell senescence in $PTEN^{-/-}$ mice. Another mechanism of MDSC-dependent tumor senescence inhibition was described in human ovarian carcinoma (108) and is based on targeting of C-terminal binding protein 2 (CtBP2) on tumor cells. CtBP2 is a transcription corepressor recruiting histone deacetylases, methylases, and demethylases on target genes, which in turn remodel chromatin condensation and gene expression. CtBP2 modulates the expression of genes involved in sphere formation in ovarian primary tumor cells. MDSCs break this equilibrium and promote cancer stemness and metastasis through induction of microRNA101 in tumor cells. The microRNA101 targeting *CtBP2* mRNA induces the expression of genes involved in sphere formation such as *OCT3/4*, *SOX2*, and *NANOG*. Accordingly, dense MDSC infiltrate, high microRNA101 expression, and low CtBP2 levels correlated in ovarian cancer patients with a worse clinical outcome (108). In pancreatic tumors, MO-MDSCs induced the proliferation of aldehyde dehydrogenase-1-positive cancer stem cells and promoted the acquisition of mesenchymal properties. The process is triggered and tuned in an autocrine loop by tumor cells, which induce recruitment of MO- and PMN-MDSCs to the tumor microenvironment and activate their immune-suppressive program through the STAT3 signaling pathway. A similar effect was observed with human $CD14^{+}HLA\text{-}DR^{-}$ MDSCs from pancreatic ductal adenocarcinoma (PDAC) patients (109).

Angiogenesis, EMT, and Metastasis

Unrestricted tumor growth is often followed by local hypoxia. To adapt to a hypoxic environment, tumor cells, which sense O_2 levels through HIF1α, release VEGF and hence stimulate sprouting and building of new vessels to increase tissue perfusion, with the purpose of fulfilling the nutrient demand and supporting the sustained proliferation. Secreted VEGF plays an additional and crucial immune-regulating role by orchestration of peripheral expansion, trafficking of MDSCs to tumor, and acquisition of immune-suppressive properties (110). Indeed, tumor hypoxia stimulates MDSCs to upregulate both ARG1 and iNOS through hypoxia response element and NF-κB (111). In addition, NO-dependent modification of HIF1α acts as positive feedback on VEGF synthesis by amplifying the reaction, and MDSCs directly reinforce this loop by producing matrix metalloproteinase-9 (MMP-9), which increases VEGF availability (112). Even though the immunoregulatory function was not specifically tested in these experiments, recruitment of cells resembling PMN-MDSCs mediated resistance to anti-VEGF antibody-mediated therapy, suggesting that PMN-MDSCs could support new vessel growth even in the presence of VEGF antibody (113). MDSCs can also mediate resistance to the tyrosine kinase inhibitor sunitinib, an antiangiogenic agent, in both preclinical models and patients with RCC (114). The presence of circulating PMN-MDSCs that produced high levels of MMP-9, MMP-8, and IL-8 inversely correlated with the clinical response to sunitinib, suggesting that MDSCs could still promote angiogenesis by different mechanisms in sunitinib-resistant tumors. Hypoxia can also affect MDSC-dependent immune dysfunctions within the premetastatic niche (115). The injection of breast cancer-preconditioned hypoxic media promoted CCL2-mediated homing of PMN-MDSCs ($Ly6C^{int}Ly6G^{hi}$) and NK cells to lungs and increased lung colonization after tumor cell injection. In this environment, NK cells lost their killing ability, thus contributing to a higher metastasis incidence (115). Another factor secreted by PMN-MDSCs during exposure to a hypoxic tumor microenvironment and involved in lung metastasis generation is *Bombina variegata* peptide 8 (116). Also, the proinflammatory proteins S100A8 and S100A9, potent chemoattractants for MDSCs, have been implicated in tumor and metastasis promotion by MDSCs (117): serum amyloid A3 induced by S100A8/A9 directly attracted MDSCs to premetastatic lungs, stimulated NF-κB signaling in a TLR4-dependent manner, and facilitated metastatic spreading (118).

After being recruited to the tumor and premetastatic niche and following the interaction of peculiar receptors—CXCR2 (PMN-MDSC), CCR2 (MO-MDSC), and CXCR4/RAGE (receptor for advanced glycation end products; both MDSC subsets)—with respective chemokines, MDSCs contribute to the generation of an immune-tolerant environment by releasing IL-6 (119) or triggering tumor cell migration (i.e., through TNF-α release) (120). MDSCs can assist the metastatic process also by inducing EMT of tumor cells, a condition in which cells acquire improved spreading skills. MDSCs attracted by CXCL5 chemokine induced EMT of melanoma cells by releasing hepatocyte growth factor and TGF-β in the primary tumor site, and the growth of primary tumor was significantly impaired by targeting PMN-MDSCs in this preclinical model (121). High-mobility group box-1 (HMGB1), a damage-associated molecular pattern protein released by tumor cells as well as leukocytes (i.e., macrophages and MDSCs) during stress and cell death, has been associated with tumor invasiveness, metastatic spreading, and EMT in colorectal carcinoma. It acts as a proinflammatory cytokine through binding to TLRs and RAGE, promoting MDSC trafficking and activation of EMT-inducing transcription factor Snail and NF-κB, which in turn activates MMP-7 (122). However, since metastatic cells show the morphology and phenotype of epithelial cells, it is conceivable that premetastatic MDSCs, by releasing the proteoglycan versican, may also control the opposite transition route (mesenchymal to epithelial) in tumor cells that reach the new district, favoring their ability to seed and colonize the organ (123). Despite these data that reinforce the concept of a role for

MDSCs in promoting metastatic tumor cell spread, some data are discordant and suggest the ability of MDSCs to inhibit metastasis by the production of a potent antiangiogenic matrix protein (i.e., thrombospondin-1) (124); this mechanism can open a new view on the relationship between MDSCs and the metastatic process.

Cachexia is a severe neoplastic syndrome characterized by body weight loss and deep metabolic changes that cannot be corrected just by increasing food intake. In digestive system cancers, cachexia has been correlated with high serum VEGF and peripheral MDSC levels, systemic inflammation, and MDSC-dependent immune dysfunctions (125). Preclinical studies on transplantable mammary tumors revealed that MDSCs can fuel some aspects of cachexia by inducing a hepatic acute-phase protein response that is either prevented by MDSC depletion or reproduced by MDSC adoptive transfer (126).

FACTORS DRIVING MDSC ACCUMULATION AND ACQUISITION OF SUPPRESSIVE FUNCTIONS

CSF-1, G-CSF, and GM-CSF are the three main regulators of proliferation and differentiation of the myeloid lineage. CSF-1 is found in many types of tumors, such as RCC (127) and about 70% of breast tumors (128), and is also implicated in macrophage trafficking. However, CSF-1R is also expressed by MO-MDSCs, and its pharmacological blockade significantly affects MO-MDSC tumor homing in melanoma and prostate tumor models (129). G-CSF induces differentiation of myeloid precursors to granulocytes and directs their recruitment to tumors (130). GM-CSF is a cytokine that can trigger myeloid differentiation toward either an immune-stimulating (DC) or an immune-suppressive (MDSC) phenotype, depending on the strength of the stimulus and on the cytokine context. Indeed, tumors promote a myeloid cell commitment toward MDSC phenotype through release of GM-CSF and IL-6, which activate an immune-suppressive, C/EBPβ-mediated program in bone marrow-derived progenitors (23). GM-CSF, with the collaboration of IL-6, can intervene in regulating MDSC function during very early stages of tumor progression. Experimental models of autochthonous PDAC, in fact, have shown progressive waves of myelomonocytic cell recruitment after initiation of the transforming program controlled by the active *Kras* oncogene, with Gr-1$^+$CD11b$^+$ cells being among the first to be recruited within the developing neoplastic lesions (131). *Kras* oncogene controls the accrual of myelomonocytic cells, and this step is mandatory for pancreatic intraepithelial neoplasia initiation and progression. *Kras* oncogene-driven inflammation at the pancreatic intraepithelial neoplasia stage critically relied on GM-CSF for both progression to PDAC and Gr-1$^+$CD11b$^+$ cell recruitment within the pancreatic stroma. This circuit was essential to alter tumor-specific CTLs, and only the blockade of either GM-CSF production or Gr-1$^+$ CD11b$^+$ cell activity restored antitumor immunity (132). Interestingly, recruited Gr-1$^+$CD11b$^+$ cells contribute with transformed epithelial cells to the local production of the cytokines IL-6 and IL-11, which activate STAT3. As discussed below, STAT3, in turn, induces antiapoptotic and proproliferative genes, fueling tumor initiation, promotion, and progression (133). GM-CSF administration was used as an adjuvant in clinical trials with inconclusive results. Indeed, both preclinical and clinical studies highlight that the cytokine concentration might be the switch regulating myeloid differentiation toward either DCs (low levels) or MDSCs (high levels) (134, 135). In the tumor microenvironment, IL-6 can be secreted by macrophages, MDSCs, monocytes, fibroblasts, and neoplastic cells. The inflammatory cytokine IL-6 drives the differentiation of CD11b$^+$Gr-1$^+$ cells into immune-suppressive cells through the activation of STAT3. IL-6's role in tumor progression has been confirmed in different tumors, such as breast, lung, ovarian, renal, and pancreatic, to inversely correlate with the clinical outcome (136). IL-4 and IL-13 act by inducing MDSC survival and activation through the immune-suppressive pathway by binding the IL-4Rα kinase subunit of the IL-4R (36). The signaling pathway downstream of IL-4Rα entails the recruitment, phosphorylation, and dimerization of STAT6, which regulates the expression of genes involved in the immune-suppressive program and survival of MDSCs. IL4-Rα-dependent STAT6 activation indeed induces TGF-β synthesis (137), ARG1 expression (138), and, together with STAT1 and STAT3, the release of ROS (8). Moreover, IFN-γ released by T cells is able to activate inflammatory monocytes (CD11b$^+$IL-4Rα$^+$) to secrete IFN-γ and IL-13, which in turn induces IL-4Rα expression and triggers in a autocrine loop the molecular processes that suppress antigen-activated CD8 T cells (36).

Interestingly, IL-4Rα genetic ablation affects MDSC-dependent immune suppression *in vivo* (36), and targeting this receptor with aptamers triggered MDSC apoptosis and delayed tumor progression (37). S100A8/A9 is a heterodimer produced mainly by circulating neutrophils and monocytes that is secreted following intracellular changes in Ca^{2+} levels. It promotes MDSC trafficking by binding to *N*-glycan-tagged plasma membrane receptors, such as RAGE. Moreover, RAGE triggering induces activation of STAT3 and NF-κB and the expression of immune-suppressive genes (117). Accordingly, S100A9 blockade decreases MDSC levels and host immune dysfunctions in tumor-bearing mice (139). Another factor that drives myeloid commitment to suppressive MDSCs is HMGB1, a structural protein located in the nucleus that, when released by necrotic cells in the extracellular space, acts as a damage-associated molecular pattern and mediator of inflammation. HMGB1 is also released by leukocytes, especially monocytes, macrophages, and DCs, as an inflammatory cytokine acting on TLRs and RAGE. In cancer, tumor-derived HMGB1 regulates MDSC levels and immune-suppressive abilities through NF-κB activation; accordingly, HMGB1 promotes differentiation of bone marrow cells toward MDSCs, contributing to suppressing antigen-activated $CD4^+$ and $CD8^+$ T cells. Finally, HMGB1 promotes IL-10 secretion in MDSCs and downregulation of CD62L on T cells (140). Osteopontin is an extracellular matrix protein produced by many cell types, such as fibroblasts, osteoblasts, osteoclasts, and bone marrow cells. Besides its role in bone remodeling, osteopontin also takes part in the immune regulatory processes, especially during cancer. Indeed, it promotes myelopoiesis and host immune suppression, and its targeting reduces immune dysfunction and tumor growth in colorectal cancer models (141). In autochthonous breast cancer models, the monocyte subset represented the highest osteopontin producer in primary tumors and lung metastases; interestingly, osteopontin genetic ablation decreased metastatic burden and altered the subset composition among MDSCs, promoting preferential PMN-MDSC development while reducing MO-MDSC-dependent immune suppression through downregulation of ARG1, IL-6, and phospho-STAT3 (142).

SIGNALING PATHWAYS REGULATING MDSC FUNCTIONS

Many transcription factors contribute to physiologic, steady-state hematopoiesis, but only some of them are associated with altered commitment of myeloid cells to MDSCs and regulation of their immune regulatory properties. In particular, members of the STAT family and C/EBPβ were shown to play a central role in the polarization of myeloid cell functions, as well as in tumor progression and alteration of immune responses to cancer. STAT1, -3, -5, and -6 can transmit polarizing signals to the nucleus (143), and each component of the family can play a distinct role in macrophage polarization and MDSC functions. A fundamental component of several signal transduction pathways associated with STAT is the activation of the JAK family. Receptor oligomerization, mostly induced by cytokine binding, triggers JAK activation by either auto- or transphosphorylation. Subsequently, activated JAKs phosphorylate receptors on target tyrosine residues, generating docking sites for STATs through the STAT Src homology 2 domain. Activated JAKs recruit and phosphorylate STATs, which leads to their dimerization and nuclear translocation, where they modulate the expression of target genes. STAT1 is a transcription factor that, after activation with type 1 and 2 IFNs, IL-1β, and IL-6, dimerizes and translocates to the nucleus, where it triggers the expression of genes involved in immune-suppressive properties of MDSCs. The role of STAT1 in mediating host immune dysfunction has been elucidated in STAT1-deficient mice. In this model, MDSCs were not able to inhibit T-cell activation due to defective iNOS and ARG1 upregulation (144). Accordingly, blocking T-cell-derived IFN-γ secretion also abrogated MDSC-mediated suppression, mainly via the block of iNOS upregulation (36). From a translational point of view, STAT1 activation in TAMs correlates with cancer progression in patients affected by follicular lymphoma (145). GM-CSF, the main factor mediating myeloid proliferation and survival, acts through STAT5 triggering (146). Activation of STAT5 and STAT3 by GM-CSF and G-CSF, respectively, induces downregulation of IFN-related factor-8 (IRF-8). This step is critical for the aberrant myelopoiesis since IRF-8-deficient mice develop myeloid cells phenotypically and functionally related to tumor-induced MDSCs (147). Accordingly, inhibition of STAT3 pathways with sunitinib prevents systemic MDSC accumulation and restores normal T-cell activation in tumor-bearing mice. However, sunitinib's efficacy was limited to the periphery and dependent on the cytokine context, since the presence of GM-CSF and activation of STAT5 at the tumor site were sufficient to confer sunitinib resistance and direct differentiation of sunitinib-sensitive PMN-MDSCs toward sunitinib-resistant MO-MDSCs (148, 149). STAT6 is a downstream transcription factor for IL-4R and IL-13R, whose role in MDSC activation is suggested by different studies, as detailed above. STAT6 deficiency prevents signaling through the type 2 IL-4R, thereby inducing enhanced immunosurveillance against primary and metastatic tumors in mice (150). The few MDSCs accumulating in $STAT6^{-/-}$ mice after physical injury showed impaired suppressive activity due to the reduction in ARG1 expression (151). Thus, it appears that STAT1, STAT5, and STAT6 play an important role in MDSC activation and that these STATs mediate the immune-suppressive function of MDSCs. Among the family of transcription factors, a particular importance has been attributed to STAT3. Several pathways downstream of STAT3 might be involved in the regulation of MDSC expansion and function. Following STAT3 activation, hematopoietic precursors release the proinflammatory proteins S100A8/A9, which in turn inhibit DC differentiation and promote the accumulation of MDSCs and their migration to the tumor site (117). STAT3-dependent upregulation of S100A8/A9 depends on NOX2 expression (139), which leads to the production of superoxide, one of the mechanisms by which MDSCs promote T-cell anergy and tolerance. STAT3 activation can directly promote NOX2 activation by upregulating the transcription of the $p47^{phox}$ and $gp91^{phox}$ NOX2 subunits (152). STAT3 can also play an indirect role in MDSC differentiation since it controls the expression of molecules such as acute-phase proteins, which assist MDSC mobilization, accumulation, and survival (153). In a model of polymicrobial sepsis, IL-6-activated STAT3 signaling in hepatocytes, through gp130 ligation, resulted in the expression of serum amyloid A and chemokine CXCL1, which cooperate in promoting the accumulation of MDSCs in the spleen (153). Finally, heat shock protein 72 (Hsp72), which is present in tumor-derived exosomes, induces suppressive activity of MDSCs via STAT3 activation. Hsp72 triggered STAT3 activation in MDSCs in a TLR2- and myeloid differentiation primary-response protein 88 (MyD88)-dependent manner through the autocrine production of IL-6 (154). Interestingly, from a therapeutic point of view, *in vitro* STAT3 inhibition abolishes the suppressive activity of MDSCs and the multitargeted tyrosine kinase inhibitor sunitinib blocked MDSC expansion in tumor-bearing mice by STAT3 signaling interference in myeloid cells (148). In summary, STAT3 activation seems to play a dominant role in MDSC biology, affecting the cells' function by different pathways.

The C/EBPβ is a basic leucine zipper transcription factor important for the differentiation of the myeloid lineage. Three different C/EBPβ isoforms are translated (starting from 3′ in-frame AUG) from the same mRNA: a 38-kDa liver-activating protein (LAP_1 or LAP*), a 36-kDa liver-activating protein (LAP_2 or LAP), and a 28-kDa liver-inhibiting protein (LIP). The upregulation of the transcriptional factor C/EPBβ induces the activation of the immune-suppressive program in myeloid progenitor cells through the activation of STAT3 (155). C/EBPβ can activate different genes by binding to their promoters, including *cmyc* (8), *IL6* (156), and the gene encoding the common signaling β-chain receptor that regulates the signal transduction for GM-CSF, IL-3, and IL-5 cytokines (157). Thus, C/EBPβ can be considered a master regulator of MDSC biology since bone marrow cells derived from C/EBPβ-deficient mice did not have the ability to differentiate *in vitro* into functional MDSCs. Moreover, $CD11b^+$ cells isolated from the spleen and the tumor of C/EBPβ-deficient mice showed an impaired immunosuppressive ability (23). C/EBPβ homologous protein (Chop) is a transcription factor induced in particular contexts, such as during endoplasmic reticulum (ER) stress. When proteins in the ER are not properly folded and cannot proceed to the Golgi apparatus for further modifications, cells activate the unfolded protein responses, which can be initiated by three different sensors: PERK (protein kinase RNA-like ER kinase), IRE1 (inositol–requiring enzyme 1), and ATF6 (activating transcription factor 6). PERK action on eIF2α induces the expression of ATF4, which in turn upregulates Chop (158). This process leads to activation of STAT3, C/EBPβ, and IL-6 release. The role of Chop was further elucidated in cancer since TDFs, through ER stress establishment, induced Chop synthesis in MDSCs and the genetic ablation of the transcription factor was able to reduce MDSC trafficking to tumor in spite of T cells, decrease their immune-suppressive properties, and critically affect tumor growth (159). Accordingly, MDSC depletion partially restored tumor growth in $Chop^{-/-}$ mice, suggesting a role for this transcription factor in orchestrating the immune-suppressive properties of MDSCs (159).

In myeloid cells, the TLR family plays an important role in NF-κB activation, primarily through MyD88. This is consistent with MDSC accumulation and activation during microbial and viral infections, as well as in trauma and sepsis. NF-κB, acting downstream of MyD88, is required for accumulation of MDSCs in a model of polymicrobial sepsis (13). TLR4 was shown to be directly involved in MDSC function (160), and lipopolysaccharide, in combination with IFN-γ, could promote MDSC expansion, probably by inhibiting differentiation of DCs (161). Moreover, $MyD88^{-/-}$ MDSCs had a substantially reduced ability to suppress T-cell activity and release cytokines compared to the wild-type counterpart both *in vitro* and *in vivo* (162).

Enzyme activity and expression (lipases, kinases, and phosphatases) may modulate myelopoiesis, contributing to tumor-promoted, MDSC-mediated host immune suppression. Phospholipase C-γ2 and Src homology 2 domain-containing inositol 5′-phosphatase-1 (SHIP-1) are two enzymes that regulate homeostasis and function of MDSCs. These genes negatively regulate MDSC biology since their ablation promotes MDSC expansion and activation in tumor-bearing mice (163, 164).

Trp53, the best-known oncosuppressor gene, found mutated in at least 50% of human tumors, is able to indirectly regulate MDSC levels. Indeed, *Trp53* ablation in mice bearing melanoma tumors promoted tumor growth through expansion of a stromal network rich in fibroblast reticular-like cells, which contributed to MDSC differentiation by releasing proinflammatory cytokines/chemokines and immunosuppressive mediators, including IL-6, IL-10, CCL3, CCL21, ARG1, and iNOS (165).

In addition to transcription factors, tumors can promote altered myelopoiesis through differential expression of microRNAs (miRs). These single-stranded, noncoding RNAs tune gene expression primarily through seed-matched sites located within the 3′ untranslated regions of the target mRNA. However, they can also bind at 5′ coding sequences of the cognate mRNA, even if with less efficacy. Different miRs are either up- or downregulated during MDSC differentiation. For example, miR-21 and miR-155 upregulation is mandatory for GM-CSF- and IL-6-mediated MDSC proliferation and differentiation (166). On the contrary, miR-142-3p affects MDSC biology through binding to mRNA coding for transcription factors and receptors involved in the activation of the immune-suppressive arsenal: through canonical binding on the 3′ untranslated region of its mRNA, miR-142-3p inhibited expression of gp130, the common subunit of the IL-6 family cytokine receptor, whereas noncanonical binding to the 5′ mRNA coding sequence altered C/EBPβ isoform expression. Accordingly, stable miR-142-3p expression in bone marrow progenitors was sufficient to reprogram TAM differentiation, improving the efficacy of cancer immunotherapy (167).

CONCLUSIONS AND FUTURE PERSPECTIVE

The strong bond linking angiogenesis with immune dysfunction and promotion of invasiveness and cancer-related morbidity is inducing clinicians to reconsider approaches targeting VEGF and blood vessel generation because of the consequent induced hypoxia. In this context, MDSC targeting could interrupt the angiogenic/immune regulatory switch, increasing the chances of therapeutic intervention, as described in a preclinical model of pancreatic adenocarcinoma (168). It is also clear that MDSCs' spectrum of protumoral actions is much broader, and MDSC targeting can open new therapeutic opportunities to control tumor progression and block metastases. Some first-generation chemotherapeutic agents, such as 5-fluorouracil (169), gemcitabine (28, 170), and docetaxel (171), as well as the combination of these drugs, such as doxorubicin plus cyclophosphamide (172), are able to control MDSC accumulation. Numerous studies in tumor-bearing mice, as well as a few clinical trials using different MDSC-targeting approaches, have shown that MDSC reduction delays tumor initiation, progression, and distal dissemination and prolongs survival of tumor-bearing hosts (Table 2). These include selective antibodies and/or aptamers (37, 113, 116, 132, 168, 173–179) against MDSC markers; molecular antagonists of essential MDSC receptors and/or molecular pathways (129, 180); molecular inhibitors of MDSC functional mechanisms used by myeloid cells to block lymphocyte reactivity and proliferation (148, 181–191); and pharmacological agents able to force MDSCs to mature into proficient APCs that can stimulate tumor-specific T cells or repolarize TAMs in proinflammatory M1 macrophages (27, 192–199).

Actually, many factors limit the characterization and analysis of MDSC subsets and thus their targeting for improving cancer immune therapy. First, MO- and PMN-MDSCs can be distinguished from monocytes and granulocytes, respectively, in terms of impairment of functional immune response and molecular signatures more than phenotypic surface markers. For example, Ly6G is a marker shared within granulocytes, PMN-MDSCs, and TANs, and the use of a Ly6G antibody may have a positive or negative effect on tumor growth depending on the targeted myeloid subset (47). Second, MDSCs are characterized by phenotypic plasticity in that they can modify their differentiation depending on the immune context (35, 51). Platforms and tools have been implemented for identifying new surface markers that allow better discrimination of MDSC subsets and can be used for *in vivo* targeting (200). However, a crucial step consists in adding to the canonic surface profiling of MDSCs a molecular signature of functional markers involved in MDSC-mediated immune regulation, such as IDO, iNOS, ARG1, and programmed cell death 1 ligand 1 (PD-L1). Integration of phenotypic and functional information could help in distinguishing between suppressive and nonsuppressive MDSCs without the need for laborious *in vitro* assays. Moreover, gene expression, proteomic, and metabolomic profiles will increase knowledge of MDSC biology and offer potential therapeutics for interrupting crucial switches of MDSC accumulation or suppression. Finally, another critical step for comparison of data produced in different laboratories relies on sample handling. For example, cryopreservation may alter not

Table 2 A synopsis of drugs targeting MDSCs[a]

Drug	Type(s) of cancer	Effect(s) on myeloid cells	Reference(s)
5-Fluorouracil	Thymoma	MDSC apoptosis	169
Gemcitabine	Sarcoma, lung, and breast cancer	MDSC apoptosis	28, 170
Doxorubicin-cyclophosphamide	Breast cancer	MDSC apoptosis	172
Docetaxel	Mammary carcinoma	MDSC apoptosis	171
Gr-1-specific antibody	Colon carcinoma	MDSC depletion	173
Ly6G-specific antibody	PDAC	MDSC depletion	168
CD124 (IL-4Rα) targeting with aptamers	Mammary cancer	MDSC and TAM depletion	37
CCL2-specific antibody	Mammary carcinoma	MDSC recruitment and angiogenesis	174
CXCR2 and CXCR4 antagonists	Breast cancer	MDSC recruitment	180
AT38	Fibrosarcoma and thymoma	Impairment of MDSC suppression and reduced recruitment	96
PROK2-specific antibody	Various mouse and human tumors	Reduced MDSC expansion and recruitment	116
Nitroaspirin	Colon carcinoma	Inhibition of MDSC-dependent immune suppression	191
Triterpenoids	Thymoma, colon, and lung carcinoma	Inhibition of MDSC-dependent immune suppression	181
Tyrosine kinase inhibitor (sunitinib)	Human RCC, fibrosarcoma, and colon, breast, lung, and kidney tumors	Low inhibition of MDSC expansion in patients	148, 182–184, 202
Cyclooxygenase-2 inhibitors	Mammary carcinoma, mesothelioma, lung carcinoma, and glioma	Inhibition of MDSC-dependent immune suppression	185–187
PDE5 inhibitors (sildenafil, tadalafil)	Breast and colon cancer, human myeloma, and HNSCC	Inhibition of MDSC-dependent immune suppression	188–190
ATRA	Sarcoma, colon carcinoma, and human RCC	MDSC differentiation in mature cells	27, 192, 193
1α25-Dihydroxyvitamin D_3	HNSCC	MDSC differentiation in mature cells	194, 195
CSF-1R antagonist	Prostate tumor and lung carcinoma	Reduced MDSC expansion and recruitment	129
CSF-1R monoclonal antibody (RG7155)	MC38 colon carcinoma and diffuse-type giant cell tumor	TAM depletion	175
GM-CSF-neutralizing antibody	Pancreatic cancer	Inhibition of proliferation	132
G-CSF-neutralizing antibody	Colon carcinoma	Inhibition of proliferation	176
VEGF-A-specific antibody (bevacizumab)	RCC	Inhibition of proliferation	113
IL-6R-specific antibody and gemcitabine	Methylcholanthrene-derived carcinoma cells	Reduced MDSC recruitment	177
Bisphosphonates	Mammary tumor	TAM depletion, inhibition of MDSC expansion	198, 199
Combined therapy with IL-12, IL-16, CpG DNA, and IL-10R-specific antibody	Lung and breast cancer	TAM reprogramming	178
CD40 agonist and gemcitabine	PDAC	TAM reprogramming	196
CD40 antibody with IL-2	RCC	TAM reprogramming in lung metastasis but not in primary tumor	179
HRG	Pancreatic and breast cancer, fibrosarcoma	TAM reprogramming	197

[a]Abbreviations: ATRA, all-*trans* retinoic acid; HNSCC, head and neck squamous cell carcinoma; HRG, histidine-rich glycoprotein; PROK2: prokineticin 2.

only MDSC phenotype but also immune-suppressive characteristics (201); thus, establishment of standard operating procedures is mandatory to avoid biased analysis. Nonetheless, despite all these technical limitations, preclinical and clinical studies indicate a pivotal role for MDSCs in immune dysfunction and promotion of cancer progression and dissemination, and support how MDSC targeting could reverse host immune dormancy, thus improving the efficacy of passive and active immunotherapies for cancer.

Acknowledgments. This work was supported by the Italian Ministry of Health; Italian Ministry of Education, Universities and Research (FIRB cup: B31J11000420001); Italian Association for Cancer Research (grants 6599, 12182, and 14103); and La Fondazione Cassa di Risparmio di Verona, Vicenza, Belluno e Ancona.

Citation. De Sanctis F, Bronte V, Ugel S. 2016. Tumor-induced myeloid-derived suppressor cells. Microbiol Spectrum 4(3):MCHD-0016-2015.

References

1. **Frank NY, Schatton T, Frank MH.** 2010. The therapeutic promise of the cancer stem cell concept. *J Clin Invest* **120:**41–50.
2. **Balkwill F, Charles KA, Mantovani A.** 2005. Smoldering and polarized inflammation in the initiation and promotion of malignant disease. *Cancer Cell* **7:**211–217.
3. **Schreiber RD, Old LJ, Smyth MJ.** 2011. Cancer immunoediting: integrating immunity's roles in cancer suppression and promotion. *Science* **331:**1565–1570.
4. **Drake CG, Jaffee E, Pardoll DM.** 2006. Mechanisms of immune evasion by tumors. *Adv Immunol* **90:**51–81.
5. **Rabinovich GA, Gabrilovich D, Sotomayor EM.** 2007. Immunosuppressive strategies that are mediated by tumor cells. *Annu Rev Immunol* **25:**267–296.
6. **Chaux P, Favre N, Martin M, Martin F.** 1997. Tumor-infiltrating dendritic cells are defective in their antigen-presenting function and inducible B7 expression in rats. *Int J Cancer* **72:**619–624.
7. **Geissmann F, Manz MG, Jung S, Sieweke MH, Merad M, Ley K.** 2010. Development of monocytes, macrophages, and dendritic cells. *Science* **327:**656–661.
8. **Gabrilovich DI, Nagaraj S.** 2009. Myeloid-derived suppressor cells as regulators of the immune system. *Nat Rev Immunol* **9:**162–174.
9. **Strober S.** 1984. Natural suppressor (NS) cells, neonatal tolerance, and total lymphoid irradiation: exploring obscure relationships. *Annu Rev Immunol* **2:**219–237.
10. **Seung LP, Rowley DA, Dubey P, Schreiber H.** 1995. Synergy between T-cell immunity and inhibition of paracrine stimulation causes tumor rejection. *Proc Natl Acad Sci U S A* **92:**6254–6258.
11. **Serafini P, Borrello I, Bronte V.** 2006. Myeloid suppressor cells in cancer: recruitment, phenotype, properties, and mechanisms of immune suppression. *Semin Cancer Biol* **16:**53–65.
12. **Kusmartsev S, Gabrilovich DI.** 2006. Role of immature myeloid cells in mechanisms of immune evasion in cancer. *Cancer Immunol Immunother* **55:**237–245.
13. **Delano MJ, Scumpia PO, Weinstein JS, Coco D, Nagaraj S, Kelly-Scumpia KM, O'Malley KA, Wynn JL, Antonenko S, Al-Quran SZ, Swan R, Chung CS, Atkinson MA, Ramphal R, Gabrilovich DI, Reeves WH, Ayala A, Phillips J, Laface D, Heyworth PG, Clare-Salzler M, Moldawer LL.** 2007. MyD88-dependent expansion of an immature GR-1$^+$CD11b$^+$ population induces T cell suppression and Th2 polarization in sepsis. *J Exp Med* **204:**1463–1474.
14. **Voisin MB, Buzoni-Gatel D, Bout D, Velge-Roussel F.** 2004. Both expansion of regulatory GR1$^+$ CD11b$^+$ myeloid cells and anergy of T lymphocytes participate in hyporesponsiveness of the lung-associated immune system during acute toxoplasmosis. *Infect Immun* **72:**5487–5492.
15. **Mencacci A, Montagnoli C, Bacci A, Cenci E, Pitzurra L, Spreca A, Kopf M, Sharpe AH, Romani L.** 2002. CD80$^+$Gr-1$^+$ myeloid cells inhibit development of antifungal Th1 immunity in mice with candidiasis. *J Immunol* **169:**3180–3190.
16. **Sunderkotter C, Nikolic T, Dillon MJ, Van Rooijen N, Stehling M, Drevets DA, Leenen PJ.** 2004. Subpopulations of mouse blood monocytes differ in maturation stage and inflammatory response. *J Immunol* **172:**4410–4417.
17. **Haile LA, von Wasielewski R, Gamrekelashvili J, Kruger C, Bachmann O, Westendorf AM, Buer J, Liblau R, Manns MP, Korangy F, Greten TF.** 2008. Myeloid-derived suppressor cells in inflammatory bowel disease: a new immunoregulatory pathway. *Gastroenterology* **135:**871–881.
18. **Makarenkova VP, Bansal V, Matta BM, Perez LA, Ochoa JB.** 2006. CD11b$^+$/Gr-1$^+$ myeloid suppressor cells cause T cell dysfunction after traumatic stress. *J Immunol* **176:**2085–2094.
19. **Verschoor CP, Johnstone J, Millar J, Dorrington MG, Habibagahi M, Lelic A, Loeb M, Bramson JL, Bowdish DM.** 2013. Blood CD33(+)HLA-DR(−) myeloid-derived suppressor cells are increased with age and a history of cancer. *J Leukoc Biol* **93:**633–637.
20. **Xiang X, Poliakov A, Liu C, Liu Y, Deng ZB, Wang J, Cheng Z, Shah SV, Wang GJ, Zhang L, Grizzle WE, Mobley J, Zhang HG.** 2009. Induction of myeloid-derived suppressor cells by tumor exosomes. *Int J Cancer* **124:**2621–2633.
21. **Talmadge JE, Gabrilovich DI.** 2013. History of myeloid-derived suppressor cells. *Nat Rev Cancer* **13:**739–752.
22. **Gabrilovich DI, Bronte V, Chen SH, Colombo MP, Ochoa A, Ostrand-Rosenberg S, Schreiber H.** 2007. The terminology issue for myeloid-derived suppressor cells. *Cancer Res* **67:**425; author reply 426.
23. **Marigo I, Bosio E, Solito S, Mesa C, Fernandez A, Dolcetti L, Ugel S, Sonda N, Bicciato S, Falisi E, Calabrese F, Basso G, Zanovello P, Cozzi E, Mandruzzato S, Bronte V.** 2010. Tumor-induced tolerance and immune suppression depend on the C/EBPβ transcription factor. *Immunity* **32:**790–802.
24. **Dolcetti L, Peranzoni E, Ugel S, Marigo I, Fernandez Gomez A, Mesa C, Geilich M, Winkels G, Traggiai E, Casati A, Grassi F, Bronte V.** 2010. Hierarchy of immunosuppressive strength among myeloid-derived suppressor cell subsets is determined by GM-CSF. *Eur J Immunol* **40:**22–35.
25. **Rossner S, Voigtlander C, Wiethe C, Hanig J, Seifarth C, Lutz MB.** 2005. Myeloid dendritic cell precursors generated from bone marrow suppress T cell responses via cell contact and nitric oxide production in vitro. *Eur J Immunol* **35:**3533–3544.

26. **Kusmartsev S, Gabrilovich DI.** 2006. Effect of tumor-derived cytokines and growth factors on differentiation and immune suppressive features of myeloid cells in cancer. *Cancer Metastasis Rev* **25:**323–331.

27. **Kusmartsev S, Cheng F, Yu B, Nefedova Y, Sotomayor E, Lush R, Gabrilovich D.** 2003. All-*trans*-retinoic acid eliminates immature myeloid cells from tumor-bearing mice and improves the effect of vaccination. *Cancer Res* **63:**4441–4449.

28. **Ugel S, Peranzoni E, Desantis G, Chioda M, Walter S, Weinschenk T, Ochando JC, Cabrelle A, Mandruzzato S, Bronte V.** 2012. Immune tolerance to tumor antigens occurs in a specialized environment of the spleen. *Cell Rep* **2:**628–639.

29. **Kusmartsev S, Gabrilovich DI.** 2003. Inhibition of myeloid cell differentiation in cancer: the role of reactive oxygen species. *J Leukoc Biol* **74:**186–196.

30. **Auffray C, Sieweke MH, Geissmann F.** 2009. Blood monocytes: development, heterogeneity, and relationship with dendritic cells. *Annu Rev Immunol* **27:**669–692.

31. **Movahedi K, Guilliams M, Van den Bossche J, Van den Bergh R, Gysemans C, Beschin A, De Baetselier P, Van Ginderachter JA.** 2008. Identification of discrete tumor-induced myeloid-derived suppressor cell subpopulations with distinct T cell-suppressive activity. *Blood* **111:**4233–4244.

32. **Youn JI, Nagaraj S, Collazo M, Gabrilovich DI.** 2008. Subsets of myeloid-derived suppressor cells in tumor-bearing mice. *J Immunol* **181:**5791–5802.

33. **Haile LA, Gamrekelashvili J, Manns MP, Korangy F, Greten TF.** 2010. CD49d is a new marker for distinct myeloid-derived suppressor cell subpopulations in mice. *J Immunol* **185:**203–210.

34. **Gabrilovich DI, Ostrand-Rosenberg S, Bronte V.** 2012. Coordinated regulation of myeloid cells by tumours. *Nat Rev Immunol* **12:**253–268.

35. **Youn JI, Kumar V, Collazo M, Nefedova Y, Condamine T, Cheng P, Villagra A, Antonia S, McCaffrey JC, Fishman M, Sarnaik A, Horna P, Sotomayor E, Gabrilovich DI.** 2013. Epigenetic silencing of retinoblastoma gene regulates pathologic differentiation of myeloid cells in cancer. *Nat Immunol* **14:**211–220.

36. **Gallina G, Dolcetti L, Serafini P, De Santo C, Marigo I, Colombo MP, Basso G, Brombacher F, Borrello I, Zanovello P, Bicciato S, Bronte V.** 2006. Tumors induce a subset of inflammatory monocytes with immunosuppressive activity on $CD8^+$ T cells. *J Clin Invest* **116:**2777–2790.

37. **Roth F, De La Fuente AC, Vella JL, Zoso A, Inverardi L, Serafini P.** 2012. Aptamer-mediated blockade of IL4Rα triggers apoptosis of MDSCs and limits tumor progression. *Cancer Res* **72:**1373–1383.

38. **Galli SJ, Borregaard N, Wynn TA.** 2011. Phenotypic and functional plasticity of cells of innate immunity: macrophages, mast cells and neutrophils. *Nat Immunol* **12:**1035–1044.

39. **Peranzoni E, Zilio S, Marigo I, Dolcetti L, Zanovello P, Mandruzzato S, Bronte V.** 2010. Myeloid-derived suppressor cell heterogeneity and subset definition. *Curr Opin Immunol* **22:**238–244.

40. **Biswas SK, Mantovani A.** 2010. Macrophage plasticity and interaction with lymphocyte subsets: cancer as a paradigm. *Nat Immunol* **11:**889–896.

41. **Murray PJ, Allen JE, Biswas SK, Fisher EA, Gilroy DW, Goerdt S, Gordon S, Hamilton JA, Ivashkiv LB, Lawrence T, Locati M, Mantovani A, Martinez FO, Mege JL, Mosser DM, Natoli G, Saeij JP, Schultze JL, Shirey KA, Sica A, Suttles J, Udalova I, van Ginderachter JA, Vogel SN, Wynn TA.** 2014. Macrophage activation and polarization: nomenclature and experimental guidelines. *Immunity* **41:**14–20.

42. **Umemura N, Saio M, Suwa T, Kitoh Y, Bai J, Nonaka K, Ouyang GF, Okada M, Balazs M, Adany R, Shibata T, Takami T.** 2008. Tumor-infiltrating myeloid-derived suppressor cells are pleiotropic-inflamed monocytes/macrophages that bear M1- and M2-type characteristics. *J Leukoc Biol* **83:**1136–1144.

43. **Movahedi K, Laoui D, Gysemans C, Baeten M, Stangé G, Van den Bossche J, Mack M, Pipeleers D, In't Veld P, De Baetselier P, Van Ginderachter JA.** 2010. Different tumor microenvironments contain functionally distinct subsets of macrophages derived from Ly6C(high) monocytes. *Cancer Res* **70:**5728–5739.

44. **Wynn TA, Chawla A, Pollard JW.** 2013. Macrophage biology in development, homeostasis and disease. *Nature* **496:**445–455.

45. **Colegio OR, Chu NQ, Szabo AL, Chu T, Rhebergen AM, Jairam V, Cyrus N, Brokowski CE, Eisenbarth SC, Phillips GM, Cline GW, Phillips AJ, Medzhitov R.** 2014. Functional polarization of tumour-associated macrophages by tumour-derived lactic acid. *Nature* **513:**559–563.

46. **Kobayashi Y.** 2008. The role of chemokines in neutrophil biology. *Front Biosci* **13:**2400–2407.

47. **Fridlender ZG, Sun J, Kim S, Kapoor V, Cheng G, Ling L, Worthen GS, Albelda SM.** 2009. Polarization of tumor-associated neutrophil phenotype by TGF-β: "N1" versus "N2" TAN. *Cancer Cell* **16:**183–194.

48. **Youn JI, Collazo M, Shalova IN, Biswas SK, Gabrilovich DI.** 2012. Characterization of the nature of granulocytic myeloid-derived suppressor cells in tumor-bearing mice. *J Leukoc Biol* **91:**167–181.

49. **Brandau S, Trellakis S, Bruderek K, Schmaltz D, Steller G, Elian M, Suttmann H, Schenck M, Welling J, Zabel P, Lang S.** 2011. Myeloid-derived suppressor cells in the peripheral blood of cancer patients contain a subset of immature neutrophils with impaired migratory properties. *J Leukoc Biol* **89:**311–317.

50. **Fridlender ZG, Sun J, Mishalian I, Singhal S, Cheng G, Kapoor V, Horng W, Fridlender G, Bayuh R, Worthen GS, Albelda SM.** 2012. Transcriptomic analysis comparing tumor-associated neutrophils with granulocytic myeloid-derived suppressor cells and normal neutrophils. *PLoS One* **7:**e31524. doi:10.1371/journal.pone.0031524.

51. **Köffel R, Meshcheryakova A, Warszawska J, Hennig A, Wagner K, Jörgl A, Gubi D, Moser D, Hladik A, Hoffmann U, Fischer MB, van den Berg W, Koenders M, Scheinecker C, Gesslbauer B, Knapp S, Strobl H.** 2014. Monocytic cell differentiation from band-stage

neutrophils under inflammatory conditions via MKK6 activation. *Blood* **124:**2713–2724.

52. **Zoso A, Mazza EM, Bicciato S, Mandruzzato S, Bronte V, Serafini P, Inverardi L.** 2014. Human fibrocytic myeloid-derived suppressor cells express IDO and promote tolerance via Treg-cell expansion. *Eur J Immunol* **44:**3307–3319.
53. **Zhang H, Maric I, DiPrima MJ, Khan J, Orentas RJ, Kaplan RN, Mackall CL.** 2013. Fibrocytes represent a novel MDSC subset circulating in patients with metastatic cancer. *Blood* **122:**1105–1113.
54. **Shi Y, Ou L, Han S, Li M, Pena MM, Pena EA, Liu C, Nagarkatti M, Fan D, Ai W.** 2014. Deficiency of Kruppel-like factor KLF4 in myeloid-derived suppressor cells inhibits tumor pulmonary metastasis in mice accompanied by decreased fibrocytes. *Oncogenesis* **3:** e129. doi:10.1038/oncsis.2014.44.
55. **Park YJ, Song B, Kim YS, Kim EK, Lee JM, Lee GE, Kim JO, Kim YJ, Chang WS, Kang CY.** 2013. Tumor microenvironmental conversion of natural killer cells into myeloid-derived suppressor cells. *Cancer Res* **73:** 5669–5681.
56. **Diaz-Montero CM, Salem ML, Nishimura MI, Garrett-Mayer E, Cole DJ, Montero AJ.** 2009. Increased circulating myeloid-derived suppressor cells correlate with clinical cancer stage, metastatic tumor burden, and doxorubicin-cyclophosphamide chemotherapy. *Cancer Immunol Immunother* **58:**49–59.
57. **Almand B, Clark JI, Nikitina E, van Beynen J, English NR, Knight SC, Carbone DP, Gabrilovich DI.** 2001. Increased production of immature myeloid cells in cancer patients: a mechanism of immunosuppression in cancer. *J Immunol* **166:**678–689.
58. **Rodriguez PC, Ernstoff MS, Hernandez C, Atkins M, Zabaleta J, Sierra R, Ochoa AC.** 2009. Arginase I-producing myeloid-derived suppressor cells in renal cell carcinoma are a subpopulation of activated granulocytes. *Cancer Res* **69:**1553–1560.
59. **Gabitass RF, Annels NE, Stocken DD, Pandha HA, Middleton GW.** 2011. Elevated myeloid-derived suppressor cells in pancreatic, esophageal and gastric cancer are an independent prognostic factor and are associated with significant elevation of the Th2 cytokine interleukin-13. *Cancer Immunol Immunother* **60:**1419–1430.
60. **Eruslanov E, Neuberger M, Daurkin I, Perrin GQ, Algood C, Dahm P, Rosser C, Vieweg J, Gilbert SM, Kusmartsev S.** 2012. Circulating and tumor-infiltrating myeloid cell subsets in patients with bladder cancer. *Int J Cancer* **130:**1109–1119.
61. **Solito S, Marigo I, Pinton L, Damuzzo V, Mandruzzato S, Bronte V.** 2014. Myeloid-derived suppressor cell heterogeneity in human cancers. *Ann N Y Acad Sci* **1319:** 47–65.
62. **Pak AS, Wright MA, Matthews JP, Collins SL, Petruzzelli GJ, Young MR.** 1995. Mechanisms of immune suppression in patients with head and neck cancer: presence of CD34$^+$ cells which suppress immune functions within cancers that secrete granulocyte-macrophage colony-stimulating factor. *Clin Cancer Res* **1:**95–103.
63. **Solito S, Falisi E, Diaz-Montero CM, Doni A, Pinton L, Rosato A, Francescato S, Basso G, Zanovello P, Onicescu G, Garrett-Mayer E, Montero AJ, Bronte V, Mandruzzato S.** 2011. A human promyelocytic-like population is responsible for the immune suppression mediated by myeloid-derived suppressor cells. *Blood* **118:**2254–2265.
64. **Walter S, Weinschenk T, Stenzl A, Zdrojowy R, Pluzanska A, Szczylik C, Staehler M, Brugger W, Dietrich PY, Mendrzyk R, Hilf N, Schoor O, Fritsche J, Mahr A, Maurer D, Vass V, Trautwein C, Lewandrowski P, Flohr C, Pohla H, Stanczak JJ, Bronte V, Mandruzzato S, Biedermann T, Pawelec G, Derhovanessian E, Yamagishi H, Miki T, Hongo F, Takaha N, Hirakawa K, Tanaka H, Stevanovic S, Frisch J, Mayer-Mokler A, Kirner A, Rammensee HG, Reinhardt C, Singh-Jasuja H.** 2012. Multipeptide immune response to cancer vaccine IMA901 after single-dose cyclophosphamide associates with longer patient survival. *Nat Med* **18:**1254–1261.
65. **Montero AJ, Diaz-Montero CM, Kyriakopoulos CE, Bronte V, Mandruzzato S.** 2012. Myeloid-derived suppressor cells in cancer patients: a clinical perspective. *J Immunother* **35:**107–115.
66. **Trellakis S, Bruderek K, Hutte J, Elian M, Hoffmann TK, Lang S, Brandau S.** 2013. Granulocytic myeloid-derived suppressor cells are cryosensitive and their frequency does not correlate with serum concentrations of colony-stimulating factors in head and neck cancer. *Innate Immun* **19:**328–336.
67. **Zea AH, Rodriguez PC, Atkins MB, Hernandez C, Signoretti S, Zabaleta J, McDermott D, Quiceno D, Youmans A, O'Neill A, Mier J, Ochoa AC.** 2005. Arginase-producing myeloid suppressor cells in renal cell carcinoma patients: a mechanism of tumor evasion. *Cancer Res* **65:**3044–3048.
68. **Idorn M, Kollgaard T, Kongsted P, Sengelov L, Thor Straten P.** 2014. Correlation between frequencies of blood monocytic myeloid-derived suppressor cells, regulatory T cells and negative prognostic markers in patients with castration-resistant metastatic prostate cancer. *Cancer Immunol Immunother* **63:**1177–1187.
69. **Gao J, Wu Y, Su Z, Amoah Barnie P, Jiao Z, Bie Q, Lu L, Wang S, Xu H.** 2014. Infiltration of alternatively activated macrophages in cancer tissue is associated with MDSC and Th2 polarization in patients with esophageal cancer. *PLoS One* **9:**e104453. doi:10.1371/journal.pone.0104453.
70. **Markowitz J, Brooks TR, Duggan MC, Paul BK, Pan X, Wei L, Abrams Z, Luedke E, Lesinski GB, Mundy-Bosse B, Bekaii-Saab T, Carson WE III.** 2014. Patients with pancreatic adenocarcinoma exhibit elevated levels of myeloid-derived suppressor cells upon progression of disease. *Cancer Immunol Immunother* **64:**149–159.
71. **Rudolph BM, Loquai C, Gerwe A, Bacher N, Steinbrink K, Grabbe S, Tuettenberg A.** 2014. Increased frequencies of CD11b$^+$CD33$^+$CD14$^+$HLA-DR^{low} myeloid-derived suppressor cells are an early event in melanoma patients. *Exp Dermatol* **23:**202–204.
72. **Weide B, Martens A, Zelba H, Stutz C, Derhovanessian E, Di Giacomo AM, Maio M, Sucker A, Schilling B,**

Schadendorf D, Büttner P, Garbe C, Pawelec G. 2014. Myeloid-derived suppressor cells predict survival of patients with advanced melanoma: comparison with regulatory T cells and NY-ESO-1- or melan-A-specific T cells. *Clin Cancer Res* **20:**1601–1609.

73. **Roca H, Varsos ZS, Sud S, Craig MJ, Ying C, Pienta KJ.** 2009. CCL2 and interleukin-6 promote survival of human CD11b+ peripheral blood mononuclear cells and induce M2-type macrophage polarization. *J Biol Chem* **284:**34342–34354.

74. **Nagaraj S, Gabrilovich DI.** 2008. Tumor escape mechanism governed by myeloid-derived suppressor cells. *Cancer Res* **68:**2561–2563.

75. **Srivastava MK, Sinha P, Clements VK, Rodriguez P, Ostrand-Rosenberg S.** 2010. Myeloid-derived suppressor cells inhibit T-cell activation by depleting cystine and cysteine. *Cancer Res* **70:**68–77.

76. **Munn DH, Sharma MD, Baban B, Harding HP, Zhang Y, Ron D, Mellor AL.** 2005. GCN2 kinase in T cells mediates proliferative arrest and anergy induction in response to indoleamine 2,3-dioxygenase. *Immunity* **22:** 633–642.

77. **Quintana FJ, Murugaiyan G, Farez MF, Mitsdoerffer M, Tukpah AM, Burns EJ, Weiner HL.** 2010. An endogenous aryl hydrocarbon receptor ligand acts on dendritic cells and T cells to suppress experimental autoimmune encephalomyelitis. *Proc Natl Acad Sci U S A* **107:**20768–20773.

78. **Rolinski J, Hus I.** 2014. Breaking immunotolerance of tumors: a new perspective for dendritic cell therapy. *J Immunotoxicol* **11:**311–318.

79. **Godin-Ethier J, Hanafi LA, Piccirillo CA, Lapointe R.** 2011. Indoleamine 2,3-dioxygenase expression in human cancers: clinical and immunologic perspectives. *Clin Cancer Res* **17:**6985–6991.

80. **Bronte V, Zanovello P.** 2005. Regulation of immune responses by L-arginine metabolism. *Nat Rev Immunol* **5:** 641–654.

81. **Mazzoni A, Bronte V, Visintin A, Spitzer JH, Apolloni E, Serafini P, Zanovello P, Segal DM.** 2002. Myeloid suppressor lines inhibit T cell responses by an NO-dependent mechanism. *J Immunol* **168:**689–695.

82. **Macphail SE, Gibney CA, Brooks BM, Booth CG, Flanagan BF, Coleman JW.** 2003. Nitric oxide regulation of human peripheral blood mononuclear cells: critical time dependence and selectivity for cytokine versus chemokine expression. *J Immunol* **171:**4809–4815.

83. **Baniyash M.** 2004. TCR ζ-chain downregulation: curtailing an excessive inflammatory immune response. *Nat Rev Immunol* **4:**675–687.

84. **Rodriguez PC, Quiceno DG, Ochoa AC.** 2007. L-Arginine availability regulates T-lymphocyte cell-cycle progression. *Blood* **109:**1568–1573.

85. **Rodriguez PC, Hernandez CP, Morrow K, Sierra R, Zabaleta J, Wyczechowska DD, Ochoa AC.** 2010. L-Arginine deprivation regulates cyclin D3 mRNA stability in human T cells by controlling HuR expression. *J Immunol* **185:**5198–5204.

86. **Raber P, Ochoa AC, Rodriguez PC.** 2012. Metabolism of L-arginine by myeloid-derived suppressor cells in cancer: mechanisms of T cell suppression and therapeutic perspectives. *Immunol Invest* **41:**614–634.

87. **Bedard K, Krause KH.** 2007. The NOX family of ROS-generating NADPH oxidases: physiology and pathophysiology. *Physiol Rev* **87:**245–313.

88. **Raad H, Paclet MH, Boussetta T, Kroviarski Y, Morel F, Quinn MT, Gougerot-Pocidalo MA, Dang PM, El-Benna J.** 2009. Regulation of the phagocyte NADPH oxidase activity: phosphorylation of gp91^{phox}/NOX2 by protein kinase C enhances its diaphorase activity and binding to Rac2, p67^{phox}, and p47^{phox}. *FASEB J* **23:** 1011–1022.

89. **Schmielau J, Nalesnik MA, Finn OJ.** 2001. Suppressed T-cell receptor zeta chain expression and cytokine production in pancreatic cancer patients. *Clin Cancer Res* **7** (3 Suppl)**:**933s–939s.

90. **Otsuji M, Kimura Y, Aoe T, Okamoto Y, Saito T.** 1996. Oxidative stress by tumor-derived macrophages suppresses the expression of CD3 ζ chain of T-cell receptor complex and antigen-specific T-cell responses. *Proc Natl Acad Sci U S A* **93:**13119–13124.

91. **Alvarez B, Radi R.** 2003. Peroxynitrite reactivity with amino acids and proteins. *Amino Acids* **25:**295–311.

92. **Hardy LL, Wick DA, Webb JR.** 2008. Conversion of tyrosine to the inflammation-associated analog 3′-nitrotyrosine at either TCR- or MHC-contact positions can profoundly affect recognition of the MHC class I-restricted epitope of lymphocytic choriomeningitis virus glycoprotein 33 by CD8 T cells. *J Immunol* **180:** 5956–5962.

93. **Nagaraj S, Schrum AG, Cho HI, Celis E, Gabrilovich DI.** 2010. Mechanism of T cell tolerance induced by myeloid-derived suppressor cells. *J Immunol* **184:**3106–3116.

94. **De Sanctis F, Sandri S, Ferrarini G, Pagliarello I, Sartoris S, Ugel S, Marigo I, Molon B, Bronte V.** 2014. The emerging immunological role of post-translational modifications by reactive nitrogen species in cancer microenvironment. *Front Immunol* **5:**69. doi:10.3389/fimmu.2014.00069.

95. **Hanson EM, Clements VK, Sinha P, Ilkovitch D, Ostrand-Rosenberg S.** 2009. Myeloid-derived suppressor cells down-regulate L-selectin expression on CD4+ and CD8+ T cells. *J Immunol* **183:**937–944.

96. **Molon B, Ugel S, Del Pozzo F, Soldani C, Zilio S, Avella D, De Palma A, Mauri P, Monegal A, Rescigno M, Savino B, Colombo P, Jonjic N, Pecanic S, Lazzarato L, Fruttero R, Gasco A, Bronte V, Viola A.** 2011. Chemokine nitration prevents intratumoral infiltration of antigen-specific T cells. *J Exp Med* **208:**1949–1962.

97. **Elkabets M, Ribeiro VS, Dinarello CA, Ostrand-Rosenberg S, Di Santo JP, Apte RN, Vosshenrich CA.** 2010. IL-1β regulates a novel myeloid-derived suppressor cell subset that impairs NK cell development and function. *Eur J Immunol* **40:**3347–3357.

98. **Wolfraim LA, Walz TM, James Z, Fernandez T, Letterio JJ.** 2004. p21^{Cip1} and p27^{Kip1} act in synergy to alter the sensitivity of naive T cells to TGF-β-mediated G_1

arrest through modulation of IL-2 responsiveness. *J Immunol* **173:**3093–3102.

99. **Brabletz T, Pfeuffer I, Schorr E, Siebelt F, Wirth T, Serfling E.** 1993. Transforming growth factor β and cyclosporin A inhibit the inducible activity of the interleukin-2 gene in T cells through a noncanonical octamer-binding site. *Mol Cell Biol* **13:**1155–1162.
100. **Becker C, Fantini MC, Neurath MF.** 2006. TGF-β as a T cell regulator in colitis and colon cancer. *Cytokine Growth Factor Rev* **17:**97–106.
101. **Serafini P, Mgebroff S, Noonan K, Borrello I.** 2008. Myeloid-derived suppressor cells promote cross-tolerance in B-cell lymphoma by expanding regulatory T cells. *Cancer Res* **68:**5439–5449.
102. **Hoechst B, Gamrekelashvili J, Manns MP, Greten TF, Korangy F.** 2011. Plasticity of human Th17 cells and iTregs is orchestrated by different subsets of myeloid cells. *Blood* **117:**6532–6541.
103. **Sinha P, Clements VK, Bunt SK, Albelda SM, Ostrand-Rosenberg S.** 2007. Cross-talk between myeloid-derived suppressor cells and macrophages subverts tumor immunity toward a type 2 response. *J Immunol* **179:**977–983.
104. **Li H, Collado M, Villasante A, Strati K, Ortega S, Cañamero M, Blasco MA, Serrano M.** 2009. The *Ink4/Arf* locus is a barrier for iPS cell reprogramming. *Nature* **460:**1136–1139.
105. **Chen Z, Trotman LC, Shaffer D, Lin HK, Dotan ZA, Niki M, Koutcher JA, Scher HI, Ludwig T, Gerald W, Cordon-Cardo C, Pandolfi PP.** 2005. Crucial role of p53-dependent cellular senescence in suppression of Pten-deficient tumorigenesis. *Nature* **436:**725–730.
106. **Collado M, Gil J, Efeyan A, Guerra C, Schuhmacher AJ, Barradas M, Benguria A, Zaballos A, Flores JM, Barbacid M, Beach D, Serrano M.** 2005. Tumour biology: senescence in premalignant tumours. *Nature* **436:**642.
107. **Di Mitri D, Toso A, Chen JJ, Sarti M, Pinton S, Jost TR, D'Antuono R, Montani E, Garcia-Escudero R, Guccini I, Da Silva-Alvarez S, Collado M, Eisenberger M, Zhang Z, Catapano C, Grassi F, Alimonti A.** 2014. Tumour-infiltrating Gr-1$^+$ myeloid cells antagonize senescence in cancer. *Nature* **515:**134–137.
108. **Cui TX, Kryczek I, Zhao L, Zhao E, Kuick R, Roh MH, Vatan L, Szeliga W, Mao Y, Thomas DG, Kotarski J, Tarkowski R, Wicha M, Cho K, Giordano T, Liu R, Zou W.** 2013. Myeloid-derived suppressor cells enhance stemness of cancer cells by inducing microRNA101 and suppressing the corepressor CtBP2. *Immunity* **39:**611–621.
109. **Panni RZ, Sanford DE, Belt BA, Mitchem JB, Worley LA, Goetz BD, Mukherjee P, Wang-Gillam A, Link DC, Denardo DG, Goedegebuure SP, Linehan DC.** 2014. Tumor-induced STAT3 activation in monocytic myeloid-derived suppressor cells enhances stemness and mesenchymal properties in human pancreatic cancer. *Cancer Immunol Immunother* **63:**513–528.
110. **Ostrand-Rosenberg S, Sinha P.** 2009. Myeloid-derived suppressor cells: linking inflammation and cancer. *J Immunol* **182:**4499–4506.
111. **Doedens AL, Stockmann C, Rubinstein MP, Liao D, Zhang N, DeNardo DG, Coussens LM, Karin M, Goldrath AW, Johnson RS.** 2010. Macrophage expression of hypoxia-inducible factor-1α suppresses T-cell function and promotes tumor progression. *Cancer Res* **70:**7465–7475.
112. **Yang L, DeBusk LM, Fukuda K, Fingleton B, Green-Jarvis B, Shyr Y, Matrisian LM, Carbone DP, Lin PC.** 2004. Expansion of myeloid immune suppressor Gr+CD11b+ cells in tumor-bearing host directly promotes tumor angiogenesis. *Cancer Cell* **6:**409–421.
113. **Shojaei F, Wu X, Malik AK, Zhong C, Baldwin ME, Schanz S, Fuh G, Gerber HP, Ferrara N.** 2007. Tumor refractoriness to anti-VEGF treatment is mediated by CD11b$^+$Gr1$^+$ myeloid cells. *Nat Biotechnol* **25:**911–920.
114. **Finke J, Ko J, Rini B, Rayman P, Ireland J, Cohen P.** 2011. MDSC as a mechanism of tumor escape from sunitinib mediated anti-angiogenic therapy. *Int Immunopharmacol* **11:**856–861.
115. **Sceneay J, Chow MT, Chen A, Halse HM, Wong CS, Andrews DM, Sloan EK, Parker BS, Bowtell DD, Smyth MJ, Möller A.** 2012. Primary tumor hypoxia recruits CD11b$^+$/Ly6C^{med}/Ly6G$^+$ immune suppressor cells and compromises NK cell cytotoxicity in the premetastatic niche. *Cancer Res* **72:**3906–3911.
116. **Shojaei F, Wu X, Zhong C, Yu L, Liang XH, Yao J, Blanchard D, Bais C, Peale FV, van Bruggen N, Ho C, Ross J, Tan M, Carano RA, Meng YG, Ferrara N.** 2007. Bv8 regulates myeloid-cell-dependent tumour angiogenesis. *Nature* **450:**825–831.
117. **Sinha P, Okoro C, Foell D, Freeze HH, Ostrand-Rosenberg S, Srikrishna G.** 2008. Proinflammatory S100 proteins regulate the accumulation of myeloid-derived suppressor cells. *J Immunol* **181:**4666–4675.
118. **Hiratsuka S, Watanabe A, Sakurai Y, Akashi-Takamura S, Ishibashi S, Miyake K, Shibuya M, Akira S, Aburatani H, Maru Y.** 2008. The S100A8-serum amyloid A3-TLR4 paracrine cascade establishes a pre-metastatic phase. *Nat Cell Biol* **10:**1349–1355.
119. **Oh K, Lee OY, Shon SY, Nam O, Ryu PM, Seo MW, Lee DS.** 2013. A mutual activation loop between breast cancer cells and myeloid-derived suppressor cells facilitates spontaneous metastasis through IL-6 trans-signaling in a murine model. *Breast Cancer Res* **15:**R79. doi:10.1186/bcr3473.
120. **Hiratsuka S, Watanabe A, Aburatani H, Maru Y.** 2006. Tumour-mediated upregulation of chemoattractants and recruitment of myeloid cells predetermines lung metastasis. *Nat Cell Biol* **8:**1369–1375.
121. **Toh B, Wang X, Keeble J, Sim WJ, Khoo K, Wong WC, Kato M, Prevost-Blondel A, Thiery JP, Abastado JP.** 2011. Mesenchymal transition and dissemination of cancer cells is driven by myeloid-derived suppressor cells infiltrating the primary tumor. *PLoS Biol* **9:** e1001162. doi:10.1371/journal.pbio.1001162.
122. **Zhu L, Li X, Chen Y, Fang J, Ge Z.** 2015. High-mobility group box 1: a novel inducer of the epithelial-mesenchymal transition in colorectal carcinoma. *Cancer Lett* **357:**527–534.
123. **Gao D, Joshi N, Choi H, Ryu S, Hahn M, Catena R, Sadik H, Argani P, Wagner P, Vahdat LT, Port JL, Stiles B, Sukumar S, Altorki NK, Rafii S, Mittal V.** 2012. Myeloid progenitor cells in the premetastatic lung

promote metastases by inducing mesenchymal to epithelial transition. *Cancer Res* **72**:1384–1394.

124. **Catena R, Bhattacharya N, El Rayes T, Wang S, Choi H, Gao D, Ryu S, Joshi N, Bielenberg D, Lee SB, Haukaas SA, Gravdal K, Halvorsen OJ, Akslen LA, Watnick RS, Mittal V.** 2013. Bone marrow-derived Gr1+ cells can generate a metastasis-resistant microenvironment via induced secretion of thrombospondin-1. *Cancer Discov* **3**: 578–589.

125. **Nakamura I, Shibata M, Gonda K, Yazawa T, Shimura T, Anazawa T, Suzuki S, Sakurai K, Koyama Y, Ohto H, Tomita R, Gotoh M, Takenoshita S.** 2013. Serum levels of vascular endothelial growth factor are increased and correlate with malnutrition, immunosuppression involving MDSCs and systemic inflammation in patients with cancer of the digestive system. *Oncol Lett* **5**:1682–1686.

126. **Cuenca AG, Cuenca AL, Winfield RD, Joiner DN, Gentile L, Delano MJ, Kelly-Scumpia KM, Scumpia PO, Matheny MK, Scarpace PJ, Vila L, Efron PA, LaFace DM, Moldawer LL.** 2014. Novel role for tumor-induced expansion of myeloid-derived cells in cancer cachexia. *J Immunol* **192**:6111–6119.

127. **Gerharz CD, Reinecke P, Schneider EM, Schmitz M, Gabbert HE.** 2001. Secretion of GM-CSF and M-CSF by human renal cell carcinomas of different histologic types. *Urology* **58**:821–827.

128. **Lin EY, Gouon-Evans V, Nguyen AV, Pollard JW.** 2002. The macrophage growth factor CSF-1 in mammary gland development and tumor progression. *J Mammary Gland Biol Neoplasia* **7**:147–162.

129. **Priceman SJ, Sung JL, Shaposhnik Z, Burton JB, Torres-Collado AX, Moughon DL, Johnson M, Lusis AJ, Cohen DA, Iruela-Arispe ML, Wu L.** 2010. Targeting distinct tumor-infiltrating myeloid cells by inhibiting CSF-1 receptor: combating tumor evasion of antiangiogenic therapy. *Blood* **115**:1461–1471.

130. **Kowanetz M, Wu X, Lee J, Tan M, Hagenbeek T, Qu X, Yu L, Ross J, Korsisaari N, Cao T, Bou-Reslan H, Kallop D, Weimer R, Ludlam MJ, Kaminker JS, Modrusan Z, van Bruggen N, Peale FV, Carano R, Meng YG, Ferrara N.** 2010. Granulocyte-colony stimulating factor promotes lung metastasis through mobilization of Ly6G+Ly6C+ granulocytes. *Proc Natl Acad Sci U S A* **107**:21248–21255.

131. **Clark CE, Hingorani SR, Mick R, Combs C, Tuveson DA, Vonderheide RH.** 2007. Dynamics of the immune reaction to pancreatic cancer from inception to invasion. *Cancer Res* **67**:9518–9527.

132. **Bayne LJ, Beatty GL, Jhala N, Clark CE, Rhim AD, Stanger BZ, Vonderheide RH.** 2012. Tumor-derived granulocyte-macrophage colony-stimulating factor regulates myeloid inflammation and T cell immunity in pancreatic cancer. *Cancer Cell* **21**:822–835.

133. **Lesina M, Kurkowski MU, Ludes K, Rose-John S, Treiber M, Klöppel G, Yoshimura A, Reindl W, Sipos B, Akira S, Schmid RM, Algül H.** 2011. Stat3/Socs3 activation by IL-6 transsignaling promotes progression of pancreatic intraepithelial neoplasia and development of pancreatic cancer. *Cancer Cell* **19**:456–469.

134. **Serafini P, Carbley R, Noonan KA, Tan G, Bronte V, Borrello I.** 2004. High-dose granulocyte-macrophage colony-stimulating factor-producing vaccines impair the immune response through the recruitment of myeloid suppressor cells. *Cancer Res* **64**:6337–6343.

135. **Parmiani G, Castelli C, Pilla L, Santinami M, Colombo MP, Rivoltini L.** 2007. Opposite immune functions of GM-CSF administered as vaccine adjuvant in cancer patients. *Ann Oncol* **18**:226–232.

136. **Trikha M, Corringham R, Klein B, Rossi JF.** 2003. Targeted anti-interleukin-6 monoclonal antibody therapy for cancer: a review of the rationale and clinical evidence. *Clin Cancer Res* **9**:4653–4665.

137. **Terabe M, Matsui S, Noben-Trauth N, Chen H, Watson C, Donaldson DD, Carbone DP, Paul WE, Berzofsky JA.** 2000. NKT cell-mediated repression of tumor immunosurveillance by IL-13 and the IL-4R-STAT6 pathway. *Nat Immunol* **1**:515–520.

138. **Bronte V, Serafini P, De Santo C, Marigo I, Tosello V, Mazzoni A, Segal DM, Staib C, Lowel M, Sutter G, Colombo MP, Zanovello P.** 2003. IL-4-induced arginase 1 suppresses alloreactive T cells in tumor-bearing mice. *J Immunol* **170**:270–278.

139. **Cheng P, Corzo CA, Luetteke N, Yu B, Nagaraj S, Bui MM, Ortiz M, Nacken W, Sorg C, Vogl T, Roth J, Gabrilovich DI.** 2008. Inhibition of dendritic cell differentiation and accumulation of myeloid-derived suppressor cells in cancer is regulated by S100A9 protein. *J Exp Med* **205**:2235–2249.

140. **Parker KH, Sinha P, Horn LA, Clements VK, Yang H, Li J, Tracey KJ, Ostrand-Rosenberg S.** 2014. HMGB1 enhances immune suppression by facilitating the differentiation and suppressive activity of myeloid-derived suppressor cells. *Cancer Res* **74**:5723–5733.

141. **Kim EK, Jeon I, Seo H, Park YJ, Song B, Lee KA, Jang Y, Chung Y, Kang CY.** 2014. Tumor-derived osteopontin suppresses antitumor immunity by promoting extramedullary myelopoiesis. *Cancer Res* **74**:6705–6716.

142. **Sangaletti S, Tripodo C, Sandri S, Torselli I, Vitali C, Ratti C, Botti L, Burocchi A, Porcasi R, Tomirotti A, Colombo MP, Chiodoni C.** 2014. Osteopontin shapes immunosuppression in the metastatic niche. *Cancer Res* **74**:4706–4719.

143. **Yoshimura A.** 2006. Signal transduction of inflammatory cytokines and tumor development. *Cancer Sci* **97**: 439–447.

144. **Kusmartsev S, Gabrilovich DI.** 2005. STAT1 signaling regulates tumor-associated macrophage-mediated T cell deletion. *J Immunol* **174**:4880–4891.

145. **Beatty GL, Paterson Y.** 2000. IFN-γ can promote tumor evasion of the immune system in vivo by down-regulating cellular levels of an endogenous tumor antigen. *J Immunol* **165**:5502–5508.

146. **Lehtonen A, Matikainen S, Miettinen M, Julkunen I.** 2002. Granulocyte-macrophage colony-stimulating factor (GM-CSF)-induced STAT5 activation and target-gene expression during human monocyte/macrophage differentiation. *J Leukoc Biol* **71**:511–519.

147. **Waight JD, Netherby C, Hensen ML, Miller A, Hu Q, Liu S, Bogner PN, Farren MR, Lee KP, Liu K, Abrams**

SI. 2013. Myeloid-derived suppressor cell development is regulated by a STAT/IRF-8 axis. *J Clin Invest* **123:** 4464–4478.

148. **Xin H, Zhang C, Herrmann A, Du Y, Figlin R, Yu H.** 2009. Sunitinib inhibition of Stat3 induces renal cell carcinoma tumor cell apoptosis and reduces immunosuppressive cells. *Cancer Res* **69:**2506–2513.

149. **Cohen PA, Ko JS, Storkus WJ, Spencer CD, Bradley JM, Gorman JE, McCurry DB, Zorro-Manrique S, Dominguez AL, Pathangey LB, Rayman PA, Rini BI, Gendler SJ, Finke JH.** 2012. Myeloid-derived suppressor cells adhere to physiologic STAT3- vs STAT5-dependent hematopoietic programming, establishing diverse tumor-mediated mechanisms of immunologic escape. *Immunol Invest* **41:**680–710.

150. **Ostrand-Rosenberg S, Clements VK, Terabe M, Park JM, Berzofsky JA, Dissanayake SK.** 2002. Resistance to metastatic disease in STAT6-deficient mice requires hemopoietic and nonhemopoietic cells and is IFN-γ dependent. *J Immunol* **169:**5796–5804.

151. **Munera V, Popovic PJ, Bryk J, Pribis J, Caba D, Matta BM, Zenati M, Ochoa JB.** 2010. Stat 6-dependent induction of myeloid derived suppressor cells after physical injury regulates nitric oxide response to endotoxin. *Ann Surg* **251:**120–126.

152. **Corzo CA, Cotter MJ, Cheng P, Cheng F, Kusmartsev S, Sotomayor E, Padhya T, McCaffrey TV, McCaffrey JC, Gabrilovich DI.** 2009. Mechanism regulating reactive oxygen species in tumor-induced myeloid-derived suppressor cells. *J Immunol* **182:**5693–5701.

153. **Sander LE, Sackett SD, Dierssen U, Beraza N, Linke RP, Muller M, Blander JM, Tacke F, Trautwein C.** 2010. Hepatic acute-phase proteins control innate immune responses during infection by promoting myeloid-derived suppressor cell function. *J Exp Med* **207:**1453–1464.

154. **Chalmin F, Ladoire S, Mignot G, Vincent J, Bruchard M, Remy-Martin JP, Boireau W, Rouleau A, Simon B, Lanneau D, De Thonel A, Multhoff G, Hamman A, Martin F, Chauffert B, Solary E, Zitvogel L, Garrido C, Ryffel B, Borg C, Apetoh L, Rébé C, Ghiringhelli F.** 2010. Membrane-associated Hsp72 from tumor-derived exosomes mediates STAT3-dependent immunosuppressive function of mouse and human myeloid-derived suppressor cells. *J Clin Invest* **120:**457–471.

155. **Zhang H, Nguyen-Jackson H, Panopoulos AD, Li HS, Murray PJ, Watowich SS.** 2010. STAT3 controls myeloid progenitor growth during emergency granulopoiesis. *Blood* **116:**2462–2471.

156. **Natsuka S, Akira S, Nishio Y, Hashimoto S, Sugita T, Isshiki H, Kishimoto T.** 1992. Macrophage differentiation-specific expression of NF-IL6, a transcription factor for interleukin-6. *Blood* **79:**460–466.

157. **van Dijk TB, Baltus B, Raaijmakers JA, Lammers JW, Koenderman L, de Groot RP.** 1999. A composite C/EBP binding site is essential for the activity of the promoter of the IL-3/IL-5/granulocyte-macrophage colony-stimulating factor receptor βc gene. *J Immunol* **163:**2674–2680.

158. **Chikka MR, McCabe DD, Tyra HM, Rutkowski DT.** 2013. C/EBP homologous protein (CHOP) contributes to suppression of metabolic genes during endoplasmic reticulum stress in the liver. *J Biol Chem* **288:**4405–4415.

159. **Thevenot PT, Sierra RA, Raber PL, Al-Khami AA, Trillo-Tinoco J, Zarreii P, Ochoa AC, Cui Y, Del Valle L, Rodriguez PC.** 2014. The stress-response sensor chop regulates the function and accumulation of myeloid-derived suppressor cells in tumors. *Immunity* **41:**389–401.

160. **Bunt SK, Clements VK, Hanson EM, Sinha P, Ostrand-Rosenberg S.** 2009. Inflammation enhances myeloid-derived suppressor cell cross-talk by signaling through Toll-like receptor 4. *J Leukoc Biol* **85:**996–1004.

161. **Greifenberg V, Ribechini E, Rössner S, Lutz MB.** 2009. Myeloid-derived suppressor cell activation by combined LPS and IFN-γ treatment impairs DC development. *Eur J Immunol* **39:**2865–2876.

162. **Liu Y, Xiang X, Zhuang X, Zhang S, Liu C, Cheng Z, Michalek S, Grizzle W, Zhang HG.** 2010. Contribution of MyD88 to the tumor exosome-mediated induction of myeloid derived suppressor cells. *Am J Pathol* **176:** 2490–2499.

163. **Capietto AH, Kim S, Sanford DE, Linehan DC, Hikida M, Kumosaki T, Novack DV, Faccio R.** 2013. Down-regulation of PLCγ2–β-catenin pathway promotes activation and expansion of myeloid-derived suppressor cells in cancer. *J Exp Med* **210:**2257–2271.

164. **Pilon-Thomas S, Nelson N, Vohra N, Jerald M, Pendleton L, Szekeres K, Ghansah T.** 2011. Murine pancreatic adenocarcinoma dampens SHIP-1 expression and alters MDSC homeostasis and function. *PLoS One* **6:**e27729. doi:10.1371/journal.pone.0027729.

165. **Guo G, Marrero L, Rodriguez P, Del Valle L, Ochoa A, Cui Y.** 2013. Trp53 inactivation in the tumor microenvironment promotes tumor progression by expanding the immunosuppressive lymphoid-like stromal network. *Cancer Res* **73:**1668–1675.

166. **Li L, Zhang J, Diao W, Wang D, Wei Y, Zhang CY, Zen K.** 2014. MicroRNA-155 and MicroRNA-21 promote the expansion of functional myeloid-derived suppressor cells. *J Immunol* **192:**1034–1043.

167. **Sonda N, Simonato F, Peranzoni E, Calì B, Bortoluzzi S, Bisognin A, Wang E, Marincola FM, Naldini L, Gentner B, Trautwein C, Sackett SD, Zanovello P, Molon B, Bronte V.** 2013. miR-142-3p prevents macrophage differentiation during cancer-induced myelopoiesis. *Immunity* **38:**1236–1249.

168. **Stromnes IM, Brockenbrough JS, Izeradjene K, Carlson MA, Cuevas C, Simmons RM, Greenberg PD, Hingorani SR.** 2014. Targeted depletion of an MDSC subset unmasks pancreatic ductal adenocarcinoma to adaptive immunity. *Gut* **63:**1769–1781.

169. **Vincent J, Mignot G, Chalmin F, Ladoire S, Bruchard M, Chevriaux A, Martin F, Apetoh L, Rebe C, Ghiringhelli F.** 2010. 5-Fluorouracil selectively kills tumor-associated myeloid-derived suppressor cells resulting in enhanced T cell-dependent antitumor immunity. *Cancer Res* **70:** 3052–3061.

170. **Suzuki E, Kapoor V, Jassar AS, Kaiser LR, Albelda SM.** 2005. Gemcitabine selectively eliminates splenic Gr-1$^+$/CD11b$^+$ myeloid suppressor cells in tumor-bearing

animals and enhances antitumor immune activity. *Clin Cancer Res* **11**:6713–6721.

171. **Kodumudi KN, Woan K, Gilvary DL, Sahakian E, Wei S, Djeu JY.** 2010. A novel chemoimmunomodulating property of docetaxel: suppression of myeloid-derived suppressor cells in tumor bearers. *Clin Cancer Res* **16**: 4583–4594.

172. **Alizadeh D, Trad M, Hanke NT, Larmonier CB, Janikashvili N, Bonnotte B, Katsanis E, Larmonier N.** 2014. Doxorubicin eliminates myeloid-derived suppressor cells and enhances the efficacy of adoptive T-cell transfer in breast cancer. *Cancer Res* **74**:104–118.

173. **Bronte V, Chappell DB, Apolloni E, Cabrelle A, Wang M, Hwu P, Restifo NP.** 1999. Unopposed production of granulocyte-macrophage colony-stimulating factor by tumors inhibits CD8$^+$ T cell responses by dysregulating antigen-presenting cell maturation. *J Immunol* **162**: 5728–5737.

174. **Qian BZ, Li J, Zhang H, Kitamura T, Zhang J, Campion LR, Kaiser EA, Snyder LA, Pollard JW.** 2011. CCL2 recruits inflammatory monocytes to facilitate breast-tumour metastasis. *Nature* **475**:222–225.

175. **Ries CH, Cannarile MA, Hoves S, Benz J, Wartha K, Runza V, Rey-Giraud F, Pradel LP, Feuerhake F, Klaman I, Jones T, Jucknischke U, Scheiblich S, Kaluza K, Gorr IH, Walz A, Abiraj K, Cassier PA, Sica A, Gomez-Roca C, de Visser KE, Italiano A, Le Tourneau C, Delord JP, Levitsky H, Blay JY, Rüttinger D.** 2014. Targeting tumor-associated macrophages with anti-CSF-1R antibody reveals a strategy for cancer therapy. *Cancer Cell* **25**:846–859.

176. **Shojaei F, Wu X, Qu X, Kowanetz M, Yu L, Tan M, Meng YG, Ferrara N.** 2009. G-CSF-initiated myeloid cell mobilization and angiogenesis mediate tumor refractoriness to anti-VEGF therapy in mouse models. *Proc Natl Acad Sci U S A* **106**:6742–6747.

177. **Sumida K, Wakita D, Narita Y, Masuko K, Terada S, Watanabe K, Satoh T, Kitamura H, Nishimura T.** 2012. Anti-IL-6 receptor mAb eliminates myeloid-derived suppressor cells and inhibits tumor growth by enhancing T-cell responses. *Eur J Immunol* **42**:2060–2072.

178. **Guiducci C, Vicari AP, Sangaletti S, Trinchieri G, Colombo MP.** 2005. Redirecting *in vivo* elicited tumor infiltrating macrophages and dendritic cells towards tumor rejection. *Cancer Res* **65**:3437–3446.

179. **Weiss JM, Ridnour LA, Back T, Hussain SP, He P, Maciag AE, Keefer LK, Murphy WJ, Harris CC, Wink DA, Wiltrout RH.** 2010. Macrophage-dependent nitric oxide expression regulates tumor cell detachment and metastasis after IL-2/anti-CD40 immunotherapy. *J Exp Med* **207**:2455–2467.

180. **Yang L, Huang J, Ren X, Gorska AE, Chytil A, Aakre M, Carbone DP, Matrisian LM, Richmond A, Lin PC, Moses HL.** 2008. Abrogation of TGFβ signaling in mammary carcinomas recruits Gr-1+CD11b+ myeloid cells that promote metastasis. *Cancer Cell* **13**:23–35.

181. **Nagaraj S, Youn JI, Weber H, Iclozan C, Lu L, Cotter MJ, Meyer C, Becerra CR, Fishman M, Antonia S, Sporn MB, Liby KT, Rawal B, Lee JH, Gabrilovich DI.** 2010. Anti-inflammatory triterpenoid blocks immune suppressive function of MDSCs and improves immune response in cancer. *Clin Cancer Res* **16**:1812–1823.

182. **Ko JS, Rayman P, Ireland J, Swaidani S, Li G, Bunting KD, Rini B, Finke JH, Cohen PA.** 2010. Direct and differential suppression of myeloid-derived suppressor cell subsets by sunitinib is compartmentally constrained. *Cancer Res* **70**:3526–3536.

183. **Ozao-Choy J, Ma G, Kao J, Wang GX, Meseck M, Sung M, Schwartz M, Divino CM, Pan PY, Chen SH.** 2009. The novel role of tyrosine kinase inhibitor in the reversal of immune suppression and modulation of tumor microenvironment for immune-based cancer therapies. *Cancer Res* **69**:2514–2522.

184. **van Cruijsen H, van der Veldt AA, Vroling L, Oosterhoff D, Broxterman HJ, Scheper RJ, Giaccone G, Haanen JB, van den Eertwegh AJ, Boven E, Hoekman K, de Gruijl TD.** 2008. Sunitinib-induced myeloid lineage redistribution in renal cell cancer patients: CD1c$^+$ dendritic cell frequency predicts progression-free survival. *Clin Cancer Res* **14**:5884–5892.

185. **Sinha P, Clements VK, Fulton AM, Ostrand-Rosenberg S.** 2007. Prostaglandin E2 promotes tumor progression by inducing myeloid-derived suppressor cells. *Cancer Res* **67**:4507–4513.

186. **Rodriguez PC, Hernandez CP, Quiceno D, Dubinett SM, Zabaleta J, Ochoa JB, Gilbert J, Ochoa AC.** 2005. Arginase I in myeloid suppressor cells is induced by COX-2 in lung carcinoma. *J Exp Med* **202**:931–939.

187. **Veltman JD, Lambers ME, van Nimwegen M, Hendriks RW, Hoogsteden HC, Aerts JG, Hegmans JP.** 2010. COX-2 inhibition improves immunotherapy and is associated with decreased numbers of myeloid-derived suppressor cells in mesothelioma. Celecoxib influences MDSC function. *BMC Cancer* **10**:464. doi:10.1186/1471-2407-10-464.

188. **Serafini P, Meckel K, Kelso M, Noonan K, Califano J, Koch W, Dolcetti L, Bronte V, Borrello I.** 2006. Phosphodiesterase-5 inhibition augments endogenous antitumor immunity by reducing myeloid-derived suppressor cell function. *J Exp Med* **203**:2691–2702.

189. **Noonan KA, Ghosh N, Rudraraju L, Bui M, Borrello I.** 2014. Targeting immune suppression with PDE5 inhibition in end-stage multiple myeloma. *Cancer Immunol Res* **2**:725–731.

190. **Weed DT, Vella JL, Reis IM, De la Fuente AC, Gomez C, Sargi Z, Nazarian R, Califano J, Borrello I, Serafini P.** 2015. Tadalafil reduces myeloid-derived suppressor cells and regulatory T cells and promotes tumor immunity in patients with head and neck squamous cell carcinoma. *Clin Cancer Res* **21**:39–48.

191. **De Santo C, Serafini P, Marigo I, Dolcetti L, Bolla M, Del Soldato P, Melani C, Guiducci C, Colombo MP, Iezzi M, Musiani P, Zanovello P, Bronte V.** 2005. Nitroaspirin corrects immune dysfunction in tumor-bearing hosts and promotes tumor eradication by cancer vaccination. *Proc Natl Acad Sci U S A* **102**:4185–4190.

192. **Nefedova Y, Fishman M, Sherman S, Wang X, Beg AA, Gabrilovich DI.** 2007. Mechanism of all-*trans* retinoic acid effect on tumor-associated myeloid-derived suppressor cells. *Cancer Res* **67**:11021–11028.

193. **Mirza N, Fishman M, Fricke I, Dunn M, Neuger AM, Frost TJ, Lush RM, Antonia S, Gabrilovich DI.** 2006. All-*trans*-retinoic acid improves differentiation of myeloid cells and immune response in cancer patients. *Cancer Res* **66:**9299–9307.

194. **Garrity T, Pandit R, Wright MA, Benefield J, Keni S, Young MR.** 1997. Increased presence of CD34[+] cells in the peripheral blood of head and neck cancer patients and their differentiation into dendritic cells. *Int J Cancer* **73:**663–669.

195. **Walsh JE, Clark AM, Day TA, Gillespie MB, Young MR.** 2010. Use of α,25-dihydroxyvitamin D_3 treatment to stimulate immune infiltration into head and neck squamous cell carcinoma. *Hum Immunol* **71:**659–665.

196. **Beatty GL, Chiorean EG, Fishman MP, Saboury B, Teitelbaum UR, Sun W, Huhn RD, Song W, Li D, Sharp LL, Torigian DA, O'Dwyer PJ, Vonderheide RH.** 2011. CD40 agonists alter tumor stroma and show efficacy against pancreatic carcinoma in mice and humans. *Science* **331:**1612–1616.

197. **Rolny C, Mazzone M, Tugues S, Laoui D, Johansson I, Coulon C, Squadrito ML, Segura I, Li X, Knevels E, Costa S, Vinckier S, Dresselaer T, Åkerud P, De Mol M, Salomäki H, Phillipson M, Wyns S, Larsson E, Buysschaert I, Botling J, Himmelreich U, Van Ginderachter JA, De Palma M, Dewerchin M, Claesson-Welsh L, Carmeliet P.** 2011. HRG inhibits tumor growth and metastasis by inducing macrophage polarization and vessel normalization through downregulation of PlGF. *Cancer Cell* **19:**31–44.

198. **Veltman JD, Lambers ME, van Nimwegen M, Hendriks RW, Hoogsteden HC, Hegmans JP, Aerts JG.** 2010. Zoledronic acid impairs myeloid differentiation to tumour-associated macrophages in mesothelioma. *Br J Cancer* **103:**629–641.

199. **Brown HK, Holen I.** 2009. Anti-tumour effects of bisphosphonates—what have we learned from in vivo models? *Curr Cancer Drug Targets* **9:**807–823.

200. **Qin H, Lerman B, Sakamaki I, Wei G, Cha SC, Rao SS, Qian J, Hailemichael Y, Nurieva R, Dwyer KC, Roth J, Yi Q, Overwijk WW, Kwak LW.** 2014. Generation of a new therapeutic peptide that depletes myeloid-derived suppressor cells in tumor-bearing mice. *Nat Med* **20:**676–681.

201. **Kotsakis A, Harasymczuk M, Schilling B, Georgoulias V, Argiris A, Whiteside TL.** 2012. Myeloid-derived suppressor cell measurements in fresh and cryopreserved blood samples. *J Immunol Methods* **381:**14–22.

202. **Ko JS, Zea AH, Rini BI, Ireland JL, Elson P, Cohen P, Golshayan A, Rayman PA, Wood L, Garcia J, Dreicer R, Bukowski R, Finke JH.** 2009. Sunitinib mediates reversal of myeloid-derived suppressor cell accumulation in renal cell carcinoma patients. *Clin Cancer Res* **15:**2148–2157.

Myeloid Cells in Health and Disease: A Synthesis
Edited by Siamon Gordon

doi:10.1128/microbiolspec.MCHD-0031-2016

Kipp Weiskopf,[1,2,3,4] Peter J. Schnorr,[2,3,4] Wendy W. Pang,[2,3,4,5]
Mark P. Chao,[2,3,4] Akanksha Chhabra,[5] Jun Seita,[2,3,4]
Mingye Feng,[2,3,4] and Irving L. Weissman[2,3,4]

50

Myeloid Cell Origins, Differentiation, and Clinical Implications

THE HEMATOPOIETIC SYSTEM

The hematopoietic stem cell (HSC) is a multipotent stem cell that resides in the bone marrow and has the ability to form all of the cells of the blood and immune system. As the quintessential stem cell, it has the ability to self-replicate and differentiate into progeny of multiple lineages. Hematopoiesis describes the process of differentiating from HSCs to mature, functional cell types of the blood lineages. The existence of HSCs was first hypothesized following early experiments that demonstrated that animals receiving lethal doses of irradiation could be rescued by transplanting unfractionated bone marrow cells (1). The transplanted cells repopulated the bone marrow of the recipients and gave rise to all the cells of the blood. In accordance with this observation, in 1961 Till and McCulloch showed that unfractionated bone marrow cells were able to generate mixed hematopoietic (myeloid and erythroid) colonies in the spleens of lethally irradiated mice (2). They subsequently demonstrated that these colonies were formed by single cells that were capable of multilineage differentiation (3). Given the limitations in technology at the time, they were unable to purify these cells further, and the experiment that showed clonal origin of spleen colonies did not include lymphoid cells (2, 3), although a later experiment did (4). Years later, with the advent of monoclonal antibodies and fluorescence-activated cell sorting, these cells could be further characterized, purified, and evaluated in functional assays. Studies have now conclusively demonstrated that HSCs are a rare population of cells that give rise to all of the cells comprising the two main branches of the hematopoietic lineage: the myeloid arm and the lymphoid arm. In mice, all long-term HSCs (LT-HSCs) are Hoxb5$^+$ (5) and located in the central marrow attached to the abluminal side of venous sinusoids.

In general, the hematopoietic lineage is organized such that HSCs sit atop the hierarchy and give rise to committed progenitor cells, which in turn give rise

[1]Department of Medicine, Brigham and Women's Hospital, Boston, MA 02115; [2]Institute for Stem Cell Biology and Regenerative Medicine; [3]Ludwig Center for Cancer Stem Cell Research and Medicine; [4]Stanford Cancer Institute; [5]Blood and Marrow Transplantation, Stanford University School of Medicine, Stanford, CA 94305.

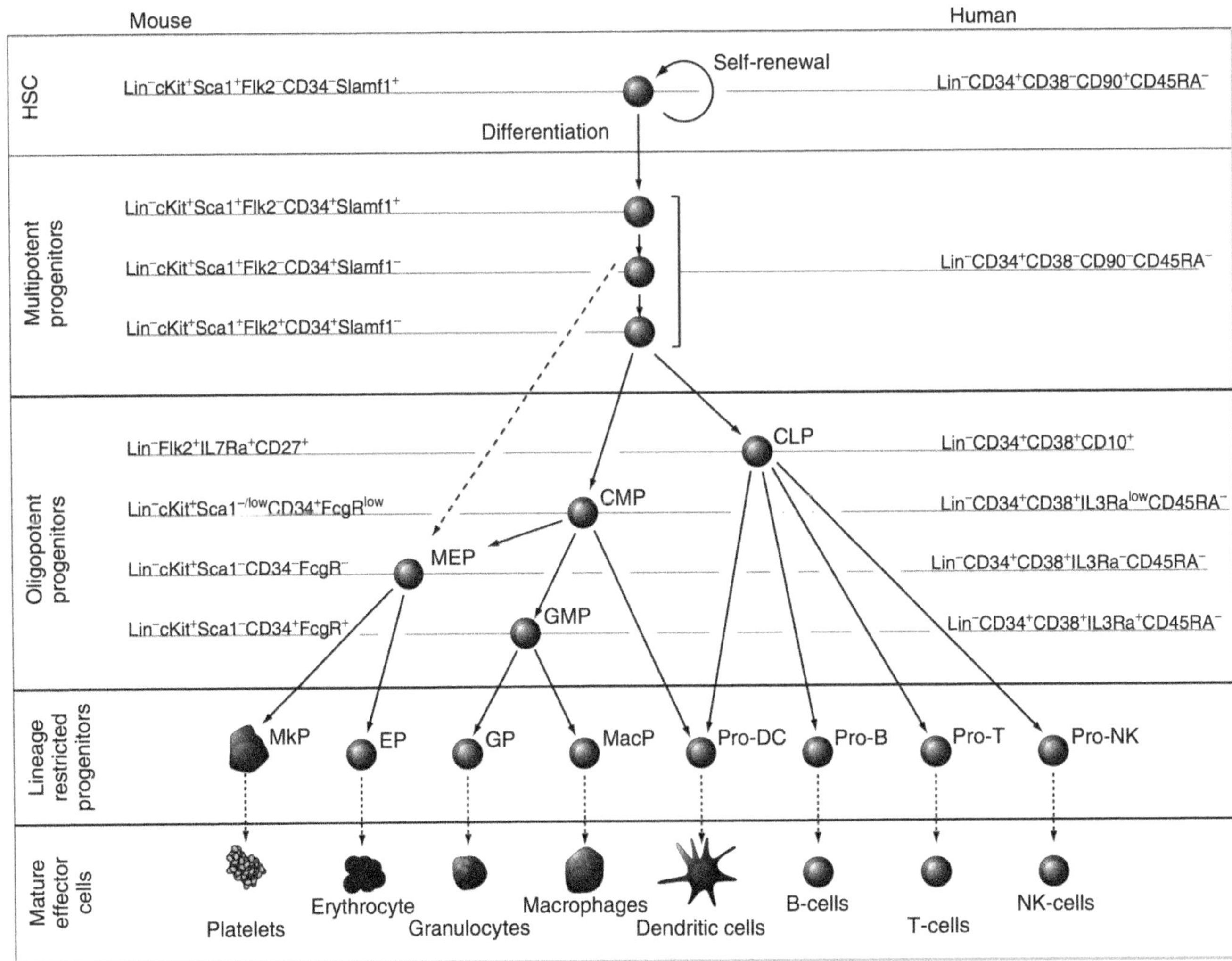

Figure 1 General organization of the hematopoietic lineage in mice and humans. The HSC can give rise to all of the cells of the blood and immune system, with multiple stepwise intermediates arising before developing into fully differentiated cells. The CMP and the CLP give rise to the two mains arms of the hematopoietic hierarchy. The CMP can give rise to all myeloid cells. Conventional surface markers for purifying each population are indicated for both mice and humans. GP, granulocyte progenitor; MacP, macrophage progenitor. Reprinted from reference 117, with permission.

to mature, differentiated cells (Fig. 1). There are two major differences between HSCs and committed progenitors: HSCs are multipotent and they have the ability to self-renew indefinitely. The transition between LT-HSC and short-term HSC (ST-HSC) is prospectively isolatable (6–8), and the poorly self-renewing ST-HSC and other multipotent progenitors (MPPs) are nevertheless fully multipotent at the single-cell level (9). Downstream of MPPs are committed progenitors that are oligopotent and have limited ability to self-renew. As general principles, when cells progress through hematopoiesis, they become more differentiated and more frequent in number. They also lose their capacity to self-renew, become more restricted in their differentiation potential, and gain expression of molecules required for functional specialization. Differentiation occurs in one direction with restriction toward a particular lineage, with no significant evidence of transdifferentiation between hematopoietic lineages under normal conditions. This chapter focuses on the characterization and isolation of the HSC in both mice and humans, as well as insight gained from the study of myelopoiesis, the development and specialization of the myeloid arm of the hematopoietic lineage.

IDENTIFYING THE HSC IN MICE

The HSC was first defined and isolated from mice. In 1988, 27 years after the work of Till and McCulloch, Spangrude et al. (1988) investigated whether distinct progenitors give rise to each cell lineage or whether a single cell could give rise to all of the cells of the hematopoietic lineage (10). Cell-sorting experiments demonstrated that only a Thy-1^{lo}Lin^{-}Sca-1^{+} population of bone marrow was able to form both myeloerythroid cells and lymphoid cells using colony-forming assays (10) (Fig. 2). As a gold standard for evaluation of HSCs, transplantation of Thy-1^{lo}Lin^{-}Sca-1^{+} cells was able to produce all hematopoietic lineages in lethally irradiated mice with ~2,000-fold enrichment over unfractionated whole bone marrow (10). While these cells were found in very small number, representing only 0.1 to 0.2% of the Thy-1^{lo}Lin^{-} cells and ~0.05% of total bone marrow, even mice injected with single Thy-1^{lo}Lin^{-}Sca-1^{+} cells were able to re-form both lymphoid and myeloid lineages (11).

The Spangrude et al. (1988) study provided the earliest definition of the HSC. Subsequent efforts were made to refine the mouse HSC population further by defining additional surface molecules that enriched for stem cell activity. C-kit, a receptor tyrosine kinase that binds stem cell factor, was discovered as another important marker that facilitates isolation of the HSC population (12). This marker would later be incorporated to define the c-Kit^{+}Thy-1^{lo}Lin^{-}Sca-1^{-} population that is often used conventionally as the basis for purifying mouse HSCs. The importance of c-Kit as an HSC marker was also demonstrated by injection of monoclonal antibodies to c-Kit *in vivo*, which caused depletion of HSCs and bone marrow ablation (13, 14). From a translational perspective, this finding supports the use of antibody-based strategies for conditioning patients for stem cell transplant without the need for radiation or chemotherapy.

Following the successful identification of the HSC population, additional studies investigated subsequent stages of hematopoietic differentiation. Morrison and Weissman (1994) further subdivided the HSC population based on low levels of the lineage markers Mac-1 and CD4. The Thy-1.1^{lo}Sca-1^{hi}Lin^{-}Mac-1^{-}CD4^{-}c-Kit^{+} population of bone marrow cells exhibited the greatest potency for LT-HSC activity (15). In secondary transplant experiments, in which the bone marrow of transplanted mice was used to reconstitute a second irradiated recipient, the LT-HSC population was the only population capable of sustaining long-term multilineage reconstitution (15). The LT-HSCs self-replicated and gave rise to ST-HSCs with more limited proliferative potential, establishing a hierarchical relationship between these two populations.

Subsequent investigations further refined the HSC population by introducing novel markers that distinguish it from downstream MPPs. As examples, Flk-2 was found to mark ST-HSCs and MPPs (6, 7), and CD48 was identified on MPPs but not HSCs (8). On the contrary, CD150 was identified as a marker expressed by nearly 50% of cells within the HSC population but <1% of the MPP population (8). The use of these additional markers allowed for the greatest purification of LT-HSC activity, such that 1 out of every 2.1 CD150^{+}CD48^{-}Sca-1^{+}Lin^{-}c-Kit^{+} cells engrafted and gave long-term multilineage reconstitution (8). This extraordinarily enriched population of HSCs represented only 0.0058% of total bone marrow cells (8). Interestingly, new studies using transposon-based clonal markers suggest that the MPP populations may make substantial contributions to hematopoiesis under normal, steady-state conditions without transplantation (16). Most recently it was found that even the most refined HSC population included both ST- and LT-HSCs (5). Pure LT-HSCs in mice have now been identified as HoxB5^{+}, and represent 20 to 50% of the cells in the CD150^{+}CD48^{-}Sca-1^{+}Lin^{-}c-Kit^{+} population. Only HoxB5^{+} cells in the subset are LT-HSCs; the rest are ST-HSCs, as best defined by secondary transplantation (5). More than 90% of these LT-HSCs are attached to the abluminal surface of the small venous endothelium mainly in marrow sinusoids, and few or none are adjacent to the endosteum or arterioles in the marrow (5). These cells represent ~1 in 100,000 to 200,000 bone marrow cells (5). No such marker systems have been found yet for pure human LT-HSCs.

DEFINING COMMITTED PROGENITOR CELLS IN THE MOUSE

With the establishment of the hematopoietic stem cell compartment, subsequent studies aimed to identify additional intermediate stages of development. In particular, researchers aimed to identify the earliest stages of hematopoietic development that would specify fate determination toward the two main branches of the hematopoietic system, the lymphoid or the myeloid arm. As the first demonstration that such a pure population of cells could exist, Kondo et al. (1997) identified a Lin^{-}IL-7Rα^{+}Thy-1^{-}Sca-1^{lo}c-Kit^{lo} population within the bone marrow that exhibited limited self-renewal relative to HSCs and was restricted in its ability to differentiate: it exclusively formed cells of the lymphoid lineage, including B, T, and NK cells (17). This popula-

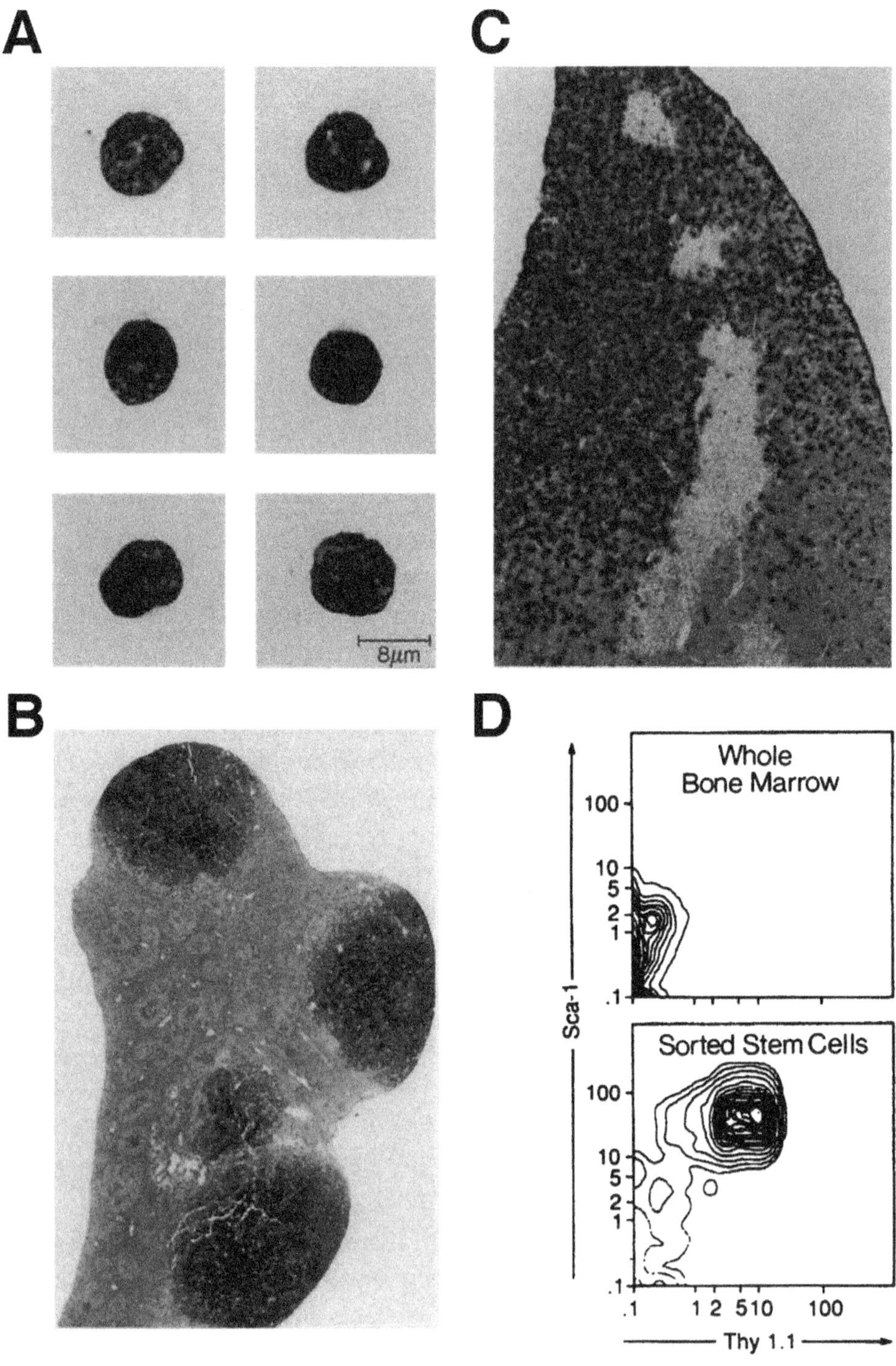

Figure 2 Purification of the first HSCs. (A) Representative examples of purified HSCs as visualized by microscopy after hematoxylin staining. (B) Myeloerythroid colonies in the spleen formed by the injection of purified HSCs into lethally irradiated mice. (C) A single lymphoid colony in the thymus formed by the injection of purified HSCs into lethally irradiated mice. (D) Fluorescence-activated cell sorting depicting HSCs as Thy-1^{lo}Sca-1^{+}. Reprinted from reference 10, with permission.

tion was therefore defined as the common lymphoid progenitor (CLP).

Investigators soon hypothesized that a counterpart to the CLP existed for the myeloid lineage. Akashi et al. (2000) identified and isolated such a cell population, deemed the common myeloid progenitor (CMP). Within an IL-7Rα$^-$Lin$^-$c-Kit$^+$Sca-1$^-$ fraction that contained myeloid progenitor activity, further subpopulations could be identified based on expression of Fcγ receptor-II/III (FcγR) and CD34 (18). The FcγR^{lo}CD34$^+$ population gave rise to the full myeloid lineage, indicating that it contained CMP activity (18). On the other hand, the FcγR^{hi}CD34$^+$ fraction primarily formed macrophages or granulocytes, and the FcγR^{lo}CD34$^-$ cells gave rise exclusively to megakaryocytes and erythroid cells (18). Therefore, these populations represented two mutually exclusive intermediates in myelopoiesis downstream of the CMP, the granulocyte/macrophage lineage-restricted progenitor (GMP), and the megakaryocyte/erythrocyte lineage-restricted progenitor (MEP). Single-cell analyses have revealed that the CMP in mice has at least two distinct subsets (19), and recent transcriptome sequencing (RNA-seq) data confirm that this population is still heterogeneous (20). Nevertheless, a hierarchical relationship was confirmed between these cell types, with CMPs giving rise to both GMPs and MEPs, and the latter cells eventually giving rise to terminally differentiated progeny (18).

IDENTIFYING MURINE MYELOID PROGENITOR CELLS WITH RESTRICTED LINEAGE POTENTIAL

The establishment of the mouse CMPs, GMPs, and MEPs suggested that murine myelopoiesis proceeds in a stepwise, linear fashion. Therefore, efforts were aimed at identifying progenitor cells that lay downstream with more restricted lineage potential, or populations that would give rise exclusively to a single lineage of cells. As an early demonstration that such a population existed, Nakorn et al. (2003) isolated and characterized the megakaryocyte-committed progenitor (MKP) by transplanting a CD9$^+$FcγR^{lo}c-Kit$^+$Sca-1$^-$IL-7Rα$^-$Lin$^-$ population into mice, which formed megakaryocytes in the spleen and platelets in the blood but no other myeloid or lymphoid cells (21). A unipotent erythrocyte-committed progenitor (EP) was also identified in a Lin$^-$c-Kit$^+$Sca-1$^-$IL-7Rα$^-$IL-3Rα$^-$TER119$^-$CD41$^-$CD71$^+$ population. *In vivo* experiments using injections into irradiated mice showed that EPs were able to form reticulocytes but lacked formation of other myeloid cells, providing evidence for their lineage restriction (22). Another unipotent progenitor, the eosinophil lineage-committed progenitor (EoP), was characterized as Lin$^-$Sca$^-$CD34$^+$c-Kit^{lo}IL-5Rα$^+$ cells, which exclusively produced eosinophils *in vitro* (23).

Fogg et al. (2006) identified a bone marrow population that expressed an established mononuclear phagocyte marker, CX_3CR1, as well as c-Kit, but was negative for other lineage markers (24). When transfused into the blood of recipient mice, the CX_3CR1^+c-Kit$^+$Lin$^-$ cells gave rise to monocytes and populations of macrophages and dendritic cells (DCs) in the spleen, identifying these cells as macrophage-DC progenitors (MDPs) (24). Specialized tissue macrophages such as peritoneal macrophages and microglia were also detected as progeny of MDPs (24). Notably, the MDP did not give rise to other cell types of the hematopoietic lineage, and the ability to form neutrophils remained within the CX_3CR1^-c-Kit$^+$Lin$^-$ population (24). Traver et al. (2000) also showed that two major populations of DCs, CD8α$^+$ and CD8α$^-$ cells, both arise downstream of the CMP (25).

Studies using β_7 integrin as a marker suggested a common origin between mast cells and basophils, cell types involved in allergic responses. Within the Lin$^-$c-Kit$^+$Sca-1$^-$ myeloid progenitor population, a β_7^+T1/ST2$^+$ subpopulation was identified that formed exclusively mast cells and was deemed the mast cell progenitor (MCP) (26). A unique Lin$^-$c-Kit$^+\beta_7^{hi}$ population in the spleen gave rise to mast cells and basophils, suggesting that this population represented a basophil-mast cell progenitor (27). In the bone marrow, a Lin$^-$CD34$^+$FcεRIα^{hi}c-Kit$^-$ population gave rise exclusively to basophils, and thus was deemed the basophil progenitor (27). A lineage hierarchy existed between these cell populations when differentiated *in vitro*, with basophil-mast cell progenitors capable of giving rise to either basophil progenitors or MCPs, although some evidence also suggests that the MCPs could arise directly from MPPs (26, 27). Recently Chen and colleagues have found that a subset of HSCs are also integrin β_7^+ and differ in their functional properties, and so the potential heterogeneity in the HSC (long-term and short-term) subsets and their eventual fates remain to be elucidated (28).

To refine other intermediate stages of myeloid development, Pronk et al. (2007) identified cell populations that represented distinct stages of development between CMPs and downstream progenitors (29). Functional studies established new populations such as pre-MEPs and pre-GMPs, which had greater proliferative potential than their respective downstream progeny: MEPs and GMPs, respectively (29). These findings helped demonstrate the multiple stepwise fashion of hematopoiesis, in which cells, as they proceed down the hematopoietic lineage and become more limited in their differentiation potential, also lose proliferative potential.

DEFINING THE HEMATOPOIETIC LINEAGE IN HUMANS

Based on the successful purification of the murine HSC, Baum et al. (1992) sought to prospectively isolate the human HSC. They identified a human fetal bone marrow Lin$^-$Thy-1$^+$CD34$^+$ population that contained multipotent hematopoietic stem cell activity (30). These candidate HSCs were highly enriched for the ability to differentiate into myeloid and lymphoid cells *in vitro* using long-term colony-forming assays and *in vivo* using immunocompromised mouse models. Hematopoietic cell transplants with purified CD34$^+$Thy-1$^-$ HSCs have been successfully performed in patients, providing clinical evidence that the human HSC activity resides in this population (31–33).

Other studies identified markers that further enriched for human HSC activity to refine the HSC population. In particular, CD38 was validated as a marker following xenotransplantation of human hematopoietic cells into immunocompromised mice, with HSC activity residing within the Lin$^-$CD34$^+$CD38$^-$ fraction (34). The full subset of CD34$^+$CD90$^+$CD38$^-$Lin$^-$ was shown by Uchida and colleagues to contain human HSCs, and these could be genetically modified with lentiviral but not retroviral vectors (35). Majeti et al. (2007) prospectively isolated human MPPs that lie downstream of HSCs. Using CD90 (Thy-1) and CD45RA as markers, they subdivided the Lin$^-$CD34$^+$CD38$^{-/lo}$ population into CD90$^+$CD45RA$^-$ and CD90$^-$CD45RA$^-$ subpopulations. Both gave rise to myeloid and lymphoid cells, but the CD90$^+$CD45RA$^-$ population exhibited 9-fold-greater human chimerism in the bone marrow (36). Thus, the CD90$^+$CD45RA$^-$ population contains the true LT-HSC in humans, while the CD90$^-$CD45RA$^-$ population represents an intermediate MPP.

HUMAN COMMITTED PROGENITORS AND RESTRICTED MYELOID DIFFERENTIATION

The first functional demonstration of a lineage-restricted human progenitor population was achieved through studies examining lymphoid development. Isolation of Lin$^-$CD34$^+$CD45RA$^+$CD10$^+$ cells yielded a human bone marrow subpopulation that was devoid of myeloid potential when transplanted into immunocompromised mice (37). However, these cells were capable of forming all of the cell types of the lymphoid arm, including T and B cells, NK cells, and lymphoid DCs, with bias to the B-cell lineage. Therefore, this oligopotent progenitor contains human CLP activity. Given this evidence, Manz et al. (2002) reasoned that counterparts to the mouse CMP, GMP, and MEP existed and sought to define and demonstrate their lineage potential by prospective isolation. The authors found that the Lin$^-$CD34$^+$CD38$^+$ population of human bone marrow has limited ability to self-renew and exhibits a high proportion of myeloid-biased differentiation (38). Expression of CD45RA and IL-3Rα (interleukin-3 receptor α) further subdivided this population, yielding three distinct subpopulations: IL-3Rα^{lo}CD45RA$^-$, IL-3Rα^{lo}CD45RA$^+$, and IL-3Rα^-CD45RA$^-$ cells. *In vitro*, the IL-3Rα^{lo}CD45RA$^-$ population gave rise to the full range of the myeloid lineage, including mixed colonies, suggesting that this population represented the human CMP (38). On the other hand, the IL-3Rα^{lo}CD45RA$^+$ population only gave rise to cells of the granulocyte and macrophage lineages, and the IL-3Rα^-CD45RA$^-$ population predominantly gave rise to cells of the erythroid and megakaryocyte lineage, thereby indicating that these populations represented the GMP and MEP, respectively (38). The same lineage hierarchy existed as with mouse myelopoiesis, with CMPs giving rise to GMPs or MEPs. Additional markers, such as the growth factor receptor TpoR, have been suggested to offer further refinement of these myeloid progenitor populations (39).

Additional studies examined unipotent committed progenitors in the human hematopoietic lineage. Some evidence suggests that a unipotent human basophil/mast cell progenitor exists within a subpopulation of CD34$^+$ cells that are recognized by an antibody to ectonucleotide pyrophosphatase/phosphodiesterase-3, or CD203c, although a unipotent population was not purified (40, 41). Mast cell potential was also identified within a CD34$^+$c-Kit$^+$CD13$^+$ fraction (42). Mori et al. (2009) identified the human EoP that lay downstream of CMP and gave rise only to eosinophils. This population was defined as IL-3Rα^{lo}CD45RA$^-$IL-5Rα^+, which also helped refine the human CMP population (38, 43). Although mature basophils also express IL-5Rα, their development was separate from the EoP and seemed to occur via CMP and GMP intermediates (43). Within the human MEP population, fractionation studies helped define the unipotent human EP as CD71$^{intermediate(int)/+}$CD105$^+$, and, when sorted to purity, gave rise exclusively to erythrocytes *in vitro* with no megakaryocyte potential (44). Additionally, an erythrocyte-biased MEP (E-MEP) was identified as $CD71^+CD105^-$ that was an intermediate between the MEP and the EP (44). Downstream stages of human erythropoiesis have also been isolated to purity, including the BFU-E and CFU-E. These populations were principally distinguished as $IL\text{-}3R^-CD34^+CD36^-$ and $IL\text{-}3R^-CD34^-CD36^+$, respectively (45). Remaining questions surrounding intermediates in human myelopoiesis include definitive isolation and characterization of additional unipotential lineages, such as a neutrophil lineage-restricted progenitor, MDP, and MKP.

COMPREHENSIVE GENE EXPRESSION PROFILING OF THE HEMATOPOIETIC SYSTEM

To better understand the complex regulation of the hematopoietic hierarchy, a robust systems approach is needed for each molecular layer (e.g., gene expression). Microarrays and RNA sequencing (RNA-seq) have become the predominant technologies for genome-wide gene expression profiling, yet they face challenges in their analysis. First, the efficiency of detection depends on a target sequence; thus each probe set on a microarray has a different sensitivity. This limits the accuracy of profiling to relative differences in gene expression based on pairwise comparisons. Many genes are scored as "not significantly different" regardless of the expression levels, and therefore the result is comparison-pair specific. Thus, conventional analysis depicts a profile of differences between cell types A and B instead of an overall profile of gene expression in cell type A. The second challenge pertains to the dynamic range of gene expression, which is different for every gene. Thus, even in microarray-based pairwise comparisons or in RNA-seq, we may misinterpret the significance when comparing differences of expression between genes. For example, if gene X has a narrow dynamic range and gene Y has a wide dynamic range, a 2-fold change in expression may be significant for gene X but not for gene Y.

To address these limitations, Seita et al. (2012) hypothesized that by examining expression of a particular gene across thousands of samples, a common reference for that gene could be established and applied to any new sample (46). Computer simulations demonstrated that if a random collection of >5,000 microarray data sets could be accumulated, meta-analysis could be performed to compute robust statistical attributes for each gene, such as the dynamic range of each gene's expression levels and a threshold between active and inactive gene expression. Subsequently, by mapping new sample data against the results of the meta-analysis, a more objective genome-wide gene expression profile could be achieved for each sample individually. Based on this result, the entire microarray data collection available through the NIH Gene Expression Omnibus (GEO) public repository was subjected to meta-analysis to obtain a dynamic range for each gene's expression level. This analysis was incorporated into a Web-based platform named the Gene Expression Commons (https://gexc.stanford.edu). Gene Expression Commons is an open platform; thus, any scientist can explore expression activity of any gene of interest and submit and analyze his or her own data. We have contributed microarray data of highly purified mouse and human hematopoietic stem and progenitor populations. The complete mouse adult hematopoiesis system containing 39 distinct populations and a partial collection of human normal and neoplastic hematopoiesis are available to the public on the Gene Expression Commons platform. This platform, designed with an intuitive interface, facilitates the identification and visualization of gene expression patterns that differ between hematopoietic populations (Fig. 3).

ORIGINS OF THE HEMATOPOIETIC SYSTEM IN THE EMBRYO-TO-FETAL TRANSITION

Based on studies by Moore and Owen using avian models, Moore and Metcalf (1970) found mouse hematopoietic cells capable of forming mixed-lineage myeloerythroid colonies *in vitro* in the embryonic day 8 (E8) mouse yolk sac, the first site of appearance of blood cells (47). Weissman et al. transplanted yolk sac blood island cells from E8 to E9 mice into the yolk sac cavity *in vivo* of the same-aged (E8 to E9) haploidentical hosts (48, 49). All such mice, when sacrificed as young adults, possessed both donor and host day 10 CFU-S, and also T cells, demonstrating HSC activity within this yolk sac population (48, 49). Medvinsky and Dzierzak challenged this finding when they cultured E10 to E12 yolk sac cells or aorta-gonad-mesonephros (AGM) dorsal aorta tissue *in vitro*, and showed higher hematopoietic activity in the AGM dorsal aorta (50). Many groups used morphological analysis to demonstrate cell clusters in the E10 ventral surface and subendothelial tissue of the dorsal aorta at those time points, consistent with an independent origin of later fetal hematopoiesis in the embryo rather than the extraembryonic yolk sacs. Keller and colleagues showed that mouse embryonic stem (ES) cells could differentiate into hematopoietic cells but also appeared to make endothelial cells in the colonies as well; they called

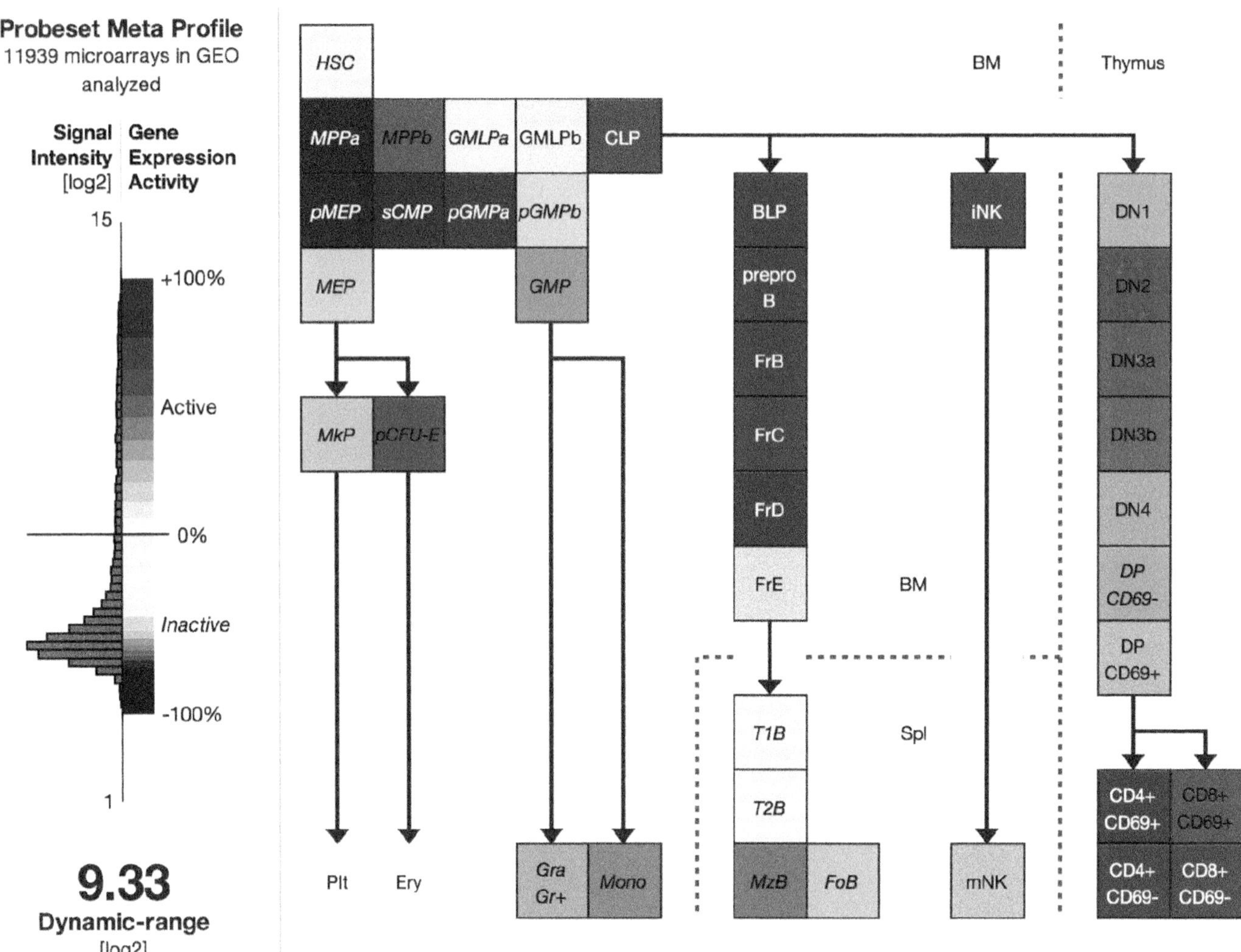

Figure 3 RNA expression pattern of IL-7Rα throughout the murine hematopoietic lineage. IL-7Rα is a critical surface molecule that helps distinguish the lymphoid arm of the hematopoietic system from HSCs, progenitors, and myeloid cells. Each box represents a different hematopoietic subpopulation. Blue indicates lower expression; pink indicates higher expression. Analysis performed using Gene Expression Commons (46). BM, bone marrow; Spl, spleen; GMLP, granulocyte/macrophage/lymphoid progenitor subset; p, pre-; s, strict; Plt, platelet; Ery, erythrocyte; Gra, granulocyte; Mono, monocyte; BLP, earliest B-lymphoid progenitor; Fr, B cell subset fraction; T1B, T1 B cell; T2B, T2 B cell; MzB, marginal zone B cell; FoB, follicular B cell; iNK, intermediate NK cell; DN, double negative T cell subset; DP, double positive T cell subset. $CD4^+$ and $CD8^+$ populations represent mature T cells.

the common precursor the hemangioblast (51). Many groups studying birds, zebrafish, and even *in vitro* dorsal aortas showed images of cells "budding" from the aortic endothelium into the blood circulation and dubbed the cells hemogenic endothelium. These cells have been proposed to be a common, clonal precursor to blood and blood vessel cells (for a review, see reference 52). However, the evidence for a required common, clonal precursor for blood and blood vessel cells in the E7.5 to E9 mouse embryo yolk sac, which appears at least 2 days before the AGM dorsal aorta has blood cells, was challenged by the analyses of tetrachimeric mice. These mice were generated by the injection of 5 red, 5 green, and 5 blue fluorescently labeled ES cells into uncolored mouse blastocysts, followed by sacrifice and analyses of the E7.5 to E9 yolk sac blood islands (53). Each blood island was generated by multiple ES donor cells. If there was a required clonal origin for blood and blood vessel cells, the colors of endothelium surrounding and blood cells within discrete blood islands should always match; however, they rarely did (53). The same result came from injecting a single-colored ES cell into uncolored mouse blastocysts; the yolk sac blood islands of these mice rarely contained blood and blood vessel cells derived from the single ES cell. This was also confirmed with the use of Flk1-cre mice crossed to our rainbow construct of three colors, with random clonal marking of each cell; again there was little or no concordance between blood and blood vessels (53). These results are consistent with the origin of blood and blood vessel cells from a common hemangioblast or hemogenic endothelium only if the cell had appeared prior to E7 (53). Lineage tracing of hematopoietic progenitors from yolk sac in *Runx1 MER-Cre-MER* mice with floxed reporters revealed that the great majority of adult blood cells of myeloerythroid and lymphoid lineages were derived from E7 to E8 mice (54). Later studies performed quantitative assessment of yolk sac versus AGM HSCs phenotypically separated and cultured *in vitro*. At all time points from E10 onward, the number of phenotypic and transplantable HSCs in the yolk sac far exceeded that in the dorsal aorta, and the appearance of large numbers of HSCs in the fetal liver could not be supported by the numbers of dorsal aorta HSCs only (55). Thus, the field has contradictory findings from these studies.

However, Merad, Geissmann, and others have demonstrated that the yolk sac blood islands are the origin of long-term resident macrophages, DCs, and microglia, distinct from the monocyte origin of such cells in postnatal mice (56–58). With the division of resident, self-renewing macrophage/DC-lineage cells and adult bone marrow HSC-derived cells of these lineages now quite clear, the clonal origin of hematopoiesis in the embryo, fetus, and postnate needs to be readdressed with much better experimental tools.

CHANGES IN MYELOPOIESIS WITH AGING

A number of studies have demonstrated a decrease in the function of the innate and adaptive immune system with age. Since HSCs give rise to the entire repertoire of blood cells, aging of HSCs directly impacts downstream myelopoiesis and lymphopoiesis. Although the number of phenotypic HSCs increases with age, the regenerative capacity of these cells decreases. The balance between lymphoid and myeloid differentiation of HSCs also changes over time, resulting in a bias toward myeloid differentiation in both mice and humans with aging (59–61). Studies have suggested that the HSC population is heterogeneous in its differentiation potential, and clonal expansion of certain subsets may contribute to changes in propensity for myeloid versus lymphoid development with increasing age (62–65). Gene expression changes and altered epigenetic landscapes have been demonstrated to contribute to these effects (60, 66, 67). Additionally, the bone marrow microenvironment can modify the degree of myeloid bias. In one study, a decrease in myeloid differentiation was observed when HSCs from aged mice were transplanted into young recipients relative to aged recipients (68). This difference was partially attributed to increases in cytokines such as RANTES within the bone marrow microenvironment of aged animals, which promotes myeloid skewing (68).

These findings have implications for HSC transplantation since the ideal selection of allogeneic donors would ensure balanced reconstitution of lymphoid and myeloid lineages in both young and aged recipients. Recent evidence indicates the small Rho GTPase Cdc42, which maintains polarity of HSCs, demonstrates an age-associated increase in expression and thereby promotes functional changes in HSCs (69). Pharmacological inhibition of Cdc42 via a novel molecule, CASIN, was sufficient to reverse the age-related changes by increasing histone H4 lysine 16 acetylation (AcH4K16) (69). This reversal of age-related changes provides a potential clinical application for HSC transplantation of autologous or aged hematopoietic cells into aged recipients.

The existence of multiple HSC populations that differ in their lineage bias may have additional implications for health and disease. Having multiple, competing HSCs certainly allows the host to be able to produce more naive T and B cells early in life and more myeloid cells later in life. This enables more appropriate generation of adaptive versus innate immune cells at different ages. But why lose lymphopoiesis later in life? We propose that another class of unipotent stem cells are present in the lymphocyte lineages—T stem/memory and B stem/memory cells. In the absence of stimulation of their antigen receptors, these cells are resting and express homing receptors that allow them to patrol lymphoid organs for the appearance of antigens, just as naive T and B cells express CD62L or integrin $\alpha_4\beta_7$ for homing to lymph nodes or Peyer's patches, respectively (for a review, see reference 70). Upon recognizing their cognate antigens, these lymphoid stem/memory cells proliferate and self-renew to create more stem/memory cells and also differentiate to effector cells. Thousands of years ago, before trains, planes, and cars, vertebrates largely were geographically limited. Thus, by a decade or two of age, individuals had encountered most of the microbial pathogens in their environments, and by that time they had developed the groups of T and B stem/memory cells required to protect them. This likely fed back to the HSC pool, shifting gradually from lymphocyte-biased to myeloid-biased stem cells. Unfortunately, our rate of evolution could not keep pace with advances in modern transportation and exposure to pathogens from around the world. Therefore, our efficient system of shifting hematopoietic biases with age has likely enabled novel pathogens to cause pandemics in the very old and very young.

ABERRANT MYELOPOIESIS AS A CAUSE OF HEMATOLOGIC DISEASE

By more rigorous characterization of the hematopoietic lineage in both mice and humans, we gain an understanding of the mechanisms that promote normal development and how to intervene when these processes contribute to disease. Myelopoiesis is perturbed in many blood disorders and diseases, including chronic myeloid leukemia (CML) and acute myeloid leukemia (AML), myelodysplastic syndrome (MDS), myeloproliferative disorders, iron or other nutrient deficiency anemias, as well as a number of inherited bone marrow failure conditions. One potential therapeutic strategy to treat these conditions is to manipulate or reverse the abnormal myelopoiesis. Understanding normal myeloid development as well as the specific aberrant defects in myeloid development found in these diseases has been and will continue to be critical in how we treat patients.

For example, in the chronic phase of CML, there is aberrant overproduction of mature myeloid cells caused by the *BCR-ABL* translocation (71). When CML advances to blast crisis phase, acquisition of additional mutations, such as in glycogen synthase kinase 3β (GSK3β), leads to the activation of β-catenin and blocks myeloid maturation at the level of the GMP (72, 73). The resulting accumulation of these progenitors, or blasts, in the bone marrow and blood severely impairs normal hematopoiesis. There are currently several highly effective tyrosine kinase inhibitors that target BCR-ABL available to treat CML in chronic phase, and these medications have been very successful in controlling this phase of disease long-term as well as reducing the frequency of transformation to blast crisis CML (74, 75). However, blast crisis CML remains a therapeutic challenge today. Targeting GSK3β or deactivating the β-catenin pathway may cause blast crisis CML to differentiate to a milder form of disease.

In acute myeloid leukemia (AML), there is a rapid, abnormal accumulation of immature myeloid progenitors in the bone marrow and blood. In one particular subtype of AML, acute promyelocytic leukemia, which is most commonly associated with a *PML-RARα* translocation, there is a differentiation block at the promyelocyte stage of myelopoiesis (76). This differentiation block can be overcome by the administration of all-*trans*-retinoic acid (ATRA), a medication that has transformed a difficult-to-treat leukemia with extremely poor prognosis into a curable disease without the need for traditional toxic chemotherapies (76). When ATRA induces differentiation of the abnormal promyelocytes, the differentiated leukemia cells exit the bone marrow and normal hematopoiesis is able to recover. Further understanding of the mechanisms behind impaired myeloid differentiation in other subtypes of AML could be important to the development of similar therapeutic strategies for these diseases.

Myelodysplastic syndrome (MDS) is an age-associated condition characterized by ineffective production of one or more mature myeloid, erythroid, or megakaryocytic lineages. We have shown that the early mutations associated with this disease lead to HSCs that outcompete normal HSCs, presumably at the proposed HSC niches (5, 77). Thus, by the time MDS is detected, >95% of the phenotypic HSCs clonally develop from a single mutated HSC. Unfortunately, this clone cannot make blood efficiently. In MDS, there is a specific and significant reduction in the number of GMPs in the bone marrow of low-risk, or early-stage, patients (77) (Fig. 4). Although there is increased apoptosis of MDS GMPs compared to normal GMPs, nonapoptotic MDS GMPs also have increased cell surface calreticulin (CRT), which targets them for programmed cell removal via phagocytosis (see below) (77). Indeed, blocking CRT can prevent the phagocytosis of MDS GMPs (77). It is possible that small molecules or other therapeutics that can inhibit phagocytosis and programmed cell removal of MDS GMPs may be able to ameliorate the often debilitating cytopenia that affects MDS patients. In high-risk MDS, oftentimes arising from low-risk MDS, the GMPs express cell surface CRT and acquire cell surface expression of CD47, which prevents phagocytosis (77). This leads to the abnormal accumulation of these progenitors in the bone marrow, which may eventually transform into AML.

In addition to BCR-ABL inhibitors and ATRA, there are other examples of our ability to successfully alter abnormal myeloid development with targeted therapies in order to treat hematopoietic disorders and diseases. Perhaps the simplest examples of this concept are iron

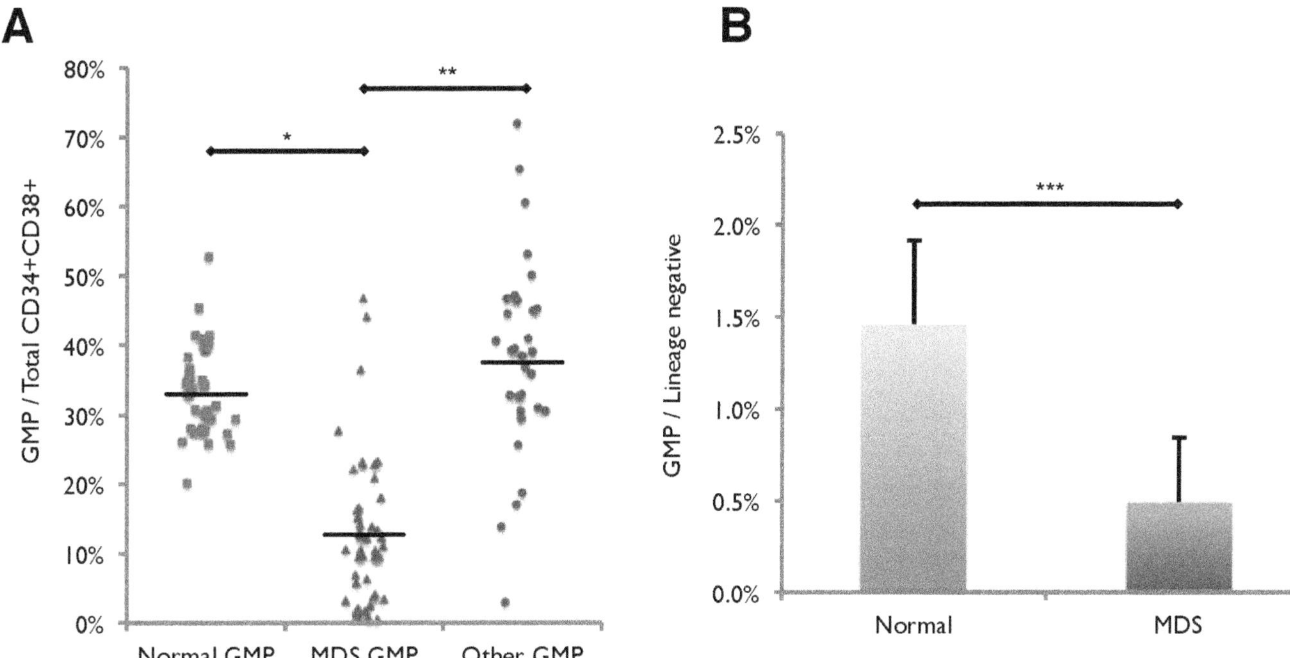

Figure 4 GMP frequency is decreased in low-risk MDS. Representative example of how hematopoietic progenitor cell populations can be altered in states of disease. (A) Frequency of GMPs out of total myeloid progenitors in normal, low-risk MDS and non-MDS diseased bone marrow samples. (B) Frequency of GMPs out of total lineage-negative bone marrow mononuclear cells in normal and low-risk MDS bone marrow samples. Asterisks indicate statistically significant differences: $*P < 10–13$, $**P < 10–10$, $***P < 0.0006$. Reprinted from reference 77, with permission.

replacement therapy for iron deficiency anemia and nutrient replacement therapy for other nutrient deficiency anemias (vitamin B12, folate, and copper). In these cases, supplementation of the deficient nutrient directly alleviates the impaired proliferation of myeloid progenitors and improves the production of mature blood cells. In polycythemia vera and essential thrombocythemia, which are myeloproliferative disorders characterized by overproduction of mature red blood cells and platelets, respectively, acquisition of a mutation in *JAK2* causes proliferation of myeloid progenitors in their respective lineages (78, 79). JAK2 inhibitors have been successfully used to treat patients with these conditions by inhibiting the proliferation of myeloid progenitors (80, 81).

Currently, there has been development of targeted therapies for only a small number of diseases that take advantage of manipulating abnormal myelopoiesis. However, the success of these agents should only inspire us to continue exploring this therapeutic strategy for other myeloid disorders, including AML, MDS, myelofibrosis, inherited bone marrow failure syndromes, and unexplained anemia of the elderly, among many other illnesses. For each of these conditions, detailed characterization of the aberrant stage in myeloid development may be crucial for the design of new targeted therapies.

CD47 AND IMMUNE SURVEILLANCE DURING NORMAL AND MALIGNANT HEMATOPOIESIS

From the study of normal hematopoiesis, we have also gained an understanding of fundamental principles that apply to other aspects of human disease. One example pertains to the CD47/SIRPα (signal regulatory protein α) axis, which is emerging as a widespread immunotherapeutic target for many cancers. CD47 is a cell surface molecule that is expressed on normal hematologic and nonhematologic tissues throughout the body, and it acts as a "marker of self" that limits macrophage phagocytosis. Oldenborg et al. (2000) provided an early functional demonstration of this concept when they transfused red blood cells from $CD47^{-/-}$ mice into wild-type mice (82). The authors observed rapid elimination of the $CD47^{-/-}$ cells from circulation by macrophages in the spleen. SIRPα, a single-pass transmembrane protein with intracellular immunoreceptor tyrosine-based inhibitory motifs, is the receptor by which CD47 transduces inhibitory signals. SIRPα is expressed on cells of the myeloid lineage, including macrophages, monocytes, granulocytes, and myeloid-derived DCs (83–87).

A role for the CD47/SIRPα axis in cancer immunoevasion was further surmised from studies on HSCs. As with red blood cells, CD47 was found to function in stem cell migration by protecting circulating HSCs from host macrophage phagocytosis (88). Mobilization of mouse HSCs from the bone marrow to the periphery led to a transient increase in CD47 expression (88). In addition, CD47 is required for *in vivo* HSC function, as CD47-deficient mouse HSCs failed to engraft in wild-type recipients, in contrast to wild-type HSC controls (88). Preservation of the CD47/SIRPα interaction was also identified as a major factor that permits HSC engraftment across xenogeneic barriers (89, 90). From a clinical perspective, downregulation of CD47 on HSCs is a possible pathogenic mechanism of hemophagocytic lymphohistiocytosis, a life-threatening condition characterized by unrestrained macrophage phagocytosis of hematopoietic cells (91).

Given its role in protecting HSCs from host macrophages during cell migration, it was hypothesized that leukemic cells, and specifically leukemic stem cells, could coopt antiphagocytic CD47 function to enable increased dissemination and immune evasion as a mechanism of tumorigenicity. In AML patients, CD47 cell surface expression was significantly increased on both leukemic blasts and leukemic stem cells compared to normal cell counterparts (92). Importantly, forced overexpression of CD47 in a human leukemia cell line enabled engraftment of fulminant leukemia *in vivo* through specific evasion of macrophage phagocytosis (88). In addition, increased CD47 expression in AML patients predicted a worse overall survival, which was an independent prognostic factor (92). Given CD47's mechanism of immune evasion in AML, it was hypothesized that blockade of the antiphagocytic CD47/SIRPα signaling axis could enable leukemia cells to become phagocytosed and eliminated. Indeed, a monoclonal antibody blocking CD47/SIRPα signaling led to robust phagocytosis of human AML cells but not normal hematopoietic cell counterparts (92). An anti-CD47 antibody rapidly eliminated leukemic disease in the peripheral blood and bone marrow in patient-derived xenografts (Fig. 5). This therapeutic effect was dependent on macrophage effector cells, as selective depletion of macrophages by liposomal clodronate completely abrogated the antileukemic effect of anti-CD47 antibodies. Furthermore, antileukemic activity was observed independent of cytogenetic and morphologic AML subtypes, suggesting that the antiphagocytic function of CD47 is a universal mechanism across

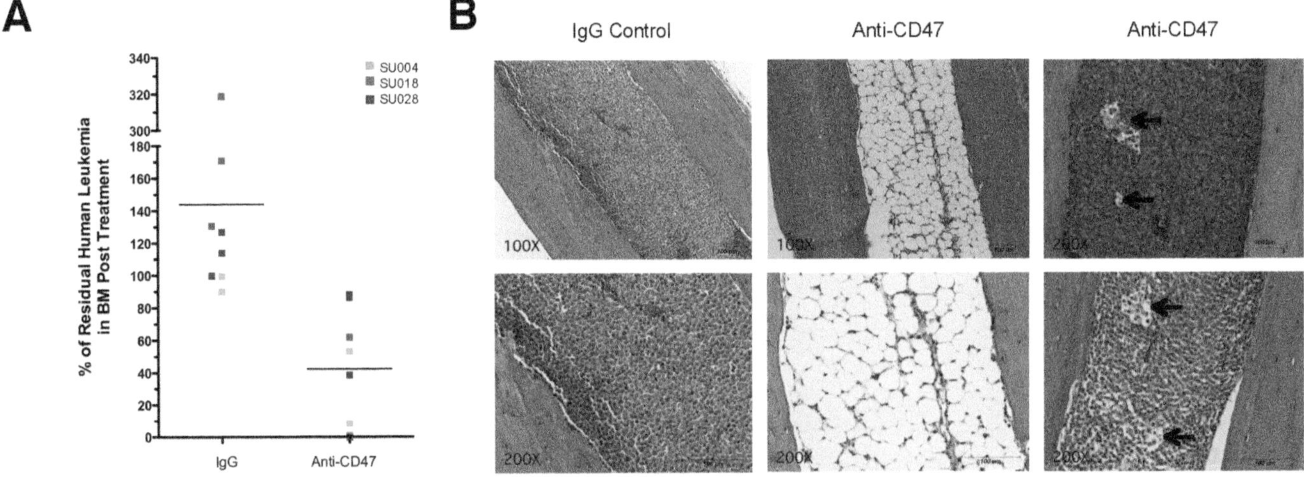

Figure 5 CD47-blocking therapies are effective in preclinical models of human cancer. Xenograft studies of mice engrafted with human AML samples that were then treated with anti-CD47 antibodies. (A) Anti-CD47 antibody treatment decreases leukemia burden, as assessed by the percent of human chimerism in the bone marrow (BM) after 14 days of treatment. (B) Bone marrow histology showing leukemia infiltration in control mice (top left, bottom left), and eradication of disease in mice treated with anti-CD47 antibodies (top middle, bottom middle). In some mice with residual tumor burden following treatment with anti-CD47 antibodies (top right, bottom right), macrophages could be seen in the bone marrow engulfing leukemia cells (black arrows). Reprinted from reference 92, with permission.

AML. Based on these preclinical data, humanized anti-CD47 antibodies have entered clinical development (93) and first-in-disease clinical trials in AML patients have begun (https://clinicaltrials.gov/ identifiers NCT02678338 and NCT02641002).

Since CD47 is normally expressed on tissues throughout the body, subsequent investigations examined whether the CD47/SIRPα axis also promoted immunoevasion in nonhematologic malignancies. Many solid tumors were found to express high levels of CD47 relative to corresponding normal tissues (94, 95). Furthermore, CD47 expression levels correlate with poor prognosis for many types of solid tumors, including ovarian cancer, gliomas, and non-small-cell lung cancer (94, 95), supporting the role of CD47 in immunoevasion. CD47-blocking therapies exhibited antitumor efficacy in a diverse range of preclinical xenograft mouse models, including ovarian, breast, colon, and bladder cancers; glioblastoma; leiomyosarcoma; pancreatic neuroendocrine tumors; small-cell lung cancer; and melanoma (94, 96–99). Furthermore, therapeutic efficacy has been observed in multiple syngeneic immunocompetent models using anti-mouse CD47 antibodies, including breast cancer, lymphoma, melanoma, and neuroendocrine tumors (94, 97, 100). Based on these data, anti-CD47 antibodies are also undergoing evaluation in clinical trials for solid tumors (https://clinicaltrials.gov/ identifiers NCT02216409 and NCT02367196).

Macrophages play a critical role as effector cells in the immune response to anticancer antibodies by performing antibody-dependent phagocytosis (101). This process is also inhibited by the CD47/SIRPα axis (102). In many cases, blockade of CD47 alone does not induce activation, but instead lowers the threshold for macrophage phagocytosis in response to an opsonizing antibody (103, 104). This principle was demonstrated using high-affinity SIRPα variants, engineered variants of the N-terminal domain of SIRPα that act as competitive antagonists to CD47 (103). When produced as high-affinity SIRPα-Fc fusion proteins, these proteins block CD47 and provide an opsonizing stimulus that induces phagocytosis, thereby yielding anticancer efficacy as a single agent (97, 100, 103). However, when used as single-domain high-affinity SIRPα "monomers," these proteins do not induce phagocytosis alone but synergize with anticancer antibodies that engage Fc receptors on macrophages (98, 103, 104). In this sense, the high-affinity SIRPα monomers may limit toxicity to normal cells expressing CD47 while stimulating robust antitumor responses by macrophages. Therapies targeting the CD47/SIRPα axis show much promise in combination with agents such as rituximab for $CD20^+$ lymphoma, trastuzumab for $HER2^+$ breast cancer, and cetuximab for epidermal growth factor receptor-positive ($EGFR^+$) solid tumors (103, 105). Blockade of CD47 can also stimulate macrophage antigen presentation after phagocytosis of cancer cells (106). This mechanism links the innate immune response to adaptive immunity. Moreover, CD47 may also enhance neutrophil antitumor responses as well as DC antigen presentation (86, 100). Thus, the CD47/SIRPα interaction acts as a myeloid-specific immune checkpoint and is a therapeutic target in many types of cancer (101).

The factors that determine the responsiveness of a tumor to CD47/SIRPα-blocking therapies are not fully understood. CD47 is expressed on both normal and malignant tissues, yet studies have found that malignant cells are more susceptible to phagocytosis. In many cases, CD47 expression levels are higher on tumors relative to normal tissues, but expression levels alone do not always predict responses (98). Some data suggest that the function of CD47 may also depend on its conformation, association with integrins, and activation state (for reviews, see references 107 and 108). It is likely that other factors such as the presence of redundant inhibitory signals or the degree of macrophage infiltration into tumors may also impact the efficacy and therapeutic window of CD47/SIRPα-blocking reagents. Further investigation of these factors in clinical studies will be valuable to identify biomarkers that predict which patients may benefit most from CD47/SIRPα-blocking therapies.

Multiple CD47/SIRPα-blocking reagents have been developed, and they differ in their therapeutic and toxicity profiles. Some anti-CD47 antagonists may exert direct cytotoxicity on cells in addition to blocking the CD47/SIRPα interaction (109–111). This process seems to occur in a caspase-independent manner (109). Based on their ability to signal directly through CD47, some reagents may also have direct effects on activating other immune subsets, such as T cells (112, 113). The principal toxicity of targeting CD47 seems to pertain to red blood cell indices, and the main side effect observed in animal models has been a moderate anemia (93, 94, 103). Administration of an initial subtherapeutic 'priming dose' may mitigate this effect (93). Agents that block the CD47/SIRPα axis without contributing an opsonizing stimulus may also avoid this toxicity (86, 103).

MYELOID CELLS AND PROGRAMMED CELL REMOVAL

Programmed cell removal (PrCR) of target cells by macrophages is determined by the balance between "eat me" and "don't eat me" signaling pathways. Blockade of "don't eat me" signals leads to the domination of "eat me" pathways in the target cells that allows them to be detected and recognized by macrophages. In 1994, we published data showing that transgenic neutrophils that overexpressed human *bcl-2* were resistant to programmed cell death, leading to their persistence *in vitro* for many days (114). In contrast, wild-type neutrophils died within hours due to apoptosis. *In vivo*, however, the number of transgenic *bcl-2* neutrophils was homeostatically regulated despite their prolonged life span. This turned out to be due to PrCR; the wild-type neutrophils expressed cell surface "eat me" signals while undergoing cell death, but the healthy *bcl-2* neutrophils expressed PrCR "eat me" signals while otherwise viable, leading to their removal by macrophages. Thus these cells, like MDS progenitors, could be removed by PrCR even if they were resistant to programmed cell death. It is of interest that the progression of MDS to AML is concurrent with the expression of CD47 by the emergent blast cells of MDS and the AML cells they generate.

We recently identified important roles of Toll-like receptor (TLR) signaling pathways in regulating phagocytic ability of macrophages against target cancer cells (115). Agonists targeting TLRs 3, 4, and 7 stimulate macrophages by triggering the phosphorylation of Bruton's tyrosine kinase (Btk), which is a member of the Tec nonreceptor protein tyrosine kinase family and is expressed along the hematopoietic system except for in T cells and NK cells. During PrCR, Btk in macrophages acts as a critical signal transducer by inducing the phosphorylation of CRT and promoting its trafficking to the cell surface. We found that CRT, a multifunctional protein that was previously identified as a bridging molecule between phagocytes and apoptotic cells to initiate phagocytosis and also an "eat me" signal on certain types of human cancers (116), is an essential effector on the surface of macrophages mediating the recognition and phagocytosis of living tumor cells (115). PrCR of tumor cells was diminished when cell surface CRT on macrophages was blocked. Further understanding of the molecular mechanisms of TLR-Btk-CRT–mediated signaling is critical for the development of novel anticancer immunotherapies through the activation of PrCR.

CONCLUSIONS

The hematopoietic hierarchy has been thoroughly defined in both mice and humans, and HSCs and their downstream progeny can now be isolated with extraordinary purity. The diseased counterparts to the normal stages of hematopoiesis are now being investigated, and these studies will continue to provide valuable information regarding the conversion to states of pathology. Furthermore, studies on aging will help us identify the changes that occur in the hematopoietic lineage over time. An understanding of hematopoiesis and myeloid cell development has broad implications for regenerative medicine, hematopoietic cell transplantation, malignancy, and infection, among many other states of illness and disease. Through these studies, we gain insight into fundamental biological processes and we gain knowledge that can be leveraged for the benefit of patients.

Acknowledgments. *We thank members of the Weissman and Shizuru labs, and R. Levy, K. C. Garcia, B. Mitchell, H. Kohrt, and J. Shizuru for helpful advice and discussions. This work was supported in part by the National Cancer Institute (F30 CA168059 to K.W., P01 CA139490 to I.L.W.), the Stanford Medical Scientist Training Program (NIH-GM07365 to K.W.), the Stanford University SPARK Program (K.W.), the Joseph & Laurie Lacob Gynecologic/Ovarian Cancer Fund (to I.L.W.), the Virginia and D.K. Ludwig Fund for Cancer Research (to I.L.W.), an Anonymous Donors Fund, the Siebel Stem Cell Institute, and the Thomas and Stacey Siebel Foundation (I.L.W.). The content of this manuscript reflects solely the views of the authors. K.W., P.J.S., M.P.C, A.C., and I.L.W. declare patents related to CD47-blocking therapies assigned to Stanford University and licensed to Alexo Therapeutics or Forty Seven, Inc. K.W. declares equity and/or consulting with Alexo Therapeutics and Forty Seven, Inc. M.P.C. and I.L.W. declare equity and/or consulting with Forty Seven, Inc. M.P.C. is an employee of Forty Seven, Inc.*

Citation. Weiskopf K, Schnorr PJ, Pang WW, Chao MP, Chhabra A, Seita J, Feng M, Weissman IL. 2016. Myeloid cell origins, differentiation, and clinical implications. Microbiol Spectrum 4(5):MCHD-0031-2016.

References

1. **Ford CE, Hamerton JL, Barnes DW, Loutit JF.** 1956. Cytological identification of radiation-chimaeras. *Nature* **177:**452–454.
2. **Till JE, McCulloch EA.** 1961. A direct measurement of the radiation sensitivity of normal mouse bone marrow cells. *Radiat Res* **14:**213–222.
3. **Becker AJ, McCulloch EA, Till JE.** 1963. Cytological demonstration of the clonal nature of spleen colonies derived from transplanted mouse marrow cells. *Nature* **197:**452–454.
4. **Wu AM, Till JE, Siminovitch L, McCulloch EA.** 1968. Cytological evidence for a relationship between normal hemotopoietic colony-forming cells and cells of the lymphoid system. *J Exp Med* **127:**455–464.
5. **Chen JY, Miyanishi M, Wang SK, Yamazaki S, Sinha R, Kao KS, Seita J, Sahoo D, Nakauchi H, Weissman IL.** 2016. Hoxb5 marks long-term haematopoietic stem

cells and reveals a homogenous perivascular niche. *Nature* 530:223–227.

6. Adolfsson J, Borge OJ, Bryder D, Theilgaard-Mönch K, Astrand-Grundström I, Sitnicka E, Sasaki Y, Jacobsen SE. 2001. Upregulation of Flt3 expression within the bone marrow Lin$^-$Sca1$^+$c-kit$^+$ stem cell compartment is accompanied by loss of self-renewal capacity. *Immunity* 15:659–669.
7. Christensen JL, Weissman IL. 2001. Flk-2 is a marker in hematopoietic stem cell differentiation: a simple method to isolate long-term stem cells. *Proc Natl Acad Sci U S A* 98:14541–14546.
8. Kiel MJ, Yilmaz OH, Iwashita T, Yilmaz OH, Terhorst C, Morrison SJ. 2005. SLAM family receptors distinguish hematopoietic stem and progenitor cells and reveal endothelial niches for stem cells. *Cell* 121:1109–1121.
9. Morrison SJ, Wandycz AM, Hemmati HD, Wright DE, Weissman IL. 1997. Identification of a lineage of multipotent hematopoietic progenitors. *Development* 124:1929–1939.
10. Spangrude GJ, Heimfeld S, Weissman IL. 1988. Purification and characterization of mouse hematopoietic stem cells. *Science* 241:58–62.
11. Smith LG, Weissman IL, Heimfeld S. 1991. Clonal analysis of hematopoietic stem-cell differentiation *in vivo*. *Proc Natl Acad Sci U S A* 88:2788–2792.
12. Ikuta K, Ingolia DE, Friedman J, Heimfeld S, Weissman IL. 1991. Mouse hematopoietic stem cells and the interaction of c-kit receptor and steel factor. *Int J Cell Cloning* 9:451–460.
13. Czechowicz A, Kraft D, Weissman IL, Bhattacharya D. 2007. Efficient transplantation via antibody-based clearance of hematopoietic stem cell niches. *Science* 318:1296–1299.
14. Chhabra A, Ring AM, Weiskopf K, Schnorr PJ, Gordon S, Le AC, Kwon HS, Ring NG, Volkmer J, Ho PY, Tseng S, Weissman IL, Shizuru JA. 2016. Hematopoietic stem cell transplantation in immunocompetent hosts without radiation or chemotherapy. *Sci Transl Med* 8:351ra105. doi:10.1126/scitranslmed.aae0501.
15. Morrison SJ, Weissman IL. 1994. The long-term repopulating subset of hematopoietic stem cells is deterministic and isolatable by phenotype. *Immunity* 1:661–673.
16. Sun J, Ramos A, Chapman B, Johnnidis JB, Le L, Ho YJ, Klein A, Hofmann O, Camargo FD. 2014. Clonal dynamics of native haematopoiesis. *Nature* 514:322–327.
17. Kondo M, Weissman IL, Akashi K. 1997. Identification of clonogenic common lymphoid progenitors in mouse bone marrow. *Cell* 91:661–672.
18. Akashi K, Traver D, Miyamoto T, Weissman IL. 2000. A clonogenic common myeloid progenitor that gives rise to all myeloid lineages. *Nature* 404:193–197.
19. Warren L, Bryder D, Weissman IL, Quake SR. 2006. Transcription factor profiling in individual hematopoietic progenitors by digital RT-PCR. *Proc Natl Acad Sci U S A* 103:17807–17812.
20. Paul F, Arkin Y, Giladi A, Jaitin DA, Kenigsberg E, Keren-Shaul H, Winter D, Lara-Astiaso D, Gury M, Weiner A, David E, Cohen N, Lauridsen FK, Haas S, Schlitzer A, Mildner A, Ginhoux F, Jung S, Trumpp A, Porse BT, Tanay A, Amit I. 2015. Transcriptional heterogeneity and lineage commitment in myeloid progenitors. *Cell* 163:1663–1677.
21. Nakorn TN, Miyamoto T, Weissman IL. 2003. Characterization of mouse clonogenic megakaryocyte progenitors. *Proc Natl Acad Sci U S A* 100:205–210.
22. Terszowski G, Waskow C, Conradt P, Lenze D, Koenigsmann J, Carstanjen D, Horak I, Rodewald HR. 2005. Prospective isolation and global gene expression analysis of the erythrocyte colony-forming unit (CFU-E). *Blood* 105:1937–1945.
23. Iwasaki H, Mizuno S, Mayfield R, Shigematsu H, Arinobu Y, Seed B, Gurish MF, Takatsu K, Akashi K. 2005. Identification of eosinophil lineage-committed progenitors in the murine bone marrow. *J Exp Med* 201:1891–1897.
24. Fogg DK, Sibon C, Miled C, Jung S, Aucouturier P, Littman DR, Cumano A, Geissmann F. 2006. A clonogenic bone marrow progenitor specific for macrophages and dendritic cells. *Science* 311:83–87.
25. Traver D, Akashi K, Manz M, Merad M, Miyamoto T, Engleman EG, Weissman IL. 2000. Development of CD8α-positive dendritic cells from a common myeloid progenitor. *Science* 290:2152–2154.
26. Chen CC, Grimbaldeston MA, Tsai M, Weissman IL, Galli SJ. 2005. Identification of mast cell progenitors in adult mice. *Proc Natl Acad Sci U S A* 102:11408–11413.
27. Arinobu Y, Iwasaki H, Gurish MF, Mizuno S, Shigematsu H, Ozawa H, Tenen DG, Austen KF, Akashi K. 2005. Developmental checkpoints of the basophil/mast cell lineages in adult murine hematopoiesis. *Proc Natl Acad Sci U S A* 102:18105–18110.
28. Murakami JL, Xu B, Franco CB, Hu X, Galli SJ, Weissman IL, Chen CC. 2016. Evidence that β7 integrin regulates hematopoietic stem cell homing and engraftment through interaction with MAdCAM-1. *Stem Cells Dev* 25:18–26.
29. Pronk CJ, Rossi DJ, Månsson R, Attema JL, Norddahl GL, Chan CK, Sigvardsson M, Weissman IL, Bryder D. 2007. Elucidation of the phenotypic, functional, and molecular topography of a myeloerythroid progenitor cell hierarchy. *Cell Stem Cell* 1:428–442.
30. Baum CM, Weissman IL, Tsukamoto AS, Buckle AM, Peault B. 1992. Isolation of a candidate human hematopoietic stem-cell population. *Proc Natl Acad Sci U S A* 89:2804–2808.
31. Michallet M, Philip T, Philip I, Godinot H, Sebban C, Salles G, Thiebaut A, Biron P, Lopez F, Mazars P, Roubi N, Leemhuis T, Hanania E, Reading C, Fine G, Atkinson K, Juttner C, Coiffier B, Fière D, Archimbaud E. 2000. Transplantation with selected autologous peripheral blood CD34$^+$Thy1$^+$ hematopoietic stem cells (HSCs) in multiple myeloma: impact of HSC dose on engraftment, safety, and immune reconstitution. *Exp Hematol* 28:858–870.

32. **Negrin RS, Atkinson K, Leemhuis T, Hanania E, Juttner C, Tierney K, Hu WW, Johnston LJ, Shizurn JA, Stockerl-Goldstein KE, Blume KG, Weissman IL, Bower S, Baynes R, Dansey R, Karanes C, Peters W, Klein J.** 2000. Transplantation of highly purified CD34[+]Thy-1[+] hematopoietic stem cells in patients with metastatic breast cancer. *Biol Blood Marrow Transplant* **6:**262–271.
33. **Muller AM, Kohrt HE, Cha S, Laport G, Klein J, Guardino AE, Johnston LJ, Stockerl-Goldstein KE, Hanania E, Juttner C, Blume KG, Negrin RS, Weissman IL, Shizuru JA.** 2012. Long-term outcome of patients with metastatic breast cancer treated with high-dose chemotherapy and transplantation of purified autologous hematopoietic stem cells. *Biol Blood Marrow Transplant* **18:**125–133.
34. **Bhatia M, Wang JC, Kapp U, Bonnet D, Dick JE.** 1997. Purification of primitive human hematopoietic cells capable of repopulating immune-deficient mice. *Proc Natl Acad Sci U S A* **94:**5320–5325.
35. **Uchida N, Sutton RE, Friera AM, He D, Reitsma MJ, Chang WC, Veres G, Scollay R, Weissman IL.** 1998. HIV, but not murine leukemia virus, vectors mediate high efficiency gene transfer into freshly isolated G0/G1 human hematopoietic stem cells. *Proc Natl Acad Sci U S A* **95:**11939–11944.
36. **Majeti R, Park CY, Weissman IL.** 2007. Identification of a hierarchy of multipotent hematopoietic progenitors in human cord blood. *Cell Stem Cell* **1:**635–645.
37. **Galy A, Travis M, Cen D, Chen B.** 1995. Human T, B, natural killer, and dendritic cells arise from a common bone marrow progenitor cell subset. *Immunity* **3:**459–473.
38. **Manz MG, Miyamoto T, Akashi K, Weissman IL.** 2002. Prospective isolation of human clonogenic common myeloid progenitors. *Proc Natl Acad Sci U S A* **99:** 11872–11877.
39. **Edvardsson L, Dykes J, Olofsson T.** 2006. Isolation and characterization of human myeloid progenitor populations—TpoR as discriminator between common myeloid and megakaryocyte/erythroid progenitors. *Exp Hematol* **34:**599–609.
40. **Bühring HJ, Simmons PJ, Pudney M, Müller R, Jarrossay D, van Agthoven A, Willheim M, Brugger W, Valent P, Kanz L.** 1999. The monoclonal antibody 97A6 defines a novel surface antigen expressed on human basophils and their multipotent and unipotent progenitors. *Blood* **94:**2343–2356.
41. **Bühring HJ, Seiffert M, Giesert C, Marxer A, Kanz L, Valent P, Sano K.** 2001. The basophil activation marker defined by antibody 97A6 is identical to the ectonucleotide pyrophosphatase/phosphodiesterase 3. *Blood* **97:** 3303–3305.
42. **Kirshenbaum AS, Goff JP, Semere T, Foster B, Scott LM, Metcalfe DD.** 1999. Demonstration that human mast cells arise from a progenitor cell population that is CD34[+], c-kit[+], and expresses aminopeptidase N (CD13). *Blood* **94:**2333–2342.
43. **Mori Y, Iwasaki H, Kohno K, Yoshimoto G, Kikushige Y, Okeda A, Uike N, Niiro H, Takenaka K, Nagafuji K, Miyamoto T, Harada M, Takatsu K, Akashi K.** 2009. Identification of the human eosinophil lineage-committed progenitor: revision of phenotypic definition of the human common myeloid progenitor. *J Exp Med* **206:**183–193.
44. **Mori Y, Chen JY, Pluvinage JV, Seita J, Weissman IL.** 2015. Prospective isolation of human erythroid lineage-committed progenitors. *Proc Natl Acad Sci U S A* **112:** 9638–9643.
45. **Li J, Hale J, Bhagia P, Xue F, Chen L, Jaffray J, Yan H, Lane J, Gallagher PG, Mohandas N, Liu J, An X.** 2014. Isolation and transcriptome analyses of human erythroid progenitors: BFU-E and CFU-E. *Blood* **124:** 3636–3645.
46. **Seita J, Sahoo D, Rossi DJ, Bhattacharya D, Serwold T, Inlay MA, Ehrlich LI, Fathman JW, Dill DL, Weissman IL.** 2012. Gene Expression Commons: an open platform for absolute gene expression profiling. *PLoS One* **7:** e40321. doi:10.1371/journal.pone.0040321.
47. **Moore MA, Metcalf D.** 1970. Ontogeny of the haemopoietic system: yolk sac origin of *in vivo* and *in vitro* colony forming cells in the developing mouse embryo. *Br J Haematol* **18:**279–296.
48. **Weissman IL, Baird S, Gardner RL, Papaioannou VE, Raschke W.** Normal and neoplastic maturation of T-lineage lymphocytes. *Cold Spring Harb Symp Quant Biol* **41:**9–21.
49. **Weissman I, Papaioannou V, Gardner R.** 1978. Fetal hematopoietic origins of the adult hematolymphoid system. *Differ Norm Neoplast Hematopoietic Cells* **5:** 33–47.
50. **Medvinsky A, Dzierzak E.** 1996. Definitive hematopoiesis is autonomously initiated by the AGM region. *Cell* **86:**897–906.
51. **Choi K, Kennedy M, Kazarov A, Papadimitriou JC, Keller G.** 1998. A common precursor for hematopoietic and endothelial cells. *Development* **125:**725–732.
52. **Adamo L, García-Cardeña G.** 2012. The vascular origin of hematopoietic cells. *Dev Biol* **362:**1–10.
53. **Ueno H, Weissman IL.** 2006. Clonal analysis of mouse development reveals a polyclonal origin for yolk sac blood islands. *Dev Cell* **11:**519–533.
54. **Samokhvalov IM, Samokhvalova NI, Nishikawa S.** 2007. Cell tracing shows the contribution of the yolk sac to adult haematopoiesis. *Nature* **446:**1056–1061.
55. **Lux CT, Yoshimoto M, McGrath K, Conway SJ, Palis J, Yoder MC.** 2008. All primitive and definitive hematopoietic progenitor cells emerging before E10 in the mouse embryo are products of the yolk sac. *Blood* **111:** 3435–3438.
56. **Ginhoux F, Greter M, Leboeuf M, Nandi S, See P, Gokhan S, Mehler MF, Conway SJ, Ng LG, Stanley ER, Samokhvalov IM, Merad M.** 2010. Fate mapping analysis reveals that adult microglia derive from primitive macrophages. *Science* **330:**841–845.
57. **Hoeffel G, Wang Y, Greter M, See P, Teo P, Malleret B, Leboeuf M, Low D, Oller G, Almeida F, Choy SH, Grisotto M, Renia L, Conway SJ, Stanley ER, Chan JK, Ng LG, Samokhvalov IM, Merad M, Ginhoux F.** 2012.

Adult Langerhans cells derive predominantly from embryonic fetal liver monocytes with a minor contribution of yolk sac-derived macrophages. *J Exp Med* **209:** 1167–1181.

58. **Gomez Perdiguero E, Klapproth K, Schulz C, Busch K, Azzoni E, Crozet L, Garner H, Trouillet C, de Bruijn MF, Geissmann F, Rodewald HR.** 2015. Tissue-resident macrophages originate from yolk-sac-derived erythro-myeloid progenitors. *Nature* **518:**547–551.
59. **Sudo K, Ema H, Morita Y, Nakauchi H.** 2000. Age-associated characteristics of murine hematopoietic stem cells. *J Exp Med* **192:**1273–1280.
60. **Rossi DJ, Bryder D, Zahn JM, Ahlenius H, Sonu R, Wagers AJ, Weissman IL.** 2005. Cell intrinsic alterations underlie hematopoietic stem cell aging. *Proc Natl Acad Sci U S A* **102:**9194–9199.
61. **Pang WW, Price EA, Sahoo D, Beerman I, Maloney WJ, Rossi DJ, Schrier SL, Weissman IL.** 2011. Human bone marrow hematopoietic stem cells are increased in frequency and myeloid-biased with age. *Proc Natl Acad Sci U S A* **108:**20012–20017.
62. **Beerman I, Bhattacharya D, Zandi S, Sigvardsson M, Weissman IL, Bryder D, Rossi DJ.** 2010. Functionally distinct hematopoietic stem cells modulate hematopoietic lineage potential during aging by a mechanism of clonal expansion. *Proc Natl Acad Sci U S A* **107:** 5465–5470.
63. **Challen GA, Boles NC, Chambers SM, Goodell MA.** 2010. Distinct hematopoietic stem cell subtypes are differentially regulated by TGF-β1. *Cell Stem Cell* **6:**265–278.
64. **Benz C, Copley MR, Kent DG, Wohrer S, Cortes A, Aghaeepour N, Ma E, Mader H, Rowe K, Day C, Treloar D, Brinkman RR, Eaves CJ.** 2012. Hematopoietic stem cell subtypes expand differentially during development and display distinct lymphopoietic programs. *Cell Stem Cell* **10:**273–283.
65. **Geiger H, de Haan G, Florian MC.** 2013. The ageing haematopoietic stem cell compartment. *Nat Rev Immunol* **13:**376–389.
66. **Chambers SM, Shaw CA, Gatza C, Fisk CJ, Donehower LA, Goodell MA.** 2007. Aging hematopoietic stem cells decline in function and exhibit epigenetic dysregulation. *PLoS Biol* **5:**e201. doi:10.1371/journal.pbio.0050201.
67. **Beerman I, Rossi DJ.** 2014. Epigenetic regulation of hematopoietic stem cell aging. *Exp Cell Res* **329:**192–199.
68. **Ergen AV, Boles NC, Goodell MA.** 2012. Rantes/Ccl5 influences hematopoietic stem cell subtypes and causes myeloid skewing. *Blood* **119:**2500–2509.
69. **Florian MC, Dörr K, Niebel A, Daria D, Schrezenmeier H, Rojewski M, Filippi MD, Hasenberg A, Gunzer M, Scharffetter-Kochanek K, Zheng Y, Geiger H.** 2012. Cdc42 activity regulates hematopoietic stem cell aging and rejuvenation. *Cell Stem Cell* **10:**520–530.
70. **Weissman IL.** 1996. From thymic lineages back to hematopoietic stem cells, sometimes using homing receptors. *J Immunol* **156:**2019–2025.
71. **Gambacorti-Passerini C, le Coutre P, Mologni L, Fanelli M, Bertazzoli C, Marchesi E, Di Nicola M, Biondi A, Corneo GM, Belotti D, Pogliani E, Lydon NB.** 1997. Inhibition of the ABL kinase activity blocks the proliferation of BCR/ABL$^+$ leukemic cells and induces apoptosis. *Blood Cells Mol Dis* **23:**380–394.
72. **Jamieson CH, Ailles LE, Dylla SJ, Muijtjens M, Jones C, Zehnder JL, Gotlib J, Li K, Manz MG, Keating A, Sawyers CL, Weissman IL.** 2004. Granulocyte-macrophage progenitors as candidate leukemic stem cells in blast-crisis CML. *N Engl J Med* **351:**657–667.
73. **Abrahamsson AE, Geron I, Gotlib J, Dao KH, Barroga CF, Newton IG, Giles FJ, Durocher J, Creusot RS, Karimi M, Jones C, Zehnder JL, Keating A, Negrin RS, Weissman IL, Jamieson CH.** 2009. Glycogen synthase kinase 3β missplicing contributes to leukemia stem cell generation. *Proc Natl Acad Sci U S A* **106:**3925–3929.
74. **Weisberg E, Manley PW, Cowan-Jacob SW, Hochhaus A, Griffin JD.** 2007. Second generation inhibitors of BCR-ABL for the treatment of imatinib-resistant chronic myeloid leukaemia. *Nat Rev Cancer* **7:**345–356.
75. **Hantschel O, Grebien F, Superti-Furga G.** 2012. The growing arsenal of ATP-competitive and allosteric inhibitors of BCR-ABL. *Cancer Res* **72:**4890–4895.
76. **Warrell RP Jr, de Thé H, Wang ZY, Degos L.** 1993. Acute promyelocytic leukemia. *N Engl J Med* **329:** 177–189.
77. **Pang WW, Pluvinage JV, Price EA, Sridhar K, Arber DA, Greenberg PL, Schrier SL, Park CY, Weissman IL.** 2013. Hematopoietic stem cell and progenitor cell mechanisms in myelodysplastic syndromes. *Proc Natl Acad Sci U S A* **110:**3011–3016.
78. **Levine RL, Wadleigh M, Cools J, Ebert BL, Wernig G, Huntly BJ, Boggon TJ, Wlodarska I, Clark JJ, Moore S, Adelsperger J, Koo S, Lee JC, Gabriel S, Mercher T, D'Andrea A, Fröhling S, Döhner K, Marynen P, Vandenberghe P, Mesa RA, Tefferi A, Griffin JD, Eck MJ, Sellers WR, Meyerson M, Golub TR, Lee SJ, Gilliland DG.** 2005. Activating mutation in the tyrosine kinase JAK2 in polycythemia vera, essential thrombocythemia, and myeloid metaplasia with myelofibrosis. *Cancer Cell* **7:**387–397.
79. **Jamieson CH, Gotlib J, Durocher JA, Chao MP, Mariappan MR, Lay M, Jones C, Zehnder JL, Lilleberg SL, Weissman IL.** 2006. The *JAK2* V617F mutation occurs in hematopoietic stem cells in polycythemia vera and predisposes toward erythroid differentiation. *Proc Natl Acad Sci U S A* **103:**6224–6229.
80. **Verstovsek S, Passamonti F, Rambaldi A, Barosi G, Rosen PJ, Rumi E, Gattoni E, Pieri L, Guglielmelli P, Elena C, He S, Contel N, Mookerjee B, Sandor V, Cazzola M, Kantarjian HM, Barbui T, Vannucchi AM.** 2014. A phase 2 study of ruxolitinib, an oral JAK1 and JAK2 inhibitor, in patients with advanced polycythemia vera who are refractory or intolerant to hydroxyurea. *Cancer* **120:**513–520.
81. **Vannucchi AM, Kiladjian JJ, Griesshammer M, Masszi T, Durrant S, Passamonti F, Harrison CN, Pane F, Zachee P, Mesa R, He S, Jones MM, Garrett W, Li J, Pirron U, Habr D, Verstovsek S.** 2015. Ruxolitinib versus standard therapy for the treatment of polycythemia vera. *N Engl J Med* **372:**426–435.

82. **Oldenborg PA, Zheleznyak A, Fang YF, Lagenaur CF, Gresham HD, Lindberg FP.** 2000. Role of CD47 as a marker of self on red blood cells. *Science* **288:**2051–2054.

83. **Adams S, van der Laan LJ, Vernon-Wilson E, Renardel de Lavalette C, Döpp EA, Dijkstra CD, Simmons DL, van den Berg TK.** 1998. Signal-regulatory protein is selectively expressed by myeloid and neuronal cells. *J Immunol* **161:**1853–1859.

84. **Seiffert M, Cant C, Chen Z, Rappold I, Brugger W, Kanz L, Brown EJ, Ullrich A, Bühring HJ.** 1999. Human signal-regulatory protein is expressed on normal, but not on subsets of leukemic myeloid cells and mediates cellular adhesion involving its counterreceptor CD47. *Blood* **94:**3633–3643.

85. **Seiffert M, Brossart P, Cant C, Cella M, Colonna M, Brugger W, Kanz L, Ullrich A, Bühring HJ.** 2001. Signal-regulatory protein α (SIRPα) but not SIRPβ is involved in T-cell activation, binds to CD47 with high affinity, and is expressed on immature $CD34^{+}CD38^{-}$ hematopoietic cells. *Blood* **97:**2741–2749.

86. **Zhao XW, van Beek EM, Schornagel K, Van der Maaden H, Van Houdt M, Otten MA, Finetti P, Van Egmond M, Matozaki T, Kraal G, Birnbaum D, van Elsas A, Kuijpers TW, Bertucci F, van den Berg TK.** 2011. CD47-signal regulatory protein-α (SIRPα) interactions form a barrier for antibody-mediated tumor cell destruction. *Proc Natl Acad Sci U S A* **108:**18342–18347.

87. **Ho CC, Guo N, Sockolosky JT, Ring AM, Weiskopf K, Özkan E, Mori Y, Weissman IL, Garcia KC.** 2015. "Velcro" engineering of high affinity CD47 ectodomain as signal regulatory protein α (SIRPα) antagonists that enhance antibody-dependent cellular phagocytosis. *J Biol Chem* **290:**12650–12663.

88. **Jaiswal S, Jamieson CH, Pang WW, Park CY, Chao MP, Majeti R, Traver D, van Rooijen N, Weissman IL.** 2009. CD47 is upregulated on circulating hematopoietic stem cells and leukemia cells to avoid phagocytosis. *Cell* **138:**271–285.

89. **Takenaka K, Prasolava TK, Wang JC, Mortin-Toth SM, Khalouei S, Gan OI, Dick JE, Danska JS.** 2007. Polymorphism in *Sirpa* modulates engraftment of human hematopoietic stem cells. *Nat Immunol* **8:**1313–1323.

90. **Yamauchi T, Takenaka K, Urata S, Shima T, Kikushige Y, Tokuyama T, Iwamoto C, Nishihara M, Iwasaki H, Miyamoto T, Honma N, Nakao M, Matozaki T, Akashi K.** 2013. Polymorphic *Sirpa* is the genetic determinant for NOD-based mouse lines to achieve efficient human cell engraftment. *Blood* **121:**1316–1325.

91. **Kuriyama T, Takenaka K, Kohno K, Yamauchi T, Daitoku S, Yoshimoto G, Kikushige Y, Kishimoto J, Abe Y, Harada N, Miyamoto T, Iwasaki H, Teshima T, Akashi K.** 2012. Engulfment of hematopoietic stem cells caused by down-regulation of CD47 is critical in the pathogenesis of hemophagocytic lymphohistiocytosis. *Blood* **120:**4058–4067.

92. **Majeti R, Chao MP, Alizadeh AA, Pang WW, Jaiswal S, Gibbs KD Jr, van Rooijen N, Weissman IL.** 2009. CD47 is an adverse prognostic factor and therapeutic antibody target on human acute myeloid leukemia stem cells. *Cell* **138:**286–299.

93. **Liu J, Wang L, Zhao F, Tseng S, Narayanan C, Shura L, Willingham S, Howard M, Prohaska S, Volkmer J, Chao M, Weissman IL, Majeti R.** 2015. Pre-clinical development of a humanized anti-CD47 antibody with anti-cancer therapeutic potential. *PLoS One* **10:** e0137345. doi:10.1371/journal.pone.0137345.

94. **Willingham SB, Volkmer JP, Gentles AJ, Sahoo D, Dalerba P, Mitra SS, Wang J, Contreras-Trujillo H, Martin R, Cohen JD, Lovelace P, Scheeren FA, Chao MP, Weiskopf K, Tang C, Volkmer AK, Naik TJ, Storm TA, Mosley AR, Edris B, Schmid SM, Sun CK, Chua MS, Murillo O, Rajendran P, Cha AC, Chin RK, Kim D, Adorno M, Raveh T, Tseng D, Jaiswal S, Enger PO, Steinberg GK, Li G, So SK, Majeti R, Harsh GR, van de Rijn M, Teng NN, Sunwoo JB, Alizadeh AA, Clarke MF, Weissman IL.** 2012. The CD47-signal regulatory protein alpha (SIRPa) interaction is a therapeutic target for human solid tumors. *Proc Natl Acad Sci U S A* **109:** 6662–6667.

95. **Zhao H, Wang J, Kong X, Li E, Liu Y, Du X, Kang Z, Tang Y, Kuang Y, Yang Z, Zhou Y, Wang Q.** 2016. CD47 promotes tumor invasion and metastasis in non-small cell lung cancer. *Sci Rep* **6:**29719. doi:10.1038/srep29719.

96. **Edris B, Weiskopf K, Volkmer AK, Volkmer JP, Willingham SB, Contreras-Trujillo H, Liu J, Majeti R, West RB, Fletcher JA, Beck AH, Weissman IL, van de Rijn M.** 2012. Antibody therapy targeting the CD47 protein is effective in a model of aggressive metastatic leiomyosarcoma. *Proc Natl Acad Sci U S A* **109:**6656–6661.

97. **Krampitz GW, George BM, Willingham SB, Volkmer JP, Weiskopf K, Jahchan N, Newman AM, Sahoo D, Zemek AJ, Yanovsky RL, Nguyen JK, Schnorr PJ, Mazur PK, Sage J, Longacre TA, Visser BC, Poultsides GA, Norton JA, Weissman IL.** 2016. Identification of tumorigenic cells and therapeutic targets in pancreatic neuroendocrine tumors. *Proc Natl Acad Sci U S A* **113:** 4464–4469.

98. **Weiskopf K, Jahchan NS, Schnorr PJ, Cristea S, Ring AM, Maute RL, Volkmer AK, Volkmer JP, Liu J, Lim JS, Yang D, Seitz G, Nguyen T, Wu D, Jude K, Guerston H, Barkal A, Trapani F, George J, Poirier JT, Gardner EE, Miles LA, de Stanchina E, Lofgren SM, Vogel H, Winslow MM, Dive C, Thomas RK, Rudin CM, van de Rijn M, Majeti R, Garcia KC, Weissman IL, Sage J.** 2016. CD47-blocking immunotherapies stimulate macrophage-mediated destruction of small-cell lung cancer. *J Clin Invest* **126:**2610–2620.

99. **Ngo M, Han A, Lakatos A, Sahoo D, Hachey SJ, Weiskopf K, Beck AH, Weissman IL, Boiko AD.** 2016. Antibody therapy targeting CD47 and CD271 effectively suppresses melanoma metastasis in patient-derived xenografts. *Cell Rep* **16:**1701–1716.

100. **Liu X, Pu Y, Cron K, Deng L, Kline J, Frazier WA, Xu H, Peng H, Fu YX, Xu MM.** 2015. CD47 blockade triggers T cell-mediated destruction of immunogenic tumors. *Nat Med* **21:**1209–1215.

101. **Weiskopf K, Weissman IL.** 2015. Macrophages are critical effectors of antibody therapies for cancer. *MAbs* **7:** 303–310.

102. **Oldenborg PA, Gresham HD, Lindberg FP.** 2001. CD47-signal regulatory protein α (SIRPα) regulates Fcγ and complement receptor-mediated phagocytosis. *J Exp Med* **193:**855–862.

103. **Weiskopf K, Ring AM, Ho CC, Volkmer JP, Levin AM, Volkmer AK, Ozkan E, Fernhoff NB, van de Rijn M, Weissman IL, Garcia KC.** 2013. Engineered SIRPα variants as immunotherapeutic adjuvants to anticancer antibodies. *Science* **341:**88–91.

104. **Weiskopf K, Ring AM, Schnorr PJ, Volkmer JP, Volkmer AK, Weissman IL, Garcia KC.** 2013. Improving macrophage responses to therapeutic antibodies by molecular engineering of SIRPα variants. *Onco Immunology* **2:**e25773. doi:10.4161/onci.25773.

105. **Chao MP, Alizadeh AA, Tang C, Myklebust JH, Varghese B, Gill S, Jan M, Cha AC, Chan CK, Tan BT, Park CY, Zhao F, Kohrt HE, Malumbres R, Briones J, Gascoyne RD, Lossos IS, Levy R, Weissman IL, Majeti R.** 2010. Anti-CD47 antibody synergizes with rituximab to promote phagocytosis and eradicate non-Hodgkin lymphoma. *Cell* **142:**699–713.

106. **Tseng D, Volkmer JP, Willingham SB, Contreras-Trujillo H, Fathman JW, Fernhoff NB, Seita J, Inlay MA, Weiskopf K, Miyanishi M, Weissman IL.** 2013. Anti-CD47 antibody-mediated phagocytosis of cancer by macrophages primes an effective antitumor T-cell response. *Proc Natl Acad Sci U S A* **110:**11103–11108.

107. **Barclay AN, Van den Berg TK.** 2014. The interaction between signal regulatory protein alpha (SIRPα) and CD47: structure, function, and therapeutic target. *Annu Rev Immunol* **32:**25–50.

108. **Soto-Pantoja DR, Kaur S, Roberts DD.** 2015. CD47 signaling pathways controlling cellular differentiation and responses to stress. *Crit Rev Biochem Mol Biol* **50:** 212–230.

109. **Mateo V, Brown EJ, Biron G, Rubio M, Fischer A, Deist FL, Sarfati M.** 2002. Mechanisms of CD47-induced caspase-independent cell death in normal and leukemic cells: link between phosphatidylserine exposure and cytoskeleton organization. *Blood* **100:**2882–2890.

110. **Kikuchi Y, Uno S, Kinoshita Y, Yoshimura Y, Iida S, Wakahara Y, Tsuchiya M, Yamada-Okabe H, Fukushima N.** 2005. Apoptosis inducing bivalent single-chain antibody fragments against CD47 showed antitumor potency for multiple myeloma. *Leuk Res* **29:**445–450.

111. **Manna PP, Frazier WA.** 2004. CD47 mediates killing of breast tumor cells via Gi-dependent inhibition of protein kinase A. *Cancer Res* **64:**1026–1036.

112. **Reinhold MI, Lindberg FP, Kersh GJ, Allen PM, Brown EJ.** 1997. Costimulation of T cell activation by integrin-associated protein (CD47) is an adhesion-dependent, CD28-independent signaling pathway. *J Exp Med* **185:** 1–11.

113. **Soto-Pantoja DR, Terabe M, Ghosh A, Ridnour LA, DeGraff WG, Wink DA, Berzofsky JA, Roberts DD.** 2014. CD47 in the tumor microenvironment limits cooperation between antitumor T-cell immunity and radiotherapy. *Cancer Res* **74:**6771–6783.

114. **Lagasse E, Weissman IL.** 1994. bcl-2 inhibits apoptosis of neutrophils but not their engulfment by macrophages. *J Exp Med* **179:**1047–1052.

115. **Feng M, Chen JY, Weissman-Tsukamoto R, Volkmer JP, Ho PY, McKenna KM, Cheshier S, Zhang M, Guo N, Gip P, Mitra SS, Weissman IL.** 2015. Macrophages eat cancer cells using their own calreticulin as a guide: roles of TLR and Btk. *Proc Natl Acad Sci U S A* **112:** 2145–2150.

116. **Chao MP, Jaiswal S, Weissman-Tsukamoto R, Alizadeh AA, Gentles AJ, Volkmer J, Weiskopf K, Willingham SB, Raveh T, Park CY, Majeti R, Weissman IL.** 2010. Calreticulin is the dominant pro-phagocytic signal on multiple human cancers and is counterbalanced by CD47. *Sci Transl Med* **2:**63ra94. doi:10.1126/scitranslmed.3001375.

117. **Seita J, Weissman IL.** 2010. Hematopoietic stem cell: self-renewal versus differentiation. *Wiley Interdiscip Rev Syst Biol Med* **2:**640–653.

Index

B

E

N

O

P

R

S

T